27th International Laser Radar Conference 2015 (ILRC 27)

EPJ Web of Conferences Volume 119 (2016)

New York City, New York, USA
5-10 July 2015

Part 1 of 2

Editors:

B. Gross
M. Arend
F. Moshary

ISBN: 978-1-5108-2531-4

Printed from e-media with permission by:

Curran Associates, Inc.
57 Morehouse Lane
Red Hook, NY 12571

Some format issues inherent in the e-media version may also appear in this print version.

Printed by Curran Associates, Inc. (2016)

For additional information, please contact EDP Sciences – Web of Conferences at the address below.

EDP Sciences – Web of Conferences
17, Avenue du Hoggar
Parc d'Activité de Courtabœuf
BP 112
F-91944 Les Ulis Cedex A
France

Phone: +33 (0) 1 69 18 75 75
Fax: +33 (0) 1 69 28 84 91

contact-edps@webofconferences.org

Additional copies of this publication are available from:

Curran Associates, Inc.
57 Morehouse Lane
Red Hook, NY 12571 USA
Phone: 845-758-0400
Fax: 845-758-2633
Email: curran@proceedings.com
Web: www.proceedings.com

TABLE OF CONTENTS

PART 1

PART 2

ILRC 27 New York City

The 27th International Laser Radar Conference
New York City, July 5-10 2015

The 27th ILRC will take place in New York City and will be hosted by the National Oceanic and Atmospheric Administration (NOAA) Cooperative Remote Sensing Science and Technology (CREST) Center and the City College of the City University of New York. The conference venue is the historic Great Hall in the Shepard Hall of the City College of New York campus. The City College of New York is located in the northern part of the Manhattan island at 138th Street and Convent Ave, and is easily accessible from midtown and the area airports by various public and private transportation systems.

MEETING WEBSITE: ILRC27.org

ILRC 27 will cover a host of lidar related topics including

- Advances in lidar techniques and new methodologies
- Atmospheric winds and turbulence, and applications in wind energy
- Atmospheric boundary layer dynamics
- Advances in aerosol monitoring and retrievals
- Cloud detection, characterization and classification
- Aerosol direct and indirect effects and radiation budget
- Lidar for atmospheric trace gas monitoring (water vapor, ozone, CO_2, CH_4, and other GHGs)
- Tropospheric and lower stratospheric dynamics and transport
- Middle and upper atmosphere physics and chemistry
- Hazard monitoring (civilian and defense applications)
- Space-based missions and validation
- Coordinated lidar networks and applications
- Combining lidar measurements and analysis with other active and passive techniques
- Assimilation of observations into forecast models
- Lidar application to marine and ecosystem monitoring/classification
- Advances in ranging and applications
- Lidar applications to land surface monitoring (elevations and terrain maps, vegetation canopy, etc.)

SteeringCommittee

- Fred Moshary -Co-Chair (City College of NY and NOAA CRE Center) moshary@ccny.cuny.edu
- M. Partick McCormick- Co-Chair (Hampton Univ. and NOAA CREST Center)*
- Xinzhao Chu (Univ of Colorado)
- Paolo Di Girolamo (Universita della Basilicata, Italy)
- R. Michael Hardesty (CIRES, U. of Colorado)
- Reza Khanbilvardi (City College of NY and NOAA CREST Center)
- Eduardo Landulfo (Instituto de Pesquisas Energéticas e Nucleares, Brazil)*
- Gelsomina Pappalardo (Institute of Methodologies for Environmental Analysis (CNR-IMAA), Italy
- Upendra Singh (NASA LARC)*
- Kevin Strawbridge (Environment Canada)*
- Nobuo Sugimoto (National Inst. for Environmental Studies, Japan)
- Ulla Wandinger (Leibniz Institute for Tropospheric Research, Germany)

Industry Liaison

Mark Arend (City College of NY and NOAA CREST Center)
marend@ccny.cuny.edu
Ruben Delgado (UMBC and NOAA CREST Center)
delgado@umbc.edu

Other Contacts

General Inquiries: ilrc27ny@gmail.com
Shakila Merchant (Administrative Director of NOAA CREST Center)
smerchant@ccny.cuny.edu

NOAA-CREST

ICLAS Executive Committee

- Upendra N. Singh., President (NASA LARC)
- Robert T. Menzies. Past President (NASA JPL)
- M. Partick McCormick- Treasurer (Hampton Univ.)

Program Committee

- Barry Gross-Chair (City College of NY and NOAA CREST Center) gross@ccny.cuny.edu
- Andreas Behrendt (Universitaet Hohenheim, Germany)*
- Sergey Bobrovnikov (Institute of Atmospheric Optics)*
- Ronald Calhoun (Arizona State U)
- Adolfo Comeron (Universitat Politècnica de Catalunya, Barcelona Spain)
- Ruben Delgado (UMBC and NOAA CREST Center)
- Belay Demoz (UMBC)
- Ferdinando DeTomasi (Universita' del Salento, Italy)*
- Edwin Eloranta (Space Science and Engineering Cener, U. Wisconsin)
- Jonathan Freedman (Arecibo National Observatory)
- Bruce Gentry (NASA GSFC)
- Gary Gimmestad (Georgia Tech.)
- Floyd Hovis (Fibertek)
- Robert Menzies (NASA JPL)
- Detlef Muller (University of Hertfordshire, UK)
- Michael Newchurch (Univ. of Alabama Huntsville)
- Doina Nicolae (National Inst. of R&D for Optoelectronics, Romania)*
- Alexandros Papayannis (Nat. Technical University of Athens, Greece)*
- Joseph Shaw (Montana State U)
- Christoph Senff (NOAA ESRL)*
- Anne Grete Straume-Lindner (ESA – ESTEC, Netherlands)
- Sara Tucker (Ball Aerospace)
- Igor Veselovskii (Physics Instrumentation Center, Troitsk, Moscow)
- Yingjian Wang (Anhui Institute of Optics and Fine Mechanics, China)*
- Stuart Young (CSIRO Marine and Atmospheric Research, Australia)

--To Propose special sessions and themes, contact Dr. Barry Gross, Chair of the Program Committee
* ICLAS member

THE CITY COLLEGE OF THE CITY UNIVERSITY OF NEW YORK

ILRC 27 New York City

CONFERENCE FORMAT

The conference will include both oral and poster sessions to be held in the Great Hall located in the historic Shepard Hall on the CCNY campus. The conference will follow the ILRC tradition of having single session oral presentations and of limiting the number of oral presentations to one per author. There are also three planned poster sessions on Monday, Tuesday, and Thursday following the oral sessions (6th, 7th, and 9th of July); and the ILRC Summer School on Sunday July 5th.

CONFERENCE REGISTRATION FEES

(All fees are payable in US Dollars)

$650.00 USD On or before May 10th, 2015
$750.00 USD On or before June 14th, 2015
$800.00 USD On or after June 15th, 2015

FULL-TIME STUDENT REGISTRATION

$350.00 USD On or before May 10th, 2015
$450.00 USD On or before June 14th, 2015
$500.00 USD On or after June 15th, 2015

PAPER SUBMISSION

Papers are limited to 2-4 pages (including references) and should be submitted in Microsoft Word (may also be submitted as a PDF document if Word submission is not possible). Formatting details are found on the Conference Website. Papers are limited to 2-4 pages (US letter size 8.5 inch x11 inch) and in Times New Roman font with 1 inch margins and two column justified format. **Papers will be peer reviewed**.

Important Dates: Regular Paper Submission is now closed. Post Deadline papers will be accepted to April 14th. Final acceptance of the papers will be communicated by May 1st. All presenters must register by May 10th (Early Registration Deadline).

LODGING

There is a wide range of lodging options in the New York City area. Hotel options near the City College on the upper west side of Manhattan will range from $120/night and up, and in the midtown Manhattan from $160/night and up. Information is available through the conference website.

There is limited dormitory space available at City College at reasonable prices. We also anticipate to have limited travel support available for students through an application process (please keep posted via the conference website).

edp sciences

EPJ Web of Conferences

Statement of Peer review

In submitting conference proceedings to EPJ Web of Conferences, the editors of the proceedings certify to the Publisher that they adhere to the **Policy on Publishing Integrity** of the journal in order to safeguard good scientific practice in publishing.

1. All articles have been subjected to peer review administered by the proceedings editors.
2. Reviews have been conducted by expert referees, who have been requested to provide unbiased and constructive comments aimed, whenever possible, at improving the work.
3. Proceedings editors have taken all reasonable steps to ensure the quality of the materials they publish and their decision to accept or reject a paper for publication has been based only on the merits of the work and the relevance to the journal.

Title, date and place of the conference

27th International Laser Radar Conference (ILRC 27)
July 5 – July 10 2015 New York City

Proceedings editor(s):

Barry Gross
Fred Moshary
Mark Arend

Date and editor's signature

5/12/16

EPJ Web of Conferences **119**, 01001 (2016)
ILRC 27
DOI: 10.1051/epjconf/201611901001

ESA'S SPACEBORNE LIDAR MISSION ADM-AEOLUS; RECENT ACHIEVEMENTS AND PREPARATIONS FOR LAUNCH

Anne Grete Straume[1*], Anders Elfving[1], Denny Wernham[1], Alain Culoma[1], Linda Mondin[1], Frank de Bruin[1], Thomas Kanitz[1], Dirk Schuettemeyer[1], Fabio Buscaglione[2], Angelika Dehn[2]

*[1]European Space Agency (ESA/ESTEC), PO Box 299, 2200 AG Noordwijk, Netherlands, *Email: anne.straume@esa.int*

[2]European Space Agency (ESA/ESRIN), Casella Postale 64, 00044 Frascati (Roma), Italy

ABSTRACT

Within ESA's Living Planet Programme, the Atmospheric Dynamics Mission (ADM-Aeolus) was chosen as the second Earth Explorer Core mission in 1999. It shall demonstrate the potential of high spectral resolution Doppler Wind lidars for operational measurements of wind profiles and their use in Numerical Weather Prediction (NWP). Spin-off products are profiles of cloud and aerosol optical properties. ADM-Aeolus carries the novel Doppler Wind lidar instrument ALADIN.

Recently the two ALADIN laser transmitters were successfully qualified and delivered for further instrument integration. The instrument delivery will follow later this year and the satellite qualification and launch readiness is scheduled for 2016.

In February 2015, an Aeolus Science and Calibration and Validation (CAL/VAL) Workshop was held in ESA-ESRIN, Frascati, Italy, bringing industry, the user community and ESA together to prepare for the Aeolus Commissioning and Operational Phases. During the Workshop the science, instrument and product status, commissioning phase planning and the extensive number of proposals submitted in response to the Aeolus CAL/VAL call in 2014 were presented and discussed. A special session was dedicated to the Aeolus CAL/VAL Implementation Plan.

In this paper, the Aeolus mission, status and launch preparation activities are described.

1. BACKGROUND AND MOTIVATON

The European Space Agency's (ESA's) Living Planet Programme includes two types of complementary user driven missions: the research oriented Earth Explorer missions and the operational service oriented Earth Watch missions. Earth Explorer missions are divided into (i) Core missions being larger missions demonstrating the capabilities of new technologies addressing issues of wide scientific interest, and (ii) Opportunity missions with a similar scope but being smaller in terms of cost to ESA. Both types of missions address the research objectives set out in the Living Planet Programme document [1]. All Earth Explorer missions are proposed, defined, evaluated and recommended by the scientific community. ADM-Aeolus was selected as the second Earth Explorer Core mission in 1999, and sets out to demonstrate the potential of space-borne Doppler Wind lidars for operational use.

2. ADM-AEOLUS MISSION

2.1 Background and objectives

The motivation for the selection of the Aeolus mission was the need for more abundant direct wind profile measurements in the current Global Observing System (GOS), which is used e.g. by NWP and climate models [2]. In the current GOS, direct wind profile measurements are obtained from radiosondes, commercial aircraft ascends and descends and ground-based wind lidars and radars. The distribution of these measurements is, however, not homogenous, with most observations taken over land in the Northern Hemisphere. Winds can also be inferred from temperature soundings when the flow is in geostrophic balance, which is only the case for large-scale wind fields in the extra-tropics. Air-Motion-Vectors deduced from cloud-tracking is limited by the difficulty in performing accurate height-assignments. It is therefore expected that the Aeolus mission will largely contribute to the

 DOI: 10.1051/epjconf/201611901001

improvement of predictions of small-scales flows and forecasts in observation-sparse regions.

The current lack of homogenous sampling of the 3-dimensional wind field in large parts of the tropics and over the major oceans leads to difficulties in the studying of key processes in coupled climate systems and further improvements of NWP. Aeolus measurements are expected to improve short-range forecasts of severe weather and modelling of tropical dynamics in particular. Recommendations from the scientific and NWP community lead to the selection of the Aeolus space-based lidar in 1999.

2.2 Instrument concept

Aeolus will embark the direct detection Doppler Wind Lidar instrument ALADIN (Atmospheric LAser Doppler INstrument). ALADIN is a pulsed UV lidar (355 nm, circularly polarized, 50 Hz), measuring continuously along the orbit (Figure 1, left panel). Its high spectral resolution concept allows for the detection of the parallel polarized molecular (Rayleigh) and particle (Mie) backscattered signals in two separate channels, each sampling the wind in 24 vertical height bins (Figure 1, right panel). This makes it possible to deliver winds both in clear and (partly) cloudy conditions down to optically thick clouds. The height of the wind measurement is detected by time-gating. A quasi-global coverage is achieved daily (~ 15 orbits per day) and the orbit repeat cycle is 7 days (109 orbits). The orbit is sun-synchronous with a local equatorial crossing-time of 6 am/pm. A detailed description of the instrument design and operation can be found in [3]. A demonstration of the measurement concept can be found in e.g. [4].

2.3 Products and product requirements

The main product from ADM-Aeolus will be horizontally projected line-of-sight (HLOS) wind profile observations (approximately zonally oriented) from the surface up to about 30 km [3]. The atmospheric backscattered signal for the individual laser pulses are averaged on-board to yield ~3 km measurements. These measurements are further averaged on-ground to observations, representing horizontal averaging on a ~87 km scale. The vertical resolution of the layer-average winds vary from 0.25 to 2 km, and can be adapted

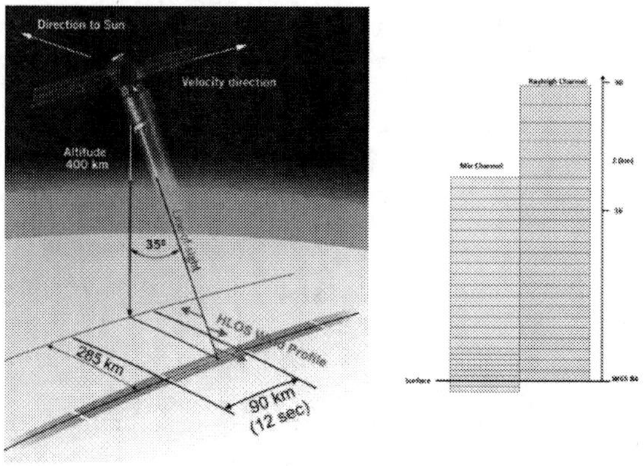

Figure 1: The Aeolus orbit, pointing and sampling characteristics (left) and an example of the vertical measurement binning (right).

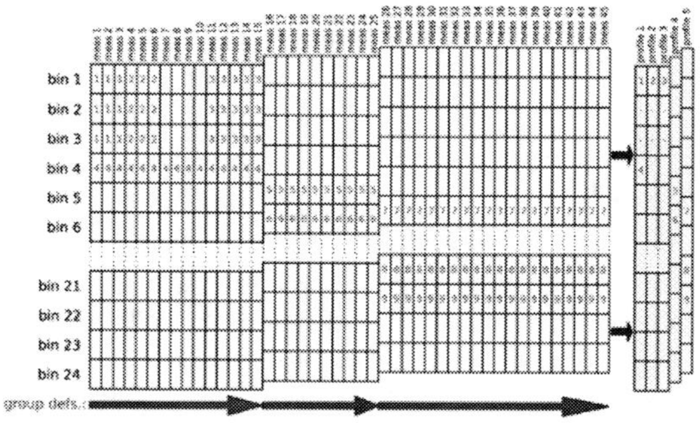

Figure 2: Schematic view of the Aeolus L2b wind observation processing. The number indicate different scene classifications (clouds versus no clouds), resulting in a number of wind profiles for an observation. These are partial or full wind profiles for the Rayleigh (cloud free) and Mie (cloud or aerosol layer winds) channels. Courtesy J. De Kloe (KNMI).

as a function of the under-laying topography and/or climate zone. The required wind accuracies are 2 m/s in the planetary boundary layer (PBL), 2-3 m/s in the free troposphere, and 3-5 m/s in the stratosphere. The wind profiles are classified into cloudy and cloud free profiles, as illustrated in Figure 2. A detailed description of the Aeolus wind profile retrievals can be found in [5].

ADM-Aeolus will also deliver height profiles of Mie and Rayleigh co-polarized backscatter and extinction coefficients, scattering ratios and lidar ratios [6] along the lidar line-of-sight. From these parameters it is possible to derive cloud and aerosol information such as layer height, multi-layer cloud and aerosol stratification, cloud and aerosol optical depths (integrated light-extinction profiles), and some information on cloud/aerosol type (lidar ratio).

2.4 Science activities in support of the user requirements and in-flight operation

During the technical development of the ALADIN laser, changes to the instrument operation strategy had to be implemented in order to meet the strict user requirements on stability and measurement accuracy. The initial strategy was to operate the instrument in so-called burst mode, where the instrument was operated at 100 Hz during 50 km and then switched off for 150 km, once every 200 km. The 200 km observation spacing assured an uncorrelated sampling scheme. A new strategy was introduced in 2010, operating the instrument in so-called continuously pulsed mode, i.e. continuous sampling along the orbit. Due to laser life-time issues and electrical power limitations, the pulse repetition frequency was reduced to 50 Hz which meant that the horizontal observation averaging had to be increased from 50 km to ~100 km to obtain sufficient SNR.

This change in the Aeolus sampling led to changes in the spatial representativity of the observations, and the Aeolus mission requirements had to be revisited to investigate whether the spatial sampling requirements could be relaxed. As a result, impact studies [7-8] were initiated to consolidate and optimize the on-ground data processing for maximum NWP impact. The studies addressed the role of NWP model background error correlations, measurement representativeness errors, and an optimized measurement processing strategy to maximize mission impact. The studies concluded that the current operational version of the ECMWF model represents the atmospheric variability well, down to horizontal scales of about 100 km (the so-called effective model resolution). Since the Aeolus observation sizes are now ~100 km, the associated observation representativeness error in the current ECMWF model would be small because the observation length is comparable to the effective model resolution. This confirmed that the current 100 km horizontal averaging strategy. The studies furthermore showed that horizontally projected line-of-sight wind profiles significantly improves ECMWF forecasts in the upper troposphere and in the tropics, and that unknown biases must be kept small and well below 1 m/s. Especially range-dependent biases are found to be detrimental for NWP. The impact is comparable to the impact of the present radiosonde network. Further recommendations to the sampling strategy were also given.

2.5 Aeolus Mission status

During the Aeolus development phase, a number of technical challenges for the Aeolus system have appeared and have been tackled. These include laser diode performance, optical mounting stability, laser induced damage on the IR and UV coatings, laser-induced contamination in vacuum, and long-term UV energy drift. The transmit lasers and optical bench assembly have been extensively retrofitted and tested, and the Aeolus transmit lasers are at the time of writing being integrated in the ALADIN instrument. The instrument delivery is scheduled for end 2015 and satellite qualification and launch readiness for end 2016. The satellite platform was ready in 2009 and has since then been in storage. The satellite Assembly, Integration and Testing teams are being reactivated. The mission Payload Data Ground Segment and Flight Operations Segment teams have also been reactivated and are preparing and running the final overall validation testing in preparation for launch. The ESA and industrial in-orbit Commissioning planning is under review and finalization.

2.6 Aeolus CAL/VAL preparations

An Announcement of Opportunity (AO) call for CAL/VAL support studies for Aeolus took place in 2007. The AO was a call for validation of the Aeolus wind and optical properties products during its commissioning and successive 3-year exploitation phase. Sixteen proposals including ground-based and airborne validation experiments were received from teams all around the globe, whereof fifteen successfully passed the review. Due to the delays of the Aeolus programme and the recent changes to the Aeolus operation, a delta-call was released in May 2014. The main objectives of the delta-call was (i) to allow confirmation of the fifteen CAL/VAL proposals from the 2007 call and (ii) to call for new proposals. In total 17 proposals were received in response of the delta-call, whereof 2 were joint national proposals and 3 included a number of institutes both national and international. Most of the proposals proposed validation of both the wind and aerosol/cloud products. Activities of special interest included tropical stratospheric balloon experiments and the inclusion of collocated measurements in polar areas. All proposals were accepted.

A Science and CAL/VAL workshop was held in ESA_ESRIN in February 2015 (http://www.aeolus-science-calval-2015.org/). The workshop brought industry, the user community and ESA together to prepare for the Aeolus Commissioning Phase and beyond. During the Workshop presentations of the science, instrument and product status, Commissioning phase planning and the extensive number of proposals submitted in response to the Aeolus CAL/VAL call in 2014 were given. Furthermore, the Aeolus CAL/VAL Implementation Plan was drafted in a plenary session. The workshop allowed the teams to start activity coordination and exchange of information.

A pre-launch campaign with the ALADIN Airborne Demonstrator (A2D) and other wind lidar instrumentation from DLR and NASA in May 2015 will, furthermore, allow for CAL/VAL rehearsal and testing of the ALADIN on-ground data processing algorithms.

3. CONCLUSIONS

With the recent start of the ALADIN instrument integration and delivery in 2015, the Aeolus Mission is nearing its completion for launch. ESA, industry and the scientific community is getting prepared for Commissioning phase, CAL/VAL and product exploitation.

REFERENCES

[1] European Space Agency, (1998) The Science and Research Elements of ESA's Living Planet Programme. ESA SP-1227, 105 pp.

[2] World Meteorological Organisation (WMO) (2004) Proc. of the 'Third WMO Workshop on the Impact of various observing systems on Numerical Weather Prediction'. WMO/TD No. 1228, 329 pp.

[3] European Space Agency, (2008) ADM-Aeolus Science Report. ESA SP-1311, 121 pp.

[4] Marksteiner, U., Airborne Lidar Observations for the validation of the ADM-Aeolus instrument, PhD Thesis and the Technical University of Munich, http://mediatum.ub.tum.de/doc/1136781/1136781.pdf.

[5] Tan, D. G. H. *et al.,* (2008) The ADM-Aeolus wind retrieval algorithms. Tellus, **60A**, pp. 191-205.

[6] Flamant, P. *et al.,* (2008) ADM-Aeolus retrieval algorithms for aerosol and cloud products. Tellus, **60A**, pp. 273–286.

[7] Horanyi, A. *et al. (2014)* The assimilation of horizontal line-of-sight wind information into the ECMWF data assimilation and forecasting system, Part I: The assessment of wind impact, Q. J. R. Meteorol. Soc. (2014) DOI:10.1002/qj.2430.

[8] Marseille, G.-J., *et al.* (2013), Final Report Vertical and Horizontal Aeolus Measurement Positioning, CCN2 to ESA contract 20940, http://www.knmi.nl/~marseill/publications/fulltexts/AE_FR_VHAMP_v1.0.pdf.

THE ATHENA-OAWL DOPPLER WIND LIDAR MISSION

Sara C. Tucker[1]*, Carl Weimer[1], R. Michael Hardesty[2]

*[1]Ball Aerospace and Technologies Corp., Boulder, Colorado, USA, *Email: stucker@ball.com,*
[2]University of Colorado, Cooperative Institute for Research in Environmental Science, Boulder, Colorado, USA

ABSTRACT

With the objective of providing tropospheric wind profile data over the mid-latitude oceans and tropics for data-starved weather forecast models, the Earth Venture Instrument (EV-I) Mission concept "Atmospheric Transport, Hurricanes, and Extratropical Numerical weAther prediction with the Optical Autocovariance Wind Lidar" (ATHENA-OAWL) was proposed in November 2013. The mission concept is described here along with a brief history of the OAWL system development and current development of an ATHENA-OAWL airborne demonstrator under NASA's Venture Technology development.

1. INTRODUCTION

The need for and feasibility of putting a Doppler Wind Lidar (DWL) into space for global wind monitoring have been covered in multiple works, spanning four decades, from Huffaker et al. 1978[1] to Baker et al 2014[2] and Atlas et al. 2015 [3]. To date, a DWL has not yet flown in space, although ESA's Aeolus mission promises to launch in late 2016. [4][5] In the US, however, technological readiness, uncertainty about what approach would provide the most impact on weather forecasting, and projected cost have attenuated space-based Doppler wind lidar system developments.

Starting in 2003, Ball Aerospace (Ball) has been developing a direct-detection method for remote measurement of winds based on the Optical Autocovariance (OA) approach first outlined in 1995 by Schwiesow and Mayor. [6] The Ball OAWL development efforts overlapped with, and still benefit from, the design and build of the CALIOP lidar at Ball for the CALIPSO mission, still in operation today. In 2007, Ball demonstrated a proof-of-concept, 3-phase OA receiver measuring aerosol-backscattered Doppler shifts verified by a sonic anemometer [7]. From 2009-2012, Ball, with additional support from NASA's Earth Science Technology Office (ESTO) built a full breadboard version of the four-phase Optical Autocovariance Wind Lidar (OAWL) and demonstrated and validated the system in ground and preliminary airborne, autonomous, flight tests.[8] Subsequent funding from ESTO and internal Ball investments led to additional development of the system, mission concepts, and space-based feasibility studies including: 1) an Instrument Design Lab (IDL) run at NASA Goddard to evaluate the flight readiness and provide parametric cost estimates of a preliminary concept and 2) OAWL-based Observing System Simulation Experiments (OSSEs, teaming with NOAA, NASA, JCSDA and Simpson Weather Associates).

2. METHODOLOGY

2.1 Optical Autocovariance

The Optical Autocovariance approach to measuring Doppler shifts works similar to a Michelson interferometer (which measures the autocovariance of a signal with itself). Rather than moving an arm of a Michelson interferometer to map out multiple fringes, however, we modify a Mach Zehnder interferometer using polarization to measure four points on a single fringe simultaneously. This quadrature Mach Zehnder direct detection approach [9][10][11] is enhanced with a reflective cats-eye design [(US Patents US7929215B1, US8077294B1] to provide field widening for use with large telescopes without costly wavefront quality requirements. Aside from minor coating losses, the reflective system is 100% transmissive, ensuring there are no return signal losses from filtering (such as observed with double-edge Fabry-Perot systems).

Through digitization and processing, the phases of Doppler shifted backscatter laser returns are estimated and compared to the estimated phase of the outgoing illumination. The corresponding phase shifts (frequency shift) linearly correspond to the line of sight wind measurements. The relatively-long 0.9 m optical path difference

(OPD) between the two arms of the interferometer in the original OAWL system, and the proposed ATHENA-OAWL mission provides increased sensitivity to Doppler shifts from narrow-band aerosol backscatter signals, enabling ground-based line-of-sight wind speed precision values of <1 m/s, similar to coherent detection system performance, but with the added benefit of being able to use photon-counting detection. A similar system operating at a shorter OPD [10] would be able to resolve Doppler shift measurements in the relatively broad-band molecular backscatter with the typical 2-3 m/s precision.

With the correct reflective coatings, the OAWL system may be built to operate at any laser wavelength. As shown in Figure 1, the existing OAWL system has demonstrated simultaneous wind measurement operation at both the 355 nm and 532 nm laser wavelengths.

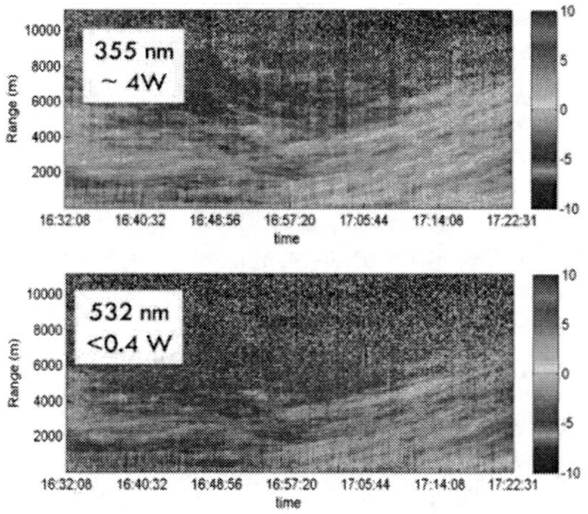

Figure 1. An example of the OAWL LOS wind speed measurements acquired simultaneously at both the 355 and 532 nm wavelengths. The color bar has units of m/s

2.2 The ATHENA-OAWL Mission

Building on the years of system development and system studies, Ball together with the University of Maryland, worked mission formulation and engineering designs for the ATHENA-OAWL mission concept, proposed to NASA's Earth Venture Instrument – 2 (EVI-2) Announcement of Opportunity in 2013.

ATHENA-OAWL is a focused science mission to measure tropospheric winds along two lines of sight, and to provide concurrent information about tropospheric aerosols. The ATHENA-OAWL Science Objectives are to:

1. Create a breakthrough in prediction and understanding of low and mid-latitude weather and climate via improved sampling of wind profiles.
2. Improve our understanding of the effects of the interaction between aerosol radiative forcing and atmospheric dynamics on the genesis and lifecycle of tropical cyclones via collocated aerosol and wind measurements in the tropic.
3. Improve our understanding of the impact of long-range dust transport and aerosol transport on global energy and water cycles, air quality, and climate via collocated aerosol and wind measurements

Based on the ISS orbit (Figure 2), ATHENA-OAWL would provide quasi-uniform sampling over >80% of the Earth's surface, focusing on the low and middle latitudes where scientific demand for additional wind profiles is strongest. Based on conservative aerosol loading profiles, measurements are expected throughout the troposphere and lower stratosphere, with sufficient returns to provide space-based line of sight wind speed estimates with < 3m/s precision. This precision and coverage meets WMO requirements for Numerical Weather Prediction (NWP) and climate reanalysis applications.

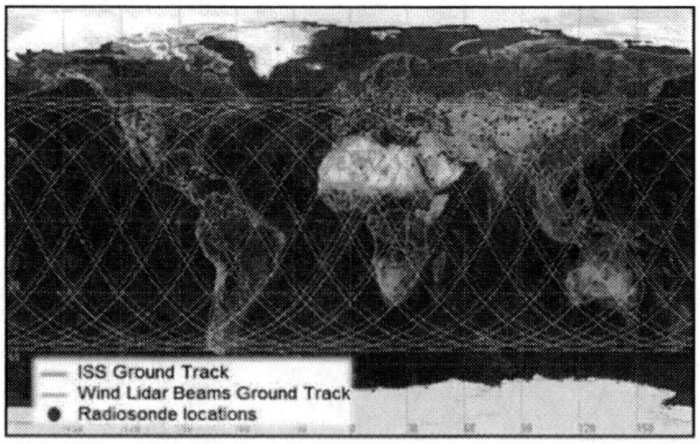

Figure 2. Current radiosonde launch locations (2 launches per day, at locations of red dots) compared to 24 hours of ISS orbit track.

By having two fixed, azimuthally-orthogonal looks at 40 degrees off nadir, the system can provide information about both components of the horizontal wind, reducing uncertainty in assimilation to NWP. Because the EVI mission opportunities are cost constrained, the team took a design to cost approach: build the system with the most science impact possible within the \$94 M cost cap (including 25% reserves). This mission approach is made possible in part by using the International Space Station (ISS), its Mission Operations, and the TDRSS downlink. ATHENA-OAWL can be a medium risk mission within the cost cap by leveraging heritage designs from CALIPSO and other space-based instruments built at Ball. The mission lifetime is 1 year (with a two year goal) and follows NASA Class C (Medium to Low Complexity) requirements.

2.3 Technological Readiness

The two critical subsystems for ATHENA-OAWL that require advanced development and space-qualification are the OA interferometer, and the narrowband 532 nm laser.

Interferometer Receiver: Prior to the proposal submission, Ball developed designs for an athermal version of the OAWL interferometer, building on heritage athermal optical designs that have already flown in space to ensure that orbital variations in temperature would not affect the interferometer alignment.

Laser: Fibertek Inc. has developed a "High Energy UV Demonstrator (HEUVD)" laser under a NASA SBIR Phase 3, and is now in final assembly. Prior to installation of the final tripling stage (used to generate the 355 nm UV light) this system first generates 532 nm light, allowing the system to be used for 532 nm testing and space-qualification. The output at 532 nm is 170 mJ at 150 Hz pulse repetition frequency for 26 W average optical power (compared to CALIPSO's 4.4 W split between 532/1064 nm). These energy levels were chosen such that, when combined with ATHENA-OAWL's beam divergence, the system achieves the same eye-safety levels at the Earth surface from ISS as achieved for CALIPSO from its higher orbit.

The laser's opto-mechanical design is traceable to CALIPSO (and to the recent ICESat II build). The contamination control and de-rating approach were developed and validated on CALIPSO which has completed over 8 Billion pulses and 9 years of continuous operation.

3. RESULTS AND ONGOING EFFORT

The ATHENA-OAWL mission was not selected for the 2013 EVI-2 award. Two awards were made: Global Ecosystem Dynamics Investigation GEDI (Univ. of Maryland), and the Ecosystem Spaceborne Thermal Radiometer Experiment on Space Station ECOSTRESS (JPL). ATHENA-OAWL was rated Category 3, however, and subsequently selected for risk reduction, Venture Technology development funding to advance its Technological Readiness Level (TRL). Under the current development effort, ATHENA-OAWL Venture-Tech (AOVT), Ball, in coordination with scientists at the University of Colorado, will demonstrate and provide validation for a two–look, 532 nm wavelength "Green-OAWL" (GrOAWL) airborne system. Under AOVT, Ball will upgrade existing OAWL instrument to match the ATHENA-OAWL two-look, two laser configuration and then perform new aircraft testing on the NASA WB-57 high altitude aircraft.

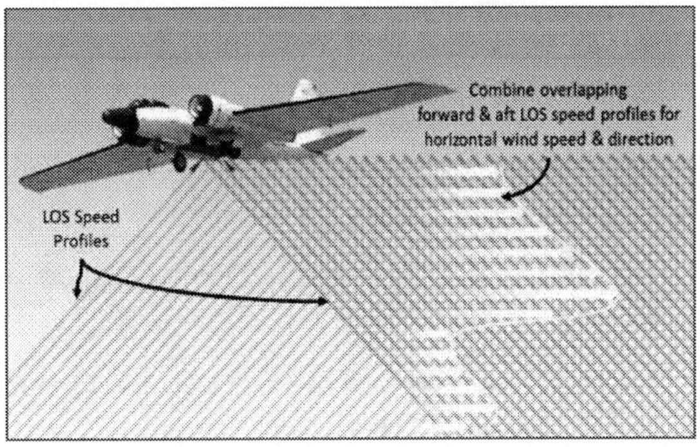

Figure 3. Example of the "GrOAWL" two-look aircraft flight configurations for measuring winds at two azimuthally separated angles. Forward beams are shown in green and aft facing beams are shown in blue. Due to the system geometry, beams have constant overlap enabling variable processing.

The system is currently in the design/build phase in preparation for demonstration flights scheduled for spring 2016. Balloon-sonde and drop sonde comparisons are planned for wind validation for different atmospheric conditions. Validation efforts will also incorporate comparisons with AERONET for aerosol calibration and radiometric performance validation.

The AOVT effort will also include space-qualification testing of key new technologies including 532 nm testing of the aforementioned

DOI: 10.1051/epjconf/201611901002

HEUVD laser at Fibertek and thermal/vacuum testing of the new athermal interferometer, to advance their TRLs.

In addition to the AOVT effort, the "HSRL for Aerosols Winds and Clouds" (HAWC-OAWL) IIP, is currently funded by NASA ESTO to develop a dual-wavelength high spectral resolution lidar + winds version of OAWL for atmospheric transport and "Chemical Weather" studies. The dual-wavelength demonstrations and validation effort will take place after the Gr-OAWL test flights.

4. CONCLUSIONS

In an effort to address the need for the missing wind profile information over the tropics and mid-latitudes, the ATHENA-OAWL Mission was proposed to the Earth Venture Instrument 2013 call. While the mission was not awarded, Venture Technology funding is enabling the build of the ATHENA-OAWL airborne GrOAWL system to prove out the two-look OAWL approach for a future space-based Doppler lidar technology demonstration. NASA WB-57 flight tests of the system with a focus on wind and performance validation, will take place in spring 2016.

ACKNOWLEDGEMENT

The authors wish to acknowledge the support of NASA's Earth Science Technology Office Instrument Incubator and Advanced Component Technology programs in making the OAWL demonstrations possible.

REFERENCES

[1] Huffaker, R.M. editor, "Feasibility Study of Satellite-Borne Lidar Global Wind Monitoring System," *NOAA Technical memorandum ERL WPL-37*, September 1978,

[2] Baker, W.E., R. Atlas, C. Cardinali, A. Clement, G. D. Emmitt, B. M. Gentry, R. M. Hardesty, E. Källén, M. J. Kavaya, R. Langland, Z. Ma, M. Masutani, W. McCarty, R. B. Pierce, Z. Pu, L. P. Riishojgaard, J. Ryan, S. Tucker, M. Weissmann, and J. G. Yoe, 2014: Lidar-Measured Wind Profiles: The Missing Link in the Global Observing System. *Bull. Amer. Meteor. Soc.*, **95**, 543–564.

[3] Atlas, R., R. N. Hoffman, Z. Ma, G. D. Emmitt, S. A. Wood, Jr., S. Greco, S. Tucker, L. Bucci, B. Annane, M. Hardesty, and S. Murillo, 2015a: Observing system simulation experiments (OSSEs) to evaluate the potential impact of an optical autocovariance wind lidar (OAWL) on numerical weather prediction. *J. Atmos. Oceanic Technol.*, submitted.

[4] Reitebuch, O., 2012a: Wind lidar for atmospheric research. In Schumann U. (Ed.): Atmospheric Physics – Background, Methods, Trends. Springer Series on Research Topics in Aerospace. ISBN 978- 916 3-642-30182-7, 487–507

[5] Reitebuch, O., 2012b: The space-borne wind lidar mission ADM-Aeolus. In Schumann U. (Ed.): 918 Atmospheric Physics – Background, Methods, Trends. *Springer Series on Research Topics in Aerospace*, 815-827

[6] Schwiesow and Mayor, 1995: Coherent Optical Signal Processing for a Doppler Lidar Using a Michelson Interferometer," Coherent Laser Radar Conference (Keystone, CO), OSA Technical Digest Series 19.

[7] Grund, C., M. Lieber, B. Pierce, M. Stephens, C. Weimer, 2008: Optical Autocovariance Wind Lidar (OAWL) for Efficient Space-Based Direct-Detection High-Resolution Aerosol Backscatter Winds, 24th *International Laser Radar Conference, Boulder, CO., 2008.*

[8] S. C. Tucker, C. Weimer, 2013: Comparing and contrasting the Optical Autocovariance Wind Lidar (OAWL) and coherent detection lidar, Coherent Laser Radar Conference, Barcelona, Spain 17 - 20 June, 2013.

[9] Liu, Z., and T. Kobayashi, 1996: Differential discrimination technique for incoherent Doppler lidar to measure atmospheric wind and backscatter ratio. *Opt. Rev.*, 3, 47–52

[10] Bruneau, D., 2001: Mach-Zehnder interferometer as a spectral analyzer for molecular Doppler wind lidar. *Appl. Opt.,* **40**, 391-399

[11] Bruneau, D., and J. Pelon, 2003: Simultaneous measurements of particle backscattering and extinction coefficients and wind velocity by lidar with a Mach-Zehnder interferometer: Principle of operation and performance assessment. *Appl. Opt.,* **42**, 1101–1114

EPJ Web of Conferences 119, 01003 (2016)
ILRC 27
DOI: 10.1051/epjconf/201611901003

OVERVIEW OF THE EARTHCARE L2 LIDAR RETRIEVAL CHAIN

Gerd-Jan van Zadelhoff[1], D. Donovan[1], U. Wandinger[2], D, Daou[1], S. Horn[2], A. Hunerbein[2], J. Fischer[3], J. von Bismarck[3], F. Filipitsch[3], N. Docter[3], M. Eisinger[4], D. Lajas[4], T. Wehr[4]

*[1] Royal Netherlands Meteorological Institute (KNMI), De Bilt, The Netherlands, *Email: zadelhof@knmi.nl*

[2] Leibniz Institute for Tropospheric Research (TROPOS), Leipzig, Germany

[3] Free University of Berlin (FUB), Institute for Space Science, Berlin, Germany

[4] European Space Agency (ESA), European Space Research and Technology Centre (ESTEC), Noordwijk, The Netherlands

ABSTRACT

In this paper an introduction to the planned L2 retrieval algorithms for the Earth Clouds and Radiation Explorer (EarthCARE) lidar ATLID is given. The ATLID instrument is a high spectral resolution lidar which will provide independent retrievals of extinction and backscatter profiles and will be launched in 2018. A short description of the intended operational ESA products is given together with the logic behind the choices made.

1. INTRODUCTION

The interactions between clouds, aerosols and solar and terrestrial radiation play key roles in the Earth's Climate. As stated in the most recent assessment report of the Intergovernmental Panel on Climate Change, clouds and aerosols are two of the largest causes of uncertainty in climate prediction: aerosols because of their uncertain direct and indirect radiative forcing of climate, and clouds because the way they change in response to a warmer world is one of the more poorly understood climate feedbacks.

It has been recognized that, despite a long history of satellite observations, novel high-quality novel observations are needed for atmospheric model evaluation and process studies. In particular, the importance of true height-resolved global observations of cloud and aerosol properties has been recognized as being essential to making progress. Currently-, elements of the A-Train satellite constellation, namely CloudSAT (carrying a 94 GHz cloud radar; [1]) and CALIPSO (carrying a 532/1064 nm lidar; [2]) continue to produce many exciting results and have demonstrated the value of multi-sensor cloud/aerosol observations including active instruments (i.e. lidar and radar).

EarthCARE[3] is an upcoming ESA/JAXA mission to fly in 2018. The instruments that will be carried onboard EarthCARE are:

- A 94 GHz, Doppler Cloud Radar supplied by Japan (CPR)
- An advanced 355 nm High-Spectral Resolution Lidar (ATLID)
- A Multispectral Cloud/Aerosol imager for narrow-band TOA radiances (MSI)
- A 3-view Broad-Band Long- and Short-Wave Radiometer for TOA radiance (BBR)

EarthCARE may be thought of as selected components of the A-Train. However, whereas CloudSAT and CALIPSO are separate platforms, EarthCARE is a single platform designed from the ground-up in order to maximize the benefit that may be realized by combining the different sensors. By combining the lidar and radar observations with the imager and radiometer radiances the production of a suite of self-consistent synergistic products will be possible.

APRIL is one of the ESA sponsored L2 algorithm development projects. The focus of APRIL is the development and implementation of ATLID and MSI L2 algorithms. Within this work the planned stand-only ATLID L2 algorithms and products are discussed.

2. The ATLID instrument

ATLID is an example of a High Spectral Resolution Lidar [4]. ATLID emits a linearly polarized spectrally narrow beam (FWHM for ATLID is around 30 MHz) and the return is detected by 3 different receiver channels. First the return is separated into components which are either the same (co-polar) or perpendicular (cross-polar) to the transmitted beams plane of polarization. Subsequently the co-polar channel is spectrally separated into 2 channels. The Mie channel is dominated by light scattered by aerosol or cloud particles whereas the Rayleigh channel is dominated by the thermally broadened molecular return from the atmosphere itself

ATLID provides atmospheric echoes with a vertical resolution of about 100 m from ground to an altitude of 20 km and 500m from 20km to 40km altitude and is designed to retrieve profiles of particle extinction and backscatter coefficients, lidar ratio, and linear depolarization ratio as well as the aerosol optical thickness (AOT) at 355 nm.

A Fabry-Pèrot (FP) etalon is used to (imperfectly) spectrally separate the two channels resulting in Rayleigh signals ending up in the Mie channel and vice versa (cross-talk). A cross-talk correction procedure is implemented as part of the L1 processing to ensure that the L2 algorithms are supplied with corrected Rayleigh and Mie signal profiles.

Due to this measurement capability, in contrast with the case of an elastic backscatter lidar, one can retrieve the aerosol/cloud extinction and backscatter profiles independently.

3. ATLID retrievals

The design of the retrieval chain is driven by expected signal-to-noise ratio [SNR] of the ATLID instrument and the EarthCARE requirements [5]. The instrument design will not provide sufficient single-shot SNR to enable an extinction retrieval using the Rayleigh signals directly, instead it is expected that at least 1 km binning is required for the thicker ice clouds and 10-150 km horizontal binning for regions consisting of aerosols only. Three separate (but interdependent) processors, as shown in Figure 1, have been proposed in order to retrieve the atmospheric state. As a first step, a feature mask algorithm (A-FM) has been created to separate regions with particle return from molecular backscatter regions only. This will enable the development of binning strategies to minimize the number of shots to bin to achieve a given SNR. The second processor will provide the best possible profile information (A-Prof) and the third derives the layer information (A-Layer), which is a direct input to the synergistic MSI-ATLID algorithms.

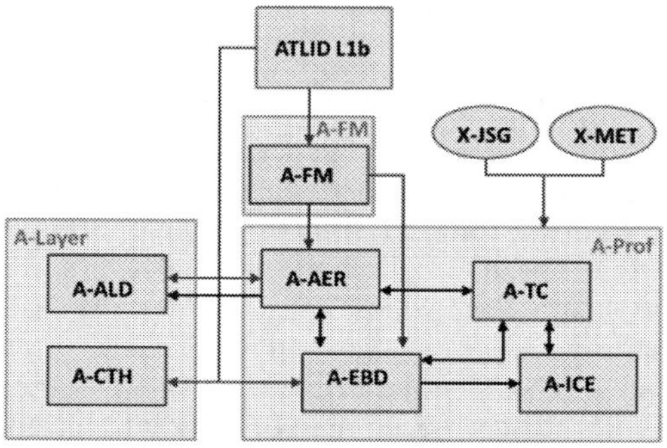

Figure 1:Flow diagram of the ATLID only algorithms. The blue boxes and arrows depict the input to the algorithms. The green boxes show the three ATLID only processors, while the yellow boxes depict the individual algorithms. Black arrows show data streams between algorithms, while green arrows indicate that all products within the processors use the same input stream. The X-MET input reflects the ECMWF meteorological information along the EarthCARE track while X-JSG (joint standard grid) defines the collocated grid, used within the synergistic algorithms, with the horizontal grid defined by the CPR and the vertical grid by ATLID.

The A-FM algorithm returns the probability of particle return, between 0 (molecular) and 10 (particles), detection at native resolution (horizontal approx. 0.2-km and vertical approx.100 m) based on the correlation of the data without focusing on a number of hard coded or input dependent thresholds. For this a combination of image reconstruction techniques has been employed. In Figure 2 an example of the Feature Mask is shown using CALIOP 1064nm data. The 1064nm data is used in preference to 532 nm data since the lower Rayleigh scattering contribution better matches ATLID's Mie channels.

The retrieval of the extinction and backscatter at a horizontal resolution of 1 km will be achieved using an optimal estimation (OE) technique [A-EBD]. This requires a reasonable a-priori input value of the local extinction to backscatter (lidar or S) ratio.

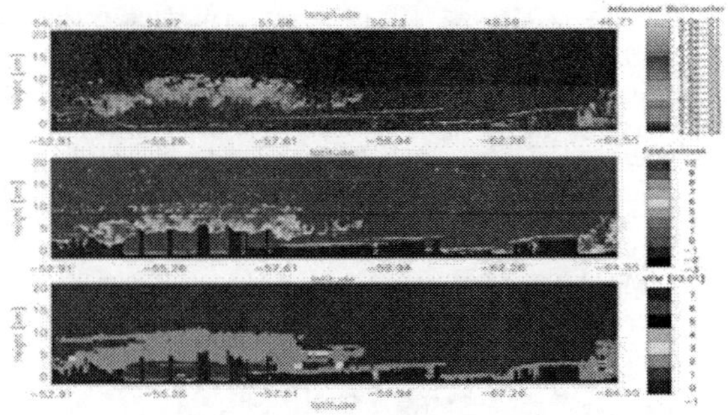

Figure 2: CALIOP night time data on 18-04-2010, the top panel represents the raw 1064nm data, the center panel the ATLID featuremask and the bottom panel the VFM (V3.01) mask.

This is in general, no problem for the ice clouds but S values can be very different for different types of aerosols. In order to deal with this issue, a separate routine will be used to derive the lidar ratio for aerosols at low horizontal resolution [A-AER]. These data will then provide the *a priori* estimate necessary for the high resolution OE procedure. The low-resolution approach will use a conventional direct retrieval using the Rayleigh signals, which requires a horizontal binning between 10 and 150 km. The OE procedure uses these retrieved lidar ratios as an *a priori* input along with both the Rayleigh and Mie channel signals to in order to produce extinction and backscatter profile retrievals at 1 km horizontal scales Both these algorithms require a target classification [A-TC] to enable the correct retrieval and binning strategies. Finally, based on the high resolution extinction, backscatter, depolarization and target classification the higher order retrieval algorithm like the synergistic classification and ice microphysical retrieval can be performed [A-ICE]. In Figure 3, an example is shown for the retrieval of the extinction at 1 km scale using a scene and modeled signals created using the EarthCARE simulator (ECSIM). The retrieved profile information can be used as a direct input to the synergetic lidar+radar retrievals.

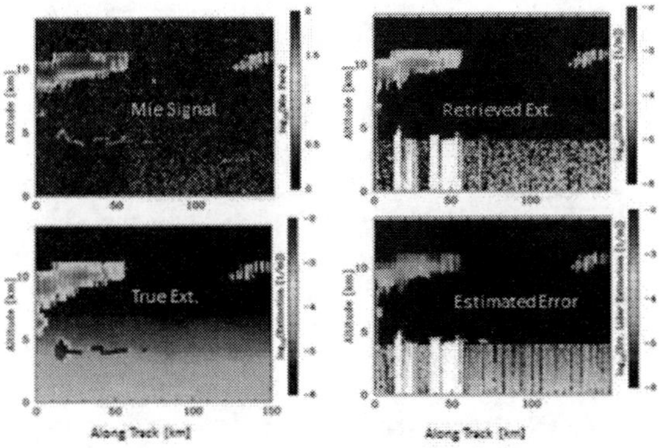

Figure 3: Simulated Mie Signals (top left) with true extinction (bottom left). The two images on the right depict the retrieved extinction and estimated errors at 1km resolution.

The third processor (A-Layer) defines the layer products for clouds and aerosols, respectively. Cloud and aerosol layers appear as regions of increased backscatter signal compared to aerosol-free and cloud-free atmospheric regions. Layers can be identified by gradients in the Mie signal measured by ATLID, or combined gradient/threshold methods. The ATLID stand-alone layer products A-CTH and A-ALD are pre-requisites for the corresponding synergistic layer products. The A-CTH algorithm makes use of a

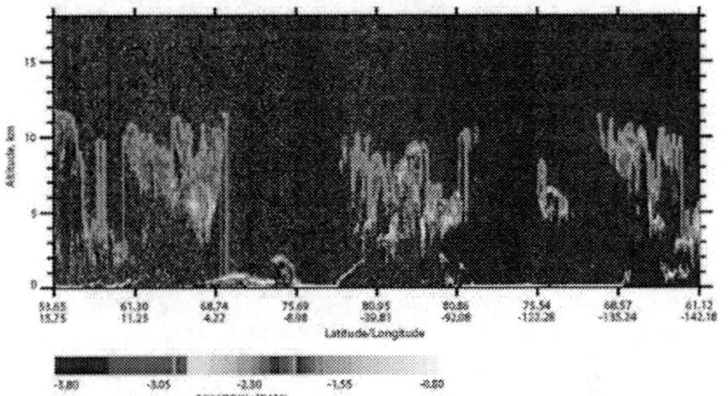

Figure 4: CTH determined with the WCT algorithm (green line) plotted on top of the 1064-nm attenuated backscatter signal along a CALIPSO daytime orbit on 1 August 2006, 1156-1213 UTC.

combined wavelet covariance transform (WCT) and threshold approach (e.g., [6], [7]) to determine the uppermost cloud top height along the ATLID track with a resolution of 1 and 10+ km. The latter, coarser resolution is needed to

 DOI: 10.1051/epjconf/201611901003

identify optically thin clouds, e.g., subvisible cirrus. An example of the CTH is shown in Figure 4 using CALIOP 1064nm data.

For the determination of aerosol layer boundaries a WCT algorithm with thresholds has been developed. The WCT is the same as for CTH determination, but the thresholds are adapted for aerosols, and a multi-layer search for base and top heights is performed. For each layer the aerosol optical depth and its error are calculated, in addition, the stratospheric AOT and the column AOT, both with errors, are computed.

4. CONCLUSIONS

With the launch of EarthCARE the active remote sensing of clouds and aerosols from space, as started by CloudSAT and Calipso, will continue for at least another three years. Since the precise instrument designs are very different from, the A-TRAIN instruments novel retrieval techniques have been defined and are currently being coded for implementation in the future ground segment. The major difference between CALIOP and ATLID is the possibility to retrieve directly and independently the cloud and aerosol extinction and backscatter profiles.

ACKNOWLEDGEMENT

The work presented in this paper is under development within the project Atmospheric Products from Imager and Lidar (APRIL) funded by ESA under contract 4000112018/14/NL/CT.

REFERENCES

[1] Stephens, G. L., and Coauthors, 2008: CloudSat mission: Performance and early science after the first year of operation. *J. Geophys. Res.*,**113**, *D009982*

[2] Winker, D. M., and Coauthors, 2010: The CALIPSO Mission: A Global 3D View of Aerosols and Clouds. *Bull. Amer. Meteor. Soc.*, **91**, *1211–1229. doi: http://dx.doi.org/10.1175/2010BAMS3009.1.*

[3] Illingworth, A., et al., 2014: THE EARTHCARE SATELLITE: The next step forward in global measurements of clouds, aerosols, precipitation and radiation. *Bull. Am. Met. Soc., doi:10.1175/BAMS-D-12-00227.1.*

[4] A. Hélière, R. Gelsthorpe, L.Le Hors, Y. Toulemont, ATLID, the backscatter lidar on board the EarthCARE satellite, *International Conference on Space Optics, 2012*

[5] EarthCARE Mission Requirement Document Version 5, EC-RS-ESA-SY-012, *ESA/JAXA, 2006*

[6] Chazette, P., J. Pelon, and G. Megie, 2001: Determination by spaceborne backscatter lidar of the structural parameters of atmospheric scattering layers, *Appl. Opt.*, **40**, *3428–3440.*

[7] Brooks, I., 2003: Finding Boundary Layer Top: Application of a wavelet covariance transform to lidar backscatter profiles, *J. Atmos. Ocean. Tech.*, **20**, *1092–1105*

HETEAC: THE AEROSOL CLASSIFICATION MODEL FOR EARTHCARE

Ulla Wandinger[1]*, Holger Baars[1], Ronny Engelmann[1], Anja Hünerbein[1], Stefan Horn[1], Thomas Kanitz[1,4], David Donovan[2], Gerd-Jan van Zadelhoff[2], David Daou[2], Jürgen Fischer[3], Jonas von Bismarck[3], Florian Filipitsch[3], Nicole Docter[3], Michael Eisinger[4], Dulce Lajas[4], Tobias Wehr[4]

[1]*Leibniz Institute for Tropospheric Research (TROPOS), Leipzig, Germany, *Email: ulla@tropos.de*

[2] *Royal Netherlands Meteorological Institute (KNMI), De Bilt, The Netherlands*

[3]*Free University of Berlin (FUB), Institute for Space Science, Berlin, Germany*

[4]*European Space Agency (ESA), European Space Research and Technology Centre (ESTEC), Noordwijk, The Netherlands*

ABSTRACT

We introduce the Hybrid End-To-End Aerosol Classification (HETEAC) model for the upcoming EarthCARE mission. The model serves as the common baseline for development, evaluation, and implementation of EarthCARE algorithms. It shall ensure the consistency of different aerosol products from the multi-instrument platform as well as facilitate the conform specification of broad-band optical properties necessary for the EarthCARE radiative closure efforts. The hybrid approach ensures the theoretical description of aerosol microphysics consistent with the optical properties of various aerosol types known from observations. The end-to-end model permits the uniform representation of aerosol types in terms of microphysical, optical and radiative properties.

1. INTRODUCTION

The Earth Clouds, Aerosols and Radiation Explorer (EarthCARE) is a joint ESA/JAXA mission planned to be launched in 2018 [1]. The multi-sensor platform comprises a cloud-profiling radar (CPR), a high-spectral-resolution cloud/aerosol lidar (ATLID), a cloud/aerosol multi-spectral imager (MSI), and a three-view broad-band radiometer (BBR). Three out of the four instruments (ATLID, MSI, and BBR) will be able to sense the global aerosol distribution and contribute to the overarching EarthCARE goals of sensor synergy and radiation closure with respect to aerosols.

The high-spectral-resolution lidar ATLID obtains profiles of particle extinction and backscatter coefficients, lidar ratio, and linear depolarization ratio as well as the aerosol optical thickness (AOT) at 355 nm along the track of the satellite. The MSI provides AOT at 670 nm (over land and ocean) and 865 nm (over ocean) across a 150-km wide swath. From combined ATLID and MSI data, the columnar Ångström exponent for the 355/670/865-nm spectral range can be inferred along track.

MSI observations will also be used to extend the two-dimensional cross-sections from lidar and radar into the three-dimensional domain, which then will serve as input to broadband radiative-transfer models [2]. In this way, fluxes, heating rates, and radiances can be calculated and top-of-atmosphere (TOA) radiances and fluxes compared to those derived from BBR measurements. The aim is to reach closure of measured and calculated TOA fluxes over each 10 km by 10 km scene with an accuracy of 10 Wm^{-2}, with the final goal to substantially decrease the uncertainties in our knowledge of global radiative forcing.

With respect to aerosols, the closure studies require a proper aerosol classification so that microphysical properties such as particle size (effective radius) or scattering and absorption efficiencies (single-scattering albedo) can be assigned to predefined aerosol types, which in turn can be identified from the measurements of their optical properties. Aerosol classification from space-borne observations is also required for quantification of anthropogenic versus natural aerosol loadings of the atmosphere, investigation of aerosol-cloud interaction, as well as assimilation purposes and validation of atmospheric transport models which carry components like dust, sea salt, smoke and pollution.

In order to facilitate a common aerosol classification and the consistency of all EarthCARE aerosol products, the Hybrid End-to-End Aerosol Classification (HETEAC) model is being developed. The model is based on a combined experimental and theoretical, i.e. hybrid, approach and allows the end-to-end simulation of aerosol properties, from microphysical to optical and radiative parameters of predefined aerosol types. In the following, we describe the basic considerations and the current development steps in more detail. Section 2 gives an overview of the classification approach. Sec. 3 shows the experimental basis for the EarthCARE typing scheme, and Sec. 4 presents a first attempt of the simulation of microphysical properties. An outlook on the next steps of HETEAC developments is given in Sec. 5.

2. AEROSOL CLASSIFICATION SCHEMES

Different approaches for aerosol classification are available in the literature. For HETEAC, the heritage of previous attempts is considered, while at the same time the specific observing capabilities of the mission are taken into account.

Previous aerosol classification schemes rely on experimental findings from passive and active remote sensing and use specific "optical footprints" primarily based on intensive, i.e. concentration-independent, optical properties but also on aerosol load, geographic location or altitude of occurrence. Such schemes have been developed in the context of the CALIPSO mission [3, 4], derived from dedicated lidar field studies [5, 6] or retrieved from passive remote-sensing observations [7].

Usually, the classification schemes distinguish three groups of aerosol types: 1) so-called pure types like marine, smoke, pollution or dust aerosols, 2) mixtures of the former types, and 3) aerosol types that occur only in specific locations like the polar region or the stratosphere. A common understanding exists to define pure types of desert dust, marine aerosol and anthropogenic pollution. These types can usually be well discriminated due to their well-defined and clearly different optical appearance. Biomass-burning aerosol or smoke is an important player as well, but differently treated due to its variable nature. The optical and microphysical properties of smoke can be quite different depending on the kind of burnt material (e.g., Savannah or boreal fires, crop burning), the kind of burning (smoldering or flaming) and the time of transport (atmospheric processing of particles). There are also different ways to account for aerosol mixtures, e.g., mixtures of dust with biomass-burning, marine or pollution aerosols or mixtures of marine and pollution aerosols.

In the CALIPSO retrievals six aerosol types, i.e. clean marine, clean continental, polluted continental, dust, polluted dust, and biomass-burning smoke, are distinguished. CALIPSO aerosol typing relies on lidar level-1 data, since the methodology has been developed primarily to select proper lidar ratios for level-2 data retrievals. Thus, selection criteria comprise integrated attenuated backscatter, approximated depolarization ratio, and vertical location of the layer as well as the kind of surface above which the observation is made.

The EarthCARE aerosol classification scheme shall preserve the aerosol types of CALIPSO as far as possible in order to allow long-term global investigations over the lifetime of both missions. However, a more robust typing based on level-2 data will be applied. Due to the high-spectral-resolution lidar approach, ATLID retrievals do not require an *a priori* estimate of the particle lidar ratio, but provide this quantity together with the particle linear depolarization ratio as an observable. EarthCARE's aerosol-type product can thus rely on measured intensive, i.e. concentration-independent, particle properties. ATLID observations are performed at 355 nm. Up to now, there is no specific aerosol typing scheme available that considers a classification based on measurements at this wavelength. Therefore, as a first step, the experimental basis of aerosol typing at 355 nm must be established.

3. EXPERIMENTAL BASIS

Figure 1 shows a collection of ground-based observations of lidar ratio and particle linear depolarization ratio at 355 nm from several field campaigns in different locations of the world. The different aerosol types (dust, smoke, pollution, marine aerosol, volcanic ash) and mixtures of them spread in this 2D-representation of intensive particle properties that will be measured with ATLID. It is obvious that a clear distinction between small, absorbing particles (high lidar ratio, small depolarization ratio = smoke or pollution), large spherical particles (small lidar ratio, small depolarization ratio = marine aerosol), and large non-spherical particles (moderate to high lidar ratio, high depolarization ratio = dust or ash) is possible. Mixtures of these types appear with intermediate values of lidar ratio and depolarization ratio. Thus, some estimate of the mixing state will be possible as well. Based on these findings, the next step is to find appropriate microphysical descriptions for the aerosol types that are consistent with the observations.

4. SIMULATION OF MICROPHYSICAL PROPERTIES

Starting with a very simple approach for the microphysical description, we assign a mono-modal particle size distribution and a (wavelength-dependent) complex refractive index to well-known basic aerosol components. Here, we follow the attempt of the aerosol project within the ESA Climate Change Initiative (Aerosol_cci) which focusses on the investigation of retrieval algorithms and products from passive satellite remote sensing. For the retrieval experiments an aerosol model based on four components is used [8]. These components comprise two fine modes, one strongly absorbing and one weakly absorbing, and two coarse modes, one with spherical particles representing sea salt and one with non-spherical particles representing dust. The mode radii and refractive

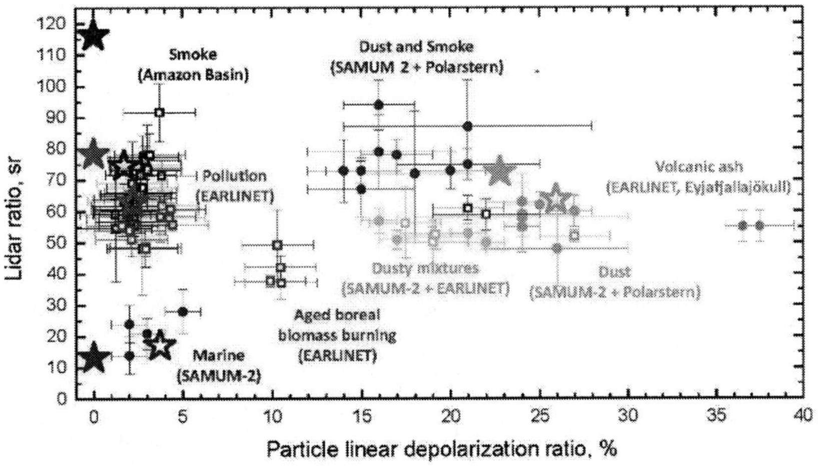

Figure 1: Ground-based observations and simulations of lidar ratio versus particle linear depolarization ratio at the ATLID wavelength of 355 nm. Measurements were performed with the Raman-polarization lidars POLIS (University of Munich, dots) and Polly^{XT} (Leibniz Institute for Tropospheric Research, open squares) during SAMUM-2 at Cape Verde [9], at the EARLINET stations of Leipzig and Munich, Germany [10], in the Amazon Basin [11], and onboard Polarstern over the North Atlantic [12]. Simulations were performed using the components of Aerosol_cci (filled stars) and variations with different refractive index and shape distribution (open stars).

indexes are typical values obtained from long-term AERONET observations (see Table 1).

Table 1: Microphysical properties of the components of the Aerosol_cci model and their optical properties at 355 nm. Italic numbers represent a variation that reproduces observational findings for Saharan dust, marine aerosol, pollution, and smoke (r_{eff}–effective radius, m_R–refractive index, real part, m_I–refractive index, imaginary part, S–lidar ratio, δ–particle linear depolarization ratio, ω_0–single-scattering albedo).

	Coarse mode: Dust	Coarse mode: Sea salt	Fine mode: Weakly absorbing	Fine mode: Strongly absorbing
r_{eff}, µm	1.94	1.94	0.14	0.14
m_R	1.56 *1.53*	1.40 *1.36*	1.40 *1.45*	1.50 *1.55*
m_I	0.005 *0.003*	0.0	0.003	0.043 *0.030*
Shape	Dubovik's spheroid distribution	Spherical *10% prolates with axis ratio 1.1*	Spherical *100% prolates with axis ratio 1.1*	Spherical *100% prolates with axis ratio 1.1*
S, sr	72.5, *63.7*	13.3 *17.1*	78.3 *63.0*	116.9 *74.6*
δ, %	22.8 *26.0*	0.0 *3.7*	0.0 *2.2*	0.0 *1.7*
ω_0	0.81 *0.87*	1.00 *1.00*	0.98 *0.98*	0.81 *0.86*

In order to describe the scattering of non-spherical particles, we apply the spheroid model and the spheroid shape distribution (mixture of prolates and oblates with axis ratios between 0.3 and 3) introduced by Dubovik et al. [13]. Figure 1 shows the values of lidar ratio and linear depolarization ratio calculated for the Aerosol_cci components at 355 nm (filled stars). It can be seen that these components already relatively well represent the properties of Saharan dust (non-spherical coarse mode), sea salt (spherical coarse mode), pollution (weakly absorbing fine mode) and smoke (strongly absorbing fine mode). Some variation of the refractive-index values and the content of non-spherical particles (see italic numbers in Table 1) shifts the theoretically calculated values close to the experimental ones (see open stars in Fig. 1).

5. NEXT STEPS

Starting from these promising results, the microphysical model will be further refined. In particular, we will combine the Aerosol_cci approach with the Optical Properties of Aerosols and Clouds (OPAC) database [14, 15]. OPAC allows the construction of aerosol types from a number of basic aerosol components with well-defined microphysical properties under consideration of hygroscopic particle growth, and it provides a comprehensive collection of refractive indexes of basic aerosol components over a wide wavelength range (0.25–40 µm). The latter is needed for the radiative-transfer calculations and the closure studies with BBR. In the new OPAC 4.0, three dust components (nucleation, accumulation, and coarse mode) are contained which are made up of non-spherical scatterers of prolate shape. A size-dependent axis-ratio distribution following experimental results obtained by Kandler et al. [16] is considered. Our first investigations show that this shape distribution better represents the experimental data than the one proposed by Dubovik et al. [13].

The experimental basis will be extended as well in order to consider the spectral behavior of extinction, backscatter and depolarization. The corresponding Ångström exponents are taken from multi-wavelength Raman lidar observations. Respective information for many of the observations shown in Fig. 1 is already available and more experimental data will be included. The consistency of the aerosol model over the entire ATLID/MSI wavelength range can be assured in this way. The HETEAC pure and mixed aerosol types shall finally be constructed from bi-modal size distributions representing the contribution of fine and coarse particles of different complex refractive index and with different shape distributions.

ACKNOWLEDGEMENT

The work presented in this paper is performed within the project Atmospheric Products from Imager and Lidar (APRIL) funded by ESA under contract 4000112018/14/NL/CT.

REFERENCES

[1] Illingworth, A., et al., 2014: THE EARTHCARE SATELLITE: The next step forward in global measurements of clouds, aerosols, precipitation and radiation. *Bull. Am. Met. Soc.*, doi:10.1175/BAMS-D-12-00227.1.

[2] Barker, H. W., et al., 2011: A 3D cloud-construction algorithm for the EarthCARE satellite mission. *Q. J. R. Meteorol. Soc.*, **137**, 1042–1058, doi: 10.1002/qj.824.

[3] Omar, A. H., et al., 2005: Development of global aerosol models using cluster analysis of Aerosol Robotic Network (AERONET) measurements, *J. Geophys. Res.*, **110**, D10S14, doi:10.1029/2004JD 004874.

[4] Omar, A. H., et al., 2009: The CALIPSO automated aerosol classification and lidar ratio selection algorithm. *J. Atmos. Oceanic Technol.*, **26**, 1994–2014. doi:10.1175/2009JTECHA1231.1.

[5] Burton, S. P., et al., 2012: Aerosol classification using airborne High Spectral Resolution Lidar measurements – methodology and examples. *Atmos. Meas. Tech.*, **5**, 73–98, doi:10.5194/amt-5-73-2012.

[6] Groß, S., et al., 2014: Towards an aerosol classification scheme for future EarthCARE lidar observations and implications for research needs. *Atmosph. Sci. Lett.*, **16**, 77–82, doi: 10.1002/asl2.524.

[7] Russell, P. B., et al., 2014: A multiparameter aerosol classification method and its application to retrievals from spaceborne polarimetry. *J. Geophys. Res. Atmos.*, **119**, 9838–9863, doi:10.1002/2013JD 021411.

[8] Holzer-Popp, T., et al., 2013: Aerosol retrieval experiments in the ESA Aerosol_cci project, *Atmos. Meas. Tech.*, **6**, 1919–1957, doi:10.5194/amt-6-1919-2013.

[9] Groß, S., et al., 2011: Characterization of Saharan dust, marine aerosols and mixtures of biomass-burning aerosols and dust by means of multi-wavelength depolarization and Raman lidar measurements during SAMUM 2. *Tellus B*, **63**, 706–724, doi:10.1111/j.1600-0889.2011.00556.x.

[10] Groß, S., et al., 2012: Dual-wavelength linear depolarization ratio of volcanic aerosols: Lidar measurements of the Eyjafjallajökull plume over Maisach, Germany. *Atm. Env.*, **48**, 85–96, doi:10.1016/j.atmosenv.2011.06.017.

[11] Baars, H., et al., 2012: Aerosol profiling with lidar in the Amazon Basin during the wet and dry season. *J. Geophys. Res.*, **117**, D21201, doi:10.1029/2012JD 018338.

[12] Kanitz, T., et al., 2013: North-south cross sections of the vertical aerosol distribution over the Atlantic Ocean from multiwavelength Raman/polarization lidar during Polarstern cruises. *J. Geophys. Res. Atmos.*, **118**, 2643–2655, doi:10.1002/jgrd.50273.

[13] Dubovik, O., et al., 2006: Application of spheroid models to account for aerosol particle nonsphericity in remote sensing of desert dust. *J. Geophys. Res.*, **111**, D11208, doi:10.1029/2005JD006619.

[14] Hess, M., et al., 1998a: Optical properties of aerosols and clouds: The software package OPAC. *Bull. Am. Met. Soc.*, **79**, 831–844.

[15] Koepke, P., et al., 2015: Technical Note: Optical properties of desert dust with non-spherical particles: data incorporated to OPAC. *Atmos. Chem. Phys. Discuss.*, **15**, 3995–4023, doi:10.5194/acpd-15-3995-2015.

[16] Kandler, K., et al., 2009: Size distributions, mass concentrations, chemical and mineralogical composition, and derived optical parameters of the boundary layer aerosol at Tinfou, Morocco, during SAMUM 2006, *Tellus B*, **61**, 32–50, doi:10.1111/j.1600-0889.2008.00385.x.

EPJ Web of Conferences 119, 01005 (2016) DOI: 10.1051/epjconf/201611901005
ILRC 27

SIMULATION OF CLOUD-AEROSOL LIDAR WITH ORTHOGONAL POLARIZATION (CALIOP) ATTENUATED BACKSCATTER PROFILES USING THE GLOBAL MODEL OF AEROSOL PROCESSES (GLOMAP)

Stuart Young[1]*, Martin Cope[1], Sunhee Lee[1], Kathryn Emmerson[1], Matthew Woodhouse[1], Nicolas Bellouin[2]

[1]*CSIRO Oceans & Atmosphere, Aspendale, VIC 3195, Australia, *Email: stuart.young@csiro.au*

[2]*Department of Meteorology, University of Reading, Reading, RG6 6BB, UK*

ABSTRACT

To permit the calculation of the radiative effects of atmospheric aerosols, we have linked our aerosol-chemical transport model (CTM-GLOMAP) to a new radiation module (UKCA-RADAER). In order to help assess and improve the accuracy of the radiation code, in particular the height dependence of the predicted scattering, we have developed a module that simulates attenuated backscatter (ABS) profiles that would be measured by the satellite-borne Cloud-Aerosol Lidar with Orthogonal Polarization (CALIOP) if it were to sample an atmosphere with the same aerosol loading as predicted by the CTM. Initial results of our comparisons of the predicted ABS profiles with actual CALIOP data are encouraging but some differences are noted, particularly in marine boundary layers where the scattering is currently under-predicted and in dust layers where it is often over-predicted. The sources of these differences are being investigated.

1. INTRODUCTION

Atmospheric aerosol is well known to lead to perturbations in the radiation balance, via direct, and indirect (i.e. changes to cloud properties) effects. Accurate representation of the aerosol size distribution and characteristics such as solubility are essential if models are to simulate correctly the coupling between aerosol loading and radiative forcing.

Because the vertical distribution of aerosol layers in the atmosphere, in addition to their optical properties, can have a significant effect on their radiative impact, we consider it necessary to assess the accuracy of our model's predictions of those vertical profiles. As we apply our model over scales ranging from continental to global, we need to compare with vertically resolved measurements made on such scales. An ideal source of such data is the CALIOP lidar on board the CALIPSO satellite [1]. While it is common to compare model predictions with measurements of such quantities as aerosol optical depth or particulate extinction coefficient, it should be remembered that CALIOP is an elastic-backscatter lidar and does not measure either of these quantities directly. Rather it retrieves these quantities from its primary measurement, the attenuated backscatter coefficient, via a combination of analysis algorithms [1,2] and an optical model of the scattering properties of the aerosol (or cloud) particles. The optical model can only describe the average properties of an identified aerosol type. A misidentification of the aerosol type, or natural variation of the optical properties from the average values, will lead to errors, of various magnitudes, in the retrieved extinction and optical depths [3]. As a result, a difference between simulated and measured profiles of a certain optical quantity could be the consequence of an incorrect prediction of aerosol concentrations, location or type, or of optical properties on the side of the chemical transport model (CTM-GLOMAP), or of errors in the optical properties used to retrieve the profiles from the measurements. In some cases, this is effectively comparing one model with another, which is neither what is intended nor is it very informative. To overcome these problems, we choose instead to compare the fundamental lidar measured quantity, the attenuated backscatter profile, with simulated profiles derived from the radiation module (UKCA-RADAER) that is used by our CTM. Attenuated backscatter is the product of the sum of the molecular and particulate backscatter coefficients at any height and the molecular and particulate two-way transmittances between that height and the lidar. It is simply the calibrated backscattered power

DOI: 10.1051/epjconf/201611901005

measured by the lidar corrected for the range-squared decrease in signal with distance from the lidar. In this way all of the uncertainties are on one side of the comparison thus permitting unambiguous assessment of the model.

2. METHODOLOGY

(a) The Models

In our modeling system, continental-scale atmospheric transport is modeled using a coupled system that comprises (1) the CSIRO Conformal-Cubic Atmospheric Model [4,5] for simulating weather, (2) a population-based anthropogenic emission inventory with natural primary emission sources such as sea salt, wind-blown dust, plus biogenic and fire emissions, and (3) a chemical transport model (CTM,[6]) for simulating the atmospheric chemical transport and subsequent fate, via wet and dry deposition, of gaseous and particulate species. It has a comprehensive chemistry incorporating the Carbon Bond 5 mechanism [7], the Volatility Basis Set approach [8] for secondary organic aerosols, and the ISORROPIA-II model [9] for secondary inorganic aerosol modeling.

The CTM is coupled with the GLObal Model of Aerosol Processes (GLOMAP, [10]), which is a comprehensive, size-resolving global aerosol model. In this work CTM-GLOMAP employs 7 modes and 5 components. Its boundary and initial conditions for aerosol components and modes are obtained from an instance of GLOMAP running in a nudged global general circulation model. The CTM-GLOMAP domain for the Australia continent has a resolution of 0.5°.

The optical properties such as aerosol optical depth (AOD), and aerosol backscatter and extinction coefficients are calculated at each model grid by the RADAER module via a Mie look-up table for the appropriate sizes and refractive indices. The lidar signals are then simulated, as described below, at each of the 19 model levels, which extend from 0.02 km to 8 km above the surface, with a vertical resolution that varies from 0.02 km near the surface to 2 km at the top of the simulation. The comparison with the measurements is then currently done off line, by extracting simulated profiles from the model output file at each model grid point along the CALIPSO ground track of interest.

(b) The Simulator

In each model grid cell, the model system predicts the aerosol components and their size modes. For each of these aerosol components, the particulate backscatter, β_P, and extinction coefficients, σ_P, are then obtained at the CALIOP wavelengths (532 nm and 1064 nm) via a pre-calculated Mie look-up table. In addition, the atmospheric temperature and pressure profiles, available from the modeled meteorology, can be used to calculate the molecular number density profile and, hence, the backscatter, β_M, and extinction coefficient, σ_M, profiles at the lidar wavelengths. The ozone number density is also available from the model allowing the ozone absorption profile, $\alpha_{O3}(z)$, to be calculated. (Ozone absorption is only significant at the shorter wavelength.) It is then a simple matter to calculate profiles of attenuated (total) backscatter:

$$\beta_T'(z) = [\beta_M(z) + \beta_P(z)]T_M^2(z_S,z)T_P^2(z_S,z),$$

where

$$T_M^2(z_S,z) = \exp\left\{-2\int_{z_S}^{z}[\sigma_M(r) + \alpha_{O_3}(r)]dr\right\},$$

and

$$T_P^2(z_S,z) = \exp\left\{-2\int_{z_S}^{z}\sigma_P(r)dr\right\}$$

are, respectively, the molecular and particulate two-way transmittance profiles. The profile of attenuated scattering ratio, $R'(z)$, is obtained from the attenuated total backscatter coefficient profile by dividing by the profile of attenuated molecular backscatter,

$$\beta_M'(z) = \beta_M(z)T_M^2(z_S,z),$$

to give

$$R'(z) = [1 + \beta_P(z)/\beta_M(z)]T_P^2(z_S,z).$$

To simulate the satellite-borne CALIOP profiles, the calculation is started at the top of the atmosphere (or the satellite altitude z_S), and proceeds down towards the surface in order to account correctly for the increasing signal attenuation with range from the satellite (decreasing height).

(c) CALIOP Attenuated Backscatter Profiles

The CALIOP data used in the comparisons here are the "Total_Attenuated_Backscatter_532" from the CALIPSO, level-1, version 3 files and the "Feature_Classification_Flags" in the corresponding, level-2, vertical feature mask (VFM) files.

Although the aim of this exercise is to compare measured and modeled profiles in an atmosphere containing air molecules and aerosol particles, in practice, the measured atmosphere often also contains clouds, which can complicate the comparison somewhat. To facilitate the comparisons, the measured data were filtered to exclude from the calculated average profiles all points in any single profile below the top of cloud layers that were identified in the corresponding VFM profiles. The filtered profiles were then averaged to produce a horizontal resolution comparable to that of the model in the domain of interest (10°S - 45°S, 110°E to 165°E). The vertical resolution of the measured data was, however, not degraded to that of the model but kept at 30 m over the complete height range.

3. RESULTS

In Figure 1 we present an example of our comparisons. The 532-nm attenuated total backscatter measured as CALIPSO passed over Australia on 7 October 2006 at approximately 1621 UTC is shown as a function of height and latitude in (a) with the simulation in (b). The aerosol subtype as identified by CALIPSO's algorithms is shown in (c). Notable features are a strongly scattering marine boundary layer extending to an altitude of 1 km to the North and 2 km to the South of the continent, and a deep moderately strongly scattering aerosol layer that extends to nearly 4 km altitude at latitudes of 20° and 30° S. Between these latitudes, the strength of the signal decreases markedly to the extent that it cannot be detected by the algorithms and does not appear in the VFM. CALIPSO's algorithms identify these aerosols as being predominantly "polluted dust" (a mixture of dust and smoke) with an indication that the weaker signal recorded between these latitudes is from pure dust.

The model generally correctly reproduces the location and heights of the marine layers, although the southern layer is slightly too low in places. The height of the continental layers is also fairly well reproduced as is the extension of the dust layer above the southern marine layer.

Although the horizontal and vertical locations of the various layers are correctly reproduced in the simulation shown, the magnitudes of the simulated signals are, in some places, rather different from what are observed.

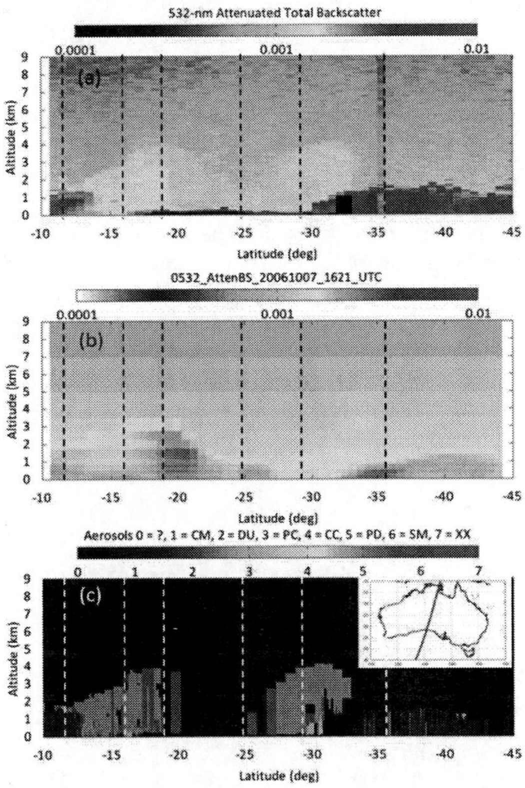

Figure 1. 532-nm Attenuated Total Backscatter (a) CALIOP, (b) Simulation. (c) Aerosol sub-type Vertical Feature Mask. CM = Clean Marine, DU = Dust, PC = Polluted Continental, CC = Clean Continental, PD = Polluted Dust, SM = Smoke, XX = not defined. Vertical, dashed lines indicate locations of profiles in Fig. 2. Inset shows CALIPSO ground track over Australia on 7 October 2006 at ~1621 UTC.

In order to study these differences in more detail, in Figure 2 we compare measured and modeled profiles of attenuated backscatter. The profiles correspond to the locations of the dashed vertical lines in Figure 1. Note that the magnitudes of the simulated profiles are larger than those of the measured profiles in the free troposphere. This is a result of the additional atmospheric attenuation that was measured between the lidar (at 705 km) and the top of the simulation (at ~ 9km) that has not been corrected for in these figures as it helps separate the profiles horizontally and, thereby, improve legibility. In the first profile, at 11.51 ° S, this difference is increased by the attenuation caused by a layer of high cloud above 9 km and not shown here. The comparison of profiles shows the marked under-prediction of scattering in the marine layers in (a) and (f), reasonable agreement in the polluted dust in (b), but less so at (e), and a significant over-estimation of the signal from dust at (c) and (d). The extent to which these differences are attributable to incorrectly predicted number densities, hydration (for the marine layers) or optical properties (e.g. the use of Mie-scattering code for dust) is being investigated and will provide useful information for the development and improvement of our model and radiation code.

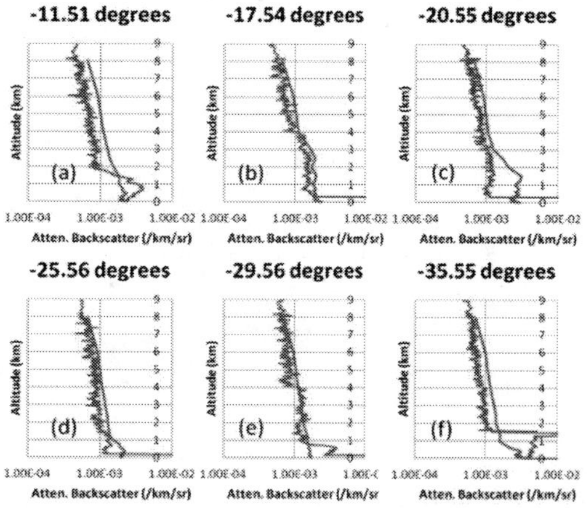

Figure 2. Profiles of 532-nm Attenuated Total Backscatter at the locations indicated by the dashed lines in Fig. 1. The red, noisy profiles are CALIOP data while the blue, smooth profiles are the simulations.

ACKNOWLEDGEMENT

The CALIPSO data were obtained from the NASA Langley Research Center Atmospheric Science Data Center.

REFERENCES

[1] Winker, D. M., et al., 2009: Overview of the CALIPSO Mission and CALIOP Data Processing Algorithms, *J. Atmos. Oceanic Technol.,* **26**, 2310–2323.

[2] Young, S. A., and M. A. Vaughan, 2009: The Retrieval of Profiles of Particulate Extinction from Cloud-Aerosol Lidar Infrared Pathfinder Satellite Observations (CALIPSO) Data: Algorithm Description, *J. Atmos. Oceanic Technol.,* **26**, 1105–1119.

[3] Young, S. A. et al., 2013: The Retrieval of Profiles of Particulate Extinction from Cloud-Aerosol Lidar Infrared Pathfinder Satellite Observations (CALIPSO) Data: Uncertainty and Error Sensitivity Analyses, *J. Atmos. Oceanic Technol.,* **30**, 395–428.

[4] McGregor, J. L., and M. R. Dix, 2001: The CSIRO conformal-cubic atmospheric GCM. In: UTAM Symposium on Advances in Mathematical Modelling of Atmosphere and Ocean Dynamics, P. F. Hodnett (Ed.), Kluwer, Dordrecht, 197-202.

[5] McGregor, J. L., and M. R. Dix, 2008: An updated description of the Conformal-Cubic Atmospheric Model, In: Hamilton K. and W. Ohfuchi (Eds.), *High Resolution Simulation of the Atmosphere and Ocean* Springer, 51–76.

[6] Cope, M. E., et al., 2009: Chemical Transport Model - Technical Description: Centre for Australian Weather and Climate Research, Technical Report 15, Melbourne, Australia.

[7] Sarwar, G., et al., 2008: Impact of an updated carbon bond mechanism on predictions from the CMAQ modeling system: Preliminary assessment, *J. Appl. Meteorol. Clim.,* **47**, 3–14.

[8] Donahue, N. M., et al., 2006: Coupled partitioning, dilution, and chemical aging of semivolatile organics, *Environ. Sci. Technol.*, **40,** 2635–2643.

[9] Fountoukis, C., and A. Nenes, 2007.: ISORROPIA II: a computationally efficient thermodynamic equilibrium model for K^+-Ca^{2+}-Mg^{2+}-NH_4^+-Na^+-SO_4^{2-}-NO_3^--Cl^--H_2O aerosols, *Atmos. Chem. Phys*, **7**, 4639–4659.

[10] Mann, G. W., et al., 2010: Description and evaluation of GLOMAP-mode: a modal global aerosol microphysics model for the UKCA composition-climate model, *Geosci. Model Dev.,* **3,** 519–551.

EPJ Web of Conferences **119**, 01006 (2016)
ILRC 27
DOI: 10.1051/epjconf/201611901006

THE EXPECTED IMPACT OF MULTIPLE SCATTERING ON ATLID SIGNALS

D.P. Donovan

Royal Netherlands Meteorological Institute (KNMI) P.O. Box 201, 3730 AE, De Bilt, The Netherlands.
Email: donovan@knmi.nl

ABSTRACT

ATLID stands for "ATmospheric LIDar" and is the lidar to be flown on the Earth Clouds and Radiation Explorer (EarthCARE) platform in 2018. ATLID is a High-Spectral Resolution (HSRL) system operating at 355nm with a narrower field-of-view and lower orbit than the CALIPSO lidar. In spite of the smaller footprint multiple-scattering (MS) will have an important impact on ATLID cloud signals and, in some aspects, the accurate treatment of MS will be more important for ATLID than CALIPSO. On the other hand, the relationship between integrated backscatter and integrated MS induced depolarization in water clouds will be similar between ATLID and CALIPSO indicating that a CALIPSO-like strategy for cloud-phase identification can be successfully applied to ATLID.

1. INTRODUCTION

It has long been known that in order to accurately model the lidar signal from clouds that the contribution of multiply scattered photons must be taken into account. This generally holds true for ground-based and aircraft-based lidar systems and is most certainly true for space-based systems. The multiply scattered return depends on both the instrument characteristics and the atmospheric optical properties. In particular, the degree of multiple-scattering (MS) depends on the cloud extinction profile, the distance from the lidar, the wavelength of the lidar, the laser divergence, the receiver field-of-view (fov) and the angular width of the forward scattering lobe associated with the particle scattering which in turn is related to the cloud/aerosol cross-sectional area. The last point is of particular importance and can lead to significant MS effects even at low optical depths. This is the case if the particle sizes are sufficiently large leading to a forward scattering lobe (which in the limit contains 1/2 the scattered energy) which is angularly narrow in comparison to the lidar fov.

ATLID is a HSRL lidar to be flown on the ESA/JAXA Explorer (EarthCARE) platform planned for launch in 2018 [1]. ATLID operates at a wavelength of 355nm with a telescope fov of 0.065 mrads (full-angle). EarthCARE's orbit will be about 400km. CALIPSO orbits at a higher altitude (about 700km) and has a large fov (0.13mads) leading to larger footprint. In spite of ATLID's smaller footprint (about 30 vs 90 meters) MS effects will still be as (or even more) important than for CALIPSO.

In this paper, we will discuss the impact of multiple-scattering on ATLID signals in ice and water clouds and how it can be accounted for and even exploited.

2. Multiple Scattering in Ice Clouds

The simplest manner in which MS effects can be modelled is by adjusting the effective particle extinction, i.e.

$$\beta_{at}(z) = \beta \exp[-2\eta \int_0^z \alpha(z')dz'] \qquad (1)$$

where z is the range from the lidar, β_{at} is the attenuated backscatter, β is the atmospheric backscatter coefficient, α is the extinction coefficient and η is the MS coefficient. This approach, originally due to Platt [2], is simple but suffers from a number of limitations. For ATLID the inability to accurately model the backscatter below semi-transparent cloud layers is a serious limitation.

Example results of MC calculations applied to idealized cirrus cloud are shown in Fig. 1. Here clouds of two different Optical Thickness (OT) comprised of hexagonal ice crystals [3] with modified-gamma particle size distributions and associated mean diameters (with diameter being identified with the crystal maximum length) of 10 and 50 microns have been used. . The corresponding effective radius ($R_{eff=}$ $0.75 < \mathrm{Vol} > / < A_c >$) values are shown in the

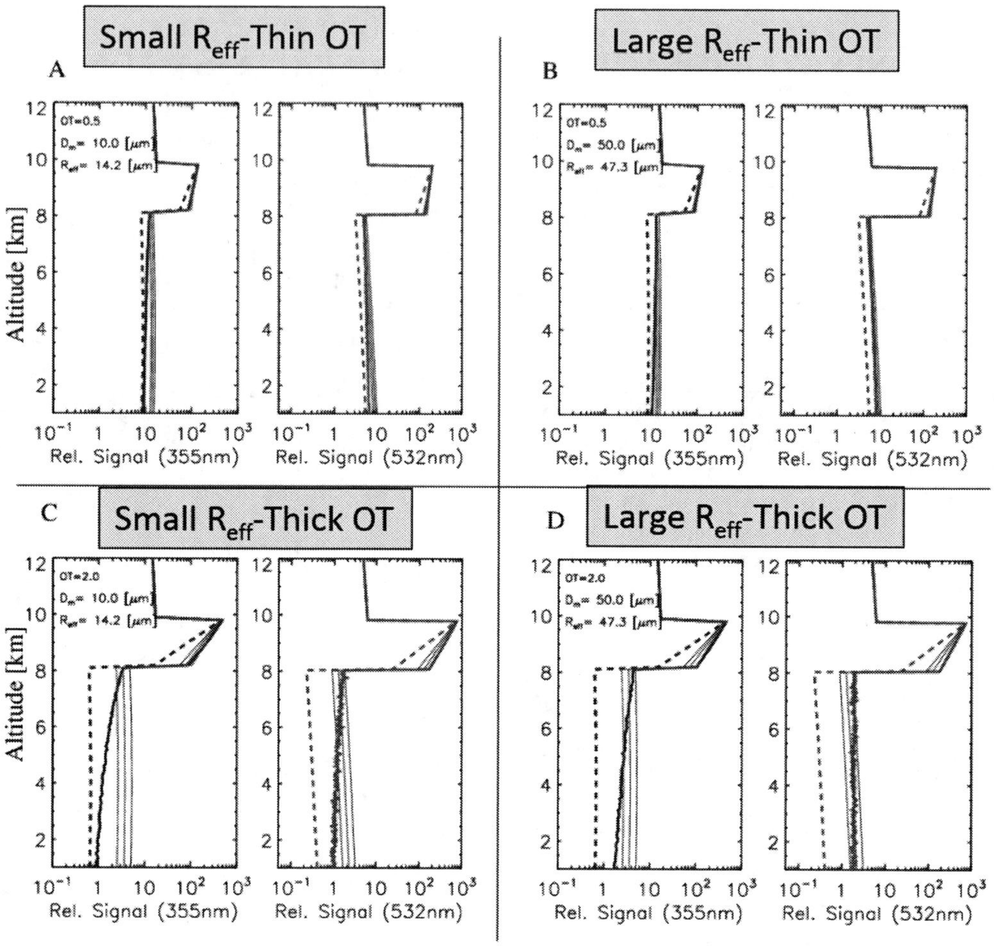

Figure 1: Example results of MC lidar radiative transfer simulations for ATLID (blue-lines), CALIPSO (Green-lines) compared to the expected Single-Scatter signal levels (Dotted Lines) and Eq. 1 with η= 0.4,0.5 and 0.6 (Red-Lines). Panel A, OT=2 and D_m=10.0 microns Panel B: Cirrus OT=2, D_m=50.0 microns and R_{eff}=47.3 microns , Panel C: Cirrus OT=0.5, D_m=10.0 microns , Panel D: Cirrus OT=0.5, D_m=50.0 microns and R_{eff}=47.3 microns.

panels. Monte-Carlo (MC) lidar radiative transfer results are shown as well as results corresponding to single-scattering only and the result of applying Eq. 1 with 3 different values of the multiple-scattering coefficient (0.4,0.5,0.6). The MC calculations were performed using the lidar component [4] of the EarthCARE simulator (ECSIM) [5]. ATLID has three channels, a co-polar `Mie' channel, a co-polar Rayleigh channel and a cross-polar Rayleigh+Mie channel. The ATLID results shown in Fig. 1 correspond to the sum of the three channels.

Referring to Fig. 1, it can be seen for this example that Eq. 1 can be used to accurately model the lidar return within the cloud for both CALIPSO and ATLID. However, in the case of ATLID, the Rayleigh signal below cloud base displays a clear "decay" towards the single-scatter signal level. The decay is not as noticeable in the CALIPSO simulations; the decay only becomes noticeable for the larger OT cases especially in the small particle case.

The behavior of the tails may be understood by considering the effective divergence of the multiple-scattered light emerging from the cirrus cloud. With reference to Fig.2, in the case of CALIPSO the instrument footprint is such that only a small about of the multiply scattered light `leaks' outside of the receiver cone leading to a long decay distance. For ATLID, the smaller fov

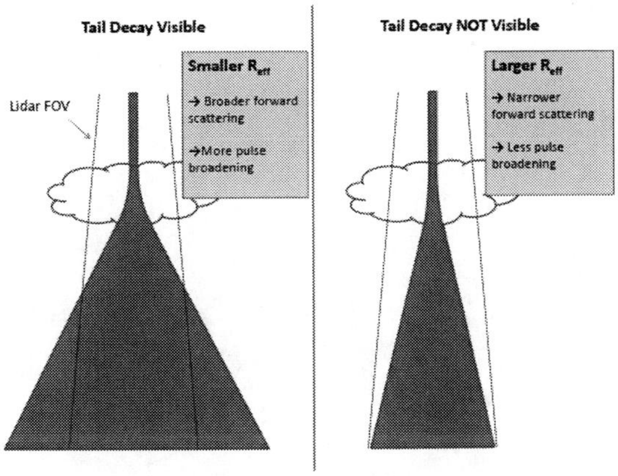

Figure 2: Schematic depiction of the beam broadening caused by MS and how it can lead to visible tails under cirrus clouds.

means that much more of the multiple-scattered light emerging from the cirrus cloud base leaves the receiver cone leading to shorter decay time.

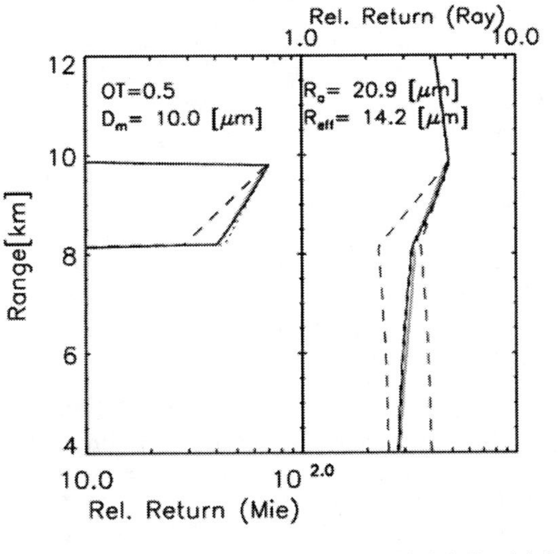

Figure 3: (Left) Mie co-polar and (Right) Rayleigh channel co-polar returns for an ice cloud of an optical thickness of 0.5 and an effective radius of 14.2 microns. Black solid: ECSIM MC results. Dashed Black: Single scattering results. Solid Grey: Hogan's model results and Red-Dashed Eq.1 with a MS factor of 0.45.

Thus, somewhat counterintuitively, for the case of targets below cirrus clouds, multiple-scattering effects will be more noticeable for ATLID than CALIPSO. In order for ATLID to retrieve accurate extinction and backscatter estimates below semi-transparent cirrus the MS tails must be accounted for.

Since Eq.1 cannot model below cloud tails, more sophisticated approaches are needed for ATLID. Fortunately, fast and accurate analytical approaches exist. One of these approaches has been developed by Hogan [6]. An example comparison between the results of an ECSIM MC calculation and the corresponding result from the fast semi-analytical model of Hogan [6] together with the result of applying Eq.1 is shown in Fig. 3. An idealized cirrus cloud with an optical thickness of 0.5 was used with a particle size mode diameter (D_m) of 10 microns giving rise to the indicated values of effective radius (R_{eff}) and mean area radius (R_a). Here it can be seen that all three approaches perform well within the confines of the cirrus cloud. However, the Rayleigh return below the ice cloud is not well-modelled by the Platt approach. The approach of Hogan, however, provides a reasonably accurate match to the MC calculations even below the cloud.

3. Depolarization in Water Clouds

The scattering properties of liquid water cloud droplets can be modelled by assuming that the droplets are perfect spheres and applying Mie theory. Under this assumption and assuming single-scattering only, no cross-polar returns from the cloud will exist. However, the occurrence of MS will give rise to a significant cross-polar return from water clouds. Fig. 4 shows the results of MC calculations for various idealized water clouds for both CALIPSO and ATLID conditions. This figure illustrates that the observed relationship between integrated attenuated backscatter and depolarization for CALIPOS can be replicated by the ECSIM MC model and that the general form of the relationship expected for ATLID will be similar to that of CALIPSO. This indicates that, similar to CALIPSO, this relationship can be used by EarthCARE as a robust means of distinguishing between ice and water clouds.

4. Conclusions

The results of extensive MC ice-cloud calculations suggest that the Platt approach will not be sufficient, in general, to model the below-

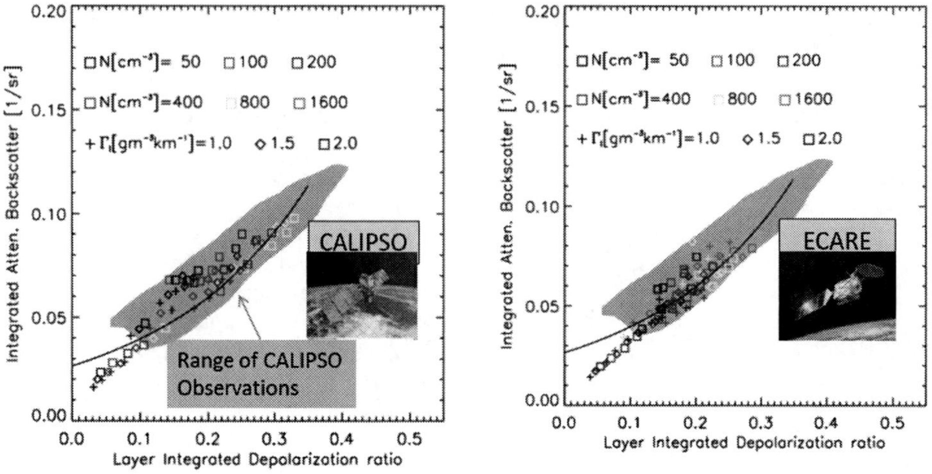

Figure 4: Integrated-layer-attenuated-backscatter vs. layer-depolarization for CALIPSO(Left) and ATLID (Right). Here the Grey-Area corresponds to the range of actual CALIPSO observations [7] and the points correspond to MC results using the indicated liquid-water lapse-rates (Γ_l) and cloud particle number concentrations (N). The solid lines correspond to Eq.(6) of [7].

cloud lidar return. This will be an important issue impacting, e.g. the retrieval of aerosol parameters below cirrus. On the other hand, the sensitivity to particle size may enable cirrus particle sizes to be estimated using the below cloud Rayleigh channel signal.

Calculations applied to idealized water clouds indicate that the layer-depolarization-vs-integrated attenuated backscatter relationship expected from ATLID will be similar to that observed by CALIPSO.

ACKNOWLEDGEMENT

Parts of this work were conducted as part of the ESA/ESTEC ATLAS project, contract No. 22638/09/NL/CT

REFERENCES

[1] Illingworth, I.J., et al., 2014: THE EARTHCARE SATELLITE: The next step forward in global measurements of clouds, aerosols, precipitation and radiation. *Bull. Amer. Meteor. Soc.*, doi:10.1175/BAMS-D-12-00227.1.

[2] Platt, C. M. R., 1981: "Remote sounding of high clouds. III: Monte Carlo calculations of multiple-scattered lidar returns." *J. Atmos.Sci.*, 38.1, 156-167.

[3] Hess M., R. Koelemeijer, P. Stammes 1998: Scattering matrices of imperfect hexagonal ice crystals, *JQRT*, doi:10.1016/S0022-4073(98)00007-7.

[4] Donovan, D.P. H. Klein Baltink, J. S. Henzing, S. R. de Roode, and A. P. Siebesma, 2015: A depolarisation lidar-based method for the determination of liquid-cloud microphysical properties, *Atmos. Meas. Tech.*, 8, 237-266 doi:10.5194/amt-8-237-2015

[5] Voors, R., D.P. Donovan, J. Acarreta , M. Eisinger,; R. Franco, et al. 2007: "ECSIM: the simulator framework for EarthCARE", Proc. SPIE 6744, Sensors, Systems, and Next-Generation Satellites XI, 67441Y (October 17, 2007); doi:10.1117/12.7377

[6] Hogan, R., 2006: "Fast approximate calculation of multiply scattered lidar returns," Appl. Opt. 45, 5984-5992.

[7] Hu, Y., M. Vaughan, Z. Liu, B. Lin, P. Yang, D. Flittner, B. Hunt, R. Kuehn, J. Huang, D. Wu, S. Rodier, K. Powell, C. Trepte, and D. Winker, 2007: "The depolarization - attenuated backscatter relation: CALIPSO lidar measurements vs. theory," Opt. Express 15, 5327-5332.

Triple-Pulsed Two-Micron Integrated Path Differential Absorption Lidar: A New Active Remote Sensing Capability with Path to Space

Upendra N. Singh*, Tamer F. Refaat, Mulugeta Petros, and Jirong Yu

*NASA Langley Research Center, Hampton, VA 23681, USA, *Email: upendra.n.singh@nasa.gov*

ABSTRACT

The two-micron wavelength is suitable for monitoring atmospheric water vapor and carbon dioxide, the two most dominant greenhouse gases. Recent advances in 2-µm laser technology paved the way for constructing state-of-the-art lidar transmitters for active remote sensing applications. In this paper, a new triple-pulsed 2-µm integrated path differential absorption lidar is presented. This lidar is capable of measuring either two species or single specie with two different weighting functions, simultaneously and independently. Development of this instrument is conducted at NASA Langley Research Center. Instrument scaling for projected future space missions will be discussed.

1. INTRODUCTION

Water vapor (H_2O) and carbon dioxide (CO_2) are the most dominant greenhouse gases that strongly contribute to the Earth's radiation budget and global warming. Extensive efforts have demonstrated successful H_2O airborne profiling using the differential absorption lidar (DIAL) technique. In the absence of similar capabilities for CO_2 profiling, integrated path differential absorption (IPDA) lidar technique is best suited for measuring the gas weighted average column dry-air volume-mixing ratio (X) [1]. Although it lacks the ranging capability, IPDA provides higher sensitivity measurements that rely on stronger hard target returns versus DIAL, which relies on weaker backscatter signals. Currently NASA is pursuing different IPDA technologies targeting CO_2 active remote sensing. Among these technologies, NASA Langley Research Center (LaRC) demonstrated a double-pulse 2-µm IPDA for CO_2 airborne measurements [2]. Nevertheless, ongoing efforts at LaRC focus on advancing such technology by enhancing the IPDA lidar instrument through triple-pulse operation [3].

A unique airborne triple-pulse, 2-µm IPDA lidar is under development at NASA LaRC. Triple-pulse IPDA lidar operation allows two simultaneous and independent measurements. With wavelength optimization, the instrument can be adjusted to either target H_2O and CO_2 measurements or target CO_2 measurements with two different weighting functions. This innovative IPDA lidar is based on a state-of-the-art, efficient, conductively cooled, and injection seeded 2-µm triple-pulse laser transmitter under development. For each laser burst, using single pump-pulse, three 2-µm pulses are generated and transmitted. These pulses are separated by 200 µsec, with adjustable pulse-energy and consequently pulse-width. Wavelength selection, tuning, switching and locking is achieved for each of the three pulses individually. The pulses short-time separation and tuning capabilities results in two simultaneous and independent measurements, respectively. This IPDA lidar instrument will leverage an existing receiver that includes a Newtonian telescope, detection, and data acquisition systems and other low-risk, commercially available components that have been validated through the previous double-pulsed operation. Compared to separate instruments for measuring two species, the triple-pulse IPDA lidar will result in reduced size, mass and power consumption while enhancing the technology for future space-based applications. With additional detection enhancements, the capabilities of this 2-µm triple-pulse IPDA lidar is comparable to a CO_2 space-based mission proposed by the European Space Agency (ESA), as presented in this paper.

2. CO_2 AND H_2O SENSING

Water vapor distribution is an essential information for converting the 2-µm IPDA measured CO_2 optical depth into X. In addition, H_2O interference is a common problem for CO_2 remote sensing in the IR spectral region, including the 2-µm wavelength. Although, this interference can limit the CO_2 measurement sensitivity, it can be significantly reduced by the proper selection of the IPDA operating wavelengths. Triple-pulse 2-

μm IPDA can further reduce such interference and provide optical depth conversion estimates by measuring H_2O simultaneously and independently while measuring CO_2. This triple-pulse IPDA measurement concept is presented in figure 1. The figure presents the vertical integrated optical depth spectral profile for CO_2 and H_2O around the CO_2 R30 line [1, 4]. The optical depths were derived using HITRAN database for absorption line parameters, assuming Voigt profile, and the US standard atmosphere model for meteorological and molecular profiles. The presented optical depth integration upper limit is based on 8 km airborne altitude, assuming a small aircraft such as the NASA B-200. Sea level is considered for the lower limit. The H_2O absorption peak, located at 2050.5322 nm, coincides close to CO_2 absorption minima between R30 and R32 lines and away from the R32 peak. The unique tuning and locking capabilities of the 2-μm laser transmitter allow proper selection of the IPDA operating wavelengths. The principle of wavelength selection for this IPDA instrument is demonstrated in the same figure. The CO_2 on- and off-line wavelengths are selected around the R30 line, so that both would have similar H_2O absorption (lower horizontal line). This minimizes the H_2O interference on the CO_2 measurements. Similarly, the H_2O on- and off-line wavelengths are selected around the absorption peak such that CO_2 interference is minimized on the H_2O measurement (upper horizontal line). Molecular interference minimization, through wavelength selection, results in independent measurement. However, CO_2 and H_2O measurements share the same wavelength (i.e., the same pulse) for on- and off-line, respectively. This allows achieving simultaneous measurement of both gases with triple pulses rather than quadruple pulses almost independently while avoiding interference from each other. Simultaneous and independent H_2O and CO_2 measurements provide an opportunity to study the two most dominant greenhouse gases variability and interaction through different environmental processes such as photosynthesis and fossil fuel burning.

3. CO_2 DOUBLE-WEIGHTING SENSING

Focusing on the CO_2 measurement, figure 2 shows the gas integrated optical depth spectra around the R30 line. At this location, H_2O optical depth, shown on the same figure, is more than order of magnitude lower than CO_2, which would contribute negligible interference, particularly in relatively dry environments. In such case, using the same 2-μm laser transmitter tuning and locking capabilities allow measuring the gas concentrations using two different weighting functions, simultaneously and independently, as marked by the vertical line in figure 2 and demonstrated in figure 3. For example, weighting function selection allows measuring CO_2 concentration near the surface for studying sources and sinks. With the same off-line and the third pulse tuned to different weighting function CO_2 concentration in free troposphere can be targeted simultaneously and independently for studying the gas transport. Other tuning for the weighting function can target the top of the boundary layer for studying CO_2 transport between the biosphere and the atmosphere. Switching between CO_2 and H_2O measurement and CO_2 double-weighting measurement, and via versa, could be achieved instantaneously through the advance wavelength seeding, tuning then locking during the IPDA lidar airborne operation. Same switching could be even applied between different, pre-defined CO_2 weighting to target different sensitivities depending on the geographical location and environment. This unique features would be attractive for space applications.

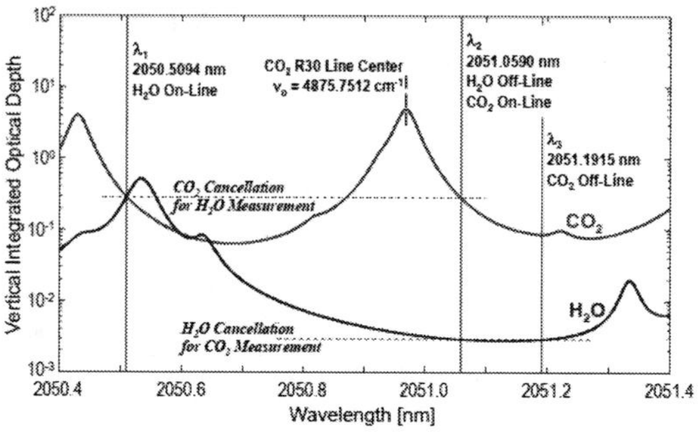

Figure 1. Comparison of the CO_2 and H_2O integrated optical depths derived using the HITRAN data base for line parameters and US Standard atmosphere model for metrological and molecular profiles. Vertical lines mark the instrument selected wavelengths for the three laser pulses, for simultaneous and independent CO_2 and H_2O measurements.

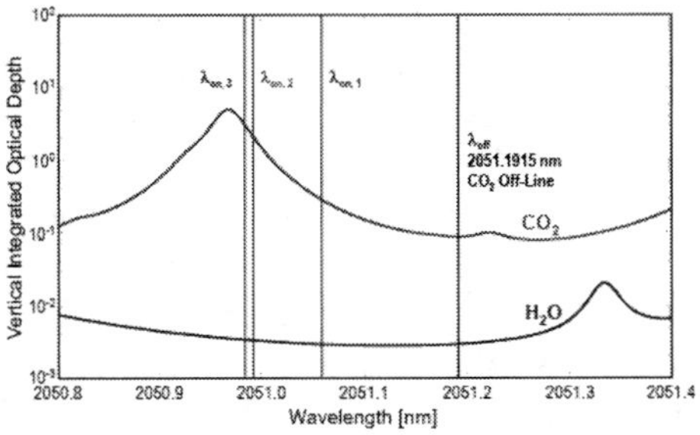

Figure 2. Focusing on the CO_2 R30 line, triple-pulse IPDA measurement can target the gas measurement with two different weighting functions, simultaneously and independently. Vertical color coded lines mark selected wavelengths examples for the three laser pulses.

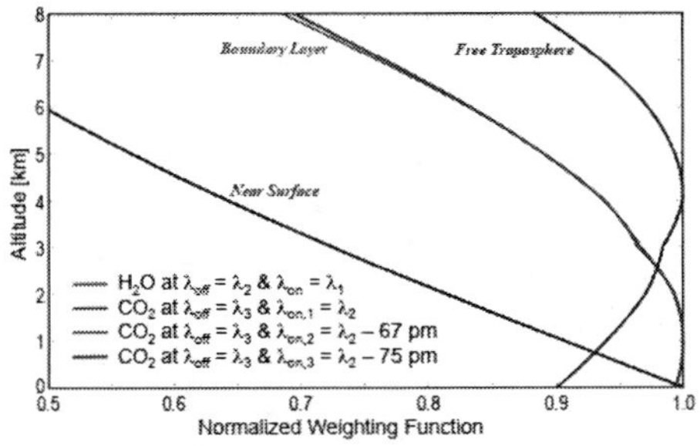

Figure 3. Corresponding color coded peak-normalized pressure-based weighting functions for H_2O and the selected CO_2 on-line wavelength.

4. 2-μm CO_2 IPDA PATH TO SPACE

The triple-pulse 2-μm IPDA lidar technology development is enabling for CO_2 measurements from space. Recent development of advance single-charge-carrier, HgCdTe electron avalanche photodiodes (e-APD) indicated a breakthrough in lidar detection technology [5]. These devices are space-qualifiable and were validated for airborne lidar operation at 1.6-μm. In co-ordination with NASA Earth Science Technology Office (ESTO), NASA LaRC is collaborating with NASA Goddard Space Flight Center (GSFC) to combine their detector with the triple-pulse transmitter towards developing a 2-μm triple-pulse IPDA lidar for airborne CO_2 measurement. Combining these technologies in a single system will result in a CO_2 IPDA lidar instrument that meets or exceed space requirement set by ESA to accomplish the same mission [6].

Table 1. CO_2 active remote sensing using the triple-pulsed 2-μm IPDA, developed at NASA LaRC, compared to ESA space requirements.

	NASA Technology	***ESA Space Requirements***
Transmitter		
Laser	Single Laser	Two Lasers
Technique	Triple-Pulse	Single-Pulse
Cooling	Conductive	Conductive
Wavelength	2.051 μm	2.051 μm
Pulse Energy	50/15/5 mJ	40 & 5 mJ
Repetition Rate	50 Hz	50 Hz
Power	3.50 W	2.25 W
Pulse Width	30-100-150	50 ns
Optical Efficiency	5.0%	5.0%
Wall-Plug Efficiency	2.1%	> 2.0%
Pulse Delay	200 μs	250 μs
Transvers Mode	TEM_{00}	TEM_{00}
Longitudinal Mode	Single Mode	Single Mode
Pulse Spectral Width	4 – 14 MHz	> 60 MHz
Beam Quality	2	< 2
Freq. Control Accuracy	0.3 MHz	0.2 MHz
Seeding Success Rate	99	99
Spectral Purity	99.9%	99.9%
Receiver		
Telescope Diameter	0.4 m	1.2 m
Detector Material	HgCdTe	---
Detector Structure	eAPD	APD
Quantum Efficiency	80%	75%
Excess-Noise-Factor	1.1	1.5
Noise-Equivalent-Power	8 $fW/Hz^{1/2}$	100 $fW/Hz^{1/2}$

Table 1 summarizes the projected parameters of the 2-μm triple-pulsed IPDA lidar and compares it to space requirements released by ESA [5]. ESA space requirements were based on a similar CO_2 IPDA lidar instrument operating at 2-μm. ESA objective is to develop future space borne active sensing mission for measuring the dry-air mixing ratio of CO_2 ***throughout*** the atmosphere with a high accuracy on the ppm level. Table 1 indicates that the triple-pulse 2-μm transmitter, in conjunction with the e-APD detector, would result in a CO_2 IPDA lidar instrument that meets or exceeds most of the transmitter and detection requirements for space-borne CO_2 measurement. The unique triple pulse capability has additional advantages. Having one laser delivering, near simultaneously, three pulses at different frequencies eliminates the complexity and need of three different lasers. This is a significant step towards reducing mass, size and power consumption of the instrument to one third and increasing the efficiency by a factor of three.

Triple-pulse transmitter eliminates the challenge and complexity in co-aligning and bore-sighting three independent beams. Concurrent to this and supported by ESTO, LaRC is investigating high repetition rate 2-μm lasers. Although operating in a single-pulse mode, pulse energy of 40 mJ was demonstrated with repetition rate as high as 200 Hz. High repetition rate is a desirable transmitter aspect allowing higher average that leads to sensitivity enhancement for space-based lidar. This early development of space qualifiable lasers and air-borne operations will reduce the risk towards future space operation.

Operating from space, adaptive targeting can be easily implemented through the 2-mm triple-pulsed IPDA lidar. The unique wavelength switching, tuning and locking capabilities of the instrument allows changing the CO_2 weighting and/or include H_2O measurement instantaneously. For example, this is useful for changing the CO_2 measurement weighting function depending on the geographical location and/or the sensing objectives. This would a valuable tool for studying the gas sources, sinks and transport.

5. CONCLUSIONS

The societal benefits of better understanding of climate change through identification of global CO_2 sources and sinks led to the desired NASA active remote sensing of the gas. The favored solutions for such goal is the IPDA lidar technique. NASA LaRC identified the desired pulsed laser attributes and wavelength regions for CO_2 active remote sensors. From foundational work in spectroscopy and modeling to airborne system validation, LaRC demonstrated the capability of CO_2 airborne measurement using double-pulse technique at 2-μm. Extending the technological capabilities, LaRC is developing a unique triple-pulsed 2-μm IPDA lidar. With wavelength optimization, this instrument is capable of achieving two simultaneous and independent measurements. The triple-pulse IPDA can either target H_2O and CO_2 measurements or target CO_2 measurements with two different weighting functions. The measurement type and/or weighting could be switched instantaneously during operation. Under NASA ESTO, and with NASA GSFC collaboration, the triple-pulse IPDA lidar capabilities can be further enhanced by including advanced e-APD based detection system. This results in an innovative CO_2 IPDA lidar that can meet or exceed space requirement set by ESA for the same objective. This 2-μm triple-pulse IPDA lidar is a possible candidate for future global CO_2 active remote sensing from space.

ACKNOWLEDGEMENTS

This work (ESTO IIP-0013, Principal Investigator: Upendra Singh, LaRC) was funded and supported by NASA Earth Science Technology Office (Program Director: George Komar and Program Manager: Parminder Ghuman).

REFERENCES

[1] U. Singh, T. Refaat, J. Yu, M. Petros, 2013: Simultaneous and independent measurement of atmospheric water vapor and carbon dioxide using a triple-pulsed, 2-micron airborne IPDA Lidar - a feasibility study, *presented at 2013 AGU Fall Meeting*, A13C-0213AGU, San Francisco, California.

[2] U. Singh, J. Yu, M. Petros, T. Refaat, R. Remus, J. Fay, K. Reithmaier, 2014: Airborne 2-micron double-pulsed integrated path differential absorption lidar for column CO_2 measurement", *Proc. SPIE* **9246**, 924602.

[3] T. Refaat, U. Singh, J. Yu, M. Petros, S. Ismail, M. Kavaya, K. Davis, 2015: Evaluation of an airborne triple-pulsed 2 μm IPDA lidar for simultaneous and independent atmospheric water vapor and carbon dioxide measurements, *Appl. Opt.* **54**, 1387-1398.

[4] U. Singh, B. Walsh, J. Yu, M. Petros, M. Kavaya, T. Refaat, N. Barnes, 2015: Twenty years of Tm:Ho:YLF and LuLiF laser development for global wind and carbon dioxide active remote sensing, *Opt. Mater. Exp.* **5**, 827-837.

[5] X. Sun, J. Abshire, J. Beck, 2014: HgCdTe e-APD detector arrays with single photon sensitivity for space lidar applications, *Proc. of SPIE* **9114**, 91140K.

[6] P. Ingmann, P. Bensi, Y. Duran, A. Griva, P. Clissold, 2008: A-Scope - advanced space carbon and climate observation of planet earth, *ESA Report for Assessment* **SP-1313/1**.

Space-Based Erbium-Doped Fiber Amplifier Transmitters for Coherent, Ranging, 3D-Imaging, Altimetry, Topology, and Carbon Dioxide Lidar and Earth and Planetary Optical Laser Communications

Mark Storm*, Doruk Engin, Brian Mathason, Rich Utano, and Shantanu Gupta

*Fibertek, Inc., 13605 Dulles Technology Drive, Herndon, VA 20171, USA, *mstorm@fibertek.com*

ABSTRACT

This paper describes Fibertek, Inc.'s progress in developing space-qualified Erbium-doped fiber amplifier (EDFA) transmitters for laser communications and ranging/topology, and CO_2 integrated path differential absorption (IPDA) lidar. High peak power (1 kW) and 6 W of average power supporting multiple communications formats has been demonstrated with 17% efficiency in a compact 3 kg package. The unit has been tested to Technology Readiness Level (TRL) 6 standards. A 20 W EDFA suitable for CO_2 lidar has been demonstrated with ~14% efficiency (electrical to optical [e-o]) and its performance optimized for 1571 nm operation.

1. INTRODUCTION

Fibertek has been developing eye-safe 1.5-um lidars and associated laser technology for numerous applications including coherent wind lidar [1,2]; 3D imaging and ranging for obstacle avoidance, docking, brownout, and rendezvous lidars [3,4]; CO_2 lidar [5]; and deep space, low/geosynchronous earth orbit (LEO/GEO) laser communications [6,7]. Many of our applications require specialized multi-stage designs to provide unique performance such as high 1-3 mJ laser pulse energies [8], kilohertz linewidth, absolute wavelength control, differential absorption lidar (DIAL), CO_2 lidar, or modulation formats for chirped pulse ranging lidar or pseudo-random noise lidar [9]. Multi-stage amplifiers are often needed to achieve 30-60 dB of net gain.

Commercial off-the-shelf (COTS) EDFA amplifiers are typically 1-10 W single stage units and used for ground-based laboratory testing and pressurized aircraft cabin lidar measurements. Systems requiring specialized multi-stage performance require customized assemblies of amplifiers and other optical components and can get large, unwieldy, and unsuitable for UAV or spaceflight platforms where small size and specialized performance are needed. Electronics and packaging can make COTS components and assemblies challenging.

This paper describes our progress in engineering EDFA transmitters into a spaceflight/UAV quality package that operates in challenging environments, has traceability, and is vibration and thermal vacuum tested toward TRL-6 qualification.

2. METHODOLOGY

Fibertek develops fiber lasers based on the performance specifications, environment of operation, and expected reliability and lifetime of the mission. After the application performance is specified Fibertek conducts an optical design and breadboard validate phase to optimize and validate the design. A team of engineers then conducts an optical, thermal, and structural design using models and finite element analysis. The design is iterated to optimize the package performance and then built and environmentally tested. Analysis products including thermal, structural finite element model (FEM), and random vibration spectrum and modal analysis are shown in **Figure 1**.

3. RESULTS

3.1 Deep Space, GEO, and LEO Laser Communications Transmitter

Fibertek is developing a space-qualifiable, highly efficient high-power (>5 W) fiber-based 1550 nm laser optical module (LOM) supporting NASA JPL. The JPL developed deep space comm system is being made available on the NASA Discovery 2014 mission instrument announcement of opportunity. The system is designed to operate from deep space, planets, and Mars and the transmitter is also suitable for high-reliability LEO/GEO communications (**Figure 2**). The transmitter achieves 6 W average and up to 1 kW peak power depending on the communications pulse format [6,7].

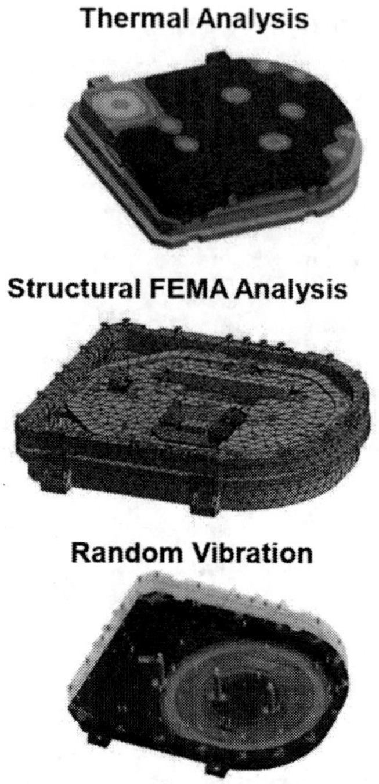

Figure 1: Thermal, Structural, and Radom Vibration Analyses Meet or Exceed Margins of Safety for Launch and Survival and Operation of Mission Temperature Ranges.

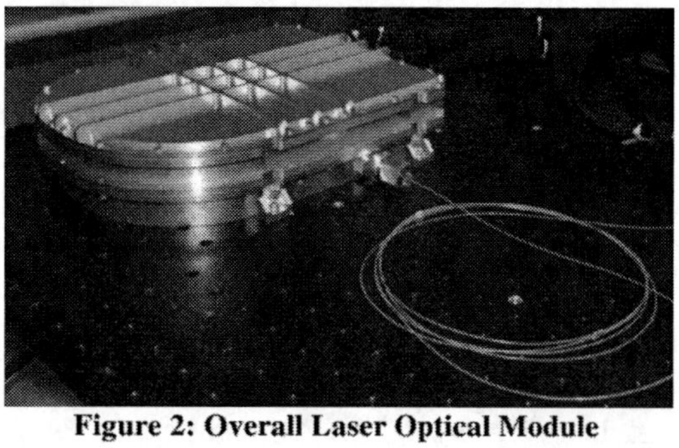

Figure 2: Overall Laser Optical Module

The transmitter is also suitable for ranging/topology because the peak powers are high. The transmitter has <6 GHz linewidth. Stimulated Brillouin scattering (SBS) is managed by precise linewidth control and use of large mode array (LMA) gain fiber in the power stage while maintaining the required diffraction limited and highly polarized (polarization extinction ratio [PER] >20 dB) output. Size and weight of the transmitter are 8″ x 10″ x 2.4″ and 3 kg, respectively. The unit includes the distributed feedback (DFB) laser diodes, EO modulators, and amplifier.

With improvements in the modulation scheme and component specification, the achieved e-o efficiency is more than 17%. Highly efficient operation is sustained for a wide range of pulse-position modulation (16 to 128-ary PPM) formats with pulsewidths varying from 8 nsec to 0.5 nsec and operation temperature 10-50°C. Pressure stress, random vibration, and thermal analyses of the designed LOM predict compliance with NASA General Environmental Verification Standard (GEVS) levels for vibration and thermal cycling in a vacuum environment. The transmitter has undergone and passed TRL-6 thermal vacuum and vibration testing to validate the design.

3.2 High-Efficiency CO_2 Amplifier Performance at 1571 nm for IPDA Lidar

Fibertek is also developing a continuous wave (CW) EDFA at 1571 nm for CO_2 IPDA (**Figure 3**). IPDA is a two-way integration of the optical-depth (OD) from platform to surface (e.g., ground or cloud top), and is measured using lock-in detection of an intensity-modulated CW fiber laser transmitter employing multiple wavelengths within a CO_2 absorption linewidth of ~1 GHz (~10 pm) [10-12].

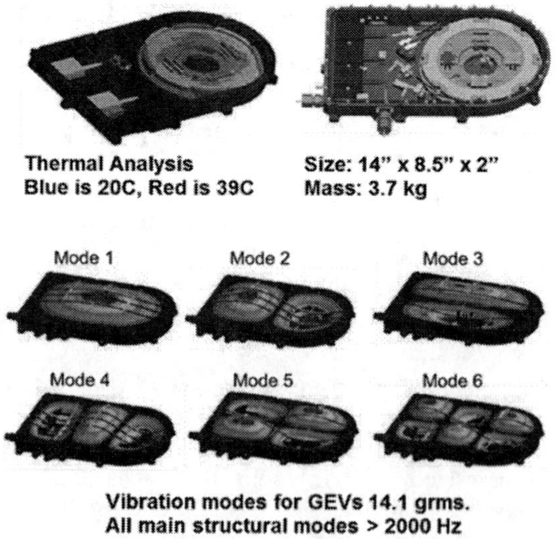

Figure 3: Solid Model of the Design and Preliminary Thermal and Structural Analysis for 20 W Average Power EDFA Amplifier

This two-stage amplifier can also support the coherent lidar DIAL method [13]. This spaceflight TRL-6 prototype transmitter is custom designed to meet NASA CO_2 lidar requirements that is far advanced compared to current demonstrated space or terrestrial lasers.

The optical transmitter diagram for the 1571 nm EDFA is a fiber laser master oscillator power amplifier (MOPA) with an ultra-low-noise master oscillator and two-stage fiber amplifiers. The MOPA seed consists of two DFB laser diodes operated CW representing the lidar on and off wavelengths. The DFB operates with a linewidth of <1 MHz and can be tuned on and off the CO_2 absorption line at 1571.11 nm.

The stage 1 EDFA is built with an Er/Ytterbium (Yb) co-doped single-mode fiber that is widely used in telecom systems. It has core and cladding diameters of 7 μm and 125 μm, and is counter-pumped with a broad area laser diode at 976 nm. The stage 1 amplifier maximum CW power is ~1 W. For a more detailed description see ref 5.

The second stage amplifier employs a commercial LMA Er/Yb co-doped fiber with cladding-pump scheme. Simple coiling of the gain fiber enables single-mode operation. We achieved maximum output power of 21 W after band-pass filter, and the optical-to-optical efficiency is about 36% as shown in **Figure 4** (left). On the right is the full spectrum after the ASE filter. The signal-to-noise ratio (SNR) is more than 30 dB. The backward spectrum (not shown) does not show any significant SBS nonlinearity.

3.3 Space-Qualified Yb MOPA for LISA and ICESat-2

Fibertek has also developed and made significant progress in qualifying a Yb MOPA transmitter. **Figure 5** shows a multi-stage Yb MOPA fiber amplifier capable of up to 20 W of power. The Yb MOPA was successfully vibration tested, thermally cycled, and radiation tested to 2x mission dosage. The unit was developed as part of the ICESat-2 design effort and designed for 400 uJ/pulse at 10 kHz repetition rate.

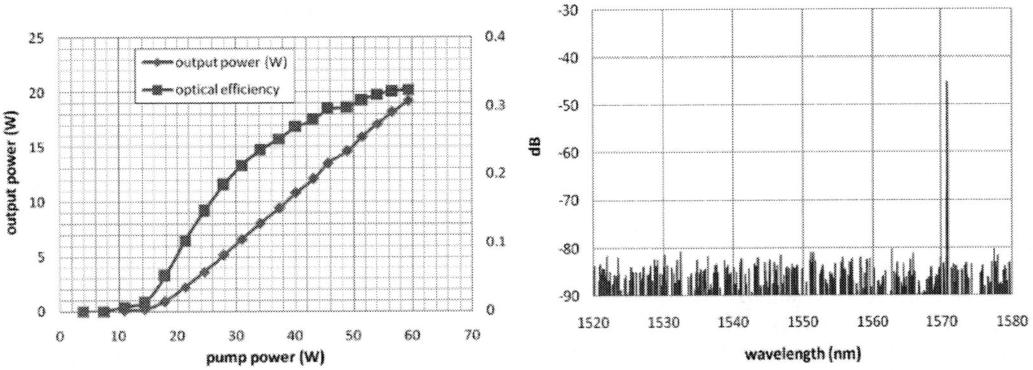

Figure 4: Final Stage Performance Measured After ASE Filter (left) and Forward Spectrum After ASE Filter with 0.01 nm Resolution (right)

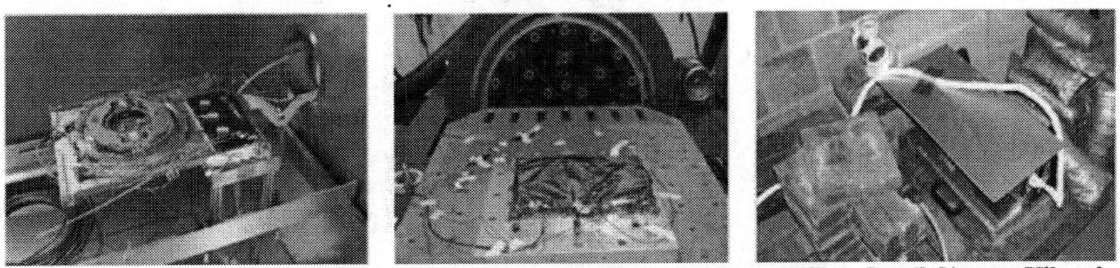

Figure 5: Space Qualification of a Yb MOPA Fiber Amplifier – in Thermal Chamber (left), on a Vibration Table (center), and Being Radiation Tested (right)

The Yb amplifier spaceflight transmitter was developed in support of the joint NASA-European Space Agency (ESA) Laser Interferometer Space Antenna (LISA) project to develop and operate a space-based gravitational wave detector.

4. CONCLUSIONS

In this paper, we demonstrate that the spaceflight EDFA MOPA and Yb MOPA fiber laser transmitter packages have been developed and environmentally tested to TRL-6 standards. The transmitters developed are high efficiency and capable of supporting high peak powers for laser communications, can be operated at 1571 nm with single-frequency transform limited performance, and are being prepared for spaceflight use. The laser communications unit supports the NASA Discovery 2014 AO. The 1571 nm EDFA supports the CO_2 IPDA lidar and the NASA ASCENDS 2007 Earth Science Decadal survey mission. The fiber laser generates 20 W CW power at 1571 nm with narrow linewidth (1-10 MHz). The optical-to-optical efficiency is 36%.

ACKNOWLEDGMENT

The authors would like to acknowledge the support of NASA. We would also like to acknowledge the work of Fibertek engineers Mehmet Akbulut and Wei Lu for their early contributions to this work.

REFERENCES

[1] Akbulut M., et al., "Pulsed coherent fiber lidar transceiver for aircraft in-flight turbulence and wake-vortex hazard detection," Proc. SPIE 8037, Laser Radar Technology and Applications XVI, 80370R (7 June 2011); doi: 10.1117/12.883990.

[2] Akbulut M., "Fiber Laser Coherent Lidar for Wake-Vortex Hazard Detection," 16th Coherent Laser Radar Conference, 2011.

[3] Gary Stevenson, Horacio R. Verdun, Peter H. Stern, Walter Koechner, "Testing the Helicopter Obstacle Avoidance System," Proc. SPIE 2472, Applied Laser Radar Technology II; doi: 10.1117/12.212025.

[4] Thomas A. DuBois, Charles A. DiPietro, Gary Stevenson, "System concept for a rotorcraft vision system to improve cargo-handling operations," SPIE Proceedings Enhanced and Synthetic Vision, Volume 2736.

[5] Lu W., "High-power, narrow linewidth 1.5-μm fiber amplifier lidar transmitter for atmospheric CO2 detection," Proc. SPIE 8961, Fiber Lasers XI: Technology, Systems, and Applications, 89610C (March 7, 2014); doi:10.1117/12.204120.

[6] Engin D., et al., "Highly reliable and efficient 1.5 μm fiber-MOPA-based, high-power laser transmitter for space communication," Proc. SPIE 9081, Laser Technology for Defense and Security X, 90810G (9 June 2014); doi: 10.1117/12.2050496.

[7] Engin D., et al., "Highly efficient and athermal 1550 nm fiber-MOPA-based high-power down link laser transmitter for deep space communication," Proc. SPIE 8610, Free-Space Laser Communication and Atmospheric Propagation XXV, 86100G (19 March 2013); doi: 10.1117/12.2005926.

[8] Torruellas, Y. Chen, B. McIntosh, J. Farroni, K. Tankala, et al., "High peak power ytterbium-doped fiber amplifiers," Proc. SPIE 6102, Fiber Lasers III: Technology, Systems, and Applications, 61020N; doi:10.1117/12.646571; http://dx.doi.org/10.1117/12.646571.

[9] Xiaoli-Xiaoli Sun and James B. Abshire, "Modified PN code laser modulation technique for laser measurements," Proc. SPIE 7199, Free-Space Laser Communication Technologies XXI, 71990P (February 24, 2009); doi:10.1117/12.817254; http://dx.doi.org/10.1117/12.817254

[10] E. Browell, K. Davis, J. Abshire, J. Dobler, G. Ehert, P. Flamant, D. Sakaizawa, U. Singh, G. Spiers, "Current Status of Active Remote Sensing of CO2 and Applications to NACP Investigations," 3rd North American Carbon Program Meeting, New Orleans, Louisiana (1-4 February 2011).

[11] James B. Abshire, et al., "A lidar approach to measure CO2 concentrations from space for the ASCENDS Mission," Proc. SPIE 7832; doi:10.1117/12.868567; http://dx.doi.org/10.1117/12.868567

[12] D. Sakaizawa, S. Kawakami, T. Tanaka, M. Nakajima, "Improvement of the 1.57 um laser absorption sensor with chirp modulation to evaluate spatial averaging carbon dioxide density," Proc. SPIE, Vol. 8182 (2011).

[13] G. D. Spiers, R. T. Menzies, J. Jacob, L. E. Christensen, M. W. Phillips, Y. Choi, E. V. Browell, "Atmospheric CO2 measurements with a 2-um airborne laser absorption spectrometer employing coherent detection," Applied Optics 50 (14), pp. 2098–2111 (May 2011).

ADVANCES IN DIODE-LASER-BASED WATER VAPOR DIFFERENTIAL ABSORPTION LIDAR

Scott Spuler*[1], Kevin Repasky[2], Bruce Morley[1], Drew Moen[2], Tammy Weckwerth[1], Matt Hayman[1], Amin Nehrir[3]

[1]*National Center for Atmospheric Research, Earth Observing Lab, Boulder, CO 80307, USA, *spuler@ucar.edu*
[2]*Montana State University, Electrical and Computer Engineering, Bozeman, MT 59717, USA*
[3]*NASA Langley Research Center, Hampton, VA 23681, USA*

ABSTRACT

An advanced diode-laser-based water vapor differential absorption lidar (WV-DIAL) has been developed. The next generation design was built on the success of previous diode-laser-based prototypes and enables accurate measurement of water vapor closer to the ground surface, in rapidly changing atmospheric conditions, and in daytime cloudy conditions up to cloud base. The lidar provides up to 1 min resolution, 150 m range resolved measurements of water vapor in a broad range of atmospheric conditions. A description of the instrument and results from its initial field test in 2014 are discussed.

1. INTRODUCTION

Water vapor is a potent greenhouse gas which is heterogeneously distributed in the lower part of atmosphere. The distribution of water vapor influences dynamical and physical processes that drive weather phenomena, general circulation patterns, radiative transfer, and the global water cycle. Terrestrial boundary layers can be especially complex given the large amplitude of diurnal cycles and surface exchange complexities. A better understanding of its spatial and temporal distribution is required to support research, and to improve forecasts of severe weather and quantitative precipitation [1, 2].

Most instrumentation capable of accurately measuring water vapor profiles have either inadequate vertical spatial resolution (e.g., satellite observation) or insufficient temporal resolution (e.g., standard, twice a day radiosonde observations) to further our understanding. Laser remote sensing – provided by either Raman lidar [3, 4] or injection seeded Ti:sapphire-based DIAL systems [5, 6, 7, 8] – are capable of high vertical range resolution and continuous measurements; however, they are traditionally expensive devices to develop and operate. The instrument plus its operational cost is especially relevant since multiple devices – a network of profilers – are required to obtain adequate horizontal spatial resolution. Research applications generally require short term deployments of a regional scale network, whereas national forecasts require national networks. For example, in the continental USA, high-resolution vertical profiles of humidity at roughly 400 locations are required [1]. Therefore, a new generation of lower cost water vapor lidar – capable of safely, accurately, continuously, and autonomously measuring water vapor in the lower troposphere with high vertical resolution – is needed to lower the barrier to investment and enable a network.

Diode-laser-based lidar, with low power consumption via electrically pumped semiconductor transmitters – capable of operation for tens of thousands of hours without the requirement for water cooling – offer inherently lower initial costs, and lower operation and maintenance costs, when compared to traditional lidar systems. The first demonstration of a diode-laser-based WV-DIAL was in 2004 by Machol et al. which was capable of nighttime operation [9]. From 2009 to 2012 Nehrir et al. significantly advanced the field, particularly with respect to transmitter design [10, 11], and subsequently demonstrated the first diode-laser-based WV-DIAL capable of daytime measurements [12]. We have built on these seminal instruments and developed them further – towards a sensor that would be suitable for a national scale network – by improving eye-safety and reliability (e.g., reducing sensitivity to environmental temperature), and enabling accurate measurement of water vapor closer to the ground surface, in rapidly changing atmospheric conditions, and in daytime cloudy conditions up to cloud base [13].

2. INSTRUMENT DESIGN

The 'next' generation diode-laser-based WV-DIAL design uses a single frequency, single mode, distributed Bragg reflector (DBR) laser diode to seed an electrically pulsed GaAs semiconductor laser diode tapered amplifier. The transmitter has rapid frequency agility to alternate between online and offline wavelengths at rates up to 100 Hz and provides spectrally pure, narrowband amplified pulses. The pulsed laser output is shaped with a pair of axicons for efficient transmission through the inner portion of a shared telescope into an opto-mechanically stable and eye-safe annular transmit beam. The outer portion of the shared telescope is used to collect backscattered photons which subsequently pass through an ultra-narrow multistage optical filtering. A critical part of the filter is a solid etalon designed to allow transmission of the online and offline radiation in adjacent cavity modes. A pair of photon-counting receiver modules – designed with narrow field-of-view for longer range, and wider field-of-view for measuring closer to the surface – are used to detect the backscattered photons. Each photon-counting module output enters a switch, which directs the signal to one of two channels on a four channel scalar card (two for each receiver) based on the transmits wavelength. The counts are gated and accumulated within the scalar card for 1 sec and then read out by a computer. A schematic is shown in Fig. 1 which appears in the Atmospheric Measurement Techniques article by Spuler et al. [13] and contains a more complete description and specifications of the system.

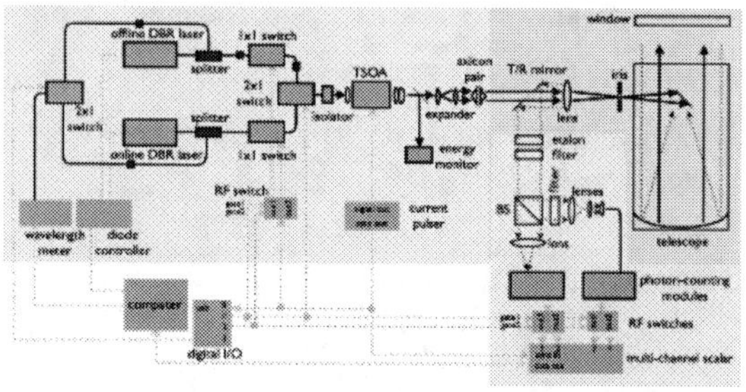

Figure 1. Schematic of the WV-DIAL system. BS = beam-splitter; T/R = transmit receive; I/O = input output. This figure appears in the Atmospheric Measurement Techniques article by Spuler et al. [13] and is used here with permission from Copernicus publications.

3. FIELD TEST RESULTS

The instrument was autonomously operated in Erie, CO, USA during the Front Range Air Pollution and Photochemistry Experiment (FRAPPE) between 1-July 2014 and 19-August 2014. The WV-DIAL operated near continuous for 50 days with an uptime of $> 95\%$. The shared telescope design proved stable – no re-alignments were made and the system had no measurable degradation in performance during the field campaign. Figure 2 shows the data from the project. The continuous observations of relative backscatter (top) illustrate the diurnal variation in the top of the aerosol layer and clearly identify cloud occurrences up to 12 km AGL. The water vapor profiles (displayed up to 7 km in the second panel) show data up to cloud base. The water vapor data illustrate substantial temporal variability and strong vertical gradients. The third panel shows background in counts per second (C s^{-1}) for the near (red) and far (black) range channels. Count rates for the far range channel remain within the linear count regime for the photon counting module (Excelitas SPCM-AQRH-13) – below 1 MC s^{-1} for all conditions encountered during the project. The maximum count rate for the near range channel exceeds the linear count rate during daytime, bright-cloudy, conditions and these periods must be filtered out of the water vapor retrieval profiles. Ongoing studies to quantify the diurnal variability and to compare with radiosondes and GPS receivers will help to characterize the system. Continuous observations from this WV-DIAL will provide opportunities for both mesoscale and climatological characterization of water vapor structure and trends.

During FRAPPE, there were 2 nighttime soundings and 43 daytime soundings. The large number of daytime sondes allowed for comparisons with the DIAL during periods of high background solar radiation – the most difficult data collection conditions for the instrument. The mean relative error was less than 10% for all of the radiosonde comparisons, from 300 m to 4 km above ground level[13]. Therefore, data from this field campaign demonstrate the capability of the WV-DIAL to provide accurate measurements of water vapor in continual operation for an extended period of time in a variety of atmospheric conditions.

A national network for water vapor may include other thermodynamic profiling instruments such as passive remote sensors that measure water vapor and temperature which are useful at low ranges, close to the surface, but in general provide low vertical resolution. During FRAPPE, the WV-DIAL was collocated with a continually-operating passive infrared radiometer (AERI: Atmospheric Emitted Radiance Interferometer.) This is a passive infrared remote sensing instrument that utilizes an interference technique to retrieve atmospheric emitted radiance [14, 15, 16, 17]. Starting with initial temperature and humidity profiles based on statistical models, an iterative solution to the radiative transfer equations is utilized to reproduce the measured atmospheric emitted radiance. This iterative solution provides the final temperature and WV profiles for clear sky conditions up to approximately 3 km. During cloudy conditions, retrievals are sensitive to the cloud properties leading to larger errors in the temperature and WV profiles, so a separate measurement of the cloud base height is often required.

A preliminary comparison of the WV DIAL and AERI is included as Fig. 3. The AERI retrievals provide continuous profiles of water vapor (along with temperature which is not shown) up to 3.5 km. The AERI profiles were retrieved every 10 min with a vertical resolution of 30 m at the surface decreasing to 600 m at 3 km. The WV-DIAL data were processed with 1 min averaging and a vertical resolution of 150 m (300 m smoothing were applied to the data above 3 km.) Note the comparable buildup in the moisture leading into 23 July. The increasing moisture values and depth are similarly observed by both instruments. The AERI and WV-DIAL are both limited in times of fog, such as that before 24 July. Both datasets show a similarly dramatic decrease in moisture just prior to 24 July. Perhaps this was caused by a frontal passage or other change in wind direction. Just prior to this abrupt change there appears to be an absolute humidity inversion in the DIAL data at 1 km AGL (note, inversions are not observable with passive remote sensors such as AERI). Again there is a comparable buildup in moisture quantity and depth at the end of the time series leading into 25 July. Further comparisons will be obtained through statistical methods for the entire FRAPPE time period. In the future we plan to investigate combining the DIAL water vapor and AERI radiances to see if the retrieved temperature profile can be improved.

4. CONCLUSIONS AND FUTURE PLANS

In summary, a next generation water vapor DIAL has been designed, constructed and tested which has the potential to enable a national scale network. The electrically-pumped semiconductor-based transmitter is

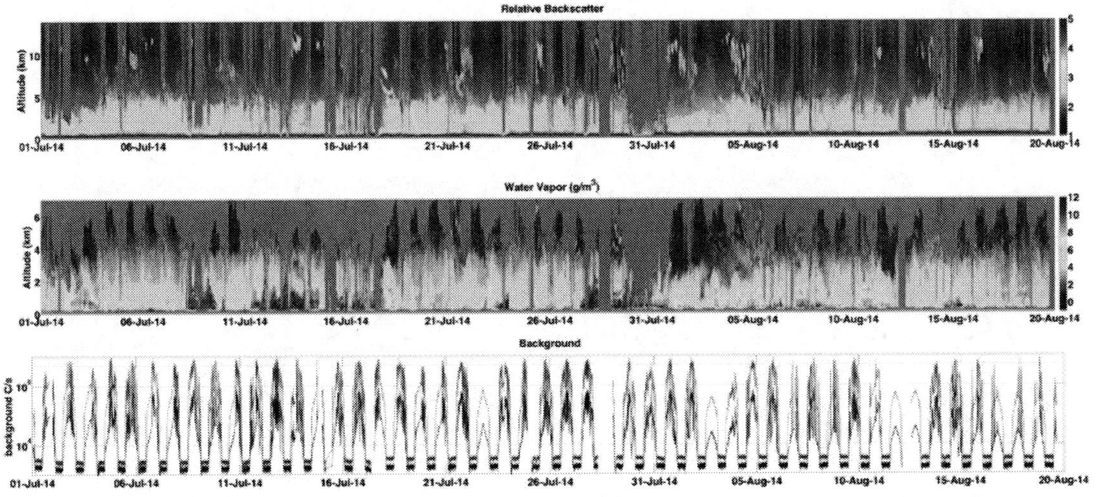

Figure 2. Fifty days of near continuous data collected 1 July to 20 August 2014. Top panel: 1 min, 150 m resolution relative backscatter from 0 to 14 km range for the far range channel. Middle panel: 10 min, 150 m resolution water vapor in g m^{-3} from 0 to 7 km range for the far range channel – 300 and 600 meter smoothing were applied to the data from 3.0-5.25 km, and above 5.25 km; respectively. Bottom panel: background in counts per second (C s^{-1}) for the near (red) and far (black) range channels.

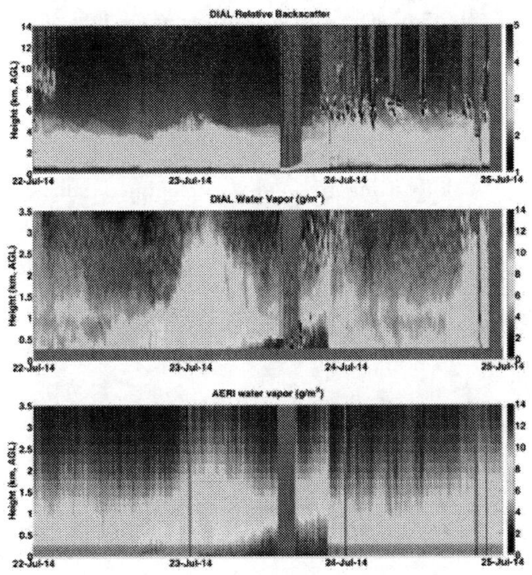

Figure 3. Comparisons between collocated WV-DIAL aerosol (top panel) and water vapor (middle panel) and AERI water vapor (bottom) 22-25 July 2014 during FRAPPE. Note the bottom 300 meters of the AERI image is grayed out for comparison purposes only.

inherently low-maintenance, low-cost, and designed to be eye-safe at the exit port. It has the required characteristics necessary for the DIAL technique; rapid frequency agility with good spectral purity, and when combined with the ultra-narrowband multistage optical filters in the receiver, the system provides continuous profiles of water vapor with complete daytime coverage – including periods of bright clouds – with better than +/- 10% relative error from 300 m above ground level to 4 km (or cloud base, whichever comes first) with 150 m nominal vertical resolution and 1 minute temporal resolution. This lower cost lidar design is a significant step towards improving our understanding of the distribution of atmospheric water vapor.

ACKNOWLEDGMENTS

We would like to acknowledge Richard Carbone for helpful discussions pertaining to the atmospheric science applications, and Richard Erickson for technical support during development. We would also like to thank to Dave Turner for processing and supplying the AERI data used for the initial inter-comparison with the WV-DIAL. NCAR is sponsored by the National Science Foundation. The authors affiliated with Montana State University would like to acknowledge the support of the National Science Foundation grant number 1206166.

REFERENCES

[1] National Research Council, 2009: *Observing Weather and Climate from the Ground Up: A Nationwide Network of Networks*. The National Academies Press, Washington, DC.

[2] National Research Council, 2010: *When Weather Matters: Science and Service to Meet Critical Societal Needs*. The National Academies Press, Washington, DC.

[3] Goldsmith, J. E., Forest, M., Blair, H., Bisson, S. E., and Turner., D. D., 1998: Turn-Key Raman Lidar for Profiling Atmospheric Water Vapor, Clouds, and Aerosols. *Applied Optics*, **37(21)**, 4979–4990.

[4] Turner, D. D., Ferrare, R. A., Brasseur, L. A. H., Feltz, W. F., and Tooman, T. P., 2002: Automated Retrievals of Water Vapor and Aerosol Profiles from an Operational Raman Lidar. *Journal of Atmospheric and Oceanic Technology*, **19(1)**, 37–50.

[5] Ertel, K., Linné, H., and Bösenberg, J., 2005: Injection-seeded pulsed Ti:sapphire laser with novel stabilization scheme and capability of dual-wavelength operation. *Applied Optics*, **44**, 5120–5126.

[6] Bösenberg, J. and Linné, H., 2006: Continuous Ground-Based Water Vapour Profiling using DIAL. In *23rd International Laser Radar Conference*, pages 679–682, Nara City, Japan.

[7] Vogelmann, H. and Trickl, T., 2008: Wide-range sounding of free-tropospheric water vapor with a differential-absorption lidar (DIAL) at a high-altitude station. *Applied Optics*, **47**, 2116-2132.

[8] Behrendt, A., Wulfmeyer, V., Riede, A., Wagner, G., Pal, S., Bauer, H., and Späth, F., 2010: Scanning differential absorption lidar for 3D observations of the atmospheric humidity field. In *25th International Laser Radar Conference*, pages 1187–1190, St. Petersburg, Russia.

[9] Machol, J. L., Ayers, T., Schwenz, K. T., Koenig, K. W., Hardesty, R. M., Senff, C., Krainak, M. A., Abshire, J. B., Bravo, H. E., and Sandberg, S. P., 2004: Preliminary Measurements with an Automated Compact Differential Absorption Lidar for the Profiling of Water Vapor. *Applied Optics*, **43(15)**, 3110–3121.

[10] Nehrir, A. R., Repasky, K. S., Carlsten, J. L., Obland, M. D., and Shaw, J. A., 2009: Water Vapor Profiling Using a Widely Tunable, Amplified Diode-Laser-Based Differential Absorption Lidar (DIAL). *Journal of Atmospheric and Oceanic Technology*, **26(4)**, 733–745.

[11] Nehrir, A. R., Repasky, K. S., and Carlsten, J. L., 2012: Micropulse water vapor differential absorption lidar: transmitter design and performance. *Optics express*, **20(22)**, 25137–51.

[12] Nehrir, A. R., Repasky, K. S., and Carlsten, J. L., 2011: Eye-Safe Diode-Laser-Based Micropulse Differential Absorption Lidar (DIAL) for Water Vapor Profiling in the Lower Troposphere. *Journal of Atmospheric and Oceanic Technology*, **28(2)**, 131–147.

[13] Spuler, S. M., Repasky, K. S., Morley, B., Moen, D., Hayman, M., and Nehrir, A. R., 2015: Field deployable diode-laser-based differential absorption lidar (DIAL) for profiling water vapor. *Atmospheric Measurement Techniques*, **8**, 1073–1087.

[14] Turner, D. D. and Löhnert, U., 2014: Information content and uncertainties in thermodynamic profiles and liquid cloud properties retrieved from the ground-based atmospheric emitted radiance interferometer (AERI). *Journal of Applied Meteorology and Climatology*, **53(3)**, 752–771.

[15] Knuteson, R., Revercomb, H., Best, F., Ciganovich, N., Dedecker, R., Dirkx, T., Ellington, S., Feltz, W., Garcia, R., Howell, H., Smith, W., Short, J., and Tobin, D., 2004: Atmospheric emitted radiance interferometer. part II: Instrument performance. *Journal of Atmospheric and Oceanic Technology*, **21**, 1777–1789.

[16] Knuteson, R., Revercomb, H., Best, F., Ciganovich, N., Dedecker, R., Dirkx, T. P., Ellington, S., Feltz, W., Garcia, R., Howell, H., Smith, W., Short, J., and Tobin, D., 2004: Atmospheric emitted radiance interferometer. Part I: Instrument Design. *Journal of Atmospheric and Oceanic Technology*, **21**, 1763–1776.

[17] Feltz, W., Smith, W., Howell, H., Knuteson, R., H., W., and Revercomb, H., 2003: Near-continuous profiling of temperature, moisture, and atmospheric stability using the atmospheric emitted radiance interferometer (AERI). *Journal of Applied Meteorology*, **42**, 584–597.

INJECTION SEEDED LASER FOR FORMALDEHYDE DIFFERENTIAL FLUORESCENCE LIDAR

G. Schwemmer [1]*, M. Yakshin[1], C. Prasad[1], T. Hanisco[2], A. R. Mylapore[3]†, I.H. Hwang[1], S. Lee[1]

[1]*Science & Engineering Services, LLC, Columbia, MD 21076, USA *Email: schwemmer@sesi-md.com*

[2]*NASA Goddard Space Flight Center, Greenbelt, MD 20771, USA*

[3]*MassTech, Inc., Columbia, MD 21076, USA*

†*currently with Aeromancer Technologies, Washington DC 20008, USA*

ABSTRACT

We describe the design and development of an injection seeded Nd:YVO_4 laser for use in a differential fluorescence lidar for measuring atmospheric formaldehyde profiles. A high repetition rate Q-switched laser is modified to accept injection seed input to spectrally narrow and tune the output. The third harmonic output is used to excite formaldehyde (HCHO) fluorescence when tuned to a HCHO absorption line. Spectral confirmation is made with the use of a photoacoustic cell and grating spectrometer.

1. INTRODUCTION

Formaldehyde (HCHO), both naturally occurring and anthropogenic, plays a key role in our understanding of short lived organic compounds in the atmospheric environment. It is a known carcinogen, respiratory irritant, and a component of smog. HCHO is a primary emission product of combustion and certain manufacturing processes, particularly for wood and paper products, but is also a degradation product of a large group of volatile organic compounds, both natural and man-made.[1] Hence, HCHO is an important target measurement for several environmental satellite programs, both existing and planned.[2]

In situ instruments can obtain precision measurements, but have limited coverage. An active in-situ sensor based on the Laser Induced Fluorescence (LIF) technique is the NASA In Situ Airborne Formaldehyde (ISAF) sensor[3], and is the foundation on which this instrument is based. Table 1 summarizes some of the current in situ instruments and techniques for formaldehyde measurements.

Besides the in-situ measurements, present remote sensing methods include long-term observations of HCHO columns from space using the Global Ozone Monitoring Experiment (GOME) satellite instrument launched in 1995.[7] GOME only provides column measurements and suffers from contaminations due to other atmospheric constituents such as O_3, NO_2, BrO and O_4. Other passive remote column measurement techniques include the Aura satellite's OMI instrument, ground-based FTIR and Multi-Axis Differential Optical Absorption Spectroscopy.[8]

Only the lidar can provide high vertical and spatial resolution simultaneously in a single instrument, making it a logical choice for HCHO science and satellite calibration and validation.

2. MEASUREMENT METHODOLOGY

HCHO fluoresces after exposure to ultraviolet (UV) radiation, appearing as a distinct series of spectral bands at specific wavelengths. The intensity of the fluorescence is proportional to the UV energy absorbed by the HCHO as well as its concentration. However, background fluorescence from other molecules and naturally occurring aerosols are also present in the same

Table 1. A comparison of some current in situ HCHO instruments and their levels of sensitivity.

Instrument	Techniques	Detection Limit (ppbv)	Time Resolution (s)
ISAF instrument[3]	LIF Technique	0.036	1
HCHO TDLAS[4]	TDLAS	0.05	2
Fourier Transform Infrared Interferometry White system – Chalmers University of Technology[5]	FTIR	0.4	5
Hantzsch AL4021 – Paul Scherrer Institute[6]	Fluorimetric Hantzsch Reaction	0.15	1.5

wavelength region. The HCHO fluorescence signal is maximized by tuning the UV transmitter to an HCHO absorption maximum (on-line). The laser is alternately switched between the on-line wavelength and an absorption minimum (off-line) in order to get a differential fluorescence measurement to remove unwanted background fluorescence from interfering gases and aerosols and determine the range resolved concentration of HCHO. A differential absorption lidar measurement can also be made from the elastic backscatter signals to get column content, but the very low concentration of HCHO makes its absorption too weak to resolve vertical structure.

The fluorescence emission spectrum of HCHO is shown in Figure 1.[9] Peaks in the spectrum correspond to vibronic transitions from the excited state to the ground state. The strongest is the 414 vibrational band located at 395.4 nm, indicated by the red line in Figure 1.

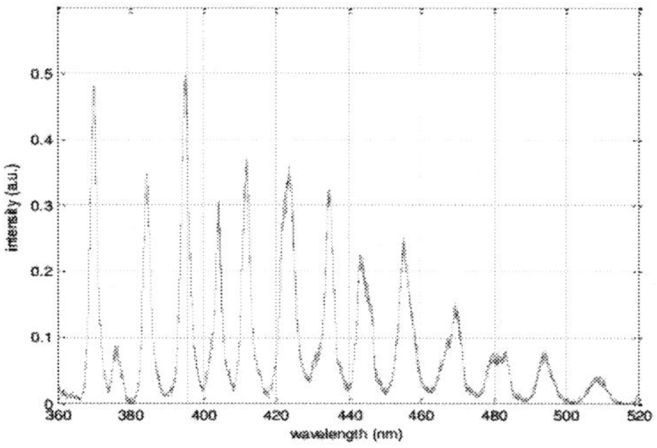

Figure 1. Fluorescence Spectrum of HCHO.

Rotationally resolved HCHO absorption measurements[10] have shown the presence of numerous strong lines in the 352 to 355 nm region, near the third harmonic wavelength of commonly used Nd:YAG and Nd:YVO_4 lasers. Figure 2 provides a plot of the portion of the HCHO absorption spectrum accessible to these lasers.

Table 2 shows the projected sensitivity of the instant lidar based upon our earlier results and the system parameters given below.[11, 12]

3. SYSTEM DESCRIPTION

We selected a diode pumped Coherent AVIA Nd:YVO_4 laser based on its reliability, high efficiency, and relatively broad (~ 0.15 nm) gain

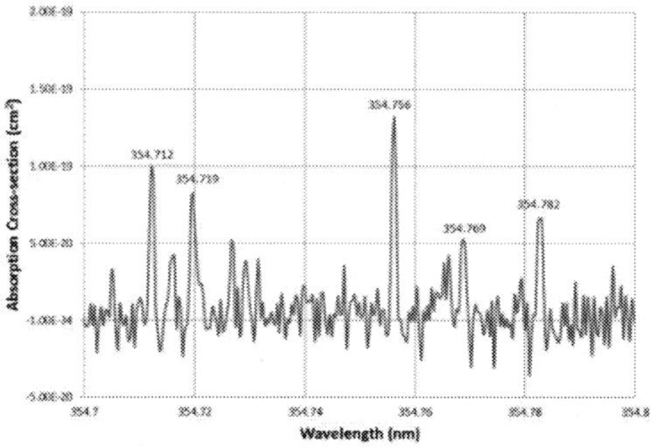

Figure 2. Rotationally resolved absorption spectrum for HCHO at 150 Torr, accessible to the third harmonic Nd:YVO_4 laser.

Table 2. Estimated sensitivity of the lidar.

Parameter	Value	
Integration Time	10 seconds	5 minutes
Range Resolution	1000 m	100 m
Sensitivity @ 1km	~85 ppt	~40 ppt

bandwidth as the starting point for the slave laser, modifying it to accommodate injection seeding. The use of a high rep-rate, low pulse energy limits this system to night time measurements only.

Figure 3 is a block diagram of the lidar system. It is configured as two assemblies, one containing a 30 cm diameter, 160 μrad field of view transceiver telescope, focal plane optics, and three photon counting photomultipliers to measure elastic backscatter, N_2 vibrational Raman backscatter, and the HCHO fluorescence signal. The fluorescence channel utilizes an interference filter produced specifically for HCHO measurements having 11-passbands between 380 and 480 nm that match the fluorescence bands shown in Figure 1. Further details of the system can be found in reference 12.

Referring again to Figure 3, the slave laser produces up to 1 W of UV Q-switched pulses at a 10 KHz rate. A narrow line (~1 MHz) single mode 1064 nm DFB diode laser is used to seed the slave laser through its output coupler. Four optical isolators (not shown) are used to prevent feedback. The slave laser head is mounted on a chill plate controlled by a small recirculating liquid chiller to maintain temperature stability.

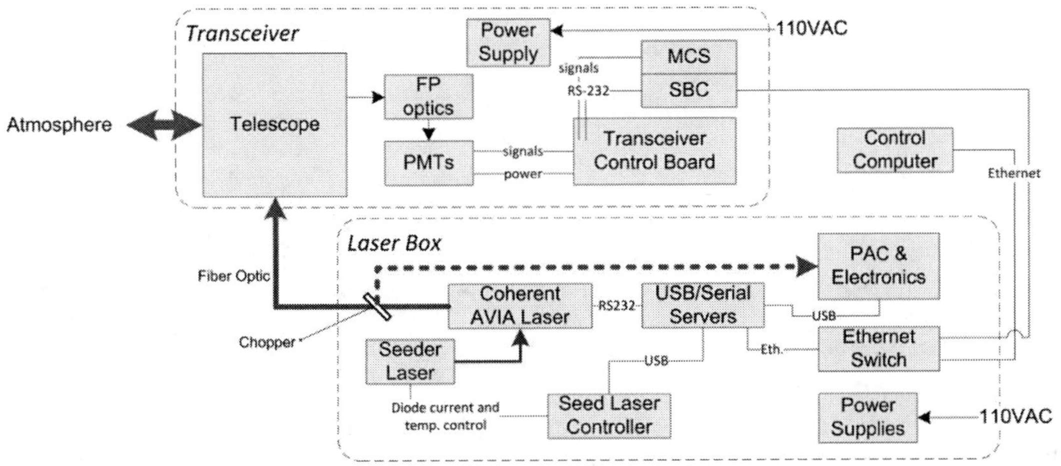

Figure 3. Block diagram of the HCHO lidar system.

4. INITIAL TEST RESULTS

A 46 Hz chopper equipped with a 4 mm wide mirror directs 0.35 ms bursts of laser pulses once per rotation to a photoacoustic spectroscopy (PAS) cell containing HCHO vapor. The PAS cell is resonant at 1.2 kHz and equipped with a microphone, whose signal is passed through a band pass amplifier centered at 1.2 kHz and digitized. We integrate the power in the PAS signal for 100 ms and normalize it by the laser power monitor signal to determine the magnitude of the absorption that takes place.

Continuous single mode seeded operation requires active control of the slave laser cavity length. However due to the monolithic nature of the construction of AVIA laser, it was not possible. Modulating the diode laser output to directly lock it to an HCHO UV absorption line is also not feasible since any rapid wavelength modulation is not conveyed to the slave laser's UV output when the cavity is not continuously seeded. Hence, we currently rely on the passive stability of the cavity to secure seeded operation. A 1-meter spectrometer equipped with a linear array detector (0.02 nm resolution in UV) is used to position the wavelength of the seed laser on or near the HCHO absorption line at 354.756 nm and also to determine the efficacy of injection seeding and its stability. When seeded, the output bandwidth shrinks from ~0.1 nm to ~0.02nm or less. The UV output power also drops significantly when the laser is well seeded, which may be due to spatial hole burning in the Nd:YVO_4 crystals when the laser is running single longitudinal mode.

Because the HCHO spectrum is fairly crowded with absorption lines, there is always a response by the PAS microphone when the laser is well seeded, partially seeded, or completely unseeded. In order to accurately locate on-line and off-line wavelengths, we perform scans of the seed laser wavelength across the gain bandwidth of the slave laser, by controlling the seed laser current. We monitor the PAS signal and the slave laser energy while also observing the spectrometer display to interpret the results. Figure 4 is a plot of the laser energy, the PAS signal amplitude, and laser power

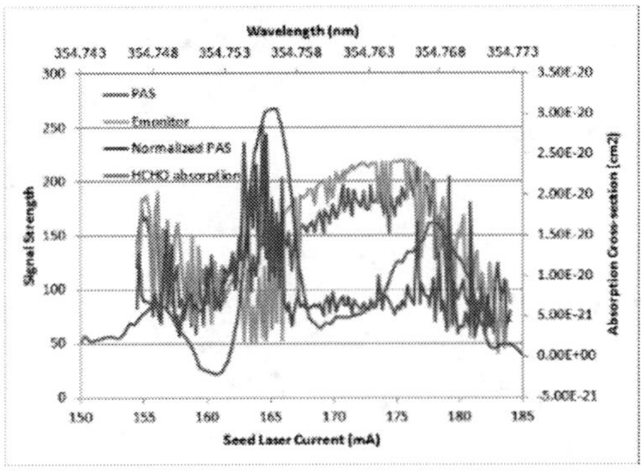

Figure 4. Current scan of the seed laser while plotting laser power (green) PAS signal power (blue), and PAS signal normalized for laser power (red). The absorption spectrum of HCHO is overlaid (violet).

normalized PAS amplitude as a function of the seed laser current. We have also overlaid a plot of the HCHO absorption cross section vs wavelength, where the correlation between diode current and wavelength was made using the spectrometer for a first order calibration and dispersion, then lining up the spectrum with features in the normalized PAS signal. Note the structure in the laser energy monitor signal (green line) across the scan as the slave laser goes in and out of seeded operation.

5. PATH FORWARD

Inconsistent seeding of the laser makes lidar data collection inefficient. This is because even when the seed laser wavelength is constant, drift in the slave laser cavity causes it to be unseeded for part of the time, and the fraction of corresponding data would need to be eliminated. Efforts are under way to achieve more stable seeded operation with the current system by improving the passive temperature stability and by using a broader line width seed laser that would overlap two to three slave cavity modes at any time, producing a multimode slave laser output that will stay seeded constantly, while still spectrally narrow with respect to the HCHO absorption line. Final integration and system level testing will follow.

ACKNOWLEDGEMENTS

We would like to acknowledge the engineering contributions of A. Achey, N. Mehta, and R. Clark of SES, and Dr. G. Li and K. Novoselov of MassTech, Inc. This work is supported by a MassTech purchase order under one of their SBIR contracts.

REFERENCES

[1] Abbot, D. S., et al., 2003: Seasonal and interannual variability of North American isoprene emissions as determined by formaldehyde column measurements from space, *Geophys. Res. Lett.*, **30 (7)**, 1886.

[2] NRC Decadal Survey 2007: Earth Science Applications from Space: National Imperatives for the next Decade and Beyond, National Research Council.

[3] M. Cazorla, *et al.*, 2015: A new airborne laser-induced fluorescence instrument for in situ detection of formaldehyde throughout the troposphere and lower stratosphere, *Atmos. Meas. Tech.*, **8**, 541-552.

[4] Wert, B. P., *et al.*, 2003: Design and performance of a tunable diode laser absorption spectrometer for airborne formaldehyde measurements, *J. Geophys. Res.*, **108**, 4350.

[5] Ritz, D., et al., 1993: An improved open-path multireflection cell for the measurement of NO2 and NO3, in: Optical methods in atmospheric chemistry, edited by: Schiff, H. I. and Platt, U., *SPIE Proceedings*, **1715**, 200–211.

[6] Kelly, T. J. and Fortune, C. R. 1994: Continuous monitoring of gaseous formaldehyde using an improved fluorescence approach, *Int. J. Environ. Anal. Chem.*, **54**, 249–263.

[7] Chance, K., *et al.*, 2000: Satellite observations of formaldehyde over North America from GOME, *Geophys. Res. Lett.*, **27(21)**, 3461–3464.

[8] Jones, N., *et al.*, 2007: Long-term tropospheric formaldehyde concentrations deduced from ground-based Fourier transform solar infrared measurements, *Atmos. Chem. Phys. Discuss.*, **7**, 14543-14568.

[9] Brackmann, C. *et al.*, 2003: Laser-induced fluorescence of formaldehyde in combustion using third harmonic Nd:YAG laser excitation, *Spectrochimica Acta Part A*, **59**, 3347

[10] Co, D. T., *et al.*, 2005: Rotationally resolved absorption cross sections of formaldehyde in the 28100-28500 cm^{-1} (351-356 nm) spectral region: Implications for in situ LIF measurements, *J. Phys. Chem. A*, **109(47)**, 10675 – 10682.

[11] Li, G., *et al.*, 2012: Measurement of atmospheric formaldehyde profiles with a laser-induced fluorescence lidar, *Laser Radar Technology and Applications* XVII *Proc. SPIE*, **8379**, Paper 18.

[12] Mylapore, A. R., *et al.*, 2014: Development of a fluorescence lidar for measurement of atmospheric formaldehyde, *Proc. SPIE*, **9080**, *Laser Radar Technology and Applications XIX; and Atmospheric Propagation XI*, 90800X (June 9, 2014); doi:10.1117/12.2053420.

IAOOS MICROLIDAR DEVELOPMENT AND FIRST RESULTS OBTAINED DURING 2014 AND 2015 ARCTIC DRIFTS

Vincent Mariage [1,3], Jacques Pelon[1], F. Blouzon[2], Stéphane Victori[3]

[1]*LATMOS, UPMC-CNRS, Paris, France*

[2]*DT/INSU, Meudon, France*

[3]*Cimel Electronique, Paris, France*

ABSTRACT

The development of a first ever autonomous aerosol and cloud backscatter lidar system for on-buoy arctic observations has been achieved in 2014, within the French EQUIPEX IAOOS project developed in collaboration with LOCEAN at UPMC. This development is part of a larger set-up designed for integrated ocean-ice-atmosphere observations. First results have been obtained from spring to autumn 2014 after the system was installed at the North Pole at the Barneo Russian camp, and in winter-spring 2015 during the Norwegian campaign N-ICE 2015. The buoys were taking observations as drifting in the high arctic region where very few measurements have been made so far. This project required the design and the conception of an all-new lidar system to fit with the numerous constraints of such a deployment. We describe here the prototype and its performance. First analyzes are presented.

1. INTRODUCTION

The arctic region is of main importance in the global climate changes, in particular because consequences of global warming are larger there. A better comprehension of this complex and fragile region goes through the analysis of the still poorly-known ocean-ice-atmosphere interactions. On the atmospheric field it means studies of vertical distribution of aerosols and clouds (type, height and optical/microphysical properties), to allow a better determination of the arctic cloud radiative forcing. To access these data vertically resolved measurements are needed, and more particularly LIDAR measurements in the high arctic region (above 75°N). Only a few campaigns have been previously performed to increase the knowledge of cloudiness arctic. Among them one can quote the SHEBA campaign (October 1997-October 1998) which provided lot of quantitative data and analysis between 75 and 78N [1]. Satellite observations provide the necessary regional and temporal coverage. CALIPSO is giving access to critical information on clouds. Still, the lower layers of the atmosphere may be occulted by upper layers, and the very high arctic regions above 82N are not sampled.

The goal of the French EQUIPEX IAOOS project is to bridge this gap by deploying a network of autonomous LIDAR set on drifting buoys, as well as others oceanographic instruments in the high arctic region. LIDAR are planned to be working during at least two years with daily few periods of measurements (up to 4 averaged profiles over 10 minutes). The opto-mechanical design of the LIDAR has been led by performances objectives of daytime molecular range of 5km in clear sky. A first buoy loaded with a simple backscattering LIDAR (i.e. no polarization) has been deployed in mid-April 2014, close to the North Pole. During about 8 months the buoy has drifted toward the north of Svalbard, providing more than 750 profiles as a whole. More recently two systems have been deployed during the first two legs of the N-ICE2015 campaign (www.npolar.no), including polarized emission for one of them.

The all-new design of the LIDAR developed within this project will be presented, in parallel with fallouts systems obtained during these drifts. Performances are also compared with simulations, and first results are given.

2. INSTRUMENT DESCRIPTION

LIDAR systems are often expensive to develop or to buy and usually require regular maintenance. Consequently a specific attention must be paid to the design for an autonomous deployment. Besides this, autonomy also means that the whole system will have to work on batteries, so with a limited available energy (few watts/hour a day).

A laser diode based system was then chosen. Such a choice would thus *a priori* make the system part of ceilometers' family, but refining its performance would however be needed to meet our requirements, allowing to better bridge the gap in LIDAR network [2]. The goal was thus to achieve a higher efficiency in molecular detection along with a smaller sensitivity to water absorption, using current laser diode technology. The choice of an optical fiber based system has been led by thermal constraints, putting the sensitive parts (electronic, detector, laser diode) in the lower part of the buoy surrounded by "warmer" arctic water. Due to the random drift of the buoy, eye safety of the system was compulsory in order to allow any deployment. Energy was therefore limited, and a "microlidar" architecture was preferred, which means high frequency (5 to 10 kHz) operation and several minutes average for each profile (10min). To optimize stability optical design was based on a bi-axial structure developed at INSU by Division Technique. Electronics and diode were provided by CIMEL Electronique. Polarized emission for the N-ICE2015 campaign has been obtained by adapting a polarizing device at the output. In order to limit satellite communication costs, each profile is spatially averaged. Data are transferred to IPEV (www.iaoos.ipev.fr) using Iridium satellite link.

First 2014 drift showed a very good behavior of the whole system, with no unexpected power consumption in arctic cold conditions, and very few profiles lost because of satellite communications. The change of frequency measurements (from 2 profiles to 4 profiles a day, and inversely) allowed some flexibility. Accelerometers are set in the buoy, which provided us with angle information concerning the tilt of the buoy due to leads and/or ridges formation. A first calibration of the transmission of the system is performed in lab. Atmospheric calibration can further be performed in the field with reference to molecular and dense water cloud scattering.

3. DATA CORRECTION AND CALIBRATION

Two main difficulties have nevertheless been met with the exploitations of the data, on this first prototype. The less disabling and rather easy to correct is an undershoot effect due to oversaturation of the APD which is the detector used for photon counting. This saturation occurs with high backscattering signals from very low (below 2km) clouds or intense haze, and prevents the APD from counting correctly during several microseconds. A polynomial fit is used to correct this perturbation, and enables the recovery of few hundred meters of useful signal. The other difficulty in processing raw data comes from random various icing of the window, despite the built-in heating system. This icing problem impacts the calibration of the IAOOS microlidar for a quantitative analysis of aerosol and cloud properties. However, as icing also increases the diffused signal measured at the very beginning of the profile, this information was used to put aside too iced profiles, and possibly correct others.

Before any analysis, raw profiles were detector corrected, since photon counting APD suffered from deadtime limitation, also corrected simultaneously with the undershoot effect.

A first analysis provided a selection of 15 clear air profiles without icing over the first four

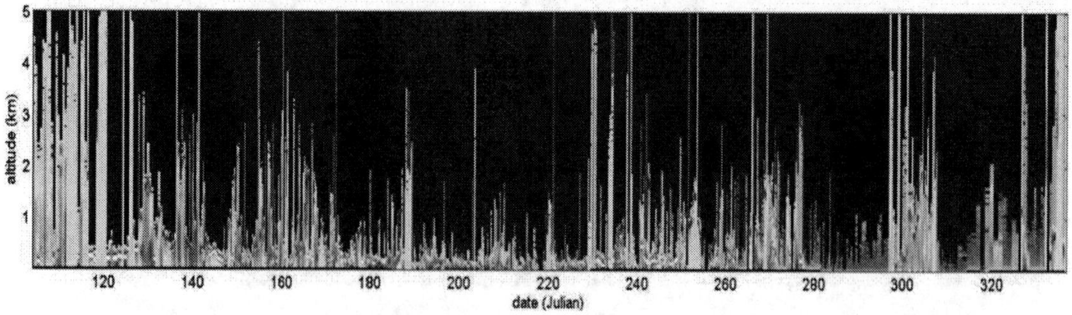

Figure 1: Altitude-time cross-section of the log(PR2) of the whole dataset of 2014 drift. Very low dense clouds, icing of the window and high tilt of the buoy limit most of the time the range of the microlidar.

month period to check calibration on molecular scattering. Results reported in Fig. 2 show no significant drift. The lidar constant is equal to 62 000, in good agreement with estimation from lab measurements and detector data sheet.

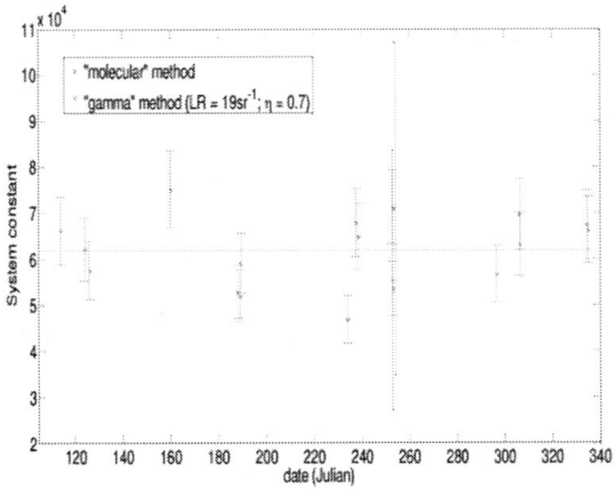

Figure 2: Values of system constants calculated either on molecular scattering (blue), or on dense water cloud (red) with the "gamma" method and adapted multiple scattering so that good agreement with molecular method was found.

Knowing the base and top of clouds (see next section), it is possible to further check the calibration by evaluating the integrated attenuated backscatter signal (usually written as γ) over dense water clouds. This method can provide another estimate of the constant system as the lidar ratio for water cloud is known to be constant or an estimate of multiple scattering knowing the clear air calibration constant [4].

The second arctic drift during the N-ICE 2015 campaign was the opportunity to test the new polarization set up of the microlidar, and to obtain more information on the system. Alignment and calibration using a polarization rotation method to balance parallel and perpendicular signals was performed in lab. From the signal measured on dense liquid water cloud, one can evaluate the multiple scattering coefficient using the relation $\eta = (1-\delta)^2/(1+\delta)^2$ [5]. A multiple scattering coefficient of 0.81 +/- 0.08 was derived after saturation correction, from test measurements over Paris. This is in good agreement with that found with the similar system from 2014 campaign. System transmission was controlled and summing up both polarization signals one obtain with this "gamma" method system constants of about 68200 +/- 8900. This is very close (better than 10%) to that obtained with the "molecular method" (73300 +/- 6800) applied to different zones of the total signal for this system.

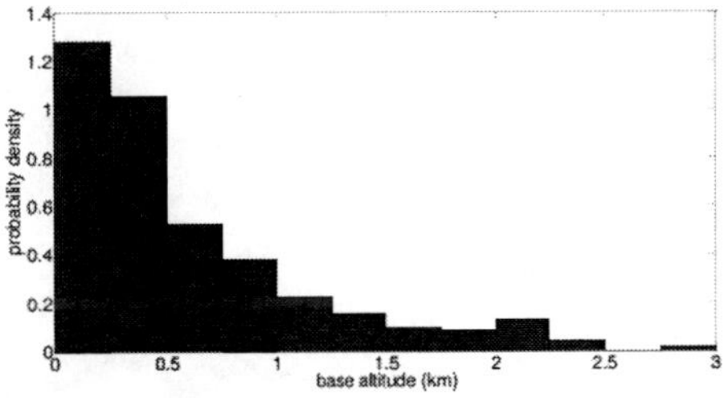

Figure 3: density of probability of the base altitude of the clouds detected between 0 and 3km.

4. RESULTS

The 2014 drift provided about 750 profiles of total backscattering profiles in the area 80-88°N and 10-25°E. More than one hundred of them were taken at the end of the drift when the window was so iced that absolutely no signal made its way through it. More than fifty percent of the remaining profiles can be considered to be totally or mostly ice- or snow-free, according to the amplitude of the diffused window signal, used as an indicator of window icing.

Using Haar wavelet and adapted threshold over vertical profiles, base and effective top of clouds have been retrieved, regardless of how high or low were the amplitude of the window diffused signal. This first analysis shows very low base of clouds (Fig. 3), frequently below1km of altitude and a high frequency of very low (under 1km) altitude cloudiness (as high as 60%), mostly due to supercooled water clouds. This is in good agreement with measurements made during SHEBA [3]. Considering profiles below half the window icing threshold, there are as high as 60% of these profiles showing signal which can be considered as a cloud.

First IAOOS polarized profiles were obtained during N-ICE2015 and normalized with the calibration constant value determined. An example is shown in Figure 4. Depolarization ratio as low as 15-22% are observed, which may mean plate crystals were prevalent [6], although

DOI: 10.1051/epjconf/201611902005

this cloud was at a temperature about –48°C according to data soundings made in Ny-Alesund. The slope below the clouds corresponds to low aerosol contribution superimposed on molecular

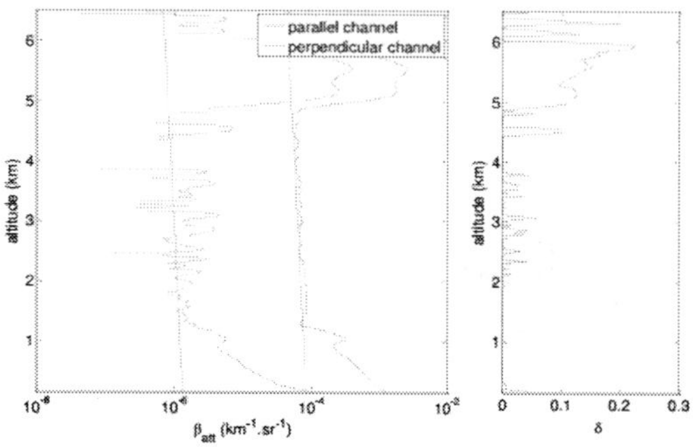

Figure 4: left: attenuated backscatter coefficient measured in both co- and cross-polarizations during the N-ICE2015 campaign with normalized molecular reference as dashed lines, and right: depolarization ratio obtained with calibration by polarization rotation method after crosstalk correction. All profiles have been filtered over 5 points.

scattering. Since the spectral bandwidth transmits only the Cabannes line, one expected about 0.36% of molecular depolarization, instead of the 1.6% observed (ratio between parallel and perpendicular β_{att} in Fig. 4). This cross-talk effect can be compensated by the previous calibration, and the depolarization ratio in right figure (Fig. 4) does not show this deviation. Depolarized peaks below the clouds are due to the very noisy perpendicular channel signal. Thus at such altitudes one can expect to detect only depolarization above 5%, this threshold increasing with the decrease of SNR. Even if there might be quite high uncertainties for some profiles, the polarization will unambiguously enable the discrimination between liquid water and ice, and help improving our knowledge concerning vertical and temporal cloud structure.

5. CONCLUSIONS

The development of first autonomous microlidar systems developed within the IAOOS project and their first deployment on buoy in the arctic has been successful. Very good first results have been obtained in terms of instrumental unattended operation. New solutions for avoiding icing of the window will be tested. Despite the icing limitation, valuable statistical analyzes could be started concerning the low level arctic clouds, and therefore IAOOS observations are expected to contribute to a better understanding of radiative budget at the surface, along with ice and ocean measurements simultaneously performed on the buoys by LOCEAN.

ACKNOWLEDGEMENT

We are thankful to the DT/INSU/CNRS for the mechanical and system control design as well as tests of the microlidar and buoy set up in the arctic. We also would like also to thank the Association Nationale de la Recherche et de la Technologie and CIMEL Electronique for supporting the PhD work of V. Mariage, and French Research Ministry EQUIPEX grant for the funding of the IAOOS project.

REFERENCES

[1] Shupe, M. D., J.M. Intrieri, 2004: Cloud radiative forcing of the Arctic surface: The influence of cloud properties, surface albedo, and solar zenith angle, *Journal of Climate*, **17**, 616-628

[2] Wiegner, M., F. Madonna, I. Binietoglou, R. Forkel, J. Gasteiger, A. Geiß, G. Pappalardo, K. Schäfer, W. Thomas, 2014: What is the benefit of ceilometers for aerosol remote sensing? An answer from EARLINET, *Atmospheric Measurement Techniques,* **7**, 2491-2543

[3] Intrieri, J. M., M. D. Shupe, T. Uttal, B. J. McCarty, 2002: An annual cycle of Arctic cloud characteristics observed by radar and lidar at SHEBA, *Journal of Geophysical Research: Oceans*, **107** (C10), SHE 5-1–SHE 5-15

[4] O'Connor, E. J., A. J. Illingworth, and R. J. Hogan, 2004: A technique for autocalibration of cloud lidar, J. Atmos. Ocean. Tech., 21, 777-786.

[5] Hu, Y., 2007: Depolarization ratio-effective lidar ratio relation: Theoretical basis for space lidar cloud phase discrimination, *Geophysical Research Letters*, **34**

[6] Sassen, K., 1991: The polarization lidar technique for cloud research: A review and current assessment, *BAMS*, **72**, 1848-1866.

EPJ Web of Conferences **119**, 02006 (2016) DOI: 10.1051/epjconf/201611902006
ILRC 27

A CONCEPT OF MULTI-MODE HIGH SPECTRAL RESOLUTION LIDAR USING MACH-ZEHNDER INTERFEROMETER

Yoshitaka Jin [1]*, Nobuo Sugimoto[1], Tomoaki Nishizawa[1], Pablo Ristori[2], and Lidia Otero[2]

[1]*National Institute for Environmental Studies, 16-2 Onogawa, Tsukuba 305-8506 Japan, *Email: jin.yoshitaka@nies.go.jp*

[2]*División Lidar, CEILAP, UNIDEF (MINDEF - CONICET), UMI-IFAECI-CNRS 3351, Juan Bautista de La Salle 4397 – B1603ALO Villa Martelli, Buenos Aires, Argentina*

ABSTRACT

In this paper, we present the design of a High Spectral Resolution Lidar (HSRL) using a laser that oscillates in a multi-longitudinal mode. Rayleigh and Mie scattering components are separated using a Mach-Zehnder Interferometer (MZI) with the same free spectral range (FSR) as the transmitted laser. The transmitted laser light is measured as a reference signal with the same MZI. By scanning the MZI periodically with a scanning range equal to the mode spacing, we can identify the maximum Mie and the maximum Rayleigh signals using the reference signal. The cross talk due to the spectral width of each laser mode can also be estimated.

1. INTRODUCTION

Observation of aerosol characterization is essential to understand the atmospheric radiation budget that affects on the climate change. High Spectral Resolution Lidar (HSRL) is an effective tool for quantitatively measuring aerosol extinction and backscattering coefficients with high sensitivity [1]. The extinction-to-backscattering ratio (lidar ratio) depends on the aerosol species, and therefore it is an important parameter for evaluating aerosol characteristics [2, 3].

HSRLs have been developed using Fabry-Perot interferometer [4], atomic filters [5], and Iodine filters [6]. Recently a method using tilted field-widened Michelson interference has been developed for an airborne HSRL [7]. HSRL can make daytime observations since the HSRL signal sensitivity is much higher than that of Raman lidar. HSRL is a key instrument for the next generation of aerosol lidar. However, traditional systems use single-longitudinal mode lasers that are costly because they need injection seeding. Moreover, traditional HSRL tends to be complicated systems since it needs to control the receiver system and tune the laser wavelength. This study presents a measurement system and some simulation results for HSRL using lasers that oscillate in a multi-longitudinal mode (hereafter referred to as MM-HSRL). The basic concept of MM-HSRL is presented by Ristori et al. [8], and a lidar using a multimode laser and MZI has been developed by Bruneau et al. [9] for wind measurement. Our MM-HSRL system is both low cost and quite simple to operate compared to traditional systems.

2. DESIGN OF A MM-HSRL SYSTEM

A MM-HSRL measurement technique is briefly described bellow. Rayleigh scattering from atmospheric molecules has a broad spectrum due to the Doppler effect. In contrast, the Mie scattering spectrum is almost the same as the transmitted laser mode. Rayleigh scattering signals are separately measured by removing Mie scattering spectrum ranges with an interferometer. For MM-HSRL, the separate measurement can be achieved by using the realistic assumption that all mode spacings of the laser resonator are the same. The interferometer should have the same mode spacing as Mie scattering (i.e., identical to the free spectral range (FSR) of the transmitted laser).

To perform the measurement described above, we design the MM-HSRL system depicted in Fig. 1. The transmitter employs a multi-mode Nd:YAG laser at a second (532nm) or third (355nm) harmonic wavelength. A part of the transmitted laser light is injected into the telescope before being emitted into the atmosphere. The reason, as described later, is that the injected laser light is measured as a reference signal for evaluating the laser spectral properties. A high-speed (> 100MHz) A/D converter is employed to get the reference signals at the lowest height (z=0).

The receiver system employs a Mach-Zehnder interferometer (MZI) composed of a roof mirror mounted on a piezo stage and two half mirrors. The received light is divided into two beams with a beam splitter. The reflected light moves on to the movable roof mirror and back in the opposite direction, and again intersects with the penetrated light at the second beam splitter (i.e., interference occurs). The superimposed light rays after the MZI are measured with two photomultiplier tubes (PMT) for opposite phases. The optical path difference (OPD) of MZI is set at the position where the MZI FSR is identical to the laser-mode spacing. Controlling the piezo actuator of the roof mirror mount, the OPD of MZI changes on the order of nanometers. The OPD changes result in changes in wavelengths of constructive and destructive interference. We scan the MZI periodically with a scanning range equal to the mode spacing within several seconds. The displacement of OPD corresponds to the laser wavelength for the mode spacing.

By scanning the MZI, we can determine the optimal OPD for separating Mie and Rayleigh scattering components. The optimal OPD is determined by envelope analysis of the reference (transmitted laser) signal during the scanning. When we regulate the reference signal to minimum, each laser-mode wavelength is blocked at the OPD. Rayleigh (Mie) signals are obtained when destructive (constructive) interference occurs at each laser-mode wavelength. Mie scattering components are removed at the optimal OPD, and the remaining Rayleigh components are measured for atmospheric signals. By setting the optimal OPD, Rayleigh and Mie scattering signals are simultaneously measured with each PMT because each MZI output is in the opposite phase.

The advantages of scanning are both to determine optimal OPD and to monitor bias caused by the laser spectral properties. Using the reference signal, we can estimate the cross talk due to the spectral width of each laser mode. The estimated cross talk is removed from the atmospheric signals. We can also monitor the displacement from optimal OPD for the reference signal and hence estimate the MZI efficiency depending on the displacement. This system does not need to know the absolute wavelength of laser modes. The important point is that the scan covers one or more mode intervals.

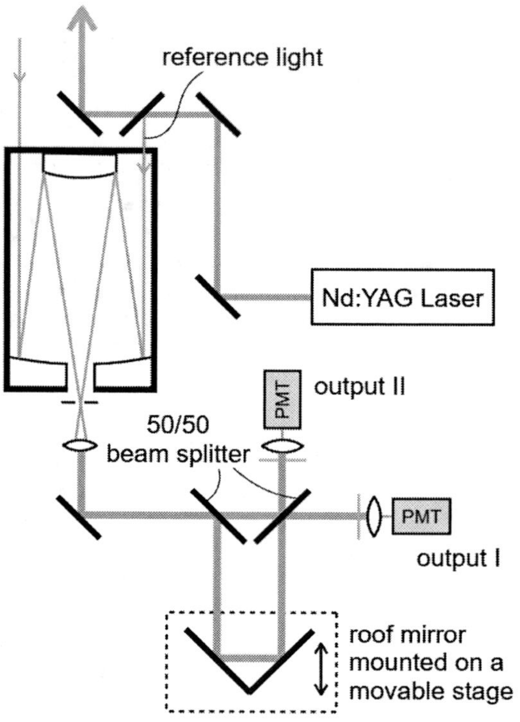

Fig. 1 System design of a MM-HSRL with a Mach-Zehnder Interferometer.

3. PRELIMINARY SIMULATION RESULTS

We will now present some simulation results for the MM-HSRL system. MZI output irradiance I is expressed by as follows:

$$I_i = 2rt \int_{-\infty}^{+\infty} I_0(k)[1 + cos(kd + \delta_i)]dk, \quad \cdots(1)$$

where subscript i denotes output 1 (parallel to the incident beam) or output 2 (perpendicular to the incident beam), $I_0(k)$ is the input irradiance, $\sqrt{r}$ and $\sqrt{t}$ are the reflectance and transmittance of the beam splitter, k is the wave number, d is the OPD, and δ is the phase shift by the beam splitters (here $\delta_1 = 0$ and $\delta_2 = \pi$). The spectral broadening of each mode is determined from the reflectivity of laser cavity. The mode spacing is identical to the laser FSR, which depends on resonator length L (i.e., $FSR = c/2L$, where c is the speed of light).

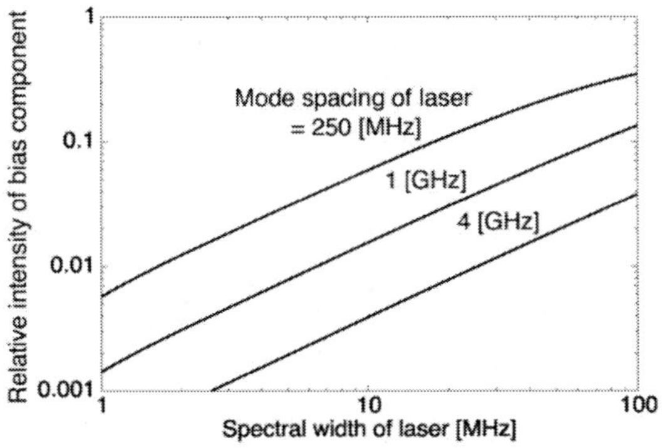

Fig. 2 Relative intensity of bias component for various spectral width and mode spacing of laser.

Using calculated output irradiance we can estimate the transmittance of aerosol and molecule signals as follows:

$$T_{i,a} = \frac{I_{i,a}}{\int_{-\infty}^{+\infty} I_{0,a}(k)dk}, \quad \cdots(2)$$

and

$$T_{i,m} = \frac{I_{i,m}}{\int_{-\infty}^{+\infty} I_{0,m}(k)dk}, \quad \cdots(3)$$

where subscripts a and m denote aerosols and molecules. We assume that aerosol signals have the same spectral properties as the laser. For molecular signals, the spectral width is set to 3GHz full-width half-maximum. The ratio T_a/T_m is an index of efficiency of HSRL [7]. A low T_a/T_m results in efficient removal of Mie signals from Mie+Rayleigh signals when detecting destructive interference.

The minimum T_a (bias component) depends on the spectral width and mode spacing of laser (Fig. 2). Large spectral widths cause difficulties to remove Mie signals and increase the bias component. Therefore the efficiency of removing Mie signals is better when using the lasers with high reflectance mirrors. A short laser cavity (large FSR) results in a wide range of MZI spectrums for the removal of Mie-signals, resulting in decrease of bias component. Laser cavity length is related to an incident angle issue as below.

Some factors in the MM-HSRL system decrease T_a/T_m. Inclining the incident beam to the interferometer is a critical factor since this system applies large OPDs to match the MZI mode spacing to the laser FSR. We can calculate OPD at the incident angle θ as follows:

$$OPD(\theta) = \frac{OPD(0°)}{cos(\theta)}. \quad \cdots(4)$$

ΔOPD ($= OPD(\theta) - OPD(0°)$) increases with the incident angle (Fig. 3). The longer $OPD(0°)$ causes a more rapid increase as expected. When ΔOPD is 0.5λ, the ratio T_a/T_m is maximal since most aerosol signals are measured with constructive interference. Therefore the incident angle should be reduced to much less than the angle at $\Delta OPD = 0.5\lambda$.

The maximum incident angle to the interferometer is determined from the field stop of the telescope and the focal length of the collimator lens. If the beam divergence of the transmitted laser is 0.1mrad (after being expanded) and the focal length of the telescope is 2m, the field stop of the telescope should exceed 0.2mm. To constrain the maximum incident angle to <0.5mrad, the focal length of the collimator lens should exceed 200mm. Therefore, this system requires lasers with very low beam divergence. Inclining the incident beam causes other errors. Tilted rays cannot overlap perfectly at the second beam splitter of the MZI. The two rays (reflected and penetrated light in MZI) overlap geometrically about 95% at the intersection point if the incident angle is 1mrad and $OPD(0°)$ is 40cm. For the area without overlap, extra signals are measured as leaking light without interference.

This decreases the efficiency removing aerosol signals. Some other sensitivity studies are provided by Ristori et al. [8].

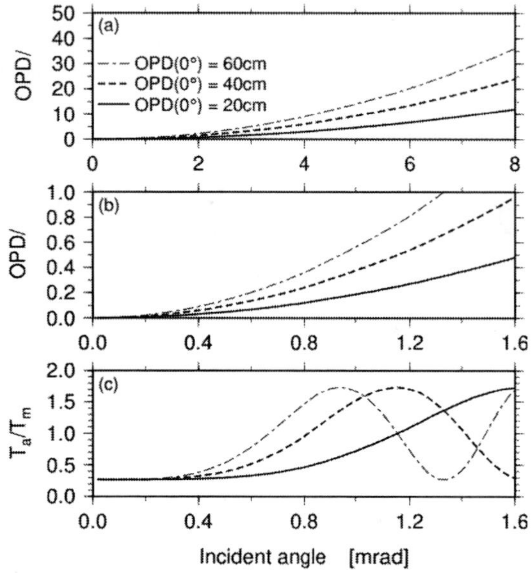

Fig. 3 Panels (a) and (b) plot OPD differences (unit laser wavelength) as a function of incident angle for three OPDs at $\theta = 0°$. Panel (b) is an expanded one of Panel (a). Panel (c) depicts the transmittance ratio as a function of the incident angle at laser finesse = 10.

4. FURTURE WORKS

As future work, we will estimate system errors caused by inhomogeneous wave front at the field stop, spherical aberration of the collimator lens, and polarization dependence of beam splitters. We will also develop a MZI control method and a signal analysis algorithm. The ratio of reference signals from two channels will be used to adjust the roof mirror position. The reference signals can also be used as bias components to correct atmospheric signals.

ACKNOWLEDGEMENT

This research was supported by the Japan Science and Technology Agency (JST) / Japan International Cooperation Agency (JICA), Science and Technology Research Partnership for Sustainable Development (SATREPS).

REFERENCES

[1] Nishizawa T., N. Sugimoto, and I. Matsui, 2010: Development of multi-wavelength high-spectral resolution lidar system (2alpha+3beta+2delta), *Proc. of 25th International Laser Radar Conference*, 903-906.

[2] Liu Z., N. Sugimoto, and T. Murayama, 2002: Extinction-to-backscatter ratio of Asian dust observed by high-spectral-resolution lidar and Raman lidar, *Applied Optics.*, **41(15)**, 2760-2767.

[3] Burton et al., 2012: Aerosol classification using airborne High Spectral Resolution Lidar measurements – methodology and example, *Atmos. Meas. Tech.*, **5**, 73-98.

[4] Fiocco G.G., G. Beneditti-Michelangeli, K. Maischberger, and E. Madonna, 1971: Measurement of Temperature and Aerosol to Molecule Ratio in the Troposphere by Optical Radar, *Nature*, **229**, 78.

[5] Shimizu H., S. A. Lee, and C. Y. She, 1983: High spectral resolution lidar system with atomic blocking filters for measuring atmospheric parameters, *Appl. Opt.*, **22**, 1373-1381.

[6] Liu Z., Ichiro Matsui, and Nobuo Sugimoto, 1999: High-Spectral-Resolution Lidar Using an Iodine Absorption Filter for Atmospheric Measurements, *Optical Engineering*, **38(10)** 1661-1670.

[7] Liu D., C. Hostetler, I. Miller, A. Cook, and J. Hair, 2012: System analysis of a tilted field-widened Michelson interferometer for high spectral resolution lidar, *Opt. Express*, **20**, 1406-1420.

[8] Ristori et al., 2015: Development of a high spectral resolution lidar using a multi-mode laser and a tunable interferometer, *Proc. of 27th International Laser Radar Conference.*

[9] Bruneau et al., 2013: Direct-detection wind lidar operating with a multimode laser, *Appl. Opt.*, **52**, 4941.

DOI: 10.1051/epjconf/201611902007

MEASURING AEROSOL OPTICAL DEPTH (AOD) AND AEROSOL PROFILES SIMULTANEOUSLY WITH A CAMERA LIDAR

John Barnes [1]*, Robert Pipes [1], Nimmi C. P. Sharma[2]

[1]*NOAA/Mauna Loa Observatory, Hilo, Hawaii, USA, *Email: john.e.barness@noaa.gov*

[2]*Department of Physics and Engineering Physics, Central Connecticut State University, New Britain, CT*

ABSTRACT

CLidar or camera lidar is a simple, inexpensive technique to measure nighttime tropospheric aerosol profiles. Stars in the raw data images used in the CLidar analysis can also be used to calculate aerosol optical depth simultaneously. A single star can be used with the Langley method or multiple star pairs can be used to reduce the error. The estimated error from data taken under clear sky conditions at Mauna Loa Observatory is approximately +/- 0.01.

1. INTRODUCTION

The simplest atmospheric aerosol lidar measurement is 180 degree backscattered light. To convert this quantity into the more useful aerosol extinction or total scatter (m^{-1}) the extinction to backscatter ratio, also known as the lidar ratio, must be assumed or measured. If the aerosol optical depth (AOD) is known, the backscatter profile can be integrated to get an average lidar ratio.

For example, this approach is used by MPLNET [2] by using daytime AOD measured by AERONET [1]. But sun-photometer measurements of AOD are limited to daytime, when many lidars cannot operate. Star-photometers have been used to get nighttime AOD [3,4] by observing single stars with a small telescope, and multiple stars have been measured simultaneously with a CCD camera [5]. But both types of photometers require an additional instrument to the lidar being used, and there can be differences in the time, location, and wavelength between the lidar and the photometer. In addition, observing single stars involves accurate pointing of the telescope, which not needed when using a wide-angle lens and camera.

This study examines the feasibility of using a novel technique of aerosol profiling, camera lidar or CLidar, to also measure AOD with no additional equipment.

2. METHODOLOGY

The CLidar technique uses a wide-angle lens on a CCD camera to image a laser beam, usually pointed vertically, from the ground to the zenith [6,7,8]. The geometry is shown in figure 1. The technique is currently limited to nighttime conditions. A continuous laser can be used since the altitude information is determined by the distance from the camera to the laser beam (usually a few 100 m), and the angle of observation. The CLidar has no overlap function as in lidar, and there is no data acquisition electronics other than the computer to operate the camera. Another difference with lidar is that light is scattered at 90 degrees at the ground and approaches 180 degree scatter as altitude increases. An important property of the fisheye lens used is that each pixel maps to a constant angle.

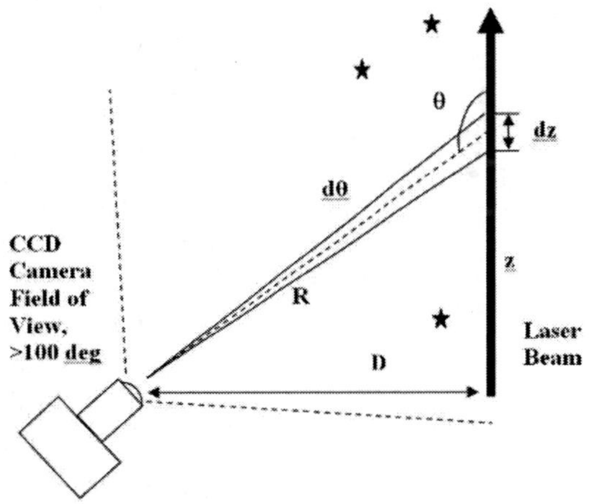

Figure 1, The geometry used for the CLidar technique along with three stars. The star brightness is a measured in the same images used for the CLidar aerosol profiles.

The integration period for the CLidar image is usually a few minutes, during which time brighter stars are also imaged. So the brightness for each

DOI: 10.1051/epjconf/201611902007

star is being measured at the same location, time, and through the same interference filter as the laser beam. Depending on the camera resolution and the lens used, a star is usually imaged onto several pixels. Pixel counts must be checked to insure there are no errors due to saturation effects.

The standard calibration method for sun-photometers, the Langley calibration, is to measure the sun as it rises or sets during which the air mass changes. An air mass change of 2 to 8 or more is often desired. The change of the log of the star brightness with air mass is linear, and the slope and zero air mass intercept are the calibration constants. A single star can be used in the same manner, although the signal will be much weaker.

The multi star method takes advantage of star brightness measured by astronomers. By using a pair of stars and the stellar magnitudes for each, the AOD can be measured without the calibration constants. This of course is not possible for sun-photometers. In this study, brighter stars in the CLidar image were identified and the Yale Bright Star Catalog Fifth edition was used for their magnitudes and relative positions to get their air masses. Interpolations were made since the magnitude data does not match the laser wavelength. The images were analyzed using software written in IDL (Interactive Data Language). The stars signals were calculated by using an IDL 2-D Gaussian plus constant non-linear least squares fit. Error estimates were made using the pixel counts for each star in the images, which were used with estimates of the other quantities needed in the error propagation calculation.

3. RESULTS

The CLidar system used a Santa Barbara Instruments Group ST-1603 CCD camera cooled to -20 C. The rectangular chip array is 1530 by 1020, 9 micron pixels. The lens was a Sigma fisheye, 10 mm focal length, F/2.8. The laser imaged is part of the NOAA/Mauna Loa Observatory lidar, which uses a doubled Nd:YAG laser operating at about 15 Watts (532 nm). Aerosol profiles were acquired but only the star-photometer data are presented here.

In figure 2 an example is shown of a Langley plot of the star Rigel in the constellation Orion on 2010/11/10. As the star rose above the horizon the integration time was 1 minute. This changed to 3 minutes as the air mass decreased and wasn't changing as rapidly. A 532 nm interference filter with a 10 FWHH bandpass was inserted between the camera and the Sigma lens. The R^2 of 0.9944 is equivalent to an AOD error of about +/-0.02, similar to other star-photometers [3,4]. There was very little aerosol over the observatory at the time, so the slope represents very clean air.

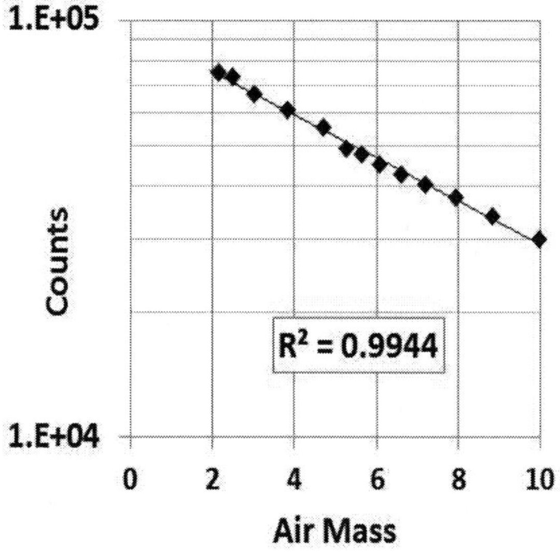

Figure 2, Langley plot using the star Rigel in Orion under very low aerosol conditions. The R^2 of 0.9944 is equivalent to an AOD error of about 0.02.

The multi-star method was also tested at Mauna Loa Observatory under very clean aerosol conditions on 2014/11/20. The exposure time was 15 seconds. When using a pair of stars to calculate AOD, the signal for each star is compared to its stellar magnitude. To minimize the error it is best to choose stars with different air masses. The best results use a star low in the sky (high air mass, high extinction) and another higher in the sky (low air mass, low extinction). In this example 10 of the brighter stars in the image were chosen which created 45 possible pairs. Many of these pairs used stars with similar air masses and generated large errors.

Figure 3 shows the results for the 5 star pairs which had a calculated AOD error of less than +/- 0.05. The AODs measured by AERONET on the daytime before and the daytime after this night were about 0.006. This should be an upper limit on the real AOD on this night since the air generally gets cleaner at night at Mauna Loa Observatory. There is currently a constant offset of -0.079 AOD generated in the analysis. This offset has been subtracted from the results shown. The source of the offset is currently under investigation, but it is not believed that it affects the AOD error estimates quoted here.

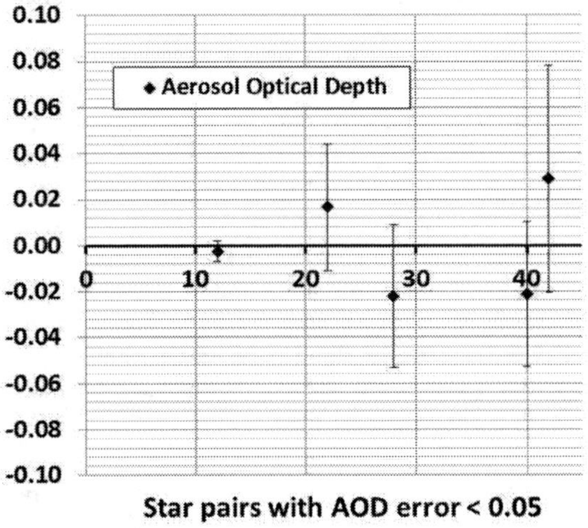

Figure 3, AOD measurements of clean sky conditions at Mauna Loa Observatory on 2014/11/10 with a 15 second exposure. The best 5 out of 45 pairs are shown.

Using all five star pairs results in an AOD error of +/- 0.0042. In figure 3 a single star pair, #12, is seen to dominate the overall error. It has a calculated error +/- 0.0044. Without this pair the error is +/- 0.018. There are other possible factors in the error calculation that may not be included such as the uniformity of the sensitivity of the CCD chip. These are currently under investigation but it is believed the error will lie between the 0.0044 and 0.018 values. The achievable error will depend on the actual operating conditions such as the number of bright stars available in the image, and background light at the observing location.

4. CONCLUSIONS

The simplicity and low cost of the CLidar technique is an attractive characteristic when measuring nighttime tropospheric aerosols. The knowledge of AOD is a strong constraint on converting the aerosol backscatter profile to an aerosol extinction profile. Using star brightness in the CLidar images to measure AOD is a promising addition with an error of +/- 0.01 possible.

ACKNOWLEDGEMENT

This work is supported by the National Oceanic and Atmospheric Administration/Earth System Research Laboratory/Global Monitoring Division.

REFERENCES

[1] http://aeronet.gsfc.nasa.gov/

[2] http://mplnet.gsfc.nasa.gov/

[3] Leiterer, U., et al., 1995: A new star photometer developed for spectral aerosol optical thickness measurements in Lindenberg, *Beitr. Phys. Atmosph.*, 68, 133-141.

[4] Perez-Ramirez, D. et al., Development and calibration of a star photometer to measure the aerosol optical depth: smoke observations at a high mountain site, *Atmos. Environ.*, 42, 2733-2738..

[5] Lanciano, O. and G. Fiocco, 2007: Nighttime measurements of atmospheric optical thickness by star photometery with a digital camera, *Applied Optics*, 46, 5176-5182.

[6] Barnes, John E., S. Bronner, Robert Beck, and N. C. Parikh, 2003: Boundary Layer Scattering Measurements with a Charge-Coupled Device Camera Lidar, *Applied Optics*, **42**, no. 15, 2647-2652.

[7] Barnes, John E., N.C. Parikh Sharma and Trevor Kaplan, Atmospheric aerosol profiling with a bistatic imaging lidar system, 2007: *Applied Optics*, **46**, no. 15, 2922-2929.

[8] Nimmi C. P. Sharma and John E. Barnes, 2015: CLidar mountain boundary layer case studies, 27th *ILRC*, New York City.

Airborne Laser Absorption Spectrometer Measurements of CO_2 Column Mixing Ratios: Source and Sink Detection in the Atmospheric Environment

Robert T. Menzies*, Gary D. Spiers, Joseph C. Jacob

Jet Propulsion Laboratory, California Institute of Technology
4800 Oak Grove Drive, Pasadena, California 91109, U.S.A.
**Email: rmenzies@jpl.nasa.gov*

ABSTRACT

The JPL airborne Laser Absorption Spectrometer instrument has been flown several times in the 2007-2011 time frame for the purpose of measuring CO2 mixing ratios in the lower atmosphere. The four most recent flight campaigns were on the NASA DC-8 research aircraft, in support of the NASA ASCENDS (Active Sensing of CO2 Emissions over Nights, Days, and Seasons) mission formulation studies. This instrument operates in the 2.05-μm spectral region. The Integrated Path Differential Absorption (IPDA) method is used to retrieve weighted CO2 column mixing ratios. We present key features of the CO2LAS signal processing, data analysis, and the calibration/validation methodology. Results from flights in various U.S. locations during the past three years include observed mid-day CO_2 drawdown in the Midwest, also cases of point-source and regional plume detection that enable the calculation of emission rates.

1. INTRODUCTION

Atmospheric CO_2 is a long-lived greenhouse gas, with sources and sinks primarily at the surface. Seeking improved capability to determine the fluxes of CO2 on various spatial scales, national space agencies have invested in Earth-orbiting passive spectrometers viewing reflected solar radiation. GOSAT (Greenhouse Gases Observing Satellite) was launched in January, 2009, and OCO-2 (Orbiting Carbon Observatory) was launched in August, 2014. These instruments are providing valuable insight into the capabilities of this technique. They have provided an unprecedented number of CO2 soundings on a global scale. Cloud and aerosol scattering, terrain complexities within the IFOV, and limited SNR (Signal to Noise Ratio) outside of daytime mid-latitudes are inherent issues that can be mitigated or eliminated using laser-based integrated path differential absorption (IPDA). An airborne laser-based approach to high-precision measurements of atmospheric CO_2 offers the potential to provide the high-accuracy mixing ratio measurements on regional scales with spatial resolution that is useful to the carbon cycle research community. The airborne environment is a good test of the component technologies and system designs. Experience with airborne IPDA lidars is also critical in assessing the instrument parameters and retrieval algorithms that are relevant to Earth-orbiting observations.

Assuming a nadir or near-nadir view, a weighted column dry air mixing ratio is obtained from the IPDA sounding because the absorption lines are pressure-broadened in the lower atmosphere. Thus the cross section at the probing on-line frequency is dependent on altitude. The weighting favors the lower troposphere when the on-line frequency is detuned one or more (surface pressure) halfwidths from line center. We can use this to advantage by detuning the on-line frequency to a location one to three halfwidths from line center results in selective probing of the CO2 in the lower troposphere, where the CO2 mixing ratio variability of interest is the highest. (Typical weighting functions for CO_2 sounding are found in references [1-4].) The JPL Laser Absorption Spectrometer (LAS) instrument probes a well characterized pressure-broadened absorption line profile at an offset of approximately two surface-pressure halfwidths in order to provide weighting functions suitable for peak response to CO2 near the surface.

DOI: 10.1051/epjconf/201611903001

Multiple flights of our airborne CO_2 LAS instrument have demonstrated the capability to observe various sources as well as large-scale CO_2 drawdown in the U.S. Midwest due to photosynthesis in the July/August time frame. Flights have also provided experience dealing with partially cloudy atmospheres, land surfaces with varied terrain and topography, and bodies of water with various degrees of surface roughness. These provide the opportunity to demonstrate the suitability of the LAS IPDA technique, to evaluate the instrument technology, and to develop and refine the high precision retrieval algorithms that are essential. Here we provide a brief instrument overview and discuss selected results from flights on the NASA DC-8 research aircraft during campaigns in July/August, 2011, February/March, 2013, and August, 2014.

2. AIRBORNE CO_2 LAS INSTRUMENT OVERVIEW

The instrument was jointly developed by JPL and Lockheed Martin Coherent Technologies (LMCT) with funding from the NASA Earth Science Technology Office Instrument Incubator Program. The transceiver approach is to utilize heterodyne detection, implementing a narrow bandwidth receiver, with frequency-stabilized narrow-linewidth laser transmitters and local oscillators. The transceiver consists of two separate transmit/receive channels for the on-line and off-line components of the IPDA measurement. Each channel has a dedicated heterodyne detector, and a cw single frequency compact Tm,Ho:YLF laser which acts both as the transmit laser and the local oscillator for heterodyne detection of the return signal. The transceiver includes a third laser locked to a frequency near line center of the R(30) CO_2 absorption line at 4875.749 cm^{-1} that provides an optical frequency reference for frequency offset-lock tuning of the other lasers. This is accomplished using a temperature controlled, hermetically sealed CO_2 absorption cell. The online and offline lasers are frequency offset-locked to this reference laser. Tunability of the on-line laser allows CO_2 measurement flexibility through on-line frequency adjustment. (The atmospheric CO2 line has a pressure-broadened FWHM of about 4 GHz near sea-level pressure.) The instrument is described in more detail in reference [5].

A frequency offset is required between the return signals and their corresponding local oscillators for heterodyne detection. By pointing the transmit beams at an offset from nadir, the return signals will experience a Doppler shift that depends on the aircraft velocity and pitch angle, thereby eliminating the need for an additional frequency shifting device in the receiver. The instrument is mounted on a frame that provides a suitable off-nadir angle. The signals from each channel are digitized with a 50 Msamples/sec, 14-bit digitizer. The samples are transformed into the spectral domain using an FFT operation followed by conversion to periodograms. The commonly used "squarer" estimator is used to determine the return power in each channel. 16K FFT's are the default in the processing scheme. The sampling duration is approximately equal to the speckle decorrelation time of the signal, τ_{decorr} , which is ~ 0.3 ms for the NASA DC-8 at nominal cruise speed. A pre-selected number of periodograms is summed, and the remainder of the signal processing steps operate on a collection of these sums.

3. ENVIRONMENTAL EFFECTS ON MEASUREMENT PRECISION

Attainment of CO_2 measurement precision of the level of ~0.3% or 1 ppm is a very challenging endeavor. We continue to incorporate into our retrieval algorithms methods for minimizing atmospheric and surface effects, particularly as potential sources of bias. Discussion of the effects of meteorological parameter uncertainties on the CO_2 measurement uncertainty can be found in Refs. [4-6]. Three additional important categories are (1) cloud detection and filtering; (2) topography; and (3) spectral reflectances of surface and above-ground scatterers. Various means of minimizing these influences on the retrievals are discussed in Ref. [7].

Methods must be developed for filtering out the scattering effects of clouds. Clouds in the field-of-view (FOV) reduce the path length, and if not recognized, bias the CO_2 retrieval. If the lidar provides time-of-flight to the backscatter source (e.g. a range-gated pulsed system, or an FM/CW system), any sources of backscatter other than that which occurs at the expected range to the surface can be set aside or filtered out. In the current implementation of our airborne system, we do not have this capability. We employ alternative methods to detect and filter out the backscatter signals that are due to clouds in the field of view. The heterodyne signals backscattered from the surface are sufficiently narrow to permit identification of cloud backscatter if there is cloud movement relative to the surface. In practice the backscatter signals from cumulus and stratocumulus are spectrally broadened also. This provides a convenient filtering method. Another filtering method depends on the ability to discern an abrupt reduction in measured DAOD (differential absorption optical depth) due to a cloud in the FOV.

We use the SRTM database along with the aircraft INS/GPS data to determine surface elevation at the laser transmitter footprint location along the ground track, with an along-track resolution of approximately 50 m. Even with sophisticated use of the SRTM database, time-dependent systematic errors of ~ 10-20 m in scattering surface elevation can occur over e.g. mountain forests. Co-aligned profiling laser altimetry with 3-5 m vertical resolution and suitably high along-track resolution would mitigate this potential error source.

4. POINT SOURCE AND URBAN PLUME DETECTION: EMISSION RATE QUANTIFICATION

Observations of mid-day summer drawdown of CO_2 over agricultural areas in the U.S. Midwest are reported in Ref. [7]. During one flight in 2011, the track was downwind of the Four Corners power plant in New Mexico. An emission rate based on the observed column CO_2 increase and MERRA wind data agreed closely with the reported annual total emission [9]. CO_2 retrievals from 2013 and 2014 flights also show evidence of urban plumes. Here we show in Fig. 1 an example of elevated CO_2 when flying over Fresno, California in March, 2013. We are comparing airborne LAS observations with total carbon emission databases including EDGAR, FFDAS, and ODIAC. Our observations indicate the potential of the airborne IPDA lidar to assess emission rates from urban areas. Results will be discussed.

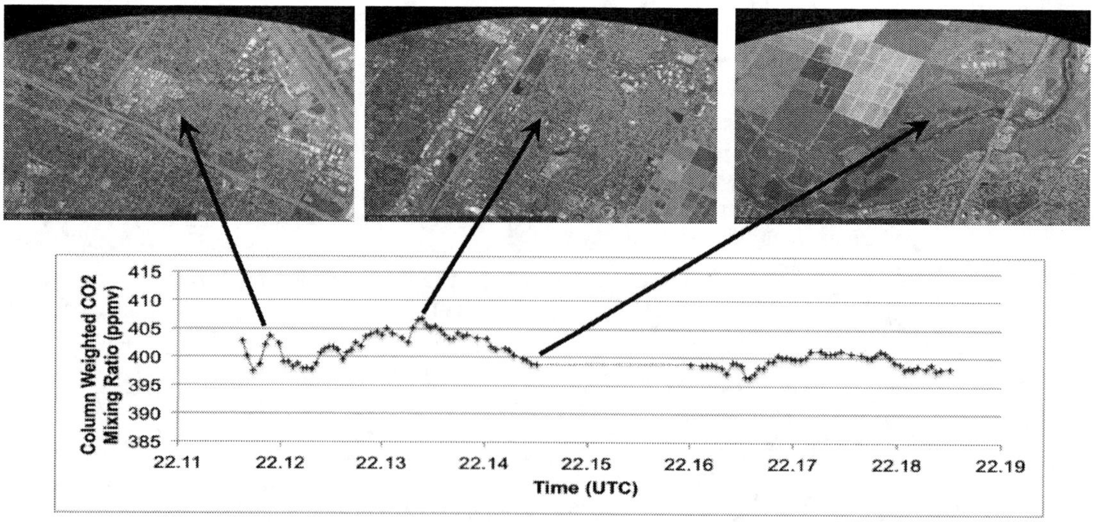

Figure 1. Elevated column CO2 over Fresno, California, March 2, 2013.

DOI: 10.1051/epjconf/201611903001

ACKNOWLEDGEMENT

This work was carried out by the Jet Propulsion Laboratory, California Institute of Technology, under contract with the National Aeronautics and Space Administration.

REFERENCES

[1] R.T. Menzies and D.M. Tratt, 2003: "Differential laser absorption spectrometry for global profiling of tropospheric carbon dioxide: selection of optimum sounding frequencies for high-precision measurements", *Appl. Opt.,* **42,** 6569-6577.

[2] G. Ehret, C. Kiemle, M. Wirth, A Amediek, A. Fix, and S. Houweling, 2008: "Spaceborne Remote Sensing of CO_2, CH_4, and N_2O by IPDA Lidar", *Appl. Phys. B,* **90**, 593-608.

[3] F. Gibert, P.H. Flamant, J. Cuesta, D. Bruneau, 2008: "Vertical 2-μm heterodyne differential absorption lidar measurements of mean CO2 mixing ratio in the troposphere", *J. Atmos. Oceanic Technol,.* doi: 10.1175_2008JTECHA1070.1.

[4] J. Caron and Y. Durand, 2009: "Operating wavelengths optimization for a spaceborne lidar measuring atmospheric CO_2 ", *Appl. Opt.***48**, 5413-5422.

[5] G.D. Spiers, R.T. Menzies, J. Jacob, L.E. Christensen, M.W. Phillips, Y. Choi, and E.V. Browell, 2011: "Atmospheric CO_2 measurements with a 2 μm airborne laser absorption spectrometer employing coherent detection", *Appl. Opt.*, **50**, 2098-2111.

[6] S. Ishii and Coauthors, 2012: "Partial CO_2 column-averaged dry-air mixing ratio from measurements by coherent 2-μm differential absorption and wind lidar with laser frequency offset locking", *J. Atmos. Oceanic Technol.,* **29**, 1169-1181.

[7] R.T. Menzies, G.D. Spiers, and J.C. Jacob, 2014: "Airborne Laser Absorption Spectrometer Measurements of Atmospheric CO_2 Column Mole Fractions: Source and Sink Detection and Environmental Impacts on Retrievals", *J. Atmos and Oceanic Tech.* **31**, 404-421.

Measurements of Atmospheric CO_2 Column in Cloudy Weather Conditions using An IM-CW Lidar at 1.57 Micron

Bing Lin[1], Michael D. Obland[1], F. Wallace Harrison[1], Amin R. Nehrir[1], Edward V. Browell[2], Syed Ismail[1], Joel Campbell[1], Jeremy Dobler[3], Byron Meadows[1], Tai-Fang Fan[4], and Susan A. Kooi[4]
[1]NASA Langley Research Center, Hampton, VA 23681
[2]STARSS II Affiliate, NASA Langley Research Center, Hampton, VA 23681
[3]Exelis Inc., Ft. Wayne, IN 46818
[4]Science System and Application, Inc, Hampton, VA 23666
Langley Research Center, Hampton, VA 23681

1. Introduction

Changes in atmospheric carbon dioxide (CO_2) concentrations are considered as one of the key factors in the net radiative heating to the Earth's climate system and in determining future global climate change [1, 2]. The U.S. National Research Council has identified the need for an Active Sensing of CO_2 Emission over Nights, Days and Seasons (ASCENDS) space mission [3] to reduce uncertainties in and to gain a better understanding of CO_2 sources and sinks.

In preparing for the ASCENDS mission, NASA Langley Research Center and Exelis Inc. have jointly developed and demonstrated the capability of CO_2 column measurements with an intensity-modulated continuous-wave (IM-CW) lidar. Results of CO_2 column measurements from aircraft flight campaigns using the prototype IM-CW lidar, Multifunctional Fiber Laser Lidar (MFLL), have been very encouraging [4, 5]. MFLL operates in the 1.57-□m CO_2 absorption band with one laser wavelength positioned on the CO_2 absorption line center (online) at 1571.112 nm, and two other laser wavelengths (offlines) positioned $\pm$50 pm on either side of the absorption line. CO_2 column differential absorption optical depth (DAOD) values are derived from combined online and offline measurements using the Integrated Path Differential Absorption (IPDA) approach. The signal-to-noise ratio (SNR) for clear sky IPDA measurements of CO_2 DAOD for a 10-s average was found to be as high as 1300, resulting in an error 0.077% or equivalent XCO_2 column precision of ~0.3 ppm [4, 6].

Measuring CO_2 column density in the presence of thin cirrus clouds, broken clouds, or cloud decks is a challenging task for both active and passive remote sensors. Passive CO_2 retrievals avoid cloudy pixels and use only clear sky measurements [7]. The purpose of this study is to extend the atmospheric CO_2 column measurements of the IM-CW lidar from clear skies to cloudy conditions. Under thin cirrus conditions, IM-CW lidar returns from the surface can be clearly separated from those from thin cirrus clouds because the received lidar returns contain vertically-resolved range information as a result of the range-encoded intensity modulation of transmitted lidar signals [4, 5, 8]. In the presence of low-level optically-thick scattered clouds or cloud decks, the lidar returns from the clouds are strong enough to measure CO_2 column DAODs from the instrument to the clouds.

2. Instrumentation and methodology

The basic architecture of MFLL for CO_2 column measurements has been discussed in detail in References 4 and 5. Basically, the seed Distributed Feedback Lasers (DFBs) are locked to the CO_2 online and offline wavelengths and fed into their corresponding intensity modulators to impart their unique ranging-codes (i.e., swept frequency) to each laser beam. A swept-frequency technique with 500-KHz bandwidth and center frequency around 350 KHz is used,

DOI: 10.1051/epjconf/201611903002

which enables the ranging measurements and the ability to separate the cloud and surface returns. The intensity modulated lasers are then simultaneously amplified using an Erbium Doped Fiber Amplifier (EDFA) to increase the transmitted power. A tiny fraction of the amplified laser power is picked off via an optical tap inside of the EDFA and sent to a detector as reference signals for transmitted laser-power normalization and calibration at the online and offline wavelengths. The lidar backscattered returns are recorded by the MFLL receiver and used as science signals for the CO_2 DAOD measurements. Digitized science and reference signals are processed by correlating these data with their corresponding range-encoded waveforms. The magnitude and position of the synthetic pulses of these correlation outputs represent the power and time-delay, respectively, of the received signals. Thus, range estimation and the discrimination of surface returns from cloud returns are achieved based on the outputs of the correlation. CO_2 DAOD values can also be retrieved from a combination of online and offline signal powers using the IPDA technique.

3. Airborne flight campaign data

The data used in this study were collected with the MFLL during the ASCENDS 2011 summer and 2013 winter airborne flight campaigns. Two flight cases are discussed: one for DAOD retrievals to the surface through optically thin cirrus clouds and the other to the tops of optically thick low clouds. In-situ measurements of the atmospheric profiles of CO_2 concentration, temperature, pressure, and humidity were obtained during aircraft spirals for comparison with the MFLL remote measurements in each case.

The thin cirrus cloud case studied here was obtained over an arid/semi-arid region around 0000 UT on 23 February 2013 near Blythe, California. The analyzed leg of this flight was maintained at an altitude of about 12.2 km, and the extended thin cirrus clouds were observed just below the aircraft. Fig. 1 shows the measured atmospheric profile as correlation power of lidar returns. Both online and offline channels clearly showed the thin cirrus clouds just below the aircraft (range close to zero). There were some changes in the range to the surface (lower panels), which resulted from small changes in the surface topography. Our estimation indicated that the average cloud optical depth of these cirrus clouds was about 0.158 with large variability ranging from near zero to as large as 0.8 with the median 0.135.

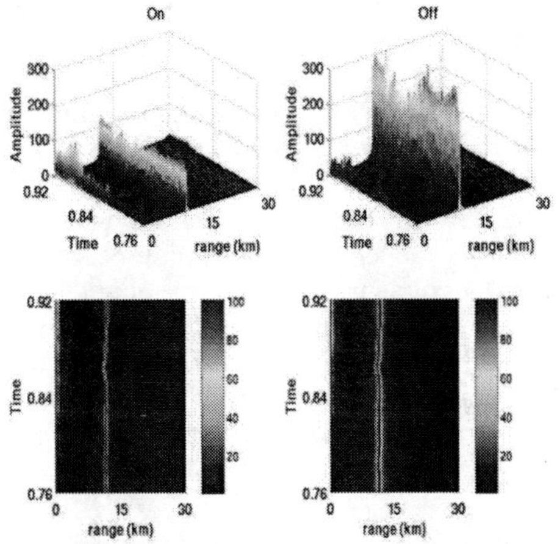

Fig.1 Lidar measured atmospheric profiles for online (left) and offline (right) observations.

The DAOD of CO_2 column to the surface and its equivalent XCO_2 values retrieved from 0.1-s integration are shown in Fig. 2. CO_2 column DAOD estimates for both clear (blue points) and cloudy (presence of intervening thin cirrus clouds; red points) conditions for this flight leg are plotted separately. Their mean and standard deviations are listed in the figure to compare with those derived from in-situ measurements (also listed). Compared to in-situ derived CO_2 DAOD column values, the lidar DAOD measurements were slightly smaller (about 0.004). Spatiotemporal differences in remotely sensed and in-situ observations are expected to account for the absolute differences between the two measurements.

The data for thick low cloud case were obtained over an agriculturally vegetated area on 10

August 2011 in the vicinity of West Branch, Iowa. Because the SNR of DAOD is proportional to the product of DAOD value and SNR of the online and offline power ratio [5], only high altitude observations of the multiple legs (1 to 7) of the flight were considered, which corresponding legs 4, 5 and 7. The flight altitudes for these legs were approximately 7.8, 9.4, and 12.5 km, respectively. The rms error of the range estimates was generally within 3 meters, which is very good and should also be adequate to meet space mission science requirements. DAOD retrievals (Fig. 3) showed consistent high precision results among the three legs although variations in DAOD retrievals to cumulus clouds were much larger than those to the surface owing to smaller DAOD values and weaker lidar power returns from the clouds. Relative precision of CO_2 DAOD column to surface was about 1.3 to 2.2% for these legs. Increasing integration time to 10-s would increase the precision to less than 0.18%. Compared to the errors of CO_2 column to surface, errors in CO_2 column to cloud tops increased due to weak cloud lidar returns.

4. Summary

This study evaluated the capability of IM-CW laser absorption lidar for CO_2 column measurements in cases of thin cirrus and thick fair weather boundary layer cumulus clouds. For thin cirrus clouds, consistent CO_2 DAOD and equivalent XCO_2 column values to surface for clear and cloudy skies were obtained in a 12-km altitude flight over an arid/semiarid region. The clear sky precision for the flight campaign case studied was about 0.72% for a 0.1-s integration. Under a very complicated multi-layer cloud environment of the mid-west vegetated area case with flight altitudes 8 to 12 km, the precision of the estimated CO_2 DAOD column to surface was as low as 1.3 – 2.2 % for 0.1-s integration. The precision of CO_2 column measurements to thick clouds was about a factor of 2 to 3 lower than that to the surface owing to much weaker lidar power returns from clouds and a smaller CO_2 column DAOD compared to those for the surface. These results indicate the potential of IM-CW lidar for CO_2 measurements over cloudy scenes in space applications.

References

[1] Intergovernmental Panel on Climate Change, "Climate Change 2007: The physical science basis, Contribution of Working Group I to the Fourth Assessment Report of the Intergovernmental Panel on Climate Change", Solomon, S., D. Qin, M. Manning, Z. Chen, M. Marquis, K. B. Averty, M. Tignor, and H. L. Miller, Eds., Cambridge University Press, Cambridge, UK, 996 pp (2007).

[2] Lin, B., L. Chambers, P. Stackhouse Jr., B. Wielicki, Y. Hu, P. Minnis, N. Loeb, W. Sun, G. Potter, Q. Min, G. Schuster, and T.-F. Fan, "Estimations of climate sensitivity based on top-of-atmosphere radiation imbalance", Atmos. Chem. Phys., 10, 1923–1930 (2010).

[3] NRC, "Earth Science and Applications from Space: National Imperatives for the Next Decade and Beyond", The National Academies Press, Washington, D.C., 2007.

[4] Dobler, J., F. Harrison, E. Browell, B. Lin, D. McGregor, S. Kooi, Y. Choi, and S. Ismail, "Atmospheric CO_2 column measurements with an airborne intensity-modulated continuous wave 1.57 μm fiber laser lidar," Appl. Opt., **52**(12), 2874-2892 (2013).

[5] Lin, B., S. Ismail, F. Harrison, E. Browell, A. Nehrir, J. Dobler, B. Moore, T. Refaat, S. Kooi, "Modeling of intensity-modulated continuous-wave laser absorption spectrometer systems for atmospheric CO2 column measurements," Appl. Opt., **50**(29), 7062-7077 (2013).

[6] Browell, E., M. E. Dobbs, J. Dobler, S. Kooi, Y. Choi, F.W. Harrison, B. Moore III, and T. S. Zaccheo, "First airborne laser remote measurements of atmospheric CO_2 for future active sensing of CO_2 from Space," Proceedings of the 8th International Carbon Dioxide Conference, Jena, Germany, 13–18 September 2009.

[7] Crisp, D., B. Fisher, C. O'Dell, C. Frankenberg, R. Basilio, and coauthors, "The ACOS CO2 retrieval algorithm – Part II: Global XCO2 data characterization", Atmos. Meas. Tech., 5, 687-707 (2012).

[8] Campbell, J., B. Lin, A. Nehrir, F.W. Harrison, M. Obland, "High Resolution CW Lidar Altimetry using Repeating Intensity Modulated Waveforms and Fourier Transform Reordering," Opt. Letters, 39, 6078-6081 (2014).

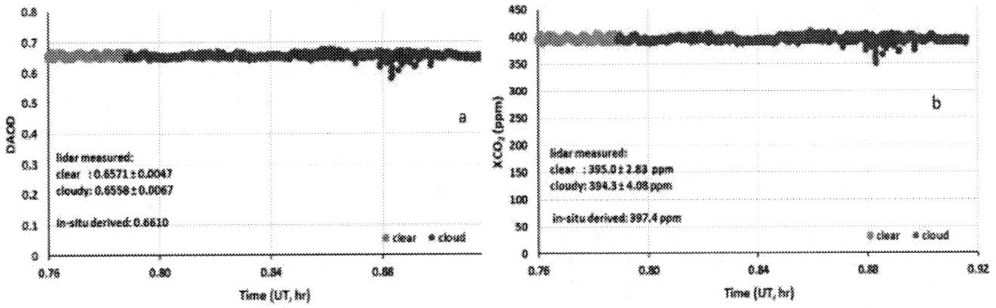

Fig. 2 Plotted are DAOD (a) of CO_2 column to the ground and its equivalent XCO_2 (b) values retrieved from 0.1-s integration of lidar measurements for both clear (blue points) and cloudy (red points) conditions. Their means and standard deviations as well as their corresponding in-situ derived values are listed.

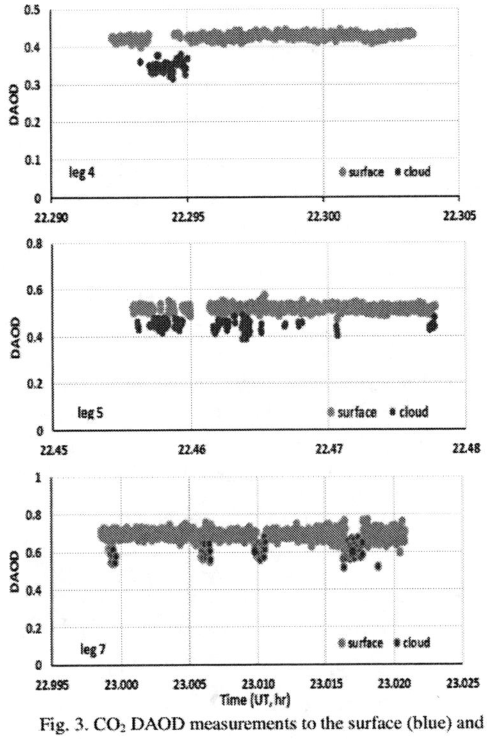

Fig. 3. CO_2 DAOD measurements to the surface (blue) and clouds (red) for legs 4, 5 and 7

DOI: 10.1051/epjconf/201611903003

MEASUREMENT OF LOWER-ATMOSPHERIC CO2 CONCENTRATION DISTRIBUTION USING A COMPACT 1.6 μm DIAL

Yasukuni Shibata*, Chikao Nagasawa, Makoto Abo

Tokyo Metropolitan University, 6-6 Asahigaoka, Hino, Tokyo 191-0065, Japan

**Email: sibata@tmu.ac.jp*

ABSTRACT

Knowledge of present carbon sources and sinks including their spatial distribution and their variation in time is one of the essential information for predicting future CO2 atmospheric concentration levels. The differential absorption lidar (DIAL) is expected to measure atmospheric CO2 profiles in the atmospheric boundary layer and lower troposphere from a ground platform.

We have succeeded to develop a compact 1.6 μm DIAL system for measuring CO2 concentration profiles in the lower atmosphere. This 1.6 μm DIAL system consists of the optical parametric generator (OPG) transmitter that excited by the LD pumped Nd:YAG laser with high repetition rate and the receiving optics that included the near-infrared photomultiplier tube operating at the analog mode and a 25 cm telescope. CO2 concentration profiles were obtained up to 2.5 km altitude.

1. INTRODUCTION

For the detailed analysis of forest carbon dynamics and CO2 fluxes of urban area, the CO2 concentration measurement techniques with high spatial and temporal resolution are required in the lower atmosphere [1]. We have developed a direct detection 1.6 μm DIAL technique to perform range-resolved measurements of vertical CO2 concentration profiles in the atmosphere. This 1.6 μm DIAL system has a 60 cm telescope for vertical measurement and a 25 cm scanning telescope for horizontal measurement. The vertical distribution of CO2 concentration from 2 km to 7 km altitude has been observed using two telescopes with different apertures [2,3].

In order to observe CO2 distribution in the tropopause, we have developed a compact light source and have improved the receiving system. This new CO2-DIAL system can observe the vertical CO2 concentration profile from 0.4 km to 2.5 km altitude.

2. COMPACT CO2-DIAL SYSTEM

In order to observe the CO2 concentration distribution in the lower altitude, the output power of transmitter and the dynamic range of the receiver of the CO2 DIAL have improved. This 1.6 μm DIAL system consists of the OPG transmitter that excited by the LD pumped Nd:YAG MOPA (Master Oscillator Power Amplifier) system [4] and a 25 cm coaxial telescope. Fig.1 shows a schematic of the 1.6 μm OPG transmitter. The MOPA system consists of an oscillator (IB laser, DiNY pQ10, 9.2 mJ, 400Hz), and a preamplifier (Cutting Edge Optronics, REA5006-3P5). A pulse energy of 23.3 mJ (1064 nm) and that of 2.1 mJ (1572 nm) are obtained at the repetition rate of 400 Hz.

Fig. 2 shows a schematic of the compact 1.6 μm CO2-DIAL system. Since the change of signal intensity is larger near the ground, we select the analog mode and use two-channel PMTs to expand the dynamic range. One is the channel for close range and another is the channel for up to the upper boundary layer. This 1.6 μm DIAL system is available to measure CO2 concentration

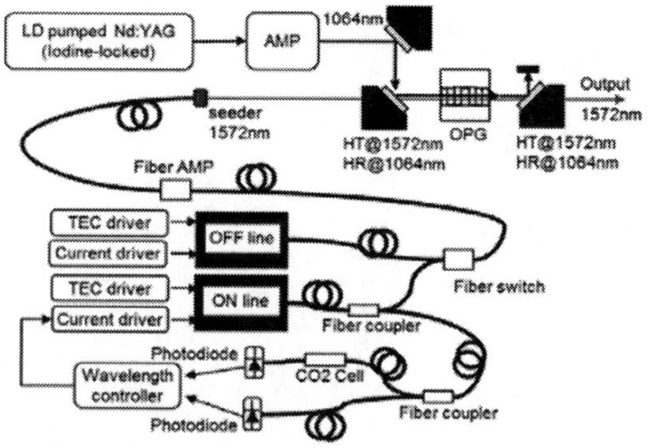

Fig. 1 Schematic of the 1.6 μm OPG transmitter.

profiles for daytime by using narrow-band interference filters (FWHM 1.0 nm). As the transmitter beam of this DIAL system is able to scan from -4 degree to 52 degree with elevation angle, the vertical distribution of lower CO2 concentration as well as the horizontal distribution from short range can be measured with high precision.

3. CO2 VERTICAL PLIFILES

The compact DIAL was achieved successfully measurements of CO2 concentration profiles for the range from 0.25 to 3 km with integration time of 30 minutes and range resolution of 300 m. Fig. 3 shows CO2 concentration profiles during a daytime segment of the measurement period. CO2 concentration increase in the lower altitude was observed at 12:00LT ~. Fig. 4 shows CO2 concentration profiles during a nighttime segment of the measurement period. CO2 concentration increase in the lower altitude was observed at 19:00LT ~.

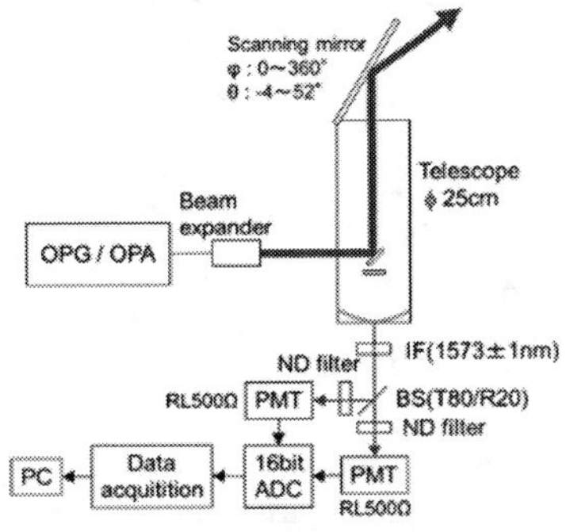

Fig. 2 Schematic of the compact 1.6 μm CO2-DIAL system.

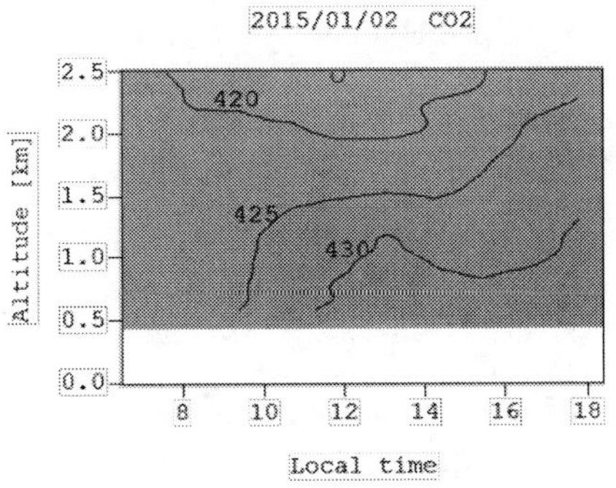

Fig.3 CO2 concentration profiles during a daytime. The contour interval is 5 ppm.

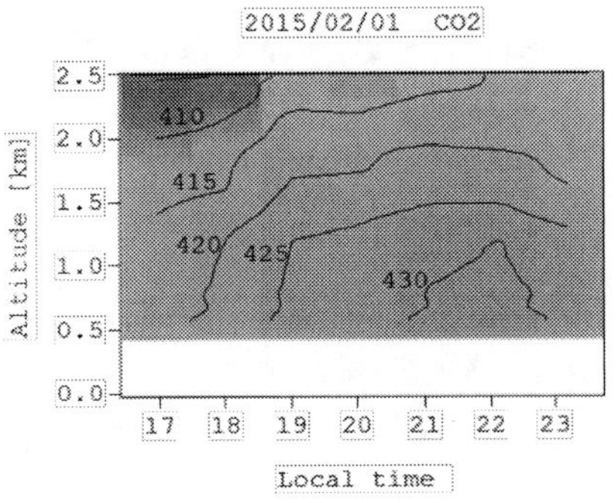

Fig. 4 CO2 concentration profiles during a nighttime. The contour interval is 5 ppm.

4. CONCLUSION

We have developed the compact 1.6 μm CO2-DIAL system for measuring in the lower atmosphere. The system operating at 1.6 μm wavelength that includes a 2 mJ QPM-OPG transmitter obtained temporal transition of vertical CO2 concentration profiles with suitable SNR in the atmospheric boundary layer during daytime and nighttime.

ACKNOWLEDGEMENT

This work was financially supported by the System Development Program for Advanced Measurement and Analysis of the Japan Science and Technology Agency.

REFERENCES

[1] B. B. Stephens et al. 2007: Science, 316 , p. 1732-

[2] Y. Shibata, C. Nagasawa, M. Abo, 25th ILRC, 243-246 (2010)

[3] Y. Shibata, C. Nagasawa, M. Abo, Proc. SPIE 8894, 889406 (2013)

[4] M. Raybauta, et al., Opt. Lett., 34, pp. 2069-2071 (2009).

EPJ Web of Conferences **119**, 03004 (2016) DOI: 10.1051/epjconf/201611903004
ILRC 27

AIRBORNE 2-MICRON DOUBLE PULSED DIRECT DETECTION IPDA LIDAR FOR ATMOSPHERIC CO_2 MEASUREMENT

Jirong Yu[1], Mulugeta Petros[1], Tamer Refaat[1], Karl Reithmaier[2], Ruben Remus[1], Upendra Singh[1], Will Johnson[3], Charlie Boyer[1], James Fay[1], Susan Johnston[2], Luke Murchison[1]

[1] *NASA Langley Research Center, MS 468, Hampton, VA 23681 USA, j.yu@nasa.gov*
[2] *Science System &Applications, Inc, One Enterprise Parkway, Hampton, Virginia 23666 USA*
[3] *Michigan Aerospace Corporation, 1777 Highland Dr. # B, Ann Arbor, MI 48108 USA*

ABSTRACT

An airborne 2-micron double-pulsed Integrated Path Differential Absorption (IPDA) lidar has been developed for atmospheric CO_2 measurements. This new instrument has been flown in spring of 2014 for a total of ten flights with 27 flight hours. This IPDA lidar provides high precision measurement capability by unambiguously eliminating contamination from aerosols and clouds that can bias the results.

1. INTRODUCTION

Active sensing of carbon dioxide concentrations will significantly increase the understanding of the gas sources, sinks, and fluxes worldwide. The mid-IR wavelength regions at 1.57μm and 2.05μm are considered suitable for atmospheric CO_2 measurements due to the existence of distinct absorption features for the gas at these particular wavelengths. NASA Langley Research Center (LaRC) developed a new approach to use double pulsed, high energy 2-micron IPDA lidar instrument to measure atmospheric CO_2. The pulsed lidar approach inherently provides means for determining range across the scattering targets. The reflected signals can be resolved between aerosols, clouds, and topographical surfaces. This lidar is operated on the long wavelength wing of CO_2 R(30) absorption line at 2050.967 nm (4875.749 cm^{-1}) in the side-line operation mode. The R(30) line is an excellent absorption line for the measurements of CO_2 in 2μm wavelength region with regard to the strength of the absorption lines, low susceptibility to atmospheric temperature variability, and freedom from problematic interference with other absorption lines [1]. This paper describes the development of a 2-micron pulsed IPDA lidar instrument that measures atmospheric CO_2 concentration between ground and airborne platform.

2. IPDA LIDAR SYSTEM

High-precision and accurate atmospheric CO_2 measurements impose stringent requirements on the lidar transmitter and receiver parameters, such as laser energy, pulse repetition rate, laser frequency control accuracy, telescope design and aperture size, high sensitivity with low noise detector and receiver design [2]. Figure 1 depicts the block diagram for the 2-micron pulsed IPDA instrument. It contains laser transmitter, thermal control unit for laser transmitter, wavelength tuning, locking and switching unit, telescope, aft optics and detector unit, signal preamplifiers, data acquisition unit, computer control and signal process unit, and data display.

The compact, rugged, highly reliable laser transmitter is based on the Ho:Tm:YLF high-energy 2-micron pulsed laser technology [3]. The laser transmitter is designed to be operated in a unique double pulse format that can produce two pulse pair in 10 Hz operation. Typically, the output energies of the laser transmitter are 100mJ and 45mJ for the first pulse and the second pulse, respectively. The first pulse is injection seeded with on-line frequency and the second pulse with off-line frequency. The double pulse operation is a unique feature of this Holmium (Ho) and Thulium (Tm) co-doped laser that has several advantages. First, the laser can be more efficient. It provides two Q-switched pulses with a single pump pulse. Second, since the time interval between the first and second pulses is typically about 150 - 200μs, the foot print overlap on the ground between the two pulses is greater than 95% for an airborne platform. This mitigates the effect of the surface reflection difference between the on-and-off line pulses on the precision of the CO_2 column density measurements. Third, the pulse pair is from the same laser. No beam combining or additional beam control is needed. The laser transmitter is

11.5 x 26.5 x 6.4 inch (29 x 67.3 x 16.5 cm) in size, and weighted less than 60lbs.

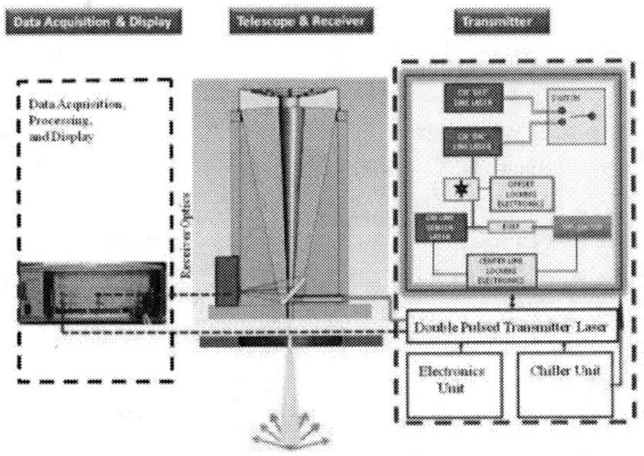

Fig. 1. 2-micron CO_2 IPDA instrument block diagram

The exact wavelengths of the pulsed laser transmitter are controlled by the wavelength control unit. The first pulse and the second pulse are injection seeded alternately by the on-line frequency and the off-line frequency. To obtain the wavelength accuracy and stability, a master wavelength reference against a sample of CO_2 in a gas cell is established. One of the CW lasers, called the center-line reference, is locked on the center of the CO_2 absorption line R(30) by a frequency modulation spectroscopic technique [4]. A second laser, called the tunable side-line, is referenced to the center-line laser by a heterodyne technique. By monitoring the heterodyne beat signal between the two lasers, the amount of detuning from the R(30) center can be determined and locked. Typically, it is locked at 2-4 GHz from the peak of R30 absorption line. A third CW laser provides the off-line wavelength. It is locked to a high precision wavemeter. The wavelength locking accuracy is less than 2 MHz and 30MHz for the on-line and off-line frequencies, respectively. The three CW lasers, an EO modulator, two detector units, an optical switch, fiber couplers and connectors are all packaged in the custom designed 3U 19 inch rack mountable box.

The telescope is a custom designed Newtonian type with 40cm diameter primary mirror size. This primary mirror is made of aluminum with diamond turning machining technique. The telescope is designed to maintain the focus point position in the temperature range between 5 and 35 °C. The lidar return signal is split into two channels, a high gain channel with 90% of signal power and a low gain channel with 10% of the signal power. A Hamamatsu InGaAs PIN photodiodes, model G5853-203, is selected and characterized for the airborne IPDA lidar application. This detector has two stage of TEC (thermo-electric cooler) and can be cooled down to -20°C. To obtain fast response and low noise required for the lidar signal detection, the diameter of the detector active area is limited to 300 micron. Thus, the NEP value is specified at $2x10^{-13}$ $W/Hz^{0.5}$ at -20°C, which is suited for the airborne IPDA lidar.

The data acquisition unit is based on two digitizers. The first is a 10-Bit, 2 GS/s digitizer (Agilent; U1065A) for laser energy monitoring and the second is a 12-bit, 420 MS/s digitizer (Agilent; U1066A) for measuring the hard target returns. Detectors are coupled to the digitizers through variable gain, high speed trans-impedance amplifiers (FEMTO; DHPCA-100). Digitizers and data storage are hosted through a personal computer that runs Microsoft XP with a 64-bit/66 MHz PCI bus. The system is capable of transferring data at sustained rates up to 400 MB/s to the host computer.

A simple real time data processing algorithm is implemented to show the lidar returning signal quality. It displays the laser energy monitor signal at on-and-off line wavelengths, the raw data from the ground returning signal, the range information from the lidar return, and the first order estimate of the Differential Absorption Optical Depth (DAOD). These signals aids to verify correct lidar operation.

3. AIRBORNE DEMONSTRATION

After the IPDA lidar has been integrated in class 10000 clean room environment, it was installed inside a mobile trailer for ground testing using calibrated hard targets before aircraft integration. Five calibrated targets with different reflections at 2-micron wavelength were set up 857 meters away from the lidar trailer. Differential optical depth measurements were obtained by calculating the natural logarithm of the off-line to the on-line return energy after normalization to the transmitted laser energy (laser energy monitor).

The theoretical differential optical depth was derived using CAPABLE meteorological data and CO_2 *in-situ* gas analyzer (LiCor; LI-840A) and US standard model [5]. In addition, the IPDA measured differential optical depth was converted to CO_2 dry mixing ratio, and was compared to the *in-situ* gas analyzer at the site. General agreement of the CO_2 temporal profiles were observed between these two measurements, verifying the 2-µm IPDA lidar operation.

Fig. 2. The double-pulsed 2-µm CO_2 IPDA lidar integrated instrument inside the NASA B-200 research aircraft

The double-pulsed 2-µm CO_2 IPDA lidar is designed for integration into a small research aircraft. The IPDA instrument size, weight and power consumption were restricted to the NASA B-200 payload requirements. This allows the system to be easily adopted in any larger airborne research platform, such as the NASA DC-8 aircraft, for future missions. In addition to the IPDA lidar, other housekeeping instruments were integrated into the B-200 aircraft. These included the CO_2 *in-situ* sensor for dry mixing ratio measurement, GPS for aircraft position, altitude and angles measurements and video recorder for target identification. Besides, aircraft built-in sensors provided altitude, pressure, temperature and relative humidity sampling at the flight position. Time stamps were adjusted to the GPS global timing. Figure 2 shows the integrated IPDA instrument inside the NASA B-200 aircraft.

The airborne IPDA lidar instrument measures the total integrated column content of CO_2 from the instrument to the ground but with weighting that can be tuned by controlling the transmitted on-line wavelengths. Therefore, the transmitter could be tuned to weight the column measurement to the surface for optimum CO_2 sources and sinks interaction studies or up to the free troposphere for optimum transport studies. In fact, it is one of the advantages that the 2-micron pulsed IPDA instrument can provide.

The 2-µm CO_2 IPDA lidar airborne testing was conducted during ten daytime flights, spanning more than 27 hours, during March 20, 2014 through April 10, 2014. Meteorological balloon radiosonde was independently lunched from LaRC site, mostly during the beginning and the end of each flight. This allows for atmospheric pressure, temperature and relative humidity vertical profiling estimates for IPDA modelling verifications. IPDA lidar airborne testing included different operating and environmental conditions. Environmental conditions included different flight altitude up to 8.3 km, different ground target conditions such as vegetation, soil, ocean, snow and sand and different cloud conditions. Two flights targeted power plant incinerators for investigating the IPDA sensitivity to CO_2 plums. Table 1 lists these flights and summarizes the conditions. The lidar instrument is robust during all of the flights. In spite of the mechanical stresses due to shaking and vibration in the small aircraft environment, the IPDA lidar did not lose alignment resulting in about 190 GB worth of raw data.

TABLE 1. Summary of the 2-micron CO_2 IPDA lidar flights during March 20, 2014 to April 10, 2014 on board the NASA B200 research aircraft

Flight #	Date	Local Time		Objective
		Start	End	
1	3/20	1:50p	3:42p	Engineering
2	3/21	2:17p	5:00p	Engineering
3	3/24	8:50a	11:50a	Engineering
4	3/27	7:15a	10:15a	Surface
5	3/27	1:35p	4:00p	Surface
6	3/31	1:30p	4:00p	Ocean/cloud
7	4/02	9:24a	11:47a	Power station
8	4/05	1:10p	4:52p	With NOAA
9	4/06	2:10p	5:10p	Power Plant
10	4/10	3:40p	5:58p	In-situ

On April 5, 2014, the NASA B-200 eighth flight coincided with another NOAA air sampling flight. Due to flight control restrictions, there was a 30 minute time lag between NOAA and NASA flights. Nevertheless, the IPDA lidar on board NASA flight sampled the same geographical location as the NOAA flask samples over the Atlantic Ocean out of the east shore of New Jersey. CO_2 flask-sampling results and meteorological data provided by NOAA were valuable for the NASA LaRC IPDA lidar instrument testing. Preliminary analysis of the airborne data indicated the IPDA lidar sensitivity to atmospheric CO_2 up to 6.6 km altitude over ocean target. Figure 3 compares the CO_2 differential optical depth obtained from the IPDA data and modelled from NOAA flask sampling data. IPDA indicated a consistent CO_2 differential optical depth offset of about 0.07.

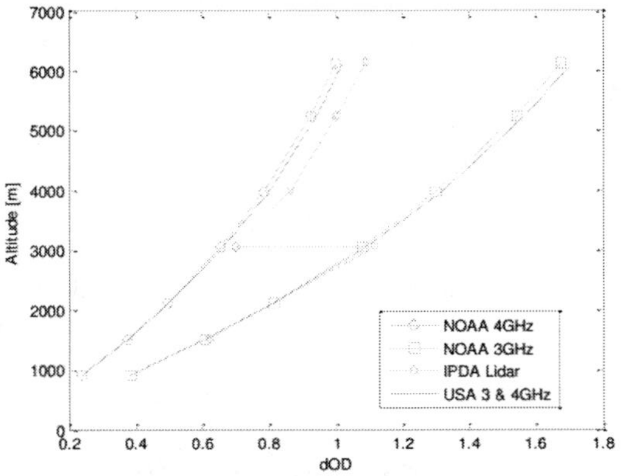

Fig. 3. CO_2 differential optical depth versus altitude calculated from NOAA flask sample data and from the IPDA lidar. Results are compared to US standard model calculations

4. CONCLUSIONS

NASA LaRC developed a double-pulse, 2-µm integrated path differential absorption (IPDA) lidar instrument for atmospheric CO_2 measurement. Advantages of the 2-micron high energy pulsed IPDA remote sensing technique include high signal-to-noise ratio measurement with accurate ranging; favorable weighting function towards ground surface to measure the source and sinks of the CO_2; and the capability to directly eliminate contaminations from aerosols and clouds to yield high accuracy CO_2 column measurements. The IPDA lidar structure is compactly and ruggedly packaged to fit in the NASA B-200 research aircraft. Ground and airborne testing of the 2-µm IPDA lidar was conducted at NASA LaRC through several validation procedures. This included instrument performance modeling through standard atmosphere and meteorological sampling. IPDA CO_2 differential optical depth measurement results agree with ground in-situ measurements and with CO_2 airborne sampling conducted by NOAA. Further detailed data processing is under work.

ACKNOWLEDGEMENT

The authors would like to thank NASA Earth Science Technology Office for funding this project. The authors acknowledge the supports from the Engineering and Research Services Directorates at NASA Langley Research Center. Thanks are also due to the dedicated efforts of the Research Systems Integration Branch personnel who made airborne flight testing possible. Acknowledgements are also due to the LaRC CAPABLE team and NOAA for providing public information that is significant for the science validation process. The authors also would like to thank Dr. Robert Menzies at JPL, Dr. Syed Ismail and Tony Notari at LaRC for very useful discussion, advice and help.

REFERENCES

[1] Ambrico, P. E., A. A. Amodeo, P. D. Gilaramo, and N. Spinelli, 2000: Sensitivity analysis of differential absorption lidar measurements in the mid-infrared region, *Appl. Opt.*, **39**, 6847-6865

[2] Ehret G., C. Kiemle, 2005: Requirement definition for future DIAL instruments, final report to ESA contract, p64

[3] Yu J., U. N. Singh, N. P. Barnes, and M. Petros, 1998: 125-mJ diode-pumped injection-seeded Ho:Tm:YLF laser, *Optics Letters* **23**, 780-782

[4] Koch, G. J., J. Y. Beyon, F. Gibert, B. W. Barnes, S. Ismail, M. Petros, P. J. Petzar, J. Yu, E. A. Modlin, K. J. Davis, and U. N. Singh, 2008: "Side-line tunable laser transmitter for Differential Absorption Lidar measurement of CO_2: Design and application to atmospheric measurement", *Appl. Opt.* **47**, 944-956

[5] G. Anderson, S. Clough, F. Kneizys, J. Chetwynd, and E. Shettle, 1986: AFGL atmospheric constituent profiles (0-120km), Air Force Geophysics Laboratory, AFGL-TR-86–0110, 954

DOI: 10.1051/epjconf/201611903005

2-µM COHERENT DIAL FOR CO_2, H_2O AND WIND FIELD PROFILING IN THE LOWER ATMOSPHERE: INSTRUMENTATION AND RESULTS

Fabien Gibert [1]*, Dimitri Edouart[1], Claire Cénac[1], Jessica Pellegrino[1], Florian Le Mounier[1], Arnaud Dumas[1]

[1]LMD-IPSL / CNRS, Ecole Polytechnique, F-91128 Palaiseau cedex, France,

**Email: gibert@lmd.polytechnique.fr*

ABSTRACT

We report on 2-µm coherent differential absorption lidar (CDIAL) measurements of carbon dioxide (CO_2), water vapour (H_2O) absorption and wind field profiling in the atmospheric boundary layer. The CDIAL uses a Tm:fiber pumped, single longitudinal mode Q-switched seeded Ho:YLF laser and a fiber-coupled coherent detection. The laser operates at a pulse repetition frequency of 2 kHz and emits an output energy of 10 mJ with a pulse width of 40 ns (FWHM). Experimental horizontal and vertical range-resolved measurements were made in the atmospheric boundary layer and compared to co-located in-situ sensor data.

1. INTRODUCTION

Water vapor (H_2O) and carbon dioxide (CO_2) are the key atmospheric gases for water and carbon cycle studies in the earth system. They are the most important greenhouse gases that rule the global warming but they are also the key components of plants biological processes (photosynthesis, respiration). Knowledge of atmospheric CO_2 and H_2O fields is of a major importance: (i) to improve our understanding of surface flux time and space heterogeneity (ii) to study the transport processes at different scales (iii) to monitor unambiguously anthropogenic emissions (for CO_2) above a given site (city, industry, carbon storage and capture experiment) for the carbon has now a price in the framework of climate and energy policy. However, knowledge of CO_2 and H_2O fields is usually useless without some information of the dynamics at different scales in the atmosphere. The wind field is of major importance to understand the diffusivity processes of these gases and to link the concentrations to surface sources and sinks.

In the instrumental point of view, CO_2 is an unprecedented challenge for optic based remote sensing techniques, as its variation is rather small in the atmosphere compared to H_2O. Typical natural variation due to its diurnal cycle (uptake by the vegetation, rectifier effect) are around 5% close to the surface but decrease to 0.1% for turbulent fluctuations in the convective boundary layer. Although differential absorption lidar experiments emerged in the last decade [1-3] the technology faced two main limitations to reach such precision: (i) a powerful and high PRF dual wavelength single mode emitter for coherent DIAL system and (ii) an efficient photodetector (i.e. avalanche photodiodes) in the near infrared for direct detection DIAL. In the framework of coherent DIAL system, a 2 µm high PRF dual-wavelength single mode pulsed Ho:YLF oscillator has been achieved and fully characterized at LMD/CNRS [4]. In this paper we report for the first time on the use of such emitter in a CDIAL system for accurate range-resolved atmospheric CO_2 absorption measurement. In addition we also demonstrate the ability of the lidar to make wind speed and H_2O absorption measurement.

2. EXPERIMENTAL SET-UP

The 2-µm domain, in addition to be an eye safe domain, is characterized by CO_2 and H_2O absorption lines that are well suited for DIAL measurements in the troposphere (Fig. 1): weak sensitivity to temperature (E" ~300 cm^{-1}), non-interference with other gas absorption lines, suited optical depth for km-range of measurement. In addition this domain corresponds to Ho based crystal laser emission band. Using this technology, the LMD/CNRS developed during the last years a 2-µm Ho:YLF emitter for DIAL application. The experimental set-up is presented in Figure 2.

DOI: 10.1051/epjconf/201611903005

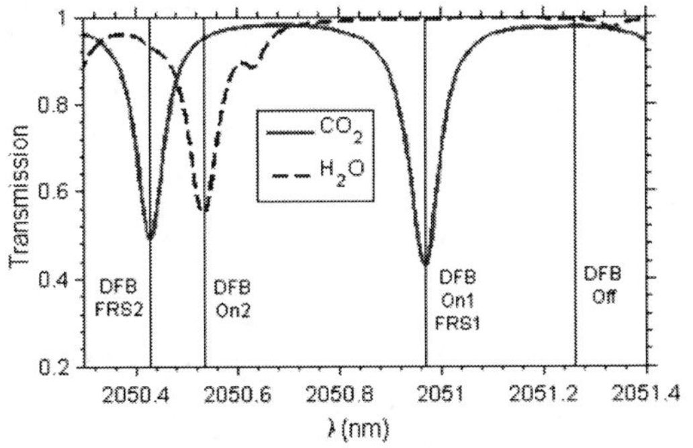

Fig. 1: 1-km horizontal single path atmospheric transmission in the 2.05-µm band and coherent DIAL wavelengths position used for CO_2 (On1) and H_2O (On2) differential absorption lidar measurements. H_2O specific humidity is 10 g/kg and CO_2 dry air mixing ratio is 380 ppm.

The Ho:YLF crystal is positioned in a ring cavity and is pumped by a commercial 100 W linearly polarized Thulium fiber laser from IPG Photonics. The advantage of such configuration compared to Nd:YAG pumped OPO emitter is (i) the compactness of the system for a given power and (ii) that no stringent characteristic is required for the pump laser, i.e. seeding, pulse repetition frequency (PRF) and beam quality [5,6]. The Ho:YLF oscillator is sequentially seeded by two fiber-coupled Distributed FeedBack laser diodes (DFB) at 2050.97 nm (On-line1) for CO_2 or 2050.53 nm (On-line2) for H_2O and 2050.26 nm (Off-line) (Fig. 1). On-line DFB is indeed frequency locked (offset locking) to a frequency reference system (FRS) using a low-pressure pure CO_2 absorption cell and Pound-Drever-Hall (PDH) technique (external frequency modulation). A similar PDH technique is used to lock the cavity length (i.e. a longitudinal mode) to the On-line seeder wavelength. Off wavelength is located in a free absorption window close to the R30 CO_2 line so that Off-line CDIAL signal serves as a reference. The laser was operated continuously and provided linearly polarized 10 mJ On-Off pulse-pair of approximately 40 ns duration at a repetition rate of 1 kHz. The beam was collimated with a built-in expanded 100-mm aperture telescope. The overall system is implemented in a container and a scanning device enables to make 3D measurements. The atmospheric molecules and particles backscattered signal is collected by the same telescope and optically mixed with the seeders on a fiber-coupled balanced InGaAs photodiode detection.

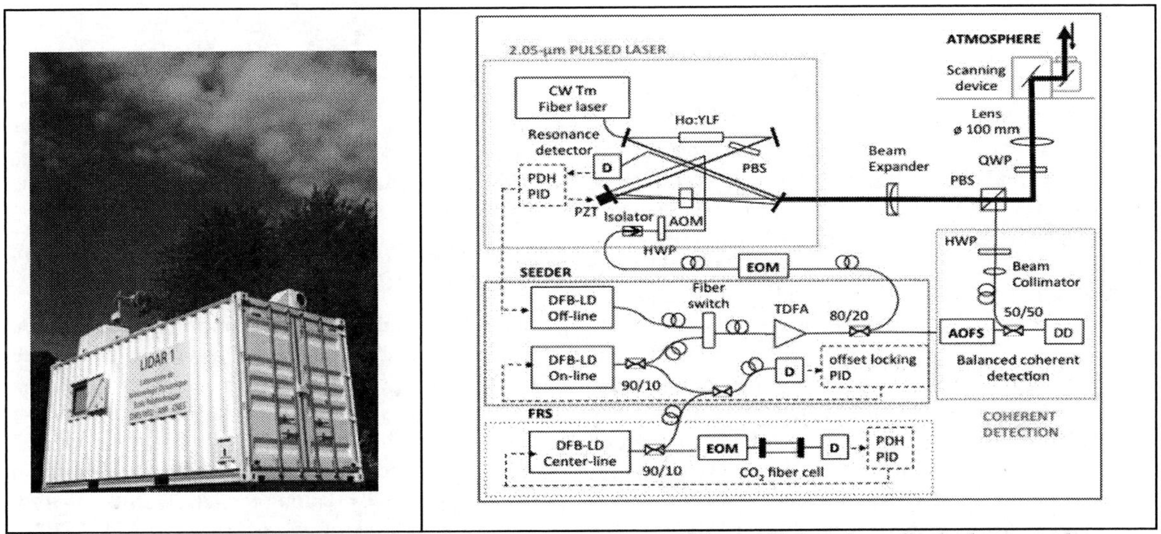

Fig. 2: Left: Coherent DIAL and in-situ mobile station for turbulence, CO_2 and H_2O cycle studies in the atmosphere. Right: 2-µm coherent DIAL experimental set-up. EOM: electro-optical modulator; AOM: acousto optical modulator; PID: proportional integral derivative; PDH: Pound Drever Hall; AOFS: acousto optic frequency shifter; TDFA: Thulium doped fiber amplifier; PBS: polarizer beam splitter; HWP: half wave plate; QWP: quarter wave plate; PZT: piezoelectric transducer.

DOI: 10.1051/epjconf/201611903005

The signal is digitized on a 400 MHz – 14-bits FPGA based high-speed data acquisition card. Real time signal processing consists in a 50-m range gate Discrete Fourier Transform accumulation over 4 s (2000 shot averaging for each wavelength). Post-processing uses Squarer and Levin-like estimators to deliver atmospheric On and Off backscattered signal power and frequency at the wanted time and space resolution. Further MATLAB based software provides a real time view of lidar reflectivity, radial velocity and CO_2 or H_2O differential absorption coefficients.

3. RESULTS

3.1 Accurate CDIAL range-resolved atmospheric CO_2 absorption measurement and mixing ratio estimates

Using the CDIAL measurements of the CO_2 absorption coefficient, one may estimates the CO_2 dry air mixing ratio with additional spectroscopic and meteorological data:

$$X_{CO2}=\frac{\alpha}{n_a\Delta\tilde{\sigma}_{CO2}}=\frac{kT\left(1+X_{H2O}\right)}{p}\frac{\alpha}{\left(\tilde{\sigma}_{CO2,On}-\tilde{\sigma}_{CO2,Off}\right)} \quad (1)$$

where α is the differential absorption coefficient due to CO_2 absorption only, n_a is the dry air density, $\Delta\sigma_{CO2}$ is the CO_2 differential absorption cross section, k the Boltzman constant, T, the temperature, p the pressure and X_{H2O} the dry air water vapor mixing ratio.

In a preliminary experiment the CDIAL was located at the second floor of the LMD building and we made horizontal measurement at ~15 m height above Ecole Polytechnique campus. In parallel, a PICARRO in-situ gas analyzer made some routine measurements. Figure 3 shows some examples of horizontal profiles of CO_2 dry air mixing ratio in late afternoon (19:00) and early morning (07:00). The time and space resolutions were 15 min and 100 m, respectively. In-situ measurements are indicated with the star markers for comparison. The experimental standard deviation for each profile is indicated in Fig.3(b). This accounts for both instrumental and natural standard deviation. The instrumental error accounting for CDIAL absorption measurements may be estimated using the high frequency data power spectrum (4 s - 50 m) and looking for the white noise level. The results indicate an instrumental standard deviation of 0.5% (2 ppm) up to a range of 500 m and 2% (8 ppm) at 1 km (solid line in Fig. 3). This error is confirmed by the CNR-based SNR calculation method [7] at least up to 700 m (dashed line in Fig.3(b)). The experimental standard deviation is nevertheless much higher (between 4 to 8 ppm up to 1 km) as a result of badly mixed anthropogenic CO_2 emissions in the nocturnal boundary layer. This is confirmed by the in-situ sensor standard deviation which shows 2 ppm at 19:00 and 6 ppm at 07:00 (typical PICARRO instrumental error is 0.2 ppm).

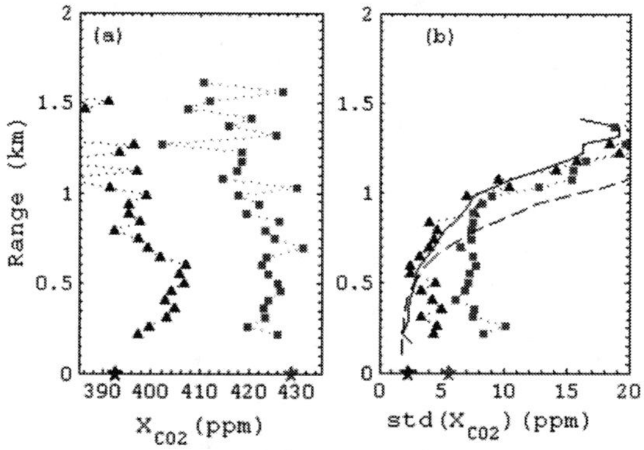

Fig. 3. (a) Examples of horizontal CDIAL CO_2 dry air mixing ratio profile. Time and space averaging are processed over 15 min and 100 m, respectively. (b) Experimental standard deviation calculated over a time gate of 3 h around each CO_2 profile. The instrumental error accounting for CDIAL absorption measurements is estimated using the high frequency data power spectrum (solid line) and using the CNR (dashed line) [7]. The different symbols are for the PICARRO in-situ gas analyzer (stars) and the CDIAL (squares) at 19:00 (black) and 07:00 (grey) in local time.

3.2 Preliminary test of CDIAL H_2O absorption measurement and specific humidity estimates

Using the nearby H_2O absorption line close to the R30 CO_2 absorption line (Fig.1), we tested the CDIAL ability for water vapor dry air mixing ratio profiling in the boundary layer. Figure 1 shows that On2 wavelength is not entirely free from CO_2 absorption. However, the low variation of atmospheric CO_2 along the CDIAL line of sight has a negligible effect of H_2O profile CDIAL retrieval. Practically, in-situ CO_2 mixing ratio measurements are usually sufficient to correct the

H_2O CDIAL differential absorption from this interference. One may estimate the dry air H_2O mixing ratio in [g/kg] by:

$$X_{H2O} = \left(\frac{\alpha - n_a X_{CO2} \Delta\tilde{\sigma}_{CO2}}{n_a \Delta\tilde{\sigma}_{H2O} - \alpha} \right) \frac{10^3 . M_{H2O}}{M_{air}} \quad (2)$$

where α is the total differential absorption coefficient due to CO_2 and H_2O absorption, $\Delta\sigma_{H2O}$ is the H_2O differential absorption cross section, M_{H2O} and M_{air} are the water and the air molar masses and X_{CO2} the dry air CO_2 mixing ratio provided by the in-situ sensor.

For the experimental measurements, the CDIAL was installed in a container and used the scanning device to look vertically. Figure 4 shows an example of a vertical profile and a comparison with a radiosonde profile. The time and space resolution were 10 s and 100 m. Although the radiosonde was not at the right time and space of the CDIAL measurement, there is a good agreement of the two profiles. The CDIAL profiles fluctuations are interpreted as both instrumental error and natural fluctuations due to turbulence. The experimental standard deviation is 10% up to 1 km.

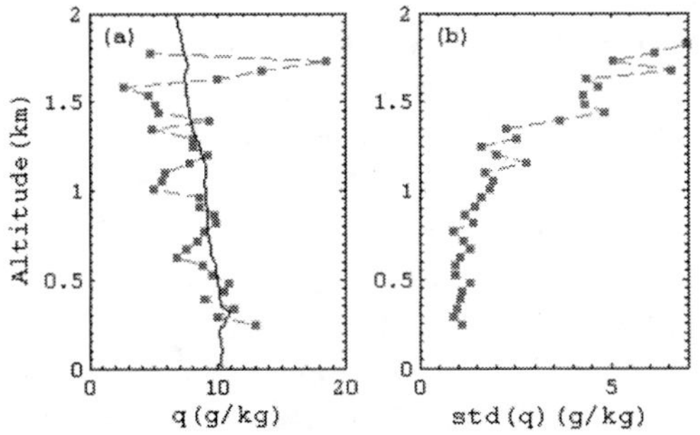

Fig. 4. (a) Example of CDIAL vertical H_2O dry air mixing ratio profile at 04:16 pm in local time. Time and space averaging are processed over 10 s and 100 m, respectively. Solid line is for the 10-km apart radiosonde profile at 01:00 pm. (b) Experimental standard deviation calculated over a time gate of 10 min.

4. CONCLUSION

To conclude, we demonstrate the 2-µm Ho emitter based coherent DIAL ability to measure CO_2 and H_2O absorption coefficient in the atmospheric boundary layer. Precision and time and space resolutions seem to be sufficient for 3-D monitoring application in the boundary layer for CO_2 and H_2O. Concerning turbulence-based fluctuations, the CDIAL may match the required performances for H_2O but not for CO_2 where both precision and time resolutions are not sufficient.

ACKNOWLEDGEMENT

The CDIAL development is supported by the French National Agency (ANR), the Centre National d'Etudes Spatiales (CNES) and the Institut National des Sciences de l'Univers (INSU-CNRS).

REFERENCES

[1] G.J. Koch, J.Y. Beyon, F. Gibert, B.W. Barnes, S. Ismail, M. Petros, P.J. Petzar, J. Yu, E.A. Modlin, K.J. Davis and U.N. Singh, 2008: Side-line tunable transmitter for differential absorption lidar measurements of CO2: design and application of atmospheric measurements, *Appl. Opt.*, **47**, 944-956

[2] F. Gibert, P.H. Flamant, J. Cuesta, D. Bruneau, 2008: Vertical 2-µm heterodyne differential absorption lidar measurements of mean CO2 mixing ratio in the troposphere, *J. Atmos. Ocean. Tech.*, **25**(9), 1477-1497

[3] D. Sakaizawa, C. Nagazawa, T. Nagai, M. Abo, Y. Shibata, M. Nakazato and T. Sakai, 2009: Development of a 1.6 µm differential absorption lidar with a quasi-matching optical parametric oscillator and photon-counting detector for the vertical CO2 profile, *Appl. Opt.*, **48**, 748-757

[4] F. Gibert, D. Edouart, C. Cénac and F. Le Mounier 2014: 2-µm high power multiple-frquency single mode Q-switched Ho:YLF laser for DIAL application, *Appl. Phys. B*, **116**, 967-976

[5] A. Amediek, A. Fix, M. Wirth and G. Ehret, 2008: Development of an OPO system at 1.57 µm for integrated path DIAL measurement of atmospheric carbon dioxide, *Appl. Phys. B*, **92**, 295-302

[6] J. Barrientos Barria, D. Mammez, E. Cadiou, J.B. Dherbecourt, M. Raybaut, T. Schmid, A. Bresson, J.M. Melkonian, A. Godard, J. Pelon and M. Lefebvre, 2014: Multispecies high-energy emitter for CO_2, CH_4, and H_2O monitoring in the 2 µm range, *Opt. Lett.*, **39**, 6719-6722

AEROSOL PRODUCTS FROM THE FUTURE SPACE LIDAR AEOLUS

Pauline Martinet[1]*, Alain Dabas[1], Vincent Lever[1], Pierre Flamant[2], Dorit Huber[3]

[1]*Météo-France, CNRM-GAME/GMEI, Toulouse, France *Email: pauline.martinet@meteo.fr*

[2]*CNRS, Laboratoire de Météorologie Dynamique, Palaiseau, France*

[3] *DoRit, Munich, Germany*

ABSTRACT

Ready for launch by the end of 2016, the Doppler lidar mission AEOLUS from the European Space Agency (ESA) will be the first High-Spectral Resolution Lidar (HSRL) in space. Operating in the UV, it implements two detection channels for aerosol and molecular backscatter. The system is primarily designed for the measurement of winds, but the HSRL capability enables the measurement of the particulate backscatter and extinction coefficients without any *a priori* assumption on the aerosol type. The level-2A (L2A) processor has been developed for these measurements and tested with synthetic data. The results show good aerosol backscatter profiles can be retrieved. Extinction coefficients are reasonable but do not reach the quality of backscatter coefficients. A precise, full, radiometric calibration of the lidar is required. A major limitation of the system is a single polarization component of the light is detected leading to an underestimation of backscatter coefficients when the atmospheric particles are depolarizing. The vertical resolution goes from 250 meters in the lowest part of the atmosphere, to 2 km in the lower stratosphere. The maximum altitude can reach above 20km. The basic horizontal averaging is 90km. Averaging on shorter distances (down to a few km) are possible but require a sufficient signal to noise ratio.

1. INTRODUCTION

The space-based Doppler mission AEOLUS of the European Space Agency should be ready for launch in the late 2016 or early 2017. It should be the first Doppler lidar in space able to measure winds over the globe throughout the entire depth of the atmosphere (from the ground up to 25km or 30km). Wind measurements will be delivered in quasi real-time and assimilated by numerical weather prediction systems of meteorological services. They should improve the quality of weather forecasts, in particular in the tropics and at high altitudes where wind observations are otherwise scarce. The single instrument aboard AEOLUS is the direct-detection Doppler lidar ALADIN. ALADIN operates a UV laser emitting 80 mJ pulses at the rate of $50s^{-1}$. It implements two sequential detection channels. The light is first passed through a narrow-band Fizeau interferometer, and the reflection on the Fizeau is transmitted through a large-band, double Fabry-Pérot. The Fizeau resolves the narrow-band spectrum of the aerosol backscatter, while the contrast of the light fluxes transmitted through the two FPs is converted into a wind Doppler shift as proposed in the late 80s by Chanin *et al.* [3]. ALADIN is thus a High-Spectral-Resolution Lidar (HSRL). Although it has been designed for wind measurements, the signals it detects bear information on the optical properties of atmospheric particles, either aerosols or hydrometeors. As demonstrated in [4] and [5], this information can be retrieved without making any a priori assumption on the aerosol Backscatter-to-Extinction Ratio (BER). Here the BER is measured by the lidar that can then not only quantify the aerosol properties of the atmosphere but also give a qualitative information on the aerosol type. This kind of information is very interesting for the scientific studies on the impact of aerosols on climate

DOI: 10.1051/epjconf/201611904001

change and could also be assimilated by numerical models predicting air quality on a global scale as it is done in the frame of the European GMES-MACC service (see http://www.gmes-atmosphere.eu).

2. METHODOLOGY

The spectrum of the light received by the lidar consists of a broad spectrum of molecular backscatter and a narrow spectrum of particle backscatter. It passes through 3 interferometers. One is a narrow band Fizeau filter, the two others are broad-band Fabry-Perot filters located on either side of the laser frequency. The power transmitted through the Fizeau (S_{Fiz}) and the sum of the powers transmitted through the two Fabry-Pérot filters (S_{FP})can be written:

$$S_{FP}(z) = \frac{E_0}{R(z)^2}[C_1\beta_{mol}(z) + C_2\beta_{aer}(z)]\, exp\big(-2\int_z^{+\infty}(\alpha_{aer}(x) + \alpha_{mol}(x))dR(x)\big)$$

$$S_{Fiz}(z) = \frac{E_0}{R(z)^2}[C_3\beta_{aer}(z) + C_4\beta_{mol}(z)]exp\big(-2\int_z^{+\infty}(\alpha_{aer}(x) + \alpha_{mol}(x))dR(x)\big)$$

Here, E_0 is the emitted laser energy, $R(z)$ is the distance from the satellite to the altitude z along the line-of-sight, β_{mol} and β_{aer} are the molecular and aerosol backscatter coefficients, α_{mol} and α_{aer} are the molecular and aerosol extinction coefficients, and C_1, C_2, C_3 and C_4 are calibration constants. The signals recorded by the lidar are integrated in 24 height-bins $[z_k, z_k + 1]$ of thicknesses varying from 250m in the lowest part of the atmosphere to 2km in the lower stratosphere:

$$S_{FP}(k) = \int_{z_k}^{z_{k+1}} S_{FP}(z)dz$$

$$S_{Fiz}(k) = \int_{z_k}^{z_{k+1}} S_{Fiz}(z)dz$$

Due to the transmission characteristics of the detectors, $C_2 < C_1$ and $C_4 < C_3$, the molecular and aerosol backscatter contributions can be separated after a crosstalk correction:

$$X(k) \approx \frac{1}{R(z_k)^2}\int_{z_k}^{z_{k+1}} \beta_{mol}(z)exp\big(-2\int_z^{+\infty}(\alpha_{aer}(x) + \alpha_{mol}(x))dR(x)\big)dz = \frac{1}{E_0}\frac{C_2S_{FP}(k)-C_3S_{Fiz}(k)}{C_2C_4-C_1C_4}$$

$$Y(k) \approx \frac{1}{R(z_k)^2}\int_{z_k}^{z_{k+1}} \beta_{aer}(z)exp\big(-2\int_z^{+\infty}(\alpha_{aer}(x) + \alpha_{mol}(x))dR(x)\big)dz = \frac{1}{E_0}\frac{C_4S_{Fiz}(k)-C_1S_{FP}(k)}{C_2C_4-C_1C_4}$$

The operation requires a good calibration of the lidar. The molecular backscatter β_{mol} and extinction α_{mol} are known functions of the air density that a numerical weather prediction model can predict with a good precision. It therefore remains two unknowns: the height-bin average aerosol backscatter $\hat{\beta}_{aer}$ and the height-bin average aerosol extinction $\hat{\alpha}_{aer}(k)$. Both can thus be determined for each height-bin.

As the receiver is transmitting a single polarization parallel to the laser emission, the backscatter coefficients above are for this single polarization.

3. RESULTS

When the molecular and aerosol backscatter contributions have been separated, the height-bin average aerosol properties can be retrieved. The L2A processor provides users with several products derived from the backscatter and the extinction coefficients: local optical depth, scattering ratio, backscatter-to-extinction ratio. In order to evaluate the performance of the chain of ADM-Aeolus processors, an end-to-end simulator (E2S) has been developed by ESA to simulate the Rayleigh and Mie signals received in the double Fabry-Pérot and the Fizeau interferometers. The L2A processor has been evaluated for different atmospheric scenarios from simple to complex scenes observed during the NASA's LITE (Lidar In-space Technology Experiment) experiment. By comparing the aerosol properties retrieved by the L2A to the input profiles used in the E2S to simulate the lidar signals, the expected accuracy of the aerosol properties can be studied. The L2A processor is expected to retrieve the backscatter coefficient and the scattering ratio with a good accuracy in cirrus clouds and boundary layers. Relative deviations from the true values are within 20 to 50 % with an average of 20 % in cirrus clouds (figure 2) and 20 to 45 % in the boundary layer. Oscillations between overestimation and underestimation are observed in the retrieval of the extinction coefficient due to the iterative algorithm used to derive this parameter. However, the overall shape of the aerosol and cloud layers is well captured by the L2A. Relative deviations for the extinction are within 20 to 60 % in cirrus layers (figure 1) and boundary layer aerosols with an average error of 45 % in layers with an extinction coefficient higher than approximately 1.10^{-4} m^{-1}. For small values of extinction, retrieval errors up to 80 % or more are observed. The retrieval inside and below opaque clouds is limited by the strong attenuation of the lidar signal (figures 1 and 2).

4. CONCLUSIONS

The on-going developments on the L2A processor have shown the potential of ADM-Aeolus to retrieve aerosol and cloud properties. With a launch scheduled in 2017, ADM-Aeolus could fill in the gap between the CALIPSO and EarthCare missions. ADM-Aeolus will also be the first satellite platform to provide information on both cloud microphysical properties and cloud dynamics. An accuracy of approximately 20 % in backscatter and 45 % in extinction can be expected in clouds and boundary layer aerosols. For the moment, these results are obtained with a horizontal averaging of 90 km. In the future, a feature finder algorithm, aiming at improving the horizontal resolution of the L2A by retrieving the aerosol properties at a smaller horizontal scale will be tested. This averaging at shorter distance requires a good signal-to-noise ratio.

ACKNOWLEDGEMENT

The work described in the paper was funded by the European Space Agency in the frame of the AEOLUS mission. We would like to thank ESA team – namely Frank de Bruin, Anne Straume, and Thomas Kanitz – and the other members of the level 1/level 2A development teams for their advice, support and feedbacks.

REFERENCES

[1] Regan, M. J., X. Wang, M. J. Osborn, 2002: Spaceborne lidar calibration from cirrus and molecular backscatter returns, *IEEE Trans. Geosci. Remote Sens.,* **40**, 2285-2290.

[2] O'Connor, E. J., A. J. Illingworth, R. J. Hogan, 2004: A technique for autocalibration of cloud lidar, *J. Atmos. Ocean Tech.,* **21(5)**, 777-778.

[3] Chanin, M. L., A. Garnier, A. Hauchecorne, J. Porteneuve, 1989: A Doppler lidar for measuring winds in the middle atmosphere. *Geophysical Research Letters*, **16**, 1273-1276.

[4] Ansmann, A., U. Wandinger, O. Le Rille, D. Lajas, A. G. Straume, 2007: Particle backscatter and extinction profiling with the spaceborne high–spectral–resolution Doppler lidar ALADIN: methodology and simulations. *Applied Optics*, **46**, 6606-6622.

[5] Flamant, P., Cuesta, J., Denneulin, M.-L., Dabas, A. and Huber, D., 2008: ADM-Aeolus retrieval algorithms for aerosol and cloud products. *Tellus*, **60A**, 273–288.

[6] Horanyi, A. *et al. (2014)* The assimilation of horizontal line-of-sight wind information into the ECMWF data assimilation and forecasting system, Part I: The assessment of wind impact, Q. J. R. Meteorol. Soc. (2014)

DOI: 10.1051/epjconf/201611904001

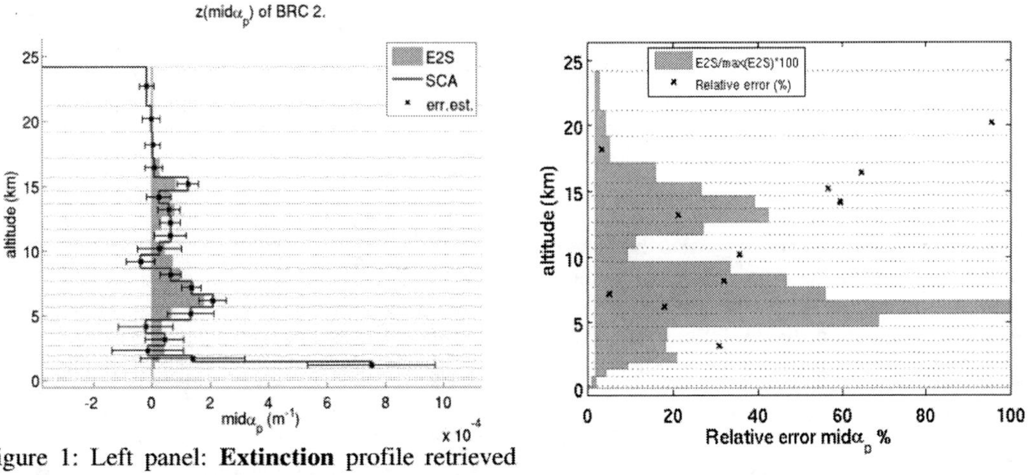

Figure 1: Left panel: **Extinction** profile retrieved by the L2A (red line) compared to the E2S input profile (gray area). The estimation of the retrieval uncertainties mainly due to the photon counting noise is shown for each layer as horizontal black line. Right panel: Relative deviation between the L2A retrieval and the E2S input profile (black crosses). The ratio of the E2S extinction value to the maximum value found over the profile is also displayed in gray area.

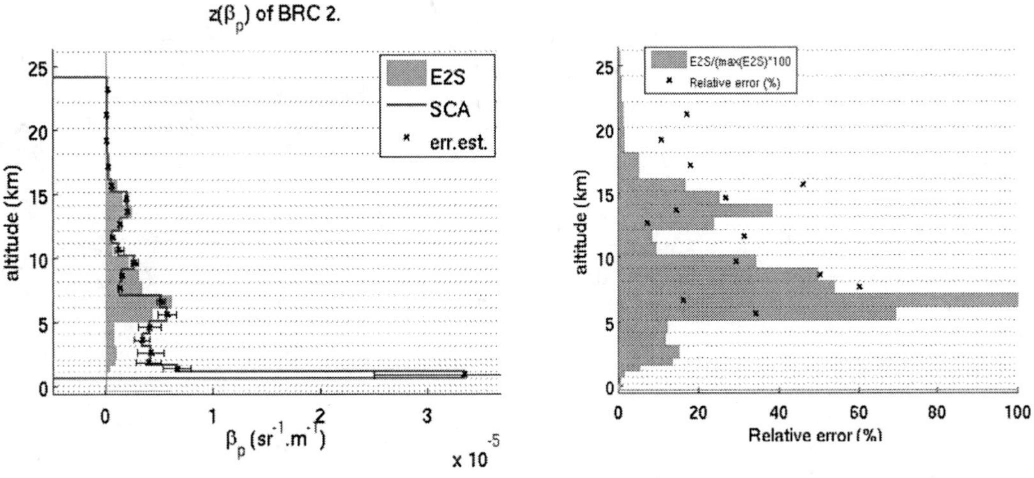

Figure 2: Left panel: Backscatter profile retrieved by the L2A (red line) compared to the E2S input profile (gray area). The estimation of the retrieval uncertainties mainly due to the photon counting noise is shown for each layer as horizontal black line. Right panel: Relative deviation between the L2A retrieval and the E2S input profile (black crosses). The ratio of the E2S backscatter value to the maximum value found over the profile is also displayed in gray area.

EPJ Web of Conferences **119**, 04002 (2016)
ILRC 27
DOI: 10.1051/epjconf/201611904002

Lidar and Laser Technology for NASA's Cloud-Aerosol Transport System (CATS) Payload on the International Space Station (JEM-EF)

Mark Storm*, Gary Stevenson, Floyd Hovis, William Gavert, Xung Dang, Abe Darab, Ti Chuang, and Patrick Burns

*Fibertek, Inc., 13605 Dulles Technology Drive, Herndon, VA 20171, USA, *mstorm@fibertek.com*

ABSTRACT

This paper describes the ISS lidar technology provided by Fibertek, Inc. in support of the NASA GSFC CATS mission and provides an assessment of the in-flight systems performance and lessons learned. During February the systems successfully operated in space for more than 300 hours using 25 W average power lasers and photon counting of aerosol atmospheric returns.

1. INTRODUCTION

Fibertek is supporting NASA's new Cloud-Aerosol Transport System (CATS) lidar for the International Space Station (ISS). CATS ISS is a Japanese Experimental Module – Exposed Facility (JEM-EF) payload launched in January 2015. The lidar has successfully completed more than 300 hours of operation during its February commissioning period. This program shows that it is possible to develop a rapid (24-month), low-cost ($14M [ref 1]) lidar instrument and successfully deploy it on ISS. The 300 hours of operation already exceeds the annual operational time for a typical aircraft remote sensing NASA lidar. This program establishes a technology heritage that lowers the risk and raises the Technology Readiness Level (TRL) for future science and technology validation remote sensing lidars.

2. METHODOLOGY

NASA's ISS CATS lidar is very unlike other space lidars costing hundreds of millions of dollars. The ISS CATS mission was really developed by a small team in a manner like a UAV mission to the space station. Goddard Space Flight Center (GSFC) and Fibertek leveraged mature heritage designs and spaceflight processes where affordable. The lasers were built to Fibertek's aircraft and flight optical standards while the lidar system and laser electronics were built using known radiation-tolerant parts at an affordable cost. Much of the systems electronics functionality was driven into high-performance Actel field programmable gate arrays (FPGAs) to minimize EEE parts count. Many EEE parts were uncertified flight parts or aerospace quality parts made from the same die as space parts. We estimate that more than 95% of the EEE parts have space-qualified plug-compatible equivalents and can be upgraded from Class D to Class C/B.

3. RESULTS

This program demonstrates that the ISS can enable a new generation of space lidars important for weather and global change research including for aerosols, methane, carbon dioxide, water vapor, ozone, and wind lidars that are currently under development (**Figure 1**).

Figure 1: CATS Lidar In-Flight Aerosol Data Showing Aerosols, Saharan Dust, Clouds, and Smoke from Fires Across Africa [2]

This paper describes the ISS lidar technology provided by Fibertek in support of the CATS mission and provides an assessment of the in-flight systems performance and lessons learned. A description of the instrument's major subsystems and photos of the systems during test and completed is shown in **Figure 2**.

CATS is a NASA GSFC lidar design that consists of the telescope, receiver optics, fiber-coupled detectors, telescope cover, boresight prisms, cover actuators, high-spectral-resolution lidar receiver, lidar controller, and lasers.

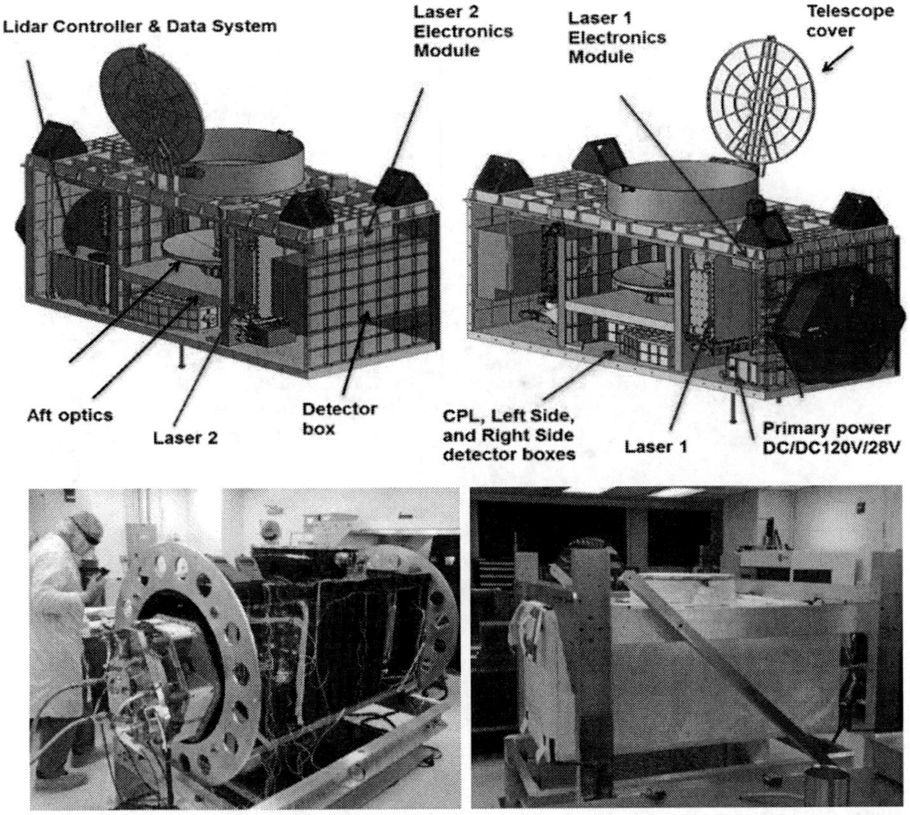

Figure 2: CATS During Integration (left) and Complete with Thermal Blankets (right)

JEM-EF payload is shown in **Figure 2** during integration at NASA GSFC. References 2-5 provide greater detail of the NASA CATS-ISS program objectives and goals, instrument design, and a more detailed description of the CATS lidar and Fibertek-provided lidar electronics and laser. CATS is a high-performance photon-counting lidar that provides aerosol lidar science data and validates new lidar technology.

Fibertek provided CATS subsystems including the lidar flight controller; two state-of-the-art 25 W average power flight lasers with operation at 1064, 532, and 355 nm; a high-power DC/DC converter (from 120V to 28V); two fault-tolerant safety interlocks; and an ISS simulator (**Figure 3**). Fibertek also developed and/or tested the ISS communications to Mil-Std-1553, Ethernet, fiber distributed data interfaces (FDDI), desktop lidar instrument controller interface, and command and control and science data communications link to the GSFC instrument Principal Investigator (PI) through the Marshall Space Flight Center (MSFC) Telescience Resource Kit (TReK) system. NASA GSFC designed, built, and integrated the overall completed JEM payload and provided the items in white: telescope, boresight prisms and cover actuators, and receiver optics and detectors.

Control Board: Lidar instrument control is provided to the CATS instrument operator through an Internet connection and the ISS downlink. Command and control functions for the instrument are implemented from a ground-based control station. The control board commands and activates up to 10 lidar operational scenarios. It provides a control interface for the two lasers, monitors and reports laser operations status and laser energies at all wavelengths, receives laser fire pulse timing information from the lasers, and triggers the receiver data collection operations. Instrument health and safety, activation of the boresighting mechanisms and the telescope cover, digital control of the secondary power supply, and receiver etalon tuning are also functions of the control board.

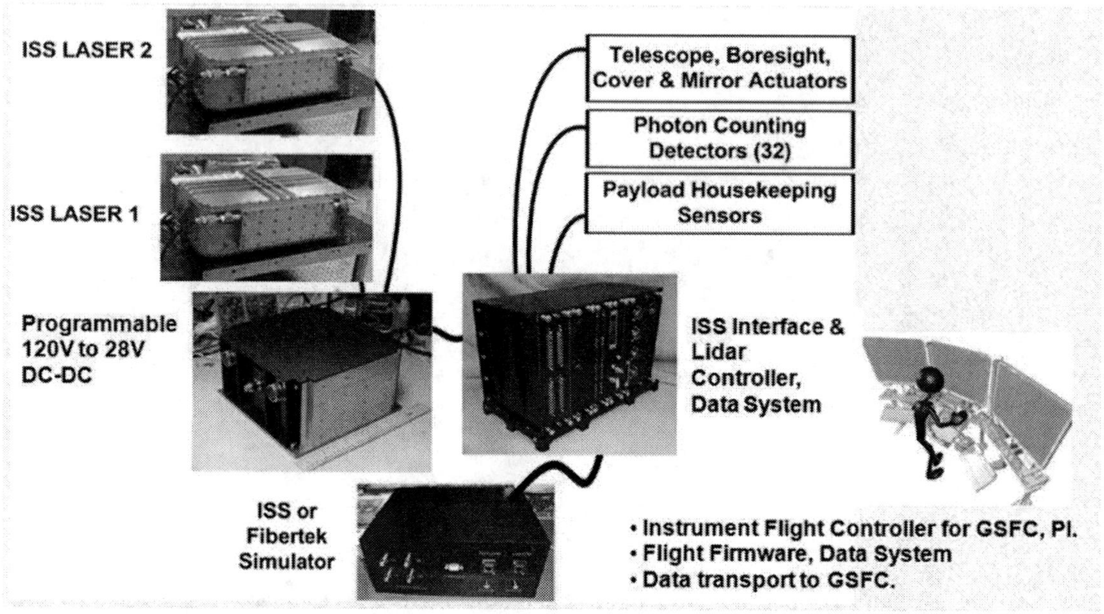

Figure 3: Description of the Fibertek-Provided Items to the ISS CATS Lidar

Data Capture Board: The data capture board supports 24 Avalanche photodetector (APD) photon-counting detector channels. The captured signals are time tagged and sent to the communications card for buffering, packetizing, and transmission to the ISS for downlink. Analysis of the data is not performed onboard the CATS instrument.

Safety: GSFC and Fibertek worked with Johnson Space Center (JSC) and MSFC to ensure all command and control functions required for instrument operation and safety. The systems can be turned off by the instrument manually or by automatic interlocks. The laser and systems can also be powered off without instrument assistance through a two-fault relay. All instrument health status, laser status, and safety functions are monitored by the control board.

ISS CATS Lasers: The CATS 25 W lasers exceed the performance of the CALISPO, ICESat-1, and upcoming ICESat-2 lasers in terms of laser power, beam quality, and spectral linewidth. Laser 1 produces 5 mJ/pulse total pulse energy with half at 1064 nm and 532 nm. Laser 2, shown in **Figure 4**, is a single-frequency transform-limited system that provides 6 mJ total energy at 1064, 532, and 355 nm.

The third harmonic generator is located outside the main laser and can be bypassed if ultraviolet (UV) lifetime issues occur. Laser 2 is not single frequency and operates at 1064 nm and 532 nm. Both lasers provide a high beam quality of <1.5 M2 at all wavelengths.

4. CONCLUSIONS

NASA's ISS CATS program has successfully demonstrated that a small team can rapidly and cost-effectively design, build, test, and deploy a space lidar capable of hundreds and thousands of hours of operation. Lidars can be developed to study focused earth science questions and to develop and validate technology for observatory class 3- to 5-year missions.

ACKNOWLEDGMENT

This work is supported by CATS Principal Investigator Dr. Mathew McGill at NASA GSFC and by Marybeth Edeen from the NASA ISS Program Research Office at the ISS National Laboratory.

DOI: 10.1051/epjconf/201611904002

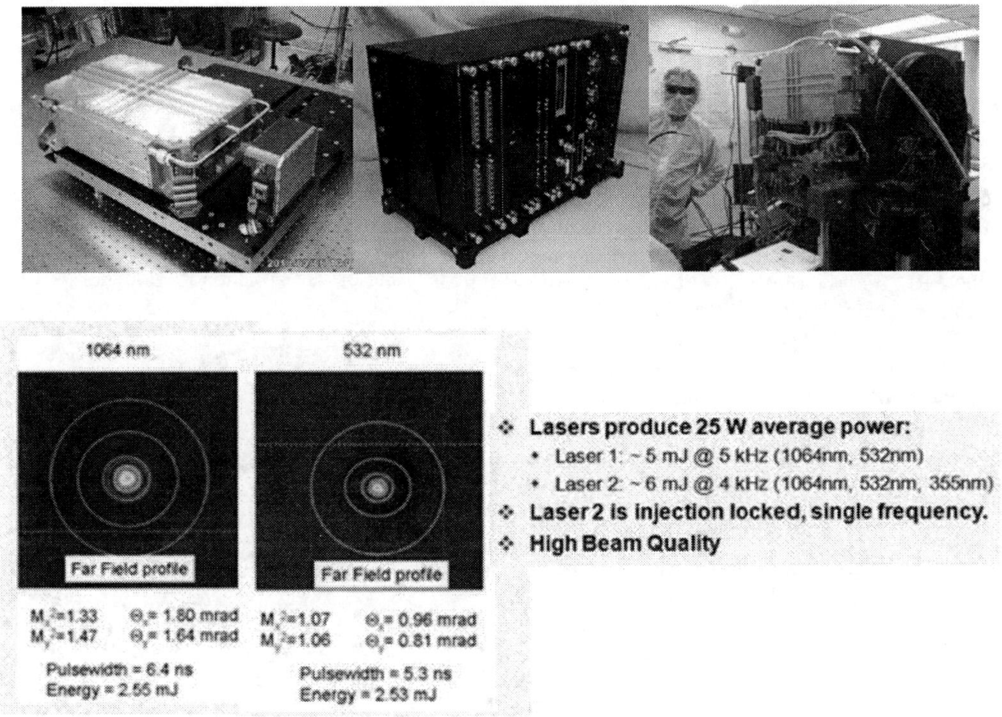

Figure 4: CATS Laser 2, Lidar Controller, and During Laser Integration (left to right)

REFERENCES

[1] http://aviationweek.com/space/station-role-earth-science-growing

[2] "Africa, From a CATS Point of View," http://www.nasa.gov/content/goddard/africa-from-a-cats-point-of-view/

[3] Mathew McGill, "The Cloud-Aerosol Transport System (CATS): A New Earth Science Capability for ISS," Proceedings from the 1st Annual ISS Research and Development Conference, June 26-27, 2012. American Astronautical Society.

[4] Mark Storm, et al., "Lidar Technology Avionics and Laser Flight Hardware for the Cloud-Aerosol Transport System (CATS) Payload on the International Space Station (JEM-EF)," Proceedings from the 1st Annual ISS Research and Development Conference, June 26-27, 2012. American Astronautical Society.

[5] Ti Chuang, et al., "Space-Based, Multi-Wavelength Solid-State Lasers for NASA's Cloud Aerosol Transport System for the International Space Station (CATS-ISS)," Proc. SPIE 8599, Solid State Lasers XXII: Technology and Devices, 85990N (March 6, 2013); DOI: 10.1117/12.2005545.

EPJ Web of Conferences 119, 04003 (2016) DOI: 10.1051/epjconf/201611904003
ILRC 27

ATLID, ESA ATMOSPHERIC LIDAR DEVELOPEMENT STATUS

João Pereira do Carmo [1], Arnaud Hélière[1], L. Le Hors[2], Y. Toulemont[2], A.Lefebvre[1]

[1]***European Space Agency**, ESTEC, NL, Email: joao.pereira.do.carmo@esa.int*

[2]***Airbus Defence and Space**, France*

ABSTRACT

The ATmospheric LIDAR ATLID[1] is part of the payload of the Earth Cloud and Aerosol Explorer[2] (EarthCARE) satellite mission, the sixth Earth Explorer Mission of the European Space Agency (ESA) Living Planet Programme. EarthCARE is a joint collaborative satellite mission conducted between ESA and the National Space Development Agency of Japan (JAXA) that delivers the Cloud Profiling Radar (CPR) instrument. The payload consists of four instruments on the same platform with the common goal to provide a picture of the 3D-dimensional spatial and the temporal structure of the radiative flux field at the top of atmosphere, within the atmosphere and at the Earth's surface. This paper is presenting an updated status of the development of the ATLID instrument and its subsystem design. The instrument has recently completed its detailed design, and most of its sub-systems are already under manufacturing of their Flight Model (FM) parts and running specific qualification activities. Clouds and aerosols are currently one of the biggest uncertainties in our understanding of the atmospheric conditions that drive the climate system. A better modelling of the relationship between clouds, aerosols and radiation is therefore amongst the highest priorities in climate research and weather prediction.

1. Introduction to ATLID, the ESA ATmospheric LIDar.

EarthCARE aims to the determination of cloud and aerosol occurrence, structure and physical properties together with collocated measurements of solar and thermal radiation at a global scale. The mission goals are to retrieve vertical profiles of clouds and aerosols, and the characteristics of their radiative and micro-physical properties, to determine flux gradients within the atmosphere and fluxes at the Earth's surface. By measuring directly the fluxes at the top of the atmosphere while clarifying the processes involved in aerosol-cloud and cloud-precipitation-convection interactions will allow to include them correctly and reliably in climate and numerical weather prediction models. The EarthCARE satellite shall achieve its mission through the operation, individual and in synergy, of its four instruments: ATLID, the CPR, the Multi-Spectral Imager (MSI) and the Broad-Band Radiometer (BBR).

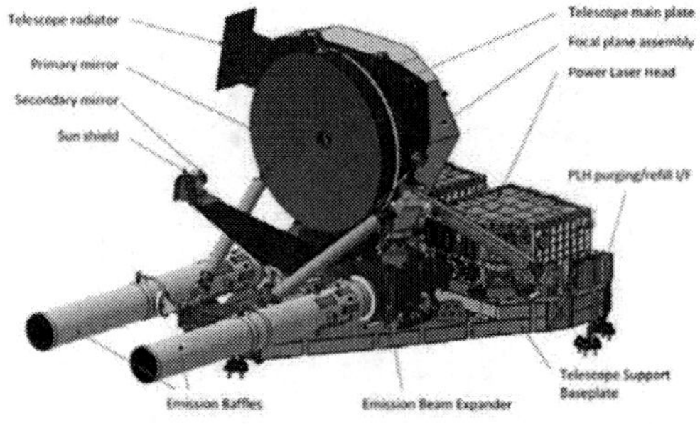

Fig. 1 – ATLID High stability assembly

The task of ATLID is to provide vertical profiles of optically thin cloud and aerosol layers, as well as the altitude of cloud boundaries. The measurements of ATLID are close to the nadir direction from a sun-synchronous orbit at 393 km altitude. These profile measurements have a vertical resolution of about 100 m from ground to an altitude of 20 km and 500m from 20km to 40km altitude. The instrument emits short laser pulses at a repetition rate of 51 Hz along the horizontal track of the satellite trajectory, so that several shots can be locally averaged to improve the signal to noise ratio. ATLID is a backscatter LIDAR instrument that uses the fact that interaction of light with molecules and aerosols leads to different spectra scattering effects. Whereas the Brownian motion of molecules induces a wide broadening of the incident light spectrum, the scattering with an aerosol does not affect the spectrum shape of the incident light. As a consequence, a simple means of separating the backscattering contributions consists in filtering the backscattered spectrum with a high spectral resolution filter centered on the emitted wavelength. This way the instrument is able to separate the relative contribution of aerosol (Mie) and molecular (Rayleigh) scattering, which allows the retrieval of the aerosol optical depth. Co-polarised and cross-polarised components of the Mie scattering contribution are also separated and measured on dedicated channels. The operating wavelength in the Ultra-Violet spectral range (355nm) was selected as the molecular scattering is high enough to measure accurate extinction profiles and aerosols/thin clouds thickness, and because laser technology (Nd:YAG laser with frequency tripling conversion) is available for operation in this spectral region.

2. ATLID Design

ATLID is designed as a self-standing instrument reducing the mechanical coupling of instrument/platform interfaces and allowing better flexibility in the satellite integration sequence. The instrument is based on a bi-static architecture consisting of two independent main sections, the emitter chain and the receiver chain. The instrument functions are shared between a 'high stability' assembly (including the telescope, focal plane optics, and the optical emission chain), and a housing structure assembly (supporting the electronic units and their radiator, the detection chain and the harness).

2.1. ATLID emitter chain design

The emitter chain includes the laser Transmitter Assembly (TxA), the Emission Beam Expander (EBEX), and the emission baffle. The TxA consist of a Reference Laser Head (RLH), a Power Laser Head (PLH) and its Transmitter Laser Electronics (TLE). The RLH provides the stable seed laser and frequency tuneability to the PLH which is a diode pumped Nd:YAG single-mode laser emitting at 355 nm. The PLH consists of a Master Oscillator, a Pump Unit and a Harmonic Conversion stage doubling and tripling the laser frequency. The laser head includes also a Beam-Steering Mechanism (BSM) finely adjusting the emission line-of-sight to continuously maintain emission / reception co-alignment in flight, based on information acquired in the reception chain by a Co-Alignment Sensor (CAS). With a power consumption of 300W the transmitter (implementing innovative mini loop heat pipes for cooling) delivers pulses of more than 38 mJ UV energy with duration of about 30 ns at a pulse repetition frequency of 51Hz. The laser transmitter has also stringent spectral requirements with a spectral linewidth below 50 MHz and a 25 GHz spectral tuneability range. The EBEX, sealed and pressurized as the TxA, is then used to expand the laser beam in order to meet the divergence requirement and minimize the laser fluence on the last diopter exposed to vacuum. This allows to mitigate the risk of Laser Induced Contamination (LIC) and the correspondent degradation of the instrument performance over its operational lifetime in space.

2.2. ATLID receiver chain design

The receiver chain includes the receiver telescope, the receiver optics, the science channels detection chain, the ATLID Control and Data Management Unit (ACDM). The ATLID telescope is an afocal Cassegrain with 620mm primary mirror diameter, made of Silicon Carbide to ensure high stability.

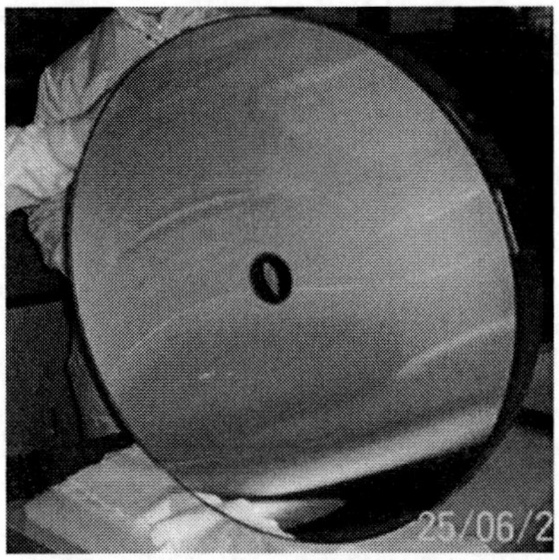

Fig.2 – ATLID Primary mirror (Airbus DS, Boostec, Safran/Reosc)

The receiver optics goes from the telescope output to the detector fiber entrances. It includes the entrance filtering optics (narrow interference filter with less than 1 nm bandwidth), the blocking filtering optics (spatial filtering with a field-stop delimiting the 65 μrad field-of-view), and two spectral filtering units: the background Fabry-Perot etalon used to finely filter the Earth background light, and the High Spectral Resolution filter (combining an optical prism assembly for the channels separation and a high resolution Fabry–Perot etalon).

Fig.3 – HSRE prisms used for channel differentiation (RUAG, TSESO)

The signal is transported to the detectors by means of fibre couplers, allowing deporting the whole detection chain on the anti-sun wall for passive cooling. Part of the flux is split at focal plane assembly entrance and imaged on the CAS which provides laser spot position information. The detection chain, composed by the three Memory CCDs and the Instrument Detection Electronics, is able to measure single photon events to meet the worst case radiometric performance requirements. The selected design provides high response together with an extremely low noise thanks to on-chip storage of the echo samples which allows delayed read-out at very low pixel frequency (typically below 50 kHz). Combined with an innovative read-out stage and sampling technique, the detection chain provides an extremely low read-out noise (< 3e- rms per sample). Finally the ACDM provides full autonomy in operation management, ensures the synchronization between laser emission and backscatter signal acquisition, the data processing and data stretching toward the spacecraft, the thermal regulation functions, the co-alignment control loop as well as the beam steering mechanism commanding, the TM/TC and observability management.

3. Industrial Consortium and Development Status

Airbus Defense and Space SAS (Airbus DS) is the prime responsible for the procurement, integration and verification of ATLID instrument and its subunits. While the ATLID system critical design review has only been recently completed, many of the subunits have already passed their qualification phases and some are already at their flight readiness status.

3.1. ATLID emitter chain development

Selex ES, as prime of the Laser Transmitter, has been developing an extensive set of hardware models as full scale breadboard used for validation and correlation with developed numerical models, and a PLH LIC test model. While the laser transmitter is largely inheriting from the Aladin instrument development for the AEOLUS mission, a significant evolution of the laser design lies in the fact that ATLID power laser head is sealed and pressurized (improving tolerance to LIC). Significant achievements have been made in the last years, including the development and the testing of a Power Laser Head Structural and Thermal Model. The model used for qualification of the housing sealing demonstrated the feasibility to meet the 3 years lifetime in orbit. The integration of a Power Laser Head Qualification Model has progressed with the assembly and testing (opto-mechanical stability confirmed after environmental testing) of the Master Oscillator section. After completing a successful qualification test campaign at Quantel Laser the Laser Amplifier has been also integrated and as soon its functional testing is successfully completed it will be followed with the integration of the UV section.

Fig.4 – PLH STM (Selex ES)

In parallel, the development of the Transmitter Electronics, supported by two engineering models, is progressing and one model was already delivered to Airbus for early function verification at instrument level. In parallel, soldering qualification is being performed, taking into account lessons learnt on Aladin electronics boards development. An extended programme for qualification of coatings, and in particular for the verification against Laser Induced Damage effects, is taking place in specialized laboratories at DLR and ESTEC. These tests are performed to coatings from optics existing throughout the entire transmitter chain (including IR and UV when relevant). The BSM assembly being developed by CEDRAT and SODERN is based on an Engineering Qualification Model (EQM) followed by the manufacturing and testing of the 2 flights models to be integrated in the Power Laser Heads. The EQM passed successfully mechanical and thermal vacuum environmental testing. Electronics coupling tests demonstrated its sub-microradian accuracy performance, just before being delivered to Airbus DS. The Emission Beam Expander by SODERN relies on two groups of lens designed as a Galilean beam expender type, inside a sealed and pressurized unit and expanding the laser beam to 120mm diameter. This requires large brazed windows, large lenses mounts development with stringent requirements in WFE, high transmission and qualification for laser irradiation. The development of the EBEX has been supported by a number of qualifications activities (large lens mount, brazed windows) successfully concluded and the FM units parts are now under integration.

3.2. ATLID receiver chain development

The receiver development is now well advanced, some Flight Models (FM) already available or ongoing manufacturing. The silicon carbide FM telescope mirrors are already coated and polished and have been assembled and aligned on its main structure. The Entrance Filtering Optics (polarisation control and narrow bandpass interference filter) and the Blocking Filter (spatial filtering) developed by Bertin Technologies in a proto-flight approach, are fully assembled and already completed their qualification test sequence. The two Fabry-Perot etalons flight models are currently under manufacturing phase at TSESO, with very challenging requirements (only a few nanometers parallelism are tolerated over a 40 mm distance), and RUAG Space will undergo their performance and environmental campaign during 2015. The Fibre Coupler Assemblies, developed by Bertin Technologies, have completed their successful qualification at QM level, and FMs assembly is ongoing. The Co-Alignment Sensor (CAS), developed by CRISA and aiming at measuring the retro-reflected laser spot with better than 1/10 pixel accuracy, completed its mechanical and thermal qualification sequence and already initiated the FM manufacturing. The ATLID Control Data and Management unit successfully passed its manufacturing review and it is now near its integration completion.

3.3. ATLID mechanical and thermal units development

ATLID mechanical and thermal functions are of particular complexity due to the large number of units, the strong dissipation from the laser source, and the interface to a carbon fibre panel. The high stiffness and high stability required for the Stable Structure Assembly (developed by APCO) supporting the telescope and the laser heads impose thick double-stages CFRP sandwich panel and strong titanium brackets. The large dissipation (above 600 W) of the units together with the inhomogeneity of materials (aluminium for electronics, carbon-fibre for interface panel) requires for the housing structure assembly a complex assembly of aluminium sandwiches, aluminium structure, titanium blades and brackets, and carbon-fibre panel designed by Airbus Defence and Space. The critical design reviews of the structures and cooling systems are now passed, and the qualification on breadboards of critical technologies (high load inserts, embedded titanium mounts) is approaching its completion.

4. CONCLUSIONS

The ATLID instrument is well progressing into its Phase C/D with the qualification and development of its subunits, many of which are already at FM level. ATLID will fly on the EarthCARE satellite which is scheduled for launch early 2018. EarthCARE, as well as the ESA wind mission AEOLUS, are paving the way for space LIDAR missions, capable of providing the scientific community with the measurements required for a better understanding of the atmospheric physics and climate evolution.

REFERENCES

[1] Hélière et al, 2014: Development of ATLID, the EarthCARE UV backscatter LIDAR, *ICSO 2014*.

[2] EarthCARE http://www.esa.int/esaLP/LPearthcare.html.

EPJ Web of Conferences **119**, 04004 (2016) DOI: 10.1051/epjconf/201611904004
ILRC 27

POTENTIALITIES AND LIMITS OF ICESAT-2 OBSERVATION FOR ATMOSPHERIC AEROSOL INVESTIGATION

L. Mona[1*], A. Amodeo[1], G. D'Amico[1]

*[1]Istituto di Metodologie per l'Analisi Ambientale CNR-IMAA, C.da S. Loja, I-85050 Tito Scalo, Potenza, Italy, *Email: lucia.mona@imaa.cnr.it*

ABSTRACT

ICESat-2(Ice, Cloud, and land Elevation Satellite-2), slated for launch in 2017, will continue the important observations of ice-sheet elevation change, sea-ice freeboard, and vegetation canopy height begun by ICESat in 2003. Among the other potential applications, ICESat-2 could provide some information about atmospheric aerosol over Polar Regions thanks to the lidar instrument. In this context, it is essential to demonstrate the ICESat-2 capability of providing vertical profiles of the aerosol backscatter coefficient and to define its potentialities and limits. First results of this investigation are reported and will be presented at the conference.

1. INTRODUCTION

Polar Regions are fragile ecosystems strongly affecting the global climate conditions. On the other hand, they are also affected by pollution coming from close regions. Because of the severe conditions in those remote regions, few measurement sites and instruments are available there. For what concerns the atmospheric aerosols, measurements from satellite are limited because passive sensors retrievals are extremely complex for several aspects: underneath bright surfaces (ice cover), expected low aerosol load and low sun angle conditions. Lidar measurements overcome all these problems. Vertical profiles of aerosol optical properties over Polar Regions are provided by CALIOP lidar since June 2006 (16-day repetition cycle). Potential ICESat-2 data could improve the temporal and spatial coverage of aerosol data on Polar Regions. In addition, the availability of aerosol optical properties vertical profiles over Polar Regions from ICESat-2 and CALIOP (and then EarthCARE, the Earth Clouds, Aerosols and Radiation Explorer) would provide a long-term data record of aerosol content over Polar Regions for air quality, climate change and climatological applications. In particular, the aerosol impact on radiation forcing and ice cover could be investigating thanks to the ICESat-2 primary products, i.e. the ice sheet heights. However, the real feasibility of aerosol study through ICESat-2 backscatter signals has to be investigated.

Three main critical issues were identified for the ICESat-2 atmospheric aerosol investigation:

a. low signal-to-noise ratio

b. laser pulses aliasing

c. needs of lidar ratio assumptions

Points a-b are strictly related to the fact that ICESat-2 lidar was designed (and optimized) for topography purposes and their effects on atmospheric aerosol measurements capability have to be investigated into details. Point c is instead common for all elastic backscatter lidars as for example CALIPSO is. In this case, methods and implications are well known, but limiting our interest to Polar Regions lidar ratio assumptions could be optimized.

Within ICESat-2 Early Adopters program, the authors investigated these critical aspects to demonstrate the utility of ICESat-2 data for atmospheric aerosol studies.

2. METHODOLOGY

This feasibility study is not starting from scratch. We built on the "giants" shoulders, taking advantage from the knowledge and measurements collected by 3 reference observational platforms: AERONET (Aerosol Robotic NETwork), the ground based sunphotometer network [1], CALIPSO the first atmospheric lidar from satellite [2], and EARLINET (European Aerosol Research Lidar Network), the first lidar network for aerosol study on continental scale [3].

In particular, CALIPSO data could be used for simulating the ICESat-2 observations. On the other hand, an estimation of the lidar ratio values at 532 nm, to be used for the backscatter retrieval, can be obtained from AERONET data. Finally, the knowledge resulting from EARLINET about lidar ratio values and CALIPSO data usage are particularly valuable for this study.

Facing the problems reported above, firstly, we limited our study to nighttime situation, because probably during the daytime, due to *a*, thin clouds and aerosols will be detectable only after a large average time (i.e. at least 40,000 lidar pulses, corresponding to about 28 km horizontal resolution).

The most critical point in ICESat-2 measurements is that the very high repetition rate of the laser (10Khz) results in the overlap of atmospheric scattering coming from altitude ranges separated by 15 km: the atmospheric scattering recorded by the instrument at height H is the sum of the scattering at height H, H+15, H-15, H+30, H-30, H+45, H-45 etc., where the numbers are in km. This has two main consequences: the calibration is limited at altitudes below15 km and there is ambiguity in aerosol/cloud height product.

To define the feasibility of using ICESat-2 lidar data for atmospheric aerosol study over Polar Regions, we define 4 main objectives:

1. perform a statistical analysis of layers on Polar regions above 15 km;
2. develop a reliability scheme for solving pulse-aliasing issue;
3. test the reliability score on CALIOP data;
4. define S values for backscatter retrieval starting from AERONET data

A statistical analysis of layers over Polar Regions was performed using 1-year of CALIPSO measurements on North (lat >66.33°N) and South regions (lat>66.33 ° S), collected for the period Jan-Dec 2012. Since ICESat-2 data will be limited in daytime conditions, only night time measurements were considered. In particular, Level 2 aerosol layer products were used. Monthly analysis of layer base and top, and layer and columnar AOD were performed.

On the base of the collected statistics, the probability to observe layers above 15 km for each polar region and month as a function of the layer thickness was evaluated.

Finally, this reliability scheme was tested on some cases as observed by CALIPSO in 2013.

For what concerns the lidar ratio values to be used into the backscatter retrieval, AERONET data collected in the Polar Regions were analysed for estimating the lidar ratio values at 532 nm to be used for the backscatter retrieval. Starting from AERONET sun-photometer data it is possible to estimate the lidar ratio S as:

$$S = 4\pi / (SSA \cdot p(180°))$$

where p(180) is the phase function at 180° and SSA the single scattering albedo [4]. Considering the low mean AOD (0.24 at 440nm) in Europe, we had to consider also AERONET Lev1.5 products in the estimation of S ratio, because all cases with AOD(440nm) < 0.4 have been masked in official Lev2.0 (quality assured) products. All available data collected in the North (lat >66.33°N) and South polar regions (lat>66.33 ° S) were considered.

3. RESULTS

The statistical analysis performed on CALIPOS data shows that layers above 15 km are observed only during Nov-Feb and May-Sept period for North and South Pole, respectively. Moreover, layers extending above 15 km are typically thicker than 1 km (see for example Figure 1). The layer thickness results to be a constraining parameter for the development of a reliability scheme to the Icesat-2 measurements.

The scheme was tested on first examples of simulated data. CALIPSO real observations for 2013 were used. The signal coming from 15-30 km altitude region was added to the 0-15 km one. Figure 2 and 3 show the real measurements and the corresponding ICESat-2 simulated data, respectively. Applying the reliability scheme to this dataset we found that the 0.8 km thick layer observed at the beginning of the time series around 2.5 km is correctly identified as true layer(probability of coming from above 15km around 5%). The tenuous but 8km deep layer observed between 3-10 km is instead identified as affected by the above 15 km region (probability greater than 99%). This is in agreement with the real observation: this layer corresponds to the layer between 17 and 25 km observed in the right-bottom of corner of figure 2.

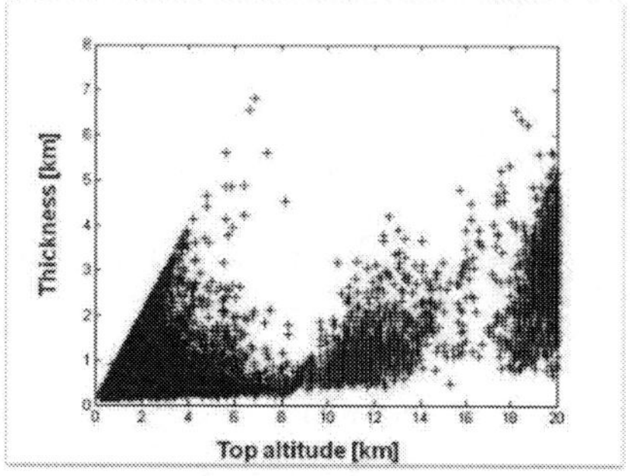

Figure 1: example of layer thickness vs top altitude values for data collected on January 2012, North Polar Region.

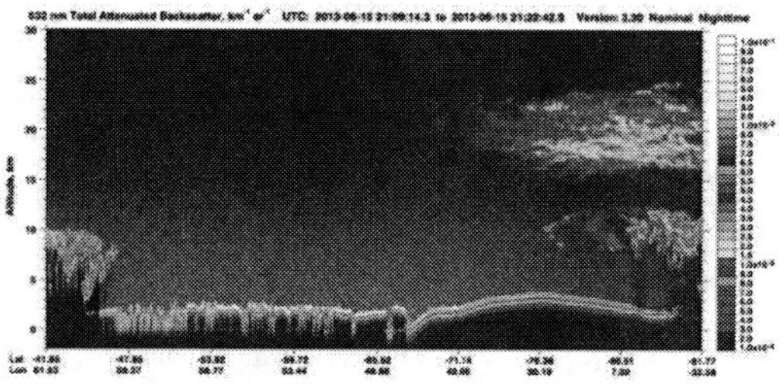

Figure 2: quick-look of CALIPSO attenuated backscatter at 532nm for 15 June 2013.

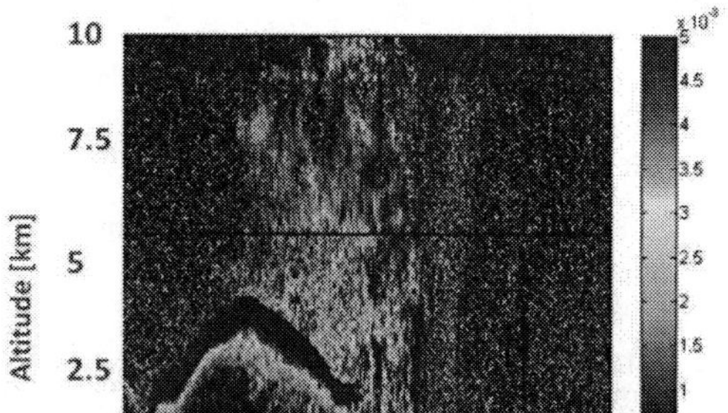

Figure 3: simulated ICESat-2 signals corresponding to North Polar Region section of figure 2.

The first examples of reliability scheme application indicate the feasibility of applying the developed reliability scheme to the Iceasat2 measurements.

For what concerns the polar lidar ratio values, first of all we found that unrealistic lidar ratio values were observed for low AOD values (< 0.025). WE found also that using median values instead of mean values, remove this aspect and gives results in agreement with the center of Gaussian fit for lidar ratio values distribution (R2>0.75) For the North Pole, data from 14 sites are available. No differences were observed for different geographical location of the sites. Grouping all North Pole lidar ratio data, we found data are Guassian distributed around a value of 58± 16 sr. Only data from 2 stations are available for South Pole, which are Gaussian distributed around 67 ± 2 sr.

These mean values are larger than what typically observed over Europe within EARLINET. However, it should be kept in mind that AERONET instrument cannot measure the backscatter coefficient but it is estimated by products f inversions algorithms. A general tendency of AERONET overestimation of EARLINET directly measured lidar ratio values was observed comparing co-located data [5]. The estimated lidar ratio values for the Polar Regions have to be therefore compared with eventually available direct S measurements.

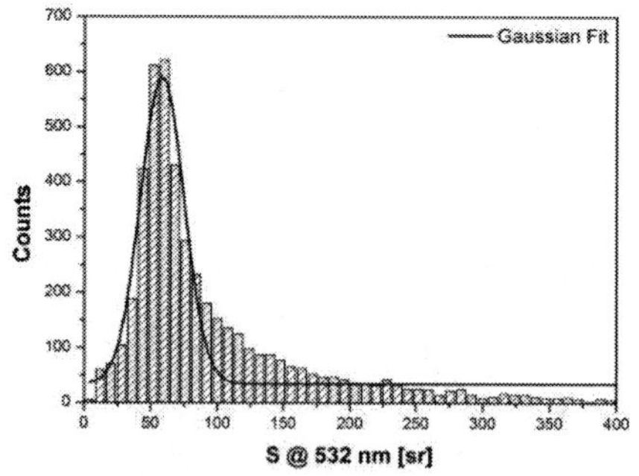

Figure 4: Counts distribution of lidar ratio values at 532nm as retrieved from AERONET inversion products (Level 15.5 data) for the North Polar Region.

4. CONCLUSIONS

A preliminary feasibility study for using ICESat-2 lidar measurements for aerosol study is in progress. First results are promising, even if they should be better assessed. Moreover, the application of the developed methodology to ICESat-2 simulated data is needed for further improvements.

ACKNOWLEDGEMENTS

CALIPSO data were obtained from the NASA Langley Research Center Atmospheric Science Data Center. We thank the AERONET PI investigators and their staff for establishing and maintaining the sites used in this investigation. The authors would like to acknowledge Stephen P. Palm (Goddard Space Flight Center) for his comments and suggestions.

REFERENCES

[1] Holben, B.N., et al., (1998) *Remote. Sens. Environ.* **66**, 1–16.

[2] Winker, D. M., et al., (2009) *J. Atmos. Oceanic Technol.*, **26**, 2310–2323.

[3] Pappalardo, G et al., (2014). *Atmos. Meas. Tech.*, **7**, 2389-2409.

[4] Welton, E.J., et al., (2002), Measurements of aerosol vertical profiles and optical properties during INDOEX 1999 using micropulse lidars, J. *Geo. Res.* **107** (19).

[5] Mona L., et al., (2009), A report on aerosol optical properties for both planetary boundary layer and free troposphere based on co-located AERONET and EARLINET data, GEOMON FP6 project deliverable D3.1.2.

Investigation on the interlink between optical properties of Dusts over Taiwan and Thar desert, India using CALIPSO data

M. Roja Raman[2,1], Wei-Nai Chen[2], Sin-Jie He[2], Li-Ting Chen[2], Kai-Hsuan Hsu[2]

[1]*UGC-SVU Centre for MST Radar Applications, S V University, Tirupati. Andhra Pradesh, India , 517502.*
E-mail: ramansvu@gmail.com
[2]*Research Center for Environmental Changes, Academia Sinica, Taipei, Taiwan, 115.*
E-mail: wnchen@rcec.sinica.edu.tw

ABSTRACT

The optical properties (Angstrom exponent and depolarization) of aerosols are analyzed to characterize the dust particles at different altitudes of the atmosphere using the backscatter signal from the space borne Lidar (CALIPSO) over Asian region with emphasis on Taiwan and Indian regions. The physical and optical properties of dust particles are found different at wet and dry weather conditions. The size distribution of dust particles at different altitudes is estimated. Finally the correlation between the particle depolarization and the angstrom exponent of dust particles at different altitudes is analyzed and the possible reasons are discussed. Further, the spatial inhomogeneity in the correlation coefficient is noticed and the possible relation with regional weather conditions is discussed.

1. INTRODUCTION

Asian dust play a major role in modulating the radiation balance by scattering and absorbing solar radiation, or modifying the cloud properties by acting as cloud condensation nuclei and ice forming nuclei in the Asia-pacific region. Asian dust is known for its complex physical and optical properties with changing weather conditions. The intensive optical properties of aerosol such as Angstrom exponent and depolarization may be considered to study the dust characteristics such as size distribution and shape of the particle respectively.

Angstrom exponent α>2 represents accumulation mode particles (Reid et al., 1964) and lower values (α near 0) represent coarse mode particles such as Saharan dusts and Asian dusts (Eck et al., 1999, Sakai et al., 2002). The depolarization ratio gives the information about the sphericity of the particle. Smaller the depolarization, more spherical the particle.

The total depolarization ratio (TDP) is define as the ratio of the return light of perpendicular to parallel polarizations, as given by the following equation:

$$TDP = \frac{P_{\perp}}{P_{||}},$$

where $P_{\perp}$ and $P_{||}$ are the integrated return power for parallel and perpendicular directions relative to the outgoing laser beam. The particle depolarization δ_p is expressed by the following equation:

$$\delta_p = \frac{R \times TDP - \delta_m}{R+1},$$

where the molecular depolarization ratio δ_m is given as 1.4% and R is backscattering ratio. The backscatter related Angstrom exponent is expressed as

$$\alpha_{back} = -\frac{d \ln \beta_p}{d \ln \lambda}.$$

2. DATA SET

In this study, level 2 data available at 5 km spatial resolution from the Cloud-Aerosol Lidar and Infrared Pathfinder Satellite Observation (CALIPSO) is employed for data analysis. Depolarization ratio at 532 nm and backscatter based Angstrom exponent are calculatef from aerosol profiles at 532 and 1064nm.

1. Study Area: (refer to Figure 1.)
   - Taiwan (21-26°N, 119-123°E).
   - Thar desert, India (20-35°N,65-80°E)
2. Period of study: 2006- 2014
3. Considering only dust cases ignoring cloud and other type of aerosol particles.
4. Typical z_{top} is 6 km but ranges from 4 to 8 km, which is depend on where is the background aerosol area. Background aerosol area is choose for an area at least 500 m is "clean".
5. Initial backscattering ratio of background aerosol in 532 nm at z_{top} is R_{532}=1.05.

6. Initial backscatter-related Angstrom Exponent at z_{top} is α_{back}=2.
7. The selection of Lidar Ratio (LR) follows the CALIPSO inversion algorithm as suggested by Omar et al. (2009).

Figure 1: Geophysical location of study areas Taiwan and That desert, India.

3. RESULTS

A typical case of 02 March 2008 showing the higher extinction coefficient values (@532nm) up to 2km at the latitudes 24-27°N that covers central and northern Taiwan is shown in the Figure 2. Since Taiwan region is known for occurrence of dust events during this season, the observed higher concentration of aerosol particles may be possibly Asian dust layers.

In order to confirm the possible dust layers, the Angstrom exponent is analysed. In general dust particles are in coarse mode, thus they result in lower angstrom exponent. However, the mixture of higher and lower values of Angstrom exponent at these latitudes show the particles are dominated by anthropogenic aerosols mixed with minor portion of coarse mode particles. The dust particles are generally non-spherical and show higher depolarisation. In this case the low depolarization ratio shows the domination of spherical particles with a mixture of few non-spherical particles.

Considering only dust cases ignoring other aerosol types, a negative correlation between particle depolarization and Angstrom exponent is expected, however this has not been reported experimentally so far. Here the observed negative correlation proves the theoretical hypothesis. It is also interesting to note that, the correlation turns likely positive during wet weather conditions that shows the presence of spherical like dust particles.

Further, to confirm the observed results, the possible dust source region (Thar Desert, India) has been considered where frequent elevated dust layers are observed. Similar analysis is repeated for Thar desert region and the similar negative correlation with more confidence is observed.

Occassionally, very poor correlaiton between depolarization and angstrom exponent is observed like in Figure.4, inspite of deep dust layers. This may be attributed to the signal saturation at 532 nm.

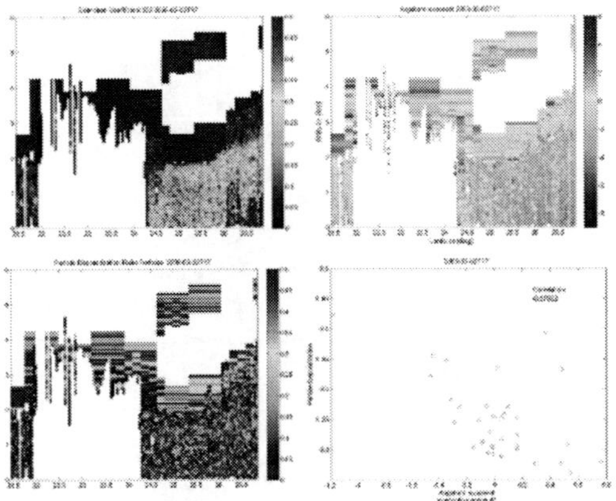

Figure 2: A typical case of CALIPSO observed Optical properties and correlation coefficient of aerosol layers over Taiwan region.

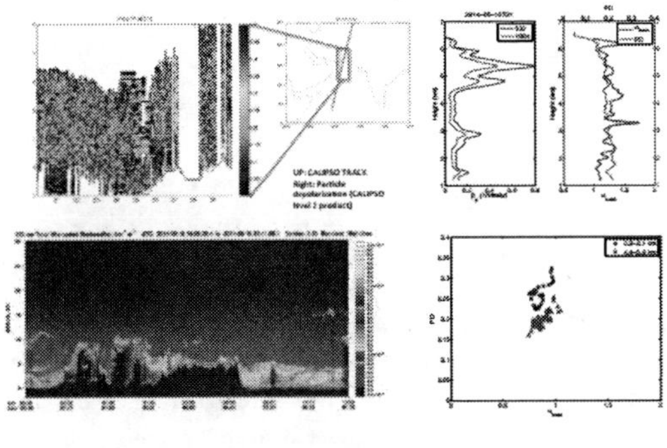

Figure 3: A dust event observed on 2014/5/16 near Thar Desert, India is select to demonstrate the calculation and interlink of dust optical properties. Attenuated backscatter is averaged from 23N to 25N. Two distinct dust layers could be observed at 2.3-3.1 km and 4.4-5.9 km. It can be noticed both layer show similar $\alpha_{back} \ll 1$ but higher PD could be found in lower layer.

Considering all the dust events over Taiwan (2006-2010) and Thar desert (2014 only), the combined correlation between Particle depolarization and Angstrom exponent shows the similar negative correlation like the individual cases however with lower correlation value.

DOI: 10.1051/epjconf/201611904005

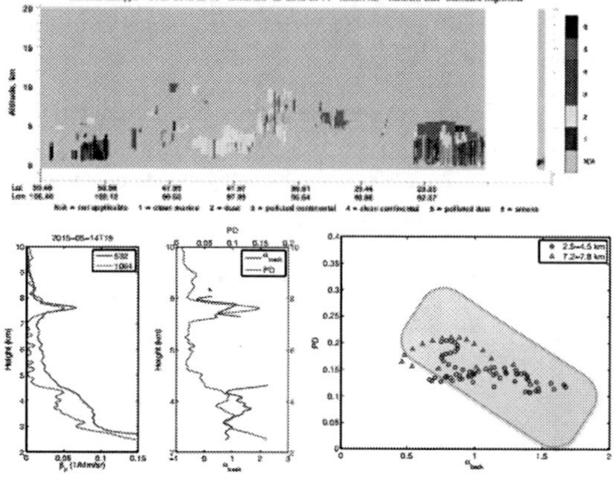

Figure 4: Similar to Fig. 3 but for Inner Mongoloa region.

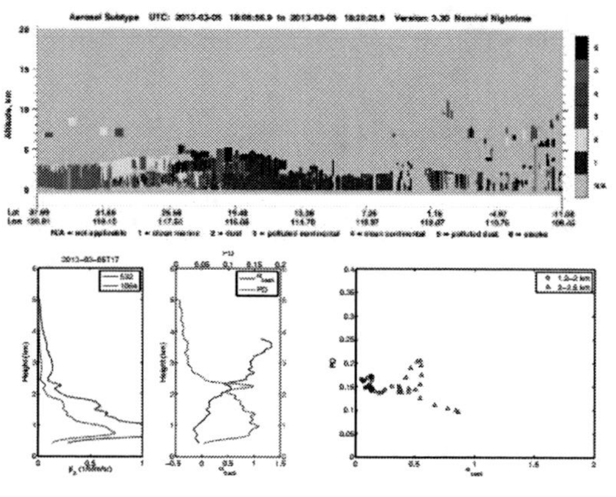

Figure 5: Similar to Fig. 3 but for IndoChina region.

This result confirms the occurrence of spherical like dust particles that may be attributed to the mixing of relatively small dust particles with the moist environment and appear like spherical particles due to moisture coating or due to external mixing with anthropogenic aerosols. More over, occassionally it is possible that the size-parameter may be independent of depolarization ratio (Mishchenko and Sassen, 1998).

4. SUMMARY

- The preliminary results shows aerosols with dust layers were observed frequently over Taiwan region using CALIPSO data supporting the earlier Lidar observed dust layers over Taipei, Taiwan.
- Observed dusts have typical PD0.2 and $\alpha_{back} \approx 1$.
- The correlation between PD and α_{back} of dusts

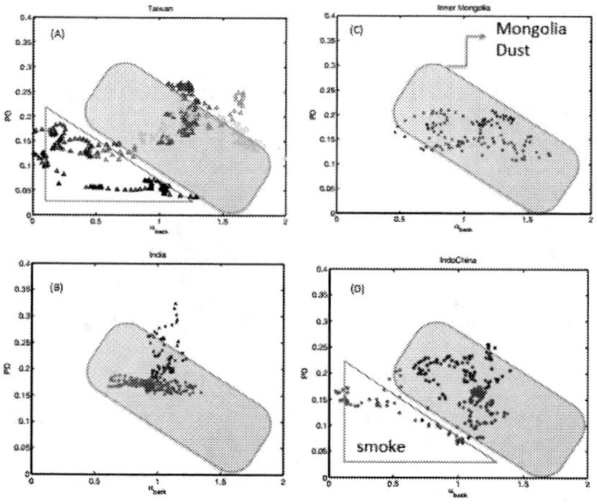

Figure 6: Results: Composed PD vs. α_{back}. (A) Taiwan: receptor of dust, smoke, and anthropogenic aerosol (B) India: dust source (C) Inner Mongolia: dust source (D) Indochina: smoke source and dust receptor.

shows a negative relation which implies that PD is dependent on particle size.

- Abnormal correlations were found especially for smokes and polluted dusts. Abnormal correlations has not been well investigated, but it might be related to mixing of dusts with water vapor according to our previous study (Chen et. al, 2007).
- The possible reason may be the condensation of water vapor on the relatively small dust particles that make the dust particle appears like spherical or small dust particles external mixing with anthropogenic aerosols.
- More detailed analysis for dusts and smokes observed by CALIPSO and ground-based lidar (include meteorological parameters) are needed to reveal the optical characteristics of aerosols.

REFERENCES

Chen, W.-N., Tsai, F.-J., Chou, C. C.-K., Chang, S.-Y., Chen, Y.-W., & Chen, J.-P. (2007). Optical properties of Asian dusts in the free atmosphere measured by Raman lidar at Taipei, Taiwan. Atmospheric Environment, 41(36), 76987714.

Eck, T. F., Holben, B. N., Reid, J. S., Dubovik, O., Smirnov, A., ONeill, N. T., Kinne, S. (1999). Wavelength dependence of the optical depth of biomass burning, urban, and desert

DOI: 10.1051/epjconf/201611904005

dust aerosols. Journal of Geophysical Research. doi:10.1029/1999JD900923.

Mishchenko, M. I., & Sassen, K. (1998). Depolarization of lidar returns by small ice crystals: An application to contrails. Geophysical Research Letters. doi:10.1029/97GL03764.

Omar, A.H. et al., 2009. The CALIPSO Automated Aerosol Classification and Lidar Ratio Selection Algorithm. JOURNAL OF ATMOSPHERIC AND OCEANIC TECHNOLOGY, 26(10), pp.19942014.

Reid, J.S., Eck, T.F., Christopher, S.A., Hobbs, P.V., Holben, B., (1999). Use of the Angstrom exponent to estimate the variability of optical and physical properties of aging smoke particles in Brazil. Journal of Geophysical Research 104, 2747327490.

Sakai, T., Shibata, T., Iwasaka, Y., Nagai, T., Nakazato, M., Matsumura, T., Hamdi, S. (2002). Case study of Raman lidar measurements of Asian dust events in 2000 and 2001 at Nagoya and Tsukuba, Japan. Atmospheric Environment, 36, 54795489. doi:10.1016/S1352-2310(02)00664-7.

Zhou, J., Yu, G., Jin, C., Qi, F., Liu, D., Hu, H., Gong, Z., Shi, G., Nakajima, T., Takamura, T., (2002). Lidar observations of Asian dust over Hefei, China, in spring 2000. Journal of Geophysical Research 107, 10.1029/2001JD000802

ACKNOWLEDGMENTS

We are thankful to the National Aeronautics and Space Administration (NASA), USA for providing the the Cloud-Aerosol Lidar and Infrared Pathfinder Satellite Observations (CALIPSO) data. One of the authors Roja Raman would like to thank Academia Sinica, Taiwan for providing Post-doctoral fellowship funded by National Science Council, Taiwan. This study was supported by Taiwan National Science Council (99-2111-M-001-008-MY3).

DOI: 10.1051/epjconf/201611904006

LOOKING INTO CALIPSO CLIMATOLOGICAL PRODUCTS: EVALUATION AND SUGGESTIONS FROM EARLINET

Nikolaos Papagiannopoulos [1]*, Lucia Mona[1], Lucas Alados-Alboledas[2], Vassilis Amiridis[3], Daniele Bortoli[4], Giuseppe D'Amico[1], Maria Joao Costa[4], Sergio Pereira[4], Nicola Spinelli[5], Ulla Wandinger[6], and Gelsomina Pappalardo[1]

*[1]Consiglio Nazionale delle Ricerche – Istituto di Metodologie per l'Analisi Ambientale (CNR-IMAA),Potenza, Italy, *Email: nikolaos.papagiannopoulos@imaa.cnr.it*

[2]Andalusian Institute for Earth System Research, University of Granada (IISTA-CEAMA), Spain

[3]Institute for Astronomy, Astrophysics, Space Applications and Remote Sensing, National Observatory of Athens, Greece

[4]Centro de Geofísica de Évora (CGE), Évora, Portugal

[5]Consorzio Nazionale Interuniversitario per le Scienze Fisiche della Materia, Napoli Research Unit, Italy

[6]Leibniz Institute for Tropospheric Research, Leipzig, Germany

ABSTRACT

CALIPSO (Cloud-Aerosol Lidar and Pathfinder Satellite Observations) Level 3 (CL3) data were compared against EARLINET (European Aerosol Research Lidar Network) monthly averages obtained by profiles during satellite overpasses. Data from EARLINET stations of Évora, Granada, Leipzig, Naples and Potenza, equipped with advanced multi-wavelength Raman lidars were used for this study. Owing to spatial and temporal differences, we reproduced the CL3 filtering rubric onto the CALIPSO Level 2 data. The CALIPSO monthly mean profiles following this approach are called CALIPSO Level 3*, CL3*. This offers the possibility to achieve direct comparable datasets. In respect to CL3 data, the agreement typically improved, in particular above the areas directly affected by the anthropogenic activities within the planetary boundary layer. However in most of the cases a subtle CALIPSO underestimation was observed with an average bias of 0.03 km^{-1}. We investigated the backscatter coefficient applying the same screening criteria, where the mean relative difference in respect to the extinction comparison improved from 15.2% to 11.4%. Lastly, the typing capabilities of CALIPSO were assessed outlining the importance of the correct aerosol type (and associated lidar ratio value) assessment to the CALIPSO aerosol properties retrieval.

1. INTRODUCTION

The CL3 product aims at characterizing the global 3-D distribution of aerosol as well as aerosol seasonal and interannual variations, and ultimately, assisting aerosol models. The main outputs are profiles of aerosol extinction coefficient at 532 nm and its vertical integral (aerosol optical depth; AOD). CL3 data are mapped onto a globally uniform 2°x5° (latitude x longitude) grid cell. The vertical resolution of the product is 60 m over the range of the heights -0.5 km to 12 km relative to the mean sea level. Monthly mean extinction profiles are computed for four conditions: nighttime/daytime all-sky and cloud-free. In addition, several quality control flags contained in the Level 2 files are used to screen the data prior to averaging. A detailed summary of the methodology used for the production of the CL3 product is provided in the Appendix of [1].

It is important to quantitatively evaluate the accuracy of any data product by comparison with independent measurements. CALIOP is firstly and foremost a profiling instrument; therefore there is nothing equivalent than comparison with profiling data. In any case, a detailed evaluation of CALIPSO extinction profiles (rather than AODs) using ground-based Raman lidars would be the ideal way to evaluate the satellite product. In the frame of the European Aerosol Research Network (EARLINET), various studies dedicated to evaluate CALIPSO Level 1 and Level 2 data have been carried out (e.g., [2]; [3]). So far, few studies

DOI: 10.1051/epjconf/201611904006

have been published regarding the CL3 dataset, such as reported from Winker et al. (2013), a comparison with High Spectral Resolution Lidar (HSRL) data given by airborne measurements during CALIPSO underflights of the NASA B200 aircraft (e.g., [4]). Moreover, a thorough assessment of CL3 product is given in [5], where the authors used multi-platform and model data to investigate its performance. Here for the first time, the performance of CL3 data is systematically compared with extinction profiles on continental scale through EARLINET correlative measurements.

2. METHODOLOGY

The CL3 data were compared against EARLINET profiles, however the correlative measurements regime and the CALIPSO mapping scheme combined with variable and unfavorable atmospheric conditions resulted in a very dissimilar number of collected observations, namely 272 CALIPSO overflights and 96 EARLINET correlative measurements. Hence, the dataset under examination posed doubts about the statistical significance and the representativeness of the aerosol monthly situation. Consequently, the standard CL3 dataset was compromised to correspond to the simultaneous EARLINET observations. Like this, we achieve to create direct comparable datasets, using only CALIPSO data segments corresponding to collected EARLINET measurements.

The CALIPSO Level 2 Version 3.01 Aerosol Profile product, which includes aerosol extinction and backscatter coefficient profiles at 532 nm is used to produce the monthly CALIPSO profiles. To ensure that only quality assured Level 2 data are included, the data were screened and low confidence retrievals were excluded, following the CL3 rubric as long as possible. More specifically each 5-km profile was processed according to screening criteria reported in the study of [6]. Differences with CL3 quality screening on the existing steps are related to the Extinction_Coefficient_Uncertainty_532 filter, which is set to a more conservative value. A further step was applied following the A6 of the Appendix of [1]. Regarding the CALIPSO monthly averaging process, a value of 0.0 km^{-1} is assigned in each profile to layers where the above criteria are invoked or no retrieval was made above 2.5 km. Moreover, the portion of extinction below the range bin that meets those conditions is excluded.

In this analysis quality screened CALIPSO extinction profiles at 532 nm are directly compared to corresponding EARLINET correlative measurements. Only the nighttime CALIPSO segment is used given the EARLINET ability of nighttime extinction measurements. For the purpose of this work the equivalent monthly mean profile is created whence at least two measurements are available within a period of a month. As far as the sky condition of the measurements, EARLINET cloud free and cirrus cloud infected data are integrated in the statistics, whilst for CALIPSO cloud free and above cloud data were employed. The reason for this disparity comes from the nature of the systems.

Typically EARLINET performs correlative measurements while the satellite ground-track lies in a 100 km radius area from the measuring site. The CL3* grid cell, therefore, corresponds to a grid which is different from CL3 one, and is dictated by the correlative measurements schedule as shown in Figure 1. The latitude boundaries of the new grid are kept the same as CL3.

Figure 1: Spatial boundaries of the CALIPSO dataset for Potenza cell. Red frame shows the CL3 grid containing Potenza EARLINET station (yellow pin), the corresponding CL3* grid is designated by the yellow frame. The blue circle has a radius of 100 km and represents the area that triggers measurements upon the satellite overflight.

3. RESULTS

This section reports as example results of the proposed methodology to the Potenza cell. The Potenza EARLINET site is a mountainous site and is located in the middle of the Mediterranean, its relatively low local aerosol content makes the observations particularly interesting for long transported particle plumes. The data used to produce the statistics consist of 33 correlative observations that generate 13 monthly profiles. Figure 2a displays the mean extinction coefficient profiles for both systems: the discrepancy below 2 km is significantly high (mean extinction bias -0.05 km^{-1}), on the other hand the differences at the upper levels are reduced (mean extinction bias less than -0.01 km^{-1}). The lower level disparity typically is weakened during summer months, and is exacerbated in winter, yet the sample size is small to quantify the periodicity of this discrepancy. Figure 2b shows the comparison of CALIPSO against ground-based profiles for the backscatter coefficient at 532 nm. The low level disparity, here, is significantly reduced. The mean relative extinction bias for the range 1-3 km is 40% while for backscatter is limited to 15%. However, this CALIPSO underestimation is in agreement with what observed for Potenza already in CALIPSO Level 1 data [7].

CALIPSO aerosol retrieval relies on the lidar ratio assignment, which in turn is based on the aerosol type classification. Therefore the lidar ratio profile for each month is estimated and directly compared to averaged unconstrained EARLINET lidar ratio profile. Figure 2c reports mean profiles of the lidar ratio measured by EARLINET station and the mean profile of values assumed by the CALIPSO algorithm for the inversion. The CALIPSO lidar ratio is for the whole altitude range $S_{aer}=49.1\pm3.3$ sr. On the other hand EARLINET measured lidar ratio exhibits higher values in the range 1.5-2.7 km, $S_{aer}=61.5\pm3.4$ sr, most likely because of the influence of more absorbing particles or the dust lidar ratio assignment that leads to CALIPSO lower values. In the height range 2.7-5 km, the agreement is satisfactory with an EARLINET mean value of $S_{aer}=50.4\pm5.2$ sr, suggesting that all the aerosol types identified compensate for this situation or the aerosol fields over the grid are frequently dust affected. The latter stems from EARLINET based studies, such as the study of [8] where the authors used a multiyear dust database and found values lidar ratio of ~50 sr for the site of Potenza.

Figure 3 reports the classification of aerosol features of the CALIPSO overpasses and the corresponding Potenza station observations. The investigated altitude range was partitioned into 1 km intervals and the percentage of layers identified within each bin is reported. Therefore, according to the boundary location, layers can be present in more than one height bin range.

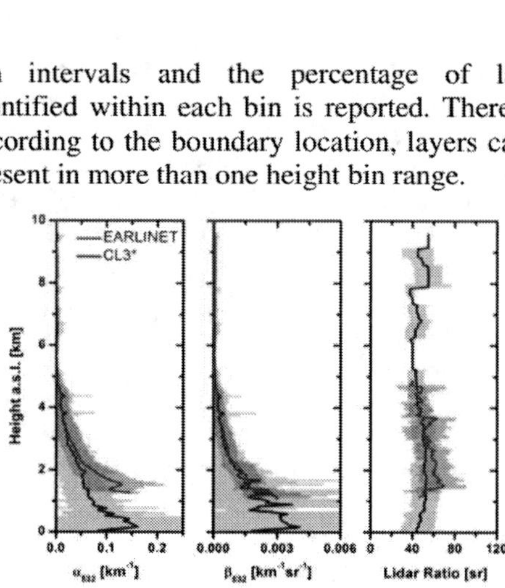

Figure 2: (a) extinction coefficient, (b) backscatter coefficient and (c) lidar ratio profiles at 532 nm.

Figure 3b gives an outlook of the aerosol types observed by CALIPSO. *Dust* and *Polluted Dust* prevail over the grid and moreover, present complementary behavior as *Polluted Dust* in the 1-2 km bin is the most important aerosol type and above that height pure *Dust* becomes the most important type. *Smoke* in the range 1-4 km keeps a constant presence. Some *Polluted Continental* cases are in the first height bin of the figure, and *Clean Continental* resides at the higher altitudes. As far as *Clean Marine* particles, they slightly affect the study area, yet below 3 km layers are identified as the ground track might be as well over the Mediterranean Sea (see Fig. 1).

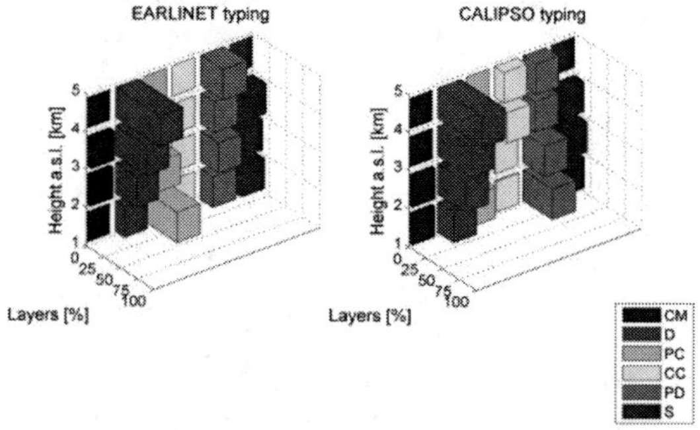

Figure 3: (a) EARLINET and (b) CALIPSO typing vertical proportioning for Potenza grid cell, the range interval is equal to 1 km. CM stands for *Clean Marine*, D for *Dust*, PC for *Polluted Continental*, CC for *Clean Continental*, PD for *Polluted Dust*, and S for *Smoke*.

Figure 3a presents the results of the typing procedure for the EARLINET correlative measurements (e.g., [9]; [10]). Again *Dust* and *Polluted Dust* affect the area around the site, although *Polluted Continental* particles in the first bin alter the aerosol situation. The latter can be ascribed to aerosols of anthropogenic origin within the nocturnal PBL and the adjoining residual layer around the area of Potenza site. *Smoke* particles, as was for the CALIPSO typing, have a range invariant character up to 4 km. *Polluted Dust* by contrast to CALIPSO, produces a rather constant influence with height. Summarizing the *Dust*, *Polluted Dust* (apart from PBL) and *Smoke* components were effectively identified by CALIPSO.

4. CONCLUSIONS

The comparison based on CALIPSO correlative measurements is essential to understand if CALIPSO measurements are representative of the corresponding station surrounding area in a climatological sense and if there are systematic deviations due to assumptions in the CALIPSO retrievals.

Besides the Potenza station, 4 more EARLINET stations were examined. In particular a subtle underestimation mainly in the PBL and its adjoining regions was found, with a relative AOD difference of -0.016. Regarding the backscatter comparison the relative difference in respect to the extinction comparison improved from 15.2% to 11.4%. Hence, using the CALIPSO backscatter coefficient, the comparison improves the relative mean biases when compared to the CALIPSO extinction coefficient. The better agreement in terms of backscatter has to be ascribed to the higher impact of lidar ratio assumption on extinction rather than on backscatter. Moreover, the realized database gives the possibility to investigate the performance of the lidar ratio assignment CALIPSO scheme.

ACKNOWLEDGEMENTS

The financial support by the European Community through the ACTRIS Research Infrastructure Action under the 7th Framework Programme under ACTRIS Grant Agreement n°262254 is gratefully acknowledged.

REFERENCES

[1] Winker, D., et al., 2013: The global 3-d distribution of tropospheric aerosols as characterized by CALIOP. *Atmos. Chem. Phys.*, **13**, 3345–3361.

[2] Pappalardo, G., et al., 2010: EARLINET correlative measurements for CALIPSO: First intercomparison results. *J. Geophys. Res.*, **115**, D00H19.

[3] Wandinger, U., et al., 2011: Aerosols and Clouds: long-term Database from Spaceborne Lidar Measurements. Technical report, ESA Publications Division. Final report, ESTEC Contract 21487/08/NL/HE, Noordwijk, the Netherlands.

[4] Burton, S., et al., 2013: Aerosol classification from airborne HSRL and comparisons with the CALIPSO vertical feature mask, *Atmos. Meas. Tech.*, **6(5)**,1397–1412.

[5] Amiridis, V., et al., 2013: Optimizing CALIPSO Saharan dust retrievals, *Atmos. Chem. Phys.*, **13**, 12089–12106.

[6] Campbell, J., et al, 2012: Evaluating night-time CALIOP 0.532 um aerosol optical depth and extinction coefficient retrievals, *Atmos. Meas. Tech.*, **5**, 2143–2160.

[7] Mona, L., et al., 2009: One year of CNR-IMAA multi-wavelength Raman lidar measurements in correspondence of CALIPSO overpass: Level 1 products comparison, *Atmos. Chem. Phys.*, **9**, 7213–7228.

[8] Mona, L., et al., 2014: EARLINET dust observations vs. BSC-DREAM8b modeled profiles: 12-year-long systematic comparison at Potenza, Italy. *Atmos. Chem. Phys.*, **14**, 8781–8793.

[9] Groß, S., et al., 2011: Characterization of Saharan dust, marine aerosols and mixtures of biomass burning aerosols and dust by means of multi-wavelength depolarization and Raman measurements during SAMUM-2, *Tellus B*, **63**, 706–724.

[10] Mona, L., et al., 2012: Multi-wavelength Raman lidar observations of the Eyjafjallajökull volcanic cloud over Potenza, southern Italy, *Atmos. Chem. Phys.*, **12**, 2229–2244.

 DOI: 10.1051/epjconf/201611904007

UTILIZING THE SYNERGY OF AIRBORNE BACKSCATTER LIDAR AND IN-SITU MEASUREMENTS FOR EVALUATING CALIPSO

Alexandra Tsekeri [1]*, Vassilis Amiridis[1], Franco Marenco[2], Eleni Marinou[1], Phil Rosenberg[2], Stavros Solomos[1], Jamie Trembath[2], James Allan[3], Asan Bacak[3] and Athanasios Nenes[4]

[1]*IAASARS, National Observatory of Athens, Athens, Greece, *Email: atsekeri@noa.gr*

[2] *Observational Based Research, Met Office, Exeter, UK*

[3]*School of Earth, Atmospheric and Environmental Sciences, University of Manchester, Manchester, UK*

[4]*School of Earth and Atmospheric Sciences and Chemical and Biomolecular Engineering, Georgia Institute of Technology, Atlanta, Georgia, USA*

ABSTRACT

Airborne campaigns dedicated to satellite validation are crucial for the effective global aerosol monitoring. CALIPSO is currently the only active remote sensing satellite mission, acquiring the vertical profiles of the aerosol backscatter and extinction coefficients. Here we present a method for CALIPSO evaluation from combining lidar and in-situ airborne measurements. The limitations of the method have to do mainly with the in-situ instrumentation capabilities and the hydration modelling. We also discuss the future implementation of our method in the ICE-D campaign (Cape Verde, August 2015).

1. INTRODUCTION

Characterizing effectively the vertical distribution of the aerosol optical and microphysical properties is very important, considering their effects on climate. The space-borne lidar CALIOP on-board the CALIPSO satellite provides profiles of the aerosol backscatter and extinction coefficients, along with information on layer-stratified types of aerosol. The validation of these products is an on-going effort employing ground-based and airborne lidars (e.g. [1]). In Figure 1 we show the validation of the CALIPSO extinction product during the South American Biomass Burning Analysis (SAMBBA) campaign [2].

In the framework of ACEMED campaign (Evaluation of CALIPSO's aerosol classification scheme over Eastern Mediterranean) we proposed a new methodology for an effective characterization of the particle vertical distribution from airborne campaigns dedicated to active remote sensing satellite validation. Our retrieval utilizes airborne backscatter lidar and in-situ measurements. The in-situ data refer to dried particle samples, thus, in order to obtain the ambient particle properties we model their hygroscopic growth with ISORROPIA II model [3].

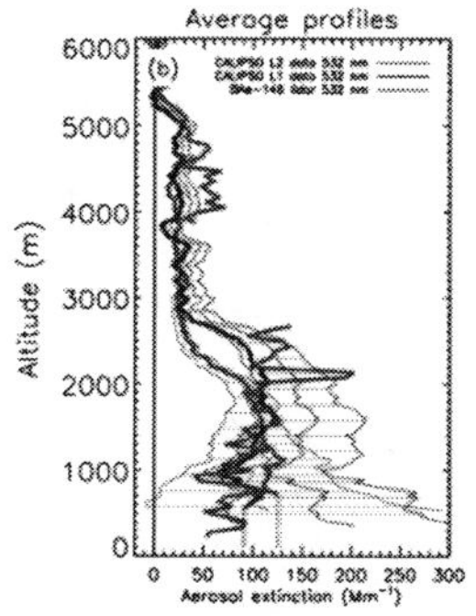

Figure 1: CALIPSO aerosol extinction coefficient at 532 nm (green), compared to the airborne lidar extinction coefficient at 532 nm (red) and the extinction coefficient retrieved from the Level 1 CALIPSO data (blue), as described in [2], during the SAMBBA campaign.(Source: [2])

2. DATA and METHODOLOGY

On 8/9/2011 the Facility for Airborne Atmospheric Measurements (FAAM) BAe-146 research aircraft performed nighttime airborne lidar and in-situ aerosol measurements at different height levels during an under-flight of CALIPSO. Central to our approach is the synergy of the acquired remote sensing and in-situ measurements: The elastic backscatter lidar signals provide information about the ambient particle properties which though is not enough to achieve a complete characterization of the ambient particles. The in-situ measurements on the other hand provide along with optical properties, the size distribution and chemical composition of the particles, from which we can derive their refractive index. Unfortunately these measurements refer to particle samples dried on-board. Moreover, they come with uncertainties, especially for the larger particles and for the imaginary part of the refractive index.

In order to calculate the aerosol optical and microphysical properties at different flight altitudes, we used the measured dry size distributions and refractive indices as a first guess in the retrieval and changed them iteratively, so as their scattering properties reproduce both the in-situ optical measurements and the lidar backscatter signals (Figure 2). The hygroscopic growth of the particles in ambient conditions was modelled with ISORROPIA II.

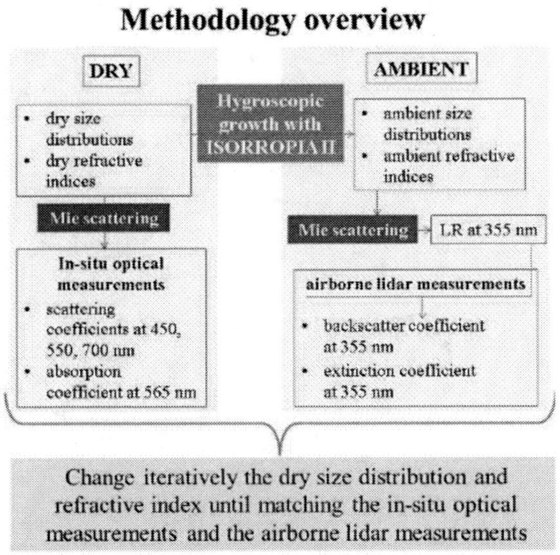

Figure 2: Methodology overview.

Airborne lidar data

The active remote sensing airborne measurements were performed with the NON-CORE Mini-Lidar Leosphere ALS450 system onboard [4] acquiring backscatter and depolarization profiles at 355 nm.

Airborne in-situ data

The in-situ instruments onboard the BAe-146 aircraft measured a large suite of the particle microphysical and optical characteristics, as the size distribution, the chemical composition, the scattering and absorption coefficients:

<u>Size distribution</u>

The number size distributions were measured with the Passive Cavity Aerosol Spectrometer Probe (PCASP) and the 1.129 Grimm Technik Sky-optical particle counter (OPC), operating at the 0.1–3 and 0.25-32 μm (nominal) diameter range, respectively. The measurements were calibrated following the methodology in [5]. As can be seen in Figure 3 the fine mode for both datasets is almost identical, but for the coarse mode this is not true, especially for particles with radius >1 μm. The comparison with the ambient size distributions provided by AERONET measurements before and after the BAe-146 aircraft flight, shows that for the fine mode there is good agreement which though deteriorates for larger particles. This can be also due to the AERONET size distribution cut-off at 15 μm.

<u>Refractive index</u>

The chemical composition provided by the Aerosol Mass Spectrometer (AMS) of FAAM can be used to estimate the particle refractive index, applying a volume mixture law to account for the contributions of the corresponding chemical groups [6]. This approach introduces some uncertainties, especially in the estimation of the imaginary part, thus in our retrieval the AMS-calculated refractive indices are used only as a first guess and the actual refractive indices are calculated from the iterative scheme described above, to achieve the best closure between all measurements. We considered different refractive indices for fine and coarse particles.

<u>Scattering and absorption</u>

The particle dry scattering coefficients at 450, 550 and 700 nm were measured with the TSI Integrating Nephelometer 3563 and the absorption coefficient at 567 nm was measured with the Radiance Research Particle Soot Absorption Photometer (PSAP). Both instruments are connected to modified Rosemount inlets, suffering from inlet enhancement/losses as well as losses along the pipelines and they do not measure the scattering properties over the whole particle size range. For this reason, in our scattering calculations we consider a sampling cut-off for particles with radius >3 μm.

Hygroscopic growth

ISORROPIA II models the thermodynamic equilibrium of aerosols in different atmospheric states [3]. The software takes as input the dried sample properties, i.e., the particle size distribution and chemical composition, along with the temperature, pressure and relative humidity (RH) of the sample, and the temperature, pressure and RH of the ambient atmosphere and calculates the hygroscopic growth of the fine and coarse modes. In our analysis a modified version of ISORROPIA II is utilized, taking in account the hygroscopic growth of the organic material as well.

3. RESULTS

Figure 4 shows the dry and the hygroscopically grown ambient size distributions at 2.7 km (the RH is 74% at this height) for the ACEMED flight of September 9th, 2011, above Thessaloniki.

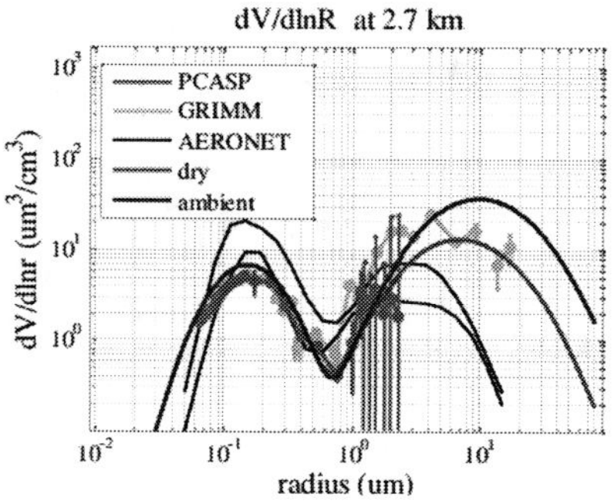

Figure 3: The retrieved size distributions for dried (red) and ambient particles (blue), at 2.7 km. The measured size distributions are from PCASP (pink) and Grimm (cyan). The ambient size distributions from AERONET before and after the flight are shown in black.

The size distributions are retrieved from our iterative scheme, so as together with the retrieved refractive indices to reproduce the lidar and in-situ measured optical properties. The fine mode is set equal to the measurements and the coarse mode is varied, due to the uncertainties in its definition. The refractive indices for fine and coarse modes are varied too.

Figure 4 shows the measured scattering coefficients versus the calculated ones from the retrieved size distributions and refractive indices. Generally, we see that our calculations reproduce the in-situ measurements well (the agreement for the absorption coefficient at 567 nm is similar).

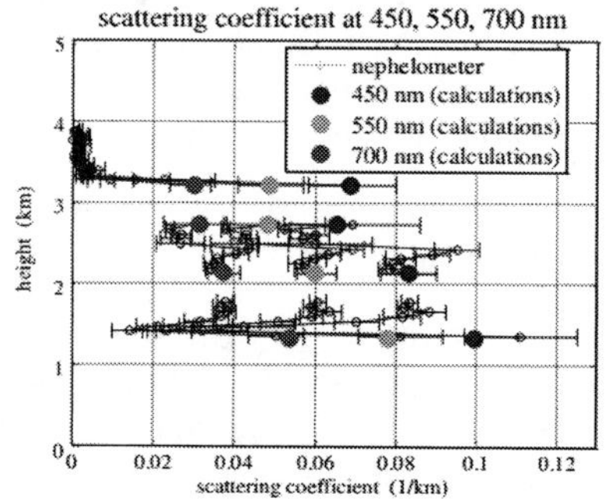

Figure 4: The calculated scattering coefficients at 450 (blue), 550 (green) and 700 nm (red), at different flight heights. The measurements from the nephelometer are shown in black.

Figure 5 shows the measured versus the calculated backscatter coefficient at 355 nm. It seems that the calculations fit the measurements well, and the same is true for the extinction coefficient at 355 nm (not shown).

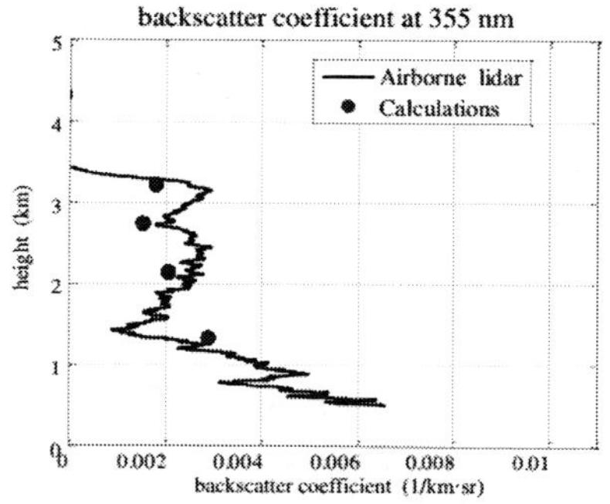

Figure 5: The backscatter coefficient at 355 nm measured by the airborne lidar (black) and calculated from the retrieved aerosol microphysichal properties (blue).

Small discrepancies between the measurements and the calculations are due to the uncertainty in the coarse mode and possibly in uncertainties in the hygroscopic modelling of the ambient particles. Moreover they can be due to changes in the atmospheric properties, since the measurements are not simultaneous, with the lidar measurements to be an average of the whole flight and the in-situ measurements to refer at different-altitude portions of the flight.

The retrieved size distributions and refractive indices are then used to calculate the backscatter and extinction coefficients at 532 nm, and to evaluate the respective CALIPSO product. As shown in Figure 6 our calculations are quite close to the CALIPSO measurements. We need to highlight here that the CALIPSO data were not used in the iterative procedure.

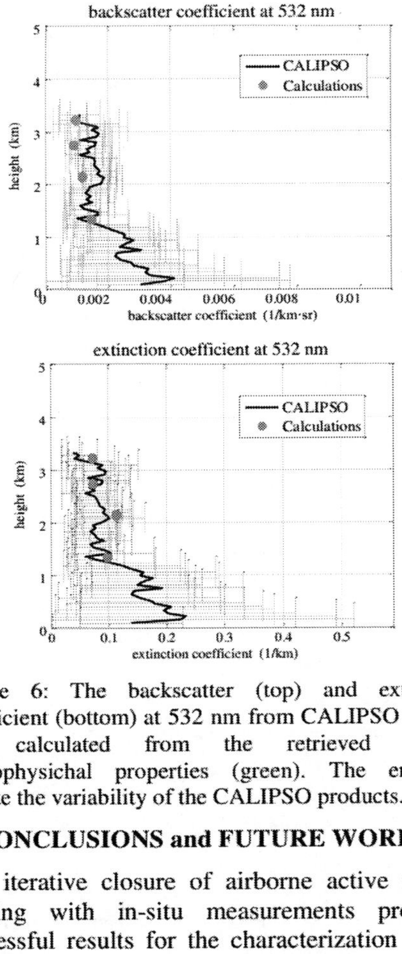

Figure 6: The backscatter (top) and extinction coefficient (bottom) at 532 nm from CALIPSO (black) and calculated from the retrieved aerosol microphysichal properties (green). The errorbars denote the variability of the CALIPSO products.

4. CONCLUSIONS and FUTURE WORK

The iterative closure of airborne active remote sensing with in-situ measurements produced successful results for the characterization of the aerosol at different flight heights for a specific case of ACEMED campaign. This method can be widely used for the validation of active remote sensing satellite products. Our aim is to utilize it in future validation campaigns, as the ICE-D campaign organized in August 2015, focused on dust monitoring.

ACKNOWLEDGEMENT

The research leading to these results has received funding from the European Community's Seventh Framework Programme (FP7/2007-2013) under grant agreement n°227159 (EUFAR: European Facility for Airborne Research in Environmental and Geo-sciences). Airborne data was obtained using the BAe-146-301 Atmospheric Research Aircraft [ARA] flown by Directflight Ltd and managed by the Facility for Airborne Atmospheric Measurements [FAAM], which is a joint entity of the Natural Environment Research Council [NERC] and the Met Office. The research leading to these results has received funding from the European Union Seventh Framework Programme ACTRIS (FP7/2007-2013) under grant agreement N°262254.

REFERENCES

[1] Burton, S. P., et al., 2013: Aerosol classification from airborne HSRL and comparisons with the CALIPSO vertical feature mask, *Atmos. Meas. Tech.*, **6**, 1397–1412.

[2] Marenco, F., et al., 2014: Airborne verification of CALIPSO products over the Amazon: a case study of daytime observations in a complex atmospheric scene, *Atmos. Chem. Phys.*, **14**, 11871-11881.

[3] Fountoukis, C. and Nenes, A., 2007: ISORROPIA II: a computationally efficient thermodynamic equilibrium model for K^+–Ca^{2+}–Mg^{2+}–NH_4^+–Na^+–SO_4^{2-}–NO_3^-–Cl^-–H_2O aerosols, *Atmos. Chem. Phys.*, **7**, 4639-4659.

[4] Marenco, F., et al., 2011: Airborne Lidar Observations of the 2010 Eyjafjallajökull Volcanic Ash Plume, *J. Geophys. Res.*, **116**, D00U05, doi:10.1029/2011JD016396.

[5] Rosenberg, P. D., et al., 2012: Particle sizing calibration with refractive index correction for light scattering optical particle counters and impacts upon PCASP and CDP data collected during the Fennec campaign, *Atmos. Meas. Tech.*, **5**, 1147-1163.

[6] Highwood, E. J., et al., 2012: Aerosol scattering and absorption during the EUCAARI-LONGREX flights of the Facility for Airborne Atmospheric Measurements (FAAM) BAe-146: can measurements and models agree?, *Atmos. Chem. Phys.*, **12**, 7251-7267.

A method for determination of cirrus extinction-to-backscatter ratio from CALIOP data

Jingbin Zhang[1*], Jan-Bai Nee[2]

[1]*Physics Department, Qufu Normal University, Qufu City 273165, Shandong Province, P. R. China*, **Email: jingbin_zhang@163.com*

[2]*Department of Physics, National Central University, Chung-Li, Taiwan 32054*

ABSTRACT

We are presenting an empirical equation to retrieve cirrus lidar ratio by using CALIOP 532 nm level 1 data for nighttime cases. Retrieval results have non-relationship with cirrus multiple scattering effects and not affected by the error of transmission. The average CALIPSO 532 nm cirrus lidar ratio over Longitude 120+/- 10 and Latitude 25+/-10 for whole year of 2008 are 21.66±0.06sr for the year of 2008 respectively, with the maximum bias of 9.25% for the year 2008, the results is fairly stable and reasonable.

1. INTRODUCTION

Cirrus clouds play a significant role in the radiative balance of the Earth's atmospheric system, a clear understanding of their optical properties is essential for climate modeling studies. Lidars are regarded as one of the leading techniques for remotely studying the properties of cirrus clouds. Generally, lidar equation has the two unknowns backscatter and extinction coefficients as well as the multiple scattering effects, in order to be able to invert the lidar equation, they have to be related either empirical or theoretical methods[1-9]. The extinction to backscatter ratio is considered to be of special importance since it is related to microphysics and ice composition of cirrus clouds, it is controlled by the complexity of ice particle shape and aspect ratio [10].Traditionally the error of retrieved cirrus lidar ratio is dependent on the errors of multiple scattering coefficient and cirrus transmission determination [11]. In order to improve the retrieval accuracy of cirrus extinction to backscatter ratio from Mie lidar return signals, an empirical equation for determination cirrus lidar ratio was employed, the retrieval results are consistent with others'.

2. METHODOLOGY

The present study uses CALIOP NASA level 1 data products for the year of 2008 was used to retrieve cirrus lidar ratio over Longitude 120+/-10 and Latitude 25+/-10. Only nighttime observations were considered in the present study, thanks to their higher signal-to-noise ratio compared to daytime observations. Lidar equation is as following:

$$P(r) = Ca\frac{[\beta_m(r) + \beta_c(r)]}{r^2}\exp\left[\int_0^r [\eta\sigma_c(r) + \sigma_m(r)]dr\right] \quad (1)$$

Were $P(r)$ is the lidar return signals from the atmosphere at the distance r; Ca is the lidar system constant; η is cloud multiple scattering coefficient; $\beta_m(r)$ and $\beta_c(r)$ are the molecular and cloud backscatter coefficients; $\sigma_m(r)$ and $\sigma_c(r)$ are the molecular and cloud extinction coefficients respectively.

The cloud backscatter coefficient $\beta_c(r)$ is as following:

$$\beta_c(r)=\frac{X(r)\exp[-2(\eta S_c-S_m)]\int_{r_0}^{r}\beta_m(r)dr}{\frac{X(r_0)}{\beta_m(r_0)}-2\eta S_c\int_{r_0}^{r}X(r)\exp[-2(\eta S_c-S_m)]\int_{r_0}^{r'}\beta_m(r')dr'}$$

(2)

Where $X(r)=r^2P(r)/C_a$ is CALIOP attenuated backscatter from the atmosphere at the distance r; $S_c=\sigma_c(r)/\beta_c(r)$ is cirrus cloud extinction to backscatter ratio and so-called lidar ratio, $S_m=8\pi/3$ is molecular extinction to backscatter ratio.

Cloud backscatter ratio BR is defined as:

$$BR(r)=\frac{\beta_c(r)+\beta_m(r)}{\beta_m(r)} \quad (3)$$

Where $\beta_c(r)$ and $\beta_m(r)$ are the backscatter coefficients of cloud and air at range r respectively.

The bottom and top height of the cloud have been considered on the following criteria: (i) for the base height of cloud, BR≥1.5 is used and this ratio should also Persists for at least 5 consecutive intervals of height range, so that the top height of cloud can be defined by considering BR=1.5 and (ii) the height of cloud base should exceed 8km. The reason for choosing the backscattering ratio of 1.5 is mainly due to the fact that the cirrus should be clearly distinguished from the other fluctuating scattering ratio [12].

According to error analyses, when aerosol optical depth is small, the relative error of retrieved cirrus optical depth has a linear relationship with the relative error of cirrus extinction to backscatter ratio [13]. An empirical equation for determination of aerosol lidar ratio S_c was developed as following [14]:

$$a_1S_c+b_2=e^{-(a_2S_c+b_2)} \quad (4)$$

Where parameters a_i and b_i (i=1,2) are least-square-fit coefficient and can be obtained by regression method, approximate solution of transcendental Eq. (4) yields an independent measurements cirrus lidar ratio S_c , the retrieved S_c has non-relationship with the multiple scattering factor η and cirrus transmission.

3. RESULTS

The retrieval results is shown in Fig.1, it is obvious that cirrus lidar ratio is nearly constant for different cirrus case occurrences, the overall mean value of cirrus lidar ratio for the area of over Longitude 120+/-10 and Latitude 25+/-10 is 21.66±0.06sr, the mean bias is 0.27% with the maximum bias of 9.25% for the year 2008. It's shown that the retrieval of cirrus lidar ratio is stable and reasonable. This result is consistent with Min Min (24.1±11.6sr), Sassen and Comstock (24.4±15.5sr) and Vaughan (23±9sr) [15-17].

4. CONCLUSION

In this paper, a new method for retrieving cirrus lidar ratio is presented, the retrieval result is consistent with others' by using traditional method. This technique is suitable for cirrus lidar retrieval.

One advantage is that this technique is not affected by the cirrus multiple scattering effects and the error of cirrus transmission determination.

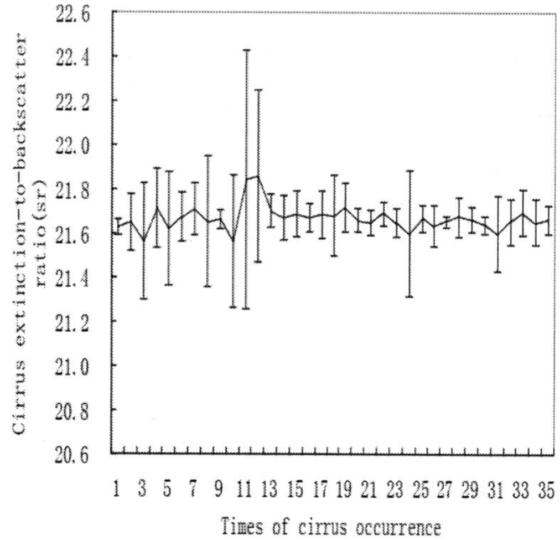

Fig.1 Retrieval results of cirrus extinction to backscatter ratio (532nm) vs. times of cirrus occurrence for the year 2008.

ACKNOWLEDGEMENT

The authors wish to thank the CALIPSO team at NASA Langley Research Center for providing the data used in our calculations.

REFERENCES

[1] Barrett, E. W. and Ben-Dov, O., 1967: Application of the lidar to air pol lution measurements, J. Appl. Meteorol., **6**, 500–515.

[2] Davis, P. A., 1969: The analysis of lidar signatures of cirrus clouds, Appl. Opt., **8**, 2099–2102.

[3] Fernald, F. G., Herman, B. M., and Reagon, J. A., 1972: Determination of aerosol height distributions by lidar, J. Appl. Meteorol., **11**, 482–489.

[4] Platt, C. M. R., 1973: Lidar and radiometric observations of cirrus clouds, J. Atmos. Sci., **30**, 1191–1204.

[5] Platt, C. M. R., 1979: Remote sounding of high clouds. I: Of visible and infrared optical properties from lidar and radiometer measurements, J. Appl. Meteorol.,**18**, 1130–1143.

[6] Platt, C. M. R., Scott, J. C., and Dilley, C., 1987: Remote sounding of high clouds.Part VI: Optical properties, J. Atmos. Sci., **44**, 729–747.

[7] Klett, J. D., 1981: Stable analytical inversion solution for processing lidar returns, Appl. Opt., **20**, 211–220.

[8] Fernald, F. G, 1984: Analysis of atmospheric lidar observations; some comments, Appl. Opt., **23**, 652–653,.

[9] Ansmann A., Wandinger U., Riebesell M., Weitkamp C. and Michaelis W., 1992: Independent measurement of extinction and backscatter profiles in cirrus clouds by using a combined Raman elastic backscatter lidar, Appl. Opt., **31**, 7113–7131.

[10] Takano, Y., and K.-N. Liou., 1995: Solar radiative transfer in cirrus clouds. Part III: Light scattering by irregular ice crystals, J. Atmos. Sci., **52**, 818-837.

[11] Chen, W. N., Chiang, C. W., and Nee, J. B., 2002: Lidar ratio and de polarization ratio for cirrus clouds, Appl. Opt., **41**, 6470–6476,.

[12] Subrata Kumar Das, Chih-Wei Chiang, and Jan-Bai Nee, 2008: Characteristics of cirrus clouds and its radiative properties based on lidar observation over Chung-Li, Taiwan, Atmos. Res. **93**, 723-735.

[13] Jinhuan Qiu, Daren Lu, 1988: A study of inversion algorithm for determining atmospheric aerosol profile from simulated space-borne Lidar signals (in Chinese), Chinese J. Atmos. Sci., **12**, 258-270.

DOI: 10.1051/epjconf/201611904008

[14] Jingbin Zhang, 2008: A method for determination of aerosol extinction-to-backscatter ratio, In 24th International Laser Radar Conference (ILRC), Boulder, CO, USA.

[15] Min Min, Wang Pucai, Zong Xuemei, 2010: Extinction to backscatter ratio of cirrus clouds retrieved by spaceborne lidar over china, Chinese Journal of Atmospheric Sciences (in Chinese), **34**, 506-512.

[16] Sassen K, Campbell J R. 2001: A midlatitude cirrus cloud climatology from the facility for atmospheric remote sensing, Part I: Macrophysical and Synoptic properties, J. Atmos. Sci., **58**, 481-496.

[17] Vaughan M, Kuehn R, Young S, et al. 2008: Validating cirrus cloud optical properties retrieved by CALIPSO. In 24th International Laser Radar Conference (ILRC), Boulder, CO, USA.

EPJ Web of Conferences **119**, 04009 (2016) DOI: 10.1051/epjconf/201611904009
ILRC 27

ANVIL PRODUCTIVITIES OF TROPICAL DEEP CONVECTIVE CLUSTERS AND THEIR REGIONAL DIFFERENCES

Min Deng[1], Gerald G. Mace[2], Zhien Wang[1],*
*[1]Department of Atmospheric Science, University of Wyoming, Laramie, Wyoming, 820072, USA, *mdeng2@uwyo.edu*
[2] University of Utah, Salt lake City, USA

ABSTRACT

Tropical deep convection's intensity and their anvil productivity are investigated and compared among 8 climatological regions with 4-year collocated and combined CloudSat and CALIPSO data. For all 8 regions, the convective clusters become deeper, while they become wider and tend to be composed of multiple rainy cores. Among 8 regions, convective clusters at the same scale over EP and AT tend to have less but wider rainy cores than those at WP, MA and IO, while those over AF, IO, MA and AM tend to have higher cloud top than those over ocean. For convective clusters less than 300 km wide over AF and MA, the rainy cores pump more ice mass of larger particles to the mid- and upper troposphere in strong updrafts.

The total anvil clouds detrained from convection counts for 0.4 to 0.8 of the cluster horizontal scale, 0.2 to 0.6 of the cluster cross section volume, and 0.05 to 0.20 of the cluster ice mass, depending on the cluster scales and height. There are two main detrainment layers. When the convective clusters is less than about 100 km, the anvil clouds are mainly detrained at about 6-8 km with a spreading ratio (ratio of maximum cluster width to convection rainy core width) less than 1.5. When convective clusters becomes 100 km or wider, it reaches the dominate detrainment layer at about 12 km, the detrainment index increase from 2 to more 6. Among 8 regions, convection clusters in MA produce the most anvil volume fraction. The more the ice mass is pumped upward in the anvil clouds till clusters are about 500 km wider. Nevertheless, the anvil ice mass pumped above 15 km is less than 0.1% of the total ice mass in the convective cluster.

1. INTRODUCTION

The tri-modal characteristics of tropical convection (Johnson et al. 1999) indicates prominent stable layers that exist over the Pacific warm pool and the tropical eastern Atlantic, which are associated with tri-modal distributions of divergence, cloud detrainment, and fractional cloudiness. Deep convection with a precipitation region~100 km in dimension is called a mesoscale convective system (MCS) (Zipser 1969, 1977; Houze 1977; Leary and Houze 1979; Houze, 2004). Sometimes frequent MCSs tend be merged or connected by their massive stratiform anvil clouds to form super clusters or mesoscale convective complex (MCC; Maddox 1980) with a scale of thousands of kilometers wide. Hydrometers produced in convective updrafts of an MCS are transported into the large mesoscale stratiform precipitating regions, where they fall out slowly or can be detrained into the environment or left aloft as stratiform anvil (Gamache and Houze 1983; Pfister et al. 2001; Mace et al. 2006; and Mullendore et al. 2009). Convective cells are capable of supplying ice to an anvil with large areal extent and thickness, which is determined by the strength and duration of the active convective updrafts as well as by the large-scale vertical motion, environmental wind shear, and moisture profile. These factors also determine how long the anvil is present after convection has ceased (Mapes and Zuidema, 1996; Mace et al. 2006; Cetrone and Houze 2009). The intensity, life cycle, and diurnal cycle of tropical deep convection have been long investigated with passive sensors and TRMM (Fu et al 1990; Chen and Houze 1997; Cetrone J. and Houze 2009; Nesbittt and. Zipser, 2002; Futyan and Del Genio 2007; Liu et al 2008) and carry a large geographic variation. For example, the diurnal fluctuation of deep convection is significant over tropical landmass. The convection over land such as central Africa, Indonesia and South America tends to overshoot tropopause frequently.

The anvil extent and its microphysical properties are related closely to the heating profiles and water vapor transport, hence it is crucial for global momentum budgets and radiative budgets, as highlighted in recent studies (e.g., Hartmann et al. 1992; Mace et al. 2011; and Klein et al. 2013). From ISCCP C2 climatology, Chou et al. (1999), Houze (1989), and Machado and Rossow (1993) showed that high cirrus fraction tends to increase strongly as deep cloud top temperature drops, which is referred as the cirrus-detrainment-temperature (CDT) relation. Such a finding stimulated further investigations about climate feedbacks associated with cirrus clouds (Ramanathan et al. 1989, 1991; Chou et al. 1999; Lindsen et al. 2002; Hartmann et al. 2002; Stephens 2005). However, besides of the horizontal extent, the vertical structure and microphysical properties of anvil clouds are very important information for further cirrus cloud radiative 3-D effect study, modeling evaluation and further climate feedback study (Klein and Jacob 1999).

The CloudSat satellite and the other A-Train satellites provide a new dimension of space-based observations of clouds and precipitation (Stephens et al 2010). Given the high sensitivity of Cloud Profiling Radar (CPR) onboard of CloudSat to sense the non-raining anvil, it has widely used to study tropical convection anvil productivity in terms of reflectivity and scales (Cetrone and Houze 2009, Yuan and Houze, 2010, 2011, 2013; Bacmeister and Stephens 2011).

The Cloud-Aerosol Lidar with Orthogonal Polarization (CALIOP) onboard of CALIPSO satellite can observe the cold top of deep convection and optically thin cirrus layers in the upper troposphere. Cirrus clouds at 14-16 km tend to be too tenuous to be detected by CloudSat, whose occurrence can be as high as ~60% (Sassen and Wang 2008).

The combination of these two active sensors provides the full cross section of deep convection system, which has been used to study the tropical convection evolution during the MJO cycle (Del Genio et al 2012). In this study we will utilize the combined and collocated range-resolved CPR and CALIOP data to characterize the horizontal and vertical structure of tropical convection clusters and their composite anvil productivities among the convective clusters in terms of horizontal scale, cross section volume. Moreover, we take advantages of ice cloud properties retrieval from the CloudSat 2C-ICE product to provide the unprecedented ice mass transport among the convection. Regional differences are compared among 8 climatological regions.

2. DATA AND METHOD

In this study we use the combined and collocated CloudSat and CALIPSO data during 2007-2010. The CALIOP and CPR provide very valuable complementary observations to cover clouds from those optically thin cirrus in the upper troposphere to precipitation near the surface. Several CloudSat standard products are used here. Cloud geometrical profile from combined CPR and CALIOP measurements (2B-GEOPROF-lidar; Mace et al., 2009) provides the cloud boundaries. Cloud scenario classification from combined CPR and CALIOP measurements (2B-CLDCLASS-Lidar, Sassen and Wang 2011) provides cloud phase (liquid, ice, or mixed-phase) determination for each cloud layers and classify it as one of eight basic cloud types, so that downstream retrieval algorithms or assumptions can be applied to the conditions for which they are considered valid.

Ice cloud microphysical properties are from CloudSat 2C-ICE product, which is a synergetic ice cloud retrieval from combining the CPR and the CALIOP measurements using a variational method to provide the profiles of extinction coefficient, ice water content (IWC) and effective radius (re) for ice particles (Deng et al., 2010, 2013). Deep convection clouds are identified as mixed-phase cloud layer by 2B-CLDCLASS-Lidar, where CALIOP signal is quickly attenuated at the cold top. Since CPR is sensitive to large ice particles due to the ~6 power law relation between reflectivity and size, 2C-ICE assumes that CPR mainly detects ice particles above the melting layer by neglecting the possible super-cool water. Therefore, 2C-ICE provides ice properties in both anvil clouds as well as the ice particles in the deep convective core.

The ECMWF-AUX product is created by the Generic-AUX Interpolate-to-Reference algorithm to provide the temperature, press and relative humidity (RH).

We use a cluster based approach to analyze the convection clusters along the A-Train track as Bacmeister and Stephens (2011). First the convection cluster (Figure 1) is defined as: a group of radar and/or lidar detected and

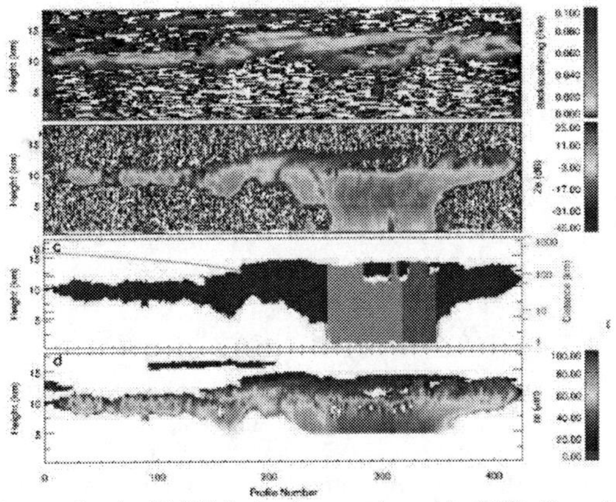

Figure 1 a) CALIOP backscattering, b) CPR Ze, c) convective cluster with cloud classification from CLDCLASS_lidar, and d) effective radius (re) from 2C-ICE. The color coded cloud types are Deep convection (Dc in red), Cumulonimbus (Ns in orange), Altostratus (As in blue) and high cirrus (Cs in purple).

horizontally or vertically connected cloud layers in GEOPROF-lidar that contains deep convective clouds classified by 2B-CLDCLASS lidar. Only the deep convective clusters with cluster top higher than about ~9 km are selected in this study, while convection in Bacmeister and Stephens (2011) includes those clusters with cloud top less than 10 km. Without using the collocated MODIS imagery, it is hard to find the horizontal 2D structure. However, as long as the convection cloud system in this study do not possess high and systematic anisotropy relative to the A-train track, the sampling pattern of A-Train should not produce major statistic biases in the spatial statistics examined here. Other studies using cloud photographs and MODIS imagery have indicated a mean cloud horizontal aspect ratio of around 2 with no systematic orientation of features (Benner and Curry, 1998; Bacmeister and Stephens 2011).

For each convective cluster (Figure 1), the convective cluster rainy core is first identified and composed of the deep convection, nimbostratus and cumulus congenstus clouds as classified in the CLDCLASS_lidar product with a width W_{cb}. Low level cumulus clouds are discriminated from this deep convective cluster. While the anvil clouds are composed of the connected altostratus and high clouds in the convective clusters. The second thin cirrus layer above 15 km is also discriminated as possible, which may not be directly associated with deep convection. The corresponding horizontal span, volume, and ice mass of the cluster and its anvil in every 480 meter are noted as $W_{clu}(z_i)$ $V_{clu}(z_i)$, $IWC_{clu}(z_i)$ and $W_{anvil}(z_i)$, $V_{anvil}(z_i)$, $IWC_{anvil}(z_i)$ respectively. The maximum horizontal span or scale and vertical depth of the convective cluster are noted as $W_{cluster}$ and D, respectively. The cross-section area (

$\sum_i W_{clu}(z_i) \times \Delta z)$ is referred as the cluster total volume (V) as in Bacmeister and Stephens (2011) to avoid confusion with the cloud horizontal area.

In this study, we define total anvil horizontal fraction, anvil volume ratio, and anvil ice mass ratio to the total of the cluster for each cluster. Then the clusters are sorted by the effective width W_{eff} (defined as the ratio of the cluster volume and maximum cloud depth: $W_{eff} = V/D$) rather than $W_{cluster}$ to produce composite results about convection clusters and their anvil productivity, avoiding the possible contamination of extensive tropospause clouds.

3. RESULT ABOUT CONVECTIVE CLUSTER

Figure 2 shows the global distribution of identified convective clusters from combined CPR and CALIOP data during 2007-2010 in a 10 x 5 degree grid resolution. ITCZ, South America, and Atlantic are the most deep convection active regions. There is a clear tendency that the most convective clusters over land or over oceans adjacent to land, in locations favors storms motion from land to ocean. This pattern is very similar to the TRMM monthly mean rainfall in Liu et al (2008). The geographical pattern of tropical deep convective cluster occurrence is well correlated with the sea surface temperature (SST) pattern. Eight tropical climatological regions are selected to represent their unique tropical convective regimes, which are similar to those in Yuan et al. (2010): Tropical Africa (AF), Indian Ocean (IO), Maritime Continent (MA), Western Pacific (WP), Eastern Pacific (EP), Southern Pacific (SP), Central American (AM), and Atlantic (AT). IO and MA are the hottest regions, while EP and AT are coldest among the 8 regions. Given the snapshots of CloudSat and CALIPSO, their observation at certain regions may be biased to certain convection development stage. The number distribution of convective clusters and probability density functions of convective cluster scales are shown in Figure 3.We can see that there is a decent sample of convection clusters for all regions, given the geographic variation for 8 regions. Convection in IO and MA, followed by AM, WP and SP, tends to be connected by the extensive anvil to form super clusters (connected MCSs or MCC) larger than 1000 km. Convection at EP and AT are very similar with a narrow scale range between 100 and 1000 km.

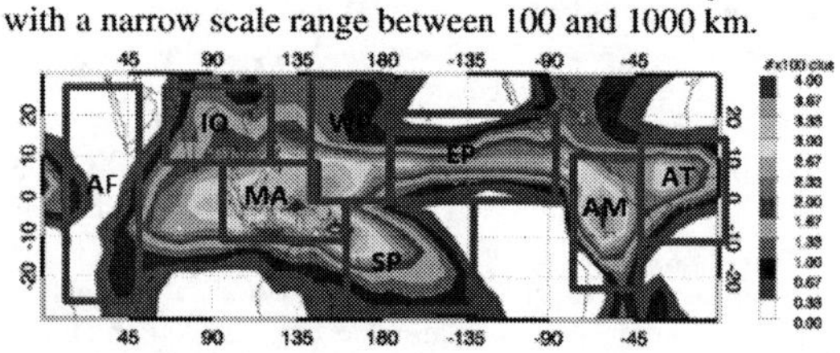

Figure 2 Maps of (a) total number of tropical convective clusters identified with CloudSat and CALIPSO There are eight regions boxed for the regional anvil cloud study: Tropical Africa (AF), Indian Ocean (IO), Maritime Continent (MA), Western Pacific (WP), Eastern Pacific (EP), Southern Pacific (SP), Central American (AM), and Atlantic (AT)

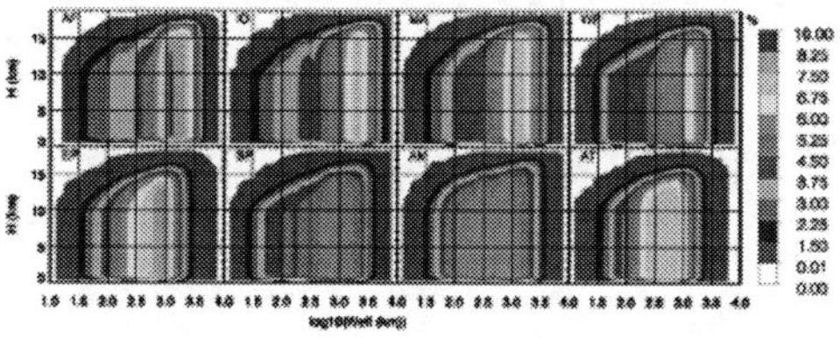

Figure 3 Composite convective clusters probable density function of occurrence by height as a function of W_{eff} for the 8 regions

Convection at AF has an outstanding scale peak at about 1000 km, while convection in AM is almost evenly distributed between 10 km to 1000 km. This is consistent with Nesbitt and Zipser (2002) that AF is revealed for its major share of moderate intense storms although IO and MA have the most extensive events. We can see that for all 8 regions, the convective clusters tend to shoot higher as their horizontal scale increases. The cluster top height increases from 12 km to more than 16 km, cluster top temperature decreases from 230 K to 190 K. At the same W_{eff} scale, the convective clusters over AF, IO, MA and AM tend to have higher cloud top than those over ocean.
The composite mean cross section volume by height is shown Figure3 for the 8 regions. First, we can see that the volume increases at all height as the cluster W_{eff} increases. Second, cross section volume has two prominent peaks vertically. The first is above the melting layer or at 6-8 km. This peak becomes less obvious as the W_{eff} increases. The second peak is above 10 km and become prominent as the W_{eff} increases. The corresponding peak height increases as W_{eff} increases. This trend would not be fully detected without the CALIPSO CALIOP observations.

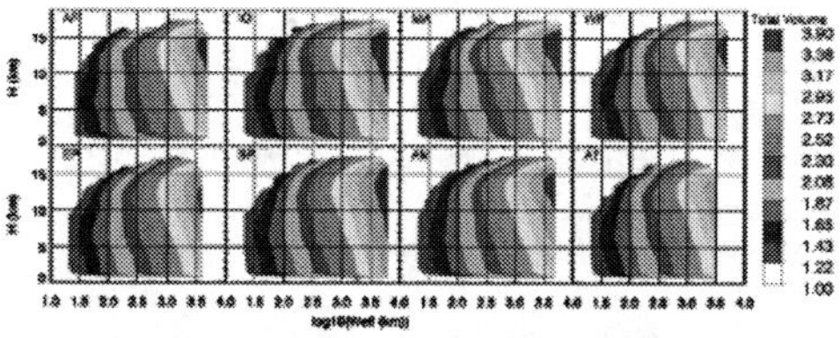

Figure 4 Composite cross section volumes (km^2 in logarithm) by height as a function of W_{eff} for the 8 regions: a) from combined CloudSat and CALIOP observations

4. DISCUSSION AND CONCLUSION

We found out that for all regions, when the convective clusters become wider, the convective clusters become deeper. The clusters tend to be composed of multiple rainy cores and each core becomes wider. Among the 8 regions, convections in EP and AT tend have less but wide rainy cores, while WP, MA and IO tend to have more but narrow rainy cores. For the convections at the same scale, AF, IO, MA and AM tend to have higher cloud top than those over ocean.

Without counting the stratiform rain, the anvil clouds detrained from convection count for 0.4 to 0.8 of the cluster horizontal fraction, 0.2 to 0.6 of the cluster cross section volume, and 0.05 to 0.20 of the cluster ice mass, depending on the cluster scales.

The vertical dependence of anvil productivity shows that there are two main detrainment layers, which is consistent with the tri-mode theory in Johnson et al. (1999). When the convection cluster is less than 100 km, the anvil cloud is mainly detrained at about 6-8 km with a detrainment index or anvil spreading ratio (ratio of maximum cluster width over convection rainy core, Yuan and Houze 2010; Bacmeister and Stephens 2011) less than 2.5. This detrainment layer is related to the stable layer near 0°C, which inhibit cloud growth, and enhance the detrainment just above it for those congestus clouds. The existence of a volume peak and anvil volume ratio near this level at times of organization of convection on the MCS (~100 km) scale suggests that processes that contribute to enhanced detrainment near that level may also contribute to restricting cloud growth to that level. When for vigorous convection clusters, undiluted or less diluted updrafts in the convection keep pump hydrometers up till it meets the higher stable level, which causes other enhanced detrainment layer between 10 and 16 km with the detainment index increasing from 2.5 to more than 6 and anvil horizontal fraction quickly increase from 0.3 to more than 0.8. Even the convective cluster with W_{eff} larger than 500 km can reach 16 km, the anvil detrainment in volumes maximize at about 14 km, which is about 1-2 km lower than the cluster top.

ACKNOWLEDGEMENT: The A-Train data are acquired from CloudSat Data center at Colorado State University. We'd like to thank the engineers, technicians, and scientists for the high-quality and well documented data. This work is supped by NASA. We also appreciate the anonymous reviewers for their suggestions and comments.

REFERENCES

Bacmeister, J. T., and G. L. Stephens (2011), Spatial statistics of likely convective clouds in CloudSat data, *J. Geophys. Res.*, 116, D04104, doi:10.1029/2010JD014444

Chen, S. S., and R. A. Houze Jr., 1997: Diurnal variation and lifecycle of deep convective systems over the Pacific warm pool. *Quart. J. Roy. Meteor. Soc.*,123, 357–388.

Chou C. J. D. Neelin (1999) Cirrus detrainment-temperature feedback. *Geophysical Research Letter,* 26, 1295-1298.

Cetrone J. and R. A. Houze Jr. 2009: anvil clouds of tropical mesoscale convective systems in monsoon regions. *Quart. J. Roy. Meteor. Soc.*135: 305–317

Deng, M., G. G. Mace, Z. Wang, and H. Okamoto, 2010: Tropical Composition, Cloud and Climate Coupling Experiment validation for cirrus cloud profiling retrieval using CloudSat radar and CALIPSO lidar, *J. Geophys. Res.,* 115, D00J15, doi:10.1029/2009JD013104 (6)

Futyan J, Del Genio A. 2007. Deep convective system evolution over Africa and the tropical Atlantic. , *J. Clim* 20: 5041–5060

Fu, R., A. D. Del Genio, and B. Rossow, 1990: Behavior of deep convective clouds in the tropical Pacific deduced from IPSSP radiance, , *J. Clim,* 3, 1129-1152. 1990.

Gamache, J. F., and R. A. Houze (1983), Water budget of a mesoscale convective system in the tropics, J. Atmos. Sci., 40(7), 1835 – 1850.

Hartmann, D. L., M. E. Ockert-Bell, and M. L. Michelsen (1992), The effect of cloud type on Earth's energy balance: Global analysis,, *J. Clim.*, 5, 1281 – 1304.

Houze, R. A., Jr., 1977: Structure and dynamics of a tropical squall line system observed during GATE. *Mon. Wea. Rev.*,105, 1540–1567.

——, 2004: Mesoscale convective systems. *Rev. Geophys.*, 42, 43

Johnson R. H. T. M. Rickenbach, S. A. Rutledge, P. E. Ciesieski, and W. H. Schubert, 1999: Trimodal characteristics of Tropical convection. *J. Clim,* 12, 2397- 2418.

Klein S. and C. Jakob 1999 Validation and sensitivities of frontal clouds simulated by the ECMWF model. *Monthly weather review,* 127, 2514-2531

Klein, S. A., Y. Zhang, M. D. Zelinka, R. Pincus, J. Boyle,and P. J. Gleckler (2013), Are climate model simulations of clouds improving? An evaluation using the ISCCP simulator, *J. Geophys. Res. Atmos.*,118, 1329–1342, doi:10.1002/jgrd.50141.

Leary, C. A., and R. A. Houze (1979), Melting and evaporation of hydrometeors in precipitation from the anvil clouds of deep tropical convection, *J. Atmos. Sci.*, 36(4), 669 – 679

Liu, C. T., E. J. Zipser, D. J. Cecil, S. W. Nesbitt, and S. Sherwood, 2008: A cloud and precipitation feature database from nine years of TRMM observations. *J. Appl. Meteor. Climatol.*,47, 2712–2728.

Lindzen, R. S., M.-D. Chou, and A. Y. Hou (2002), Comments on "No evidence for iris", *Bull. Am. Meteorol. Soc.*, 83, 1345 – 1348.

Maddox RA. 1980. Mesoscale convective complexes. Bull. Am. Meteorol. Soc.61: 1374–1387.Machado, L. A. T., and W. B. Rossow, Structural characteristics and radiative properties of tropical cloud clusters, *Mon. Weather Rev.*, 121, 3234-3260, 1993.

Mapes, B. E., and P. Zuidema, 1996: Radiative-dynamical consequences of dry tongues in the tropical troposphere. *J. Atmos.Sci.*, 53,620–638.

Mace GG, Deng M, Soden B, Zipser E. 2006. Association of tropical cirrus in the 10–15-km layer with deep convective sources: An observational study combining millimeter radar data and satellite derived trajectories. *J. Atmos. Sci.* 63: 480–503.

Mullendore, G. L., A. J. Homann, K. Bevers, and C. Schumacher (2009), Radar reflectivity as a proxy for convective mass transport, *J. Geophys. Res.*, 114, D16103, doi:10.1029/2008JD0114

Nesbittt S. W. E. Zipser, 2002: the diurnal cycle of rainfall and connective intensity according to three years of TRMM measurement. , *J. Clim,* 16, 1456-1475

Pfister, L., et al., Aircraft observations of thin cirrus clouds near the tropical tropopause, *J. Geophys. Res.*, 106, 9765 – 9786, 2001.

Ramanathan, V., R. D. Cess, E. F. Harrison, P. Minnis, B. R. Barkstrom, E. Ahmad, and D. Hartmann, Cloud-radiative forcing and climate: results from the earth radiation budget experiment, *Science* 243, 57-63, 1989

Stephens, G. L. (2005), Cloud feedbacks in the climate system: A critical review, *J. Clim.*, 18, 237 – 273

Yuan J. and R. A. Houze Jr. 2010: Global variability of mesoscale convective system anvil structure from A-train satellite data , *J. Clim,* 23, 5864-5888

Yuan J. and R. A. Houze Jr. 2011: vertical structure of anvil clouds of tropical mesoscale convective systems observed by CloudSat. J. Atmos. Sci. 68, 1653-1674.

Yuan J. and R. A. Houze Jr. 2013: Deep convective systems observed by A-Train in the tropical Indo-Pacific regions affected by the MJO. J. Atmos. Sci. 70, 465-486

Zipser, E. J., 1969: The role of organized unsaturated downdrafts in the structure and rapid decay of an equatorial disturbance. *J. Appl. Meteor.,* 8,799–814

EPJ Web of Conferences **119**, 04010 (2016) DOI: 10.1051/epjconf/201611904010
ILRC 27

AEROSOL OPTICAL PROPERTIES ABOVE OPAQUE WATER CLOUDS DERIVED FROM THE CALIOP VERSION 4 LEVEL 1 DATA

Zhaoyan Liu [1,2]*, David Winker [2], Ali Omar [2], Mark Vaughan [2], Jayanta Kar [1,2], Charles Trepte [2], Yongxiang Hu [2], Gregory Schuster [2], and Stuart Young [3]

[1]*Science Systems and Applications Inc. Hampton, VA23666, USA, *Email: Zhaoyan.liu@nasa.gov*

[2]*NASA Langley Research Center, Hampton, VA23681, USA*

[3] *CSIRO Oceans and Atmosphere, Aspendale, Victoria, 3195, Australia*

ABSTRACT

In a previous study we evaluated the above-cloud aerosol optical depth (AOD) retrieval at 532 nm in the CALIOP version 3 (V3) data products for two selected spatial domains along the Saharan dust transport pathway and African smoke transport pathway. In that study we rescaled the V3 level-1 (L1) data to compensate for known V3 calibration biases, and then derived aerosol intrinsic properties such as lidar ratio (S_a) and particulate depolarization ratio (PDR) for comparison with the CALIOP dust and smoke aerosol models. The calibration of the recently released version 4 (V4) CALIPSO L1 data product is significantly improved over V3. So in this paper we repeat our previous analysis using the new V4 L1 data. A comparison shows that our rescaled V3 and the new V4 data are different only by ~1% in the two selected spatial domains. The retrieved AOD values decrease by ~2.6% in both domains from V3 to V4. When the data is screened to exclude weakly scattering layers, the median S_a retrieved from the V4 L1 data in the dust transport region is reduced by 1.4 sr to 43.0±8.3 sr. The median S_a value in the smoke transport region is increased by 0.8 sr to 71.2±15.1 sr. The PDR values remain almost unchanged for the screened data.

1. INTRODUCTION

Beginning in June 2006, a global data set of height-resolved measurements of aerosols and clouds has been continuously acquired by the Cloud-Aerosol Lidar with Orthogonal Polarization (CALIOP), deployed aboard the Cloud-Aerosol Lidar and Infrared Pathfinder Satellite Observations (CALIPSO) platform. These active sensor data offer a new and unique opportunity to characterize the global three-dimensional (3-D) distribution of aerosols [1]. Aerosol extinction profiles and AOD can be derived from the CALIOP measurements even for aerosols located over clouds or other bright surfaces. CALIOP's ability to quantify the spatial distribution and optical properties of above-cloud aerosols represents an important step forward, as this information is required to more accurately assess aerosol intercontinental transport and radiative and climate impacts.

In a previous study [2] we evaluated the above-cloud aerosol extinction and optical depth retrieval at 532 nm in the CALIOP V3 data products for two selected spatial domains along the Saharan dust transport pathway and African smoke transport pathway (refer to Fig. 1). In that study, we also derived the aerosol intrinsic properties such as S_a and PDR and compared them with the CALIOP dust and smoke aerosol models. Six years of the V3 nighttime data were analyzed. However, in the V3 data release, stratospheric aerosols in the calibration region (30-34 km) introduced a low bias of a few percent into the L1 532-nm attenuated backscatter profiles [3]. For this reason, all the L1 profiles were recalibrated using a scaling factor determined by comparing the backscatter signals from a higher altitude range (36 – 40 km) to the reference molecular scattering signal [4].

The CALIPSO project recently released a new version (V4) of L1 data in which the calibrations have been improved significantly. Major changes at 532 nm include moving the molecular normalization region up to 36–39 km (from 30–34 km in V3), thus greatly reducing calibration biases due to stratospheric aerosols. In this paper, we repeat our earlier analyses using the new V4 L1 data in order to further evaluate the dust and smoke aerosol intrinsic properties reported in our previous study.

2. METHODOLOGY

Hu et al. [5] developed a technique that uses CALIOP measurements of opaque water clouds (OWCs) as a reference to determine the optical depth of overlying transparent aerosol or cirrus layers. This approach takes advantage of the well-behaved relationship between the layer-integrated depolarization ratio and the multiple scattering in water clouds. The optical depth for aerosol layers above an opaque water cloud can be derived using

$$\tau_{aerosol} = -\frac{1}{2}\ln\left(\frac{\mathrm{H}\,\gamma'_{WC,TS}}{\gamma'_{WC,SS,NA}}\right), \qquad (1)$$

where $\gamma'_{WC,TS} = \int_{r_{WC,top}}^{r_{WC,base}} B'(r)dr$ is the integrated attenuated backscatter within the underlying OWC, *B'* is the attenuated backscatter corrected for the attenuation due to the molecular scattering and ozone absorption, and H is the OWC layer effective multiple scattering factor [5] defined as

$$\mathrm{H} = \left(\frac{1-\delta_I}{1+\delta_I}\right)^2. \qquad (2)$$

In equations (1) and (2), δ_I is the layer-integrated volume depolarization ratio within the OWC and *B'* is the attenuated backscatter coefficient corrected for molecular and ozone attenuation, and the subscripts WC, SS and TS represent, respectively, water clouds, single scattering and total scattering (i.e., single scattering and multiple scattering together).

The denominator in Eq. (1), $\gamma'_{WC,SS,NA}$, is obtained from a climatology of the single-scattering integrated attenuated backscatter for OWCs that have no aerosol (NA) or any other features detected above. A location dependent climatology of $\gamma'_{WC,SS,NA}$ was developed from six years (2007–2012) of the CALIOP data to account for the spatial variability in the microphysical properties of water clouds [2].

Retrievals using the OWC technique are performed on CALIOP L1 attenuated backscatter profiles, averaged horizontally to 5 km. The OWCs selected for constrained retrievals are (1) single layered with (2) top heights less than 2 km for which (3) opaque water clouds are detected in all 15 single-shot profiles within each 5-km average, and the standard deviation of these 15 single shot top heights is less than 50 m. More details can be found in Liu et al. [2].

The AOD derived using the OWC technique can subsequently be used as a constraint to retrieve S_a, PDR, extinction and backscatter coefficients for the overlying aerosol layer using a two-component solution of the lidar equation [6].

3. RESULTS

Six years (2007–2012) of the CALIOP V4 L1 data have been analyzed. To select OWCs, the V3 L2 vertical feature mask (VFM) data are used, because the V4 L2 data products have not yet been created. As in our previous study, we focus our analysis in this paper on the two selected regions shown in Fig. 1. North Africa is the largest source of dust emissions in the world, and transport of Saharan dust across the tropical North Atlantic reaches a maximum during the summer. For this reason, the months of June – August are selected for analyzing the dust transport region. Outbreaks of smoke due to biomass burning in southern Africa are most frequent in the months of July – September, and we therefore select these months for analyzing the smoke transport region.

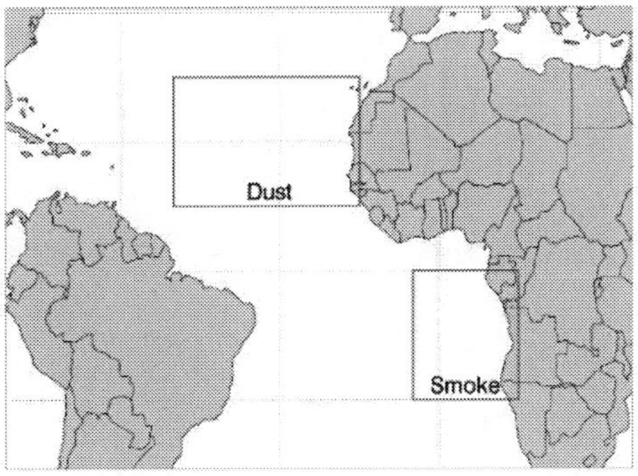

Figure 1 Spatial domains analyzed (red boxes). The northern region (10°N-30°N, 50°W-15°W) is along the Saharan dust transport pathway over the tropical North Atlantic, while the southern region (20°S-0°, 5°W-15°E) is along the smoke transport pathway over the tropical South Atlantic.

Figures 2 and 3 present results from the spatial domains indicated by the red boxes in Fig. 1. The mean value of the dust S_a distribution is 47.1 sr, with a median of 43.5 sr, a mode of 42.5 sr, and a standard deviation of 24.5 sr. For the PDR distribution the mean is 0.255, the median is 0.275, the mode is 0.279, and the standard deviation is 1.64. For the screened data using attenuated scattering ratio (ASR= $\int_{WC,top}^{8km} B'dr / \int_{WC,top}^{8km} \beta_m dr - 1$) > 0.3, the mean/median/mode values are 43.4/43.0/42.5 ±8.3 for S_a, and 0.280/0.281/ 0.282±0.045 for PDR.

DOI: 10.1051/epjconf/201611904010

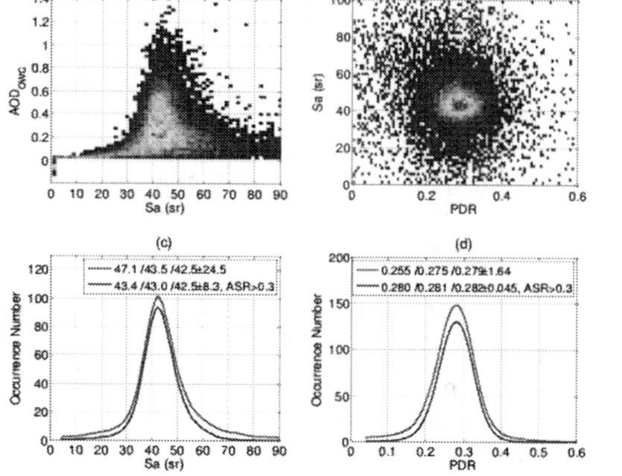

Figure 2 Analysis results for the dust transport region as shown in Fig. 1. The upper row shows 2-D distributions of (a) AOD_{OWC} vs. S_a retrieved using AOD_{OWC} as a constraint and (b) S_a vs. PDR, while the lower row shows histograms of (c) S_a and (d) PDR occurrence frequencies. The red curves in (c) and (d) include all data and the blue curves are screened data using ASR>0.3. The numbers in the legends of (c) and (d) show mean/median/mode ± standard deviation.

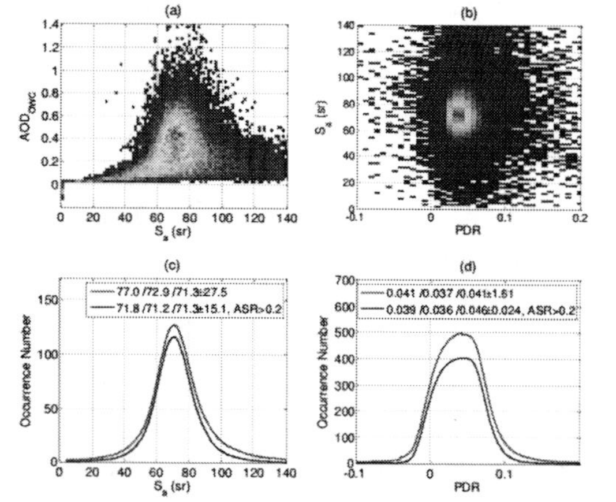

Figure 3 Analysis results for the smoke transport region as shown in Fig. 1. The upper row shows 2-D distributions of (a) AOD_{OWC} vs. S_a retrieved using AOD_{OWC} as a constraint and (b) S_a vs. PDR, while the lower row shows histograms of (c) S_a and (d) PDR occurrence frequencies. The red curves in (c) and (d) include all data and the blue curves are screened data using ASR>0.2. The numbers in the legends of (c) and (d) show mean/median/mode ± standard deviation.

The S_a values retrieved in the smoke transport region have mean/median/mode values of 77.0/72.9/71.3±27.5 sr for all the data and 71.8/71.2/71.3±15.1 sr for the screened data using ASR>0.2. The PDR values retrieved in the smoke region are typically smaller than 0.1, with mean/median/mode values of 0.041/0.037/0.041 ±1.61 for all smoke layers analyzed and 0.039/0.036/0.046 ±0.024 for the layers with ASR>0.2.

4. COMPARISONS

Figure 4 presents a comparison of the S_a and PDR retrieved in the dust and smoke transport regions from the rescaled V3 and standard V4 L1 data using the OWC constrained retrieval. In this comparison, weakly scattering aerosol layers were screened out using the ASR test. In general, the retrieved PDR values from the two L1 datasets have an almost identical distribution, because the PDR retrieval is largely insensitive to calibration errors [7]. Small difference is seen in the median retrieved S_a value in the dust transport region (decreased by 1.4 sr, or 3.2%), while the S_a difference for the smoke transport region is even smaller (increased by 0.8 sr, or 1.1%).

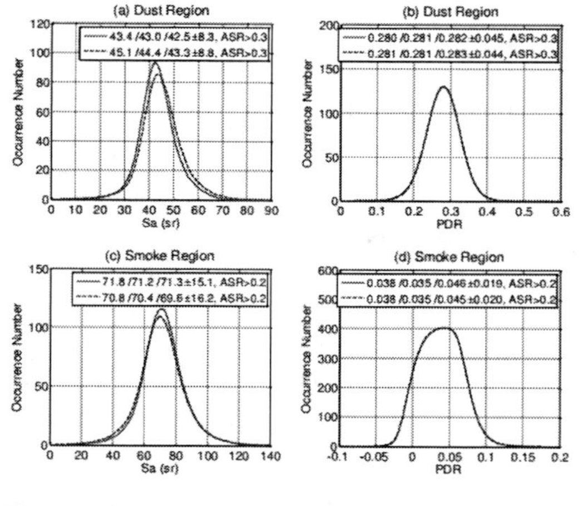

Figure 4 S_a and PDR retrieved using the OWC constrained technique in the dust (a and b) and smoke (c and d) transport regions from the rescaled V3 (blue) and V4 (red) L1 data, with weakly scattering aerosol layers screened out.

We examine the calibration in the two L1 datasets. As shown in Fig. 5, the rescaled V3 L1 data is, on average, 0.8% smaller than the V4 L1 data in the dust transport region and 1.1% larger in the smoke transport region. These differences are small and within the estimated calibration uncertainties [3]. On the other hand, the retrieved AOD from the V4 data is smaller than the V3 data by 0.006 (2.6%) in

DOI: 10.1051/epjconf/201611904010

the dust transport region and by 0.08 (2.6%) in the smoke transport region.

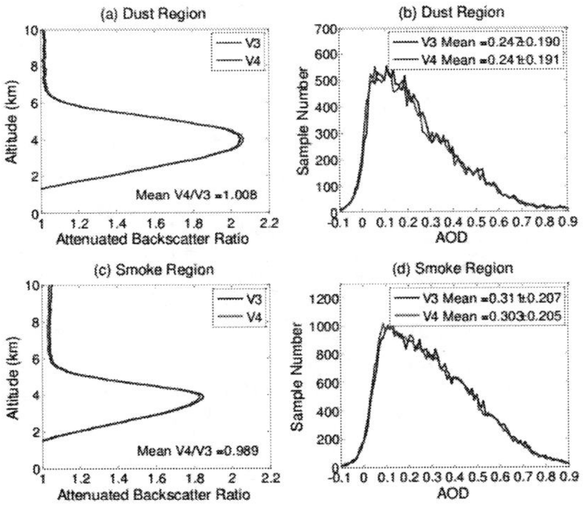

Figure 5 Average attenuated backscatter ratio (a and c) and histograms of AOD (b and d) retrieved from the V3 (blue) and V4 (red) L1 data using the OWC technique.

5. CONCLUSIONS

Following on our previous study that analyzed six years of the CALIOP V3 data to retrieve the aerosol optical depth above opaque water clouds and dust and smoke lidar ratio, in this paper we repeated the same analysis using the recently released V4 L1 data. The rescaled V3 L1 data that was used in the previous study is, on average, different from the V4 L1 data by ~1% in the two selected spatial domains. The retrieved above-cloud aerosol optical depths are 2.6% smaller using the V4 L1 data than the V3 L1 data in both domains, while the particular depolarization ratio values using both versions have an almost identical distribution for moderately scattering layers because the retrieval of this parameter is largely insensitive to the data calibration. Lidar ratio is an intrinsic aerosol property and a key parameter that limits the accuracy of the extinction retrieval. The retrieved median lidar ratio value is decreased by 1.4 sr using the V4 L1 data in the dust transport region. This is mainly because the V4 L1 data in this region is ~1% larger than the rescaled V3 L1 data while the above-cloud aerosol optical depths retrieved from the V4 data are ~2.6% smaller than those retrieved from the rescaled V3 data. In the smoke transport region the lidar ratio is increased by 0.8 sr. Unlike in the dust transport region, in the smoke transport region the attenuated backscatter and the retrieved aerosol optical depths are both slightly smaller in the V4 analysis than in the V3 analysis. Because these differences have mutually compensating effects, the difference in the retrieved smoke lidar ratios from the two data versions is smaller than that in the retrieved dust lidar ratios.

This is an on-going study, and it would be useful to perform the same analysis once again when the V4 L2 data products become available.

REFERENCES

[1] Winker, D. M., J. L. Tackett, B. J. Getzewich, et al.: The global 3-D distribution of tropospheric aerosols as characterized by CALIOP, Atmos. Chem. Phys., **13**, 3345–3361, 2013.

[2] Liu, Z., D. Winker, A. Omar, M. Vaughan, et al.: Evaluation of CALIOP 532-nm aerosol optical depth over opaque water clouds, Atmos. Chem. Phys., **15**, 1265-1288, 2015.

[3] Rogers, R. R., C. A. Hostetler, J. W. Hair, et al.: Assessment of the CALIPSO Lidar 532 nm Attenuated Backscatter Calibration Using the NASA LaRC Airborne High Spectral Resolution Lidar, Atmos. Chem. Phys., 11, 1295-1311, 2011.

[4] Vernier, J. P., J. P. Pommereau, A. Garnier, et al.: The tropical stratospheric aerosol layer from CALIPSO lidar observations, J. Geophys. Res., 114, D00H10, 2009.

[5] Hu, Y., M. Vaughan, Z. Liu, et al.: Retrieving optical depths and lidar ratios for transparent layers above opaque water clouds from CALIPSO lidar measurements, IEEE Geosci. Remote Sens. Lett., **4**, 523-526, 2007.

[6] Young, S. A., and M. A. Vaughan: The retrieval of profiles of particulate extinction from Cloud Aerosol Lidar Infrared Pathfinder Satellite Observations (CALIPSO) data: Algorithm description, J. Atmos. Oceanic Technol., **26**, 1105-1119, 2009

[7] Liu, Z., T. D. Fairlie, I. Uno, et al.: Transpacific Transport and Evolution of the Optical Properties of Asian Dust, J. Quant. Spectrosc. Ra., **116**, 24-33, 2013.

STATISTICAL CHARACTERISTICS OF AEROSOL EXTINCTION COEFFICIENT PROFILE IN EAST ASIA FROM CALIPSO

Xuejin Sun, Junhao Zhou, Yongbo Zhou*

*College of Meteorology and Oceanpgraphy, PLA University of Science & Technology, Nanjing 211101, China, *Email: dingboluo@yahoo.com*

ABSTRACT

Aerosol extinction coefficient profile (ECP) is important in radiative transfer modeling, however, knowledge of ECP in some area has not been clearly recognized. To get a full understanding of statistical characteristics of ECP in three Asian regions: the Mongolian Plateau, the North China Plain and the Yellow Sea, CALIPSO aerosol product in 2012 is processed by conventional statistical methods. Orbit averaged ECP turns out to be mainly exponential and Gaussian patterns. Curve fitting shows that the two ECP patterns account for more than 50 percent of all the samples, especially in the Yellow Sea where the frequency of occurrence even reaches over 80 percent. Parameters determining fitting curves are provided consequently. To be specific, Gaussian pattern is the main ECP distribution in the Mongolian Plateau and the Yellow Sea, and exponential pattern predominates in the North China Plain. Besides, aerosol scale height reaches its maximum in summer and in the Mongolian Plateau. Meanwhile, the uplifting and deposition of dust during transportation are potentially explanations to the occurrence of Gaussian ECP. The results have certain representativeness, and contribute to reducing uncertainties of aerosol model in relevant researches.

1. INTRODUCTION

Aerosol extinction coefficient profile (ECP) is one of the most uncertain factors in radiation transfer modeling. With the deployment of lidar in atmosphere detection, measurement of ECP becomes feasible, and achievements have been made in characterizing statistical characteristics of ECP in certain areas[1]. However, existing researches are usually confined to depict ECP qualitatively[2~3]. Besides, knowledge of ECP in certain areas remains scarce. In this study, three Asian regions are selected: the Mongolian Plateau (38~44°N; 98~110°E), the North China Plain (33~44°N; 111~119°E) and the Yellow Sea (31~38°N; 121~127°E), which are denoted as A, B and C (black boxes in Fig. 1). They are regions where ground-based lidar station is sparse, therefore, relevant researches are rare. The results provide a full understanding of ECP in the three regions. More details are given below.

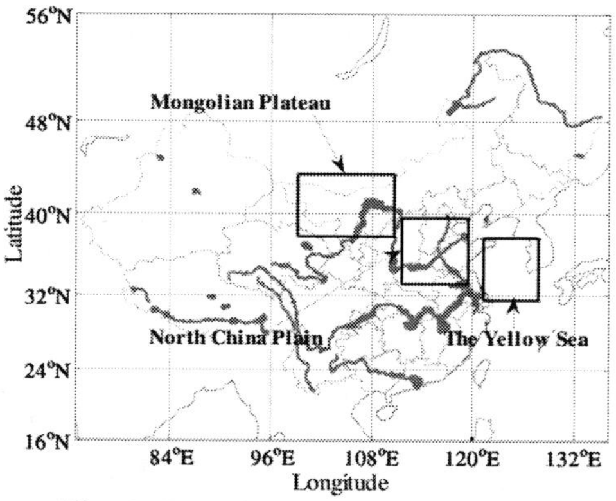

Fig. 1 Locations of selected regions

2. METHODOLOGY

2.1 CALIPSO DATA

CALIPSO aerosol product in 2012 is used to explore the statistical properties of ECP in the selected regions. CALIPSO is a member of the A-Train constellation of satellites which make global observation of ECP feasible[4]. Borne on CALIPSO, the Cloud-Aerosol Lidar with Orthogonal Polarization (CALIOP) provides aerosol product containing information on ECP and aerosol type. The product is gridded at a vertical resolution of 30 m below 8.2 km[5], and correlates well with ground-based lidar[6].

2.2 STATISTICAL METHODS

Orbit averaged ECP is generated by averaging valid data per layer in the selected regions. Besides, aerosol scale height and aerosol type are investigated. Aerosol scale height is defined as

DOI: 10.1051/epjconf/201611904011

the height below which AOD reaches 63 percent of the whole-layer AOD[2]. In CALIPSO aerosol product, aerosol is classified into six types: clean marine, dust, polluted continental, clean continental, polluted dust and smoke. Aerosol per layer is interpreted as one of the six types, and the frequency of occurrence for each aerosol type is generated

3. RESULTS

Polynomial curve fitting is performed to the orbit averaged ECP whose valid layers surpass 50 percent of the whole layers below 10 km. If the fitting rate of orbit averaged ECP to (1) or (2) is larger than 0.6, orbit averaged ECP is deemed as the pattern corresponding to the larger fitting rate.

$$k(z) = a \cdot \exp(-b \cdot z) + c \quad (1)$$

$$k(z) = a \cdot \exp[-(\frac{b-z}{c})^2] + d \quad (2)$$

where $k(z)$ denotes ECP. Curve fitting shows that the two ECP patterns account for more than 50 percent of all the ECP samples. For the Yellow Sea, the percentage even reaches over 80 percent (Fig.2).

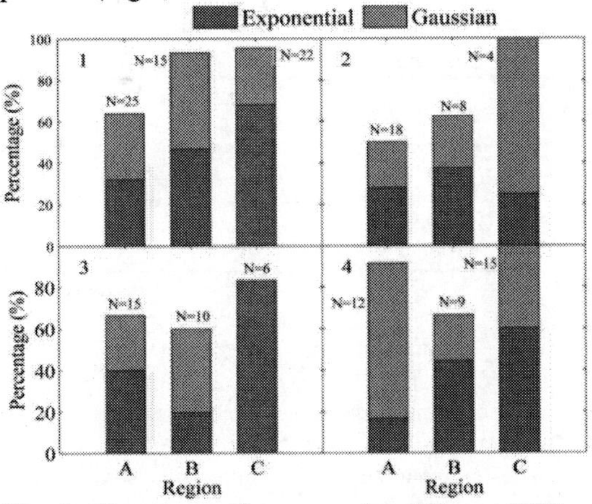

Fig. 2 Occurrence frequency for different ECP patterns. 1-4 in the top left corner denotes season from spring to winter. N is the the total number of profiles

Orbit averaged ECP with AOD larger than 0.4 only occurs in spring and winter in the Mongolian Plateau, while can be found in spring, summer and winter in the Yellow Sea and all year round in the North China Plain. In the Mongolian Plateau, scale height reaches its maximum in summer. In the North China Plain, aerosol scale height remains 2~3 km. Scale height reaches its maximum and minimum in summer and winter for AOD between 0.6~0.8. For orbit averaged ECP with extremely large AOD (AOD > 0.8), variation of scale height with season is contrary to that of AOD within 0.6~0.8. In the Yellow Sea, scale height generally reaches its maximum in summer (Fig. 3).

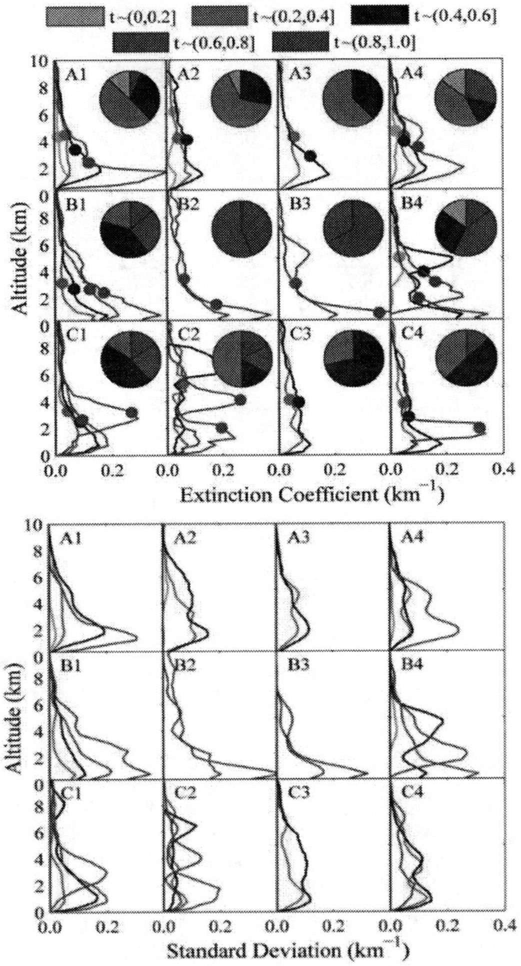

Fig. 3 The upper panel denotes orbit averaged ECP at 532-nm, where the inner pie charts indicate occurrence frequency for different AOD bins, and the colored rounds represent the scale height. The lower panel denotes standard deviation of orbit averaged ECP shown in upper panel. 1-4 in the top left corner denotes season from spring to winter, and A-C represents the three selected regions

According to (1) and (2), Gaussian and exponential patterns are shaped by 4 and 3 parameters, which are presented as G(a,b,c,d) and E(a,b,c) in Table 1-5.

Table 1 ECP parameters for AOD below 0.2

	Spring	Summer	Autumn	winter
A	G(0.03,1.6,1.2,0.01)	——	No data	——
B	No data	No data	No data	——
C	No data	No data	No data	No data

Table 2 ECP parameters for AOD between 0.2 and 0.4

	Spring	Summer	Autumn	winter
A	G(0.04,3.1,2.9,0.01)	G(0.06,2.3,2.6,0.02)	G(0.05,2.8,2.7,0.01)	G(0.06,2.8,2.9,0.01)
B	E(0.20,1.1,0.0)	No data	No data	No data
C	G(0.06,1.6,1.6,0.02)	——	——	E(0.09,0.2,0)

Table 3 ECP parameters for AOD between 0.4 and 0.6

	Spring	Summer	Autumn	winter
A	G(0.13,2.2,1.2,0.03)	G(0.08,1.7,0.8,0.05)	G(0.15,1.9,1.5,0.02)	G(0.09,2.4,1.5,0.03)
B	E(0.19,0.4,0.0)	No data	No data	——
C	E(0.17,0.3,0.0)	——	——	G(0.18,0.2,0)

Table 4 ECP parameters for AOD between 0.6 and 0.8

	Spring	Summer	Autumn	winter
A	No data	No data	No data	G(0.18,2.3,2.1,0.02)
B	E(0.29,0.2,0)	E(0.21,0.5,0.0)	E(0.18,0.5,0.0)	E(0.42,0.7,0.0)
C	E(0.21,0.3,0)	G(0.20,1.4,1.1,0.03)	No data	No data

Table 5 ECP parameters for AOD between 0.8 and 1.0

	Spring	Summer	Autumn	winter
A	G(0.33,1.4,0.9,0.04)	No data	No data	No data
B	E(0.42,0.3,0)	E(0.41,0.6,0.0)	G(0.71,0.9,0.0)	G(0.21,2.5,1.4,0.03)
C	——	——	No data	G(0.33,1.8,0.7,0.04)

4. CONCLUSIONS AND DISCUSSIONS

In general, Gaussian pattern is the predominant ECP pattern in the Mongolian Plateau and the Yellow Sea, and exponential pattern predominates in the North China Plain.

In the Mongolian Plateau, dust and polluted dust are the main aerosol types in spring and winter, during sand-dust weather conditions, dust aerosol can stretch to upper troposphere[7], the uplifting of dust aerosol may concentrate and forms a single peak of extinction coefficient at a certain layer, which is the typical characteristic of Gaussian distribution, which may be potential explanations to the prevalence of Gaussian pattern. In the Yellow Sea, aerosol is a mixture of clean marine, dust, polluted dust and smoke, of which clean marine aerosol only concentrates in lower layer in summer and autumn, namely, aerosol in this region is mainly transported from outer place. In summer and autumn, clean marine aerosol frequently occurs due to sea water evaporation. Clean marine aerosol is generally smaller than dust in particle size, which weakens extinction effect. Apart from these, aerosol in lower layer could be deposited into the ocean during long-range transportation, the remaining part may concentrates at a high layer, which is also typical characteristics of Gaussian pattern. The North China Plain is usually under the effect of smoke, dust and polluted dust, besides, seasonal variation of aerosol type is complex, accordingly, factors contributing to the predominance of Exponential distribution remain unclear. More details are shown in Fig. 4.

DOI: 10.1051/epjconf/201611904011

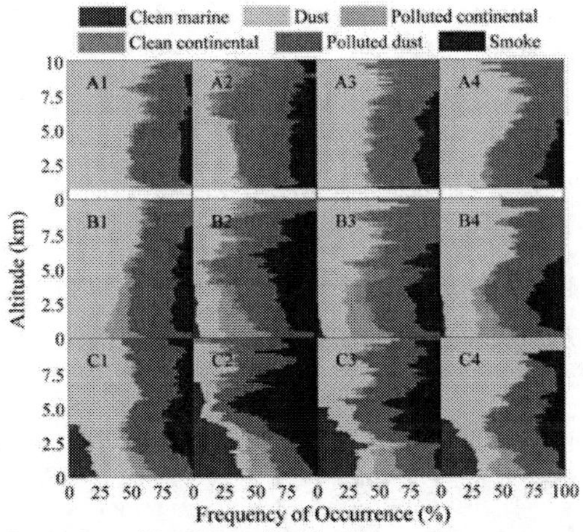

Fig. 4 Vertical distribution of occurrence frequency for each aerosol type. Characters in the top left corner have the same meanings as that of Fig. 3

The predominant ECP patterns in the three selected regions have certain representativeness, and they help to reduce uncertainties in aerosol model in relevant researches. However, it is noteworthy that ECP is a comprehensive function of many factors including geography, synoptic circulation et al. Therefore, further researches should focus on validation of ECP patterns through systematical analyses and measurements.

ACKNOWLEDGEMENT

We owe our thanks to the NASA Langley Atmospheric Science Data Center (ASDC) for its help in elaborating and providing the CALIPSO aerosol product.

REFERENCES

[1] Xu, C., Ge, J, Huang, J., Fu, Q., Liu, H., Chen, B., 2014: Observations of dust aerosol over China based on CALIPSO spaceborne lidar, *J. Desert Res.*, **34(5)**, 1253-1362.

[2] Turner, D. D., Ferrare, R. A., Brasseur, L. A, 2001: Average aerosol extinction and water vapor profiles over the Southern Great Plains, *Geophys. Res. Lett.*, **28(23)**, 4441-4444.

[3] Zhang, J., Li, X., 2012: Vertical distribution of sand-dust aerosols and the relationships with atmospheric environment, *J. Arid Land,* **4(4)**, 357-368.

[4] Yu, H., Chin, M., Winker, D. M., Omar, A. H., Liu, Z., Kittaka, C., Diehl, T. ,2010): Global view of aerosol vertical distributions from CALIPSO lidar measurements and GOCART simulations: Regional and seasonal variations, *J. Geophys. Res.: Atmospheres,* 115(D4): 1984-2012.

[5] Winker, D. M., Vaughan, M. A., Omar, A., Hu, Y., Powell, K. A., Liu, Z., Hunt, W. H., Young, S. A., 2009: Overview of the CALIPSO mission and CALIOP data processing algorithms, *J. Atmos. Ocean. Tech.,* **26(11)**, 2310-2323.

[6] Wu, Y., Cordero, L., Gross, B., Moshary, F., Ahmed, S., 2014: Assessment of CALIPSO attenuated backscatter and aerosol retrievals with a combined ground-based multi-wavelength lidar and sunphotometer measurement, *Atmos. Environ.*, **84**, 44-53.

[7] Huang, J., Minnis, P., Chen, B., Huang, Z., Liu, Z., Zhao, Y., Yi, Y., Ayers J. K., 2008: Long-range transport and vertical structure of Asian dust from CALIPSO and surface measurements during PACDEX. *J. Geophys. Res.*, **113(D23212)**, 1-13

EPJ Web of Conferences **119**, 04012 (2016) DOI: 10.1051/epjconf/201611904012
ILRC 27

Laser Remote Sensing from ISS: CATS Cloud and Aerosol Level 2 Data Products (Heritage Edition)

Sharon Rodier[1,3*], Steve Palm[2,4], Mark Vaughan[3], John Yorks[2,4], Matt McGill[4], Mike Jensen[1,3], Tim Murray[1,3], Chip Trepte[3]

[1]*Science Systems and Applications Inc., Hampton, VA USA *Email: sharon.d.rodier@nasa.gov*

[2]*Science Systems and Applications Inc., Greenbelt, MD, USA*

[3]*NASA, Langley Research Center, Hampton, VA, USA*

[4]*NASA, Goddard Space Flight Center, Greenbelt, MD USA*

ABSTRACT

With the recent launch of the Cloud-Aerosol Transport System (CATS) we have the opportunity to acquire a continuous record of space-based lidar measurements spanning from the Cloud-Aerosol Lidar and Infrared Pathfinder Satellite Observations (CALIPSO) era to the start of the EarthCARE mission. Utilizing existing well-validated science algorithms from the CALIPSO mission, we will ingest the CATS data stream and deliver high-quality lidar data sets to the user community at the earliest possible opportunity. In this paper we present an overview of procedures necessary to generate CALIPSO-like lidar level 2 data products from the CATS level 1 data products.

1. INTRODUCTION

The Cloud-Aerosol Transport System (CATS) instrument was developed at NASA's Goddard Space Flight Center (GSFC) and deployed to the International Space Station (ISS) on 10 January 2015. CATS is mounted on the Japanese Experiment Module's Exposed Facility (JEM_EF) and will provide near-continuous, altitude-resolved measurements of clouds and aerosols in the Earth's atmosphere. The CATS ISS orbit path provides a unique opportunity to capture the full diurnal cycle of cloud and aerosol development and transport, allowing for studies that are not possible with the lidar aboard the CALIPSO platform, which flies in the sun-synchronous A-Train orbit. The CATS instrument consists of two high repetition rate lasers operating at two wavelengths (1064 nm and 532 nm) with four configurable fields of view (FOV). By combining a specific laser and FOV(s) the CATS instrument can be configured for three primary operational modes (see [1] for details). In science mode 1, laser 1 measures elastic backscatter and volume depolarization in a dual FOV mode (left FOV and right FOV) at 1064 nm and 532 nm. In science mode 2, laser 2 measures elastic backscatter and volume depolarization at 1064 nm and makes High Spectral Resolution Lidar (HSRL) measurements at 532 nm via the forward FOV. The final configuration is science mode 3, which uses laser 2 to measure elastic backscatter and volume depolarization at 532 and 1064 nm via the aft FOV.

One of the primary science objectives of CATS is to continue the CALIPSO aerosol and cloud profile data record to provide continuity of lidar climate observations during the transition from CALIPSO to EarthCARE. To accomplish this, the CATS project at NASA's Goddard Space Flight Center (GSFC) and the CALIPSO project at NASA's Langley Research Center (LaRC) are closely collaborating to develop and deliver a full suite of CALIPSO-like level 2 data products that will be produced using the newly acquired CATS level 1B data whenever CATS is operating in science modes 1 or 3. The CALIPSO mission is now well into its ninth year of on-orbit operations, and has developed a robust set of mature and well-validated science algorithms to retrieve the spatial and optical properties of clouds and aerosols from multi-wavelength lidar backscatter signals [2]. By leveraging both new and existing NASA technical resources, this joint effort by the CATS and CALIPSO teams will deliver validated lidar data sets to the user community at the earliest possible opportunity. The science community will have access to two sets of CATS Level 2 data products. The "Operational" data products will be produced by the GSFC CATS

DOI: 10.1051/epjconf/201611904012

team utilizing the new instrument capabilities (e.g., multiple FOVs and 1064 nm depolarization), while the "Heritage" data products created using the existing CALIPSO algorithms and the CATS 532 nm channels and the total 1064 nm channel.

Throughout this abstract we outline the development of the CATS Level 2 Heritage (L2H) software and data products and describe the modifications made to the ingested CATS Level 1B data stream and the CALIPSO Level 2 processing algorithms in order to successfully interface two disparate data processing systems. Simulated CATS Level 1B data provided an initial assessment of CATS Heritage Level 2 data quality.

2. METHODOLOGY

Producing high quality CALIPSO-like Level 2 products requires addressing differences between the instrument sampling rates, vertical resolution, and signal-to-noise (SNR) characteristics. To ingest the CATS data, the 532 nm and 1064 nm attenuated backscatter channels must be converted from single shot profiles sampled at ~350 m horizontally with a uniform 60 m vertical resolution spanning a total of 30 km (28 km to -2 km), to the CALIPSO 5 km downlinked major frame structure that consists of 15 profiles sampled at ~330 m horizontally that span 42 km (40 km to -2 km) vertically and are averaged onboard both vertically and horizontally to different spatial resolutions that vary with range from the lidar. Additional required parameters such as molecular and ozone number density are also calculated and ancillary instrument parameters are extracted to create required CALIPSO parameters such as QC Flags, noise scale factors, and RMS background measurements. A set of four software modules (called Product Generation Executables or PGEs) were constructed to accomplish these tasks:

1) Ingest/Archive: receives CATS data files from GSFC and archives them for ordering by the public via the LaRC Atmospheric Sciences Data Center.

2) L1B Pre-Processor: converts CATS L1B into CALIPSO-like L1B by applying CALIPSO's onboard averaging scheme to the CATS data.

3) L2 FOV Merge: while each CATS FOV is processed separately by the Heritage software, the results from both FOVs are reported in single L2H data files. This is consistent with the CATS L1B files, which also report all FOVs in a single file for each orbit segment.

4) CATS Lidar Browse Image creation: generates images of the CATS L1B and L2H data products for display on the CATS web pages.

Figure 1 illustrates the Heritage processing flow for Science Mode 1 from ingest to pre-processing, execution of the L2 algorithms, to post-processing and finally archival of the CALIPSO-like data sets. Science Mode 3 follows a similar flow, but does not require the FOV Merge PGE since CATS data is acquired using only a single FOV.

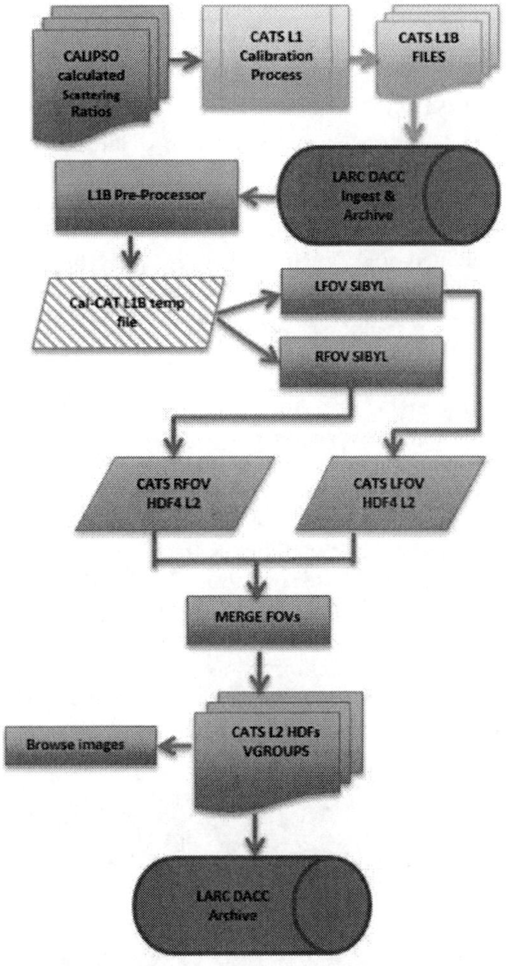

Figure 1: CATS Heritage Processing Flow for Science Mode 1

Due to SNR differences between the two instruments, turning of the detection levels threshold will be required to accurately perform feature classification of the clouds and aerosols in each scene. Preliminary results from the first CATS downlinks show that the CATS SNR is higher than the CALIPSO SNR at night, especially at 1064 nm. However, the CATS daytime SNR is lower than CALIPSO and, for reasons not yet fully understood, somewhat lower than prelaunch predictions. With respect to the Level 2 Heritage products, these differences are expected to have minimal impact on the quality of the layer detection and optical properties results reported in the L2H products, largely due to the extensive spatial averaging done by the CALIPSO retrieval algorithms.

3. RESULTS

The development and validation of the Heritage data products benefited substantially from high quality CATS L1B simulation data. Multiple orbits created from the meteorological products generated by NASA's Global Modeling and Assimilation Office (GMAO) provided realistic scenes of clouds and aerosol at varying heights and optical depths throughout the atmosphere. These simulations provided the baseline examples of day-night SNR differences that were used to determine the prelaunch configuration (e.g., layer detection thresholds) of the CATS Heritage retrieval system. Figures 2a, 2b and 2c show an extended cirrus cloud deck above boundary layer aerosols during a continuous transition from daytime (left side) to nighttime (right side) data acquisition. While the differences in SNR are plainly evident (Figures 2a and 2b), the retrieval algorithms are uniformly successful in properly locating and identifying the layers, irrespective of lighting conditions (Figure 2c).

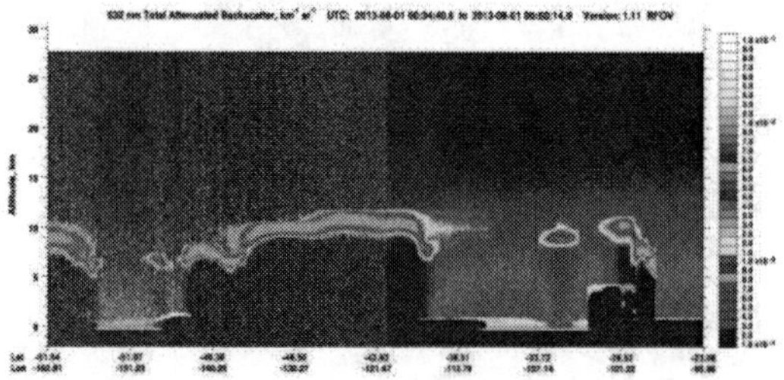

Figure 2a: Simulated CATS L1B 532 nm total attenuated backscatter showing cirrus clouds above boundary layer aerosols.

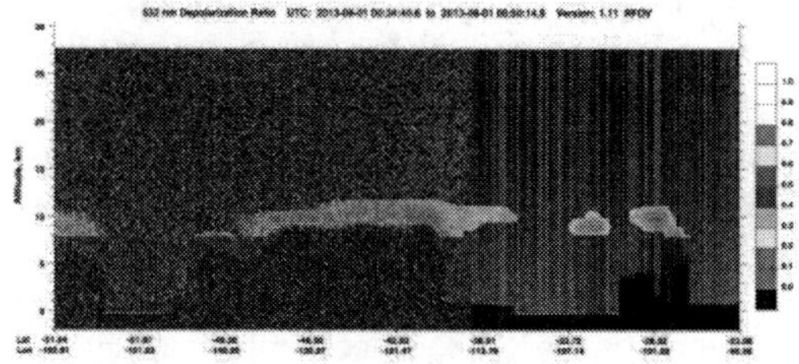

Figure 2b: CATS L1B 532 nm volume depolarization ratios

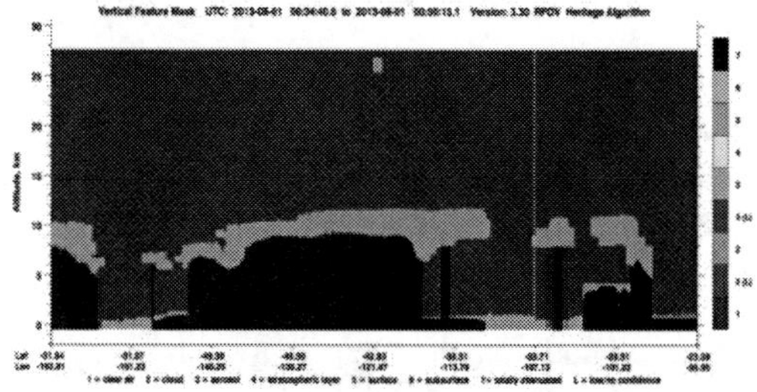

Figure 2c: CATS L2H vertical feature mask showing locations of cloud (light blue) and aerosol (orange) layers identified in the CATS. The high altitude (~26 km) layer at ~43°S is a false positive introduced by high levels of solar background noise.

The CATS Lidar Browse Image PGE will produce all the standard L1 and L2 images generated by the CALIPSO analyses. In additional, some CATS specific images such as plots of the 1064 nm volume depolarization will also be produced. Figures 3a–3e show a subset of the CATS L1 and L2 browse products that will be available to the science community.

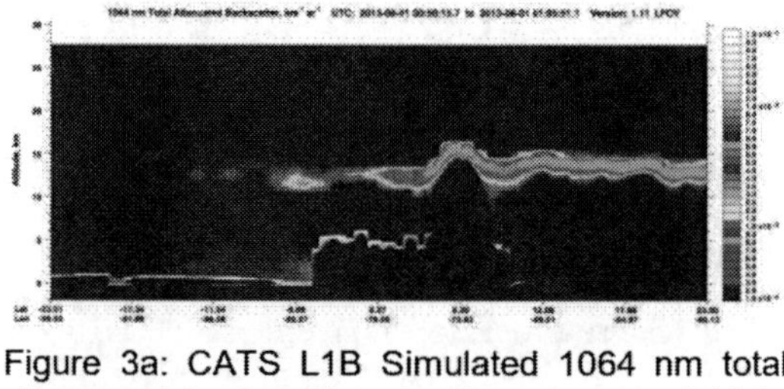

Figure 3a: CATS L1B Simulated 1064 nm total attenuated backscatter showing cirrus and high-altitude aerosol lying over (left) a stratus deck embedded in a height-varying aerosol layer and (center) midlevel opaque water clouds.

DOI: 10.1051/epjconf/201611904012

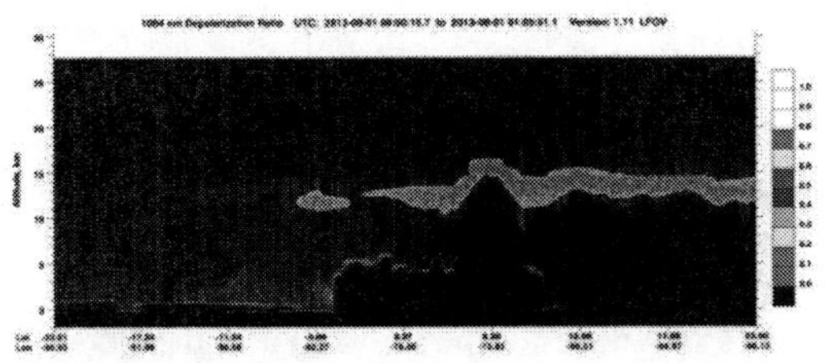

Figure 3b: CATS L1B 1064 nm volume depolarization ratios

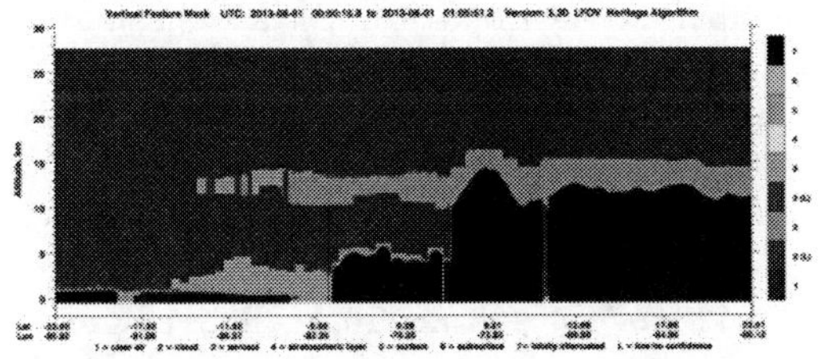

Figure 3c: CATS L2H vertical feature mask for the data shown in Figures 3a and 3c

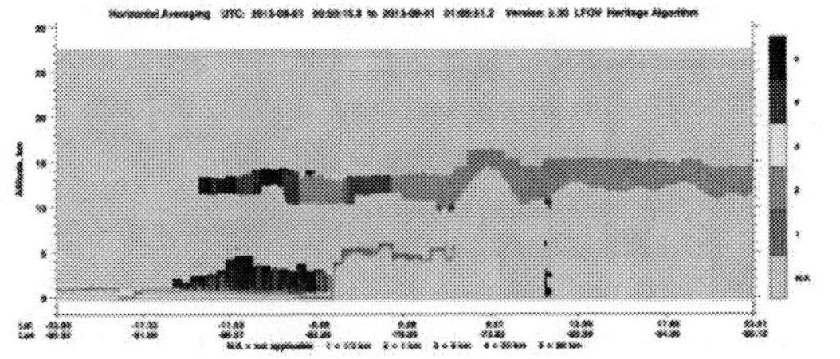

Figure 3d: CATS L2H horizontal averaging required for layer detection

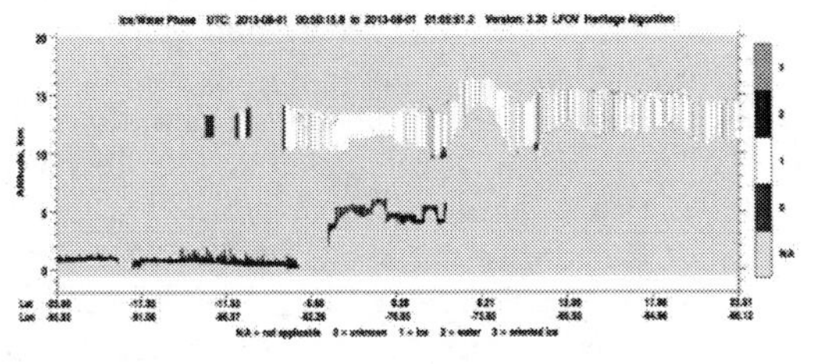

Figure 3e: CATS L2H ice-water phase

4. CONCLUSIONS

The first CATS data was telemetered back to Earth on 10 February 2015, and as of this writing the GSFC CATS team is completing their initial post-launch verifications of their data handling, geolocation and instrument calibration algorithms and code. Once these quality assessment procedures have been successfully completed, routine transfer of the CATS L1B files from GSFC to LaRC will begin.

The CATS L2H data products shown in the images above were generated with a wholly unmodified version of the CALIPSO Level 2 Version 3.30 analysis software. While this software is quite capable of handing the CATS simulated data, some minor modifications may be required to best accommodate the real world data stream. Furthermore, a certain amount of configuration and quality assurance testing remains to be done to optimize L2H retrieval performance for the actual (as opposed to simulated) CATS data. Upon the conclusion of these activities, production of the CATS L2H data products will begin. At present, the LaRC CATS team is anticipating public release of the L2H products by early-to-mid summer of 2015.

REFERENCES

[1] Yorks, J. E., M. J. McGill, S. P. Palm, D. L. Hlavka, P. A. Selmer, E. P. Nowottnick, M. A. Vaughan and S. D. Rodier, 2015: An Overview of the Cloud-Aerosol Transport System (CATS) Processing Algorithms and Data Products, *Proceedings of the 27th International Laser Radar Conference (ILRC 27)*

[2] Winker, D. M., M. A. Vaughan, A. H. Omar, Y. Hu, K. A. Powell, Z. Liu, W. H. Hunt, and S. A. Young, 2009: "Overview of the CALIPSO Mission and CALIOP Data Processing Algorithms", *J. Atmos. Oceanic Technol.*, **26**, 2310–2323, doi:10.1175/2009JTECHA1281.1.

CALIOP CALIBRATION: VERSION 4.0 ALGORITHM UPDATES

Brian J. Getzewich[1*], Jason L. Tackett[1], Jay Kar[1], Anne Garnier[1], Mark A. Vaughan[2], Bill Hunt[1]

*[1]SSAI, 1 Enterprise Parkway, Suite 200, Hampton, VA 23666, USA, *Email: brian.j.getzewich@nasa.gov*
[2]NASA, MS 475, NASA Langley Research Center, Hampton, VA 23681, USA

ABSTRACT

The Cloud-Aerosol Lidar with Orthogonal Polarization (CALIOP) lidar, onboard the Cloud-Aerosol Lidar and Infrared Pathfinder Satellite Observation (CALIPSO) satellite, has been providing a near continuous record of high-resolution vertical profiles of clouds and aerosols properties since the summer of 2006. Key to the generation of these vertical profiles is proper calibration of the 532 nm and 1064 nm channels. This abstract summarizes improvements to the calibration techniques used to calibrate the 532 nm and 1064 nm signals for the recent version 4 (V4) Lidar Level 1 data release.

1. INTRODUCTION

Because of differences in signal-to-noise ratios and intra-orbit thermal stability, CALIOP uses different techniques to calibrate the 1064 nm measurements and the 532 nm daytime and nighttime measurements. The 532 nm nighttime calibration procedure [1] uses a high-altitude molecular normalization technique [2] in which the calibration is determined by taking the ratio of the measured signal to the expected signal computed using an atmospheric model. This approach assumes that all constituents of the nighttime normalization region (i.e., including aerosol loading) can be accurately modeled or characterized. Unfortunately, the same technique cannot be used during daytime, because the signal-to-noise ratio (SNR) is not sufficient due to the influence of background solar radiation. Instead, the 532 nm daytime calibration coefficients are derived by matching the daytime 'clear air' scattering ratios (i.e., the ratio between measured backscatter and modeled molecular signal) to previously calibrated nighttime clear air scattering ratios measured at the same latitude. For the version 3 (V3) data, the clear air calibration transfer region was located between 8 km and 12 km globally [3]. Similarly, the 1064 nm channel does not have sufficient molecular backscatter to allow referencing to clear air at high altitudes, so instead, a scale factor is computed to match the 1064 nm signals from dense cirrus clouds to the corresponding 532 nm signals, with the assumption that the 1064 nm/532 nm particulate backscatter ratio should be close to 1. The 1064 nm calibration coefficient is then the product of this 1064 nm scale factor and the previously determined 532 nm calibration coefficient [4].

The new calibration techniques used for V4 retain the underlying scientific basis established in V3, but with several key modifications. For the 532 nm nighttime calibration, the height of the calibration altitude was increased from 30-34 km to 36-39 km in order to mitigate the influence of stratospheric (volcanic) aerosols [5]. The 532 nm daytime continues to match day to night scattering ratios in a calibration transfer region, but now this region has been raised to a higher altitude in V4 to avoid clouds and increase the number of samples used in calculating daytime calibration coefficients. The V4the transfer region is still 4 km deep, but it now follows the 400K isentropic surface which is typically between 13-18 km, always above the tropopause and above the meteorologically active part of the atmosphere [6], thus better satisfying the assumption that there is no diurnal variability of the scattering ratios in the transfer region. For the 1064 nm calibration, the scale factors are in V4 calculated and applied as a function of the granule elapsed time, to better represent the relative variations of the 532 nm and 1064 nm signals along daytime or nighttime orbit segments. The criteria for selecting the dense cirrus clouds that are used in the scale factor calculation have also been changed significantly.

2. AVERAGING & ERROR MITIGATION

To compensate for the significant reduction in the SNR at the higher calibration altitude regions, the number of orbits required for calibration was increased. To keep the random uncertainty at or below standards established in the V3 release requires averaging nighttime data over 11 orbits and daytime data over 105 orbits (~7 days). The increase in the number of orbits applied also compensates for nighttime signal spikes due to the South Atlantic Anomaly.

3. 532 NM NIGHTTIME RESULTS

Figure 1 shows nighttime monthly mean 532 nm calibration coefficients for October 2010 computed for version 3 (left) and for the V4 algorithm (right). In the V3 the presence of the stratospheric aerosol layer in the tropics is clearly visible as a region of high calibration coefficients located between the equator and 20°N. The South Atlantic Anomaly (SAA) is also visible as an oval of reduced calibration coefficients centered at ~20°S and ~40°W. By contrast, neither of these geophysical features appears in the V4 data. In both images, the diminution of the calibration coefficients in the southern hemisphere, as the satellite approaches the night-to-day terminator, indicates changes in the on-board thermal environment that perturb the alignment between the laser transmitter and receiver.

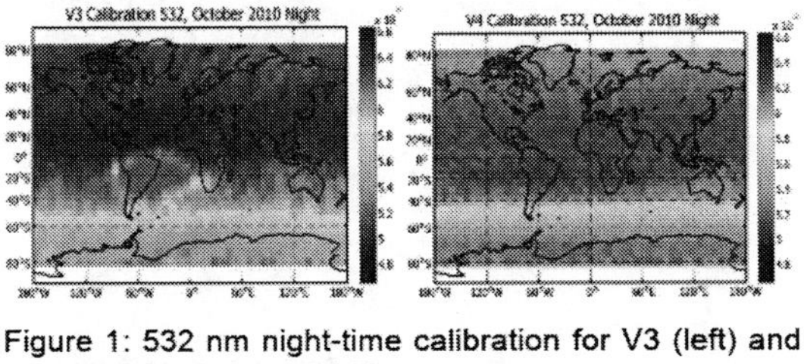

Figure 1: 532 nm night-time calibration for V3 (left) and V4 (right) for October 2010.

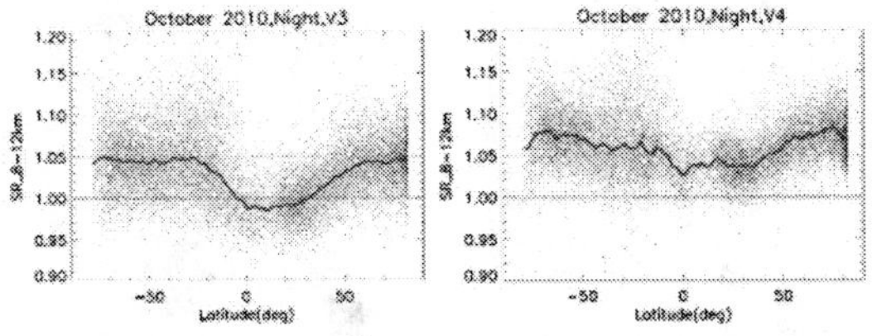

Figure 2: 532 nm night-time scattering ratios derived from clear-air at 8-12km for version 3 (left) and version 4 (right) for October 2010.

The median reduction in the calibration coefficients from V3 to V4 is ~3%. This in turn leads to an increase in the total attenuated backscatter coefficients by the same amount. The attenuated scattering ratios in the 30–34km region are now increased by up to 5% over previous versions, indicating seasonal variations which are consistent with the independent predictions [4]. Stratospheric aerosol loading at these altitudes is now clearly captured by CALIOP measurements, showing up as spatial structures which are consistent with stratospheric dynamics. Similarly, the clear-air scattering ratios at 8–12 km showed an anomalous dip (scattering ratios < 1) in the tropics in V3 (figure 2, left), and are all correctly greater than one in V4 (figure 2, right).

4. 532 NM DAYTIME RESULTS

In the V3 release, the agreement between day and night clear-air attenuated scattering ratios at high altitudes was poor and varied with latitude, possibly due to diurnal variability in aerosol loading and clouds reducing the potential number of clear-air samples in the 8-12km V3 calibration transfer region (figure 3, top). V4 (figure 3, bottom) shows a marked improvement in the agreement between day and night clear-air attenuated scattering ratios above the V4 calibration transfer region (to within 1 ± 3%), demonstrating (1) the validity of assuming that there is no diurnal variability in stratospheric aerosol loading at these altitudes, (2) that the day/night agreement of clear-air attenuated scattering ratios holds even above the night to day calibration transfer region where they are forced to agree, and (3) the V4 calibration procedure adjusts for differences in day and night instrument behavior.

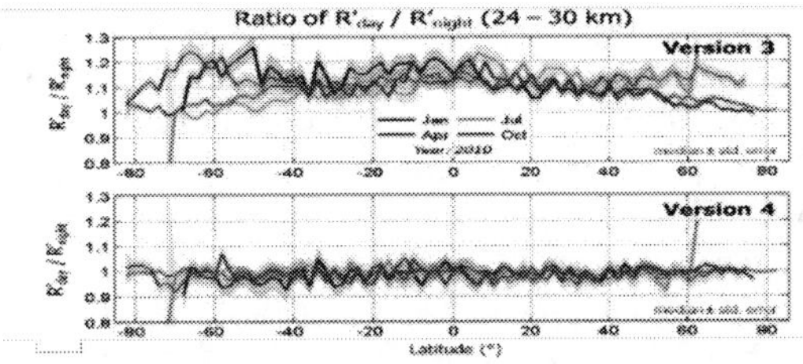

Figure 3: ratio of day to night clear-air attenuated scattering ratios at 24-30km for V3 (top) and V4 (bottom) for January, April, July and October 2010. The South Atlantic Anomaly has been excluded.

Figure 4 shows the 532 nm daytime calibration (top) and ratio between V3 and V4 (bottom) over 79 months. Discontinuities in the data reflect those instances when calibration could not be performed due to a change of state of the instrument (e.g., bore-sight alignments) or when there was an appreciable gap in the data record (e.g., solar flares). Seasonal differences in V3/V4 calibration at high latitudes are attributable to differences in anchoring neighboring day-to-night orbits.

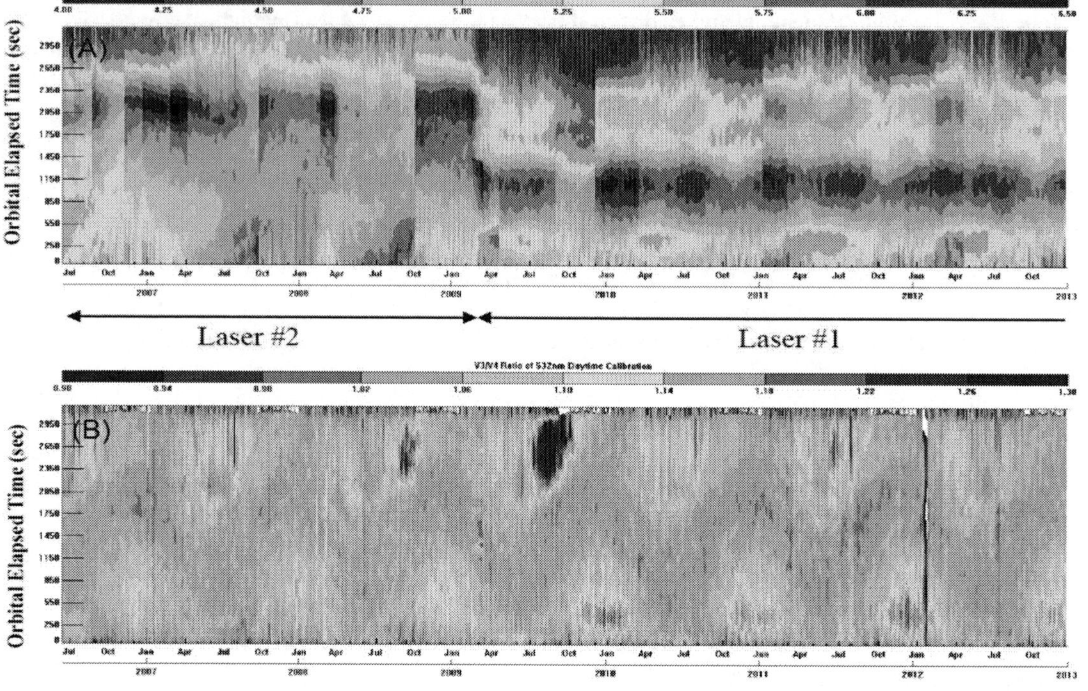

Figure 4: 532 nm daytime calibration (top) and Version 3 / Version 4 ratio of 532 nm daytime calibration from June 6, 2006 to December 31, 2012.

5. 1064 NM RESULTS

Figure 5 illustrates the changes in the 1064 nm scale factor between V3 (left) and V4 (right) for both nighttime data (top row) and daytime data (bottom row) for the month of October 2010. The V3 scale factors show a large discontinuity between the very high night-time values and the very low day-time values, along with abrupt changes in value between adjacent orbit segments. The V4 scale factors, on the other hand, vary smoothly as a function of orbit elapsed time, with continuous values at both terminators. The magnitudes of the changes from V3 to V4 in the nighttime mean scale factors range between 20% lower (northern hemisphere) to 7% higher (southern hemisphere), with the median nighttime difference being 6% lower. The daytime mean scale factors range between 14% lower (northern hemisphere) to 19% higher (southern hemisphere), with the median daytime difference being 3% lower. Figure 6 illustrates the changes in the 1064 nm calibration coefficients between V3 (left) and V4 (right) for both nighttime data (top row) and daytime data (bottom row) for the month of October 2010.

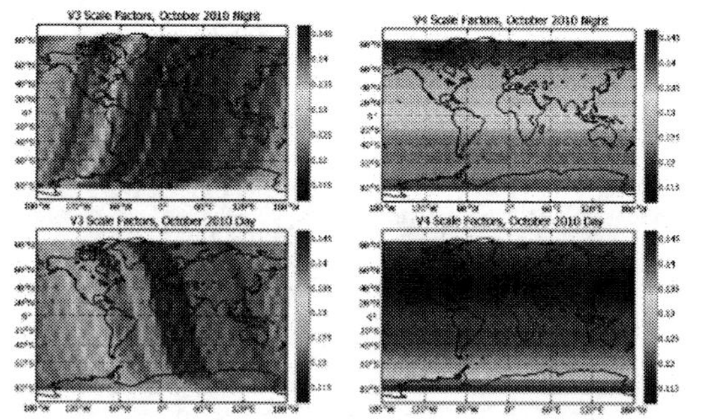

Figure 5: Monthly mean 1064 nm calibration scale factors for October 2010, with V3 shown on the left and V4 on the right. Upper panels show nighttime data; lower panels show daytime data.

The data in the top row of Figure 6 represents the 532 nm nighttime calibration coefficients shown in Figure 1 multiplied by the nighttime 1064 nm calibration scale factors shown in the top row of Figure 5. Similarly, the data in the bottom row of Figure 6 represents the 532 nm daytime calibration coefficients shown in Figure 1 multiplied by the daytime 1064 nm calibration scale factors shown in the bottom row of Figure 5.

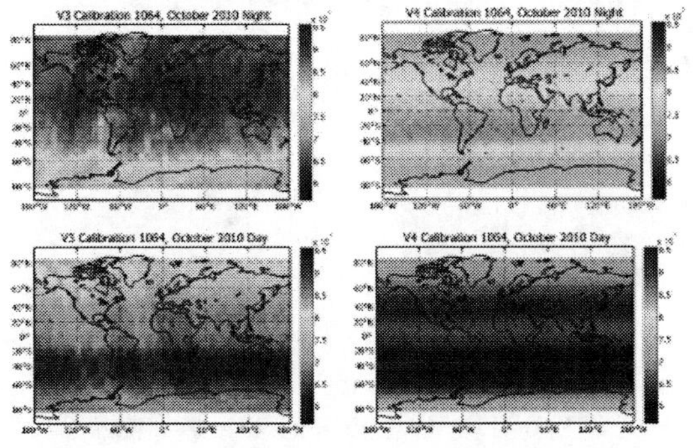

Figure 6: Monthly mean 1064 nm calibration coefficients for October 2010. V3 calibration coefficients are shown on the left; V4 calibration coefficients are shown on the right. The upper panels show nighttime data and the lower panels show daytime data

The V3 1064 nm calibration coefficients exhibit spatial artifacts and irregularities that arise from both the V3 532 nm calibration coefficients and from the V3 1064 nm scale factors. In contrast to the V3 data, the V4 1064 nm calibration coefficients produced by the new calibration and scale factor algorithms are seen to vary smoothly both latitudinal and longitudinally. The magnitudes of the changes from V3 to V4 in the nighttime mean 1064 nm calibration coefficients range between 22% lower (northern hemisphere) to 8% higher (southern hemisphere), with the median nighttime difference being 8% lower. The daytime mean 1064 nm calibration coefficients range between 20% lower (northern hemisphere) to 10% higher (southern hemisphere), with the median daytime difference again being 8% lower.

REFERENCES

[1] Russell, P. B. et al., 1979: "Methodology for error analysis and simulation of lidar aerosol measurements. *Appl. Opt.,* **18**, 3783-3797.

[2] Powell, K. A. et al., 2009: "CALIPSO Lidar Calibration Algorithms: Part I - Nighttime 532 nm Parallel Channel and 532 nm Perpendicular Channel", *J. Atmos. Oceanic Technol.*, **26**, 2015–2033, doi:10.1175/2009JTECHA1242.1.

[3] Powell, K. A. et al., 2010: "The CALIOP 532-nm Channel Daytime Calibration: Version 3 Algorithm", Proceedings of the 25th International Laser Radar Conference, 5–9 July 2010, St.-Petersburg, Russia, 1367-1370, ISBN 978-5-94458-109-9.

[4] Vaughan, M. A. et al., 2012: "Chaos, Consternation and CALIPSO Calibration: New Strategies for Calibrating the CALIOP 1064 nm Channel", in Reviewed and Revised Papers Presented at the 26th International Laser Radar Conference (ILRC 2012), 25-29 June 2012, Porto Heli, Greece, 655–658.

[5] Vernier, J.-P. et al., 2009: "Tropical stratospheric aerosol layer from CALIPSO lidar observations", *J. Geophys. Res.*, **114**, D00H10, doi:10.1029/2009JD011946.

[6] Hoskins, B. J., 1991: "Towards a PV-θ view of the general circulation", *Tellus*, 43AB, 27–35, doi:10.1034/j.1600-0889.1991.t01-3-00005.x.

EPJ Web of Conferences 119, 05001 (2016)
ILRC 27
DOI: 10.1051/epjconf/201611905001

CONTINUOUS TIME SERIES OF WATER VAPOR PROFILES FROM A COMBINATION OF RAMAN LIDAR AND MICROWAVE RADIOMETER

Andreas Foth[1], Holger Baars[2], Paolo Di Girolamo[3], Bernhard Pospichal[1]

[1]*Leipzig Institute for Meteorology (LIM), University of Leipzig, Leipzig, Germany,*
Email: andreas.foth@uni-leipzig.de
[2]*Leibniz Institute for Tropospheric Research (Tropos), Leipzig, Germany*
[3]*Scuola di Ingegneria, Università degli Studi della Basilicata, Potenza Italy*

ABSTRACT

In this paper, we present a method to retrieve continuous water vapor profiles from a combination of a Raman lidar and a microwave radiometer. The integrated water vapor from the microwave radiometer is used to calibrate the Raman lidar operationally resulting in small biases compared to radiosondes. The height limitations for Raman lidars (cloud base and daylight contamination) can be well compensated by the application of a two–step algorithm combining the Raman lidars mass mixing ratio and the microwave radiometers brightness temperatures.

1. INTRODUCTION

Water vapor plays a key role in the energy budget and in the description of the thermodynamic state. Moreover, it is the most important green house gas. Besides its high spatio-temporal variability, water vapor is involved in many atmospheric processes e.g. cloud formation or the transport of latent heat. This complicates the implementation of water vapor in climate or weather prediction models. To achieve an extensive comprehension spatially and temporally resolved measurements are necessary. In the framework of the German research project High Definition Clouds and Precipitation for advancing Climate Prediction "HD(CP)^2" the intense observation campaign HD(CP)^2 Observational Prototype Experiment (HOPE) was performed in western Germany [1]. During HOPE, both active and passive remote sensing systems observed the vertical water vapor distribution. An active method is given by the Raman lidar technique enabling high vertical resolution measurements of water vapor. However, these measurements are highly disturbed by clouds and daylight. Furthermore Raman lidars need to be calibrated with a simultaneously measuring instrument. All these three problems can be compensated by the use of a microwave radiometer (MWR). Passive microwave radiometry provides atmospheric water vapor observations with high temporal, but limited vertical resolution during both day and nighttime. However, the integrated water vapor (IWV) can be retrieved very accurately and can be used to calibrate the Raman lidar. In addition, clouds are semitransparent in the microwave region which allows an operation during all weather conditions except for precipitation.

The synergy of these instruments provides complementary information on the water vapor structure. Thus, when both Raman lidar and MWR are measuring collocated and simultaneously, continuous water vapor profiles can be obtained on a routine basis also during cloudy conditions. This paper has two major objectives which are presented after a brief introduction in the instrumentation and the lidar methodology. Firstly, the application of a calibration method which is only based on IWV from MWR in a very straightforward way to offer a broad application [2]. Secondly, the development of a two-step algorithm combining the Raman lidar mass mixing ratio and the MWR brightness temperature. Applying both steps, the Kalman filter and the optimal estimation method, water vapor from within and above a cloud can be retrieved.

2. INSTRUMENTATION

During HOPE ground–based remote sensing observations were conducted around Jülich in western Germany in April and May 2013. Besides the Jülich Observatory for Cloud Evolution (JOYCE) two further observation sites were temporarily set up 4 km around Jülich, the Leipzig Aerosol and Cloud Remote Observations System (LACROS) and the Karlsruhe Institute for Technologies (KIT) site. The instruments used in this study will be briefly explained in the following subsections.

2.1. POLLY^{XT}

The lidar measurements at the LACROS site were performed with the fully automatic portable multiwavelength Raman and polarization lidar Polly^{XT} by the Leibniz Institute for Tropospheric Research (TROPOS). The system receives backscattered light at 355, 532 and 1064 nm and Raman scattered light at 387, 407 and 607 nm wavelengths. Further informations are given in [3]. No water vapor observations were performed during daytime due to the high daylight background. The temporal and vertical resolutions amount to 30 s and 30 m, respectively.

2.2. BASIL

The University of Basilicata lidar system (BASIL) observations were conducted at the JOYCE site during the HOPE campaign. In contrast to Polly^{XT} BASIL performs high resolution and accurate measurements of atmospheric water vapor, in both day and nighttime due to its more powerful laser. A thorough description of the technical characteristics, measurement capabilities and performances is given in [4]. For water vapor measurements BASIL uses the same wavelengths as Polly^{XT}. The maximum resolutions are 1 s in time and 7.5 m in height, respectively, and can be traded-off to improve the measurement precision.

2.3. MICROWAVE RADIOMETER HATPRO

The humidity and temperature profiler (HATPRO) is a passive remote sensing instrument that measures atmospheric emission at two frequency bands in the microwave spectrum. Vertical humidity informations can be retrieved by observing at seven frequencies along the 22.235 GHz H_2O absorption line. Along the O_2 absorption complex from 51 to 58 GHz vertical temperature informations can be retrieved. To obtain atmospheric quantities such as IWV, statistical retrievals were used. Such retrievals are based on a multi-linear regression between atmospheric profiles and modeled brightness temperatures. Both MWRs from JOYCE and LACROS utilize the same retrieval algorithms based on a long term dataset of radiosondes.

3. AUTOMATED CALIBRATION

The following section dealing with the calibration of the Raman lidar mass–mixing ratio is based on [2]. Using the inelastic backscatter from molecular nitrogen (N_2) at 387 nm and from water vapor (H_2O) at 407 nm, the Raman lidar technique enables the determination of water vapor mixing ratio profiles [5]. The mass mixing ratio can be obtained by forming

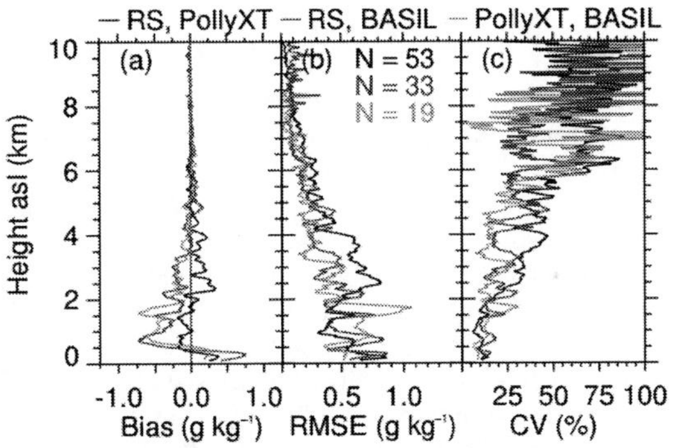

Figure 1: Figure 8. (a) Absolute bias between the radiosonde (RS) and Polly^{XT} (black), between RS and BASIL (red) and between Polly^{XT} and BASIL (green). (b) Root-mean-square error (RMSE) of the water vapour mixing ratio. The numbers indicate the sample size. (c) Coefficient of variation (CV, relative RMSE).

and rearranging the ratio of the Raman lidar equations for both signals:

$$m_{H_2O}(z) = C\,\frac{P_{\lambda_{H_2O}}(z)}{P_{\lambda_{N_2}}(z)}\,\frac{T_{\lambda_{N_2}}(z)}{T_{\lambda_{H_2O}}(z)}. \qquad (1)$$

In Eq. (1) identical overlap factors were assumed. The last term in Eq. (1) describes the differential transmission ratio at λ_{N_2} and λ_{H_2O} which is not explained in detail here. The second term indicates the signal ratio which is directly measured. Differences in the range–independent Raman backscatter cross sections for both channels are adsorbed within the calibration factor C which can be determined by comparison with a simultaneous measurement from a reference instrument. Usually, a linear regression between the water vapor mixing ratio from a radiosonde and the signal ratio P_{H_2O}/P_{N_2} from the lidar is used [6] and the calibration factor is defined as the slope of the regression line. However, in previous experiments radiosondes showed a significant sonde–to–sonde variability [7] as well as a dry bias [8]. Therefore, Raman lidars are often calibrated by using the IWV retrieved from a MWR. Using this method C is defined as the ratio between the IWV measured with the MWR and the integrated signal ratio from the lidar. C can be calculated from the mean of each 20 min clear sky interval.

The calibration factor from Polly^{XT} based on the IWV from MWR varies only slightly with a relative error of 5 % during the two month of HOPE. Such high stability of the calibration factor allows deriving Raman lidar water vapor profiles also during cloudy conditions. In these cases the calibration factor from the last 20 min clear sky interval can be applied. With this technique continuous water vapor profiles up to cloud base can be retrieved during all non–precipitating night cases. A statistical analysis of the absolute bias, the root mean square error (RMSE) and the coefficient of variation (CV) between the radiosonde and the two lidars is given in Fig. 1. For the comparisons with the radiosonde only clear sky nighttime measurements within less than 2 h before or after the radiosonde launch time are considered. The profiles are averaged over 20 min and are interpolated to the lidars height grid. For the comparison between both lidars only simultaneous 20 min averages are investigated. One has to consider that several lidar profiles were compared to one RS profile (e.g. lidar profiles from 21:20, 21:40 and 22:00 UTC to the radiosonde from 23:00 UTC). The biases are largest in the planetary boundary layer with maximum values values of about 0.75 g kg^{-1} mostly induced by inhomogeneous conditions at the different measurement locations (distances about 4 km). Above the planetary boundary layer the bias converges to zero. The RMSE decreases with height and the CV increases with height due to the decreasing amount of water vapor. Finally, these low biases confirm the application of the MWR to calibrate the Raman lidar resulting in rather accurate profiles.

4. RETRIEVAL

After automatically calibrating the lidar measurement, the two–step algorithm is applied which is sketched in Fig. 2. If the calibrated profiles are truncated at the cloud base (determined from the lidar 1064 nm channel) they need to be expanded to the full height range. This is done by the Kalman filter [10] using a combination of the current calibrated lidar measurement and an estimated state which is the time projection of a previous analyzed state. The resulting filtered state serves as input to the one–dimensional variational (1D–VAR) algorithm. The optimum estimation of the atmospheric state by a given MWR measurement of brightness temperatures ($\mathbf{z}$) and the filtered state as a priori ($\mathbf{x}^{\mathrm{F}}$) can be found by minimizing a cost function of the form [10]:

$$\begin{aligned} \mathrm{J}(\hat{\mathbf{x}}) &= [\hat{\mathbf{x}} - \mathbf{x}^{\mathrm{F}}]^{T}\,\mathbf{P}^{\mathrm{F}}\,[\hat{\mathbf{x}} - \mathbf{x}^{\mathrm{F}}] \qquad (2)\\ &+ [\mathbf{z} - \mathbf{F}(\hat{\mathbf{x}})]^{T}\,\mathbf{P}_{\mathrm{z}}^{-1}\,[\mathbf{z} - \mathbf{F}(\hat{\mathbf{x}})] + \mathbf{J}_{\mathrm{p}}. \end{aligned}$$

The time index k is omitted here for clarity. $\hat{\mathbf{x}}$ is the optimal estimate of the atmospheric state. $\mathbf{P}^{\mathrm{F}}$ and $\mathbf{P}_{\mathrm{z}}$ denote the error covariance matrices of the filtered state and the MWR measurement, respectively. $\mathbf{J}_{\mathrm{p}}$ indicates a penalty term to avoid supersaturation. The forward model operator $\mathbf{F}(\mathbf{x})$ calculates the brightness temperatures from a given state $\mathbf{x}$. For the modeling of the brightness temperatures from a given profile a non–scattering microwave radiative transfer model is used for gas absorption [11] and liquid water absorption [12]. For the calculation of liquid water a modified adiabatic assumption is used [13]. $\mathrm{J}(\hat{\mathbf{x}})$ is minimized iteratively using the Levenberg-Marquardt method [10]. The iteration starts with a first guess which is based on the average of 225 radiosonde profiles (during HOPE) and stops after the convergence criterion is reached. In summary, the filtered profiles are modified such that the modeled brightness temperatures match those measured with the MWR. Figure 3 illustrates the retrieved mixing ratio profiles (a) and associated errors (b) from a Polly^{XT} observation on 22 April 2013. The time resolution amounts to 5 min. The solid white line indicates the truncation height. While only using the lidar no water vapor observation from above the truncation height is available due to daylight contamination between 18:00 and about 19:30 UTC and strong signal attenuation within clouds (dotted white line). But in

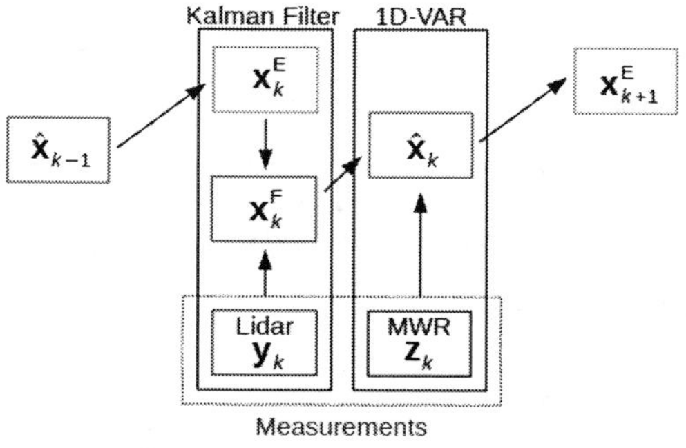

Figure 2: Sketch of the retrieval scheme. Starting with an analyzed state $\hat{\mathbf{x}}_{k-1}$ which is projected in time to the estimated state $\mathbf{x}_k^{\mathrm{E}}$. This state is then combined with the current lidar measurement $\mathbf{y}_k$ to the filtered state $\mathbf{x}_k^{\mathrm{F}}$ using the Kalman filter. $\mathbf{x}_k^{\mathrm{F}}$ is then used as the a priori input to the one–dimensional variational (1D–VAR) assimilation scheme. The a priori profile is modified such that the modeled brightness temperature match those measured with the microwave radiometer (MWR) $\mathbf{z}_k$ resulting in the most probable estimated state $\hat{\mathbf{x}}_k$ which is again projected in time in the consecutive step. This figure is adapted from [9].

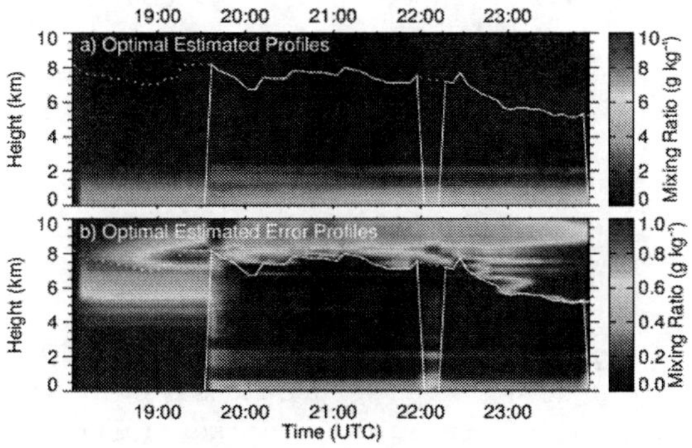

Figure 3: (a) Height–time display of the optimal estimated mixing ratio profiles derived from $Polly^{XT}$ observations on 22 April 2013. (b) Associated Errors. The solid white line indicates the truncation height and the dotted line indicates the cloud base.

combination with a microwave radiometer and by applying the two–step algorithm water vapor information can also be retrieved from within or above the cloud and during strong daylight contamination. Between 18:00 and 19:30 UTC the retrieved profiles are driven by the first guess. Therefore, no precise layers can be retrieved. At around 22 UTC the profiles are truncated due to the routine depolarization calibration of the lidar. The larger errors near the surface are caused by the manual increase of the lidar measurement error to account for the overlap problem. Above the cloud base an enhanced a priori error is applied.

5. OUTLOOK

With the presented automatic algorithm continuous water vapor profiles can be retrieved during all non-precipitating conditions on a routine basis but with limited maximum lidar measurement height during daytime. In the next steps we will evaluate the retrieval with radiosonde launches. We plan to apply the retrieval to the data sets of HOPE and LACROS in Leipzig. This technique in combination with Doppler lidar data enables the investigation of turbulent entrainment of water vapor at cloud boundaries. In the future this algorithm can be implemented in the Cloudnet data processing [14].

REFERENCES

[1] Macke, A. 2014, HOPE—A German intensive field campaign to capture the spatiotemporal variability of the thermodynamics, energetics, and microphysics of the cloudy troposphere with high resolution, *14th Conference on Atmospheric Radiation*, J6.4, Boston, MA, USA, July 07 - 11, 2014.

[2] Foth, A. et al. 2015, Water vapour profiles from Raman lidar automatically calibrated by microwave radiometer data during HOPE, *Atmos. Chem. Phys. Discuss.* **15**, 6567–6599.

[3] Althausen, D. et al. 2009, Portable Raman lidar PollyXT for automated profiling of aerosol backscatter, extinction, and depolarization, *J. Atmos. Oceanic Technol.* **26**, 2366–2378.

[4] Di Girolamo, P. et al. 2009, Multiparameter Raman lidar measurements for the characterization of a dry stratospheric intrusion event, *J. Atmos. Oceanic Technol.* **26**, 1742–1762.

[5] Whiteman, D. N. 2003, Examination of the traditional Raman lidar technique. II. Evaluating the ratios for water vapor and aerosols, *Appl. Opt.* **42**, 2593–2608.

[6] England, M. N. et al. 1992, Atmospheric water vapor measurements: Comparison of microwave radiometry and lidar, *J. Geophys. Res.* **97**, 899–916.

[7] Nash, J. et al. 2005, WMO intercomparison of high quality radiosonde systems: Final report, *WMO Rep.* p. 118 pp.

[8] Turner, D. D. et al. 2003, Dry Bias and Variability in Vaisala RS80-H Radiosondes: The ARM Experience, *J. Atmos. Oceanic Technol.* **20**, 117–132.

[9] Schneebeli, M. 2009, Advancements in Ground-Based Microwave Remote Sensing of the Troposphere - Calibration, Data Retrieval and Applications, PhD thesis, Institute of Applied Physics, University of Bern.

[10] Rodgers, C. D. 2000, *Inverse Methods for Atmospheric Sounding - Theory and Practice*, **2**, Wolrd Scientific Publishing.

[11] Rosenkranz, P. W. 1998, Water vapor microwave continuum absorption: A comparison of measurements and models, *Radio Science* **33**, 919–928.

[12] Liebe, H. J. et al. 1991, A model for the complex permittivity of water at frequencies below 1 THz, *I. J. Infrared Milli.* **12**, 659–675.

[13] Karstens, U. et al. 1994, Remote sensing of cloud liquid water, *Meteorol. Atmos. Phys.* **54**, 157–171.

[14] Illingworth A. J. et al. 2007, Cloudnet, *Bull. Amer. Meteorol. Soc.* **88**, 883–898.

EPJ Web of Conferences **119**, 05002 (2016) DOI: 10.1051/epjconf/201611905002
ILRC 27

RECONSTRUCTION OF N_2O AND CH_4 CONTENT BY DIAL MEASUREMENTS AT WAVELENGTHS OF OVERTONE CO LASER

O. A. Romanovskii[1,2*], G.G. Matvienko[1,2], O. V. Kharchenko[1] and S. V. Yakovlev[1,2]

[1] *V.E. Zuev Institute of Atmospheric Optics SB RAS, 1 Zuev Square, Tomsk 634021, Russia,* *roa@iao.ru*
[2] *National Research Tomsk State University, 36 Lenina Pr., Tomsk 634050, Russia*

ABSTRACT

The paper presents the results of laboratory experiments on measurement of absorption and extinction of radiation of the overtone CO laser at wavelengths used for sensing of methane and N_2O in the mid-IR spectral range with the differential absorption (DIAL) method, as well as the concentrations of the studied gases reconstructed from the analysis of experimentally obtained absorption coefficients.

1. INTRODUCTION

In connection with the growing pollution of the atmosphere, the problem of real-time monitoring of gas concentrations becomes increasingly urgent. This problem can be solved with the aid of remote laser sensing of the atmosphere. The maximal information in remote determination of the atmospheric composition with high temporal resolution can be obtained only by the optical method with the use of lasers, that is, by lidar method [1, 2]. The development of laser remote IR spectroscopy requires the development and implementation of new laser sources in the mid-IR spectral region generating radiation in the as wide as possible spectral range with a small frequency step. The tuning range of the overtone CO laser (2.5-4.2 μm) [3, 4] allows investigations in the field of laser spectroscopy, as was demonstrated in [5 - 11]. In addition to the laboratory laser spectroscopy, the overtone CO laser is interesting for some applications, which require the transport to long distances, because the range of the overtone radiation covers the atmospheric transparency window in the spectral range 3.3-4.2 μm [1]. The absorption of radiation of the overtone CO laser in the atmosphere was studied in [12, 13], where both the linear and nonlinear absorption, including induced absorption and "bleaching" by laser radiation, were taken into account in the calculations. The list of 50 spectral lines of the overtone CO laser characterized by the lowest absorption in the atmosphere is given in [12]. The minimal absorption coefficient at such lines in the atmosphere is about 0.01 km^{-1} (H_2O and CO_2 continuum absorption taken into account). At the same time, the overtone CO laser can find application in remote sensing of the atmosphere. The weak absorption of laser radiation in the atmosphere allows spectroscopic measurements of concentrations of various substances at long distances [14]. It was demonstrated in [15, 16] that lines of the overtone CO laser are promising for remote analysis of minor gas constituents (MGCs) of the atmosphere.

The aim of this study was to conduct model laboratory experiments on remote measurement of absorption and total extinction of radiation of the overtone CO laser in methane and N_2O with the following reconstruction of concentrations of the minor gas constituents under study.

2. EXPERIMENT

Laboratory experiments on lidar sensing of methane and N_2O by the path DIAL scheme were carried out at the wavelengths selected by the technique developed in [16] for sensing of methane (3.440 μm) and N_2O (3.877 μm). The experiments were conducted at the laser setup developed in the Laboratory of Gas Lasers of the Lebedev Physical Institute RAS [17].

The absorption and extinction of radiation scattered from a topographic target and passed through the medium with the studied gas were measured with OPHIR 3A-SH calorimeters and FSG-22-3A1 microcryogenic Ge:Au photodetectors (SVOD photoresistors). The optical arrangement of experiments on measurement of the absorption and extinction coefficients of radiation of the overtone CO laser in methane and N_2O is shown in Fig. 1.

The laser cavity of the overtone CO laser was formed by the M1 spherical mirror (R~20 m) and the diffraction grating (420 lines/mm, blazing angle of 27°) operating in the autocollimation mode in the first diffraction order and outputting the radiation in the zero diffraction order. The laser beam aperture was determined by the D1 intracavity diaphragm 25 mm in diameter. The He-Ne laser was used for alignment of the laser cavity. The spectral tuning was performed by turning the diffraction grating. The grating turning angle was controlled with an auxiliary semiconductor laser (LTs-1, radiation wavelength of 0.65 μm).

In this configuration, a 10-cm long cell was filled with the gas mixture at a pressure of 1 atm: studied gas constituents with nitrogen (N_2O:N_2, CH_4:N_2) in proportion 1:24 at the 4% concentration of the absorbing gas. To measure the laser pulse energy E_0, a part of the laser radiation passed through the diaphragm D2 20 mm in diameter was directed to the first calorimeter through reflection from the ZnSe plane-parallel plate and the spherical mirror M2 (R~0.25 m).

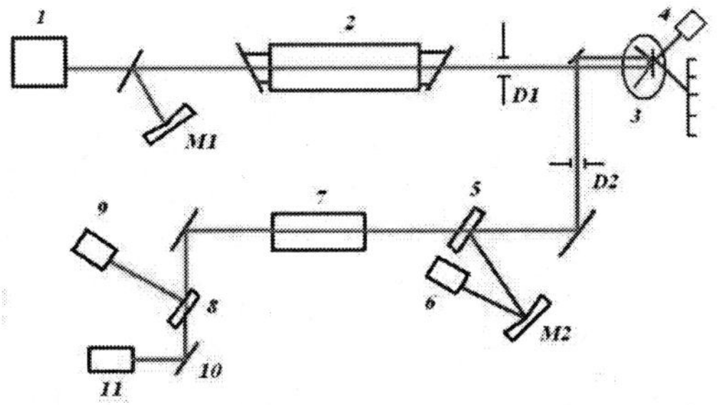

Figure 1. Experimental setup for measurement of absorption and extinction of radiation scattered from the topographic target and passed through the medium with the studied gas: (1) He-Ne laser, (2) CO laser, (3) diffraction grating, (4) semiconductor laser, (5) ZnSe plate, (6) Ophir 3A-SH calorimeter, (7) absorbing cell (L=10 cm), (8) CaF_2 plate, (9) IRS/ Ophir 3A-SH calorimeter, (10) topographic target, and (11) SVOD photoresistor

As the laser radiation passed through the cell, a part of the radiation was directed to the second calorimeter /IRS (infrared spectrometer) through reflection from the CaF_2 plane-parallel plate. The second calorimeter measured the energy of the laser beam E passed through the absorbing cell. The absorbed radiation was calculated in accordance with the Bouguer—Lambert—Beer law. Another part of the laser radiation was directed to the topographic target behind the CaF_2 plate. The topographic target was represented by the diffusely scattering reflector with an albedo of 0.8. The SVOD photoresistors measured the energy of radiation passed through the medium with the studied gas and the radiation part lost at reflection from the topographic target. The radiation extinction was measured by the technique described in [16]. The calculated and measured absorption and extinction coefficients for N_2O and methane are summarized in Table 1.

Table 1. Calculated and measured absorption and extinction for N_2O and methane

λ in air, μm	abs. coeff., cm^{-1} (calculation)	abs. coeff., cm^{-1} (experiment)	extinction coeff., cm^{-1} (experiment)
		N_2O	
3.841	0.0014	0.0095	-
3.852	0.0067	0.0004	0.0129
3.877	0.1178	0.0953	0.1202
3.892	0.0209	0.0502	-
3.897	0.0102	0.0072	-
3.907	0.0508	0.0262	0.0029
3.918	0.0307	0.0223	0.0066
3.930	0.0061	0.0014	0.0185
		CH_4	
3.328	0.0025	-	0.0090
3.336	0.0005	0.00462	0.0151
3.345	0.0381	0.0199	0.0276
3.432	0.0003	0.0029	-
3.440	0.0247	0.0078	0.0564
3.444	0.0011	-	0.0563
3.453	0.1159	0.1040	0.1733

It can be seen from the tabulated experimental data on absorption and extinction that the measured (absorption and extinction) and calculated (absorption) values are in a good agreement in the entire spectral range, in which the measurements were conducted. Some discrepancies between the calculated and measured methane absorption coefficients can be explained by the influence of residual interfering absorption of water vapor in the measuring cell or (for calculation of the extinction coefficient) by the inhomogeneity of the laser radiation scattered from the topographic target and passed through the medium with the gas mixture under study. The error of energy measurement by the OPHIR 3A-SH calorimeter, which did not exceed 5%, should be taken into account as well.

3. RECONSTRUCTION OF THE GAS CONCENTRATIONS

For the wavelengths selected for sensing of methane and N_2O at the operation with the topographic target, the inverse problem of reconstruction of the concentrations of this minor atmospheric gases in the 10-cm long absorbing cell with the aid of the overtone CO laser was solved from the analysis of the experimentally determined absorption coefficients. The reconstructed concentrations in comparison with the concentrations in the calibration mixture of the studied gases in the cell are shown in Fig. 2.

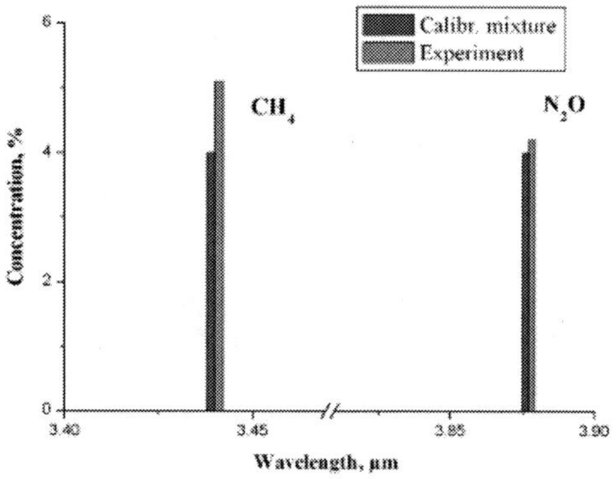

Figure 2. Reconstructed concentrations of N_2O and methane in the case of overtone CO laser radiation at the selected sensing wavelengths

It can be seen from the figure that the accuracy of measurement of the absorption coefficients of the gases under study influences significantly the solution of the inverse problem of concentration reconstruction.

4. CONCLUSIONS

The test laboratory experiments (based on numerical simulation) on measurement of absorption and extinction of the overtone CO laser radiation in mixtures with the studied gases yield the acceptable agreement between the calculated and measured values except for few noninformative wavelengths, which allows the developed technique of wavelength selection to be applied along with the DIAL method. The reliability of the obtained results is confirmed by the solution of the inverse problem on reconstruction of the concentrations of the studied gas constituents from analysis of the experimentally obtained absorption coefficients.

ACKNOWLEDGMENTS

This work was supported in part by the Russian Foundation for Basic Research (Grant No. 13-05-98074 - r_sibir-a), the Russian Scientific Foundation (Grant No. 14-27-00022), the President of the Russian Federation (Grant NS-4714.2014.5 in support of leading scientific schools), and the Tomsk State University (Tomsk State University Competitiveness Improvement Program).

REFERENCES

[1] R. M. Measures, Laser Remote Sensing: Fundamentals and Applications [Russian translation], *Mir, Moscow* 1987. 550 pp.

[2] P. P. Geiko, V. E. Privalov, O. A. Romanovskii and O.V. Kharchenko Femtosecond Laser Radiation Frequency Converters for Lidar Monitoring of the Atmosphere *Technical Physics Letters*. 2009. V. **35**. N 8. P. 733-736.

[3] N. G. Basov, A. A. Ionin, A. A. Kotkov, A. K. Kurnosov, J. E. McCord, A. P. Napartovich, L. V. Seleznev, N. G. Turkin and G. D. Hager Pulsed laser operating on the first vibrational overtone of the CO molecule in the 2.5—4.2-μm range: 1. Multifrequency lasing *QUANTUM ELECTRON*. 2000. **30** (9), 771–777.

[4] N. G. Basov, A. A. Ionin, A. A. Kotkov, A. K. Kurnosov, J. E. McCord, A. P. Napartovich, L. V. Seleznev, N. G. Turkin and G. D. Hager Pulsed laser operating on the first overtone of the CO molecule in the 2.5—4.2-μm range. II. Frequency-selective lasing *QUANTUM ELECTRON*. 2000. **30** (10), 859–866.

[5] M. Mürtz, T. Kayser, D. Kleine, S. Stry, P. Hering, and W. Urban Recent developments in cavity ring-down spectroscopy with tunable cw lasers in the mid-infrared *Proc. SPIE*. 1999. V. **3758**. P. 53 - 61.

[6] H. Danke, G. von Basum, K. Kleinermanns, P. Hering and M. Mürtz Rapid formaldehyde monitoring in ambient air by means of mid-infrared cavity leak-out spectroscopy *Appl. Phys. B*. 2002. V. **75**. P. 311 -316.

[7] H. Dahnke, D. Kleine, W. Urban, P. Hering and M. Mürtz Isotopic ratio measurement of methane in ambient air using mid-infrared cavity leak-out spectroscopy *Appl. Phys. B*. 2001. V. **72**. P. 121 - 125.

[8] H. Danke, J. Kahl , G. Schuler, W. Boland, W. Urban and F. Kühnemann On-line monitoring of biogenic isoprene emissions using photoacoustic spectroscopy *Appl. Phys. B*. 2000.

V. **70** P. 275 - 280.

[9] H. Danke, D. Klaine, P. Hering and M. Mürtz Real-time monitoring of ethane in human breath using mid-infrared cavity leak-out spectroscopy *Appl. Phys. B*. 2001. V. **72**. P. 971 - 975.

[10] I. E. Santosa., L. J. J. Laarhovan, J. Harbinson, S. Driscoll and F. J. M. Harren Laser-based trace gas detection of ethane as a result of photooxidative damage in chilled cucumber leaves *Rev. Scientific Instruments*. 2003. № **74**. P. 680 - 683.

[11] H. Schmitz, M. Mürtz and H. Bleckmann Responses of the infrared sensilla of Melanophila acuminate (Coleoptera: Buprestidae) to monochromatic infrared stimulation *J. Comparative Physiology A*. 2000. № **186**. P. 543 - 549.

[12] O. G. Buzykin, S.V. Ivanov, A.A. Ionin, A.Yu. Kozlov, A.A. Kotkov and L.V. Seleznev Linear and nonlinear absorption of the overtone CO laser radiation in the atmosphere *Atmospheric and oceanic optics*. 2001. V.**14**. No. 05. P. 361-367

[13] O. G. Buzykin, A. A. Ionin, S. V. Ivanov, A.A. Kotkov, L.V. Seleznev and A.V. Shustov Resonant absorption of first-overtone CO laser radiation by atmospheric water vapor and pollutants *Laser and Particle Beams*. 2000. № **18**. P. 697 - 713.

[14] O. G. Buzykin, S. V. Ivanov, A. A. Ionin, A.A. Kotkov and A.Yu. Kozlov Spectroscopic Detection of Sulfur Oxides in the Aircraft Wake *J. Russian Laser Research*. 2005. № **26**. P. 402 - 426.

[15] A. A. Ionin, Yu. M. Klimachev, A. Yu. Kozlov, A. A. Kotkov, O. A. Romanovskii, L. V. Seleznev, D. V. Sinitsyn, O. V. Kharchenko, A. V. Shelestovich and S. V. Yakovlev Wideband CO laser in problems of laser sensing of minor gaseous components in the atmosphere *Russian Physics Journal*, vol. **51** issue 11., p. 1200 – 1207.

[16] O.A. Romanovskii, O.V. Kharchenko and S.V. Yakovlev Methodological aspects of lidar ranging of trace gases in the atmosphere by differential absorption *JOURNAL OF APPLIED SPECTROSCOPY*. 2012. V. 79, Issue: 5, Pages: 793-800.

[17] A. Ionin, Yu. Klimachev, A. Kotkov, A. Kozlov, O. Rulev, L. Seleznev, D. Sinitsyn, S. Vetoshkin, "Multiline Laser Probing for Active Media CO:He, CO:N_2, and CO:O_2 in Wide-Aperture Pulsed Amplifier", *Journal of Russian Laser Research*, **27**(1), p. 33-69 (2006).

EPJ Web of Conferences **119**, 05003 (2016) DOI: 10.1051/epjconf/201611905003
ILRC 27

NUMERICAL SIMULATIONS OF A 2.05 μm Q-SWITCHED Ho:YLF LASER FOR CO_2 IPDA SPACE REMOTE SENSING

Jessica Barrientos Pellegrino*, Dimitri Edouart, Fabien Gibert, Claire Cenac

*Laboratoire de Météorologie Dynamique, Route de Saclay, 91128 Palaiseau Cedex, *Email: jessica.pellegrino@lmd.polytechnique.fr*

ABSTRACT

We report on numerical simulations of the performances of a 2.05 μm double pulse Q-switched Ho:YLF laser for the monitoring of CO_2 from space. A Q-switched Holmium laser set-up based on a MOPA configuration is proposed to fulfill the requirements of a IPDA space-borne measurement. Double pulse operation is considered to obtain a 250 μs delay time between the ON and OFF pulse emissions. Numerical simulations results show that up to 40 mJ ON pulse can be extracted from the Ho:YLF laser at a repetition rate of 350 Hz with an optical efficiency of 17 %.

1. INTRODUCTION

A particular interest is focused on space borne active remote sensing of greenhouse gases. Indeed, greenhouse gases (GHGs) measurements on a global scale are necessary to identify, locate and quantify GHGs sources and sinks in order to improve the comprehension of climate change. In this context, instruments able to detect, identify and quantify atmospheric trace gases such as CO_2, H_2O or CH_4 from space are required. Previous studies enable to identify the emitter specifications, in terms of emitted wavelength to address the most important greenhouse gases, output energy, frequency stability, and beam quality, were derived from the overall instrument error budget for such space borne measurements [1, 2[2]]. A peculiar aspect of the space borne monitoring of the atmospheric CO_2 dry-air mixing ratio, is a high accuracy on the ppm level or 0.25 % assuming a mean concentration of 400 ppm. To meet this stringent need, we have to address challenging technical requirements such as: (1) a transmitter delivering high energy pulses (higher than the mJ scale) at repetition rate (PRF) higher than hundreds of Hertz, (2) with a good spectral and spatial quality of the emitted radiation, (3) multiple wavelengths emission capability in single mode operation and (4) double pulse emission with a delay time between ON and OFF emissions of 250 μs [3]. Several approaches were investigated to fulfill such stringent requirements for space applications. Some of them are based on injection-seeded laser oscillators [4, 5] or on single mode optical parametric oscillator with optical parametric amplifier (OPO-OPA) source [6] at 2.05 μm for space CO_2 monitoring. Others approaches are based on injection-seeded optical parametric oscillators (OPOs) emitting around 1.57 and 1.6 μm, for CO_2 or CH_4 monitoring, respectively [7 - 9]. These examples are operating in a single pulse mode. However, to fulfill the need on the delay time between ON and OFF emissions, there are also developments and assessments studies on emitter operating in double pulse mode [10] or triple-pulsed mode [11] for DiAL measurement based on codoped Ho:Tm laser. These systems deliver high energy pulses but at a repetition rate lower than 50 Hz. Here, we propose a numerical study of the performances of a double pulse Q-switched Holmium laser with a repetition rate higher than 100 Hz dedicated to the atmospheric CO_2 monitoring from space.

2. Q-switched Ho:YLF laser SET-UP

As you can see on the Figure 1, the suggested Ho-doped fluorides laser set-up is based on a master oscillator-power amplifier (MOPA) configuration. This configuration enables to produce high energy level pulses (> 10 mJ) while maintaining a high spectral and spatial beam qualities as required for LIDAR applications. Ho-doped fluorides are more attractive laser materials than Ho:YAG as they have much longer upper laser level lifetimes (~ 14 ms) and higher peak emission cross-sections (1.6×10^{-20} cm^2 versus 1.2×10^{-20} cm^2) [12]. The typical emission bands of Ho-doped fluorides are more adapted to the monitoring of CO_2 from space. In addition, YLF and LLF thermal lens are weaker than YAG which enables to generate diffraction-limited beams even under intense end pumping.

DOI: 10.1051/epjconf/201611905003

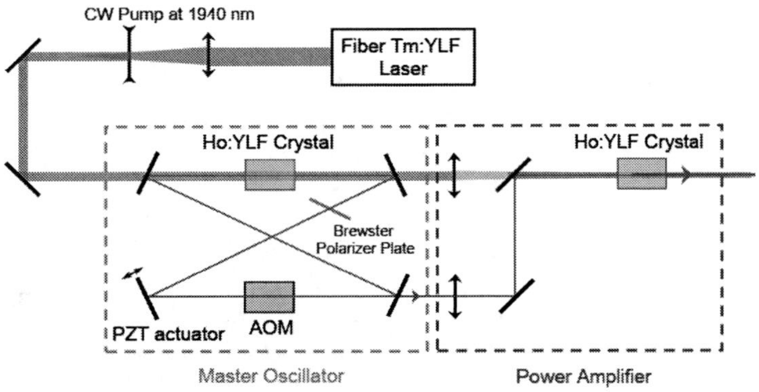

Figure 1: Ho-doped fluorides laser set-up proposed for double-pulse operation.

However, Ho:YLF and Ho:LLF have a stronger quasi-three-level nature than Ho:YAG [13].

The Ho:YLF master oscillator (MO) relies on a CW Tm fiber laser pump which has the main advantage to provide a simple and robust pumping architecture. A 0.5 at. % doped Ho:YLF crystal is put in a 1-m long ring cavity to limit up-conversion processes. This cavity configuration for the MO is expected to ease injection-seeding operation and avoid spatial hole burning to reach higher frequency stability. The PRF and the double pulse operation of the MO is controlled by the Q-switching rate of the acousto-optic modulator (AOM). Careful design and operation has to be done to avoid laser damage and reach the specified Ho:YLF emission wavelength around 2051 nm. This latter band is in competition with a second emission band around 2065 nm. It has been demonstrated that the laser emission frequency shifts with optical characteristics of the output coupler and could be further tuned to the specified wavelength by both accurate control of crystal temperature and selective spectral components inside the cavity [14]. YLF host crystal is birefringent and produces laser emission on both π and σ polarisations whatever the Tm fiber laser beam polarization is. Nevertheless, pumping on the π axis will be searched for more efficiency. The polarization of the emitted beam is chosen by inserting a Brewster polarizer plate inside the cavity. The Q-switched master oscillator is sequentially injection seeded to produce specific ON- and OFF- wavelengths, line width, pulse width, beam quality and adaptable pulse repetition rate [15]. To achieve space energy requirement, a power amplifier (PA) composed of a Ho:YLF crystal amplified the MO laser beam to energy level > 10 mJ.

3. NUMERICAL SIMULATION RESULTS

Double pulse operation, consisting in Q-switching the laser cavity two times in a row, has been simulated. We use a numerical model based on rate equations describing the dynamics of the laser manifolds involved in 2051 nm laser emission and quasi resonant pumping around 1940 nm [16]. Re-pumping during the delay time between ON and OFF pulses is considered in the simulations.

PRF (Hz)	ON pulse energy (mJ)	OFF pulse energy (mJ)	Optical efficiency (%)	Max. Fluence (J/cm^2)
100	14,5	4	3,5	5
250	14	3	8,5	5
350	13	3	11,5	5
400	12,5	3,5	13	4,5
500	12	4,5	15,5	4,5

Table 1 : Simulated MO parameter sets in double pulse operation to achieve asymmetric pulse energies. Length crystal : 50 mm.

Double-pulse operation assessment of the MO

As two pulses must be emitted in a row, high energy storage is required and an available pump power of 50 W is considered for the MO. To limit the maximal fluence to 5 J/cm^2, a 500 µm beam size is optimal in the MO. The simulation results show that MO laser performances slightly depend on the crystal length. As you can see in Table 1, High PRF operation is necessary to obtain high optical efficiency. Respective ON and OFF pulse energies depend on the intermediate cavity loss level after the first Q-switch. Assuming OFF pulse energies higher than 3 mJ to ensure good pulse energy stability, the maximal achievable ON pulse energy is limited to 15 mJ.

Most of the crystal gain is used to emit the high energy ON pulse (11-15 mJ). As a consequence, the OFF pulse is built with a low gain crystal and its duration is much longer than the other one.

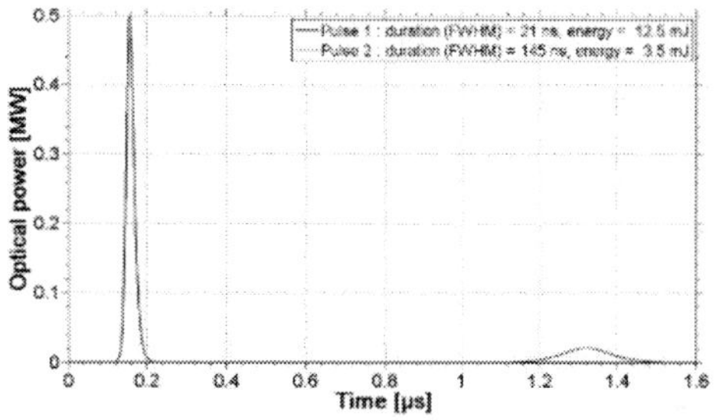

Figure 2 : Simulated pulse temporal profiles in double pulse operation.

ON pulse durations range between 17 and 27 ns and OFF pulse durations between 135 and 185 ns. Figure 2 displays an example of simulated pulse duration power profiles in double pulse operation. The first pulse is a high energy (12.5 mJ) and short pulse of 21 ns. Indeed, as it is generated with a high crystal gain, its built-up time is shorter than 200 ns. The second pulse is much longer (145 ns), has a low energy (3.5 mJ) and shows up after more than 1 µs delay time. This inability to generate short OFF pulses is a drawback of the double pulse operation. The AOM can be used to artificially shorten the OFF pulse duration but this will reduce the expected pulse energy and may affect the pulse spectral line width.

Double-pulse operation assessment of the MOPA set-up

Moreover, assuming these energy levels, a pulse amplifier gain between 2 and 3 is needed to answer the requirements for a space-borne measurement. So, MOPA set-up in double pulse operation has been simulated as a whole. Amplifying up to 40 mJ energy and keeping maximal fluence as low as 5 J/cm^2 is very challenging.

PA pump power (W)	PRF (Hz)	PA beam waist (µm)	ON pulse energy (mJ)	OFF pulse energy (mJ)	MOPA optical efficiency (%)
40	100	800	43	15.5	5.5
50	250	800	41	14	12
50	350	800	41	13.5	17
50	400	800	39.5	12.5	19

Table 2 : Simulated MOPA performances in double pulse operation.

All the MOPA parameters and resulting simulated performances that fulfill the pulse energy requirement and achieve the lowest maximal fluence (~ 6 J/cm^2) are gathered in Table 2. With 50 W pump for the PA stage and 800 µm beam size, the simulation shows that 40 mJ pulse energy is achieved at 350 Hz PRF with 50 mm long PA crystal length and 17 % optical efficiency. At 500 Hz PRF, 40 mJ pulse energy is obtained with 700 µm beam size but the maximal fluence reaches 8 J/cm^2. Pulse energies are not higher than 35 mJ with 800 µm beam size.

4. CONCLUSIONS

To fulfill the requirements of space-borne CO_2 monitoring, we propose a double pulse Q-switched Holmium laser based on a MOPA configuration. Double pulse operation is achieved by Q-switching the laser cavity two times in a row. The drawback of this method is the difference between the ON and OFF pulse duration that would probably affects LIDAR measurement. Nevertheless numerical simulations results are very promising. Indeed, they show that up to 40 mJ can be extracted from the Ho:YLF laser at a repetition rate larger than 100 Hz on the ON pulse while maintaining a maximal fluence around 5 J/cm^2. At 350 Hz PRF, an optical efficiency of 17 % is achieved. The development and the characterization of this emitter will start soon.

ACKNOWLEDGEMENT

This work is supported by the European Space Agency (ESA) through the contract 4000113667/15/NL/PA, "2.05 µm Pulsed Holmium Laser for Atmospheric CO_2 monitoring".

REFERENCES

[1] J. Caron and Y. Durand, "Operating wavelengths optimization for a spaceborne lidar measuring atmospheric CO2", *Appl. Opt.*, 48(28):5413–5422, 2009.

[2] G. Ehret, C. Kiemle, M. Wirth, A. Amediek, A. Fix, and S. Houweling, "Space-borne remote sensing of CO2, CH4, and N2O by integrated path differential absorption lidar: a sensitivity analysis", *Appl. Physics B*, 90(3-4):593–608, 2008.

[3] A-SCOPE (Adances Space Carbon and Climate Observation of Planet Earth), ESA Report for Assessment, SP-1313/1, 2008.

[4] F. Gibert, P. H. Flamant, D. Bruneau, and C. Loth, "Two-micrometer heterodyne differential absorption lidar measurements of the atmospheric CO2 mixing ratio in the boundary layer", *Appl. Opt.*, 45(18):4448–4458, 2006.

[5] G. J. Koch, J. Y. Beyon, F. Gibert, B. W. Barnes, S. Ismail, M. Petros, P. J. Petzar, J. Yu, E. A. Modlin, K. J. Davis, and U. N. Singh, "Side-line tunable laser transmitter for differential absorption lidar measurements of CO2: design and application to atmospheric measurements", *Appl. Opt.*, 47(7):944–956, 2008.

[6] J. Barrientos Barria, D. Mammez, E. Cadiou, J. B. Dherbecourt, M. Raybaut, T. Schmid, A. Bresson, J. M. Melkonian, A. Godard, J. Pelon, and M. Lefebvre. "Multispecies high-energy emitter for CO2, CH4, and H2O monitoring in the 2 μm range", *Opt. Lett.*, 39(23):6719–6722, 2014.

[7] A. Fix, C. Büdenbender, M. Wirth, M. Quatrevalet, A. Amediek, C. Kiemle, and G. Ehret, "Optical parametric oscillators and amplifiers for airborne and spaceborne active remote sensing of CO2 and CH4", 2011.

[8] K. Numata, S. Wu, and H. Riris, "Fast-switching methane lidar transmitter based on a seeded optical parametric oscillator," *Appl.Physics B*, 116(4):959–966, 2014.

[9] D. Sakaizawa, S. Kawakami, M. Nakajima, Y. Sawa, and H. Matsueda, "Ground-based demonstration of a CO2 remote sensor using a 1.57μm differential laser absorption spectrometer with direct detection", *Journal of Applied Remote Sensing*, 4(1):043548–043548-17, 2010.

[10] U. N. Singh, J. Yu, M. Petros, T. F. Refaat, R. Remus, J. Fay, K. Reithmaier, "Column CO2 measurement from an airborne solid-state double-pulsed 2-micron integrated path differential absorption lidar", Proceedings of the International Conference on Space Optics, 2014.

[11] T. F. Refaat, U. N. Singh, J. Yu, M. Petros, S. Ismail, M. J. Kavaya, and K. J. Davis, "Evaluation of an airborne triple-pulsed 2μm IPDA lidar for simultaneous and independent atmospheric water vapor and carbon dioxide measurements", Appl. Opt., 54(6):1387–1398, 2015.

[12] B. M. Walsh, N. P. Barnes, and B. Di Bartolo, "On the distribution of energy between the tm 3f4 and ho 5i7 manifolds in tm-sensitized ho luminescence", *Journal of Luminescence*, 75(2):89 – 98, 1997.

[13] M. Eichhorn, "Quasi-three-level solid-state lasers in the near and mid infrared based on trivalent rare earth ions", *Appl. Physics B*, 93(2-3):269–316, 2008.

[14] D. Edouart, F. Gibert, F. Le Mounier, C. Cénac, P. H. Flamant, "2-μm high repetition rate laser transmitter for CO2 and Wind Lidar (COWI)", 11-14, Proceedings of the 26th International Laser Radar Conference, 2012.

[15] F. Gibert, D. Edouart, C. Cénac, and F. Le Mounier, "2-μm high-power multiple-frequency single-mode Q-switched Ho:YLF laser for dial application", *Appl. Physics B*, 116(4):967–976, 2014.

[16] N. P. Barnes, B. M. Walsh, and E. D. Filer, "Ho:ho upconversion: applications to ho lasers", *J. Opt. Soc. Am. B*, 20(6):1212–1219, 2003.

EPJ Web of Conferences **119**, 05004 (2016) DOI: 10.1051/epjconf/201611905004
ILRC 27

WATER VAPOR PROFILES UP TO THE UT/LS FROM RAMAN LIDAR AT REUNION ISLAND (21°S, 55°E) : TECHNICAL DESCRIPTION, DATA PROCESSING AND COMPARISON WITH SONDES

Hélène Vérèmes[(1,2)]*, Philippe Keckhut[3], Jean-Luc Baray[4], Jean-Pierre Cammas[(1,2)], Davide Dionisi[5], Guillaume Payen[2], Valentin Duflot[(1,2)], Franck Gabarrot[2], Jimmy Leclair De Bellevue[1], Françoise Posny[1], Stéphanie Evan[1], Susanne Meier[6], Holger Vömel[7] and Ruud Dirksen[6]

[1]*Laboratoire de l'Atmosphère et des Cyclones, UMR8105, Saint-Denis de La Réunion, France,*
**Email: helene.veremes@univ-reunion.fr*
[2]*Observatoire des Sciences de l'Univers de La Réunion, UMS3365, Saint-Denis de la Réunion, France,*
[3]*Laboratoire ATmosphères, Milieux, Observations Spatiales-IPSL, UMR8190, Guyancourt, France,*
[4]*Laboratoire de Météorologie Physique, UMR6016, Clermont-Ferrand, France,*
[5]*Institute of Atmospheric Sciences and Climate-CNR, Roma, Italy,*
[6]*Deutscher Wetterdienst, Meteorological Observatory Lindenberg, Lindenberg, Germany,*
[7]*National Center for Atmospheric Research, Boulder, CO, USA*

ABSTRACT

The Maïdo high-altitude observatory located in Reunion Island (21°S, 55°E) is equipped with an innovative lidar designed to monitor the water vapor in the whole troposphere up to the lower stratosphere with a Raman system and to monitor, simultaneously, the temperature in the stratosphere and in the mesosphere based on a Rayleigh scattering technique. Several improvements have been performed on the new instrument to optimize the water vapor mixing ratio measurements thanks to the experience of the previous system. The choice of the operational configuration of the system and the calibration methodology were realized during the campaign MALICCA-1 (MAïdo LIdar Calibration CAmpaign) which provided simultaneous measurements of water vapor and ozone in April 2013. The lidar water vapor profiles are calibrated with water vapor columns obtained from a collocated GNSS receiver. By comparing CFH and Vaisala radiosondes and satellites water vapor mixing ratio profiles with the Raman lidar profiles, the performances of the lidar are shown to be good in the troposphere. With a suitable integration time period, the ability of measuring quantities of a few ppmv in the lower stratosphere is demonstrated. This Raman lidar will provide regular measurements to international networks with high vertical resolution profiles of water vapor in order to document various studies and to insure a long-term survey of the troposphere and of the lower stratosphere.

1. INTRODUCTION

A Raman water vapor lidar has been designed, upgrading a former system running in Reunion Island, in order to measure the water vapor up to the (sub)tropical UT/LS (Upper Troposphere / Lower Stratosphere). This system is located at the Maïdo observatory (2160 m asl) and is operating since October 2012 [1]. The main goal of the MALICCA campaign in April 2013 (Maïdo facility) was to test the potential configurations and to determine the best one for routine measurements. Choices have also been consequently made for the routine data processing code. The present paper will introduce the configuration of the system and of the retrieval code, the first pieces of information on the vertical resolution and the evaluation of the performances, and finally a brief overview of the scientific studies initiated around the first two years data supplied by this lidar.

2. THE MAIDO RAMAN WATER VAPOR LIDAR

Similarly to the ozone DIAL system, the Raman water vapor system operated at Saint-Denis (Reunion Island) before being transferred to the Maïdo observatory [2], enhancing very significantly the reception and emission parts. The new Raman lidar has been designed to monitor water vapor until the lower stratosphere and, simultaneously, the temperature in the stratosphere and in the mesosphere. The system is conceived to work at two wavelengths, 532 and 355 nm, generated by two Quanta Ray Nd:Yag

DOI: 10.1051/epjconf/201611905004

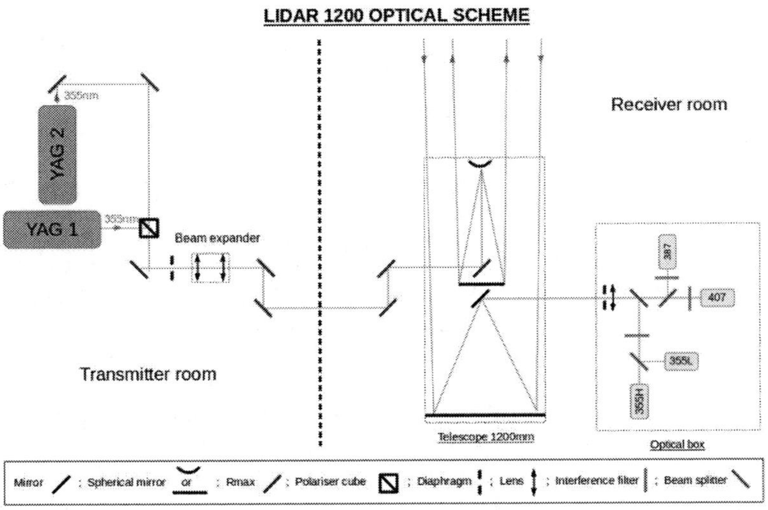

Figure 1. Optical scheme of the Maïdo Raman lidar on a routine basis

lasers, in order to double the power, with a repetition rate of 30 Hz. The emitting pulse of each laser, 532 and 355 nm (that can be synchronized and coupled through polarization cubes), has an energy of, respectively, about 800 and 375 mJ.pulse^{-1}, and a duration pulse of 9 ns. A coaxial geometry for emission and reception has been designed to avoid parallax effects, to extend measurements down to the ground and to facilitate the alignment. The backscattered signal (collected by a 1.2 m diameter telescope) is transferred to the optical ensemble by a set of lenses and mirrors instead of fiber-optic cables. Indeed, the fluorescence in fiber-optic cables has been shown to cause systematic biases. This is an important issue when collecting small Raman scattering by water vapor compared with the large elastic scattering able to generate fluorescence. The light coming from the primary mirror is reflected by a secondary flat mirror, tilted at 45°, directing the light in the optical box unit used to separate the backscattered Raman and Rayleigh signals (figure 1). The field of view (FOV) of the system can be adjusted from 3.0 to 0.5 mrad thanks to a diaphragm field stop at the entrance of this box. The spectral separation of the light is realized by dichroic beam splitters and interference filters (band pass or high pass). Regarding the photon detection and data acquisition, Hamamatsu miniature photomultiplier tubes (PMT) and Licel transient recorders are used.

In April 2013, the different configurations of the system have been tested and the configuration chosen for operational use is : two lasers, the 355 nm wavelength and a 2 mm (0.5 mrad) FOV (figure 1). Such FOV induces a geometrical obturation that is similar for both channels and allows to reduce PMT saturation and to retrieve water vapor mixing ratio down to the ground. Details on the different tests and the final choice of the optical configuration are given in [3].

3. WATER VAPOR PROFILES RETRIEVAL

Uncalibrated profiles are retrieved with a vertical resolution whose values increase with the altitude to compensate the decrease of the signal to noise ratio, with a time selection depending on the water vapor variability at several levels and with the calibration coefficient (calculated afterward from ancillary data column of water vapor).

The vertical resolution of the initial data is 15 m. For retrieving the profiles on a routine basis, data are smoothed with a Blackman filter whose number of points changes with the altitude. The vertical resolution of the profiles with this filter has been calculated by using the cutoff frequency of the transfer function of the filter. The vertical resolution in the lowest layers, in the mid-troposphere, in the upper troposphere and in the lower stratosphere is around 200-400 m, 1, 1.2-1.3 and 1.5 km, respectively.

Profiles are calibrated using integrated water vapor (IWV) columns obtained from GNSS measurements. The recovering altitude of the lidar is very low, and thus, the use of collocated GNSS total columns for the calibration of its profile is appropriate. The receiver is able to collect raw data from GPS and GLONASS satellites constellations. One of the basic GPS atmospheric products is the tropospheric delay which is a measurement of the GPS signal delay that has traveled between a GPS satellite and a ground-based receiver with respect to propagation in a vacuum. The GPS network used in our typical differential simulation includes twenty one other local stations mainly located around the Reunion volcano massif and about fifteen overseas stations to ensure efficiently high numbers of baselines.

The total absolute error (ΔWVMR) on the calibrated profiles (WVMR) is calculated by :

$$\frac{\Delta WVMR}{WVMR} = \sqrt{\underbrace{\left(\frac{dw}{w}\right)^2}_{(a)} + \underbrace{\left(\frac{dC}{C}\right)^2}_{(b)} + \underbrace{\left(\frac{d\Gamma_\Delta}{\Gamma_\Delta}\right)^2}_{(c)}},$$

where *(a)* is the relative error associated to the statistical error of the detectors (PMT), *(b)* is the error linked to the calibration coefficient and *(c)* linked to the extinction coefficient term.

$$\frac{dw}{w} = \frac{\sqrt{S_N^2 \times \left(N_H + \sigma_{BH}^2\right) + S_H^2 \times \left(N_N + \sigma_{BN}^2\right)}}{S_N S_H}$$

where dw is the lidar statistical error and w the uncalibrated water vapor mixing ratio, N=S-B with S the number of photons counted by the PMT per shot, B the associated background and σ_B the error on B. *N* and *H* subscripts stand for N_2 and H_2O Raman wavelengths.

4. DATA VALIDATION PROCEDURE

The calibration coefficient is supposed to be almost constant if no instrumental change occurs. In order to determine this coefficient, ratios between integrated lidar measurements and GNSS IWV on same temporal extents are checked on the whole dataset. Periods with a constant ratio have been identified on the 2013 and 2014 databases. The average ratio of each period can be considered as the calibration coefficient of the associated profiles. For the operational data processing, the former validated coefficient will be applied to the profile but it will be necessary to identify instrumental changes on a few month overview in order to check possible alteration of the calibration coefficient and thus to recalculate it if necessary. In addition to the known changes, systematic lamp measurements are made at the beginning of each night of measurements in order to survey and quickly identify potential sudden instrumental changes. Once the coefficients are validated, data can be calibrated and used for geophysical purposes.

5. PERFORMANCE EVALUATION

The MALICCA-1 campaign data set shows a good agreement between water vapor lidar measurements and other almost simultaneous and collocated measurements (Vaisala RS92, MLS satellite) [3]. The relative difference between lidar profiles and fifteen radiosoundings was lower than 10% for the low and mid-troposphere and between 10 and 20% for the upper troposphere. Two integration methods has been tested : 240 minutes for an absolute error of approximately 2 ppmv from 17 to 20 km, and 1 ppmv at 20 km for a monthly integration, which demonstrates the ability of the lidar to measure quantities of only few ppm in the UT/LS [3]. The monthly water vapor average profile measured by MLS matches well in the lower stratosphere with the lidar average profile of MALICCA-1 [3].

In November 2014, a first CFH sonde was launched during a 120 minutes water vapor lidar operation. The lidar signal was workable up to

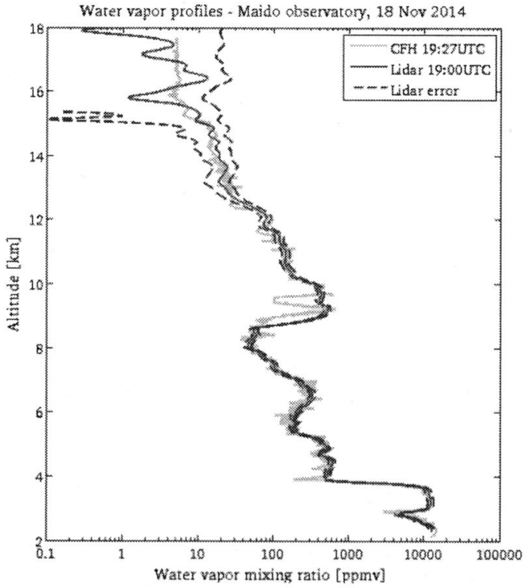

Figure 2. Water vapor mixing ratio profiles on 18 November 2014, in blue : CFH measurement (19:27UTC), and in green : lidar measurement (19:00UTC). Red dotted lines represent the lidar total absolute error on the profile.

DOI: 10.1051/epjconf/201611905004

approximately 15 km. The CFH profile reached 18 km. The comparison between both profiles shows a good agreement up to the upper troposphere (tropopause at 17 km) considering the lidar total error (figure 2). A non negligible difference occurs in the 10 km wet layer sampled by the sonde and has to be investigated. The instrumental conditions on November 18th allowed the lidar to work with only one laser. This result opens good perspectives for future lidar-CFH comparisons, especially those of the MORGANE (Maïdo ObseRvatory Gas and Aerosols Ndacc Experiment) campaign in May 2015.

6. CAMPAIGN AND ROUTINE MODES

Some intensive measurement campaigns provided a detailed data set of water vapor profiles on short periods in April 2013, November 2013 and November 2014. These campaigns allowed scientific studies on various topics : stratosphere-troposphere exchanges, moistening of the lower stratosphere, cirrus clouds... [4]. Besides, measurements are performed once or twice a week. 58 night measurements were made in 2013, 81 in 2014. With the specific subtropical location of the Maïdo observatory, this instrument will participate to a long-term survey of climate changes.

7. CONCLUSIONS AND PERSPECTIVES

The optimal routine system configuration and data processing code have been determined for the Raman water vapor lidar of the Maïdo facility. The first evaluation of the performances of this lidar shows a good accuracy in the troposphere. Its ability of measuring quantity of few ppm in the lower stratosphere has also been demonstrated, but requires adapted integration time periods. The MORGANE campaign will supply suitable data sets (lidars and lidar-CFH sondes comparisons) to go further in this investigation and to make the Raman lidar granted for the NDACC network label. Future data sets of this lidar will represent opportunities to validate some satellites water vapor measurements [4]. The Maïdo observatory offers a large panel of data thanks to the diversity of its remote sensing instruments. The profiles of collocated and simultaneous lidar measurements of water vapor and ozone will offer opportunities to further investigate tropical UT/LS transport and chemical processes.

ACKNOWLEDGEMENTS

To the ACTRIS project and support of the European Community (Research Infrastructure Action under the FP7 "Capacities" specific programme for Integrating Activities, ACTRIS Grant Agreement no. 262254). To the European Commission (FEDER program), Région Réunion and CNRS for their strong support in the building of the Maïdo facility. To the Université de La Réunion and CNRS for their strong support of the OPAR station (Observatoire de Physique de l'Atmosphère de La Réunion) and the OSU-R activities. To E. Golubic, P. Hernandez and L. Mottet who are deeply involved in the routine lidar observations at OPAR.

REFERENCES

[1] Baray J.-L et al., 2013 : Maïdo observatory: a new high-altitude station facility at Reunion Island (21° S, 55° E) for long-term atmospheric remote sensing and in-situ measurements, *Atmos. Meas. Tech.*, **6**, 2865-2877.

[2] Hoareau C. et al., 2012 : A Raman lidar at La Reunion (20.8°S, 55.5°E) for monitoring water vapour and cirrus distributions in the subtropical upper troposphere : preliminary analyses and description of the future system, *Atmos. Meas. Tech.*, **5(6)**, 1333-1348.

[3] Dionisi D. et al., 2015 : Water vapor observations up to the lower stratosphere through the Raman lidar during the Maïdo Lidar Calibration Campaign", *Atmos. Meas. Tech.*, **8**, 1425-1445.

[4] Keckhut P. et al., 2015 : Introduction to the MALICCA campaign dedicated to the validation of Upper Air Meteorological parameters, *J. Appl. Remote Sens*, **9(1)**, 094099.

EPJ Web of Conferences **119**, 05005 (2016) DOI: 10.1051/epjconf/201611905005
ILRC 27

Ozone profiles obtained by DIAL technique at Maïdo Observatory in La Reunion Island: comparisons with ECC ozone-sondes, ground-based FTIR spectrometer and microwave radiometer measurements.

T. Portafaix (1)*, S. Godin-Beekmann (3), G. Payen (2), M. de Mazière (4), B. Langerock (4), S. Fernandez (5), F. Posny (1), J.P. Cammas (2), J. M. Metzger (2), H. Bencherif (1), C. Vigouroux (4), N. Marquestaut (2),

*(1) LACy (UMR 8105) – Université de la Réunion, Saint Denis, France, * thierry.portafaix@univ-reunion.fr*
(2) OSU-Réunion - Université de La Réunion & CNRS (UMS 3365), Saint Denis, France
(3) LATMOS-IPSL, Guyancourt, France
(4) BIRA-IASB, Brussels, Belgium
(5) IAP, University of Bern, Switzerland

ABSTRACT

A DIAL lidar system performing stratospheric ozone profile measurements from 15 to 45 km is installed at Reunion Island (southwest of Indian Ocean). The purpose of this communication is to present this DIAL system mounted now at the new Maïdo Observatory since February 2013, and the ozone profile retrieval. The first stratospheric ozone profiles obtained during 2013 and 2014 will be presented and discussed. Inter-comparison and differences observed with other high vertical resolution ozone profiles performed by ECC ozonesonde will be shown. Finally, comparisons with low vertical resolution ozone profiles retrieved from microwave and FTIR remote sensing measurements performed at Maïdo will be carried out, making appropriate use of the associated averaging kernels

1. INTRODUCTION

Reunion Island is situated in the Southwest of Indian Ocean (55° East, 21° South). This location is particularly well suited for monitoring stratospheric ozone in the tropical belt of the southern hemisphere but also for studying quasi-horizontal (isentropic) and vertical (stratosphere-troposphere) exchange processes, variability and composition change of the tropical stratosphere in relationship with large-scale dynamics.

In support of these scientific topics, a new high-altitude observatory was built on the west coast of the island at 2200 m asl (Baray et al. 2013). It has been operating since October 2012 (Baray et al. 2013). The main objective of this station consists in observing the long-term chemical composition of the atmosphere in this region of the world where very few measurements are operational.

This facility called Maïdo Observatory hosts many instruments for monitoring the atmosphere by various techniques: in-situ measurements, radiosoundings, active and passive remote sensing experiments...

Stratospheric ozone measurements are regularly performed, on a routine basis following 3 experiments:

- a DIAL system to perform high-resolution vertical profiles of ozone in the 15-45km altitude range

- a high spectral resolution Fourier- Transform infrared spectrometer (FTIR) operated by the Belgian Institute for Space Aeronomy ozone is measured between 10_40km

- balloonborn soundings with ECC component to measure ozone profiles from ground to 30km

In addition, a microwave radiometer of University of Bern, has operated between late 2013 and early 2015.

2. STRATOSPHERIC DIAL SYSTEM AT REUNION ISLAND

This LIDAR was installed at Reunion Island in 2000 and moved to Maïdo facility in 2013 after instrumental updates.

Like any DIAL system, it requires the use of a pair of emitted wavelengths.

Laser sources are a tripled Nd:Yag laser (Spectra-Physics Lab 150) and a XeCl excimer laser (Lumonics PM 844). The Nd:Yag provides the non-absorbed beam at 355 nm with a pulse rate of 30 Hz and a power of 5W, and the excimer provides the absorbed beam at 308 nm with a pulse rate of 40 Hz and a power larger than 9W. An afocal optical system is used to reduce the divergence of the beam to 0.5 mrad.

The receiving telescope is composed of 4 parabolic mirrors (diameter: 500 mm). The backscattered signal is collected by 4 optical fibers located at the focal point of each mirror. The spectrometer used for the separation of the wavelengths is a Jobin Yvon holographic grating (3600 linesmm−1, resolution 3 Å.mm−1, efficiency >25 %).

The two Rayleigh beams at 308 and 355 nm are separated initially by the holographic grating and separated again at the output of the spectrometer by a lens system in the proportion 8% and 92 %, respectively, in order to adapt the signal to the non-saturation range of the photon-counting system. The optical signals are detected by 6 Hamamatsu non-cooled photomultipliers (PM). A mechanical chopper is used to cadence the laser shots and cut the signal in the lower altitude range where PM are saturated. This chopper consists of a steel blade rotating at 24 000 rpm in primary vacuum.

6 acquisition channels are recorded simultaneously: 2 channels at 355 nm corresponding to the lower and upper parts of the profile, 2 channels at 308 nm (lower and upper parts) and 2 Nitrogen Raman channels at 332 and 387 nm. In addition to the mechanical gating, both upper Rayleigh channels at 355 nm and 308 nm, are equipped with an electronic gating in order to cut the signals for the altitudes below 16 km and prevent signal-induced noise.

The system was moved to Maïdo Observatory by the end of 2012, after the update of the electronic system (now LICEL TR and PR transient recorders) and of the XeCl excimer laser. This new configuration allows us to obtain ozone profiles in the 15–45 km altitude range.

The lidar signals are recorded in a 3 min time file but averaged over the whole night acquisition (2 to 3h time integration per night) to increase the signal-to-noise ratio.

It is necessary to apply different corrections to the signal. The background signal is estimated and removed using an average or a linear regression in the high altitude range where the useful lidar signal is negligible (over 80 km). Another correction of the photomultiplier saturation for low layers is also required and applied.

The ozone number density is retrieved from the slope of signals after derivation (Godin-Beekmann et al., 2003).

The lidar signals are corrected the Rayleigh extinction using a composite pressure temperature profile computed from nearby meteorological soundings performed daily at Reunion Airport and the Arletty model (based on meteorological data from the European Centre)

It is also necessary in the DIAL technique to use a low-pass filter. The logarithm of each signal is fitted to a 2nd order polynomial and the ozone number density is computed from the difference of the derivative of the fitted polynomial. Varying the number of points on which the signals are fitted completes the filtering.

3. OTHER STRATOSPHERIC OZONE INSTRUMENTS AT MAIDO FACILITY.

A ground-based microwave radiometer (GROMOS-C) designed to measure middle atmospheric ozone profiles has been installed at Maïdo Observatory in 2014 and removed in early 2015. It has been specifically designed for campaigns and is remotely controlled and operated continuously under all weather conditions. It measures the pressure broadened line at 110.836 GHz and can also measure the CO line at 115.271 GHz. The vertical profiles are retrieved by optimal estimation method (Fernandez at al., 2015).

DOI: 10.1051/epjconf/201611905005

FTIR solar absorption measurements at high spectral resolution (from 0.0110 to 0.0035 cm^{-1} for ozone spectra) are performed by a Bruker 125HR spectrometer installed in 2013. This instrument is dedicated to NDACC measurements in the mid-infrared, covering the spectral range 600 to 6500 cm^{-1} (1.5 to 16 μm),and particularly the ozone retrievals are performed using the 1000-1005 cm^{-1} window in the 600-1400 cm^{-1} spectra (MCT detector, KBr beam-splitter). From the measured absorption spectrum, an inverse method (optimal estimation method) is used to trace back the vertical abundance profiles of gases present in the atmosphere. For ozone, information on about four independent layers in the atmosphere can be retrieved, roughly one in the troposphere and three in the stratosphere, up to about 45 km (Vigouroux et al., 2008). This instrument is operated remotely and automatically with an updated version of the BARCOS system (Neefs et al., 2007). In addition to the continuous monitoring of the atmospheric chemical composition and transport processes, the intention is also to participate to dedicated observations campaigns.

In addition, ECC ozone soundings are performed weekly at Reunion Island since 1998. The ozonesonde currently used is of ECC Z Ensci type with a 0.5% KI buffered solution from Droplet Measurement Technology [DMT]. It is coupled to a meteorological radiosonde M10 from MeteoModem. The effective vertical resolution of the ozone data is between 50 and 100 m [Thompson et al., 2003a,b, 2007]. The ozone measurement accuracy is around ±4% in the stratosphere below 10 mbar pressure level and the precision in total ozone column measured by the ECC sonde is around 5%. These ozone measurements are part of the SHADOZ (Thompson et al., 2003a, 2003b) and NDACC networks.

4. INTER- COMPARISONS

The first comparisons with ECC simultaneous sounding are very encouraging (Baray et al. 2013) with differences less than 10% throughout the profile. Figure 1 presents an example for June 23, 2014.

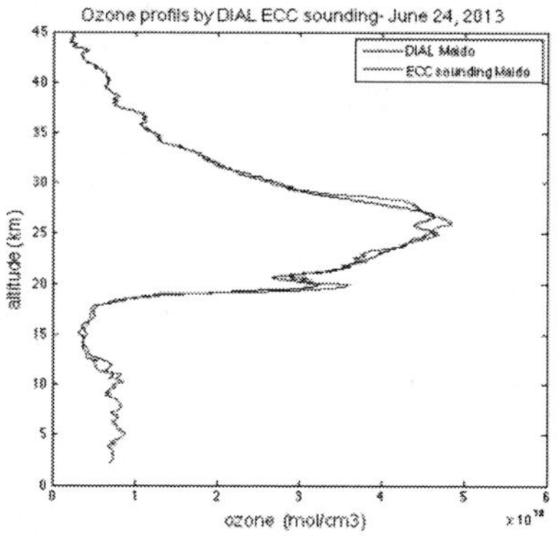

Fig 1: Ozone profiles on 24 June 2013 by stratospheric DIAL (black line) and ECC-ozonesonde at Maïdo Observatory (blue).

Other comparison between DIAL and GROMOS –C ozone profiles after applying the averaging kernel show a very good agreement in the layer between 5 and 20 hPa, with differences less than 5 %. But differences are more important on the lower or upper layers. It can reach more than 15 % in the 20-100 hPa layer.

These comparisons with the microwave radiometer were made using the DIAL "Rapid Delivery" profiles for the NDACC network, using average parameters for photomultiplier desaturation and background signal removing.

These medium parameters can introduce some additional error in the lower or upper part of the resulting profiles. It will be important for the final version of this paper to make these comparisons from consolidated lidar profiles using refined parameters.

The stratospheric ozone LIDAR is already NDACC qualified. It should be noted however that an inter-comparison campaign of all the NDACC lidar systems (water vapor, temperature, ozone) installed the Maïdo Observatory with the mobile system of NASA-GSFC (McGee et al., 1995) is planned for May 2015.

Comparisons with FTIR will be performed for the three layers between 15 and 45 km. The FTIR measurements in the ozone spectral range will be intensified during this 2015 intercomparison campaign.

ACKNOWLEDGEMENT

The present work is supported by LACy, OSU-Réunion and the FP7 European NORS project. The authors acknowledge the European Community, the Région Réunion, the CNRS, and the University of La Réunion for their support and contribution in the construction phase of the research infrastructure OPAR (Observatoire de Physique de l'Atmosphère à La Réunion). OPAR and LACy are presently funded by CNRS (INSU) and Université de La Réunion, and managed by OSU-R (Observatoire des Sciences de l'Univers à la Réunion, UMS 3365). We acknowledge Anne M. Thompson (NASA/GSFC, USA) the SHADOZ network principal Investigator, and E. Golubic, P. Hernandez and L. Mottet who are deeply involved in the routine lidar measurements at Maïdo facility.

REFERENCES

[1] Baray, J.-L., Courcoux, Y., Keckhut, P., Portafaix, T., Tulet, P., Cammas, J.-P., Hauchecorne, A., Godin Beekmann, S., De Mazière, M., Hermans, C., Desmet, F., Sellegri, K., Colomb, A., Ramonet, M., Sciare, J., Vuillemin, C., Hoareau, C., Dionisi, D., Duflot, V., Vérèmes, H., Porteneuve, J., Gabarrot, F., Gaudo, T., Metzger, J.-M., Payen, G., Leclair de Bellevue, J., Barthe, C., Posny, F., Ricaud, P., Abchiche, A., and Delmas, R.: Maïdo observatory: a new high-altitude station facility at Reunion Island (21° S, 55° E) for long-term atmospheric remote sensing and in situ measurements, Atmos. Meas. Tech., 6, 2865-2877, doi:10.5194/amt-6-2865-2013, 2013.

[2] Fernandez, S., A. Murk, and N. Kämpfer, GROMOS-C, a novel ground based microwave radiometer for ozone measurement campaigns, Atmos. Meas. Tech. Discuss., 8, 3001–3048, 2015 www.atmos-meas-tech-discuss.net/8/3001/2015/ doi:10.5194/amtd-8-3001-2015

[3] Godin-Beekmann, S., J. Porteneuve, A. Garnier, Systematic DIAL ozone measurements at Observatoire de Haute-Provence, J. Env. Monitoring, 5, 57-67, 2003

[4] McGee, T. J., Gross, M. R., Singh, U. N., Butler, J. J., and Kimvilakani, P. E.: Improved stratospheric ozone lidar, Opt. Eng., 34,1421–1430, 1995.

[5] Neefs, E.; de Mazière, M.; Scolas, F.; Hermans, C.; Hawat, T.: BARCOS, an automation and remote control system for atmospheric observations with a Bruker interferometer, Rev. Sci. Instrum., 78, 035109:1–035109:8, 2007.

[6] Thompson, A. M., et al. (2003a), Southern Hemisphere Additional Ozonesondes (SHADOZ) 1998–2000 tropical ozone climatology: 1. Comparison with Total Ozone Mapping Spectrometer (TOMS) and ground based measurements, J. Geophys. Res., 108(D2), 8238, doi:10.1029/ 2001JD000967.

[7] Thompson, A. M., et al. (2003b), Southern Hemisphere Additional Ozonesondes (SHADOZ) 1998–2000 tropical ozone climatology: 2. Tropospheric variability and the zonal wave one, J. Geophys. Res., 108(D2), 8241, doi: 10.1029/ 2002JD002241.

[8] Vigouroux, C., De Mazie`re, M., Demoulin, P., Servais, C., Hase, F., Blumenstock, T., Kramer, I., Schneider, M., Mellqvist, J., Strand-berg, A., Velazco, V., Notholt, J., Sussmann, R., Stremme, W., Rockmann, A., Gardiner, T., Coleman, M., and Woods, P.: Evaluation of tropospheric and stratospheric ozone trends over Western Europe from ground-based FTIR network observations, At- mos. Chem. Phys., 8, 6865–6886, doi:10.5194/acp-8-6865-2008, 2008.

Double-Pulse Two-Micron IPDA Lidar Simulation for Airborne Carbon Dioxide Measurements

Tamer F. Refaat*, Upendra N. Singh, Jirong Yu, and Mulugeta Petros

NASA Langley Research Center, Hampton, VA 23681, USA,

**Email: tamer.f.refaat@nasa.gov*

ABSTRACT

An advanced double-pulse 2-µm integrated path differential absorption lidar has been developed at NASA Langley Research Center for measuring atmospheric carbon dioxide. The instrument utilizes a state-of-the-art 2-µm laser transmitter with tunable on-line wavelength and advanced receiver. Instrument modeling and airborne simulations are presented in this paper. Focusing on random errors, results demonstrate instrument capabilities of performing precise carbon dioxide differential optical depth measurement with less than 3% random error for single-shot operation up to 11 km altitude. This study is useful for defining CO_2 measurement weighting function for adaptive targeting, instrument setting, validation and sensitivity trade-offs.

1. INTRODUCTION

Carbon dioxide (CO_2) is an important greenhouse gas, which critically affects Earth's climate. NASA has being developing several passive and active remote sensing instruments targeting precise monitoring of atmospheric CO_2 [1-2]. Recent efforts at NASA Langley Research Center (LaRC) focused on developing, integrating and validating a unique double-pulse 2-µm integrated path differential absorption (IPDA) lidar for CO_2 remote sensing [3-4]. This instrument utilizes a state-of-the-art, compact, hardened and highly efficient laser transmitter. The laser transmitter is based on solid state Ho:Tm:YLF crystal technology. The transmitter produces two Q-switched laser pulses at 2-µm wavelength for every 792 nm pump pulse. Targeting the CO_2 R30 line, the first 2-µm pulse is set to the on-line wavelength. The on-line wavelength can be tuned and locked to any position from 1 GHz to 6 GHz with 1 GHz step relative to the CO_2 R30 line peak toward longer wavelength. The transmitted on-line pulse energy is set to 90 mJ with 200 nsec pulse width. The second 2-µm pulse, produced within 200 µsec from the first, is set to a fixed off-line wavelength with 45 mJ pulse energy and 350 nsec pulse width. The laser repetition rate is set to 10 Hz. The IPDA receiver consists of a high quality 0.4 m diameter Newtonian telescope, which focuses the radiation, reflected from a hard target, onto 300-µm diameter detectors through field-imaging aft-optics. The detectors of choice are the extended range InGaAs pin photodiodes (Hamamatsu; G5853-23), the outputs of which are applied to trans-impedance amplifiers (TIA) (Femto; DHPCA-100). Similar detector and TIA combination is integrated inside the transmitter enclosure for laser pulse energy monitoring [5]. Signals from the hard target and the laser energy monitor are digitized and recorded using 12-Bit, 200 MS/s (Agilent; U1066A) and 10-Bit, 500 MS/s (Agilent; U1065A) digitizers, respectively.

CO_2 double-path differential optical depth, *dOD*, is the main IPDA product. Modeling the double-pulse 2-µm IPDA lidar instrument is the focus of this paper. IPDA lidar modeling is critical for instrument validation and simulation. Instrument validation is conducted by comparing and evaluating the IPDA measurement versus model calculations. Besides, modeling provides the necessary tools for converting *dOD* into CO_2 weighted average column dry-air volume-mixing ratio that can be compared to the gas *in-situ* sensors. Instrument simulations are necessary to define the measurement sensitivity and limits against the various IPDA parameters settings, such as on-line wavelength and gain selections, and operating conditions, such as airborne platform altitude, ground conditions and environment. In the presented model, six different altitudes are assumed, spanning from 1 km to 11 km, with 2 km step. This allows fast scaling to any altitude assuming a small aircraft platform, such as the NASA B-200. This IPDA model is based on the US standard atmosphere model which provides typical meteorological profiles

DOI: 10.1051/epjconf/201611905006

[1]. End products include operation charts, which produce estimates of return signal strength, *dOD*, signal-to-noise ratio (SNR) and random error.

2. SPECTROSCOPY

Figure 1 shows the composite absorption spectra of CO_2 around the R30 line, located at 2050.9670 nm. The figure also compares similar spectra for water vapor (H_2O), which is the dominant interfering gas for CO_2 measurement at 2-µm. The spectra were derived using HITRAN for line parameters, assuming Voigt line profile at two altitudes [1]. In this calculation, 5550 and 1816 lines were included for CO_2 and H_2O, respectively. The vertical lines mark the different IPDA on-line tuning and off-line positions. Due to selected wavelength positions, H_2O differential cross section is negligible with respect to CO_2, which results in minimal molecular interference error. Focusing on the selected operational wavelengths, the variation of the absorption cross section, σ, versus altitude is shown in figure 2.

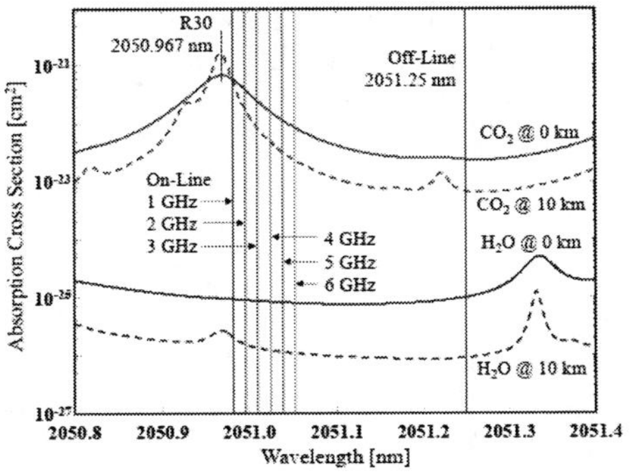

Figure 1. CO_2 and H_2O absorption spectra and IPDA operational wavelength settings.

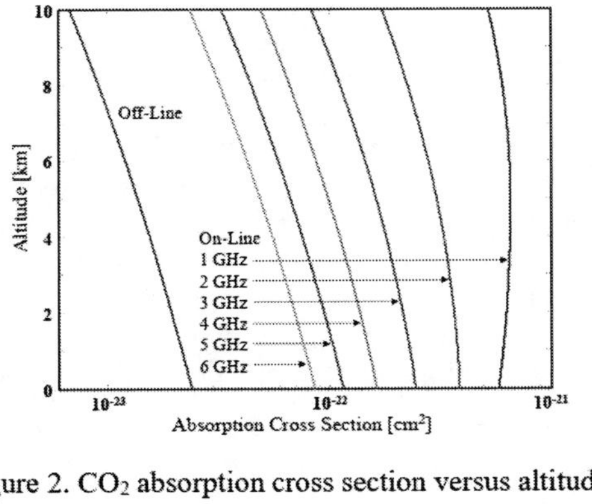

Figure 2. CO_2 absorption cross section versus altitude.

3. IPDA LIDAR SIMULATION

IPDA lidar operation relies on the hard target lidar equation, where the return power, P (in Watt) is given by the equation

$$P = \eta_r \cdot \varphi_r \cdot \frac{A}{(R_A - R_G)^2} \cdot \frac{E}{t} \cdot \frac{\rho}{\pi} \cdot T \tag{1}$$

for a nadir airborne instrument at a fixed altitude, R_A, and ground elevation, R_G, above sea level. In this equation, η_r is the receiver optical efficiency, φ_r is the overlap function, A is the telescope area, E is the transmitted energy, t is the transmitted pulse width, ρ is the surface reflectivity and T is the total transmission, defined at the operating wavelength, λ. For a given wavelength, the total transmission was calculated according to

$$T = \exp\left(-OD_{cd} - OD_{wv} - OD_A\right) \tag{2}$$

where OD_{cd}, OD_{wv} and OD_A are the CO_2, H_2O and aerosol optical depths, respectively. Aerosol optical depth was obtained through standard model [1]. Appling US standard atmosphere model, the CO_2 double-path optical depth is calculated from

$$OD_{cd} = 2\int_{R_G}^{R_A} \sigma_{cd} \cdot N_{cd} \cdot dr \tag{3}$$

where N_{cd} is the CO_2 number density. N_{cd} vertical profile was updated from the US standard model to a recent surface dry mixing ratio of 397 ppm. H_2O double-path optical depth was calculated using the same equation. Figure 3 shows the total transmission profiles obtained for the operating wavelengths from different airborne altitudes.

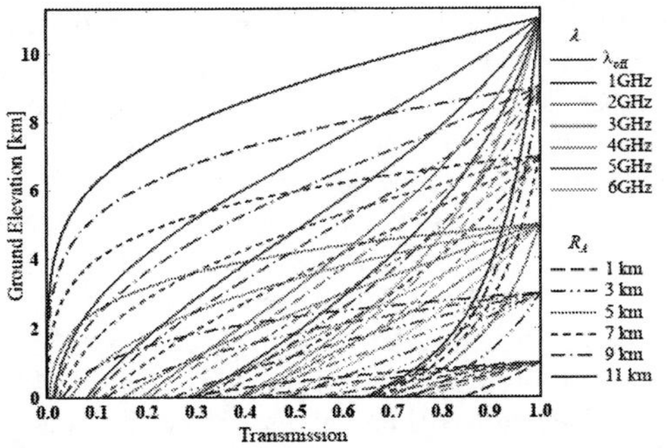

Figure 3. IPDA lidar nadir transmission at different operating on-line settings and altitudes.

DOI: 10.1051/epjconf/201611905006

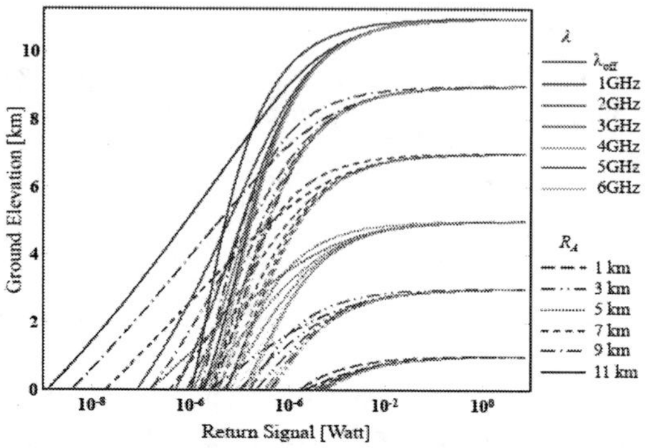

Figure 4. IPDA lidar nadir return power from ocean surface at different operating conditions.

The presented transmissions were substituted in equation (1) to estimate the hard target return power, as shown in figure 4. For this calculation 65% receiver optical efficiency, with full overlap (φ_r=1), and moderate ocean reflectivity were considered (i.e., R_G=0 and ρ=0.08). With detector responsivity set to 1.15 A/W, the return power chart is useful to set proper TIA gains that match the output voltage to the digitizer input limits. Besides, it gives a quick estimate for critical IPDA performance parameters, such as SNR and random error.

4. IPDA LIDAR PERFORMANCE

Figure 5 shows the 2-µm IPDA lidar, single-shot power SNR variation with ground elevation for different airborne altitude and wavelength settings. SNR were calculated as the ratio of the return power, shown in figure 4, to the total noise power. The total noise power was obtained by combining the instrument fixed noise and signal dependent shot noise. Fixed noises include electronic noises and background radiation. Dominant electronic noises sources, such as detector dark current and TIA feedback Johnson noise, input current and voltage noises and coupling noise, were considered. Other electronic noise sources, such as digitizer noise and interference, were found weak and neglected. The presented SNR is based on the IPDA lidar receiver assuming ideal performance. SNR could be deteriorated due other factors such as speckles and pulse energy fluctuations of the transmitted laser. Both of these effects were minimized in the system design and thus neglected in this analysis.

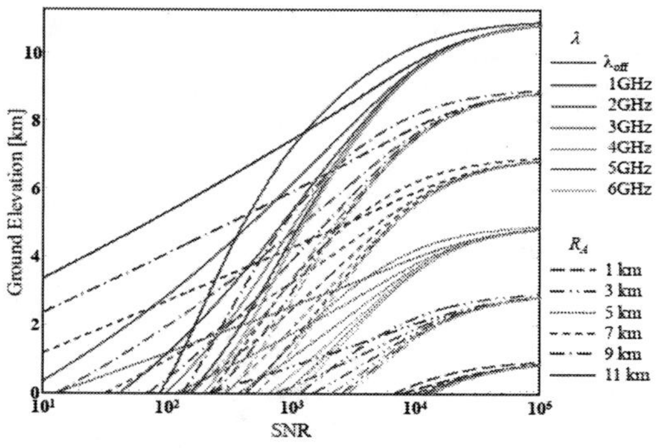

Figure 5. IPDA lidar SNR estimates at different operating conditions.

For example, speckle SNR could be enhanced by the transmitted laser beam divergence and the sampling time. Besides, laser pulse fluctuation effects could be minimized by proper energy monitoring and normalization per shot [5].

During the 2-µm IPDA operation the selected on-line and off-line return pulses are collected per shot. During data analysis, both on- and off-lines pulses are reduced to obtain the CO_2 *dOD* measurement, which according to equation 3, can be modeled as

$$dOD_{cd} = OD_{cd}(\lambda_{on}) - OD_{cd}(\lambda_{off}) \tag{4}$$

where λ_{on} and λ_{off} are the on-line and off-line wavelengths, respectively. The uncertainty associated with *dOD* measurement, $\delta(dOD)$, can be estimated from the SNR of the return signals using the equation [1]

$$\delta(dOD_{cd}) = \sqrt{\frac{SNR_{on}^{-2} + SNR_{off}^{-2}}{N}} \tag{5}$$

This *dOD* uncertainty represents the measurement random error caused by the receiver. In the above equation SNR_{on} and SNR_{off} are the on- and off-line return signal SNR, respectively, presented in figure 5, and N is the number of shot averages. Figure 6 shows the variation of the CO_2 double-path differential optical depth with ground elevation for different flight altitude and operating on-line settings. The corresponding random errors are shown in figure 7, for single-shot operation.

The random error was calculated as the ratio of the measurement uncertainty, from equation (5) to differential optical depth, from equation (4).

Results indicate the possibility of measuring CO_2 optical depth accurately with less than 3% single-shot random error (i.e., without averaging). From equation (5), shot averaging would reduce the error further. For example, operating the IPDA instrument from a platform at 1 km altitude, a minimum error of 0.02% can be achieved with 1 GHz on-line setting. At this setting, both on- and off-line return powers would be within 100 μW with *dOD* around unity and 10^4 SNR. Operating the instrument from a platform at 11 km altitude, 4 GHz would result in a minimum error of 1.2% at sea level that is dominated by the off-line SNR. This error would reduce to 0.12% with 10 sec shot average. For the same altitude, 3 GHz would result in lower errors than 4 GHz for ground elevation above 800 m.

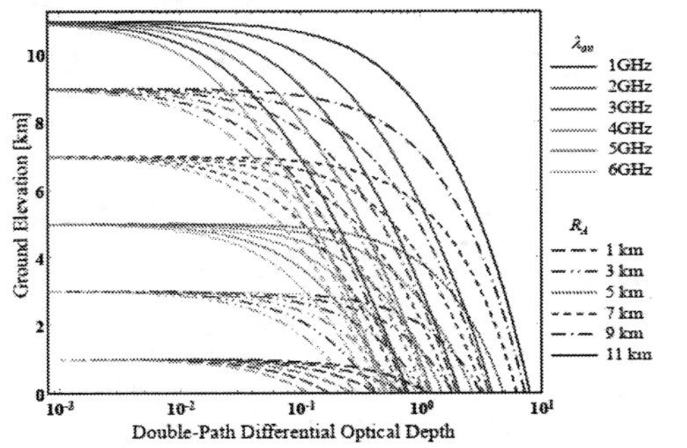

Figure 6. Double-pulse 2-μm CO_2 IPDA lidar double-path differential optical depth.

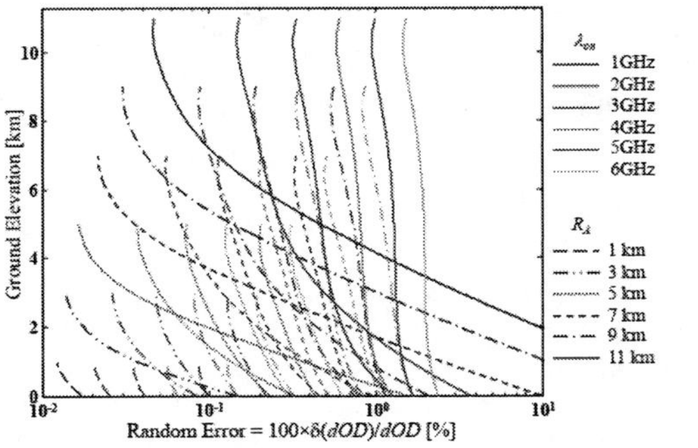

Figure 7. CO_2 differential optical depth random error, calculated assuming single-shot operation.

4. CONCLUSIONS

An advanced double-pulse 2-μm IPDA lidar has been developed at NASA LaRC for measuring atmospheric CO_2. The instrument utilizes a state-of-the-art 2-μm tunable laser transmitter and advanced receiver. IPDA lidar modeling is useful for defining the instrument setting and measurement validation and sensitivity trade-offs. Airborne simulations resulted in less than 3% random error for single-shot operation up to 11 km altitude. This demonstrates the 2-μm IPDA lidar capability of performing precise CO_2 measurement from airborne platforms. This model has been used to validate the CO_2 2-μm IPDA lidar instrument performance from ground and airborne platforms as will be the presented in future studies.

ACKNOWLEDGEMENT

This work (ESTO ATI-QRS-2012, Principal Investigator: Upendra Singh, LaRC) was funded and supported by NASA Earth Science Technology Office (Program Director: George Komar and Program Manager: Parminder Ghuman).

REFERENCES

[1] T. Refaat, U. Singh, J. Yu, M. Petros, S. Ismail, M. Kavaya, K. Davis, 2015: Evaluation of an airborne triple-pulsed 2 μm IPDA lidar for simultaneous and independent atmospheric water vapor and carbon dioxide measurements, *Appl. Opt.* **54**, 1387-1398.

[2] U. Singh, B. Walsh, J. Yu, M. Petros, M. Kavaya, T. Refaat, N. Barnes, 2015: Twenty years of Tm:Ho:YLF and LuLiF laser development for global wind and carbon dioxide active remote sensing, *Opt. Mater. Exp.* **5**, 827-837.

[3] J. Yu, M. Petros, K. Reithmaier, Y. Bai, B. Trieu, T. Refaat, M. Kavaya, U. Singh, S. Ismail, 2012: A 2-micron pulsed integrated path differential absorption lidar development for atmospheric CO_2 concentration measurements," *in 26th ILRC*, Porto Heli, Greece.

[4]. U. Singh, J. Yu, M. Petros, T. Refaat, R. Remus, J. Fay, K. Reithmaier, 2014: Airborne 2-micron double-pulsed integrated path differential absorption lidar for column CO_2 measurement", *Proc. SPIE* **9246**, 924602.

[5] T. Refaat, M. Petros, R. Remus, J. Yu, U. Singh, 2014: Laser energy monitor for double-pulsed 2-μm IPDA lidar application, *Proc. SPIE* **9246**, 924606.

EPJ Web of Conferences **119**, 05007 (2016)
ILRC 27
DOI: 10.1051/epjconf/201611905007

Remote Sensing of Greenhouse gases by combining Lidar and Optical Correlation Spectroscopy

C. Anselmo[1*], B. Thomas[2], A. Miffre[1], M. Francis[1], J.P. Cariou[3], P. Rairoux[1]

[1]*Institut Lumière Matière, CNRS, UMR 5306 Université Lyon 1, 69622 Villeurbanne, France, *Email : christophe.anselmo@univ-lyon1.fr*

[2]*NOAA CREST Center and Remote Sensing Laboratory, Department of Electrical Engineering, City College, NY, 10031, US*

[3]*Leosphere France, 14-16 rue Jean Rostand, 91400 Orsay, France*

ABSTRACT

In this contribution, we present recent work on the ability to achieve range-resolved greenhouse gases concentration measurements in the Earth's atmosphere (CH_4, H_20) by combining broadband optical correlation spectroscopy (OCS) with lidar. We show that OCS-Lidar is a robust methodology, allowing trace gases remote sensing with a low dependence on the temperature and pressure-variation absorption cross section. Moreover, we evaluate, as an experimental proof, the water vapor profile in the planetary boundary layer using the 4ν 720 nm absorption band.

1. INTRODUCTION

As underlined in the latest IPPC report [1], there is a need for determining the spatial and temporal concentrations of atmospheric constituents. Therefore, Lidar is an efficient monitoring device to probe trace species [2] and also particulate matter [3] coupled with complex atmospheric processes [4]. We have recently developed the OCS-Lidar, a novel instrument that is based on coupling a Lidar with Optical Correlation Spectroscopy (OCS). Its principle and the theoretical approach have then been established [5] leading to a comprehensive study on OCS-Lidar errors budget in the NIR [6]. Moreover OCS-Lidar based on a femtosecond broadband laser source has been successfully applied to retrieve water vapor in the lower atmosphere using the 4ν 720 nm absorption band [5].

This abstract is organized as follows. The OCS-Lidar methodology and the formalism aimed at retrieved the TG concentration are presented in Section 2. Numerical simulations of [CH_4] remote sensing for extreme pressure and temperature conditions are presented in Section 3. Finally, a field measurement of OCS-Lidar dedicated to water vapor in the planetary boundary layer is presented. The abstract ends with a conclusion.

2. METHODOLOGY

OCS-Lidar is an active optical remote sensing method able to monitor atmospheric trace using broadband laser pulse. Its principle is displayed in figure 1.

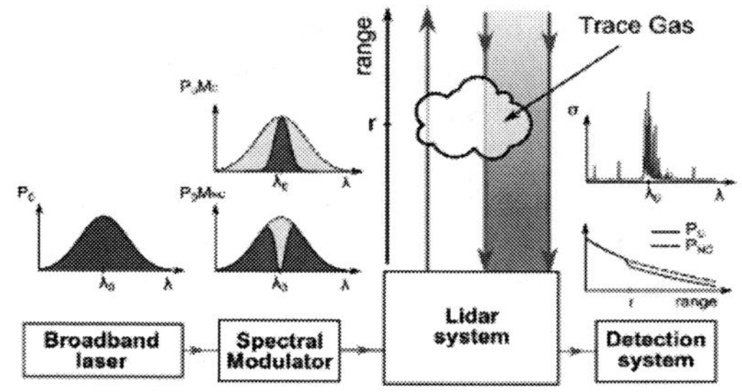

Fig. 1: Scheme of the OCS-Lidar principle.

The broadband laser pulse power density $P_0(\lambda)$ is spectrally shaped by alternatively applying two different amplitude modulations M_C and M_{NC}. A spectral modulator is used to custom one emission laser pulse to be spectrally correlated (C) with the trace gas absorption spectrum while a second one, which is a non-correlated (NC) laser pulse, serves as OCS-Lidar reference signal.

The target gas (TG) concentration profile is retrieved from the signals without the requirement of a permanent calibration gas cell. However a calibration cell could serve to reveal experimental biases. The backscattered OCS-Lidar signal P_i(r) can be written as follows:

$$P_i = \frac{K(r)}{r^2} \int_{\Delta\lambda} P_0(\lambda).M_i(\lambda).\beta(r,\lambda).T^2(r,\lambda).\eta(\lambda)\, d\lambda \quad (1)$$

where M_i is the modulator transmission which ensures the correlation (i = C) or not (i = NC). Δλ is the spectral bandwidth of the optical set-up. $\beta(r, \lambda)$ is the atmosphere backscattering coefficient, while $K(r)$ depends on the receiving optics geometry and $\eta(\lambda)$ is the optical set-up efficiency. Finally, T(r,λ) is the transmission of the atmosphere which includes the trace gas (TG) concentration profile. From the ratio of OCS-Lidar signals P_C(r) and P_{NC}(r), we retrieved an analytical expression of [TG](r) by approximating the TG transmission present in T(r,λ) with a Taylor expansion series of the Lambert Beer equation. The mathematical description can be found elsewhere [5].

 DOI: 10.1051/epjconf/201611905007

3. EXPERIMENT

3.1 Methane

A numerical model has been developed to generate NIR OCS-Lidar signals and to evaluate the errors budget associated to [CH_4] remote sensing. It includes detector noise and variation of the Methane absorption spectral lines with the atmosphere thermodynamic changes. Fig. 2 shows, in the form of a color plot, the relative error on the retrieved [CH_4] by considering the temperature and pressure variations in the troposphere. There, for the input of the model, we consider a 20 ppm Methane peak concentration located at a 1000 meters distance with a 100 meters width range. This numerical simulation [6] shows that the error on [CH_4] does not overcome 14 % when the temperature and pressure respectively vary by 200 K and 900 mbar.

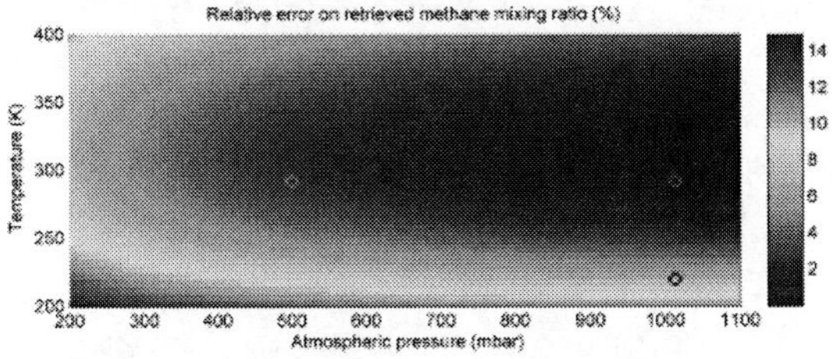

Fig. 2: Relative error in % on retrieved Methane mixing ratio assuming initial atmosphere temperature and pressure of 292 K and 1013 mbar respectively.

3.2 Water Vapor

The first experimental proof of OCS-Lidar has been performed on the measurement of the water vapor content in the PBL [5].

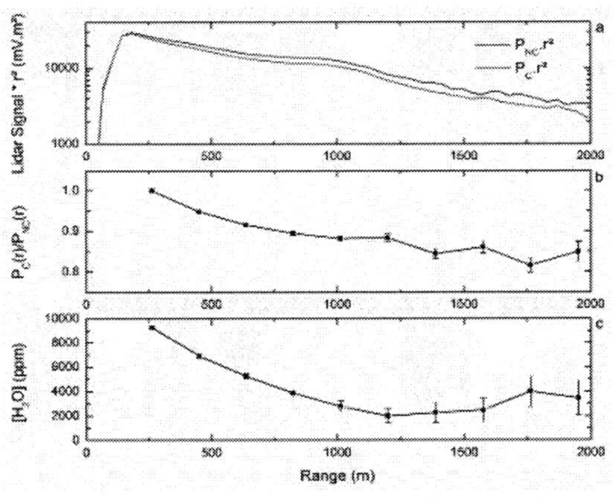

Fig. 3: OCS-lidar water vapor measurement in the PBL. (a) Range-corrected OCS-lidar signals. (b) The ratio P_C /P_{NC} (c) Retrieved water vapor mixing-ratio profile.

We also demonstrate the ability to realize the experiment using two different configurations of M_i (see Eq. 1): on the laser emission or on the reception part. Fig. 3 shows the results of the field experiment when achieving the amplitude modulation $M_i(\lambda)$ on $P_0(\lambda)$. The water vapor concentration profile in the PBL could be retrieved up to a 2000 meters distance with a 200-m range resolution. The retrieved water vapor mixing-ratio profile starts at (7,850 ± 55) ppm at a 260-m range and reaches (3,500 ± 1,000) ppm at the maximum range of 1900 meters. These first experimental results show water vapor mixing-ratio values in agreement with standard atmosphere. Moreover, near ground level (30 m), in situ hydrometer measurement shows similar values within 15 %.

4. CONCLUSIONS

We here present the OCS-Lidar based broadband laser source, which is a novel optical methodology to remotely retrieve the concentration of atmospheric greenhouse gases. We demonstrate the ability of this methodology to achieve water vapor filed measurements. Moreover, based on a numerical model, we demonstrate that the OCS-Lidar applied to [CH_4] detection is almost insensitive to changes in absorption cross-section due to extreme temperature and pressure variations. Future work on [CH_4] and on other traces gases remote sensing are in progress.

REFERENCES

[1] IPCC, Climate Change 2013: The Physical Science Basis. Working Group I Contribution to the Fifth Assessment Report of the Intergovernmental Panel on Climate Change (Cambridge University Press, (2013).

[2] C. Weitkamp, "Lidar. Range-Resolved Optical Remote Sensing of the Atmosphere". Springer Ed., (2005).

[3] G. David et al., "Retrieving simulated volcanic, desert dust, and sea-salt particle properties from two / three-component particle mixtures using UV-VIS polarization Lidar and Tmatrix," Atmos. Chem. Phys. **13** (14), 6757–6776 (2013).

[4] David, et al., "UV polarization lidar for remote sensing new particles formation in the atmosphere", *Opt. Exp.*, **22**, A1009-A1022, (2014).

[5] B. Thomas et al., "Remote sensing of atmospheric gases with optical correlation spectroscopy and Lidar: first experimental results on water vapor profile measurements," Appl. Phys. B **113**(2), 265–275 (2013).

[6] B. Thomas et al., "Remote sensing of methane with broadband laser and optical correlation spectroscopy on the Q-branch of the 2v3 band," J. of Mol. Spec., **291**, 3–8, (2013).

THE ZUGSPITZE RAMAN LIDAR: SYSTEM TESTING

Katharina Höveler, Lisa Klanner, Thomas Trickl, Hannes Vogelmann

Karlsruher Institut für Techologie, IMK-IFU, Kreuzeckbahnstr. 19, 82467 Garmisch-Partenkirchen Germany; e mail: thomas.trickl@kit.edu

ABSTRACT

A high-power Raman lidar system has been installed at the high-altitude research station Schneefernerhaus (Garmisch-Partenkirchen, Germany) at 2675 m a.s.l., at the side of the existing wide-range differential-absorption lidar. An industrial XeCl laser was modified for polarized single-line operation at an average power of about 175 W. This high power and a 1.5-m-diameter receiver are expected to allow us to extend the operating range for water-vapour sounding to more than 25 km, at an accuracy level of the order of 10 %. In addition, temperature measurements in the free troposphere and to altitudes beyond 80 km are planned. The system is currently thoroughly tested and exhibits an excellent performance up to the lowermost stratosphere. We expect that results for higher altitudes can be presented at the meeting.

1. INTRODUCTION

The high spatial and temporal variability of water vapour in the climate-relevant upper troposphere (UT) and lower stratosphere (LS) (e.g., [1]), together with LS mixing ratios of the order of 5 ppm, impose tough boundary conditions for vertical sounding. Lidar measurements fulfil the resolution requirements, but due to the very low mixing ratio of about 5 ppm lidar sounding of H_2O in the stratosphere is a particularly demanding task for ground-based systems. Considerable efforts for extending lidar measurements into the LS have been reported by groups of the Network for the Detection of Atmospheric Composition Change (NDACC, http://www.ndsc.ncep.noaa.gov/). The measurements, carried out during night-time with Raman lidar systems, have been extended to about 20 km with uncertainties of the order of 20 % [1, 2]. This is an important step towards filling the existing gap in achieving frequent accurate, vertically resolved measurements between 10 and 20 km.

In order to improve these specifications we have developed a Raman lidar system that should provide roughly one hundred times more backscatter signal than existing systems. More reliable results can be obtained in a substantially shorter data-acquisition time. The new lidar system yields an ideal extension of the measurements with our differential-absorption lidar (DIAL) that provides accurate water-vapour profiles in most of the free troposphere [1,4-7]. Both systems are located in the same laboratory at the Schneefernerhaus high-altitude station (UFS) at an altitude of 2675 m. The DIAL has been in routine operation since 2007.

Due to the considerable light absorption by the tropospheric water vapour a range extension of the DIAL measurements into the stratosphere would require a research platform located at an unrealistic altitude of about 7.5 km [4]. Therefore, stratospheric DIAL measurements are restricted to airborne systems (e.g., [8]). Despite their considerably lower sensitivity Raman lidar systems offer a higher potential for ground-based routine measurements of stratospheric water vapour since the tropospheric radiation losses are substantially less significant. In contrast to the DIAL method, Raman backscattering is, after suppressing light from minor sources such as stars, background free during night-time and the concentrations can be directly related to the signal level. The quality of the H_2O data grows with the number of photons collected.

2. LASER SYSTEM

A rather explicit system description was given in our previous ILRC contribution [9]. Here, we just concentrate on the most important properties.

The radiation source of the Zugspitze Raman lidar is a Lambda SX XeCl laser (Coherent, formerly Lambda Physik) with up to 1.2 J pulse energy at 308 nm and 350 Hz repetition rate. Stabilized operation at 300 W can be maintained over at least 55 h. Since this laser system is normally used for industrial applications it was not fully suitable for the lidar application. We had to modify the optical layout to obtain linearly polarized single-line output with strongly reduced beam divergence.

The cavity was extended as shown in Fig. 1. A custom-made large intra-cavity etalon with 0.1 mm plate spacing and 70 mm diameter (SLS Ltd., compromise: T = 0.9634, R = 0.55) and a thin-film polarizer (Laseroptik G.m.b.H., T = 0.94 %) were installed. The narrow beam was expanded to square shape (about 35×35 mm^2) with a cylindrical telescope for reducing the intensity impinging on both etalon and polarizer. Apertures were introduced in order to avoid damage due to strong reflections or stray light. The etalon is mounted on a rotation stage that can be rotated manually or under computer control. The etalon is the main source of loss that strongly grows at high repetition rates, limiting the output energy to about 0.5 J at 350 Hz.

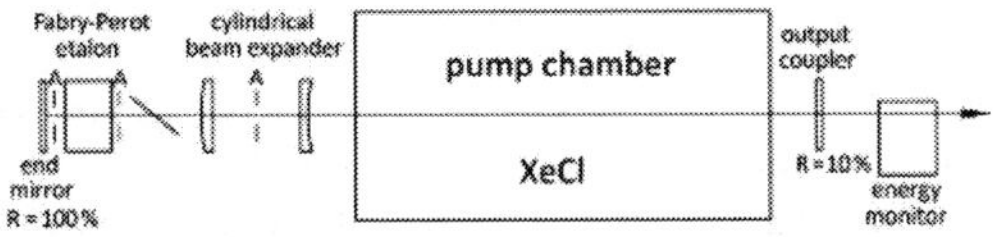

Fig. 1: Schematic diagram of the modified laser system; A: sand-blasted aluminium apertures

The laser has been operated on its high-frequency component (λ = 307.955 nm). All other contributions are very small. The spectral impurity can be as low as 0.5 %, but is mostly around 1 % at the highest repetition rates. The residual contamination, which cannot be avoided, is in agreement with the unfiltered direct forward emission of the laser of a few millijoules. The wavelength is controlled to within 0.025 nm by a calibrated grating spectrometer (Ocean Optics, model HR4000).

Linear polarization of the radiation is important for achieving spectrally clean stimulated Raman shifting with best efficiency (see below) [9] and for efficiently reducing the strong Cabannes component in the rotational Raman channels of the detection system. After inserting the T = 94-% thin-film polarizer a power loss of just 4 % was observed. A degree of linear polarization of 99.5 % was obtained and can be further improved.

The laser beam is focussed into a Raman shifter 3.6 m long with an f = 2 m lens. The Raman cell is filled with about 30 bar of hydrogen for generating a reference emission at 353.1 nm needed for ozone corrections. The long focal length is chosen to avoid radiation losses by optical breakdown and the generation of higher Stokes orders. At low repetition rates a conversion up to 20 % was observed at 20 bar. For higher repetition rates the conversion diminishes since the energy gradually decreases to about 0.5 J which seems to be the threshold for Raman emission. The observation of 353-nm radiation is extremely sensitive to the laser alignment. Therefore, we plan to reduce the focal length to 1.75 m.

From the focal point the beam expands towards an f = 10 m concave mirror used for collimation (Fig. 2). This means an overall beam expansion by a factor of five and a reduction of the beam divergence to less than 0.5 mrad, necessary for sufficiently tight focusing of the backscattered light in the two receivers.

3. RECEIVERS

Due to the expected nine-decade dynamic range of the system [9] two separate Newtonian receivers for near-field and far-field detection are used (d = 0.38 m, f = 2 m, and d = 1.5 m, f = 5 m, respectively). The entire set-up is shown in Fig. 2. Because of the long focal length of the large mirror the receivers are mounted in a tower on the terrace above the lidar laboratory, covered by a dome suitable for arctic conditions (Baader).

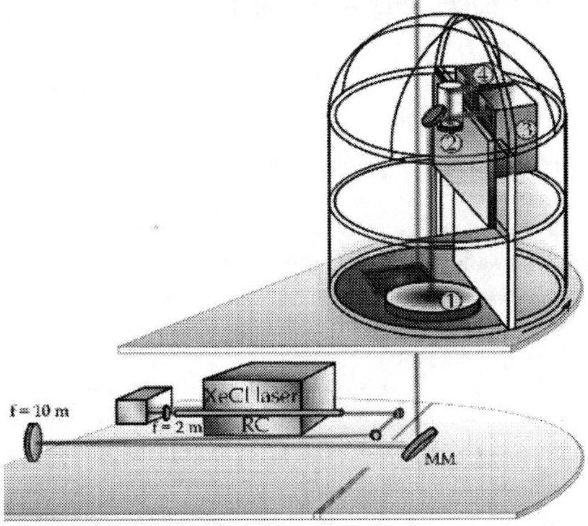

Fig. 2: Sketch of the Raman lidar system; abbreviations: RC: Raman cell, filled with hydrogen; MM: motorized beam steering mirror; 1: principal mirror of the far-field telescope; 2: near-field telescope; 3: far-field polychromator; 4: near-field polychromator. The entire receiver tower on the roof is rotated counterclockwise by more than 90° in the drawing for better visibility.

DOI: 10.1051/epjconf/201611905008

The radiation from both telescopes is focussed into two six-channel polychromators as shown in Ref. 9. A combination of polarization-sensitive optics and interference filters (Laseroptik G.m.b.H. and Materion Barr). The spectral widths are 0.25 nm (f.w. h.m.) for all channels except for H_2O where a 0.75-nm filter with 70 % transmittance is used.

The radiation is detected with Hamamatsu R7400 P-03 photomultiplier tubes (PMTs) with actively stabilized sockets yielding single-photon pulses without ringing (Romanski Sensors). The signal is processed by Licel 12-bit/20-MHz transient digitizers with a new ground-free input stage and a 5-GHz photon counting system (FAST ComTec).

4. LIDAR TESTING

Rayleigh Signals

The 308-nm backscattered radiation collected by both receivers had to be attenuated by a factor of 1000 for matching the useful voltage range of the detectors. Cutting off some of the near-field portion of the signal by shifting a blade in the focal plane in front of the PMT was not ultimately successful due to the finite divergence of the laser beam, and the operating range could not be extended to much more than 60 km. The operating range for 353 nm will be determined after improving the Raman conversion.

Water Vapour

The first water vapour measurement took place on December 12, 2012. A peak analogue signal of about 10 mV was observed at a distance of about 1 km. The analogue signal covered an astonishing dynamic range of 5 decades after a very small exponential background correction (2×10^5 laser shots). The range-corrected signal is shown in Fig. 3, together with the 24:00 UTC Munich radiosonde data, scaled to match the lidar data, and an NCEP (National Centers for Environmental Prediction) interpolation or the position of our site. Due to a situation with stratospheric layers during that day some differences exist. However, the general behaviour looks reasonable, even with some indication of a humidity drop above the tropopause. During the measurement the polychromator box was not closed to allow for optimizing the alignments. Thus, there was a certain level of residual background light in the detection system that led to elevated noise.

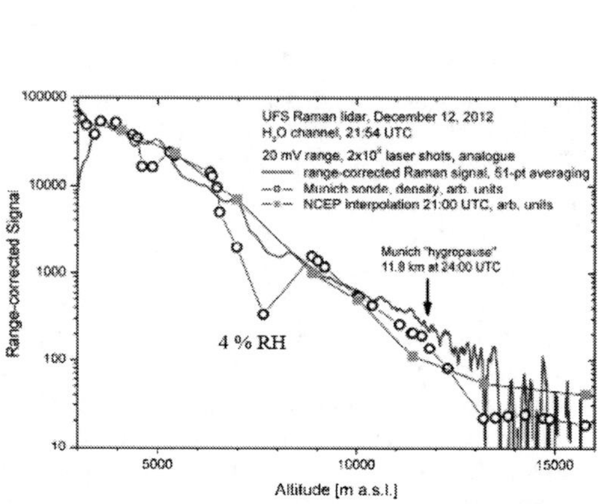

Fig. 3: First water-vapour profile of the large receiver on December 12, 2012 (range-corrected signal); the data are smoothed with a 51-point running average (corresponding to a 187.5-m vertical resolution [10]).

In Fig. 4 the first successful test including single-photon counting is shown for a measurement time of about 1 h (10^6 laser shots). The humidity can be traced to about 13.5 km, the mixing ratio between 10 and 12 km being already as low as 10-15 ppm. The agreement of photon-counting and analogue data is excellent (starting at 4.2 km) despite a slight exponential background correction of the digitizer output. The photon-counting data are only slightly less noisy, in surprising agreement with the larger 15-m range interval. In the upper troposphere the shape of the Munich sonde humidity profile is almost identical. In absence of better information we calibrated our data in this altitude range. There are pronounced differences between lidar and sonde below 6.7 km (100 km distance).

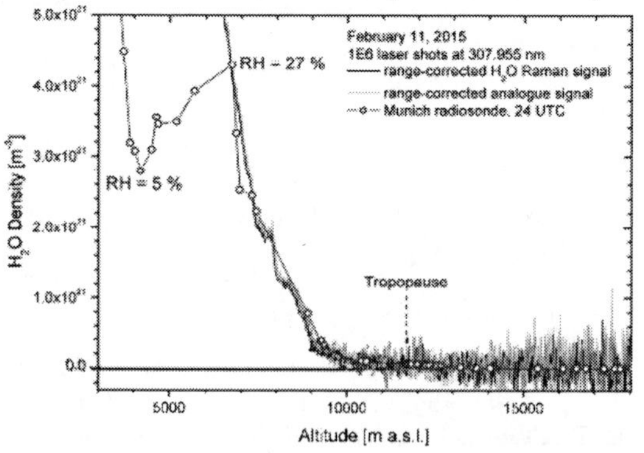

Fig. 4: Comparison of photon-counting (black) and analogue results; the noise is caused by about 7500 photon counts in the chosen 100-ns bins due to insufficient shielding during the test phase.

During the following week the near-field telescope could be successfully tested. There was good overlap for distances above 0.3 km. The water-vapour profile extended to 9 km.

Temperature

The temperature measurements in the stratosphere will most likely be based on the 353-nm signal and the rather strong N_2 vibrational Raman signals. First successful temperature retrievals from the two rotational Raman channels were made for the small telescope and agree with the sonde and NCEP temperatures within 1 K in a major part of the free troposphere. The results are currently examined in more detail. We plan to try also alternative retrieval algorithms.

5. CONCLUSIONS

After a major break the test phase of the Zugspitze Raman lidar was recently resumed. The results show that, after reducing the background light by 2-3 decades, a substantial gain in the H_2O operating range can be achieved. The outcome of this effort will be presented at the meeting.

The results clearly demonstrate the necessity of using the H_2O DIAL, operated in the same laboratory, for calibration and long-term quality assurance. One successful comparison was already made during a complex stratospheric intrusion event with several extremely dry layers seen by both systems.

More uncertainty is associated with the density-related temperature measurements that were predicted to cover a range up to more than 80 km. At this time our hope in the 308-nm Rayleigh channel is strongly diminished. Unless a small-size chopper could be used an optimization of the 353-nm signal is necessary. In the troposphere the 307.36-nm interference filter insufficiently blocks interference from the elastic channel. Several options will be tested.

ACKNOWLEDGEMENTS

The authors thank H. P. Schmid for his support and the UFS team for their great help. The system development was funded by the Bavarian State Ministry of Environment and Health.

REFERENCES

[1] Vogelmann, H., R. Sussmann, T. Trickl, A. Reichert, 2015: Spatiotemporal variability of water vapor investigated using lidar and FTIR vertical soundings above the Zugspitze, *Atmos. Chem. Phys.*, **15**, 3135-3148

[2] Whiteman D., K. Rush, S. Rabenhorst, W. Welch, W. Cadirola, G. McIntryre, F. Russo, M. Adam, D. Venable, R. Connell, I. Veselovskii, R. Forno, B. Mielke, B. Stein, T. Leblanc, S. McDermid, H. Vömel, 2010: Airborne and Ground-Based Measurements Using a High-Performance Raman Lidar, *J. Atmos. Oceanic Technol.*, **27**, 1781-1801

[3] Leblanc T., I. S. McDermid, T. D. Walsh, 2012: Ground-based water vapor raman lidar measurements up to upper troposphere and lower stratosphere for long-term monitoring, *Atmos. Meas. Technol.*, **5**, 17-36

[4] Vogelmann H., T. Trickl, 2008: Wide-range sounding of free-tropospheric water vapor with a differential-absorption lidar (DIAL) at a high-altitude station, *Appl. Opt.*, **47**, 2116-2132

[5] Wirth M., A. Fix, G. Ehret, J. Reichardt, R. Begie, D. Engelbart, H. Vömel, B. Calpini, G. Romanens, A. Apituley, K. M. Wilson, H. Vogelmann, T. Trickl, 2009: Intercomparison of Airborne Water Vapour DIAL Measurements with Ground Based Remote Sensing and Radiosondes within the Framework of LUAMI 2008, Contribution S07-P01-1 (3 pp.) in: *Proceedings of the 8th International Symposium on Tropospheric Profiling*, A. Apituley, H. W. J. Russchenberg, W. A. A. Monna, Eds., RIVM (The Netherlands), ISBN 978-90-6960-233-2

[6] Vogelmann H., R. Sussmann, T. Trickl, T. Borsdorff, 2011: Intercomparison of atmospheric water vapor soundings from the differential absorption lidar (DIAL) and the solar FTIR system on Mt. Zugspitze, *Atmos. Meas. Technol.*, **4**, 835-841

[7] Trickl, T., H. Vogelmann, H. Giehl, H. E. Scheel, M. Sprenger, A. Stohl, 2014: How stratospheric are deep stratospheric intrusions? *Atmos. Chem. Phys.* **14**, 9941-9961

[8] Kiemle, C., M. Wirth, A. Fix, G. Ehret, U. Schumann, T. Gardiner, C. Schiller, N. Sitnikov, G. Stiller, 2008: First airborne water vapor measurements in the tropical upper troposphere and mid-latitudes lower stratosphere: accuracy evaluation and intercomparisons with other instruments, *Atmos. Chem. Phys.*, **8**, 5245-5261

[9] L. Klanner, T. Trickl, H. Vogelmann, 2012: On the Way to Combined DIAL and Raman-Lidar Sounding of Water Vapour on Mt. Zugspitze – a Progress Report, pp. 853-826 in: *Reviewed and Revised Papers Presented at the 26th International Laser Radar Conference (ILRC 2012)*, Porto Heli (Greece), A. Papayannis, D. Balis, V. Amiridis, Eds., published in Greece on behalf of ICLAS

[10] VDI guideline 4210, Part 1, 1999: *Remote Sensing: Atmospheric measurements with LIDAR, Measuring gaseous air pollution with DAS LIDAR*, Verein Deutscher Ingenieure (VDI), (Düsseldorf, Germany), 47 pp.

DOI: 10.1051/epjconf/201611905009

REMOTE SENSING OF ATMOSPHERIC METHANE WITH SIMULTANEOUS RANGING USING CHIRPED LASER DISPERSION SPECTROSCOPY

Genevieve Plant[1], Andreas Hangauer[1,2], Gerard Wysocki[1*]

[1]*Electrical Engineering, Princeton University, NJ 08544, USA, *Email: gwysocki@princeton.edu*
[2]*Currently: Siemens, Munich, Germany*

ABSTRACT

We present a new sensing technology that allows for simultaneous sensing and ranging using chirped laser dispersion spectroscopy (CLaDS). In conjunction with previous works demonstrating the effectiveness of CLaDS for remote sensing, this new configuration yields spectroscopic and ranging information from a single measurement, and is implemented for continuous, multi-path detection of atmospheric methane.

1. INTRODUCTION

In remote sensing of atmospheric trace gases, knowledge of the path length is crucial to accurate extraction of the species concentration. However, in most case, this optical path length must be known beforehand or measured separately. Currently, the standard technique for atmospheric ranging and sensing is lidar (light radar), where pulsed light probes the target environment. The timing of the returned backscatter is used to extract the path length, and the received intensity is analyzed to determine the path-integrated concentration. We present an alternative, continuous wave (CW), approach based on the fundamentals of chirped laser dispersion spectroscopy (CLaDS). A proof-of-concept study has shown that both species concentration and path length can be extracted from a single, CW measurement. Development of a system that combines this CW sensing and ranging feature, with a CLaDS system that monitors atmospheric methane will be presented.

2. CHIRPED LASER DISPERSION SPECTRSOCOPY FOR REMOTE ATMOSHPERIC SENSING

Chirped laser dispersion spectroscopy is a new technique that probes the optical dispersion induced when probe laser light interacts with a targeted species [1]. When the laser frequency coincides with a molecular transition, light is absorbed, however, in addition the light also experiences dispersion which is related to resonant absorption via the Kramers-Kronig relationship. Although the sensed gas parameters are extracted from the spectral measurement in a way very similar to that of absorption-based techniques, CLaDS has unique properties that make it particularly well suited for remote sensing configurations. These include CLaDS signal linearity with concentration when implemented in a single-sideband configuration [2], immunity to intensity fluctuations [1], and possible harmonic detection using chirp-modulated CLaDS (CM-CLaDS) [3]. We have already demonstrated the operation of a CLaDS instruments for the application of atmospheric methane monitoring [4]. The developed instrument was tested in a wetland to target natural emissions of methane. Multiple paths were probed using a motorized mirror to cycle through the various path locations (defined by retro-reflector placement) [5]. However, like most systems, the optical length for each path was initially measured separately and kept constant throughout the campaign. Therefore the new capability of simultaneous ranging adds an important functionality that will enable dynamic variable-path chemical sensing.

3. SIMULTANEOUS SENSING AND RANGING

We have recently demonstrated an upgraded CLaDS technique that performs simultaneous chemical sensing and ranging [6]. The ranging measurement is performed using a CW radar method implemented in the form of the frequency chirped modulation sidebands (chirp 2 of frequency Ω in Fig. 1), which are used in conventional CLaDS to probe the target molecular transition. As a result of the propagation delay between the transmitter and the receiver, this additional chirp causes a frequency shift to the 0th harmonic signal equal to $\tau_g \cdot S_{11}$ (where τ_g [s] is the total group delay and S_{11} [Hz/s] is the carrier chirp rate), while the molecular signals (CM-CLaDS spectra) demodulated at higher harmonics are not affected. Thus extraction of the path-integrated concentration is performed using the same algorithms as in conventional CLaDS systems.

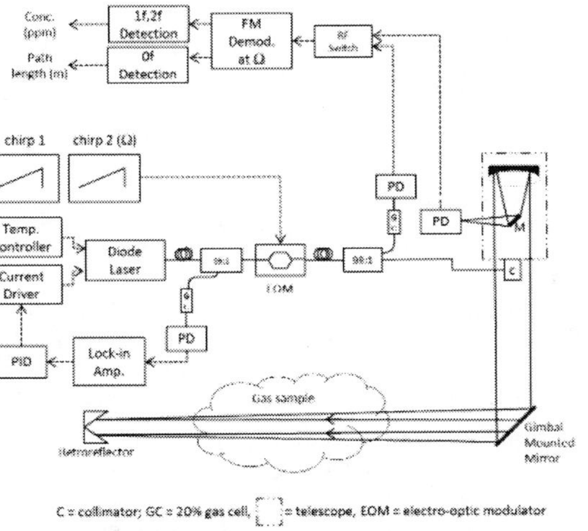

Figure 1. System diagram for remote detection of atmospheric methane with simultaneous ranging information. The major difference with respect to a conventional CLaDS is the addition chirp of modulation sidebands ("chirp2").

Results from proof-of-concept testing show the path length and gas signal measured for multiple path lengths (Fig.2). The path length, L, is determined using the frequency offset, Δf (shown in the top plot of Fig.2, and the carrier chirp rate.

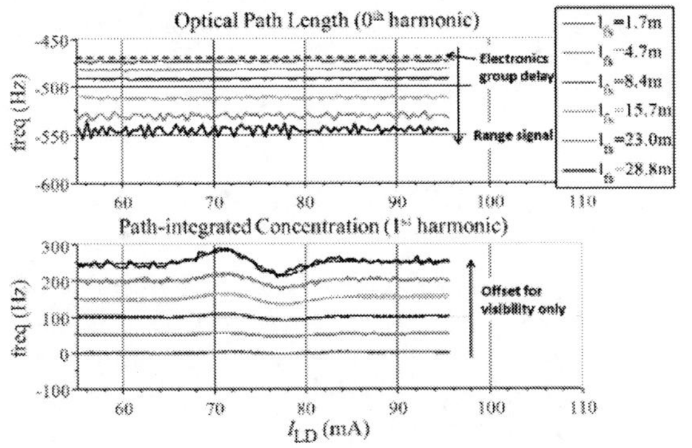

Figure 2. Proof of concept results demonstrating the offset of the 0th harmonic signals containing the pathlength information, while the 1st harmonic provided the CLaDS signal used for methane concentration retrieval [6].

The relationship follows as $L = (\Delta f\text{-}\Delta f_{ofs})/(S_{||}\cdot c)$, where Δf_{ofs} and c are the frequency offset produced by the delay of the generation electronics (a constant) and the speed of light, respectively. Integration of this range-retrieval method into the second generation instrument for atmospheric methane sensing allows for increased flexibility on path selection as well as the possibility for implementation of moving targets or hard targets at unknown distances.

4. MULTI-PATH MEASUREMENTS

Due to the chirp of the carrier frequency, | |, different beatnote frequencies will be generated for different path lengths. Thus, multiple paths can be sensed simultaneously while the corresponding signals are extracted using straightforward filtering. This method of simultaneous multi-path sensing has been demonstrated so far only in a fiber-based multi-node sensor systems [7], and in this work a first open-path implementation into our remote methane sensor will be discussed.

5. CONCLUSIONS

Using the foundation of chirped laser dispersion spectroscopy, a simultaneous ranging and sensing technique has been developed using a continuous-wave laser source. Implementation into a methane sensor for multi-path atmospheric monitoring applications will be discussed in details.

ACKNOWLEDGEMENTS

The authors would like to acknowledge financial support by the NSF CAREER award CMMI-0954897 and by the NSF ERC MIRTHE award EEC-0540832

REFERENCES

[1] G. Wysocki and D. Weidmann, "Molecular dispersion spectroscopy for chemical sensing using chirped mid-infrared quantum cascade laser," Optics Express **18**, 26123-26140 (2010).
[2] M. Nikodem and G. Wysocki, "Measuring optically thick molecular samples using chirped laser dispersion spectroscopy," Optics Letters **38**, 3834-3837 (2013).
[3] M. Nikodem, D. Weidmann, and G. Wysocki, "Chirped laser dispersion spectroscopy with harmonic detection of molecular spectra," Applied Physics B **109**, 477-483 (2012).
[4] M. Nikodem, G. Plant, D. Sonnenfroh, and G. Wysocki, "Open-path sensor for atmospheric methane based on chirped laser dispersion spectroscopy," Applied Physics B, 1-7 (2014).
[5] G. Plant, Y. Tian, T. Wang, and G. Wysocki, "Improved Signal Processing for Distributed Sensing Network based on Chirped Laser Dispersion Spectroscopy," in *CLEO: 2014*, (Optical Society of America, 2014), SM3E.2.
[6] A. Hangauer and G. Wysocki, "Simultaneous ranging and remote sensing utilizing chirped laser dispersion spectroscopy," in *CLEO: 2014 Postdeadline Paper Digest*, OSA Technical Digest (online) (Optical Society of America, 2014), STh5C.5.
[7] G. Plant, A. Hangauer, Y. Tian, T. Wang, and G. Wysocki, "Methane Leak Detection Sensor Network based on Fiber-coupled Chirped Laser Dispersion Spectroscopy," in *Light, Energy and the Environment*, (Optical Society of America, 2014), ETu4A.2.

EPJ Web of Conferences **119**, 05010 (2016) DOI: 10.1051/epjconf/201611905010
ILRC 27

1.6 MICRON FIBER LASER SOURCE FOR CH_4 GAS LEAK DETECTION

Nicolas Cézard*, Philippe Benoit, Guillaume Canat[1]

[1]*ONERA – The French Aerospace Lab – F-91761 Palaiseau, France,*

**Email:nicolas.cezard@onera.fr*

ABSTRACT

We report on the development of a new pulsed fiber laser source at 1645.5 nm, based on stimulated Raman amplification. This laser source is intended to be used in a future lidar system, dedicated to methane gas leak monitoring in the vicinity of industrial facilities. In this paper we discuss reasons for choosing the 1645.5 nm wavelength, and then we present the two-stage amplification architecture of our fiber laser source under development. Recent experimental results are provided and perspectives are drawn.

1. INTRODUCTION

Methane is an important atmospheric greenhouse gas, as well as a major energy source. Methane gas leaks in industrial plants can cause human injuries, environmental concerns, and financial losses. It is then of primary interest to develop remote sensing methods to characterize methane gas leaks in the vicinity of industrial facilities, typically at a few hundred meters, up to 1 km range.

Integrated-Path Differential Absorption Lidars (IP-DIAL) for CH_4 detection at various ranges have been demonstrated in the past in the 1.6μm band, using semiconductor lasers [1] or solid-state Optical Parametric Oscillators (OPO) [2-4]. An OPO-based spaceborne IPDA lidar is currently under development [5]. Range-Resolved systems (RR-DIAL) have also been developed using OPOs in the 1.6μm band or 3.3μm band [6-7]. These systems are efficient but complex and cumbersome (trucks).

In this work, we consider another approach and wish to investigate CH_4 IP-DIAL and RR-DIAL solutions based on silica-fiber-based components for both lidar source and lidar receiver. The main advantages of silica-fiber-based technologies are (i) high potential for lidar compactness, robustness and portability relying on telecom-type components (ii) design versatility for shaping spectral and temporal pulse properties and (iii) possibility of using fiber coherent detection for simultaneous or sequential range-resolved wind and CH_4 measurements with a single lidar. This last property is particularly attractive in the context of gas leak monitoring, since evaluating the leak rate implies to measure not only the gas concentration but also the gas speed [8].

To our knowledge, so far fiber laser based detection of CH_4 has been reported only in short-range (≈10m) IP-DIAL arrangement [9]. The fiber laser used Raman stimulated amplification and delivered 1.2 W cw radiation at 1651 nm. To address range-resolved lidar measurement, our objective is to address pulsed operation (typ.100 ns) and reach high peak powers.

2. WAVELENGTH SELECTION FOR CH4 DETECTION WITH A FIBER LIDAR

Though it is well-known that silica fibers do not transmit light beyond 2.4 μm, it is interesting to first consider the wavelength selection problem on a more general basis.

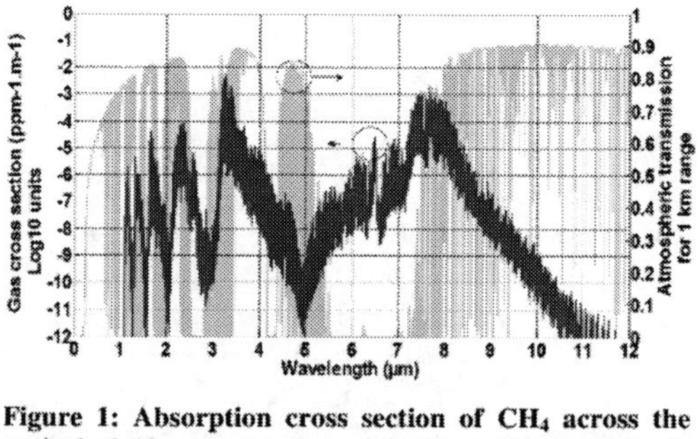

Figure 1: Absorption cross section of CH_4 across the optical 1-12 μm spectrum (red), and atmospheric transparency for 1 km propagation (blue). Both are calculated from Hitran 2012 database.

DOI: 10.1051/epjconf/201611905010

The four spectral bands that feature highest CH_4 absorption cross-sections in the 1-12 µm wavelength range are (from highest to lowest cross-section) 3.3 µm, 8 µm, 2.4 µm, and 1.6 µm (see fig.1). For 1 km lidar range, choosing a high-transparency atmospheric window is necessary. At 3.3 µm, CH_4 cross-section is so high (close to 5.10^{-3} ppm^{-1}.m^{-1}) that the natural background methane concentration (~1.8 ppm) would completely extinct the laser light after a few hundred meters propagation. The 8µm window can be discarded because it is extremely challenging for current high power laser systems. Consequently, near-infrared bands at 2.4 µm and 1.6 µm, with moderate absorption cross-sections close to 5.10^{-5} ppm^{-1}.m^{-1}, are the best suited bands for CH_4 leaks monitoring at 1 km range. Choosing silica-based fiber technology further excludes the 2.3-2.4 µm band, which stands at the very edge of the silica transparency. Therefore, the 1.6 µm band chosen for this work is optimal both for technical reasons (fiber transparency) and with respect to the present application (kilometer range lidar).

A close-up of fig.1 around the 1.63-1.67 µm band shows four high cross-section transitions at 1645.54 nm, 1650.96 nm, 1653.72 nm, and 1648.23 nm (figure 2). All these transitions are actually made of multiple fundamental lines that merge into single ones under the atmospheric pressure broadening effect. Two important criteria must be evaluated in order to choose the best transition to work with: the water-vapor interference level and the cross-section temperature sensitivity.

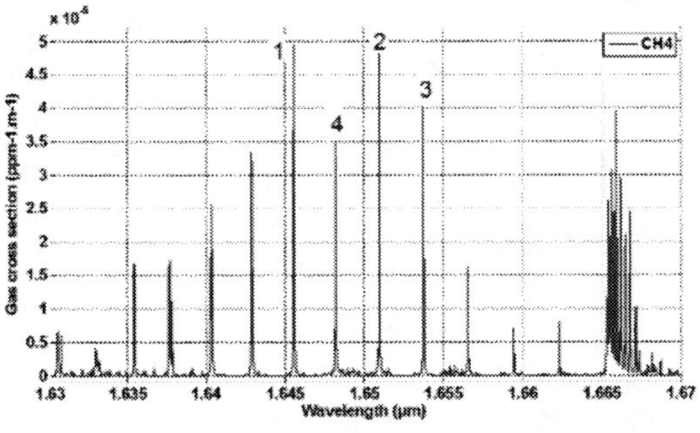

Figure 2: Absorption cross section of CH_4: close-up on the 1.63-1.67 µm range. Four candidate transitions with high cross-sections are numbered.

A calculation of the water vapor optical depth for 1 km range in a worse-case extremely humid atmosphere (30 000 ppm of water vapor) reveals that (i) transitions 2, 3 and 4 are well isolated (more than two linewidths apart) from the first significant H_2O neighboring line and consequently shall not be disturbed (ii) transition 1 is separated by about one linewidth from the first significant H_2O line. It is therefore a little more sensitive to H_2O interference, but only on the short-wavelength side of the transition, and with a small expected impact since the optical depth of this H_2O line reaches only 0.1 after 1 km in this very high humidity scenario.

The Relative Systematic Error (RSE) expected to occur on gas concentration retrieval because of temperature misreading can be evaluated by the following formula, derived for lorentzian lines with negligible OFF-line cross-section:

$$RSE = \frac{d\sigma}{\sigma} = G_1(\nu_{ON})\frac{dT}{T} \quad \text{with}$$

$$G_1(\nu_{ON},T) = \left[(1+x_{TIPS}) - \frac{100hc\Delta E}{kT} - n_{air}\frac{1-[(\nu_{ON}-\nu_c)/\gamma_L]^2}{1+[(\nu_{ON}-\nu_c)/\gamma_L]^2}\right]$$

Where σ is the gas cross section (ppm^{-1}.m^{-1}), T the temperature, x_{TIPS} the power exponent of the molecule Total Internal Partition Sum function (x_{TIPS} = 1.54 for CH_4 at 296 K, as computed from a power function fit of the TIPS data provided by HITRAN 2012 database), ΔE the fundamental energy level of the absorption transition in cm^{-1}, n_{air} the power exponent of the transition intensity temperature dependence, ν_c the central wavenumber in cm^{-1} and γ_L the line HWHM in cm^{-1}. However, here the lineshapes are not strictly lorentzians, because each line is made of merged multiplets. Therefore, temperature sensitivities have also been assessed using a line-by-line atmospheric propagation code at temperature T=296K and T+1K, and by calculating:

G_2=T.[σ(T+1)−σ(T)]/σ(T).

A significant discrepancy between G_1 and G_2 is found only for line 1, as shown in Table 1. For line 1, a 3K error (dT/T=1% at 296 K) turns into approximately 1.2% RSE for CH_4 concentration retrieval when the gas is probed on line center. Lines 2,3 and 4 are a little more sensitive than line 1. For all lines, sensitivities are currently considered acceptable values for our purpose.

In conclusion, because it exhibits both the highest cross-section and the lowest temperature sensitivity, we have chosen to begin with the 1645.54 nm transition. Water-vapor interference is expected to remain very small, especially if the Off-line wavelength is chosen along the long-wavelength side of the line center.

Line N°	ΔE (cm^{-1})	G_1 (ν_c,296K)	G_2 (ν_c,296K)
1	220	0.8	1.2
2	105	1.3	1.3
3	63	1.5	1.4
4	157	1.1	1.2

Table 1: Spectroscopic parameters and temperature sensitivity coefficients for the four selected lines. Coefficients n_{air}=0.7 and x_{TIPS}=1.54 are the same for all lines. G_1 (lorentzian model) and G_2 (line-by-line calculation) are given here for $\nu_{ON}=\nu_c$

3. FIBER LASER SOURCE AT 1645.5 nm

Radiation at 1645 nm cannot be efficiently amplified in conventional Erbium-doped silica fibers, because the Erbium gain band stretches from 1530 nm up to 1620 nm only. Instead, our fiber laser design relies on Raman stimulated amplification. When a powerful laser pump beam propagates into a silica fiber, it interacts non-linearly with the silica matrix and experiences a 13 THz down-frequency (spontaneous) Raman shift. If the pump wavelength is 1535 nm, the Raman shift leads to the desired 1645 nm wavelength. Consequently, if a low-power seed at 1645 nm (signal wavelength) and a high-power beam at 1535 nm (pump wavelength) propagate into the fiber at the same time, stimulated Raman amplification can occur and transfers energy from the 1535 nm pump to the 1645 nm signal. In practice, the Raman gain is spectrally quite large and pump wavelength within 10-15 THz from the signal wavelength are acceptable (13 THz being the optimum).

Raman amplifiers are known for their simplicity and efficiency. The main drawback is the Raman gain weakness (only 8.10^{-14} m/W), which implies using a long fiber length and/or a high power pump to provide high signal amplification. Because long fiber lengths are required, the pump wavelength is quickly prone to experience Stimulated Brillouin Scattering (SBS), which limits the usable pump power. SBS mitigation strategies are therefore required to increase the pump power. Moreover, in order to generate high signal power at 1645 nm for lidar application, SBS mitigation techniques must also be used for the signal wavelength.

The fiber laser architecture is shown on figure 3. It is a two-stage MOFPA (Master Oscillator Fiber Power amplifier). The seeding master oscillator is provided by a DFB (Distributed FeedBack) laser diode tuned at 1645.54 nm (typically 20 mW of cw power). In the first amplifier stage, the 1645 nm cw seed and a pulsed high peak-power pump at 1563 nm are injected in co-propagation into a highly non-linear fiber (HNLF) through a fiber multiplexer. The pump is spectrally broadened to mitigate SBS at 1563 nm and to enable increased pump power. At the exit, the signal power at 1645 nm typically reaches 1W peak-power (+17 dB gain) with 80 ns pulses at 20 kHz. This could be improved in the future by optimizing the pump wavelength (1563 nm is far from the optimal pump wavelength).

In the second Raman amplifier stage, a high peak-power 1545 nm broad linewidth pump laser is coupled in a second HNLF in contra-propagation with the 1645 nm signal. Using a broadband pump laser allows here again mitigating SBS detrimental effects, and increasing the pump power (through two successive Erbium-Doped fiber amplifiers). A drawback is that this configuration also transfers and amplifies the pump intensity noise toward signal intensity noise. However, in lidar systems, signal intensity noise can usually be compensated for by pulse-to-pulse power reference measurements if needed.

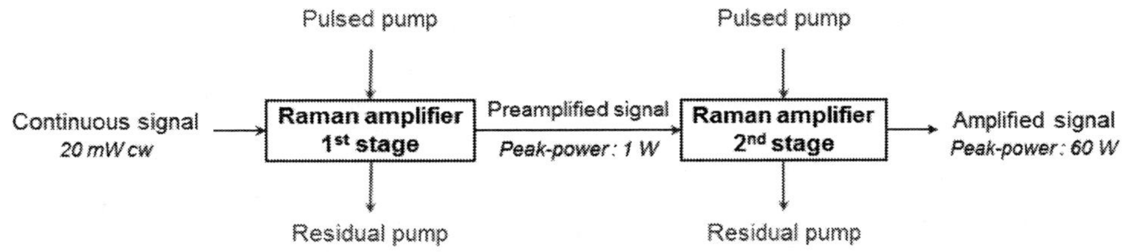

Figure 3 : Architecture of the fiber laser source with two amplification stages.

We have recently obtained 60 W peak power at 1645 nm at the 2nd stage exit (+18 dB) with 30 ns pulses (signal pulses experience time compression during the amplification process) at 20 kHz (see figure 4). These are encouraging results, though they call for many further investigations. So far the gain is believed to be mainly limited by an observed 2nd-stage pump power depletion induced by spontaneous Raman emission at superior orders. This effect must then be properly investigated. Pulses duration also need to be reshaped in the first and second stage in order to obtain the desired 100-ns duration at 2nd stage exit. Finally, the 1645 nm signal spectral width will have to be measured carefully, since frequency chirp effects are expected to occur (due to the optical Kerr effect that takes place in the HNLF along with Raman scattering).

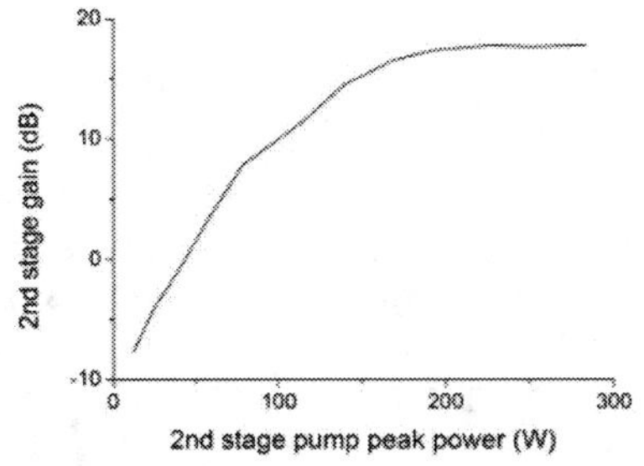

Figure 4 : Gain of the second Raman amplifier vs. pump peak power. The gain is 'negative' at low pump power because the amplifier has its own internal losses.

4. CONCLUSIONS

A 60 W peak-power fiber laser source relying on Raman stimulated amplification at 1645.5 nm has been reported for the first time to our knowledge. This wavelength has been identified as a good choice (high - but not too high - cross-section, low temperature sensitivity, low water-vapor interference level) for future lidar remote sensing of CH_4 gas leaks up to 1 km range. The obtained power may already allow performing IP-DIAL or short-range RR-DIAL measurements of methane. Another perspective is to keep improving the laser power and optimize the spectral and temporal parameters to allow both CH_4 and wind measurements to be made with a single lidar system.

ACKNOWLEDGEMENT

This work is a result of TOTAL and ONERA collaboration under NAOMI contract.

REFERENCES

[1] T. Iseki et al. Meas. Sci. Technol. 11, 594–602 (2000)
[2] A. Fix et al., Proc. of 22nd International Laser Radar Conference (2004)
[3] J.Barrientos Barria et al. Appl.Phys.B. 117, pp 509-518 (2014)
[4] K. Numata et al. Applied Physics B 116, 959-966 (2014)
[5] G.Ehret et al., Proc. of 13th Coherent Laser Radar Conference (2013)
[6] www.npl.co.uk
[7] www.ineris.fr
[8] P.Weibring et al., Appl.Phys. B 66, 383-388 (1998)
[9] D. Mitchell et al., Proc. SPIE 7503, 20th International Conference on Optical Fibre Sensors, 750350 (2009)

EPJ Web of Conferences **119**, 05011 (2016) DOI: 10.1051/epjconf/201611905011
ILRC 27

DEVELOPMENT OF FIELD-DEPLOYABLE DIODE-LASER-BASED WATER VAPOR DIAL

Phong Pham Le Hoai[1]*, Makoto Abo[1], Tetsu Sakai[2]

[1]*Graduate School of System Design, Tokyo Metropolitan University, Tokyo 191-0065, Japan,*
**Email: phamlehoaiphong@gmail.com*

[2]*Meteorological Research Institute, Tsukuba, Ibaraki 305-0052, Japan*

ABSTRACT

In this paper, a field-deployable diode-laser-based differential absorption lidar (DIAL) has been developed for lower-tropospheric water vapor observation in Tokyo, Japan. A photoacoustic cell is used for spectroscopy experiment around absorption peaks of 829.022 nm and 829.054 nm. The water vapor density extracted from the observational data agrees with the referenced radiosonde data. Furthermore, we applied modulated pulse technique for DIAL transmitter. It enables DIAL to measure water vapor profile for both low and high altitude regions.

1. INTRODUCTION

Water vapor mostly exists in the planetary boundary layer, a bottom part of the troposphere. The water vapor variation is one of the major factors causing the Earth's weather phenomena, climate change, and the other atmospheric activities. Hence, a precise and high-resolution water vapor profile is necessary for the weather forecast and climate study. Lidar, an active optical remote sensing has been utilized for water vapor monitoring. There are two typical techniques of lidar including DIAL and Raman lidar.

Raman lidar technique can measure high-resolution vertical profile of multiple gases simultaneously. The scattering molecule is basically identified from the frequency shift caused by the inelastic Raman scattering. It requires a high power laser pulse transmitter and a large aperture telescope to be able to receive small scattering cross section. The Raman lidar also needs periodic calibration device.

DIAL technique utilizes a pair of lasers tuned on and off the gas absorption peak, referred as online and offline laser signal. The molecular number density is basically obtained from the ratio of return online and offline signals as well as the prior knowledge of molecular absorption coefficients at online and offline wavelengths. The DIAL is able to cancel the calibration issue in Raman lidar. Furthermore, a diode laser based DIAL with the characteristics of low laser pulse energy and high pulse repetition is eye-safe, low cost, compact, and deployable.

A field-deployable diode-laser-based water vapor DIAL has been developed by the researchers at Montana State University through several generation systems. The DIAL prototype is able for continuous and autonomous operation during long-period measurement test [1]. Based on this initial work, we aim at developing a water vapor DIAL for heavy rain prediction during the summer season in urban areas of Japan.

2. DIAL INSTRUMENT

2.1 Laser transmitter

The block diagram of the laser transmitter is shown in Fig. 1. The output from a pair of distributed Bragg reflector (DBR) diode lasers are collimated using aspheric lens. The laser beam profile is reshaped by an anamorphic prism pairs (AP). Then the laser beam passes a Faraday isolator to prevent feedback and a half-wave plate ($\lambda/2$) to align polarization. Afterward, the laser beam is coupled in a single mode polarization maintaining fiber.

Both laser beams, then direct to a 2x1 solid-state high-speed fiber optical switch (SW) driven by a pulse generator. The seeding laser, switching between the online and offline lasers, passes through a half-wave plate and a Faraday isolator before injected to a tapered semiconductor optical amplifier (TA). The output laser beam of TA is circularly shaped by spherical and cylindrical

DOI: 10.1051/epjconf/201611905011

lenses. Then, the laser beam is expanded for eye safety before transmitting to the air.

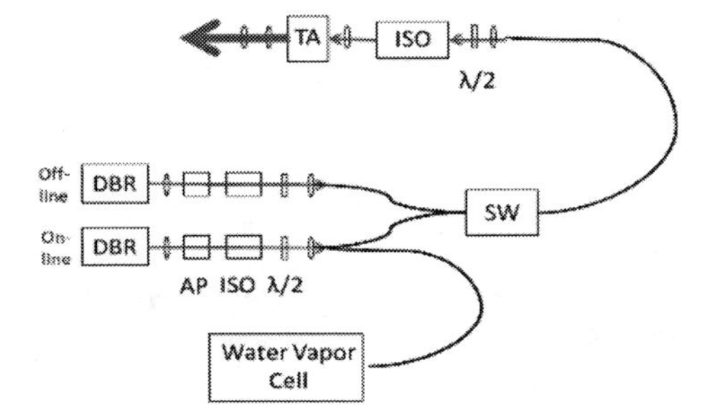

Figure 1. Block diagram of the laser transmitter.

2.2 Absorption spectroscopy experiment

In the scope of this work, we conducted absorption spectroscopy experiment utilizing the photoacoustic cell to measure water vapor absorption around operating wavelengths. A strong water vapor absorption peak at 829.022 nm and a weaker absorption peak at 829.054 nm used for this experiment are shown in Fig. 2.

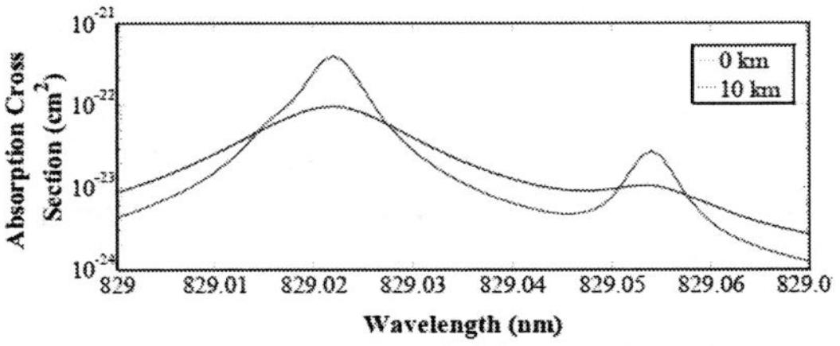

Figure 2. Voigt profile of the water vapor absorption cross section around the operational lines.

Photoacoustic spectroscopy is a sensitive technique for small absorption measurement. Fundamentally, a laser beam is directed to the absorption cell and the pressure variation in the cell is detected by a sensitive microphone, as shown in Fig. 3. The output signal of the microphone is proportional to the absorption cross section in the wavelength that the incident laser beam is tuned to. Then the output signal of microphone passes through an amplifier, a 1.5kHz-to-2.5kHz band pass filter, and a lock-in amplifier. Fig. 4 and Fig. 5 present the normalized absorption cross section from spectroscopy experiment using the photoacoustic cell and the referenced Voigt profiles around 829.022 nm 829.054 nm, respectively. The measured absorption peaks are shifted by 1 pm compared to the theoretical absorption peaks in both cases. The shifted wavelengths may relate to the calibration issue of the wavemeter.

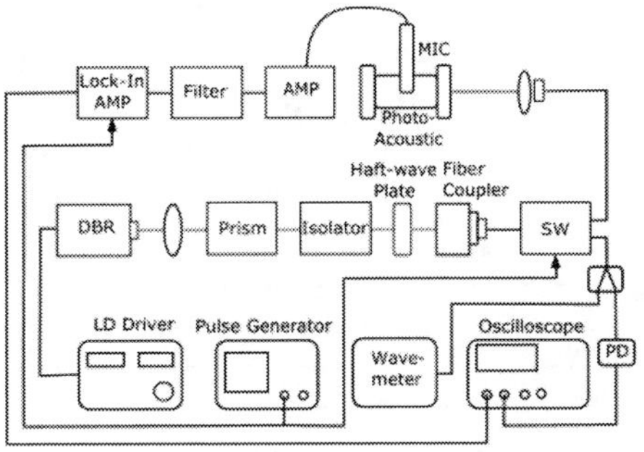

Figure 3. Schematic of spectroscopy experiment setup using the photoacoustic cell.

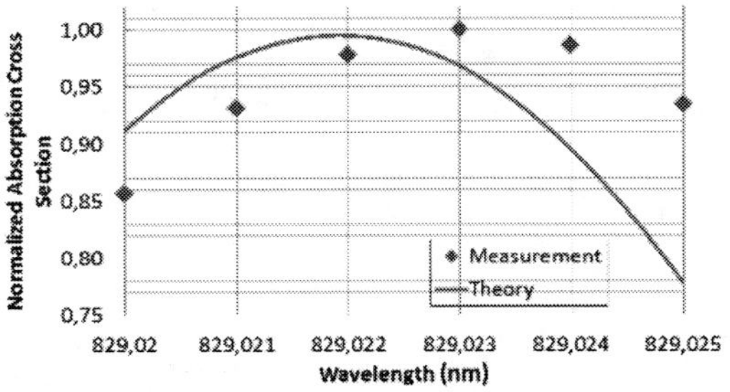

Figure 4. Normalized absorption cross section from the photoacoustic cell experiment and Voigt profile around 829.022 nm.

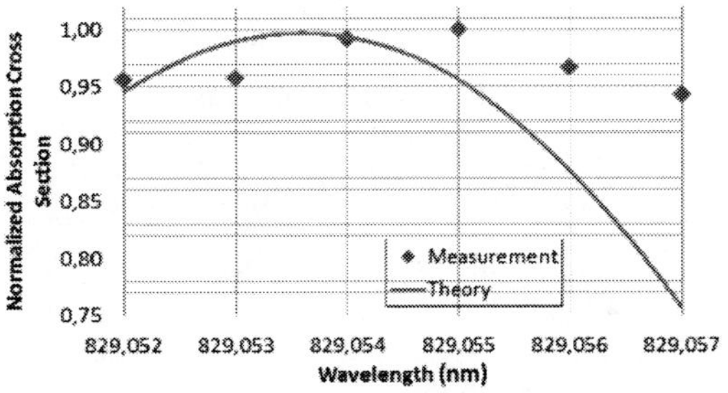

Figure 5. Normalized absorption cross section from the photoacoustic cell experiment and Voigt profile around 829.054 nm.

3. TEST OBSERVATION

The DIAL receiver includes a 50 cm diameter telescope, a photomultiplier tube (PMT), and a lidar transient recorder. The DIAL parameters for testing observation are described in Table 1. The photon-count numbers of the online and offline signals after subtracting background are plotted in

Fig. 6. The water vapor density with a range resolution of 450m is compared to the referenced radiosonde data, as shown in Fig. 7. As a result, the trend of lidar data coincides with the radiosonde data. However, the recent DIAL can not observe low region water vapor density due to the limit of driving pulse width.

Table 1. The H_2O DIAL experimental parameters

Online wavelength	829.054 nm
Offline wavelength	829.124 nm
Pulse energy	0.24 μJ
Pulse width	3 μs
Repetition rate	1kHz
Accumulation time	1 hour
Optical efficiency of filter	50%
Quantum efficiency	10%
Telescope diameter	50 cm
Field of view	3 mrad

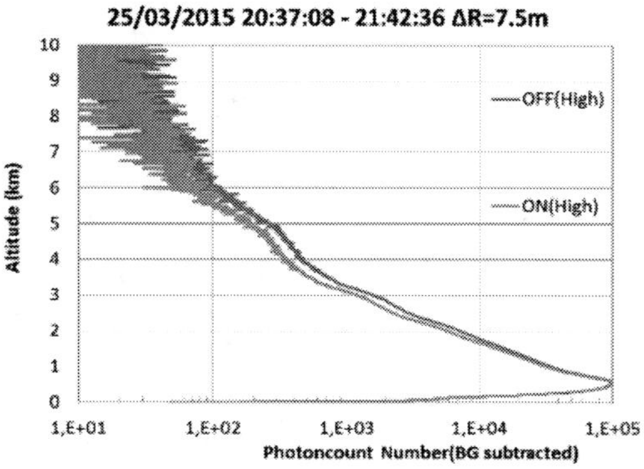

Figure 6. The photon-count number of the online and offline signals after subtracting background.

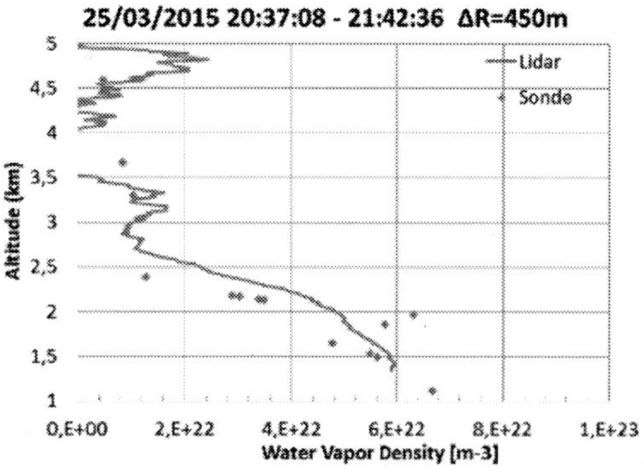

Figure 7. Lidar water vapor density with a range-resolution of 450m and the referenced radiosonde data.

4. MODULATED PULSE TECHNIQUE

The concept of modulated pulse technique is illustrated in Fig. 8 [2]. A single macropulse, transmitting N subpulses, is repeated every period of t_{rep}. The width of a single macropulse is given as

$$t_{macro} = Nt_c = Nt_b, . \quad (1)$$

where subpulse width t_c equals to bit width t_b. The waveform of a single macropulse laser is determined by

$$x(t) = \sum_{i=0}^{N-1} a_i p(t - it_b) , \quad (2)$$

where a_i, defined by pseudorandom sequence, equals 1 or 0, and $p(t)$, the envelope of a subpulse, is described as

$$p(t) = \begin{cases} 1 & (0 \le t \le t_b) \\ 0 & otherwise \end{cases} . \quad (3)$$

Similarly, the envelope of a macropulse is described as

$$m(t) = \begin{cases} 1 & (0 \le t \le t_{macro}) \\ 0 & otherwise \end{cases} . \quad (4)$$

Consequently, the complete laser waveform is given as

$$s(t) = \sum_{j=0}^{\infty} x(t) m(t - jt_{rep}) . \quad (5)$$

At the receiver, the return signal is a convolution form of modulated laser pulse, the medium response impulse $h(\tau)$ (i.e., backscattering response), and background noise $n(t)$ (i.e., Poisson noise). The corresponding return waveform $y(t)$ of a single modulated macropulse $x(t)$ is given as

$$y(t) = \int_0^{t_{rep}} x(t-\tau) h(\tau) d\tau + n(t) . \quad (6)$$

Finally, the return signal $z(t)$ is demodulated by cross-correlation of $x(t)$ and $y(t)$ as

$$z(t) = R_{yx}(\tau) = \int_0^{t_{rep}} y(t) x(t-\tau) dt . \quad (7)$$

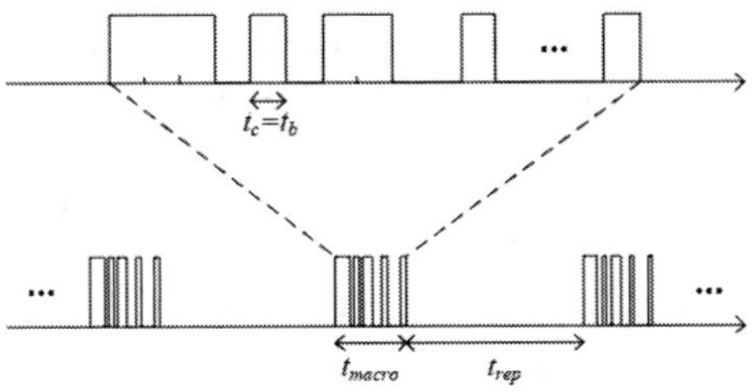

Figure 8. The concept of modulated pulse laser technique.

Theoretically, a periodic cross-correlation $z(t)$ of the continuous and periodic return signal $y(t)$ leads to a neglected sidelobe result. However, we can not keep the transmitted signal continuously in reality. By operating simultaneously a pair of online and offline wavelengths, the sidelobe can be canceled in the water vapor density result.

Single pulse modulation, 7-bit pulse modulation, and 15-bit pulse modulation are used for simulation. Specifications of the DIAL are given in Table 2. The HITRAN database of the water vapor absorption line is utilized for water vapor profile calculation. The aerosol profile is assumed as a U.S. standard model with 23 km ground visibility. The DIAL error is represented by total effective error including random error such as signal detection error, and systematic error such as temperature relative error and wavelength tuning error.

Table 2. The H_2O DIAL simulation parameters

Online wavelength	829.054 nm
Offline wavelength	829.124 nm
Pulse energy	1 μJ
Pulse width	100 ns
Repetition rate	10 kHz
Accumulation time	10 min
Sky spectral radiance	12.1 mW/m^2/nm/sr
Total opt. efficiency-Receiver	29% (day-time)
Quantum efficiency	45%
Telescope diameter	35 cm
Field of view	224 μrad

The total effective errors of three modulation models are plotted in Fig. 9. It is evident that the total effective of single pulse modulation is lowest in the low region. However, within 10% error, 7-bit pulse modulation, and 15-bit pulse modulation reach higher altitude compared to single pulse modulation. In detail, the maximum reached altitude within 10% error of single pulse modulation, 7-bit pulse modulation, and 15-bit pulse modulation are 1.5 km, 2.2 km, and 2.8 km, respectively. Though 7-bit pulse modulation is not optimum for lower/ higher region measurement compared to single/ 15-bit pulse modulation, it still maintains appropriate error through the whole region of the lower troposphere.

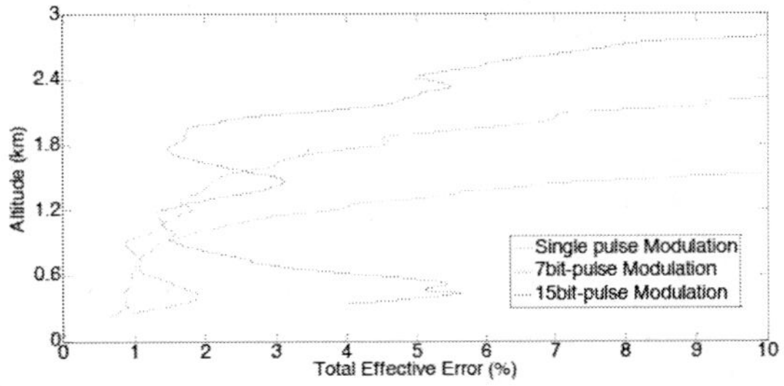

Figure 9. The total effective errors of single pulse modulation, 7-bit pulse modulation, and 15-bit pulse modulation.

5. CONCLUSIONS

In this research, we are developing a field-deployable diode-laser-based water vapor DIAL. Photoacoustic absorption cell can be utilized for wavelength locking of DBR laser to online by spectroscopy experiment. The water vapor profile retrieved from observational data agrees with the referenced radiosonde data. In future, we will apply modulation pulse technique and wavelength locking system as well as improve the TA efficiency for the DIAL transmitter.

ACKNOWLEDGEMENT

This work was supported by JSPS KAKENHI Grant Number 26282115.

REFERENCES

[1] Repasky, K. S., Moen, D., Spuler, S., Nehrir, A. R., and Carlsten, J. L., 2013: Progress towards an Autonomous Field Deployable Diode-Laser-Based Differential Absorption Lidar (DIAL) for Profiling Water Vapor in the Lower Troposphere, *Remote Sens.*, **5(12)**, 6241-6259.

[2] Cochenour, B., Mullen, L., and Muth, J., 2011: Modulated pulse laser with pseudorandom coding capabilities for underwater ranging, detection, and imaging, *Appl. Opt.*, **50(33)**, 6168-6178.

EXPERIMENTAL EVALUATION OF THE UV RAMAN LIDAR SENSITIVITY IN DETECTION OF TRACES OF CHEMICAL COMPOUNDS

Sergey Bobrovnikov [1,2]*, Evgeny Gorlov [1,2], Viktor Zharkov [1]

[1] *V.E. Zuev Institute of Atmospheric Optics, SB RAS, Tomsk, Russia, *Email: bsm@iao.ru*

[2] *National Research Tomsk State University, Tomsk, Russia.*

ABSTRACT

Experimental results are presented on the remote detection of traces of some chemical compounds on the surface with the help of Raman lidar built on the basis of an excimer KrF laser with a narrow line emission and multi-channel spectrum analyzer based on the diffraction spectrograph and a time gated ICCD camera. The sensitivity of the system is evaluated for the sensing range of 10 m. At the accumulation of the signal over 1000 laser pulses, the detection threshold of the nitrogen-containing chemical compounds of about of units of μg/cm^2 has been reached. The effect of the substrate material on the sensitivity of the Raman-lidar method for detecting traces of chemicals on the surface is analyzed.

1. INTRODUCTION

As previously shown in [1, 2], the lidar method based on the effect of Raman scattering in the ultraviolet (UV) wavelength range allows to achieve the sensitivity of about several ppm in detection of gaseous substances in the atmosphere. At the same time, the effect of Raman scattering allows to detect and identify substances not only in the gaseous phase, but also in the liquid and solid phases. Solution of the problem of development of the lidar method for the remote detection and identification of traces of chemical compounds on the surfaces of various items is of great practical and scientific interest and should be studied to assess its sensitivity, noise-stability, and the application field [3–5]. For this purpose, the work was done to upgrade and use the Raman-lidar method for the detection of traces of high-energy nitrogen-containing chemical compounds on the surface.

2. UV RAMAN LIDAR

The UV Raman lidar is built on the coaxial scheme. An excimer KrF laser with a narrow lasing line is used as the radiation source.

The scattered radiation reaches the receiving telescope aperture that creates an image of the scattering volume on the entrance slit of the spectrometer (Shamrock 500i). At the output of the spectrograph in the localization area of the spectrum, a cathode of a time gated ICCD camera is set that allows to detect the signal in the photon counting mode and in the mode of accumulation of charge. At the entrance of the spectrograph, an edge filter blocking the laser line of 248.3 nm is installed.

3. EXPERIMENTAL SCHEME AND EVALUATION OF THE MAXIMUM LIDAR SENSITIVITY

The Raman lidar for the detection of chemical compounds on the surface was tested in the laboratory room under standard conditions. The lidar transceiver system was focused at a distance of 10 m, where a bracket with fixed samples possessing different surface concentrations of traces of chemical compounds was installed.

The width of the entrance slit of the spectrograph was 125 μm, which corresponds to the spectral resolution of ~ 15 cm^{-1}. The signal along the slit of the spectrograph was integrated. The signal was recorded in the photon counting mode at the accumulation over 1000 laser pulses. At the initial stage of the experiment, the Raman-lidar spectrum response of the substrate was recorded. Figure 1 (curve 1) shows the experimental Raman spectrum recorded from the substrate (a clean microscope slide).

In all the recorded spectra, broad Raman bands of the glass components (600 cm^{-1} and 1150 cm^{-1}) [6], as well as the vibrational-rotational Raman band of the atmospheric oxygen molecules (1556 cm^{-1}) are observed.

DOI: 10.1051/epjconf/201611905012

After the identification of the Raman spectra bands obtained in the experiments and subtraction of the substrate spectrum and Raman spectra of major atmospheric gases, the lidar response spectrum of traces of chemical substance deposited on the surface was selected.

The limit of the detectable concentration of traces was determined in the process of detecting the traces with a predetermined threshold surface concentration at a threshold value of the signal/noise ratio equal to 2.

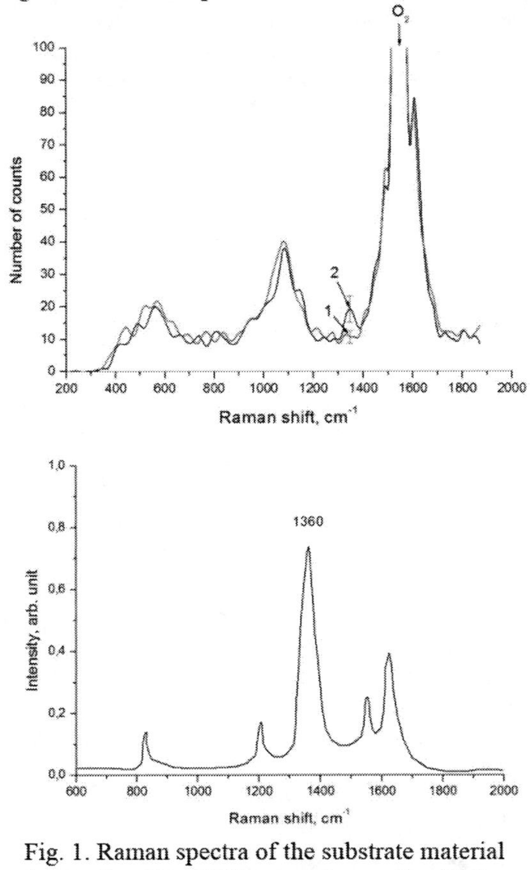

Fig. 1. Raman spectra of the substrate material (curve 1), of the TNT traces (curve 2) with the surface concentration of 500 ng/cm^2 (upper figure) and the reference spectrum of TNT [7] (lower figure).

For the experiments, some high-energy nitrogen-containing substances with low volatility contained in the database of Raman spectra were used. When choosing the substances, the urgency of their remote detection was taken into account.

Figure 1 shows the results of experiments on the remote spectrometry of traces of trinitrotoluene (TNT) by the Raman method: curve 1 - Raman spectrum of the glass slide material; curve 2 - Raman spectrum of the TNT traces deposited on a glass slide.

The signal observed in the region of the Raman shifts of 1500-1700 cm^{-1} is due to the vibrational-rotational Raman band of the atmospheric oxygen molecules.

It follows from the experiments on determining the sensitivity of the Raman lidar using the data on the detection of traces on the silica glass substrates (the glass microscope slides) that the limits of the detectable concentrations are 0.5 μg/cm^2 for TNT, 25 μg/cm^2 for pentaerythritol tetranitrate, and 20 μg/cm^2 for ammonium nitrate.

It is obvious that the possible superposition of Raman bands of the substrate material on the Raman bands of the detected substances can significantly limit the sensitivity of the method. That is why, it is very important to know the spectroscopic properties of expected substrates for a more accurate assessment of the real limit of the detectable concentration and determination of the Raman-method application field.

4. INFLUENCE OF THE SUBSTRATE ON THE SENSITIVITY OF THE RAMAN-LIDAR METHOD FOR DETECTING TRACES

To investigate the noise level of the substrates, a series of experiments on the assessment of the luminescence intensity of the substrate surfaces, when exposed to KrF-laser radiation, was performed.

An analysis of experimental data showed that the noise level in the recorded region of the spectrum depends on the optical properties of the substrate material. For example, substrates with negligible absorption of the exciting radiation (248 nm) give the Raman bands, whose intensities are 3-5 orders of magnitude greater than those of the Raman signals of the traces (depending on the ratio between the substrate thickness and that of the substance layer under study).

Obviously, this effect can be attributed to a large depth of the radiation penetration into the substrate material, and consequently, to a large

DOI: 10.1051/epjconf/201611905012

total volume of interaction. Figure 2 shows the Raman spectrum of an optical quartz glass KU-1.

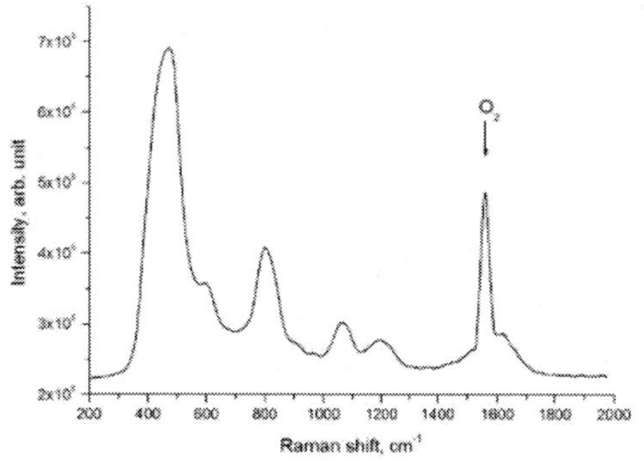

Fig. 2. Raman spectrum of the quartz glass KU-1.

In the weakly absorbing substrates, fluorescence is observed, whose spectrum covers the area of the frequency shifts and extends beyond this area. In this case, a substantial resonant rise of the Raman scattering intensity in the substrate material can be observed (Figure 3).

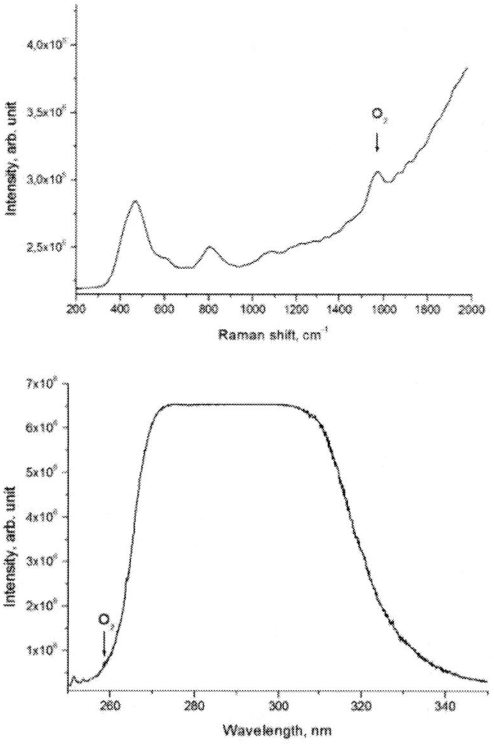

Fig. 3. Raman (upper figure) and fluorescence (lower figure) spectra in the quartz glass KV.

In contrast, in the highly absorbing substrates, a decrease in the intensity of Raman bands of the substrate material is observed due to the decrease in the penetration depth. As shown by experiments, in this case, the intensity of Raman signals in the substrate material is comparable with the intensity of Raman signals in the traces of the tested substances (Fig. 1).

In the strongly absorbing substrates, an intense fluorescence was observed due to the high level of absorption of the exciting radiation in the substrate material (Fig. 4). However, the maximum of the fluorescence band intensity is shifted to the red region of the spectrum. The spectral region of the Raman shifts is free of noise.

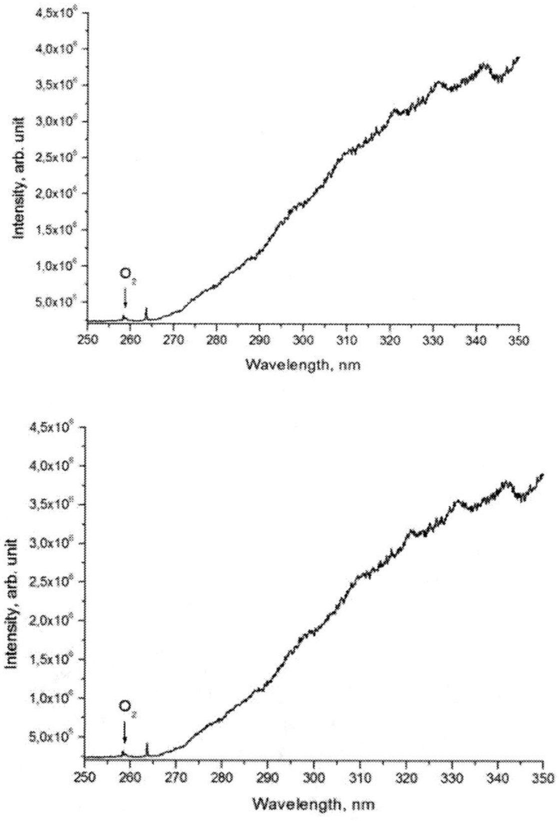

Fig. 4. Fluorescence spectra of the blue cotton (upper figure) and white cotton (lower figure).

5. CONCLUSIONS

As a result of the statistical analysis of the data, it was found that the UV Raman lidar based on an excimer KrF laser provides reliable detection of traces of chemical compounds with the surface density of about several µg/cm^2, depending on the Raman cross section of the detected molecules.

It is found that the sensitivity of the method depends on the optical properties of the substrate material. Substrates transparent to the exciting radiation, as a rule, limit the sensitivity of the method due to the high efficiency of the Raman processes and broadband fluorescence.

ACKNOWLEDGEMENT

This work was supported in part by the RFBR under Grant No. 13-05-00096a and No. 13-08-98013, the RF President grant No. NSH-4714.2014.5 and No. MK-6286.2015.10, and the Russian Science Foundation (Agreement No. 14-27-00022).

REFERENCES

[1] Bobrovnikov, S. M., E. V. Gorlov, V. I. Zharkov, 2012: Deep UV Raman lidar gas analyzer for environment monitoring, Proc. of the 26th ILRC 2012, Porto Heli, Greece, pp. 215–217.

[2] Bobrovnikov, S. M., E. V. Gorlov, V. I. Zharkov, 2013: Experimental estimation of the sensitivity of the UV Raman lidar, Atmos. Ocean. Opt., **26 (01)**, pp. 70-74.

[3] Carter, J. C., S. M. Angel, M. Lawrence-Snyder, J. Scaffidi, R. E. Whipple, and J. G. Reynolds, 2005: Standoff Detection of High Explosive Materials at 50 Meters in Ambient Light Conditions Using a Small Raman Instrument, Appl. Spectrosc., **59(6)**, pp. 769–775.

[4] Jander, P. and R. Noll, 2009: Automated Detection of Fingerprint Traces of High Explosives Using Ultraviolet Raman Spectroscopy. Appl. Spectrosc. **63(5)**, pp. 559–563.

[5] Moros, J., J. A. Lorenzo, K. Novotný and J. J. Laserna, 2013: Fundamentals of stand-off Raman scattering spectroscopy for explosive fingerprinting, J. Raman Spectrosc., **44**, pp. 121–130.

[6] Seuthe, T., M. Grehn, A. Mermillod-Blondin, H. J. Eichler, J. Bonse, and M. Eberstein, 2013: Structural modifications of binary lithium silicate glasses upon femtosecond laser pulse irradiation probed by micro-Raman spectroscopy, Opt. mater. express, **3(6)**, pp. 755–764.

[7] Gaft, M., L. Nagli, 2008: UV gated Raman spectroscopy for standoff detection of explosives, Opt. Mater., **30**, pp. 1739–1746.

ACTIVE STAND-OFF DETECTION OF GAS LEAKS USING A SHORT RANGE HARD-TARGET BACKSCATTER DIFFERENTIAL OPTICAL ABSORPTION SYSTEM BASED ON A QUANTUM CASCADE LASER TRANSMITTER

Adrian Diaz[1], Benjamin Thomas[1], Paulo Castillo[2], Barry Gross[1] and Fred Moshary[1]

[1] *NOAA-CREST, City College of New York, New York, NY, 10031, United States*

[2] *Brookhaven National Lab., New York, NY, 11973, United States*

ABSTRACT

Fugitive gas emissions from agricultural or industrial plants and gas pipelines are an important environmental concern as they can contribute to the global increase of greenhouse gas concentration. Moreover, they are also a security and safety concern because of possible risk of fire/explosion or toxicity. This study presents gas concentration measurements using a quantum cascade laser open path system (QCLOPS). The system retrieves the path-averaged concentration of N_2O and CH_4 by collecting the backscattered light from a scattering target. The gas concentration measurements have a high temporal resolution (68 ms) and are achieved at sufficient range (up to 40 m, ~ 130 feet) with a detection limit of 2.6 ppm CH_4 and 0.4 ppm for N_2O. Given these characteristics, this system is promising for mobile/multidirectional remote detection and evaluation of gas leaks. The instrument is monostatic with a tunable QCL emitting at ~ 7.7 μm wavelength range. The backscattered radiation is collected by a Newtonian telescope and focused on an infrared light detector. Puffs of N_2O and CH_4 are released along the optical path to simulate a gas leak. The measured absorption spectrum is obtained using the thermal intra-pulse frequency chirped DFB QCL and is analyzed to obtain path averaged gas concentrations.

1. INTRODUCTION

In order to reduce greenhouse gas concentrations, there is a need for tools able to detect fugitive gas emissions from agricultural, industrial plants, or natural gas pipeline infrastructure. This is of concern for the global warming [1] as well as possible toxicity or fire and explosive risks. Optical methods to measure fugitive emission flux has been developed in order to detect greenhouse (or hazardous) gases [2-5], these methods are appropriate for fugitive emissions as they cover larger area than point sensors.

In this contribution, we present a potential system for fast gas concentration measurements based on a distributed feedback quantum cascade laser (DFB - QCL), its central wavelength of emission when used in a long pulsed (200+ nanoseconds) mode is rapidly chirped during the course of the pulse resulting in a well defined frequency sweep allowing this type of laser to be use for intra-pulse spectroscopy [6]. Using the backscattered light of a hard target, this system is promising for mobile/multidirectional remote detection and quantification of fugitive gas leak. Laboratory measurements have been performed on methane (CH_4) and nitrous oxide (N_2O), respectively the second and third most potent anthropogenic greenhouse gases [1] over a distance of 40 meters (~ 130 ft) and with a high temporal resolution (68 ms).

2. METHODOLOGY

The system is based on differential optical absorption spectroscopy (DOAS) [7]. A quantum cascade laser (QCL) emits a laser pulse also referred to as the reference signal $I_0(\upsilon)$. This pulse is pointed towards a topographic scattering target (in this demonstration, a well-characterized near-Lambertian scattering target). A coaxial optical Newtonian telescope is used to collect the backscattered light which is then focused on an optical detector resulting in a measured signal $I(\upsilon)$. During propagation, the laser pulse undergoes optical extinction due to the presence of the gas leak.

A remarkable feature of the QCLOPS system is that it does not use a spectrally resolved detector. In order to measure an absorption spectrum, we use the intra-pulse frequency chirp of a pulsed quantum cascade laser (1 cm^{-1} in ~ 200 nanoseconds in our case), which is remarkably stable and repeatable [6, 8]. Therefore, the time resolved intensity measurements I(t) can be converted to a spectrally resolved intensity $I(\upsilon)$ using the experimentally-determined chirping rate of the QC laser.

The negative logarithm of the ratio between the reference signal $I_0(\upsilon)$ and the measured signal $I(\upsilon)$ is equal to the gas absorbance times a constant, from which the target gas concentration can be retrieved using the least-squares fitting technique [9-11]. The range to the target is measured using time of flight of the hard target backscatter (echo). The intensity of the signal $I(\upsilon)$ measured by the system may vary due to laser intensity fluctuations or changes in scattering efficiency of the target, as a consequence only the relative absorbance can be measured. However, using the least-squares fitting technique there is no need to measure the absolute absorbance and therefore no need to calibrate the signal intensity.

3. EXPERIMENTAL RESULTS

Measurements of N_2O and CH_4 have been performed illuminating a scattering target 40 m away (~ 130 feet) in the mid-infrared spectral range on rotational absorption bands located between 1295 and 1296 cm^{-1} (~ 7.7 μm). This spectral range has been chosen since N_2O and CH_4 can be addressed within the tunable range of the QC laser used. The repetition rate is 20 kHz and the average power reaches 0.20 mW.

Figure 1 shows an example of the recorded signals using our system for the N_2O leak through the laser optical path. The reference signal $I_0(t)$ is plotted as a function of time, represented by a green line (laser pulse in absence of a leak); the measurement signal I(t), is displayed as a red line, showing the strong absorption effects on the laser pulse when a puff of N_2O is released from a tank.

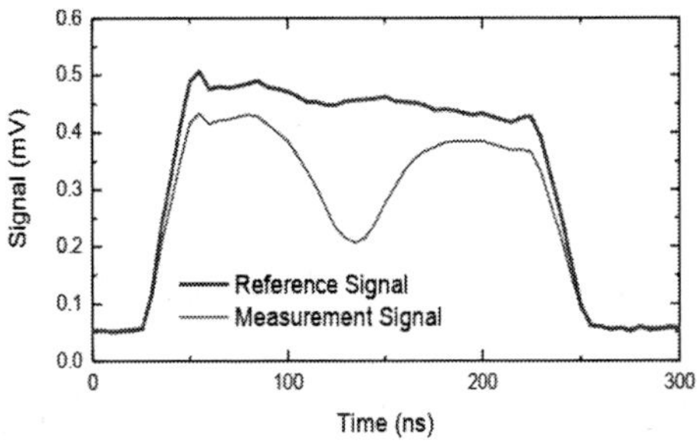

Fig. 1: Reference signal $I_0(t)$ (grey line) and measurement signal I(t) (black line) displayed as functions of time. The measurement signal shows the absorption line due to the presence of the N_2O gas along the optical path.

Figure 2 presents the path-averaged concentration for methane and nitrous oxide, retrieved by the least-squares analysis, with a 68 ms temporal resolution. In both cases, a small amount of pure N_2O (~ mL) or CH_4 (~ 10^2 mL) have been released near the laser beam.

The average N_2O ambient concentration is about 0.22 ppm with strong fluctuations (± 0.2 ppm).

When the gas puff is released, the retrieved concentration peaks and reaches 11 ± 1 ppm. The methane concentration reaches 34 ± 3 ppm then oscillate for approximately 20 seconds. The concentration stabilized to a concentration slightly higher than ambient level (5 ± 1.5 ppm) and eventually returns to value ambient level (1.8 ± 1.3 ppm)

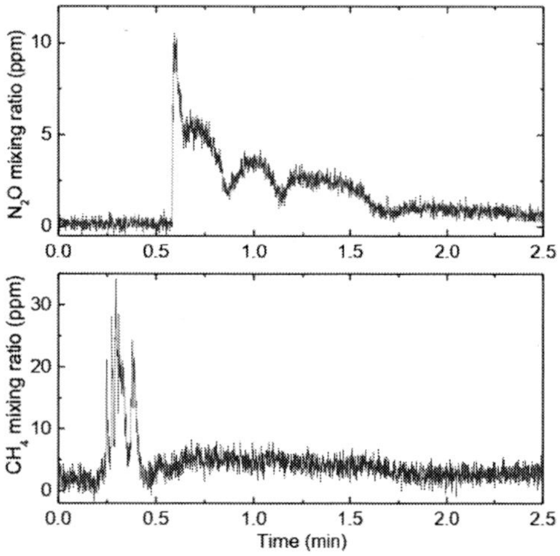

Fig. 2: Time evolution of the path-averaged mixing ratio of N_2O (top graph) and CH_4 (bottom graph).

Both measurements show that the system is able to detect small leaks of either methane or nitrous oxide with a high time response, as well as evaluating the integrate path mixing ratio with uncertainty better than 10 % while measuring the leak. The detection limit (at 2σ) of the instrument in this configuration (40 m range, 68 ms time resolution) is 0.4 ppm for N_2O and 2.6 ppm for CH_4, which is very close to their respective ambient mixing ratio. The detection limit can easily be improved by reducing the time resolution or deploying a more powerful QCL.

4. SUMMARY AND CONCLUSION

In order to perform mobile or multidirectional remote detection of fugitive gas emission, we demonstrate in this contribution a backscatter Quantum Cascade Laser Open Path System (QCLOPS) to measure trace gas with a high time resolution using topographic scattering. The trace gas concentration measurement is based on differential absorption spectroscopy, with the distance to the target determined by the hard target lidar return. The use of pulsed quantum cascade laser to carry out intra-pulse spectroscopy for lead detection in a backscatter configuration is, to our knowledge, new. Using the intra-pulse frequency chirp of the QCL, the absorbance along the optical path is measured, allowing to retrieve the target gas path-averaged concentration using a least-squares analysis. As experimental proof, we presented the detection of two gases, methane and nitrous oxide. Small puffs of gas have been released along the laser optical path and measured by the QCLOPS. With the backscattering target at 40 m (~ 130 feet) range, the QCLOPS has a detection limit of 2.6 ppm for CH_4 and 0.4 ppm for N_2O.

This experiment shows feasibility of this methodology. The use of a diffuse-reflectance target is an experimental proof that the gas sensor does not require setting a retro-reflector and can be directly pointed towards topographic targets such as pipes, gas tanks, buildings or walls. However, topographic targets present a wide range of backscattering efficiencies, leading to a variable optical signal and therefore various uncertainties and detection limits. Our current work is now focused on using the high time resolution and compactness of the instrument to achieve measurements to be taken from a moving platform. As an example, if set on a platform moving at 50 km/h, the instrument could perform one concentration measurement every meter. Thus, it makes possible the localization of gas leaks over large distances in a short time frame and with an excellent spatial resolution.

 DOI: 10.1051/epjconf/201611905013

ACKNOWLEDGEMENT

This research was supported by the Mid Infrared technologies for Health and the Environment (MIRTHE) Center and National Science Foundation (NSF MIRTHE Center: EEC-0540832 NOAA CREST CENTER: NA11SEC4810004).

REFERENCES

[1] Stocker, T. F., Qin, D., Plattner, G. K., Tignor, M., Allen, S. K., Boschung, J., & Midgley, B. M. (2013). IPCC, 2013: Climate change 2013: the physical science basis. Contribution of working group I to the fifth assessment report of the intergovernmental panel on climate change.

[2] Demtröder, W. (2008). Laser spectroscopy 2. Experimental techniques. *Springer Berlin Heidelberg*.

[3] Weitkamp C., 2005: Lidar, Range-Resolved Optical Remote Sensing of the Atmosphere, *Springer, Berlin, NewYork.*

[4] Fujii T., T.Fukuchi, 2005: Laser Remote Sensing, *CRCPress, BocaRaton.*

[5] Hodgkinson J., R. P. Tatam, 2013: Optical gas sensing: a review, *Meas. Sci. Technol.,* **24**, doi:10.1088/0957-0233/24/1/012004.

[6] Duxbury, G., Langford, N., McCulloch, M. T., & Wright, S., 2005: Quantum cascade semiconductor infrared and far-infrared lasers: from trace gas sensing to non-linear optics. *Chemical Society Reviews*, **34** (11), 921-934.

[7] Platt U., and Jochen Stutz, 2008: Differential optical absorption spectroscopy, *Springer Berlin Heidelberg.*

[8] Kosterev, A., 2002: Chemical sensors based on quantum cascade lasers, *Quantum Electronics*, IEEE Journal of 38.6, 582-591.

[9] Bevington, P. R., & Robinson, D. K., 2003: Data reduction and error analysis. *McGraw–Hill, New York.*

[10] Chapra S., and R. Canale, 2010: Numerical Methods for Engineers, 6th ed., *McGraw-Hill Higher Education, Boston.*

[11] Press, W. H., 2007: Numerical recipes 3rd edition: The art of scientific computing, *Cambridge University press.*

CHARACTERIZING THE VERTICAL PROCESSES OF OZONE IN COLORADO'S FRONT RANGE USING THE GSFC OZONE DIAL

John T. Sullivan [1,2]*, Thomas J. McGee[3], Raymond M. Hoff[1,2], Grant Sumnicht[4], Laurence Twigg[4]

[1]Department of Physics, University of Maryland, Baltimore County, Baltimore, MD 21250, USA,

**Email: johnsullivan@umbc.edu*

[2]Joint Center for Earth Systems Technology, Baltimore, MD 21250, USA,

[3]Atmospheric Chemistry and Dynamics Laboratory, NASA GSFC, Greenbelt, MD 20771, USA

[4]Science Systems and Applications Inc., Lanham, MD 20706, USA

ABSTRACT

Although characterizing the interactions of ozone throughout the entire troposphere are important for health and climate processes, there is a lack of routine measurements of vertical profiles within the United States. In order to monitor this lower ozone more effectively, the National Aeronautics and Space Administration (NASA) Goddard Space Flight Center TROPospheric OZone DIfferential Absorption Lidar (GSFC TROPOZ DIAL) has been developed and validated within the Tropospheric Ozone Lidar Network (TOLNet). Two scientifically interesting ozone episodes are presented that were observed during the 2014 Deriving Information on Surface Conditions from Column and Vertically Resolved Observations Relevant to Air Quality (DISCOVER AQ) campaign at Ft. Collins, Colorado.

The first case study, occurring between 22-23 July 2014, indicates enhanced concentrations of ozone at Ft. Collins during nighttime hours, which was due to the complex recirculation of ozone within the foothills of the Rocky Mountain region. Although quantifying the ozone increase aloft during recirculation episodes has been historically difficult, results indicate that an increase of 20 - 30 ppbv of ozone at the Ft. Collins site has been attributed to this recirculation. The second case, occurring between Aug 4-8th 2014, characterizes a dynamical exchange of ozone between the stratosphere and the troposphere. This case, along with seasonal model parameters from previous years, is used to estimate the stratospheric contribution to the Rocky Mountain region. Results suggest that a large amount of stratospheric air is residing in the troposphere in the summertime near Ft. Collins, CO. The results also indicate that warmer tropopauses are correlated with an increase in stratospheric air below the tropopause in the Rocky Mountain Region.

1. INTRODUCTION

Tropospheric ozone is an important trace gas to characterize because in abundance it can be harmful to humans and vegetation. Above the ground level, the enhancement of ozone has been historically difficult to measure directly due to its diurnal and season variability, relatively short lifetime and nonlinear formation. Tropospheric ozone concentrations within the planetary boundary layer in urban areas, such as the Denver, CO region, are typically dominated by photochemistry when local ozone precursor emissions, such as nitrogen oxides and certain volatile organic compounds, are concentrated near the surface in the presence of near ultraviolet sunlight. At higher altitudes, within the free troposphere, ozone is a key greenhouse gas and the sources can be classified as 1) dynamic Stratospheric-Tropospheric Exchange (STE), 2) advection from upwind sources and 3) regional redistribution from convection and/or lightning [1].

Although the interactions and processes of ozone throughout the entire troposphere are crucial for air quality and climate analyses, the uncertainty is rather high due to a lack of extended routine measurements of vertical profiles within the United States. To help aid in the current lack of tropospheric ozone profiles, the ground-based Tropospheric Ozone Lidar NETwork (TOLNet), which currently consists of five stations across the

United States has been developed. The primary purpose of the instruments within TOLNet is to provide regular, high fidelity profile measurements of ozone within the troposphere, particularly the changing ozone dynamics and laminae inside the PBL.

As part of this network, the NASA Goddard Space Flight Center TROPospheric OZone DIfferential Absorption Lidar (GSFC TROPOZ DIAL) has been developed in a transportable trailer to take routine measurements of tropospheric ozone near the Baltimore - Washington D.C. area and various campaign locations. Similar ground based instruments have been developed within TOLNet [2][3][4] but the GSFC TROPOZ DIAL is the first to measure ozone profiles near sea level directly in an urban environment within the mid-Atlantic U.S.

2. THE DIAL METHOD

For the DIAL technique, at least two different wavelengths are transmitted into the atmosphere. These wavelengths are chosen based on the spectral properties of ozone, the atmospheric constituent of interest. For this work [5], the TROPOZ transmitted wavelengths are generated by focusing the output of two frequency-quadrupled Neodymium-doped Yttrium Aluminum Garnet (Nd:YAG) laser beams (266 nm), each into a pair of high pressure Raman cells, one filled with hydrogen and the other deuterium. Both cells have been optimized using helium as a buffer gas. The more absorbing wavelength, at 288.9 nm, is called the "on" wavelength and the less absorbing wavelength, at 299.1 nm, is the "off" wavelength. Because the inelastic Raman conversion is done prior to transmission into the atmosphere, and detection of atmospheric scattering is performed at the same transmitted wavelengths, the TROPOZ retrieval uses an elastic technique.

3. OZONE RECIRCULATION CASE STUDY

Over the last several decades, the population growth, complex terrain, and mesoscale meteorological conditions throughout Colorado's Front Range have been shown to favor severe pollution episodes [6]. An elevated ozone pollution episode was observed between 22-23 July 2014 during the DISCOVER AQ campaign, in which the Ft. Collins (Figure 1), NREL - Golden, and BAO Tower sites exceeded the EPA National Ambient Air Quality Standard (NAAQS) 8-hour average of 75 ppbv. Several other ground sites also observed enhanced levels of ozone. However, the 8-hour exceedance that occurred at the Ft. Collins - West site was delayed by nearly two hours with respect to the other sites. For this reason, the elevated ozone was likely attributed to the additional recirculation and advection of ozone throughout the domain.

To understand the synoptic meteorological conditions for the domain during the test case, several NOAA national meteorological maps and surface ozone maps from the EPA AIRNOW network are presented. The on-board in situ ozone monitor and wind measurements from the NASA P3B, which performed several spiral down overpasses at multiple sites, are also presented. Lidar measurements of ozone, wind speed/direction, and aerosols are presented as well as several ozonesondes. A back trajectory of the polluted air mass is shown with the HYSPLIT trajectory model. These results are also compared to the meteorological and chemical output from a regional air quality model that was being initialized daily during the DISCOVER AQ campaign.

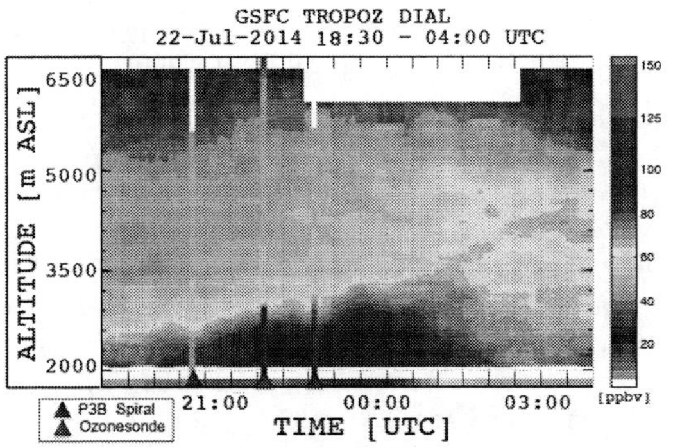

Figure 1: TROPOZ time series observations from Ft. Collins, CO during the DISCOVER AQ Campaign from 22-23 July 2014.

With the suite of models, surface monitors, profilers, and aircraft observations during the event, it is possible to show that the complex interactions of winds aloft and near surface winds were the coupled driving mechanism for the enhancement of ozone over the most of the campaign sites. More specifically, evening observations occurring after a very photochemically active ozone production day, indicate that 1) upslope winds advected elevated concentrations of ozone to high elevation mountain sites and during the upslope to downslope lower level wind reversal, ventilation was deterred which subsequently elevated ozone concentrations by 20-30 ppbv at sites near the eastern ridge of the Rocky Mountains, and 2) a reversal in the aloft winds caused by a quickly moving cold front advected a southeasterly flow of polluted ozone-rich air from the Denver metropolitan area towards the Rocky Mountains, which helped to maintain elevated ozone at sites near the eastern ridge of the Rocky Mountains.

4. STRATOSPHERIC TROPOSPHERIC EXHCAHNGE (STE) CASE STUDY

Intrusions of stratospheric air into the troposphere and tropopause folds are typically prompted by upper level synoptic meteorology such as baroclinic waves and upper level frontogenesis in the jet stream region [7]. The frontogenetically induced ageostrophic cyclonic circulation forces stratospheric air down into the troposphere resulting in a folding of the dynamical tropopause [8]. For the mid-latitude Rocky Mountain region, stratosphere-troposphere exchanges (STE) and tropopause folds are typically induced by the formation of an upper level cut-off low pressure system and its associated trough in the presence of the polar front or subtropical jet stream [9].

Measurements made in instances directly involved in STE events or tropopause folds in the northern Hemisphere have been previously described in winter, spring, and fall. However, it is also important to characterize summertime STE events because the upper level meteorology favors the summertime occurrence of cut off low pressure systems and STE events. An interesting example of a summertime STE event occurred during the 2014 DISCOVER AQ mission. Because of the campaign, high-resolution lidar and ozonesonde observations were available to quantify the upper level ozone laminae and resolve the dynamics surrounding the event, which can be seen in Figure . One unique feature of this STE event is that it likely entrained regional wildfires during the subsidence and advection near the Rocky Mountain region. Along with detailed supporting observations, the intrusion and advection of a stratospheric air mass over the Rocky Mountain region from 04-08 August 2014 will be characterized.

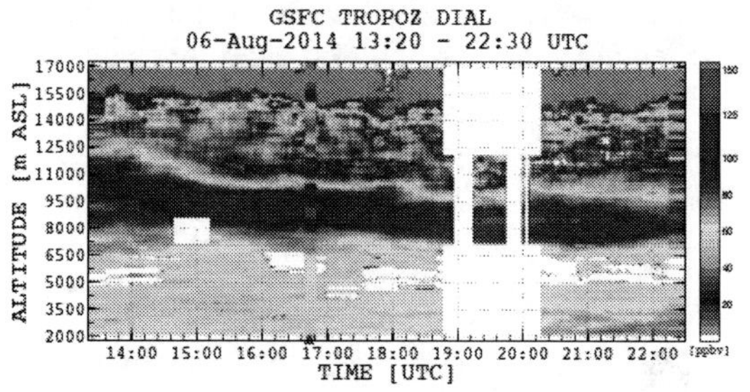

Figure 2: TROPOZ time series observations from Ft. Collins, CO during the DISCOVER AQ Campaign on 06 August 2014.

To emphasize the many similarities of the upper level dynamics that commonly induces STE events in the Rocky Mountain region, regardless of season, meteorological maps from a case study in each of the remaining seasons of 2014 are presented. In each STE event, the transfer of the dry, cold, and ozone-rich stratospheric air mass was associated with the presence of an Aleutian Low pressure system and developing cut off low near the Pacific coast of southern California. The air masses were quickly advected over the Pacific Southwest United States by a persistent southwesterly upper level tropospheric jet stream and eventually transported over the Rocky Mountain region.

During the summertime STE event, the Real-time Air Quality Modeling System (RAQMS) ozone output produced a very similar ozone filament in altitude and concentration to the TROPOZ and ozonesonde observations. Because of this agreement, the latter portion of this presentation will use RAQMS model output to analyze the stratospheric contribution to Ft. Collins and the Rocky Mountain region. To do this, a methodology of defining a correlation between the RAQMS ozone to potential vorticity ratio in the summertime STE event at Ft. Collins is determined. From this relationship, results suggest that 1) the occurrence of stratospheric air in the upper free troposphere from 2012-2014 is largely controlled by the temperature of the tropopause, in which warmer temperatures correlate well with larger occurrences of STE events and vice versa, 2) about one third of stratospheric air in Summer 2014 has resided below the tropopause for less than 12 hours, while another third has lasted 24 - 48 hours, implying some STE events are rather shallow, but nearly the same amount penetrate deeper in the troposphere, and observations during the Summer 2014 DISCOVER AQ campaign indicate that 3) a semi-permanent stratospheric reservoir exists in the free troposphere that is likely affecting the chemical and radiative budget at Ft. Collins.

5. CONCLUSIONS

Tropospheric ozone is an important trace gas to characterize because it can be harmful to humans and vegetation. The enhancement of lower tropospheric, or boundary layer ozone, typically occurs when local ozone precursors, such as nitrogen oxides and volatile organic compounds, are concentrated near the surface in the presence of near ultraviolet sunlight. At higher altitudes, in the free troposphere, ozone is a key greenhouse gas. Intrusions and layers of ozone occurring in the free troposphere are typically advected from local and regional sources or dynamically introduced from the stratosphere. Although the interactions and processes of ozone throughout the entire troposphere are crucial for air quality and climate analyses, there is a lack of routine measurements of vertical profiles within the United States.

The GSFC TROPOZ DIAL is a new and unique tool for the air quality community because it can provide a continuous, high resolution and novel perspective of ozone pollution episodes during the daytime from near the surface to the tropopause. Depending on the atmospheric event or scientific question the TROPOZ is intended to investigate, averaging of the data may be minimized to detect natural variability and small ozone fluctuations or maximized to reach higher altitudes for climatology purposes. The TROPOZ has been utilized, along with a suite of other instruments and models, to help better characterize the vertical processes of ozone.

ACKNOWLEDGEMENT

This work was supported by UMBC/JCET (Task #374, Project 8306), the Maryland Department of the Environment (MDE, Contract #U00P4400079), NOAA-CREST CCNY Foundation (Sub-Contract 740 #49173B-02) and the National Aeronautics and Space Administration. The authors wish to also gratefully acknowledge the support for this study provided by Jack Kaye (NASA HQ), the NASA Tropospheric Chemistry Program and the Tropospheric Ozone Lidar Network (TOLNet).

REFERENCES

[1] Anne M. Thompson et al, 2003: Southern Hemisphere Additional Ozonesondes (SHADOZ) 1998–2000 tropical ozone climatology 2. Tropospheric variability and the zonal wave-one. J. Geo. Research: Atmospheres, **108(D2),** 8241.

[2] I. Stuart McDermid, Georg Beyerle, David A. Haner, and Thierry Leblanc, 2002: Redesign and improved performance of the tropospheric ozone lidar at the Jet Propulsion Laboratory Table Mountain Facility. Appl. Opt., **41(36),** 7550-7555.

[3] R. J. Alvarez et al., 2011: Development and application of a compact, tunable, solid-state airborne ozone lidar system for boundary layer profiling. J. Atmos. Ocean. Tech., **28(10)**, 1258.

[4] Shi Kuang, Michael J. Newchurch, John Burris, and Xiong Liu, 2013: Ground- based lidar for atmospheric boundary layer ozone measurements. Appl. Opt., **52(15)**, 3557–3566.

[5] J. T. Sullivan, T. J. McGee, G. K. Sumnicht, L. W. Twigg, and R. M. Hoff, 2014: A mobile differential absorption lidar to measure sub-hourly fluctuation of tropospheric ozone profiles in the Baltimore - Washington, D.C. region. Atmos. Meas. Techniques, **7(10)**, 3529–3548.

[6] William D Neff. 1997: The Denver Brown Cloud studies from the perspective of model assessment needs and the role of meteorology. J. of Air & Waste Mgmt. Assoc., **47(3)**, 269–285.

[7] Edwin F Danielsen, 1968: Stratospheric-tropospheric exchange based on radioactivity, ozone and potential vorticity. J. of the Atmospheric Sciences, **25(3)**, 502–518.

[8] M Beekman et al., 1997: Regional and global tropopause fold occurrence and related ozone flux across the tropopause. J. of At. Chem., **28**, 29.

[9] AO Langford, CD Masters, MH Proffitt, E-Y Hsie, and AF Tuck, 1996: Ozone measurements in a tropopause fold associated with a cut-off low system. Geophy. Res. Letters, **23(18)**, 2501–2504.

RESULTS OF A LONGER TERM NDACC MEASUREMENTS COMPARISON CAMPAIGN AT MAUNA LOA OBSERVATORY

Thomas J. McGee [1]*, Laurence W. Twigg[2], Grant K. Sumnicht[2], Thierry Leblanc[3] and John Barnes[4]

[1]*Laboratory for Atmospheric Chemistry and Dynamics, NASA Goddard Space Flight Center, Greenbelt, MD 20771, USA, *Email: Thomas.j.mcgee@nasa.gov*

[2]*Science Systems and Applications, Inc. Lanham MD, USA*

[3]*Table Mountain Facility, Jet Propulsion Laboratory, Wrightwood, CA USA*

[4]*NOAA, Earth Science Research Laboratory, Mauna Loa Observatory, Global Monitoring Division, Hilo, HI USA*

ABSTRACT

Between November, 2015 and January, 2015, the Goddard Space Flight Center operated a pair of lidar instruments at the NOAA facility at Mauna Loa on the Big Island of Hawaii (Lat. 19.5N, Lon. 155.5 W, Altitude 3.397 km). Measurements were made during six different four week periods during this time period by both the NASA GSFC Stratospheric Ozone Lidar (STROZ) and the Aerosol and Temperature (ATL) lidar. Also making measurements were the JPL Stratospheric Ozone Lidar and the NOAA Aerosol and Water Vapor Lidar. All instruments participate and archive data with the Network for the Detection of Atmospheric Composition Change. Measurement comparisons were made among various instruments in accordance with the standard intercomparison protocols of the NDACC.

1. INTRODUCTION

In August of 2012, the GSFC Stratospheric Ozone Lidar (STROZ) was shipped to Mauna Loa Observatory from a previous campaign at Lauder, New Zealand, for the purpose of an NDACC Measurement Validation Campaign. The STROZ Lidar makes measurements of the vertical profiles of temperature, aerosols and ozone in the stratosphere. The instrument has also been modified to make measurements of water vapor, from near the ground to the upper troposphere. This instrument was designed and built for the purpose of making such periodic comparisons of instruments making similar vertical profile measurements as part of the NDACC Validation Protocols which were established at the very beginning of the Network. These types of campaigns are extremely important to the credibility of measurements which are used for the purpose of determining the existence and the magnitude of observed or suspected trends. Goddard Space Flight Center has provided these services within NDACC since immediately prior to the official founding of NDACC (then NDSC) in 1991. Shortly after the STROZ lidar arrived and was set up at the observatory, the Aerosol and Temperature Lidar (ATL) arrived at the site from its' home base at the Goddard Space Flight Center. The AT Lidar is capable of making vertical measurements of numerous aerosol properties, including aerosol backscatter at three wavelengths (1064, 532 and 355 nm) , as well as extinction at two wavelengths (532 and 355 nm). Aerosol depolarization is recorded at 532 nm at the same time. The instrument make two separate temperature measurements, one a rotational Raman retrieval at 532 nm which measures the temperature vertical profile between .5 km above the instrument to about 22 km, and the second measurement is a Rayleigh/Vibrational Raman measurement between about 12 and 80 km above sea level. The instrument also uses the UV Raman technique to retrieve a water vapor profile from approximately 0.2 km above the site to about 22 km above sea level. The JPL Stratospheric Ozone Lidar has very similar ozone and stratospheric temperature capabilities to the GSFC STROZ lidar, and has been operated at MLO since 1992. This instrument also retrieves an aerosol backscatter profile, but does not retrieve water vapor. The NOAA Aerosol Lidar has been operating at MLO since the early 1970's, with Temperature and Water Vapor lidar measurements added subsequently. The original NOAA lidar utilized a Ruby laser (694 nm) but this was later converted to a Nd-YAG laser transmitting at 532 and 1064 nm. The NOAA results will also be presented. In all there were approximately 45 coincident ozone profile measurements made by the two lidar systems. Figure 1. Below shows the results of one such period, February, 2013.

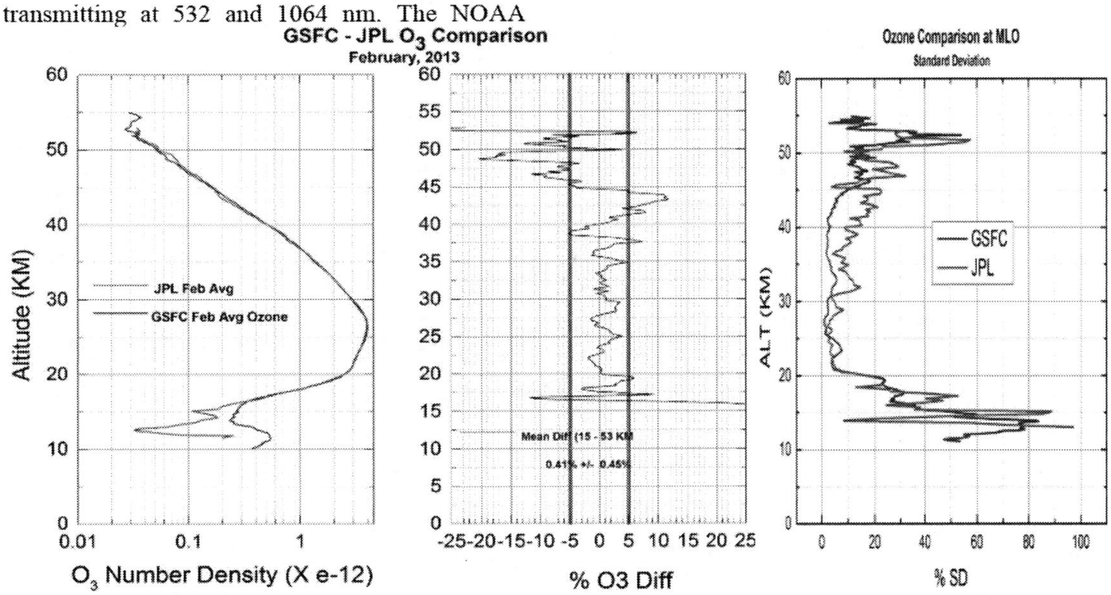

Figure 1. Results of comparisons between the GSFC and the JPL ozone measurements during February, 2013. There were a total of six coincident measurements during that period, largely because of weather.

MLO aerosol data record is one of the longest records in the world. The primary purpose for the Goddard AT lidar being at MLO for this extended period was to determine the added benefits that the AT lidar measurements could provide to the lengthy aerosol data record at MLO.

2. RESULTS

Measurements by the GSFC lidar instruments began in November of 2012. Early in this campaign the YAG laser within the STROZ instrument failed and parts from the AT Lidar YAG laser were shifted to the STROZ YAG laser so that an ozone intercomparison could begin. The AT YAG was repaired after the November measurement period. The poster being described will largely present data from the Ozone comparisons between the JPL Ozone lidar and the STROZ, although temperature and water vapor

There were six coincident measurements between the JPL and GSFC instruments and the overall results are presented in the figure. Between about 16 km (where the tropopause is located) and 45 km (where signal to noise ratio is getting small) the agreement is shown to be within +/- 5%, generally considered very good. Below the tropopause, the ozone concentration is small and absorption at the wavelengths used (308, 355 nm) is insufficient for an accurate measurement.

The poster will present data from all of the periods measurements were made, and will also include comparisons between the NOAA water vapor and the GSFC water vapor from both the STROZ and AT lidars, as well as temperature measurements from the JPL and GSFC lidars.

LIDAR OBSERVATION OF TROPOPAUSE OZONE PROFILES IN THE EQUATORIAL REGION

Yasukuni Shibata*, Chikao Nagasawa, Makoto Abo

Tokyo Metropolitan University, 6-6 Asahigaoka, Hino, Tokyo 191-0065, Japan

**Email: sibata@tmu.ac.jp*

ABSTRACT

Tropospheric ozone in the tropics zone is significant in terms of the oxidizing efficiency and greenhouse effect. However, in the upper troposphere, the ozone budget in the tropics has not been fully understood yet because of the sparsity of the range-resolved observations of vertical ozone concentration profiles. A DIAL (differential absorption lidar) system for vertical ozone profiles have been installed in the equatorial tropopause region over Kototabang, Indonesia (100.3E, 0.2S). We have observed large ozone enhancement in the upper troposphere, altitude of 13 - 17 km, concurring with a zonal wind oscillation associated with the equatorial Kelvin wave around the tropopause at equatorial region.

1. INTRODUCTION

Recently, the significance of tropospheric ozone in the tropics has been recognized in terms of the oxidizing efficiency and greenhouse effect, and various efforts have been made to investigate its spatial distribution and temporal variation. However, especially in the upper troposphere, the equatorial ozone has not been fully understood yet because of the lack of the ground-based and long-term observations. Some papers of the equatorial ozonesonde observation report that a phenomenon of large ozone enhancement in the upper troposphere, altitudes of 12 - 16 km, sometimes occurred [1,2]. Fujiwara et al. reports related to the equatorial Kelvin wave around the tropopause and the Madden-Julian oscillation (MJO) in the troposphere. The downward motion associated with the Kelvin wave and the MJO transported the stratospheric ozone into the troposphere, and the air mixing due to the Kelvin wave breaking at the tropopause also caused stratosphere-troposphere exchange.

We have constructed the lidar facility for survey of atmospheric structure over troposphere, stratosphere, mesosphere and low thermosphere over Kototabang, Indonesia in the equatorial region [3]. The lidar observations started from 2004, and routine observations of clouds and aerosol in the troposphere and stratosphere are continued now. We installed a DIAL (differential absorption lidar) system for high-resolution measurements of vertical ozone profiles in the equatorial tropopause region over Kototabang, Indonesia in 2014. This ozone lidar is possible to investigate the phenomenon of increasing ozone concentration of the equatorial upper troposphere in detail.

2. OZONE DIAL SYSTEM

We have installed DIAL system for high-resolution measurements of vertical ozone profiles in the equatorial tropopause region over Kototabang. There were many ozone DIAL systems in the world, but their systems are almost optimized for stratospheric ozone layer measurement or tropospheric ozone measurement [4, 5]. Because of deep ozone absorption in the UV region, the wavelength selection is important [6]. Over the equatorial region, the tropopause height is almost 17 km. So we use 314 nm for on-line and 355nm for off-line.

A schematic diagram of the integrated DIAL system is shown in Fig. 1. The radiation source of the lidar was a Q-switched frequency-tripled Nd:YAG laser (355 nm) with a 35 mJ output energy and a 10 Hz repetition rate. The dye laser consisted of a dual-wavelength generator. In case of the ozone measurements, the dye laser provided an on-line wavelength of 314 nm with a 4 mJ output energy. A harmonic generator equipped with a BBO crystal is used. The backscattered radiation from the atmosphere is collected by a 45 cm diameter telescope and a dichroic mirror divide the optical signal into two channels (for 314 nm and 355 nm). The field of view of the receiving telescope is 2 mrad, and the distance between the optical axes of the telescope and the radiation beam is 2.5 m.

The DIAL system was achieved successfully test observation of O3 concentration profiles around equatorial troposphere with integration time of 6 hours and range resolution of 450 m on June 19, 2014. Because of the high frequency of occurrence of equatorial cirrus cloud, we need enough integration time for the observation ozone concentration over the tropopause. Fig. 2 shows the ozone concentration profile and radiosonde temperature profile. The radiosonde launched at the Tabing airport, Padang (100.2E, 0.5S) approximately 90 km from the lidar site. A large ozone enhancement in the upper troposphere, altitude of 13 - 17 km, was observed. Fig. 3 shows the variation of temperature profile at Padang in June, 2014. The tropopause height moved downward from 18 km on June 7 to 16 km on June 17. After June 18, the tropopause height jumped up to 18 km concurring with a zonal wind oscillation associated with the equatorial Kelvin wave around the tropopause at equatorial region. Therefore, the downward motion associated with the Kelvin wave transported the stratospheric ozone into the troposphere.

3. CONCLUSION

We have conducted an observation with the ozone DIAL in Indonesia in June 1995 to investigate the

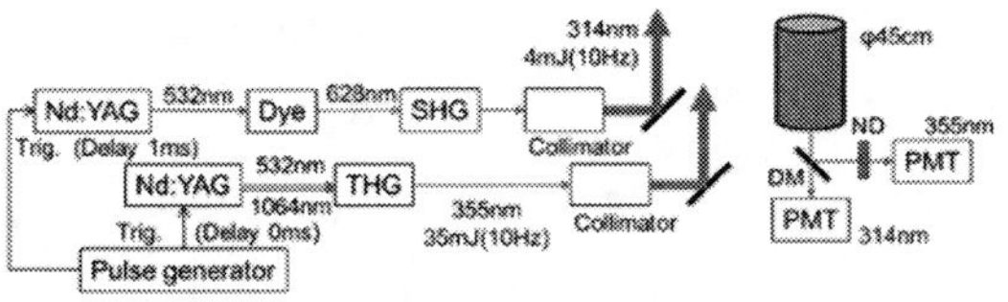

Fig. 1 Block diagram of the ozone DIAL system.

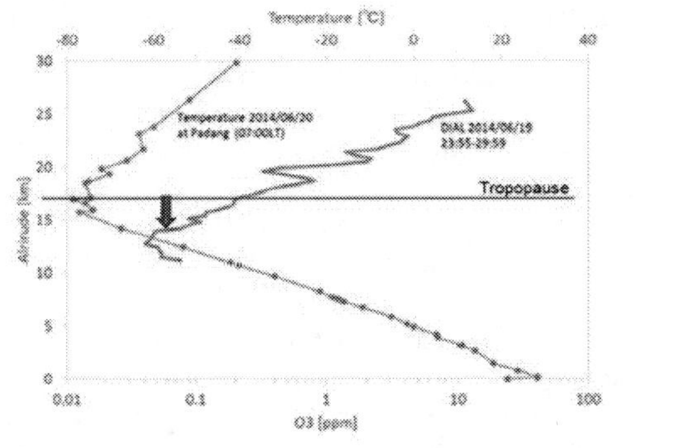

Fig.2 The ozone concentration profile and radiosonde temperature profile at Padang.

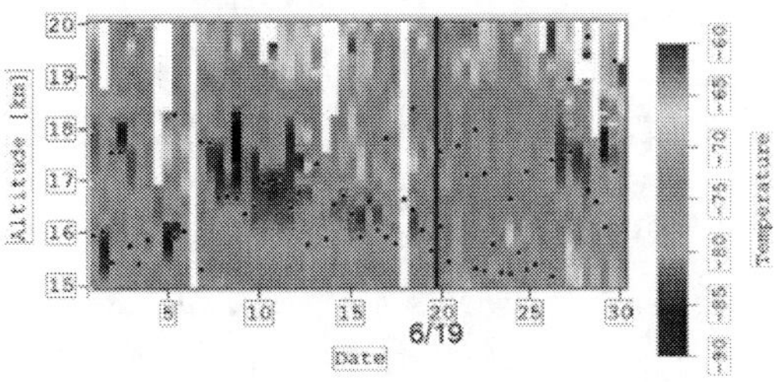

Fig. 3 Variations of vertical distribution of atmospheric temperature at Padang from June 1 to June 30, 2014. The tropopause height is indicated by asterisks (*).

phenomenon of ozone enhancement in the tropical upper troposphere. A large ozone enhancement in the upper troposphere was observed. The O3 DIAL with high temporal resolution would clarify the relationship of the time variation of the ozone enhancement and the Kelvin wave. We will also contribute to the elucidation of the climate change by getting observational information about high resolution ozone density profiles, and the wave-propagation and material transportation using ozone as a tracer from the troposphere to the lower stratosphere over the equator.

ACKNOWLEDGEMENT

This work was financially supported by Grants-in-Aid for Scientific Research (No. 233401043).

REFERENCES

[1] Komala, N., S. Saraspriya, K. Kita, and T Ogawa, Atmos. Environ., 30, 1851-1856, 1996.

[2] Masatomo Fujiwara, Kazuyuki Kita, Toshihiro Ogawa, J. Geophys. Res., Vol. 103, No. D15, 19173-19182, 1998.

[4] C. Nagasawa, M. Abo, Y. Shibata, 23rd International Laser Radar Conference, No.2O-8, pp.43 - 46, 2006.

[5] Gerard J. Megie, Gerard Ancellet, and Jacques Pelon, Appl. Opt. 24, 3454-3463, 1985.

[6] Masahisa Nakazato, Tomohiro Nagai, Tetsu Sakai, and Yasuo Hirose, Appl. Opt. 46, 2269-2279, 2007.

[7] M. Abo, C. Nagasawa, Y. Shibata, 26th International Laser Radar Conference, No.S1P-44, 2012.

WATER VAPOUR MIXING RATIO MEASUREMENTS IN POTENZA IN THE FRAME OF THE INTERNATIONAL NETWORK FOR THE DETECTION OF ATMOSPHERIC COMPOSITION CHANGE - NDACC

Benedetto De Rosa*, Paolo Di Girolamo*, Donato Summa, Dario Stelitano, Ignazio Mancini

*Scuola di Ingegneria, Università degli Studi della Basilicata, Viale dell'Ateneo Lucano n. 10, 85100 Potenza – Italy , *Emails: bundit@hotmail.it; digirolamo@unibas.it*

ABSTRACT

In November 2012 the University of BASILicata Raman Lidar system (*BASIL*) was approved to enter the International Network for the Detection of Atmospheric Composition Change (NDACC). This network includes more than 70 high-quality, remote-sensing research stations for observing and understanding the physical and chemical state of the upper troposphere and stratosphere and for assessing the impact of stratosphere changes on the underlying troposphere and on global climate. As part of this network, more than thirty ground-based Lidars deployed worldwide are routinely operated to monitor atmospheric ozone, temperature, aerosols, water vapour, and polar stratospheric clouds.

In the frame of NDACC, BASIL performs measurements on a routine basis each Thursday, typically from local noon to midnight, covering a large portion of the daily cycle. Measurements from BASIL are included in the NDACC database both in terms of water vapour mixing ratio and temperature. This paper illustrates some measurement examples from BASIL, with a specific focus on water vapour measurements, with the goal to try and characterize the system performances.

1. INTRODUCTION

Water vapour is a key atmospheric trace gas controlling weather and climate. The effects of water vapour are large in the upper troposphere and lower stratosphere, but there are few measurements of water vapour concentrations and its long-term variation in this altitude region. Measurements in this altitude region have been traditionally guaranteed by the use of operational radiosondes or balloon-borne frost-point hygrometers, the latter remaining the best source of high quality water vapour measurements in the in the UTLS region[1], but being yet too expensive to be used on operational basis.

For the above motivations, in the early 2000s, the international Network for the Detection of Stratospheric Change (NCSC), now Network for the Detection of Atmospheric Composition Change (NDACC), decided to include water vapour Raman lidars among its instruments.

In fact, while NDACC remains committed to long-term monitoring of changes in the stratosphere, with an specific emphasis on the evolution of the ozone layer, its priorities have broadened to include issues such as the detection of trends in the overall atmospheric composition and understanding their impacts on the stratosphere and troposphere, and establishing links between climate change and atmospheric composition.

The Raman Lidar system *BASIL* joined NDACC in November 2012, with the primary goal of providing accurate routine measurements of the vertical profile of water vapour mixing ratio and temperature, these two parameters being of primary importance for the purposes of NDACC, with valuable contribution to the NDACC primarily expected in terms of water vapour measurements.

2. BASIL

Lidar measurements discussed in this paper were performed by the Raman Lidar system *BASIL* in Potenza, Italy (40°38'45" N, 15°48'29" E, 730 m above mean sea level). The system is hosted in a sea-tainer located on the roof of the Scuola di Ingegneria at the Università degli Studi della Basilicata. The major feature of *BASIL* is represented by its capability to perform high-resolution and accurate measurements of atmospheric water vapour and temperature, both in daytime and nighttime, based on the application of the vibrational and rotational Raman lidar techniques in the UV [2, 3, 4, 5, 6]. Besides water

DOI: 10.1051/epjconf/201611905017

vapour and temperature, *BASIL*, in its NDACC configuration, also provides measurements of particle backscatter at 355 and 532 nm, particle extinction at 355 nm and particle depolarization at 355 nm. and water vapor mixing ratio. All lidar-based measurements of relative humidity can also be obtained from the simultaneous water vapor mixing ratio and temperature measurements.

As stated in the affiliation letter a valuable contribution to the NDACC from BASIL is primarily expected in terms of water vapour mixing ratio measurements. For this reason, a major emphasis has been put on this parameter, especially in terms of calibration and validation efforts.

In the frame of NDACC, BASIL operates on a routine basis each Thursday, typically from local noon to midnight, covering a large portion of the daily cycle. Water vapour mixing ratio measurements routinely performed by BASIL in the frame of the NDACC program, besides being included in the long-term measurement activities NDACC database, are also used for the purpose of validating the water vapor measurements from other sensors, especially satellite-borne sensors, of studying upper tropospheric humidity variability at timescales varying from minutes to days.

BASIL makes use of a Neodymium-doped Yttrium Aluminium Garnet (Nd:YAG) laser source, equipped with second and third harmonic generation crystals and capable of emitting pulses at 355 and 532 nm, with single pulse energies of 500 and 300 mJ, respectively, a pulse repetition rate of 20 Hz and a pulse duration of 5–6 ns. Laser beams at 355 and 532 nm are simultaneously transmitted in the atmosphere along the zenith.

The receiver is built around a telescope in Newtonian configuration (40-cm diameter primary mirror). Collected radiation is split into eight separate portions by means of dichroic or partially reflecting mirrors. Specifically, two portions are fed into the detection channels used for temperature measurements; two other portions are sent to the water vapour and nitrogen Raman channels; another two portions are fed into the 355 and 532 nm elastic channels; and a fraction of the signal entering the 355-nm channel is split into two additional portions to allow the detection of the parallel and cross-polarized elastic signals used for the determination of particle depolarization.

Signal selection is performed by means of narrowband interference filters, while signal detection is accomplished by means of photomultipliers located in cascade with each interference filter; detected signals are finally sampled by means of photon-counting units. Data are acquired with a vertical and temporal resolution of 30m and 1min, respectively. A block diagram of the system is given in Fig.1.

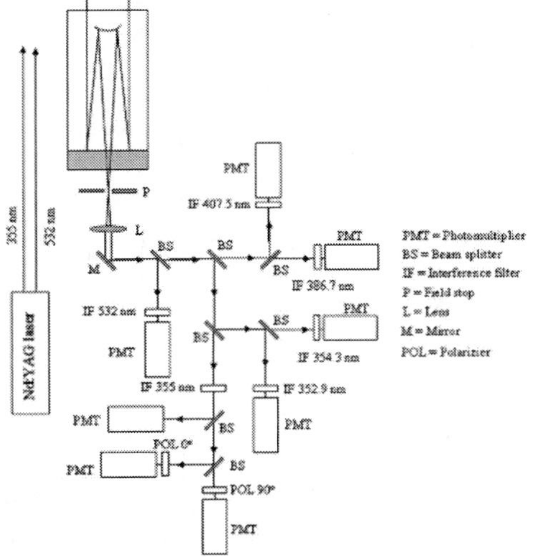

Fig.1. Block diagram of the system setup.

3. METHODOLOGY

The Raman lidar technique for the determination of the water vapor mixing ratio profile has been extensively discussed in literature (for a review see Whiteman 2003a and 2003b). The approach is based on the collection of the vibrational Raman backscattered signals from water vapour and nitrogen molecules, i.e.:

$$P_{H_2O}(R) = P_0 \frac{c\Delta t}{2} \frac{A_{tel}}{R^2} O(R) n_{H_2O}(R) \sigma_{H_2O} T_{\lambda_0}(R) T_{\lambda_{H_2O}}(R)$$

$$P_{N_2}(R) = P_0 \frac{c\Delta t}{2} \frac{A_{tel}}{R^2} O(R) n_{N_2}(R) \sigma_{N_2} T_{\lambda_0}(R) T_{\lambda_{N_2}}(R)$$

The water vapour mixing ratio $x_{H_2O}(R)$ can be obtained from the power ratio of water vapour to molecular nitrogen vibrational Raman signals through the application of the expression:

$$x_{H_2O}(R) = k \frac{P_{H_2O}(R)}{P_{N_2}(R)} \Delta T_{\lambda_{H_2O}, \lambda_{N_2}}$$

with k being the lidar system calibration coefficient and $\Delta T_{\lambda_{H_2O},\lambda_{N_2}} = T_{\lambda_{N_2}}(R)/T_{\lambda_{H_2O}}(R)$ being the differential transmission term, which accounts for the different atmospheric transmission at the two Raman wavelengths λ_{H_2O} and λ_{N_2}. When stimulated at 354.7 nm the water vapour and molecular nitrogen vibrational Raman signals are located at λ_{H_2O}=407.5 nm and λ_{N_2}=386.7 nm, respectively. $\Delta T_{\lambda_{H_2O},\lambda_{N_2}}$ is primarily caused by Rayleigh scattering and can easily be corrected for by the use of radiosonde or standard-atmospheric profiles of number density. An additional component of $\Delta T_{\lambda_{H_2O},\lambda_{N_2}}$ is associated with the wavelength dependence of particle extinction, which is usually very small (1-2 %) and is easily determinable from lidar measurements of particle extinction at 354.7 nm (Whiteman, 2003b).For the purposes of NDACC the calibration coefficient k was determined based on the comparison between simultaneous and co-located water vapour mixing ratio profiles from the Raman lidar and radiosondes for an extended measurement sample. The comparison is performed over an altitude region with an extent of 1-3 km in order to have a statistical significant number of data points per profile to be compared. The selection of this altitude region accounts for the existing distance between the lidar site and the radiosonde launching facility.

The statistical error is calculated by means the analytical expression :

$$\Delta\chi_{H2O}(z) = \chi_{H2O}(z)\sqrt{\frac{P_{H2O}(z)+b_{H2O}}{[P_{H2O}(z)]^2}+\frac{P_{N2}(z)+b_{N2}}{[P_{N2}(z)]^2}}$$

4. RESULTS

As an example of the measurements carried out in the frame of NDACC, figure 2 illustrates the time evolution of the water vapor mixing ratio over a 7 h period of from 16:36 UTC to 23:22 UTC on 09 October 2014. Figure 2 is plotted as a succession of 1 min averaged consecutive profiles. To reduce signal statistical fluctuations, vertical smoothing is applied to the data to achieve an overall vertical resolution of 150 m. The figure clearly highlights the day-to-night transition, with the final part of the daytime portion clearly distinguishable on the left side of the figure with noisy data above 4-5 km.

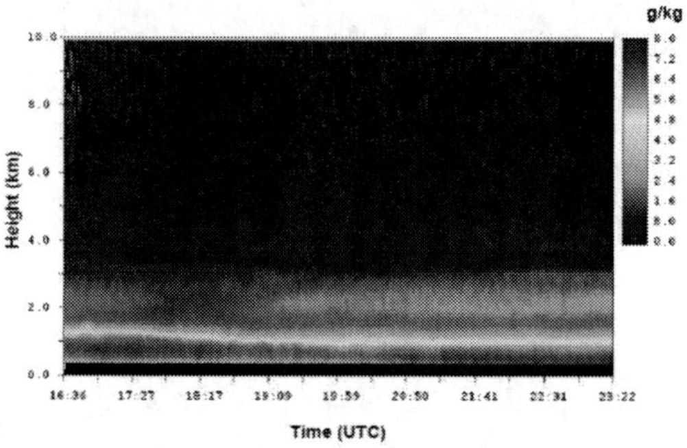

Fig. 2. Time evolution of the water vapor mixing ratio over a 7 h period of from 16:36 UTC to 23:22 UTC on 09 October 2014.

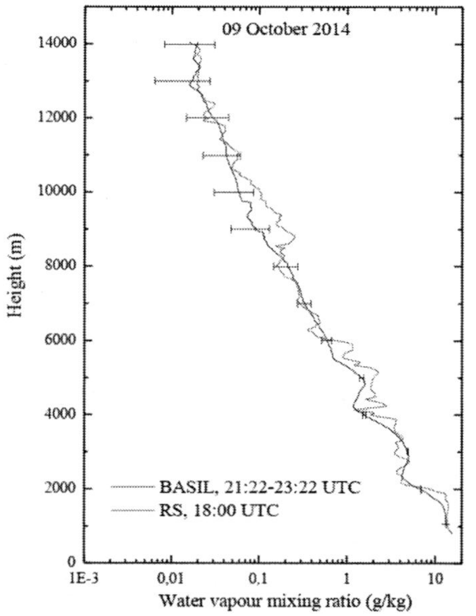

Fig. 3. Vertical profile of the water vapour mixing ratio with the error bars as obtained considering a 2 h time integration between 21:22 and 23:22 UTC on 09 October 2014, together with the simultaneous profile as measured by a radiosonde launched from the nearby site of Tito Scalo at 18:00 UTC on 09 October 2014.

Figure 3 illustrates the vertical profile of the water vapour mixing ratio as obtained for the same day considering a 2 h time integration between 21:22 and 23:22 UTC. The vertical resolution is 300 m up to 6 km, 600 m up to 9 km and 1200 m above. For comparison the almost simultaneous water vapour mixing ratio profile as measured by a radiosonde launched from the nearby site of Tito Scalo (7 km away) at 18:00 UTC on 09 October 2014. The agreement between BASIL and the radiosonde is quite good, especially when considering the distance between the two sites. The lidar profile from BASIL is found to reach approx. 14 km, with a capability to go down to 0.02 g/kg. The capability to reach 14-15 km and a detection level of 0.01-0.02 g/kg has been observed in most of the 2 h averaged clear sky measurements from BASIL in the frame of NDACC. In the case of measurements integrated over all night, the lidar profile from BASIL is found to reach approx. 18-20 km. More results from this measurement effort, also focused on a statistical analysis on the collected data, will be illustrated and discussed at the Conference.

ACKNOWLEDGEMENT

We wish to warmly thank Istituto di Metodologie per l'Analisi Ambientale, Consiglio Nazionale delle Ricerche, in Tito Scalo, for the provision of the radiosnde data.

REFERENCES

[1] Vomel, H., H. Selkirk, L. Miloshevich, J. Valverde-Canossa, J. Valdes, E. Kyro, R. Kivi, W. Stolz, G. Peng, and J. A. Diaz,2007: Radiation dry bias of the Vaisala RS92 humidity sensor. *J. Atmos. Oceanic Technol.*, **24**, 953-963.

[2] Whiteman, D. N., 2003: Examination of the traditional Raman lidar technique. I. Evaluating the temperature-dependent lidar equations, *Appl. Opt.*, **42**, 2571-2592.

[3] Behrendt, A. and Reichardt, J.: Atmospheric temperature profiling in the presence of clouds with a pure rotational Raman lidar by use of an interference-filter-based polychromator, *Appl. Optics*, **39**, 1372–1378, 2000.

[4] Di Girolamo, P., R. Marchese, D. N. Whiteman, B. B. Demoz, 2004: Rotational Raman Lidar measurements of atmospheric temperature in the UV. *Geophys. Res. Lett.*, **31**, doi: 10.1029/2003GL018342.

[5] Di Girolamo, P., A. Behrendt, and V. Wulfmeyer, 2006. Spaceborne profiling of atmospheric temperature and particle extinction with pure rotational Raman lidar and of relative humidity in combination with differential absorption lidar: performance simulations, *Appl. Opt.,* **45**, 2474-2494, doi:10.1364/AO.45.002474.

[6] Bhawar, R., P. Di Girolamo, D. Summa, C. Flamant, D. Althausen, A. Behrendt, C. Kiemle, P. Bosser, M. Cacciani, C. Champollion, T. Di Iorio, R. Engelmann, C. Herold, Müller, D., S. Pal, M. Wirth, V. Wulfmeyer, 2011: The Water Vapour Intercomparison Effort in the Framework of the Convective and Orographically-Induced Precipitation Study: Airborne-to-Ground-based and airborne-to-airborne Lidar Systems, *Quarterly Journal of the Royal Meteorological Society*, **137**, 325–348.

EPJ Web of Conferences 119, 05018 (2016) DOI: 10.1051/epjconf/201611905018
ILRC 27

LATE SUMMER OZONE VARIABILITY IN THE LOWER TROPOSPHERE OF THE EASTERN MEDITERRANEAN

Christoforos Tsamalis [1,3]*, Alexandros Papayannis[2], Gerard Ancellet[1], François Ravetta[1]

[1]*Sorbonne Universités, UPMC Univ. Paris 06, Université Versailles St-Quentin, CNRS/INSU, LATMOS-IPSL, 75252, Paris, France*

[2]*Physics Department, National Technical University of Athens, 15780, Zografou, Greece*

[3]*Now at Met Office, Exeter, EX1 3PB, UK, *Email: christoforos.tsamalis@metoffice.gov.uk*

ABSTRACT

During the STAAARTE 96 Hellen aircraft campaign the lidar ALTO together with in situ measurements examined the ozone and aerosols variability in the lower troposphere over the Eastern Mediterranean region. Ozone mean value in the free troposphere (FT) measured from ALTO was 48 ppb, while it was 45 ppb from in situ observations; the aerosols mean scattering coefficient (550 nm) was 31 Mm^{-1}. The FT ozone distributions of the two instruments are significantly different to distinct sampling. Air masses origin examination using the FLEXPART model for low and high FT ozone observations during late summer indicate that low ozone masses come from Mediterranean/North Africa regions travelling at low altitude, while high ozone masses emanate from Europe's middle or upper troposphere.

1. INTRODUCTION

In the Mediterranean region air masses with different origin are found creating an atmospheric crossroads. This favours the transport of atmospheric constituents in the region like ozone, its precursors and aerosols, at high variable concentrations, which under the high insolation, especially during the summer period, create an ideal environment for photochemical reactions to take place [1]. Indeed, during summertime an ozone pool is observed over the Eastern Mediterranean free troposphere (FT), which is one of the regions with the highest background levels of tropospheric ozone worldwide [2]. The number of the studies focusing in the FT ozone in the area is limited (e.g. [3]), mainly using either models, which are affected by the emissions uncertainties or satellites, which do not provide the required vertical resolution to observe ozone layers e.g. [4].

2. THE STAAARTE HELLEN 96 CAMPAIGN

The STAAARTE (Scientific Training and Access to Aircraft for Atmospheric Research Throughout Europe) Hellen 96 airborne campaign took place during the period 19-26 September 1996 over Greece (Fig. 1, Tab. 1). The main purpose of the campaign was to characterise the ozone variability in the lower troposphere over the region during late summer and to identify the processes that contribute to this variability. The French Fokker 27 medium-range research aircraft (ARAT) realised in total 10 flights flying at an altitude of 3-4.3 km a.s.l. (Fig. 1, Tab. 1).

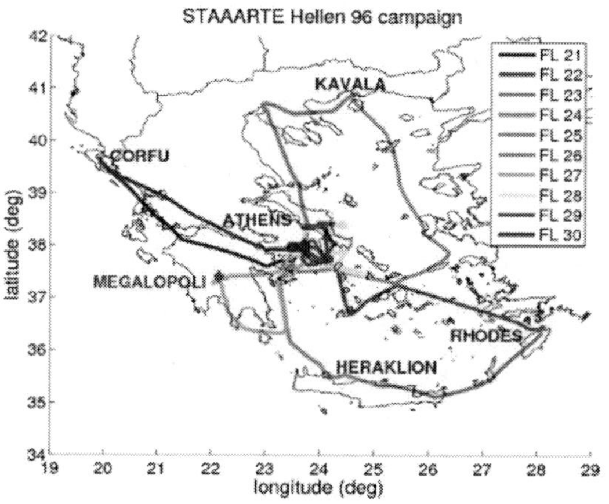

Fig. 1: Flights tracks.

The main instrument on board the aircraft was the lidar ALTO, a UV differential absorption lidar (DIAL) system based on Nd:YAG laser and stimulated Raman scattering in deuterium to generate three wavelengths (266, 289 and 316 nm) in order to measure ozone and aerosols profiles [5]. The ALTO optical receiver includes a 40 cm Cassegrain telescope coupled to a grating

Flight	Date	Time (UTC)	Altitude (km)	Path
21	19/9/1996	13:22-14:37	3.33±0.02	Corfu – Athens
22	21/9/1996	07:03-08:48	3.93±0.02	Athens – Rhodes
23	21/9/1996	10:30-11:54	4.23±0.04	Rhodes – Heraklion
24	21/9/1996	13:15-14:48	3.02±0.02	Heraklion – Athens
25	22/9/1996	07:12-09:02	3.64±0.02	Athens – Kavala
26	22/9/1996	10:26-12:45	3.77±0.15	Kavala – Athens
27	24/9/1996	08:02-10:06	3.63±0.03	Athens – Megalopoli – Athens
28	24/9/1996	12:57-15:12	4.17±0.19	Athens – Euboea – Athens
29	25/9/1996	07:57-10:02	3.79±0.03	Athens – Euboea – Athens
30	26/9/1996	06:56-08:22	4.25±0.03	Athens – Corfu

Tab. 1: Information (flight number, date, flight time, mean altitude with standard deviation and path) about the flights during the STAAARTE Hellen 96 campaign.

spectrometer by a 1.5 mm diameter fused-silica optical fiber. Two wavelength pairs are used for the ozone observation 266-289 nm and 289-316 nm, the first for short distances (0.5-2 km) while the second for longer distances (1.5-5 km). At near range (<0.5 km) where the overlap function between the laser beam and the telescope field of view is still range dependant the observations are excluded. ALTO is able to characterise narrow layers (<0.5 km) with good accuracy (<5-10 ppb). At 316 nm the ozone absorption can be neglected in the troposphere and the aerosol profile is expressed through the non-dimensional scattering ratio, which is the ratio of total (molecules and aerosols) backscattering to molecular backscattering. Aboard the aircraft there were also in situ instruments measuring ozone (Thermo-Electron 49PS with accuracy 1 ppb) and aerosols scattering coefficient at 550 nm (MRI 1550B with uncertainty 10%).

3. RESULTS

Tab. 2 presents the mean values and the range of ozone and aerosols in the FT measured by the in situ instruments. The lower ozone values (<30 ppb) encountered during flights 25 and 26, while the higher (>70 ppb) during Flight 24. Regarding aerosols the lower values were found during flights 23, 28 and 29 (just above 20 Mm-1) and the higher (>60 Mm-1) during flight 21 due to Saharan dust intrusion (not shown). The distributions of ozone measured in situ and from ALTO in the FT and the aerosols are depicted in Fig. 2.

According to ALTO the ozone mean value is 48 ppb and slightly lower for the in situ at 45 ppb, while the mean aerosols scattering coefficient is 31 Mm-1 (with the high value tail being mainly due to the dust event). Interestingly the Ozone distributions are different, with the ALTO one being bimodal and having significantly more ozone measurements in the range 55 to 70 ppb. This is mainly due to different sampling of air masses between the two instruments as it can be seen in Fig. 3. During flight 29 ALTO is able to detect an ozone layer at 2.5-3 km, which is not affected by aerosols or clouds as it can be seen by the scattering ratio being almost 1 indicative of molecular scattering only. However, this layer is missed by the in situ instrument measuring higher in the troposphere.

Flight	Ozone (ppb)	Aerosols (Mm^{-1})
21	48±9 [34, 65]	40±9 [29, 66]
22	45±11 [29, 63]	35±1 [32, 39]
23	46±3 [42, 51]	25±1 [22, 29]
24	61±9 [42, 76]	31±1 [27, 34]
25	38±5 [27, 46]	35±5 [29, 56]
26	36±4 [28, 47]	30±3 [24, 39]
27	47±3 [40, 53]	29±2 [26, 37]
28	48±2 [43, 53]	27±4 [21, 38]
29	44±2 [38, 48]	31±4 [21, 41]
30	39±2 [34, 43]	29±3 [23, 40]

Tab. 2: Mean, standard deviation and range [minimum, maximum] of FT values for ozone and aerosols.

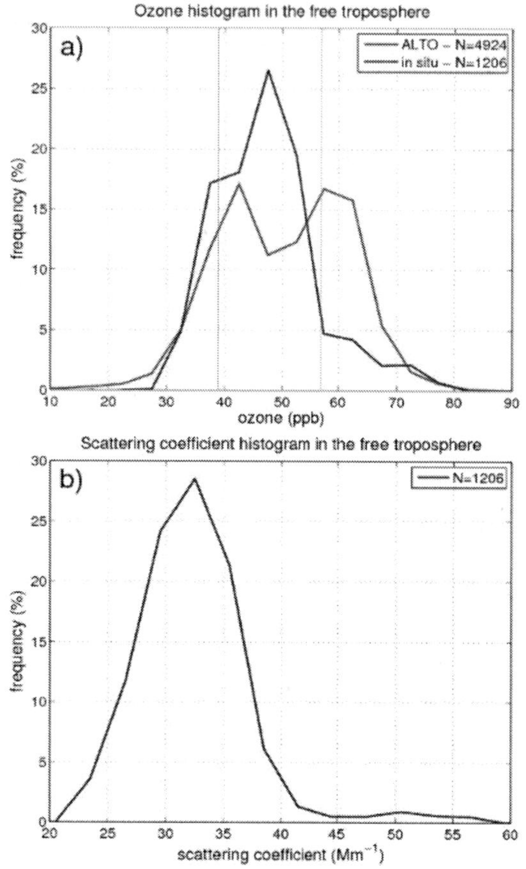

Fig. 2: Histograms of free tropospheric (a) ozone and (b) aerosols from in situ (blue) and ALTO (red) observations.

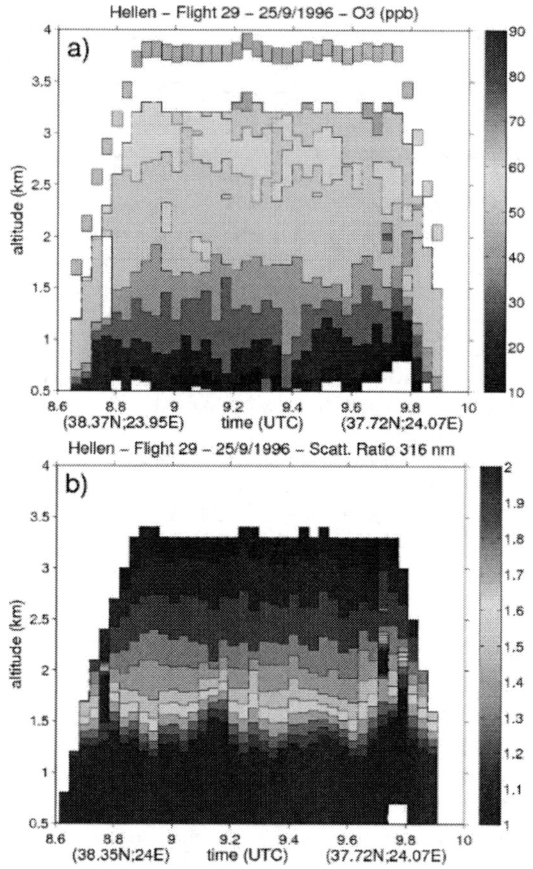

Fig. 3: ALTO measurements of (a) ozone (with the in situ above) and (b) aerosols.

The origin of the low (<39 ppb) and high (>57 ppb) ozone air masses based on the ALTO distribution of Fig. 2a and corresponding to the first and the third quartiles are examined using the FLEXPART model [6]. Fig. 4 shows only air masses at every location when the percentage at this location is larger than 0.1%, in order to have a minimum statistical significance. It can be observed that the air masses characterized by low ozone have either maritime origin (Mediterranean Sea and Atlantic Ocean) or coming from northern Africa while always their mean altitude is below 3 km. This reflects the export of clean maritime or depleted by dust particles air from the local boundary layer.

On the other hand, the air masses with high ozone are coming from Europe and their mean altitude is above 4 km reaching 8 km 4-5 days before their observation. Although, there is no indication from FLEXPART of a stratospheric intrusion within this time interval, their relatively increased potential vorticity points towards old stratospheric masses.

ACKNOWLEDGEMENT

The STAAARTE Hellen 96 Campaign was financed by the European Union through the Training and Mobility of Researchers (TMR) Programme - Access to Large-scale Facilities/EU DG XH, which is gratefully acknowledged.

DOI: 10.1051/epjconf/201611905018

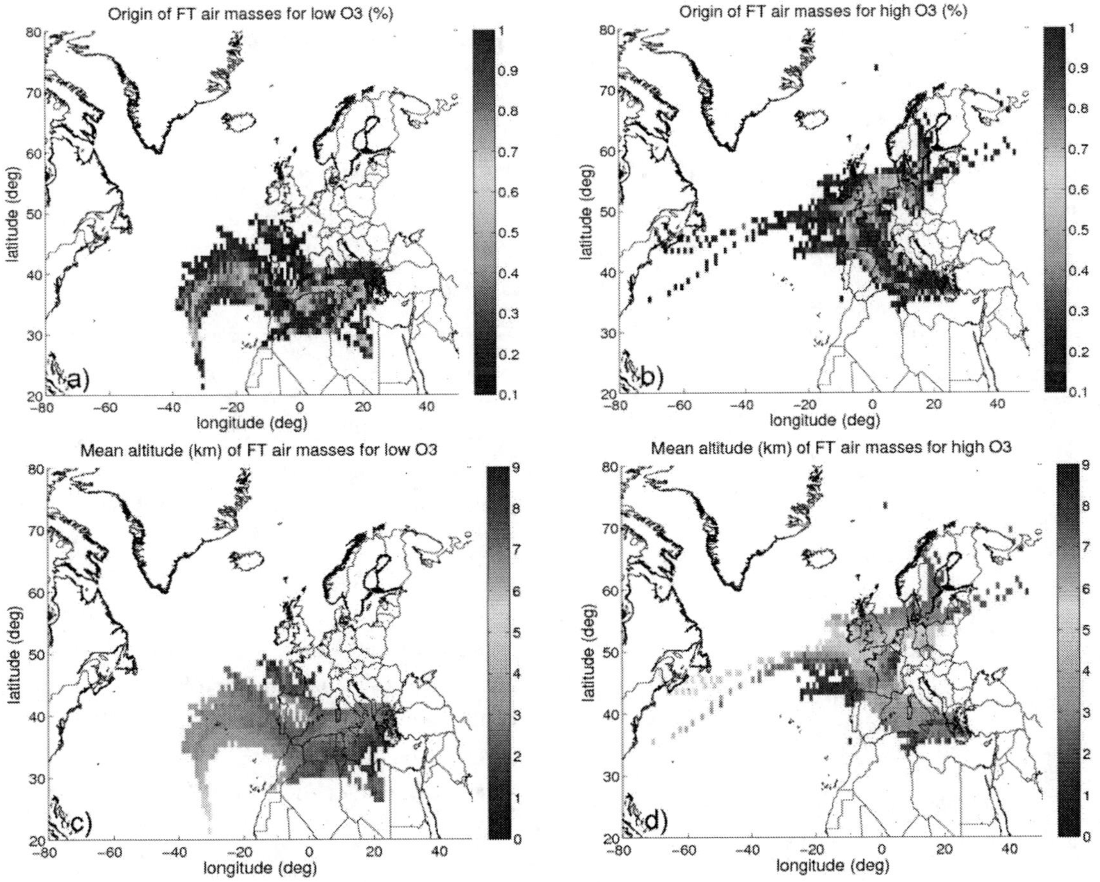

Fig. 4: (a, b) Origin of Free Tropospheric air masses during the campaign and (c, d) mean altitude for low (<39 ppb) and high (>57 ppb) ozone concentrations.

C. Tsamalis is grateful to Eugenides Foundation for financial support.

REFERENCES

[1] Lelieveld J. et al., 2002: Global air pollution crossroads over the Mediterranean, *Science,* **298,** 794-799.

[2] Zanis P. et al., 2014: Summertime free-tropospheric ozone pool over the eastern Mediterranean/Middle East, *Atmos. Chem. Phys.,* **14,** 115-132.

[3] Kalabokas P. D. et al., 2013: Examination of the atmospheric conditions associated with high and low summer ozone levels in the lower troposphere over the eastern Mediterranean, *Atmos. Chem. Phys.,* **13,** 10339-10352.

[4] Safieddine S. et al., 2014: Summertime tropospheric ozone assessment over the Mediterranean region using the thermal infrared IASI/MetOp sounder and the WRF-Chem model, *Atmos. Chem. Phys.,* **14,** 10119-10131.

[5] Ancellet G. and Ravetta F., 1998: Compact airborne lidar for tropospheric ozone: description and field measurements, *Appl. Opt.,* **37(24),** 5509-5521.

[6] Stohl A. C. et al., 2005: Technical note: The Lagrangian particle dispersion model FLEXPART version 6.2, *Atmos. Chem. Phys.,* **5,** 2461-2474.

DOI: 10.1051/epjconf/201611905019

A Case Study on Observed and Simulated CO_2 Concentration Profiles in Hefei based on Raman Lidar and GEOS-Chem Model

Yinan Wang[1*], Daren Lü[1], Weilin Pan[1], Kee Yuan[2]

[1]*Key Laboratory of Middle Atmosphere and Global Environment Observation, Institute of Atmospheric Physics, Chinese Academy of Sciences, Beijing,100029,China, *Email: wangyinan@mail.iap.ac.cn*

[2]*Anhui Institute of Optics and Fine Mechanics, Chinese Academy of Sciences, Hefei 230031, China*

ABSTRACT

Observations of atmospheric CO_2 concentration profiles provide significative constraints on the global/regional inversions of carbon sources and sinks. Anhui Institute of Optics and Fine Mechanics of Chinese Academy of Sciences developed a Raman Lidar system to detect the vertical distribution of atmospheric CO_2. This paper compared the observations with the modeled results from a three-dimensional global chemistry transport model-GEOS-Chem, which showed a good agreement in the trend of change with lidar measurements. The case study indicated a potential for better simulating vertical distribution of atmospheric CO_2 by combining with lidar measurements.

1. INTRODUCTION

Observations of atmospheric CO_2 concentration profiles provide significative constraints on the global/regional inversion of carbon sources and sinks. Raman Lidar has possessed the capability of conducting high temporal and vertical resolution measurements of CO_2 concentration profiles [**1-3**]. By assimilating those informations of vertical distribution of CO_2 concentration in atmosphere under synthetic inversion framework, the uncertainty of the regional carbon budget prediction will be reduced. In this paper, we proceeded a case study on CO_2 concentration profiles observed by a complicated Raman lidar system, developed by Anhui Institute of Optics and Fine Mechanics of Chinese Academy of Sciences, and compared observations with simulated results from a global three-dimensional chemistry transport model-GEOS-Chem, which could provide a valuable basis for better using lidar data to improve the simulation of atmospheric CO_2 concentration.

2. DATA AND METHOD

2.1 GEOS-CHEM MODEL

GEOS-Chem is a global 3-D chemical transport model of atmospheric composition driven by assimilated meteorological observations from the Goddard Earth Observing System (GEOS) of the National Aeronautics and Space Administration (NASA) Global Modeling and Assimilation Office (GMAO). We used the model (v9-01-02) to simulate the atmospheric CO_2 concentrations on a horizontal resolution of 2° latitude × 2.5° longitude and on a vertical resolution of 47 layers from the surface to 0.01 hPa. The model was integrated with GEOS-5 meteorology for eight years (2004–2011) from an initial condition of uniform CO_2 (375.0 ppmv) as a "spin up" period [**4**]. The sources (or sinks) for the simulations in

DOI: 10.1051/epjconf/201611905019

this study included emissions from fossil-fuel burning and cement production, biomass burning, biofuel burning, balanced biosphere (net ecosystem production), ocean, net terrestrial exchange, aircraft, and marine.

2.2 RAMAN LIDAR SYSTEM

The Raman lidar system applied 354.7nm third harmonic of Nd:YAG laser, which was transmitted with 350 mJ pulse energy and repetition rate of 20 Hz. The receiver employed a photomultiplier tube with quantum efficiency of 25% and 200 MHz photon counter, and detected Raman backscattering 371.66 nm (shift 1285cm(-1)) signal [detailed lidar description can be found in **5**, **6**]. The principle of the system is based on the Raman scattering by CO_2 molecules in the atmosphere. The laser beam simultaneously motivated Raman scattering of atmospheric CO_2 and N_2, and the number of received photons is in proportion to the molecular density of the two gases. By detecting the backscattering Raman signal ratio of CO_2 to N_2 at different altitude, we can obtain the relative concentration of atmospheric CO_2 as the following formula:

$$\frac{n_1(z)}{n_2(z)} = \frac{\beta_{CO2}}{\beta_{N2}} \frac{\eta_1}{\eta_2} \frac{N_1(z)}{N_2(z)} \exp\left[\int_0^z [\alpha(\lambda_1, z) - \alpha(\lambda_2, z)] dz\right]$$

Where $N_1(z)$ and $N_2(z)$ represent backscattering Raman signal ratio of CO_2 and N_2 received by lidar system, η_1 and η_2 denote the instrument efficiency constants of the two channels, β indicates the scattering cross section.

3. RESULTS

To deduce the atmospheric CO_2 concentration at different altitude, a Li-7500 H_2O/CO_2 analyzer installed on a tower at 30m height above the ground was employed to calibrate the observation results (as shown in Fig.1). The vertical resolution of the lidar system was 30m, and the signals were accumulated by 20000 pulses which generated a CO_2 concentration profile every 17 minutes. Fig.1 showed that the lidar observed CO_2 concentration (i.e. the backscattering Raman signal ratio of CO_2 to N_2 during the night of May 2, 2012 was in reasonable accordance with observations from Li-7500 except for the first measurement which was likely due to the instability of the lidar system at the beginning. The signals were then linear fitted with the CO_2 concentration observed by Li-7500 to obtain the profiles.

After the inversions of atmospheric CO_2 vertical

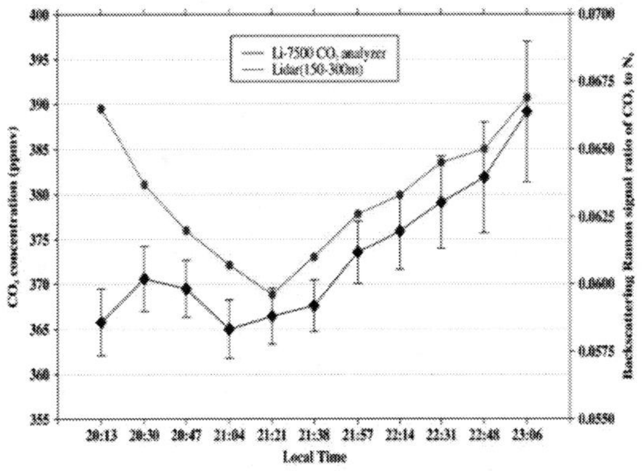

Fig.1. Comparisons between Li-7500 CO_2 analyzer and Raman lidar on May 2, 2012

profiles, we compared the average results with the model simulations (Fig.2). The nearest grid of GEOS-Chem was selected to represent the observed CO_2 concentrations in HeFei (39°58′N, 116°22′E), and the vertical layers under 1.05 km were chosen and linear interpolated to the resolution of the observations. The simulations mainly captured the observed vertical variations of CO_2 concentration with a lower changing amplitude (~15 ppmv). However, the observed CO_2 concentration decreased from ~430 ppmv at 0.03 km to ~385 ppmv at 0.09 km. This indicated that the modeled CO_2 concentration profile could not preferably reflect the coupling between the regional carbon sources/sinks and synoptic atmospheric transport, which could be

DOI: 10.1051/epjconf/201611905019

resolved by improving the resolution of the model and optimizing the subgrid parameterization scheme.

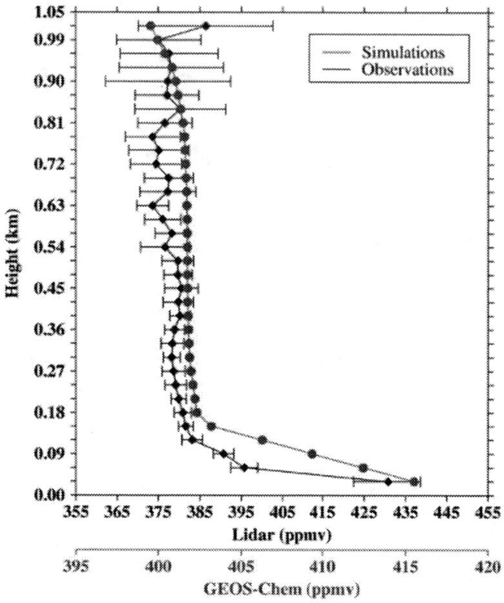

Fig.2. The mean lidar observed and simulated CO_2 concentration profiles on May 2, 2012. The error bars denote the spread of the observations.

4. CONCLUSIONS

The observations from Raman lidar system have shown a good agreement in the trend of change with the Li-7500 H_2O/CO_2 analyzer. Although the model simulations mainly captured the characteristics of atmospheric CO_2 vertical distribution, further in-depth studies need to be carried out to improve the capability of the model combining with lidar observations.

ACKNOWLEDGMENTS

This work was supported by the National natural Science Foundation of China (NSFC) program (Grant No. 41127901).

REFERENCES

1. Sugimoto N, Minato A, 1993: Long-path absorption measurement of CO_2 with a Raman-shifted tunable dye laser, Applied optics, **32(33)**: 6827-6833.

2. Zhao P, Zhang Y, Wang L, et al., 2008: Measurement of tropospheric CO_2 and aerosol extinction profiles with Raman lidar, Chinese Optics Letters, **6(3)**: 157-160.

3. Hu S. X., Zhang S. G., Yuan K. E., et al., 2013: Atmospheric CO_2 uncertainty in Raman lidar measurements, Liangzi Dianzi Xuebao/ Chinese Journal of Quantum Electronics, 2013, **30(1)**.

4. Nassar R, Jones DBA, Suntharalingam P et al., 2010: Modeling global atmospheric CO_2 with improved emission inventories and CO_2 production from the oxidation of other carbon species, Geosci Model Dev, **3**:889–948

5. Hong G. L., Zhao Y. C., Zhao Y. F., et al., 2006: Raman lidar for profiling atmospheric CO_2, ACTA PHYSICA SINICA, **55(2)**: 983-987.

6. Zhao P, Zhao Y. C, Wang L, et al., 2007: Analysis of influence of atmosphere extinction to Raman lidar monitoring CO_2 concentration profile, Chinese Physics, **16(8)**: 2486.

IMPLEMENTATION OF MICROPULSE LIDAR AT 4.5 μM AND 1.5 μM FOR AEROSOL AND CLOUD STUDY

Morann Dagan [1]*, Benjamin Thomas[1], Barry Gross[1], and Fred Moshary[1]

*[1]NOAA CREST Center and Remote Sensing Laboratory, Department of Electrical Engineering, City College, NY, 10031, US, *Email: morann.dagan@gmail.com*

ABSTRACT

Identifying and quantifying ambient aerosols and their interactions with clouds are important for air-quality and climate studies. Advances in infrared technologies on fiber lasers, quantum cascade lasers and IR detectors have made developing micro-pulse (low energy) IR lidar systems operating in the infrared spectral range feasible. We present in this contribution a micropulse dual channel (IR wavelength) lidar system for studying aerosol and cloud optical properties. The system operates at 1.545 μm (6472.5 cm-1) and at 4.55 μm (2197.8 cm-1) with high repetition rates and microjoule pulses. The system is intended to be coupled with an existing UV, visible, near infrared lidar system at the city college of New York, part of the CREST lidar network Preliminary backscattered signals from this system are here presented and compared to SNR simulation.

1. INTRODUCTION

The study of aerosols is important in order to better understand the effects on health and climate. PM2.5 aerosols have been linked to health issues including lung cancer [1], and cardiopulmonary syndromes which can lead to mortality [2]. Recent studies [3] show that particle matter suspended in air is responsible for 370,000 premature deaths every year in Europe. In addition, atmospheric aerosols play a key role in the Earth's atmosphere radiative balance both directly, by light extinction, and indirectly, through complex processes, with cloud formations for example, involving physical and chemical properties of aerosols [4]. However, their contribution to the Earth's radiative budget is still subject to large uncertainties [5] and in need of further studies.

In this context, the lidar is a very efficient tool to remotely provide range-resolved measurements of the optical properties of aerosols, such as absorption and scattering [6]. Advances in technologies, such as IR fiber lasers, Quantum Cascade (QC) lasers and improvement in IR light detector sensibility and operating temperatures, make new IR lidar systems possible for research and development. These innovations make way for building a low energy IR lidar that is compact, lightweight and runs at room temperature. Despite being of much less energy as typical YAG:Nd based lidars, these lasers offer much higher repetition rates making aerosol or cloud backscattering measurements possible with sufficient SNR.

This project outlines the design of a two channel micropulse IR backscatter lidar system and presents results obtained using this lidar to study aerosol and cloud base observations. This IR system adds two additional channels to the NOAA CREST Center lidar facility at the Optical Remote Sensing Lab at CCNY which currently runs at 3 elastic channels; 1064 nm, 532 nm and 355 nm and 2 Raman channels [7]. Larger spectral coverage can improve the separation of fine and coarse aerosols, as well as advance our knowledge of aerosol and cloud interactions.

2. METHODOLOGY

The system runs with two channels at wavelengths of 1.545 μm (6472.5 cm-1) and at 4.55 μm (2197.8 cm-1). The 1.545 μm laser source is a 20 kHz Keopsys fiber laser with a peak power of 4 kW and pulse width of 7 ns (30 μJ/pulse). The laser source for the 4.55 μm channel is a 100 kHz Pranalytica Inc. QC laser which can deliver a peak power of 4.5 watts with a pulse width of 202 ns (0.9 μJ/pulse). Quantum-Cascade lasers offer several Watts of pulsed peak optical power, while retaining a good far field pattern as required for laser remote sensing techniques in atmospheric research. Mid-IR QC lasers produce pulsed signal at room temperature and are compact in size.

DOI: 10.1051/epjconf/201611906001

The receiver consists of an f/3 10" primary mirror with a focal length of 762 mm focusing on a Vigo PVI-4TE-5 detector for the 4.55 μm channel and a Thorlabs APD110C for the 1.545 μm channel.

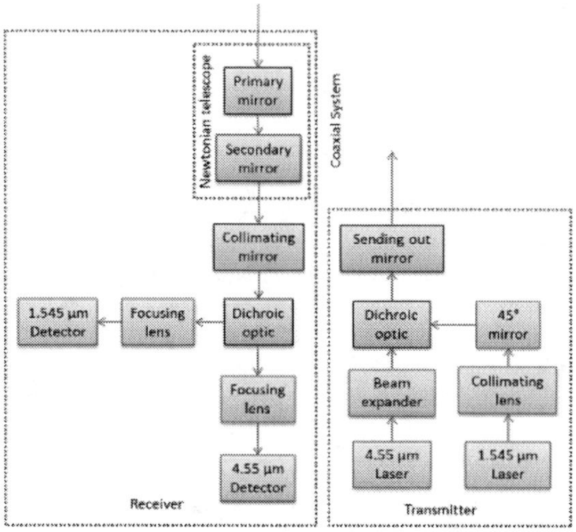

Figure 1: System Design

The signal is then recorded by a 16 bit Gage digitizer with an onboard FPGA card for a maximum of 1024 hardware averages. This acquisition system has a maximum sampling rate of 200 MS/s.

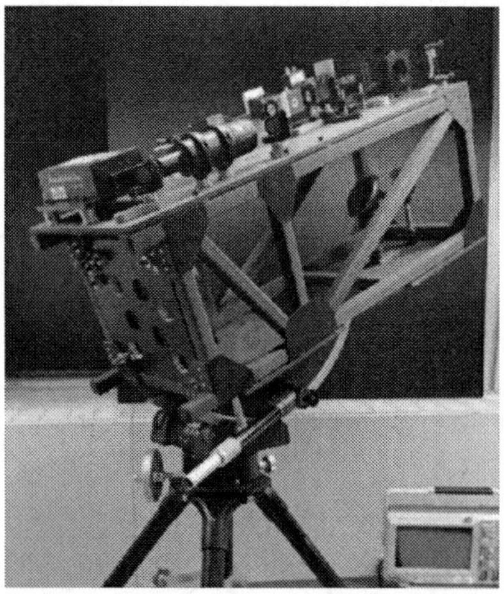

Figure 2: IR lidar

The dual channel IR lidar was designed to be light and compact, weighing less than 30 lbs; therefore making it portable. The lidar structure has a carbon fiber frame making it ideal for transporting and running in outdoor conditions.

3. RESULTS

At 1.545 μm, backscattered aerosol signal was collected on a clear day with the lidar pointing at a slant angle of 30° from horizontal within the planetary boundary layer. Figure 3 shows the processed aerosol data as a function of range. Pulses were averaged in the time domain for approximately 15 minutes or 15 x 10^6 pulses, and then averaged over 30 m in the spatial domain to further improve the SNR. Results show a good agreement between the simulated and measured SNR (assuming a reasonable aerosol backscattering coefficient of 10^{-7} $m^{-1}.sr^{-1}$).

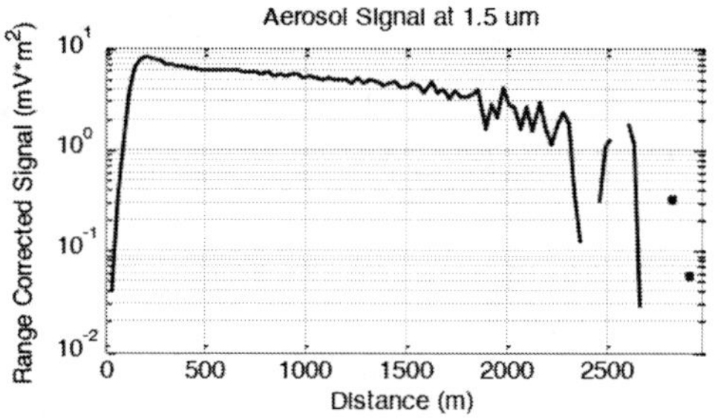

Fig. 3: Aerosol backscattered signal from the 1.5 μm wavelength channel

At 4.55 μm, backscattered cloud signal was collected with the lidar also pointing at an angle of 30°. Pulses were averaged in the time domain with a 60 second time resolution and then averaged in the spatial domain with a resolution of 36 m. Figure 4 displays the cloud peak at approximately 900 m. As for the 1.5 μm channel, a simulation of the SNR was implemented for comparison using the backscattering coefficient of 2 x 10^{-4} $m^{-1}.sr^{-1}$ retrieved from a ceilometer. At 900 m distance the simulated SNR is ~500 and the SNR from the collected run is 270.

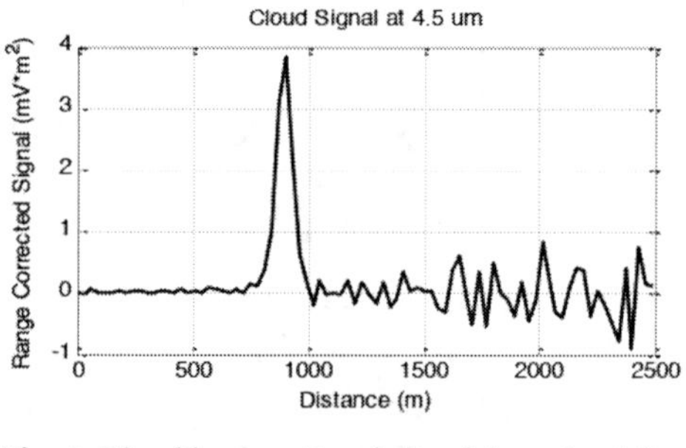

Fig. 4: Cloud backscattered signal from the 4.5 μm wavelength channel

DOI: 10.1051/epjconf/201611906001

4. CONCLUSIONS

We present in this contribution a micropulse dual channel infrared lidar operating at 1.545 μm (fiber laser) and at 4.55 μm (quantum cascade laser). The two lasers have low energy (resp. 30 μJ and 0.9 μJ) with high repetition rate (resp. 20 kHz and 100 kHz). Results showing aerosol backscattering as well as cloud backscattered signal are presented showing that these laser types can be suitable for atmospheric study in the infrared spectral range. The signal to noise ratio obtained on measured signals is in good agreement with the simulated SNR. The system is planned to be implemented with an existing Raman-Mie lidar operating in the UV, visible and near infrared wavelength for further studies on urban aerosols as well as cloud base dynamic.

ACKNOWLEDGEMENT

This study was supported by National Oceanic and Atmospheric Administration (NOAA) under the NOAA CREST Grant # NA11SEC4810004. *The statements, findings, conclusions, and recommendations presented in this manuscript are those of the authors and do not necessarily reflect the views of the National Oceanic and Atmospheric Administration or the Department of Commerce.* This research was also supported by the Mid Infrared technologies for Health and the Environment (MIRTHE) Center and the National Science Foundation (NSF) under grant # EEC-0540832.

REFERENCES

[1] Raaschou-Nielsen, O., et al., 2013: Air pollution and lung cancer incidence in 17 European cohorts: Prospective analyses from the European Study of Cohorts for Air Pollution Effects (ESCAPE), *The Lancet Oncology*, **14(9)**, 813.

[2] Cohen, A. J., H. Ross Anderson, B. Ostro, K. D. Pandey, M. Krzyzanowski, N. Künzli, K. Gutschmidt, A. Pope, I. Romieu, J. M. Samet, K. Smith, 2005: *The Global burden of disease due to outdoor air pollution, Journal of Toxicology and Environmental Health*, **68**, 1301.

[3] Mokdad, A. H., J. S. Marks, D. F. Stroup, J. L. Gerberding, 2004: Actual causes of death in the United States, 2000, *J. Amer. Med. Assoc.*, **291**, 1238.

[4] Twomey S., 1974: Pollution and the planetary albedo, *Atmos. Environ.*, **8**, 1251.

[5] Stocker, T. F., D. Qin, G. K. Plattner, M. Tignor, S. K. Allen, J. Boschung, B. M. Midgley, 2013: Climate change 2013: the physical science basis. Contribution of working group I to the fifth assessment report of the intergovernmental panel on climate change, IPCC.

[6] Weitkamp C., 2005: Lidar, Range-Resolved Optical Remote Sensing of the Atmosphere, Springer, Berlin, New York.

[7] Wu, Y., S. Chaw, B. Gross, F. Moshary, S. Ahmed, 2009: Low and optically thin cloud measurements using a Raman-Mie lidar, *Applied Optics*, **48(6)**, 1218-27.

PULSED SINGLE FREQUENCY FIBER LASERS

Shibin Jiang

AdValue Photonics Inc

3440 E. Britannia Drive, Suite 190, Tucson, AZ 85706, USA

sjiang@advaluephotonics.com

ABSTRACT

Pulsed single frequency fiber lasers with mJ level near 1 micron, 1.55 micron and 2 micron wavelengths were demonstrated by using our proprietary highly doped fibers. These fiber lasers exhibit excellent long term stable operation with M^2<1.2.

1. INTRODUCTION

Coherent lidar and atmospheric sensing require compact coherent pulsed narrow linewidth single frequency laser sources. High pulse energy and high peak power is preferred, especially for airborne and space-borne applications. In order to achieve high pulse energy and high peak power, free-space solid-state lasers are typically used in order to avoid the optical nonlinearity. Free-space solid-state solid state lasers are challenging for packaging. All-fiber monolithic laser sources are highly desirable for these applications, especially for airborne and space borne applications, because fiber laser sources offer a much more compact and robust solution. With the advancement of fiber lasers in the recent years fiber based component become readily available. Single-frequency pulse radiation can simply be generated from intensity-modulated distributed feedback (DFB) semiconductor laser and/or single-frequency fiber oscillators in combination with fiber based acoustic optic modulator (AOM). The pulse energy can be increased by following fiber amplifiers. However, the pulse energy and peak power of the fiber amplifiers are limited because of the stimulated Brillouin scattering (SBS). The pulse energy is generally limited to near 0.1mJ level for most commercially available fiber lasers although much higher values were reported by using large mode-area fibers and photonic crystal fibers. In this paper we report single frequency pulsed fiber laser with mJ level near 1 micron, 1.55 micron and 2 micron laser wavelengths.

2. RARE-EARTH DOPED GLASSES AND FIBERS

We developed multi-component non-silica glasses as the host of the rare-earth ions, which permit a higher doping concentration because of their less-defined glass network as compared to silica glass. Yb, Er, Er/Yb, Tm, and Tm/Ho doped glasses were fabricated as the core glasses using high purity chemicals with less than 1ppm of iron and copper. Glass melting was processed in platinum crucibles. The hydroxyl ion OH- concentration in these glasses are controlled in order to eliminate the non-radiative energy transfer from the excited level of rare-earth ions to OH-. We carefully control the glass melting process in order to prevent other contaminations from ceramic insulation materials and heating elements, ensuring a low propagation loss. The absorption spectra are recorded on a UV-Vis-Nir Spectrophotometer. Refractive index of core glasses at 632nm, 980nm and 1550nm are characterized. The glass transition temperature, softening temperature, and thermal expansions coefficient are measured.

Undoped glasses for the inner and external cladding layers are designed and fabricated. Refractive index of undoped cladding glasses are characterized for the calculation of the numerical aperture (NA).

Rod-in-tube fiber drawing technique is used for our fiber fabrication. The core glass rod will be drilled from a bulk rare-earth doped glasses, and the barrel of the rod will be polished to a high surface quality. Cladding glass tubes will be fabricated from undoped glasses. Both inside and outside surfaces of the glass tubes made from the bulk cladding glasses are polished to a high surface quality. The inside diameter of the inner cladding tube should be perfectly matched with the diameter of the core glass. The outside diameter of the inner cladding tube should be perfectly matched with the inside diameter of the outer cladding tube. Fiber drawing process is carried our in-house using a fiber drawing tower that is optimized for non-silica fiber drawing.

The high doping concentration of our non-silica fibers enables high pump absorption and high gain per unit length, which is particularly essential for high pulse energy single frequency laser amplification. We used these fibers demonstrated single frequency fiber laser, Q-switched all-fiber fiber lasers, as well as mode-locked fiber lasers.

3. RESULTS

We directly modulate DFB single frequency diode lasers for a few nanoseconds pulse generation and AOM modulation for a few hundred nanoseconds pulse generation. When the pulse width is relatively long, pulse shaping is necessary in order to generate Gaussian pulse after the amplification. Single-frequency pulse radiation can simply be generated from intensity-modulated distributed feedback (DFB) semiconductor laser and/or single-frequency fiber oscillators in combination with fiber based acoustic optic modulator (AOM).

For 1030nm fiber laser, pulse energy of 1mJ and pulse width of 200ns to 300ns was demonstrated. When the pulse width is 4.5ns, a peak power of 88kW was demonstrated.

A 1550nm single frequency fiber laser with an average power of 8.4W, pulse energy of 0.42mJ and peak power of 2.1kW (200ns, 20kHz) were successfully demonstrated by using an 50cm long fiber. An Er fiber laser at 1532nm was built as the pump source in order to improve the efficiency. A single-frequency DFB diode laser was used as the seed laser. Three-stage low-noise fiber pre-amplifiers were used to boost the power of the seed pulses. Narrow-band filter is used to filter ASE before it is launched into the last stage amplifier. The slope efficiency is measured to be greater than 56%.

A single frequency fiber laser at 1950nm was built and AOM was used to modulate the pulse. Pulse shaping was applied in order to generate Gaussian like pulse after the amplification. A Tm-doped fiber with a length of 30cm was used for amplification. The pump laser is multi-mode 793nm laser diode. With 45W pump laser an output of near 16W was demonstrated. The pulse energy is greater than 0.3mJ.

4. CONCLUSIONS

We demonstrated and commercialized single frequency pulsed fiber lasers with mJ level near 1 micron, 1.55 micron and 2 micron wavelengths by using our proprietary highly doped single mode fibers. Since these fibers are truly singe mode fiber, these fiber lasers exhibit stable long term operation with excellent beam quality. We are in the processing of further scale up the pulse energy.

DITHER CAVITY LENGTH CONTROLLER WITH IODINE LOCKING

Marty Lawson [1]*, Ed Eloranta[2]

[1]*UW SSEC Lidar Group, UW Madison, WI 53706, USA, *Email: marty.lawson@ssec.wisc.edu*

[2] *UW SSEC Lidar Group, UW Madison, WI 53706, USA*

ABSTRACT

A cavity length controller for a seeded Q-switched frequency doubled Nd:YAG laser is constructed. The cavity length controller uses a piezo-mirror dither voltage to find the optimum length for the seeded cavity. The piezo-mirror dither also dithers the optical frequency of the output pulse. [1]. This dither in optical frequency is then used to lock to an Iodine absorption line.

1. INTRODUCTION

Many lidar systems use seeded Q-switched lasers for their combination of pulse energy and narrow spectrum. Seeded Q-switched lasers generally need a controller to keep the main cavity modes locked to the exact frequency of the seed laser. We constructed a custom cavity length controller that uses end mirror dither to estimate the first derivative of the pulse build up time versus end mirror voltage. This first derivative estimate is then used to control the average end mirror voltage towards the optimum buildup time.

The optical pulse form the main cavity emerges with a frequency centered on the main cavity mode nearest to the seed laser frequency. [1] Thus the end mirror position dither also appears as a dither in output optical frequency. If a sample of the output optical pulse is then sent through a molecular absorption cell, the cavity length controller can also estimate the first derivative of the molecular absorption. The molecular absorption derivative is then used to lock the center frequency of the seed laser to the center of the chosen absorption line.

2. METHODOLOGY

A new cavity length controller (CLC) for a Photonics seeded Q-switched frequency doubled solid state laser was constructed. Figure 1 shows how the new CLC is connected to the laser and lidar control PC. It consists of a microcontroller board with inputs for the laser sync pulse, two energy monitors, and an optional input for the main cavity light pulse. For output it has an analog voltage for driving the piezo-mirror, and a logic output to signal the lidar data system if each laser pulse is good. The logic output is named Laser Pulse OK (LPOK). For PC communications it has a USB to serial connection that is used for data, commands, and firmware updates. See figure 1.

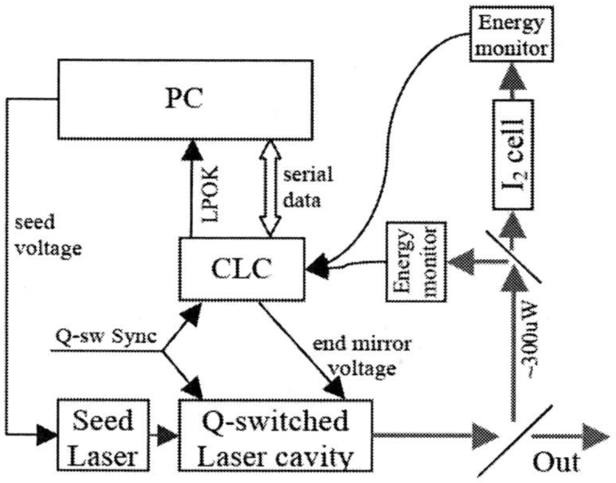

Figure 1: Block diagram of combined cavity length and Iodine locking controller.

The two energy monitors are configured to measure the absorption of a short Iodine cell. A small sample (<300uW average) of the output pulse is sent towards a beam splitter. One leg of the split beam is sent to the first energy monitor, while the second beam is passed through a short Iodine cell before detection by a second energy monitor.

Control of cavity length is done by minimizing the estimate of the first derivative of cavity build up time versus end mirror voltage. This derivative is estimated by adding a small (0.5-1V) dither voltage to the piezo end mirror voltage and averaging the difference between the cavity buildup time when the cavity is longer than average and shorter than average. An integral, double-integral controller is then used to adjust the average piezo end mirror voltage such that the buildup time is minimized.

The controller includes a laser pulse quality flagging function. (LPOK) The LPOK function improves lidar data-quality while also improving the robustness of the cavity length controller. The LPOK function rejects laser pulses that take too long to build up. The LPOK threshold is dynamically set. First minimum buildup time over 128 shots is recorded (the CLC lost-lock scan insures that there are always several seeded shots in this interval). This local minimum buildup time is then time averaged using a digital low-pass filter. Finally the LPOK threshold is set 50-100ns longer than the average of the minimum buildup time. The CLC directly uses the LPOK signal to detect when it has lost cavity length lock. When loss of cavity length lock is detected, the CLC starts a rapid scan for another seeded main cavity mode.

The cavity length controller is also linked to two energy monitors and a short Iodine cell configured to allow pulse by pulse measurement of Iodine absorption. Due to the end mirror position dither, the laser pulses also have an optical frequency dither [1]. This frequency dither in the pulses can be used with the Iodine cell to estimate the first derivative of the Iodine absorption.

Figure 2 shows the Iodine absorption and first derivative estimates during a frequency scan. Finally, the Iodine slope estimate is transmitted to the PC for use in a PID frequency locking loop.

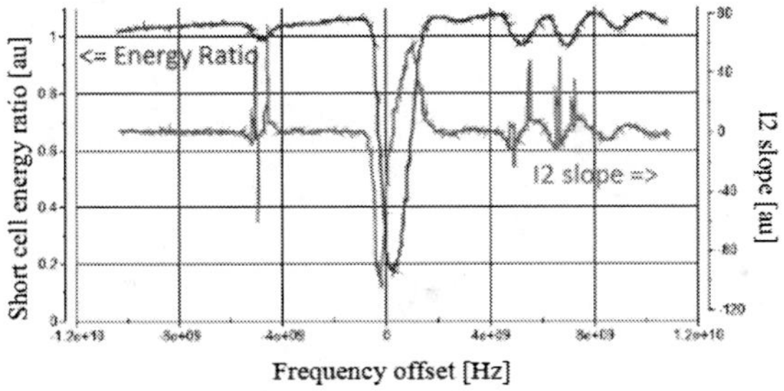

Figure 2: Transmission through the I2 cell and slope estimate versus frequency offset from I2 1109.

3. RESULTS

Figure 3 shows the ARM NSHSRL lidar system at Barrow Alaska recovering from several jumps in seed laser operating point. Recovery from the 500-600MHz jumps is prompt and well damped. (Note: the PC control is limiting how quickly the I_2 locking recovers) The controller will also recover from larger jumps in frequency as long as the jumps stay within the base of the Iodine 1109 line.

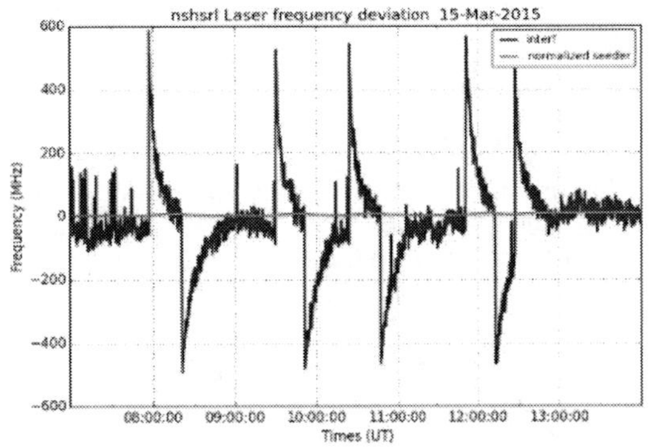

Figure 3: Slope estimate Iodine locking dealing with DPSS seed laser operating point jumps. (measured with a fiber interferometer)

Figure 4 shows the frequency locking error recorded by the ARM MF2 HSRL system while deployed on the NOAA Ship Ronald H. Brown during eight hours of typical operation in dock.

Frequency errors of the average slope estimate are 10 MHz peak to peak. The constant DC offset in frequency comes from the Iodine slope estimate calibration procedure (minimum and maximum represent the shot-shot variation in the Iodine slope estimate).

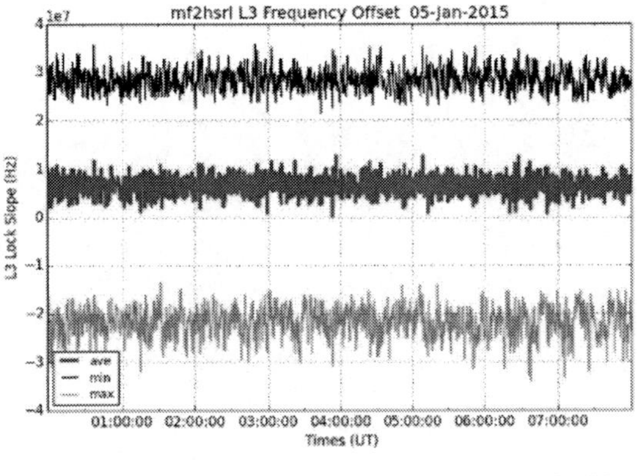

Figure 4: Frequency error over 8 hours with I_2 locking active. (calibrated I_2 slope)

This cavity length controller with I2 locking has been deployed to all five of the currently operating HSRLs built by the SSEC Lidar Group. The dither CLC replaced a Brillion shifted edge-locking Iodine locking system. [2] The dither CLC provides the same short term Iodine locking frequency error. The dither CLC also operates without periodic calibration, needs less optical power, and has less stringent alignment requirements.

4. CONCLUSIONS

With minimal additional hardware, a cavity length controller for a seeded laser can leverage the parasitic optical frequency modulation from dithering the piezoelectric end mirror to lock the average frequency of the laser to a molecular absorption filter or other frequency standard.

REFERENCES

[1] Siegman, A. E., 1986: Lasers, ISBN-10-0935702113, 1129 1170, University Science Books, California, USA.

[2] Eloranta, E. W. and I. A. Razenkov, 2006: Frequency locking to the center of a 532 nm iodine absorption line by using stimulated Brillouin scattering from a single-mode fiber, *Optics Letters*, **31**, 234-237.

DOI: 10.1051/epjconf/201611906004

PRELIMINARY LIDAR EXPERIMENT TO STUDY THE BACKSCATTER AMPLIFICATION

Igor A. Razenkov*, Victor A. Banakh

*V.E. Zuev Institute of Atmospheric Optics, Tomsk, 634055, RUSSIA, *Email: lidaroff@iao.ru*

ABSTRACT

Long-term continuous measurements for detection relative backscatter amplification on a horizontal path of 2 km long are performed by using a specific micro pulse lidar. The laser beam path is limited by a solid obstacle. The lidar is located next to an ultrasonic anemometer that measures 3D wind velocity and temperature; the laser spot on the obstacle is observed by using a telephoto lens. The results showed that the backscatter amplification has a clear diurnal variation. Moreover, the backscatter amplification was completely absent in the morning and evening under neutral stratification in the atmospheric surface layer. At night and in the daytime there was a significant increase of the backscatter amplification coefficient.

1. INTRODUCTION

The backscatter amplification effect occurs in the turbulent atmosphere, when the probe laser beam and scattered light propagate along the same trajectory. The increase of the scattered light intensity fluctuations at the point of reception is the result of the correlation of direct and backward light waves. The point of reception must be located on the axis of the probe beam [1-3].

To register the relative backscatter amplification (RBSA) was build the lidar with beam expansion via receiving telescope [4]. RBSA-lidar has one transmitting channel and two receiving channels: on-axis receiving channel registers echo exactly on the axis of the laser beam, the off-axis channel registers the signal 200 mm apart from the probing beam axis. The system continuously works in Tomsk since the end of August 2014.

2. EXPERIMENT

Figure 1 shows an arrangement of the experiment for the lidar test. Lidar operates at the elevation angle close to horizontal. The laser beam is directed to the wall at the distance of about 2 km. A modified web camera mounted on the Maksutov telescope with a focal length of one meter is located next to the lidar. The mast with the ultrasonic anemometer [5,6] is located above the RBSA-lidar on the roof of the building.

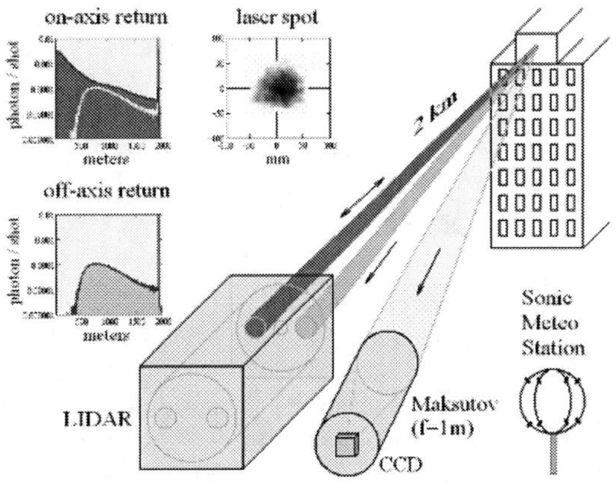

Fig. 1. Experimental setup.

Figure 1 also shows on-axis return, off-axis return, and laser spot on the wall. Experiments are done in the city Tomsk. Figure 2 shows a Google map. The lidar is located in the building of the Institute of Atmospheric Optics. Sounding path (straight yellow line) passed over the city (roads, small river, buildings, and trees).

Fig. 2. A map of the experiment site (Tomsk, Russia).

DOI: 10.1051/epjconf/201611906004

The lidar operates in the photon counting mode, the counts accumulation time interval is ten minutes. Data processing for the instrument is described in [4]. Fiber laser pulse repetition rate is 50 KHz. Each measurement cycle contains 30 million shots. Just an average was taken only a couple signal photons per shot. To improve statistics the data are averaged over a 50 m range intervals. The statistical error for the returns from the farthest range bin is less than 1%. Note that off-axis return is equal to zero up to ranges of approximately 500 m due to lidar's geometry. The data acquired from ranges from 0.9 km to 2 km are considered to be reliable.

The coefficient N of relative backscatter amplification is determined from the ratio of on-axis and off-axis returns. The instrument allows measure the coefficient N with an accuracy of ±5%, even when the changes of the backscattering coefficient along the sounding path are >100%.

3. DATA EXAMPLES

The RBSA-lidar continuously operated during a half-year since August 2014. Figure 3 shows example of a two-day recording of the amplification coefficient *N(2km)* at the distance up to 2 km in clear atmosphere in September 2014.

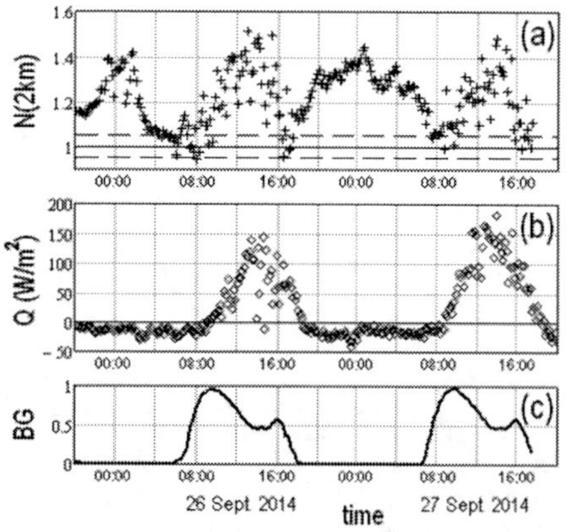

Fig. 3. The typical two days time series of (a) relative backscatter amplification N in clear conditions. (b) turbulent heat flux Q, and (c) the magnitude of the normalized background BG. Data are recorded on 26 and 27 September 2014. Hereinafter, time is local.

The diurnal variation of the N is clearly seen. Coefficient N is equal to unit twice a day within a 10% error of the accepted value. This happens in the mornings at 8 am and in the evenings at 5 pm of local time. It follows from Fig. 3(b) that at about the same time the turbulent heat flux Q crossed the zero line. The third graph (c) shows the normalized values of the background BG, which was defined as an off-set of the lidar returns. The appearance of a second "evening" of the maximum in the curve BG is due to the fact that the lidar path was directed to the North-West.

At day time and night time turbulent heat fluxes have opposite signs and in absolute value differ from each by the factor of about six. In this case, the amplification coefficient at the day and night increased to approximately the same value of 1.4. Also Fig. 3 shows that the data spread in the daytime is larger than that at night time.

Figure 4 shows another example of data obtained in 4 days period from 1 to 4 October 2014. Clear day is only on October 2. We can see two clear maxima in the behavior of the N which is similar to the data in Fig. 3. Some clouds appear on October 3, and the coefficient N and turbulent flux Q are smaller. It was cloudy on October 1 and 4; the heat flux Q and the amplification of N are even smaller. Note that we should not expect a very high correlation between RBSA-lidar data with the data of the anemometer, because it was located at the lidar site, but away from measuring volume.

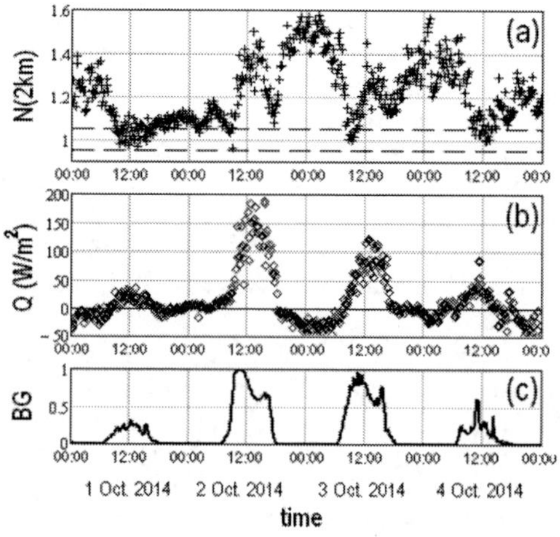

Fig. 4. Four days time series of (a) backscatter amplification N, (b) turbulent heat flux Q, and (c) the magnitude of the background BG when changing cloud amount. Data are recorded starting on October 1 2014.

Figure 5 shows the correlation between the coefficient N and the turbulent heat flux Q for the data in Fig. 4. A sharp increase of the amplification factor N is seen under the stable temperature stratification for $Q<0$ (night time). With the increase of the heat flux for $Q>0$ (day time) there is a slow increase of N.

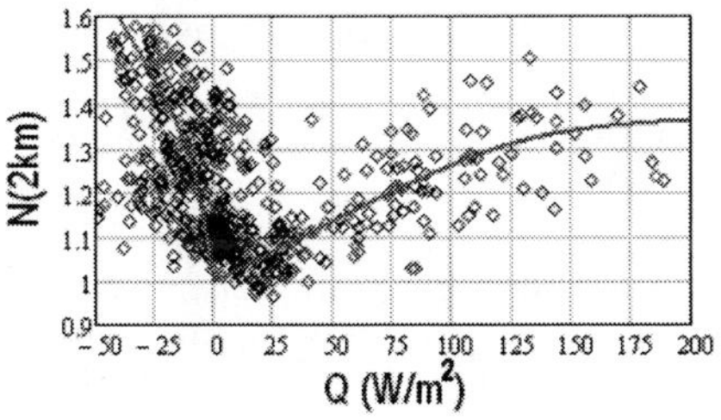

Fig. 5. Correlation plot of the amplification coefficient N vs. the turbulent heat flux Q. Interpolation is shown by the solid blue line. Was used data from the Fig. 4. Data were obtained from 1 to 4 October 2014.

For explanation of this we can assume that in a day time the radius of coherence of the laser beam is smaller than the Fresnel zone and less than the size of the aperture of the receiver respectively [3,7]. As a result, the averaging effect of the receiving aperture is stronger under larger turbulence intensity; this reduces the resulting estimate of the coefficient N. In the RBSA-lidar data analysis, it may make sense to consider separately "night" and "day" turbulence.

Figure 6 presents the data obtained in winter time. The atmosphere was not very clean on these days, but according to our lidar data, the visibility was more than 6 km in the presence of some clouds. Fig. 6 shows (a) the amplification factor N, (b) standard deviation of the centre of gravity of the laser spot (*RMS CG*) at a distance of 2 km, (b) the structural constant of refractive index C_n^2 and (d) the magnitude of the background *BG*. *RMS CG* data were measurable only in the dark time by recording the jitter of the image of the laser spot by the camera sensor (see Fig. 1). C_n^2 was calculated from anemometer's data [6].

Figure 7 shows (a) the correlation between the factor N and the beam jitter *RMS CG*, and (b) between factor N and the structure constant of refractive index C_n^2 for the data in Fig. 6.

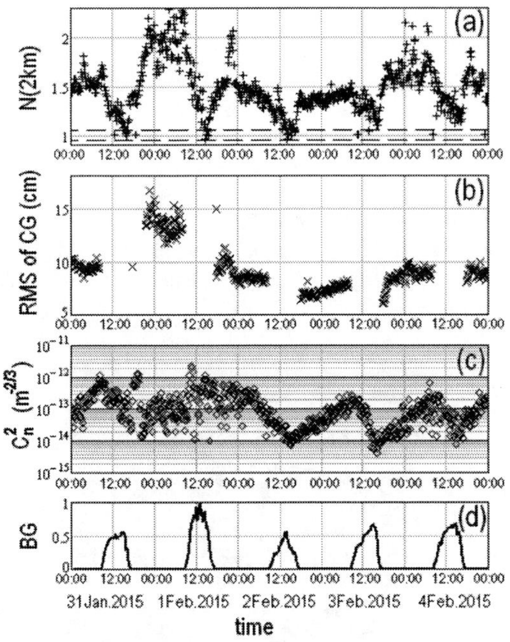

Fig. 6. Five days time series of (a) relative backscatter amplification *N(2km)*, (b) mean square displacement of the centre of gravity of the laser spot *RMS CG* on the wall at a distance of 2 km, (c) the structure constant of the refractive index C_n^2 and (d) the magnitude of the normalized background *BG*. Data were obtained from 31 January to 4 February 2015.

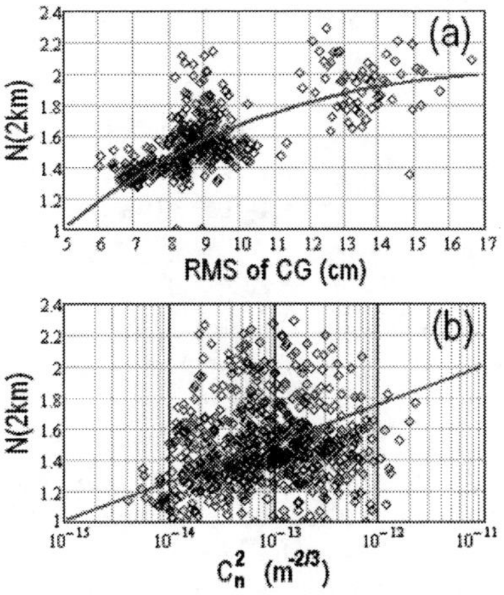

Fig. 7. Correlation plots of (a) factor N vs. laser spot *RMS CG*, and (b) factor N vs. the structure constant of refractive index C_n^2. Fitting curves is shown with the solid blue line. Was used data from the Fig. 6. Data were obtained from 31 January to 4 February 2015.

A positive correlation for both graphs in Fig. 7 should be noted. Fig. 7(b) shows a large spread of points due to all day time and night time data were putted on the one correlation plot. Was mentioned above that day time and night time behavior of the RBSA measurements is different.

Figure 8 presents laser spots at 2 km and shows beam's intensity, shape, and position changes. When turbulence is higher (a) the motion of laser spot is larger than when turbulence in lower (b, c).

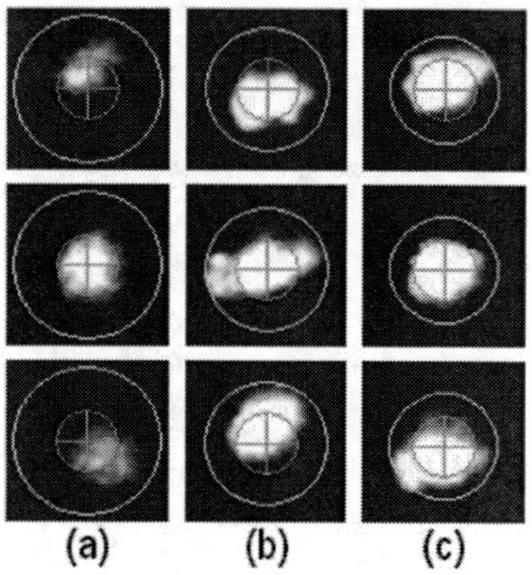

Fig. 8. The beam spot at 2 km. Sets of three random images from data were obtained on the night (a) 1 February 2015, (b) 2 February 2015, and (c) 3 February 2015. Big blue circles are *RMS CG* from Fig. 6(b) plus beam size 100 mm (small green circle with cross).

4. CONCLUSION

The two-channel micro pulse RBSA-lidar has been demonstrated continuous measurements of relative backscatter amplification in a turbulent atmosphere. Now it is sure that the diurnal variation of the received series of lidar data is consistent with the data from the meteorological observations and does not contradict to the atmospheric surface layer physics. For the stable temperature stratification at night time and for unstable stratification in the day time the relative backscatter amplification can have a fairly large and approximately the same values. It can be argued that the RBSA coefficient depends on the state of the atmosphere and reflects somehow this state. At the same time, this coefficient depends on the system parameters of the transceiver such as the beam size and the receiver size. We hope that the theory of RBSA-lidar will be developed.

ACKNOWLEDGMENTS

The authors many thank Dr. Sergei Odintsov and Dr. Vladimir Gladkikh for kindly providing meteo data of the ultrasonic anemometer.

This study was supported by the Russian Scientific Foundation for Maintenance and Development (Project: 14-17-00386).

REFERENCES

[1] Vinogradov A. G., Kravtsov Yu. A., and Tatarskii V. I., "The Effect of Intensification of Back Scattering by Bodies That Are Situated in a Medium Having Random Inhomogeneities," Radiophys. Quant. Electr. **16** (7), 818–823 (1973).

[2] Kravtsov Yu.A. and Saichev A.I., "Effects of Double Passage of Waves in Randomly Inhomogeneous Media," Physics-Uspekhi **25** (7), 494–508 (1982).

[3] Banakh V. A. and Mironov V. L., *Lidar in a Turbulent Atmosphere* (Artech House, Boston and London, 1987).

[4] Razenkov, I. A., Banakh V. A., Nadeev A.I. Aerosol Lidar for the Relative Backscatter Amplification Measurements, 27th International Laser Radar Conference, July 5-10, 2015, present proceedings, New York, USA.

[5] Gladkikh V.A, Makienko A.E. Digital ultrasonic meteostation. // Devices. 2009. V.109. No. 7. P. 21-25 (in Russian).

[6] Gladkikh V.A, Mamyshev V.P., Odintsov S.L. Experimental evaluation of structural constant of refractive index in the atmospheric surface layer. // Atmospheric and oceanic optics. 2015. V. 28. No. 4. P. 309-318 (in Russian).

[7] *Laser Beam Propagation in the Atmosphere.* Edited by J. W. Strohbehn. Springer Series Topics in Applied Physics, Vol. 25. (Berlin, Heidelberg, New York: Springer-Verlag, 1978).

DOI: 10.1051/epjconf/201611906005

DEVELOPMENT OF A HIGH SPECTRAL RESOLUTION LIDAR USING A MULTI-MODE LASER AND A TUNABLE INTERFEROMETER

Pablo Ristori [1*], Lidia Otero [1-3], Yoshitaka Jin [2], Nobuo Sugimoto [2], Tomoaki Nishizawa [2], Eduardo Quel [1-3]

[1]*División Lidar, CEILAP, UNIDEF (MINDEF - CONICET), UMI-IFAECI-CNRS 3351, Juan Bautista de La Salle 4397 – B1603ALO Villa Martelli, Buenos Aires, Argentina, *Email: pablo.ristori@gmail.com*

[2]*National Institute for Environmental Studies, 16-2 Onogawa, Tsukuba 305-8506 Japan*

[3]*IUE – Escuela Superior Técnica, Facultad de Ingeniería, Av. Cabildo 15 – C1426AAA CABA, Argentina*

ABSTRACT

A High Spectral Resolution Lidar (HSRL) using an unseeded laser is designed to separate Mie from total atmospheric backscatter by means of a tunable interferometer. The separation is achieved by tuning the interferometer's free spectral range (FSR) to longitudinal mode separation of the laser. When this condition is achieved, the interferometer transmits aerosol and molecular backscatter with different efficiencies due to their spectral properties. We estimate those transmissions for the perfect tuning case and introducing accuracy or precision errors on the interferometer alignment.

1. INTRODUCTION

Conventional aerosol lidar systems collect the atmospheric backscattered radiation from aerosols and molecules at the same wavelength of the laser emission. Many inversion methods [1-5] intend to determine aerosol backscatter and extinction modelling the molecular profile by using the atmospheric temperature and pressure height distribution. This method needs to estimate the aerosol backscatter-to-extinction ratio.

Rotational-vibrational nitrogen and oxygen Raman shifted backscatter returns are not affected by aerosol backscatter and can be combined to the elastic return to determine aerosol backscatter and extinction. However, the Raman cross section is small and the Raman shift may be important. This method needs the extinction spectral dependence (Ångström coefficient) which can be imposed or calculated from other profiles [6]. Pure rotational Raman returns may provide a stronger signal which is spectrally closer to the elastic emission but the signal is very weak.

As the spectral broadening of the laser emission due to the molecular backscatter is much wider than the one from aerosol backscatter, lidars can use this property to separate molecules from aerosol. The laser emission must be such that the broadened returns from molecules and aerosols are spectrally different. Traditional HSRL uses a single mode laser which linewidth is narrower that the molecular backscatter broadening. The collected light passes through a Fabry Perot interferometer (acting as a narrow band-pass filter) or an atomic filter (acting as a notch filter) [7-8] before detection. On this respect, a 60 cm cavity Nd:YAG seeded laser can provide a linewidth 200 times smaller than an unseeded laser with multiple oscillating modes creating a linewidth of 1 cm^{-1}.

A different HSRL technique using an unseeded multimode laser is proposed on this paper.

2. METHODOLOGY

This method intends to separate the aerosol and molecular backscatter returns from the atmospheric backscattered generated by the fine spectral structure of the emission modes of an unseeded laser. The laser cavity generates longitudinal modes equally spaced in the spectrum modulated by the laser gain curve. For example, a 60 cm effective length cavity will have 120 modes in 1 cm^{-1}. The aerosol backscattered return will have similar spectral signature than the unseeded laser but the molecular backscatter return will be pressure and temperature broadened.

Let's consider a filter placed at the lidar detection unit which spectral transmission function is periodic and which local maximums (or minimums) coincide with the longitudinal modes of the unseeded laser source. This filter will affect considerably the aerosol return, but not the atmospheric-broadened molecular return which will be constant and will have an efficiency equal to the spectrally-integrated filter transmission. The interferometer (Fabry Perot, Michelson, Mach Zehnder) path difference must be equal or multiple (or sub-multiple) of the laser effective cavity length.

The signal obtained for the fundamental or laser harmonics by this channel is represented by eq. 1:

$$S_{new}(r) = a\,S_{mol}(r) + b\,S_{aer}(r) \qquad (1)$$

where S_{mol} and S_{aer} are the range r dependent molecular and aerosol attenuated backscatter returns respectively; the variables a and b are the molecular and aerosol detection efficiencies for this channel. This is a major change from a standard channel, that collects the molecular and aerosol signals with the same efficiency, and makes possible the calculation of separated molecular and aerosol attenuated backscatter independently. If a second detector collecting the complementary light, the system will improve its performance. This can be done using Michelson or Mach Zehnder interferometers. The first consequence of this improvement is that the channel sum provides the total backscattered signal making useless the presence of an additional elastic channel. Furthermore, analyzing equations (2) and (3)

$$S_{new_1}(r) = \frac{1}{2} S_{mol}(r) + \left(\frac{1+c}{2}\right) S_{aer}(r) \qquad (2)$$

$$S_{new_2}(r) = \frac{1}{2} S_{mol}(r) + \left(\frac{1-c}{2}\right) S_{aer}(r) \qquad (3)$$

it can be seen that the subtraction of (2) from (3) (after compensating for different detection efficiencies) gives the aerosol backscatter return as seen on equation (4)

$$c\,S_{aer}(r) = S_{new_1}(r) - S_{new_2}(r) \qquad (4)$$

This is the basic concept of the multi-mode High Spectral Resolution Lidar (MM-HSRL).

Reference [10] shows a direct detection wind lidar that uses a multimode laser and a quadri-channel Mach-Zehnder interferometer to measure aerosol wind speed extracted from its Doppler-shifted return.

3. ANALYSIS

To start the analysis prior to the construction we must take into account several conditions. This work study the interferometer misalignment effects while [9] studies comes other issues like angular alignment and some laser emission properties.

This system is very sensible to the interferometer path difference. Solving the aerosol equation like in (4) when this path is not the optimum can lead to an ill posed problem adding noise or making impossible to reach to a solution. We use the matrix condition number (maximum to minimum singular value ratio) as an indicator of the stability of this inversion.

We will study the alignment problem considering first the effect of the precision analyzing the signal behavior close to the absolute maximum, and then the effect of the accuracy analyzing the local maximum intensities as a function of its distance to the absolute maximum.

For this analysis a simulation was done using the following parameters:

- The laser considered was a Nd:YAG with maximum gain set at 1064.15 nm
- Laser gain curve was simulated to have Gaussian shape of 1 cm^{-1} FWHM
- The cavity effective length was considered to be 48.5 cm (FSR=0.0104 cm^{-1}) with a front mirror reflectivity equal to 16%
- The atmospheric molecular broadening was set 3 GHz
- No broadening nor wavelength shift was considered for the aerosol backscatter
- The interferometers simulated were Fabry Perot (reflectivity 40%) and Michelson (or Mach Zehnder)

4. RESULTS

4.1 System Precision

If the interferometer is close to its best alignment position (accurate alignment) then it has reached a region at which minor movements create an important change in the intensity of the transmitted aerosol backscatter radiation. Then a precise alignment is required. For the analysis simulation the ratio from the maximum to the minimum is about 2.33 and 1.91 for Michelson and Fabry Perot respectively. Figure 1 shows the percentage of the backscatter intensity transmitted by the interferometers from molecular and aerosol backscatter returns. Figure 2 show the condition number (maximal to minimal singular value ratio) where larger is worst.

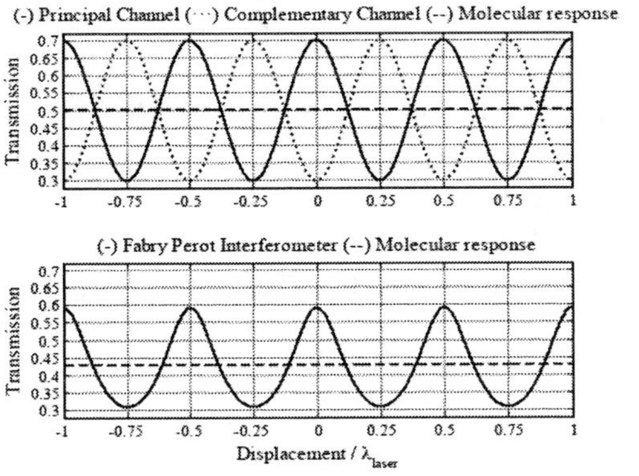

Figure 1: Transmission efficiency: (upper panel) for the Principal and Complementary aerosol channels of Michelson / Mach–Zehnder Interferometer compared to the molecular transmission efficiency which in both cases is 0.5; (Lower panel) for Fabry Perot aerosol and molecular channels

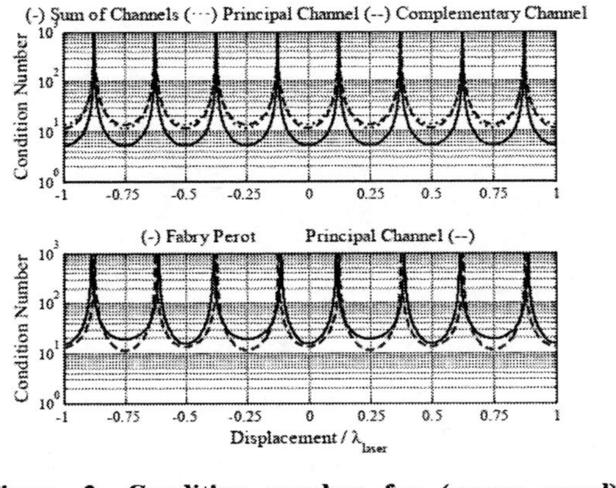

Figure 2: Condition number for (upper panel) Michelson/Mach-Zehnder Interferometer: Principal and Complementary Channel in dashed and dotted lines; combination in continuous line; (lower panel) Condition number comparison between Principal Channel above and Fabry Perot channel

4.2 System Accuracy

Even if the interferometer may track a local maximum, it may be still far from the absolute maximum (lack of accuracy). As for the system precision study, Figure 3 shows the percentage of the backscatter intensity from molecules and particles in a system which is precisely aligned to its local maximum as a function of its displacement to the absolute maximum.

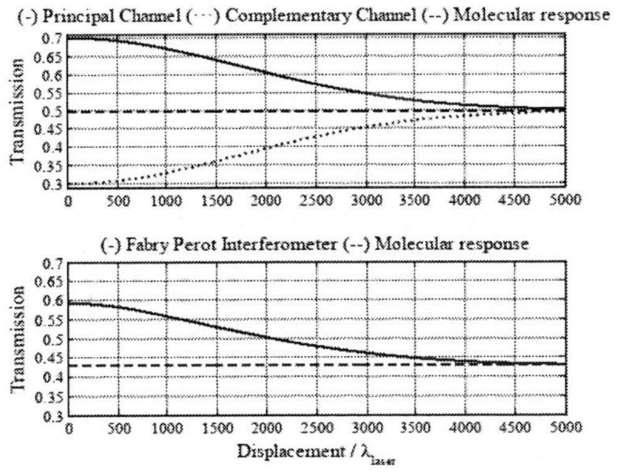

Figure 3: Same as Figure 1 but following the maximum transmission up to 5000 λ_{laser}

From these curves it is clear that the accuracy of the simulated system should be better that ±0.1% of the cavity length (±1 mm) to have a 10% reduction of the aerosol signal.

Figure 4 shows the condition number. Using both channels of the Michelson interferometer leads always to better accuracy in the calculation.

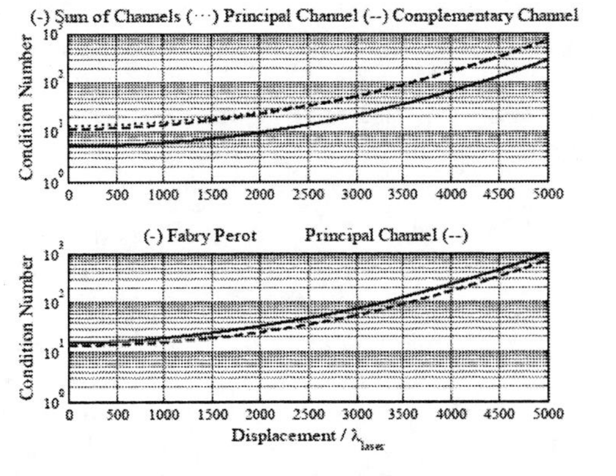

Figure 4: Same as Figure 2 but following the maximum transmission up to 5000 λ_{laser}

5. CONCLUSIONS

This work shows to the authors' knowledge the first study of a MM-HSRL for aerosol observations. It important to notice that the emission complexity is similar to traditional aerosol lidar systems and that the detection is not so complex as the standard Fabry Perot based HSRL. To use this technique, the alignment to reach a local maximum and to stay closer as possible to the absolute maximum are critical issues to address.

ACKNOWLEDGEMENT

The researchers want to acknowledge the Japan Science and Technology Agency (JST) / Japan International Cooperation Agency (JICA), Science and Technology Research Partnership for Sustainable Development (SATREPS).

REFERENCES

[1] Klett, J. D., 1981: Stable analytical inversion solution for processing lidar returns. *App. Opt.*, **20**(2), 211-220.

[2] Fernald, F. G., 1984: Analysis of atmospheric lidar observations- Some comments. *App. Opt.*, **23**(5), 652-653.

[3] Wagner, J., A. Ansmann, U. Wandinger, P. Seifert, A. Schwarz, Matthias Tesche, A. Chaikovsky, and O. Dubovik, 2013: Evaluation of the Lidar/Radiometer Inversion Code (LIRIC) to determine microphysical properties of volcanic and desert dust. *Atm. Meas. Tech.* **6** (7), 1707-1724.

[4] Ansmann, A., P. Seifert, M. Tesche, and U. Wandinger, 2012: Profiling of fine and coarse particle mass: case studies of Saharan dust and Eyjafjallajökull/Grimsvötn volcanic plumes, *Atmos. Chem. Phys.*, **12**, 9399–9415, doi:10.5194/acp-12-9399-2012.

[5] A. Shimizu, N. Sugimoto, I. Matsui, K. Arao, I. Uno, T. Murayama, N. Kagawa, K. Aoki, A. Uchiyama, and A. Yamazaki, 2004: Continuous observations of Asian dust and other aerosols by polarization lidar in China and Japan during ACE-Asia, *J. Geophys. Res.*, **109**, D19S17, doi:10.1029/2002JD003253.

[6] Otero L., P. Ristori, E. Quel, 2008: Multiwavelength Aerosol and Water Vapor Raman Lidar System at CEILAP, Buenos Aires, Argentina. Case Study: November 10, 2006. *Opt. Pura Apl.*, **41** (2) 171 - 175,

[7] Shimizu H., S. A. Lee, and C. Y. She, 1983: High spectral resolution lidar system with atomic blocking filters for measuring atmospheric parameters, *Appl. Opt.*, **22**, 1373-1381.

[8] Nishizawa T., N. Sugimoto, and I. Matsui, 2010: Development of multi-wavelength high-spectral resolution lidar system ($2\alpha+3\beta+2\delta$), *Proc. of 25th International Laser Radar Conference*, 903-906.

[9] Jin Y., N. Sugimoto, T. Nishizawa, P. Ristori, and L. Otero, 2015: A concept of multi-mode high spectral resolution lidar using Mach-Zehnder interferometer, *Proc. of 27th International Laser Radar Conference*.

[10] Bruneau, D., F. Blouzon, J. Spatazza, F. Montmessin, J. Pelon, B. Faure, 2013: Direct-detection wind lidar operating with a multimode laser, *Appl. Opt.*, 52, 4941.

DESIGN OF A LOW COST DIODE-LASER-BASED HIGH SPECTRAL RESOLUTION LIDAR (HSRL)

Matthew Hayman[1], Scott Spuler[1], Bruce Morley[1], Edwin W. Eloranta[2]

[1]*Earth Observing Laboratory, National Center for Atmospheric Research, PO Box 3000, Boulder, CO, 80307, USA, mhayman@ucar.edu*

[2]*Space Science and Engineering Center, University of Wisconsin-Madison, 1225 W. Dayton St., Madison, WI, 53706, USA*

ABSTRACT

A concept for an eye-safe, semiconductor-based high spectral resolution lidar has been developed at the National Center for Atmospheric Research. The lidar operates at a wavelength of 780 nm near several rubidium absorption peaks. A rubidium vapor cell is used to block aerosol backscatter in one channel to provide a molecular backscatter measurement for calculating extinction and backscatter ratio (calibrated backscatter).

Laser and optical components around 780 nm are widely developed due to the large growth in atomic cooling and trapping of rubidium. Thus this instrument can be built largely using mature commercial-off-the-shelf parts. The simulation of the conceptual design shown here uses known commercial products and suggests that such an instrument could be used for quantitative profiling of the lower troposphere.

1. INTRODUCTION

Commercially available backscatter lidar (e.g. Vaisala Ceilometer, Sigma Space Micropulse Lidar) are a major source of lidar data used in studies relating to clouds and aerosols. The availability, mobility, cost and ease of use associated with these instruments is often sufficiently attractive for scientific researchers to accept the relatively limited supply of quantitative results provided by such systems. The directly measured data product provided by these lidar is attenuated backscatter, in which separating particle backscatter and atmospheric optical depth present an intractable problem.

By contrast, quantitative optical parameters can be accurately and reliably retrieved using Raman and high spectral resolution lidars. Both techniques rely on measuring a molecular backscatter profile to calibrate aerosol backscatter and infer an extinction profile. Raman lidar relies on the unique vibrational Raman spectra of nitrogen or oxygen molecules to separate molecular backscatter from the total aerosol and molecular backscatter signals [1]. Raman lidars are often appealing for their all-in-one ability to simultaneously profile clouds and aerosols as well as other important atmospheric state parameters such as temperature and water vapor. However, Raman scattering is several orders of magnitude weaker than Rayleigh scattering so such systems typically require high power lasers which are both expensive, require regular maintenance, and make it difficult to achieve routine unattended deployment.

High spectral resolution lidar (HSRL) provides a second option for quantitative profiling of aerosols and clouds which can operate at eye-safe intensities. An HSRL measures molecular backscatter by blocking the spectrally narrower aerosol backscatter signal with an optical notch filter. Thus, the molecular channel only sees the spectral wings of the molecular backscatter spectrum.

Originally, Fabry-Pérot etalons were used to perform this spectral filtering [2]. These typically come with some level of complexity, requiring tuning, temperature control, limited optical throughput (where etalon size often limits the lidar's primary aperture or field-of-view), and insufficient optical depth at single passes.

Iodine absorption lines have been used to perform spectral filtering [3]. There are several lines with high optical depth near the common 532 nm doubled Nd:YAG wavelength. Iodine cells require much less temperature control and have no sensitivity to angle of incidence. However, obtaining reliable, low maintenance narrow band lasers at this wavelength continues to be a challenge. Also the highly visible green light presents problems for unattended operation, where the light may serve as a distraction to overhead aircraft. This is particularly problematic in for night operation and instances where slant transmit angles are needed.

Recent developments in water vapor DIAL technology at Montana State University (MSU) and the National Center for Atmospheric Research (NCAR) have shown that semiconductor lasers can be used to build reliable, eye-safe, low cost water vapor profiling systems[4; 5]. We will show here that similar development of a semiconductor laser HSRL, using rubidium absorption lines for spectral filtering, presents a useful opportunity to perform quantitative microphysical measurements of the atmosphere with a reliable, low cost system. Optical elements such a narrow linewith semiconductor lasers and amplifiers are available as mature commercial off the shelf (COTS) components due to their heavy use in atomic cooling and trapping where often times the requirements on the optical elements are much more stringent than what is needed for an HSRL.

The use of lower power lasers and longer wavelengths means the proposed rubidium HSRL (Rb-HSRL) will have lower performance compared to its existing HSRL

DOI: 10.1051/epjconf/201611906006

and Raman lidar counterparts. However, compared to existing low cost commercial systems, the Rb-HSRL would be a great leap forward in providing quantitative optical measurements and present new opportunities for deployment on platforms previously incapable of housing such devices.

2. SIMULATION

The simulated design of a Rb-HSRL is a derivative of the current design of water vapor DIAL at NCAR [5]. A distributed Bragg reflector semiconductor laser at 780 nm is coupled into a tapered semiconductor optical amplifier (TSOA). This amplifies the laser and provides the pulse of $15\mu J$ over a 900 ns pulse. The beam is expanded to fill the inner part of the transmit telescope with a radius of $10cm$. The laser repetition rate is dictated by an assumed maximum permissible energy (MPE) of $0.47mW/cm^2$. That is, direct exposure to the laser is within eye-safe limits. The resulting pulse repetition rate is just below 10 kHz. This repetition rate limits the maximum altitude of the lidar system to less than 15 km, however solar background levels will result in lower resolvable altitudes during daytime operation.

The construction of the system assumes a transceiver configuration where the aperture stop and therefore the telescope is divided into a transmit and receive portion [5]. This is accomplished by placing a mirror with a hole at a plane conjugate to the telescope primary (See Fig. 1). The transmit laser is passed through the hole, while the received light is reflected off the mirror into the detection path. This provides the stability of a transceiver lidar with higher isolation between transmitter and receiver at the cost of reduced receiver area. The receiver telescope has a diameter of 40 cm where the inner 20 cm diameter is dedicated to the transmit laser.

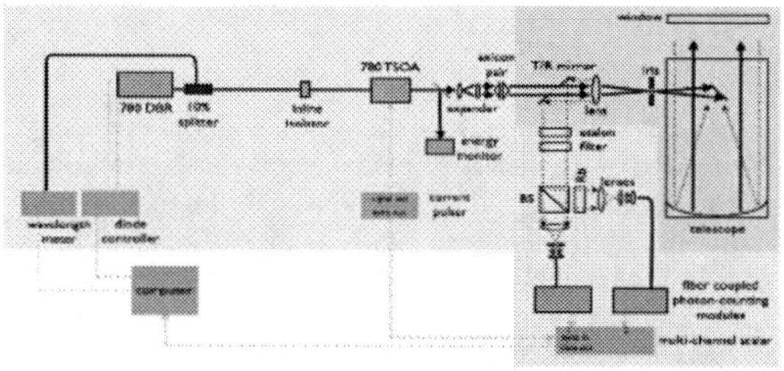

Figure 1: Proposed layout of the Rb-HSRL based on the water vapor DIAL system designed at NCAR/MSU [5]. The instrument uses a shared aperture transceiver for pointing stability with good isolation. A rubidium cell is heated to 320K to block aerosol backscatter in the molecular channel.

The receiver system consists of a solid etalon for daylight rejection with a full width half max (FWHM) of 8 GHz and a free spectral range (FSR) of 50 GHz. The light then passes through a 90:10 beam splitter. Because it is generally the weaker signal, 90% of the received light is directed to the molecular channel through a heated rubidium cell, while the other 10% is directed onto a detector for the total backscatter measurement. It should be noted that a polarizing beam splitter could easily be installed in the total channel to provide polarization measurements. This comes at the expense of adding another detector and data acquisition channel. It may also require a polarizer in the transmitter.

The Doppler broadened rubidium spectra is shown for naturally occurring isotopic abundance in Fig. 2. In addition, Thorlabs offers an Rb 87 cell which is 98% pure (Rb 85 is the other more common isotope of rubidium). For simplicity, we assume in this simulation that we are using the Rb 87 cell provided by Thorlabs (7.2 cm long) as well as their corresponding cell oven. Typically aerosol blocking with optical depth greater than 6 is desirable for accurate liquid cloud characterization. This requires this rubidium cell to be heated to greater than 320 K. The resulting optical depth spectrum is shown in Fig. 2.

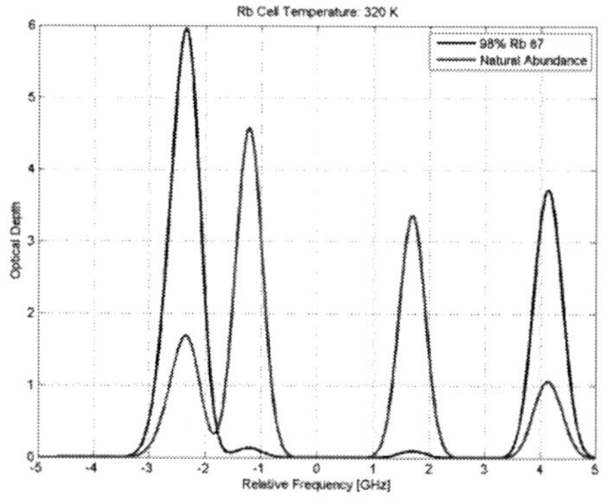

Figure 2: Rubidium absorption spectra for a 7.2 cm cell at 320 K. The Rb-HSRL design considered here assumes the use of the Rb 87 cell available from Thorlabs.

This simulation uses a U.S. standard atmosphere with aerosol loading chosen to be the lower-decile aerosol backscatter coefficients from page 944, Table 4 in [6], scaled from 10.6 um wavelength to 780 nm. To obtain extinction figures, we assume a constant lidar ratio of 60. Also an overlap function is roughly approximated where full overlap approximately occurs at

$$z = \frac{D_{telescope}}{\theta_{FOV}}, \tag{1}$$

where $D_{telescope}$ is the telescope diameter and θ_{FOV} is the full angle field-of-view (FOV) of the receiver.

The profile calculated for the Rb-HSRL was normalized for component, power and wavelength and compared to a similarly normalized profile measured by the NCAR/NSF GV-HSRL. This allows us to better estimate efficiency

DOI: 10.1051/epjconf/201611906006

Table 1: Simulation parameters for Rubidium HSRL

Transmitted pulse width	900 ns
Transmitted pulse energy	15 μJ
Pulse repetition rate	9.8 kHz
Laser wavelength	780 nm
Transmitted beam diameter	20 cm
Telescope diameter	40 cm
Effective receiver collection area	940 cm^2
Field-of-view	150 μrad
Receiver efficiency	0.006
Receiver optical bandwidth	8 GHz (FWHM)
Rubidium cell length	7.2 cm
Rubidium cell temperature	320 K
Range resolution	150 m
Solar background	1.15 $\mathrm{mWcm^{-2}\mu m^{-1}sr^{-1}}$

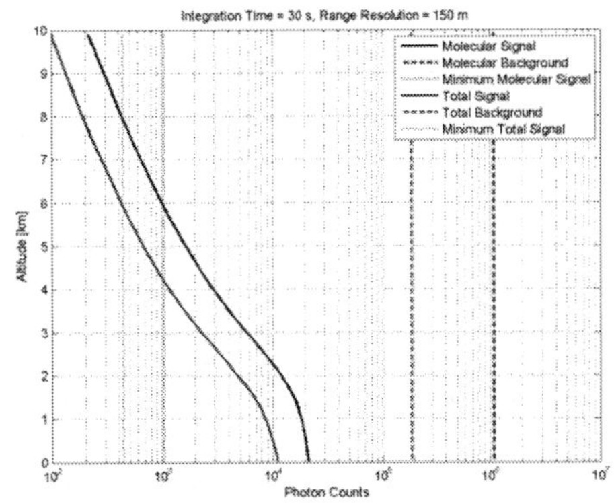

Figure 3: Backscatter (solid), solar background (course dashed) and noise floor (fine dashed) levels of the total (red) and molecular (blue) channels of the Rb-HSRL for a U.S. standard atmosphere [6] with 30 seconds of integration.

terms in the lidar system and develop a conservative performance estimate. In this case, we added a multiple of 0.1 to Rb-HSRL simulations to obtain agreement between the two normalized profiles.

The proposed Rb-HSRL is simulated by calculating all relevant range dependent quantities and applying them to a normalized backscatter spectrum for aerosols (which are assumed to be 1000 times more massive than air molecules), and a temperature and pressure dependent molecular spectrum [7]. The subsequent backscatter spectrum is then multiplied by the relevant receiver filter function (etalon for the total channel, and etalon with rubidium spectra for the molecular channel). The simulation parameters are summarized in Table 1.

The pure backscatter signals are shown in Fig. 3. The course dashed lines indicate the solar background levels and the fine dashed lines indicate the signal level where SNR (signal divided by variance) is 1. Below an SNR of 1, integration will not improve the resolution of the signal. Thus, for this particular simulation, the maximum resolvable altitude is approximately 6 km. Cloud backscatter information is still likely to be available at higher altitudes. While not enough to provide full tropospheric profiles, this still represents an promising development where the Rb-HSRL can provide quantitative backscatter data over approximately half of the troposphere. The extinction resolution is shown in Fig. 4. The resolution of the system is too course to resolve aerosols but may be useful for estimating cloud extinction.

Additional improvements to Rb-HSRL may be possible as technology continues to advance. For example, higher energy tapered semi-conductor amplifiers may become available as fabrication of longer chips becomes possible. These would increase the maximum altitude of the system and reduce integration times.

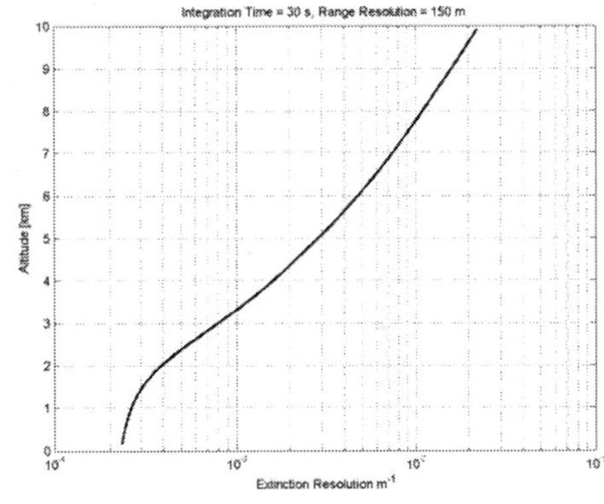

Figure 4: Extinction resolution of the Rb-HSRL for a U.S. standard atmosphere [6].

3. CONCLUSION

We have demonstrated that a diode based, eye safe HSRL is feasible for making calibrated backscatter measurements. The system relies on existing commercial components that have largely been proven in the NCAR/MSU water vapor DIAL. It uses a heated rubidium cell as a spectral filter to block aerosol returns. The instrument could retrieve backscatter ratio measurements up to approximately 6 km during daytime operation. Daytime extinction measurements would largely be limited to clouds. This instrument offers a unique pathway forward in cloud and aerosol profiling due to its potential for unattended operation and low cost.

ACKNOWLEDGMENTS

The National Center for Atmospheric Research is sponsored by the National Science Foundation.

REFERENCES

1. Whiteman, D. N., "Examination of the traditional Raman lidar technique. i. evaluating the temperature-dependent lidar equation," Appl. Opt., **42**(15), pp. 2571–2592 (2003).
2. Eloranta, E., Chapter 5: High Spectral Resolution Lidar in *Lidar: Range-Resolved Optical Remote Sensing of the Atmosphere*, Springer, New York, U.S.A. (2005).
3. Piironen, P., Eloranta, E., "Deomonstration of a high-spectral-resolution lidar based on an iodine absorption filter," Opt. Lett. **19**, 234–236 (1994).
4. Nehrir, A. R., Repasky, K. S., Carlsten, J. L., Obland, M. D. and Shaw, J. A., "Water vapor profiling using a widely tunable, amplified diode-laser-based differential absorption lidar (DIAL)," J. Atmos. Ocean. Technol., **26**, pp. 733–745 (2009).
5. Spuler, S. M., Repasky, K. S., Morley, B., Moen, D., Hayman, M., Nehrir, A. R., "Field-deployable diode-laser-based differential absorption lidar (DIAL) for profiling water vapor," Atmps. Meas. Tech, **8**, pp. 1073–1087 (2015).
6. Pike, E. R., Sabatier, P.C., *Scattering, Two-Volume Set: Scattering and inverse scattering in Pure and Applied Science*, Elsevier Science, 2001)
7. Boley, C. D., Desai, R. C., Tenti, G., "Kinetic models and brillouin scattering in a molecular gas," Canadian Journal of Physics, **50**(18), pp. 2158–2173, 1972.

ATMOSPHERIC TEMPERATURE PROFILE MEASUREMENTS USING MOBILE HIGH SPECTRAL RESOLUTION LIDAR

Ilya I. Razenkov, Edwin W. Eloranta

University of Wisconsin, 1225 W. Dayton St., Madison, WI, 53706, USA, razenkov@wisc.edu

ABSTRACT

The High Spectral Resolution Lidar (HSRL) designed at the University of Wisconsin-Madison discriminates between Mie and Rayleigh backscattering [1]. It exploits the Doppler effect caused by thermal motion of molecules, which broadens the spectrum of the transmitted laser light. That allows for absolute calibration of the lidar and measurements of the aerosol volume backscatter coefficient. Two iodine absorption filters with different absorption line widths (a regular iodine vapor filter and Argon buffered iodine filter) allow for atmospheric temperature profile measurements. The sensitivity of the measured signal-to-air temperature ratio is around 0.14%/K. The instrument uses a shared telescope transmitter-receiver design and operates in eye-safe mode (the product of laser average power and telescope aperture equals 0.1 Wm^2 at 532 nm).

1. INTRODUCTION

The atmospheric temperature is an important parameter that describes the atmospheric state. It plays an important role in many atmospheric processes. Passive remote sensors lack the spatial resolution necessary for many applications. Radiosonde measurements are usually available only twice a day. Some types of lidars are capable of measuring the atmospheric temperature profile with a relatively large spatial resolution and with more frequent measurement.

Several investigators have made atmospheric temperature profile measurements using a High Spectral Resolution Lidar (HSRL) [2-5]. These techniques suffer from low signal-to-noise ratio, low sensitivity to air temperature, and/or operate highly exceeding the eye-safe limit on the transmitted laser power; and also suffer from limitations in filter and laser frequency stability.

The standard HSRL technique requires the atmospheric temperature and pressure profile for instrument calibration. In this paper, we report the atmospheric temperature profile measurements performed by the University of Wisconsin – Madison High Spectral Resolution Lidar. Two iodine absorption filters with different absorption line widths enable atmospheric temperature profile measurements. A molecular iodine I2 vapor filter provides standard HSRL measurements. A second channel using an argon buffered iodine filter provides additional information for atmospheric temperature measurements. The instrument's capability to measure the atmospheric temperature profile makes it intrinsically calibrated. The presented temperature measurement technique requires a superior instrument performance in order to achieve good measurement accuracy. Work done in attempt to perfect this technique have improved instrument performance and data quality for standard HSRL measurements.

2. HIGH SPECTRAL RESOLUTION LIDAR SYSTEM

The simplified schematic of the HSRL lidar is shown in Figure 1. The transmitter uses an injection seeded diode pumped Nd:YAG laser. The laser operates at 4kHz repetition rate. The single mode diode DFB seed laser (1064 nm) is used to injection seed the host laser cavity. The seed laser is frequency locked to the center of iodine absorption line #1109 using a frequency locking setup with active control loop. The setup uses a 100mm long iodine absorption cell and two energy monitors. The computer controlled DAQ applies voltage to the seed laser controller and does a slow frequency tuning of the laser. The laser cavity controller dithers the host cavity end mirror changing the output frequency with around 12MHz amplitude and measures the absorption line slope at the frequency. The zero slope value corresponds to the bottom of the absorption line. The slope information is transferred to the computer providing an active feedback for laser frequency control. That provides the frequency locking accuracy of around 12MHz with a frequency jitter of around 12MHz amplitude. Some of the lidar transceiver properties are summarized in Table 1.

DOI: 10.1051/epjconf/201611906007

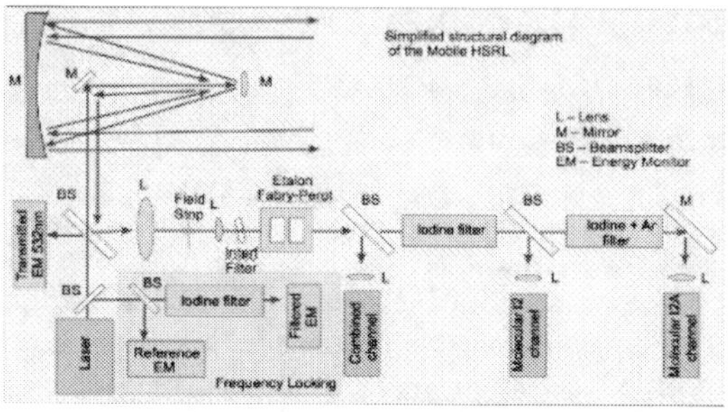

Figure 1. A simplified structural diagram of HSRL system.

Table 1: Lidar parameters.

Wavelength	532 nm
Transmitted eye-safe power	0.3W @532nm
Transmitter Spectral width	< 60 MHz
Spectral purity	>5000:1
Scanning telescope (Dall-Kirkham)	20x, Diam 400 mm
Receiver Spectral bandwidth	8 GHz
Iodine blocking filters' bandwidth	1.8 GHz (narrow filter), 2.85 GHz (wide filter)

Some portion of the laser light is picked off from the beam by a wedged beam pick-off for frequency locking setup, laser spectrum quality control and the interferometer. The broadband beam splitter (R/T=65/35S) is used as a transceiver switch. Then, the beam is further expanded by 20x Dall-Kirkham telescope (400 mm aperture) and transmitted into the atmosphere. The telescope is shared between transmitter and receiver. The sky background noise is rejected by the interference filter (0.35 nm FWHM) and Fabry-Perot etalon (FWHM=7.5 GHz, FSR=375 GHz, peak transmission is 76%).

An argon-broadened filter allows ~0.1% minimum transmission in the center of the 1109 line and cannot provide sufficient rejection of aerosol signal. For that reason a serial arrangement of the filters is used.

3. ARGON BUFFERED IODINE FILTER EXPERIMENTS

A buffer gas such as argon added to the filter can broaden the absorption line of the iodine filter. A series of experiments were performed in order to find the optimum pressure of the argon gas in the iodine cell. The filter is optimized to provide sensitivity to the atmospheric temperature and transmission of molecular signal. The experimental setup for iodine absorption line broadening with argon buffer gas is shown in Figure 2. The filter transmission curves are shown in Figure 3.

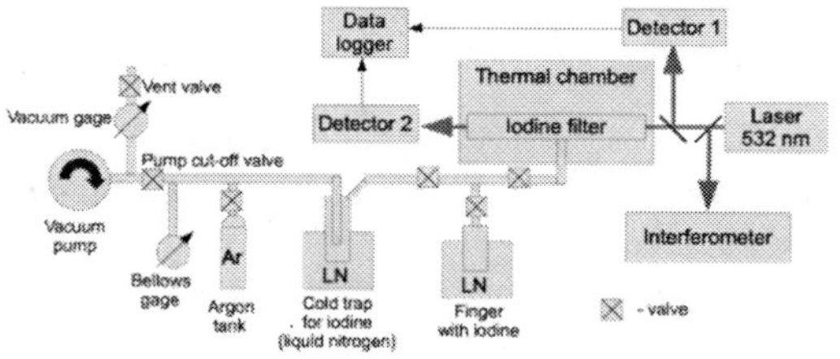

Figure 2. Experimental setup for iodine filter broadening using Argon gas.

The molecular iodine absorption spectrum changes as the argon pressure increases: the line width increases, peak absorption and continuum transmission decrease.

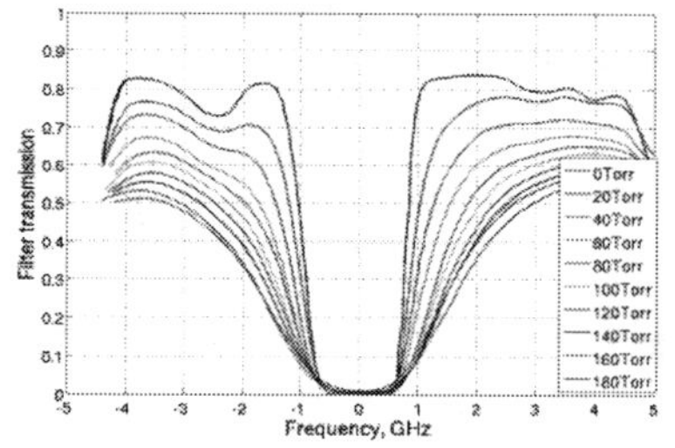

Figure 3. A set of experimentally measured molecular iodine filter transmission curves for different argon gas pressures ranging from 0 Torr to 180 Torr.

The absorption filters contain a fixed amount of iodine to provide stable filter properties. Heaters maintain all the iodine in the gaseous phase and ensure that no iodine condensation occurs on the windows of the cell.

4. CALCULATION OF OPTIMUM ARGON PRESSURE

The selection of the optimum pressure for the broadened iodine filter is done by modeling the lidar temperature measurements. The real lidar returns from the operating instrument are used to perform the analysis that accounts for the actual instrument efficiency. The ratio R of the signals of the two molecular channels is a function of the atmospheric temperature and pressure. The received signal in the channels *Molecular I2* ($N_{m,1}$) and *Molecular I2A* ($N_{m,2}$) are:

$$N_{m,1}(z) = N_0(z)(1 - T_{bs})f_{m,1}(z),$$

$$N_{m,2}(z) = N_0(z)\,T_{bs}f_{m,2}(z),$$

where $f_{m,1}(z) = \int F_1(\nu)\mathcal{R}(T(z), P(z), \nu)d\nu$,

$f_{m,2}(z) = \int F_1(\nu)F_2(\nu)\mathcal{R}(T(z), P(z), \nu)d\nu$,

and $\int \mathcal{R}(T(z), P(z), \nu)d\nu = 1$, $N_0(z)$ – number of photons received from a distance z (actual lidar counts); T_{bs} – beamsplitter transmission; ν – frequency; $F_1(\nu), F_2(\nu)$ – transmission spectrums of a standard iodine and broadened iodine filters; $\mathcal{R}(T, P, \nu)$ – spectrum of molecular return from Tenti's s6 model; $T(z)$, $P(z)$ – atmospheric temperature and pressure; $T(z)$, $P(z)$ are the temperature and pressure measured by radiosonde.

The ratio of the signals in two channels is the function of temperature and pressure for a given value of T_{bs}:

$$R(z) = \frac{N_{m,1}(z)}{N_{m,2}(z)} = \frac{1 - T_{bs}}{T_{bs}} \frac{f_{m,1}(T(z), P(z))}{f_{m,2}(T(z), P(z))}$$

The instrument temperature sensitivity coefficient $\frac{dR}{dT}$ at each altitude can be calculated by using the expression above.

The photon counting error is calculated by using the following relationship:

$$dR_{er} = \sqrt{\frac{N_{m,1}}{N_{m,2}^2} + \frac{N_{m,1}^2}{N_{m,2}^3}}.$$

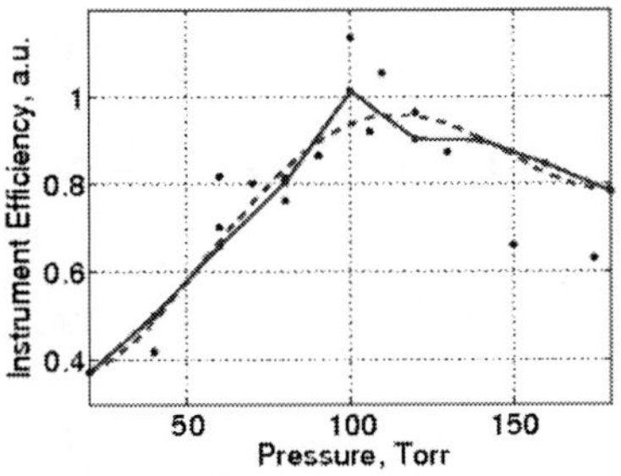

Figure 4. Modeled instrument temperature efficiency at 5 km as a function of argon pressure. Scattered blue dots are the efficiencies for different tests with various amount of iodine in the cell, the solid green line is for a test with a fixed amount of iodine and the dashed green line is the third order polynomial fit to all the tests.

The ratio of the sensitivity coefficient (signal-to-air temperature ratio) to the counting error (instrument T efficiency) is used to find the optimum argon gas pressure for broadened filter. The curve for the instrument T efficiency as a function of buffer gas pressure calculated at 5 km is shown in Figure 4. The maximum of the ratio is found to be at around 110 Torr of the argon gas pressure. The instrument temperature sensitivity coefficient $\frac{dR}{dT}$ for broadened filter (110 Torr) is around 0.14%/K.

5. ATMOSPHERIC TEMPERATURE MEASUREMENT RESULTS

The atmospheric temperature profile measurement requires the instrument calibration. The channels' transmission spectrum functions $\xi_1(\nu)F_1(\nu)$, $\xi_2(\nu)F_1(\nu)F_2(\nu)$ (where ξ_1, ξ_2 are the channel gains) are measured by performing a laser frequency scans across the receiver bandwidth. This provides the filter spectral bandpass and the relative sensitivity of the channels.

To obtain the atmospheric temperature we assume that the pressure at a starting range height is known. Then the temperature follows from the ratio of the received photon counts of the two molecular channels. When the temperature and pressure at one altitude are know, the pressure at the next range bin can be obtained by using the relationships for hydrostatic balance and ideal gas law. By repeating this, the temperature and pressure profiles are obtained.

The atmospheric temperature profile measured on the March 11, 2015 from 800 to 1500 UTC Madison, WI is shown in the Figure 5. The vertical resolution is 300 meters. The temperature profile measured by the HSRL still exhibits an offset between -5 and +5 K when compared with the radiosonde sounding result. A correction on the relative gain of the two molecular channels must be applied for its correction. This correction factor is determined by matching the lidar temperature and balloon temperature at one range height.

The difference in optical alignment of the two channels introduces a range dependent differential geometry error in the temperature measurements. The iodine absorption filters in the instrument are mounted in the motorized mechanical mounts and can be remotely removed from the received signal beam path without disturbing the optics. The correction for the range dependent alignment error

 DOI: 10.1051/epjconf/201611906007

is performed by measuring the profile of the ratio of the two molecular channels when the broadened absorption filter is removed from the optical path. This range dependent correction is applied to the relative gain of the two channels. The measured differential geometry correction profile and the temperature error associated with it are shown in the Figure 6.

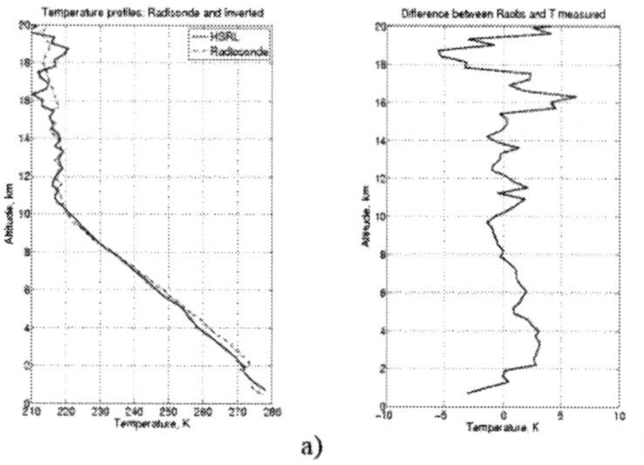

a) b)

Figure 5. a) Measured atmospheric temperature profile (blue, solid) March 11, 2015 from 800 to 1500 UTC Madison, WI and radiosonde temperature profile (green, dashed) 12UTC Green Bay, WI sounding; b) The difference of lidar and radiosonde measured atmospheric temperature profiles.

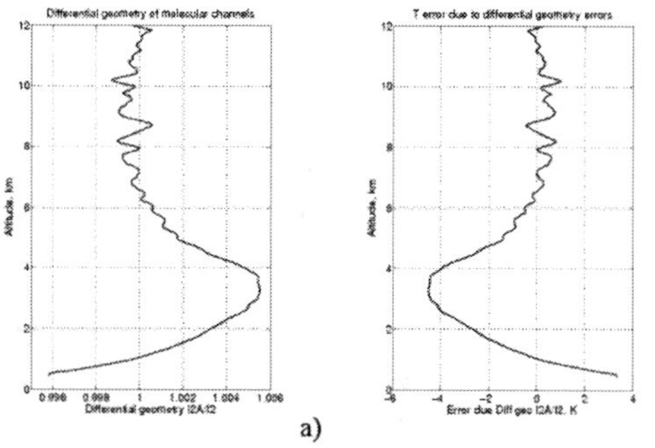

a) b)

Figure 6. a) Measured differential geometry correction; b) Correction for differential geometry error.

The difference in geometry of the two channels is the largest source of errors. The inverted profile mismatches to the radiosonde observations in the lower 3 km. We assume that it is caused by the changes in the differential geometry and nonlinear detector response. The sensitivity of the instrument is relatively low, which requires a precise measurement (long averaging time), accurate correction of the channel relative overlaps (differential geometry), long term mechanical stability for optics, accurate calibration and frequency locking, stability of detector gains and linearity of detector response.

6. CONCLUSIONS

We have reported the atmospheric temperature profile measurements performed by our HSRL lidar. This is made possible by using two iodine absorption filters with different absorption line widths. The instrument's signal-to-air temperature sensitivity is 0.14 %/K. The instrument has an advantage of moderate temperature sensitivity, relatively high efficiency, and eye-safe operation over previously reported works.

The temperature measurement accuracy currently lacks the long-term mechanical stability of optics, stability of detector gains and linearity of detector response. The instrument requires future improvement for robust routine measurements that will be useful for meteorological applications.

ACKNOWLEDGEMENT

This work is supported by the Space Science and Engineering Center University of Wisconsin – Madison.

REFERENCES

[1] Eloranta, E. W.: High Spectral Resolution Lidar, *in Lidar: Range-Resolved Optical Remote Sensing of the Atmosphere,* Klaus Weitkamp ed., Springer, NY, pp. 25-30.

[2] Shimizu H. et al, 1986: Atmospheric temperature measurement by a high spectral resolution lidar, *Applied Optics*, Vol. 25, Issue 9, pp. 1460-1466 (1986).

[3] Dengxin H., 2004: Ultraviolet high-spectral resolution Rayleigh-Mie lidar with a dual-pass Fabry-Perot etalon for measuring atmospheric temperature profiles of the troposphere, *Applied Optics*, Vol. 44 Issue 7, pp.1305-1314 (2005).

[4] Hair J.W. et al, 2001: "High spectral resolution lidar with iodine-vapor filters: measurement of atmospheric-state and aerosol profiles", *Applied Optics*, Vol. 40, Issue 30, pp. 5280-5294 (2001).

[5] Liu Z-S. et al 2009: Iodine-filter-based high spectral resolution lidar for atmospheric temperature measurements, *Optics Letters*, Vol. 34, Issue 18, pp. 2712-2714 (2009).

EPJ Web of Conferences 119, 06008 (2016)
ILRC 27
DOI: 10.1051/epjconf/201611906008

An innovative rotational Raman lidar to measure the temperature profile from the surface to 30 km altitude

Alain Hauchecorne[1]*, Philippe Keckhut[1], Jean-François Mariscal[1], Eric d'Almeida[1], Pierre-Richard Dahoo[1], Jacques Porteneuve[1]

*[1]LATMOS-IPSL, Université Versailles St-Quentin, CNRS, Guyancourt, France, *Email: alain.hauchecorne@latmos.ipsl.fr*

ABSTRACT

A concept of innovative rotational Raman lidar with daylight measurement capability is proposed to measure the vertical profile of temperature from the ground to the middle stratosphere. The optical filtering is made using a Fabry-Pérot Interferometer with line spacing equal to the line spacing of the Raman spectrum. The detection is made using a linear PMT array operated in photon counting mode. We plan to build a prototype and to test it at the Haute-Provence Observatory lidar facility. to achieve a time resolution permitting the observation of small-scale atmospheric processes playing a role in the troposphere-stratosphere interaction as gravity waves. If successful, this project will open the possibility to consider a Raman space lidar for the global observation of atmospheric temperature profiles.

1. INTRODUCTION

The temperature is a fundamental parameter for meteorology and climatology. It is measured, (1) twice a day using radiosondes with good vertical resolution but in a limited number of stations in the world and, (2) from space but often with a vertical resolution limited by the weighting function of the instrument. Lidars provide a good vertical resolution and continuity of measurements. Currently the temperature is measured by Rayleigh lidar at NDACC ground stations but only above 30 km due to the presence of aerosols below this altitude [1] [2]. There is currently no space lidar in flight or in project measuring the temperature. The rotational Raman lidar technology allows access to the evolution of the temperature profile in the troposphere and lower stratosphere. It requires developing an efficient optical system with separation between the Rayleigh and the rotational Raman signal, the first being 2 to 3 orders of magnitude stronger than the second. It requires acquiring the full rotational Raman spectrum of molecular nitrogen, the temperature being deduced from the evolution of the relationship between the spectral lines.

2. METHODOLOGY

In the Rayleigh lidar, monochromatic laser pulses are sent vertically into the atmosphere and a temporal analysis of the backscattered light provides information about the vertical structure and composition of the atmosphere. Above the top of the stratospheric aerosol layer (about 30 km), Mie scattering is negligible and the received signal due only to Rayleigh scattering is directly proportional to the atmospheric density. The temperature profile is deduced from the density profile assuming that the density is in hydrostatic equilibrium and obeys the perfect gas law. The profile is limited upwards by the signal to noise ratio at about 80-90 km, depending on the power of the lidar.

It is obvious that below about 30km, depending on the height and abundance of aerosols, it is not possible to use the Rayleigh scattering technique to measure atmospheric density and temperature because of the contribution of Mie scattering to the signal. It is, however, possible to use the Raman vibrational and rotational components of the scattering by the main gases of the atmosphere, O_2 and N_2. The vibrational Raman scattering is about 3 orders of magnitude smaller than the Rayleigh scattering. In order to achieve a useful accuracy in the lower stratosphere with this technique, it is then necessary to have a lidar with high energy-surface product.

As for the Rayleigh scattering, the vibrational Raman signal is directly proportional to the atmospheric density and the temperature profile is computed from the density profile assuming the hydrostatic equilibrium. It is necessary to correct the signal for the atmospheric transmission, which cannot be assumed constant below 30 km. This technique is very well adapted for observations in the lower stratosphere during periods with low background aerosol loading during which an estimation of the optical thickness of the aerosol layer can be done with enough accuracy [3].

The vibrational Raman technique is limited to the altitude above the tropopause due to the frequent presence of thin cirrus with unknown optical thickness in the upper troposphere. Furthermore, after a major volcanic eruption, like the Pinatubo in June 1991, it is no more possible to use the vibrational Raman technique in the lower stratosphere due to the uncertainty on the very high optical thickness of the volcanic aerosol layer. In this case an alternative technique exists consisting in the direct measurement of the temperature using the rotational Raman lidar technique.

The Raman rotational spectrum of a diatomic molecule such as N_2 or O_2 is composed of two branches, Stokes and anti-Stokes. Each branch is composed of a series of lines with an intensity depending on temperature. Lines close to the Rayleigh line vary inversely to the temperature when liens far to the Rayleigh line vary in phase with the temperature. If we measure the ratio between two lines with an opposite dependence in temperature, we can obtain a direct measurement of the atmospheric temperature. This is generally done with two anti-Stokes lines near 529 nm and 530.5 nm [4] [5] [6]. Anti-Stokes lines are preferred to the Stokes lines that may be perturbed by the fluorescence of aerosols. The difficulty of these measurements is due to the very small shift of wavelength between the laser excitation line (532nm) and the observed lines. The needed rejection factor at the wavelength of the laser line is of the order of 10^{-7} and is obtained by using two interference filters in series. This method gives a relative temperature profile and has to be calibrated once with a radiosonde profile to obtain the two parameters of the calibration curve. This curve can be used as long as the characteristics of the lidar are not changed.

Figure 1 presents a profile obtained at Haute-Provence Observatory (OHP, 44°N, 6°E) integrated during 10h with a 1-km vertical resolution, showing a good agreement with the coincident radiosonde. This technique has been installed in the ALOMAR lidar facility (Andoya Space Center, 69°N, 16°E) in 1995 and is operated regularly since this date. It has been used to study the properties of polar stratospheric clouds, showing the capability to obtain temperature profiles inside the clouds [6]. However the low level of the signal after filtering by the two band-pass filters limits the use of the rotational Raman technique for dynamics studies in the lower stratosphere that need e better time resolution. For instance for the study of gravity waves a time resolution less than 30 minutes is needed.

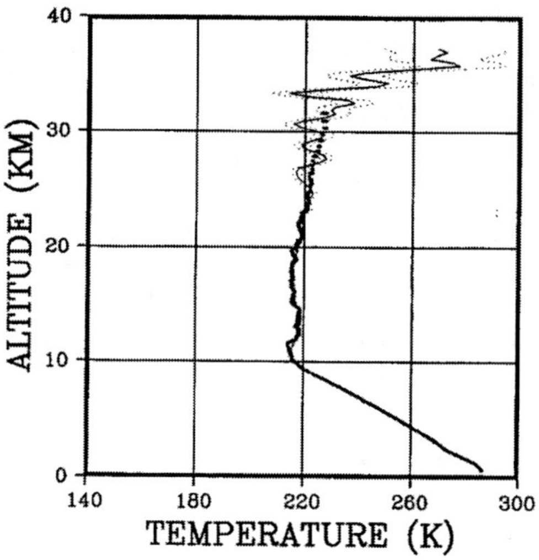

Figure 1: Rotational Raman profile at OHP on April 9, 1991: lidar (thin line) with 1σ uncertainty (dotted line) compared with a radiosonde (thick line). From [4].

We propose to develop an innovative rotational Raman lidar using the signal recorded in the full rotational Raman spectrum. The integrated intensity of the full N_2 rotational Raman spectrum represents 2% of the intensity of the Rayleigh line, which is not at all negligible. In order to do that, it is needed to acquire the very low backscatter signal at the frequency required to achieve the expected vertical resolution (1 μs for 150 m). New array detectors, like the Hamamatsu 32-channel linear array PMT assembly H7260 operated in photon counting mode, are well adapted for this objective.

Another challenge is the high dynamics between the Rayleigh-Mie line and rotational Raman lines that are individually 3 to 4 orders of magnitude weaker. An elegant solution to solve this problem proposed by [7] is to transmit the optical signal through a Fabry-Pérot interferometer (FPI) with a line spectral spacing exactly equal to the spacing between the lines of the N_2 rotational Raman spectrum (7.958cm-1 [6]). The Rayleigh line is placed exactly in the middle of two FPI lines, corresponding to a minimum of transmission. For a FPI with finesse 30, the intensity of the Rayleigh-Mie line is attenuated by a factor 160 and is only 6 times larger than the stronger Raman lines. The use of a FPI has another advantage. It decreases the sky background light by a factor approximately equal to its finesse, giving to the Raman lidar a daylight measurement capability.

We started the definition of the rotational Raman detection system. The objective is to build a prototype and to test it in the NDACC station at Haute-Provence Observatory to benefit from the facilities available on site (laser source, telescopes, ...). Such a system will be able to perform a continuous survey of the temperature profile from the surface to the middle stratosphere, providing a powerful tool to study the variability of the temperature profiles due to atmospheric processes and could serve as a reference to define a future space lidar for measuring the temperature in the troposphere and lower stratosphere.

3. CONCLUSION

We propose an innovative rotational Raman lidar with daylight measurement capability to measure continuously the vertical profile of temperature from the ground to the middle stratosphere. The optical filtering is made using a FPI with line spacing equal to the line spacing of the N_2 rotational Raman spectrum. This allows benefiting from the signal in the whole N_2 rotational Raman spectrum and achieving a time resolution permitting the observation of small-scale atmospheric processes playing a role in the troposphere-stratosphere interaction as upward propagating gravity waves. If successful, this project will open the possibility to consider a rotational Raman space lidar for the global observation of atmospheric temperature profiles with a high vertical resolution.

ACKNOWLEDGEMENT

This work is made with the support from CNRS-INSU, CNES and UVSQ.

REFERENCES

[1] Hauchecorne A., M.L. Chanin, Density and temperature profiles obtained by lidar between 35 and 70 km, Geophys. Res. Lett., 7, 565-568, 1980.

[2] Keckhut P., A. Hauchecorne, M.L. Chanin, A critical review of the data base acquired for the long term surveillance of the middle atmosphere by Rayleigh lidar, J. Atm. Ocean. Tech., 10, 850-867, 1993.

[3] Keckhut P., M.L. Chanin, A. Hauchecorne, Stratospheric temperature measurements using Raman lidar, Applied Optics, 29, 5182-5186, 1990.

[4] Hauchecorne A., M.L. Chanin, P.Keckhut, and D. Nedeljkovic, Lidar monitoring of the temperature in the middle atmosphere, Appl. Phys. B, 55, 29-34, 1992.

[5] Nedeljkovic N., A. Hauchecorne, M.L. Chanin, Rotational Raman lidar to measure the atmospheric temperature from the ground to 30 km, IEEE Transactions on Geoscience and Remote Sensing, 31, 91-101, 1993.

[6] Behrendt, A., J. Reichardt, Atmospheric temperature profiling in the presence of clouds with a pure rotational Raman lidar by use of an interference-filter-based polychromator, Applied Optics, 39 (9), 1372-1378, 2000.

[7] Fierli, F., A. Hauchecorne and B. Knudsen, Analysis of polar stratospheric clouds formation using temperature and aerosols measured by ALOMAR RMR lidar, J. Geophys. Res., 106, 24127-24141.2001.

[8] Arshinov, Y., S/ Bobrovnikov, I. Serikov, A. Ansmann, U. Wandinger, D. Althausen, I. Mattis, D. Müller Daytime operation of a pure rotational Raman lidar by use of a Fabry–Perot interferometer, Applied Optics, 2005, 44, 3593-3603, 2005.

EPJ Web of Conferences **119**, 06009 (2016) DOI: 10.1051/epjconf/201611906009
ILRC 27

COMPUTATIONAL INVESTIGATIONS OF THZ TRANSMITTANCE IN THE ATMOSPHERE

Xingyu Yang[1], Casey Calhoun[1], and Ronald Calhoun[2]

[1]*BASIS Scottsdale, Scottsdale, AZ 85259*
[2]*Arizona State University, Tempe, AZ 85281*

ABSTRACT

With the recent scientific advancements in Terahertz (THz) wave propagation and reception technology, there has been significant development in new possibilities for using THz waves – offering new possibilities in THz detection and ranging. A first foundational step toward this goal is to better understand THz transmittance in the turbulent atmosphere. In this project, a frequency modulation pattern of THz waves was created by utilizing a system of shifting frequency based on temperature, air humidity, and distance of transmission. The total path loss of the wave in air, based on the wave spread and molecular absorption, was then modeled using radiative transfer theory, onto a set of JavaHAWK filtered-HITRAN data representative of an air sample. This data was used to generate a path loss matrix, which was then used to optimize frequency of transmission for the specific conditions. The concept to be evaluated is whether adaptive frequency modulated THz might usefully decrease transmission losses by adjusting to atmospheric conditions (such as local variations in temperature and humidity).

I. INTRODUCTION

One significant challenge to the development of terahertz detection and ranging is the high level of molecular absorption associated with these frequencies **[1]**. The presence of water vapor in the atmosphere provides a particularly significant contribution to this absorption, with potential path losses of up to 10 dB **[2]**. To avoid this path loss, the THz band is divided into smaller usable segments of approximately 10-100 GHz, windows which are found and characterized by modeling the path loss, or reduction in power density of the molecule **[2]**.

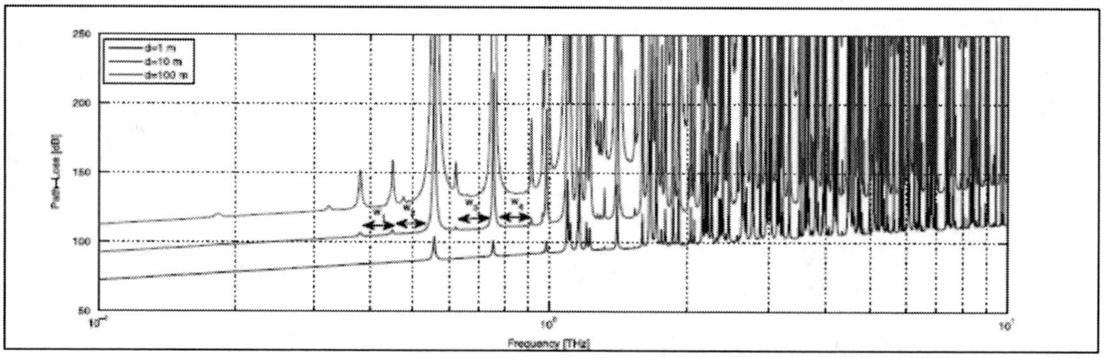

Figure 1 - Path Loss in THz Band for Different Transmission Distances (Image credit to Kurner)

The purpose of this project is to characterize the path loss in THz atmospheric transmission by taking into consideration the spreading loss and the molecular absorption loss based on conditions such as air temperature, air humidity, and transmission distance, and then minimize this level of path loss in order to create better possibilities for detection and ranging applications.

A path loss matrix for a given ambient temperature, humidity level, and transmission distance, was determined for variables of frequency and distance. The data extrapolated from this matrix allows for the optimization of a frequency band to minimize path loss. The project uses the gas properties of oxygen and H2O based on data extrapolated from JavaHAWK(JAVA). The code is then used to investigate the aspects that affect the total path loss.

II. METHODOLOGY

In order to minimize the amount of wave attenuation, the highest sources of path loss in a sample of air were characterized, and key factors affecting this amount of path loss were identified. Using the properties of THz waves, it was found that, in a sample of air, absorption by water molecules created the highest source of path loss. Thus, the unique characterizing factor of path loss was found to be the molecular absorption coefficient (units of M/cm-1), also known as the molecular attenuation coefficient. This coefficient was modeled, and found to depend on variable factors such as the pressure, temperature, humidity of the environment, and properties of transmission of EM wave such as distance of transmission, and, especially, the frequency of the THz wave transmitted. Thus, it was established that if the optimal frequency for a certain ambient factors could be determined, that the amount of path loss could be minimized, which would result in minimal power input for the device.

We used a line-by-line model to simulate the various cases of molecular absorption based on path loss, generating a power matrix of total path loss with variables of distance and frequency. From this matrix, the ideal frequency can be extrapolated. Our envisioned design for use of this modulation scheme consists of a transmitter that changes frequency transmission rate based on perceived factors of ambient temperature, pressure, and humidity, and detected distance of transmission to reduce the path loss of the signal.

HITRAN (high resolution transmission), source of molecular information used, is a database that contains the parameters necessary simulate atmospheric molecular transmission and radiance from the microwave through ultraviolet region of the spectrum. This database is available online, for free, upon request. JavaHAWK is an open source software included with the database that performs simple functions on the information, such as filtering by wavelength. Since HITRAN, itself, contains a distribution of isotopologues - isotope molecule - of a certain gas g, the software JavaHAWK (JAVA HITRAN ATMOSPHERIC WORKSTATION) is used to filter and compile the molecules of resonance within the Terahertz range, specifically within air. In the line-by-line model, the total path loss was found by referencing the parameters of each isotopologue within the JavaHAWK filtered HITRAN data, and finding the sum of the path loss. To create a frequency-optimization modulation design for a THz

DOI: 10.1051/epjconf/201611906009

model, a power matrix was generated, detailing the total path loss in dB for a specific set of atmospheric conditions, using a model for the amount of attenuation of the molecule.

This was created by using radiative transfer theory, and the Beer-Lambert law to model the molecule's properties of absorbance - the amount of the EM wave absorbed by the molecule - and spreading - the free-space loss due to the attenuation of molecules moving through a medium. The variable medium absorbance is found as the sum of the individual medium absorbance coefficients of each individual isotopologue. For a gas, its own molecular absorbance coefficient is found by taking the weighted sums of the isotopologues its isotopologues. Within the HITRAN database, for each gas type (i.e. - H2O, O2, etc.), there is a distribution of isotopologues that models the actual distribution of each in a random sample of that gas. Thus, taking into consideration the percentage of that gas, the total molecular absorption coefficient can be found by simply taking the sum of the individual molecular absorption coefficients.

III. RESULTS

We created a path loss matrix for atmospheric conditions of 1 atm and 298 K; the distribution of this matrix is represented in Figure 2. In this figure, the 'windows' of H2O absorption can be seen in dark red, indicating sections of absorbance that result in significant power loss. This allows us to determine the frequency ranges that should be avoided.

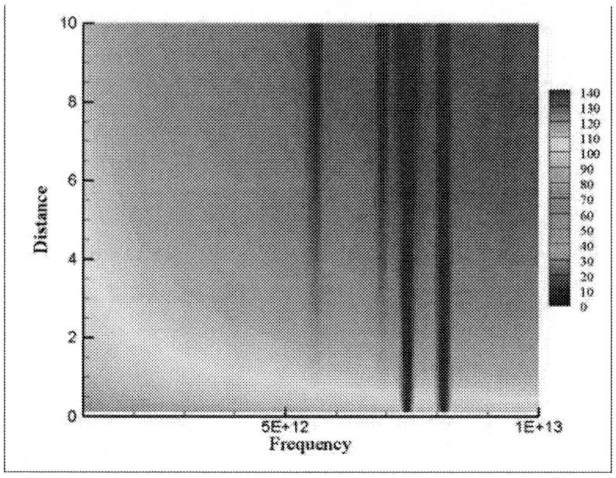

Figure 2 - A color-intensity distribution found for sample power matrix Image Credit to Authors

IV. CONCLUSION

We developed a method of finding the optimal THz frequency for a certain set of parameters within a short transmittance distance by creating and optimizing a power matrix.. The information contained in this power matrix can help shape future designs of THz detection and ranging devices by directing the specific frequencies at which power loss is minimized **[3]**. Future research should investigate the application of this power matrix to specific devices, and how they will minimize path loss based on these specific conditions. In addition, this can be applied, specifically, to LIDAR detections systems within the THz range to allow for optimization of frequency and power **[4]**, **[5]**.

REFERENCES

[1] C. Han, I.F. Akyildiz. "Distance-aware multi-carrier (DAMC) modulation in terahertz band communication." *Proc. IEEE International Conference on Communications.* 2014.

[2] T. Kürner, "THz Communications – Approaching Wireless 100 Gbit/s", in Proc. International Symposium on Future THz Technology FTT 2012, Nara, Japan, November 2012.

[3] Jornet, Josep Miquel, and Ian F. Akyildiz. "Channel modeling and capacity analysis for electromagnetic wireless nanonetworks in the terahertz band."*Wireless Communications, IEEE Transactions on* 10.10 (2011): 3211-3221.

[4] Ian F. Akyildiz, Josep Miquel Jornet, Chong Han. "Terahertz band: Next frontier for wireless communications." *Physical Communication* on 12(2014) 16-32.Holger, Karl. "Wireless Transmission." *Mobile Communications.* 1.3st ed. Computer Networks Group Universität Paderborn, 2008.

[5] Ho-Jin Song; Nagatsuma, T.: "Present and Future of Terahertz Communications," IEEE Transactions on Terahertz Science and Technology, vol.1, no.1, pp.256-263, September 2011.

EPJ Web of Conferences **119**, 06010 (2016)
ILRC 27
DOI: 10.1051/epjconf/201611906010

TIME-CORRELATED SINGLE-PHOTON COUNTING RANGE PROFILING OF MOVING OBJECTS

Julia Hedborg*, Per Jonsson, Markus Henriksson, Lars Sjöqvist

*Dept. of Electro-optical Systems, Swedish Defence Research Agency, FOI, PO Box 1165, SE-581 11 Linköping, Sweden. *Email: julia.hedborg@foi.se*

ABSTRACT

Time-correlated single-photon counting (TCSPC) is a laser radar technique that can provide range profiling with very high resolution. Range profiles of multiple surface objects and geometrical shapes are revealed using multiple laser pulses with very low pulse energy. The method relies on accurate time measurements between a laser pulse sync signal and the registration of a single-photon event of reflected photons from a target.

TCSPC is a statistic method that requires an acquisition time and therefore the range profile of a non-stationary object (target) may be corrupted. Here, we present results showing that it is possible to reconstruct the range profile of a moving target and calculate the velocity of the target.

1. INTRODUCTION

The TCSPC technique offers the possibility to measure range profiles of stationary targets with resolution in the order of centimeters [1]. These profiles can be used for recognition and classification of remote optical systems or other reflecting objects.

The principle of TCSPC range profiling is based on time-of-flight measurements between the emitted laser pulse and the detection of a single photon. The registered photons from detection events are sorted into time bins according to their arrival time relative to the latest emitted laser pulse. These are typically presented as histograms with time bins on the x-axis and counts on the y-axis. The number of detected laser pulses should be less than 5% in order to avoid detector "pile-up" effects and fulfil the Poisson statistic requirements [2]. This enables usage of very low laser power that may be an advantage for some applications and facilitates meeting eye-safety requirements.

A measurement of a stationary target, like a mirror or an optical system, will give a characteristic range profile (or signature), where multiple closely spaced surfaces are resolved. However, if the target is moving, the profile is blurred over the time bins. Since the resolution is very high, a small movement is enough to corrupt the profile.

The aim of this work was to investigate the possibility of measuring range profiles of moving targets with a TCSPC system. By using the time information provided by the timing acquisition hardware we can reconstruct the characteristic range profile and also measure the velocity of the moving target.

2. SYSTEM DESCRIPTION

The main components of the experimental system are a picosecond pulsed laser, a transceiver unit for emittance of laser pulses and collection of reflected photons, a single-photon silicon detector and the timing acquisition hardware (Figure 1).

The laser is a super continuum (SC) fiber laser (Fianium Ltd. UK) emitting radiation from 450 to 1900 nm. It operates at 40 MHz pulse repetition frequency (or 25 ns pulse intervals) and has a pulse duration of 4 ps. An acousto-optical (AO) filter is used to select the wavelength and reduce the emitted laser power. A fixed wavelength of 832±5 nm was used in this work.

The detector is a SPAD (Single Photon Avalanche Diode) detector from Micro Photon Devices S.r.l. Italy, with a 100 µm pixel diameter, <250 dark counts/s, 87 ns detector dead time and <50 ps timing jitter. A band pass filter (λ_c=834 nm, $\Delta\lambda$=3 nm) is mounted in front of the detector to reduce the background radiation.

The timing acquisition hardware (PicoHarp300 from PicoQuant GmbH, Germany) registers the time between every photon detected by the SPAD device and the last electrical synchronisation pulse from the SC fiber laser.

A transceiver module with a wide field of view (FOV) adapted for short range laboratory measurements was used. The module is monostatic with the same optics for transmitting and collecting photons having a combined FOV of approximately 5 mrad. A CCD camera mounted next to the transceiver helped in aiming the system on the target.

The laser pulse frequency of 40 MHz, gives a range interval of 7.5 m between pulses and an ambiguity distance of 3.75 m for the system. The absolute distance to an object can be determined e.g. by using pseudo-random modulation of the laser pulse train [3, 4], but that feature is not implemented in the current set-up.

A more thorough description of the system is found in [5], except for the new monostatic wide FOV transceiver head.

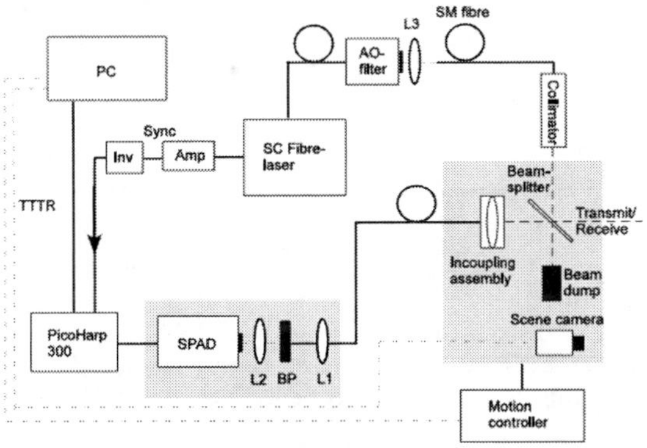

Figure 1. Overview of the TCSPS system used for the experiments. The transceiver module was mounted on a turn table. Abbreviations: SM (single mode), Amp (amplifier), Inv (inverter), TTTR (time-tagged time-resolved signal), Sync (laser pulse sync signal), L1 (collimating lens), L2/L3 (focusing lenses).

3. EXPERIMENTAL METHOD

For the experiments an indoor measurement facility was used. At a distance of about 50 m from the TCSPC transceiver, two different rails were positioned on which the targets were placed.

The first rail is 50 cm long with a motor driving step with constant velocity at the speed of a few cm/s. The second rail is 5 m long (longer than the ambiguity distance of 3.75 m) and the targets were moved by pulling a line tied to a wagon comprising the mounted targets. The velocity was not constant and the rail not completely straight giving a slightly wobbly movement of the target.

The targets used for measurements were corner cube retro reflectors (CCRR) that provide strong reflections and a diffuse cone that gives a much weaker signal. First, static measurements were performed to provide reference range profiles. Subsequently measurements during movement of the targets on the two different rails were carried out.

The PicoHarp software was used to collect experimental data and in-house developed MATLAB routines were used off-line to analyse the results. The length of each time bin was set to 8 ps.

By using the PicoHarp TTTR (time-tagged time-resolved signal) mode for data recording both the time relative to the latest sync pulse and the number of sync pulses from the start of the measurement are recorded for every detected photon. The scheme enables reconstruction of the range profile of the moving object. This was done by sorting the signals into time intervals much shorter than the total acquisition time, and creating a histogram for every time interval (Figure 2).

By cross correlating the histogram from each time interval with the first one, the number of time bins the target has moved compared to the first time interval was calculated. This gives the distance as a function of time and by fitting a polynomic function to these values, a distance correction of every single detection event is calculated.

Though the number of counts for every time interval is low and the histograms noisy, a range profile that is very similar to a static measurement profile can be achieved. The reconstruction of the time intervals shown in Figure 2 is presented in Figure 3.

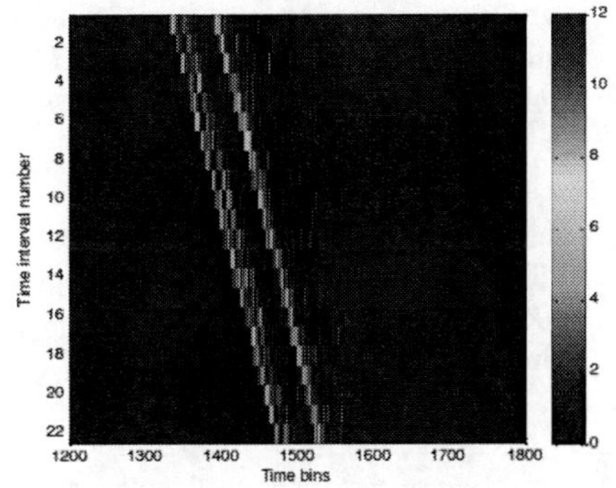

Figure 2. Histograms of every time interval (y-axis) plotted as a picture. The color scale represents number of counts for every time bin (x-axis). The diagonal shape is a result of the movement of the object, in this case three CCRRs.

4. RESULTS

To illustrate the reconstruction and the resolution of the system three different CCRRs were placed close to each other and moved on the short rail at a velocity of 0.025 m/s. CCRRs give strong reflections, but the signals were filtered to resemble longer distance or higher speed than was used. This results in few counts/s and it is not possible to achieve a good range profile by shortening the acquisition time because of the low count rate or by averaging over longer time intervals because of the movement. After the reconstruction, the range profile appears and is almost identical to a static measurement of the profile. This is illustrated in Figure 3 where one of the time interval histograms from Figure 2 is plotted together with the reconstructed and a static profile. The distance between the first two peaks (closest two CCRRs) of the range profile is 1.7 cm.

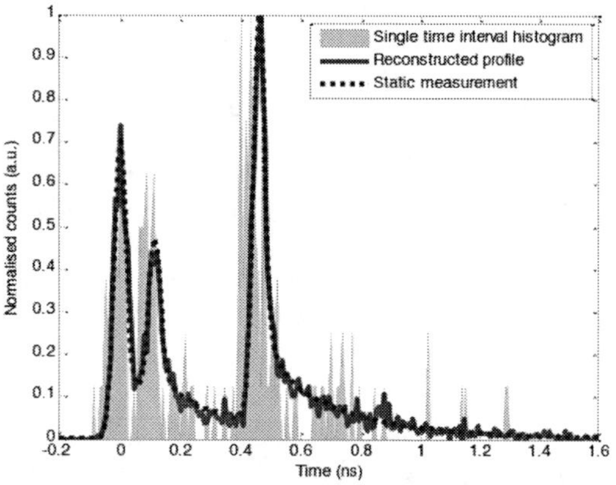

Figure 3. Three corner cubes closely placed on the short rail measured during movement. The gray area is one of the time interval histograms from Figure 2.

A static measurement of a single CCRR determines the system response and its FWHM (full width at half maximum), that is a measure of the system resolution, was 49 ps. This corresponds well to the specification of the detector, which gives the largest time jitter of the system components. For a more thorough characterisation of the system response, see [5].

The CCRR was subsequently mounted on the short and long rail, respectively, and studied during movement. The range profiles after reconstruction compared with the static profile are shown in Figure 4. The non-compensated histogram from this measurement before the reconstruction is also shown illustrating the broadening from 0.2 m movement during the measurement.

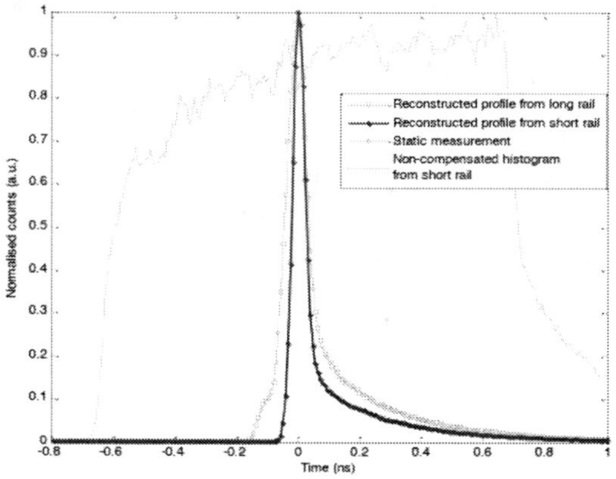

Figure 4. Range profile of the CCRRs, from static measurement and reconstructed range profiles from the measurement during movement on the long and short rails. The dotted line is the histogram of a moving CCRR as it looks before reconstruction.

The reconstructed range profile of the CCRR measured during movement on the short rail has the same FWHM as the static measurement. The velocity was calculated to 0.025 and 0.050 m/s for two measurements at different speed settings. The reconstruction works excellent for both cases. Because of the strong signal a sampling time less than 0.05 s is required to achieve a characteristic range profile studying the CCRR.

On the longer rail, the movement was not controlled and nonlinear. However, the signals were successfully reconstructed for the CCRR. The FWHM was 81 ps which is broader than the static measurement, but the range profile is still recognisable (see Figure 4). The broadening is due to the non-linear movement that was not included in the fitted curve used for the reconstruction. The difference of fitted curve and the real velocity is illustrated in Figure 5. The average velocity for this measurement was 0.32 m/s.

The studied cone was 30 cm high and had a base diameter of 30 cm with a white diffuse surface. Since the reflection of the cone is weak compared to the CCRR, longer time intervals (0.2-0.3 s) was required for the reconstruction in

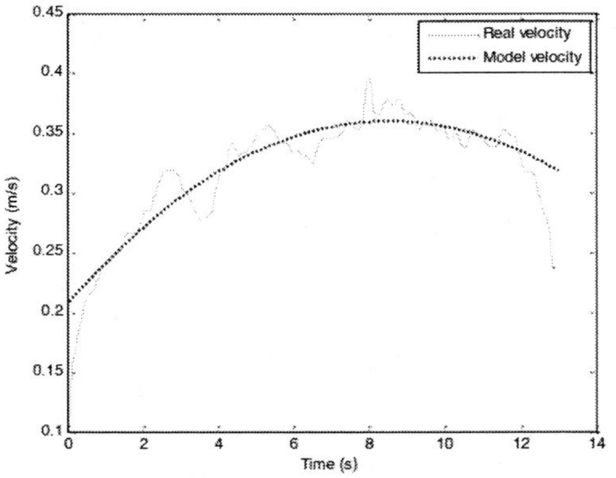

Figure 5. Speed as a function of time for the corner cube traveling on the long rail. The dotted line represents the model curve used for the reconstruction.

order to obtain histograms useable for the cross-correlation. Figure 6 shows the cone profile from the static measurement and the reconstructed profile from the measurement during movement on the short rail.

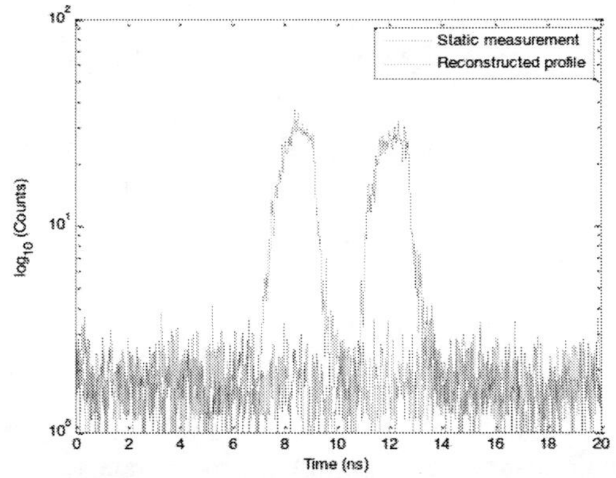

Figure 6. Range profiles for the cone from the static measurement (blue) and the reconstructed range profile from measurement during movement (green). The curves were filtered with a gaussian filter to reduce some noise.

There is a slight difference between the two profiles, which are time shifted for better visualisation. The FWHM of the static cone profile was 1.47 ns and about the same for the reconstructed signal. The average velocity was calculated to 0.027 m/s, a little higher than the speed calculated from the cube measurement which should be more correct. A measurement at 0.05 m/s was also performed with similar result.

5. CONCLUSIONS

The TCSPC system enables high-resolution characterisation of remote, retro reflecting or reflecting objects using low power eye-safe laser illumination. The presented results show that it is possible to reconstruct range profiles of moving object with this technique. This technique works even when the necessary integration time is longer than the period for which the movement is negligible, i.e. for objects with low signal strength compared to the velocity. Reconstruction of the signal requires no prior information of the original range profile and the instantaneous and average velocity of the object can be calculated. Objects moving longer distances than the ambiguity distances can also be compensated with respect to velocity induced broadening.

With further developed reconstruction algorithms, objects moving at greater velocity or in more nonlinear paths should be possible to identify or classify.

ACKNOWLEDGEMENT

This work has been supported by the Swedish Armed Forces research program and the Swedish Defence Materiel Administration (FMV).

REFERENCES

[1] G. Buller, A. Wallace, 2007: Ranging and Three-Dimensional Imaging Using Time-Correlated Single-Photon Counting and Point-by-Point Acquisition, *IEEE J. Sel. Top. in Quant. Electr.*,**13**(4), 1006-1015.

[2] G. Buller, R. Collins, 2010: Single-photon generation and detection, *Meas. Sci. Technol*, **21**(1), 012002.

[3] N. J. Krichel, A. McCarthy, G. S. Buller, 2010: Resolving range ambiguity in a photon counting depth imager operating at kilometer distances, *Opt. Express*, **18**(9), 9192-9206.

[4] P. Hiskett, C. S. Parry, A. McCarthy, G. S. Buller, 2008: A photon-counting time-of-flight ranging technique developed for the avoidance of range ambiguity at gigahertz clock rates, *Opt. Express*, **16**(18), 13685-13698.

[5] L. Sjöqvist, M. Henriksson, P. Jonsson, O. Steinvall, 2014: Time-correlated single-photon counting range profiling and reflectance tomographic imaging, *Adv. Opt. Techn.*, **3**(2), 187-197.

EPJ Web of Conferences **119**, 06011 (2016) DOI: 10.1051/epjconf/201611906011
ILRC 27

HIGH FIDELITY IMAGING ALGORITHM FOR THE UNDIQUE IMAGING MONTE CARLO SIMULATOR

Grégoire Tremblay [1]*, Gilles Roy [2]

[1]AEREX Avionics, 324 St-Augustin av., Lévis, Qc, Canada, G6Z 8G8, gtremblay@aerex.ca
[2]DRDC Valcatier, 2459 de la Bravoure, Québec, Qc, Canada, G3J 1X5

ABSTRACT

The Undique imaging Monte Carlo simulator (Undique hereafter) was developed to reproduce the behavior of 3D imaging devices. This paper describes its high fidelity imaging algorithm.

1. INTRODUCTION

The field of 3D imaging is evolving rapidly. Advance in Flash ladar [1-2], simplified the task of laser 3D imaging. These technologies can be used for observation of targets through foliage [1], aerosol [3], or netting [2].

We developed Undique in an effort to understand the influence of these phenomena on Flash lidar image quality. The simulator reproduces not only the scattering of the photon in the propagation range but also 3D solid targets and basic receiver behavior.

In section 2 we give a brief description of Undique capabilities. Section 3 describes the high fidelity imaging algorithm. Finally, section 4 presents some examples of simulated results.

2. UNDIQUE

Undique (*un-di-kwe*) is a latin word meaning “from all parts”, a name quit adequate when propagating photons in aerosols. It is a versatile tool developed to reproduce the behaviour of different types of active imaging system such as classical lidar system, passive 2D imager, active imager, or range gated 3D imager. To simulate these systems, we need a 2D gated detection system, a source, and a propagation range. In the case of 3D range gated imager, a solid 3D target is also simulated. Fig. 1 shows a schematic view of the multiple components emulated by the system.

2.1 Simulated sub systems characteristics

Undique generates individual photons. The source is monochromatic but its wavelength can be adjusted before simulations. The photons polarization state is defined using the Stokes parameters. To simulate a gated camera, we keep a record of the total distance travelled by the photons.

The source can be located anywhere in the simulated environment. Multiple sources or pre-defined source patterns such as bar patterns of radial patterns can be used.

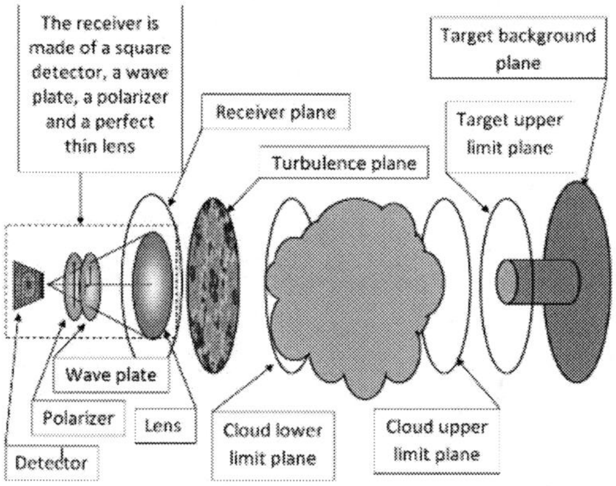

Fig. 1. The Undique Monte Carlo simulation environment. We show the various components of the system.

The gated camera is made of a perfect thin lens and a square detector. The aperture diameter, the focal length and the focus of the camera can be changed. To provide polarization analysis capability, a polarizer and a retarder are located in front of the detector.

The propagation range is the distance between the camera and the solid target. It is divided in a series of propagation planes. Each plane generates a specific behavior. The simulation basic planes are the camera’s aperture plane, the solid target upper limit, and the target background plane. We also have planes to define the obscurant limits and for turbulence effects (still under development). Photons propagate from one plane to the next. When it arrives at a plane, actions are taken and propagation is modified accordingly.

The solid target is at the end of the propagation range. It is a 3D structure that rests on an infinite solid background. We can have up to 20 individual targets. These targets can be cylindrical, rectangular, or have specific regular patterns. Albedo and BRDF of solid targets can be changed.

Undique relies on phase function information to scatter the photons. The first aerosols included in the simulator were spherical water droplets described by the Mie theory. The aerosols can be monodispersed or have a normal, log normal, bi-log normal, or gamma distribution.

The simulator can also use Rayleigh scattering, Henyey Greenstein phase function and user defined phase function for atmospheric propagation. In water, Fournier Forand and pure water phase functions are available.

Aerosols can be located anywhere in the propagation range. It can span from the camera to the target or be concentrated in a determined position. We can simulate up to three clouds having either uniform, linearly changing, or parabolic extinction profile.

2.2 Direct propagation and variance reduction

When used in *direct propagation* mode, the simulator propagates ballistic photons until they are captured by the camera or reach a predetermined delay. In order to increase the throughput of the simulator, the *variance reduction* mode generates events every time an interaction occurs within the system Field Of View (FOV). Three different variance reduction methods are used. The first method is the well-known semi-analytical method [4] also known as the expected value method [5]. The second is an importance sampling method used to increase the amount of events occurring in front of the receiver. The third method is splitting and Russian roulette used to produce a defocusing effect.

3. HIGH-FIDELITY IMAGE GENERATION

Conventional atmospheric Monte Carlo [6] are usually coaxial with the source, focused at infinity, and observe symmetrical system. The camera aperture is not considered in the imaging process. The errors caused by these approximations are usually less important than the intrinsic variability of the simulation.

In recent years, active systems have been integrated in various industrial processes [7]. The aperture diameter can represent a few percent of the total range. In these conditions, the photon distribution over the camera's aperture can have a significant impact. When simulating engulfed system, the aperture's contribution is paramount.

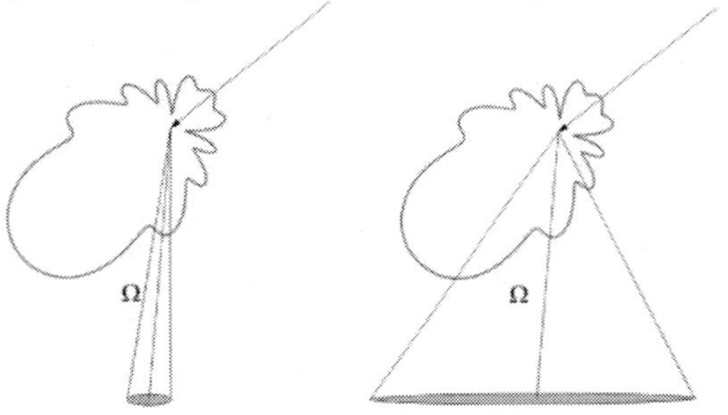

Fig. 2. Phase function projection over a small (left) and a large (right) solid angle Ω.

In *direct propagation* mode, ballistic photons touching the camera aperture are focused on the detector respecting the optical properties of thin lenses. In *variance reduction* mode, a sub photon having the probability to reach the receiver is send toward it. This probability is provided by the phase function. Fig.2 shows the two possible cases. On the left, the range is large and the camera's aperture sustains a small solid angle Ω. On the right the solid angle sustained by the aperture is large and the projection of the phase function changes considerably over it. When the camera's aperture is small compared to the distance to travel, the probability to go toward the small solid angle Ω is

$$P_\Omega = \frac{F(\theta,\varphi)}{d^2}\pi r^2_{aperture} \quad , \qquad (1)$$

where $F(\theta, \varphi)$ is the phase function as a function of elevation θ and azimuth φ, d is the range and r the aperture radius. When the camera's aperture is larger, the phase function projection over the aperture varies. The scattered light can go toward a larger Ω and the new probability is

$$P_\Omega = \frac{1}{d^2}\int_{\varphi=\varphi_0}^{\varphi_1}\int_{\theta=\theta_0}^{\theta_1} F(\theta,\varphi)\sin\theta d\theta d\varphi \ . \qquad (2)$$

This calculation can be perform faster by putting $F(\theta, \varphi)\sin\theta d\theta$ in a look-up-table (LUT).

We can either produce an exact mapping of the aperture's probability distribution to locate the photon or use an alternate, less computer time consuming strategy. We chose the latter and we explain how we proceeded in the next section.

3.1 The photon pointing into the receiver

In Fig. 3, an interaction occurs at point A inside the camera FOV, and the photon travels along line AB; point B is inside the aperture. The circumference of the aperture is divided in 24 sections of 15°. They are used to define 24 slices having point B as vertex. Knowing the interaction coordinate in A, the contact point coordinate in B, and the edge coordinate of a slice in E, we can find the elevation angle θ defined by these three points. Knowing the azimuthal angle $d\varphi$ sustained by the slice, we can find the probability to be scattered in that slice using Equ. 2. The operation is repeated for all the slices and we get a probability distribution. A random number is drawn to determine the slice touched by the photon. Using the LUT as a probability distribution along elevation, a second random number is used to find the photon location in θ and a third is used to determine φ.

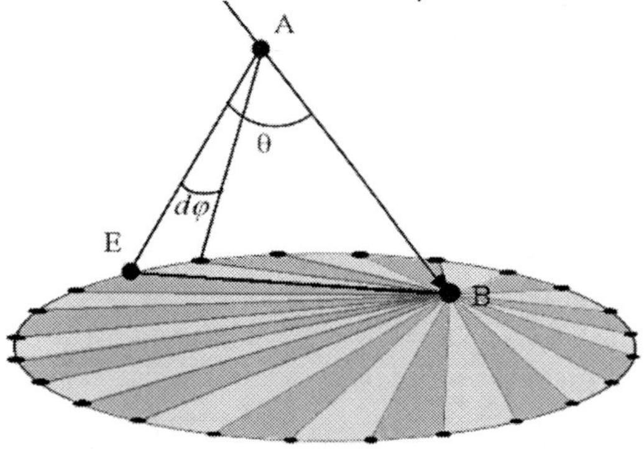

Fig. 3. Schematic view of the photon distribution on the camera's aperture in variance reduction: the photon pointing inside case. The slices are shown in yellow and blue.

3.2 The photon pointing outside the receiver

This situation is shown in Fig. 4. The interaction between the photon and the aerosol happens at point A but this time the projected location of the photon is outside the aperture (point B). The aperture is divided in 24 slices having the same angular width $d\varphi$. A line is traced between B and E, the far edge of the aperture. The probability to be scattered between point D and E is computed for every slice using θ_1, θ_2, and $d\varphi$ in Equ. 2. From that point, the distribution of the photons on the aperture is similar to the previous case.

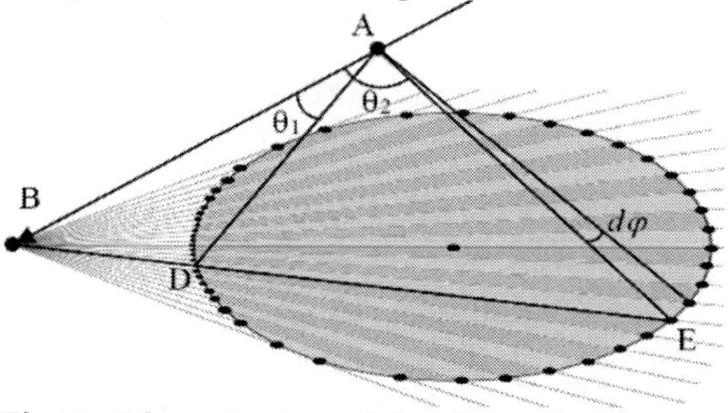

Fig. 4. *Schematic view of the photon distribution on the camera's aperture in variance reduction: the photon pointing outside the aperture case. The slices are shown in yellow and blue.*

4. SOME EXAMPLES

In this section we present simulations made using Undique to demonstrate its versatility. The first example in Fig. 5 represents a solid target made of concentric cylinders seen through aerosol. This was done to assess the impact of aerosol on our capacity to detect 3D shapes through obscuration [8].

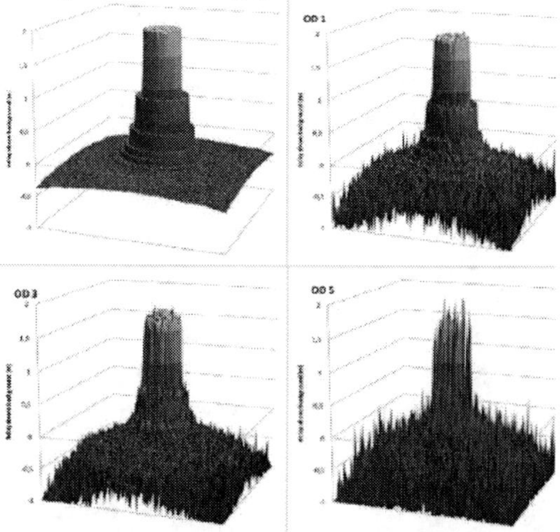

Fig. 5. *Simulation of a solid target seen through an aerosol made of 10 µm water droplets. Top left: OD0. Top right: OD1. Bottom left: OD3. Bottom right: OD5 (from [8]).*

In Fig. 6 we look at a radial source through an F/2 system. The source-camera distance is equal to 100 times the aperture diameter and the FOV is 50 mrad. On the left, the source is focused; on the right, defocus generates the expected phase inversion in the radial pattern. This is only possible if the photons are distributed adequately on the aperture.

In Fig. 7 we show the case of a polarised signal propagating through a test case. The case is chosen because it can be calculated analytically and because it is highly sensitive to small errors in polarization management.

Fig. 6. Phase inversion of a radial pattern generated by defocus of the optical system.

Fig. 7. Polarized source reflected by aerosols in a test case. We see the total reflected signal (top left), the principal polarization(top right), the 45° polarization (bottom left) and the secondary polarization (bottom right).

5. CONCLUSION

The Undique Monte Carlo simulator is a versatile tool that can be used to reproduce most passive and active imager observing a scene under obscuration. Its unique imaging algorithm enables it to reproduce the behavior of actual systems under a wide range of conditions.

REFERENCES

[1] Murray, J.T., Moran, S.E., Roddier, N., Vercillo, R., Bridges, R., Austin, W., "Advanced 3D polarimetric flash ladar imaging through foliage", Proc. SPIE, 5086, 84-95 (2003).

[2] Halmos, M.J., "Eyesafe 3D flash Ladar for target under obscuration", Proc. SPIE, 5086, 70-83 (2003).

[3] Trickey, E., Church, P., Cao, X., "Characterization of the OPAL Obscurant Penetrating LiDAR in various Degraded Visual Environments", Proc. SPIE, 8737, 87370E (2013).

[4] Poole, L.R., Venable, D.D., Cambell, J.W., Semianalytic Monte carlo radiative transfer model for oceanographic lidar systems, Appl. Opt. 20, 3653-3656 (1981).

[5] Spanier, J., Gelbard, E.M., "Monte Carlo Principles and Neutron Transport Problems", Dover (1969).

[6] Marchuk, G.I., Mikhailov, G.A., Nazaraliev, M.A., Darbinjan, R.A., Kargin, B.A., Elepov, B.S., The Monte Carlo Methods in Atmospheric Optics, Springer series in Opical Science, (1980).

[7] Berrocal, E., Meglinski, I., Jermy, M., "New model for light propagation in highly inhomogeneous polydisperse turbid media with applications in spray diagnostics", Opt. Express, 13, 9181-9195 (2005).

[8] Tremblay, g., Cao, X., Roy, G., "The effect of dense aerosol cloud on the 3D information content of flash Lidar", Proc. SPIE, 7828, 78280C (2010).

DOI: 10.1051/epjconf/201611906012

CHALLENGES AND SOLUTIONS FOR FREQUENCY AND ENERGY REFERENCES FOR SPACEBORNE AND AIRBORNE INTEGRATED PATH DIFFERENTIAL ABSORPTION LIDARS

Andreas Fix*, Mathieu Quatrevalet, Benjamin Witschas, Martin Wirth, Christian Büdenbender, Axel Amediek, and Gerhard Ehret

*Deutsches Zentrum für Luft- und Raumfahrt (DLR), Institut für Physik der Atmosphäre, Oberpfaffenhofen, Germany, *Email: Andreas.Fix@dlr.de*

ABSTRACT

The stringent requirements for both the frequency stability and power reference represent a challenging task for Integrated Path Differential Absorption Lidars (IPDA) to measure greenhouse gas columns from satellite or aircraft. Currently, the German-French methane mission MERLIN (Methan Remote Lidar Mission) is prepared. At the same time CHARM-F, an aircraft installed system has been developed at DLR as an airborne demonstrator for a spaceborne greenhouse gas mission. The concepts and realization of these important sub-systems are discussed.

1. INTRODUCTION

The Integrated Path Differential Absorption Lidar (IPDA) technique using hard target reflection in the near IR has the potential to deliver methane and carbon dioxide column measurements from satellite or aircraft with unprecedented accuracy [1].

Carbon dioxide and methane are the two most important anthropogenic greenhouse gases contributing to global radiative forcing and have significantly increased since the beginning of the industrial era. CO2 has grown from a pre-industrial average atmospheric mole fraction of about 280 parts per million volume (ppm) to 396 ppm in 2013 which is an increase of 42 %. The increase of methane was even higher than for CO2: it rose by 153% from ~700 parts per billion volume (ppb) in the pre-industrial period to 1824 ppb in 2013 [2]. CH4 has an estimated global warming potential per molecule 25 times greater than CO2 over a 100 year horizon and, despite its much lower abundance, thus is the second most significant anthropogenic greenhouse gas.

Large uncertainties in their budget, however, and feedback mechanisms which are, if at all, only partly understood, limit the accuracy of climate change projections. In order to reliably predict the climate of our planet, and to help constrain political conventions on greenhouse gas avoidance, adequate knowledge of the sources and sinks of these greenhouse gases and their feedbacks is mandatory. In spite of the recognized importance of this issue, our current understanding about sources and sinks of the gases CO2 and CH4 is still inadequate. In order to improve our current knowledge about methane fluxes the German-French methane mission MERLIN has been initiated. The goal of MERLIN is to measure the spatial and temporal gradients of atmospheric CH4 columns with high precision and low bias. MERLIN, which is currently in its definition phase and whose launch is scheduled for 2019 will be the first IPDA lidar for greenhouse gas monitoring in space. At the same time DLR has accomplished to develop CHARM-F a state-of-the-art airborne demonstrator which is able to simultaneously measure both greenhouse gases, CH4 (at ~1645 nm) and CO2 (at ~1571 nm), onboard the German research aircraft HALO (Fig. 1). Recently, its first test flights have successfully been performed.

Figure 1. Artist's view of MERLIN (source: CNES) and photograph of HALO starting.

In many aspects and sub-systems the instruments resemble each other. Both use e.g. optical parametric oscillators (OPOs) in a double-pulse mode as the transmitter [3]. Of particular importance for both instruments are the sub-modules required for the frequency stabilization of the transmitter wavelength and, since the IPDA technique, in contrast to DIAL, requires the exact knowledge of the energy ratio of outgoing on-line and off-line pulses, the energy calibration chain. Wavelength accuracy and energy calibration represent a potential source of systematic error contribution to the measurement of XCH4 and XCO2, the weighted averages of the greenhouse gas dry-air volume mixing ratio which is the primary data product.

2. REQUIREMENTS

In preparation of the MERLIN mission, feasibility studies have been undertaken (e.g. [4]). The target, breakthrough and threshold mission requirements for the accuracy of XCH4 have been set to 1 ppb, 2 ppb, and 3 ppb, respectively, for an observation averaged over 50 km along track. The allowed target, breakthrough and threshold values for the random error are a factor of 8-12 higher, thus 8 ppb, 18 ppb, and 36 ppb, respectively. Using inverse modelling, such error margins would allow constraining the CH4 observations to the point where regional to continental scale CH4 fluxes can be estimated to generally within less than 20% of their mean values on average [5].

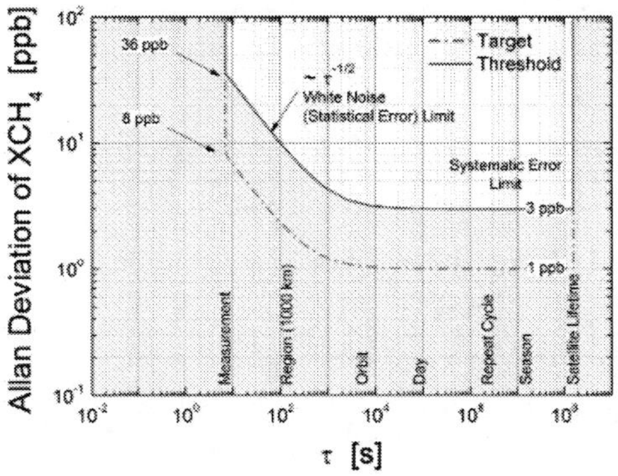

Figure 2. Template expressing MERLIN random and systematic error requirements in terms of Allan deviation.

The concepts of "random error" and "systematic error" may be interpreted in terms of uncorrelated and correlated measurement noise, respectively. It is, however, important on which timescales these specifications have to be met. Therefore, a methodology is implemented that gives rise to the individual timescales involved and serves as a benchmark to assess the observing of requirements. It uses the well-known overlapping Allan variance which is a powerful tool to identify and quantify these different types of noise sources. Figure 2 gives an example of how the requirements are transferred into a template.

3. ENERGY REFERENCE

For the monitoring of the on-line to off-line ratio of outgoing pulse energies an accuracy of the order of $2 \cdot 10^{-4}$ to $8 \cdot 10^{-4}$ is pursued to meet the above given requirements. The preferred set-up would use an identical detector for both lidar signal and pulse energy monitor to circumvent potential detector ageing effects. It is thus important to employ pick-up, attenuation, transmit and detection schemes that do not alter the energy ratio. For this purpose, the use of integrating spheres appear to be beneficial since those are insensitive to beam pointing and intensity profile variations and can be fibre-coupled. Integrating spheres also provide an adequate means for wavelength-independent attenuation. Using established theory [6], the attenuation can be accurately predicted also for double integrating sphere arrangements.

However, given the coherence of the lidar transmitter, speckle effects need to be considered in detail [7]. Therefore, the OPO was used within a laboratory testbed to investigate possible monitoring schemes.

One of the main challenges when determining the accuracy of the energy ratio measurement is the lack of a reference system that may serve to provide the "true" value against which the experimental set-up can be referenced. Therefore the energy ratios had to be simultaneously measured twice using the same or different measurement technique and analysed using statistical methods. Again, the Allan variance is used to identify uncorrelated and correlated measurement noise. Since thermal detectors are generally to slow to resolve individual pulses, only detectors based on the pyroelectric or photoelectric effect are to be considered.

In a series of experiments it could be shown that the speckle patterns at the exit port of integrating spheres of two subsequent pulses within a double-pulse sequence of ~ 360 μs appear to be partly correlated. This is exemplarily depicted in Fig. 3. While the Allan deviation of the energy ratio of single pulses with a separation time of 10 ms results in an almost ideal -0.5 slope, this is not true for the double pulse mode.

In order to destroy this correlation, several methods have been evaluated. Reasonable results were achieved by means of reducing the speckles by angular diversity, i.e. the bare fibre ends were mechanically vibrated (Fig. 3). Unfortunately this is no viable option for space use. An alternative approach uses oscillating diffusors that rapidly vary the speckle pattern and which are currently tested within the CHARM-F system.

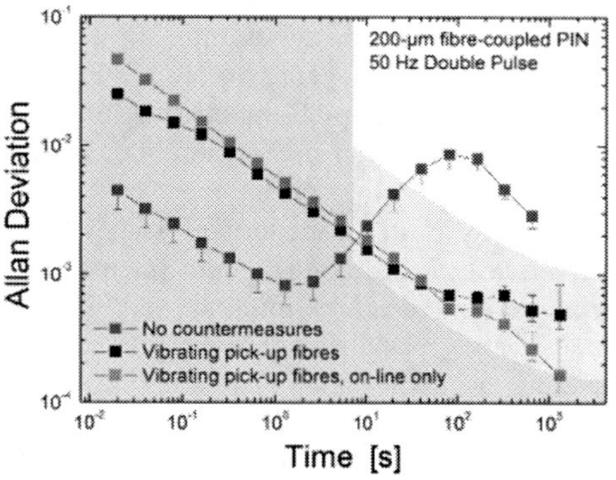

Figure 3. Allan deviations of the pulse energy ratio of two subsequent pulses divided by the same ratio recorded with a second detector. Shown are three cases: without countermeasures against the speckle correlation, and employing vibrating pick-up fibers for the nominal on-/offline sequence and for on-line only, respectively

4. FREQUENCY REFERENCE

Concerning frequency stabilization CHARM-F and MERLIN are following different concepts. CHARM-F's frequency stabilization has been developed consisting of various DFB lasers. In general, two reference lasers are stabilized onto absorption features of CH4 and CO2, respectively, using a multipass absorption cell filled with defined quantities of both gases. The on-line and off-line wavelengths are generated by individual seed lasers locked with a defined offset to their respective references using the offset-locking technique.

The frequency stability for the seed lasers have been verified using a frequency comb. For CH4 and CO2 less than 3 MHz p-p and 70 kHz p-p was achieved, respectively, which fully meets the requirements which are approximately 10 MHz for CH4 and 100 kHz for CO2, respectively.

The OPO cavity is matched to the seed wavelengths using a heterodyne technique whereby the frequency difference between the outgoing OPO pulse and the incoming cw seed radiation is measured on a shot-by-shot basis [8]. This is done by mixing part of the seed light which is acousto-optically frequency shifted with the outgoing OPO pulse by using a fast photo diode. From the beat spectrum of each pulse, its power spectrum is calculated by fast Fourier transform and the error signal for the loop control generated from the difference between the two frequencies.

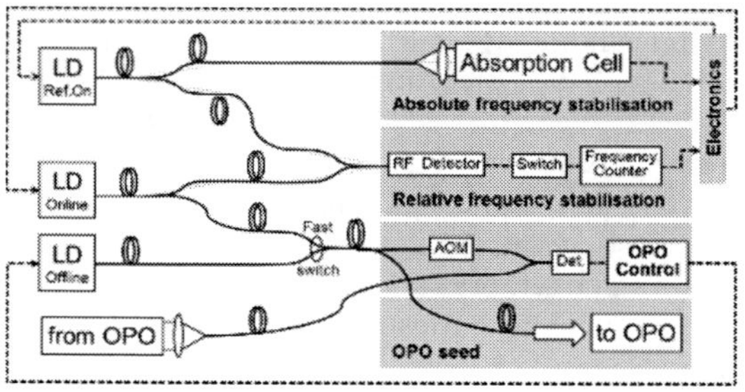

Figure 4. Schematic set-up of the frequency reference of the CHARM-F system. For simplicity only the branch for one greenhouse gas is shown.

Figure 4 shows the beat spectrum for the CH4 transmitter. An OPO bandwidth of ~35 MHz is derived. The offset frequency from the nominal 500 MHz value (given by the AOM frequency) is used for the OPO control.

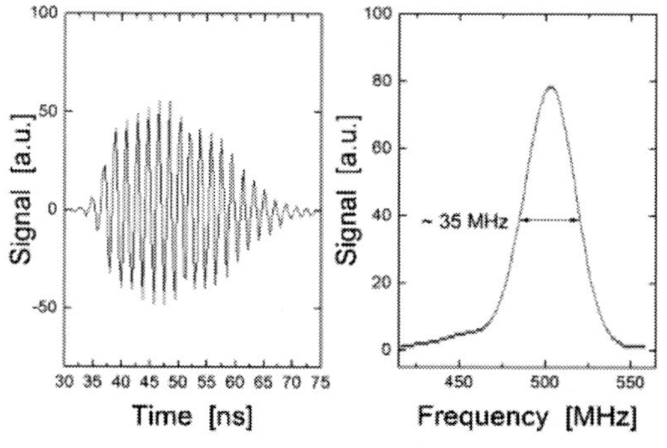

Figure 5. Beat spectrum (left) and power spectrum (right) of the OPO at a wavelength of 1645nm.

For MERLIN, however, a different baseline was chosen that seems more favorable in terms of power demands and recurrent parts for the sake of space qualification. Here, the reference is likewise stabilized to an absorption cell, but relative frequency stabilization and OPO control will be performed using a Fizeau interferometer [9]. The respective fringes are imaged onto a linear

DOI: 10.1051/epjconf/201611906012

InGaAs array. Using a laboratory set-up, first experimental tests have been performed directed to optimize the parameters of the interferometer. Using a frequency comb and changing the wavelength in defined steps the response of the Fizeau interferometer was analyzed. After pre-processing, a long-term stability of < 5 MHz was achieved (Fig. 6).

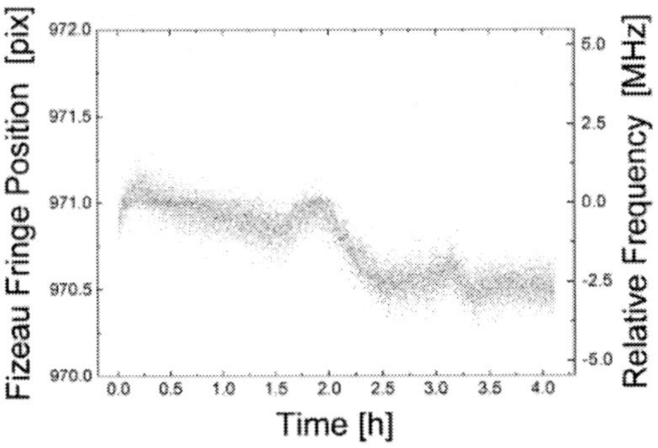

Figure 6. Time series of the peak position of the Fizeau fringe. The left axis shows the respective frequency deviation.

These investigations are to be continued within the design phase of MERLIN.

5. CONCLUSIONS

As a conclusion, frequency and energy reference are fundamental to both airborne and spaceborne IPDA to measure the most important anthropogenic greenhouse gases. Detailed investigations have been performed on both issues within the development and design phase of CHARM-F and MERLIN. The coherence of the lidar transmitter gives rise to speckle effects which have to be considered for the monitoring of the energy ratio of outgoing on- and off-line pulses.

For the frequency reference of CHARM-F, a very successful stabilization scheme has been developed which will also serve as the reference for MERLIN, for which a differing variant was chosen due to space qualification issues. Preliminary results show good prospects. Further investigations are in progress and will be reported.

ACKNOWLEDGEMENT

We would like to H.R. Khalesifard for experimentally supporting the energy monitoring experiments. This work was partly supported by Airbus D+S GmbH and OHB Systems AG on behalf of the German Space Administration, as well as the European Space Agency.

REFERENCES

[1] Ehret, G., A. Amediek, C. Kiemle, M. Wirth, A. Fix, and S. Houweling, 2008: Space-borne remote sensing of the greenhouse gases CO2, CH4, N2O by integrated path differential absorption lidar: a sensitivity analysis, *Appl. Phys.* **B90**, 593–608.

[2] WMO 2014: The State of Greenhouse Gases in the Atmosphere Based on Global Observations through 2013, Greenhouse Gas Bulletin No. 10.

[3] Fix A., 2012: Tunable Light Sources for Lidar Applications, in: U. Schumann (ed.), Atmospheric Physics, Research Topics in Aerospace, Springer, Berlin Heidelberg 2012.

[4] Kiemle, C., M. Quatrevalet, G. Ehret, A. Amediek, A. Fix, and M. Wirth, M., 2011: Sensitivity studies for a space-based methane lidar mission, *Atmos. Meas. Tech.* **4**, 2195-2211.

[5] Ehret, G., P.H. Flamant et al. 2011: MERLIN User Requirements Document, issue: 1-8.

[6] Pickering, J.W., S.A. Prahl, N. van Wieringen, J.F. Beek, H.J.C.M. Sterenborg, and M.J.C. van Gemert, 1993: Double-integrating-sphere system for measuring the optical properties of tissue, *Appl. Opt.* **32**, pp. 399-410.

[7] Boreman, G.D., A.B. Centore and Y. Sun, 1990: Generation of laser speckle with an integrating sphere, *Opt. Eng.* **29**, pp. 339-342.

[8] Fix, A., C. Büdenbender, M. Wirth, M. Quatrevalet, A. Amediek, C. Kiemle, and G. Ehret 2011: Optical parametric oscillators and amplifiers for airborne and spaceborne active remote sensing of CO2 and CH4, *Proc. SPIE* **8182**, pp. 818206.

[9] Cahen, C., J. P. Jegou, J. Pelon, P. Gildwarg, J. Porteneuve, 1981: Wavelength stabilization and control of the emission of pulsed dye lasers by means of a multibeam Fizeau interferometer. *Revue de Physique Appliquée,* **16 (6)**, 353-358.

DOI: 10.1051/epjconf/201611906013

HO:YLF LASER PUMPED BY TM:FIBER LASER

Kohei Mizutani[1]*, Shoken Ishii[1], Toshikazu Itabe[1], Kazuhiro Asai[2], Atsushi Sato[2]

[1]*National Institute of Information and Communications Technology, 4-2-1 Nukui-Kita, Koganei, Tokyo 184-8795, Japan, *Email: mizutani@nict.go.jp*

[2]*Tohoku Institute of Technology, 35-1 Yagiyama-Kasumi, Taihaku-ku, Sendai, Miyagi 982-8577, Japan*

ABSTRACT

A 2-micron Ho:YLF laser end-pumped by 1.94-micron Tm:fiber laser is described. A ring resonator of 3m length is adopted for the oscillator. The laser is a master oscillator and an amplifier system. It is operated at high repetition rate of 200-5000 Hz in room temperature. The laser outputs were about 9W in CW and more than 6W in Q-switched operation. This laser was developed to be used for wind and CO_2 measurements.

1. INTRODUCTION

2-micron solid lasers doped with Tm and/or Ho are suitable for remote sensing of wind, CO_2 and H_2O[1]. We have developed 100mJ pulse energy class laser diode(LD)-side pumped Tm,Ho:YLF oscillator and 460mJ pulse energy Tm,Ho:YLF amplifier [2,3]. The pumping module of these lasers used composite rod and 28 LD. It was heavy and high cost. We cooled the rod conductively to -80C in the vacuum container for effective laser output. For easy operation and accessibility, we developed moderate output compact lasers with a pumping module using non-composite rod and 12 LD. The pumping module was also cooled to -80C. The output pulse energy is 50-100mJ at 30Hz and it was used for a CO_2 DIAL/Wind Doppler lidar system (Co2DiaWiL) [4]. Another method to get high average output power is to use high power CW Tm:fiber lasers emitting around 1940nm as a pumping source. Efficient pumping of Ho:YLF rod by Tm:fiber laser is possible[5,6]. High pulse repetition 2.05-2.06 micron Ho:YLF laser end-pumped by Tm:fiber laser were developed for CO_2 and wind measurements at [7,8,9]. Advantages of 1.9-micron pumping are low thermal load and possibility of the use of room temperature rod. Commercial pumping sources are available. High repetition is an advantage for CO_2 coherent DIAL measurements. Then, we are developing such Ho:YLF laser for coherent measurements of CO_2 and wind. We describe the status of the development of Ho:YLF laser end-pumped by Tm:fiber laser here.

2. EXPERIMENTAL SET UP

A random polarized Tm:fiber laser of 50W maximum output at 1940nm (IPG Photonics) is used to end-pump the laser rod of a Ho(0.5%):YLF crystal of 4mm diameter x 40mm length. The rod in the pumping module is conductive-cooled from three heat Cu sinks which are cooled by water (Figure 1).

Figure 1 Pumping module of Ho:YLF

A long laser pulse width is needed to achieve narrow line width for coherent lidar applications. Adoption of long resonator is an easy way to get a long pulse. Ring resonator is also needed to avoid spatial hole burning. Then, we adopted a 3m ring resonator for the laser oscillator. The system configuration of the ring oscillator and also the amplifier is shown in Figure 2. S-polarized beam of the Tm:fiber laser is reduced to the size of about 1.2mm diameter by a telescope of two lenses and enters to the resonator. The pump beam passes through a dichroic mirror (DM1) with high transmission for the pump light and high reflectivity for the laser light. After the pump beam is absorbed by the rod, the residual pump light passes through another dichroic mirror (DM2). P-polarized pump beam after a PBS is rotated by 90 degree by a half wave-plate. It is used to pump the amplifier rod in the pumping module of the same design. The pump beam for the amplifier is also reduced in size to match with the oscillator laser output by a telescope of two lenses and passes DM3. The laser output of the oscillator is connected to the amplifier. The unidirectional operation is achieved by a high reflection mirror put after the output coupler of 90%. An acousto-optic modulator is inserted for Q-switched operation. A dichroic mirror (DM3) reflected the oscillator laser light to the amplifier rod. The amplified laser light is reflected by a dichroic mirror (DM4), and the residual pump beam energy passes through DM4 and enters a dumper.

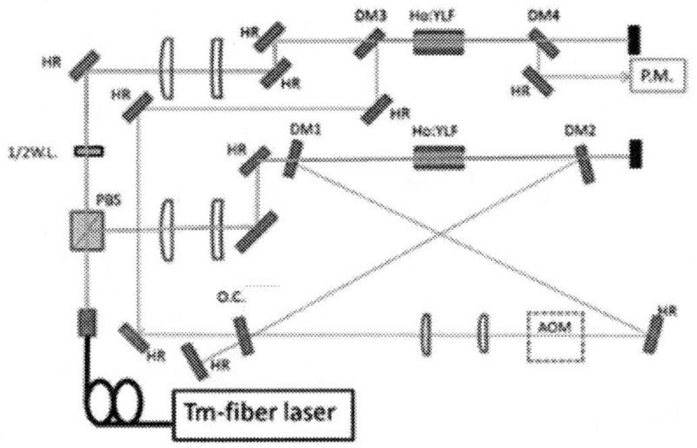

Figure 2 Oscillator and amplifier configuration

3. RESULTS

We show the CW output characteristics in the ring oscillator and also the amplifier in Figure 3. Lenses with focal length of 1000mm and 4000mm were used in the rig resonator. In the maximum input, the oscillator output was 5.6W and amplifier output was 9.1W. The gain by the amplifier is about 1.6. Figure 4 shows oscillator and amplifier output characteristics at CW and 5 kHz Q-switch operations. Lenses with focal length of 1000mm and 5000mm were used in the ring resonator. The maximum output at 5 kHz was 7.5W, compared with 7.9W at CW.

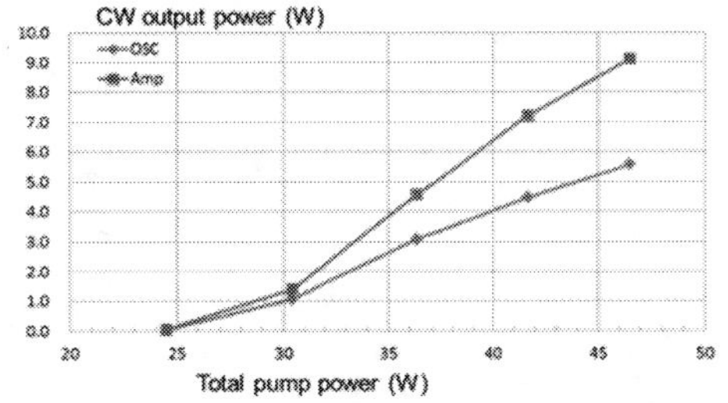

Figure 3 CW output characteristics with lenses of 1000mm and 4000mm focal lengths

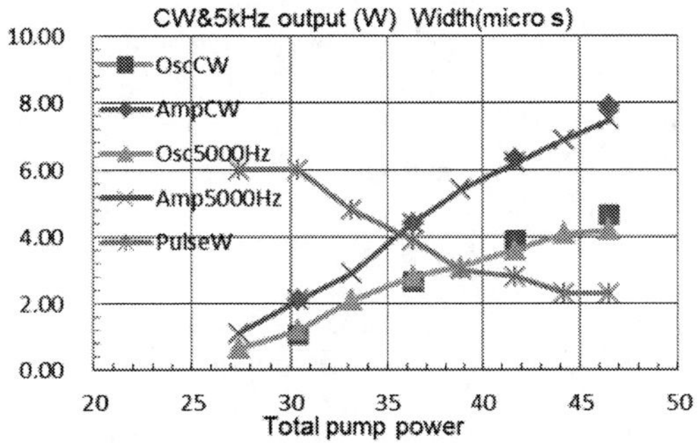

Figure 4 Oscillator and amplifier output characteristics at CW, and 5 kHz Q-SW operation with lenses of 1000mm and 5000mm focal lengths

Slope efficiency at 5 kHz pulse operation was about 33%. Pulse width was measured for the oscillator and decreased according to pump power. The minimum pulse width of 2.3 μs at 5 kHz was obtained at the maximum total pump power of 46.5W. The amplifier output characteristics at 200 Hz to 5 kHz are shown in Figure 5. The output characteristics were similar in more than 1 kHz. According to the decrease of the repetition, output became smaller in 200 – 500 Hz.

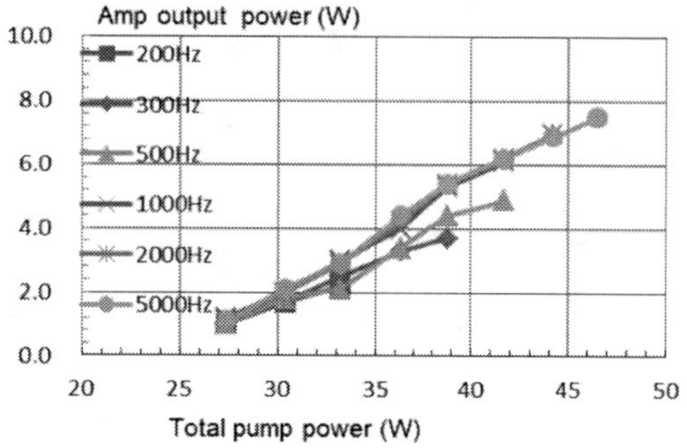

Figure 5. Amplifier output characteristics at 200 Hz to 5 kHz.

We did not make experiment in the pump condition where pulse energy seemed to be too large in view of damage concern. The maximum pulse energy of 12mJ in this experiment was obtained at the repetition of 300 Hz and the total pump power of 38.8W. The oscillator and amplifier pulse energy dependences on repetition rate between 500Hz and 5KHz are shown in Figure 6 for the total pump power of 41.7W. The pulse width is also shown in the Figure 6.

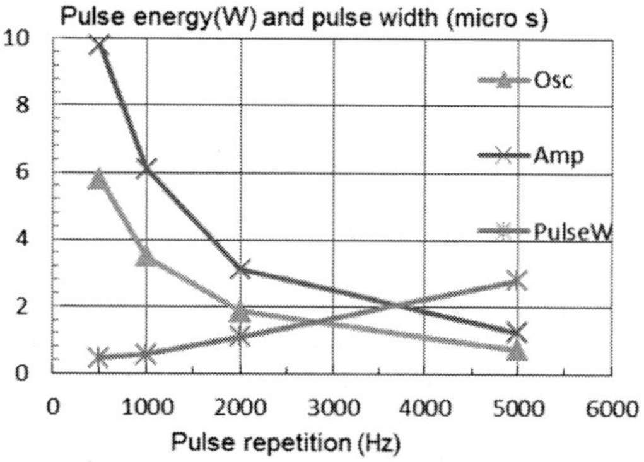

Figure 6. Oscillator and Amplifier output and width dependences on pulse repetition at 500 Hz to 5 kHz for a pump power of 41.7W.

According to the decrease of the repetition, the pulse energy became larger and the pulse width did shorter. The amplifier pulse energy at 500 Hz is 9.8mJ and the pulse width is about 0.44μs. According to increase of pulse energy, the pulse width became short. High peak power may damage some of optical components, especially dichroic mirror DM1 and DM2. We need more consideration about damage in the dichroic mirror to obtain higher pulse energy in the lower repetition. Then we will use a lower reflectance output coupler than 90% for the oscillator.

We will use a ramp-and-fire mechanism [10] for injection-seeding. HR mirror will be removed and the CW seed source will be introduced to the resonator through the output coupler. In this case, unidirectional operation could be induced without the HR mirror after the output coupler.

4. CONCLUSIONS

We made an experiments of a Ho:YLF laser end-pumped by Tm:fiber laser. The ring oscillator and amplifier system was operated at CW and high repetition rate of 200-5000 Hz. The MOPA system showed maximum outputs of 8 - 9W at CW. The average output at repetition more than 1 kHz was similar to that at CW. Pulse outputs of more than 6W were obtained at more than 1 kHz. However efficiency was smaller at 200 – 500 Hz. The maximum pulse energy of about 12mJ was obtained at the repetition of 300 Hz and the total pump power of 38.8W. We could expected higher pulse energy in the repetition rata of 200 – 1000 Hz, if we used maximum pump power. The use of lower reflectance output coupler will permit higher pulse energy. More consideration about damage concern is important to achieve higher pulse energy. This laser will be used for wind and CO_2 concentration measurements. High repetition is especially suitable to CO_2 coherent DIAL observations.

REFERENCES

[1] Koch G.J., M. Petros, J. Yu, 2001: Coherent DIAL with Ho:Tm:YLF Laser for measurement of wind, water vaper, and carbon dioxide, Proc. 11th Coherent Laser Radar Conference, 130-133.

[2] Mizutani K., et al., 2005: Development of conductive cooled 2micron lasers, Proc. 13th Coherent Laser Radar Conference, 32-35.

[3] Mizutani K., et al., 2006: Development of conductively cooled 2micron laser oscillators, Proc. 23rd International Laser Radar Coherent Conference, 1037-1038 .

[4] Ishii S., et al., 2012: Development of coherent 2-μm differential absorption and wind lidar with laser frequency offset locking technique, Proc.

SPIE 7860, 786004-1-6 .

[5] Dergachev A., P. F. Moulton, T. E. Drake, 2005: High-power, high-energy Ho:YLF Laser Pumped with Tm:fiber laser, in Advanced Solid-State Photonics, OSA Trends in Optics and Photonics , 98, 608-612.

[6] Koen, W., C. Bollig, H. Strauss, M. Schellhorn C. Jacobs, M. J. D. Esser, 2010: Compact Fiber-laser-pumped Ho:YLF oscillator-amplifier system, Applied Physics B 99, 101-106.

[7] Singh, U.N., Y. Bai, J. Yu, M. Petros, P. Petzar, B. Trieu, H. Lee, 2009: High repetition rate pulsed 2-micron laser transmitter for coherent CO2 DIAL measurement, Proc. 15th Coherent Laser Radar Conference, 77-80.

[8] Edouart D. ,Gibert F., Mounier F. Le, Bruneau D., Flamant P., 2010: 2-Micron high-repetition rate laser transmitter for biosphere-atmosphere flux measurements, Proc. 25th ILRC ISBN 978-5-94458-109-9, 31-34.

[9] Gibert F., Edouart D. , Cenac, C., Mounier F. Le, 2014: 2-μm high frequency single-mode Q-switched Ho:YLF laser for DIAL application, Appl. Phys. B116, 967-976.

[10] Henderson S.W., Yuen E. H., Fry E. S., 1986: Fast Resonance-detection Technique for Single-frequency Operation of Injection-seeded Nd:YAG Lasers, Optics Letters, 11, 715-717.

LIDAR ARCHITECTURE FOR HARSH ENVIRONMENT APPLICATIONS

Philip Church[1]*

*[1]Neptec Technologies Corp. Ottawa, Ontario, Canada, *Email:pchurch@neptec.com*

ABSTRACT

An overview is provided of the obscurant-penetrating OPAL lidar sensor developed for harsh environments, including poor visibility conditions. The underlying technology, hardware and software architecture of the sensor are presented along with some examples of its software modules' applications. The paper also discusses the performance of the OPAL in the presence of various types of obscurants.

1. INTRODUCTION

Automation is part of a rapid technological evolution in many industries. Several benefits include productivity increase, allowing for an improvement in workplace safety, and facilitating significant gains through reduced maintenance requirements and managing potential labour shortfalls [1], [2]. Automation in heavy machinery industries, such as mining and construction, will require vision sensors capable of operating in harsh environments and see through obscurants. Neptec Technologies Corp. has developed the OPAL lidar product line in fulfillment of this requirement.

2. OPAL Lidar Architecture

The OPAL lidar sensors are designed as compact, rugged, multi-purpose 3D sensors, specifically built for harsh environments running real-time applications. OPAL overcomes two major limitations of traditional lidar when operating in harsh environments; object perception in the presence of obscurants and the ability to withstand severe environmental conditions in the field.

The lidar sensor is environmentally sealed (IP67) with no external moving parts, has an operating temperature range of -40° to +65°C, with no fans or heaters for improved reliability, and is ruggedized to withstand significant vibration and shock levels while operating. The OPAL comes in two distinct designs for the scanning field-of-view (FOV): a panoramic 360° FOV and a conical 60° or 90° FOV.

2.1 Lidar Hardware

Figure 1 illustrates an OPAL-120 with the conical FOV along with its main hardware architecture. The lidar operates at a wavelength of 1540nm and achieves a Pulse Repetition Frequency (PRF) of up to 200kHz.

The OPAL lidars produce scan patterns using Risley prisms under independent motor control. Figure 2 shows an example of increased pattern density coverage for scanning periods ranging from 100ms to one second. The quality of the pattern coverage is highly dependent upon the selection of the motor speeds. A given motor speed ratio will uniquely define a particular scan pattern. Integer speed ratios will typically cause the patterns to repeat at the same locations from one scan to the next. Patterns that incorporate phase variations between scans provide better uniform coverage as repetitious patterns are minimized. This is shown quantitatively in Figure 3 where a Figure of Merit (FOM) is developed to measure the uniformity of a specific scan pattern; the larger the FOM, the less uniform the pattern. It can be interpreted from the graph that scan patterns corresponding to motor speed ratios that are integers will not provide good coverage uniformity.

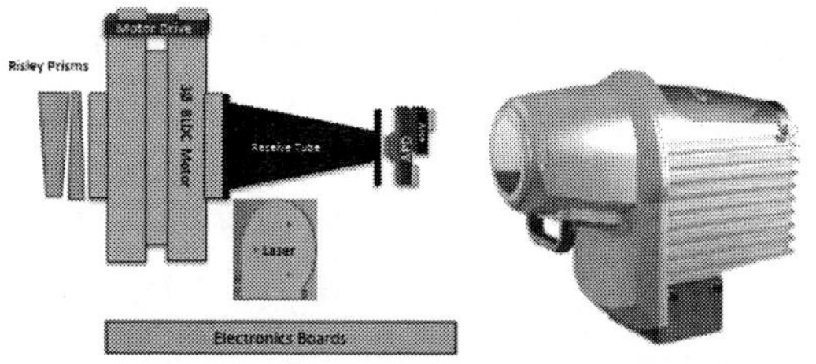

Figure 1: OPAL-120 Hardware Architecture

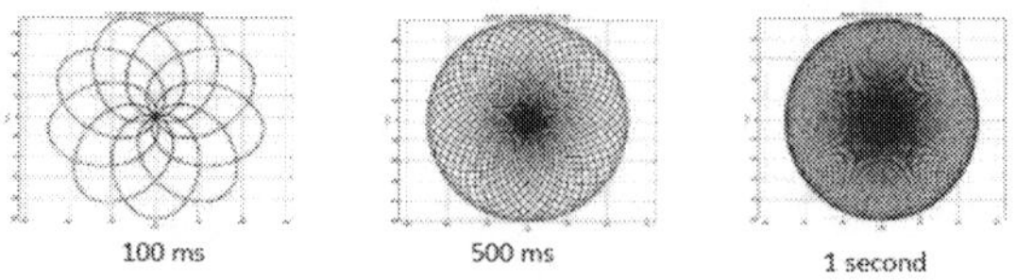

Figure 2: Sample Scan Patterns

DOI: 10.1051/epjconf/201611906014

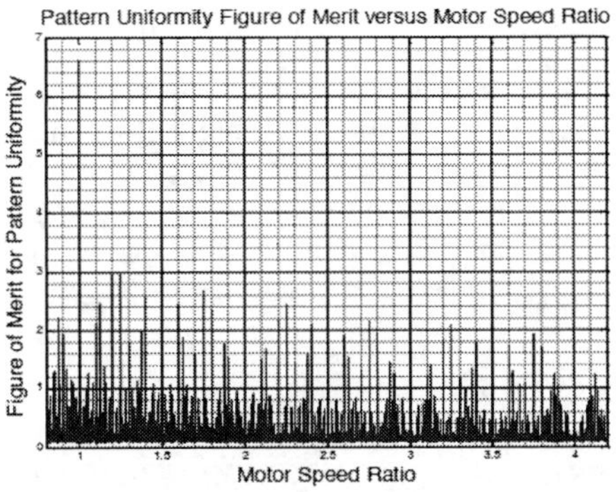

Figure 3: Figure of Merit (FOM) of the scan uniformity versus the motor speed ratio.

OPAL sensors incorporate Neptec's patented real-time obscurant penetrating lidar technology. Originally developed for helicopters landing in brownout conditions in the desert, OPAL is able to penetrate obscurants such as dust, fog, rain and snow to dynamically image objects engulfed within those obscurants. This capability is accomplished, in part, through a unique hardware design where features of the return waveform are evaluated in real-time without the need for full processor-intensive digitization. A real-time filter is also implemented in software to filter out the returns coming from obscurant particles. An example of dust penetration by the OPAL in an open-pit mining area is shown in Figure 4, 5 and 6 during a shoveling operation. The bottom image, acquired with OPAL, clearly shows the mine face behind the haul truck compared to an obscured image captured by conventional lidar.

Figure 4: Mining shovel loading a hauling truck

Figure 5: Imaging with a conventional lidar

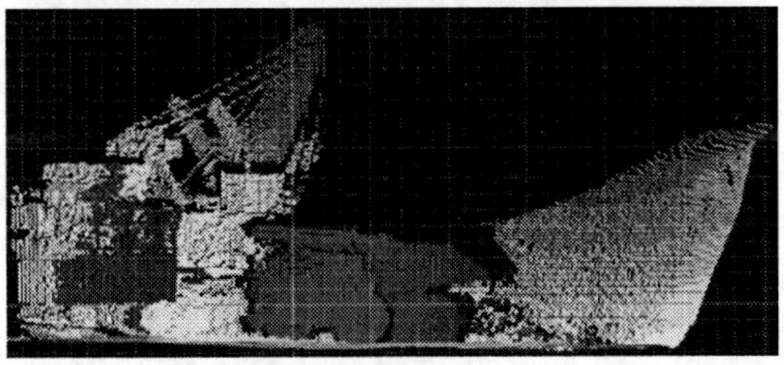

Figure 6: Imaging with OPAL lidar

2.2 Software Architecture

3DRi (3D Real-time intelligence) is a collection of highly efficient algorithms and software to extract actionable information from 3D sensor data in real-time and from moving platforms. This technology, when combined with the OPAL sensors, provides an autonomous machine the ability to perceive its environment, dynamically respond to changes, and identify and track objects in its field of view in real-time and while in motion. 3DRi makes it easy to develop, integrate and support intelligent 3D machine vision systems for machine automation and other industrial applications. The 3DRi modular "plug-ins" are shown in Figure 7 and can be implemented based upon the application requirements.

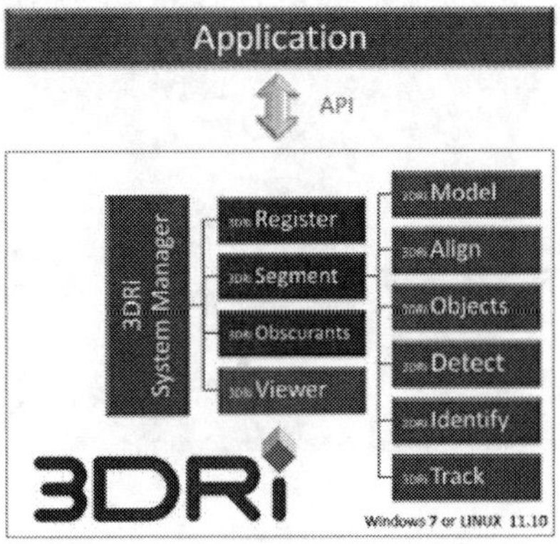

Figure 7: 3DRi software architecture

An API is usually customized to interface with a specific application target such as a vehicle or heavy machinery that requires control input.

Some examples of 3DRi processing are shown below. Figure 8 shows raw lidar data of a landing zone where dust from the soil has just been stirred up by an approaching helicopter. Figure 9 shows how the 'Obscurant' plug-in, filters out the dust to reveal objects hidden in the dust. Finally, Figure 10 shows how the 'Segment' plug-in separates out the ground plane to leave only the above ground objects that represent a potential hazard to the helicopter.

Figure 8: Landing zone raw lidar data including dust and above ground objects.

Figure 9: Dust removed leaving only ground and above ground objects.

Figure 10: Ground segmented out, leaving only above ground objects.

3. Obscurants' Penetration Results

It is important for a lidar sensor operating in harsh environments to be capable of penetrating obscurants to some degree. Dust and smoke are often present in mining operations. Airborne platforms often face rain, fog and snow. Landing helicopters can also generate large severe brownouts and whiteouts when done in unprepared areas.

A series of evaluations on the OPAL were conducted using the aerosol chamber facility at Defence Research and Development Canada (DRDC) Valcartier in Quebec, Canada [3]. The chamber has a length of 22m. The optical depth (defined as $-ln(T)$, T being the transmittance) of the aerosol clouds generated in the chamber ranges from 0.01 to over 10; the range that can be monitored by a transmissometer is limited to optical depths of 0.01 to 4. The aerosol chamber was used to evaluate laser propagation in dust and water fog.

Figure 11 illustrates a few examples of obscurants' measured penetration ranges compared with the predicted penetration ranges using the single scattering lidar equation. The mass extinction coefficients were derived from the Mie scattering model for a laser wavelength of 1540nm and using particles' properties, such as their expected size and refraction index.

For the case of dust, a particle size of 20μm is used for the model, with the sensor at a standoff distance of 43m from the dust cloud. Experiments were conducted with the MIL-810 dust that has an average particle size of 20μm. The agreement between the actual and predicted ranges is fairly good; however, in reality there is a particle size distribution that will affect the penetration characteristics and this is not accounted for by the current model. In addition, the model assumes an obscurant cloud with a constant density, which is rarely the case in reality.

For dense water fog cases, a particle size of 10μm is used for the model, with the sensor at a standoff distance of 46m from the fog cloud. The agreement between the model and the field measurements is good; however, in reality the water fog also has a particle size distribution that will affect the penetration characteristics, and this is not accounted for by the current model.

In the case of snow as an obscurant, the data was gathered using a helicopter hovering over a fresh snow field to create a whiteout scenario [4]. Independent measurements of the snow generation rate were not available. For the model, a snowfall rate of 230mm/hr was used, corresponding to an extremely heavy snowfall observed by weather services. It can be noted that there is a significant discrepancy between the model and the measurements, due to the fact that a true whiteout model is unavailable.

Finally, a measured detection range in clear conditions of wires with a diameter size of half inch (½") is shown with a close agreement with the model.

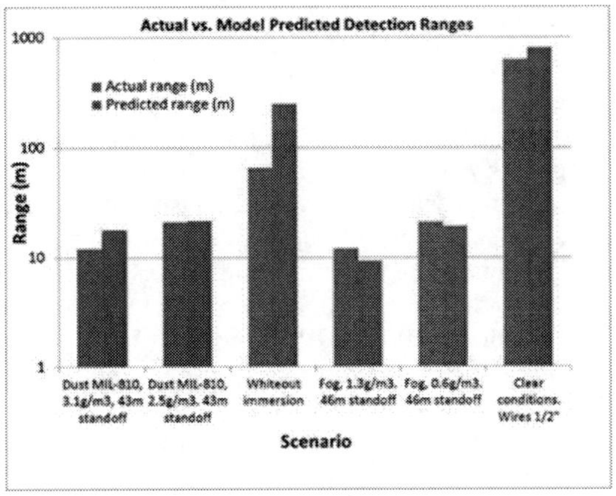

Figure 11: Comparison of predicted detection ranges in obscurants with measured detection ranges.

4. CONCLUSIONS

The hardware and software architectures of the OPAL lidar sensor have been described. The sensor architecture is well optimized for harsh and poor visibility environments. The OPAL lidar is currently successfully employed to acquire 3D vision data for machinery control in environments such as open-pit mining, heavy construction, ship loading, and military helicopter landing in unprepared areas. Obscurant penetration performances of the sensor are in reasonable agreement with predictions from a single scattering model. This holds true despite the fact that detected light returns will include multiple scattering events and the obscurants' density and particle sizes are not uniform in real situations.

Further tests with obscurants in controlled environments are planned to better characterize the sensor resolution and accuracy as a function of the obscurant type and density.

ACKNOWLEDGEMENT

The author would like to thank Dr. Gilles Roy and Dr. Simon Roy from DRDC Valcartier. Part of this work was supported by DRDC Valcartier under contract W7701-10-3768 awarded to Neptec by Public Works and Government Services Canada.

REFERENCES

[1] Roberts, J.M. et al (1999). Robot Control of a 3,500 Tonne Mining Machine. *Proceedings of the 1999 IEEE International Workshop on Robot and Human Interaction* (pp. 213-218). Pisa, Italy.

[2] Deloitte Touche Tohmatsu Ltd. (2012). *Tracking the trends 2013: The top 10 issues mining companies may face in the coming year* (Tech. Rep. No. 12-3013). Canada: Deloitte Design Studio.

[3] Evan Trickey, Philip M. Church (2013), Xiaoying Cao (2013), "Characterization of the OPAL obscurant penetrating lidar in various degraded visual environments", *Proc. SPIE 8737, Degraded Visual Environments: Enhanced, Synthetic, and External Vision Solutions, 87370E* May 16, 2013

[4] Roy, S., Cao X., Roy, G., Trickey E., "Evaluation of an obscurant-penetrating 3D-imaging lidar in whiteout conditions", *26th International Laser Radar Conference (ILRC), Porto Heli, Greece,* June (2012).

EPJ Web of Conferences **119**, 06015 (2016) DOI: 10.1051/epjconf/201611906015
ILRC 27

NEAR-RANGE RECEIVER UNIT OF NEXT GENERATION POLLY^{XT} USED WITH KOLDEWAY AEROSOL RAMAN LIDAR IN ARCTIC

Iwona S. Stachlewska[1*], Krzysztof M. Markowicz[1], Christoph Ritter[2], Roland Neuber[2], Birgit Heese[3], Ronny Engelmann[3], Holger Linne[4]

[1]*Institute of Geophysics, Faculty of Physics, University of Warsaw, (IG FUW) Warsaw, Poland, *E-mail: iwona.stachlewska@igf.fuw.edu.pl*

[2]*Alfred-Wegener Institute for Polar and Marine Research (AWI), Potsdam, Germany.*

[3]*Institute for Tropospheric Research (TROPOS), Leipzig, Germany.*

[4]*Max-Planck Institute for Meteorology (MPI), Hamburg, Germany.*

ABSTRACT

The Near-range Aerosol Raman lidar (NARLa) receiver unit, that was designed to enhance the detection range of the NeXT generation Polly^{XT} Aerosol-Depolarization-Raman (ADR) lidar of the University of Warsaw, was employed next the Koldeway Aerosol Raman Lidar (KARL) at the AWI-IPEV German-French station in Arctic during Spring 2015. Here we introduce shortly design of both lidars, the scheme of their installation next to each other, and preliminary results of observations aiming at arctic haze investigation by the lidars and the iCAP a set of particle counter and aethalometer installed under a tethered balloon.

1. INTRODUCTION

Presence of the absorbing aerosols in troposphere causes locally warming effect at the altitude of their occurrence and cooling effect below, at the ground. Measurements have shown that black carbon could be a minor, but due to its absorption, an important component found in arctic haze advection events over Ny-Alesund [1], what has been confirmed by chemical in-situ measurements from Zeppelin station [2].

In the frame of the iAREA project, an arctic field campaign was performed during 26 March to 7 May 2015 at Ny-Alesund, Spitsbergen, Arctic. We aimed at assessing a possibility of profiling absorbing particles during arctic haze event, using a set of a compact particle counter and an aethalometer installed under a tethered balloon, in combination with ground-based observations of a near-range aerosol elastic and Raman lidar signals measured at UV and VIS channels, that completed measurements of an advanced multi-wavelength multi-channel aerosol-raman-depolarization lidar.

The Near-range Aerosol Raman lidar (NARLa), was initially developed for the Radiative Transfer Labolatory (RTLab) of the IG FUW in Warsaw to serve along with the advanced 8-channel ($2\alpha + 3\beta + 2\delta$ + wv) aerosol-depolarization-Raman lidar (ADR lidar) of Polly^{XT}-type, that was developed withinin the SONATA-BIS project in a scientific cooperation with TROPOS in Leipzig [3]. The ADR lidar has a completed overlap at about 600 m, which was not sufficiently low to measure Winter boundary layer structures, that are expected to be often below that altitude over Warsaw [4]. Since our long-term goal at RT Lab is to measure aerosols within the boundary layer and in the free troposphere, to investigate exchange processes between them, we required an additional receiving/detecting unit (possibly independent on the ADR lidar receiving/detecting unit), that could be applied for the near range measurements starting from almost a ground level (NARLa).

The compact design of NARLa and its operation (independent on the ADR lidar), allowed us to use this near-range unit with other lidar systems, for example with the KARL lidar during the arctic field campaign in Spring 2015 on Spitsbergen. Preliminary analysis of the data taken on April 3rd and 7th, 2015 shows that on the first day background aerosol was measured while on the other a possible arctic have event occurred.

2. INSTRUMENTATION

The NARLa is dedicated to near-range detection with a full overlap around 120m above the ground for a configuration with the ADR lidar, and thus it lowers the overlap-affected height range of the overall system of about 480 m and allows to improve the detection within the boundary layer. It and comprises 4 channels ($2\alpha + 2\beta$) at 355 and 532 nm. The channels for detection of polarization were not implemented as this unit is normally used with the ADR lidar that collects the polarization signals at both 532 and 355nm channels, and thus provides particle linear depolarization profiles almost down to the ground. The photon counting data acquisition system for the NARLa and the ADR lidar were adopted after the MPI-Hamburg design to provide a small and relatively low-priced data acquisition for scientific lidar measurements characterized by an excellent response to the high dynamic range of lidar data.

The NARLa comprises the Newtonian type telescope of a diameter of 5cm and focal length of 20cm, a fiber core of 0,4 mm resulting in FOV of 2mrad. A fiber-optic scrambler consisting of a ball lens of 2 mm diameter links two fiber patch cables to clean angular distortions. After, the light passes a set of beam splitters and interference filters before the elastic signals are detected at 355 and 532nm, and the Nitrogen Raman channels collect the light at 387 and 607nm (Fig.1). For all channels the Hamamatsu P10721-110 are used in photon counting mode with low overall dead-time of about 2ns at each channel.

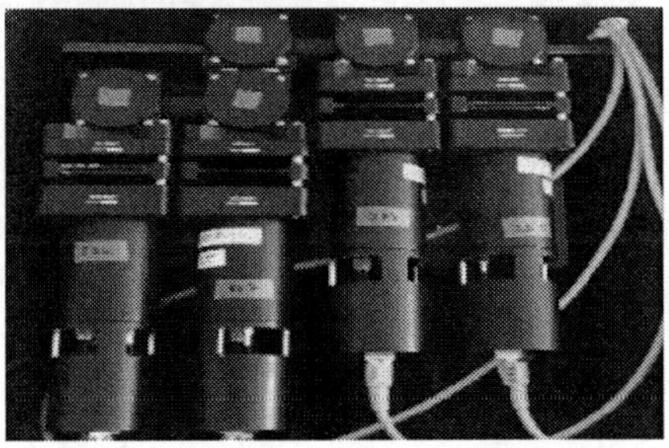

Fig.1: The NARLa detector unit consisting of elastic and Raman channels at 355nm and 532nm.

The acquisition is triggered with scattered laser light and obtained with 800 MHz photon counters with 7.5m spatial resolution. The first 256 signal bins are allocated for pre-trigger and signal is obtained up to ~40km.

The operation of NARLa was tested initially in February 2015 in an installation at the ACTRIS-EARLINET Supersite, next to a primary receiver of the MARTHA lidar at TROPOS in Leipzig. Then, at the beginning of March 2015 it was tested in an installation next to the ADR lidar at IG FUW in Warsaw. The installation of NARLa naxt to the KARL at AWI-IPEV station in Ny-Alesund was done from 27 March till 6 May 2015 (Fig.2). In all of the mentioned configurations we used the powerful lasers of the three advanced lidar systems as a source of radiation for NARLa.

Fig.2: Installation of the NARLa receiver/detection unit next to the KARL lidar receiver at Ny-Alesund.

The Koldewey Aerosol Raman Lidar (KARL) [5] is a coaxial ($2\alpha + 3\beta + 2\delta + 2wv$) Nd:Yag based system with ~10W / color, with 0.6 mrad laser divergence and 70 cm diameter receiving mirror. An aperture stop that is movable in position and diameter allows measurements in the troposphere (overlap range above 700m) and stratosphere. The KARL lidar is member of the NOAA - Network for the Detection of Atmospheric Composition Change (NDACC).

Table 1. Detection channels used by the three lidars ADR, KARL, and NARLa.

Detected Wavelength (nm)	Detection Type	ADR lidar Warsaw Poland	KARL Warsaw Spitsbergen Arctic	NARLa mobile detection module
1064	Elastic total	x	x	
660	Raman H_2O		x	
607	Vibrational Raman N2	x	x	x
532	Elastic total	x		x
532	Elastic cross	x	x	
532	Elastic parallel		x	
407	Raman H_2O	x	x	
387	Vibrational Raman N_2	x	x	x
355	Elastic total	x		x
355	Elastic cross	x	x	
355	Elastic parallel		x	

Additionally, profiling of absorbing particles was performed with the iCAP set, that comprises two in-situ sensors – the AE51 aethalometer and the HandiLaz aerosol particle counter together with the GoPro camera. The iCAP set (Fig.3) was installed on the tethered balloon that can be released to the maximum altitude of ~1600m. During the campaign in Spring 2015 the iCAP measurements were performed up to 5 times a day in favorable weather conditions (not too strong wins speed and not too thick clouds). Duration of a single ascent/profile is of ~30 min.

The quick-looks a vieved online via the website of the polish aerosol observational network Poland AOD (http://www.polandaod.pl).

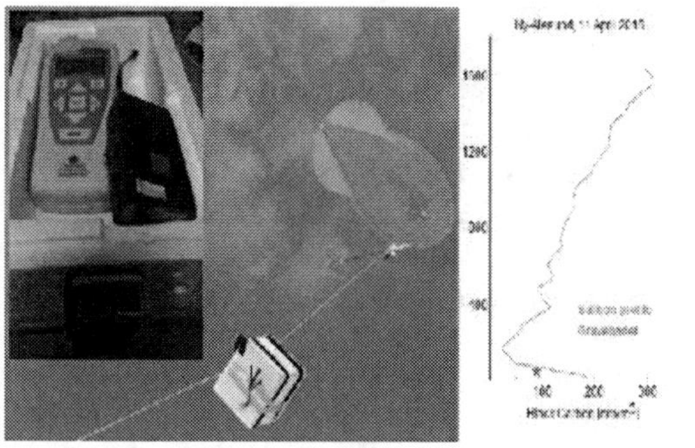

Fig.3: The iCAP set installed on tethered balloon for absorbing particles profiling. On the right an ascent with black carbon measurement taken at Ny-Alesund on 11/04/ 2015 compared with in-situ measurement at Gruvebadet station.

3. RESULTS

The layered structure of troposphere obtained with NARLa receiver at 532nm attenuated backscatter signals on 03/04/2015 and 07/04/2015 is depicted in Fig.4.

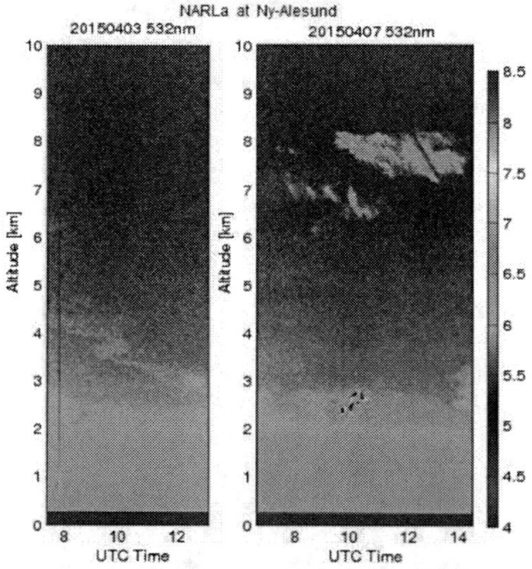

Fig.4: Attenuated backscattering coefficient profiles at 532 nm [1/m/sr] measured by NARLa on 03/04/2015 and 07/04/2015 in Ny-Alesund, Arctic. The signals are available for evaluation in the range between about 300m up to 10km. Quick-looks of the data set are given at http://www.polandaod.pl.

The first day was characterized by clear-sky weather conditions (background aerosol). On the second day the aerosol load in lower troposphere was significantly enhanced (arctic haze) and in the upper atmosphere Cirrus clouds appeared.

A preliminary backscatter coefficient profiles calculated for KARL and NARLa observations at 12 UTC on 03/04/2015 at 355, 532 and 1064nm are compared in Fig.5. All profiles were averaged over 1h (no smoothing was applied and range resolution was kept at 7.5m), prior to applying the Klett's evaluation scheme with lidar ratio of 30sr and calibration value set at 3.75km (NARLa) and 12km (KARL). The agreement of VIS channels (green) is clear but reason for profiles shift at UV channels (blue) need to be further investigated.

As a next step, an evaluation of entire data set of NARLa and KARL obtained during Spring 2015 campaign using the Ansmann evaluation scheme to obtain independently extinction coefficient and lidar ratio profiles will be conducted together with analysis of absorption profiles obtained using the iCAP set.

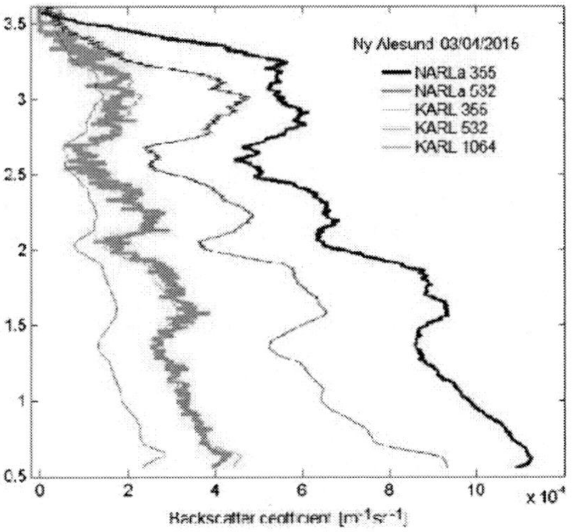

Fig. 2. The backscatter profiles obtained around 12 UTC on 03/04/2015 by NARLa and the KARL receivers in Ny-Alesund, Arctic. The disagreement of the UV profiles still need to be investigated.

4. CONCLUSIONS

The NARLa proved robust and compact enough for easy installation next to the different lidars (MARTHA in Leipzig, ADR in WARSAW, KARL in Ny-Alesund). The latter configuration show that it is capable of performing aerosol measurements even up to 10 km at UV and VIS channels in tough Arctic conditions.

ACKNOWLEDGEMENT

The development of NARLa was conducted in the frame of the SONATA-BIS Project: *Integrated study of climate processes involving absorbing aerosols,* funded by the National Science Centre of Poland (2012/05/E/ST10/01578).

The arctic campaign in 2015 was financed in the frame of the iAREA Project: *Impact of absorbing aerosols on radiative forcing in the European Arctic* funded by the Polish-Norwegian Research Programme (Pol-Nor/196911/38/2013).

REFERENCES

[1] Yamanouchi T., et al. 2005: *Arctic Study of Tropospheric Aerosol and Radiation (ASTAR) 2000: Arctic haze case study,* Tellus 57B, 141–152

[2] Tunved P., et al. 2013: *Arctic aerosol life cycle: linking aerosol size distributions observed between 2000 and 2010 with air mass transport and precipitation at Zeppelin station, Ny-Alesund, Svalbard,* Atmos. Chem. Phys., 13, 3643-3660

[3] Engelmann R., et al. 2015: *EARLINET Raman lidar Polly^{XT}: The NeXT Generation*, Atmos. Meas. Techn. Discuss. (submitted April 2015)

[4] Stachlewska I.S., et al. 2012: *Ceilometer observations of the boundary layer over Warsaw,* Poland, Acta Geophys. 60, 1386-1412, **2012**

[5] Hoffmann A., et al. 2010: A redesigned Raman lidar for cloud and aerosol profiling in the Arctic, iILRC 25, St. Petersburg.

WATER VAPOR, CLOUD AND AEROSOL PROPERTIES ON THE TIBETAN PLATEAU USING MULTI-LIDAR MEASUREMENTS

Songhua WU*, Guangyao DAI, Dongxiang WANG, Xiaochun ZHAI, Xiaoquan SONG

Ocean Remote Sensing Institute, Ocean University of China, Qingdao 266100, China.

**Email: wush@ouc.edu.cn*

ABSTRACT

The 3rd Tibetan Plateau atmospheric expedition experiment campaign were operated in the Tibetan Plateau during July and August 2014 by utilizing the Water vapor, Cloud and Aerosol Lidar (WVCAL), Coherent Doppler Wind Lidar and ceilometer VAISALA CL31. The observation was carried out in Nagqu area (31.5°N, 92.05°E), which is 4508 meters above the mean sea level. Water vapor mixing ratio, cloud height, vertical wind speed and vertical water vapor flux was measured by these lidars. The inversion methods of data products of lidars are described in details in this paper. Furthermore, the clouds heights measured by lidar and ceilometer were compared to verify the performance of the lidar. Finally, the case studies of water vapor mixing ratio, water vapor flux and cloud height and statistics were provided.

1. INTRODUCTION

The Tibetan Plateau is located in a critical and sensitive junction of four climatic systems: the Westerlies, the East Asian Monsoon, the Siberian cold polar airflow, and the Indian monsoon. In turn, the Tibetan Plateau influences the atmosphere in East Asia area and even the whole northern hemisphere.

Water vapor is the most active element and has great impact on the change of the climate and weather. It influences atmospheric circulation and temperature structure by condensation and evaporation processes. Together with water vapor, aerosol and clouds play an important role in the global energy budget and radiative forcing both directly and indirectly. Due to the important influence and the rareness of measurement, the observation of these compositions at the Tibetan Plateau becomes significant and urgent. Aiming at the detection of water vapor mixing ratio, Raman technique is applied. Raman Lidar technique depends on the detection of Raman backscattered radiation from atmospheric molecules [1,2]. The process of Raman scattering is characterized by a wavelength shift of the scattered radiation in respect to the exciting wavelength. The shift is uniquely associated with the internal transitions between the rotational-vibrational energy levels of the molecules [3]. This technique was firstly used by Melfi and Cooney [1] and the profiles of water vapor mixing ratio was retrieved. Due to the development of the high power laser source, the vertical operation range of the Raman Lidars can be extended up to 7 km [4] and throughout the troposphere [5]. Raman Lidars with high power source can also measure atmospheric boundary layer for a complete diurnal cycle with high resolution spatial and temporal resolutions [6].

When come to the measurement and classification of the aerosol and clouds, the polarization Lidar technique is a well-established method [7,8]. The polarization Lidar technique bases on the depolarization character of light and the particles scattered radiation. This technique is widely used in the identification of clouds types, desert dust, volcanic ash [9]. It is also helpful to improve the estimation of the microphysical properties of dust from optical measurements [10].

In order to study the boundary layer dynamics on plateau, the wind field needs to be observed. A pulsed coherent Doppler wind Lidar was used to measure the wind profile and vertical velocity in the boundary layer accurately to understand the unique property of atmospheric circulation and hydrological cycling on the Tibetan Plateau.

2. METHODOLOGY

2.1 WATER VAPOR MIXING RATIO

The Lidar equation can be described as

$$P(z,\lambda_R)+P_{BR}= P_0(\lambda_L)\Delta z\frac{A_0 O(z)}{z^2}\xi(\lambda_R)\beta_R^{\pi}(z,\lambda_R)T^{up}(z,\lambda_L)T^{down}(z,\lambda_R) \quad (1)$$

$$T^{up}(z,\lambda_L)=\exp[-\int_{z_0}^{z}\alpha(z',\lambda_L)dz']$$
$$T^{down}(z,\lambda)=\exp[-\int_{z}^{z_0}\alpha(z',\lambda_L)dz'] \tag{2}$$

Where $P_0(\lambda_L)$ is the laser pulse energy, Δz is the range resolution, A_0 is the aperture of the telescope, $\xi(\lambda_R)$ is the receiving efficiency at a given wavelength, $\beta_R^{\pi}(z,\lambda_R)$ is the backscatter coefficient at λ_R, $\alpha(z,\lambda_L)$ is the extinction coefficient, $T^{up}(z,\lambda_L)$ and $T^{down}(z,\lambda_R)$ are the atmospheric transmission at λ_L and λ_R respectively.

According to equation (1), the backscatter signal of N_2 and H_2O are obtained as $P(z,\lambda_{N_2})$ and $P(z,\lambda_{H_2O})$. The water vapor mixing ratio can be calculated by equation (3):

$$w(z)=C\frac{P(z,\lambda_{H_2O})}{P(z,\lambda_{N_2})}\Delta T(\lambda_{N_2},\lambda_{H_2O},z) \tag{3}$$

Where: C is Raman channel constant and can be obtained by contrast of Lidar data and radiosonde data, $\Delta T(\lambda_{N_2},\lambda_{H_2O},z)$ is the differential atmospheric transmission at nitrogen and water vapor wavelengths and is calculated by equation (4):

$$\Delta T(\lambda_{N_2},\lambda_{H_2O},z)=\exp\left(-\int_{z_0}^{z}[\alpha(z',\lambda_{N_2})-\alpha(z',\lambda_{H_2O})]dz'\right) \tag{4}$$

We assume that the relationship between Lidar data $W_{Lidar}=\frac{P(z,\lambda_{H_2O})}{P(z,\lambda_{N_2})}\Delta T(\lambda_{N_2},\lambda_{H_2O},z)$ and radiosonde data W_{Sonde} as equation (5):

$$W_{Sonde}=C*W_{Lidar}+D \tag{5}$$

After the calibration and validation using the radiosonde data, the water vapor mixing ratio can be rewritten as

$$W_{Lidar}^{Cal}=218.9976*W_{Lidar}-0.33976 \tag{6}$$

2.2 VERTICAL VELOCITY AND VERTICAL WATER VAPOR FLUX

The Coherent Doppler lidar takes advantage of the fact that the frequency of the echo signal is shifted compared to the local-oscillator light because of the Doppler effect which occurs from backscattering of aerosols. The Doppler shift in the frequency of the backscattered signal is analyzed to calculate the line-of-sight (LOS) velocity component of the air motion. The Doppler shift f_D can be obtained by equation (7):

$$f_D=\frac{2|\vec{V}_{LOS}|}{\lambda} \tag{7}$$

Where $\vec{V}_{LOS}$ is the line-of-sight (LOS) velocity, λ is the laser wavelength and is equal to 1550 nm in this lidar system.

When the LOS velocities in four directions $\vec{V}_{LOS,E}$, $\vec{V}_{LOS,W}$, $\vec{V}_{LOS,S}$ and $\vec{V}_{LOS,N}$ were measured, the vertical wind speed can be calculated by equation (8):[11]

$$\vec{V}_{ver}=\frac{1}{4\sin\theta}(\vec{V}_{LOS,E}+\vec{V}_{LOS,W}+\vec{V}_{LOS,S}+\vec{V}_{LOS,N}) \tag{8}$$

Where θ is the elevation.

With the synchronized observations of the water mixing ratio and vertical velocity, the vertical water vapor flux $Flux_{WV,ver}$ can be calculated by equation (9):

$$Flux_{WV,ver}=\rho_{WV}*\left|\vec{V}_{ver}\right| \tag{9}$$

Where ρ_{WV} is the absolute humidity that can be obtained from the water vapor mixing ratio.

2.3 CLOUD HEIGHT MEASUREMENT AND CLOUD STATISTICS

WVCAL is capable of detecting of cloud height. For purpose of determining the cloud height, the algorithm that combines "differential zero-crossing algorithm" [12] and "threshold algorithm" [13] is taken into account. Generally cloud height can be determined directly from these zero crossings of the first derivative of backscatter intensity $dP(z)/dz$. However, in order to exclude interference of spurious zero crossings, The $dP(z)/dz$ is determined from a least-squares fit by using a multipoint window that slides through successive points from z_0 to the end of the lidar profile. Besides, a range-dependent threshold is defined, based on the noise in the recorded lidar signal. Signal excursion above the threshold value is identified as a cloud. By this algorithm, cloud layers apparent in the return signal are identified. However, if the lower clouds were too thick to be penetrated, the higher clouds will not be detected and cannot be identified.

3. RESULTS

Here we provide case studies of water vapor mixing ratio in Fig. 1. Several intercomparisons of lidar derived vertical profiles with radiosondes are presented. In this figure, the green line indicates the water vapor mixing ratio measured by lidar and the pink line shows the error bar of the data. Meanwhile the blue line shows the data of radiosonde. From these figures, one dry layer could be seen at about 2.8km to 3km in left figure and one distinct wet layer could be seen at about 2.5km to 3km in the right figure. All of the water vapor mixing ratio profiles was averaged every 60 minutes and the range resolution was 75 m.

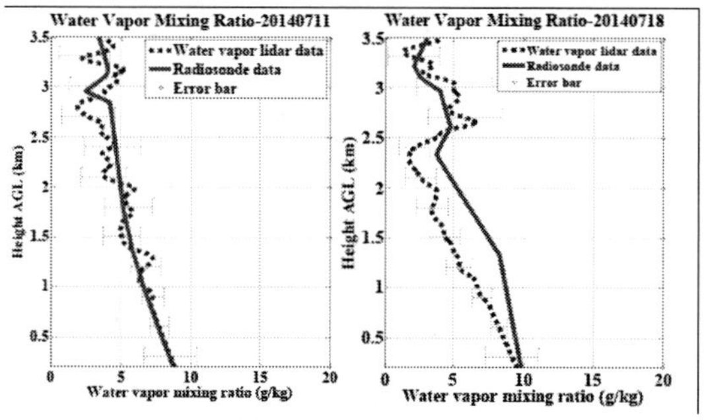

Fig.1. Water vapor mixing ratio study cases in Nagqu on 11and 18 July 2014.

In term of the vertical velocity and vertical water vapor flux, one case study on 15 August 2014 is presented below. Fig. 2 is the time serials of the vertical velocity profile of 164 minutes obtained from the Coherent Doppler Wind Lidar. By combining absolute humidity and vertical velocity data, the vertical water vapor flux can be calculated and the temporal development is shown in Fig. 3.

Fig.2. Time serials of vertical velocity profile.

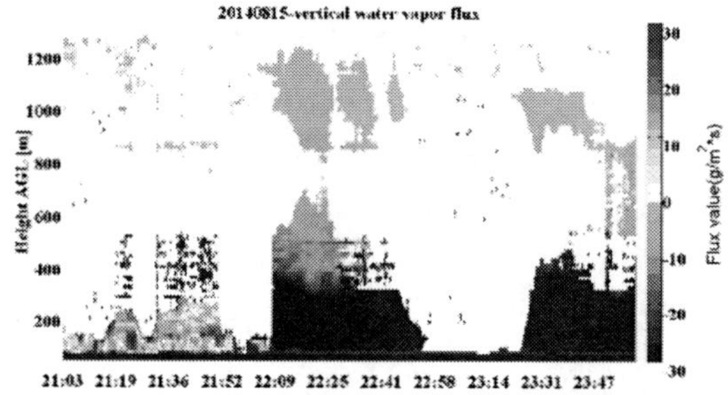

Fig.3. time serials of vertical water vapor flux.

The cloud heights measured by WVCAL and VAISALA CL31 ceilometer are compared. One case study on 18 July 2014 is presented. From the results on that day, it can be figured out that the correlation coefficient and standard deviation of cloud heights measured by polarization channel and infrared channel of Lidar were 91.87% and 0.28km respectively. Meanwhile, the correlation coefficient and standard deviation of cloud heights measured by polarization channel of Lidar and VAISALA were 80% and 0.37km respectively.

Lastly, the cloud statistics in Nagqu is shown in Fig. 4. The number and frequency of clouds layers are analyzed based on the 3rd Tibetan Plateau atmospheric expedition experiment campaign from July to Augest, 2014. The frequency of cloud appearance was about 59.88% in summer at the observatory.

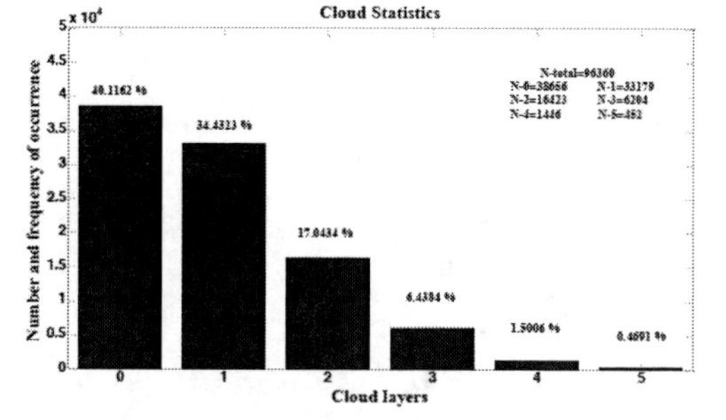

Fig.4. Cloud statistics in Nagqu from Jul 8, 2014 to August 16, 2014

4. CONCLUSIONS

By utilizing WVCAL, information about water vapor was acquired. Result from the abundant vegetation, lakes effect and wet Indian monsoon and East Asia monsoon, water vapor mixing ratio in Tibetan plateau in summer was rich.

With the help of the WVCAL and the Coherent Doppler Wind Lidar, the vertical wind speed and vertical water vapor flux were calculated. The upwelling and deposition of the water vapor were monitored.

The cloud base height measured by 532 nm channel, 1064 nm channel of Lidar and VAISALA CL31 can match well and the correlation coefficient of each other was larger than 80%. However, because of the difference in emission wavelengths, deviation of cloud base height exists. The standard deviation was about 0.2 to 0.4 km and was acceptable.
Cloud statistics was provided during 3rd Tibetan Plateau atmospheric expedition experiment campaign in Nagqu, 2014. The frequency of cloud appearance was about 59.88%. Consequently, clouds in Nagqu were abundant in summer.

ACKNOWLEDGEMENT

This work was funded by CAMS/LAWS (Chinese Academy of Meteorological Science/Lab oratory of Severe Weather)

REFERENCES

[1] Dinoev, T., 2009: Automated Raman lidar for day and night operational observation of tropospheric water vapor for meteorological applications(Doctoral dissertation, ÉCOLE POLYTECHNIQUE FÉDÉRALE DE LAUSANNE).

[2] Renaut, D., & Capitini, R., 1988: Boundary-layer water vapor probing with a solar-blind Raman Lidar: validations, meteorological observations and prospects, *Journal of Atmospheric and Oceanic Technology*, **5(5)**, 585-601.

[3] Demtroder, W., & Demtroder, W., 2005: Molecular physics: theoretical principles and experimental methods, *Wiley VCH.*

[4] Whiteman, D. N., Melfi, S. H., & Ferrare, R. A., 1992: Raman Lidar system for the measurement of water vapor and aerosols in the Earth's atmosphere, *Applied Optics*, **31(16)**, 3068-3082.

[5] Goldsmith, J. E. M., Blair, F. H., Bisson, S. E., & Turner, D. D., 1998: Turn-key Raman Lidar for profiling atmospheric water vapor, clouds, and aerosols, *Applied Optics*, **37(21)**, 4979-4990.

[6] Froidevaux, M., Higgins, C. W., Simeonov, V., Ristori, P., Pardyjak, E., Serikov, I., ... & Parlange, M. B., 2013: A Raman lidar to measure water vapor in the atmospheric boundary layer. *Advances in Water Resources*, **51**, 345-356.

[7] Ansmann A., Tesche M., Althausen D., et al., 2008: Influence of Saharan dust on cloud glaciation in southern Morocco during the Saharan Mineral Dust Experiment, *Journal of Geophysical Research: Atmospheres (1984–2012)*, **113(D4)**.

[8] Freudenthaler, V., Esselborn, M., Wiegner, M., Heese, B., Tesche, M., Ansmann, A., ... & Seefeldner, M., 2009: Depolarization ratio profiling at several wavelengths in pure Saharan dust during SAMUM 2006. *Tellus B*,**61(1)**, 165-179.

[9] Sassen, K., Zhu, J., Webley, P., Dean, K., & Cobb, P., 2007: Volcanic ash plume identification using polarization lidar: Augustine eruption, Alaska. *Geophysical Research Letters*, **34(8)**.

[10] Wiegner M, Gasteiger J, Kandler K, et al., 2008: Numerical simulations of optical properties of Saharan dust aerosols with emphasis on lidar applications. *Tellus B*, **61(1)**, 180-194.

[11] Cariou, J. P., & Boquet, M., 2011: Leosphere pulsed lidar principles. *Leosphere, Orsay (FR)*, 1-32.

[12] Pal, S. R., Steinbrecht, W., & Carswell, A. I., 1992: Automated method for lidar determination of cloud-base height and vertical extent, *Applied optics*, **31(10)**, 1488-1494.

[13] Winker, D. M., & Vaughan, M. A., 1994: Vertical distribution of clouds over Hampton, Virginia observed by lidar under the ECLIPS and FIRE ETO programs, *Atmospheric research*, **34(1)**, 117-133.

OBSERVATIONS OF WIND PROFILE OF MARINE ATMOSPHERE BOUNDARY LAYER BY SHIPBORNE COHERENT DOPPLER LIDAR

Songhua Wu*, Jiaping Yin, Bingyi Liu, Jintao Liu, Hongwei Zhang, Xiaoquan Song, Kailin Zhang
*Ocean University of China, Qingdao 266100, China, *Email: wush@ouc.edu.cn*

ABSTRACT

Pulsed Coherent Doppler Lidar (CDL) system is so good as to prove the feasibility of the marine atmosphere boundary layer detection. A ship-mounted Coherent Doppler lidar was used to measure the wind profile and vertical velocity in the boundary layer over the Yellow sea in 2014. Furthermore, for the purpose of reducing the impact of vibration during movement and correcting the LOS velocity, the paper introduces the attitude correction algorithm and comparison results.

1. INTRODUCTION

Marine accounted for 70% of the surface of the earth. Marine atmosphere Boundary Layer (MABL) is the connection between sea surface and free atmosphere. The flux of air-sea interface is a major parameter to description the matter-energy interaction. The detection of wind field of MABL has important significance to the flux calculation.

Researchers perform a lot of experiments to detect the wind field in MABL. At present, Buoy, wind tower, ocean surveying vessel, radiosonde and weather radar are used in wind observation. Comparing to the techniques above, the sensitivity is typically lower than lidar. Conventional measuring method is difficult to provide high-precision wind velocity, especially vertical speed, which is demanding for the measurement of air-sea boundary layer turbulence flux. The CDL system is an available tool which has high spatial and temporal resolution in the clear atmosphere and 3D scanning mode are used for measurements of the wind field and the atmospheric turbulent flow [1] feasible. But the ship motion should be considered or else wind speed especially the vertical velocity, which impact on the calculation of the flux, would be inaccuracy [5].

This paper introduces the experimental investigation which was carried out to measure the wind field of MABL of the Yellow Sea and the Bohai Gulf by pulsed CDL in 2014.The Yellow Sea is the northern part of the East China Sea, which is a marginal sea of the Pacific Ocean. It is located between mainland China and the Korean Peninsula. It is linked with the Bohai Sea (the innermost gulf of the Yellow Sea on the coast of Northeastern and North China) by the Bohai Strait. The Yellow Sea is effected by temperate monsoon climate, so in winter the weather is cold and dry while in summer it's warm and moist. In winter (from October to next March), the wind of the North Yellow Sea is from the north-west and the average wind speed is 6~7m/s. At the same time, the wind speed of the South Yellow Sea is 8~9m/s. In April, these waters experience the change of monsoons which lead to the instability of the wind direction. In May, due to the southerly monsoon, the average wind speed is 4~6m/s. In summer (from June to August), the wind is always from the south, and the average speed is 4~6m/s. The lidar MABL experiment was carried out in May 2015 while the wind is mainly coming from the south. Wind profile and the validation with radiosonde during the cruise test are presented in this paper.

2. METHODOLOGY

Lidar is one of the most accurate remote sensing techniques that transmit a laser beam into the atmosphere and detect the backscattering of velocities the characteristic of the wind field can be deduced. The 3D beam scanner and detection range of 3 km enable the system to monitor the whole marine boundary layer. Because of the all-fiber structure adoption, the system is stable, reliable and high-integrated. The diagram of pulsed CDL setup is shown in Fig. 1. The system specifications are summarized in Table 1.

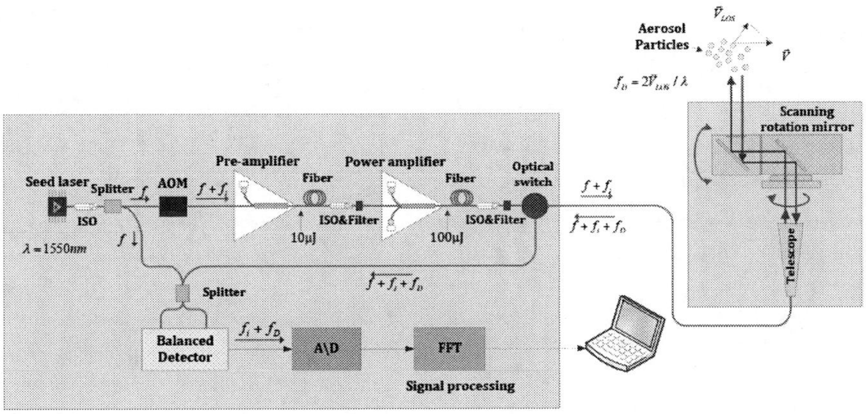

Fig.1.The pulsed CDL setup. A 1.5-μm semi-conductor single frequency laser and an Erbium doped fiber amplifier is used as the transmitter. A fiber optical circulator and a telescope are used as the optical transceiver.

Table 1.Specifications of the CDL.

CDL	Specification
Wavelength	1.5μm
Repetition rate	10kHz
Pulse energy	150μJ
Speed measurement range	±50m/s
Speed accuracy	≤0.3 m/s
Scanner pointing accuracy	0.1°
Scan speed	1-55 °/s
Spatial resolution	30m
Measurement range	30m to 3000m
Power dissipation	<150W
weight	<50kg

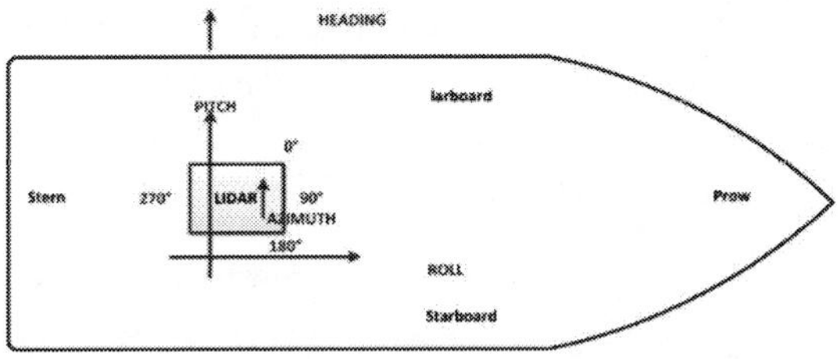

Fig.2. The installation Diagram of the ship-mounted CDL.

A ship-mounted Coherent Doppler lidar was used which causes the errors of wind profile retrieval. At the same time, the speed of the ship itself and ocean wave will be stacked to the line-of-sight velocity. The impact on the vertical speed is much more obvious. So the correction of the direction of the transmitting laser beam in geographic coordinate system is very essential. It can be precisely calculated from the attitude data during movement by an algorithm of attitude transformation. In order to acquire the speed, position and attitude angles of the vessel during movement, a two-antenna differential GPS and an inertial navigation system were installed on the ship-mounted CDL. The inertial navigation system was fixed on the base of the scanner to keep constant relative angles with reference to the transmitting laser beam. After installation, a calibration is performed to determine the initial orientation of the laser beam.

Radiosonde is an accurate meteorological measuring means of MABL profile. But the balloon is unrecyclable and the temporal resolution is low. The CDL system which has high

 DOI: 10.1051/epjconf/201611906017

radiosonde were conducted for 17 days and comparison results were obtained to verify the feasibility of attitude calibration.

3. RESULTS

To test the modification results, selecting the data detected by CDL at the corresponding time points when the radiosonde was set up and averaging the data for 1 min. This corresponded to a time series with 20 values. The spatial resolution of CDL and radiosonde was 30 m and 10 m, respectively. Measurement range was 3 km and 9 km, respectively.

The data of fixed measurement at15:00 (UTC+8h), May 9, 2014 and the data of underway measurement at 7:00 (UTC+8h) May 13, 2014 were chosen as modification instances. We compared the wind velocity and direction in Fig.3~6.The disparity of wind velocity between before and after the modification was small when the ship was anchored, while the disparity was bigger when the ship was voyaging.

Fig.3 and Fig.4 show the results of the anchored measurement at 15:52 (UTC+8h), corresponding to the longitude 37.00°N and latitude 122.86°E. The heading angle recorded by inertial navigation system was 269.30°, the practical heading angle was 359.30°. In the coordinate system of the ship, the North azimuth was 0° and the pitch angle was 59.98° recorded by sever-system. In the geographic coordinate, the North azimuth was 267.94° and the average value of pitch angle was 59.19°. The standard deviation of the pitch angle was 0.04° so that the swing of the ship was not obvious. The difference of wind direction between before and after attitude modification was 90° due to the different coordinate definition.

The wind velocity measured by CDL were consistent with that of radiosonde with the RMS error better than 0.90 m/s before modification and 0.84 m/s after modification below 2.4 km. Low Signal Noise Ratio (SNR) makes the data unavailable higher than 2.4 km,

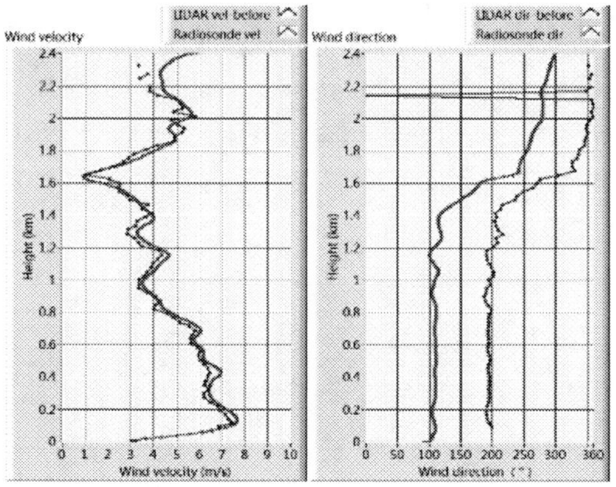

(a) (b)

Fig.3.Anchored observation below 2.4 km (before attitude modification): Wind velocity and wind direction measured by radiosonde (red line) and CDL (blue line) at 15:52 (UTC+8h), May 9, 2014. (a)Wind velocity. (b)Wind direction.

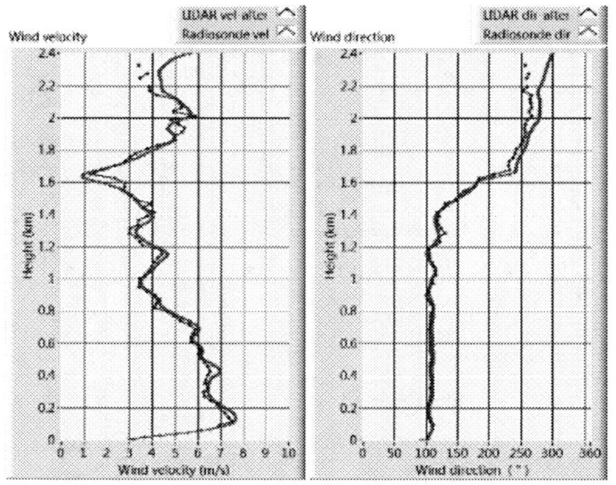

(a) (b)

Fig.4.Anchored observation (after attitude modification): Wind velocity and wind direction measured by radiosonde (red line) and CDL (blue line) at 15:52 (UTC+8h), May 9, 2014. (a)Wind velocity. (b)Wind direction.

Fig.5 and Fig.6 show the results of the voyage measurement at 07:44 (UTC+8h), corresponding to the coordinate (38.93°N, 119.75°E). The heading angle recorded by inertial navigation system was 299.40°, the practical heading angle was 29.40°. The North azimuth was 0° and the

 DOI: 10.1051/epjconf/201611906017

was a slight swing of the ship.

The comparison of the results between CDL and radiosonde were better below 1.4 km. The east component of ship's speed was 3.2 m/s and the north component was 3.7 m/s recorded by inertial navigation system. So the speed of the ship was 4.9 m/s. The wind velocity was shown in Fig.6 after the ship speed was calibrated. The results up to 1.8 km altitude of the wind velocity measured by CDL were inconsistent with that of radiosonde with a large RMS error of 3.66 m/s before attitude correction and were consistent with that of radiosonde with a RMS error of 0.63 m/s after correction.

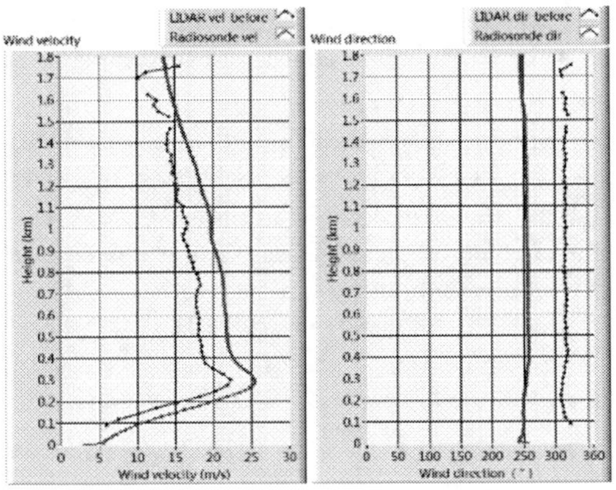

(a) (b)

Fig.5.Voyage observation (before attitude modification): Wind velocity and wind direction measured by radiosonde (red line) and CDL (blue line) at 07:44 (UTC+8h), May 13, 2014. (a)Wind velocity. (b)Wind direction.

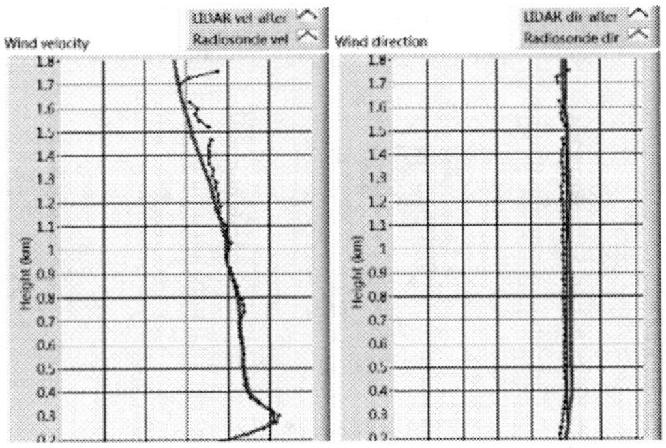

(UTC+8h), May 13, 2014. (a) Comparison of wind velocity. (b) Comparison of wind direction.

4. CONCLUSIONS

The observations of MABL wind profile by 1.5-μm all-fiber pulsed CDL system were conducted successfully. Our research team calibrated the speed and direction of the wind field and verified the possibility of attitude correction through the comparison experiment with radiosonde. The CDL is a practical tool for MABL wind field research and a powerful technique for the wind flux. The subsequent significance is to calculate the turbulence flux by vertical and horizontal wind velocity.

ACKNOWLEDGEMENT

This work was supported by NSFC under grant 41375016 and 41471309.

REFERENCES

[1] Hawley, J. G., Tang, R., Henderson, S. W., Hale C. P., Kavaya M. J., and Moerder D., 1993: Coherent launch-site atmospheric wind sounder: Theory and experiment, *Appl. Opt.*, **32**, 4557–4567.

[2]Eberhard, W. L., Cupp, R. E., Healy, K. R., 1989: Doppler lidar measurements of profiles of turbulence and momentum flux, *J. Atmos. Oceanic Technol.*, **6**, 809-819.

[3]Banta, R. M., Pichugina, Y. L., Brewer, W. A., 2006: Turbulent velocity-variance profiles in the stable boundary layer generated by a nocturnal low-level jet, *J. Atmos. Oceanic Technol.*, **62**, 2700-2719.

[4]Smalikho, I. N., Köpp, F., Rahm, S., 2005: Measurement of atmospheric turbulence by 2-μm Doppler lidar, *J. Atmos. Oceanic Technol.*, **22 (11)**, 1733-1747.

[5]Mitsuta, Y., Fujitani, T., 1974: Direct measurement of turbulent fluxes on a cruising ship, *Bound –Layer. Meteor.*, **6(1)**, 203-217.

NONUNIFORM ATMOSPHERIC SCATTER OF SHORT-PULSE OPTICAL RADIATION

I.V. Zaloznaya, E.D. Zaloznaya

V.E. Zuev Institute of Atmospheric Optics SB RAS 1 Akademik Zuev Square, Tomsk, 634021, Russia

iya@iao.ru

ABSTRACT

The results of calculation of the mean intensity of spatially partially coherent pulsed optical beam scattered by an atmospheric layer are presented. It is shown that in contrast to cw radiation, which scatters uniformly, the scatter of pulsed beam becomes nonuniform for shorter pulses. For the femtosecond pulse duration the backscattered radiation is localized near the strictly backward direction in the area with transverse dimensions comparable with the initial size of the sensing beam.

1. INTRODUCTION

The diffractive broadening of broadband pulsed optical beams as a function of the pulse duration has been studied in [1-5], and it has been shown that as the pulse duration decreases, the diffractive broadening of the beam becomes smaller than that of continuous wave (cw) beams. In the limit of zero pulse duration (δ - pulse), the beam diffractive broadening does not occur at all [3-5]. In lidar problems the atmospheric scatter of laser radiation is one of the key elements [6, 7]. In this paper, the scatter of short pulse optical beams by an atmospheric layer is analyzed.

2. METHODOLOGY

Let a laser beam of pulsed radiation with the initial complex amplitude distribution $U(0,\boldsymbol{\rho}',t)$ after propagation of the path L in the direction of the Z axis in the atmosphere is scattered by an atmospheric layer. The backscattered radiation is received by a photodetector in the plane $Z=0$.

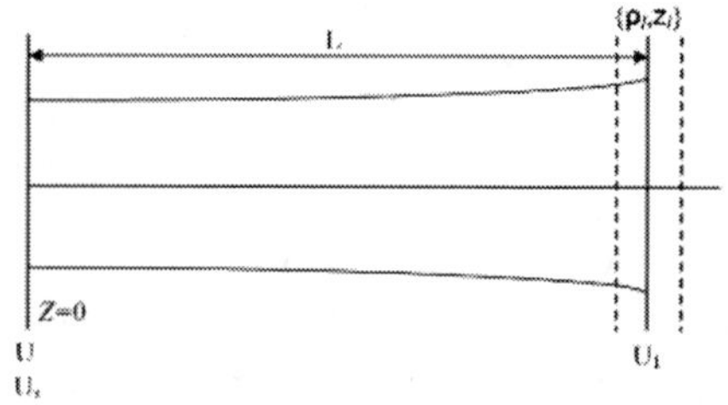

Fig. 1. Propagation and scattering geometry.

The complex amplitude of the field incident onto a scattering particle can be written in the form

$$U_1(z_l,\boldsymbol{\rho}_l,t)=\int d^2\rho' U\left(0,\boldsymbol{\rho}',t-\frac{z_l}{c}\right)G(0,\boldsymbol{\rho}';z_l,\boldsymbol{\rho}_l), \quad (1)$$

where

$$G(0,\boldsymbol{\rho}';z_l,\boldsymbol{\rho}_l)=\frac{ke^{ikz_l}}{2\pi i z_l}\exp\left\{i\frac{k}{2z_l}(\boldsymbol{\rho}_l-\boldsymbol{\rho}')^2\right\}\exp(i\psi_1)$$

is the Green's function of the parabolic equation for the complex amplitude of the field propagating in free space from point $(0,\boldsymbol{\rho}')$ to point $(z_l,\boldsymbol{\rho}_l)$, ψ_1 – is a random phase of the field due to turbulence, z_l is the longitudinal coordinate of the scattering particle, $\boldsymbol{\rho}_l=\{x,y\}$ is the radius vector determining its position in the transverse plane, $k=2\pi/\lambda=\omega/c$, λ is wavelength, ω is frequency, c is the speed of light, t is time, i is imaginary unit. The initial field of the sensing pulsed beam satisfies the condition

$$U(0,\boldsymbol{\rho}',t)=U_0(\boldsymbol{\rho}')A(t). \quad (2)$$

The complex amplitude of the field scattered by the particle, in the plane $Z=0$ can be written as

$$U_{sl}(0,\boldsymbol{\rho},t)=\alpha_l U_1\left(z_l,\boldsymbol{\rho}_l,t-\frac{2z_l}{c}\right)G(z_l,\boldsymbol{\rho}_l;0,\boldsymbol{\rho})e^{i\psi_2}, (3)$$

where $G(z_l,\boldsymbol{\rho}_l;0,\boldsymbol{\rho})$ is the Green's function and ψ_2 is a turbulent phase for the backward propagation from the scattering particle with coordinates $(z_l,\boldsymbol{\rho}_l)$ to point $(0,\boldsymbol{\rho})$ in the receiving plane, α_l is the backscattering amplitude.

The total field in the plane $(0,\boldsymbol{\rho})$ can be found through summation of the right-hand side of Eq.(3) over all scattering particles N_s in the layer

$$U_s=\sum_{l=1}^{N_s}U_{sl}\,. \quad (4)$$

According to [3-5], the strength of the electric field $E_1(z_l,\boldsymbol{\rho}_l,t)=e^{i\omega_0 t}U_1(z_l,\boldsymbol{\rho}_l,t)$ incident on the particle can be written in the form

$$E_1(z_l,\boldsymbol{\rho}_l,t)=\int d\omega\, U(z_l,\boldsymbol{\rho}_l,\omega)\exp\{-i(\omega-\omega_0)t\}, \quad (5)$$

where

$$U(z_l,\boldsymbol{\rho}_l,\omega)=\int d\rho'\, U(0,\boldsymbol{\rho}')G(0,\boldsymbol{\rho}';z_l,\boldsymbol{\rho}_l), \quad (6)$$

$$U(0,\mathbf{\rho}',\omega) = \frac{1}{2\pi} U_0(\mathbf{\rho}') \int dt\, e^{i\omega t} A(t) = P(\omega) U_0(\mathbf{\rho}'), \quad (7)$$

ω_0 is the carrier frequency.

$$A(t) = \int d\omega\, P(\omega) e^{-i\omega t}. \quad (8)$$

Upon the use of Eqs. (3) and (5)-(7) in Eq. (4), we obtain the following equation for the strength of the electric field scattered by an atmospheric layer $E_s(0,\mathbf{\rho},t) = e^{i\omega_0 t} U_s(0,\mathbf{\rho},t)$ in the plane $Z = 0$:

$$E_s(0,\mathbf{\rho},t) = \sum_{l=1}^{N_z} \alpha_l \int d\omega\, P(\omega) \exp\{-i(\omega - \omega_0)t\}$$
$$\times \int d\mathbf{\rho}' U_0(\mathbf{\rho}') G(0,\mathbf{\rho}'; z_l, \mathbf{\rho}_l) G(z_l,\mathbf{\rho}_l; 0,\mathbf{\rho}) e^{i\psi_1 + i\psi_2}. \quad (9)$$

For a thin scattering layer (short pulse) $|z_l - L| << L$, we can take $z_l \approx L$ in Eq. (3), keeping only the dependence on z_l in the fast oscillating factor e^{ikz_l} [8]. Then the Green's functions in Eq. (9) take the form

$$G(0,\mathbf{\rho}; z_l,\mathbf{\rho}_l) = \frac{k e^{ikz_l}}{2\pi i L} \exp\left\{ i \frac{k}{2L} (\mathbf{\rho} - \mathbf{\rho}_l)^2 \right\}. \quad (10)$$

Equation (9) allows us to write the following equation for the function of mutual coherence of the electric field strength of the partially coherent pulsed radiation $\Gamma_2(0,\mathbf{\rho}_1,\mathbf{\rho}_2,t_1,t_2) = \langle E_s(0,\mathbf{\rho}_1,t_1) E_s^*(0,\mathbf{\rho}_2,t_2) \rangle$ scattered by an atmospheric layer:

$$\Gamma_2(0,\mathbf{\rho}_{1,2}; t_{1,2}) = \sum_{l=1}^{N_z} \sum_{l'=1}^{N_z} \alpha_l \alpha_{l'} \int d\omega_{1,2} \langle P(\omega_1) P^*(\omega_2) \rangle$$
$$\times \exp\{-i(\omega_1 - \omega_0)t_1 + i(\omega_2 - \omega_0)t_2\} \quad (11)$$
$$\times \int d\mathbf{\rho}'_{1,2} \langle U_0(\mathbf{\rho}'_1) U_0^*(\mathbf{\rho}'_2) \rangle G(0,\mathbf{\rho}'_1; z_l,\mathbf{\rho}_l) G(z_l,\mathbf{\rho}_l; 0,\mathbf{\rho}_1)$$
$$\times G^*(0,\mathbf{\rho}'_2; z_{l'},\mathbf{\rho}_{l'}) G^*(z_{l'},\mathbf{\rho}_{l'}; 0,\mathbf{\rho}_2) \langle e^{i(\psi_1 + \psi_2 - \psi_3 - \psi_4)} \rangle,$$

where according to Eq. (7)

$$\langle P(\omega_1) P^*(\omega_2) \rangle = \frac{1}{(2\pi)^2} \int dt_{1,2} \langle A(t_1) A^*(t_2) \rangle$$
$$\times \exp\{ i[(\omega_1 t_1) - (\omega_2 t_2)] \}, \quad (12)$$

the angular brackets $\langle ... \rangle$ denote averaging over an ensemble of random realizations. At the equal coordinates $\mathbf{\rho}_1 = \mathbf{\rho}_2 = \mathbf{R}$, Eq. (11) describes the intensity distribution of the scattered radiation in the source plane $Z = 0$.

According to [9], for the arbitrary probability distribution of the initial position of scattering particles in space, the terms with $l \neq l'$ in Eq. (11) can be neglected because of the fast oscillating terms e^{ikz_1} in the Green's functions (10), if the scale of variation of the probability density function far exceeds the wavelength of the scattered field. The scales of turbulent inhomogeneities of the velocity field in the atmosphere far exceed the wavelength of optical waves, so we can use this approximation and restrict our consideration to only the terms with $l = l'$ in summation in Eq. (11). That is, the right-hand side of Eq. (11) $\sum_{l=1}^{N_z} \sum_{l'=1}^{N_z} \alpha_l \alpha_{l'} \Phi(z_l, z_{l'}; \mathbf{\rho}_l, \mathbf{\rho}_{l'})$ takes the form $\sum_{l=1}^{N_z} |\alpha_l|^2 \Phi(z_l, \mathbf{\rho}_l)$.

Then, for the averaging over microphysical factors, we can use the relation [10]

$$\left\langle \sum_{l=1}^{N_z} |\alpha_l|^2 \Phi(z_l,\mathbf{\rho}_l) \right\rangle_a =$$
$$= \int_0^\infty dz_l \int d\mathbf{\rho}_l \rho_c(z_l) \sigma_\pi(z_l) \Phi(z_l,\mathbf{\rho}_l), \quad (13)$$

where $\rho_c(z_l)$ is the concentration of scattering particles, $\sigma_\pi(z_l) = \langle |\alpha_l|^2 \rangle$ is the differential scattering cross section. Averaging over turbulent phase is carried out in accordance with known rules [6, 11]. But, as to impact of turbulence, we will neglect it. In what follows we will consider limiting case of "far zone", where the Fresnel number $\Omega_0 = (k_0 a^2)/L$ of transmitting aperture with radius a for the carrier frequency $k_0 = \omega_0/c$ is much less than unity and turbulent broadening of sensing beam, at least, on inclinated and height paths is less than diffractive one [11]. As a result, Eq. (11) acquires the form

$$\Gamma_2(0,\mathbf{\rho}_{1,2}; t_{1,2}) = \int_0^\infty dz_l \rho_c(z_l) \sigma_\pi(z_l) \int d\omega_{1,2} \langle P(\omega_1) P^*(\omega_2) \rangle$$
$$\times \exp\{ i(\omega_1 - \omega_0)t_1 + i(\omega_2 - \omega_0)t_2 \} \int d\mathbf{\rho}_l$$
$$\times \int d\mathbf{\rho}'_{1,2} \langle U_0(\mathbf{\rho}'_1,\omega_1) U_0^*(\mathbf{\rho}'_2,\omega_2) \rangle G(0,\mathbf{\rho}'_1; z_l,\mathbf{\rho}_l)$$
$$\times G(z_l,\mathbf{\rho}_l; 0,\mathbf{\rho}_1) G^*(0,\mathbf{\rho}'_2; z_{l'},\mathbf{\rho}_{l'}) G^*(z_{l'},\mathbf{\rho}_{l'}; 0,\mathbf{\rho}_2). \quad (14)$$

In Eq. (14), we can use the Gauss models for the functions of spatial and temporal correlation of the initial field [12, 13]

$$\langle A(t_1) A(t_2) \rangle = A_0^2 \exp\left\{ -\frac{(t_1 + t_2)^2}{2T^2} - \frac{(t_1 - t_2)^2}{T_0^2} \right\}, \quad (15)$$

$$\langle U_0(\mathbf{\rho}'_1) U_0^*(\mathbf{\rho}'_2) \rangle = U_0^2 \exp\left\{ -\frac{\mathbf{\rho}_1'^2 + \mathbf{\rho}_2'^2}{2a^2} - \frac{(\mathbf{\rho}'_1 - \mathbf{\rho}'_2)^2}{\rho_0^2} \right\}. \quad (16)$$

where $\frac{1}{T_0^2} = \frac{1}{4T^2} + \frac{1}{T_c^2}$, T is the pulse duration, T_c is the coherence time of the source field, a is radius, F is the wavefront curvature radius, and

DOI: 10.1051/epjconf/201611906018

ρ_0 is the radius of spatial coherence of the beam field in the initial plane.

Upon the integration over the thickness of the scattering layer [8], coordinates of the initial plane $\boldsymbol{\rho}_1'$ and $\boldsymbol{\rho}_2'$, and the plane of the scattering layer $\boldsymbol{\rho}_l$ in Eq.(14), we obtain for $\Gamma_2(\mathbf{R},\boldsymbol{\rho},t,\tau)$, where $\mathbf{R} = \frac{\boldsymbol{\rho}_1 + \boldsymbol{\rho}_2}{2}$, $\boldsymbol{\rho} = \boldsymbol{\rho}_1 - \boldsymbol{\rho}_2$, $t = (t_1 + t_2)/2$, $\tau = t_1 - t_2$, the expression in a form of a 2-D integral.

This expression allows one to obtain a formula for the mean intensity of the pulsed radiation scattered by an atmospheric layer $I(\mathbf{R})$, if put $\boldsymbol{\rho}_1 = \boldsymbol{\rho}_2$. In particular, for sensing beam with $T_0 = T$ in a far zone $\Omega_0 << 1$ we obtain the following equation for the mean intensity

$$I(\mathbf{R}) = I(R) = \Gamma_2\left(R, 0, t = \frac{L}{c}, 0\right) =$$

$$C\,\omega_0 T \int d\Omega_{1,2} \frac{\Omega_1^2 - \Omega_2^2}{Q_{\Omega F}} \times \qquad (17)$$

$$\times \exp\left\{-(\omega_0 T)^2\left[\Omega_2^2 + \frac{1}{4}(\Omega_1 + 1)^2\right] + \frac{R^2}{a_g^2}\left[i\frac{\Omega_2}{\Omega_0} - \frac{\Omega_2^2 a_\rho}{\Omega_0^2 Q_{\Omega F}}\right]\right\},$$

where $C = \frac{1}{2}\beta_\pi (a/L)^4 U_0^2 A_0^2$, $\beta_\pi = \rho_c \sigma_\pi$ denotes the backscattering coefficient, $Q_{\Omega F} = \Omega_1^2 + \Omega_2^2 a_\rho - i2\Omega_2 a_\rho \Omega_0^{-1}$, $a_\rho = 4(a/\rho_0)^2 + 1$, $a_g = L/(k_0 a)$.

Equation (17) allows us to calculate the mean intensity of the field of scattered beam for any pulse duration. At $\omega_0 T \to \infty$, it transforms into the equation corresponding to cw radiation $I = \beta_\pi \left(a^4/L^4\right) U_0 A_0/2$, where there is no dependence on the vector $\mathbf{R}$, that is, the intensity of the scattered cw radiation is distributed uniformly over the transverse plane. In contrast to the cw radiation, the integrand (17) at the finite values of the parameter $\omega_0 T$, which characterizes the pulse duration, keeps the dependence on the spatial coordinate $\mathbf{R}$. That is, the uniform scattering of the cw radiation becomes nonuniform in the case of the pulsed sensing radiation.

3. RESULTS

Figure 2 shows the intensity distributions of the beam with $\Omega_0 = 0.1$ scattered by the atmospheric layer. The calculations were based on Eq.(17) without regard for the constant C. It follows from the figure that the diffuse uniform scattering of cw optical radiation ($\omega_0 T = 10^8$) becomes nonuniform as the pulse duration decreases.

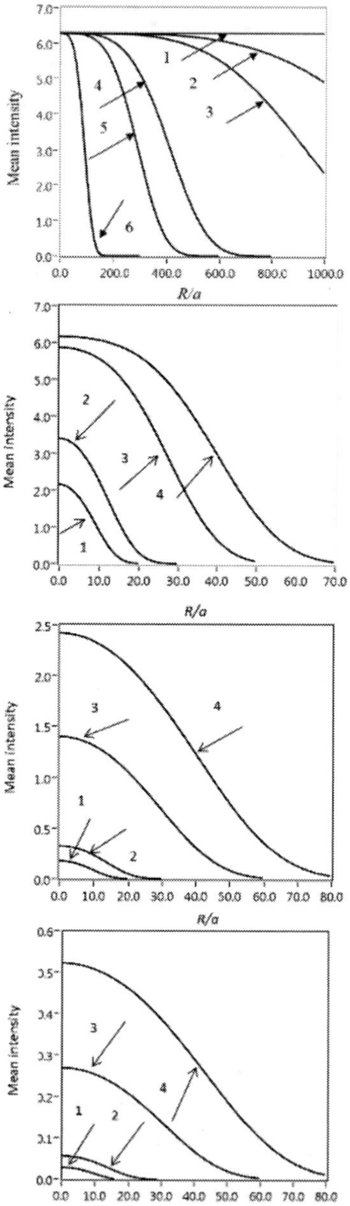

Fig. 2. Intensity distribution of the scattered collimated beam, Ω_0=0.1: a) $\omega_0 T$=10^8 (1), 10^5 (2), $5\cdot10^4$ (3), 10^4 (4), $5\cdot10^3$ (5), 500(6); b), c), d) $\omega_0 T$=5 (1), 10 (2), 50 (3), 10^2 (4); $a/\rho_0 = 0$(a, b), 2(c), 5(d).

As the parameter $\omega_0 T$ decreases, the scattered radiation localizes in the smaller vicinity of the strictly backward direction (compare Figs.2a and 2b), and for femtosecond pulses $\omega_0 T = 5$ the scattered radiation is localized in the transverse plane near the axis of the sensing beam within a few its initial dimensions.

Deterioration of the initial spatial coherence of the pulsed sensing beam has no effect on the localization of the scattered radiation near the beam axis (Figs. 2c, 2d). In the full accordance with Eq. (19), as the radius of spatial coherence of the initial field ρ_0 decreases, only the absolute value of the mean intensity decreases, while the scale of its decrease in the transverse plane remains unchanged.

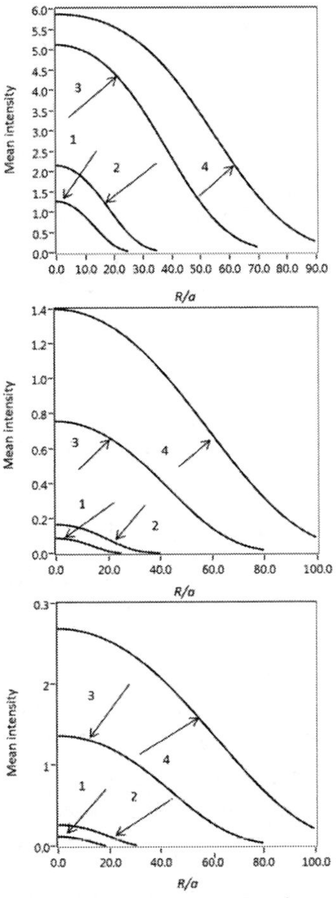

Fig. 3. Intensity distribution of the scattered collimated beam, $\Omega_0 = 0.05$: $\omega_0 T = 5$ (1), 10 (2), 50 (3), 10^2 (4); $a/\rho_0 = 0$(a), 2(b), 5(c).

The decrease of the Fresnel number of the transmitting aperture Ω_0 diminishes the effect of nonuniformity of scattering of the pulsed radiation (Fig.3).

It can be seen from Fig. 3 and Fig. 2 that for $\Omega_0 = 0.05$ and $\omega_0 T = 5$ the scattered radiation is localized near the axis of the sensing beam in the area with transverse dimensions approximately two times greater than that for $\Omega_0 = 0.1$.

Deterioration of the initial spatial coherence of sensing beam does not lead to changes of transverse dimensions of this area (Figs. 3b and 3c). This leads to only the significant decrease in the intensity of the scattered radiation.

ACKNOWLEDGEMENT

This study was supported by the Russian Scientific Foundation for Maintenance and Development (Project 14-17-00386).

REFERENCES

[1] I.V. Zaloznaya, A.V. Falits, Atmospheric and Oceanic Optics **22**, pp.590-594 (2009).
[2] L.O. Gerasimova, I.V. Zaloznaya, Atmospheric and Oceanic Optics **24**, pp.185-189 (2011) [in Russian]
[3] V.A. Banakh, Opt. Lett. **36**, pp.4554-4556 (2011).
[4] V.A. Banakh, L.O. Gerasimova, I.V. Zaloznaya, O.V. Tikhomirova, Atmospheric and Oceanic Optics **26**, pp.185-189 (2013).
[5] V.A. Banakh, L.O. Gerasimova, Atmospheric and Oceanic Optics **26**, pp.5-10 (2013) [in Russian]
[6] V. A. Banakh, V.L. Mironov, "Lidar in a turbulent atmosphere", Artech House. Boston & London, 1987, 185 p.
[7] V. Banakh, I. Smalikho, "Coherent Doppler wind lidars in a turbulent atmosphere", Artech House. Boston-London. 2013. 250 p.
[8] V.A. Banakh, Atmospheric and Oceanic Optics **20**, pp.271-274 (2007).
[9] B. Crosignani, P. Di Porto, M. Bertolotti, "Statistical Properties of Scattered Light", Academic Press, New York-San Francisco-London, 1975, 206 p.
[10] A. Ishimaru, "Wave Propagation and Scattering in Random Media", Academic Press, 1978, 572 pp.
[11] V.E. Zuev, V.A. Banakh, V.V. Pokasov, "Optics of turbulent atmosphere", Leningrad, Gidrometeozdat, 1988, 270 p. [in Russian].
[12] I.P. Christov, Optica Acta **33**, pp.63-72 (1986).
[13] A.I. Kon, V.I. Tatarskii, Radiophysics and Quantum Electronics. **15**, pp.1187-1192 (1972).

EPJ Web of Conferences **119**, 07001 (2016) DOI: 10.1051/epjconf/201611907001
ILRC 27

TOWARD AN OBSERVATION OF VOLCANIC ASH: WHICH KIND OF OBSERVATION CAN BE MADE BY DIFFERENT INSTRUMENTS

Florence Besson [1]*, Jean-Luc Lampin [1]*

[1]Météo-France,Upper Air Observation Department,42 Av Coriolis 31057 Toulouse Cedex, France,
**Email: florence.besson@meteo.fr, jean-luc.lampin@meteo.fr*

ABSTRACT

In the first part, this presentation describes how an intercomparison was conducted between several lidars and ceilometers, the processing algorithms used, the difficulties encountered and the method used to design an optimum volcanic ash detection network.

In the second part, we describe the work progress of a feasibility study concerning a WMO intercomparison for volcanic ash detection that is beginning with experts and colleagues from many others countries.

1. INTRODUCTION

The 2010 eruptions of the Eyjafjallajökull (Icelandic volcano) resulted in one of the greatest air traffic disruption in Western Europe with 107,000 flights cancellations during an 8-day period, accounting for 48% of total air traffic and roughly 10 million passengers. The estimated cost of this European controlled airspace shutdown is about € 1.3 billion according to the International Air Transport Association (IATA).

London Volcanic Ash Advisory Center (VAAC) was responsible for providing information about the ash plume to the relevant civil aviation authorities in the form of Volcanic Ash Advisories (VAA). On this basis, the authorities made decisions about when and where airspace should be closed due to the safety issues.

It was difficult for the VAAC to provide accurate information about the size and concentration of the volcanic particles. As a result, Toulouse VAAC asked Météo-France to find solutions to detect volcanic ashes from the ground to a 12km altitude and to assess their concentration.

2. METHODOLOGY

During summer 2012, Météo-France conducted an intercomparison campaign between several lidars and ceilometers with in-situ measurements (balloon-borne particle counters and aircraft sensor) in order to assess their ability to detect Saharan dust for lack of volcanic ash. The observations produced by the instruments are obtained thanks to several algorithms: STRAT from SIRTA and BASIC from the LOA (both French laboratories). It turns out that it was difficult to compare the data one among other for several reasons: there is no aerosol measurement reference, instruments have different wavelengths, they can be dual-polarized. Results show the aerosols lidar technology may be a good mean to meet the VAAC requirements. Moreover, having the desire to built an efficient network in terms of number and location of sites over metropolitan France, Météo-France has run a model of pollutant dispersion, named MOCAGE, in retro-plume mode with different configurations in order to define the network which provides the best coverage.

3. RESULTS

Important differences appeared between ceilometers and lidars concerning Saharan dust observations. The ceilometers were not able to see the aerosols layer because of the noise while it was clearly visible with lidar.

The structure of the plume was well described by the lidars and was coherent with the forecast of the pollutant dispersion model.

It's hard to compare instruments with each other because of their very different technology (wavelength, laser energy, detection mode…).

We also observed that double polarization is interesting by giving information about sphericity of the particle and that the Raman N2 channel is promising, but we didn't have our own algorithm able to deal with N2 Raman channel.

4. CONCLUSIONS

Lidars have shown their great potential for detecting sub-visible atmospheric particles and providing accurately their location in altitude with an accuracy of around 10-100 meters. Lidar technology has been improved and commercial systems are now available. Such instruments can be the best candidate for detecting from the ground, particles coming from volcanic eruptions, pollution, biomass burning, mineral sources, … or can be used for dynamic information such as height of the boundary layer. For operational uses existing lidar systems need to be evaluated with international standards. This is the reason why a WMO intercomparaison would be appreciated but many issues remain and solving them is the aim of the feasibility study that is beginning.

DOI: 10.1051/epjconf/201611907002

SIGNAL TO NOISE RATIO ESTIMATIONS FOR A VOLCANIC ASH DETECTION LIDAR. CASE STUDY: THE MET OFFICE.

George Georgoussis [1], Mariana Adam[2], George Avdikos[1*]

[1]*Raymetrics S.A., Spartis 32,Metamorfosis, Greece *Email: gavdikos@raymetrics.gr*

[2]*Observations, Met Office, Exeter, Devon EX1 3PB, United Kingdom*

ABSTRACT

In this paper we calculate the Signal-to-Noise (*SNR*) ratio of a 3-channel commercial (Raymetics) volcanic ash detection system, (LR111-D300), already operating under Met Office organization. The methodology for the accurate estimation is presented for day and nighttime conditions. The results show that *SNR* values are higher than 10 for ranges up to 13 km for both nighttime and daytime conditions. This is a quite good result compared with other values presented in bibliography and proves that such system is able to detect volcanic ash over a range of 20 km.

1. INTRODUCTION

The Eyjafjallajokull eruption started on 20 March 2010, for almost 30 days resulted in large amounts of volcanic ash being injected into the atmosphere [1]. This event strongly affected the European and global air transport industry since volcanic ash (VA) plumes were subjected to long-range transport and carried over large areas in Central Europe due to strong westerly winds. Although volcanic ash is composed of materials with very small dimensions, the damages it may cause can be significant because, after expelled in the atmosphere to a very high altitude (kilometers), it is scattered and usually falls on large surfaces. In all cases, volcanic ash at high altitudes in the atmosphere can pose a serious hazard to aircraft engines [2]. VA has a highly corrosive effect, primarily on gas turbines of the aircrafts as they get temporarily or definitively blocked after aspirating the dust, which melts at high temperatures inside the combustion chamber, and then cools down and solidifies on the cooler parts of the engine. Such a situation can cause sudden and permanent engine failure. VA poses a hazard also to gas turbine blades and causes friction scratches on the windshield, resulting in poor visibility [3]. Monitoring and detecting VA particles remains a difficult task since aerosol particles are highly inhomogeneous and vary in time and space. Consequently, aerosol observations and measurements have to be global, continuous and systematic. Ground-based aerosol remote sensing instruments such as lidar can contribute significantly to the understanding of aerosols properties and potential associated impacts, thanks to their distribution. Among lidar systems, those which have the capability of detecting aerosols' type (depolarization lidar systems) are appropriate for detecting and distinguishing volcanic ash [4]. In several European countries, meteorological agencies decided to develop lidar networks (e.g. Met Office) in order to support decision-making in the case of volcanic ash events. In our paper we aim to show that automatic, depolarization lidar systems with routine operation can be a quite powerful tool for detecting volcanic over a range of 20 km or more. The paper is structured as follows: Sect. 2 is dedicated to the description of the Raymetrics lidar system and to the methodology for the estimation of *SNR* values including the error propagation. In Sect. 3 we present the results of the *SNR* analysis for both nighttime and daytime conditions. In the last section, we conclude with the most important results arising from the *SNR* values.

2. METHODOLOGY

The Met Office lidar network consists of two Raymetrics depolarization Raman lidars, located in Camborne. In the near future, 8 new systems will be dispersed at several locations over UK and operate as a lidar network of 10 systems. For more information please see Adam et al. [5]. The Raymetrics lidar system is a portable, 3-channel (355p, 355s and 387nm) *eye-safe depolarization Raman lidar system* suitable for detection of any volcanic ash, fully automated including a lot of sensors, electronics sub-systems and interlocks (which block the operation at emergency

DOI: 10.1051/epjconf/201611907002

situations). It can work 24 h per day outdoors in an unattended mode under almost any weather conditions (Fig. 1). A pulsed laser beam at 355 nm is emitted into the atmosphere. The energy per emitted pulse is 53 mJ, while the pulse duration is ~ 10 ns. A beam expander is used at the emission unit in order to expand the laser beam by a factor of 7, so that the eye safety is completely fulfilled. The repetition rate is 20 Hz. The backscattered radiation is collected by a Cassegrain telescope of 300mm in diameter and having an f-number of 5 (focal length f = 1500 mm). The collected radiation is spectrally analyzed (using beam splitters), filtered (using narrow band interference filters) and focused on photomultiplier tubes (PMTs) which are used to detect the received lidar signals in the analog and the photon counting mode. The corresponding raw signal spatial resolution is 7.50 m. For this study the raw temporal resolution of the retrieved aerosol profiles was set at 1 min, while the spatial resolution was 15 m. The full overlap of the instrument is achieved for altitudes higher than ~250m.

The basis of the methodology for the *SNR* estimation is similar to that described in [6]. At instrumental level, data processing includes correction such as background sky irradiance, which is measured by averaging lidar signal at high altitude range (above 49 km). Finally, lidar profiles within 10 min were accumulated and signal-to-noise ratio is estimated according to the procedure that follows.

For *N* laser shots, the counts per bin number

$$C'^{n}_{m} = \sum_{n=0}^{Nshots} C^{n}_{m} \text{ m} \quad (1)$$

(m memory bin from 0 to 8000). In case that we have a resolution of 15 meters it is like sampling every 10 MHz, so

$$M^{n}_{m}(Hz) = \frac{10*10^{6}*C^{n}_{m}}{Nshots} \Rightarrow M^{n}_{m}(MHz) = \frac{10*\sum_{n=0}^{Nshots} C^{n}_{m}}{Nshots} \quad (2)$$

Thus, if we want convert back to Counts per bin (which is the sum of counts for *Nshots*) we simply reverse the above equation:

$$C'^{n}_{m} = \frac{Nshots * M^{n}_{m}(MHz)}{10} \quad (3)$$

Figure 1. Raymetrics volcanic ash detection system installed and operated by Met Office

If we assume that an electronic system can average 12000 shots (e.g. 10min data at 20Hz) then we should get

$$M^{n}_{m} = \frac{10*C'^{n}_{m}}{12000} = \frac{10*\sum_{n=1}^{12000} C^{n}_{m}}{12000} \quad (4)$$

In that case, if we like to go back to Counts we should get

$$C'^{n}_{m} = \frac{12000*M^{n}_{m}}{10} \quad (5)$$

During nighttime, the background for photon counting is almost zero such that it can be neglected. However, in case there is some noise (e.g. during daytime) we proceed as following

$$C'^{n}_{m} \equiv C_{total} = C_{sig} + C_{bg} \Rightarrow C_{sig} = C_{total} - C_{bg} \quad (6)$$

From error propagation law, we obtain:

$$\Delta C_{sig} = \sqrt{(\Delta C_{total})^2 + (\Delta C_{bg})^2} \quad (7)$$

leading to

$$SNR = \frac{C_{sig}}{\Delta C_{sig}} = \frac{C_{sig}}{\sqrt{(\Delta C_{total})^2 + (\Delta C_{bg})^2}} \quad (8)$$

For photon counting the signal noise follows the Poisson statistics. The error of a counting rate is equal to its square-root. Hence, we get:

$\Delta C = \sqrt{C}$ and then $\Delta C_{total} = \sqrt{C_{total}}$ =>

$(\Delta C_{total})^2 = C_{total}$ and similar $(\Delta C_{bg})^2 = C_{bg}$

Finally,

$$SNR = \frac{C_{sig}}{\Delta C_{sig}} =$$

$$\frac{C_{sig}}{\sqrt{(\Delta C_{total})^2 + (\Delta C_{bg})^2}} = \frac{C_{sig}}{\sqrt{C_{sig} + 2C_{bg}}} \quad (9)$$

3. RESULTS

In this section we present the results of the methodology previously described. The analysis involves data processing for both daytime (noisy) and nighttime (noiseless) conditions, depicted in Figs 2 and 3 respectively. First, we processed the data for daytime conditions. To do this, we averaged 10 profiles, each one corresponding to 1 minute data. In all cases, the alignment was checked by fitting with molecular atmosphere. The Range Corrected Signal (*RCS*) for 355-parallel channel (analog mode) during daytime is depicted at Fig. 2 (upper plot). In order to carry out a more accurate analysis, we performed gluing analysis (by plotting photon counting versus analog signal with background correction in order to convert the analog to photon counting) (Fig. 2, middle plot). Finally, the *SNR* values were calculated by using equation (9) from the previous section (Fig. 2, bottom plot).

The black curve corresponds to 355-parallel signal (glued signal) whilst the red curve to 355-perpendicular signal (glued signal). As it can be easily seen, the *SNR* is equal to 10 at about 11.5 km for the 355-perpendicluar signal and 13 km for the parallel signal, which is a quite good result as compared with other references [6]. For comparison reasons, we made the same analysis for nighttime conditions (Fig. 3). The value of *SNR* of 10 was found to be at ~ 17.8 km and 18.7 km for parallel and perpendicular channel respectively. The green curve (Figure 3, bottom plot) corresponds to *SNR* values for the Raman channel (387 nm).

4. CONCLUSIONS

By processing daytime and nighttime routine lidar measurements, calculation for the Signal-to-Noise Ratio was feasible. According to our analysis, the

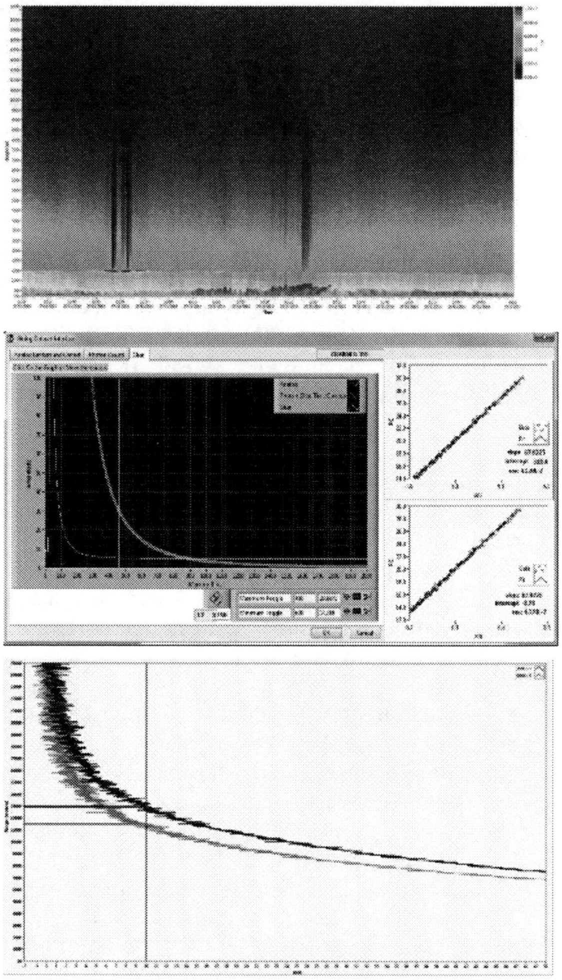

Figure 2. Daytime conditions (27/11/2014): *RCS* for 355nm (Analog) (top). GUI interface of Gluing procedure for 355-parallel signal (middle). *SNR* values versus range (bottom).

SNR can reach the value of 10 at the range of 11.5 km and 13.0 km for both 355 nm channels leading to a very promising result. We showed that the system is capable of detecting volcanic ash (VA) aerosols for ranges up to 20 km or more. A network of 10 lidar systems across UK could provide very useful information in case that an event similar to that of 2010 takes place.

ACKNOWLEDGEMENT

The authors would like to thank Met Office for giving us the permission to use the system picture and for providing the data for processing.

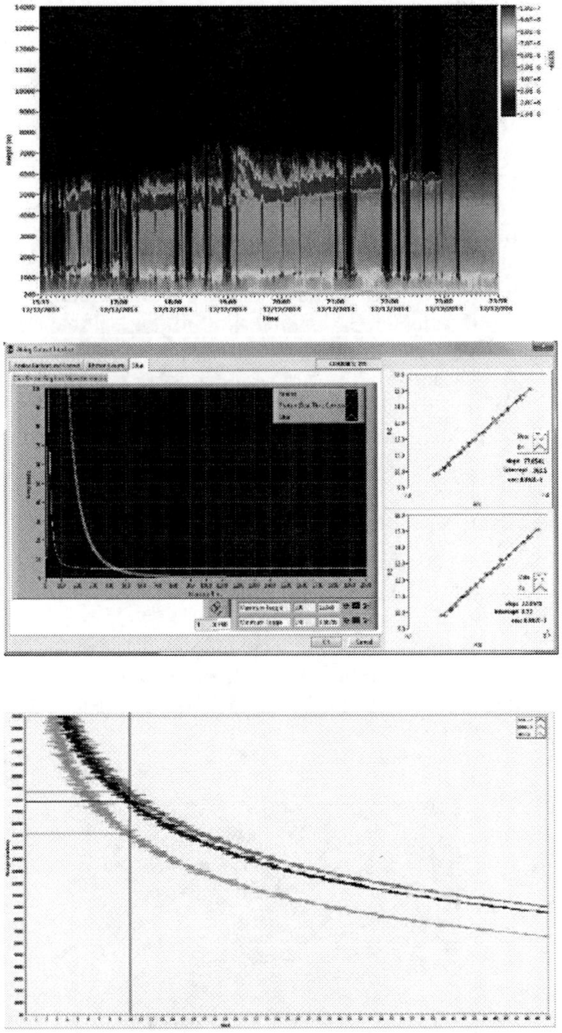

Figure 3. Nighttime conditions (12/12/2014): *RCS* for 355nm (Analog) (top). GUI interface of Gluing procedure for 355-parallel signal (middle). *SNR* values versus range (bottom).

REFERENCES

[1] Sigmundsson, F., Hreinsdottir, S., Hooper, A., Arnadottir, T., Pedersen, R., Roberts, M. J., O'skarsson, N., Auriac, A., Decriem, J., Einarsson, P., Geirsson, H.,Hensch, M., O' feigsson, B. G., Sturkell, E., Sveinbjornsson, H., and Feigl, K. L. 2010: Intrusion triggering of the 2010 Eyjafjallajokull explosive eruption, *Nature*, **468**, 426–430.

[2] A. Mortier, P. Goloub, T. Podvin, C. Deroo, A. Chaikovsky, N. Ajtai, L. Blarel, D. Tanre1, and Y. Derimian 2013: Detection and characterization of volcanic ash plumes over Lille during the Eyjafjallajokull eruption, *Atmos. Chem. Phys.*, **13,** 3705–3720.

[3] Ajtai, N., Stefanie, H., Stoian, L. C., and Oprea, M. G.: The volcanic ash and its impact on European air transport industry 2010: A case study on the detection and impact of the the Eyjafjallajokull volcanic ash plume over North-Western Europe between 14 and 21 April 2010, *AES Bioflux,* **2**, 57–68.

[4] Muller, D., Ansmann, A., Mattis, I., Tesche, M., Wandinger, U., Althausen, D., and Pisani, G. 2007: Aerosol-type-dependent lidar ratios observed with Raman lidar, *J. Geophys. Res.*, **112,** D16202, 1-11.

[5] Adam, M. et al. 2015: From operational ceilometer network to operational lidar network, *27 ILRC* – Coordinated lidar networks and applications.

[6] B. Heese, H. Flentje, D. Althausen, A. Ansmann, and S. Frey 2010: Ceilometer lidar comparison: backscatter coefficient retrieval and signal-to-noise ratio determination, *Atmos. Meas. Tech.*, **3**, 1763–1770.

EPJ Web of Conferences **119**, 07003 (2016) DOI: 10.1051/epjconf/201611907003
ILRC 27

REMOTE SENSING OF VOLCANIC ASH AT THE MET OFFICE

F. Marenco*, J. Kent, M. Adam, J. Buxmann, P. Francis and J. Haywood

*Met Office, Exeter EX1 3PB, United Kingdom; *Email: franco.marenco@metoffice.gov.uk*

ABSTRACT

The eruption of Eyjafjallajökull in 2010 has triggered the rapid development of volcanic ash remote sensing activities at the Met Office. Volcanic ash qualitative and quantitative mapping have been achieved using lidar on board the Facility for Airborne Atmospheric Measurements (FAAM) research aircraft, and using improved satellite retrieval algorithms. After the eruption, a new aircraft facility, the Met Office Civil Contingencies Aircraft (MOCCA), has been set up to enable a rapid response, and a network of ground-based remote sensing sites with lidars and sunphotometers is currently being developed. Thanks to these efforts, the United Kingdom (UK) will be much better equipped to deal with such a crisis, should it happen in the future.

1. INTRODUCTION

The eruption of Eyjafjallajökull in 2010 has put the Met Office at the centre of public attention, due to its responsibilities as Volcanic Ash Advisory Centre (VAAC). The policy followed by aviation authorities until then had been to avoid flying in any volcanic ash, but it was rapidly updated into a number of concentration thresholds. The need to estimate concentrations has since then occurred for forecasters, and this has triggered advances in terms of research, and improvements of the observing infrastructure.

2. AIRCRAFT OBSERVATIONS

In 2010, a detailed dataset of volcanic ash concentrations has been compiled from airborne lidar measurements [1]. The data were collected over the UK and surrounding seas, during six flights of the FAAM research aircraft (Fig. 1). In addition to the remote sensing, in-situ data were collected during plume penetrations, and yielded information on the size-distribution. This information has been exploited to evaluate the ash specific extinction, and hence to convert lidar-derived extinction to concentration (Fig. 2).

This dataset has been successfully used for the improvement of dispersion models [see e.g. Ref. 2 and 3] and for the validation of satellite products based on the Infrared Atmospheric Sounding Interferometer (IASI) and the Spinning Enhanced Visible and Infrared Imager (SEVIRI) [see e.g. Ref. 4 and 5]. The volcanic ash was observed to be in patches extending 85-550 km in the horizontal and 0.5-3 km in the vertical, with typical concentrations of 300-650 $\mu g/m^3$ (maxima of 800-1900 $\mu g/m^3$). The altitude of the volcanic layers was found between 2 and 8 km.

To be able to face future eruptions, the MOCCA aircraft has been set up (Fig. 1). This Cessna 421, driven by piston engines, carries an SO_2 analyser, a three-wavelength nephelometer, a Cloud, Aerosol, and Precipitation Spectrometer (CAPS), and a Leosphere ALS450 backscatter lidar operating at 355 nm.

Figure 1. The FAAM (top) and MOCCA (bottom) research aircraft.

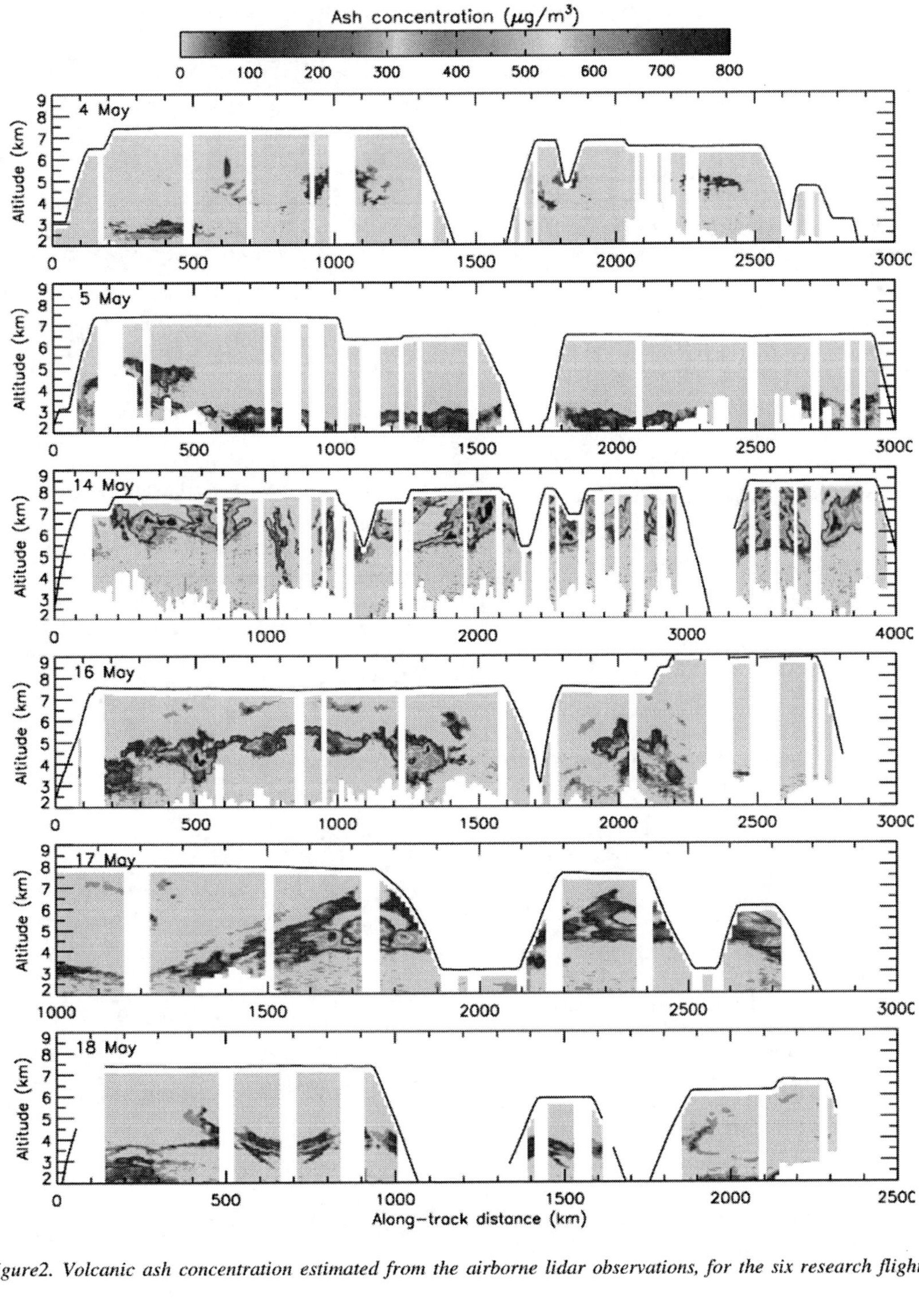

Figure2. Volcanic ash concentration estimated from the airborne lidar observations, for the six research flights.

3. QUALITATIVE AND QUANTITATIVE SATELLITE RETRIEVALS

A robust method to detect volcanic ash, using data from the infrared channels of SEVIRI has been investigated [4]. Moreover, the simultaneous retrieval of quantitative volcanic ash physical properties using a one-dimensional variational analysis framework has been developed. Sensitivity experiments have shown that the retrieved quantities are strongly dependent on the choice of ash refractive index. The retrievals compare well with the airborne lidar results (see Fig. 3).

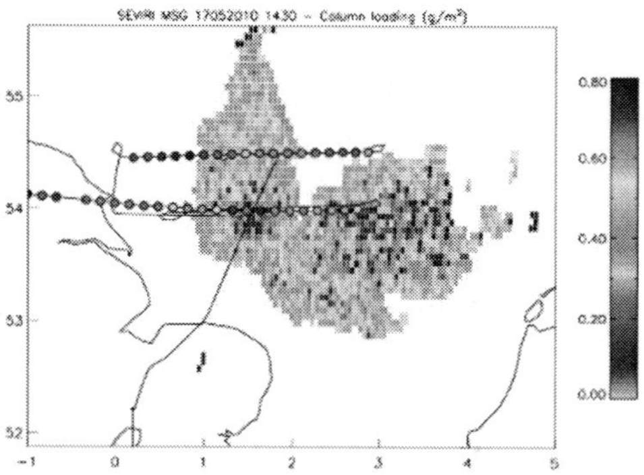

Figure 3. Volcanic ash column loading over the North Sea, determined from Meteosat products for 17/5/2010 at 14:30. Overlaid is the flight track of the FAAM aircraft, and the coloured circles represent the column loadings determined with the aircraft lidar.

Moreover, simulated satellite imagery is being produced every time the Numerical Atmospheric-dispersion Modelling Environment (NAME) dispersion model runs [6]. The simulated imagery (see Fig. 4) permits likewise comparison with the actual imagery, and is a useful tool for assessing the quality of the forecast and for understanding how the ash layer properties affect the imagery itself.

4. IMPROVED GROUND-BASED REMOTE SENSING NETWORKS

For the purpose of detecting volcanic ash plumes, nine ground-based sites distributed over the UK will be equipped with single-wavelength Raymetrics LR111-D300 Raman/depolarization

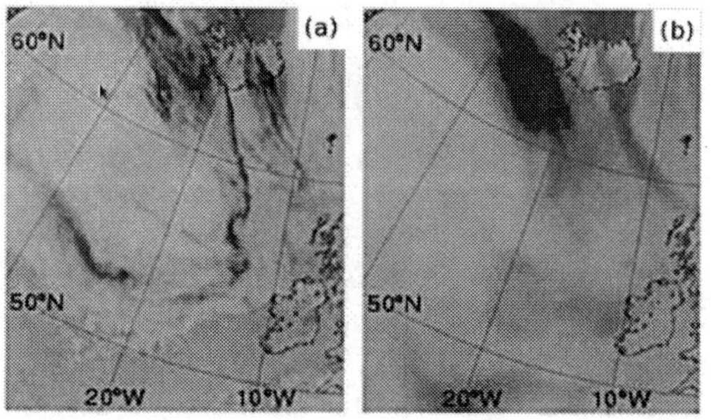

Figure 4. (a) "Dust RGB" product from SEVIRI/MSG for 9/5/2010 at 12:00 and (b) imagery simulated from the NAME dispersion model output for the same time. Courtesy of Sarah Millington.

lidar systems operating at 355 nm and 8-wavelength CIMEL CE318 NE DPS9 sunphotometers (Fig. 5). In addition, one mobile unit will be also available, having the same equipment. Currently, one fixed site and the mobile unit are installed. These developments represent a great improvement of coverage over the country for ground-based remote sensing observations, and are complemented by the observations of the operational ceilometers network [7].

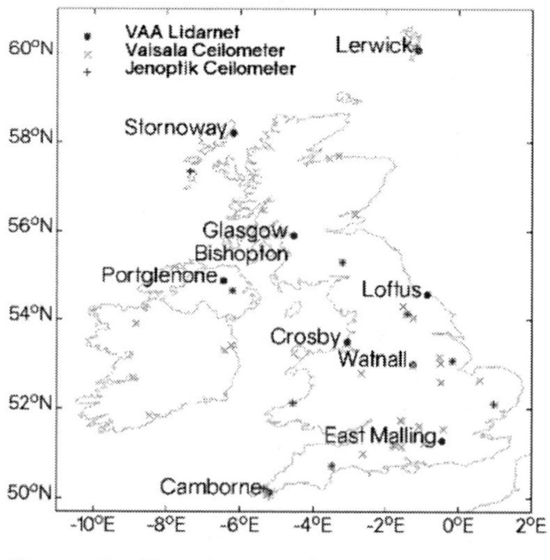

Figure 5. The future volcanic ash lidar and sunphotometer network, and the operational ceilometers network.

ACKNOWLEDGEMENTS

Airborne data was obtained using the BAe-146 Atmospheric Research Aircraft flown by Directflight Ltd. and managed by the Facility for Airborne Atmospheric Measurements (FAAM), which a joint entity of the Natural Environment Research Council (NERC) and the Met Office. The MOCCA aircraft and the ground-based lidar network are funded by the Civil Aviation Authority and the Department for Transport.

REFERENCES

[1] Marenco et al, Airborne lidar observations of the 2010 Eyjafjallajökull volcanic ash plume, JGR 2011.

[2] Webster et al, Operational prediction of ash concentrations in the distal volcanic cloud from the 2010 Eyjafjallajökull eruption, JGR 2012.

[3] Kristiansen et al, Performance assessment of a volcanic ash transport model mini-ensemble used for inverse modeling of the 2010 Eyjafjallajökull eruption, JGR 2012.

[4] Francis et al, Retrieval of physical properties of volcanic ash using Meteosat: A case study from the 2010 Eyjafjallajökull eruption, JGR 2012.

[5] Newman et al, A case study of observations of volcanic ash from the Eyjafjallajökull eruption: 2. Airborne and satellite radiative measurements, JGR 2012.

[6] Millington et al, Simulated volcanic ash imagery: A method to compare NAME ash concentration forecasts with SEVIRI imagery for the Eyjafjallajökull eruption in 2010, JGR 2012.

[7]Adam et al, From operational ceilometer network to operational lidar network, 27th ILRC – Coordinated lidar networks and applications, 2015.

EPJ Web of Conferences **119**, 07004 (2016) DOI: 10.1051/epjconf/201611907004
ILRC 27

CASE STUDY ON COMBINED LIDAR-PHOTOMETER RETRIEVAL OF VOLCANIC ASH PROPERTIES

Josef Gasteiger[1]*, Matthias Wiegner[1], Carlos Toledano[2], Silke Groß[3], Volker Freudenthaler[1]

[1]*Meteorologisches Institut, Ludwig-Maximilians-Universität, 80333 München, Germany*
**Email: josef.gasteiger@lmu.de*

[2]*Grupo de Óptica Atmosférica, Universidad de Valladolid, Valladolid, Spain*

[3]*Institut für Physik der Atmosphäre, Deutsches Zentrum für Luft- und Raumfahrt, Oberpfaffenhofen, Germany*

ABSTRACT

We study the benefit of considering sun-/sky-photometer measurements in a microphysical lidar retrieval. Furthermore, to assess the importance of the aerosol model employed by the retrieval, we compare results obtained using a spheroid aerosol model with results using an advanced aerosol model that considers irregular particle shapes. Preliminary results are shown for the mass-extinction conversion factor and the single scattering albedo during a measurement case of long-range-transported volcanic ash.

1. INTRODUCTION

It had been shown that multi-wavelengths Raman- and depolarization-lidar-systems can provide aerosol data that is sufficient for the retrieval of microphysical properties of volcanic ash [1]. The uncertainties of such retrievals are quite large because of limitations of the information content of the lidar data and also because of assumptions required in the aerosol model employed by the retrieval. Combining vertically-resolving lidar measurements with coincident ground-based sun-/sky-photometer measurements seems to be a promising approach to reduce the retrieval uncertainties because of increased information content.

Lopatin et al. [2] developed a combined lidar-photometer retrieval based on the AERONET photometer algorithm extended by consideration of multi-wavelength backscatter coefficients from lidar. They found that the lidar ratio can be retrieved with improved accuracy if the lidar data is considered in addition to the photometer data.

In the present study, we compare retrieval uncertainties of a lidar-only retrieval with uncertainties when photometer data is considered in addition by the retrieval. Furthermore, we compare the retrieval results when a spheroid aerosol model is used with results when a more complex model is used. For these comparisons, we consider the mass-extinction conversion factor $\eta = M / \alpha_{ext}$, which allows one to calculate the particle mass concentration M from extinction coefficients α_{ext} (e.g. provided by advanced lidar systems), and the single scattering albedo, which is relevant for radiative transfer. We use measurement data of long-range transported volcanic ash at Maisach/Munich (Germany) in April 2010 for this study.

2. METHODOLOGY

Our retrieval approach is based on the lidar retrieval scheme described by [1] and is extended by an improved model for aerosol ensembles and consideration of photometer measurements. It uses the Monte-Carlo method for sampling the microphysical parameters of aerosol ensembles. By comparing modeled results with measurement data, solutions of the retrieval are found.

In the present study we consider two aerosol models. The first model **(ash model A)** describes particle ensembles consisting of prolate and oblate spheroids. The particle size distributions are mono-modal log-normal distributions. The refractive index is wavelength-independent and does not vary within an ensemble. The particle aspect ratio distributions are described by modified log-normal distributions. Model A is described in detail in Section 3 of [1].

The second model **(ash model B)** describes ensembles of irregularly-shaped particles, including deformed spheroids, aggregates, and edged particles as shown in Fig. 1 of [3]. The optical properties of these particles were calculated using the Discrete Dipole Approximation [4] up to a size parameter x (=

$2\pi r/\lambda$) of about 20. For larger particles ($x > 20$), it is assumed that the lidar ratio and linear depolarization ratio is the same as for $x = 20$. For the sky radiance simulations required for the photometer retrieval part, large irregularly-shaped particles ($x > 20$) are substituted by prolate spheroids with a prescribed wide aspect ratio distribution (with median around 2.0 and maximum 5.0). Model B ensembles consist of two independent modes. The size distribution of each mode is a log-normal distribution. The refractive index is wavelength-dependent as given by the refractive indices of the mineral components of OPAC [5]. To account for the natural variability, the real part of the refractive index is allowed to vary by ±0.04 from OPAC mineral, and the imaginary part is multiplied with a random number in the range from 0 to 1. Furthermore, absorbing and non-absorbing particles can be mixed within each mode with the mode-average imaginary part not changed and the relative number of non-absorbing particles in the range from 0 to 0.5.

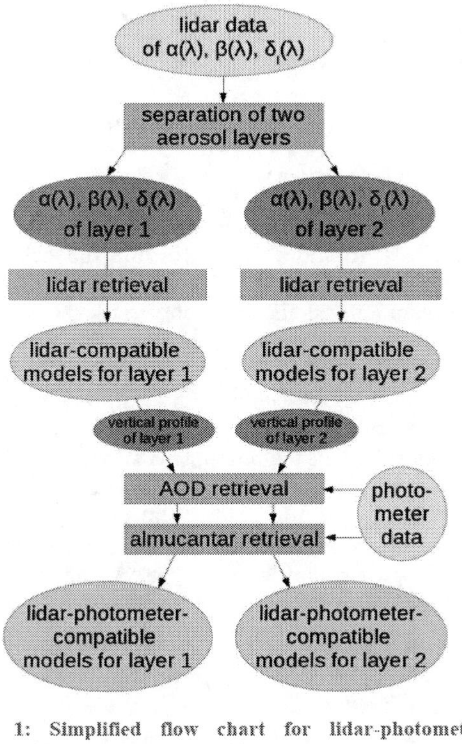

Fig. 1: Simplified flow chart for lidar-photometer retrieval in case of two aerosol layers.

A simplified flow chart of the combined lidar-photometer-retrieval is shown in Fig. 1. As the first step, the lidar profile can be separated into layers based on the intensive aerosol properties. Each aerosol layer is then determined by its (height-independent) intensive optical properties and its vertical structure. For each aerosol layer, the lidar retrieval is applied independently. The aerosol models that are compatible with the lidar measurements are stored for each layer. As the next step, the lidar-compatible ensembles of each layer are combined to atmospheric setups using the extinction coefficient of each layer at a reference wavelength. The wavelength dependence of the extinction is modeled for each ensemble to calculate the wavelength-dependent aerosol optical thickness (AOD) of each atmospheric setup. The AODs are compared with the spectral AOD measured by the photometer. For the setups that agree with the AOD measurements within the measurement uncertainties, radiative transfer calculations are started. The DISORT solver with intensity correction [6], which is included in the libRadtran toolbox [7], is used to model sky radiances in the almucantar geometry at the angles and wavelengths of the photometer. The ground albedo is varied between 0.0 and 0.4. The simulated and measured sky radiances are compared and the setups that agree within the measurement uncertainties are accepted as solutions of the combined lidar-photometer retrieval. As a result, the AOD and the sky radiance measurements are used for refining the set of solutions from the lidar part.

For the present study measurement uncertainties of 5% for the aerosol optical depth and 10% for the sky radiances are assumed.

3. RESULTS

We applied this approach to lidar measurements of MULIS and POLIS in Maisach around 2 UTC on 17 April 2010 in combination with AOD and almucantar measurements of a CIMEL photometer in Munich (about 25 km from the lidar) around 8:12 UTC on the same day. Two aerosol types were detected by the lidar, one type in the boundary layer up to about 1.7 km above ground and volcanic ash in a separate layer above. The intensive aerosol properties did not change

 DOI: 10.1051/epjconf/201611907004

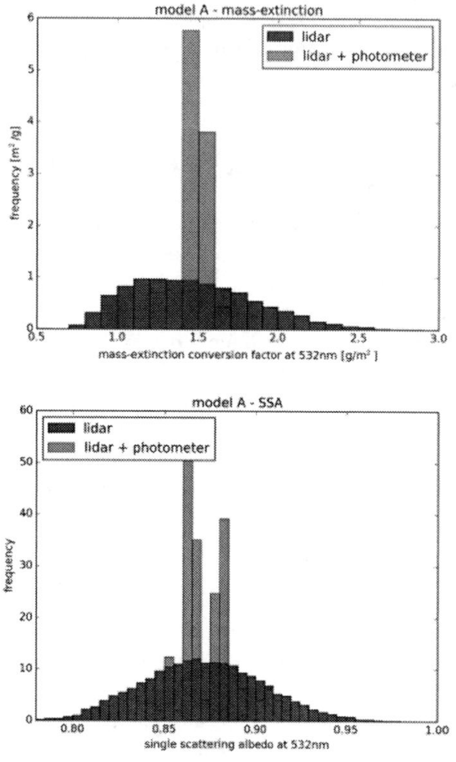

Fig. 2: Mass-extinction conversion factor (upper panel) and single scattering albedo (lower panel) of the solutions of the retrievals using model A.

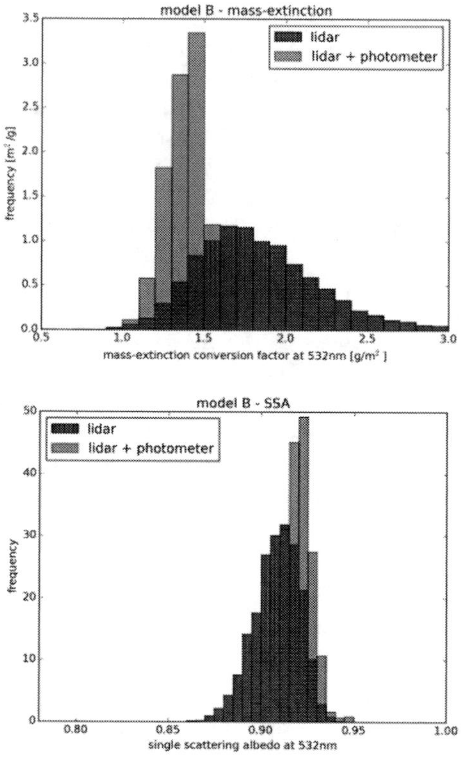

Fig. 3: Mass-extinction conversion factor (upper panel) and single scattering albedo (lower panel) of the solutions of the retrievals using model B.

during that time period, but the aerosol loading was variable.

As a consequence, the aerosol extinction coefficients during the photometer measurement are not known from lidar. To overcome this gap, measurements of a ceilometers located a few meters from the photometer are used to estimate the contributions of each layer. An optical depth of 0.30 at λ = 532 nm for the boundary layer aerosol and 0.25 for the volcanic ash aerosol provide good consistency with the ceilometer measurements in Munich. These layer optical depths were assumed subsequently, together with height-independent extinction coefficients within each layer. The boundary layer aerosol was characterized by low depolarization ratios, thus spherical particles were assumed for the retrieval of the boundary layer aerosol properties. 10000 lidar-compatible solution ensembles were retrieved for each layer (and each model).

Fig. 2 shows the frequency distributions for the retrieved solution ensembles for the ash layer when ash model A is used. The red bars illustrate the solutions of the lidar-only retrieval, whereas the green bars show the solutions of the combined lidar-photometer retrieval. The upper panel of Fig. 2 shows a significant reduction of the uncertainty of the mass-extinction conversion factor by about a factor 8 when the photometer measurements are considered (95% uncertainty range shrinked from 0.87 - 2.28 g/m^2 to 1.44 - 1.60 g/m^2). The lower panel illustrates a reduction of the uncertainty of the single scattering albedo at 532 nm by about 50%.

Fig. 3 is analogous to Fig. 2, however with the more complex ash model B used instead of model A. The width of the uncertainty range of the mass-extinction conversion factor is reduced by about 75% (from 1.21 - 2.82 g/m^2 to 1.15 - 1.56 g/m^2) if the additional photometer data is considered in case of model B. For the single scattering albedo a reduction of the uncertainties by 30% is found (from 0.882 - 0.929 to 0.901 - 0.934).

Comparison of the results using ash model A with results using model B reveals that the retrieved mass-extinction conversion factor is not strongly sensitive to the aerosol model employed. Only in the lidar-only case, a small shift towards larger mass-extinction conversion factors is found when the simpler model A is replaced by the complex and more realistic ash model B. The uncertainties of the mass-extinction conversion factor are somewhat larger when using model B which can be understood in view of the increased number of model parameters of model B. For the single scattering albedo a significant shift towards larger values and a reduction of its uncertainties occurs if model A is replaced by model B.

The AOD data was found not to reduce the uncertainties in all investigated cases, thus the uncertainty reduction can be attributed mainly to the consideration of the sky radiance data. Only a very low fraction of the sampled atmospheric setups (about 10^{-5}) were compatible with the AOD and sky radiance measurements in case of model A. The sampling frequency of solutions of the lidar-photometer retrieval was significantly higher (about $1.8 \cdot 10^{-2}$) in case of model B.

4. CONCLUSIONS

We investigated the benefit of photometer measurements and advanced aerosol models for a lidar-based aerosol retrieval. Our preliminary results show significantly reduced uncertainties for the retrieval of the mass-extinction conversion factor, and thus the aerosol mass concentration, when lidar measurements are complemented by photometer measurements. Improvements were found also for the single scattering albedo, but they were less significant than for the mass-extinction conversion factor.

Usage of the advanced model instead of the spheroid model leads to small changes of average retrieved parameters and the widths of their uncertainty ranges, indicating that the spheroid model is sufficient for volcanic ash retrievals. However, as the frequency of sampling retrieval solutions is much higher when the advanced model is used, a deeper look into the effect of the aerosol model is required for final conclusions.

ACKNOWLEDGEMENT

This work was funded by the LMUexcellent project EVAeNT.

REFERENCES

[1] Gasteiger, J., S. Groß, V. Freudenthaler, M. Wiegner, 2011: Volcanic ash from Iceland over Munich: mass concentration retrieved from ground-based remote sensing measurements, *Atmos. Chem. Phys.*, **11**, 2209-2223.

[2] Lopatin, A., O. Dubovik, A. Chaikovsky, P. Goloub, T. Lapyonok, D. Tanré, P. Litvinov, 2013: Enhancement of aerosol characterization using synergy of lidar and sun-photometer coincident observations: the GARRLiC algorithm, *Atmos. Meas. Tech.*, **6**, 2065-2088.

[3] Gasteiger, J., M. Wiegner, S. Groß, V. Freudenthaler, C. Toledano, M. Tesche, K. Kandler, 2011: Modeling lidar-relevant optical properties of complex mineral dust aerosols, *Tellus B*, **63**, 725-741.

[4] Yurkin, M.A, A.G. Hoekstra, 2011: The discrete-dipole-approximation code ADDA: capabilities and known limitations, *J. Quant. Spectrosc. Radiat. Transfer*, **112**, 2234–2247.

[5] Hess, M., P. Koepke, I. Schult, 1998: Optical Properties of Aerosols and clouds: The software package OPAC, *Bull. Am. Met. Soc.*, **79**, 831-844.

[6] Buras, R., T. Dowling, C. Emde, 2011: New secondary-scattering correction in DISORT with increased efficiency for forward scattering, *J. Quant. Spectrosc. Radiat. Transfer*, **112**, 2028–2034.

[7] Mayer, B., A. Kylling, 2005: Technical note: The libRadtran software package for radiative transfer calculations – description and examples of use, *Atmos. Chem. Phys.*, **5**, 1855–1877.

EPJ Web of Conferences **119**, 07005 (2016) DOI: 10.1051/epjconf/201611907005
ILRC 27

LIDAR OBSERVATION OF THE 2014 KELUT VOLCANIC STRATOSPHERIC AEROSOLS AT KOTOTABANG, INDONESIA

Makoto Abo*, Yasukuni Shibata, Chikao Nagasawa

*Tokyo Metropolitan University, Tokyo 191-0065, Japan, *Email: abo@tmu.ac.jp*

ABSTRACT

The Kelut (Kelud) volcano (7.9S, 112.3E) in the Java island of Indonesia erupted on 13 February 2014. The CALIOP observed that the eruption cloud reached 26km above sea level. We have observed this stratospheric aerosol from 28 February 2014 at equatorial lidar site located in the Sumatra island of Indonesia (0.2S, 100.3E). We observed the depolarization maximum to be up to 2km below the backscatter maximum in April 2014. We also observed the vertical transportation process of stratospheric aerosol to troposphere by equatorial Kelvin wave.

1. INTRODUCTION

Stratospheric aerosols play important roles in climate regulation and atmospheric chemistry. The volcanic eruption of Mt. Pinatubo (15.14N, 120.35E) on 15 June 1991 injected huge amounts of SO2 and ash into the stratosphere. The Volcanic Explosivity Index (VEI) was 6. The eruption injected into the stratosphere an amount of SO2 estimated to be about 20 Tg [1]. The injected SO2 was oxidized to sulfuric acid particles through homogeneous nucleation [2]. Based on some satellite data, the stratospheric aerosol optical thickness (AOT) increased after 2000 as the result of a series of moderate but increasingly intense volcanic eruptions [3].

At 16:15 UTC, Mt. Kelut in East Java, Indonesia, erupted a mixture of gas, ash and water. The eruption was short lived and no subsequent explosions have been reported. VEI was 4. From CALIOP data, the inertial plume ejected material up to 26 km in the tropical stratosphere, but most of the plume remained at 19-20 km [4]. A number of sensors continued to track the plume over the next few days. As the plume continued to move west over the Indian Ocean, it thinned and spread out, following local atmospheric circulation. Aerosol index returns show that much of the ash injected into the atmosphere was removed by three days after the eruption. The SO2 plume remained more coherent and could still be seen clearly in OMI and OMPS data 4 days after the eruption. OMI estimates suggest that SO2 is being removed from the plume at a rate of roughly 0.01 Tg per day. IASI data suggests that the removal is happening more rapidly, at a rate of 0.02 - 0.03 Tg per day, however IASI reported higher initial concentrations of SO2 than OMI [5]. Although most of the SO2 was injected into the stratosphere, the total SO2 mass measured was ~0.2 Tg (confirmed by multiple sensors including OMI, AIRS and IASI), making this eruption fairly modest in terms of SO2 release [6].

In this paper we report observations of stratospheric aerosols following the volcanic eruption of Mt. Kelut (7.9S, 112.3E) in February 2014 at lidar site in Kototabang (0.2S, 100.3E), Indonesia.

2. LIDAR SYSTEM

We had constructed the lidar facility for survey of atmospheric structure over troposphere, stratosphere, mesosphere and low thermosphere over Kototabang (100.3E, 0.2S), Indonesia in the equatorial region [7]. The lidar system consists of the Mie and Raman lidars for tropospheric aerosol, water vapor and cirrus cloud measurements, the Rayleigh lidar for stratospheric and mesospheric temperature measurements and the Resonance lidar for metallic species in the mesopause region. The most parts of this lidar system are remotely controlled via the Internet from Japan. The full lidar observations started from 2004. The routine observations of clouds and aerosol in the troposphere and stratosphere are continued now. We have installed DIAL (differential absorption lidar) system for high-resolution measurements of vertical ozone profiles in the equatorial tropopause region in 2013 and

532nm polarization lidar system for stratospheric aerosol monitoring in 2014.

A polarizer divided photons at 2 into components parallel (P) and perpendicular (S) to the transmitted laser polarization plane. The received photons were converted to electrical signals by photomultiplier tubes (PMT). A photon counter is used to process the output signals of the PMTs.

We used the nearest operational radiosonde data to calculate the atmospheric molecular density. The lidar backscattered signal was interactively normalized to unity around 27–32 km, where aerosol-free conditions could be assumed. The total linear depolarization ratio d is defined as

$$d = S/(P + S) \times 100(\%)$$

where P and S are the parallel and perpendicular components of the backscattered signals [8].

3. RESULTS

Over Kototabang, stratospheric aerosol layers were observed on 28 February 2014 about 15 days after the eruption (Figure 1(a)). The peak values of scattering ratio was 4 at 19 km and the values of depolarization ratio was 4% at 19 km. Non-spherical ash particles were probably included in the layers with sulfuric acid particles that were produced from SO2 through chemical reactions. Non-spherical particles were also present in the lower region (13-17km) of the aerosol layer.

On 29 April 2014 (2 months after the eruption), the peak altitude of scattering ratio move upward to 21 km but the peak altitude of depolarization ratio was still at 19 km (Figure 1(b)). Light spherical sulfuric acid particles should be moved upward by the upwelling tropical branch of the Brewer-Dobson circulation but the non-spherical ash should not be moved upward.

On 26 June 2014 (4 months after the eruption), the peaked stratospheric aerosol layers could not seen (Figure 1(c)). On 20-26 June 2014, aerosol transportation from the stratosphere to the troposphere by Kelvin-Helmholtz instability in the tropical tropopause layer should be occurred over the Kototabang.

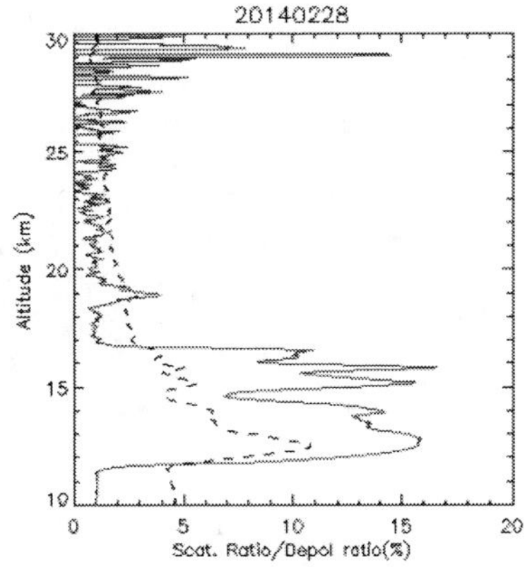

(a)

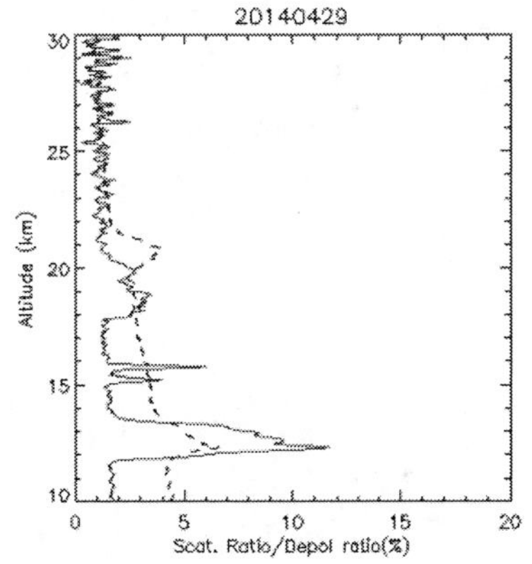

(b)

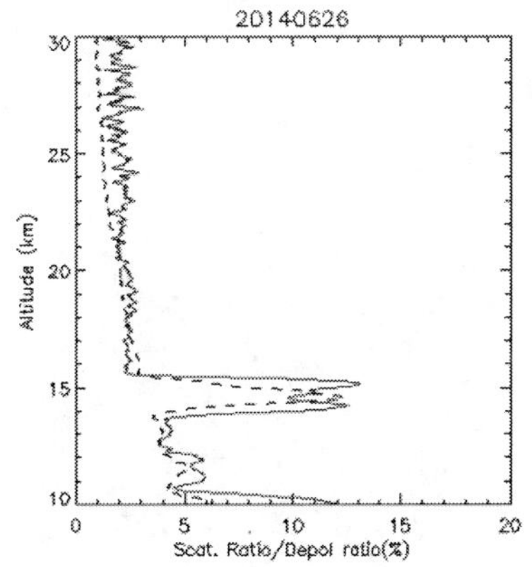

(c)

Figure 1 Profiles of scattering ratio (dotted) and depolarization ratio (solid) on (a) 15 days after the eruption, (b) 2 months after the eruption, (c) 4 months after the eruption.

DOI: 10.1051/epjconf/201611907005

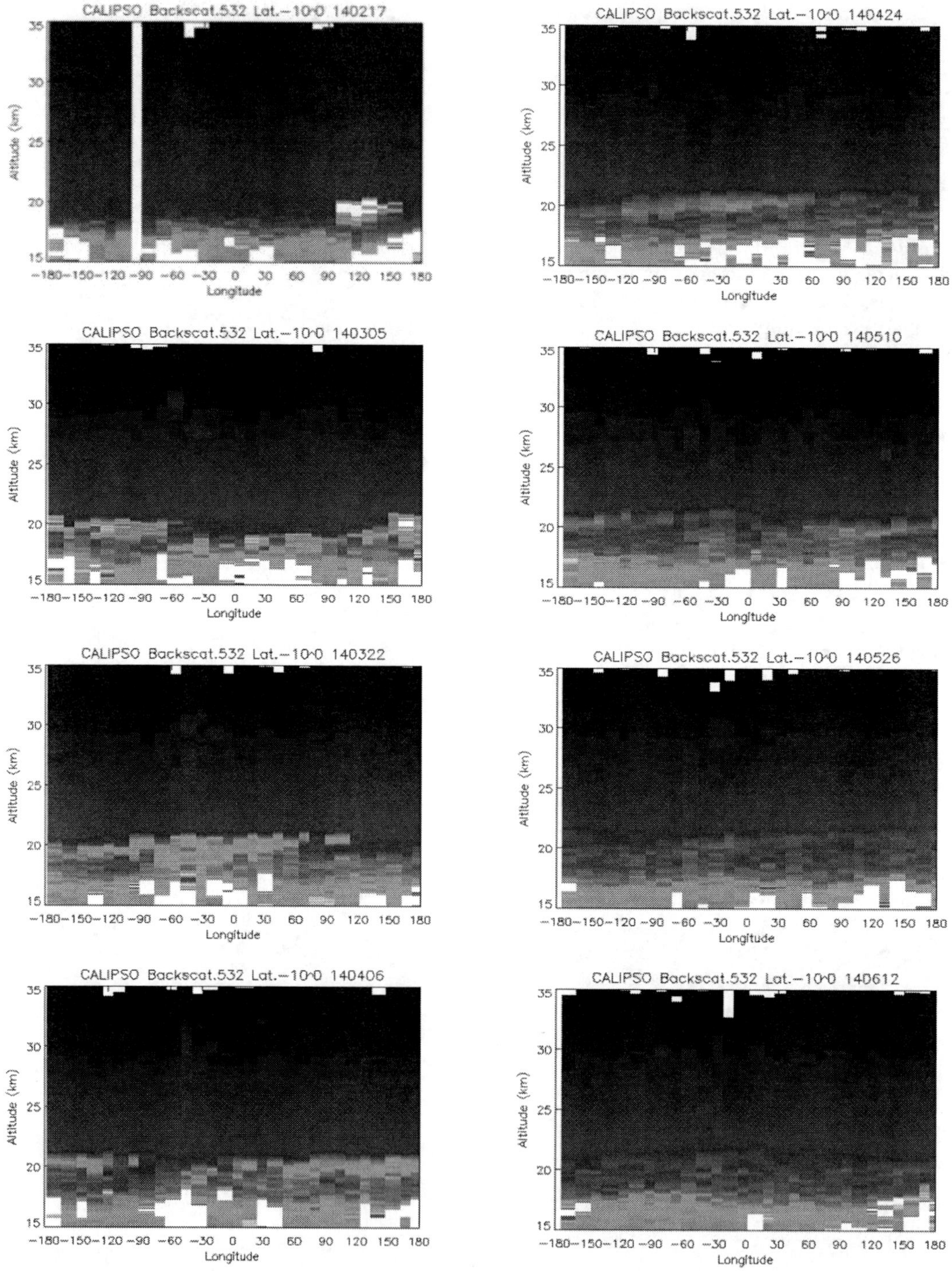

Figure 2 Two days composit of the CALIOP meridional mean [0-10S] of 532nm total attenuated Backscattering ratio every 16 days from 17 February 2014.

Figure 2 shows two days composit of the CALIOP meridional mean [0-10S] of 532nm total attenuated Backscattering ratio every 16 days from 17 February 2014. Stratospheric aelosol cloud around 20km exist over 100-120E longitude on 17-18 February 2014. It begun to move eastward and did one lap of earth in approximately one month. Aelosol layer was spread in the longitude direction with the lapse of time. Aerosol shold be carried by a equatorial Kelvin wave. The spread in the latitude direction was slower than in the longitude direction

4. CONCLUSIONS

We have observed 2014 Kelut volcanic stratospheric aerosol from 28 February 2014 at equatorial lidar site located in the Sumatra island of Indonesia (0.2S, 100.3E). We observed the depolarization maximum to be up to 2km below the backscatter maximum in April 2014. We also observed the vertical transportation process of stratospheric aerosol to troposphere by equatorial Kelvin wave instability.

From CALIOP data, Kelut aelosol cloud movement of eastward by equatorial Kelvin wave was cleary seen. Aelosol layer was spread in the longitude direction with the lapse of time but the spread in the latitude direction is slower.

As the SO2 mass injected into the stratosphere by Kelut eruption is moderate, these observations are good example of transportation of materials from the equatorial stratosphere to the troposphere.

ACKNOWLEDGEMENT

This work was supported by Collaborative Research based on MU Radar and Equatorial Atmosphere Radar.

REFERENCES

[1] Bluth, G. J. S., Doiron, S. D., Schnetzler, C. C., Krueger, A. J., and Walter, L. S., 1992: Global tracking of the SO2 clouds from the June, 1991 Mount Pinatubo eruptions, Geophys. Res. Lett., 19, 151–154, doi:10.1029/91GL02792,.

[2] Wu, P.-M., Okada, K., Tanaka, T., Sasaki, T., Nagai, T., Fujimoto, T., and Uchino, O., 1994: Balloon observation of stratospheric aerosols over Tsukuba, Japan Two years after the Pinatubo volcanic eruption, J. Meteor. Soc. Jpn., 72, 475–480,.

[3] Vernier, J.-P. et al., 2011: Major influence of tropical volcanic eruptions on the stratospheric aerosol layer during the last decade, Geophys. Res. Lett., 38, L12807, doi:10.1029/2011GL047563.

[4] Global Volcanism Program (GVP) 2014: Smithsonian Institution, Kelut Eruption Page. accessed Mar 2015. [Available at http://www. volcano.si.edu/volcano.cfm?vn=263280.]

[5] Carn, S., and Telling, J., 2014, Kelut 2014, IAVCEI (International Association of Volcanology and Chemistry of the Earth's Interior) Remote Sensing Commission (RSC) (URL: https://sites.google.com/site/iavceirscweb /eruptions/kelut-2014)

[6] National Aeronautics and Space Administration (NASA) Goddard Space Flight Center, 2014 (13 February), Kelut (Kelud) Eruption- February 13, 2014 (URL: http://so2.gsfc.nasa.gov/pix/special/2014/kelut/Kelut_summary_Feb14_2014.html)

[7] Nagasawa, C., Abo, M., Shibata, Y., 2006: Lidar System for Observations of Equatorial Lower and Upper Atmosphere, Reviewed and revised papers presented at the 23rd International Laser Radar Conference, No.2O-8, 43-46.

[8] Uchino, O. et al., 2012: On recent (2008–2012) stratospheric aerosols observed by lidar over Japan, Atmos. Chem. Phys., 12, 11975–11984.

[9] Nagai, T., Uchino, O., Fujimoto, T., Sai, Y., Tamashiro, K., Nomura, R., and Sunagawa, T., 1993: Lidar observation of the stratospheric aerosol layer over Okinawa, Japan, after the Mt. Pinatubo volcanic eruption, J. Meteor. Soc. Jpn., 71, 749–755.

[10] Mega, T., Yamamoto, M. K., Luce, H., Tabata, Y., Hashiguchi, H., Yamamoto, M., Yamanaka, M. D., and Fukao, S., 2010: Turbulence generation by Kelvin - Helmholtz instability in the tropical tropopause layer observed with a 47 MHz range imaging radar, J. Geophys. Res., 115, D18115, doi:10.1029/ 2010JD013864.

VALIDATION OF ASH OPTICAL DEPTH AND LAYER HEIGHT FROM IASI USING EARLINET LIDAR DATA

D. Balis[1], N.Siomos[1], M. Koukouli[1], L. Clarisse[2], E. Carboni[3], L.Ventress[3], R. Grainger[3] , L. Mona[4] and G. Pappalardo[4]

[1]*Aristotle University of Thessaloniki ,Greece, *Email:balis@auth.gr*
[2]*Université Libre de Bruxelles, Brussels, Belgium*
[3]*University of Oxford, Atmospheric, Oceanic & Planetary Physics Clarendon Laboratory, United Kingdom*
[4]*Consiglio Nazionale delle Ricerche, Istituto di Metodologie per l'Analisi Ambientale (CNR-IMAA), Tito Scalo, Potenza, Italy*

ABSTRACT

The 2010 eruptions of the Icelandic volcano Eyjafjallajökull attracted the attention of the public and the scientific community to the vulnerability of the European airspace to volcanic eruptions. The European Space Agency project "Satellite Monitoring of Ash and Sulphur Dioxide for the mitigation of Aviation Hazards", called for the creation of an optimal End-to-End System for Volcanic Ash Plume Monitoring and Prediction. This system is based on improved and dedicated satellite-derived ash plume and sulphur dioxide level assessments, as well as an extensive validation, using among others ground-based measurements (Koukouli et al., 2014). The validation of volcanic ash levels and height extracted from IASI/MetopA is presented in this work with emphasis on the ash plume height and ash optical depth levels. European Aerosol Research Lidar Network [EARLINET] lidar measurements are compared to different satellite estimates for two eruptive episodes. The validation results are extremely promising within the estimated uncertainties of each of the comparative datasets.

1. SATELLITE DATA

Satellite data from IASI (Clerbaux et al, 2009) were provided by both the Université Libre de Bruxelles, hereafter, ULB, and Oxford University, hereafter OXF. The algorithmic processing of the ULB institute (Clarisse et al., 2010; 2013) resulted in two different output types depending on the assumed volcanic ash characteristics whereas the algorithmic processing of OXF (Thomas et. al, 2009) resulted in four different output types, three of which assumed a fixed volcanic ash layer height analysis.

2. VALIDATION DATA

For the validation of the ash plume height as well as the optical depth of the ash plume, LIDAR data from the EARLINET network will were used [http://www.earlinet.org/]. EARLINET is the first aerosol lidar network on a continental scale with the main goal to provide data for the aerosol distribution over Europe at 25 participating stations. The available aerosol extinction and backscatter profiles make feasible the derivation of the ash plume height and optical thickness. A relational database, containing the output of the 4-D analysis of EARLINET data related to the Eyjafjallajökull volcanic eruptions of 2010, has been set up and is freely available on request at http://www.earlinet.org (Pappalardo et al., 2013).

3. RESULTS

The values of each satellite product have been restricted within an area of variable radius, dependent of the satellite, around each EARLINET station. Out of those values the closest spatially and chronically to each station has been selected and has been compared with the respective EARLINET layer for each day. The spatial filtering has been applied before the temporal filtering. For all the satellite products a comparison of the Aerosol Optical Depth (AOD) has taken place. For the few satellite products that provided volcanic ash layer height information a comparison of volcanic ash layer height was also performed. The AOD of the EARLINET layers was derived by the layers' integrated backscatter

DOI: 10.1051/epjconf/201611907006

coefficient multiplied by a constant conversion factor (lidar ratio) with a value of $50sr^{-1}$. An estimated 20% uncertainty on the EARLINET AOD was employed due to the uncertainty on the lidar ratio [typically between 40 and 60 sr^{-1}]. Concerning the layer height comparisons, the height of the center of the layers' mass was used and as estimated uncertainty, the distance between the mass center from the top and base of the later was employed. As far as the wavelength from which the AOD values were extracted is concerned; all the satellite AOD products were calculated using a wavelength of 550nm. Correspondingly, only the 532nm AOD LIDAR measurements were used in the following comparisons.

The ash characteristics extracted by the IASI/MetopA instrument by the Oxford team have a double providence, since the team operates two different algorithms; one, henceforth called "nominal" applied to an optimal estimation technique which permits also the retrieval of ash plume height however is a lot more cost/time effective, and a second one, henceforth called "fast", is less time effective however requires a fixed plume height as input. As we can see in FIGURE 1, the ash AOD extracted from the IASI/MetopA Oxford nominal algorithm is quite low, with values rarely rising above the 0.2 level. Depending on the spatiotemporal choices imposed (only one is shown here), the number of common points for the entire European domain may rise from 18 to 36, with little effect on the accordance with the ground-based AOD assessment. This accordance provides quite promising correlation coefficients between 0.6 and 0.85 depending on the restrictions on the correlative dataset. Great care is needed though when interpreting these findings, due to the small amount of data points.

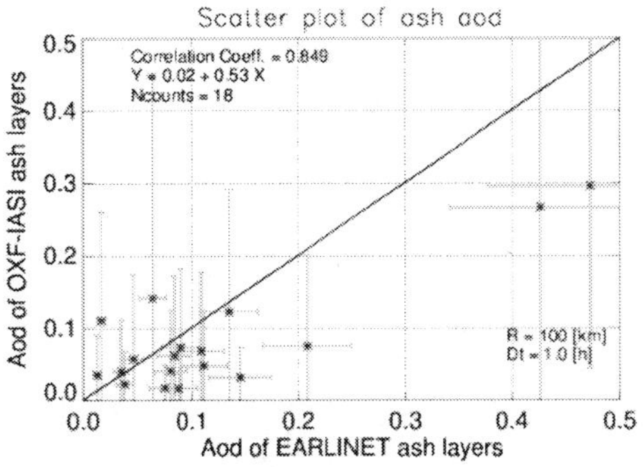

FIGURE 1. Scatter diagram between Oxford IASI nominal algorithm ash AOD and the co-located EARLINET stations for the Eyjafjallajökull eruptions in 2010 (coincidences within 100km and 1h).

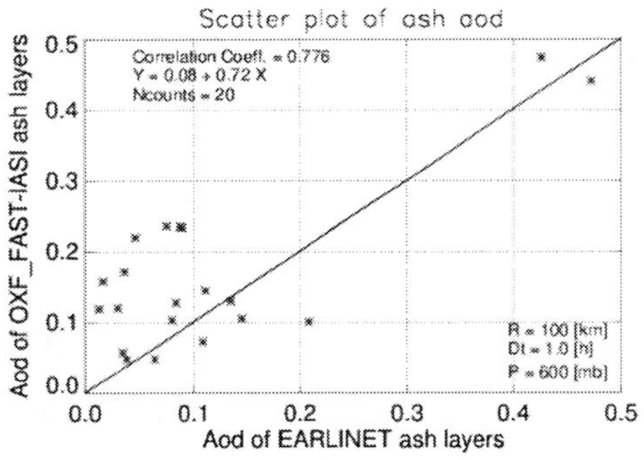

FIGURE 2 Scatter diagrams between the Oxford IASI fast algorithm ash AOD and the co-located EARLINET stations for the Eyjafjallajökull eruptions in 2010.

Concering the validation of the "fast" algorithm we present in FIGURE 2 the comparisons that show the best correlation with EARLINET data, which correspond to fixed ash layer height at 600hPa.

Two different refractive indices were used as input to the direct fitting of the ULB IASI algorithm, resulting in two very similar ash AOD datasets. The comparisons when using the refractive index according to the Eyjafjallajökull ash [FIGURE 3] perform better with a correlation coefficient of 0.94 for the narrower spatiotemporal choice. Even though the sampling is poor, this product shows very promising results.

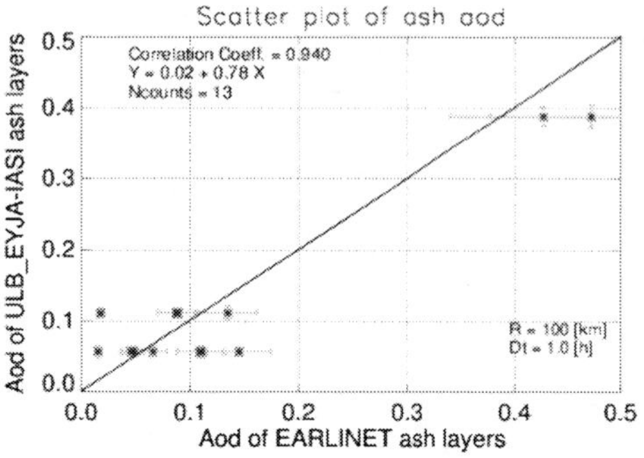

FIGURE 3. Scatter diagrams between the ULB IASI algorithm ash AOD and the co-located EARLINET stations for the Eyjafjallajökull eruptions in 2010 with assumed refractive indices according to Eyjafjallajökull ash.

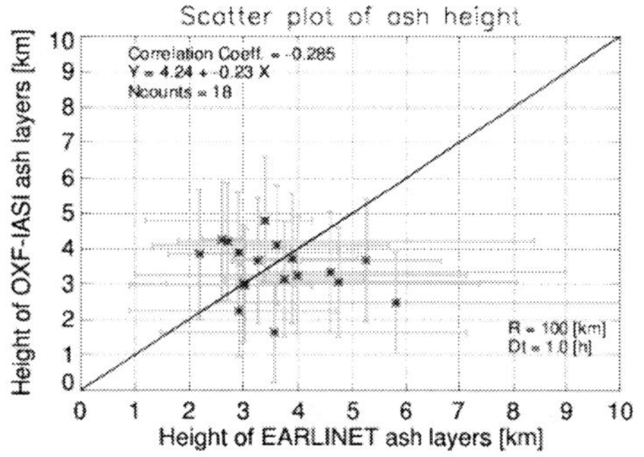

FIGURE 4. Scatter diagrams between the Oxford IASI nominal algorithm ash plume height and the co-located EARLINET stations for the Eyjafjallajökull eruptions in 2010.

In FIGURE 4 the comparison of the ash plume height extracted from the IASI/MetopA Oxford nominal algorithm and the one observed by the EARLINET network is shown. We note than the spread of plume heights found by the EARLINET network is higher than those found by the Oxford nominal IASI algorithm leading to rather poor comparisons. The scatter bars provide an indication of the amount of uncertainty inherent in both sets of observations.

4. CONCLUSIONS

In this paper we presented comparisons between volcanic ash AOD and ash layer height as these are estimated by IASI with ground based lidar measurements performed within EARLINET. Concerning the ash AOD the Oxford/IASI optimal estimation algorithm shows an acceptable correlation with the ground values, with coefficients ranging between 0.6 and 0.85 depending on the spatiotemporal co-location criteria used. Even though this comparisons provides a rather low amount of common values, these are of the same order of magnitude as the LIDAR ones. The Oxford/IASI Fast algorithm also provides AOD estimates with the same order of magnitude as the LIDAR, with correlations ranging between 0.7 and 0.8 for different algorithm setups. The ULB/IASI AOD estimates are also quite promising, with correlations ranging between 0.74 and 0.94, depending on the assumed refractive index. All comparisons however suffer from the small number of co-locations found. As far as the ash plume height is concerned, the Oxford nominal IASI algorithm ash plume height comparisons are quite satisfactory with similar mean estimated ash layer height and spread.

ACKNOWLEDGEMENT

The authors would like to acknowledge the European Space Agency SACS-2/SMASH project. The ground-based EARLINET LIDAR database has been downloaded from www.earlinet.org. The financial support for EARLINET by the European Union under grant RICA 025991 in the Sixth Framework Program is gratefully acknowledged. Since 2011 EARLINET has been integrated in the ACTRIS Research Infrastructure Project supported by the European Union Seventh Framework Program
(FP7/2007-2013) under grant agreement no. 262254

REFERENCES

Clarisse, L., F. Prata, J.-L. Lacour, et al. (2010), A correlation method for volcanic ash detection using hyperspectral infrared measurements, Geophys. Res. Lett., 37, L19806, doi:10.1029/2010GL044828.

Clarisse, L., Coheur, P.-F., Prata, F., et al., (2013), A unified approach to infrared aerosol remote sensing and type specification, Atmos. Chem. Phys., 13, 2195-2221, doi:10.5194/acp-13-2195-2013.

Clerbaux, C., Boynard, A., Clarisse, L., et al., (2009), Monitoring of atmospheric composition using the thermal infrared IASI/MetOp sounder, Atmos. Chem. Phys., 9, 6041-6054, doi:10.5194/acp-9-6041-2009.

Koukouli, M. E., Balis, D. S., Dimopoulos, S., et al., SACS-2/SMASH validation report on the Eyjafjallajökull and Grimsvotn eruptions, http://sacs.aeronomie.be/ [last accessed, March 27th, 2015], 2014.

Pappalardo et al., Four-dimensional distribution of the 2010 Eyjafjallajokull volcanic cloud over Europe observed by EARLINET, Atmos. Chem. Phys., 13, 4429-4450, doi:10.5194/acp-13-4429-2013

Thomas, G.E., E. Carboni, A.M. Sayer, et al., (2009), Oxford-RAL Aerosol and Cloud (ORAC): aerosol retrievals from satellite radiometers in Satellite Aerosol Remote Sensing Over Land (Eds: A.A. Kokhanovsky and G. de Leeuw), Springer.

VERTICAL VARIATION OF OPTICAL PROPERTIES OF MIXED ASIAN DUST/POLLUTION PLUMES ACCORDING TO PATHWAY OF AIRMASS TRANSPORT OVER EAST ASIA

Sung-Kyun Shin [1]*, Detlef Müller[2], K.H. Lee[3], D. Shin[4], Y. J. Kim[1], Y. M. Noh[5]

[1]*Gwangju Institute of Science and Technology (GIST), Republic of Korea, *Email: skyun@gist.ac.kr*

[2]*Uiversity of Hertfordshire, United Kingdom*

[3]*Ganneung-Wonju National University, Republic of Korea*

[4]*National Institute of Environmental Research, Republic of Korea*

[5]*Ineternational Environmental Research Center (IERC), GIST, Republic of Korea*

ABSTRACT

We use five years (2009 – 2013) of multiwavelength Raman lidar measurements at Gwangju, Korea (35.10° N, 126.53° E) for the identification of changes of optical properties of East Asian dust in dependence of its transport path over China. Profiles of backscatter and extinction coefficients, lidar ratios, and backscatter-related Ångström exponents (wavelength pair 355/532nm) were measured at Gwangju. Linear particle depolarization ratios were used to identify East Asian dust layers. We used backward trajectory modelling to identify the pathway and the vertical position of dust-laden air masses over China during long-range transport. Most cases of Asian dust events can be described by the emission of dust in desert areas and subsequent transport over highly polluted regions of China. The Asian dust plumes could be categorized into two classes according to the height above ground in which these plumes were transported: (I) the dust layers passed over China at high altitude levels until arrival over Gwangju, and (II) the Asian dust layers were transported near the surface and the lower troposphere over industrialized areas before they arrived over Gwangju. We find that the optical characteristics of these mixed Asian dust layers over Gwangju differ in dependence of their vertical position above ground over China and the change of height above ground during transport. The mean linear particle depolarization ratio was 0.21±0.06 (at 532 nm), the mean lidar ratios were 52±7 sr at 355 nm and 53±8 sr at 532 nm, and the mean Ångström exponent was 0.74±0.31 in case I. In contrast, plumes transported at lower altitudes (case II) showed low depolarization ratios, and higher lidar ratio and Ångström exponents. The mean linear particle depolarization ratio was 0.13 ± 0.04, the mean lidar ratios were 63±9 sr at 355 nm and 62±8 sr at 532 nm, respectively, and the mean Ångström exponent was 0.98±0.51. These numbers show that the optical characteristics of mixed Asian plumes are more similar to optical characteristics of urban pollution.

1. INTRODUCTION

East Asian dust is particularly complicated as it usually travels over densely populated and highly industrialized areas of China before it moves out over Pacific Ocean. During transport over East Asia dust mixes with pollutants such as industrial soot, toxic material, and acidic gases (Sun et al., 2005). However, there exist few studies on the degree of mixing that occurs between dust and pollution during transport, the effect of the direction of dust transport across China, and the vertical distribution of Asian dust layers during long-range transport over China. There still is a lack of understanding of how much of the mixing of dust with pollutants depends on the vertical distribution over East Asia when dust passes over source regions of anthropogenic pollution. One reason of our limited knowledge is that there are only few vertically-resolved, long-term observations of pollution over East Asia. In this study we use Raman lidar data taken at Gwangju, South Korea, between 2009 and 2013. In our study we focus specifically on lidar observations of Asian dust layers as they passed over China. The main objective of this study is to investigate the variation of optical properties of mixtures of Asian dust with anthropogenic pollution in dependence of the pathways and vertical distributions of these mixed dust layers during long-range transport. We identify these dust layers by the linear particle depolarization ratio. We present vertically-resolved optical properties such as lidar ratio and Ångström exponent. We also categorize the optical properties of these pollution plumes according to their transport pathway and their vertical distribution. Section 2 presents the methods used in this study. Section 3 presents our results. We discuss our results and summarize our findings in section 4.

2. METHODOLOGY

The lidar station, dubbed MRS.LEA (Multi-wavelength Raman Spectrometer Lidar in East Asia) of the Gwangju Institute of Science and Technology (GIST) is located at 35.10° N, 126.53° E in the west-south-western part of the Korean peninsula. A description of the lidar system is given by Noh et al. (2008). The system allows us to retrieve vertical profiles of the particle backscatter coefficients at 355, 532, and 1064 nm, the particle extinction coefficients at 355 and 532 nm, the linear particle depolarization ratio at 532 nm, the water-vapor mixing-ratio, and profiles of silicon-dioxide (Müller et al., 2010). Profiles of silicon-dioxide (quartz) can be used as tracer of the concentration of mineral dust. In this contribution we use the signals needed for measuring particle backscatter and extinction coefficients at 355 and 532 nm and the linear particle depolarization ratio. The profiles of particle backscatter coefficients (β_p) at 355 and 532 nm were calculated with the Raman method. The overlap effect which describes the incomplete overlap between outgoing laser beam and field of view of the receiver telescope is cancelled out for the case of profiles of the backscatter coefficient because the ratios of two signals (elastic signals from particles and molecules and the nitrogen Raman signals) are. In that way we can retrieve vertical profiles of the backscatter coefficient to 400 m above ground. The vertical profiles of the aerosol extinction coefficients (α_p) at 355 and 532 nm were derived with the use of the nitrogen vibration Raman signals at 387 and 607 nm, respectively. We derive particle extinction-to-backscatter ratios (lidar ratios, denoted as S in this contribution) at 355 and 532 nm from the profiles of β_p and α_p. The backscatter-related Ångström exponent for the wavelength pair of 355/532 nm (denoted as $\mathring{A}_\beta$) is computed, too. The measurements were carried out at nighttime under cloud-free conditions. In this contribution we use the linear particle depolarization ratio according to the definition by as below:

$$\delta_p = \frac{\delta(z)R_B(z) - \delta_m}{R_B(z) - 1}.$$

The term δ_m is the linear depolarization ratio of air molecules at the wavelength and bandwidth of the emitted laser wavelength. We used the value δ_m=0.0044 (Behrendt and Nakamura, 2002). This value takes account of our interference filter which have a full width at half maximum of 1.0 nm. $R_B(z)$ is the backscatter ratio, expressed as $(\beta_p+\beta_m)/\beta_m$ at altitude z. β_m denotes the backscatter coefficient of atmospheric molecules. The calibration of the polarization channels was carried out by using rotating polarizers following the methodology explained by Tesche et al. (2009).

3. RESULTS

Figure 1 shows the transport pathway and the change of the vertical position of the dust plumes during transport to our lidar site. It is clear that backward trajectories cannot provide us with information on the concentration of dust and anthropogenic pollution in the air masses prior to observation over Korea. Still, backward trajectories show if the air masses originated from or nearby the desert regions, and whether the air masses passed over densely populated/industrialized regions. Case I includes those Asian dust plumes that passed over industrialized areas in China at high altitude level (> 3km height above ground) as shown in figure 6a. The Asian dust plumes were classified as Case II when they were transported through the near surface/lower troposphere (< 3km height above ground) over industrialized areas in China, i.e., longitude range between 110° E and 125° E; the locations of industrialized and densely populated regions in China are shown in figure 1. The depolarization ratios of the Asian dust plumes we observed are lower compared to depolarization ratios of pure dust particles. For example,

Freudenthaler et al. (2009) report a value of δ_p = 0.31 at 532 nm for pure Saharan dust observed during SAMUM 2006. The values of δ_p and the corresponding values of $\mathring{A}_\beta$ and S at 355 nm and 532 nm for the Cases I and II are also shown in figure 1.

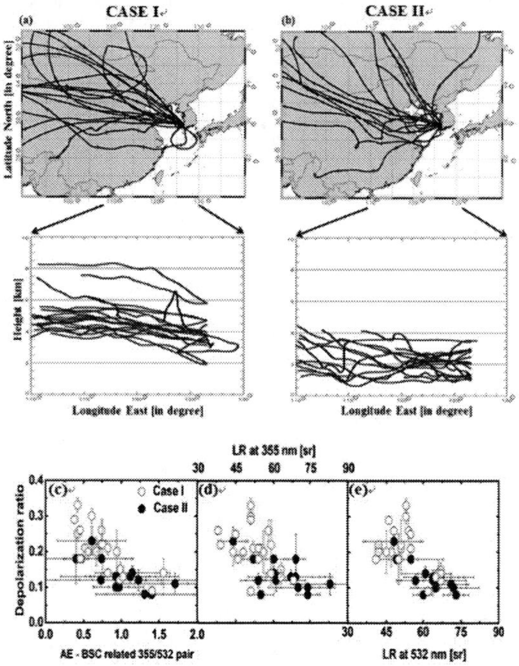

Figure 1. (top panel) Transport pattern of the dust plumes that originated in the desert regions and passed over industrialized/populated regions of China before arrival over the Korean peninsula. (middle panel) Vertical position of the dust layers during transport: (a) Dust layers passed over China at high altitude (Case I) (b) dust layers were transported over China through the near surface/lower troposphere (Case II). (bottom panel) Scatter diagram of the linear particle depolarization at 532 nm versus (c) the backscatter-related Ångström exponent (355/532 nm wavelength pair), and the (d), (e) lidar ratio (at 355 nm and at 532 nm) with respect to the Case I and Case II. The two categories I, II are denoted by different colors.

We find different clusters of the optical properties of the dust layers when we take into consideration their vertical position during transport. Cases I show larger values of δ_p compared to case II. On average $\mathring{A}_\beta$ of case I is smaller than $\mathring{A}_\beta$ of case II. The average values of δ_p and $\mathring{A}_\beta$ are 0.21±0.06 and 0.74±0.31, respectively, for case I. In contrast, δ_p and $\mathring{A}_\beta$ are 0.13±0.04 and 0.98±0.35, respectively, for case II. The lowest values of S at 355 nm and 532 nm are also measured for high values of δ_p (0.21±0.06). We find values of 52±7 sr at 355 nm and 53±8 sr at 532 nm, respectively, for case I. Comparably high values of S were found for case II, i.e. 63±9 at 355 nm and 62±8sr at 532 nm. In that case the value of δ_p is 0.13±0.04. There are several previous studies that report on depolarization ratios of polluted dust after long-range transport. According to these studies the observed dust particles were partly/completely mixed with anthropogenic pollution (Shimizu et al., 2004; Chen et al., 2007). As a result of the mixing of dust with anthropogenic pollution, the values of δ_p were lower than the values of pure dust, which is estimated to be at 0.3 - 0.35 (Murayama et al., 2004; Freudenthaler et al., 2009). Likewise, the values of $\mathring{A}_\beta$ and S also differ compared to the values of $\mathring{A}_\beta$ and S of pure dust. We assume that the dust particles carried more anthropogenic pollution in cases where the air masses travelled near the surface. Consequently, the optical characteristics of the dust/pollution layers of case II are dominated by the optical properties of anthropogenic pollutants. In contrast, the optical properties of dust layers that travelled at high altitudes (Case I) are less influenced by urban/industrial pollutants. Thus, the optical properties of these dust layers are more likely to be those of pure dust. The clusters denoted Case I and Case II were classified according to the altitude the dust-laden air masses passed over industrialized/populated regions of China. The differences of the optical properties of the dust layers are shown in figure 2. The difference of the optical characteristics of East Asian dust layers that travelled in surface-near heights and at high altitudes is obvious. The values of δ_p, $\mathring{A}_\beta$ and S are 0.12±0.01, 1.00±0.43, and 63±7 sr at 355 nm and 64±6 sr at 532 nm, respectively, when Asian dust passed over China below 1 km height above ground. These values reflect the fact that the optical properties of the dust/pollution plumes are dominated by the anthropogenic part of the particles in these plumes. In contrast, the values of δ_p, $\mathring{A}_\beta$, and S were 0.23±0.02, 0.60±027, and 50±7 sr at 355 nm and 49±8 sr at 532 nm, respectively, when the dust layers passed over China at high altitudes, i.e., above 4 km. These values more likely reflect the optical characteristic of Asian dust particles that are less affected by the contribution of anthropogenic pollution. The altitude in which the Asian dust layers passed over China have significant influence on their optical characteristics. In our study, we took 3 km above ground as threshold value as we observed a notable change of optical properties of the dust/pollution plumes if they travelled below or below 3 km height above ground. Pollution particles below 3 km could mix and interact with Asian dust particles (more influence). In contrast, we assume that optical properties of dust particles above 3 km are not that much influenced by anthropogenic pollution as the mixing of pollution into these heights is less intense. This height of 3 km is also in relatively good agreement with the average height of planetary boundary layers. Pollutants emitted at the surface predominantly stay in this height (Basha et al., 2009).

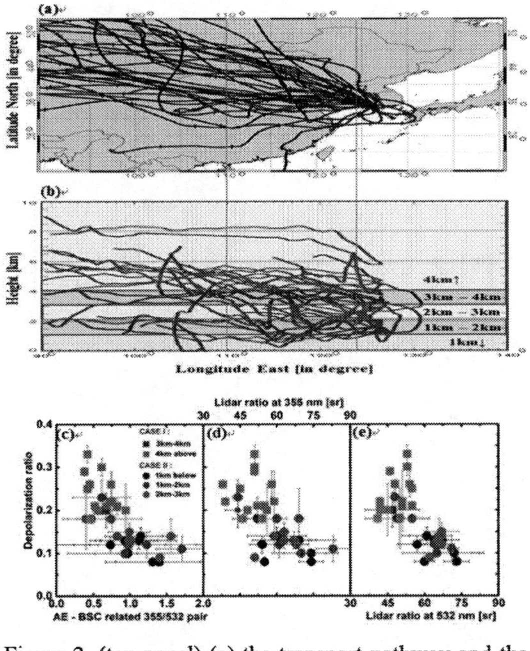

Figure 2. (top panel) (a) the transport pathway and the classification of East Asian dust layers with respect to (b) their altitude above ground when they passed over industrial regions of China. (bottom panel) Scatter plots of the linear particle depolarization at 532 nm versus (c) the backscatter-related Ångström exponent (355/532 nm wavelength pair), versus (d) the lidar ratio at 355 nm, respectively (e) the lidar ratio at 532 nm in dependence of the 5 altitude categories. The height of the Asian dust layers above ground is denoted by different colors and symbols. Case I included the layers from 3 – 4 km and above 4 km (blue and pink colored squares). Case II includes the layers from 0 - 1 km, from 1-2km, and from 2-3km height above ground (black, green and violet circles).

4. CONCLUSIONS

Our results suggest that the transport pathway as well as the vertical position of Asian dust during long-range transport may have significant impact on the optical properties of mixed Asian dust layers.

ACKNOWLEDGEMENT

This work was supported by the "Advanced Research on Applied Meteorology" of National Institute of Meteorological Research (NIMR) funded by the Korea Meteorological Administration (KMA) and the Eco-Innovation Project 2012000160004 from Korean Ministry of Environment through the Advanced Environmental Monitoring Research Center.

REFERENCES

[1] Basha, G., and Ratnam, M. V., 2009: Identification of atmospheric boundary layer height over a tropical station using high-resolution radiosonde refractivity profiles: Comparison with GPS radio occultation measurements, Journal of Geophysical Research: Atmospheres (1984–2012), **114**.

[2] Behrendt, A., and Nakamura, T., 2002: Calculation of the calibration constant of polarization lidar and its dependency on atmospheric temperature, Optics express, **10**, 805-817

[3] Chen, W.-N., Chen, Y.-W., Chou, C. C., Chang, S.-Y., Lin, P.-H., and Chen, J.-P., 2009: Columnar optical properties of tropospheric aerosol by combined lidar and sunphotometer measurements at Taipei, Taiwan, Atmospheric Environment, **43**, 2700-2708.

[4] Freudenthaler, V., Esselborn, M., Wiegner, M., Heese, B., Tesche, M., Ansmann, A., Müller, D., Althausen, D., Wirth, M., and Fix, A., 2009: Depolarization ratio profiling at several

DOI: 10.1051/epjconf/201611908001

wavelengths in pure Saharan dust during SAMUM 2006, Tellus B, **61**, 165-179.

[5] Müller, D., Mattis, I., Tatarov, B., Noh, Y., Shin, D., Shin, S., Lee, K., Kim, Y., and Sugimoto, N., 2010: Mineral quartz concentration measurements of mixed mineral dust/urban haze pollution plumes over Korea with multiwavelength aerosol Raman-quartz lidar, Geophysical Research Letters, **37**.

[6] Murayama, T., Müller, D., Wada, K., Shimizu, A., Sekiguchi, M., and Tsukamoto, T., 2004: Characterization of Asian dust and Siberian smoke with multi-wavelength Raman lidar over Tokyo, Japan in spring 2003, Geophysical Research Letters, **31**.

[7] Noh, Y. M., Kim, Y. J., and Müller, D., 2008: Seasonal characteristics of lidar ratios measured with a Raman lidar at Gwangju, Korea in spring and autumn, Atmospheric Environment, **42**, 2208-2224.

[8] Shimizu, A., Sugimoto, N., Matsui, I., Arao, K., Uno, I., Murayama, T., Kagawa, N., Aoki, K., Uchiyama, A., and Yamazaki, A., 2004: Continuous observations of Asian dust and other aerosols by polarization lidars in China and Japan during ACE-Asia, Journal of Geophysical Research: Atmospheres (1984–2012), **109**.

[9] Sun, Y., Zhuang, G., Wang, Y., Zhao, X., Li, J., Wang, Z., and An, Z., 2005: Chemical composition of dust storms in Beijing and implications for the mixing of mineral aerosol with pollution aerosol on the pathway, Journal of Geophysical Research: Atmospheres (1984–2012), **110**.

[10] Tesche, M., Ansmann, A., Müller, D., Althausen, D., Mattis, I., Heese, B., Freudenthaler, V., Wiegner, M., Esselborn, M., and Pisani, G., 2009: Vertical profiling of Saharan dust with Raman lidars and airborne HSRL in southern Morocco during SAMUM, Tellus B, **61**, 144-164.

EPJ Web of Conferences **119**, 08002 (2016)
ILRC 27
DOI: 10.1051/epjconf/201611908002

SAHARAN AND ARABIAN DUST AEROSOLS: A COMPARATIVE CASE STUDY OF LIDAR RATIO

Carmen Córdoba-Jabonero[1]*, Ismail Sabbah[2], Mar Sorribas[1,3], José Antonio Adame[1], Emilio Cuevas[4], Faisal Al Sharifi[5], Manuel Gil-Ojeda[1]

[1]*Instituto Nacional de Técnica Aeroespacial (INTA), Atmospheric Research and Instrumentation Branch, Torrejón de Ardoz, 28850-Madrid, Spain, *Email: cordobajc@inta.es*

[2]*Department of Natural Sciences, College of Health Sciences, the Public Authority for Applied Education and Training, Kuwait*

[3]*Instituto Interuniversitario de Investigación del Sistema Tierra en Andalucía-Centro Andaluz de Medio Ambiente (IISTA-CEAMA), Universidad de Granada (UGR), Granada, Spain*

[4]*Agencia Estatal de Meteorología (AEMET), Izaña Atmospheric Research Centre, Santa Cruz de Tenerife, Spain*

[5]*Department of Environmental Sciences, College of Health Sciences, the Public Authority for Applied Education and Training, Kuwait*

ABSTRACT

This work presents a first comparative study of the Lidar Ratio (LR) values obtained for dust particles in two singular dust-influenced regions: the Canary Islands (Spain, close to the African coast in the North Atlantic Ocean), frequently affected by Saharan dust intrusions, and the Kuwait area (Arabian Peninsula) as usually influenced by Arabian dust storms. Synergetic lidar and sun-photometry measurements are carried out in two stations located in these particular regions for that purpose. Several dusty cases were observed during 2014 in both stations and, just for illustration, two specific dusty case studies have been selected and analyzed to be shown in this work. In general, mean LR values of 54 sr and 40 sr were obtained in these studies cases for Saharan and Arabian dust particles, respectively. Indeed, these results are in agreement with other studies performed for dust particles arriving from similar desert areas. In particular, the disparity found in Saharan and Arabian dust LR values can be based on the singular composition of the suspended dust aerosols over each station. These results can be useful for CALIPSO extinction retrievals, where a single LR value (40 sr) is assumed for pure dust particles independently on the dust source region.

1. INTRODUCTION

The extinction-to-backscattering coefficient ratio (Lidar Ratio, LR) is an aerosol-type dependent parameter associated to the chemical composition, particle shape and aerosol size distribution. Hence, the LR estimation from lidar measurements is a crucial task in aerosol research, since this parameter can provide valuable information on optical/microphysical properties of aerosols. A special emphasis is devoted to dust, as shown by diverse published works [i.e., 1-11, among others]. Moreover, data from diverse observational platforms and techniques, as space-borne instrumentation (CALIPSO/CALIOP and Aqua/MODIS) as ground-based aerosol networks (EARLINET and AERONET), and analysis from several models (BSC-DREAM, NAAPS, ...) have been used for assessment of the Aerosol Optical Depth (AOD), mainly, regarding the LR value used for dust aerosols. Hence, the single LR value assumed by CALIOP/CALIPSO (www-calipso.larc.nasa.gov) for pure dust (40 sr) is under discussion. Instead, a higher value (58 sr) is proposed by EARLINET (European Aerosol Lidar NETwork, www.earlinet.org) based on multi-year lidar measurements of Saharan dust intrusions over Europe [7, and references herein]. However, lower LR values are reported for Arabian dust by only a few works [i.e., 6; 8]. Currently, lidar measurements carried out in the Arabian Peninsula and surroundings are still scarce.

The aim of this work is to show a first comparative analysis on LR values retrieved for dust particles present over two different dust-influenced regions: the Canary Islands (Spain, close to the African coast in the North Atlantic Ocean), frequently affected by the arrival of Saharan dust intrusions, and the Kuwait area (Arabian Peninsula) as usually influenced by Arabian dust storms. Hence, dust observations were performed in particularly located lidar stations under both Saharan and Arabian dust influence in order to establish a long-term dust aerosol monitoring over these stations in the future. In this first step, two case studies of these dust observations are presented. Results can be useful for retrieving AOD from CALIPSO observations, despite a single LR value (40 sr) is assumed for pure dust particles independently on the dust source region.

2. METHODOLOGY

Lidar measurements in synergy with sun-photometry data are performed in two aerosol stations:

1) the Izaña observatory (IZO, 28.3ºN, 16.5ºW, 2400 m a.s.l.), belonging to the Spanish Meteorological Agency (AEMET), is located at the Tenerife Island, in a pristine environment and representative of the Free Troposphere (FT) stations; hence, only pure dust particles are present under Saharan dust intrusion occurrence; and
2) the Kuwait station (KW, 29.3ºN, 47.9ºE, 40 m a.s.l.), belonging to the College of Health Sciences of Kuwait, is located in the Arabian Peninsula and usually influenced by dust storms. Their impact is particularly relevant in the Boundary Layer (BL) since they are transported by surface level northwesterly "Shamal" winds to Kuwait from Syria, Iraq and Jordan areas [12].

Synergetic measurements of both height-resolved lidar and columnar-integrated sun-photometry observations were performed in both areas. LR values are retrieved by using an elastic inversion procedure based on the AOD constraint and assuming the presence of pure dust particles. Details of this method are reported in [11].

2.1. Tenerife instrumentation

Lidar measurements for IZO site were carried out by a Micropulse Lidar v. 3 (MPL-3), within NASA/MPLNET (Micro Pulse Lidar NETwork, mplnet.gsfc.nasa.gov), located in the AEMET/Santa Cruz de Tenerife observatory (SCO, 40 km far from IZO, at 52 m a.s.l.). The MPL-3 is an elastic lidar, operated by the Instituto Nacional de Técnica Aeroespacial (INTA), with the following main features: highly-pulsed (2500 Hz) and low-energy (10 µJ, max.) laser at 523 nm, and Cassegrain-type telescope receiving system. It is a small, easy-handle system with coaxial configuration, high autonomy and low attendance, operational in full-time continuous mode (24 h day-1/365 days year-1). Routine measurements (MPLNET settings) are 75-m vertical resolution and 1-minute integrating time.

IZO is also a NASA/AERONET (AErosol RObotic NETwork, aeronet.gsfc.nasa.gov) site. Retrieved inversion products level 1.5 (Cloud Screened) and 2.0 (Quality-assured) used are the spectrally resolved Aerosol Optical Depth (AOD) and Ångström Exponent (AEx). In particular, the Cimel-318 500-nm channel data are used in correspondence to the closer wavelength to that of the MPL-3 laser (523 nm).

2.2. Kuwait instrumentation

A polarized Micropulse Lidar (P-MPL) has been recently installed in the Kuwait station, and routine measurements are performed from November 2013 to date for dust research. This system is similar to that MPL-3 together to polarization capabilities included. The main features are the same as those for the MPL-3 system, but the laser is at 532 nm, and the receiving system is composed by a Matsukov-Cassegrain telescope with two polarization channels. Routine measurements are 30-m vertical resolution and 30-second integrating time.

Sun-photometer data, mainly AOD values, are obtained with a portable sun-photometer (Microtops II).

3. RESULTS

 DOI: 10.1051/epjconf/201611908002

Several dusty cases were observed during 2014 in both stations. In particular, one case study for each station was selected: 29 June for IZO site and 09 August for KW station, reporting AOD values of 0.35 and 0.46, respectively (Note that IZO is placed at around 2400 m a.s.l.). LR values were retrieved for these dusty days, and preliminary results were analyzed.

Extinction profiles for Saharan and Arabian dust retrieved, respectively, from IZO MPL-3 and KW P-MPL measurements are shown in Figures 1 and 2, respectively.

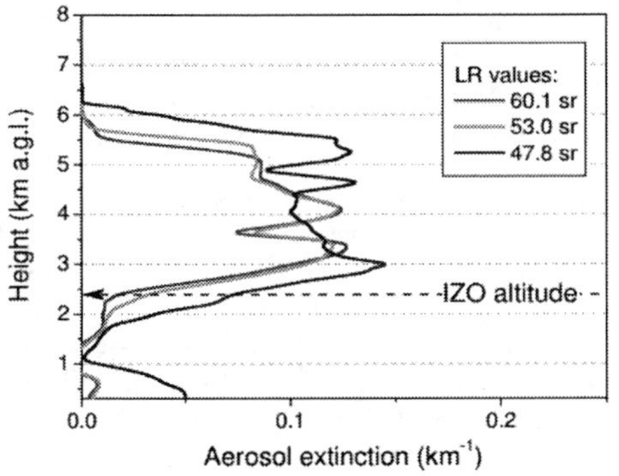

Figure 1. Saharan dust observed at IZO station: Aerosol extinction profiles derived from MPL-3 measurements performed on 29 June 2014 at several times (color lines).

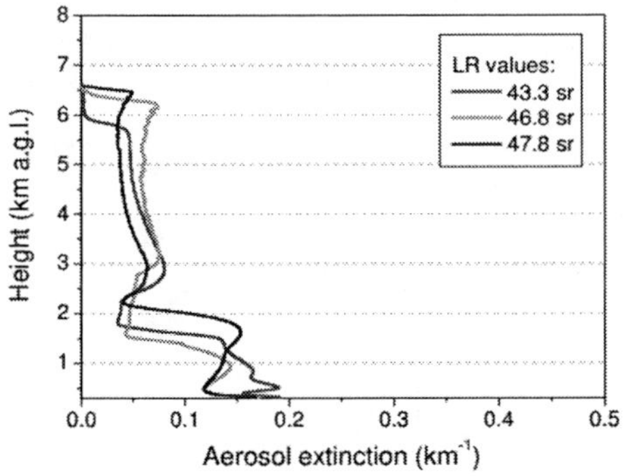

Figure 2. Arabian dust observed at KW station: Aerosol extinction profiles derived from P-MPL measurements performed on 09 August 2014 at several times (color lines).

Mean LR values of 54 ± 5 sr and 46 ± 2 sr are obtained for Saharan and Arabian dust particles, respectively, for these days (see Figs. 1 and 2). These preliminary results are in agreement with other LR values reported for dust in previous works [i.e., 6-8; 11] dependent on the desert dust source region. In addition, that observed disparity can be based on the singular composition of the suspended dust aerosols arriving from different desert areas, as reported in [6]. In that work an inverse relationship was found between LR and real refractive index, based on AERONET LR retrievals; in particular, Saharan dust presented a lower value of the real refractive index than that found for dust from the Arabian Middle East area; correspondingly, they obtained a higher LR (55 sr) for the Saharan dust than for Arabian dust (43 sr) [6]. Similar LR values were also derived by lidar measurements performed in Cyprus region for Arabian dust particles [8]. In general, it should be noted that results obtained in this work are specific for those particular cases analyzed.

4. CONCLUSIONS

A novel comparative analysis has been performed between LR values retrieved from MPL measurements carried out in two different dust-influenced stations: SCO and KW sites, as affected, respectively, by Saharan and Arabian dust intrusions.

In this first study, one dusty case was selected for each station. Their derived LR values are in agreement with those obtained in other regions affected by both Saharan and Arabian dust arrivals, highlighting this way the diversity found in LR for dust particles arriving from different dust source regions. In particular, the LR assumed by CALIPSO for pure dust (40 sr) could be a correct value for dust over regions as Arabian Peninsula and surroundings. In addition, results can be relevant, indeed, for extinction retrievals from observations performed by future space-borne elastic lidar instrumentation, as that aboard next ESA Sentinel and EarthCARE satellites.

Finally, a more extended dataset is to be analyzed in order to establish a long-term dust aerosol monitoring over these stations in the future.

Hence, similarities and differences might be plausibly obtained in LR retrievals for dust by using a larger dataset for each station.

ACKNOWLEDGEMENTS

This work has been supported by both the Spanish Ministerio de Economía y Competitividad (MINECO) under grant CGL2011-24891 (AMISOC project) and the Kuwait Foundation for the Advancement of Sciences under project 2011-1401-01. We acknowledge NOAA Air Resources Laboratory (ARL) for the provision of the HYSPLIT model. Authors also thank the staff of both Tenerife and Kuwait stations for providing maintenance and support of all the instrumentation involved in this work.

REFERENCES

[1] Cattrall et al., 2005: Variability of aerosol and spectral lidar and backscatter and extinction ratios of key aerosol types derived from selected Aerosol Robotic Network locations, *J. Geophys. Res.*, **110**, D10S11, doi:10.1029/2004JD005124.

[2] Mona et al., 2006: Saharan dust intrusions in the Mediterranean area: Three years of Raman lidar measurements, *J. Geophys. Res.*, **111**, D16203, doi:10.1029/2005JD006569.

[3] Müller et al., 2007: Aerosol-type-dependent lidar ratios observed with Raman lidar, *J. Geophys. Res.*, **112**, D16202, doi:10.1029/2006JD008292.

[4] Córdoba-Jabonero et al., 2011: Synergetic monitoring of Saharan dust plumes and potential impact on surface: a case study of dust transport from Canary Islands to Iberian Peninsula, *Atmos. Chem. Phys.*, **11**, 3067-3091, doi:10.5194/acp-11-3067-2011.

[5] Campbell et al., 2012: Evaluating nighttime CALIOP 0.532 μm aerosol optical depth and extinction coefficient retrievals, *Atmos. Meas. Tech.*, **5**, 2143–2160, doi:10.5194/amt-5-2143-2012.

[6] Schuster et al., 2012: Comparison of CALIPSO aerosol optical depth retrievals to AERONET measurements, and a climatology for the lidar ratio of dust, *Atmos. Chem. Phys.*, **12**, 7431–7452, doi:10.5194/acp-12-7431-2012.

[7] Amiridis et al., 2013: Optimizing CALIPSO Saharan dust retrievals, *Atmos. Chem. Phys.*, **13**, 12089-12106, doi:10.5194/acp-13-12089-2013.

[8] Mamouri et al., 2013: Low Arabian dust extinction-to-backscatter ratio, *Geophys. Res. Lett.*, **40**, 4762–4766, doi:10.1002/grl.50898.

[9] Omar et al., 2013: CALIOP and AERONET aerosol optical depth comparisons: One size fits none, *J. Geophys. Res. Atmos.*, **118**, 1–19, doi:10.1002/jgrd.50330.

[10] Tesche et al., 2013: Ground-based validation of CALIPSO observations of dust and smoke in the Cape Verde region, *J. Geophys. Res. Atmos.*, **118**, 2889–2902–, doi:10.1002/jgrd.50248.

[11] Córdoba-Jabonero, C., J. A. Adame, D. Grau, E. Cuevas and M. Gil-Ojeda, 2014: Lidar Ratio discrimination retrieval in a two-layer aerosol system from elastic lidar measurements in synergy with sun-photometry data, *Proc. International Conference in Atmospheric Dust* (DUST 2014), *ProScience*, **1**, 243-248, published by Digilabs, DOI:10.14644/dust.2014.040.

[12] Sabbah, 2010: Impact of aerosol on air temperature in Kuwait, *Atmospheric Res.*, **97**, 303-314.

STUDY OF AFRICAN DUST WITH MULTI-WAVELENGTH RAMAN LIDAR DURING "SHADOW" CAMPAIGN IN SENEGAL

Igor Veselovskii[1], Philippe Goloub[2], Thierry Podvin[2], Valentyn Bovchaliuk[2], Didier Tanre[2], Yevgeny Derimian[2], Mikhail Korenskiy[1], Oleg Dubovik[2]

[1]*Physics Instrumentation Center, Troitsk, Moscow Region, 142190, Russia*
E-mail: igorv@pic.troitsk.ru
[2]*Laboratoire d'Optique Atmosphérie, Université de Lille-CNRS, 59650, Villeneuve d'Ascq*

ABSTRACT

West Africa and the adjacent oceanic regions are very important locations for studying dust properties and their influence on weather and climate. The SHADOW (Study of SaHAran Dust Over West Africa) campaign is performing a multi-scale and multi-laboratory study of aerosol properties and dynamics using a set of in situ and remote sensing instruments at an observation site located at IRD (Institute for Research and Development) Center, Mbour, Senegal (14^{0}N, 17^{0}W). In this paper, we present the results of lidar measurements performed during the first phase of SHADOW which occurred in March-April, 2015. The multiwavelength Mie-Raman lidar acquired 3β+2α+1δ measurements during this period. This set of measurements has permitted particle intensive properties such as extinction and backscattering Ångström exponents (BAE) for 355/532 nm wavelengths corresponding lidar ratios and depolarization ratio at 532 nm to be determined. The backscattering Ångström exponent during the dust episodes decreased to ~-0.7, while the extinction Ångström exponent though being negative, was greater than -0.2. Low values of BAE can likely be explained by an increase in the imaginary part of the dust refractive index at 355 nm compared to 532 nm.

1. INTRODUCTION

Africa is one of the major sources of atmospheric dust and numerous campaigns were organized to study the dust properties near the origin source [1,2,3]. The SHADOW (Study of SaHAran Dust Over West Africa) campaign is performing a multi-scale and multi-laboratory study of aerosol properties and dynamics using a set of in situ and remote sensing instrumentation (multi-wavelength Raman LIDAR, Wind-LIDAR,, nephelometer, aethalometer, sun/lunar photometer, airborne sunphotometer, optical particle counter) in the framework of the CaPPA (Chemical and Physical Processed in The Atmosphere) project (http://www.labex-cappa.fr/). The site is located at IRD (Institue for Research and Development) Center, Mbour, Senegal (14^{0}N, 17^{0}W). The objective of the experiment is to report the optical, chemical and physical properties of the aerosols as well as the source apportionment in a location where aerosol loading can be very large and aerosol type depends on the season. Two enhanced observing periods of 7 weeks are considered: March-April 2015 when dust originating from the Sahara/Sahel region is dominant and December 2015-January 2016 when dust and carbonaceous aerosols resulting from fire activities are in variable proportion and transported at different altitudes. Other types of aerosols can also be present such as sulfates from nearby urban areas or maritime aerosols depending on the air mass flow. The mixed state of these various chemical components results in different radiative properties of the aerosols.

We hereinafter focus our study on multiwavelength Mie-Raman lidar measurements performed during the first phase of the SHADOW campaign for the period 8 March - 24 April. During this period approximately 40 day- and night-time measurement sessions were performed and numerous strong dust episodes were observed.

2. LIDAR SYSTEM DESCRIPTION

The results presented were obtained with LILAS multi-wavelength Mie-Raman lidar based on a tripled Nd:YAG Spectra Physics INDI laser with a 20 Hz repetition rate. The laser output power at λ=355 nm is about 2 W. The backscattered light is collected by a 40-cm aperture Newton telescope operated at 47 dg to horizon. The system is capable of detecting three backscattered signals at the laser wavelengths and two Raman signals, thus three backscattering and two extinction coefficients (so called 3β+2α set) can be calculated. The outputs of the detectors are recorded at 7.5 m range resolution using Licel transient recorders that incorporate both analog and photon-counting electronics. The full geometrical overlap of the laser beam and the telescope FOV is achieved at ~1500 m range, which determines the lower limit of the full set of our 3β+2α measurements due to the difficulty of calculating aerosol extinction in the overlap region. To improve the system operational capabilities, we detected rotational Raman scattering at 530 nm instead traditionally used vibrational nitrogen backscatter at 608 nm. [4]. For each profile 4000 laser pulses were accumulated so the temporal resolution of the measurements was about 3 minutes. The measurements were performed both in day and night time. The system has also capability for depolarization measurements at 355 nm and 532 nm wavelengths.

3. DUST PARTICLE PROPERTIES DERIVED FROM RAMAN LIDAR OBSERVATIONS

The vertical distribution of particle intensive properties is strongly influenced by the origin of the air masses which during the SHADOW measurement period were arriving either from ocean or continental regions. In this section, we present the results of three measurements sessions on 13 and 29 March which were characterized by different types of air masses.

13 March

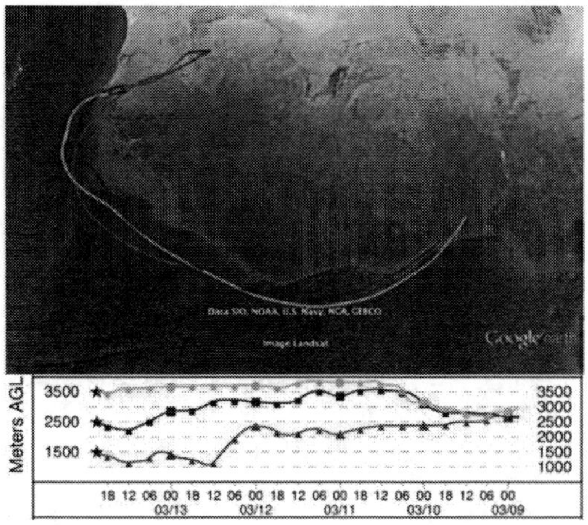

Fig.1. Five-day backward trajectories for the air mass in Mbour at altitudes 1500 m, 2500 m, 3500 m, on 13 March 2015 at 21:00 UTC.

As follows from fig.1, on 13 March the air masses at all heights were transported mainly over the ocean, but the back trajectory for 1500 m presents a "loop" over continent, so the corresponding air masses may contain more dust compared to other heights. Fig.2 shows the vertical profiles of 3β+2α measurements together with lidar ratios LR_{355}, LR_{532}, depolarization ratio δ_{532}, and Angstrom exponents $A^{\alpha}_{355/532}$, $A^{\beta}_{355/532}$ on 13 March 2015. The aerosol layer extended up to 3500 m but the extinction coefficient α was relatively small; at both 355 and 532 nm wavelengths it did not exceed 0.16 km^{-1}. The particle depolarization ratio at 532 nm was approximately 31±4.5% inside the dust layer (up to ~2750 m) and decreased to less than 15% at 3250 m. Likewise, the $A^{\alpha}_{355/532}$ and $A^{\beta}_{355/532}$ are close to zero up to 2750 m, but above that height both EAE and BAE start to increase indicating the presence of smaller particles. The lidar ratios LR_{355} and LR_{532} are approximately 53±8 sr inside the dust layer. Above 2750 m uncertainties are larger but the mean values of LR do not seem to change above the 2750 m height.

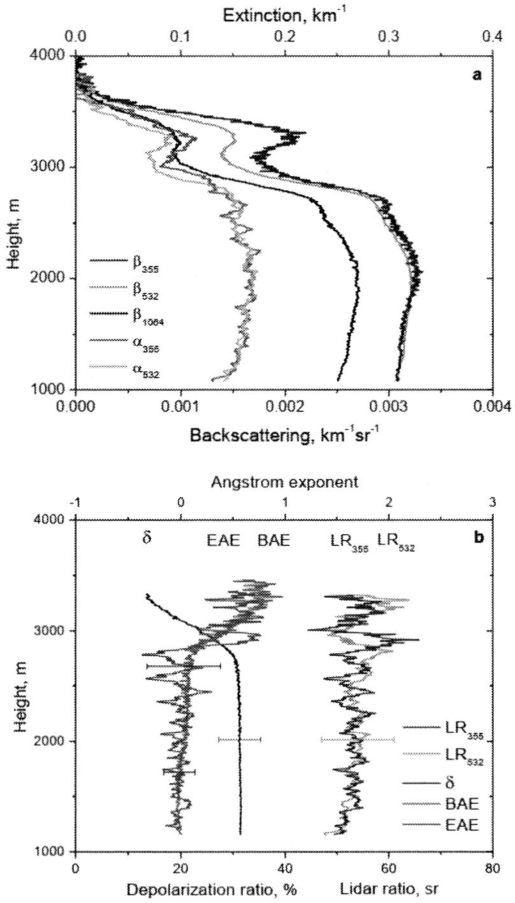

Fig.2. Vertical profiles of (a) backscattering and extinction coefficients and (b) lidar ratios, depolarization ratio, backscattering and extinction Angstrom exponents at 355/532 nm measured on 13 March 2015 for period 20:30-21:30 UTC.

29 March

The backtrajectories from the night of 29-30 March indicating a strong dust case are shown in Fig.3. The air masses at low altitude were transported over the continent and were strongly loaded with dust. Fig.4 presents the height profiles of the same particle parameters as in fig.2 but for 29 March. The extinction coefficient inside the dust layer (below 1500 m) is greater than 0.6 km^{-1} for both wavelengths. The backscattering coefficient at 355 nm inside the dust layer is lower than β_{532} which is consistent with the lidar ratio at 355 nm being larger than that at 532 nm with values as large as 65 sr.

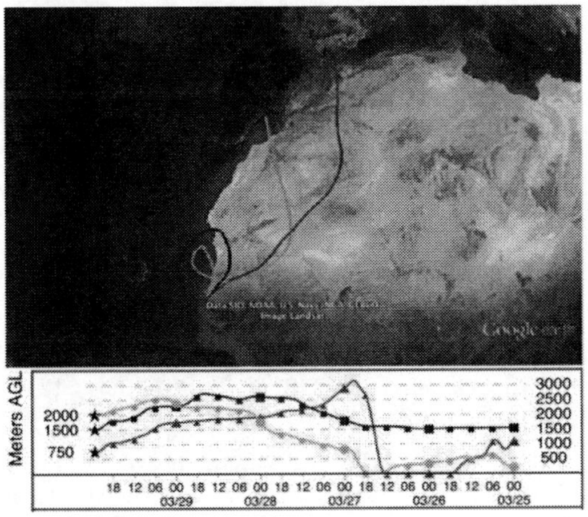

Fig.3. Five-day backward trajectories for the air mass in Mbour at altitudes 750 m, 1500 m, 2000 m on 29 March 2015 at 23:00 UTC.

The $A^{\beta}_{355/532}$ (BAE) is negative with a minimum value of less than -0.8, while EAE is still close to 0 as was observed on 13 March. The negative values of BAE observed on 29 March can be the result of the spectral dependence of the imaginary part of the dust refractive index (RI). It is well known that the imaginary part of RI of dust increases in the UV spectral region compared to that at 532 nm wavelength [3]. The ground based measurements performed during the SAMUM campaign demonstrated that the imaginary part of the dust RI could vary from m_I=0.005 at 532 nm to m_I=0.02 at 355 nm [3]. Such a strong enhancement of m_I may lead to a decrease of the backscattering coefficient [5].

To estimate the impact of the m_I enhancement at 355 nm on the values of EAE and BAE at 355/532 nm wavelengths, numerical simulations were performed. Extinction and backscattering Ångström exponents were calculated using the model of randomly oriented spheroids as described in [5] for a bimodal particle size distribution:

$$\frac{dn(r)}{d\ln(r)} = \sum_{i=f,c} \frac{N_i}{(2\pi)^{1/2}\ln\sigma_i}\exp\left[-\frac{(\ln r - \ln r_i)^2}{2(\ln\sigma_i)^2}\right]. \quad (1)$$

where $N_{f,c}$ is particle number density in the fine (*f*) and the coarse (*c*) mode. Each mode is represented by a lognormal distribution with modal radius $r_{f,c}$ and dispersion $\ln\sigma_{f,c}$. For the fine mode the values r_f=0.1 µm, $\ln\sigma_f$ =0.4 were used. For the coarse mode r_c=1.0 µm and three values $\ln\sigma_c$ =0.4, 0.5, 0.6 were considered.

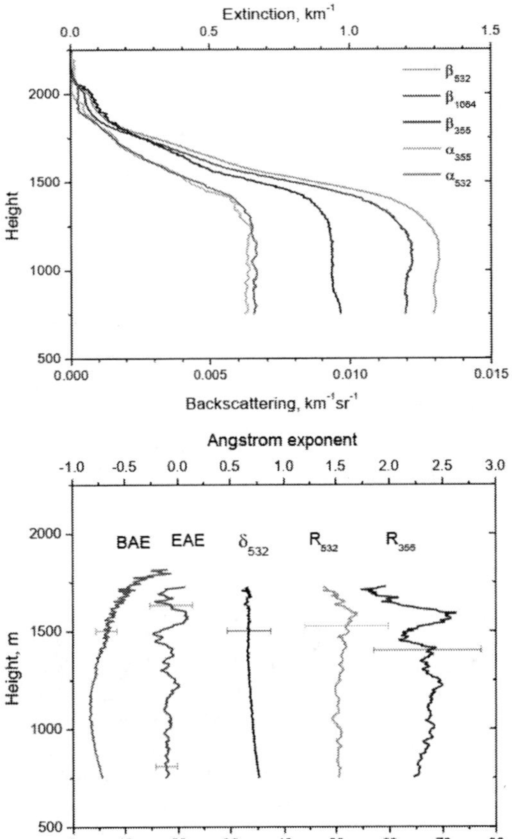

Fig.4. Vertical profiles of (a) backscattering and extinction coefficients and (b) lidar ratios, depolarization ratio, backscattering and extinction Angstrom exponents measured on 29 March 2015 for period 22:00-23:30 UTC.

Three size distributions used in computations are shown as insert in fig.5. The ratio N_c/N_f in all cases was 0.01, and the real part of CRI was 1.55 for all wavelengths. The imaginary part was fixed at 0.005 for 532 nm while it varied within the 0.005 – 0.05 range at 355 nm. Values of EAE and BAE as a function of m_I at 355 nm are given by fig.5. The EAE shows no significant sensitivity to changes in m_I, but BAE decreases rapidly with an increase of m_I at 355 nm. Although the present sensitivity study is limited, it illustrates the importance of accounting for the spectral dependence of $m_I(\lambda)$ in the analyses such as this.

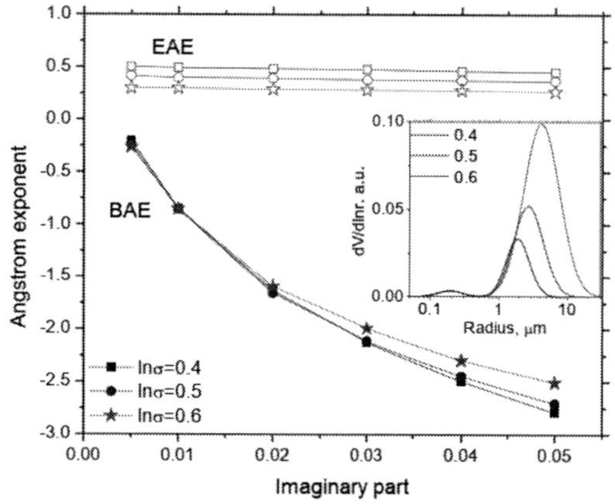

Fig.5. Extinction and backscattering Angstrom exponent for 355/532nm wavelengths as a function of the imaginary part of the refractive index at 355 nm. The CRI at 532 nm was kept m=1.55-i.005. Computations were performed using the model of randomly oriented spheroids for three bimodal PSDs shown in the insert.

CONCLUSION

The lidar measurements performed in March-April 2015 during the first phase of the SHADOW campaign in Senegal has provided a significant amount of information about dust particle parameters. The use of rotational Raman scattering in the LILAS for 532 nm observations improved the α_{532} measurements and, as a result, the calculation of lidar ratio and extinction Angstrom exponent were improved as well. The mean values of lidar ratios of pure dust were about 53±8 sr for both 532 nm and 355 nm wavelengths, which agrees with the values observed during SAMUM 1 (Morocco) and SAMUM 2 (Cape Verde) campaigns [3]. The mean value of particle depolarization ratio at 532 nm was 30±4.5%, however during strong dust episodes this ratio increased up to 35±5%, which is also in agreement with the results of SAMUM campaigns. The backscattering Angstrom exponent at 355/532 nm wavelengths during the dust episodes decreased to ~-0.7, while the EAE values, though being negative, were higher than -0.2. Low values of BAE may be a result of enhanced dust absorption at 355 nm.

Acknowledgments: Authors thank labex CaPPA for supporting this campaign and are very grateful to IRD-Dakar (Institut de Recherche pour le Développement) and specially A. Diallo and T. Ndiaye for their welcome and efficient support.

References

[1]. Reid, J. S. and Maring, H. B: Foreword to special section on the Puerto Rico Dust Experiment (PRIDE), J. Geophys. Res. 108, 8585, doi:10.1029/2003JD003510, 2003.

[2] Tanre, D., Haywood, J., Pelon, J., Leon, J.-F., Chatenet, B., Formenti, P., Francis, P., Goloub, P., Highwood, E. J., Myhre, G.: Measurements and modeling of the Saharan dust radiative impact: overview of the Saharan Dust Experiment (SHADE). J. Geophys. Res. 108, 8574, doi:10.1029/2002JD003273, 2003.

[3]. Ansmann, A., Petzold, A., Kandler, K., Tegen, I., Wendisch, M., Müller, D., Weinzierl, B., Müller, T., Heintzenberg, J.: Saharan Mineral Dust Experiments SAMUM–1 and SAMUM–2: what have we learned?, Tellus, 63B, 403–429, 2011.

[4]. Veselovskii, I., Whiteman, D. N., Korenskiy, M., Suvorina, A., Perez-Ramirez, D.: Use of rotational Raman measurements in multiwavelength aerosol lidar for evaluation of particle backscattering and extinction, Atmos. Meas. Tech., 8, 4111–4122, 2015.

[5]. Veselovskii I., O. Dubovik, A. Kolgotin, T. Lapyonok, P. Di Girolamo, D. Summa, D. N. Whiteman, M. Mishchenko, and D. Tanré, 2010: Application Of Randomly Oriented Spheroids For Retrieval Of Dust Particle Parameters From Multiwavelength Lidar Measurements, J. Geophys. Res., **115**, D21203, doi:10.1029/2010JD014139, 2010.

DOI: 10.1051/epjconf/201611908004

VERTICAL RESOLVED DUST MASS CONCENTRATION AND BACKSCATTER COEFFICIENT RETRIEVAL OF ASIAN DUST PLUME USING QUARTZ RAMAN CHANNEL IN LIDAR MEASUREMENTS

Young M. Noh[1]*, Detlef Mueller[2], Sungkyun Shin[3]

[1]*International Environmental Analysis and Education Center, Gwangju Institute of Science and Technology, 1 Oryong dong, Buk-gu, Gwangju 500-712, Republic of Korea,* *Email:nym@gist.ac.kr
[2]*Institute University of Hertfordshire, College Lane, Hatfield, Herts AL 10 9AB, United Kingdom*
[3]Advanced Environmental Monitoring Research Center, Department of Environmental Science and Engineering, Gwangju Institute of Science and Technology

ABSTRACT

In this work, we present a method for estimating vertical resolved mass concentration of dust immersed in Asian dust plume using Raman scattering of quartz (silicon dioxide, silica). During the Asian dust period of March 15, 16, and 21 in 2010, Raman lidar measurements detected the presence of quartz, and successfully showed the vertical profiles of the quartz backscatter coefficient. Since the Raman backscatter coefficient was connected with the Raman backscatter differential cross section and the number density of quartz molecules, the mass concentration of quartz in the atmosphere can be estimated from the quartz backscatter coefficient. The weight percentage from 40 to 70 % for quartz in the Asian dust was estimated from references. The vertical resolved mass concentration of dust was estimated by quartz mass concentration and weight percentage. We also present a retrieval method to obtain dust backscatter coefficient from the mixed Asian dust and pollutant layer. OPAC (Optical Properties of Aerosol and Clouds) simulations were conducted to calculate dust backscatter coefficient. The retrieved dust mass concentration was used as an input parameter for the OPAC calculations. These approaches in the study will be useful for characterizing the quartz dominated in the atmospheric aerosols and estimating vertical resolved mass concentration of dust. It will be especially applicable for optically distinguishing the dust and non-dust aerosols in studies on the mixing state of Asian dust plume. Additionally, the presented method combined with satellite observations is enable qualitative and quantitative monitoring for Asian dust.

1. INTRODUCTION

Dust has considerable influence on the Earth's radiative budget (Schwartz and Andreae, 2002). Next to the Sahara, an equally important source of dust is Central Asia from where dust is transported over East Asia to the adjacent Pacific Ocean. In Asia the problem of dust in the atmosphere is much more severe than in other areas of the world, as dust often is mixed with urban haze. Scientific knowledge on the mixing state of dust with anthropogenic pollution is very limited. Thus there is strong need the measurement technologies that allow us to monitor the vertical distribution of mineral dust immersed in continental pollution plumes. In this work, we present a method for estimating vertical resolved mass concentration of dust immersed in Asian dust plume using Raman scattering of quartz (silicon dioxide, silica). We also present a retrieval method to obtain dust backscatter coefficient from the mixed Asian dust and pollutant layer.

2. METHODOLOGY

We collected data during the Asian dust period of March 15, 16, and 21 in 2010 with multi-wavelength Raman lidar system at the Gwangju Institute of Science and Technology (GIST; 35.2° N, 126° E), Republic of Korea (South Korea). From the signals detected at 532 nm (elastic backscattering) and 607 nm (molecular Raman backscattering signals) we derive profiles of particle backscatter and extinction coefficients and the linear volume (particle plus molecule) depolarization ratio at 532 nm. The linear particle depolarization ratio follows according to the methods described by Sakai et al. (2003) and Noh et al. (2013). We installed Raman channel at 546 nm for inferring the mineral quartz concentration. The channels utilize the Raman return signals from silicon dioxide which is one main component of mineral quartz. The measurement technology is described by Tatarov and Sugimoto (2005) who use Raman return signals at 546 nm. Instrument setup, optical components, and details of the operation mode are described in Tatarov et al. (2011).

The weight percentage range from 40 to 70 % for quartz in the Asian dust was estimated from references (Feng et al., 2002; Ganzei and Razahigaeva, 2006). The vertical resolved mass concentration of dust was estimated by quartz mass concentration and weight percentage.

We also retrieved dust backscatter coefficient from the Asian dust mixed with pollutants by OPAC (Optical Properties of Aerosol and Clouds) simulations. The retrieved dust mass concentration was used as an input parameter for the OPAC calculations. It was assumed that the Asian dust is composed of nuclei, accumulation, and transported mode with mass mixing ratio of 0.110, 0.747, 0.153, respectively. The concentration of Asian dust by quartz measurements was converted as number concentration of each mode by following equation (1) :

$$N_i = \frac{D_z R_i}{M_i^*} \quad (1)$$

Where N_i represent the number concentration of Asian dust mode i, D_z denotes mass concentration of Asian dust at height z, M_i^* is weight per unit of grain. The number concentration of each component was used as an input parameter for the OPAC calculations to retrieve extinction coefficient.

3. RESULTS

3.1. Vertical resolved mass concentration of Asian dust

Figure 1 shows the vertical profiles of the aerosol extinction coefficient at 532 and quartz backscatter coefficient at 546 nm measured on three days, i.e., on 15, 16 and 21 March 2010.

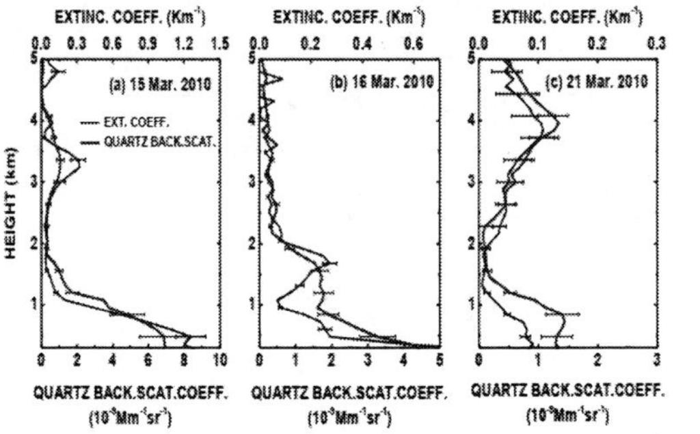

Figure 1. Extinction coefficient (black) and quartz backscatter coefficient (Blue).

Figure 2 shows the profiles of the mineral quartz and dust concentration. The profiles were obtained from the profiles of the quartz-backscatter coefficient and the values of the Raman scattering cross-section for quartz. We find 5-100 μg/m^3 mineral quartz. Dust concentration was calculated by assuming the ratio of quartz in the Asian dust as 40 – 70 %.

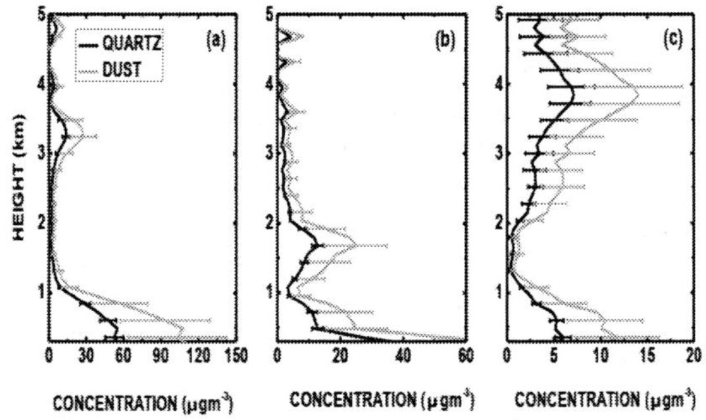

Figure 2. Vertical resolved mass concentration of quartz (black) and dust(gray).

Measurements of PM-10 are done on a routine basis by the Korea Meteorological Administration (KMA) in Gwangju. Table 1 shows PM-10 concentration observed at the surface, mineral quartz and dust concentration at the height of 360 m. The surface-level observations were 210 ± 110, 124 ± 6, and 42 ± 3 μg/m^3 as an average value during lidar measurement period. The concentration of dusts is 49.8, 55.6 and 19.0 % in Asian dust plumes on 15, 16, and 21 March 2010, respectively. This means we can double the dust concentration values we derived from our lidar observations. Our lidar derived dust concentration means pure dust originated from dust source regions. However, the surface measured PM-10 concentrations can include urban anthropogenic particle or local dust particle. Another reason for the higher in-situ values may be mixed with anthropogenic aerosols originated from industrial region of China during long-range transport. Since the dust plumes originated from dust source regions have to pass over industrial areas of China, anthropogenic aerosols can be mixed with dust plumes.

Table 1. PM-10 by in-situ measurement and quartz and dust concentrations by lidar measurements.

Date (2010)	PM_{10} (μg m^{-3})	SiO2 (μg m^{-3})	Dust (μg m^{-3})	Dust/ PM_{10}
03.15	215 ± 110	70 ± 25	107 ± 30	0.498
03.16	124 ± 6	45 ± 10	69 ± 16	0.556
03. 21	42 ± 3	5 ± 2	8 ± 3	0.190

3.2. Retrieval of dust backscatter coefficient

The dust backscatter coefficient was retrieved from vertical resolved dust concentration by OPAC. Figure 3 shows the backscatter coefficients of dust, non-dust, and total dust plumes. The backscatter coefficient of non-dust particle was calculated by subtracting dust backscatter coefficient from backscatter coefficient of total dust plume.

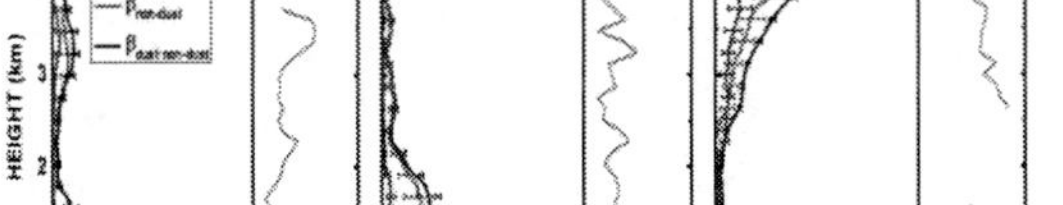

Figure 3. Backscatter coefficient of dust (blue), non-dust (red) and all (dust + non-dust) particles. The ratio of dust to the total aerosol backscatter coefficient is shown as gray line.

Figure 3 also shows the ratio of the dust backscatter coefficient to the total aerosol backscatter coefficient. The average ratios are 0.19 ± 0.07, 0.14 ± 0.06 and 0.31 ± 0.10 on 15, 16 and 21 March 2010, respectively. The ratios at the lowest lidar observation height (300 m) are 0.22, 0.17 and 0.15 on 15, 16 and 21 March 2010, respectively. These values are lower than the ratio of dust/PM_{10} in Table 1. The difference of the mass-scattering efficiency of dust and non-dust particles can be one reason for the low value of the dust backscatter coefficient. The aerosol mass-scattering efficiency (m^2g^{-1}), which is also referred to as mass-scattering coefficient or specific light-scattering is determined using concurrent measurements of the aerosol light-scattering coefficient and some estimate of particle mass (Charlson et al., 1999). Andreae et al. (2002) find that the fine-mode aerosols play a dominant role in the optical characteristics of the atmosphere in desert regions in spite of the desert location and the large contribution of mineral dust to the aerosol mass burden. Titos et al. (2012) report from one year of measurements (from March 2006 to February 2007) performed at an urban site in Southern Spain (Granada) that sulfate exhibits the largest mass scattering efficiency (7 ± 1 m^2g^{-1}) and dust aerosols present the lowest mass-scattering efficiency (0.2 ± 0.1 m^2g^{-1}). Hand and Malm (2007) reviewed the values of mass-scattering efficiencies of common aerosol species obtained from 60 studies carried out since 1990. The data represent a variety of time periods and locations on Earth. Hand and Malm (2007) recommend a value of 0.7 m^2g^{-1} of the mass-scattering efficiency for coarse mode dust while the mass-scattering efficiency of other anthropogenic aerosols is 2 – 6 m^2g^{-1}.

4. CONCLUSIONS

This research presents, for the first time, a new retrieval method to obtain dust backscatter coefficient from the mixed Asian dust and pollutant layer by combining quartz Raman measurement and OPAC simulation. OPAC (Optical Properties of Aerosol and Clouds) simulations were conducted to calculate dust backscatter coefficient. The retrieved dust mass concentration was used as an input parameter for the OPAC calculations.

Our research confirmed that the quratz Raman measurement is useful for estimating vertical resolved mass concentration of dust. It will be also especially applicable for optically distinguishing the dust and non-dust aerosols in studies on the mixing state of Asian dust plume. Additionally, the presented method combined with satellite observations is enable qualitative and quantitative monitoring for Asian dust.

ACKNOWLEDGEMENT

This work was supported by the "Advanced Research on Applied Meteorology" of National Institute of Meteorological Research (NIMR) funded by the Korea Meteorological Administration (KMA). This research was also supported by International Environmental Research Center (IERC).

REFERENCES

[1] Andreae, T.W., Andreae, M.O., Ichoku, C., Andreae, Maenhaut, W., Cafmeyer, J., Karnieli, A., Orlovsly, L., 2002. Light scattering by dust and anthropogenic aerosol at a remote site in the Negev desert, Israel. *J. Geophys. Res.* **107**, 252–290.

[2] Charlson, R. J., Anderson, T. L., and Rodhe, H., 1999: Direct climate forcing by anthropogenic aerosols: Quantifying the link between atmospheric sulfate and radiation, *Contrib. Atmos. Phys*. **72**, 79– 94.

[3] Feng, Q., K.N. Endo, and G.D. Cheng, 2002: Dust storms in China: A case study of dust storm variation and dust characteristics, *Bull. Eng. Geol. Environ.*, **61**, 253-261.

[4] Ganzei, L.A. and N.G. Razzhigaeva, 2006: Composition of sand storm particles in the southern far east, *Lithol. Miner. Resour.*, **41(3)**, 215-221.

[5] Hand, J.L., and Malm, W.C., 2007. Review of aerosol mass scattering efficiencies from ground-based measurements since 1990, *J. Geophys. Res.* **112**, D16203, doi:10.1029/2007JD008484.

[6] Tatarov, B., and N. Sugimoto, 2005: Estimation of quartz concentration in the tropospheric mineral aerosols using combined Raman and highspectral-resolution lidars, *Opt. Lett.*, **30**, 3407–3409.

[7] Tatarov. B., et al. 2011: Lidar measurements of Raman scattering at ultraviolet wavelength from mineral dust over East Asia, *Opt. Exp.*, **19**. 1569-1581.

[8] Sakai, T., Nagai, T., Nakazato, M., Mano, Y., and Matsumura, T., 2003: Ice clouds and Asian dust studied with lidar measurements of particle extinction-to-backscatter ratio, particle depolarization, and water-vapor mixing ratio over Tsukuba, *Appl. Opt.*, **42**, 7103-7116.

[9] Schwartz, S. E., and M. O. Andreae. 2002: Uncertainty in climate change caused by aerosols, **272**, 1121–1122, doi:10.1126/science.272.5265.112.

[10] Noh, Y.M., Lee, H., Müller, D., Lee, K., Shin, D., Shin, S., Choi, T.J., Choi, Y.J., Kim, K.R., 2013: Investigation of the diurnal pattern of the vertical distribution of pollen in the lower troposphere using LIDAR, *Atmos. Chem. Phys.* **13**, 7619-7629.

[11] Titos, G., Foyo-Moreno, I., Lyamani, H., Querol, X., Alastuey, A., and Alados-Arboledas, L., 2012. Optical properties and chemical composition of aerosol particles at an urban location: An estimation of the aerosol mass scattering and absorption efficiencies, *J. Geophys. Res.*, **117**, D04206, doi:10.1029/2011JD016671.

MINERAL DUST IMPACT ON SHORT- AND LONG-WAVE RADIATION AND COMPARISON WITH CERES MEASUREMENTS

Salvatore Romano*, Maria Rita Perrone

Department of Mathematics and Physics, Università del Salento, Lecce, 73100, Italy

**Email: salvatore.romano@unisalento.it*

ABSTRACT

Clear-sky downward and upward radiative flux measurements both in the short- and in the long-wave spectral range have been used to estimate and analyze the radiation changes at the surface due to the mineral dust advection at a Central Mediterranean site. Then, short- and long-wave radiative fluxes retrieved from the CERES (Clouds and the Earth's Radiant Energy System) radiometer sensors operating on board the EOS (Earth Observing System) AQUA and TERRA platforms have been used to evaluate the mineral dust radiative impact at the top of the atmosphere. Satellite-derived radiative fluxes at the surface have been compared with corresponding ground-based flux measurements, collocated in space and time, to better support and understand the desert dust radiative impact. Results referring to the year 2012 are reported.

1. INTRODUCTION

Mineral dust aerosols play an important role in the Earth's radiation budget influencing significantly the climate system through direct (scattering and absorption of solar and terrestrial radiation) and indirect effects (acting as cloud and ice nuclei of condensation). In a more detail, mineral aerosols cause the decrease of the incoming short-wave (SW) radiation at the surface (cooling effect) as a consequence of their scattering and absorption properties. Conversely, they enhance the green-house effect in the long-wave (LW) spectral range by trapping the outgoing terrestrial radiation at the surface (warming effect). The Mediterranean basin represents one of the most important areas for the study of the mineral aerosol impacts due to its proximity to the Sahara desert. Mineral dust outbreaks affect the Mediterranean area mostly in summer and contribute to one of the highest radiative effects in the world [1].

The ground-based radiative flux measurements analyzed in this study have been performed throughout the year 2012, at the Mathematics and Physics Department of the Università del Salento, Lecce (Italy) (40.33°N; 18.11°E), a coastal site of the Central Mediterranean area. Two Kipp & Zonen pyranometers (CMP 21 model) and two Kipp & Zonen pyrgeometers (CGR 3 model) have been used to retrieve upward and downward irradiance (F_{UP} and F_{DN}, respectively) at the surface in the SW (0.31-2.8 μm) and in the LW (4.5-42 μm) spectral range, respectively. Then, the CERES radiometers have allowed estimating directly F_{UP} at the top of the atmosphere (TOA) and through an algorithm F_{DN} and F_{UP} at the surface in the SW (0.3-5 μm) and in the LW (5-100 μm) spectral range, respectively.

2. METHODOLOGY

The Barcelona Supercomputing Center Dust REgional Atmospheric Model (BSC-DREAM8b v2.0, http://www.bsc.es) and the Ångstrom exponent (Å) calculated from the Aerosol Optical Depth (AOD) at 440 and 870 nm retrieved from AERONET (AErosol RObotic NETwork) sun/sky photometer measurements (http://aeronet.gsfc.nasa.gov) have been used to identify dusty days. In particular, we have considered as dusty days the ones characterized by a daily mineral dust loading greater than 0.05 g m^{-2} and a daily Å mean value lower than 0.9, in accordance with BSC-DREAM8b and AERONET data, respectively. The true-color images from MODIS (MODerate resolution Image Spectroradiometer) on board AQUA and/or TERRA satellites (http://modis.gsfc.nasa.gov) and the analytical back-trajectories from the HYSPLIT (Hybrid Single Particle Lagrangian Integrated Trajectory) model (http://ready.arl.noaa.gov) have also been used to support BSC-DREAM8b and AERONET results.

Clear-sky conditions have been defined for CERES footprints with cloud cover percentages less than 5%, according to the method adopted in [2]. We have analyzed the SW and LW radiative fluxes at the surface and at the top of the atmosphere available in the Single Scanner Footprint (SSF) Level2 products of the CERES sensors (http://ceres.larc.nasa.gov). In particular, we have considered only CERES radiative flux values related to a field of view (FOV) center within an area of 0.5° latitude by 0.5° longitude around the study site.

3. RESULTS

Fig. 1 provides the monthly evolution of the AOD at 440 nm and the Å(440,870) mean values during the analyzed year 2012. The AOD monthly mean values ranged from 0.10 on December to 0.26 on August, while the Å monthly mean values varied from 1.82 on February to 0.77 on November. Then, Fig. 2 shows the absolute frequency of the clear-sky dusty and dust-free days, identified by using the previously described methodology, for each month of the studied year. No clear-sky dusty days were found in January, February, October, November, and December of the year 2012. The number of clear-sky dust-free days ranged from 4 on January and December to 23 on August, while the number of clear-sky dusty days varied from 3 on March to 6 on July. In particular, we have found 30 dusty days that represent the 19% of the total clear-sky days from March to September 2012. These results confirm that desert dust events are more frequent in spring/summer over the Mediterranean area. In fact, note that July was the month characterized by one of the largest (0.25) and lowest (1.27) monthly mean values of AOD and Å, respectively, as reported in Fig. 1. Observe that a combination of large AOD with small Å may indicate a significant contribution of coarse mode mineral dust to the whole aerosol load. In addition, it is worth noting that the AOD mean value increased from 0.20 on dust-free days to 0.28 on dusty days, while the Å mean value decreased from 1.40 on dust-free days to 0.84 on dusty days during the period affected by the desert dust outbreaks (March-September 2012). Daytime mean values of clear-sky radiative fluxes have been divided into dust-free and dusty days mean values both in the SW and in the LW spectral domain, respectively, in order to quantify and investigate the monthly-based impact of mineral dust on the radiation measurements. Ground-based flux measurements have revealed that the monthly mean value of SW-F_{DN} decreased from 2.2% (April) to 4.2% (May) and the monthly mean value of LW-F_{DN} increased from 2.3% (May) to 6.1% (March) as a consequence of the mineral dust advection. In addition, the mean value of SW-F_{DN} decreased from 590 W m^{-2} on dust-free days to 570 W m^{-2} on dusty days, considering the period affected by the mineral dust intrusion (March-September 2012). Conversely, in the same period, the mean value of LW-F_{DN} increased from 360 W m^{-2} on dust-free days to 370 W m^{-2} on dusty days. These flux changes were related to the previously reported variations of the analyzed aerosol properties (AOD and Å).

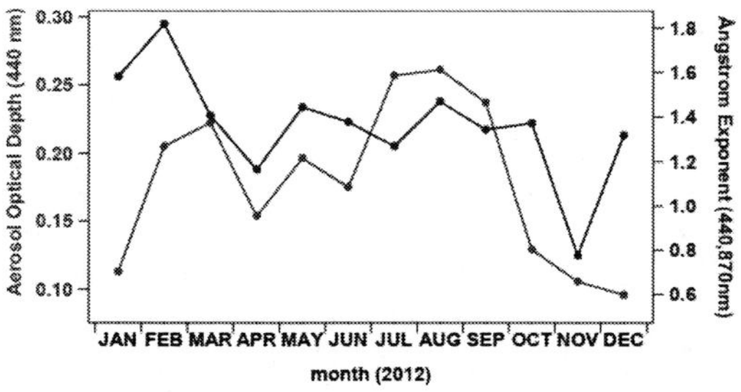

Fig. 1 Temporal evolution of the monthly mean values of the AOD at 440 nm (in red, on the left axis) and of the Å(440,870) (in blue, on the right axis) during the analyzed year 2012.

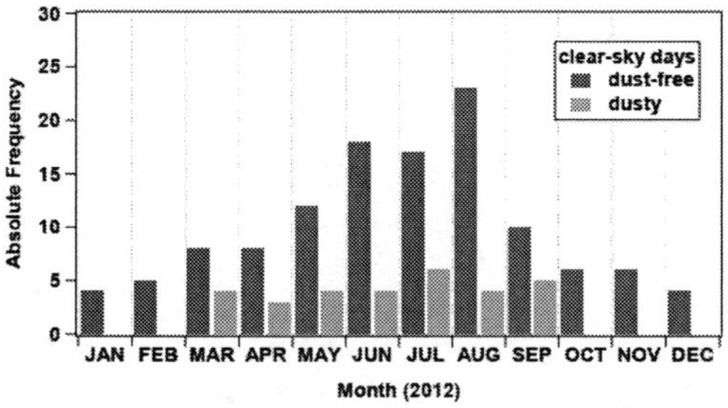

Fig. 2 Monthly-based absolute frequency of the clear-sky dust-free (in blue) and dusty (in orange) days during the analyzed year 2012.

A moderate Saharan dust episode observed over south-eastern Italy from 9 to 13 July 2012 is analyzed in the following as a case study. Fig. 3 shows the temporal evolution of the AOD at 440 nm and of the Å(440,870) at the study site from 8 to 14 July 2012. Note that July 10 and 12 were the days mostly affected by the desert dust advection during the analyzed period. In fact, the AOD, which was equal to 0.12 at midday of July 8 (dust-free day), reached the largest instantaneous values on July 10 and 12 (0.50 and 0.45, respectively), according to Fig.3. In addition, the instantaneous Å values decreased from 1.65 at midday of July 8 to 0.56 and 0.54 at midday of July 10 and 12, respectively. Then, the mineral dust loading that was equal to 0.02 g m^{-2} at 12:00 UTC of July 8 increased up to 0.12 and 0.15 g m^{-2} at 12:00 UTC of July 10 and 12, respectively, as a consequence of the Saharan dust advection, in accordance with the BSC-DREAM8b results. In particular, these AOD, Å, and mineral dust loading changes have been responsible for a surface SW-F_{DN} (SW-F_{UP}) decrease of 86 and 83 W m^{-2} (19 and 16 W m^{-2}) on July 10 and 12, respectively. In the same time interval, the surface LW-F_{DN} increased of 36 and 55 W m^{-2}, while this effect was less evident for LW-F_{UP} (increase of 9 and 18 W m^{-2}) since the latter is more affected by the ground properties. It is worth observing that the F_{UP} evolution found at the top of the atmosphere is opposite to the one observed at the surface. The CERES-based TOA SW-F_{UP} increased from 145 W m^{-2} at midday of July 8 to 238 W m^{-2} at midday of July 12, as a consequence of the desert dust intrusion. On the contrary, the TOA LW-F_{UP} decreased from 327 to 309 W m^{-2} in the same time interval.

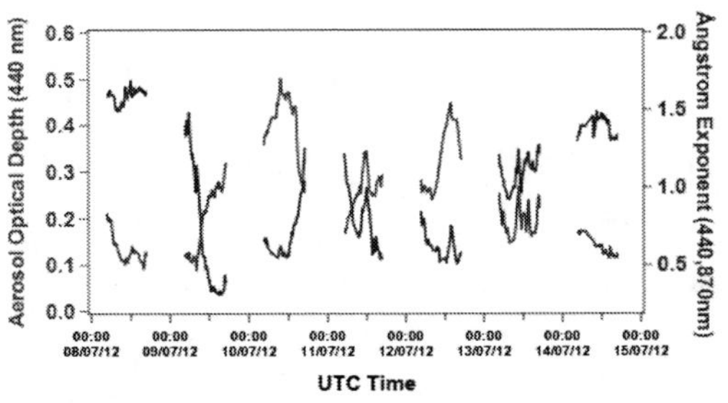

Fig. 3 Temporal evolution of the AOD at 440 nm (in red, on the left axis) and of the Å(440,870) (in blue, on the right axis) during the analyzed period of July 8-14, 2012.

During the analyzed dust outbreak, the clear-sky and satellite-derived surface F_{DN} values have been compared with the collocated in time and space ground-based measurements in order to better support and understand the mineral dust radiative impact. Satellite-derived and ground-based F_{DN} at the surface presented a high linear correlation coefficient (0.87) in the SW spectral range. In a more detail, the SW-F_{DN} values measured at the surface on average slightly overestimated of 23 W m^{-2} (2.6%) the CERES-derived results during the studied period. This percentage difference resulted slightly greater than the total uncertainty related to the SW flux measurements (2%). Yan et al. [2] also found ground-based SW-F_{DN} at the surface on average greater than corresponding CERES-derived values during summer. A different relationship between satellite- and ground-based LW-F_{DN} values at the surface was found for daytime and nighttime. A very low mean percentage difference (1.5%) was observed for the nighttime period, resulting within the experimental error of 3%. Conversely, all the LW-F_{DN} values measured at the surface resulted larger than the related CERES-derived values for the daytime period, with a greater mean percentage difference (6.8%), indicating the presence of a systematic error.

4. CONCLUSIONS

The mineral dust impact on ground- and satellite-based irradiance measurements both in the SW and in the LW range have been evaluated during the Saharan dust outbreaks that have affected south-eastern Italy throughout the year 2012. We have found that the mean value of the dusty days SW-F_{DN} at the surface was 20 W m^{-2} smaller than the dust-free days one from March to September 2012, which was the period affected by the desert dust intrusions. Conversely, in the same period, the mean value of the dusty days LW-F_{DN} at the surface was 10 W m^{-2} greater than the dust-free days one. In addition, we have found that the surface SW- and LW-F_{DN} decreased up to 9% and increased up to 13%, respectively, compared to a pristine dust-free day during the studied mineral dust outbreak of July 9-13, 2012. These outcomes are due to the cooling and warming effect by mineral dust at the surface in the SW and LW spectral range, respectively. On the contrary, we have found that the mineral dust advection was responsible for an increase and a decrease of the F_{UP} at the top of the atmosphere in the SW and in the LW spectral range, respectively. Finally, we have observed that the ground-based SW-F_{DN} was on average 2.6% greater than the corresponding ones retrieved from CERES radiometer sensors resulting within the experimental uncertainty. The mean percentage difference between ground- and satellite-based LW-F_{DN} at the surface, which was equal to 1.5% on nighttime and 6.8% on daytime, resulted within and without the experimental error, respectively.

ACKNOWLEDGEMENT

S. Romano has carried out this work with support of a PhD fellowship from the Mathematics and Physics Department of the Università del Salento. This study has been supported by the European Community through the ACTRIS Research Infrastructure Action under the 7th Framework Programme, ACTRIS grant agreement no. 262254. The authors gratefully acknowledge the NASA/ Goddard Space Flight Center for the provision of CERES data and satellite images. The NOAA Air Resources Laboratory (ARL) and the Barcelona Super-Computing Centre are also acknowledged for their contribution of HYSPLIT and DREAM model results, respectively.

REFERENCES

[1] Lelieveld, J., H. Berresheim, S. Bormann, P.J. Crutzen, F.J. Dentener and coauthors, 2002: Global air pollution crossroads over the Mediterranean, *Science*, **298**, 794-799.

[2] Yan, H., J. Huang, P. Minnis, T. Wang, J. Bi, 2011: Comparison of CERES surface radiation fluxes with surface observations over Loess Plateau, *Remote Sens. Environ.*, **115**, 1489-1500.

EPJ Web of Conferences **119**, 08006 (2016) DOI: 10.1051/epjconf/201611908006
ILRC 27

STRONG SAHARAN DUST EVENT DETECTED AT LALINET LOA-UNAL STATION, OVER MEDELLÍN, COLOMBIA BY ACTIVE AND PASSIVE REMOTE SENSING

Andrés Bedoya*[(1)], Daniel Nisperuza [(1)], Dairo Alegría[(1)], Mauricio Múnera[(1)], Juan Luis Guerrero-Rascado[(2,3)], Carmen E. Zapata[(4)] , Jose F. Jiménez[(4)], Eduardo Landulfo[(5)] and Álvaro Bastidas[(1)]

[1]*Lasers and Spectroscopy Group, Facultad de ciencias, Universidad Nacional de Colombia, Medellín 050034, Colombia, *Email: aebedoyav@unal.edu.co*

[2]*Dpt. Applied Physics, Faculty of Sciences, University of Granada, Fuentenueva s/n, 18071, Granada, Spain*

[3]*Andalusian Institute for Earth System Research (IISTA-CEAMA), Avda. del Mediterráneo s/n, 18006, Granada, Spain*

[4] *Laboratorio de Calidad del Aire (CALAIRE), Facultad de Geociencias y Medio Ambiente, Universidad Nacional de Colombia, Medellín 050041, Colombia.*

[5] *Centro de Lasers e Aplicações, Instituto de Pesquisas Energéticas e Nucleares (IPEN), Avd. Prof. Lineu Prestes2242, 05508-000, São Paulo, Brazil 4*

ABSTRACT

Passive and active remote sensing techniques are well used for understanding optical and microphysical characteristics of aerosol layers. Lidar has the ability to resolve stratifications of the complex vertical structures in the atmosphere and determine the existence of aerosols which has been transported for long-ranges through the evaluation of the optical properties such as particle backscatter and extinction coefficients, among others. CIMEL sunphotometer data (AERONET network) give information about optical properties such as Aerosol Optical Depth (AOD), Single Scattering Albedo (SSA), and Angström Exponent (AE) and microphysical properties such as size distribution. The LOA-UNAL lidar station as part of the LALINET (Latin American LIdar NETwork) [1], involves an elastic coaxial system operating in zenith mode used for monitoring the atmosphere at Medellín-Colombia (6.26°N, 75.58°W, 1470 m asl). This work presents a Saharan dust even over Medellín, Colombia, 27th June, 2014, observed simultaneously with lidar, sun-photometer and complementary global mass transport model HYSPLIT.

1. INTRODUCTION

Anthropogenic and natural radiative forcing is a very important topic which has many uncertainties (e.g. mineral dust particles). The magnitude of dust emissions to the atmosphere depends on the surface wind speed and many soil-related factors such as texture, moisture and vegetation cover [2]. Mainly sources of mineral dust particles on global scale are Sahara and Gobi deserts which are located on the African continent. Their principal effect is by scattering and absorption (direct effect), and their interaction with clouds (indirect effect) [3].

Mineral dust aerosols from North Africa represent one of the largest sources of aerosols available to the atmosphere, and their generation and transport are thought to be modulated by African easterly waves. The Sahara desert as the world's most important dust source adjoins directly to the Atlantic Ocean. A major part of the Saharan mineral dust injected to the atmosphere is exported towards the northern tropical Atlantic [4-6].

In this paper, we report an event of dust particles advected over Medellín city in Colombia, see figure 1, which was registered by first time in this region, using remote sensing instrumentation: lidar and sun-photometer.

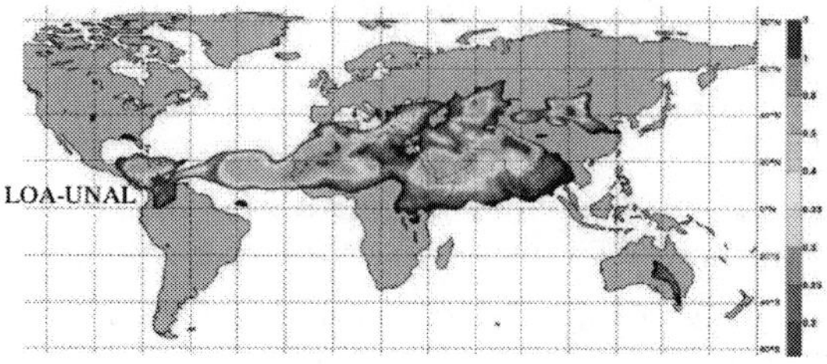

Figure 1 MACC model for representation of Dust Optical Depth transport until LALINET LOA-UNAL station.

2. METHODOLOGY

The available infrastructure of the LALINET-LOA-UNAL station is used for monitoring the air quality, as well as in-situ (namely $PM_{2.5}$ and PM_{10} impactors), and remote sensing (coaxial elastic lidar), operating at 532 nm and 355 nm (see figure 2), which derives aerosol optical properties trough the Klett-Fernald-Sasano algorithm [6] and the evolution of atmospheric boundary layer [7]. Also, it was used a CIMEL sun photometer of NASA-AERONET which allowing us to getting direct products (AOD and AE, and inversion products such as particle size distribution and SSA, with uncertainties about 0.01 at wavelengths larger than 440 nm and about 0.02 at wavelengths shorter that this value [8].

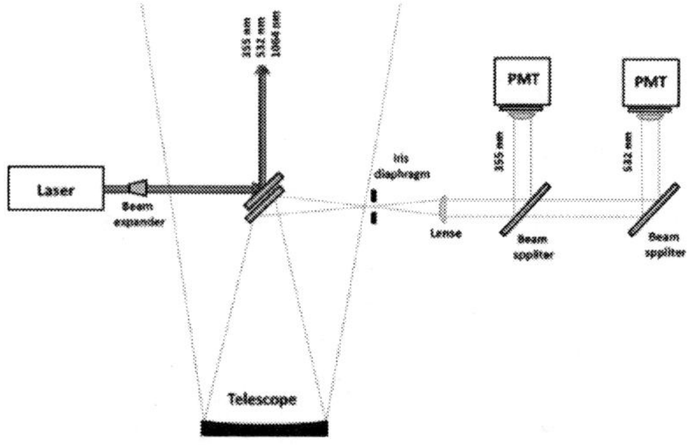

Figure 2 Lidar LOA-UNAL coonfiguration

3. RESULTS

The analysis of this event involved 5-day backtrajectories, calculated from HYSPLIT model, to evaluate the origin of the mineral particle advected over Medellín. The backtrajectories were computed for six altitudes (since 1000 to 6000 m agl) using GDAS database (Global data assimilation (System,ftp://www.arl.noaa.gov/pub/archives/gdas1/). Figure 3 presents the backtrajectories on 27th June 2014.

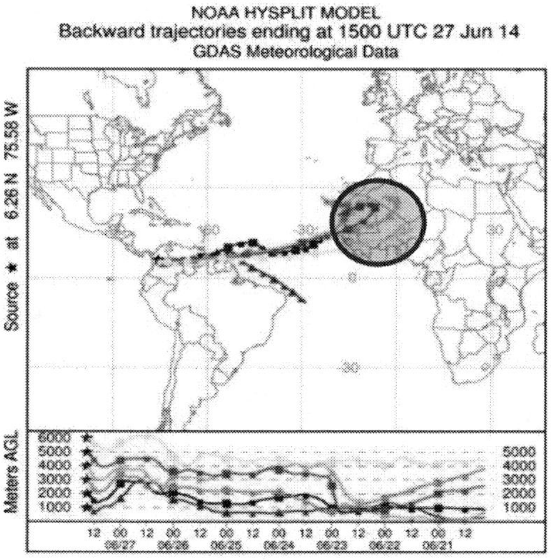

Figure 3 Hysplit 5-day backtrajectories on 27th June 2014 ending at Medellín

The mineral particulate directly comes from the North region of Africa (Sahara desert), traveled cross the Atlantic Ocean supported by typical easterly wave. The calculated backtrajectories for this event show an injection of particles which reaches this city at medium height levels (3000-5000 m agl).

On the other hand, the event was characterized by the increasing of the optical parameters as well as AOD_{500}, (0.2-0.8) $AE_{440-870}$ (0.1-0.3), and SSA obtained by the sun-photometer at Medellín. Also, the size distribution shows a predominance of the coarse mode over fine. [Figure 4] [9].

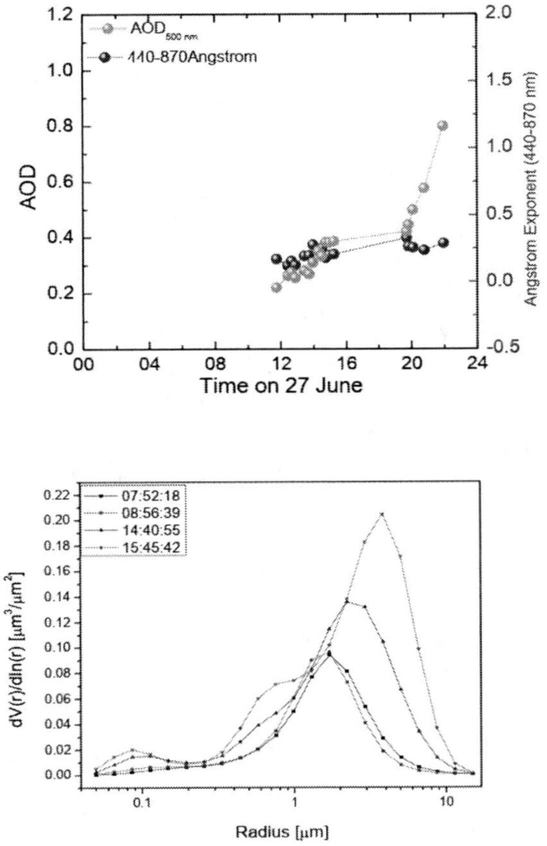

Figure 4 Sun-photometer data on 27th June: (a) AOD_{500} and AE440-870 ; (b) Particle size distribution.

Above of the size distribution, the modal radius coarse (~8 µm), was compared with PM_{10} (140 µg/m^3) data in the surface, which shown also an important increase.

In addition, lidar products were obtained through the inversion of Klett-Fernald- Sasano algorithm, in particular, the particle backscatter coefficient evolution at 532 nm (using a constant lidar ratio of 55 sr.) during all day, (Figure 5), involving the corresponding high accumulation.

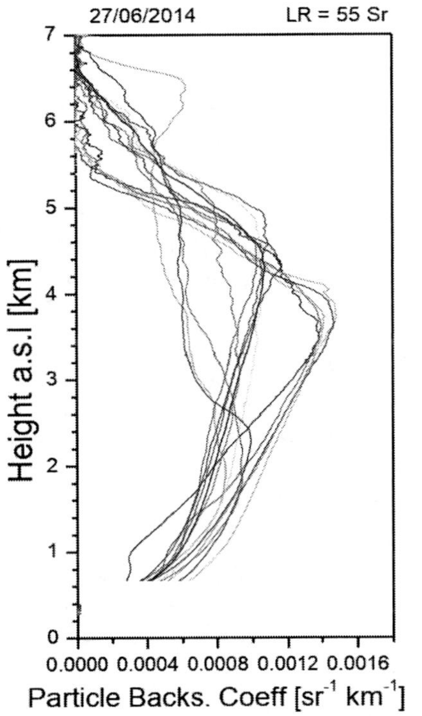

Figure 5 30-min averaged particle backscatter coefficient at 532 nm, computed with a lidar ratio of 55 sr.

4. CONCLUSIONS

The methodology used in this study allowed us to characterizing for the first time a Saharan dust event over this unexplored Colombian Andean region. The synergy of passive and active remote sensing technical with the calculus of some backtrajectories air mass and in-situ measurements, were the support for obtain this results associated with Saharan dust event over Medellín, Colombia.

It was possible to see the high accumulation of particles on the backscatter coefficient, (about 3 to 5 km agl), during all time measurements.

This phenomenon will be studied about the local atmospheric dynamic, air quality, and its health impact.

ACKNOWLEDGEMENT

This work has been supported by Universidad Nacional de Colombia through the LOA-UNAL station of LALINET, also by the air quality laboratory (CALAIRE) and by the University of Granada through the contract "Plan Propio. Programa 9. Convocatoria 2013". We acknowledge the use of NOAA- HYSPLIT, NAAPS and MACC models.

REFERENCES

[1] Towards an instrumental harmonization in the framework of LALINET: dataset of technical specifications, vol 9246 (924600). September 2014. Journal of applied Remote Sensing.

[2] IPCC 2013, Climate change 2013: the physical Science Basis, chapter 7

[3] Mahowald, N.M., and Kiehl, L.M. (2003). Mineral aerosol and cloud interactions. Geophysical Research Letters, 30.

[4] Wong, S. and Dessler, A.E. (2005). Suppression of deep convection over tropical North Atlantic by Saharan Air Layer. Geophysical Research Letters, 32.

[5] Middleton, N.J and Goudie, A.S. (2001). Saharan Dust: sources and trajectories, vol 26, 165-181. Royal Geographical Society.

[6] Huang, J., Zhang, C., Prospero, M.J. (2010). African outbreaks: A satellite perspective of temporal and spatial variability over the tropical Atlantic Ocean, vol 115, 1-20. Journal of Geophysical Research.

[7] Klett, J. D. (1985). Lidar inversion with stable backscattering to extinction ratios. Applied Optics, 24, 1638-1643.

[8] Nisperuza, D.J., Bedoya, A.E., Alegria, D.L., Múnera, M., Jiménez, J.F., Zapata, C.E. and Bastidas, A.E. Lidar measurements and wavelet covariance transform method to estimate the atmospheric boundary layer heights in Medellín, Colombia, Vol 47, 123-130. June 2014. Sociedad Española de Óptica (OPA).

[9] Holben, B.N., Eck, T.F., Slutsker, I., Tenré, D., Buis, J. P., Setzer, A., Vermote, E., Reagan, J. A., Kaufman, Y. J., Nakajima, T., Lavenu, F., Jankowiak, I., Smirnov, A. AERONET-A Federated Instrument Network and Data Archive for Aerosol Characterization, 1998.

[10] Dubovik, O. H. (2000). Variability of absorption and optical properties of key aerosol types observed in worldwire locations. Journal of the atmospheric sciences, 59, 590-608.

INTEGRATED STUDY OF AD-NET MIE-LIDAR NETWORK AND DATA ASSIMILATED CTM FOR ASIAN DUST EPIDEMIOLOGY IN JAPAN

Atsushi Shimizu[1*], Tomoaki Nishizawa[1], Nobuo Sugimoto[1], Ichiro Matsui[1], Hiroshi Kobayashi[2], Takashi Maki[3], Thomas T. Sekiyama[3], Kumiko Kanatani[4]

[1]*National Institute for Environmental Studies, 16-2 Onogawa, Tsukuba 305-8506 Japan*

**Email: shimizua@nies.go.jp*

[2]*University of Yamanashi, 4-4-37 Takeda, Kofu 400-8510 Japan*

[3]*Meteorological Research Institute, 1-1 Nagamine, Tsukuba 305-0052 Japan*

[4]*Kyoto University, Yoshida-konoe-cho, Sakyo-ku, Kyoto 606-8501 Japan*

ABSTRACT

Mie-scattering lidar data are going to be incorporated into data assimilation in chemical transport model to represent more reliable / useful horizontal distribution of Asian dust over Japan. The result is utilized in an epidemiology which surveys effect of Asian dust to human health. Wider application of AD-Net is expected in the field of environmental researches.

1. INTRODUCTION

Recent days Mie-scattering lidar data is utilized in Japan for environmental epidemiology on Asian dust [1]. Lidar is able to estimate concentrations of dust particles in fine time resolution using backscatter and depolarization ratio. It is compared with various medical dataset obtained in hospitals, or in domestic questionnaire based reports. Although several case studies were published, some limitations are recognized in these days. Especially, insufficient horizontal coverage is a problem in the epidemiology because patients are usually widely distributed. In this study, a data assimilation is introduced in a chemical transport model (CTM) with dust extinction coefficient by lidars, then the output on the grid points in CTM is supplied for comparison with dataset of health. Also, lidar system is updated to observe lower altitude to catch the actual air which is directly affect human beings. In addition, introduction of polarization optical particle counter (POPC) is planned. Finally, dust exposure estimated by lidar/POPC is distributed for public in real time.

2. CURRENT STATUS OF AD-NET

Dual wavelength polarization lidars are currently operated in 17 sites in East Asia. Each lidar measures backscatter at 532 / 1064 nm and depolarization ratio at 532 m, with 30 m vertical resolution and 15 minutes time resolution. As operated automatically regardless of weather conditions, classification of scatterers including aerosols, clouds, rain drops, snow, and so on are practically important. For clear air region, Fernald's method is applied at 532 nm to retrieve aerosol backscatter (or extinction) coefficients and particulate depolarization ratio. Finally dust extinction and spherical particle extinction coefficients are estimated [2]. Results are uploaded to NIES WWW server, and distributed to the Ministry of Environment Japan every hour. Currently the lowest altitude of valid data is fixed at 120 m from surface.

3. LIDAR DATA TREATMENT FOR HIND CAST EPIDEMIOLOGY

Usually information related to human health are collected daily basis. Actual atmospheric condition changes in a day, thus a well-designed exposure data from atmospheric observation is needed. Lidar can measure atmosphere in high time resolution, but statistical approach is indispensable for epidemiology. Fig 1 is an example of time-height sections of backscatter coefficients and depolarization ratio accompanied with dust / spherical particle extinctions. These quantities are estimated in the time resolution of 15 minutes (96 times per day) in AD-Net, but just averaging over time in a day is not suitable because some kind of contamination from extremely large extinction by cloud / fog / snow are not excluded perfectly. Thus we introduced median value during one day in the initial stage of collaboration with epidemiology. However, this is not the best solution for drastically changing atmospheric conditions. For example, a heavy dust passed rapidly over lidar and clear air mass arrived just after dust, median cannot depict the short lived dust event at all. We have proposed two methods to conquer this difficulty.

(1) Averaging of segmented median

Median is obtained every 3 hours, then 8 values are averaged with weighting.

(2) Level counting by hourly basis

Hours in which dust extinction exceeds threshold value are counted in a day.

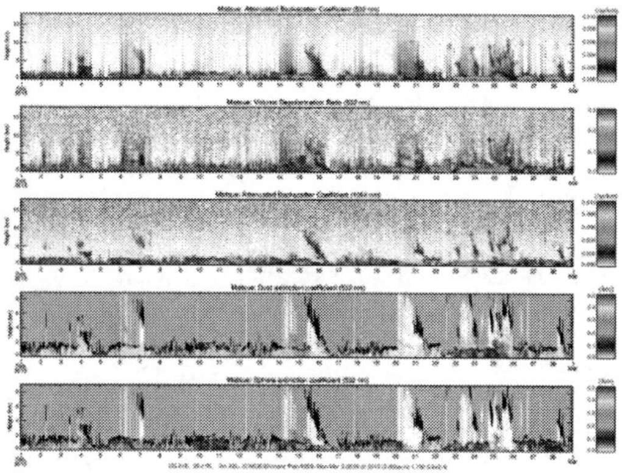

Figure 1. Time height indications of (from top) attenuated backscatter coefficient at 532 nm, volume depolarization ratio at 532 nm, attenuated backscatter coefficient at 1064nm, dust extinction coefficient, and spherical particle extinction coefficient observed in February 2015 at Matsue (133E, 35N).

4. PLANS FOR OBTAINING MORE USEFUL DATA

(1) Data assimilated CTM

Dust concentration can be calculated by chemical transport model (CTM) with appropriate meteorological data. If the horizontal grid scale of the model is small enough to represent the local exposure level in (for example) each city, the problem about distance between lidar location and colony of patients can be defeated. Then the accuracy of calculated concentration of dust is the key factor. Recently data assimilation technique is combined with CTM, and lidar data can be included in this system [3,4]. Thus, dust extinction estimated by the lidar network will raise the reliability of concentration predicted by CTM, and such results are the most suitable for epidemiology. We are now considering to make a long period assimilated dust concentration data in East Asian region.

(2) POPC

POPC measures size and depolarization ratio at 120 degree of independent particle at 780 nm. As this equipment is operated at the surface, the results can be directly compared with information related to the human health. Challenging is the relationship with the number concentration by POPC and dust extinction by lidar or mass concentration (TSP etc.) of dust.

(3) Smaller telescope for lidar

Currently AD-Net does not estimate optical properties below 120 m altitude due to insufficient overlap of laser and FOV of the telescope. For epidemiology, information from lower altitude is more important because human breathes air at the surface. We additionally attach a smaller telescope with wider FOV for each lidar in order to retrieve optical properties around 60 m from the surface.

5. CONCLUSIONS

A plan to make lidar data more useful for an epidemiology related to Asia dust is explained. A closer application of lidar data in the civic life is demanded in Japan, and continuous improvements of equipment / data treatment are indespensable.

ACKNOWLEDGEMENT

This work is supported by the Environment Research and Technology Development Fund (5-1502) of the Ministry of Environment Japan.

REFERENCES

[1] Ueda, K., A. Shimizu, H. Nitta, and K. Inoue (2012), Long-range transported Asian Dust and emergency ambulance dispatches, *Inhalation Toxicology*, *24* (12), 858-867, doi:10.3109/08958378.2012.724729.

DOI: 10.1051/epjconf/201611908007

[2] Shimizu, A., N. Sugimoto, I. Matsui, K. Arao, I. Uno, T. Murayama, N. Kagawa, K. Aoki, A. Uchiyama, and A. Yamazaki (2004), Continuous observations of Asian dust and other aerosols by polarization lidars in China and Japan during ACE-Asia, *J. Geophys. Res.*, *109* (D19), D19S17, doi:10.1029/2002JD003253.

[3] Yumimoto, K., I. Uno, N. Sugimoto, A. Shimizu, and S. Satake (2007), Adjoint inverse modeling of dust emission and transport over East Asia, *Geophys. Res. Lett.*, *34*, L08806, doi:10.1029/2006GL028551.

[4] Sekiyama, T. T., T. Y. Tanaka, A. Shimizu, and T. Miyoshi (2010), Data assimilation of CALIPSO aerosol observations, *Atmos. Chem. Phys.*, *10* (1), 39-49, doi:10.5194/acp-10-39-2010.

EPJ Web of Conferences **119**, 08008 (2016) DOI: 10.1051/epjconf/201611908008
ILRC 27

STUDY CASE OF AIR-MASS MODIFICATION OVER POLAND AND ROMANIA OBSERVED BY THE MEANS OF MULTIWAVELENGTH RAMAN DEPOLARIZATION LIDARS

Montserrat Costa-Surós[1], Lucja Janicka[1], Iwona S. Stachlewska[1*], Anca Nemuc[2], Camelia Talianu[2], Birgit Heese[3], Ronny Engelmann[3]

*[1]Institute of Geophysics, Faculty of Physics, University of Warsaw, Warsaw 02-093, Poland, *Email: iwona.stachlewska@igf.fuw.edu.pl*

[2]National Institute of R&D for Optoelectronics Romania (INOE), Magurele, Romania

[3]Institute for Tropospheric Research (TROPOS), Leipzig, Germany

ABSTRACT

An air-mass modification, on its way from Poland to Romania, observed between 19-21 July 2014 is discussed. The air-mass was investigated using data of two multi-wavelength lidars capable of performing regular elastic, depolarization and Raman measurements in Warsaw, Poland, and in Magurele, Romania. The analysis was focused on evaluating optical properties of aerosol in order to search for similarities and differences in the vertical profiles describing the atmospheric layers above the two stations within given period.

1. INTRODUCTION

There are a few multi-wavelengths lidar systems in Eastern Europe. According to EARLINET [1] there are lidar stations in Magurele (Romania), Minsk (Belarus), Belsk (Poland). The second Polish lidar site in Warsaw only recently joined the EARLINET lidar network. The low number of remote sensing stations is one of the reasons why little studies of air-mass modification have been done over Eastern Europe, in comparison with western part of Europe.

The air-mass modification studies are important to assess how particular modification influences boundary layer composition and dynamics, clouds properties, radiative transfer, and its effect on weather and regional climate. Therefore, we frequently search in the available lidar data sets for special events occurring over the Eastern Europe, e.g. biomass burning from forest fires, desert dust advection, episodes of cold air masses descending from Arctic. This resulted in a few case studies using lidar data from Warsaw and Magurele, e.g. optical properties of long-range transported volcanic ash during Eyjafjallajökull eruption in April 2010 were analyzed over Poland and Romania [2]. A recent study of the Canadian forest fire event observed over Warsaw during July 2013 revealed another air-mass path transporting mineral dust particles form Sahara [3]. Here we discuss a case of an air-mass flow from Poland to Romania, over the NW-SE transect that allows for assessing the role of Carpathian Mountains in this air-mass modification.

2. METHODOLOGY

The measurements, taken by two multi-wavelength Raman lidars, one based at the Radiative Transfer Laboratory (RT-Lab) of the Institute of Geophysics, Faculty of Physics, University of Warsaw (Poland) and the second installed at the RADO site of the National Institute of R&D for Optoelectronics (INOE) in Magurele (Romania), have been evaluated. The study is complemented with meteorological data collected at two other sites: the SolarAOT in Strzyżów (Poland) equipped, among other instrumentation, with the AERONET photometer and the CHM15k ceilometer, and the Observatory of Cluj (Romania). Along with these data we used information obtained from satellite imagery to describe the local and regional meteorological situation development and to further characterize the aerosols properties. As the four mentioned above stations are located along a north-west to south-east transect, the main objective of the study was to evaluate properties of the aerosol transported by the air flow over Poland, Slovakia, further to Romania.

The RT-Lab site at Warsaw (52.21° N, 20.98° E, 96 m a.s.l.) joined the EARLINET in 2015, contributing to the network activities with the 8-channel (2α+3β+2δ+VW) Aerosol-Depolarization -Raman (ADR) lidar, that is a NeXT generation Polly^{XT} system developed in a scientific cooperation with TROPOS, Germany [4]. Moreover, the RT-Lab and the SolarAOT sites are part of the Poland-AOD Consortium network (*http://www.polandaod.pl*).

The RADO site at Magurele (44.05° N, 26.03° E, 93 m a.s.l.) near Bucharest, equipped among other instrumentation with the 7-channel (2α+3β+1δ+WV) Raman Lidar (RALi), that was developed by RAYMETRICS, Greece, is recognized as one of advanced EARLINET stations since 2007 [1]. The RADO is also an AERONET site. A list of differences in the detection scheme of the RALi lidar with respect to the ADR lidar are given in Table 1.

<table>
<tr><td rowspan="2"></td><td colspan="2">Detected Wavelength [nm]</td><td colspan="2">Detection type</td></tr>
<tr><td>RALi lidar</td><td>ADR lidar</td><td>RALi lidar</td><td>ADR lidar</td></tr>
<tr><td rowspan="2">Elastic total</td><td>1064</td><td>1064</td><td>A</td><td rowspan="6">PC</td></tr>
<tr><td>355</td><td>355
532</td><td>A & PC</td></tr>
<tr><td>Vibrational Raman N_2</td><td>607
387</td><td>607
387</td><td>A & PC</td></tr>
<tr><td>Elastic parallel</td><td>532</td><td></td><td>A & PC</td></tr>
<tr><td>Elastic cross-parallel</td><td>532</td><td>532
355</td><td>A & PC</td></tr>
<tr><td>Raman H_2O</td><td>408</td><td>407</td><td>PC</td></tr>
</table>

Table 1. Detection channels for the multi-wavelength lidars at the RADO in Magurele and the RT-Lab in Warsaw (A: analog, PC: photon counting).

In order to assess the evolution and modification of the air flow properties during the overpass over the two countries between 19-21 July 2014, the particle backscatter coefficient profiles at 355, 532 and 1064 nm and the particle extinction profiles at 355 and 532 nm were obtained at both sites [5]. Also the calibrated depolarization profiles at 532 nm at RADO, as well as at 532 and 355 nm at RT-Lab were calculated [6]. If feasible, the retrieval of aerosol microphysical properties for found characteristic layers will follow, in a near future [7]. The ceilometer measurements at SolarAOT site were evaluated using routines by [8, 9, 10] for retrieval of both aerosol content and PBL/cloud layers structure.

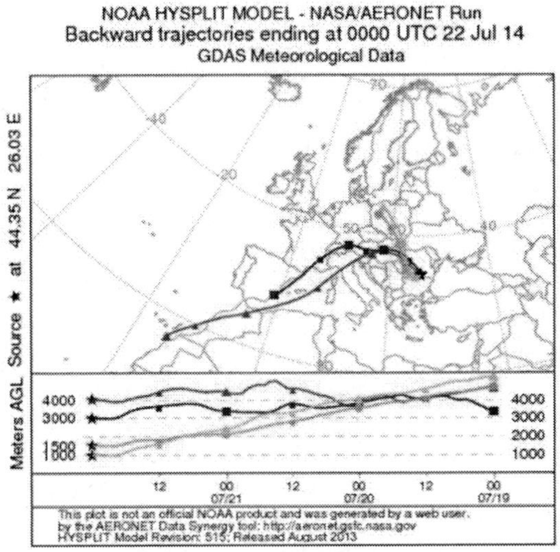

Figure 1. NOAA Hysplit backward trajectories ending at 00 UTC on 22nd July 2014 in Magurele at 1, 1.5, 3 and 4 km. Note that the lowermost air-masses crossed Poland on July 19th, 2014 and reached Romania on July 21st.

3. RESULTS

Accordingly to NOAA Hysplit 4-days backward trajectories (Fig.1), ending up at Magurele at 1, 1.5, 3 and 4 km, during the identified period an air-mass (Fig.1, in green, magenta) that crossed Poland on July 19th, 2014 descended on the following day over the SolarAOT station in Strzyżów, then further over Cluj site in Romania, and ended up in Magurele on July 21st, 2014. The air-mass that ended at 3 km in Magurele

(Fig.1, blue), originated in south-eastern coast of Spain in Europe. The air-mass that advected at 4 km over Magurele (Fig.1, red), came from the west coast of Morocco in Africa. Neither of the three identified air masses were mixed into each other as their paths were well separated in time and horizontal/vertical space.

At both sites the backscatter coefficient profiles at (Fig. 2) show distinctly layered structure in the atmosphere. Note that the profiles from Warsaw have resolution of 7.5 m and were smoothed using Savitzky-Golay filter, while the profiles from Magurele were smoothed with an adaptive smoothing filter and their effective resolution is linearly increasing from 7.5 m to 1500 m.

Both lidars reveal multi-layer aerosol structures during this period. The lidar data from 21 to 22 UTC collected at RT-Lab on July 19th, 2014 show backscatter peaks roughly at 2 km and at 2.7 km for VIS and UV channels (Fig. 2, left). Additional layers are visible at about 3 km (only in VIS) and at 7.2 km (only in UV). Whereas, the lidar data at RADO on July 21st, 2014 reveals layered structure with peaks at roughly 2 km, 3.2 km, and 4 km (Fig. 2, right).

Here we focus on two distinct air-masses, i.e. the air-mass from Africa that was directly advected at 4 km over Magurele, and the air-mass from over Poland, that reached Magurele at 1.5 km. Not only the origin of these air-masses is different, they also have not been mixed and thus, one can study their properties depending on where they came from.

The latter air mass was likely modified by the Carpathian Mountains on its way, as it descended following the terrain between the two countries. The backscatter values at 1.5 km over Magurele are $1.4\cdot10^{-6}$ at 532 nm and $3.1\cdot10^{-6}$ $m^{-1}sr^{-1}$ at 355 nm; while the same air-mass which was detected at about 3.5 km over Warsaw show backscatter coefficients of $5\cdot10^{-6}$ at 532 nm and $2.2\cdot10^{-6}$ $m^{-1}sr^{-1}$ at 355 nm. The linear depolarization ratio (not shown here) for the air-mass advected at an altitude of 1.5 km over Magurele on July 21st, 2014 is of about 5% at 532 nm, while the same air-mass was at 3.5 km over Warsaw on July 19th, 2014 where it had only 1% at 532 and 2.4% at 355 nm. Higher depolarization value of 9% at 355 nm was calculated at 8.5 km indicating Cirrus.

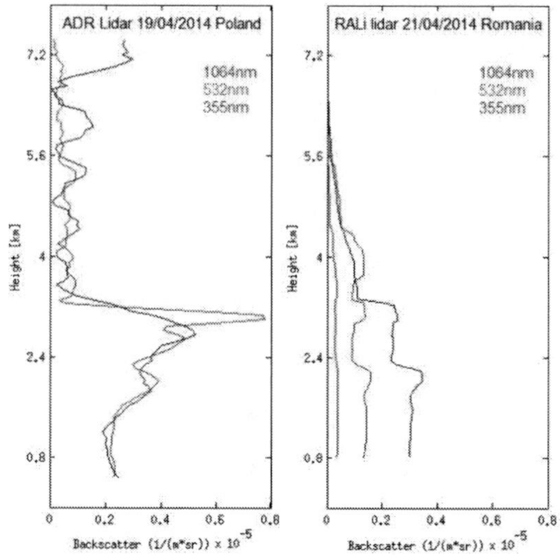

Figure 2. Backscatter coefficient profiles obtained by Raman method from 21 to 22 UTC on 19/07/2014 over Warsaw (left) and from 18:50 to 19:50 UTC on 21/07/2014 over Magurele (right).

The modificanion af the air mass can be followed by investigating the CHM15k ceilometer retrievals at 1064 nm over Strzyżów (quicklooks are available online at www.polandaod.pl), that show structures of enhanced aerosol load at an altitude of about 1 km between 19 UTC on July 19th, 2014 and 5 UTC on July 20th, 2014.

The backward trajectories (Fig. 1) indicate that the air located at the lower heights (between 1 and 1.5 km) over Magurele on July 21st, 2014 was roughly at about 4 km over Warsaw 2 days earlier. The aerosol layer present over Warsaw at height about 3 km for 532 nm and those present between 2-2.7 km for 532 nm and 355 nm (Fig.2) was likely modified by Carpathian Mountains prior to its arrival at Magurele, since it is not showing the same characteristics over Magurele which it had over Warsaw. The second air-mass of interest, i.e. a well defined layer detected at about 4 km over Magurele on July 21st, 2014 was directly transported at that altitude from Morocco (Fig.1). Thus it is likely that it contained the mineral dust particles that originated form Saharan Desert, what is also confirmed by enhanced linear depolarization ratio of 24% (not shown here) obtained by RALi at this altitude. The backscatter over Magurele for the layer at 4 km is of about $0.3\cdot10^{-6}$ at 1064 nm, $1.3\cdot10^{-6}$ at 532 nm and

DOI: 10.1051/epjconf/201611908008

$0.9 \cdot 10^{-6}$ $m^{-1}sr^{-1}$ at 355 nm, that indicates the dust particles were aged.

4. CONCLUSIONS

Meridional air-flow, coming from Warsaw to Magurele, is the point of interest in this study. The results have been interpreted on basis of long-range aerosol transport analysis and lidar derived backscatter coefficients and depolarization ratios measured over Warsaw and Magurele. Based on the optical properties derived from the lidar signals at the two sites we are able to assess the air-mass modification between 2 days period and analyze the role of Carpathian Mountains. In this case study is shown how the air arriving to Magurele at height about 1.5 km, coming from high altitudes >4.5km over north-western Poland, has changed its characteristics along the way. Moreover, a dust layer coming from Africa has been identified over Magurele, showing the typical depolarization characteristics for such aerosol type [6].

ACKNOWLEDGEMENT

We want to acknowledge Aerosols, Clouds, and Trace gases Research Infrastructure Network (ACTRIS) for the Trans National Access (TNA) training activity we were able to perform at RADO installations in December 2014.

The Polish lidar development was financed by FNITP, Poland (Grant No.519/FNITP/115/2010).

This work has been partially supported by the Romanian Programme STAR–ESA Project No.39/2012- SIAFIM

We acknowledge the NOAA Air Resources Laboratory (ARL) for the provision of the HYSPLIT transport model (http://www.arl.noaa.gov/ready.php) used for the interpretation for the results obtained in this publication.

REFERENCES

[1] Pappalardo, et al., 2014: EARLINET towards an advanced sustainable European aerosol lidar network, *Atmos. Meas. Tech.*, **7**, 2389–2409.

[2] Nemuc, A., I.S. Stachlewska, J. Vasilescu, A. Gorska, D. Nicolae, and C. Talianu, 2014: Optical properties of long-range transported volcanic ash over Romania and Poland during Eyjafjallajökull eruption in 2010, Acta Geophys., 62(2), 350-366.

[3] Janicka, L., I.S. Stachlewska, K.M. Markowicz, H. Baars, R. Engelmann, and B. Heese, 2015: Lidar measurements of Canadian forest fire smoke episode observed in July 2013 over Warsaw, Poland, *International Laser Radar Conference* (ILRC 27th), 5-10 July 20215, New York, USA, Paper ID: 246.

[4] Althausen, D., R. Engelmann, H. Baars, B. Heese, A. Ansmann, D. Müller, M. Komppula, 2009: *Portable Raman Lidar PollyXT for Automated Profiling of Aerosol Backscatter, Extinction, and Depolarization.* J. Atmos. Oceanic Technol., **47**, 2366-2378.

[5] Ansmann A., Riebesell M., and C. Weitkamp, 1990: Measurement of atmospheric aerosol extinction profiles with a Raman lidar, *Opt. Lett.*, **15(13)**, 746-748.

[6] Freudenthaler, V., et al., 2009: Depolarization ratio profiling at several wavelengths in pure Saharan dust during SAMUM 2006, *Tellus B*, **61(1)**, 165-179.

[7] Veselovskii, I., et al., 2002: Inversion with regularization for the retrieval of tropospheric aerosol parameters from multiwavelength lidar sounding, Appl Optics 41(18), 3685-3699.

[8] Stachlewska I.S., Piądłowski M., Migacz S., Szkop A., Zielińska A.J, and Swaczyna P.L., 2012; Ceilometer observations of the boundary layer over Warsaw, Poland, *Acta Geophys*. **60(5)**, 1386-1412.

[9] Sokół P., I.S. Stachlewska I. Ungureanu, S. Stefan, 2013: Evaluation of the Boundary Layer Morning Transition Using the CL-31 Ceilometer Signals. *Acta Geophys.*, **60(2)**, 367-380.

[10] Costa-Surós M., J. Calbó, J.A. González, and J. Martin-Vide, 2014: Behavior of cloud base height from ceilometer measurements. *Atmos. Res.*, **127**, 64-76.

DIRECT ESTIMATION OF FINE AND COARSE MODE PARTICLE PARAMETERS FROM MULTIWAVELENGTH LIDAR MEASUREMENTS

Alexei Kolgotin[1], Mikhail Korenskiy[1], Igor Veselovskii[1], David N. Whiteman[2]

[1]*Physics Instrumentation Center, Troitsk, Moscow Region, 142190, Russia,*
E-mails: alexeift@yahoo.com, *miklekor@ pic.troitsk.ru, igorv@ pic.troitsk.ru*
[2]*NASA Goddard Space Flight Center, Greenbelt, MD 20771, USA,*
E-mail: david.n.whiteman@nasa.gov

ABSTRACT

An approach for the direct estimation (DE) of particle parameters in the fine and coarse mode from multiwavelength lidar measurements is presented. Particle size distributions in both modes are approximated by rectangular functions, so the particle density is estimated directly without solving the inverse problem. The numerical simulation demonstrates that the particle volume in both modes can be estimated from 3β+2α lidar measurements with uncertainty of ~25% for a wide range of size distributions. The technique developed was applied to the observations of NASA GSFC Raman lidar. Comparison of the results obtained with DE and regularization approach applied to the same set of data demonstrates agreement between these two techniques.

1. INTRODUCTION

Retrieval of particle microphysical properties from multiwavelength Mie-Raman lidar observations has become a widely discussed problem during the last decade. Numerous approaches were suggested for inversion of lidar data to particle parameters [e.g.1-3]. However, even application of complicated mathematical methods does not guarantee that the acceptable solutions can be found without additional constraints on solution space. The main reason is that the inverse problem is strongly underdetermined: from lidar measurements normally only three backscattering (β) and two extinction (α) coefficients are available in a limited spectral interval 355-1064 nm (so called 3β+2α set). All reported results are related mainly to the retrieval of total particle volume and effective radius attributed also to the total size distribution. At the same time in many applications it is very important to separate the contributions of the fine and the coarse modes. Here we present an approach for direct estimation of the fine and the coarse mode parameters without solving the inverse problem.

2. METHODOLOGY

The particle extinction and backscattering coefficients are related to the volume size distribution $v(r)$ through the integral equation

$$\int_{r_{min}}^{r_{max}} k_i^V(m,r)v(r)dr = g_i, \qquad (1)$$

where g_i are lidar measured data (α or β) at different wavelengths, r is the particle radius, $m=m_R - im_I$ - complex refractive index (CRI) and $k_i^V(m,r)$ are the kernel functions calculated on a base of Mie theory. Eq. (1) can be also rewritten for number $n(r)$ and surface $s(r)$ density using the corresponding kernel functions $k_i^N(m,r)$ and $k_i^S(m,r)$.

The size distribution $v(r)$ can be approximated by a rectangular function on some interval $[r_{min}, r_{max}]$ with maximal value $\frac{V}{r_{max}-r_{min}}$ such that for g_i data

$$\frac{V_i}{r_{min}-r_{max}}\int_{r_{min}}^{r_{max}} k_i^V(m,r)dr = g_i \qquad (2)$$

Then the total volume can be found as

$$V_i = \frac{g_i(r_{max}-r_{min})}{K_i^V(m,r_{min},r_{max})} \qquad (3)$$

where

$$K_i^V(m,r_{min},r_{max}) = \int_{r_{min}}^{r_{max}} k_i^V(m,r)dr \qquad (4)$$

is the integrated kernel function.

The volume found in this way is not unique and depends on the choice of the interval $[r_{min}, r_{max}]$ and refractive index. So if the volume is obtained by using several optical data, it is obvious that for "true" solution the results obtained with different g_i should coincide.

Eq. (3) is an algebraic equation containing five unknown parameters V, m_R, m_I, r_{min}, r_{max}. In contrast to eq. (1) it is *parametric* problem. In principle, the eq. (3) can be solved if 3β+2α measurements are available. To solve this equation we predefine the set of input parameters r_{min}, r_{max}, m_R, m_I the same way as we did in [2].

If we have N_0 data (N_0=5 for 3β+2α set) we obtain N_0 solutions $V_1,\ldots, V_{N_0}$ for each set of input parameters. The mean value of volume $\bar{V}$ is

$$\bar{V} = \frac{1}{N_0}\sum_{i=1}^{N_0} V_i \quad (5)$$

and the relative deviation from mean value Δ_V is

$$\Delta_V = \frac{100\%}{\bar{V}}\sqrt{\frac{1}{N_0}\sum_{i=1}^{N_0}(V_i - \bar{V})^2} \quad (6)$$

The deviation Δ_V can be treated as discrepancy in solution space $\rho_V = \Delta_V$. Since all solutions $V_1,\ldots, V_{N_0}$ should coincide we should look for minimum of discrepancy ρ_V to determine input parameters m_R, m_I, r_{min}, r_{max}.

We can also introduce the discrepancy ρ_g in the optical data space

$$\rho_g = \sqrt{\frac{1}{N_0}\sum_{i=1}^{N_0}\left(\frac{g_i - \bar{g}_i}{g_i}\right)^2} \quad (7)$$

where back-calculated optical data are found as

$$\bar{g}_i = \frac{\bar{V}}{r_{max} - r_{min}} K_i^V(m, r_{min}, r_{max}),\ i=1,\ldots,N_0 \quad (8)$$

The numerical simulations performed demonstrate, that best results are obtained when both discrepancies are used, so in our algorithm we minimize the sum

$$\rho = \rho_g + \rho_V \quad (9)$$

Minimization of discrepancy provides the estimation of particle volume.

Similar consideration can be performed for the particle number *n(r)* and surface *s(r)* concentrations and effective radius can be calculated as $r_{eff} = 3\frac{V}{S}$.

The approach presented can be extended to the case of bimodal PSD. The $v(r)$ is represented by a combination of two independent distributions related to the fine $v^f(r)$ and course $v^c(r)$ modes

$$v(r) = v^f(r) + v^c(r) \quad (10)$$

The size distributions for the fine and coarse mode can be approximated by a rectangular functions on interval $[r_{min}^f, r_{max}^f]$ and $[r_{min}^c, r_{max}^c]$ respectively with maximal values $\frac{V^f}{r_{max}^f - r_{min}^f}$ and $\frac{V^c}{r_{max}^c - r_{min}^c}$, as shown in fig.1.

For a bimodal size distribution the eq.(3) is transformed to :

$$V^f \frac{K_i^V(m, r_{min}^f, r_{max}^f)}{r_{min}^f - r_{max}^f} + V^c \frac{K_i^V(m, r_{min}^c, r_{max}^c)}{r_{min}^c - r_{max}^c} = g_i \quad (11)$$

and for every pair of g_i, g_j data the volumes V^f, V^c can be found as:

$$\begin{cases} V_{j,i}^f = \dfrac{A_i^{V^c}(m)g_j - A_j^{V^c}(m)g_i}{A_j^{V^f}(m)A_i^{V^c}(m) - A_j^{V^c}(m)A_i^{V^f}(m)} \\ V_{j,i}^c = \dfrac{A_j^{V^f}(m)g_i - A_i^{V^f}(m)g_j}{A_j^{V^f}(m)A_i^{V^c}(m) - A_j^{V^c}(m)A_i^{V^f}(m)} \end{cases} \quad (12)$$

where

$$A_l^{V^c}(m) = \frac{K_l^V(m, r_{min}^c, r_{max}^c)}{r_{min}^c - r_{max}^c};$$
$$A_l^{V^f}(m) = \frac{K_l^V(m, r_{min}^f, r_{max}^f)}{r_{min}^f - r_{max}^f} \quad (13)$$

and $l=i, j$. Thus for 5 input data we have ten equations (11) for volume density. To solve these we minimize the discrepancy (9) as was explained above. Since solutions V^f, V^c are expressed explicitly we call this approach the *direct* estimation (DE). The important distinction from regularization retrieval is that DE does not contain an ill-posedness.

We should recall, that problem (12) contains eight unknown parameters V^f, V^c, m_R, m_I, r_{min}^f, r_{max}^f, r_{min}^c, r_{max}^c, while only five input data are available, so the problem is strongly underdetermined. Such situation is typical, when lidar measurements are inverted to particle parameters: instead a unique solution a family of solutions is obtained. The limitations we put on the "search space" of predefined parameters work as a constrain and help to decrease the number of solutions in the family. The found solutions are ranged in accordance with discrepancy ρ. Normally we are searching for m_R and m_I inside the intervals 1.3-1.6 and 0-0.015 respectively. The particle radii are taken on interval 0.075 - 6.0 μm, so the total number of parameters sets considered is of order 10^5. Usually we average 0.1% of total amount of solutions.

3. NUMERICAL SIMULATION

A numerical simulation was performed to estimate the uncertainties of DE approach for retrieval the particle volume, effective radius and refractive index. The particle bimodal size distribution was represented as:

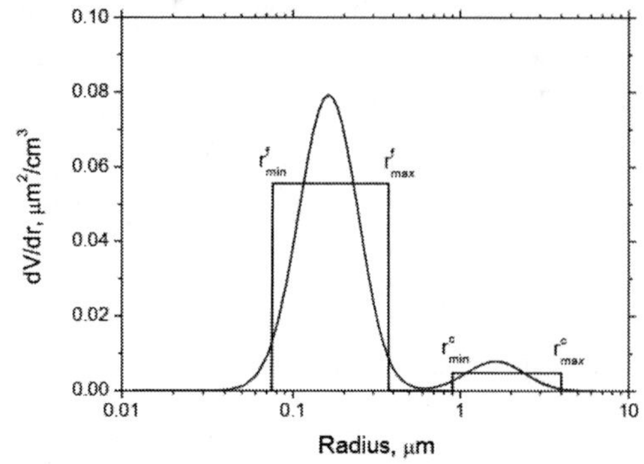

Fig.1. The model (blue) and retrieved (black) with DE approach the particle volume size distribution.

$$\frac{dn(r)}{d\ln(r)} = \sum_{i=f,c} \frac{N^i}{(2\pi)^{1/2}\cdot \ln\sigma^i}\cdot \exp\left[\frac{\left(\ln r - \ln r_0^i\right)^2}{2\left(\ln\sigma^i\right)^2}\right] \quad (14)$$

Here N^i is the total particle number of the i-th mode, r_0^i describes the mode radius and $\ln\sigma^i$ is the dispersion. The index $i = f, c$ corresponds to the fine and coarse mode, respectively. For all computations the values r_0^f=0.1μm, r_0^c=1 μm and $\ln\sigma^{f,c}$=0.4 were used and the complex refractive index (CRI) was assumed to be the same for both modes. To account for different PSDs, three ratios N^c/N^f=10^{-2}, 10^{-3}, 10^{-4} were considered. For ratio 10^{-2}, the coarse mode is predominant, while for ratio 10^{-4} the main part of the volume is contained in the fine mode.

The 3β+2α lidar measurements are not very sensitive to the coarse mode contribution when the fine mode is predominant, because the maximal wavelength available is only 1064 nm. This is illustrated by Fig.2, showing the contribution of the coarse mode to the optical data for different ratios V^c/V^t, where V^t=V^f+V^c. Keeping in mind that typical uncertainty for optical data is 5-10%, we cannot expect retrieval of the coarse mode parameters for V^c/V^t<0.1. Basing on results of numerous simulations we introduced the limitations in the algorithm: for retrieved V^c/V^t<0.2 the parameters of the coarse mode are not considered.

Fig.3a shows the uncertainties of estimation of V^f, V^c, V^t for these three N^c/N^f ratios. For N^c/N^f=10^{-4} the contribution of the coarse mode is too low, so the program doesn't consider it. For the intermediate case, when contributions of the fine and coarse mode are comparable, the uncertainty of volume estimation is about 20%. For the coarse mode predominance the errors for V^f are still about 20%, while for V^c it is below 5%. Errors for effective radius are higher and may reach 40% for N^c/N^f=10^{-3} as shown in Fig.2b. The uncertainty of estimation of the real part of refractive index is estimated as ±0.05.

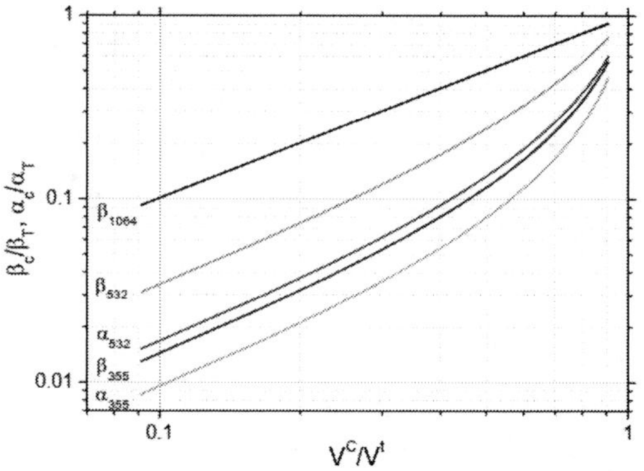

Fig.2. Contribution of the coarse mode to the optical data depending on the coarse mode volume fraction. Computations were performed for m=1.45-i0.005.

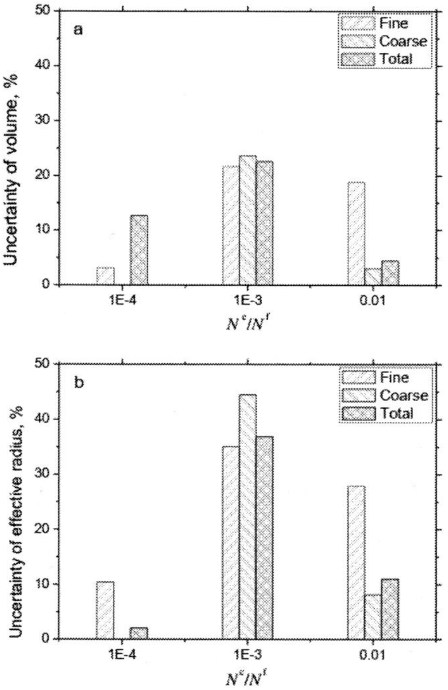

Fig.3. Uncertainties of (a) volume and (b) effective radius retrieval for different ratios of particle number density in the coarse and fine mode. Simulations were performed for m=1.45-i0.005.

4. APPLICATION TO LIDAR MEASUREMENTS

To test the developed algorithm the measurements performed by NASA GSFC multiwavelength Raman lidar [4] on 22 August 2006 were used. The measured 3β+2α lidar data were inverted to the particle microphysics by using regularization [2] and DE algorithms. Fig.4.shows the vertical profiles of the particle volume density obtained with two approaches.

Contribution of the coarse mode for this day was too low, so DE algorithm automatically put it to zero, as described in previous section. The volume densities V^f obtained with the two approaches are close, while the effective radii differ significantly for the heights above 1000 m. DE leads to higher values of r_{eff} and to lower values of the real part of CRI above 1000 m height, as shown in Fig.5. This agrees with results of numerical simulations, showing that DE approach is most suitable for estimation of particle volume.

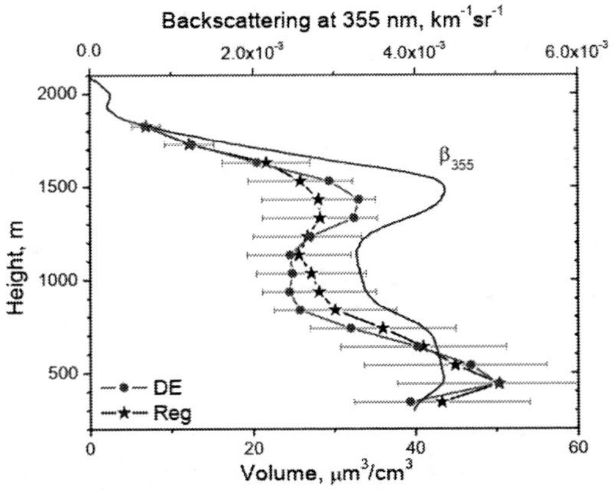

Fig.4. Vertical profiles of the volume density obtained with regularization and DE approach. Blue solid line shows the backscattering coefficient at 355 nm.

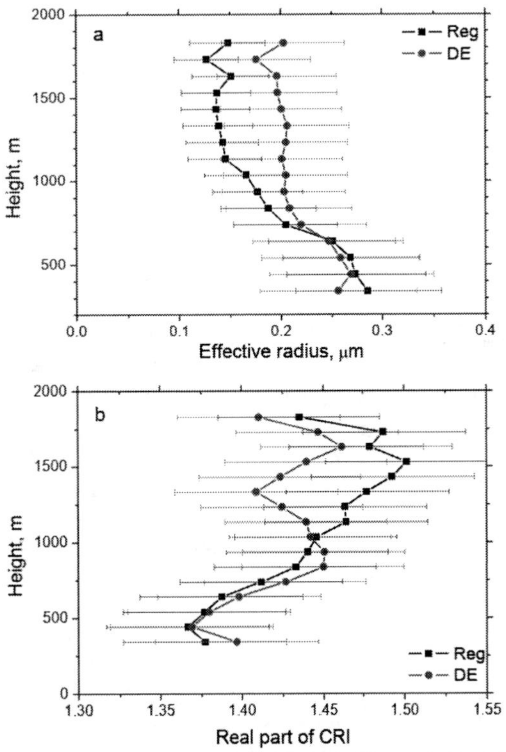

Fig.5. Vertical profiles of (a) effective radius and (b) real part of CRI obtained with regularization and DE approach.

5. CONCLUSION

An approach for estimation of the particle volume in the fine and in the coarse mode is presented. The numerical simulations performed demonstrate that particle volume in both modes can be derived with uncertainty about 25% for a wide range of PSDs. To evaluate this approach DE and regularization algorithms were applied to the same multiwavelength measurements. The comparison demonstrates an agreement between these two approaches. In all measurements considered the fine mode was predominant and contribution of the coarse mode was too small to be retrieved with the DE technique. In the future we plan to consider desert dust measurements, characterized by strong coarse mode. To treat the dust particles the spheroids mode will be incorporated in the algorithm in the nearest future.

REFERENCES

[1] Ansmann, A. and Müller, D.: Lidar and atmospheric aerosol particles, in "Lidar. Range-Resolved Optical Remote Sensing of the Atmosphere", Weitkamp, C. ed., Springer, New York, 2005, pp. 105-141.

[2] Veselovskii I., et al.,: Inversion with regularization for the retrieval of tropospheric aerosol parameters from multiwavelength lidar sounding, *Appl. Opt.*, 41, pp. 3685 – 3699, 2002.

[3] Chemyakin E., et al,: "Arrange & average" algorithm for the retrieval of aerosols parameters from multiwavelength HSRL/Raman lidar data, *Appl. Opt.*, 41, pp. 3685 – 3699, 2014.

[4] Veselovskii I., et al.: Demonstration of aerosol property profiling by multi-wavelength lidar under varying relative humidity conditions, J. of Atmospheric and Oceanic Tech. 26, 1543-1557, 2009.

EPJ Web of Conferences **119**, 08010 (2016) DOI: 10.1051/epjconf/201611908010
ILRC 27

GLOBAL DUST TRANSPORT AS OBSERVED BY A-TRAIN SATELLITES

Tao Luo [1]*, Zhien Wang[1], Damao Zhang[1]

[1] *University of Wyoming, Dept. Atmospheric Science, Laramie, WY, 82070,USA, *Email: tluo@uwyo.edu*

ABSTRACT

This paper presented a new height-resolved view of global dust transport based on 2007-2010 A-train observations. First, a new dust identification methodology was developed to improve optically thin dust layer detection. Second a new dust partition methodology was developed and applied to CALIPSO lidar measurements to derive dust partitions in external mixed aerosols. These new approaches allow a new view of global dust distribution from dense dust layers near the strong source regions to the optically thin, but significant dust layers from the point of view of aerosol–cloud interactions, over the weak source regions, the transport areas, and the upper troposphere. The results will not only help us to better understand global dust transport and dust-cloud interactions, but also provide critical information for model evaluations and improvements.

1. INTRODUCTION

Dust aerosols are well known for the role in modulating the climate system at local and global scales [1]. The airborne dust mass is of great importance to models in aspects of radiations and cloud microphysics [1,2]. However, current models still have large uncertainties in simulating dust optical depth, vertical extinction profiles and seasonal variations [3], which stresses the need of better dust observations to improve dust related processes in the models, especially in the free troposphere where the dust loading is lower [4, 5]. Reliable height-resolved dust identification and mass estimation approaches are needed to provide more accurate dust loading estimations, and thus to improve our understanding of the mixing of the climate-relevant aerosol components, the long range transport of the dust, and the impact of aerosols on regional climate [5,6,7].

This paper aims to provide a new and complete global height-resolved view of dust distribution by improving thin dust identification based on Cloud-Aerosol Lidar and Infrared Pathfinder Satellite Observations (CALIPSO) Cloud-Aerosol Lidar with Orthogonal Polarization (CALIOP) measurements. Section 2 introduces the data and section 3 introduces the new dust identification and retrieval methodology. Section 4 presents the results and conclusions are in the section 5.

2. Data

Nighttime CloudSat and CALIPSO measurements and operational meteorology datasets for the period of January 2007 to December 2010 were used in this study.

CALIOP provides 532-nm total attenuated backscatter (β_{532}) and perpendicular polarization component (β_{532p}) [8], with 333-m along-track and 30-m vertical resolution below 8.2 km. The CALIOP level 1B data are calibrated and geo-located. The $\beta_{532}$and β_{532p}data were averaged to a 60-m vertical resolution and were used to identify dust aerosols.

CloudSat carries a 94-GHz cloud profiling radar (CPR), which provides an instantaneous footprint of approximately 1.3 km (at mean sea level) and has 125 vertical bins with a bin size of about 240 m. The 2B-GEOPROF product provides hydrometeor identification of cloud mask [9].

Temperature and pressure profiles from the European Center for Medium range Weather Forecasting AUX-algorithm (ECMWF-AUX) were used to model molecular backscattering.

3. METHODOLOGY

The well-used volume depolarization ratio (VDR) method often misses the thin dust layers [10]. The β_{532p} is more sensitive to the presence of a weak dust layer than the β_{532} or VDR, which shows more peaks in weak or dense dust layer signals than in clear sky signals or other types of aerosols or clouds. A new method was developed based on this dust signal characteristic in β_{532p} channel, as following steps.

Firstly, the cloud information was determined by combining CALIPSO and CloudSat products to better detect optically thin ice clouds with large irregular-shaped particles [11]; Then, for cloud-free regions, peak index was built by identifying and setting the peaks in β_{532p} to be 1 (non-peaks to be 0) and were averaged by 6 × 3 bins to generate the peak mask. The possible dust layer was identified in the peak index by the threshold of 0.04, and then refined by a further three-time iterative to set any peaks to zero when the 11 × 3 and 21 × 3 moving averages of the peak mask are smaller than the clear sky value. The so-derived quick dust mask was further refined by the layer-integrated depolarization ratio (LPDR) to remove other types of aerosols with spherical particles or misidentified clouds. The details can be found in [10].

Then, a new vertically resolved dust separation method based on depolarization measurements was developed to better determine dust loading within a polluted dust layer (dust mixed with other aerosols) by

$$\beta'_{Total}(z) = (\beta_a(z) + \beta_m(z))e^{-2\int_0^z (S_{eff}\beta_{all}(z')+S_m\beta_m(z'))dz'} \quad (1)$$

$$\beta_a(z) = (1+c)\beta_d(z) \quad (2)$$

$$S_{eff} = \frac{S_d + cS_{nd}}{1+c} \quad (3)$$

$$c(z) = \frac{\delta_p'(z) - \delta_d'(z)}{\delta_{nd}'(z) - \delta_d'(z)} \quad (4)$$

Here, $\beta'_{Total}(z)$ is the measured backscattering coefficient; $\delta_p' = \frac{\delta_p}{\delta_p+1}$; $\delta_d' = \frac{\delta_d}{\delta_d+1}$; $\delta_{nd}' = \frac{\delta_{nd}}{\delta_{nd}+1}$; δ_p, δ_d, and δ_{nd} are depolarization ratios for mixed aerosol, dust and non-dust aerosol respectively; S_d, S_{nd} and S_m are lidar ratios for dust, non-dust and molecules respectively.

The particle depolarization ratio is less affected by the choice of lidar ratio or retrieval method, which could act as a good constrain to separate dust and non-dust aerosols in polluted dust [12]. Therefore, the first-guess δ_p was retrieved from lidar equation with lidar ratio of 65. Then, with assumed S_d, S_{nd}, D_d and D_{nd}, height-resolved S_{eff} can be calculated and δ_p can be retrieve with the updated S_{eff}. Several iterations were performed to retrieve δ_p and update S_{eff}, until the δ_p converges. Finally, the dust backscattering coefficient was separated from the β_a. The sensitive tests showed that the new method retrieves dust backscattering from polluted dusts with uncertainties less than ~11%[12]. More details of this new vertical separation method can be found in [12].

With the retrieved dust extinctions, the polluted dust in dust mask was then further refined to remove any dust mask without dust extinction retrievals. Then, the retrieved dust extinctions were converted to dust mixing ratio with specific extinction coefficient of 0.5×10^3 m^2 kg^{-1} at 532 nm [13].

Lidar signals over South America are strongly affected by the South Atlantic Anomaly (SAA), which results in more noisy spikes in measured signals. This part of the data was removed from the data processing and analysis [11], shown as the blank area over Southern America in Fig. 1.

4. RESULTS

Figure 1 shows the 2007-2010 global 2.5° × 2.5° mean dust mixing-ratio distributions in December, January, and February (DJF); March, April, and May (MAM); June, July and August (JJA); and September, October, November (SON). The mean mixing ratio was calculated by (total dust mixing ratio)/(total observation number) in each grid box at a given height range. The total observation number includes all sky conditions.

The new dust identification methodology provides a full view of global dust distributions by detecting more dust cases and providing more accurate dust layer boundaries [11], as highlighted in Fig. 1 in terms of dust loading. A main dust band lies from the Sahara dust region to the Asia dust region with strong dust loading at 0-3km ranges near the sources and with the dust loading gradually decreasing upward and far away from the source regions.

Over the Sahara and mid-east region (longitude of 20°W to 60°E, latitude of 5°N to 33°N), the mean dust loadings with the lower 6 km AGL are 2.62×10^{-8}, 4.46×10^{-8}, 4.18×10^{-8}, and 2.79×10^{-8} kg/kg for DJF, MAM, JJA and SON respectively. The upward transportation of those dust are mainly within lower 6 km AGL except for MAM (up to 12 km AGL) and JJA (up to 9 km). At lower levels (<6km AGL), the western transportation of those dusts is the weakest in DJF. But in other seasons, significant amounts of dust can be transported to the Caribbean Sea and influence North America especially in MAM and JJA. Small amount of those dust can be transported northerly to European, except in MAM, during which dust is transported to cover most of the Northern Hemisphere. African dust can be significantly transported eastward over the Indian Ocean to India at the order of 10^{-8} kg/kg.

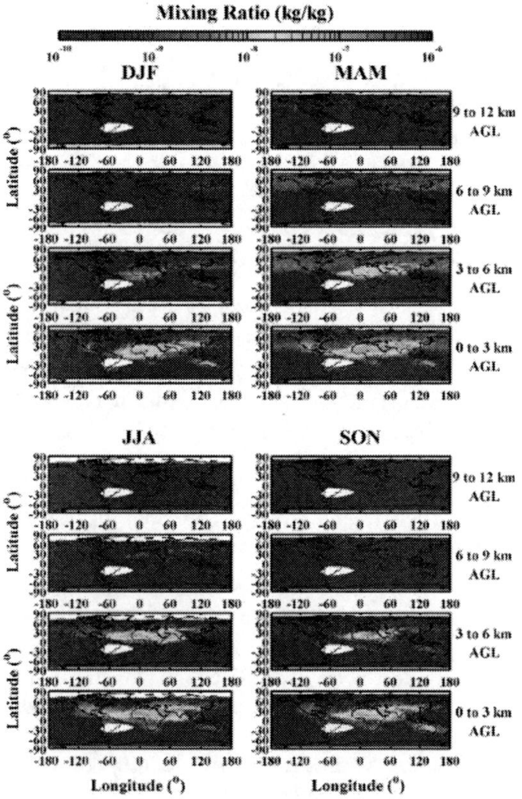

Figure 1. Four-year mean height-resolved global distribution of dust mixing-ratios in different seasons. 3-km vertical bins are used to illustrate dust vertical structure. The blank areas over Arctic and Antarctic indicate the boundaries of satellite observations.

For the Asia dust region, the mean dust mixing ratio in each season over the Taklimakan and Gobi desert those values with the lower 6 km AGL are 1.29×10^{-8}, 3.6×10^{-8}, 2.23×10^{-8}, and 1.88×10^{-8} kg/kg for DJF, MAM, JJA and SON respectively. The upward transportation of those dusts are main lower than 6km, and can reach up to 12km in MAM, and 9km AGL in JJA. The Asia dust can be transported to the east over the Pacific during every season and reaches North America. In MAM, Asian dust can be transported northward over Russia to Arctic.

Over North America, the mean dust loadings are 0.14×10^{-8}, 0.30×10^{-8}, 0.32×10^{-8}, and 0.24×10^{-8} kg/kg for DJF, MAM, JJA and SON respectively. In DFJ and SON, the dust layers are mainly within lower 6km with low dust loadings contributed mainly by local sources. The dust loading in MAM and JJA are stronger than DJF and SON, and the dust layer can reach up to 12 km AGL in MAM and 9 km AGL in JJA due to long-range dust transportations from Sahara and Asia dust source regions.

Over the Australia dust region, the dusts are mainly from local sources. The dust layers are mainly below 6km in DJF and SON, and below 3 km in MAM and JJA. The dust loadings in the lower 3km are 0.25×10^{-8}, 0.25×10^{-8}, 0.20×10^{-8}, and 0.4×10^{-8} kg/kg for DJF, MAM, JJA and SON respectively. Those dusts are transported westward to Indian Ocean year round. Furthermore, there are relatively stronger eastward transportations in DFJ and SON than the other seasons.

5. CONCLUSIONS

With the newly developed dust identification methodology for improving optically thin dust layer detection, and the newly developed dust partition methodology for externally mixed aerosols, this paper presents a new global height-resolved view of dust transportation based on multi-year A-train satellite observations. The new results allow us to better understand processes controlling dust concentrations over weak source regions, transport areas, and the upper troposphere. One of our ongoing work is to evaluate and improve the model simulations with the dataset.

ACKNOWLEDGEMENTS

The CALIPSO dataset was obtained from the NASA Langley Research Center Atmospheric Science Data Center (eosweb.larc.nasa.gov). The CloudSat 2B-GEOPROF and ECMWF-AUX products were obtained from CloudSat Data Processing Center (cloudsat.cira.colostate.edu). This research was funded by NASA grant NNX13AQ41G and a contract from NASA/JPL.

REFERENCES

[1] IPCC (2007). In: Solomon, S., et al. (Eds.), Climate Change 2007: The Physical Science Basis, Contribution of Working Group I to the Fourth Assessment Report of the Intergovernmental Panel on Climate Change, Cambridge University Press, Cambridge and New York, p. 996.

[2] Zhang, D., Z. Wang, A. Heymsfield, J. Fan, D. Liu, and M. Zhao (2012), Quantifying the Impact of Dust on Heterogeneous Ice Generation in Midlevel Supercooled Stratiform Clouds, Geophys. Res. Lett. (39), L18805.

[3] Evan, A. T., C. Flamant, S. Fiedler, and O. Doherty (2014), An Analysis of Aeolian Dust in Climate Models, Geophys. Res. Lett., 41, 5996–6001.

[4] Liu, X., X. Shi, K. Zhang, E.J. Jensen, A. Gettelman, D. Barahona, A. Nenes, and P. Lawson (2012b), Sensitivity Studies of Dust Ice Nuclei Effect on Cirrus Clouds with the Community Atmosphere Model CAM5, Atmos. Chem. Phys. (12), 12061-12079.

[5] Uno, I., Z. Wang, M. Chiba, Y. S. Chun, S. L. Gong, Y. Hara, E. Jung, S.-S. Lee, M. Liu, M. Mikami, S. Music, S. Nickovic, S. Satake, Y. Shao, Z. Song, N. Sugimoto, T. Tanaka, and D. L. Westphal (2006), Dust Model Intercomparison (DMIP) Study over Asia: Overview, J. Geophys. Res. (111), D12213.

[6] Johnson, M. S., N., Meskhidze, and V. P., Kiliyanpilakkil (2012), A Global Comparison of GEOS-Chem-predicted and Remotely-sensed Mineral Dust Aerosol Optical Depth and Extinction Profiles. J. Adv. Model. Earth Syst., 4(3).

[7] Zhang, D., D. Liu, T. Luo, Z. Wang, and Y. Yin (2015), Aerosol Impacts on Cloud Thermodynamic Phase Change over East Asia Observed with CALIPSO and CloudSat Measurements, J. Geophys. Res. Atmos., 120.

[8] Winker, D. M., M. A. Vaughan, A. Omar, Y. Hu, K. A. Powell, Z. Liu, William H. Hunt, and S. A. Young (2009), Overview of the CALIPSO Mission and CALIOP Data Processing Algorithms, J. Atmos. Oceanic Technol. (26), 2310-2323.

[9] Mace, G. (2007), Level 2 GEOPROF Product Process Description and Interface Control Document Algorithm Version 5.3, http://www.cloudsat.cira.colostate.edu/ICD/2B-GEOPROF/2B-GEOPROF_PDICD_5.3.doc (last access 08/11/2014).

[10] Luo, T., Z. Wang, D. Zhang, X. Liu, Y. Wang, and R. Yuan (2015), Global Dust Distribution from Improved Thin Dust Layer Detection using A-train Satellite Lidar Observations, Geophys. Res. Lett., 42, 620–628.

[11] Wang, Z., G. Stephens, and T. Deshler (2008), Association of Antarctic Polar Stratospheric Cloud Formation on Tropospheric Cloud Systems, Geophys. Res. Lett. (35), L13806.

[12] Luo, T., Z. Wang, F. A. Richard, C. A. Hostetler, R. Yuan, and D. Zhang, (2015). Vertically Resolved Separation of Dust and other Aerosol Types by a New Depolarization Method. Optics Express 23(11), 14095-14107.

[13] Ansmann, A., P. Seifert, M. Tesche, and U. Wandinger, (2012). Profiling of Fine and Coarse Particle Mass: Case Studies of Saharan dust and Eyjafjallajökull/Grimsvötn Volcanic Plumes. Atmos. Chem. Phys. (12), 9399-9415.

LIDAR CHARACTERIZATION OF BOUNDARY LAYER TRANSPORT AND MIXING FOR ESTIMATING URBAN-SCALE GREENHOUSE GAS EMISSIONS

R. Michael Hardesty[1], W. Alan Brewer[2], Scott P. Sandberg[2], Ann M. Weickmann[1], Paul B. Shepson[3], Maria Cambaliza[3], Alexie Heimburger[3], Kenneth J. Davis[4], Thomas Lauvaux[4], Natasha L. Miles[4], Daniel P. Sarmiento[4], A. J. Deng[4], Brian Gaudet[4], Anna Karion[1], Colm Sweeney[1], James Whetstone[5]

[1]*Cooperative Institute for Research in Environmental Sciences, University of Colorado/NOAA, Boulder CO, 80305 USA, *Email: mike.hardesty@noaa.gov*

[2]*NOAA Earth System Research Laboratory, CSD, Boulder, CO 80305*

[3]*Purdue University, Lafayette, IN USA*

[4]*Pennsylvania State University[3], College Station, PA USA*

[5]*National Institute of Standards and Technology[7], Gaithersburg, MD, USA*

ABSTRACT

A compact commercial Doppler lidar has been deployed in Indianapolis for two years to measure wind profiles and mixing layer properties as part of project to improve greenhouse measurements from large area sources. The lidar uses vertical velocity variance and aerosol structure to measure mixing layer depth. Comparisons with aircraft and the NOAA HRDL lidar generally indicate good performance, although sensitivity might be an issue under low aerosol conditions.

1. INTRODUCTION

The Indianapolis Flux Experiment (INFLUX) is aimed at improving methods for estimation of greenhouse gas emissions at urban scales. Current estimates rely on "bottoms-up" techniques, in which emissions from individual sources within the area of interest are summed to produce an overall estimate. The INFLUX experiment incorporates aircraft and surface-based observations, as well as chemical transport models, to investigate potential improvement in regional scale emissions gained by applying a top-down approach. INFLUX observational components include several-times-per-month aircraft measurements of gas concentrations and meteorological parameters, as well as a number of towers observing CO_2, CH_4, and CO and a single continuously operating Doppler lidar to estimate wind, turbulence and aerosol structure in the boundary layer. Here we discuss the role and utility of Doppler lidar observations to characterize the boundary layer during INFLUX.

2. ROLE OF DOPPLER LIDAR IN INFLUX

INFLUX focuses on developing and assessing model and observational techniques for estimating greenhouse gas emissions. An important element is the application of mass balance techniques [1] in which an aircraft equipped with chemical sensors samples greenhouse gas concentrations upwind and downwind of the city, as shone in Fig. 1. The Doppler lidar is a critical element, providing the needed estimates of wind speed and direction, boundary layer depth, and mixing needed for calculation of the upwind and downwind flux, from which the emission rates can be computed.

3. INDIANAPOLIS DEPLOYMENT

Utility of Doppler lidar for mass balance studies was demonstrated during the Uintah ozone study [1] where the NOAA High Resolution Doppler Lidar (HRDL, [2]) was used in conjunction with aircraft measurements to assess leakage from gas wells in the Uintah basin oil and gas field. Because INFLUX was designed to be a multi-year study, we opted to deploy a commercial, low-pulse-energy, high pulse rate lidar for the study [3]. The lidar, purchased from Halo Photonics, was deployed on the roof of a classroom building at Ivy Tech Community College northeast of

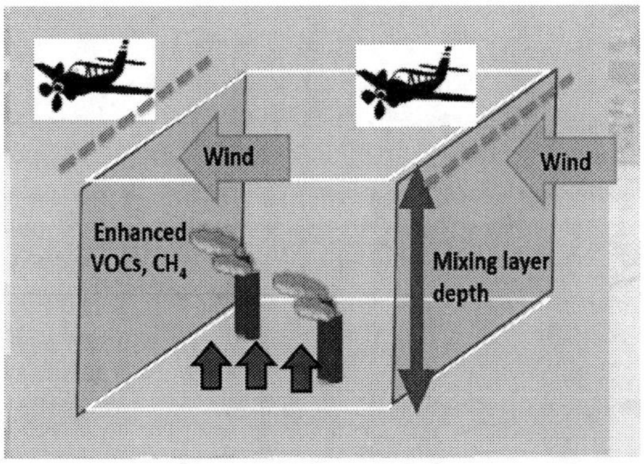

Figure 1: Schematic showing mass-balance technique for estimating emissions.

Indianapolis in April, 2013, and operated, with the exception of a maintenance interval in summer 2013, continuously until spring, 2015, when it was sent back to the manufacturer for maintenance and performance upgrades. A fixed scan pattern, repeated every 20-30 minutes, was implemented to measure the needed boundary layer parameters. The pattern is comprised of several conical scans at different elevation angles for measurement of horizontal wind profiles, vertical scans at orthogonal azimuths for observations of low level winds under stable conditions, and fixed vertical stares to observe vertical turbulence and mixing. Estimates of vertical velocity variance and aerosol backscatter profiles are computed from the vertical stares to characterize atmospheric mixing.

Figure 2 shows a 12-hour time height cross-section of vertical velocity variance and aerosol backscattered signal observed during the daytime (12Z-00Z) on September 13, 2013. The growth of the mixing layer to about 1500 m is clearly seen in both figures. However, the presence of a residual aerosol layer extending from the surface to ~ 1 km above the surface between 12Z and 16Z is also observed, complicating the ability to measure mixing layer depth during that period. For this reason we use both aerosol and vertical velocity variance for estimation of mixing layer depth.

The vertical velocity variance profile in Fig. 2 also shows evidence of uncertainties due to sampling error. Because staring is typically carried out for 10-15 minutes of every scan sequence, the inability to statistically represent large eddies under certain conditions is manifested as variability in the estimated mixing height.

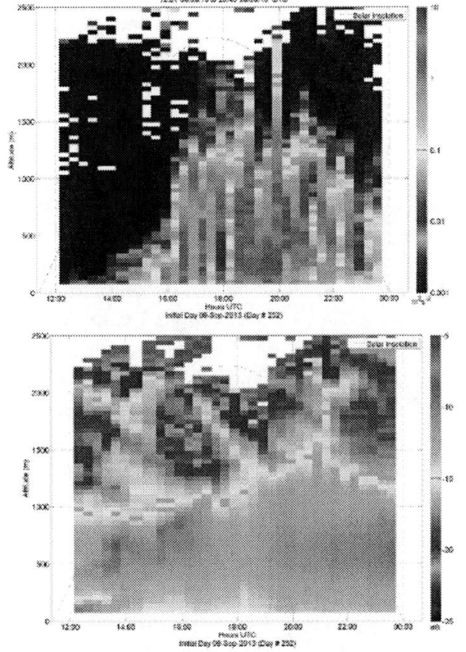

Figure 2: Time-height cross-sections showing vertical velocity variance (top) and aerosol signal (bottom) over a 12 hour daytime period.

Figure 3 shows wind profiles computed from conical and vertical scans from the same period. Note that this scanning strategy, which includes several low elevation scans, provides high-resolution measurements near the surface where

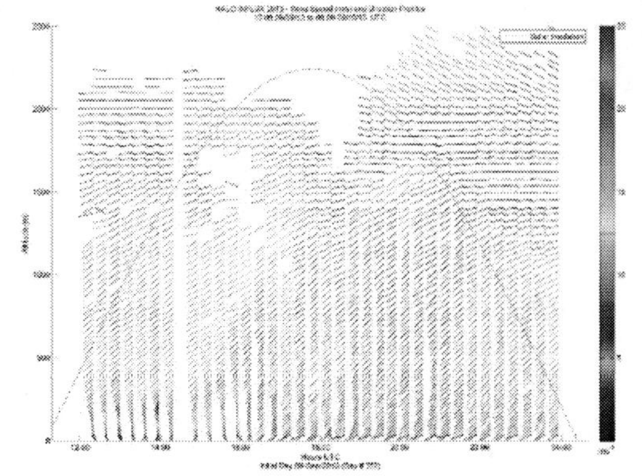

Figure 3: 12-hour time-height cross section of wind profiles computed from lidar conical scans.

wind structure can be quite detailed.

4. TWO-LIDAR OBSERVATION PERIOD

During spring 2014 we deployed our HRDL lidar in Indianapolis for comparison studies with the INFLUX Halo instrument and to investigate spatial variability of the lidar-measured properties. For studying spatial variability, the HRDL was situated at Indianapolis International Airport, southwest of the city. Measurements of boundary layer properties were obtained by both lidars in conjunction with mass-balance studies incorporating observations from the Purdue aircraft ([4] The flights typically included spirals above the lidar site(s), thus providing opportunities to compare aircraft and lidar estimates of boundary layer properties.

Figure 4 shows lidar measurements of vertical velocity variance, along with aircraft measurements of CO_2, methane, potential temperature and water vapor observed during aircraft spirals over the lidar sites at the airport and Ivy Tech. The spirals are displaced in time by slightly less than one hour. From the figures, it can be seen that the mixing layer is growing steadily during this period. Mixing layer depths at the Ivy Tech site observed by the lidar are deeper than those observed at the airport- this is also seen in the aircraft data. It is worth noting that estimating the mixing layer depth from the aircraft observations in this case is not straightforward – the profiles of both the greenhouse gases and meteorological variables show significant structure, perhaps because the aircraft is sampling within several distinct convective plumes. Over the course of the 2-year deployment a number of similar aircraft-lidar comparisons exist and are being analyzed to assess different techniques for estimating mixing layer depth as it pertains to mass-flux measurements.

The lidar observations are also of significant use in assessing and improving the performance of chemical models to estimate emissions. Figure 5 shows the improvement obtained by assimilating 12 hours of lidar wind data into a WRF chemical model. The model readily accepts the lidar data

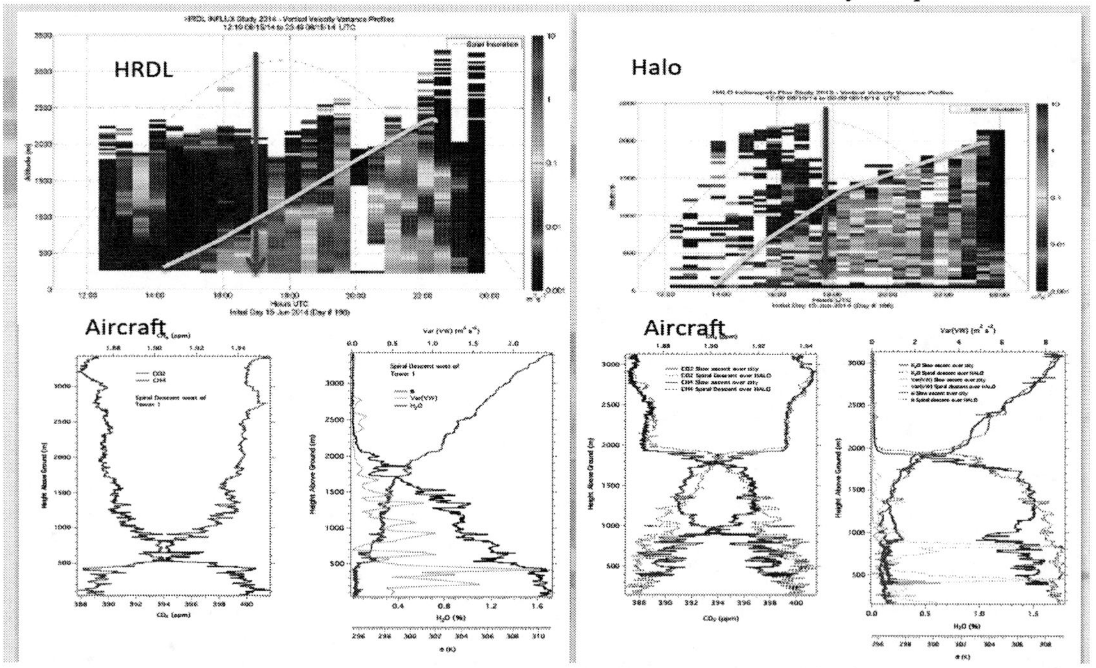

Figure 4: Observations of vertical velocity variance from HRDL (top, left) and Halo (top, right) compared with aircraft in situ sampling (bottom) of CO2 (red), methane (blue,) water vapor (black) and potential temperature (green).

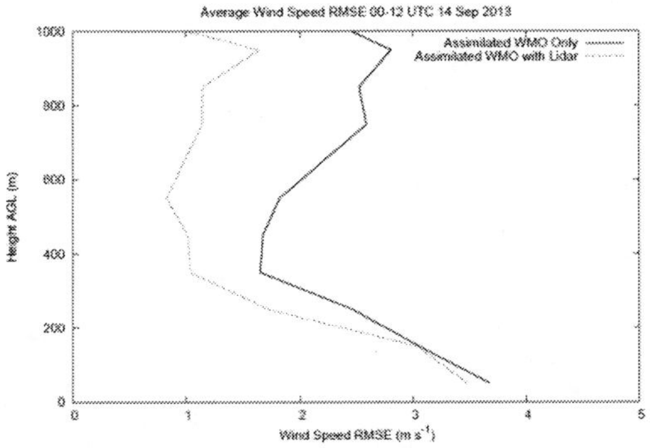

Figure 5: Improvement in model estimates of wind profile based on assimilation of lidar wind profiles during INFLUX.

and uses it to improve the wind estimate, indirectly verifying the utility of the observations.

5. INVESTIGATING SENSITIVITY

A key requirement for a lidar deployed for long-term measurements of boundary layer properties is the ability to reach the top of the mixing layer under a variety of conditions. During periods of low aerosol loading, we have seen indications that, unlike HRDL, the Halo instrument does not always see the full mixing layer. We have compared measurements from the HRDL and Halo instruments in side-by-side comparisons. Generally, it is seen that the measurements from the Halo instrument are noisier than those from HRDL, which is not unexpected given the two order-of-magnitude difference in pulse energy. HRDL always sees the terrestrial mixed layer and typically penetrates further into the free troposphere. It is also observed that the direct wind speed comparisons are generally in good agreement. We are currently testing algorithms to determine how often the Halo instrument is not seeing the full boundary layer in order to assess whether this is significant for future deployments.

6. FUTURE PLANS

We are continuing deployment of the Halo lidar at INFLUX. However, the lidar is scheduled to be returned to the manufacturer in spring 2015 for modification to improve sensitivity. Prior to redeployment in Indianapolis, the instrument will be installed next to HRDL in Boulder for additional sensitivity tests. We also intend to include a micro-pulse aerosol lidar for studying the relative capability to estimate mixing layer depth.

ACKNOWLEDGEMENTS

We appreciate the assistance of INFLUX colleagues Jocelyn Turnbull and Kevin Gurney in providing insights on greenhouse gas emissions. This research was partially supported by the National Institute of Standards and Technology.

REFERENCES

[1] Petron, G., A. Karion, C. Sweeney, et al, 2014: A new look at methane and nonmethane hydrocarbon emissions from oil and natural gas operations in the Colorado Denver-Julesburg Basin, *J. Geophys. Res.*, ***119**, 6836-6852.*

[2] Grund, C. J.,R. M. Banta, J. L. George et al, 2001: High resolution Doppler lidar for boundary layer and cloud research, *J. Atmos. Ocean. Tech.*, ***18**, 376-393.*

[3] Hogan, R. J., A. L. Grant, A. J. Illingworth, et al., 2009: Vertical velocity variance and skewness in clear and cloud-topped boundary layers as revealed by Doppler lidar. *Quart. J. Roy. Meteor. Soc.*, ***135**, 635-643.*

[4] Cambaliza, M. O. L., P. B. Shepson, D. R. Caulton, et al, 2014: Assessment of uncertainties of an aircraft-based mass balance approach for quantifying urban greenhouse gas emissions, *Atmos. Chem. Phys.*, ***14**, 9029-9050.*

AIRBORNE RAMAN LIDAR AND ITS APPLICATIONS FOR ATMOSPHERIC PROCESS STUDIES

Zhien Wang[1], Perry J. Wechsler[1], Nick Mahon[1], Decheng Wu[1], Bo Liu[1], Matthew Burkhart[1], Brent Glover[1], William Kuestner[1], Wayne Welch[2], and Andrew Thomson[2]*

*[1]Department of Atmospheric Science, University of Wyoming, Laramie, Wyoming, 820072, USA, *zwang@uwyo.edu*

[2]Welch Mechanical Designs, LLC, Maryland, USA

ABSTRACT

Although ground-base Raman lidars are widely used for atmospheric observations, the capabilities of airborne Raman lidar is not fully explored. Here we presented two recently developed airborne Raman lidar systems for the studies of atmospheric boundary layer process, aerosols, and clouds. The systems are briefly introduced. Observation examples are presented to illustrate the unique observational capabilities of airborne Raman lidar and their applications for atmospheric process studies.

1. INTRODUCTION

Water vapor and aerosol are the two most important ingredients for the weather and climate system [1]. Aerosols can change energy distributions in the atmosphere and at the surface and the top of atmosphere through both direct and indirect effects. Water vapor and their transportation control the global water cycle. Meanwhile, water vapor is the most important greenhouse gas in the atmosphere. Because the sources and sinks for water vapor and primary aerosols are from the surface, the turbulence mixing and convective vertical transportation in the atmospheric boundary layer (ABL) control air-sea interactions and land-air interactions. Therefore, ABL processes and surface inhomogenity significantly influence the spatial distributions of water vapor and aerosols. These present a great challenge in observing and modeling aerosol and water vapor.

The spatial variability of water vapor has impacts on many other important atmospheric processes. Weckwerth discussed the effect of small-scale moisture variability on thunderstorm initiation and showed that the typical radiosonde profile has difficulties to capture the small-scale water vapor variability [2]. One important mesoscale water vapor inhomogeneity is the dry line, which is a subject of increasing interest due to its frequent association with convection initiation and subsequent development of severe weather in the southern Great Plains and other locations [3]. Through entrainment and detrainment, small-scale moisture variability also control cloud evolutions.

Therefore, many studies are designed to better understand the water vapor and aerosol variability in the atmosphere, which presents a high demand on new observation capabilities [4]. The spatial variability of aerosol properties can be observed with a range of different methods from airborne sampling and ground-based lidar measurements to space-borne passive and active measurements. However, the spatial variability of water vapor on small scales is still mainly dependent on airborne in situ observations, but the in situ data are limited at the flight level data, which are hard to interpret sometimes due to strong vertical and horizontal water vapor variability. Airborne DIAL systems have demonstrated to be able to provide accurate water vapor measurements with a vertical resolution ~200-400 m and a horizontal resolution few hundred meters to few km [5]. Similar measurements are demonstrated by a large airborne Raman lidar focusing on profiling tropospheric aerosols and water vapor [6].

To provide required high temporal and spatial resolution aerosol, water vapor and temperature measurements for improving processed-oriented study, we developed two airborne Raman lidars to integrate remote sensing with in situ sampling from a small airborne platform, such as University of Wyoming King Air (UWKA).

This paper first provides an overview of Wyoming Compact Raman Lidar (WCRL) and Multi-function Airborne Raman Lidar (MARLi) and their capabilities. Then, WCRL observation examples from the King Air Planetary boundary layer Exploratory Experiment (KAPEE), which was conducted in 2010 to test WCRL, are provided to illustrate the airborne Raman lidar measurements capabilities and applications. Finally, a summary is presented.

2. AIRBORNE RAMAN LIDAR SYSTEMS

2.1 WCRL specification

Aimed at working together with in situ instruments in UWKA, the WCRL used a compact box-style design to integrate laser and transmitting optics, telescope, and receiving optics with a T-shape optical bench (see Fig. 1). The two other units of the WCRL are the laser power supply and the data acquisition and control system. The compact and integrated design not only provides the needed stability to maintain the alignment, but also make it easy to be integrated into different aircraft. The mechanical design of the system is provided by Welch Mechanical Designs, LLC.

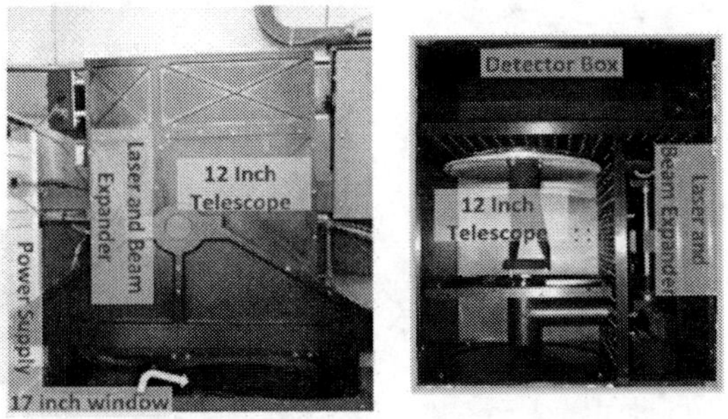

Figure 1. Photos show the installation of WCRL on the UWKA and its internal structure.

WCRL uses a Big Sky CRF 400 laser, which has a compact laser head and stable cavity design. After a 5X beam expender, the laser beam is transmitted out coaxially. The receiver consists a 12-inch diamond turning telescope providing an excellent thermal stability. The detection system has four channels: 355 nm perpendicular and parallel depolarization channels, 387nm nitrogen Raman channel, and 407 nm water vapor Raman channel. The system specification is listed in Table 1.

Table 1: Main specification of WCRL

Transmitter	
Laser Wavelength	355 nm Nd:YAG
Pulse Repetition Frequency	30 Hz
Pulse width and Energy	~8 ns and 50 mJ
Receiver	
Telescope	Cassegrain, 12 inch, 0.5-2 mrad
Detectors	PMTs
Data acquisition system	Analog-only, 200 MHz 4 channel A/D

In order to characterize fine structure of water vapor and aerosols, the system mainly use short-range strong signals by taking advantage of strong range dependent of lidar signals. Thus, an A/D data acquisition system rather then a photon counting system is used to provide high spatially resolved rdata. For single shot data from UWKA (cruising speed ~100 m/s), the system can provide 0.75 m vertical, ~3.5 m horizontal raw data. Different post-averaging can be done according to atmospheric features and applications [7].

2.2 MARLi specification

With the low laser power and Raman scattering cross-section, the daytime measurements range of WCRL is limited to ~500 m below the flight level for the summertime ABL, which is not enough for many ABL process studies. Furthermore, the vertical structure of temperature in the ABL is critical for ABL process studies. For these reasons and with the NSF MRI support, we are developing MARLi to profile temperature, water vapor, aerosol, and clouds simultaneously.

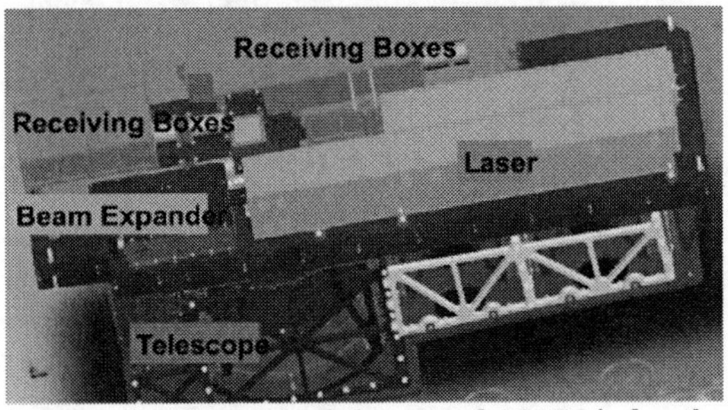

Figure 2. The general layout of MARLi for the installation on UWKA.

To improve the measurement range, MARLi used a powerful laser with ~14W power and 30Hz at

355nm. Together with a 16-inch telescope, MARLi is expecting to provide a factor 10 stronger atmospheric Raman signals than those from WCRL. To improve MARLi's capability to profile water vapor over high albedo targets, such as above clouds during daytime, MARLi is capable to transmit 266 nm laser to provide the solar-blind measurements of water vapor. Such measurement capability will be critical to study cloud-environment interactions. MARLi will have total 13 detection channels to cover needed elastic and Raman channels. Based on the performance of WCRL, MARLi is expected to achieve 0.5 K and 5% or better random errors for daytime temperature and water vapor measurements at 90m vertical and 1000m horizontal resolutions in the summer time ABL when aircraft flies at 2 km AGL and 100m/s.

3. WCRL OBSERVATION EXAMPLES

KAPEE was conducted during June 2010 to explore WCRL's capabilities, especially in conjunction with in situ sampling of aerosols and water vapor. During KAPEE, in situ water vapor measurements were available from both a LI-COR CO_2/H_2O analyzer and a chilled mirror hygrometer (EdgeTech). The chilled mirror hygrometer can provide reliable water vapor measurements with dew/frost point accuracy better than ±0.2°, but with a relatively slow temporal response. On the other hand, LI-COR analyzer can provide 1% accuracy water vapor measurements at a faster temporal resolution. Other than offering flight level measurements, these two instruments are used to calibrate WCRL water vapor measurements. The calibration of airborne Raman lidar can be done in two easy ways, which are more convenient than the calibration of ground-based Raman lidar systems. First, water vapor profiles within a small cloud free region collected from aircraft spiral descending or ascending can be use to compare with WCRL profile and to exam the overlap differences of between the water vapor and nitrogen Raman channels and to calibrate the system constant [7]. After knowing the short-range overlap difference, we can simply calibrate WCRL water vapor measurements or monitor calibration variations by comparing WCRL near aircraft measurements with in situ measurements. KAPEE results show that good agreements among these measurements can be achieved. Such a comparison is provided in Fig. 3c, where WCRL and LI-COR measurements follow with each other very well and the chilled mirror measurements showed a slow response crossing the dryline.

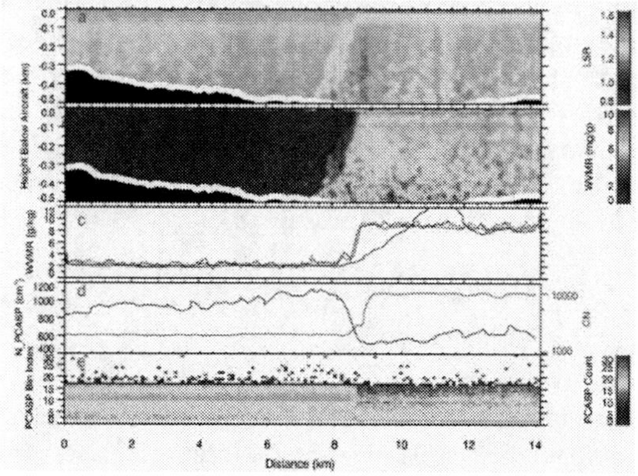

Figure 3. A dryline cross sections observed by WCRL on 22 June 2010: (a) lidar scattering ratio (LSR) and (b) water vapor mixing ratio (WVMR). (c) Comparison of WVMR measured by in situ probes (solid lines, red-Li-COR; black-chilled mirror) and by Raman lidar at 40m below aircraft (dashed line). (d) CN and PCASP total aerosol concentrations. (e) Aerosol size distributions measured by PCASP. The white lines in (a) and (b) indicate the surface.

The vertical structure of the dryline measured by WCRL is presented in Figs. 3a and 3b. In order to show the fine structure across the dryline, a horizontal running averaging with a less than 50m window was done in Fig. 2b. It is clear that the sharp transition from the dry side to the moisture side occurred within 100 m. With WRCL data recorded for each laser shot, different post averaging can be done to better characterize different atmospheric features. For this case, an extended horizontal averaging can be done for either the dry or the moisture side to reduce the random noise in water vapor measurements.

Due to the low power laser of WCRL, its daytime measurement range is limited to ~500 m below the aircraft. Thus, it took two horizontal legs to reveal the complete dryline structure in the ABL. The spatial distribution of water vapor shows a complex dynamic interaction on the moist side of the dryline, which pushes a 5-km wide water vapor plume into the free troposphere. A deep convection was triggered along this dryline near the time of the observations and developed into a severe storm causing significant damages along the path [8].

Figure 3 also shows WCRL aerosol (3a) and in situ PCASP aerosol size distributions and condensation nuclei (CN) number concentrations (3d and 3e). As expected, aerosol measurements showed similar coherent structures as water vapor across the dryline. The synergy of lidar and in situ aerosol measurements allow for more effective characterizations of aerosol spatial variations.

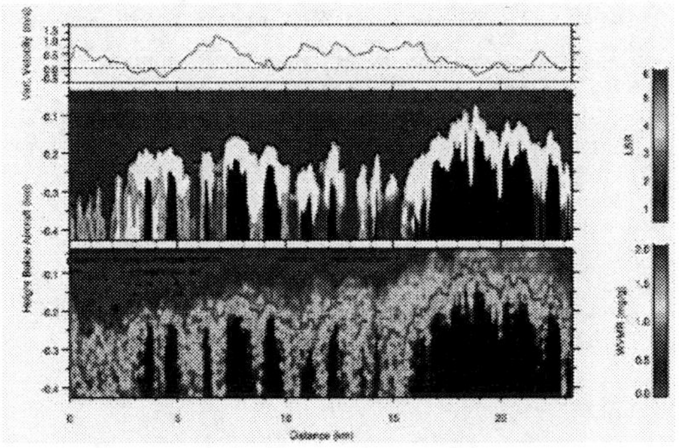

Figure 4. Nighttime measurements of water vapor structure above altocumulus clouds on 25 June 2010. Top: air vertical velocity at the flight level; middle: LSR; bottom: WVMR. The black dots in the LSR plot and the red dots in the WVMR plot are identified cloud top heights.

Another important application of high spatially resolved water vapor measurements is to study cloud and environment interactions, which strongly control cloud evolutions and are poorly understood and simulated in models. Figure 4 provide an example to show altocumulus and environment interactions. Although there is a weak temperature inversion at the cloud top height based on in situ measurements, strong water mixing above cloud top is clearly illustrated with WVMR distributions. Above a few convective cells, the mixing zones are more than 100 m deep. These measurements offer new data not only to evaluate large eddy simulations, but also to better study related cloud entrainment and detrainment processes.

4. CONCLUSIONS

Examples presented here clearly show that airborne Raman lidar is a powerful tool to measure multiple atmospheric parameters simultaneously to support many important atmospheric studies, such as, convective initiation, cloud-environment interactions, atmospheric boundary layer over land and oceans, and mesoscale dynamics. Such studies are fundamental to improving climate simulations and our capability to forecast weather and air quality.

WCRL will be deployed in summer 2015 for the PECAN (Plains Elevated Convection at Night) campaign, which is a large project supported NSF, NOAA, NASA, DOE to advance the understanding of continental, nocturnal, warm-season precipitation. MARLi will be tested in the laboratory in summer 2015 and on NSF/NCAR C-130 in fall 2015. Thus more results from these systems will be presented during the meeting.

ACKNOWLEDGEMENT

This research is supported by NSF under award of AGS-0645644 and AGS-1337599 and University of Wyoming. The UWKA is supported by NSF award ATM 0334908.

REFERENCES

[1] IPCC, 2007: Climate Change 2007: The Physical Science Basis. Solomon, S., et al. (eds.). Cambridge University Press, Cambridge, United Kingdom and New York, NY, USA.

[2] Weckwerth, T.M., 2000: The effect of small-scale moisture variability on thunderstorm initiation. *Mon. Wea. Rev.*, **128**, 4017-4030.

[3] Bluestein, H. B., and S. S. Parker, 1993: Modes of isolated, severe convective storm formation along the dryline, *Mon. Wea. Rev.,* **121,** 1354–1372

[4]Weckwerth, T. M., and D. B. Parsons, 2006: A review of convection initiation and motivation for IHOP_2002, *Mon. Wea. Rev.*, **134**, 5–22.

[5] B Ismail, S., E. V. Browell, R. A. Ferrare, S. A. Kooi, M. B. Clayton, V. G. Brackett, and P. B. Russell, 2000: LASE measurements of aerosol and water vapor profiles during TARFOX, *J. Geophys. Res.*, **105**, 9903-9916

[6] Whiteman, D. N., et al., 2010: Airborne and ground-based measurements using a high-performance Raman lidar, *J. Atmos. Ocean. Technol.,* **27**, 1781–1801.

[7] Liu B., et al., 2014: Compact airborne Raman lidar for profiling aerosol, water vapor and clouds, *Optics express*, **22** (17), 20613-20621.

[8] Bergmaier, P. T., et al., 2014: A Dryline in Southeast Wyoming. Part II: Airborne In Situ and Raman Lidar Observations, *Mon. Wea. Rev.*, **142**, 2961–2977.

APPLICATION OF CABAUW LIDAR DATA FOR CAMPAIGNS, NEW METHODOLOGY DEVELOPMENT AND SATELLITE VALIDATION ACTIVITIES

Arnoud Apituley* (1), Dave Donovan (1), Marco de Bruine (2), Bram Sanders (1), Marijn de Haij (1), Dimitra Mamali (3), Lukas Pfitzenmaier (3)

[1]*Royal Netherlands Meteorological Institute (KNMI), *Email: apituley@knmi.nl*

[2]*Utrecht University, The Netherlands,* [3]*Delft University of Technology, The Netherlands,*

ABSTRACT

Cabauw lidar data were used for the development of several new methods, as well as in the validation of new techniques based on other sensor data. The potential of the site that is equipped with a suite of in-situ and remote sensing equipment provides the possibility to develop new methods, and test them using independent observations.

Examples are shown for several recent campaigns conducted at the site, new methods developed using lidar data, and for satellite validation, including preparation for future missions.

1. INTRODUCTION

The Cesar Observatory in Cabauw [1], located in the western part of The Netherlands (51.971° N, 4.927° E) in a polder 0.7 m below mean sealevel. At the site a large set of instruments is operated to study the atmosphere and its interaction with the land surface. The site is currently equipped with a range of lidar systems (Fig. 1, Tab. 1). Caeli, the Cabauw Water Vapour, Aerosol and Cloud lidar, is a high-performance multi-wavelength Raman lidar and is part of EARLINET. It is one of the key instruments installed at the site. Also operated at the site are a UV-backscatter lidar with depolarisation and a ceilometer. The CHM15k was operated at the site for a test campaign.

Table 1. *Cabauw lidar sytems and capabilities. The right column 'Oper' indicates round-the-clock operation. Polarisation sensitivity is indicated by (δ).*

	Wavelength (nm)							Oper
Lidar	1064	904	607	532	407	387	355	24/7
Caeli	✔		✔	✔ (δ)	✔	✔	✔	
UV-lidar							✔ (δ)	✔
LD40		✔						✔
CHM15k	✔							✔

2. CAMPAIGNS

2.1 PEGASOS (2012)

A campaign with the PEGASOS Zeppelin [2] in Cabauw was held in 2012 to quantify the magnitude of regional to global feedbacks between atmospheric chemistry and a changing climate and to reduce the corresponding uncertainty of the major ones. During the campaign in Cabauw, the lidar measurements were used to identify atmospheric layers, traced by the aerosol content, while in-situ trace gas measurements were carried out from the Zeppelin platform and balloon borne NO_2 sondes [3].

2.2 ACCEPT (2014)

In the ACCEPT campaign – Analysis of the Composition of mixed-phase Clouds with Extended Polarization Techniques – a novel set of instruments to retrieve ice crystal microphysical properties in a large size range from micrometer to cm-scales was investigated. Co-located measurements

Figure 1. *Overview of the CESAR site in Cabauw, The Netherlands (left), Caeli (middle) and the UV-lidar (ALS450) with the CHM15k (right).*

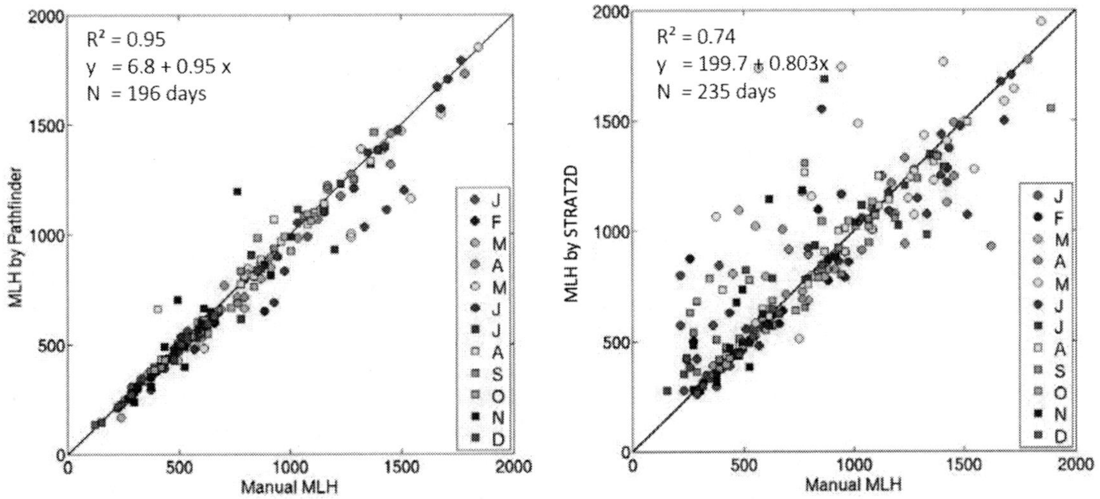

Figure 2. *Full year analysis of MLH retrievals at midday. (a) Pathfinder, (b) STRAT2D [8]. The reduced spread of the Pathfinder results indicates better consistency of the MLH retrieval over STRAT2D.*

of the off-zenith pointing polarization-lidar Polly-XT of TROPOS was exploited to detect layers of horizontally aligned planar crystals, causing specular reflections which affect only the signal measured with zenith-pointing Caeli [4].

2.3 Ceilometer tests (2014)

The Raman lidar Caeli as well as the UV-lidar and the LD40 acted as reference instruments in the selection procedure for a new ceilometer for the Dutch ceilometer network [5]. During the tests in November 2014 – January 2015, a wide range of conditions were observed from which the performance of the CHM15k under test could be verified against the requirements for the ceilometer network.

3. NEW METHODOLOGY DEVELOPMENT

3.1 Volcanic Ash

A new practical method for estimations of the mass load of volcanic ash based on UV-lidar depolarization lidar data was developed. The key point of this method is to estimate the extinction based on the measured aerosol (ash) depolarization ratio. The extinction can subsequently be used for mass estimations. The method was tested using the Raman lidar extinction data using observations from the Eyjafjallajökull eruption in 2010 [6].

Also, a method was introduced to derive integral properties of the aerosol size distribution, e.g., aerosol mass, from tropospheric multi-wavelength Raman lidar aerosol extinction and backscatter data, using an adapted form of the principal component analysis (PCA) technique. [7] Since the refractive index of general tropospheric aerosols is variable and aerosol types can vary within one profile, an inversion technique applied in the troposphere should account for varying aerosol refractive indices. Using PCA, provided a sufficiently complete set of appropriate refractive index dependent kernels is used, no a priori information about the aerosol type is necessary for the inversion of integral properties. This method was tested and applied to Caeli Raman lidar observations.

3.2 Pathfinder Mixed Layer Height Tracking

A new method for tracking the height of the mixed layer (MLH) using graph theory was developed. For independent verification, the collocated data from the wind profiler were used, as well as manually derived MLH. It was found that the new method outperformed other known methods for MLH tracking, especially in cases where other methods tend to jump back and forth between different layers [9] as shown in Fig.2.

3.3 Aerosols from smartphone spectropolarimetric measurements

A novel approach was developed actively involving the general public using their smartphones to form an atmospheric measurement network for aerosols at high spatial and temporal resolution.

This was established using iSPEX [10], a low-cost, mass-producible optical add-on for smartphones with a corresponding app. The aerosol optical thickness (AOT) maps derived from iSPEX measurements are in good agreement with the spatial AOT structure derived from satellite imagery and temporal AOT variations derived from ground-based precision photometry. Since relating the column integrated AOT measurements to (ground level) air quality is not straightforward, the Raman lidar measurements were used to separate the tropospheric AOT contribution from the total column.

4. SATELLITE VALIDATION ACTIVITIES

4.1 Aerosol Layer Height (ALH) from the O_2 A band

The lidar data was used for testing a new technique to derive the aerosol layer height from passive satellite observations in the O_2 A-band, that can be applied in future operational earth observation platforms [11]. For a number of selected cases, Cabauw lidar data was used to assess the performance of the new ALH algorithm. The ALH algorithm will be implemented in the Sentinel-5 precursor processor suite [12]. Validation activities are planned for the Sentinel-5p/TROPOMI validation.

4.2 ESA ADM-Aeolus aerosol and cloud product and ESA EarthCare aerosol and cloud product validation

Future work will include validation activities for the upcoming European satellite missions Aeolus [13], and EarthCare [14] using the lidars, radars and other profiling instruments at the Cabauw site as core instruments.

5. CONCLUSIONS

A range of examples has been given of the application of lidar data in campaigns in the recent past for atmospheric studies as well as new method development. Lidar data has also been used for validation of new satellite retrieval methods and will be used in the near future for the validation of new satellite instruments.

For a number of applications, the simultaneous availability of an advanced, operator controlled lidar running side by side with automated lidars has been very beneficial for method development and validation.

ACKNOWLEDGEMENTS

This work was partially supported by the Netherlands Institute for Space Research, project number GO-2005/075, the ESA project AO/1-7017/11/NL/MP, and the financial support for EARLINET in the ACTRIS Research Infrastructure Project by the European Union under grant agreement n. 262254 in the 7th Framework Programme (FP7/2007-2013) is gratefully acknowledged.

REFERENCES

[1] Apituley, A., H. Russchenberg, H. van der Marel, F.C. Bosveld, R. Boers, H. ten Brink, G. de Leeuw, R. Uijlenhoet, B. Arbresser-Rastburg and T. Röckmann, Overview of Research and Networking With Ground Based Remote Sensing for Atmospheric Profiling at the Cabauw Experimental Site for Atmospheric Research (CESAR) – The Netherlands, 2008, IEEE International Geoscience & Remote Sensing Symposium, 6/7/2008-11/7/2008, Boston, USA.

[2] http://pegasos.iceht.forth.gr/the-project.html

[3] Sluis, W. W., Allaart, M. A. F., Piters, A. J. M., and Gast, L. F. L.: The development of a nitrogen dioxide sonde, *Atmos. Meas. Tech.*, 3, 1753-1762, doi:10.5194/amt-3-1753-2010, 2010.

[4] http://atmos.weblog.tudelft.nl/accept-2/

[5] Haij, M.J. de, W.M.F. Wauben, H. Klein Baltink and A. Apituley, Determination of the mixing layer height by a ceilometer, Proceedings of the 8th International Symposium on Tropospheric Profiling, ISTP-2009, 19/10/2009-23/10/2009, A. Apituley, H.W.J. Russchenberg, W.A.A. Monna (Eds.), 2009, Delft, ISBN 978-90-6960-2.

[6] Donovan, D.P. and A. Apituley, A Practical Depolarization Ratio Based Inversion Procedure: Lidar Measurements of the Eyjafjallajo˙kull Ash cloud over the Netherlands, *Appl. Opt.*, 2013, 52, 11, 2394-2415, doi:10.1364/AO.52.002394.

[7] Graaf, M. de, A. Apituley and D.P. Donovan, Feasibility study of Integral property retrieval for tropospheric aerosol from Raman lidar data using principle component analysis. *Appl. Opt.*, 2013, 52, 10, 2173-2186, doi:10.1364/AO.52.002173.

[8] Morille, Y., M. Haeffelin, P. Drobinski, and J. Pelon (2007) STRAT: An automated algorithm to retrieve the vertical structure of the atmosphere from single-channel lidar data. *JTECH*, 24, 761-775.

[9] Bruine, M. de, A. Apituley, H. Klein Baltink, D. Donovan, M. de Haij, Applying graph theory for consistent tracking of the mixed layer height over the day from backscatter lidar measurements, AMT, in preparation, 2015.

[10] Frans Snik, Jeroen H. H. Rietjens, Arnoud Apituley, Hester Volten, Bas Mijling, Antonio Di Noia, Stephanie Heikamp, Ritse C. Heinsbroek, Otto P. Hasekamp, J. Martijn Smit, Jan Vonk, Daphne M. Stam, Gerard van Harten, Jozua de Boer, Christoph U. Keller and 3187 iSPEX citizen scientists (2014), Mapping atmospheric aerosols with a citizen science network of smartphone spectropolarimeters, *Geophys. Res. Lett.*, 41, 7351–7358, doi:10.1002/2014GL061462.

[11] Sanders, A.F.J., J.F. de Haan, M. Sneep, A. Apituley, P. Stammes and J.P. Veefkind, Evaluation of an aerosol layer height retrieval algorithm for Sentinel-5 Precursor and Sentinel-4: Application to O2 A band observations from GOME-2, Presentation: EUMETSAT Meteorological Satellite Conference, 22/9/2014-26/9/2014, Geneva, EUMETSAT.

[12] J.P. Veefkind, I. Aben, K. McMullan, H. Förster, J. de Vries, G. Otter, J. Claas, H.J. Eskes, J.F. de Haan, Q. Kleipool, M. van Weele, O. Hasekamp, R. Hoogeveen, J. Landgraf, R. Snel, P. Tol, P. Ingmann, R. Voors, B. Kruizinga, R. Vink, H. Visser, P.F. Levelt, TROPOMI on the ESA Sentinel-5 Precursor: A GMES mission for global observations of the atmospheric composition for climate, air quality and ozone layer applications, Rem. Sens. Env., 120, 2012, pp. 70-83, http://dx.doi.org/10.1016/j.rse.2011.09.027.

[13] http://www.esa.int/Our_Activities/Observing_the_Earth/The_Living_Planet_Programme/Earth_Explorers/ADM-Aeolus/ESA_s_wind_mission

[14] http://www.esa.int/Our_Activities/Observing_the_Earth/The_Living_Planet_Programme/Earth_Explorers/EarthCARE/Overview2

EPJ Web of Conferences **119**, 09004 (2016) DOI: 10.1051/epjconf/201611909004
ILRC 27

Initial Results from the Experimental Measurement Campaign (XMC) for Planetary Boundary Layer (PBL) Instrument Assessment (XPIA) Experiment

W.A. Brewer[1], A. Choukulkar[2,] S. Sandberg[1], A. Weickmann[2], J Lundquist[3], V. Iungo[4], R. Newsom[5], R. Delgado[6]

[1]*Chemical Sciences Division, National Oceanic and Atmospheric Administration, Boulder, CO*

[2]*Cooperative Institute for Research in Environmental Sciences, Boulder, CO*

[3]*University of Colorado, Boulder, CO*

[4]*University of Texas at Dallas, Dallas, TX*

[5]*Pacific Northwest National Laboratory, Richland, WA*

[5]*University of Maryland Baltimore County, Baltimore, MD*

ABSTRACT

The Experimental Measurement Campaign (XMC) for Planetary Boundary Layer (PBL) Instrument Assessment (XPIA) is a DOE funded study to develop and validate methods of making three dimensional measurements of wind fields. These techniques are of interest to study wind farm inflows and wake flows using remote sensing instrumentation. The portion of the experiment described in this presentation utilizes observations from multiple Doppler wind lidars, soundings, and an instrumented 300m tower, the Boulder Atmospheric Observatory (BAO) in Erie, Colorado.

1. INTRODUCTION

Knowledge of the time evolving, three dimensional wind field is essential in developing and evaluating large eddy simulation (LES) models, the study of turbine wake effects, and optimization of turbine layout for wind farms.

A network of scanning, pulsed, Doppler wind lidars can be used to measure the wind speed over a common volume or along a common vertical line (virtual towers (Calhoun, 2006)). These observations, projected onto multiple line-of-site (LOS) look angles, are combined to estimate the three dimensional wind field.

The goal of this study was to develop an understanding of the trade-off between:

(1) Synchronous vs asynchronous observations,

(2) The spatial resolution of the common observational points within the measurement volume vs the temporal update rate of entire volume

(3) To optimize the geometric layout of the observational network given maximum range of the lidars, the rate at which the measurements are made, the size of the measurement volume, and the need to have sufficiently large relative look angles to maintain precision in the 3-D wind retrieval.

Utilizing a network of commercially available, low-pulse-energy, high-pulse-repetition frequency (PRF), Doppler wind lidars, the Experimental Measurement Campaign (XMC) for Planetary Boundary Layer (PBL) Instrument Assessment (XPIA) Experiment was conducted at the Boulder Atmospheric Observatory (BAO) from March 3rd – May 1st 2015. It was funded by the Department of Energy (DOE) and was designed to develop and evaluate methods for making three dimensional measurements of the wind field using remote sensing techniques.

2. METHODOLOGY

The XPIA study was conducted at the NOAA BAO, the site of a 300m meteorological tower outfitted with a set of three-axis sonic anemometers positioned on the NW and SE sides of the tower, every 50m in height. The portion of the study described in this presentation consisted of eight Doppler wind lidars: five commercial scanning units (either Leosphere 200S or Halo

DOI: 10.1051/epjconf/201611909004

Photonics Streamline), one scanning research system (NOAA High Resolution Doppler Lidar - HRDL), and two vertically profiling systems (Leosphere V1).

The locations of the instruments are shown in Figure 1 and described in Table 1. The red rings in the figure have 500m and 1000m radii, centered on the BAO tower. The HALO lidar operated primarily in a vertically staring mode and was located at the super site, just south of the tower, along with two Leopshere V1 lidar wind profilers.

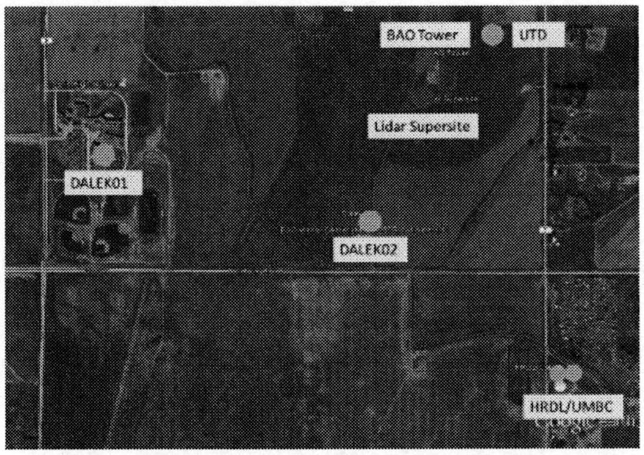

Figure 1. Instrument locations during XPIA field campaign. The inner red circle is of radius 500 m centered on the tower while the outer circle is 1000 m radius centered on the BAO tower.

Designation	Model	Affiliation
DALEK01 (D1)	200S	NOAA/ESRL
DALEK02 (D2)	200S	NOAA/ESRL
UTD	200S	U of Texas, Dallas
UMBC	200S	U of Maryland, BC
HRDL	Research	NOAA/ESRL
At Super Site:		
UCA	V1	Univ of Colorado
UCB	V1	Univ of Colorado
HALO	Streamline	DOE/PNNL

Table 1 The instrumentation deployed during the part of the XPIA experiment described in the presentation

The location of the four scanning Leosphere 200s units were chosen to be within 1.1km of the tower (in order to provide measurement overlap in the case of moderately clean conditions) and positioned such that relative azimuths of the beams, when intersecting near the tower, would not degrade the precision of the 3-D wind retrieval due to geometry.

Projecting the D2 line of site to the tower through the tower and measuring the relative azimuth look angles to that projection, the relative angles of the [D1, D2(Proj), HRDL/UMBC] line of sites at the tower are [97,128,135] degrees (where 120 degrees would be equally spaced). The location of the UTD system is closer to the tower than the other units and, while its azimuthal line of site to the tower is nearly parallel with D1, it is closer and provides a steeper elevation angle than is possible from the other systems.

The co-located HRDL and UMBC 200S have nearly identical look angles. During the latter stages of the study, when the UMBC system was present, HRDL was operated in a survey mode to provide contextual information about the horizontal wind field over the entire measurement area while the four 200S units did synchronized scans.

The overlapping measurement volumes used in this study were typically either centered on the instrumented tower or over the lidar supersite (where continuous vertical velocity data from the HALO unit and continuous horizontal wind profiles from the V1 units could be used for evaluation purposes). Several measurement volumes and scanning approaches were evaluated: asynchronous scanning/horizontal plane, asynchronous scanning/vertical line (virtual tower), synchronous staring, small and large synchronous staring/horizontal plane array.

Initially, the lidars performed scans to determine the location of the booms holding the sonic anemometers on the tower and then performed synchronized stares at each level. This was done to determine the pointing angle offsets of each lidar, verify the alignment of the sonic anemometers, and evaluate the accuracy of software tools used to facilitate coordinated pointing of the multiple lidars.

The next step was to probe various measurement volumes with multiple lidars. Figure 2 shows the

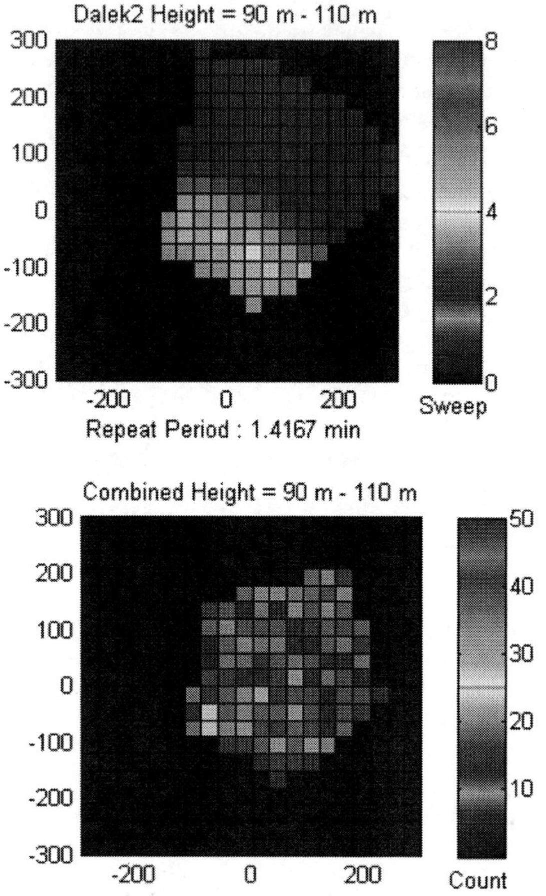

Figure 2. Simulated scan geometry required to cover a horizontal plane centered at the tower extending +/- 100m horizontally at a height of 100m AGL. The top plot shows that it takes 5 sweeps to provide coverage from D2. The bottom plot shows the combined number of measurements from three systems (D1, D2, HRDL) in each 30mx30m grid cell between the heights of 90 and 110m AGL.

simulated measurement density and scan geometry required to cover a plane extending +/- 100m horizontally around the tower at a height of 100m. The data are gridded into 30mx30m cells extending from 90m-110m vertically. The D2 unit, at a range of 600m, requires a 5 elevation plane azimuthal sector scan to completely fill the measurement volume (top plot). The lower plot shows the combined measurement count for three lidars (D1, D2, HRDL), asynchronously scanning, with a pattern repeat time of 1.4 minutes. Several versions of this pattern that were designed to either minimize the pattern repeat period or maximize the horizontal resolution were attempted. The lidars also performed a set of asynchronous elevation (RHI) scans that were designed to intersect (virtual towers) over the tower and the supersite. This pattern was repeated for several days through a range of atmospheric conditions.

Two sets of horizontal-plane, synchronized stares were designed to probe a 3x3 or 4x4 measurement grid at a height of 100m. The smaller 3x3 grid extended roughly 200m on a side and the larger 4x4 grid extended 1800m. The lidar scans were programmed to allow roughly 10 seconds dwell at each location and an attempt was made to keep the pattern repeat time shorter than 2 minutes.

In order to study the effect of the geometry of the look angles on the 3 dimensional fit (especially the vertical component). A series of 10 minute synchronous stares were programmed to have all four 200S units staring above the super site at a range of increasing altitudes (and elevation angles) over a period of several days. During these scans, the HRDL lidar performed a set of two plane azimuthal sector scans designed to provide contextual information on the horizontal wind field.

3. RESULTS

At the time of the submission of this abstract, the XPIA study is still underway. Significant effort has gone into developing tools to aid in multiple lidar coordinated scan design and implementation. Programing scans on different manufacturer's scanner control software (or different versions of the same software) so that they remain synchronous has proven to be challenging and techniques to do this have been developed. The pointing accuracy and stability of the commercial lidar systems have been evaluated. Preliminary results from the operations described in this abstract will be presented.

ACKNOWLEDGEMENT

The authors would like to thank Jim Wilczak and Daniel Wolfe (Physical Sciences Division, NOAA) for their assistance in the experimental design and deployment of the lidars at the BAO facility. We thank the Lefthand water district and Erie High School for allowing us to place lidars on their property. In addition the authors would like to acknowledge financial support from DOE and NOAA.

REFERENCES

[1] R Calhoun, R Heap, M Princevac, R Newsom, H Fernando, D Ligon (2006): Virtual towers using coherent Doppler lidar during the Joint Urban 2003 dispersion experiment. *Journal of Applied meteorology and climatology* **45** (8), 1116-1126

DOI: 10.1051/epjconf/201611910001

DOPPLER LIDAR IN THE WIND FORECAST IMPROVEMENT PROJECTS

Yelena Pichugina[1, 2*], Robert Banta[2], Alan Brewer[2], Aditya Choukulkar[1, 2], Melinda Marquis[2], Joe Olson[1, 2], and Mike Hardesty[1, 2]

[1]*Cooperative Institute for Research in Environmental Sciences, Boulder, CO*

[2]*National Oceanic and Atmospheric Administration, Boulder, CO*

ABSTRACT

This paper will provide an overview of some projects in support of Wind Energy development involving Doppler lidar measurement of wind flow profiles. The high temporal and vertical resolution of these profiles allows the uncertainty of Numerical Weather Prediction models to be evaluated in forecasting dynamic processes and wind flow phenomena in the layer of rotor-blade operation.

1. INTRODUCTION

Intelligent development of wind-energy (WE) power plants requires high-quality measurements in the turbine rotor layer of the atmosphere to estimate the wind resources and understand meteorological processes controlling the Boundary Layer (BL) for the optimal design and operation of wind turbines, wind plants, and other components of the electrical grid. A potentially important tool in providing characteristics and behavior of the BL in response to various atmospheric conditions, stability, seasonality, and diurnal cycle is the numerical weather prediction (NWP) model, but without measurements in this layer for verification, the accuracy and fidelity of model output is unknown. To address the need for wind flow measurements at turbine-rotor heights, Doppler-lidar technologies are an attractive option. These technologies are aimed at providing cost effective data through the layer swept by modern turbine rotor blades at needed temporal and vertical resolutions. In recent years the number of research projects in WE involving lidars has significantly increased. Lidar measurements are used to study wind flow phenomena, including wake effects [1, 2] and Low Level Jets [3], and also for evaluating numerical model forecasting capabilities, and for finding ways to improve model skill and reduce the errors. Measurements from several offshore, ship-borne campaigns are available that use National Oceanic and Atmospheric Administration/Earth System Research Laboratory (NOAA/ESRL)'s High-Resolution Doppler Lidar (HRDL), which is a remote-sensing lidar system equipped with a motion-compensation system that has been shown to produce high-quality profile measurements [4]. Recently the HRDL dataset from a field campaign, the New England Air Quality Study (NEAQS-2004), designed to study air quality off the New England coast, was used to investigate properties of the flow in the rotor layer in the offshore environment, and to validate performance of NOAA NWP models.

2. PREDICTION OF OFFSHORE WINDS

The NOAA study to inform requirements for an observation network for WE, sponsored by the U.S. Department of Energy, was conducted in 2013. In the study, called the Position of Offshore Wind Energy Resources, or POWER, project, the unique measurements from ship-based lidar were used to verify NOAA NWP models and obtain information on key atmospheric phenomena that impact offshore wind forecasts. The verification of model skill in predicting hub-height winds over the ocean was performed by comparing hourly averaged lidar wind profiles to these produced by model. The model verification performed here is an evaluation of model skill in predicting hub-height winds over the ocean, using a total of 12 model configurations based on two modeling systems, the hourly-updated Rapid Refresh (RAP) system and a new hourly-updated version of North America Mesoscale (NAMRR) forecast system. The two models are run at both normal resolution ("parent" model) and at high-resolution ("nested" model) called High-Resolution Rapid Refresh or HRRR and NAMRR CONUS-Nest respectively. The short description of these models is given in Table 1.

Table 1. Description of models used in the POWER study, horizontal grid resolution and NOAA facility running these models.

Model	Grid	Facility
Rapid Refresh (RAP)	13 km	OAR/ESRL
High Resolution Rapid Refresh (HRRR)	3 km	OAR/ESRL
North American Mesoscale RR (NAMRR)	12 km	NWS/NCEP
NAMRR Contiguous United State (ConusNest)	4 km	NWS/NCEP

Retrospective runs of all models were performed for two intensive measurement-modeling study periods of the POWER analysis, one week in July and one week in August 2004, selected to represent various wind-flow conditions as the research ship cruised around the Gulf of Maine. During both periods the research ship was covering a wide 241 x 250 km (150 x 155 mi) area as shown in Figure 1 (right panels). Examples of 15-min average wind profiles as time-height cross sections are given in Figure 1 (left panels) to illustrate the high vertical resolution of the data that allowed fast changing dynamic processes such as LLJs or sudden changes in wind speed or direction to be sampled.

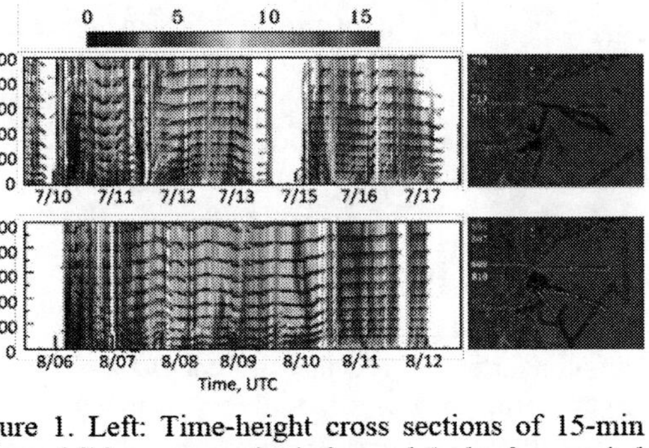

Figure 1. Left: Time-height cross sections of 15-min averaged lidar-measured wind speed (color bar, scaled from 0 to 16 ms^{-1}) and direction (arrows), computed from HRDL conical scans during two intensive measurement-modeling study periods of the POWER analysis (top: July period, bottom: August period). Vertical axis is height above sea level (m), and horizontal axis is time in UTC. Right: Google maps illustrating color coded ship tracks for each day during selected periods in (top) July and (bottom) August.

Due to the heavy rain and dense fog observed from the evening of 13 July until mid-afternoon of 15 July, HRDL was not operated during these hours and measurement hours during several other days were also shortened by synoptic conditions (blank area on the left panels). For comparison with modeled winds, lidar data were averaged over 1-hr intervals and interpolated to the model heights. A visual overview of time-height cross sections of the HRDL-measured wind-speed data and model-initialization data shows that the overall features and trends are represented in the models. However, a quantitative analysis showed differences in flow strength and timing between modeled and observed wind profiles. Sample comparisons of (black lines) measured and (color lines) modeled wind profiles for the initialization time are shown in Figure 2 to illustrate the general findings. The largest discrepancies were observed for stronger winds and LLJ-like profiles, which are more frequent during nighttime hours as demonstrated in the top four profiles of 16 July.

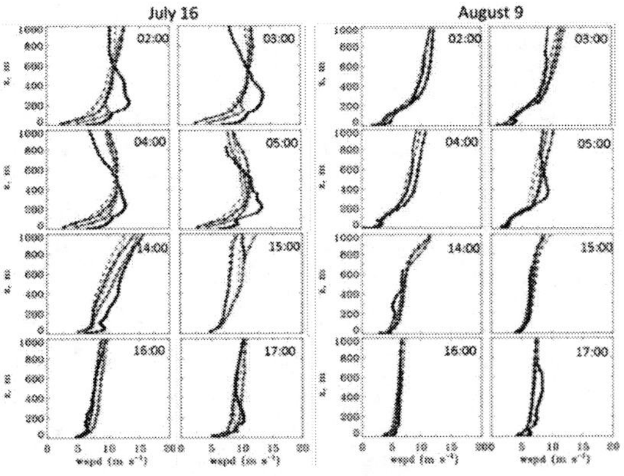

Figure 2. Examples of lidar-measured and modeled wind profiles (left two columns) July 16; (right two columns) August 9. Lidar profiles are shown in black. Red and blue lines show profiles modeled by NAMRR model in parent (NAMRR) and nested (NAMRR ConusNest) modes correspondently. Solid and dashed lines indicate model runs with different sets of data assimilated into model. Discussion of the difference of the assimilated sets of data and its impact on the results of the comparisons is beyond the scope of this paper.

Overall, better agreement was observed for weak and moderate wind speeds as shown in all profiles for 9 August. Similar results were observed for all days in both study periods showing larger discrepancies below 200 m, for stronger (>10 ms^{-1}) winds, and in the presence of LLJ events in period-averaged comparisons of wind profiles. Figure 3 shows mean error statistics for each study period.

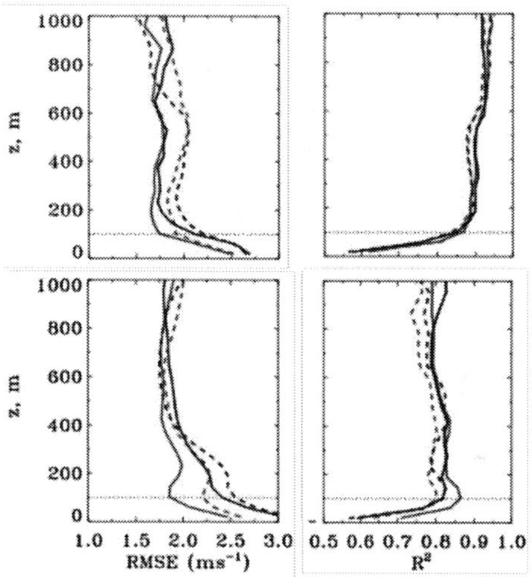

Figure 3. Profiles of the Root-mean Square error (RMSE) and correlation coefficients (R^2) between observed and initial-condition model wind speed averaged over (top) August and (bottom) July study periods. Red lines on all panels are for NAMRR-CONUS and blue lines are for NAMRR parent mode of the model runs. Solid and dashed lines are for different sets of data assimilated into each model as described in Figure 2.

HRDL wind data were used to verify model forecasts out to 12 h lead time. To test model performance, HRDL data were averaged over the layer from 20 to 500 m ASL. Figure 4 shows the RMSE for the eight model configuration runs for the August period. Overall, a quantitative statistical assessment of the mean-wind accuracy of the models showed the RMSE between observed and modeled winds of 2-2.5 m/s for the first 2-3 lead hours, increasing for the later forecast hours. Such errors could be explained partially by the use of an archive dataset from an experiment that was designed to trace plume dispersion over the water surface (www.esrl.noaa.gov/csd/ICARTT/.1), and these objectives are different from model validation purposes. Another part of the error could be induced by the using older versions of models in the study and the availability of historical datasets to assimilate into the retrospective model runs. The overall results from the POWER project demonstrated the importance of observational data

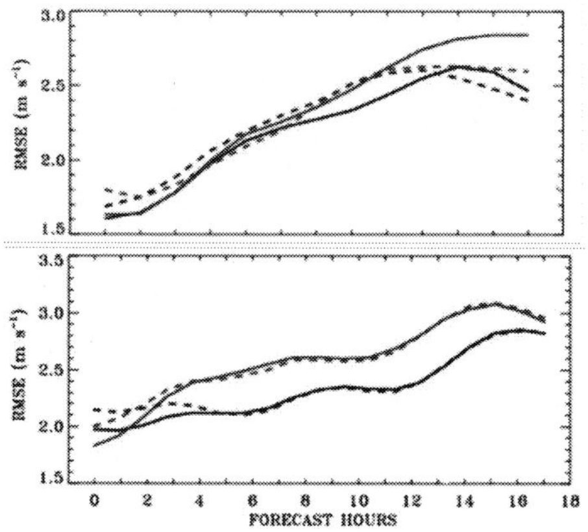

Figure 4. RMSE between observed and modeled wind in August 6-12 for data averaged over 20-500m layer. Top panel: (red) HRRR and (blue) RAP models. Bottom panel: (red) NAMRR ConusNest and (blue) NAMRR models. Solid and dashed lines are for different sets of data assimilated into each model as described in Fig. 2.

to validate NWP models, and to determine the uncertainty of wind resource assessment in one of the US offshore areas projected for wind plant development. Comparisons of model ability to simulate diurnal cycles, strong and weak winds, LLJ-like profiles, and other time- and space-dependent features of the flow, indicated areas of most needed model improvement. These areas will also be of primary interest for the validation and improvement of model skill in prediction of the wind flow over a complex terrain.

3. WIND FORECAST IMPROVEMENT PROJECT (WFIP II)

The WFIP-II, aimed at developing a better understanding of the factors that shape wind flow over complex terrain, will be conducted in the Columbia River Gorge area near the border of Oregon and Washington states, east of the Cascade Mountains. The project, sponsored by the Department of Energy, is a collaborative effort to improve model performance in predicting wind flow and BL phenomena in the mountain region occupied by numerous wind farms. The main objectives of this experiment are to characterize complex terrain atmospheric phenomena that impact model accuracy and to validate and improve weather prediction models, including the NOAA HRRR operational model and other, finer scale models such as WRF or LES. Participating organizations will deploy a variety of instruments in the study area for an 18-month period to provide real-time measurements including several scanning and wind-profiling lidars. Two NOAA scanning Doppler lidars, obtained commercially and updated for remote operation and data acquisition to better fit project objectives, will also operate for the duration of the experiment. Two proposed sites, in the vicinity of Wasco and Arlington airports (Figure 5), were selected for possible location of these instruments. These lidars will either be operating from each site, providing profiles of wind flow parameters, or will operate from one (Wasco) site in the Dual-Doppler scanning mode providing measurements of turbulence as well as mean-wind profiles. I either case lidar data will be used for the comparison with model output.

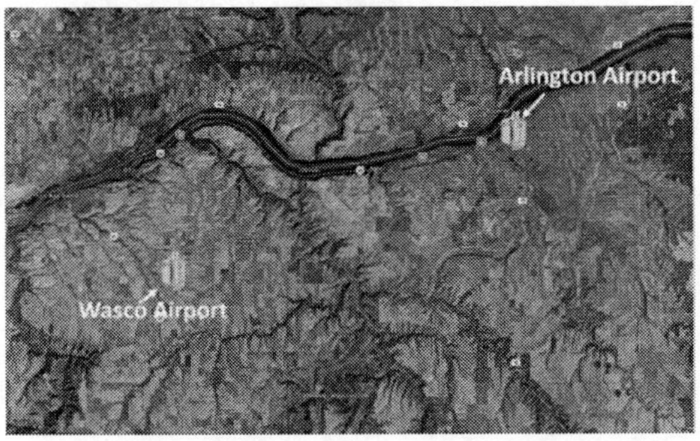

Figure 5. The USGS animated map shows proposed locations for the NOAA scanning Doppler lidars. Red dots on the map indicate wind turbines.

WFIP-II will take advantage of the results from two prior field programs designed to test the uncertainty of measurements from various types of lidar, and their coordinated scanning strategies. Both of these short-term experiments were conducted at Erie, Colorado. One of the experiment, called the **L**idar **U**ncertainty **M**easurement **EX**periment **(LUMEX)** was executed in in the summer of 2014 and another, called the **EX**perimental Measurement Campaign for **P**lanetary Boundary Layer **I**nstrument **A**ssessment **(XPIA)** was executed in the spring of 2015. An overview these experiments will be presented at this conference by Brewer et al. (2015) and Choukulkar et al. (2015).

4. CONCLUSIONS

The offshore Doppler lidar dataset allowed detailed

verification of NWP model output in the atmospheric layer, occupied by turbine rotors. The scientific and measuring capabilities of the WFIP-II project, represented by the collaborating organizations, will allow an unprecedented characterization of the key atmospheric processes in complex terrain leading to the better BL parametrization and improved wind flow forecasts.

ACKNOWLEDGEMENT

The POWER and WFIP-II projects are sponsored by the U.S. Department of Energy (DOE), Energy Efficiency and Renewable Energy office, Wind and Water Power program. We want to thank all our colleagues for the collaboration and their tremendous contribution during the POWER project and in the preparation of the forthcoming WFIP-II project: Stan Benjamin, Eric James, and Jaymes Kenyon - all from the NOAA/ ESRL Global Science Division; James Wilczak, Laura Bianco and Irina Djalalova - all from the NOAA/ ESRL Physical Science Division Jacob Carley and Geoff DiMego from the NOAA National Weather Service (NWS). We also want to thank Joel Cline from DOE for the leadership in preparation of both projects; management, execution and reporting results from the POWER project.

REFERENCES

[1] Aitken, M.L., J.K. Lundquist, R.M. Banta, and Y.L. Pichugina, 2014: Quantifying wind turbine wake characteristics from scanning remote sensor data. *J. Atmos. Ocean. Technol.*, **31**, 765–787.

[2] Pichugina Y.L., R. M. Banta, W.A. Brewer, J.K. Lundquist, R. M. Hardesty, M.L. Aitken, R.J. Alvarez, S.P. Sandberg, A M. Weickmann N.D. Kelley, and J.D. Mirocha, 2012: Estimation of wind turbine wake characteristics from Doppler lidar measurements, *26th ILRC, Proceedings*, 4pp

DOI: 10.1051/epjconf/201611910001

[3] Banta, R.M., Y.L. Pichugina, N.D. Kelley, R.M. Hardesty, W.A. Brewer, (2013): Wind energy meteorology: Insight into wind properties in the turbine-rotor layer of the atmosphere from high resolution Doppler lidar. *Bulletin of the American Meteorological Society*, 94, **6**, 883-902.

[4] Pichugina, Y.L., R. M. Banta, W. A. Brewer, S. P. Sandberg, and R. M. Hardesty, 2012: Doppler-lidar-based wind-profile measurement system for offshore wind-energy and other marine-boundary-layer applications. J. Appl. Meteor. Climatol. 51, 8, pp 327-349.

EPJ Web of Conferences **119**, 10002 (2016) DOI: 10.1051/epjconf/201611910002
ILRC 27

ALL-FIBER AIRBORNE COHERENT DOPPLER LIDAR TO MEASURE WIND PROFILES

Jiqiao Liu[*1,2], Xiaopeng Zhu[1,2], Weifeng Diao[1,2], Xin Zhang[1,2], Yuan Liu[1,2], Decang Bi[1,2], Liyuan Jiang[1,2], Wei Shi[1,2],Xiaolei Zhu[1,2],Weibiao Chen[1,2]

[1]*Key Laboratory of Space Laser Communication and Detection Technology, CAS,*

[2]*Shanghai Institute of Optics & Fine Mechanics, CAS, Shanghai, China,*

**Email: x_qiao@siom.ac.cn*

ABSTRACT

An all-fiber airborne pulsed coherent Doppler lidar (CDL) prototype at 1.54 μ m is developed to measure wind profiles in the lower troposphere layer. The all-fiber single frequency pulsed laser is operated with pulse energy of 300μJ, pulse width of 400ns and pulse repetition rate of 10kHz. To the best of our knowledge, it is the highest pulse energy of all-fiber eye-safe single frequency laser that is used in airborne coherent wind lidar. The telescope optical diameter of monostatic lidar is 100 mm. Velocity-Azimuth-Display (VAD) scanning is implemented with 20 degrees elevation angle in 8 different azimuths. Real-time signal processing board is developed to acquire and process the heterodyne mixing signal with 10000 pulses spectra accumulated every second. Wind profiles are obtained every 20 seconds. Several experiments are implemented to evaluate the performance of the lidar. We have carried out airborne wind lidar experiments successfully, and the wind profiles are compared with aerological theodolite and ground based wind lidar. Wind speed standard error of less than 0.4m/s is shown between airborne wind lidar and balloon aerological theodolite.

1. INTRODUCTION

A coherent Doppler lidar (CDL) is proved to be an effective tool for wind profiles measurement[1] , windshear[2] and wake vortices detection and warning[3] in clear air with high temporal and spatial resolution and high accuracy in lower atmosphere layer. The laser source is the most important part in a CDL system. With the development of single frequency laser technology, 10 μ m CO_2 laser[4] ,1 μ m solid laser[5], 2 μ m solid laser[6], 1.6 μ m solid laser[7] and 1.5 μ m fiber laser[8] are utilized as the lidar transmitter source successively. Over the past decade, single frequency pulsed fiber lasers are studied widely. All-fiber CDL system at 1.5 μ m has attracted much attentions because of its eye safety, compact size, flexible arrangement, and mature fiber components technology from telecommunication industry[9]. It is also suitable for airborne wind profile measurement because of its stable performance in harsh environment. Several companies have developed commercial all-fiber CDL systems since 2004[10]. In this paper, an all-fiber pulsed CDL prototype system for wind sensing and some experiment results are presented. In the airborne field campaign, the CDL measured wind profiles are compared with balloon aerological theodolite and ground-based wind lidar , and good agreements are shown.

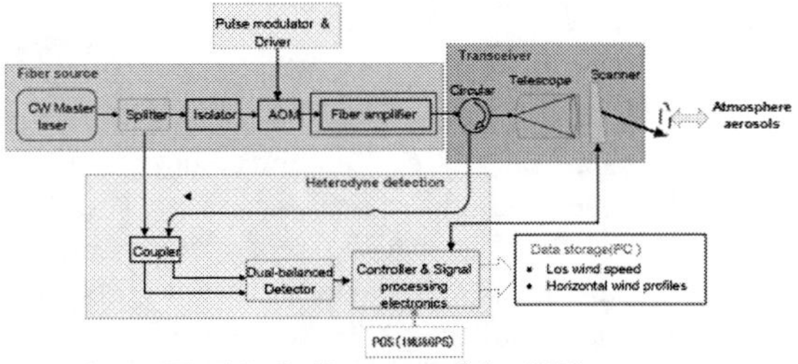

Fig.1. The block diagrams of the CDL system

2. SYSTEM CONFIGURATION

The block diagrams of the CDL system is shown in Fig 1. The CDL system consists of all-fiber laser source and optical transceiver and heterodyne detection unit with real-time signal processor. The CDL system is compact with size of 450mm×400mm×400mm and mass of 50kg. A laptop or PC is used to control the lidar and store measured data in every azimuth angle. Further estimate of the 3-dimensions wind profiles could be implemented in the computer.

DOI: 10.1051/epjconf/201611910002

An eye-safe, narrow-linewidth, polarized, pulsed, fiber laser based on master oscillator power amplifier (MOPA) configuration is acted as the lidar source[11]. A commercial continuous single frequency Distributed Feed Back Laser Diode (DFB-LD) with 10 kHz linewidth and 20mw power is used as the seeder laser. The seeder is divided into two parts by the 1×2 optical coupler. One part (about 0.8mW) is sent to the balanced detectors as the local oscillator for heterodyne detection. The other part is pulsed modulated and frequency shifted by the AOM at repetition rate of 10 kHz, pulse duration of 400ns. The IF (Intermediate frequency - IF) frequency is 160MHz. Through multi-stages fiber amplifiers, fiber laser with pulse energy of 300 μJ, linewidth of 1.1MHz at wavelength of 1540nm is obtained. The pulse shape of the output pulse is shown in Fig.2 .

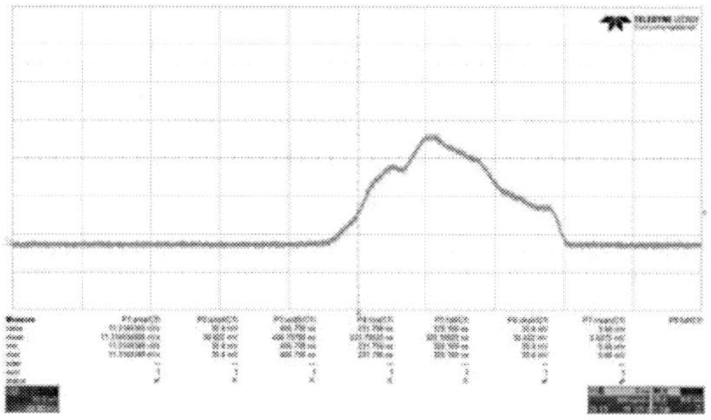

Fig.2. The shape of the output laser pulse

The optical transceiver combines a circulator and a telescope and an optical scanner. The high power circulator is used to separate transmitting fiber laser and signal backscattered by aerosols. The optical diameter of the telescope is 100mm. The output beam from telescope is conically scanned by a compact refractive scanner with 20° deflection angle. VAD scanning could be implemented in less than 20 seconds with 8 different azimuth angles.

In the heterodyne detection unit the local oscillator from seeder is mixed with signal backscattered by aerosols from transceiver, then it is received by the InGaAs balanced detector. The heterodyne signal from the balanced receiver is sampled by the high speed AD converter with 1G bps rate. The sampling time is 40 μs and 30m range resolution is realized with 200ns range gate.

Fast Fourier Transform (FFT) of 1024 points and power spectrum accumulation with 10000 pulses are implemented in every range gate. Accumulated power spectra in every azimuth are sent to the computer through RS422 interface. Then wind profiles could be calculated.The specifications of the system are summarized in the table 1.

Table1 The specifications of the CDL system

Parameters	value
Wavelength	1540nm
Pulse energy	300 μJ
Pulse width	400ns
Pulse repetition rate	10 kHz
Power of LO	0.8mW
Range resolution	30m
Telescope aperture	Φ100mm
Scanner	Φ100mm 20° deflection angle
IF	160MHz
AD	1G/s ,10bit
Accumulated pulses	10000

3. RESULTS

3.1 Ground experiments

Ground validation experiments were carried out for the CDL system compared with standard radiosonde balloon. Measured wind profiles results for both instruments were shown in Fig.3 and good agreements were demonstrated. The radiosonde balloon was released at 8:00 pm on Dec 24 in Shanghai upper atmospheric sounding site (121.46E, 31.40N), where was about 22 kilometers from the CDL system experiment site (121.23E, 31.38N). The measured wind profiles below 0.5km were ignored for comparison because of larger horizontal space between two sites. Coefficients of correlation for the CDL system and radiosonde from 0.5 km to 2.9 km were proved to be 0.93 for wind direction and 0.91 for wind speed. The standard errors of wind speed and wind direction were 0.85 m/s and 4.0°, respectively.

DOI: 10.1051/epjconf/201611910002

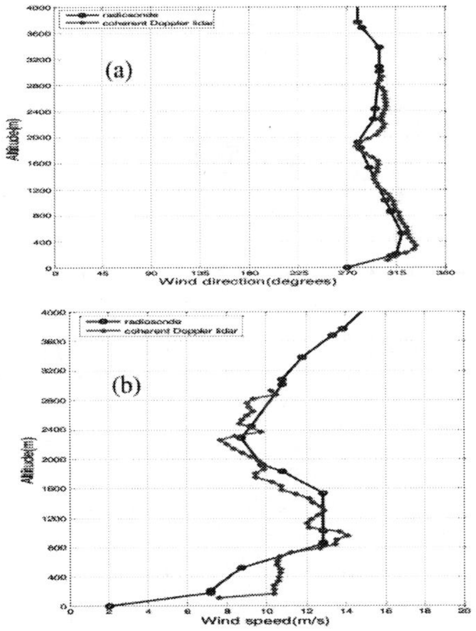

Fig.3. Comparison of wind profiles between CDL and radiosonde: (a) horizontal wind speed direction, (b) horizontal wind speed.

3.2 Airborne experiments

Airborne CDL validation experiment is carried in Roncheng city, Shangdong province, China on Jan 19, 2015. The CDL system is installed in a fixed wing aircraft with almost 50m/s speed and 2000m flight height. For airborne CDL, the speed and attitude angles should be corrected to calculate accurate wind profiles. So the POS system composed of Inertial Navigation System (INS) and global positioning system(GPS) is installed in the aircraft with the lidar together to record aircraft speed and attitude angles (yaw ,pitch and roll). Wind profiles measured from airborne CDL are compared with ground-based CDL and aerological theodolite in the campaign simultaneously to assess the airborne CDL performance. The sites of the ground-based lidar and the balloon aerological theodolite are shown in Fig.4, and the distances between the aircraft ground track and the ground campaign stations are about 500 m. The whole flight time is lasted about 5 hours, and the horizontal wind profiles with time are shown in Fig.5. Between 13:40 (Local time) and 15:40 the aircraft is completely flying above the Yellow Sea.

Fig.4. Aircraft ground track and ground campaign stations

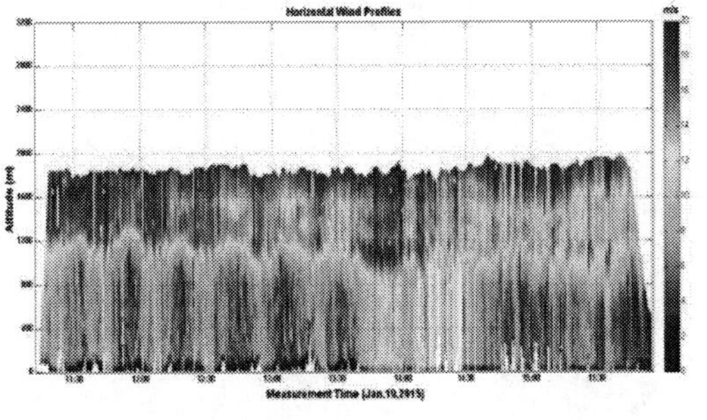

Fig.5. The horizontal wind profiles with time

For one case on 12:22 airborne lidar shows good agreements with ground-based lidar and the balloon aerological theodolite. The wind speed standard error between the three system are less 1m/s, and the wind direction differences are about 10° ,as shown in Fig.6 and Fig.7. Especially, wind speed standard error between airborne lidar and balloon aerological theodolite is less 0.4m/s, with wind speed coefficient of correlation of 0.9867 , which is shown in Fig.8.

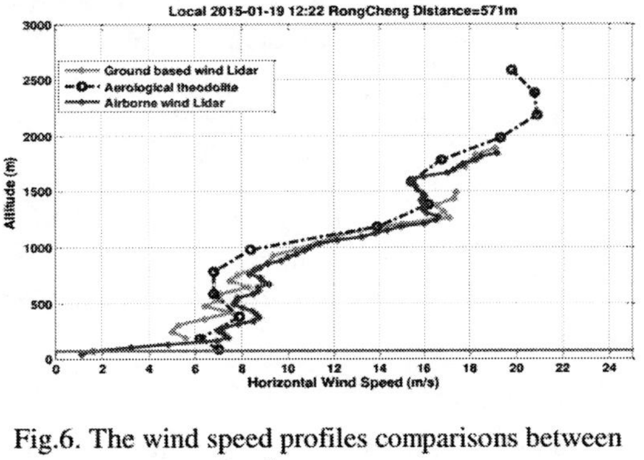

Fig.6. The wind speed profiles comparisons between the three systems.

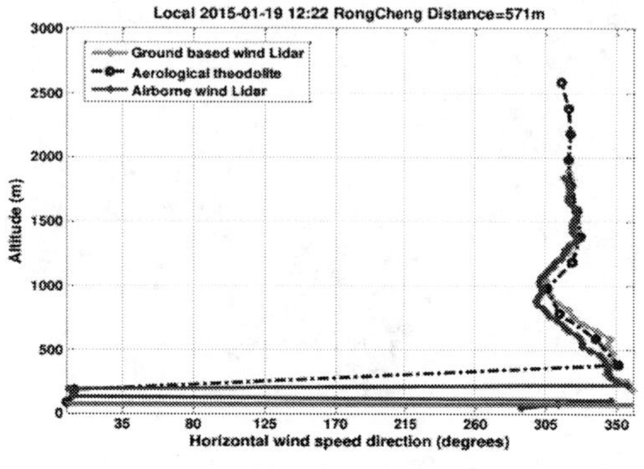

Fig.7. The wind directions profiles comparisons between the three systems.

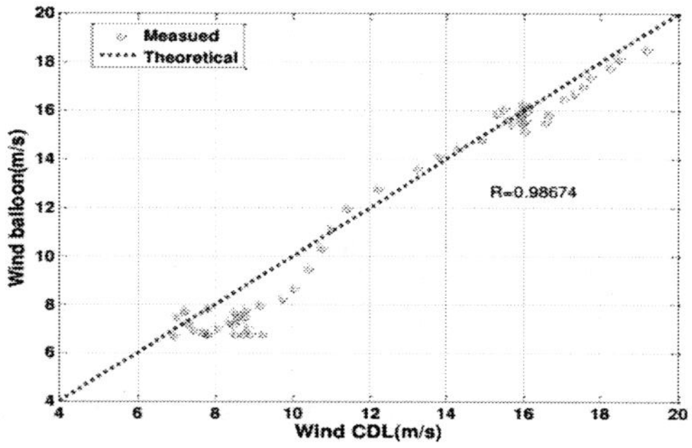

Fig.8. Linear correlation of wind speeds measured by the CDL and aerological theodolite

4. CONCLUSIONS

An all-fiber airborne pulsed CDL prototype with high laser pulse energy is developed to measure wind profiles. Ground and airborne campaign experiments are implemented to validate the CDL performance. In the ground campaign, the standard errors of wind speed and wind direction are shown to be 0.85 m/s and 4.0° respectively between CDL and radiosonde. In the airborne campaign, the airborne CDL measured wind profiles are compared with the ground-based CDL and the aerological theodolite. Wind speed standard error between airborne CDL and aerological theodolite are less 0.4m/s is validated. This lidar will be used for long-range wind profiles measurements in future in ground-based or airborne platform.

ACKNOWLEDGEMENT

This work is supported by National Natural Science Foundation of China (No. 60908036) and the Youth Innovation Found of China High Resolution Earth Observation.

REFERENCES

[1] Asaka, K., T. Yanagisawa, and Y. Hirano, 2001:1.5-μm eye-safe coherent lidar system for wind velocity measurement. SPIE. 4153: p. 321-328.

[2] Frehlich, R., 1994:Coherent Doppler lidar signal covariance including wind shear and wind turbulence. Appl. Opt. 33(27): p. 6472-6481.

[3] Dolfi-Bouteyre, A., et al., 2007:1.5 μm all fiber pulsed lidar for wake vortex monitoring, in 14th Coherent Laser Radar Conference.

[4] Huffaker, R.M. and R.M. Hardesty, 1996:Remote sensing of atmospheric wind velocities using solid-state and CO2 coherent laser systems. Proceedings of the IEEE. 84(2): p. 181-204.

[5] Kane, T.J., et al., 1987:Coherent laser radar at 1.06 um using Nd:YAG lasers. Opt. Lett. 12(4): p. 239-241.

[6] Jirong, Y., et al. An all solid-state 2um laser system for space coherent wind lidar. in Aerospace Conference Proceedings, 2000 IEEE. 2000.

[7] Hannon, S.M., S.R. Vetorino, and J.V. Pelk, 2007:Next generation Doppler lidar sensor at 1.6μm, in 14th Coherent Laser Radar Conference: Colorado.

[8] Ando, T., et al., 2008:All-fiber Coherent Doppler LIDAR for wind sensing. Materials Research Society. 1076: p. 1076-K04-05.

[9] Kameyama, S., et al., 2007:Compact all-fiber pulsed coherent Doppler lidar system for wind sensing. Applied Optics. 46(11): p. 1953-1962.

[10] Ando, T., S. Kameyama, and Y. Hirano, 2008:All-fiber Coherent Doppler LIDAR technologies at Mitsubishi Electric Corporation, in 14th International Symposium for the Advancement of Boundary Layer Remote Sensing IOP.

[11] Zhang, X., et al., 2014:Single-frequency polarized eye-safe all-fiber laser with peak power over kilowatt. Applied Physics B. 115(1): p. 123-127.

CURRENT APPLICATIONS OF SCANNING COHERENT DOPPLER LIDAR IN WIND ENERGY INDUSTRY

R Krishnamurthy[1], M Boquet[1], E Osler [2*]
[1] LEOSPHERE SAS, 14 Rue Jean Rostand, 91400, France
[2] Renewable NRG Systems, Vermont, USA, *Email: evan@nrgsystems.com

ABSTRACT

Scanning Doppler Lidars have become more prominent in the wind energy industry for a variety of applications. Scanning Lidar's provide spatial variation of winds and direction over a large area, which can be used to assess the spatial uncertainty of winds and analyze complex flows. Due to the recent growth in wind energy, wind farms are being built in complex terrain areas and fine tuning of the existing wind farms for optimized performance have gained significant interest. Scanning Lidar is an ideal tool for improved assessment of flow over complex terrains and wake characterization of large wind farms. In this article, the various applications of Lidar in the wind industry are discussed and results from several campaigns conducted in US and Europe is presented. The conglomeration of results provided in this article would assist wind energy developers and researchers in making improved decisions about their wind farm operations and pre-construction analysis using scanning Lidar's.

1. EMERGING OPPORTUNITIES IN WIND ENERGY

Wind energy is one of the leading contenders for providing clean renewable energy. In the United States, 35% of new installed power over the last five years was from wind energy (AWEA 2012). By the end of 2012, the U.S. had 60 GW ($120 billion investment) of installed wind power. Because of this large and increasing national investment, cutting-edge developments in the remote sensing industry have gained significant interest to optimize the performance of existing and future wind farms. Traditionally, wind measurements have been carried out using meteorological towers which provide points measurements at hub heights. With the increase in rotor swept area and hub heights, the impact of wind shear and turbulence intensity on turbines become dominant and point measurements from meteorological towers are no longer a good representation of the wind interacting with the turbine [1]. In addition, wakes from large wind turbine designs have introduced several challenges in wind farm design. Remote sensing techniques – especially scanning coherent Doppler LIDAR (LIght Detection And Ranging), can bridge the gap by providing high-resolution velocity measurements in three-dimensional (3D) space.

The ability of scanning Lidar to produce data rich 3D measurements over a wind farm site helps improve our understanding of the fluid dynamical processes occurring over a wide range of scales. This leads to accurate characterization of wind turbine inflow parameters and allows better understanding different atmospheric conditions and wind events which influence wind power output. In addition, accurate measurements of the boundary layer at a wind farm site is required to properly characterize local boundary layer events such as the low-level jet, wind speed and direction shear, which are important for effective wind resource assessment. Coherent Doppler lidars have been used extensively for boundary layer studies [2], mountain meteorology [3, 4], wind resource assessment [5] and dispersion modeling [6]. This paper outlines the key results from various scanning Lidar deployments for wind energy applications. These results showcase the ability of the Lidar remote sensing platform to play a pivotal role in various aspects of wind farm design.

2. CURRENT APPLICATION OF SCANNING LIDARS

2.1 ONSHORE WIND RESOURCE ASSESSMENT

Accurate wind resource assessment requires spatial measurements of wind speed

 DOI: 10.1051/epjconf/201611910003

and direction at hub-height over the complete wind farm domain and the vertical structure over the turbine swept area. Wind farm resource assessment was carried out using 3D scanning lidar measurements performed at a wind development site in USA. The site-setup and initial results are shown below [7].

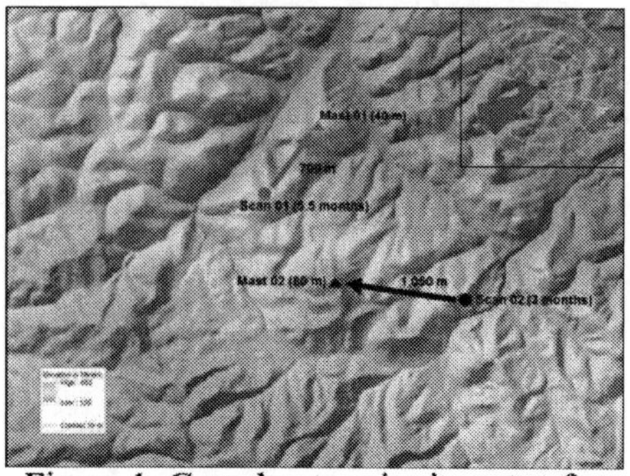

Figure 1. Complex terrain site setup for scanning Lidar site assessment.

The lidar measurements for several month were helpful in understanding the vertical and spatial structure of the wind field. Coherent Doppler lidar enables calculations of the wind speed distributions as a function of space and height which can be used to estimate the wind energy density for a given site.

Table 1. Wind speed correlation and bias comparison with tower measurements

Wind speed at 40 m	**Slope**	**R^2**	**Bias**
	0.99	94%	-1.8%

2.2 OFFSHORE WIND RESOURCE ASSESSMENT

The relatively long range of the lidar and the wind farm location along the coast allows characterization of land-sea interactions. The ability of the Doppler lidar to make measurements over the sea surface provides a strategic advantage due to the prohibitive cost of setting up meteorological towers offshore (over $1Million). Spatial variation of wind speed and direction were measured using a three-dimensional scanning Doppler lidar near a confidential coast in Europe. The analysis demonstrates the ability of the 3D scanning lidars to perform effective resource assessment for near-shore wind farms through measurement of the horizontal and vertical flow structure. Figure 2 prior onshore verification of the scanning Lidar being conducted in Germany by DNV GL. Initial results of the verification process are also shown in Table 2.

The results from the verification study conducted are shown in the tables below.

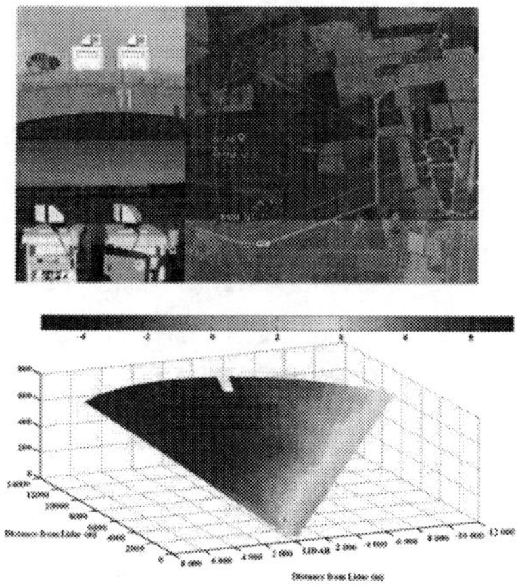

Figure 2. Site setup of the onsite verification of scanning Lidar by DNV GL.

U12 PLC4 - Wind Direction all WS > 3 m/s	Slope	R^{2**}	Mean WD Difference
(i) Along (+/- 45°)	0.999	0.995	-2.742
(ii) Orthogonal (+/- 45°)	1.014	0.998	-1.924
(iii) All directions	1.003	0.998	-2.325

U12 PLC4 - Wind Speed all > 2 m/s	Mean error / [m/s]	STD error / [m/s]	Slope	R^{2**}
(i) Along (+/- 45°)	0.24	0.32	0.991	0.976
(ii) Orthogonal (+/- 45°)	0.32	0.40	0.982	0.966
(iii) All directions	0.28	0.36	0.986	0.971

Table 2. Direction (top) and wind speed (bottom) comparison to tower measurements on site at 80 m.

2.3 WIND TURBINE/WIND FARM WAKE

Large wind farms have been experiencing lower Annual Energy Production (AEP), due to lack of complete information of wind farm or wind turbine wakes. Scanning Lidar's provide 3D assessment of wakes and help fine tune the wind farm for optimal performance. Several campaigns have been conducted in offshore and onshore wind farms to assess the wake behavior. OWA (Offshore Wind Accelerator) is an industry funded program in the United Kingdom aiming to reduce the cost of offshore wind energy by 10%. This project includes 4 major research areas, one of which is characterizing wake effects. Scanning Lidar's were used to:

1. Measure turbine inflow and outflow parameters
2. Validate advanced wake models
3. Characterize single turbine wakes
4. Characterize wind farm wakes

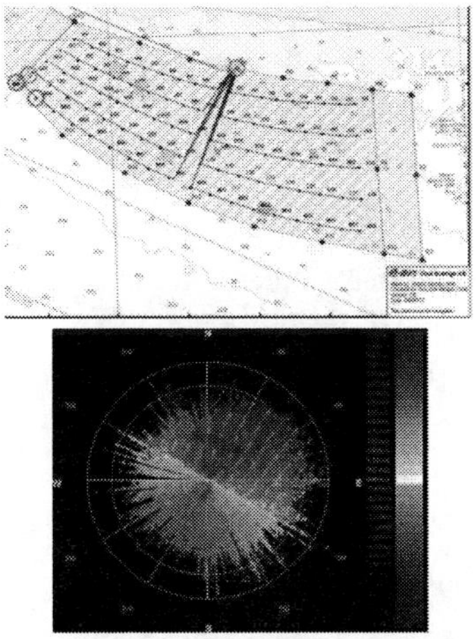

Figure 3. Site setup and measurements observed from scanning Lidar. Wind turbine wakes can be observed in the right figure.

3 CONCLUSIONS

Scanning Lidar's provide a new and attractive tool for wind energy applications. The frequent use of scanning Lidar's for wind energy applications for WRA, Power Curve and wake measurements have improve the AEP of wind farms. In this presentation, details on the impact of AEP for the above mentioned projects would be discussed.

REFERENCES

[1] Wagner et al., Rotor equivalent wind speed for power curve measurements, 2009, Journal of Physics Conference.

[2] Pichugina et al., Doppler Lidar-Based Wind-Profile Measurement System for Offshore Wind-Energy and Other Marine Boundary Layer Applications, 2010, Journal of Applied Meteorology and Climatology.

[3] Krishnamurthy et al., Wind Turbulence estimates in a complex terrain using Scanning Doppler lidar, 2011, Meteorological Applications

[4] Choukulkar et al., Investigation of a complex nocturnal flow in Owens valley, California using coherent Doppler lidar, 2012, Boundary Layer Meteorology.

[5] Krishnamurthy et al., 2012, Wind farm characterization using scanning Doppler lidar, 2013, Wind Energy.

[6] Collier et al., Dual-Doppler lidar measurements for improving dispersion models, 2005, B AM METEOROL SOC, 86, pp.825.

[7] Seig et al., Scanning Doppler lidar for reduced uncertainty, 2014, AWEA Wind Resource Assessment conference.

LIDAR UNCERTAINTY MEASUREMENT EXPERIMENT (LUMEX) – UNDERSTANDING SAMPLING ERRORS

A. Choukulkar[1*], W.A. Brewer[2], R.M. Banta[2], M. Hardesty[1], Y. Pichugina[1], Christoph Senff[1], S. Sandberg[2], A. Weickmann[1], B. Carroll[3], R. Delgado[3], A. Muschinski[4]

[1]*Cooperative Institute for Research in Environmental Sciences, Boulder*
[2]*Chemical Sciences Division, National Oceanic and Atmospheric Administration, Boulder*
[3]*University of Maryland Baltimore County, Baltimore County*
[4]*North-West Research Associates, Boulder*

ABSTRACT

Coherent Doppler LIDAR (Light Detection and Ranging) has been widely used to provide measurements of several boundary layer parameters such as profiles of wind speed, wind direction, vertical velocity statistics, mixing layer heights and turbulent kinetic energy (TKE). An important aspect of providing this wide range of meteorological data is to properly characterize the uncertainty associated with these measurements.

With the above intent in mind, the Lidar Uncertainty Measurement Experiment (LUMEX) was conducted at Erie, Colorado during the period June 23rd to July 13th, 2014. The major goals of this experiment were the following:

1. Characterize sampling error for vertical velocity statistics
2. Analyze sensitivities of different Doppler lidar systems
3. Compare various single and dual Doppler retrieval techniques
4. Characterize error of spatial representativeness for separation distances up to 3 km
5. Validate turbulence analysis techniques and retrievals from Doppler lidars

This experiment brought together 5 Doppler lidars, both commercial and research grade, for a period of three weeks for a comprehensive inter-comparison study. The Doppler lidars were deployed at the Boulder Atmospheric Observatory (BAO) site in Erie, site of a 300 m meteorological tower. This tower was instrumented with six sonic anemometers at levels from 50 m to 300 m with 50 m vertical spacing.

A brief overview of the experiment outline and deployment will be presented. Results from the sampling error analysis and its implications on scanning strategy will be discussed.

1. INTRODUCTION

Every measurement system has uncertainties associated with it and the coherent Doppler lidar is no exception. Understanding these uncertainties is essential not only to improve the measurement technique, but also to ensure proper interpretation of the observations.

The uncertainties associated with Doppler lidar measurements can be split in to the following components:

1. Instrument error
2. Sampling error
3. Error of spatial representativeness
4. Vector retrieval error

Each of these errors are explain briefly below.

Instrument Error:

Instrument error or instrument noise can be described as the error associated with making the radial velocity measurement. Instrument error arises due to imperfections of the instrument such as limited spectral resolution of the detector, frequency chirp in the transmitted laser signal, and detector noise, among others (for further details, see: Rye and Hardesty 1993; Dabas et al., 1998; Frehlich 2004).

Sampling Error:

The sampling error can be divided into two parts: (a) Instrument sampling error and (b) Statistical sampling error. Instrument sampling error occurs due to the finite spatial and temporal resolution of the measurement which results in some scales of motion being misrepresented. Statistical sampling error occurs when statistics estimated from measurements do not represent true ensemble estimates.

Error of Representativeness:

Doppler lidar measurements represent a spatial and temporal average depending on the scanning strategy and retrieval technique used. However, correct interpretation of this spatial and temporal footprint is not trivial and hence introduces an error of representativeness in the measurements.

Vector Retrieval Error:

Coherent Doppler lidars measure the projection of the wind along the line of sight of the beam known as the line of sight (LOS) velocity or radial velocity. Various single and dual Doppler techniques can be used to estimate the two-dimensional wind vector from this radial velocity measurement. The error associated with making this estimation is termed vector retrieval error.

In this presentation, measurements made during LUMEX will be used to present an analysis of the sampling error commonly introduced when estimating vertical velocity variance. This quantity is an important variable in studies related to mass flux transport and pollutant dispersion.

2. METHODOLOGY

The Lidar Uncertainty Measurement Experiment (LUMEX) was conducted at the Boulder Atmospheric Observatory (BAO) in Erie CO. The BAO site is the location of a 300 m met mast which was instrumented with three-axis sonic anemometers from 50 m to 300 m above ground at 50 m intervals. In addition, two scanning Doppler lidars (one NOAA's High Resolution Doppler Lidar (HRDL) and one University of Maryland Baltimore County (UMBC), Leosphere 200S) and three profiling Doppler lidars were also deployed at various location in and around the BAO facility. The location of the instrumentation is shown in Figure 1.

This experiment had two stages. The first stage was calibration and sampling error evaluation and the second stage was vector retrieval uncertainty and error of spatial representativeness evaluation.

Figure 1. Instrument locations during LUMEX field campaign. The inner red circle is of radius 500 m centered on the tower while the outer circle is 1000 m radius centered on the tower.

During the first stage of the experiment, the two scanning Doppler lidars (HRDL and UMBC 200S) were set to stare vertically for 2 full days. In this presentation, the data from these two days will be used to present an analysis on the sampling error for the vertical velocity variance.

3. RESULTS

Figure 2 shows the vertical velocity measurements over a three hour period on June 29th, 2014. The period is from 1900 UTC to 2200 UTC (1300 hrs to 1600 hrs local time). We see that during this period, the boundary layer is dominated by convective motions and could be called quasi-stationary state.

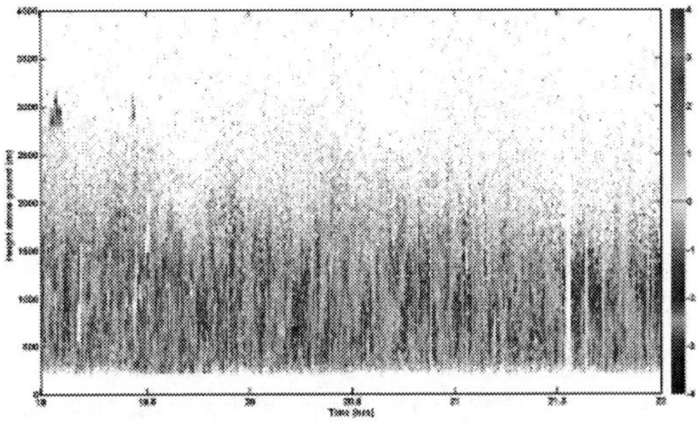

Figure 2. Vertical velocity measurements from June 29th, 2014. Warmer colors represent updrafts and cooler colors represent downdrafts.

From this data, vertical velocity variance is estimated for different averaging periods. The sampling error (Lenschow et al., 1994; 2000) for the different variance values is also estimated and plotted in Figure 3.

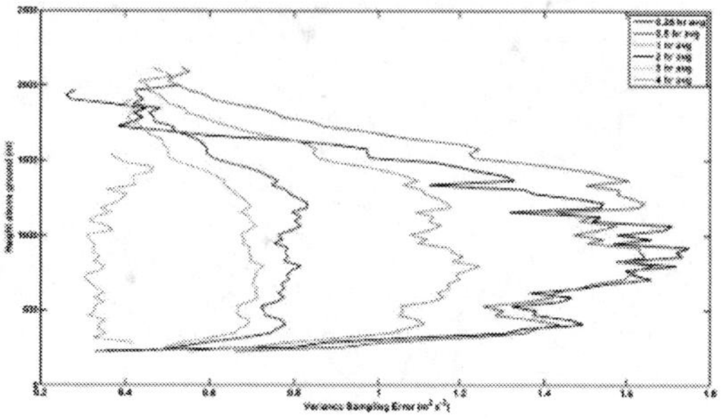

Figure 3. Vertical velocity variance sampling error for different averaging periods ranging from 15 minutes to 4 hours.

As seen from Figure 3, the sampling error reduced steadily as the averaging time is increased. This is consistent with the definition of the ensemble average for a stationary process. However, cases of stationary processes do not always occur in the atmospheric boundary layer. It is common to see large scale features like convective rolls or fronts that have time scales on the order of hours. One such large scale feature is shown in Figure 4.

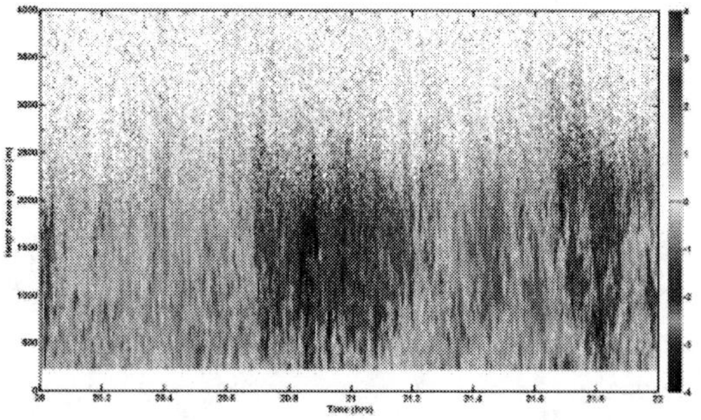

Figure 4. Vertical winds measured on June 30th, 2014 from 2000 UTC to 2200 UTC (1400 hrs to 1600 hrs local time)

The sampling errors associated with the vertical velocity variance are shown in Figure 5. It is seen that the smaller time averages have a smaller sampling error associated with them. But this is due to the smaller time averages not capturing the large scale feature completely. It is only when we do a one hour average see the magnitude of our sampling error. Therefore, it is important to note that smaller sampling errors do not always imply that the process has been sampled correctly.

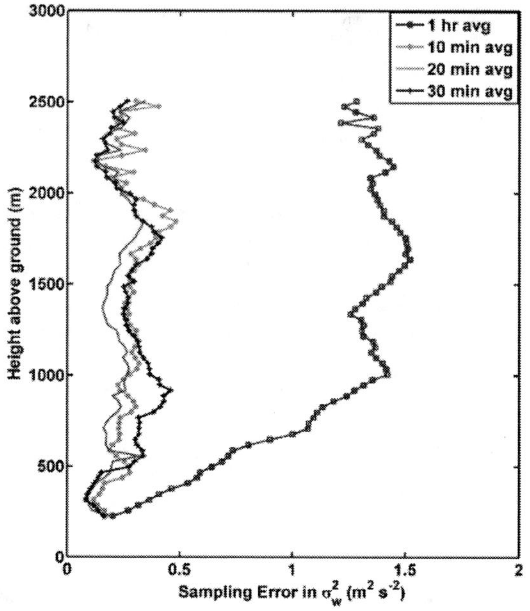

Figure 5. Vertical velocity sampling errors for data collected on June 30th, 2014 from 2000 UTC to 2200 UTC.

4. CONCLUSIONS

The LUMEX field campaign allows a unique opportunity to evaluate the different sources of uncertainty in Doppler lidar measurements. In this presentation, an overview of the sampling error analysis will be presented. In addition, ongoing work to incorporate this analysis in real-time measurements will be presented.

This sampling error analysis is being extended to horizontal velocity retrievals as well in a subsequent field deployment Experimental Measurement Campaign for Planetary Boundary Layer Instrumentation Analysis (XPIA) conducted in March-April, 2015.

ACKNOWLEDGEMENT

The authors would like to thank Daniel Wolfe (Physical Sciences Division, NOAA) for his assistance in deployment of this instrumentation at the BAO facility. In addition, the authors would like to thank our collaborators in this field campaign, Julie Lundquist (University of Colorado Boulder) and Keith Barr (Lockheed Martin Coherent Technologies) for bringing their

DOI: 10.1051/epjconf/201611910004

instrumentation and lending their expertise in this study.

REFERENCES

[1] Rye B.J., Hardesty R.M. (1993): Discrete spectral peak estimation in incoherent backscatter heterodyne lidar. II. Correlogram accumulation. *IEEE Transactions Geoscience and Remote Sensing*, **31**, 28–35.

[2] Dabas A.M., Drobinski P., Flamant P.H. (1998): Chirp-induced bias in velocity measurements by a coherent Doppler CO2 lidar. *Journal of Atmospheric and Oceanic Technology*, **15**, 407–415.

[3] Frehlich R. (2004): Velocity Error for Coherent Doppler Lidar with Pulse Accumulation. *Journal of Atmospheric and Oceanic Technology*, **21**, 905-920.

[4] Lenschow D., V. Wulfmeyer, and C. Senff, (2000): Measuring second through fourth-order moments in noisy data. *J. Atmos. Oceanic Technol.*, **17**, 1330–1347.

CHARACTERIZATION OF TURBULENT PROCESSES BY THE RAMAN LIDAR SYSTEM BASIL IN THE FRAME OF THE HD(CP)^2 OBSERVATIONAL PROTOTYPE EXPERIMENT – HOPE

Paolo Di Girolamo[1]*, Donato Summa[1], Dario Stelitano[1], Marco Cacciani[2], Andrea Scoccione[2], Andreas Behrendt[3], Volker Wulfmeyer[3]

[1]*Scuola di Ingegneria, Università degli Studi della Basilicata, Viale dell'Ateneo Lucano n. 10, 85100 Potenza – Italy , *Email: digirolamo@unibas.it*

[2]*Dipartimento di Fisica, Università di Roma "Sapienza", Piazzale Aldo Moro, n. 2, 00100 Roma – Italy*

[3]*Institut fuer Physik und Meteorologie, Universitaet Hohenheim, Hohenheim – Germany*

ABSTRACT

Measurements carried out by the Raman lidar system *BASIL* are reported to demonstrate the capability of this instrument to characterize turbulent processes within the Convective Boundary Layer (CBL). In order to resolve the vertical profiles of turbulent variables, high resolution water vapour and temperature measurements, with a temporal resolution of 10 sec and a vertical resolution of 90 and 210 m, respectively, are considered. Measurements of higher-order moments of the turbulent fluctuations of water vapour mixing ratio and temperature are obtained based on the application of spectral and auto-covariance analyses to the water vapour mixing ratio and temperature time series. The algorithms are applied to a case study (IOP 5, 20 April 2013) from the HD(CP)2 Observational Prototype Experiment (HOPE), held in Central Germany in the spring 2013. The noise errors are demonstrated to be small enough to allow the derivation of up to fourth-order moments for both water vapour mixing ratio and temperature fluctuations with sufficient accuracy.

1. INTRODUCTION

Measurements of higher-order moments of moisture and temperature fluctuations provide unique and essential information for the characterization of turbulent processes within the convective boundary layer (CBL). Water vapour variance is a key parameter in many turbulence, convection, and cloud parameterizations. Within the CBL, water vapour variance increases with height, achieving a maximum at the top of the CBL due to the mixing of moist air in the updrafts with the drier air from above the CBL ([1];[2]). Moreover, water vapour skewness and kurtosis are found to be characterized by an appreciable vertical variability within the CBL, which changes patterns during different phases of the CBL evolution.

Lidar systems, based on their capability to provide high space and time resolution and accurate measurements of atmospheric water vapour and temperature, have nowadays reached the level of maturity needed to investigate the relevant atmospheric processes and enable measurements of turbulent variables within the CBL (among others, [3] [4]). The major advantage of the lidar techniques is represented by its capability to characterize turbulent variables from the proximity of the surface up to interfacial layer and above.

In the present paper we report what we believe are the first simultaneous and co-located measurements of water vapour and temperature turbulent fluctuations performed by a single lidar system. These measurements have been carried out by the Raman lidar system *BASIL* exploiting its capability to perform high-resolution and accurate measurements of atmospheric temperature and water vapour, both in daytime and night-time, based on the application of the rotational and vibrational Raman lidar techniques in the UV, respectively ([5];[6]). Measurements of turbulent variables throughout the CBL by water vapour Raman lidar had been demonstrated to be possible by Wulfmeyer *et al.* [7] based on the use of the data from the Atmospheric Radiation Measurement (ARM) Raman lidar. However, these authors came to the conclusion that the noise errors affecting the ARM Raman lidar water vapour mixing ratio measurements are too large to derive fourth-order moments with sufficient accuracy. Thus, to the best of our knowledge, *BASIL* is the first Raman lidar with a demonstrated capability to retrieve daytime profiles of water vapour turbulent fluctuations up to the fourth order throughout the atmospheric CBL, this capability being combined with the one to also measure daytime profiles of temperature fluctuations up to the fourth order.

2. METHODOLOGY

BASIL is a ground-based Raman Lidar hosted in a transportable sea-tainer. The major feature of *BASIL* is represented by its capability to perform high-resolution and accurate measurements of atmospheric temperature and water vapour, both in daytime and night-time, based on the application of the rotational and vibrational Raman lidar techniques in the UV. Besides temperature and water vapour, *BASIL* is also capable of providing measurements of particle backscatter at 355, 532 and 1064 nm, particle extinction at 355 and 532 nm and particle depolarization at 355 and 532 nm ([8]).

The accuracy of the measurements performed by BASIL is high enough to allow the retrieval of vertical profiles throughout the atmospheric CBL of the two turbulent variables (namely, water vapour mixing ratio and temperature) up to the fourth order during daytime. Results from this system are obtained based on the application of the approach introduced by Lenschow et al. [9], which allows to estimate higher-order moments of turbulent variables in the presence of noisy data.

As the atmospheric variance and the noise variance a measured atmospheric variable are uncorrelated, the total variance Var_m can be expressed as [9] $Var_m = Var_a + Var_n$, with Var_a being the atmospheric variance and Var_n being the noise variance.

Different procedures may be considered to separate atmospheric variance from the noise variance in the total measured variance. The auto-covariance method is probably the most effective and straightforward among these procedures. This method is based of the consideration that atmospheric fluctuations are correlated in time, while instrumental noise fluctuations are uncorrelated [9]. This approach determines the atmospheric variance based on the computation of the auto-covariance function for the considered atmospheric variable and then extrapolating this function to zero lag based on the application of a power-law fit. As specified in Lenschow *et al.* [9], the auto-covariance function at zero lag represents the total measured variance and, consequently, the noise variance can be determined as the difference between the auto-covariance function extrapolated to zero lag and its value at zero leg.

3. RESULTS

In the selection of the case study to be included in this work, attention was paid on identifying weather conditions characterized by the presence of a consolidated, well-mixed and quasi-stationary CBL. Typically time segments with a duration of 1-2 hours are considered as in fact for longer periods the CBL can no longer be considered as being quasi-stationary, while the consideration of shorter periods would reduce the number of sampled thermals and thus increase the sampling errors.

Figure 1 illustrates the time-height plot of the particle backscatter coefficient at 1064 nm, □β_{par}, between 11:30 and 13:30 UTC, on 20 April 2013, revealing the presence of a significant boundary layer aerosol loading (with values of □β_{par} in excess of $2x10^{-5}$ m^{-1} sr^{-1}), which testifies the presence of a well-mixed and quasi-stationary CBL at this time of the day, extending up to a height of approximately 1300 m. The figure reveals the alternation of updrafts associated with thermals of warm air rising from the ground and downdrafts associated with thermals of cool air sinking from the free troposphere. The evolution of z_i (black line in figure 1), determined with a conventional approach based on the detection of the strongest gradient of the aerosol backscatter signal, is found to be characterized by a limited time variability, with a mean value during the observation period of 1290 m a.g.l and a standard deviation of 77 m.

Figure 2 illustrates the time-height cross section of the water vapour mixing ratio (upper panel) and temperature (lower panel) in the same time interval considered in figure 1. The upper panel of figure 3 reveals the variability of the water vapour mixing ratio within the CBL associated with the presence of alternating updrafts and downdrafts. The largest variability is observed in the interfacial layer, which is characterized by the

DOI: 10.1051/epjconf/201611910005

penetration of the warm humid air rising from the ground and the entrainment of cool dry air sinking from the free troposphere. As for water vapour, the largest variability temperature profile (lower panel of figure 2) is observed in the interfacial layer, possibly as a results of sensible heat entrainment fluxes.

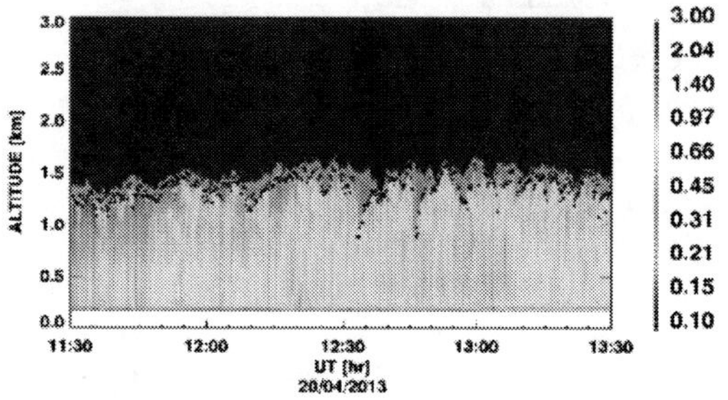

Figure 1: Time-height cross-section of β_{par} between 11:30 and 13:30 UTC on 20 April 2013. The black line in the figure identifies the CBL height z_i.

In order to characterize the quality of water vapour mixing ratio and temperature measurements, an accurate assessment of noise error profiles is necessary. Profiles of noise error affecting water vapour mixing ratio and temperature measurements are illustrated in figure 3, this being quantified as the root-square of the noise variance. For the vertical and temporal resolution selection considered for the turbulence measurements, i.e. 90 m/200m and 10 s, respectively, the statistical error affecting water vapour mixing ratio measurements is smaller than 0.06 g/kg up to 1.4 km, while the statistical error affecting temperature measurements is smaller than 2 K up to 1.3 km.

Figure 4 shows the vertical profiles of atmospheric and total variance for water vapour mixing ratio (upper panel) and temperature (lower panel). A maximum of the water vapour mixing ratio variance profile is observed in the interfacial layer at 1260 m, i.e. 0.29 g^2/kg^2 with 0.013 g^2/kg^2 and 0.032 g^2/kg^2 for the sampling error and noise error, respectively. The near-zero values in the lower CBL indicate weak surface forcing. In the interfacial layer, the variance reaches a maximum as a result of the large variability produced by the vertical exchange associated with the strong updrafts and downdrafts [9]. For what concerns the temperature variance, a maximum of 0.4 K^2 is observed at 1310 m. Larger values of the temperature variance in the interfacial layer are the result of entrainment [1].

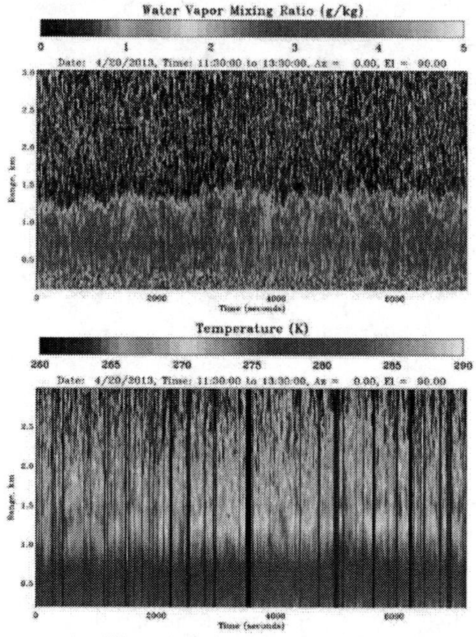

Figure 2: Time-height cross section of water vapour mixing ratio (upper panel) and temperature (lower panel) in the same time interval considered in figure 1.

Vertical profiles of the third-order moment for water vapour mixing ratio and temperature have also been determined, but are not shown here and will be illustrated at the Conference. Values of the third order moment of water vapour mixing ratio fluctuations are found to be negative between 900 and 1290 m, with a negative peak value of -0.029±0.005 g^3/kg^3 at 1140 m, while a large positive peak is observed just above CBL top, with a maximum of 0.156±0.009 g^3/kg^3 at 1380 m. The third order moment of temperature fluctuations is characterized by a peak of -0.11±0.01 K^3 around the top of the CBL at 1280 m. Water vapour mixing ratio fourth-order moment, also determined but not shown here, is almost zero up to ≅ 750 m (i.e. z/z_i =0.6), is characterized by a maximum of 0.28±0.13 g^4/kg^4 around the top of the CBL (at 1350 m), while temperature fourth-order moment, also not shown, shows a maximum of 0.36±0.10 K^4 around the top of the CBL (at 1310 m).

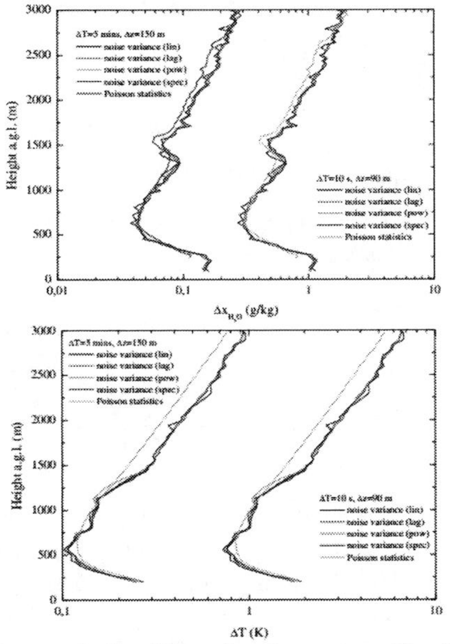

Figure 3: Profiles of noise error affecting water vapour mixing ratio (upper panel) and temperature (lower panel) measurements.

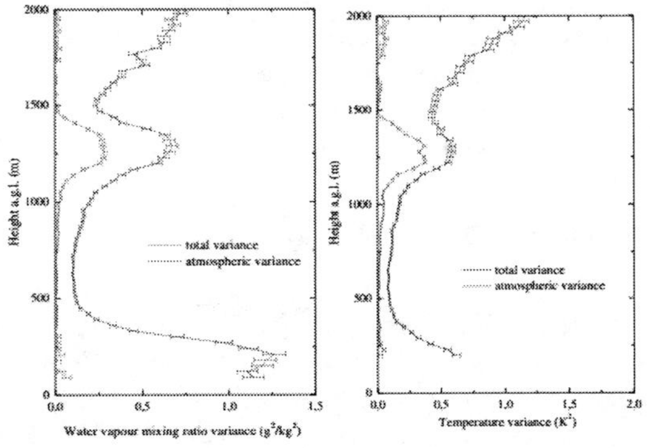

Figure 4: Vertical profiles of atmospheric and total variance for water vapour mixing ratio (left panel) and temperature (right panel).

4. CONCLUSIONS

This paper provides a detailed characterization of the performances of the Raman lidar *BASIL* and demonstrates that profiles of turbulent variables can be determined throughout the CBL with sufficient accuracy. For this purpose measurements from the HD(CP)2 Observational Prototype Experiment (HOPE), held in Central Germany in the spring 2013, are considered.

REFERENCES

[1] Wulfmeyer, V., 1999: Investigation of turbulent processes in the lower troposphere with water-vapour DIAL and radar-RASS, *J. Atm. Sci.*, **56**, 1055–1076.

[2] Wulfmeyer, V., 1999: Investigations of humidity skewness and variance profiles in the convective boundary layer and comparison of the latter with large eddy simulation results, *J. Atm. Sci.*, **56**, 1077–1087.

[3] Frehlich, R., L. Cornman, 2002: Estimating spatial velocity statistics with coherent Doppler lidar, *J. Atmos. Ocean. Tech.*, **19**, 355–366.

[4] Behrendt, A., V. Wulfmeyer, E. Hammann, S. K. Muppa, and S. Pal: 2014, Profiles of second- to third-order moments of turbulent temperature fluctuations in the convective boundary layer: first measurements with Rotational Raman Lidar", *Atmos. Chem. Phys. Discuss.*, **14**, 29019-29055.

[5] Di Girolamo, P., R. Marchese, D. N. Whiteman, B. B. Demoz, 2004: Rotational Raman Lidar measurements of atmospheric temperature in the UV. *Geophys. Res. Lett.*, **31**, doi: 10.1029/2003GL018342.

[6] Di Girolamo, P., D. Summa, R. Ferretti, 2009: Multiparameter Raman Lidar Measurements for the Characterization of a Dry Stratospheric Intrusion Event, *J. Atmos. Ocean. Tech.*, **26**, 1742-1762.

[7] Wulfmeyer, V., D. D. Turner, S. Pal, E. Wagner, 2010: Can water vapour raman lidar resolve profiles of turbulent variables in the convective boundary layer ?, *Bound.-Lay. Meteorol.*, doi:10.1007/s10546-010-9494-z.

[8] Di Girolamo, P., D. Summa, R. Bhawar, T. Di Iorio, M. Cacciani, I. Veselovskii, O. Dubovik, A. Kolgotin: 2012, Raman lidar observations of a Saharan dust outbreak event: Characterization of the dust optical properties and determination of particle size and microphysical parameters, *Atm. Env.*, **50**, 66-78.

[9] Lenschow, D. H., V. Wulfmeyer., C. Senff, 2000: Measuring second-through fourth-order moments in noisy data. *J. Atmos. Ocean. Tech.*, **17**, 1330–1347.

DOI: 10.1051/epjconf/201611911001

Cloud-Aerosol Interactions: Retrieving Aerosol Ångström Exponents from CALIPSO Measurements of Opaque Water Clouds

Mark Vaughan[1]*, Zhaoyan Liu[1,2], Yong-Xiang Hu[1], Kathleen Powell[1], Ali Omar[1], Sharon Rodier[1,2], William Hunt[1,2], Jayanta Kar[1,2], Jason Tackett[1,2], Brian Getzewich[1,2], Kam-Pui Lee[1,2]

*[1]NASA Langley Research Center, Hampton, VA 23681, USA, *Email: mark.a.vaughan@nasa.gov*

[2]Science Systems and Applications, Inc. (SSAI), Hampton, VA 23666, USA

ABSTRACT

Backscatter and extinction from water clouds are well-understood, both theoretically and experimentally, and thus changes to the expected measurement of layer-integrated attenuated backscatter can be used to infer the optical properties of overlying layers. In this paper we offer a first look at a new retrieval technique that uses CALIPSO measurements of opaque water clouds to derive optical depths and Ångström exponents for overlying aerosol layers.

1. INTRODUCTION

Hu et al., 2007 [1] introduced a technique to retrieve optical depths for aerosol layers lying above opaque water clouds (OWC). For a nadir-viewing, polarization-sensitive lidar operating at wavelength λ, the layer integrated attenuated backscatter for water clouds, $\gamma'_{w,\lambda}$, is given by [1]

$$\gamma'_{w,\lambda} = \exp\left(-2\int_0^{z_{top}} \sigma_{a,\lambda}(r)\,dr\right)\int_{z_{top}}^{z_{base}} B_\lambda(r)\,dr$$
$$= T^2_{a,\lambda}(0, z_{top})\frac{1 - T^2_{w,\lambda}(z_{top}, z_{base})}{2\eta_{w,\lambda}S_{w,\lambda}}. \quad (1)$$

In this expression, $B_\lambda(r)$ is the measured attenuated total backscatter coefficient at range r with corrections applied for molecular and ozone attenuation. $T^2_{a,\lambda}(0, z_{top})$ is the aerosol two-way transmittance between the lidar and cloud top. Similarly $T^2_{w,\lambda}(z_{top}, z_{base})$ is the two-way transmittance of the water cloud between z_{top} and z_{base}. $S_{w,\lambda}$ is the water cloud lidar ratio, which can either be established empirically or derived theoretically from Mie calculations, and $\eta_{w,\lambda}$ is the water cloud layer-effective multiple scattering factor, which can be obtained from the cloud layer-integrated volume depolarization ratio, $\delta_{w,\lambda}$. For OWCs, $T^2_{w,\lambda}(z_{top}, z_{base}) = 0$, and thus the aerosol optical depth (AOD) for aerosol layers lying above OWCs can be retrieved using [1]

$$\tau_{a,\lambda} = -\frac{1}{2}\ln\left(2S_{w,\lambda}\gamma'_{w,\lambda}\eta_{w,\lambda}\right)$$
$$= -\frac{1}{2}\ln\left(2S_{w,\lambda}\gamma'_{w,\lambda}\left(\frac{1-\delta_{w,\lambda}}{1+\delta_{w,\lambda}}\right)^2\right) \quad (2)$$

Since its introduction, the OWC retrieval technique has been used extensively with CALIPSO data to, among other things, investigate "direct climate forcing by absorbing aerosols located above boundary layer clouds" [2] and to characterize lidar ratios of dust and smoke transported from Africa out over the Atlantic Ocean [3]. However, while the CALIPSO lidar makes backscatter measurements at both 532 nm and 1064 nm, depolarization is only measured in the 532 nm channel [4], and thus the Hu OWC technique cannot be applied to CALIPSO 1064 nm data.

2. 1064 nm OWC RETRIEVAL TECHNIQUE

Figure 1 shows CALIPSO measurements acquired off the southwest coast of Africa on 12 July 2008. On the left side of the images a smoke plume is seen lying over a largely opaque stratus deck. The smoke layer is not present on the right side. The presence/absence of the smoke layer is clearly reflected in the magnitude of the OWC attenuated backscatter color ratio (1064 nm/ 532 nm) shown in the lower panel. In the clear air regions south of ~22°S, the OWC color ratios lie between 1.1 and 1.2, whereas beneath the smoke layer (e.g., at ~9.7°S) the OWC color ratios can exceed 1.6. The higher OWC color ratios occur because smoke attenuates the 532 nm signal much more rapidly than it does the 1064 nm signal. We can quantify the attenuation differences using layer-integrated attenuated color ratio, χ'; i.e.,

DOI: 10.1051/epjconf/201611911001

$$\chi' = \frac{\gamma'_{w,1064}}{\gamma'_{w,532}} = \frac{T^2_{a,1064}(0,z_{top})\int_{z_{top}}^{z_{base}} B_{1064}(z)dz}{T^2_{a,532}(0,z_{top})\int_{z_{top}}^{z_{base}} B_{532}(z)dz}. \quad (3)$$

In clear skies having negligible particulate loading above an OWC (e.g., as in the right side of Figure 1), $T^2_{p,532}(0, z_{top}) = T^2_{p,1064}(0, z_{top}) = 1$ and thus

$$\chi'_{clear} = \int_{z_{top}}^{z_{base}} B_{1064}(z)dz \Big/ \int_{z_{top}}^{z_{base}} B_{532}(z)dz. \quad (4)$$

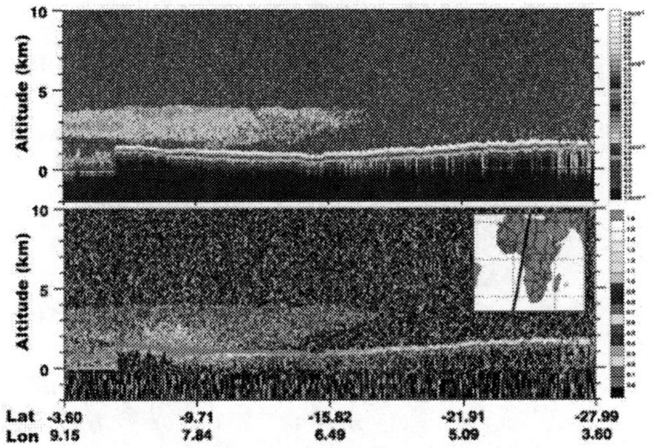

Figure 1: Version 4.00 CALIPSO nighttime data acquired 12 July 2008; upper panel shows 532 nm attenuated backscatter coefficients while the lower panel shows the attenuated backscatter color ratios (i.e., the 1064 nm attenuated backscatter coefficients divided by the 532 nm attenuated backscatter coefficients).

When aerosol layers lie above, χ'_{clear} is rescaled by the ratio of the aerosol two-way transmittances; that is

$$\chi' = \left(\frac{T^2_{a,1064}(0,z_{top})}{T^2_{a,532}(0,z_{top})}\right)\chi'_{clear}. \quad (5)$$

Using equation (5) we can derive estimates of the AOD at 1064 nm:

$$\tau_{1064} = -0.5\ln\left(T^2_{a,532}(0,z_{top})\frac{\chi'}{\chi'_{clear}}\right). \quad (6)$$

Note that the parameters on the right hand side of equation (6) are all potentially known values. χ' is a direct measurement; $\tau_{a,532}$ and hence $T^2_{a,\lambda}(0, z_{top})$ can be retrieved using the Hu OWC technique; and, as described in the following section, high quality estimates of χ'_{clear} can be obtained empirically from the 9-year CALIPSO data set.

3. CHARACTERIZING χ'_{clear} FOR OWC

Obtaining reliable measurements of χ'_{clear} depends on several factors: cloud top and base must be accurately detected and $\gamma'_{w,532}$ must be corrected for contributions from molecular scattering (e.g., as described in [5]); even faint aerosol loading of the atmosphere above OWC must be confidently identified and rigorously excluded; and, most importantly, the signals in all channels must be accurately calibrated. To best satisfy the final requirement, in this study we use the CALIPSO version 4.00 level 1 nighttime data, which delivers improved calibrations, especially at 1064 nm [6]. To achieve the requirements imposed by the penultimate condition, we used the CALIPSO simulator [7] to generate "known to be clear" test data, from which we established the mean and standard deviation of the 532 nm integrated attenuated backscatter within a purely molecular atmosphere as a function of altitude. To be included in the data set from which χ'_{clear} was calculated

(a) a candidate cloud must have a top below 3 km and be the only layer detected in a 5-km horizontal average;

(b) all candidate clouds must be classified as opaque, and have minimum values of $\gamma'_{w,\lambda} \geq 0.04$ sr^{-1} at both 532 nm and 1064 nm;

(c) cloud top temperature must be greater than 0°C; and

(d) the integrated attenuated backscatter of the "clear air" above cloud top must lie within ±1 standard deviation of the corresponding mean value obtained from Figure 2.

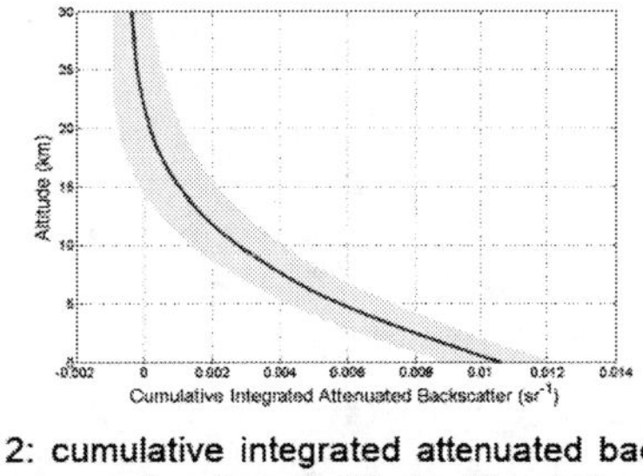

Figure 2: cumulative integrated attenuated backscatter at 532 nm as a function of altitude. The dark green line represents the mean calculated over 3404 simulated profiles. A one standard deviation uncertainty envelope is shown in pale green.

The resulting distribution of χ'_{clear} values obtained from the V4 CALIPSO data acquired globally during all of 2008 is shown in Figure 3. The distribution is both tight and symmetric: the median is essentially identical to the mean, and the standard deviation is ~4% of the mean.

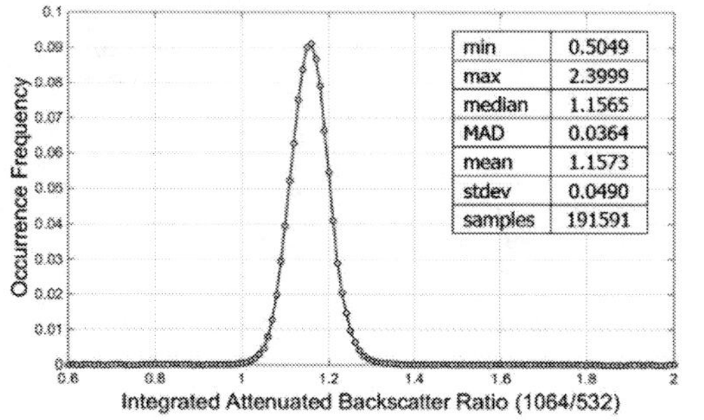

Figure 3: distribution of χ'_{clear} for opaque water clouds measured by CALIPSO during all of 2008.

Absent multiple scattering effects, χ'_{clear} is simply the ratio of water cloud lidar ratios at the two wavelengths; i.e., $\chi'_{clear} = S_{w,532} / S_{w,1064}$. However, multiple scattering can never be neglected in the analysis of CALIPSO water cloud measurements, and thus in practice

$$\chi'_{clear} = \frac{\eta_{w,532}\, S_{w,532}}{\eta_{w,1064}\, S_{w,1064}} = \left(\frac{\eta_{w,532}}{\eta_{w,1064}}\right)\left(\frac{S_{w,532}}{S_{w,1064}}\right). \quad (7)$$

4. AEROSOL ÅNGSTRÖM EXPONENTS

As a first test, we applied our retrieval scheme to nighttime data acquired during June–September 2008 over the same smoke transport region off the southwest coast of Africa (0°–20°S, 5°W–15°E) that was analyzed in [3]. Aerosol above cloud (AOC) samples were identified by applying criteria (a) and (b) used to establish χ'_{clear}. However, for the AOC data set no restrictions were placed on the cloud integrated attenuated backscatters. Furthermore, to assure the presence of a reasonable amount of aerosol loading, the 532 nm integrated attenuated backscatter above cloud top was required to exceed the mean plus four standard deviations of the corresponding mean value obtained from Figure 2.

Choosing a region largely dominated by smoke allows us to make an initial assessment of the effectiveness of the method by comparing the retrieved aerosol extinction Ångström exponents to an expected value of ~2 (e.g., see [2] and references therein). We compute the Ångström exponents, α, using the 532 nm AOD derived from the Hu OWC technique (i.e., equation (2)) and the 1064 nm AOD from equation (6):

$$\alpha = \frac{-\log\left(\tau_{1064}/\tau_{532}\right)}{\log\left(1064/532\right)}. \quad (8)$$

The Ångström exponent distribution obtained from the AOC data set is shown in Figure 4. From a proof of concept point of view, the most important characteristics of this distribution are that the mode and mean values are both ~2, which is wonderfully consistent with our initial expectations. On the other hand, the distribution is substantially broader than the distribution of χ'_{clear}, with a standard deviation approximately equal to ~60% of the mean. However, this should not be too surprising, given that the signal-to-noise ratios (SNR) within opaque water clouds (i.e., where χ'_{clear} is measured) can be two orders of magnitude larger than the typical SNR for CALIPSO aerosol measurements. Furthermore, the fact that the median absolute distance (MAD) is much smaller (at ~26% of the median) suggests that the relatively large standard deviation is influenced more by large outliers than by an overall dispersion of the values.

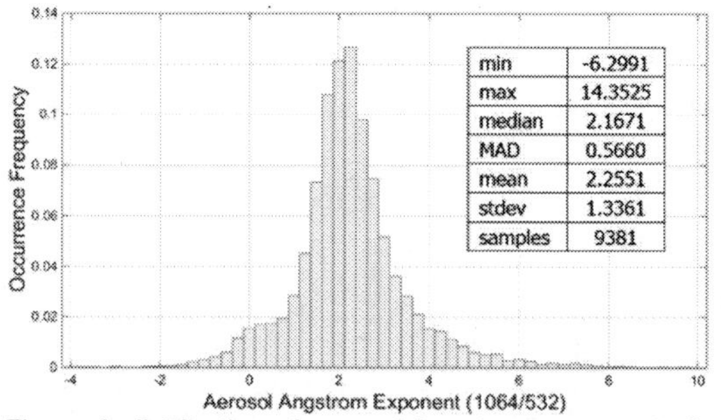

Figure 4: distribution of aerosol Angstrom exponents for June-September 2008

5. NEXT STEPS

Clearly the next step in the development of this technique will be to use the 1064 nm AODs obtained from equation (6) as constraints to retrieve estimates of 1064 nm lidar ratios (e.g., as is done at 532 nm in [3]). However, whether this can be done reliably with the standard CALIPSO data products remains to be seen, as the above-cloud 1064 nm signals will likely require considerably more spatial averaging than the 532 nm signals.

In closing, we note that this work used a research version of the CALIPSO level 2 products created by applying the CALIPSO V3.30 level 2 analysis software to the newly released V4.00 level 1 data. However, while a full suite of level 2 products was generated, this investigation did not use the CALIPSO aerosol data products. Instead, the results presented here were derived from information reported in the CALIPSO level 1B products and level 2 5-km cloud layer products: aerosol optical depths and Ångström exponents are derived solely from changes to the measured optical properties of clouds that occur in the presence of above-cloud aerosols (e.g., as seen in Figure 1).

REFERENCES

[1] Hu, Y., M. Vaughan, Z. Liu, K. Powell, and S. Rodier, 2007: Retrieving Optical Depths and Lidar Ratios for Transparent Layers Above Opaque Water Clouds From CALIPSO Lidar Measurements, *IEEE Geosci. Remote Sens. Lett.*, **4**, 523–526, doi:10.1109/LGRS.2007.901085.

[2] Chand, D., T. L. Anderson, R. Wood, R. J. Charlson, Y. Hu, Z. Liu, and M. Vaughan, 2008: Quantifying above-cloud aerosol using spaceborne lidar for improved understanding of cloudy-sky direct climate forcing, *J. Geophys. Res.*, **113**, D13206, doi:10.1029/ 007JD009433.

[3] Liu, Z., D. Winker, A. Omar, M. Vaughan, J. Kar, C. Trepte, Y. Hu and G. Schuster, 2015: Evaluation of CALIOP 532-nm AOD over Opaque Water Clouds, *Atmos. Chem. Phys.*, **15**, 1265–1288, doi:10.5194/acp-15-1265-2015.

[4] Hunt, W. H., D. M. Winker, M. A. Vaughan, K. A. Powell, P. L. Lucker, and C. Weimer, 2009: "CALIPSO Lidar Description and Performance Assessment", *J. Atmos. Oceanic Technol.*, **26**, 1214–1228, doi:10.1175/2009JTECHA1223.1.

[5] Vaughan, M. A., Z. Liu, M. J. McGill, Y. Hu, and M. D. Obland, 2010: On the Spectral Dependence of Backscatter from Cirrus Clouds: Assessing CALIOP's 1064 nm Calibration Assumptions Using Cloud Physics Lidar Measurements", *J. Geophys. Res.*, **115**, D14206, doi:10.1029/2009JD013086.

[6] Vaughan, M., A. Garnier, Z. Liu, D. Josset, Y. Hu, K.-P. Lee, W. Hunt, J.-P. Vernier, S. Rodier, J. Pelon, and D. Winker, 2012: Chaos, Consternation and CALIPSO Calibration: New Strategies for Calibrating the CALIOP 1064 nm Channel, 26th International Laser Radar Conference (ILRC), Porto Heli, Greece.

[7] Powell, K. A., Liu, Z., and W. H. Hunt, 2006: Simulation of Random Electron Multiplication in CALIPSO Lidar Photomultipliers, 23rd International Laser Radar Conference (ILRC), Nara, Japan.

EPJ Web of Conferences 119, 11002 (2016) DOI: 10.1051/epjconf/201611911002
ILRC 27

AEROSOL/CLOUD MEASUREMENTS USING COHERENT WIND DOPPLER LIDARS

Philippe Royer [1]*, Matthieu Boquet[1], Jean-Pierre Cariou[1], Laurent Sauvage[1], Rémy Parmentier[1]

*[1]LEOSPHERE, 14-16 rue Jean Rostand, 91400, Orsay, FRANCE, *Email: proyer@leosphere.com*

ABSTRACT

The accurate localization and characterization of aerosol and cloud layers is crucial for climate studies (aerosol indirect effect), meteorology (Planetary Boundary Layer PBL height), site monitoring (industrial emissions, mining,...) and natural hazards (thunderstorms, volcanic eruptions).

LEOSPHERE has recently developed aerosol/cloud detection and characterization on WINDCUBE long range Coherent Wind Doppler Lidars (CWDL). These new features combine wind and backscatter intensity informations (Carrier-to-Noise Ratio CNR) in order to detect (aerosol/cloud base and top, PBL height) and to characterize atmospheric structures (attenuated backscatter, depolarization ratio). For each aerosol/cloud functionality the method is described, limitations are discussed and examples are given to illustrate the performances.

1. INTRODUCTION

Coherent Lidars are nowadays well-recognized for their accurate measurements of wind speed profiles based on the Doppler shift effect on atmospheric particles and clouds. The signal reflectivity (CNR for Carrier-to-Noise Ratio) profiles can also be retrieved from the signal strength of the detected Doppler peak.

Based on the CNR and wind information, specific aerosol and cloud functionalities were developed on WINDCUBE long range scanning Doppler lidar at 1.54 microns. They consist in 6 aerosol/cloud features:

1) Aerosol/cloud detection
2) Aerosol/cloud discrimination
3) Attenuated backscatter coefficient
4) Planetary Boundary Layers height detection
5) Depolarization ratio
6) Speciation of atmospheric structures

The general flowchart of the aerosol/cloud measurements is given on figure 1. Each function is described and discussed in the next section

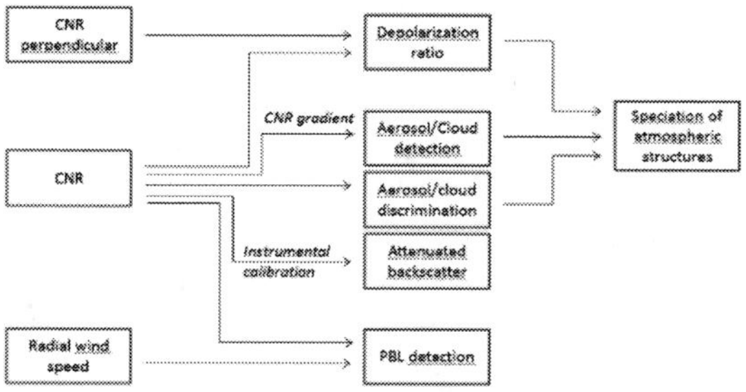

Figure 1: General flowchart of the aerosol/cloud measurements on WINDCUBE lidars

2. RESULTS

2.1 AEROSOL/CLOUD DETECTION

The layer detection is based on automated detection of gradient on CNR signal. Figure 2 shows the three steps to detect base and top of each atmospheric layer:

- Step1 : CNR signal is filtered and the gradient is computed using a Sobel mask
- Step 2 : The gradient are automatically segmented using an OTSU method [1]
- Step 3 : Base and top of atmospheric layers are identified

CWDL aerosol/cloud detection has been compared the aerosol lidar (R-MAN510) and showed quite similar results (see figure 3).

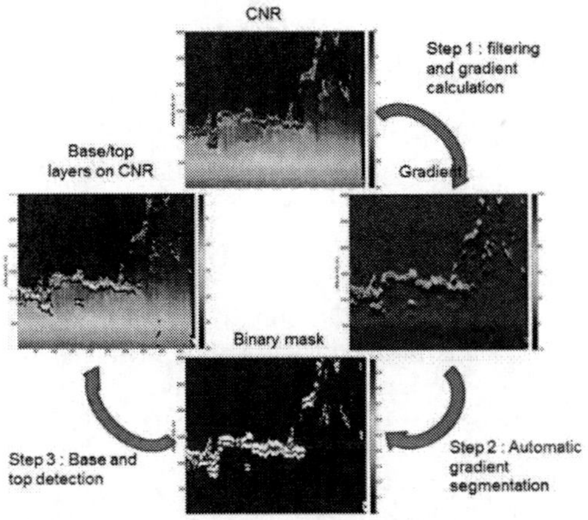

Figure 2: Automatic detection of base and top of atmospheric layers based on CNR signal

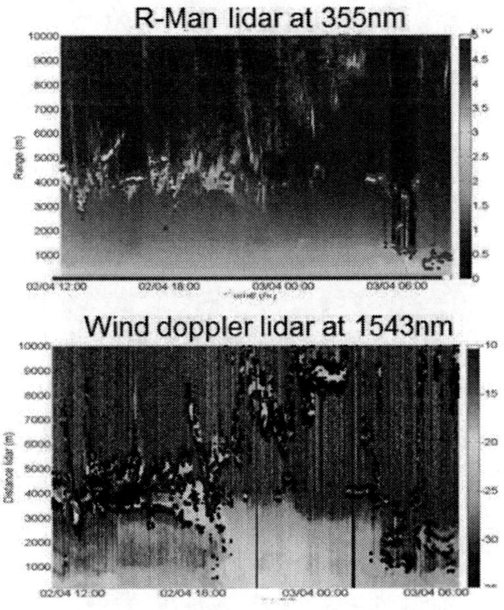

Figure 3: Comparison of base and top of atmospheric layers detected with an aerosol lidar at 355nm (red dots on the top panel) and a CWDL at 1543 nm (black dots on the bottom panel)

2.2 AEROSOL/CLOUD DISCRIMINATION

Each detected structure is discriminate in aerosol of cloud based on CNR peak intensity. The probability of aerosol and cloud is estimated with the following steps:

- compute average CNR gradient in layer between base and top
- estimate probability of presence aerosol and cloud using value of gradient and statistical results from WINDCUBE200S campaigns
- discriminate layer into aerosol or cloud using probability of presence and neighborhood information

Examples of aerosol/cloud discrimination are given on figure 4 in PPI (Plan Position Indicator) and fixed LOS (Line Of Sight) measurements.

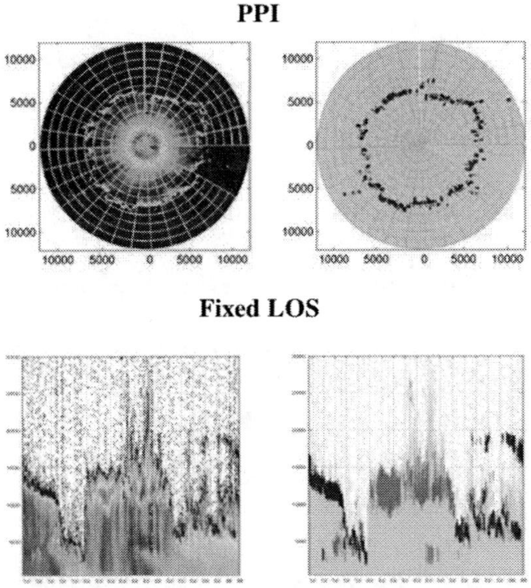

Figure 4: Examples of CNR signals (left panels) and aerosol (orange) and cloud (brown) discrimination (right panels) in PPI (top panel) and fixed LOS (bottom panel)

2.3 ATTENUATED BACKSCATTER

The attenuated backscatter coefficient is a function of the aerosol type and concentration. This parameter can be used to highlight aerosol and cloud layer in the atmosphere. An automatic calibration is performed every hour to compute and remove the instrumental function of the instrument.

DOI: 10.1051/epjconf/201611911002

Figure 5 shows the comparison of attenuated backscatter coefficient retrieved with a WINDCUBE200S lidar at 1543 nm and the absolute backscatter coefficient at 355 nm from the R-MAN510 aerosol lidar.

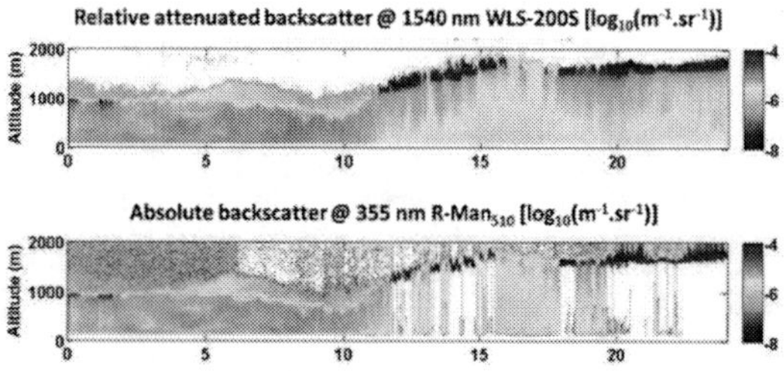

Figure 5: Comparison of attenuated backscatter coefficient retrieved from a CWDL at 1543 nm (top panel) and an aerosol lidar (R-MAN510) at 355 nm (bottom panel)

2.4 PLANETARY BOUNDARY LAYER HEIGHT

Residual and mixing layers detections are described on figure 6.

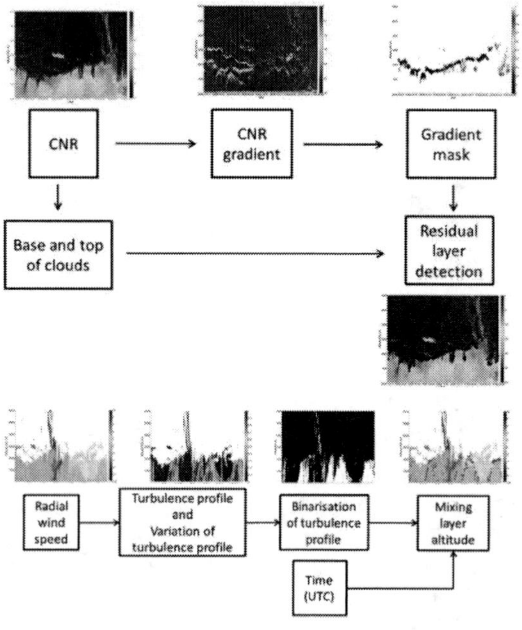

Figure 6: Residual (top panel) and mixing (bottom panel) layer detections on WINDCUBE long range lidars

The residual layer is the transition between aerosol in the boundary layer and the free troposphere. The residual layer is bounded above by a capping inversion. The residual layer is retrieved using CNR gradient (top panel on figure 7).

Mixing layer is the layer where positive buoyancy flux at the surface creates a thermal instability and thus generates additional or even major turbulence. Mixing layer is thus retrieved using turbulence information determined by vertical wind dispersion (bottom panel on figure 7, [2]).

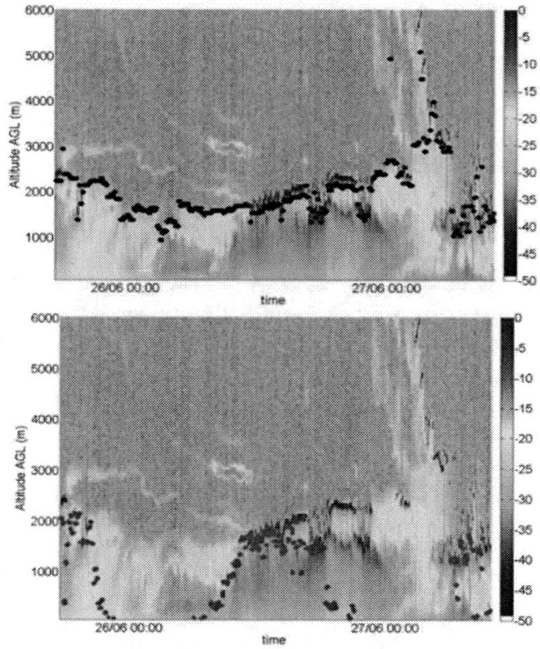

Figure 7: Example of residual layer (black dots on top panel) and mixing layer detection (red dots on bottom panel)

2.5 DEPOLARIZATION RATIO

Depolarization ratio gives information on particle shape. High depolarization ratio are expected for non-spherical scatterers (ice clouds, dust, ashes) whereas low depolarization are expected from spherical particles (sea salt, pollution, biomass burning) or water-clouds. Depolarization ratio is computed at 1543nm using information from co-polar and cross-polar CNR signals.

DOI: 10.1051/epjconf/201611911002

Dust event has been measured with high depolarization ratio in the boundary layer (near to 30%, figure 8). This case is referenced in the operational website of the BSC-DREAM8b model [3] with a maximum concentration of 50µg/m^3 over Paris (figure 8). Water clouds and ice clouds have been also been observed at the same time at the top and above the PBL.

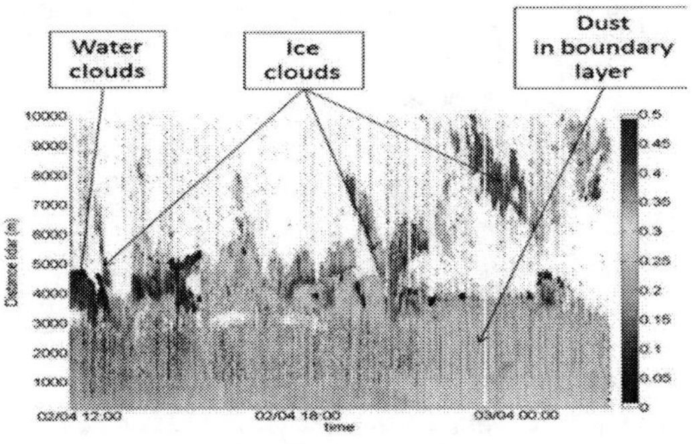

Figure 8: Depolarization measurement during a dust event observed over Paris the 2nd of April 2014

2.6 SPECIATION OF ATMOSPHERIC STRUCTURES

The mean value of depolarization ratio gives information on shape and type of each aerosol and cloud layer detected (figure 5). Combining information from aerosol/cloud detection and discrimination and depolarization measurement the atmospheric structures are classified into 6 types: aspherical aerosols (ash, dusts), spherical aerosols (sea salt, biomass burning, continental pollution), water clouds (water droplets), ice clouds (e.g. cirrus clouds), unclassified aerosol and unclassified clouds.

An example of atmospheric structure speciation from WINDCUBE200S measurements is given on figure 9.

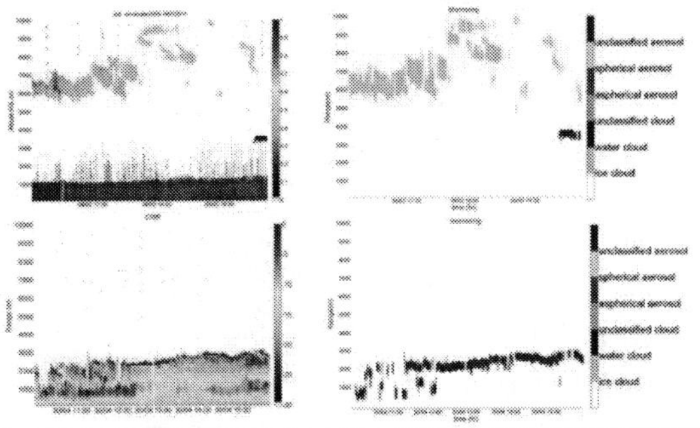

Figure 9: Example of atmospheric structure speciation for ice (top) and water clouds (bottom)

3. CONCLUSIONS

Based on wind and CNR information, aerosol and cloud capabilities have been developed on CWDL for the detection and the characterization of atmospheric structures. The results have been compared with aerosol lidar retrievals and showed similar performances.

The main benefit of CWDL technology is the ability to retrieve both wind (direction and intensity) and aerosol/cloud information. This ability is particularly important for the monitoring of the dispersion on industrial sites, the understanding of atmospheric hazard (thunderstorms formation) and the air traffic survey (identification and localization of volcanic ashes).

REFERENCES

[1] Otsu, N., 1979: A threshold selection method from gray-level histograms, *IEEE Trans. Sys., Man., Cyber.*, vol. 9, 1979, p. 62–66

[2] Bianco, L., Wilczak, J. M., White, A. B., 2008: Convective Boundary Layer Depth Estimation from Wind Profilers: Statistical Comparison between an Automated Algorithm and Expert Estimations, *J. Tech. A,* DOI: 10.1175/2008JTECHA981.1, 1397-1413.

[3] Basart et al., 2012: Development and evaluation of the BSC-DREAM8b dust regional model over Northern Africa, the Mediterranean and the Middle East, Tellus B, 64, 18539, doi:10.3402/tellusb.v64i0.18539.

AIRBORNE DIFFERENTIAL ABSORPTION AND HIGH SPECTRAL RESOLUTION LIDAR MEASUREMENTS FOR CIRRUS CLOUD STUDIES

Silke Gross [1]*, Andreas Schaefler[1], Martin Wirth[1], Andreas Fix[1]

[1]*Institute of Atmospheric Physics, German Aerospace Center (DLR), 82234 Oberpfaffenhofen, Germany,*
**Email: silke.gross@dlr.de*

ABSTRACT

Aerosol and water vapor measurements were performed with the lidar system WALES of the German Aerospace Center (DLR) onboard the German research aircraft G550-HALO during the HALO Techno-Mission in October and November 2010 and during the ML-Cirrus mission in March and April 2014 over Central Europe and the North Atlantic region. Curtains composed of lidar profiles beneath the aircraft show the water vapor mixing ratio and the backscatter ratio. Temperature data from ECMWF model analysis are used to calculate the relative humidity above ice (RHi) in the 2-D field along the flight track to study the RHi distribution inside and outside of cirrus clouds at different stages of cloud evolution.

1. INTRODUCTION

Although cirrus clouds have a large impact on the Earth's climate system [1] [2], especially their radiative properties are still only insufficiently understood [3][4]. The radiative properties of cirrus clouds highly depend on their microphysical composition [5], e.g. the ice crystal shape, the size distribution and the number concentration, which are strongly dependent on ambient vertical movement [6]. To improve our knowledge of cirrus clouds a better understanding of the micro- and macro-physical properties of cirrus clouds is mandatory. Airborne differential absorption lidar (DIAL) measure the 2-dimensional atmospheric structure along the flight path and thus provide valuable information to study the macro-physical properties as well as the fine-structure of cirrus clouds. Furthermore the combination of DIAL measurements with additional temperature information (e.g. from ECMWF model analysis) enable the investigation of relative humidity distribution and ice-supersaturation inside and outside of cirrus clouds, which is a crucial measure for cirrus cloud formation and evolution processes [7].

We present measurements of cirrus cloud properties performed with the DLR differential absorption and high spectral resolution lidar (HSRL) system WALES during the HALO Techno-Mission in autumn 2010 and during the ML_Cirrus mission in spring 2014. We provide RHi curtains beneath the HALO flight path from combined lidar water vapor measurements and temperature data from ECMWF model analysis, and investigate the RHi distribution inside and outside of cirrus clouds. A special focus of our studies is on the analysis of differences in RHi distribution for different types of cirrus clouds and cirrus cloud evolution stages.

2. METHODOLOGY

2.1 Differential Absorption Lidar WALES

The multi-wavelength lidar system WALES of the DLR-Institute for Atmospheric Physics is capable to emit simultaneously four wavelengths, three online and one offline, in the water vapor absorption band around 935 nm. Additionally to the 935 nm channel, the system is equipped with polarization sensitive aerosol channels at 532 nm and 1064 nm, the first with high spectral resolution capabilities using an iodine filter in the detection path [8]. A detailed technical description of the system was done by [9]. An analysis of the systems accuracy can be found in [10].

2.2 HALO

The new German research aircraft HALO (High Altitude and LOng range) is based on a modified Gulfstream G550. It is operated by the Flight Department of the DLR, Oberpfaffenhofen. HALO has a flight range of more than 8000 km, a flight time of more than 10 h, and a maximum cruising height of more than 15 km. Therefore measurements with WALES on HALO achieve the necessary distance to cirrus cloud tops to

DOI: 10.1051/epjconf/201611911003

prevent saturation of the detector units due to strong backscattered light from the clouds.

2.3 ML_Cirrus mission

The Mid-Latitude Cirrus mission ML_Cirrus is a joint activity of different German research centers and universities led by the DLR-Institute of Atmospheric Physics, and took place in March and April 2014. Measurements were performed with a set of remote sensing and in-situ measurements onboard HALO to study aviation induced cloudiness as well as natural mid-latitude cirrus clouds in Central Europe and the North Atlantic. The operational base of HALO during ML_Cirrus was Oberpfaffenhofen. Altogether 12 research flights over Germany, the North Atlantic region, and South-Western Europe were carried out. Additional data for cirrus cloud studies are available from the HALO Techno-Mission.

3. RESULTS - CASE STUDY ON 11 APRIL 2014

The main focus of the research flight on 11 April was to measure cirrus cloud properties in different stages of cloud evolution. Model simulations indicated rapidly ascending air masses over the extra tropical North-Atlantic region and high cirrus cloud evolution probability north-westerly of Scotland followed by cirrus cloud resolving areas in the region of the English Channel.

To achieve the research aims of this flight, four consecutive flight lags at different altitudes (three for in-situ measurements, and one at high altitude for lidar measurements) were planned parallel to the 500 hPa isobars (see Figure 1).

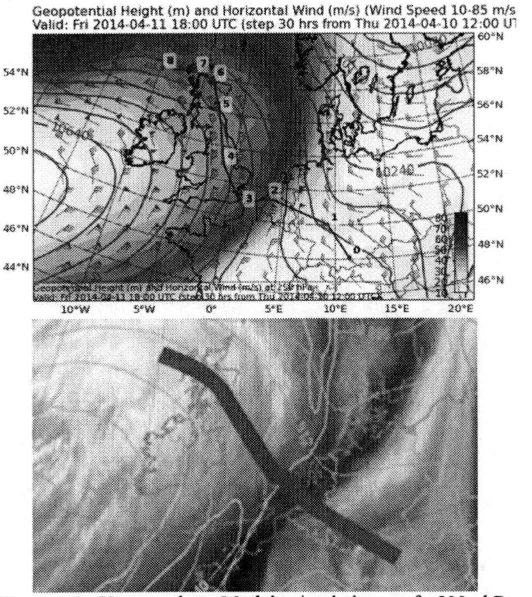

Figure 1 Upper plot: Model simulations of 500 hPa geopotential height and horizontal wind for 11 April 2014, 18 UTC and planned HALO flight path for the fourth flight leg (blue line). Lower plot: water vapor satellite measurements on 11 April 2014 in combination with 500 hPa geopotential height (cyan lines) and HALO flight path for the fourth flight leg (red line).

Figure 2 gives an overview over the lidar measurements during the fourth flight leg of this research flight. It shows the lidar backscatter ratio cross-section at 532 nm. The measurements extend from the formation region of the cirrus cloud at Northern Scotland to the dissolving area over the English Channel.

Figure 2 Extinction corrected backscatter ratio at 532 nm as measured with WALES on HALO on 11 April 2013.

To study RHi distribution and supersaturation in- and outside the cirrus cloud the RHi field along the flight track was calculated using a combination of WALES' water vapor measurements and interpolated ECMWF model temperature data. The applicability of ECMWF model temperature to study cirrus cloud properties was recently investigated by [11]. The RHi cross section along the fourth flight leg is shown in Figure 3.

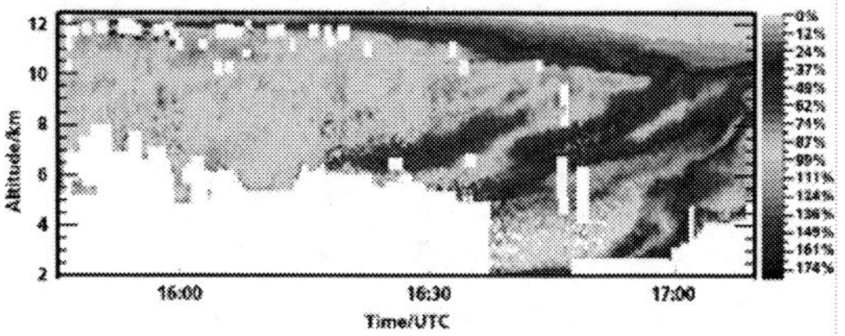

Figure 3 Relative humidity over along the fourth flight leg on 11 April 2014 ice calculated from combined lidar water vapor mixing ratio and ECMWF model analysis temperature data.

The threshold for the decision cloud/no-cloud was set to an empirical value of the backscatter ratio at 532 nm of 2. Taking the whole cloud measured on 11 April 2014 into account, the RHi distribution inside the cloud (not shown here) shows highest frequencies for RHi values around 95 %, a strong broadening to RHi values below 80 % and a slight broadening to RHi values larger than 120 %. To determine the height dependence of RHi within the cirrus cloud, the cloud was divided in height levels of 500 m.

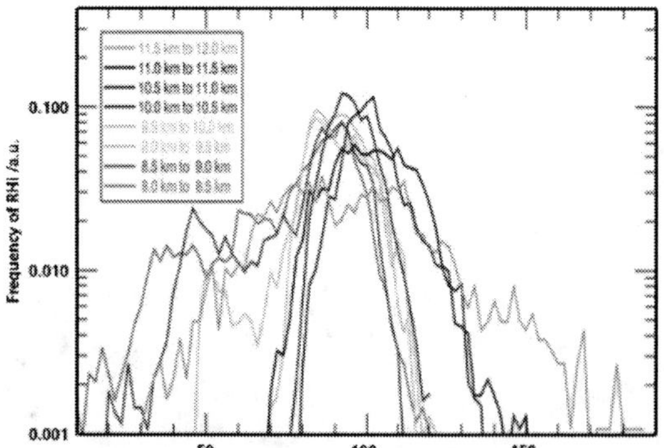

Figure 4 Frequency distribution of the relative humidity over ice dependent on the height level inside the fully developed cirrus cloud.

The RHi values in the mid-part of the cirrus cloud show a narrow distribution with a mean value of about 95% while the RHi distribution of the lowest cloud layers show a strong broadening towards lower values. Even RHi values smaller 50 % were measured in the lowermost cirrus cloud layers. The highest RHi values were found in the uppermost cloud layers (see Figure 4).

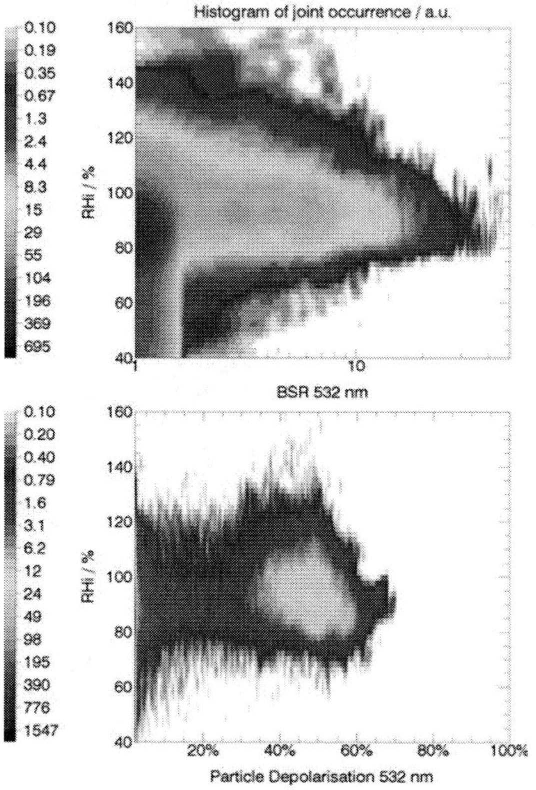

Figure 5 Histogram of joint occurrence of the relative humidity over ice (RHi) and the extinction corrected backscatter ratio (BSR) at 532 nm (upper plot) and of the RHi and the particle depolarization ratio at 532 nm (lower plot).

Additional to the analysis of the RHi field along the flight track, correlated measurements of the 2-D distribution of the RHi and the backscatter ratio (BSR), as well as of the RHi and the particle depolarization ratio (PDR) at 532 nm were performed (Figure 5), showing RHi values near saturation for large BSRs. Outside the cloud high ice-supersaturation up to more than 150 % was found. The PDR inside the cloud mainly ranged between 0.3 and 0.6 with most frequent values

DOI: 10.1051/epjconf/201611911003

around 0.5. High RHi values inside the cloud are correlated with quite high PDRs of 0.4 - 0.5.

4. SUMMARY AND OUTLOOK

Combined airborne DIAL and HSRL measurements, performed during the HALO Techno-Mission and the ML_Cirrus mission, are used to study the fine-structure and the macro-physical properties of cirrus clouds. RHi fields beneath the flight track, calculated from combined lidar water vapor and ECMWF model analysis temperature data, show large differences in the height dependent RHi distribution within the cirrus cloud for cirrus clouds at different stages of evolution and formation mechanisms, as well as in analysis of ice-supersaturation inside and outside the cirrus clouds.

The measured dataset provides an ideal basis to investigate the properties and differences of cirrus clouds in different stages of evolution and under different meteorological conditions. Therefore the WALES measurements during the ML_Cirrus mission will be systematically analyzed under consideration of the meteorological situation and the age / stage of evolution of the cirrus cloud. Besides our studies concerning ice-supersaturated regions in the vicinity of cirrus clouds, we will investigate the link of RHi and backscatter intensity to other lidar parameters like the depolarization or extinction-to-backscatter ratio, and we will examine dominant horizontal and vertical scales of the cirrus cloud properties. Our results will be combined with model analyses and in-situ measurements to answer some general questions related to altitude, age, and long term history dependence of the cirrus cloud properties.

ACKNOWLEDGEMENT

The authors like to thank the staff members of the HALO aircraft of DLR Flight Experiments for preparing and performing the measurement flights.

REFERENCES

[1] Cox, S. K., 1971: Cirrus clouds and the climate, *J. Atmos. Sci.*, **28**, 1513-1515.

[2] Liou, K.-N., 1986: Influence of cirrus clouds on weather and climate processes: a global perspective, *Mon. Wea. Rev.*, **114**, 1167-1199.

[3] IPCC, climate change 2007: The scientific basis, Cambridge University Press.

[4] IPCC, climate change 2013: The Physical Science Basis.

[5] Fusina, F., U. Lohmann, and P. Spichtinger (2007). Impact of ice supersaturated regions and thin cirrus on radiation in the midlatitudes, *J. Geophys. Res.*, **112**, D24S14.

[6] Heymsfield, A, 1977: Preciptitaion development in stratiform ice clouds: A microphysical and dynamical study, *J. Atmos. Sci.*, **34**, 367-381.

[7] Heymsfield, A. J., and L. M. Miloshevich, 1995: Relative humidity and temperature influences on cirrus formation and evolution: Observations from wave clouds and FIRE II, *J. Atmos. Sci.*, **52**, 4302-4326.

[8] Esselborn, M., M. Wirth, A. Fix, M. Tesche, and G. Ehret, 2008: Airborne high spectral resolution lidar for measuring aerosol extinction and backscatter coefficients, *Appl. Opt.*, **47(3)**, 346-358, doi:10.1364/AO.47.000346.

[9] Wirth, M., A. Fix, P. Mahnke, H. Schwarzer, F. Schrandt, and G. Ehret, 2009: The airborne multi-wavelength h2o-dial wales: system design and performance, *Appl. Phys. B: Lasers Opt.*, **96(1)**, 201-213, doi:10.1007/s00340-009-3365-7.

[10] Kiemle, C., M. Wirth, A. Fix, G. Ehret, U. Schumann, T. Gardiner, C. Schiller, N. Sitnikow, and G. Stiller, 2008: First airborne water vapor lidar measurements in the tropical upper troposphere and mid-latitudes lower stratosphere: accuracy evaluation and intercomparisons with other instruments, *Atmos. Chem. Phys.*, **8**, 5245-5261.

[11] Groß, S., Wirth, M., Schäfler, A., Fix, A., Kaufinann, S., and Voigt, C., 2014: Potential of airborne lidar measurements for cirrus cloud studies, *Atmos. Meas. Tech.*, **7**, 2745-2755, doi:10.5194/amt-7-2745-2014.

UNDERSTANDING SEASONAL VARIABILITY IN THIN CIRRUS CLOUDS FROM CONTINUOUS MPLNET OBSERVATIONS AT GSFC IN 2012

Simone Lolli [1]*, Jasper R. Lewis[1], Ellsworth J. Welton[2], James R. Campbell[3], Y. Gu[4]

[1]*NASA-JCET, Code 612, Greenbelt, MD 20771, USA, *Email: simone.lolli@nasa.gov*

[2]*NASA, Code 612, Greenbelt, MD 20771, USA*

[3] *Naval Research Laboratory, Monterey, CA 93940, USA*

[4] *UCLA, Los Angeles, CA 90095, USA*

ABSTRACT

Optically thin cirrus cloud (optical depth < 0.3) net radiative forcing represents one of the primary uncertainties in climate feedback, as sub-visible clouds play a fundamental role in atmospheric radiation balance and climate change. A lidar is a very sensitive optical device to detect clouds with an optical depth as low as 10^{-4}. In this paper we assess the daytime net radiative forcing of sub-visible cirrus clouds detected at Goddard Space Flight Center, a permanent observational site of the NASA Micro Pulse Lidar Network in 2012. Depending on their height, season and hour of the day, the solar albedo effect can outweigh the infrared greenhouse effect, cooling the earth-atmosphere system rather than warming it exclusively. As result, based on latitude, the net forcing of sub-visible cirrus clouds can be more accurately parameterized in climate models.

1. INTRODUCTION

Cirrus clouds, and especially optically thin ice crystal clouds with an optical depth below 0.03 exist globally [1], but are mostly found in the tropical upper troposphere, where the temperature is very low [2]. These sub-visible cirrus clouds (SVC) have been investigated over several decades both from in-situ measurements through airborne probes [15] and through ground-based lidar profiling [1]. SVC clouds are responsible for a large proportion of the uncertainties of climate change projections by General Circulation Models (GCM). This comes from poorly understood and represented interactions and feedbacks between dynamic, microphysical and radiative processes affecting cirrus clouds ([3], [4]). A recent IPCC report (AR4) shows that different climate models parameterize ice formation in a very simplified way, leading to a factor of ten in difference between zonally averaged Ice Water Content (IWC; [5]). As a consequence, climate modeling is very sensitive to small changes in cirrus coverage or ice microphysics. Large uncertainties in climate prediction caused by processes involving cirrus clouds highlight the importance of more quantitative information by observation, especially on SVC.

SVC are primary regulators of water vapor concentrations in the upper troposphere-lower stratosphere region, which plays a fundamental role in atmospheric radiation balance [15] and then climate change effects seen at the surface. Lidars are ideally sensitive to detecting clouds with an optical depth (COD) as low as 10^{-4} The main purpose of this study is to assess properties and the seasonal variability of SVC net radiative effect and statistical properties, detected at Goddard Space Flight Center (GSFC), a permanent observational site of the NASA Micro Pulse Lidar NETwork (MPLNET) in 2012.

2. METHODOLOGY

MPLNET [6], [7], [12], [13] is a global federated lidar network, begun in 1999, which counts more than twenty stations (deployed worldwide, from tropics, mid-latitudes, equator, Arctic and Antarctic regions) equipped with homogenous single wavelength elastic commercially available lidars. Project instruments sample the atmosphere continuously with a spatial resolution of 75m and a temporal resolution of 1 min in all different meteorological conditions. Measurements are centrally processed in near-real time at GSFC, and are made publicly available online to the scientific community (http://mplnet.gsfc.nasa.gov).

The new Version 3 (V3) MPLNET cloud algorithm (Lewis et al., 2015, *in progress*) is applied in this study to detect single-layer optically-thin clouds profiled at the GSFC site in 2012. The cloud extinction profile is retrieved simultaneously at 20 and 30 sr, respectively, to replicate the potential variance within this applied solution. All cloud samples are included, depending on the temporal sampling resolution used to resolve them in the V3 product (1, 5 or 20 min averaging). SVC microphysical properties are retrieved using [8] parameterization. *IWC* and average ice crystal dimensions (*Dge*) profiles are retrieved by integrating lidar cloud extinction at both 20/30 sr settings with WMO radiosonde temperature and pressure measurements from nearby Sterling, Virginia (~ 50 km).

Radiative forcing parameters are calculated with the Fu-Liou-Gu (FLG) radiative transfer code [9], [10]. FLG is a δ-four stream transfer scheme for flux calculations in both solar and infrared spectra, which are divided into six and twelve bands, respectively. The K-distribution method is used to parameterize non-gray gaseous absorption by CH_4, O_3, CO_2 and H_2O. Parameterization of cirrus clouds assumes hexagonal ice crystals, randomly-oriented, and combines the extinction coefficient α with D_{ge} and IWC at each range bin r as:

$$\alpha(r) = \frac{4(3)^{1/3}}{3\rho_i}\frac{IWC(r)}{D_{ge}(r)} \quad (1)$$

where ρ_i is the ice density. The vertically resolved ice crystal mean effective size D_{ge} (in µm) is calculated with an empirical formula that links crystal size and cloud top temperature [8].

3. RESULTS

At GSFC in 2012, more than 1e4 SVC cases were profiled. Each cloud is characterized by several parameters, including radiative forcing, cloud optical depth, cloud thickness, cloud position, etc., making a visualization of n-dimensional space impossible.

Principal Component Analysis (PCA) is a common multivariate statistical technique that reveals what variables in a given dataset form a coherent subset that is relatively independent of one another. PCA extracts the maximum variance from a data set with each component. The first principal component is the linear combination of observed variables that efficiently separate subjects by maximizing the variance of their component scores. The second component is computed from the residual correlation. It is the linear combination of observed variables that extracts maximum variability. This variability is uncorrelated with the first component. The subsequent components also extract maximum variability from the residual correlations and are independent from all the other components. The extracted components represent most of the variance of the original data set and can be used in further analysis, like for characterizing the most common cirrus cloud features and how they change with season. PCA is applied here in this analysis seasonally. Each cirrus cloud is represented as a point the 10-dimensional space with parameters reported in Table 1.

PCA parameters	Units/Values
Net cloud forcing (TOA)	W/m^2
COD	--
Day/Night	1 Day – 0 Night
Cloud Top	Km
Dge	µm
pIWC	g/m^2
Ground-Cloud Top Diff.	K
SZA	Deg
Cloud Top Temp	K
DOY	--

Table 1 Cirrus cloud retrieved and measured parameters.

Figure 1 shows how SVC cluster together. The two primary clusters represent daytime (upper strip) and nighttime clouds (lower strip). The first component (PC1) is related to cloud top temperature, cloud top altitude, crystal average size and the temperature difference between the ground and cloud top (higher factor loading). This component is characterized as the thermodynamic properties of SVC.

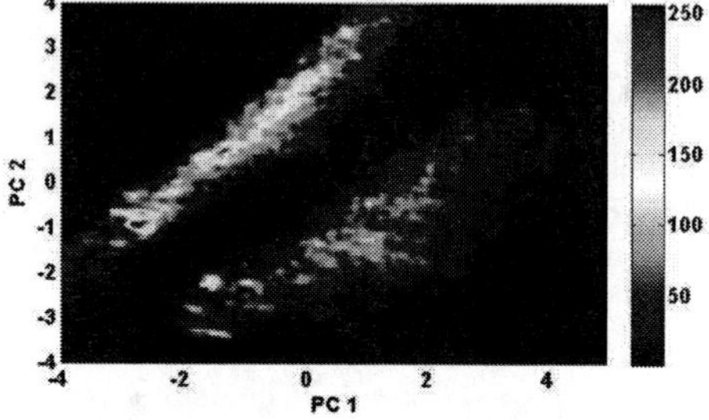

Figure 1 SVC PCA analysis at Goddard in 2012. The two distinct patterns represent day(up) and night (down) clouds. PC 1 and PC2 are the scores related to the first and second component, linear combination of parameters in Table 1.

SVC with higher values on PC1 are those clouds with warmer cloud tops, lower in altitude, with bigger crystals and *IWC* but with a smaller difference between ground temperature and cloud top.

The second component (PC2) relates to net TOA cloud radiative forcing (CRF), day/night flag and solar zenith angle. PC2 is thus characterized by the radiative properties of the cloud, and higher PC2 scores in Fig. 1 represent SVC with lesser CRF (in absolute value) and lower SZA. While nighttime clouds are always warming clouds, daytime SVC can both cool and warm the earth-atmosphere system depending on SZA and cloud top altitude (the former is related to the IWC and the latter to the crystal average size; [11]). The main daytime cluster (case density > 100) shows that SVC are cooling the atmosphere when cloud top is generally lower than 11.5 km.

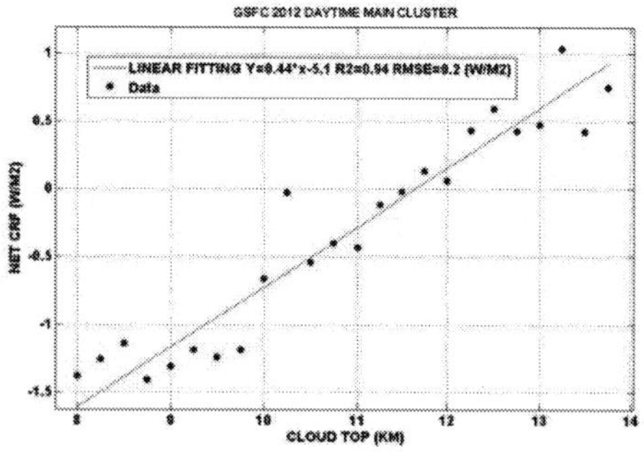

Figure 2 Daytime main cluster (density>100) showing Net Cloud Radiative Forcing (CRF) vs. cloud top altitude (Km). 11.5Km cut-off point from cooling to warming

Conversely, SVC warm the atmosphere when the SZA is generally less than 54 degrees.

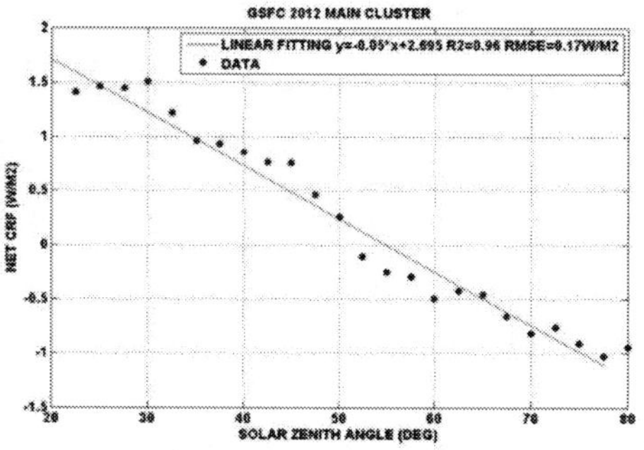

Figure 3 Daytime main cluster (density>100) showing Net Cloud Radiative Forcing (CRF) vs. SZA (deg). For SZA= 54 deg transition from warming to cooling.

By performing this same analysis seasonally, we derive approximate threshold SZA and SVC separating daytime cooling and warming (Table 2)

	CLOUD TOP	SZA
DJF	12.25 KM	37.5
MAM	11.1 KM	61
JJA	10.8 KM	64
SON	11.4 KM	58

Table 2 Seasonal variability of transition point between cooling and warming (Cloud Top) and warming and cooling (SZA).

During winter, daytime SVC are more prone to cool the atmosphere, as threshold points are 12.25 km for cloud top height and 37.5 degrees for SZA. Summer SVC are more keen to warm the earth-atmosphere, as threshold top is 10.8 Km and SZA 64 deg. Keep in mind that this analysis relates both to common seasonal distributions of SVC heights and the duration of the solar day relative to SZA.

4. CONCLUSIONS

From new Version 3 Micro Pulse Lidar cloud products, 1e4 optically thin cirrus clouds (COD <0.03, or sub-visual cirrus; SVC) were profiled in 2012 at NASA GSFC permanent MPLNET observation site. For each cloud, net radiative forcing was calculated with Fu-Liou-Gu code, where retrieved cloud extinction is related to both IWC and the average crystal size (Dge) after integrating local radiosonde temperature profiles. Principal components analysis (PCA) is used to reduce dimensions and visualize the main cloud clusters by their thermodynamic and radiative properties. Daytime PCA clusters are evaluated to determine when the solar albedo effect is outweighing the infrared greenhouse effect, producing a net cooling effect.

Surprisingly, thin cirrus clouds, depending on position and hour of the day, are primarily atmospheric coolers, contradicting previous results in literature. Seasonally, the variability of threshold cloud top and solar zenith angle (SZA) values indicate that SVC during winter are more prone to cool the atmosphere.

This study is a first step toward reducing the uncertainty in parameterization of the net radiative forcing of thin cirrus cloud in global climate models. Further analysis will be performed on other years and MPLNET sites.

ACKNOWLEDGEMENT

The Micro Pulse Lidar Network is supported by the NASA Earth Observing System and Radiation Sciences Programs.

REFERENCES

[1] Sassen, K., et al., 1989: Optical scattering and microphysical properties of subvisual cirrus clouds, and climate implications. *Journal of Applied Meteorology*, 28.

[2] Jensen, E.J., O et al., 1996: On the formation and persistence of subvisual cirrus clouds near the tropical tropopause. *Journal of Geophysical Research*, 101, 95JD03575

[3] IPCC: Climate Change 2013 – The Physical Science Basis, Work- ing Group I Contribution to the Fifth Assessment Report of the Intergovernmental Panel on Climate Change, edited by: Inter- governmental Panel on Climate Change, Cambridge University Press, Cambridge, UK and New York, NY, USA, 2014.

[4]Spang, R.,et al.,, S. 2015: Satellite observations of cirrus clouds in the Northern Hemisphere lowermost stratosphere, *Atmos. Chem. Phys.*, 15, 927-950, doi:10.5194/acp-15-927-2015.

[5] Immler, F., Treffeisen, R., Engelbart, D., Krüger, K., and Schrems, O. 2008: Cirrus, contrails, and ice supersaturated regions in high pressure systems at northern mid latitudes, *Atmos. Chem. Phys.*, 8, 1689-1699, doi:10.5194/acp-8-1689-2008.

[6] Welton, et al, 2001, Global monitoring of clouds and aerosols using a network of micro-pulse lidar systems, in Lidar Remote Sensing for Industry and Environmental Monitoring. *Proc. SPIE*, 4153, 151-158, 2001

[7] Lolli S., Welton E. J., Campbell J. R., 2013: Evaluating Light Rain Drop Size Estimates from Multiwavelength Micropulse Lidar Network Profiling. *J. Atmos. Oceanic Technol.*, **30**, 2798–2807.

[8] Heymsfield, A. et al., 2014: Relationships between Ice Water Content and Volume Extinction Coefficient from In Situ Observations for Temperatures from 0° to −86°C: Implications for Spaceborne Lidar Retrievals*. *J. Appl. Meteor. Climatol.*, **53**, 479–505.

[9] Fu, Q. and Liou, K. N. 1992: On the correlated k-distribution method for radiative transfer in nonhomogeneous atmospheres, *J. Atmos. Sci.*, 49, 2139–2156.

[10] Gu, Y. et al.,: 2003: Parameterization of cloud-radiation processes in the UCLA general circulation model, J. Climate, 16, 3357–3370.

[11] Grenfell, T. C., and S. G. Warren, 1999: Representation of a nonspherical ice particle by a collection of independent spheres for scattering and absorption of radiation, *J. Geophys. Res.*, 104(D24), 31,697–31,709.

[12] Spinhirne, J.D., et al., 1995 Compact Eye Safe Lidar *Sys., Rev. Laser Eng.*, 23, 112-118,

[13] Lolli, S., et al, 2011. EZ Lidar: A new compact autonomous eye-safe scanning aerosol lidar for extinction measurements and PBL height detection. Validation of the performances against other instruments and intercomparison campaigns. *Ópt.Pura y Apl.*, 44 (1), 33-41.

[14] Reverdy, M., et al, 2012: On the origin of subvisible cirrus clouds in the tropical upper troposphere, Atmos. Chem. Phys., 12, 12081-12101, doi:10.5194/acp-12-12081-2012

[15] Solomon, S., et al., 2010: Contributions of stratospheric water vapor to decadal changes in the rate of global warming, Science, 327, 1219-33

DOI: 10.1051/epjconf/201611911005

RESEARCH ON THE RELATIONSHIP BETWEEN CLOUD TEMPERATURE AND OPTICAL DEPTH USING ROTATIONAL AND VIBRATIONAL RAMAN LIDAR

Jia Su, M. Patrick McCormick*, and Liqiao Lei

*Center for Atmospheric Sciences, Department of Atmospheric and Planetary Sciences, Hampton University, Hampton, Virginia 23668, USA, *Email: pat.mccormick@hamptonu.edu*

ABSTRACT

Clouds play a key role in the climate system, for they can result in a warming or a cooling effect according to their characteristics and altitudes. Raman Lidars have been proven to be a very useful remote sensing tool to characterize cloud properties and locations. In this paper, cloud temperature and optical depth are obtained using rotational Raman (RR) and vibrational Raman techniques. Results of cloud temperature and optical depth (OD) observed by the Hampton University (HU) Rotational-Vibrational Raman Lidar are presented. The paper discusses the influence of cloud OD on temperature of the cloud base and top. From these measurements, the relation of low-altitude cloud OD and temperature is summarized. These analyses are unique in that they combine simultaneous measurements of these quantities that can lead to an improvement in the understanding of cloud radiation transfer and effects.

1. INTRODUCTION

The study of clouds is very important in the understanding climate change. Reliable measurements of cloud temperature and optical depth are important for improving our understanding of cloud physics, cloud dynamics, and for validating cloud-resolving models [1, 2]. Raman Lidars have been proven to be a very useful remote sensing tool to measure cloud properties. Atmospheric temperature can be measured using RR lidars [3-4]. Moreover, the capabilities of RR technique are extended to the measurement of cloud temperature. Behrendt et al. developed the multi-cavity polychromator to measure cloud temperature using RR lidar when cloud backscatter ratio is smaller than 45 [3]. Su et. al present a technique to obtain cloud temperature by correcting for the influence of the residual elastic-signal on RR channels [4]. And cloud OD can be independently obtained from a nitrogen vibrational Raman signal. In this paper, cloud temperature and OD are both obtained using the RR and vibrational Raman techniques. Results of cloud temperature and OD observed by the Hampton University (HU) Rotational-Vibrational Raman Lidar are presented. The paper discusses the influence of cloud OD on temperature of cloud base and top. From these measurements, the relation of low-altitude cloud OD and temperature is summarized. These analyses are unique in that they combine simultaneous measurements of quantities that can lead to an improvement in the understanding of cloud radiation transfer and effects.

2. METHODOLOGY

2.1. Rotational Raman method to obtain temperature

Cloud temperature can be derived from two RR signals and written as[4]:

$$T(z) = \frac{a}{In\left(\frac{X_{r1}^{L}(z,\lambda_{r1}) - k_1 X_e(z,\lambda_e)}{X_{r2}^{L}(z,\lambda_{r2}) - k_2 X_e(z,\lambda_e)}\right) + b} \quad (1)$$

where the subscripts of X_e, X^L_{r1} and X^L_{r2} refer to one elatic and two RR backscattering signals; λ is wavelength; z is altitude; T is temperature; a and b are calibrated coefficients obtained using a balloon-sounding's temperature; k_1 and k_2 are two RR channels' transmission for elastic backscatter signals from HU lidar elastic channel.

2.2. Vibrational Raman method to obtain OD

Cloud OD can be derived from Nitrogen vibrational Raman signals and written as:

$$OD = \frac{In\left[\frac{X_r(z_{top})}{X_r(z_{base})} \times \frac{m(z_{base})}{m(z_{top})}\right]}{-2} - \sum_{z=base}^{z=top} T_m(z)dz \quad (2)$$

Where the subscripts of *r, m, top* and *base* refer to Raman, molecular, cloud top and base; *X* is the lidar signal; *z* is altitude; *OD* is optical depth; *T* is molecular extinction; *dz* is lidar range resolution.

3. Equipment

The HU lidar system includes three parts: laser system, receiver telescope and detector system, and data acquisition and analysis system. The Nd:YAG laser emits three laser beams with wavelengths of 1064, 532 and 355 nm. The lidar system can measure high-resolution (7.5 m) vertical profiles of elastic backscatter at the three operational wavelengths, and Raman scattering in the near-UV that are collected by a Cassegrain-configured telescope with a 48 inch diameter and 1~4 mrad field-of-view. The lidar has a non-coaxial configuration and detects return signals between the ranges of 1 km and approximately 30 km. Four photomultiplier tubes are used for the detection of the backscattered signals (532 nm and 355 nm) and Raman backscattered signals, while an avalanche photodiode is used to detect the backscattered 1064 nm signal. This provides the ability to retrieve aerosol backscatter and extinction coefficients [5-7] from the elastic measurements, and temperature [4] and H_2O from the rotational and vibrational Raman measurements, respectively, with a vertical range that extends from the near-surface to the top of the troposphere (~15km) depending on atmospheric conditions. Fig.1 shows the structure and filter system of HU lidar.

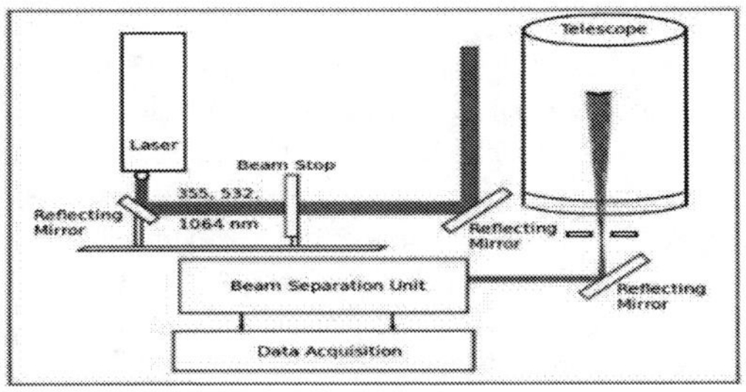

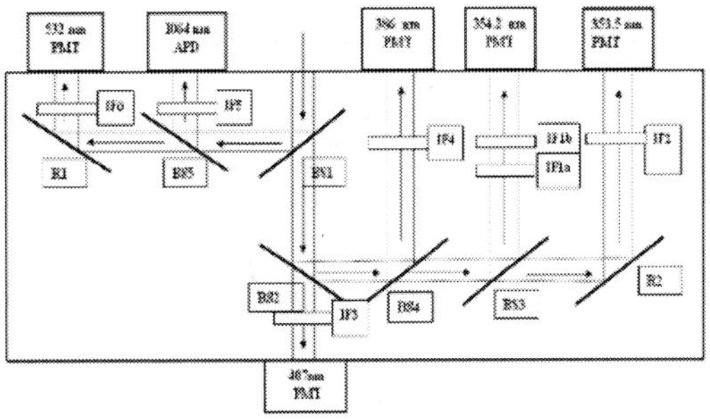

Fig.1 The detector system.

4. RESULTS

The example for a cumulus cloud is demonstrated with this technique. Fig. 2(a) shows a 5-min average of elastic backscatter returns at 354.7 nm, as well as two channel RR returns at 22:10 (local time) on October 26, 2011. Fig. 2(a) shows that the cumulus clouds appear at approximately 4.0 to 4.5 km altitude; 2(b) presents the HU lidar temperature results derived with the RR-signals (red line) and balloon-sounding's temperature results (square symbols). The blue bars are error bars of temperature retrieved using RR signals. Clearly, the lidar-derived cloud temperatures agree well with balloon-sounding's results. Presented in Fig. 2(c) is the temporal variation of HU lidar temperature results for a time period of 18:00–22:00 on October 26, 2011. Fig. 2(d) shows HU lidar range-corrected signals for the same time period., For this case, the clouds show stable heights, depths and a few hours life-time at 18:30 – 20:00 and 20:20 – 22:00. Fig 2. (e) shows corresponding cloud OD variation obtained by HU lidar. Fig. 2(f) shows the corresponding results of temperature of cloud base and top and cloud OD. It is found that the temperature of cloud base (red line) has a slight increasing trend. It means night cloud can reflect Earth radiation and warms the air below the cloud. Fig. 2(g) shows the corresponding results of temperature difference of cloud base and top and cloud OD. It is found that the higher the cloud OD is, the higher the difference of temperature of the cloud base and top is. These results indicated that the technique is feasible.

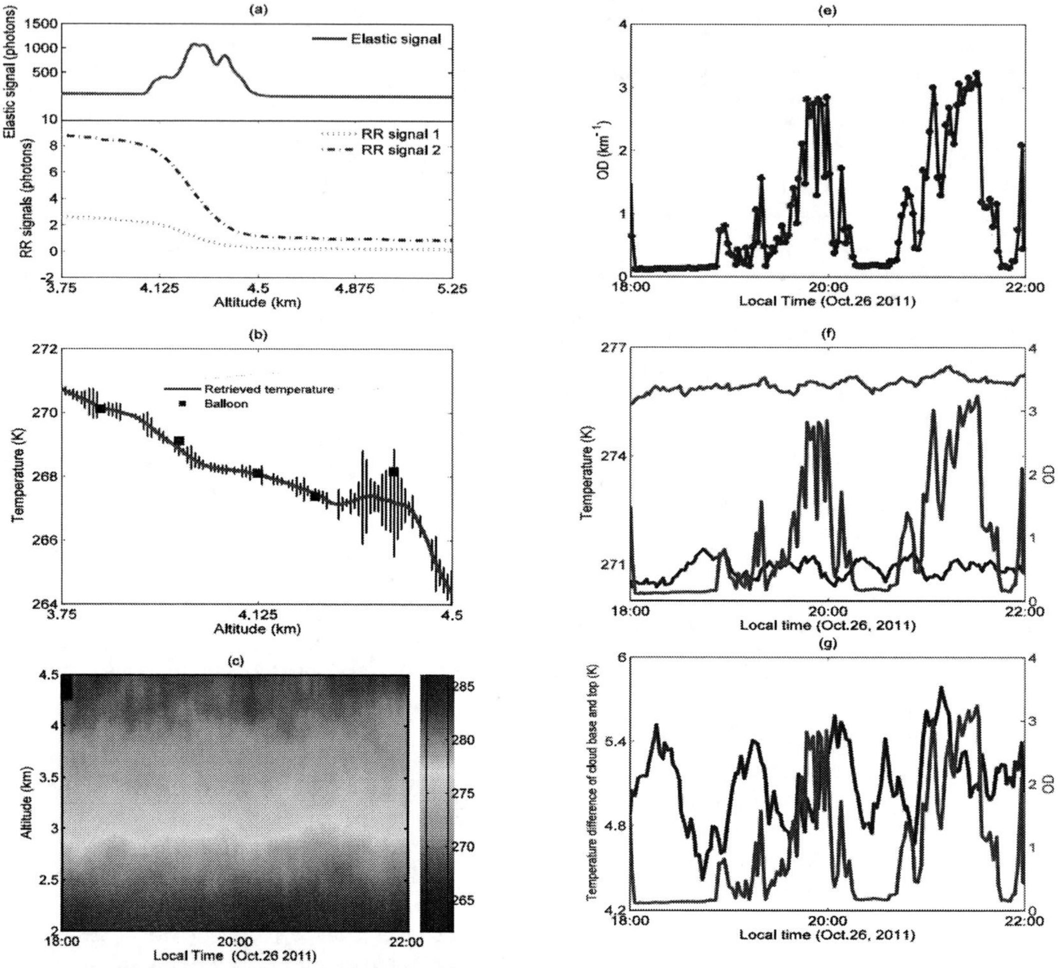

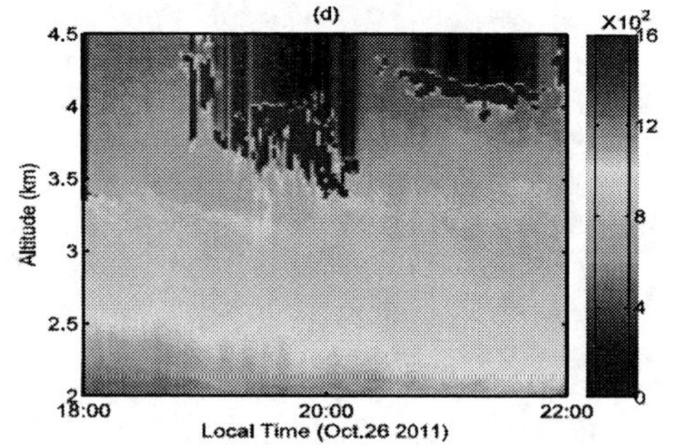

Fig.2 (a) elastic backscatter returns at 354.7 nm and two channel RR returns at 22:10 (local time) on October 26, 2011; (b) HU lidar temperature results (solid line) and a balloon-sounding's temperature results (square symbols) launched at HU; (c) Temporal variation of HU lidar temperature for a time period of 18:00–22:00 22:00 on October 26, 2011; (d) HU lidar range-corrected signals for a time period of 18:00–06:00 22:00 on October 26, 2011; (e) cloud OD variation measured by the HU lidar for a time period of 18:00–22:00 on October 26, 2011; (f) Temperatures retrieved of cloud base (red line) and cloud top (blue line), and cloud OD (green line) over the same period; (g) Temperature difference between cloud base and cloud top, as well as the cloud OD.

5. CONCLUSIONS

This work has developed a new method to research the relationship between cloud temperature and cloud OD using a Raman Lidar. Cloud temperature and optical depth are obtained using the rotational Raman and vibrational Raman techniques. A nocturnal result of cloud temperature and optical depth observed by the HU Rotational-Vibrational Raman Lidar are presented. The results of cloud temperature base and top, and cloud OD, are analyzed. . It is found that the higher the cloud OD, the higher the difference of temperature of the cloud base and top. . We will continue studying this outcome in the future.

ACKNOWLEDGEMENT

This study was supported by the National Oceanic and Atmospheric Administration (NOAA) under Grant - CREST Grant # NA06OAR4810162, and by the US Army Research, Development and Engineering Command (AQC) Center (DOD) under HU PIRT Award # 551150-211150).

REFERENCES

[1] Sinkevich AA, and Lawson RP. A Survey of Temperature Measurements in Convective Clouds. J. Appl. Meteor 2005; 44: 1133-1136.

[2] Albrecht BA, Cox SK, and Schubert WH. Radiometric measurements of in-cloud temperature fluctuations. J. Appl. Meteor 1979; 18: 1066-1068.

[3] Behrendt A, and Reichardt J. Atmospheric temperature profiling in the presence of clouds with a pure RR lidar by use of an interference-filter-based polychromator. Appl. Opt 2000; 39: 1372-1376.

[4] Su. J., McCormick, M. P., Wu, Y., Lee III, R. B., Lei, L., Liu, Z., & Leavor, K. R. (2013). Cloud temperature measurement using rotational Raman lidar, Journal of Quantitative Spectroscopy & Radiative Transfer, 125, 45-50.

[5] Tao, Z., Liu, Z., Wu, D., and McCormick, M. P. (2008a). Determination of aerosol extinction-to-backscatter ratios from simultaneous ground-base and spaceborne lidar measurements, Optics Letters, 33, 2986-2988.

[6] Su, J., M. P. McCormick,, Z. Liu,, K. Leavor,, R. B. Lee III,, J. Lewis, M. T. Hill, 2010: Obtaining ground-based lidar geometrical form factors using coincident spaceborne lidar measurements, *Appl. Opt*, **49**, 108-113.

[7] Su, J., McCormick, M. P., Liu, Z., Lee III, R. B., Leavor, K. R., & Lei, L. (2012). Transmittance ratio constrained retrieval technique for lidar cirrus measurements, Optics Letters, 37, 1595-1597.

DOI: 10.1051/epjconf/201611911006

IMPORTANCE OF RAMAN LIDAR AEROSOL EXTINCTION MEASUREMENTS FOR AEROSOL-CLOUD INTERACTION STUDIES

Zaw Han[1], Yonghua Wu[1], Fred Moshary[1], Barry Gross[1*], Alex Gilerson[1]

[1]*Optical Remote Sensing Lab, City College of New York, USA, * gross@ccny.cuny.edu*

ABSTRACT

Using a UV Raman Lidar for aerosol extinction, and combining Microwave Radiometer derived Liquid Water Path (LWP) with Multifilter Rotating Shadowband Radiometer derived Cloud Optical depth, to get cloud effective radius (R_{eff}), we observe under certain specialized conditions, clear signatures of the Twomey Aerosol Indirect effect on cloud droplet properties which are consistent with the theoretical bounds. We also show that the measurement is very sensitive to how far the aerosol layer is from the cloud base and demonstrate that surface PM25 is far less useful. Measurements from both the DOE ARM site and new results at CCNY are presented.

1. INTRODUCTION

One of the outstanding issues regarding the earth's energy balance and resultant forcing on the climate are the indirect effects that aerosols have on cloud radiative properties. In particular Twomey [1] described a mechanism in which increased aerosol loading results in higher concentration of cloud condensation nuclei which ultimately lead to increased cloud droplet number concentration and smaller effective cloud droplets which however have to be compared for constant liquid water paths. Unlike direct cloud effects on the climate, the aerosol-cloud indirect interactions are very hard to measure from satellites directly since simultaneous aerosol loadings (especially below cloud base) and cloud properties such as COD and/or Reff are not possible and therefore efforts at quantifying these effects are limited to statistical trend studied over large domains.

While such statistical studies have provided some useful information, poor temporal statistics makes it particularly difficult to establish underlying causal mechanisms. In addition, an equally difficult issue is to retrieve the true aerosol loading near the cloud base. In particular, it is quite evident that column measures of aerosol optical depth do not provide the vertical information needed to quantify the pollution level at the cloud base. Furthermore, the relationship between optical depth and particle number depends on the microphysical distribution models and does not discriminate between aerosol types below cloud base that impact cloud micro-physics and aerosol particles located higher than cloud top, that only interact with the radiative transfer above clouds. Therefore, it is clearly important to develop short term measurements with high temporal resolution to eliminate many of the above cited ambiguities. In particular, under such an approach, simultaneous measurements of clouds and aerosol loading below cloud base along with important vertical updraft data are achievable. Most important, within the measurement cycle, the exact nature of the cloud within the cloud lifecycle including general meso-scale processes is less important for these short term inter-cloud measurements.

Various efforts have been made to develop retrieval algorithms to infer COD and Reff from ground based passive radiometric measurements. In one approach, a Microwave

Radiometer (MWR) is combined with Millimetre-wavelength Cloud Radar (MMCR) to infer droplet diameters. In this case, the radar reflectivity and the MWR LWP provide the information to retrieve R_{eff}. However, the cost of the MMCR can be quite substantial. On the other hand, significant success in combining COD from the Multifilter Rotating Shadowband Radiometer (MFRSR) together with LWP from MWR is a reasonable cost-effective solution [2]. One drawback in this method is obviously the limitation that observations must be made during daytime. However, significant aerosol-cloud interaction signals should exist during daytime especially under conditions of strong convective heating which enhances vertical uptake. The relative simplicity and low cost of this method has motivated the deployment of such an approach at the CCNY

The main signature we wish to observe is the change in droplet effective radius (R_{eff}) as a function of aerosol loading. While a direct evaluation of aerosol number density is optimal, such a measurement is not realistic near cloud base. To this end, we make use of the conventional Raman Lidar technique to estimate aerosol extinction which is closely related to number density. In particular, to quantify the changes between droplets and aerosol loading, the Aerosol Cloud Index (ACI) was defined [3] as,

$$ACI = -\left(\frac{d(\log(R_{eff}))}{d(\log(\alpha))}\right) \quad (1)$$

which represents the relative change in mean cloud droplet effective radius (Reff) for a relative change in aerosol extinction, α_{ext} for clouds having the same LWP. It is useful to note that an upper theoretical limit of ACI < 1/3 exists.

2. METHODOLOGY

We make use of the methods described in [12]. Briefly, the MWR provides a direct retrieval of liquid water path (LWP) while the MFRSR measures the diffuse transmittance (T_{diff}). From these 2 measurements, we can simultaneously solve for cloud optical depth (τ_{cod}) and cloud droplet effective radius (R_{eff}) using the two independent relationships for LWP and T_{diff}

$$LWP(\tau_{cod}, R_{eff}) \approx \frac{2}{3}\tau_{cod}R_{eff} \quad (2)$$

$$T_{diff}(\tau_{cod}, R_{eff}) \quad (3)$$

The complexity of the measurements and the strict requirements to observe a meaningful Aerosol-Cloud signal severely limit the number of potential observations. The factors that must be accounted for include

1) Fine aerosol modes with angstrom coefficient with high Single Scattering Albedo (SSA). This allows us to restrict our cases to sulfate dominated aerosols
2) Low altitude water phase clouds with reasonably small Liquid Water path (50 < LWP < 90) to remove complications of mixed ice phase and maximize the observation of aerosol influence. In particular, higher LWP clouds will clearly be less affected by the aerosol fields.
3) Reasonably strong aerosol loadings to reduce the noise in the aerosol retrieval using the Raman Lidar approach
4) A significant vertical wind uptake which allows for a better mixing of the aerosol CCN's with the cloud LW. (use of HYSPLIT to insure vertical uptake)
5) Sufficiently homogeneous cloud decks that allow us to approximate the cloud diffuse radiance with a 1D radiative transfer model.

3. RESULTS

To explore the aerosol-cloud interaction and illustrate the need for a complete measurement testbed including lidar, we plot in figure 1, the Raman lidar derived extinction profile for a stable water phase cloud case from the ARM site [4].

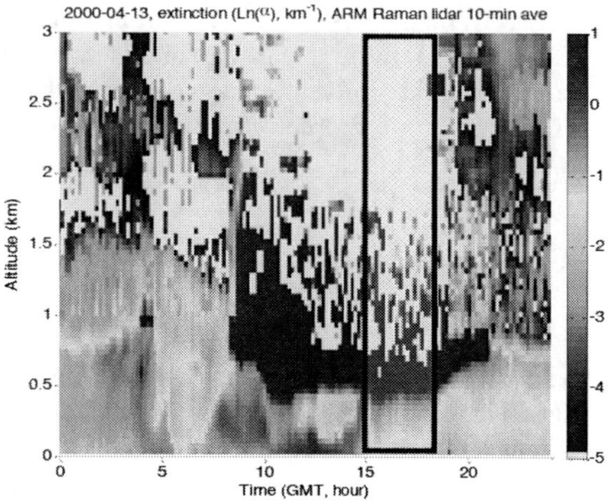

Fig. 1: Raman Lidar derived extinction profile and marked cloud boundary for April 13, 2000 at ARM SGP site

In figure 2, we show the relationship between the effective radius and aerosol extinction where we consider both the surface extinction (lidar ground bin = 30 m) as well as the extinction 150 m below the cloud base. Here the cloud base was determined using a simple extinction coefficient threshold of 1 m^{-1} and is displayed as black circles in figure 1. We note that the aloft aerosol measurement is critical in observing the AIE. In the example above, we have an ACI index of 0.18 which is reasonable based on the theoretical limits and other observational results.

The ARM site measurements motivated the deployment of a similar MFRSR / MWR / Raman Lidar testbed at CCNY in the hopes of being able to not only see AIE cases but to identify quantitative differences between them. However, as we pointed out in section 2, the difficulties in getting sufficient cases is the result of the strict conditions that a positive measurement must have. One interesting issue which we did not explore with the ARM site was the accuracy of the LWP measurement from the MWR. In order to explore this, we make use of 2 methods of calculating the LWP. 1) The NN approach which requires training at the local site 2) a simplified dual channel method [15].

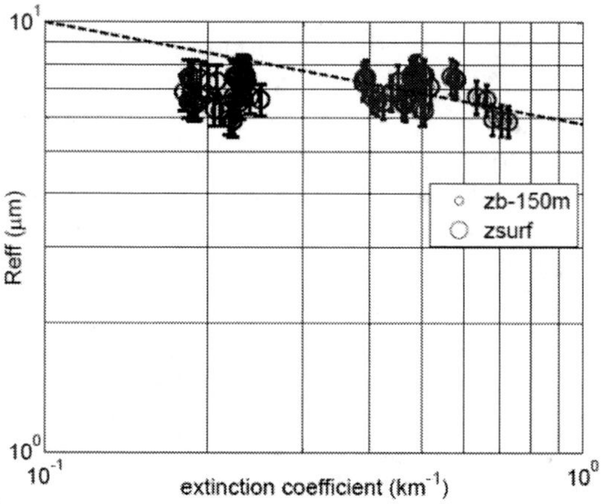

Fig. 2: Log-Log plot of -Raman lidar extinction below cloud against the cloud droplet effective radius. 'zb-150m' (blue circles) indicates the upper air atmospheric column layer which was used in the comparison (i.e 150 m below the cloud base altitude denoted by zb), while 'zsurf' refers to the atmospheric layer closest to the surface (~ 30m).

In figure 3, we see that strong agreement occurs for all moderate LWP cases.

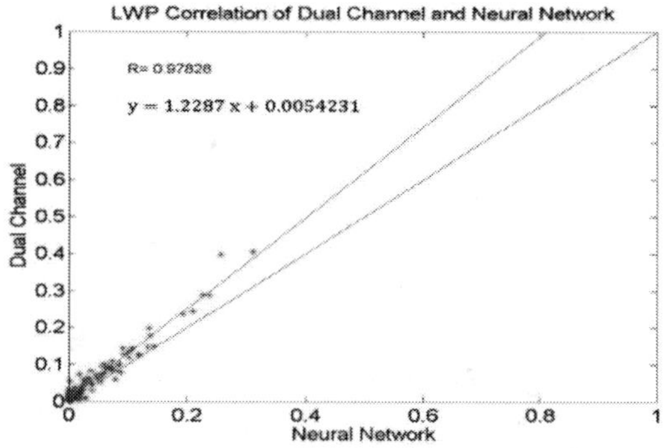

Fig. 3: Integrated liquid water path LWP (mm) correlation between two methods

We applied both the DC and NN method to a case where all conditions were met. (Other cases will be presented at the conference). The results are shown in figure 4. The main observation is that the DC method actually allows for more retrievals without any degradation. In fact, the larger number of retrievals allows improved regression analysis in this case. In addition, the ACI magnitude is very comparable between the 2 methods. In addition, this example, just like the ARM site demonstrates that the results degrade if we use aerosol layers too far below the cloud. In fact, strong degradation of the ACI signal occurs if we are further than 300 meters below cloud base (not shown here).

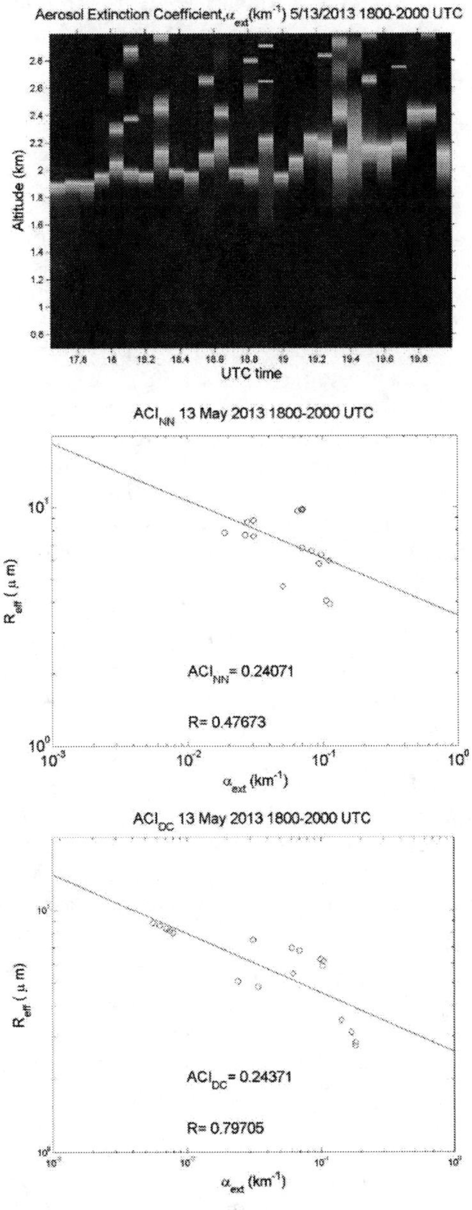

Fig. 4: Example of ACI for CCNY system. a) 3D lidar image b) ACI using NN c) using DC

4. CONCLUSIONS

We have made a preliminary investigation of the potential of quantifying and generally observing Aerosol Cloud Interactions at both the DOE ARM site and CCNY. We have demonstrated that accurate ACI measurements require LIDAR observations that can probe the aerosol loading sufficiently near cloud base. We also note (not shown here) that using surface $PM_{2.5}$ mass as a proxy for particle number.

In addition, to provide some internal consistency checks, we use 2 different MWR retrieval approaches and show good consistency between the methods. In addition, we will present matchups with MODIS cloud retrievals that demonstrate the general soundness of the ground based approach.

ACKNOWLEDGEMENT

This work is partially supported by the NOAA CREST Center under contract NOAA #NA17AE1625. *The statements, findings, conclusions, and recommendations presented in this manuscript are those of the authors and do not necessarily reflect the views of the National Oceanic and Atmospheric Administration or the Department of Commerce.*

REFERENCES

[1] Twomey, S., Influence of pollution on shortwave albedo of clouds. *J Atmos Sci* **1977**, *34*, 1149-1152.

[2] Min, Q.L.; Harrison, L.C., Cloud properties derived from surface mfrsr measurements and comparison with goes results at the ARM SGP site. *Geophys Res Lett* **1996**, *23*, 1641-1644

[3] Feingold, G.; Remer, L.A.; Ramaprasad, J.; Kaufman, Y.J., Analysis of smoke impact on clouds in Brazilian biomass burning regions: An extension of Twomey's approach. *J Geophys Res-Atmos* **2001**, *106*, 22907-22922

[4] Madhavan, B.L.; He, Y.; Wu, Y.; Gross, B.; Moshary, F.; Ahmed, S. Development of a Ground Based Remote Sensing Approach for Direct Evaluation of Aerosol-Cloud Interaction. Atmosphere, 3, 468-494 (2012).

DEPOLARIZATION RATIO OF CLOUDS MEASURED BY MULTIPLE-FIELD OF VIEW MULTIPLE SCATTERING POLARIZATION LIDAR

Hajime Okamoto[1]*, Kaori Sato[1], Toshiyuki Makino[1], Tomoaki Nishizawa[2], Nobuo Sugimoto[2] Yoshitaka Jin[2], Atsushi Shimizu[2]

[1]*Research Institute for Applied Mechanics, Kyushu University, Kasuga, Fukuoka 816-8580, Japan,*
**Email: okamoto@riam.kyushu-u.ac.jp*

[2]*National Institute for Environmental Studies, Tsukuba, Ibaraki 305-8506, Japan*

ABSTRACT

We have developed the Multiple Field of view Multiple Scattering Polarization Lidar (MFMSPL) system for the study of optically thick low-level clouds. It has 8 telescopes; 4 telescopes for parallel channels and another 4 for perpendicular channels. The MFMSPL is the first lidar system that can measure depolarization ratio for optically thick clouds where multiple scattering is dominant. Field of view of each channel was 10mrad and was mounted with different angles ranging from 0 mrad (vertical) to 30mrad. And footprint size from the total FOV was achieved to be close to that of Cloud-Aerosol Lidar and Infrared Pathfinder Satellite Observation (CALIPSO) lidar at the altitude of 1km in order to reproduce similar degree of multiple scattering effects as observed from space. The MFMSPL has started observations since June 2014 and has been continuously operated at National Institute for Environmental Studies (NIES) in Tsukuba, Japan. Observations proved expected performance such that measured depolarization ratio was comparable to the one observed by CALIPSO lidar.

1. INTRODUCTION

Conventional ground-based lidar has a limited capability to detect low-level clouds due to strong attenuation of lidar signals and maximum optical thickness is about 3. On the other hand, space-borne lidar such as CALIPSO lidar can measure signals from deeper part of the clouds. And the depolarization ratio for low-level clouds measured by the CALIPSO lidar often exceeded 30% while ground-based lidar usually showed smaller than 10% of depolarization ratio [1]. The difference in depolarization ratio between ground-based conventional lidar and space-borne lidar attributed to the different foot print size between the two. That is, the footprint size of conventional ground-based lidar with FOV of 1 mrad is about 1m at the altitude of 1km while that of CALIPSO lidar was 90m. Analysis of CALIPSO lidar data showed depolarization ratio of water clouds tended to increase as estimated extinction of the layer increased [2]. The characteristics were reproduced by the results of numerical simulations by Monte Carlo method. And the depolarization ratio of optically thick water clouds becomes comparable to that of randomly oriented ice particles. Due to these characteristics, the discrimination of cloud particle phase was not possible by depolarization information alone in case of space borne lidar observations. Thus it was necessary to take into account the combination of attenuation of backscattered lidar signals and depolarization to discriminate between randomly oriented ice, horizontally oriented ice particles and water particles [2]. It is also worth to note that there is no global product of three-dimensional distribution of water cloud microphysics using CALIPSO data.

There were several attempts to overcome the limitation of detectability of optically thick clouds in the ground-based lidar measurements. Off-beam lidar was introduced to detect lidar signals that were strongly affected by multiple scattering [3,4 and 5]. These have much wider field of view than the conventional lidar and were especially suited to detect multiple scattering signals from water clouds. It is also noted that no ground-based lidar that can detect depolarization ratio of clouds comparable to the observed values by CALIPSO lidar have been built.

Main aims of the study are to develop a new lidar system than can detect depolarization ratio from optically thick parts of the clouds. Multi-field of view lidar system was introduced with each field of view of 10 mrad and resultant total FOV is 40mrad. The foot print size at the altitude of 1km was close to that of CALIPSO so that similar degree of multiple scattering effects was expected in the new lidar system.

2. Description of the Multiple-Field of view Multiple Scattering Polarization Lidar

ND:YAG laser was used in the system and only the second harmonic wavelength (532nm) of the laser is detected. And the new lidar consists of 8 telescopes to achieve total FOV of 30mrad and FOV of each telescope is chosen to be 10mrad to assure the same polarization signatures. Four telescopes for parallel channels are located in a line and angles from the vertical direction are 0,10,20 and 30 mrad for channels 1, 3, 5 and 7, respectively. Another four telescopes for perpendicular channels (channels 2, 4, 6 and 8) are also located in a line parallel to the previous line and the angles were chosen to be the same as parallel channels. Each channel consists of polarizer, interference filter, lens and detector where photo multiplier tubes (PMT) are used. The temporal and vertical resolutions of original data were 10 seconds and 6m, respectively.

Fig.1. The Multiple Field of view Multiple Scattering Polarization Lidar operated at National Institute for Environmental Studies in Tsukuba since June 2014.

Signal to noise ratio of the MFMSPL system was not sensitive enough to detect signals from upper layers for the calibration target. Therefore calibration of the each channel of MFMSPL was conducted as follows. There were two steps for the calibration; the first one was the absolute calibration of channel 1 (vertical) by the comparison of the collocated National Institute for Environmental Studies (NIES) dual wavelength polarization lidar. Prior to the absolute calibration of the MFMSPL, the NIES lidar was calibrated using signals of upper layers than 6km. The data for aerosol region observed in June 5, 2014 were chosen for the absolute calibration because there is in generally negligible impact of multiple scattering on the lidar backscattering signals from aerosols at 532nm. The second calibration procedure is about relative calibration of all channels of the MFMSPL. All 8 telescopes were set to be parallel to vertical direction and polarizers for even channels were not used for the calibration purpose. Then data for aerosol regions were used, too. The calibration was conducted relative to channel 1.

Cloud mask methods for the off-beam channels (from channel 3 to 8) were different from that for the on-beam channels (channel 1 and 2). Cloud mask scheme for on-beam channels were developed similar to the one for ground-based lidar and also for CALIOP lidar [6,7] where discrimination of clouds, aerosols, molecules and noise were performed. And the cloud mask for off-beam channels were done by discrimination between noise and clouds since signals from off-beam channels were not influenced by aerosols and molecules.

3. OBSERVATIONAL RESULTS

The lidar has been continuously operated since June 2014. We showed the preliminary results from observations of low-level clouds on 9 June 2014. Apparent (attenuated) backscattering coefficient coefficient for each parallel channel were shown in Fig. 2.

DOI: 10.1051/epjconf/201611911007

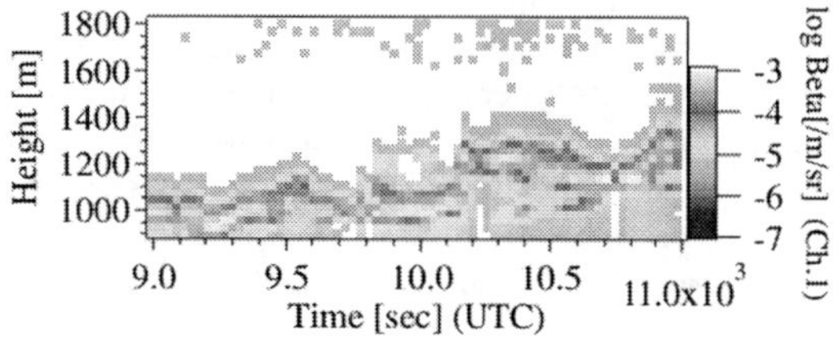

(a)

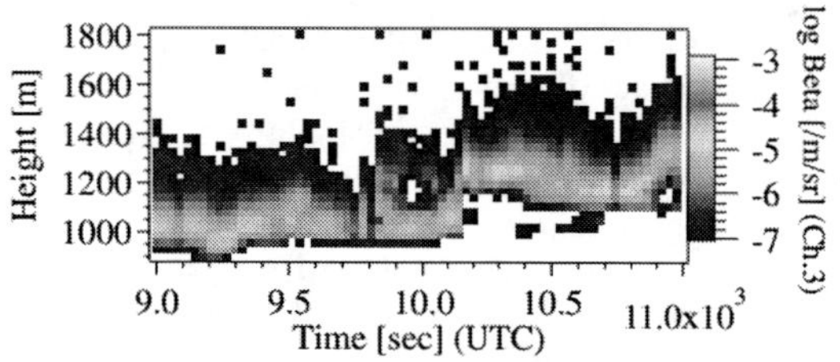

(b)

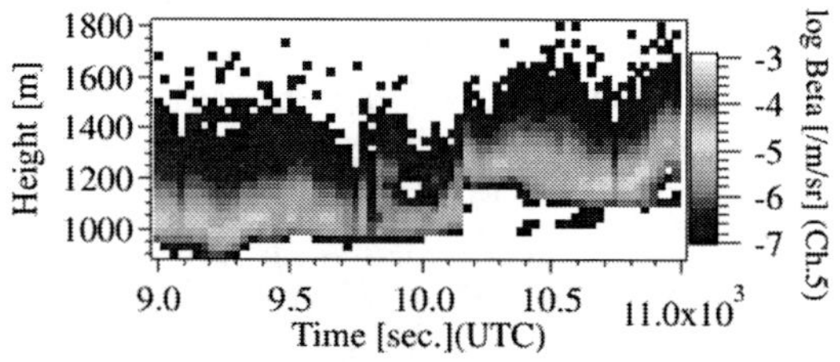

(c)

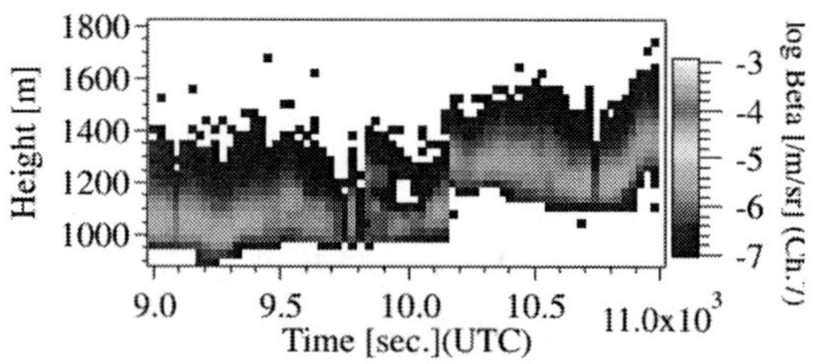

(d)

Fig.2. Time-height plot of apparent backscattering coefficient for parallel channels 1,3,5 and 7. As channel number becomes larger, the direction of the receiver deviated from the direction of laser beam.

Cloud top height (CTH) tended to be higher for outer channels as expected. The CTH for on-beam channel (Ch.1) was lower than 1.2km at about 9000 seconds and it increased to 1.4km at 10500 seconds. The CTH of the same clouds detected by channel 3 was generally higher than that detected by channel 1, e.g., 1.4km at 9000 seconds and 1.6km at 10500 seconds.

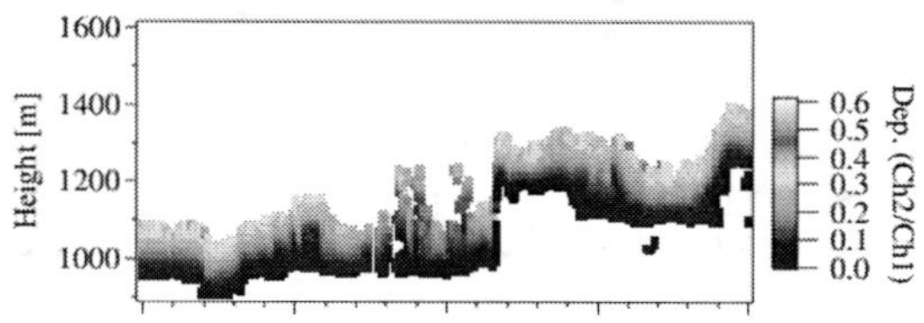

(a)

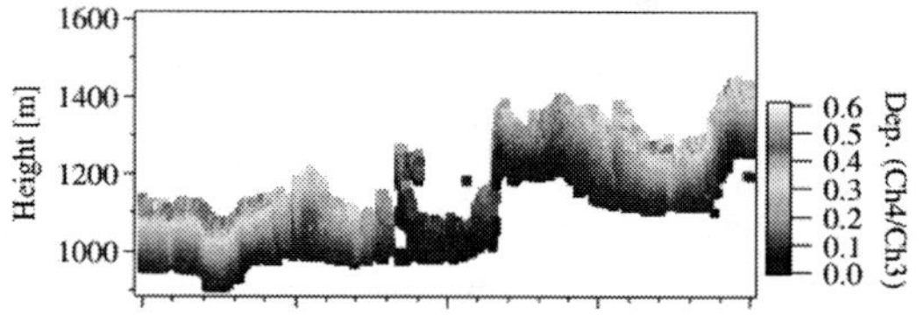

(b)

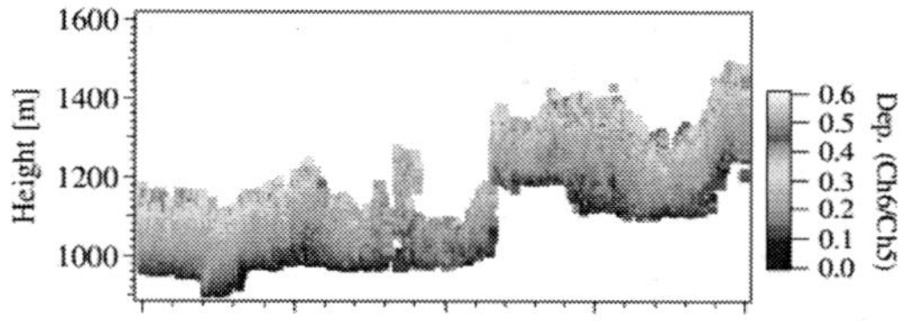

(c)

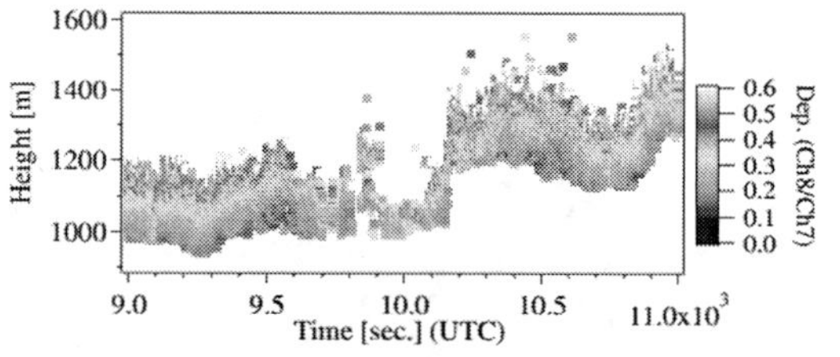

(d)

Fig.3. Same as Fig.3 but for depolarization ratio. Dep. (Ch1/Ch2) in the figure denotes depolarization ratio was estimated by using channels 1 and 2.

The depolarization ratio was also estimated for each set of parallel and perpendicular channels (Fig. 3). The depolarization ratio increased as altitude increased. The value estimated using channel 7 and 8, corresponding to the angle of 30 mrad, often reached 50%. Therefore, it was suggested that the MFMSPL have a potential to capture the similar feature found in the CALIPSO lidar observations of optically thick clouds where multiple scattering contribution was significant.

4. SUMMARY

We have developed the new type of ground-based lidar, MFMSPL, that can observe depolarization signals from optically thick clouds. It has 8 telescopes for detectors; 4 for parallel and 4 for perpendicular channels. The total FOV was about 30mrad. The lidar has started observations since June 2014. The MFMSPL was designed to achieve similar footprint size of CALIPSO lidar at altitude of 1km. The MFMSPL successfully detected much higher cloud top compared with that by conventional lidar at the same place. The depolarization ratio from off-beam channel was comparable to the values detected by the CALIPSO lidar. Co-located observations of the MFMSPL and 95GHz cloud radar have been conducted from February and April in 2015. Analysis of the MFMSPL and the cloud radar would provide a unique opportunity to develop and evaluate the performance of the system. And it is also expected that the system can help to develop the retrieval algorithm for water microphysics using CALIPSO and CloudSat data as previously conducted for the ice microphysics [8]. The JAXA-ESA mission EarthCARE was scheduled to be launched in 2018 and the EarthCARE will carry high spectral resolution lidar at 355nm and 94GHz Doppler cloud radar. Using the MFMSPL data, similar development should be possible for the new mission. The MFMSPL can also provide the fundamental data for the study of interaction between clouds and aerosols.

ACKNOWLEDGEMENT

This work was supported in part by the JSPS Kakenhi Grant Numbers 25247078 and 15K17762 and also supported by the Collaborated Research Program of Research Institute for Applied Mechanics, Kyushu University.

REFERENCES

[1] Hu, Y., D. Winker, M. Vaughan, B. Lin, A. Omar, C. Trepte, D. Flittner, P. Yang, S. L. Nasiri, B. Baum, W. Sun, Z. Liu, Z. Wang, S. Young, K. Stamnes, J. Huang, R. Kuehn and R. Holz, 2009: CALIPSO/CALIOP Cloud Phase Discrimination Algorithm, *J. Atmos. Oceanic Technol.*, **26**, 2293-2309, doi: /10.1175/2009JTECHA1280.1.

[2] Yoshida, R., H. Okamoto, Y. Hagihara and H. Ishimoto, 2010: Global analysis of cloud phase and ice crystal orientation from Cloiud-Aerosol Lidar and Infrared Pathfinder Satellite Observation (CALIPSO) data using attenuated backscattering and depolarization ratio, *J. Geophys. Atmos.*, **115**, D00H32, doi:1029/2009JD012334.

[3] Polonsky, N. I., S. P. Love and A. B. Davis, 2005: Wide-Angle Imaging Lidar development at the ARM Southern Great Plains site: Intercomparison of cloud property retrievals, *J. Atmos. Oceanic Technol.*, **22**, 628-648.

[4] Davis, B. A. 2008: Multiple-scattering lidar from both sides of the clouds: Addressing internal strudcture, *J. Geophys. Res.*, **113, D14S10**, doi:10.1029/2007JD009666.

[5] Caharan, R. F., M. McGill, and J. Kolansinski, 2005: THOR-Cloud thickness from offbeam lidar returns, *J. Atmos. Oceanic Technol.*, **22**, 605-627.

[6] Okamoto, H., T. Nishizawa, T. Takemura, H. Kumagai, H. Kuroiwa, N. Sugimoto, I. Matsui, A. Shimizu, S. Emori, A. Kamei and T. Nakajima, 2007: Vertical cloud structure observed from shipborne radar and lidar: Midlatitude case study during the MR01/K02 cruise of the research vessel Mirai, *J. Geophys. Res.*, **112**, D08216, doi:10.1029/2006JD007628.

[7] Hagihara, Y., H. Okamoto and R. Yoshida, 2010: Development of a combined CloudSat-CALIPSO cloud mask to show global cloud distribution, *J. Geophys. Res.*, **115**, D00H33, doi:10.1029/2009JD012344.

[8] Okamoto, H., K. Sato and Y. Hagihara, 2010: Global analysis of ice microphysics from CloudSat and CALIPSO: intercomparison of specular reflection in lidar signals, *J. Geophys. Res.*, **115**, D22209, doi:10.1029/2009JD013383.

FROM ANTARCTICA LIDAR DISCOVERIES TO OASIS EXPLORATION

Xinzhao Chu[1]*, Zhibin Yu[1], Weichun Fong[1], Cao Chen[1], Jian Zhao[1], Ian F. Barry[1], John A. Smith[1], Xian Lu[1], Wentao Huang[1], and Chester S. Gardner[2]

*[1]Cooperative Institute for Research in Environmental Sciences & Department of Aerospace Engineering Sciences, University of Colorado Boulder, CO 80309, USA, *Email: Xinzhao.Chu@colorado.edu*

[2]University of Illinois at Urbana-Champaign, Urbana, IL 61801, USA

ABSTRACT

Stunning new science discoveries including neutral thermospheric metal layers in the 100–200 km altitude from McMurdo lidar campaign and other world lidar observations have led to a new initiative in the middle and upper atmosphere science community—the very large-aperture lidar Observatory for Atmosphere Space Interaction Studies (OASIS). These discoveries and the recent technology breakthroughs in Fe and Na Doppler lidars are presented to illustrate the science drivers and technology foundations forming the basis for OASIS.

1. INTRODUCTION

Can lidar observations and resulting discoveries help advance humanity's exploration of the universe, especially in the search for exoplanets? This is a question to which all lidar researchers wish to answer a big "YES" but wondered "how" in the past. In this paper we present a pathway, enabled by the new lidar discoveries from Antarctica and then from the rest of the world, which can make good contributions to this question through exploring the Earth's Space-Atmosphere Interaction Region (SAIR) with lidar.

The pathway starts at understanding the Earth—our home in the universe. Earth has remained habitable during the last 3.5 billion years because of the effects of a set of key universal processes that govern the evolution of planetary atmospheres everywhere. For example, the Earth's magnetic field shields our atmosphere from the energetic particles in the solar wind that are partially responsible for stripping Mars of its primordial atmosphere. The Earth's carbonate-silicate cycle provides a feedback mechanism that controls atmospheric CO_2 levels and climate over long time scales, preventing the runaway greenhouse effect that is responsible for the high temperatures on Venus today. Many coupled feedback processes are at play in a planetary atmosphere system. But most of these are unknown and cannot be addressed without a complete description of the atmosphere, from lower interaction with land/ocean to upper interaction with space [1].

The natural upward extension of a planet's atmosphere ultimately leads to its interaction with space, where atmospheric neutral gases entwine with the dynamic plasma of space. This SAIR is common to all planetary systems, yet its properties, and the processes that govern them, are not described sufficiently to fully understand its role in an atmosphere's development and evolution. However, the SAIR is known to be essential for sustaining life on Earth by absorbing extreme solar radiation, ablating meteoric materials, regulating gaseous escape, dissipating energetic particles and fields from space, while balancing influences from the planet itself in the form of wave energy and momentum originating from the lower atmosphere.

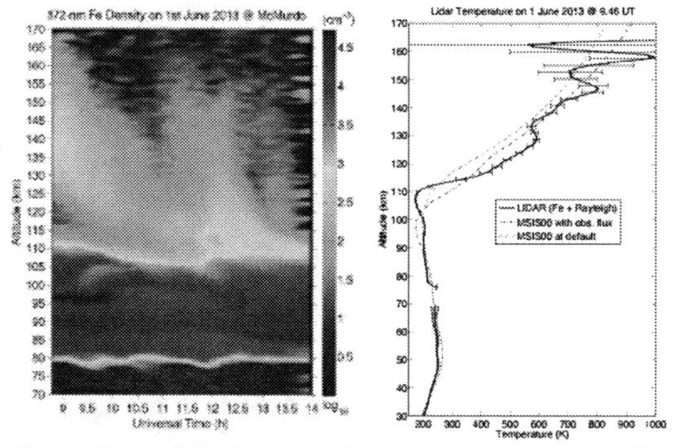

Fig. 1. Neutral Fe layers and temperatures observed by Fe lidar reaching 170 km at McMurdo, Antarctica.

The overarching goal of the pathway is to significantly advance our understanding of the fundamental, universal processes in the Earth's SAIR and how they shape the atmospheres of Earth-like planets throughout our galaxy. By doing so with advanced lidars and cluster instruments, the world's lidar researchers provide spectroscopic signatures for Earth-like planets!

This pathway remained a dream in the past, because no lidars could detect the atmosphere or space above ~110 km, until the discoveries of neutral metal layers in the altitude range of 100–200 km made by an Fe Boltzmann lidar at McMurdo, Antarctica [2] and then followed by other lidar measurements in the rest of the world. Such neutral metal layers with an example shown in Figure 1a provide excellent tracers for lidar to observe the neutral atmosphere into the thermosphere and their interaction with the ionosphere. This stunning lidar observation at McMurdo was made with a telescope of 40-cm diameter. To turn the dream into reality, scientists from the atmosphere and space science community began to consider 100 m^2 telescopes with advanced next-generation lidars. We name this pathway OASIS—The Observatory for Atmosphere Space Interaction Studies [1].

In this paper several new science discoveries during the McMurdo lidar campaign from 2011 to 2015, along with the most recent lidar technology breakthroughs, are presented to illustrate the science drivers and technology foundations forming the basis for OASIS.

2. MCMURDO LIDAR CAMPAIGN

The McMurdo lidar campaign started in Dec 2010 when an Fe Boltzmann temperature lidar was installed at the Arrival Heights observatory (77.8°S, 166.7°E) by the University of Colorado via collaboration between United States Antarctic Program and Antarctica New Zealand [3]. The lidar has been operated there ever since. The lidar observations are all year round and cover full diurnal cycle, weather permitting. Four graduate students (Zhibin Yu, Brendan Roberts, Weichun Fong, and Cao Chen) have braved through the harsh Antarctic winters of 2011–2014, and the 5th winter-over student (Jian Zhao) is operating the lidar to collect data in 2015. The rich lidar data from McMurdo cover from ~10 km all the way to over 170 km.

McMurdo lidar data have enabled many new science studies, in combination with co-located radar and satellite observations as well as numerical modeling, e.g., the discoveries of thermospheric Fe layers, fast gravity waves and elevated temperatures under geomagnetic storms [2], polar mesospheric clouds at 78°S [3], diurnal variations of Fe layers and solar effects on the layer bottomside [4], two simultaneous inertia-gravity waves in Antarctica [5], large eastward planetary waves in the stratosphere [6], temperature tides from 30–110 km [7], and vertical evolution of potential energy density and vertical wavenumber spectrum of Antarctic gravity waves from 35 to 105 km [8]. The lidar temperature data led to a discovery of fast amplitude growth with altitude of diurnal temperature tides in the winter lower thermosphere and dependence on geomagnetic activity, which was then studied with CTIPe model and attributed to the Hall-ion-drag induced adiabatic cooling and heating [9].

These discoveries and observations provide unique opportunities to study the complex physical, chemical, dynamical, energetic and electro-dynamical processes in the Earth's atmosphere and space environment. Here we choose a few new studies to present.

3. THERMOSPHERIC METAL LAYERS OPEN A NEW DOOR TO EXPLORE SAIR

The most intriguing discovery is perhaps the neutral Fe layers with clear gravity wave signatures observed well into the thermosphere of 100–200 km altitude. Such discovery challenges the current understandings of the upper atmosphere composition, chemistry, dynamics and energetics, while the neutral metal layers open the door for lidars to measure neutral temperatures and winds in the thermosphere directly. Measurements of the neutral thermosphere are woefully incomplete and in critical need to advance our understanding of and ability to predict the SAIR. Figure 1b demonstrates one of the first lidar measurements of neutral temperatures from 30 to 170 km on 1 June 2013 at McMurdo. Modern Fe Doppler lidars can certainly use such metal layers to measure the neutral winds directly.

An earlier example on 28 May 2011 exhibits clear gravity wave features from 35 to 155 km with periods of 1.5–2 h (see Figure 2a) [2]. *Chu et al.* [2] hypothesize that the thermospheric Fe layers are formed through the neutralization of vertically converged Fe^+ layers that descend in height following the gravity wave downward phase progression. Conversion from Fe^+ to Fe in the thermosphere is most likely through the direct recombination of Fe^+ with electron: $Fe^+ + e^- \rightarrow Fe + h\nu$. Except for this qualitative explanation, we know very little about these layers, e.g., why, when, and where these thermospheric layers occur, how they vary, and to what altitude range they extend. Although the main metal layers between ~75 and ~105 km come from cosmic dust entering the Earth's atmosphere, it is puzzling how these heavy neutral metals are transported to such high altitudes.

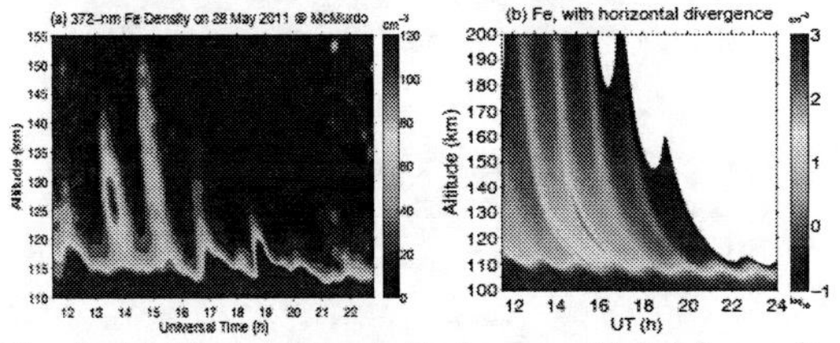

Fig. 2. (a) Thermospheric Fe layers with 1.5-h gravity waves up to 155 km observed by lidar on 28 May 2011 at McMurdo [2]. (b) Numerical simulation of Fe layers by the University of Colorado Fe/Fe^+ model.

To quantitatively explain the observations and advance the understanding, we have developed a thermospheric Fe/Fe^+ model for the polar region. Details of this model can be found in *Yu and Chu* [10] and in a PhD dissertation of Dr. Zhibin Yu. Here we illustrate one of the model results in Figure 2b and briefly introduce some important model findings. Figure 2b represents a nice reproduction of the Fe layer observations, except the modeled contrast is not as high as the observations.

Numerical simulations confirm the hypothesis by *Chu et al.* [2] that the thermospheric Fe layers are produced through the neutralization of vertically converged Fe^+

DOI: 10.1051/epjconf/201611912001

layers. The Fe/Fe^+ model quantitatively demonstrates that the conversion from Fe^+ to Fe above 120 km is mainly through the direct electron-Fe^+ recombination. While the meteor injection rate of iron species is negligible above 120 km, we find that the polar electric field is sufficient in transporting Fe^+ ions from their main deposition region (~80–110 km) to the E–F regions, therefore supplying the source for Fe atoms in the thermosphere (see Figure 3).

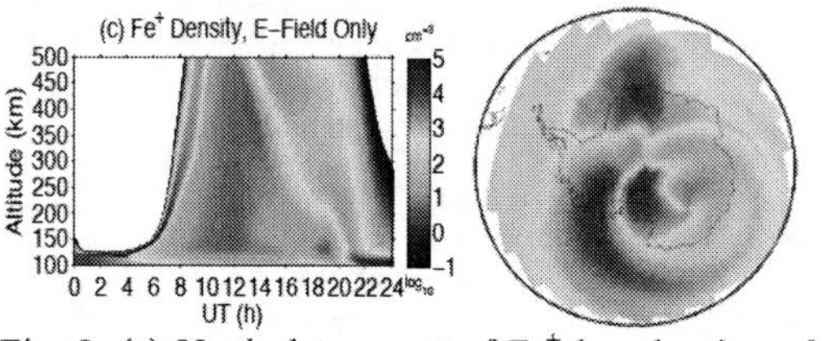

Fig. 3. (a) Vertical transport of Fe^+ ions by the polar electric field on 28 May 2011 at McMurdo. (b) A map of vertical ion drift velocity at 117 km: Ion upward transport (red) patch is over McMurdo for many hours.

Furthermore, the model predicts that diurnal variations of the convection electric field (Figure 3b) should lead to diurnal occurrence of thermospheric Fe layers. This has been proven by the 50 hours of continuous Fe observations taken by Mr. Cao Chen in June 2014 at McMurdo (see Figure 4). Fe layers occurred three times with the layer height peaking around 8 UT.

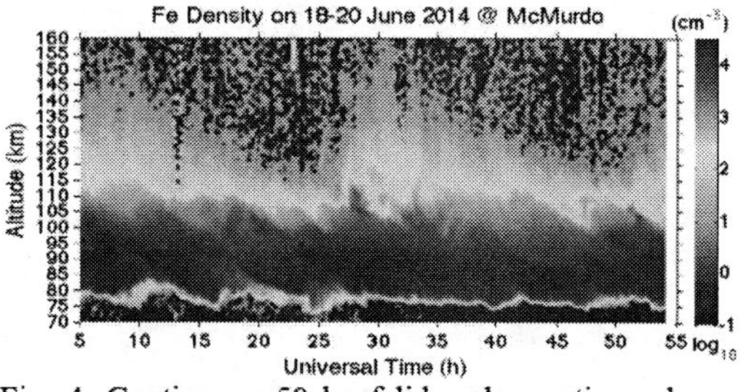

Fig. 4. Continuous 50 h of lidar observations show a diurnal period occurrence of thermospheric Fe layers.

The thermospheric metal layers are not limited to McMurdo but appear to be a global phenomenon. Indeed, similar layers have been observed at other Antarctic stations, e.g., Fe layers at Davis [11] and Na layers at Syowa. Similar metal layers have also been observed at mid- and low-latitudes following the McMurdo discoveries, e.g., K layers up to 155 km at Arecibo, Puerto Richo, and Na layers up to 170 km at Lijiang, China and Cerro Pachon, Chile [12]. In principle, as long as metal ions like Fe^+, Na^+ and K^+ can be transported upward, neutral metal atoms can be formed through direct recombination so they are detectable by lidars with sufficient detection sensitivity. These metals provide tracers for the world's lidar researchers to measure the SAIR worldwide!

4. POLAR WAVES SHAPING THE SAIR

The recent McMurdo lidar data in March 2015 indicate the Fe layers reaching 180 km with ~1–3 h wave period. Such waves frequently occur and survive into the thermosphere as shown in Figures 1, 2 and 4. They have gravity wave nature. However, the 4–9 h waves demonstrated in Figure 5 are so persistent that they have been observed lasting throughout 3 days in winter and even 7 days of a lidar marathon in summer. Are they relatively localized inertia-gravity waves or planetary-scale inertia gravity waves (forced Class I gravity mode)? This remains a mystery–for details see *Chen et al.* [13]. Considering such waves easily cause ±30 K temperature perturbations near 100 km, it is important to identify its nature, sources and impacts.

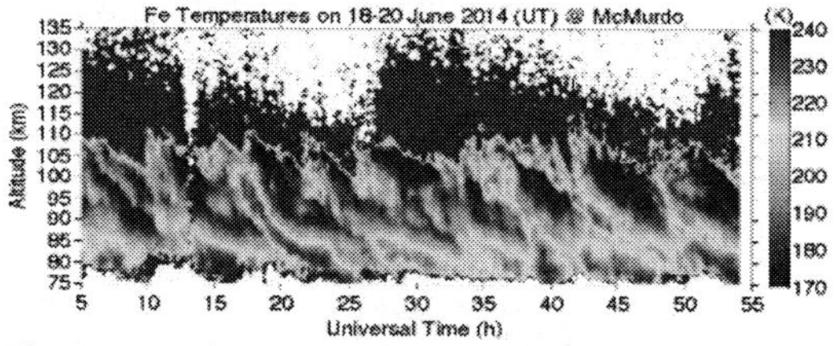

Fig. 5. Continuous 50-h temperature measurements show persistent 4–9 h waves on 18–20 June 2014.

In addition to the several studies on gravity, tidal and planetary waves mentioned earlier, the extensive datasets collected in May 2014 show the long-period temperature variations over 5–6 days (Figure 6), indicating planetary wave influences in the mesopause region. Atmospheric waves across various spatial and temporal scales transport energy, momentum and constituents among different atmospheric regions, so shape the SAIR significantly.

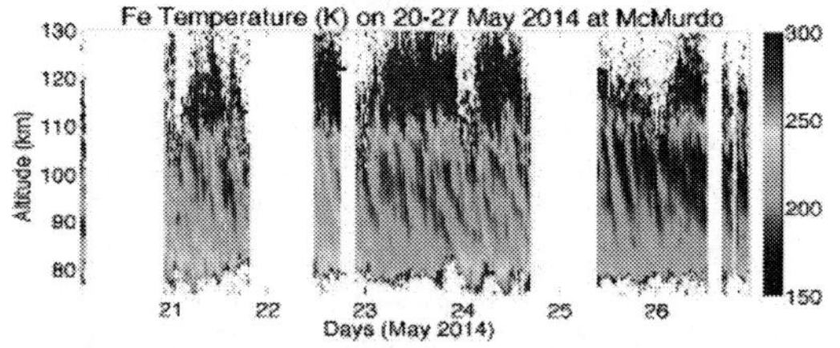

Fig. 6. Temperatures measured over 6 days in May 2014 indicate planetary wave influences.

5. RECENT LIDAR BREAKTHROUGHS

Several new powerful Fe and Na Doppler lidars with simultaneous wind, temperature & density capabilities have been developed recently. They represent some significant advancement in lidar technologies. One is high-efficiency lidar receiver architecture developed at the University of Colorado Boulder that allows the Na Doppler lidars to achieve very high resolutions with ~80-cm diameter telescopes, enabling new science inquiries from turbulence to the middle thermosphere [12]. Another is the Major Research Instrumentation (MRI) Fe/Rayleigh/Mie Doppler lidar that has revolutionized the resonance lidar transmitter and receiver technologies. The advanced concepts of modular resonance Doppler lidar are integrated with a unique Pulsed Alexandrite Ring Laser (PARL) and a new acousto-optic modulator (AOM) path design, along with absolute frequency stabilization of laser pulses using Doppler-free Fe spectroscopy and optical heterodyne detection (OHD) techniques. Combined with high-efficiency receiver and data acquisition system, the MRI Fe Doppler lidar has achieved superb Fe signal levels at Table Mountain in 2014, making this lidar a new workhorse for SAIR exploration. Details will be given in the full paper.

6. CONCLUSIONS

Three basic questions have commanded the attention of atmosphere and space scientists for generations [1] –
1) What are the fundamental processes that shape the Earth's upper atmosphere and govern its evolution?
2) How do these processes affect weather and climate?
3) What roles do they play on other planets?
These questions are especially relevant today as rising concentrations of greenhouse gases are changing Earth's climate and increasing numbers of extra-solar planets are being discovered in our galaxy.

Using the recently discovered neutral metal layers in the thermosphere, the modern Doppler lidars can directly measure the neutral winds and temperatures along with metal species densities to address key SAIR science issues. Because the densities of thermospheric metal layers are usually 50–1000 times lower than the main layer peak at ~90 km, the very large-aperture and high-power lidars as envisioned by the OASIS are required to realize the world lidar researchers' dreams to answer these key questions.

The OASIS pathway will enable lidar researchers to explore the plasma-neutral atmosphere coupling in SAIR, investigate transport and turbulence processes, and study influences of space processes (e.g., cosmic dust, energetic particle precipitation, and radiation) and planetary processes (e.g., compositions and waves). Understanding these universal processes will certainly help advance the exploration of the universe.

ACKNOWLEDGEMENT

We sincerely acknowledge the staff of United States Antarctic Program, McMurdo station, Antarctica New Zealand, and Scott Base for their superb support. This work was supported by NSF grants ANT-0839091, PLR-1246405, MRI-0723229, and AGS-1136272.

REFERENCES

[1] OASIS, 2014: Exploring the interaction of Earth's atmosphere with space, *Report to NSF*, available at http://rsss.csl.illinois.edu/workshop.
[2] Chu, X., Z. Yu, C. S. Gardner, C. Chen, and W. Fong, 2011: Lidar observations of neutral Fe layers and fast gravity waves in the thermosphere (110–155 km) at McMurdo (77.8°S, 166.7°E), Antarctica, *Geophys. Res. Lett.*, **38**, L23807, doi:10.1029/2011GL050016.
[3] Chu, X., et al., 2011: First lidar observations of polar mesospheric clouds and Fe temperatures at McMurdo (77.8°S, 166.7°E), Antarctica, *Geophys. Res. Lett.*, **38**, L16810, doi:10.1029/2011GL048373.
[4] Yu, Z., et al., 2012: Diurnal variations of the Fe layer in the mesosphere and lower thermosphere: Four season variability and solar effects on the layer bottomside at McMurdo (77.8°S, 166.7°E), Antarctica, *J. Geophys. Res.*, **117**, doi:10.1029/2012JD018079.
[5] Chen, C., et al., 2013: Inertia-gravity waves in Antarctica: A case study using simultaneous lidar and radar measurements at McMurdo/Scott Base, *J. Geophys. Res.*, **118**, doi:10.1002/jgrd.50318.
[6] Lu, X., et al., 2013: Eastward propagating planetary waves with periods of 1-5 days in the winter Antarctic stratosphere as revealed by MERRA and lidar, *J. Geophys. Res.*, **118**, doi:10.1002/jgrd.50717.
[7] Fong, W., et al., 2014: Winter temperature tides from 30 to 110 km at McMurdo, Antarctica: Lidar observations and comparisons with WAM, *J. Geophys. Res.*, **119**, doi:10.1002/ 2013JD020784.
[8] Lu, X., et al., 2015: Vertical evolution of potential energy density and vertical wave number spectrum of Antarctic gravity waves from 35 to 105 km at McMurdo (77.8°S, 166.7°E), *J. Geophys. Res.*, **120**, doi:10.1002/2014JD022751.
[9] Fong, W., X. Chu, X. Lu, C. Chen, T. J. Fuller-Rowell, M. Codrescu, and A. D. Richmond, 2015: Lidar and CTIPe model studies of the fast amplitude growth with altitude of the diurnal temperature "tides" in the Antarctic winter lower thermosphere and dependence on geomagnetic activity, *Geophys. Res. Lett.*, **42**, doi:10.1002/2014GL062784.
[10] Yu, Z., and X. Chu, 2015: Source, formation and evolution of thermospheric Fe layers: Exploration with a thermospheric Fe/Fe^+ model, *J. Geophys. Res.*, manuscript in preparation.
[11] Lübken, F.-J., et al., 2011: First measurements of thermal tides in the summer mesopause region at Antarctic latitudes, *Geophys. Res. Lett.*, **38**, L24806, doi:10.1029/2011GL050045.
[12] Smith, J. A., and X. Chu, 2015: High-efficiency receiver architecture for resonance-fluorescence and Doppler lidars, *App. Opt.*, **54(11)**, pp. 3173-3184, doi: 10.1364/AO.54.003173.
[13] Chen, C., et al., 2015: Wave dynamics mystery discovered by lidar in Antarctica, *Proceeding of 27th International Laser Radar Conference*, New York.

THE TECHNIQUES AND PROGRESS OF WIND AND TEMPERATURE LIDAR IN WIPM

Faquan Li[1*], Yong Yang[1,2], Xuewu Cheng[1], Yajuan Li[1,3], Xin Lin[1], Yuan Xia[1,3], Linmei Liu[1], Shalei Song[1], Zhenwei Chen[1], Jun Xiong[1,3], Kuijun Wu[1,3], Shunsheng Gong[1]

[1] State Key Laboratory of Magnetic Resonance and Atomic and Molecular Physics, Wuhan Institute of Physics and Mathematics, Chinese Academy of Sciences, Wuhan 430071, China,

**Email: lifaquan@wipm.ac.cn*

2 Department of Atmospheric and Planetary Sciences, Center for Atmospheric Sciences, Hampton University, Hampton, Virginia 23668, USA,

3 University of Chinese Academy of Sciences, Beijing 100049, China

ABSTRACT

In this paper, a lidar system in Wuhan Institute of Physics and Mathematics (WIPM, 30.5°N,114.5°E) for atmospheric density, temperature and wind observation was reported. The design and techniques of this lidar were described in detail. The atmospheric temperature of the troposphere, stratosphere and lower mesosphere were measured by the Raman, Rayleigh and sodium channel of this lidar system, respectively.

1. INTRODUCTION

The vertical profiles of atmospheric density, temperature and wind describe the variation of atmospheric phenomena directly. They are key parameters for the research and application of atmospheric dynamics and thermodynamics[1, 2].

The lidar (light detection and range) was developed to observe the atmospheric density, temperature and wind with high spatial resolution, high temporal resolution and high spectral resolution. The Rayleigh lidar measure the temperature of the stratosphere and mesosphere over altitudes of ~35 km to ~80 km[3]. The altitude lower limit for a Rayleigh temperature lidar is decided by the aerosol in the lower atmosphere. The vibrational and pure rotational Raman lidar which is based on the pure molecular scattering without Mie scattering by aerosol particles, is developed to measure the atmosphere from near ground to the stratosphere[4-7]. The altitude upper limit for the Rayleigh temperature lidar depends on the atmospheric molecular number density and lidar parameters, especially the laser power and effective area of the optical receivers. The metal resonance fluorescence lidar is employed to observe the mesosphere and lower thermosphere, since the effective cross-section of the metal atoms in the mesopause is several orders of magnitude larger than that of the Rayleigh scattering[8-10]. The sodium lidar is one of the most popular metal resonance fluorescence lidars due to its high population and large cross section[11, 12]. The Rayleigh lidar, Raman lidar and potassium resonance fluorescence lidar were combined to observe the whole profile of the atmospheric temperature from 1 to 105 km[13].

In this paper, a lidar system for atmospheric observation was reported. This lidar system was designed to observe the vertical profiles of the atmospheric density, temperature and wind. The Rotational Raman channel, Rayleigh temperature channel and sodium resonance fluorescence channel of this lidar system were employed to measure the temperature of the troposphere, stratosphere and mesopause respectively. The design and progress of this system were described in section 2. In section 3, the observational results of temperature profile was discussed.

2. METHODOLOGY

As a part of the APSOS (Atmospheric Profiling Synthetic Observation System) project granted by the NSFC (National natural Science Foundation of China), a new lidar system is designed and under construction aiming to investigate the neutral atmospheric dynamics and thermodynamic progress, especially in the tropopause, stratopause and mesopause area. Compared with the old lidar system in WIPM (Wuhan Institute of Physics and Mathematics)[14], the pulsed lasers of the transmitter in the new lidar are narrow band with single mode, and therefore more atmospheric parameters will be measured by this new system, especially the wind and the temperature in the troposphere and mesopause. This lidar system includes the transmitter, optical receiver and signal and data platform, as shown in figure 1.

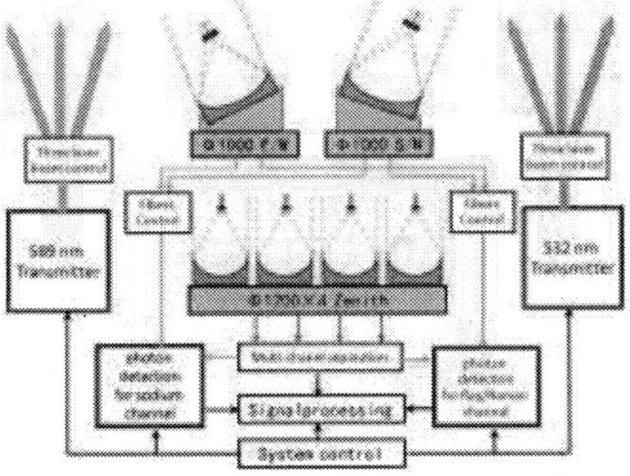

Figure 1 the schematic of the lidar system

The transmitter includes a 589 nm yellow laser and a 532 nm green laser, as shown in figure 2. The frequency of the 532 nm laser is locked by the absorption spectrum of the iodine line 1109.

Figure 2 589 nm and 532 nm laser beams of the transmitter

The frequency of 589 nm seeder laser is locked by Doppler-free spectroscopy of the sodium D2 line. Then, two acousto-optic modulators are employed to generate three frequencies with a shift of 630 MHz, 0 and -630 MHz to the seeder laser, respectively[11]. The cw seeder laser is amplified by a pulsed dye amplifier. These three frequencies are changed one by one with a period of 30 seconds to measure the number of neutral sodium atoms with three kinds of different velocity, which are sensitive to the wind and temperature.

The optical receiver is combined by three telescopes, two for west and north direction observing with an effective area of about 0.78 m^2 each other, and one combined telescope for zenith observing with an effective area of about 1.13 m^2. The combined telescope is combined by four telescopes with a diameter of 1.2 m each other, and will be shared for multi channels. The lidar echoes excited by multi laser beams are focused at different positions of the focal plane of the telescope and received by their own fibers, as shown in figure 3. This technique has been verified in the optical receiver of the old lidar system in WIPM[14].

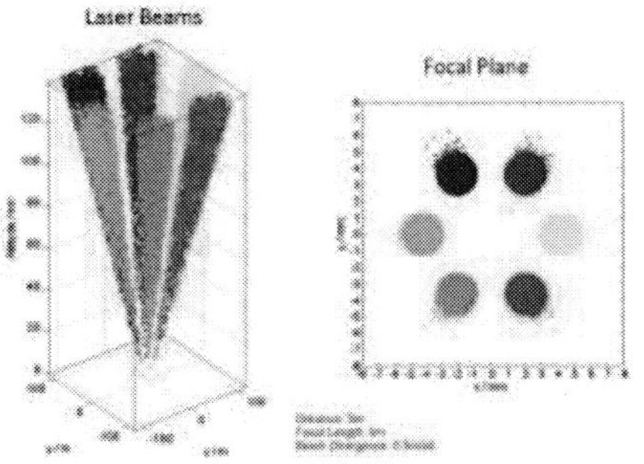

Figure 3 laser beams (left) and their image (right) in the focal plane of the telescope

The signal processing and data platform records the lidar echoes and processes 16 data sets. One of them is for the Rayleigh temperature channel, two for the rotational Raman channel, four for the Doppler wind channel and nine for the sodium channel. An iodine filter is employed in the Doppler wind channel to discriminate the Doppler shift induced by the atmospheric wind. An interference-filter-based polychromator is used in the rotational Raman channel to block the Rayleigh and Mie scattering and split the Raman scattering echoes with high and low quantum number J into two photomultiplier tubes. Based on lidar echoes, the data platform retrieves the aerosol backscatter coefficient, atmospheric density in the stratosphere and mesosphere and sodium density in the mesopause. The temperature for altitudes between 5~40 km is retrieved within the pure rotational Raman scattering methodology. The temperature in the mesosphere is retrieved based on the ideal gas law and hydrostatic equilibrium assumption. The echoes from the sodium atoms with different velocities of thermal motion are used to calculate the temperature ratio and wind ratio, and therefore to obtain the wind and temperature. The relationships between the temperature, wind and their ratios are show in figure 4.

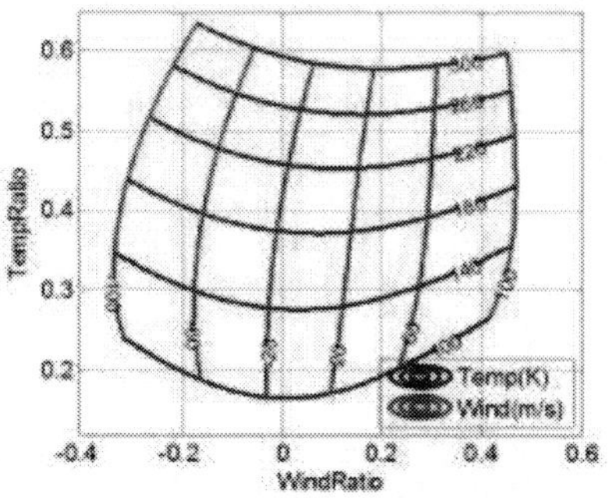

Figure 4 The temperature ratio and wind ratio are used to calculate the temperature and wind

3. RESULTS

The new lidar system is still under construction in the laboratory and the telescopes have not been installed so far. However, a small telescope has been used as a substitute to tune and adjust the lidar system. The lidar echoes of all channels have been obtained, and some of them are shown in figure 5 and 6.

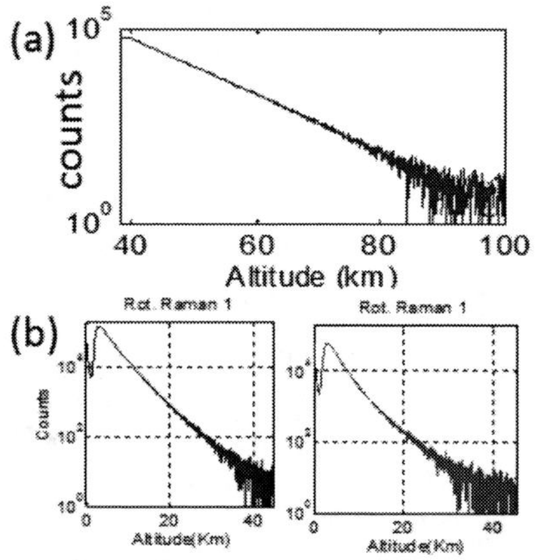

Figure 5 lidar echoes of Rayleigh channel(a) and Raman channel(b)

The atmospheric density, temperature and wind are measured based on the observation of the rotational Raman channel, Rayleigh temperature channel, sodium channel and Doppler wind channel. The vertical profiles of temperature will be shown in this section. Figure 7 shows the temperature in the troposphere and stratosphere observed by the rotational Raman channel with a 55 min temporal resolution and 75m spatial resolution. The temperature measured by the lidar (blue solid line in the left panel of figure 7) agrees well with the radiosonde measurements (red circular in the left panel of the figure 7) provided by the Wuhan Weather Station[7].

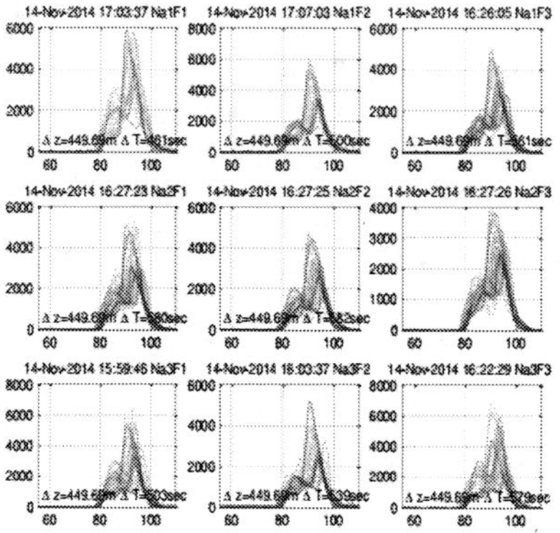

Figure 6 lidar echoes of sodium channel

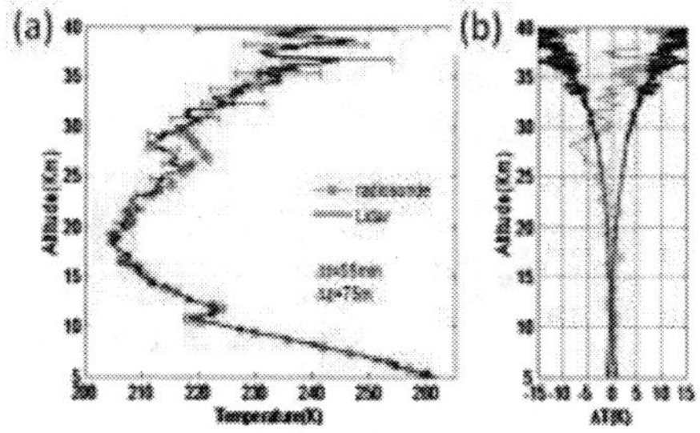

Figure 7 Temperature profile (left) and its error (right) measured by the rotational Raman channel

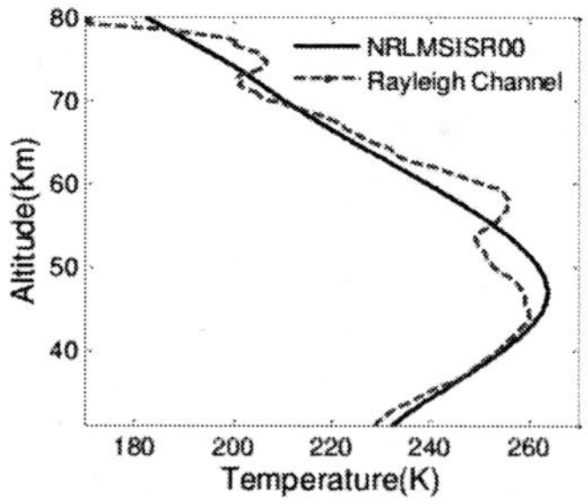

Figure 8 Temperature in the stratosphere and mesosphere measured by the Rayleigh channel

The temperature profile of the stratosphere and mesosphere retrieved from the Rayleigh signal is shown in Figure 8. Figure 9 shows the temperature profile of the mesopause measured by the sodium channel. The temporal and spatial resolution are 30 min and 900 m, respectively. The lidar observed temperature is consistent with the atmospheric model. The measurement precision will be scaled in further data process.

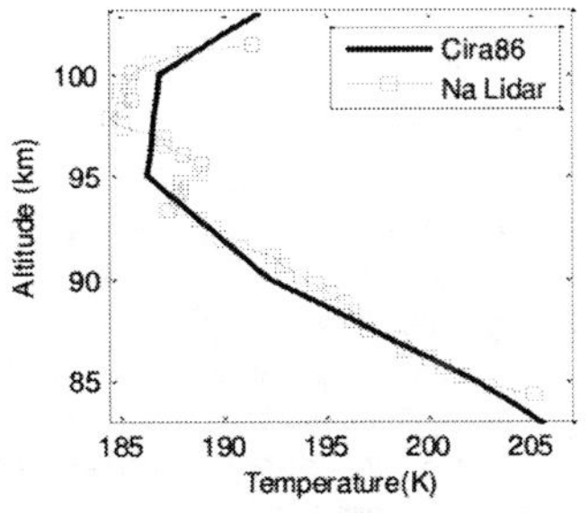

Figure 9 The temperature profile of the mesopause measured by the sodium channel

The temperature results from Raman, Rayleigh and sodium channel will be merged to obtain the whole atmosphere thermodynamic structure. The temperature profiles from Raman and Rayleigh channel have a ~5-km overlap, which could verify the measurement for each other. The combined telescope will be used to enhance the effective area of the optical receiver in the next step. Thus, there would also be an overlap for the temperature profiles measured by the Rayleigh and sodium channel. This lidar system will be installed in two chambers and take part in the field observation.

4. CONCLUSIONS

A new lidar system was developed in Wuhan Institute of Physics and Mathematics to observe the atmospheric density, temperature and wind. The design and progress of this lidar system was reported. The atmospheric temperature of the troposphere, stratosphere and mesopause were measured by the Raman, Rayleigh and sodium channel of this lidar respectively.

ACKNOWLEDGEMENT

The radiosonde data was supported by the Wuhan Weather Station. This work was supported by the National Natural Science Foundation of China, under the Grant nos. 11403085, 41127901, 41101334.

REFERENCES

[1]. Dunker T, Hoppe U-P, Feng W, Plane JMC, Marsh DR. 2015: Mesospheric temperatures and sodium properties measured with the ALOMAR Na lidar compared with WACCM.

[2]. Whiteway JA, Carswell AI. 1994: Rayleigh lidar observations of thermal structure and gravity wave activity in the high arctic during a stratospheric warming, *J Atmos Sci*, **51(21)**,3122-36.

[3]. Hauchecorne A, Chanin ML, Keckhut P. 1991: Climatology and trends of the middle atmospheric temperature (33–87 km) as seen by Rayleigh lidar over the south of France, *Journal of Geophysical Research: Atmospheres*, **96(D8)**,15297-309.

[4]. Cooney J. 1972: Measurement of atmospheric temperature profiles by Raman backscatter, *J Appl Meteorol*, **11(1)**,108-12.

[5]. Keckhut P, Chanin M, Hauchecorne A. 1990: Stratosphere temperature measurement using Raman lidar, *Applied optics*, **29(34)**,5182-6.

[6]. Di Girolamo P. 2004: Rotational Raman Lidar measurements of atmospheric temperature in the UV, *Geophysical Research Letters*, **31(1)**.

[7]. YaJuan L, ShaLei S, Faquan L, Xuewu C, Zhenwei C, Linmei L, et al. 2015: High Precision Measurements of lower atmospheric temperature based on Pure Rotational Raman Lidar, submitted to *Chinese Journal of Geophysics*.

[8]. Chu X, Pan W, Papen GC, Gardner CS, Gelbwachs JA. 2002: Fe Boltzmann temperature lidar: design, error analysis, and initial results at the North and South Poles, *Applied Optics*, **41(21)**,4400-10.

[9]. Höffner J, Lautenbach J. 2009: Daylight measurements of mesopause temperature and vertical wind with the mobile scanning iron lidar, *Optics Letters*, **34(9)**,1351-3.

[10]. Gibson AJ, Thomas L, Bhattachacharyya SK. 1979: Laser observations of the ground-state hyperfine structure of sodium and of temperatures in the upper atmosphere, *Nature*, **281**.

[11]. She CY, Sherman J, Yuan T, Williams BP, Arnold K, Kawahara TD, et al. 2003: The first 80-hour continuous lidar campaign for simultaneous observation of mesopause region temperature and wind, *Geophysical Research Letters*, **30(6)**,1319.

[12]. Hu XO, Yan ZA, Guo SY, Cheng YQ, Gong JC. 2011: Sodium fluorescence Doppler lidar to measure atmospheric temperature in the mesopause region, *Chinese Sci Bull*, **56(4-5)**,417-23.

[13]. Alpers M, Eixmann R, Fricke-Begemann C, Gerding M, Hoeffner J. 2004: Temperature lidar measurements from 1 to 105 km altitude using resonance, Rayleigh, and Rotational Raman scattering, *Atmospheric Chemistry and Physics*, **4(3)**,793.

[14]. Cheng XW, Gong SS, Li FQ, Dai Y, Song J, Wang JM, et al. 2007: 24 h continuous observation of sodium layer over Wuhan by lidar, *Sci China Ser G*, **50(3)**,287-93.

EPJ Web of Conferences **119**, 12003 (2016) DOI: 10.1051/epjconf/201611912003
ILRC 27

WINTER TEMPERATURE AND TIDAL STRUCTURES FROM 2011 TO 2014 AT MCMURDO STATION: OBSERVATIONS FROM FE BOLTZMANN TEMPERATURE AND RAYLEIGH LIDAR

Weichun Fong[1,2], Xinzhao Chu[1,2], Xian Lu[1], Timothy J. Fuller-Rowell[1,3], Mihail Codrescu[3], Arthur D. Richmond[4], Zhibin Yu[1,2], Brendan Roberts[1,2], and Cao Chen[1,2]

[1] University of Colorado at Boulder, 216 UCB, CIRES, Boulder, CO 80309, USA.
Email: weichun.fong@colorado.edu and Xinzhao.Chu@colorado.edu
[2]Aerospace Engineering Sciences, University of Colorado, Boulder, CO, USA.
[3] SWPC, NOAA, Boulder, CO, USA. [4] HAO, NCAR, Boulder, CO, USA

ABSTRACT

McMurdo station (77.8°S, 166.7°E), locating at the poleward edge of the auroral oval, provides great opportunities for researchers to study the interactions among neutral atmosphere, ionosphere and magnetosphere. More than four years of valuable data have been collected, leading to several new discoveries from the McMurdo lidar campaign. Presented here are the winter temperature tides and their responses to the magnetospheric sources. Winter temperature structures from the lidar observations are also presented for this high southern latitude.

1. INTRODUCTION

Polar middle and upper atmosphere is complex and is one of the least understood regions on the Earth. Between 100 and 200 km, the ion-neutral collisions have apparent altitude dependence due to the exponentially decay of the neutral density, therefore, the neutral constituents and their dynamics have large impacts on the plasma dynamics [1]. Meanwhile, the open geomagnetic field lines allow the energetic particles penetrate into the lower thermosphere or even mesosphere under strong geomagnetic disturbance, and the energy carried by these energetic particles eventually transfer into the ionosphere-thermosphere (IT) system. Such strong coupling between the polar IT system and the magnetosphere complicates the situation when considering the energy budget and dynamics, and it is still unclear that how magnetosphere energy is dissipated and distributed spatially and temporally in this region.

Atmospheric temperature, as one of the important parameters, dominates numerous geophysical phenomena and is also a key variable to climate change studies. Temperature and its variations in the polar IT region are not only affected by the energy propagating from the lower atmosphere but also strongly related to the magnetospheric energy dissipation. However, temperature measurements of the high latitude regions are rare comparing with those of middle and low latitudes. Very few ground-based instruments, especially active remote sensing ones, have made measurements at such high southern latitude so far due to the difficulty to access the harsh environment, and the spaceborne satellites are usually limited by their viewing geometry and cannot provide high resolution measurements.

McMurdo station (77.8°S, 166.7°E), sitting at the poleward edge of the auroral oval (Figure 1), provides a great opportunity for atmospheric scientists to study how neutral atmosphere interacts with the ionosphere and is affected by the magnetosphere. University of Colorado (CU) Lidar group has deployed an Fe Boltzmann lidar system at McMurdo since December 2010. The McMurdo lidar campaign is aiming to address these questions with high-quality observations for new science discoveries. The lidar system can be operated during both day and night, leading to full diurnal coverage. The lidar was operated at almost every clear sky since the day of installation, and over 4000 hours of data has been collected so far.

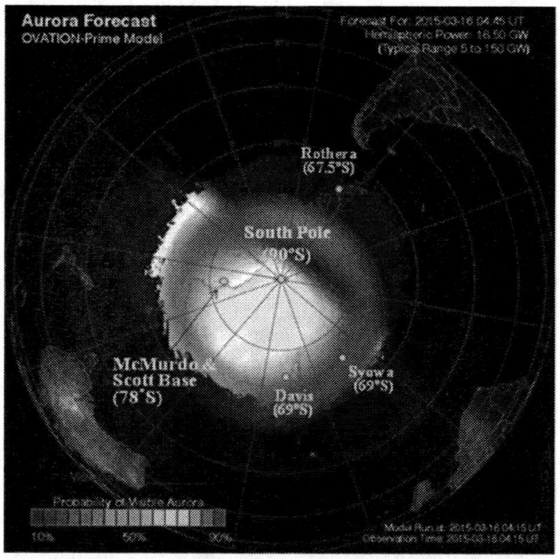

Figure 1. Map of Antarctica with aurora forecast, taken and modified from: http://www.swpc.noaa.gov/products/30-minute-aurora-forecast

In this paper, we introduce the lidar observational campaign in winter, and present the results of winter temperature and tidal structures at McMurdo.

2. LIDAR CAMPAIGN AT MCMURDO

2.1 Fe Boltzmann Lidar System

The lidar system was originally developed by Chu, Gardner and co-workers in 1998 at University of Illinois. It was deployed to the North Pole and the Arctic on board NSF/NCAR Electra aircraft in 1999, to the South Pole (1999–2001), and to Rothera (2002–2005). From 2009 to

2010, the lidar system was refurbished and upgraded by the University of Colorado lidar group. It was tested and collected data simultaneously with a Na Doppler lidar system co-located at Table Mountain Lidar Observatory, Boulder, Colorado in the summer of 2010. After that, the whole Fe system was shipped to McMurdo station, and has the lidar campaign started since December 16, 2010. One of the main purposes of the McMurdo lidar campaign is to fill in the data gap in between the South Pole (90°S) and Rothera (67.5°S) stations.

The lidar system consists of two independent channels of 372 and 374 nm, corresponding to the two absorption lines of atomic Fe. At the same time, each channel works individually as a Rayleigh lidar. The details of the refurbishment and upgrade of the lidar system could be found in [2]. Temperatures in the MLT region (80–110 km) can be derived from both channels using the Fe Boltzmann techniques based on the Maxwell-Boltzmann distribution law as shown in equation (1) [3], and from 30–70 km using the Rayleigh integration technique [4].

$$\frac{P_{374}}{P_{372}} = \frac{g_2}{g_1}\exp\left(-\Delta E / k_B T\right) \quad (1)$$

P is the population of the state, g is the degeneracy factor, ΔE is the energy difference between the two states, k_B is the Boltzmann constant, and T is the temperature.

2.2 Winter Operation of Fe Lidar system

This Fe lidar system is a sophisticated system. For each winter, one member of CU lidar team will be deployed to McMurdo and stay 8 months through whole Antarctic winter. Most of the major maintenances and system tests are usually done in the summer, while the largest human power and access of the resources guarantee that the lidar system is kept at the best status for the coming winter. While entering the winter season, the operation of the system switches to the "campaign mode", which means only the essential maintenances will be performed on the system. The goal of the winter is to collect data as much as possible, weather permitting. Table 1 shows the statistics of the lidar data collected from April to September from the past four years. In average, over 60 hours/month of data were collected in 2011 and 2012, and over 80 hours/month in 2013 and 2014.

Table 1. Winter lidar observation statistics (2011–2014)

Month	April	May	June	July	Aug	Sept
2011	55(5)	59(6)	70(4)	48(3)	72(5)	31(4)
2012	70(7)	57(5)	85(5)	32(4)	32(3)	87(6)
2013	118(8)	57(5)	85(6)	90(5)	72(5)	76(5)
2014	54(5)	127(4)	147(5)	77(3)	25(4)	60(5)
Total	297(25)	300(20)	387(20)	247(15)	201(17)	254(20)

*The numbers indicate the total hours (episodes) of data collected in the corresponding month.

During winter, for most of the time, the system is operated in darkness (May to August) or has day and night alternation (April and September). In the total dark night, the solar background level is very low, and the photomultiplier tubes (PMT) receive near zero photon counts at background altitudes. Under such circumstance, three changes are made in the receiver chain for better and more stable signal levels: 1) the daytime etalon filter in between chopper and PMT is removed, 2) the interference filter is changed from 0.3 nm (pass-band width) to 4.2 nm, and 3) the 2-mm pinhole in front of the chopper is replaced with a 4-mm one to increase the equivalent field of view of the telescope. With the etalon removed and wider bandwidth of interference filter is used, the receiver optical efficiency increases by more than 3 times [5] and the larger pinhole decreases the fluctuation due to the transmitted beam swinging during lidar run.

Figure 2 shows an example of a 65-hour continuous lidar run by Cao Chen from June 28–30, 2014. In Figure 2, strong temperature perturbations between 85 and ~110 spanned the whole observation period. The peak-to-peak amplitude of perturbation is about 40 K at 90 km, and the oscillating periods are between 7 and 9 hours. Such strong temperature perturbations with periods between 4 and 10 hours were frequent observed during winter at McMurdo. The sources of these strong waves are still remaining unclear, but a strong candidate could be the inertia gravity wave (IGW) generated at stratosphere jet near 40 km. *Chen et al.* [6] reports the first coincident observation of IGWs by this Fe lidar and the co-located Scott Base medium frequency (MF) radar. Contrary to the high frequency oscillation structure in the MLT region, the Rayleigh temperature appears to have longer oscillating period, for which the period is greater than 2 days.

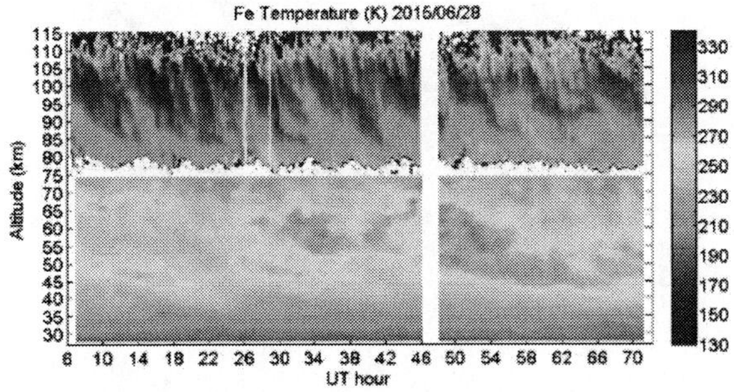

Figure 2. Temperature contour from 30 to 115 km at McMurdo from June 28–30, 2014. The lidar was run continuously for 65 hours.

From Figure 2, the altitude span of the temperature contour in the MLT region also indicates the distribution of the Fe layer in the MLT region. In order to derive robust temperature, the Fe density has to reach certain level. In the wintertime, the Fe layer usually spans from as low as 75 km to over 110 km. In some occasions, the Fe layer can extend as high as over 155 km [7].

3. WINTER TEMPERATURES AND TIDES

3.1 Winter Temperature Structures

Four years of temperature measurements enable us to investigate the winter temperature structure from 30 to 110 km. Figure 3 shows the lidar measurement of monthly mean temperatures from May to August in 2011-14. For each month, the temperature data were first binned to resolution of 1 km in altitude and 1 hour in time. Then the monthly average profile was calculated by taking the mean of the diurnal composite contour, which was formed by averaging the measurements with the same UT hour in the same month. Finally, the four-year monthly mean temperature (bold black in Figure 3) was obtained by simply averaging the four profiles from each year.

It can be seen that the four years lidar observations show similar winter mesopause and stratopause heights. The mean mesopause is at about 96–100 km and the stratopause is at about 51–53 km from May through August. The mean mesopause temperatures are between 180 and 190 K and the stratopause temperatures are between 260 and 275 K through four months. Comparing the monthly mean profiles of four years, it is clear that large inter-annual variations of temperature exist. Especially in the MLT region, for example, the mesopause temperature can have 20 K interannual variability in July.

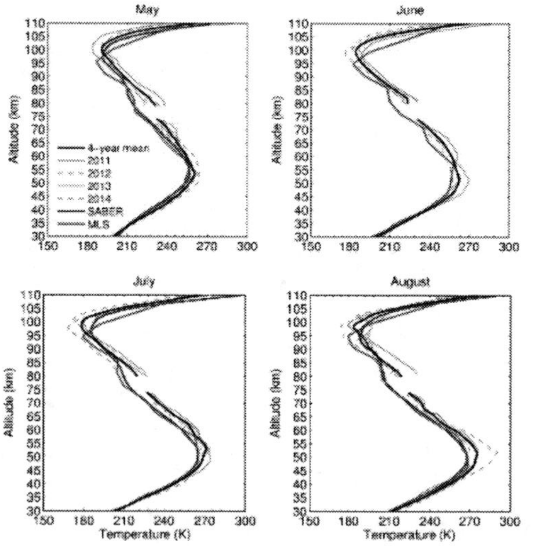

Figure 3. Monthly mean temperature profiles from 30 to 110 km at McMurdo from 2011 to 2014. The same periods of satellite measurements by SABER and MLS are plotted for comparison (no SABER data in June).

The lidar results are also compared to two satellite measurements, the Sounding of the Atmosphere using Broadband Emission Radiometry (SABER) experiment on TIMED satellite and the Microwave Limb Sounder (MLS) on the EOS Aura satellite. Both satellite measurements in each month are the mean of temperature profiles retrieved from 2011 to 2014 near McMurdo station. For most of the months, SABER has colder stratopause and warmer mesopause, but the heights of stratopause and mesopause are similar to the lidar measurements. Furthermore, above 100 km, SABER agrees with lidar measurement quite well. It is worth mentioning that SABER does not have measurements in June because of the yaw maneuver of the TIMED and the viewing geometry of SABER. MLS also exhibits colder stratopause than lidar. However, the vertical resolution of MLS in the MLT region is reduced to about 4 km, which could not provide trustable mesopause height and temperature. But generally speaking, the MLS appears to have similar mesopause temperatures as lidar.

3.2 Winter Temperature Tides

The winter temperature tides from the four years of lidar measurements are also investigated. Figure 4a shows the composite contour of temperature perturbations from 16 months lidar data of May through August through 2011 to 2014. Clear diurnal and semidiurnal features can be seen in the Rayleigh (30–70 km) and MLT (80–110 km) region, respectively. A harmonic fitting model comprising a mean temperature plus diurnal and semidiurnal tides was used to quantify the tidal structure. From Figure 5a, it can be seen that the diurnal tidal amplitudes are about 1–3 K below 100 km, but swiftly increase above 100 km. Semidiurnal tide also has small amplitudes (< 3 K) below 100 km (Figure not shown). The reconstructed composite contour from the diurnal and semidiurnal tides, as shown in Figure 4b, captures the major features of Figure 4a, indicating that the tidal information in Figure 5a and 5b are reasonable. The detailed data process procedures and discussions can be found in [8].

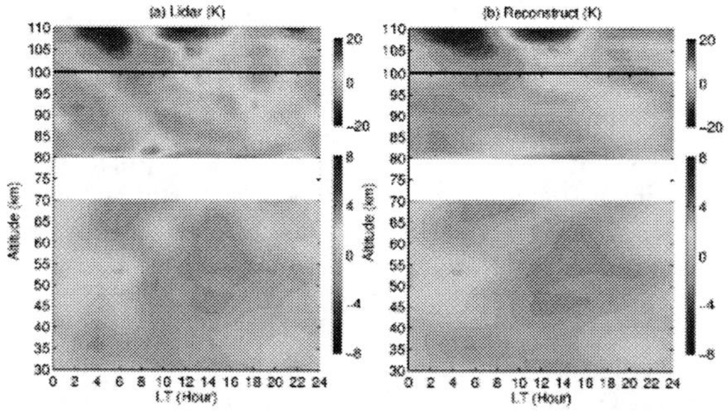

Figure 4. (a) Composite and (b) reconstructed contours of temperature perturbations from 2013 to 2014 data. The temperatures below and above 100 km are indicated by the bottom and top color bars, respectively.

The further investigation of the fast increase with altitude of the diurnal tidal amplitudes above 100 km reveals that the tidal sources are from magnetosphere instead of the source from lower atmosphere [9]. By separating the temperature data under different Kp index conditions, we found that the tidal amplitude increases in strength with the Kp magnetic activity index, as shown in Figure 5c and 5d. A global circulation model, the Coupled Thermosphere

Ionosphere Plasmasphere Electrodynamics (CTIPe) model, was used for the mechanistic studies of this tidal behavior.

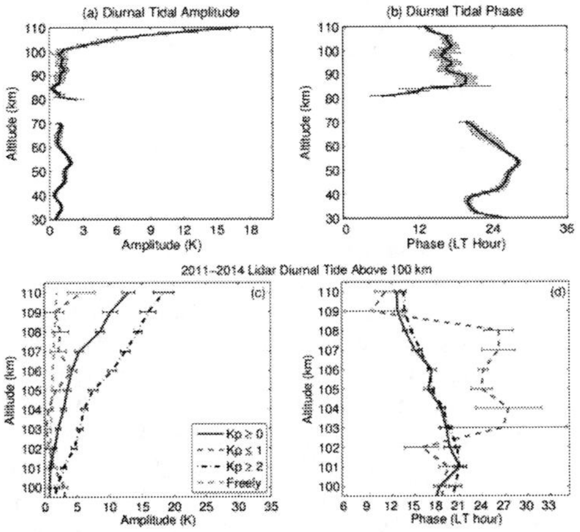

Figure 5. (a-b) Lidar observations of winter temperature diurnal tidal structures from 30 to 110 km and (c-d) tidal amplitudes and phases above 100 km under different Kp conditions at McMurdo from 2011 to 2014.

An interesting finding is that, between 100 and ~116 km, the model results show that the adiabatic cooling/heating associated with Hall ion drag is the dominant source of this feature. As shown in Figure 6, with the real-time run of the CTIPe model, the tidal amplitudes form a concentric ring structure circling the geomagnetic pole and overlap the auroral zone, which point to the magnetospheric source origin. The simulated tidal amplitudes are also comparable with those of the lidar observations. A test run was then launched with the Hall ion drag removed from the model. The results plotted in Figure 6 show that the tidal amplitudes significantly decrease, indicating that the Hall ion drag plays an important role in the lower thermosphere. The detailed discussion can be found in [9].

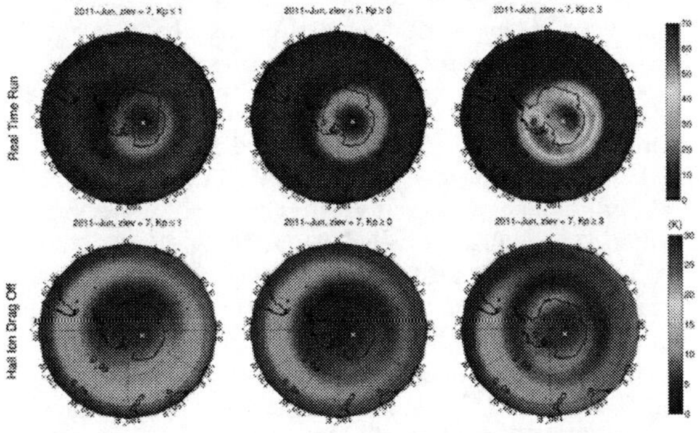

Figure 6. CTIPe model output of diurnal temperature tidal amplitudes at level 7 (~116 km). Upper row is the real-time run and lower row is without Hall ion drag in the model.

4. CONCLUSIONS

The McMurdo lidar campaign has provided invaluable data of the polar middle and upper atmosphere. The temperature measurements during winter not only fill in the data gap in between the South Pole and Rothera stations, but also can provide crucial information to the climate models and space weather forecast models. The measurements and studies of temperature tides above 100 km also help advance our understandings of the relation between IT system and the magnetosphere.

ACKNOWLEDGEMENTS

We acknowledge Wentao Huang, Zhangjun Wang, Chester Gardner, John Smith, Jian Zhao, and Ian Barry for their contributions to the project. We also thank the staff of United States Antarctica Program, McMurdo Station, Antarctica New Zealand, and Scott Base for their support. This project was supported by the USA National Science Foundation OPP grants ANT-0839091 and PLR-1246405. W. F. was partially supported by NSF grant AGS-1136272.

REFERENCES

[1] Richmond, A. D., and J. P. Thayer, 2000: Ionospheric electrodynamics: A tutorial, in *Magnetospheric Current Systems, Geophys. Monogr. Ser., 118*, pp. 131–146, edited by S.-I. Ohtani et al., AGU, Washington, D.C.
[2] Wang, Z. et al., 2012: Refurbishment and upgrade of Fe Boltzmann/Rayleigh temperature lidar at Boulder for a McMurdo lidar campaign in Antarctica, *Proc. of the 26th ILRC*, pp. 207–210, Porto Heli, Greece, June 2012.
[3] Gelbwachs, J. A., 1994: Iron Boltzmann factor LIDAR: proposed new remote-sensing technique for mesospheric temperature, *Applied Optics*, 33, 7151–7156.
[4] Hauchecorne, A. and M. Chanin, 1980: Density and temperature profiles obtained by lidar between 35 and 70 km, *Geophys. Res. Lett.*, 7(8), 565–568.
[5] Chu, X. et al., 2002: Fe Boltzmann temperature lidar: design, error analysis, and initial results at the North and South Poles. *Applied Optics 41* (21), 4400–4410.
[6] Chen, C. et al., 2013: Inertia-gravity waves in Antarctica: A case study using simultaneous lidar and radar measurements at McMurdo/Scott Base (77.8°S, 166.7°E), J. Geophys. Res. Atmos., 118, 2794–2808.
[7] Chu, X. et al., 2011: Lidar observations of neutral Fe layers and fast gravity waves in the thermosphere (110–155 km) at McMurdo (77.8°S, 166.7°E), Antarctica, *Geophys. Res. Lett., 38*, L23807.
[8] Fong, W. et al., 2014: Winter temperature tides from 30 to 110 km at McMurdo (77.8°S, 166.7°E), Antarctica: Lidar observations and comparisons with WAM, *J. Geophys. Res. Atmos., 119*.
[9] Fong, W. et al., 2015: Lidar and CTIPe model studies of the fast amplitude growth with altitude of the diurnal temperature "tides" in the Antarctic winter lower thermosphere and dependence on geomagnetic activity, *Geophys. Res. Lett., 42*, doi:10.1002/2014GL062784.

EPJ Web of Conferences **119**, 12004 (2016) DOI: 10.1051/epjconf/201611912004
ILRC 27

EXPLORATION OF WHOLE ATMOSPHERE LIDAR: MACH-ZEHNDER RECEIVER TO EXTEND FE DOPPLER LIDAR WIND MEASUREMENTS FROM THE THERMOSPHERE TO THE GROUND

John A. Smith [1]* and Xinzhao Chu[1]

[1]*University of Colorado Boulder, 216 UCB, CIRES, Boulder, CO 80309, USA,*

**Email: John.A.Smith@colorado.edu and Xinzhao.Chu@colorado.edu*

ABSTRACT

A receiver employing a field-widened Mach-Zehnder interferometer (MZI) is investigated for extending the wind measurement range of a narrowband Fe Doppler (372 nm) lidar from its existing measurement range in the mesosphere and lower thermosphere (MLT) down to the ground. This design uses the multiple transmitted frequencies available from the base Fe Doppler lidar in combination with an MZI receiver to make a measurement of the Doppler shift from Rayleigh-Mie scattering that is independent of aerosol backscatter ratio, temperature and pressure of the lidar volume and also independent of geometric overlap, the chopper function and any other factor affecting the signal in both MZI channels equally. A ratio is constructed from the three frequencies and two channels of the interferometer that exhibits a measurement performance of 1.75 times the Cramer-Rao lower bound, which is comparable to the dual MZI (DMZ) while preserving the insensitivity to backscatter spectrum of the quad MZI (QMZ). Using actual data obtained recently from the Fe Doppler lidar, we show the expected measurement performance of this whole atmosphere lidar instrument concept.

1. INTRODUCTION

The Fe Doppler lidar is one of a class of lidars referred to as "resonance Doppler" lidars that exploit the exceptionally high backscatter cross-section of metal atoms deposited by meteor ablation in the MLT region, typically Fe, Na or K, to sense wind and temperature in this region. By probing the backscatter cross-section at three fixed frequencies distributed about the resonance line, these lidars infer the line's Doppler broadening and bulk Doppler shift, thereby making measurements of the temperature and wind of the backscatter volume, respectively [1]. In light of the recent discoveries of neutral metal layers well into the thermosphere of 100-200 km [2], the Fe Doppler lidar is now thought to be capable of measuring wind and temperature in the 75 to ~ 200 km region.

To extend resonance Doppler lidar wind measurements to the lower mesosphere and stratosphere, we propose the addition of a Mach-Zehnder interferometer as a spectral discriminator in the receiver system of the Fe Doppler lidar to discern Doppler shifts from Rayleigh and Mie scattering (see Figure 1). An Fe Doppler lidar equipped with such an instrument would be especially well-suited to fulfilling the need for wind measurement in the so-called "radar gap" in the statically stable region 30-60 km, which lacks sufficient turbulence or ionization for routine radar measurement. Therefore, a mobile "whole-atmosphere" lidar that implements such a receiver could present a much-needed capability to the atmospheric science community. Despite the potential scientific benefits, routine wind profiling through the complete atmospheric column remains elusive for a single system and usually requires campaigns spanning multiple earth observing platforms incorporating balloons, MST radars and rockets. For example, the ability to measure wind and temperature from 35 to 150+ km with a single, mobile lidar would help to discover the source(s) of short-period inertia gravity waves recently observed in Antarctica [2], [3].

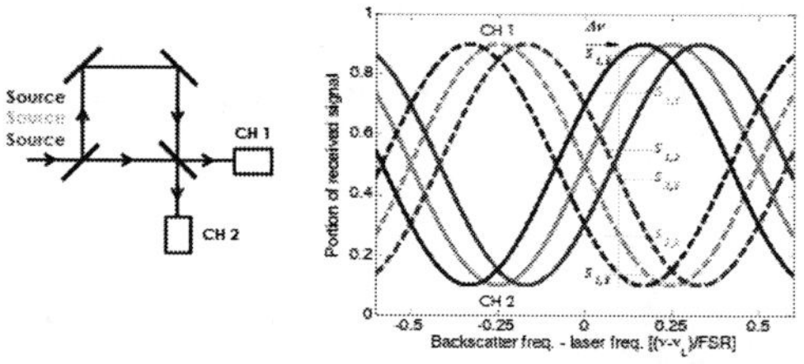

Figure 1. Response functions of the XMZ apparatus introduced in this work. Signals from all channels and frequencies exhibit sensitivity to the Doppler shift.

Atomic line filters have been implemented in Na-Doppler lidars [4], [5] to infer wind from Rayleigh scattering, but the low power attainable from current 589 nm transmitters and relatively weak molecular scattering at this wavelength severely limits the attainable precision of wind and temperature in the radar gap. Fe lidars, by contrast, operating at 372 nm with higher power transmitters are much more capable in the Rayleigh region and can attain a comparable signal in the MLT region due to the much higher abundance of Fe compared to Na [1].

Liu and Kobayashi [6] were the first to propose the use of a Mach-Zehnder interferometer (MZI) as a frequency discriminator for direct-detection Doppler

 DOI: 10.1051/epjconf/201611912004

lidar. Their proposed transmitter employed the Nd:YAG fundamental (1.064 μm) and inferred wind predominantly from aerosol, rather than molecular scattering. Bruneau [7] optimized their MZI for molecular scattering and demonstrated wind measurements up to 40 km [8]. Our approach is based on the MZI tuned to molecular scattering since our interests are predominantly in the stratosphere and higher where scattering from aerosols, which are relatively sparse in this region, is much weaker than that from molecules. In addition, the shorter optical path difference of an MZI optimized for molecular scattering, versus an MZI optimized for aerosol scattering, allows more extensive field-widening [9], which is crucial for resonance lidars that require a large transmitter beam divergence to avoid saturation of the layer and large collecting areas to detect single photon signal levels from the upper atmospheres [10].

2. METHODOLOGY

Fe lidars are unique in that the high backscatter cross-section of molecules at the UV Fe transition wavelength, and typically high power of lasers at this wavelength, allows them to serve as highly effective Rayleigh lidars also. Laser technology at the time limited the first Fe lidars to broadband systems that were capable only of temperature measurement. Recent developments have introduced narrowband systems capable of measuring wind also, the Fe-Doppler lidar at Boulder, Colorado [11] being the first example of such a lidar. This lidar is capable of measuring temperature and wind in the MLT region and temperature below the MLT region down to 30 km with a transmitter locked to the Fe a^5D_4-z^5F_5 transition at 372.0995 nm and shifted alternately in time by 742 MHz above and below the precise transition wavelength. Its reliance on spectroscopy to derive spectral information from species in the MLT region restricts the routine wind measurement range to regions of persistent Fe, or about 75–115 km. Therefore, to measure wind below this range requires a spectrally sensitive receiver.

In this work, we analyze a field-widened dual Mach-Zehnder (DMZ) for the task of combining, using the same resonance Fe-Doppler beam, Rayleigh Doppler lidar measurements from the troposphere to mesosphere with resonance Doppler measurements that extend upwards from the mesosphere to the lower thermosphere. The concept is then to marry a Rayleigh-Mie Doppler direct-detection lidar technique with the resonance Doppler lidar technique to cover wind measurement throughout the whole atmosphere. Rather than operating at a single transmitted frequency, as previous works [6], [7] have investigated, we investigate the benefits of using the three available transmitted frequencies of this resonance Doppler lidar system in combination with the DMZ system. We will refer to this scheme as the X-frequency Mach-Zehnder or "XMZ" since, in principle, an arbitrary number of spectral response phases/frequencies is possible in this scheme **(Error! Reference source not found.)**. By using the information from three frequencies, a Doppler shift estimate can be made independent of the atmospheric contrast factor in addition to instrument defects. The benefits of the XMZ technique are possible due to a reduction in the number of variables compared to the QMZ technique (**Figure**) presented in previous works [6], [7], which required a waveplate, two polarizers and two additional detectors, each with associated uncertainties.

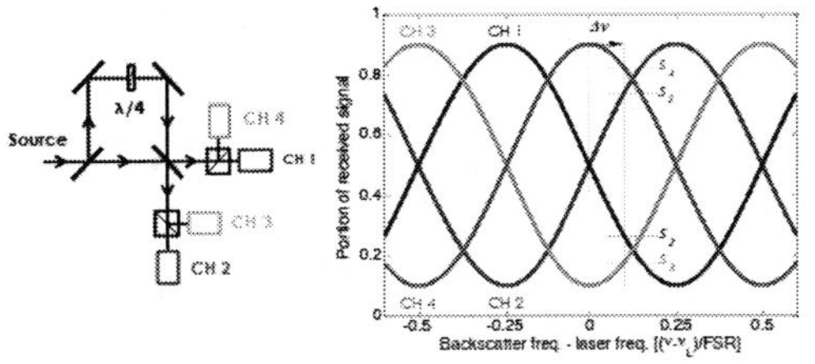

Figure 2. Response functions of the QMZ apparatus introduced by previous authors. Two of the four channels are insensitive to Doppler shift.

Another key benefit of the Mach-Zehnder techniques, including the XMZ technique described here, is that all of the signal is recovered and summation of the signals in both channels results in a quantity free of interference effects, provided the detectors are balanced. Therefore, after summation, the same resonance Doppler lidar techniques for retrieving wind and temperature in the MLT region are still valid. Unlike many other techniques that may discard a portion of the backscatter [4], [5], [12], the MZI techniques, by recovering the complete spectrum, enable a single receiver to profile wind and temperature through the complete atmospheric column with nearly the same precision as the original instrument and without modification to any existing retrieval techniques.

3. RESULTS

An excellent treatment of the theory of the measurement of Doppler shift using the DMZ and QMZ techniques is provided in Bruneau [7]. Discriminator signals for the XMZ and QMZ configurations can be defined that are both insensitive to temperature, aerosol and pressure variations of the lidar volume. The XMZ and QMZ configurations are therefore our preferred methods for retrieval of wind in the stratosphere and upper troposphere. One additional advantage of the XMZ configuration is that since all three frequencies and associated discriminator signals are sensitive to the Doppler shift, one can then define a discriminator signal that yields a measurement performance similar to the DMZ, but, unlike the DMZ, is insensitive to backscatter spectrum. The shot-noise limited performance of the XMZ, DMZ and QMZ techniques is compared here.

The frequency of the AOM shift is 742 MHz, reference temperature is 250 K and the wavelength is 372 nm. The measurement performance of all three techniques is given here, with performance for the DMZ and QMZ techniques given by Bruneau [7] (1). We assume that the only source of noise is shot noise and that the background is negligible.

$$\Delta u'_{QMZ} = \Delta Q \cdot \left(\frac{dQ}{du'}\right)^{-1} \sim \frac{2.33}{\sqrt{\sum_i S_i}} \qquad (1)$$
$$\Delta u'_{DMZ} \sim \frac{1.65}{\sqrt{\sum_i S_i}}$$
$$\Delta u'_{XMZ} \sim \frac{1.75}{\sqrt{\sum_i S_i}}$$

The XMZ possesses a measurement performance that is 25% better than the QMZ while preserving the QMZ's key feature, which is the insensitivity to aerosol, temperature and pressure effects.

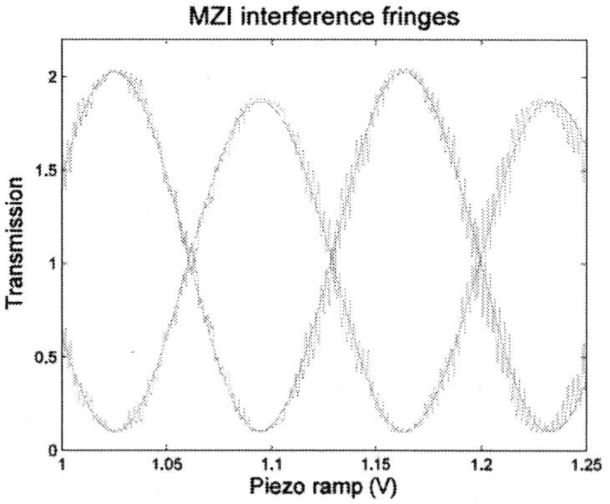

Figure 3. Mach-Zehnder fringes in both channels with sinusoidal fits indicated by the black curves to determine the instrumental contrast coefficients.

Field-widening using the compensating plate has been demonstrated at the University of Colorado Boulder and an instrument-limited fringe contrast of between 0.89 and 0.93 has been obtained with the apparatus (Figure). One leg of the interferometer is varied by ramping a piezo-actuated stage to view the interference. The mirror, beamsplitters and field compensating plate were polished and coated by LightMachinery to obtain transmitted and reflected wavefront errors of better than λ/30. A fiber with a core diameter of 1000 μm and NA of 0.22 was filled on the input side with light from a narrow linewidth, Toptica DL-100 372-nm external cavity diode laser. Illumination from this fiber provided the range of angles representing the optical extent of the true receiver source. Frequencies matching resonant acoustic modes of the enclosure were evident in the scans from insufficient damping of ambient acoustic noise, but the long-term drift is small due to the thermally compensated nature of this design.

Based on data obtained from the Fe Doppler lidar to date, we have simulated the uncertainties. This figure includes data that was taken with the transmitter operating at a reduced power level of 1.45 W due to optical damage concerns. An upgrade to the optics will allow the lidar to operate at power levels of up to 6 W. Once the design power is attained, the wind measurement precision will be within 4 m/s for 15 minute and 3 km resolutions up to 75 km, above which the wind measurement is accomplished from the resonance 3-frequency technique. For the present power levels, we still expect a precision of 1 m/s up to 50 km and a maximum uncertainty of 8 m/s at 76 km at a resolution of 15 minutes horizontally and 2.9 km vertically (Figure 4). The parameters of the Fe Doppler lidar transmitter and receiver are listed in **Error! Reference source not found.**.

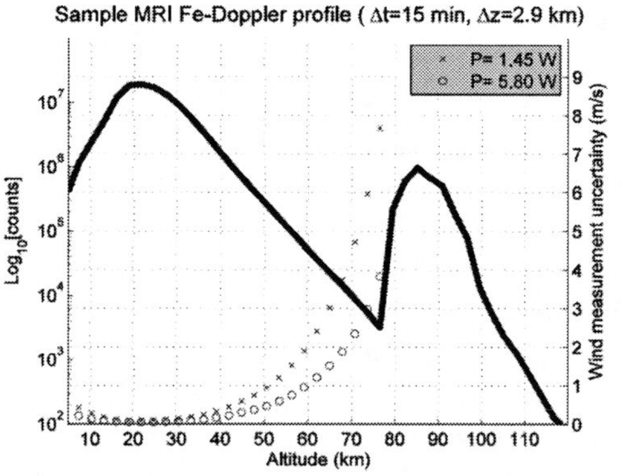

Figure 4. Based on data obtained from the Fe-Doppler lidar to date, we predict the line-of-sight wind measurement uncertainty.

4. CONCLUSIONS

We have described a method of extending the wind measurement range of an Fe-Doppler lidar from the MLT region down to the surface or near surface using a tilted, field-widened Mach-Zehnder interferometer. By detecting the signal at both channels of the interferometer, and at three distinct frequencies within a quarter of the free spectral range, the measurements can be made independent of the aerosol backscatter ratio, temperature and Brillouin scattering contributions to the backscatter spectrum. In addition, the uncertainty of the measurements is limited to only 1.75 times the Cramer-Rao lower bound and there are only two detectors to balance.

Table 1. Parameters of the Fe-Doppler lidar transmitter and receiver

Transmitter	
Power	1.45 W
Wavelength	372.0995 nm
Spectral width	35-55 MHz
Rate	35 Hz
Pulse width	250 ns FWHM
Divergence	250 μrad
Receiver	
FOV	500 μrad
Aperture	0.5 m^2
PMT QE	43%

A significant benefit of the Mach-Zehnder-receiver-based Fe Doppler lidar techniques described here is that all of the signal is recovered and no portion of the backscatter is discarded. Consequently, summation of the signals in both channels results in a quantity free of interference effects, provided the detectors are balanced. The same resonance Doppler lidar techniques for retrieving wind and temperature in the MLT region can be applied to the summation of signals from both channels. The Fe Doppler lidar with the multiple phase MZI techniques, by recovering the complete spectrum, enables a single lidar receiver to profile wind and temperature through the complete atmospheric column with nearly the same precision as the original instrument and without modification to any existing retrieval techniques. We believe that the successful implementation of such new technologies represents significant progress towards a whole atmosphere lidar that will enable new science avenues.

ACKNOWLEDGEMENT

We acknowledge the support of a NSF Major Research Instrumentation award (MRI-0723229), CRRL grant (AGS-1136272), a NASA Earth and Space Science Fellowship award (NNX09AO35H), and an internal CIRES Innovative Research Proposal award.

REFERENCES

[1] Chu, X. and G. C. Papen, 2005: Resonance Fluorescence Lidar for Measurements of the Middle and Upper Atmosphere, in *Laser Remote Sensing*, Fujii and Fukuchi, Eds. Taylor & Francis Group, 179–432.

[2] Chu, X., Z. Yu, C. S. Gardner, C. Chen, and W. Fong, 2011: Lidar observations of neutral Fe layers and fast gravity waves in the thermosphere (110-155 km) at McMurdo (77.8°S, 166.7°E), Antarctica, *Geophys. Res. Lett.*, **38**, L23807, doi: 10.1029/2011GL050016.

[3] Chen, C., X. Chu, A. J. McDonald, S. L. Vadas, Z. Yu, W. Fong, and X. Lu, 2013: Inertia-gravity waves in Antarctica: A case study using simultaneous lidar and radar measurements at McMurdo/Scott Base, *J. Geophys. Res. Atmos.*, **118**, doi: 10.1002/jgrd.50318.

[4] Huang, W., X. Chu, B. Williams, S. Harrell, and C.-Y. She, 2009: Na double-edge magneto-optic filter for Na lidar profiling of wind and temperature in the lower atmosphere, *Opt. Lett.*, **34(2)**, 199–201.

[5] Huang, W., X. Chu, J. Wiig, B. Tan, C. Yamashita, T. Yuan, J. Yue, S. D. Harrell, C.-Y. She, B. P. Williams, J. S. Friedman, and R. M. Hardesty, 2009: Field demonstration of simultaneous wind and temperature measurements from 5 to 50 km with a Na double-edge magneto-optic filter in a multi-frequency Doppler lidar, *Opt. Lett.*, **34(10)**, 1552–1554.

[6] Liu, Z. and T. Kobayashi, 1996: Differential discrimination technique for incoherent Doppler lidar to measure atmospheric wind and backscatter ratio, *Opt. Rev.*, **3(I)**, 47–52.

[7] Bruneau, D., 2001: Mach-Zehnder interferometer as a spectral analyzer for molecular Doppler wind lidar, *Appl. Opt.*, **40(3)**, 391–399.

[8] Bruneau, D., A. Garnier, A. Hertzog, and J. Porteneuve, 2004: Wind-velocity lidar measurements by use of a Mach-Zehnder interferometer, comparison with a Fabry-Perot interferometer, *Appl. Opt.*, 43(1), 173–182.

[9] Hilliard, R. and G. Shepherd, 1966: Wide-Angle Michelson Interferometer for Measuring Doppler Line Widths, *J. Opt. Soc. Am.*, 56(3), 362–369.

[10] Smith, J. A., and X. Chu, 2015: High-efficiency receiver architecture for resonance-fluorescence and Doppler lidars, *App. Opt.*, **54(13)**, in press.

[11] Chu, X., 2006: Alexandrite-ring-laser-based Fe Doppler lidar for mobile/airborne deployment, Proc. *23rd International Laser Radar Conference*, 385-388.

[12] Hair, J., L. Caldwell, D. Krueger, and C.-Y. She, 2001: High-spectral-resolution lidar with iodine-vapor filters: measurement of atmospheric-state and aerosol profiles, *Appl. Opt.*, **40(30)**, 5280–5294.

EPJ Web of Conferences **119**, 13001 (2016) DOI: 10.1051/epjconf/201611913001
ILRC 27

LIDAR SOUNDINGS BETWEEN 30 AND 100 KM ALTITUDE DURING DAY AND NIGHT FOR OBSERVATION OF TEMPERATURES, GRAVITY WAVES AND TIDES

Michael Gerding*, Kathrin Baumgarten, Josef Höffner, Franz-Josef Lübken

Leibniz-Institute of Atmospheric Physics at the Rostock University, Kühlungsborn, Germany
**Email: gerding@iap-kborn.de*

ABSTRACT

Ground-based temperature measurements by lidar are an important tool for the understanding of long-term temperature changes as well as the propagation of gravity waves and tides. Though, mesospheric soundings are often limited to nighttime conditions due to the low signal-to-noise ratio during the day. We developed a daylight-capable RMR lidar for temperature soundings in the middle atmosphere. The influences of the narrowband detector on the calculated hydrostatic temperatures as well as their correction are described. The RMR lidar is complemented by a co-located resonance lidar. We present an example for tidal analyses and short-term variability of tidal amplitudes.

1. INTRODUCTION

Temperature is one of the key parameters needed for the understanding of dynamical and chemical processes in the whole atmosphere. Lidar is the only ground-based technique providing continuous temperature profiles from the troposphere up to the lower thermosphere (e.g. [1]). Depending on altitude, rotational and vibrational Raman scattering, Rayleigh and/or resonance scattering are used. Long-term data sets of precise lidar-derived temperature profiles are important for the understanding of climate change in the whole middle atmosphere [2]. Temporally resolved temperature data can be used to derive (altitude resolved) amplitudes of gravity waves, tidal and planetary waves (e.g. [3]). Unfortunately, especially Rayleigh lidars are often limited to nighttime conditions, as the solar background radiation typically exceeds the laser backscatter from the mesosphere by several orders of magnitude. By this, observations of tides and planetary waves are inhibited. Observations of absolute temperatures and trends are restricted to nighttime only, impeding comparisons with other data sets.

In 2010 we developed a new RMR lidar at our site at Kühlungsborn/Germany (54°N, 12°E), subsequently replacing an old RMR lidar observing middle atmospheric temperatures since 2002 [1,4]. The old lidar is lacking effective daylight suppression and was operated during nighttime only. The new RMR lidar is designed for observations of noctilucent clouds at 532 nm [5] and for measurements of temperature profiles up to ~75 km (~90 km) during day (night), respectively. Until autumn 2012 the lidar has been complemented by a daylight-capable potassium resonance lidar for temperature observations between ~85 and 105 km. The old RMR lidar was operated in parallel during nighttime until August 2013.

A first study of tidal temperature variations has been published by [6]. Here we provide an overview of the instrumental setup and the correction of filter effects (Section 2). In Section 3 we show an example of tidal temperature variations.

2. METHODOLOGY

The new RMR lidar is based on an injection-seeded Nd:YAG laser used at second harmonic (532 nm). The external seedlaser is locked to an iodine absorption line. Detailed information is given in Table 1. The laser is emitted co-axially with the receiving telescope by means of different steerable mirrors. A beam-stabilization based on a Piezo-coupled mirror is used to keep the laser beam within the 62 µrad field of view of the telescope [7]. Input for beam-stabilization is provided by a CCD camera observing the beam in real-time through the same telescope and a 90:10 beam splitter. The received signal is guided to a detection bench consisting essentially of a narrow-band optical filter (interference filter and double-etalon) and an avalanche photodiode.

DOI: 10.1051/epjconf/201611913001

Table 1: Summary of instrument parameters of the new RMR lidar at Kühlungsborn

wavelength (vac.) [nm]	532.24
power @ 532 nm [W]	~ 22
repetition rate [pps]	30
beam diameter after expansion [m]	0.09
beam divergence after expansion [μrad]	50
field of view [μrad]	62
telescope diameter [m]	0.81
interference filter width (FWHM) [pm]	130
etalon 1 free spectral range [pm]	120
etalon 1 observed finesse	27
etalon 2 free spectral range [pm]	140
etalon 2 observed finesse	20

The bandwidth of the etalons is only about 4 pm, i.e. comparable to the Doppler-broadened Rayleigh backscatter (~2.5 pm). Due to the Airy shape of the transmission functions of both etalons, part of the Rayleigh backscatter is suppressed by the etalons. The overall transmission is reduced to 80 – 90% of the transmission of the unbroadened light. Figure 1 shows the transmission function of the first etalon and the spectrum of the Doppler-broadened backscatter before and after the etalon.

Obviously, the transmission of the etalon depends on the atmospheric temperature at the particular

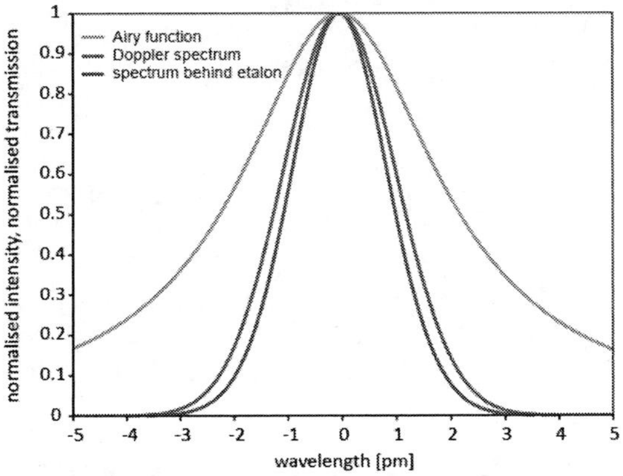

Fig. 1: Transmission of first etalon and relative intensity. Doppler broadening calculated for 300 K.

scattering altitude. By this, an altitude dependent factor must be added to the lidar equation. This factor needs to be accounted for calculation of a relative density profile from the photon count profile. The relative density profile is a pre-requisite for temperature inversion [8]. We have implemented an iterative transmission correction to get the correct density profile. The algorithm first calculates temperatures ignoring the transmission effect. Based on this temperature profile the nominal transmission is retrieved and used for correction of the photon count profile. The corrected backscatter profile is again inverted into temperatures. Figure 2 shows the temperature profiles with and without transmission correction in comparison with the data of the old RMR lidar being operated simultaneously. Please note that both lidars are completely independent systems, using different lasers, telescopes, detectors etc. Above the Rayleigh temperature profile we show the K lidar data that is used for seeding the inversion of the data from the old lidar. Below 43 km no correction is necessary for the new RMR lidar, because in this range a separate detector in front of the double etalon is used. The systematic error of the uncorrected profile increases from 85 km down to ~60 km due to increasing temperature, i.e. decreasing transmission of the etalons. Below 60 km the effect is reverse due to the changing temperature

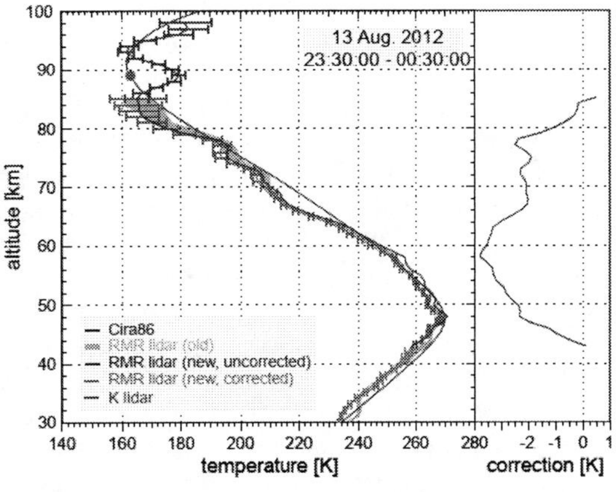

Fig. 2: Temperature profile of 14 August 2008, 0 UT obtained by old and new RMR lidar, and K lidar. The right profile shows the difference between uncorrected and true Rayleigh temperature profile.

gradient. The differences between temperatures of both RMR lidars are within the statistical uncertainty after correction. The inversion of the new lidar data is independently seeded by CIRA-86 temperatures [9] at 89 km.

We do not show data below 30 km altitude due to the increasing effect of aerosols. An aerosol correction is currently under development.

3. RESULTS

Between June 2010 and December 2014 the new RMR lidar was in operation for ~5200 hours. Soundings continue since whenever weather allows. Figure 3 shows a sounding running for ~16 h during 13/14 August 2012 (cf. profile presented in Fig. 2). Temperatures are calculated with a running average of 2 h. The solar elevation during the sounding was up to 50°. Please note that local solar time (LST) at Kühlungsborn is ~45 min ahead of UT. During the day the uppermost altitude for RMR temperature retrieval decreases due to increasing noise. Simultaneously the available height range from the K lidar shrinks. During the night (until ~4 UT) the gap at 80 km occurs due to the use of CIRA-86 instead of K lidar data for Rayleigh temperature retrieval. Fig. 3 shows some temporal variation of temperatures due to gravity waves and tides.

For a quantitative analysis of wave amplitudes the deviations of temperatures from the daily mean profile are examined. As described by [6], typically ~100 h of lidar data are used to calculate amplitudes and phases of 24 h, 12 h, and 8 h waves for each available altitude bin. Depending on weather conditions, each harmonic fit covers a period of ~4 – 30 d. Variability of tidal amplitudes can be large in the middle atmosphere due to changing transmission conditions for the particular wave modes. Figure 4 shows tidal amplitudes for different periods in September 2014 together with the monthly mean. Here only the diurnal and semidiurnal variations are considered due to possible confusion of the terdiurnal tide and long-period gravity waves on short time scales. Mean amplitudes of the 24 h tide are around 1 K up to 65 km and partly up to 2 K between 45 and 50 km. Above 65 km the amplitude increases strongly. The semidiurnal tide is somewhat smaller in the stratopause region (0.5 – 1 K), increasing steadily above 55 km.

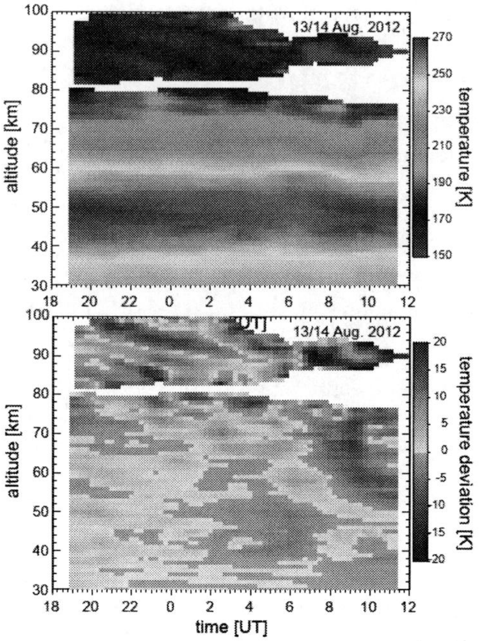

Figure 3: Temporal variation of temperatures during night and day. Solar elevation was up to 50°. Upper plot: absolute temperatures, lower plot: deviation from the daily mean profile.

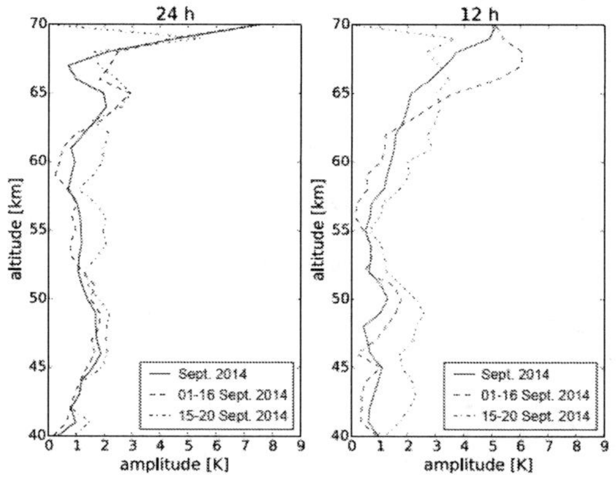

Figure 4: Amplitudes of diurnal and semidiurnal variations for different periods in September 2014 (period 1: 96 h, period 2: 101 h) and for the monthly mean.

Depending on altitude, amplitudes in the shorter sounding periods often vary by 100% or even more. The monthly mean amplitude in some altitudes ranges is smaller than the average of the individual periods, e.g. around 50 km and 65 km for both harmonics. This indicates a phase shift of the tide between both sounding periods due to, e.g., changing propagation conditions. Further analysis of the propagation conditions of individual tidal modes is outside the scope of this paper.

4. CONCLUSIONS

We have shown that effective daylight suppression by means of a narrow field of view and narrowband optical filters allows Rayleigh temperature soundings in the middle atmosphere up to ~75 km during the day. We perform regular RMR lidar temperature soundings during day and night at Kühlungsborn/Germany since 2010, complemented until autumn 2012 by a K resonance lidar. The transmission of the etalons depends on the Doppler broadening of the backscatter signal. This temperature (or altitude) dependent effect results in a systematic temperature bias of up to ~4 K. We correct the backscatter profile in an iterative process making use of an initially calculated temperature profile. The final temperature profile agrees well with the simultaneous, co-located data of the old RMR lidar. Continuous lidar soundings allow the calculation of tidal amplitudes and phases. While many other instruments like space-based profilers require about 2 months for tide observations, lidar requires only about 4 days of sounding, if weather allows. Tidal variability can be at least 100% in amplitude, if short-term data are compared with, e.g., a monthly average. Altitude ranges where the average amplitude is lower than short-term data indicate a change in phase progression, e.g., due to changing propagation conditions. Future soundings will allow further studies of tidal variability above our mid-latitude station, as well as the separation of gravity waves and tides.

ACKNOWLEDGEMENT

The authors gratefully acknowledge Maren Kopp for her help in the installation of the new RMR lidar. We thank Torsten Köpnick and Michael Priester for the maintenance and operation of the lidar systems at IAP. We also acknowledge all our students for their numerous hours of lidar operation. This research has been partly supported by the Deutsche Forschungsgemeinschaft (DFG) under Grant GE 1625/2-1.

REFERENCES

[1] Alpers, M., R. Eixmann, C. Fricke-Begemann, M. Gerding, and J. Höffner, 2004: Temperature lidar measurements from 1 to 105 km altitude using resonance, Rayleigh, and Rotational Raman scattering, *Atmos. Chem. Phys.*, ***4***, 793-800

[2] Angot, G., P. Keckhut, A. Hauchecorne, and C. Claud, 2012: Contribution of stratospheric warmings to temperature trends in the middle atmosphere from the lidar series obtained at Haute-Provence Observatory (44°N), *J. Geophys. Res.*, **117,** D21102, doi: 10.1029/2012JD017631

[3] Gerrard, A. J., Y. Bhattacharya, and J. P. Thayer, 2011, Observations of in-situ generated gravity waves during a stratospheric temperature enhancement (STE) event, *Atmos. Chem. Phys.*, **11**, 11913–11917

[4] Gerding, M., J. Höffner, J. Lautenbach, M. Rauthe, and F.-J. Lübken, 2008: Seasonal variation of temperatures between 1 and 105 km altitude at 54°N observed by lidar, *Atmos. Chem. Phys.*, **8**, 7465-7482

[5] Gerding, M., M. Kopp, P. Hoffmann, J. Höffner, and F.-J. Lübken, 2013: Diurnal variation of mid-latitude NLC parameters observed by daylight-capable lidar and their relation to ambient parameters, *Geophys. Res. Lett.*, **40**, 6390-6394

[6] Kopp, M., M. Gerding, J. Höffner, and F.-J. Lübken, 2015: Tidal signatures in temperatures derived from daylight lidar soundings above Kühlungsborn (54°N, 12°E), *J. Atmos. Solar-Terr. Phys.*, *127*, 37-50, doi:10.1016/j.jastp.2014.09.002

[7] Eixmann, R., M. Gerding, J. Höffner, and M. Kopp, 2015: Lidars with narrow FOV for spectral and daylight measurements, *IEEE Transactions on Geoscience and Remote Sensing*, in print

[8] Hauchecorne, A., and M.-L. Chanin, 1980: Density and temperature profiles obtained by lidar between 35 and 70 km, *Geophys. Res. Lett.*, **7**, 565–568

[9] Fleming, E. L., S. Chandra, J. J. Barnett, and M. Corney, 1990: Zonal mean temperature, pressure, zonal wind, and geopotential height as functions of latitude, *Adv. Space Res.*, **10**, 11–59

EPJ Web of Conferences **119**, 13002 (2016)
ILRC 27
DOI: 10.1051/epjconf/201611913002

INTERCOMPARISON OF OZONE AND TEMPERATURE PROFILES DURING OZITOS+ 2014 CAMPAIGN IN RÍO GALLEGOS, ARGENTINA

Jacobo Salvador [1,2]*, Elian Wolfram [1], Facundo Orte[1], Raúl D'Elia[1], Jonathan Quiroga[1], Eduardo Quel[1], Felix Zamorano[3], Raúl Pérez[3], Israel Villa[3], Hirofumi Oyama[4], Akira Mizuno[4].

[1] *Centro de Investigaciones en Láseres y Aplicaciones, CEILAP-UNIDEF (MINDEF-CONICET), UMI-IFAECI-CNRS-3351,UMI3351, Villa Martelli, Argentina, *Email: jacosalvador@gmail.com*

[2]*Universidad Nacional de la Patagonia Austral, UNPA-UARG, Sta Cruz, Argentina*

[3]*LICA Atmospheric Research Laboratory of Magallanes University, Punta Arenas Chile*

[4]*ISEE Institute for Space-Earth Environmental Research, Nagoya University, Japan.*

ABSTRACT

In the framework of SAVER-Net project, the **OZ**one profIle a**T** Río Galleg**OS** (OZITOS+) campaign was held in the city of Río Gallegos, Argentina (51.5 S; 69.1 W). This experiment was conducted on October 14 -18, 2014 and its main goal was to compare the ozone and temperature profiles using three different measurement techniques such as Differential Absorption Lidar (DIAL), ozonesonde and Millimeter Wave Radiometer (MWR). Also other ground-based and satellite-based instruments were included in the experiment but in this work we only present preliminary results from ground-based instruments deployed in the site. The DIAL instrument is part of Network Data for Atmospheric Composition Change (NDACC) network, and the usual protocols of quality assurance imposed for the network involve regular validation/comparisons experiments. The lidar ozone profiles measured with the lidar are compared with ozone profiles obtained with independent techniques, usually with higher or same resolution as lidar. The experiment are made collocated spatial and temporally. For that reason the Chilean team joined to Japanese and Argentine team at Río Gallegos to develop the experiment.

On October 2014, the Río Gallegos Observatory station was inside the polar vortex during first two weeks and after that polar vortex have moved far away from Río Gallegos during the 3rd week of October, when the intercomparison campaign was held.

In this paper we are present a preliminary results of the campaign, computing the ozone and temperature profiles from DIAL with ozonesondes and MWR.

1. INTRODUCTION

The assessments of change induced by human activities on Earth atmosphere composition and thus on global climate requires a long term and regular survey of the troposphere and stratosphere layers.

In that frame, in 1990, the International Network for the Detection of Stratospheric Change (NDSC, http://www.ndsc.ncep.noaa.gov) was formed to provide a consistent, standardized set of long-term measurements of stratospheric trace gases, particles, and physical parameters via a suite of globally distributed sites. Within the NDSC, which was expanded and renamed Network for the Detection of Atmospheric Composition Change (NDACC) in 2006, lidar (laser radars) are key instruments for measuring the stratospheric ozone and temperature profile. Lidar measurements of ozone and temperature are self-calibrating [1,2]. This makes them especially suited for long-term monitoring.

The Atmospheric Observatory of Southern Patagonia (OAPA) at Río Gallegos Argentina (51.5 S; 69.1 W) is an extension of Laser and Application Research Department (CEILAP) in Buenos Aires. Since June 2005, a shelter with a differential absorption lidar developed in collaboration with Service d'Aéronomie (CNRS) was deployed in this Patagonian city, 2600 km far away Buenos Aires. This instrument was endorsed to NDACC in December 2008 and since this date several intercomparison campaigns were held as validation/control performance of the DIAL instrument.

During 2005 and 2007, with the financial support of JICA (Japanese International Cooperation Agency), was held the SOLAR campaign (www.division-lidar.com.ar). The principal objective of this campaign was study the ozone layer when the polar vortex crosses over the continental part of Argentina, in South America. Río Gallegos is located in the surf zone of polar vortex, making very interesting place to observe the evolution and perturbation that ozone hole produces on the stratospheric ozone profile.

In the framework of the Project "Development of Atmospheric Environmental Risk Management System in South America", (SAVER-Net, 2013-2017) financed by JICA-JST through Science and Technology Research Partnership for Sustainable Development (SATREPS) was held the OZITOS+ campaign during October 2014.

In this work the preliminary results of this experiment are presented. The paper is organized as follow: in Section 2 are described the different instruments involved in this campaign and the section 3 the results of comparisons are presented followed with a short conclusions.

2. METHODOLOGY

There are three instruments involved in this OZITOS+ campaign: the Differential Absorption Lidar (DIAL), the ozone sondes and a millimeter-wave radiometer (MWR). All of them measure the vertical profile of ozone content. However, they have different vertical resolution and dynamic range, thus they are complementary measurements that permit to obtain a vertical ozone profile from ground base to mesosphere (around 80 km).

In the following paragraph we present a brief description of these three instruments used during the OZITOS+ campaign:

2.1 Ozone Balloon-borne

A Lockheed Martin LMG6 ground system unit, LMS6 meteorological radiosonde and electrochemical concentration cell (ECC) ozonesondes from EN-SCI Corporation have been used with 1 % solution of KI. The ozone sensor of the ECC ozonesonde is an iodine/iodide redox concentration cell composed of two platinum electrodes immersed in neutral buffered iodide solutions of different concentration in the anode and cathode chambers. An electric current is generated when air containing ozone is pumped into the cathode. Both ECC and LMS6 are carried aloft on balloons from the ground to the burst altitude of the ballon, normally near 28–30 km.

The ozone sondes launched in OZITOS+ campaign were managed by the personal of Atmospheric Research Laboratory of Magallanes University, that is the Chilean counterpart of SAVER-Net project.

2.2 Differential Absorption Lidar

This instrument was developed at the CEILAP laboratory and is an improved version of the previous one. The DIAL technique requires two emitted wavelengths that in this instrument are generated by two laser emitter: an excimer (XeCl) laser emitting at 308 nm with 30 Hz repetition rate, and maximum energy per pulse of 300 mJ is used for the wavelength absorbed by ozone. The reference wavelength corresponds to the third harmonic of Nd-YAG laser emission, at 355 nm, 30 Hz repetition rate and 130 mJ maximum energy. The optical receiver of the DIAL instrument that collects the backscattered photons is made up of four Newtonian (f/2) telescopes. Each of them has a 50 cm diameter with parabolic aluminized surfaces of 48 cm diameter. A typical experiment lasts 3–4 h, leading to a horizontal spatial resolution of 100–200 km. The measurements are taken during the night and with clear skies. The vertical resolution ranges from 0.7 km at 14 km to 4 km at 35 km, with a corresponding total uncertainty varying from 3 to 15 % in the same altitude range. This instrument contributes to NDACC and is one of only five instruments of this type in the Southern Hemisphere. The instrument is inside a shelter that was deployed inside an area of the Air Force Military Base of Río Gallegos in 2005, where today is emplaced the Observatorio Atmosférico de la Patagonia Austral (OAPA). The full description of DIAL instrument can be found in reference [3].

2.3 Millimeter Wave Radiometer

The MWR was developed and belongs to ISEE group from Nagoya University, Japan. In the millimeter measurements, molecular emission lines due to thermally excited rotational transitions at 110.836 GHz are observed. We can measure the emission spectra day and night

DOI: 10.1051/epjconf/201611913002

continuously because this method does not require any background light sources unlike absorption and scattering measurements. This is a great advantage. Such continuous measurements provide us valuable information on variations of the vertical distributions of ozone in the middle atmosphere with various timescale. In addition, the millimeter wave radiation is not affected by aerosols. Therefore, the vertical profile of ozone mixing ratio can be calculated exactly from the equations of radiative transfer at the millimeter wavelength. Typical altitudinal coverage and resolution are ~ 20-80 km and ~ 10-20 km, respectively.

3. EXPERIMENT AND RESULTS

The OZITOS+ campaign was held during October 14 – 17, 2014 (Julian Day 287 to 291) in the Río Gallegos, Argentina. A total of six Electrochemical Chamber Cell (ECC) ozone sonde were launched during this campaign. In general the launch time was near midnight local time (3:00 UTC) and a second sounding after 3 hs of departure time of the first one. The lidar measurement always needs clear skies conditions to get successful measurement output and for this reason this instrument is very weather dependent. During the week of the experiment, two main windows of clear skies were used, on the night of October 15 and the night of October 18.

Table I: Schedule of the observation days of the DIAL. Measurements coincident with the ozonesonde.

Day of Oct, 2014	Start Time UTC	End Time UTC	Integral Time
15	00:33:00	03:12:51	148 min
15	03:13:52	05:43:50	148 min
15	05:44:51	08:16:10	150 min
16	05:31:30	07:34:58	116 min
16	No obs.	No obs.	No obs.
18	00:10:15	02:47:36	148 min
18	02:48:37	05:18:35	148 min
18	05:19:37	7:48:59	148 min

Ozonesondes used during the campaign are small, lightweight, balloon-borne instruments capable of making measurements from the surface to near 35 km in altitude. In the Table I, we shown a schedule of the observation days of the DIAL measurements correlated with the ozonesonde flights.

In figure 1 we represents the total ozone column around the period of the campaign at Río Gallegos represented as vertical dashed line.

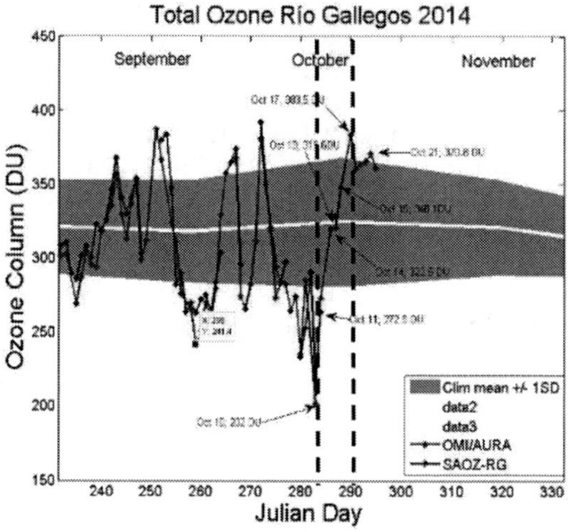

Figure 1. Total ozone column for Río Gallegos from OMI/AURA instrument and SAOZ instrument. White line is the climatological ozone mean (1978-2013) +/- 1SD (shadow gray zone). We show OMI/AURA and SAOZ data at Río Gallegos Station.

We have included in the representation the total ozone column provided by OMI/AURA and ground-based SAOZ instrument installed in the OAPA station. Both measurements are contrasted with the climatological data obtained from the Multi Sensor Reanalysis (MRS) database in the period 1978 to 2013 (white line) and its corresponding standard deviation (shadow gray zone) [4].

Figure 2 shows the ozone profile measured by the three techniques (DIAL-Balloon-borne-MWR) between October 15-18, 2014. Plots are represented in ozone number density. Shadow rows from Table I. are the days compared in the figure below.

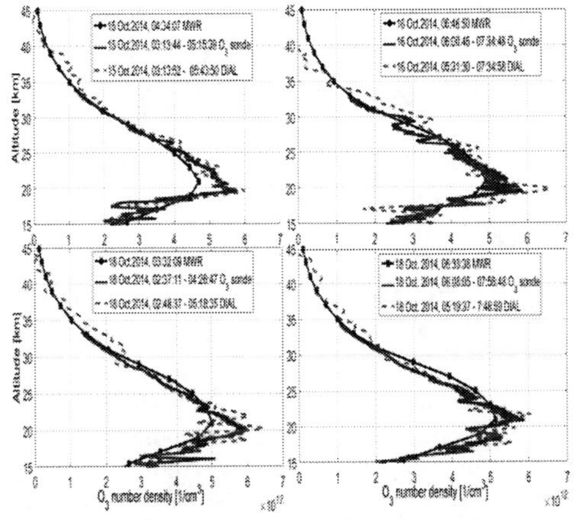

Figure 2. Example of comparisons with multi-instruments observing ozone profiles measured by DIAL (dashed line), ozonesonde (continuous line) and MWR (square line) on Oct 15-18, 2014.

The four measurements above shows good agreement between ~ 20 and 30km.

Figure 3 represents the temperature profiles computed from DIAL system during the campaign period, one profile per day since Oct. 14 to 18, compared with NCEP reanalysis at Río Gallegos.

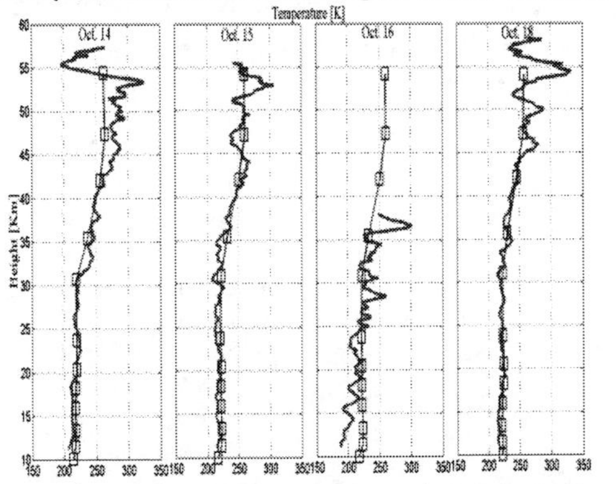

Figure 3. Evolution temperature profiles derived from Rayleigh Lidar (continuous line) and compared with Data reanalysis NCEP (square line) for Río Gallegos during the campaign days Oct 14-18 2014.

Applying signal processing over the Rayleigh lidar signals we can obtain the temperature profiles [5]. As the measurements obtained by DIAL system are longer than two hours integration time (IT) we have computed the temperature profile using an IT around 60 min. coincident at middle period of time of flight of the ozone sounding

4. CONCLUSIONS

This first joined activity between Argentina, Chile and Japan for the ozone observation profiles at Patagonia in the framework of SAVER-Net project has been very successful from scientific and social point of view, given the opportunity to researchers of three countries to share their experiences and work together for a common goal. The preliminary analysis of the results obtained in this activity shows good agreement between different instruments and puts in evidence the high quality observation capability that SAVER-Net project produce for the monitoring of ozone hole and related issues associated with this.

ACKNOWLEDGEMENT

We would like to make special thanks to the Project "Development of Atmospheric Environmental Risk Management System in South America" (2013-2017) financed by JICA-JST through Science and Technology Research Partnership for Sustainable Development (SATREPS). Also we would like to thank a SAOZ Network and FAA (Fuerza Aérea Argentina) for allowing us to use the Antarctic Storehouse during the campaign.

REFERENCES

1. Hauchecorne, A. and Chanin, M.-L.: Density and temperature profiles obtained by lidar between 35 and 70 km, Geophys. Res. Lett., 7, 565–568, 1980.
2. Megie, G. J., Ancellet, G., and Pelon, J.: Lidar measurements of ozone vertical profiles, Appl. Opt., 24, 3454–3463, 1985.
3. Wolfram E.A., J Salvador, R D'Elia, C Casiccia, N Paes Leme, A Pazmiño, J Porteneuve, S Godin-Beekman, H Nakane and E J Quel. New differential absorption lidar for stratospheric ozone monitoring in Patagonia, South Argentina *J. Opt. A: Pure Appl. Opt.* **10** (2008) 104021 (7pp). doi:10.1088/1464-4258/10/10/104021.
4. Van der A, R. J., M. A. F. Allaart, and H. J. Eskes (2010), Multi sensor reanalysis of total ozone, Atmos. Chem. Phys. Discuss., 10, 11,401–11,448, doi:10.5194/acpd-10-11401-2010
5. Jacobo O. Salvador, Elian Wolfram, Raul D'Elia, Eduardo Quel, Temperature lidar retrieval using a Rayleigh lidar in Río Gallegos, Argentina. Óptica Pura y Aplicada 44 (2) 367-372 (2011).

DOI: 10.1051/epjconf/201611913003

Simultaneous Observations of Mesoscale Gravity Waves Over the Central US with CRRL Na Doppler Lidars and USU Temperature Mapper

Xian Lu[1*], Cao Chen[1], Wentao Huang[1], John A. Smith[1], Jian Zhao[1], Xinzhao Chu[1*],

Tao Yuan[2], Pierre-Dominique Pautet[2], Mike J. Taylor[2]

[1]*University of Colorado Boulder, 216 UCB, CIRES, Boulder, CO 80309, USA,*

**Email: xian.lu@colorado.edu and Xinzhao.Chu@colorado.edu*

[2]*Utah State University, Logan, UT, USA*

ABSTRACT

We present the first coordinated study of a 1-h mesoscale gravity wave event detected simultaneously by a Na Doppler lidar at Boulder, CO (40.1°N, 105.2°W), and a Na Doppler lidar and an airglow temperature mapper (AMTM) at Logan, UT (41.7°N, 111.8°W) in the mesopause region on 27 Nov. 2013. The vertical and horizontal wavelengths are ~16.0±0.3 and 342.0±10.4 km, corresponding to vertical and horizontal phase speeds of ~4.4±0.1 and 95.0±3.0 m/s, respectively. The wave propagates from Logan to Boulder with an azimuth angle of ~138.1±1.7° clockwise from North. A uniqueness of this study is that the 1-h wave amplitudes on vertical winds have been quantified for the first time by the STAR Na lidar at Boulder. The GW polarization relation between vertical wind and temperature is evaluated. The intrinsic period of the wave is Doppler shifted to ~100 min by a background wind of 40 m/s, which is confirmed by USU lidar wind observations. This study illustrates a great potential of combining multiple instruments to fully characterize mesoscale gravity waves and inspect their intrinsic properties.

1. INTRODUCTION

Atmospheric gravity waves (GWs) are ubiquitous in the Earth's atmosphere and play irreplaceable roles in its energetics, dynamics and chemistry [1]. Among them, mesoscale GWs with horizontal wavelengths of 50–500 km and periods of 1–4 h, have important effects on weather systems by affecting mesoscale cloud and precipitation patterns [2]. They also play a significant role in the transport of momentum in the mesosphere and lower thermosphere (MLT) region, and affect ionospheric variability by seeding medium-scale travelling ionospheric disturbances. Physical and dynamical mechanisms capable of generating mesoscale GWs are largely related to lower-atmosphere activity, such as convection, topographic forcing, geostrophic adjustment, and diabatic heating. Thus, the mesoscale GWs serve as important media for terrestrial weather to impact space variability.

Numerous techniques have been developed to detect and characterize GWs, while the coordinated studies that can fully resolve high to medium frequency and mesoscale gravity waves are relatively sparse. Almost all techniques have their own limitations and suffer the problem of observational filtering as each of them is only sensitive to certain portion of wave spectra and can capture certain part of wave characteristics [3]. Nevertheless, it is likely that multiple techniques have certain overlapping wave spectra that can be fully or marginally resolved and in this case, coordinated studies become meaningful.

This study aims at combining the three ground-based instruments (two Na Doppler lidars and an airglow imager) for the first case study of a 1-h mesoscale GW event over the central US and providing direct wave characterizations in the MLT region. In addition to temperatures (T) and horizontal winds (u, v), we provide the direct measurements of GW amplitudes in vertical winds (w), which have been rarely reported in previous studies due to their small magnitudes and technical challenges to measure them. Such case study is important as it lays the foundation for statistical studies, which benefit the future investigations of wave sources and wave impacts upon the upper atmosphere.

2. COORDINATED OBSERVATIONS

This study was enabled by the simultaneous observations with two Na Doppler lidars of the Consortium of Resonance and Rayleigh Lidars (CRRL) at Boulder, CO [4], and at Logan, UT [5], combining with an Advanced Mesospheric Temperature Mapper (AMTM) [6] located at Bear Lake Observatory (41.9°N, 111.4°W) nearby Logan. The observations were made from ~1–13 UT in the night on 27 November 2013. Under the CRRL umbrella, the Consortium Technology Center (CTC), hosted by the University of Colorado Boulder, operates a STAR Na Doppler lidar from the Table Mountain Lidar Observatory north of Boulder while the Utah State University (USU) lidar group operates a Na Doppler lidar from the USU campus in Logan. Here "STAR" stands for Student Training and Atmospheric Research.

The STAR lidar signal levels reached over 1000 counts/shot from 80 to 115 km with the averaged laser power of ~500 mW at 30 Hz repetition rate and a telescope primary mirror of ~80 cm in diameter. The raw lidar data were collected in the temporal and spatial resolutions of 3 s and 24 m. After binning to 5 min and 0.96 km, the measurement uncertainties in w and T are ~0.5 m/s and ~1 K near the Na layer peak. The USU lidar was operating in a two-beam setup with one pointing 30° north and the other one pointing 20° east. For 5 min×1 km resolution, the uncertainty near the Na peak was ~3–6 K in T, and varied between 2–8 m/s in the Line-Of-Sight (LOS) wind.

The AMTM is an infrared digital imaging system that measures selected emission lines in the mesospheric OH (3,1) band to create intensity and temperature maps of the mesosphere around 87 km. The raw temperature data are obtained with a spatial resolution of ~0.5 km and a temporal resolution of ~0.5 min, with an uncertainty of +/-1K in temperature. Two temperature keograms, one for South-North (S-N) direction and one for West-East (W-E) direction, are composed of the S-N and W-E slices cutting across the center of each raw image and proceeding with time. The horizontal ranges over the S-N and W-E directions are 120 and 150 km, respectively. The AMTM data are also binned into 5 min to have the same temporal interval as lidar data.

3. RESULTS

3.1. Salient 1-h Wave Signatures

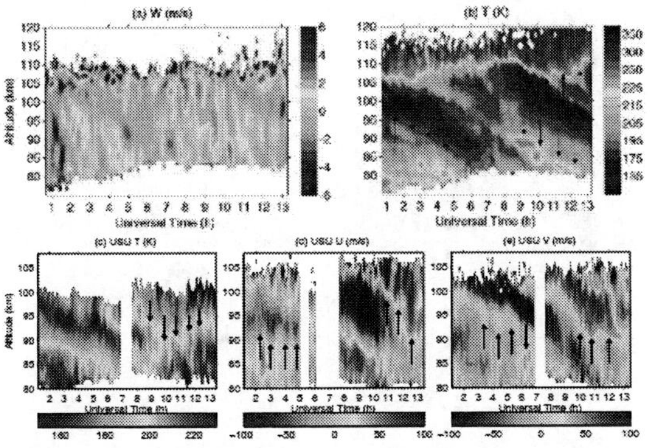

Figure 1. (a) Vertical winds and (b) temperatures from the STAR lidar at Boulder, (c)-(e) temperature and horizontal winds from USU lidar at Logan.

This study was stimulated by the signatures of short-period waves in the vertical winds measured by the STAR Na Doppler lidar. The most salient wave signatures in vertical winds are those with periods of ~30–60 min (Figure 1a). The downward phase progressions indicate that the wave signatures are authentic and correspond to upward energy propagation. The signatures of longer period waves are visible in vertical winds as a background to these short-period waves. Such longer period waves or possibly tides become the prominent features in temperatures and horizontal winds (e.g., Figure 1b-e), from which higher-frequency waves are difficult to detect. Even so, the 1-h waves are noticeable features (highlighted by black arrows in Figure 1b-e).

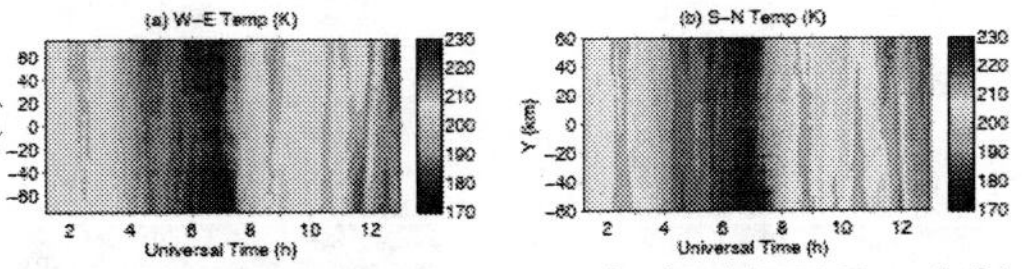

Figure 2. Raw AMTM keograms in the (a) W-E and (b) S-N directions, respectively.

Figure 2 shows the raw AMTM measurements of OH airglow temperatures at Logan without any filtering; however, obvious GW signatures with a period of ~1 h are visible after ~8 UT in both directions, indicating that the presence of this wave is a robust feature.

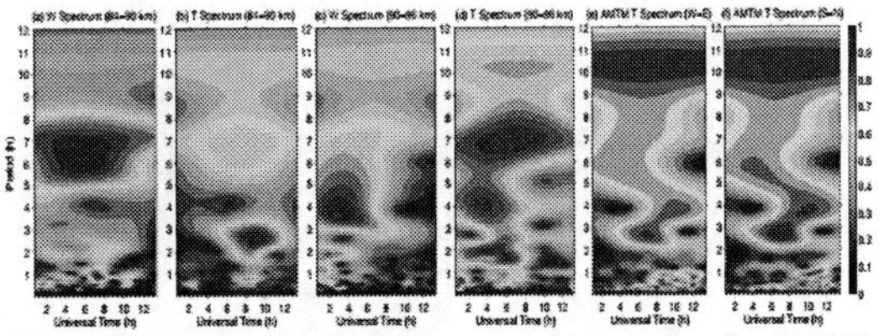

Figure 3. Normalized Morlet wavelet spectra for STAR (a) w and (b) T'/T_0 averaged in 84–90 km, (c) w and (d) T'/T_0 averaged in 90–96 km, (e) for AMTM T'/T_0 in W-E and (f) S-N directions, respectively.

To obtain perturbation fields, we subtract the nightly mean at each altitude for lidar data and at each range for AMTM. Relative temperature perturbations (T'/T_0) are further derived by dividing with nightly mean temperatures. GW Polarization relation predicts that the amplitude ratio of w to T'/T_0 is proportionally to the intrinsic frequency [7]. Thus, the T'/T_0 wavelet spectra are weighted by multiplying frequency, in order to be comparable with w and highlight high frequency waves. Figure 3a-d show the normalized wavelet spectra for STAR lidar, and 3e-f for the AMTM.

A rich wave spectra can be seen from the wavelet analyses. In addition to a wave component of 10–11 h which is likely the semidiurnal tide, we see strong waves with periods of 2–3 and 4–8 h. The 1-h waves persistently occur through the night while strengths vary with time. We choose a narrow passband of 0.5–1.5 h to extract the quasi 1-h waves due to the existence of the 2–3 h waves. The filtered perturbations are shown in Figure 4 for lidar data and Figure 5a-b for

AMTM. The 1-h waves are obvious through 82–107 km and in the W-E and S-N directions near 87 km.

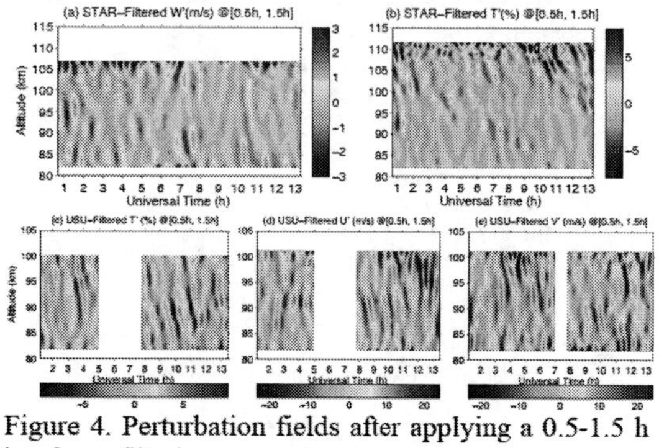

Figure 4. Perturbation fields after applying a 0.5-1.5 h bandpass filtering.

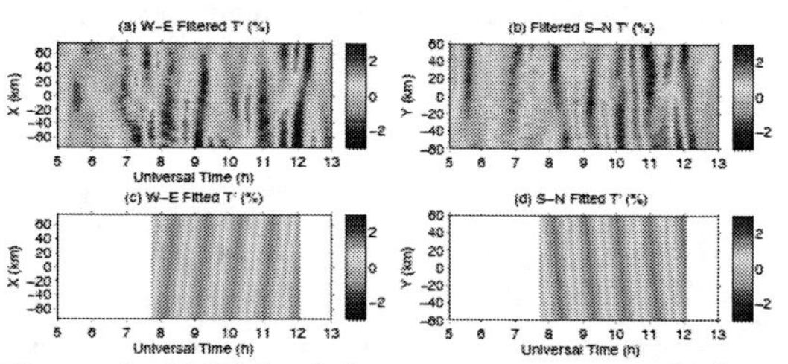

Figure 5. AMTM relative temperature perturbations filtered at 0.5–1.5 h, and the corresponding 2D fittings.

3.2. Observed Wave Characteristics

The AMTM keograms contain the information of time versus distance, from which wave period and wavelength along either W-E or S-N direction can be inferred from the following 2D fitting:

$$\frac{T^{'}(x,t)}{\overline{T}(x)} = A\cos\left(\frac{2\pi}{\tau}t - \frac{2\pi}{\lambda}x - \varphi\right) \qquad (1)$$

where τ and λ are period and wavelength along the W-E or S-N direction, respectively. Both wavelet spectra (Figure 3) and filtered perturbations (Figure 5) illustrate strong 1-h waves between ~7.7 and 12.1 UT for AMTM, which are chosen for the fitting. The fitted periods in both directions are ~0.96 h, indicating that the same wave is identified. The wavelengths along the W-E and S-N directions from the fittings are 511.9±21.8 km and -459.6±19.8 km, respectively. The positive sign means waves propagate eastward and the negative sign means they propagate southward.

We project such horizontal structure onto a US map centered at the states of Colorado and Utah in Figure 6. The horizontal wavelength is determined to be 342.0±10.4 km, and the azimuth angle is 138.1°±1.7, clockwise from North. Under such scenario, for one cycle of wave propagation, the projection onto the Logan-Boulder line would be ~395.2 km. Since the distance between these two locations is ~580 km, the 1-h wave will take ~1.47 h to propagate from Logan to Boulder. Thus, the phase shift between the two stations is: $\Delta\varphi = 2\pi(1.47-1) \approx 2.95$ rad at the same altitude.

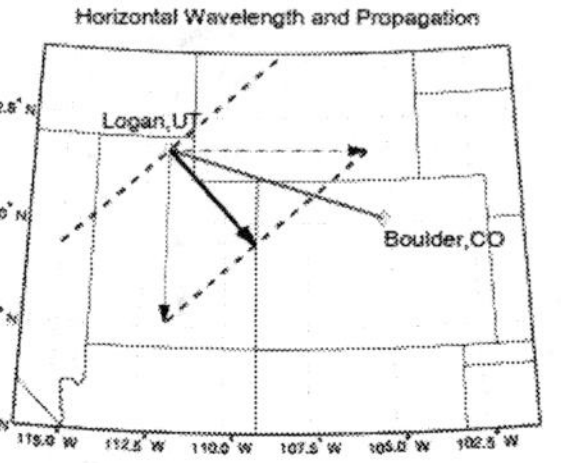

Figure 6. Geographic projection of the horizontal structure of the 1-h wave onto the US map. Thick black line with an arrow denotes the wave propagation vector ($\vec{k_h}$). Dashed blue lines are two wave fronts.

Lidar measurements surpass in providing range-resolved information of wave amplitudes and phases, and from vertical variations of phase, we can derive vertical wavelengths. Figure 7 shows the amplitudes and phases of the 1-h waves, derived from the sinusoidal fittings of the filtered perturbations [8] (Figure 4). The time window for the fitting is 7.7–10.7 UT for Logan and 9–12 UT for Boulder, which considers the time durations of relatively strong waves and the time delay caused by wave propagation.

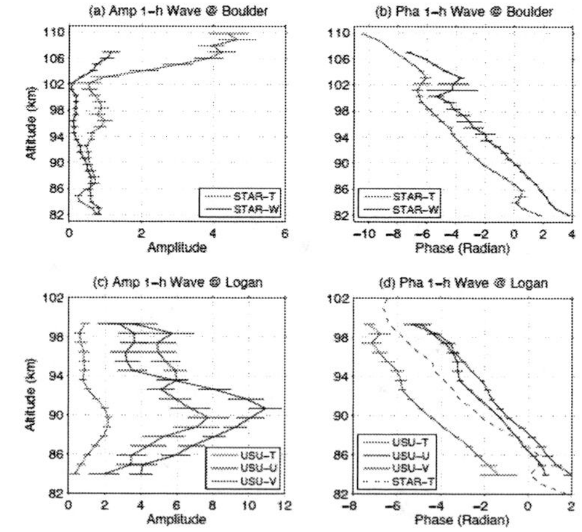

Figure 7. Amplitude and phase of the 1-h wave. The amplitude unit of T'/T_0 is % and that of winds is ms^{-1}.

The mean amplitudes in T'/T_0 and w at Boulder are 1.04% and 0.44 m/s, and are 1.16%, 6.04 and 5.15 m/s in T'/T_0, u and v at Logan, respectively. The vertical wavelengths are similar among these components. The mean value is ~16.0±0.3 km. The phase difference in temperature between Logan and Boulder is about 3 rad near 87 km, which is consistent with the AMTM results. This confirms that the same wave is observed and its horizontal structure has been properly retrieved.

3.3. Intrinsic Wave Properties

According to the linear wave theory, the non-dissipative GWs should obey the following dispersion and polarization relations [7]:

$$\hat{\omega}^2 = \frac{N^2 k_h^2 + f^2\left(m^2 + \frac{1}{4H^2}\right)}{k_h^2 + m^2 + \frac{1}{4H^2}} \quad (2)$$

$$\hat{T} = \frac{N^2\left(im + \frac{1}{2H}\right) - \frac{\hat{\omega}^2}{\gamma H}(\gamma - 1)}{g\hat{\omega}\left(-m - \frac{i}{2H} + \frac{i}{\gamma H}\right)}\hat{w} \quad (3)$$

where $\hat{\omega}$, m, and k_h are the wave intrinsic frequency, vertical and horizontal wavenumbers. N and H are buoyance frequency and scale height of the background atmosphere and equal to ~0.022 s^{-1} and 6.1 km for the MLT region [9]. f is the inerital frequency and γ is the ratio of specific heats. By substituting horizontal and vertical wavelengths into Eq (2), the intrinsic period is calculated as ~101.4 min. Since $\hat{\omega} = \omega - \vec{k} \cdot \vec{U}$, the mean wind along the wave propagating direction should be ~38.5 m/s, which is in accordance with Figure 1d-e showing that near the peak of the airglow layer (~87km) and from 7.7 to 12 UT, eastward zonal and southward meridional winds would have a positive projection onto the wave direction.

The polarization relation predicts that the ratio of wave amplitude in relative temperature ($\hat{T}$) to vertical wind ($\hat{w}$) is ~4.5% and the phase difference is ~1.69 rad. From Figure 7b, we derive such amplitude ratio and phase difference at each altitude and obtain their means and standard deviations as 3.4±1.9%, and 1.88±0.4 rad. The theoretical predictions are within these ranges. It is expected that wave dissipations may alter the polarization relation and be responsible for the biases.

4. CONCLUSIONS

Simultaneous observations of two Na Doppler lidars and a temperature mapper allow us to unambigiously derive the characteristics of the 1-h mesoscale wave, shown as a dominant feature in the vertical winds of MLT region. The horizontal and vertical wavelengths of this wave are ~340.0 and 16.0 km, corresponding to phase speeds of ~95.0 and 4.4 m/s. It propagates southeastward from Logan, UT to Boulder, CO with an azimuth angle of ~138°. Due to the Doppler effect of background winds, its intrinsic period is shifted to ~100 min. The mean amplitude of this 1-h wave in vertical wind is about 0.44 m/s.

A preliminary inspection of the Na lidar data at Boulder, CO has frequently revealed 0.5-2 h waves in vertical wind, temperature and Na density variations. The statistical study of their characteristics are needed to fully understand the distribution of their scales, magnitudes, and intermittency. The current study has shown a great capability of using multiple data sources to resolve the characteristics of such mesoscale and high-to-medium-frequency GWs. Future studies will be performed to link the lower-atmosphere wave sources with the observed waves in the middle atmosphere, and their coupling effects on ionospheric variabilities.

ACKNOWLEDGEMENT

The STAR lidar work was supported by NSF grants AGS-1136272 (CRRL) and AGS-1115224. X.L. was also supported by AGS-1343106 (CEDAR).

REFERENCES

[1] Holton, J. R., 1983: The influence of gravity wave breaking on the circulation of the middle atmosphere, *J. Atmos. Sci.*, **40**, 2497-2507.

[2] Zhang, F., 2004: Generation of mesoscale gravity waves in the upper-tropospheric jet-front systems, *J. Atmos. Sci.*, **61**, 440–457.

[3] Alexander, M. J., 1998: Interpretations of observed climatological patterns in stratospheric gravity wave variance, *J. Geophys. Res.*, **103(D8)**, 8627–8640.

[4] Smith, J. A., W. Fong, B. Roberts, W. Huang and X. Chu, 2012: Very high-resolution Na Doppler lidar at Boulder, Colorado, *Proc. 26th International Laser Radar Conference*, 151-154.

[5] Yuan, T., et al., 2013: Coordinated investigation of summer time mid-latitude descending E layer (Es) perturbations using Na lidar, ionosonde, and meteor wind radar observations over Logan, Utah (41.7°N, 111.8°W), *J. Geophys. Res. Atmos.*, **118**, 1734–1746.

[6] Pautet, P.-D., et al., 2014: An advanced mesospheric temperature mapper for high-latitude airglow studies, *Appl. Opt.*, **53(26)**, 5934–5943.

[7] Vadas, S. L., 2013: Compressible f-plane solutions to body forces, heatings, and coolings, and application to the primary and secondary gravity waves generated by a deep convective plume, *J. Geophys. Res. Space Physics*, **118**, 2377–2397, doi:10.1002/jgra.50163.

[8] Lu, X., et al., 2009: Gravity wave propagation and dissipation from the stratosphere to the lower thermosphere, *J. Geophys. Res.*, **114**, D11101.

[9] Lu, X., et al., 2015: Vertical evolution of potential energy density and vertical wave number spectrum of Antarctic gravity waves from 35 to 105 km at McMurdo (77.8°S, 166.7°E), *J. Geophys. Res. Atmos.*, **120**, doi:10.1002/2014JD022751.

DOI: 10.1051/epjconf/201611913004

ANTARCTIC WAVE DYNAMICS MYSTERY DISCOVERED BY LIDAR, RADAR AND IMAGER

Cao Chen [1]*, Xinzhao Chu[1], Weichun Fong[1], Xian Lu[1], Adrian J. McDonald[2],

Dominique Pautet[3], and Mike Taylor[3]

[1]*University of Colorado at Boulder, 216 UCB, CIRES, Boulder, CO 80309, USA, Email: Cao.Chen@Colorado.edu*
[2]*University of Canterbury, Department of Physics and Astronomy, Christchurch, NZ*
[3]*Center for Atmospheric and Space Sciences, Utah State University, Logan, UT, USA*

ABSTRACT

Since the start of the McMurdo Fe lidar campaign, large-amplitude (~±30 K), long-period (4 to 9 h) waves with upward energy propagating signatures are frequently observed in the MLT temperatures. Despite its frequent appearance, such type of wave was neither widely observed nor well understood in the past. At McMurdo (77.8°S, 166.7°E), the simultaneous observations of such waves using lidar, radar and airglow imager can provide 3-D intrinsic wave-propagation properties, which are greatly needed for understanding their sources and potential impacts. This study presents the first coincident observation of these 4–9 h waves by lidar, radar and airglow imager in the Antarctic mesopause region.

1. INTRODUCTION

Although observations in the Antarctic middle and upper atmosphere started very early, there were no range-resolved temperature measurements at McMurdo until the start of Chu group's lidar campaign [1]. Since then, many new discoveries have been made by lidar observations in Antarctica, e.g., the thermospheric neutral iron (Fe) layers [1], solar effects on Fe layer bottomside, two simultaneous Inertia Gravity Waves (IGWs) [2], large eastward planetary waves and super-exponential growth of thermal tide amplitude above 100 km. These discoveries are only made possible by the robustness and powerfulness of the lidar system. To achieve this, Chu group members have upgraded the lidar several times, and have been constantly maintaining the system well. Therefore, this lidar is capable of running continuously for several days with high performance, which enables the statistical study of wave occurrence and properties.

Benefited from the capabilities of the Fe Boltzmann lidar, a completely new wave phenomenon has been discovered at McMurdo, i.e., strong and persistent 4 to 9 h period waves with short vertical wavelengths (λ_z) from the stratosphere all the way to the lower thermosphere that occur all year round, e.g., [1,2]. However, after its discovery, the true identity of these waves still remains a mystery to the community: Are these waves ordinary IGWs, or short-period Atmospheric Normal Modes (ANMs), or something unknown before? Initial case studies have suggested that some of these waves are ordinary IGW [2]. However, persistent appearance of these waves over McMurdo contradicts the current understanding of Gravity Waves (GWs) being intermittent. Furthermore, what could be the persistent sources for these waves around Antarctica? There are no commonly known sources, especially in summertime when polar vortex has disappeared. On the other hand, ANMs, natural resonant modes of atmospheric free oscillations [3] or 8-h/6-h tides can be possible causes. However, these explanations cannot solve the mystery because the observed vertical wavelength λ_z (~20 km) are much shorter than those predicted for ANMs or tides according to both theory and simulations. The lack of a convincing theory for these 4-9 h waves in Antarctica has put forth a big mystery.

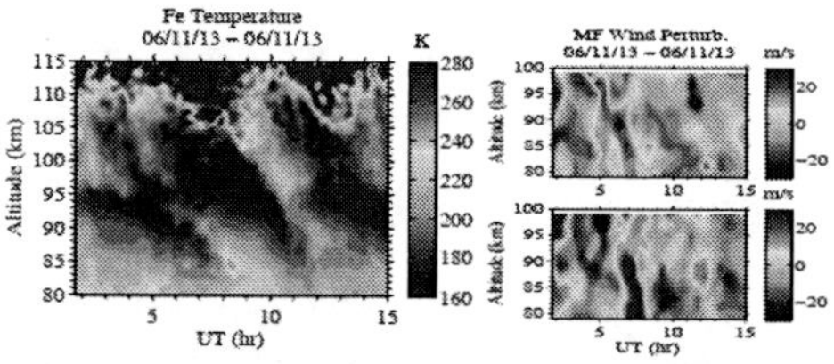

Figure 1. (a) Raw temperatures from 1.5 to 15 UT on 11 June 2013 by lidar. (b) Zonal wind (top) and meridional wind (bottom) data by MF radar at the same time with the 12-h and 24-h tides removed.

Solving this mystery is crucial to improving the current weather and climate models and to understanding climate changes especially in Polar Regions. Currently, models have simulated large-scale waves such as planetary waves and tides well, but most of them cannot directly resolve small-scale waves such as GWs and ANMs. These small-scaled waves, although smaller in scales, transport significant amounts of energy and momentum up to the MLT region where they strongly influence the mean wind and alter the temperature structure. Therefore, this incapability has become the biggest uncertainty in weather and climate forecasting models and has greatly undermined the models' predictive capability.

To resolve this mystery, we need the 3-dimensional (3-D) intrinsic wave properties, i.e., observed and intrinsic periods, horizontal propagation direction, wavelengths λ_h and λ_z. Fortunately, two other instruments collocated with our Fe Boltzmann lidar enable such study. One is Scott Base MF radar, which has continuous temporal coverage and measures MLT winds between 70 and 100 km.

Combining our lidar with this radar, intrinsic wave properties can be derived, e.g., [2]. Another is Utah State University (USU) all-sky infrared imager provides measurements on the intensity of infrared OH emission layer (~87 km). The OH imager data can provide horizontal propagation information directly.

2. CASE STUDY

2.1 Simultaneous and common volume observations of the 4–9 h waves by lidar, radar and imager

On 11 June 2013, strong wave perturbations with period $\tau \sim 5$ h were observed in Fe lidar temperatures as shown in Figure 1a. This wave has downward phase progression and λ_z of ~20 km. The raw MLT temperature data have resolutions of 0.1 h and 0.1 km. The data we used were smoothed temporally and vertically with Hamming windows of 0.5 h and 1 km FWHM. Therefore, only the waves with $\tau \geq 1$ h, and $\lambda_z \geq 2$ km were resolved. MF wind data also shows a 5-h wave as in Figure 1b, after removing the semidiurnal and diurnal tides, we can see large wind perturbations in both the zonal and meridional directions. We first assume this wave as IGW, and then test it with a linear GW theory. At the same time, the 5-h wave was clearly observed by the OH imager. Figure 2 shows two keograms, one for the South-North (S-N) direction and another for West-East (W-E), which are composed of the S-N and W-E slices at the center of each raw image proceeding with time. The horizontal ranges over the S-N and W-E directions are 280 and 350 km, while the temporal and spatial resolutions are 0.5 min and ~ 1 km, respectively.

Data Analysis

In order to minimize contamination of tidal and planetary wave oscillations, the temperatures, winds and OH intensities are all band-pass filtered by a 6th-order Butterworth filter with pass band between 2.5–7 h. We then fit two OH intensity keograms to a non-linear monochromatic wave model to extract the horizontal propagation properties of the waves. The model *I(t,x,y)* which takes two forms for each keogram, is written explicitly as follows.

$$I(t,x,0) = AI_x \cos(2\pi\omega t - 2\pi k x + \phi_0) + I_{x0} \quad (1)$$

$$I(t,0,y) = AI_y \cos(2\pi\omega t - 2\pi l y + \phi_0) + I_{y0} \quad (2)$$

In this model, ω is observed frequency, k is the zonal wavenumber, and l is meridional wavenumber. The fitting process is to minimize the chi-square fitting error.

$$\chi^2 = \sum_i \frac{\left(I(t_i,x_i,0) - \bar{I}(t_i,x_i,0)\right)^2}{uI(t_i,x_i,0)^2} + \sum_j \frac{\left(I(t_j,0,y_j) - \bar{I}(t_j,0,y_j)\right)^2}{uI(t_j,0,y_j)^2} \quad (3)$$

We utilize a Monte-Carlo sampling method to do the fitting [4]. Basically, for each step of the fitting, a set of random parameters around the current model parameters was generated. Then, a new χ^2 is calculated. If the new χ^2 is smaller than the pervious one, then this new set of model parameters was accepted. We continue this search until the fitting error is below our tolerance, and the parameters are converged. This method avoids the entrapment in local likelihood maxima, and therefore a useful way to solve non-linear fitting problems. λ_h and propagation direction are then derived:

$$\lambda_h = \frac{1}{\sqrt{k^2+l^2}} \quad (4)$$

$$\theta = \tan^{-1}\left(\frac{l}{k}\right) \quad (5)$$

The fitted results are λ_h~1760 km, θ~180°.

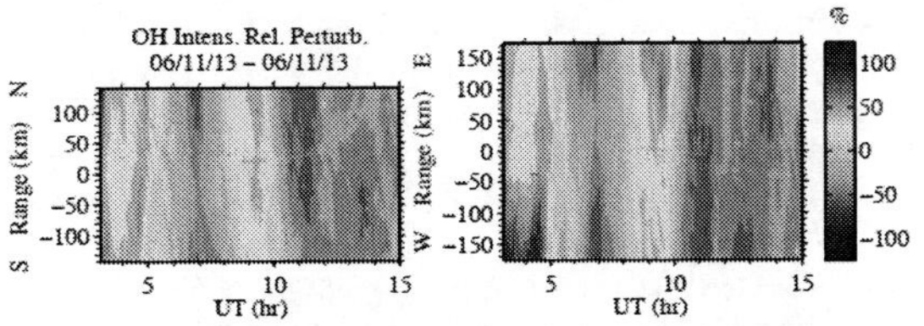

Figure 2. Perturbations of OH intensities on 11 June 2013.

As shown in [2], combining lidar temperatures and radar winds, we can calculate the intrinsic wave properties for IGW. The radar wind vector at each altitude can tell the wave propagation direction but with a 180° ambiguity. Then by comparing the phase difference between the wind and temperature perturbations, we can resolve this ambiguity. We take a new approach to obtain this phase difference, which is to fit the temperatures and winds together using the following wave model, see [5].

$$T(t) = |T|\cos(2\pi\omega(t - t_0)) + T_0 \quad (6)$$

$$U(t) = A\cos\varphi\cos(2\pi\omega(t - t_1)) - B\sin\varphi\sin(2\pi\omega(t - t_1)) + U_0 \quad (7)$$

$$V(t) = A\sin\varphi\cos(2\pi\omega(t - t_1)) + B\cos\varphi\sin(2\pi\omega(t - t_1)) + V_0 \quad (8)$$

where, t_0 is the phase of temperature perturbation, and t_1 is the phase of the horizontal parallel wind perturbation. We find the phase difference is ~297°, which is very close to that predicted by linear theory for ordinary IGW, (~270°). This means that our assumption of this wave being ordinary IGW is valid. After the 180° ambiguity was resolved, we can then use the method in [2] to solve the intrinsic properties. The results are shown in Figure 3.

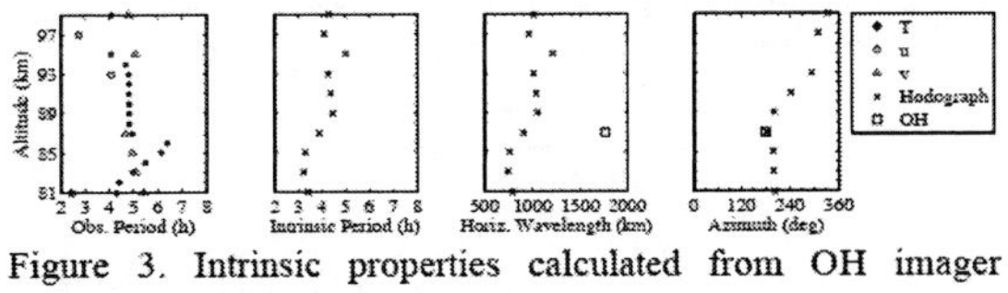

Figure 3. Intrinsic properties calculated from OH imager keograms and lidar/radar data.

As noticed, the directly fitted OH keograms yields much longer λ_h than that calculated from lidar and radar. The discrepancy could be due to the fact that OH density variations do not have the same response as the temperature variations, due to the chemistry involved.

3. STATISTICS

Although we have reported several case studies of these 4–9 h waves identified with IGWs [2,6], we believe a statistical study of the wave properties will be more helpful to obtain a comprehensive interpretation of the waves. In this section, we utilize the winter (May, June, and July) temperature data from 2011 to 2014, to derive several statistical wave properties of these 4–9 h waves. We select observation episodes longer than 12 h to do the study because this allows us to accurately determine the wave periods. A total of 135, 267, 172 hours of data were used in this study for May, June and July, respectively. Figure 4a shows an example of the persistence of these 4–9 h waves in our observations. During nearly 50 hours of observations on 18 June 2014, clear downward phase progressions are always presented. Figure 4b shows the corresponding frequency spectra at 85 km, 90 km and 95 km in which the waves peak at ~5.5 h, ~6.5 h, and 8 h. Persistent presence and discrete distribution of the dominant periods of the 4–9 h waves as shown in Figure 4 are very common in our observations.

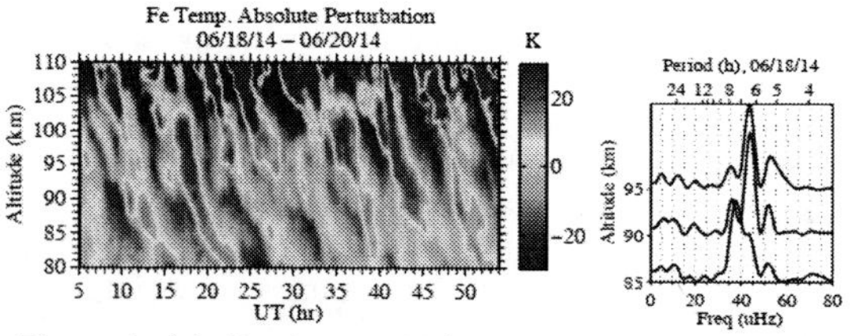

Figure 4. (a) Continuous 50-h temperature measurements show persistent 4–9 h waves on 18–20 June 2014. (b) The frequency spectra at 85 km, 90 km and 95 km.

Wave Frequency Spectra

Observations have shown that ordinary GWs tend to have a so-called "canonical" wave spectrum, i.e., in the spectral region where waves are regarded as being saturated; the power spectral density (PSD) follows a form of m^{-p} or ω^{-q} (m and ω are the vertical wavenumber and frequency, respectively; $-p$ and $-q$ are the corresponding slope of the power) [7]. According to diffusive filtering theory [7], $p=2q-1$. The studies done by *Lu et al.* [8] showed that, $-p$ is around -2.26 at McMurdo, therefore, $-q$ should be around -1.6.

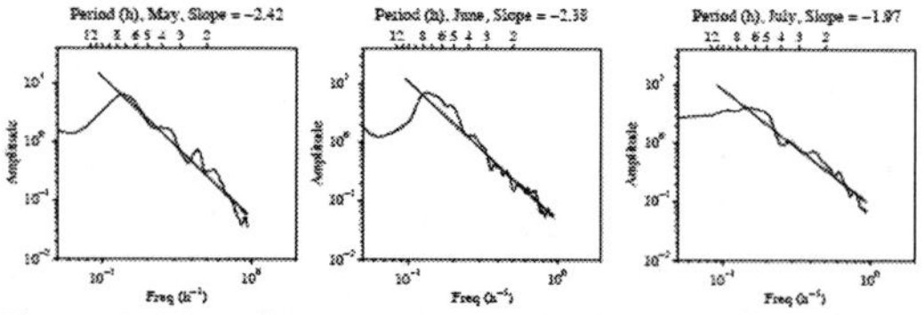

Figure 5. Mean frequency spectra in the MLT for May, June and July. Black dashed lines are fittings of the power law shapes in the range of ~1–11 h. The data used to calculate the spectra have 0.5 h and 1 km intervals in time and space.

We plot the mean frequency spectra for each month in Figure 5. Two things are different from the predicted universal power law. First, the slopes in all three months are much steeper than -1.6. Second, the spectral energies in periods range 4–9 h are higher than the power law fitting suggesting that these waves do not follow power law and therefore are different from the ordinary GWs.

Occurrence frequency

Ordinary GWs are believed to be intermittent, and therefore the occurrence frequency of these waves is critical when identifying whether they are ordinary GWs. Because multiple waves were often present at the same time, wavelet techniques were used in analyzing temperature time series at each altitude as described in [9]. Multiple peaks were identified in wavelet power as a function of periods and time, each of which is considered as one wave event. Then the time span of each wave event is determined from the full-width-half-maximum (FWHM) of the corresponding wavelet power peak. Usually, 2–6 wave events can be identified at each altitude during each episode of observation using this method. Figure 6 shows the percentage of the time span of a wave event over entire observation time as a function of the wave period in May, June, July averaged over all altitudes through 4 years of lidar campaign.

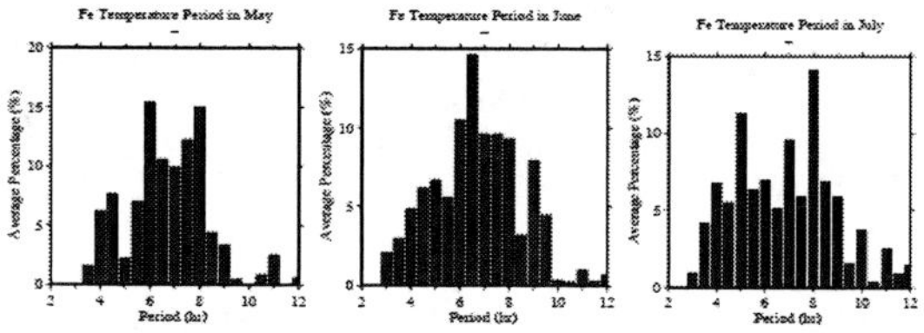

Figure 6. Percentage of the time span of a wave event over entire observation time as a function of the wave period in May, June, July through 4 years of lidar campaign. Results are averaged between 81 and 105 km.

Clearly shown in this Figure 6 is the prevalence of 4–9 h waves, which are detected by the algorithm ~92%, ~87% and ~84% of the entire observation time in May, June, and July respectively. This high occurrence frequency contradicts the current understanding of GWs being intermittent. Another distinct feature is that the discrete distribution of the dominant periods. For example, in July, the occurrence frequency of waves with τ~8 h is two-thirds higher than the waves in the neighboring period bins. If these are ordinary GWs, this means that the sources of these waves have preference over certain periodicity.

Dominant vertical wavelengths

To determine the λ_z of these waves, we first filter the wave perturbations with a 6-order Butterworth band-pass filter (pass bands between 3 to 12 h), and then calculate vertical spatial spectra for each episode, followed by averaging over each month. Figure 7 shows the mean vertical spatial spectra. The dominant λ_z~22 km for all three months.

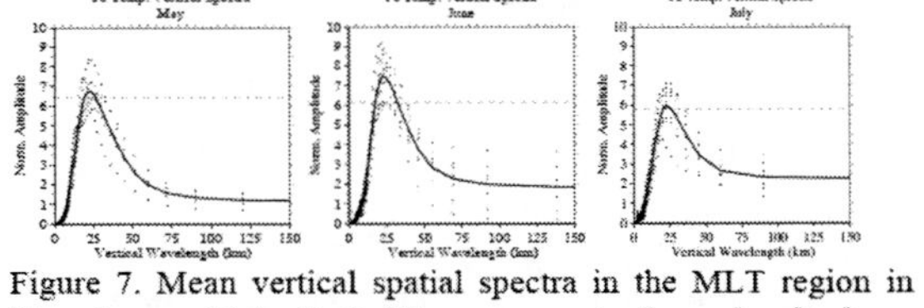

Figure 7. Mean vertical spatial spectra in the MLT region in May, June and July. Dotted lines are spectra for each episode.

Amplitude

We then plot the wave amplitude (standard deviation) with altitudes in Figure 8. The waves have been categorized into three groups: 3–5 h, 5–9 h, and 9–11 h. At least in May and June the 5–9 h wave amplitude clearly has a faster growth rate than the other two groups of waves, suggesting less damping encountered, or less saturated.

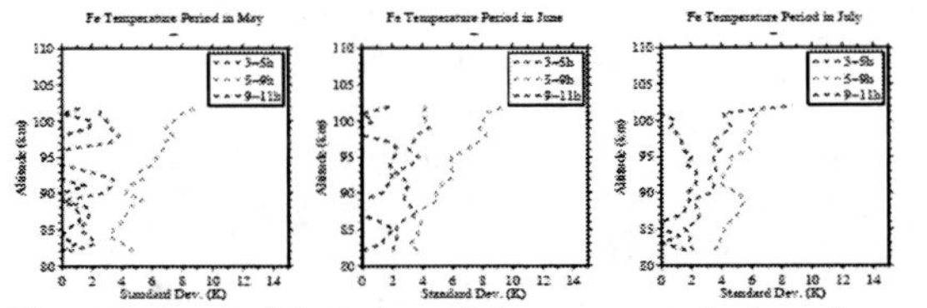

Figure 8. Standard deviations of the wave perturbations change with altitude for waves periods between 3-5 h, 5-9 h, 9-11 h in May, June, July.

4. CONCLUSIONS AND DISCUSSION

Large-amplitude (~±30 K), long-period (4 to 9 h) upward propagating waves are frequently observed in the MLT temperatures at McMurdo. Simultaneous observations of such waves using lidar, radar and airglow imager has revealed that one of the wave events on 11 June 2013 has the following properties: τ ~ 5 h, λ_z~20 km, λ_h~1100 km, propagating from north to south at azimuth θ~180°.

Statistical study has shown that, unlike ordinary GWs, these waves have higher energy than the predicted universal power-law spectra, and very high occurrence frequencies. The discrete distribution of the dominant periods is also a distinct feature for these waves. Short-period ANM, 6-h or terdiurnal tides cannot be the cause of these waves because of the short vertical wavelength (λ_z~22 km). Planetary scale Inertio-gravity waves simulated by *Mayr et al.* [10] at high latitudes have τ ~ 10 h, λ_z~20 km, is one possible candidate. These waves are related to the Class I gravity mode discussed in the classical literature. However, *Mayr et al.* only showed waves with τ ~ 10 h. Whether their simulations also have periodicity between 4 and 9 h is unknown. The planetary-scale IGWs by *Mayr et al.* have zonal wave number 0–4, and long horizontal wavelength λ_h~4000 km. Multiple-station observation network is required to examine the zonal structure of the 4–9 h waves.

ACKNOWLEDGEMENTS

We sincerely acknowledge Jeff Forbes and Chet Gardner for invaluable discussion. We appreciate the staff of United States Antarctic Program, McMurdo station, Antarctica New Zealand, and Scott Base for their superb support. This work was supported by NSF grants ANT-0839091 and PLR-1246405. C.C. was partially supported by AGS-1136272.

REFERENCES

[1] Chu, X., Z. Yu, C. S. Gardner, C. Chen, and W. Fong, 2011: Lidar observations of neutral Fe layers and fast gravity waves in the thermosphere (110–155 km) at McMurdo (77.8°S, 166.7°E), Antarctica, *Geophys. Res. Lett.*, **38**, L23807, doi:10.1029/2011GL050016.

[2] Chen, C., et al., 2013: Inertia-gravity waves in Antarctica: A case study using simultaneous lidar and radar measurements at McMurdo/Scott Base, *J. Geophys. Res.*, **118**, doi:10.1002/jgrd.50318.

[3] Forbes, J. et al., 1999: Lamb waves in the lower thermosphere: Observational evidence and global consequences, *J. Geophys. Res.*, **104**, doi: 10.1029/1999JA900044

[4] Mosegaard, K., & Tarantola, A., 1995: Monte Carlo sampling of solutions to inverse problems. *J. Geophys. Res.*, **100**, doi:10.1029/94JB03097

[5] Nakamura, T. et al., 1993: Characteristics of gravity waves in the mesosphere observed with the middle and upper atmosphere radar 2. Propagation direction, *J. Geophys. Res.*, **98**, doi:10.1029/92JD03030.

[6] Chen, C. et al., 2012: Lidar and radar investigation of inertia gravity wave intrinsic properties at McMurdo, Antarctica, *Proceedings of the 26th ILRC*, Greece.

[7] Gardner, C. S., 1994: Diffusive filtering theory of gravity wave spectra in the atmosphere, *J. Geophys. Res.*, **99**, doi:10.1029/94JD00819.

[8] Lu et al., 2015: Vertical evolution of potential energy density and vertical wave number spectrum of Antarctic gravity waves from 35 to 105 km at McMurdo (77.8°S, 166.7°E), *J. Geophys. Res.*, accepted.

[9] Zink, F., & Vincent, R. A., 2001: Wavelet analysis of stratospheric gravity wave packets over Macquarie Island: 1. Wave parameters, *J. Geophys. Res.*, **106**, doi: 10.1029/2000JD900847.

[10] Mayr, H. G. et al., 2004: Properties of internal planetary-scale inertio gravity waves in the mesosphere. *Ann. Geophys.*, **22**, doi:10.5194/angeo-22-3421-2004.

GROUND-BASED RAYLEIGH-MIE DOPPLER LIDAR FOR WIND MEASUREMENTS IN THE MIDDLE ATMOSPHERE

S. M. Khaykin[1]*, A. Hauchecorne[1], J. Porteneuve[1], J.-F. Mariscal[1], E. D'Almeida[1], J.-P. Cammas[2], G. Payen[2], S. Evan[2], and P. Keckhut[1]

[1]LATMOS-IPSL, Université Versailles St.-Quenitn, CNRS/INSU, Guyancourt, France,
**Email: sergey.khaykin@latmos.ipsl.fr*
[2]OSUR, UMS3365, Université de la Réunion, Saint-Denis, La Réunion, France

ABSTRACT

A unique Rayleigh-Mie Doppler wind lidar, measuring Doppler shift between the emitted and backscattered light using direct-detection technique is deployed at Observatory of Haute Provence Southern France) and at Reunion island (tropical Indian Ocean). The instrument was shown capable of wind measurements between 5 and 50 km with accuracy better than 1 m/s up to 30 km. The system consists of a monomode Nd:Yag laser operating at 532 nm, three telescopes and a double-edge Fabry-Perot interferometer. The laser light is sent alternatively in the vertical as well as zonal and meridional directions at 40° from the zenith using a rotating mirror. The two components of the horizontal wind are obtained from the measurement of the Doppler shift of the return signal spectrally filtered by the Fabry-Perot etalon. After demonstration of the method in 1989 the measurements were used for studying stratospheric dynamics as well as for constructing wind climatology up to 50 km altitude. A new system, featuring a more compact design was installed at Maïdo observatory at Reunion Island (21° S). The design of the instrument, results of observations and comparison against GPS radiosondes are presented. Application of Doppler lidar for validation of the future ADM-Aeolus satellite mission is discussed as well.

1. INTRODUCTION

Wind measurements above 25 km are rare. The only information available on a regular basis is zonal mean winds inferred from satellite-borne pressure measurements assuming geostrophic balance. However this does not allow understanding the propagation of gravity waves, whose filtering, dissipation and/or breaking strongly depend on wind speed and direction. Observations of high vertical resolution (~ 1 km) are required for detecting wind shears around the stratopause, which play a crucial role in reflecting or filtering of the waves. Following this requirement, a direct- detection Doppler lidar for wind observations in the middle atmosphere has been developed in France in 1989 [1,2]. After deployment of the Doppler lidar at Observatory of Haute Provence (OHP) in Southern France (44 °N, 6 °E, 600 m a.s.l.) in 1993 the technique was shown capable of wind measurements between 5 and 50 km independent of the presence of aerosols [3,4]. The wind lidar was operated on a regular based during 1994-1999 and occasionally since then. The observations were used to construct wind climatology in the middle atmosphere [3] and to study gravity wave spectra and small-scale wind fluctuations [5].

In 2012 the Rayleigh-Mie Doppler lidar was deployed at a newly established tropical observatory located on mount Maïdo (21 °S, 55 °E, 2200 m a.s.l.) at Reunion island [6]. The wind measurements above Reunion Island represent a substantial value for studying the dynamics of the tropical middle atmosphere – a region where observations are particularly sparse. They are also of big relevance for validation of the upcoming European Space Agency (ESA) ADM-Aeolus satellite mission for global observations of wind. ADM-Aeolus mission will carry a Rayleigh-Mie Doppler lidar, based on the same principle as the French Doppler lidars.

2. METHODOLOGY

The horizontal wind components are obtained by measuring Doppler shift between emitted and backscattered light by means of a double-edge Fabry-Perot interferometer (direct-detection technique). The Doppler shift is due to projection of molecules or particles velocities onto the line of sight of the laser inclined off zenith. The laser beam is sent alternatively to zonal and meridional direction for obtaining both wind components, whereas the vertical pointing is used for measuring the zero Doppler shift reference (in assumption that the vertical wind is negligible). A wind velocity of 1 m/s corresponds to the Doppler shift of about 2 fm with the off-zenith angle of 40°.

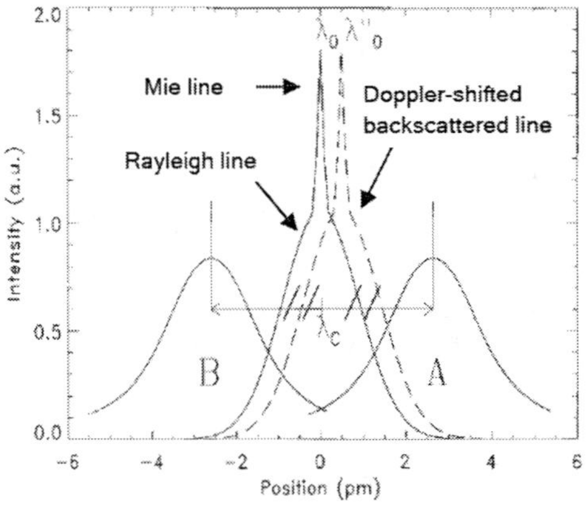

Fig. 1. Shape of the backscattered line embedded between FPI bandpasses. The A and B labels refer to the different areas of FPI [3].

After elastic scattering from air molecules and possibly other particles, the spectral shape of the backscattered laser line shows two distinct components (Fig. 1). The wide part is the Rayleigh component that is broadened because of the random motion of the air molecules. The spectral width can take values between 2 and 2.4 pm for atmospheric temperature from 200 to 280 K [3]. The fine Mie component is assumed to have the same spectral width as the laser source and to follow the same Doppler shift as the Rayleigh component. The Doppler-shift determination is provided by a double-edge FPI with two half-disks areas whose air gap is slightly different, resulting in two bandpasses, named A and B. Each area intersects a half-disk of the incoming light. Both bandpasses have a FWHM of 2.03 pm and are spaced 5.24 pm apart. The FPI is set in a sealed chamber ended by a pressure-changing piston driven by a stepper motor. The FPI bandpasses are spectrally tuned symmetrically on both sides of the Rayleigh–Mie backscattered line (see Fig. 1) by air pressure adjustment, i.e., through the adaptation of the air-gap refractive index. For a given line of sight, at an angle θ from the zenith, we calculate the response R(z, θ) at each altitude z, defined as

$$R(z,\theta) = \frac{CN_A(z,\theta) - N_B(z,\theta)}{CN_A(z,\theta) + N_B(z,\theta)} \quad (1)$$

where $N_A(z, \theta)$ and $N_B(z, \theta)$ are the number of photons transmitted through the bandpasses A and B respectively; and C is a corrective factor accounting for the possible imbalance between channels A and B. The response $R(z, \theta)$ is representative of the backscattered line position with respect to the FPI bandpasses.

3. DESCRIPTION OF THE INSTRUMENT

The Doppler lidar uses a Q-switched, injection-seeded, frequency-doubled Nd:YAG laser operating in monomode as a light source. The pulse repetition rate of the laser is 30 Hz, and the energy per pulse is ~330 mJ at 532 nm (24 W mean energy). The emitted spectral shape is assumed to be near Gaussian, with a full width at half-maximum (FWHM) of ~ 0.14 pm.

The OHP system includes three fixed subassemblies: one is pointed toward the zenith while the two others are tilted at 40° off zenith to the North and East directions (Fig. 2). The different subassemblies, i.e. different line of sights are used successfully during the acquisition. The unique laser beam is steered by a rotating mirror with three predefined positions. Each mirror position sends the beam to the chosen subassembly, composed of a central afocal transmitter and four receiving parabolic mirrors providing a total collective area of 0.78 m^2.

Fig. 2. Schematic view of the lidar optical setup at OHP. Four telescope assemblies for Northward, Eastward, Westward and vertical pointing are shown. The Westward pointing assembly is no longer used.

The photons are collected by means of four multimode optical fibers located at each focal point. A 1-nm bandpass is used as a spectral pre-filter before FPI to reduce the sky background light. The photons N_A and N_B transmitted through the FPI's bandpasses A and B are detected with R9880U-20 Hamamatsu photomultipliers (PMT's) in photon-counting mode. To extend the altitude range of measurements, there are two pairs of PMT's collecting 90% (high-altitude channels) and 10% (low-altitude channels) of the signals. The acquisition is done with a Licel transient recorder providing 16384 time bins of 100 ns width (i.e. 15 m radial resolution). Electronic obturation is used to protect the PMTs from the strong signals coming from low altitudes. Measurements are limited to nighttime and absence of optically thick clouds. A narrow-band interference filter with 0.1 nm bandpass, limiting the vertical range and resolution, can be used to extend the operation time frame to twilight conditions.

The Doppler lidar system at Maïdo observatory makes use of the same laser and acquisition system as those at OHP but features a more compact design of the reception set-up. The telescope is composed of a single rotating mirror, which serves for both the emission and reception. The collective area of the telescope is 0.3 m^2. The smaller collective area compared to OHP lidar is compensated by the high elevation of the station and cleaner atmosphere above the Indian Ocean.

4. RESULTS

The OHP Doppler lidar has been operated on a regular basis during 1994-1998 and on a campaign basis since then. The wind observations were used to build the climatology of wind velocity components in the 12 – 50 km height range [3] and for retrieving the gravity wave vertical wavenumber and apparent spectra as well as for constructing the time series of wave kinetic energy [5]. The error estimates showed the 1-σ uncertainty varying significantly with altitude and integration time; for 115 m vertical resolution the uncertainty ranges from 0.8 m/s at 13 km for 1 hour integration time to 6.8 m/s at 20 km for 5 min integration. For an entire night of integration (7 hours) the uncertainty stays below 1 m/s up to 30 km altitude [5]. These estimates are confirmed by intercomparison with GPS-equipped radiosondes [3].

The new Maïdo Doppler lidar was deployed in 2012 and brought to a nominal operation in 2014. It is currently operated on a campaign basis [7]. Fig. 3 shows the results of comparison of wind profiles obtained with Doppler lidar (2 h integration time) against the radiosounding data and European Center for Medium-Range Weather Forecast (ECMWF) operational analysis. The mean of two radiosoundings launched at 15 h local time from a location ~30 km away was used for intercomparison with the lidar, whose operation was started around midnight. The results of comparison show good agreement of lidar profiles (low- and high-altitude channels) of both zonal and meridional wind components with those of radiosondes above 7 km altitude. The ECMWF zonal and meridional wind profiles agree well with the radiosonde throughout the entire altitude range and confirm the vertical structure observed by the lidar between 30 and 43 km, where the error in lidar profile becomes substantial. For better accuracy in the upper stratosphere a longer than 2 hours integration time is required.

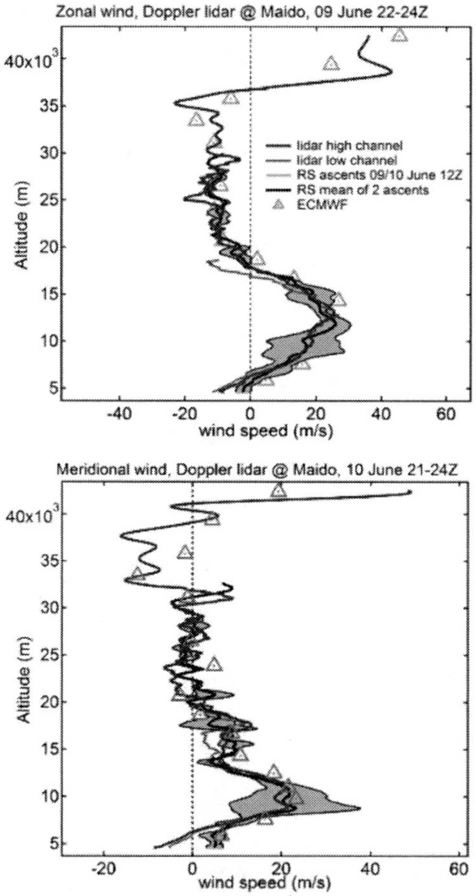

Fig. 3. Comparison of zonal (top) and meridional (bottom) wind velocity profiles obtained using Maïdo Doppler lidar on 09 and 10 June 2014 with the corresponding M10 GPS radiosondes and ECMWF operational analysis. Grey-shaded area indicates the range of values from radiosonde profiles.

5. VALIDATION OF ADM-AEOLUS USING DOPPLER LIDARS

The ESA ADM-Aeolus is the first mission

DOI: 10.1051/epjconf/201611913005

to provide global observations of wind from the lower troposphere up to about 30 km altitude. The satellite is expected to be launched and set into sun-synchronous orbit in late 2016. The Atmospheric LAser Doppler INstrument (ALADIN) instrument onboard Aeolus is composed of a laser emitting at 355 nm and a 1.5-m telescope. The Doppler shift is detected using Fizeau and Fabry-Perot interferometers for Mie and Rayleigh scattering analysis respectively. A similarity of the measurement techniques exploited by ALADIN (Rayleigh channel) and by the French ground-based Doppler lidars makes the latter an attractive mean for Aeolus validation. We present a proposal for ADM-Aeolus CAL/VAL activities based on the operations of two wind lidars operating at mid-latitude and tropical sites.

The validation program includes a pre-CAL/VAL phase aimed at development of a spatial-temporal collocation criteria for ground-based/satellite measurement match. An advantage of the Doppler lidar with respect to conventional GPS radiosondes is that the lidar provides continuous high-resolution wind measurements that can be used to determine temporal collocation criteria for validation of ADM-Aeolus. The spatial collocation criteria will be defined using a dense radiosonde network in Europe.

6. SUMMARY

Two direct-detection Rayleigh-Mie Doppler wind lidars are deployed at mid-latitude (OHP) and tropical (Maïdo) sites. The lidar is shown capable of wind measurements between 5 and 50 km with accuracy better than 1 m/s up to 30 km. Observations are used for studying mesoscale wind fluctuations and gravity waves in the mid-stratosphere as well as for constructing wind climatology up to 50 km altitude. The results of comparison with GPS radiosondes suggest good performance of the Doppler lidars and confirm the error estimates. A validation program for ADM-Aeolus space mission is described. The validation activities include the development of spatiotemporal collocation criteria based on continuous measurements by ground-based lidar and radiosonde network.

ACKWNOLEDGEMENTS

The work was supported by the French National Space Agency (CNES), CNRS-INSU and by the EU FP7 ARISE project.

REFERENCES

[1] Chanin, M. L, A. Garnier, A. Hauchecorne, J. Porteneuve (1989): A Doppler lidar for measuring winds in the middle atmosphere, *Geophys. Res. Lett.*, 16(11), 1273–1276.

[2] Garnier A., M. L. Chanin, 1992: Description of a Doppler Rayleigh lidar for measuring winds in the middle atmosphere, *Appl. Phys., B*, 55, 35-40.

[3] Souprayen, C., A. Garnier, A. Hertzog, A. Hauchecorne, J. Porteneuve (1999): Rayleigh–Mie Doppler wind lidar for atmospheric measurements. I. Instrumental setup, validation, and first climatological results, *Appl. Opt.*, 38, 12.

[4] C. Souprayen, A. Garnier, and A. Hertzog (1998): Rayleigh–Mie Doppler wind lidar for atmospheric measurements. II. Mie scattering effect, theory, and calibration, *Appl. Opt.*, 2422- 2431.

[5] Hertzog, A., C. Souprayen and A. Hertzog (2001): Measurements of gravity wave activity in the lower stratosphere by Doppler lidar, *J. Geophys. Res.,* 106, D8, 7879-7890.

[6] Baray, J.-L., Courcoux, Y., Keckhut, P., Portafaix, T., Tulet, P., Cammas, J.-P., Hauchecorne, A., Godin Beekmann, S., De Mazière, M., Hermans, C., Desmet, F., Sellegri, K., Colomb, A., Ramonet, M., Sciare, J., Vuillemin, C., Hoareau, C., Dionisi, D., Duflot, V., Vérèmes, H., Porteneuve, J., Gabarrot, F., Gaudo, T., Metzger, J.-M., Payen, G., Leclair de Bellevue, J., Barthe, C., Posny, F., Ricaud, P., Abchiche, A., and Delmas, R.: Maïdo observatory: a new high-altitude station facility at Reunion Island (21° S, 55° E) for long-term atmospheric remote sensing and in situ measurements, *Atmos. Meas. Tech.*, 6, 2865-2877, doi:10.5194/amt-6-2865-2013, 2013.

[7] Keckhut, P., Y. Courcoux, J-L. Baray, J. Porteneuve, H. Vérèmes, A. Hauchecorne, D. Dionisi, F. Posny, J-P. Cammas, G. Payen, F. Gabarrot, S. Evan, S. Khaykin, R. Rüfenacht, B.Tschanz, N. Kämpfer, Ph. Ricaud, A. Abchiche, J.Leclair-de-Bellevue, V. Duflot (2015): Introduction to the MALICCA campaign dedicated to the validation of Upper Air meteorological parameters, *J. Appl. Rem. Sens.*, in press.

DOI: 10.1051/epjconf/201611913006

VARIATIONS IN MESOSPHERIC NEUTRAL DENSITIES FROM RAYLEIGH LIDAR OBSERVATIONS AT UTAH STATE UNIVERSITY

David L. Barton[1], Vincent B. Wickwar[1]*, Joshua P. Herron[2], Leda Sox[1], Luis A. Navarro[1]

[1]*Physics and CASS, Utah State University, Logan, UT 84322, USA, *Email: vincent.wickwar@gmail.com*

[2]*Space Dynamics Lab, Utah State University, Logan, UT 84322, USA*

ABSTRACT

A Rayleigh lidar was operated from 1993 to 2004, at the Atmospheric Lidar Observatory (ALO; 41.7°N, 111.8°W) at the Center for Atmospheric and Space Sciences (CASS) on the campus of Utah State University (USU). Observations were carried out on over 900 nights, 729 of which had good data starting at 45 km and going upward toward 90 km. They were reduced for absolute temperatures and relative neutral number densities. The latter at 45 km can be put on an absolute basis by using atmospheric models that go up to at least 45 km. The models' absolute number densities at 45 km are used to normalize the lidar observations, thereby providing absolute densities from 45 to 90 km. We examine these absolute density profiles for differences from the overall mean density profile to show altitudinal structure and seasonal variations.

1. INTRODUCTION

Observing neutral density throughout the mesosphere is challenging, as has been discussed in relation to the development of the NRLMSISe00 empirical model [1]. The available instrumentation, such as balloons, reach up to ~30 km, while rockets are launched infrequently. For the thermosphere, densities are obtained from satellites and incoherent scatter radars.

However, ground-based, Rayleigh-scatter lidars, such as the ALO-USU Rayleigh-scatter lidar, can provide a solution. Their mesospheric observations can provide absolute temperatures and relative neutral number densities in the 35 km to 115 km or so altitude region [e.g., 2, 3].

An absolute scale can be applied to these relative densities by using empirical and reanalysis models that overlap with the Rayleigh data. The next section, Section 2, describes both the methodology for creating a pseudo-density climatology for a composite year from the relative densities and for using four models to apply absolute scales to this pseudo-density climatology. Section 3 describes the results found for the altitudinal and seasonal variations in the neutral number densities about an absolute mean profile during the composite year. It provides some conclusions about these variations and indicates future directions.

2. METHODOLOGY

There are two steps to finding the mesospheric neutral number density climatology. The first is to use the Rayleigh lidar observations to make a climatology of relative number density profiles over a composite year. For this, all the densities at 45 km are normalized to a constant. The second is to apply estimates of absolute densities at 45 km to scale the observed relative densities.

2.1 USU Relative Density Variations

The ratio of major mesospheric neutral constituents, N_2 and O_2, is assumed constant, giving rise to a constant Rayleigh backscatter coefficient in the mesosphere. Above 45 km, the atmospheric transmission is taken to be 1.00. Under these conditions, the lidar return multiplied by range-squared gives the relative neutral number density. The densities from each of the 729 nights, for which the data extend down to 45 km, were included in this work. The all-night averaged density profiles were then scaled to a constant at 45 km.

These observations occurred over 11 years. However, what is most meaningful is a climatology that shows how the densities vary over the course of a year. Accordingly, a composite year is created. For each day in this composite year, the nighttime densities at each altitude are averaged over all the nights within a 31-day window, centered on that night in each of the 11 years. In early January, the 31-day window includes nights from the previous December and,

 DOI: 10.1051/epjconf/201611913006

in late December, it includes nights from the next January. This averaging, applied to the norma-

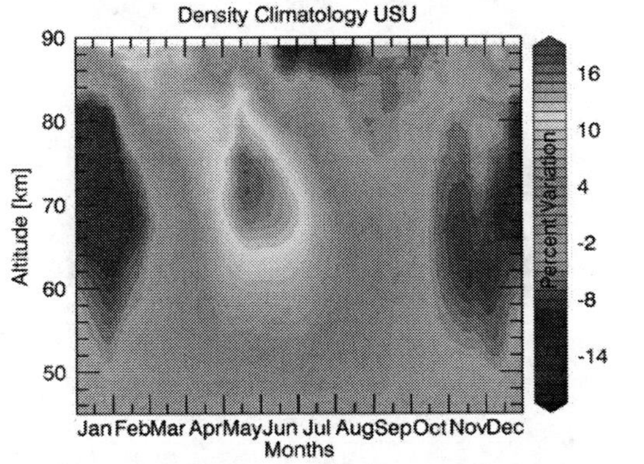

Figure 1. Climatology of relative density variations for a composite year derived from the ALO Rayleigh lidar observations. For each day, it shows the percentage variation at each height from the average annual density.

lized densities, produces a pseudo-density climatology (not shown) for the composite year.

With densities varying by over a factor of 600 between 45 and 90 km, they would have to be displayed on a semilog plot, which would make it extremely difficult to see seasonal variations that are on the order of only 30–45%. To better display these small variations, all 365 nights in the composite year are averaged together to find a mean or reference profile, $\langle n(h) \rangle$. Then, at each altitude, h, the percentage density difference $N(h)$ for each night's composite density $n_{31}(h)$ from the mean annual density $\langle n(h) \rangle$ is calculated as

$$N(h) = 100 \times \left[n_{31}(h) - \langle n(h) \rangle \right] / \langle n(h) \rangle . \quad (1)$$

These percentage differences are shown in Figure 1.

2.2 USU Absolute Densities Variations

While Figure 1 does show significant variations in the observed relative densities with both altitude and season, these variations would be more meaningful if they were tied to an absolute density at 45 km. This can be done by referring to several appropriate atmospheric models that go up to at least 45 km altitude. Four models, instead of one, were used to give an indication of the variability in this normalization. They include an empirical model, the NRLMSISe00 model (MSIS) [1], which is used extensively as a reference, three retrospective, reanalysis models that combine both first principle calculations and observations. The Climate Prediction Center model (CPC) [4,5] is the oldest of those being considered. It goes up to 1 hPa, which is just above 45 km. The ECMWF model (ERA Interim) [6] is more recent. It, too, goes to 1 hPa. NASA's Modern-Era Retrospective analysis for Research and Applications model (MERRA) [7] goes to 0.1 hPa, ~65 km.

For each model, the number density was found at 45 km for ALO for each of the 729 observed days. For the MSIS model, this was straight forward as the densities above any point on any day are calculated as a function of altitude. For the reanalysis models, this was more involved. Profiles of pressure and temperature are given in their respective databases as a function of geopotential altitude on a global grid for every day during a period that includes the ALO observations. These geopotential altitudes were converted to geometric altitudes and interpolated to find the temperatures and pressures at 45 km. The ideal gas law then gave the number densities.

As previously, a composite year was created for the four model-dependent sets of 45-km density values, which are shown in Figure 2. The four curves have similar shapes, showing ~30% greater density in summer than in winter. They also have a spread, with the highest densities being 20–30% greater than the lowest in all seasons. This spread is impressively small, considering that these are independent models, two of which are at their upper altitude limit.

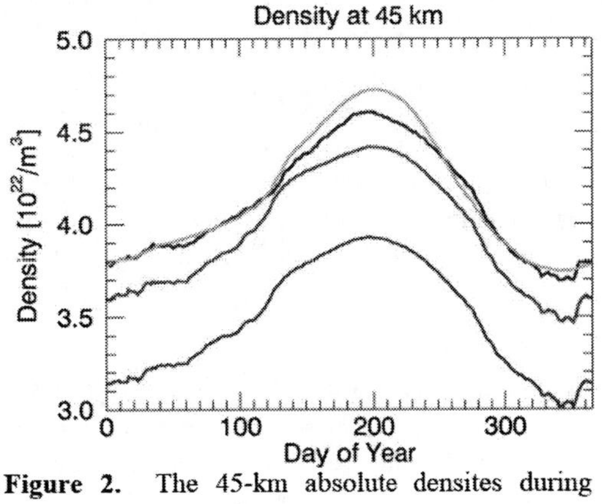

Figure 2. The 45-km absolute densites during a

DOI: 10.1051/epjconf/201611913006

composite year for the MSIS (Yellow), CPC (Blue), MERRA (Red), and ERA Interim (Green) models.

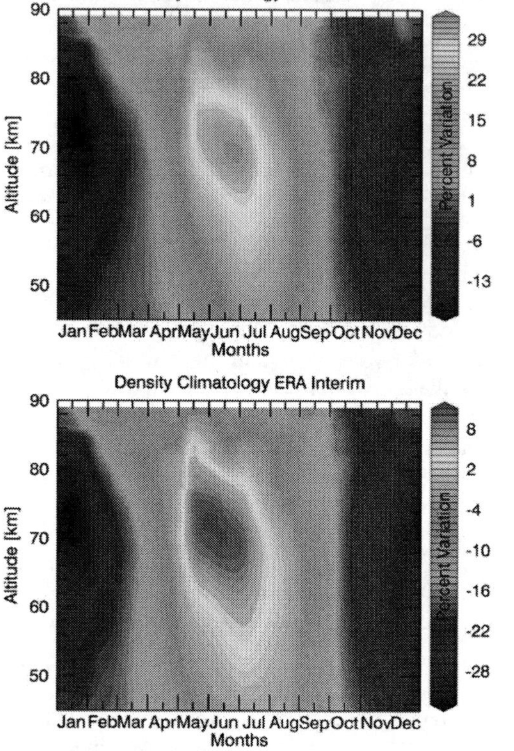

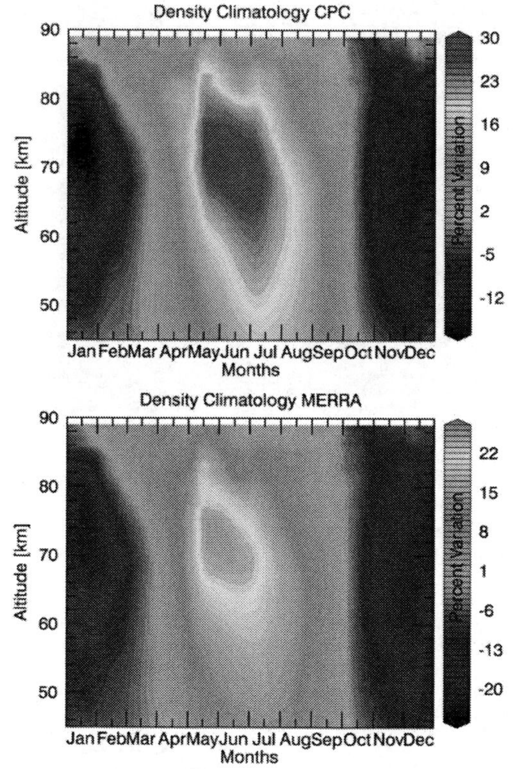

Figure 3. Climatologies of density variations from a mean absolute density profile for a composite year derived from the ALO Rayleigh lidar observations normalized to the four models at 45 km. The models are MSIS (upper left), CPC (upper right), ERA Interim (lower left) and MERRA (lower right). (The color scale is different for each model.)

By applying these 45-km density values to the observed relative density profiles, four sets of absolute number-density profiles are created. From these, four climatologies are formed (but not shown), each giving absolute number densities between 45 and 90 km over a composite year.

The four absolute densities at each altitude for each day of the composite year are averaged together to find an overall mean absolute density profile. The differences from this overall mean are then calculated for each model using the equivalent of Equation 1, displaying the results as percentage variations in the four parts of Figures 3.

3. RESULTS AND CONCLUSIONS

The four absolute number density curves in Figure 2 have similar shapes, showing an annual variation at 45 km, over the composite year, that is ~30% greater in summer than in winter. The maximum is near mid July (day 200); the minimum near mid December (day 350). This asymmetry in the rise and fall of the density suggests contributions of semiannual and perhaps higher order variations, even in this region dominated by solar radiation. The curves have a small spread. The ordering from highest to lowest number density is MSIS, CPC, MERRA, and ERA Interim, with MSIS being 20–30% greater than ERA Interim. That they are not closer together reflects the limitations in our current state of knowledge of the absolute densities at 45 km.

The four sets of density values at 45 km, applied to the relative densities from the lidar, provide absolute number density profiles throughout most of the mesosphere, 45–90 km. For reference, a mean profile was created from the four sets of density profiles in the composite year. For each model, the differences from the mean are shown as percentages in Figure 3.

In the region above 65 km, which is usually thought to be dominated by dynamics, the difference profiles show a summer, density maximum that extends from May well into July. It is centered at 72 km, but extends from 45 to 85 km. Above 80 km, the maximum is in mid May, whereas below 62 km it is in early July. They also show a winter density minimum above 65 km in mid January that is centered near 72 km. It runs from December through January. At 70 km, winter extends about 6 months, from a week into October until the beginning of April. In addition, below 65 km, there is another deep minimum in late November and early December. This minimum extends to a much lower altitude than the January minimum.

The summer-to-winter transition in the fall is marked by a very sharp decrease in density over the same two weeks at the beginning of October at all altitudes. No such sharp increase occurs in the spring at a given time. Instead, the winter-to-spring transition appears to propagate down from 90 to 70 km from early January to early April. At 70 km, the transition takes place over about two weeks. At lower altitudes, it occurs earlier and more gradually. At 45 km it starts in January and takes 3 months.

This study suggests some future work. Given these new, absolute densities from 45–90 km, they should be compared to those from the MSIS model, which covers this whole region, and to the MERRA model, which go up to 0.1 hPa, ~65 km.

It also suggests significant observational improvements at USU. One is to extend the lidar observations downward to overlap better with absolute values from the reanalysis models or even directly with balloons. Rayleigh observations could be extended to between 30 and 35 km. Raman-scatter observations could extend the observations to about 15 km. A second is a more sensitive Rayleigh system to extend the observations upwards into the lower thermosphere. Initial steps in this direction have been taken [2].

Going beyond USU, a major improvement would be to develop a simpler and more affordable Rayleigh system, which would cover this extended altitude range, and could be acquired and operated by many smaller research groups around the world to form a global observing network.

ACKNOWLEDGMENTS

David Barton would like to acknowledge financial support from USU, the Physics Dept, and the College of Science. Leda Sox would like to acknowledge the Physics Dept's Blood and Gene Adams Scholarships. Many thanks to Dr. Doug Drob of NRL for the MSIS algorithm and to Dr. Bill Randel of NCAR for the CPC data and discussions. Many thanks to ECMWF for permitting access to their database and use of the ERA Interim model, and to similarly to NASA for the MERRA reanalysis model and data. The lidar data were acquired with support from several grants from the Atmospheric Sciences Division of the National Science Foundation.

REFERENCES

[1] Picone, J. M., A. E. Hedin, D. P. Drob, and A. C. Aikin, 2002: NRLMSISE-00 empirical model of the atmosphere: Statistical comparisons and scientific issues, *J. Geophys. Res.*, **107(A12)**, 1468–1483, doi:10.1029/2002JA009430.

[2] Wickwar, V. B., L. Sox, M. T. Emerick, J. P. Herron, D. L. Barton, 2015: Early temperatures observed with the extremely sensitive Rayleigh lidar at Utah State University, *ILRC27*.

[3] Herron, J.P., 2007: *Rayleigh-Scatter Lidar Observations at USU's Atmospheric Lidar Observatory (Logan, UT) — Temperature Climatology, Temperature Comparisons with MSIS, and Noctilucent Clouds*, PhD Dissertation, 156 pp, Utah State University, Logan, UT.

[4] Gelman, M. E., A. J. Miller, K. W. Jihnson, and R. N. Nagatani, 1986: Detection of long term trends in global stratospheric temperature from NMC analyses derived from NOAA satellite data. *Adv. Space Res.* **6**, 17–26.

[5] Randel, W. J., et al., 2004: The SPARC Intercomparison of middle-atmosphere climatelogies, *J. Climate,* **17**, 986-1003.

[6] Berrisford, P., et al., 2011: *The ERA-Interim Archive, Version 2.0.* ERA report series, 1. Technical Report, 23 pp, ECMWF.

DOI: 10.1051/epjconf/201611913006

[7] Rienecker, M. M., et al., 2011: MERRA: NASA's modern-era retrospective analysis for research and applications, *J. Climate*, **24(14)**, 3624-3648.

EARLY TEMPERATURES OBSERVED WITH THE EXTREMELY SENSITIVE RAYLEIGH LIDAR AT UTAH STATE UNIVERSITY

Vincent B. Wickwar [1]*, Leda Sox[1], Matthew T. Emerick[1], Joshua P. Herron[2], David L. Barton[1]

[1]*Physics and CASS, Utah State University, Logan, UT 84322, USA, *Email: vincent.wickwar@gmail.com*

[2]*Space Dynamics Lab, Utah State University, Logan, UT 84322, USA*

ABSTRACT

Rayleigh-scatter lidar observations were made at the Atmospheric Lidar Observatory (ALO) at Utah State University (USU) from 1993–2004 from 45–90 km. The lidar operated at 532 nm with a power-aperture-product (PAP) of ~3.1 Wm^2. The sensitivity of the lidar has since been increased by a factor of 66 to 205 Wm^2, extending the maximum altitude into new territory, the lower thermosphere. Observations have been extended up to 115 km, almost to the 120 km goal. Early temperatures from four ~4-week periods starting in June 2014 are presented and discussed. They are compared to each other, to the ALO climatology from the original lidar [1], and to temperatures from the NRLMSISe00 empirical model [2].

1. INTRODUCTION

The region between the maximum weather balloon altitude of ~30 km and the minimum radar, airglow, or resonance lidar altitude of ~80 km can only be regularly observed from the ground by Rayleigh lidar [3]. The original ALO lidar, mentioned above, covered much of this region. A similar difficulty in making ground-based observations exists in the lower thermosphere above ~100 km. The intent of the upgraded ALO lidar is to extend Rayleigh observations up to ~120 km. The upgrade has involved a big increase in both laser power and telescope collecting area. So far the upgraded system has reached 115 km.

One general point of making ground-based Rayleigh observations is to observe the time evolution of various aspects of the middle atmosphere. Another is to observe the coupling among regions: between the stratosphere and mesosphere and between the mesosphere and thermosphere.

A more specific and detailed statement about the science to which Rayleigh lidar contributes comes from the list of past and ongoing studies carried out using data from the original ALO Rayleigh lidar. Studies taking advantage of much of the 11-year data set include a midlatitude, mesospheric temperature climatology [1], an extensive gravity wave study [4], a detailed examination of possible climate or solar cycle changes [5], a study of the midlatitude mesosphere during Sudden Stratospheric Warmings [6], and an exploration of neutral density variations [7].

In addition there are unexpected results that arise from observing as frequently as possible. These include midlatitude observations of noctilucent clouds [8,9] and their association with large-amplitude waves, and an all-night observation of a thin layer descending from the mesosphere into the stratosphere.

Furthermore, Rayleigh lidar observations are complementary to observations from satellites, such as those from SABER on TIMED [10]. Whereas the lidar can measure time evolution, the satellite can measure latitudinal and longitudinal variability. Rayleigh lidars are also complementary to resonance lidars, which have a more limited altitude range, roughly 80–100 km, though they have better altitude and time resolution. At USU, either on campus or at the nearby Bear Lake Observatory, the ALO Rayleigh lidar observations overlap with those from a cluster of other instruments including a Na lidar; all-sky cameras imaging OH, $O(^1S)$, and O_2 Atmospheric bands; a meteor wind radar; an ionosonde (for measuring D and E region ionization); and a magnetometer (for measuring E-region currents).

It is expected that the extremely sensitive ALO Rayleigh lidar will continue the types of studies described above, extending them in time and from the mesosphere into the lower thermosphere.

Section 2, on methodology, discusses the acquisition and temperature reduction of the early data from the upgraded system. Section 3, on results, shows these early temperatures and their relation to climatological and model temperatures.

Section 4 concludes by comparing these temperatures and indicates future directions.

2. METHODOLOGY

Considerable data were acquired with the upgraded system during summer and fall 2014. The data were acquired during periods ranging, typically, from 4 to 10 hours. The data were averaged in time over these whole periods and 3 km in altitude. The analysis used the standard Chanin-Hauchecorne method involving hydrostatic equilibrium and the ideal gas law [3]. For this early work, the mesospheric composition was extended upward unaltered, giving rise to a constant backscatter coefficient and mean molecular mass. The initial value for the temperature integration at the highest altitude for each night was taken from the NRLMSISe00 (MSIS) empirical model. With this technique, the temperatures are highly accurate 10–20 km below the highest altitude. That altitude for each night is taken to be where the ratio of signal-to-standard deviation drops below 16. For a given night, the temperature uncertainties derived from Poisson statistics are far less than 1 K at 70 km and ~12 K at the highest altitude. Each of the four periods had between 10 and 16 nights of observations.

3. RESULTS

Figure 1 shows 13 temperature profiles, in various colors, from the mid summer period 12 June–13 July extending upwards from 70 km. The highest two nights reached 114 km. Also shown in Figure 1 is a red curve representing the average of the observed lidar temperature profiles. At 70 km, it is the average of all the nights in the observed period. At the highest altitudes, the top 5 km or so, the number of nights in the average decreases rapidly to only 1 or 2. Because of the effects of geophysical variability and the decreasing number of profiles at the highest altitudes, this red average curve becomes noisier towards the top. Also shown in Figure 1 is a green curve between 70 and ~90 km that represents the climatological average found with the earlier Rayleigh lidar [1]. That value is an average over a 31-night window, applied to each of the 11 years. The initial temperature value, at the highest altitude, for the data reductions for these curves was taken from the 8-year temperature climatology from the nearby Na lidar at Colorado State University (CSU) [11]. Finally, a black curve is shown between 70 and 115 km. It is calculated from the MSIS model [2] for the center day, for 6:00 UT.

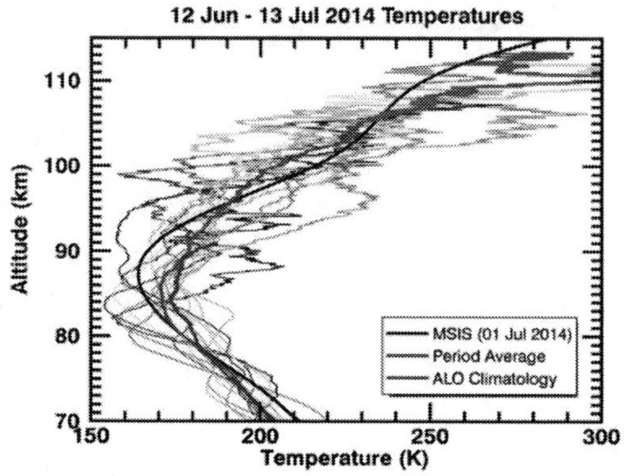

Figure 1. Nighttime temperatures in mid summer on 13 nights between 12 June and 13 July 2014, in the upper mesosphere and lower thermosphere.

Figure 2(a) also refers to the mid-summer period, 12 June–13 July. For clarity, it reproduces only the red, green and black curves from Figure 1. Again, these three curves represent, respectively, the average temperatures for the new lidar observations, the climatological temperatures from the earlier lidar observations, and the MSIS empirical model temperatures for the center of the night, at 6:00 UT. Figures 2(b-d) show the corresponding red, green, and black curves for late summer, 14 July–17 August; early fall, 12 September–06 October; and late fall, 29 October–17 November; respectively.

4. CONCLUSIONS

Two contributions to the nighttime profiles stand out in Figure 1: a core of days with a 10–20 K band of temperatures with many small amplitude oscillations; a significant number of days with large amplitude (≥ 20 K) and long wavelength (≥ 10 km) waves below ~100 km.

A minimum exists in the average temperatures in Figures 2(a-d) at 80–85 km throughout this time period. In summer the minimum at 85 km is the summer mesopause. It is at ~170 K in mid summer and at ~180 K in late summer. In fall this feature becomes a relative minimum at ~80 km. It is at ~190 in early fall and ~210 K in late fall.

DOI: 10.1051/epjconf/201611913007

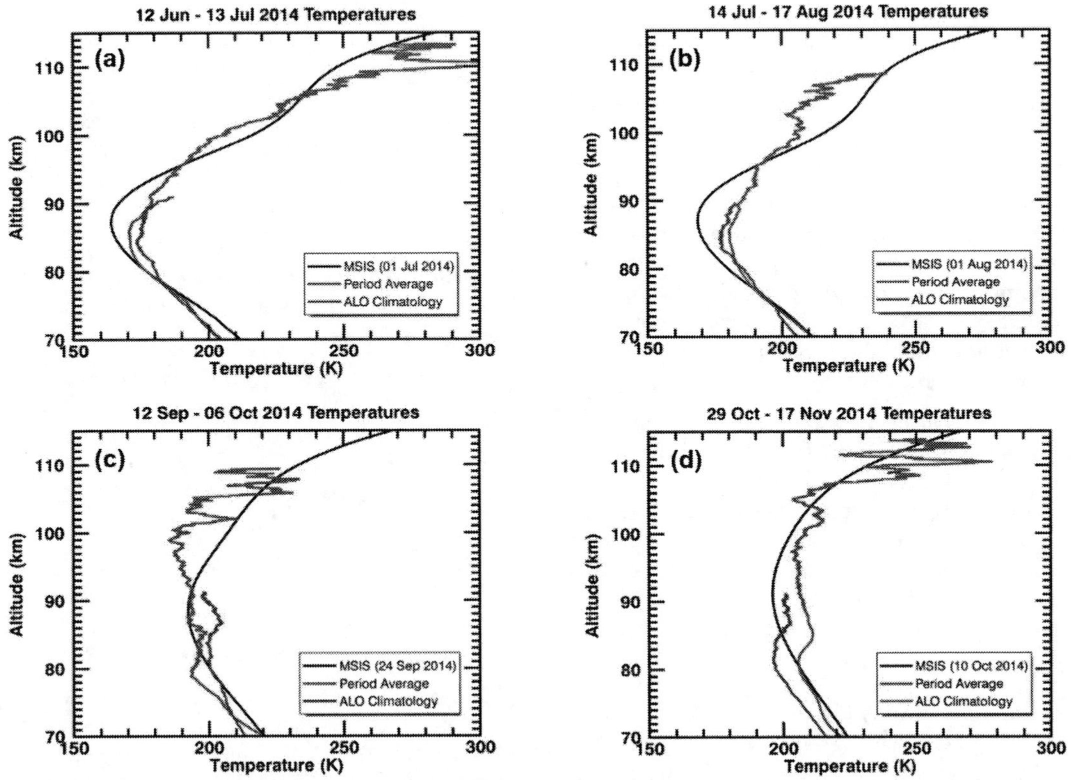

Figure 2. Average nighttime temperatures from the upper mesosphere and lower thermosphere for four periods between mid summer and late fall 2014: (a) 12 June–13 July [13 nights], (b) 14 July–17 August [16 nights], (c) 12 September–06 October [11 nights], and (d) 29 October–17 November [10 nights]. The red curves are for the current observations. The green curves are from the earlier ALO climatology. The black curves are from the MSIS model.

A second minimum appears near 100 km in both the early and late fall. It is the higher altitude, winter mesopause. It is at ~185 K in early fall and ~205 K in late fall. The transition between summer and winter behaviors occurs very rapidly in early fall.

From approximately 85–100 km, the average new profiles follow a straight line with a slight positive slope in the two summer periods and a slight negative slope in the two fall periods. The negative slope in fall gives rise to the winter mesopause near 100 km.

Furthermore, this 15-km region of almost straight temperature profiles differs greatly from the MSIS profiles. The latter are curved with a minimum between 88 and 92 km. In mid and late summer, the MSIS profiles have a mesopause ~5 km higher and 10–15 K colder than the observations. In the early fall the MSIS mesopause only moves up to 89 km, making its temperatures ~20 K warmer than the observed mesopause temperatures near 100 km. In going from early to late fall, the MSIS mesopause barely moves and its temperatures increase less than the observed ones, with the result that the MSIS temperatures are now slightly colder than the observed mesopause temperatures.

Agreement of these 2014 averages and the ALO climatological temperatures from 70–90 km is remarkably close in summer, especially considering that the new averages are from only a few days in one year, while the climatology is from many days over 11 years. In fall the differences approach 10 K. They are within the geophysical variability found for the climatology. This means these new results support the original ALO climatology that, with the same caveats, means they also support reasonable agreement of Rayleigh and Na temperatures at 90 km during these four periods. At that altitude, the new temperatures are good because the downward integration started ~20 km higher. The initial values near 90 km for the ALO climatology, in contrast, are based on the CSU Na climatology [11].

In the future, assuming that funding is available, the top altitude will be raised and regular observations will continue. The bottom altitude will be lowered with continuing system improvements, including more detector channels. Extensive comparisons will also be carried out with observations from the other instruments in the USU mesospheric cluster.

While this lidar opens the mesosphere and lower thermosphere to ground-based observations at one mid-latitude location, it suggests the future importance of developing more affordable and simpler Rayleigh systems with comparable sensitivity that can be acquired and operated by small research groups as part of a global network at many locations around the world.

ACKNOWLEDGMENTS

We gratefully acknowledge support for this work from a Space Dynamics Lab IR&D grant, Utah NASA Space Grant Consortium, Howard L. Blood and Gene Adams Graduate Scholarships, USU, USU College of Science, USU Physics Department, personal contributions, and the work of many students: Thomas Amely, Marcus Bingham, Ryan Martineau, Luis Navarro, and Lance Petersen. Many of the observations were carried out by a dedicated group of students: Jordan Burns, Preston Hooser, David Moser, Rebecca Petrick, Warren Schweigert, Patrick Sharp, Joe Slansky, and Bryant Ward. The observatory, telescope, and laser laboratory were built with funds from NSF, AFOSR, and USU.

REFERENCES

[1] Herron, J.P., 2007: *Rayleigh-Scatter Lidar Observations at USU's Atmospheric Lidar Observatory (Logan, UT) — Temperature Climatology, Temperature Comparisons with MSIS, and Noctilucent Clouds*, PhD Dissertation, 156 pp, Utah State University, Logan, UT.

[2] Picone, J. M., A. E. Hedin, D. P. Drob, and A. C. Aikin, 2002: NRLMSISE-00 empirical model of the atmosphere: Statistical comparisons and scientific issues, *J. Geophys. Res.*, **107(A12)**, 1468–1483, doi:10.1029/2002JA009430.

[3] Hauchecorne, A., and M. L. Chanin, 1980: Density and temperature profiles obtained by lidar between 35 and 70 km, *Geophys. Res. Lett.* **7**, 565–568.

[4] Kafle, D. N., 2009: *Rayleigh-lidar observations of mesospheric gravity wave activity above Logan, Utah*, PhD Dissertation, 209 pp, Utah State University, Logan, UT.

[5] Wynn, T. A., 2010: *Statistical analysis of the USU lidar data set with reference to mesospheric solar response and cooling rate calculation, with analysis of statistical issues affecting the regression coefficients*, PhD Dissertation, 232 pp, Utah State University, Logan, UT.

[6] Sox, L, V. B. Wickwar, C. Fish, J. P. Herron, 2015: Temperature deviations in the midlatitude mesosphere during stratospheric warmings as measured with Rayleigh-scatter lidar, *ILRC27*.

[7] Barton, D. L., V. B. Wickwar, J. P. Herron, L. Sox, L. A. Navarro, 2015: Variations in Mesospheric neutral densities from Rayleigh lidar observations at Utah State University, *ILRC27*.

[8] Wickwar, V. B., M. J. Taylor, J. P. Herron, B.A. Martineau, 2002: Visual and lidar observations of noctilucent clouds above Logan, Utah, at 41.7°N, *J. Geophys. Res.*, **107(D7)**, 4054–4059, 10.1029/2001JD001180.

[9] Herron, J. P., V. B. Wickwar, P. J. Espy, and J. W. Meriwether, 2007: Observation of a noctilucent cloud above Logan, Utah (41.7°N, 111.8°W) in 1995, *J. Geophys. Res.*, ***112* (D19203)**, 10.1029/2006JD007158.

[10] Mertens, C. J., et al., 2004: SABER observations of mesospheric temperatures and comparisons with falling sphere measurements taken during the 2002 summer MaCWAVE campaign, *Geophys. Res. Lett.*, **31 (L03105),** doi:10.1029/2003GL018605.

[11] She, C. Y., et al., 2000: Eight-year climatology of nocturnal temperature and sodium density in the mesopause region (80 to 105 km) over Fort Collins, CO (41°N, 105°W), *Geophys. Res. Lett.*, **27(20)**, 3289-3292.

Temperature deviations in the midlatitude mesosphere during stratospheric warmings as measured with Rayleigh-scatter lidar

Leda Sox [1]*, Vincent Wickwar[1], Chad Fish[2], Joshua P. Herron[3]

[1]*Center for Atmospheric and Space Sciences and Physics Department, Utah State University, UT 84322, USA, *leda.sox@gmail.com*

[2] *Atmospheric & Space Technology Research Associates (ASTRA), Boulder, CO USA*

[3]*Space Dynamics Laboratory, Utah State University, Logan, UT 84322, USA*

ABSTRACT

While mesospheric temperature anomalies associated with Sudden Stratospheric Warmings (SSWs) have been observed extensively in the polar regions, observations of these anomalies at midlatitudes are sparse. The original Rayleigh-scatter lidar that operated at the Atmospheric Lidar Observatory (ALO; 41.7°N, 111.8°W) in the Center for Atmospheric and Space Sciences (CASS) on the campus of Utah State University (USU) collected an extensive set of temperature data for 11 years in the 45–90 km altitude range. This work focuses on the extensive Rayleigh lidar observations made during six major SSW events that occurred between 1993 and 2004, providing a climatological study of the midlatitude mesospheric temperatures during these SSW events. An overall disturbance pattern was observed in the mesospheric temperatures during these SSWs. It included coolings in the upper mesosphere, comparable to those seen in the polar regions during SSW events, and warmings in the lower mesosphere.

1. INTRODUCTION

SSWs are major disturbances in the polar region of the winter hemisphere that are defined by major changes in stratospheric temperature and circulation. SSWs are characterized by a temperature increase of tens of degrees Kelvin, averaged over 60°–90° latitude at 10 hPa (roughly 32 km), and a weakening of the polar vortex that persists for the order of a week at 60° and 10 hPa level [1]. Polar vortices are cyclones centered on both of the Earth's poles that are present throughout the stratosphere in the winter hemisphere. Strong eastward zonal winds define the polar vortices. Increased planetary wave (PW) activity in the winter hemisphere leads to increased PW breaking in the polar stratosphere and the deposition of the PW's westward momentum in the polar vortex. This weakens the polar vortex, and in the case of major SSWs, can reverse the zonal wind direction to westward. The reversal of the stratospheric jet allows more eastward propagating gravity waves (GWs) to travel up into the mesosphere where, under normal winter conditions, westward propagating GWs dominate. The atypical wintertime GW filtering and the resulting dominance of eastward GWs induce an equatorward circulation in the mesosphere, similar to what it is in summer, which leads to the cooling of the upper mesosphere. While these mesospheric coolings have been observed in the polar regions for several decades [2] they have not been observed at midlatitudes often [3], [4].

2. METHODOLOGY

2.1 Classifying SSWs from 1993-2004

In this initial study of the mesosphere's response to northern hemisphere SSWs above Logan, UT, we will focus on periods when there were major SSW events during the USU Rayleigh-Scatter Lidar's (RSL's) original 11-year operational run. Here, a major SSW is defined by both a stratospheric temperature increase averaged over the latitudes 60° to 90° N at 10 hPa and a complete reversal of the zonal-mean winds from eastward to westward at 60° N and 10 hPa, as seen in NASA's Modern-Era Retrospective Analysis for Research and Applications (MERRA) reanalysis dataset [5]. This creates a complete change in the circulation, or a breakdown, of the polar vortex [1]. Two major SSWs (one in mid-December and one in late February), at northern latitudes, can be identified in Figures 1 (a) and (b). The blue vertical lines mark the wind reversal, or onset as it is referred to in this study.

2.2 Calculating temperature deviations

The original RSL system ran at a midlatitude site, on the campus of Utah State University from

1993–2004. It was comprised of a Nd:YAG laser operating at 30 Hz and either 18 or 24 W at a wavelength of 532 nm. The receiving telescope had a diameter of 44 cm, and thus a receiving aperture of 0.15 m^2. The USU RSL measured relative densities that were then used to determine absolute temperatures using a slight variation of the Chanin-Hauchecorne method [6], which assumes hydrostatic equilibrium and employs the ideal gas law. The initial temperature values used in this analysis came from either the Colorado State University Sodium Lidar climatology [7] or the Mass Spectrometer and Incoherent Scatter Radar (MSISE90) model [8]. A composite year temperature climatology was created by averaging over a window 31 nights wide centered on each night and 11 years deep, for the USU RSL data [9].

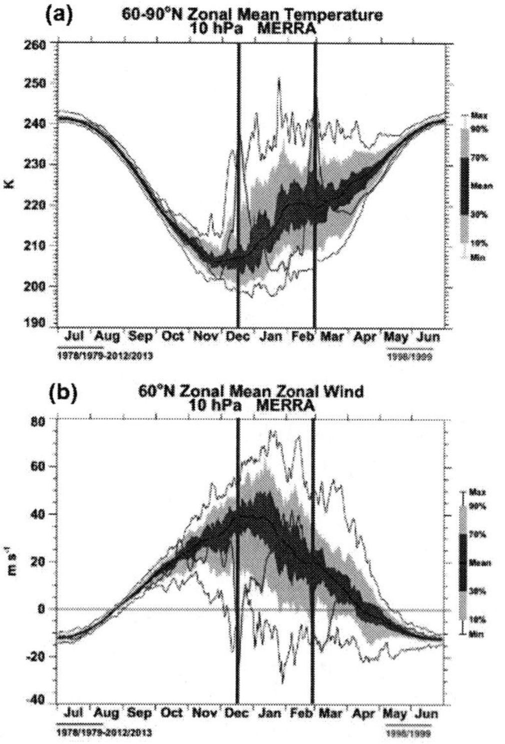

Figure 1. (a) Zonal mean temperatures at 10 hPa, averaged over 60°–90° N and (b) zonal mean zonal winds at 10 hPa and 60° N. Vertical blue lines indicate major SSWs

In order to define coolings and warmings during an SSW event, relative to the RSL climatology, temperature difference plots [Figs 2 (a-f)] were created by subtracting each night of the SSW event's temperatures by the temperature profile from February 3rd of the RSL climatology.

Between the six SSW events, the annual range of these events spans December 15th to April 9th. February 3rd was selected because it was in the middle of this range and because the climatological days about each SSW also included the SSW temperatures. In Figure 2 (a-f), the red vertical lines indicate the SSW onset date. The SSW onset date is defined here as the date when the zonal mean zonal winds switch from eastward to westward [Figure 1 (b)]. In other words, this is the date on which the SSW is distinguished as a major SSW, rather than a minor one. The temperature differences were then smoothed in time with a window of 5 days.

3. RESULTS

To characterize these warmings, temperature difference, or deviation, plots are given in Figure 2 (a-f). While the data coverage and deviations from the climatological background temperatures vary from event to event, in general, a pattern of warmings in the lower mesosphere and coolings in the upper mesosphere becomes apparent after the onset date of the SSW. In Figure 2, the red vertical lines indicate the onset date for each SSW event period. The upper mesospheric coolings are located from ~70–95 km and the lower mesospheric warmings are located from ~45-70 km. Both the observed coolings and warmings are roughly one order of magnitude higher than those predicted in Liu and Roble [2002] for midlatitudes. They are more comparable to the coolings and warmings that have been found in the polar mesosphere [2] and range from less than −50 K (coolings) to more than +50 K (warmings). The warming/cooling pattern shown here also indicates that the midlatitude lower mesospheric warmings extend higher in altitude and the coolings start higher in altitude than their polar counterparts [2], [10].

During a major SSW event, the polar stratosphere switches from typical winter conditions to summer conditions, e.g. Figure 1, for the order of a week. At approximately the same time, we see that the midlatitude mesosphere, shown here, undergoes a similar abrupt, season-like switch in

 DOI: 10.1051/epjconf/201611913008

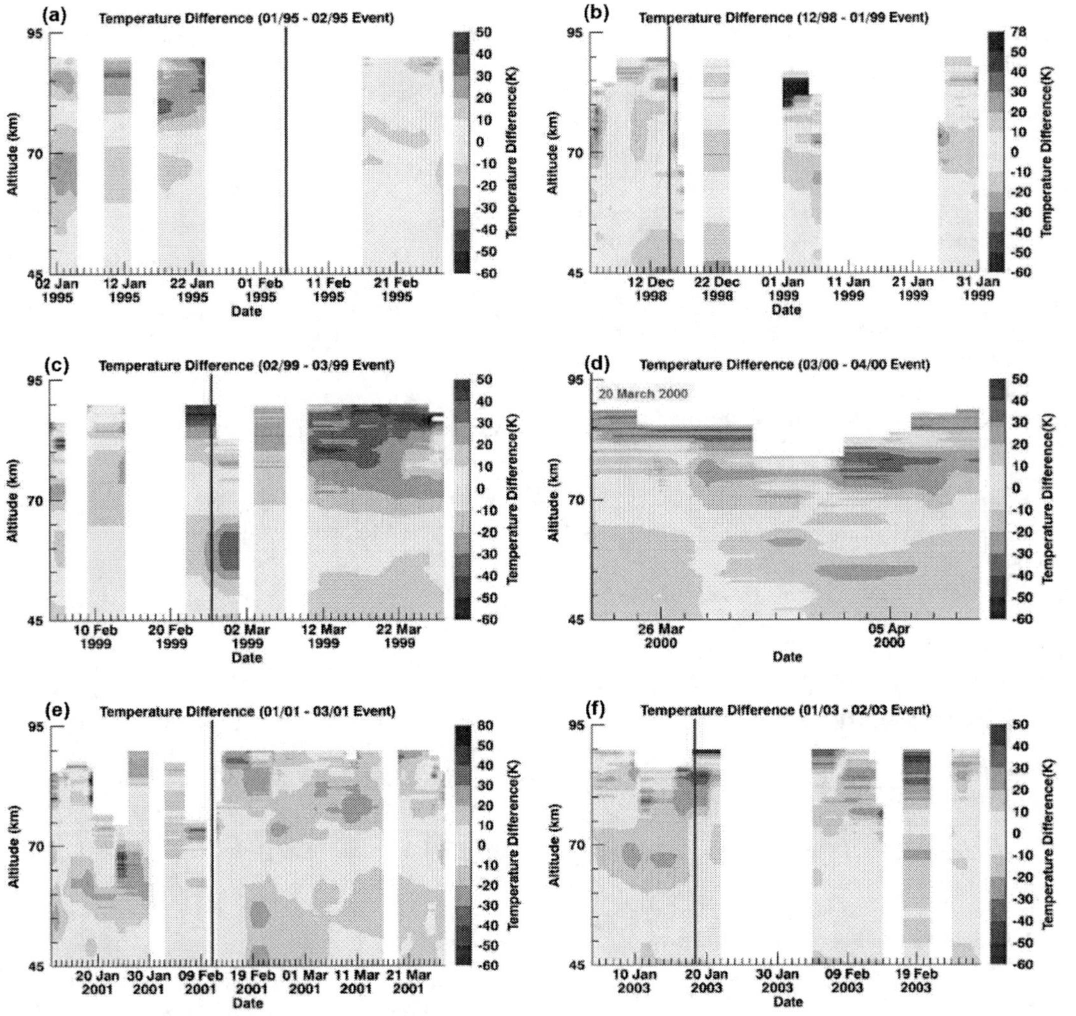

Figure 2. Temperature difference plots from the six SSW event periods in which RSL data was obtained between 1993 and 2004. Temperature differences were calculated by subtracting the climatological February 3rd temperature profile from the temperature profiles from each night of the SSW period. Negative values represent coolings and positive values represent warmings. Red vertical lines indicate SSW onset date. Note that some plots have different color contour ranges.

its temperatures. For example, the range of temperatures, between 45 and 90 km, is ~265–180 K in the undisturbed winter and ~280–160 K in the summer and during major SSWs [not shown]. The transition from winter to summer temperatures in the mesosphere results in a cooling of the upper mesosphere and a warming of the lower mesosphere [9]. These temperature deviations can be seen in all six event periods given in Figure 2. This suggests that the coupling of the stratosphere and mesosphere during SSWs extends to midlatitudes for *most* major SSW events and that this coupling at lower latitudes is not a rare occurrence.

4. CONCLUSIONS

In this study, we explored the temperature structure of the midlatitude mesosphere during the time periods of six major sudden stratosphere warmings that occurred in the Northern

Hemisphere from 1993–2004. Data were used from both the NASA MERRA reanalysis database and from the dense temperature dataset acquired by the USU Rayleigh lidar that overlapped significantly with the six SSWs. One important finding was that the whole mesosphere tends to switch from the climatological temperature range of winter to that of summer from the time of the stratospheric zonal wind reversal at 60°N. A second important observation was that the mesospheric temperature deviations from the RSL winter climatology, i.e. coolings in the upper mesosphere and warmings in the lower mesosphere, are roughly the same magnitude at midlatitudes as they are in the polar regions. The data presented here also suggests that the extension of SSW-related anomalies to midlatitudes is not a rare occurrence. Rather, these anomalies are seen during every major SSW observational period. These results imply that sudden stratospheric warmings are not only coupled to different atmospheric regions, but are coupled to different latitudinal regions, as well.

ACKNOWLEDGEMENT

Leda Sox would like to acknowledge financial support from USU, the Utah NASA Space Grant Consortium, and the USU Physics Department's Blood Scholarship and Gene Adams Scholarship. The data were acquired with support from several grants from the Atmospheric Sciences Division of the National Science Foundation.

REFERENCES

[1] Labitzke, K., and B. Naujokat, 2000: The lower arctic stratosphere in winter since 1952, *SPARC Newsl.*, **15**, 11–14.

[2] Labitzke, K., 1972: Temperature changes in the mesosphere and stratosphere connected with circulation changes in winter, *J. Atmos. Sci.*, **29**, 756– 766.

[3] Hauchecorne, A. and M. L. Chanin, 1983: Mid-Latitude Lidar Observations of Planetary Waves in the Middle Atmosphere during the Winter of 1981-1982, *J. Geophys. Res.*, **88 (C6),** 3843-3849.

[4] Yuan, T., B. Thurairajah, C.-Y. She, A. Chandran, R. L. Collins, and D. A. Krueger, 2012: Wind and temperature response of midlatitude mesopause region to the 2009 Sudden Stratospheric Warming, *J. Geophys. Res.,* **117**, D09114, doi:10.1029/2011JD017142.

[5] Rienecker, M. M., M. J. Suarez, R. Gelaro, R. Todling, J. Bacmeister, E. Liu, M. G. Bosilovich, S. D. Schubert, L. Takacs, G.-K. Kim, S. Bloom, J. Chen, D. Collins, A. Conaty, A. da Silva, W. Gu, J. Joiner, R. D. Koster, R. Lucchesi, A. Molod, T. Owens, S. Pawson, P. Pegion, C. R. Redder, R. Reichle, F. R. Robertson, A. G. Ruddick, M. Sienkiewicz, and J. Woollen, 2011: MERRA: NASA's Modern-Era Retrospective Analysis for Research and Applications. *J. Climate,* **24**, 3624–3648.

[6] Hauchecorne, A. and M. L. Chanin, 1980: Density and temperature profiles obtained by lidar between 35 and 70 km, *Geophys. Res. Lett.* **7**, 565–568.

[7] She, C. Y., S. Chen, Z. Hu, J. Sherman, J. D. Vance, V. Vasoli, M. A. White, J. R. Yu, and D. A. Krueger, 2000: Eight-year climatology of nocturnal temperature and sodium density in the mesopause region (80 to 105 km) over Fort Collins, CO (41 N, 105 W), *Geophys Res. Lett.*, **27**, 3289-3292.

[8] Hedin, A.E., 1991: Extension of the MSIS thermosphere model into the middle and lower atmosphere, *J. Geophys. Res.*, **96 (A2),** 1159-1172.

[9] Herron, J. P., 2007: *Rayleigh-Scatter Lidar Observations at USU's Atmospheric Lidar Observatory (Logan, UT) — Temperature Climatology Comparisons with MSIS, and Noctilucent Clouds*, PhD Dissertation, Utah State University, Logan, UT.

[10] Liu, H.-L., and R. G. Roble, 2002: A study of a self-generated stratospheric sudden warming and its mesospheric-lower thermospheric impacts using the coupled TIME-GCM/CCM3, *J. Geophys. Res.*, **107 (D23),** 4695, doi:10.1029/2001JD001533.

EPJ Web of Conferences **119**, 13009 (2016) DOI: 10.1051/epjconf/201611913009
ILRC 27

TEMPERATURE CLIMATOLOGY WITH RAYLEIGH LIDAR ABOVE OBSERVATORY OF HAUTE-PROVENCE: DYNAMICAL FEEDBACK

Philippe Keckhut[1*], Alain Hauchecorne[1], Beatriz Funatsu[2], Serguey Khaykin[1], Nahouda Mze[1], Chantal Claud[3], Guillaume Angot[1]

[1] LATMOS-IPSL, UVSQ, UPMC, CNRS/INSU, UMR 8190, Guyancourt, France,
**Email:* keckhut@latmos.ipsl.fr
[2]LETG-Rennes COSTEL, CNRS UMR 6554, Université Rennes 2, Rennes, France
[3]Laboratoire de Météorologie Dynamique, CNRS UMR 8539, IPSL, Ecole Polytechnique, Palaiseau, France

ABSTRACT

Rayleigh lidar in synergy with satellite observations (SSU and AMSU) allow insuring an efficient monitoring and showing that cooling has continued. New approach for trend detection has been developed allowing a better estimate of changes due to radiative forcing. Stratospheric Warmings and gravity waves contribute to insure a dynamical feedback of the long-term changes.

1. INTRODUCTION

Rayleigh lidar measurements were performed continuously since 1979 at Observatory of Haute-Provence in south of France (44°N) allowing detecting temperature trends in the stratosphere and mesosphere.

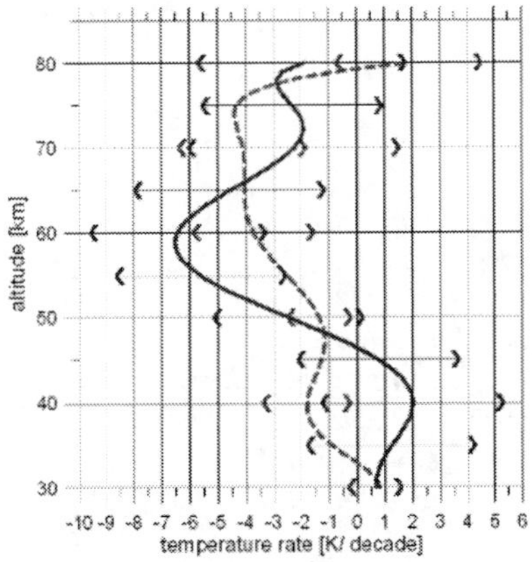

Figure 1. Temperature trends derived from the OHP lidar for winter (blue) and summer (red) periods with a multi-function linear regression.

Time evolution reveals a general cooling during the 1995-2005 period, and a slowdown of the cooling that is not yet understood and coincides with the end of SSU series with a questionable continuity between the successive instruments [1]. Some differences with other lidar series have been reported [2] as well as differences between summer and winter estimates (Figure 1).

In addition to radiative forcing induced by ozone decrease and greenhouse gas increase, dynamic feedbacks are investigated while the variability of the middle atmosphere is driven by waves propagation and their interaction with the mean flow mainly through Sudden Stratospheric Warming and mesospheric dissipation.

2. SSU AND AMSU CONTINUITY

The SSU experiment has provided the unique temperature series from space. While the continuity suffers from tidal effect due to the orbit changes from the successive satellites and their time drifting [2], the continuity is now insured by AMSU. Comparisons between lidar and AMSU reveal very good agreements when data obtained on the same days are considered and partial sampling appear critical [3].

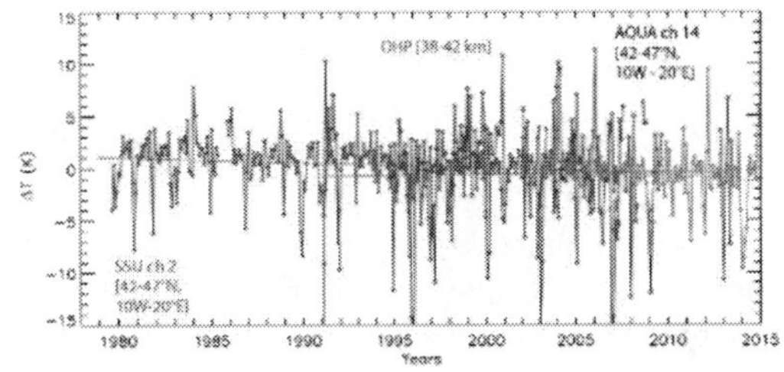

Figure 2. Temperature series in the upper stratosphere given by SSU (blue), OHP lidar (red), and AQUA-AMSU (grey).

For some AMSU similar drifts and tide effects occur [4], however some platforms like METOP do not exhibit drift and the combinations of some of them avoid the tidal effects. Lidars and SSU series have shown a temperature plateau [2] during the period 1995-2005 and AMSU series have confirmed that cooling has continued and that the plateau was a transitory effect (Figure 2).

3. EFFECT OF STRATOSPHERIC WARMINGS

Stratospheric warmings are the main cause of temperature variability during winter-time (Figure 3). The breakdown of the vortex induces a large warming in the upper stratosphere associated with a cooling in the mesosphere.

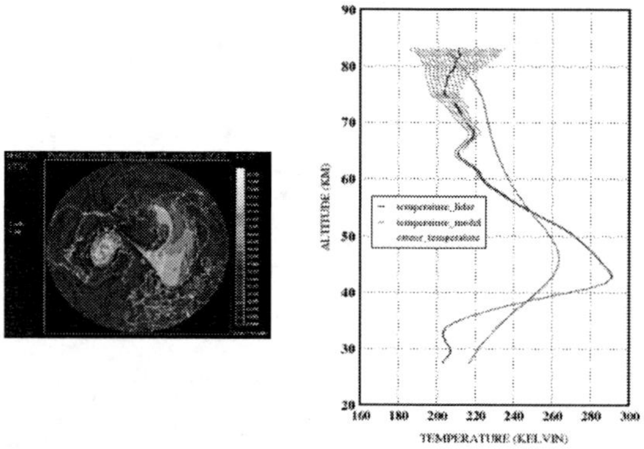

Figure 3. Temperature profile during a stratospheric warming on January 2013 (right panel) and the corresponding PV map in the stratosphere at 475 K.

While their occurrence can be forced by the mean state of the atmosphere and could be modified with climate change, it is important to estimate temperature change due to radiative effect alone. This estimate has been performed in tracking the main mode of the temperature distribution instead of the mean (Figure 4). In applying such methodology the warming observed in December-February period due to stratospheric warming seasonal cycle have been removed [4] and trends during winter and summer reduced. The impact of different sampling is probably the main cause of differences between the different lidar and rocket series with satellite data while the impact of stratospheric warming is not zonal.

Stratospheric warmings were associated with cooling at ground and some mesospheric effect more than a month before the breaking is also noticed suggesting a potential role of gravity waves.

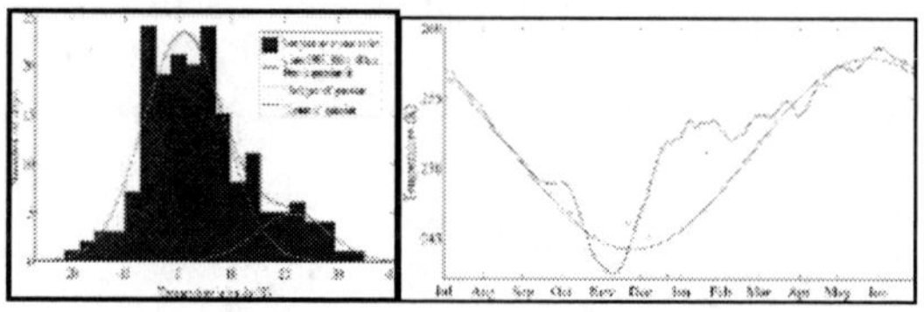

Figure 4. Histogram of stratopause temperature (left panel) and seasonal cycle (right panel) estimated from the mean temperature and the maximum of the main mode.

4. EFFECT OF GRAVITY WAVES

Gravity waves potential energy has been derived from COSMIC GPS and Rayleigh lidar providing consistent information between the both methods. The observed seasonal variations of stratospheric gravity waves activity is found mainly driven by the wind filtering process [6].

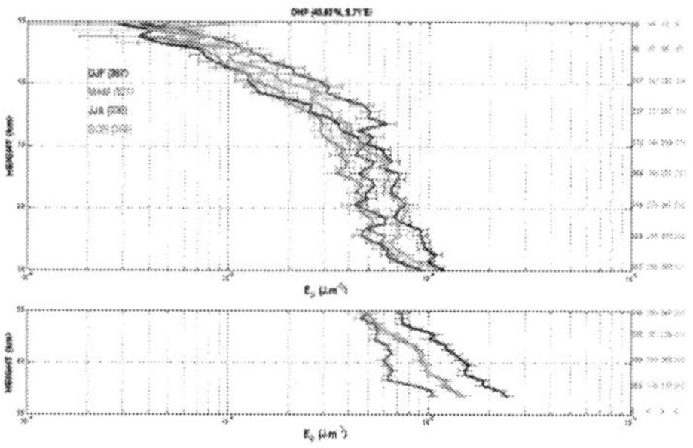

Figure 5. Vertical profiles of potential energy per unit mass (in J.kg top) in the upper stratosphere and mesosphere averaged over winter (December-January-February, blue line), spring (March-April-May, green line), summer (June-July-August, red line), and autumn (September-October-November, cyan line). The horizontal error bars indicate the ±1σ uncertainty of the median.

Dramatic increases of gravity waves activity in the upper stratosphere above OHP have been observed before and during the SSW event in early January 2013. This is attributed to planetary wave activity, which strongly perturbs the zonal winds in the middle atmosphere, resulting in wave generation by adjustment of the unbalanced flow. The gravity wave potential energy has been derived using a variance method on the raw signal [7]. We observed night-to-night variations in the lidar measurements, and inter-annual variability in gravity wave activity based on 16 years of Rayleigh lidar data (30–85 km). This method provides additional information on the vertical structure and on the interactions of the gravity waves with the mean flow. Because the energy is constant with altitude when there is no interaction with the mean flow, it is possible to investigate the altitude where the largest energy exchange occurs. This analysis shows that wave energy dissipates in the background atmosphere depending on the altitude and the season above OHP (Figure 5). Gravity waves are dissipating above ~70 km during all seasons and a little bit higher during summer, but there is relatively little dissipation at lower altitudes. Below 40 km there are also some interactions that have been also confirm by the COSMIC observations when the flow is highly perturbed as it can happen during stratospheric warmings.

5. CONCLUSIONS

Lidar temperature series obtained since 1979 allow monitoring long-term changes due to anthropogenic change. Global estimate can be performed in synergy with satellite series. The plateau observed during 1995-2005 period was a temporary anomaly and our investigation shows that the cooling has continued with similar amplitude as during the 1980s. The new approach to analyze temperature trends reveals the dynamical feedback associated with stratospheric warmings and may explain differences between estimates from different lidar sites and between winter and summer. Gravity waves analysis allows identifying altitude where interactions with the mean flow are the largest. Their potential feedback on the long-term still needs to be estimated. Synergy of lidar and GPS technique could be used.

ACKNOWLEDGEMENT

The work was supported by EU FP7 ARISE project (no. 284387) and the French ANR StraDyVariUS project. Lidar measurements at Haute Provence observatory were obtained in the frame of NDACC (Network for detection of Atmospheric Composition Change) with the support from CNRS-INSU, CNES and UVSQ and are publicaly available on NOAA NDACC data portal (http://www.ndsc.ncep.noaa.gov) and the French Data Center ESPRI-ETHER (http://www.pole-ether.fr).

REFERENCES

[1] Thompson D.W.J et al., 2014: The mystery of recent stratospheric temperature trends, doi:10.1038/nature11579.

[2] Keckhut, P. et al., 2011: An evaluation of uncertainties in monitoring middle atmosphere temperatures with the lidar network in support of space observation, J. Atmos. Sol. Terr. Phys, 73(5-6), 627-642.

[3] Funatsu, B. M. et al., 2011: Investigations of stratospheric temperature regional variability with lidar and Advanced Microwave Sounding Unit, J. Geophys. Res., 11(D8).

[4] Keckhut P. et al., 2015: Tidal effects on stratospheric temperature series derived from successive Advanced Microwave Sounding Units *Q.J.R. Meteorol. Soc.,* 141, 477–483.

[5] Angot, G. et al., 2012: Contribution of stratospheric warmings on temperature trends in the middle atmosphere from the lidar series obtained at observatory of Haute-Provence (44°N), J. Geophys. Res., 117(D21).

[6] Khaykin, S. et al., 2015: seasonal variation of gravity wave activity at mid-latitudes from 7 years of cosmic gps and Rayleigh lidar temperature observations, *Geophys. Res. Lett.*, 42(4), 1251-1258.

[7] Mzé, N., et al., 2014, Vertical distribution of gravity wave potential energy from long-term Rayleigh lidar data at a northern middle latitude site, *J. Geophys. Res.,* 119(21), 12069-12083.

EPJ Web of Conferences **119**, 13010 (2016)
ILRC 27
DOI: 10.1051/epjconf/201611913010

Winter Mesospheric Thermal Structure over Tibetan Plateau

Shuai Qiao[1,2], Weilin Pan[1]*, Daren Lü[1,]
[1]*Key Laboratory of Middle Atmosphere and Global Environment Observation, Institute of Atmospheric Physics, Chinese Academy of Sciences, Beijing,100029,China, *Email: panweilin@mail.iap.ac.cn*
[2]*University of Chinese Academy of Sciences, Beijing, 100049, China*

ABSTRACT

A mobile Rayleigh temperature lidar was deployed in Golmud (36.25°N 94.54°E), Qinghai in China for making measurements of mesosphere temperature from 55 up to 90 km. The mesospheric thermal structure was obtained during the winter seasons of Year 2013 and Year 2014. At the altitude of 57~79 km, lidar temperatures were much colder than the MSIS-00 model predictions. However, in the lower mesosphere region, lidar measurements were in reasonable agreement compared with the TIMED-SABER satellite data.

1. INTRODUCTION

Temperature is a key parameter for understanding the chemical, dynamic, and radiative processes in the atmosphere. Middle atmosphere is sensitive to solar activity as well as human being activities. In the stratosphere and mesosphere, rising CO_2 concentration increases the effects of radiative cooling. Modeling studies have predicted that mesosphere will cool in response to the increase of CO_2 [1]. On the other hand, atmospheric temperature is affected by waves (gravity waves, tides, and planetary waves) as well as atmospheric circulation [2]. While extensive temperature observations of middle atmosphere have been conducted at several critical sites, including the South Pole and the North Pole [3,4], few data were available in the Tibetan Plateau area, which is called the Third Pole with its great importance for monitoring global climate change.

Lidar has been used for decades and proved to be a powerful instrument for detecting the atmosphere. Rayleigh lidar, with its high spatial and temporal resolution, is an effective means for measuring vertical temperature profiles in the middle atmosphere [5].

The MARMOT (Middle Atmosphere Remote Mobile Observatory in Tibet) lidar was recently developed by the Institute of Atmospheric Physics (IAP), Chinese Academy of Sciences (CAS). From August 2013 until December 2014, MARMOT lidar has been deployed in Lhasa, Tibet and in Golmud, Qinghai. In this paper, we report winter thermal structure derived from ~400 h of lidar measurements at Golmud during the winter seasons in 2013 and 2014.

2. INSTRUMENTATION

The MARMOT lidar consisted of three parts: laser transmitter, optical receiver, and signal acquisition & control unit (as shown in Fig.1). Lidar transmitter was mainly a Nd:YAG laser working at the wavelength of 532 nm. Optical receiver was a prime focus telescope of Φ1000 mm diameter. Rayleigh backscattered signals of 532 nm were collected by the telescope and then detected by a PMT. A mechanical chopper in the receiver chain blocked any signals below 50 km in order to prevent PMT saturation caused by strong lidar return signals from lower altitudes. A raw photon-count profile is obtained for every 1800 laser shots accumulated. The temporal and spatial resolutions are 1 min and 30 m,

DOI: 10.1051/epjconf/201611913010

Table 1. Lidar observational time during the winters of 2013 and 2014

Date(DEC/2013)	*1*	*2*	*3*	*4*	*5*	*6*	*7*					
Time(Hours)	**9**	**12**	**12**	**12**	**12**	**12**	**12**					
Date(NOV/2014)	*12*	*13*	*14*	*16*	*17*	*18*	*19*	*21*	*24*	*25*	*27*	*30*	
Time(Hours)	**7**	**11**	**10**	**11**	**11**	**11**	**10**	**2**	**5**	**4**	**10**	**4**	
Date(DEC/2014)	*1*	*2*	*3*	*4*	*6*	*7*	*9*	*10*	*11*	*12*	*13*	*14*	*15*
Time(Hours)	**12**	**12**	**12**	**12**	**13**	**5**	**7**	**9**	**13**	**7**	**10**	**8**	**13**
Date(DEC/2014)	*16*	*17*	*18*	*19*	*20*	*21*	*22*	*23*	*24*	*25*	*26*	*27*	
Time(Hours)	**13**	**13**	**13**	**4**	**8**	**3**	**2**	**8**	**11**	**10**	**2**	**13**	

respectively.

Fig.1. The MARMOT lidar system, including the laser transmitter (upper figure), the optical receiver (middle figure), and the signal acquisition part (lower figure).

The main principle of Rayleigh temperature derivation is as follows. Considering that the aerosol content is extremely low above 30 km, the lidar backscattered signals mainly come from the Rayleigh scattering by gaseous molecules. By calculating the atmospheric backscattered signals, we can obtain the relative density profiles in the atmosphere. If we assume that the atmospheric temperature at an initial altitude can be taken from the MSIS-00 model [6], then by combining the ideal gas law and the hydrostatic equation, atmospheric temperature profiles can be derived downward using the relative density profiles [7].

3. OBSERVATIONAL RESULTS

Lidar data from November 12 to December 27, 2014 were used to calculate the mesospheric mean temperature profile in winter. The lengths of observation period during these days were listed in Table.1. In Fig.2, the red solid line represents the mean temperature profile in winter 2014, and the blue dashed line represents the mean temperature profile predicted by the MSIS-00 model in the same time period. The lidar temperatures in lower mesosphere between 57 km and 79 km were considerably lower than the MSIS-00 predictions. In Fig.3, the red line shows the difference between lidar temperatures and MSIS-00 temperatures. The greatest difference is ~16 K at the height of ~66 km. For comparison purpose, we chose another period of lidar data from December 1 to December 7, 2013, which were the only available lidar measurements during that winter season. The cyan solid line in Fig.2 represents the mean temperature profile in

DOI: 10.1051/epjconf/201611913010

winter 2013. Although a little warmer than the winter of 2014 at the altitude between 57 km and 79 km, it is still considerably colder than the MSIS-00 results.

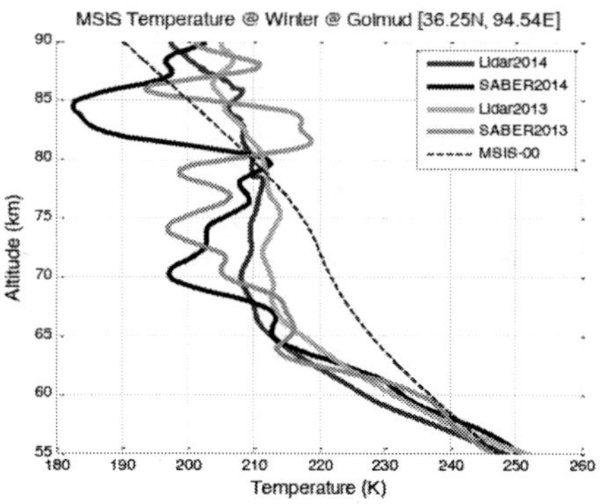

Fig.2. Comparison among the lidar measured temperatures (cyan solid line for Year 2013 and red solid line for Year 2014), the MSIS-00 model temperatures (blue dashed line for Year 2014), and the TIMED-SABER measured temperatures (green solid line for Year 2013 and black solid line for Year 2014) in the winter seasons over Golmud.

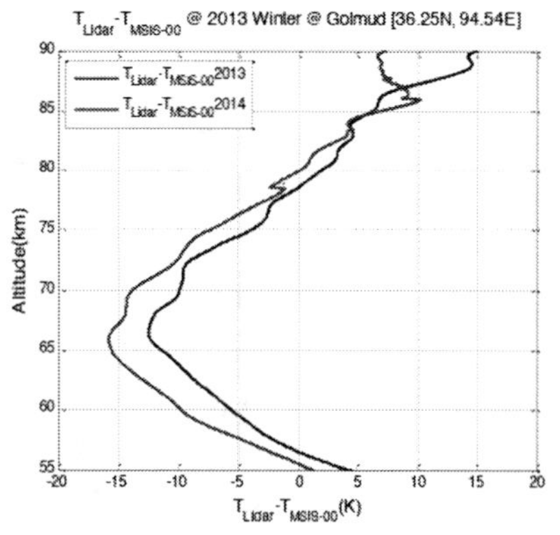

Fig.3. The difference between the lidar measured temperatures and the MSIS-00 temperatures in Year 2013 (blue line) and in Year 2014 (red line).

In addition, as plotted in Fig.2 are temperature data obtained from the Sounding of the Atmosphere using the Broadband Emission Radiometry (SABER) instrument [8] onboard the Thermosphere-Ionosphere-Mesosphere-Energetics and Dynamics (TIMED) satellite, which was launched on December 7, 2001. The black solid line in Fig.2 shows the temperature profile obtained by SABER at 36.20°N, 94.61°E on December 5, 2014. The green solid line in Fig.2 represents the temperature profile obtained by SABER at 37.01°N, 111.16°E on December 4, 2013. Both SABER measured temperatures show significant inversion layers in the upper mesosphere. In the lower mesosphere region, the satellite data were in reasonable agreement with the lidar temperatures, both considerably colder than the MSIS temperatures.

4. DISCUSSIONS

The MARMOT lidar measurements obtained over Tibet Plateau has been used for studying the mesospheric thermal structure during two winter seasons. The discrepancy between lidar and MSIS-00 results have suggested some limitations of the MSIS model over the Tibetan Plateau, since the model was developed under the circumstances that not much middle atmosphere observational data were available in that area. Therefore, lidar measurements could provide an effective tool for improving current atmospheric modeling, and for better understanding the middle atmosphere thermal structure, the waves, and the processes of momentum and energy exchange within layers.

ACKNOWLEDGMENTS

This work was supported by National natural Science Foundation of China (NSFC) program (Grant No. 41127901) and the "100 Technical Talents" Program of the Chinese Academy of Sciences (CAS).

REFERENCES

[1] Portmann, R. W., G. E. Thomas, S.

Solomon, and R. R. Garcia, 1995: The importance of dynamical feedbacks on doubled CO2-induced changes in the thermal structure of the mesosphere, Geophys. Res. Lett., **22**, 1733–1736.

[2] Lü, D., and H. Chen, 2003: Advances in middle atmosphere physics research, Chinese J. Atmos. Sci. (in Chinese), **27(4)**, 750–769.

[3] Gardner C. S., G. C.Papen, X. Chu, and W. Pan, 2001: First lidar observations of middle atmosphere temperatures, Fe densities, and polar mesospheric clouds over the North and South Poles, Geophys. Res. Lett., **28(7)**, 1199–1202.

[4] Pan W., C. S. Gardner, and R. G. Roble, 2002: The temperature structure of the winter atmosphere at South Pole, Geophys. Res. Lett., **29(16)**, 49-1-49-4.

[5] Gardner C. S., D. C. Senft, and T. J. Beatty, 1989: Rayleigh and sodium lidar techniques for measuring middle atmosphere density, temperature and wind perturbations and their spectra [M]. In World ionosphere/thermosphere study.

[6] Picone, J. M., A. E. Hedin, D. P. Drob, and A. C. Aikin, 2002: NRLMSISE-00 empirical model of the atmosphere: Statistical comparisons and scientific issues, J. Geophy. Res., **107(A12)**, SIA 15-1–SIA 15-16.

[7] Chanin, M. L., 1984: Review of lidar contributions to the description and understanding of the middle atmosphere, J. Atmos. Solar-Terr. Phys., **46**, 987–993.

[8] Russell III, J. M., M. G. Mlynczak, L. L. Gordley, et al., 1999: Overview of the SABER experiment and preliminary calibration results, Proc. SPIE, 277-288.

DOI: 10.1051/epjconf/201611914001

REAL TIME TURBULENCE ESTIMATION USING DOPPLER LIDAR MEASUREMENTS

Lucie Rottner [1*], Christophe Baehr [1]

[1]*Météo-France, CNRM-GAME, Toulouse, France.* **lucie.rottner@meteo.fr*

ABSTRACT

A preliminary work on a new way to estimate atmospheric turbulence using high-frequency Doppler lidar measurements is presented. The turbulence estimations are based on wind reconstruction using 3D Doppler lidar observations and a particle filter. The suggested reconstruction algorithm links the lidar observations to numerical particles to obtain turbulence estimations every time new observations are available. The high frequency of the estimations is a new point which is detailed and discussed. Moreover, the presented algorithm ables to reconstruct the wind in three dimensions in the observed volume. We have thus locally access to the spatial variability of the turbulent atmosphere. The suggested algorithm is applied to a set of real observations. The obtained results are very encouraging : they show significant improvements on turbulent parameter estimations.

1. INTRODUCTION

Turbulent phenomena in the Atmospheric Boundary Layer (ABL) are characterized by small spatial and temporal scales which make them difficult to observe and to model. As Doppler lidar provide fine and high-frequency observations, they are instruments of predilection to study wind and turbulence in the ABL. With the development of wind farms and the constant growth of traffic on airports, the need for turbulence estimations is growing. Lot of different parameters have been studied to estimate turbulence. In this paper, the novelty is not about the turbulent parameters themselves. Instead of working on these parameters we suggest a new way to reconstruct the wind from the observations. The reconstruction algorithm is a nonlinear Bayesian estimation method based on wind observations. Using the algorithm, the 3D wind is reconstructed and atmospheric turbulent parameters are estimated in a conic volume. The estimation method is based on a particle filter. Using this filter, the wind observations are associated to particle systems driven by a local Lagrangian turbulence model. The particles have both fluid and stochastic properties. Here is the innovation of our method : spatial averages and covariances may be deduced from the particles themselves. Thus wind gusts and turbulent structures may be detected in real time. The theoretical framework for turbulence estimations have been defined by Baehr [1] and then applied to vertical lidar [2]. A lidar WindCube is used to get 3D estimations. The WindCube geometry is described. This description leads to one assumption : the atmosphere is assumed to be locally homogeneous. Next the particle filter is introduced. Then the turbulent kinetic energy computation is discussed. Finally an application of the reconstruction algorithm to real WindCube observations is presented. The results show good wind reconstructions. For the turbulent parameter estimations, the main new point is the frequency of the estimations : turbulent parameters are estimated at the observation frequency.

2. WINDCUBE GEOMETRY

Wind reconstruction and turbulent parameter estimation are based on high-frequency and high-quality wind observations. To get such observations, a lidar WindCube provided by LEOSPHERE is used. It is a heterodyne pulsed Doppler lidar which has a specific geometry. It has five different lines of sight. There is one vertical line surrounded by four oblique lines with an oblique angle of 28° (see figure 1). After each emission of the lidar beam, the lidar gets back 10 observations with a spacing of 20 meters. The lidar performs measures one line after an other. A full revolution takes 4 seconds. The lidar has a maximum range of 220 meters covering the low ABL. Thanks to these characteristics, the lidar instrument allows to

DOI: 10.1051/epjconf/201611914001

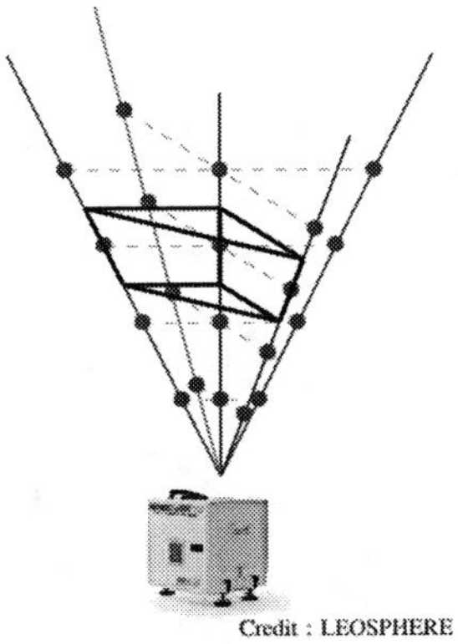

FIGURE 1: Lidar WindCube : 1 vertical line of sight and 4 oblique lines. The observed volume is split in boxes. One of them is drawn is black.

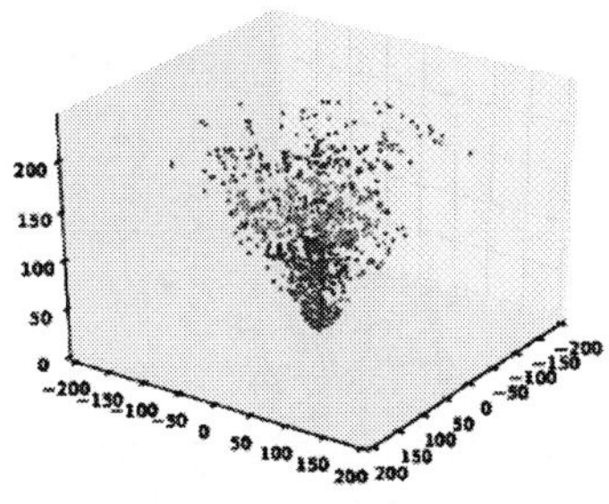

FIGURE 2: Illustration of a particle system filling a conic volume. The colored dots are the numerical particles.

observe inertial turbulent scales in the ABL.

We may notice that as the lidar uses the Doppler effect to measure wind speed, it only observes the projection of the wind following the line of sight. So, to reconstruct the wind we have to use three observations following three different lines. We logically choose two consecutive observation points on a same vertical level and the vertical observation point related to this level. It makes appear an hypothesis : in order to reconstruct the turbulent atmosphere, we assume that the atmosphere is homogeneous between these three points and during a pattern repeat period. It is the assumption of locally homogenous medium. The observed volume is split in boxes. Each box contains three observations of radial wind (see fig. 1). Thus the atmosphere is considered homogenous in each box. This partition is usefull to compute turbulence estimations.

3. PARTICLE FILTER

When a set of data is collected, the usual question is to estimate the real state of a system using one or more observations at different time steps. It is the principle of data assimilation. The reconstruction of the turbulent atmosphere may be seen as an assimilation process : the wind observation is assimilated in the particle model. For each time step k, the state vector of the system ξ_k is associated to the observation vector y_k, where the dimension of the vector y_k is often smaller than the dimension of the state vector ξ_k. Then some coordinates of the vector ξ_k are hidden for the observation y_k. Assuming that ξ_k is a Markov process and the observational noise is a random variable, the classical problem consists in evaluating the probability density function (pdf) :

$$p(\xi_k|y_{1:k}) = \frac{p(y_k|\xi_k).p(\xi_k|y_{1:k-1})}{p(y_k|y_{1:k-1})}$$

with the following time recursion due to the Markov property

$$p(\xi_k|y_{1:k-1}) = \int p(\xi_k|\xi_{k-1})p(\xi_{k-1}|y_{1:k-1})d\xi_{k-1}.$$

For usual data assimilation, the evaluation of the expectation of $p(\xi_k|y_{1:k})$ is based on variational methods and a meteorological model is used for the time prediction. Here the whole pdf is described using particle filters (PF) designed to study nonlinear processes. A PF uses a set of numerical particles to describe the probability density function (pdf) of interest. The observed volume is filled with the numerical particles. They follow the air flow and carry the local characteristics of the atmosphere. The PF is made of two sequential steps. First there is the evaluation of the pdf $p(\xi_k|y_{1:k})$. The particle velocities are updated using the last observations : these updated particles are samples of the pdf $p(\xi_k|y_{1:k})$. This resampling is called "importance sampling". Then there is a Markovian time prediction. Here, a physical model is used to predict the state at the next time step. The model is locally consistent with the Kolmogorov K41 laws and has been first introduced for turbulence estimation by [1]. Thus using the particle representation at the time step k, the pdf is approximated by :

$$p(\xi|y_{1:k}) \sim p^N(\xi|y_{1:k}) = \frac{1}{N}\sum_{i=1}^{N}\delta(\xi - \xi_k^i)$$

where N is the number of particles and ξ_k^i is the particle i at the time step k.

4. MEASURING TURBULENCE

To estimate atmospheric turbulence, a usual parameter is the turbulent kinetic energy (TKE). The TKE is calculated using the variance of the wind velocity fluctuation. Let $\mathbf{V} = (u, v, w)$ be the 3D velocity vector. Following the Reynolds decomposition, $\mathbf{V}$ may be split in a mean component $\overline{\mathbf{V}}$ and a fluctuation $\mathbf{V}'$, $\mathbf{V} = \overline{\mathbf{V}} + \mathbf{V}'$. From this split, we can define the turbulent kinetic energy :

$$\text{TKE} = \frac{1}{2}(\overline{u'^2} + \overline{v'^2} + \overline{w'^2})$$

Ideally the averaging operator should be determined by ensemble averaging. As the ensemble average is unreachable in usual pratical situations, time averaging is used instead. To use time averaging, two hypotheses are made : the turbulent flow is assumed ergodic and stationary. In this framework, time average and spatial average are equilvalent. Therefore, the TKE is obtained using the variance of the 3D wind speed $\mathbf{V}$ over a period T. At each observation point, the classical TKE is given in discrete time by :

$$\text{TKE} = \tfrac{1}{T}\textstyle\sum_{k=1}^{T}(\mathbf{V}_k - <\mathbf{V}>_T)^2$$

where $< . >_T$ is the time average. To mesure the TKE using a lidar, the integration time T has to be specified. Pichugina and O'Connor suggest methods to estimate the TKE and the eddy dissipation rate using a Doppler lidar [4] [3]. Following their works, to compute the TKE with a time average we choose T = 10min. This TKE is used as a reference in this paper. Here using the particle system, average and variance can be computed using the particle values. It leads that the suggested method overcomes the ergodicity and stationarity hypotheses. At any time, the particles represent the wind and its spatial variability. Thus the local wind speed variance is computed at each time step k. For N particles close to the observation point, the TKE is given by :

$$\text{TKE}(k) = \tfrac{1}{N}\textstyle\sum_{i=1}^{N}(\mathbf{V}_k^i - <\mathbf{V}_k>_N)^2$$

where $< . >_N$ is the spatial average computed using the updated particles. At each time step k, the TKE is estimated in each box used to reconstruct the wind. We have thus access to the spatial and time variability of the turbulence.

5. RESULTS

To get relevant turbulent estimations, the first step is to assess the wind reconstruction. The aim is to reconstruct the 3D wind in the observed volume using the particle filter. The WindCube data have been provided by LEOSPHERE. The results are given in the classical frame. The observed wind is obtained by geometric construction using the three observations. We illustrate the results only in one box with the horizontal wind following the first coordinate u. The figure 3 shows the time series of the observed wind and the reconstructed wind over a 30min period with a 4s time step. The reconstructed serie can barely be separated from the observed one. The power spectrum densities (PSD) are presented on figure 4. They confirm that the reconstructed wind follows the K41 Kolmogorov laws for the inertial sub-range turbulence. The K41 energy cascade is represented by a -5/3 slope drawn in green. The estimated wind is thus coherent with the observations. Its realistic shape is validated by the PSD which follows the energy cascade. The validation of the 3D wind estimations leads us to the estimations of turbulent parameters. The graph 5 illustrates the contribution of the particle method for turbulence estimation. For one box, it shows over a 2h period with a 4s time step the classical TKE calculated on 10 minutes and the TKE estimated at each time step with the particle method. The usual TKE is made of 10-minute long steps, whereas the particle system gives back a TKE made of irregular peaks. It seems that particles capture a spatial and time variability of the atmospheric turbulence which is clearly unreachable so far. However, even if the particle TKE shape is coherent with the intermittent nature of the turbulence, this estimation method has still to be validated. The natural approach to validate the particle TKE estimations is to average them on 10-minute period and then to compare average particle TKE with classical TKE. In average, the particle

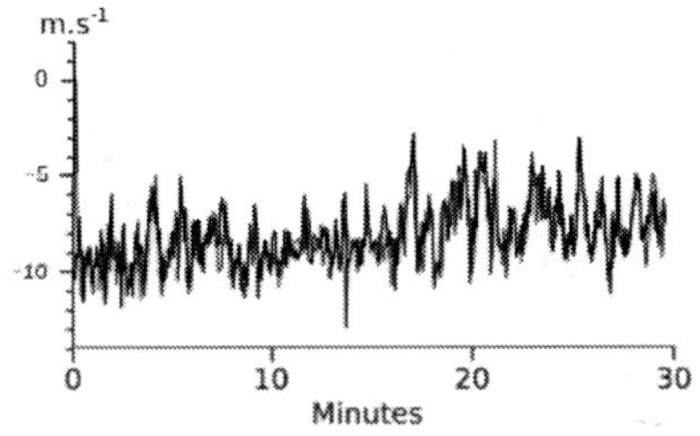

FIGURE 3: Time series of the horizontal reconstructed wind (in red) and the horizontal observed (by lidar) wind (in black), with a 4s time step.

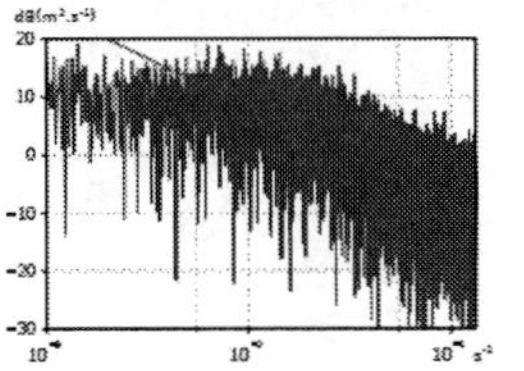

FIGURE 4: Power spectrum densities of the horizontal reconstructed wind. The Kolmogorov energy cascade is drawn in green.

TKE estimations are coherent with the classical estimations (not shown). However, an average validation is not enough to validate the peaky shape of the instantaneous particle TKE. As this variability is unreachable with classical methods, we have no way to provide further validations of instentaneous turbulence estimations.

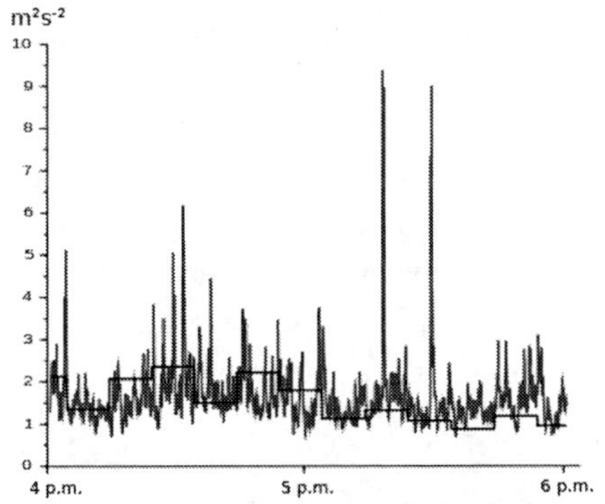

FIGURE 5: Usual TKE (black) versus particle TKE (red), for the 08/26/2010 from a test site in Greece.

6. CONCLUSION

A new way to estimate turbulent parameters has been introduced. Using our method the wind and the turbulent structure advection may be visualized in real time. The reconstruction algorithm have been applied on lidar WindCube measurements. The WindCube has been used because of its conic configuration. This configuration allows to reconstruct the wind in 3D and to estimate turbulent parameters in the observed volume. To validate the algorithm we have compared the estimated wind to the observed wind. The results show that the algorithm is able to reconstruct the 3D wind. We have then worked on TKE estimations. We have seen that the reconstruction algorithm brings completely new information on turbulence. Thanks to the association of high-frequency observations with particle methods, we have now access to local reconstruction of the turbulent atmosphere. Fine turbulence estimations are thus available. Experiences on real data show very good results. However the validation of the high-frequency turbulence estimations is still an open question as there is no way to assess the time variability of turbulent parameters so far.

ACKNOWLEDGEMENT

The author would like to thanks LEOSPHERE for providing the WindCube data.

REFERENCES

[1] C. Baehr. Nonlinear filtering for observations on a random vector field along a random path. application to atmospheric turbulent velocities. *ESAIM : Mathematical Modelling and Numerical Analysis*, 44(05) :921–945, 2010.

[2] C. Baehr, B. Claire, C. Fleur, D. Alain, and P. Bruno. Retrieval of the turbulent and backscattering properties using a nonlinear filtering technique applied to doppler lidar observation. In *ISARS*, 2012.

[3] E. J. O'Connor, A. J. Illingworth, I. M. Brooks, C. D. Westbrook, R. J. Hogan, F. Davies, and B. J. Brooks. A method for estimating the turbulent kinetic energy dissipation rate from a vertically pointing doppler lidar, and independent evaluation from balloon-borne in situ measurements. *Journal of Atmospheric and Oceanic Technology*, 27(10) :1652–1664, 2010.

[4] Y. L. Pichugina, S. C. Tucker, R. M. Banta, W. A. Brewer, N. D. Kelley, B. J. Jonkman, and R. K. Newsom. Horizontal-velocity and variance measurements in the stable boundary layer using doppler lidar : Sensitivity to averaging procedures. *Journal of Atmospheric and Oceanic Technology*, 25(8) :1307–1327, 2008.

EPJ Web of Conferences **119**, 14002 (2016) DOI: 10.1051/epjconf/201611914002
ILRC 27

AIRCRAFT WAKE VORTEX PARAMETRIZATION BASED ON 1.5-μm COHERENT DOPPLER LIDAR DATA

V.A. Banakh, I.N. Smalikho

V.E. Zuev Institute of Atmospheric Optics SB RAS 1 Akademik Zuev Square, Tomsk, 634021, Russia
banakh@iao.ru, smalikho@iao.ru

ABSTRACT

A strategy of measurement by a 1.5-μm pulsed coherent Doppler lidar "Stream Line" has been developed, and a method for estimation of aircraft wake vortices from the lidar data has been proposed. The principal possibility of obtaining the information about the vortex situation over an airport airfield with the Stream Line lidar has been demonstrated.

1. INTRODUCTION

An aircraft flight is always accompanied by the formation of a pair of vortices. An aircraft wake can extend to many kilometers and constitute a danger for other aircrafts. A pulsed coherent Doppler lidar (PCDL) is now the best tool for the experimental study of wake vortices. The 2-μm PCDL is used most widely for measurement of wake vortex parameters (vortex axis coordinates and circulation) [1]. These parameters were estimated from lidar data by different methods, which are described in [2-5].

The necessary condition for lidar measurements of aircraft vortex parameters is the rather high spatiotemporal resolution of raw data and the high signal-to-noise ratio (SNR), which can be provided by the 2-μm PCDL in the atmospheric boundary layer. In contrast to the 2-μm PCDL [6], the Stream Line 1.5-μm PCDL developed by HALO Photonics [7] has the 20 times lower probing pulse energy [8]. Therefore, for the Stream Line lidar, the SNR defined as the ratio of the mean power of lidar return signal to the mean power of noise in the frequency band of 50 MHz at the same concentration of aerosol particles is at least several tens times lower than SNR of the 2-μm PCDL.

In this paper, we study the possibility of obtaining the information about an aircraft wake from data measured by the Stream Line lidar. A measurement strategy is developed, and a method for estimation of aircraft vortex axis coordinates and circulation is proposed. The results of test measurements by the Stream Line lidar at the air field of the Tomsk Airport are presented.

2. METHODOLOGY

To obtain the information about aircraft wake vortices from data of PCDL, the scanning of the probing beam in the vertical plane perpendicularly to the aircraft flight line is used during the measurements. Figure 1 shows the measurement geometry for the case of up-down beam scanning, that is, the elevation angle φ variable in time.

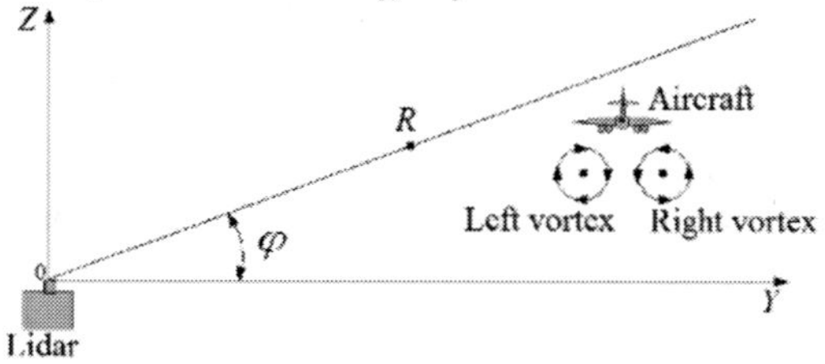

Fig. 1. Geometry of measurement by PCDL scanning in the vertical plane.

The raw data measured by the Stream Line lidar are an array of estimates of correlation functions of the normalized complex lidar signal $\hat{C}(lT_s, R_k, \varphi_m; n')$, where $l = 0, 1, ..., 6$, $T_s = B^{-1}$ is the sampling interval, $B = 50$ MHz is the frequency bandwidth, $R_k = R_0 + k\Delta R$ is the distance from the lidar, $k = 0, 1, ..., K-1$, $\Delta R = cT_s/2 = 3$ m, c is the speed of light, $\varphi_m = m\Delta\varphi$ is the elevation angle, $m = 0, 1, ..., M-1$, $\Delta\varphi = \omega_s \Delta T$ is the scanning angle resolution for the rate ω_s, $\Delta T = N_a / f_P$ is the time of measurement, N_a is the number of laser pulses used for accumulation [9], $f_P = 15$ kHz is the pulse repetition frequency, and $n' = 1, 2, 3, ...$ is the scan number. If the lidar signal is normalized to the square root of the mean noise power, then the estimate of the signal-to-noise ratio can be determined as $\hat{\mathrm{SNR}}(R_k, \varphi_m; n') = \hat{C}(0, R_k, \varphi_m; n') - 1$.

Using Eq. (5) from [9], we can obtain the array of estimates of Doppler spectrum $\hat{S}_D(V_{l'}, R_k, \varphi_m; n')$

from $\hat{C}(lT_s, R_k, \varphi_m; n')$, where $V_{l'} = (l' - L'/2)\Delta V$, $l' = 0, 1, ..., L'-1$, L' is the number of spectral channels, which can be taken arbitrarily large by adding $L'-7$ complex zeros to the array $\hat{C}(lT_s, R_k, \varphi_m; n')$, $\delta V = (\lambda/2)B/L'$, and λ = 1.5 µm is the wavelength. Since SNR has very low values in the case of the Stream Line lidar, the estimates of spectral moments [2] and velocity envelopes [3] cannot be obtained from $\hat{S}_D(V_{l'}, R_k, \varphi_m; n)$ with the acceptable accuracy. Therefore, to extract the information about aircraft wake vortices from the data measured by this lidar, we obtained estimates of radial velocity $\hat{V}_r(R_k, \varphi_m; n')$ at the point of maximum of the Doppler spectrum, that is, $\max\{\hat{S}_D(V_{l'}, R_k, \varphi_m; n')\} = \hat{S}_D(\hat{V}_r, R_k, \varphi_m; n')$.

The focusing of the probing beam to some distance F allows the SNR in the vicinity of the focus to be increased significantly. Our numerous experiments for the case of the Stream Line lidar have shown that at F = 300 m the SNR(R_k) is equal, on average, to 6 dB at the point of maximum $R_k = F$, to 14 dB at R_k = 100 m, and to –16 dB at R_k = 600 m. If in this case N_a = 1500 pulses are used for the accumulation of data, then for distances 100 m $\le R_k \le$ 600 m the probability P_b of the bad estimate of the radial velocity does not exceed 10^{-4} and the bias of $\hat{V}_r(R_k, \varphi_m; n')$ is very close to zero.

Our studies have allowed us to determine the parameters optimal for measurements of vortices generated by a landing aircraft at a height lower 50 m by the Stream Line lidar: the distance between the lidar and the runway of about 300 m, F = 300 m, N_a = 1500, $\omega_s = 2^0$/s, and M = 76. Taking into account that f_P = 15 kHz, the measurement time is ΔT = 0.1 s, the scanning angle resolution is $\Delta\varphi = 0.2^0$, the maximum elevation angle is $\varphi_{M-1} = (M-1)\Delta\varphi = 15^0$, and the scan duration is $T_{scan} = M\Delta T = 7.6$ s.

Let the array of lidar estimates of the radial velocity $\hat{V}_r(R_k, \varphi_m; n')$ contain the information about aircraft wake vortices starting from the scan number $n' = n_0 + 1$ and up to $n' = n_0 + N$. To avoid the influence of the background wind, we obtain the data array

$$\tilde{V}_r(R_k, \varphi_m; n) = \hat{V}_r(R_k, \varphi_m; n_0 + n) - \hat{V}_r(R_k, \varphi_m; n_0), \quad (1)$$

where $n = n' - n_0 = 1, 2, ..., N$. Figure 2(a) exemplifies the distribution $\tilde{V}_r(R_k, \varphi_m; 2)$ obtained from the data measured by the Stream Line lidar at the air field of the Tomsk Airport, as a B737 aircraft crossed the scanning plane at a height of 37 m. We can see that at distances from 250 to 330 m from the lidar, the pair of wake vortices affects the radial velocity.

To estimate coordinates of axes of the right and left wake vortices from the data analogous to those shown in Fig. 2(a), we introduce the functions $E(R_k; n)$, $\varphi_{\max}(R_k; n)$, and $\varphi_{\min}(R_k; n)$ determined by the following algorithm:

$$\max_{\varphi}\{\tilde{V}_r(R_k, \varphi_m; n)\} = \tilde{V}_r(R_k, \varphi_{\max}; n), \quad (2)$$

$$\min_{\varphi}\{\tilde{V}_r(R_k, \varphi_m; n)\} = \tilde{V}_r(R_k, \varphi_{\min}; n), \quad (3)$$

$$E(R_k; n)\} = [\tilde{V}_r(R_k, \varphi_{\max}; n)]^2 + [\tilde{V}_r(R_k, \varphi_{\min}; n)]^2, \quad (4)$$

where the maxima and minima fall within the angle range φ_m for every distance R_k.

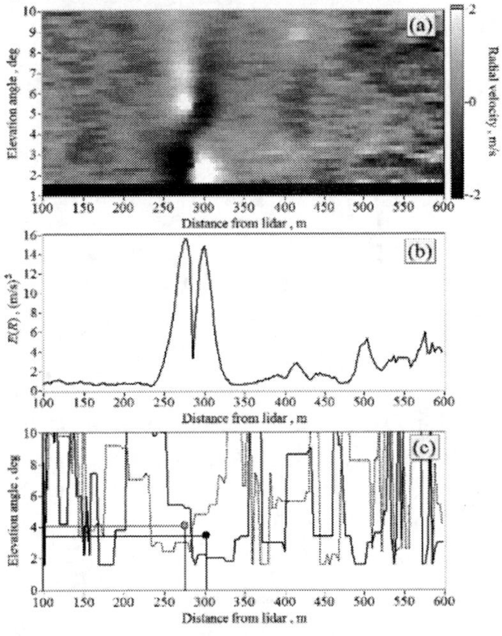

Fig.2. Distribution of the difference of estimates of radial velocities $\tilde{V}_r(R_k, \varphi_m; 2)$ (a) and the corresponding functions (b): $E(R_k; 2)$, (c): $\varphi_{\max}(R_k; 2)$ (solid curve) and $\varphi_{\min}(R_k; 2)$ (dashed curve).

 DOI: 10.1051/epjconf/201611914002

Figures 2(b) and 2(c) show $E(R_k;2)$, $\varphi_{\max}(R_k;2)$, and $\varphi_{\min}(R_k;2)$ obtained from the data depicted in Fig. 2(a). We can see that $E(R_k;2)$ has two pronounced peaks. The results of our numerical investigation for the Stream Line lidar (when the longitudinal size of the sensing volume is around 30 m) indicate that the positions of the far and near peaks are almost identical with the distances between the lidar and the axes of the right $R_{C1}(n)$ and left $R_{C2}(n)$ aircraft vortices, while $[\varphi_{\max}(R_{C1};n)+\varphi_{\min}(R_{C1};n)]/2$ and $[\varphi_{\max}(R_{C2};n)+\varphi_{\min}(R_{C2};\ n)]/2$ are very close to the angular coordinates of the axes of the right $\varphi_{C1}(n)$ and the left $\varphi_{C2}(n)$ vortices, respectively.

The algorithm for estimation of coordinates of the axis of the i-th aircraft vortex ($i=1$ for the right vortex and $i=2$ for the left one) can be the following. For the n-th scan, we first find the position of absolute maximum of the function $E(R_k;n)$ in the entire range of distances $R_k \in [R_0\,,(K-1)\Delta R]$, that is, we use the procedure

$$\max_R\{E(R_k;n)\} = E(R_{\max};n)\,. \qquad (5)$$

Two vortices generated by an aircraft rotate in the opposite directions (see Fig.1). Consequently, if $\varphi_{\max}(R_{\max};n') > \varphi_{\min}(R_{\max};n')$, then $R_{\max} = R_{C2}$. Other-wise, $R_{\max} = R_{C1}$.

Assume that the condition $\varphi_{\max}(R_{\max};n) > \varphi_{\min}(R_{\max};n)$ is true. Then the coordinates of the axis of the left vortex $\{R_{C2}(n)\,,\varphi_{C2}(n)\}$ in the polar coordinate system are determined as

$$\{R_{C2}(n) = R_{\max}\,,\varphi_{C2}(n) = \\ = [\varphi_{\max}(R_{\max};n)+\varphi_{\min}(R_{\max};n)]/2\}. \qquad (6)$$

The position of the second peak $R'_{\max}$, by analogy with Eq. (5), should be sought for in the range of distances $R_k \in [R_{\max}+b_0/2\,,(K-1)\Delta R]$, where $b_0=(\pi/4)B_A$ is the theoretical initial distance between the vortex axes, and B_A is the aircraft wingspan [10], and then

$$\{R_{C1}(n) = R'_{\max}\,,\varphi_{C1}(n) = \\ = [\varphi_{\max}(R'_{\max};n)+\varphi_{\min}(R'_{\max};n)]/2\}. \qquad (7)$$

If $\varphi_{\max}(R_{\max};n) < \varphi_{\min}(R_{\max};n)$ and $R_{\max} = R_{C1}(n)$, then $R'_{\max} = R_{C2}(n)$ is sought for in the range $R_k \in [R_0\,,R_{\max}-b_0/2\,]$. In this case, $\varphi_{C1}(n) = [\varphi_{\max}(R_{\max};\ n)+\varphi_{\min}(R_{\max};n)]/2$ and $\varphi_{C2}(n) = [\varphi_{\max}(R'_{\max};n)+\varphi_{\min}(R'_{\max};n)+\varphi_{\min}(R'_{\max};n)]/2$. In Figure 2(c), the black and grey circles show the coordinates of the axes of, respectively, the right and left aircraft vortices as obtained from the data of Figure 2(a).

The elevation angle $\varphi_m^{(n)}$ is a function of time t measured from the instant, when the aircraft crosses the scanning plane $\varphi_m^{(n)} \equiv \varphi(t)$. From the equality $\varphi(t_n^{(i)}) = \varphi_{Ci}(n)$, we can determine the time, when the sensing beam crosses the axis of the i-th aircraft vortex for the n-th scan. The value of $t_n^{(i)}$ is the age of the i-th vortex. Using the obtained arrays of $R_{Ci}(n)$ and $\varphi_{Ci}(n)$, we can represent the coordinates of the vortex axes as functions of time in the Cartesian coordinate system

$$\mathbf{r}_{Ci}(t_n^{(i)}) = \{Z_{Ci}(t_n^{(i)}) = \\ = R_{Ci}(n)\sin[\varphi_{Ci}(n)]\,, Y_{Ci}(t_n^{(i)}) = R_{Ci}(n)\cos[\varphi_{Ci}(n)]\}\,.(8)$$

The next step in processing of lidar data is obtaining estimates of the circulation of the right $\hat{\Gamma}_1(t_n^{(1)})$ and left $\hat{\Gamma}_2(t_n^{(2)})$ vortices. We propose to determine the vortex circulation through minimization of the functionals $\rho(\Gamma_1)$ and $\rho(\Gamma_2)$, that is, using the procedure

$$\min\{\rho(\Gamma_i)\} = \rho(\hat{\Gamma}_i), \qquad (9)$$

where

$$\rho(\Gamma_i) = \sum_m [\hat{V}_r(R_{Ci},\varphi_m;n) - \bar{V}_r(R_{Ci},\varphi_m\,|\,\Gamma_1,\Gamma_2)]^2, \ (10)$$

$\bar{V}_r(R_{Ci},\varphi_m\,|\,\Gamma_1,\Gamma_2)$ is the radial velocity calculated theoretically at arbitrary values of circulation of the right Γ_1 and left Γ_2 vortices. Since the estimates of the radial velocity are obtained from the position of maximum of the estimates of the Doppler spectrum, the analytical formula for the model function $\bar{V}_r(R_{Ci},\varphi_m\,|\,\Gamma_1,\Gamma_2)$ cannot be derived. This problem can be solved in the following way.

Based on the theory developed in [4] and taking into account that the temporal profile of the probing pulse power is close to the Gaussian one, we have derived the equation for the correlation function of the complex lidar signal $C(lT_s,R_{Ci},\varphi_m)$. Based on calculation of $C(lT_s,R_{Ci},\varphi_m)$ at different values of Γ_1 and Γ_2, we obtain the Doppler spectra $S_D(V_{r'},R_{Ci},\varphi_m)$ and then, using the procedure $\max\{S_D(V_{r'},R_{Ci},\varphi_m)\} = S_D(\bar{V}_r,R_{Ci},\varphi_m)$, we find $\bar{V}_r(R_{Ci},\varphi_m\,|\,\Gamma_1,\Gamma_2)$.

The estimates $\hat{\Gamma}_1$ and $\hat{\Gamma}_2$ are obtained by the iterative procedure.

3. RESULTS

To test the described method for estimation of the wake vortex parameters from data measured by the Stream Line lidar, we have conducted an experiment at the airfield of the Tomsk Airport on 9 July 2014. The lidar was installed 315 m far from the runway. For the time of continuous measurement by this lidar from 05:00 to 07:10 LT, four B737-800 aircraft crossed the plane of scanning by the probing beam at a height of about 37 m.

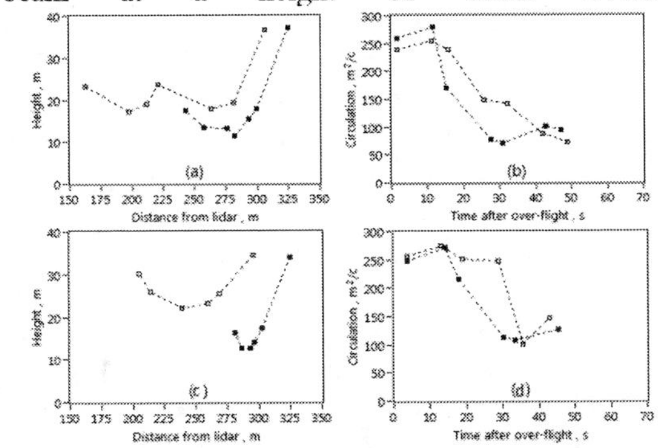

Fig.3. Trajectories of axes (a, c) and evolution of circulation (b, d) of the left (open squares) and right (closed squares) vortices generated by B737 aircraft over the airfield of the Tomsk Airport on 9 July 2014 at 05:09:21 (a, b) and 07:02:20 (c, d).

Figure 3 shows the final results of processing of the data measured by the Stream Line lidar. According to the data of Fig. 3, the estimated initial distance between the axes of the right and left vortices $\hat{b}_0 = |\mathbf{r}_{C1}(t_1^{(1)}) - \mathbf{r}_{C2}(t_1^{(2)})|$ is, on average, 25 m, which is close to the theoretical value $b_0 = 27$ m for this type of aircraft, while the estimate of the initial vortex circulation $\hat{\Gamma}(t_1^{(i)})$ is, on average, 250 m^2/s, which is close to the theoretical value of circulation obtained from calculation by the equation for the initial circulation $\Gamma_0 = M_A g/(\rho_a b_0 V_A)$ [10], where M_A is aircraft mass, and V_A is the true velocity of aircraft, g is gravitation acceleration, and ρ_a is air density.

The trajectories and circulations of vortices shown in Figure 3 are characteristic in the case of vortex generation at low altitudes [4]. The direction and speed of vortex transport by the cross wind correspond to the weather conditions of the experiment. The lifetime of the wake generated by aircraft of this type is about 1 min.

4. CONCLUDIONS

Thus, in this work we have developed the strategy of measurement by the Stream Line lidar and proposed the method for processing of data measured by this lidar in order to estimate the wake vortex parameters. The method has been tested in numerical and field experiments. The experimental results depicted in Figure 3 indicate the principal possibility of obtaining the information about the presence of aircraft wake vortices in the vicinity of a runway and the vortex strength (circulation) from data measured by the Stream Line lidar.

ACKNOWLEDGEMENT

This study was supported by the Russian Scientific Foundation for Maintenance and Development (Project 14-17-00386).

REFERENCES

[1] S.W. Henderson, P.J. Suni, C.P. Hale, S.M. Hannon, J.R. Magee, D.L. Bruns, and E.H. Yuen, IEEE Trans. Geosci. Remote Sens. **31**, 4 (1993).

[2] S.M. Hannon and J.A. Thomson, *J. Mod. Opt.* **41**, 2175 (1994).

[3] F. Köpp, S. Rahm, and I. Smalikho, *J. Atmos. Ocean. Technol.* **21**, 194 (2004).

[4] V.A. Banakh and I.N. Smalikho I.N. *Coherent Doppler wind lidars in a turbulent atmosphere.* (Artech, 2013) p. 248.

[5] S. Rahm and I. Smalikho, *J. Aircraft.* **45**, 1148 (2008).

[6] I. Smalikho, F. Köpp, and S. Rahm, *J. Atmos. Ocean. Technol.* **22**, 1733 (2005).

[7] G. Pierson, F. Davies, and C. Collier, *J. Atmos. Ocean. Technol.* **22**, 240 (2009).

[8] V.A. Banakh, I.N. Smalikho, A.V. Falits, B.D. Belan, M.Yu. Arshinov, and P.N. Antokhin, Atmos. Ocean. Opt. **27**, 911 (2014).

[9] V.A. Banakh, A. Brewer, Y.L. Pichugina, and I.N. Smalikho, *Atmos. Ocean. Opt.* **23**, 333 (2010).

[10] T. Gerz, F. Holzäpfel, and D. Darracq, *Progress in Aerospace Sciences*, **38**, 181 (2002).

FLIGHT TESTS OF THE DELICAT AIRBORNE LIDAR SYSTEM FOR REMOTE CLEAR AIR TURBULENCE DETECTION

Patrick Vrancken[1*], Martin Wirth[1], Gerhard Ehret[1], Benjamin Witschas[1], Henk Veerman[2], Robert Tump[2], Hervé Barny[3], Philippe Rondeau[3], Agnès Dolfi-Bouteyre[4], Laurent Lombard[4]

[1] *Deutsches Zentrum für Luft- und Raumfahrt (DLR) – German Aerospace Center, Institute of Atmospheric Physics, Oberpfaffenhofen, 82234 Weßling, Germany, *Email: patrick.vrancken@dlr.de*

[2] *National Aerospace Laboratory (NLR), Anthony Fokkerweg 2, 1059CM Amsterdam, The Netherlands*

[3] *THALES Avionics, 25 Rue Jules Védrines, 26027 Valence Cedex, France*

[4] *ONERA - The French Aerospace Lab, 91123 Palaiseau, France*

ABSTRACT

An important aeronautics application of lidar is the airborne remote detection of Clear Air Turbulence which cannot be performed with onboard radar. We report on a DLR-developed lidar system for the remote detection of such turbulent areas in the flight path of an aircraft. The lidar, consisting of a high-power UV laser transmitter and a direct detection system, was installed on a Dutch research aircraft. Flight tests executed in 2013 demonstrated the performance of the lidar system to detect local subtle variations in the molecular backscatter coefficient indicating the turbulence some 10 to 15 km ahead.

1. INTRODUCTION

In commercial aviation, Clear Air Turbulence (CAT) encounter is a leading cause for injuries to cabin crew and passengers causing M$ per year to airlines. Occurring at cruise altitudes in clear air, CAT may not be detected by radar due to the absence of hydrometeors. Active optical sensing with lidar is the only possible means of remotely detecting CAT. Following detection, a warning to cabin crew and passengers may be issued or a mitigation action initiated (deceleration or evasion maneuver, e.g.). Generally, lidar detection is possible with different methods, by either measuring wind speed fluctuations with coherent Doppler wind lidar or air density fluctuations, the latter method described herein. Despite its apparent suitability, coherent Doppler lidar has the disadvantage of relying on backscatter from aerosols that are, at cruise flight altitudes, not sufficiently present for a generally reliable and long-range detection of CAT. Second, it measures wind speed in the direction of line-of-sight whereas the hazardous turbulent winds occur vertically, thus perpendicular to the flight path. Due to anisotropy of the turbulence, these horizontal and vertical wind speed fluctuations are not the same, and thus the turbulence strength may not be determined reliably.

Another method to measure CAT is proposed in [1], and a technology demonstration activity, funded by the European Commission aeronautics research programme, is described in [2]. The detection principle relies on tiny fluctuations of air temperature and thus density when air parcels undergo the up- and downwelling motion in the buoyancy range of the turbulence. Under CAT conditions, a simple relationship between vertical wind speed, air density ρ_{air} and hence backscatter coefficient $\beta_{\pi,mol}$ fluctuation may be derived [2]:

$$\sigma_w = \frac{g}{N}\cdot\left(\frac{\partial\rho_{air}}{\rho_{air}}\right) = \frac{g}{N}\cdot\left(\frac{\partial\beta_{\pi,mol}}{\beta_{\pi,mol}}\right)$$

where σ_w is the vertical wind speed turbulent fluctuation, g the gravity acceleration, and N the stratification stability that may be derived sufficiently well from numerical weather forecast (or determined in-situ). For a practical lidar application, a strong molecular backscatter is thus crucial and may be achieved with a UV wavelength. In the described approach within the European DELICAT project [2], a high power UV laser beam is aimed into the flight direction and the backscattered radiation is sufficiently averaged in order to attain the necessary level of resolution, which is on the percent level for the "moderate" turbulence. In the following, this lidar system is described, an overview of the flight campaign is given and the measurement results are discussed.

DOI: 10.1051/epjconf/201611914003

2. LIDAR SYSTEM

The lidar system for the detection of the small turbulence-induced air density fluctuations was designed by DLR and is comprised of the following elements (a synopsis is given in Fig. 1): The transmitter is based on the high-power Nd:YAG system of the WALES lidar [3].

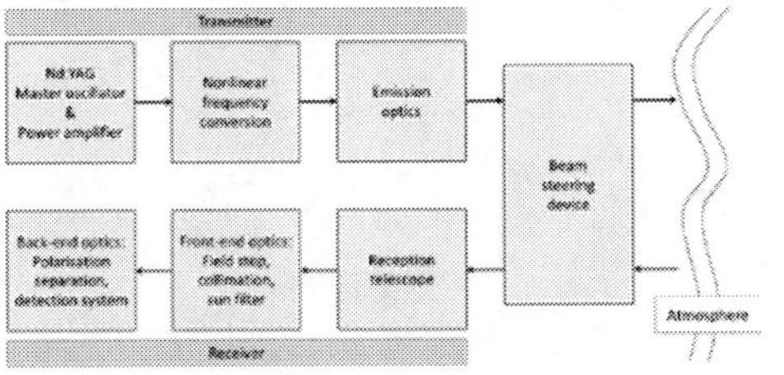

Figure 1: Synopsis of the DELICAT lidar system

A single-mode MOPA monolithic ring laser is locked to an Iodine absorption cell and subsequently amplified in three linear power amplifiers. Its linearly polarized short Q-switched infrared pulses are frequency converted to green and subsequently to UV by sum-frequency generation. Technical details are given in Table 1.

Figure 2: Transmitter: ND:YAG laser and THG.

The transmit laser beam is then guided onto a beam steering system, developed by THALES. This system ensures that the laser beam is always pointed into flight direction, no matter what the attitude of the aircraft (due to angle of attack for a chosen flight speed, mass, etc.). The system, comprised of two moving and one fixed mirror (the latter on the outside of the aircraft fuselage) tracks the aircraft movements by reading the ARINC data from the aircraft's inertial platform (IRS) and a measurement boom. The laser beam is thus guided, over the mirrors and some windows to the outside of the aircraft and directed into the flight path direction.

The receiver system (monostatic architecture regarding the transmit beam), consists of a 140 mm Newtonian telescope, spatial and frequency filters, a polarisation separation assembly and two PMT detection devices.

Figure 3: Receiver system.

The polarisation separation assembly allows the measurement of the depolarisation due to aerosols (the molecular depolarisation amounts to approximately 0.5% under the respective conditions). This depolarisation would give an indication of spurious aerosol backscatter, admittedly not systematically, but at least for Cirrus ice crystals and the like. A spectral separation with a high resolution spectral filter (such as a Fabry-Perot interferometer, e.g.) is planned for an upcoming upgraded version of the lidar. The present receiving system further allows, due to its modular layout, the versatile testing of various detection schemes for other aeronautics applications, such as volcanic ash or hazardous crystalline icing conditions detection.

The PMT detectors' read-out is digitized with a selectable sample rate, 30 MHz (i.e. a range resolution of 5 m) has been chosen for the present tests. The detectors, as well as the data acquisition system (also controlling the transmitter) are likewise issued from the WALES lidar [3].

The lidar system elements, together with the beam steering system are integrated into a stiff rack which is in turn mounted onto the test aircraft's seat rails. Apart from the lidar, a two-circuit cooling system has been developed, evacuating the laser thermal discharge to the outside air and thus preventing the heating-up of the small cabin of the research aircraft.

DOI: 10.1051/epjconf/201611914003

Table 1: Lidar performance parameters

Parameter	Value
Repetition rate	100 Hz
Laser pulse energy (IR)	400 mJ
Laser pulse energy (UV)	85 mJ
Line width (IR)	54 Mhz
Frequency stability (IR)	≤ 1 MHz
Pulse length	8 ns
Beam quality M^2 (UV)	4.3
Beam divergence	150 µrad
Telescope diameter	140 mm
Field of view	530 µrad
Sun filter bandwidth (FWHM)	0.5 nm
System depolarisation	0.9 %
Detector NEP	0.25 fW/√Hz
Analog/digital converter	14 bit
Sampling rate	30 MHz (5 m)
Beam steering dynamic	1 Hz (sine)
Pointing accuracy	0.1°

3. TEST AIRCRAFT AND FLIGHT CAMPAIGN

After ground test at Oberpfaffenhofen, Germany, the lidar system was installed on the NLR's Cessna Citation 2 aircraft. It offers a maximum ceiling of 43,000 ft at maximum cruise speed of 200 m/s. The aircraft was fitted with an aerodynamic fairing covering the external mirror that bends the laser beam to the front (see Fig. 4). Further, it was equipped with an external liquid cooling system (jointly developed by NLR and DLR), an inertial reference system (IRS), a noseboom measuring side-slip angle, and a fast temperature sensor. As a truth for lidar-detected turbulent zones ahead could thus be used the vertical acceleration (from IRS) – the response to vertical wind speed fluctuation σ_w – and the temperature fluctuations, both when the aircraft is travelling through the probed area some seconds after lidar detection.

All modifications on the aircraft, on the fuselage, in the cabin, related to procedures and to laser safety within and outside the aircraft have been subject to airworthiness certification. An STC (Supplemental Type Certificate) was issued to the aircraft PH-LAB with the DELICAT modifications in July, 2013 by Dutch CAA.

The flight test campaign [4] was performed in July and August 2013 from Amsterdam Schiphol airport. The flight test crew of NLR, together with lidar operators of DLR have been assisted by Warsaw University and Météo France, procuring aeronautical weather forecast and different CAT forecast metrics. Another task was to identify areas "devoid" of aerosols and Cirrus (i.e. aerosol mass concentration below a certain threshold) by exploiting the MOCAGE aerosol forecast model.

Figure 4: Left: View on the outside starboard fuselage with the fairing for laser beam transmission and reception. Right: Complete lidar and beam steering system with operator HMI.

In total, 11 flights with over 30 flight hours have been performed whereof 15 hours of lidar operation (≈175 Gb of data). Over the whole campaign, the lidar system has been working reliably without failures, constantly delivering high output power and strong, low-noise backscatter signal beyond ranges of 15 km.

4. FLIGHT TEST RESULTS

As a first step in data analysis, all data (i.e. aircraft, beam steering and lidar data) have been combined in order to identify the sections usable for scientific exploitation. For this purpose, ONERA developed a series of filters [4] to be applied first on the aircraft data (for maneuver exclusion, e.g.) and then on the lidar data (for aerosol contamination exclusion). The validated data were distributed to the concerned project partners for further evaluation. Generally, the CAT yield was rather low (as confirmed by aircraft in-situ measurements) exhibiting only "light" intensity. This encountered intensity was below the designed detection threshold – nevertheless, with advanced processing techniques, respective correlations between in-situ measured turbulence (by vertical acceleration, e.g.) and lidar measurements could be shown. Going beyond the scope of this communication, we will here focus on the lidar performance itself.

DOI: 10.1051/epjconf/201611914003

This performance evaluation is carried out by analyzing "zero" measurements, i.e. backscatter measurements in clear, smooth (non-turbulent), and aerosol-devoid air. We should recall here that the key for resolving the turbulence-induced air density fluctuations is a substantial averaging process. Therefore, the noise characteristics of the receiving and detection chain have to be evaluated. Only random "white" noise leads to a reduction of the dispersion of the mean of a measurement with the square root of the number of averaged data points.

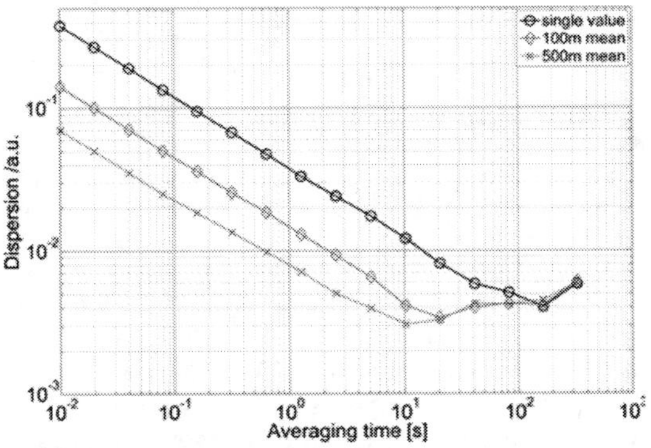

Figure 5: Stability of the backscatter measurement at 10 km distance for different integration times (at 100 Hz pulse rate) and spatial averages (at FL250).

From the campaign data, different "zero" domains have been selected (temporally close to light CAT encounters) and analyzed by employing the "time variance", a modified Allan variance. This tool basically acts as a robust and conservative estimator of the dispersion of an average – over the whole set of data. Fig. 5 shows such a plot for a 10 km distance: The sub-percent level, needed for moderate turbulence detection, is attained by averaging 1400 single data points (5 m resolution), i.e. 14 s long, or by 64 merged 500 m - bins, integrating less than a second. Since CAT is effective on aircraft on rather medium (tens to hundreds meters) scales, short integration times are possible. Further, longer ranges than 10 or 15 km may be reached since the detection system exhibits white noise behavior up to some tens of seconds. This analysis shows the suitability of such a lidar system to effectively retrieve the density fluctuations occurring in clear air turbulence.

5. Conclusion

Within the European DELICAT project, a lidar system for the airborne remote detection of CAT was designed, built, tested and evaluated by DLR and partners. The system showed superior signal to noise and stability characteristics in order to remotely resolve the air density fluctuations that are caused by CAT. Moderate turbulence at cruise altitudes may thus be detected at least 10 to 15 km ahead with the present system. This activity validates the application of lidar technology for improving aeronautics safety by the mitigation of CAT encounter.

Acknowledgements

The authors would like to thank the numerous colleagues from NLR, THALES, Warsaw University, Méteo France and DLR that contributed to the preparation and execution of the DELICAT flight tests. The research leading to these results has received funding from the European Union Seventh Framework Programme FP7/2007-2013 under grant agreement n° 233801.

References

[1] Feneyrou, P., J. Lehureau, and H. Barny, "Performance evaluation for long-range turbulence detection using ultraviolaet lidar," *Applied Optics*, **48**, no. 19, pp. 3750-3759, 2009.

[2] Vrancken, P., M. Wirth, D. Rempel, G. Ehret, A. Dolfi-Bouteyre, L. Lombard, L., T. Gaudo, D. Rees, H. Barny, Ph. Rondeau, and F. Dalaudier, "Clear air turbulence detection and characterisation in the DELICAT airborne lidar project." In *25th International Laser Radar Conference (ILRC),* 2010.

[3] Wirth, M., A. Fix, P. Mahnke, H. Schwarzer, F. Schrandt, and G. Ehret, "The airborne multiwavelength water vapor differential absorption lidar WALES: system design and performance," *Applied Physics B: Lasers and Optics*, **96**, no. 1, pp. 201–213, 2009.

[4] Veerman, H.P., P. Vrancken, and L. Lombard, "Flight Testing DELICAT – A Promise for Medium-Range Clear Air Turbulence Protection", In *European SETP and SFTE Flight Test Symposium*, 2014.

DOI: 10.1051/epjconf/201611914004

IMPROVING MARYLAND'S OFFSHORE WIND ENERGY RESOURCE ESTIMATE USING DOPPLER WIND LIDAR TECHNOLOGY TO ASSESS MICROTMETEOROLOGY CONTROLS

Alexandra St.Pé [1]*, Daniel Wesloh[1], Graham Antoszewski[1], Farrah Daham[1], Navid Goudarzi[2], Scott Rabenhorst[3], Ruben Delgado[4]

[1]*University of Maryland, Baltimore County, Baltimore, MD 21250, USA, *Email: astpe@umbc.edu*

[2]*University of Maryland, College Park, MD 20742, USA*

[3]*Naval Research Laboratory, Washington D.C., 20375, USA*

[4]*Joint Center for Earth Systems Technology, Baltimore, MD, 21250, USA*

ABSTRACT

There is enormous potential to harness the kinetic energy of offshore wind and produce power. However significant uncertainties are introduced in the offshore wind resource assessment process, due in part to limited observational networks and a poor understanding of the marine atmosphere's complexity. Given the cubic relationship between a turbine's power output and wind speed, a relatively small error in the wind speed estimate translates to a significant error in expected power production. The University of Maryland Baltimore County (UMBC) collected in-situ measurements offshore, within Maryland's Wind Energy Area (WEA) from July-August 2013. This research demonstrates the ability of Doppler wind lidar technology to reduce uncertainty in estimating an offshore wind resource, compared to traditional resource assessment techniques, by providing a more accurate representation of the wind profile and associated hub-height wind speed variability. The second objective of this research is to elucidate the impact of offshore micrometeorology controls (stability, wind shear, turbulence) on a turbine's ability to produce power.

Compared to lidar measurements, power law extrapolation estimates and operational National Weather Service models *underestimated* hub-height wind speeds in the WEA. In addition, lidar observations suggest the frequent development of a low-level wind maximum (LLWM), with high turbine-layer wind shear and low turbulence intensity within a turbine's rotor layer (40m-160m). Results elucidate the advantages of using Doppler wind lidar technology to improve offshore wind resource estimates and its ability to monitor under-sampled offshore meteorological controls impact on a potential turbine's ability to produce power.

1. INTRODUCTION

Offshore wind energy promises to be a significant domestic renewable energy source for coastal electricity loads. Maryland is engaging in groundbreaking work to inform the planning and development of a utility-scale offshore wind farm. In 2013, the Offshore Wind Energy Act was passed and outlined Maryland's plan to lead in United States offshore wind energy development, incentivizing the development of 500 MW of offshore wind capacity at least 10 nautical miles (NM) off the coast of Ocean City, MD. Although there is enormous potential to produce offshore wind energy in Maryland, significant uncertainties, and associated high costs, exist when developing offshore wind energy projects.

In part, wind project uncertainties exist due to the natural intermittency of wind. Unlike traditional energy resources, the wind resource, and therefore its power output is highly variable, making it particularly challenging to compete and integrate with traditional power sources, which supply a relatively steady flow of power. The variability wind is driven by local meteorological conditions. Monitoring and predicting the relationship between micrometeorology controls (stability, wind shear, turbulence) and a wind turbine's power curve performance at a distinct location, is critical for reducing project uncertainties, in particular, uncertainties that arise during an offshore wind resource assessment [1].

Unfortunately, due to limited high spatial and temporal offshore met-ocean measurements at a proposed offshore wind farm location, characterizing, with a high degree of certainty, the meteorological conditions that impact a potential wind turbine's performance, thus the wind resource, is challenging. To overcome data limitations, available offshore surface wind data is often extrapolated, using the power law equation (1) to estimate the wind speed at hub-height:

$$\frac{u(z_2)}{u(z_1)} = \left(\frac{z_2}{z_1}\right)^{\alpha} \quad (1)$$

where z_1 and z_2 are the measurement height and the desired height to estimate the wind speed, respectively. The theoretical shear exponent, α, varies by site, and is estimated to be 0.11 for an offshore environments [2]. Although the power law method provides a first estimate, the technique is limited as it assumes a logarithmic or neutral stability wind profile. Similarly, turbine manufacturer estimates of expected power curves, known as the Manufacturers' Power Curve (MPC), also assume neutral stability and only mechanical turbulence [2]. In reality, depending on local meteorological conditions, offshore wind deviates from convective, neutral, and stable profiles. The ramifications of not accounting for 'non-standard' turbine inflow conditions during an initial offshore wind resource estimate is significant, since expected power output is proportional to the cube of wind speed at hub-height. Therefore, a relatively small wind speed error introduced from a non-neutral conditions, translates to significant error in expected power production. Many efforts have been made to quantify the error introduced from the inherent atmospheric assumptions within traditional wind resource assessment techniques [3-6]. The objective of this research is to use the results of a two-month (July 17-August 17, 2013) offshore measurement campaign to demonstrate the advantages of deploying remote sensing technologies to quantify the uncertainty introduced from traditional wind resource assessment techniques and elucidate the impact of micrometeorology on a turbine's ability to produce power. A case study is presented to illustrate the impact of atmospheric stability on a turbine's power output and demonstrate the importance of high spatial and temporal measurements to accurately estimate the potential capacity of an offshore wind farm.

2. METHODOLOGY

The University of Maryland Baltimore County (UMBC) deployed Doppler wind lidar technology and other in-situ instruments within Maryland's offshore Wind Energy Area (WEA) during the State's sponsored geophysical survey (July-August 2013). Elastic and Doppler wind lidars (Leosphere ALS-450 355 nm and Windcube V2 Offshore) collected vertical profiles of aerosols and wind (speed and direction) on board the *Scarlet Isabella* research vessel. The group also launched weather balloons equipped with radiosonde instruments that measured vertical profiles of temperature, moisture, and wind speed and direction. Since the ship was in motion during most of the campaign, the translational motion was accounted for during data post-processing.

Buoy wind speed data were obtained from the closest the station to Maryland's WEA (38.461°N, 74.703°W). The anemometer height of Buoy 44009 data is 5m above the site elevation (sea level) [7]. Surface wind data are converted to wind speed at hub height (100m) using the Power Law extrapolation technique (1).

3. RESULTS

The July-August 2013 offshore Doppler wind Lidar data set was compared to available[1] buoy data to assess the relationship between empirically measured hub-height winds and power law extrapolated[2] estimates. Overall, both datasets suggest greatest wind speeds occurred during late day or overnight (between 06:00 pm - 06:00 am local time). However, compared to lidar measurements, the buoy power law extrapolations significantly *underestimated* hub-height (100 m) wind speeds. Hub-height mean wind speed values are 4.723 m/s and 8.143 m/s from buoy and Doppler wind data, respectively. Similar results were found when comparing offshore lidar measurements to operational Numerical Weather Prediction (NWP) models. Using a suite of NWP models, results indicate an *underestimation* of the offshore wind speed from 1.63-3.15 m/s, depending on model spatial resolution. Large model error in wind direction is also demonstrated, ranging from 26.9 -72.9 degrees. Finally, offshore lidar measurements were compared to the Department of Energy's National Renewable Energy Laboratory (NREL) estimate of the offshore wind resource within Maryland's WEA. The average July-August 2013 lidar hub-height wind speed is 7.4 m/s, while the NREL *annual* estimate is 8.2 m/s. Given the Mid-Atlantic wind resource is climatologically weakest during the summer season [8], results suggest that the *annual* mean wind speeds within Maryland's WEA may be higher than the initial NREL estimate.

Collectively, the results suggest significant differences between empirical hub-height wind measurements and power law/model estimates. The discrepancies are the result of 'non-standard' (neutral) turbine inflow conditions offshore. For example, the mean diurnal

[1]Due to buoy data availability, buoy-Doppler wind Lidar comparison was calculated from July 17-August 17, 2013.

[2] α =0.11 (offshore)

variability of lidar wind speeds demonstrated the frequent development of a low-level wind maximum (LLWM), with associated high turbine-layer wind shear and low turbulence intensity. High wind shear exponents were observed across a turbine's typical rotor layer (40m-160m), and 21% of the time values were negative. A case study event on August 30-31, 2013 was analyzed to investigate possible drivers of measurement versus estimate discrepancies during the campaign. August 30-31 offshore radiosonde data indicated an increase in atmospheric stability led to the development of a late afternoon/nocturnal LLWM near 100m. Further, the onset of the LLWM was marked by a sharp change in wind direction (northeast to southwest). To determine whether offshore conditions were similar to onshore, August 30-31 offshore data sets were compared to the nearest Maryland Department of Environment wind profiler, in Cambridge, MD. Although the wind profiler suggests a nocturnal low-level jet developed on land overnight on August 31, the event superseded an initial increase in offshore stability, wind speed, and change in wind direction within Maryland's WEA. The physical processes driving August 30-31 offshore conditions may be related to a variety of factors, such as coastal upwelling in the WEA or the development of a sea-land breeze circulation. More measurements and investigation is needed to determine the cause of the 'non-standard' turbine inflow conditions observed on August 30-31 as well as throughout the measurement campaign within Maryland's WEA.

4. CONCLUSIONS

Significant uncertainties are introduced when estimating a turbine's ability to produce power, due in part to limited met-ocean data networks and atmospheric assumptions inherent in traditional offshore wind resource assessment techniques. More measurements of site-specific micrometeorology and investigation of its impact on a turbine's power curve performance are needed to reduce offshore wind resource uncertainties. Results from a two month offshore data collection campaign in Maryland's WEA demonstrate significant error between measured hub-height wind speeds and traditional techniques used to estimate these values. The results suggest 'non-standard' inflow conditions occurred within Maryland's WEA, and elucidate the advantages of Doppler wind lidar technology for accurately assessing the offshore wind resource, as well as quantifying micrometeorological controls that impact a turbine's power curve performance.

ACKNOWLEDGEMENT

Funding for this work was provided from the Maryland Energy Administration Clean Energy Program (Contract Number U00P3400325) and the Joint Center for Earth System Technology Fellowship Program. The authors would also like to acknowledge the DOD/Army Research Laboratory for their Partnership in Research Transition (PIRT) Program and NOAA-CREST funding. Finally, a special thank you to Leosphere/Renewable NRG System and CB&I/Coastal Planning and Engineering Environmental and Infrastructure for instrument and data post-processing support.

REFERENCES

[1] Archer, C. L., Colle, B. A., Monache, L., 2014: Meteorology for coastal/offshore wind energy in the United States: recommendations and research needs for the next 10 years. *Bulletin of the American Meteorological Society*.

[2] Emeis, S. 2013: Wind Energy Meteorology: Atmospheric Physics for Wind Power Generation. Springer. ISBN 978-3-642-30523-8, pp. 34.

[3] Wharton and Lundquist, 2012: Atmospheric stability affects wind turbine power collection. *Environmental Research Letters*, **7**, 014005, doi: 10.1088/1748-9326/7/1/01400

[4] Vanderwende and Lundquist 2012, The modification of wind turbine performance by statistically significant distinct atmospheric regimes, *Environmental Research Letters*, **7**,034035, doi:1088/1748-9326/7/3/034035

[5] Barthelmie et al. 2013, Meteorological controls on wind turbine wakes, Proceedings of the IEEE, Vol. **101**, No.4

[6] Honrubia et al. 2012: The influence of turbulence and vertical wind profile in wind turbine power curve, *Progress in Turbulence and Wind Energy IV*, SPPHY, **141**, pp.251-254

[7] Station 44009, available at : http://www.ndbc.noaa.gov/station_page.php?station=44009

[8] Zhang, D. L., Zhang, S., & Weaver, S. J. 2006: Low-level jets over the mid-Atlantic states: Warm-season climatology and a case study. *Journal of applied meteorology and climatology*, **45**(1), 194-209.

EPJ Web of Conferences **119**, 14005 (2016) DOI: 10.1051/epjconf/201611914005
ILRC 27

Lidar Measurements of Wind Flow through a Rotor-swept Area in Tehachapi Pass

Kai Zhou, Ronald Calhoun*

Mechanical & Aerospace Engineering, Arizona State University, Tempe, AZ85287-9809, USA
**Email: Ronald.Calhoun@asu.edu*

1. INTRODUCTION

Retrieval of wind fields from a single-Doppler lidar with small azimuthal sectors is challenging. Both simpler retrieval algorithms such as Velocity Azimuth Display (VAD) or Volume Velocity Processing (VVP), and more sophisticated techniques such as Four Dimensional Variational Analysis (4DVAR) using VAD for mean velocities have limitations on small sectors of data. Narrow datasets may result in non-orthogonality and collinearity of basis functions, which causes numerical instability. Although careful inspection of data can reduce the instability, large conditional number computations which use data over small sectors may make results unacceptable [1].

Since estimates of wind velocities retrieved from single Doppler lidar over limited sectors may be of interest (i.e., winds through a single wind turbine or wakes behind buildings), a method to retrieve wind velocity in these situations would be valuable. In this work, we experiment with an algorithm which is able to provide these velocity estimates under the assumption that the wind direction is relatively unchanging or varies linearly over a larger sector, and quasi-2D ridge can have a tendency to homogenize wind direction. The wind direction of the larger neighborhood is obtained from a sector-VAD-like method while magnitudes are allowed to vary fully locally.

2. EXPERIMENTAL SETUP

Data presented in the paper were collected during a field experiment in Tehachapi Pass in California during the period from February 21st to 23rd, 2015. The objective of the campaign was to capture the behavior of the wind flowing through the rotor-swept area of a wind turbine. The measured wind field will be used as a forcing function in a wind farm simulation using data assimilation.

According to the prevailing wind direction of Tehachapi Pass, the lidar was placed northwest (upwind) of a turbine with distance of 560 m at 35°5'39.00"N, 118°20'27.60"W, with the goal of capturing the wake downwind of the turbine.

The lidar is a 1.6 μm *WindTracer* Doppler with a pulse repetition frequency of 750 Hz. It was configured to average over 100 pulses resulting in a product rate of 7.5 Hz. The range-gate length for the lidar was set at 100 m.

A "stack" of Plan Position Indicator (PPI) scans was executed at elevation angles ranging from 5.5 deg to 10.6 deg with step of 0.255 deg. In each PPI scan, the lidar scanned with different rates in the three regions (shown in Fig. 1) with scanning rates of 5 deg/sec in region I (119.9 deg to 130.4 deg, measured clockwise from north) and region III (136.4 deg to 146.4 deg), and 0.5 deg/sec in region II (130.4 deg to 136.4 deg). Region II corresponded to the rotor-swept area. The three regions cover not only the rotor-swept area but also the space in the vicinity as shown in Fig. 1. With this scanning pattern, region I and III have fewer data points than the region II. Each stack of PPI took approximately 1 minute.

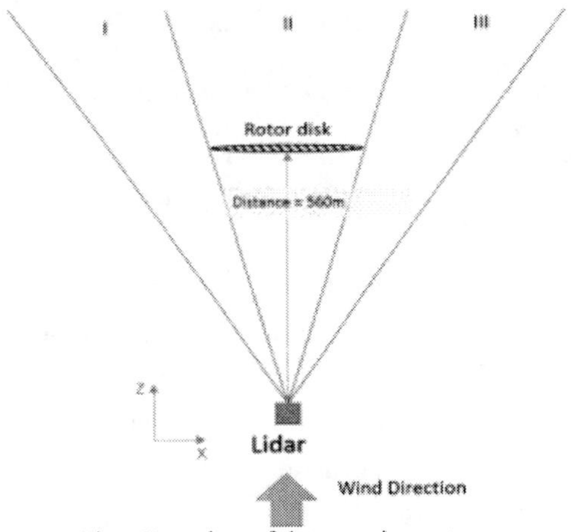

Fig.1 Top view of the scanning pattern

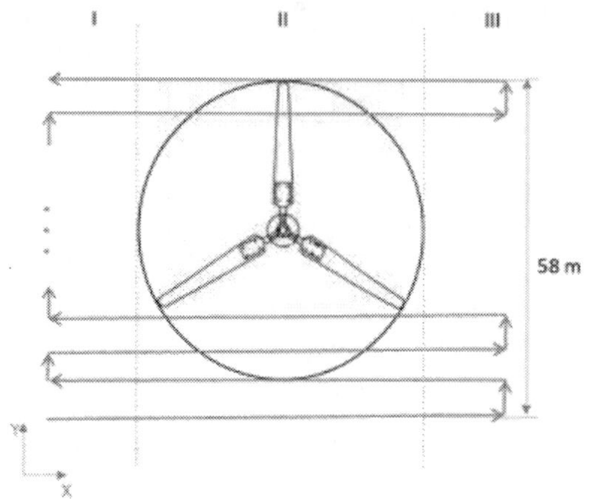

Fig.2 Front view of the scanning pattern

3. ALGORITHM

This algorithm is based on the assumption that the change of the magnitude of wind velocity in region II due to the rotor is more important than that due to the variation of wind direction so that the wind direction in each range gate in region II is treated as uniform or varying linearly over each elevation level. We assume that the wind direction in the rotor-swept area can be reasonably estimated using a sector-VAD fit over the larger sector formed from regions I and III. Note typical wind turbine operation presumes that efficient operation can be obtained with the yaw tracking only running-means of wind direction. Under these assumptions and the relatively rapid scanning of sectors I and III, the wind direction in the rotor-swept area can be obtained from observations in region I and III without large increases in cost, and the data points measured over the wider sector are sufficiently dispersed for the viability of sector-VAD method.

As given in [2], the equation for VAD is given as:

$$V_r = [V_x \cos(\beta) \;\; V_y \cos(\beta)] \cdot [\sin(a) \;\; \cos(a)]^{\mathrm{T}} \quad (1)$$

where V_r is the lidar radial velocity observation; $[\sin(a) \;\; \cos(a)]^{\mathrm{T}}$ is the basis function matrix and $[V_x \cos(\beta) \;\; V_y \cos(\beta)]$ is the parameter vector; a and β are the azimuth and the elevation angles, respectively; V_x and V_y are the velocity in the x and y directions, respectively. Knowing the radial velocity and the look direction, uniform wind direction at each range gate can be obtained from V_x and V_y computed by least square regression to equation (1) with careful outlier rejection suggested by Boccippio [1].

In region II, as the radial velocity is the projection of wind vector on the look direction of the lidar, the wind speed $|\mathbf{V}|$ can be solved from the equation (2)

$$V_r = |\mathbf{V}| \cdot \mathbf{r} \cdot \mathbf{L} \quad (2)$$

Here, V_r is the radial velocity in region II, $\mathbf{r}$ is a unit wind direction vector calculated in region I and III, $\mathbf{L}$ is a unit look direction vector of the lidar which is a function of azimuth and elevation angles. Therefore, the mean velocity vector $\mathbf{V} = |\mathbf{V}| \cdot \mathbf{r}$ in the rotor-swept area is retrieved at each data point.

4. RESULTS

The slow scanning rate gives good spatial resolutions in the rotor-swept area. The horizontal resolution between successive measurements is approximately from 0.7 m at the turbine, to 1 m at five diameters (5D), 290 m downstream the turbine, due to diverging beams. The vertical resolution between successive elevation angles is approximately from 2.5 m at the turbine to 3.8 m at 5D. The radial (along the beam, not the rotor) resolution is 100 m due to the range-gate length. The fast scanning rate in region I and III consumes little time, resulting in high temporal resolution without reducing the quality of the results.

Two examples of radial velocity measured at 19:33:43 UTC (11:33:43 local time) on Feb.21, 2015 at elevation angle 8.05 deg, the angle for hub height and 8.56 deg are shown in Fig. 3a and Fig. 4a. The black line indicates the wind turbine being scanned. White regions in Fig.3 and Fig.4 are missing or contaminated data. However, the overall velocity deficits behind the turbine are clearly observed in the plots. Resulting retrieved wind speeds from the two datasets are shown in Fig.3b and Fig.4b, respectively. The retrieved wind velocities apparently recover the velocity deficit in the wakes.

5. CONCLUSIONS AND FUTURE WORK

A new algorithm of retrieving wind velocity from small sectors of lidar observations is proposed. It is demonstrated by retrieving the wind flow through a rotor-swept area of a wind turbine located in Tehachapi Pass. Preliminary results show the retrieval method appears to recover the wind field at least for the current cases.

The given assumptions clearly have limitations, but may have usefulness for certain applied problems such as wind flowing through a rotor swept area of a wind turbine. Future work will be to test the level of error involved in these assumptions as a function of conditions. Verification of the algorithm with more experimental observations under various wind conditions, and with numerical simulation of wind field with a single wind turbine is another possibility for further study.

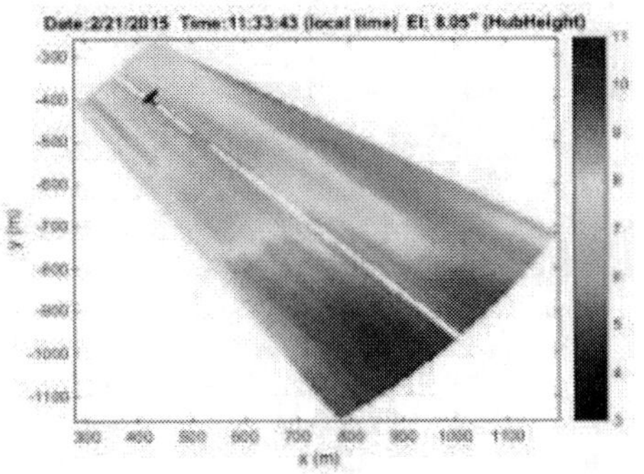

Fig.3a Radial velocity from the PPI scan at elevation angle of 8.05 deg

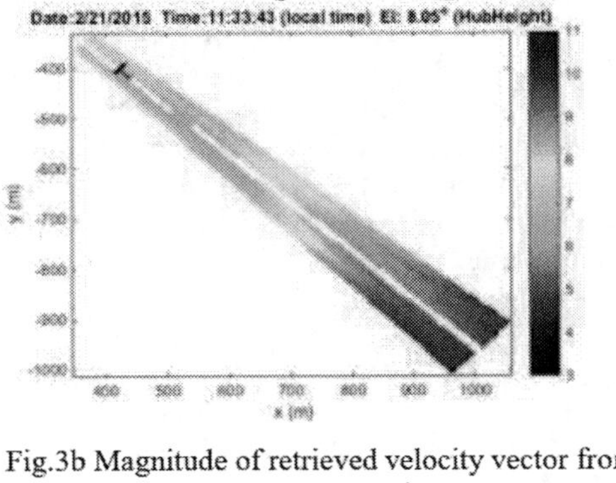

Fig.3b Magnitude of retrieved velocity vector from the PPI scan at 8.05 deg

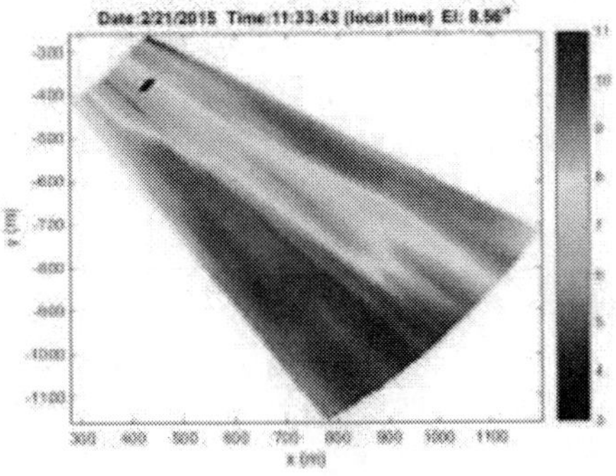

Fig.4a Same as Fig. 3a, but at 8.65 deg

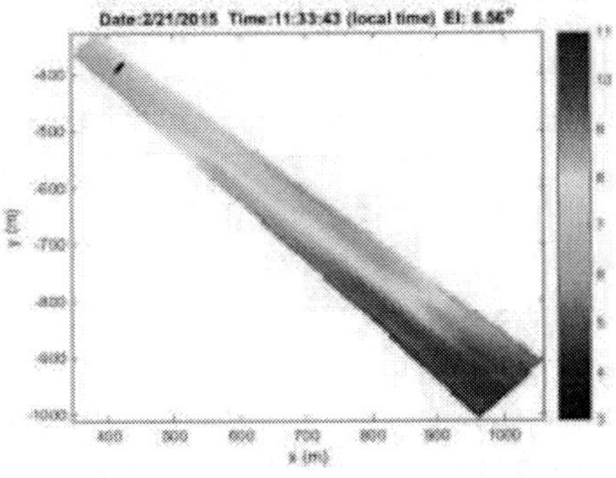

Fig.4b Same as Fig. 3b, but at 8.65 deg

ACKNOWLEDGEMENT

Funding from Electric Power Research Institute (EPRI, program manager Dr. Aidan Tuohy) and NSF (program manager Dr. Gregory Rorrer) are gratefully acknowledged (grants 00-10001389 and 1335868, respectively). We would especially like to thank *WindLand* for site permissions and help with facility setup. Valuable guidance was provided by James Blatchford of CAISO.

REFERENCES

[1] Boccippio, D. J. 1995: A diagnostic analysis of the VVP single-Doppler retrieval technique, *Journal of Atmospheric and Oceanic technology*, **12(2),** 230-248.

[2] Browning, K. A., & Wexler, R. 1968: The determination of kinematic properties of a wind field using Doppler radar. *Journal of Applied Meteorology*, **7(1),** 105-113.

EPJ Web of Conferences **119**, 14006 (2016) DOI: 10.1051/epjconf/201611914006
ILRC 27

AUGMENTED REALITY BASED DOPPLER LIDAR DATA VISUALIZATION: PROMISES AND CHALLENGES

Cherukuru, N. W., [1]* and Calhoun, R.[1]

[1]*Environmental Remote Sensing Group, Arizona State University, AZ 85281, USA, *Email: ncheruku@asu.edu*

ABSTRACT

Augmented reality (AR) is a technology in which the enables the user to view virtual content as if it existed in real world. We are exploring the possibility of using this technology to view radial velocities or processed wind vectors from a Doppler wind lidar, thus giving the user an ability to see the wind in a literal sense. This approach could find possible applications in aviation safety, atmospheric data visualization as well as in weather education and public outreach.

As a proof of concept, we used the lidar data from a recent field campaign and developed a smartphone application to view the lidar scan in augmented reality. In this paper, we give a brief methodology of this feasibility study, present the challenges and promises of using AR technology in conjunction with Doppler wind lidars.

1. INTRODUCTION

Augmented reality (AR) is a technology in which 3-D virtual content is integrated with visual content and presented to the user. The virtual content could be anything that the user cannot directly perceive with his own senses [1]. Based on the technology used to present the virtual content, AR technologies have been classified into two categories. In the first, the virtual content is presented as an overlay on the live camera feed, and displayed on a screen. The second category uses see through Head mounted displays (HMD) in which the virtual content is placed in front of the user's eyes with the help of optical combiners [1]. In the current study, we explore the potential applications of AR technology in conjunction with Doppler lidar which could enable the user to view the surrounding wind fields in an intuitive manner.

There have been a number of studies discussing the capabilities of Doppler lidars in detection of wind shear events, wake vortices, microbursts and other hazardous situations to aircraft and rotorcrafts to name a few [2][3][4]. Previous flight simulation studies on helicopter pilots have shown that visualization of airflow improved the helicopter pilot's landing performance under stressful operational conditions [5].

To test the feasibility of AR technology with Doppler lidars we first implemented this method with a smartphone. Section 2 gives a brief description of the methodology. The results are discussed in section 3 along with the challenges identified in the process. Finally, we conclude with the description of the advantages of using this method on alternate platforms.

2. METHODOLOGY

There are four important components in the AR system that we implemented for viewing Lidar data in this test study: a) Location, attitude (roll, pitch and yaw) and heading sensors, b) a processor c) camera and d) display.

Almost all smartphones come equipped with MEMS gyroscope (Micro-Electromechanical systems) to obtain the attitude data, solid-state compass for heading information and a GPS receiver.

First, the location and attitude sensors determine the GPS location as well as the orientation of the device being used to view the lidar data. The processor then acquires the wind data from the lidar and renders the scan as a Layer object. In future we could use standard graphics libraries like OpenGL to construct the lidar scan data for a more efficient implementation. The image of the lidar scan is then scaled and proper perspective transform is applied based on the orientation of the viewing device. Next, the image of the scan is translated based on the heading information obtained from the compass. The image transformation parameters depend on the focal length of the camera lens, field of view of the camera as well as the resolution of the final

Figure-1. Screenshot of the AR application. The red double headed arrow at the base of the lidar was added to the screenshot after it was captured to show the effect of sensor drift.

display. At this step, the rendered and transformed wind data from the lidar is displayed as an overlay on the live camera feed on the device's screen. The virtual image of the lidar scan is transformed continuously based on the sensor data giving the impression that the scan is fixed at a location in space. This whole system was implemented in Objective C.

The wind data was obtained from a 1.6μm LMCT windtracer lidar. The lidar was virtually placed on a nearby hill that was about 11 km away from the observation point. The difference in elevations between the location of the virtual lidar and the observation point was 390 m.

3. RESULTS

A screen shot of the smartphone was taken while using this application (see figure 1). The RHI (range-height indicator) scan shows the radial velocity at each range gate. 'R', 'P', 'Y' correspond to the roll, pitch and yaw of the device while the screenshot was captured. The colors represent the radial velocity in m/s. The black circle with a green line and a green dot shows the orientation of the viewer with respect to the lidar scan, i.e., currently, the viewer is looking at the scan from the south and the RHI scan is oriented, NE to SW.

3.1 Challenges:

The major challenge was with the accuracy of the sensors used. Although a smartphone can use different methods to obtain the GPS location that may affect the accuracy of the GPS significantly, our observed accuracy seem to conform to the specified ±10m range. However, the accuracy can be improved with a commercially available external GPS receiver. The second challenge is with the drift in MEMS gyroscope. All inertial sensors have the issue with drift causing the lidar scan to drift while using the application. Figure 1 also shows the effects of drift and location error of the GPS (red double headed arrow). There have been a few studies to reduce this and correct for the drift programmatically, which could be used in our application [6]. The solid-state compass was sensitive to the proximity of metallic surfaces. While not a major issue, it required calibration from time to time. Although latency measurement between the completion of scan and rendering of the image was not possible with the current setup, we observed that the response of the image to user movements was almost instantaneous. Thus we expect the latency to come from the data transfer between the devices, which can be reduced significantly if the visualization is done on the same computer that is analyzing the lidar data.

4. Promise on alternate platforms:

4.1 Video based approach:

Smartphone will be the most accessible devices for all users but the issue with latency discussed in the challenges section needs to be studied with lidar data collected in real time with data transfer between the lidar and the smartphone

Oculus rift with stereo camera rig could be a feasible option. The main advantage of using oculus rift setup is that it provides an immersive environment and it can also be directly connected with the computer acquiring the data from the lidar. This may help in resolving the latency issue.

4.2 Optical Technology:

A well-known device in this category is Google Glass. Although Google is by far the most portable AR glasses, their biggest disadvantage for our lidar application is lack of full screen capabilities. This shortcoming can be met with the recently announced Microsoft HoloLens, a see through head mounted display by Microsoft.

With regard to military applications of the AR system, this technology can be easily integrated with the new head mounted displays like JEDEYE™ (from Elbit Systems Ltd.) that currently use the AR technology to help pilots see through brownouts, fog and other low vision conditions [7].

ACKNOWLEDGEMENTS

The authors would like to thank ASU Lightworks for their support.

REFERENCES

[1] R. Azuma, August 1995: A survey of augmented reality, *Computer Graphics, SIG-GRAPH Proceedings*, **pp. 1–8**.

[2] C. M. Shun and P. W. Chan, 2008: Applications of an Infrared Doppler Lidar in Detection of Wind Shear. *J. Atmos. Oceanic Technol.*, **25**, 637–655.

[3] Menzies, Robert T., and R. Michael Hardesty, 1989: Coherent Doppler lidar for measurements of wind fields, *Proceedings of the IEEE* 77.3: 449-462.

[4] Köpp, F., Rahm, S., & Smalikho, I., 2004: Characterization of Aircraft Wake Vortices by 2-μ m Pulsed Doppler Lidar. *Journal of Atmospheric and Oceanic Technology,* 2**1(2)**, 194-206.

[5] Aragon, C. R., & Long, K. R., 2005: Airflow hazard visualization for helicopter pilots: Flight simulation study results. *Annual forum proceedings – American Helicopter Society* **(Vol. 61, No.1, p.1)**. AMERICAN HELICOPTER SOCIETY, INC.

[6] Hol, J. D., 2011: Sensor fusion and calibration of inertial sensors, vision, ultra-wideband and GPS. *Linköping studies in science and technology. dissertations*, (1368).

[7] H.-U. Doehler ; Sven Schmerwitz and Thomas Lueken, 2014: *Visual-conformal display format for helicopter guidance* ", *Proc. SPIE* 9087, Degraded Visual Environments: Enhanced, Synthetic, and External Vision Solutions.

EPJ Web of Conferences **119**, 14007 (2016) DOI: 10.1051/epjconf/201611914007
ILRC 27

OBSERVATIONS AND ANALYSIS OF TURBULENT WAKE OF WIND TURBINE BY COHERENT DOPPLER LIDAR

Songhua Wu[1*], Jiaping Yin[1], Rongzhong Li[2], Xitao Wang[2], Bingyi Liu[1], Jintao Liu[1]
1. Ocean Remote Sensing Inst., Ocean University of China, Qingdao 266100, China,
2. Seaglet Environmental Technology Inc. Qingdao 266100, China,
**Email: wush@ouc.edu.cn*

ABSTRACT

Turbulent wake of wind turbine will reduce the power output of wind farm. The access to real turbulent wake of wind turbine blades with different spatial and temporal scales is provided by the pulsed Coherent Doppler Lidar (CDL) which operates by transmitting a laser beam and detecting the radiation backscattered by atmospheric aerosol particles. In this paper, the authors discuss the possibility of using lidar measurements to characterize the complicated wind field, specifically wind velocity deficit by the turbine wake.

1. INTRODUCTION

Concentration of global energy shortage is closely linked with our living and has become a concern to scientists because of the "Greenhouse Effect". Wind energy industry is growing fast as a sustainable energy source. Because of the obstacle of turbine blades, when the turbine extracts power from the wind, a turbulent wake evolves downwind of the turbine. The power output of the whole wind farm operating within the turbulent wake will be decreased comparing to the turbine operating in the free wind. According to the study by Barthelmie et al., with spacing between the turbines of 4-8 D (Diameter of turbine scanning plane), power dissipation due to wind turbine wakes can be 5%–15%. Spacing of turbines beyond 8-10D is unlikely due to the high cost of installing cables [1]. Furthermore, the wake effect can lead increase of turbulence induced stress and reduce the life time of wind turbines and their components in wind farm. Studies on the turbulent wake are significant for optimizing the turbine arrangement and power output in a wind farm [2].

Turbulent wake characteristics of wind turbines have been studied since 1980s. Current industry standard for wind turbine wake measurements are in-situ meteorological masts with wind cup and vane anemometers on it [3]. But it is difficult to build on the sea due to high cost and complexity. Because of the global growth of wind energy industry, more techniques, like ultrasonic anemometer, profilers and SODAR (Sonic Detection and Ranging) are increasingly being used to measure wind wake [4]. The measurement range based on relative air humidity, and the sensitivity is typically lower than lidar leading to much reduced data rates.

Comparing to the techniques above, the CDL has produced high spatial and temporal resolution measurements of wind fields [5]. The research of wind turbine wake detected by a continuous-wave CDL was done by Bingöl et al. [6] and Trujillo et al. [7]. CDL was put into turbine control study by QinetiQ to examine the requirements of a lidar system at Postlow in 2005[8].

Experimental investigations of the wind field in vicinity of wind turbine with the use of CDL were carried on in 2014. This paper describes the measurement process for the pulsed CDL that was performed to obtain wake parameters of the wind turbine wake, and presents some results of the field test under various atmospheric conditions.

2. METHODOLOGY

Lidar is one of the most accurate remote sensing techniques that transmit a laser beam into the atmosphere and the backscattered signal is detected. The1.5-μm all-fiber pulsed CDL with high resolution, which was utilized for the tests at mountain area, takes advantage of the fact that the frequency of the echo signal is shifted compared to the local-oscillator light because of the Doppler effect which occurs from backscattering of aerosols. The Doppler shift in the frequency of the backscattered signal is analyzed to obtain the line-of-sight (LOS) velocity component of the air motion. From the LOS velocities the characteristic of the turbulent wake can be deduced.

The OUC/ORSI CDL system is based on all-fiber laser technology and fast digital-signal-processing technology. Because of the all-fiber structure adoption, the system is stable, reliable and high-integrated. The spread of optical fiber communication caused drop in prices and mass production [9]. The Fast Fourier Transform (FFT) spectral estimates are real time processed with FGPA. The 2-axis beam scanner and detection range of 3000m enable the system to monitor the whole wind farming filed. The diagram of pulsed CDL setup is shown in Fig. 1. The main system specifications are summarized in Table 1.

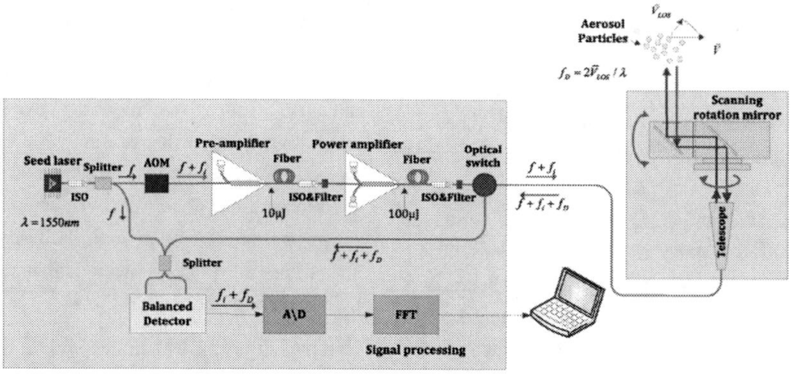

Fig. 1.The pulsed CDL setup. A 1.5-μm semi-conductor single frequency laser and an Erbium doped fiber amplifier is used as the transmitter. A fiber optical circulator and a telescope are used as the optical transceiver.

Table1.Specifications of the CDL.

CDL	Specification
Wavelength	1.5μm
Repetition rate	10kHz
Pulse energy	150μJ
Speed measurement range	±50m/s
Speed accuracy	≤0.3 m/s
Scanner pointing accuracy	0.1°
Scan speed	1-55 °/s
Spatial resolution	30m
Measurement range	30m to 3000m
Power dissipation	<150W
weight	<50kg

3. RESULTS

The turbulent wake of wind turbine blades with different sizes from 50m to 100m had been observed by using the CDL in the detecting experiment. The lidar was located at the center of the area of interest. To measure the LOS component of the wind, two different scanning modes were applied: Plane-Position Indicator (PPI) scanning mode and Range-Height Indicator (RHI) scanning mode. For the PPI scanning, the elevation of the laser beam is constant and the azimuth angle is continuously changed. For the RHI scanning, the azimuth angle is not changed but the elevation is increased continuously.

In the field campaigns, the lidar was located from a sufficient distance from wind turbines so as to observe the upwind and leeward of the turbine. Taking into account the high accuracy of the spatialresolution of CDL, LOS velocity measurements of the wind turbines are shown below: one PPI scan (Fig.2) on January 3rd, 2014 and one RHI scan (Fig.3) on December 26th, 2013. The position of the wind turbines and the wind direction are indicated in the plot. The red and blue colors respectively indicate positive (away from lidar) and negative (towards lidar) movement of the particles along the laser beam. Fig. 2 shows a PPI scan through the rotor blades at mountain area. The width of lidar range gates is 60m.The scanning speed is 1°/s. The elevation is 2.00° for this measurement. Because of the terrain effects, the turbulence caused by hills can be seen in Fig. 2. Low Signal Noise Ratio (SNR) makes the data unavailable near 3 km. Hard target returns (when the laser pulse hits terrain) also decrease data availability. Fig. 3 shows a RHI scanning result. The scanning speed is 1°/s. The elevation varied from 0° to 20° and the azimuth is 150° for this measurement.

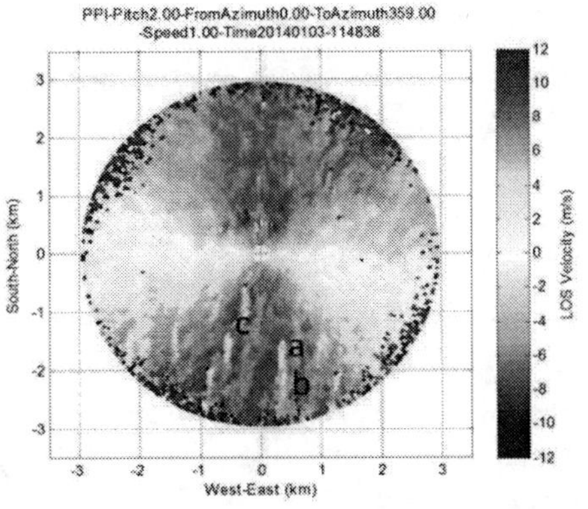

Fig.2. LOS wind field of wind farm by PPI scanning. The background wind is blowing from north to south. Wind direction and position of the different wind turbines are indicated in the figure. Turbine b submerged under the turbulent wake of turbine a.

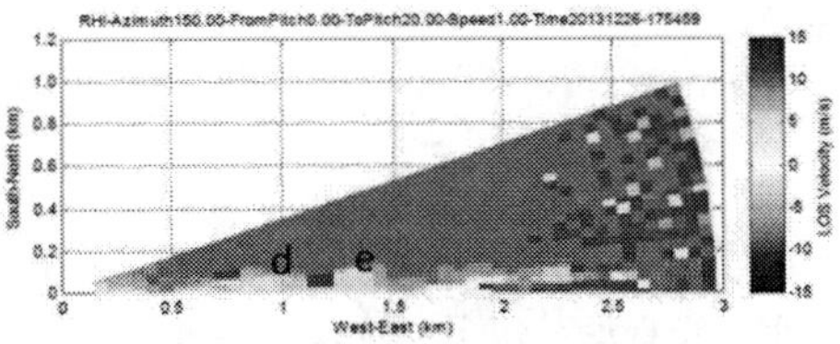

Fig.3. LOS velocity detected by RHI scanning mode through the rotor blades (two wind turbines). The wind is blowing from the northwest parallel to the laser beam. The position of the wind turbines d and e are at a range of 700 m and 1200m respectively.

The rotor diameter is 80m (1D) and hub height is 110m in this case. The width of downwind turbulence is about 2D at the distance of 1D downwind of the rotor. In this observation, turbine wake lengths are from 3D to 10D depends on atmospheric conditions. The velocity deficit behind the wind turbine c, e due to the wake vortices of turbine blades (derived from the PPI scanning mode in Fig.2 and RHI scanning mode in Fig.3) are shown in Fig.4 and Fig.5, respectively. For wind turbine c, two rotor diameter range downwind of the LOS velocity at 50m high is reduced about 54% in comparison to the maximum LOS velocity values. After 9D downwind is reduced about 14%. For wind turbine e, one rotor diameter range downwind of the LOS velocity at height of 80m is reduced about 52%. After 5D downwind (10% reduction of the LOS velocity at 80m high), the wind field recovers slowly. For wind turbine d, because of the limited distance between two turbines, the maximum range of wake of turbine d recorded in this measurement was only 3D. Find the vertical profile of LOS velocity in upwind and 1D range downwind respectively and calculate the velocity deficit. The velocity deficit at height of 40~80m is about 55% as shown in Fig. 6.

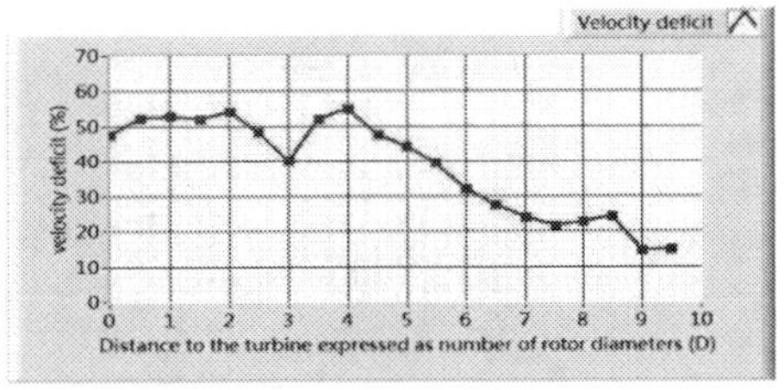

Fig.4.Velocity deficit downwind behind the wind turbine c as the function of the blade diameter D due to the wake vortices of turbine blade (derived from the PPI scanning mode in Fig.2).

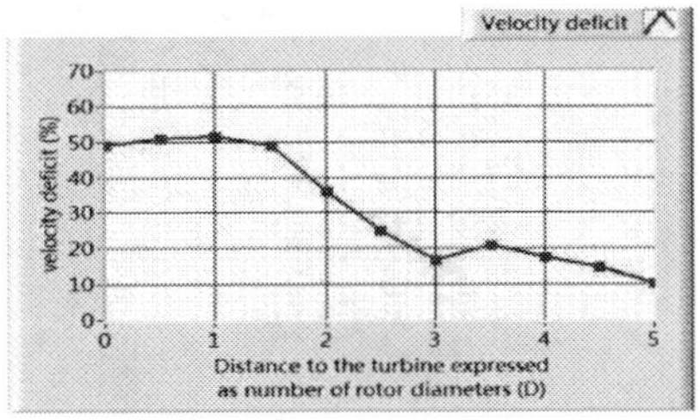

Fig.5.The velocity deficit downwind behind the wind turbine e due to the wake vortices of turbine blade (derived from the RHI scanning mode in Fig.3).

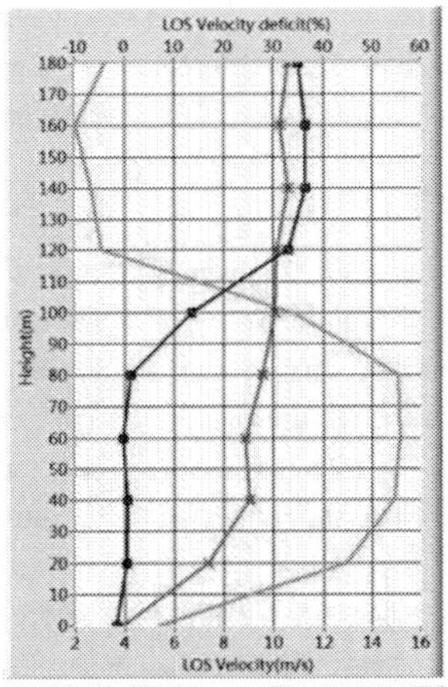

Fig.6.The vertical profile of velocity in upwind (red line) and downward (blue line) of turbine d. Green line is the vertical profile of velocity deficit up to 180m.

The purpose for Yaw control of a turbine is to regulate the aerodynamic power. Here we focus on the turbulence wake dynamic when the yaw control happens. Fig.7 shows the change of turbulence wake before and after the yaw control. CDL can be used to measure velocity and take action to control wind turbine in advance.

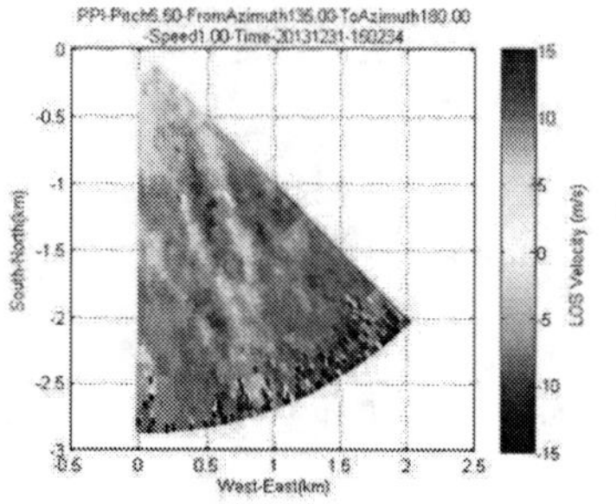

(a)

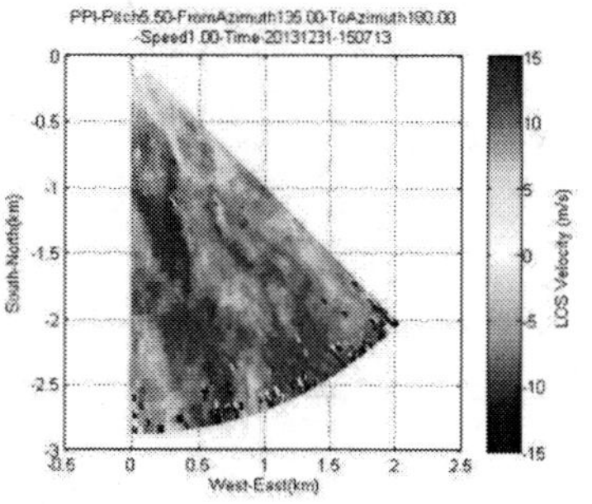

(b)

Fig. 7.The PPI scanning mode of LOS velocity shows that the scale of turbulence wake is much small after the yaw control. The width of lidar range gates is 30m. The scanning speed is 1°/s. The elevation is 3.00°.

4. CONCLUSIONS

During the wind farming experiments in 2014, OUC/ORSI lidar team measured the wind speed and direction of a wind field and the wake parameters of wind turbines through LOS velocity obtained by the 1.5-μm all-fiber pulsed CDL system. The paper shows that the1.5-μm CDL is a practical tool for wind energy research and a powerful technique for the wind energy industry. The subsequent significance is to improve the power output efficiency of the wind turbines and reduce dissipation. The CDL system can detect the interaction of wind turbines and will provide a new basis for wind farm site selection, design and optimization.

ACKNOWLEDGEMENT

This work was supported by NSFC under grant 41375016 and 41471309. The authors wish to thank Dr. Hongdong Zhu for valuable discussion on wind turbine technology. Thanks to the whole lidar group of Ocean University of China.

REFERENCES

[1]Frandsen, S., Barthelmie, R. J., Pryor, S., Rathmann, O., et al., 2006: Analytical modelling of wind speed deficit in large offshore wind farms, *Wind Energy*, **9**, 39-53.

[2]Smalikho, I. N., et al., 2013: Lidar investigation of atmosphere effect on a wind turbine wake, *J. Atmos. Oceanic Technol.*, **30**, 2554–2570.

[3] Amar, F. B., Elamouri, M., Dhifaoui, R., 2008: Energy assessment of the first wind farm section of Sidi Daoud, Tunisia, *Renewable Energy*, **33**, 2311–2321.

[4] Barthelmie, R., Folkerts, L., Larsen, G., et al., 2006: Comparison of wake model simulations with offshore wind turbine wake profiles measured by sodar, *J. Atmos. Oceanic Technol.*, **23(7)**, 888–901.

[5]Högström, U., Asimakopoulos, D. N., Kambezidis, H., et al., 1987: A field study of the wake behind a 2MW wind turbine, *Atmos. Environ.*, **22**, 803–820.

[6]Bingöl, F., Mann, J., Larsen, G. C., 2010: Light detection and ranging measurements of wake dynamics, part I: one-dimensional scanning, *Wind Energy*, **13**, 51–61.

[7] Trujillo, J.-J., Bingöl, F., Larsen, G. C., et al., 2011: Light detection and ranging measurements of wake dynamics. Part II: Two-dimensional scanning, *Wind Energy*, **14**, 61–75.

[8]Jørgensen, H., Mikkelsen, T., Mann, J., Bryce, D., et al., 2004: Site wind field determination using a CW Doppler Lidar- comparison with cup anemometers at Risø, *Proc. EWEA Special Topic Conference: The Science of Making Torque from Wind, The Netherlands*, 261-266.

[9] Kameyama, S., Ando, T., Asaka, K., Hirano, Y., Wadaka, S., 2007: Compact all-fiber pulsed coherent Doppler lidar system for wind sensing, *Appl. Opt.* ,**46(11)**, 1953-1962.

EPJ Web of Conferences **119**, 14008 (2016) DOI: 10.1051/epjconf/201611914008
ILRC 27

AIRCRAFT WAKE VORTEX MEASUREMENT WITH COHERENT DOPPLER LIDAR

Songhua Wu*, Bingyi Liu, Jintao Liu
Ocean Remote Sensing Institute, Ocean University of China, Qingdao 266100, China,
**Email: wush@ouc.edu.cn*

ABSTRACT

Aircraft vortices are generated by the lift-producing surfaces of the aircraft. The variability of near-surface conditions can change the drop rate and cause the cell of the wake vortex to twist and contort unpredictably. The pulsed Coherent Doppler Lidar Detection and Ranging is an indispensable access to real aircraft vortices behavior which transmitting a laser beam and detecting the radiation backscattered by atmospheric aerosol particles. Experiments for Coherent Doppler Lidar measurement of aircraft wake vortices has been successfully carried out at the Beijing Capital International Airport (BCIA). In this paper, the authors discuss the Lidar system, the observation modes carried out in the measurements at BCIA and the characteristics of vortices.

1. INTRODUCTION

Wake vortices are generated by the lift-producing surfaces of the aircraft[1]. The lifting surfaces of all aircraft produce wake vortices to some extent[2,4]. The vortex created by a large aircraft can have a catastrophic effect on a small airplane following closely behind[3].The variability of near-surface conditions (temperature, humidity, wind shear, etc.) can change the drop rate and cause the vortex cell to twist and contort unpredictably[4]. Wake Vortex behaviors very close to the ground have very complex causes that could change abruptly and need for real time monitoring[5].

The CDL(Coherent Doppler Lidar) system is a useful tool that can measure wind, turbulence, aircraft wake vortices, and so forth[5-10].The pulsed Coherent Doppler Lidar Detection and Ranging is an indispensable access to real aircraft vortices behavior which transmitting a laser beam and detecting the radiation backscattered by atmospheric aerosol particles[11]. The Line-Of-Sight (LOS) velocity component of the air motion could get by analyzing the Doppler shift in the frequency of the backscattered signal[12]. From the LOS velocities and the broadening process of the backscattering spectrum the characteristic of the aircraft vortices can be deduced.

The pulsed CDL system (Fig.1) is based on all-fiber laser technology and fast digital signal processing technology. The laser equipped in the CDL system is 1.5-μm eye-safe fiber laser which has an adjustable pulse length of 100ns to 800 ns and a pulse repetition frequency of 10 kHz. The measurement range of the CDL system is ±50m/s, while the speed measurement uncertainty is 0.1 m/s. The scanner and detection range of 3000m enable the system to detect the aircraft wake vortices. The fiber-based optical circuit works stable even under vibrating environment and temperature variation. The Fast Fourier Transform (FFT) spectral estimates are processed with FPGA in real time. All these design of the pulsed CDL system make it stable, reliable and high-integrated [13].

2. METHODOLOGY

Fig.1. the Coherent Doppler Lidar at BICA airport

The OUC Lidar team developed and carried out aircraft wake vortex field observation on the final approach to test out the performance of LOS velocity measurement in RHI (Range Height Indicator) and in ATOM (Along Track Observation Mode) configurations.

For tracking the wake vortices of a landing aircraft in RHI mode, the laser beam scanned in a plane perpendicular to the flight path of the aircraft. Fig. 2 shows the transverse observation mode of the pulsed coherent Doppler Lidar. However, the spatial precision of wake vortex core is limited due to the pulse duration and the data sampling rate.

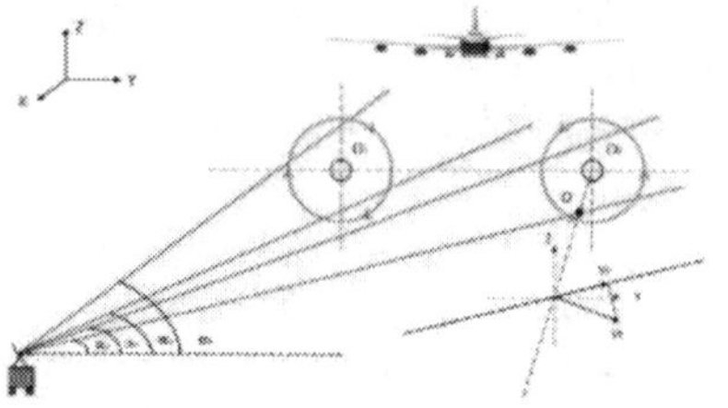

Fig.2. Transverse Observation Mode by Lidar.

In the axial detection or so-called ATOM setup, the Lidar was installed below the taking-off/landing slope in front of the runway. The aircraft were flying towards the Lidar beam passing the detection volume. The Lidar made a number of scan measurements to form a probing area covering the wake vortex corridors. (As shown in Fig.3)ATOM was designed to evaluate the feasibility of improving the localization accuracy of the vortex core.

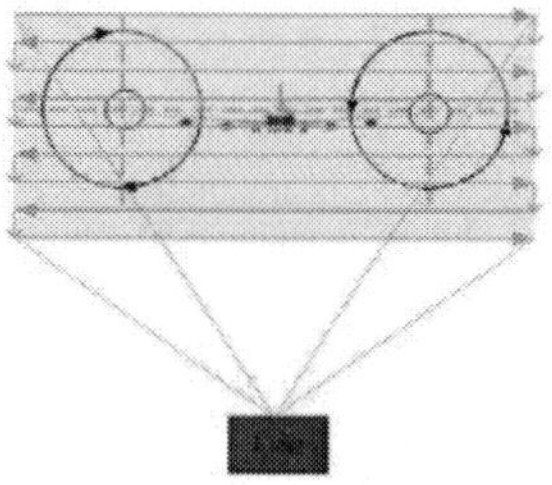

Fig. 3. Along Track Observation Mode (ATOM).

3. RESULTS

During the trial field test, a simple Doppler spectrum subtract method was applied to confirm the Doppler broadening behavior by wake vortex and to define an initial architecture which has been adopted in the BICA field experiment.

For each elevation angle in a scan measurement, Lidar obtained range-resolved signal with a resolution of 30 m. In each spatial sampling point, the backscattered signal is detected and spectrum is then calculated by a Fourier Transform.Fig.4 shows the frequency spectrum of one range bin with a LOS pointing through the vortex in core vicinity. It consists of three adjacent elevations. The blue plot shows a symmetrical and Gaussian-like spectrum from background wind, whereas the red and green plot indicate the spectral broadening up to around 10 point, approximately 10 MHz due to the wake vortices when Lidar probe beam cross the fraction of it. The difference between each two adjacent plots (shown as magenta and olive dashed line) had a zero-crossing pattern which could be used to locate the vortex.

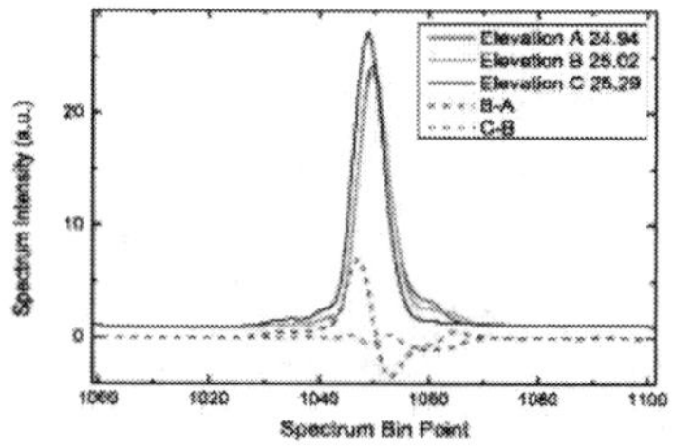

Fig.4. Power spectrum from three adjacent elevation scan in one range bin.

In the Lidar scan patterns such as Fig.5, the red and blue colors indicate positive (away from Lidar) and negative (towards Lidar) movement of the air mass along the laser beam, respectively. It should be noted that the color of the plot is the volume speed of the detected air mass. If there is no wake vortex generated, the color indicates the background atmospheric wind speed projection on the direction of laser beam. If there is wake vortex, the color shows the weighted sum of wake vortex and background wind, which means the color is shifted by the wake vortex, so the wake vortex could be extracted from the background.

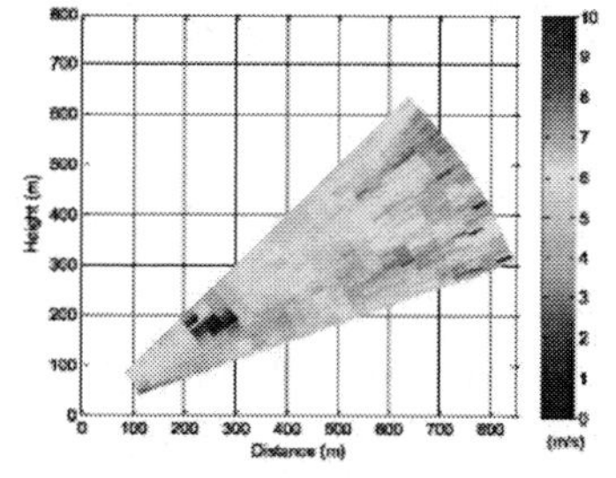

Fig.5. Volume averaged LOS velocity shows positive and negative velocity pattern of vortices.

Fig.5 shows the volume averaged LOS velocity, from which a pair of wake vortices were observed with characteristics of positive and negative velocity pattern at a height of about 200 m and a horizontal distance of 200m to 300m. The color inside the vortex region has a physical meaning of averaged tangential speed in one detection volume. The center of the pair of positive and negative velocity shows the core position of each wake vortex.

In order to investigate the characteristics of wake vortex, the envelopes of both positive and negative velocity have to be extracted. To obtain the envelope representing the sum of the tangential velocity of the vortex and the radial wind velocity, the threshold has to be chosen in accordance with the measured Doppler spectrum.

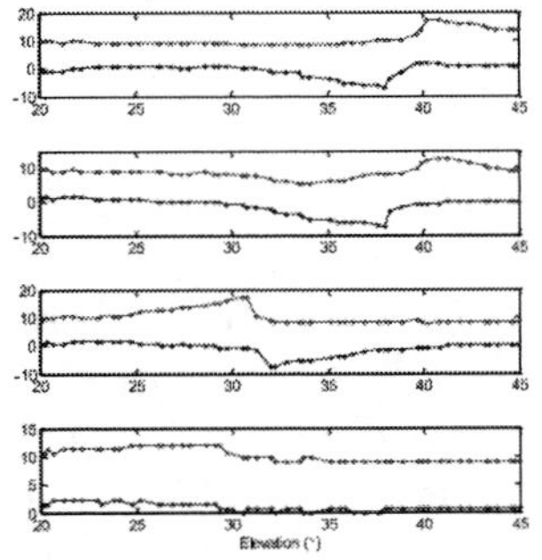

Fig.6. Envelops of velocity along elevation within 9-12 (from top to bottom) range bins. The red and blue curves show the maximum and minimum velocities in each bin.

Fig.6 shows the velocity envelops along four range bins where wake vortices can be captured. The characteristics of a pair of wake vortices can be found in the top figure (bin 9) with elevation around 39 and in the third figure (bin 12) with elevation around 31.In addition, time dependence of the vortex circulation can be achieved as shown in Fig.7.

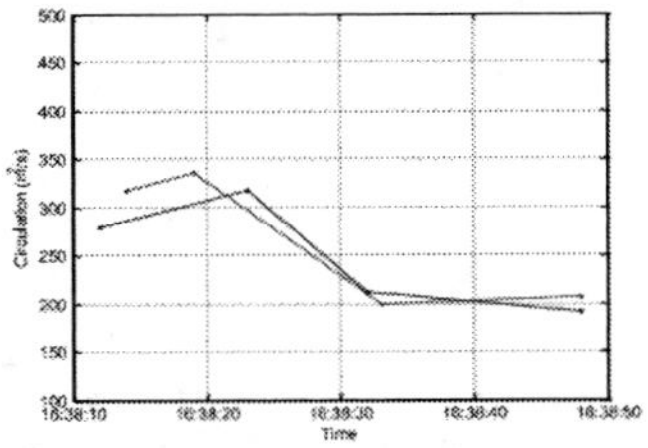

Fig.7. Circulation of a pair of vortices.

As the vortices usually move downwards, an ascending laser beam scan with increasing elevation goes across the pair of vortices faster, which means the vortices are compressed. So the detected radiuses of the vortices are less than their real values. Consequently, the value of vortex circulation is underestimated. On the contrary, when the Lidar scans downwards in a descending scan, the feature of the vortices are expanded and the vortex circulation is overestimated.

Since the movement of the core position can be obtained by a series of descending and ascending scans, the trajectory and speed of vortices can be estimated. Consequently, this information can be used to correct the expanded wake vortex to obtain its exact dimension and circulation. Fig.8 shows the corrected distribution of vortex velocity with radial distance.

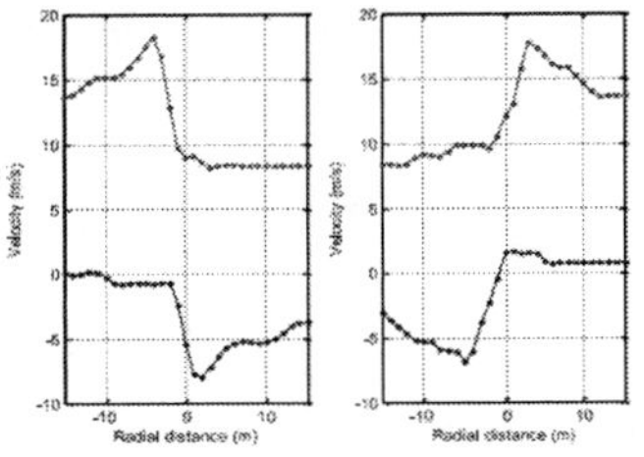

Fig.8. Corrected velocity envelops around vortices core.

Consequently, the time series of the circulation can be calculated from the corrected velocity distribution as shown in Fig.9 Compared to Fig.7, the expansion and compression of circulation value due to alternate scanning direction was calibrated.

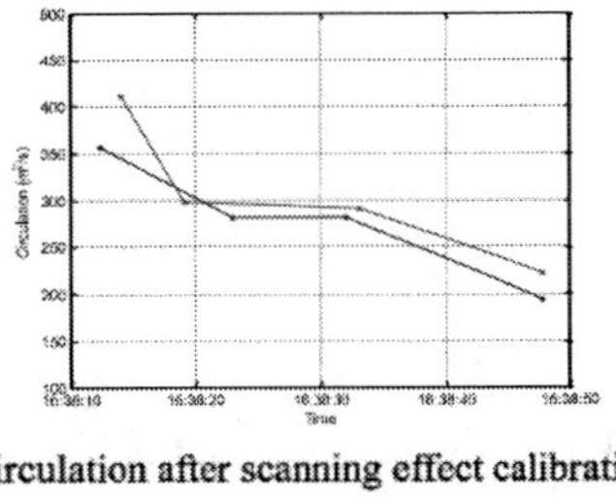

Fig.9. Circulation after scanning effect calibration.

In the ATOM, for tracking the wake vortices of a landing aircraft, the laser beam scanned in a volume along the flight path of the aircraft. To find out the feature of wake vortex, quick-look figures of radical velocity and spectrum widening were generated by processing all data measured in ATOM. Compared to the spectrum of background wind, the intensity of wake vortex is weaker. The wake vortex will generate Doppler frequency shift and extend the width of frequency spectrum. Fig.10 shows the LOS velocity and velocity dispersion measured in ATOM.

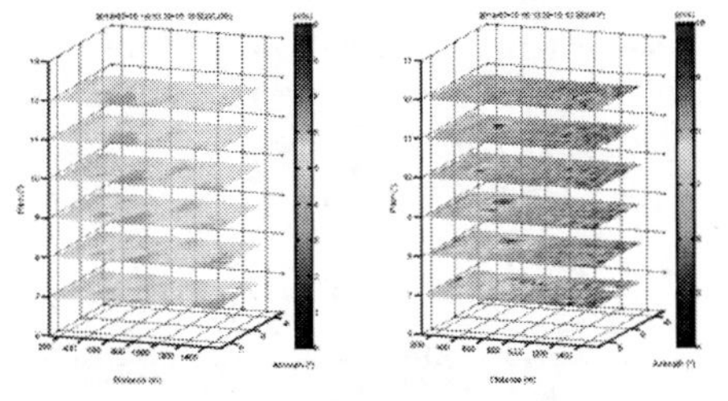

Fig.10. LOS velocity (left) and velocity dispersion (right) measured by Along-Track-Observation-Mode.

For each aircraft flew over, a time series of LOS velocity and velocity dispersion volume scans can be obtained by Lidar system. Analyzing the velocity dispersion is an indispensable access to study the evolution of the wake vortices generated by aircraft.

4. CONCLUSIONS

Wake vortex in clear air can only be derived from Lidar measurements up to now. It is possible to detect the wake vortices generated by aircraft as a consequence of lift with different Lidar observation mode. From series of field tests at BCIA and TAO, our research team estimated the circulation, the position of wake vortex core and the velocity dispersion.

The indispensable access to real wake vortex behavior is provided by Lidar measurement that traces full-scale wake vortices in the free atmosphere. The paper shows that the 1.5-μm pulsed CDL is a practical tool for aircraft wake vortex research. The system can detect the intensity of wake vortex which has impact on setting efficient interval and securing the lifting aircraft.

ACKNOWLEDGEMENT

This work was supported by the Boeing-COMAC Aviation Energy Conservation and Emissions Reduction Technology Center (AECER). Thanks to the whole Lidar Group of Ocean University of China.

REFERENCES

1. Khatwa, R., & Helmreich, R. L., 1998: Analysis of critical factors during approach and landing in accidents and normal flight, *Flight Safety Digest*, **17(11-12)**, 1-2.
2. Gandhi, N. T., 2012: Automatic Dependent Surveillance-Broadcast Enabled, Wake Vortex Mitigation Using Cockpit Display, *Doctoral dissertation, Ohio University.*
3. Connell, L. J., Taube, E. A., Drew, C. R., & Barclay, T. E., 2010: ASRS Reports on Wake Vortex Encounters.
4. Air Bus, 2005: Wake Turbulence Awareness / Avoidance.
5. Harris, M., Young, R. I., Köpp, F., Dolfi, A., & Cariou, J. P., 2002: Wake vortex detection and monitoring, *Aerospace science and Technology*, **6(5)**, 325-331.
6. Agnes Dolfi-Bouteyre, et al., 2009: 1.55μm Pulsed Fiber Lidar for wake vortex detection (axial or transverse), *Wakenet3-Europe 1st major workshop.*
7. K-ograve, F., pp, Smalikho, I., Rahm, S., Agn-ccedil, Dolfi, S. & Gordon, N., 2003: Characterization of aircraft wake vortices by multiple-Lidar triangulation, *AIAA Journal*, **41(6)**, 1081-1088.
8. Barbaresco, F., Brovelli, P., Currier, P., Garrouste, O., Klein, M., Juge, P., & Schneider, J. Y., 2012: Radar sensors for wind & wake-vortex monitoring on airport: first results of SESAR P12. 2.2 XP0 trials campaign at Paris CDG Airport, *In proceeding of the seventh European conference on radar in meteorology and hydrology (ERAD 2012), Toulouse.*
9. Moneuse, J.F., 2010: SESAR Project 12.2.2: Runway Wake Vortex Detection, Prediction and decision support tools, *2nd Wakenet3-Europe Major Workshop, Toulouse.*
10. Köpp, F., Rahm, S., & Smalikho, I., 2004: Characterization of Aircraft Wake Vortices by 2-μm Pulsed Doppler Lidar, *Journal of Atmospheric and Oceanic Technology*, **21(2)**, 194-206.
11. Darracq, D., Corjon, A., Fré, dé, Ducros, R., Keane, M, & Redfern, M. 2000: Simulation of wake vortex detection with airborne Doppler lidar, *Journal of aircraft*, **37(6)**, 984-993.
12. Barbaresco, F., Juge, P., Klein, M., Ricci, Y., Schneider, J., & Moneuse, J., 2011: Optimising runway throughput through wake vortex detection, prediction and decision support tools, *In Digital Communications-Enhanced Surveillance of Aircraft and Vehicles (TIWDC/ESAV), 2011 Tyrrhenian International Workshop on. IEEE*, 27-32.
13. Wu, S., Yin, J., Liu, B., Liu, J., Li, R., Wang, X & Zhang, K., 2014: Characterization of turbulent wake of wind turbine by coherent Doppler Lidar, *In SPIE Asia Pacific Remote Sensing.*

DOI: 10.1051/epjconf/201611914009

DESIGN OF A NON-SCANNING LIDAR FOR WIND VELOCITY AND DIRECTION MEASUREMENT

Bo Liu [1]*, Zhangxian Peng [1]

[1] *Institute of Optics and Electronic, CAS, CHINA, *Email: boliu@ioe.ac.cn*

ABSTRACT

A Doppler lidar system for wind velocity and direction measurement is presented. The lidar use a wide field of view (FOV) objective lens as an optical antenna for both beam transmitting and signal receiving. By four fibers coupled on different position on the focal plane, the lidar can implement wind vector measurement without any scanning movement.

1. INTRODUCTION

Lidars were employed for wind sensing since early 1970s [1]. Recently, following advances in fiber optic communications, laser wavelengths close to 1550 nm are used extensively in all-fiber coherent doppler wind lidars (CDL), and many commercial systems are available [2-4]. The fiber optic technology, used in optical communications industry, is employed for the generation, amplification, transmission, and manipulation of the laser beam in all-fiber CDLs [5]. For a doppler wind lidar, only radial velocity can be measured by one beam direction, to get the real wind vector, at least three beam directions are required. Most of wind lidars scan the laser beam to different directions to complete the vector measurement with a scanning device. In some cases, the mechanical scanning is not a good choice due to its slow movement, especially for those time resolution sensitive systems, such as airborne and spaceborne systems. In this paper, we discussed a design of non-scanning all-fiber wind lidar with a wide filed view optical antenna.

2. System design

The configuration of the system is shown in Fig. 1. The system consists of a transceiver, an optical antenna, and a signal processor. The transceiver consists of a laser source, optical couplers, frequency shifter, and pulsed fiber amplifier. The optical antenna, which consists of an optical circulator, a 1x4 fiber switch and a wide FOV objective lens. Balanced amplified photo detectors and A/D convertors are used for signal receiving. The outgoing laser pulse can be switched to one of four fibers which are coupled on four different points on the focal plane of the objective lens. The four points are on the field of (0°, 12.5°), (0°,-12.5°), (12.5°, 0°) and (-12.5°,0°) respectively. By this design, the lidar can measure four different radial velocities on four directions by a fiber switch within 2ms. (Switching speed 0.5ms)

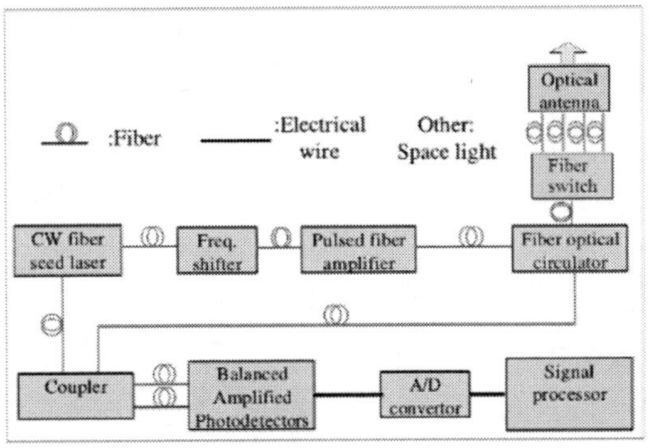

Fig.1. System configuration

Figure 2 shows the configuration of the lidar beam direction and the optics design. In a vertically setting, the four beams are pointed to the East, North, West and South at 12.5° zenith angle. The objective lens consists of five pieces lens and optimized by fiber coupling efficiency. In the optical design, except for the fiber coupling efficiency, the effective receiving aperture of each point also should be taken into account. The aperture of the first lens is 76mm, and each beam use 91% of this aperture which means the effective receiving aperture of each point is about 69mm.

Figure 3 shows the system wavefront function and spot diagram of focal plane. The RMS wavefront error is 0.05 waves which is good enough for coherent detection. The RMS radius of the four points is 2.47 μm and the coupling efficiency of the signal mode fiber is 53%.

DOI: 10.1051/epjconf/201611914009

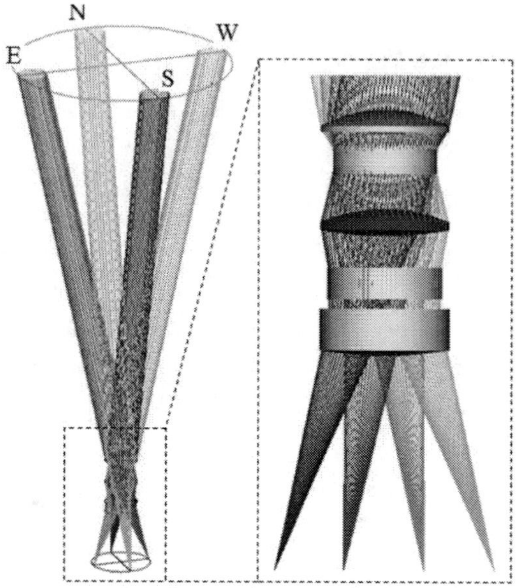

Fig.2. lidar beam direction and the optics design

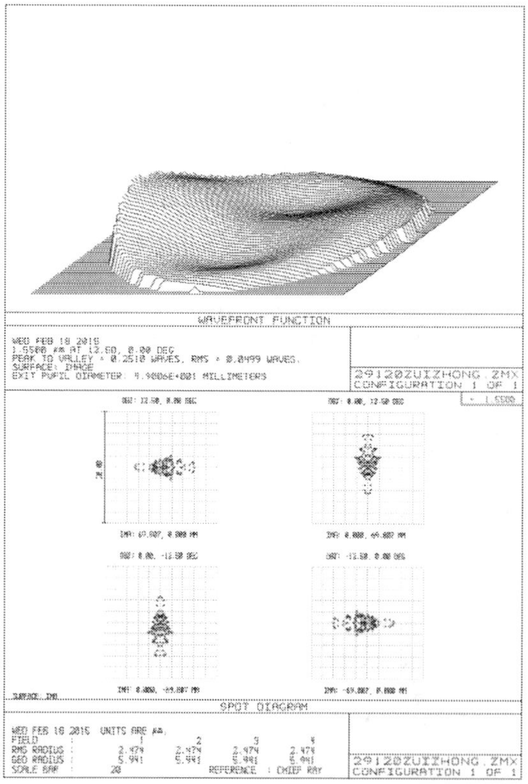

Fig.3. System wavefront function and spot diagram·

3. Summary

A non-scanning all-fiber pulsed wind doppler lidar was designed. The system use a wide filed view optical antenna for beam transmitting and signal receiving. Four fibers were coupled with the objective lens on four symmetrical points, and a fiber switch was used to switch the lidar beam to four directions which allow the lidar complete wind vector measurement without any scanning movement. The optics design and system configuration were presented.

REFERENCES

[1] R. M. Huffaker, A. V. Jelalian, and J. A. L. Thompson, “Laser- Doppler system for detection of aircraft trailing vortices,” Proc. IEEE **58,** 322–326 (1970).

[2] ZephIR 300 technical specifications, (ZephIR Lidar, 2014).

[3] WINDCUBE V2, a 200m vertical wind Doppler lidar, (Leosphere, 2014).

[4] WINDAR functional specifications, (WINDAR PHOTONICS, 2014).

[5] C.F.Abari, A. T. Pedersen, and JakobMann, "An all-fiber image-reject homodyne coherent Doppler wind lidar" Opt. Express. **22**(21), 25880–25894 (2014).

DOI: 10.1051/epjconf/201611915001

CLIDAR MOUNTAIN BOUNDARY LAYER CASE STUDIES

Nimmi C. P. Sharma [1]*, John E. Barnes[2]

[1]*Department of Physics and Engineering Physics, Central Connecticut State University, New Britain, CT 06050 USA, *Email: sharmanim@ccsu.edu*

[2]*National Oceanic and Atmospheric Administration, Mauna Loa Observatory, Hilo, HI USA*

ABSTRACT

A CCD Camera Lidar system called the CLidar system images a vertically pointing laser from the side with a spatially separated CCD camera and wide angle optics. The system has been used to investigate case studies of aerosols in mountain boundary layers in in the times following sunset. The aerosols detected by the system demonstrate the wide variation of near ground aerosol structure and capabilities of the CLidar system.

1. INTRODUCTION

The CCD Camera Lidar (CLidar) system is a bistatic laser radar system designed for near-ground aerosol studies [1][2][3]. It has applications in aerosol profiling and boundary layer studies [4]. With the CLidar system, a laser (pulsed or continuous) is transmitted vertically into the atmosphere. The detector consists of a CCD camera equipped with a 10 nm laser line interference filter and wide angle optics. The detector is located far (typically 100 m or more) from the laser transmitter and images the entire laser beam from ground to zenith at once from the side. The signal is integrated on the CCD for several minutes. The configuration allows for sub-meter altitude resolution near the ground. Data are collected all the way to ground level, with no need for overlap corrections. Altitude from which the side-scatter is returned is determined by geometry from the laser and detector locations, eliminating the need for sophisticated timing electronics. The signal is influenced by laser polarization angle. Since the system is bistatic, the scattering angle varies for each altitude. Thus retrieval of aerosol extinction from the signal requires assumptions or measurements of the aerosol scattering phase function. The instrument operates only at night.

Mountain boundary layer measurements are of interest for many sites. Airflows upslope and downslope vary with time. Mauna Loa Observatory (MLO), located on the flank of Mauna Loa volcano on the big Island of Hawaii at an altitude of 3396 m above sea level, is a designated Atmospheric Baseline Observatory of the National Oceanic and Atmospheric Administration's (NOAA) Global Monitoring Division (GMD), where long term atmospheric parameters are measured to determine global climate parameters. Air at this location is usually clean tropospheric air, however daytime upslope flows occur, followed by evening downslope flows. To study the atmospheric structure in the near ground region at this location, the CLidar instrument was used to map vertical aerosol distributions in the near ground region just after sunset. CLidar data were combined with data from Aerosol Robotic Network (AERONET) sunphotometers [5].

2. METHODOLOGY

CLidar measurements were conducted at MLO during the post-sunset period of the upslope to downslope flow transition to gain insights on variations of aerosol distributions present at these times near ground level. A linearly polarized Spectra Physics Nd:YAG model GCR-6 with 300-600 mJ/pulse operating at a wavelength of 532 nm, a pulse width of 8 ns and a pulse repetition rate of 30 Hz was transmitted vertically into the atmosphere. A 10 nm (FWHH) laser line filter between the camera and the lens was employed to reject background light. The detector used for these experiments was a Santa Barbara Instruments Group ST-237A camera. The beam was imaged with a Kinoptic lens with a 120 degree field of view and a focal length of 1.98 mm. The detector was located 139 m from the transmitter at a polarization angle of 89.4 degrees. The returned scatter signal was integrated on the camera in exposures typically of 5 minutes 32 seconds duration. CLidar experimental runs usually lasted one to two hours. Data for the case studies presented here span the months from February through August.

DOI: 10.1051/epjconf/201611915001

Data images were analyzed using a beam finding routine which fits a function consisting of a Gaussian plus a constant to the data in the direction normal to the beam at each altitude. The constant is taken as representative of the background at each altitude. Data are normalized to match molecular scattering in a high-altitude fitting range, just as in typical backscatter lidar analysis. An aerosol phase function is then selected. The phase function shows the amount of light scattered into each angular direction. An AERONET sunphotometer [5] is available at Mauna Loa. This sunphotometer performs almucantar scans from which the overall column average aerosol scattering phase function may be derived. In the cases presented here, the AERONET-derived phase function for the daytime hours preceding the CLidar measurements was used. In the event that AERONET data are not available, phase functions from the CALIPSO phase function classes [6] could also employed. CLidar data were corrected for two way transmission, and the aerosol to molecular side scatter ratios were calculated. Lastly the aerosol extinction as a function of altitude and time were derived. This was done by using the aerosol phase function to convert from the measured side scatter to total scatter and an assumed single scattering albedo to finally derive extinction.

3. RESULTS

Three sample cases are presented here to demonstrate the breadth of near ground aerosol conditions encountered in this mountain boundary layer regime. Figures 1, 2 and 3 show the time averaged CLidar-derived aerosol extinction as a function of altitude for each of the three nights. Averaging times were typically 1.5 to two hours. Figures 4, 5 and 6 show the CLidar-derived aerosol extinction contour plots for the three case studies. Here the horizontal axis is the time in UTC, the vertical axis is altitude above ground level (where the instrument ground level at MLO is 3.402 km above sea level at the bottom of the plot), and the color map represents the derived aerosol extinction in Mm^{-1}.

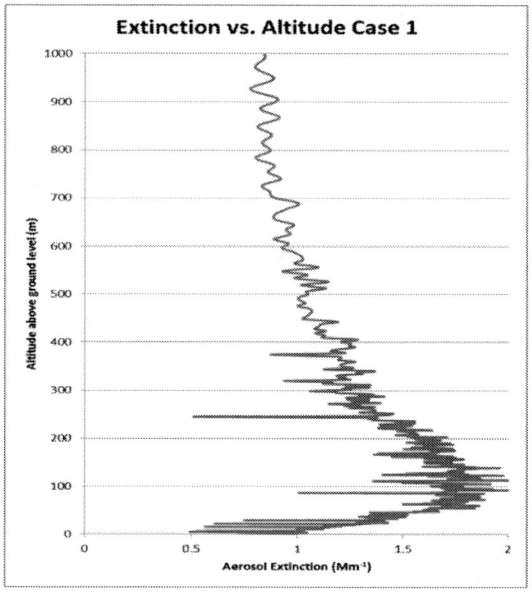

Figure 1: MLO 2/13/07 Aerosol Extinction vs. Altitude

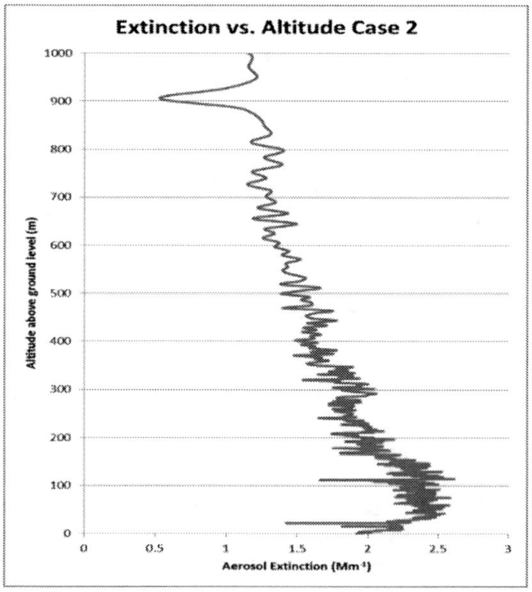

Figure 2: MLO 4/12/07 Aerosol Extinction vs. Altitude

DOI: 10.1051/epjconf/201611915001

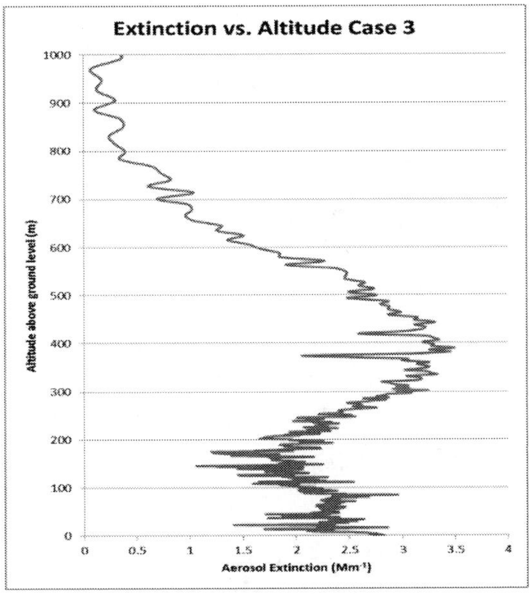

Figure 3: MLO 8/27/07 Aerosol Extinction vs. Altitude

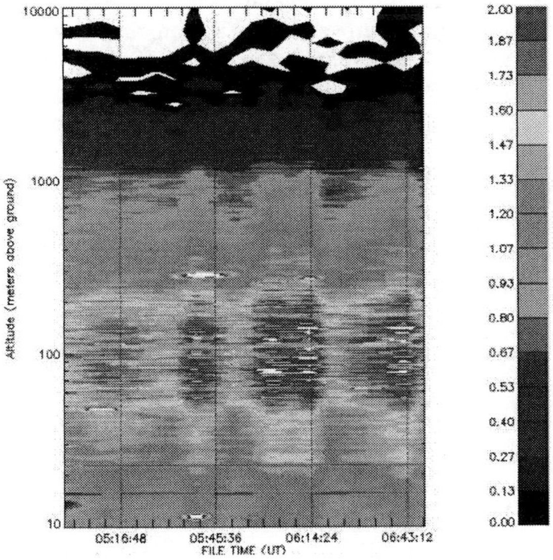

Figure 4: MLO Case 1 Aerosol Extinction in Mm^{-1}

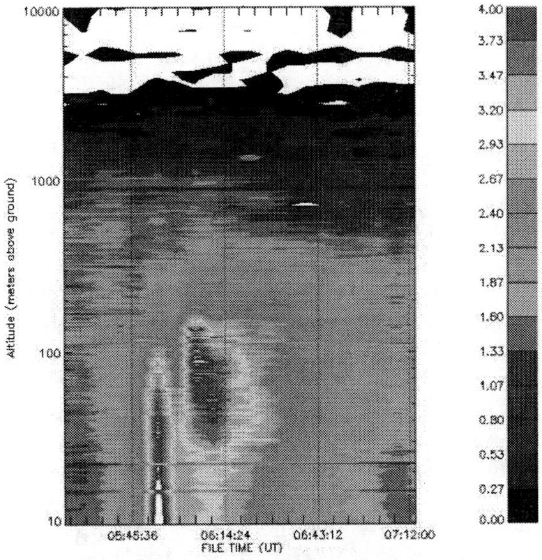

Figure 5: MLO Case 2 Aerosol Extinction in Mm^{-1}

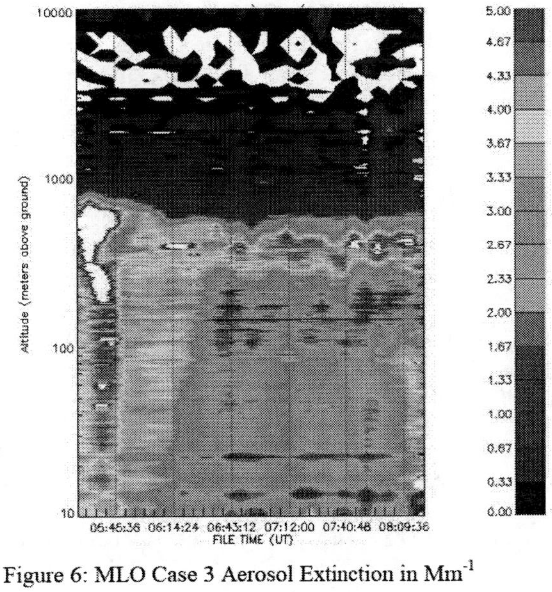

Figure 6: MLO Case 3 Aerosol Extinction in Mm^{-1}

4. DISCUSSION AND CONCLUSION

All three cases presented here show low to moderate level aerosol present at Mauna Loa during the evening flow transition time. Cases of no detectable aerosol and much higher aerosol were also observed. Case 1 demonstrates persistent low-level aerosol in the near ground region. Aerosol scattering is highest at altitudes between 50 and 150 meters above ground level. The aerosol declines gradually at higher altitudes. Although not visible on the plot in Fig.1, the aerosol persists to an altitude of almost 3500 m above ground level over the experiment time frame. Case 2 shows a low passing plume of aerosol that enters the measurement region and then begins to dissipate by the end of the experimental time. The aerosol extinction at ground level changed by a factor of 3 to 4 over the 1-2 hour time frame. The aerosol varied strongly over very small distance scales as well, in some instances changing by a factor of 2 over less than 50 m altitude. Case 3 demonstrates an interesting example of near ground aerosol that would appear to clear out if measured only by surface instruments, while a lofted layer persists at approximately 400 m, out of range of direct sampling instrumentation. The variation in these cases highlights the need for instrumentation capable of characterizing aerosol altitude profiles all the way to the ground. The CLidar instrument maps aerosols with high altitude resolution in this near ground region. Often the information obtained is not accessible from any other instruments. It is expected that the CLidar, combined with additional instrumentation, will be a useful tool for helping to characterize boundary layer dynamics on mountain and other sites.

ACKNOWLEDGEMENT

This material is based upon work supported by the National Science Foundation under Grant 0320800. Any opinions, findings, and conclusions or recommendations expressed in this material are the opinions of the author(s) and do not necessarily reflect the views of the National Science Foundation. The work was also supported by NOAA. The authors also acknowledge AERONET for the use of the AERONET phase function data.

REFERENCES

[1] Barnes, John E., S. Bronner, Robert Beck, and N. C. Parikh, 2003: Boundary Layer Scattering Measurements with a Charge-Coupled Device Camera Lidar, *Applied Optics*, **42**, no. 15, 2647-2652.

[2] Barnes, John E., N.C. Parikh Sharma and Trevor Kaplan, Atmospheric aerosol profiling with a bistatic imaging lidar system, 2007: *Applied Optics*, **46**, no. 15, 2922-2929.

[3] Barnes, John E. and Nimmi C. P. Sharma, 2012: An inexpensive active remote sensing instrument for assessing aerosol distributions, *Journal of Air and Waste Management Association*, **62**, no. 2, 198-203.

[4] Sharma, N. C. Parikh, John E. Barnes, Trevor B. Kaplan and Antony D. Clarke, 2011: Coastal Aerosol Profiling With A Camera Lidar and Nephelometer, *Journal of Atmospheric and Oceanic Technology (JTECH)*, **28**, No. 3. 418-425.

[5] http://aeronet.gsfc.nasa.gov/

[6] Omar, Ali H. et al., 2009: The CALIPSO automated aerosol classification and lidar ratio selection algorithm, *Journal of Atmospheric and Oceanic Technology*, **26**, 1994-2014.

LIDAR OBSERVATIONS OF LOW-LEVEL WIND REVERSALS OVER THE GULF OF LION AND CHARACTERIZATION OF THEIR IMPACT ON THE WATER VAPOUR VARIABILITY

Paolo Di Girolamo[1]*, Cyrille Flamant[2], Marco Cacciani[3], Donato Summa[1], Dario Stelitano[1], Ignazio Mancini[1], Evelyne Richard[4], Véronique Ducrocq[5], Mathieu Nuret[5], Frédérique Said[4]

*[1]Scuola di Ingegneria, Università degli Studi della Basilicata, Viale dell'Ateneo Lucano n. 10, 85100 Potenza – Italy , *Email: digirolamo@unibas.it*

[2]Laboratoire Atmosphères, Milieux, Observations Spatiales,UMR 8190 CNRS &UPMC &UVSQ, Paris-France

[3]Dipartimento di Fisica, Università di Roma "Sapienza", Piazzale Aldo Moro, n. 2, 00100 Roma – Italy

[4]Laboratoire d'Aérologie, UMR 5560 CNRS, Université de Toulouse, Toulouse, France

[5]CNRM-GAME, UMR3589 Météo-France & CNRS, Toulouse, France

ABSTRACT

Water vapour measurements from a ground-based Raman lidar and an airborne differential absorption lidar, complemented by high resolution numerical simulations from two mesoscale models (Arome-WMED and MESO-NH), are considered to investigate transition events from Mistral/Tramontane to southerly marine flow taking place over the Gulf of Lion in Southern France in the time frame September-October 2012, during the Hydrological Cycle in the Mediterranean Experiment (HyMeX) Special Observation Period 1 (SOP1). Low-level wind reversals associated with these transitions are found to have a strong impact on water vapour transport, leading to a large variability of the water vapour vertical and horizontal distribution. The high spatial and temporal resolution of the lidar data allow to monitor the time evolution of the three-dimensional water vapour field during these transitions from predominantly northerly Mistral/Tramontane flow to a predominantly southerly flow, allowing to identify the quite sharp separation between these flows, which is also quite well captured by the mesoscale models.

1. INTRODUCTION

Mistral and Tramontane are two strong dry northerly winds which are often found to blow in the Mediterranean basin. They usually blow in winter or spring, though they occur in all seasons. Mistral and Tramontane sometimes last only one or two days, frequently lasts several days, and sometimes lasts more than a week. They bring cold and dry continental air over the sea and are frequently observed to extend few hundred kilometres off the coast (among others, [1] Jansa, 1987). In the Gulf of Lion, Southern France, the occurrence of Mistral and Tramontane at the coast are often observed to alternate with southerly marine flows. Marine flows advect moist air onshore, its inland penetration being strongly dependent on the large scale flow [2]. During *HyMeX-SOP1* (September-November 2012, Ducrocq *et al.*, 2014) several events when the southerly marine flow broke through and overcame the Mistral/Tramontane flow were observed. The complex interaction between the southerly marine flow and the Mistral/Tramontane events may lead to a rapid transition between the two wind regimes along the coastline [3]. This research effort is dedicated to the measurements performed by the ground-based University of BASILicata Raman Lidar system (*BASIL*,[4]) and the airborne differential absoption lidar (DIAL) system LEANDRE 2 [5], which are used to illustrate the high time and space variability of the water vapour field associated with the occurrence of transition events from Mistral/Tramontane to southerly marine flow in Southern France during *HyMeX-SOP 1*. Measurements from these two systems are also used to validate the numerical simulations from the mesoscale model MESO-NH and the mesoscale numerical weather prediction model AROME-WMED.

2. METHODOLOGY

BASIL is a ground-based Raman Lidar hosted in a transportable sea-tainer. The major feature of *BASIL* is represented by its capability to perform high-resolution and accurate measurements of atmospheric temperature and water vapour, both in daytime and night-time, based on the application of the rotational and vibrational Raman lidar techniques in the UV. Besides temperature and water vapour, *BASIL* is also capable of providing measurements of particle backscatter at 355, 532 and 1064 nm, particle extinction at 355 and 532 nm and particle depolarization at 355 and 532 nm [6].

In the frame of *HyMeX-SOP1*, *BASIL* was deployed in Candillargues and operated between 5 September and 5 November 2012, collecting more than 600 hours of measurements, distributed over 51 measurement days and 19 intensive observation periods (IOPs).

During *HyMeX-SOP1*, the CNRS DIAL-LEANDRE2 was located onboard the French ATR42, operated by SAFIRE. The system was operated in zenith and nadir-pointing mode with the primary mission of characterizing the water vapour and aerosol inflow feeding heavy precipitating systems. The system includes a tunable laser whose emission is tuned upon a water vapour absorption line selected from two rotation–vibration bands in the near infrared (727–770 nm). The system allows water vapour mixing ratio measurements to be performed with a precision ranging from less than 0.1 g kg^{-1} at 4.5 km above sea level to less than 0.4 g kg^{-1} near the surface for an along-beam resolution of 150 m and accumulation of 100 individual profiles, which, for an ART42 flying speed of 100 m/s, corresponds to an along-track resolution of nearly 1 km. During *HyMeX SOP1*, the CNRS DIAL performed over 60 flight hours: 8 of which supported by the *European Facility For Airborne Research* Project *"WaLiTemp"* and in coordination with the Raman lidar BASIL, and the reminder as part of the *HyMeX SOP1* endowment supported by the MISTRAL program.

Numerical simulations considered in this paper are performed with two non-hydrostatic models: Méso-NH and AROME-WMED. The Méso-NH has been jointly developed by the Laboratoire d'Aérologie and by CNRM-GAME. It makes use of two interactively nested model domains with a horizontal mesh size of 9 km and 3 km for the coarse and the fine domain, respectively. AROME-WMED is a convection-permitting model with a domain covering the entire western Mediterranean basin, with a forecast interval up to 48 h. In addition to the operational observations used in the 3-h data assimilation cycle, the model assimilates some of the observations made available in real-time during SOP1 (i.e. surface observations over Spain, radiosoundings from the research sites, radar data (Doppler winds and reflectivity) and satellite data).

3. RESULTS

Figure 1illustrates the time evolution of the wind direction and speed and relative humidity for the period from 00:00 UT on 19 September 2012 to 00:00 UT on 21 September 2012 from a meteorological surface station located in Candillargues. Gradients of surface wind direction and speed and relative humidity are considered to identify the transitions from Mistral/Tramontane to predominantly southerly marine flow and vice versa.

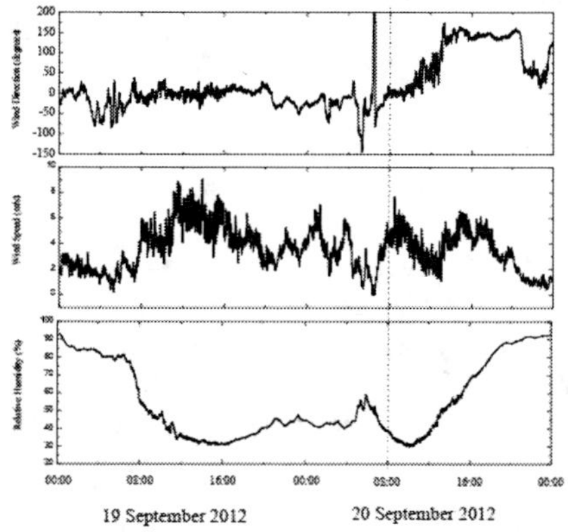

Figure 1: Time evolution of the wind direction and speed and relative humidity for the period 19-21 September 2012 as measured by the surface meteorological station located in Candillargues. The vertical dashed line identifies the time when the chance in wind direction and speed takes place.

The observed variability of surface wind direction and speed and relative humidity which is depicted in figure 1 clearly identifies the southerly marine flow onset from a previous weak

Mistral/Tramontane event around 08:00 UT on 20 September 2012, with wind direction progressively changing from about 0 to about 150 degrees (from Northerly to South-easterly), wind speed progressively decreasing from approx. 6 to 2 m/s and relative humidity increasing from 30 to 90 %.

Figure 2a illustrates the time evolution of the water vapour mixing ratio as measured by BASIL over a time period of approx. 24 h from 16:49 UT on 19 September 2012 to 16:33 UT on 20 September 2012. This figure reveals the arrival of the moist air in Candillargues associated with the south-easterly marine flow starting from approx. 08:00 UT. A strong gradient in the water vapour field is visible starting from 09:30 UT. Before the arrival of the humid layer, during the Mistral/Tramontane flow, the mixing ratio values in the boundary layer are not exceeding 5-6 g/kg. The humid layer associated with the marine flow is found to extend up to the boundary layer top, with mixing ratio values as large as 15 g/kg. Above the boundary layer top the water vapour mixing ratio variability is very limited, with mixing ratio values not exceeding 3 g/kg, both before and after the arrival of the flow, mainly because the wind reversal is taking place only in the lower levels (primarily in the boundary layer) and it is not extending throughout the free troposphere. The humid layer is first revealed in the upper portion of the boundary layer and then it is found to progressively fill the boundary layer extending down to the surface. In the presence of the south-easterly flow the earlier arrival of the humid layer in the upper portion of the boundary layer and its progressive later arrival at lower levels is caused by the presence of higher horizontal wind speed values (7-8 m/s) in the upper portion of the boundary layer with respect to those found (2-5 m/s) in the lower portion, as can be easily revealed in figure 2c from the measurements performed by the UHF wind profiler located in Candillargues.

Figure 2b illustrates the time evolution of the range corrected signal at 532 nm, RCS_{532}, as measured by BASIL over a time period of 18 h from 00:00 UT to 18:00 UT on 20 September 2012. As the range corrected signal is expressed in arbitrary units no colour legend was introduced for this panel. The figure reveals a sensitive change in particle backscatter, which is caused by the arrival of an aerosol-loaded air mass associated with the south-easterly marine flow shortly after the low-level wind reversal. This change in particle backscatter is superimposed to the particle backscatter variability associated with the development of the boundary layer, which starts building up around the same time when the wind reversal takes place. The warmer and more humid air advected by the marine flow is characterized by a larger aerosol loading, predominantly of maritime origin, which is replacing the colder and drier air advected by the Mistral/Tramontane flow, predominantly loaded with continental aerosol. RCS_{532} is found to increase by approx. 50 % after the marine flow onset.

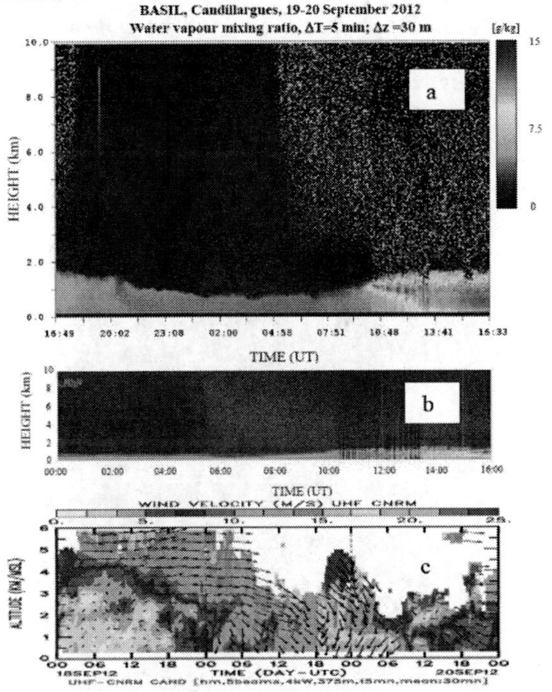

Figure 2: a) Water vapour mixing ratio as measured by BASIL over a the time period from 16:49 UT on 19 September 2012 to 16:33 UT on 20 September 2012. b) RCS_{532} as measured by BASIL over the time period from 00:00 UT to 16:00 UT on 20 September 2012. c) Wind speed and directions on 18-19-20 September 2012 as measured by the UHF wind profiler located in Candillargues.

Figure 3 illustrates the time evolution of the water vapour mixing ratio as simulated by the mesoscale models MESO-NH (lower panel) and AROME WMED (upper panel) over the time period of approx. 24 h from 17:00 UT on 19 September 2012 to 17:00 UT on 20 September 2012. MESO-NH and AROME-WMED outputs are provided with a time step of 1 h. Model analysis from both MESO-NH and AROME-WMED properly simulate the time-height structure of the humidity field observed by BASIL, reproducing the arrival of the humid air in the boundary layer with similar mixing ratio values (8-12 g/kg) for AROME-WMED and slightly lower values (in the range 6-8 g/kg) for MESO-NH. The depth of the moist layer also seems to be slightly different for the two models, with AROME-WMED being in better agreement with BASIL (moist layer extending for both up to approx. 1500 m) than the one from MESO-NH (moist layer extending up to approx. 1300 m).

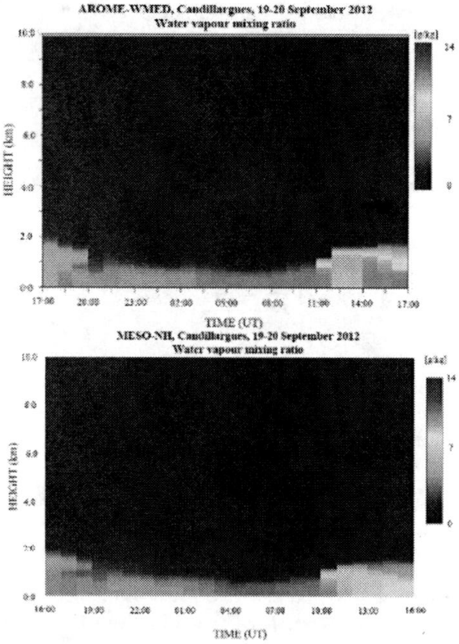

Figure 3: Time evolution of the water vapour mixing ratio as simulated by the mesoscale models AROME WMED (upper panel) and MESO-NH (lower panel) over the 24 h time period from 16:00 UT (17:00 UT for AROME WMED) on 19 September 2012 to 16:00 UT (17:00 UT for AROME WMED) on 20 September 2012.

However, both models well capture the elevated humid layer present in the free troposphere which is visible in BASIL data, this layer having a vertical extent of approx. 2 km, with a base at 4 km around 16:00 UT and progressively descending down to 3 km around 06:00 UT. The bias and RMS deviation between BASIL and the two mesoscale models have been computed and will be discussed at the Conference. Unfortunately, the CNRS DIAL-LEANDRE2 was not flying during this period. More cases describing low-level wind reversals associated with transition events from Mistral/Tramontane to southerly marine flow taking place in Southern France, complemented by the availability of simultaneous measurements by CNRS DIAL-LEANDRE2, will be discussed at the Conference.

REFERENCES

[1] Jansa, A., 1987: Distribution of the Mistral: A satellite observation. *Meteorol. Atmos.Phys.*, **36**, 201-214.

[2] Arritt, R.W., 1993: Effects of the large scale flow on characteristics features of the sea breeze, *J. Appl. Meteor.*, **32**, 116-125.

[3] Bastin, S., C. Champollion, O. Bock, P. Drobinski, F. Masson, 2005: On the use of GPS tomography to investigate water vapor variability during a Mistral/sea breeze event in southeastern France, *Geophys. Res. Lett.*, 32, L05808.

[4] Di Girolamo, P., D. Summa, R. Ferretti, 2009: Multiparameter Raman Lidar Measurements for the Characterization of a Dry Stratospheric Intrusion Event, *J. Atmos. Ocean. Tech.*, **26**, 1742-1762.

[5] Bruneau D, P. Quaglia, C. Flamant, M. Meissonnier, J. Pelon, 2001: Airborne lidar LEANDRE II for water-vapor profiling in the troposphere. I: System description. *Appl. Opt.* **40**: 3450–3461.

[6] Di Girolamo, P., D. Summa, R. Bhawar, T. Di Iorio, M. Cacciani, I. Veselovskii, O. Dubovik, A. Kolgotin: 2012, Raman lidar observations of a Saharan dust outbreak event: Characterization of the dust optical properties and determination of particle size and microphysical parameters, *Atm. Env.*, **50**, 66-78.

Multiwavelength Lidar observation of the atmospheric response to the 20th March 2015 partial solar eclipse in Rome Tor Vergata: preliminary results.

Gian Luigi Liberti[1]*, Davide Dionisi[1], Stefano Federico[1], Fernando Congeduti[1]

*[1]Institute of Atmospheric Sciences and Climate-CNR, Via del Fosso del Cavaliere 100, 00133 Roma, Italy, *Email: g.liberti@isac.cnr.it*

ABSTRACT

This study reports some preliminary analyses of multichannel lidar measurements taken in Rome Tor Vergata (Italy) during the 20th March 2015 partial solar eclipse. The objective is assessing the capability of the instrument to document the effect of the eclipse in the lower troposphere, with a particular emphasis on the information content at relatively small temporal and spatial scales.

1. INTRODUCTION

Solar eclipses are a unique opportunity for studying the atmospheric response to the large scale and sharply variation of the incoming solar radiation. Due to the rarity of the event, only few studies by the means of lidar have been conducted to characterize the phenomenon. One of the main observed effects is on the formation and the evolution of the Planetary Boundary Layer (PBL) during the solar eclipse. An aerosol backscatter lidar was used by Kolev et al. [1] to investigate the dynamics of the PBL during the solar eclipse of 11 August 1999, over Bulgaria. This study demonstrated that the solar eclipse affects the meteorological parameters of the atmosphere near the ground, the ozone concentration and the height of the mixing layer. During the solar eclipse of 29 March 2006, lidar data from two sites were used to characterize the Planetary Boundary Layer (PBL) height evolution over Greece, during the solar eclipse of 29 March 2006 [2]. In this case the solar eclipse induced a decrease of the PBL height, indicating a suppression of turbulence activity similar to that during the sunset hours.

According with astronomic data the eclipse in Rome started at about 8:24 UTC and ended at 10:43 UTC reaching the maximum (53% totality) at about 9:32 UTC. The aim of this paper is to document its effect in the troposphere through the multi-wavelength Rayleigh-Mie-Raman lidar located at Rome Tor Vergata (Italy). The analysis of the atmospheric response at different temporal and spatial scales has been performed.

2. METHODS

2.1 METEOROLOGICAL DATA

Figure 1 shows 1-minute sampled surface meteorological data from the automatic weather station on the site of the lidar system. As expected during the eclipse there is a cooling of the air temperature caused by the decrease of the solar short-wave irradiance. The perturbation to the solar irradiance caused by the eclipse is larger than 300 W/m^2, while the surface temperature decrease is greater than 1° C. There is a time shift between the effective air temperature estimated by the downward LW radiation at the air temperature at the surface during the eclipse (the minimum of the latter lasts that of the former). This difference is likely caused by the different heat capacity of the two systems, the air temperature at the surface being influenced by the ground temperature. The maximum of the pressure is partially due to the semidiurnal tides. The 2m air density was estimated from pressure and temperature. The air density shows an increase at the surface caused by the stratification of the atmosphere during the eclipse. It is also important to note that the curve of the solar irradiance, if we exclude the perturbation of the eclipse, shows the typical behavior of a clear sky day. Also, the surface winds are small and their evolution shows an increase during the afternoon, caused by the intensification of the sea-breeze and valley-mountains flow. Thus, with the exception of the eclipse perturbation, the evolution of the PBL follows that of a clear sky wind calm evolution.

2.2 DESCRIPTION OF THE LIDAR SYSTEM AND DATA PROCESSING

The Rayleigh–Mie–Raman (RMR) lidar of ISAC CNR is part of the experimental site in the

DOI: 10.1051/epjconf/201611915003

suburban hilly area of Tor Vergata (41.88°N, 12.68°E, 107 m ASL).

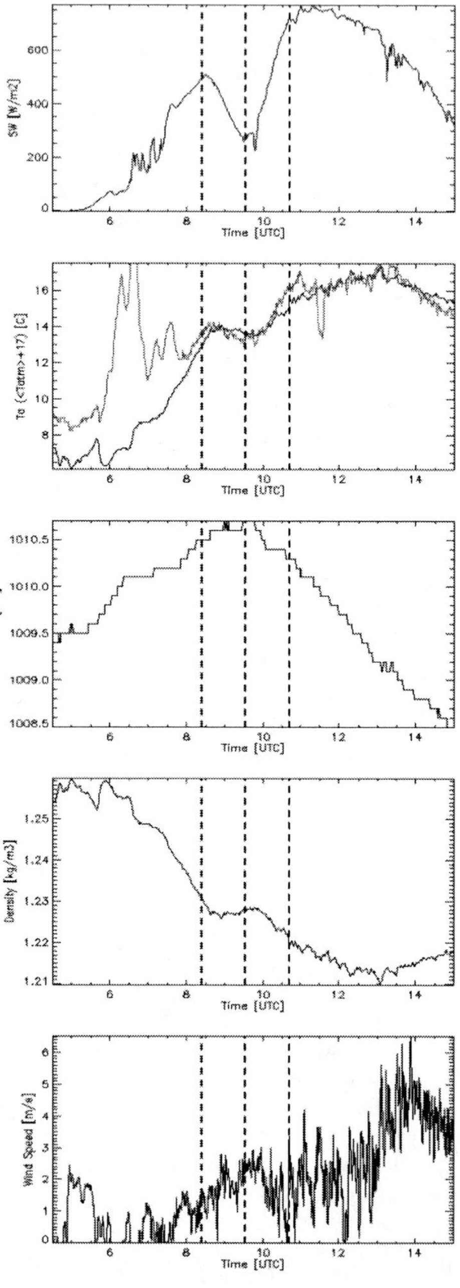

Figure 1. 2m meteorological data: Red line in the air temperature panel is the effective sky temperature (<Tatm>) as estimated from the downward LW radiation (to match the 2m air temperature range the plotted line is <Tatm>+17°C).

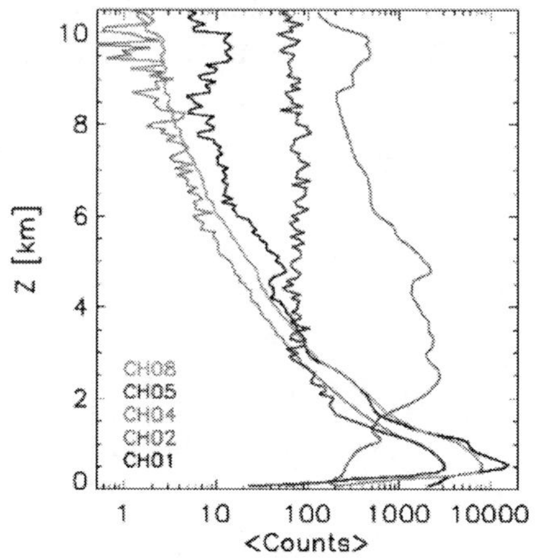

Figure 2. Mean profiles of dead-time corrected counts.

The transmitter is based on an Nd:YAG laser with 2nd (532 nm) and 3rd (355 nm) harmonic generators. Backscattered radiation is collected and analyzed at 4 wavelengths of interest: 355 and 532 nm for the elastic backscatter and 386.7 and 407.5 nm for Raman scattering of N_2 and H_2O molecules, respectively. The receiver is in a multiple-telescope configuration [3], allowing the sensing of a wide-altitude atmospheric interval. For the Raman and elastic backscatter, two different collection channels (CHs) are employed: one for the lower range (0.1–5-km altitude, using a 30-cm telescope and an optical fiber) and one for the upper range (2–13-km altitude, using an array of nine 50-cm aperture telescopes and an optical fiber bundle). An additional 15-cm telescope is used to collect the 532 nm elastic backscatter from the lower atmosphere (0.5–8 km). During the eclipse, the following signals were acquired with the 30 cm telescope: the 532 and 355 nm elastic backscattering (CH02, CH08); the N2 (CH04 at 387 nm) and WV (CH05 at 407 nm) Raman backscatter. The 532 nm elastic backscattering was also acquired with the 15 cm telescope (CH01). In the following, we analyze the signals corrected for the background value and the dead-time. During daytime the background is a function of the intensity of solar backscattered radiation and therefore should vary consistently during the eclipse.

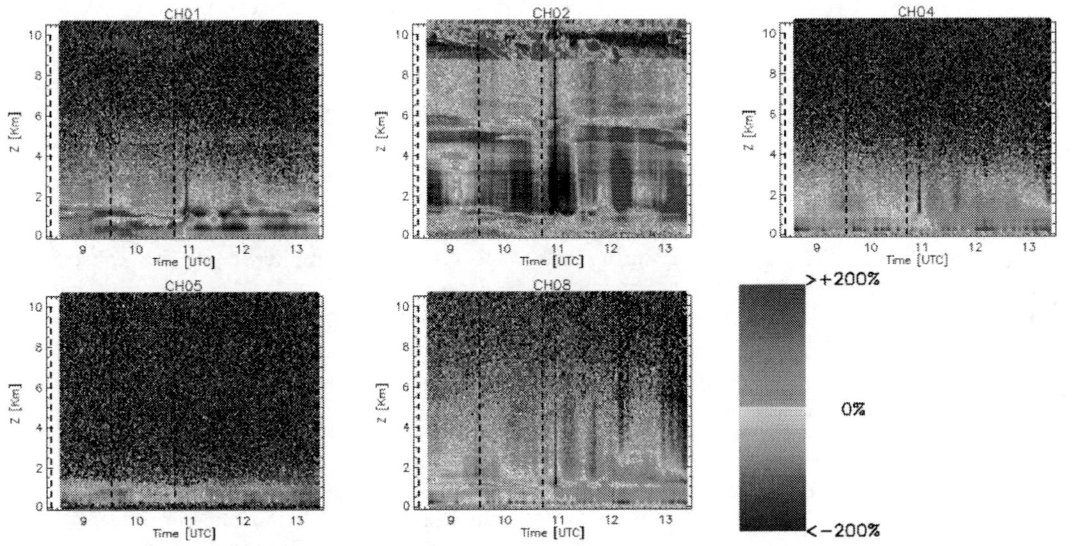

Figure 3. Time series of the profiles of the anomalies of the signals [%], with respect to the average (Fig.2). Dashed vertical bars are plotted in correspondence of beginning, maximum and ending time of the eclipse.

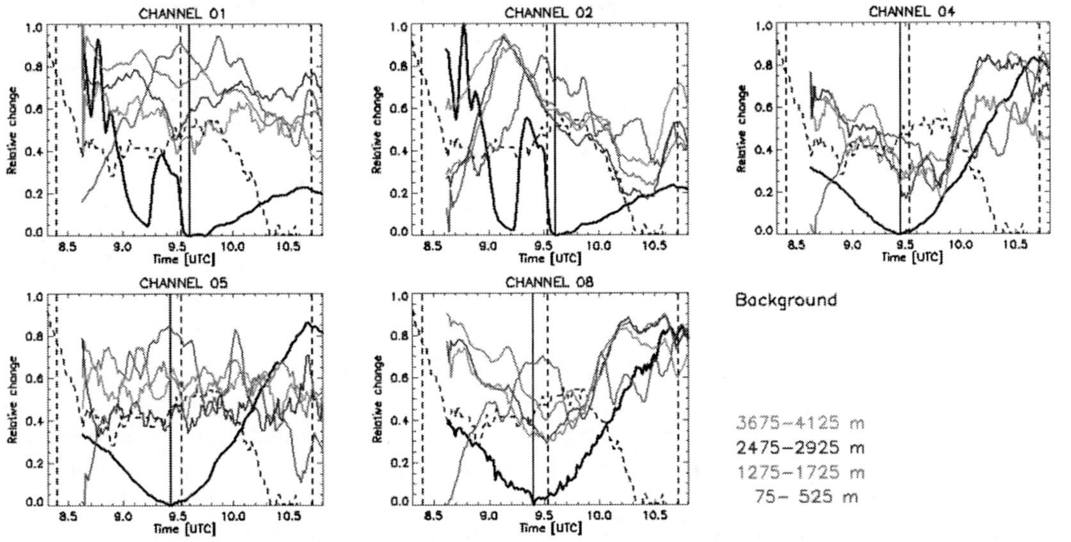

Figure 4. Time series of corrected signals for different atmospheric layers averaged over a 7' by 450 m moving box and normalized between 0 and 1 during the eclipse (dashed vertical bars in correspondence of beginning, maximum and ending time of the eclipse). Also plotted the normalized background value (black line) and the time of its minimum (black continuous vertical line).

3. RESULTS

Figure 2 shows the mean profiles of the dead-time corrected counts acquired by the different channels between 8:38-13:28 UTC. During the eclipse, the main permanent features on the signals are the presence of the aerosols in the PBL up to 2 km and a cirrus between 8.7 and 10 km.

Figure 3 shows the time series of single 1-min profiles during the same time window represented as anomaly [%] with respected to the averaged profiles. Channels 4 and 8 are the most sensitive to the atmospheric density. Considering the anomaly of the signals (Figure 3), during the eclipse it is apparent a decrease of the density

DOI: 10.1051/epjconf/201611915003

between 500 m and 2000 m. This feature is clear also in Figure 4, which shows the relative change of the signals averaged for different atmospheric layers. However, for channels 4 and 8, in addition to the decrease of the signal for the levels above 525 m, there is an increase of the signals for the layer between 75 m and 525 m, especially during the first part of the eclipse. This behavior shows the stratification of the atmosphere during the eclipse, due to the decrease of the solar irradiance. After the eclipse, the PBL shows the behavior of a clear sky, calm-wind day with the increase of the density between 500 m and 2-3 km (above those levels the Lidar signal becomes noisy) and a decrease at the lowest levels, as a consequence of the mixing of the PBL.

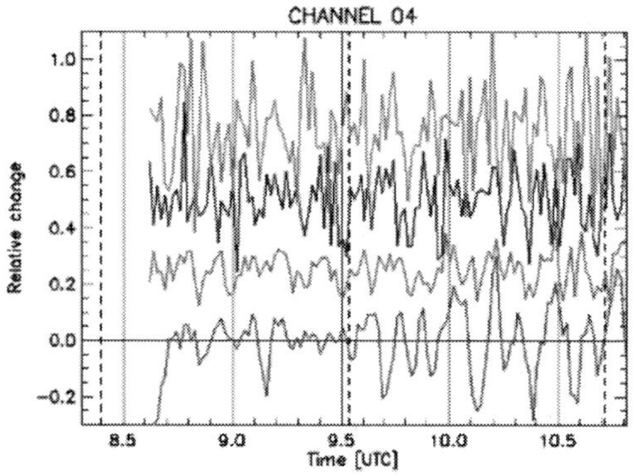

Figure.5 Time series of vertically averaged 1-min anomalies (with respect to the time averaged values Fig.4) for CH04 at different atmospheric levels. Same legend as Fig.4 for the altitude of the level. Dashed vertical lines are the beginning, the maximum and the ending time of the eclipse.

Figure 5 shows the time series of vertically averaged 1-min anomalies (with respect to the time averaged values Fig.4) for CH04 at different atmospheric levels. The anomalies seem to show some periodic behavior. Assuming that within the acquisition cycle of a single laser shot there are no changes of periodic nature in the performances of the whole acquisition chain, a periodic behavior in the signal could be generated by a periodic variation of the power of the emitting laser as well as of the properties of the atmosphere. The latters can modulate the measured signal in two ways: through the variability of the signal extinction due to the atmosphere below the backscattering level, through the variability of the properties of the backscattering level. With the above assumption, a periodicity in the signal due uniquely to a possible variation of the laser power should appear with the same frequency at all levels. Because of the different behavior with the altitude we may conclude that the observed fluctuations with period between 5' and 10' are due to changes in the atmospheric properties. Figure 5 shows also a decrease of the amplitude of the gravity waves, whose characteristics differs at different levels. This damping could be caused by the decrease of the surface heat forcing and we are currently investigating this issue.

4. CONCLUSIONS

During the eclipse the system observed an increase of the stratification of the atmosphere. The delay in the atmospheric response, of few minutes is consistent with what observed at the surface with the 2-min meteorological data. The occurrence of high frequency (period less than about 10 min) variations has been shown. From this preliminary analyses we exclude that the observed high frequency variability is due uniquely to instrument induced periodic signals. From a very qualitative analysis it seems that the altitude of the base of the Cirrus observed in CH2 data during the whole morning could be correlated to the occurrence of the eclipse. Further investigations need to be conducted on this point. Following this signal-based preliminary analysis we plan to perform the study of the corresponding retrieved geophysical parameters and to compare with other available observations during the event.

REFERENCES

[1] Kolev, N., et al., 2005: Aerosol Lidar and in situ ozone observations of the planetary boundary layer over Bulgaria during the solar eclipse of 11 August 1999, *Int. J. Remote Sens.*, **26**, 3567–3584.

[2] Amiridis, V et al., 2007. Aerosol Lidar observations and model calculations of the Planetary Boundary Layer evolution over Greece, during the March 2006 Total Solar Eclipse. *Atmos. Chem. Phys.*, **7**, 6181–6189.

[3] Congeduti, F., et al., 1999: The multiple-mirror lidar '9-eyes.' *J. Opt.*, **1A**, 185–191.

The National Weather Service Ceilometer Planetary Boundary Layer Project

M. Hicks[1], D. Atkinson[1], B. Demoz[3], K. Vermeesch[2,3], R. Delgado[3]

[1] *NWS, Silver Spring, MD, USA;* [2] *NASA/GSFC, Greenbelt, MD, USA;* [3] *UMBC, Baltimore, MD, USA*

ABSTRACT

The National Weather Service (NWS) is investigating the potential of utilizing the Automatic Surface Observing System's (ASOS) cloud base height indicator, the Vaisala CL31 ceilometer, to profile aerosols in the atmosphere. Field test sites of stand-alone CL31 ceilometers have been established, primarily, around the Washington DC metropolitan area, with additional systems in southwest USA and Puerto Rico. The CL31 PBL project examines the CL31 data collected for data quality, mixing height retrieval applicability, and its compliment to satellite data. This paper reviews the topics of the CL31 data quality and mixing height retrieval applicability.

1. INTRODUCTION

The National Weather Service (NWS) along with the Federal Aviation Administration (FAA) and the Department of Defense (DOD) owns an extensive suite of Automatic Surface Observing System (ASOS) stations. These stations span the USA with close to 1,000 sites and monitor atmospheric surface conditions (temperature, dewpoint, wind, etc.) and cloud base levels. The NWS, FAA and DOD are in the process of updating acquisition hardware and software for these stations as part of a Service Life Extension Program (SLEP). In this regard, the NWS Office of Science and Technology (OS&T) and Office of Operational Systems (OOS) are investigating the potential of adding functionality to the ASOS cloud base height indicator, the Vaisala CL31 ceilometer.

The OS&T NextGen Program and the OOS Sterling Field Support Center (SFSC) are collaborating with the NOAA Center of Atmospheric Sciences (NCAS) - Howard University Beltsville, MD Research Campus (HUBC), and University of Maryland Baltimore County (UMBC) to investigate the CL31 applicability to deriving boundary layer mixing heights and its compliment to satellite data. This project is known as the CL31 PBL Project. Of these topics, this paper discusses a review of status of the project and evaluation of the CL31 data quality and mixing height applicability.

2. METHODOLOGY

The CL31 PBL Project team has established data collection points in several locations: (1) on the west and northeast of Washington, DC metropolitan area in Sterling, VA (SFSC), College Park, MD (NCWCP), and Beltsville, MD (HUBC) since March 2013; (2) southwest Baltimore, MD at UMBC since October 2014; (3) University of Puerto Rico in Mayaguez since October 2014; and (4) west Texas alone Mexico's border at the University of El Paso since February 2015. The CL31 systems collect data at a 15 s frequency with a spatial resolution of 20 m from surface to 7.5 km.

LIDARs (Light Detecting and Ranging) and radiosoundings are used to evaluate the CL31's data quality and retrieved mixing height products, respectively. At SFSC, HUBC and UMBC micro-pulse LIDARs (MPL) are used to evaluate the CL31's data quality. HUBC also has a newer generation Vaisala ceilometer, CL51, used to further evaluate the CL31's data quality and Vaisala BL-View mixing height software. The MPL's normalized relative backscatter (NRB) profile data is used to build a relation with the CL31's backscattered intensity; in addition, their profile gradient intensities are related (Campbell et al. 2002).

The SFSC, HUBC, and UMBC sites facilitate routine radiosonde soundings. Their datasets are used to evaluate LIDAR mixing height algorithms applied to CL31 profile data. The radiosonde-based mixing heights are derived using the Bulk Richardson method. In addition, the mixing height comparison results are categorized by stability conditions; with the mean surface Bulk Richardson values (below 0.2 km) less than -1 indicating unstable conditions and those greater or equal to -1 stable conditions.

 DOI: 10.1051/epjconf/201611915004

3. RESULTS

This section includes preliminary results of this ongoing study. So far, the collected CL31 data at SFSC have been qualitatively compared to that of the MPL. All major atmospheric features sensed by the MPL system were sensed with the CL31 system, up to 7.5 km, as shown in Figure 1. More quantitative comparisons are currently being conducted. For example, the height of the gradient maximum and minimum up to 7.5 km in 3 hr averaged backscatter profile periods are being correlated between the two systems. In addition, the difference of daily averaged NRB to that of the backscatter intensity of the CL31 is being correlated. This qualitative and quantitative analysis will be conducted at all LIDAR hosting data collection sites.

The CL31 PBL Project team has also evaluated Compton et al. 2013, Hicks et al. 2015 and Vaisala BL-View mixing height algorithms. So far, this analysis has been done for SFSC and HUBC. These algorithms, relative to radiosonde-based derivations, showed the CL31 to be capable of detecting mixing heights, with a bias ranging from -176 m to 527 m for unstable atmospheric conditions and 173 m to 1158 m for stable atmospheric conditions.

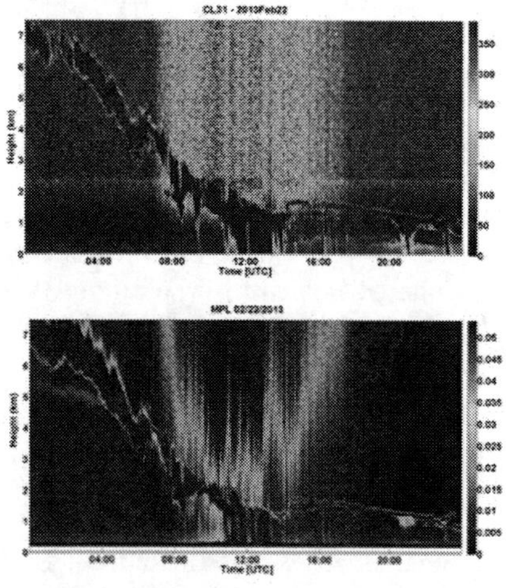

Figure 1. Comparison of time series of CL31 backscatter observations in top panel and MPL NRB observations in bottom panel at SFSC on February 22, 2013.

4. CONCLUSION

The CL31 ceilometer has a strong potential to be used for more than detecting cloud base heights in the ASOS network of almost 1000 stations across the USA. The profile data from these systems reflect real atmospheric features including frontal passages, low level jets, volcanic ash, smoke, thin clouds, and pollutants. In addition, the CL31 allows for the observation of mixing height which is lacking operationally.

The CL31 PBL team is pleased with the preliminary results and look forward to completing the analysis. The final goal of this project will be to provide concrete information on the impact of the CL31 profile data on volcanic ash, frontal passage, pollutant dispersion, and bay breeze monitoring/tracking and its overall impact on weather and dispersion forecast modeling.

REFERENCES

[1] Campbell, J. R., D. L. Hlavka, E. J. Welton, C. J. Flynn, D. D. Turner, J. D. Spinhirne, V. S. Scott III, I. H. Hwang, 2002: Full-time, Eye-Safe Cloud and Aerosol Lidar Observation at Atmospheric Radiation Measurement Program Sites: Instruments and Data Processing, *J. Atmos. Ocean Tech.*, **19**, 431-442.

[2] Compton, J. C., R. Delgado, T. A. Berkoff, R. M. Hoff, 2013: Determination of planetary boundary layer height on short spatial and temporal scales: A demonstration of the Covariance Wavelet Transform in ground based wind profiler and lidar measurements, *J. Atmos. Ocean Tech.*, **30**, 1566-1575.

[3] Hicks, M., S. Ricardo, E. Joseph, 2015: *The* Evaluation of a New Atmospheric Boundary Layer (ABL) Height Detection Method for Light Detecting and Ranging (LIDAR) Observations, Manuscript submitted for publication, *J. Atmos. Ocean Tech.,* Manuscript submitted for publication.

EPJ Web of Conferences **119**, 15005 (2016) DOI: 10.1051/epjconf/201611915005
ILRC 27

EXPLORING THE TURBULENT URBAN BOUNDARY BY USE OF LIDARS AND MICROWAVE RADIOMETERS

Mark Arend[1*], Ivan Valerio[1], Stephen Neufeld[1], Raymond Bishir[2], Younghu Wu[1]
Fred Moshary[1], David Melecio-Vazquez[2], Jorge Gonzalez[2]

[1] *Optical Remote Sensing Laboratory, Electrical Engineering Department and NOAA CREST CENTER, The City College of New York, New York, NY 10031*EMAIL: marend@ccny.cuny.edu*

[2]*Mechanical Engineering Department and NOAA CREST CENTER, The City College of New York, New York, NY 10031*

ABSTRACT

A Doppler lidar has been developed using fiber optic based technologies and advanced signal processing techniques. Although this system has been operated in a scanning mode in the past, for this application, the system is operated in a vertically pointing mode and delivers a time series of vertical velocity profiles. By co-operating the Doppler lidar with other instruments, including a back scatter lidar, and a microwave radiometer, models of atmospheric stability can be tested, opening up an exciting path for researchers, applied scientists and engineers to discover unique phenomena related to fundamental atmospheric science processes. A consistent set of retrievals between each of these instruments emphasizes the utility for such a network of instruments to better characterize the turbulent atmospheric urban boundary layers which is expected to offer a useful capability for assessing and improving models that are in great need of such ground truth.

1. INTRODUCTION

Research in turbulent boundary layer dynamics has been and will likely continue to be of interest to scientists and engineers because of the wide applications across many disciplines. However, defining turbulence and deriving closed form physical solutions that are based on strict axiomatic and physical principles continues to be an unsolved challenge even though much advancement in modern science and mathematics has been made [1]. Experimental methods that utilize Doppler lidars are very useful for investigating flows since the motion as well as the constituent concentration can be tracked simultaneously, thus providing additional metrics for assessing the flow dynamics. When applied to real world situations such as the urban atmospheric boundary layer, the possibility exists for there being relevant applications that depend on robust and field-able instruments and networks. A fiber based Doppler lidar system [2-7], a back scatter Raman lidar system [8] and a network of other Ground based remote sensing instruments (including a microwave radiometer for measuring temperature and water vapor profiles) are included in a set of instruments comprising the NYCMetNet [9], developed by members of the Optical Remote Sensing Lab at the City College of New York. This suite of instruments are sometimes (but not always) co-operating and when there is overlap, there is an opportunity to look closely at the consistencies between the instruments and expose any inconsistencies by following a time series of the vertical profiles and comparing these visualizations with one another.

2. METHODOLOGY

COHERENT DOPPLER LIDAR: The Coherent Doppler lidar has been constructed and tested at CCNY. The system consists of a 1542 nm CW laser fiber laser that is split in to two paths such that one path is used for a local oscillator and the other path is amplified by a single mode erbium doped fiber amplifier EDF and then modulated by an acousto-optic modulator, producing 200 nsec pulses at 20 kHz. These pulse are amplified further with a final amplifier and sent to an optical circulator before be launched in to the 4 inch clear aperture telescope from the single mode fiber. The back scattered light from the atmosphere is gathered by the telescope, passed through the circulator and mixed with the local oscillator with a 2x2 coupler. A 12h MHZ bandwidth balanced photodetector followed by TIA amplifiers detect the signal which is then sampled by a 400 MS/sec 12 but ADC. These signals are processed with a FPGA that calculates gated Fourier transforms (or autocorrelations depending on the chosen mode) and accumulate power spectra for further processing on a host computer.

DIRECT DETECTION BACKSCATTER LIDAR : The direct detection backscatter lidar was made using a Q-switched Spectra-Physics Quanta Ray Pro230 Nd-YAG with variable output power up to 475 mJ @ 532 nm, 950 mJ @ 1064 nm and 300 mJ @ 355 nm. This system currently has five channels with wavelengths at 1064 nm, 532 nm, 355 nm, including a Raman channel at 407 nm for water with a second Raman channel at 387 for nitrogen. The repetition rate is 30 Hz with 1-2 ns pulse duration @ 532 nm; <0.5 mrad beam divergence. The beam is on axis to the telescope. See ref. WU for more information.

MICROWAVE RADIOMETER: The Radiometrics Profiling Radiometer, model MP-3000A, is also operated at CCNY in Manhattan and produces vertical profiles from the surface to 10 km, producing high-resolution temperature, relative humidity and water vapor profiles, and low-resolution liquid profiles. The MP-3000A incorporates two radio frequency (RF) subsystems in the same cabinet. These RF subsystems share the same antenna and antenna pointing system. The temperature profiling subsystem utilizes sky brightness temperature observations at selected frequencies between 51 and 59 GHz. The water vapor profiling subsystem utilizes sky brightness temperature observations at selected frequencies between 22 and 30 GHz. Atmospheric stability is related to the actual sounding relative to a critical sounding which depends on how much water vapor is in the air. Fig. 1 shows an illustration of the dry adiabatic sounding and shows how stability is referenced to the dry adiabatic laps rate Γ_d.

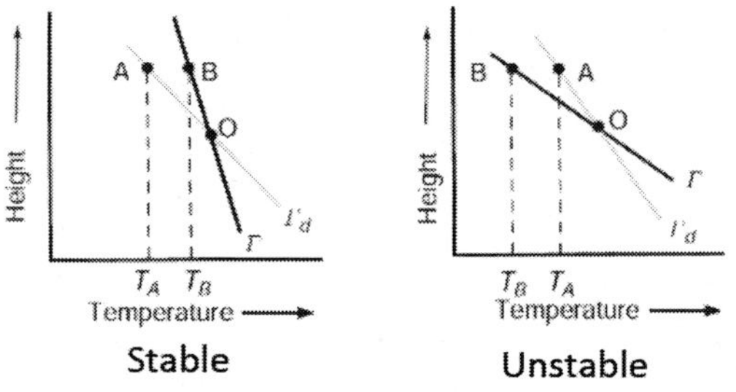

Fig. 1 Stable and unstable sounding relative to the dry adiabatic laps rate Γ_d

The virtual potential temperature is a convenient measurement for the air since it takes into account the variability in density between dry air and moist air ("virtual"), and difference in temperatures due to pressure variation.

The Radiometer is used to derive the virtual potential temperature from the analytical profiles of temperature, T, and relative humidity, RH. From the observed temperature profiles, the potential temperature and the saturation pressure are calculated. From the observed relative humidity profiles, the saturation mixing ratio and the actual mixing ratio values are generated. These values are then used to derive virtual potential temperature generates a vertical profile that can be used for classifying the stability of the atmosphere. If the virtual potential temperature increases with elevation the atmosphere is considered stable. An unstable (or convective) atmosphere is defined for conditions in which the virtual potential temperature decreases with height. Neutral stability is defined when there is no change in virtual potential temperature with height.

RESULTS: Figs. 2-7 shows the processed signals from each of these three instruments for a particular day (10/31/2014). Colored boxes represent the time and height where the data from each instrument is sampled since both the lidars are not continuously operated. Magenta color represents the coherent Doppler Lidar sampling and the Green color represents the direct detection lidar sampling.

DOI: 10.1051/epjconf/201611915005

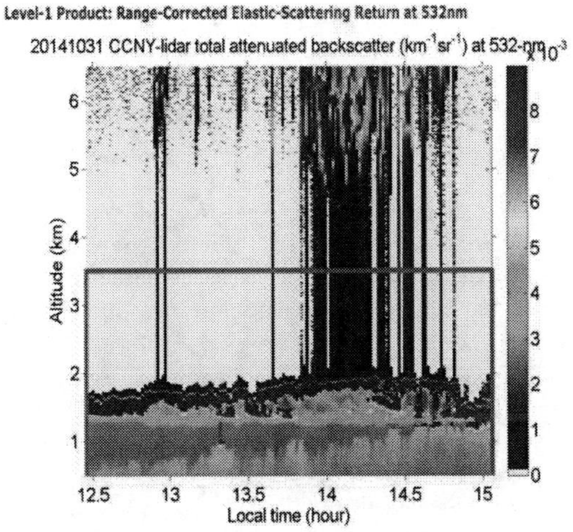

Fig. 2 Direct detection lidar at 532 nm

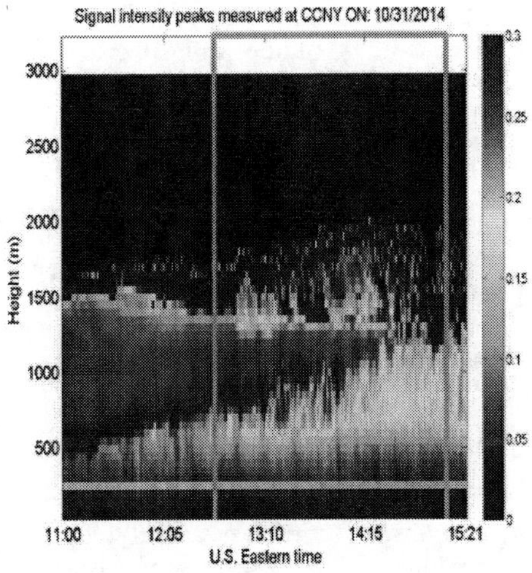

Fig. 3 Coherent Detection Doppler lidar Signal Intensity

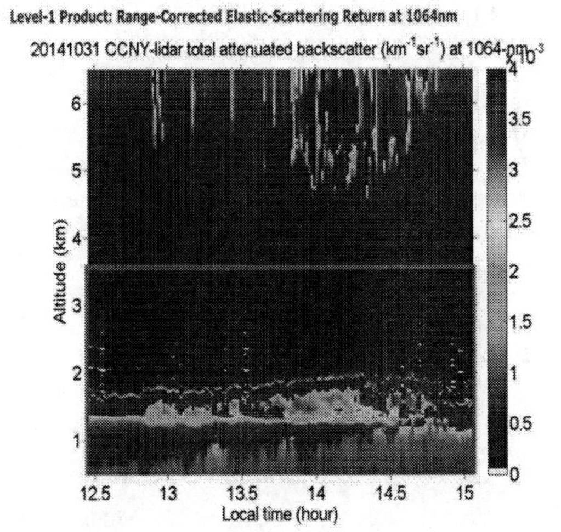

Fig. 4 Direct detection Lidar at 1064 nm

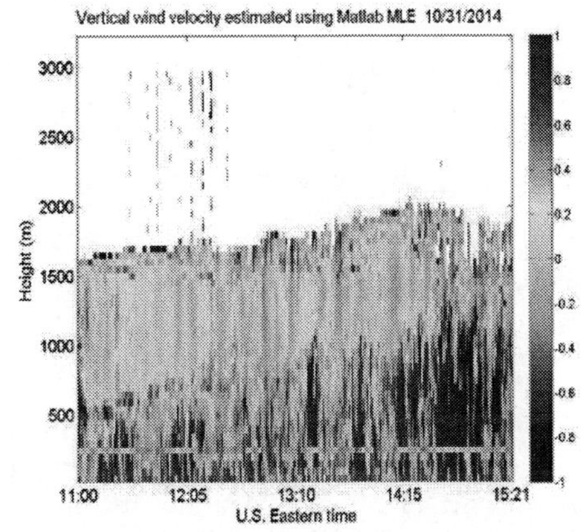

Fig. 5 Coherent Detection Lidar Vertical Velocity (m/s)

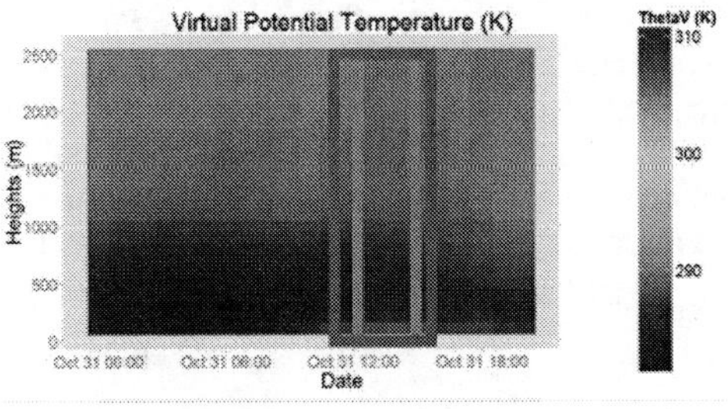

Fig. 6 Virtual Potential Temperature

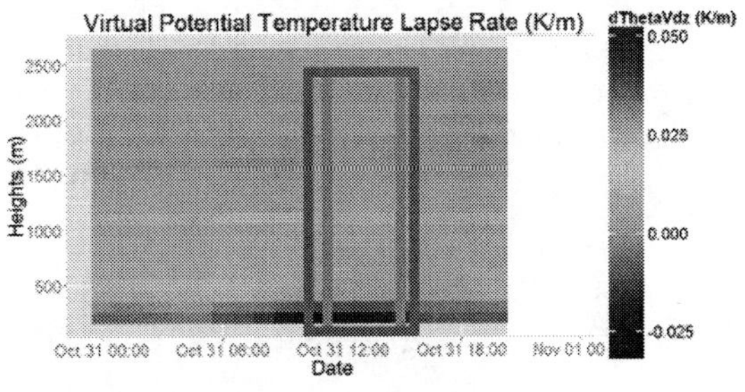

Fig. 7 Virtual Potential Temperature Laps rate

DISCUSSION: On this particular day (Oct 31, 2014) the stability of the atmosphere was not being forced by any significant mean surface winds and the upper air winds were sufficiently quite so as to not cause significant transport of aerosols horizontally as observed by other instruments in the CCNY NYCMetNet. This then simplifies the interpretation of the temporal evolution of the three instruments as follows: The large thermal gradient of the virtual potential temperature (Fig. 6) a few hours before and after noon shows strong instability which would be expected to cause a significant uplifting. This gradient is shown even more dramatically in the laps rate (Fig. 7, the negative of the derivative of virtual potential temperature vs. height) plot. This episode is consistent with the strong vertical motion that begins to evolve during the same time period (Fig 5) which dramatically increases during the latter part of this time period. The growth of the boundary layer and the increased presence of aerosols throughout this time period is very evident in Figs. 2, 3 and 4. Figs. 2, 3 and 4 are also consistent with each other in depicting the presence of the aerosol plume that lifts up under the clouds at 1.5 to 2 km.

CONCLUSION: A coherent Doppler lidar, a direct detection back scatter lidar and a microwave radiometer have been operated co-operatively to demonstrate that each of these instruments respond to a dynamically changing atmospheric boundary layer stability condition in a a manner that is consistent across the suite of instruments. This inspires further studies to demonstrate these principles under other dynamically changing conditions so that a stability climatology can be documented which should have an impact on model development.

ACKNOWLEDGEMENTS: Supported in part by NOAA Cooperative Agreement No: NA11SEC4810004

REFERENCES

[1] Gorban A.N.; Karlin, I. "Hilbert's 6th Problem: Exact and Approximate Hydrodynamic Manifolds for Kinetic Equations" Bull. Amer. Math. Soc. 51 (2014), 187-246

[2] M. Arend, S. Abdelazim, M. Lopez and F. Moshary, "A comparison of two embedded programming techniques for high rep rate coherent Doppler lidars," in *SPIE*, 2013.

[3] S. Abdelazim, D. Santoro, M. Arend, F. Moshary and S. Ahmed, "Field programmable gate array processing of eye-safe all-fiber coherent wind Doppler lidar return signals," in *SPIE, Remote Sensing*, Prague, 2011.

[4] S. Abdelazim, D. Santoro, M. Arend, F. Moshary and S. Ahmed, "Wind Velocity Estimate and Signal to Noise Ratio Analysis of an All Fiber Coherent Doppler Lidar System," in *16th Coherent Laser Radar Conference*, Long Beach, California, June, 2011.

[5] S. Abdelazim, D. Santoro, M. Arend, F. Moshary and S. Ahmed, "All-fiber Coherent Doppler LIDAR for Wind Sensing," in *5th Symposium on Lidar Atmospheric Applications, 91st American Meteorological Society Annual Meeting*, Seattle, Washington, 2011.

[6] S. Abdelazim, D. Santoro, M. Arend, F. Moshary and S. Ahmed, "Development of All-fiber Coherent Doppler LIDAR System for Wind Sensing," in *15th Symposium on Meteorological Observation and Instrumentation, 90th AMS Annual Meeting*, Atlanta, GA, 2010.

[7] S. Abdelazim, D. Santoro, M. Arend, F. Moshary and S. Ahmed, "All-fiber Coherent Doppler LIDAR for Wind Sensing," in *15th Annual Coherent Laser Radar Conference*, Toulouse, France, 2009.

[8] Wu, Y., Chaw, S. Gross, B., Moshary, F., Ahmed, S. "Low and optically thin cloud measurements using a Raman- Mie lidar" Applied Opt. 48, 1218-1227, 2009.

[9] Committee on Urban Meteorology: Scoping the Problem, Defining the Needs; Board on Atmospheric Sciences and Climate; Division on Earth and Life Sciences, National Research Council, 2012, Urban Meteorology: Forecasting, Monitoring, and Meeting Users' Needs, Washington, DC: The National Academies Press, p54, http://nycmetnet.ccny.cuny.edu

DOI: 10.1051/epjconf/201611915006

ATMOSPHERIC BOUNDARY LAYER HEIGHT EVOLUTION WITH LIDAR IN BUENOS AIRES FROM 2008 TO 2011

Ezequiel Eduardo Pawelko[1], Jacobo Omar Salvador[1], Pablo Roberto Ristori[1], Juan Vicente Pallotta[1], Lidia Ana Otero[1], Eduardo Jaime Quel[1]

[1] *CEILAP (CITEDEF-CONICET), UMI-IFAECI-CNRS 3351, Juan B. de La Salle 4397, B1603ALO, Villa Martelli, Buenos Aires, Argentina, E-mail: epawelko@citedef.gob.ar*

ABSTRACT

The analysis of the atmospheric boundary layer top height evolution is obtained from 2008 to 2011 in Buenos Aires using the multiwavelength lidar located at CEILAP (CITEDEF-CONICET) (34°33' S; 58°30' W; 17 m asl). Algorithms recognition based on covariance wavelet transform are applied to obtain seasonal statistics. This method is being evaluated for use in the Lidar Network in Argentina and it is being deployed in Patagonia region currently. The technique operates in real time in both low and high aerosol loads and with almost no human supervision.

1. INTRODUCTION

The Atmospheric Boundary Layer (ABL) is the lowest layer that is directly influenced by the Earth´s surface. It responds to surface forcing by frictional drag, evaporation and transpiration, and sensible heat transfer with a timescale of one hour or less [1]. Lidar, light detection and ranging, is a valuable tool to study the atmospheric boundary layer. This instrument works with high spectral and spatial resolution that allows identifying the ABL height. Aerosol vertical structure is considered as a good tracer to determine the ABL height. Lidar backscatter profiles show a strong negative gradient at the top of this layer which can be recognized with different mathematic methods. Algorithms based on threshold values [2], minimum vertical gradient in the lidar profile [3], maximum in variances of signals [4] and a method that fits an idealized profile [5] are widely used. It is recognized that these approaches are sensitive to noise and small-scale structure in the lidar profiles.

A robust algorithm based on CWT (Covariance Wavelet Transform) is proposed by [6, 7]. The method can recognize the ABL height in profiles with lower signal to noise ratio than other algorithms.

2. METHOD AND RESULTS

2.a. Lidar setup

The tropospheric multiwavelength lidar system was developed to monitor range-resolved aerosol optical properties. Its dynamic range covers a region from 200 m to 10 km in coaxial overlap (from 2009 up to present) and from 600 m to 10 km in biaxial overlap setup. The lidar emission system is based on a Nd:YAG laser (Continuum Surelite III P-IV) with 3 wavelengths at 10 Hz (650 mJ at 1064 nm). First and second harmonic are also generated. The IR wavelength is normally used for studying the ABL. The backscattered radiation is collected by an f/2, 1 m focal length Newtonian telescope and focused into a 1.8 mm diameter optical fiber that works as a field stop. The data acquisition system had a 20 MHz sampling rate and a 7.5 m vertical resolution [8].

2.b.- Method

The ABL height recognition is obtained applying a CWT (Covariance Wavelet Transform) and Haar function based algorithm [7]. CWT and Haar function are shown in equations 1 and 2 respectively. In those part, *a* and *b* are the dilatation and translation factors respectively, *z* is the height vector and *f(z)* is the lidar profile. The atmospheric boundary layer height is presented as an abrupt decrease of the lidar backscatter.

$$W_f(a,b) = \frac{1}{a}\int_{z_1}^{z_2} f(z)h\left(\frac{z-b}{a}\right)dz \qquad (1)$$

$$h\left(\frac{z-b}{a}\right) = \begin{cases} +1: b-a/2 \le z \le b \\ -1: b \le z \le b+a/2 \\ 0: \text{elsewhere} \end{cases} \qquad (2)$$

Equation 3 is the spectral variance proposed by [7, 9] to calculate the wavelet dilation factor. The traslation *b* that produces the maximum value in CWT is the atmospheric boundary layer top height.

 DOI: 10.1051/epjconf/201611915006

$$\left[D^2(a)=\int_{z_1}^{z_2}\left[W_f(a,b)\right]^2 db\right]_{Max} = a \qquad (3)$$

2.c.- Results

The results of a seasonal study of the ABL height evolution with lidar between 2008 and 2011 are presented. 197 days of measurements are processed. The ABL height is obtained from 1064 nm backscatter wavelength. Measurements with clouds supported on the ABL are not processed.

Figures 1 to 4 show the ABL height evolution in the summer, fall, winter and spring.

(a)

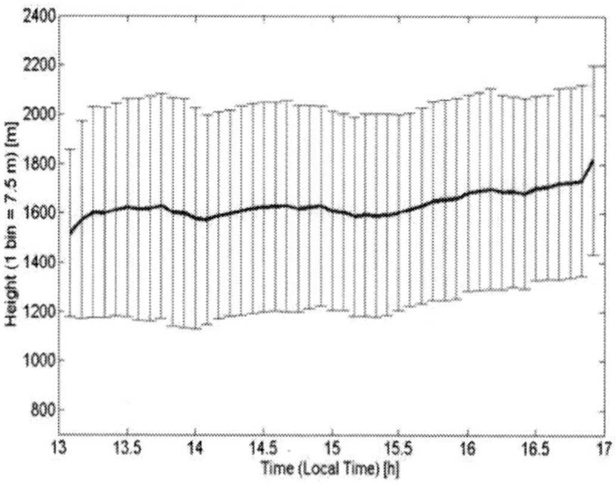

(b)

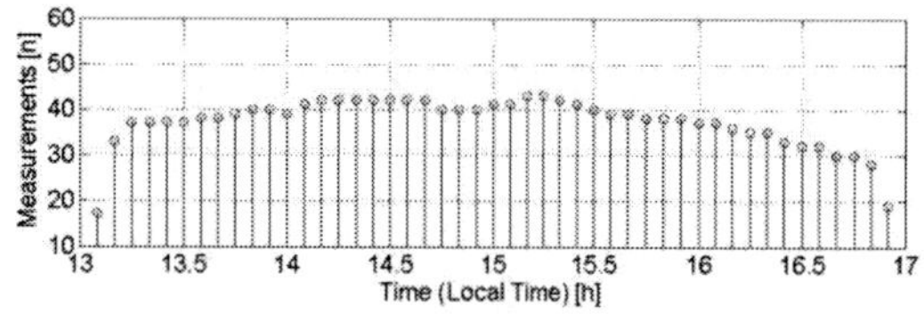

Figure 1. (a) Daytime ABL height evolution (mean value +- one standard deviation) in the summer between 2008 and 2011 (1064 nm, 7.5 m spatial resolution, 5 min temporal resolution) (50 measurements: 03/1/2008, 03/25/2008, 01/05/2009, 01/09/2009, 01/20/2009, 01/21/2009, 01/22/2009, 01/23/2009, 01/26/2009, 02/02/2009, 02/06/2009, 02/12/2009, 02/17/2009, 02/18/2009, 02/19/2009, 02/25/2009, 02/26/2009, 12/30/2009, 01/05/2010, 01/08/2010, 01/12/2010, 01/13/2010, 01/21/2010, 01/22/2010, 01/24/2010, 01/25/2010, 01/28/2010, 01/29/2010, 02/08/2010, 02/11/2010, 02/12/2010, 03/15/2010, 03/16/2010, 12/21/2010, 12/27/2010, 01/06/2011, 01/18/2011, 01/20/2011, 01/21/2011, 02/02/2011, 02/03/2011, 02/04/2011, 02/14/2011, 02/15/2011, 03/01/2011, 03/02/2011, 03/14/2011, 03/18/2011, 12/27/2011, 12/29/2011). (b) Number of measurements processed at each time interval.

(a)

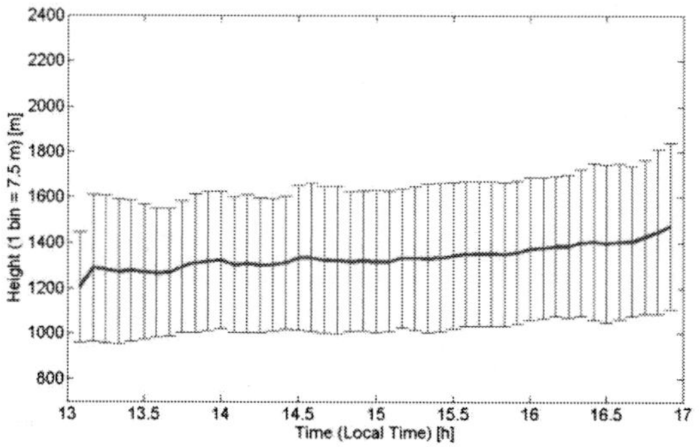

(b)

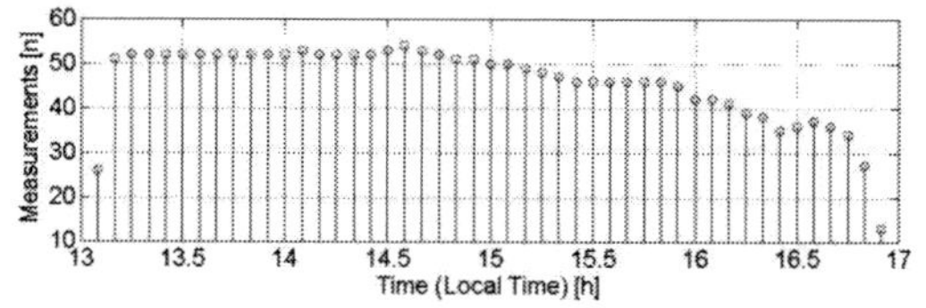

Figure 2. (a) Daytime ABL height evolution (mean value +- one standard deviation) in the fall between 2008 and 2011 (1064 nm, 7.5 m spatial resolution, 5 min temporal resolution) (57 measurements: 04/08/2008, 04/10/2008, 04/14/2008, 04/17/2008, 04/18/2008, 04/21/2008, 04/22/2008, 04/24/2008, 04/25/2008, 05/01/2008, 05/05/2008, 05/06/2008, 05/09/2008, 05/10/2008, 05/29/2008, 05/30/2008, 06/06/2008, 06/10/2008, 06/12/2008, 06/13/2008, 06/17/2008, 03/23/2010, 03/25/2010, 03/29/2010, 04/05/2010, 04/06/2010, 04/16/2010, 04/22/2010, 04/23/2010, 04/26/2010, 04/28/2010, 04/29/2010, 04/30/2010, 05/07/2010, 05/10/2010, 05/12/2010, 06/01/2010, 06/07/2010, 03/21/2011, 03/28/2011, 03/30/2011, 04/01/2011, 04/06/2011, 04/26/2011, 04/27/2011, 05/02/2011, 05/03/2011, 05/04/2011, 05/06/2011, 05/09/2011, 05/19/2011, 06/07/2011, 06/08/2011, 06/09/2011, 06/10/2011, 06/13/2011, 06/14/2011). (b) Number of measurements processed at each time interval.

(a)

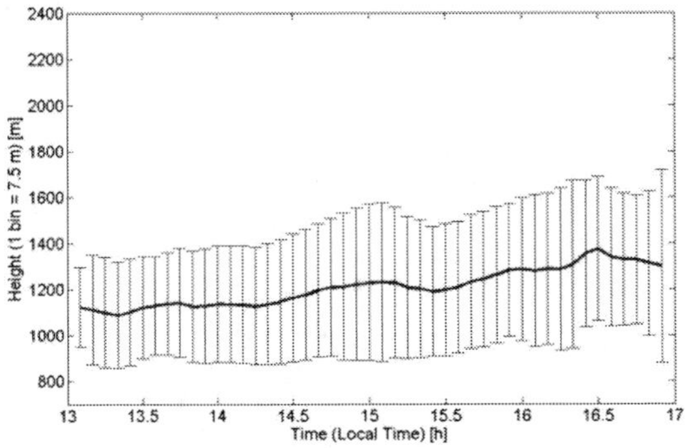

(b)

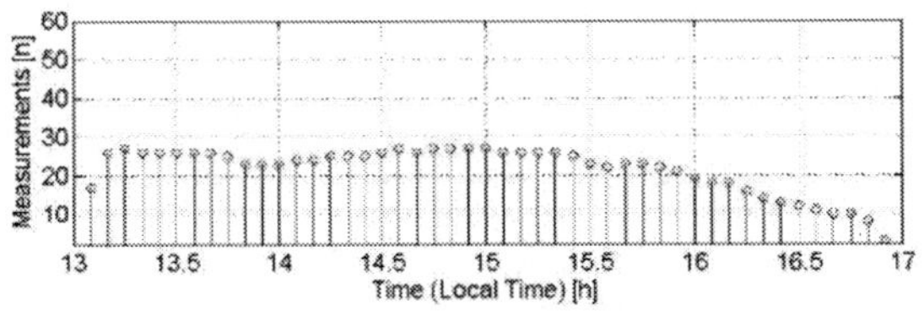

Figure 3. (a) Daytime ABL height evolution (mean value +- one standard deviation) in the winter between 2008 and 2011 (1064 nm, 7.5 m spatial resolution, 5 min temporal resolution)(38 measurements: 09/12/2008, 09/17/2008, 08/11/2009, 08/20/2009, 08/26/2009, 09/09/2009, 06/28/2010, 07/15/2010, 07/16/2010, 07/23/2010, 08/092010/, 08/20/2010, 08/24/2010, 08/25/2010, 09/06/2010, 09/07/2010, 09/09/2010, 09/14/2010, 06/27/2011, 06/28/2011, 06/29/2011, 07/01/2011, 07/02/2011, 07/06/2011, 07/12/2011, 07/28/2011, 07/29/2011, 08/03/2011, 08/04/2011, 08/05/2011, 08/11/2011, 08/23/2011, 08/25/2011, 08/30/2011, 09/02/2011, 09/12/2011, 09/14/2011, 06/29/2011). (b) Number of measurements processed at each time interval.

(a)

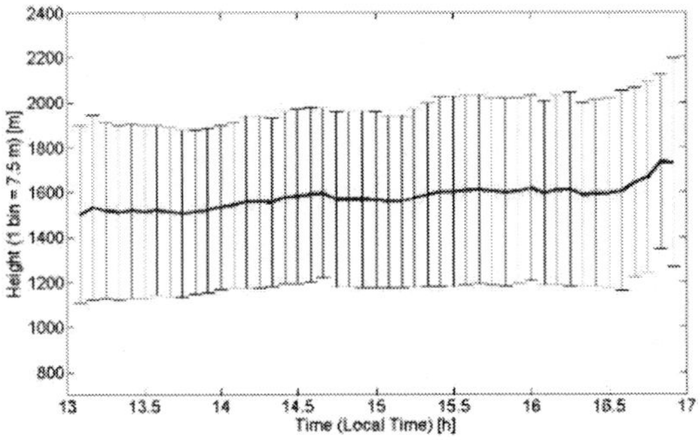

(b)

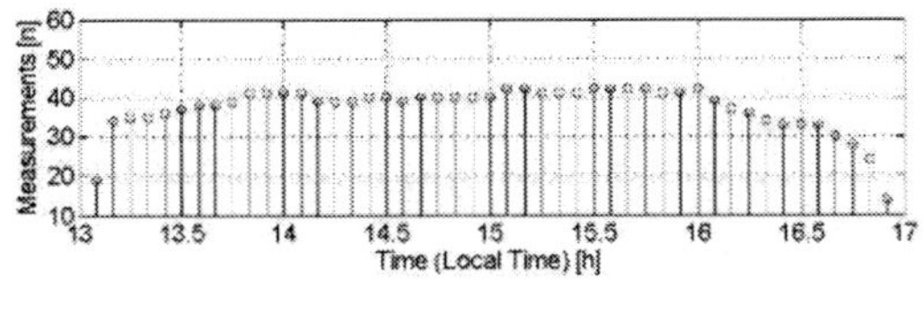

Figure 4. (a) Daytime ABL height evolution (mean value +- one standard deviation) in the spring between 2008 and 2011(1064 nm, 7.5 m spatial resolution, 5 min temporal resolution)(52 measurements: 09/22/2008, 09/25/2008, 09/26/2008, 10/02/2008, 10/06/2008, 10/16/2008, 10/20/2008, 10/24/2008, 10/29/2008, 10/30/2008, 10/31/2008, 11/05/2008, 11/06/2008, 11/13/2008, 11/17/2008, 11/21/2008, 12/03/2008, 12/04/2008, 12/05/2008, 12/15/2008, 12/16/2008, 11/05/2009, 11/16/2009, 10/04/2010, 10/12/2010, 11/24/2010, 11/29/2010, 12/01/2010, 12/02/2010, 12/03/2010, 12/06/2010, 12/16/2010, 09/27/2011, 10/03/2011, 10/06/2011, 10/11/2011, 10/17/2011, 10/18/2011, 10/20/2011, 10/21/2011, 10/26/2011, 10/27/2011, 10/31/2011, 11/01/2011, 11/11/2011, 11/16/2011, 11/22/2011, 11/23/2011, 11/29/2011, 12/01/2011, 12/07/2011, 12/12/2011). (b) Number of measurements processed at each time interval.

Table I presents results of ABL height in Buenos Aires.

Table I. seasonal atmospheric boundary layer height between 2008 and 2011.

Seasons	Measurements [n]	ABL top height [m]	
		Mean	Standard deviation
Summer	50	1634.9	411.6
Fall	57	1336.5	314.6
Winter	38	1208.0	285.7
Spring	52	1577.9	399.4

3. CONCLUSIONS

A preliminary seasonal study of the daytime atmospheric boundary layer height evolution is performed with aerosol lidar and an algorithm based on covariance wavelet transform. A database of 197 lidar measurements in Buenos Aires is processed. The results show that this layer varies depending on the season, between 1223 and 2046 m in the summer, 1021 and 1651 m in the fall, 922 and 1493 m in the winter and between 1178 and 1977 m in the spring, using mean value with one standard deviation computed from 13 to 17 h local time.

ACKNOWLEDGMENTS

Authors would like to thank JICA (Japan International Cooperation Agency) and JST (Japan Science and Technology Agency) through the Science and Technology Research Partnership for Sustainable Development (SATREPS) for supporting this work under the project: "Development of Atmospheric Environmental Risk Management System in South America".

REFERENCES

[1] Stull R. 1998: "*An Introduction to Boundary Layer Meteorology*", Kluwer Academic Publishers, Dordrecht, Boston, London.

[2] Boers, R., E. Eloranta, 1986: Lidar measurements of the atmospheric entrainment zone and the potential temperature jump across the top of the mixed layer; Bound.-Lay. Meteorol., **34**, pp. 357–375.

[3] Flamant, C., J. Pelon, P. Flamant, P. Durand, 1997: Lidar determination of the entrainment zone thickness at the top of the unstable marine atmospheric boundary layer, Bound.-Lay. Meteorol., **83**, pp. 247–284.

[4] Hooper, W., E. Eloranta, 1986: Lidar measurements of wind in the planetary boundary layer: the method, accuracy and results from joint measurements with radiosonde and kytoon, J. Clim. Appl. Meteor., **25**, pp. 990–1001.

[5] Steyn, D., M. Baldi, R. Hoff, 1999: The detection of mixed layer depth and entrainment zone thickness from lidar backscatter profiles, J. Atmos. Ocean Tech., **16**, pp. 953–959.

[6] Cohn, S., W. Angevine, 2000: Boundary layer height and entrainment zone thickness measured by lidars and wind-profiling radars, J. Appl. Meteor., **39**, pp. 1233–1247

[7] Davis K., N. Gamage, C. Hagelberg, C. Kiemle, D. Lenschow, P. Sullivan, 2000: An objective method for deriving atmospheric structure from airborne lidar observations, J. Atmos. Oceanic. Technol., **17**, pp. 1455-1468.

[8] Otero L., P. Ristori, E. Pawelko, J. Pallotta, E. Quel, 2011: Six-Year Evolution of Multiwavelength Lidar System at CEILAP, V Workshop on Lidar Measurements in Latin America, Opt. Pura Apl., **44**, pp. 13-18.

[9] Gamage N., C. Hagelberg, 1993: Detection and Analysis of Microfronts and Associated Coherent Events Using Localized Transforms. J. Atmos. Sci., 50, pp. 750–756.

LIDAR MEASUREMENTS OF OZONE IN THE UPPER TROPOSPHERE – LOWER STRATOSPHERE AT SIBERIAN LIDAR STATION IN TOMSK

O. A. Romanovskii[1,2*], S.I. Dolgii[1], V. D. Burlakov[1], A.A. Nevzorov[1] and A. V. Nevzorov[1]

[1] *V.E. Zuev Institute of Atmospheric Optics SB RAS, 1 Zuev Square, Tomsk 634021, Russia, roa@iao.ru*

[2] *National Research Tomsk State University, 36 Lenina Pr., Tomsk 634050, Russia*

ABSTRACT

The paper presents the results of DIAL measurements of the vertical ozone distribution at the Siberian lidar station. Sensing is performed according to the method of differential absorption and scattering at wavelength pair of 299/341 nm, which are, respectively, the first and second Stokes components of SRS conversion of 4th harmonic of Nd:YAG laser (266 nm) in hydrogen. Lidar with receiving mirror 0.5 m in diameter is used to implement sensing of vertical ozone distribution in altitude range of 6-16 km. The temperature correction of zone absorption coefficients is introduced in the software to reduce the retrieval errors.

1. INTRODUCTION

Ozone plays a key role as a shield against hard UV solar radiation for all living on our planet. Also, it is an important climate-forcing agent, which plays a significant role in thermal balance of the planet. Activation of the processes of destruction of stratospheric ozone layer, formation and spread of ozone anomalies ("ozone holes") over Antarctica, Europe, and Siberia necessitated arrangement of planetary-scale monitoring. This has led to creation of network of ground-based ozonometer stations, which measure the total ozone (TO) content with the help of different spectrophotometers (foreign instruments such as Dobson, Junker, and Brewer spectrophotometers, and national instrument such as M – 124 ozonometer). In addition, information on TO is inferred from satellite (TOMS and other) measurements. Laser sensing of ozonosphere had been regular at a number of observatories since the second half of 1980s. It provides information on vertical ozone distribution (VOD), successfully complicating a similar information obtained *in situ* with the help of ozonesondes, as well through "onion-peeling" approach from satellites (SAGE-II, Terra-Aqua, etc.) [1]. The multiyear lidar observations of stratospheric ozone had made it possible to obtain information on climatology of ozonosphere, especially above 30 km, where ozonesonde data become unrepresentative. The lidar measurements of VOD are performed on the basis of method of differential absorption of backscattered energy of laser radiation in UV wavelength range of 240-360 nm (the so-called Hartley-Huggins band). As a rule, for sensing the stratospheric ozone, the fundamental frequency of excimer XeCl laser (308 nm) is used as λ_{on}, where absorption of sensing radiation by ozone is strong. At the same time, as a reference frequency (with weak absorption) λ_{off}, researchers use either the first component of its SRS conversion in hydrogen (353 nm), with efficiency of conversion reaching 40%, or the third harmonic of Nd:YAG laser (355 nm), the energy of which may reach more than 100 mJ.

Long-term period of lidar observations of stratospheric ozone (since 1989) at the measurement complex Siberian Lidar Station (SLS), V.E. Zuev Institute of Atmospheric Optics, Siberian Branch, Russian Academy of Sciences (Tomsk: 56.5°N; 85.0°E) [2] showed that the part, most significant for studying the ozonosphere, is located in the lower stratosphere, where ozone is subject to the effect of dynamic factor. This part totally determines the TO variations in atmospheric column. Poorly studied scenarios of stratosphere-troposphere exchange, deformation of ozonosphere by jet streams, and formation of fine tongue-like structure of ozone layer also develop in this part. However, higher concentration sensitivity of lidar measurements is required to perform these studies in altitude ranges of upper troposphere – lower stratosphere, thus necessitating passage to shorter-wavelength region of UV spectrum, where ozone absorption cross section is greater.

2. SELECTION OF WAVELENGTHS

Excimer KrF laser (248 nm) or the fourth harmonic of Nd:YAG laser (266 nm) in combination with technique of SRS conversion in H_2, D_2, CO_2, and other gases [3-6] are usually used for tropospheric measurements of ozone. Hydrogen and deuterium are most widespread, in this regard. A possible set of wavelengths, corresponding to the 1st, 2nd, and 3rd Stokes (C) frequencies of SRS conversion in H_2, D_2, and CO_2, is presented in Table.

Pumping radiation	Wavelengths (nm) corresponding to the Stokes frequencies (*C*) of the SRS conversion		
	in H_2 C1 C2	In D_2 C1 C2 C3	in CO_2 C2 C3
Nd:YAG, 266 nm KrF, 248 nm	299 341 277 313	289 316 268 291 319	287 299

Diverse wavelength combinations are used in different altitude ranges of the troposphere and lower stratosphere. For instance, the wavelength pairs 289/316 and 287/299 nm make it possible to obtain ozone profile up to the heights of about 10 km [3,4]; the pair 292/319 nm can be used up to the heights of 14-16 km [5]; and the pairs 277/313 and 292/313 nm can be used up the heights of 8-12 and 15 km, respectively [6].

The wavelength λ_{on}=299 nm lies in the region of ozone absorption band with absorption cross section $\sigma_{299}=4.4\cdot10^{-19}$ cm^2, a factor of 3 larger than the absorption cross section at the wavelength of 308 nm: $\sigma_{308}=1.4\cdot10^{-19}$ cm^2. The maximal height of sensing is determined primarily by the range from which the signal at λ_{on} is recorded, which is always shorter than the range from which the signal at λ_{off} is recorded, due to stronger absorption by ozone. From this viewpoint, λ_{on}=299 nm is more preferable than 277 or 292 nm. Wavelengths 299 and 341 are implemented in one sensing beam (in one SRS cell), in contrast to, e.g., pair 292/313 nm. It is noteworthy that system on the basis of hydrogen-filled SRS cell is cheaper than deuterium-filled cell. It is also important to remember about technical feasibility of spectral separation, during reception, of signals at closely lying wavelengths.

To study the vertical ozone distribution in the upper troposphere – lower stratosphere, we developed and put into measurement mode a lidar, as part of SLS, for measuring the ozone concentration in the upper troposphere – lower stratosphere [7]. Sensing is performed according to the method of differential absorption and scattering at wavelength pair of 299/341 nm, which are, respectively, the first and second Stokes components of SRS conversion of 4th harmonic of Nd:YAG laser (266 nm) in hydrogen. Lidar with receiving mirror 0.5 m in diameter is used to implement sensing of vertical ozone distribution in altitude range of 6-16 km.

3. LIDAR FOR SENSING THE VERTICAL OZONE DISTRIBUTION IN THE UPPER TROPOSPHERE – LOWER STRATOSPHERE

Figure 1 presents the block-diagram of the developed lidar.

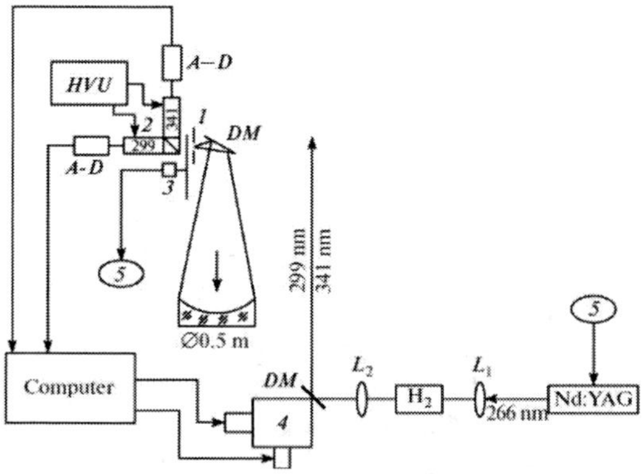

Fig. 1. Block-diagram of the lidar: (*1*) field diaphragm, (*2*) cell for spectral selection with a PMT, (*3*) mechanical shutter, (*4*) automated adjustment unit of the rotating output mirror, (*5*) system for synchronizing the shutter operation time and the moment of emission of laser pulses, (*DM*) deflecting mirrors, (Nd:YAG) solid-state laser, (H_2) SRS conversion cell with H_2, (*A–D*) amplifiers–discriminators, (*HVU*) high-voltage power supply units for the PMT, and (L_1, L_2) lenses.

As the source of laser radiation, we use the 4th harmonic (266 nm) of fundamental frequency of Nd:YAG laser (model LS-2134UT, LOTIS TII firm, Minsk) with its subsequent SRS conversion in hydrogen to the first (299 nm) and second (341 nm) Stokes components.The vertical ozone profiles are retrieved from lidar signals with the help of universal software [8], which allows the altitude profiles of ozone concentration to be calculated according to the method of differential absorption and scattering for three wavelength pairs 272/289 nm, 299/341 nm, and 308/353 nm. The software applies linear smoothing for both input lidar data and for retrieval results. Linear smoothing (smoothing by sliding average) is a well-known procedure, widely used for processing the experimental data in different natural science regions. Linear smoothing is a particular case of digital filtering, possessing random signal error, filter with rectangular window, and unit weighting coefficients.

The software permits us: to read off lidar data and save the retrieval results in ASCII format; smooth the lidar signals and retrieval results using sliding mean. The software incorporates temperature correction of ozone absorption coefficients to reduce the retrieval errors. When lidar signals are retrieved at sensing wavelengths 272/289 nm and 299/341 nm, large aerosol concentrations in the altitude range of 0-20 km should be taken into consideration; therefore, aerosol correction is accounted for in the software. Seasonal model values of altitude distribution of temperature and molecular backscattering coefficient for midlatitude winter and summer are introduced in the software for the calculations.

4. RESULTS OF MEASUREMENTS

Examples of retrieved profiles of vertical ozone distribution are presented in Figure 2 (a - November 2014, b - January-February 2015).

Figure 2 presents the retrieved profiles of ozone concentration for November 2014 and January-February 2015 and compares them with Krueger midlatitude model [9] and IASI satellite data. The retrieved and model profiles quite well coincide, while deviations from IASI satellite data are quite natural for ozone dynamics in separate observation days.

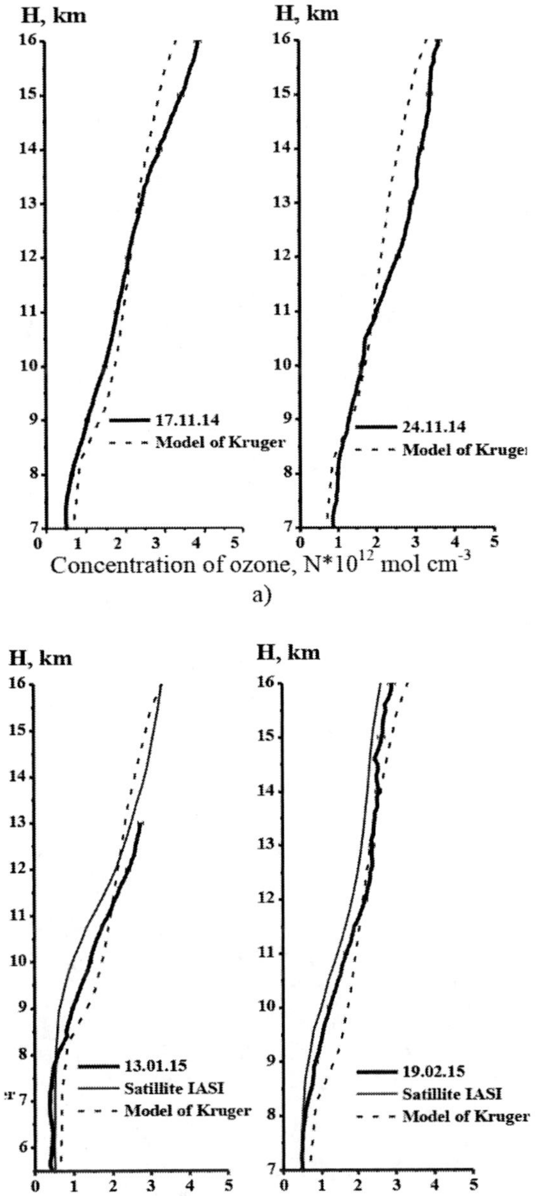

Figure 2 – Retrieved vertical profiles of ozone concentration at wavelengths 299/341 nm (a - November 2014, b - January-February 2015) compares them with Krueger midlatitude model and IASI satellite data.

5. CONCLUSIONS

Lidar measurements at wavelengths 299 and 341 nm agree with model estimates and satellite (IASI) observations, which indicate acceptable accuracies of ozone sensing in the altitude range of about 6-16 kmкм. We can also note that, for ozone sensing in the altitude range of 5-20 km, lidar on the basis of Nd:YAG laser is more preferable than lidar on the basis of excimer KrF laser which is more costly, more complex in exploitation, needs especially pure gases for working mixture, frequent purification or replacement of resonator optics.

ACKNOWLEDGEMENTS

This work was supported in part by the Ministry of Science and Education of the Russian Federation (Agreement No. 14.604.21.0100 (RFMEFI60414X0100)), the Russian Science Foundation (Agreement no. 14-27-00022), The President of the Russian Federation (Grant NSh-4714.2014.5).

REFERENCES

[1] http://sage.nasa.gov/SAGE2/

[2] Burlakov, V.D., Dolgii, S.I., Nevzorov, A.V., 2004, Modification of the measuring complex at the Siberian Lidar Station. *Atmospheric and oceanic optics*. **17** (10), 756-762.

[3] Galani, E., Balis, D., Zanis, P., Zerefos, C., Papayannis, A., Wernli, H., and Gerasopoulos, E., 2003, Observations of stratosphere-to-troposphere transport events over the eastern Mediterranean using a ground-based lidar system. *J. Geophys. Res.* **108**, D12. P.STA12/1-STA12/10.

[4] Masahisa Nakazato, Tomohiro Nagai, Tetsu Sakai, and Yasuo Hirose, 2007, Tropospheric ozone differential-absorption lidar using stimulated Raman scattering in carbon dioxide. *Appl. Opt.* **46** (12), 2269-2279.

[5] Bukreev, V.S., Vartapetov, S.K., Veselovskii, I.A., Galustov, A.S., Kovalev, Yu.M., Prokhorov, A.M.,. Svetogorov, E.S., Khmelevtsov, S.S., Li, Ch.Kh., 1994, Lidar system for sensing the stratospheric and tropospheric ozone on the basis of excimer lasers. *Quantum Electronics*. **21** (6), 591- 596.

[6] Eisele, H., H.E. Scheel, R. Sladkovic, and T. Trickl, 1999, High resolution lidar measurements of stratosphere-troposphere exchange. *J. Atmos. Sci.,* **56** (3), 319-330.

[7] Burlakov, V.D., Dolgii, S.I., Makeev, A.P., Nevzorov, A.V., Romanovskii, O.A., Kharchenko, O.V, 2010, A differential-absorption lidar for ozone sensing in the upper atmosphere-lower stratosphere. *Instruments and Experimental Techniques*. 2010. **53** (6). 886-889.

[8] Nevzorov, A.V., Nevzorov, A.A., Romanovskii, O.A., 2014, Software for retrieving the ozone altitude profiles from data of atmospheric laser sensing. *Proc. SPIE.* **9292**, 92923L. doi:10.1117/12.2074718.

[9] Krueger A.J. , Minzner R.A. , 1976, A mid-latitude ozone model for the 1976 U.S. standard atmosphere. *J. Geophys. Res.,* 81 (D24) 4477-4481.

AEROSOL ACTIVITY AND HYGROSCOPICITY COMBINED WITH LIDAR DATA IN THE URBAN ATMOSPHERE OF ATHENS, GREECE IN THE FRAME OF THE HYGRA_CD CAMPAIGN

Aikaterini Bougiatioti[1,2]*, Alexandros Papayannis[1], Stergios Vratolis[3], Athina Argyrouli[1], Nikolaos Mihalopoulos[4,5], Maria Tsagkaraki[5], Athanasios Nenes[2,6] and Konstantinos Eleftheriadis[3]

[1]Laser Remote Sensing Laboratory, National Technical University of Athens, 15780 Athens, GREECE

**Email: kbougioti@gmail.com*

[2]School of Earth & Atmospheric Sciences and Chemical & Biomolecular Engineering, Georgia Institute of Technology, Atlanta 30332, GA, USA

[3]ERL, INRSTES, N.C.S.R. Demokritos, 15310 Agia Paraskevi, Attiki, GREECE

[4]IERSD, National Observatory of Athens, Palaia Penteli, GREECE

[5]ECPL, Department of Chemistry, University of Crete, 71003 Heraklion, GREECE

[6]ICE-HT, Foundation for Research and Technology, Hellas, 26504 Patras, GREECE

ABSTRACT

Measurements of cloud condensation nuclei (CCN) concentrations between 0.2-1.0% supersaturation and aerosol size distribution were performed at an urban background site of Athens during HygrA-CD. The site is affected by local and long-range transported emissions as portrayed by the external mixing of the particles, as the larger ones appear to be more hygroscopic and more CCN-active than smaller ones. Activation fractions at all supersaturations exhibit a diurnal variability with minimum values around noon, which are considerably lower than unity. This reinforces the conclusion that the aerosol is mostly externally mixed between "fresher", less hygroscopic components with more aged, CCN active constituents.

1. INTRODUCTION

The international HygrA-CD field campaign took place between 15 May and 22 June, 2014, in the frame of the Initial Training on Atmospheric Remote Sensing (ITARS) project. Hygra-CD brought together different instruments and expertise aiming to enhance our understanding on the impact of aerosols and clouds on weather and climate (http://hygracd.impworks.gr). It was based on the synergy between remote sensing and in-situ instrumentation, making also use of numerical weather prediction and atmospheric modeling.

Aerosol particles, especially the anthropogenic ones, are important components of climate change, via absorption and scattering of incoming radiation [1]. They can also act as CCN and "indirectly" force climate through modification of cloud radiative properties and precipitation efficiency, according to the Twomey effect [2]. The main physico-chemical principles involved in the transformation ("activation") of CCN into a cloud droplet involve the effects of curvature and solute on the equilibrium water vapor pressure, represented by the Köhler theory [3]. It constitutes the theoretical basis for linking aerosol to CCN activity, by determining the characteristic level of ambient water vapor supersaturation (S_c), required for particles to activate into cloud droplets. For a given aerosol particle, S depends on the particle dry size and chemical composition.

Simple representations of solute hygroscopicity in Köhler theory are fitted to a one-parameter model, where the hygroscopicity parameter (κ) represents a quantitative measure of aerosol water uptake characteristics and CCN activity [4]. Typical values of κ vary between 0, for insoluble wettable materials that do not significantly affect water activity via adsorption, ~0.1-0.2 for oxidized secondary organic aerosol (SOA) and ~0.6 for pure ammonium sulfate and nitrate.

Continental and marine aerosols tend to cluster around κ~0.3-0.7, respectively, with a fair amount of variability across space and time, depending mainly on the organics mass fraction, relative abundance of ammonium sulfate, as well as particle size [5].

2. METHODOLOGY

The measuring site in Athens was located in Demokritos WMO-GAW station (37.995° N, 23.816° E, 250 m a.s.l.), at a distance of ~8 km from the city center. The sampling location was influenced by aerosol particles (PM) emitted from the city under S and NW wind flow regimes. The instrumentation setup consisted of a Droplet Measurement Technologies (DMT) streamwise thermal-gradient CCN counter [6] and a scanning mobility particle sizer (SMPS) composed of a custom-made differential mobility analyzer and a condensation particle counter (TSI Model 3772) used to measure the dry aerosol size distribution. Throughout the campaign, the instrument was operated at a total flow rate of 0.5 L min^{-1}, with a sheath-to-aerosol flow ratio of 10:1, and a top-bottom column difference, ΔT between 4 and 15 K. Concentrations were measured at each supersaturation for 10 min, yielding a CCN spectrum consisting of 5 different supersaturations every 50 min.

To support the aerosol ground measurements the AIAS mobile one-wavelength (532 nm) elastic backscatter depolarization lidar system [7] was deployed in the field to provide information about the dynamic activity of the Planetary Boundary Layer (PBL) (from ~ 300 m above ground level to 1500 m a.sl.) and to strengthen the link between CCN activity and cloud formation.

Aerosol activate in the CCNc when S_c > than the instantaneous supersaturation *(S),* in the CCNc column. This S_c corresponds to a critical diameter ($d_{p,c}$) above which all particles activate. These parameters are used to determine the parameter *κ:*

$$\kappa = \frac{4A^3}{27 d_{p,c}^3 S_c^2},$$ where $A=(4M_w\sigma_w)/(RT\rho_w)$, R is the universal gas constant and M_w, σ_w and ρ_w are the molar mass, surface tension and density, respectively, of water at the average mid-column temperature, T, in the CCNc. The $d_{p,c}$ is obtained by matching the concentration of CCN activated at a given *S* with the backwards integrated SMPS number distribution.

3. RESULTS

In this paper we focus on the period of 18-22 June 2014. In Figure 1 (upper and lower graphs) we present the available data of the spatio-temporal evolution of the range-corrected aerosol backscattered lidar signal at 532 nm obtained by AIAS system on 18 and 21 June 2014 from near ground (0.4 km) to 5 km height.

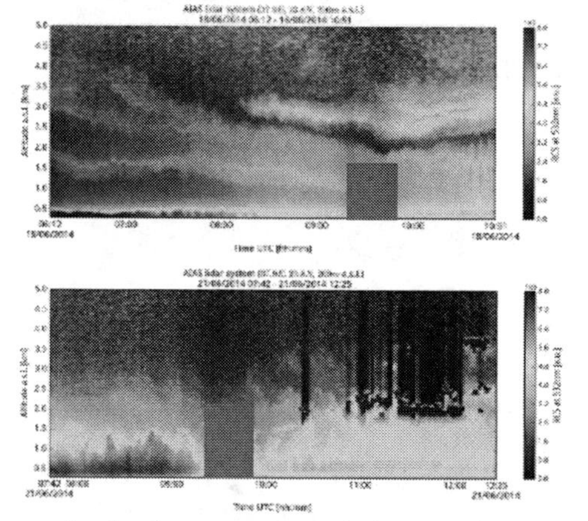

Figure 1: Spatio-temporal evolution of the range-corrected aerosol backscattered lidar signal at 532 nm obtained by AIAS system on 18 and 21 June 2014.

In Fig.1 (upper graph) we see that the PBL is limited to 500 m a.s.l. Taking in account the Athens basin topography, being surrounded by mountains higher than 1 km, this extremely shallow PBL leads to very poor ventilation of the basin and thus to very low air pollution dispersion; therefore high concentrations of aerosols are to be expected near ground. On 21 June (Fig. 1, lower graph) the PBL is significantly higher (~2 km) and strong convective conditions drive ground aerosols to higher heights leading to their dispersion and dilution, so forming a cloud-topped PBL.

Figure 2 shows the CCN concentrations measured from 17 to 22 June. As seen by AIAS, the highest concentrations are observed on 18 June, linked to very low PBL heights. As the PBL height is gradually increased during the following days, vertical dispersion leads to concurrent decrease of PM concentrations as well. As expected, CCN concentrations were higher with increasing supersaturation. CCN at ss >0.6% were generally similar, showing that most of the hygroscopic particles are activated by this supersaturation. The correlation of increased CCN concentrations with similar trends in NO_2 (Fig. 2) is an indication that anthropogenic pollution may also lead to increased CCN numbers, possibly through atmospheric processes. CCN activity can be also represented by the activation fraction (AF), which is the ratio between the measured CCN and the total condensation nuclei (CN), calculated by the integration of the aerosol size distribution.

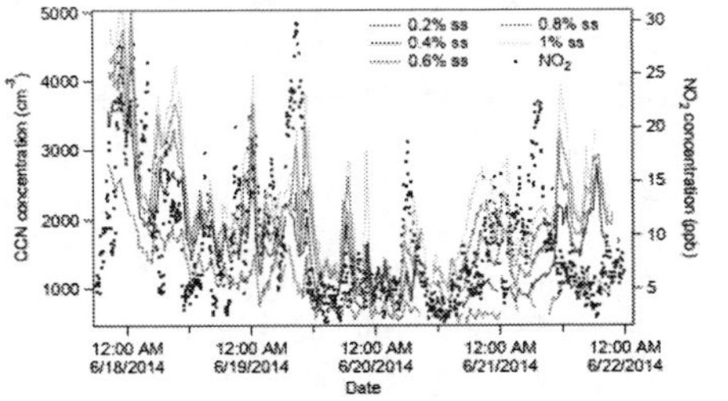

Figure 2: CCN concentration levels at the measured supersaturations during the measurement period.

Figure 3 shows the calculated AFs at the different measured supersaturations. The average values of the AFs for the five different supersaturations are 0.24±0.13, 0.31±0.14, 0.35±0.15, 0.39±0.16 and 0.43±0.18, from lowest to highest *S*, respectively.

At the lowest supersaturation the measured CCN accounted only for a quarter of the total CN, probably because only a minor fraction of the CN was larger than the activation diameter. It can be seen that generally there is a diurnal variability with ratios having the highest values during nighttime and lower around noon, with values being substantially lower than unity. AFs between 0.6 and 1% ss do not differ a lot, showing that the CCN-active particles that are present activate at this range of supersaturation.

Also the fact that generally AFs are substantially higher during nighttime denotes the fact that during night the aerosol is more aged and more hygroscopic, compared to the aerosol during day. This might be associated to fresh, local sources such as traffic, emitting principally smaller, non-hygroscopic particles that can form external mixtures and lower the CCN activity of the sampled aerosol. Similar to our observations, Rose et al. [8] observed at the mega-city Guangzhou in China, that at the lowest *S*, the maximum activated fraction remained well below one, indicating substantial proportions of externally mixed CCN-inactive particles with much lower hygroscopicity-most likely soot particles.

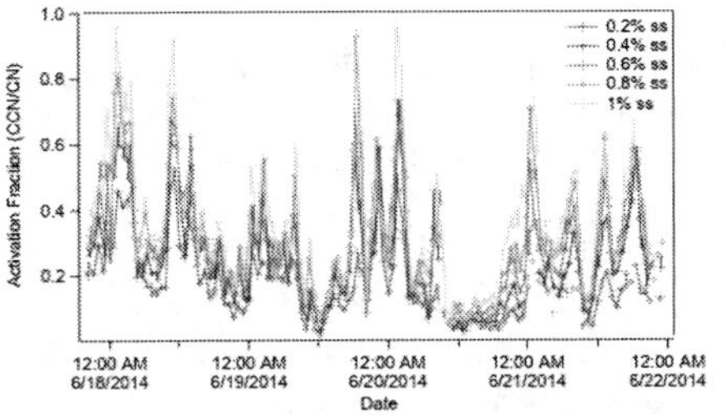

Figure 3: Activation fractions (CCN/CN) calculated for the different supersaturations.

Finally, the κ parameter was calculated for all supersaturations, based on the methodology already presented in the precedent section (Fig. 4). The median values for the five supersaturations were 0.36, 0.19, 0.13, 0.1 and 0.09, respectively. It can be seen that κ values are higher at the lowest supersaturation and diminish with increasing *S*. This is expected, as the lowest supersaturation corresponds to particles with larger critical diameters. These particles are usually also more hygroscopic, containing higher amounts of inorganic salts. Smaller particles tend to exhibit reduced hygroscopicity.

The observed values are closer to typical values observed for oxidized organics [9] therefore this external mixing causes the overall drop of total hygroscopicity of the sampled aerosol. Once more, the values for the three higher supersaturations are fairly close, indicating the similar chemical composition and mixing state aerosols.

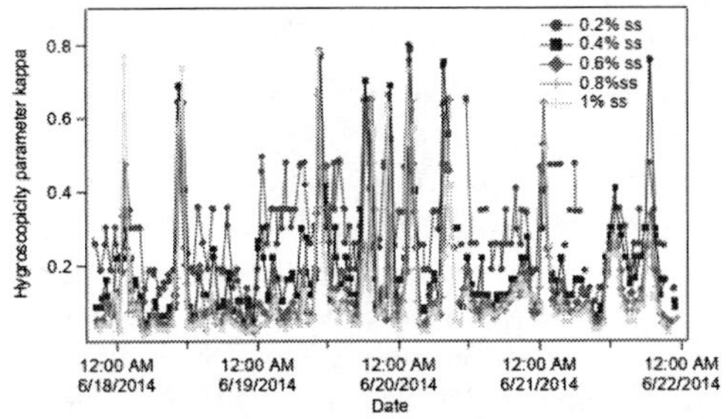

Figure 4: Hygroscopicity parameter, κ, calculated based on the size distributions for all supersaturations.

DOI: 10.1051/epjconf/201611915008

Finally, these values are very close to the measured κ values at the same site during summer 2012, where median κ for 60 nm and 80 nm particles was 0.29 and 0.43, respectively. This reinforces the conclusion that in this environment there are mostly smaller, less hygroscopic particles present. Our future plans include implementing ISORROPIA II to infer the PM chemical properties aloft (~500 m asl.) [10] and compare them with those obtained at ground level.

4. CONCLUSIONS

In this paper we presented the combination of CCN and lidar measurements which were conducted during HygrA-CD in 2014. During the first days of the CCN measurements the highest concentrations were observed, verified by the structure of the PBL seen by the lidar. This extremely shallow PBL, characterized by poor ventilation and limited dispersion, led to high CCN concentrations. With the increase of the PBL height on the following days, the CCN concentrations decreased. AFs exhibited diurnal variability with higher values during nighttime, which remained well below unity during daytime, probably because of external mixing with non-hygroscopic components. Finally, κ was calculated for all supersaturations, having lower values with increasing supersaturation. As lower supersaturation values are characterized by larger critical diameters, this means that larger particles are more CCN-active. As supersaturation increases, there is a decrease in critical diameter, which means that smaller particles start to activate, as well. κ-values at the lowest supersaturation are close to values observed for continental aerosol, while those of the smaller particles are more representative for organic aerosol.

ACKNOWLEDGEMENTS

KB, AP were supported by the MACAVE research project which is implemented within the framework of the action Supporting of Postdoctoral Researchers of the Operational Program Education and Lifelong Learning (action's beneficiary: General Secretariat for Research and Technology), and is co-financed by the European Social Fund (ESF) and the Greek State. The work of AA was supported by the EU's 7th Framework Programme (FP7/2007-2013) for Research, Technological Development and Demonstration under grant agreements no 289923 (MC ITN-ITARS). SV and KE acknowledge EnTeC FP7-REGPOT-2012-2013-1, no 316173.

REFERENCES

[1] Intergovernmental Panel on Climate Change: Climate Change 2013: *Synthesis Report*, 2013.

[2] Twomey, S., 1977: The influence of pollution on the shortwave albedo of clouds, *J. Atmos. Sci.*, **34**, 1149-1152.

[3] Köhler, H., 1936: The nucleus in the growth of hygroscopic droplets, *T. Faraday Soc.*, **32**, 1152.

[4] Petters, M. D. and S.M. Kreidenweis, 2007: A single parameter representation of hygroscopic growth and cloud condensation nucleus activity, *Atmos. Chem. Phys.*, **7**, 1961-1971.

[5] Bougiatioti A., et al., 2011: Size-resolved CCN distributions and activation kinetics of aged continental and marine aerosol, *Atmos. Chem. Phys.*, **11**, 8791-8808.

[6] Roberts, G. and A. Nenes, 2005: A continuous-flow streamwise thermal gradient CCN chamber for atmospheric measurements, *Aerosol Sci. Tech.*, **39**, 206-221.

[7] A. Papayannis, et al., 2012: ARIADNE: The Greek Lidar Network, *Proc. 26th ILRC,* Porto Heli, Greece, 349-352.

[8] Rose, D., et al., 2010: Cloud condensation nuclei in polluted air and biomass burning smoke near the megacity Guangzhou, China-Part 1: Size-resolved measurements and implications for the modeling of aerosol particle hygroscopicity and CCN activity, *Atmos. Chem. Phys.*, **10**, 3365-3383.

[9] Engelhart, G. J., et al., 2011: CCN Activity of Isoprene Secondary Organic Aerosol, *J. Geophys. Res.*, **116**, D02207, doi:10.1029/2010JD014706.

[10] Mamouri, R.E., et al., 2012: Multi-wavelength Raman lidar, sun photometric and aircraft measurements in combination with inversion models for the estimation of the aerosol optical and physico-chemical properties over Athens, Greece, *Atmos. Meas. Tech.*, **5**, 1793-1808.

EPJ Web of Conferences **119**, 16001 (2016) DOI: 10.1051/epjconf/201611916001
ILRC 27

BACKSCATTERING PROPERTIES OF NONSPHERICAL ICE PARTICLES CALCULATED BY GEOMETRICAL-OPTICS-INTEGRAL-EQUATION METHOD

Kazuhiko Masuda[1], Hiroshi Ishimoto[2]*, Tetsu Sakai[2], Hajime Okamoto[3]

[1] *Meteorological Satellite Center, Nakakiyoto 3-235, Kiyose, Tokyo 204-0012, JAPAN,*

[2] *Meteorological Research Institute, Nagamine 1-1, Tsukuba, Ibaraki 305-0052, JAPAN,*

**Email: hiroishi@mri-jma.go.jp*

[3] *Kyushu University, Research Institute for Applied Mechanics, 6-1 Kasuga-koen Kasuga, Fukuoka 816-8580, JAPAN*

ABSTRACT

Backscattering properties of ice crystal models (Voronoi aggregates (VA), hexagonal columns (COL), and six-branched bullet rosettes (BR6)) are calculated by using geometrical-optics-integral-equation (GOIE) method. Characteristics of depolarization ratio (δ) and lidar ratio (L) of the crystal models are examined. δ (L) values are 0.2~0.3 (4~50), 0.3~0.4 (10~25), and 0.5~0.6 (50~100) for COL, BR6, and VA, respectively, at wavelength λ=0.532μm. It is found that small deformation of COL model could produce significant changes in δ and L.

1. INTRODUCTION

Optical properties of nonspherical ice crystals influence the accuracy of retrieval of cirrus cloud information from remote sensing measurements. Improvements in modeling of ice crystal shapes and numerical calculation schemes are of increasing importance. Optical properties of three kinds of crystal models of VA (irregular particles), COL and BR6 (regular particles) have been investigated from visible to infrared wavelength regions by using the finite-difference time-domain (FDTD) and the GOIE methods [1]. In this work, the depolarization ratio (δ) and lidar ratio (L) for VA, BR6, and COL models are examined to study a feasibility for applying theoretical calculations to lidar measurements.

2. METHODOLOGY

Ice crystal models and computational scheme used in this work are described in [1]. A short summary is presented here. Size of ice crystals is expressed in terms of the radius of volume equivalent sphere (R_{eq}). Eleven (eight) kinds of R_{eq} is considered for VA and COL (BR6) from 2.8 μm (28.0 μm) to 202.0μm. The aspect ratio of COL and elements of BR6 are changed depending on the crystal size. Seven sizes of VA model are considered, for which the number of faces are changed from 92 to 2667. In GOIE, the reflected and refracted near fields are calculated by the ray-tracing technique, and then, the scattered far field can be obtained by integrating the near fields over the particle's surface [2]. Preliminary results of the backscattering phase function $F_{11}(\pi)$ and δ, defined by the perpendicular-to-parallel polarization components of backscattered light, have been presented in [1] with a note that "the values may include large errors". Note that, in [1], number of particle orientations is determined from the convergence check of average phase function, not from the convergence of backward radiation. In this respect, it is demonstrated that large numbers of random orientations (N_{ori}) are necessary to calculate the radiation of backward direction accurately [3-4]. After some trial and error, N_{ori} for COL and BR6 are chosen to be 24000~96000 and 48000, respectively, whereas N_{ori} for VA is changed from 500 to 24000 depending on the number of faces.

3. RESULTS

As a preliminary study, we compared, in Figure 1, the M_{11} element of Mueller matrix calculated by GOIE for quasi-horizontally oriented hexagonal plates with those calculated by the physical optics approximation method [5] that is essentially the same as GOIE. Except for some fluctuations at tilt angles less than 10 degree, GOIE calculation shows close agreement with the results by Borovoi et al. that is open to the public via World

Wide Web [5].

$F_{11}(\pi)$ and δ at λ=0.565μm are calculated in a similar manner to [1] except for using the above-mentioned N_{ori} values (Fig.2). Ice crystals are assumed to be randomly oriented in three-dimensional space. It has been confirmed that the fluctuations in $F_{11}(\pi)$ and δ with respect to size parameter X_{eq} are much reduced by increasing numbers of random orientations.

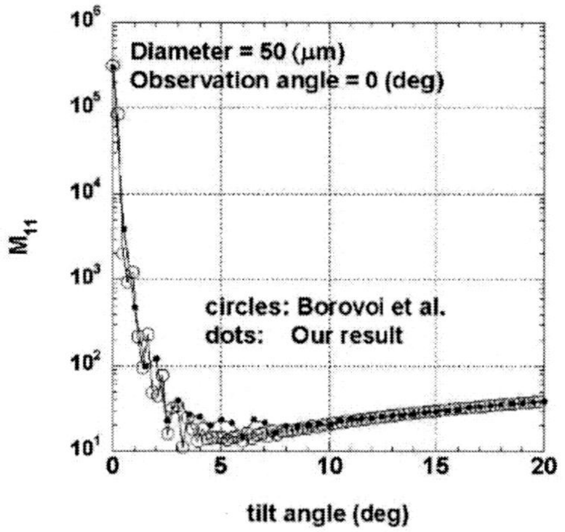

Fig. 1: Backscattering Mueller matrix element (M_{11}) at λ=0.532μm with observation angle of 0 degree for quasi-horizontally oriented ice plates of 50.0μm diameter and 11.7μm height as a function of plate tilt angle. M_{11} is defined as $F_{11}\sigma_{sca}/4\pi$, where F_{11} and σ_{sca} are the phase function and scattering cross section, respectively.

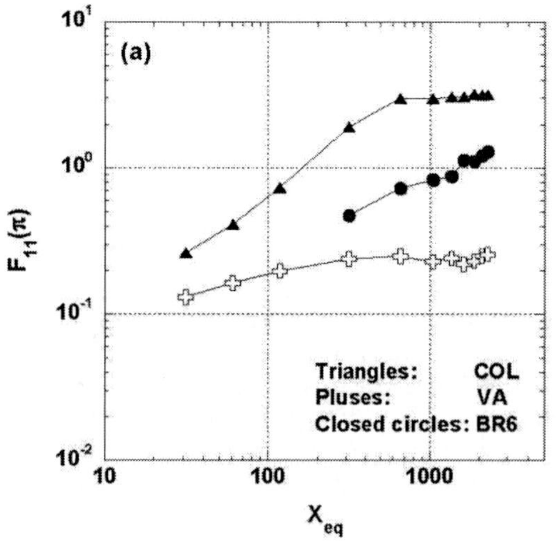

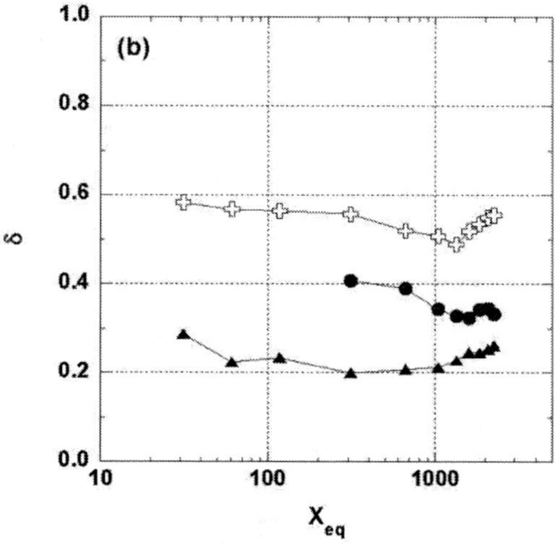

Fig. 2: Backscattering phase function $F_{11}(\pi)$, (a) and depolarization ratio δ, (b) with respect to the size parameters at wavelength λ=0.565μm for the Voronoi aggregates (VA), hexagonal columns (COL), and six-branched bullet rosettes (BR6).

Figure 3 shows the relation between δ and L predicted from theoretical calculations at λ= 0.532μm. L is defined as σ_e/σ_π where σ_e and σ_π are the extinction and backscatter cross sections, respectively. To examine a feasibility for applying theoretical calculations to lidar measurements, the range of (L, δ) for upper clouds over Tsukuba, Japan observed with the lidar is also presented in Fig.3.

Because the columnar particles with length of 80~150μm are dominant in the microscopic image of the cloud particles taken using the hydrometer videosonde over the lidar observational site [6], from now on, we concentrate on the analysis of lidar measurements by using COL model.

The (L, δ) calculations for COL of comparable sizes to the observational data are indicated with the symbols A and B. Although the results are within the observational range of (L, δ), most measurements of (L, δ) produced by the lidar are inconsistent with the calculations for COL model.

It should be noted that the COL model is assumed to be composed of perfect hexagonal ice crystals without any roughness or deformations. On the contrary, the microscopic image of the ice cloud particles shows deformation from the perfect hexagonal shapes to some extent. We did sensitivity tests to investigate the effect of deformation of COL on L and δ.

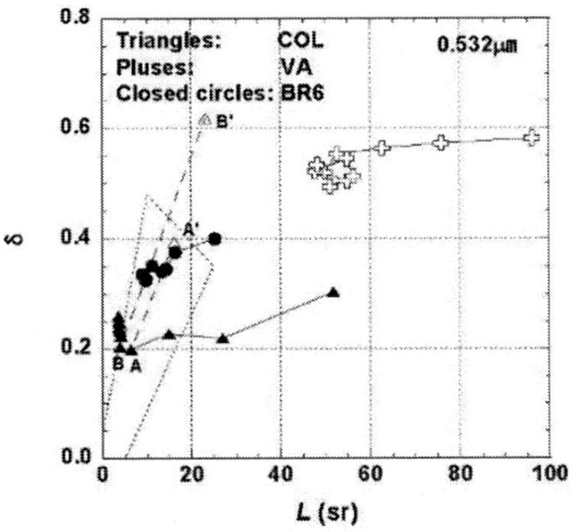

Fig. 3: Diagram showing the relation between the lidar ratio (L) and depolarization ratio (δ) for randomly oriented ice crystal models at wavelength λ=0.532μm calculated by GOIE. The area surrounded by dotted lines roughly shows (L, δ) of upper clouds measured with the lidar over Tsukuba, Japan on 29 March 2004 [6]. The symbol A shows the results for COL of R_{eq}=28.0μm with the radius (a) and length (l) of 23.1μm and 66.1μm, respectively; symbol B for R_{eq}=59.4μm, a=45.0μm, l=167.0μm. The symbols A' and B' indicate the results for modified COL models (See text in detail).

Center of each hexagonal face of COL is moved towards the column center by 0.5% of the column length. Thus, shallow hollows are created on the hexagonal faces. The (L, δ) for the modified COL model are shown by open triangles indicated with the symbols A' and B'. It is noted that L and δ values significantly changes by the small deformation of COL.

This changes in L and δ can be explained as follows. The backscattered light by hexagonal plates and columns are composed of specular and corner-reflection terms [7]. The corner-reflection term originates from refraction/reflection events at orthogonal faces of the hexagonal crystals. For particles randomly oriented in three-dimensional space, the corner-reflection term is dominant in backscattering radiations. Thus, the L and δ values could be significantly influenced by disturbing the orthogonality between neighboring faces, particularly for large sizes of particles.

4. CONCLUSIONS

Backscattering properties of nonspherical ice particles are calculated for three kinds of ice crystal models (VA, BR6, COL) by use of the GOIE method. Particularly, δ and L are examined for their applications to the lidar measurements. The calculated L and δ are compared with measurements by the lidar. It is difficult to explain all observed δ and L values by theoretical calculations with the above-mentioned crystal models. However, small deformation of the COL model produce significant changes in δ and L. In consequence, wide ranges of L and δ values observed by the lidar could be explained by some deformation of crystal models.

ACKNOWLEDGEMENT

This work was supported by the Global Change Observation Mission Climate (GCOM-C1) research project of Japan Aerospace Exploration Agency (JAXA). We thank Drs. Borovoi and Konoshonkin for valuable comments on calculation of backscattering radiation of quasi-horizontally oriented hexagonal plates.

REFERENCES

[1] Ishimoto, H., K. Masuda, Y. Mano, N. Orikasa, A. Uchiyama, 2012: Irregularly shaped ice aggregates in optical modeling of convectively generated ice clouds, *J. Quant. Spectrosc. Radiat. Transfer*, **113,** 632-643.

[2] Masuda, K., H. Ishimoto, Y. Mano, 2012: Efficient method of computing a geometric optics integral for light scattering by nonspherical particles, *Pap. Meteor. Geophys.*, **63,** 15-19.

[3] Okada, Y., 2008: Efficient numerical orientation averaging of light scattering properties with a quasi-Monte-Carlo method, *J. Quant. Spectrosc. Radiat. Transfer*, **109,** 1719-1742.

[4] Um, J., G. M. McFarquhar, 2013: Optimal numerical methods for determining the orientation averages of single-scattering properties of atmospheric ice crystals, *J. Quant. Spectrosc. Radiat. Transfer,* **127,** 207-223.

[5] Borovoi, A., A. Konoshonkin, N. Kustova, H. Okamoto, 2012: Backscattering Mueller matrix for quasi-horizontally oriented ice plates of cirrus clouds: application to CALIPSO signals, *Opt. Express,* **20(27),** 28222-28233.

[6] Sakai, T., N. Orikasa, T. Nagai, M. Murakami, K. Kusunoki, K. Mori, A. Hashimoto, T. Matsumura, T. Shibata, 2006: Optical and microphysical properties of upper clouds measured with the Raman lidar and hydrometer videosonde: A case study on 29 March 2004 over Tsukuba, Japan, *J. Atmos. Sci.,* **63,** 2156-2166.

[7] Borovoi, A., A. Konoshonkin, N. Kustova, 2013: Backscattering by hexagonal ice crystals of cirrus clouds, *Opt. Lett.,* **38(15),** 2881-2884.

ON DEPOLARIZATION LIDAR-BASED METHOD FOR THE DETERMINATION OF LIQUID-CLOUD MICROPHYSICAL PROPERTIES

Gilles Roy[1*], Xiaoying Cao[2], Grégoire Tremblay[3] and Robert Bernier[4]

[1]*RDDC Valcartier 2459 de la Bravoure, Québec, Qc, Canada, gilles.roy@drdc-rddc.gc.ca*
[2]*Lidar Consultant, Ottawa, Canada;* [3]*AEREX Avionic, Levis, Canada;* [4]*IOSL, Mirabel, Canada,*

ABSTRACT

Under single scattering conditions, water droplets clouds do not depolarize the backscattered light. However, backscattered light from multiple scattering will be depolarized. The level of depolarization is a function of the droplets size, the cloud extinction coefficient value and profile; it has also an important dependency on the lidar field-of-view (FOV). The use of depolarization information to retrieve cloud microphysical properties, using Multiple-FOV has been the object of studies, [1], [2]. Recently the use of the depolarization, at a single FOV, has been studied for cloud with linear liquid water content profiles, [3], [4]. In this paper we present the mechanism leading to depolarization and identify the FOV values for which the information on particle size is high. Also Monte Carlo simulations for cloud with constant and ramp up profiles are presented. The degree of linear depolarization as a function of cloud penetration is significantly different for both cloud profiles. This suggests that the use of the degree of linear depolarization at a single FOV should be used with caution to determine clouds micro-physical parameters.

1. DEPOLARIZATION RATIOS

Backscattered light from spherical particles does not change the polarization orientation state of a linearly polarized incident light. This particularity allows the discrimination of spherical particles from non-spherical particles. However, under multiple scattering conditions, even spherical particles generate depolarized signals. The linear depolarization ratio (δ_{lin}) at a scattering angle β is defined as the ratio of the power integrated over the azimuthal angles φ in the plane of polarization orthogonal ($I_{\perp}$) and parallel ($I_{//}$) to the linearly polarized source:

$$\delta_{lin}(\beta) = \frac{<I_{\perp}>}{<I_{//}>} = \frac{\int_0^{2\pi} I_y(\beta,\varphi)d\varphi}{\int_0^{2\pi} I_x(\beta,\varphi)d\varphi}, \qquad [1]$$

Using Mie theory, it can be shown [1] that:

$$\delta_{lin}(\beta) = \frac{(P_2 \cos^2\beta - 2P_3\cos\beta + P_1)}{(3P_2\cos^2\beta + 2P_3\cos\beta + 3P_1)} \qquad (2)$$

Where in P_1, P_2 and P_3 are the Mie scattering phase matrix elements defined by Deirmendjian.

Figures 1 shows the phase functions and the depolarization ratios at a wavelength of 0.532 µm for linear incident waves on water-droplets of diameters 4 µm and 20 µm. The water droplets size distribution is log-normal with a geometric standard deviation of 0.1; the distribution was truncated for values smaller than 0.02 the maximum of the distribution.

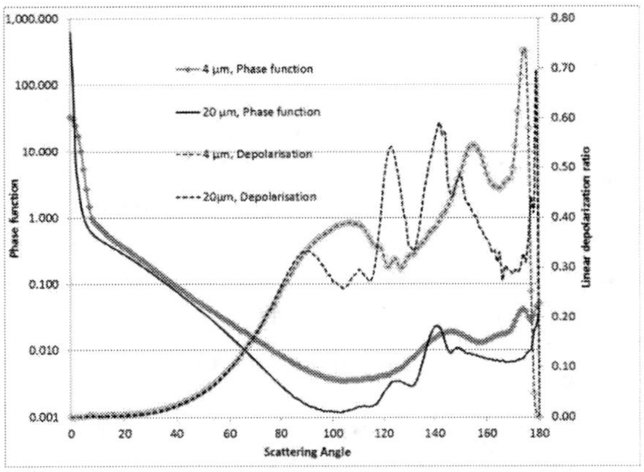

Figure 1. Phase functions and linear depolarization ratios for 4 and 20 µm in diameter water droplets.

For depolarization, the major observations are:

- The light is not depolarized in the forward scattering direction for angles as large as 20°.
- The depolarization is equal to 0 at 180°.
- The depolarization increases very sharply around 180°.

Figure 2 shows the depolarization as a function of scattering angles ranging from 170° to 180° for water droplets of 4, 8, 12, 16 and 20 μm (same distribution as for figure 1). For smaller particles, the depolarization peak shifts toward smaller scattering angles. However the maximum values keep very close to each other.

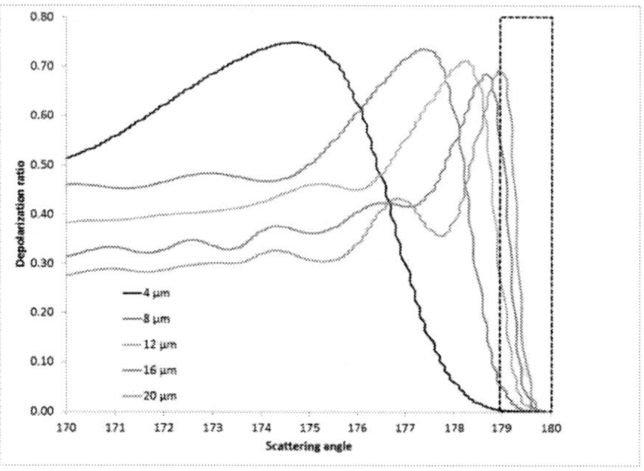

Figure 2. Depolarization ratios for water droplets of 4, 8, 12, 16 and 20 μm in diameter. The rectangular box represents a 1° width.

2. SCATTERING AND DEPOLARIZATION

Double scattering events and higher scattering events leading to depolarization are illustrated in Figure 3. For clarity, only a single backscattering has been illustrated at distance z_c from the lidar system. This backscattering is preceded by a forward scattering event that can come from different distances z_a. A lidar with a given FOV (θ_i) will measure the backscattered light without any distinction regarding the forward scattering event position in space. The second order scattering signal detected will be the sum of all those signals and it will be given by:

$$P_D(z_c,\theta_i) = P(z_c)\int_0^{z_c}\int_0^{\beta_i} F(z,\beta)\sin\beta d\beta dz,$$

where $P(z_c) = P_0 \exp[-2\int_0^{z_c}\alpha(z)dz\ ]\frac{c\tau}{2}\frac{4\pi A}{z_c^2}$, and

$$F(z,\beta) = [\alpha_s(z)p(r,z,\beta)][\alpha_s(z_c)p(r,z_c,\beta_b)],$$

where, c, τ and A are respectively the speed of light, the pulse width and the collecting optics area; α_s is the scattering extinction coefficient; $p(r,z,\beta)$ and $p(r,z_c,\beta_b)$ are the values of the phase function for the forward (β) and backward scattering angles ($\beta_b = \pi - \beta + \theta$) for a particle of radius r; z_a is the distance of the beginning of the cloud; z_c is the distance where the scattered radiation is measured; the quantity $[\alpha_s(z)p(r,z,\beta)]$ represents the forward scattering coefficient while $[\alpha_s(z_c)p(r,z_c,\beta_b)]$ represents the backscattering coefficient. The FOV θ is easily related to the scattering angle β in the simple geometric relation: $\tan\beta = \frac{z_c \tan\theta}{z_c - z}$.

For compactness of the equations, the particle size dependence of the functions F and δ_{lin} has been dropped out.

The depolarization ratio for scattering angle within β_i is calculated as follows:

$$\bar{\delta}_{lin}(z_c,\beta_i) = \frac{\int_0^{z_c}\int_0^{\beta_i} F(z,\beta)\delta_{lin}(z_c,\beta_b)\sin\beta d\beta dz}{\int_0^{z_c}\int_{0_i}^{\beta_i} F(z,\beta)\sin\beta d\beta dz}$$

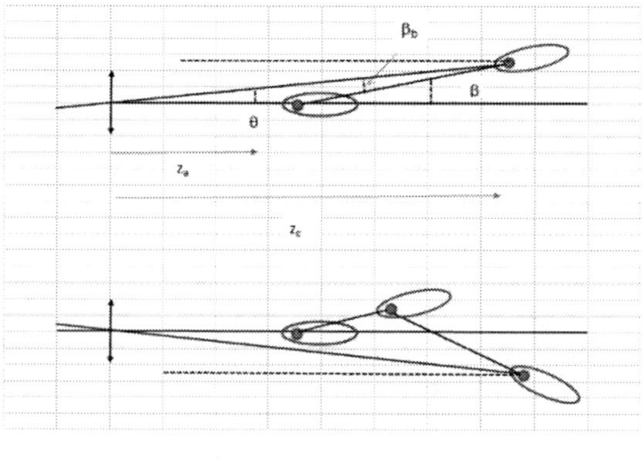

Figure 3. Double scattering processes and higher scattering processes leading to backscattering depolarization.

Figure 1 clearly shows that the depolarization ratios for scattering angles smaller than 20° are very close to zero. Considering the small field of view covered by a typical lidar detector, the measured depolarization effect must come from the backscattering. Figure 4 shows the function $F(z,\beta)\delta_{lin}(z,\beta)$ as a function of the scattering angle β.

DOI: 10.1051/epjconf/201611916002

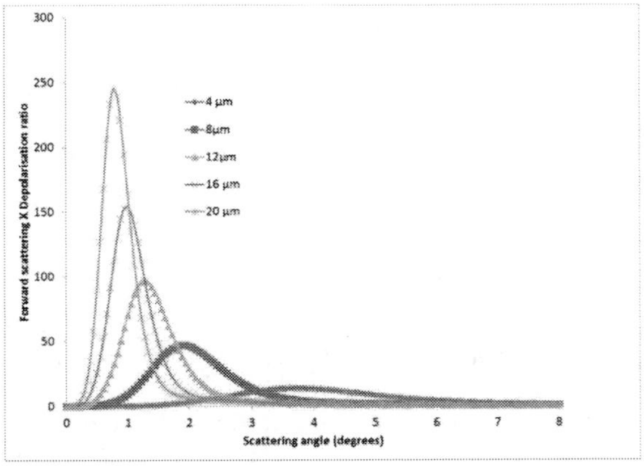

Figure 4. The product of the forward scattering phase function at scattering angle β and the depolarization ratio at angle β'= β - π as a function of the scattering angle β.

Clearly the product of the forward scattering phase function at scattering angle β and the depolarization ratio at angle β'= β - π as a function of scattering angle β contains information on the size of the particle and a Multiple FOV lidar will be able to extract that information under some conditions. Can a single FOV depolarization measurement provide information on droplet size? Is there an optimal FOV to obtain information on droplet size?

In view of a better understanding of the problem, Figure 5 shows the product of normalized cumulative forward scattering and depolarization ratio as a function of scattering angle. The cumulative signal at a specific scattering angle is the signal which would be recorded by a lidar with that specific angle as FOV. The different arrows are for different cloud geometries and lidar FOVs. For small scattering angles, the average scattering angle, <β> over a penetration depth Δz ($\Delta z = z_c - z_a$) is equal to: $<\beta>=\theta\frac{2z_c+\Delta z}{\Delta z}$.

From Fig. 5 and Table 1, it is clear that for average scattering angles greater than 2.5 degrees the product of cumulative forward scattering and depolarization ratio reach values very close to each other (except for the 4 µm droplets). A small FOV would thus show better differentiation capability, which reinforces the interest of the depolarization single FOV lidar based method, but at the cost of an often significantly reduced signal.

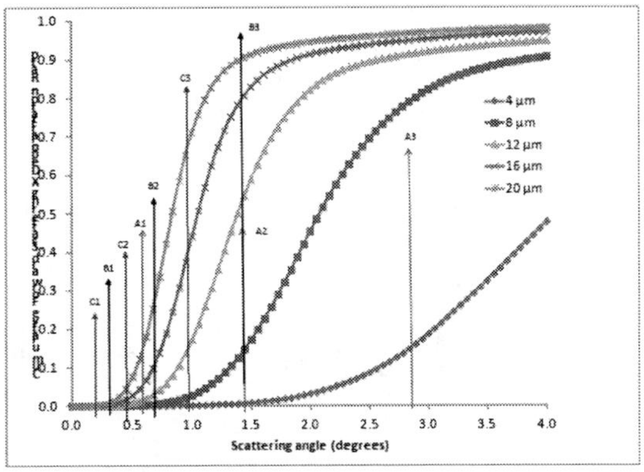

Figure 5. Normalized product of cumulative forward scattering and depolarization ratio as a function of scattering angle. The different identified arrows are for different cloud geometries and lidar FOVs.

Table 1 Average scattering angle within lidar FOV of 0.5 and 2 mrad for different cloud base and cloud penetration depth.

		FOV: 0.5 (mrad)	FOV: 2.0 (mrad)
Cloud base (m)	τ z (m)	τ (µ(Deg)	τ ε(Deg)
500	10	0.72 (A1)	2.89
1000	10	1.44 (A2)	5.76
2000	10	2.87 (A3)	11.49
500	20	0.37 (B1)	1.46
1000	20	0.72 (B2)	2.89
2000	20	1.44 (B3)	5.76
500	30	0.25 (C1)	0.98
1000	30	0.48 (C2)	1.94
2000	30	0.96 (C3)	3.85

So far we have focused our effort on understanding the physics of the depolarization effect caused by multiple scattering and its relation to the lidar FOV and did not address the effect of the cloud extinction profile. In the following and last section, the effect of cloud extinction profile will be studied using Monte Carlo simulation.

3. MONTE CARLO SIMULATION

The Undique Monte Carlo (MC) simulator is a multithread software based on the Bohren and Huffman Mie scattering computation routine. The simulator is made to reproduce the characteristics of an actual Flash Lidar system. It consists of an emitter/receiver system, a target and a propagation range including aerosols of various properties. The particularity of the Undique MC is its capability to image the scattered light on a detector array. See [6] for details.

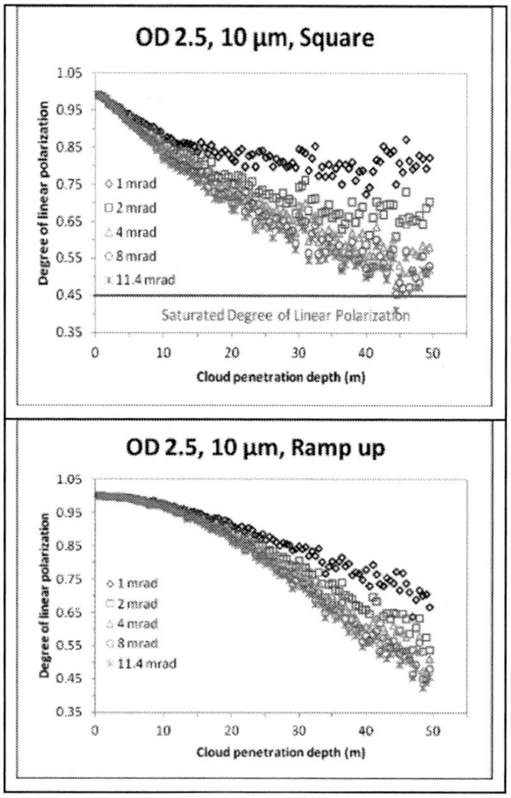

Figure 6 Degree of linear polarization as a function of cloud penetration depth.

The different parameters for the MC simulation are listed; *Emitter:* 532nm, 0.5 mrad beam divergence, Pulse width and shape: 10 ns, Gaussian; *Receiver:* Receiver diameter and focal length; 0.2m and 0.527m; *Cloud* base: 1000m, depth: 50 m, optical depth: 2.5, profiles: Square and Ramp up.

Figure 6 shows the degree of linear polarization as a function of cloud penetration. Degree of linear polarization was used to facilitate comparison with [3]. The cloud extinction profile has a strong influence on the degree of linear polarization (or depolarization ratio) and such an effect is only partly accounted for in [3] and [4]. In absence of knowledge of cloud extinction profile or working hypothesis [4], the simple measurement of depolarization for a given FOV will most likely not be enough to provide information on droplet size.

CONCLUSION

Droplet size information is contained in the lidar depolarization signal, especially at small FOVs where the higher order scattering effects are less important and where the product of cumulative forward scattering and depolarization ratio shows distinct values as a function of droplet size for constant cloud extinction profile. A non-constant cloud profile will have an effect on the product of cumulative forward scattering and depolarization ratio and will most likely make droplet size recovery unreliable.

REFERENCES

[1] G. Roy, L. R. Bissonnette, C. Bastille, and G. Vallée, 1999: Retrieval of droplet-size density distribution from multiple field-of-view cross-polarized lidar signals, *Appl. Opt.* **38**, pp.5202-5211.

[2] I. Veselovskii, M. Korenskii, V. Griaznov, D. N. Whiteman, M. McGill, G. Roy, L. Bissonnette "Information contain of data measured with multiple-field-of-view lidar," *Appl. Opt.* **45,** 6839–6848(2006)

[3] K. Dukhyeon and J. Lee, 2014: Measuring cloud droplet effective radius and liquid water content using changes in degree of linear polarization along cloud depth, *Opt. Lett., 39,* pp. 3378-3381.

[4] D.P. Donovan, H. K. Baltink, , J.S. Henzing, S.R. de Roode and A.P. Siebesma, 2015,: A depolarization lidar-based method for the determination of liquid-cloud microphysical properties, *Atmos. Meas. Tech..* **8**, pp.237-266.

[5] K. Sassen, H. Zhao, 1995: Lidar multiple scattering in water droplet clouds: Toward an improved treatment, *Opt. Rev., 2,* pp. 394-400.

[6] G. Tremblay, G. Roy, 2015: High fidelity imaging algorithm for the Undique Monte Carlo simulator. *27rd International Laser Radar Conference,* New York, USA.

EPJ Web of Conferences **119**, 16003 (2016) DOI: 10.1051/epjconf/201611916003
ILRC 27

AN AUTONOMOUS POLARIZED RAMAN LIDAR SYSTEM DESIGNED FOR SUMMIT CAMP, GREENLAND

Robert A. Stillwell[1]*, Ryan R. Neely III[2,3], Peter Pilewskie[4], Michael O'Neill[3], Jeffrey P. Thayer[1], Matthew Hayman[5]

*[1]Aerospace Engineering Sciences, University of Colorado, *Presenting Author, [2]School of Earth and Environment, University of Leeds, [3]Cooperative Institute for Research in Environmental Science, University of Colorado, [4]Atmospheric and Oceanic Sciences, University of Colorado, [5]Research Aviation Facility, National Center for Atmospheric Research*

ABSTRACT

A dearth of high-spatial and temporal resolution measurements of atmospheric state variables in the Arctic directly inhibits scientific understanding of radiative and precipitation impacts on the changing surface environment. More reliable and frequent measurements are needed to better understand Arctic weather processes and constrain model predictions. To partially address the lack of Artic observations, a new autonomous Raman lidar system, which will measure water vapor mixing ratio, temperature, extinction, and cloud phase profiles through the troposphere, is designed for deployment to Summit Camp, Greenland (72° 36' N, 38° 25' W, 3250 [m]). This high-altitude Arctic field site has co-located ancillary equipment such as a Doppler millimeter cloud radar, microwave radiometers, depolarization lidars, ceiliometer, an infrared interferometer and twice-daily radiosondes. The current suite of instruments allows for a near comprehensive picture of the atmospheric state above Summit but increased spatial and temporal resolution of water vapor and temperature will reveal detailed microphysical information. A system description will be provided with an emphasis on the Monte Carlo safety analysis done to ensure eye safety in all relevant weather conditions.

1. INTRODUCTION

Unattended and automated lidar systems are becoming more common as atmospheric observational capabilities of lidar systems grow. Of particular interest for expanding observing capabilities are locations, which are historically undersampled due to the logistical challenges of maintaining an observational presence. The Greenland Ice Sheet is one such location due to its remote and extreme environment. Precipitative and radiative effects of tropospheric clouds and aerosols remain uncertain due to a lack of observations. High spatial and temporal resolution observations of temperature, water vapor mixing ratio, and cloud properties are needed. The Summit Polarized Raman Lidar (SuPR) is under construction to fill a vital gap in the current observational abilities of the Arctic Observing Network.

Using the Raman scattering technique, the unique rotational and vibrational structure of atmospheric constituents can be leveraged to uniquely identify species of interest. Combining that technique with polarization sensitivity yields more relevant cloud information but creates certain design challenges related to the trade of requirements and priorities.

2. SYSTEM DESIGN

The measurements that SuPR will make are split into three equally competing scientific goals: measurement of temperature, water vapor mixing ratio, and polarization properties of clouds and aerosols. Great care has been taken in the design phase to balance these three goals. By designing a Raman lidar system using the full Stokes Vector polarization treatment, one can accommodate the unique requirements of the three measurements. A system diagram of SuPR is given in Figure 1 with some relevant transmitter and receiver specifications given in Table 1.

Care has been taken with the system's polarization and physical alignment to allow for partially automated alignment procedures; this is helpful when considering the limited seasonal availability of the system. Extra optics as well as motorized mounts are included to perform automated polarization alignments.

By calculating the Mueller matrix of each element in the system and using it instead of scalar values, one could simulate the performance not only of the Raman returns but the systematic effects on the polarization measurements. Photon returns were calculated, using the Stokes Vector Lidar Equation, given in Equation 1.

 DOI: 10.1051/epjconf/201611916003

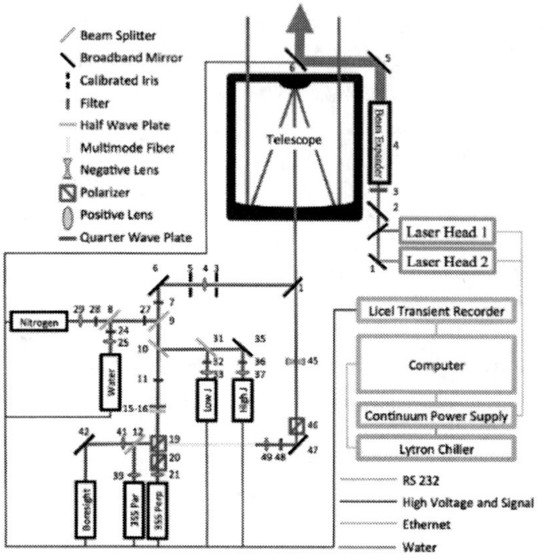

Figure 1: A functional block diagram of the SuPR system. This design contains 4 Raman channels and two polarization sensitive channels.

Table 1: Relevant design specifications for SuPR.

Transmitter	Spec	Receiver	Spec
Divergence	166 [μrad]	Aperture	61 [cm]
Pulse Energy	400 [mJ]	FOV	252 [μrad]
Pulse Rate	30 [Hz]	Filter Width	0.3 [nm]
Polarization	Any	Resolution	7.5 [m]

$$\bar{N} = \bar{\bar{O}}\,\bar{\bar{M}}_{R_x}\left[\left(\bar{\bar{G}}\frac{A}{R^2}\Delta R\right)\bar{\bar{T}}_{atm}\bar{\bar{F}}\bar{\bar{T}}_{atm}\bar{\bar{M}}_{T_x}\bar{S}_{T_x} + \bar{S}_B\right] \quad (1)$$

Here, $\bar{S}_{T_x}$, $\bar{\bar{M}}_{T_x}$, $\bar{\bar{T}}_{atm}$, $\bar{\bar{F}}$, $\bar{\bar{M}}_{R_x}$, $\bar{\bar{G}}$, $\bar{\bar{O}}$, and $\bar{N}$ are the transmitted Stokes Vector, the Mueller matrix of the transmitter, Mueller matrices of atmospheric propagation, the Mueller matrix of scattering, the Mueller matrix of the receiver, the channel dependent overlap function, the matrix defining the observational channels, and the photon counts in those channels respectively. With those photon counts, retrievals were performed to estimate the measurement sensitivity. In the interest of brevity, only the retrievals are shown here. Because the atmosphere must be prescribed to determine raw photon counts, the exact value is known. This can be used to compare theoretical to retrieved values to determine measurement performance. This is done in Figure 2 for a 15-minute integrating period with temperature and water vapor profiles taken from Summit, measured by Vaisala radiosonde. These profiles are typical of what has been observed at Summit in early spring from a twice-daily radiosonde program. The background is taken from a UV spectrometer at Summit. Finally, a Poisson random number is used to simulate shot noise from photon counting.

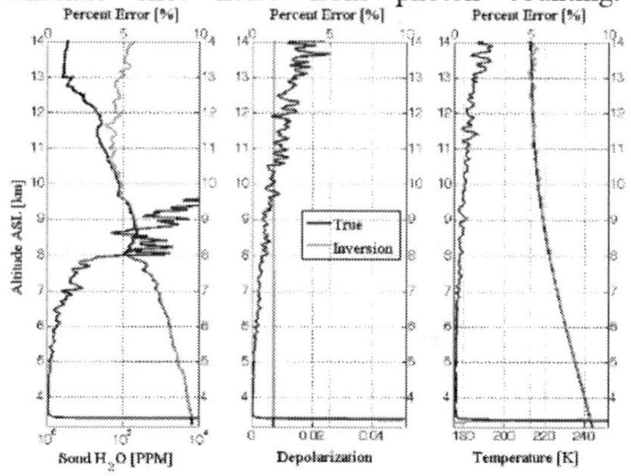

Figure 2: Simulated retrievals for a 15-minute integration under background conditions observed at Summit during the summer-time (i.e. day time). The absolute anomaly is the percent error between the atmosphere modeled to simulate the raw signal and retrieved values. The resolution limit seen is a function of the high background modeled; the same simulation run during the winter-time (i.e. night time) yields similar resolution for polarization and temperature but increases the ability to retrieve water vapor mixing ratio to approximately 5 [PPM].

3. EYE SAFETY

Due to the high energy density of pulsed lasers for Raman applications, burns are possible due to the fast heating of live tissue where the energy is accumulated much quicker than can be dissipated. To avoid this, the American National Standards Institute (ANSI) defines maximum permissible energy (MPE) levels for safe laser operation. These levels are based on pulse duration, peak power, repetition rate, and wavelength [1].

One major factor in the design effort for SuPR was the concern for eye safety. As the system is not designed to be deployed to a populated area, the accepted risks can be different than more traditional lidars. For example, the beam energy need not be below the MPE everywhere but the beam does need to have eye safe scattering in all conditions. There are many methods to meet the ANSI standards; among them are the use of low-power/high-repetition rate lasers, beam expansion [4], and the use of collocated surveillance radar [5]. For high-powered Raman lidar systems, the first is impractical, the second introduces cost and complexity due to the large expansions that are often needed (possibly on the order of 10-20 times), and the third introduces complexity in incorporating such an ancillary system. Here, an accompanying method will be introduced which accounts for all relevant atmospheric absorption and scattering to ensure beam safety without introducing extra complexity or expense associated with the methods stated.

A Monte Carlo scattering model was created which accounts for atmospheric absorption and scattering to determine where a lidar beam is safe after propagation/scattering. This model will be applied to SuPR with weather conditions common at Summit but can be applied to other systems with different characteristics. In this case a fog example is given but the same simulations were run for clear air and blowing snow.

3.1 Monte Carlo Coordinate System

The method for defining scattering is in spherical coordinates where the phase function of a scatterer can be calculated directly from Mie theory for spheres and using ray tracing methods for non-spherical particles [3, 6]. The rotation matrix used to connect the propagation direction before and after scattering is given in Equation 2 where θ_i is the angle from zenith of scattering event i, ϕ_i is the azimuth angle of scattering event i, and C and S are cosine and sine respectively [2].

$$\boldsymbol{M}_i = \begin{bmatrix} C(\theta_i)\,C(\phi_i) & C(\theta_i)\,S(\phi_i) & -S(\theta_i) \\ -S(\phi_i) & C(\phi_i) & 0 \\ S(\theta_i)\,C(\phi_i) & S(\theta_i)\,S(\phi_i) & C(\theta_i) \end{bmatrix} \quad (2)$$

The propagation direction, $\bar{\Omega}$, is:

$$\bar{\Omega}_i = [S(\theta_i)C(\phi_i) \quad S(\theta_i)S(\phi_i) \quad C(\phi_i)]^T \quad (3)$$

The position of the photon, $\bar{\tau}$, is a function of the scattering directions θ_i and ϕ_i for each event, the number of total scattering events in the region of interest, n, and the optical depth traveled before the scattering event, τ_i, and is given in Equation 4.

Using this method, the photon location can be tracked through a region of interest to find its exit position. This position can be summed into a discretized boundary to create a map of photon locations in optical depth space.

$$\bar{\tau} = \sum_{i=1}^{n} \tau_i \prod_{j=0}^{i-1} \boldsymbol{M}_j^T(\theta_j, \phi_j)\bar{\Omega}\,(\theta_i, \phi_i) \quad (4)$$

3.2 Scattering Characteristics

The three cases chosen for analysis represent three different methods of determining scattering characteristics as well as the three prominent weather patterns observed at Summit. Due to the coordinate systems defined, what is needed is the azimuthal and angle from zenith of any scattering event. The phase functions can be calculated exactly using Mie theory and ray tracing. Here, a geometric optics approach is assumed for ice. These exact phase functions can be converted using their normalized cumulative distribution function into a weighted random number. Using a random number and then comparing it to a cumulative distribution, a weighted random number can be created which is used in the Monte Carlo analysis to define scattering directions. This method is completely general, not limited to the phase functions used. An example is given for the fog case in Figure 3 which assumes a wavelength of 354.71 [nm] with scatterers defined by a gamma distribution of width parameter $\alpha = 7$ and effective radius of 10 [µm]. Similar steps can be completed for both the blowing snow case and clear air. For the blowing snow case, the phase function is from an assumed mix of roughened crystal habits of effective radius 30 µm [6], typical of what has been observed at Summit; for the clear air case, the phase function is derived directly from dipole (molecular) scattering.

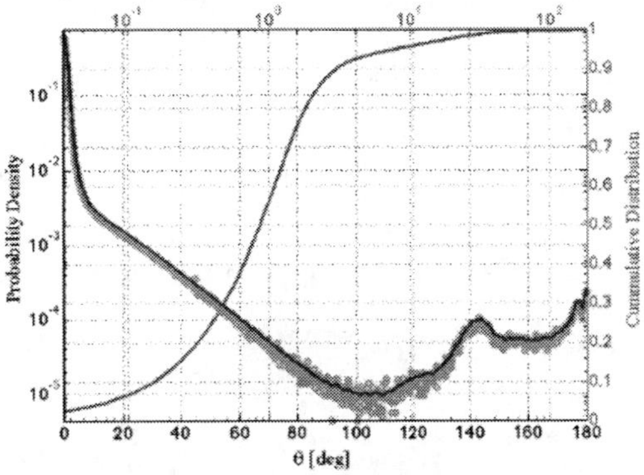

Figure 3: A phase function for spherical fog droplets. The exact solution from Mie theory is the solid blue line, the Monte Carlo approximation using 10^7 particles with 1000 bin discretization in green dots, and the cumulative distribution function used to convert to weighted random numbers in red. Note the top abscissa is logarithmic.

3.3 Energy Maps

Given a region of interest around the lidar beam, a map of energy density is calculated which is then compared to relevant MPE values. The region of interest is defined as a cylinder around the beam with a set radius, r, and height, h. That region is then converted from units of distance to optical depth. This requires knowledge of the optical depth per meter, which is different for all three cases. For air, a simple Rayleigh scattering extinction model suffices. For water and ice particle scattering, optical depth can be calculated from the knowledge of particle size distribution, $n(r)$, density, ρ, effective radius, r_e, and liquid water path, LWP. This conversion is given in Equation 5.

$$r_e = \frac{\int_0^\infty r^3 n(r) dr}{\int_0^\infty r^2 n(r) dr} = \frac{3}{2} \frac{LWP}{\rho \tau} \quad (5)$$

The following results are for the phase function in Figure 3. The system parameters given in Table 1 are all simulated which yields a multipulse MPE of 1.23 [mJ/cm^2]. This map is given in Figure 4.

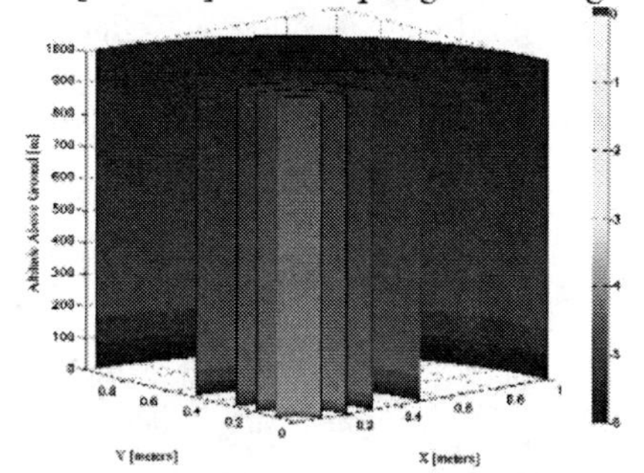

Figure 4: Log$_{10}$ energy density in [mJ/cm^2] caused by fog scattering. This map shows the intensity as a function of radius for 10, 20, 30, 50, and 100 cm. Red is in excess of the MPE; the direct beam is the only source intense enough to exceed the MPE at 355 [nm]. However, the energy is very near the MPE for 532 [nm].

The impact of the direct beam and forward scattering is evident in Figure 5.

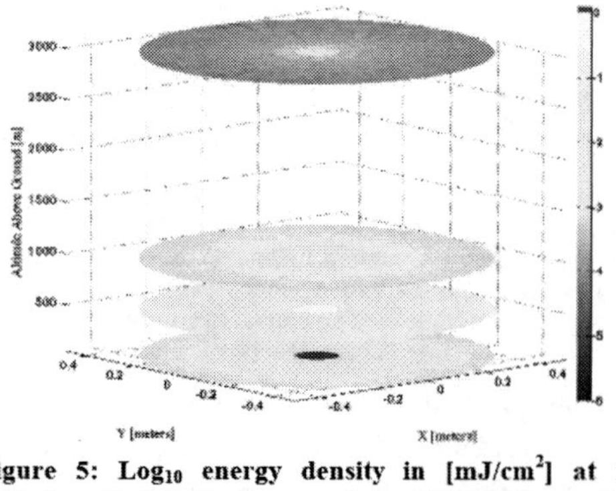

Figure 5: Log$_{10}$ energy density in [mJ/cm^2] at four altitudes. Red is in excess of the MPE. Monte Carlo simulations are run with 10^7 photons and show no beam hazards outside of the direct beam below aircraft flight altitude in foggy conditions.

4. CONCLUSIONS

The work shown here provides a design for an Arctic Raman/Polarization lidar system. The system designed meets its three pronged scientific requirements of measuring water vapor, temperature and polarization properties through the troposphere in all background conditions. Furthermore, a detailed atmospheric scattering simulation was performed to test eye safety concerns in inclement weather. Shown were the results from diurnally forced fog events, which show that the only hazard is the direct beam. Other simulations that include blowing snow and clear air scattering also yielded similar results. The beam is safe at 355 [nm] but given the same system at 532 [nm], scattered energy in excess of the MPE is readily observable with this design. It should also be noted that these results could also be applied to address lidar observations under multiple scattering conditions.

ACKNOWLEDGEMENT

This material is based upon work supported by the National Science Foundation Graduate Research Fellowship under Grant No. DGE 1144083 and National Science Foundation Grant No. AON 1303864.

REFERENCES

[1] American National Standard for Safe Use of Lasers (ANSI Z136.1). Laser Institute of America, 2013.

[2] Bohren, C. F. and E. E. Clothiaux. Fundamentals of Atmospheric Radiation. Wiley-VCH Verlag GmbH and Co. KGaA, 2005.

[3] Bohren, C. F. and D.R. Huffman. Absorption and Scattering of Light by Small Particles. Wiley-VCH Verlag GmbH and Co. KGaA, 1998.

[4] Radlach, M. et al. Scanning rotational Raman lidar at 355 nm for the measurement of tropospheric temperature fields. Atmos. Chem. Phys., 8, 159-169, 2008.

[5] Thayer, J. P. et al. Rayleigh lidar system for middle atmosphere research in the arctic. Opt. Eng. 36, 2045-2061. 1997.

[6]Yang, P., L. Bi, B. A. Baum, K.-N. Liou, G. Kattawar, and M. Mishchenko, 2013: Spectrally consistent scattering, absorption, and polarization properties of atmospheric ice crystals at wavelengths from 0.2 µm to 100 µm. *J. Atmos. Sci.*, 70, 330-347.

EPJ Web of Conferences **119**, 16004 (2016)
ILRC 27
DOI: 10.1051/epjconf/201611916004

MEASUREMENTS OF TERMINAL VELOCITIES OF CIRRUS CLOUDS IN THE UPPER TROSPHERE

Jan Bai Nee, W. N. Chen, C. W. Chiang, S.K. Das

Department of Physics, National Central University, Chungli, Taiwan 32001

jbnee@phy.ncu.edu.tw

ABSTRACT

Cirrus clouds are composed of ice crystals condensed from humidity due to low temperature condition in the upper atmosphere. The microphysics of cirrus clouds including sizes and shapes of ice particles are not well understood but are important in climate modeling. Ice crystal will fall under gravitational sedimentation to reach terminal velocities which depend on the size, mass, and ice habit. We studied here the terminal velocity of cirrus clouds by using lidar observations at Chungli (25N, 121E). The terminal velocities for a few cases of stable cirrus clouds are measured to determine the ice particle sizes and processes in the upper atmosphere.

1. INTRODUCTION

Cirrus cloud is an important element in understanding the climate and radiation of the earth' s atmosphere. However, modeling efforts are hampered by the lack of many important parameters which include ice particle size, habit, terminal velocity, composition and optical properties and related parameters. Terminal velocity is caused by the gravitational settling of ice crystals and depends on the size, mass, and shape of the ice crystals. Terminal velocity is important not only for understanding the microphysics of cirrus clouds but also for understanding the dehydration process in the upper tropospheric (UT) region.

In this paper, we study, terminal velocity by using lidar observations at Chungli (25 N, 121E), Taiwan. As previously reported, cirrus clouds in the UT are often observed [1]. For stable clouds, the terminal velocities can be measured and used for deriving the ice crystal properties [1-2].

2. Instrument and methodology: Lidar system for cirrus research

The lidar facility used to detect the cirrus cloud has been reported before [1-2]. Briefly, a laser at 532 nm operated at 30 Hz frequency is sent into the atmosphere in the zenith direction. A Newtonian type telescope of 18 inch diameter is employed for the collection of scattering light. The backscattering light is detected by a photomultiplier tube combined with an interference filter centered at 532 nm with FWHM 1 nm. The photon counting signals are analyzed by a fast multichannel analyzer (SR430) and stored in a computer. Polarization measurements are carried out also to derive particle nature. The lidar signals are analyzed according to the Fernald integration method in terms of backscattering and depolarization ratios. For normal cirrus clouds, the average backscattering ratio is about 10 and depolarization ratio <0.5.

3. Observation of falling cirrus clouds

Cirrus clouds are often observed in the upper troposphere. Factors affecting the height, composition, and stability of cirrus cloud include wind, waves, temperature, and transport of humidity. However, descending stable cirrus clouds are frequently observed. Figures (1a) and (1b) show the contours for a cirrus clouds at about 16 km in the tropopause region. Fig. (1a) shows results for the backscattering ratio (BR) and Fig (1b) for the depolarization ratio (DR). As we can see in Figure 1, falling cirrus cloud is observed in both plots and terminal velocity can be measured as shown in Fig.2. The cloud position is defined by using a cloud centroid position given as: $h=\int Rzdz/\int Rdz$ where R is the backscattering ratio and z is the height position. Fig. 2 shows the centroid of the cloud descended 0.4 km in about 5 hours corresponding to a speed about 2.5 cm/sec.

Fig. 3 shows a case of rising and lowering of high cloud observed. In the top panel, we observe the rising of cirrus cloud and a growth process. The process will freezes out the humidity and causes dehydration in the region. The upper plot of Fig 3 gives a slow rising speed of 3 cm/s and the bottom part shows varied speed depending on the cloud positions. We can see the cloud are separated into three parts each descending with different speeds ranging between 6 to 18 cm/sec depending on the height. By assuming ice particles shapes such as hexagonal plate and cylinder, the speeds can be determined.

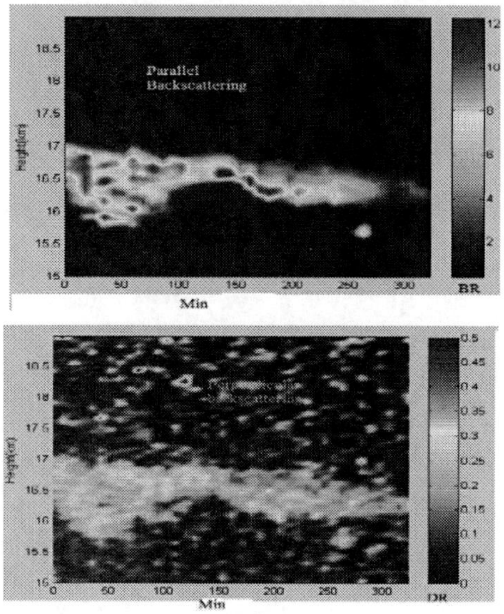

Figure 1. Contour diagram of a 16 km cirrus clouds observed on 2004 June 29. (1a) top: backscattering ratio(BR) for a cirrus cloud at 16 km (1b) bottom: the depolarization ratio (DR).

4. Summary

Terminal velocity of cirrus clouds can be measured by using lidar observations. A few cases of high cirrus cloud showing steadily descending are presented here. We found terminal velocities of clouds ranged 2-20 cm /s corresponding to various ice crystals of sizes. Based on ice habit model, terminal velocity can be used to derive sizes of different types of ices. For hexagonal plate and cylinder, the size of ices of 1-100 μm are determined as will be shown.

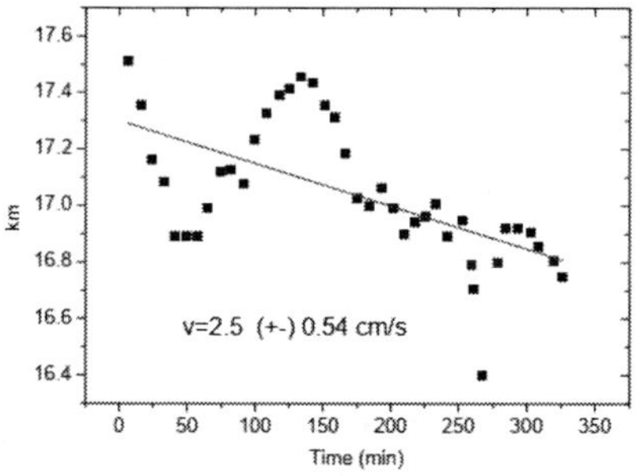

Figure 2. The cloud centroid position as a function of time.

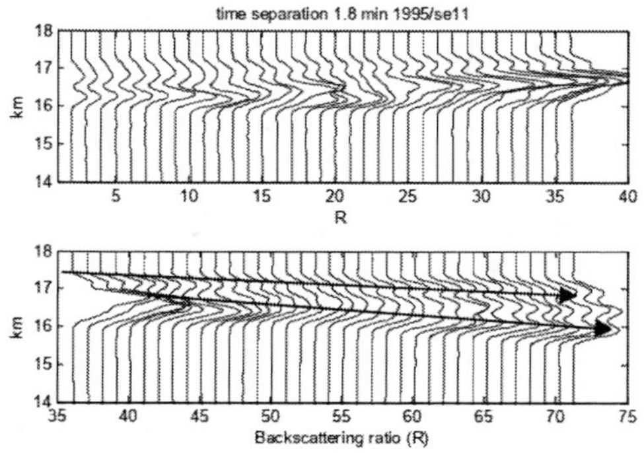

Figure 3. A case of tropopause thin cirrus ascending (upper plot) and descending (lower plot). The profiles are backscattering ratio R progressively offset by a constant in 1.8 min time interval.

REFERENCES

[1] Nee,J. B., W. N. Chen, C. N. Lien, C. I . Lin, 1998: Lidar detection of cirrus cloud in Chung-Li (25 N, 121 E) , J. Atmos. Sci.55, 2249-2257.

[2] Das, S.K., C.-W. Chiang, and J. B. Nee, , 2009: Characteristics of cirrus cloud and its radiative properties based on lidar observation over Chung-Li, Taiwan, Atmos. Res., **93**,723-735.

[3] Das, S. K., J. B. Nee, and C. W. Chiang, A LIDAR study of the effective size of cirrus ice crystals over Chung-Li, Taiwan, J. Atmos. Sol. Terr. Phys. 72,781–788, 2010.

AUTHOR INDEX

AUTHOR INDEX

AUTHOR INDEX

AUTHOR INDEX

AUTHOR INDEX

AUTHOR INDEX

AUTHOR INDEX

AUTHOR INDEX

AUTHOR INDEX

AUTHOR INDEX

27th International Laser Radar Conference 2015 (ILRC 27)

EPJ Web of Conferences Volume 119 (2016)

New York City, New York, USA
5-10 July 2015

Part 2 of 2

Editors:

B. Gross
M. Arend
F. Moshary

ISBN: 978-1-5108-2531-4

Printed from e-media with permission by:

Curran Associates, Inc.
57 Morehouse Lane
Red Hook, NY 12571

proceedings.com

Some format issues inherent in the e-media version may also appear in this print version.

Printed by Curran Associates, Inc. (2016)

For additional information, please contact EDP Sciences – Web of Conferences at the address below.

EDP Sciences – Web of Conferences
17, Avenue du Hoggar
Parc d'Activité de Courtabœuf
BP 112
F-91944 Les Ulis Cedex A
France

Phone: +33 (0) 1 69 18 75 75
Fax: +33 (0) 1 69 28 84 91

contact-edps@webofconferences.org

Additional copies of this publication are available from:

Curran Associates, Inc.
57 Morehouse Lane
Red Hook, NY 12571 USA
Phone: 845-758-0400
Fax: 845-758-2633
Email: curran@proceedings.com
Web: www.proceedings.com

TABLE OF CONTENTS

PART 1

PART 2

DOI: 10.1051/epjconf/201611916005

COMPARISON OF CIRRUS CLOUD CHARACTERISTICS AS ESTIMATED BY A MICROPULSE GROUND-BASED LIDAR AND A SPACEBORNE LIDAR CALIOP DATASETS OVER LILLE, FRANCE (50.60 °N, 3.14 ° E)

Nohra Rita [1]*, Parol Frédéric[1], Dubuisson Philippe[1]

[1]*Laboratoire d'Optique Atmosphérique, Lille University, CNRS* 59655 Villeneuve d'Ascq Cedex, France,
* rita.nohra@ed.univ-lille1.fr

ABSTRACT

Our goal is to establish a climatology of cirrus cloud properties over Lille, France (50.60°N, 3.14 °E) using a ground-based lidar. A statistical analysis of mid-latitude cirrus clouds from lidar data in Lille over the period 2008-2013 is presented and discussed. The macrophysical properties (cloud base altitude, cloud top altitude, geometrical thickness, mid-cloud temperature) and optical properties (cloud optical thickness and lidar ratio) are evaluated and compared between the ground-based and the spaceborne lidar CALIOP (Cloud-Aerosol Lidar with Orthogonal Polarization) measurements for the period 2008-2013. We found similar results of macrophysical properties derived from both lidars. In addition a difference in the optical properties results is due to the multiple scattering and the heterogeneity of the observed scenes.

1. INTRODUCTION

Cirrus clouds are classified by the World Meteorological Organization (WMO) as high altitude clouds, detached in the form of delicate filaments and composed of ice crystals. Because of their high altitudes, cirrus clouds play an important role in the climate system with two opposite effects. They trap the outgoing infrared radiation, but at the same time, they can reflect the incoming solar radiation back out to space. The balance between these two radiative processes, the greenhouse and albedo effects, determines the net impact of cirrus on our climate system. Each contribution depends on the cirrus clouds microphysical and macrophysical properties (Stephens et al., 1990) [1]. Many active remote sensors are used to study high level clouds such as lidars and radars. Moreover several methods have been developed over time to determine and characterize geometrical and optical properties of cirrus clouds using ground-based or spaceborne lidars measurements. The objectives of this study are: (1) to establish a methodology of detection and characterisation of cirrus clouds using a micropulse lidar (MPL) data, (2) to retrieve cirrus clouds properties using the MPL database from 2008-2013, in order to construct a cirrus cloud climatology over Lille, (3) to compare the retrievals properties with CALIOP measurements, and finally (4) to evaluate the potential of the MPL to study high level clouds.

2. METHODOLOGY

First, we present the instrumentation and datasets characteristics used in this work. Then a basic description of the methodology used in this study to determine the macrophysical and optical properties of cirrus clouds using the MPL in Lille and CALIOP, is presented.

2.1. Ground-based lidar

A micropulse lidar has been installed in the Laboratoire d'Optique Atmosphérique (LOA), at University of Lille (France) since 2006. This lidar is operating in automatic mode 24 hours per day, 7 days per week. This MPL uses a Qswitch active ND: yag laser which emits a light pulse of ~ 15 ns at 532 nm. The repetition rate is 4.7 kHz with a pulse energy of 14 μJ. The photons are collected by a 200 mm diameter telescope with an altitude resolution of 15 m. A 1 minute integration time is used. This MPL is not equipped with a polarizing detector; hence the information of ice crystals size or phase of cirrus clouds cannot be obtained.

2.2 Spaceborne lidar CALIOP

The CALIPSO mission (Cloud Aerosol and Infrared Pathfinder Satellite Observations) was launched in April 2006 to measure global profiling of clouds and aerosols. The CALIOP lidar is a two-wavelength (532 nm and 1064 nm) polarization-sensitive lidar set on CALIPSO satellite. CALIOP is a three-channel elastic backscatter lidar. The vertical resolution of

 DOI: 10.1051/epjconf/201611916005

CALIPSO data is 30 m from the surface to 8.2 km and 60 m from 8.2 km to 20.2 km. The layer products are generated at three different spatial resolutions: 1/3, 1 and 5km. The temperature used in CALIOP products is derived from the GEOS-5 data product. In order to compare cirrus clouds properties obtained from CALIOP measurements to those obtained from the ground-based lidar over Lille from 2008 to 2013, we used the CALIOP Level 2 v3.01, 3.02 and 3.30, 5km cloud layer products with a spatial domain of ± 1.5 ° in latitude and longitude round the ground site. We obtained 725 overpasses in this zone (day & night).

2.3. Methodology basic description

The first step of the methodology is to identify the presence of cirrus clouds, by using the method proposed by Platt et al., (1994) [2]. Hence the following geometrical properties are found: cloud base altitude, cloud top altitude and geometrical thickness. The second step is to determine the cirrus cloud optical parameters, which are the Cloud Optical Thickness (COT) and the lidar ratio (LR). The LR corresponds to extinction-to-backscatter ratio, assumed constant within the cloud. This ratio depends on crystal shape, size and orientation. Hence the effective optical depth of cirrus cloud is determined by using the transmittance method (Chen et al., 2002) [3]. This method consists to apply two linear fits of backscattering signals just below and above the cloud layer. The derived effective lidar ratio implies an iterative process, for the COT obtained by applying klett inversion forward method (Klett, 1985) [4] coincide with the COT obtained by the transmittance method. These two effectives parameters are corrected by the multiple scattering parameter η. It is still poorly known, as well as it depends on factors including cloud optical and microphysical properties and lidar configuration (field of view of the receiver, laser beam divergence).

3. RESULTS

In our study, only clouds with cloud base temperature colder than -25 °C are considered to ensure the majority composition of ice crystals. In Lille, this threshold corresponds to a typical cloud base altitude above 6300 m. The atmospheric temperature measurements are obtained from radiosondes launched in Brussels (50.75°N, 4.77°E) at 00:00 UTC, while the MPL lidar in Lille is approximately 115 km from Brussels station.

3.1. Geometrical and temperature characteristic

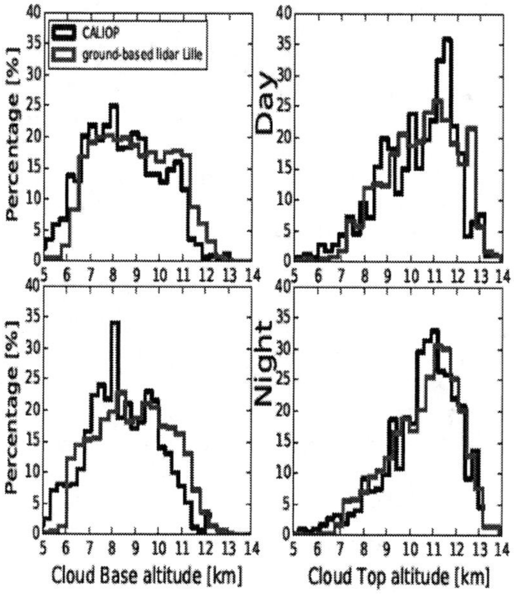

Figure 1: *Distribution of cloud base and cloud top altitude. Black and red lines correspond respectively to CALIOP and ground-based lidar data. The upper two figures correspond to the results obtained during the night and the lower ones to these obtained during the day.*

The MPL is very sensitive to the solar background noise, therefore we separate between nighttime and daytime lidar measurements. Figure 1 shows the distribution of cloud base altitude, and cloud top altitude obtained by the MPL and CALIOP. Cirrus clouds base altitudes are generally observed from 7 to 11 km, and top cloud altitudes from 8 to 14 km with a peak at 11 km. Similar results are obtained from both lidars. There is generally close agreement between results obtained for the cloud top altitudes during night. Also, 87% of total cases studied have a geometrical thickness between 0.5 and 2.5 km. Figure 2 shows the mid-cloud temperature, i.e. the temperature at the geometric midpoint of the cloud layer. Mid-cloud temperatures range from -30 to -70 °C, with a maximum of cases between -45 and -55 °C.

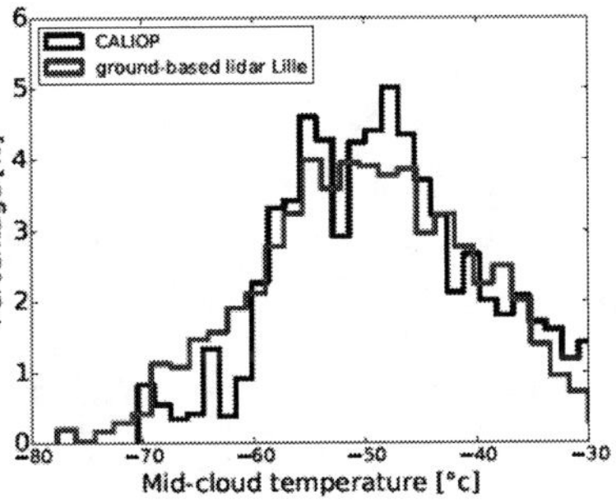

Figure 2: *Distribution of mid-cloud temperature for nighttime measurements.*

The average retrievals of macrophysical properties are presented in table 1, with a mean difference varying between 0.1 and 0.5 km. Table 1 shows that the cirrus clouds detected during night have slightly higher cloud top altitude, colder mid-cloud temperature, and are geometrically thicker, than cirrus clouds detected during day. Besides no remarkable diurnal difference for the cloud base altitudes is found.

Parameters	Day	night
Cloud base altitude (km)	8.92 ± 1.65 (8.39 ± 1.65)	8.91±1.60 (8.38 ± 1.57)
Cloud top altitude (km)	10.46±1.59 (10.25 ± 1.62)	10.62±1.50 (10.55 ± 1.51)
Mid-cloud temperature(°C)	-49±10 (-47 ± 10)	-50±9 (-48 ± 9)
Geometrical thickness (km)	1.54±0.91 (1.86 ±1.55)	1.71±0.93 (2.17 ± 2.01)

Table 1. *Average and standard deviation of macrophysical properties of cirrus clouds over Lille obtained from the MPL lidar and CALIOP (in parentheses) during daytime and nighttime measurements.*

3.2. Optical properties of cirrus clouds

The CALIOP cloud products use a multiple scattering factor of 0.6 and a fixed lidar ratio (LR) of 25 sr, which is one of the causing differences in the optical parameters obtained by the MPL and CALIOP. Sassen and Comstock (2001) [5] assumed a multiple scattering factor of 0.9 for sub-visible cirrus clouds, of 0.8 for relatively thick clouds and 0.6 to 0.7 for optically thick clouds. Because the multiple scattering factor depends on many factors as cited before and in order to find this factor for the MPL lidar, we determine η as the ratio of the COT obtained by CALIOP and the COT obtained by the MPL before correction, i.e. effective COT; we found a variable multiple scattering factor with a mean value equal to 0.75. Figure 3 shows the distribution of different classes of cirrus clouds according to their optical thickness. Visible cirrus clouds, characterized by their COT between 0.03 and 0.3 as defined by Sassen and Cho, (1992) [6] reach 67 % of the total observed cirrus clouds by the MPL before the correction of multiple scattering effect. It is reduced to 56 %, after the correction (figure 3).

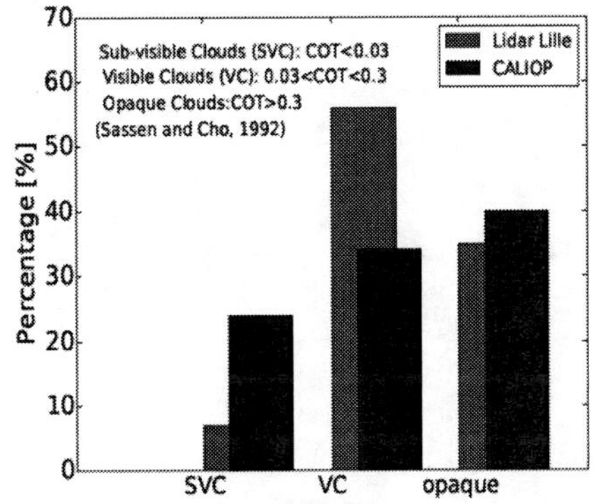

Figure 3: *Different classes of cirrus clouds obtained by MPL and CALIOP. Red histograms correspond to the results obtained by the MPL after the correction of the multiple scattering effect (η=0.75).*

3.3. Monthly variation of cirrus clouds

Figure 4 shows the monthly variation of altitudes of cirrus clouds over Lille cumulated for the 6 years. It shows a similar shape variation obtained from both lidars. Lille has a temperate oceanic climate; furthermore four seasons are generally recognized. Cirrus clouds are more frequently observed in autumn (September till November) with 37 % of total cases (figure 4, a). The mean value of the tropopause during the six years is at 11.3 ± 1.4 km with a mean temperature of -59 ± 6 °C. The cloud top altitudes tend to track the tropopause (figure 4, b). However cirrus clouds are higher during summer (June till August) than during winter (December till February).

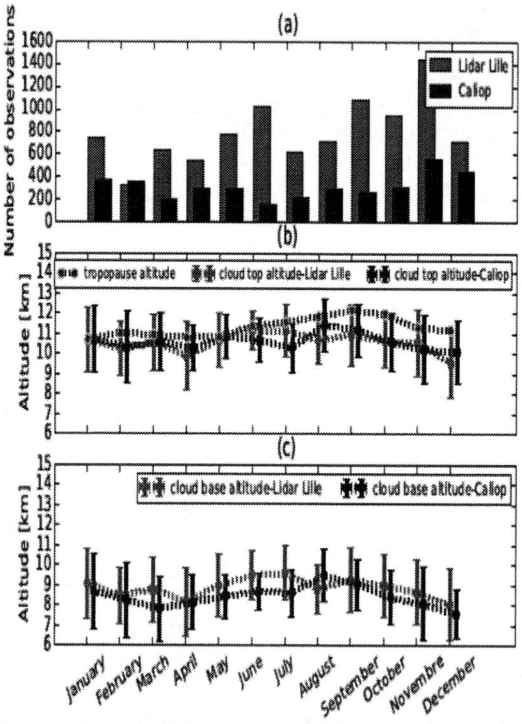

Figure 4: *(a) Histogram of the number of observations of cirrus. (b) Mean and standard deviation of cirrus top altitude and mean tropopause altitude. (c) Mean and standard deviation of base cirrus altitude obtained from ground-based and CALIOP lidar.*

4. CONCLUSIONS

This study provides an analysis of macrophysical and optical properties of cirrus clouds over Lille. An agreement was found between the results obtained by the ground-based and spaceborne lidar especially for the macrophysical properties. A slightly difference is obtained for the optical properties due to the multiple scattering and the impact of cirrus clouds heterogeneities on the retrieved COT. This study shows the potential of the micropulse lidar to study high clouds and to create a climatology of mid-latitude cirrus clouds. However, with some limits: a maximum value of retrieved cirrus COT is 1.2, the solar background noise and the presence of thick lowest cloud creates difficulties in the detection of cirrus clouds.

ACKNOWLEDGEMENT

The authors would like to thank the Lille University and the Regional Council of Nord-Pas-de Calais for financial support of this work. The authors also thankful T.Podvin and computer team at Laboratore d'Optique Atmosphérique (LOA) for their assistance.

REFERENCES

[1] Stephens, G.L., S. Tsay, P.W. Stackhouse, Jr., and P.J Flatau, 1990: The relevance of the microphysical and radiative properties of cirrus clouds to climate and climate feedback, *J. Atmos. Sci.*, **47**, 1742-1753.

[2] Platt, C., Young, S., Carswell, A., Pal, S., McCormick, M., Winker, D., DelGuasta, M., Stefanutti, L., Eberhard, W., Hardesty, M., Kolev, I., Kapreolev, B., Sassen, K., Da-ren, LU., Shamanaev, V., Uchino, O., Mizuno, Y., Wandinger, U., Weitkamp, C., Ansmann, A. et Wooldridge, C, 1994: The Experimental Cloud Lidar Pilot Srudy (ECLIPS) for Cloud-Radiation Research, *J .Am. Meteorolo. Soc*, **75**, 1635-1654.

[3] Chen, W. N., Chiang, C. W et Nee, J. B, 2002: The lidar ratio and depolarization ratio for cirrus clouds, *J. Appl. Opt*, **41**, 6470-6497.

[4] Klett, J. D, 1985: Lidar inversion with variable backscatter/extinction ratios, *J. Appl. Opt,* **24**, 1638-1643.

[5] Sassen, K. and Comstock, J. 2001: A midaltitude cirrus cloud climatology from the facility atmospheric remote sensing. Part III: Radiative properties, *J. Atmos.Sci,* **58**, 2113-2127.

[6] Sassen, K. and Cho, B. S, 1992: Subvisual-thin cirrus lidar dataset for satellite verification and climatological research, *J. Appl. Met.* **31**, 1275-1285.

DOI: 10.1051/epjconf/201611916006

MICROPHYSICAL PROPERTIES OF WARM CLOUDS DURING THE AIRCRAFT TAKE-OFF AND LANDING OVER BUCHAREST, ROMANIA

[1*]Sabina Stefan, [1,2]Sorin Nicolae Vajaiac, [1,2]Andreea Boscornea

[1] *University of Bucharest, Faculty of Physics, P.O.BOX, MG-11, Magurele, Bucharest, Romania,*

**Email: sabina_stefan@yahoo.com*

[2] *National Institute for Aerospace Research "Elie Carafoli" (INCAS), Bucharest, Romania*

ABSTRACT

This paper is focused on airborne measurements of microphysical parameters into warm clouds when the aircraft penetrates the cloud, both during take-off and landing. The experiment was conducted during the aircraft flight between Bucharest and Craiova, in the southern part of Romania. The duration of the experimental flight was 2 hours and 35 minutes in October 7th, 2014, but the present study is dealing solely with the analysis of cloud microphysical properties at the beginning of the experiment (during the aircraft take-off) and at the end, when it got finalized by the aircraft landing procedure. The processing and interpretation of the measurements showed the differences between microphysical parameters, emphasizing that the type of cloud over Bucharest changed, as it was expected. In addition, the results showed that it is important to take into account both the synoptic context and the cloud perturbation due to the velocity of the aircraft, in such cases.

1. INTRODUCTION

Cloud systems are the most important components of the atmosphere. The microphysical properties of clouds play an important role in weather forecasting, in the radiative and energetic budget of the earth, and, as such, in climatological studies [1]. The parameterization of cloud radiative properties, which is important for climatic models, depends on the parameterization of microphysical parameters like effective radius, drop size distributions and liquid water content (LWC) [2].
The drop/droplets concentrations and their size distributions are influenced by environment conditions, especially by temperature, pressure and atmospheric aerosol particles concentrations [3].

This study was designed to analyze the microphysical parameters during a take-off and landing procedures over the Bucharest airport, in time of an experimental flight between Bucharest and Craiova, which are two cities in southern part of Romania.
The details about the flight, the instruments of the Airborne Laboratory and the methods used for measurement processing are included in Section 2, namely "Methodology". Section 3 of the study discusses the results and their interpretation, which confirms the importance of microphysical proprieties knowledge. The paper ends with some conclusions.

2. METHODOLOGY

The flight started with the profiling of a cloud layer above Bucharest. Following this, the flight was in line toward Craiova, and close to Craiova, the aircraft made a vertical sounding. After one hour, the aircraft flew back toward Bucharest, and it finished with the landing procedure, which occurred into a thick cloud over Bucharest. In our study, we found it necessary to examine the properties of the clouds during the procedures of take-off and landing. The measurements were carried out with CAPS system mounted on the left wing of aircraft Beechcraft C90 GTx, which is part of the Airborne Laboratory for Environmental Atmospheric Research in the National Institute for Aerospace Research "Elie Carafoli" (INCAS), in Bucharest, Romania.
The CAPS (Cloud Aerosol Spectrometer Precipitation) was used for measurements. The CAPS system has two components: CAS (Cloud and Aerosol Spectrometer) for small particles/droplets (0.51 to 50 μm) and CIP (Cloud Imagine Particles) for big droplets (15 to 930 um with optional 15-micron resolution). This system provides information about the effective diameter of cloud/aerosol particles, their concentrations, LWC (liquid water content), the atmospheric

temperature and pressure. The CAS is based on light-scattering by smaller particles [4]. CAS records just droplets which are suspended in the air indefinitely, and, in addition to this, CIP recorded the falling precipitation drops. The number of particles of different sizes is stored in Bin channels and the effective diameters are differently defined.

3. RESULTS

The first part of the flight, from Bucharest, is a standard take-off procedure and is then continued with a horizontal line. The lowest altitude of the first part of the flight was observed at the beginning, namely at 672 m, and the highest occurred when the aircraft moved to horizontal position, at 3300 m. During this first part of our analysis, the temperature started from 9°C and did not go beyond the negative temperature of -3°C, at the highest altitude. However, the cloud layer got up to 1°C and its thickness had a value of 1732 m. These macroscopic characteristics indicate a warm/water cloud with cloud top at 2360 m. During the slow cloud penetration by the aircraft (which lasted for 4 minutes), the variations of the microphysical parameters were determined using CAS and CIP measurements (Fig. 1 upper panel).

In the case of aircraft landing, the temperatures have positive values that increased from 3°C to 10°C. It is known [5] that the radii values of the particles/cloud droplets are the result of a few important processes in warm clouds, as follows: a) the heterogeneous nucleation of cloud droplets on CCN (cloud condensation nuclei), r~1µm; b) the growth by diffusion of water vapor up to r~ 20 µm; c) growth by collision and coalescence up to r~100 µm; and, d) precipitation r≥100 µm. In Fig. 1 upper panel one can note the variation with the height of the microphysical parameter values inside of the warm cloud. Thus, the variation with the height of effective diameters show that there are aerosol particles at cloud base, small cloud droplets due to CCN presence and large cloud droplets (cloud drops radius between 30 and 100 µm) due to collision-coalescence or accretion processes. Toward the cloud top, the radius of the droplets increases and the radius of the cloud drops decreases.

The high droplets' concentration in the cloud is associated with small size particles (Fig.1 upper panel).

The measurements with CIP show low drop concentrations: max 20cm^{-3} at cloud base to 140 cm^{-3} close to the cloud top. These results are similar to Pruppacher and Klett (1997) observations for continental water clouds.

The microphysical parameters in the time of aircraft landing are displayed in Fig.1 lower panel. The microphysical parameters are very different from those observed at the time of aircraft take-off (Fig.1 lower panel). This means that the meteorological conditions changed over Bucharest, as well as the cloudy system. The increasing of the cloud drops radii determined the decreasing of their number concentrations, which is explained by the enhancement of collision – coalescence processes (Fig.1 lower panel).

In the last part of the flight, during the landing procedure, the dynamics of processes inside the cloud were accelerated due to the aircraft velocity and the temperature and pressure variations.

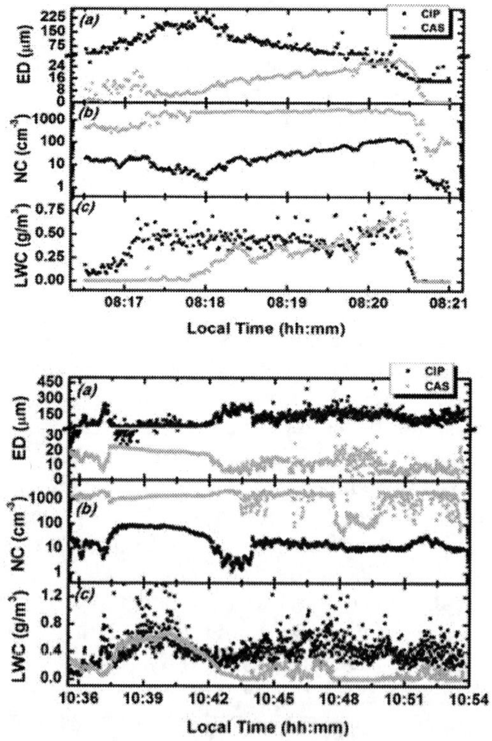

Fig.1 Effective diameter, number concentration and LWC measured by CAS and CIP in clouds

 DOI: 10.1051/epjconf/201611916006

over Bucharest during take-off (upper panel) and during landing (lower panel)

The microphysical parameters are very different than those from the time of aircraft take-off.

The liquid water content (LWC) (Figs.1 and 2) varies as per the number concentrations of drops.

It is noteworthy that in the first meters, the values of LWC are small (0.05g m^{-3}), as was expected due to the aerosol dominance. LWC increases with height above the cloud base (at 628m), up to 0.7g m^{-3}.

The variation of LWC with height into cloud during aircraft ascension, and the variation during the landing procedure are different, which emphasizes the difference between the cloud types. At the beginning of the flight, it was stratus and, after two hours, the nimbostratus cloud covered the sky airport area over Bucharest.

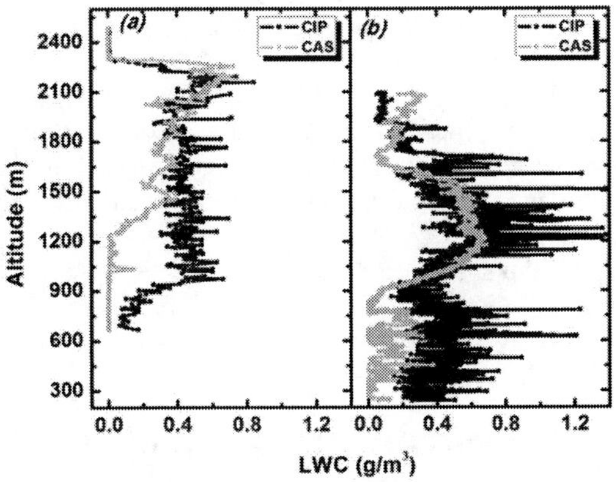

Fig.2 Variation of LWC with height during aircraft take-off (a) and during landing (b)

4. CONCLUSIONS

The aircraft measurements during the flight on October 07 2014 were used to investigate the properties of the clouds.

The discrepancy or the similarity of effective diameters for cases with significantly different number concentrations, and in consequence, with different LWC, can be attributed to both meteorological (temperature, dynamics) and microphysical factors, associated to differences in the cloud types.

The cases presented here highlight the microphysical parameters values (according to Pruppacher and Klett (1997) data) presence of warm clouds as follows: the aircraft take-off was into stratus cloud, and, lastly, the cloud system over Bucharest was a precipitating nimbostratus.

Combined analyses of the CAS and CIP microphysical measurements lead to better insights into the key processes within clouds.

We have shown that the remote sensing measurements provide an important context for the interpretation of the microphysics measurements, and extend them along trajectories for interpreting the evolution of hydrometeors within clouds.

ACKNOWLEDGEMENT

This work is supported by the project Clouds and Aerosols Profiling in support of ESA's sentinel 4 and 5 missions (CAPESA), which take place under program Technology Space and Advanced Research (STAR), Romanian Space Agency (ROSA).The work of the first author, Sabina Stefan, was supported by the national project PNII-PT-PCCA-2013-4-2074.

REFERENCES

[1] Intergovernmental Panel on Climate Change (IPCC), Climate Change 2007: The Physical Science Basis. UK: *Cambridge University Press*, 2007.

[2] Pawlowska, Hanna, J. L. Brenguier, and Frederic Burnet, (2000): Microphysical properties of stratocumulus clouds, *Atmospheric research* **55.1**, 15-33.

[3] Sharon, T. M., B. A. Albrecht, H. H. Jonsson, P. Minnis, M. M. Khaiyer, T. M. van Reken, J. Seinfeld, and R. Flagan, 2006: Aerosol and cloud microphysical characteristics of rifts and gradients in maritime stratocumulus clouds. *Journal of the atmospheric sciences*, **63**, 983-997.

[4] D. Baumgardner, H. Jonsson, W. Dawson, D. O'Connor, and R. Newton, 2001, The cloud, aerosol and precipitation spectrometer: A new instrument for cloud investigations, *Atmospheric research*, **59**, 251-264.

[5] Pruppacher, H. R., and J. D. Klett, 1997: *Microphysics of Clouds and Precipitation.* Published by Kluwer Academic Publisher, 714 pp.

EVALUATION OF RETRIEVAL ALGORITHMS FOR ICE MICROPHYSICS USING CALIPSO /CLOUDSAT AND EARTHCARE

Hajime Okamoto[1]*, Kaori Sato[1], Yuichiro Hagihara[1], Hiroshi Ishimoto[2], Anatoli Borovoi[3], Alexander Konoshonkin[3] , Natalia Kustova[3]

[1]*Research Institute for Applied Mechanics, Kyushu University, Kasuga, Fukuoka 816-8580, Japan, *Email: okamoto@riam.kyushu-u.ac.jp*

[2]*Meteological Research Institute, Japan Meteorological Agency, Tsukuba, Ibaraki 305-0052, Japan*

[3]*V.E. Zuev Institute of Atmospheric Optics, Rus. Acad. Sci., Tomsk 634021, Russia*

ABSTRACT

We developed lidar-radar algorithms that can be applied to Cloud-Aerosol Lidar and Infrared Pathfinder Satellite Observation (CALIPSO) lidar and CloudSat data to retrieve ice microphysics. The algorithms were the extended version of previously reported algorithm [1] and can treat both of nadir pointing of CALIPSO lidar period and 3°-off-nadir pointing one. We used the scattering data bank produced by the physical optics methods [2] and created lidar look-up tables of quasi-horizontally oriented ice plates (Q2D-plate) for nadir- and off-nadir lidar pointing periods. Then LUTs were implemented in the ice retrieval algorithms.

We performed several sensitivity studies to evaluate uncertainties in the retrieved ice microphysics due to ice particle orientation and shape. It was found that the implementation of orientation of horizontally oriented ice plate model in the algorithm drastically improved the retrieval results in both for nadir- and off-nadir lidar pointing periods. Differences in the retrieved microphysics between only randomly oriented ice model (3D-ice) and mixture of 3D-ice and Q2D-plate model were large especially in off-nadir period, e.g., 100% in effective radius and one order in ice water content, respectively. And differences in the retrieved ice microphysics among different mixture models were smaller than about 50% for effective radius in nadir period.

1. INTRODUCTION

Ice clouds play an essential role in the climate system through their radiative effects. Combination of 94GHz cloud radar on CloudSat and lidar on CALIPSO was proved to be effective to derive global distribution of ice microphysical properties [1]. Since the lidar wavelength was much smaller than the ice particle size, the lidar backscattering properties (backscattering coefficient and depolarization ratio) depend on particle shape, orientation, size, ice water content (IWC) and laser tilt angle (=off-nadir angle). Currently details of horizontally oriented ice particle models have not been taken into account in the analysis of CloudSat and CALIPSO for the off-nadir period and lidar backscattering properties of oriented ice were assumed to be similar to that of randomly oriented ice. In this study, we developed the new lidar-radar algorithms that can be applied to both periods. And uncertainties of the retrieved ice microphysics were investigated.

2. ANALYSIS OF CALIPSO LIDAR DATA

CALIPSO lidar observations showed that specular returns, i.e., large backscattering coefficient (ß), from ice clouds were frequently observed from December 2006 to November 2007 and the period corresponds to the one when the laser direction was close to nadir (off-nadir angle was 0.3°) (Fig. 1a) [3]. And fraction of large ß (>10^{-4} [1/m/sr]) drastically reduced from December 2007 to November 2008, corresponding to the off-nadir period (3°) (Fig.1b). In these analyses, ice clouds were first identified by the phase discrimination scheme from attenuation and depolarization ratio of CALIPSO lidar observations [3] and only night time data was used.

When the data was sorted by values of ß at 532nm, the depolarization ratio of ice particles for the largest ß category (>10^{-4} [1/m/sr]) was minimum (<5%) when temperature (T) >-20°C for the nadir pointing period (Fig. 2a). And smaller ß categories generally showed larger depolarization ratio in the same T range and the depolarization ratio was smaller than 15%.

The depolarization ratio below -20°C during off-nadir period was similar to nadir period, while the value above -20°C in the off-nadir period was about 20%, which was larger compared with the value in the nadir period,. These differences between the two periods might attribute to the change in the backscattering properties of oriented ice particles between both periods.

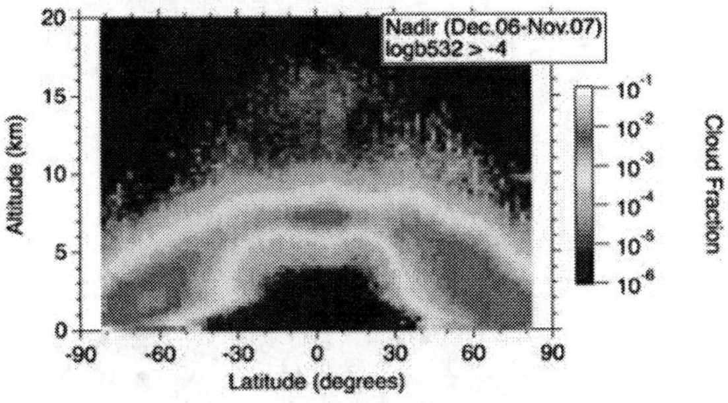

(a)

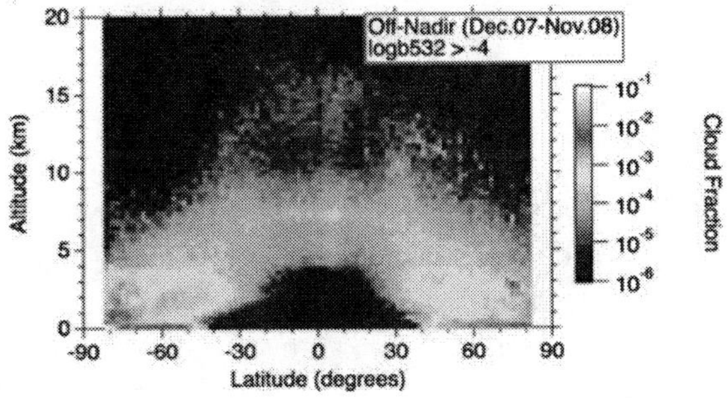

(b)

Fig.1 Height-latitude plot of annual mean cloud fraction for attenuated backscattering coefficient at 532nm >10^{-4} [1/m/sr] from CALIPSO lidar observations. (a) Nadir pointing period (0.3°), from December 2006 to December 2007. (b) Off-nadir pointing period (3°) from December 2007 to December 2008.

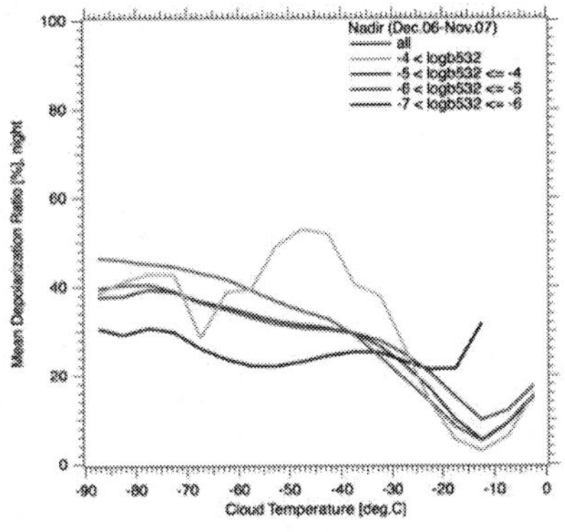

(a)

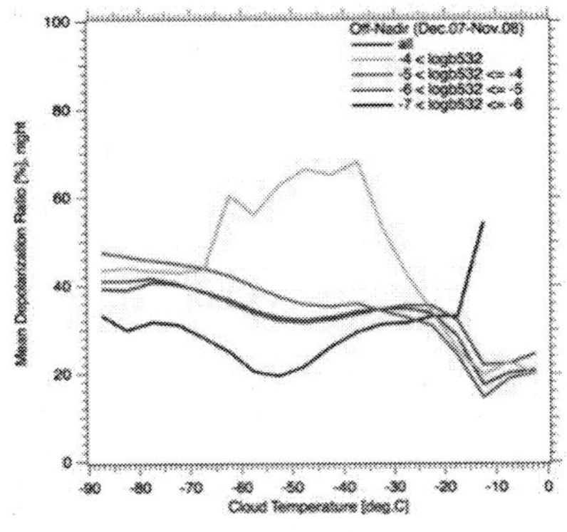

(b)

Fig.2 Annual mean depolarization ratio of ice clouds by CALIPSO lidar. Dependence of temperature on the depolarization ratio sorted by ß was shown. (a) Nadir period. (b) Off-nadir period.

Next, we examined lidar backscattering properties of quasi-horizontally oriented ice plates with different orientation models and that for randomly oriented ice particles by using physical optics method [2]. The size distribution was assumed to be modified Gamma distribution function with variation of the characteristic radius [1]. The state of orientation of ice plate was characterized by Gamma distribution function with effective distribution width θ_{eff} that denotes standard deviation of canting angle of the particles. We considered three values of θ_{eff}, i.e., 0.1°, 0.5° and 1° [4]. ß of Q2D-plate was generally one to three orders of magnitude larger compared with that of 3D-ice for nadir pointing period (Fig.3). ß for

Q2D-plate for the nadir period differed by a factor of 4 among the three θ_{eff} models.

There were larger differences among Q2D-plate models in off-nadir period compared with those in nadir period. Q2D-plate with $\theta_{eff} = 1°$ showed the largest ß value than other types and Q2D-plate models with 0.1° and 0.5° showed smaller ß than that of 3D-ice in the period. That suggests uncertainties in the ice models might lead to larger uncertainties in the ice microphysical retrievals in off-nadir period.

Since depolarization ratio of Q2D-plate was close to 0% and that for 3D-ice was 20-40%, mixture of Q2D-plate and 3D-ice model can potentially explain the observed depolarization ratio shown in Fig.2. Therefore the mixture of Q2D-plate and 3D-ice model with variation of the mixing ratio of Q2D-plate was considered in the retrieval algorithms as in [1].

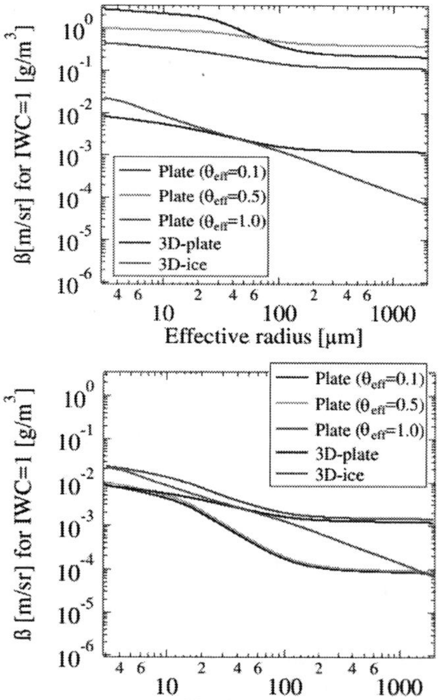

Fig.3 ß for quasi-horizontally oriented ice plates and randomly oriented ice models as function of effective radius. (a) ß for Nadir period and (b) ß for off-nadir period. Ice water content was set to be 1g/m^3.

3. RETRIEVAL RESULTS OF ICE MICROPHYSCS FROM CALIPSO AND CLOUDSAT

We estimated look up tables for CALIPSO lidar by the physical optics method described in Section 2 and LUTs for CloudSat radar by the discrete dipole approximation. Then lidar-radar algorithm [1] for ice microphysics with the new look up tables was applied to the CALIPSO and CloudSat data. ß and depolarization ratio at 532nm and radar reflectivity factor were used and effective radius, IWC and mass mixing ratio of Q2D-plate to the total IWC were retrieved. The results were shown for off-nadir period (Fig 4).

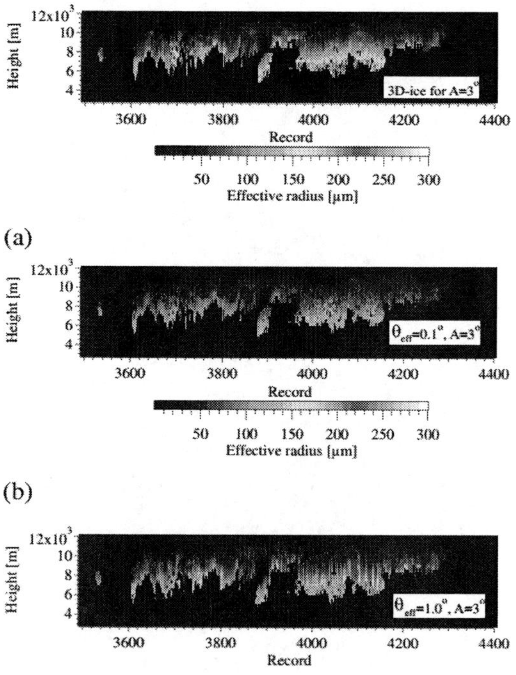

(c)

Fig. 4 Time-height plot of retrieved ice effective radius for θ_{eff} to be 0.1°, 0.5° and 1° by lidar-radar algorithms for off-nadir period. (a) results of pure 3D-ice model, (b) results for mixture of 3D-ice and Q2D-plate with θ_{eff} of 0.1° and (c) results for mixture of 3D-ice and Q2D-plate with θ_{eff} of 1°.

When only 3D-ice model was considered and Q2D-plate model was not included in the

DOI: 10.1051/epjconf/201611916007

retrievals, effective radius was overestimated in the lower part of cloud layers compared with the results for the mixture of 3D-ice and Q2D-plate models. The differences in effective radius and IWC were more than a factor and one order, respectively. This demonstrated the need to take into account the Q2D-plate and the orientation of Q2D-plate in off-nadir period. Differences between the models with θ_{eff} to be 0.1° and 1° were also large, e.g., about 50% in effective radius and about one order. Thus we considered that the ice orientation would be the major source of retrieval uncertainties in off-nadir period.

Currently there are no retrieval algorithms that can treat ß at 1064nm in addition to ß at 532nm. The backscattered color ratio defined as ratio of ß at 1064nm to ß at 532nm was investigated. First we used lidar signals from cloud top part of water clouds without dense aerosols or cirrus clouds above to re-calibrate ß at 1064nm for CALIPSO lidar. The color ratio of water clouds was assumed to be unity for the re-calibration of ß at 1064nm. Then global analysis of the color ratio of ice particles was conducted. It was shown that the color ratio of ice clouds is about 0.9 with standard deviation of about 0.3. Calculation by physical optics indicated that the color ratio for Q2D-plate was smaller than 1 when effective radius is smaller than 40μm and the value was about 1 for larger particles except for θ_{eff} to be 0.1° where the value was much larger than 1 in the nadir period. That for 3D-plates was from 0.7 to 0.8 for the same period. The color ratio for Q2D-plate were larger than 1 for smaller particles and smaller than 1 for large size when θ_{eff} was assumed to be 0.1° or 0.5° for off-nadir period. Therefore the backscatter color ratio might be able to be used to discriminate different orientation models and this could reduce uncertainties in the retrieved ice microphysics.

4. SUMMARY

We implemented the new look up tables for quasi-horizontally oriented ice plates into the lidar-radar algorithm to retrieve ice microphysics by CALIPSO lidar and CloudSat data. The new algorithms can treat both data obtained in the nadir and the off-nadir laser pointing periods of CALIPSO lidar. We considered standard deviation of canting angle of ice plates from 0.1 to 1 to estimate ß by the physical optics method for CALIPSO lidar signals. We found mixture of 3D-ice and Q2D-plate models could reproduce observed ß and depolarization ratio for nadir and off-nadir periods.

Sensitivity studies showed that larger differences in the retrieved ice microphysics among Q2D-plate models existed for off-nadir period than those for nadir-period. To reduce uncertainties in the retrieved microphysics, the information of backscattered color ratio of ice might help. JAXA-ESA joint mission EarthCARE will carry high spectral resolution lidar (ATLID) and Doppler 94GHz cloud radar in space in 2018. Since ATLID will use off-nadir pointing similar to CALIPSO, it is expected to have similar effects considered in this studies.

ACKNOWLEDGEMENT

This work was supported in part by the JSPS Kakenhi Grant Number 25247078 and 15K17762 and also supported by the Collaborated Research Program of Research Institute for Applied Mechanics, Kyushu University.

REFERENCES

[1] Okamoto, H., K. Sato and Y. Hagihara, 2010: Global analysis of ice microphysics from CloudSat and CALIPSO: intercomparison of specular reflection in lidar signals, *J. Geophys. Res.*, **115**, D22209, doi:10.1029/2009JD013383.

[2] Borovoi, A., A. Konoshonkin, N. Kustova, H. Okamoto, 2012: Backscattering Mueller matrix for quasi-horizontally oriented ice plates of cirrus clouds: application to CALIPSO signals, Opt. Express, **20**(27), 28222-28233,

[3] Yoshida, R., H. Okamoto, Y. Hagihara and H. Ishimoto, 2010: Global analysis of cloud phase and ice crystal orientation from Cloiud-Aerosol Lidar and Infrared Pathfinder Satellite Observation (CALIPSO) data using attenuated backscattering and depolarization ratio, *J. Geophys. Atmos.*, **115**, D00H32, doi:1029/2009JD012334.

[4] Sassen, K., 1980: Remote sensing of planar ice crystal fall altitudes, *J. Meteor. Soc. Japan*, **58**, 422-429.

EPJ Web of Conferences **119**, 16008 (2016) DOI: 10.1051/epjconf/201611916008
ILRC 27

STUDY OF DROPLET ACTIVATION IN THIN CLOUDS USING GROUND-BASED RAMAN LIDAR AND ANCILLARY REMOTE SENSORS

Marco Rosoldi [1]*, Fabio Madonna[1], Pilar Gumà Claramunt[1] and Gelsomina Pappalardo[1]

[1]*Consiglio Nazionale delle Ricerche (CNR), Istituto di Metodologie per l'Analisi Ambientale (IMAA), C.da S. Loia, I-85050, Tito Scalo, Potenza, Italy, *Email: marco.rosoldi@imaa.cnr.it*

ABSTRACT

A methodology for the study of cloud droplet activation based on the measurements performed with ground-based multi-wavelength Raman lidars and ancillary remote sensors collected at CNR-IMAA observatory, Potenza, South Italy, is presented. The study is focused on the observation of thin warm clouds. Thin clouds are often also optically thin: this allows the cloud top detection and the full profiling of cloud layers using ground-based Raman lidar. Moreover, broken clouds are inspected to take advantage of their discontinuous structure in order to study the variability of optical properties and water vapor content in the transition from cloudy regions to cloudless regions close to the cloud boundaries. A statistical study of this variability leads to identify threshold values for the optical properties, enabling the discrimination between clouds and cloudless regions. These values can be used to evaluate and improve parameterizations of droplet activation within numerical models. A statistical study of the co-located Doppler radar moments allows to retrieve droplet size and vertical velocities close to the cloud base. First evidences of a correlation between droplet vertical velocities measured at the cloud base and the aerosol effective radius observed in the cloud-free regions of the broken clouds are found.

1. INTRODUCTION

Studies on global climate change show that the effects of aerosol-cloud interactions (ACI) on the Earth's climate and radiative balance, also known as indirect aerosol effects, are the most uncertain among all the effects involving the atmospheric constituents and processes [1]. The activation of cloud droplets occurs through the water vapor condensation when super-saturation conditions are reached in an air mass. It represents the microphysical link between aerosols and clouds and it is the basis towards a better understanding of ACI.

The droplet activation and the ACI are not yet accurately quantified and parameterized in the weather and climate numerical models due to the complexity of the processes which take place far away from the ground, where the availability of accurate in situ measurements is limited in time and space sampling. Therefore, the study of droplet activation and ACI is strongly needed to evaluate and improve the ability of numerical models to forecast clouds and their microphysical properties, as well as to better quantify ACI in global climate models. This investigation needs an accurate estimation of optical and microphysical properties of clouds and aerosols close to the cloud boundaries, that can be achieved using ground-based remote sensing techniques.

This work describes a methodology to improve the knowledge of the processes underlying the droplet activation. It is based on the observations performed with multi-wavelength (355, 532 and 1064 nm) Raman lidars, a Ka-band cloud Doppler radar and a microwave radiometer operative at CIAO (CNR-IMAA Atmospheric Observatory), located in Potenza, Southern Italy (40.60N, 15.72E, 760 m above sea level) and on the study of thin warm clouds. These clouds are low or midlevel super-cooled clouds characterized by a liquid water path (LWP) less than about 100 gm^{-2} [2]. Thin warm clouds are often optically thin, i.e they allow the Raman lidar to measure the backscattered signal through the entire cloud structure and above, enabling the cloud top detection. Optically thin broken warm clouds are observed, where broken refers to single layer clouds with a not homogeneous horizontal structure, characterized by cloud fields separated by, cloud-free regions. The discontinuous structure of broken clouds allows to study the variability of optical properties and water vapor content in the transition from cloudy regions to cloudless regions close to cloud boundaries.

2. METHODOLOGY

Night time observations of low and mid-level optically thin broken clouds are performed simultaneously with the multi-wavelength Raman lidars, the cloud Doppler radar and the microwave radiometer. The cloud phase is identified using the radar linear depolarization ratio (LDR) and atmospheric temperature profiles obtained by the microwave radiometer in the time range when the cloud is observed.

Taking advantage of the discontinuous structure of the broken clouds, the vertical profiles of the extensive aerosol optical properties (extinction coefficient α at 355 and 532 nm and backscattering coefficients β at 355, 532 and 1064 nm) and of the water vapor content (water vapor mixing ratio and relative humidity) are retrieved using the combined Raman-elastic lidar technique in two different ways: "with clouds", by averaging only the lidar signals containing clouds, in a selected time window; "without clouds", by averaging only the cloud-free signals before and after the clouds (skipping the signals containing clouds), in the same time window. The lidar signals contaminated by clouds are identified by visual inspection of the nitrogen Raman lidar signals at 387 nm, assuming that a strong negative slope in these signals is due to the extinction by liquid water droplets. The profiles with clouds provide the cloud optical properties and the water vapor content in cloudy regions, while the profiles without clouds provide the aerosol optical properties and water vapor content in cloudless regions close to the cloud boundaries.

For each observed cloud layer, the vertical profiles of optical properties and water vapor content are calculated using the same effective vertical resolution. The selected time windows for lidar analysis are in the range 15 - 45 minutes, while the used effective vertical resolutions are in the range 160-350 m. These resolutions have been optimized in a way to obtain lidar profiles at the highest vertical resolution with a random uncertainty lower than 30%. The profiles of extinction coefficient with clouds are corrected for multiple scattering [3,4].

For each observed cloud, the vertical regions where extinction coefficients at 355 nm with and without clouds do not show significant differences, within their error bars, represent atmospheric regions out of the cloud layer, where aerosols are not activated into droplets; the altitude range where the extinction coefficient at 355 nm with clouds is significantly higher than the corresponding extinction coefficient without clouds constitutes the region where aerosol particles grow and droplet activation occurs, identified as the "activation region".

The cloud base and top are determined by using the nitrogen Raman lidar signals at 387 nm with and without clouds, calibrated on the same molecular lidar signal at 387 nm derived from a standard atmosphere model. The cloud base height is identified by an inflection point in the Raman signal with clouds, corresponding to a local minimum of the first derivative of the signal; specifically it is identified as the height at which the derivative starts to decrease on the decreasing edge of the local minimum. The cloud top height is identified as the height where the Raman signal with clouds starts to overlap the corresponding Raman signal without clouds.

The cloud base and cloud top temperatures as well as cloud the LWP are estimated by averaging the microwave radiometer products in the same time window of the lidar analysis.

The cloud optical depth (COD) is obtained by integrating the extinction coefficient profile with clouds, corrected for multiple scattering, over the altitude range between the base and the top of the cloud.

The source and type of aerosols in cloud-free regions close to the cloud boundaries are identified from the analysis of the air mass back-trajectories provided by NOAA Hysplit model [5], along with the values of multi-wavelength aerosol lidar ratios (S) and the Ångström exponents (å) [6], averaged over the altitude range between the cloud base and the cloud top. In the same region, the aerosol size distribution close to the clouds is also retrieved from multi-wavelength aerosol extensive optical properties, averaged over the altitude range between the base and the top of the cloudy region, using the algorithm developed by Veselovskii [7].

DOI: 10.1051/epjconf/201611916008

3. RESULTS

Twenty cloud layers have been studied. Using the methodology described in section 2, it is found that these clouds are characterized by a vertical extension of the activation region in the range 290-1040 m, the cloud base height in the range 1440-4550 m above sea level (a.s.l.), the cloud top height in the range 1640-4730 m a.s.l., the cloud thickness in the range 90-500 m, the cloud base and the cloud top temperatures in the range 270-286 K and 268-284 K, respectively, the COD at 355 nm in the range 0.06-1.20. All the observed clouds are liquid or mixed phase clouds with a LWP lower than 180 gm^{-2}. In the cloud-free regions close to the cloud boundaries, the aerosol type is originated from different sources (dust, continental, urban-industrial, marine, mixed).

Figure 1 shows the probability density function (pdf) of α (panel a), β (panel b) and S (panel c) at 355 nm, calculated from profiles with clouds (blue columns) and without clouds (red columns) in the altitude range between the base and the top of the activation region. These pdfs lead to identify threshold values enabling the discrimination between clouds and cloud-free atmosphere. In particular, values of α and β higher than 3.5×10^{-4} m^{-1} and 1.0×10^{-5} m^{-1} sr^{-1}, respectively, and values of S lower than 25 sr (blue areas in fig.1) show a very high probability of droplet activation and cloud formation and are classified as cloudy regions. Values of α and β lower than 1.0×10^{-4} m^{-1} and 0.6×10^{-5} m^{-1} sr^{-1}, respectively, and values of S higher than 55 sr (red areas in fig.1) show a very low probability of droplet activation and cloud formation and are classified as cloudless regions. Values of α, β, and S between the above threshold values, i.e. α in the range 1.0×10^{-4} - 3.5×10^{-4} m^{-1}, β in the range 0.6×10^{-5} - 1.0×10^{-5} m^{-1} sr^{-1} and S in the range 25 - 55 sr (green areas in fig.1) show comparable probabilities of droplet activation or not; these values do not allow to discriminate between cloudy and cloudless regions and might correspond to the transition region between clouds and cloud-free atmosphere also called "Twilight Zone" [8]. These results can be used to directly evaluate and improve the ability of cloud numerical models to predict the formation of liquid water clouds, by ground-based Raman lidar measurements of extinction and backscattering coefficients. This evaluation also implies an indirect validation of parameterizations used in numerical models for the droplet activation.

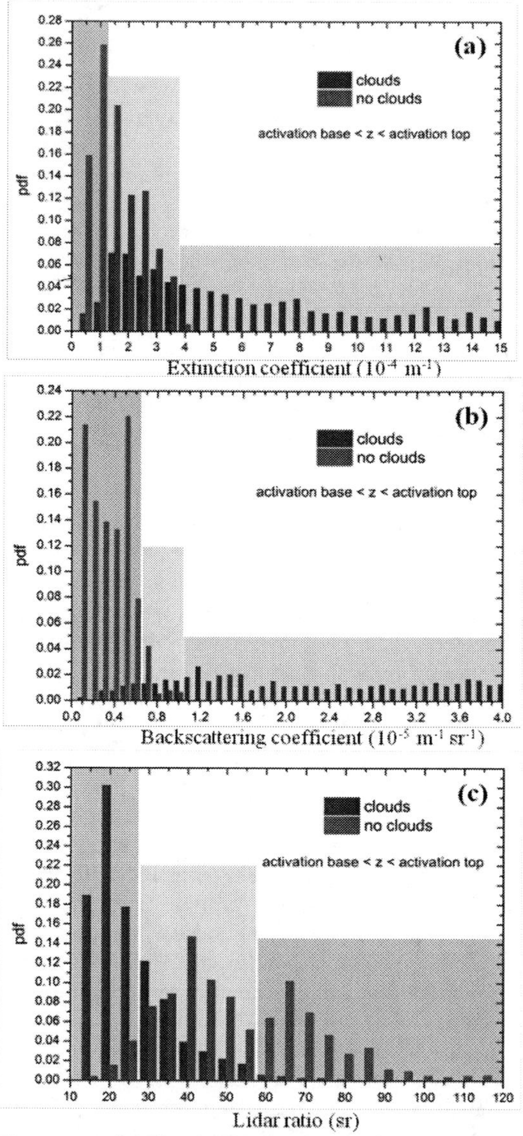

Figure 1. Probability density function (pdf) of the aerosol extinction coefficient (a), aerosol backscatter coefficient (b) and lidar ratio (c) at 355 nm, calculated from the Raman lidar profiles "with clouds" (blue columns) and "without clouds" (red columns) in the altitude ranges between the base and the top of the activation region.

Only in ten of the considered cases the radar provides a Doppler spectrum. The pdfs of mean Doppler velocity and reflectivity measured by the radar at the closest available altitude to the cloud base, are calculated. The Doppler velocity pdf shows that close to the cloud base droplet vertical velocity is typically dominated by downdrafts, with velocities ranging within - 1.5 m s^{-1} and + 0.5 m s^{-1}, with a median value around - 0.25 m s^{-1}. Using the radar reflectivity and scattering simulations [9], it is possible to relate the pdf of the radar reflectivity to droplet size: the two main modes of the pdf are corresponding to droplet sizes around 10 µm and 15 µm, respectively. Finally, the scatter plot of droplet vertical velocities, measured at the cloud base and averaged over the time window of lidar analysis for each cloud layer, versus the effective radius of aerosol particles in cloud-free regions close to the cloud boundaries is considered. The scatter plot indicates that aerosols with an effective radius smaller than 0.8 µm are activated to form droplets moving downward, while aerosols with larger effective radius tend to form droplets moving upward.

4. CONCLUSIONS

A methodology for the experimental investigation of cloud droplet activation is presented. The study is focused on optically thin warm clouds with a broken structure. It is based on the observations performed with ground-based multi-wavelength Raman lidars, a cloud Doppler radar and a microwave radiometer. Geometrical and microphysical properties for twenty cases characterized by this cloud type are reported. Type, source, and microphysical properties of aerosols in cloud-free regions close to the cloud boundaries are also retrieved. To enable an efficient discrimination between clouds and cloudless regions, threshold values of α, β and S at 355 nm are identified. These values can be used to evaluate and improve numerical models and their parameterizations of cloud droplet activation. Finally, droplet sizes and vertical velocities close to the cloud base have been correlated with the size of aerosol particles, revealing that droplet activation for aerosol particles smaller than 0.8 µm generates strong downdrafts, while activation for particles larger than 0.8 µm generates moderate updrafts.

ACKNOWLEDGEMENT

This work was supported by the European Union (FP7/2007-2013) for the ACTRIS Research Infrastructure under grant n. 262254.

REFERENCES

[1] Stocker, T.F. et al., 2013: Contribution of Working Group I to the Fifth Assessment Report of the Intergovernmental Panel on Climate Change, *Cambridge University Press.*

[2] Turner, D. D. et al., 2007: Thin liquid water clouds: Their importance and our challenge, *Bull. Amer. Meteor. Soc., 88, 177–190.*

[3] Eloranta, E.E., 1998: Practical model for the calculation of multiply scattered lidar returns, *Applied Optics, Vol.37, N.12, 2464 – 2472.*

[4]Wandinger, U., 1998: Multiple scattering influence on extinction and backscatter coefficient measurements with Raman and high spectral resolution lidars, *Applied Optics, Vol.37, N.3, 417-427.*

[5]Draxler, R.R., G.D. Rolph, 2014: HYSPLIT (Hybrid Single-Particle Lagrangian Integrated Trajectory) Model access via NOAA ARL READY Website, *NOAA Air Resources Laboratory, Silver Spring, MD.*

[6]Müller, D. et al., 2007: Aerosol-type-dependent lidar ratios observed with Raman lidar, *Journal of Geophysical Research: Atmospheres, Vol.112, D16202.*

[7] Veselovskii, I. et al., 2004: Inversion of multi wavelength Raman lidar data for retrieval of bimodal aerosol size distribution, *Applied Optics, Vol.43, N.5, 1180 – 1195.*

[8] Koren, I. et al., 2007: On the twilight zone between clouds and aerosols, *Geophysical Research Letters, Vol. 34, L08805.*

[9] Carey, L. D., S. A. Rutledge, 1998: Electrical and multiparameter radar observations of a severe hailstorm, *J. Geophys. Res., 103, 13979–14000.*

DEVELOPING A CLIMATOLOGY OF CIRRUS LIDAR RATIOS USING UNIVERISTY OF WISCONSIN HSRL OBSERVATIONS

Ralph Kuehn*[1], Robert Holz[1], Edwin Eloranta[1], Mark Vaughan[2], and Johnathan Hair[2]

[1]*Cooperative Institute for Meteorological Satellite Studies (CIMSS), University of Wisconsin-Madison, WI 53706, USA,*

[2]*NASA Langely Research Center, Hampton, VA 23681-2199*

**Email: ralph.kuehn@ssec.wisc.edu*

ABSTRACT

Measurements of ice cloud lidar ratio from the University of Wisconsin High Spectral Resolution Lidar System (UW HSRL) are shown, for the period 2013/06/18-2013/11/04 in Hunstville, AL. These data were acquired as part of the SEAC4RS campaign. The layer-averaged median lidar ratio ,at 532 nm, for non-precipitating ice-clouds, 0.3 < OD < 3.0, observed during the experiment was determined to be 25.7 +/- 10.6 sr. As part of this work we've also developed an automated cloud and precipitation classification and detection algorithm.

1. INTRODUCTION

Our ability to detect and probe the vertical extent of cirrus was hugely improved with the launch of the NASA- CNES (CALIPSO) mission in April 2006. However, our skill at retrieving the optical properties of the cirrus detected by CALIPSO is not yet commensurate with our detection abilities. As with any new observing system, CALIPSO faces challenges and uncertainties in the retrieval of the geophysical parameters from its fundamental measurements. The data acquired during NASA's many intensive airborne and ground-based field campaigns can provide new insights into these knowledge gaps that can be used to improve retrieval accuracies and, equally important, help constrain overall retrieval uncertainties using more detailed and higher signal-to- noise measurements. To improve CALIPSO optical properties retrievals, we show High Spectral Resolution Lidar (HSRL) measurements acquired during the SEAC4RS campaign from Huntsville, AL, USA. Because the extinction and optical depth retrievals for elastic backscatter lidars like CALIOP typically rely on a priori assumptions about layer-mean extinction-to- backscatter ratios (AKA lidar ratios), uncertainties are high. In contrast, HSRLs can directly determine ice cloud extinction and lidar ratio by separately measuring the molecular and particulate components of the total backscattered signal, thus largely eliminating many of the uncertainties inherent in elastic backscatter retrievals. During the SEAC4RS campaign, three HSRL systems were deployed: an aircraft system on the DC8 and two ground-based systems, this paper focuses on measurements of lidar ratios acquired during the period 2013/6/18-2013/11/4 in Hunstville, AL.

2. METHODOLOGY

The HSRL technique (Shipley et al., 1983, Piironen and Eloranta 1994) relies on spectral discrimination between particle and molecular backscatter in the receiver using optical filters which enables independent retrievals of backscatter and extinction without reliance on uncertain assumptions about the particle microphysics. By design the HSRL technique provides measurements that are self-calibrated requiring only the current atmospheric density profile. The University of Wisconsin's (UW) HSRL instruments are specifically designed and optimized for cirrus cloud measurements in that they have a very narrow receiver field of view (FOV) (~100urad) which minimizes the effects of multiple scattering. The UW-HSRL system deployed in Huntsville, measured the particulate backscatter, extinction, and linear depolarization at a single wavelength (532 nm). The nominal resolution for the backscatter and depolarization (extinction) is 15 meters (75 meters) in the vertical, for 120 seconds of integration. The vertical coverage is from ~300 m AGL to 30 km.

 DOI: 10.1051/epjconf/201611916009

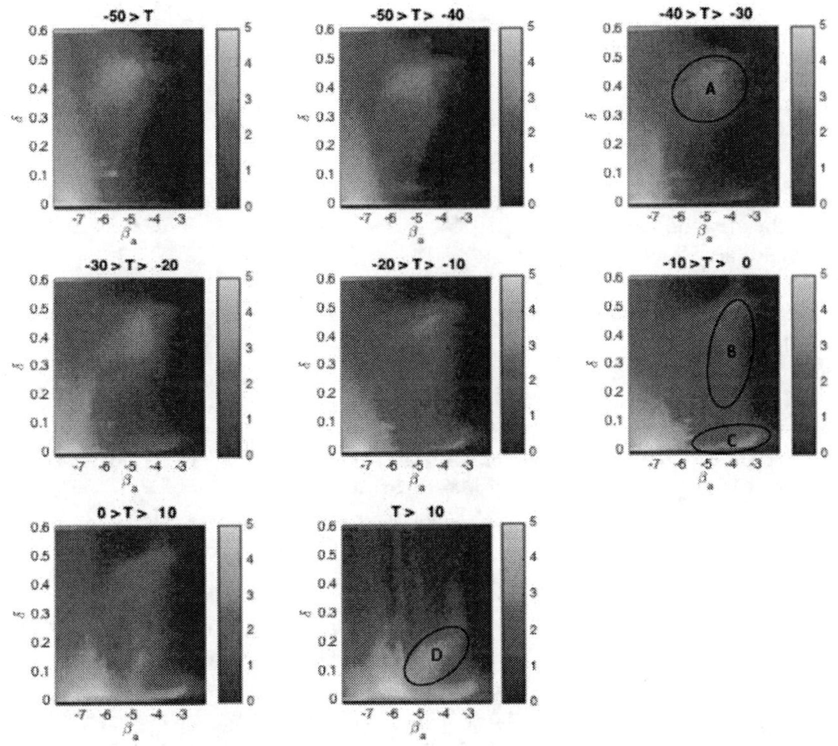

Figure 1 Two dimensional histogram of depolarization vs backscatter cross-section (log10) for different temperature regimes (T is in Celsius). Number concentration, i.e. intensity, is on a log10 scale. Annotated regions correspond to ice-clouds (A), melting ice-cloud that is precipitating (B), water clouds (C) and liquid precipitation (D). See text for further information.

For the SEAC4RS campaign a new set of QA flags, based upon expert inspection of the data, were incorporated into the data processing stream to provide data users additional information on data quality. In conjunction with the existing QC flags, which are based upon molecular-channel SNR and other system health and status information, processing and analysis of these large data sets is now tractable. Using this new quality controlled dataset we have implemented routines to classify aerosols, clouds, and precipitation in HSRL measurements.

The classification of 'features' in HSRL data is done on a range bin-by-range bin basis rather than by scanning each profile. We've found that with the typically high SNR and up-looking orientation of the HSRL this allows for more flexibility in both finding and classifying features. Depolarization ratio vs backscatter cross-section for a range of atmospheric temperature regimes, that are used to classify features as shown in Figure 1. Annotated ellipses are drawn to help the discussion and but do not necessarily indicate thresholds of the classification algorithm. Data were averaged to one minute intervals with a 30 m vertical range bin from 400 to 15 km range, QA and QC filters, in particular the molecular signal lost filter was applied. The molecular signal lost is valuable for filtering out low SNR data from above fully attenuating features.

In figure 1, non-precipitating ice clouds dominate the region indicated by 'A', these clouds are not found below the freezing level and depolarization and backscatter cross-sections are tightly constrained around 0.4 and 1x10^(-5) (m*sr-1), respectively. Depolarizing aerosols at higher altitude typically have backscatter cross-sections below 1x10^(-6) (m*sr-1). Melting ice clouds are found in the temperature range -20 to 0 C. Backscatter cross-sections are high ~ 1.0x10^(-4) (m*sr-1) and depolarization are quite variable, from 0.1 to 0.55. The variability in depolarization can be explained by the contribution from multiple scattering at high depolarizations, to large quantities of liquid drops or melting ice at lower depolarizations. Almost all of these clouds are associated with moderate to high precipitation rates. Water clouds dominate the area indicated by 'C', low to no depolarization and large backscatter cross-section values. Note that the depolarization increases at very high backscatter values from the increase in multiple scattering at high extinction values. Also note that none of these water clouds are found below -40 C. Finally, area 'D' denotes liquid precipitation, these are all 'warm' features, above 0 C, and depolarization increases exponentially with backscatter because of the increase in multiple scattering from large spherical drops.

These regions are used to separate the point measurements into 4 different feature masks for the different types: water clouds, ice clouds, melting ice clouds, and precipitation. From the feature mask, top and bases of the features are determined and the layer properties of the features are calculated. Threshold values are shown below:

Water cloud: Backscatter cross-section > 1.45x10^(-5), depolarization < line defined by y=m*x+b, where m = 0.0545, b= 0.2936, b is the log10 of backscatter cross section.

Ice cloud: Backscatter cross-section > 1.87x10^(-5), 0.3 < depolarization < 0.58, T < -5 C.

Ice cloud, precipitating: Backscatter cross-section > 1x10^(-5), 0.26 < depolarization < 0.53, T > 0 C

Precipitation: Backscatter cross-section > 1x10^(-5), 0.08 < depolarization < 0.3, T > 0 C

3. RESULTS

Using the classification scheme above, we've calculated the lidar ratio for ice clouds (no predicating) with optical depths greater than 0.3 and with a base height above 4 km. These results are shown in Figure 2. The mean and median lidar ratios are 26.7, and 25.7 +/- 10.6 sr respectively.

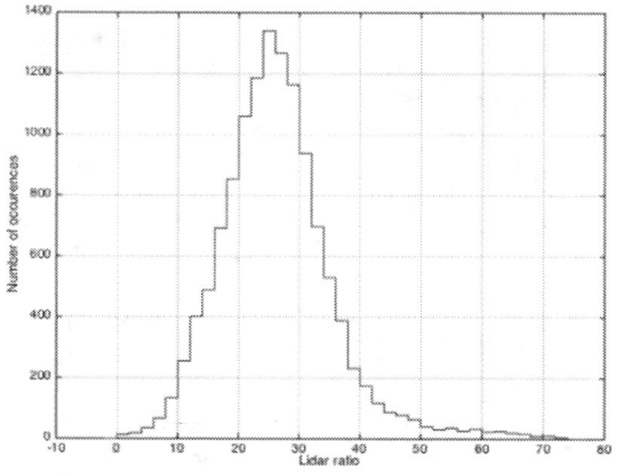

Figure 2 Measurements of lidar ratio in cirrus clouds from the UW-HSRL in Huntsville, AL, 2013/06/18-2013/11/04.

4. SUMMARY

In this paper we have shown first results of ice cloud lidar ratios measured with the UW HSRL lidar system in Huntsville, AL during SEAC4RS. These results are the first part of the analyses of the data set. With the addition of improvements to the feature classification scheme, adding corrections for multiple scattering, and a detailed error analyses we can then look at profiles of lidar ratio measurements, and investigate the relationships between the lidar ratio and other optical parameters measured by the HSRL.

ACKNOWLEDGEMENT

This work is supported by NASA grant NNX15AD08A.

REFERENCES

[1] Shipley, S. T., D. H. Tracy, E. W. Eloranta, J. T. Trauger, J. T. Sroga, F. L. Roesler and J. A. Weinman, 1983: A High Spectral Resolution Lidar to Measure Optical Scattering Properties of Atmospheric Aerosols, Part I: Instrumentation and Theory, *Appl. Optics*, **23**, 3716–3724, doi:10.1364/AO.22.003716.

[2] Piironen, P., and E.W. Eloranta 1994: Demonstration of a high-spectral-resolution lidar based on an iodine absorption filter, *Optics Letters*, **19**, 234-236.

EPJ Web of Conferences **119**, 16010 (2016) DOI: 10.1051/epjconf/201611916010
ILRC 27

DEPOLARIZATION LIDAR DETERMINATION OF CLOUD-BASE MICROPHYSICAL PROPERTIES

D.P. Donovan[1], H. Klein Baltink[1], J.S. Henzing[2], S. de Roode[3] and A.P. Siebesma[1,3]

1 Royal Netherlands Meteorological Institute (KNMI). PO Box 201, 3730 AE De Bilt, The Netherlands. donovan@knmi.nl

2 Netherlands Organisation for Applied Scientific Research (TNO), Utrecht, The Netherlands.

3 Technical University of Delft (TUD), Delft, The Netherlands.

ABSTRACT

The links between multiple-scattering induced depolarization and cloud microphysical properties (e.g. cloud particle number density, effective radius, water content) have long been recognised. Previous efforts to use depolarization information in a quantitative manner to retrieve cloud microphysical cloud properties have also been undertaken but with limited scope and, arguably, success. In this work we present a retrieval procedure applicable to liquid stratus clouds with (quasi-)linear LWC profiles and (quasi-)constant number density profiles in the cloud-base region. This set of assumptions allows us to employ a fast and robust inversion procedure based on a look-up-table approach applied to extensive lidar Monte-Carlo multiple-scattering calculations. An example validation case is presented where the results of the inversion procedure are compared with simultaneous cloud radar observations. In non-drizzling conditions it was found, in general, that the lidar- only inversion results can be used to predict the radar reflectivity within the radar calibration uncertainty (2-3 dBZ). Results of a comparison between ground-based aerosol number concentration and lidar-derived cloud base number considerations are also presented. The observed relationship between the two quantities is seen to be consistent with the results of previous studies based on aircraft-based in situ measurements.

1. INTRODUCTION

The fact that a linear polarization lidar will detect a cross-polarized signal due to the occurrence of multiple-scattering in liquid water clouds has been recognised since at least 1970 [1]. Extensive field and laboratory observations [2] of the depolarization of laser radiation in water clouds have been made and various theoretical approaches have been developed ranging from Monte Carlo-based (MC-based) models to semi-analytic approaches; see [4] for a review.

In spite of the long history and the increasing understanding of the relevant phenomenon, the use of depolarization measurements to retrieve cloud extinction and microphysical information appears to not have seen widespread implementation. This may be due to the fact that much of the theoretical work has either used homogeneous clouds (over simplified) or on the general problem of retrieving arbitrary cloud property profiles (too ambitious). Yet another reason may be the shift in attention to multi-field-of-view (FOV) lidars.

2. METHODOLOGY

In this work we present a retrieval procedure using single FOV depolarization lidars. The retrieval is based on assuming that the cloud-base region can be characterised by a quasi-linear (with height) LWC profile (i.e. constant LWC lapse rate) and constant cloud particle number density. This set of assumptions allows us to reduce the cloud variables to two parameters. In turn, this enables the development of a fast and robust inversion procedure using a look-up-table approach based on stored results from accurate lidar Monte-Carlo (MC) simulations.

The procedure does not rely on the absolute calibration of the lidar signals. However, the lidar depolarization calibration including cross-talk effects must be accounted for. In order to take into account any errors in the depolarization and cross-talk correction procedures the relevant instrument parameters taken into account using an Optimal

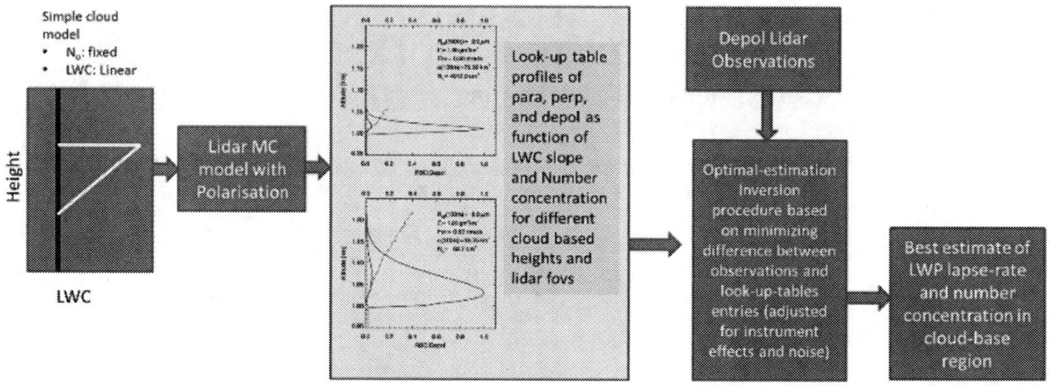

Figure 1: Schematic depiction of the inversion process.

Estimation (OE) approach. This is depicted in Fig.1. Extensive details may be found in [2].

Development and testing of the inversion procedure was facilitated by end-to-end testing of the procedure using Large-Eddy-Simulation (LES) model data. An example is presented in Fig.2. Here, data from the DALES model [6] driven by observed conditions over the Dutch CESAR measurement site at Cabauw in the central Netherlands were used. The simulations

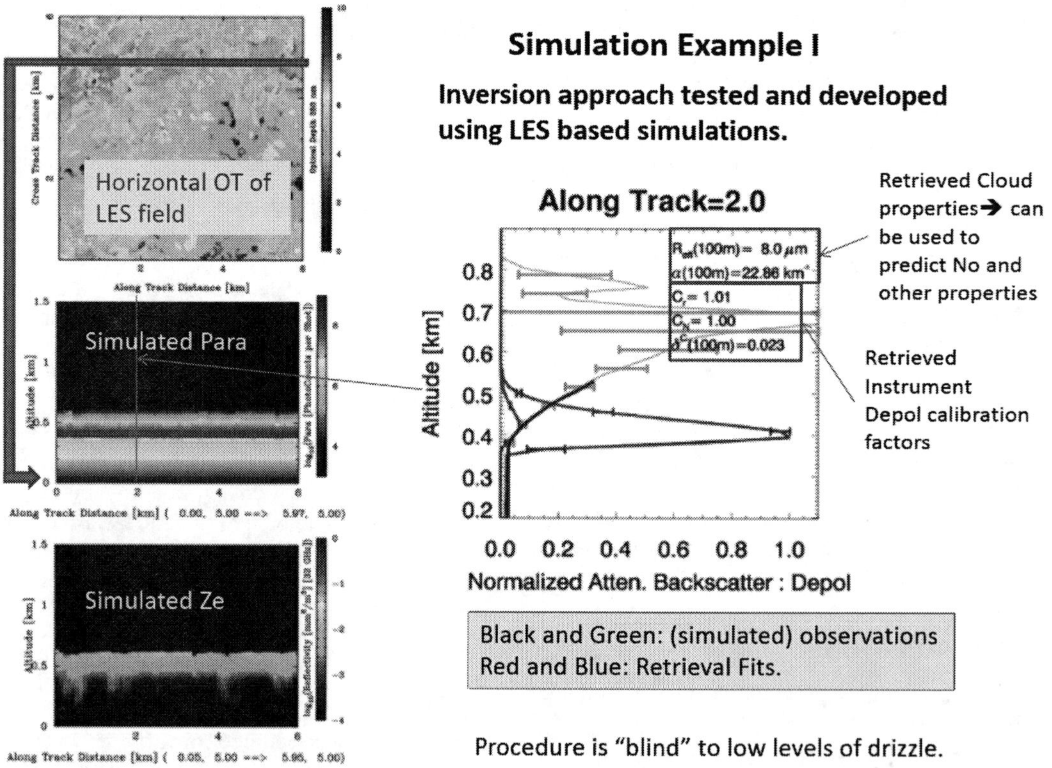

Figure 2: Example end-to-end simulation results (for one profile) using LES data.

 DOI: 10.1051/epjconf/201611916010

indicate that the procedure should produce robust results in the presence of non-ideal cloud macrostructure and instrument noise.

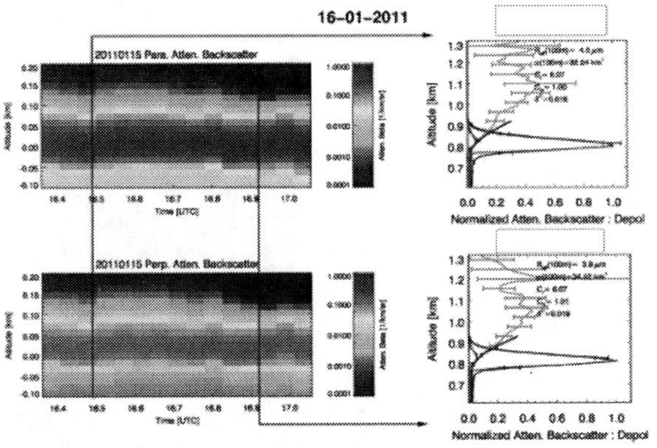

Figure 3: Results of the retrieval applied to lidar data for two periods on 16/01/2011. The Black-lines correspond to the normalized para and perp attenuated backscatters while the observed depol. ratio is given by the Green-line The Red-lines are the fits to the observed attenuated backscatters and the Blue-line is the corresponding fit depolarization ratio. Γ_l is the liquid water lapse rate, while Cr, C and δ^c are instrument parameters associated with the polarization calibration and cross-talk.

It is also interesting to note that, unlike, techniques involving the quantitate use of radar reflectivity that the procedure is robust against the presence of drizzle.

3. RESULTS

An example application of the depolarization lidar inversion to real observations is shown in Fig. 3. The lidar observations were made with a 355 nm Leosphere ALS-450 backscatter lidar. The system has been in operation at Cabauw since mid-2007 with breaks in the record ranging from weeks to several months. The data was acquired with a vertical resolution of 15 m and a temporal resolution of about 30 s. Here it can be seen that very good fits to actual data are obtained.

As well as the lidar measurements, we also make use of the 35 GHz lidar observations at Cabauw. The cloud radar is a vertically pointing Doppler radar with a vertical resolution of 89 m and a temporal resolution of approximately15 s. In Fig.4, time series of lidar-derived parameters for several hours on Jan 15, 2011 are shown. A comparison between the `cloud-base' radar reflectivity predicted on the basis of the lidar retrieval results and the observed values is also presented. The agreement between the lidar

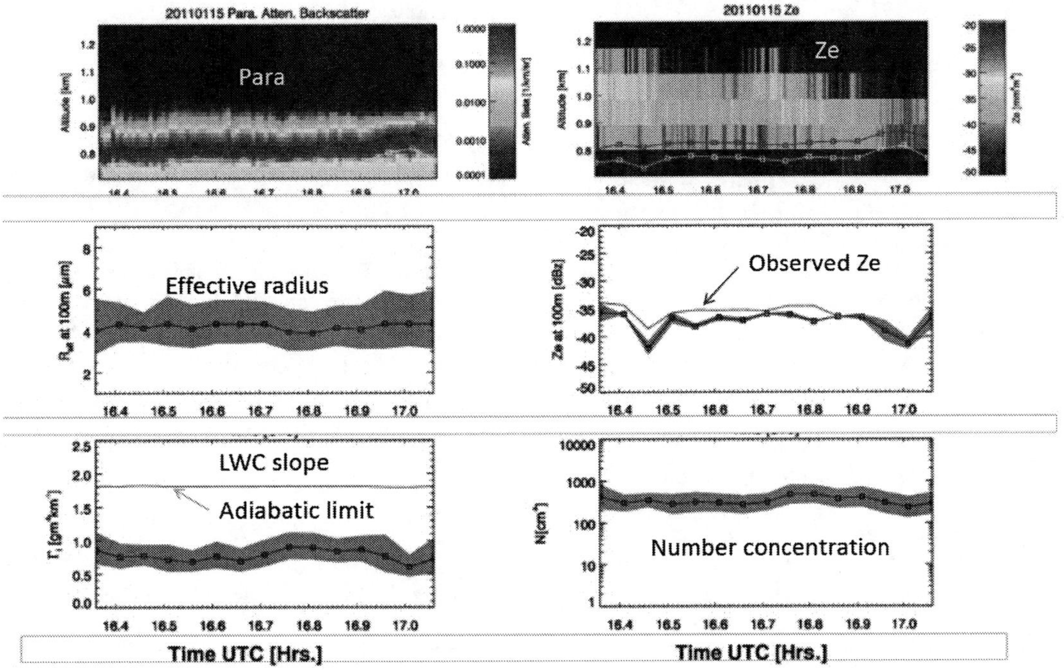

Figure 4: Example time series inversion results including observed radar reflectivities averaged 100 m above cloud base and reflectivities calculated on the basis of the lidar inversion results.

derived estimates and the observations agree within the 2-3 dBz calibration uncertainty of the radar. This was generally observed to be the case in non-drizzling conditions and strongly supports the validity of the lidar inversion results.

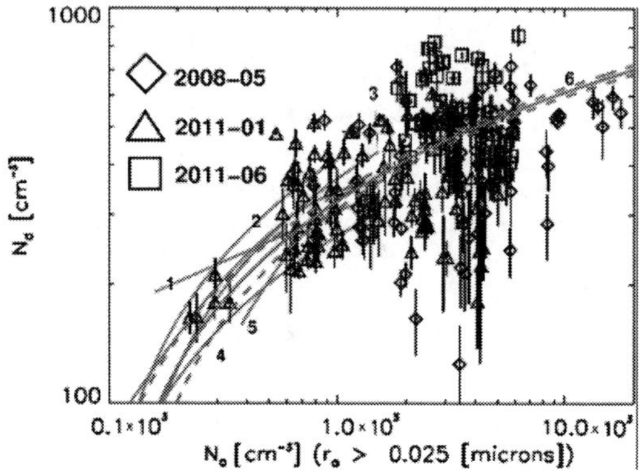

Figure 5: Lidar-derived cloud-base number density (Nd) and SMPS ground-based number density of (Na). The Light-Blue lines represent previous empirical relationships based on in-situ aircraft measurements. Each point represents a 30 min. average result. The data spans about 3 months of observations of boundary layer stratus clouds.

As well as the remote-sensing equipment, Cabauw also hosts a number of in situ probes including a Scanning Mobility Particle Sizer (SMPS) instrument which measures aerosol size distributions between diameters of 10 and 470 nm. As described in [7], the SMPS instrument is connected to a laminar flow sampling tube with an inlet at 60 m.

A comparison between lidar derived cloud number density and in-situ results are shown in Fig. 5. Here a range of boundary-layer cases was considered. It can be seen that the comparison is consistent with a range of previously derived empirical relationships based on in-situ aircraft observations e.g. [8].

4. CONCLUSIONS

A new method for retrieving liquid stratus cloud-based microphysical properties using common depolarization lidar signals has been developed. The method appears to produce accurate robust results. This conclusion is based on simulations driven by LES data, consistency with radar observations in non-drizzling clouds and consistency with the results of previous aircraft in-situ based investigations into the correlation between cloud and aerosol number densities. For more information see [2]

ACKNOWLEDGEMENTS

The aerosol size distribution measurements used in this paper were partly funded by European Union Seventh Framework Program (FP7/2007–2013) under grant agreement no. 262254.The LES simulations used in this work were sponsored by the Dutch National Computing Facilities Foundation (NCF) who provided the use of supercomputer facilities.

REFERENCES

[1] Liou, K.-N. and Schotland, R. M., 1971: Multiple backscattering and depolarization from water clouds for a pulsed lidar system, *J. Atmos. Sci.*, **28**,772–784.

[2] Sassen, K., 2005: Polarization in lidar, in: Lidar, edited by: Weitkamp, C., vol. 102 of Springer Series in Optical Sciences, 19–42, Springer, New York, doi:10.1007/0-387-25101-4_2, 2005.

[3] Chaikovskaya, L., 2008: Remote sensing of clouds using linearly and circularly polarized laser beams: techniques to compute signal polarization, in: Light Scattering Reviews 3, edited by: Kokhanovsky, A. A., Springer Praxis Books, 191–228, Springer, Berlin Heidelberg.

[5] Donovan, D.P. et al., 2015: A depolarization lidar-based method for the determination of liquid-cloud microphysical properties, *Atmos. Meas. Tech.*, **8**, 237-266, 2015 doi:10.5194/amt-8-237-2015

[6] Heus, T., van Heerwaarden, et al., 2010: Formulation of the Dutch Atmospheric Large-Eddy Simulation (DALES) and overview of its applications, *Geosci. Model Dev.*, **3**, 415–444.

[7] Mensah, A. A. et al., 2012: Aerosol chemical composition at Cabauw, The Netherlands as observed in two intensive periods in May 2008 and March 2009, *Atmos. Chem. Phys.*, **12**, 4723 4742,doi:10.5194/acp-12-4723-2012.

[8] Leaitch, W. R., et al., 1992: The relationship between cloud droplet number concentrations and anthropogenic pollution: observations and climatic implications, *J. Geophys. Res.*, **97**, 2463–

DOI: 10.1051/epjconf/201611916010

2474, doi:10.1029/91JD02739.

DOI: 10.1051/epjconf/201611916011

MPLNET V3 CLOUD AND PLANETARY BOUNDARY LAYER DETECTION

Jasper R. Lewis [1,2]*, Ellsworth J. Welton[2], James R. Campbell[3], Phillip C. Haftings[2,4]

[1]*Joint Center for Earth Systems Technology, University of Maryland Baltimore County, Baltimore, MD 21250, USA, *Email: jasper.r.lewis@nasa.gov*

[2]*NASA Goddard Space Flight Center, Greenbelt, MD 20771, USA*

[3]*Naval Research Laboratory, Monterey, CA 93943, USA*

[4]*Science System and Applications, Inc. Lanham, MD 20706, USA*

ABSTRACT

The NASA Micropulse Lidar Network Version 3 algorithms for planetary boundary layer and cloud detection are described and differences relative to the previous Version 2 algorithms are highlighted. A year of data from the Goddard Space Flight Center site in Greenbelt, MD consisting of diurnal and seasonal trends is used to demonstrate the results. Both the planetary boundary layer and cloud algorithms show significant improvement of the previous version.

1. INTRODUCTION

The National Aeronautics and Space Administration (NASA) Micropulse Lidar Network (MPLNET) (Welton et al. 2001, http://mplnet.gsfc.nasa.gov) is a federated network of micropulse lidar (MPL) systems deployed worldwide in support of basic science and the NASA Earth Observing System program (Wielicki et al. 1995). Vertical profiles of aerosols and clouds are measured continuously both day and night at polar, midlatitude, and tropical sites using a standardized instrument and data processing algorithm. Thorough uncertainty characterization makes for straightforward comparisons between sites.

Long-tem, continuous measurements of the planetary boundary layer (PBL) depth are rare, but necessary for ascertaining the diurnal and seasonal variation for studies of air quality, weather, and climate. Similarly, it is important to have highly accurate, quantitative data records of cloud properties that span several years and geographic regions. Ground-based lidar measurements, like those provided by MPLNET, are crucial for verification and validation of satellite missions, as well as modeling processes.

2. ALGORITHM DESCRIPTIONS

The raw MPL signal is processed to produce the Level 1 normalized relative backscatter (NRB) which is a function of atmospheric parameters and only one instrument parameter, the lidar calibration constant. The NRB is reported at one-minute temporal resolution up to 30-km, typically at 75-m vertical resolution (although some sites operate at 30-m vertical resolution). The NRB is defined as:

$$\mathrm{NRB}(z) = C\beta(z)T^2(z) \quad (1)$$

where C is the calibration constant, β is the backscatter coefficient from molecules and particles, T^2 is the atmospheric two-way transmittance from molecules and particles, and z is the altitude. The NRB and the calculation of its uncertainty are discussed by Campbell et al. (2002) and Welton and Campbell (2002). Higher level MPLNET products, including PBL and cloud retrievals, are derived from the Level 1 NRB.

2.1 PBL Retrievals

The Version 2 (V2) PBL depth is retrieved using a the wavelet covariance transform and a Haar wavelet as described by Davis et al. (2000) and Brooks (2003). PBL depths are determined using five-minute averages and regridded to a one-minute temporal grid. The V2 PBL products were plagued with contamination from low-level (less than 5 km altitude) clouds and residual layers which resulted in an unrealistic representation of the PBL.

The Version 3 (V3) PBL retrieval uses the procedure outlined by Lewis et al. (2013). The new algorithm uses the first derivative of a Gaussian wavelet in place of the Haar wavelet because it more closely resembles the gradient in the lidar profile and therefore results in less noise in the retrieval. Three features representing possible choices for the PBL depth are retrieved using Canny edge detection from the wavelet covariance transform. A fuzzy logic scheme is used as a means for layer attribution to select the most appropriate choice of PBL depth from the three possible features. The choice of PBL depth depends on a combination of factors including: the time of day, altitude of the retrieved feature, variance in the lidar profile, and most recently retrieved features. A continuity scheme is used to suppress unphysical fluctuations in the retrieved PBL depth.

In addition to the best estimate of the PBL depth, the altitudes of all three retrieved features are also included in the PBL products as they can be useful for identification of aerosol layers and the residual layer. Data flags are also provided to indicate day and night retrievals.

Lewis et al. (2013) demonstrated the V3 method of PBL retrieval produced shallower PBL depths which better exhibited the diurnal growth and collapse expected (see Figure 1).

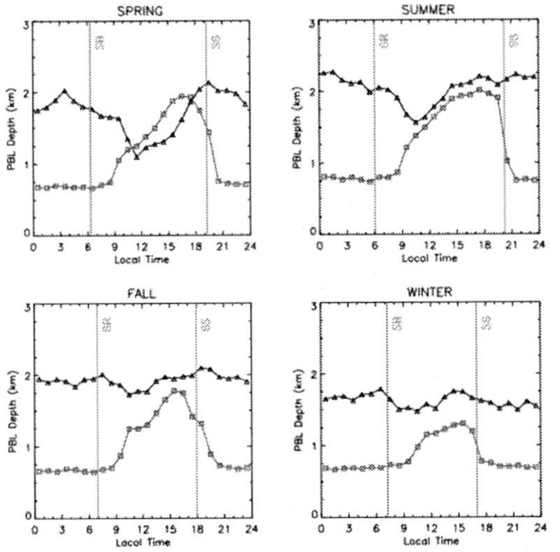

Figure 1. Comparison of V2 (black) and V3 (red) PBL depths at GSFC in 2010 (from Lewis et al. 2013)

2.2 Cloud Retrievals

The V2 cloud retrievals are obtained using a combination of two methods; one relying on gradients in the lidar backscatter profile, and the other based on uncertainty in the lidar signal. The first method is best suited for detection of low-level (primarily liquid phase) clouds and is referred to as the gradient-based cloud detection method (GCDM). The second method is designed for use in situations of low signal to noise (high solar background) and best suited for elevated clouds (i.e. cirrus). This second method is referred to as the uncertainty-based cloud detection method (UCDM). The UCDM was initially introduced by Campbell et al. (2008) to detect polar stratospheric clouds at the South Pole and uses two tunable thresholds and one objective threshold in order to resolve cloud boundaries.

The V3 cloud algorithm is based on the V2 algorithm with several meaningful changes described by Lewis et al. (2015, in preparation). The GCDM and UCDM are used in the V3 algorithm; however, a more synergistic merger of the two algorithms is used which allows the best suited method of cloud detection to be used in both day and night measurements. Previously, the UCDM was used primarily for daytime measurements retrievals when the signal-to-noise ratio was lowest.

The objective threshold used in the V3 cloud detection algorithm is attenuated within cloud layers which allows for possible better retrievals of the cloud top height and the detection of some cloud layers missed by the V2 algorithm.

The V3 cloud detection algorithm also employs a multi-temporal resolution averaging scheme to improve retrievals in conditions with high solar background. Cloud retrievals at the base one-minute temporal resolution are combined with retrievals at intermediate (five-minute) and long (twenty-minute) averages to produce a combined cloud scene.

Cloud phase is determined using a cloud top temperature threshold suggested by Sassen and Campbell (2001) and Campbell et al. (2015) in order to identify cirrus based on the ice nucleation threshold. An estimated cloud optical depth (COD) is provided for clouds identified as cirrus using the procedure described by Chew et al. (2011). The fidelity of the COD estimate is limited by the choice of extinction-to-backscatter ratio and accuracy of cloud boundaries retrieved.

Lidar signals are unable to penetrate through optically thick clouds which causes uncertainty in the value of the cloud top. In these cases, the estimated COD will be lower than the true optical depth. Similarly, attenuation from the bottom-most cloud layer leads to uncertainty in retrievals of higher clouds for lidar profiles containing multiple cloud layers. As a result, the estimated COD will be most reliable for single-layer, optically thin clouds.

Data flags are provided to indicate the method of cloud retrieval (GCDM or UCDM), beam attenuation, and the temporal resolutions used to identify cloud layers.

Figures 2-4 show comparisons of the V2 and V3 cloud retrievals. Because the majority of changes are related to the UCDM, there are only slight differences are seen with low clouds (less than 2 km) and middle clouds (between 2 and 5 km). The largest differences are seen in relation to high clouds (above 5-km).

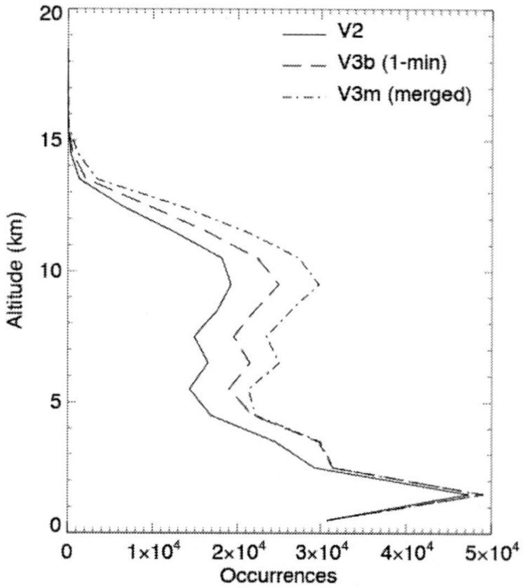

Figure 2. Vertical comparison of V2 (solid line) and V3 (dashed line and dash-dotted line for 1-minute and multi-temporal resolutions, respectively) cloud retrievals.

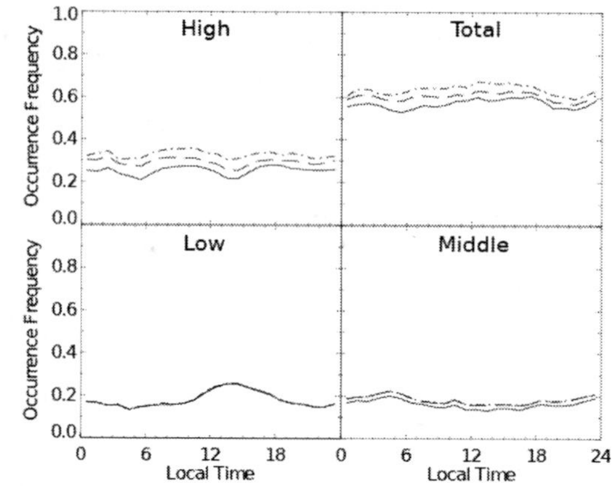

Figure 3. Diurnal comparison of V2 and V3 cloud retrievals.

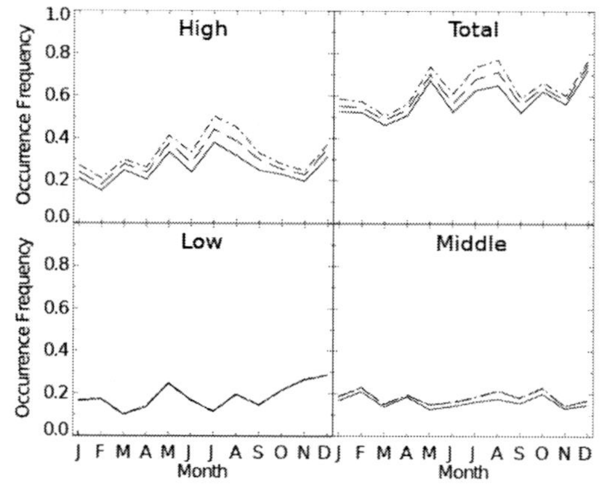

Figure 4. Monthly comparison of V2 and V3 cloud retrievals.

3. SUMMARY

The Version 3 planetary boundary layer and cloud algorithms are presented. The new algorithms are compared to the Version 2 algorithms to show improvements to diurnal and seasonal retrievals. Long-term continuous measurements like those provided by the Micropulse Lidar network are essential for studies of climate involving satellites and modeling processes.

ACKNOWLEDGEMENT

We are thankful to the MPLNET PIs and staff for their efforts in establishing and maintaining the GSFC site.

REFERENCES

Campbell, J. R., D. L. Hlavka, E. J. Welton, C. J. Flynn, D. D. Turner, J. D. Spinhirne, V. S. Scott, and I. H. Hwang, 2002: Full-time, eye-safe cloud and aerosol lidar observation at Atmosphere Radiation Measurement program sites: Instrument and data processing. J. Atmos. Oceanic Technol., 19, 431-442.

Campbell, J. R., K. Sassen, and E. J. Welton, 2008: Elevated cloud and aerosol layer retrievals from micropulse lidar signal profiles. J. Atmos. Oceanic Technol., 25, 685-700.

Campbell, J. R., M. A. Vaughan, M. Oo, R. E. Holz, J. R. Lewis, E. J. Welton, 2015: Distinguishing cirrus cloud presence in autonomous lidar measurements. Atmos. Meas. Tech., 8, 435-449, doi:10.5194/amt-8-435-2015.

Chew, B. N., J. R. Campbell, J. S. Reid, D. M. Giles, E. J. Welton, S. V. Salinas, and S. C. Liew, 2011: Tropical cirrus cloud contamination in sun photometer data. Atmos. Environ., 45, 6724-6731.

Lewis, J. R., E. J. Welton, A. M. Molod, and E. Joseph (2013), Improved boundary layer depth retrievals from MPLNET, J. Geophys. Res. Atmos., 118, 9870–9879, doi:10.1002/jgrd.50570.

Sassen, K. and J. R. Campbell, 2001: A midlatitude cirrus cloud climatology from the Facility for Atmospheric Remote Sensing. Part I: Macrophysical and synoptic properties. J. Atmos. Sci., 58, 481-496.

Welton, E. J. and J. R. Campbell, 2002: Micropulse lidar signals: Uncertainty analysis. J. Atmos. Oceanic Technol., 19, 2089-2094.

EPJ Web of Conferences **119**, 16012 (2016) DOI: 10.1051/epjconf/201611916012
ILRC 27

SUBTROPICAL AND POLAR CIRRUS CLOUDS CHARACTERIZED BY GROUND-BASED LIDARS AND CALIPSO/CALIOP OBSERVATIONS

Carmen Córdoba-Jabonero[1*], Fabio J. S. Lopes[2], Eduardo Landulfo[2], Héctor Ochoa[3], Manuel Gil-Ojeda[1]

[1]*Instituto Nacional de Técnica Aeroespacial (INTA), Atmospheric Research and Instrumentation Branch, Torrejón de Ardoz, 28850-Madrid, Spain, *Email: cordobajc@inta.es*

[2]*Instituto de Pesquisas Energéticas e Nucleares (IPEN), Center for Lasers and Applications, São Paulo, Brazil*

[3]*Instituto Antártico Argentino/Dirección Nacional del Antártico (IAA/DNA), Buenos Aires, Argentina*

ABSTRACT

Cirrus clouds are product of weather processes, and then their occurrence and macrophysical/optical properties can vary significantly over different regions of the world. Lidars can provide height-resolved measurements with a relatively good both vertical and temporal resolutions, making them the most suitable instrumentation for high-cloud observations. The aim of this work is to show the potential of lidar observations on Cirrus clouds detection in combination with a recently proposed methodology to retrieve the Cirrus clouds macrophysical and optical features. In this sense, a few case studies of cirrus clouds observed at both subtropical and polar latitudes are examined and compared to CALIPSO/CALIOP observations. Lidar measurements are carried out in two stations: the Metropolitan city of Sao Paulo (MSP, Brazil, 23.3°S 46.4°W), located at subtropical latitudes, and the Belgrano II base (BEL, Argentina, 78°S 35°W) in the Antarctic continent. Optical (COD-cloud optical depth and LR-Lidar Ratio) and macrophysical (top/base heights and thickness) properties of both the subtropical and polar cirrus clouds are reported. In general, subtropical Cirrus clouds present lower LR values and are found at higher altitudes than those detected at polar latitudes. In general, Cirrus clouds are detected at similar altitudes by CALIOP. However, a poor agreement is achieved in the LR retrieved between ground-based lidars and space-borne CALIOP measurements, likely due to the use of a fixed (or low-variable) LR value in CALIOP inversion procedures.

1. INTRODUCTION

The influence of Cirrus clouds on weather and climate is actually an evident fact. Indeed, they can act as modulators in the radiation balance of the Earth-atmosphere system, and their heating or cooling effects can be observed at both regional and global scales [1]. However, the sensitivity of Cirrus clouds to factors associated with human-induced climate changes, i.e. greenhouse effect [2] and contamination of the upper troposphere from increasing aircraft traffic [3], is still poorly investigated. The predominance of infrared greenhouse warming versus solar albedo cooling depends sensitively on both the altitudes and microphysical compositions of the Cirrus clouds [4]. Indeed, cloud height has an evident impact. Hence, high tropical Cirrus can be particularly effective greenhouse modulators. Conversely, lower Cirrus over Polar Regions could be more efficient for albedo effects. Moreover, Cirrus clouds are product of weather processes, and then their occurrence and macrophysical/optical properties can vary significantly over different regions of the world. Since Cirrus clouds usually are located from 7 km height up to the tropopause, active remote sensing techniques, as lidars, are usually used for detection of Cirrus clouds from ground-based [5] and space [6] observations.

The aim of this work is to show the potential of ground-based lidar observations on Cirrus clouds detection in combination with a recently proposed methodology [7] to retrieve the Cirrus clouds macrophysical and optical features. In this sense, Cirrus clouds observations are performed in particularly located lidar stations at both subtropical and polar latitudes in order to establish a long-term Cirrus clouds monitoring over these

DOI: 10.1051/epjconf/201611916012

stations in the future. In this first step, a case study of Cirrus clouds observed over these subtropical and polar stations are presented. Their both macrophysical (vertical extent, top/base heights and thickness), and optical (COD-Cloud Optical Depth, LR-Lidar Ratio=extinction-to-backscatter ratio) properties are reported. A comparative analysis is also performed between ground-based lidar measurements and space-borne CALIPSO/CALIOP (www-calipso.larc.nasa.gov) observations [8]. Similarities and discrepancies are also discussed.

2. METHODOLOGY

2.1 Ground-based lidar measurements

Cirrus (Ci) clouds were detected over both subtropical and polar stations: a) the Metropolitan city of Sao Paulo station (MSP, Brazil, 23.3°S 46.4°W, 740 m a.s.l.), located at subtropical latitudes and managed by the Instituto de Pesquisas Energéticas e Nucleares (IPEN), and b) the Antarctic Belgrano II base (BEL, Argentina, 78°S 35°W, 256 m a.s.l.), managed by the Instituto Antártico Argentino (IAA).

A Micropulse Lidar v.4 (MPL-4) is deployed in routine operation at the BEL site; MPL-4 is a standard system of those are within NASA/MPLNET (MicroPulse Lidar NETwork, mplnet.gsfc.nasa.gov), belonging to the Spanish Institute for Aerospace Technology (INTA) in collaboration with the IAA (Argentina). The lidar system operating in the MSP station (SPL system) belongs to the IPEN/Centro de Lasers e Aplicaçoes (IPEN/CLA), and is within LALINET a.k.a. ALINE (Latin America LIdar NETwork, lalinet.org). The main ground-based lidar features are shown in Table 1.

In this work, a case study is presented for each station: Cirrus clouds were detected on 11 June 2007 (early austral winter) over MSP station and on 20 May 2010 (late austral autumn) over BEL site. Both macrophysical and optical properties of the Cirrus clouds observed in those days are retrieved from lidar measurements by using the methodology reported in [7]. Among the macrophysical features, the vertical extent, the top and base heights and the thickness are obtained. The optical properties retrieved, considering also multiple scattering effects, are the Backscattering Ratio (BSR) profiles, Cloud Optical Depth (COD) and Lidar Ratio (LR, extinction-to-backscatter ratio). In addition, Cirrus layers detected along the day are classified according to their COD (τ) into: sub-visual (svCi, τ<0.03), semi-transparent (stCi, τ: 0.03-0.3), and opaque (opCi, τ>0.3). All those parameters were examined for the selected case studies, and compared to CALIPSO/CALIOP Cirrus retrievals.

Table 1. Main information about the two G-B lidar systems and the space-borne CALIOP.

Lidar system	SPL	MPL-4	CALIOP
Wavelength (nm)	532 (*)	527	532 (*)
Pulse energy pulse (mJ)	150 - 330	0.010 (max.)	110
Pulse repetition freq. (Hz)	10 – 20	2500	20.25
Eye-safe	No	Yes	No
Depolarization	No	Yes	Yes
Raman capability	Yes	No	No
Acquisition settings	Vert. res.: 3.75 m Int. time: 1-2 min	Vert. res.: 75 m Int. time: 1 min	Vert. res.: 30 m (< 8.3 km), 60 m (8.3-20.5 km) Horiz. res.: 333 m

(*) Used in this work.

2.2 Space-borne CALIOP observations

CALIOP (Cloud-Aerosol Lidar with Orthogonal Polarization) is a two-wavelength elastic lidar system aboard CALIPSO satellite [8]. Its principal features are shown in Table 1. CALIPSO Level 1 data were analyzed in order to retrieve the total attenuated backscatter product and calculate the BSR profiles for each station. A total of 300 profiles were selected in order to obtain a total horizontal range of 100 km. Each 15 consecutive profiles of the total attenuated backscatter were averaged in order to obtain a BSR with a horizontal resolution of 5 km. Hence, 20 BSR profiles centered in the coordinates of each lidar station were examined. In particular, CALIPSO overpasses were performed at around 17:08 UTC and 01:34 UTC, respectively, over MSP and BEL stations for those selected days; the closest

DOI: 10.1051/epjconf/201611916012

distance between the CALIPSO ground-track and MSP and BEL sites was 72.5 km and 75.3 km, respectively. COD values were obtained from CALIPSO Level 2 Cloud Layer products with 5-km horizontal resolution, being retrieved the number of cloud layers detected and their base and top altitudes, as well as their Cloud-Aerosol Discrimination score (CAD) [9]. COD values were also used to classify the Cirrus clouds, and the LR for each cloud profile was also provided.

3. RESULTS

3.1 Subtropical MSP case

The evolution of the Cirrus clouds observed over MSP station on 11 June 2007 based on G-B SPL measurements presented a layered structure composed of one or two layers with alternating COD values between stCi and opCi clouds, but no svCi were detected. Figure 1 shows the BSR profiles obtained from SPL (left panel) and CALIOP (right panel) observations as classified depending on the COD. Both stCi and opCi occurrence is found between 8 and 11 km height, at similar altitudes as those found for CALIOP detected Cirrus clouds. In addition, G-B retrieved LR values are higher for stCi (20-30 sr) than for opCi (10-20 sr) clouds. However, a fixed value of 25 sr is used in CALIOP LR retrievals independently on the Ci-type category.

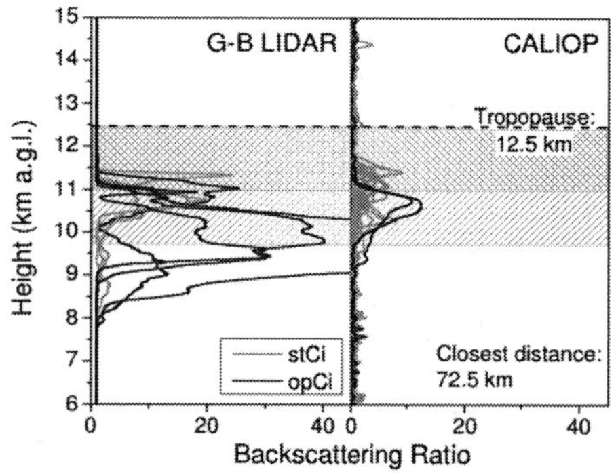

Fig. 1. BSR profiles for the subtropical Cirrus clouds observed over MSP station on 11 June 2007 depending on COD (svCi: red lines, stCi: green lines, and opCi: blue lines) obtained from SPL (left) and CALIOP (right) measurements.

3.2. Polar BEL case

In this case, opCi are observed over BEL station on 20 May 2010 based on G-B MPL-4 measurements for the first two hours, followed by svCi and stCi (CODs for these stCi are close to 0.08, values nearly within svCi category). Cirrus persist at similar heights along the day (unfortunately, no data available from 16:00 UTC on). Figure 2 shows the BSR profiles obtained from MPL-4 (left panel) and CALIOP (right panel) observations as classified depending on the COD. In general, polar Cirrus are found at lower altitudes than those for the subtropical case, as expected at polar latitudes. Cirrus clouds detected by CALIOP are only within opCi category, but their altitudes are coincident with those observed by G-B lidars. In addition, LR values derived from MPL-4 measurements are between 15 and 30 sr for svCi, and 30 sr for both stCi and opCi clouds. These LR values are dissimilar in comparison to those obtained for the Ci-type categories found in the subtropical case. In addition, CALIOP-retrieved LR are slightly variable (22±2 sr) for the only opCi clouds detected, being values relatively lower than those derived by G-B lidar measurements for these opCi clouds.

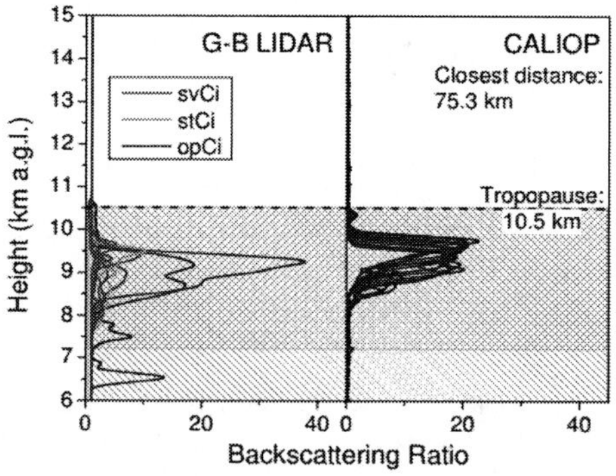

Fig. 2. The same as for Fig. 1, but for the polar Cirrus clouds observed over BEL station on 20 May 2010 obtained from MPL-4 (left) and CALIOP (right) measurements.

4. CONCLUSIONS

Cirrus clouds have been analyzed in terms of their macrophysical and optical properties as retrieved by a recently developed methodology applied to lidar measurements. A case study of Cirrus clouds observed at both subtropical and polar latitudes are presented in this work. These cases correspond to the Cirrus detected over the subtropical MSP station and the Antarctic BEL base. Their BSR profiles together with their altitude levels, COD and LR values have been reported. This study has been focused on the classification of the daily cloud features into three Ci-type categories related to their COD (τ): svCi (τ<0.03), stCi (τ: 0.03-0.3), and opCi (τ>0.3).

Based on G-B lidar measurements, subtropical svCi and opCi clouds present LR values lower than 10 sr and 10-20 sr, respectively, while they are higher for polar Cirrus clouds (svCi: 15-30 sr, and opCi: ~ 30 sr). However, svCi were not detected by CALIOP, and a larger fixed (25 sr) and a smaller low-variable (22±2 sr) LR values were derived/used for subtropical and polar opCi clouds, respectively. stCi clouds represent a mixed state of those svCi and opCi clouds. Regarding their altitudes, subtropical svCi were located higher than opCi clouds (from 10 km and 8 km height, respectively, up to tropopause height-levels). Polar Cirrus were observed, in general, at lower altitudes, as expected in Polar Regions. For comparison, Ci clouds were detected by CALIOP at similar altitudes. It should be noted that these results are specific for those study cases analyzed. However, in general, subtropical Cirrus clouds present lower LR values and are found at higher altitudes than those Ci detected at polar latitudes. Similarities and differences can be plausibly found, as long as a larger dataset can be available to be analyzed in each station.

These results can be into consideration to improve the treatment of the different latitude-types of Cirrus clouds to be introduced in large-scale models for climate-related issues, as well as to be useful in validation assessments of the retrieval of their properties from future satellites (i.e., next ESA/EarthCARE and Sentinels missions).

ACKNOWLEDGEMENTS

This work has been supported by the Spanish Ministerio de Economía y Competitividad (MINECO, grant CGL2011-24891) and the Brazilian Fundação de Amparo à Pesquisa do Estado de São Paulo (FAPESP, grant 2011/14365-5). C. C.-J. also thanks FAPESP for their support (grant 2013/11836-2) during her research stay in the Instituto de Pesquisas Energéticas e Nucleares (IPEN, Brazil). Authors are grateful to the staff of all the stations responsible for instrumentation maintenance and support.

REFERENCES

[1] Liou, 1986: Influence of cirrus clouds on weather and climate processes: A global perspective, *Mon. Weather Rev.* **114**, 1167-1200.

[2] Iacono et al., 2008: Radiative forcing by long-lived greenhouse gases: Calculations with the AER radiative transfer models, *J. Geophys. Res. Atmos.* **113**, D13103.

[3] Boucher, 1999: Air traffic may increase cirrus cloudiness, *Nature* **397**, 30-31.

[4] Myhre et al., 2013: Anthropogenic and Natural Radiative Forcing. In: *Climate Change 2013: The Physical Science Basis. Contribution of Working Group I to the Fifth Assessment Report of the Intergovernmental Panel on Climate Change* [Stocker, T.F., D. Qin, G.-K. Plattner, M. Tignor, S.K. Allen, J. Boschung, A. Nauels, Y. Xia, V. Bex and P.M. Midgley (eds.)]. Cambridge University Press, Cambridge, United Kingdom and New York, NY, USA.

[5] Sassen and Campbell, 2001: A midlatitude cirrus cloud climatology from the Facility for Atmospheric Remote Sensing. Part I: Macrophysical and synoptic properties, *J. Atmos. Sci.* **58**, 481-496.

[6] Sassen et al., 2008: Global distribution of cirrus clouds from CloudSat/Cloud-Aerosol Lidar and Infrared Pathfinder Satellite Observations (CALIPSO) measurements, *J. Geophys. Res.* **113**, D00A12, doi:10.1029/2008JD009972.

[7] Larroza et al., 2013: Towards an automatic lidar cirrus cloud retrieval for climate studies, *Atmos. Meas. Tech.* **6**, 3197-3210.

[8] Winker et al., 2009: Overview of the CALIPSO mission and CALIOP data processing algorithms, *J. Atmos. Ocean. Tech.* **26**, 2310-2323.

[9] Liu et al., 2009: The CALIPSO Lidar cloud and aerosol discrimination: Version 2 algorithm and initial assessment of performance, *J. Atmos. Ocean. Tech.* **26**, 1198-1213.

CLOUD AND AEROSOL INTERACTION OBSERVED IN SKYNET HEFEI SITE IN CHINA

Yingjian Wang[1,3], Lu Sun[1,2], Dong Liu[1], Zhien Wang[1,4], Zhenzhu Wang[1], Chenbo Xie[1]

[1]*Key Laboratory of Atmospheric Composition and Optical Radiation, Anhui Institute of Optics and Fine Mechanics, Chinese Academy of Sciences, Hefei, Anhui 230031, China*
[2]*University of Chinese Academy of Sciences, Beijing 100049, China*
[3]*University of Science and Technology of China, Hefei, Anhui 230026, China*
[4] *Department of Atmospheric Sciences, University of Wyoming, WY 82070, USA*
**Email: wyj@aiofm.ac.cn*

ABSTRACT

The interaction relationship between cloud and aerosol is studied via their optical depth and cloud effective radius based on ground-based remote sensors. By attenuated backscatter obtained by lidar the optical depth of cloud and aerosol is retrieved. Combing with liquid water path observed by microwave radiometer, the cloud number concentration and cloud effective radius is also retrieved based on the adiabatic hypothesis Cases studies shows that during the stable stratocumulus with interval precipitation period and aerosol with vertical motion, the cloud effective radius shows both negative and positive relationship with aerosol optical depth. It may be due to the difference of liquid water path of the cloud properties and shows complex interaction with aerosol.

1. INTRODUCTION

Cloud and aerosol interaction and their climate effect have been at cutting edge of global climate change research since nineteen nineties. The 5th assessment report of Intergovernmental Panel on Climate Change (IPCC 2013) still consider that aerosol-cloud-precipitation interaction and its climate effect is one of the most uncertain factors in climate changing study. As the physical process of aerosol-cloud-radiation-precipitation interaction is rater complex, there are still many unknown issues and misunderstandings which are needed to make clear. Therefore, most process and parametrization cannot be presented in the numerical model exactly. If we hope to improve the parameterization schemes, we need to learn the 3D structures of clouds and aerosol, their microphysical properties and their interaction. Because of lacking relevant observational data, most of the parameterization schemes used in our country are not well-developed. It has become barrier of the research of climate change.

Combined observation by lidar and microwave radiometer from 16th to 17th, November, 2012 and from 13th to 14th, December, 2012 are chosen to retrieve the cloud microphysical properties and study their interaction with local aerosol.

2. INSTRUMENTS AND DATASET

SKYNET Hefei site (31.89° N, 117.17° E) is one of the super observation sites in SKYNET program. This super site consists of a NIES two-wavelength depolarization lidar at the wavelengths of 532nm and 1064nm and depolarization at 532nm, a sky radiometer, a microwave radiometer, a sky camera and some other radiation instruments with data archived from October, 2002 till now. The photos and main specifications of NIES two-wavelength depolarization lidar and two-band microwave radiometer is shown in fig 1,2 and table1, 2.

Fig. 1 Two-wavelength depolarization lidar initialized in SKYNET Hefei site by NIES, Japan

DOI: 10.1051/epjconf/201611916013

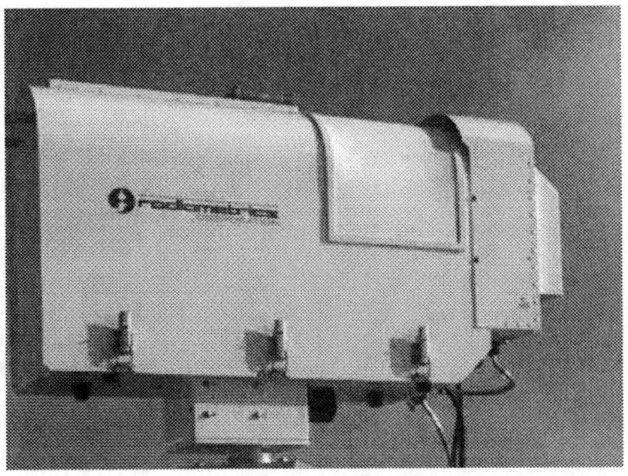

Fig.2 Two-band microwave radiometer installed in Hefei by SKYNET

Table1. The specifications of NIES two-wavelength depolarization lidar

Laser	Nd:YAG(532,1064nm)
Repetition	10Hz
Power	20mJ
Diameter of Telescope	20cm
Direction	Zenith(fixed)
Detector	PMT, APD
Range Resolution	7.5m
Signal Accumulation	4000 shots （5min）
Operation	4 profiles/hour

Table2. The specification of two-band microwave radiometer

Frequency/GHz	23.8	31.4
Beam width/°	5.8	4.5
Gain/dB	31	33
Sampling Time	<1min	
Resolution	0.25K	
Accuracy	0.5K	
Angle Scanning Range	Whole Sky	
Scanning Speed	Azimuth 3° /s; Elevating angle >90° /s	

3. ALGORITHMS AND RESULTS

3.1 Cloud vertical structure retrieval

There exists a plenty of publications referred to the cloud vertical structure retrieval. We develop a universal algorithm which is suitable for most cases of high aerosol loading condition in China and our lidar system. Samples are shown that it works very well with our lidar system.

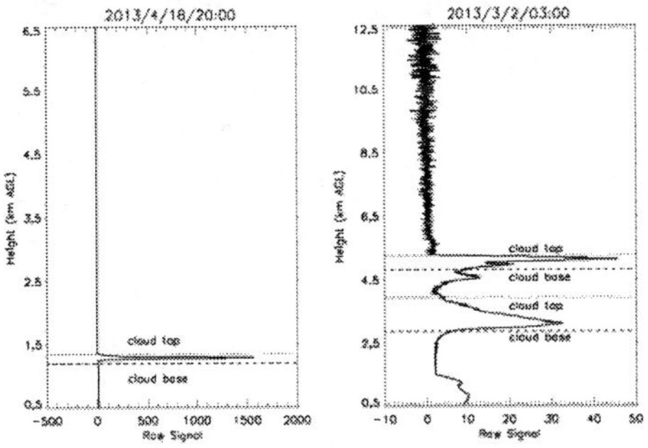

Fig.3. some cases of cloud vertical profile retrieval (left. low cloud ; right. multiple-layered cloud)

These are two vertical profiles that are typical cases including low cloud and multiple-layered cloud in Fig.3, one can see clearly that how it works faced with different kinds of cloud vertical profiles, single-layered and multiple-layered clouds. It shows that we can get exact cloud base and cloud top under different circumstance. Whereas, when it comes to multiple-layered cloud, it is hard for our lidar to detect because of the laser power and signal-noise-ratio. The right plot in Fig.3 shows that the algorithm is suitable for detecting multiple-layered cloud.

3.2 Cloud microphysical properties retrieval

Based on the Robert Woods' field campaigns, the relationship between liquid water path and cloud optical depth as shown in the following is proposed:

$$q_L(z) = f_{ad}\Gamma_{ad}z \quad (1)$$

$$LWP = \frac{5}{9}\rho_w \tau r_e(h) \quad (2)$$

$$LWP = \int_{z=0}^{h} q_L dz = \frac{1}{2} f_{ad}\Gamma_{ad}h^2 \quad (3)$$

q_L is liquid water content (g/cm^2), f_{ad} is the degree of adiabatic degree. It usually equals to 1 near cloud base. Γ_{ad} is related to pressure and temperature. r_e is cloud droplet effective radius. τ is cloud optical depth. Based on that and adiabatic assumption, cloud droplet effective radius could be calculated. Combining the aerosol optical depth observed by lidar and cloud droplet effective radius, we find in some typical cases, there exists a negative relationship between them. In the top panel Fig 4, it shows that the results match the Twomey effect (indirect effect of aerosol). However, in some cases, there exists opposite relationship between them as shown in the bottom panel in Fig 4. Comparing with others' observation results, it is possible because of the lack of liquid water path. If liquid water path is high enough for aerosol to become cloud condensation nuclei, there is negative relationship. If not, there is positive relationship because aerosol competes for liquid water to become cloud condensation nuclei.

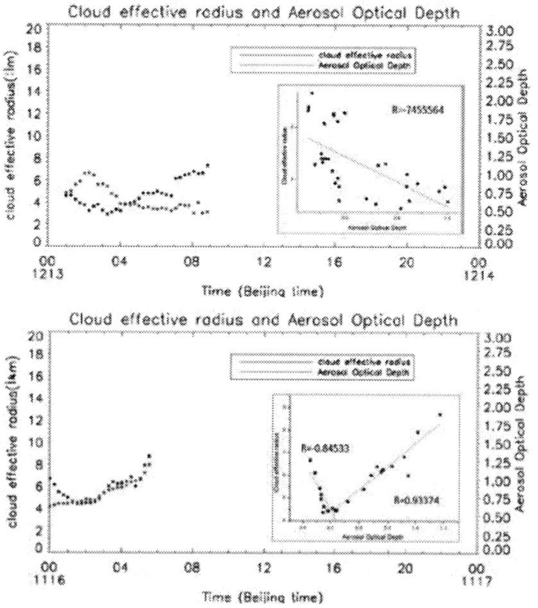

Fig 4 the relationship between cloud droplet effective radius and aerosol optical depth (top: negative relationship; bottom: positive relationship)

4. CONCLUSION

Due to the lack of observation instruments and temperature and humidity vertical profiles, it is difficult to derive exact cloud vertical structure. Moreover, aerosol retrieval method has been limited because there exists opaque thick cloud. If High Spectral Resolution Lidar has been adopted, aerosol extinction coefficient could be obtained without the assumption of lidar ratio. Based on multiple wavelength lidar, aerosol spectral distribution could be got. With Raman lidar, we could get water vapor and temperature vertical profiles. Combining lidar and Radar, cloud boundary could be acknowledged more exactly. So there is still possiblity for improvement in lidar-based cloud physical properties retrieval with state-of-art technology and long-term observation data.

Acknowledgement: This research has been supported by national basic research program of China (973 project) No. 2013CB955802, the project of Chinese Academy of Science No. XDA05100302, and NSCF No. 41075016. The authors would like to also thank Dr. Nobuo Sugimoto's group to initialize the lidar system in Hefei, China.

References

[1] Zhien Wang and Sassen, "Cloud Type and Macrophysical Property Retrieval Using Multiple Remote Sensors," Appl. Meteor, 40(10), 1665-1683(2001).

[2] Yongxiang Hu, Mark Vaughan, Zhaoyan Liu et al. Retrieving Optical Depths and Lidar Ratios for Transparent Layers Above Opaque Water Clouds From CALIPSO Lidar Measurements[J]. IEEE Geoscience and Remote Sensing Letters, 2007(4):523-526

[3] Jiming Li, Yongxiang Hu, Jianping Huang, et al. A New Method for Retrieval of the Extinction Coefficient of Water Clouds by Using the Tail of the CALIOP Signal [J]. Atmospheric Chemistry and Physics Discuss, 2010, 10: 28151-28181

[4] Robert Woods. Relationships between optical depth, liquid water path, droplet concentration and effective radius in an adiabatic layer cloud

TOWARDS IMPROVED CIRRUS CLOUD OPTICAL DEPTHS FROM CALIPSO

Anne Garnier[1,2]*, Mark Vaughan[2], Jacques Pelon[3], David Winker[2], Chip Trepte[2], and Stuart Young[4]

[1]*Science Systems and Applications, Inc., Hampton, VA, 23666, USA, *Email: anne.garnier@latmos.ipsl.fr*

[2]*NASA Langley Research Center, Hampton, VA, 23681, USA*

[3]*Université Pierre et Marie Curie, CNRS, IPSL, LATMOS, Paris, 75252, FRANCE*

[4]*CSIRO Oceans & Atmosphere, Aspendale, Victoria, 3195, AUSTRALIA*

ABSTRACT

This paper reviews recent advances regarding the retrieval of optical depths of semi-transparent cirrus clouds using synergetic analyses of perfectly collocated observations from the CALIOP lidar and the IIR infrared radiometer aboard the CALIPSO satellite.

1. INTRODUCTION

The CALIPSO satellite, launched nearly 9 years ago, carries the CALIOP (Cloud and Aerosol Lidar with Orthogonal Polarization) lidar operating at two wavelengths (532 and 1064 nm), the 3-channel Imaging Infrared Radiometer (IIR) operating in the 8-12 μm thermal infrared spectral range, and the Wide Field-of-view Camera (WFC) operating in the visible domain [1]. Observations from the three instruments are almost perfectly temporally and spatially collocated along the lidar track, allowing detailed synergetic analyses. In particular, visible (532 nm) extinction optical depths retrieved from CALIOP [2] have been extensively compared with absorption optical depths retrieved at 12.05 μm from the IIR [3]. Indeed, CALIOP and IIR retrieval techniques are based on independent physical principles, and are subject to different sources of uncertainty. To ensure meaningful comparisons between the active and passive observations, this study examines semi-transparent clouds that do not fully attenuate the CALIOP beam. In addition, only atmospheric columns containing one single cloud layer are considered. At night, CALIOP retrievals use one of two different techniques, which are briefly reviewed in section 2, and the consistency between the three independent retrievals is initially investigated in section 3. The CALIOP multiple scattering correction factor inferred from refined comparisons between CALIOP and IIR is then presented in section 4, which is followed by our conclusions in section 5.

2. CALIOP CIRRUS OPTICAL DEPTHS

Visible optical depths, τ_{vis}, for semi-transparent cirrus measured by CALIOP are retrieved using one of two totally independent techniques. The first consists in directly measuring the layer apparent 2-way transmittance, T_c^2, on those occasions when clear air is found immediately above and below the layer [2]. For these layers, T_c^2 is obtained directly from the ratio of the mean attenuated scattering ratios in clear air regions above and below the cloud, and hence is insensitive to calibration errors. This technique allows a direct determination of the layer "apparent" optical depth, $\eta.\tau_{vis}$, where η represents the required correction for the contribution from multiple scattering. Thus, τ_{vis} can be derived as:

$$\tau_{vis} = \frac{-\ln\left(T_c^2\right)}{2\eta} \quad (1)$$

In the version 3 algorithm, η is assumed constant throughout the layer, identical in-cloud and below cloud, and equal to 0.6. The lidar ratio, S_c, can then be retrieved from the particulate layer-integrated attenuated backscatter, γ', as [4]:

$$S_c = \frac{1 - e^{-2\eta\tau_{vis}}}{2\eta\gamma'} \quad (2)$$

Note that in practice, this initial estimate of S_C is refined iteratively until the integral of the retrieved extinction profile exactly matches the directly-determined optical depth [2]. In the version 3 algorithm, this so-called 'constrained retrieval' is applied when the uncertainty in S_c is estimated to be less than 40%. This condition, which requires relatively large signal-to-noise ratios in the clear air regions, is fulfilled almost exclusively by nighttime measurements. Furthermore, because the signal is noisy, the measured optical depth distribution is increasingly truncated as true optical depth decreases, because a larger fraction of these small optical depths may become negative or do not satisfy the estimated relative uncertainty requirement [5]. Consequently, both the optical depths and the lidar ratios obtained from constrained retrievals exhibit a high bias at optical depths smaller than about 0.6 that increases as optical depth decreases [6]. For consistency, optical depths reported in the CALIOP 5-km cloud layer product are derived from the measured apparent 2-way transmittance (Eq. 1) only when the conditions imposed by the constrained technique are satisfied.

When the constrained technique is deemed not possible, τ_{vis} reported in the products is retrieved using an initial lidar ratio which is set to 25 sr in version 3 data products. Within the retrieval code, this initial lidar ratio can be increased or reduced to prevent extinction retrievals from diverging [2]. For semi-transparent cirrus clouds, most of the retrievals are 'unconstrained', with final lidar ratios identical to the initial ones. Nevertheless, the apparent 2-way transmittance derived for those layers is also reported in the products, from which the layer optical depth can be derived using Eq. 1, and compared with optical depths retrieved using $S_c = 25$ sr.

3. COMPARISONS WITH IIR CIRRUS OPTICAL DEPTHS

The IIR optical depth retrieved at 12.05 μm, τ_a, is essentially an absorption optical depth derived from the cloud emissivity, ε, as [3]:

$$\tau_a = -\ln(1-\varepsilon) \qquad (3)$$

where

$$\varepsilon = \frac{R_m - R_{BG}}{R_{BB} - R_{BG}} \qquad (4)$$

In Eq. 4, R_m is the measured radiance. R_{BB} is the radiance at the top of the atmosphere (TOA) of a blackbody source located at the centroid altitude of the attenuated backscatter profile whose thermodynamic temperature, T_c, is retrieved from ancillary meteorological data. R_{BG} represents the correction for the TOA background radiance which would be observed had the cloud not been present. R_{BG} is preferably constrained by observations in neighboring pixels. Otherwise, R_{BG} is computed in real-time using the FASRAD radiative transfer model [7] and ancillary atmospheric and surface data.

Fig. 1 shows 2D-histograms of nighttime τ_{vis} and $2\tau_a$ for single-layered semi-transparent cirrus. All clouds were measured over oceans and classified (with high confidence) as being composed of randomly oriented ice (ROI). In order to minimize possible biases, IIR τ_a is taken from R_{BG} measured in neighboring pixels. The τ_{vis}/τ_a ratio is relatively insensitive to cirrus cloud microphysical properties. For this data set, it is expected to be mostly around 2 [6]. In Fig.1a, CALIOP τ_{vis} is as reported in the version 3 products. Two ensembles of samples are clearly seen. The larger values of τ_{vis} are from the constrained retrievals and hence the 2-way transmittance (Eq. 1), whereas the smaller values of τ_{vis} are from unconstrained retrievals and the default lidar ratio $S_c = 25$ sr. The clear discontinuity between both ensembles in terms of comparison with IIR is explained by the fact that the default lidar ratio is not consistent with the lidar ratios derived from the constrained retrievals, and is too small by about 25%, as discussed in [8]. In Fig. 1b, τ_{vis} of clouds for which unconstrained retrievals are chosen in the version 3 algorithm is instead obtained from the apparent 2-way transmittance (Eq. 1). Here, no discontinuity between the two ensembles of retrievals is seen, which confirms that the default lidar ratio is too small in the mean for the unconstrained retrievals. Thus, it appears that the constrained technique could be extended to smaller optical depths, and likely to more daytime measurements, by relaxing the threshold in the relative lidar ratio uncertainties used in the version 3 algorithm, which, in turn, would substantially decrease the number of unconstrained retrievals. Analyses in [6] indicate that constrained retrievals could thus be extended down to $2\tau_a = 0.3$ with no evidence of systematic high bias. It is noted that below this limit, the

 DOI: 10.1051/epjconf/201611916014

number of non-physical negative values of τ_{vis} increases rapidly due to the unavoidable increasing noise.

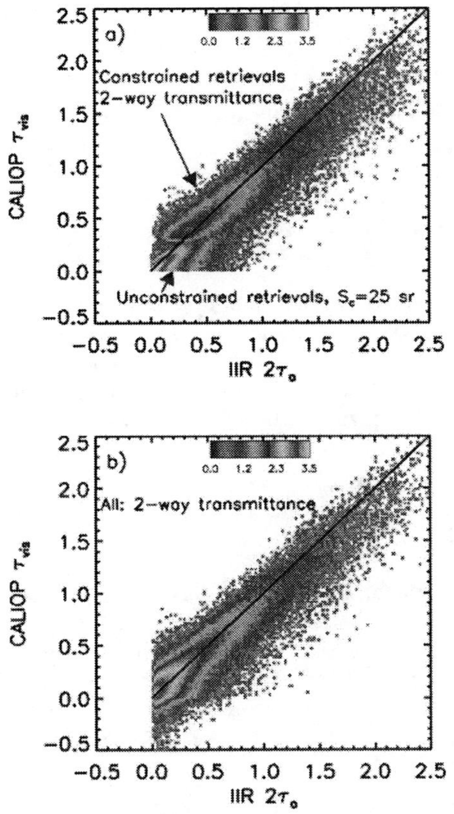

Fig. 1. 2D-histogram of nighttime CALIOP visible optical depth (τ_{vis}) and 2xIIR absorption optical depth at 12.05 µm ($2\tau_a$) with R_{BG} constrained by neighboring observations for single-layered semi-transparent cirrus composed of ROI (high confidence) over oceans during 2008. **a)** CALIOP version 3 products; **b)** for unconstrained retrievals, τ_{vis} is from the apparent 2-way transmittance and not from S_c = 25 sr. The black line represents τ_{vis}/τ_a = 2. The color code is the decimal logarithm of the number of samples.

4. MULTIPLE SCATTERING FACTOR

Fig.1b shows that the relationship between the CALIOP τ_{vis} derived from the apparent 2-way transmittance and the IIR τ_a is overall close to the expectations. However, detailed analyses show that the τ_{vis}/τ_a ratio increases as the temperature T_c decreases, which is in disagreement with theoretical simulations and microphysical properties derived from the IIR microphysical algorithm [6]. As the CALIOP multiple scattering correction factor η is the only free parameter in these analyses, the proposed explanation is a temperature-dependent multiple scattering factor, η_T. Fig. 2 shows η_T against T_c inferred from these comparisons [6] by following the approach initially introduced in [4].

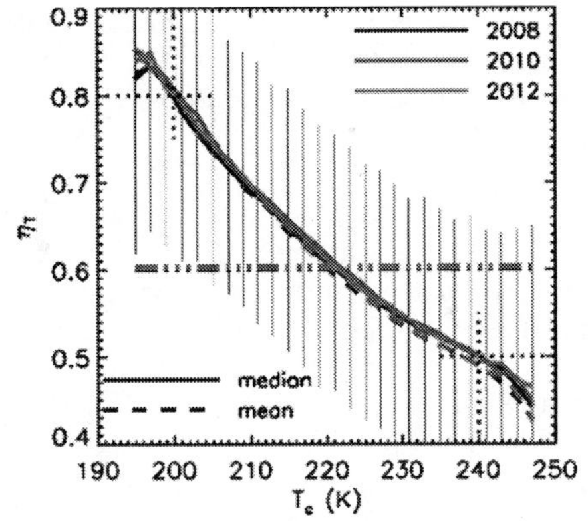

Fig. 2: CALIOP median (solid) and mean (dashed) multiple scattering factor ± standard deviation inferred from the ratios of CALIOP visible optical depth τ_{vis} derived from the apparent 2-way transmittance to IIR absorption optical depth at 12.05 µm (τ_a). The horizontal dashed dotted lines are the mean values overall. See text for details.

Clouds are composed of ROI (high confidence), and those clouds with base temperatures warmer than -20°C are discarded to minimize possible contamination from mixed phase clouds. IIR background references are from both measurements in neighboring pixels and computations to maximize the number of samples. Analyses are over oceans, and $2\tau_a$ is limited to values larger than 0.3 to minimize residual biases at small optical depth for both CALIOP and IIR. τ_a is corrected for biases in R_{BB} that have been quantified using CALIOP extinction profiles [6], and is on average increased by 1% at $2\tau_a$ = 0.3 up to 7% at $2\tau_a$ = 2. The median (solid lines) and mean (dashed lines) values of η_T are shown for 2008 (black), 2010 (red), and 2012 (blue), demonstrating a remarkable repeatability from one year to the other. The median η_T is found to decrease from $\eta_T = 0.8$ at 200 K to $\eta_T = 0.5$ at 240 K. The horizontal dashed dotted lines show the mean values overall, which are 0.601, 0.604, and 0.602 respectively in 2008, 2010, and 2012. These values are found to be essentially identical to the constant value η=0.6 taken in the version 3 operational algorithm. S_c is found to be weakly varying with T_c when η is taken constant. When using η_T from Fig. 2, S_c is found to increase by 50% from T_c = 200 K to T_c = 228 K and to be mostly constant at warmer temperatures.

5. CONCLUSIONS

Previous analyses showed that CALIOP constrained and unconstrained retrieval techniques are in good agreement when a lidar ratio derived from the statistical analysis of constrained retrievals is chosen as a default lidar ratio for CALIOP unconstrained retrievals. Here, we report that constrained retrievals can be improved, in the mean, by eliminating the threshold on the estimated relative lidar ratio uncertainties used in the version 3 algorithm. At the same time, this significantly increases the number of optical depths obtained directly from measurements rather than by initializing the retrievals with a default lidar ratio. Furthermore, analyses of the relationship between visible extinction optical depths derived from the apparent 2-way transmittances and infrared absorption optical depths indicate that the CALIOP multiple scattering correction factor, taken equal to 0.6 in the version 3 algorithm, varies from 0.8 at 200 K to 0.5 at 240 K. The mean value overall is found essentially equal to 0.6, so that on average, CALIOP visible optical depths compare as expected with IIR effective optical depths at 12.05 µm. These findings will be used as guidance for the definition of the future version 4 of the products.

ACKNOWLEDGEMENTS

The authors are thankful to the Data Management Team at NASA/LaRC and to ICARE data center in Lille. CNES is acknowledged for its support.

REFERENCES

[1] Winker, D. M., J. Pelon, J. A. Coakley, Jr., S. A. Ackerman, R. J. Charlson, P. R. Colarco, P. Flamant, Q. Fu, R. Hoff, C. Kittaka, T. L. Kubar, H. LeTreut, M. P. McCormick, G. Megie, L. Poole, K. Powell, C. Trepte, M. A. Vaughan, B. A. Wielicki, 2010: The CALIPSO Mission: A global 3D view of aerosols and clouds, *Bull. Am. Meteorol. Soc.*, **91**, 1211-1229.

[2] Young, S. A. and M. A. Vaughan, 2009: The retrieval of profiles of particulate extinction from Cloud Aerosol Lidar Infrared Pathfinder Satellite Observations (CALIPSO) data: Algorithm description, *J. Atmos. Oceanic Technol.*, **26**, 1105–1119.

[3] Garnier, A., J. Pelon, P. Dubuisson, M. Faivre, O. Chomette, N. Pascal, D. P. Kratz, 2012: Retrieval of cloud properties using CALIPSO Imaging Infrared Radiometer. Part I: effective emissivity and optical depth, *J. Appl. Meteor. Climatol.*, **51**, 1407–1425.

[4] Platt, C. M. R., 1973: Lidar and radiometer observations of cirrus clouds, *J. Atmos. Sci.*, **30**, 1191-1204.

[5] Young, S. A., M. A. Vaughan, R. E. Kuehn, D. M. Winker, 2013: The retrieval of profiles of particulate extinction from Cloud-Aerosol Lidar Infrared Pathfinder Satellite Observations (CALIPSO) data: uncertainty and error sensitivity analyses, *J. Atmos. Oceanic Technol.*, **30**, 395-428.

[6] Garnier, A., J. Pelon, M. A. Vaughan, D. M. Winker, C. R. Trepte, P. Dubuisson, 2015: Optical depths of semi-transparent cirrus clouds over oceans from CALIPSO infrared radiometer and lidar measurements, and an evaluation of the lidar multiple scattering factor, *Atmos. Meas. Tech. Discuss.*, **8**, 2143–2189.

[7] Dubuisson P., V. Giraud, O. Chomette, H. Chepfer, J. Pelon, 2005: Fast radiative transfer modeling for infrared imaging radiometry, *J. Quant. Spectr. Rad. Tr.*, **95**, Issue 2, 1 October 2005, 201-220.

[8] Garnier, A., M. A. Vaughan, P. Dubuisson, D. Josset, J. Pelon, D. M. Winker, 2012: Multi-sensor cirrus optical depth estimates from CALIPSO, in: *Reviewed & Revised Papers Presented at the 26th International Laser Radar Conference*, Porto Heli, Greece, 25-29 June 2012, 691-694.

EPJ Web of Conferences **119**, 16015 (2016) DOI: 10.1051/epjconf/201611916015
ILRC 27

BACKSCATTER MATRIX OBSERVATIONS BY THE GV-HSRL

Matthew Hayman, Scott Spuler, Bruce Morley

Earth Observing Laboratory, National Center for Atmospheric Research, PO Box 3000, Boulder, CO, 80307, USA, mhayman@ucar.edu

ABSTRACT

The GV-HSRL is a high spectral resolution lidar capable of measuring calibrated backscatter, extinction and circular depolarization from the ground or NSF Gulfstream V platform. In the spring of 2012, the instrument was modified to measure the full backscatter matrix of atmospheric scatterers. This modification enabled us to investigate the polarization properties of oriented particles and further understand where particles orient and how they may impact depolarization lidar data. Observations were performed from the ground at different times with the lidar's tilt angle at 4°, 22° and 32° degrees off zenith. This instrument found oriented ice crystals only produced observable polarization effects at the 32° lidar tilt angle. By contrast, a significant fraction of rain observations have oriented scattering matrices at all three lidar tilt angles. Thus conventional depolarization lidar is generally well suited for characterizing ice crystals but not liquid precipitation.

Oriented ice crystals are most commonly investigated by looking for specular scatter from horizontally oriented plates. While this method offers excellent sensitivity to small populations of oriented plats, it has very little capability to determine the fraction of oriented plates if the population is larger than about 1%. We show here that observing f_{12} scattering matrix element at 32° off zenith may be more effective for estimating the oriented fraction of particles in a volume.

1. INTRODUCTION

A typical depolarization lidar system intrinsically assumes that the particles under investigation are randomly oriented, which, by symmetry and reciprocity must have a diagonal backscatter matrix with three degrees of freedom (backscatter, depolarization and where asymmetric particles exist, circular diattenuation)[1; 2]

$$\mathbf{F}(\pi) = \beta \begin{bmatrix} 1 & 0 & 0 & f_{14} \\ 0 & 1-d & 0 & 0 \\ 0 & 0 & d-1 & 0 \\ f_{14} & 0 & 0 & 2d-1 \end{bmatrix}, \quad (1)$$

where β is the volume backscatter coefficient and d is the propensity of the medium to depolarize. The term f_{14} is generally assumed to be zero in atmospheric scattering [3; 4]. The scattering angle argument π indicates this is a backscattering phase matrix independent of incident direction.

When this assumption is valid, the standard depolarization ratio is sufficient to fully describe the backscatter properties of the volume. It is known, however, that particles do sometimes orient in the atmosphere. In such cases, the backscatter matrix does not necessarily reduce to the form in Eq. (1) and generally must take the form [3]:

$$\mathbf{F}(\vec{k}_i, -\vec{k}_i) = \begin{bmatrix} \beta & f_{12} & 0 & f_{14} \\ f_{12} & f_{22} & 0 & 0 \\ 0 & 0 & f_{33} & f_{34} \\ f_{14} & 0 & -f_{34} & f_{44} \end{bmatrix}. \quad (2)$$

where now there are six independent terms in the scattering matrix and the argument $\vec{k}_i$ is the incident wave vector, indicating that the scattering matrix is dependent on direction of incidence. The resulting depolarization measurements will depend on several factors, such as the lidar's specific pointing angle, and the polarization transmitted and measured by the lidar. These potential errors are not well quantified due to the very limited number of oriented scattering matrix observations. Thus it is not clear if and when the oriented scattering matrix terms described in Eq. (2) will impact polarization lidar observations.

For a backscatter matrix to appear in the form shown in Eq. (2), three conditions must be met [4]. Oriented particles must be present, the oriented particles must not exhibit symmetry about the incident and backscatter angle and finally the backscatter signal from the oriented particles must be large enough that it is not overwhelmed by the signal from randomly oriented particles. There are very few observations that indicate typical values for oriented scattering matrix terms in the atmosphere. Furthermore there are very few observations to indicate where all three conditions are satisfied for observing these unique scattering matrices.

There has been substantial scientific interest in occurrence of oriented ice crystals [5; 6; 7]. Lidar observations typically take the approach of looking for strong specular reflections from plates by directing the instrument very close to zenith or nadir. This technique has the benefit of being highly sensitive to the presence of even a small population of ice crystals, but comes at the expense of diluting the signatures of all other particles in the volume.

In Hirakata et al. [5], a number of statistics are used to characterize the where 2D (horizontally oriented) and 3D (randomly oriented) plates exist. They notably define a region classified as a "mix" of 2D and 3D plates. However, it should be noted that anywhere 2D plates are observed, it is likely 3D plates are also present. Because the backscatter from the 2D plates dominates the signal,

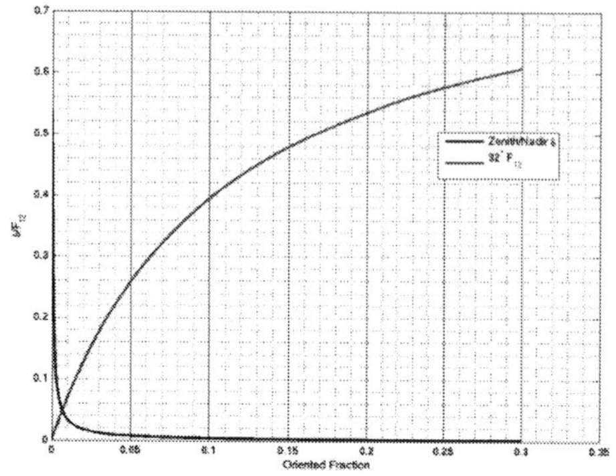

Figure 1: Depolarization ratio (blue) or diattenuation (green) for different oriented fractions.

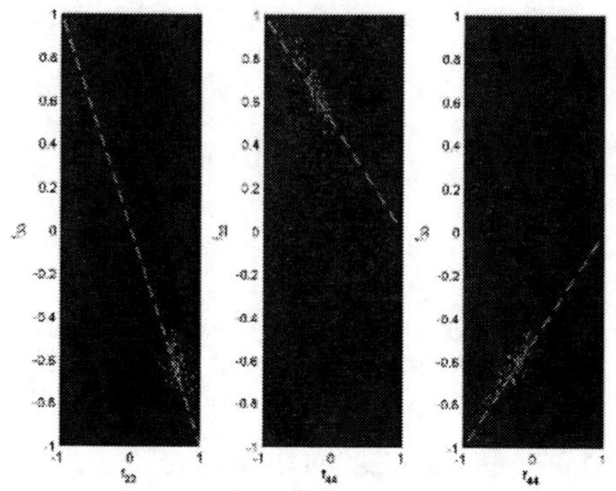

Figure 2: Histograms of the f_{22}, f_{33} and f_{44} of oriented ice crystal observations. The dashed green line shows the expected relationship for randomly oriented ice crystals.

it is almost impossible to observe the 3D plates or any of their corresponding properties. Furthermore, any attempt to determine the relative mixing of oriented and randomly oriented particles is only practical for very small oriented particle fractions.

Consider a volume consisting of a combination of randomly oriented and oriented plates. The total backscatter is given by the sum of the two backscatter terms

$$\beta_T = \beta_R + \beta_O, \tag{3}$$

where the subscript T corresponds to the observed total volume, R corresponds to randomly oriented terms and O corresponds to oriented terms. The total depolarization is the backscatter weighted mean of the two backscatter terms

$$\beta_T d_T = \beta_R d_R + \beta_O d_O. \tag{4}$$

The volume backscatter coefficient is given by the number of particles in the volume multiplied by their backscatter cross section. For the randomly oriented particles this is given by

$$\beta_R = \sigma_R n_R, \tag{5}$$

and the oriented particles are given by a similar equation, but their backscatter cross section is much larger than that of their randomly oriented counterparts, so let $\sigma_O = A\sigma_R$ making the oriented volume backscatter coefficient

$$\beta_O = \sigma_O n_O = A\sigma_R n_O. \tag{6}$$

The presence of oriented ice crystals are identified by their low depolarization, often inside an ice cloud. The depolarization of a volume consisting of oriented and randomly oriented particles is

$$d_T = \frac{d_R + (Ad_O - d_R)F_O}{(A-1)F_O + 1}, \tag{7}$$

where d_T is the total depolarization, d_R is the depolarization of the randomly oriented particles, d_O is the depolarization of the oriented particles (typically very close to zero), A is the oriented particle enhancement factor and F_O is the fraction of oriented particles in the population and is given by

$$F_O = \frac{n_O}{n_O + n_R}. \tag{8}$$

By the same principles that we obtain Eq. (7) we can also obtain a metric for the total diattenuation of the volume (the f_{12} element of the scattering matrix).

$$f_{12}^{(T)} = \frac{Af_{12}^{O} F_O}{(A-1)F_O + 1}, \tag{9}$$

Note that $f_{12} = 0$ for randomly oriented particles.

A scattering simulation performed by Lei Bi and Ping Yang of $100\mu m$ plates has been used to determine that for specular backscatter, $A \approx 1000$. It is assumed that $d_R = 0.7$ which is near the maximum value we observe with the GV-HSRL, giving maximum sensitivity to oriented fraction. By contrast, for a lidar operating at 32° off zenith, $A \approx 10$ and $|f_{12}| \approx 0.75$.

We plot the expected total volume depolarization and diattenuation in Figure 1 as a function of the oriented fraction F_O. While measuring specular backscatter is highly sensitive to small populations of oriented particles, it is virtually impossible to make distinctions between oriented fractions larger than 1%. By contrast, diattenuation maintains a more consistent sensitivity over a large range of F_O, making it a more useful parameter for estimating the oriented particle fraction.

2. OBSERVATIONS

Observations of scattering matrices presented here are from summer in 2012 and 2013 at lidar tilt angles of 4°,

 DOI: 10.1051/epjconf/201611916015

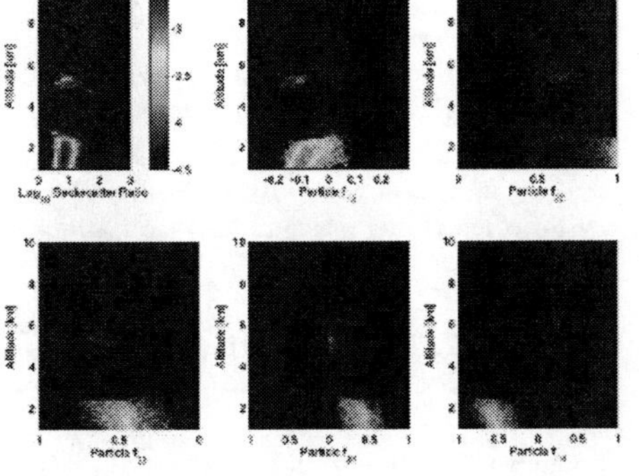

Figure 3: Altitude resolved histograms of scattering matrix elements when observed at a lidar tilt angle of 32° during June 2012. This figure is originally published in Optics Express in [4] and is used here with permission.

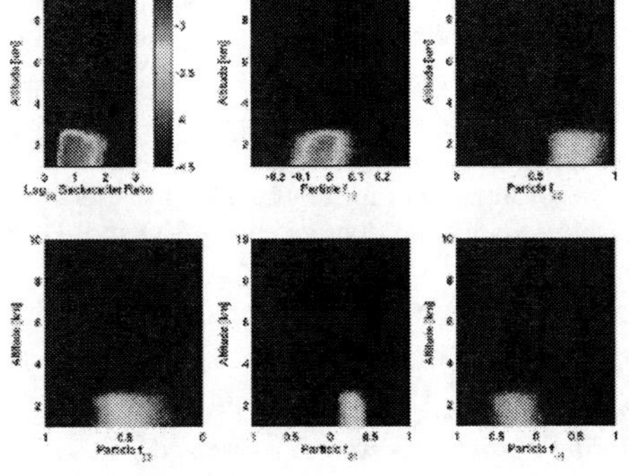

Figure 4: Altitude resolved histograms of scattering matrix elements when observed at a lidar tilt angle of 22° during July 2012. This figure is originally published in Optics Express in [4] and is used here with permission.

22° and 32° off zenith. All observations were made in Boulder, CO, USA and the lidar technique used for measuring the backscatter matrix is described in [8].

In nearly all cases, oriented ice crystals were only identifiable by significant diattenuation (f_{12}) terms. In fact, the other matrix terms typically conformed quite well to Eq. (1). Figure 2 shows the relationship between the diagonal terms of oriented ice crystal observations when the lidar was tilted 32° off zenith. The green dashed line indicates the relationship expected for randomly oriented particles given by

$$f_{44} = 1 - 2f_{22} = 1 + 2f_{33}. \qquad (10)$$

In [3] the principle parameter used to identify the presence of oriented ice crystals is the orientation parameter χ which they define as

$$\chi = \frac{f_{22} + f_{33}}{f_{11} + f_{44}}. \qquad (11)$$

Under the assumption that the single particle backscatter matrix has $f_{33} = f_{44}$ and $f_{22} = 1$, $\chi \to 1$ as the particle orientation distribution narrows, and $\chi = 0$ for a completely randomly oriented population.

The results from the GV-HSRL in Figure 2 conflict with the use of χ in [3]. In our observations, matrix diagonals do not significantly deviate from the randomly oriented form in Eq. (1) so χ is always near zero. This may be the result of a significant population of randomly oriented ice crystals also present in the volume. By contrast, [3] uses χ to identify the presence of oriented ice crystals and finds significant populations with a near zenith directed lidar.

Our processing routines use non-zero off diagonal elements to indicate the presence of oriented particles in a volume. In particular, we look for non-zero diattenuation (f_{12}) and retardance (f_{34}). The altitude resolved histograms of all oriented scattering matrix elements are shown in Figures 3, 4 and 5 for lidar tilt angles of 32°, 22° and 4° respectively. These data sets were recorded in the summer and observations of oriented particles above 3 km are generally regarded as oriented ice crystals where observations below 3 km are rain.

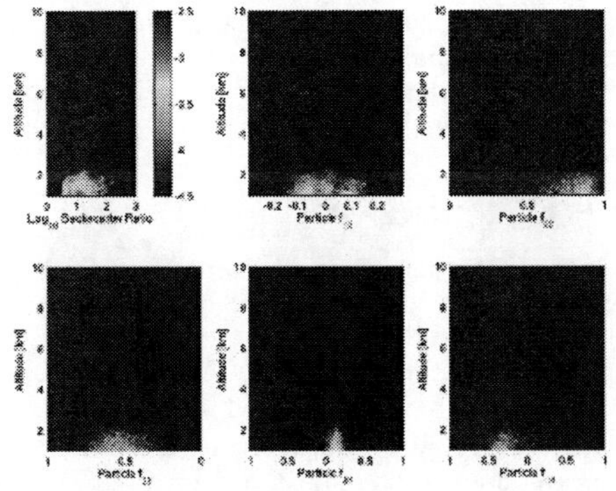

Figure 5: Altitude resolved histograms of scattering matrix elements when observed at a lidar tilt angle of 4° during August 2013. This figure is originally published in Optics Express in [4] and is used here with permission.

The oriented ice observations only appear when the lidar is tilted 32° off zenith. In the other two cases, it appears that either the off diagonal terms are small and/or the oriented backscatter signal is overwhelmed by randomly oriented particles. Even at 32° tilt angle, only the f_{12} term shows significant contribution from oriented particles. Noting that the maximum measured $|f_{12}|$ is approximately 0.2 we can use Figure 1 to roughly estimate that the oriented fraction of particles are typically between 0 and 3.5%.

In contrast to ice observations, the rain scattering matrices appears to contain significant oriented scattering terms at all three lidar tilt angles. Rain scattering matrices exhibit significant diattenuation (f_{12}) and retardance (f_{34}) and the diagonal terms generally do not conform to Eq. (10).

3. CONCLUSION

Observations performed by the GV-HSRL of oriented scattering matrices showed that oriented polarization effects are quite rare in ice clouds. Their peak occurrence was found to be of the order of 1% at altitudes of 5 km and they were only observed in significant numbers when the lidar was tilted at 32° off zenith. Even when oriented particle polarization effects were observed, they were only significant in the f_{12} parameter and diagonal terms seemed to conform relatively well to the functional form of randomly oriented particles. It is notable that these observations differ significantly from [3] where near zenith pointing lidar observed and characterized oriented ice crystals using deviations from the expected diagonal relationship of randomly oriented particles.

Our observations also show that rain has significant oriented scattering terms and cannot be reliably described by the linear depolarization ratio. Also, when circular depolarization is reported for rain, the angle of incidence should be specified. It may be possible to improve characterization of liquid precipitation using scattering matrix observations of rain. In a number of profiles, we see size sorting common to rain storms, where rain is highly oriented at the beginning and becomes less so as the storm passes.

ACKNOWLEDGMENTS

The National Center for Atmospheric Research is sponsored by the National Science Foundation.

REFERENCES

1. Flynn, C. J., Mendoza, A., Zheng, Y., Mathur, S. 2007: Novel polarization-sensitive micropulse lidar measurement technique. *Opt. Express* **15**, pp. 2785-2790.
2. Gimmestad, G. 2008: Re-examination of depolarization in lidar measurements. *Appl. Opt.* **47**, pp. 3795-3802.
3. Kaul, B. V., Samokhvalov, I. V., Volkov, S. N. 2004: Investigating particle orientation in cirrus clouds by measuring backscattering phase matrices with lidar. *Appl. Opt.*, **43**, pp. 6620-6628.
4. Hayman, M., Spuler, S., Morley, B. 2014: Polarization lidar observations of backscatter phase matrices from oriented ice crystals and rain. *Opt. Express*, **22**, pp. 16976-16990.
5. Hirakata, M., Okamoto, H., Hagihara, Y., Hayasaka, T., Oki, R. 2014: Comparison of global and seasonal characteristics of cloud phase and horizontal ice plates derived from CALIPSO with MODIS and ECMWF. *J. Atmos. Oceanic Technol.*, **31**, pp. 2114-2130.
6. Noel. V. and Sassen K. 2005: Study of planar ice crystal orientations in ice clouds from scanning polarization lidar observations. *J. Appl. Meteor.*, **44**, pp. 653-664.
7. Wesbrook, C. D. 2008: The fall speeds of sub-100μm ice crystals. *Q. J. R. Meteorol. Soc.*, **134**, pp. 1243-1251.
8. Hayman, M., Spuler, S., Morley, B., VanAndel, J. 2012: Polarization lidar operation for measuring backscatter phase matrices of oriented scatterers. *Opt. Express*, **20**, pp. 29553-29567.

DOI: 10.1051/epjconf/201611916016

USE OF LIDAR DERIVED OPTICAL EXTINCTION AND BACKSCATTERING COEFFICIENTS NEAR CLOUD BASE TO EXPLORE AEROSOL-CLOUD INTERACTIONS

Zaw Han [1*], Yonhgua Wu[1], Barry Gross[1], Fred Moshary[1]

[1] *Optical Remote Sensing Lab, City College of New York, USA, *Email: zhan@ccny.cuny.edu*

ABSTRACT

Combination of microwave radiometer (MWR) and mutlifilter rotating shadowband radiometer (MFRSR) measurement data together with SBDART radiative transfer model to compute cloud optical depth (COD) and cloud droplet effective radius (R_{eff}). Quantify the first aerosol indirect effect using calculated R_{eff} and aerosol extinction from Raman lidar measurement in urban coastal region. Illustrate comparison between ground-based and satellite retrievals. Demonstrate relationship between surface aerosol ($PM_{2.5}$) loading and R_{eff}. We also explain the sensitivity of aerosol-cloud-index (ACI) depend on the aerosol layer from cloud base height. Potential used of less noisy elastic backscattering to calculate the ACI instead of using Raman extinction. We also present comparison of elastic backscattering and Raman extinction correlation to R_{eff}.

1. INTRODUCTION

It is generally understood that recent changes in climate have impacted the Earth's bio-sphere. Greenhouse gases play a significant role in the overall global energy balance, but other effect such as aerosols and the interaction of aerosols with clouds are much less understood. These uncertainties are the main source of uncertainty in the overall energy balance. In particular, the aerosol direct effect that is defined as the solar radiation, which is either absorbed or scattered by aerosol particles is estimated to partially compensate for the global warming due to greenhouse gas forcing but the magnitude of this effect is difficult to estimate due to the variety of aerosol properties that can either act as a cooling mechanism or a warming mechanism.

Far more complex are the aerosol-cloud interactions in which aerosols are thought to modify the microphysical and therefore the radiative properties and lifetime of clouds. Generally, the aerosol can affect on cloud directly and indirectly, and consequences of the effects that lead to alter the climate condition in the local, regional, and global scale. Feingold quantified the aerosol indirect effect using models and radar reflectivity for cloud mask, Raman LIDAR for aerosol extinction, microwave radiometer for integrated liquid water path, combination of radar radiometer for cloud droplet effective radius at ARM CART site [1]. Additionally, Min propose the new technique for simultaneously retrieving cloud optic properties (τ_{cod},R_{eff}) and LWP using radiative transfer model simulation [2]. In this paper, we provide the first results of aerosol-cloud interaction in an urban coastal environment (NYC).

2. METHODOLOGY

In order to measure the cloud and aerosol properties simultaneously, we make use of the technique discussed in [3]. In particular, a combination of a microwave radiometer (MWR) and the multi filter rotating shadow band radiometer (MFRSR) are needed for retrievals of cloud optical depth and cloud droplet effective radius. Basically, we used SBDART model to create look-up-table (LUT), HYSPLIT model to determine origin of air particles, and ground based instruments retrieval to calculate aerosol extinction coefficients and cloud optic properties. SBDART is a FORTRAN source code run on LINUX/UNIX that computes plane-parallel radiative transfer in clear and cloudy sky condition within the Earth's atmosphere and at the surface.

In this study, atmospheric transmittance is the quantity that can be measured and depends critically on cloud optical properties (τ_{cod}, R_{eff}) as well as the observed solar zenith angle. Figure 1&2 illustrate that the behavior of cloud optical depth and inverse transmittance with varying cloud droplet effective radius with fixed solar zenith angle and varying solar zenith angle with fixed cloud droplet effective radius.

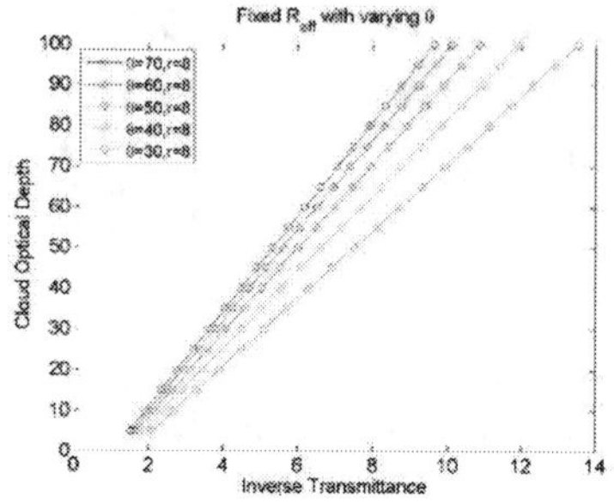

Figure 1. Fixed R_{eff} with varying solar zenith angle.

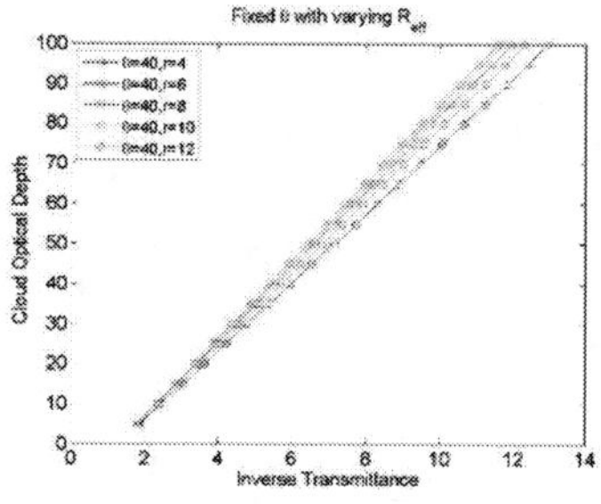

Figure 2. Fixed solar zenith angle with varying R_{eff}.

The retrieval of cloud properties requires the simultaneous retrieval of cloud diffuse transmittance and liquid water path. The cloud diffuse transmittance is obtained by eq. 1 using the 415 nm channel of MFRSR. In order to use MFRSR, an estimation of the Top of Atmosphere (TOA) irradiance or calibration constant (I_0) must be made. In particular, the atmospheric transmittance is evaluated using 415 nm channel because it is insensitive to absorption by ozone and atmospheric gases, suitable for natural surface, and low surface albedo.

$$T_{diff} = \frac{I_{diff}}{I_0 \cos(\theta)} \quad (1)$$

In order to calculate the cloud optical depth and cloud droplet effective radius, the other crucial parameter is integrated liquid water path that can be obtained from MWR. In preliminary calculations, we made using of the default setting from the Radiometric Corporation on their Neural Network (NN) method.

Concisely, microwave radiometer (MP3000A) used four points nonlinear model for brightness temperature and surface meteorological data which is defined as level 1 product. Then train by neural network to retrieve level 2 product. Level 2 product consist of profiling and column integrated liquid water path.

Inauspiciously, since no other independent ground based measurements exist for integrated liquid water path validation, we decided to compare against a significantly different dual channel method [4] to retrieve integrated liquid water path and water vapor so that we compare two retrievals method to assess the accuracy of liquid water path.

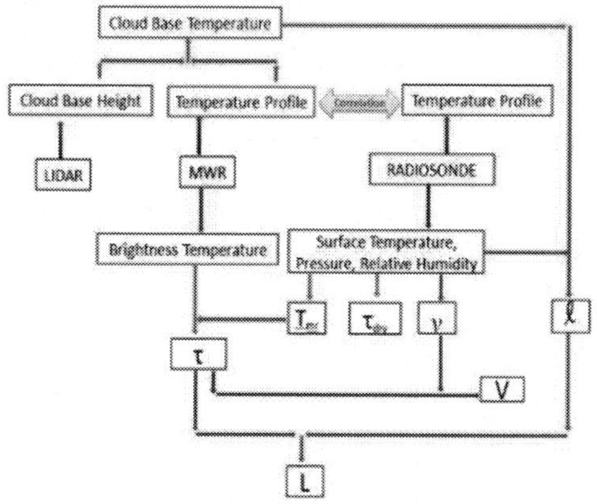

Figure 3. Dual-Channel LWP retrieval algorithm.

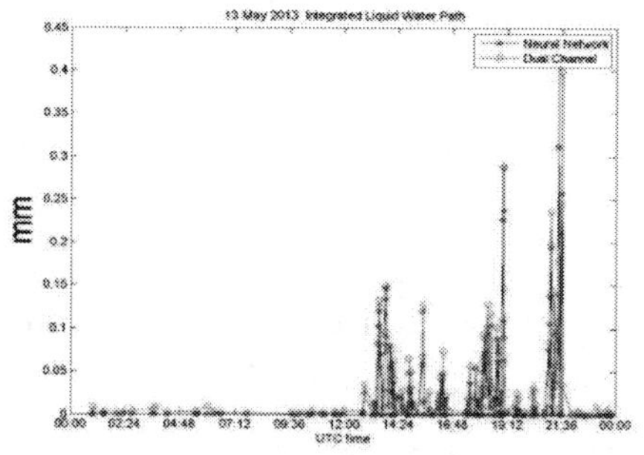

Figure 4. Integrated Liquid Water Path (mm) retrieved by dual channel and default neural network algorithms.

Once the cloud properties are obtained, we use Raman LIDAR to derive aerosol extinction coefficients (α_{ext}) near cloud base which serve as a proxy to the aerosol number density. In particular, the following Raman LIDAR eq. (2) is used to compute α_{ext} below cloud.

$$\alpha_{ext}(\lambda_L, z) = \frac{\frac{d}{dz}\left[\ln\frac{N(z)}{z^2 P(z)}\right] - \alpha_{mol}(\lambda_L, z) - \alpha_{mol}(\lambda_R, z)}{1 + \frac{\lambda_L}{\lambda_R}} \quad (2)$$

3. RESULTS

This result is clearly consistent with the aerosol indirect effect. In this example, we also note that the DC method actually seems to outperform the NN approach. This can be attributed to the strict requirement of Microwave Brightness Channels which results in a smaller number of overall LWP retrievals and therefore less Cloud R_{eff} measurements. Even so, the overall consistency between the two approaches can be seen by the general good agreement with the ACI values.

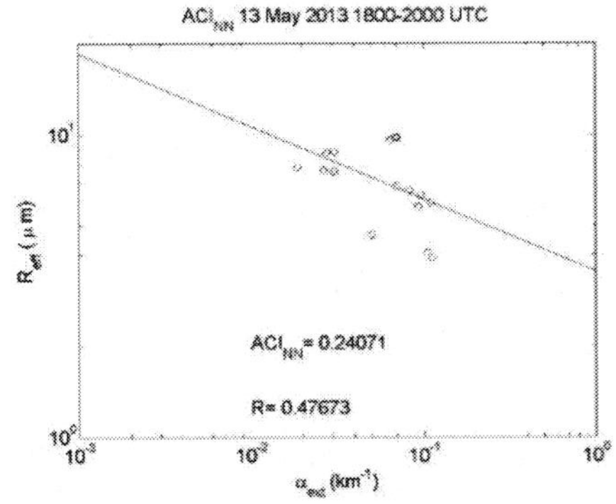

Figure 5. ACI using NN method.

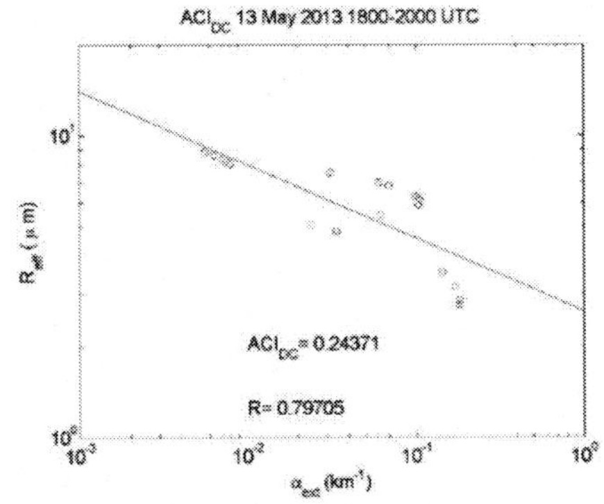

Figure 6. ACI using DC method.

In getting these results, care was taken to look at the underlying aerosol as close to the cloud base as possible without including any of the cloud fields. In this case, we see that at least 100-150 meters gap was necessary to guarantee that no cloud contamination was made in the aerosol estimates. The sensitivity of this measurement to the vertical layer chosen can be seen in figure 7&8 where both the ACI coefficient and correlation (for both DC and NN approaches) are plotted as a function of the aerosol reference height.

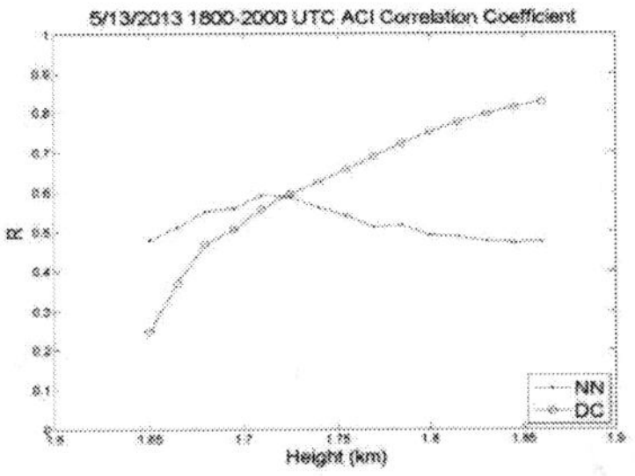

Figure 7. Variation of correlation coefficient (R) of ACI as a function of the aerosol loading (height).

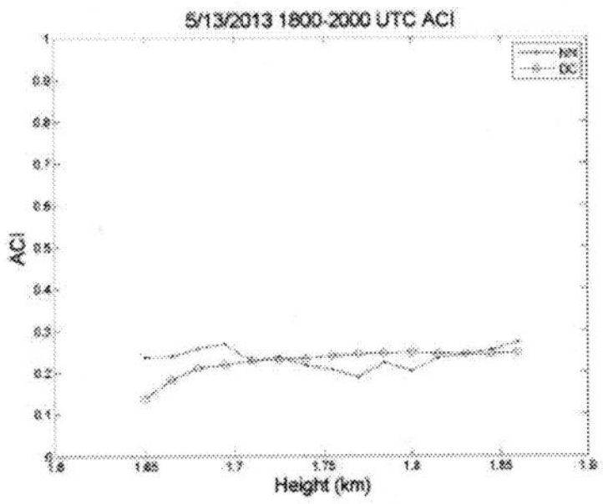

Figure 8. ACI within the theoretical limit as a function of height (aerosol range).

For using elastic backscattering at 1064 nm, we first explore the relationship between lidar derived aerosol extinction and number concentration from AERONET. Figure 9&10 show that behavior between number concentration against extinction and backscattering. To see the results using backscatter at 1064nm, we re-examine the two cases for both NN and DC cloud droplet effective radius retrievals using extinction and backscattering. Table 1 represent correlation coefficient comparison of extinction (α) and backscattering (β) using neural network and dual channel retrieval algorithm.

Retrieval method	Morning		Evening	
	α	β	α	β
NN	0.83	0.99	0.47	0.62
DC	0.79	0.79	0.91	0.98

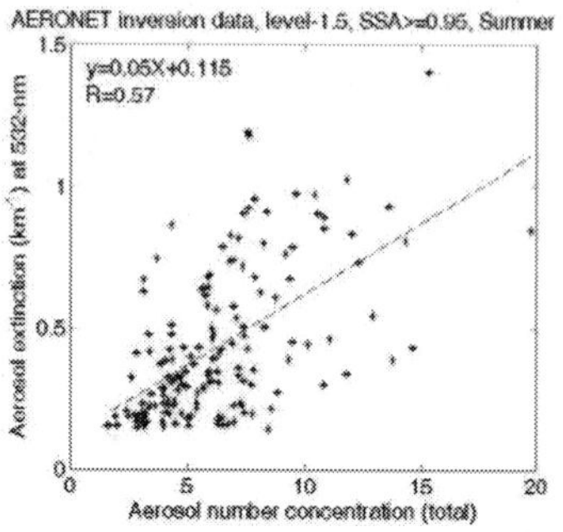

Figure 9. Aerosol extinction and number concentration.

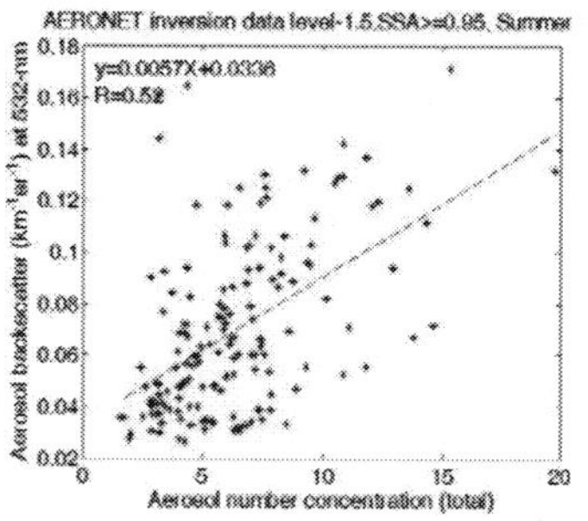

Figure 10. Aerosol backscattering and number concentration.

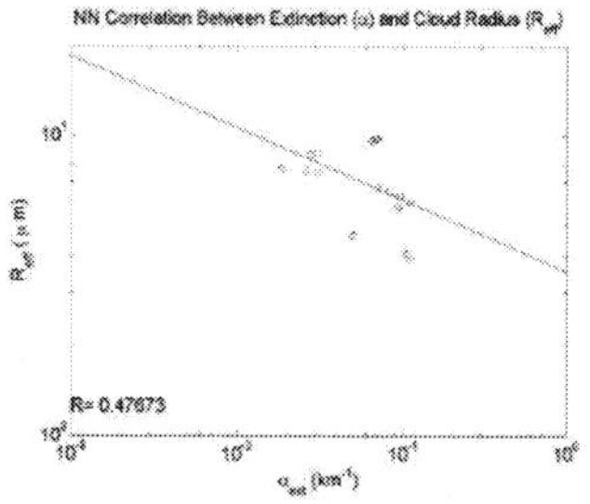

Figure 11. NN retrieval (1800-2000 case) of cloud droplet effective radius changed by aerosol extinction coefficient

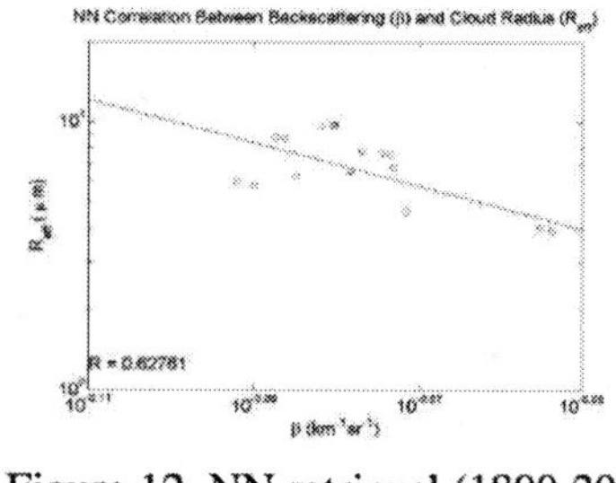

Figure 12. NN retrieval (1800-2000 case) of cloud droplet effective radius modified by aerosol.

4. CONCLUSIONS

In this paper, we have made a preliminary investigation of the potential of quantifying and generally observing Aerosol Cloud Interactions. In particular, the LWP retrieval using DC agrees well with the NN when both data are available, but allows for more retrievals which improves the overall statistics when using allowing for a larger number of data retrievals. We also demonstrated that the observation of the ACI effect is very sensitive to the distance from the cloud base. We found the ACI correlation will persist for as much as 200 meters below the cloud.

Finally, we explored the potential of using the backscattering coefficient at 1064 nm as a proxy for fine mode particulates. Then, for all the cases where the extinction was shown to be a useful indicator of ACI, the backscatter approach was used and we found that the backscatter signal has a generally better correlation.

ACKNOWLEDGEMENT

This work is partially supported by the NOAA CREST Center under contract NOAA #NA17AE1625.

REFERENCES

[1] Feingold G., Eberhard W.L., Veron D.E., Previdi M., "First measurements of the Twomey indirect effect using ground-based remote sensors", *Geophys. Res. Lett.,* (2003).

[2] Min Q., and M. Duan, "Simultaneously retrieving cloud optical depth and effective radius for optically thin clouds", *J. Geophys. Res.,* **110**, D21201 (2005).

[3] Madhavan B.L., He Y., Wu Y., Gross B., Moshary F., and Ahmed S., "Development of a ground based remote sensing approach for direct evaluation of aerosol-cloud interaction", *Atmosphere*, **3**, 468-494 (2012).

[4] Liljegren J.C., E.E. Clothiaux, G.G. Mace, S. Kato, and X. Dong, "A new retrieval for cloud liquid water path using a ground-based microwave radiometer and measurements of cloud temperature", *J. Geophys. Res.,* **10** (D13), 14485-14500, July 16 (2001).

EPJ Web of Conferences **119**, 17001 (2016) DOI: 10.1051/epjconf/201611917001
ILRC 27

FIVE YEARS LIDAR RESEARCH ON BOARD THE FACILITY FOR AIRBORNE ATMOSPHERIC MEASUREMENTS (FAAM)

Franco Marenco

Met Office, Observations Based Research, Exeter EX1 3PB, United Kingdom.

Email: franco.marenco@metoffice.gov.uk

ABSTRACT

I will present a summary of the results obtained with the backscatter lidar on-board the FAAM research aircraft. This simple instrument has been used in several campaigns, and has contributed successfully to the characterization of volcanic ash, mineral dust, biomass burning aerosols, clouds, and the boundary layer structure. Its datasets have been used in many applications, from numerical weather predictions to the validation of satellite remote sensing.

1. INTRODUCTION

The Facility for Airborne Atmospheric Measurements (FAAM) BAe-146 research aircraft (www.faam.ac.uk) is a major asset for the British scientific community. Its purpose is to provide independent observations of atmospheric fields and to permit simultaneous and complementary observations from a range of in situ and remote sensing instruments. The flying programme is primarily aimed at improving the Met Office weather prediction model and its parametrisations, and to verify/validate satellite retrievals. The aircraft may also be occasionally used operationally, e.g. to face civil contingencies.

The aircraft is equipped with a simple ALS450 backscatter lidar manufactured by Leosphere, operating at 355 nm and pointing at nadir. The success of this simple instrument relies on several factors: (1) it is complemented by full suite of measurements on board FAAM, which provide in situ and passive remote sensing information; (2) the campaigns of the FAAM aircraft are supported by multi-disciplinary teams capable of exploiting the data for a wide spectrum of scientific objectives; and (3) the instrument's ease of operation and relative robustness allow intensive use for campaigns in different parts of the world, targeting different objectives. In this presentation I shall present some of the major applications of the instrument.

2. AEROSOL APPLICATIONS

The airborne lidar has been extensively used for aerosol measurements, and its data have been combined with in situ measurements obtained by nephelometer and optical particle counters.

In 2010, the lidar provided a detailed dataset of spatially-resolved volcanic ash concentrations [1]. Thanks to the on-board in situ optical particle counters and to passive remote sensing with the ARIES infrared spectrometer, a closure of the observations has been possible [2,3].

In 2011 and 2012 the aircraft has taken part in the Fennec experiment, targeted at studying the meteorology of the Saharan Heat Low region [4]. The lidar measurements have been very successful at mapping the turbulent structure of the Saharan atmospheric boundary layer[5].

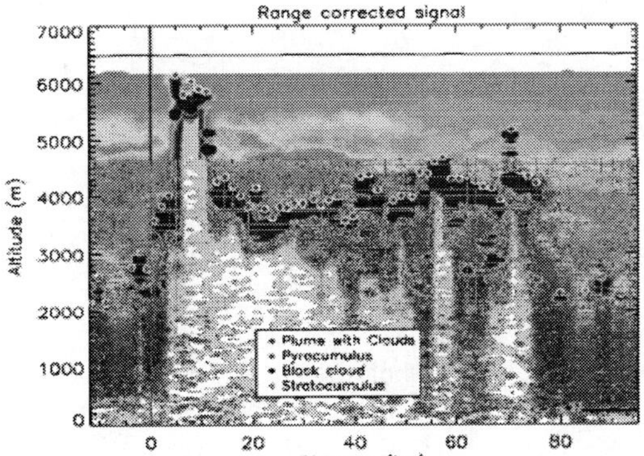

Figure 1. Mapping single-fire smoke plumes by lidar in the Amazon.

Later in 2012, a major aerosol campaign has been the South American Biomass Burning Analysis (SAMMBA). Aerosols in the Brazilian rainforest have been extensively studied during the biomass burning season using ground-based [6] and airborne measurements. The airborne lidar has produced the detailed mapping of the regional haze and a thorough analysis of its vertical distribution. The geometry of single fire plumes has also been studied (see Fig. 1).

3. CLOUD APPLICATIONS

The airborne lidar has been used in research flights around the United Kingdom, targeting the properties of cirrus. Here, the lidar quantitative measurements, together with other observations, helped validating a microphysical model of the cloud properties, capable of predicting the scattering properties across the electromagnetic spectrum from UV to radar frequencies (see Fig. 2 and Ref. [7]) and the relationship between the phase function and the atmospheric state [8].

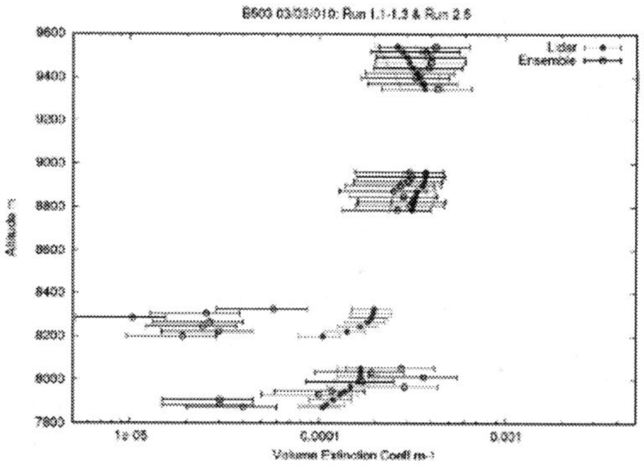

Figure 2. Cirrus extinction coefficient profile, determined by lidar (red) and a microphysical model (black).

A simple threshold-based approach to cloud detection and the measurement of cloud top height has been established [9]. It has enabled the mapping of the evolution of stratus and stratocumulus, when advected over land. It has been successfully exploited in a comprehensive study, aimed at understanding the strengths and weaknesses of the Met Office numerical weather prediction model [10].

For a study on cloud cover over the Sahara desert, the lidar cloud-detection algorithm has been reinforced using co-located broadband passive infrared measurements, yielding a robust product. Along-track cloud maps at ~300 m resolution have been established during 25 research flights (see Fig. 3), with the purpose of verifying model predictions and Meteosat observations (analysis is ongoing).

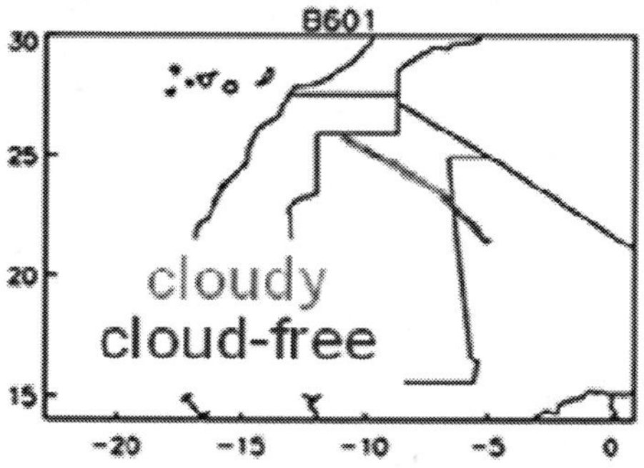

Figure 3. Cloud map in the Sahara, established by lidar and infrared radiometer.

4. THE ICE-D CAMPAIGN

In August 2015, the FAAM aircraft will be detached to Cape Verde to take part in the ICE-D campaign. The campaign objectives are focused on the ice nucleation properties of mineral dust, aerosol mapping for the assessment of dust predictions, and the verification of dust remote sensing with Meteosat and CALIPSO.

Two additional components of the campaign will be the validation of aerosol ground-based remote sensing retrievals (pending funding application). The SAVEX-D component will assess the retrievals of size-distribution, single-scattering albedo, and refractive index obtained from AERONET and SKYRAD sunphotometer retrievals. The ACCURATE component will the evaluation of the GARRLiC algorithm using multi-wavelength Raman lidar and AERONET [11].

DOI: 10.1051/epjconf/201611917001

ACKNOWLEDGEMENTS

Airborne data was obtained using the BAe-146 Atmospheric Research Aircraft flown by Directflight Ltd. and managed by the Facility for Airborne Atmospheric Measurements (FAAM), which a joint entity of the Natural Environment Research Council (NERC) and the Met Office. Fennec was funded by NERC (grant NE/G017166/). SAMBBA was funded by the Met Office and NERC (grant NE/J009822/1). This research has been carried partly at the University of Leeds, within the framework of the Met Office Academic Partnership scheme.

REFERENCES

[1] Marenco et al, Airborne lidar observations of the 2010 Eyjafjallajökull volcanic ash plume, J. Geophys. Res. 116, 10.1029/2011JD016396, 2011.

[2] Turnbull et al, A case study of observations of volcanic ash from the Eyjafjallajökull eruption: 1. In situ airborne observation, J. Geophys. Res. 117, 10.1029/2011JD016688, 2012.

[3] Newman et al, A case study of observations of volcanic ash from the Eyjafjallajökull eruption: 2. Airborne and satellite radiative measurements, J. Geophys. Res. 117, 10.1029/2011JD016780, 2012.

[4] Ryder et al, Advances in understanding mineral dust and boundary layer processes over the Sahara from Fennec aircraft observations, Atmos. Chem. Phys. Discuss. 15, 199-290, 2015.

[5] Garcia-Carreras et al, Turbulent structure and diurnal growth of the Saharan atmospheric boundary layer, J. Atmos. Sci., 10.1175/JAS-D-13-0384.1, 2014.

[6] Brito et al, Ground-based aerosol characterization during the South American Biomass Burning Analysis (SAMBBA) field experiment, Atmos. Chem. Phys. 14, 12069-12083, 2014.

[7] Baran et al, A self-consistent scattering model for cirrus. II: The high- and low-frequencies, Q. J. Roy. Met. Soc. 140, 1039-1057, 2014.

[8] Baran et al, On the relationship between the scattering phase function of cirrus and the atmospheric state, Atmos. Chem. Phys. 15, 1105-1127, 2015.

[9] Allen et al, Atmospheric composition and thermodynamic retrievals from the ARIES airborne TIR-FTS system -- Part 2: Validation and results from aircraft campaigns, Atmos. Meas. Tech. 7, 4401-4416, 2014.

[10] Osborne et al, Evolution of stratocumulus over land: comparison of ground and aircraft observations with numerical weather prediction simulations, Bound. Lay Meteorol. 153, 165-193, 2014.

[11] Wagner et al, Evaluation of the Lidar/Radiometer Inversion Code (LIRIC) to determine microphysical properties of volcanic and desert dust, Atmos. Meas. Tech. 6, 1707-1724, 2013.

IMPLEMENTATION OF ROTATIONAL RAMAN CHANNEL IN MULTIWAVELENGTH AEROSOL LIDAR TO IMPROVE MEASUREMENTS OF PARTICLE EXTINCTION AND BACKSCATTERING AT 532 NM

Igor Veselovskii[1], David N. Whiteman[2], Michael Korenskiy[1], A.Suvorina[1], Daniel Perez-Ramirez[2,3]

[1]*Physics Instrumentation Center, Troitsk, Moscow Region, 142190, Russia, E-mail:*
E-mail: igorv@pic.troitsk.ru
[2]*Mesoscale Atmospheric Processes Laboratory, NASA Goddard Space Flight Center, Greenbelt, MD 20771, USA,*
E-mail: david.n.whiteman@nasa.gov
[3]*Applied Physics Department, University of Granada, 18071, Granada, Spain.*

ABSTRACT

We describe a practical implementation of rotational Raman (RR) measurements in an existing Mie-Raman lidar to obtain measurements of aerosol extinction and backscattering at 532 nm. A 2.3 nm width interference filter was used to select a spectral range characterized by low temperature sensitivity within the anti-Stokes branch of the RR spectrum. Simulations demonstrate that the temperature dependence of the scattering cross section does not exceed 1.0% in the 230-300K range making accurate correction for this dependence quite easy. With this upgrade, the NASA/GSFC multiwavelength Raman lidar has demonstrated useful α_{532} measurements and was used for regular observations. Examples of lidar measurements and inversion of optical data to the particle microphysics will be given in presentation.

1. INTRODUCTION

Mie-Raman lidars acquiring elastic backscatters simultaneously with nitrogen vibrational Raman signal at 387 nm and 608 nm are widely used for independent mesurements of particle backscattering (β) and extinction (α) coefficients. While high signal-to-noise measurements of nitrogen vibrational scattering at 387 nm during both daytime and nighttime have been possible, the use of vibrational Raman scattering at 608 nm has posed certain problems. First of all, cross section of vibrational Raman scattering is almost three orders lower comparing to Raleigh one, which limits range and temporal resolution of the measurements. Another issue related to Raman nitrogen measurements at both 387 nm and 608 nm is the significant wavelength shift of the nitrogen Raman component with respect to the laser line. As a result, information about the spectral dependence of aerosol extinction is needed when α and β are computed [1], which can be an additional source of uncertainty. A well-known solution for both of these issues is the use of High Spectral Resolution Lidar (HSRL). The recent results provided by the NASA/LaRC airborne lidar demonstrate the possibility of reliable HSRL operation at both 355 nm and 532 nm [2]. However such lidars, due to their complexity and cost, are still unique instruments. At the same time, there are numerous existing Raman lidars which can potentially be upgraded to improve Raman measurements at 608 nm. One of the ways this can be accomplished is to use pure rotational Raman (RR) scattering instead vibrational scattering. The small spectral separation of the elastic and RR components of the signal implies that aerosol backscatter and extinction can be obtained without information about the Ångström exponent. Furthermore, the backscatter cross section of anti-Stokes component of pure RR nitrogen scattering exceeds that of vibrational scattering by approximately a factor of 15-20, when all lines for nitrogen and oxygen are integrated [3].

Pure rotational Raman scattering is widely used in laser remote sensing for the measurement of atmospheric temperature [e.g. 4,5], where the pure RR and elastic signals are generally separated by the use of narrow-band interference filters or diffraction-grating- spectrometers. Rotational Raman scattering can also be used for the particle extinction and backscattering calculation. However, an important consideration in the use of RR measurements for calculating α and β is the temperature dependence of the RR spectra of N_2 and O_2 molecules. This issue in principle can be resolved by using the sum of lidar returns from the Stokes and anti-Stokes branches, which is essentially temperature independent. However, measuring all Stokes and anti-Stokes lines with equal efficiency is technologically difficult. The alternate approach considered in this paper is to select a portion of the RR spectrum characterized by a low temperature sensitivity inside one of the spectral branches. Such an approach simplifies the implementation of a RR channel in conventional multiwavelength Raman lidar systems without significant modification of the receiver design. In the approach presented here, selection of the RR lines is done by use of a wide-band interference filter which transmits the desired portion of the anti-Stokes branch. Centering the filter transmission band near the temperature insensitive lines of the RR spectrum decreases the temperature sensitivity of the RR scattering cross section to a level appropriate for tropospheric measurements.

2. NUMERICAL SIMULATIONS

Fig.1 shows RR spectra of nitrogen and oxygen weighted with corresponding number density fractions at T=270 K for a laser wavelength of 532.12 nm. In our implementation, the filter transmitted only portions of the anti-Stokes branch in order to have the possibility to shift the filter transmission band toward shorter wavelengths by filter tilting, if necessary. The differential cross section of RR scattering can be calculated as:

$$\sigma^{RR} = \eta_{N_2} \sum_i \xi_R(\lambda_i)\sigma^{RR}_{N_2}(\lambda_i) + \eta_{O_2} \sum_k \xi_R(\lambda_k)\sigma^{RR}_{O_2}(\lambda_k) \quad (1)$$

Here $\sigma^{RR}_{N_2}$, $\sigma^{RR}_{O_2}$, are the differential scattering cross sections of nitrogen and oxygen, $\xi_R(\lambda_i)$ are the filter transmittances of the Raman channel at wavelengths λ_i corresponding the lines of RR spectrum, η_{N_2}, η_{O_2} are the fractions of nitrogen and oxygen concentration.

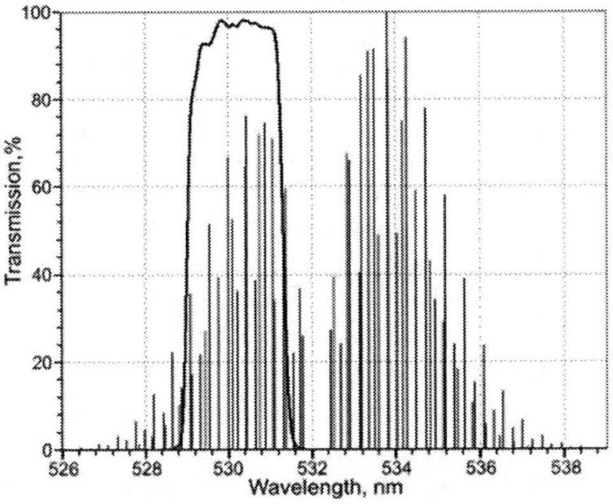

Fig.1. Spectrum of RR scattering by nitrogen (blue) and oxygen (red) for T=300 K. Black line shows transmittance of the interference filter used in the experiment. Laser line is at 532.12 nm.

The frequency shift of rotational lines with respect to the frequency of the laser radiation is small compared with vibrational scattering, so in practice the spectral dependence of particle extinction can be neglected. The intensities of the individual lines in the RR spectrum have a strong dependence on temperature, so the scattering cross section is temperature (and thus range) dependent. For RR the standard expressions for calculation of extinction and backscattering coefficients [1] is modified as:

$$\alpha^a(z) = \frac{1}{2}\frac{d}{dz}\ln\left[\frac{N(z)}{P_R(z)z^2}\right] - \alpha^m(z) + \Delta\alpha^T \quad (2)$$

$$\beta^a(z) = -\beta^m(z) + \beta^m(z_0)\cdot\frac{P_R(z_0)P_L(z)N(z)}{P_L(z_0)P_R(z)N(z_0)}F(z) \quad (3)$$

Where $P_{L,R}(z)$ is the power of elastic and Raman backscatters, *N(z)* is the density of air molecules. The reference height z_0 is chosen in the free troposphere where $\beta^a_L(z_0) << \beta^m_L(z_0)$. Indices "a" and "m" correspond to aerosol and molecular components of backscattering and extinction coefficients.

$\beta^T = \beta^a + \beta^m$, $\alpha^T = \alpha^a + \alpha^m$

The expression for extinction (2) contains the term $\Delta\alpha^T = \frac{1}{2}\frac{d}{dz}\ln\left[\sigma^{RR}(z)\right]$ arising from the height variation of atmospheric temperature and, as a result, variation of the scattering cross section. This term can be significant in the presence of strong temperature gradients.

The expression (3) for backscattering coefficient contains the factor $F(z) = \frac{\sigma_{RR}(z)}{\sigma_{RR}(z_0)}$ responsible for the temperature dependence of β^a, arising from temperature variation from T_0 to T at heights z_0 and z respectively. The error in backscattering ε^T_β due to neglecting of this temperature dependence (*F*=1) can be estimated from (3) as:

$$\varepsilon^T_\beta = \frac{\beta^a - \beta^a_{F=1}}{\beta^a} = \frac{\beta^T}{\beta^T - \beta^m}\frac{F-1}{F} = \frac{R}{R-1}\frac{F-1}{F} \quad (4)$$

where R is the scattering ratio $\frac{\beta^T}{\beta^m}$. For $\beta^a >> \beta^m$ the error ε^T_β becomes $\frac{F-1}{F}$, while for low aerosol loading there is an enhancement factor $\frac{R}{R-1}$.

To quantify the possible uncertainties in the evaluation of backscattering and extinction coefficients arising from the temperature dependence of σ^{RR}, numerical simulation studies were performed. The RR line with least temperature sensitivity is in the vicinity 530.4 nm (J=9) for O_2 and in vicinity of 530.2 nm (J=7) for N_2, so centering the filter transmission band near these values will minimize the sensitivity of σ^{RR} to changes in temperature. When choosing filter parameters the desire is to maximize the cross section σ^{RR} by including only the strongest rotational lines since unnecessarily wide transmission band will lead to an increase of the sky background noise and an overall decrease in signal to noise under daytime conditions.

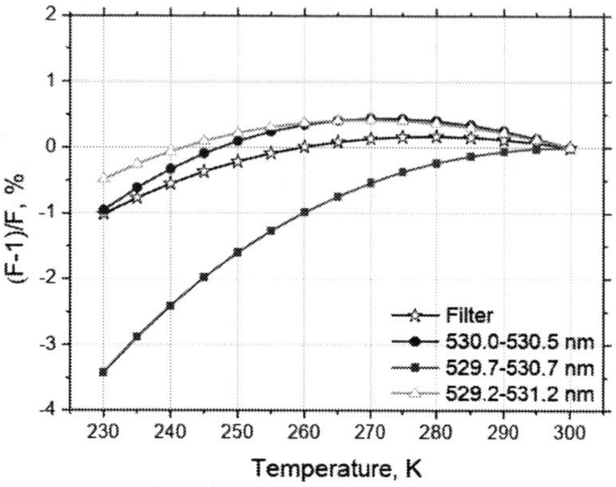

Fig.2. The relative change of RR scattering cross section (F-1)/F as a function of temperature for spectral range 530.0-530.5 nm, 529.7-530.7 nm, 529.2-531.2 nm. Open stars show results for the interference filter used in the experiment.

Fig.2 shows the change of $\frac{F-1}{F}$ as a function of temperature for the spectral intervals of Raman spectrum 530.0-530.5 nm, 529.7 - 530.7 nm and 529.2-531.2 nm. Computations were performed with T varying in the range 230-300K where the reference temperature T_0 was taken to be T_0=300K. For all intervals the relative change of RR scattering cross section $\frac{F-1}{F}=\frac{\sigma^{RR}(T)-\sigma^{RR}(T_0)}{\sigma^{RR}(T)}$ does not exceed 3.5%. The ratios of $\sigma^{RR}(T_0)$ to the total cross section of anti-Stokes branch for these three intervals are 0.2, 0.37 and 0.7. Of these intervals, that of 529.2-531.2 nm is the most attractive, as it contains approximately 70% of the Raman signal provided by the entire anti-Stokes band and possesses a small temperature dependence (the variation of σ^{RR} is below 1.0%). Based on these considerations, we chose this interval as the target for the interference filter design.

The transmission spectrum of the interference filter used in our experiments, taken from the manufacturer's data sheet (Alluxa, CA, USA), is shown in fig.1. The filter has transmission greater than 95% at the peak with bandwidth of approximately 2.3 nm FWHM. The suppression of the elastic scattering at 532.12 nm is performed with OD>4. The variation of $\frac{F-1}{F}$ for this filter in 230-300 K temperature range doesn't exceed 1%, so in many practical cases F=1 in (3) can be used.

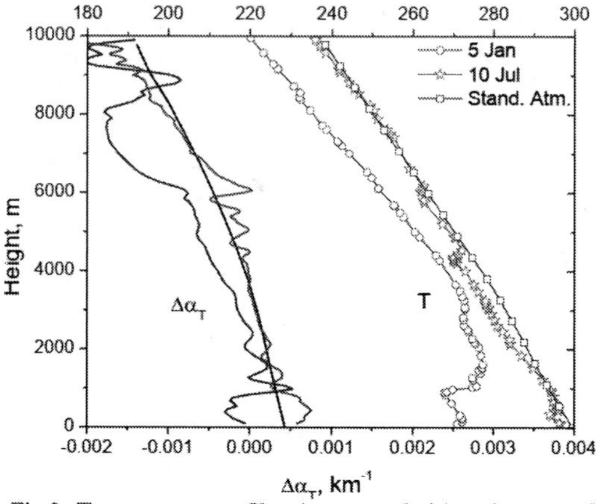

Fig.3. Temperature profiles (open symbols) and errors of extinction coefficient calculation $\Delta\alpha_T$ due to height variation of temperature (solid lines). Computations were performed for US Standard Atmosphere temperature profile (black) and for temperature measured by radiosonde on 5 January (blue) and 10 July 2013 (red).

The error of the extinction coefficient calculation $\Delta\alpha_T$ induced by the temperature height variation depends, as it follows from (2), on the gradient of the temperature profile. To estimate this error we have tested the temperature profiles from the US Standard Atmosphere model and from radiosonde launches performed at Dulles airport, VA in the vicinity of Washington DC. The results of computations are shown in fig.3. To calculate the range derivation from discrete temperature measurements the cubic spline over 7 points was used. For the model temperature profile the decrease of the temperature with height provides a negative contribution to the calculated particle extinction. The error $\Delta\alpha_T$ changes with height but it is below 0.002 km^{-1} for the heights up to 10 km. The same figure shows the results for radiosonde launches on 5 January and 10 July 2013. Corresponding temperature profiles present low-scale oscillations, still for both days $\Delta\alpha_T$<0.002 km^{-1}. Computations performed for days at different seasons have demonstrated that, in general, temperature variations contribute an error in calculated extinction of less than 2% for extinction values of 0.1 km^{-1} and greater.

Using the scattering cross sections provided by Fenner [3] we estimate that the scattered power in the RR channel is a factor of 15 higher than that of vibrational nitrogen Raman channel at 608 nm. Thus, under skylight limited conditions, the RR channel would provide greater signal-to-noise if the width of the filter used for the vibrational Raman measurement is greater than approximately 0.15 nm..

3. MODIFICATION OF GSFC RAMAN LIDAR

A description of the NASA/GSFC multiwavelengh Raman lidar is presented in [6]. To incorporate the RR channel at 530 nm the lidar receiving module was modified. The optical configuration permitted simultaneous measurements of the vibrational and rotational Raman signals. The spectral components at 608 nm and 532 nm were separated by a dichroic mirror. The additional beamsplitter then reflected approximately 10% portion of the 532 signal into the Mie channel with more than 85% being transmitted into the RR channel. Two RR filters each possessing the characteristics shown in fig.1 were used together to provide more than 8 orders of magnitude rejection of the elastic signal. To compare the scattered power in 608 nm and RR 530 nm channels, we made two sequential measurements using the same R1924 PMT for both channels. After correction for difference in quantum efficiency of the photocathode at 608 nm and 530 nm, we estimate the enhancement of scattered power in RR channel compared with the vibrational one to be more than a factor 10.

The initial measurements were taken under cloudy conditions to test the blocking of the Mie scattering by the RR filters. Those tests did not reveal any enhancement of the signal measured in the RR channel due to backscatter from the base of the cloud. Fig.4 shows backscattering and extinction coefficients at 532 nm calculated from vibrational and rotational Raman signals. The maximal difference between backscattering profiles was observed in the maximum of scattering layer at approximately 1750 m height and did not exceed 2%.This figure also clearly demonstrates the improvement of the extinction calculation when the RR signal at 530 nm is used instead of that of vibrational scattering at 608 nm.

4. CONCLUSION

The use of rotational Raman scattering, instead of vibrational, removes the concern due to wavelength scaling, while simultaneously permitting measurements to take advantage of a much larger cross section. We have presented a practical approach to implementing rotational Raman measurements in multiwavelength aerosol lidar using sharp edge interference filters to separate the RR and Mie signals. During the design phase of the development, numerical simulations were performed demonstrating that a measurement of aerosol extinction could be made with low temperature dependence using a portion of the either anti-Stokes of Stokes components. This selection also achieved a signal increase of more than a factor or 10 when compared with vibrational nitrogen. Upgraded this way the NASA/GSFC multi-wavelength Raman lidar has demonstrated improved capabilities for α_{532} measurements which in turn has improved the inversion of $3\beta+2\alpha$ optical data to particle microphysics. The examples of regular lidar observations will be given in presentation.

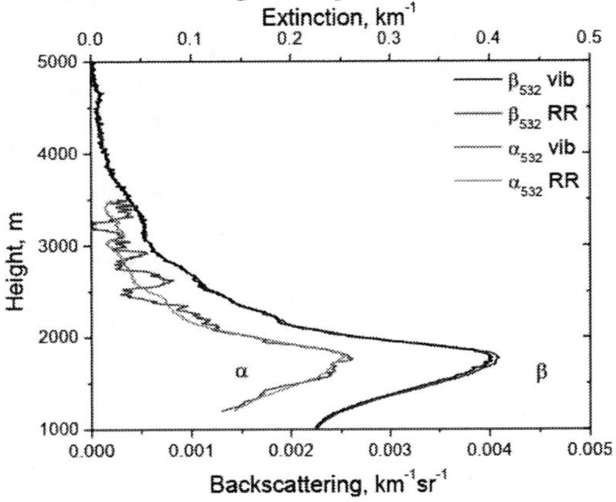

Fig.4. Extinction and backscattering coefficients at 532 nm computed from nitrogen vibrational Raman and from RR signal

References

[1]. Ansmann, A., et al.: Combined Raman elastic-backscatter lidar for vertical profiling of moisture, aerosols extinction, backscatter, and lidar ratio, Appl.Phys.B, *55*, 18-28, 1992.

[2] Müller, D., et al.: Airborne Multiwavelength High Spectral Resolution Lidar (HSRL-2) observations during TCAP 2012: vertical profiles of optical and microphysical properties of a smoke/urban haze plume over the northeastern coast of the US, Atmos. Meas. Tech., 7, 3487–3496, 2014.

[3] Fenner, W. R., et al.: Raman cross section of some simple gases, J.Opt.Soc.Amer. 63, 73-77, 1973.

[4] Behrendt, A. and Reichardt, J.: Atmospheric temperature profiling in the presence of clouds with a pure rotational Raman lidar by use of an interference-filter-based polychromator, Appl. Optics, 39, 1372–1378, 2000.

[5] Arshinov,Y., et al.: Daytime operation of a pure rotational Raman lidar by use of a Fabry–Perot interferometer, Appl.Opt. 44, 3593-3603, 2005.

[6] I. Veselovskii, et al. "Characterization of forest fire smoke event near Washington, D.C. in Summer 2013 with multi-wavelength lidar", Atmos. Chem. Phys. 15, 1647–1660, 2015.

DOI: 10.1051/epjconf/201611917003

DEVELOPMENT OF THREE-WAVELENGTH POLARIZATION-RAMAN LIDAR AND APPLICATION TO SHIPBORNE MEASUREMENTS

Zhangjun Wang [1]*, Libin Du[1], Xianxin Li[1], Bin Zhou[2], Xiangqian Meng[1], Chao Chen[1], Qiaojun Liu[1], Xingtao Liu[1]

[1]*Shandong Academy of Sciences Institute of Oceanographic Instrumentation, Qingdao, 266001, China, *Email: zhangjun.wang@hotmail.com*

[2]*Marine Engineering Prospecting Institute of North China Sea, SOA, Qingdao, 266033, China*

ABSTRACT

A Three-Wavelength Polarization-Raman Lidar (TWPRL) system for aerosol and clouds was developed. This lidar system provides α at 532 and 355 nm, β at 355, 532 and 1064 nm, and σ at 532 nm as well as water vapor content using Raman lidar techniques. The temporal and vertical variation of aerosols and clouds could be determined. We conducted shipborne TWPRL measurements over Yellow Sea of China from August to September in 2014. The derived aerosol optical properties indicate that the developed lidar system worked very well. 24-hour continuous measurements with the shipborne TWPRL during the cruise are presented.

1. INTRODUCTION

Atmospheric aerosols have many effects on global warming and cooling, through altering the radiative properties of the atmosphere [1]. Knowledge of the vertical structure of cloud and aerosol scattering properties or layers from varying climate regimes is fundamental to understand their role in the atmospheric processes as well as on human health and environment.

Lidar measurements of atmospheric aerosol profiles have become important for atmospheric researchers during last decades. Since 2001, The National Institute for Environmental Studies (NIES) has been constructing ground-based lidar networks with automated dual wavelength/polarization Mie-Lidar systems to observe air quality in the Asian region [2]. From June 2006, CALIOP (Cloud-Aerosol Lidar with Orthogonal Polarization) has been measuring global aerosol and cloud vertical distributions continuously with very high spatial resolution [3]. Aerosol optical and microphysical properties by ground-based sunphotometer measurements and aerosol vertical profiles by lidar measurements, performed within AERONET[4] and MPLNET[5], are used to analyze aerosols effects on radiative budget and atmospheric heating rate profiles. Within the European Aerosol Lidar Research Network (EARLINET) [6], now part of the Aerosols, Clouds, and Trace gases Research InfraStructure Network (ACTRIS), more than 30 lidar systems were used to monitor the aerosol vertical distribution over Europe.

In addition, ships provide a mobile platform in oceans allowing the cover of large geographical regions. Single-wavelength lidars were used during field campaign like INDEOX[7], ACE2[8], ACE-Asia[9]. Dual-wavelength Mie/polarization lidar measured the aerosol in the northern west Pacific region [10]. Kanitz et al. [11] deployed a multiwavelength Raman lidar (Polly^{XT}) with a polarization-sensitive receiver channel on RV Polarsten. The cruise went from Bremerhaven, Germany to Punta Aernas, Chile from 16 October 2009 to 25 November 2009.

A Three-Wavelength Polarization-Raman Lidar (TWPRL) was designed and developed in Shandong Academy of Sciences Institute of Oceanographic Instrumentation (SDIOI) for aerosol and clouds measurement. The lidar system was tested and performed aboard the research vessel Xiangyanghong No. 8 over the Yellow Sea and Jiaozhou Bay of China from 29 Aug. to 2 Sept. 2014. The data presented here give information on the vertical distribution and optical properties of the Jiaozhou Bay and Yellow Sea of China during the lidar campaign. 24-hour continuous measurements with the shipborne TWPRL during the cruise are also presented.

2. LIDAR SYSTEM

The configuration and specifications of the

TWPRL system are presented at Figure 1 and Table 1. This is a typical lidar instrument containing Nd:YAG laser as a transmitter, Schmidt-Cassegrain telescope to collect scattered laser radiation, dichroic beam splitter to split beam of light from the telescope and set of photodetectors connected to the PC computer.

The lidar is conceived to perform vertical measurements. The transmitter emits laser pulses at 355, 532, and 1064 nm with a repetition rate of 20 Hz. All three laser beams are aligned on two optical axes: one is the 355 nm channel; another one is the 532 and 1064 nm channel. The outgoing laser beams are expanded by a 5 X beam expander to reduce the divergence angle, and then directed into the sky through a set of mirrors.

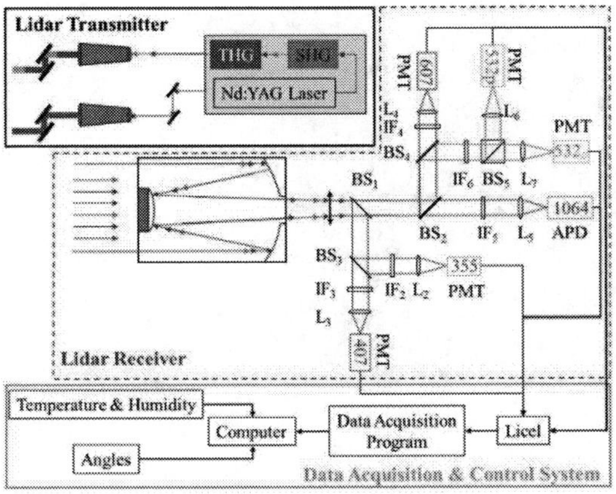

Figure 1. The three-wavelength Polarization-Raman lidar system configuration.

The laser beam is coupled from side into the field-of-view (FOV) of the receiver telescope. Behind the pinhole, a collimating lens (L) collimates and transmits the light to six channels. The beam is separated according to wavelength with beam splitters and transmitted to photomultipliers. The Dichroic Beam Splitters (DBSs) were selected to reflect light with shorter wavelengths and transmit light with longer wavelengths because it is easier to produce DBS with higher transmission values at the longer wavelengths than at the shorter wavelengths. The signals are acquired using Licel transient recorder with 6 modules. Narrow band interference filter is used to suppress the sky background radiation in order to increase the signal-to-noise ratio (SNR) in each channel.

Table 1. Specifications of the TWPRL system

Size	1.7 m (H)*1.25 m (W)*0.7 m (D) (Container: 2 m *2 m*2 m)
Weight	300 kg (without container)
Transmitter	
Laser type	Flash-Lamp pumped Nd:YAG laser
Wavelength	355, 532 and 1064 nm
Single Pulse Energy	80 mJ@355 nm, 50 mJ@532 nm, 100 mJ@1064 nm
Repetition Rate	20 Hz
Divergence Angle	0.15 mrad (using a 5x expander)
Receiver	
Telescope	Schmidt-Cassegrain, D=30 cm
Field of View	0.3 mrad
Interference Filter	1 nm (FWHM) for each wavelength
Detectors	APD for 1064 nm
	PMT for 355, 532, 407 and 607 nm
Data Acquisition and Control system	
Data acquisition card	Licel Transient recorder (40 MHz, 12 bit)
Height resolution	3.75 m
Maximum altitude	40 km
Angles of the ship	MTi-G-700 (0.2°)

From the elastically backscattered signals, and signals inelastically (Raman) scattered from nitrogen and water vapor molecules, we determine particle volume backscatter coefficients at 355, 532, and 1064 nm (3β), and particle volume extinction coefficients at 355 nm and 532 nm (2α), depolarization ratio (σ), and water vapor mixing ratio at 407 nm. In order to deploy it on the Xiangyanghong No.8, the lidar system was housed in a constant temperature and constant humidity container.

3. RESULTS

Figure 2 shows the cruise line of the Xiangyanghong No.8 during the campaign, which started on 29 Aug. and ended on 2 Sept. 2014.

The cruises went from the Black Point (120.2967° E, 36.0510° N) at 8:00 to Blue Point (120.5652° E, 35.9601° N) at 10:00 on 30 August 2014. Then we did measurement more than 24 hours in Blue Point. The second cruise started in the Blue Point in the Yellow Sea of China at 18:00 on 31 August 2014 towards the Purple Point in the Jiaozhou Bay and a further day-night observation was made.

The signals are recorded up to 50 km with 3.75 m resolutions and every 1000 pulse. In the analysis, the signals are averaged over 2.5 minute in time and 30 m in range. We conducted 24-hour continuous measurements and fixed position observations with the developed TWPRL during the cruise.

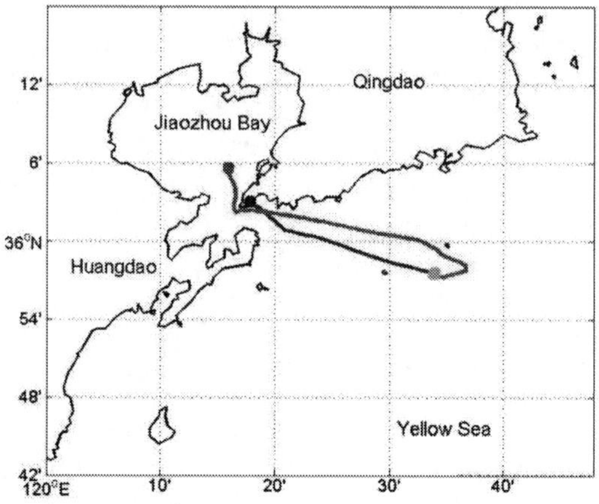

Figure 2. The cruise of the Xiangyanghong No.8

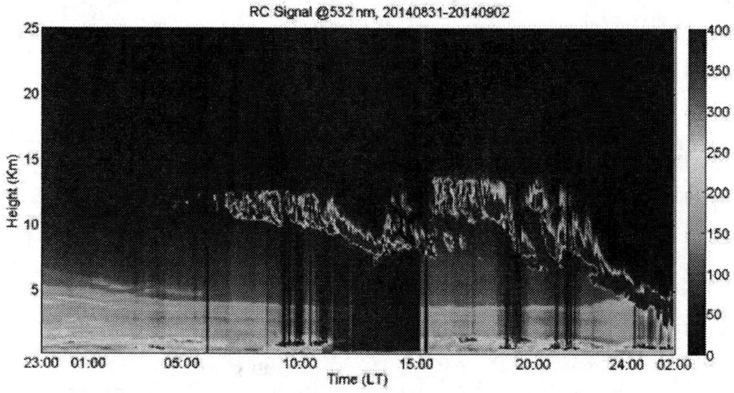

Figure 3. Time-Height display of the range-corrected signal measured in fixed position.

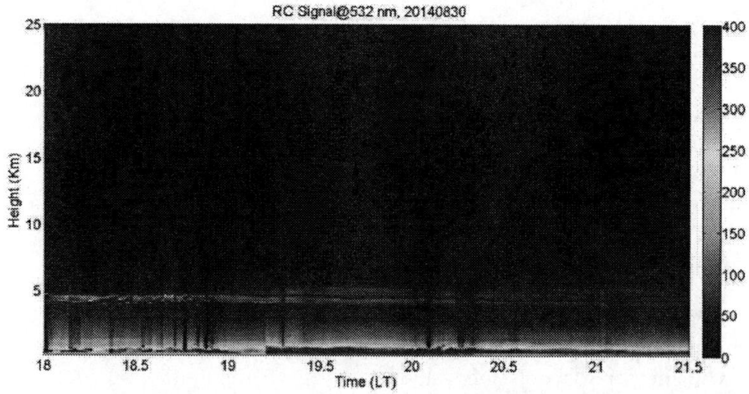

Figure 4. Time-Height display of the range-corrected signal measured along the sea route.

Two examples of the measurement results are depicted in Figure 3 and Figure 4. Figure 3 shows the 27-hour continuous measurements from 23:00 on 31 Aug. to 02:00 on 02 Sept. over the Jiaozhou Bay (120.2649° E, 36.0939° N). The height of clouds decreases from 13 km at 16:00 to 3.5 km at 02:00 on 02 Sept., then it is raining. Figure 4 presents the latitude-longitude-dependent aerosol vertical distribution observed at 532 nm along the sea route from (120.5652° E, 35.9601° N) to (120.2649 °E, 36.0939° N).

4. CONCLUSIONS

A three-wavelength Polarization-Raman lidar was developed and conducted aboard the Xiangyanghong No. 8 from August to September 2014. The lidar system works very well during this campaign. The vertical distribution of marine aerosol over Yellow Sea was measured in fixed position for more than 24 hours. Furthermore, it is planning to do another two lidar campaign in May 2015 and September 2015 over the Yellow Sea of China.

ACKNOWLEDGEMENT

The study is funded by the Public science and technology research funds projects of ocean under grant No. 201205036, the International Science & Technology Cooperation Program of China under grant No. 2011DRF20090, the Progress Program of Science and Technology of Sh andong with grant No. 2014GSF117006, and grant No. 13-1-4-183-jch supported by Qingdao Municipal Science and Technology Commission.

REFERENCES

[1]Leon J. F., Chazette P., Pelon J., Dulac F., and Randriamiarisoa H., 2002: Aerosol direct radiative impact over the INDOEX area based on passive and active remote sensing, *Journal of Geophysical Research*, **107 (D19)**, 8006.

[2]Shimizu A., Sugimoto N., Matsui I., Arao K., Uno I., Murayama T., Kagawa N., Aoki K., Uchiyama A., Yamazaki A., 2004: Continuous observations of Asian dust and other aerosols by polarization lidars in China and Japan during ACE-Asia, *Journal of Geophysical Research: Atmospheres*, **109 (D19)**, S17.

[3]Winker D. M., Hunt W. H., and McGill M. J., 2007: Initial performance assessment of CALIOP, *Geophysical Research Letters*, **34(19)**, L19803.

[4]Holben B. N., Eck T. F., Slutsker I., Tanre D., Buis J. P., Setzer A., Vermote E., Reagan J. A., Kaufman Y. J., Nakajima J., Lavenu F., Jankowiak I., and Smirnov A., 1998: AERONET-A federate instrument network and data archive for aerosol characterization, *Remote Sens. Environment.*, **6**, 1-16.

[5]Welton E. J., Berkoff, T. A., et al., 2010: The NASA micropulse lidar network (MPLNET): summary of the last 10 years, current status, and future plans, *25th International Laser Radar Conference*, 875-878.

[6]Matthais V., et al., 2004: Aerosol lidar intercomparison in the framework of the EARLINET project. 1, Instruments, *Appl. Opt.*, **43(4)**, 961-976.

[7]Welton E. J., Voss. K. J., Quinn. P. K., Flatau. P. J., Markowicz K., Campbell. J. R., Spinhirne. J. D., Gordon H. R., Johnson J. E., 2002: Measurements of aerosol vertical profiles and optical properties during INDOEX 1999 using micropulse lidars, *Journal of Geophysical Research*, **107(D19)**, 8019.

[8]Welton E. J., Voss K. J., Gordon H. R., Maring H., Smirnov A., Holben B., Schmid B., Livingston J. M., Russell P. B., Durkee P. A., Formenti P., and Andreae M. O., 2000: Ground-based lidar measurements of aerosols during ACE-2: instrument description, results, and comparisons with other ground-based and airborne measurements, *Tellus B.*, **52**, 636-651.

[9]Schmid B., Hegg D. A., Wang J., Bates D., Redemann J., Russell P. B., Livingston J. M., Jonsson H. H., Welton E. J., Seinfeld J. H., Flagan R. C., Covert D. S., Dubovik O., and Jefferson A., 2003: Column closure studies of lower tropospheric aerosol and water vapor during ACE-Asia using airborne Sun photometer and airborne in situ and ship-based lidar measurements, *Journal of Geophysical Research: Atmospheres*, **108(D23)**, 8656.

[10]Sugimoto N., Matsui I., Liu Z., Shimizu A., Asai K., Yoneyama K., and Katsumata M., 2001: Latitudinal distribution of aerosols and clouds in the western Pacific observed with a lidar on board the research vessel Mirai, *Geophys. Res. Lett.*, **28**, 4187-4190.

[11]Kanitz. T., Ansmann A., Englemann R., and Althausen D., 2013: North-south cross sections of the vertical aerosol distribution over the Atlantic Ocean from multiwavelength Raman/polarization lidar during Polarsten cruises, *Journal of Geophysical Research: Atmospheres*, **118**, 2643-2655.

DOI: 10.1051/epjconf/201611917004

PERFORMANCE ASSESSMENT OF MOBILE RAYLEIGH DOPPLER LIDARS FOR MIDDLE ATMOSPHERE RESEARCH

Yuli Han[1], Ruocan Zhao[1*], Dongsong Sun[1, 2]

[1] *CAS Key Laboratory of Geospace Environment, Department of Geophysics and Planetary Sciences, University of Science and Technology of China, Hefei, 230026, China, *Email: canlan@mail.ustc.edu.cn*

[2] *Mengcheng National Geophysical Observatory, School of Earth and Space Sciences, University of Science and Technology of China, Hefei, 230026, China*

ABSTRACT

Recently, two sets of mobile Rayleigh Doppler lidars were implemented in University of Science and Technology of China (USTC) for atmospheric gravity waves research. One of them works in a step stare scanning mode with azimuths corresponding to four cardinal points, while the other one consists of three fixed subassemblies: one points to the zenith and the two others are titled at 30° from the zenith with east and north pointings, respectively. They both operate at eye-safe wavelength 354.7 nm and adopt a triple Fabry-Perot interferometer (FPI) as frequency discriminator. In order to assess the performance of the Doppler lidars, comparison experiments were performed between them. Perhaps, it is the first time to make direct comparison between scanning and non-scanning Rayleigh Doppler lidars.

1. INTRODUCTION

Study of atmospheric gravity waves has been an intense activity in recent years because of their myriad effects and contributions to atmospheric circulation, structure, and variability. In the middle atmosphere, between ~10 and 110 km altitudes, one of the most intriguing aspects of the atmosphere is the role gravity waves play in influencing its state by transporting energy and momentum between widely separated regions and generating turbulence [1]. In situ, ground-based, and space-based observation instruments have contributed great to our knowledge of gravity wave scales, amplitudes, vertical propagation and seasonal and geographic variability. Now in USTC, we have developed Mie Doppler lidar, mobile Rayleigh Doppler lidar and sodium lidar, providing profiles of wind and temperature from ground up to lower thermosphere (with wind sounding gap between 60 and 80 km) with 0.2-1 km vertical resolution and a precision of 1-10 m/s depending on altitude. Combined observations using these instruments would greatly advance our knowledge of middle atmospheric dynamics and gravity waves. In this paper, we present our first wind field measurements by combination of two sets of mobile Rayleigh Doppler lidar, as shown in Fig. 1. Meanwhile, the comparison of two lidars' data is performed to assess the reliability and the stability of them.

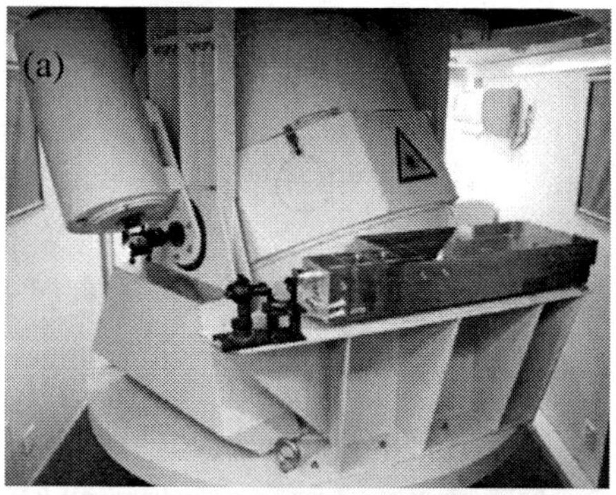

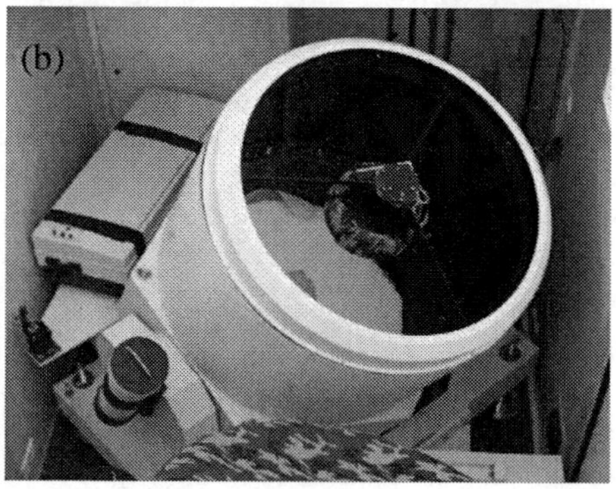

Fig. 1 Photos of the scanning (a) and non-scanning (b) mobile Rayleigh Doppler lidar system in USTC.

2. Mobile Wind Lidar Instruments

Two sets of mobile Rayleigh Doppler lidars are both based on the double-edge technique to realize direct-detection. The principle, optical layout, specifications, and solutions of system-level optical frequency stabilization for the scanning wind lidar is discussed in detail in [2]. The scanning wind lidar works in a step stare scanning mode at a fixed zenith angle of 30° with azimuths corresponding to four cardinal points. Therefore the systematic error due to uncalibrated zero-wind reference can be eliminated by differential measurement of two opposite pointings. However, such superiority cannot be found in another wind lidar system, the non-scanning one. Instead of using opposite pointing to compensate the systematic error, we proposed to determine the zero wind shift by scanning the temperature-stabilized FPI with vertical atmospheric backscatter. A detailed description of the wind lidar system calibration and stabilization is presented in [3, 4].

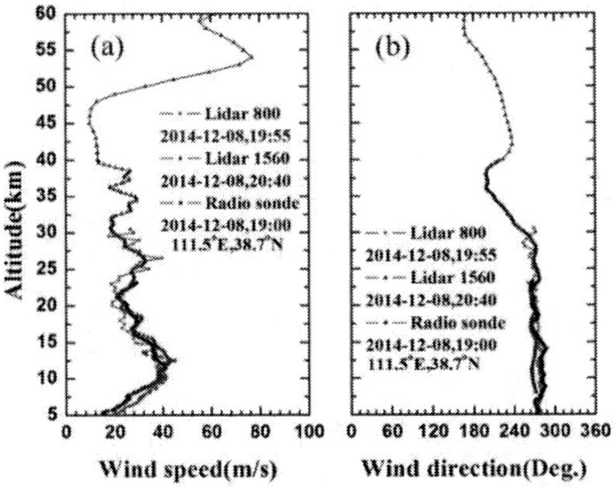

Fig. 2 Wind speed (a) and wind direction (b) of two lidars compared with radiosonde data. Lidar 800 refers to the scanning one and lidar 1560 refers to the non-scanning one.

3. Comparison Experiments

Observations of two lidars were performed from 6:00 pm to 7:00 am on clear weather nights. Radiosonde observations were operated twice at the same place on 7:00 pm and 7:00 am. 53 days of valid wind field data was acquired during three month of observations. Wind profiles of two lidars close to the radiosonde observation time are compared with the radiosonde data, as shown in Fig. 2. An accumulating time of 30 min is adopted during the wind inversion. The resolution of the scanning lidar and the non-scanning lidar changes from 100m to 500m at 15km and from 200m to 1000m at 40km, respectively. Fig. 2 shows that the profiles of two lidars have good agreement with each other and both have good agreement with the radiosonde. It should be noted that the data is regard as invalid when the SNR (signal to noise ratio) is less than 17. Therefore, the wind profile of the scanning lidar has valid data up to 30km in Fig. 2.

In order to have a further study of the reliability of the data at high altitude (from 30km to 40km) where the signal of the scanning lidar is weak, we adjust the observation directions of two lidars to the same direction and compares the LOS (line of sight) wind velocity of two lidars at the same time, as shown in Fig. 3(a). The LOS wind profiles have good agreement below 35km. However, the difference between two profiles increases as altitude grows above 35km. This difference is caused by the weak signal of the scanning lidar above 35km. Fig. 3(b) shows the SNR of two lidars corresponding to these two LOS wind profiles. The SNR of the non-scanning lidar is greater than 17 below 50km (see the blue dashed line). However the SNR of the scanning lidar is less than 17 above 35km, which is the main reason why the difference between two LOS wind profiles increases above 35km. Meanwhile, six continuous LOS wind profiles of two lidars are plotted together in Fig. 4 to illustrate the reliability of two lidars at different altitude. As shown in this figure, the differences of the profiles increase as SNR decreases and the detection limit of the scanning lidar is about 35km, which is conform with the discussion above.

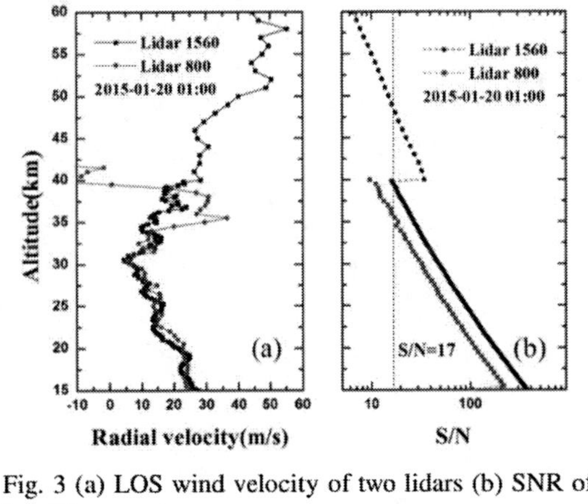

Fig. 3 (a) LOS wind velocity of two lidars (b) SNR of two lidars

DOI: 10.1051/epjconf/201611917004

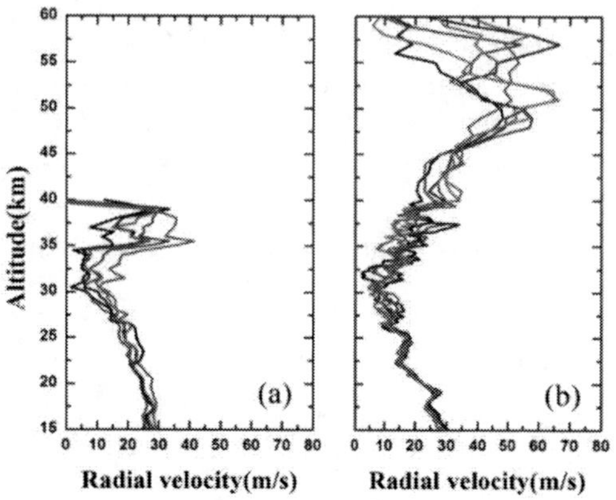

Fig. 4 Six continuous LOS wind profiles of the scanning (a) and non-scanning (b) lidar

4. CONCLUSIONS

Two sets of mobile Rayleigh Doppler lidars are implemented in USTC for atmospheric dynamical research. Comparison experiments are performed to assess the performance of the lidars, and the result demonstrates that their combination could provide simultaneous wind data reliable and precision enough for atmospheric gravity waves research. To fulfill the observation gap between 60 km to 80 km in the mesosphere would be our future work.

ACKNOWLEDGEMENT

This work was supported by National Natural Science Foundation of China 41174130, 41174131 and 41304123.

REFERENCES

[1] Fritts, D. C., M. J. Alexander, 2003: Gravity wave dynamics and effects in the middle atmosphere, *Rev. Geophys.*, **40(1)**.

[2] Xia, H., X. Dou, D. Sun, Z. Shu, X. Xue, Y. Han, D. Hu, Y. Han, T. Cheng, 2012: Mid-altitude wind measurements with mobile Rayleigh Doppler lidar incorporating system-level optical frequency control method, *Opt. Exp.*, **21(14)**, 15286-15300.

[3] Dou, X., Y. Han, D. Sun, H. Xia, Z. Shu, R. Zhao, M. Shangguan, J. Guo, 2014: Mobile Rayleigh Doppler lidar for wind and temperature measurements in the stratosphere and lower mesosphere, *Opt. Exp.*, **22(105)**, A1203-A1221.

[4] Ruo-can, Z., X. Hai-Yun, D. Xian-Kang, S. Dong-Song, H. Yu-Li, S. Ming-Jia, G. Jie, S. Zhi-Feng, 2015: Correction of temperature influence on the wind retrieval from a mobile Rayleigh Doppler lidar, *Chin. Phys. B*, **24(2)**, 024218.

Extending and Merging the Purple Crow Lidar Temperature Climatologies Using the Inversion Method

Ali Jalali[1], R.J.Sica[1,2,3], P. S. Argall[1,4]

[1]Department of Physics and Astronomy, University of Western Ontario, London, Canada

**Email: ajalali3@uwo.ca*

[2]Visiting Scientist, MétéoSuisse, Payerne, Switzierland

[3]Visiting Scientist, Environmental Fluid Dynamics Laboratory, École Polytechnique Fédérale de Lausanne, Switzerland

[4]School of Applied Science and Technology, Fanshawe College, London, Ontario

ABSTRACT

Rayleigh and Raman scatter measurements from The University of Western Ontario Purple Crow Lidar (PCL) have been used to develop temperature climatologies for the stratosphere, mesosphere, and thermosphere using data from 1994 to 2013 (Rayleigh system) and from 1999 to 2013 (vibrational Raman system). Temperature retrievals from Rayleigh-scattering lidar measurements have been performed using the methods by Hauchecorne and Chanin (1980; henceforth HC) and Khanna et al. (2012). Argall and Sica (2007) used the HC method to compute a climatology of the PCL measurements from 1994 to 2004 for 35 to 110 km, while Iserhienrhien et al. (2013) applied the same technique from 1999 to 2007 for 10 to 35 km. Khanna et al. (2012) used the inversion technique to retrieve atmospheric temperature profiles and found that it had advantages over the HC method. This paper presents an extension of the PCL climatologies created by Argall and Sica (2007) and Iserhienrhien et al. (2013). Both the inversion and HC methods were used to form the Rayleigh climatology, while only the latter was adopted for the Raman climatology. Then, two different approaches were used to merge the climatologies from 10 to 110 km. Among four different functional identities, a trigonometric hyperbolic relation results in the best choice for merging temperature profiles between the Raman and Low level Rayleigh channels, with an estimated uncertainty of 0.9 K for merging temperatures. Also, error function produces best result with uncertainty of 0.7 K between the Low Level Rayleigh and High Level Rayleigh channels. The results show that the temperature climatologies produced by the HC method when using a seed pressure are comparable to the climatologies produced by the inversion method. The Rayleigh extended climatology is slightly warmer below 80 km and slightly colder above 80 km. There are no significant differences in temperature between the extended and the previous Raman channel climatologies. Through out this study, we discuss the data collected by PCL for a long term (1994–2013) up to 110 km. Moreover, different merging functions in various methods were used to merge the temperature climatologies for different channels.

1. INTRODUCTION

Rayleigh and Raman lidar systems are designed to detect photon backscatter from the atmosphere. These systems provide high temporal and vertical resolutions, making them suitable candidates to study the climatology of the atmosphere. The Purple Crow Lidar (PCL)(Sica et al. [1995, 2001]) is currently located at the Environmental Science Western Field Station (43.07 N, 81.33W, 275 m) near The University of Western Ontario in London, Canada and has been operating since 1992. Now, a Nd:YAG solid state laser is used to produce 532 nm beam with a repetition rate of 30 Hz with 1000 mJ energy per pulse for Rayleigh and Raman lidars. The conventional method for retrieving temperature from lidar measurements was first proposed by Hauchecorne and Chanin in 1980. In previous PCL climatologies, this method was used for obtaining temperature from the PCL measurements. One important difference in this study is a grid search optimization (inversion) technique, which was used to analyze Rayleigh channel data in order to improve the deficiencies of the HC method. The PCL is comprised of three channels: Raman, analog Rayleigh and digital Rayleigh. The Raman channel measures the temperature of the upper troposphere and lower stratosphere with data collected from 1999 to the present from 10 to 45 km. The digital Rayleigh and analog Rayleigh channels measure temperature in the upper mesosphere and lower thermosphere, which the measurements have been conducted since 1994. Currently, the PCL has two PMTs for the Rayleigh channel. The new one, labeled as Rayleigh Licel (High level Rayleigh), and old Rayleigh channel labeled as the Rayleigh Hamamatsu (Low level Rayleigh). The old Rayleigh channel was used for calibration and alignment while testing the linearity of the new Licel system. These three channels were used to perform PCL climatology.

2. METHODOLOGY

Different techniques have been suggested to retrieve temperature from lidar measurements. There are two methods for retrieving temperature from lidar measurements, these methods are called the conventional method which was first proposed by Hauchecorne and Chanin in 1980 and the inversion method presented by Khanna et al., 2012. Each approach has its own deficiencies and benefits. In the HC method, using hydrostatic equilibrium, the ideal gas law and the lidar equation, we can find a relation between observed lidar output and the temperature at each altitude in the lidar range. The details of this technique are described in Hauchecorne and Chanin (1980). In order to integrate the pressure relation from the top to bottom, it is necessary to know the value of the seed pressure or seed temperature at the top. Because the temperature and density measurements in the thermosphere have large fluctuations, the initial guess of the pressure or temperature at the top has a large uncertainty, and this uncertainty will be carried through the computation process of retrieved temperatures in lower altitudes. The starting point at the top for the Rayleigh channel is set between 100 to 110 km, and the corresponding pressure is obtained from the CIRA model. Because of the uncertainty in the value of the seed pressure or temperature, the uncertainty of the retrieved temperatures is large for the first 15 km below the highest altitude. Therefore, it is common to remove the first 10 to 15 km from the temperature profiles in order to achieve more accurate results. However, we lose valuable information in the top layers-the disadvantage of this method. One solution for this deficiency of the conventional method is the inversion method. The inversion method is an optimization scheme that uses an iterative approach to estimate the true value of parameters by starting from an initial guessed value of the parameter vector and proceeding by updating the parameters with their improved values step by step until it reaches the best estimate for temperature. In the inversion method, still we use the seed pressure but contrary to the conventional method, the seed pressure is at the bottom and the integration is upward. In the upward integration method, the seed pressure and temperature, which are defined to be between 30 to 35 km, can be measured accurately using a variety of methods. Therefore, using this method will decrease the uncertainties. Since the inversion method does not provide statistical model or model parameter uncertainties, the uncertainties of its results can be estimated using Monte Carlo techniques, whereas in the conventional method the uncertainties are calculated analytically. Details of the inversion method are explained in Khanna et al., (2012). In order to extend our results to include an extra 10 km on top, the inversion method is used to perform temperature climatology for the PCL High Level Rayleigh channel. In this study, the Argall and Sica methodology was followed to form the temperature climatologies where the methodology is explained in detail in Argall and Sica (2007). The temperature climatologies were calculated for separately for the High Level Rayleigh, Low Level Rayleigh and Raman channels.

3. RESULTS

The temperature climatologies were performed for High Level Rayleigh, Low Level Rayleigh and Raman channel. The PCL channels overlap in some regions in altitude. In overlapping regions, the temperature should be identical but there are differences between them. Therefore, the different ranges are combined to have an individual profile for the entire atmosphere from 10 km up to 110 km and are known as "merged" profile. Retrieved temperatures for each channel have more statistical accuracy at the bottom rather than the top of each profile. As a result, some regions for each channel are more reliable in a common area. Therefore, instead of using simple averages between temperatures, four functions were chosen to merge the temperature profiles between different channels: sine/cosine, simple linear, error function (erf), and hyperbolic functions. PCL measurements have been obtained with better or worse precision because each lidar channel measurement in a common area has different precision. In this case, the data points are expressed by assuming the same distribution with the same mean but with different standard deviations. Therefore the weighted average of the measurements is the most probable value. The weighted averages for the Raman and LLR channel between 30 to 35 km, as well as LLR and HLR channel between 35 to 40 km, were calculated. The weighted average result was chosen as a reference for evaluating merging functions. The related errors for the PCL are a combination of statistical errors due to the photon counting and systematic errors due to applied corrections. Although, the applied corrections in merging area are fairly small. Consequently, the weighted average function was chosen as a reference for the evaluation of the merging functions. Two different approaches were used to evaluate the merging functions and are explained in following sections. In the first approach, in order to extend the range of lidar measurements, temperature climatologies from different channels were merged together. First, merging altitudes were found by considering several factors, including: photon count linearity, temperature standard deviation, and temperature differences between different channels. The best altitude range between the Raman and LLR channels was 30 to 35 km and between the LLR and HLR was 35 to 40 km. Then temperature climatologies were merged together using each merging function between the Raman and LLR channels, as well as LLR and HLR channels. The differences between weighted average temperatures and merged parts are considered as matrices, where these matrices represent the average uncertainty for each function for an entire year. In order to evaluate each function, the standard deviations of these matrices were calculated. The results show that the hyperbolic function is the best function for merging the Raman and LLR channel with uncertainty of 0.9 K and the error function is a better function for the LLR and HLR with an uncertainty of 0.7 K. After finding the best merging functions between the different channels, the combined PCL climatology was found.

The second approach involved merging the nightly temperature profiles together first and then creating the temperature climatology. In the first step the common dates among the Raman, LLR and HLR channels were found a total of 297 nights. The second method was unlike the last approach, where the merged area was constant. In the second approach, the critical height was calculated for each night, and then the merged area was defined as the critical height plus 5 km, if this summation was less than 50 km for merging HLR-LLR. In the case that the result of the critical height + 5 km was larger than 50 km for the HLR-LLR merging, the merged area was set between 45 to 50 km. The LLR-Raman merged area was chosen between 30 and 35 km. Temperature climatology differences between the weighted average result and each function was calculated. Finally, the standard deviation of each difference matrix was determined to be the uncertainty for each function. The results for this approach are consistent with the first approach. Figure 1 shows the PCL merged temperature climatology. The white area in this figure is related to the lack of data.

Finally, the new extended climatology was compared to previous climatology. The results revel the increase in temperature between 35 to 80 km and decrease above 80 km. Also there are no significant differences in temperature between the extended and the previous Raman channel climatologies.

4. CONCLUSIONS

The inversion method method showed a significant improvement in the integration of temperature profiles using the lidar equation when compared to the conventional technique. In this study, temperature climatologies were calculated for all available PCL Rayleigh measurements in order to investigate and evaluate the new method, as well as update the PCL temperature climatology using inversion method. The comparison between the conventional and inversion method results revealed that given the same uncertainty level, the inversion method had the advantage of gaining approximately an about extra 10 km in retrieved temperatures. Between the tested functions for merging the different temperature channels, the hyperbolic function had the best result for merging the Raman and LLR channels, with an uncertainty of 0.9 K. The error function yielded a better result for merging temperatures between the LLR and HLR channels, with a 0.7 K uncertainty. The different approaches for evaluating merging functions confirmed their results.

ACKNOWLEDGEMENT

Ali Jalali would like to thank the NSERC CREATE Training Program in Arctic Atmospheric Science and the Canadian Network for the Detection of Atmospheric Change (CANDAC) for their support. We would also like to acknowledge the support of the Canadian National Science and Engineering Research Council.

REFERENCES

[1] Argall, P. S. and R. J. Sica. A comparison of rayleigh and sodium lidar temperature climatologies. Ann. Geophys., 25:27–35, 2007.

[2] Hauchecorne, A., M. Chanin, and P. Keckhut. Climatology and trends of the middle atmospheric temperature (33-87 km) as seen by rayleigh lidar over the south of france. J. Geophys. Res., 96:5169–5183, 1991.

[3] Iserhienrhien, B., R. J. Sica, and P. S. Argall. A 7-year lidar temperature climatology of the mid-latitude upper troposphere and stratosphere. Atmosphere-Ocean, 51:532–540, 2013.

[4] Khanna, Jaya, Justin Bandoro, R. J. Sica, and C. Thomas McElroy. New technique for retrieval of atmospheric temperature profiles from rayleigh-scatter lidar measurements using nonlinear inversion. Applied Optics, 51:7945–52, 2012.

[4] Sica, R. J., S. Sargoytchev, P. S. Argall, E. F. Borra, L. Girard, C. T. Sparrow, and S. Flatt. Lidar measurements taken with a large aperture liquid mirror. 1. rayleigh scatter system. Applied Optics, 34:6925–6936, 1995.

[5] Sica, R. J., Z. A. Zylawy, and P.S. Argall. Ozone corrections for rayleigh-scatter temperature determinations in the middle atmosphere. Journal of Atmospheric and Oceanic Technology, 18:1223–1228, 2001.

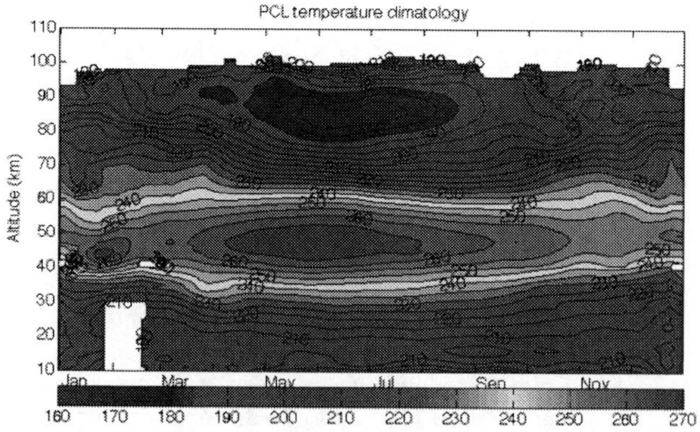

Figure 1- The PCL temperature climatology.

EPJ Web of Conferences **119**, 17006 (2016)
ILRC 27
DOI: 10.1051/epjconf/201611917006

AUTOMATING THE PURPLE CROW LIDAR

Shannon Hicks [1]*, R.J. Sica[1], P.S. Argall[2]
[1]*Dept. of Physics and Astronomy, University of Western Ontario, London, Canada,*
**Email: shicks26@uwo.ca*
[2]*School of Applied Science and Technology, Faculty of Technology, Fanshawe College, London, Canada*

ABSTRACT

The Purple Crow LiDAR (PCL) was built to measure short and long term coupling between the lower, middle, and upper atmosphere. The initial component of my MSc. project is to automate two key elements of the PCL: the rotating liquid mercury mirror and the Zaber alignment mirror. In addition to the automation of the Zaber alignment mirror, it is also necessary to describe the mirror's movement and positioning errors. Its properties will then be added into the alignment software. Once the alignment software has been completed, we will compare the new alignment method with the previous manual procedure. This is the first among several projects that will culminate in a fully-automated lidar. Eventually, we will be able to work remotely, thereby increasing the amount of data we collect. This paper will describe the motivation for automation, the methods we propose, preliminary results for the Zaber alignment error analysis, and future work.

1. INTRODUCTION

The Purple Crow Lidar (PCL) is located 10km north of the University of Western Ontario at the Environmental Science Field Station. It uses a Neodymium: Yttrium-Aluminium-Garnet (Nd:YAG) solid state laser to produce a 532 nm beam, with an average power of 950 mJ and a repetition rate of 30 Hz [1]. The initial beam diameter is 12mm, but it is quickly passed through a 5X beam expander, to produce a 60mm beam. It is then reflected off of a series of three mirrors, the last of which is the motorized Zaber mirror that controls the alignment of the system (Figure 1). The laser is directed skyward and backscattered photons are collected and focused by the 2.6 m liquid mercury primary mirror onto 4 photomultiplier tubes (PMTs). Two of the channels simultaneously measure analog and digital photocounts and are used for the Rayleigh-scatter and vibrational nitrogen channels, as well as aligning the laser. The other two channels are digital; one channel is used with an attenuator for the elastic return, the other for water vapour. With these channels, the PCL is capable of measuring temperature, density, pressure, and water vapour mixing ratio profiles[1]. In time, we may increase the altitude range of the lidar as well as expand our detection capabilities to include particulates.

In the past, the PCL alignment has been conducted manually, whenever the observer notices a drop in signal. This is not only inefficient, but also compromises the integrity of sections of the data throughout an observing night. The focus of my project is to use MATLAB to program the Zaber mirror and lidar control system to automatically align the laser beam, and keep it aligned throughout the course of the night. The process for this will be described in the Methodology section. We will then compare the new data with previous observations and look for any significant improvement in data quality.

2. METHODOLOGY

There are two concerns that our automation method must address: 1) staying within the field of view (FOV) of our telescope, and 2) small, stray clouds confusing the Zaber mirror program. To avoid both, we will oscillate the Zaber mirror between two positions over the course of a minute, or "dither", thereby gaining two profiles, one from each position(Z1 and Z2). We then pick the position that has the best signal, move to it, and then begin dithering again in a direction orthogonal to the previous movement. Dithering will always keep us in the FOV of the telescope and will also remove cloud bias.

Clouds move across the FOV of the telescope on the scale of minutes. If we do not dither, and instead integrate the signal over the course of a minute at each position (the current method), the cloud would cause a dramatic increase in counts in the resulting profile. We would be unable to determine if the increase in counts was due to a cloud or if the signal at that position was valid.

DOI: 10.1051/epjconf/201611917006

However, dithering the Zaber mirror on the order of seconds (much faster than the speed of the cloud) should provide reliable information.

Now that the alignment algorithm is established, we must determine the Zaber mirror's positioning error. If the error is significant, the alignment algorithm will need to compensate. I will investigate three aspects of the mirror: its movement step size, the relevance of its movement speed, and how accurately the mirror's actuators return to the same position. The smallest increment the mirror can move is 1 microstep, which translates to 1.48 μrads of tilt [2]. In order to document any change in position, the mirror movement must be visible to the human eye, so we have created a set-up so that 50 microstep increments are measurable. A small, red, laser is shone on the Zaber mirror and is directed down a hallway of 13.44 meters. The mirror is then commanded to move according to each test's parameters. The movement is recorded on graph sheets at the end of the hallway. Preliminary results from the first tests are located in the preliminary results section.

3. RESULTS

Before conducting the first error test, it was necessary to calculate the required accuracy of the PCL system at our highest observational altitude. First, we considered the full width FOV of the PCL telescope, which is 0.39 mrad [3]. The highest observable altitude, using the Inversion Method[4], is 105 km [5]. Therefore, the full FOV at that altitude is 43.55 m. Next, this must be compared to the laser beam size at the same altitude. This calculation is much less trivial, as one must know the divergence of the beam and the distances between each mirror that directs the beam. With a beam divergence of 17 μrads [6], a path length of 13.418m [1], and an initial size of 60mm after expansion, we can calculate the final diameter of the beam to be 1.85 m at an altitude of 105 km. The maximum positional error of the Zaber mirror must be the difference of the radius of the field of view and the laser radius, which is 21 m. This corresponds to 198.5 μrads of movement of the Zaber mirror, or 134 microsteps. The Zaber mirror has a positional error of roughly 10 microsteps, therefore it is not necessary to do a complete analysis on the Zaber movement at small command values.

However, it *is* necessary to measure the error of the second two scenarios. The motor speed's effect has not yet been analyzed, and will be left for future work. The last scenario is crucial, as it models the "dithering" technique. Preliminary tests have shown that there is no significant drift when the mirror oscillates, however, we have not tested if oscillation step-size is a factor. A longer hallway is also needed in order to produce statistically significant results.

4. CONCLUSIONS

The goal of this project is to automate the alignment procedure of the PCL in order to improve data collection and integrity. Future work will include testing the Zaber motor speed and its effect on positional error, and conducting a more comprehensive test of the dithering error. We will also be comparing the lidar signal received using the new alignment method against that which is received while using a manual alignment. Automation of the alignment will result in more complete climatologies and increased accuracy of water vapour profiles. As part of the Network for Detection of Atmospheric Composition Change (NDACC), the new data will be added to data from 70 other lidars, and used to study long term changes in atmospheric composition.

ACKNOWLEDGEMENT

I would like to thank the NSERC CREATE Training Program in Arctic Atmospheric Science and the Canadian Network for the Detection of Atmospheric Change (CANDAC) for their support. We would also like to acknowledge the support of the Canadian National Science and Engineering Research Council.

REFERENCES

[1]Wing, R., 2012: Multi-sensor calibration and validation of the UWO-PCL Water Vapour Lidar. *Master's Thesis.*

[2]Zaber Technologies, 2015: Actuator Characterization, *Internet.*

[3]R.J. Sica, S. Sargoytchev, P.S. Argall, E.F. Borra, L. Girard, C.T. Sparrow, S. Flatt, 1995: Lidar measurements taken with a large-aperture liquid mirror. 1. Rayleigh-scatter system. *Applied Optics*, **34(30)**:6925–6936.

[4]R.J. Sica, A. Haefele, 2015: Retrieval of temperature from a multiple-channel Rayleigh-scatter lidar using an optimal estimation method. *Applied Optics,* **54**, 1872-1889

[5]Jalali, A., 2015: Extending and merging the Purple Crow Lidar temperature rayleigh and vibrational raman climatologies, *Master's Thesis.*

[6]Melles Griot, 2015: *Laser Accessory Data Sheet.*

FIGURES

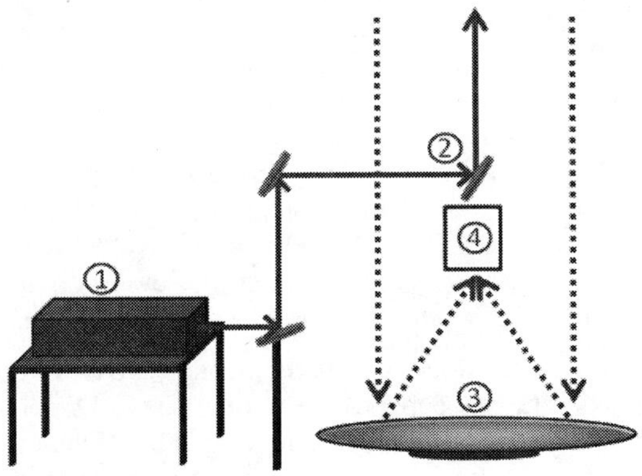

Figure 1: In numerical order: The Nd:YAG 532 nm Laser, Zaber Alignment mirror, 2.6 m Liquid Hg mirror, Detector Box

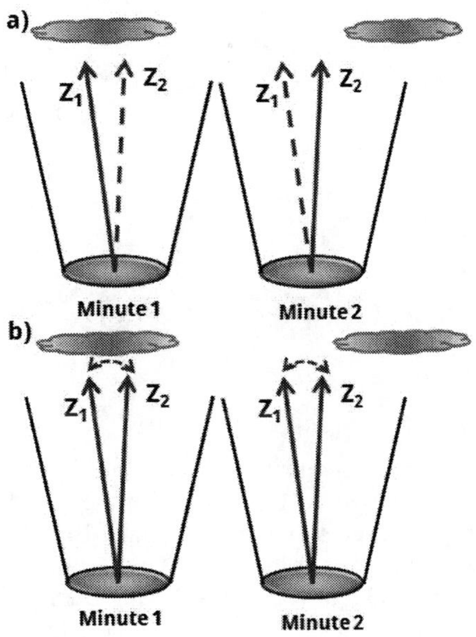

Figure 2: a) No Dithering: Integrating for one minute in each position can result in a false positive/better position at Z2. We are unable to tell if Z1 or Z2 is better due to cloud interference. b) Dithering: Both Z1 and Z2 are affected simultaneously by the cloud in the first minute. Both are open in the second – we will get a clear picture of the better position at both times.

DOI: 10.1051/epjconf/201611917007

HIGH SPECTRAL RESOLUTION LIDAR BASED ON A POTASSIUM FARADAY DISPERSIVE FILTER FOR DAYTIME TEMPERATURE MEASUREMENT

Makoto Abo*, Phong Pham Le Hoai, Kouki Aruga, Chikao Nagasawa, Yasukuni Shibata

Graduate School of System Design, Tokyo Metropolitan University, Tokyo 191-0065, Japan,
**Email: abo@tmu.ac.jp*

ABSTRACT

In this paper, a new high-spectral-resolution lidar technique is proposed for measuring the profiles of atmospheric temperature in daytime. Based on the theory of high resolution Rayleigh scattering, the feasibility and advantages of using potassium (K) Faraday dispersive optical filters as blocking filters for measuring atmospheric temperature are demonstrated with a numerical simulation. It was found that temperature profiles could be measured within 1K error for the height of 9 km with a 500 m range resolution in 60 min by using laser pulses with 1mJ/pulse and 1 kHz, and a 50 cm diameter telescope. Furthermore, we are developing compact pulsed laser system for temperature lidar transmitter.

1. INTRODUCTION

The Earth's weather is driven by the convective processes associated with the atmospheric temperature, water vapor density, and wind, which mostly occur in lower troposphere. Hence, vertical profiles of temperature, humidity, and wind are important data for weather forecasting and prediction of meteorological disaster. Lidar is a preferable instrument to measure these parameters due to the high spatial and temporal resolution profile. The lidar technique for measuring water vapor and wind has been already established. However, a compact temperature lidar for daytime observation has not been realized.

Recently, there are various lidar techniques for temperature measurement including Rayleigh/ Raman Lidar, differential absorption lidar (DIAL), rotational Raman lidar, and high spectral resolution lidar (HSRL). Rayleigh and Raman lidar technique is based on molecular lidar signal that is proportional to molecular number density. This technique requires a hydrostatic equilibrium assumption and an aerosol free altitude region for measuring temperature. DIAL technique is based on the temperature dependence of molecular absorption line. This technique needs a precise tuning laser and three operating wavelengths. Rotational Raman lidar is based on temperature dependence of the population of molecular rotation level. The characteristic of this technique includes narrow linewidth laser, narrow filter, and small scattering cross section.

HSRL technique is fundamentally based on temperature dependence of the Doppler broadening of Rayleigh scattering. This technique required a precise tuning laser and very narrow filter. HSRL performance mostly depends on the stability of filters and laser wavelength. Hence, it is only utilized for measuring nighttime temperature. In this project, we aim at developing a new HSRL system with K Faraday dispersive optical filters to measure atmospheric temperature profile in daytime.

2. HIGH SPECTRAL RESOLUTION LIDAR FOR TEMPERATURE MEASUREMENT

2.1 High spectral resolution lidar with absorption filters

The schematic of temperature measurement by the HSRL with absorption cell is shown in Fig. 1. The left figure shows the spectral profile of Rayleigh/Mie backscattering containing a narrow aerosol peak near the laser transmitter frequency and a much broader spectrum of molecular scattering. The backscattering signal passing through two absorption cell filters with different temperature results in different residual scattering spectra. Consequently, the atmosphere temperature can be calculated from the ratio of two residual scattering signals [1].

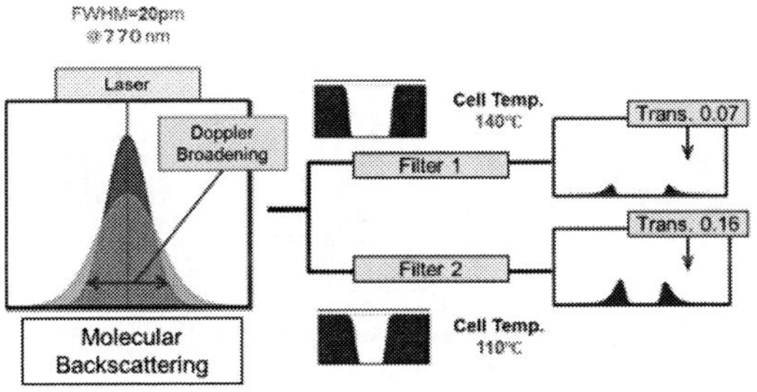

Figure 1. The schematic of temperature measurement by the HSRL with absorption cell.

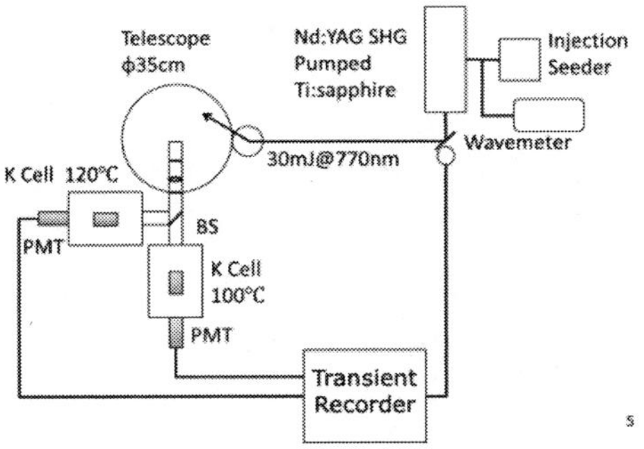

Figure 2. Diagram of HSRL experimental setup with K absorption filters.

Diagram of the HSRL experimental setup with K absorption filters is shown in Fig. 2. The transmitter utilizes a Nd:YAG- SHG laser as a pump source for Ti: sapphire laser. The output laser with 30 mJ pulse energy operates at wavelength of 770 nm. The receiver contains a 35 cm telescope, two K absorption cells at 100 °C and 120 °C, photomultiplier tubes (PMT), and a transient recorder. The temperature profile, measured at Hino, Tokyo with 150 m range resolution is compared to the radiosonde data at Tateno, as shown in Fig. 3. The lidar result coincides with radiosonde data.

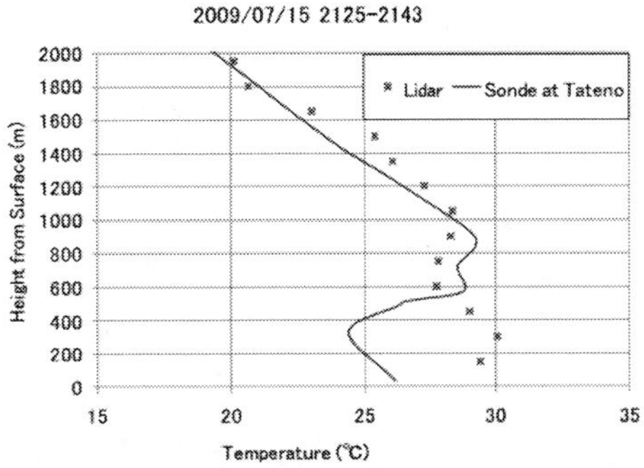

Figure 3. Lidar temperature profile measured at Hino, Tokyo and radiosonde data at Tateno.

2.2 Faraday Anomalous Dispersion Optical Filter

A Faraday anomalous dispersion optical filter (FADOF) includes an atomic vapor cell placed between 90° cross polarizers, as shown in Fig. 4. Magnetic field along the propagation direction induces circular birefringence in the dispersive atomic vapor. Due to the Faraday rotation, only narrow range of wavelengths can be transmitted through filter with high efficiency [2].

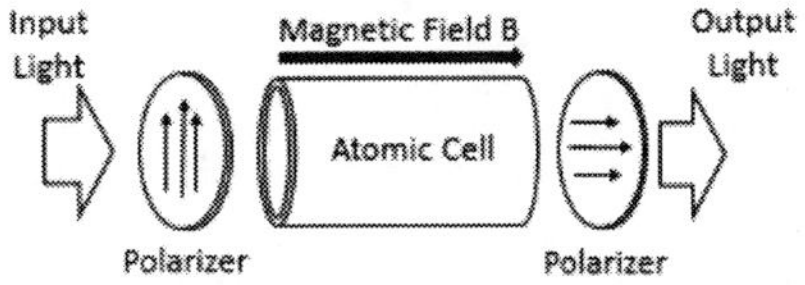

Figure 4. Faraday Anomalous Dispersion Optical Filter.

The transmission profile of FADOF depends on cell temperature, magnetic field, and cell length. Examples of band pass filter and band stop filter are plotted in Fig. 5. The advantage of FADOF is narrow bandwidth and rejected background. While the background rejection of a conventional absorption filter depends on additional interference filter. Hence, we proposed a new HSRL with FADOF for daytime temperature measurement. We also replace a tunable laser of transmitter by compact and simple laser system such as a vapor laser and pulsed tapered semiconductor optical amplifier system.

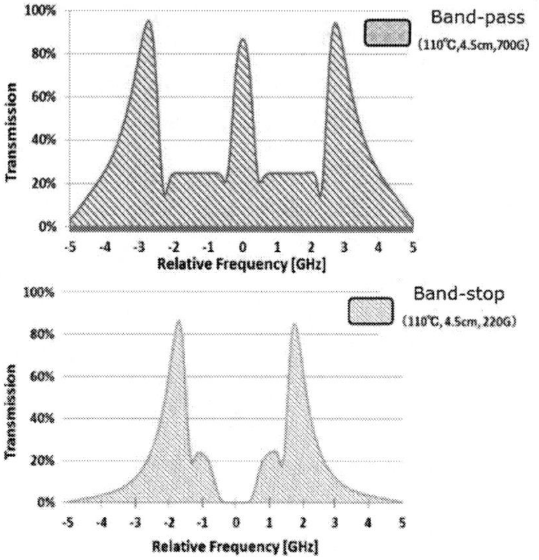

Figure 5. Examples of band pass filter and band stop filter of FADOF.

2.3. Proposed HSRL with K FADOF

The schematic of transmission measurement of the K FADOF is shown in Fig. 6. The laser beam from an external cavity diode laser (ECDL) passes through ND filter and Faraday filter, and measured by a photodiode. Measured and calculated transmission profiles of the K FADOF under different conditions are plotted in Fig. 7 The cell temperature is 115 °C, cell length is 7.5 cm, and magnetic field is 330 G and 115 G respectively. Consequently, the experimental results agree well with theoretical results.

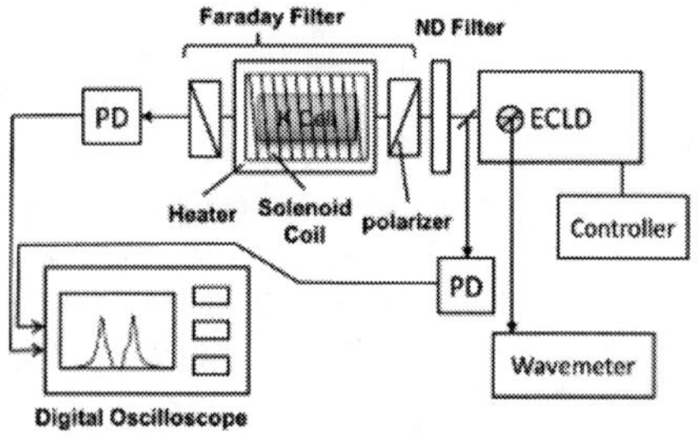

Figure 6. Schematic of transmission measurement of the K FADOF.

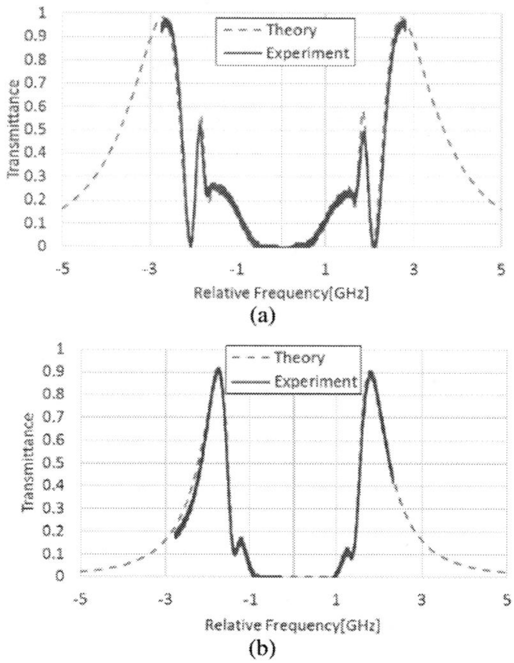

Figure 7. Measured (solid line) and calculated (dashed line) transmission profiles of the K FADOF. The cell temperature is 115 °C, cell length is 7.5 cm, and magnetic field is (a)330 G and (b)115 G.

The HSRL parameters for temperature error simulation are given in Table 1. A laser beam with pulse energy of 1 mJ and repetition rate of 1 kHz is transmitted to low altitude region from ground to 6 km and high altitude region from 6 km to 20 km. The receiver contains 20 cm / 50 cm telescope for low / high altitude system. The K FADOF specifications include cell length of 7.5 cm, cell temperature of 115 °C, and magnetic field of 115 G / 330 G for low / high altitude region.

Table 1. The HSRL parameters for simulation.

	Low Altitude	High Altitude
Target Altitude	0 km~6 km	6 km~20 km
Pulse energy	1 mJ	
Repetition rate	1kHz	
Telescope Aperture	20 cm	50 cm
Range Resolution	200 m	500 m
Accumulation time	10 min	60 min

The simulated temperature errors of conventional HSRL and proposed HSRL are shown in Fig. 8. Our proposed HSRL using K FADOF shows

DOI: 10.1051/epjconf/201611917007

smaller error than the conventional HSRL using absorption filter during daytime and nighttime as well as low and high altitude region. For low altitude system, the proposed HSRL can measure the temperature up to 3km within 1K error for daytime with a 200 m range resolution in 10 minutes. For high altitude system, the proposed HSRL can measure the temperature up to 9km within 1K error for daytime with a 500 m range resolution in 60 minutes. Hence, the proposed HSRL is appropriate for daytime temperature measurement.

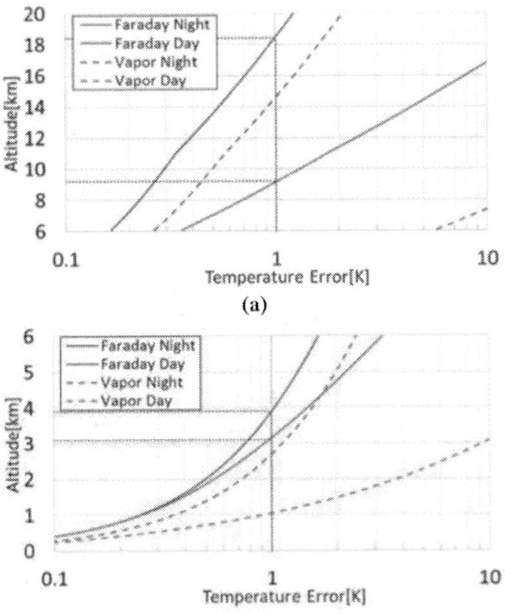

Figure 8. The simulated temperature errors of absorption-filter HSRL (dashed line) and proposed Faraday-filter HSRL (solid line) for nighttime (blue) and daytime (red) observations with (a) high altitude system and (b) low altitude system.

3. TRANSMITTER

We are developing two type of transmitter for the compact and simple temperature HSRL. One laser is a pulsed K vapor laser. The K vapor laser can easily realize narrow-linewidth and precise tuning. A simple configuration for experimental setup is shown in Fig. 9. A pumped laser beam passes through haft wave plate, a polarized beam splitter and injected to a K vapor cell in oven with AR coating. A high reflector mirror placed behind the K cell to reflect the laser beam back. Then the output laser beam directs to the polarized beam splitter and to be transmitted to the air.

Another laser is tapered amplifier diode system by seeding with continuous-wave DBR laser [3]. The DBR laser is easily tuned to K resonance line using wavelength-locking system, and tapered amplifier diode can provide high repletion rate pulse laser.

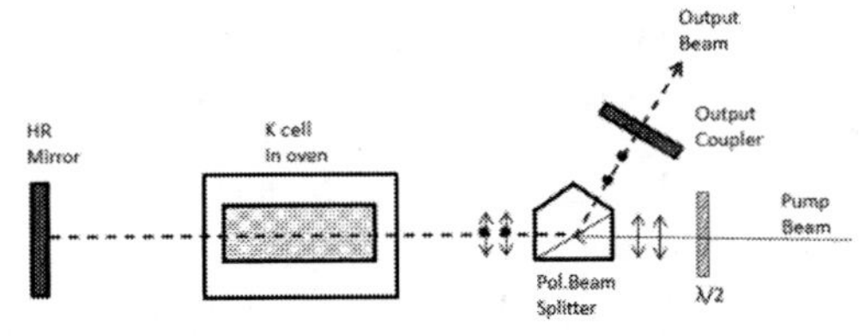

Figure 9. Experimental Setup of a Pulsed Laser Pumped K Vapor Laser

4. CONCLUSIONS

In this paper, we propose a new high-spectral-resolution lidar for measuring the profiles of atmospheric temperature in daytime. The K Faraday dispersive optical filters act as blocking filters for measuring atmospheric temperature. The simulation results show that the proposed HSRL is suited for daytime temperature measurement. Furthermore, we are developing compact pulsed laser system for temperature lidar transmitter.

ACKNOWLEDGEMENT

This work was supported by JSPS KAKENHI Grant Number 26550014.

REFERENCES

[1] Shimizu, H., Lee, S. A., and She, C. Y., 1983: High spectral resolution lidar system with atomic blocking filters for measuring atmospheric parameters, *Appl. Opt.*, **22**, 1373-1381.

[2] Zhang, Y., Jia, X., Ma, Z., Wang, Q., 2001: Optical filtering characteristic of potassium Faraday optical filter, *IEEE J. of Quantum Electronics*, **37**, 372-375.

[3] Xiong, Y., Murphy, S., Carlsten, J. L., Repasky, K., S., 2006: Design and characteristics of a tapered amplifier diode system by seeding with continuous-wave and mode-locked external cavity diode laser, Opt. Eng. **45**(12), 124205.

EPJ Web of Conferences **119**, 17008 (2016) DOI: 10.1051/epjconf/201611917008
ILRC 27

LIDAR MEASUREMENTS SUPPORTING THE OCULAR HAZARD DISTANCE CALCULATION USING ATMOSPHERIC ATTENUATION

K. Ove S. Gustafsson[1]*, Rolf Persson[1], Frank Gustafsson[1], Folke Berglund[1], Julia Hedborg[1], Jonas Malmquist[2]

[1]Swedish Defence Research Agency, P.O. Box 1165, SE-581 11 Linköping, Sweden,
**Email: ove.gustafsson@foi.se*

[2] Swedish Defence Materiel Administration, Vidsel Test Range, Box 74, SE-94 295 Vidsel, Sweden

ABSTRACT

A series of lidar measurements has been performed at the Vidsel Test Range, Vidsel, situated in the inland of the very northern part of Sweden, as a part of an assessment of reducing the laser hazard distance using atmospheric attenuation within the calculations of nominal ocular hazard distance (NOHD). The question was "How low is the atmospheric attenuation as function of height in this area, using a wavelength of 1064 nm?" The work included building a ground based backscatter lidar, performing a series of measurements and analyzing the results. The measurements were performed during June to November, 2014, with the objective to measure at clear air and good weather situations.

The lidar measurements at 1064 nm showed a very low atmospheric attenuation as a function of height to altitudes of at least 10 km at several occasions. The lowest limit of backscatter coefficient possible to measure with this instrument is $0.3 \cdot 10^{-7}$ m^{-1} sr^{-1}. Assuming a lidar ratio varying between 30 – 100 sr, this was leading to an extinction coefficient of about 0.9 - $3 \cdot 10^{-6}$ m^{-1}. The atmospheric attenuation reduces the laser hazard distance with about 50 – 56 % depending on the lidar ratio. A recommendation is to monitor the atmospheric attenuation at the occasions when the method to the reduced laser hazard distance using atmospheric attenuation is used.

1. INTRODUCTION

Lidar measurements were performed in an attempt to provide values on atmospheric attenuation as function of altitude for usage of powerful lasers at wavelength 1064 nm (Nd:YAG). The Lidar measured backscatter coefficient as function of height and standard values of lidar ratio was used to estimate the extinction coefficients. The ideal situation would have been using for instance three laser wavelengths measuring the backscatter coefficient and adding the information from two different Raman lidar signals, as suggested by Müller at al. [1] to calculate the extinction coefficient. Such system was not available and only the single backscatter was measured.

The distance where risk for injuries or damage of an eye when laser irradiance level exceeded the Maximum Permissible Exposure (MPE) is equal to the Nominal Ocular Hazard Distance (NOHD). The NOHD for lasers can be calculated using standardized methods as given by [2] and [3]. The NOHD is given, see [4], in the standard case by equation:

$$NOHD = \frac{1}{\theta}\sqrt{\frac{4P}{\pi\, MPE} - a^2}$$

where P is the laser power, θ the laser beam divergence and a is the laser beam radius at the sender.

The NOHD for a typical designator laser with small divergence can be several tens of kilometers. As an example a designator laser, assuming a Nd:YAG laser, with wavelength 1064 nm, power 0.180 W, pulse length 15 ns, PRF 11.5 Hz and beam divergence 0.08 mrad, will have NOHD of 48 km (multiple pulses during 10 sec).

One way of handling those risk distances, which might be too long for ordinary firing ranges or embargoed areas, are by probabilistic calculations of danger and by including the atmospheric attenuation. For such a long laser beam path the atmospheric transmission attenuation will be significant, depending on the level of attenuation. The suppression of the risk distance can be substantial even for moderate extinction coefficient if the atmospheric attenuation is included within the calculations of Ocular Hazard Distance (OHD).

The common way of calculating the NOHD abandons the use of atmospheric attenuation lowering the laser safety ranges in the civil society. The military community of interest has proposed

 DOI: 10.1051/epjconf/201611917008

several methods for exercises or testing taking into account the atmospheric attenuation under special conditions see [5] and [6]. A method to calculate the influence of a homogenous atmosphere on the laser safety range is given by [2]

$$OHD_{\mu} = \frac{2W\left(\frac{\mu(a+\theta r)\cdot \exp\left(\frac{\mu(a+\theta r)}{2\theta}\right)}{2\theta}\right)}{\mu} - \frac{a}{\theta}$$

where *W* is the Lambert W function, [7], μ is the extinction coefficient and *r* is the beam radius.

The introduction of atmospheric attenuation in the calculation of the laser hazard distance could reduce the risk distance considerably, see Figure 1. The figure show the laser hazard distance as a function of visibility where a second scale of the abscissa gives the extinction coefficient of 1064 nm radiation. The dashed line show the NOHD for this particular laser system calculated with the standard method and the solid line shows the OHD when the atmospheric attenuation is taken into the account.

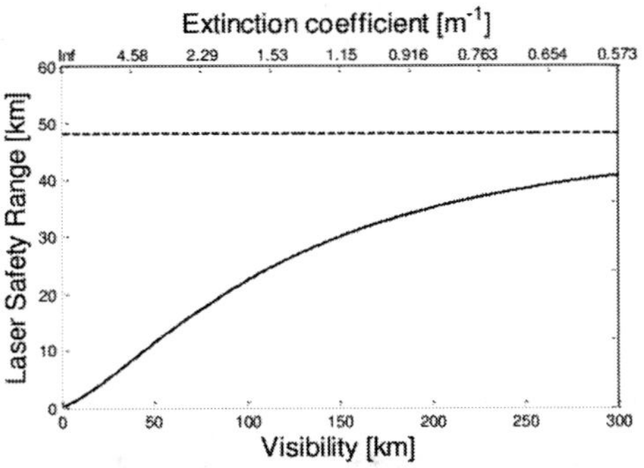

Figure 1. NOHD calculated by standard methods (dashed line) and OHD as function of visibility (solid line) for a designator laser using Nd:YAG laser.

The lidar system data, methodology of the lidar measurement and analysis of the signal is presented within chapter 2 together with a comparison between the 1064 nm lidar and our old 355 nm lidar system. In the comparison both system where measuring from the same place, within a 3m distance. In chapter 3 examples of the results from the 2014 campaign is given. In the concluding chapter are the results discussed.

2. METHODOLOGY

A new lidar system was built up mostly from used equipment. A new detector with power supply unit and cooler support electronics was purchased from Licel GmbH.

The laser is a Quantel Brilliant with pulse energy 230 mJ at 1064 nm, pulse length 5 ns, PRF 10 Hz, beam divergence 0.5 mrad and beam diameter 3 mm. The laser beam is directed with a 45° mirror to zenith. The beam is protected up to over eye level. See Figure 2.

The receiver telescope is constructed with hyperbolic mirrors, the primary mirror is 200 mm and the secondary mirror is 82 mm in diameter. The focal length is 615 mm. Its axis is located 130 mm from the laser beam which gives full overlap well below 500 m from ground. The detector is an APD module from Licel GmbH (S11518-10) with a sensor diameter of 1 mm which gives a FOV of 1.6 mrad. The detector is thermoelectric cooled. A narrow filter with FWHM 12 nm decreases the background noise level.

Figure 2. The 1064 nm lidar system was located in a small house with moveable roof for theodolite measurements.

Data Acquisition is performed with a Licel TR40 unit where 4000 bins are digitized at 40 MHz sample rate. Usually 600 returns are averaged and saved as individual data files on a Windows computer during 30 minutes measurements. Measurement with covered laser beam during 15 minutes (laser warm up) gives 15 background files. The files were sent by the Lidar operator from Vidsel to FOI with email for analysis.

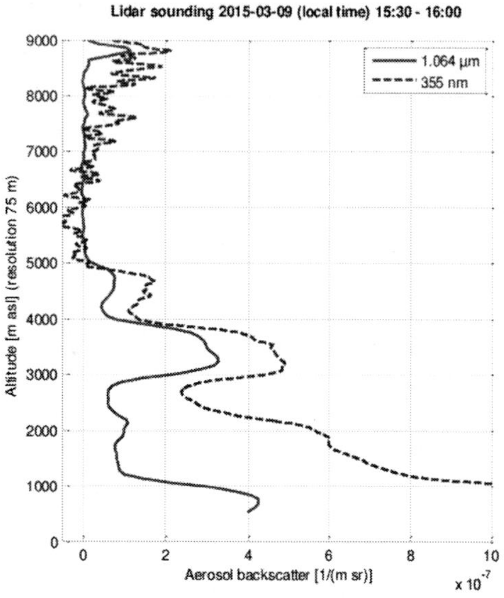

Figure 3. Backscatter levels from low levels up to 9 km. Comparison between the new 1064 nm lidar and our old 355 nm lidar system.

The data files were viewed as "quicklook" graphs for data control and to make it convenient to eliminate bad measurements, e.g. clouds. Quick-looks are reachable at http://earlinet.foi.se/ (together with results from our previous engagements in Earlinet). Files sorted for further usage were analyzed with Klett's & Fernald's methods [8] and [9]. Meteorological data for this analysis were achieved from SMHI (Swedish Meteorological and Hydrological Institute) as 1D-HIRLAM prognosis, calculated for the Lidar sites at Vidsel and Linköping four times every day.

The objective of the lidar measurements was to perform measurement at least once a week during six months, but only at occasions with very clean air, that means low amount of aerosol. The goal was not accomplished mainly since the weather and cloudiness did not fulfill the requirement of the campaign. Since the measurement was performed inside a test range, the range administration demanded that all measurement should be reported to the control tower in advance and the area had to be closed for flight operations. The laser risk range of the lidar laser is 11 km.

3. RESULTS

During this period about 25 measurements were performed but only 16 could be completed and analyzed. At several occasions a measurement was started but too few profiles (due to clouds) could be collected to give a significant measurement. It was desired that at least 15 minutes of measurement, resulting in 15 profiles, were performed before the analysis was done. This was not always the case.

An example of measurement is shown in figure 4, where the number of profiles measured was lower than 15 but the number of aerosol layers looked very interesting. At least two layers are shown, one (at least) below 3 km and one above 7 km. In between there is an altitude interval with very clean air.

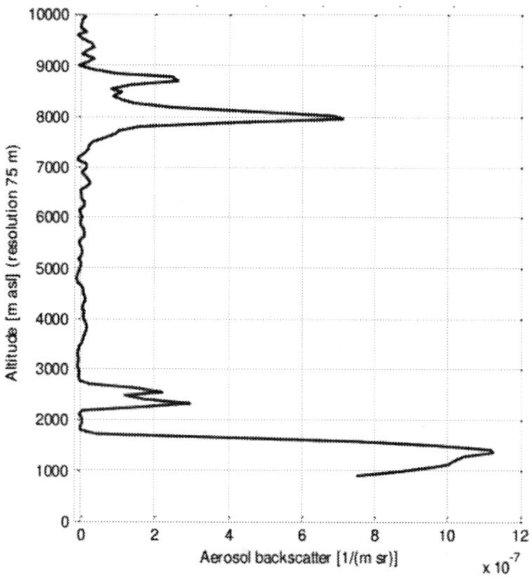

Figure 4. Lidar measurements at 09:30 UTC the 2014-09-10.

Since the purpose was different during this campaign than otherwise (looking for the situation with the lowest of aerosol content, lowest attenuation) the interest was focused on the clean days. Days with particular clean air were the June 30, September 1, and September 17. A lidar measurement from the September 17 is shown in figure 5.

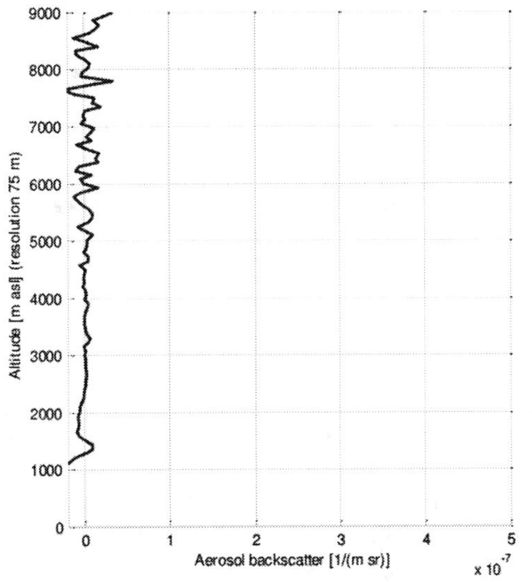

Figure 5. Lidar measurements at about 12:30 UTC the 2014-09-17.

The figure shows that the backscatter coefficient is lower than $0.3 \cdot 10^{-7}$ $m^{-1} \cdot sr^{-1}$. The backscatter extinction below zero is a consequence of background elimination and noise since values below zero does not exist in principle.

4. CONCLUSIONS

The objective was to find a lowest level of aerosol backscatter coefficient during his campaign, since the implication of low levels of extinction for this particular wavelength has very long laser hazards distances. The noise level did not give the opportunity setting a definitive lowest level of the aerosol extinction. The lowest possible measureable backscatter coefficient with this instrument was concluded to be less than $0.3 \cdot 10^{-7}$ $m^{-1} \cdot sr^{-1}$. This value does give an extinction value that depend of the type of aerosol to be < 0.9 - $3 \cdot 10^{-6}$ m^{-1}, assuming a lidar ratio in the interval of 30 – 100 sr. This leads to OHD values between 21 – 24 km. This is a reduction of the NOHD with 50 - 56 %, depending on the lidar ratio assumption. The molecule extinction coefficient was assumed to be $5 \cdot 10^{-6}$ m^{-1}, calculated by Modtran (program from Ontar Corporation) assuming no aerosol within the atmosphere.

The highest aerosol backscatter coefficient measured during this campaign was just above $2 \cdot 10^{-6}$ $m^{-1} \cdot sr^{-1}$, but again the objective was to find out the lowest values.

Reducing the NODH with atmospheric attenuation should be carried through with care and it is recommended to monitor the extinction coefficient during the activities.

ACKNOWLEDGEMENT

The author acknowledges the support from the Swedish Armed Force (FM) and Swedish Defence Material Administration (FMV), through Magnus Weideryd.

REFERENCES

[1] Müller D., Wandinger U., and Ansmann A., 1999: Microphysical particle parameters from extinction and backscatter lidar data by inversion with regularization: theory, Applied Optics, **38(12)**, 2346-2357.

[2] ANSI Z136.1-2014, 2014, [American National Standard for safe use of lasers].

[3] IEC 60825-1:2014, 2014 [Safety of laser products – Part 1: Equipment classification and requirements],.

[4] Gustafsson K. O. S., 2013: Questions about using of atmospheric attenuation calculating the nominal ocular hazard distance, Proc. of SPIE Vol. **8897**, 88970M-1-8.

[5] JSP390 – "Military Laser Safety", 2005, UK Ministry of Defence – Ordnance Board

[6] STANAG 3606 – Edition 5, 1991, "Evaluation and Control of Laser Hazards".

[7] Steinvall O. K., 2009: Laser system range calculations and the Lambert W function, Applied Optics, **48(4)**, 1-7.

[8] Klett J. D., 1981: Stable analytical inversion solution for processing lidar returns, Applied Optics **20**, 211–220.

[9] Fernald F. G., 1984: Analysis of atmospheric lidar observations: some comments, Applied Optics **23(5)**, 652–653.

EPJ Web of Conferences 119, 17009 (2016) DOI: 10.1051/epjconf/201611917009
ILRC 27

PERFORMANCE CHARACTERISTICS OF COMPACT MOBILE LIFS (LASER-INDUCED FLUORESCENCE SPECTRUM) LIDAR

Takayuki Tomida[1]*, Naoto Nishizawa[2], Kosuke Sakurai[3], Hikaru Suganumata[2], Shodai Tsukada[3], Sung-Moo Song[1], Ho-Dong Park[4] and Yasunori Saito [1]

[1]Institute of Engineering, Academic Assembly, [2]Graduate School of Science and Technology, [3]Faculty of Engineering, Shinshu University, 4-17-1 Wakasato, Nagano City, Nagano 380-8553, Japan, [4]Institute of Science, Academic Assembly. 3-1-1 Asahi, Matsumoto City, Nagaano 390-8621, Japan
**Email: tomida@cs.shinshu-u.ac.jp*

ABSTRACT

We developed a compact but versatile laser-induced fluorescence spectrum (LIFS) lidar that has potential use for material or aerosol identification outside experimental rooms. The compactness and mobility of the LIFS lidar means observations can be more freely conducted at any place and any time. Its performance characteristics were validated by three-dimensional fluorescence imaging of targets and remote detection of quasi bio/organic aerosols.

1. INTRODUCTION

Because of their unique characteristics, lidars have been widely using in environmental monitoring. Most lidars are best suited to experimental rooms. If lidars could move around a field freely at the time needed, their usefulness would expand. Field observations are important because the need might be sudden. Sudden events include thunder storms, volcano eruptions, and other natural events, and human-related activities/accidents such as chemical factory fires, dispersion of agricultural chemicals, gas leakages of buried pipe, and so on. The outbreaks of these events are hard to predict. Sometimes the events are so localized that we have to approach the area as closely as possible and yet avoid danger.

Using a mobile lidar is one of the best solutions for that purpose. Mobile lidars have already been developed, but most of them require a vehicle that is exclusively manufactured for the lidar. In addition, the size of the vehicles tends to be large because conventional laser systems and measurement instruments are big. Moreover, the power consumption is also large. Traditional mobile lidars were mainly based on Mie scattering and/or an absorption process with filter detection. Recent technologies such as microelectromechanical systems (MEMS) are capable of downsizing laser systems and spectroscopic instruments and decreasing the power consumption, while still maintaining the performance. So, lidars can be smaller even if we use a complicated spectroscopic instrument in addition to filter detection. Then, compact but versatile mobile lidars will be available for the abovementioned outbreaks and events. In other words, just-in-time lidar observation will be possible.

The Shinshu University lidar group has been interested in several phenomena occurring in the "livingsphere", where strong relations exist between human life and the natural environment [1]. Of many various materials in the livingsphere, we focused on biological and organic materials because they affect human health and life most directly. A quick response to such accidents with these materials is important.

In this paper, we describe a compact but versatile mobile laser-induced fluorescence spectrum (LIFS) lidar that is applicable to multi-observation.

2. METHODOLOGY

With the abovementioned events taken into account, a new lidar was designed to have multipurpose use and mobility.

2. 1 Laser-induced fluorescence spectroscopy

To detect biological and organic materials in the livingsphere, we introduced laser-induced fluorescence (LIF) spectroscopy through lidar signal detection. Since almost all materials exhibit their own unique auto-fluorescence spectrum, it is possible to judge the types of material by

DOI: 10.1051/epjconf/201611917009

comparing lidar observation data and the LIF spectrum database. We can easily find materials such as pollens, yellow-sands, minerals, detergents, and agricultural chemicals. Both natural materials and artificial products exist everywhere in our livingsphere.

If we set up the spectrometer used for fluorescence detection to match a specific observation target, LIFS lidar can be easily transformed for many uses, for example, Raman lidar and high-resolution Mie-Rayleigh lidar.

2.2 Compact vehicles

One simple idea behind lidar mobility is that lidar can be loaded into a vehicle. Another idea is the intended space of observation is limited. However, the actual roadways to approach the target space may be narrow and winding, and so the vehicles must be compact and commercially available. Fortunately, we can easily find such vehicles in our daily life. Figure 1 shows an example. The standard carrier space that can transport the lidar is about 1725 mm (L) x 1240 mm (W) x 1200 mm (H) [2].

3. Development of LIFS lidar

Figure 1 shows the LIFS lidar that we developed. Our lidar consists of a THG YAG-laser (Quantel, Ultra100-355-20), the energy of which at 355 nm is 20 mJ, a cassegrainian telescope with a diameter of 254 mm (Meade, LX200-25), and a photonic multichannel analyzer with spectral sensitivity of 200–900 nm (Hamamatsu, PMA-12). The analyzer is optically connected to the telescope with an optical bandle fiber. Synchronized detection of the fluorescence with laser oscillation [3] was achieved by changing the gate opening time of the charged-coupled device (CCD) of the analyzer by using a digital delay/pulse generator with a delay resolution of 5 ps (Stanford Research Systems, RG535). A laptop computer (Fujitsu, FMV-BIBLO) controlled the whole lidar operation.

A small power generator with output power of 900 VA (Yanmar, G900is) was prepared for observation at the aimed areas, where no power lines existed.

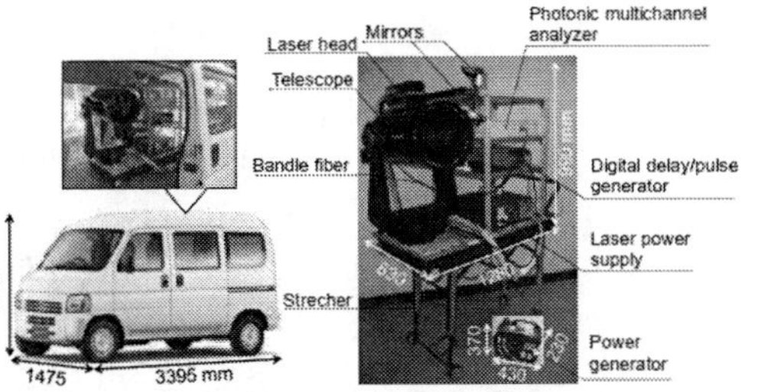

Fig. 1. Photograph of developed LIFS lidar and the lidar loaded inside a compact vehicle.

The LIFS lidar was set on a stretcher so that we could easily carry it from our laboratory and load it inside a car.

4. RESULTS

4.1 Database of LIF spectrum

We prepared powdered materials for the data collection experiment to make the LIF spectrum database. Some materials were made from solids that are candidates for aerosols. They were crushed by a cryogenic sample crusher (Japan Analytical Industry Co., Ltd., JFC-300) in the following procedure. First, materials that had dropped on the ground were frozen by liquid nitrogen. Then the frozen materials were crashed by shaking them with a stainless ball inside a stainless container. LIF spectra of the powdered materials were measured with the same instruments used for the LIFS lidar.

Because each of the LIF spectra of the powdered materials has a different shape (see Fig. 2 & 4), the LIF spectra can be used to classify aerosol species. We have collected LIF spectra of over 20 different materials so far. Some of the materials are fallen leaves, dried insects, bird feathers, and pieces of a road sign block. They are classified into vegetation, animal matter, and artificial materials.

4.2 Performance characteristics of LIFS lidar

1) Three-dimensional LIF spectrum imaging

In this paper, the targets were a plant and a wooden board. They were arranged side by side, at 15 m and 20 m away from the lidar, respectively. The laser beam scanned the target

DOI: 10.1051/epjconf/201611917009

area by rotating the telescope on which the laser was set, and at the same time the telescope collected the LIF spectrum from each target. This experiment was conducted under natural light conditions inside a gymnasium.

LIF spectra from the plant leaf and the wooden board are shown in Fig. 2. A LIF spectrum of the plant leaf has three peaks; a large 425 nm peak with a small swell around 500 nm, and two peaks around 685 nm and 740 nm of chlorophyll. It is well known that chlorophyll is one of the most important biomolecules for plants. Thus, the LIFS lidar has the potential to provide the physiological status of a plant non-destructively and remotely. For example, the relative intensities of the two chlorophyll LIF spectra had a strong relation to the chlorophyll concentration, which varied depending on the month and season [4].

A LIF spectrum from the wooden board has only a smooth curve with a peak around 425 nm. The result confirmed that plants and wooden boards can be distinguished by comparing their spectra.

The laser scan made a fluorescence image. Depth information was also obtained by changing the detection time of the photonic multichannel analyzer because the actual gate opening time of the CCD detector was increased by steps of 10 ns.

Figure 3 is a three-dimensional image of the chlorophyll LIF spectra, in which a wavelength region from 730 nm to 740 nm was extracted from the whole spectrum shown in Fig. 2. In Fig. 3, the 730–740 nm fluorescence image appears at the position of the plant and disappears where the plant is not present. Also, the image can be found

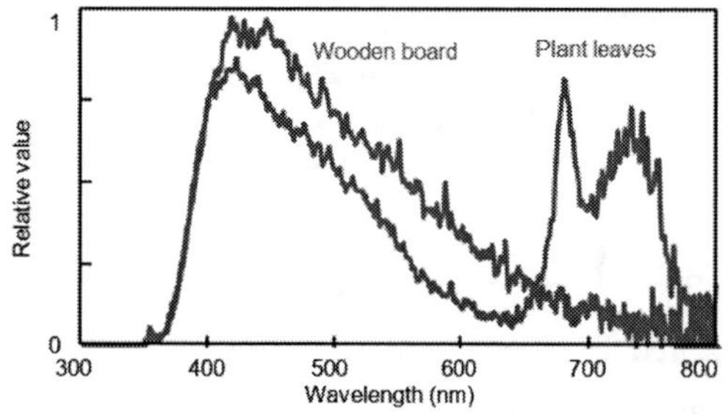

Fig. 2. LIF spectra obtained from a plant leaf and a wooden board by the LIFS lidar.

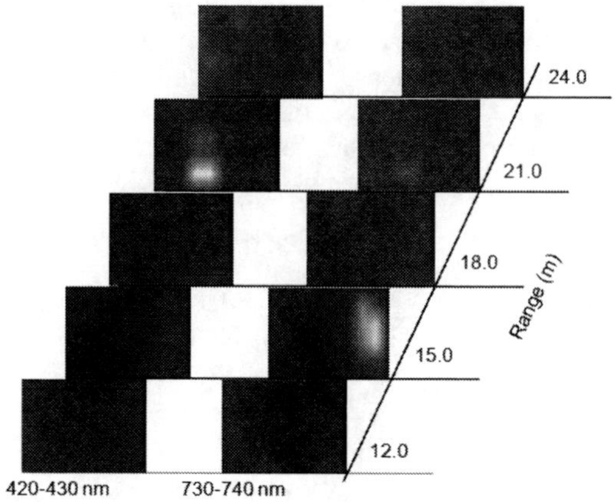

Fig. 3. Three-dimensional images of the 730–740 nm and the 420–430 nm LIF lidar spectrum.

at the position where the wooden board is placed, because the board has a very little fluorescence at the 730–740 nm range. It has been reported that ultraviolet-induced fluorescence of leaves provides us with information, for example, on photosynthesis, primary photochemical reactions, the presence and accumulation of the product [5]. This shows that the distribution of the plant living status can be remotely monitored by the LIFS lidar.

Figure 3 also shows an image having a wavelength region extracted from 420 nm to 430 nm, which are the peaks of both targets. At 15.0 m only the plant image appears but it is a quite weak fluorescence image, at 21.0 m an image of the wooden board appears clearly. The image appearing at 24.0 m is the back wall of the gym, which cannot be seen in the 730-740 nm image.

2) Remote detection of LIF spectrum of quasi bio/organic aerosols in daytime

Powdered materials were prepared as quasi aerosols. The powder was held by two non-fluorescence quartz plates located 20 m away from the lidar. An experiment of LIFS lidar fluorescence detection was made outside in the daytime. The whole system could be operated by the power generator.

Figure 4 is a spectrum of the powdered frass of fall webworm larvae. The frass is easily crushed by stepping on it and is subsequently blown into the air by the wind, becoming a bioaerosol. The surface of the frass itself looked brown, but the LIF spectrum of the powdered one resembled that of the plant leaf shown in Fig. 2. This is because pieces of leaves that the larva ate were included in the frass. The difference between the surface color and the powdered LIF spectrum is interesting because it helps to understand the process of the degradation of material.

A LIF spectrum of the powdered road corn/pylons which were wasted on the construction ground is also shown in Fig. 4. Plastic is one of the materials that becomes artificial aerosol (i.e., human made and human activity related), for example, by being crushed by a passing car. Artificial aerosols do not easily decompose in nature, so the LIF spectrum will be the same for a long time. The crushed or powdered material is carried everywhere through human activities or by natural means. As most plastics show a stronger fluorescence characteristic than do natural materials, they are a good marker to track anthropogenic activities for a very long time, as Corcorn et al. reported [6]. The LIFS lidar observation can give provide more information.

Fluorescence monitoring is usually made under dark conditions because (auto-)fluorescence intensity is very weak. In our case, the synchronous detection during a very short period of 10 ns (CCD gate width) could reduce the background sunlight noise effectively. This success offers daytime information that is fundamental and indispensable data for understanding the dynamical phenomena driven by solar energy.

5. CONCLUSIONS

Mobile laser-induced fluorescence spectrum lidar, LIFS lidar, was developed. The versatile functions made possible by the lidar mobility were confirmed by field observations. The results showed the possibility of three-dimensional monitoring of biological and organic aerosols, plant health status, the degradation process of materials, and long-time tracking of anthropogenic activities.

The compact and mobile LIFS lidar is confirmed to make field observations easier.

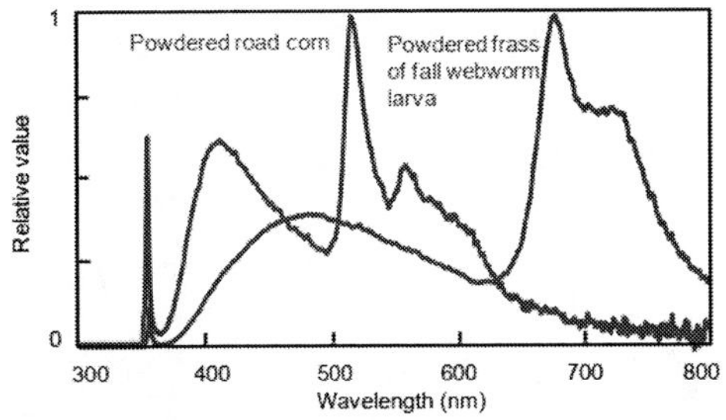

Fig. 4. Remote detection of LIF specta of powdered frass of fall webworm larvae and powdered road corn by the LIFS lidar.

ACKNOWLEDGEMENT

This work is supported by Japan Society for the Promotion of Science (JSPS), KAKENHI Grant Number 25289122, and also partially supported by Korea Institute of Civil Engineering and Building Technology (KICT), Project Number 2015-0123.

REFERENCES

[1] Saito, Y. et al., 2012: Multi-element monitoring of livingsphere with laser-induced fluorescence spectrum (LIFS) lidar, Papers of the 26th ILRC25 (29 June 2012, Porto Heli, Greece), pp. 301-304.

[2] For example, catalogue of HONDA ACTYVAN, http://www.honda.co.jp/ACTY/van/webcatalog/spec/.

[3] Yasunori Saito, 2007: Laser-induced fluorescence spectroscopy/technique as a tool for field monitoring of physiological status of living plants, Proc. SPIE, Vol. 6604, pp. 66041W-1 -11.

[4] Saito, Y. et al., 2002: Remote estimation of the Chlorophyll concentration of living trees using laser-induced fluorescence imaging lidar, Optical Review, 9, pp. 37-39.

[5] Cerovic, Z. G. et al., 1999: Ultraviolet-induced fluorescence for plant monitoring: present status and prospects, Agronomie, 19, pp. 543-578.

[6] Corcorn, P. L. et al., 2014: An anthropogenic marker horizon in the future rock record, GSA Today, 24, pp. 4-8, doi: 10.1130/GSAT-G198A.1.

1579 NM FIBER LASER SOURCE FOR SPACEBORNE MONITORING OF CARBON DIOXIDE

Nicolas Cézard*[1], Laurent Lombard, Julien Le Gouët, Didier Goular, Alexandre Bresson, Agnès Dolfi-Bouteyre, Guillaume Canat

[1]*ONERA – The French Aerospace Lab – F-91761 Palaiseau, France,*

**Email:nicolas.cezard@onera.fr*

ABSTRACT

We report on the development of a 1579 nm pulsed fiber laser source with high peak-power, intended to be used as a lidar source for CO_2 monitoring from space. We first discuss water-vapor sensitivity of spaceborne CO_2 measurements by lidar and point the interest of the 1579 nm wavelength with that respect. Then we detail the current development status of the pulsed fiber laser source.

1. INTRODUCTION

Integrated-Path Differential Absorption Lidar (IPDA) has been proposed for next-generation spaceborne CO_2 monitoring systems [1]. Two wavelengths windows, at 2 µm and 1.5 µm, were identified as suitable for this task. We will focus here on the 1.5µm band, which shows a less favorable weighting function in the troposphere but benefits of more mature detector technologies. A detailed multi-criteria wavelength selection was conducted in early studies that led to identify the 1572 nm CO_2 absorption transition as a good candidate [1]. Several laser transmitters have been proposed to address this wavelength and fulfill all the desired requirements for space operation. An OPO-based pulsed transmitter has been developed through A-SCOPE project [2], while fiber-based pulsed lasers are developed for ASCENDS[3-4]. A fiber-based, frequency modulated cw laser has also been proposed at 1573 nm [5].

In 2009, it was suggested by Caron and Durand [6] that using the CO_2 absorption transition at 1579 nm could dramatically reduce the measurement sensitivity to water vapor uncertainties. ONERA was contracted in 2011 by ESA to develop a high energy pulsed fiber laser emitting around 1579 nm, through HEPILAS project. In this paper, we first recall how water-vapor sensitivity can be evaluated for a spaceborne CO_2 IPDA system, and we confirm, with detailed conditions, that this sensitivity can be minimized by an appropriate choice of ON and OFF wavelengths around the 1579 nm transition. Then we report on the current status of the 1579 nm pulsed fiber laser source under development in HEPILAS project.

2. WATER-VAPOR SENSITIVITY OF A CO_2 IPDA LIDAR AT 1579 NM

The IPDA lidar aims at measuring the CO_2 column-integrated dry air volume mixing ratio, noted ρ_{CO2}. This quantity can be estimated from the measured Differential Absorption Optical Depth (DAOD) noted $\Delta\tau_{measured}=\tau(\lambda_{ON})-\tau(\lambda_{OFF})$, using the following relationship:

$$\rho_{CO_2} = \frac{1}{IWF_{assumed}}\left[\Delta\tau_{measured} - \Delta\tau_{H_2O\_assumed}\right]$$

The retrieval thus requires ancillary assumptions concerning values of the water-vapor DAOD and of the so-called Integrated Weighting Function (IWF) defined as:

$$IWF_{assumed} = \int_0^{P_G} \frac{\Delta\sigma_{CO_2}(P)}{g m_{dry}\left(1+k\rho_{H_2O\_assumed}(P)\right)} dP$$

Where P_G is the ground pressure, $\Delta\sigma_{CO2}(P)$ is the pressure-dependent difference between ON-line and OFF-line absorption cross-sections (in m²), g=9.8 m/s² is the Earth gravity, $k=m_{dry}/m_{H2O}$ is a constant ratio, m_{dry} and m_{H2O} are the average masses of one molecule of dry air and water vapor respectively, and $\rho_{H2O\_assumed}(P)$ is the

DOI: 10.1051/epjconf/201611917010

assumed vertical profile of water vapor dry air volume mixing ratio.

Any error concerning the water vapor profile induces errors on both IWF and $\Delta\tau_{H2O}$, that eventually translate into a Relative Systematic Error (RSE) on the retrieved value of ρ_{CO2}. However, as pointed out by Caron and Durand [6], biases induced by both terms have opposite signs and can mutually cancel under favorable conditions. A useful figure of merit for water-vapor sensitivity can be obtained assuming that the water vapor profile is known up to a scaling constant $(1+\varepsilon)$, that is :

$$\rho_{H_2O\_assumed}(P) = (1+\varepsilon)\rho_{H_2O\_real}(P)$$

Following Eqs. 18-22 in ref.6, it can then be shown that the RSE induced on ρ_{CO2} by an ε error on the water vapor profile reads:

$$RSE = -\frac{\varepsilon}{\Delta\tau_{CO_2}} \int_0^{P_G} \frac{\rho_{H_2O}(P)\left(\Delta\sigma_{H_2O}(P) - k\rho_{CO_2}\Delta\sigma_{CO_2}(P)\right)}{g m_{dry}\left[1+k\rho_{H_2O}(P)\right]^2} dP$$

This induced RSE is a bi-dimensional function of ON and OFF wavelengths. It can cancel if the water vapor DAOD is slightly positive (not null), as can be seen from the right hand side term's numerator. Figure 1 shows the RSE (expressed in dB) induced on ρ_{CO2} by an ε=5% error on the water vapor vertical profile, as a function of ON-line and OFF-line frequencies (expressed as frequencies relative to the 1579nm line center). For a worst-case assessment of humidity sensitivity, we have assumed a tropical profile for H_2O, with a 20 g/kg specific humidity on ground. A constant 380 ppm profile was assumed for CO_2. Voigt lineshapes and HITRAN 2008 database were used to compute H_2O and CO_2 cross sections.

Adapting Table.1 of ref.2 to the current wavelength, the threshold RSE (maximum tolerable) and target RSE (scientific objective) are computed to be approximately 0.02% (-37 dB) and 0.006 % (-42 dB) respectively. Figure 1 clearly shows a line were the water-vapor-induced RSE is virtually cancelled (below -60 dB). For example, if the ON-line wavelength is chosen to be +0.7 GHz, then the OFF-line wavelength should be tuned at -11.3 GHz (red point on figure 1). Though not plotted here, it can also be shown that the RSE cannot be cancelled if the ON-line frequency is chosen negative with respect to the line center.

However, beyond the water-vapor insensitivity criterion, selecting the lidar ON-line wavelength requires trading off many other figures of merit. Indeed, measurement RSE can also be induced by errors on the temperature profile, ground pressure, ground altitude, laser spectral stability, spectral linewidth, spectral purity…The Relative Random Errors (RRE) and the allowed RRE & RSE budgets are also ON-line-wavelength-dependent. But if a positive ON-line frequency can be selected as a good tradeoff, then an OFF-line wavelength can also be found so that the water-vapor induced RSE is cancelled or at least minimized. To our knowledge, this possibility was not numerically assessed in the literature so far, and such a behavior is not found in the vicinity of the A-SCOPE transition at 1572 nm. This justifies the interest of the 1579 nm wavelength for a spaceborne IPDA CO_2 lidar.

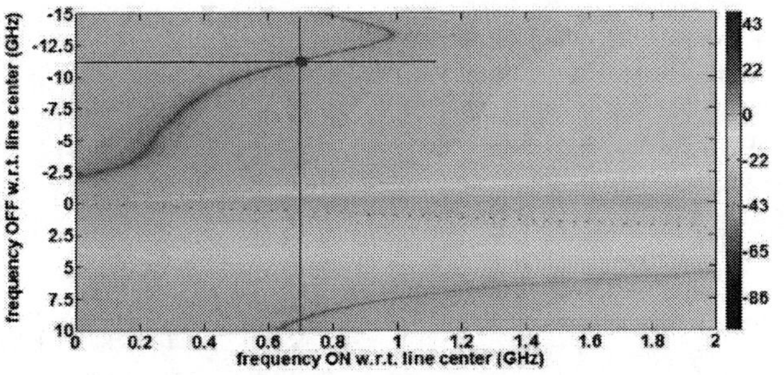

Figure 1: RSE in dB arising from a 5% error on a tropical water-vapor profile (20 g/kg at ground). The blue line in the up-left corner indicates a line of virtually cancelled water vapor sensitivity (very low induced RSE).

3. HIGH ENERGY PULSED FIBER LASER SOURCE AT 1579 NM

A high energy, pulsed fiber laser source at 1579 nm is under development at ONERA through ESA-funded HEPILAS project. The expected performances are the following: pulse energy 260µJ, pulse duration 150ns, beam quality M^2<2, pulse linewidth<60 MHz, laser stability 200 kHz over 10 seconds.

Laser amplification is challenging here because i) Erbium gain is lower at 1579 nm than at the more 'classical' 1572 nm wavelength, and ii) since short pulses are demanded, the Stimulated Brillouin Scattering (SBS) threshold peak power is quickly reached. It is worth noticing, for comparison, that the 1572 nm fiber laser under development for ASCENDS project [4] delivers much longer pulses (1.5 µs).

The HEPILAS laser system architecture is a three stage master oscillator power amplifier (MOPA) seeded by tunable and well-referenced oscillators. The seed lasers are Telecom DFB laser diodes that are stabilized by a master/slave architecture. At the output of the stabilized oscillator, 2 mW peak power with low noise are available with a time pattern composed of 'ON' and 'OFF' pulses controlled by an optical switch. The pattern pulse repetition frequency is 2 kHz.

We measured the short term stability of the beatnote of two identical free running seeders using a high speed counter. An Allan standard deviation of ~ 220 kHz over 10 s was measured for the beatnote. Assuming identical stability for both diodes, a stability of 155 kHz over 10 s can be attributed to each DFB seeder. For stable spaceborne operation, it is planned to lock a master oscillator at the center of the 1579 nm transition using a CO_2 multipass absorption cell and a modulation/lock-in technique, and then reference ON-line and OFF-line oscillators to the master using offset beatnote stabilization techniques. We also measured the spectral purity of seed lasers using a high resolution OSA (for broadband measurement) and heterodyning on an electrical spectrum analyzer (for high resolution measurement). A single longitudinal mode behavior with 50 dB of Side-Mode Suppression Ratio and a spectral purity better than 99.95% within 1 GHz were found. These parameters meet requirements for spaceborne operation.

Starting from a 2 mW peak power seed, the objective is to reach about 2 kW peak power at the exit of a three-stage amplification chain. However, amplification of single frequency DFB lasers at high peak power leads to Stimulated Brillouin Scattering. The threshold for this nonlinear effect is of the order of 80 W per meter of fiber in standard single-mode fibers. Therefore, in order to reach the desired high peak power at the third stage output, we considered simultaneous implementation of various well-known SBS mitigation techniques, such as the use of large mode area fibers, composition tuning, and strain gradient generation along the fiber.

We have built a preliminary version of the fiber laser source, sketched in figure 2.

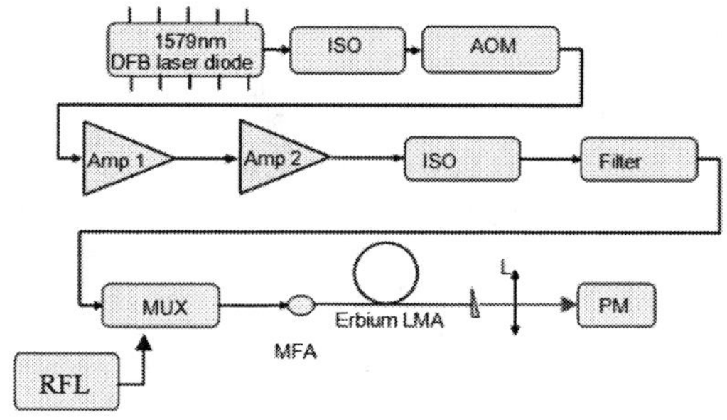

Figure 2: Preliminary setup of HEPILAS fiber laser source

A 30 mW single frequency 1579 nm DFB laser diode is connected, after optical isolation (ISO), to an Acousto-Optic Modulator (AOM) used to generate gaussian pulse shape with pulse duration 150 ns at 4 kHz pulse repetition frequency. The pulses enter a double pass preamplifier (Amp 1) built using a standard single-mode L-band fiber core-pumped at 1480 nm. The amplified pulses enter a second stage of amplification (Amp 2), built using a 12 µm-core Erbium-doped fiber, core-pumped at 1480 nm. After isolation and spectral filtering to remove the Amplified Spontaneous Emission (ASE), 31 W peak power pulses enter in the third power amplifier stage. The latter is built using a 40 µm-core nLight Er40-60 Large Mode Area (LMA) fiber. It is core-pumped in co-propagating configuration by a 4W Raman Fiber Laser (RFL) (through a single-mode wavelength multiplexer (MUX) and a Mode Field Adaptation (MFA)). The peak power is monitored by a power-meter (PM) at the third stage exit.

Using this setup, the peak power was found to be quickly limited by SBS to 550 W. We then coiled the fiber with a triangular strain distribution. It is indeed known that the central frequency of the Brillouin gain curve can be shifted by applying strain. Applying an optimal strain distribution for a given signal distribution in the fiber will result in a flat Brillouin spectrum, and consequently, an increased SBS threshold peak power. We have shown that a triangular distribution can be used in this case. With this improvement, we have obtained up to 260 µJ per pulse, corresponding to 1.7 kW peak power. The peak power was thus increased by a factor almost 4 using fiber strain distribution. The slope efficiency reaches 23%. So far the output 1579 nm peak power is only limited by the available pump power. There was some concern that strain could modify the mode repartition in the LMA fiber, but we measured an excellent beam quality with $M^2<1.2$.

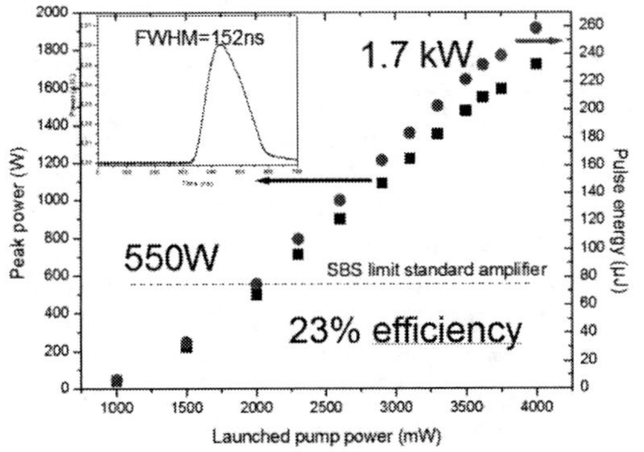

Fig. 3. Peak power and pulse energy at the exit of the third amplification stage, as a function of pump power. A triangular strain profile is applied to the LMA fiber, allowing overcoming the 550 W limit (obtained without strain).

4. CONCLUSIONS

For CO_2 monitoring by a spaceborne IPDA Lidar, working around 1579 nm could offer a very low sensitivity to water-vapor profile errors, with an appropriate choice of ON-line and OFF-line wavelengths. It can thus be seen as an interesting alternative to the initially selected 1572 nm wavelength. A fiber laser is under development at ONERA to further assess this possibility. Though the 1579 nm wavelength is less favorable than 1572 nm for Erbium-doped fiber amplification, we have obtained 1.7 kW peak power for 150 ns pulses (pulse energy 260 µJ), using SBS mitigation techniques (especially strain distribution). Further work will focus on increasing the peak power further, and assessing strained-fiber lifetime and radiation handling capabilities.

ACKNOWLEDGEMENT

The fiber laser development is funded by the European Space Agency under HEPILAS project (contract #4000104022/11/NL/CP).

REFERENCES

[1] G.Ehret et al, Appl. Phys.B 90, 593-608 (2008)

[2] J.Caron et al, SPIE Proc. 7479, 74790E, 1-15 (2009)

[3] K.Numata et al, Appl.Opt. 50, 1047-1056 (2011)

[4] W. Lu et al, SPIE Proc. 8961, 89610C, 1-10 (2014)

[5] S.Kameyama et al. Opt. Lett. 34, 1513-1515 (2009)

[6] J.Caron et Y.Durand, Appl.Opt. 48, 5413-5422 (2009)

EPJ Web of Conferences **119**, 17011 (2016) DOI: 10.1051/epjconf/201611917011
ILRC 27

CHARACTERISTICS OF THE OPG SYSTEM USIG QUASIPHASE-MATCHED NONLINEAR CRYSTALS FOR 1.6 μm CO2 DIAL

Yasukuni Shibata*, Chikao Nagasawa, Makoto Abo

Tokyo Metropolitan University, 6-6 Asahigaoka, Hino, Tokyo 191-0065, Japan

**Email: sibata@tmu.ac.jp*

ABSTRACT

We have developed a direct detection 1.6 μm differential absorption lidar (DIAL) technique to perform range-resolved measurements of vertical CO2 concentration profiles in the atmosphere. Our 1.6 μm DIAL system consists of the optical parametric generator (OPG) and amplifier (OPA) transmitter that excited by the LD pumped Nd:YAG laser with high repetition rate (500 Hz). The OPG system consists of a quasi-phase-matched (QPM) crystal and does not need a cavity. The output power of the OPA system is 6 mJ, the full width at half maximum (FWHM) of the spectrum is about 280 MHz and spectrum purity is 91.0 +- 0.2 ~ 0.5%. CO2 concentration error from fluctuation of the spectrum purity is 0.3% at 6 km altitude and 0.4 % at 10 km altitude.

1. INTRODUCTION

In recent years, there have been significant advances in a QPM nonlinear optical frequency conversion [1, 2]. The QPM condition is produced to use periodically poled ferroelectric crystals. Optical parametric oscillator (OPO), amplifier (OPA), and generator (OPG) devices are widely recognized as versatile coherent tunable spectroscopic sources. Many applications of PPLN-parametric radiation sources, such as laser remote sensing and molecular spectroscopy, require broadly tunable and narrow linewidth operation in the infrared region [3, 4].

We developed an optical parametric oscillator (OPO) laser system for 1.6 μm CO2 DIAL [5]. In order to improve the measurement accuracy of CO2 profiles, development of high power and wavelength stabilized laser system has been conducted. We have developed a new high-power 1.6 μm laser transmitter based on a parametric master oscillator-power amplifier (MOPA) system pumped by a LD-pumped Q-switched Nd:YAG laser which has the injection seed laser locked to the iodine absorption line. The master oscillator is the OPG system and the amplifier is the OPA system [6]. Since the OPO system has a cavity mirror, running the system without mode hopping requires complex control of cavity length. By contrast, the OPG system has no cavity mirror, so there is no need to control cavity length. In this paper, we describe characteristics of the OPG system for the 1.6 μm CO2-DIAL.

2. CHARACTERISTICS OF THE OPG SYSTEM

2.1 Experimental setup

The OPO system requires either active control of the cavity length or slight misalignment of the cavity. On the other hand, the OPG system does not need a cavity and instead rely on sufficient conversion efficiency to be obtained with a single pass through the crystal. Figure 1 shows a schematic diagram of the LD pumped Q-switched Nd:YAG pumped OPG/OPA system.

The OPG consisted of a QPM crystal (35×3×3 mm3) with a domain period of 30.9 μm with the duty ratio close to the ideal value of 0.5. The end facets were coated with an anti-reflection material to prevent feedback effects induced by residual reflection. The pump source was a Nd:YAG laser, that was injection seeded by the iodine locked cw Nd:YAG laser. The pulse energy of the pumping laser is 100 mJ and repetition rate is 500 Hz. This pump laser is used less than 40 mJ for each QPM crystal to avoid damaging during high-energy pumping. Each QPM crystal was mounted on a copper holder, and the temperature was maintained at 36 °C using a Peltier module. The holder' s temperature was stabilized to within 0.01 °C when the copper holder was covered with a plastic case. Single-frequency oscillation of the QPM-OPG was achieved by injection seeding, as

DOI: 10.1051/epjconf/201611917011

described in the following. Two injection seeder were DFB lasers having maximum power of 40mW with a 1MHz oscillation spectrum Their oscillation wavelength was coarse tuned by temperature and fine tuned by adjusting injection currents. The partial power of the on-line wavelength was split in the wavelength control unit. The DFB laser was locked to the on-line wavelength (1572.992 nm) by referencing the fiber coupled multipath gas cell (path length 800 mm) containing pure CO2 at a pressure of 700 Torr by monitoring the feedback signal of a wavelength-controlled unit. Stabilization was estimated to within 4.0 MHz rms of the line center of the CO2 absorption spectra. The DFB laser as an injection seeder of the off-line wavelength is operated by the free run mode with the stability of 0.5 pm/hour. Both on-line and off-line DFB lasers were connected to an optical fiber switch. Although the previous switching speed was 250 Hz. The OPG/OPA output power is 6 mJ and the full width at half maximum (FWHM) of the pulse width was about 20 ns.

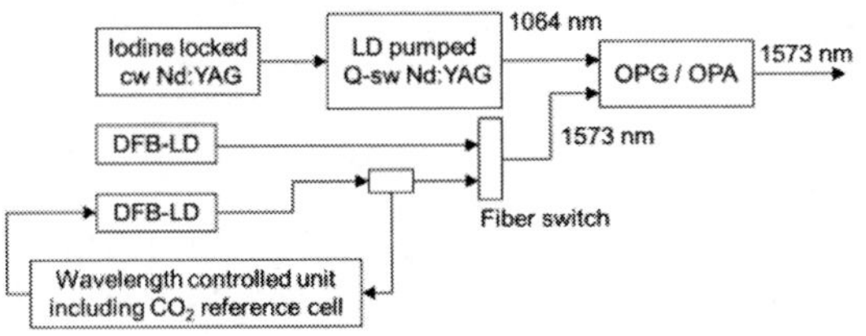

Fig. 1 Schematic diagram of the 1.6 μm transmitter for CO2-DIAL.

2.2 Line narrowing

The threshold power of the injection seeder radiation for successive injection seeding was about 1.0 mW. As the power of the injection seeder radiation was increased, the linewidth of OPA output narrowed and the peak power increased. The seeded OPA radiation was easily tuned by changing the wavelength of the injection seeder radiation, and was mode-hop free in the range of the parametric radiation spectrum of the unseeded OPA. The seeded OPA signal can be continuously tuned within about 10 nm by changing the crystal temperature using the wavelength of the injection seeder radiation. Fig. 2 shows the etalon fringe pattern (FSR = 2.0 GHz) and OPA spectra with seeded and unseeded. The linewidth of the OPA radiation was reduced to ~280 MHz or less by injection seeding. The spectral width of unseeded OPA is about 1.4 nm. Fig. 3 shows OPA spectra with on-line seeded and off-line seeded. Spectral purities of on-line seeded and off-line seeded were 91.0 % respectively. In a long term measurement, shown in Fig. 4, fluctuations of spectral purity were 0.47 % of on-line and 0.23 % of off-line. We calculate the CO2 concentration error given by this purity fluctuations. Fig. 5 shows CO2 distribution error as functions of purity error at 3 km altitude and 6 km altitude. Purity of on-line and off-line are respectively 91.0%. Influence on CO2 concentration error of offline purity fluctuation is less than the online purity fluctuation. This is due to the difference in the Influence on the absorption cross-section by prity fluctuation. Fig. 6 shows CO2 density error as a function of altitude with on-line purity 91.0±0.5% and off-line purity 91.0±0.2%. From this result, CO2 concentration error is 0.3% at 6 km altitude and 0.4 % at 10 km altitude.

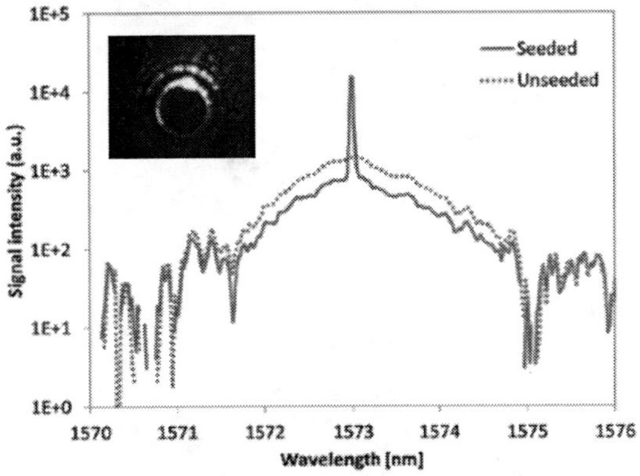

Fig. 2 Etalon fringe pattern and signal spectra of injection-seeded and unseeded. (FSR = 2.0 GHz)

DOI: 10.1051/epjconf/201611917011

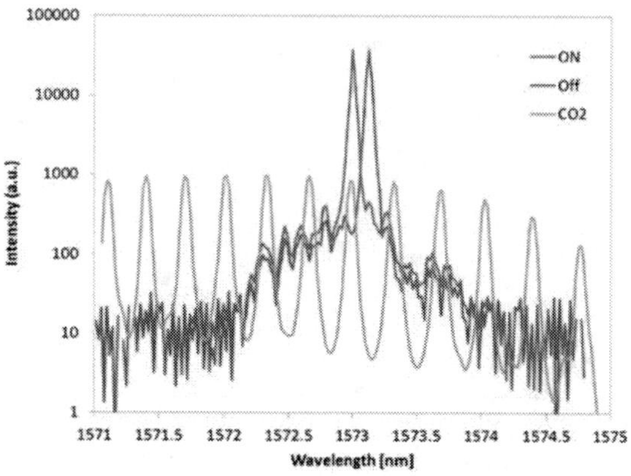

Fig. 3 Spectra of the OPA radiation measured by means of an optical spectrum analyzer: narrow-band online and offline radiation during seeded operation and calculated CO2 absorption lines.

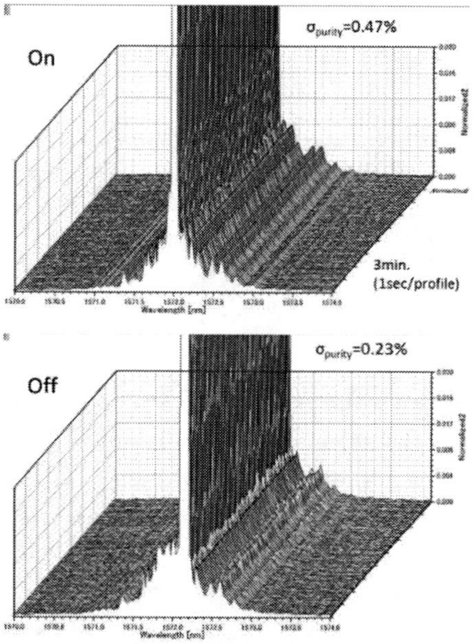

Fig. 4 Long term spectra measurements (3 min.) of on-line and off-line.

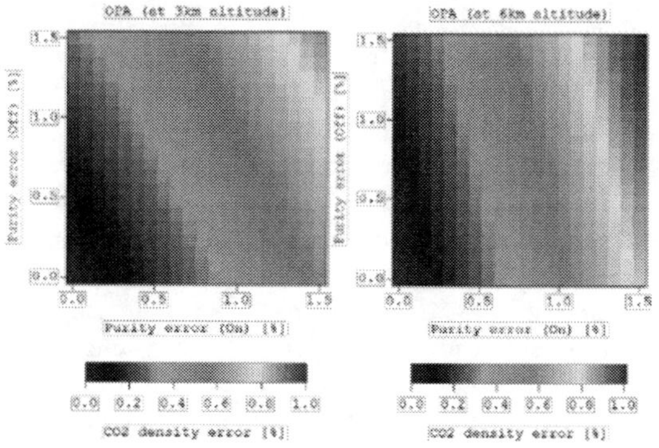

Fig. 5 CO2 density error as functions of purity error at 3 km altitude and 6 km altitude. Purity of on-line and off-line are respectively 91.0%.

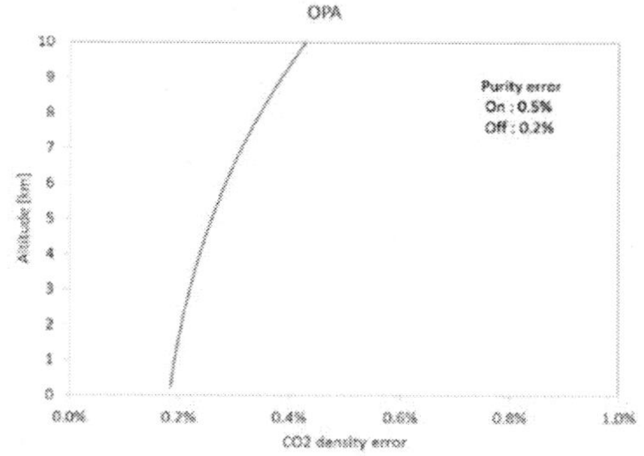

Fig. 6 CO2 density error as a function of altitude with on-line purity 91.0±0.5% and off-line purity 91.0±0.2%.

3. CONCLUSION

Injection seeding using DFB-lasers was performed with a OPG / OPA system for the 1.6 μm CO2 DIAL. The linewidth of the OPA radiation was reduced to ~280 MHz or less by injection seeding. The OPG radiation was easily tuned by changing the wavelength of the injection seeder radiation, and was modehop free. The mode-hop free performance of the OPG radiation made CO2 DIAL easy to perform. The output power of the OPA system is 6 mJ and the full width at half maximum (FWHM) of the spectrum is about 280 MHz. The spectrum purity of the OPA system is 91.0 % and fluctuations of spectral purity were 0.47 % of on-line and 0.23 % of off-line. +- 0.2 ~ 0.5%. CO2 concentration error from fluctuation of the spectrum purity is 0.3% at 6 km altitude and 0.4 % at 10 km altitude.

ACKNOWLEDGEMENT

This work was financially supported by the System Development Program for Advanced Measurement and Analysis of the Japan Science and Technology Agency.

REFERENCES

[1] J. A. Armstrong, N. Bleombergen, J. Ducuing, and P. S. Pershen, Interactions between Light Waves in a Nonlinear Dielectric, Phys. Rev. 127, 1918 (1962)

[2] J. Webjorn, F. Laurell, and G. Arvidsson, Blue light generated by frequency doubling of laser diode light in a lithium niobate channel waveguide, IEEE Photon. Tech. Lett. 1, 316 (1989)

[3] K. Suizu, H. Minamide, T. Usami, K. Nakamura, K. Kawase, H. Ito, Line-width narrowing for nanosecond PPLN optical parametric generation (OPG) by injection seeding, Lasers and Electro-Optics, 2002. CLEO '02. Technical Digest. Summaries of Papers Presented at the Conference on Lasers and Electro-Optics, 329-330 (2002)

[4] M. Rahm, G. Anstett, J. Bartschke, T. Bauer, R. Beigang and R.Wallenstein, Widely tunable narrow-linewidth nanosecond optical parametric generator with self-injection seeding, 2005 Conference on Lasers & Electro-Optics (CLEO) , 1300-1302 (2005)

[4] D. Sakaizawa, C. Nagasawa, T. Nagai, M. Abo, Y. Shibata, M. Nakazato and T. Sakai, Development of a 1.6 μm differential absorption lidar with a quasi-phase-matching optical parametric oscillator and photon-counting detector for the vertical CO2 profileAppl. Opt. Vol.48, 748-757 (2009)

[5] Y. Shibata, C. Nagasawa, M. Abo, QPM-OPG based high power 1.6 μm laser transmitter for CO2-DIAL, 25th ILRC, 243-246 (2010)

EPJ Web of Conferences **119**, 17012 (2016) DOI: 10.1051/epjconf/201611917012
ILRC 27

EMPIRICAL DETERMINATION OF OPTIMAL PARAMETERS FOR SODIUM DOUBLE-EDGE MAGNETO-OPTIC FILTERS

Ian F. Barry [1]*, Wentao Huang[1], John A. Smith[1], and Xinzhao Chu[1]

[1]*Cooperative Institute for Research in Environmental Sciences & Department of Aerospace Engineering Sciences, University of Colorado Boulder, CO80309, USA, *Email: Ian.Barry@colorado.edu*

ABSTRACT

A method is proposed for determining the optimal temperature and magnetic field strength used to condition a sodium vapor cell for use in a sodium Double-Edge Magneto-Optic Filter (Na-DEMOF). The desirable characteristics of these filters are first defined and then analyzed over a range of temperatures and magnetic field strengths, using an IDL Faraday filter simulation adapted for the Na-DEMOF. This simulation is then compared to real behavior of a Na-DEMOF constructed for use with the Chu Research Group's STAR Na Doppler resonance-fluorescence lidar for lower atmospheric observations.

1. INTRODUCTION

The Na-DEMOF has previously demonstrated a capacity for high-resolution temperature and wind measurements in the stratosphere and lower mesosphere [3]. However, processing these measurements requires calibration that assumes vertical winds that average to zero, resulting in a set of calibration constants that vary on a nightly or even an hourly basis. Figure 1 shows the result of deriving temperature and vertical wind after calibrating the system every 30 minutes and integrating over the 4-hour collection period. Using the average calibration over the entire collection period results in a negative 3-10 K bias in temperature. Reliably calibrated data can only be obtained by reducing this instability, which in turn necessitates the identification and stabilization of the factors involved. To this end, the theoretical and practical variables that are known to affect Na-DEMOF performance will be described in terms of their stability and effects on the filter function.

The four theoretical factors that have been identified as influential on vapor cell filter performance are: vapor density, vapor temperature, optical thickness, and magnetic field strength. The simulation developed by S. D. Harrell and modified by Wentao Huang to calculate the absorption and dispersion of orthogonal circular polarizations of light within a Na vapor cell allows for calculation of the DEMOF filter function by specifying each of these variables [1, 2]. Vapor density is determined by controlling the temperature at which a reservoir of solid Na is held, and optical thickness is determined by this vapor density and the physical length of the cell.

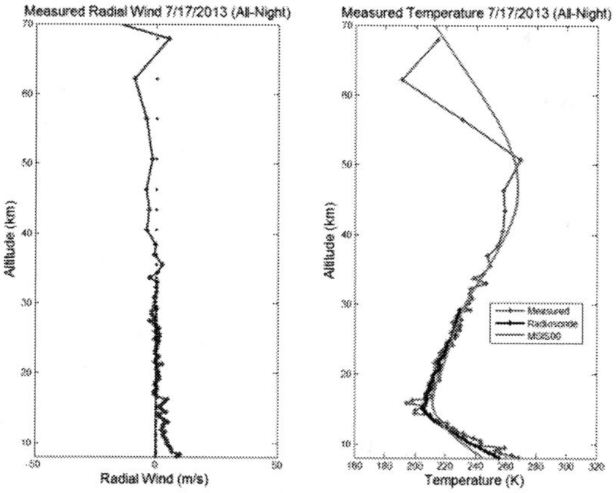

Fig. 1. Measurements of temperature and vertical wind achieved from 10 to 65 km using 30-minute calibration scheme.

The additional practical considerations that have been found to affect Na-DEMOF performance since its construction are: vapor cell window depolarization caused by stress creating scattered areas of disorganized birefringence, Na condensation on the vapor cell windows, and oxidation of the solid Na reservoir due to imperfections in vapor cell vacuum. The window depolarization results in a constant efficiency reduction but can be avoided by loosening the structure holding the windows, and the window condensation has been mitigated by locating the heaters near the windows and occasionally evaporating the condensate from the outside using a heat gun. Oxidation causes a progressive reduction of Na vapor availability requiring higher temperatures to produce the desired amount of vapor, and new manufacturing methods are being employed to improve the vacuum and resultant vapor cell lifetimes.

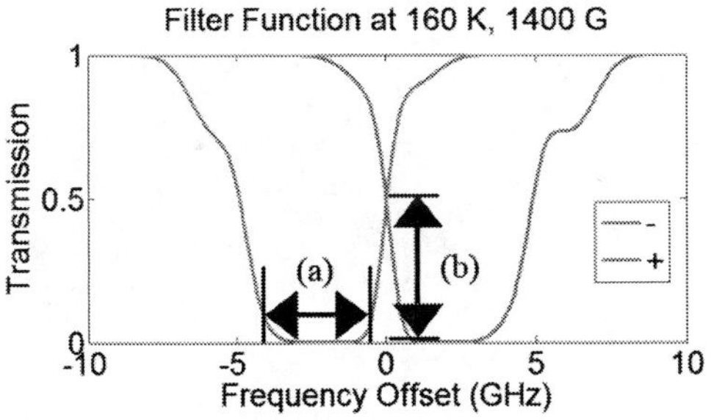

Fig. 2. Example filter function with temperature at 160 K and magnetic field strength at 1400 G. (a) shows the trough width and (b) shows the crossover height. The edge slope is determined by the slopes in the region within 0.5 GHz of the crossover point.

Figure 2 provides an example of an ideal simulated Na-DEMOF filter function for lower altitude wind and temperature derivations. In order to provide the optimal sensitivities for useful data acquisition, the transmitted laser frequencies must lie within the region shown in the center of the figure, where the two channels of the DEMOF act as sloped edge filters. This is best achieved by tuning the temperature and magnetic field used to condition the vapor cell such that the crossover point occurs near 50% transmission, the edge of the filter are sloped gradually enough to accommodate the transmitted laser wavelengths, and the transmission runs from near 100% outside the filter region to near zero within the trough. To determine the performance of a Na-DEMOF filter, these characteristics have been defined as crossover height, edge slope, and trough width.

Ideally, a Na-DEMOF should also maintain stable operation for extended periods of time, requiring calibration no more frequently than once per night and replacement after no less than a month for reasonable operability. The final paper will demonstrate these stabilities and the derived atmospheric winds and temperatures, including analysis of the sensitivity of the derived results to slight errors in calibration.

Later on we discovered that the hourly or half-hourly variations of calibration factors were caused by the operators' periodic rotation of a quarter wave plate used in the Na-DEMOF receiver setup. Such manual rotation of the quarter wave plate could not ensure a stable calibration of polarization states, resulting in varied calibration factors depending on the orientation angle of the quarter wave plate. Once the quarter wave plate was fixed to an angle (supposedly the optimum orientation), the calibration factor became constant throughout the night observations.

2. METHODOLOGY

Design considerations limit the difference that may be achieved between the vapor temperature and the "tip temperature" for the solid Na reservoir controlling the vapor density, and in the cell designed by Wentao Huang the same temperature controller is used for both. This simplifies the theoretical problem to one dependent only on temperature, magnetic field strength, and vapor cell length. The temperatures of the Na reservoir and vapor within the cell are not determined directly, but due to the insulation around the vapor cell can be assumed to lie steadily within 2-5 K of the control temperature, with heaters located near the windows and temperature measured by an RTD near the center of the cell. The control temperature can be set and obtained within 0.1 K, and length is fixed but can be determined accurately to within a fraction of a millimeter. However, the magnetic field is neither constant, as assumed by the simulation, nor easily characterized. Similarly, the vapor density is affected by the age of the vapor cell, the mass of solid Na present in the reservoir, and the insulation-dependent temperature distribution around the cell. Therefore, the filter function must be tested over a range of temperatures in order to empirically determine the operating temperature which produces a crossover height of 50% transmission.

In order to guide this testing, it is important to have an idea of how temperature and magnetic field strength affect the filter function. To this end, the Na-DEMOF simulation code was used to determine the trough widths, edge slopes, and crossover heights of filter functions for ideal vapor cells held at various temperatures within various strengths of magnetic field.

3. RESULTS

Figures 3-5 show the dependence of the three key filter function characteristics on temperature and magnetic field strength. Trough width was found to increase with increasing temperature, edge slope was found to decrease with increasing temperature except at a magnetic field strength range between 1350-1400 Gauss, and crossover height was found to decrease with increasing temperature. Figure 5 can also be used to determine the effective magnetic field strength of a Na-DEMOF filter once the optimal temperature producing 50% transmission crossover height has been achieved.

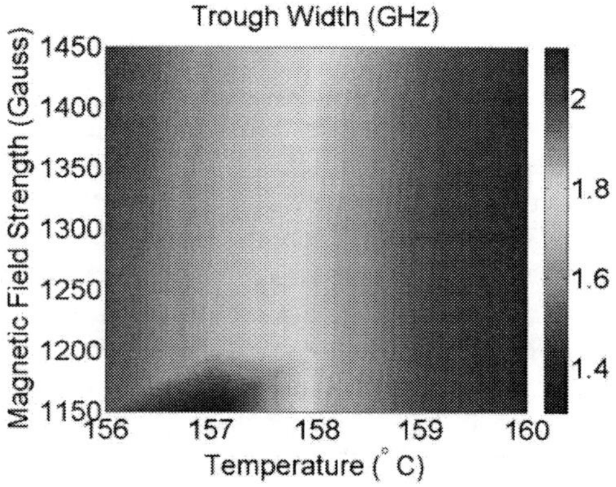

Fig. 3. Dependence of trough width on temperature and magnetic field strength.

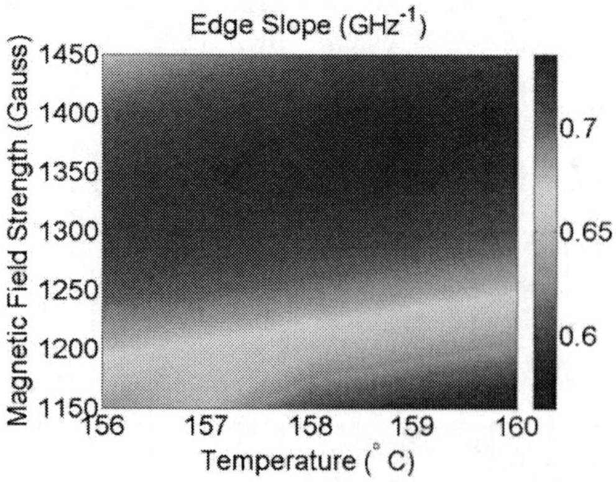

Fig. 4. Dependence of edge slope on temperature and magnetic field strength.

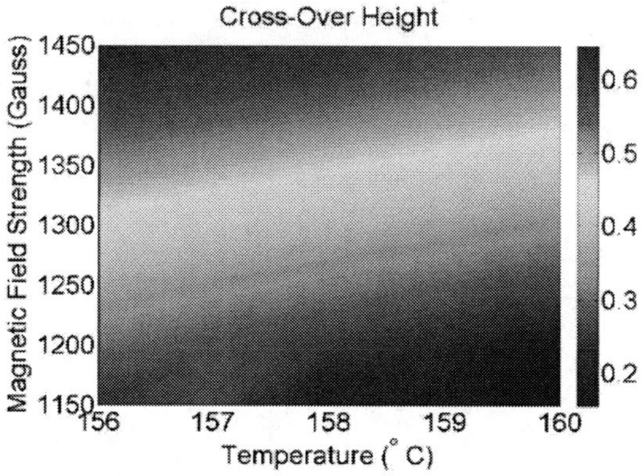

Fig. 5. Dependence of crossover height on temperature and magnetic field strength.

4. CONCLUSIONS

For the greatest sensitivity determined by edge slope, a temperature of 160° C or hotter and a magnetic field strength of around 1400 Gauss is ideal. However, for a broader edge filter function allowing the widest range of transmitted laser wavelengths, lower magnetic field strengths or higher temperatures may be required. In either case, higher temperatures provide more complete rejection within the function trough and thus more effective return signal frequency dependence. However, sensitivity from high temperatures must be traded off with cell lifetime degradation from oxidation and condensation, so the lowest possible temperature for a given application will grant the longest lifetime. The results of a filter optimized using this method will be presented in the final paper, demonstrating derivation of temperature and zonal and meridional winds from the stratosphere throughout the mesosphere and lower thermosphere.

ACKNOWLEDGEMENT

We acknowledge the support of a NSF CRRL grant (AGS-1136272) and CEDAR grant (AGS-1042257). I.B. was partially supported by PLR-1246405.

REFERENCES

[1] Harrell, S. D., C.-Y. She, T. Yuan, D. A. Krueger, H. Chen, S. S. Chen, Z. L. Hu, 2009: Sodium and potassium vapor Faraday filters revisited: theory and applications, J. Opt. Soc. Am., **26(4)**, 659-670.

[2] Huang, W., X. Chu, B. P. Williams, S. D. Harrell, J. Wiig, C. Y. She, 2009: Na double-edge magneto-optic filter for Na lidar profiling of wind and temperature in the lower atmosphere, Optics Letters, **34(2)**, 199-201.

[3] Huang, W., X. Chu, J. Wiig, B. Tan, C. Yamashita, T. Yuan, J. Yue, S. D. Harrell, C.-Y. She, B. P. Williams, J. S. Friedman, R. M. Hardesty, 2009: Field demonstration of simultaneous wind and temperature measurements from 5 to 50 km with a Na double-edge magneto-optic filter in a multi-frequency Doppler lidar, Optics Letters, **34(10)**, 1552-1554.

[4] Schreier, F., 1992: The Voigt and Complex Error Function: A Comparison of Computational Methods, J. Quant. Spectrosc. Radiat. Transfer, **48(5/6)**, 743-763. O'Connor, E. J., A. J. Illingworth, R. J. Hogan, 2004: A technique for autocalibration of cloud lidar, *J. Atmos. Ocean Tech.,* **21(5)**, 777-778.

PULSE-SHAPE CONTROL IN AN ALL FIBER MULTI-WAVELENGTH DOPPLER LIDAR

Albert Töws*, Jan Lehmann and Alfred Kurtz
Institute of Physics, Cologne University of Applied Sciences, 51643 Gummersbach, Germany,
**Email: albert.toews@fh-koeln.de*

ABSTRACT

Pulse distortion during amplification in fiber amplifiers due to gain saturation and cross talk in a multi-wavelength Doppler lidar are discussed. We present a feedback control technique which is capable of adjusting any predefined pulse shape and show some examples of feedback controlled pulse shapes.

1. INTRODUCTION

Using fiber amplifiers in MOPA configuration in all-fiber coherent Doppler lidar systems has become more attractive due to many advantages of such configurations. [1]

Erbium doped fiber amplifiers (EDFA) are pumped with laser diodes to achieve the population inversion which is needed for amplification of the injected seed pulses. During the amplification of the seed pulses the population inversion decreases as the pump process is rather slow compared to the pulse duration. This leads to a drop in gain along the pulse duration and characteristic pulse shapes with a large spike on the leading edge of the output pulses are generated. Therefore, all-fiber lidar systems show strong pulse distortion due to gain saturation at typical pulse lengths and repetition frequencies. [2]

The main pulse energy limiting effect is the stimulated Brillouin scattering (SBS). Due to the large spike on the leading edge of the output pulse, the SBS threshold is reached at very low possible pulse energy. Therefore, a certain preshaping of the seed pulse is necessary to prevent SBS at very low energy. [3]

In this multi-wavelength lidar system four pulses with different wavelengths are simultaneously amplified in one EDFA. Every wavelength reduces the density of excited Erbium ions in its specific sublevel. By means of fast phononic relaxation processes between energy sublevels the reduction of population inversion is balanced. This leads to a mutual gain dependence of simultaneously amplified wavelengths due to crosstalk between those wavelength-channels. [4]

Furthermore, pulse shape and peak power strongly depend on pump power, temperature changes, and ageing of the EDFA.

Due to pulse distortion caused by the mentioned processes, a careful control of pulse shapes is essential for stable and reliable lidar system operation. One important advantage of fiber amplifiers is that the output shape can be affected by the shape of the seed pulse. Different compensation techniques were presented in [5,6]. In this work we present a pulse shape feedback control technique which is capable of adjusting any given pulse shape for each channel of the four-wavelength Doppler lidar system.

2. METHODOLOGY

The schematic setup of the four-wavelength Doppler lidar system with a feedback controlled pulse shaping unit is shown in figure 1. The four-wavelength lidar consists of five overall units: Master oscillator unit (MO), pulse-shaping unit, amplifier and transceiver unit, detector unit, and signal processing unit. The MO unit consists of four external cavity diode lasers. The wavelengths are chosen from the ITU-grid near 1.55 μm. All four channels are multiplexed by a wavelength division multiplexer (WDM) and amplified by one EDFA. 80% of the MO laser light is directed to the pulse shaping unit, where the laser light is demultiplexed, in order to shape the pulses of each channel with electro-optic modulators (EOM) separately. Then all of the pulses are frequency shifted and preshaped by an acousto-optic modulator (AOM). The shifted and shaped pulses are amplified with a two-stage EDFA. Via a circulator the amplified pulses are directed to the

DOI: 10.1051/epjconf/201611917013

telescope. In the detection unit the wavelengths of the local oscillator and the backscattered light are demultiplexed, thus, every channel can be coupled separately to the balanced detector. The amplified differential signal eliminates the DC and amplifies the AC component of the heterodyne signal. The signal processing unit extracts the information about wind velocity and signal strength. More detailed information about the four-wavelength Doppler-lidar is found in [7], including advantages using multiple wavelengths.

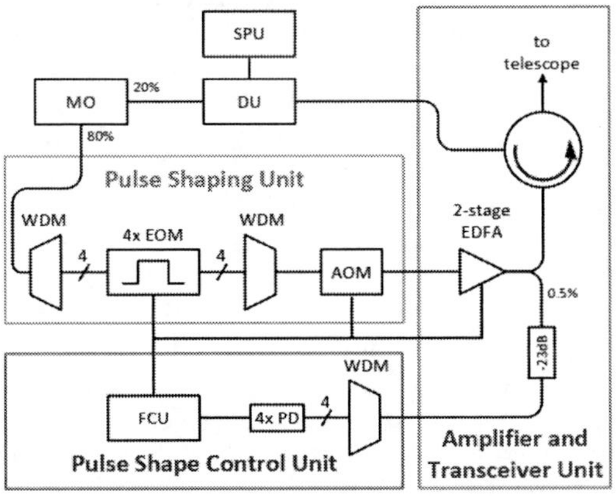

Figure 1: Schematic setup of the four-wavelength lidar system with feedback controlled pulse shaping (WDM – wavelength division multiplexer; EDFA – erbium doped fiber amplifier; EOM – electro-optical modulator; AOM – acousto-optical modulator; MO – master oscillator, DU – detector unit; SPU – signal processing unit; FCU – feedback control unit)

The system described in [7] was extended by a pulse shape control unit (PSCU). 0.5 % of the output pulse power of the EDFA is used to monitor the pulse shape. After attenuation, the wavelength-channels are demultiplexed. Four fast photodiodes measure the shape of the pulses for each wavelength separately. The pulses are digitized with a sampling rate of 100 MHz by an analog-to-digital converter within the feedback control unit (FCU). The FCU is connected to a personal computer (PC), where the pulse shape and the SBS threshold can be predefined. By means of a digital-to-analog converter, the FCU controls the EOMs, the AOM, and the pump power of the EDFA. To achieve a larger control range it is useful to preshape the pulses with the AOM.

Starting the PSCU, the EDFA is turned on at low pump power to avoid excess of the SBS level. Then the feedback control unit calculates the correction needed with an integral controller and it changes the seed pulse shape with the EOMs for each channel. Simultaneously, the EDFA pump power is controlled to adjust the output power of the EDFA close to SBS threshold.

The pulse control is performed by a PC software and consists of two nested control loops as illustrated in figure 2.

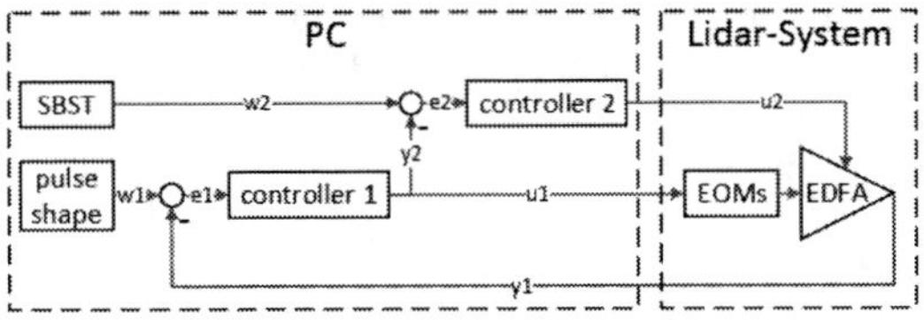

Figure 2: Schematic layout of the pulse control loops (EDFA – erbium doped fiber amplifier; EOM – electro-optical modulator; SBST – stimulated Brillouin scattering threshold)

Control loop one (blue) is responsible for pulse shape control. It stabilizes user defined pulse shapes by regulating the value for each sample of the pulses. The EOMs represent the actuator of this control loop, while the EDFA is the control path.

Control loop two (red) controls the pump power of the EDFA. The pump power is automatically adjusted to the lowest level at which the pulses can be completely developed, depending on the preset pulse amplitude. In most cases this amplitude is chosen close to the SBS threshold. There are two benefits resulting from this approach. On the one hand the amplified spontaneous emission (ASE) of the EDFA is reduced. On the other hand the amplitude resolution of the pulses is maximized by using the full dynamic range of the EOMs.

The controlled variable of control loop two is the maximum amplitude value of the EOMs used for pulse shaping. The EDFA is used as an actuator by manipulating the pump power, which affects control loop one as a disturbance variable. Furthermore, control loop one acts as a control path for control loop two. For those reasons the regulation of the pump power is designed to work much slower than the regulation of the pulse shape. This enables an adjustment of the pump power and pulse shape at the same time.

3. RESULTS

Amplifying four shifted pulses simultaneously by one EDFA, all pulses are distorted through gain saturation without preshaping and feedback control (figure 3). The first pulse (green) is much stronger than the following pulses. The dip following the leading edge of the green pulse shows that the SBS threshold is already reached at very low pulse energy due to the large spike.

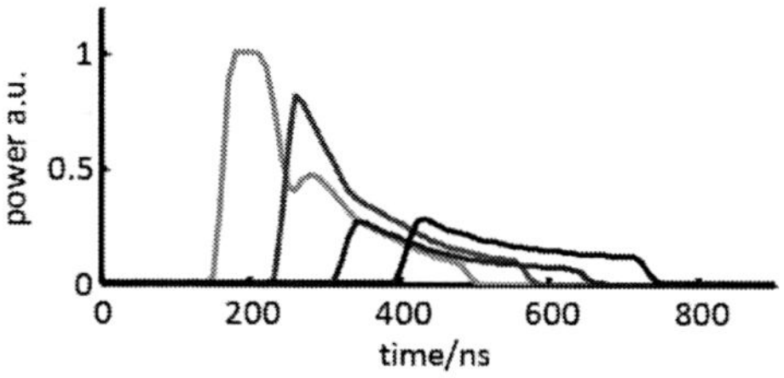

Figure 3: Four shifted pulses simultaneously amplified without preshaping and feedback control

For best preshaping using the AOM, a second degree polynomial input signal was chosen. Figure 4a shows the preshaped seed pulses. As the square pulses created by the EOMs pass a single AOM, as shown in figure 1, they are all affected in the same manner.

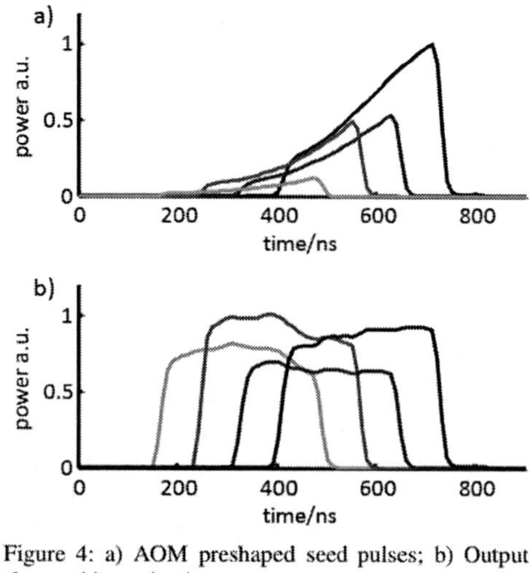

Figure 4: a) AOM preshaped seed pulses; b) Output pulse shape with preshaping

Using this preshaping with the AOM, the effect of gain saturation is significantly reduced. Figure 4b demonstrates the positive effect of preshaping. However, the pulse shapes of the wavelength-channels are disturbed by crosstalk between those pulses.

To achieve user defined and long-term stable pulse shapes, preshaping is not sufficient. The remaining distortions can be compensated by feedback controlling of pulse shapes using the EOMs. Figure 5 shows four shifted square pulses which are amplified in one EDFA with preshaping and feedback control by the PSCU. The corrected seed pulses by the EOMs and AOM can be seen in figure 5a. These corrections were necessary to generate a pulse train of four delayed square pulses (figure 5b). Obviously, all distortions are well regulated. To work stable at the falling and rising edge, the control algorithm needs at least two or three points on the edges. Since the sampling rate is 100 MHz the rise-time and trailing-edge time of the controlled pulses is limited to the range of 20 ns to 30 ns. This limitation can be avoided by applying a higher sampling rate. However, in general the main limitation is due to the rise-time and trailing-edge time of the AOM and EOMs, these are in the range of 20 ns in this system.

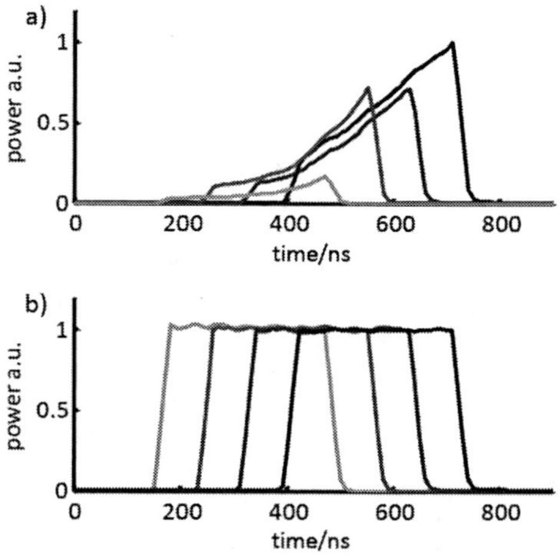

Figure 5: a) Seed pulses with feedback control; b) Output pulse shape of four simultaneously amplified and shifted pulses

Another advantage of the feedback control is that the chosen pulse shapes can be different on each channel as every channel is controlled separately. As an example, a Gauß-shaped and a square pulse are controlled at the same time. This capability enables us to compare the response of both pulse shapes under the same atmospheric conditions with the four-wavelength lidar system. Both regulated pulse shapes with a full-width half-maximum (FWHM) of 300 ns are shown with their corresponding seed pulse in figure 6.

Due to the stronger gain applied to the leading edge of the pulses, the output pulse (figure 6b) seems to rise earlier than the seed pulse.

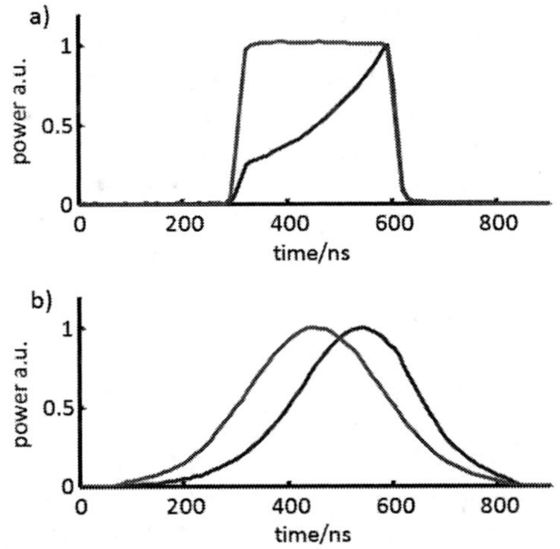

Figure 6: a) Channel with square pulse; b) Channel with Gauß-shaped pulse (blue: seed pulse, red: output pulse)

This system is capable of adjusting any given pulse shapes with a pulse duration from 100 ns to 1200 ns. The only limitation for the sake of stability is the rise-time and falling-edge time of at least 20 ns.

In general, the feedback control unit needs about five minutes from start-up to completely adjusting the predefined pulse shapes. Short disturbances are compensated within seconds by control loop one. Long term changes like temperature influences are compensated mainly by control loop two.

4. CONCLUSIONS

We present a feedback control technique which is capable of adjusting any given pulse shape for each channel of a multi-wavelength Doppler lidar system.

In future we will increase the control speed by applying models for gain saturation. In addition, we will use the backscattered SBS-Stokes wave to detect the SBS threshold for each wavelength-channel.

ACKNOWLEDGEMENT

The authors acknowledge the Federal Ministry of Education and Research (BMBF) for financial support under FKZ 17055X10.

REFERENCES

[1] S. Kameyama, T. Ando, K. Asaka, Y. Hirano and S. Wadaka, Compact all-fiber pulsed coherent Doppler lidar system for wind sensing, *Appl. Opt.*, **46**, 1953-1962.

[2] A.W. Naji, et al., Review of erbium-doped fiber amplifier, *International Journal of the Physical Sciences*, **6(20)**, 4674-4689.

[3] R. Su, et al., Dependence of stimulated Brillouin scattering in pulsed fiber amplifier on signal linewidth, pulse duration, and repetition rate, *COL*, **10(11)**.

[4] M. Dasan, et al., An investigation on transient effects in EDFA with variable duty cycle and wavelength multiplexed signals, *Microwave and Optical Technology Letters*, **57(2)**, 352-358.

[5] M. Michalska, J. Swiderski and M. Marcin, Arbitrary pulse shaping in Er-doped fiber amplifiers - Possibilities and limitations, *Optics & Laser Technology*, **60**, 8-13.

[6] G. Sobon, et al., Pulsed dual-stage fiber MOPA source operating at 1550 nm with arbitrarily shaped output pulses, *Appl. Phys. B*, **105**, 721–727.

[7] A. Töws, A. Kurtz: A multi-wavelength lidar system based on an erbium-doped fiber MOPA-system, *Proc. SPIE*, **9246**.

EPJ Web of Conferences **119**, 17014 (2016)
ILRC 27

DOI: 10.1051/epjconf/201611917014

SIGNAL TO NOISE RATIO CHARACTERIZATION OF COHERENT DOPPLER LIDAR BACKSCATTERED SIGNALS

Sameh Abdelazim[1*], David Santoro[2], Mark Arend[2], Fred Moshary[2], Sam Ahmed[2]
[1]*Fairleigh Dickinson University, Teaneck, NJ, 07666, USA, *EMAIL: azim@fdu.edu*
[2]*The City College of New York, New York, NY 10031*

ABSTRACT

An eye-safe coherent Doppler Lidar (CDL) system for wind measurement was developed and tested at the Remote Sensing Laboratory of the City College of New York (CCNY). The system employs a 1542 nm fiber laser to leverage components' availability and affordability of the telecommunication industry. A balanced detector with a bandwidth extending from dc to 125 MHz is used to eliminate the common mode relative intensity noise (RIN). The system is shot noise limited i.e., the dominant component of received signals' noise is the shot noise. Wind velocity can be measured under nominal aerosol loading and atmospheric turbulence conditions for ranges up to 3 km while pointing vertically with 0.08 m/s precision.

1. INTRODUCTION

Coherent Doppler Lidar (CDL) is widely used for remote sensing of the atmosphere. The first CDL wind-sensing system was reported by Huffaker *et al.* [1] in 1970, where a 10.6 µm cw CO_2 laser was used. Since the late 1980s, researchers started to deploy CDL systems with newly developed solid-state lasers which lead to advantages of size, weight, reliability, and lifetime. Operation at shorter wavelengths allows for higher spectral resolution, which means higher velocity resolution. Much effort has been put into 2 µm pulsed systems mainly intended for wind measurements [2] [3]. Hawley *et al.* [4] developed a 1.06 µm pulsed CDL system for wind sensing. Karlsson *et al.* [5] reported a 1.5 µm cw all-fiber wind sensing CDL system. An all-fiber coherent Doppler lidar for wind sensing was developed and tested at the Remote Sensing Laboratory of the City College of New York (CCNY) that utilizes 1.5 µm fiber laser pulsed at 20 KHz [6] [7] [8].

2. SYSTEM'S CONFIGURATION

The system's configuration is shown in Fig. 1 and consists of a 1542 nm fiber laser with two outputs; one output is used as a local oscillator (LO) whereas the other output is pulse shaped and then amplified before being transmitted into the atmosphere. For pulse shaping, the system uses two acousto-optic modulators (AOM) with 42 MHz frequency shift each, resulting in an 84 MHz total frequency shift of the input laser frequency. An optical amplifier is then deployed to increase the pulse energy to 14 µj at a repetition rate of 20 kHz before transmitting the pulse into the atmosphere. The laser pulse is coupled to the telescope (100 mm Diameter) with an optical circulator. The telescope also collects the backscattered atmospheric signals and via the same circulator directs the signals to an optical coupler. The Local oscillator and atmospheric backscattered signals are optically combined using a 2x2 variable coupler. The two outputs of the coupler are fed to a balanced detector to eliminate the common mode relative intensity noise (RIN) of the laser source that mixed the local oscillator with the atmospheric signal. Received signals are sampled at 400 MHz using an analog to digital converter card (ADC) equipped with an FPGA for signal pre-processing. The balanced detector has a bandwidth extending from dc to 125 MHz and provides transimpedance amplification and a second stage voltage amplification. Anti-aliasing filters precede the ADC of the receiver.

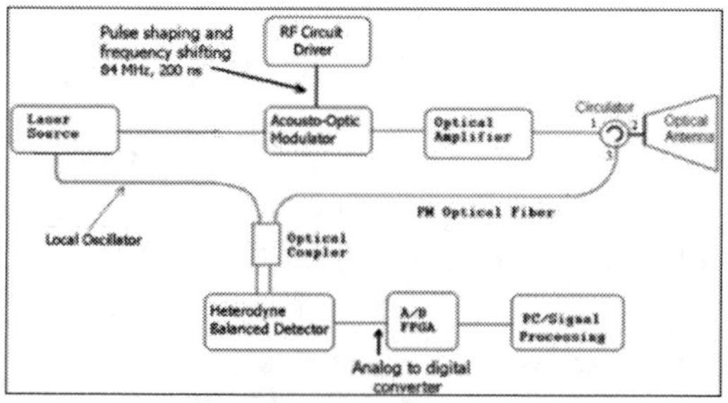

Fig. 1 Coherent Doppler lidar system's configuration.

3. THE SYSTEM'S SNR ANALYSIS

The system's receiver has a non-flat frequency response. Figure 2 shows the response of the receiver due only to shot noise from the local oscillator. The power spectra of received signals need to be corrected for the detector's non-flat gain shape. This is done by dividing the power spectrum of received signals by power spectrum of reference (shot noise) signal.

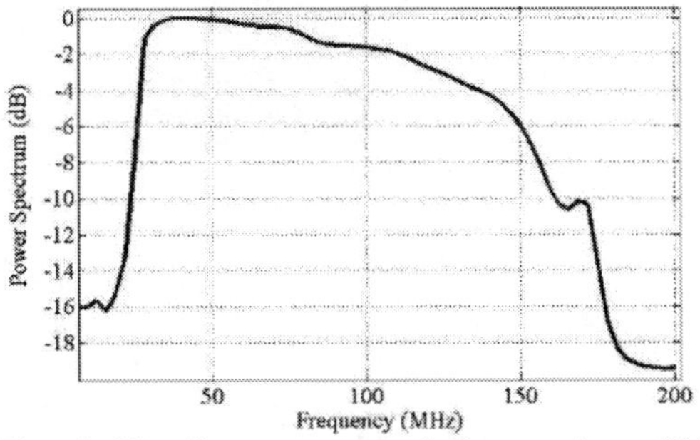

Fig. 2 Non-flat response of the receiver. This response was obtained by applying only the (LO) to the balanced detector while turning off and removing the input to the final amplifier. The vertical axis is in dB and normalized to the peak.

The reference signal is obtained using the balanced detector while LO is turned on and no atmospheric received signals are present (final amplifier is turned off), i.e. this power spectrum only represents shot noise. The SNR can be presented as follows:

$$SNR = \frac{P_{sig}^{(N)}}{P_{nois}} \qquad (1)$$

where, P_{nois}, and $P_{sig}^{(N)}$ are the noise and signal power spectra accumulated over N pulses, respectively. For a shot noise limited receiver system, the noise power is due to shot noise: $P_{shot}^{(N)}$ which can be written as:

$$P_{shot}^{(N)} = < P_{shot} > \pm \sigma_{shot}^{(N)} \qquad (2)$$

where; $< P_{shot} >$ is the average shot noise power, which is a fixed level characterized by the laser source, and $\sigma_{shot}^{(N)}$ is the shot noise variation around its fixed level when averaged N times (standard deviation of the shot noise for N accumulation). The mean shot noise fixed power level is given by:

$$P_{shot} = 2eiB \qquad (3)$$

where; e is the electronic charge, i is the photo-current, and B is the detector's bandwidth. Since shot noise follows Poisson's distribution then the averaged shot noise variation $\sigma_{shot}^{(N)}$ can be expressed as:

$$\sigma_{shot}^{(N)} = \frac{P_{shot}}{\sqrt{N}} \qquad (4)$$

Therefore, SNR, equation (1), can be re-written as:

$$SNR = \sqrt{N}\left(\frac{P_{sig}^{(N)}}{P_{shot}}\right) \qquad (5)$$

Correcting for the non-flat detector's gain involves dividing accumulated measured signals' power spectrum by a reference signal's power spectrum, which can be expressed as:

$$\frac{P_{measured}}{P_{ref}} \qquad (6)$$

where; $P_{measured}$ is the measured signals' power, which includes both signals and noise powers, and can be expressed as:

$$P_{measured} = P_{sig}^{(N)} + P_{Noise} \qquad (7)$$

$$= P_{sig}^{(N)} + P_{shot}^{(N)} \qquad (8)$$

P_{ref} is the power spectrum of LO when no received signals are input to the detector, i.e. only represents shot noise. As a result, the ratio between measured power and ref. power (equation 6) can be written as:

$$\frac{P_{measured}}{P_{ref}} = \frac{P_{sig}^{(N)}}{P_{shot}^{\prime(N)}} + \frac{P_{shot}^{(N)}}{P_{shot}^{\prime(N)}} \qquad (9)$$

$$= \frac{P_{sig}}{P_{shot}} + 1 \qquad (10)$$

This quantity has an average value $< \frac{P_{measured}}{P_{ref}} >$ and a (fluctuation) standard deviation value σ that can be expressed as follows:

$$\sigma = \sqrt{\frac{2}{N}} \qquad (11)$$

In our analysis, we calculate the following parameter: $\frac{P_{measured}}{P_{ref}} - 1$ by dividing signals power spectrum by a reference signals power spectrum and then subtracting one, which results to:

$$\frac{P_{measured}}{P_{ref}} - 1 = \frac{(SNR)}{\sqrt{N}} + \sqrt{\frac{2}{N}} \qquad (12)$$

Therefore, when accumulating 10,000 pulses and in order to extract signal out of noise, the signal power should be at least equal to noise power, i.e. SNR =1. As a result, the value we calculate is equal to: $\frac{1}{\sqrt{10{,}000}} + \frac{\sqrt{2}}{\sqrt{10{,}000}} = 0.024$, which is the value we set as a threshold below which received signals are ignored. It is worth noting that the signal power we report is not really the signal power, but it's the signal power normalized to the shot noise, in other words, it is the SNR for a single shot. To verify that the standard deviation (fluctuation) of the result of equation 10 (measured signals divided by reference signals) follows: $\sqrt{\frac{2}{N}}$ relationship, the ratio of measured noise to a reference shot noise was characterized by calculating the standard deviation of approximately 250k points. Each point is a measured noise power accumulated 10k times divided by a reference shot noise power also accumulated 10k times. The standard deviation is found to be 0.0155, approximately 10% more than what we estimated in equation 13. This difference can be a result of an effect of other noise sources such as thermal noise, i.e. the system is not 100% shot noise limited. In other words, the total noise in the receiver is not solely due to shot noise, but other noises have a small contribution as well. The histogram of the power ratio of the 250k analyzed points is shown in figure 3.

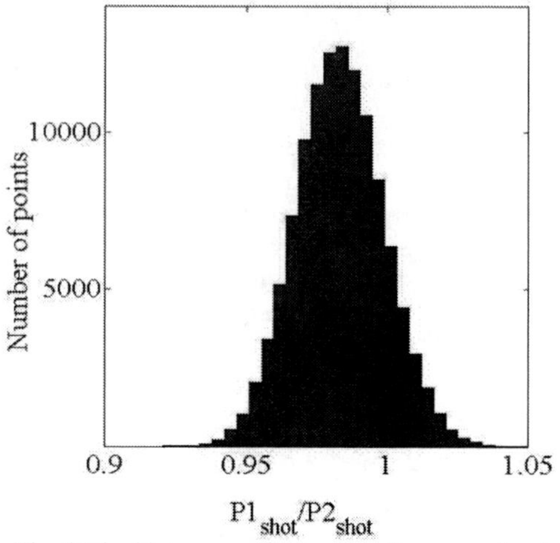

Fig. 3 The histogram of the ratio of measured noise power (10k accumulations) to a reference shot noise power (also 10k accumulations)

4. MEASUREMENTS

The SNR of backscattered signals and vertical wind velocity vs. time and height measured at the Remote Sensing Laboratory of the CCNY between 14:00-1800 (EDT) on July 12th, 2012 are presented in Fig. 4 (A) and (B) respectively. Over the course of four hours backscattered signals were measured up to approximately 2.5km vertically. Initially the SNR peaks between 0.5km and 1km. Cloud patterns appear at approximately 14:20 and 14:40 at 2km where the SNR is significantly greater than that of lower altitude. The SNR continues to peak at 1km throughout the measurement period. Fig. 4(B) shows the vertical wind velocity vs. time and height. Downdrafts are represented by red colors whereas updrafts are represented by blue colors. Vertical velocities are dominated by downdrafts during the first hour of measurements followed by updrafts dominance for the following hour. It is worth noting that the laser beam was focused at approximately 2km.

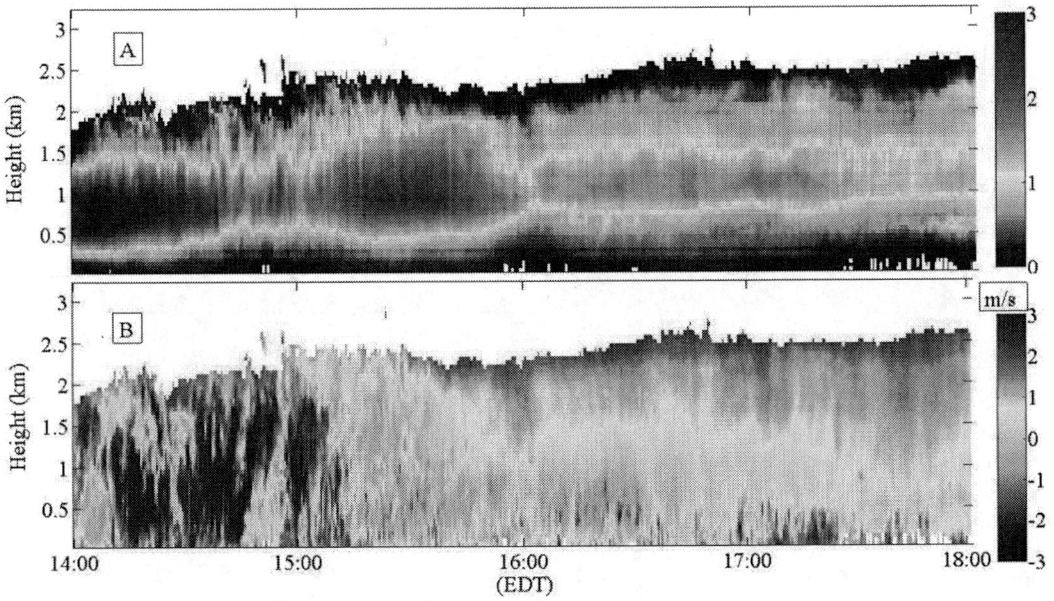

Fig. 4 SNR of backscattered signals(A), and Vertical wind velocity (m/s) (B) vs. time and height measured at CCNY Remote Sensing Laboratory between 14:00 – 18:00 EDT on July. 12th, 2012.

5. CONCLUSION

The SNR of backscattered signals of a shot noise limited CDL system for wind measurement was analyzed and presented. The measured signals power spectrum is divided by a reference signals power spectrum to remove the effect of a non-flat receiver's frequency response. In order to obtain a measurement realization, the ratio of received signals power to the reference signals power has to be greater than 1.025. The measured fluctuation of the signal power divided by a reference power is 0.015 whereas the calculated shot noise fluctuation is 0.014 for 10,000 accumulated pulses. A 10% difference between measured and calculated shot noise shows that the system is not 100% shot noise limited.

REFERENCES:

[1] R. M. Huffaker, A. V. Jelalian and J. A. L. Thompson, "Laser-Doppler system for detection of aircraft trailing vortices," *IEEE,* vol. 58, p. 322–326, 1970.

[2] M. J. Kavaya, S. W. Henderson, J. R. Magee, C. P. Hale and R. M. Huffaker, "Remote wind profiling with a solid-state Nd:YAG coherent lidar system," *Opt. Lett.,* vol. 14, no. 15, p. 776–778, 1989.

[3] J.-P. Cariou, B. Augere and M. Valla, "Laser Source requirements for coherent lidars based on fiber technology," *C. R. Physique,* vol. 7, p. 213–223, 2006.

[4] J. G. Hawley, R. Targ, S. W. Henderson, C. P. Hale, M. J. Kavaya and D. Moerder, "Coherent launch-site atmospheric wind sounder: theory and experiment," *Applied Optics,* vol. 32, no. 24, pp. 4557-4568, 1993.

[5] C. J. Karlsson, F. A. A. Olsson, D. Letalick and M. Harris, "All-Fiber Multifunction Continuous-Wave Coherent Laser Radar at 1.55 vm for Range, Speed, Vibration, and Wind Measurements," *Appl. Opt.,* vol. 39, pp. 3716-3726, 2000.

[6] S. Abdelazim, D. Santoro, M. Arend, F. Moshary and S. Ahmed, "All-fiber Coherent Doppler LIDAR for Wind Sensing," in *5th Symposium on Lidar Atmospheric Applications, 91st American Meteorological Society Annual Meeting,* Seattle, Washington, 2011.

[7] S. Abdelazim, D. Santoro, M. Arend, F. Moshary and S. Ahmed, "Wind Velocity Estimate and Signal to Noise Ratio Analysis of an All Fiber Coherent Doppler Lidar System," in *16th Coherent Laser Radar Conference,* Long Beach, California, June, 2011.

[8] S. Abdelazim, D. Santoro, M. Arend, F. Moshary and S. Ahmed, "Field programmable gate array processing of eye-safe all-fiber coherent wind Doppler lidar return signals," in *SPIE, Remote Sensing,* Prague, 2011.

EPJ Web of Conferences **119**, 17015 (2016) DOI: 10.1051/epjconf/201611917015
ILRC 27

APPROACHES FOR IMPROVED DOPPLER ESTIMATION IN LIDAR REMOTE SENSING OF ATMOSPHERIC DYNAMICS

Sreevatsan Bhaskaran and Ronald Calhoun

*Environmental Remote Sensing Group, Arizona State University, AZ 85281, USA, *Email: sbhaska4@asu.edu*

ABSTRACT

Laser radar (Lidar) has been used extensively for remote sensing of wind patterns, turbulence in the atmospheric boundary layer and other important atmospheric transport phenomenon. As in most narrowband radar application, radial velocity of remote objects is encoded in the Doppler shift of the backscattered signal relative to the transmitted signal. In contrast to many applications, however, the backscattered signal in atmospheric Lidar sensing arises from a multitude of moving particles in a spatial cell under examination rather than from a few prominent "target" scattering features. This complicates the process of extracting a single Doppler value and corresponding radial velocity figure to associate with the cell. This paper summarizes the prevalent methods for Doppler estimation in atmospheric Lidar applications and proposes a computationally efficient scheme for improving Doppler estimation by exploiting the local structure of spectral density estimates near spectral peaks.

1. INTRODUCTION

Estimation of range rate by measuring Doppler shift of scattered signals is a foundational capability in radar remote sensing of dynamic scenes. In the context of atmospheric sensing with laser radar, the backscattered signal from a single range cell results from interaction of the transmitted light with a large number of particles in the cell. Each of these will have its own velocity, and a typical objective is to deduce a mean range rate from the superposition of these backscattered signals to describe the aggregate atmospheric motion within the cell.

Most current-generation commercial lidars estimate Doppler using straightforward digital signal processing methods such as the Periodogram Discrete spectral Peak Estimator [1] or the Maximum Likelihood Estimator described in [2]. Both of these methods begin by forming a discrete spectral density estimate from digitally sampled temporal data representing the range cell of interest and then identifying a single dominant peak in this spectrum estimate. The (discrete) frequency of this peak is accepted as the Doppler frequency from which a singly radial velocity representing the entire cell is calculated.

In view of the high data flow rate in Lidar systems, it is important that the signal processing algorithms be computationally efficient as well as accurate. For Doppler frequency estimation, accuracy of an unbiased estimator may be characterized by its variance, which is desirable to bring as close as possible to the Cramér-Rao lower bound [2].

In this paper we examine methods for Doppler estimation, inspired by [4], that exploit a small neighborhood of values around the maximal peak in the spectral density estimate to improve estimation accuracy relative to accepting the frequency of the single peak as the estimate. Tradeoffs between additional computational complexity entailed in these methods and the improvement in estimation accuracy they provide are quantified, and their performance is compared to performance descriptors for other algorithms that appear in [2].

2. METHODOLOGY

We use a standard periodogram spectrum estimator based on the Discrete Fourier Transform (DFT) to estimate the spectrum from multiple blocks of time series data corresponding to a given range cell. This approach intrinsically produces spectral density estimates at a discrete set of frequencies ("bins"), the spacing of which (in Hertz) is determined by the duration (in seconds) of the time series segments of which DFTs are taken. Choosing a single peak from this discrete spectrum estimate to represent the Doppler frequency inherently limits the frequency resolution to be no finer than the bin spacing of the DFT. The method presented here affords the possibility of a Doppler estimate that falls between two DFT bins by performing an interpolation that uses a small number of periodogram values on each side of the peak and considers the spectral shape of the DFT bins in a mathematically appropriate way. The computational complexity of this approach is seen to exceed the single-peak method by an amount that increases with the number of neighboring values used in the interpolation; it is small when small local neighborhoods of the peak are used.

It is noteworthy that the Doppler estimation problem in Lidar sensing of atmospheric dynamics differs significantly from the classical problem of estimating the frequency of a tone in noise. This is observed in actual Lidar data and also in data simulated from standard models [5,6], such as shown in Figure 1. The Doppler signal is not tonal, but has numerous frequency components arising from the diversity of scatterers within the range bin. When the aggregate velocity of these particles is well defined, the spectral energy is correspondingly concentrated, but rarely within a single bin of the periodogram. There is also uncertainty associated with the time-frequency bandwidth limitation, which adds to the spread of the peak power over multiple spectral bins.

This characteristic of the data provides further rationale for considering more than a single peak in arriving at a figure to represent the aggregate radial velocity of the flow in a range cell. Our approach builds on the method for estimating the frequency of a tone in noise given in [4] and shown to yield estimation error variance on the same order as the Cramér-Rao bound. We adapt from the tonal setting to accommodate the frequency spread observed in lidar Doppler estimation and we describe the performance in terms of Cramer-Rao efficiency assuming a mathematical model of the frequency spread.

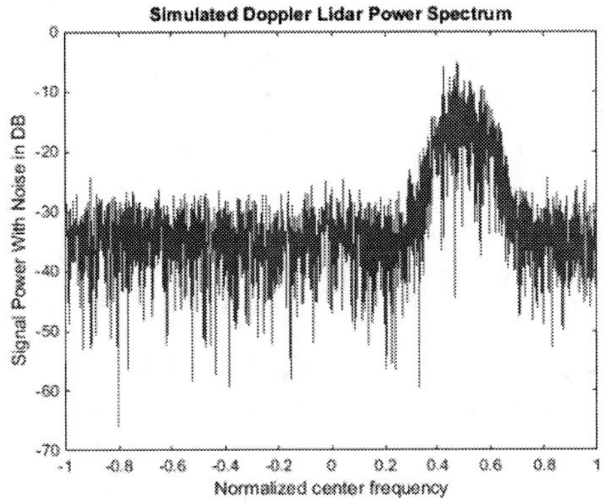

Fig.1 Simulated[5,6] Doppler Lidar Spectrum, Sampling time T = 0.5sec, SNR = 10 dB, Spectral Width of signal $\sigma = 0.625$. Demonstrates the smeared peak frequency observed on the frequency spectrum, over multiple channels.

3. RESULTS

The paper aims at presenting an exploratory technique, tested on Lidar returns. The localized point complex interpolation technique, is analyzed both in terms of its estimation fidelity and computational complexity. A detailed feasibility study of this technique, and comparison with the two most prominent estimator algorithms specified in the introduction, is required to completely justify the use of the new estimator algorithm for Lidar returns.

CONCLUSION

The technique suggested here is a promising motivation to improve the uncertainty associated with the peak estimation problem in a Lidar. Simultaneously, we would also need to address the necessity to process the huge data flow rate in a Lidar. A possible contender in this area is using neural network approaches. This would be an exploratory future direction, to address the huge data flow rate problem and to optimize the speed of the signal processing algorithms used in a Lidar system

DOI: 10.1051/epjconf/201611917015

REFERENCES

[1] B. J. Rye and R. M. Hardesty, "Discrete spectral peak estimation in incoherent heterodyne backscatter Lidar I – Spectral estimation and Cramér-Rao lower bound," *IEEE Trans. Geosci. Remote Sens.,* **31**, no. 1, January 1993.

[2] R. G. Frehlich and M. J. Yadlowsky, "Performance of mean frequency estimators for Doppler radar and Lidar," *J. Atmos. Ocean Tech., Vol. 11.*

[3] D. S. Zrinc, "Simulation of weather like Doppler spectra:," *Journal of Applied Meteorology*, **14**, June 1975.

[4] B. G. Quinn, "Estimating frequency by interpolation using Fourier coefficients," *IEEE Trans. Signal Processing,* **42**, no. 5, May 1994.

DOI: 10.1051/epjconf/201611917016

PERSPECTIVES OF THE EXPLICIT RETRIEVAL OF THE COMPLEX REFRACTIVE INDEX OF AEROSOLS FROM OPTICAL DATA TAKEN WITH LIDAR

Alexei Kolgotin[1], Detlef Müller[2,3], Eduard Chemyakin[2], Anton Romanov[4]

(1)Physics Instrumentation Center, Troitsk, Moscow Region, 142190, Russia, alexeift@yahoo.com
(2)Science Systems and Applications, Inc. (SSAI), NASA LaRC, Hampton, VA, 23666, USA, and
(3)University of Hertfordshire, Hatfield, Hertfordshire, AL10 9AB, UK
(4)The National University of Science and Technology, Leninskii av. 4, Moscow, 119049

ABSTRACT

We developed an explicit approach for the retrieval of the complex refractive index from optical data, i.e. backscatter and extinction coefficients measured with lidar. In this approach we assume that we know the particle size distributions as well as the optical data. On the basis of this approach we carried out numerical simulations in order to test the uncertainty of the retrieval of the complex refractive index in dependence of the combination of extinction and backscatter coefficients, the measurement wavelengths and measurement errors.

1. INTRODUCTION

A number of methods for the retrieval of particle microphysical parameters from multiwavelength lidar measurements was developed in the past years [1-5]. The methods require us to apply a grid of complex refractive indices (CRI) as input; for example, real parts in this grid may range from 1.3 to 1.8 and the imaginary part may range from 0 to 0.05. The final solutions of the CRI grid then are defined *implicitly* in the sense that the solutions fulfill a specific set of constraints and mathematical rules. For example, the constraint is applied that the optical data that follow from these solutions agree to the input optical data that were used in the retrieval. Besides that, the rule of averaging the solutions in the vicinity of the modified discrepancy minimum can be exploited as well [3]. In this contribution we explored an approach which allows us to derive the CRI *explicitly* from optical data (OD) taken with lidar. The particle size distribution (PSD) in our approach is assumed to be known.
In section 2 we present the methodology. In section 3 we carry out numerical simulations with synthetic optical data. We assume that the PSDs can be described by the log-normal law. Section 4 summarizes our results.

2. METHODOLOGY

OD such as extinction $\alpha(\lambda)$ and backscatter $\beta(\lambda)$ coefficients measured with lidar at wavelength λ are related to the PSD $f(r)$ via integral equations of the form

$$\int_0^\infty K_g(\lambda,m,r)f(r)dr = g(\lambda), \qquad g=\alpha,\beta \quad (1)$$

where the kernels of the integral equation $K_g(\lambda,m,r)$ are described by Mie theory in case of spherical particles of radius r and CRI $m=m_R\text{-}im_I$. In our previous work we investigated Eq. (1) to estimate the PSD $f(r)$ on the basis of input data $\alpha(\lambda)$ and $\beta(\lambda)$. If we assume that there is some estimation of $f(r)$ [2-3] then we can rewrite eq. (1) and use this estimation of $f(r)$ for the retrieval of m, i.e. we can write

$$g(m)=g \qquad \text{where } g=\alpha,\beta \quad (2)$$

and for fixed wavelength λ we obtain

$$g(m) = \int_0^\infty K_g(\lambda,m,r)f(r)dr, \qquad g=\alpha,\beta \quad (3)$$

This equation is a typical nonlinear algebraic equation and we define it on the domain

$$m_R\in[1.3, 1.8], \; m_I\in[0, 0.099] \qquad (4)$$

Since we deal with the unknown parameter m that depends on pairs of values (m_R,m_I) we need to consider Eq. (2) two times, i.e. for the different data types α and β or at the different wavelengths λ_1 and λ_2 in order to make the problem determinate. As a result we obtain two unknown parameters m_R and m_I for two nonlinear algebraic equations which theoretically can be solved.
Unfortunately, an analytical solution of the system (2) is not possible. For that reason we reformulate this problem in another way. Let us compute three different kinds of discrepancies:

$$\rho_g^2(\mathbf{m})=[g-g(\mathbf{m})]^2/g^2 \qquad g=\alpha,\beta \,(5)$$

$$\rho_t^2(\mathbf{m})=[\rho_\alpha^2(\mathbf{m})+\rho_\beta^2(\mathbf{m})]/2 \qquad (6)$$

The vector **m** denotes the pair (m_R,m_I). In that case, the solution of the system (2) can be treated as a minimization problem for the discrepancies

$\rho_j^2(\mathbf{m})=\rho_j^2(m_R, m_I) \rightarrow \min \quad j=\alpha,\beta,t$ (7)

where we consider the constraint used in (4).

There are plenty of approaches to minimize the functions $\rho_j(\mathbf{m})$: gradient method, random search, method of cyclic descent (MCD) and so on [6]. In view of the information we have about the kernel functions we use the MCD in our investigation.

This method of function optimization does not require us to calculate the derivative and one of the modifications of the MCD used here can be described by the following algorithm:

1. We define an arbitrary initial value, for example, $\mathbf{m}^0=(1.3;0)$
2. One of the two unknown parameters is fixed, for example, m_R^0 and we consider the one-dimensional minimization problem for ξ:

$$\rho_j^2(m_R^0, \xi) \rightarrow \min \quad j=\alpha,\beta,t \quad (8)$$

on the domain $\xi\in[0, 0.1]$. This problem can be solved, for example, with the golden section method [6]. After this step the 1st approximation for the parameter m_I is found as

$$m_I^1=\xi^*, \quad (9)$$

where ξ^* is the solution of problem (8).

3. Another parameter is fixed, in our case m_I^1, and the one-dimensional minimization problem is considered again for ξ,

$$\rho_j^2(\xi, m_I^1) \rightarrow \min \quad j=\alpha,\beta,t \quad (10)$$

on the domain $\xi\in[1.3, 1.8]$. After this step the 1st approximation for the parameter m_R is found, i.e.,

$$m_R^1=\xi^*, \quad (11)$$

where ξ^* is the solution of problem (10).

4. We compare the two points $\rho_j(m_R^0, m_I^0)$ and $\rho_j(m_R^1, m_I^1)$. If the deviation $\Delta\rho_j$

$$\Delta\rho_j=|\rho_j(m_R^1, m_I^1)-\rho_j(m_R^0, m_I^0)|<\varepsilon \quad j=\alpha,\beta,t (12)$$

is fulfilled at the preliminarily initialized threshold ε the iteration process is stopped. Otherwise steps 1-4 should be repeated.

5. With the last iteration n we obtain the solution $\mathbf{m}^*=\mathbf{m}^n$ from problem (7).

The minimization procedure including the steps 1-5 formalizes our explicit approach to the retrieval of the CRI from a set of OD measured with lidar.

3. NUMERICAL SIMULATION

We used our approach of retrieving the CRI with synthetic OD that were computed directly with Eq. (1) for known CRI and PSD $f(r)$ which is described by the lognormal law. We used a range of values for the mean radius $r_0\in[0.1, 0.3]$ μm and the mean width $\sigma\in[1.3;2.0]$ (defined according to the lognormal law) and we computed synthetic OD for different λ. Here we present some results of the CRI retrievals on the basis of a lognormal PSD with r_0=0.22 μm, σ=1.65 and m=1.4-i0.0019. It is not a very special case. However it illustrates quite well the major particularities of the problem of CRI retrievals.

First, we tested OD sets that consist of 1-3 backscatter coefficients (corresponding to measurement wavelengths of 355, 532, and 1064 nm) and 1-2 extinction coefficients (corresponding to measurement wavelengths of 355 and 532 nm). We show a few examples of our studies.

As we discussed in section 2 we need at least two coefficients to make our problem determinate. We considered in our first simulation the input OD that contain only backscatter (2β) or only extinction (2α) coefficients at λ_1=355 nm and λ_2= 532 nm. Unfortunately, the minimization procedure for the functionals $\rho_\beta(\mathbf{m})$ and $\rho_\alpha(\mathbf{m})$ failed for these combinations of input OD. This result agrees well with previous studies [2, 3]. If we include the backscatter coefficient at λ_3=1064nm we do not find a significant improvement of the results based on this set of 3β.

The retrieval results for the input set 1β+1α (λ=532 nm) are shown in Fig. 1: the discrepancies ρ_t (a), ρ_β (b) and ρ_α (c) are shown in dependence of the parameters m_R and m_I. The minimum values close to 0 % correspond to the true CRI (red circle) on these maps. The maximum values are 22.6 % for ρ_α and 1180 % for ρ_β. A very important and typical property of the two surfaces shown in Fig. 1b and 1c is that they contain the light areas ("anti-gradient" direction) with smallest values (0%-1%) distributed from the point m=1.4-i0.0019 to the point m=1.8-i0.099 for ρ_β and from the point m=1.4-i0.0019 to the point m=1.55-i0.099 for ρ_α. Apparently it results from the properties of the kernel functions $K_\beta(\lambda,m,r)$ and $K_\alpha(\lambda,m,r)$.

To extract in a better way the behavior of the three surfaces near the minimum values we connected the 5 points with solid lines along the smallest discrepancies. Unfortunately, the lines are spread out along all anti-gradient directions. It means that even a small averaging interval of the discrepancy [3] can lead to the large variations of the CRI in the range (4).

Fig. 1 shows the CRI (big blue close circle) retrieved with our minimization procedure described in section 2. In that procedure we used the initial values $\mathbf{m}^0=(1.8;0.0)$ and $\varepsilon=10^{-13}$. After

n=22 iterations we obtained the solution $\mathbf{m}^*=\mathbf{m}^{22}$=(1.4;0.001898). If we ask for more accuracy it can be improved. For example, for $\varepsilon=10^{-14}$ we obtain $\mathbf{m}^*=\mathbf{m}^{28}$=(1.4;0.001899). The convergence speed depends on the behavior of the discrepancy surface: the more pronounced the optimum is the faster the convergence is. It is also defined by the PSD and the wavelength.

The next important question concerns the conditionality of our problem (7); in other words, what is the stability of this solution scheme with regard to different errors? Here we consider this question just with respect to the OD measurement error δg.

We could employ our minimization procedure for different realizations of that error for a fixed OD set in order to investigate the uncertainty of the CRI retrieval δm. However it can be seen immediately from the maps for ρ_t, ρ_β and ρ_α. For that reason we plotted in Fig. 1 the small open circles which show the area where the discrepancy $\rho_j \leq 10\%$ reflects a measurement error of 10%.

As we can see solutions corresponding to that area are spread along all anti-gradient directions from m=1.4-i0.0019 to m=1.8-i0.099. The results do not improve if we take into account other combinations of OD that can be measured by lidar, for example, $3\beta+2\alpha$. Recent studies [7] also demonstrate that extra constraints are necessary for accurate estimations of the CRI.

We therefore investigated options that would allow us to improve the CRI retrieval. We considered measurement scenarios in which we might have another extinction value at a wavelength above 532 nm. We are not aware of lidars that measure extinction at 1064 nm. However, we could think of a measurement configuration in which we combine lidar with a passive remote sensor that measures at around 1064 nm or a longer wavelength. For example AERONET sunphotometer provides optical depth at 1640 nm. We could use lidar profiles to derive profiles of extinction from optical depth measured at this longer wavelength. Previous feasibility studies have shown how this could be done, e.g., Balis et al. [9]. We show the results for the scenario in which we have extinction at 1064 nm.

Fig. 2 shows results for $1\beta+1\alpha$ at λ=1064 nm. The extinction anti-gradient changes the direction significantly, and for 10% measurement error the CRI retrieval is still accurate to $\pm i0.013$ for the imaginary part and to-±0.025 for the real part (see small open circles in Fig. 2a). If we use extinction at 2 μm we obtain the imaginary part to an accuracy of $i\pm i0.004$ and the real part to an accuracy of ±0.

We note one more result of our analysis. A recent publication [8] demonstrated a specially developed approach to extracting the real part of the CRI. Our study shows that if we can estimate either the real or the imaginary part of the CRI we can derive immediately the other part of the CRI on the basis of the anti-gradient direction area that is shown in the ρ_t map. For example, if in our simulation case we would constrain the real part within the range [1.40;1.45] the imaginary part range could be constrained to the interval [0;0.015] (see Fig. 1a).

4. CONCLUSION AND SUMMARY

We present an approach for the retrieval of the CRI in an explicit way from OD taken with lidar. We carried out numerical simulations in order to test the capabilities of this method.

Our approach allows us to estimate the CRI with any pre-assigned accuracy if the OD and the PSD are known. However the method can be employed only for the case of combinations of backscatter and extinction coefficients.

We investigated the uncertainty of the CRI retrievals in dependence of the measurement error. We showed that we can achieve an accuracy of $\pm i0.05$ for the imaginary part and ±0.3 for the real part if the uncertainty of the OD is 10% and if we do not apply any other constraints to the investigated parameters.

The accuracy of the CRI retrieval can be improved significantly if, for example, extinction coefficients from optical depth measurements at longer wavelengths are used. Data at longer wavelengths could for instance be provided by a combination of lidar with passive remote sensing. The example for 1064 nm shows that the uncertainties of the real and imaginary parts decrease to ±0.025 and $\pm i0.013$, respectively, even for measurement errors of 10%.

We will develop our approach further in order to investigate the influence of incorrectly given PSD on the uncertainty of the CRI retrieval.

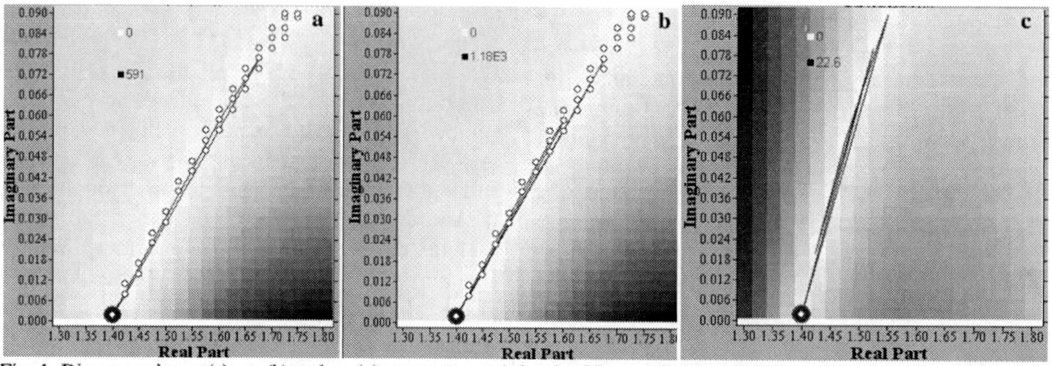

Fig. 1. Discrepancies ρ_t (a), ρ_β (b) and ρ_α (c) versus (m_R, m_I) for the OD set $1\beta+1\alpha$ at λ=532 nm. Big close red and blue circles show the true values and the retrieved values of the CRI, respectively. The solid line connects the five points that are obtained from the smallest discrepancy values which are less than 1%. The small open circles are the retrieved CRIs if the OD error varies from 0 to 10 % (not shown for ρ_α). Legend's black square means 591 % (a), 1180% (b) and 22.8% (c)

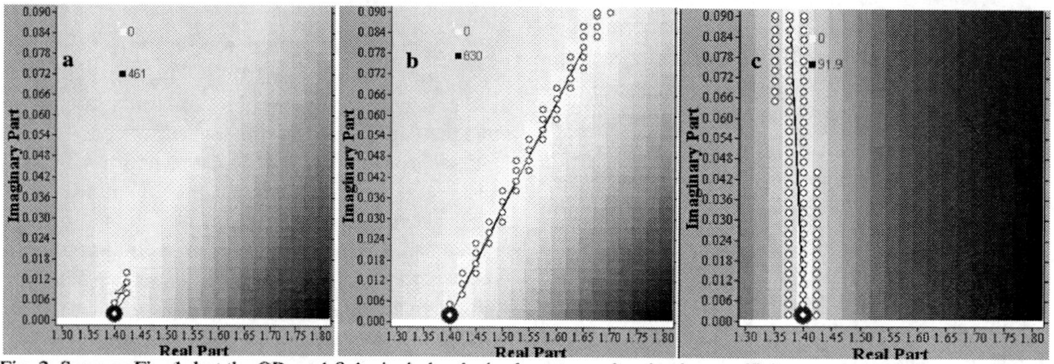

Fig. 2. Same as Fig. 1. but the OD set $1\beta+1\alpha$ includes the backscatter and extinction coefficients at λ=1064 nm.

REFERENCES

1. Donovan D., A. Carswell, 1997: Principal component analysis applied to multiwavelength lidar aerosol backscatter and extinction measurements, *Appl.Opt.*36, pp. 9406-9424.
2. Müller D., et al., 1999: Microphysical particle parameters from extinction and backscatter lidar data by inversion with regularization: theory, *Appl. Opt.*, 38, pp. 2346 – 2357.
3. Veselovskii I., et al., 2002: Inversion with regularization for the retrieval of tropospheric aerosol parameters from multiwavelength lidar sounding, *Appl. Opt.*, 41, pp. 3685 – 3699.
4. Böckmann C., et al, 2005: Microphysical aerosol parameters from multiwavelength lidar, *Journal of Optical Society of America*, A, 22, pp. 518–528.
5. Chemyakin E., et al, 2014: "Arrange & average" algorithm for the retrieval of aerosols parameters from multiwavelength HSRL/Raman lidar data, *Appl. Opt.*, 53, pp. 7252 – 7266.
6. P. E. Gill, W. Murray and M. H. Wright, 1982: Practical Optimization. Emerald Publishing).
7. . Veselovskii I., et al, 2005: Information Content of Multiwavelength Lidar Data With Respect to Microphysical Particle Properties Derived From Eigenvalue Analysis, *Appl. Opt.* 44, 5292-5303.
8. Veselovskii I., et al, 2012: Linear estimation of particle bulk parameters from multi-wavelength lidar measurements, *Atm. Meas. Tech.*, 5, 1–30.
9. Balis, D., E. Giannakaki, D. Müller, V. Amiridis, K. Kelektsoglou, S. Rapsomanikis and A. Bais (2010). "Estimation of the microphysical aerosol properties over Thessaloniki, Greece, during the Scout-O3 campaign with the synergy of Raman lidar and sunphotometer data." Journal Geophysical Research 115: D08202, doi:10.1029/2009JD013088.

THE CARBON AEROSOL / PARTICLES NUCLEATION WITH A LIDAR: NUMERICAL SIMULATIONS AND FIELD STUDIES.

Alain Miffre[1], Christophe Anselmo[1], Mirvatte Francis[1], Gregory David[2] and Patrick Rairoux[1*]

[1]Institute of Light and Matter, Lyon University, 10 rue Ada Byron, France,
[2]Currently with ETH Zürich, Laboratory of Physical Chemistry, W.Pauli Strasse 10, Zürich, Switzerland
**Email: patrick.rairoux@univ-lyon1.fr*

ABSTRACT

In this contribution, we present the results of two recent papers [1,2] published in Optics Express, dedicated to the development of two new lidar methodologies. In [1], while the carbon aerosol (for example, soot particles) is recognized as a major uncertainty on climate and public health, we couple lidar remote sensing with Laser-Induced-Incandescence (LII) to allow retrieving the vertical profile of very low thermal radiation emitted by the carbon aerosol, in agreement with Planck's law, in an urban atmosphere over several hundred meters altitude. In paper [2], awarded as June 2014 OSA Spotlight, we identify the optical requirements ensuring an elastic lidar to be sensitive to new particles formation events (NPF-events) in the atmosphere, while, in the literature, all the ingredients initiating nucleation are still being unrevealed [3]. Both papers proceed with the same methodology by identifying the optical requirements from numerical simulation (Planck and Kirchhoff's laws in [1], Mie and T-matrix numerical codes in [2]), then presenting lidar field application case studies. We believe these new lidar methodologies may be useful for climate, geophysical, as well as fundamental purposes.

1. INTRODUCTION

Recent advances in lidar concern both atmospheric traces [4] and aerosols up to three-component particle mixtures involving volcanic ash, desert dust, sea-salt or sulfate particles [5]. However, up to now to our knowledge, no lidar methodology exists that allows a range-resolved specific detection of the carbon aerosol in the atmosphere over several hundreds of meters, despite its key role on human health, as being carcinogenic in urban polluted areas [6], and in radiative forcing, black carbon being the second contributor to global warming after CO_2 [7]. In the literature, the terminology related to the carbon aerosol involves soot particles, carbonaceous particles, and biomass burning, as well as secondary organic aerosols, to quote only a few. The carbon aerosol light absorbing properties have been addressed through laboratory measurements on soot particles [8] and accurate numerical simulations [9]. While determining the spatial and temporal behavior of the carbon aerosol is essential for developing carbon reduction emission strategies, in this contribution, we perform a field experiment, supported by a numerical simulation, based on coupling the lidar and LII-techniques, to retrieve a range-resolved profile of LII-thermal radiation emitted by the carbon of an urban atmosphere over several hundred meters.

For paper [2], the starting point is the definition of the lidar backscattering coefficient, showing that small-sized particles may strongly contribute to the backscattering coefficient when present in high concentrations, as implied by the detection of molecular backscattering with a lidar. Hence, and as a second point, we here discuss on the optical requirements ensuring a lidar backscattering device to remotely observe new particle formation, as we published in [2]. Considering a dust NPF-event, we applied the Mie and T-matrix numerical codes to simulate the backscattering coefficient of spherical freshly nucleated particles and non-spherical desert dust particles. We hence showed that, to remotely observe such nucleation events promoted by non-spherical particles with an elastic lidar, it should operate in the UV spectral range and be polarization-resolved. Two atmospheric case studies are then proposed, on nucleation events promoted by desert dust or volcanic ash particles.

The paper is organized as follows. In Section 2, we present the numerical simulations and the methodology of our two papers [1, 2] by starting with the carbon aerosol lidar methodology. We then present lidar field application case studies for both papers in Section 3 and then conclude.

2. NUMERICAL SIMULATIONS AND METHODOLOGY

We here focus on the optical requirements ensuring a lidar device to be sensitive to the carbon aerosol (Section 2.1, [1]), then to particles nucleation (Section 2.2, [2]).

2.1 LII-lidar formalism and numerical simulation

Starting from Planck and Kirchhoff's laws, we address the thermal radiation intensity emitted by a laser-illuminated particle volume at a range R from a lidar laser source. For incandescence occurs during a characteristic time τ, the LII-lidar signal takes the mathematical form of a convolution product in which the Planck's distribution is involved. We then addressed the main requirements for operating the atmospheric LII-lidar methodology in the low troposphere by numerically simulating the LII-lidar equation. To observe an LII-lidar signal in the low troposphere, the laser beam divergence should be very low (in the range of 0.05 mrad). The numerical simulation of the LII-lidar signal (see Fig. 1) shows that, even when the carbon aerosol has a simple flat-top vertical profile, the interpretation of the LII-lidar signal is not straightforward. More details are given in [1] where the LII-lidar formalism is set and the LII-lidar equation is established.

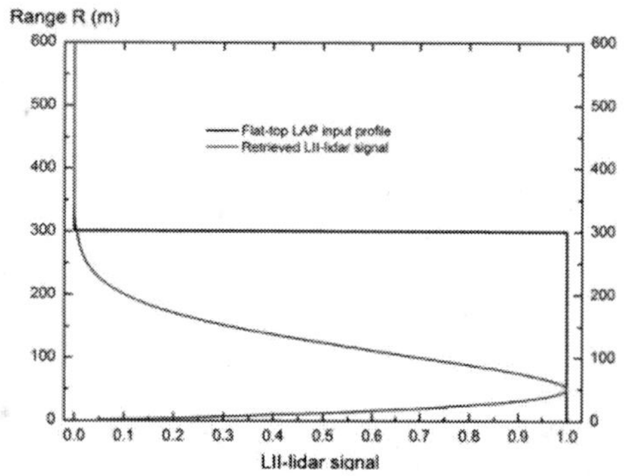

Fig. 1 Numerical simulation of the LII-lidar signal for an input flat-top carbon aerosol vertical profile [1].

2.2 Methodology for remotely sensing a NPF-event with a lidar backscattering device

Starting from an observed NPF-event promoted by desert dust particles [10], we numerically simulated the corresponding lidar particles backscattering coefficient β_{NPF}, by applying the Mie theory to freshly nucleated sulfuric acid particles and the T-matrix numerical code [9] for non-spherical desert dust particles. We hence retrieved the optical requirement for detecting such a nucleation event (NPF-event) promoted by desert dust particles, for which the lidar backscattering coefficient β_{NPF} is equal to 0.7 $Mm^{-1}.sr^{-1}$ at the onset of the nucleation process. To highlight the contribution of each particle size to β_{NPF}, we plotted the integrand of β_{NPF} in Fig. 2. In the UV spectral range, particles for diameters between 50 and 800 nm contribute to β_{NPF}, which is hence interestingly increased (β_{NPF} is four times higher than in the IR-spectral range).

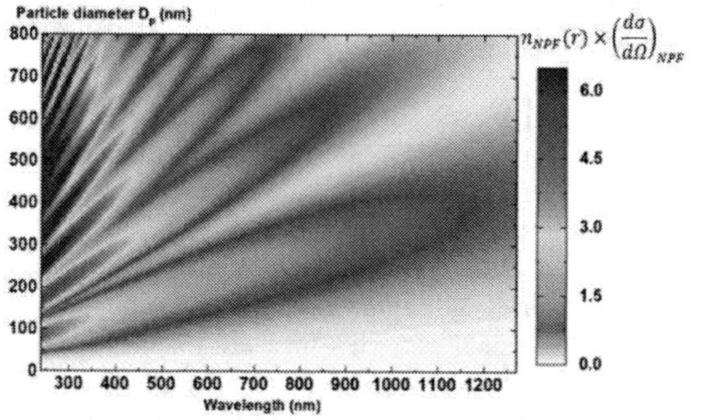

Fig. 2 Numerical simulation of β_{NPF}'s integrand [2].

Since nucleation events promoted by desert dust particles appear at low dust concentrations [11], the correlative behavior of β_{NPF} with that of β_{dust} can be used as optical tracers for remote sensing such NPF-events. Hence, β_{NPF} can be addressed by using a sensitive and accurate UV polarization lidar. The polarization acts as a particle shape discriminator, allowing backscattering separation of spherical particles from any other non-spherical particle, whatever its size. Besides (see Fig.2), the wavelength acts as a particle size discriminator: the shorter the wavelength is, the more sensitive to fine and ultrafine particles backscattering is.

3. FIELD APPLICATION CASE STUDIES

In this section, we report on the first experimental observation [1, 2] of LII-lidar thermal radiation in an urban atmosphere (Section 3.1), of lidar remote sensing of particles nucleation (Section 3.2).

3.1 Observation of a LII-lidar vertical profile

Two experimental detector configurations are proposed, allowing either spectrally-resolved LII-lidar measurements over a range interval or range-resolved LII-lidar measurements at a single detection wavelength (LII-lidar vertical profile). The latter configuration is coupled with an elastic backscattering lidar to address all atmospheric aerosols, either refractory, which may incandesce, or not. Fig. 3a presents the observed LII-lidar signal as a function of the wavelength of the detected radiation. Interestingly, apart from observed Raman vibrational sidebands of water vapor and nitrogen, a continuous signal increase can be adjusted in nice agreement with Planck's law [1]. Fig. 3b presents the corresponding range-resolved LII-lidar signal measured in the Lyon urban troposphere, over several hundred of meters, by using a 1064 nm lidar laser, to avoid superposition with Raman sidebands (down-conversion process).

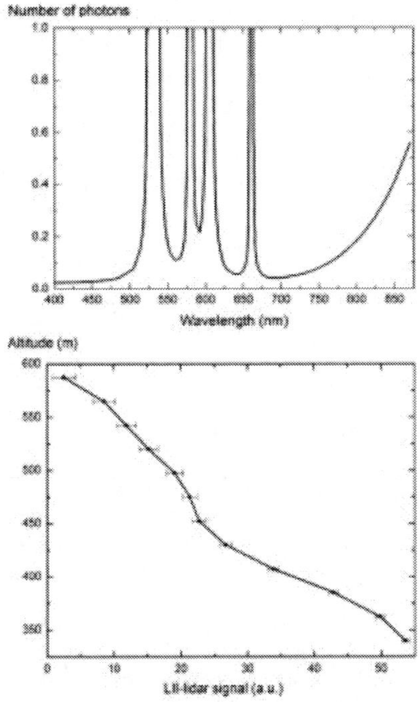

Fig. 3 (a): Measurement of LII-lidar signal spectrum [1], in agreement with Planck's law. (b): Vertical profile of LII-lidar signal in Lyon urban city's troposphere [1].

3.2 Lidar observation of a nucleation event

To apply Section 2.2's methodology, a sensitive and accurate UV-polarization lidar is required since the more accurate the polarization analysis is, the more sensitive to β_{NPF} the lidar is. We hence operated our home-built UV-polarization lidar [11]. By applying the methodology we published in GRL [12], we obtained the Figure 4 time-altitude maps for the spherical (β_s) and non-spherical desert dust (β_{ns}) lidar backscattering coefficients.

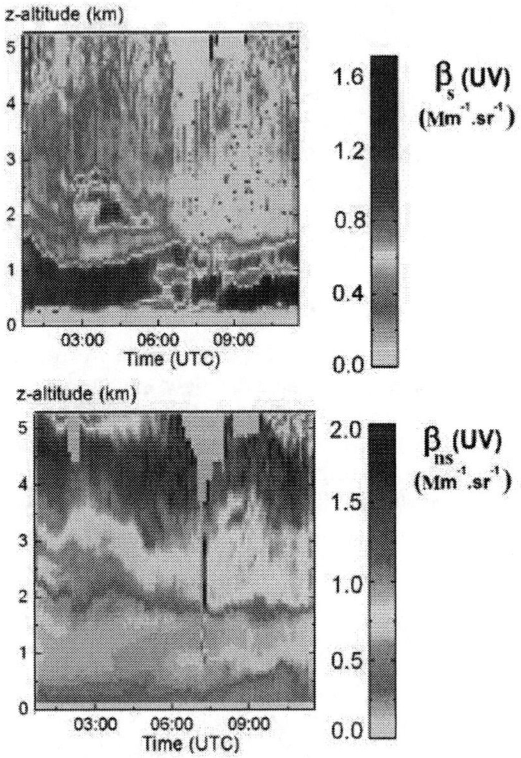

Fig. 4 Time-altitude maps of the spherical and non-spherical lidar particles backscattering coefficients [2].

Interestingly, s-particles are only observed at the border of the dust ns-layer, i.e. a β_s-enhancement only occurs at low dust particles backscattering, in agreement with our numerical simulation [2]. This behavior can a priori result from nucleation, coagulation or condensation on pre-existing surfaces. However, nucleation is the only one leading to such an s-particles backscattering enhancement, occurring only at low dust particles backscattering. Condensation may also occur and for this reason, it is only when β_{ns} is low that the β_s-increase is indeed observed. Hygroscopic growth may also be responsible for the observed β_s-enhancement but in our case study, the relative humidity remains constant (ECMWF-analysis).

Coagulation might be also invoked, but in this case, photochemistry would not be implied, in contradiction with Dupart et al.'s laboratory experiments [10]. As a result, our numerical simulation and lidar experiments provide strong arguments in favor of the remote lidar observation of a nucleation event promoted by dust particles. Actually, the state-of-the-art atmospheric science does not provide other processes that may induce such an optical behavior of a β_s-enhancement occurring only at low dust concentrations. More details on our newly developed methodology and its limitations are discussed in [2]. In addition, a second case study is proposed on the lidar observation of nucleation events promoted by ashes, released from the Icelandic volcano.

4. CONCLUSIONS

In this contribution, we presented the results of two recently published lidar new methodologies, to detect the carbon aerosol [1] or particles nucleation [2], the latter being awarded as June 2014 OSA Spotlight. More details and outlooks are proposed in the corresponding papers. The oral presentation will focus on the implications of these new methodologies for the lidar community.

ACKNOWLEDGEMENT

The authors thank their former PhD students, Grégory David (ETH Zürich) and Benjamin Thomas (NYCC) for their help. CNRS and INERIS are thanked for partly funding this work.

REFERENCES

[1] Miffre, A., C. Anselmo, S. Geffroy, E. Fréjafon and P. Rairoux: Lidar remote sensing of laser-induced incandescence on light absorbing particles in the atmosphere, Optics Express, Vol. 23, Issue 3, pp. 2347-2360 (2015).

[2] David, G., B. Thomas, Y. Dupart, B. D'Anna, C. George, A. Miffre and P. Rairoux, UV polarization lidar for remote sensing new particles formation in the atmosphere, *Opt. Exp.*, **22**, A1009-A1022, (2014).

[3] J. Almeida et al., Molecular understanding of sulphuric acid-amine particle nucleation in the atmosphere, Nature **502**(7471), 359–363 (2013).

[4] B. Thomas, G. David, C. Anselmo, J.-P. Cariou, A. Miffre, and P. Rairoux, Remote sensing of atmospheric gases with optical correlation spectroscopy and Lidar: first experimental results on water vapor profile measurements, *Appl. Phys. B* **113**(2), 265–275 (2013).

[5] G. David, B. Thomas, T. Nousiainen, A. Miffre, and P. Rairoux, Retrieving simulated volcanic, desert dust, and sea-salt particle properties from two / three-component particle mixtures using UV-VIS polarization Lidar and Tmatrix, *Atmos. Chem. Phys.* **13** (14), 6757–6776 (2013).

[6] J. Löndhal, E. Switelicki, E. Lindgren, and S. Loft, Aerosol exposure versus aerosol cooling of climate: what is the optimal reduction strategy for human health ?, *Atmos. Chem. Phys.* **10**(19), 9441–9449 (2010).

[7] IPCC, Climate Change 2013: The Physical Science Basis. Working Group I Contribution to the Fifth Assessment Report of the Intergovernmental Panel on Climate Change (Cambridge University Press, (2013).

[8] M. Schnaiter, H. Horvath, O. Möhler, K.-H. Naumann, H. Saathoff, and O. W. Schöck, UV-VIS-NIR spectral optical properties of soot and soot-containing aerosols, *J. Aerosol Sci.* **34**(10), 1421–1444 (2003).

[9] M. Mishchenko, L. Liu, and D. W. Mackowski, T-matrix modeling of linear depolarization by morphologicallycomplex soot and soot-containing particles, *J. Quant.Spectrosc.Radiat.Trans.* **123**, 135–144 (2013).

[10] Y. Dupart, S. M. King, B. Nekat, A. Nowak, A. Wiedensohler, H. Herrmann, G. David, B. Thomas, A. Miffre, P. Rairoux, B. D'Anna, and C. George, Mineral dust photochemistry induces nucleation events in the presence of SO2., *Proc. Natl. Acad. Sci.* U.S.A. **109**(51), 20842–20847 (2012).

[11] G. David, A. Miffre, B. Thomas, and P. Rairoux, Sensitive and accurate dual-wavelength UV-VIS polarization detector for optical remote sensing of tropospheric aerosols, *Appl. Phys. B* **108**(1), 197–216 (2012).

[12] A. Miffre, G. David, B. Thomas, and P. Rairoux, Atmospheric non-spherical particles optical properties from UV-polarization lidar and scattering matrix, *Geophys. Res. Lett.* **38**(16), L16804 (2011).

EPJ Web of Conferences 119, 18002 (2016) DOI: 10.1051/epjconf/201611918002
ILRC 27

INDEPENDENT RETRIEVAL OF AEROSOL TYPE FROM LIDAR

Doina Nicolae[1]*, Jeni Vasilescu[1], Camelia Talianu[1], Alexandru Dandocsi[1,2]

*[1]National Institute of R&D for Optoelectronics, Romania, *Email: nnicol@inoe.ro*
[2]University Politehnica of Bucharest, Faculty of Applied Sciences, Bucharest, Romania

ABSTRACT

This paper presents an algorithm for aerosol typing from multiwavelength lidar data, based on Artificial Neural Networks. The aerosol model used to simulate optical properties for the training of the network is described. The algorithm is tested on real observations from ESA-CALIPSO database.

1 INTRODUCTION

A major step forward in Earth Observation has been made by the successful implementation of active remote sensing from space (CALIPSO, ADM-AEOLUS, ATLID). Although having multiple advantages (high dynamic range, high spatial and temporal resolution), lidar data suffer by a limited physical content. Both the problems of retrieving optical profiles and microphysical properties from lidar data are ill-posed, needing advanced mathematical algorithms [1]-[3].

Our work refers to the use of Artificial Neural Networks (ANN) combined with an extended learning process in order to retrieve the aerosol class from lidar optical data. The algorithm relies on ANN with tens to hundreds of neurons, which is trained to identify the most probable aerosol type by using aerosol intensive optical parameters as input. The advantage of ANNs compared to other methods is that can work with significant noisy data, as long as a pattern is present.

An ANN represents a mathematical projection of the human neural network. It is based on neurons, axons and synapses, the information being propagated as a neural influx. The analysis process takes part during the propagation of information through synapses, where it is adjusted to properly interpret the given data. The network complexity, architecture and dimensions depend on input data complexity and could contain hundreds of neurons.

The ANN is guided through a learning process (supervised or unsupervised) to recognize the pattern of the data. The main goal of the training process is to find the optimal value of weights and to have the minimum errors. [4].

2 METHODOLOGY

More than 22 multiwavelength Raman lidars are currently operating in Europe, as part of EARLINET (the European Aerosol Research LIdar NETwork). Typical configuration includes 3 elastic backscatter channels (1064, 532 and 355 nm), 2 vibrational Nitrogen Raman channels (607 and 387 nm) and 1 depolarization channel (532 nm). Using the combination of these channels, 8 aerosol intensive parameters can be obtained: 2 lidar ratios (532 and 355 nm), 1 Ängström exponent (355/532 nm), 2 color ratios (355/532 and 532/1064 nm), 2 color indexes (355/532 and 532/1064 nm), and 1 particle depolarization ratio (532 nm). The ultimate goal of our algorithm is to be able to retrieve the aerosol type using these 8 parameters as input, without any complementary information (backtrajectory, column data, source sensitivity).

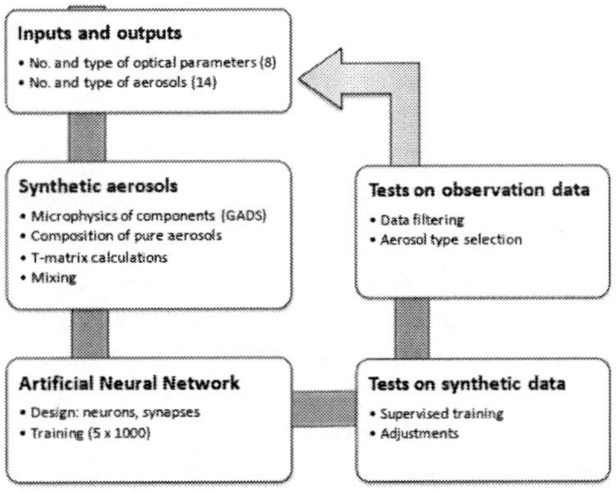

Fig. 1 Schematics of the method

The algorithm relies on a custom-designed ANN with 8 inputs and a number of outputs corresponding to the number of aerosol types to be recognized. The success of the retrieval depends on the information content of the inputs, as well as on the extent of the training program. A high number of good training cases is required to obtain a reasonable level of confidence. Such validated database does not exist at this moment, therefore we developed an aerosol model to provide the necessary inputs for the ANN. Observations were used to adjust the aerosol model and to test the performance of the ANN.

2.1 Synthetic data

The synthetic database is generated by simulating the optical properties of various aerosol types based on available information on the microphysics of basic components. The algorithm combines the GADS database (Global Aerosol DataSet) with OPAC model (Optical Properties of Aerosol and Clouds) and T-Matrix code in order to compute, in an iterative way, the intensive optical properties of each aerosol type. Four classes of humidity (50%, 70%, 80%, 90%) are considered, out of eight classes in OPAC, high humidity values being excluded in order to avoid ambiguous results related to activation of the hygroscopic particles. Each aerosol type (pure) was built as an internal mixture of typical components which do not interact physically or chemically, having different mixing ratios. Basic components were picked up from OPAC [5], and GADS database is used for the microphysical properties of each component.

The database is built in two stages: a) microphysical and optical properties of pure aerosols, as described before; b) optical properties of mixed aerosols, using the properties obtained in stage a). Six classes of pure aerosol are defined: Continental, Continental Polluted, Dust, Marine, Smoke, and Volcanic.

Particles were considered spheroids with different axis ratios, and the T-matrix code [6] was used to generate the optical properties. Axis ratios were picked up from the literature [7]. Extensive (backscatter and extinction coefficients) and intensive optical parameters (Ängström exponents, lidar ratios, color ratios, particle depolarization ratios) were computed.

Measurements have associated uncertainties (systematic and statistical errors), either due to the instrument - biases, linearity, calibration), or due to the data treatment– algorithms applied to correct / average raw signals and to calculate data products. In order to simulate values as close as possible to the ones obtained from lidar measurements, we considered for each parameter an associated relative error of maximum 20%, and calculated the possible range based on the initially simulated value. Basically, within this range any value of the parameter is accepted.

Eight classes of mixed aerosols were considered, corresponding to the most frequent aerosol types observed by EARLINET (the European Aerosol Research LIdar NETwork): Continental Dust, Marine Mineral, Continental Smoke, Dust Polluted, Coastal, Coastal Polluted, Mixed Dust, and Mixed Smoke. Optical properties of mixed aerosols were computed as linear combination of corresponding optical properties of pure types.

2.2 ESA-CALIPSO database

The ESA-CALIPSO database contains EARLINET ground-based retrievals which are used for the derivation of conversion factors of pure aerosol types and typical mixtures of aerosols [8]. The data are grouped in aerosol categories, more or less related to CALIPSO aerosol types. Measurements were used to identify aerosol layers and calculate extensive and intensive optical parameters. For each layer, mean lidar ratios and Ångström exponents were calculated from the backscatter and extinction profiles provided by the stations. The aerosol type and source region for each aerosol layer was identified through a number of auxiliary data, e.g. FLEXPART [9].

ESA-CALIPSO database includes a total number of 718 cases and 21 aerosol and cloud types, out of which only 13 contains all the necessary parameters (3 backscatters, 2 extinctions and 1 depolarization). To increase the number of cases to test the ANN, we added the particle depolarization, assuming typical values (from the literature) for the corresponding aerosol type. We obtained 105 cases, out of which a large number represents mixture of 3 or more pure types. In order to compromise between the performance of the algorithm and the number of output aerosol types, we made a selection of 14 representative types, including mixtures of 3.

2.3 Artificial Neural Networks

A complex Jordan /Elman ANN with 6 hidden layers and 50 processing elements per layer was built to check the capability of the algorithm to

DOI: 10.1051/epjconf/201611918002

identify the aerosol type. Eight inputs were fed into the algorithm, corresponding to the aerosol intensive optical parameters which can be calculated from a $3\beta+2\alpha+1\delta$ lidar. Fourteen outputs were considered, corresponding to the 14 aerosol types (6 pure and 8 mixed).

About 70% of the synthetic data described above were used to train the network in 5 steps, with 1000 iterations per training set. Measured data were not used to train the ANN because even a small error in the initial classification can lead to improper operation of the algorithm. Testing was performed on both synthetic and measured data. The relative error of each parameter was used to generate 19 additional values, and combination of all obtained parameters was considered for the synthetic and the observational database.

3 RESULTS

Comparative results of the optical parameters from synthetic data and ESA-CALIPSO are shown below.

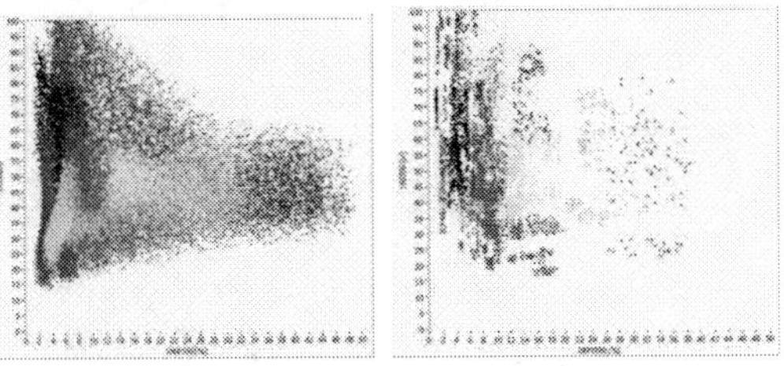

Fig. 2 Lidar ratio versus particle depolarization at 532nm for Continental (brown), ContinentalPolluted (black), Marine (blue), Dust (orange), Smoke (green), Volcanic (dark red), DustPolluted (Red), ContinentalDust (yellow), Coastal (cyan), CoastalPolluted (violet), ContinentalSmoke (olive), MixedDust (magenta), MixedSmoke (light green) and MarineMineral (steel) aerosols: left - synthetic aerosols; right – ESA-CALIPSO database [8]

As shown in Fig. 2, synthetic data agree quite well with the ESA-CALIPSO, for each of the aerosol type. With the considered accuracy and definition of types, clusters are not completely distinct in the synthetic data (left). This is also true for the observational data (right). Additionally, wide spreading of the measured parameters (right panel) is due to the presence of more complex mixtures (not considered in our simulations), incorrect calibration or underestimation of the measurement uncertainty. Although several aerosol types are frequently observed in Europe, the observational cases available are not sufficient to describe all possible combinations of optical parameters, leading to apparently incomplete clusters, which must not be seen as a failure of the aerosol model to simulate real measurements. For example, marine aerosols are rarely present in the ESA-CALIPSO database, and the uncertainty is higher than 20% therefore are not represented in the right panel. Coastal aerosols are also poorly represented. For such types, no conclusion can be drawn.

The ANN was tested after five training cycles with synthetic data. Tests were performed considering three degrees of discrimination: 55%, 65% and 75%.

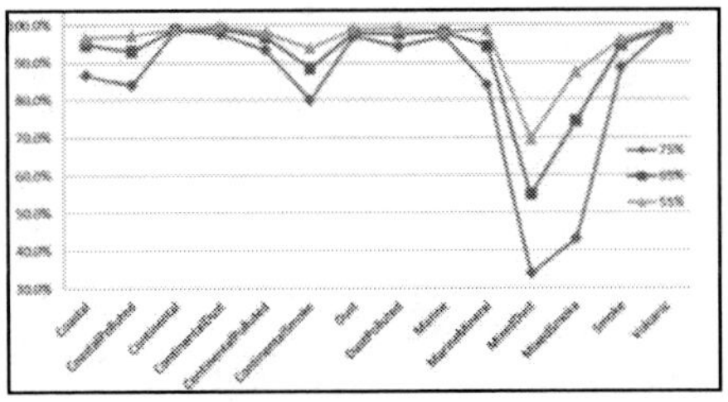

Fig. 3 Performance of the ANN after training on synthetic data; different degrees of discrimination

Twelve aerosols types are identified in more than 80% of the cases, Mixed Dust and Mixed Smoke being recognized in only 30% cases when the highest discrimination level is applied. All aerosols types are well classified in more than 70% cases when 55% discrimination level is applied (Fig. 3). Blind test was applied on ESA-CALIPSO database, with a discrimination level of 75%, and results are shown in Fig. 4.

Type ESA-CALIPSO \ Type ANN	Coastal Polluted	Continental	Continental Polluted	Continental Smoke	Dust	Mixed Dust	Smoke	Volcanic	Unknown
clean continent. + dust + marine aerosol						1			
clean continental		2							
clean continental + dust aerosol		1							
clean continental + marine aerosol		2							1
dust					1				3
dust + marine aerosol (= mixed dust)		1							4
poll. continent. + dust + marine aerosol		1							2
polluted contin. + dust + marine + smoke				1					
polluted continental		1	6	2			1		3
polluted continental + marine + smoke	1								
polluted continental + marine aerosol	5		1						
polluted continental + smoke aerosol			2				2		
polluted continental aerosol + dust		1					2		2
polluted dust (dust + smoke)				1					1
smoke				1			1		
volcanic aerosol					2		1	1	

Fig. 4 Comparative results of aerosol types from ESA-CALIPSO and the ANN

Note that, although our algorithm is not always in perfect agreement with the classification in ESA-CALIPSO (marked in green), there are only small differences which are due to a different definition of mixtures (marked in blue). For example, 1 case of polluted continental is classified by the ANN as Continental, 2 are classified as Continental Smoke, and 1 as Smoke. Similarly, 2 volcanic cases are classified by the ANN as Dust, and 1 as Smoke, only 1 being classified as Volcanic. However, the type could not be retrieved for a significant number of cases (16) when applying a discrimination level of 75%. A possible cause is the underestimation of the measurement uncertainties, and/or too demanding constraints for the level of confidence of the ANN.

4 CONCLUSIONS

This paper presents an algorithm for aerosol typing based on Artificial Neural Networks and lidar-retrieved aerosol intensive optical properties. Synthetic data used to train the network was generated by combining OPAC and T-matrix calculations and considering both pure and mixed aerosols. Values and ranges are within the ones published in the literature and generally agree with ESA-CALIPSO. Observations were used to adjust the aerosol model by direct comparison, and to test the performance of the algorithm.

We applied the algorithm on ESA-CALIPSO data with accuracy better than 20%. A large number of cases were correctly classified by the algorithm, although the distinction between Dust and Volcanic is still to be improved. In this case, the retrieval depends strongly on the quality of the spectral parameters (particle depolarization is similar, as well as lidar ratios), which is limited by the signal-to-noise ratio and calibration of the lidar. In case of observation data with accuracy lower than 20%, complex mixtures of 3 aerosol types cannot be retrieved without accepting a lower confidence level for the ANN.

ACKNOWLEDGEMENT

This work has been supported by the ESA contract no. 4000110671/14/I-LG, NATALI and by the STAR–ESA Programme Project No.55/2013- CARESSE.

References

[1] Müller, D., Wandinger, U., Ansmann, A., 1999: Microphysical particle parameters from extinction and backscatter lidar data by inversion with regularization: theory. *Appl. Opt.* **38** (12), 2346-2357.

[2] Osterloh, L., Böckmann, C., Mamouri, R. E., Papayannis, A., 2011: An adaptive base point algorithm for the retrieval of aerosol microphysical properties. *Open Atmospheric Science Journal* **5**, 61-73.

[3] Veselovskii, I., Kolgotin, A., Griaznov, V., Müller, D., Franke, K., Whiteman, D. N., 2004: Inversion of multiwavelength raman lidar data for retrieval of bimodal aerosol size distribution, *Appl. Opt.* **43** (5), 1180-1195.

[4] Gupta P. and Christopher S. A., 2009: Particulate matter air quality assessment using integrated surface, satellite, and meteorological products: 2. A neural network approach, *J. Geophys. Res.*, **114**, D20205

[5] Hess, M., P. Koepke, and I. Schult, 1998: Optical properties of aerosols and clouds: The software package OPAC, *Bull. Am. Meteorol. Soc.*, **79**, 831 – 844

[6] Mishchenko, M. I., and L. D. Travis, 2000: Light Scattering by Nonspherical Particles, *Academic Press*, San-Diego, 690pp.

[7] Dubovik, O., Sinyak, A., Lapyonok, T., Holben, B. N., Mishchenko, M., Yang, P., Eck, T. F., Volten, H., Munoz, O., Veihelmann, B., van der Zande, W. J., Leon, J.-F., Sorokin, M., and Slutsker, I., 2006: Application of spheroid models to account for aerosol particle nonsphericity in remote sensing of desert dust, *J. Geophys. Res.*, **111**, D11208, doi:10.1029/2005JD006619.

[8] Wandinger U., A. Hiebsch, I. Mattis, G. Pappalardo, L. Mona, and F. Madonna, 2011: Aerosols and Clouds, Long-term Database from Spaceborne Lidar Measurements, *Final report*, ESTEC Contract 21487/08/NL/HE.

[9] Stohl, A., M. Hittenberger, and G. Wotawa, 1998: Validation of the Lagrangian particle dispersion model FLEXPART against large scale tracer experiment data, *Atmos. Env.*, **32**, 4245-4264.

EPJ Web of Conferences **119**, 18003 (2016) DOI: 10.1051/epjconf/201611918003
ILRC 27

VERTICAL PROFILES OF AEROSOL OPTICAL AND MICROPHYSICAL PROPERTIES DURING A RARE CASE OF LONG-RANGE TRANSPORT OF MIXED BIOMASS BURNING-POLLUTED DUST AEROSOLS FROM THE RUSSIAN FEDERATION-KAZAKHSTAN TO ATHENS, GREECE

Alexandros Papayannis[1]*, Athina Argyrouli[1], Panayotis Kokkalis[1], Georgios Tsaknakis[1], Ioannis Binietoglou[2], Stavros Solomos[3], Stylianos Kazadzis[4], Stefanos Samaras[5], Christine Böckmann[5], Panagiotis Raptis[3] and Vassilis Amiridis[3]

[1]*Laser Remote Sensing Laboratory, National Technical University of Athens, 15780 Athens, Greece*
**Email: apdlidar@central.ntua.gr*

[2]*INOE, National R&D Institute for Optoelectronics, Magurele, Romania*

[3]*National Observatory of Athens, Palaia Penteli, Greece*

[4]*Physikalisch-Meteorologisches Observatorium Davos, World Radiation Center, Switzerland*

[5]*University of Potsdam, Institute of Mathematics, Germany*

ABSTRACT

Multi-wavelength aerosol Raman lidar measurements with elastic depolarization at 532 nm were combined with sun photometry during the HYGRA-CD campaign over Athens, Greece, on May-June 2014. We retrieved the aerosol optical [3 aerosol backscatter profiles (b_{aer}) at 355-532-1064 nm, 2 aerosol extinction (a_{aer}) profiles at 355-532 nm and the aerosol linear depolarization ratio (δ) at 532 nm] and microphysical properties [effective radius (r_{eff}), complex refractive index (m), single scattering albedo (ω)]. We present a case study of a long distance transport (~3.500-4.000 km) of biomass burning particles mixed with dust from the Russian Federation-Kazakhstan regions arriving over Athens on 21-23 May 2014 (1.7-3.5 km height). On 23 May, between 2-2.75 km we measured mean lidar ratios (LR) of 35 sr (355 nm) and 42 sr (532 nm), while the mean Ångström exponent (AE) aerosol backscatter-related values (355nm/532nm and 532nm/1064nm) were 2.05 and 1.22, respectively; the mean value of δ at 532 nm was measured to be 9%. For that day the retrieved mean aerosol microphysical properties at 2-2.75 km height were: r_{eff}=0.26 μm (fine mode), r_{eff}=2.15 μm (coarse mode), m=1.36+0.00024i, ω=0.999 (355 nm, fine mode), ω=0.992(355 nm, coarse mode), ω=0.997 (532 nm, fine mode), and ω=0.980 (532 nm, coarse mode).

1. INTRODUCTION

The HygrA-CD campaign was an international field campaign which took place in Athens, Greece between 15 May and 22 June 2014, within the Initial Training on Atmospheric Remote Sensing-ITARS project (www.itars.net). HYGRA-CD (http://hygracd.impworks.gr) brought together a suite of different instruments and expertise aiming to enhance our understanding on the impact of aerosols and clouds on weather and climate. It was based on the synergy between remote sensing and in-situ instrumentation, making also use of numerical weather prediction and atmospheric modeling.

Tropospheric aerosols play a crucial role in climate change through scattering (cooling effect) and absorbing (warming effect) incoming solar and outgoing thermal radiation [1]. Despite recent progress documented in the latest Intergovernmental Panel for Climate Change (IPCC) [1], the uncertainty about the current level of radiative forcing due to aerosols (0.5 Wm^{-2}) is still relatively large compared to that of global warming (greenhouse) gases (0.25 Wm^{-2}).

In order to clarify the mechanisms of aerosol radiative forcing and reduce the respective uncertainties, detailed knowledge of the vertical profiles of the particle optical (b_{aer}, a_{aer}, AE and LR) together with microphysical (r_{eff}, m, ω) and chemical properties (water content, dry chemical composition), as well as their mass concentration

EPJ Web of Conferences **119**, 18003 (2016) DOI: 10.1051/epjconf/201611918003
ILRC 27

are required [1-3]. Raman lidars have proven to be the most adapted tools in aerosol characterization experiments since they can provide the vertical profiles of b_{aer} and a_{aer} with very high spatial and temporal resolution [3 and references therein].

2. METHODOLOGY

In this study a synergy of a multi-wavelength Raman lidar, a 532 nm depolarization lidar and a sun photometer was used to derive the aerosol optical and microphysical properties aloft. The Raman lidar system (EOLE) [4] together with the mobile single-wavelength depolarization lidar (AIAS) are part of the ARIADNE Greek lidar network [5]. EOLE and AIAS are based in the Laser Remote Sensing Unit (LRSU) of the National Technical University of Athens (NTUA) (37.9°N, 23.6°E, 200 m a.s.l.). The data processing was based on the Single Calculus Chain (SCC) developed within EARLINET.

The AERONET CIMEL sun photometer located at the National Observatory of Athens (NOA) (www.aeronet.gsfc.nasa.gov) was used to derive the daytime evolution of the columnar aerosol optical properties, like the aerosol optical depth (AOD), the Ångström exponent (AE) at 440 nm/870 nm and the aerosol size distribution. The microphysical particle properties inside specific atmospheric layers were retrieved using the inversion code provided by Böckmann et al. 2005 [6] and Osterloh et al., 2013 [7].

3. RESULTS

In this paper we will focus on a case study of long-range transport of mixed aerosols arriving over Athens on 21-23 May 2014 in the height range 1.7-3.5 km height a.s.l. The prevailing synoptic meteorological conditions and air mass trajectories based on WRF-FLEXPART model simulations [8] showed that during the studied period a low pressure system was established over the Balkans and a high pressure system north of the Caspian and Aral Seas. The relative position of these two synoptic systems was governing the air mass flow over the Eastern Europe and the Balkans. On 21 May the synoptic situation resulted in transportation of elevated smoke particles from the persisting fires in S. Russian Federation and Kazakhstan towards the Baltic Sea. Displacement of the trough axis towards the east on 22 May, resulted in wind convergence along the frontal area between the two systems and in the formation of a secondary smoke plume, this time over the Black Sea, together with transportation of Asian dust. On 23 May these particles merged with the older smoke particles coming from the North. According to our model, the establishment of a northern flow over Greece, mainly on 22-23 May, favored the transportation of smoke particles from Russia-Kazakhstan towards Athens, as seen in the model results (Fig. 1) where elevated smoke layers arrived between 3.5-8 km on 22 May, while on 23 May smoke should also be found near ground.

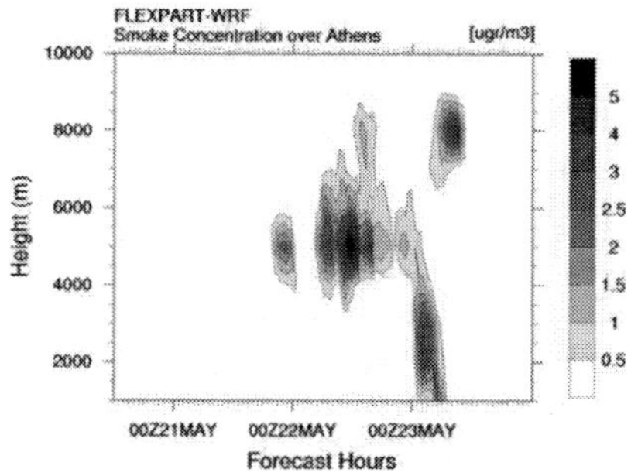

Figure 1. Time-height plot of smoke concentration (in μg m^{-3}) over Athens from 21 to 23 May 2014.

WRF-FLEXPART model simulations agree with the 8-day air mass back trajectories calculated by the HYSPLIT model. For instance on the evening of 23 May the air masses arriving over Athens between 2-2.75 km height originated from the southern Russian-northern Kazakhstan biomass burning regions and were mixed with desert dust during their overpass over central Kazakhstan and NW Iran desert areas, resulting in air masses with a mixture of dust and smoke particles.

The evolution of the columnar daily mean AOD values obtained at 340-500-1020 nm and the AE values at 440 nm/870 nm and 380nm/500nm, as derived by the NOA CIMEL sun photometer, are depicted in Fig. 2 (upper plot) for the period 21-23 May 2014. In that figure (lower plot) we show the respective columnar particle size distribution (in um). From Fig. 2 we can see that the AOD increases with the arrival of the air masses originating from the biomass (Russian-Kazakhstan) and desert (S. Kazakhstan-NW Iran)

DOI: 10.1051/epjconf/201611918003

regions on 22 and 23 May, as previously explained. On 22 May (Fig. 2-lower plot) the fine (r=0.15 μm) particles (smoke) dominate (AE~1.7-1.85) while on 23 May the coarse ones (r=1-4 μm) (mostly mixed desert dust and smoke) dominate (AE~0.65-1.2).

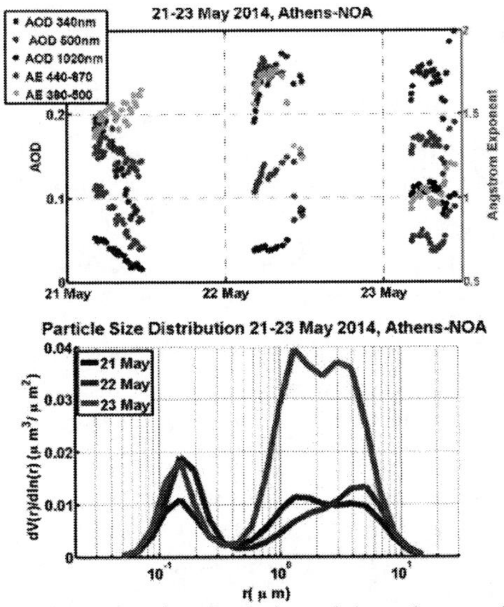

Figure 2. *upper plot:* Evolution of the columnar daily mean AOD obtained over Athens at 340-500-1020 nm and AE values at 440nm/870nm and 380nm/500nm, as derived by the NOA CIMEL sun photometer; *lower plot:* the particle size distribution (in μm) of the aerosol column (21-23 May 2014).

The corresponding EOLE Raman lidar measurements, expressed in terms of the temporal evolution of the range-corrected lidar signals (RCS) at 1064 nm, on 21-23 May (Fig. 3) showed discrete aerosol layers between 0.5-3.5 km heights over Athens, extending up to 3.5-4 km height. In Figure 4 we present the aerosol optical properties (backscatter and extinction coefficients, AE and Ångström backscatter and extinction-related exponent) retrieved by EOLE lidar between 1-5 km height on 23 May 2014 (20:30-21:40 UTC): b_{aer} at 355-532-1064 nm, a_{aer} and LR at 355-532 nm and Ångström backscatter (355nm/532nm and 532nm/1064nm) and extinction-related (355nm/532nm) exponents.

In Fig. 4 we observe a strong aerosol layer between 2.2-3 km extending up to nearly 3.5-4 km height. In the 2-2.75 km height region we measured mean LR values of 35 sr (355 nm) and 42 sr (532 nm), with mean particle linear depolarization ratio of 9% (532 nm), and AE aerosol backscattered-related values (355nm /532nm and 532nm/1064nm) of 1.22 and 2.05, respectively; these values are higher than the 0.95-1.2 measured at the end of the day by CIMEL in the atmospheric column. The columnar microphysical properties from CIMEL for 21-23 May were: r_{eff}=0.32-0.59 μm, m=(1.48-1.51)+(0.0018-0.0080)i, ω=0.93-0.96. These columnar values are compared with the retrieved mean aerosol microphysical properties from EOLE data on 23 May (2-2.75 km asl.): r_{eff}=0.26 μm (fine mode), r_{eff}=2.15 μm (coarse mode), m=1.36+0.00024i, ω=0.999 (355 nm, fine mode), ω=0.992 (355 nm, coarse mode), ω=0.997 (532 nm, fine mode), ω=0.980(532 nm, coarse mode).

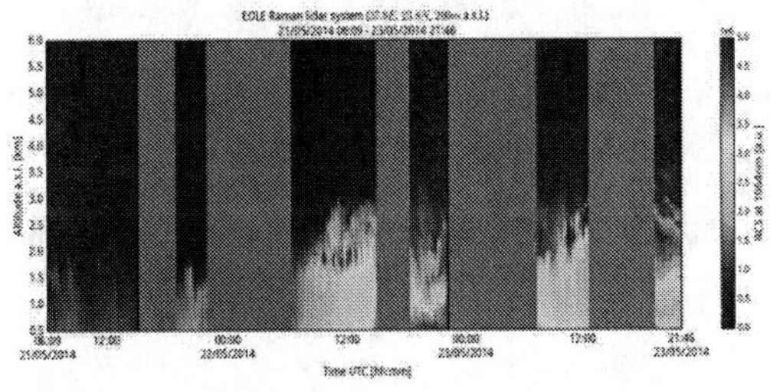

Figure 3. Spatio-temporal evolution of the range-corrected backscattered lidar signals (RCS) obtained by the EOLE lidar over Athens at 1064 nm between 21 (07:06 UTC) and 23 (21:45 UTC) May 2014.

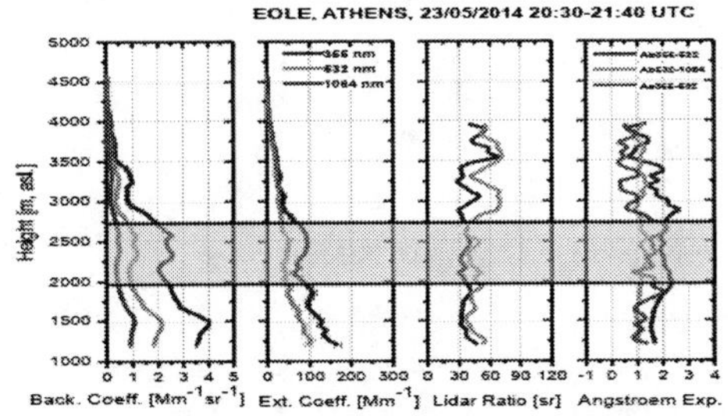

Figure 4. Vertical profiles of the aerosol optical parameters (extinction and backscatter coefficients, LR, AE) retrieved from the EOLE Raman lidar measurements on 23 May 2014 (20:30–21:40 UTC).

All these aerosol microphysical values compare well with those measured under similar atmospheric conditions regarding mixed smoke aerosols with polluted dust [9,10]. Finally, in Fig. 5 we present the forward FLEXPART simulations of the fire smoke integrated column (μg/m^2) for 23 May 2014 (21:00 and 22:00 UTC).

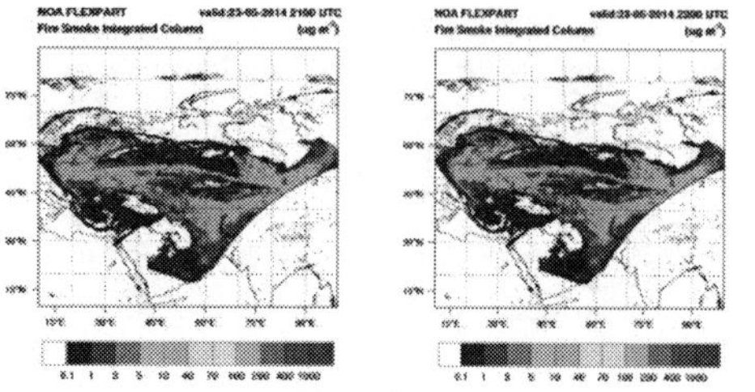

Figure 5. Forward FLEXPART simulations of the fire smoke integrated column (μg/m^2) for 23 May 2014 (*left:* 21:00 UTC and *right:* 22:00 UTC).

4. CONCLUSIONS

In this paper we presented a rare case of long-range transport of mixed aerosols (biomass burning particles mixed with dust) arriving from the S. Russian-Kazakhstan-NW Iran regions over Athens on 21-23 May 2014. In the 2-2.75 km height region we measured on 23 May mean LRs of 35 sr (355 nm) and 42 sr (532 nm) and mean AE aerosol backscatter-related values (355nm/532nm and 532nm/1064 nm) of 2.05 and 1.22, respectively; the mean value of δ at 532 nm was 9%. In this height region the mean retrieved aerosol microphysical properties r_{eff}=0.26 μm (fine mode), r_{eff}=2.15 μm (coarse mode), m=1.36+0.00024i, ω=0.999 (355 nm, fine mode) ω=0.992 (355 nm, coarse mode), ω=0.997 (532 nm, fine mode), ω=0.980(532 nm, coarse mode) are compared with the columnar microphysical properties from CIMEL for 21-23 May: r_{eff}=0.32-0.59 μm, m=(1.48-1.51)+(0.0018-0.0080)i and ω=0.93-0.96.

ACKNOWLEDGEMENTS

AP, PK, GT were supported by the MACAVE research project which is implemented within the framework of the action Supporting of Postdoctoral Researchers of the Operational Program Education and Lifelong Learning (action's beneficiary: General Secretariat for Research and Technology), and is co-financed by the European Social Fund (ESF) and the Greek State. The work of AA, IB and SS has received funding from the European Union 7th Framework Programme (FP7/2007-2013): People, ITN Marie Curie Actions Programme (2012-2016) in the frame of ITARS under grant agreement n°289923. SS, SK and VA were supported by the European Union 7th Framework Programme (FP7-REGPOT-2012-2013-1) in the framework of the project BEYOND, under grant agreement n°316210.

REFERENCES

[1] Intergovernmental Panel on Climate Change: Climate Change 2013, 2013: Synthesis Report.

[2] Bierwirth, E., et al., 2009: Spectral surface albedo over Morocco and its impact on radiative forcing of Saharan dust, *Tellus,* **B61,** 252–269.

[3] Mamouri, R.E., et al., 2012: Multi-wavelength Raman lidar, sun photometric and aircraft measurements in combination with inversion models for the estimation of the aerosol optical and physico-chemical properties over Athens, Greece, *Atmos. Meas. Tech.*, **5**, 1793-1808.

[4] Kokkalis, P., et al., 2012: The EOLE lidar system of the National Technical University of Athens, *Proc. 26th Intern. Laser Radar Conference,* Porto Heli, Greece, 629-632.

[5] Papayannis, A., et al., 2012: ARIADNE: The Greek Lidar Network, *Proc. 26th Intern. Laser Radar Conference,* Porto Heli, Greece, 349-352.

[6] Böckmann, C., et al., 2005: Microphysical aerosol parameters from multi-wavelength lidar, *J. Opt. Soc. Am.,* **A22**, 518-528.

[7] Osterloh, L., et al., 2013: Regularized inversion of microphysical atmospheric particle parameters: theory and application, *J. Comp. Phys.*, **237**, 79-94.

[8] Brioude, J., et al, 2013: The Lagrangian particle dispersion model FLEXPART-WRF version 3.1, *Geosci. Model Dev.*, **6**, 1889-1904, doi:10.5194/gmd-6-1889-2013.

[9] Papayannis, A., et al., 2012: Optical-micro-physical properties of Saharan dust aerosols and composition relationship using a multiwave-length Raman lidar, in situ sensors and modelling: a case study analysis, *Atmos. Chem. Phys.*, **12**, 4011–4032.

[10] Müller, D., et al., 2007: Aerosol-type-dependent-lidar ratios observed with Raman lidar, *J. Geophys. Res.*, **112**, D16202, doi: 10.1029/2006JD008292.

INTEGRATED OBSERVATION OF AEROSOL PLUMES TRANSPORT AND IMPACTS ON THE AIR QUALITY REMOTE SENSING IN THE NORTHEAST U.S.

Yonghua Wu*, Chowdhury Nazmi, Zaw Han, Cuiya Li, Barry Gross, Fred Moshary

Optical Remote Sensing Lab and NOAA-CREST, City College of New York (CCNY), NY 10031, USA.
**Email: yhwu@ccny.cuny.edu*

ABSTRACT

In this paper, we present a cluster analysis of plume transport paths to New York City (NYC, 40.821°N, 73.949°W) for the 8-year period during 2006-2013. We also show cases of such aloft aerosol plumes intrusion and mixing into the boundary layer (PBL) and the impact on local air quality. Range-resolved monthly occurrence frequency and modification of local aerosol optical properties are presented. The NOAA-HYSPLIT cluster analysis indicates 6 main transport paths; and the optical properties (optical depth-AOD, Angstrom exponent-AE and single scatter albedo-SSA) of aerosol for each cluster are characterized. We further illustrate the impact of these aloft plumes on the satellite MODIS estimate of ground $PM_{2.5}$ levels and observe that when the aloft plumes-layer AODs are filtered out using lidar, the correlation of MODIS AOD-$PM_{2.5}$ can be much improved.

1. INTRODUCTION

Aloft aerosol plumes from both forest-fires and dust storms events are often transported over long distance, affecting both climate radiation and air quality on regional/continental scales [1, 2]. They often result in significant contribution to the column aerosol optical depths (AOD) which can bias the estimate of $PM_{2.5}$ from the satellite column observation [3, 4]. The optical properties and time-height statistics of these transported plumes can be characterized by long-term ground lidar observations. In this study, we present the climatology of aloft aerosol plumes using synergistic observations from a co-located multi-wavelength lidar and AERONET-sunphotometer. The potential impacts on the local aerosol optical properties, air quality and the MODIS AOD-$PM_{2.5}$ correlation are analyzed.

2. INSTRUMENTS AND METHODOLOGY

At CCNY, a suite of ground-based remote sensing instruments are deployed on a building roof, near a standard surface air quality monitoring station deployed by the New York State Department of Environment Conservation for $PM_{2.5}$ mass concentration, ozone and CO measurement. The instruments include a 3-wavelength elastic-Raman lidar, a ceilometer (Vaisala CL-31), an AERONET Cimel sun/sky radiometer, a multi-filter shadow band radiometer (YES MFR-7) and a microwave radiometer (Radiometrics MP-3000A). The lidar emits three wavelengths (355-, 532- and 1064-nm) with a flashlamp-pumped Nd: YAG laser at a repetition rate of 30 Hz (Spectra-physics Quanta-Ray PRO-320) [5]. Three elastic-scattering and two Raman-scattering returns by nitrogen and water vapor molecules excited by 355-nm laser output are collected by a receiver telescope (Ø50-cm). The signals are acquired by LICEL transient recorders (TR40-160) and recorded with 1-min average. AERONET sun/sky radiometer provide column AOD, Angstrom exponent (AE), microphysical parameters (volume size distribution and refractive index, effective radius-R_{eff}), and SSA [6].

The lidar is generally operated under the clear or partly cloudy sky at 10:00-18:00 local time, 3-day/week. An aloft-aerosol-layer event is identified via the lidar imagery of range-corrected returns at 1064-nm, with the aerosol-layer above PBL, duration time > 3 hr and geometric depth ≥ 300 m. Due to general SNR limit of Raman-channel in the daytime, we derive aerosol extinction profiles with the elastic returns by using SP-measured AOD to constrain lidar-ratio, and the resultant Angstrom exponent profile is derived to classify aerosol types in the plumes [7].

3. OBSERVATIONS

3.1 A case of Canadian smoke plume transport

A heavy smoke transport event was observed by the CCNY-lidar on Aug. 8, 2014. Figure 1 (a) shows the time-height cross section of the aerosol extinction coefficient, Angstrom exponent, constrained lidar-ratio and segmented AOD. Lidar data illustrate one dense aerosol plume layer at 4~8 km altitude with the corresponding Angstrom exponent of 1.6-2.0, which indicates that the plume is fine mode aerosol dominated. The column averaged lidar-ratios were then obtained with values of 71.7 ± 3.5 sr at 532-nm and 39.8 ± 4.9 sr at 1064-nm, respectively. The column AOD measurements are mostly due to the aloft-aerosol-plumes. The NOAA-HYSPLIT backward trajectory and NOAA-HMS (Hazard Map System) satellite analysis indicate that these plumes are transported from Canadian forest fire events, and traveled to the eastern US within 3-day by the high-level jets. This demonstrates the significant effect long range transport can have on local conditions and illustrates the value of multi-wavelength techniques to identify the plume characteristics.

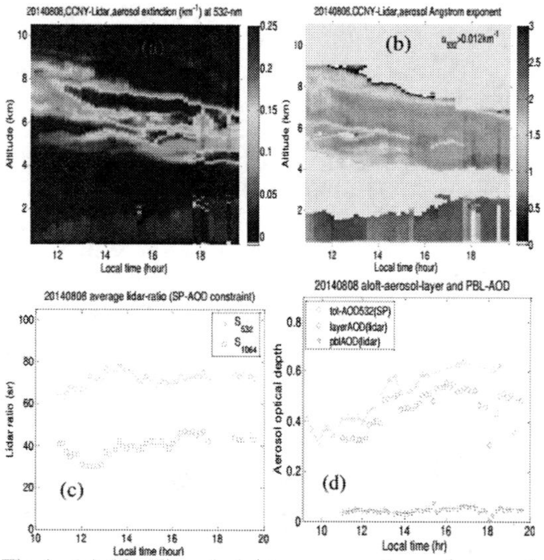

Fig.1. (a)-(d) Time-height cross section of aerosol extinction, Angstrom exponent at 532-1064-nm, lidar-ratio and aloft-layer AOD on Aug.8, 2014.

3.2 Range-resolved aloft aerosol layer occurrence and modification of local aerosol optical properties

Our routine lidar observations frequently detect aloft aerosol layers over NYC. Our long term datasets allow us to build climatology of aerosol plume transport and properties. The range-resolved monthly average occurrence of aloft aerosol layers is estimated from the identified aloft-plume days over the total days of lidar observation during 2006 – 2013. The result is given in Fig. 2a. Most episodes occur mainly from March to September, with higher altitude in spring and lower in summer, making the summer months most susceptible to smoke intrusions into the PBL. The monthly average of sunphotometer (SP) AOD and Angstrom exponent on the aloft-aerosol-plume days and clear days are compared in Fig.2b-c. Generally, significant enhancement of AOD occurs on the aloft-plume-days during April to September. Angstrom exponents become much lower in April on plumes days indicating more coarse-mode particles that might be associated with the Asian dust transport. In summer, on plumes days, the higher Angstrom exponents indicate fine-mode dominant particles that might be associated with the aged smoke transport, but they are only slightly lower than the ones on the clear day with the urban aerosols.

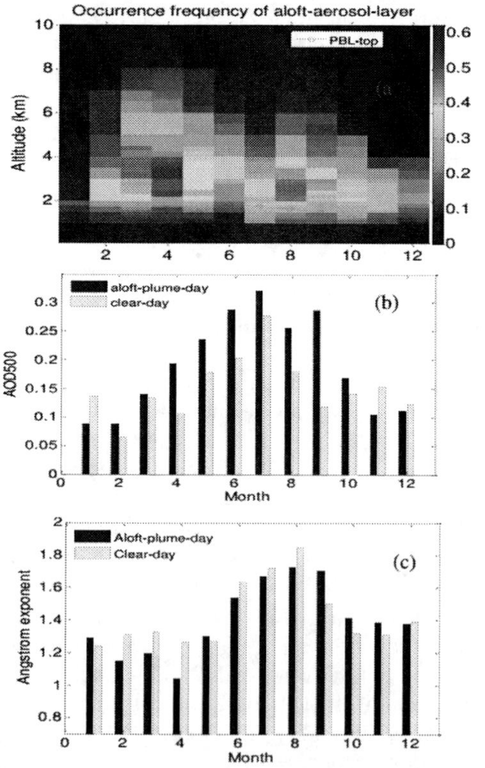

Fig.2 (a)-(c) Monthly occurrence of aloft-aerosol-plume, column AOD, Angstrom exponent in NYC.

DOI: 10.1051/epjconf/201611918004

3.3 Cluster analysis of aerosol transport paths

With the identified plumes events from the lidar images, the transport pathways of elevated aerosol plumes are classified with the cluster analysis of HYSPLIT-trajectories. A total of 6 clusters are determined from the total spatial variance (TSV) of 283 trajectories (ending points at 15:00 UTC, 4-km altitude and duration of 72-hr long) as shown in Fig. 3a. The mean aerosol properties (AOD, AE, SSA and R_{eff}) are analyzed for each cluster shown in Fig.3. Cluster-2 shows the air mass transported from the Great Lakes area (shortest range travel, 22% fraction), with aerosols that show both larger AOD and Angstrom exponent with SSA=0.91. This is possibly associated with the transport of industrial emissions. Cluster-6 (longest distance travel, 9% fraction) shows the air mass transported from the Pacific Ocean via the northwest US; with aerosols that have smaller Angstrom exponent and large effective radius with the mean SSA≈0.95. This is likely linked with the trans-Pacific Asian dust transport. Other clusters show similar optical properties of aerosols with the mean AOD of 0.2 and Angstrom exponent of 1.5, the smaller SSA and effective radius may indicate the transport of aged biomass burning aerosols.

3.4 Impact on the local air quality

The aloft-aerosol-plumes can be mixed down to the surface via the processes of PBL entrainment and/or air subsidence, thus potentially affecting the surface air quality. The mixing-down events can be visually identified from the time-height lidar imagery. In Fig. 4, the average surface $PM_{2.5}$ concentration is compared for the days where aloft-plume mixing-down is observed to clear days or days where no-mixing is observed. It can be seen that the $PM_{2.5}$ concentrations are clearly higher on the mixing-down days than for the other days. In particular, during 12:00-16:00 local time, the $PM_{2.5}$ shows a dramatic increasing trend on the mixing-down days but is stable and low level on the clear days due to the PBL development during mid-day.

3.5 Plume impact on MODIS-AOD/Surface $PM_{2.5}$ correlation

As shown in Fig.1, the lidar data can be used to study the effect of aloft plumes on potential correlation between the MODIS-AOD and surface $PM_{2.5}$. We compare their linear regression results before and after filtering out the aloft-plume-layer data in summer, for 2010 to 2013. As shown in Fig.5, when filtering out the aloft plume cases, a better linear correlation coefficient and a bigger slope are observed as well as a reduction of the DC offset which is indicative of removal of aloft contamination As expected, poor correlation is seen when only the smoke plume cases are studied. We observe similar effects in spring with the dust transport; the linear slope increases from 20.6 to 42.3 and the correlation coefficient is improved from 0.39 to 0.51 when filtering out the aloft-aerosol-plumes data (not shown here).

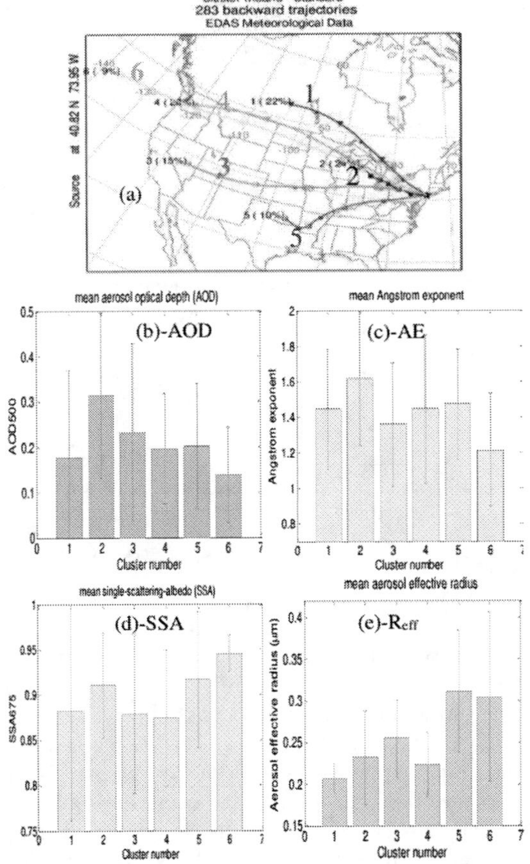

Fig.3. (a) Cluster analysis of aloft-aerosol-plume transport paths, (b) column average AOD, (c) Angstrom exponent, (d) single-scattering-albedo, and (e) effective radius for each cluster.

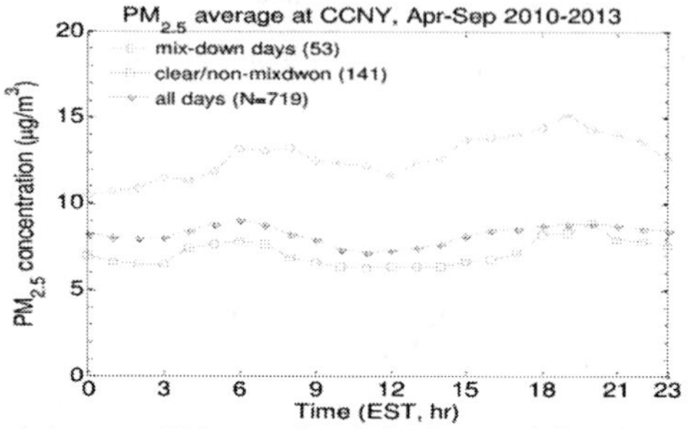

Fig.4 Average $PM_{2.5}$ on the aloft-plume mixing-down days and clear days.

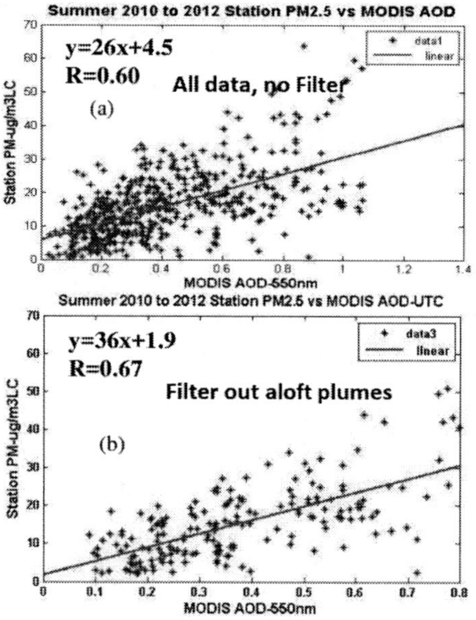

Fig.5. Correlation of MODIS-AOD and $PM_{2.5}$ in summer, 2010-2012, NYC. (a) All lidar days, (b) lidar days after filtering out aloft-plumes.

4. SUMMARY

Extensive aloft-aerosol-plume events and regional transport to the northeastern US are observed with the ground-based multi-wavelength lidar and sunphotometer in New York City. Range-resolved monthly average occurrence of aloft aerosol layers indicates the main transport to occur during March to September which results in the significant enhancement of AOD. Lower Angstrom exponents on the plumes days in April indicate the coarse-mode dust aerosol transport while in summer the higher Angstrom exponents indicate the fine-mode aged smoke transport. Meanwhile, the cluster analysis of transport trajectories indicates the primary 6 transport paths at 4-km altitude. Cluster-2 air-mass transported from the Great Lakes area shows larger AOD and Angstrom exponent while Cluster-6 air-mass traveling from the Pacific Ocean shows the smaller Angstrom exponent and AOD. Other clusters may represent the transport of aged biomass burning aerosols. Additionally, impact of aloft-aerosol-layer on the MODIS column-AOD and ground $PM_{2.5}$ correlation is demonstrated. Both correlation coefficient and linear slope increase while a reduction in offset occurs when filtering out the aloft aerosol plumes. Finally, we note a significant increase in surface $PM_{2.5}$ during plume intrusion cases into the PBL which points to mix-down of the aerosols.

ACKNOWLEDGEMENT. This study was supported by National Oceanic and Atmospheric Administration (NOAA) under the CREST Grant # NA11SEC4810004. Authors greatly appreciate the data from NASA-MODIS and AERONET, EPA, NOAA-HMS and HYSPLIT projects.

REFERENCES

[1]. Kaufman, Y. J., et al., 2005: The effect of smoke, dust, and pollution aerosol on shallow cloud development over the Atlantic Ocean, *Proc. Nat. Acad. Sci.*, **102**, 11207-11212.

[2]. Colarco, P R., et al., 2004: Transport of smoke from Canadian forest fire to the surface near Washington, D.C.: Injection height, entrainment, and optical properties, 2004, *J. Geophy. Res.*, **109**, D06203, doi: 10.1029/2003JD004248.

[3]. Engel-Cox, Hoff, R. M., et al., 2006: Integrating lidar and satellite optical depth with ambient monitoring for 3-dimensional particulate characterization. *Atmos. Environ.* **40**, 8056– 8067.

[4]. Hoff, R. M., Christopher, S. A., 2009: Remote sensing of Particulate Pollution from Space: Have we reached the Promised Land, *J. Air Waste Manage. Assoc.*, **59**, 645–675.

[5]. Wu, Y., S. Chaw, B. Gross, et al., 2009: Low and optically thin cloud measurements using a Raman-Mie lidar, *Appl. Opt.*, **48**, 1218-1227.

[6]. Holben, B.N., et al., 1998: AERONET A federated instrument network and data archive for aerosol characterization, *Rem. Sens. Environ.*, **66**, 1-16.

[7]. Eck, T. F., B. N. Holben, J. S. Reid, et al., 1999: Wavelength dependence of the optical depth of biomass burning, urban, ad desert dust aerosols, *J. Geophys. Res.*, **104**, 31333–31349.

LIDAR MEASUREMENTS OF CANADIAN FOREST FIRE SMOKE EPISODE OBSERVED IN JULY 2013 OVER WARSAW, POLAND

Lucja Janicka[1]*, Iwona S. Stachlewska[1], Krzysztof M. Markowicz[1], Holger Baars[2], Ronny Engelmann[2] and Birgit Heese[2]

[1]*Institute of Geophysics, Faculty of Physics, University of Warsaw (IGFUW), Pasteura 7, 02-093 Warsaw, Poland, *Email: lucja.janicka@student.uw.edu.pl*

[2]*Leibniz Institute for Tropospheric Research, Leipzig, Germany*

ABSTRACT

This paper presents a preliminary study of aerosol optical properties of air-mass advected on 10th July 2013 from Canada above Warsaw, Poland, during the forest fire event that occurred in Quebec at the beginning of July 2013. The observations were conducted with use of the modern version of 8-channel Polly^{XT} lidar capable of measuring at 3β+2α+2δ+VW and interpreted with available information from the MACC model, the CALIPSO and MODIS satellite sensors, the AERONET data products and the data gathered within the Poland-AOD network.

1. INTRODUCTION

Biomass burning aerosols released into the atmosphere during wildfires events have a significant impact on climate [1]. Smoke particles are composed mainly of organic carbon and black carbon which efficiently scatter and absorb solar radiation. Biomass smoke particles can act also as convenient cloud nuclei. Smoke properties depend on variety of factors such as a fuel type, moisture, combustion phase, or wind direction. Properties of biomass burning aerosols change rapidly with time. In comparison to the fresh smoke, larger size and more spherical shape of the aged smoke particles are observed. Favorable weather conditions can spread contaminants on a large area what can cause a worldwide influence on the radiative transfer.

In this study we focus on the long-range transported Canadian biomass burning smoke, that reached and was observed over Warsaw between 8-10 July 2013. Summer 2013, was the driest one in the past 40 years in Quebec, Canada. Thunderstorms in May 2013 caused ignition which finally led to great forest fires in July 2013, when over 2 800 km2 of boreal forests were burned.

The event was observed, among others, by the multi-wavelength depolarization-Raman lidar located at the Radiative Transfer Laboratory (RT-Lab) in Warsaw, Poland. Here, we discuss some of the mentioned above observations.

2. METHODOLOGY

The Radiative Transfer Laboratory (RT-Lab) of the Institute of Geophysics, Faculty of Physics, University of Warsaw is part of the Poland-AOD research network (*http://www.polandaod.pl*). Measurements of the optical properties of aerosols, along with determination of related to them radiative effects in the troposphere are the main tasks of the laboratory. At the end of June 2013 measurement station was enriched by an installation of the automated multi-wavelength-Raman-polarization lidar of a Polly^{XT} type [2]. The lidar started operating in the first week of July 2013, thus the advection of air masses containing smoke from Canadian fires was the first major event observed by the lidar.

In the first step lidar measurements were used to estimate altitudes of aerosol layers that could possibly contain biomass burning aerosols. This information was used as an input to calculate air-mass backward trajectories using the HYSPLIT trajectory model [3].

Optical properties (backscatter and extinction coefficients, lidar ratios, depolarization ratios) were retrieved using standard approaches applied at the lidar sites of EARLINET [4].

The particle backscatter coefficient profiles at 355, 532 nm and the particle extinction profiles at 355 and 532 nm were obtained independently using Raman channels [5]. If it is not possible to use it, the assumption of the lidar ratio is taken [6], [7]. Also the calibrated depolarization profiles at 355 and 532nm were calculated [8]. In the near future, we plan to perform also the retrieval of micro-physical properties of aerosols for characteristic layers found for different periods of the event [9].

Finally, the data were interpreted with other available means: the model MACC, the CALIPSO and MODIS satellite sensors, the AERONET data products and using the data gathered within the Poland-AOD network [10].

3. RESULTS

An aerosol layer is visible at a height of about 2.5 km from the beginning of measurements at 08:30UTC on 8 July 2013 to 05:00 UTC on 10 July 2013 (not shown here; quicklooks of the signals are available via the Polly^{XT} Net Website at *http://polly.tropos.de*). For this period, six-days backward trajectories indicate inflow from over Canada that was uniform at all altitudes. Thus, we assume that this layer contained biomass burning aerosols. From about 18:00 UTC on 8 July to 06:00UTC on 9 July there is visible increased backscattered signal, especially at 1064 nm, at this height. It is probable that the biomass burning aerosol served as cloud condensation nuclei for the clouds that formed at 2.5 km.

From 12:00 UTC on 10 July, aerosol layers appear at higher levels. Optical properties were calculated using averaged data from 18:45 UTC to until 19:45 UTC on that day. Fig. 1. shows the temporal development of the range-corrected signal at 1064 nm. Aerosol layers at about 3 and 5 km are clearly visible. At a height of about 1.5 km, clouds associated with the weather front began to form.

Backscatter and extinction coefficient profiles up to 7 km are shown in Fig. 2. At altitudes of circa 1.5 km, 3 km and 5.5 km three aerosol layers are discernible. The lowest one likely reflects boundary layer and cloud layer forming a weather front. The other two are likely to be due to the advection of air-masses containing aerosols. Maximum values for the extinction of about 0.18 km^{-1} for 355 nm and 0.1 km^{-1} for 532 nm were observed in layers at altitudes 1.5 and 5.5 km. At the same time the backscatter coefficient was lower in the highest layer than in the lowest one and reached 0.003 and 0.002 km^{-1}sr^{-1} at 532 nm, respectively, for lowermost and uppermost layer. For 355 nm backscatter coefficient values were 0.0015 and 0.0008 km^{-1}sr^{-1}, respectively.

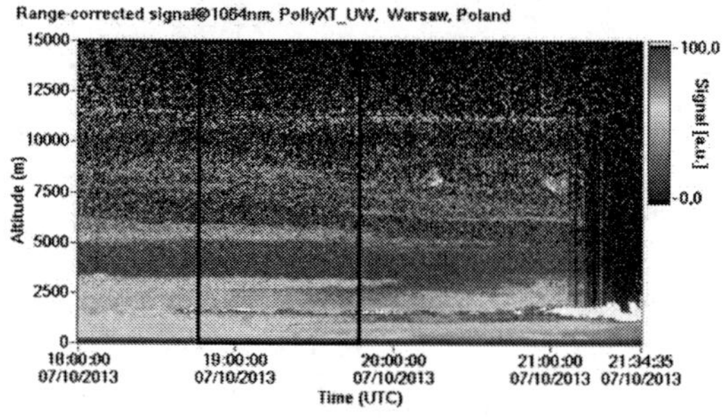

Figure 1. Range-corrected signal at 1064 nm between 18:00UTC and 21:34UTC on 10 July 2013. The black rectangle indicates the period chosen for the data analysis.

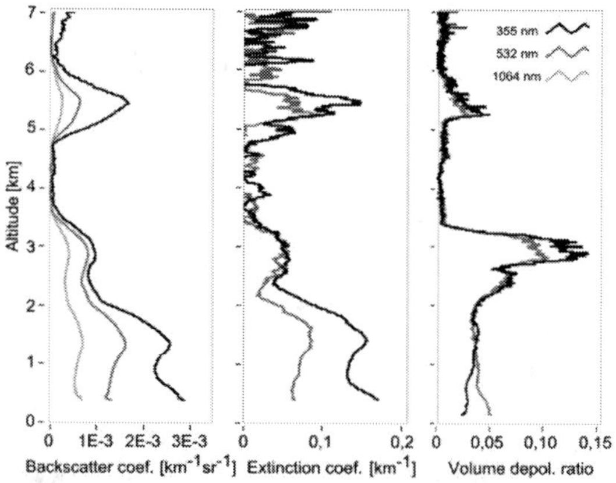

Figure 2. Vertical profiles of particle backscatter coefficient, particle extinction coefficient, linear volume depolarization ratio on 10 July 2013, averaged over 18:45UTC to 19:45UTC.

The aerosol layer that was located at 3 km strongly differed from that at 5.5 km. Here the backscatter and extinction coefficients values for 355 and 532 nm are very similar to each other, which is in contrast to the upper layer where

DOI: 10.1051/epjconf/201611918005

values for 355 nm are larger. Backscatter and extinction coefficients for 532 nm did not vary very much between the layers 5.5 km and 3 km. However, coefficients for 355 nm are significantly higher for the upper layer. It can be concluded that it is more likely that in the top layer strongly absorbing particles characteristic for the biomass burning aerosols could be present but the lower layer had rather different origin. This hypothesis is also confirmed by the profile of volume depolarization ratio shown in Fig. 2. The high values of about 15% for 355 nm and 10% for 532 nm were observed at an altitude of 3 km. At a height of 5.5 km the depolarization ratio was less than 5% for both wavelengths. A low depolarization is characteristic for aged biomass burning aerosol [11], [12]. Relatively high values of depolarization can indicate mineral dust particles [13]. Given in Fig. 3, twelve-days HYSPLIT backward trajectories suggest that air masses incoming at 3 km could origin from Sahara. The trajectory ending at 5 km shows advection of biomass burning aerosol from Canada which is about six days old.

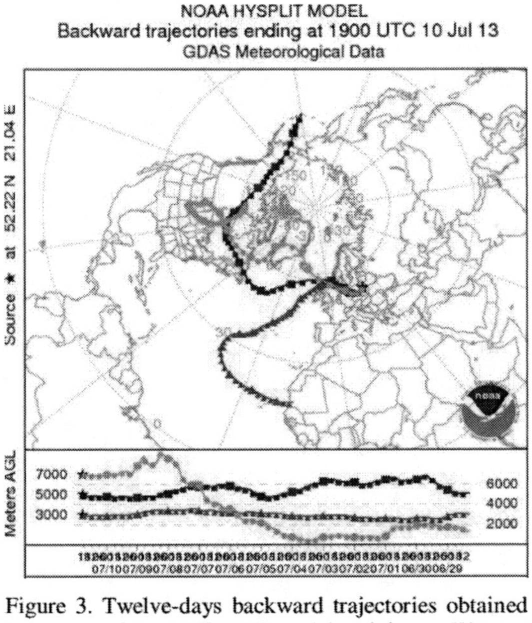

Figure 3. Twelve-days backward trajectories obtained by means of the HYSPLIT model arriving at Warsaw on 10th July 2013.

The mean lidar ratio (not shown here) for this two layers is about 55 and 75 sr for 355 nm and 70 and 95 sr for 532 nm at 3 km and 5.5 km height, respectively. This values are consistent with lidar ratio values found during SAMUM campaign [13]. Lidar ratios at 532 nm are larger than the ones at 355 nm. This property was noticed for aged biomass burning aerosols [14]. The mean color ratio β(532/1064) was about 1.3 at both layers. The mean color ratio β(355/532) at height 3 km was lower and reached about 0.5, in the opposite of the upper layer, where this value was about 2 and was higher than β(532/1064). The Ångström exponent for extinction at 3 km was below 0.5 which means large aerosol particle size.

4. CONCLUSIONS

The new Polly^{XT} lidar operating at RT-Lab in Warsaw since June 2013 allowed for the detection of biomass burning aerosols incoming over Europe from Canadian wildfires in July 2013. Profiles of particle backscatter and particle extinction coefficients retrieved from one hour measurements from 18:45 UTC on 10 July 2013 showed, next to the boundary layer at about 1.5 km, two distinct layers of aerosol at 3 and 5.5 km. Wavelength independent backscatter and extinction coefficients were observed in the lower one, whereas in the upper one this properties highly depended on wavelength. High extinction, especially for 355 nm, and low depolarization was observed in the top layer. It can be inferred that layer at 5.5 km contained biomass burning aerosols. The high lidar ratios obtained in this layer also confirm this hypothesis. However, the color ratio β(355/532) for this layer was high, what is not typical for aged biomass burning aerosol where aging processes lead to increased particle size.

The HYSPLIT backward trajectories and high volume depolarization ratio indicate that lower layer of 3 km can include mineral dust particles from Sahara. Advection of Saharan dust particles over Warsaw is not unusual (another recent example of such event is discussed in [15]), however they occur most frequently in May and June.

ACKNOWLEDGEMENTS

Part of the research was done within the SONATA BIS Project "Integrated study of climate processes involving absorbing aerosols" funded by NCN, Poland (Grant No. 2012/05/E/ST10/01578).

The lidar development was financed by FNITP, Poland (Grant No.519/FNITP/115/2010).

We acknowledge the NOAA Air Resources Laboratory (ARL) for the provision of the HYSPLIT transport model (http://www.arl.noaa.gov/ready.php) used for the interpretation for the results obtained in this publication.

REFERENCES

[1] Reid, J.S., R. Koppmann, T.F. Eck, D.P. Eleuterio, 2005: A review of biomass burning emissions part ii: intensive physical properties of biomass burning particles, *Atmospheric Chemistry and Physics*, **5**, pp. 799-825.

[2] Althausen et al. 2009: Portable Raman Lidar Polly^{XT} for Automated Profiling of Aerosol Backscatter, Extinction, and Depolarization. *J. Atmos. Oceanic Technol.*, **26**, 2366-2378.

[3] Draxler, R.R. and Rolph, G.D., 2012, HYSPLIT (Hybrid Single-Particle Lagrangian Integrated Trajectory) Model access via NOAA ARL READY Website (http://ready.arl.noaa.gov/HYSPLIT.php). NOAA Air Resources Laboratory, Silver Spring, MD

[4] Pappalardo, et al., 2014: EARLINET towards an advanced sustainable European aerosol lidar network, Atmos. Meas. Tech., 7, 2389–2409.

[5] Ansmann A., Riebesell M., and C. Weitkamp, 1990: Measurement of atmospheric aerosol extinction profiles with a Raman lidar, *Opt. Lett.*, **15(13)**, 746-748.

[6] Klett, J. D., 1981: Stable analytical inversion solution for processing lidar returns. *Appl. Opt.*, **20**, 211–220.

[7] Fernald, F. G., 1984: Analysis of atmospheric lidar observations: Some comments. *Appl. Opt.*, **23**, 652–653.

[8] Freudenthaler, V., et al., 2009: Depolarization ratio profiling at several wavelengths in pure Saharan dust during SAMUM 2006, *Tellus B*, **61(1)**, 165-179.

[9] Veselovskii, I., Kolgotin, A., Griaznov, V., Muller, D., Wandinger, U., and Whiteman, D.N., 2002: Inversion with regularization for the retrieval of tropospheric aerosol parameters from multiwavelength lidar sounding, *Appl Optics* **41(18)**, 3685-3699.

[10] Chiliński M.T. et al., 2014: Study of Aerosol Radiative Properties During Long-range Transport of Biomass Burning from Canada to Central Europe in July 2013, *14th Conference on Atmospheric Radiation*, conference poster, https://ams.confex.com/ams/14CLOUD14ATRAD/webprogram/Paper250309.html

[11] Baars, H., et al., 2012: Aerosol profiling with lidar in the Amazon Basin during the wet and dry season, *Journal of Geophysical Research*, **117**, D21201, doi:10.1029/2012JD018338.

[12] Nicolae D. et al. 2012: Change of Optical and Microphysical Properties of Biomass Burning Aerosol During Transport in the Free Troposphere, *Reviewed and Revised Papers 26th International Laser Radar Conference*, 463-466.

[13] Groß, S., et al., 2011: Characterization of Saharan dust, marine aerosols and mixtures of biomass-burning aerosols and dust by means of multi-wavelength depolarization Raman lidar measurements during SAMUM 2. *Tellus B*, **63**, pp. 706-724.

[14] Muller D., et al, 2005: Raman lidar observations of aged Siberian and Canadian forest fire smoke in the free troposphere over Germany in 2003: Microphysical particle characterization, *Journal of Geophysical Research*, **110**, D17201.

[15] Costa-Surós M., I.S. Stachlewska, A. Nemuc, C. Talianu, B. Heese, and R. Engelmann, 2015: Study case of air-mass modification over Poland and Romania observed by the means of multi-wavelength Raman depolarization lidars, *International Laser Radar Conference* (ILRC 27th), 5-10 July 2015, New York, USA, Paper ID: 315.

CENTRAL ASIAN DUST EXPERIMENT (CADEX): MULTIWAVELENGTH POLARIZATION RAMAN LIDAR OBSERVATIONS IN TAJIKISTAN

Julian Hofer [1]*, Dietrich Althausen [1], Sabur F. Abdullaev [2], Ronny Engelmann [1], Holger Baars [1]

[1] *Leibniz Institute for Tropospheric Research, Leipzig, Germany, *Email: hofer@tropos.de*

[2] *Physical Technical Institute of the Academy of Sciences of Tajikistan, Dushanbe, Tajikistan*

ABSTRACT

For the first time lidar measurements of vertical aerosol profiles are conducted in Tajikistan/ Central Asia. These measurements just started on March 17th, 2015. They are performed within the Central Asian Dust Experiment (CADEX) in Dushanbe and they will last at least one year. The deployed system for these observations is an updated version of the multiwavelength polarization Raman lidar Polly^{XT}. Vertical profiles of the backscatter coefficient, the extinction coefficient, and the particle depolarization ratio are measured by this instrument. A first and preliminary measurement example of an aerosol layer over Dushanbe is shown.

1. INTRODUCTION

Tajikistan (Fig. 1) lies in the global dust belt in close proximity of some major dust sources like the Taklimakan desert, the Aralkum desert of the desiccating Aral Sea, the Kyzylkum and Karakum deserts east of the Caspian Sea, and the Iranian Dasht-e-Kavir and Dasht-e-Lut deserts. Therefore Tajikistan is frequently affected by severe dust events. Furthermore, Central Asia and especially Tajikistan are highly affected by climate change. For example, dramatic glacier shrinking took place in the last decades [1], which has also an effect on the water resources of Tajikistan and the whole Central Asian area, since trans-regionally important rivers as Amu Darya and Syr Darya are fed by glacier meltwater.

Aerosol observations in Tajikistan are therefore highly important to understand regional and global transport of mineral dust and its effects on radiation budget, cloud formation etc.

Up to now, only few experiments were performed to characterize the aerosol over Central Asia. So, a joint Soviet-American research group did dust research in Tajikistan in 1989 [2]. They characterized the chemical and physical properties of mineral dust collected in 2 dust storms [3, 4]. During these dust storms in 1989 optical depths of up to 3.5 (at 550 nm) were measured. Furthermore a few vertical profiles of the dust distribution were measured by aircraft [5]. They found well mixed dust aerosols up to a height of 3 km.

Nevertheless, the knowledge of the vertical aerosol distribution over Tajikistan and especially the transport of mineral dust over Central Asia is still insufficient. Therefore CADEX aims to provide long-term data on vertical profiles of the backscatter coefficient, the extinction coefficient, and the particle depolarization ratio.

2. FIELD SITE AND INSTRUMENTATION

The measurement site is located in an urban area on a slightly elevated hill near the Physical Technical Institute of the Academy of Science of Tajikistan in Dushanbe. Its coordinates are 38°33'34" N, 68°51'22" E, and its altitude is 864 m a.s.l..

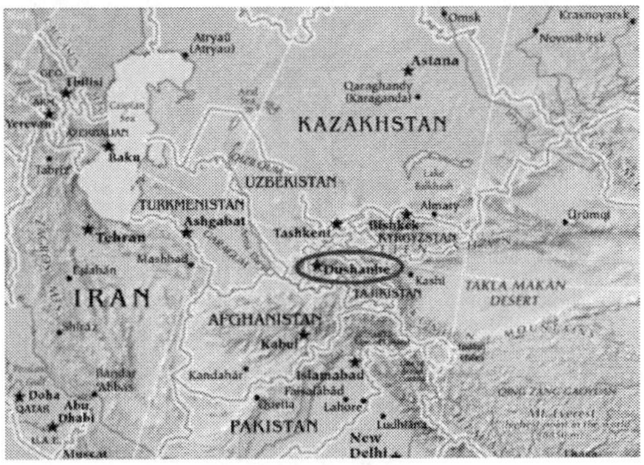

Fig. 1: Physical map of Central Asia with Dushanbe/Tajikistan (red circle). (The World Factbook 2013-14. Washington, DC: Central Intelligence Agency, 2013)

The updated version of the multiwavelength polarization Raman lidar Polly^{XT} [6, original design] is used for the lidar measurements. The

DOI: 10.1051/epjconf/201611918006

last improvements were the installation of an additional depolarization channel at 355 nm for this campaign (Fig. 2) and the change of most of the electronic components to the current Polly^{XT} systems standard parts to ensure higher counting rates (see contribution Engelmann et al. [7] at this conference). All Polly^{XT} systems contain a laser that emits light at 355 nm, 532 nm, and 1064 nm. The receiver of the used system has 8 channels. With this system, vertical profiles of the backscatter coefficient are determined for all three emitted wavelengths. The channels at 387 nm, 607 nm, and 407 nm allow to detect Raman scattering at night time to determine extinction coefficient and water vapor profiles. Another two channels are installed to detect cross-polarized light at 355 nm and 532 nm. This allows to measure the particle depolarization ratio at these wavelengths. The system has a vertical resolution of 7.5 m and temporal resolution of 30 s.

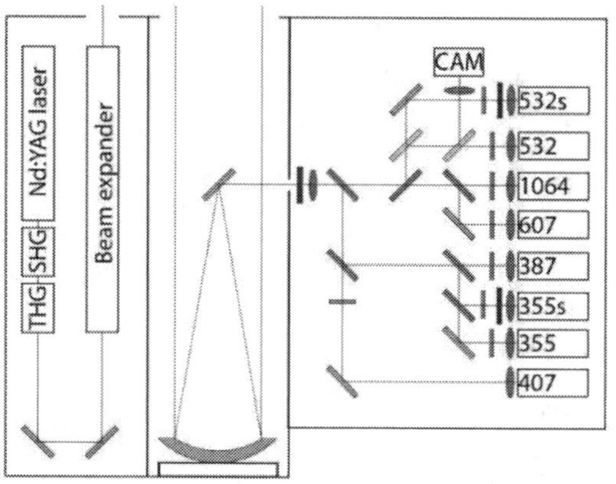

Fig. 2: Scheme of Polly^{XT}. 1064 nm Nd:YAG laser with second (SHG) and third harmonic generation (THG). Orange: Mirrors, including parabolic 30 cm main mirror. Blue: Lenses. Green: Dichroic beam splitters. Yellow: Non-polarizing 50% beam splitters. Magenta: Polarizers. Grey: Neutral density filters. Numbers: Photomultiplier tubes for the different wavelengths. CAM: Camera.

Additionally, a CIMEL sun photometer from the AERONET station in Dushanbe is collocated with the Polly^{XT} lidar to measure the aerosol optical depth and retrieve further particle properties, like size distribution, fine and coarse mode fraction.

There are also devices measuring ground level aerosols at the measurement site in Dushanbe. A Digitel DHA-80 High Volume Sampler is collecting PM10 particulates on quartz fiber filters for subsequent chemical analysis. A Grimm Environmental Dust Monitor EDM-180 is measuring particle size distribution of ground level aerosols without any initial cut off size.

Later, simulations with the regional dust model COSMO-MUSCAT [8] will be performed to access the transport and radiative effects of the dust during the time of the field campaign. The properties measured with the Polly^{XT} serve as input parameters for the model.

3. FIRST MEASUREMENTS

An example of one of the first measurements during CADEX is shown in Fig. 3. The temporal evolution of the range-corrected signal at 1064 nm on March 19th, 2015 reveals a lofted aerosol layer, located between the planetary boundary layer and the high-level cirrus clouds. The layer is settling down from about 6 km to about 4.5 km from 00:04 to 06:00 UTC. Between 03:00 and 04:00 UTC a cirrus cloud has formed in the altitude of about 6 km where the layer was located before.

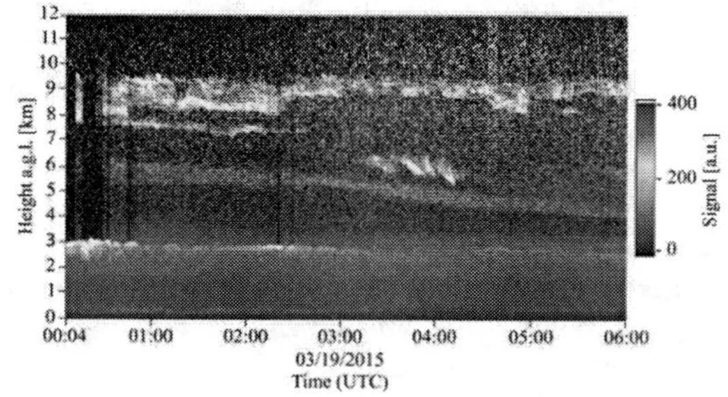

Fig. 3: Temporal evolution of the range-corrected signal at 1064 nm on March 19th, 2015.

HYSPLIT [9, 10] back trajectories indicate that the air, carrying this aerosol layer at 6 km height, was transported over the Middle East to Tajikistan (Fig. 4). Despite being transported along the global dust belt the air did not reach into the planetary boundary layer.

Fig. 5 shows vertical profiles of the backscatter coefficient (a), the extinction coefficient (b), the depolarization ratio (c) and the lidar ratio (d) averaged between 00:50 and 01:05 UTC on

March 19th, 2015. The analysis was conducted with the Raman method [11]. The depolarization ratio was calculated based on the internal depolarization calibration which is automatically

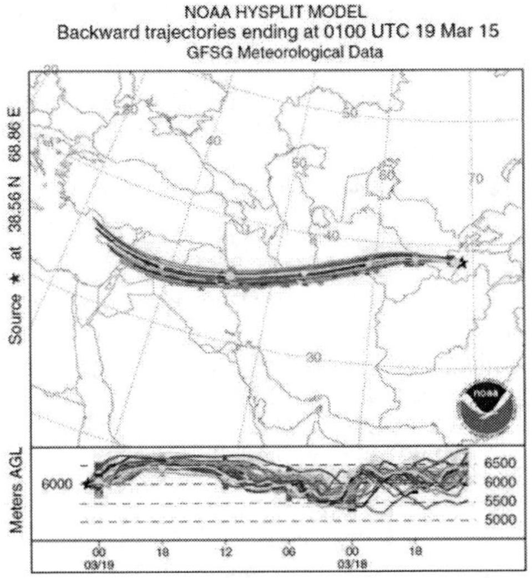

Fig. 4: HYSPLIT ensemble backward trajectories arriving on 6 km over Dushanbe at 01:00 UTC on March 19th, 2015.

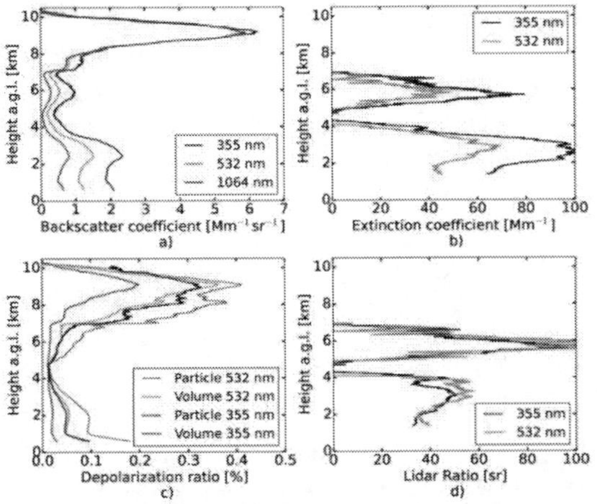

Fig. 5: Averaged profiles from 0050 to 0105 UTC on March 19th, 2015, of a) backscatter coefficients for 355 nm, 532 nm, and 1064 nm, b) extinction coefficients for 355 nm and 532 nm, c) volume and particle depolarization ratios for 355 nm and 532 nm, and d) lidar ratios for 355 nm and 532 nm. The backscatter coefficient and the depolarization ratio are smoothed over 161, the extinction coefficient and the lidar ratio over 231 vertical bins.

Performed by the system three times a day [12]. Low signal-to-noise ratios did not allow the calculation of extinction coefficients and lidar ratios above 7 km. The lofted aerosol layer between about 5 to 6 km depolarizes only weakly (Fig. 4c)) but has larger extinction values (Fig. 4b)). Also it shows rather large lidar ratios of > 60 sr. So, bringing this information together leads to the conclusion that this layer is a particle layer originating mainly from pollution and/or burning processes and it is most likely not mineral dust. However, the shown data have to be considered yet as very preliminary. Ongoing overviews of the measurements in Dushanbe are available under http:\\polly.rsd.tropos.de and www.tropos.de/en/research/projects-infrastructures-technology/research-projects/cadex-home/.

4. OUTLOOK

The Polly^{XT} lidar is measuring since March 17th, 2015 in Dushanbe, Tajikistan. It is planned to operate it continuously until at least April 2016. The obtained backscatter, extinction and depolarization profiles will be used to characterize the aerosol types [e.g. 13, 14]. Analyses of HYSPLIT back trajectory clusters have indicated that the air masses arriving in Tajikistan can carry desert dust from as far as the Sahara and the Middle East. Hence we will statistically analyze the aerosol characterization and especially the measured aerosol optical properties according to the air masses' origin. Also anthropogenic pollution from local and distant sources is likely to form a complex aerosol mixture together with desert dust over Tajikistan. This needs approaches to separate optical profiles of dust from non-dust like [15]. Furthermore a separation of coarse mode and fine mode dust like [16] will be done with the performed polarization measurements.

ACKNOWLEDGEMENT

We want to thank Dr. K. W. Fomba, Dr. K. Müller, and Dr. B. Heinold from TROPOS Leipzig, and Dr. G. Schettler from GFZ Potsdam for their contributions to CADEX. Special thanks go to the Physical Technical Institute of the Academy of Sciences of Tajikistan for their collaboration and for providing the measurement site and maintaining the instrumentations. Further, we would like to thank the German Ministry of Education and Research for funding this project.

REFERENCES

[1] Kayumov, A.: Glaciers Resources of Tajikistan in Condition of the Climate Change, State Agency for Hydrometeorology of Committee for Environmental Protection under the Government of the Republic of Tajikistan.

[2] Golitsyn, G., Gillette, D. A., 1993: Introduction: A joint Soviet-American experiment for the study of Asian desert dust and its impact on local meteorological conditions and climate. *Atmos. Environ.*, **27(16)**, 2539-2544.

[3] Gomes L., Gillette D. A., 1993: A comparison of characteristics of aerosol from dust storms in Central Asia with soil-derived dust from other regions. *Atmos. Environ.*, **27(16)**, 2539-25.

[4] Sokolik, I. N., Andronova, A. V., Johnson, T. C., 1993: Complex refractive index of atmospheric dust aerosols. *Atmos. Environ.*, **27(16)**, 2495-2502.

[5] Pachenko, M. V., Terpugova, S. A., Bodhaine, B. A., Isakov, A. A., Sviridenkov, M. A., Sokolik, I. N., Romashova, E. V., Nazarov, B. I., Shukurov, A. K., Chistyakova, E. I., Johnson, T. C., 1993: Optical investigations of dust storms during U.S.S.R.-U.S. experiments in Tadzhikistan, 1989. *Atmos. Environ.*, **27(16)**, 2503-2508.

[6] Althausen, D., Engelmann, R., Baars, H., Heese, B., Ansmann, A., Müller, D., Komppula, M., 2009: Portable Raman Lidar Polly^{XT} for Automated Profiling of Aerosol Backscatter, Extinction, and Depolarization. *J. Atmos. Oceanic Technol.*, **26(11)**, 2366-2378.

[7] Engelmann, R., Ansmann, A., Bühl, J., Heese, B., Baars, H., Althausen, D., Marinou, E., Amiridis, V., Mamouri, R., Vrekoussis, M., 2015: Observation of Arabian and Saharan dust in Cyprus with a new generation of the smart Raman lidar Polly. *Proceedings of the 27th ILRC*, New York City, 2015.

[8] Heinold, B., Helmert, J., Hellmuth, O., Wolke, R., Ansmann, A., Marticorena, B., Laurent, B., and Tegen, I., 2007: Regional modeling of Saharan dust events using LM-MUSCAT: Model description and case studies. *J. Geophys. Res. - Atmos.*, **112(D11)**.

[9] Draxler, R. R., Rolph, G. D., 2015: HYSPLIT (HYbrid Single-Particle Lagrangian Integrated Trajectory) Model access via NOAA ARL READY Website (http://ready.arl.noaa.gov/HYSPLIT.php). NOAA Air Resources Laboratory, Silver Spring, MD.

[10] Rolph, G. D., 2015: Real-time Environmental Applications and Display sYstem (READY) Website (http://ready.arl.noaa.gov). NOAA Air Resources Laboratory, Silver Spring, MD.

[11] Ansmann, A., Wandinger, U., Riebesell, M., Weitkamp, C., Michaelis W., 1992: Independent measurement of extinction and backscatter profiles in cirrus clouds by using a combined Raman elastic-backscatter lidar. *Appl Opt.*, **31(33)**, 7113-7131.

[12] Freudenthaler, V., Esselborn, M., Wiegner, M., Heese, B., Tesche, M., Ansmann, A., Müller, D., Althausen, D., Wirth, M., Fix, A., Ehret, G., Knippertz, P., Toledano, C., Gasteiger, J., Garhammer, M., Seefeldner, M. 2009: Depolarization ratio profiling at several wavelengths in pure Saharan dust during SAMUM 2006. *Tellus*, **61B**, 165-179.

[13] Groß, S., Esselborn, M., Weinzierl, B., Wirth, M., Fix, A., Petzold, A., 2013: Aerosol classification by airborne high spectral resolution lidar observations. *Atmos. Chem. Phys.*, **13**, 2487-2505.

[14] Burton, S. P., Ferrare, R. A., Hostetler, C. A., Hair, J. W., Rogers, R. R., Obland, M. D., Butler, C. F., Cook, A. L., Harper, D. B., Froyd, K. D., 2012: Aerosol classification using airborne High Spectral Resolution Lidar measurements – methodology and examples. *Atmos. Meas. Tech.*, **5**, 73-98.

[15] Tesche, M., Ansmann, A., Müller, D., Althausen, D., Engelmann, R., Freudenthaler, V., Groß, S., 2009: Vertically resolved separation of dust and smoke over Cape Verde by using multiwavelength Raman and polarization lidars during Saharan Mineral Dust Experiment 2008. *J. Geophys. Res. – Atmos.*, **114(D13)**.

[16] Mamouri, R., Ansmann, A., 2014: Fine and coarse dust separation with polarization lidar. *Atmos. Meas. Tech.*, **7**, 3717-3735.

DOI: 10.1051/epjconf/201611918007

3D STRUCTURE OF SAHARAN DUST TRANSPORT TOWARDS EUROPE AS SEEN BY CALIPSO

Eleni Marinou[1*], Vassilis Amiridis[1], Alexandra Tsekeri[1], Stavros Solomos[1], Panos Kokkalis[1], Emmanouil Proestakis[1], Michael Kottas[1], Ioannis Binietoglou[2], Prodromos Zanis[3], Stelios Kazadzis[4], Ulla Wandinger[5] and Albert Ansmann[5]

*[1]IAASARS, National Observatory of Athens, Athens, Greece, *Email: elmarinou@noa.gr*

[2] National Institute of R&D for Optoelectronics, Magurele, Romania

[3] Department of Meteorology and Climatology, School of Geology, Aristotle University of Thessaloniki, Thessaloniki, Greece

[4] Physikalisch-Meteorologisches Observatorium Davos, World Radiation Center, Switzerland

[5]Leibniz Institute for Tropospheric Research, Leipzig, Germany

ABSTRACT

We present a 3D multi-year monthly mean climatology of Saharan dust advection over Europe using an area-optimized pure dust CALIPSO product. The product has been developed by applying EARLINET-measured dust lidar ratios and depolarization-based dust discrimination methods and it is shown to have a very good agreement in terms of AOD when compared to AERONET over Europe/North Africa and MODIS over Mediterranean. The processing of such purely observational data reveals the certain seasonal patterns of dust transportation towards Europe and the Atlantic Ocean. The physical and optical properties of the dust layer are identified for several areas near the Saharan sources, over the Mediterranean and over continental Europe.

1. INTRODUCTION

CALIPSO is capable of providing a multi-year, robust 4D dust climatology, a task that cannot easily be achieved by passive sensors, especially over bright surfaces like deserts. However, limitations on retrieval performance using CALIPSO exist, especially regarding the classification of dust and its mixtures based on the approximate particle depolarization ratio and the retrieval of the dust extinction that is based on the Lidar Ratio (LR) assumption.

Amiridis et al. (2013) [1] demonstrated a method for the retrieval of pure dust extinction from CALIPSO based on a depolarization-based separation of pure dust backscatter fraction in dusty mixtures (Tesche et al., 2009) and the application of a regionally uniform LR for calculating dust extinction. This LR value (equal to 58 sr) has been taken from climatological values derived by long-term EARLINET measurements. It has been shown in [1] that the final dust product is consistent with AERONET collocated measurements over Sahara and Europe and also consistent with MODIS over the Mediterranean for collocated cells with low cloudiness.

Here, we present a climatology of the Saharan dust 3D structure and load over Europe and North Africa based on the product introduced and validated in [1]. We briefly present the methodology in section 2, demonstrate the first results in section 3 and conclusions in section 4.

2. METHODOLOGY

We produce a monthly mean dust L3 product based on averaging CALIPSO L2 product following the methodology introduced in [1]. In brief, the product is developed using the filtering CALIPSO techniques introduced in [2]. The product aggregation is performed in a 1x1 degree spatial resolution for the backscatter coefficient at 532 nm. Prior averaging, we obtain the dust backscatter coefficient for pure dust by applying the method introduced in [3]. This method makes use of the particle backscatter coefficient and the particle depolarization ratio at 532 nm in order to separate the backscatter contributions of the weakly depolarizing aerosol components ("other type") from the contribution of strongly depolarizing particles (pure dust). We avoid using the polluted dust and pure dust CALIPSO types in order to eliminate possible misclassifications introduced due to the use of L1 estimation of the depolarization ratio. Instead, we use the L2 linear particle depolarization ratio (recalculated from L2 total and perpendicular backscatter at 532 nm) to classify dust and polluted dust components following the definitions given in [1]. After discriminating the pure dust from its mixtures with other types, we apply the EARLINET-based climatological dust LR of 58 sr on each backscatter profile prior averaging. Finally, we apply a slightly different concept on averaging the dust extinction on the 1x1 degree resolution, by acknowledging zero extinction contribution from the "other" non-dust aerosol types. This alteration would result on a product that is representative of the dust extinction and correspondingly the dust columnar AOD in regard to the total AOD. In other words, we calculate the dust AOD contribution to the total AOD.

The methodology introduced in [1] and applied here on CALIPSO record between the years 2007 and 2014, has been validated against AERONET over land and compared to MODIS over the Mediterranean. A demonstration of those evaluations is given in the following figures.

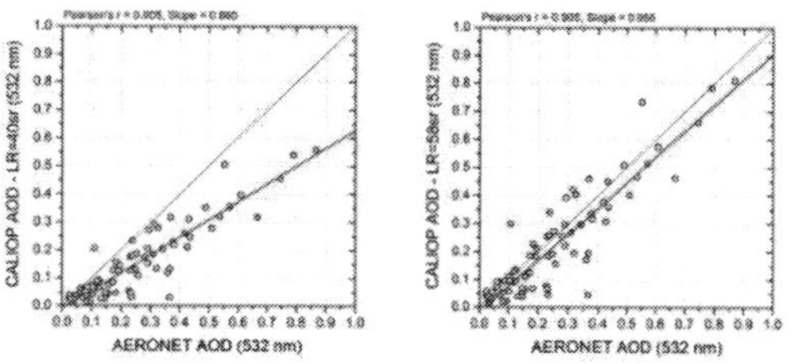

Figure 1: Scatter plot comparison of CALIPSO AOD versus collocated AERONET measurements when LR is equal to 40 sr (left) and when LR is equal to 58 sr (right).

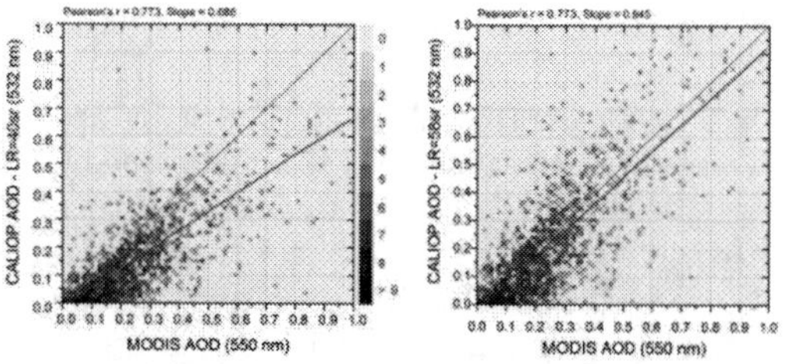

Figure 2: Comparison of CALIPSO AODs (1x1 degree) versus collocated MODIS-Aqua Level 3 product using LR equal to 40 sr (left) and LR equal to 58 sr (right). The 2D scatterplots represent the number of cases found for each CALIPSO/MODIS AOD bin between 0 and 1.0 (bin step equal to 0.0125). MODIS data are not filtered, while CALIPSO data are filtered according to CAL-L3 specifications. Only CALIPSO overpasses that are cloud-free and for which the aerosol classification scheme reveals only dust presence are considered.

3. RESULTS

From the CALIPSO dust product we provide here a climatology for the domain between 20 and 60 in latitude and between -20 and 30 in longitude which includes Europe and North Africa and where we believe that the EARLINET-derived LR value of 58 sr can be applied. In Figure 3 we present a seasonal distribution of the dust AOD in the Region of Interest (RoI).The four seasons are defined based on the dominant dust transportation patterns. During January to March (J-F-M) limited dust activity is observed almost uniformly over Sahara and the AOD remains below 0.2 with the exception of specific dust hot-spots. The formation of mid-latitude low pressure systems during this period results in transportation of dust from the deserts of Libya towards Greece and a mean seasonal AOD of 0.15-0.2 is found over eastern Mediterranean. For the period April-June (A-M-J) dust production is evident over the entire desert and AOD values exceeding 0.5 are found especially at lower latitudes (20°N-30°N). Over central and eastern Mediterranean, South Italy, Greece and Turkey the mean CALIPSO dust AOD is close to 0.15. During July-September (J-A-S) dust activity is mostly limited to the western part of the Sahara and the AOD near the sources is up to 0.6. The migration of the ITCZ (Intertropical Convergence Zone) towards higher latitudes and the dominance of trade wind patterns (easterlies) favor the transportation of dust towards the Atlantic Ocean. Increased AOD values up to 2 are also found over western Mediterranean, south Italy and the south coast of Spain. For October up to December (O-N-D) dust activity is mostly suppressed except from the south-west desert areas close to the Sahel zone where AOD remains close to 0.3-0.4.

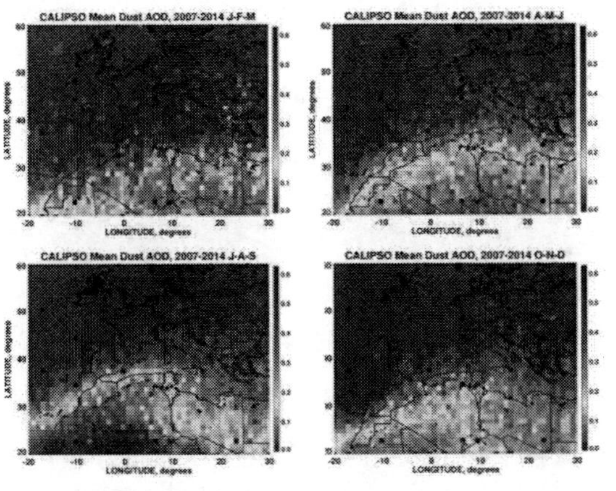

Figure 3: Dust AOD of 1x1 degree spatial resolution averaged for the four seasons over RoI using CALIPSO observations from 2007 to 2014.

In Figure 4 we present the Top Layer Height (TLH) of the dust layer produced from the averaged profiles within the 1x1 degree cells. During the first period (J-F-M) dust remains in general below 3-4 km over land areas. Over the sea, several transport paths are evident and dust extends higher than 3 km. In A-M-J the dust layer top is up to 6 km over Africa and up to 3-4 km over the sea and south Europe. The highest top layer values (>8 km) are found for the period (J-A-S) over west Africa and Atlantic Ocean. Intrusion of the lower tropospheric Atlantic monsoon south of the ITCZ and the development of MCS (mesoscale-convective systems) favor the elevation of dust at this area and the transatlantic transportation of dust along the easterlies. During the same period, the dominance of the strong Saharan high favors mobilization of dust from the western part of Sahara towards west Mediterranean and Europe. This pattern leads to elevated dust up to 3-4 km over south European countries and the Balkans. In O-N-D the situation is similar to A-M-J however with relatively lower heights.

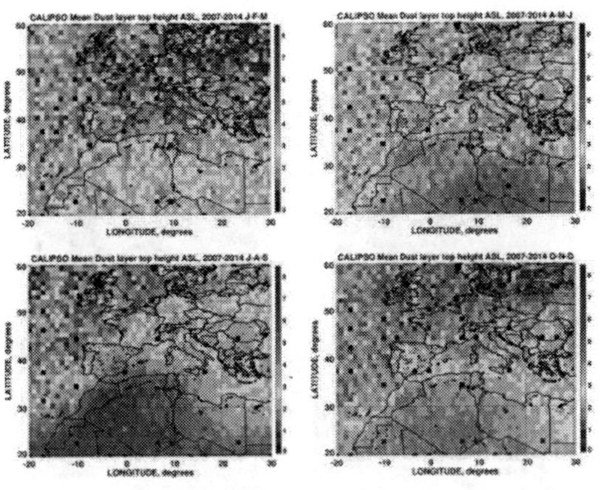

Figure 4: Seasonal maps of dust Top Layer Height (km) over the RoI using CALIPSO observations from 2007 to 2014.

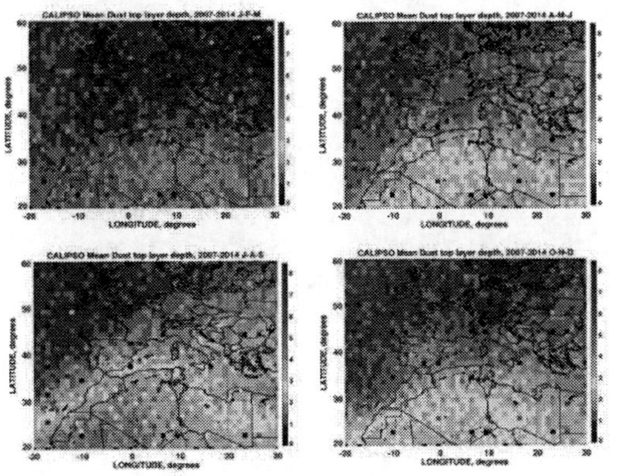

Figure 5: Seasonal maps of the Dust Layer Depth over the RoI using CALIPSO observations from 2007 to 2014.

The dust layer thickness (Figure 5) is an important parameter for determining the severity of dust episodes over certain areas and also the vertical structure of dust plumes. This last parameter is of particular importance for radiative transfer calculations and dust-cloud interactions. As seen in this figure, the dust plumes that affect Europe are at the range of 1-2.5 km in depth while dust plumes off the west African coast can be 5-6 km thick.

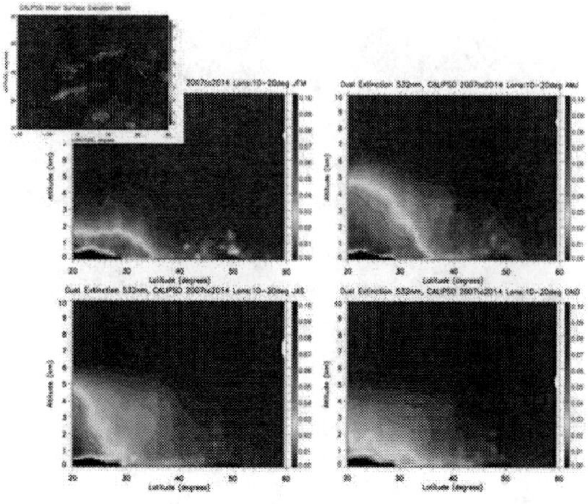

Figure 6: Seasonal maps of dust extinction cross sections for the longitudinal belt between 10 and 20 degrees using CALIPSO observations from 2007 to 2014.

A very important asset of this particular CALIPSO dust product is the three-dimensional information on dust extinction that is seen for example at the vertical South-to-North cross-sections of Figure 6 for the area between 10°E-20°E. This region includes the dust sources of Libya and central Sahara, central Mediterranean, Italy, the Alps and part of North Europe. It is evident from this figure that dust extinction over central Mediterranean (30°N-45°N) is higher than 0.02 throughout the year. Moreover, elevated dust is seen over the Alps (47°N-52°N) and significantly high values (>0.03) are especially found for the period J-F-M.

4. CONCLUSIONS and FUTURE WORK

CALIPSO is capable of providing a multi-year, robust 4D dust climatology, a task that cannot easily be achieved by passive sensors, especially over deserts. However, limitations on retrieval performance using CALIPSO exist, especially regarding the classification of dust and its mixtures based on the approximate particle depolarization ratio and the LR assumption. In this paper, we brief our previous work on the potential improvement of CALIPSO dust retrievals over Europe and North Africa by using a dust LR of 58 sr, demonstrating that a regional correction is feasible when using a universal and spatially constant LR. Then, we demonstrate the potential of this product to provide pure dust 2D and 3D distributions over the RoI. Future work includes the application of the methodology in similar studies over the deserts in the Middle East and China in order to optimize CALIPSO dust retrievals over these areas as well. Ground-based measurements of the dust LR and particle depolarization ratio for these regions will be vital for the success of implementing similar improvements.

ACKNOWLEDGEMENT

The publication was supported by the European Union Seventh Framework Programme (FP7-REGPOT-2012-2013-1), in the framework of the project BEYOND, under Grant Agreement no. 316210 (BEYOND – Building Capacity for a Centre of Excellence for EO-based monitoring of Natural Disasters). The research leading to these results has received funding from the European Union Seventh Framework Programme (FP7/2007-2013) under grant agreement no. 262254 (ACTRIS). The research leading to these results has received funding from the European Union Seventh Framework Programme (FP7/2007-2013) under grant agreement n° 606953 (MarcoPolo). CALIPSO data were provided by NASA. We thank the ICARE Data and Services Center for providing access to the data used in this study.

REFERENCES

[1] Amiridis, V., et al., 2013: Optimizing CALIPSO Saharan dust retrievals, *Atmos. Chem. Phys.*, **13**, 12089-12106, doi:10.5194/acp-13-12089-2013.

[2] Winker, D. M., et al., 2013: The global 3-D distribution of tropospheric aerosols as characterized by CALIOP, *Atmos. Chem. Phys.*, **13**, 3345-3361, doi:10.5194/acp-13-3345-2013.

[3] Tesche, M., et al., 2009: Vertically resolved separation of dust and smoke over Cape Verde using multiwavelength Raman and polarization lidars during Saharan Mineral Dust Experiment 2008, *Journal of Geophysical Research D: Atmospheres*, **114** (13), D13202, doi: 10.1029/2009JD011862.

DUST TRANSPORT ACROSS THE ATLANTIC STUDIED BY AIRBORNE DOPPLER WIND LIDAR DURING THE SALTRACE EXPERIMENT IN 2013

Fernando Chouza [1]*, Oliver Reitebuch[1], Stephan Rahm[1], Bernadett Weinzierl[1,2]

[1] *Institut für Physik der Atmosphäre, Deutsches Zentrum für Luft- und Raumfahrt (DLR), Oberpfaffenhofen, Germany, *Email: fernando.chouza@dlr.de*

[2] *Meteorologisches Institut, Ludwig-Maximilians-Universität München (LMU), München, Germany*

ABSTRACT

During the SALTRACE field experiment, conducted during June/July 2013, the Saharan dust transport across the Atlantic was analyzed by a set of ground based, in-situ and airborne instruments, including a 2-μm coherent DWL (Doppler wind lidar) mounted onboard the DLR Falcon 20 research aircraft. An overview of the measurements of aerosol backscatter and extinction, horizontal and vertical winds retrieved from the DWL are presented together with a brief description of the applied methods. The retrieved measurements provide direct observation of Saharan dust transport mechanisms across the Atlantic as well as island induced lee waves in the Barbados region.

1. INTRODUCTION

Mineral dust is a key component of the climate system and the most mass abundant type of aerosol, accounting for about half of the total aerosol mass emissions (e.g. Hinds 1999), acting as cloud and ice nuclei and playing an important role in the radiation budget of the earth.

The Saharan desert has been identified as the world largest source of mineral dust (e.g. Mahowald et al., 2005). Each year, large amounts of Saharan dust are transported across the Atlantic into the Caribbean region (e.g. Prospero, 1999). Despite the progress made during the last decades, the transport mechanisms of the Saharan dust over the Atlantic and their relative importance remain under discussion (e. g. Tsamalis et al., 2013; Karyampudi et al., 1999).

The Saharan Aerosol Long-range Transport and Aerosol-Cloud-Interaction Experiment (SALTRACE: http://www.pa.op.dlr.de/saltrace) conducted during June/July 2013 provided a whole new set of measurements, including in-situ, airborne and ground based observations. For the first time, an airborne coherent DWL was deployed onboard the DLR Falcon 20 research aircraft to study the Saharan dust transport over the Atlantic in combination with an extensive set of in-situ instruments to measure aerosol properties.

The airborne coherent DWLs usually provide the horizontal wind vector and qualitative aerosol measurements. Quantitative aerosol backscatter and extinction profiles were retrieved from the SALTRACE data set by means of a new calibration method (Sec. 3) for the first time. In addition the retrieval of the vertical wind from airborne wind lidar was largely refined (Sec. 4). The simultaneous retrieval of aerosol properties and, horizontal and vertical winds (Sec. 4), enhance the capabilities of the DWL for aerosol transport studies.

2. INTRUMENTAL DESCRIPTION

The coherent DWL deployed on the DLR Falcon 20 research aircraft during SALTRACE is based on a CLR Photonics instrument (Henderson et al., 1993) combined with a two wedge scanner and acquisition system developed at DLR (Köpp et al., 2004; Reitebuch, 2012). The system operates at a wavelength of 2.02254 μm, with a pulse full width at half maximum (FWHM) of 400 ns, a pulse energy of 1-2 mJ, and a repetition frequency of 500 Hz.

When mounted on an aircraft, the system can be operated in two scanning modes: the conical scanning and the nadir pointing mode. The conical scanning mode allows the retrieval of horizontal wind speed vector with a horizontal resolution of around 6 km (depending on the aircraft speed) and a vertical resolution of 100 m, while the nadir pointing mode can be used to retrieve vertical wind measurements with a horizontal resolution of approximately 200 m (Sec. 4). Both modes allow, by means of an adequate calibration (Sec. 3), the retrieval of backscatter and extinction coefficients with a horizontal resolution of approximately 200 m and a vertical resolution of 100 m.

3. BACKSCATTER CALIBRATION

In order to retrieve calibrated backscatter and extinction coefficients from the DWL, a novel calibration method was developed and validated (Chouza et al., 2015).

Based on sun-photometer measurements of the extinction coefficient wavelength dependency and the concurrent measurements of the DWL and a ground based aerosol lidar, calibration constants corresponding to different aerosol types were calculated. Those constants can then be applied to retrieve calibrated backscatter and extinction coefficient profiles at a wavelength of 532 nm from the coherent DWL measurements during other flight periods.

4. VERTICAL WIND RETRIEVAL

Vertical wind can be derived from the DWL measurements. Due to the relatively low magnitude of the vertical winds $\vec{w}$ with respect to the horizontal winds $\vec{V}$ and the Falcon flight speed $\vec{v}_{ac}$, different instrumental corrections have to be applied for the retrieval of the vertical wind in order to reduce the systematic errors introduced by the uncertainties in the DWL pointing direction $\vec{L}$ and the projection of the horizontal winds (Fig. 1).

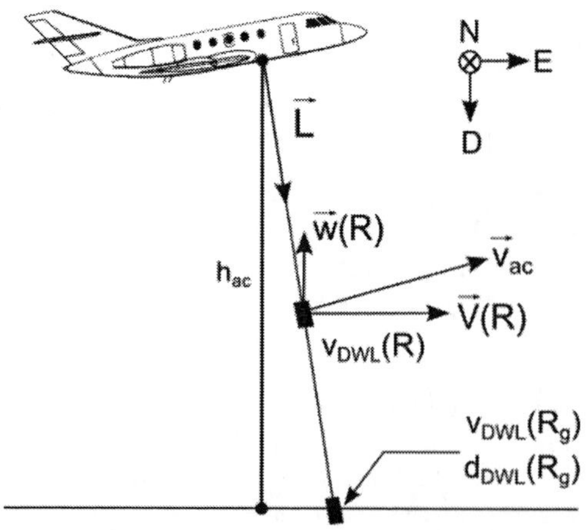

Fig. 1. Sketch of the vertical wind measurements and its various contributors for correction.

The proposed retrieval method is based on a 3-steps approach. First, an estimation of the DWL mounting angles and pointing offsets is calculated based on ground return speed $v_{DWL}(R_g)$ and distance $d_{DWL}(R_g)$ from all measurement flights. Then, the residual ground speed error, when available, is subtracted from the LOS speed $v_{DWL}(R)$ for each flight leg. Finally, if horizontal wind speed profiles are available, its projection is calculated and subtracted from the LOS speed.

5. RESULTS

Figure 3 shows an example of the retrieved backscatter and extinction coefficients using the method introduced in Sec. 3. The measurements, conducted during a flight on 11 June 2013 from La Palma to Sal (Fig. 2, a), illustrate the northern boundary of the SAL (Saharan Air Layer), its internal structure and the formation of clouds on its top (between 14:22 and 14:43 UTC). As it can be seen from these measurements, the SAL has its lower boundary at around 1 km and extends vertically up to 7 km in the Cape Verde region.

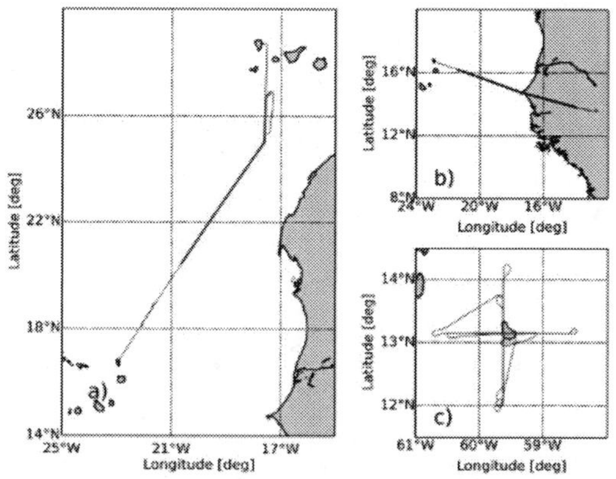

Fig. 2. Flight tracks (black) corresponding to the flights on 11 June between La Palma and Sal (a), 12 June between Sal and Dakar (b) and 26 June around Barbados (c). The colored flight track segments correspond to the measurement sections displayed in Figs. 3-5.

On top of the marine boundary layer (below 1000-500 m) clouds are present, which restrict the possibility of backscatter coefficients retrieval below the clouds.

Figure 4 shows the retrieved horizontal wind speed and direction for a flight conducted on 12

June 2013 between Sal and Dakar (Fig. 2, b). Between 3 and 6 km altitude a region with high wind speeds (more than 20 m s^{-1}) is visible, which corresponds to the African Easterly Jet (AEJ). The AEJ is considered to be the main dust advection mechanism associated with the SAL (Karyampudi et al., 1999). In addition to the AEJ, a land-sea breeze system over Dakar (14.72 N, 17.40 W, ~ 9:50 UTC) can also be observed.

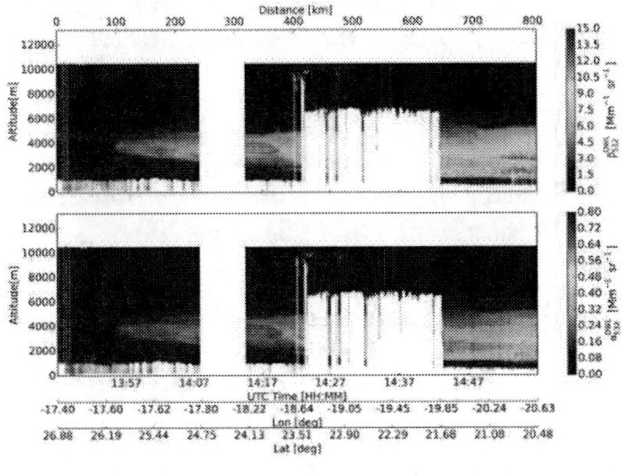

Fig. 3. Retrieved backscatter (upper panel) and extinction coefficients (lower panel) for a flight on 11 June 2013 along the West coast of Africa. The latitude and longitude of the track are shown in the bottom of the figure. An estimated distance scale is shown in the top of the figure. The white color indicates regions were no atmospheric signal is available (e.g. below clouds, low laser energy).

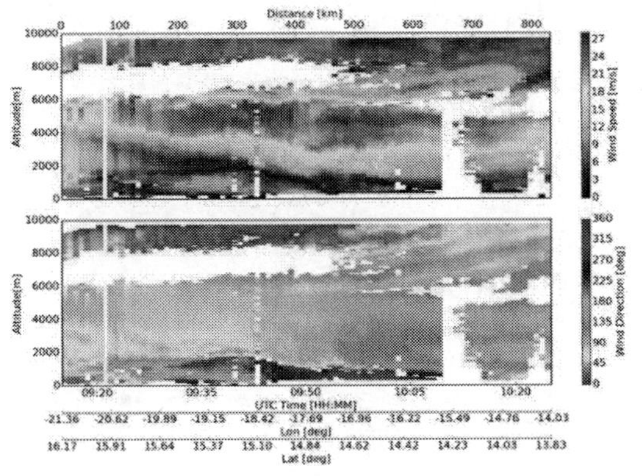

Fig. 4. Retrieved horizontal wind speed (upper panel) and direction (lower panel) for a flight on 12 June 2013 from Sal to Dakar.

Besides the simultaneous retrieval of backscatter coefficients and horizontal winds described in Sec. 2 and Sec. 3, also vertical winds can be obtained from the DWL for flight sections during which the system was operating in nadir pointing mode (Sec. 4). Figure 5 shows an example of the retrieved vertical wind speed and backscatter coefficient for a section (flown eastward) of the flight performed on the 26-27 June 2013 in the Barbados region.

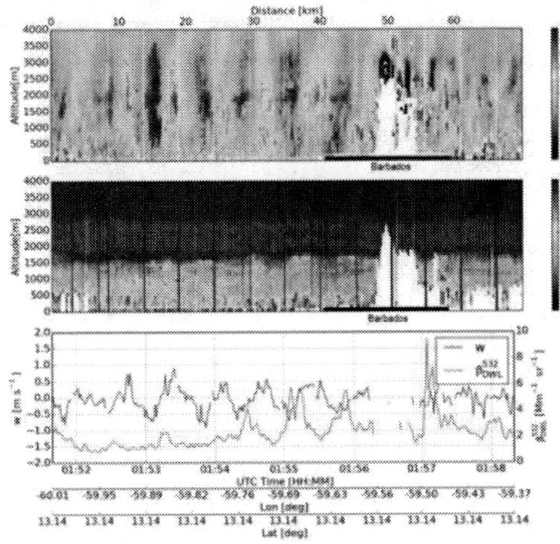

Fig. 5. Retrieved vertical wind speed (upper panel) and backscatter coefficient (mid panel) sections together with the backscatter coefficient and vertical wind for an altitude of 1.5 km (lower panel) on the lee side of Barbados. Positive vertical winds (red) correspond to upward winds, while negative (blue) winds correspond to downward winds.

Between 1:52 and 1:56 UTC waves with amplitudes of up to 1 m s^{-1} on the lee side of Barbados (overflown between 1:56 and 1:57 UTC) (Fig. 2, c) were observed. The structure of the waves can be also seen in the backscatter coefficient, especially at the boundary between the mixed layer (1000 m to 1500 m) and the SAL (1500 m up to 4000 m), at 1.5 km (Fig. 4, lower panel). A 90-degrees phase shift between the aerosol backscatter coefficient and the vertical wind velocity is present, which is typical of gravity waves (Durran, 1990).

6. SUMMARY

An overview of the DWL measurement dataset retrieved during SALTRACE was presented. The backscatter and extinction profiles retrieved on the north boundary of the SAL revealed its complex internal structure, vertical extension and shape, while the horizontal wind speed measurements provided a direct observation of the AEJ, associated with the long range transport of Saharan dust. Simultaneous backscatter and vertical wind retrievals exposed the existence of island induced lee waves in the Barbados region.

Further studies using simultaneous backscatter and wind measurements will be performed in order to give us a better understanding of the long range and vertical transport mechanisms associated with the Saharan dust transport over the Atlantic.

ACKNOWLEDGEMENT

This work was funded by the Helmholtz Association under grant number VH-NG-606 (Helmholtz-Hochschul-Nachwuchsforschergruppe AerCARE). The SALTRACE campaign was mainly funded by the Helmholtz Association, DLR, LMU and TROPOS.

REFERENCES

[1] Hinds, W.C., 1999: Aerosol Technology: Properties, Behaviour and Measurement of Airborne Particles, John Wiley & Sons, Inc., New York.

[2] Mahowald, N. M., A.R. Baker, G. Bergametti, N. Brooks, R.A. Duce, T.D. Jickells, N. Kubilay, J.M. Prospero, I. Tegen, 2005: Atmospheric global dust cycle and iron inputs to the ocean, *Global Biogeochemical Cycles*, **19**, 1–15.

[3] Prospero, J. M., 1999: Long-range transport of mineral dust in the global atmosphere: Impact of African dust on the environment of the southeastern United States, *Proc. Natl. Acad. Sci. U. S. A.*, **96**, 3396-3403.

[4] Tsamalis, C., A. Chédin, J. Pelon, V. Capelle, 2013: The seasonal vertical distribution of the Saharan Air Layer and its modulation by the wind, *Atmos. Chem. Phys.*, **13**, 11235-11257.

[5] Karyampudi, V. M., S.P. Palm, J.A. Reagen, H. Fang, W.B. Grant, R.M. Hoff, C. Moulin, H.F. Pierce, O. Torres, E.V. Browell, S.H. Melfi, 1999: Validation of the Saharan dust plume conceptual model using lidar, Meteosat, and ECMWF data, *B. Am. Meteorol. Soc.*, **80**, 1045–1075.

[6] Henderson, S.W., P. Suni, C.P. Hale, S.M. Hannon, J.R. Magee, D.L. Bruns, E.H. Yuen, 1993: Coherent laser radar at 2 μm using solid-state lasers, *IEEE Trans. Geosci. Remote Sens.*, **31(1)**, 4-15.

[7] Köpp, F., S. Rahm, I. Smalikho, 2004: Characterization of Aircraft Wake Vortices by 2-μm Pulsed Doppler Lidar, *Journal of Atmospheric and Oceanic Technology*, **21**, 194–206.

[8] Reitebuch, O., 2012: Wind Lidar for Atmospheric Research, in: Atmospheric Physics, Schumann, U., Springer, Berlin Heidelberg, 487-507.

[9] Chouza, F., O. Reitebuch, S. Groß, S. Rahm, V. Freudenthaler, C. Toledano, B. Weinzierl, 2015: Retrieval of aerosol backscatter and extinction from airborne coherent Doppler wind lidar measurements, *Atmos. Meas. Tech. Discuss.*, **8**, 1935-1986.

[10] Durran, D.R., 1990: Mountain waves and downslope winds. Atmospheric Processes Over Complex Terrain (ed. W.Blumen), Am. Meteorol. Soc., 59–81.

EPJ Web of Conferences **119**, 18009 (2016) DOI: 10.1051/epjconf/201611918009
ILRC 27

MEASUREMENT OF THE LINEAR DEPOLARIZATION RATIO OF AGED DUST AT THREE WAVELENGTHS (355, 532 AND 1064 nm) SIMULTANEOUSLY OVER BARBADOS

Moritz Haarig[1]*, Dietrich Althausen[1], Albert Ansmann[1], André Klepel[1], Holger Baars[1], Ronny Engelmann[1], Silke Groß[2], Volker Freudenthaler[2]

[1]Leibniz Institute for Tropospheric Research (TROPOS), Permoserstr. 15, 04318 Leipzig, Germany,
**Email: haarig@tropos.de*

[2]Meteorological Institute, Ludwig-Maximilians-University, Theresienstr. 37, 80333 Munich, Germany

ABSTRACT

A ground-based polarization Raman lidar is presented, that is able to measure the depolarization ratio at three wavelengths (355, 532 and 1064 nm) simultaneously. This new feature is implemented for the first time in a Raman lidar. It provides a full dataset of 3 backscatter coefficients, two extinction coefficients and 3 depolarization ratios (3+2+3 lidar system). To ensure the data quality, it has been compared to the well characterized two-wavelength polarization lidar POLIS. Measurements of long-range transported dust have been performed in the framework of the Saharan Aerosol Long-Range Transport and Aerosol-Cloud-Interaction Experiment (SALTRACE) in the Caribbean.

1. INTRODUCTION

Aerosols affect the radiative budget of the Earth. Mineral dust is one major component of the global aerosol load. It demonstrates a high temporally and spatially variability and furthermore the dust may change its properties and size distribution during transport or gets mixed with other species of aerosols. So observation and modelling of the properties and distribution of dust is a challenging task.

Lidar measurements provide vertically resolved profiles of the optical properties of the aerosols, such as the particle backscatter and extinction coefficient and the depolarization ratio. In order to get most detailed information about the dust particles, several wavelengths are used. With this additional spectral information, it will be easier for inversion algorithms to retrieve microphysical properties from the measured optical properties. Especially for dust investigations the aerosol linear depolarization ratio at 1064 nm is of great benefit (Gasteiger and Freudenthaler, 2014).

In order to provide such an extensive data set, the lidar system BERTHA of the TROPOS has been newly set up. Additionally to the backscatter at three wavelengths and the extinction at two wavelengths, the simultaneously measurement of the linear depolarization ratio at three wavelengths is now implemented.

Long-range transported dust has been observed the this improved lidar system in the Saharan Aerosol Long-Range Transport and Aerosol-Cloud-Interaction Experiment (SALTRACE) on Barbados (13°N, 59°W) in 2013 (summer) and in 2014 (winter and summer). The campaigns aim to analyze the properties of long-range transported Saharan dust, taking advantage of the fact, that the dust is transported about 5000 km only over the ocean without pollution or continental influence. In line with the previous campaigns SAMUM-1 (Heintzenberg, 2008) in southern Morocco (2006) and SAMUM-2 (Ansmann et al., 2011) at the Cape Verde (2008), the SALTRACE campaign completes the investigation of the transport way of the dust from the source region in the Sahara to the Caribbean. A measurement example of a dust outbreak will be shown.

2. LIDAR SYSTEM

The multi-wavelength polarization Raman lidar BERTHA (Backscatter Extinction lidar-Ratio Temperature Humidity profiling Apparatus) of the Leibniz Institute of Tropospheric Research (TROPOS) is a container-based mobile lidar system. It has been renewed after the SAMUM campaigns and contains a new receiver unit enabling the measurement of the depolarization at three wavelengths simultaneously. Furthermore a high spectral resolution channel (HSRL) at 532 nm was added.

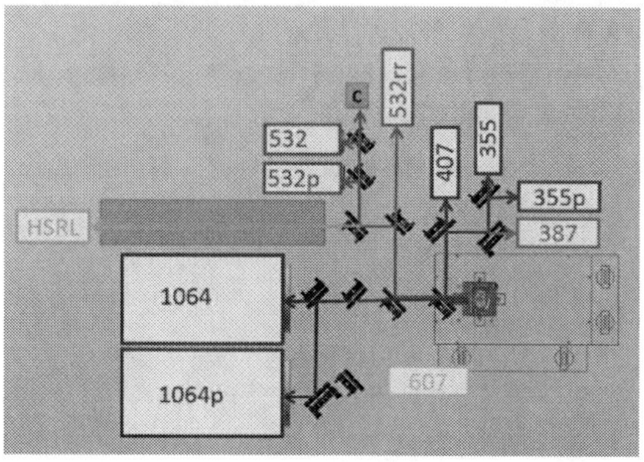

Figure 1 Sketch of the receiver unit of lidar system BERTHA with the three depolarization channels, labeled p (courteously of A. Klepel).

Two Nd:YAG laser (Continuum, Powerlite) transmit linearly polarized laser pulses at 355, 532 and 1064 nm. Two additional linear polarizers clean the polarization of the outgoing light. The repetition rate is 30 Hz and the pulse length 50 ns. The pulse energy is 1.6 J (for 1064 nm), 800 mJ (for 532 nm) and 150 mJ (for 355 nm). The beam is pointed to the sky under a zenith angle of 5°.

A 53 cm Cassegrain telescope collects the backscattered light. The receiver unit (see Fig. 1) separates the light into 13 detection channels: three elastic backscatter channels (355, 532 and 1064 nm) and three cross polarized elastic backscatter channels (355p, 532p and 1064p), two Raman channels (387 and 607 nm), three rotational Raman channels (532rr), a water vapor channel at 407 nm and a HSRL channel at 532 nm. All detectors are photomultiplier tubes (PMT) from Hamamatsu (H10721P-110). For the 1064 nm channels the PMTs R3236 from Hamamatsu are used. To reduce the signal noise, they are cooled down to lower than -30°C. In front of each cross polarized channel a linear polarizer perpendicular to the polarization of the laser beam is placed. To ensure the good quality of the depolarization measurements a Δ90° calibration (Freudenthaler et al., 2009) is performed for every measurement.

The signals are detected with a range resolution of 7.5 m and a time resolution of 5 to 30 s. Table 1 summarizes the output of vertical profiles of the BERTHA system.

Table 1 Output of vertical profiles of the BERTHA system.

Vertical profiles	**Wavelength (nm)**
Particle backscatter coeff.	355, 532, 1064
Particle extinction coeff.	355, 532
Lidar ratio	355, 532
Extinction and backscatter Ångström exponent	355/532 532/1064
Volume and Particle linear depolarization ratio	355, 532, 1064

A second lidar system POLIS (Portable Lidar System, contribution of V. Freudenthaler at the 27th ILRC) from the Ludwig-Maximilians-University, Munich, was deployed in the same campaign. It is a small portable 6-channel Raman and depolarization lidar system measuring the co- and cross-polarized backscattered light at 355 nm and 532 nm (day- and night-time) and the N_2-Raman shifted backscattered light at 387 nm and 607 nm (night-time only).

3. MEASUREMENT EXAMPLE

To give an example for the measurements of the depolarization ratio at three wavelengths the dust case of June 30, 2013 will be shown. It is a typical example for a strong dust event over the Western Caribbean. The Saharan dust plume was measured in parallel with the lidar systems BERTHA and POLIS, placed 30 m apart from each other, on Barbados during the SALTRACE campaign.

Figure 2 shows the range-corrected signal for the total and the cross polarized 1064 nm channel measured by the BERTHA system. The aerosol layer reaches up to 4.1 km about ground level. Combining the information of both channels the dust layer can be determined between about 1.8 km and 4.1 km. It is characterized by a relative humidity of around 30%. The layers beneath the dust layer are significantly more humid (around 80%). The relative humidity is obtained by a radiosonde launched at 0000 UTC at the measurement site.

DOI: 10.1051/epjconf/201611918009

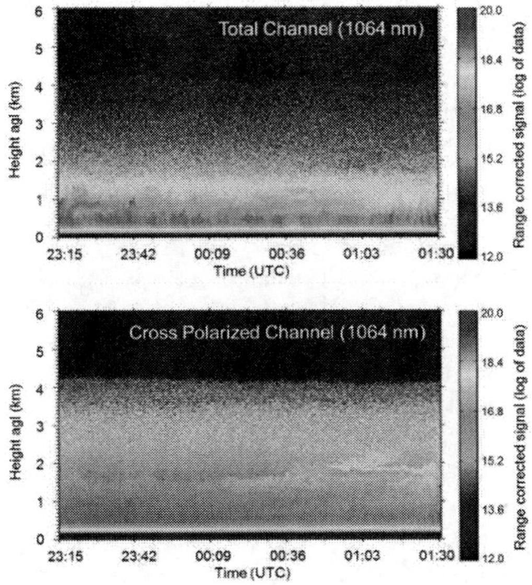

Figure 2 BERTHA measurement of 30 June 2013, 23:15, till 1 July 2013, 01:30 UTC, range-corrected signal for the 1064 nm total channel (upper part) and 1064 nm cross polarized channel (lower part). The first 500 m in these plots are note reliable due to high uncertainties in the overlap correction.

The atmospheric conditions remained almost constant over more than two hours (as it can be seen in Fig. 2). The retrieved profiles are averaged over this measurement period and shown in Fig. 3 for both lidar systems. The particle backscatter at 355 and 532 nm from BERTHA can be compared to the measurements obtained by POLIS and they agree very well. BERTHA measured additionally the particle backscatter at 1064 nm, which reaches in the dust layer only about 70% of the particle backscatter at the other wavelengths. The particle extinction coefficient for BERTHA is determined via the Raman method (Raman channels at 387 and 607 nm), for POLIS via the Fernald/Klett inversion with a fixed lidar ratio within the dust layer.

The measured lidar ratio within the transported dust layer (1.8 to 4.1 km) is approximately 61 sr and 71 sr for the 355 nm and 532 nm, respectively. For the Fernald-Klett algorithm a typical lidar ratio for dust of 55 sr is taken.

The Ångström exponent yields to different values for backscatter and extinction, caused by different influence of the particle size distribution on these quantities. A lesser amount of large particles could be responsible for the lower backscatter at 1064 nm.

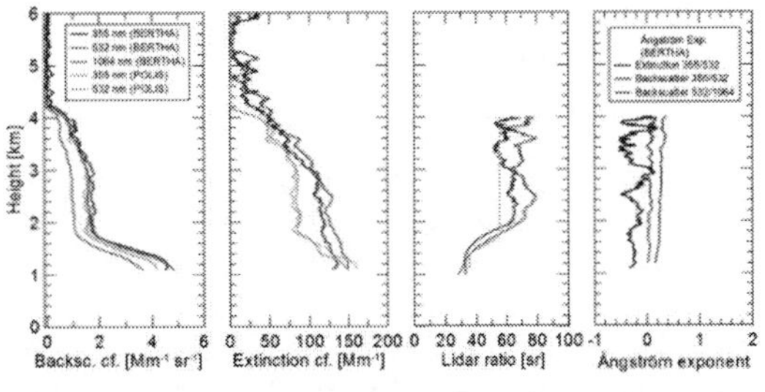

Figure 3 Vertical profiles for 30 June 2013, 2315 - 0130 UTC for the two lidar systems BERTHA and POLIS: the particle backscatter coefficient (100 m vertical smoothing), the particle extinction coefficient (750 m vertical smoothing), the lidar ratio (750 m vertical smoothing) and the Ångström exponent. BERTHA measurements are indicated in blue (355 nm), dark green (532 nm) and red (1064 nm), POLIS measurements in cyan (355 nm) and light green (532 nm).

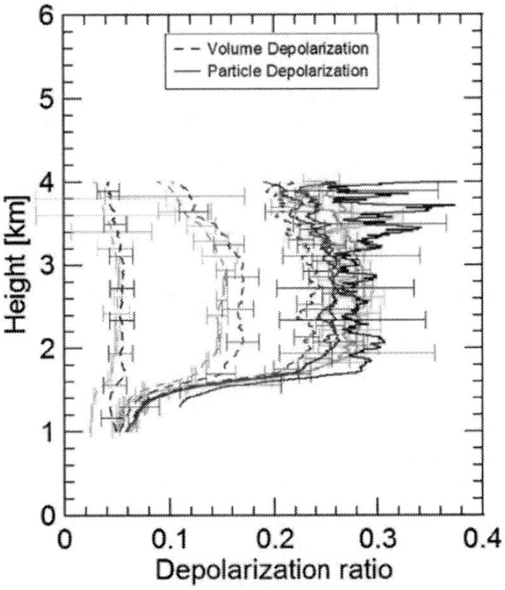

Figure 4 Vertical profiles of the depolarization ratio for 30 June 2013, 2315-0130 UTC. Colors for BERTHA and POLIS are the same as in Fig. 3, dashed lines correspond to the volume linear depolarization ratio, continuous lines to the particle linear depolarization ratio. The vertical smoothing length for the BERTHA measurements is 100 m, for POLIS 190 m.

BERTHA measured a particle linear depolarization ratio in the dust layer (1.8 to 4.1 km height) of 0.25±0.02 for 532 nm. This is slightly lower than the POLIS measurement (0.27±0.01). For 355 nm the particle linear depolarization ratio agrees very well between BERTHA (0.27±0.07) and POLIS (0.27±0.02). For 1064 nm a particle linear depolarization ratio of 0.24±0.1 can be reported.

4. CONCLUSIONS AND OUTLOOK

A method to measure the depolarization at three wavelengths simultaneously has been implemented and a unique data set of long-range transported Saharan dust has been recorded in two periods in summer 2013 and summer 2014 within the SALTRACE campaign. Furthermore pure marine conditions and dust, that had been mixed with biomass burning smoke over Western Africa and transported over the Atlantic Ocean, was measured in winter 2014 at Barbados. Dust-smoke mixtures will be shown at the conference as well.

The information, that can be obtained from the lidar ratio at two wavelengths (355 and 532 nm), the Ångström exponent for extinction and two backscatter ratios as well as the linear depolarization ratio at three wavelengths (355, 532 and 1064 nm), consists of 8 vertically resolved parameters, which may serve as an input for more sophisticated aerosol classification schemes (Burton et al., 2012; Groß et al., 2013) and furthermore for the retrieval of microphysical properties of the aerosol particles. It will be also used to improve the modelling of dust particles (Gasteiger et al., 2011) and the separation of non-dust, fine mode and coarse mode dust (Mamouri and Ansmann, 2014).

The measurement of the depolarization ratio at three wavelengths combined with Raman lidar measurements reveals a deeper look into the aerosol properties by means of remote sensing.

REFERENCES

[1] Heintzenberg, J, 2008: The SAMUM-1 Experiment over southern Morocco: Overview and introduction, *Tellus Ser. B*, **61**, 2-11, doi: 10.1111/j.1600-0889.2008.00403.x.

[2] Ansmann, A., A. Petzold, K. Kandler, I. Tegen, M. Wendisch, D. Müller, B. Weinzierl, T. Müller, and J. Heintzenberg 2011: Saharan mineral dust experiments SAMUM-1 and SAMUM-2: What have we learned?, *Tellus Ser. B*, **63**, 403–429, doi:10.1111/j.1600-0889.2011.00555.x.

[3] Gasteiger, J and Freudenthaler, V, 2014: Benefit of depolarization ratio at λ = 1064 nm for the retrieval of the aerosol microphysics from lidar measurements, *Atmos. Meas. Tech.*, **7**, 3773-3781, doi: 10.5194/amt-7-3773-2014.

[4] Freudenthaler, V., Esselborn, M., Wiegner, M., Heese, B., Tesche, M., Ansmann, A., Müller, D., Althausen, D., Wirth, M., Fix, A., Ehret, G., Knippertz, P., Toledano, C., Gasteiger, J., Garhammer, M., Seefeldner, M., 2009. Depolarization ratio profiling at several wavelengths in pure saharan dust during samum 2006. *Tellus* **61B**, 165–179.

[5] Freudenthaler, V., Seefeldner, M., Groß, S., Wandinger, U., 2015: Accuracy of linear depolarization ratios in clean air ranges measured with POLIS-6 at 355 and 532 nm. *Contribution to the 27th ILRC*, New York

[6] Burton, S. P., Ferrare, R. A., Hostetler, C. A., Hair, J. W., Rogers,R. R., Obland, M. D., Butler, C. F., Cook, A. L., Harper, D. B. and Froyd, K. D., 2012: Aerosol Classification of Airborne High Spectral Resolution Lidar Measurements – Methodology and Examples, *Atmos. Meas. Tech.*, **5**, 73–98, doi:10.5194/amt-5-73-2012.

[7] Groß, S., Esselborn, M., Weinzierl, B., Wirth, M., Fix, A., and Petzold,A., 2013: Aerosol classification by airborne high spectral resolution lidar observations, *Atmos. Chem. Phys.*, **13**, 2487–2505, doi:10.5194/acp-13-2487-2013.

[8] Gasteiger, J., Wiegner, M., Groß, S., Freudenthaler, V., Toledano, C., Tesche, M., and Kandler, K., 2011: Modelling lidar-relevant optical properties of complex mineral dust aerosols, *Tellus B*, **63**, 725– 741, doi:10.1111/j.1600-0889.2011.00559.x.

[9] Mamouri, R.E. and Ansmann, A., 2014: Fine and coarse dust separation with polarization lidar, *Atmos. Meas. Tech.*, **7**, 3717–3735, doi:10.5194/amt-7-3717-2014

EPJ Web of Conferences **119**, 19001 (2016) DOI: 10.1051/epjconf/201611919001
ILRC 27

THE ASIAN DUST AND AEROSOL LIDAR OBSERVATION NETWORK (AD-NET): STRATEGY AND PROGRESS

Tomoaki Nishizawa, Nobuo Sugimoto, Ichiro Matsui, Atsushi Shimizu, Akiko Higurashi, and Yoshitaka Jin

National Institute for Environmental Studies, 16-2 Onogawa, Tsukuba, 305-0052, Japan, nisizawa@nies.go.jp

ABSTRACT

We have operated a ground-based lidar network AD-Net using dual wavelength (532, 1064nm) depolarization Mie lidar continuously and observed movement of Asian dust and air pollution aerosols in East Asia since 2001. This lidar network observation contributed to understanding of the occurrence and transport mechanisms of Asian dust, validation of chemical transport models, data assimilation and epidemiologic studies. To better understand the optical and microphysical properties, externally and internally mixing states, and the movements of Asian dust and air-pollution aerosols, we go forward with introducing a multi-wavelength Raman lidar to the AD-Net and developing a multi-wavelength technique of HSRL in order to evaluate optical concentrations of more aerosol components. We will use this evolving AD-Net for validation of Earth-CARE satellite observation and data assimilation to evaluate emissions of air pollution and dust aerosols in East Asia. We go forward with deploying an in-situ instrument polarization optical particle counter (POPC), which can measure size distributions and non-sphericity of aerosols, to several main AD-Net sites and conducting simultaneous observation of POPC and lidar to clarify internally mixed state of Asian dust and air pollution aerosols transported from the Asian continent to Japan.

1. INTRODUCTION

Since 2001, National Institute for Environmental Studies (NIES) has continuously operated a ground-based lidar network, the Asian Dust and aerosol lidar observation Network (AD-Net), and observed movement of Asian dust and air pollution aerosols in East Asia in cooperation with various research organizations and universities under several research programs [e.g., 1,2]. The AD-Net uses a compact two-wavelength (532 and 1064nm) backscatter (β) and one-wavelength (532nm) polarization (δ) Mie lidar system (i.e., 2β+1δ Mie lidar system) and contributed to understanding of the occurrence and transport mechanisms of Asian dust, validation of chemical transport models, and data assimilation study [e.g., 3,4]. Furthermore, concentrations of Asian dust and air pollution aerosols evaluated from the AD-Net observation has been used in epidemiologic studies (e.g., effect on a human body) [5].

To better understand the optical and microphysical properties and the movements of Asian dust and air-pollution aerosols, we go forward with attaining the multichannel of the lidars settled at several main sites in AD-Net, using a Raman lidar technique, a high spectral resolution lidar (HSRL) technique, and their combined techniques. The independent extinction and multichannel measurements can provide useful data for characterizing aerosol optical properties as demonstrated by multi-wavelength Raman lidar measurements in EARLINET [6]. In addition, the independent multichannel measurements are essential for classifying individual aerosol components (e.g., mineral dust, sea-salt, and black carbon) [7]. Derived extinction coefficients, backscatter coefficients, and depolarization ratio for total aerosols at multi-wavelengths and optical properties for individual aerosol components are useful for validations of chemical aerosol transport models and satellite measurements and data assimilation as well as aerosol characterization. HSRL is a highly sensitive method and helps us to measure signals with sufficient signal-to-noise ratio in daytime as well as nighttime. Thus, we have developed a compact and automatically operated two-wavelength (355, 532nm) HSRL using a injection seeded Nd:YAG laser, a Fabry-Perot etalon interference at 355nm and iodine absorption filter at 532nm [12]. We start developing a novel HSRL using a multi-mode Nd:YAG laser and Mach-Zender interferometer to realize a robuster and lower-cost multi-wavelength HSRL [8]. To clarify internally mixed state of Asian dust and air pollution aerosols transported from the Asian continent to Japan, we go forward with introduction of an in-situ instrument optical particle counter with a function of depolarization measurement to several main sites of AD-Net [9,10].

In this paper, we report the latest status of the AD-Net observation (Section 2), and denote future strategy and ongoing projects of the AD-Net (Section 3).

2. LATEST STATUS OF AD-NET

Fig. 1 indicates the locations of AD-NET observation sites. Table 1 summarizes information for the individual observation sites. AD-Net has 20 observation sites in East Asia and is recognized as one of the contributing network of WMO/GAW. A part of the network is operated as the Asian dust monitoring network of Ministry of the Environment (MOE), Japan, and a part of the network is operated in the radiation network SKYNET [11].

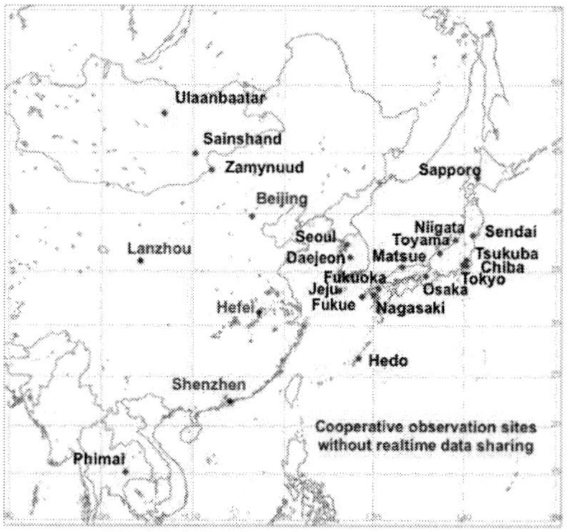

Fig. 1 AD-Net observation sites

2.1 Standard AD-Net lidars

Lidars are settled in a container or a building with a glass window on the roof and operated regardless of weather conditions continuously. $2\beta(532,1064)+1\delta(532)$ Mie lidars are settled at 13 sites of the AD-Net. A commercial Nd:YAG laser is employed as a light source and emitted into zenith direction in 10Hz repetition, 20mJ/pulse for both wavelengths. We use a cassegrainian telescope of 20cm diameter, two photo multiplier tubes (PMTs) in analogue mode for copol and crosspol signal measurements at 532nm, an avalanche photo diode (APD) for 1064nm signal measurements. They are operated automatically, and the 5-minute averaged lidar profiles with 6m vertical resolution are recorded every 15minutes in the continuous observations. The recorded data are transferred to NIES in real time except for the cooperative stations in China (i.e., Lanzhou, Hefei, and Shenzhen). The observed data are analyzed automatically, and attenuated backscatter coefficients at 532 and 1064nm, total depolarization ratio at 532nm, extinction coefficient, backscatter coefficient, and depolarization ratio of aerosols at 532nm, mixing layer height, and extinction coefficients of mineral dust and spherical aerosols at 532nm are distributed in public in quasi real time as the standard product of AD-Net [http://www-lidar.nies.go.jp/AD-Net/]. We use the Fenrald's method with S1=50sr to estimate aerosol extinction and backscatter coefficients. We then apply the method using the particle depolarization ratio to estimate the dust and spherical aerosol extinction coefficients [1]. It should be noted that the AD-Net standard product are retrieved for all the 20 sites of AD-Net and those are distributed in public.

Table 1 Information for the AD-Net sites

Site	Organization	Lon. (E)	Lat. (N)	Height (m)	Start of Observation	Lidar system*2
Tsukuba, Japan*1	NIES	140.12	36.05	30	1996	B, D
Toyama, Japan*1	MOE/Toyama	137.10	36.70	28	2004	A
Matsue, Japan	MOE/Shimane	133.01	35.48	5	2005	B
Nagasaki, Japan	MOE/Nagasaki	129.98	32.94	206	2001	A
Niigata, Japan	MOE/Niigata	138.94	37.84	1	2007	A
Tokyo, Japan	MOE	139.71	35.69	42	2008	A
Sapporo, Japan*1	Hokkaido Univ.	141.33	43.06	30	2003	A
Sendai, Japan*1	Tohoku Univ.	140.83	38.26	154	2005	A
Chiba, Japan*1	Chiba Univ.	140.12	35.62	20	2007	A
Osaka, Japan	Kinki Univ.	135.59	34.65	19	2008	A
Fukue, Japan*1	Chiba Univ. / NIES	128.68	32.75	50	2002	B
Hedo, Japan*1	NIES	128.25	26.87	60	2005	C
Fukuoka, Japan*1	Kyushu Univ.	130.48	33.52	30	2012	C
Seoul, Japan*1	Seoul Nat. Univ.	126.95	37.46	116	2006	B
Gosan Jeju, Korea	Seoul Nat. Univ.	126.16	33.29	35	2010	A
Daejeon, Korea	Mokwon Univ.	127.34	36.33	80	2002	A
Ulaanbaatar, Mongolia	IMH	106.90	47.92	1320	2007	A
Sainshand, Mongolia	IMH	110.12	44.87	937	2007	A
Zamyuud, Mongolia	IMH	111.90	43.72	962	2007	A
Phimai, Thailand*1	Chulalongkorn Univ.	102.57	15.18	212	2002	B

*1 SKYNET. *2 A indicates 2β (532,1064) + 1δ(532) Mie lidar system; B indicates 1α(532) + 2β(532,1064) + 1δ (532) Mie-Raman lidar system; C indicates 2α (355, 532) + 3β (355, 532,1064) + 2δ (355, 532) Raman lidar system; and D indicates 2α (355, 532) + 3β (355, 532,1064) + 2δ (355, 532) HSRL-Raman lidar system.

2.2 Introduction of Raman lidar technique

We have improved the AD-Net Mie lidars at six main sites (i.e., Tsukuba, Fukue, Hedo, Matsue, Seoul, and Phimai) by adding a nitrogen Raman scatter measurement channel since 2009 to better understand aerosol optical properties and realize more advanced aerosol classification [12,13]. The improved lidars (i.e., $1\alpha+2\beta+1\delta$ Mie-Raman lidar) provide extinction (α) at 532nm, backscatter (β) at 532 and 1064nm, and depolarization ratio (δ) at 532nm. We use a PMT in photon-counting mode for the Raman backscatter measurements. The recorded data are transferred to NIES in real time except for the cooperative station, Beijing in China. The observed data are automatically analyzed [13]. We apply wavelet transform analysis to reduce signal noise; then average the observed data every hour and 120m range; and retrieve the 1α+2β+1δ data by the conventional direct method using nighttime data observed with appropriate signal-to-noise ratio. To evaluate aerosol components using these 1α+2β+1δ data, we developed an algorithm to retrieve extinction coefficients at 532nm for four aerosol components: strong light absorption aerosols such as black carbon (BC), nonspherical aerosols such as mineral dust (dust), large-size and spherical aerosols such as sea-salt (sea-salt), and small-size, weak-absorption aerosols such as air-pollution aerosols excepting for black carbon (air pollution) using the derived 1α+2β+1δ data (1α+2β+1δ method), and we demonstrated the algorithm performance by applying to the observed data [13]. This method assumes the external mixture of the four aerosol components (i.e., BC, dust, sea-salt, and air pollution) at each layer. Size distributions, refractive indices, particle shapes, and optical properties of the aerosol components are modeled based on previous studies. The vertical profiles of the extinction coefficients at 532nm for the four aerosol components are estimated by a look-up table method using their modeled optical properties.

3. STRATEGY AND ONGOING PROJECT

3.1 Introduction of multi-wavelength Raman lidars for more advanced data assimilation

We deployed a two wavelength Raman lidar at Hedo and Fukuoka sites of AD-Net as well as University of Toyama, Japan (36.70N, 140.12E) and started continuous measurements using the lidars in collaboration with Kyushu university and University of Toyama. The main objective of this study is to estimate extinction coefficients of aerosol components, conduct data assimilation analysis using the retrieved concentrations of aerosol components, and finally evaluate emission of dust and air pollution aerosols and build aerosol assimilated data in each region of East Asia. The multi-wavelength Raman lidar system provides 2α(355,532) + 3β(355,532,1064) + 2δ(355,532) data. The lidar uses a commercial Nd:YAG laser and emit the laser into zenith direction in 10Hz repetition and laser energies of 150, 100 and 60mJ/pulse for 1064, 532, 355nm. We use a cassegrain telescope of 20cm diameter, and PMTs for elastic channels at 532 and 355nm in analog mode and for nitrogen Raman channels at 607 and 387nm in photon-counting mode. Depolarization measurement functions at 355 and 532nm are implemented. We will retrieve extinction coefficients of dust, sea-salt, BC, and air pollution aerosols and size distributions of dust, sea-salt, and air pollution aerosols using the 2α+3β+2δ data in a similar to the 1α+2β+1δ method. This algorithm assumes external mixture of the four aerosol components; the size distributions, refractive indices, and particle shapes and optical properties except for the size distributions of dust, sea-salt, and air pollution aerosols are modeled based on previous studies [14].

3.2 Development of multi-wavelength HSRL for ground-based lidar network observations

We have developed a compact and automatically operated two-wavelength (355, 532nm) HSRL using a single-longitudinal mode laser, a Fabry-Perot etalon interference at 355nm and iodine absorption filter at 532nm [15]. Using this iodine absorption filter technique for 532nm and a nitrogen Raman lidar technique for 355nm, we constructed a 2α(355,532) + 3β(355,532,1064) + 2δ(355,532) HSRL-Raman lidar at the Tsukuba site of AD-Net and started continuous measurement from 2014 (Table 1). Since the multi-wavelength HSRL system has to fix the laser wavelength to an iodine absorption line and further control the etalon transmittance line to the fixed laser wavelength, and keep controlling the wavelength of the laser and the etalon during continuous measurement, this system has technical complexity and difficulty. To reduce the complexity and difficulty, we start developing a new HSRL technique using a multi-longitudinal mode Nd:YAG laser and Mach-Zender interferometer [8]. The key feature of this system is not to need controlling the laser wavelength though the interferometer has to be controlled. Further, this system is applicable to any laser wavelength, different from HSRL using an iodine absorption filter. In addition, this system is lower-cost since it does not use a single-longitudinal mode laser. Thus, we expect that the multi-wavelength HSRL using this developed HSRL technique will be more effective and appropriate for ground-based lidar network observations.

3.3 Validation observation of the EarthCARE

An international aerosol, cloud, and radiation observation mission from space "EarthCARE" is ongoing by ESA, JAXA, and NICT. A 355nm high-spectral resolution lidar ATLID as well as a multi spectral imager MSI, a cloud profiling radar CPR, and a broad band radiometer BBR are aboard the EarthCARE satellite [16]. We are developing an algorithm to estimate $1\alpha+1\beta+1\delta$ at 355nm, mixing layer height, aerosol/cloud discriminator, and extinction coefficients of BC, sea-salt, dust, and air pollution aerosols from the ATLID and MSI measurements in collaboration with JAXA to distribute those derived parameters as JAXA level 2 products in public after the EarthCARE will be launched. We are ongoing to introduce multi-wavelength Raman lidars and HSRLs with measurement function at 355nm to AD-Net towards the EarthCARE validation observation. It should be noted that the data analysis of the multi-channel lidar with more channels than the ATLID is effective to validate the ATLID-derived extinction coefficients of aerosol components because the algorithm using the multi-channel lidar data needs less assumption and so more accurate (see subsection 3.1).

3.4 Introduction of polarization optical particle counter (POPC) to study internal mixture of aerosols

The POPC is an in-situ instrument to measure the forward scattering and the two polarization components of backscattering for single particles at 780nm and provides particle size and depolarization ratio for single particles. We conducted simultaneous observations using the POPC and the lidar at the AD-Net site in Seoul and captured the characteristics of pure Asian dust and internally mixed polluted Asian dust; the optical characteristics measured using the lidar were consistent with the POPC measurements [10]. This indicates that the simultaneous measurements of POPC and lidar are effective to clarify and characterize internally mixing state of Asian dust and air pollution aerosols. We will deploy the POPC to the other AD-Net sites.

ACKNOWLEDGEMENT

This study was partly supported by the Environmental Research and Technology Development Fund of the MOE, Japan (5B-1202), Grant-in-Aid for Scientific Research (25220101) from Japan Society for the Promotion of Science (JSPS), and EarthCARE first algorithm and validation research announcement by JAXA.

REFERENCES

[1] Shimizu, A., et al., 2004: *J. Geophys. Res.*, **109**, D19S17, doi:10.1029/2002JD003253.

[2] Sugimoto, N., et al., 2005: *Water, Air, and Soil Pollution: Focus*, **5**, 145-157.

[3] Yumimoto, K., et al., 2008: *Atmos. Chem. Phys.*, **8**, 2869-2884.

[4] Uno, I., et al., 2008: *Geophys. Res. Lett.*, 35, L06803, doi:10.1029/2007GL032329.

[5] Onishi, K., et al., 2012: *Atmos. Env.*, **49**, 94-102, doi:10.1016/j.atmosenv.2011.12.018.

[6] URL: http://www.earlinet.org/

[7] Nishizawa, T., et al., 2008: *Trans. Geosci. Rem. Sens.*, **41**, 1743-54.

[8] Jin, Y., et al., 2015: *Proc. 27th ILRC*, to be submitted.

[9] Kobayashi, H., et al., 2014: Atmos. Env., http://dx.doi.org/10.1016/j.atmosenv.2014.05.006.

[10] Sugimoto, N., et al., 2015: *J. Quant. Spec. Rad. Trans.*, **150**, 107-113.

[11]http://atmos.cr.chiba-u.ac.jp/index_atmos.html

[12] Xie, C., et al., 2008: *Appl. Opt.*, **47**, 4945-4951.

[13] Nishizawa, T., et al., 2012: *Proc of SPIE*, 8526, 8526Q, doi:10.1117/12.977215.

[14] Nishizawa, T., et al., 2008: *Proc. 24th ILRC*, 361-364.

[15] Nishizawa, T., et al., 2010: *Proc. 25th ILRC*, 903-906.

[16] Illingworth, A. J., et al., 2016: *BAMS*, doi:10.1175/BAMS-D-12-00227.1

DOI: 10.1051/epjconf/201611919002

EARLINET: 12-YEAR OF AEROSOL PROFILING OVER EUROPE

L. Mona[1], L. Alados Arboledas[2], V. Amiridis[3], A. Amodeo[1], A. Apituley[4], D. Balis[5], A. Comeron[6], M. Iarlori[7], H. Linné[8], D. Nicolae[9], A. Papayannis[10], M.R. Perrone[11], V. Rizi[7], N. Siomos[5], U. Wandinger[12], X. Wang[13] and G. Pappalardo[1]

[1]*Istituto di Metodologie per l'Analisi Ambientale CNR-IMAA, C.da S. Loja, I-85050 Tito Scalo, Potenza, Italy, *Email: lucia.mona@imaa.cnr.it*

[2] *Andalusian Institute for Earth System Research. IISTA-CEAMA, Universidad de Granada, Spain*

[3] *National Observatory of Athens, IAASARS, Athens, Greece*

[4]*KNMI - Royal Netherlands Meteorological Institute, The Bilt, The Netherlands*

[5]*Aristotle University of Thessaloniki, 54124 Thessaloniki, Greece*

[6] *Universitat Politècnica de Catalunya, Barcelona, Spain*

[7]*CETEMPS/Dipartimento di Scienze Fisiche e Chimiche - Università Degli Studi dell'Aquila, Italy*

[8]*Max-Planck-Institut für Meteorologie, Hamburg, Germany*

[9] *National Institute of R&D for Optoelectronics, Magurele-Bucharest, Romania*

[10]*National Technical University of Athens, Physics Department, Athens, Greece*

[11]*Università del Salento, Department of Mathematics and Physics, Lecce, Italy*

[12] *Leibniz-Institut für Troposphärenforschung, Leipzig, Germany*

[13]*Consorzio Nazionale Interuniversitario per le Scienze Fisiche della Materia, Napoli, Italy*

ABSTRACT

EARLINET has been collecting high quality aerosol optical profiles over Europe since 2000. The comparison with automatic collected dataset of aerosol optical depth (AOD) from AERONET and MODIS demonstrates the effectiveness of EARLINET regular measurement schedule for climatological studies. The analysis of optical properties in the local boundary layer indicates that the general decrease of AOD observed by different platforms over Europe in the last decade could be due to the modification of aerosol properties (towards less absorbing and larger particles) in the lower troposphere.

1. INTRODUCTION

The importance of aerosol, and tropospheric aerosol in particular, on the atmospheric radiation budget and air quality is nowadays well recognized. The latest IPCC report states that the main cause of the uncertainty of the current estimation of aerosol direct and indirect effects on climate is the high variability in time and space of the distribution and microphysical properties of tropospheric aerosols [1]. The vertical distribution of aerosol is of particular importance because it is a combined signature of atmospheric transport patterns, residence time in the atmosphere, and the efficiency of the vertical exchange.

EARLINET (European Aerosol Research Lidar Network) [2,3] has been providing aerosol optical properties vertical profiles over Europe since May 2000. As a unique lidar network mainly based on Raman aerosol systems, EARLINET is the only "tool" able to provide a climatology of aerosol profiles. Here for the first time a long-term climatological study is performed using the quality assured database, in terms of AOD, free troposphere contribution to the AOD and intensive properties. The accuracy and representativeness of the measurements is addressed combining EARLINET with external datasets typically used as reference in literature such as AERONET and MODIS AOD data.

2. METHODOLOGY

The long-term aerosol observations collected within EARLINET allow a climatological study of aerosol properties over Europe. Here data published in the first volumes of EARLINET data were used [4-8]. EARLINET assures high quality products in terms of aerosol optical properties profiles thanks to quality assurance and quality check protocols. Every 2 years EARLINET quality checked data are published on the CERA database of the world data center for climate (WDCC) (www.dkrz.de/WDCC/). Data are freely available, as soon as stations PI finalize them, through EARLINET (www.earlinet.org) and ACTRIS data portal (actris.nilu.no). Additional datasets and products tailored for specific interests are moreover freely available on request.

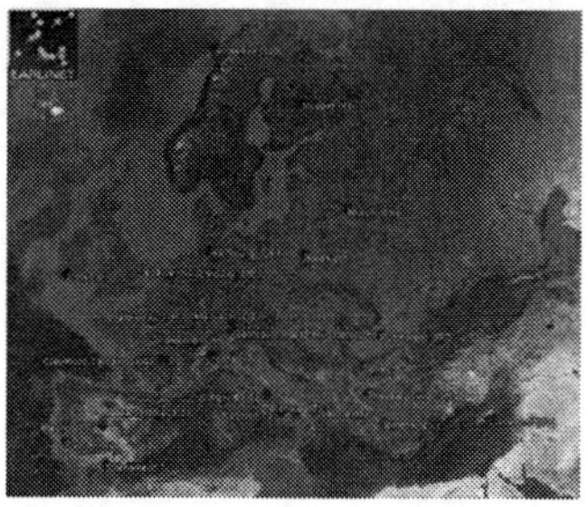

Figure1. Map of the EARLINET stations updated on June 2014.

In particular, only profiles related to measurements performed following the climatological schedule established within EARLINET were used for climatological purposes in order to avoid potential biases due to specific intensive measurements campaign related to events. The columnar AOD is calculated from each extinction profile of the EARLINET database, integrating the aerosol extinction profile over the whole column. In the AOD calculation from lidar measurements it has been taken into account that extinction values are not provided down to the ground because of the laser-telescope overlap region. For calculating the total column aerosol optical depth, it is considered that the aerosol extinction is constant in the lowest part of the atmosphere because of the well mixed conditions typically observed in the first night-time hours.

From the extinction profile, it is also possible to provide an estimation of the free troposphere contribution (FT) to the total aerosol load. FT can be calculated in percentage to the total AOD starting from information provided by lidar about the PBL top [9] and the extinction profile directly measured by lidars.

As reference point and for representativeness considerations, AOD values provided by the closest Aerosol Robotic NETwork (AERONET) site and by MODIS (MODerate Resolution Imaging Spectroradiometer) at 1° x1° resolution were used. The Ångström exponent as measured by AERONET are used for scaling AOD with the wavelength when needed. In particular we used only quality assured AERONET data (namely Level 2 data) and MODIS collection 5 data. The temporal coincidence between EARLINET and AERONET/MODIS data cannot be obtained at a lower scale than 1 day, because the AOD EARLINET data are measured in nighttime conditions, while AERONET and MODIS measurements are possible only in daytime. Starting from these considerations, the comparison between the different datasets is carried out comparing EARLINET data to AERONET and MODIS daily average.

This permits, first of all, to measure directly the AOD, a key parameter for understanding the aerosol role on radiation budget. The comparison of EARLINET AOD measurements vs. MODIS and AERONET ones, allows on one hand to confirm the high quality of EARLINET data and on the other hand to investigate the representativeness of our network regular scheduling for climatological studies.

3. RESULTS

The comparison between EARLINET AOD and the corresponding values from AERONET collocated stations and MODIS 1°x1° confirms the reliability of EARLIENT columnar AOD, even under the assumption of well-mixed conditions below the first sounded point. In particular the one-to-one comparison with collocated AERONET data shows that, for all the stations, difference in AOD follows a Gaussian distribution centered around values in agreement with 0 within 1 std deviation (on average centered around 0.07). Differences are distributed according to a Gaussian distribution even for the MODIS comparison. The agreement in this case is also improved (center of the distribution on average is 0.02) probably because the larger horizontal area could smooth out the differences due instead to the non simultaneous measurements.

DOI: 10.1051/epjconf/201611919002

Once the good reliability of EARLINET AOD columnar values is proved, the larger temporal coverage of both AERONET and MODIS can be used to understand if the regular schedule is sufficient for climatological purposes and at which extend. Temporal series of monthly, seasonal and annual averages resulting from the 3 datasets are compared in terms of root mean square (RMS) difference and correlation coefficient. Figure 2 and 3 report the RMS as a function of the correlation coefficient for EARLINET vs. AERONET and MODIS, repsectively. We also considered as interesting parameter the number of cases N/N_{tot} for each time series for which the observed difference is within one standard deviation: i.e. cases for which $t = |AOD_{ear} - AOD_{aer}| / \sigma_{AER}$ is less than 1.

In the AERONET comparison, it is clear that a pretty good agreement is found already for the monthly averages (small RMS and correlation above 0.4 and colors corresponding to N/N_{tot} greater than 50%). The seasonal average improves the situation moving points towards the bottom right corner and cold colors. Even negative correlations are recorded for annual averages because of too few simultaneous data between EARLINET and AERONET. The presence of negative correlation between time series when few points are compared, is observed also in the comparison with MODIS (Figure 3). Apart from this, a very good agreement is found for both seasonal and annual averages.

This analysis shows that the EARLINET climatological schedule, taking also into account that typically up to half of the scheduled measurements are performed because of weather and/or /man power limitations, is significant for climatological purposes if at least seasonal averages are considered and they could be representative also for 1° x1° for these time scales.

Consequently, the EARLINET dataset can be used to investigate continental and regional modification of aerosol properties over Europe. The last IPCC report [1] shows a general tendency of a reduction of the AOD over Europe, but little research has been done to date to identify the reason of it.

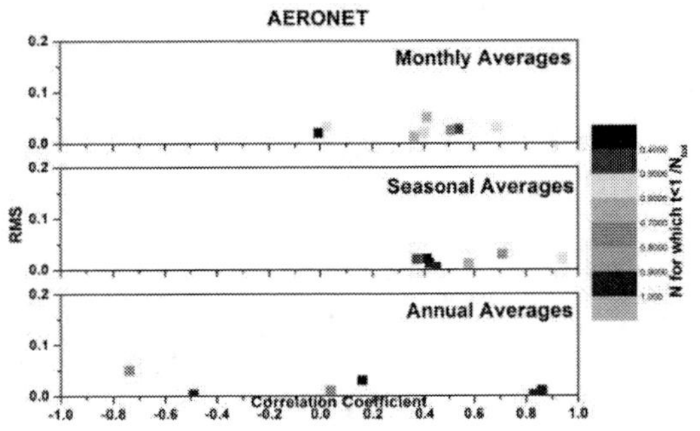

Figure 2: RMS as a function of correlation coefficient between EARLINET and AERONET time series of monthly, seasonal and annual averages (from top to bottom). Colors correspond to percentage of cases with observed differences in 1 std deviation.

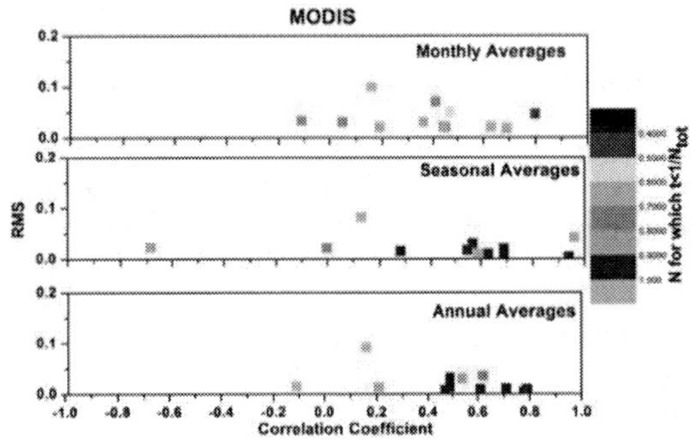

Figure 3: as Fig. 2 but for EARLINET vs. MODIS series.

The EARLINET measurements of course show this decrease. EARLINET observations add to this general framework another important point: in 2000-2010 a general tendency for decreasing of PBL contribution to the columnar AOD is observed (not shown). To understand this point, intensive properties as lidar ratio (S) at 355 nm and backscatter-related Angstrom exponent (δ) at 355/532 nm averaged in the PBL were investigated. The analysis was limited to these parameters because they are the ones with the largest number of available measurements. Figure 4 reports the time series of S and δ for one station in the Mediterranean and one in Central Europe with the largest number of measurements at the present time (Potenza and Leipzig, respectively).

DOI: 10.1051/epjconf/201611919002

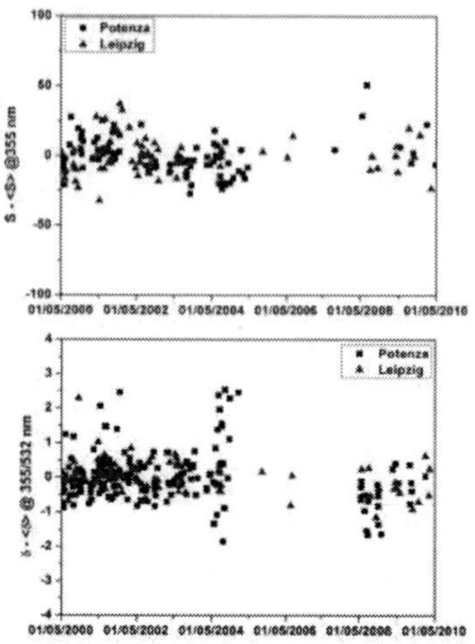

Figure 4. Temporal behavior of PBL mean values of lidar ratio at 355 nm (S) and backscatter-related Ångström exponent at 355/532 nm (δ) for Potenza and Leipzig stations. Data are normalized in respect to each site mean value.

Data are normalized respect to the specific site mean values for removing differences due to the specific location and emissions. It can be observed an increase in both S and δ for the 2000-2002 period and a slight decrease afterwards. This would indicate a possible modifications towards less absorbing and slightly larger particles in the PBL over Europe.

4. CONCLUSIONS

A detailed study about the significance of EARLINET measurements has been carried out. The high quality dataset of aerosol optical properties vertical profiles has been compared with well recognized datasets of columnar measurements. This allowed a quantification of reliability of EARLINET columnar investigation. The efficiency of EARLINET climatological schedule for climatological purpose has been demonstrated. The representativeness on average of EARLINET punctual measurements at 1°x1° scale over seasonal /annual scale was proved.

EARLINET columnar observations in agreement with other platforms show a reduction of AOD over Europe. The EARLINET capability to investigate the vertical profiles shows that this reduction is due to the decrease of the PBL contribution to the total column AOD. Through investigation of aerosol intensive properties, we found that this could be ascribed to a modification of aerosol optical characteristics towards less absorbing and slightly larger particles. This could be related to the efficiency of pollution reduction actions over EU. Further investigations about vertical profiles and differences among the European regions could be carried out using the EARLINET database. In addition a combined study with near the surface measurements could add a further insight in this context.

ACKNOWLEDGEMENT

The financial support for EARLINET in the ACTRIS Research Infrastructure Project by the European Union under grant agreement n. 262254 in the 7th Framework Programme (FP7/2007-2013) is gratefully acknowledged.

REFERENCES

[1] Boucher, O., et al., (2013), Fifth Assessment Report of the Intergovernmental Panel on Climate Change Cambridge University Press, Cambridge

[2] Bosenberg J. et al, (2001) in *Advances in Laser Remote Sensing,* 155–158.

[3] Pappalardo, G et al., (2014). *Atmos. Meas. Tech.,* **7**, 2389-2409

[4] The EARLINET publishing group 2000–2010, (2014)doi:10.1594/WDCC/EN_all_measurements_2000-2010.

[5] The EARLINET publishing group 2000–2010, (2014b)doi:10.1594/WDCC/EN_Climatology_2000-2010.

[6] The EARLINET publishing group 2000–2010, (2014)doi:10.1594/WDCC/EN_Calipso_2006-2010.

[7] The EARLINET publishing group 2000–2010, (2014)doi:10.1594/WDCC/EN_VolcanicEruption_2000-2010.

[8] The EARLINET publishing group 2000–2010, (2014)doi:10.1594/WDCC/EARLINET_SaharanDust_2000-2010.

[9] Matthias V. et al., (2014), *J. Geophys. Res.* **109.**

 DOI: 10.1051/epjconf/201611919003

LIDAR CALIBRATION CENTRE

Gelsomina Pappalardo[1], Volker Freudenthaler[2], Doina Nicolae[3], Lucia Mona[1], Livio Belegante[3], Giuseppe D'Amico[1]

[1]Consiglio Nazionale delle Ricerche, Istituto di Metodologie per L'Analisi Ambientale (CNR-IMAA), Tito Scalo, Potenza, Italy

[2]Ludwig Maximilians University, Meteorological Institute Munich in German

[3]National Institute of R&D for Optoelectronics, Magurele, Romania

ABSTRACT

This paper presents the newly established Lidar Calibration Centre, a distributed infrastructure in Europe, whose goal is to offer services for complete characterization and calibration of lidars and ceilometers. Mobile reference lidars, laboratories for testing and characterization of optics and electronics, facilities for inspection and debugging of instruments, as well as for training in good practices are open to users from the scientific community, operational services and private sector. The Lidar Calibration Centre offers support for trans-national access through the EC HORIZON2020 project ACTRIS-2.

1. INTRODUCTION

Lidars and ceilometers are based on the same principle, but significantly different from the engineering point of view, as they are meant for different applications. However, calibration is one common challenge. Lidars are more complex in terms of laser power, number of detection channels and therefore of the optics and electronics. This helps to obtain more quantitative data products [1] (e.g. backscatter and extinction profiles at several wavelengths, lidar ratios, Angstrom exponents and particle depolarization ratios), however with an uncertainty that has to be well documented. An important contribution to the data uncertainty comes from the instrument's imperfections, which are generally difficult to quantify. Each optical or electronic component has an impact, but they also interact at a certain level, and depend on the environmental conditions. For example, the transmission of the interference filters at the laser wavelength changes with temperature, while the dynamic range of the whole system varies with the laser-telescope alignment, also depending on the temperature.

Lidars operate in different configurations and designs. The quality of the data products depends strongly on the instrument performance and calibration, but a direct calibration is not possible because they measure remotely in ambient conditions. Coupling of the system with the atmosphere is also a disturbing factor. Due to the complexity of the instrument and large variety of designs, systematic errors are difficult to be objectively assessed. On the other hand, data uncertainty is underestimated if systematic errors are not included.

Worldwide, there are several lidar networks, regional or global in scope. EARLINET (the European Aerosol Research Lidar NETwork) was founded in 2000 as a research project for establishing a quantitative, comprehensive, and statistically significant database for the horizontal, vertical, and temporal distribution of aerosols on a continental scale [2]. Since then, EARLINET has collected and published more than 45000 vertical profiles of aerosols [3],[4],[5],[6],[7]. GALION (the Global Aerosol LIdar Network) strives to determine the vertical component of the spatio-temporal distribution of aerosol properties related to climate forcing and air quality, in a network of ground-based lidar stations [8]. To achieve this goal, homogeneous data products are required. This homogeneity can only be achieved by implementing common calibration procedures and making use of the existing calibration facilities.

Calibration is possible by direct comparison with reference instruments off-and-on, but has to be regularly checked by means of quality assurance (QA) tests applied by properly trained staff.

2. LICAL INFRASTRUCTURE

The Lidar Calibration Centre (LiCal) is a multi-installation facility, offering a wide range of possibilities to test and calibrate lidars and ceilometers, starting from the characterization and optimization of single components to the assessment of the whole system's performance. It combines a series of direct and remote techniques, involving both the equipment and the operators, to ensure a proper evaluation, maintenance, calibration and operation. Each of the seven installations has a specific role in this process.

The Lidar Calibration Training Laboratory (LiCalTrain) is operated by the National Institute of R&D for Optoelectronics in Bucharest, Romania. The infrastructure consists of a multiwavelength depolarization Raman lidar, a scanning depolarization UV lidar, a tropospheric ozone lidar and complementary instrumentation (Scanning Microwave radiometer, C-ToF Aerosol Mass Spectrometer, Aerosol Chemical Speciation Monitor, Aethalometer, Aerodynamic Particle Sizer, Integrating nephelometer), as well as dedicated software tools to test, characterize, calibrate and validate lidar measurements. It is ideal to train the operators how to apply regular check-ups and maintain the calibration.

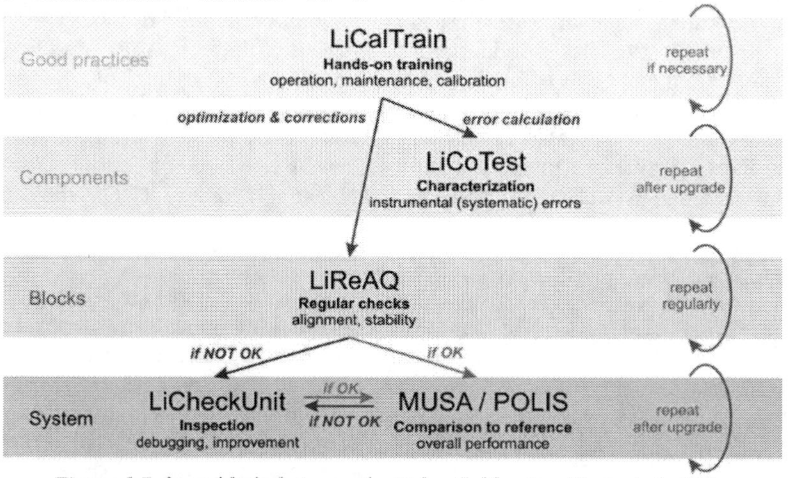

Figure 1 Role and links between the Lidar Calibration Centre's facilities

The mobile reference lidar POLIS operated by the Meteorological Institute at Ludwig Maximilians University in Munich, Germany, measures at two elastic wavelengths - 355 nm and 532 nm and their N_2-Raman shifted wavelengths - 387 nm and 607 nm, and linear depolarisation ratio at 532 nm and 355 nm. It is portable and can be easily moved to the user's site by air cargo to perform side-by-side comparison. It is the only reference lidar with depolarization at 355 nm.

The reference lidar MUSA operated by the Institute of Methodologies for Environmental Analysis - National Research Council, in Potenza, Italy is a multi-wavelength depolarization Raman with 3 elastic – 1064, 532, 355 nm, 2 Raman – 607, 387 nm, and linear depolarization ratio at 532 nm. It is the only reference instrument operating simultaneously 3 elastic and 2 Raman channels. The system can be moved to the user site (if the user's lidar is fixed) or can be operated at the host for longer intercomparisons.

The availability of the two reference systems provides a good geographical coverage and reduces time delays between demand and supply.

The Lidar Components Testing Laboratory (LiCoTest) hosted by the National Institute of R&D for Optoelectronics has a long expertise in characterizing laser beams (pulse width, Gaussian parameters, beam shape) and optical components (angles between plane surfaces, refractive index, surface characterization), using power and energy meters, Tektronics oscillioscopes, Beam Profiler with M^2 option, BeamScope, goniometer and interferometer. LiCoTest also covers spectroscopic ellipsometric measurements and spectral characterization of interferential filters (bandwidth, transmission, blocking), dichroic beam splitters and mirrors, and characterization of the aging process with the Perkin Elmer UV-VIS NIR Spectrometer. Electronic modules are characterized and tested with Agilent and Tektronics pulse/form generators and analysers.

This installation delivers full characterization of the optical and electronic components of the lidar transmitter and receiver, which is essential when assessing the instrumental systematic errors.

The Lidar Remote Quality Assurance (LiReQA) and the Lidar Check-up Unit (LiCheckUnit) are operated by the Meteorological Institute at Ludwig Maximilians University

The Lidar Remote Quality Assurance (LiReQA) operates dedicated software tools (raytracing software for analyzing the transmitter and receiver optics, lidar polarization analysis software, lidar radiation power budget analysis software, analysis of test measurements specialized for individual lidar setups) to perform fault diagnosis of any lidar system.

The Lidar Check-up Unit (LiCheckUnit) is a portable unit (laser power meter, polarization analysis instruments for inspection of transmitter and receiver optics, adapted instruments for checking the alignment of the lidar system, e.g. CCD camera & special optics adapted to the lidar system, lidar pulse simulator for inspection of the detection electronics) which can be used at the user site to inspect and debug the instrument.

LiCal has been set up following a long process for standardization and harmonization of calibration procedures within EARLINET community. Mature techniques were picked up by LiCal and are now offered to the users as services, while research and technological development continues in collaboration with EARLINET and ACTRIS. The techniques and the methodologies used by LiCal emerged from EARLINET's scientific and technological activities.

The services offered by the LiCal infrastructure are part of a complete characterization and optimization of lidar instruments. While the characterization of basic components (laser, optical and electronic parts) is performed at LiCoTest (laser beam characterization, optical and spectral characterization of lenses, filters, mirrors and fiber optics, spectroscopic ellipsometry measurements, polarization measurements, paraxial analysis of optical systems, electronic modules synchronization, stability, response time/speed), the assessment of the overall system's performance is made by a side-by-side comparison with POLIS or MUSA, which are both EARLINET reference mobile lidar systems.

Systematic quality assurance of the optical and electronic chains and fault diagnosis is performed by LiReQA, while users are trained at LiCalTrain to apply the specific tests. Inspection, debugging and optimization of lidars at their site is performed by LiCheckUnit.

3. RESULTS

The various installations of LiCal have offered in the last years a significant number of accesses within EARLINET, ACTRIS (Aerosols, Clouds and Trace gases Research Infrastructure Network), ESA-CEOS, Initial Training for atmospheric Remote Sensing network (ITARS), as well as for the master and doctoral programs at the Romanian, Italian and German Universities, and national research programs. On average, LiCal installations have hosted about 48 projects (intercomparisons, sample analysis, training, system inspection) and 1260 access days per year.

Within the frame of EARLINET and ACTRIS, LiCal has been serving the scientific community (about 20 active lidar stations, regular access), the user community (several lidar/ceilometer operators belonging to weather services and aviation) and the private sector (about 3 companies / developers and integrators).

LiCal was also involved in intercomparison (SLiCE2000, EARLI09, SPALI10, SOLI10, ROLI10, AQUILI12, NALI13, LELI13) and scientific campaigns (ACTRIS Summer 2012 Campaign, Charmex EARLINET 72h exercise-Jul. 2012, EMEP summer and winter campaign-Jun.-Jul. 2012 and Jan.-Feb. 2013, Charmex 2013, CHARADMExp 2014).

This activity strongly contributed to increase the number of available quality assured EARLINET data as demonstrated by the large number of publications collected in the last years (a complete list is reported at www.earlinet.org). In particular, EARLINET data have been recently used for aerosol characterization in terms of optical and microphysical properties in conjunction with in situ measurements and transport models (e.g. [9],[10],[11],[12]).

The long-term EARLINET quality assured database is the basis for climatological studies, systematic comparison with models and satellite observations ([13],[14]). New products were derived as further outcome of this calibration activity, e.g the calibrated linear volume depolarization ratio [15], the particle linear depolarization ratio [16]. At the same time, more and more lidars were upgraded with Raman and depolarization channels, so the number of accesses to LiCal increased as well.

4. CONCLUSIONS

A Lidar Calibration Centre was recently set-up in Europe. Its task is to provide complete calibration of lidars and ceilometers, and to provide proper tools to assess the instrumental biases. It is a distributed infrastructure, relying on state-of-the-art laboratories for testing and characterization of optical and electronic components and blocks, on mobile reference instruments, and on experienced experts for each topic. Training in good practices in lidar operation and maintenance is also provided. The Lidar Calibration Centre serves ACTRIS (Aerosol, Clouds and Trace gases Research InfraStructure), but also other world-wide networks and operators, system and components manufacturers. Open access and support to the users is provided through the HORIZON2020 contract ACTRIS-2.

ACKNOWLEDGEMENT

This work is supported by the EC HORIZON2020 project ACTRIS-2, contract INFRAIA-1-2014-2015 no. 654109.

REFERENCES

[1] Gasteiger, J. et al., 2011: Volcanic ash from Iceland over Munich: mass concentration retrieved from ground-based remote sensing measurements *Atmos Chem Phys*, **11**, 2209-2223.

[2] Pappalardo et al., 2014: EARLINET: towards an advanced sustainable European aerosol lidar network, *Atmos. Meas. Tech.*, **7**, 2389–2409.

[3] The EARLINET publishing group 2000–2010, (2014)doi:10.1594/WDCC/EN_all_measurements_2000-2010.

[4] The EARLINET publishing group 2000–2010, (2014b)doi:10.1594/WDCC/EN_Climatology_2000-2010.

[5] The EARLINET publishing group 2000–2010, (2014)doi:10.1594/WDCC/EN_Calipso_2006-2010.

[6] The EARLINET publishing group 2000–2010, (2014)doi:10.1594/WDCC/EN_VolcanicEruption_2000-2010.

[7] The EARLINET publishing group 2000–2010, (2014)doi:10.1594/WDCC/EARLINET_SaharanDust_2000-2010.

[8] Bösenberg J., Hoff R. et al., 2007: Plan for the implementation of the GAW Aerosol Lidar Observation Network GALION, *GAW Report No. 178*, Hamburg, Germnay

[9] Pappalardo, G. et al., 2013.: Four-dimensional distribution of the 2010 Eyjafjallajökull volcanic cloud over Europe observed by EARLINET, *Atmos. Chem. Phys.*, **13**, 4429-4450.

[10] Sicard, M. et al., 2012: Monitoring of the Eyjafjallajökull volcanic aerosol plume over the Iberian Peninsula by means of four EARLINET lidar stations, *Atmos. Chem. Phys.*, **12**, 3115-31130.

[11] Wagner, J. et al., 2013: Evaluation of the Lidar/Radiometer Inversion Code (LIRIC) to determine microphysical properties of volcanic and desert dust, *Atmos. Meas. Tech.*, **6**, 1707-1724.

[12] Nicolae, D. et al., 2013: Characterization of fresh and aged biomass burning events using multiwavelength Raman lidar and mass spectrometry, *J. Geophys. Res. Atmos.*, **118**, 2956–2965.

[13] Mona, L. et al., 2012: Lidar Measurements for Desert Dust Characterization: An Overview, *Advances in Meteorology*, 2012, 356265, 1-36.

[14] Pappalardo, G. et al., 2010: EARLINET correlative measurements for CALIPSO: first intercomparison results, *J. Geophys. Res.*, doi:10.1029/2009JD012147.

[15] Bravo-Aranda, J. A. et al., 2013: Analysis of lidar depolarization calibration procedure and application to the atmospheric aerosol characterization, *Int. J. Rem. Sens.*, **34** (9-10), 3543-3560.

[16] Freudenthaler, V. et al., 2009: Depolarization-ratio profiling at several wavelengths in pure Saharan dust during SAMUM 2006, *Tellus B*, **61**, 165-179.

ALINE/LALINET NETWORK STATUS

Eduardo Landulfo [1]*, Fabio Juliano da Silva Lopes[1,2], Gregori de Arruda Moreira[1], Márcia Talita Amorim Marques[1], Marcelo Osneide[1], Juan Carlos Antuña[3] , René Estevan Arredondo[3] , Juan Luiz Guerrero Rascado[4,5], Lucas Alados-Arboledas[4,5] Alvaro Bastidas[6], Daniel Nisperuza[6], Andrés Bedoya[6], Mauricio Múnera[6], Dairo Alegría[6], Ricardo N. Forno[7], Maria Fernanda Sánchez[7], Oscar Lazcano[7] , Elena Montilla-Rosero[8,9], Antonieta Silva[8,9], Cristofer Jimenez[8,9], Eduardo Quel[10], Pablo Ristori[10] , Lidia Otero[10], Henrique M.J. Barbosa[11], Diego A. Gouveia[11], Boris Barja[3,11].

[1]*Laser Environmental Application Laboratory, Instituto de Pesquisas Energéticas e Nucleares, Av. Prof. Lineu Prestes 2242, São Paulo, Brazil 01258-010, Brazil, *Email: elandulf@ipen.br*

[2]*Universidade de São Paulo, Instituto de Astronomia, Geofísica e Ciências Atmosféricas, Rua do Matão, 1226, Cidade Universitária, CEP 05508-000, São Paulo, Brasil*

[3]*Centro Meteorológico de Camagüey, Instituto de Meteorología de Cuba, Cuba*

[4]*Instituto Interuniversitario de Investigación del Sistema Tierra en Andalucía (IISTA-CEAMA), Av. De Mediterráneo, 18006, Granada, España*

[5]*Dpto. Física Aplicada, Universidad de Granada, Fuentenueva s/n, 18071, Granada, España*

[6]*Escuela de Física, Universidad Nacional de Colombia, Calle 59ª Nº 63-20, Medellín, Colombia*

[7]*Laboratorio de Física de la Atmósfera, Universidad Mayor de San Andrés, Casilla 8635, La Paz, Bolivia*

[8]*Centro de Óptica y Fotónica CEFOP, Universidad de Concepción, Casilla 4016, Concepción, Chile*

[9]*Dpto. Física, Universidad de Concepción, Casilla 160-C, Concepción, Chile*

[10]*División Lidar, CEILAP (UNIDEF-CONICET), San Juan Bautista de La Salle 4397 (B1603ALO), Villa Martelli, Buenos Aires, Argentina*

[11]*Universidade de São Paulo, Instituto de Física, Rua do Matão, Travessa R, 187, CEP 05508-090, São Paulo, Brasil*

ABSTRACT

The Latin American Lidar Network, ALINE a.k.a LALINET is a federation lidar network established in 2008 which became a member of GALION/GAW program in 2013. Currently the network consists of 9 operational stations with the perspective of two more stations to be included. The network today covers more than 18 million Km^2 and spans in latitude from -52° to 21° and in longitude from -78° to -47°. It should cover a larger area in the future as planned with the inclusion of more active stations.

1. INTRODUCTION

LALINET is a federation lidar network established in 2013 as a member of GALION for 4D aerosol observation related to topics such as long-term climatic trends, identification of aerosols in the atmosphere and their optical properties and monitoring of long range transport, such as dust, volcanic ashes and biomass burning onto South America. Initially before its establishment the network organized itself via biennial meetings that have been taking place every two years since 2001. These meetings were initially created to nucleate the laser remote sensing groups in South and Central America. Among the participants we highlight the presence of experts and observers from other groups in the world. With the engagement of the groups and setup of informal agreements new insights and scientific activities have been released such as pre-workshop introductory courses on atmospheric science and lidar fundamentals. The scientific drives for the network are those connected to characterize the temporal-spatial distribution of aerosols and their variations on a regional and continental scale. On a local scale the aerosol plumes are used to study the atmospheric dynamics acting as planetary boundary layer tracers and also to help in validating models and satellite observation products [1]. LALINET's most recent efforts are focused in establishing a standard measurement protocol and in improving a dataset common algorithm for data processing to be shared among all participating stations and adding more stations to the present network.

Table 1. ALINE/LALINET lidar stations.

Lidar system	City/Country	Coordinates
ba-BA-AR	Bariloche / Argentina	41.15°S 71.16°W
not applicable	Camagüey / Cuba	21.4°N 77.8°W
co-CEFOP-UDEC	Concepción/Chile	36.84°S 73.02°W
cr-CR-AR	Comodoro Rivadavia/ Argentina	43.24°S 65.33°W
ma-MA	Manaus / Brazil	2.89°S 59.97°W
me-LOA-UNAL	Medellin/ Colombia	6.26°N 75.58°W
ne-NE-AR	Neuquén/ Argentina	38.59°S 68.15°W
pa-LIPAZ	La Paz / Bolivia	16.54°S 68.07°W
sp-CLA-IPEN-MSP-LIDAR-I	São Paulo / Brazil	23.56°S 46.74°W
sp-CLA-IPEN-II	São Paulo / Brazil	Variable

2. NETWORK WORKSHOPS

The biennial meetings that have been taking place every two years since 2001, when the workshops were initially created to nucleate the laser remote sensing groups in South and Central Americas. To achieve such goal, experts and observers from other groups in the world were invited at each edition. With the engagement of the groups and informal agreements being made new insights and scientific activities have been implemented, as pre-workshop introductory courses on atmospheric science and lidar fundamentals.

The network workshops have been held at most of the sites of the original teams (except Arecibo). Altogether 7 editions of this meeting have happened – Cuba(2001,2003); Colombia(2005); Brazil (2007); Argentina (2009); Bolivia (2011); Chile (2013) – with a scheduled workshop to happen in Cuba, in April 2015. These meetings have steered most of the LALINET activities until the last edition when additional technical workshops were created to help in establishing measurement and data analysis protocols.

The I Workshop on Lidar Inversion Algorithms for LALINET were held in Concepción, Chile, from 10 to 13 March 2014. The main goal of this first workshop was to compare the inversion algorithms for elastic backscatter lidars from each Latin-American Lidar group in order to develop a uniform, unified and improved algorithm.

The algorithm evaluation and improvement was based on the analysis of three simulated lidar datasets and a comparison with the expected results. The first dataset was provided by the Institute for Tropospheric Research (iFT), Leipzig – Germany, and corresponds to the data used in the EARLINET. The second dataset was a modification of the first one, to include different levels of noise. The third dataset was another modification to include a cloud. As a result of the comparison, a standard for the retrieval of the particle backscatter from elastic lidar data was defined. Matlab and Mathematica routines for doing these calculations were prepared, which are being implemented to beused by all stations in the network.

3. INSTRUMENTAL INVENTORY

Most of the LALINET systems are not series-produced instruments and, therefore, present large differences in configuration and capabilities, in that condition efforts are being made to provide a network harmonization and, as a result, it is mandatory to have means of system optimization. A review of the current instrumental status of all LALINET systems was done and analyzed in detail in order to assess the potential performance of the network and to detect networking weaknesses. In this process it was detected that procedures must be done to achieve as much as possible unattended systems to improve temporal coverage and to reduce the manpower needed, which in turn should have repercussions on the disposition devoted on data analysis. The number of emitted wavelengths should increase in order to achieve the triad 355+532+1064 nm, with the aim to improve the spectral capabilities of LALINET and be able to detect these channels that are the minimum lidar wavelengths to derive particle microphysical properties by combination with sun-photometer data. Besides, no 3+2 systems are available to extract microphysical properties by inversion method with regularization. At present, the Nd:YAG laser is the workhorse for the lidar community but there is no preference in the laser model (almost any LALINET lidar is based on a different laser models). Because almost 90% of LALINET systems emit laser radiation at 355 nm, almost 90% at 532 nm and 80% at 1064 nm, only 80% of LALINET systems emit simultaneously at these three wavelengths therefore limiting the number final products that can be derived from part of the lidar systems. It is expected that the results of this diagnostic will bring benefits to both the existing stations and to the new groups, and will help the improvement of the overall quality of the aerosol data products derived from LALINET for contributing to the scientific knowledge.

4. PILOT CAMPAIGN

A pilot campaign was carried out during Sep 10-14th 2012, which was representative as many biomass-burning activities occur during this period of the year. This pilot campaign was the first coordinated effort to perform simultaneous lidar measurements in Latin America. The difficulties involved in the coordination of such simultaneous measurements were mainly due to (1) reduced manpower, (2) manually operated lidar stations, and (3) weather permitting conditions. This effort allowed us to establish some common network standards such as molecular cross-section and the same methodology for retrieving particle backscatter from elastic lidar data.

5. CONCLUSIONS

LALINET (http://lalinet.org) is a Latin American coordinated lidar network focused on the vertically-resolved monitoring of the particle optical properties distribution (i.e. particle backscatter and extinction profiles) over Latin America, as well as other atmospheric species such as ozone and water vapor. This federative lidar network started on voluntary basis in 2001 and was officially recognized by WMO in 2013.

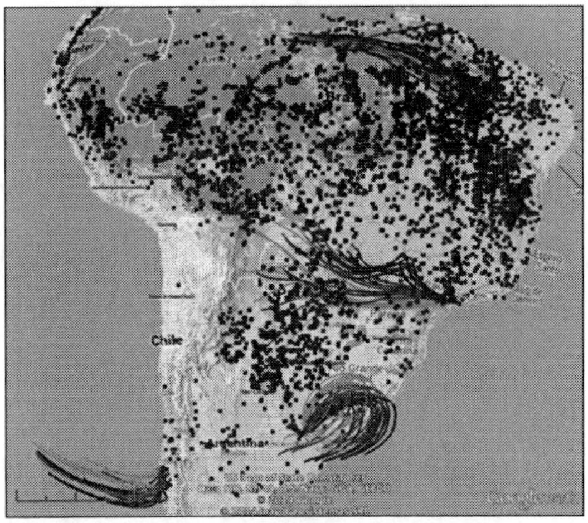

Figure 1 Fires (dark blue) detected by INPE algorithm (http://www.inpe.br/queimadas/) using AQUA/MODIS satellite data from September 10th to 14th 2012 are shown. Lines correspond to 48 h HYSPLIT back trajectories (http://ready.arl.noaa.gov/HYSPLIT.php) starting at the time selected for each station and altitude of maximum backscatter (Fig. 3). Trajectories were calculated in ensemble mode using GDAS meteorological data.

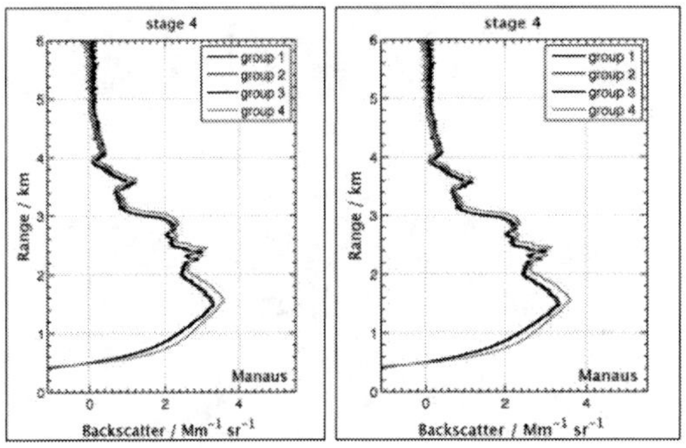

DOI: 10.1051/epjconf/201611919004

Figure 2 Particle backscatter coefficients (Mm-1 sr-1) obtained by each group participating in the analysis of data from the Pilot Campaign. Left: São Paulo and right: Manaus. No overlap correction was applied at this stage.

The network is developing measurement and data analysis protocols and aims to establish some seasonal intensive campaigns in order to provide representative dataset from a region in the globe which still is underrepresented in terms of atmospheric and climatic data available with special interest in aerosol probing and its specification though active remote sensing.

ACKNOWLEDGEMENTS

The authors would like to thank the supporting agencies in special: - co-CEFOP-UDEC Lidar station is supported by the CONICYT grants PIA PFB0824 and FONDECYT 11110126 .

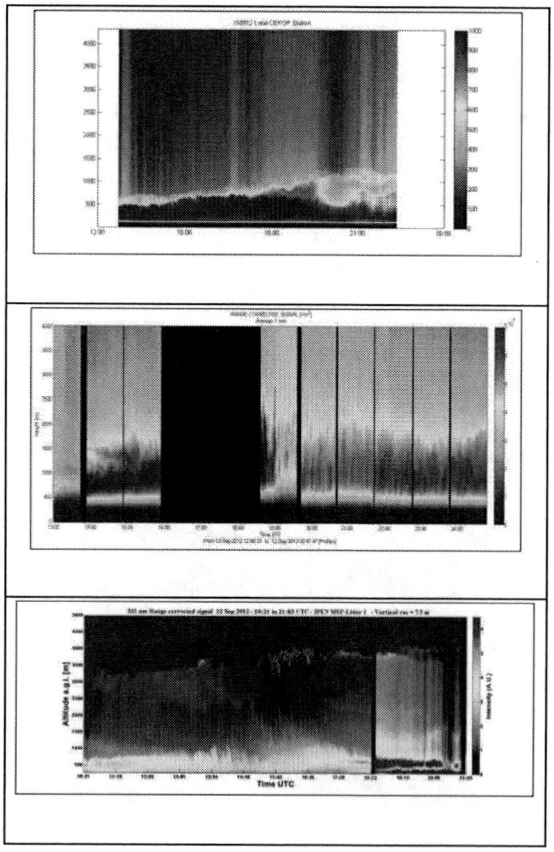

Figure 3 – Quicklook panel for some of the stations running a coordinated campaign in 2012. From top to bottom: co-CEFOP-UDEC, and sp-CLA-IPEN-MSP-LIDAR-I.

REFERENCES

[1] Lopes, F. J. S., E. Landulfo, M. A. Vaughan, 2013: Evaluating CALIPSO's 532 nm lidar ratio selection algorithm using AERONET sunphotometers in Brazil, *Atmos. Meas. Tech.*, **6(11)**, 3281-3299.

[2] Guerrero-Rascado, J.L., E. Landulfo, J. C. Antuña, H. M. J. Barbosa, B. Barja, A. E. Bastidas, A. E. Bedoya, R. da Costa, R. Estevan, R. N. Forno, D. A. Gouveia, C. Jiménez, E. G. Larroza, F. J. S. Lopes, E. Montilla-Rosero, G. A. Moreira, W. M. Nakaema, D. Nisperuza, L. Otero, J. V. Pallotta, S. Papandrea, E. Pawelko, E. J. Quel, P. Ristori, P. F. Rodrigues, J. Salvador, M. F. Sánchez, and A. Silva: Towards an instrumental harmonization in the framework of LALINET: dataset of technical specifications, Proceedings of SPIE 2014, vol. 9246, 92460O-1—92460O-14 (2014) doi: 10.1117/12.2066873

[3] Barbosa, H.M.J., F.J.S. Lopes, A. Silva, D. Nisperuza, B. Barja, P. Ristori, D.A. Gouveia, C. Jimenez, E. Montilla, G.L. Mariano, E. Landulfo, A. Bastidas, E.J. Quel, 2014: The first ALINE measurements and intercomparison exercise on lidar inversion algorithms, *Opt. Pura Apl.*, **47(2)**, 99-108.

[4] E. Montilla-Rosero, A. Silva, C. Jimenez, R. Hernandez, C. Saavedra: Optical Characterization of low tropospheric aerosols by the Southern East Pacific Lidar station (Concepcion, Chile). Paper submitted to Journal of Aerosol Science, March, 2015.

[5] Silva, A., Montilla-Rosero,E., Jimenez C., Saavedra C. Hernandez R. (2012). Tropospheric Measurements of Aerosol Optical Properties at Concepcion (36S, 73W), Proc. 26th International Laser Radar Conference (26th ILRC), Vol I, 521-524, Porto Heli, Greece.

EPJ Web of Conferences **119**, 19005 (2016) DOI: 10.1051/epjconf/201611919005
ILRC 27

NEW RESULTS FROM THE NOAA CREST LIDAR NETWORK (CLN) OBSERVATIONS IN THE US EASTCOAST

***Fred Moshary [a], Zaw Han [a], Yonghua Wu [a], Barry Gross [a]**
[a]*Optical Remote Sensing Lab., City College of New York, NY, 10031, USA *moshary@ccny.cuny.edu*
Daniel Wesloh[b], Raymond M. Hoff[b], Ruben Delgado[b]
[b]*Physics Department and Joint Center for Earth Systems Technology, University of Maryland, Baltimore County, MD USA 21250*
Jia Su[c], Liqiao Lei [c], Robert B. Lee III[c], M. Pat McCormick[c]
[c]*Department of Atmospheric and Planetary Sciences, Hampton University, Hampton, VA, 23668 USA*
Jesus Diaz[d], Carlos Cruz[d], Hamed Parsiani [d]
[d]*Department of Electrical and Computer Engineer, University of Puerto Rico, Mayagüez (UPRM) Campus, PO Box 9000, Mayagüez , P.R. 00681-9000*

ABSTRACT
This paper presents coordinated ground-based observations by the NOAA-CREST Lidar Network (CLN) for profiling of aerosols, cloud, water vapor, and wind along the US east coast including Caribbean region at Puerto Rico. The instrumentation, methodology and observation capability are reviewed. The applications to continental and intercontinental-scale transport of smoke and dust plumes, and their large scale regional impact are discussed.

1. INTRODUCTION
Intra-continental and intercontinental transport of aerosol plumes (e.g. Asian dust, African dust, and biomass burning) play an important role in air quality and climate radiation studies. Global and regional chemical transport models and satellite sensors have been used to assess plume transport, but we still lack detailed knowledge of the vertical stratification of aerosol and their corresponding optical and microphysical changes during transport [1]. Multi-wavelength lidars are able to measure profiles of aerosol extinction and backscatter coefficient and thus the Angstrom exponent which contains information about particle size. In particular, a ground-based lidar network has the capability to observe the aerosol temporal-spatial distribution and variation in the regional-continental scale, such as EARLINET in Europe, MPL-Net in USA, AD-net in East Asia, and GALION on the global scale [2], thus providing the opportunities to verify the models and satellite products. In this study, we review the ground-based NOAA-CREST (Cooperative Remote Sensing Science and Technology Center) Lidar Network (CLN) in the US east coast, the observation capability and applications on the smoke and dust transport are presented.

2. LIDAR SYSTEM AND METHODOLOGY
CLN is comprised of lidar facilities at mainly 4 sites (campuses) supported by the NOAA CREST Center (see Fig.1) [3-4]: City College of New York (CCNY) in New York City (40.821°N/73.95°W; the University of Maryland, Baltimore County (UMBC, 39.255°N/76.71°W); Hampton University, Hampton, VA (HU, 37.02°N/76.337°W), and University of Puerto Rico at Mayaguez, PR (UPRM, 18.30°N/67.20°W). The CLN sites primarily deploy multi-wavelength elastic-Raman lidar for aerosol, cloud, PBL, water vapor observations. The elastic-Raman lidars are based on solid state Nd:YAG lasers with multiple wavelengths (355, 532, 1064-nm) [5,6,7]. The high pulse powers allow the observation and quantification of multi-layer plumes; and the multi-wavelength capacity enables us to classify aloft aerosol type (smoke or dust). In addition, the use of Raman channels excited by the 355-nm laser beam allows for a direct measure of water vapor and aerosol extinction profile. Observations from select CLN sites are available for over 8 years. More recently, rotational Raman temperature lidar has been deployed in the HU site and wind Doppler lidars have been deployed to characterize wind variability within the boundary layer at CCNY, UMBC and HU, with the applications for PBL dynamic, low-level jet, wind resource assessment and validation and assimilation of winds in weather forecast models.

DOI: 10.1051/epjconf/201611919005

Existing CLN lidars are either collocated with AERONET CIMEL sun/sky radiometers or are in near proximity. Simultaneous column aerosol optical depth (AOD) measured by AERONET-radiometer is used to constrain the lidar-derived aerosol extinction profile and column lidar-ratio, this avoids an arbitrary assumption on lidar ratio that may result in the large retrieval error. In using multiple lidar systems, consistency checks between the different systems are warranted. In particular, we have made cross-validation checks of the CLN-lidar algorithms for retrieving aerosol extinction and backscatter profiles. Relative differences are generally less than 5% below 7-km. CLN provides near real-time lidar images to local Environmental Agencies in MD, NY, NJ, CT, PR states.

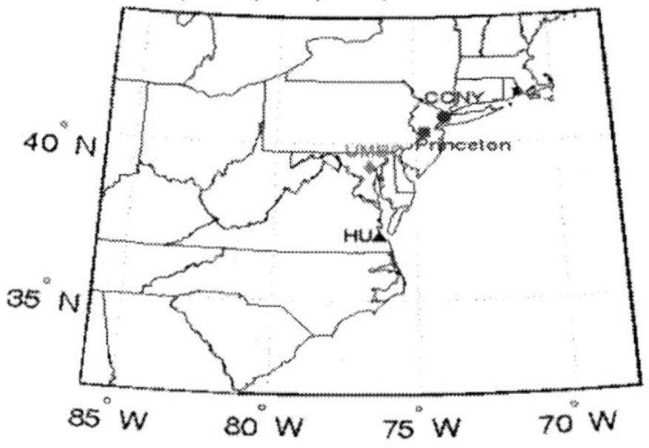

Fig. 1. NOAA-CREST lidar network sites.

3. RESULTS and APPLICATIONS

3.1 Regional transport of Canadian smoke

As an example, Figure 2 shows the time-height variation of aerosol by CLN on Aug.8, 2014. Aloft aerosol plumes can be clearly seen at CCNY, UMBC and HU sites. Both satellite GOES-AOD image and NOAA-HYSPLIT model indicate that the plumes are transported from Canada as shown in Fig. 3. Large Angstrom exponents (≥1.8) from ARONET and MODIS satellite indicate that they are fine-mode dominant particles and thus associated with the smoke plumes transport. At CCNY-site, the column averaged lidar-ratios were obtained with values of 60~80 sr at 532-nm and 40~45 sr at 1064-nm, respectively. The aloft plumes contribute 80~90% fraction to the total-AODs. Importantly, the aloft plume heights show a decreasing trend with time at CCNY and UMBC sites, and even get close to PBL-top and mix downward after 16:00 at HU site. Such air subsidence and PBL entrainment mix the plumes down to the surface, thus resulting in air quality degradation.

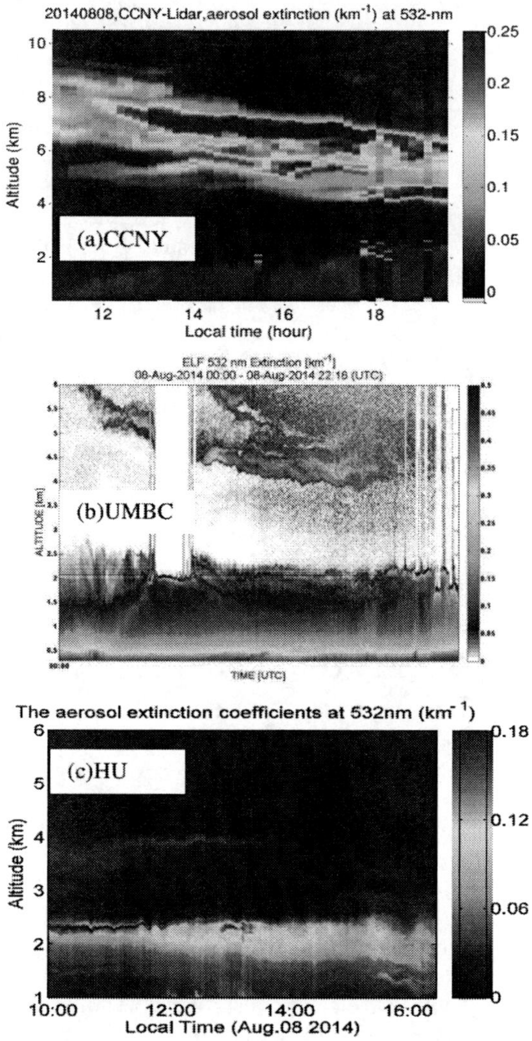

Fig.2. Smoke plumes transport to US east coast from Canada on Aug. 8, 2014. (a) CCNY, (b) UMBC, (c) HamptonU.

3.2 African and Asian dust transport

Figure 4 shows the time-height distribution of range-corrected lidar returns at Puerto Rico on July 1, 2014. The elevated aerosol layers can be seen at 2~4 km altitude. The co-located AERONET-sunphotometer data indicates the coarse-mode dominant aerosols as shown in Fig.5. The HYSPLIT model shows that the air mass at 4-km altitude is transported from Africa for about 9 days.

 DOI: 10.1051/epjconf/201611919005

The high AOD and small Angstrom exponent from the MODIS/Aqua image indicate this event is African dust transport and the observed aloft coarse-model aerosols are the transported dust.

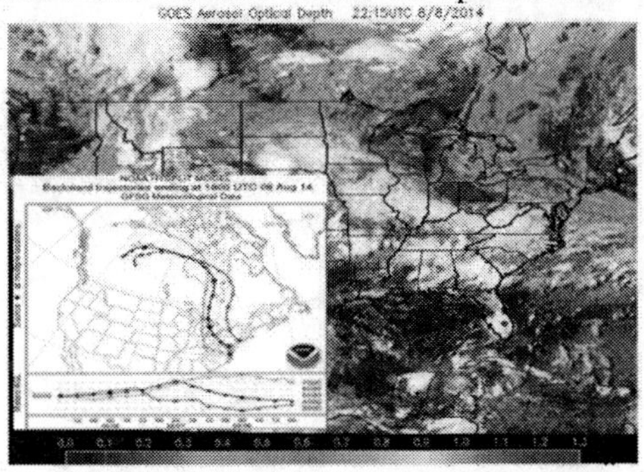

Fig.3 NOAA-GOES AOD and NOAA-HYSPLIT backward trajectories ending at 5-km, 18:00 UTC, Aug. 8, 2014.

In addition, the trans-Pacific transports of Asian dust in spring and mixture with other aerosols are also observed and characterized by CLN in the US East Coast [4]. The potential influence on the air quality and validation of model profiling prediction are verified [8]. A large scale transport episode observed by CLN in March 2012 is shown in Fig 6. Trajectory analysis and satellite data indicate that the multilayer plumes are mixture of trans-pacific dust transport to US (>3 km) together with mixed smoke and dust from south-southwest US.

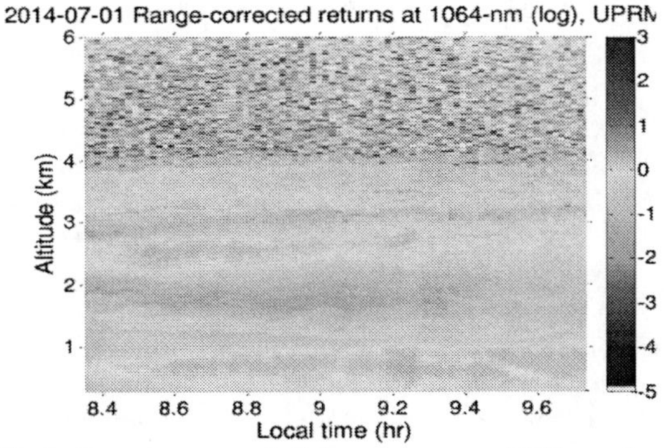

Fig.4. Range-corrected lidar returns at 1064-nm by UPRM lidar on July 1, 2014.

4. CONCLUSIONS

In summary, we present a coordinated ground-based NOAA-CREST Lidar Network (CLN) along the US east coast and the Caribbean region. The lidar systems have being used to profile aerosols, cloud, water vapor wind, and atmospheric structure demonstrate large-scale aloft aerosol plumes / process for climate, weather and air quality transport to the eastern United States that cause a

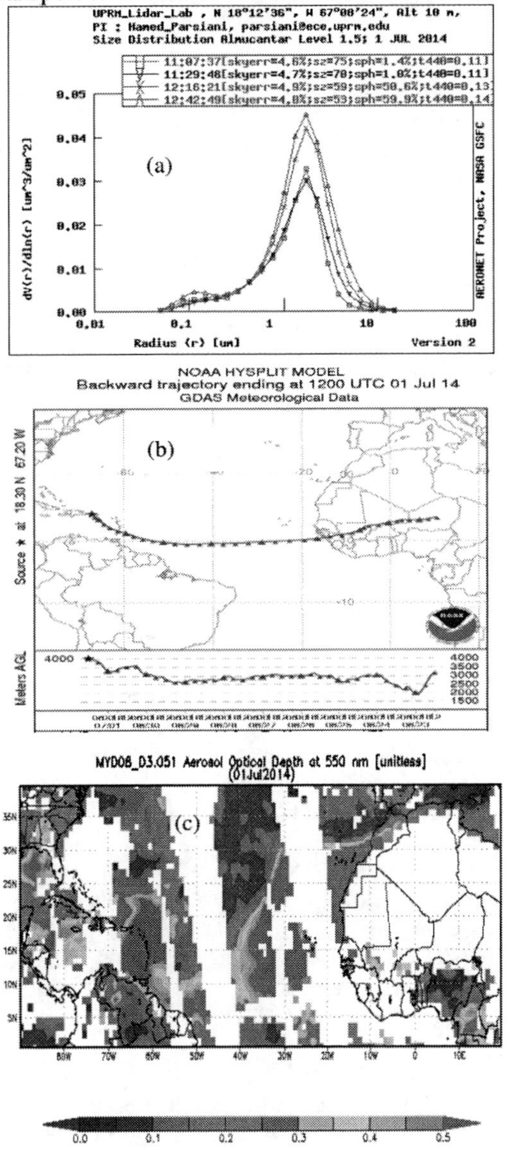

Fig.5 (a) AERONET aerosol size distribution, (b) HYSPLIT transport trajectory at 4-km altitude, (c) MODIS/Auqa AOD on July 1, 2014.

significant change in aerosol optical properties and potential influences on air quality. CLN provides near real-time images to local Environmental

DOI: 10.1051/epjconf/201611919005

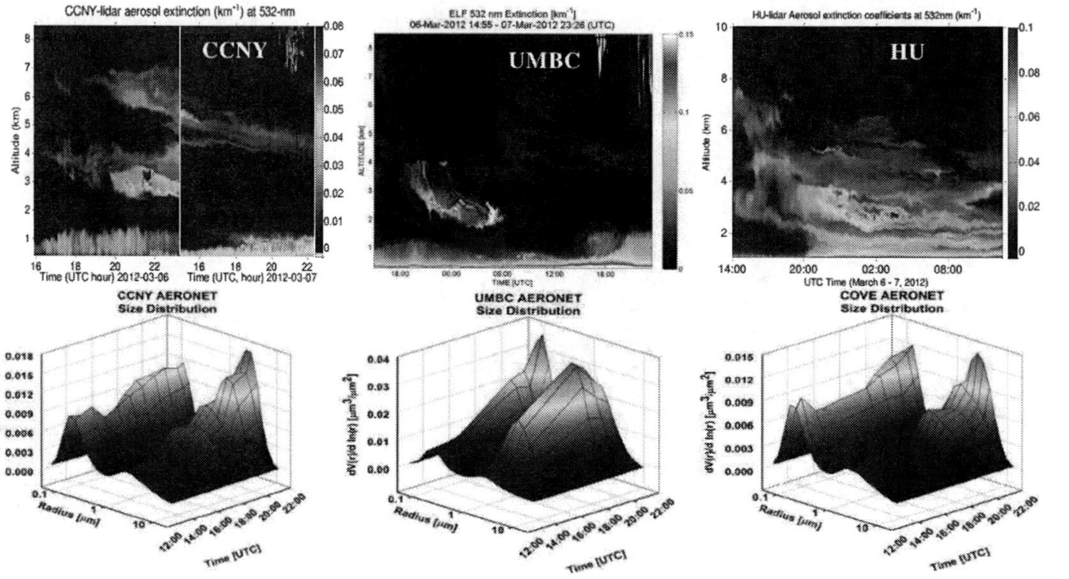

Fig. 6 Large scale plume transport observed by CLN sites at CCNY, UMBC and Hampton U on 03/06/2012 together with the corresponding NASA AERONET inversions at these sites.

Agencies in MD, NY, NJ, CT, PR states. CLN has been expanding its capabilities wind lidar observations at select sites, and plans are underway to expand the number of sites by including observations from other US lidar and ceilometers sites. Overall, the CLN observations help improve understanding of atmospheric physical and chemical processes/dynamics; validation of climate, weather and air quality forecast models.

ACKNOWLEDGEMENT

This study was supported by National Oceanic and Atmospheric Administration (NOAA) under the CREST Grant # NA11SEC4810004. Authors greatly appreciate the data from NASA-AERONET and NOAA-HYSPLIT projects.

REFERENCES

[1] National Research Council, Global Sources of Local Pollution: An Assessment of Long-Range Transport of Key Air Pollutants to and from the United States, National Academies Press, Washington, DC, 2009.

[2] Bosenberg, J., Hoff, R., Plan for the implementation of the GAW Aerosol lidar observation Network GALION, GAW report No.178, WMO/TD-No.1443, 2007.

[3] Hoff, R., F. Moshary, S. Ahmed, B. Gross, P. McCormick, H. Parsiani, CREST Lidar Network (CLN), CREST Publications Series, Vol.7: No.01 2009.

[4] Moshary, F., Cordero, L., Wu, Y., et al., Assessment of long-scale plume transport to the U.S. East coast using coordinated CREST lidar network and synergistic AERONET and satellite measurements, Proc. SPIE 8894, p. 88940K, 2013.

[5] Wu, Y., Chaw, S., Gross, B., Moshary, F., Ahmed, S., Low and optically thin cloud measurements using a Raman-Mie lidar, Appl. Opt., 48, 1218-1227, 2009.

[6] Su, J., McCormick, M. P., Wu, Y., Lee, R. B., Lei, L., Liu, Z., Leavor, K., Cloud temperature measurement using rotational Raman lidar, J.Quant.Spectro.Rad.Trans.,125,45-50,2013.

[7] Compton, J. C., Delgado, R., Berkoff, T. A, Hoff, R. M., Determination of planetary boundary layer height on short spatial and temporal scales: a demonstration of the covariance wavelet transform in ground-based wind profiler and lidar measurements, J. Atmos. Oceanic Technol. 30 (7), pp. 1566–1575, 2013.

[8] Wu, Y., Han, Z., Nazmi, C., Gross, B., Moshary, F., A trans-Pacific Asian dust episode and its impacts to air quality in the east coast of U.S., Atmos. Environ.,106, 358–368, 2015.

AEROSOLS MONITORING NETWORK TO CREATE A VOLCANIC ASH RISK MANAGEMENT SYSTEM IN ARGENTINA AND CHILE

Eduardo Quel [1-3*], Nobuo Sugimoto [2], Lidia Otero [1-3], Yoshitaka Jin [2], Pablo Ristori [1], Tomoaki Nishizawa [2], Francisco González [1], Sebastián Papandrea [1], Atsushi Shimizu [2], Akira Mizuno [4]

[1]*División Lidar, CEILAP, UNIDEF (MINDEF - CONICET), UMI-IFAECI-CNRS 3351, Juan Bautista de La Salle 4397 – B1603ALO Villa Martelli, Buenos Aires, Argentina, *Email: pablo.ristori@gmail.com*

[2]*National Institue for Environmental Studies, 16-2 Onogawa, Tsukuba 305-8506 Japan*

[3]*IUE – Escuela Superior Técnica, Facultad de Ingeniería, Av. Cabildo 15 – C1426AAA CABA, Argentina*

[4]*Atmospheric Environment Division, Solar-Terrestrial Environment Laboratory, Nagoya University, Furo-cho, Chikusa-ku, Nagoya 464-8601 Japan*

ABSTRACT

Two main decisions were made in Argentina to mitigate the impact of the recent volcanic activity in de country basically affected by the presence of volcanic ash in the air and deposited over the Argentinean territory. The first one was to create a risk management commission were this risk between others were studied, and second to develop new ground based remote sensing technologies to be able to identify and inform the risk close to the airports. In addition the Japanese government program for Science and Technology joint Research Partnership between Argentina, Chile and Japan for Sustainable Development (SATREPS) accepted to fund this cooperation due to the potential future utilization of the research outcomes to the benefit of the society. This work present the actual achievements and expected advance of these projects that try to joint efforts between national and international agencies as well as countries on behalf of a better understanding of the risks and a joint collaboration on the mitigation of suspended ashes impact over the aerial navigation.

1. INTRODUCTION

The intense volcanic activity in the Andes Mountain Range affecting Argentina and Chile as for example Chaitén (2008), Puyehue Cordón Caulle (2011), Copahue (2014), Villarrica (2015), had an important impact in the Argentinean and Chilean society by means of the activation of diverse risk management protocols at both sides of the Andes Mountain Range, like evacuation of vulnerable regions, and in some cases the suspension for several days to months of the aerial navigation as mentioned on the previous ILRC meeting [1]. At that meeting only Bariloche Station was installed (Feb 2, 2012). Eyjafjallajökull eruption major consequence in the Argentinean region was the introduction of quantitative evaluation criteria based on threshold concentrations to decide which aerial space to close. Then several studies and developments were done in Argentina to model [2] and to detect volcanic ashes. Actually the network has new stations: Comodoro Rivadavia (CDR) on Oct 12, 2012; Neuquén (NQN) Dec 12, 2013, Río Gallegos (RGL) Jul 1, 2014 and Aeroparque (AEP) Feb 2, 2015. All these stations, are installed inside the Airports or very close to it (<1km). A last station was also installed at the Lidar Division (CEILAP) which is intended to be not only a transportable station but also a station where to test the updates before being sent to the sites and a training place for operators and researchers.

2. THE SPECIAL PROJECT MD 31554/11

After the eruption of Puyehue Cordón Caulle volcano the Ministry of the Defense in Argentina proposed to finance the construction of an aerosol monitoring network at the Lidar Division to be operated by the National Weather Service (NWS) (dependent of this Ministry) which is responsible of the regional Volcanic Ash Advisory Center (VAAC) at Buenos Aires covering the region (South of 10°S and between 10°W and 90°W).

This project consisted on designing upgradable and transportable aerosol monitoring stations to be placed at different places to be designed. They were built in a 20 feet maritime container (figure 1) adapted to work as a laboratory. Half of the inner space has a multi-wavelength lidar (355 nm, 532 nm and 1064 nm) and the second half contains a room were the UPS, computers and acquisition systems of the complementary instruments were placed. Those instruments are installed at the rooftop of the container. They were initially a CIMEL Sunphotometer part of the AERONET network, a nephelometer to measure TSP, PM_{10}, $PM_{2.5}$ and PM_1; a weather station, a broadband, UVA and UVB pyranometers, and a 7 channel Radiometer. These systems are part to a single and indivisible network. Further updates of the Project were done to include vibro-rotational Raman channels (384 nm, 408 nm and 607 nm) and depolarization channels for both UV and visible elastic wavelengths (except for the actual fiber-based design at Bariloche Station as seen on [3]). Other complementary equipment is also planned to be added is a multi-axes DOAS unit.

Figure 1: Aerosol Monitoring Station: Left: Lidar Container. Right: Lidar unit

Part of this project consist on creating the lidar operation capacity at the NWS stations were the lidars are placed and at the NWS sites were part data will be processed and analyzed.

3. THE SATREPS "SAVER.Net" PROJECT

This project complements in several ways the MD31554/11 project. The previous project intended to transfer technology from one institution to another in one country. In this case those institutions are partners on a common research involving three different countries in strong collaboration. It must also be noticed that social point of view is the most important one. The most ambitious goal is to contribute the global community developing a social management system for environmental risk in South America, concerning suspended volcanic ash risk. From the scientific and technological point of view the project increase the number of monitoring stations adding a third multi-wavelength and depolarization Raman lidar station at Punta Arenas in southernmost region of Chilean Patagonia as seen on figure 2, as well as two new monitoring sites containing iodine cell-based HSRL's and based on [4]. These last stations are an important contribution from NIES to the Lidar Division at CEILAP and the whole members and Institutions integrating this project. The complementing instrumentation is similar to the one mentioned in the previous project. The lidar and container design will be similar but it is intended to improve some construction details based on the current acquired experience.

From the point of view of data analysis SAVER.Net will start learning from AD-NET experience implementing their algorithms [5] but it is expected a fast and growing collaboration by learning from other inversion algorithms developed in the lidar community [6] and also creating new capabilities by its own focused on local aerosol conditions and volcanic ash behavior. I must be noticed at this point that one of the most important problems with volcanic ash in Patagonia is ash deposits after a main eruption being lifted and transported by the wind several kilometers away from their origin.

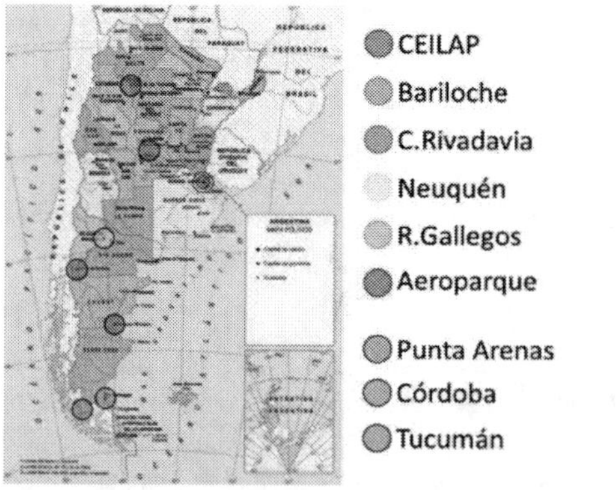

Figure 2: Aerosol Monitoring Station locations. From CEILAP to Aeroparque the stations are already installed along the period Feb/2012 to Feb/2015. The new stations are being built at CEILAP, except Punta Arenas which will be partially built at Chile

DOI: 10.1051/epjconf/201611919006

A second aspect of this project is the development of a UV risk management system but this exceeds the scope of this paper.

4. THE LIDAR SYSTEMS AT THE STATIONS

All the lidar systems were built inside an aluminum frame as seen on figure 3. All the parts that need to be perfectly aligned (laser, telescope and spectroscopic box) are interconnected using a metal frame that is at the time attached to the aluminum frame via shock absorber units (figure 4-a). Inside this, the telescope and polychromator units are air coupled using a diaphragm at the telescope focal plane (figure 4-b). The light is then collimated and the wavelengths sent using hi-pass beamsplitters to the photomultipliers (Hamamatsu H10721-110 or -20). The optical paths are identical. The infrared radiation passes unaltered to the detection unit (Licel APD).

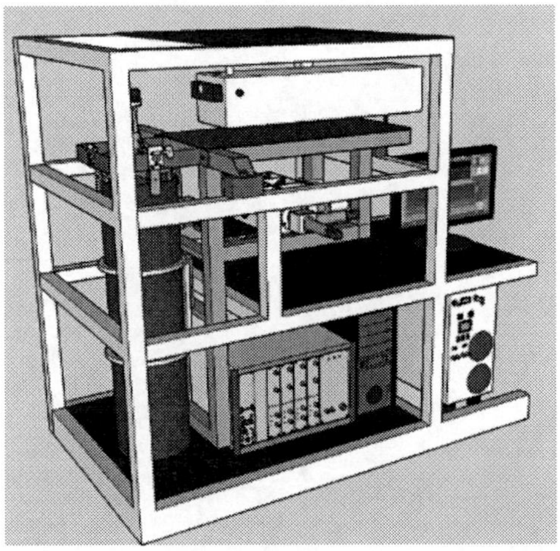

Figure 3: Diagram for Multi-wavelength Raman Depolarization Aerosol lidar.

Figure 4-c shows the aluminum frame were the laser (Continuum Surelite I – 30Hz), the energy source, the acquisition system (Licel operating at 20 MSPS) and the computer are placed. The output alignment unit (figure 4-d) has a coaxial configuration. The beams are not expanded since low divergence is not desired for this instrument. The light is oriented by a fused silica prism. The manual alignment is quite precise (angular sensitivity of the prism below 1.67 mrad per turn) and no telescope obstruction is done by the alignment procedure.

The observation protocol consists on acquiring 3 minutes of lidar signals every 15 minutes on a 24 h/day basis. The data is actually being sent for analysis and fast visualization to CEILAP and NIES.

Basic changes on the HSRL version consists on changing the lasers to the seeded version (NP-Photonics) of Surelite II-10Hz spectrally tuned to match a specific absorption line of a heated Iodine cell using acousto-optical modulators to improve the wavelength tuning [4].

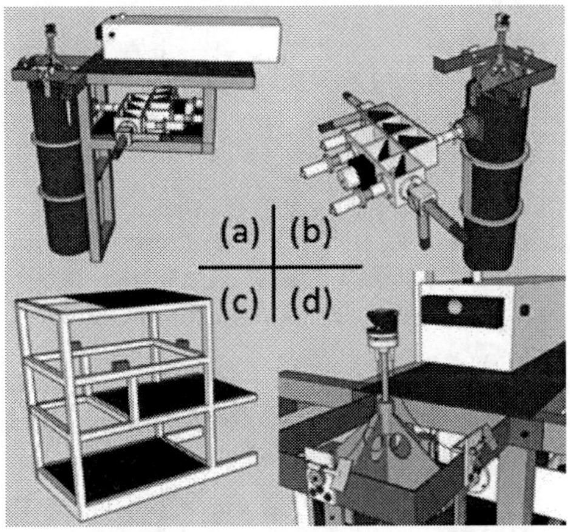

Figure 4: Parts of the Multi-wavelength Raman Depolarization Aerosol lidar. (a) Iron Frame, (b) Telescope and spectroscopic box (c) Aluminum frame, (d) laser alignment unit

5. CONCLUSIONS

This work shows the evolution of the Aerosol monitoring network based on lidars and other instrumentation to provide risk management capabilities inside the Argentina and the southern region of Chile. This network is intended to contribute to ALINE/LALINET and GALION. The synergy of the projects MD 31554/11 and SAVER.Net made possible a fast evolution of the network in a short period of time however this network is still growing and will need some time before reaching the intended maturity.

ACKNOWLEDGEMENT

The researchers want to acknowledge the National Weather Service Lidar Operators, Japan Science and Technology Agency (JST) / Japan International Cooperation Agency (JICA), Science and Technology Research Partnership for

DOI: 10.1051/epjconf/201611919006

Sustainable Development (SATREPS), CITEDEF and the Ministry of Defense.

REFERENCES

[1] Ristori P., Otero L., Pawelko E., Pallotta J., D'Elía R., Chouza F., González F., Dworniczak J. C., Vilar O., Pereyra A., Fernández M., Lema S., Sugimoto N., Quel E., 2012, Development of an Argentinean Lidar Nework to Monitor the Volcanic Plume and Dust in Patagonia, Proc of the 26th International Laser Radar Conference: 357-360

[2] Collini, E., Osores M. S., Folch A., Viramonte J. G., Villarosa G., Salmuni G., 2013, Volcanic ash forecast during the June 2011 Cordón Caulle eruption. Natural hazards **66** (2): 389-412.

[3] Pawelko E., Ristori P., Otero L., D'Elía R., ., Pereyra A., Vilar O, Chouza F., Pallotta J., González F., Fernández M., Lema S., Sugimoto N., Quel E., 2012, Multiwavelength Raman Lidar Construction to Monitor Volcanic Ash and Aerosols in Bariloche International Airport, Argentina, Proc of the 26th International Laser Radar Conference: 285-288

[4] Shimizu H., S. A. Lee, and C. Y. She, 1983: High spectral resolution lidar system with atomic blocking filters for measuring atmospheric parameters, *Appl. Opt.*, **22**, 1373-1381.

[5] A. Shimizu, N. Sugimoto, I. Matsui, K. Arao, I. Uno, T. Murayama, N. Kagawa, K. Aoki, A. Uchiyama, and A. Yamazaki, 2004. Continuous observations of Asian dust and other aerosols by polarization lidar in China and Japan during ACE-Asia, J. Geophys. Res., **109**, D19S17, doi:10.1029/2002JD003253.

[6] Ansmann, A., Seifert, P., Tesche, M., and Wandinger, U., 2012. Profiling of fine and coarse particle mass: case studies of Saharan dust and Eyjafjallajökull/Grimsvötn volcanic plumes, Atmos. Chem. Phys., **12**, 9399–9415, doi:10.5194/acp-12-9399-2012.

EPJ Web of Conferences **119**, 20001 (2016)
ILRC 27
DOI: 10.1051/epjconf/201611920001

TOLNet – A Tropospheric Ozone Lidar Profiling Network for Satellite Continuity and Process Studies

Michael J. Newchurch[1*], Shi Kuang[1], Thierry Leblanc[2], Raul J. Alvarez II[3], Andrew O. Langford[3], Christoph J. Senff[3,4], John, F. Burris[5], Thomas J. McGee[5], John T. Sullivan[5,6], Russell J. DeYoung[7], Jassim Al-Saadi[7], Matthew Johnson[8], Alex Pszenny[9]

[1]Atmospheric Science Department, University of Alabama in Huntsville, Huntsville, AL 35805, USA

[2]NASA Jet Propulsion Laboratory, Wrightwood, CA 92397, USA

[3]NOAA Earth System Research Laboratory, Boulder, CO 80305, USA

[4]Cooperative Institute for Research in the Environmental Sciences, University of Colorado at Boulder, Boulder, CO 80309, USA

[5]NASA Goddard Space Flight Center, Greenbelt, MD 20771, USA

[6]Dept. of Atmospheric Physics, University of Maryland Baltimore County, Baltimore, MD 21250, USA

[7]NASA Langley Research Center, Hampton, VA 23681, USA

[8]NASA Ames Research Center, Moffett Field, CA 94031, USA

[9]NASA Headquarters, Washington, DC 20546, USA

**Corresponding author:* mike@nsstc.uah.edu

Abstract

Ozone lidars measure continuous, high-resolution ozone profiles critical for process studies and for satellite validation in the lower troposphere. However, the effectiveness of lidar validation by using single-station data is limited. Recently, NASA initiated an interagency ozone lidar observation network under the name TOLNet to promote cooperative multiple-station ozone-lidar observations to provide highly time-resolved (few minutes) tropospheric-ozone vertical profiles useful for air-quality studies, model evaluation, and satellite validation. This article briefly describes the concept, stations, major specifications of the TOLNet instruments, and data archiving.

Background

Although the U.S. EPA (Environmental Protection Agency) reports hourly surface ozone mixing ratios from a dense national ground-level monitoring network, upper-air ozone observations are extremely sparse. Ozone can be generated throughout the troposphere as a secondary product of industrial or mobile emissions and it can also be transported from the stratosphere [1]. Upper-air observations are particularly important because the ozone produced aloft and transported over appreciable distances can significantly affect surface air quality.

Lidar offers a number of significant advantages including continuous high spatial and temporal resolution vertical profiles, automated, high-quality retrievals under both day and nighttime conditions, and the ability to generate consistent, long-term, data sets. The lidar observations are critically important when a requirement exists for continuous ozone retrievals with high temporal and vertical resolution such as in the PBL where chemical and physical variations are both rapid and complicated. Ground-based lidars, including mobile and fixed-station configurations, have lower operation cost than airborne lidars but have limited geographic coverage. This limitation motivates the development of a ground-based ozone lidar network to make simultaneous ozone profiling at multiple locations for process studies, model evaluation and satellite validation at reasonable cost.

TOLNet, the Tropospheric Ozone Lidar Network, is an interagency initiative started by NASA, NOAA, and USEPA in 2011 to provide highly time resolved measurements of

DOI: 10.1051/epjconf/201611920001

tropospheric ozone vertical profiles to address fundamental air-quality science questions in cooperation with national, state, and local organizations. TOLNet also contributes to intensive field campaigns such as DISCOVER-AQ, and development of future satellite missions GEOstationary Coastal and Air Pollution Events (GEO-CAPE), and Tropospheric Emissions: Monitoring of Pollution (TEMPO).

Objectives and Recent Activity

The primary scientific objective of TOLNet is to provide high spatio-temporal measurements of ozone from near surface to the top of the troposphere. Such high-fidelity lidar measurements can provide accurate representations of the PBL and free-tropospheric ozone structure. With these observations of the detailed ozone structure, satellite science teams and the modeling community can study the character of lower-atmospheric ozone and also assess the accuracy and vertical resolution with which a geosynchronous instrument could retrieve the observed laminar ozone structures.

A second objective is to identify an ozone lidar instrument design that would be suitable to populate an expanded network to address the needs of NASA/EPA/NOAA air-quality scientists and managers who increasingly express a desire for ozone vertical profiles.

The third motivation for TOLNet is basic scientific research into the processes creating and destroying the ubiquitously observed ozone laminae and other ozone features in the troposphere. The sources and mechanisms controlling ozone distribution and variations are being investigated by using high-resolution lidar data, synergistic meteorological measurements, and atmospheric chemistry model simulations.

TOLNet recently contributed simultaneous observations of ozone profiles to the summer 2014 DISCOVER-AQ Colorado campaign with all three lidars first deployed together at the NOAA Boulder Atmospheric Observatory (BAO), and then separately at Ft. Collins, Golden, and the BAO. The individual lidars have also conducted measurements at various sites from California to Maryland. The scientific foci of these observations include the following: Los Angeles pollution outflow; stratosphere-troposphere exchange (STE) and transport from East Asia; wintertime ozone exceedances resulting from oil and gas development in Utah; Southeast US ozone structure and variability resulting from lightning, STE, laminar evolution, and residual layer transport and entrainment; Chesapeake Bay effects on PBL ozone; and correlative measurement with the NSF-sponsored Wintertime Investigation of Transport, Emission, and Reactivity (WINTER) campaign in early 2015.

Stations and Instruments

TOLNet currently comprises five lidars, including three mobile systems located at the NOAA Earth System Research Laboratory (ESRL, Boulder, CO), NASA Langley Research Center (LaRC, Hampton, VA), and NASA Goddard Space Flight Center (GSFC, Greenbelt, MD), and two fixed-base systems located at the NASA Jet Propulsion Laboratory, Table Mountain, CA), and the University of Alabama in Huntsville (UAH, Huntsville, AL). Some of these lidar stations already had a long observation history before becoming part of the TOLNet.

Table 1 summarizes the major specifications of the TOLNet lidars. Currently, Nd:YAG pumped Raman-shifted lasers at 289 and 299 nm are popular for ozone measurement due to their easy installation, robustness, good conversion efficiency, and good output beam quality [2]. However, the Nd:YLF pumped Ce:LiCAF laser offer flexibility on wavelength choice. Near-surface ozone retrieval has been difficult for ozone lidars due to the required altitude for overlap between laser beam and telescope caused by the bistatic configuration. Recently, some TOLNet stations have extended the ozone lidar observations down to 125 m AGL by modifying the lidar receiving system [3] to better characterize the interactions between surface and upper air.

Lidar data QA/QC and archiving

The uncertainty of the ozone lidar retrievals depends strongly on the signal integration time and the vertical (or spatial) resolution. The TOLNet lidars generally retrieve ozone at an uncertainty of ~10% at near-range and ~20% at far range with an integration time of few minutes and vertical resolution of hundreds of meters [4,6,7,8]. Most of the TOLNet lidars validate their ozone retrievals using collocated ozonesonde measurements although occasional intercomparison between TOLNet lidars was made (e.g., during the summer 2014 DISCOVER-AQ Boulder campaign). Standardization of the retrieval algorithm and definition of the vertical resolution for all TOLNet lidars is an ongoing project.

The TOLNet data are accessible at http://www-air.larc.nasa.gov/missions/TOLNet/ although consulting with the instrument principal or co-investigators before usage is highly encouraged. We provide the digital data to users with both the widely used ICARTT format(http://www-ir.larc.nasa.gov/missions/etc/IcarttDataFormat.htm) and the TOLNet format which is a new data protocol recently formed by the TOLNet working group. Figure 1 shows the TOLNet data webpage and few lidar curtain examples.

Summary

NASA's TOLNet has made significant progress toward enhancing the capability for vertical ozone profiling and fostering corporation between five member stations including NASA/JPL, NOAA/ESRL, UAH, NASA/GSFC, and NASA/LaRC. All these stations have made considerable progress concerning lidar platform construction, receiver or transmitter upgrades, and near-surface channel development since the initiation of this project. A vehicle for the public archiving and dissemination of TOLNet data using two well-defined standards has been developed to promote research and collaboration among TOLNet investigators and the broader scientific community. A main objective of TOLNet is to provide high-resolution lidar data at multiple stations to modeling and satellite teams for validating and improving the fidelity of tropospheric ozone measurements by NASA's next-generation geostationary instruments.

References

[1] Kuang, S., M. J. Newchurch, J. Burris, L. Wang, K. Knupp, G. Huang, 2012: Stratosphere-to-troposphere transport revealed by ground-based lidar and ozonesonde at a midlatitude site, *J. Geophys. Res.*, **117**, D18305.

[2] Schoulepnikoff, L. de, V. Mitev, V. Simeonov, B. Calpini, H. van den Bergh, 1997: Experimental investigation of high-power single-pass Raman shifters in the ultraviolet with Nd:YAG and KrF lasers, *Appl. Opt.*, **36**, 5026-5043.

[3] Kuang, S., M. J. Newchurch, J. Burris, X. Liu, 2013: Ground-based lidar for atmospheric boundary layer ozone measurements, *Appl. Opt.*, **52**, 3557-3566.

[4] McDermid, I., G. Beyerle, D. Haner, T. Leblanc, 2002: Redesign and improved performance of the tropospheric ozone lidar at the jet propulsion laboratory table mountain facility, *Appl. Opt.*, **41**, 7550-7555.

[5] Alvarez, R. J., et al, 2011: Development and application of a compact, tunable, solid-state airborne ozone lidar system for boundary layer profiling, *J. Atmos. Ocean. Technol.*, **28**, 1258-1271.

[6] Kuang, S., J. F. Burris, M. J. Newchurch, S. Johnson, S. Long, 2011: Differential Absorption Lidar to measure subhourly variation of tropospheric ozone profiles, *IEEE Trans. Geosci. Remote Sens.*, **49**, 557-571.

[7] Sullivan, J. T., McGee, T. J., Sumnicht, G. K., Twigg, L. W., and Hoff, R. M., 2014: A mobile differential absorption lidar to measure sub-hourly fluctuation of tropospheric ozone profiles in the Baltimore–Washington, D.C. region, *Atmos. Meas. Tech.*, **7**, 3529-3548.

[8] De Young, R.J., Carrion, W., Pliutau, D.: A Compact Mobile Ozone Lidar for Atmospheric Ozone and Aerosol Profiling, *Proceedings of SPIE Remote Sensing Conf.*, 9246-7, Amsterdam, Netherlands, September 2014.

DOI: 10.1051/epjconf/201611920001

Table 1. Summary of the TOLNet lidars.

Name	JPL TMO (Table Mountain tropospheric Ozone) DIAL	TOPAZ (Tunable Optical Profiler for Aerosol and oZone) lidar	RO_3QET (Rocket-city O3 Quality Evaluation in theTroposphere) lidar	GSFC TROPOZ (TROPospheric OZone) DIAL	LMOL (Langley Mobile Ozone Lidar)
Affiliation	NASA/JPL	NOAA/ESRL	UAH	NASA/GSFC	NASA/LaRC
Host location	Wrightwood, CA	Boulder, CO	Huntsville, AL	Greenbelt, MD	Hampton, VA
Set-up	Fixed-location	Mobile	Fixed-location	Mobile	Mobile
Transmitter type	Quadruple Nd:YAG pumped Raman laser	Quadruple Nd:YLF pumped Ce:LiCAF laser (tunable wavelength)	Quadruple Nd:YAG pumped Raman laser	Quadruple Nd:YAG pumped Raman laser	Quadruple Nd:YLF pumped Ce:LiCAF laser (tunable wavelength)
Wavelength (nm)	289, 299	Typically 287, 291, 294	283, 289, 299	289, 299	Typically 285, 291 (527 for aerosol)
Receiver size (cm)	91, 5, 5	50	40, 10, 2.5	41, 2.5, 2.5	40
Measurable range (km AGL)	0.1-23	0-3	0.1-12	0.2-12 (day) 0.2-19 (night)	0.1-4
Reference	[4]	[5]	[3, 6]	[7]	[8]

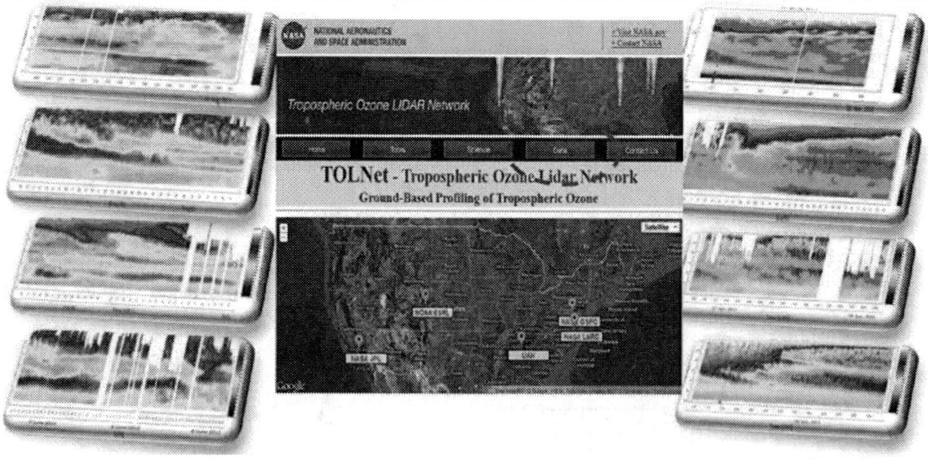

Figure 1. TOLNet webpage and lidar measurement examples. Left ozone curtains are measured in summer and the right curtains are measured in spring/fall.

DOI: 10.1051/epjconf/201611920002

COMPARISON OF LONG TERM TROPOSPHERIC OZONE TRENDS MEASURED BY LIDAR AND ECC OZONESONDES FROM 1991 TO 2010 IN SOUTHERN FRANCE

G. Ancellet [1*], A. Gaudel[1], S. Godin-Beekmann[1]

*[1]LATMOS, UPMC/CNRS/UVSQ, Paris, France *Email:* gerard.ancellet@latmos.ipsl.fr

ABSTRACT

ECC (Electrochemical Concentration Cell) ozonesondes and UV DIAL (DIfferential Absorption Lidar) measurements have been carried out simultaneously at OHP (Observatoire de Haute Provence, 44°N, 6.7°E, 690 m) since 1991. A unique long-term trend assessment by two different instruments operated routinely at the same location is possible. Air mass trajectories have been calculated for all the ozone observations available at OHP. The bias between the seasonal mean calculated with lidar and ECC ozone vertical profiles for 4 time-periods of 5 years is 0.6 ppbv in the free troposphere (4-8 km). Larger differences (> 10 ppbv) are explained by the need for clear sky conditions during lidar observations. The measurements of both instruments have been combined to decrease the impact of short-term atmospheric variability on the trend estimate.

1. INTRODUCTION

Considering the spatial and temporal variations of the tropospheric ozone trend calculated in Northern and central Europe **[1]**, analysis of ozone vertical profiles in Southern Europe is missing. Regular lidar and ECC (Electrochemical Concentration Cell) ozonesondes are available since 1991 and can be used for a long-term trend analysis at OHP (Observatoire de Haute Provence, 44°N, 6.7°E, 690 m)). A comparison of ozone inter-annual variability for these two instruments has not been made previously using such long time series. Furthermore, OHP is a good station to characterize the Mediterranean basin where there is a significant spatial positive anomaly of tropospheric ozone, especially during summer. In this contribution the tropospheric ozone data sets are described including a description of the transport patterns influencing OHP for both ECC and lidar data in section 1. Results of the comparison in terms of ozone inter-annual variability are then given in section 3.

2. METHODOLOGY

Electrochemical ozonesondes have been launched every week since 1991 to measure the ozone vertical profile at OHP using Science Pump Corporation (SPC) for the period 01/1991-03/1997 and ENSCI (Z type) thereafter. Changing ozonesonde manufacturers may have introduced an ozone overestimate of 3% according to the BESOS 2004 campaign **[2]**. Therefore, from March 1997, we have applied a correction of 3\% to the ENSCI data. The time evolution of the sonde normalization factor the DOBSON total ozone measurements at OHP shows a general decrease of the order of 5\% during the 20-year period which may correspond to the use of the ENSCI sondes after 03/1997 The precision of ECC measurement was estimated to be in the range of 5-10\% which corresponds to accuracy better than 5 ppbv **[3]**.

Ozone is also measured between 3 and 14 km with the UV DIAL (DIfferential Absorption Lidar) technique using two wavelengths: 289-299 nm from 1990 to 1993 and 289-316 nm from 1993 until now **[4]**. The latter wavelength pair is generated with a single Raman cell filled with Deuterium and pumped with the 4th harmonic of a Nd-YAG laser. Data are usually recorded during at least one hour just after sunset. The ozone absorption cross sections are taken from Bass and Paur **[5]**. According to several studies, the uncertainty on the absorption cross section is within 1.5\% near 290 nm **[6]**. The upper bound of the lidar measurement range corresponds to a statistical error less than 20\%. In this work, we do not consider data above 8 km and the statistical error always remains less than 10%. The lower bound of the measurement range is chosen to minimize the effect of the overlap function between the laser beam and the telescope field of view. Regarding the

 DOI: 10.1051/epjconf/201611920002

systematic error due to atmospheric interferences, only the effect of Rayleigh extinction is corrected using radiosoundings from the nearby WMO station. The systematic error due to the aerosol interference is not corrected in the free troposphere and it is the major term limiting the lidar accuracy **[7]**. Using error analysis and results from inter-comparison campaigns, the precision of the lidar measurement was estimated to be better than 9\%. The accuracy is of the order of 5 ± 5 ppbv.

The seasonal means are calculated for four periods of 5 years: 1991-1995, 1996-2000, 2001-2005 and 2006-2010. At least 40 profiles by period are used to calculate a seasonal average for each instrument. For both instruments, ozone concentrations are first averaged in two layers of 2 km width in the free troposphere from 4 to 8 km. The top boundary at 8 km avoids taking into account observations above the tropopause. Because lidar data are less reliable below 3 km before 2010, the comparison in the range 2-4 km is not considered.

For each ozone vertical profiles, the transport variability is characterized by three days backward trajectories calculated with FLEXTRA **[8]** for three altitude levels: 700 hPa, 500 hPa, 400 hPa. FEXTRA is initialized with wind fields of ECMWF ERA-interim with a horizontal resolution of 1° and a vertical resolution of 60 model levels. Three day backward trajectories are also calculated at the same level for every day of the 20-year time series in order to characterize the representativeness of our data sampling.

3. RESULTS

The differences between lidar and ECC measurements will be discussed in this section, using seasonal averages and aiming at distinguishing the influence of measurement sampling from differences related to the measurement technique.

According to Figure 1, a maximum in spring/summer and a minimum in winter/fall are seen for both instruments, for the four selected periods. The inter-annual variability of the seasonal means is generally the same for lidar and ECC except in winter/spring at 4-6 km and, in spring and fall at 6-8 km with lidar showing larger variability. It is a bit surprising that lidar with more observations indicates larger inter-annual variability. It may only be explained by some inter-annual variability of the meteorological conditions necessary for the lidar observations. Looking at the seasonal differences of ozone ECC minus lidar, the average ozone difference is about 0.9 ppbv in both layers. The error bars on the differences ECC minus lidar correspond to 2σ confidence interval. Seasonal differences fluctuate between -5 ppbv and +5 ppbv but they can reach 8 or 12 ppbv in 1996-2000 at both levels and in 2006-2009 at 6-8 km. The lowest biases are observed for 1991-1995 and 2001-2005 in both layers.

The analysis of the air mass trajectories for both data records show that the largest differences between the geographical air mass origins sampled by lidar and ECC are found in 1996-1999, in 2001-2005 and in 2006-2009, with ECC being launched more often in air masses coming from the South Western sector. In 1996-2000 ECC database include also more ascent cases than lidar at 400 hPa. Less events with ascending air masses and southerly flow bias low the lidar ozone concentrations in the upper troposphere especially in spring.

Significant transport differences observed in 1996-2000 and 2006-2009 are consistent with the larger ozone seasonal differences (> 10 ppbv) measured by both instruments. It means that a bias is introduced in the lidar database because of the specific meteorological situations required for the lidar measurement. There is also some kind of bias in the ECC database when using weekly observations. Saunois **[9]** show that the sampling uncertainty when using 4 profiles per month must be considered as it is often higher than the measurement accuracy. In 2001-2005, differences of air mass origins between both instruments seems to not influence ozone at OHP. In 1991-1995, no bias of transport is observed. Therefore the ozone bias of 0.6 ppbv found in 1991-1995 can be a good proxy for the impact of the instrumental bias on the 5-year seasonal average.

DOI: 10.1051/epjconf/201611920002

4. CONCLUSIONS

For the first time, ECC and lidar data, available at OHP since 1991, have been compared over a 20 year period. The comparison conducted using this long time series reveals a bias between both measurement types (ECC - lidar) of the order of 0.6 ppbv. This bias is calculated using the seasonal differences for the 5-year period 1991-1995 (50-80 profiles in average per season) for which the air mass transport differences are lowest. This bias agrees with single profile differences of the order of 8\% observed during ECC/lidar inter-comparison campaigns **[3]**. Ozone differences larger than 10 ppbv sometimes found between the two datasets can be explained by meteorological biases resulting from a lack of lidar observations for air masses transported from the Low-latitude North Atlantic. In order to have a data base more representative of the various types of air masses observed at OHP, measurements of both instruments have been combined to calculate the seasonal and annual averages.

ACKNOWLEDGEMENT

The CNRS/INSU, CNES and ADEME are gratefully acknowledged for funding the tropospheric ozone observations at the OHP NDACC station. The engineers operating the lidar and launching ozonesondes at OHP are also acknowledged.

REFERENCES

[1] Logan, J.A., Staehelin, J., Megretskaia, I.A., Cammas, J.P., Thouret, V., Claude, H., De Backer, H., Steinbacher, M., Scheel, H.E., Stübi, R., Fröhlich, M. Derwent, R., 2012: Changes in ozone over europe: Analysis of ozone measurements from sondes, regular aircraft (MOZAIC) and alpine surface sites. J. Geophys. Res. 117.

[2] Deshler, T., Mercer, J.L., Smit, H.G.J., Stubi, R., Levrat, G., Johnson, B.J., Oltmans, S.J., Kivi, R., Thompson, A.M., Witte, J., Davies, J., Schmidlin, F.J., Brothers, G., Sasaki, T., 2008: Atmospheric comparison of electrochemical cell ozonesondes from different manufacturers, and with different cathode solution strengths: The balloon experiment on standards for ozonesondes, *J. Geophys. Research*, 113, 10.1029/2007JD008975.

[3] Smit, H., Straeter, W., Johnson, B., Oltmans, S., Davies, J., Tarasick, D., Hoegger, B., Stubi, R., Schmidlin, F., Northam, T., 2007: Assessment of the performance of ecc-ozonesondes under quasi-flight conditions in the environmental simulation chamber: Insights from the Jülich ozone sonde intercomparison experiment (JOSIE). J. Geophys. Res., 112, D19306.

[4] Ancellet, G. Beekmann, M., 1997: Evidence for changes in the ozone concentrations in the free troposphere over southern france from 1976 to 1995. Atmos. Environ., **31,** 2835-2851.

[5] Bass A.M. And Paur R.J., 1985: The ultraviolet cross-sections of ozone. i. The measurements. ii - results and temperature dependence, *Atmospheric ozone: Proceedings of the Quadrennial Ozone Symposium*, Ghazi (Ed.), 606- 616.

[6] Yoshino,K., Freeman, D.E., Esmond, J.R., Parkinson, W.H., 1988: Absolute absorption cross-section measurements of ozone in the wavelength region 238–335 nm and the temperature dependence, *Planetary and Space Science*, **36(4)**, 395-398.

[7] Ancellet, G., Ravetta, F., 2003: On the usefulness of an airborne lidar for O3 layer analysis in the free troposphere and the planetary boundary layer, *J. Environ. Monit*, **5**, 47-56 ?

[8] Stohl, A., and Koffi., N.E., 1998: Evaluation of trajectories calculated from ECMWF data against constant volume balloon flights during ETEX. *Atmos. Environ.*, **24**, 4151-4156

[9] Saunois, M. and Emmons, L. and Lamarque, J.-F. and Tilmes, S. and Wespes, C. and Thouret, V. and Schultz, M., 2012: Impact of sampling frequency in the analysis of tropospheric ozone observations, *Atmos. Chemistry and Physics*, **12,** 6757-6773.

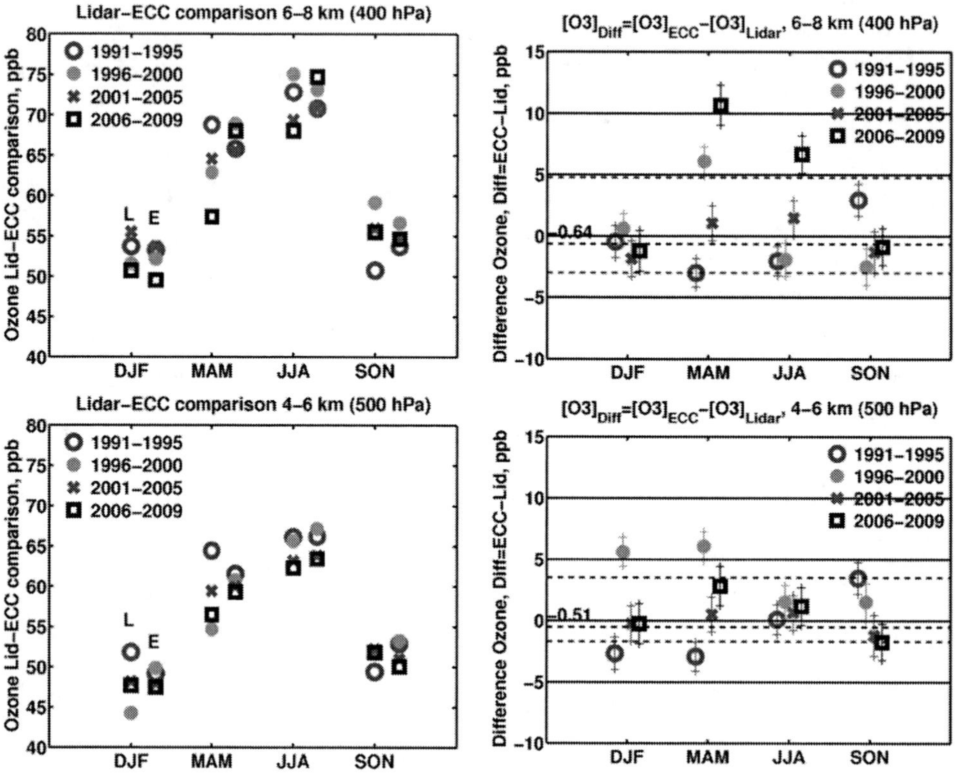

Fig. 1 Seasonal variability of the ozone 5-year average in ppbv for ECC (E) and lidar (L) (left column) and seasonal differences of ozone (ECC minus Lidar) (right column) at 6-8 km (400hPa) (top rank) and 4-6km (500hPa) (bottom rank). Black dotted lines represent the annual average of the seasonal differences of ozone in 1991-1995 taken as the best period to represent instrument differences. Blue dotted lines correspond to the standard deviation (2σ) of all the differences.}

DOI: 10.1051/epjconf/201611920003

LIDAR MEASUREMENTS OF TROPOSPHERIC OZONE IN THE ARCTIC

Jeffrey Seabrook and James Whiteway

Centre of Research in Earth and Space Science, York University, Toronto, Ontario, M3J 1P3
Email: whiteway@yorku.ca

ABSTRACT

This paper reports on differential absorption lidar (DIAL) measurements of tropospheric ozone in the Canadian Arctic during springtime. Measurements at Eureka Weather Station revealed that mountains have a significant effect on the vertical structure of ozone above Ellesmere Island. Ozone depletion events were observed when air that had spent significant time near to the frozen surface of the Arctic Ocean reached Eureka. This air arrived at Eureka by flowing over the surrounding mountains. Surface level ozone depletions were not observed during periods when the flow of air from over the sea ice was blocked by mountains. In the case of blocking there was an enhancement in the amount of ozone near the surface as air from the mid troposphere descended in the lee of the mountains. Three case studies will be shown in the presentation, while one is described in this paper.

1. INTRODUCTION

Lidar systems for tropospheric ozone measurements have been developed at York University for field deployment. One was installed on the Canadian Coast Guard Icebreaker Ship Amundsen [1] and on the Polar-5 (DC-3) aircraft [2] for studies of air chemistry over Arctic sea ice. Another lidar for ground-based measurements of tropospheric ozone was constructed for the Canadian Network for the Detection of Atmospheric Change (CANDAC) and installed as part of the Polar Environment Atmospheric Research Laboratory (PEARL) at Eureka Weather Station on Ellesmere Island in the Canadian Arctic. The lidar was installed in 2007 and operated for several intensive periods until 2012.

The measurement campaigns over sea ice have focused on surface level ozone depletion events (ODEs) that are well known to occur during springtime in the Arctic [3,4,5]. The ozone depletion has been explained as being caused by Bromine atoms that originate from photochemical reactions on snow and fresh sea ice surfaces [6]. The aircraft and ship based lidar measurements over the sea ice have found that the ODEs extend from the surface to heights of a few hundred meters [1,2]. Analyses of air trajectories revealed that ozone depleted air spent significant time in the surface layer over the frozen ocean surface and open leads [1,2].

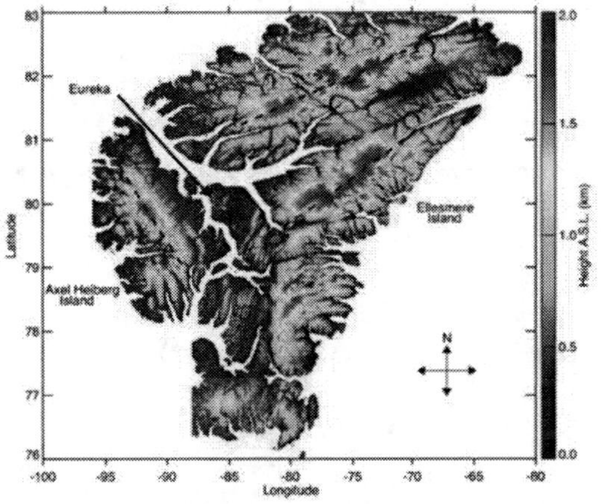

Figure 1. Map of topography in the vicinity of Eureka Weather Station on Ellesmere Island.

The current study primarily concerns ground based lidar measurements of tropospheric ozone over Eureka Weather Station on Ellesmere Island. Figure 1 shows the topography of Ellesmere Island and Axel Heiberg Island, with Eureka indicated on the map at 80° N, 85° W. Air from over the sea ice interacts with various mountain ranges to the west, southeast, east and northeast before it reaches Eureka. It was found that traversal of these mountain ranges had a significant effect on the vertical structure of ozone measured above Eureka. One case study will be presented that involved significant changes in the height distribution of ozone in the lower troposphere over a 6-day period. The observations are interpreted using back trajectory analysis and this has provided an illustration of the effect of mountains on the amount of near surface ozone over land in the Arctic.

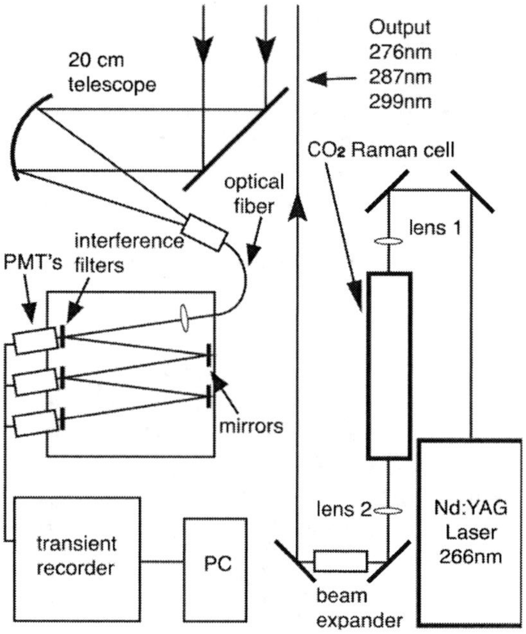

Figure 2. Schematic diagram of the differential Absorption Lidar (DIAL) at Eureka

2. METHOD

The ground based Differential Absorption Lidar (DIAL) instrument used for this study had a design that was essentially equivalent to the system employed for previous ship and aircraft based studies of ozone over sea ice [1,2]. A schematic diagram of the lidar is shown in Fig. 2. The fourth harmonic of a Q-switched Nd:YAG laser (266 nm, 70 mJ per pulse, 20 Hz repetition rate) was focused into a gas cell filled with 140 PSI of CO_2 to generate pulsed light at wavelengths of 276 nm, 287 nm and 299 nm by stimulated Raman scattering [7] and then directed into the atmosphere along the zenith. Backscattered light was collected with a 20 cm off axis parabolic mirror and a 1.5 mm diameter optical fiber positioned in the focal plane, 500 mm from the mirror, to form a receiver field of view of 3.0 mrad. Full overlap between the emitted laser pulse and the telescope field of view occurred at a range of approximately 250 m and the signal received from below that height was not used in the ozone analysis. The three wavelengths (276, 287, and 299 nm) were separated in the receiver using the transmittance and reflectance from interference filters having a bandwidth of 1 nm and tilted at an angle of 5 degrees. The backscatter signals were detected using photomultipliers to generate electrical signals that were recorded using photon counting for the weak signals from distances greater than 1.8 km (typically) and using analog to digital conversion for the strong signals in the near range. The raw data was recorded with a range gate of 7.5 m and an integration period of 1200 laser shots (1 min). During data processing the recorded backscatter signal profiles were averaged both spatially and temporally in order to reduce the measurement uncertainty. The 276/287 nm analog signals were used below a height of 1.8 km and the 287/299 nm photon counting signals were used at greater heights (depending on signal strength).

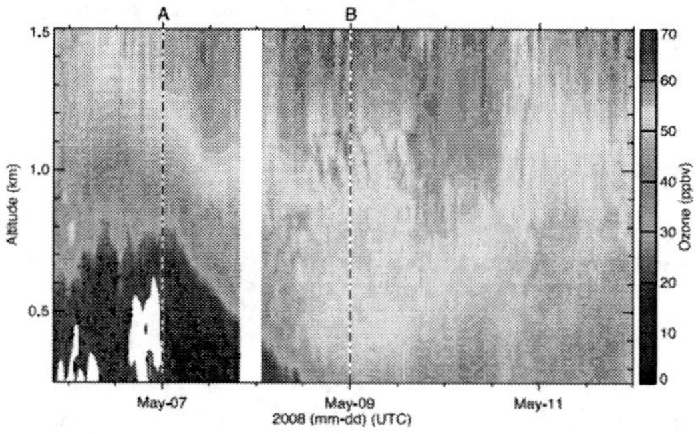

Figure 3. Contour plot of lidar ozone measurements during the period from May 6th to 12th, 2008.

3. OBSERVATIONS AND ANALYSIS

One case will be presented here that illustrates the influence of the surrounding topography on the vertical distribution of ozone. Figure 3 shows a contour of the lidar measurements of ozone vertical distribution during the 6-day period from the 6th to 12th of May in 2008. There was a surface level ozone depletion event (ODE) that extended to a height of 800 m on May 7th when the volume mixing ratio of ozone decreased to less than 10 ppb. The ODE was then replaced after 9 May by air with enhanced ozone content of 60 ppb (in comparison with the background of 40 ppbv) that descended from greater heights.

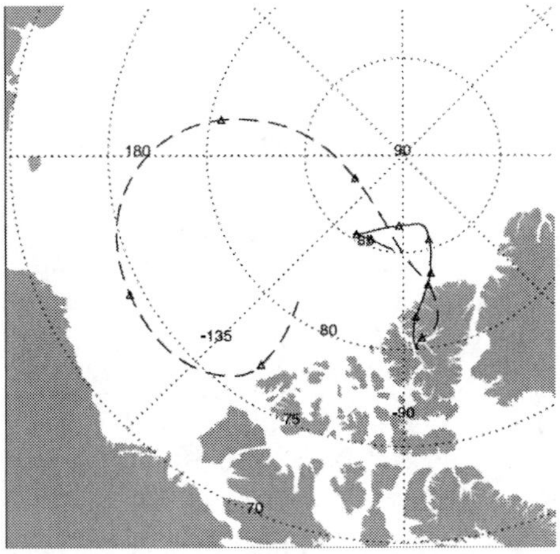

Figure 4. Back trajectories of air arriving at Eureka at a height of 300m above sea level on May 7th (solid) and May 9th (dashed).

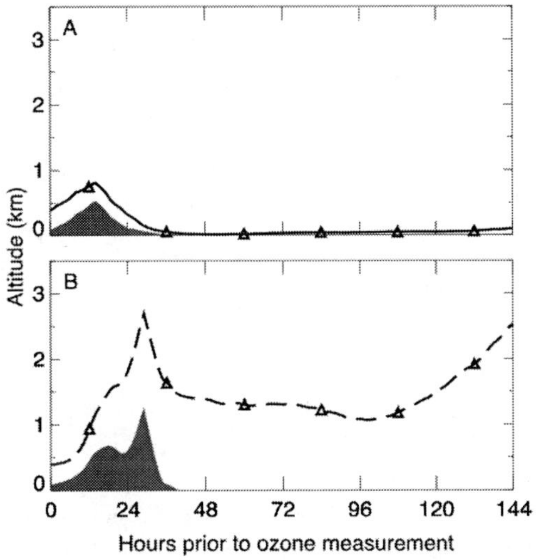

Figure 5. Height along the back trajectories shown in Fig 4 for May 7th (solid) and May 9th (dashed). The shaded region indicates the height of the topography beneath the air trajectory.

The back trajectories for air arriving above Eureka were calculated with the HYSPLIT model [8]. These are shown in terms of geographical position in Fig. 4 for May 7th, when there was an ozone depletion, and for May 9th, when there was an ozone enhancement (points A and B in Fig. 3). In each case the air originated from over the frozen Arctic Ocean. The reason for the transition from ozone depleted air to the ozone enhancement is apparent in Fig. 5, which shows the height along the calculated back trajectories. The air arriving over Eureka on May 7th had spent significant time in contact with sea ice during the previous 6 days. It traversed the mountains on Ellesmere Island prior to arriving at Eureka. The air that was arriving above Eureka on May 9th was not in contact with the surface of the frozen Arctic Ocean during the previous 6 days. This air had descended from heights above 2 km in the lee of the mountains northeast of Eureka.

4. INTERPRETATION

When air near the surface blows inland from the ocean it can flow over a mountain range if the wind speed is sufficient that there is enough kinetic energy to overcome the force of buoyancy and the associated change in potential energy. If the wind speed is too slow it will not be able to flow over the mountain. For an extended mountain range the flow will be blocked and replaced by air flowing down from greater heights on the lee side of the mountains. In this simplified analysis the conditions for flow over the mountain, or blocking, can be differentiated with the Froude number, $Fr = U/Nh$, where U is the wind speed perpendicular to the axis of the mountain range, N is the buoyancy frequency, and h is the change in height required for air to flow over the mountains. The Froude number is the square root of the ratio of kinetic energy to the change in potential energy. When $Fr < 1$ the airflow will be blocked by the mountains, and when $Fr > 1$ the flow will pass over the mountains [9]. The flow conditions determine whether the air at the surface in the lee of the mountains is originating from the sea ice surface, and thus ozone depleted, or from the free troposphere with greater ozone concentration.

The Froude number was calculated using the wind and temperature profiles from the HYSPLIT model at a position on the back trajectories that was upwind of the mountains. For air that was arriving at Eureka during the ozone depletion on May 7th the Froude number upwind of the mountains was 1.30. As this is greater than unity, the wind speed near the sea ice surface was large enough to flow over the mountains. Previous observations have found that air that spends significant time in contact with the sea ice surface is depleted in ozone. This explains why the air that arrived at Eureka on May 7th was depleted in ozone.

The Froude number upwind of the mountains had a value of 0.11 for the air arriving over Eureka on May 9th. With a Froude number less than unity, the flow from over the sea ice was blocked by the mountains. The compensating descent in the lee of the mountains brought air from heights above 2 km with greater ozone content.

5. CONCLUSION

A differential absorption lidar (DIAL) for measurements of tropospheric ozone was operated at Eureka Weather Station on Ellesmere Island in the Canadian Arctic. Large variations in the vertical distribution of ozone were observed during spring that are associated with surface level ozone depletion events over sea ice and the effects of mountains on atmospheric flow patterns.

ACKNOWLEDGEMENTS

This work was carried out as part of the Canadian Network for the Detection of Atmospheric Change (CANDAC). Funding was provided by the Canadian Natural Sciences and Engineering Research Council (NSERC) Discovery program, the NSERC International Polar Year (IPY) Program, the Canadian Foundation for Climate and Atmospheric Sciences (CFCAS), and the Canadian Foundation for Innovation (CFI).

REFERENCES

[1] Seabrook, J., J. Whiteway, J., R. Staebler, J. Bottenheim, L. Komguem, L. Gray, D. Barber, and M. Asplin, 2011: LIDAR measurements of Arctic boundary layer ozone depletion events over the frozen Arctic Ocean, *J. Geophys. Res.*, **116,** doi:10.1029/2011JD016335.

[2] Seabrook, J., J. Whiteway, L. Gray, R. Staebler, A. Herber, 2013: Airborne lidar measurements or surface ozone depletion over Arctic sea ice, *Atmos. Chem. Phys.,* **13**, 1 – 7.

[3] Oltmans, S. and W. Komhyr, 1986: Surface ozone distributions and variations from 1973-1984 measurements at the NOAA geophysical monitoring for climatic change baseline observatories, *J. Geophys. Res.*, **91**, 5229 – 5236.

[4] Bottenheim, J., A. Gallant, and K. Brice, 1986: Measurements of NOy species and O3 at 82 N latitude, *Geophys. Res. Lett.*, **13**, 113 – 116.

[5] Barrie, L., J. Bottenheim, R. Schnell, P. Crutzen, and R. Rasmussen, 1988: Ozone destruction and photochemical reactions at polar sunrise in the lower Arctic atmosphere, *Nature*, 334, 138 – 141.

[6] Simpson, W.R. et al., 2007: Halogens and their role in polar boundary-layer ozone depletion, *Atmos. Chem. Phys.*, **7**, 4375 – 4418.

[7] Nakazato, M., T. Nagai, T. Sakai, Y. Hirose, 2007: Tropospheric ozone differential-absorption lidar using stimulated Raman scattering in carbon dioxide, *Appl. Opt.*, **46**, 2269 – 2279, doi:10.1364/AO.46.002269.

[8] Draxler, R. and G. Hess, 1998: An overview of the HYSPLIT 4 modeling system for trajectories, dispersion, and deposition, *Australian Meteorological Magazine*, **47**, 295 – 308.

[9] Wang, C., G. Chen, 2003: On the formation of leeside mesolows under different Froude number flow regime in TAMEX, *J. Met. Soc. Japan,* **81**, 339 – 365.

DOI: 10.1051/epjconf/201611920004

AIRBORNE LIDAR MEASUREMENTS OF POLLUTION ABOVE THE OIL SANDS REGION IN NORTHERN ALBERTA

Monika Aggarwal[1]*, James Whiteway[1], Jeffrey Seabrook[1], Lawrence Gray[1], Kevin B. Strawbridge[2]

[1]*Centre for Research in Earth and Space Science, York University, Toronto, Ontario, M3J 1P3,*
**Email: aggarwal@yorku.ca*
[2]*Air Quality Processes Research Section, Environment Canada, Toronto, Ontario, M3H 5T4*

ABSTRACT

Lidar measurements of ozone and aerosol were conducted from a Twin Otter aircraft above the oil sands region of northern Alberta. For the majority of the flights, significant amounts of aerosol were observed within the boundary layer, up to an altitude of 2.0 km above sea level (ASL), while the ozone concentration remained at background levels (30-45 ppb) downwind of the industry. On August 24th the lidar measured a separated layer of aerosol above the boundary layer, at a height of 2.0 km ASL, in which the ozone mixing ratio increased to 70 ppb. Backward trajectory calculations revealed that the air containing this separated aerosol layer had passed over an area of forest fires. Directly below the layer of forest fire smoke, pollution from the oil sands industry was observed. Measurements of the backscatter linear depolarization ratio were obtained with a ground based lidar operated by Environment Canada within the oil sands region. The depolarization measurements aided in discriminating between the separate sources of pollution from industry and forest fires. The depolarization ratio was 5-6% in forest fire smoke and 7-10% in the industrial pollution.

1. INTRODUCTION

A lidar system for tropospheric ozone and aerosol measurements was developed at York University for installation on aircraft [1,2]. This was installed on a Twin Otter aircraft in August 2013 for participation in a field campaign that was designed to study the air pollution from the oil sands extraction activities north of Fort McMurray, Alberta.

The chemical species of interest for the measurements was Ozone (O_3). Ozone is a secondary pollutant as it is produced from the photochemical reactions between NO_x and VOCs in the presence of sunlight. Typically the background level of ozone in the troposphere is between 30 and 40 ppb. The exposure of crops and living organisms to ozone concentrations higher than the background levels can result in damage to biological tissue and decrease the rate of photosynthesis in plants [3].

The Twin Otter aircraft provided a platform for nadir directed lidar ozone and aerosol measurements above and downwind from the oil sands industry. The field campaign consisted of a total of five flights out of Fort McMurray during the period between August 22 and August 26, 2013. The flights were planned using meteorological analysis so that each flight path intersected the trajectory of air that had passed over the oil sands. The results from flight segments on August 23 and 24, 2013 are shown in this paper.

2. MEASUREMENT TECHNIQUE

A schematic diagram of the lidar is shown in Fig. 1. The fourth harmonic of a Q-switched Nd:YAG laser (266 nm, 70 mJ per pulse, 20 Hz repetition rate) was focused into a gas cell filled with 140 PSI of CO_2 to generate pulsed light at wavelengths of 276 nm, 287 nm and 299 nm by stimulated Raman scattering [4] and then directed into the atmosphere along the zenith. The second harmonic output from the laser at a wavelength of 532 nm was also directed into the atmosphere. Backscattered light was collected with a 15 cm off axis parabolic mirror and two separate 1.5 mm diameter optical fibers positioned in the focal plane, 500 mm from the mirror, to form two

DOI: 10.1051/epjconf/201611920004

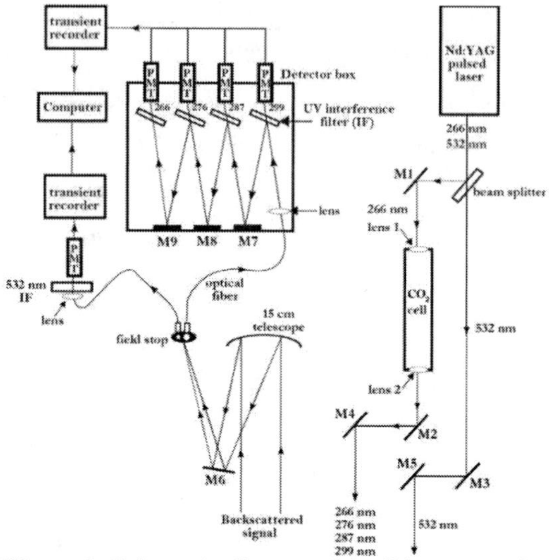

Figure 1: Schematic diagram of the lidar system for aircraft based measurements of aerosol and ozone.

separate receiver fields of view, each of width 3.0 mrad. The 532 nm signal was aligned to one of the optical fibers, while the UV wavelengths were aligned to the other fiber. The four UV wavelengths (266, 276, 287, and 299 nm) were separated in the receiver using the transmittance and reflectance from interference filters having a bandwidth of 1 nm and tilted at an angle of 5 degrees. The backscatter signals were detected using photomultipliers to generate electrical signals that were recorded using photon counting for the weak signals from distances greater than 1.8 km and using analog to digital conversion for the strong signals in the near range. The raw data was recorded with a range gate of 3.75 m for the 532 nm signal and 7.5 m for the UV signals. The temporal integration was 10 seconds. During data processing the recorded backscatter signal profiles were averaged both spatially and temporally in order to reduce the measurement uncertainty.

The ozone concentration was derived by using the Differential Absorption Lidar (DIAL) technique [5]. For the lidar derived ozone measurements presented in this paper, the analog measurements of the 276/299 wavelength pair were used closer to the altitude of the aircraft and were combined with the photon counting measurements of the 266/299 wavelength pair for altitudes closer to the ground. Regions where extreme aerosol loading or clouds interfered with the ozone retrieval have been omitted in the analysis.

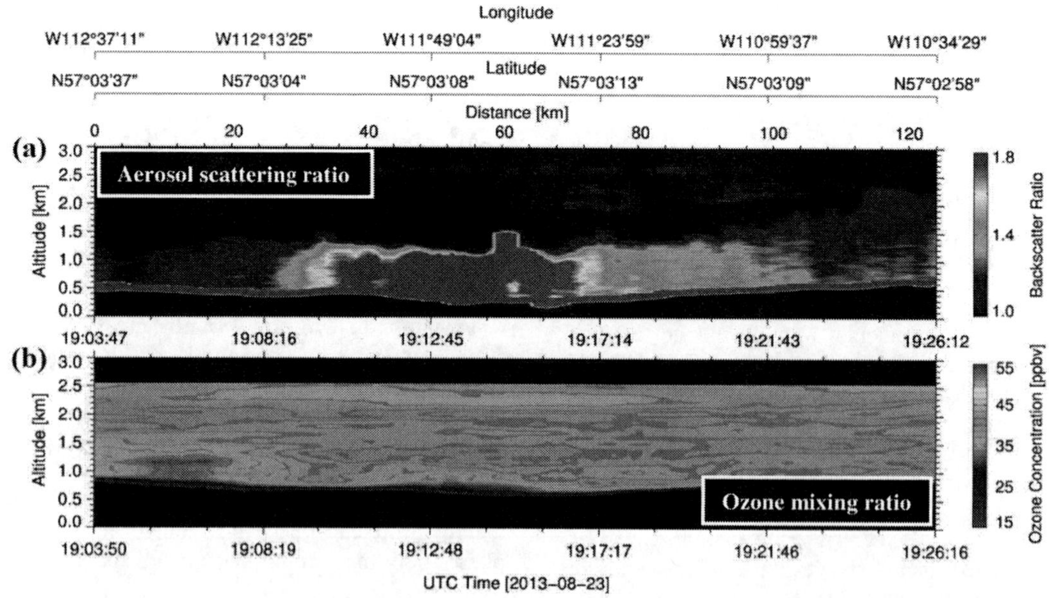

Figure 2: **(a)** The aerosol backscatter ratio and **(b)** the ozone mixing ratio derived from the lidar measurements for a flight segment on August 23, 2013. The height is above sea level.

DOI: 10.1051/epjconf/201611920004

3. OBSERVATIONS

3.1 Flight 2: August 23, 2013

Measurements acquired during one leg on the flight of 23 August 2013 are shown in Fig 2. The flight path passed across the oil sands extraction industry from downwind (east) to upwind (west) at a constant latitude of N57°03'. In Fig. 2(a), significant amounts of aerosol were observed over the industry and the height of the boundary layer is 1.5 km ASL. The corresponding ozone concentration along this flight segment is shown in Fig. 2(b). The ozone concentration remained consistent with background levels throughout the flight leg, with values ranging between 35 and 45 ppb. An increase in ozone concentration was not observed downwind from the industry.

3.2 Flight 4: August 24, 2013

Airborne lidar measurements to the north of Fort McMurray were made on August 24th and the corresponding flight path is shown in Fig. 3.

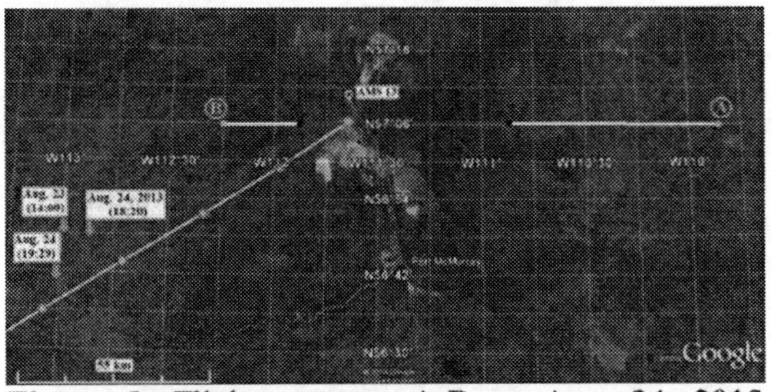

Figure 3: Flight segment A-B on Aug. 24, 2013 (produced on Google Earth). The red section along a flight leg represents the region where forest fire smoke was observed. The forest fires are depicted by orange arrows with the estimated start time beside each fire. A backward trajectory was initiated from an altitude of 2.0 km ASL at a time of 21:00 UTC on Aug. 24, 2013. The round marks along the trajectory represent a time interval of one hour. The green flag indicates the location of Environment Canada's ground based lidar.

In this flight leg, the Twin Otter started at point A (a distance of ~90 km to the east of the oil sands industry) and travelled westbound to point B along a constant latitude of N57°06'. Along this flight segment, the lidar observed a separated layer of material at an altitude of 2.0 km ASL and pollution underneath as shown in Fig.4.

By using the HYSPLIT model [6], a backward air trajectory was initiated from an altitude of 2.0 km ASL at a time and location the lidar observed the separated aerosol layer. Figure 3 shows that the measured air passed over forest fires to the southwest of the flight track. In Fig. 4(a), the concentration of ozone in the forest fire smoke layer was measured to be up to 70 ppb. This is consistent with previous observations of in situ ozone measurements in North American forest fires [7].

Environment Canada's ground based lidar system observed the same layer of forest fire smoke at an altitude of ~1.8 km ASL. In Fig. 4(c), the depolarization ratio in the forest fire smoke layer was measured to be 5-6%. The depolarization ratio in the pollution from the oil sands industry at heights below 1.5 km had a larger depolarization ratio with values between 7% and 10%. This provides further evidence that the layer with enhanced ozone had originated from forest fires rather than industrial pollution.

4. CONCLUSION

Air pollution above the Alberta oil sands region was studied with an airborne lidar system. Significant amounts of aerosol were observed up to an altitude of 2.0 km ASL, in which the ozone concentration remained at background levels (35–45 ppb). On the afternoon of August 24th, a layer of forest fire smoke was observed and this contained ozone concentrations of up to 70 ppb, which is significantly greater than the background.

ACKNOWLEDGEMENT

Financial support for this study was provided by FedDev Ontario, Environment Canada, the Natural Sciences and Engineering Research Council of Canada (NSERC), and the Canadian Foundation for Innovation (CFI). The authors would like to thank Mr. Cordy Tymstra from the Fort McMurray wildfire branch (under the Environment and Sustainable Resource Development of Alberta) for providing the forest fire locations.

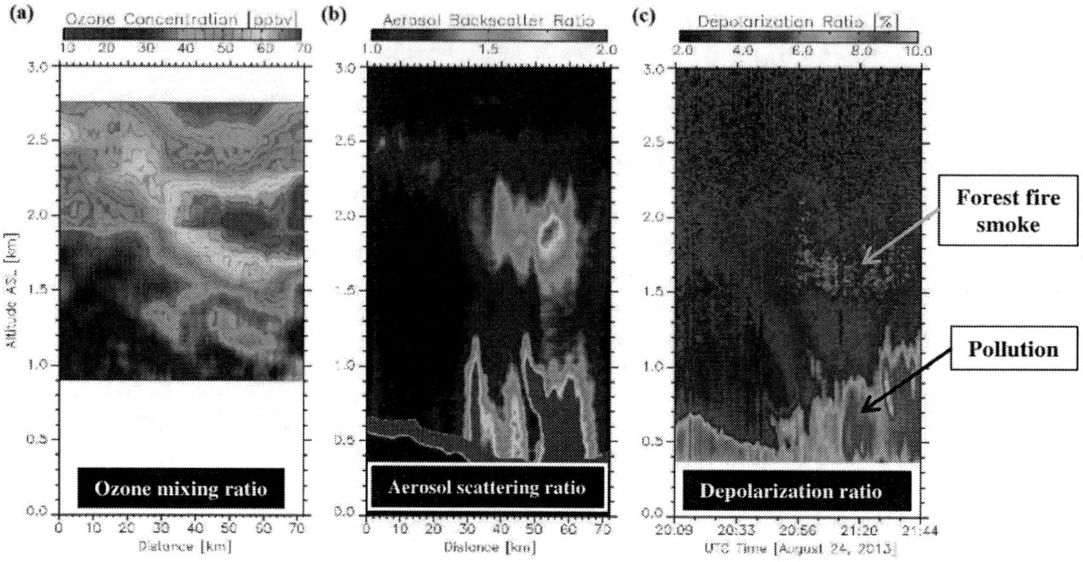

Figure 4: (a) The ozone mixing ratio and **(b)** the aerosol backscatter ratio derived from the lidar measurements along a section of the flight segment A-B (as shown in Fig. 3) where forest fire smoke was observed. Distance is along the red section of the flight segment in Fig. 3. **(c)** Depolarization ratio measurements taken by Environment Canada's ground based lidar system located at 57.15° N, 111.6° W. This site is indicated as AMS 13 in Fig. 3.

REFERENCES

[1] Seabrook, J., J. Whiteway, J., R. Staebler, J. Bottenheim, L. Komguem, L. Gray, D. Barber, and M. Asplin, 2011: LIDAR measurements of Arctic boundary layer ozone depletion events over the frozen Arctic Ocean, *J. Geophys. Res.*, **116,** doi:10.1029/2011JD016335.

[2] Seabrook, J., J. Whiteway, L. Gray, R. Staebler, A. Herber, 2013: Airborne lidar measurements or surface ozone depletion over Arctic sea ice, *Atmos. Chem. Phys.*, **13**, 1 – 7.

[3] Morgan, P. B., E. A. Ainsworth, S. P. Long, 2003: How does elevated ozone impact soybean? A meta-analysis of photosynthesis, growth and yield, *Plant, Cell and Environ.*, **26**, 1317 – 1328.

[4] Nakazato, M., T. Nagai, T. Sakai, Y. Hirose, 2007: Tropospheric ozone differential-absorption lidar using stimulated Raman scattering in carbon dioxide, *Appl. Opt.*, **46**, 2269 – 2279, doi:10.1364/AO.46.002269.

[5] Kovalev V. A., and W. E. Eichinger, 2004: *Elastic Lidar: Theory, Practice, and Analysis Methods*, New Jersey, USA: John Wiley & Sons.

[6] Draxler, R.R. and G. D. Rolph. HYSPLIT (HYbrid Single-Particle Lagrangian Integrated Trajectory) Model access via NOAA ARL READY Website (http://www.arl.noaa.gov/HYSPLIT.php). NOAA Air Resources Laboratory, College Park, MD

[7] Aggarwal, M., 2011: *Lidar measurements of forest fire smoke and stratospheric aerosol*, Master of Science thesis, York University, Toronto, Canada.

USING OZONE LIDAR TO INVESTIGATE SOURCES OF HIGH OZONE CONCENTRATIONS IN THE WESTERN UNITED STATES

C. J. Senff[1,2], A. O. Langford[2], R. J. Alvarez II[2], Wm. A. Brewer[2], R. M. Banta[2], R. D. Marchbanks[1,2], S. P. Sandberg[2], A. M. Weickmann[1,2], J. S. Holloway[1,2], E. J. Williams[2]

[1]*CIRES, University of Colorado, Boulder, CO 80309, USA, *Email: christoph.senff@noaa.gov*

[2]*NOAA/ESRL/CSD, Boulder CO 80305, USA*

ABSTRACT

We have used NOAA's Tunable Optical Profiler for Aerosol and oZone (TOPAZ) ozone lidar to investigate the sources of high surface ozone concentrations in two different regions of the western United States (US): the Uintah Basin in northeast Utah and Clark County in southern Nevada, which includes the city of Las Vegas. The Uintah Basin is a booming oil and gas producing region that often suffers from very high wintertime ozone concentrations. Clark County experiences violations of the US ozone standard primarily in spring and early summer despite a lack of any major local pollution sources. TOPAZ lidar observations, in conjunction with surface in situ measurements and model results, provided strong evidence that the high wintertime ozone concentrations in the Uintah Basin are primarily driven by local emissions associated with oil and gas exploration, whereas the Clark County ozone exceedances are often caused by ozone-rich air that is transported from the lower stratosphere all the way down to the earth's surface.

1. INTRODUCTION

Atmospheric ozone concentrations in the US have generally decreased over the last two decades, especially in large cities, due to effective controls of ozone precursor emissions. However, the recent boom in oil and natural gas exploration in the US has led to an increase in regional ozone precursor emissions. These emissions are often poorly regulated and have been implicated in raising ambient ozone levels, in some cases well above the US ozone National Ambient Air Quality Standard (NAAQS), which is currently set at 75 ppbv for 8-hour average ozone. Exceedances of the ozone NAAQS have also been linked to stratosphere-to-troposphere transport (STT) of ozone and, to a lesser degree, transport of ozone from Asia. Observations and modeling studies have indicated that high-ozone layers of stratospheric or Asian origin can be transported from the free troposphere to the surface and cause a deterioration of air quality. This is especially true for the US Intermountain West, where elevated terrain and deep atmospheric boundary layers facilitate transport of free tropospheric air to the surface. These processes are bound to cause more ozone exceedances in the future as springtime ozone background concentrations over the western US are slowly increasing and the ozone NAAQS will likely be lowered to between 65 and 70 ppbv. This will effectively shrink the margin of ozone than can be added to the background before an exceedance occurs.

We deployed the TOPAZ ozone lidar, together with NOAA's High Resolution Doppler Lidar (HRDL), a large array of in situ instruments, and other remote sensors, at the Uintah Basin Wintertime Ozone Study (UBWOS) in January and February 2013 to investigate the sources of high surface ozone concentrations that had been observed during the winter months in the Uintah Basin. The main objective of the lidar deployment was to determine whether long-range transport of ozone, STT, or emissions from a nearby power plant could cause the high surface ozone levels, or if emissions from the local oil and gas extraction operations were the main source. Another goal was to characterize the vertical extent of the high ozone concentrations.

We also used TOPAZ, along with a smaller complement of in situ sensors, during the Las Vegas Ozone Study (LVOS) in May and June 2013 to investigate the causes for ozone NAAQS exceedances that occur in Clark County, Nevada during spring and early summer [1]. The primary objective was to assess the seasonal influence of STT and long-range transport from Asia on surface ozone concentrations. Secondary goals included the characterization of local photochemical production and regional transport of ozone from the Los Angeles Basin, and

 DOI: 10.1051/epjconf/201611920005

understanding the impacts of wildland fires on ozone levels in the Las Vegas area.

2. METHODOLOGY

The TOPAZ ozone lidar, deployed during both UBWOS and LVOS, is based on an all-solid-state transmitter and emits three tunable ultraviolet (UV) wavelengths between 285 and 300 nm [2]. Originally conceived as an airborne, nadir-pointing instrument, TOPAZ was converted to an up-looking, truck-based system in early 2012 [3] (Fig. 1). A roof-mounted two-axis scanner permits pointing of the laser beam at several elevation angles at a fixed but changeable azimuth angle. Zenith operation is achieved by moving the scanner mirror out of the laser beam path. A typical scan sequence consists of observations at elevation angles of 2, 6, 10, and 90 degrees. The ozone profiles from the various elevation angles are spliced together to create composite vertical profiles extending from 15 m up to about 3 km AGL (Fig. 2). The time resolution of the composite ozone profiles is typically 5 – 15 minutes and their effective vertical resolution increases with altitude from 3 to 90 m. The TOPAZ lidar is unique because it allows profiling of ozone very close to the surface at high spatial resolutions, well below the minimum range of comparable zenith-looking lidars.

Figure 1: TOPAZ truck with roof-mounted scanner.

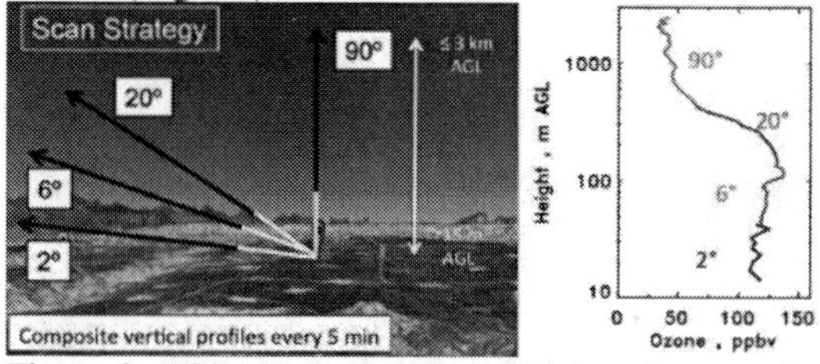

Figure 2: TOPAZ scan pattern (right) and composite vertical ozone profile (left).

The HRDL Doppler wind lidar, deployed during the UBWOS campaign, provides profiles of horizontal wind speed and direction, horizontal and vertical turbulence, and boundary layer height typically at a time resolution of 20 minutes [4]. Wind profiles retrieved with HRDL in scanning mode are measured to within 5 meters of the surface and with 5-meter vertical resolution.

During LVOS, surface ozone and carbon monoxide (CO) concentrations were determined at 1-min resolution using a commercial UV-absorbance monitor and a modified commercial vacuum UV-fluorescence monitor, respectively. A GPS-enabled weather station recorded 5-min averaged winds, temperature, and relative humidity from a 3-m mast. The LVOS measurement activities were guided by forecasts and analyses from the NOAA/ESRL/CSD FLEXPART particle dispersion model and the NOAA/NESDIS/CIMSS Real Time Air Quality Modeling System (RAQMS). Interpretation of the measurements was aided by support from the NOAA/ESRL/GFDL AM3 chemistry-climate model.

3. RESULTS

3.1 UBWOS study

The Uintah Basin in northeast Utah is about 120 x 80 kilometers in size, almost completely surrounded by mountain ranges, and contains more than 10,000 active oil and gas wells (Fig. 3).

Figure 3: Map with the topography of the Uintah Basin in northeast Utah.

The TOPAZ and HRDL lidars, along with a comprehensive suite of chemistry and aerosol in situ sensors were situated at the Horse Pool site in the eastern part of the Basin. Normally, the ground in the Uintah Basin is snow covered throughout the winter and a persistent, shallow cold-pool layer of air forms in the Basin. A strong temperature inversion capping this cold-pool layer coupled with very light winds causes a buildup of pollutants in the Basin over several days, which can lead to very high surface ozone concentrations of up to 150 ppbv. These ozone episodes come to an end when high winds and strong mixing associated with passing storm systems dilute and flush out the high ozone concentrations. Figure 4 shows an ozone time-height cross section

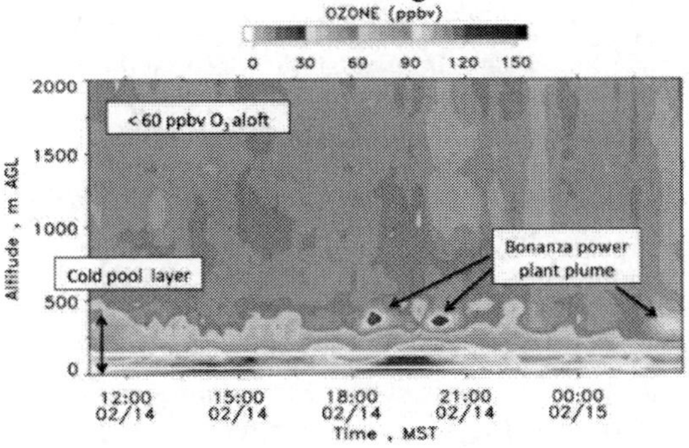

Figure 4: Ozone structure observed with the TOPAZ lidar on February 14-15, 2013.

observed with the TOPAZ lidar over a 16-hour period on February 14-15 2013. The ozone structure observed on that day was typical for all high ozone days encountered during the UBWOS study. The lidar observations show that the high ozone concentrations were confined to the cold-pool layer, which was about 400 m deep. The plume from the nearby Bonanza power plant (Fig. 3) manifests itself in the lidar observations as low-ozone patches (Fig. 4) because of titration of ambient ozone by power plant NOx. The power plant plume stayed above the temperature inversion at the top of the cold-pool layer and did not get mixed down to the surface. Therefore, the Bonanza power plant emissions were not part of the surface ozone precursor mix. Ozone concentrations above the cold-pool layer were below 60 ppbv at all times, which is consistent with free-troposphere background ozone concentrations. Thus, the lidar observations showed no evidence of long-range transport of ozone aloft or STT that could have caused the high surface ozone levels.

3.2 LVOS study

During the LVOS study, the TOPAZ lidar and the ozone, CO, and water vapor in situ instruments were located about 45 km west of Las Vegas on top of Angel Peak at about 2700 m MSL (Fig. 5).

Figure 5: Map of LVOS study area.

TOPAZ collected approximately 200 hours of data on 26 days during May/June 2013. Fig. 6 depicts the ozone lidar and surface in situ observations on May 24-25. The lidar data show three high-ozone filaments sloping downward toward the surface, two of which reached the

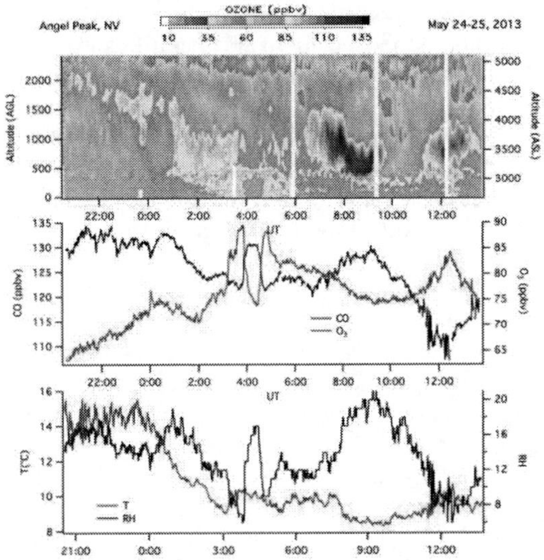

Figure 6: Vertical ozone distribution observed with TOPAZ on May 24-25, 2013 and coincident surface ozone, carbon monoxide, relative humidity, and air temperature measurements.

ground. At the same time, the in situ sensors show a simultaneous increase in surface ozone and a decrease in CO and relative humidity, indicating that these ozone filaments originated in the lower stratosphere. These measurements show the first ever example of a stratospheric intrusion being

 DOI: 10.1051/epjconf/201611920005

tracked continuously to the surface by lidar and then directly detected on the ground by in situ measurements. These observations were corroborated by FLEXPART back trajectory calculations and AM3 as well as RAQMS model analyses. Nocturnal drainage flows transported the high-ozone air into the Las Vegas valley, leading to exceedances of the ozone NAAQS on the following day. During LVOS, we observed two other cases where high-ozone air of stratospheric origin reached the surface at Angel Peak and subsequently caused violations of the ozone standard in Clark County. Contributions to elevated surface ozone from wildland fires and pollution plumes from the Los Angeles Basin were of secondary importance during the LVOS campaign and did not result in any ozone NAAQS exceedances.

4. CONCLUSIONS

TOPAZ lidar measurements from the UBWOS campaign revealed that neither STT, nor long-range transport of ozone, nor power plant emissions contributed in any significant way to the high wintertime surface ozone concentrations in the Uintah Basin. Our observations strongly support the conclusion, also reached by the other UBWOS investigators, that the local emissions associated with the oil and gas exploration are the primary driver of the high surface ozone levels in the Uintah Basin.

During LVOS, TOPAZ lidar observations, in conjunction with the collocated in situ measurements and supported by model results, provided compelling evidence that STT significantly increased surface ozone concentrations in the Las Vegas area during the late spring and early summer of 2013. Both measurements and model analyses suggest that ozone of stratospheric origin directly contributed to the exceedances of the ozone NAAQS in Clark County on three days in May and June 2013.

ACKNOWLEDGEMENT

We would like to acknowledge funding support from the NOAA Health of the Atmosphere and Climate Programs, the NASA Tropospheric Ozone Lidar Network, the Utah Department of Environmental Quality, the Western Energy Alliance, and the Clark County Department of Air Quality.

REFERENCES

[1] Langford, A. O., C. J. Senff, C. J., R. J. Alvarez II., J. Brioude, J., O. R. Cooper, J. S. Holloway, M. Y. Lin, R. D. Marchbanks, R. B. Pierce, S. P. Sandberg, A. M. Weickmann, E. J. Williams, 2014: An overview of the 2013 Las Vegas Ozone Study (LVOS): Impact of stratospheric intrusions and long-range transport on surface air quality, Atmospheric Environment (2014), doi: 10.1016/j.atmosenv.2014.08.040.

[2] Alvarez, R. J., et al, 2011: Development and application of a compact, tunable, solid-state airborne ozone lidar system for boundary layer profiling, *J. Atmos. Ocean. Technol.*, doi: 10.1175/JTECH-D-10-05044.1

[3] Alvarez, R. J., Senff, C. J., A. M. Weickmann, S. P. Sandberg, A. O. Langford, R. D. Marchbanks, W. A. Brewer, and R. M. Hardesty, 2012: Reconfiguration of the NOAA TOPAZ Lidar for Ground-based Measurement of Ozone and Aerosol Backscatter, Proceedings of the 26th International Laser Radar Conference, Porto Heli, Greece, 25-29 June 2012, pp. 249-252.

[4] Grund, C. J., R. M. Banta, J. L. George, J. N. Howell, M. J. Post, R. A. Richter, and A. M. Weickmann, 2000: High-resolution Doppler lidar for boundary layer and cloud research, J. Atmos. Ocean. Technol., **18**, 376-393.

EPJ Web of Conferences **119**, 21001 (2016)
ILRC 27
DOI: 10.1051/epjconf/201611921001

GLOBAL AEROSOL DIRECT RADIATIVE EFFECT FROM CALIOP AND C3M

Dave Winker[1*], Seiji Kato[1], and Jason Tackett[2]

*[1]NASA Langley Research Center, Hampton, VA, 23681 *email: david.m.winker@nasa.gov*

[2]SSAI, Hampton, VA, 23666

ABSTRACT

Aerosols are responsible for the largest uncertainties in current estimates of climate forcing. These uncertainties are due in part to the limited abilities of passive sensors to retrieve aerosols in cloudy skies. We use a dataset which merges CALIOP observations together with other A-train observations to estimate aerosol radiative effects in cloudy skies as well as in cloud-free skies. The results can be used to quantify the reduction of aerosol radiative effects in cloudy skies relative to clear skies and to reduce current uncertainties in aerosol radiative effects.

1. INTRODUCTION

Aerosols affect the Earth's radiation budget by scattering sunlight back to space and by absorbing sunlight before it reaches the surface. Model estimates of global aerosol radiative effects in cloudy-skies depend in part on cloud albedo and on the relative vertical distribution of aerosol and cloud. This represents a major source of uncertainty in model estimates of aerosol forcing, as the aerosol vertical distribution in models is poorly constrained and there is a large diversity between models[1]. Additionally, global models do a poor job at predicting the cloud cover and top height of shallow warm clouds[2]. On the other hand, published estimates of global aerosol radiative effects based on satellite observations (e.g. [3]) have been hampered by a reliance on passive satellite sensors, which are largely restricted to aerosol retrievals in clear-sky conditions. Assumptions or model estimates are then required to extend the clear-sky estimates to all-sky conditions. This limitation is particularly important for aerosol located above bright, low clouds. These estimates based on satellite observations consistently predict larger aerosol cooling than model based estimates. There are many factors contributing to this discrepancy, but a large part remains unexplained. These uncertainties can be reduced using observations from the CALIOP lidar, on the CALIPSO satellite, which retrieves the vertical distribution of aerosol extinction in both clear and cloudy skies[4]. Chand et al. [5] and have used observations from CALIOP and MODIS in a regional study of the radiative effects of aerosol above cloud. Oikawa et al. [6], using a different approach, produced the first global all-sky estimate. Here, using a different approach from [5] and [6], we combine observations from CALIPSO and the A-train to estimate global aerosol radiative effects in both clear and cloudy skies.

2. DIRECT RADIATIVE EFFECT

Direct aerosol radiative forcing (DARF), as it is usually defined, refers to the radiative effect of anthropogenic aerosol in the present data relative to pre-industrial conditions. This inherently requires the use of models to estimate aerosol radiative effects in pre-industrial conditions. Here, we estimate the aerosol direct radiative effect (DRE), defined as the perturbation in diurnally averaged radiative fluxes due to both natural and anthropogenic aerosol. This quantity is more directly observable and is also more relevant to the radiative impact of aerosol on the present-day Earth radiation budget.

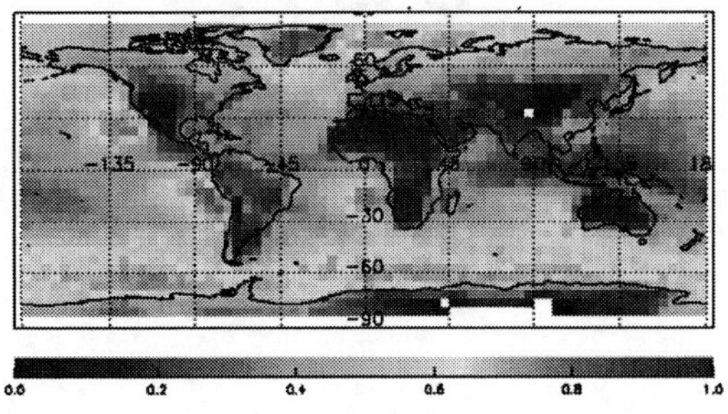

Figure 1. Mean low cloud amount from CALIOP, 2008 annual average.

Significant amounts of aerosol are located above low clouds in certain regions. Figure 1 shows the global distribution of low clouds, having tops typically less than 2 km. Figure 2 shows aerosol scale heights from the CALIPSO Level 3 aerosol product[7] and Figure 3 shows AOD of aerosol below thin clouds and above low clouds.

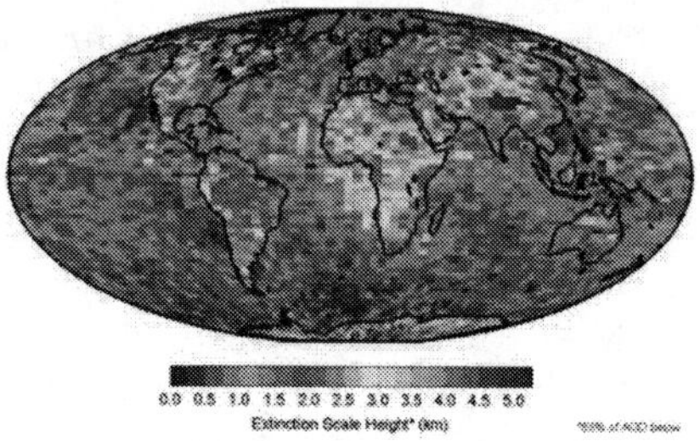

Figure 2. Aerosol scale height, defined as the height at which 63% of the AOD lies below, for August 2008.

Figure 3 shows regions where significant aerosol is located in cloudy columns. Three primary regions are seen in August 2008 where significant AOD is located above low opaque clouds: the Atlantic dust transport region, west of the coast of southern Africa, and the Arabian sea.

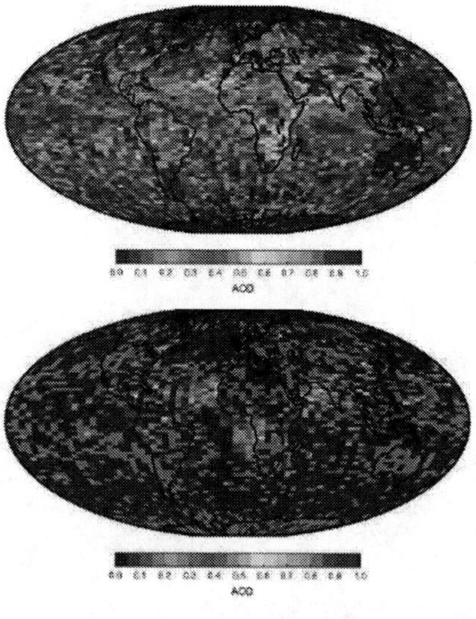

Figure 3. Upper panel: AOD in columns containing optically thin cloud. Lower panel, AOD above low opaque cloud

In addition to aerosol extinction at 532 nm, calculation of aerosol DRE requires aerosol spectral properties, cloud properties, surface albedo, and radiative transfer calculations. Our estimates are based on the CERES-MODIS-CALIPSO-CloudSat (C3M) product[7]. The C3M product contains profiles of SW and LW irradiance computed from instantaneous, collocated MODIS aerosol/cloud data and profiles from CALIPSO and CloudSat, all matched to CERES footprints. Aerosol extinction profiles in both clear and cloudy skies come from CALIOP. Aerosol absorption is estimated using information on aerosol type and optical properties from a combination of CALIOP and the MATCH global aerosol model. Aerosol absorption for the various aerosol species is based on the OPAC optical properties database[8]. Diurnally averaged fluxes are derived and used to estimate annual and seasonal mean DRE.

3. RESULTS

Diurnally averaged fluxes have been derived from C3M and used to estimate annual and seasonal mean DRE. Table 1 gives annual mean shortwave (SW) aerosol DRE at the top of the atmosphere (TOA). Cloudy-sky DRE is derived from all-sky and clear-sky DRE using:

$$DRE_{cldy} = DRE_{all} - (1 - A_c)\, DRE_{clear} / A_c$$

where A_c is the cloud fraction.

Table 1. 2008 global annual mean SW TOA aerosol DRE.

	Aerosol DRE, W/m^2
All-sky	-2.34
Clear-sky	-3.30
Cloudy-sky	-1.93

It can be seen that cloudy-sky aerosol DRE is less negative than in clear-skies. This is due to two effects: masking of aerosol radiative effects by overlying cloud, and warming effects of absorbing aerosol over clouds and other bright surfaces.

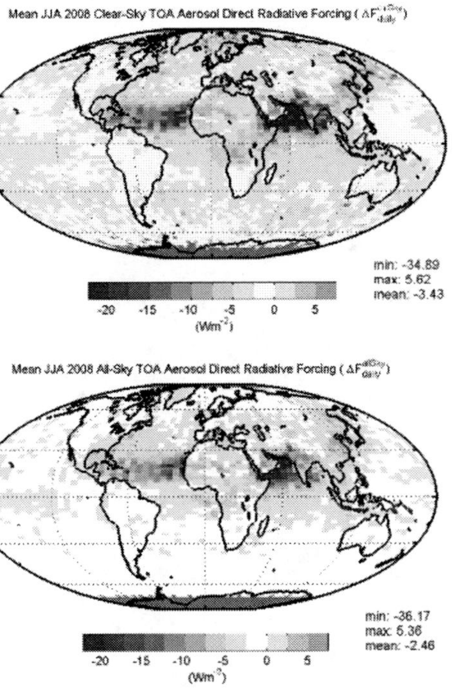

Figure 4. Shortwave TOA aerosol DRE, JJA 2008. Top: Clear-sky only. Bottom: All-sky.

Figure 4 compares the global distribution of aerosol DRE in clear and cloudy skies. Blue shades indicate net aerosol cooling at TOA. Reddish shades indicate net aerosol warming. For absorbing aerosol, there is a critical value of aerosol single scatter albedo where the aerosol effect goes from cooling to warming. The magnitude of the critical value depends on the magnitude of the underlying column (Figure 5).

In the top panel of Figure 4, showing clear-sky effects only, a net warming effect is only seen over the relatively bright surfaces of the Sahara and middle Eastern deserts and the snow-covered surface of Greenland. The differences seen between the upper and lower panels are due to absorbing aerosol above bright clouds. Figure 4 shows results for JJA, which is the burning season in Brasil and southern Africa. The net aerosol warming in these regions offsets part of the clear-sky cooling resulting in a less negative global mean aerosol DRE.

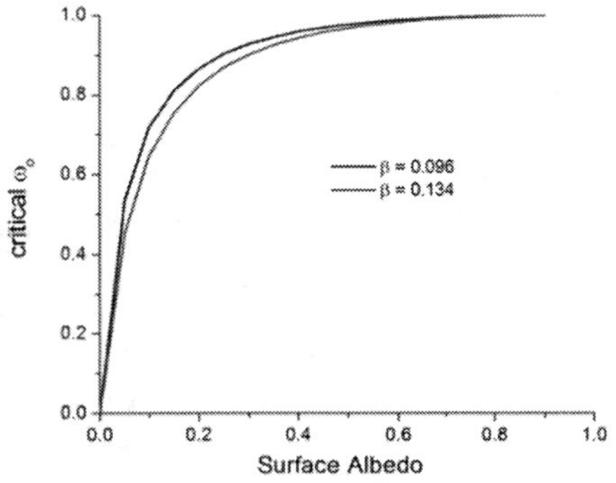

Figure 5. Critical value of aerosol single scatter albedo as a function of surface albedo. The two curves correspond to fine mode and coarse mode aerosol particles having upscatter fraction, β.

4. SUMMARY

An approach combing passive and active observations with model data has been used to estimate aerosol direct radiative effects in both clear and cloudy skies. The results show that all-sky DRE is less negative than in clear skies, due both to the masking effect of clouds and the warming effect of absorbing aerosol. This result helps to explain the discrepancy between model and observational estimates of aerosol radiative effects. Next steps include an analysis of the variability of DRE at regional scales, and an analysis to estimate uncertainties of the derived DRE values.

REFERENCES

[1] Koffi, B., M. Schulz, F.-M. Bréon, J. Griesfeller, D. M. Winker, Y. Balkanski, S. Bauer, T. Berntsen, M. Chin, W. D. Collins, F. Dentener, T. Diehl, R. C. Easter, S. J. Ghan, P. A. Ginoux, S. Gong, L. W. Horowitz, T. Iversen, A. Kirkevag, D. M. Koch, M. Krol, G. Myhre, P. Stier, and T. Takemura, 2012: "Application of the CALIOP Layer Product to evaluate the vertical

distribution of aerosols estimated by global models: Part 1. AeroCom phase I results", *J. Geophys. Res.*, **117**, D10201, doi:10.1029/2011JD016858.

[2] Nam, C. C. W., S. Bony, J.-L. Dufresne, and H. Chepfer, 2012: "The 'too few, too bright' tropical low-cloud problem in CMIP5 models, *Geophys. Res. Lett.* **39**, doi:10.1029/2012GL053421.

[3] Bellouin, N., A. Jones, J. Haywood, and S. A. Christopher, 2008: Updated estimate of aerosol direct radiative forcing from satellite observations and comparison against the Hadley centre climate model. *J. Geophys. Res.*, **113** (D10205).

[4] Winker, D. M., M. A. Vaughan, A. H. Omar, Y. Hu, K. A. Powell, Z. Liu, W. H. Hunt, and S. A. Young, 2009: "Overview of the CALIPSO Mission and CALIOP Data Processing Algorithms", *J. Atmos. Oceanic Technol.*, **26**, 2310–2323, doi:10.1175/2009JTECHA1281.1.

[5] Chand, D., R. Wood, T. L. Anderson, S. K. Satheesh, and R. J. Charlson, 2009: "Satellite-derived direct radiative effect of aerosols dependent on cloud cover", *Nat. Geosci.* **2**, 181–184, doi:10.1038/ngeo437.

[6] Oikawa, E., T. Nakajima, T. Inoue and D. Winker, 2013: "A study of the shortwave direct aerosol forcing using ESSP/CALIPSO observation and GCM simulation", *J. Geophys. Res.*, **118**, 3687–3708, doi:10.1002/jgrd.50227.

[7] Winker, D. M., J. L. Tackett, B. Getzewich, Z. Liu, M. A. Vaughan, and R. R. Rogers, 2013: "The global 3-D distribution of tropospheric aerosols as characterized by CALIOP", *Atmos. Chem. Phys.*, **13**, 3345–3361, doi:10.5194/acp-13-3345-2013

[7] Kato, S., S. Sun-Mack, W. F. Miller, F. G. Rose, Y. Chen, P. Minnis, and B. A. Wielicki, 2010: "Relation of cloud occurrence frequency, overlap, and effective thickness derived from CALIPSO and CloudSat merged cloud vertical profiles", *J. Geophys. Res.*, **115**, D00H28, doi:10.1029/2009JD012277.

[8] Hess, M., P. Koepke, and I. Schult, 1998: Optical Properties of Aerosols and Clouds: The Software Package OPAC. *Bull. Amer. Meteor. Soc.* **79**, 831-844.

DIRECT, LONGWAVE RADIATIVE FORCING OF MINERAL DUST: IMPROVEMENT OF ITS ESTIMATION BY MEANS OF TOOLS RECENTLY DEVELOPED BY THE EARLINET COMMUNITY

Michaël Sicard[1,2]*, Constantino Muñoz-Porcar[1], Adolfo Comerón[1], Alejandro Rodriguez[1], Francesc Rocadenbosch[1,2], Ruben Barragan[1,2]

[1] *Remote Sensing Laboratory, Universitat Politècnica de Catalunya, Barcelona, Spain, *Email: msicard@tsc.upc.edu*

[2]*Ciències i Tecnologies de l'Espai - Centre de Recerca de l'Aeronàutica i de l'Espai / Institut d'Estudis Espacials de Catalunya (CTE-CRAE / IEEC), Universitat Politècnica de Catalunya, Barcelona, Spain*

ABSTRACT

The objective of this work is to investigate how the knowledge of the vertically-resolved fine and coarse mode aerosol optical depth modifies the longwave radiative forcing. Since relatively little the EARLINET (European Aerosol Research Lidar Network) community has developed codes that combine sun-photometer and lidar data to retrieve a set of parameters vertically-resolved related to the size distribution (fine and coarse mode extinction coefficients, fine and coarse mode volumetric concentrations, etc.). We concentrate on the case of mineral dust whose size distribution is often dominated by the coarse mode. This work demonstrates that the knowledge of the vertically-resolved fine and coarse mode aerosol optical depth modifies the LW RF as compared to the classical approach with a unique profile of total aerosol optical depth. The results show that when the coarse mode predominates the classical approach underestimates the dust longwave radiative forcing by 10 to 20 % at the surface. The effect at the top of the atmosphere is not systematic because of the predominance of fine particles near the top of the dust layer.

1. INTRODUCTION

Atmospheric aerosols have a remarkable effect on the Earth-atmosphere radiative budget. Indeed, aerosols and their interactions with clouds contribute to the largest uncertainties in the estimation of the Earth's changing energy budget [1]. Nowadays many radiative transfer models have been developed to locally estimate the aerosol direct radiative forcing (RF) [2][3]. In the longwave (LW) spectral range, the aerosol radiative properties are usually estimated theoretically with a Mie code. The parameter that contains the absorption and scattering quantities, the extinction coefficient, is normalized to the extinction coefficient in the shortwave (SW) spectral range, most of the time in the visible spectral range [4][5], or to the number concentration [6]. As measurements of the extinction coefficient or of its integral, the aerosol optical depth (AOD), are available in the shortwave spectral range, the equivalent extinction coefficient or optical depth in the LW spectral range can be deduced thanks to that normalized, theoretical extinction coefficient.

In this work we use codes recently developed by the EARLINET community to separate the contributions of the small and the large particles. It is known that for a constant value of extinction in the visible spectral range, the theoretical extinction value in the LW spectral range (calculated with a Mie code) will vary with particle size. It is then expected that considering small and large particles separately will give extinction values in the LW different than considering the total bimodal distribution as a whole. This work investigates how the knowledge of the vertically-resolved fine and coarse mode AOD modifies the LW RF usually estimated with a single profile of total AOD.

2. METHODOLOGY

Irradiances in the longwave spectral region are calculated using the GAME code [7]. The generic parameters used as input in the Mie code (particles' shape, refractive index, size distribution and density) and in GAME (atmospheric pressure, temperature and relative humidity, and surface albedo) are described in [5] and [8].

To estimate the aerosol RF in the LW spectral range the profile of the LW AOD is calculated by multiplying the SW AOD profile to the normalized LW extinction coefficient (previously calculated with a Mie code). The single scattering albedo and the asymmetry factor are intensive parameters and are estimated in the LW with a Mie code. In the following, the altitude dependency has been omitted for the sake of clarity.

In the classical approach, a bimodal size distribution (e.g. measured by a sun-photometer) is considered to calculate the normalized LW extinction coefficient, $\bar{\alpha}_{LW,bm}$, and a profile of total AOD at 532 nm, $AOD_{532,t}$, (e.g. measured by a lidar) is considered for the SW AOD profile. The profile of the LW AOD is calculated as:

$$AOD_{LW,class} = \bar{\alpha}_{LW,bm} AOD_{532,t} .$$

In the new approach, called the Mode Separation Profiling (MSP) method like in [8], two mono-modal size distributions (e.g. measured by a sun-photometer) are considered to calculate the coarse and fine normalized LW extinction coefficients, $\bar{\alpha}_{LW,c}$ and $\bar{\alpha}_{LW,f}$, respectively. Two profiles of coarse and fine AOD at 532 nm, $AOD_{532,c}$ and $AOD_{532,f}$, (e.g. measured by combining lidar and sun-photometer, see next Section) are considered for the SW AOD profile. The profile of the LW AOD is calculated as:

$$AOD_{LW,MSP} = \bar{\alpha}_{LW,c} AOD_{532,c} + \bar{\alpha}_{LW,f} AOD_{532,f} .$$

As $\bar{\alpha}_{LW,c}$ can be larger than $\bar{\alpha}_{LW,bm}$, in cases predominated by large particles it can occur that:

$$AOD_{LW,class} < AOD_{LW,MSP} .$$

For instance, in the extreme case $AOD_{532,f} = 0$, the latter equation is satisfied if $\bar{\alpha}_{LW,bm} < \bar{\alpha}_{LW,c}$.

In Fig. 1 we plot the LW normalized extinction for the three size distributions: the bimodal and both coarse and fine mono-modal distributions. The input data are the ones of the first case study (see next Section) given in Table 1. It is clearly visible that $\bar{\alpha}_{LW,bm} < \bar{\alpha}_{LW,c}$ (by a factor of 2 approximately) in the entire spectral range even though over the whole column the fraction of coarse mode AOD (47 %, see Table 1) is lower than the one of fine mode AOD. It results that, even though $AOD_{532,t} > AOD_{532,c}$, there might be some altitude ranges where the large particles predominate where $AOD_{LW,class}$ could be lower than $AOD_{LW,MSP}$, implying a modification in the LW RF.

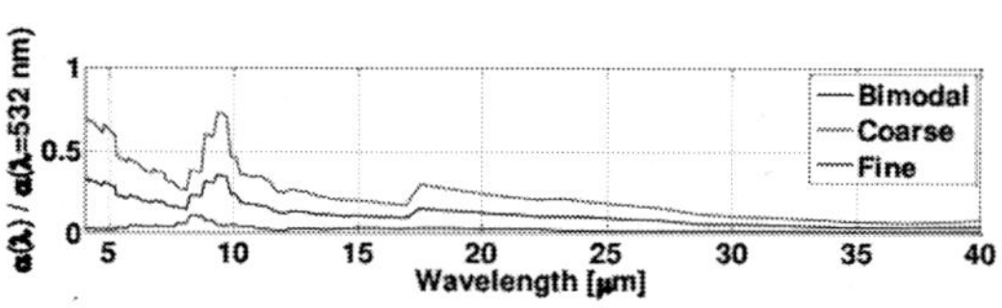

Fig. 1. Normalized extinction coefficient in the LW spectral range for Case 1.

3. RESULTS

The code used to estimate separately the coarse and fine mode fractions of the extinction coefficient and volumetric concentration is LIRIC (LIdar-Radiometer Inversion Code) [9]. LIRIC calculates the fine and coarse particle concentration profiles, using the backscattered lidar signals at three wavelengths (355, 532 and 1064 nm) and the column averaged aerosol microphysical properties retrieved from a sun photometer. More information about its performances can be found in [10] and [11].

We use a total of four lidar and AERONET sun-photometer measurements in Barcelona with different loads and vertical distributions of mineral dust. Table 1 gives some description of the four cases in terms of AOD and size distribution, as well as the results of our radiative forcing calculations with both the MSP and the classical methods. More details about the mineral dust vertical distribution of those four cases can be found in [8].

At the surface ΔF^{MSP} ranges from +3.1 to +11.0 Wm^{-2}. All values of ΔF^{class} are 10 to 20 % lower than ΔF^{MSP}. The largest difference, 20 %, is not reached for the case with the highest column-averaged coarse-to-total AOD fraction but for the case with the highest difference between $\bar{\alpha}_{LW,bm}$ and $\bar{\alpha}_{LW,c}$ in the bottom part of the mineral dust layer ($\bar{\alpha}_{LW,c}$ is approximately double, see [8]).

At the top of the atmosphere (TOA) the subestimation of ΔF^{class} is not systematic. The differences compared to ΔF^{MSP} range between -16 and +7 %. The overestimation of ΔF^{class} of +7 % occurs for the case with the highest difference between $\bar{\alpha}_{LW,bm}$ and $\bar{\alpha}_{LW,c}$ in the top part of the mineral dust layer ($\bar{\alpha}_{LW,bm}$ is approximately triple, see [8]).

Case	AOD_{532}	$\frac{AOD_{532,c}}{AOD_{532}}$	Coarse mode	Fine mode	LW RF at surface	LW RF at TOA
			$r_{V,c}$; $\sigma_{V,c}$; $C_{V,c}$	$r_{V,f}$; $\sigma_{V,f}$; $C_{V,f}$	ΔF^{MSP}; ΔF^{class}	ΔF^{MSP}; ΔF^{class}
	[ ]	[%]	[μm]; [ ]; [μm³μm⁻²]	[μm]; [ ]; [μm³μm⁻²]	[Wm⁻²]	[Wm⁻²]
1	0.19	47	1.708; 0.631; 0.066	0.152; 0.547; 0.018	+3.1; +2.8 (-10%)	+0.6; +0.6 (0%)
2	0.23	54	1.953; 0.588; 0.101	0.131; 0.503; 0.026	+5.4; +4.3 (-20%)	+2.3; +2.2 (-4%)
3	0.23	57	2.206; 0.629; 0.112	0.143; 0.520; 0.020	+4.4; +3.7 (-16%)	+1.5; +1.6 (+7%)
4	0.48	72	1.809; 0.545; 0.257	0.117; 0.647; 0.042	+11.0; +9.3 (-15%)	+5.0; +4.2 (-16%)

Table 1. LW RF calculated by the MSP and the classical methods, ΔF^{MSP} and ΔF^{class} respectively. The percentage after ΔF^{class} is its relative difference with respect to ΔF^{MSP}. r_V, σ_V and C_V represent the column-averaged volumetric radius, standard deviation and volumetric concentration. The suffixes c and f stand for coarse and fine, respectively.

4. CONCLUSIONS

The calculation of mineral dust, and aerosols dominated by large particles in general, direct radiative forcing in the longwave spectral range can be improved when both profiles of coarse and fine mode extinction coefficients can be retrieved separately. Our work demonstrates that the classical method underestimates the LW RF at the surface by 10 to 20 %. As the predominance of the large particles decreases with increasing height, this subestimation is not systematic at the TOA. However the strong predominance of large particles at the surface has also an effect on the forcing in the upper layers.

In presence of large particles, the improved estimation of LW RF suggests that the total (SW+LW) forcing is reduced (since both forcings are usually of opposite sign), and that aerosol impact on regional climate near dust sources could be lower than the actual estimations.

ACKNOWLEDGEMENT

This work is supported by the 7th Framework Programme project Aerosols, Clouds, and Trace Gases Research Infrastructure Network (ACTRIS) (grant agreement no. 262254). Lidar measurements in Barcelona are supported by the Spanish Ministry of Science and Innovation and FEDER funds under the projects TEC2012-34575, UNPC10-4E-442, CGL2011-13580-E/CLI and CGL2011-16124-E/CLI.

REFERENCES

[1] IPCC, "Climate Change 2013, The Physical Science Basis", Working Group I Contribution to the Fifth Assessment Report of the Intergovernmental Panel on Climate Change, T.F. Stocker, D. Qin, G.-K. Plattner, M. Tignor, S.K. Allen, J. Doschung, A. Nauels, Y. Xia, V. Bex, and P.M. Midgley (eds.), Chapter 7: Clouds and Aerosols, 571-658, doi: http://dx.doi.org/10.1017/CBO9781107415324.016, Cambridge University Press, New York, 2014.

[2] Ricchiazzi, P., S. Yang, C. Gautier, D. Sowle, 1998: SBDART: A Research and Teaching Software Tool for Plane-Parallel Radiative Transfer in the Earth's Atmosphere, *Bull. Am. Meteor. Soc.*, **79(10)**, 2101–2114.

[3] Key, J., A. J. Schweiger, 1998: Tools for atmospheric radiative transfer: Streamer and FluxNet, *Computers & Geosciences*, **24(5)**, 443–451.

[4] Dufresne, J., C. Gautier, P. Ricchiazzi, 2002: Longwave scattering effects of mineral aerosols, *American Meteorological Society*, **59**, 1959-1966.

[5] Sicard, M., S. Bertolín, M. Mallet, P. Dubuisson, A. Comerón, 2014: Estimation of mineral dust longwave radiative forcing: sensitivity study to particle properties and application to real cases in the region of Barcelona, *Atmos. Chem. Phys.*, **14**, 9213-9231.

[6] Sokolik. I. N., O. B. Toon, R. W. Bergstrom, 1998: Modeling the radiative characteristics of airborne mineral aerosols at infrared wavelengths, *J. Geophys. Res.*, **103**, 8813-8826.

[7] Dubuisson, P., J. Roger, M. Mallet, O. Dubovik, 2006: A Code to Compute the Direct Solar Radiative Forcing: Application to Anthropogenic Aerosols during the Escompte Experiment, Proc. International Radiation Symposium (IRS 2004) on Current Problems in Atmospheric Radiation, edited by H. Fischer, B.-J. Sohn and A. Deepak, Hampton, 127-130.

[8] Sicard M., S. Bertolín, C. Muñoz, A. Rodríguez, F. Rocadenbosch, A. Comerón, 2014: Separation of aerosol fine- and coarse-mode radiative properties: Effect on the mineral dust longwave, direct radiative forcing. *Geophys. Res. Lett.*, **41**, doi:10.1002/2014GL060946.

[9] Chaikovsky, A., Dubovik, O., Goloub, P., Tanré, D., Pappalardo, G., Wandinger, U., Chaikovskaya, L., Denisov, S., Grudo, Y., Lopatsin, A., Karol, Y., Lapyonok, T., Korol, M., Osipenko, F., Savitski, D., Slesar, A., Apituley, A., Arboledas, L. A., Binietoglou, I., Kokkalis, P., Granados Muñoz, M. J., Papayannis, A., Perrone, M. R., Pietruczuk, A., Pisani, G., Rocadenbosch, F., Sicard, M., De Tomasi, F., Wagner, J., Wang, X., 2012: Algorithm and software for the retrieval of vertical aerosol properties using combined lidar/ radiometer data: Dissemination in EARLINET, in: Proceedings of the 26th International Laser and Radar Conference, vol. 1, Porto Heli, Greece, 25–29 June 2012, 399–402.

[10] Wagner, J., A. Ansmann, U. Wandinger, P. Seifert, A. Schwarz, M. Tesche, A. Chaikovsky, O. Dubovik, 2013: Evaluation of the Lidar/Radiometer Inversion Code (LIRIC) to determine microphysical properties of volcanic and desert dust, *Atmos. Meas. Tech.*, **6**, 1707–1724.

[11] Granados-Muñoz, M. J., Guerrero-Rascado, J. L., Bravo-Aranda, J. A., Navas-Guzmán, F., Valenzuela, A., Lyamani, H., Chaikovsky, A., Wandinger, U., Ansmann, A., Dubovik, O., Grudo, J. O., and Alados-Arboledas, L., 2014: Retrieving aerosol microphysical properties by Lidar-Radiometer Inversion Code (LIRIC) for different aerosol types, *J. Geophys. Res.*, **119**, 4836-4858.

DOI: 10.1051/epjconf/201611921003

DRIZZLE MEASUREMENTS USING HIGH SPECTRAL RESOLUTION LIDAR AND RADAR DATA

Edwin W. Eloranta

*University of Wisconsin, Madison, WI 53706, USA, *Email: eloranta@ssec.wisc.edu*

ABSTRACT

The ratio of millimeter radar and High Spectral Resolution Lidar (HSRL) backscatter are used to determine drizzle rates which are compared to conventional ground based measurements. The robustly calibrated HSRL backscatter cross section provides advantages over measurements made with traditional lidars.

1. INTRODUCTION

Marine stratus clouds are an important feature of the global climate system. Cloud lifetime is sensitive to drizzle rates. Drizzle not only removes water from the cloud but evaporation of the falling droplets cools the sub-cloud layer acting to suppress convection. Accurate measurements of drizzle rates will improve our understanding of cloud maintenance.

Several investigators have used simultaneous lidar and radar observations to determine particle size [1,2,3]. However, measurements made with conventional lidar are hampered by: 1) changes in the transmission of the output window caused by water accumulation, 2) the difficulty of correcting the backscatter signal for atmospheric extinction, 3) the effects of multiple scattering, and 4) the need to convert backscatter measurements to extinction. The use of High Spectral Resolution Lidar (HSRL) data avoids many of these problems. HSRL backscatter measurements are referenced to the known molecular scattering cross-section at each point in the profile and are thus independent of changes in window and atmospheric transmission.

This study uses data collected during the US Department of Energy Atmospheric Sciences program MAGIC campaign. Instruments including a suite of conventional precipitation gauges, a High Spectral Resolution Lidar, along with 3.2 mm wavelength WACR and a 8.6 mm wavelength KAZR radars, were installed on the container ship *Horizon Spirit* as it made repeated trips between Long Beach, CA and Honolulu, HI. Figures 1 though 4 provide an example of the measurements used in our retrieval of drizzle rates.

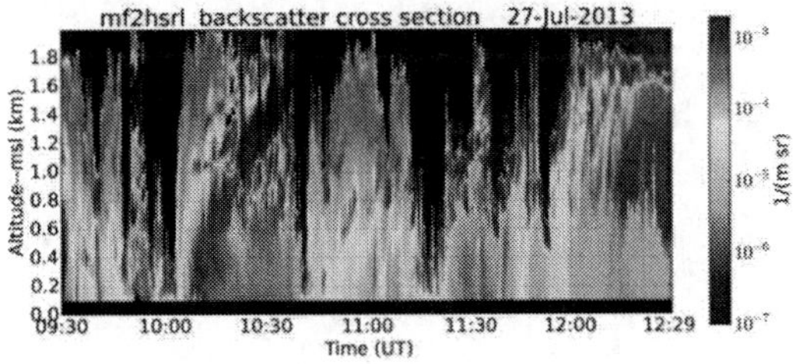

Figure 1: HSRL measured 532nm backscatter cross-section on 27-Jul-2013. Water clouds produce the areas with backscatter cross-sections near 1e-3 1/(m sr). Drizzle shafts near the surface appear with backscatter cross-sections between 1e-5 and 1e-4 1/(m sr).

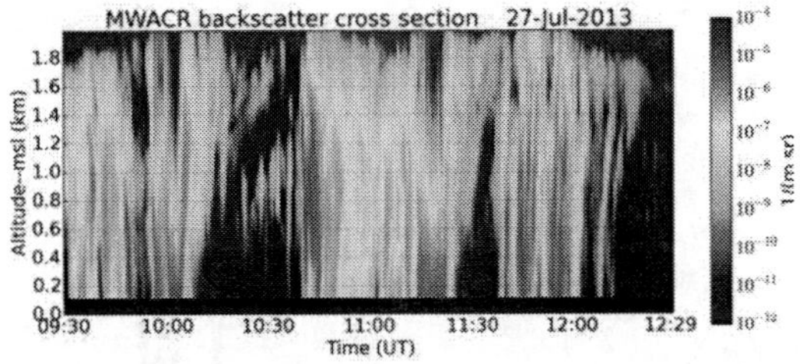

Figure 2: Radar backscatter cross section measured with the WACR 3.2 mm wavelength radar. (note: cross section is 4 π times usual radar definition).

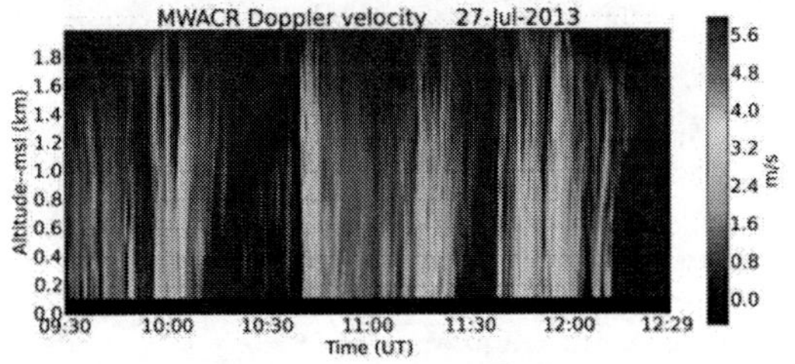

Figure 3: Doppler velocity measured with the WCAR radar on 27-Jul-2013.

DOI: 10.1051/epjconf/201611921003

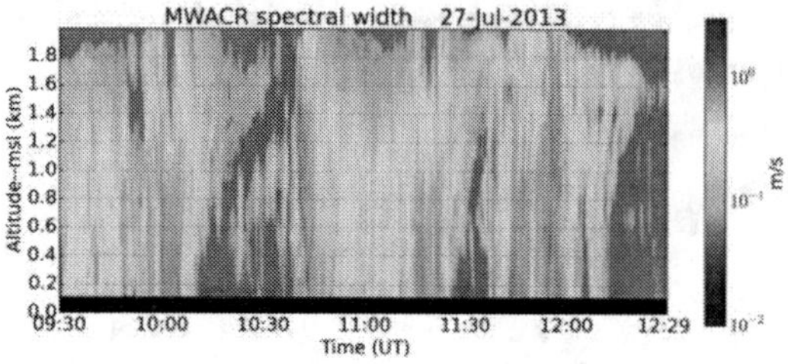

Figure 4: Spectral width measured with the WACR radar on 27-Jul-2013.

2. DIZZLE PARTICLE SIZE RETRIEVAL

We assume a modified gamma distribution [4] of particle diameters:

$$\frac{dN}{dD} \sim D^{\alpha} \cdot exp(-\frac{\alpha}{\gamma} \cdot (\frac{D}{D_m})^{\gamma})$$

Where: N = number of particles

D = particle diameter

D_m = mode diameter

α,γ = shape parameters

Using Mie theory we compute the backscsatter cross-section as a function of size parameter for both 532nm and 8.6 mm wavelengths using the appropriate indices of refraction. Calculations extend over the range of diameters between 1 micron and 5 mm. The optical cross-sections were carried out at size parameter intervals of 0.01 and averaged to a spacing of 1 size parameter.

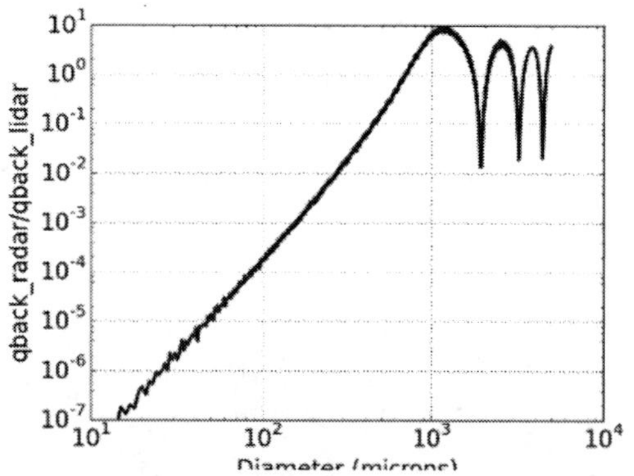

Figure 5: Ratio of radar (λ=3.2 mm) to lidar (λ=532 nm) backscatter efficiencies for monodisperse particle size distribution.

It is convenient to define a diameter D* as a ratio of the radar and lidar backscatter cross-sections:

$$D* = \lambda_{radar} \cdot (\frac{bsRadar}{bsLidar})^{\frac{1}{4}}$$

This ratio can be computed for our size distribution using Mie theory backscatter efficiencies $Q_{bsRadar}$ for radar and $Q_{bsLidar}$ for lidar scattering along with the radar wavelength, λ_{radar}:

$$D* = \lambda_{radar} \cdot (\frac{\int D^2 \cdot N(D) \cdot Q_{bsRadar} \cdot dD}{\int D^2 \cdot N(D) \cdot Q_{bsLidar} \cdot dD})^{\frac{1}{4}}$$

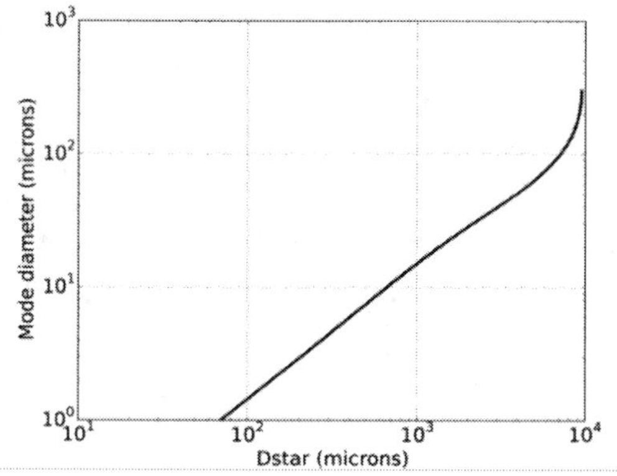

Figure 6: The relationship between D* and mode diameter for a modified gamma distribution calculated with γ=0.7 and α=1 for lidar and radar wavelengths of 532nm and 3.2 mm respectively at a temperature of 285K.

This relationship allows mode diameters to be derived from D* measurements. Formulating the relationship in terms of the backscatter cross-section takes advantage of the robustly calibrated HSRL measurement. This measurement does not require corrections for window transmission or atmospheric extinction below the scattering volume. In addition, multiple scattering effects largely cancel out in the HSRL backscatter retrieval.

Using Mie theory for the radar avoids errors caused by the Rayleigh approximation. However, the maximum mode diameter which can be retrieved is limited because the relationship between D* and the mode diameter becomes multi-valued when the particle sizes increase into the Mie regime .

Because the size distribution is known the mode diameter can be easily converted to effective diameter (effective diameter = ratio of particle volume to area).

DOI: 10.1051/epjconf/201611921003

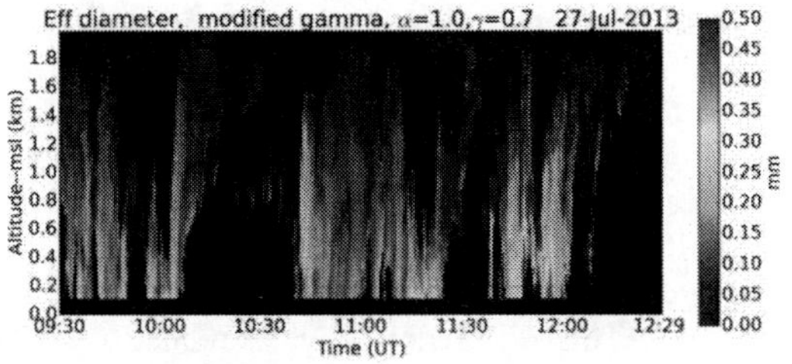

Figure 7: Effective diameter for the data shown with ξ=1 and γ=0.7.

Beard's parameterization [5] of fall velocity is used to compute the radar cross-section weighted fall velocity. The radar measured Doppler velocity and spectral width are independent measurements that can be compared to the modeled radar weighted velocities. Values of the shape parameters, α and γ can be adjusted to improve the comparison between the measured and modeled velocities.

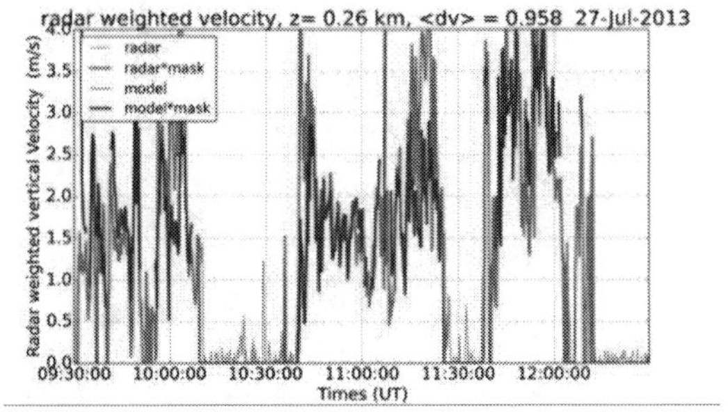

Figure 8: Comparison between measured Doppler velocities and the modeled radar weighted fall velocity at an altitude of 130 m with α = 1, γ = 0.7.

Even though we do not expect exact agreement because the Doppler velocity includes contributions due to vertical air motion, figures 8 and 9 show good agreement between the retrieved radar weighted velocities and the measured Doppler velocities.

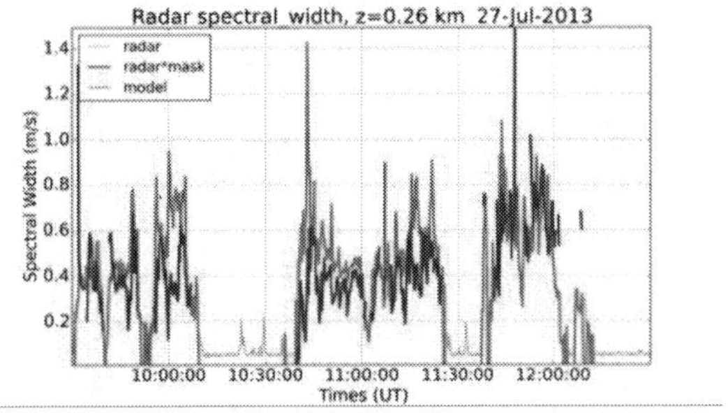

Figure 9: Comparison of measured and modeled spectral width as a function of time at an altitude of 130 m assuming α = 1 and γ = 0.7.

3. DRIZZLE RATE RETRIEVAL

The extinction cross-section, of particles that are large compared to the wavelength is twice the geometric area [6]:

$$\beta_{ext} = 2 \cdot N < A >$$

Where:

N = particle number density

<A> = average geometric area of particles

Because the effective diameter is $<D^3>/<D^2>$, the liquid water content(LWC) can be computed using the density of water:

$$LWC = \frac{1}{3} \cdot \rho_w \cdot D_{eff} \cdot \beta_{ext}$$

The lidar extinction cross-section is derived from the lidar backscatter cross-section using:

$$\beta_{ext} = bsLidar \cdot \frac{\int D^2 \cdot N(D) \cdot Q_{ext}(D) \cdot dD}{\int D^2 \cdot N(D) \cdot Q_{bs}(D) \cdot dD}$$

Where :

Q_{ext} and Q_{bs} are extinction and backscatter efficiencies calculated from Mie theory.

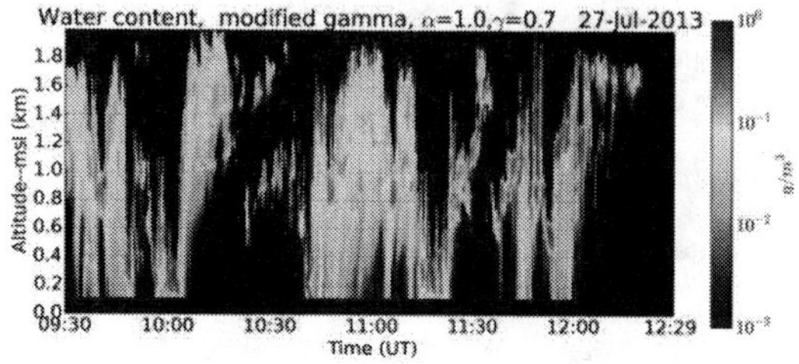

Figure 10: Liquid water content.

The derived liquid water content is not valid where clouds and drizzle are present together because this retrieval has assumed a mono-modal size distribution.

With the particles size distribution known, the mass weighted fall velocity can be computed using the Beard parameterization. The precipitation rate is computed from the product of the mass weighted fall velocity and the liquid water content.

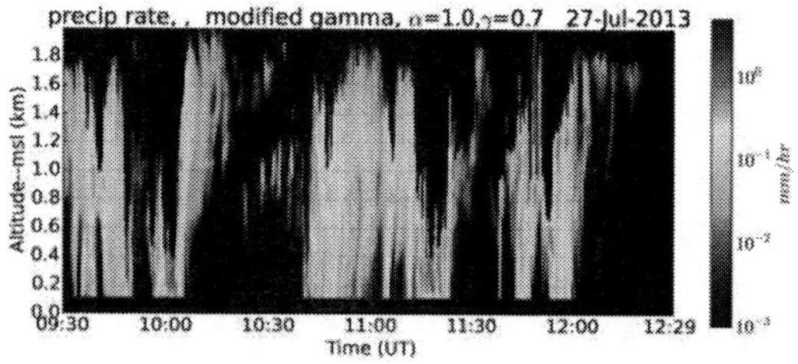

Figure 11: Precipitation rate image for 27-Jul-2013 derived using $\alpha = 1, \gamma = 0.7$.

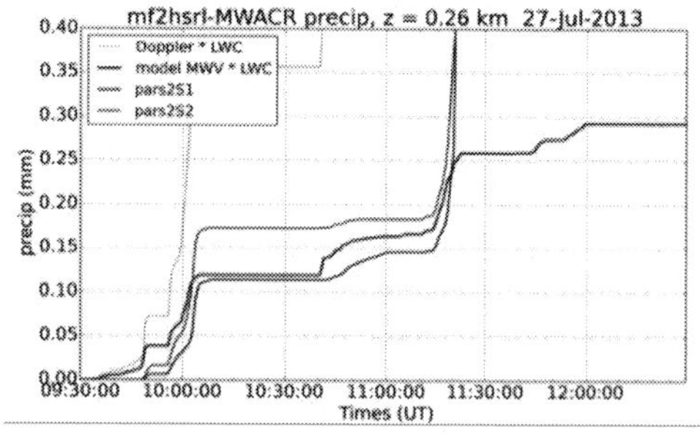

Figure 12: The accumulated precipitation measured by two shipboard distrometers (pars2S1 and pars2S2) and the lidar-radar retrieval. Also shown is the derived liquid water content multiplied by the radar Doppler velocity.

Figure 11 highlights the difficulty of making conventional precipitation measurements on a moving ship. Lidar-radar retrievals agree with PAR2S1 measurements prior to 11:45 UT, but do not agree as well with PARS2S2 measurements. After 11:45UT, the two distrometers give very different precipitation rates. Two conventional rain gauges on the ship also provide wildly different precipitation rates. Using the 8.6mm KAZR radar data in the retrieval indicated ~10% less precipitation with a best fit to velocities using $\gamma \sim 0.5$. This difference may be related to radar calibrations differences or because the WACR antenna was gyro-stabilized, while KZAR pointing angle rolled with the ship.

4. CONCLUSIONS

Lidar-Radar drizzle retrievals using HSRL data show considerable promise. The technique is sensitive to the functional form of the assumed size distribution and needs to be tested in a setting where more accurate conventional measurements are possible.

ACKNOWLEDGEMENT

This research was support by Department of Energy Grant DE-SC0006933.

REFERENCES

[1] Interi J.M., J. L. Stephens, W. L. Eberhard and, T. Uttal, 1993: A Method for Determining Cirrus Cloud Particle Sizes Using Lidar and Radar Backscatter Technique, *J. Applied Meteor.*, **32**, 1074-1082.

[2] Donovan, D.P., and A.C.A.P. van Lammeren, 2001: Cloud effective particle size and water content profile retrievals using combined lidar and radar observations: 1. Theory and examples, *J. Geophy. Res. Atmos.*,**106**, 27425-27448.

[3] O'Connor, E. J., A. J. Illingworth, R. J. Hogan, 2004: Retrieving Statocumulus Drizzle Parameters Using Doppler Radar and Lidar, *J. Atmos. Ocean Tech.*, **21(5)**, 777-778.

[4] Deirmendjain, D., 1969: *Electromagnetic Scattering on Spherical Polydispersions.* American Elsevier, 290 pp.

[5] Beard, K. V., 1976: Terminal velocity and shape of cloud and precipitation drops aloft. *J. Atmos. Sci.*, **33,** 851–864.

[6] van de Hulst, H. C., 1957: *Light Scattering by Small Particles.* John Wiley and Sons, 470 pp.

DOI: 10.1051/epjconf/201611921004

COMBINED DIAL SOUNDING OF OZONE, WATER VAPOUR AND AEROSOL

Thomas Trickl*, Hannes Vogelmann

*Karlsruher Institut für Technologie, IMK-IFU, Kreuzeckbahnstr. 19, 82476 Garmisch-Partenkirchen, Gernany, *e mail: thomas.trickl@kit.edu*

ABSTRACT

Routine high-quality lidar measurements of ozone, water vapour and aerosol at Garmisch-Partenkirchen since 2007 have made possible more comprehensive atmospheric studies and lead to a growing insight concerning the most frequently occurring long-range transport pathways. In this contribution we present as examples results on stratospheric layers travelling in the free troposphere for extended periods of time without eroding. In particular, we present a case of an intrusion layer that subsided over as many as fifteen days and survived the interference by strong Canadian fires. These results impose a challenge on atmospheric modelling that grossly overestimates free-tropospheric mixing.

1. INTRODUCTION

Differential absorption lidar (DIAL) methods have been applied in numerous atmospheric process and air-quality studies [1]. The most important species are ozone and water vapour, the free-tropospheric concentrations of other chemically relevant species such as NO or NO_2 mostly being too low for. In the Garmisch-Partenkirchen area (Bavarian Alps, Germany) IMK-IFU has developed three DIAL systems for ozone and water vapour [2,3,4,5] that have been extensively used in focussed atmospheric investigations. In 2007 a programme of routine measurements of ozone, water vapour and aerosol was started. By combining the measurements for these three species enhanced information for the atmospheric studies has been obtained. In this contribution we give examples of the scientific work related to these measurements.

2. METHODS

The ozone DIAL is located at IMK-IFU (Garmisch-Partenkirchen) at an altitude of 740 m a.s.l. It is a three-wavelength system based on stimulated Raman shifting the 248.50 nm output of a KrF laser (Lambda Physik, LPX250; operated at 400 mJ/pulse and 99 Hz repetition rate) in H_2 and D_2. Ozone is calculated for the wavelength combinations 277 nm – 313 nm, 292 nm – 313 nm and, in the presence of aerosol [5], 277 nm – 292 nm. This allows internal quality assurance. Two Newtonian receivers (0.13 m and 0.50 m diameter, respectively) have been used for dividing the eight-decade dynamic range [2]. For both the near-field and the far-field detection systems 1.1-m Wadsworth-type grating spectrographs with a spectral resolution of 0.35 nm are used. The H5783P-06 Hamamatsu photomultiplier tubes used since 1996 were recently replaced by the successor model R7400U-03 with actively stabilized sockets provided by Romanski Sensors. Licel 12-bit, 20 MHz transient digitizers with custom-made ground-free input are used for all six channels. With just minor smoothing (50 m range resolution [7]) the signals exhibit a noise level of about 1×10^{-6} of the input voltage range, a major step forward in signal quality, making single photon counting almost unnecessary. The uncertainty of the ozone mixing ratios is 2–3 ppb in the lower troposphere and 5–10 ppb in the upper troposphere due to the use of 292 nm as the "on" wavelength. A detailed system publication is in preparation.

Because of the excellent signal quality aerosol structures corresponding to backscatter coefficients of less than 1×10^{-7} m^{-1} sr^{-1} can be resolved at 313 nm within most of the troposphere. In this way at least some substitution of our non-funded high-spectral-resolution aerosol lidar [8] has been possible.

The water-vapour DIAL [5] is located at 2675 m a.s.l. in the top laboratory of the nearby Schneefernerhaus high-altitude research station (UFS). It is operated with typically 100 mJ from a 250-mJ flashlamp-pumped Ti:sapphire laser with a repetition rate of 20 Hz determined by the pump laser of the two oscillators (Continuum, Powerlite 8020). The oscillators are widely tunable single-longitudinal-mode optical parametric oscillators (Continuum, Mirage). Several wavelength pairs next to 817 nm are used, depending on the humidity conditions. The absorption cross sections of Ponsardin

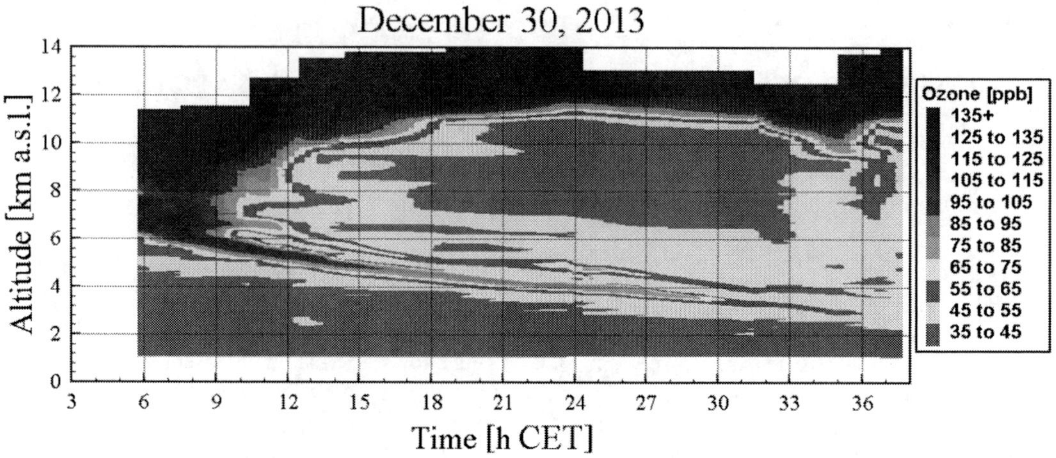

Fig. 1: Example of two parallel intrusion layers registered with the ozone DIAL; no H_2O measurements were possible because UFS was closed on these days. The minimum relative humidity of the Munich radiosonde (station 10868) in both layers was 1 % which, for unknown reasons, is the lowest value listed.

and Browell [9] have been used and verified to be better than the 2008 HITRAN values (e.g., [10,11]). A 0.65-m-diameter Newtonian receiver collects the backscattered radiation that is separated into near- and far-field channels by a combination of a 1-% beam splitter and an edge in a focus of the far-field channel. Avalanche photodiodes (APDs; Licel) and the same transient digitizer type as in the case of the ozone DIAL are used. The performance is currently limited by the noise from the APD modules and a laser power below the principal capability of the system [5]. A vertical range up to the tropopause region is achieved even during daytime. Up to about 7 km a.s.l. an uncertainty of 5 % and no bias exceeding 1 % have been demonstrated.

3. RESULTS

The measurements of ozone, water vapour and aerosol allow distinguishing between air from a remote boundary layer, characterized by the presence of particles, or from the stratosphere, characterized by low humidity and enhanced ozone. It is interesting to note that we have seen much less high-ozone structures in the free troposphere related to intercontinental transport than in the late 1990s and slightly afterwards. By contrast, most pronounced ozone peaks could be traced back to the stratosphere with travel times between two and fifteen days.

Meanwhile, stratospheric air intrusions have been observed on more than 300 days with one of the two DIAL systems or both. The structures and layer widths look quite different. One spectacular example is shown in Fig. 1. On about 60 % of the measurement days stratospheric layers could be observed demonstrating that the ozone import from the stratosphere is by no means negligible as frequently claimed. Most astonishingly the humidity level in intrusion layers reaching the lower troposphere has been found to range mostly between 0 and 2 %, much lower than recorded by the in-situ instrumentation at Alpine summit levels [12]. Our H_2O results clearly demonstrate that ozone mixing ratios of sometimes just 60 ppb are not caused by mixing of the surrounding tropospheric air into these layers.

A summary of the analysis of about 80 intrusion days with the water-vapour DIAL is given in Fig. 2. Three ranges of travel times are distinguished by estimates from four-day forward trajectory plots daily received for stratospheric intrusion forecasts [12,13,14] and from 315-h backward runs with the HYSPLIT trajectory model (http://ready.arl.noaa.gov/HYSPLIT_traj.php). The results show that the probability of tropospheric admixture grows with the travel time in the troposphere. However, even for the longest travel times some layers do not change their lower-stratospheric characteristics.

A particularly intriguing example has been observed on July 16, 2013. July 2013 was characterized by very strong smoke from boreal North America sometimes filling a major part of the troposphere. On July 16, a thin smoke layer directly hit the Zugspitze summit (2962 m) and, therefore, could be observed at UFS (2670 m). The 313-nm aerosol backscatter coefficients reached 1.2×10^{-5} m^{-1} sr^{-1} (Fig. 3), in agreement with the observed horizontal visual range of less than 10 km. Ozone was just slightly enhanced, as confirmed by the UFS measurements, but the humidity values revealed by the H_2O DIAL were almost zero (Fig. 4), which strongly suggests a stratospheric origin.

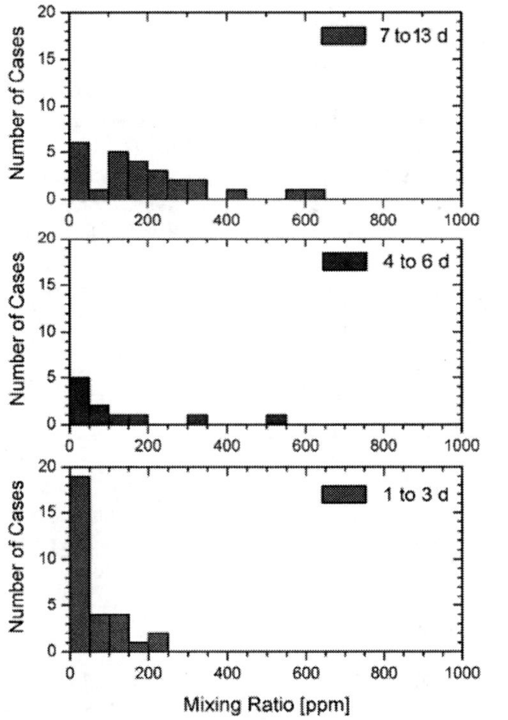

Fig. 2: Water-vapour mixing ratio at the centre of stratospheric air intrusion layers for different travel times [12] (for details see text)

HYSPLIT calculations extended to more than the maximum 315 h for a single run revealed a very slow subsidence from about 9 km over at least 15 d, from Siberia via parts of the "road of fire" that had formed between Alaska and Eastern Canada to Europe. The backward trajectory bundle stayed unbelievably coherent on the way around the northern hemisphere.

It is difficult understand the extreme dryness and narrow structure of the smoke layer at 3.2 km on July 16. A strong perturbation of a thin atmospheric layer by a hot ascending air mass would be expected to occur. Thus, it is more likely that the stratospheric air tongue intersected a smoke column in a later, rather stationary phase. Low-humidity conditions are a prerequisite of pronounced forest fires. Dry air would not significantly change the H_2O content of the intrusion layer. Alternatively, sedimentation of heavy particles into the layer could be thought of.

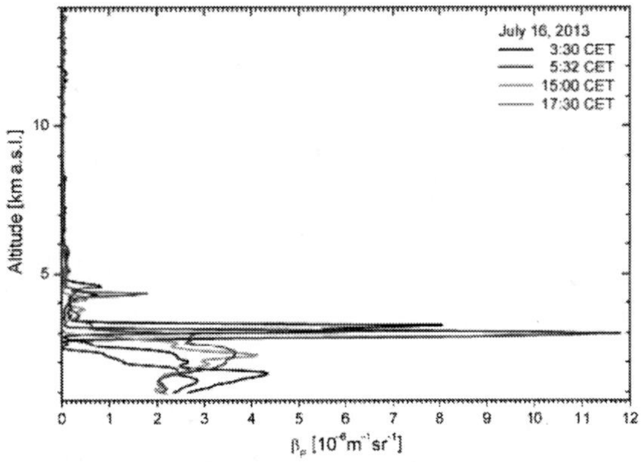

Fig. 3: 313-nm aerosol backscatter coefficients on July 16, 2013, showing a pronounced Canadian smoke plume at the altitude of the Zugspitze peak (2962 m a.s.l.) next to Garmisch-Partenkirchen

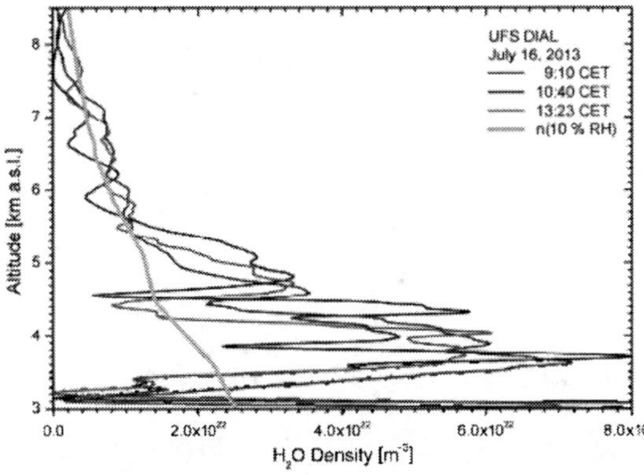

Fig. 4: Three water-vapour profiles from the H_2O DIAL at UFS on July 16, 2013, showing the extreme dryness in the lower smoke layer in Fig. 3

ACKNOWLEDGEMENTS

The authors thank H. P. Schmid for his support. L. Ries made available in-situ data from the Global Atmosphere Watch laboratory at UFS, and M. Sprenger has maintained the invaluable intrusion forecast service over many years. The support by the UFS team is gratefully acknowledged. This work contributes to the European Aerosol Research Lidar Network, recently partly supported by the European Union within the ACTRIS project.

REFERENCES

[1] Trickl, T., 2010: Tropospheric trace-gas measurements with the differential-absorption lidar technique, pp. 87-147 in: Recent Advances in Atmospheric Lidars, L. Fiorani, V. Mitev, Eds., INOE Publishing House, Bucharest (Romania), *Series on Optoelectronic Materials and Devices*, Vol. 7, ISSN 1584-5508, ISSN 978-973-88109-6-9; a revised version can be obtained from the author (thomas.trickl@kit.edu).

[2] Kempfer, U., W. Carnuth, R. Lotz, T. Trickl, 1994: A wide-range UV lidar system for tropospheric ozone measurements: Development and application, *Rev. Sci. Instrum.* **65**, 3145-3164

[3] Eisele, H., T. Trickl, 1997: Second Generation of the IFU Stationary Tropospheric Ozone Lidar, pp. 379-382 in: *Advances in Atmospheric Remote Sensing with Lidar,* Selected Papers of the 18th International Laser Radar Conference, A. Ansmann, R. Neuber, P. Rairoux, U. Wandinger, Eds., Springer (Berlin, Heidelberg, New York)

[4] Brenner, P., O. Reitebuch, K. Schäfer, T. Trickl, A. Stichternath, 1997: A Novel Mobile Vertical-sounding System for Ozone Studies in the Lower Troposphere, pp. 383-386 in: *Advances in Atmospheric Remote Sensing with Lidar,* see [3].

[5] Vogelmann, H., T. Trickl, 2008: Wide-range sounding of free-tropospheric water vapor with a differential-absorption lidar (DIAL) at a high-altitude station, *Appl. Opt.* **47**, 2116-2132

[6] Eisele, H., T. Trickl, 2005: Improvements of the aerosol algorithm in ozone-lidar data processing by use of evolutionary strategies, *Appl. Opt.* **44**, 2638-2651

[7] VDI guideline 4210, Part 1, 1999: *Remote Sensing: Atmospheric measurements with LIDAR, Measuring gaseous air pollution with DAS LIDAR*, Verein Deutscher Ingenieure (VDI), (Düsseldorf, Germany), 47 pp.

[8] Giehl, H., T. Trickl, 2010: Testing the IFU High-Spectral-Resolution Lidar at the 2009 Leipzig Field Campaign, pp. 920-923 in: *Proceedings of the 25th International Laser Radar Conference*, G. Matvienko, A. Zemlyanov, Eds., published by V. E. Zuev Institute of Optics (Tomsk, Russia)

[9] Ponsardin, P. L., E. V. Browell, 1997: Measurements of $H_2{}^{16}O$ linestrengths and air-induced broadenings and shifts in the 815 nm spectral region, *J. Mol. Spectrosc.* 185, 58–70.

[10] Wirth, M., A. Fix, G. Ehret, J. Reichardt, R. Begie, D. Engelbart, H. Vömel, B. Calpini, G. Romanens, A. Apituley, K. M. Wilson, H. Vogelmann, T. Trickl, 2009: Intercomparison of Airborne Water Vapour DIAL Measurements with Ground Based Remote Sensing and Radiosondes within the Framework of LUAMI 2008, Contribution S07-P01-1 (3 pp.) in: *Proceedings of the 8th International Symposium on Tropospheric Profiling (ISTP2009)*, Delft (The Netherlands), A. Apituley, H. W. J. Russchenberg, W. A. A. Monna, Eds., RIVM (The Netherlands, 2009), http://www.knmi.nl/~apituley/files/istp8/, ISBN 978-90-6960-233-2

[11] Vogelmann, H., R. Sussmann, T. Trickl, T. Borsdorff, 2011: Intercomparison of atmospheric water vapor soundings from the differential absorption lidar (DIAL) and the solar FTIR system on Mt. Zugspitze, *Atmos. Meas. Technol.* **4**, 835-841

[12] Trickl, T., H. Vogelmann, H. Giehl, H. E. Scheel, M. Sprenger, A. Stohl, 2014: How stratospheric are deep stratospheric intrusions? *Atmos. Chem. Phys.* **14**, 9941-9961

[13] Zanis, P., T. Trickl, A. Stohl, H. Wernli, O. Cooper, C. Zerefos, H. Gaeggeler, A. Priller, C. Schnabel, H. E. Scheel, H. J. Kanter, L. Tobler, P. W. Kubik, P. Cristofanelli, C. Forster, P. James, E. Gerasopoulos, A. Delcloo, A. Papayannis, H. Claude, 2003: Forecast, observation and modelling of a deep stratospheric intrusion event over Europe, *Atmos. Chem. Phys.* **3**, 763-777

[14] Trickl, T., H. Feldmann, H.-J. Kanter, H. E. Scheel, M. Sprenger, A. Stohl, H. Wernli, 2010: Deep stratospheric intrusions over Central Europe: case studies and climatological aspects, *Atmos. Chem. Phys.* **10**, 499-524

EPJ Web of Conferences **119**, 21005 (2016)
ILRC 27
DOI: 10.1051/epjconf/201611921005

Modeling lidar multiple scattering

Kaori Sato [1]*, Hajime Okamoto[1], Hiroshi Ishimoto[2]

[1] *Research Institute for Applied Mechanics, Kyushu University, Kasuga, Kasuga koen 6-1, 816-8580, JAPAN*, *Email: sato@riam.kyushu-u.ac.jp

[2] *Meteorological Research Institute, Japan Meteorological Agency, Nagamine 1-1, Tsukuba 305-0052, JAPAN*

ABSTRACT

A practical model to simulate multiply scattered lidar returns from inhomogeneous cloud layers are developed based on Backward Monte Carlo (BMC) simulations. The estimated time delay of the backscattered intensities returning from different vertical grids by the developed model agreed well with that directly obtained from BMC calculations. The method was applied to the Cloud-Aerosol Lidar and Infrared Pathfinder Satellite Observations (CALIPSO) satellite data to improve the synergetic retrieval of cloud microphysics with CloudSat radar data at optically thick cloud grids. Preliminary results for retrieving mass fraction of co-existing cloud particles and drizzle size particles within low-level clouds are demonstrated.

1. INTRODUCTION

The combined use of attenuated backscatter coefficient (β) and depolarization ratio (δ) is effective to reveal detailed microphysical structures within clouds. From space-borne lidars, they have been widely used to identify cloud phase [1][2], or further combined with cloud radar to determine mass concentration of non-spherical ice particles [3][4]. However, there are still areas of improvements in order to qualitatively retrieve vertically resolved cloud microphysics or to detect cloud base where multiple scattering effects are dominant as in optically thick low-level clouds [5]. In these situations, Monte Carlo methods are reliable for simulating the effects on lidar returns but are inefficient for inversion purposes [6]. The main objective of this paper is to improve lidar-only and radar-lidar synergy retrievals of cloud microphysics by updating the treatment for multiple scattering of depolarized lidar returns that are capable of handling global data.

2. METHODOLOGY

The concept of the method is to combine the parallel and perpendicular components of the backscattered intensities calculated by BMC for vertically homogeneous cloud cases to construct lidar returns from any inhomogeneous cloud layers (fig.1). To reflect the effect of inhomogeneity on the time delayed lidar returns, the excessive distances traveled within each vertical layer had been parameterized based on BMC simulation to determine the effective extinction σ_{eff} of the inhomogeneous media. Here, σ_{eff} is defined as the extinction including multiple scattering. In this way, repeated Monte Carlo simulations can be avoided for estimating the lidar returns in inversion schemes and the numerical cost in multiple scattering regimes can be reduced to that equivalent to assuming single scattering.

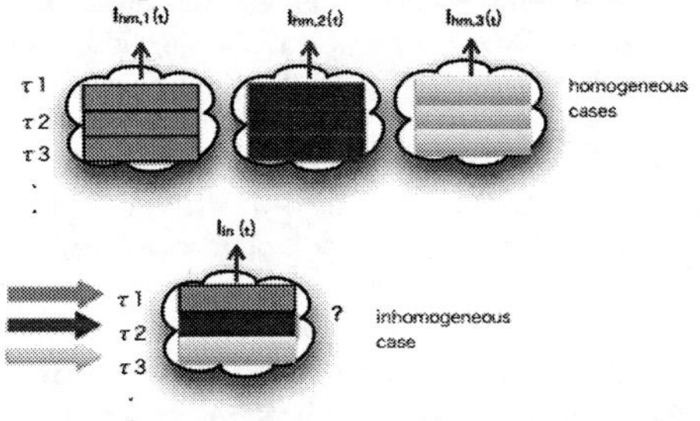

Fig.1 Basic concept. $I_{hm}(t)$ and $I_{in}(t)$ are the observed backscattered intensity at time t for homogenous and inhomogeneous cases, respectively, and τ_i is the optical thickness of the i-th layer from cloud top.

To test the validity of the developed method, BMC simulations were performed considering observation of liquid phase clouds with vertically varying microphysic from the CALIOP platform at 532 nm wavelength. For convenience, the microphysical properties were varied at 240m vertical resolutions corresponding to that of CloudSat radar, and result for an ideal cloud with three consecutive layers are provided as a representative case. The BMC method used in this study to simulate the lidar returns is known to compare well with other Monte Carlo codes tested against the same simulation settings [7].

3. RESULTS

Comparison between the estimated backscattered intensities for both parallel and perpendicular component and those obtained from direct BMC calculation had good agreement with each other. Figure 2 shows the result for the total backscatter coefficient. On the other hand, mere combination of the backscattered intensities estimated from BMC for homogeneous profiles and neglecting inhomogeneity effects largely deviated from the truth.

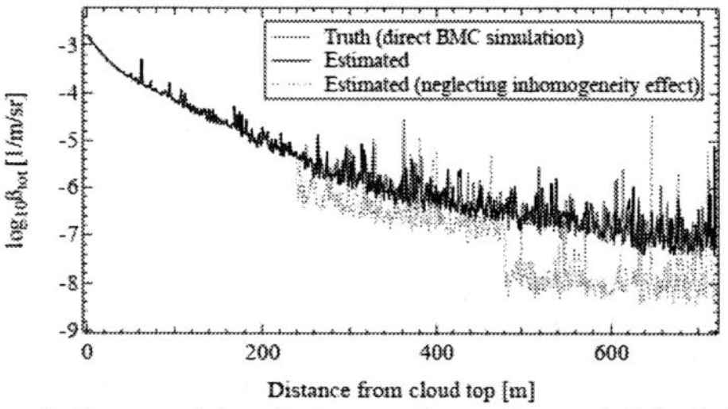

Fig. 2 Comparision between the estimated (black line) and simulated truth (blue line) of the total lidar backscatter coefficient for CALIOP at 532 nm wavelength. Extinction coefficients of the cloud layers were varied from 30.0, 15.0, 7.5 per kilometer at 240m resolutions from cloud top. Also shown are the estimated profile connecting BMC calculated backscattered intensities for homogeneous cases at layer boundaries without taking in to account the effect of inhomogeneity (gray dash line).

The backscattered intensities apparently returning from each vertical lidar grid were decomposed into the contributions from other grids. Comparison between those estimated and the truth implied that potentially the method is capable of determining the information the lidar returns contain about the microphysics of a particular cloud layer (Fig.3).

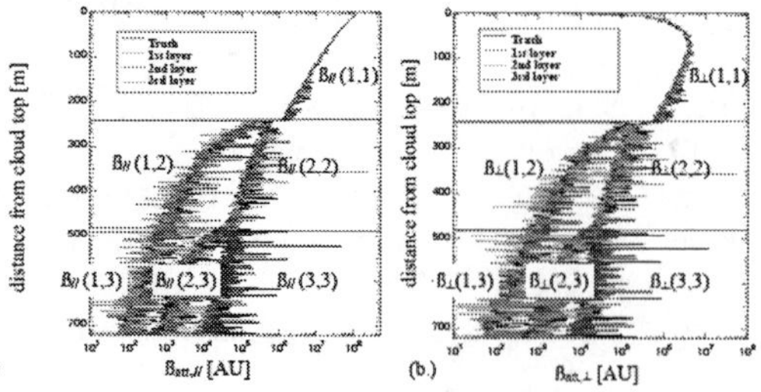

Fig. 3 (a.) Parallel and (b.) perpendicular component of the backscattered intensity for the same case as in fig.2. The estimated profile (colored line) and the truth via direct BMC simulation (black line) are compared. $\beta_{//\perp}(i,j)$ indicates the contribution of the microphysics at the i-th layer to the apparent backscattered intensity from the j-th layer.

4. APPLICATION

The developed lidar forward model was further tested against radar reflectivity (dBZ_e) obtained by CloudSat radar and β and δ obtained from CALIOP lidar to demonstrate its usability for cloud studies. Here, both sensors were combined to infer microphysical processes occurring within water clouds (Fig.4). To find solutions that satisfy each observable having different size dependence, the shape of the size distribution was assumed to be capable of detecting up to two different modes (i.e, mono-modal or bi-modal) within each vertical radar/lidar grid. The particle sizes and mass ratio were retrieved without fixing the mode radius. The retrieval results show that the synergy seems to be adequately functioning to separate drizzle (and precipitation) from cloud particles (Fig.5).

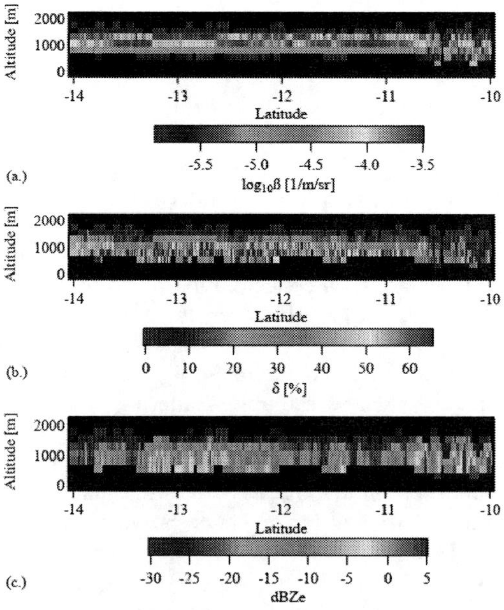

Fig.4 Latitude-height cross section of (a.) β and (b.) δ from CALIOP and (c.) dBZe from Cloudsat for low level cloud observed in October 2009. The CALIOP data are matched to the resolution of the CloudSat data and hydrometeor detection [4] has been performed.

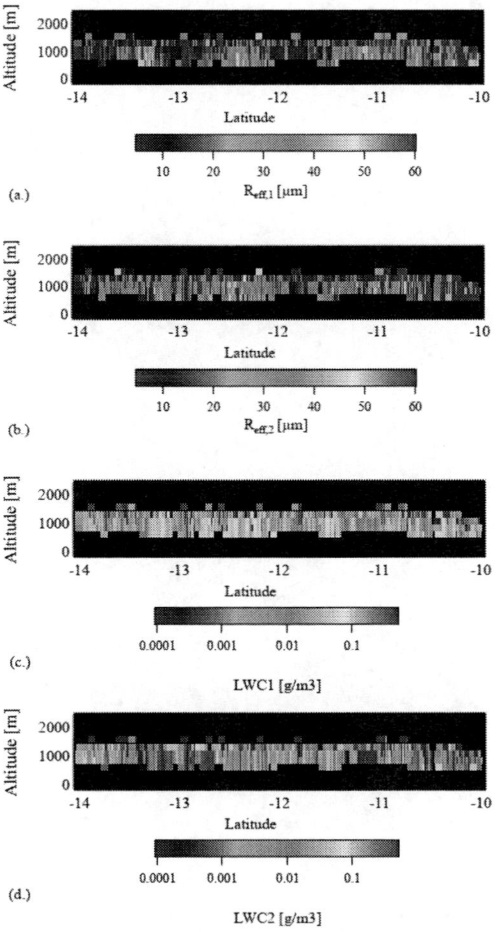

Fig.5 The same as fig. 4 but for the retrieved effective radius (R_{eff}) of (a.) mode 1 and (b.) mode 2, and (c.)/(d.) the liquid water contents (LWC) corresponding to (a.)/(b.). When the lidar/radar grid have bi-modal microphysical property, the smaller/larger mode is automatically identified as mode 1/mode 2, and therefore mode 1 and mode 2 does not necessarily correspond to cloud particle and drizzle (or precipitation), respectively.

5. CONCLUSIONS

Practical modeling of Monte Carlo simulations of multiple scattering effects from clouds was performed to analyze depolarized lidar returns from arbitral inhomogeneous cloud layers. For the cases tested, the method showed good performance. The developed forward model was used to inversely derive low-level cloud microphysics observed by CALIOP lidar. Especially, its potential for the synergy with CloudSat radar at heavily attenuated regions was focused. Preliminary results showed the effectiveness of the synergy to study the transition region, from cloud particles to drizzle and precipitation, which the representation in climate models is largely uncertain. In future, further validation of the methods is planned using collocated ground-base observation of Multiple Field of view Multiple Scattering Polarization Lidar (MFMSPL) and cloud radar, which realizes low-level cloud observations comparable to that of space-borne active sensors from ground.

ACKNOWLEDGEMENT

This study was supported by JSPS KAKENHI Grant Numbers 15K17762 and 25247078. K. Sato was also supported in part by the Kyushu University Interdisciplinary Programs in Education and Projects in Research Development, and by the program of Special Coordination Funds for Promoting Science and Technology, “Supporting positive activities for female researchers”.

REFERENCES

[1] Hu Y., M. Vaughan, Z. Liu, B. Lin, P. Yang, D.Flittner, B. Hunt, R. Kuehn, J. Huang, D. Wu, S. Rodier, K. Powell, C. Trepte, and D. Winker, 2007: The depolarization–attenuated backscatter relation: CALIPSO lidar measurements vs. theory, *Opt. Express*, 15(9), 5327–5332, doi:10.1364/OE.15.005327.

[2] Yoshida, R., H. Okamoto, Y. Hagihara, and H. Ishimoto, 2010: Global analysis of cloud phase and ice crystal orientation from Cloud-Aerosol Lidar and Infrared Pathfinder Satellite Observation (CALIPSO) data using attenuated backscattering and depolarization ratio, *J. Geophys. Res.,* 115, D00H32, doi:10.1029/2009JD012334.

[3] Okamoto, H., K. Sato, and Y. Hagihara, 2010: Global analysis of ice microphysics from CloudSat and CALIPSO: Incorporation of specular reflection in lidar signals, *J. Geophys. Res.*, 115, D22209, doi:10.1029/2009JD013383.

[4] Sato, K., and H. Okamoto, 2011: Refinement of global ice microphysics using spaceborne active sensors, *J. Geophys. Res.*, 116, D20202, doi:10.1029/2011JD015885.

[5] Illingworth A., H. Okamoto, and others, 2015: THE EARTHCARE SATELLITE: THE NEXT STEP FORWARD IN GLOBAL MEASUREMENTS OF CLOUDS, AEROSOLS, PRECIPITATION AND RADIATION, *Bull. Amer. Meteor. Soc.*, 96, 1311-1332, doi: http://dx.doi.org/10.1175/BAMS-D-12-00227.1
[6] Bissonnette, L. R., Lidar multiple scattering. In: C. Weitkamp (Ed.), 2005: Lidar: Range-Resolved Optical Remote Sensing of the Atmosphere, Springer Series in Optical Sciences, Vol. 102, ISBN: 0-387-40075-3, Springer, New York, 273-305.
[7] Ishimoto, H., and K. Masuda, 2002: A Monte Carlo approach for the calculation of polarized light: Application to an incident narrow beam, *J. Quant. Spectrosc. Radiat. Transfer,* 72, 467–483, doi:10.1016/ S0022-4073(01)00136-4

COMBINED ATMOSPHERIC AND OCEAN PROFILING FROM AN AIRBORNE HIGH SPECTRAL RESOLUTION LIDAR

Johnathan Hair [1]*, Chris Hostetler[1], Yongxiang Hu[1], Michael Behrenfeld[2], Carolyn Butler[3], David Harper[1], Rich Hare[1], Timothy Berkoff[1], Antony Cook[1], James Collins[3], Nicole Stockley[4], Michael Twardowski[5], Ivona Cetinić[6+], Richard Ferrare[1], and Terry Mack[7]

*[1]NASA Langley Research Center, Hampton VA 23681, USA, *Email: johnathan.w.hair@nasa.gov, [2]Oregon State University, Corvallis, OR, 97331, USA, [3]Science Systems and Applications Inc., Hampton, VA 23666, USA, [4]WET Labs, Inc., Narragansett, RI 02882, [4]Harbor Branch Oceanographic Institute, Florida Atlantic University, Ft. Pierce, FL* 34946, *[6]University of Maine, Walpole, ME 04573,* [+] currently NASA Goddard Space Flight Center/USRA, Greenbelt, MD 20771, *[7]Analytical Mechanics Associates, Hampton, VA 23666*

ABSTRACT

First of its kind combined atmospheric and ocean profile data were collected by the recently upgraded NASA Langley Research Center's (LaRC) High Spectral Resolution Lidar (HSRL-1) during the 17 July – 7 August 2014 Ship-Aircraft Bio-Optical Research Experiment (SABOR). This mission sampled over a region that covered the Gulf of Maine, open-ocean near Bermuda, and coastal waters from Virginia to Rhode Island. The HSRL-1 and the Research Scanning Polarimeter from NASA Goddard Institute for Space Studies collected data onboard the NASA LaRC King Air aircraft and flight operations were closely coordinated with the Research Vessel Endeavor that made in situ ocean optical measurements. The lidar measurements provided profiles of atmospheric backscatter and particulate depolarization at 532nm, 1064nm, and extinction (532nm) from approximately 9km altitude. In addition, for the first time HSRL seawater backscatter, depolarization, and diffuse attenuation data at 532nm were collected and compared to both the ship measurements and the Moderate Resolution Imaging Spectrometer (NASA MODIS-Aqua) satellite ocean retrievals.

1. INTRODUCTION

A combined ocean and atmospheric lidar has the potential to provide key measurements needed to better understand both the ocean optical properties, and hence the biogeochemical properties of the global oceans, and air-sea exchange processes important for climate studies. For example, it has been shown that the particle organic carbon stocks in the ocean are better estimated from particle backscatter than chlorophyll concentrations [e.g. 1]. In addition, lidar has the ability to provide vertical distributions of the optical properties within the ocean that are not possible from current passive satellite measurements. Recent results using the CALIPSO satellite measurements have revealed that global measurements [2] of vertically integrated backscatter within the ocean are possible, even from an instrument not designed for ocean profiling. As noted in [2], using the High Spectral Resolution Lidar and a passive ocean color sensor could result in a 3-D distribution of the upper ocean ecosystem.

To demonstrate the quantitative capabilities of such a system, the airborne HSRL that has flown on the King Air aircraft during many field campaign since 2006 [3] was recently configured to provide high vertical profiling to 3 optical depths below the ocean surface. The atmospheric HSRL measurements have been compared to both in situ and other remote sensors [5] and continues to provide key validation data for CALIOP sensor on the CALIPSO satellite [6].

2. METHODOLOGY

The HSRL technique relies on the difference between the spectral distributions of 180° backscatter from particles and molecules [2]: backscatter from particles (e.g., plankton or atmospheric aerosols) is at the same frequency as the transmitted laser pulse, whereas molecular backscatter is shifted in frequency. The basic diagram of the lidar system is provided in Fig 1A. As shown, an iodine vapor cell functions as a spectral filter in the receiver and frequency discrimination is achieved by tuning the narrowband laser to an absorption line of the iodine gas. The spectral distribution of the backscatter is shown in Fig. 1B for both the atmosphere and the ocean. The Co-polarized and Cross-polarized Channels measure both molecular

 DOI: 10.1051/epjconf/201611922001

and particulate backscatter, while the Molecular Channel measures only the molecular backscatter. Backscatter from water molecules in the ocean (blue) is dominated by Brillouin scattering, which is shifted both up and down in frequency from the source laser by ~7-8 GHz while backscatter from air molecules (red) is both Brillouin and Doppler shifted (Cabannes) scattering, resulting in broadening of about 3 GHz. Molecular scattering is therefore transmitted by the iodine vapor filter, the shaded regions shown in Fig. 1B, while backscatter from particles in the atmosphere and ocean is effectively blocked. The choice of the particular iodine line can be configured such that the ocean Brillouin scattering is insensitive to changes due to changes in ocean temperature and salinity.

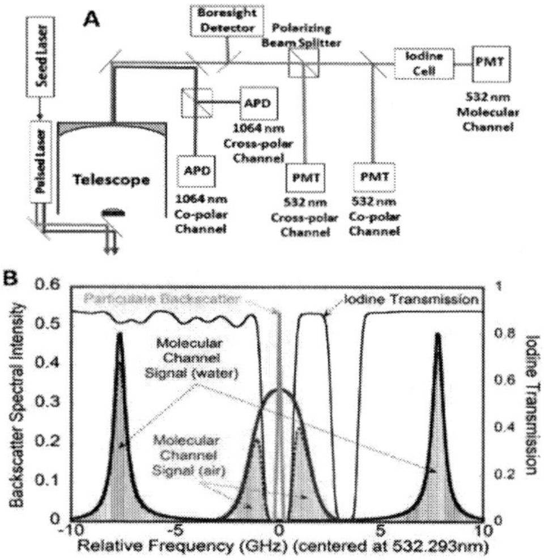

Figure 1. (A) Block diagram of the HSRL-1 instrument and (B) the spectra of backscatter signals at 532 nm relative to the iodine vapor filter absorption features.

The instrument, measurement technique, and calibration methods for the HSRL instrument are outlined in [3] for the atmospheric measurements and are performed similarly for the ocean measurements. As noted above, the instrument was recently upgraded to provide high resolution sampling of the ocean from near the surface to approximately three optical depths (40-45m in the open clear ocean) at a resolution of 1m in the ocean (120MHz sampling). New detection electronics were developed along with integrating new Hybrid Photodetectors (HPD) to reduce the effects of after-pulsing. In addition, the laser was upgraded to narrow the pulse width (6ns) and increase the pulse repetition rate (4000Hz) with slightly higher energies (2.5mJ).

3. RESULTS

During SABOR we conducted 24 science flights split evenly between bases in Portsmouth NH, St. George's Bermuda, and Hampton, VA onboard the NASA LaRC King Air aircraft which flew at an altitude of 8-9km. The flights included overpasses of the RV Endeavor research vessel during the campaign, thus providing coordinated measurements made remotely and in situ of the seawater.

The flight track of the first science flight on 18 July 2014 is shown in Fig. 2 (inset); the track is color-coded with aerosol optical depth (532nm). The combined atmosphere and ocean backscatter for the region in the highlighted red box is also shown. This segment shows a region that covered a large dynamic range in aerosol optical depth (~0.04 – 0.14) and a large dynamic range in the particulate backscatter in the ocean (0.3-16.7 $km^{-1}sr^{-1}$). Note that the altitude scale for the atmosphere is in kilometers and meters for the ocean. The large atmospheric backscatter above 4km in the northern portion of the flight was dominated by smoke advected into this region from forest fires in the Western US. This condition was prevalent during most of the flights conducted from Portsmouth NH. This provides an ideal case to compare the atmospheric corrections required for the ocean color retrievals from MODIS (AQUA satellite) and VIIRS (NPP satellite). Also note that the backscatter in the ocean increases around 19UT as the flight track approaches the George's Bank region off the coast of Massachusetts, which is known for high phytoplankton stocks and productivity. The Gulf of Maine also has higher backscatter values that vary with depth compared to the open ocean.

This particular flight also offered a first-of-kind opportunity to compare airborne lidar data with satellite retrievals of the ocean optical properties. It is expected that for the flight altitude and instrument parameters, the lidar attenuation is approximately equal to the exponential decrease of the downwelling irradiance with depth, known as the diffuse attenuation coefficient [8], which is

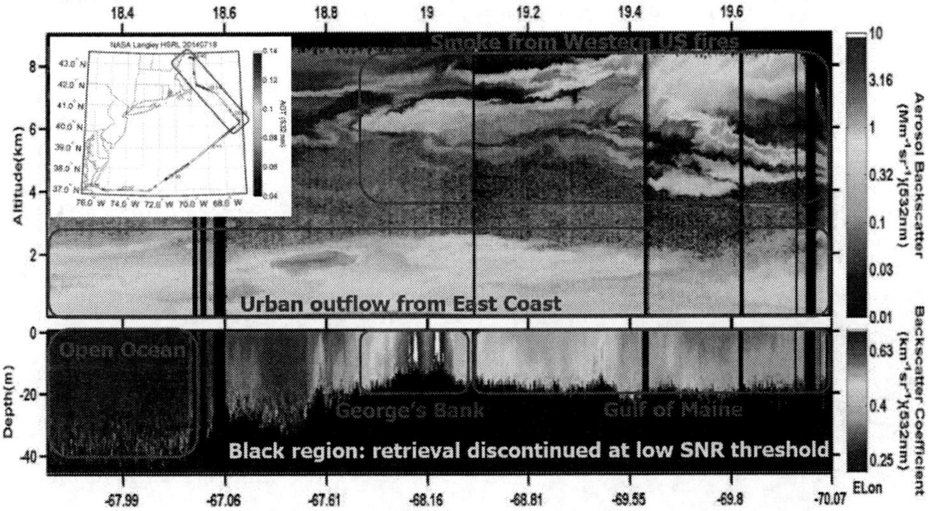

Figure 2. Particulate backscatter coefficient for the atmosphere (upper) and ocean (lower). The black regions in the ocean denote where no retrieval is performed at low SNR or is during a calibration period or aircraft turn.

retrieved by MODIS. The fundamental backscatter measurement from the lidar is the aerosol volume backscatter coefficient, $\beta(\pi)$ [3]. The fundamental retrieval from the satellite is the hemispherical particulate backscatter, (b_{bp}). The lidar hemispherical backscatter is retrieved through a similar approach to that of Boss and Pegau [7] which assumes the following relationship; $b_{bp}=2\pi*\chi(\theta)*\beta(\theta)$. For the lidar, It is noted that the conversion factor, $\chi(\pi)$, between the lidar $\beta(\pi)$ and the hemispherical backscatter can vary depending on size, shapes, and composition of the particle assemblage.

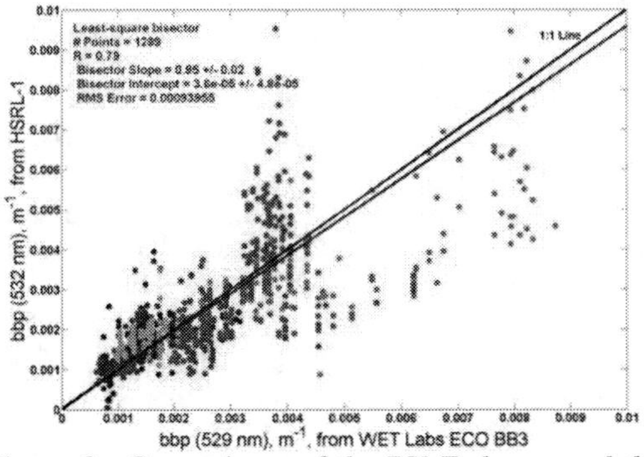

Figure 3. Comparisons of the RV Endeavor and the lidar retrieved hemispherical particulate backscatter. Colors on the graph indicate different sampling locations.

However, during the SABOR campaign we evaluated the correlation between the in situ measurements of b_{bp}, as outlined in [7], and the corresponding values derived from the lidar data and found a strong correlation when using $\chi(\pi)=0.5$ for the lidar as shown in Fig 3. The slope of the correlation was 0.95 with an $R^2=0.79$ using all profile data including the complex coastal waters and the open ocean. The data shown in this figure are in situ profile data collected during specific locations when the aircraft overflew these locations and the aircraft was within 1km. This result using an extensive number of comparisons is encouraging and is unprecedented in terms of using quantitative lidar measurements enabled by the HSRL technique to compare retrievals of backscatter with passive remote sensor retrievals.

Comparisons of lidar- and satellite-based retrievals of diffuse attenuation coefficients are shown if Figs. 4&5. The flight track, color coded to show the lidar retrieval of diffuse attenuation, is shown with the MODIS QAA retrievals (4-km resolution) of diffuse attenuation [9]. A correction of the water absorption between 532nm, the wavelength the lidar measures, and the MODIS retrievals at 488nm is applied. No conversion is made for the particulate spectral dependence in this case. For this flight there was a high correlation (R=0.94, slope=0.97) between the

DOI: 10.1051/epjconf/201611922001

interpolated MODIS and lidar attenuation values (not shown). During the campaign, we composited the comparisons for all flights that had measurements from the two instruments and the correlation was impressive over a range of approximately 0.03-0.23 m^{-1} with an R=0.96, slope=1.0 as shown in Fig 5. Similar comparisons were made with the MODIS GSM retrievals with an R=0.9 and a slope=0.75.

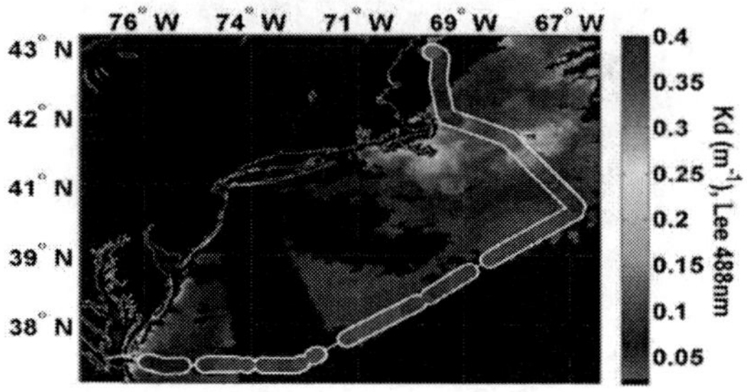

Figure 4. MODIS QAA retrieval of diffuse attenuation compared with the HSRL retrieved attenuation.

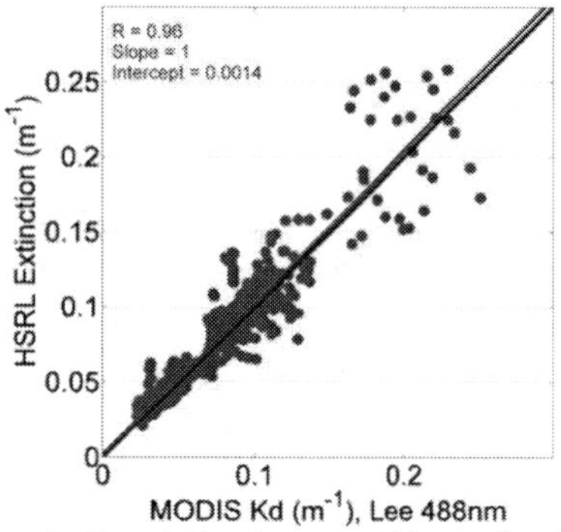

Figure 5. Correlation between lidar extinction and the MODIS QAA diffuse attenuation retrievals.

Similar to the diffuse attenuation results, comparisons were also made between estimates of particulate hemispherical backscatter (b_{bp}) with no spectral corrections. Although the correlations are not as high for the backscatter data, the results offer an unprecedented set of measurements to evaluate the various ocean retrieval models. For example, the correlations of backscatter between the lidar and the satellite retrievals varied by R=0.76, 0.88, and 0.82 and slope= 1.1, 1.3, and 1.5 between the different retrieval algorithms, QAA, GSM, and GIOP [9], respectively.

4. SUMMARY

Results described here demonstrate that an HSRL lidar can provide both vertical profiling and direct measurements of ocean optical properties that can be used to assess and potentially improve retrievals from current and future passive ocean color sensors. The combination of these sensors will provide significant improvements in our understanding and ability to assess global ocean measurements such as particulate organic carbon and net primary productivity. It is also clear that the full benefit of lidar measurements to ocean science community has not yet been fully realized.

ACKNOWLEDGEMENT

We acknowledge the Ocean Biology and Biogeochemistry and Radiation Sciences Programs at NASA Headquarters for support. We also acknowledge the Research Services Directorate at LaRC for supporting the King Air flights.

REFERENCES

[1] Behrenfeld, M.J. and E. Boss, Beam attenuation and chlorophyll concentration as alternative optical indices of phytoplankton biomass. Journal of Marine Research, 2006. 64(3): p. 431-451.

[2] Behrenfeld, Michael J., et al. "Space-based lidar measurements of global ocean carbon stocks." Geophysical Research Letters 40.16 (2013): 4355-4360.

[3] Hair, J.W., et al., Airborne High Spectral Resolution Lidar for profiling aerosol optical properties. Applied Optics, 2008. 47(36): p. 6734-6752.

[5] Rogers, R. R., et al. NASA LaRC airborne high spectral resolution lidar aerosol measurements during MILAGRO: observations and validation. Atmospheric Chemistry and Physics 9.14 (2009): 4811-4826.

[6] Rogers, Raymond R., et al. Assessment of the CALIPSO Lidar 532 nm attenuated backscatter calibration using the NASA LaRC airborne High Spectral Resolution Lidar. Atmospheric Chemistry and Physics 11.3 (2011): 1295-1311.

[7] Boss, Emmanuel, and W. Scott Pegau. Relationship of light scattering at an angle in the backward direction to the backscattering coefficient. Applied Optics 40.30 (2001): 5503-5507.

[8] Gordon, Howard R. "Interpretation of airborne oceanic lidar: effects of multiple scattering." Applied Optics 21.16 (1982): 2996-3001.

[9] NASA Ocean Biology (OB.DAAC). Moderate Resolution Imaging Spectrometer (MODIS) Ocean Color Data, 2014.0 Reprocessing. NASA OB.DAAC, Greenbelt, MD, USA.

DOI: 10.1051/epjconf/201611922002

DEVELOPMENT OF FLUORESCENCE IMAGING LIDAR FOR BOAT-BASED CORAL OBSERVATION

Masahiko Sasano[1]*, Motonobu Imasato[1], Hiroya Yamano[2], Hiroyuki Oguma[3]

[1]Underwater Technology Center, National Maritime Research Institute,

*Shinkawa 6-38-1, Mitaka, Tokyo 181-0004, Japan, *Email: sasano@nmri.go.jp*

[2]Center for Environmental Biology and Ecosystem, National Institute for Environmental Studies,

[3]Centerfor Environmental Measurement and Analysis, National Institute for Environmental Studies,

Onogawa 16-2, Tsukuba, Ibaraki 305-8506, Japan

ABSTRACT

A fluorescence imaging lidar system installed in a boat-towable buoy has been developed for the observation of reef-building corals. Long-range fluorescent images of the sea bed can be recorded in the daytime with this system. The viability of corals is clear in these fluorescent images because of the innate fluorescent proteins. In this study, the specifications and performance of the system are shown.

1. INTRODUCTION

Reef-buildings Corals (hereafter corals) are mainly distributed in tropical shallow water (0-30 m in depth) and play a role in primary producers of coral reefs. Despite the importance as a species, it has been reported that coral distribution areas are rapidly diminishing because of various marine environmental factors [1]. In addition, it is predicted that global climate change (such as ocean warming and ocean acidification) will affect decline of corals [2]. Therefore, understanding their current status is important. Coral observation is considered an urgent requirement.

As corals are marine organisms, wide-area observations are technically difficult. In air surveillance, the waves on the sea surface and the absorption and scattering of light in seawater interfere with coral observation. In underwater surveillance, the field of view (FOV) is narrow, and navigation is slow. In this study, a boat-based surveillance technique has been developed as an improved method of coral observation beneath the sea surface.

Ordinary boat-based video observation has several weaknesses:

- The clarity of video images depends on the amount of solar radiation (cloudy conditions) because of passive observation;
- Video image resolution is affected by image blurring due to boat propulsion and motion;
- Sometimes checking coral viability is difficult because of short observation times and insufficient information.

In this study, a new active coral observation method has been developed to address the above mentioned problems.

Most corals have innate fluorescent proteins in the surface of their flesh, and they emit fluorescence with colors ranging from blue to yellow-green by ultraviolet (UV) excitation. Their fluorescence lifetime is approximately 1 to 3 ns. When corals die, the fluorescent proteins are degraded and no longer emit fluorescence. Thus, the detection of UV excited fluorescence is a sign of live corals.

2. METHODOLOGY

Observing UV excited fluorescence of corals is preferred in the nighttime in order to avoid background sunlight. However, surveying in the daytime is necessary because the boat operator must confirm the sea traffic safety and the safety of operating the boat with the shallow sea bottom in the observed coral reef area. Therefore, a fluorescence imaging lidar system has been developed for operating in the daytime.

In this system, the wavelength of the UV pulsed laser is 355 nm, the pulse width is approximately 9 ns, and the exposure time of the ICCD camera is approximately 100 ns. The effect of background sunlight can be suppressed by setting a very short exposure time. In addition, the image blurring caused by the boat motion does not occur because of the very short duration of the laser irradiation. Thus, clear fluorescent sea bed images would be obtained in the daytime.

Table 1 Specifications of the lidar.

Laser	wavelength	355 nm
	pulse width	9 ns
	power	90mJ / pulse
	spread angle	1~100 mrad (adjustable)
ICCD	gate width	100 ns
	gain	10^9 (max)
DAQ	repetition	5 Hz

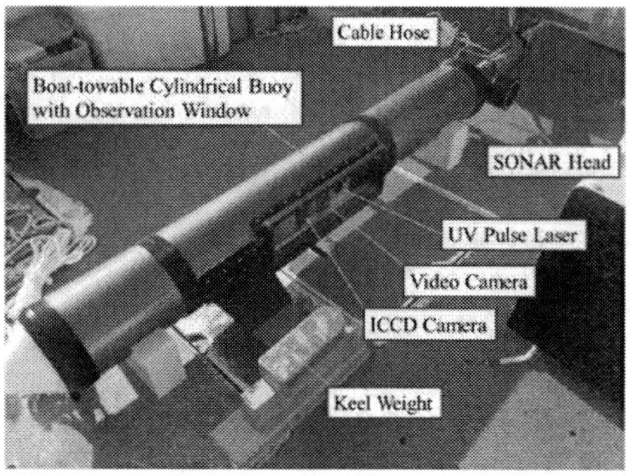

Fig.1 Fluorescence imaging lidar system installed in a boat-towable cylindrical buoy.

3. RESULTS

Coral observations have been operated using this lidar system in Taketomi Island, Okinawa, Japan (24.33N, 124.09E). A large number of fluorescent images of the sea bed in shallow coral reef sea area have been successfully obtained. In addition, passive sea bed images, boat position data with an accuracy of 1 m, and bathymetry data with an accuracy of 0.1 m were also simultaneously recorded using a video camera, DGPS, and SONAR, respectively. Sample images from the video and the lidar are shown in Fig. 2.

Estimating the coral viability is much easier using fluorescent images than video images because of the intensity of the coral fluorescence. The boat-based lidar surveillance around the Taketomi Island (boat track ~14 km long) was recorded with a repetition of 5 Hz, and the total observation time was approximately 3 h.

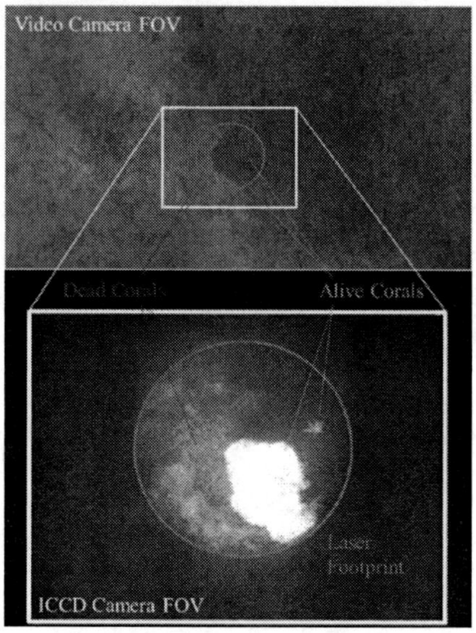

Fig.2 Sample image from recorded video (upper) and fluorescence imaging lidar (lower).

4. CONCLUSIONS

A fluorescence imaging lidar system installed in a boat-towable buoy has been developed, and sea bed fluorescent images were successfully obtained in the daytime. This study indicates that the observed lidar data are useful in surveying coral distributions.

ACKNOWLEDGMENT

This study was funded by the Global Environment Research Account for National Institute, Japan. The authors express their appreciation to all those who participated in this study, and thank the Ministry of Land, Infrastructure, Transport and Tourism of Japan (MLIT) and the Ministry of the Environment of Japan (MOE) for permitting publication of this paper.

REFERENCES

[1] Clive Wilkinson, et al., “Status of Coral Reefs of the World: 2008”, Global Coral Reef Monitoring Network report (2008).

[2] IPCC Working Group II 5th Assessment Report (2014).

EPJ Web of Conferences **119**, 22003 (2016) DOI: 10.1051/epjconf/201611922003
ILRC 27

OCEAN LIDAR MEASUREMENTS OF BEAM ATTENUATION AND A ROADMAP TO ACCURATE PHYTOPLANKTON BIOMASS ESTIMATES

Yongxiang Hu[1]*, Mike Behrenfeld[2], Chris Hostetler[1], Jacques Pelon[3], Charles Trepte[1], John Hair[1], Wayne Slade[4], Ivona Cetinic[5], Mark Vaughan[1], Xiaomei Lu[1], Pengwang Zhai[6], Carl Weimer[7], David Winker[1], Carolus C. Verhappen[1], Carolyn Butler[1], Zhaoyan Liu[1], Bill Hunt[1], Ali Omar[1], Sharon Rodier[1], Anne Lifermann[2], Damien Josset[8], Weilin Hou[8], David MacDonnell[1], Ray Rhew[1]

*[1]NASA Langley Research Center, Hampton, VA 23681, USA, *Email: Yongxiang.hu-1@nasa.gov*
[2]Oregon State University, Corvalis, OR, USA; [3]Centre National d'Etudes Spatiales, France
[4]Sequoia Sci. Inc., USA; [5]University of Maine, Walpole, ME 04573; [6]UMBC, MD, USA;
[7]Ball Aerospace Corp., CO, USA; [8]NRL Stennis, MS, USA

ABSTRACT

Beam attenuation coefficient, *c*, provides an important optical index of plankton standing stocks, such as phytoplankton biomass and total particulate carbon concentration. Unfortunately, *c* has proven difficult to quantify through remote sensing. Here, we introduce an innovative approach for estimating *c* using lidar depolarization measurements and diffuse attenuation coefficients from ocean color products or lidar measurements of Brillouin scattering. The new approach is based on a theoretical formula established from Monte Carlo simulations that links the depolarization ratio of sea water to the ratio of diffuse attenuation K_d and beam attenuation C (i.e., a multiple scattering factor).

On July 17, 2014, the CALIPSO satellite was tilted 30° off-nadir for one nighttime orbit in order to minimize ocean surface backscatter and demonstrate the lidar ocean subsurface measurement concept from space. Depolarization ratios of ocean subsurface backscatter are measured accurately. Beam attenuation coefficients computed from the depolarization ratio measurements compare well with empirical estimates from ocean color measurements. We further verify the beam attenuation coefficient retrievals using aircraft-based high spectral resolution lidar (HSRL) data that are collocated with in-water optical measurements.

1. INTRODUCTION OF THE CONCEPT

Though originally designed for retrieving spatial and optical properties of clouds and aerosols, new applications of CALIOP measurements suggest that space-based lidars can provide physical properties of ocean surface [1] and subsurface [2][3]. Lidars can be used for retrievals of particulate backscattering, diffuse attenuation coefficients, the size spectrum and vertical distribution of ocean particles. These retrievals complement products derived from passive ocean color sensors and can contribute to reduced uncertainties in global ocean plankton stocks, primary productivity, and carbon export estimates. Here, we introduce an innovative approach for retrieving the beam attenuation coefficient from the subsurface depolarization ratio measured with space-based and/or aircraft-based lidars.

The beam attenuation coefficient, *c*, has proven an elusive ocean property to retrieve from remote sensing measurements [4]. One approach has been to estimate *C* from chlorophyll concentration [5], but this approach can suffer from chlorophyll (1) being influenced by physiological processes (i.e., intracellular changes in pigmentation in response to light and nutrient conditions) [6] and (2) not providing a robust index of the non-phytoplankton particle populations. Due to the highly forward peaked scattering phase function in water (with asymmetry factor around 0.95), we can only measure the effective attenuation coefficient (K_d), which is linked to *C* through the so-called multiple scattering factor, η ($\eta = K_d/c$). Accurately quantifying the magnitude and effects of multiple scattering is the primary obstacle in obtaining reliable measurements of *c* [4].

Multiple scattering can cause depolarization [7]. Thus for non-absorbing media with spherical particles, η can be estimated accurately from lidar depolarization measurements [8][9][10]. Monte Carlo simulations of ocean lidar backscatter suggest that a similar relationship between multiple scattering factor and depolarization ratio (δ) exists for absorbing media as well, i.e.,

$$\eta = \left(\frac{\varpi^2 - \delta}{\varpi^2 + \delta} \right)^2, \qquad (1)$$

where ϖ is the ratio of scattering and extinction coefficients for the water and its constituents. For open ocean at 532 nm, $\varpi \approx 1 - \eta$ [11] and thus,

$$1 - \varpi \approx \left(\frac{\varpi^2 - \delta}{\varpi^2 + \delta} \right)^2. \qquad (2)$$

Solving Eq. 2, ω and η (=1-ω) can be derived from depolarization, δ (red line in Fig. 1),

$$K_d / c = 1 - \varpi = e^{-f(\delta)}. \qquad (3)$$

When $\delta > 0.002$, $f(\delta) \approx 0.222 + 19.46\delta - 188.39\delta^2 + 1288.1\delta^3 - 3974.9\delta^4 + 4684.2\delta^5$ and

$$c = K_d e^{f(\delta)}. \qquad (4)$$

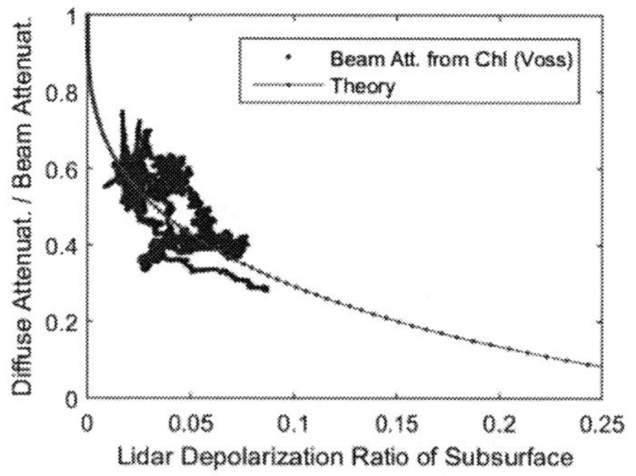

Figure 1. Solution of Equation (2) (red line). Comparison with collocated MODIS *c* (Voss, 1992).

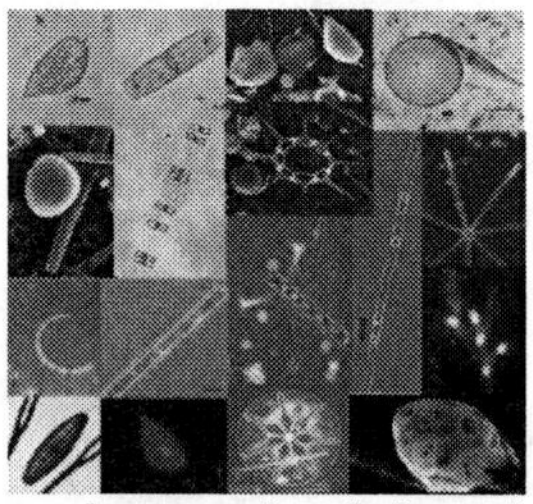

Figure 2. Microphotographs of phytoplankton cells demonstrating high diversity in morphological structures.

Eq. 2 is valid for sea waters with relatively small depolarization in backscatter direction. Depolarization ratios are near zero for backscatter by density fluctuation (Brillouin scattering) and for small soft particles with small relative refractice index. It is also likely valid for sea waters that may include some larger, non-spherical particulates, since the contribution of larger particulate to backscatter is relatively small and its bulk scattering properties are determined primarily by the tiny structures within the particles (Fig. 2) that have single scattering properties similar to spherical particles.

2. CALIPSO 30° Tilt: DEMONSTRATING OCEAN LIDAR IN SPACE

CALIOP's vertical resolution at 532 nm is 30 m. At average open ocean surface wind speeds (~6 m/s), the attenuated backscatter (532 nm parallel channel) from the ocean surface is about 30 times stronger than the subsurface backscatter. This makes it difficult to estimate the depolarization ratio of light backscattered by the ocean subsurface and to show clear ocean subsurface signals from CALIOP measurements.

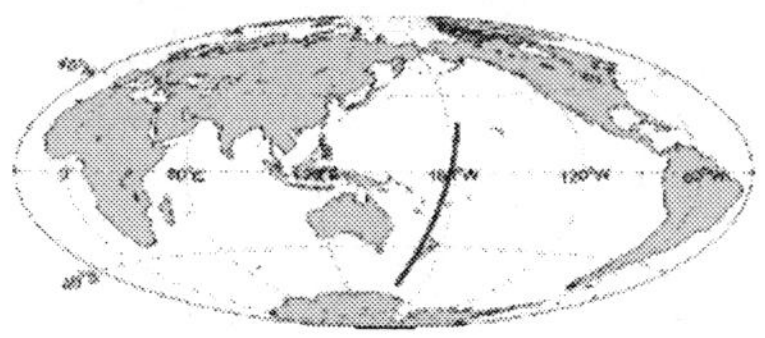

Figure 3. Orbit track when CALIPSO spacecraft is tilted 30° backward on July 17, 2014.

If the lidar is pointed 30^0 off-nadir, the ocean surface signal is reduced by more than two orders of magnitude [4], and CALIOP can then accurately measure ocean subsurface backscatter. The small surface contribution to 532 nm subsurfacebackscatter can be removed using 1064 nm measurements as its subsurface signals are near zero due to stronger absorption by water. Thus a 30° tilt of the CALIPSO satellite (and thus the CALIOP lidar) can help demonstrate our space-based ocean lidar concept.

CNES and NASA tilted the CALIPSO satellite 30^0 forward on July 17, 2014 (Fig. 3) in order to make accurate ocean subsurface backscatter measurements. During this special operation, CALIOP clearly detected ocean subsurface signals from both 532 nm parallel (upper panel, Fig. 4) and perpendicular (middle panel, Fig. 4) channels. Very little backscatter is seen in the 1064 nm channel (lower panel of Figure 4) near the ocean surface. It suggests that the ocean surface does not contribute to the subsurface signal in the 532 nm channels, because ocean surface backscatter at 532 nm is about 30% less than ocean surface backscatter at 1064 nm. Both the parallel and perpendicular components of 532nm

DOI: 10.1051/epjconf/201611922003

backscatter are measured accurately by CALIOP at 30^0 off-nadir. The column integrated depolarization ratio of ocean subsurface backscatter can be accurately measured.

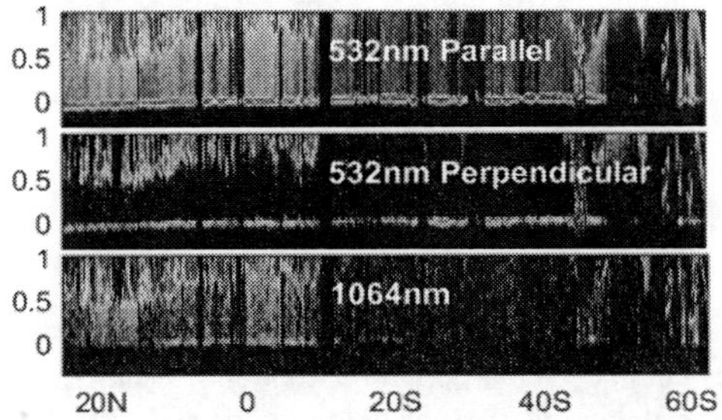

Figure 4. The lowest 1 km CALIOP backscatter profiles. Upper panel: 532 nm parallel; lower panel: 1064 nm total; middle panel: 532nm perpendicular.

3. COMPARISON OF BEAM C: CALIOP vs MODIS

Figure 5 shows beam c (red line) derived from CALIOP depolarization measurements (blue line) together with collocated MODIS diffuse attenuation coefficient estimates scaled to 532 nm (black line). Difference between CALIOP's *c* estimates and *c* based on MODIS chlorophyll measurements (green line) [7] are mostly within 30%.

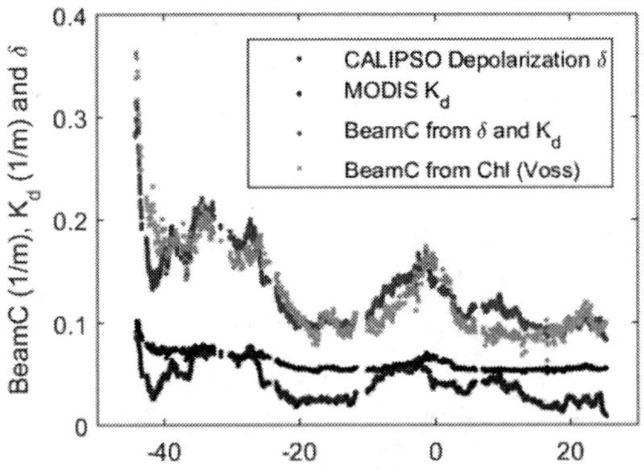

Figure 5. Beam attenuation coefficient comparisons between the lidar (red) method and the chlorophyll (green) method.

4. COMPARISONS BETWEEN AIRCRAFT LIDAR AND IN WATER MEASUREMENTS

During July 2014, NASA's Ship-Aircraft Bio-Optical Research project (SABOR) acquired both aircaft HSRL measurements [12] and in-water optical measurements along the track of an ocean-going research vessel [13]. Here we compare the beam attenuation coefficients derived from the lidar and the in-water measurements for the aircraft flight on July 26, where the flight track (red line in Figure 6) is close to the ship track (green).

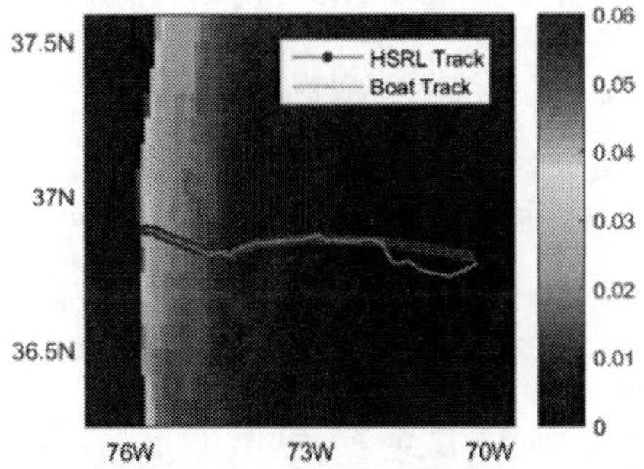

Figure 6. Aircraft track (red line) and track of the research vessel (green). The background is CDOM absorption coefficient (m^{-1}) estimated rom MODIS.

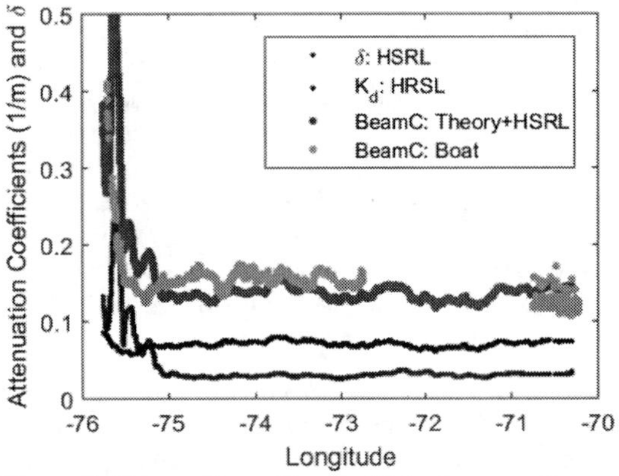

Figure 7. Beam attenuation coefficient comparisons between the HSRL lidar (red) and in water (green) measurements.

Because Brillouin backscatter is frequency shifted, HSRL can make separate measurements of particulate backscatter $\beta_P(z)$ and Brillouin scattering $\beta_B(z)$ profiles. HSRL provides accurate subsurface depolarization measurements (blue line in Figure 7), and can measure K_d directly from the vertical Brillouin backscatter profile as $K_d = -\Delta\log[\beta_B(z)]/\Delta z$. K_d can also be computed from the column integrated Brillouin signal, which is inversely proportional to diffuse attenuation (black line in Figure 7),

$$K_d = a\beta_m / \beta_B \ , \qquad (5)$$

where β_m is the molecular backscatter signal of the air right above the ocean surface, and a is a constant related to instrument filter characteristics and theoretical molecular backscatter coefficients of the air and the water near ocean surface.

Using Eq. 4, beam attenuation (red line in Fig. 7) was computed from the HSRL ocean surface depolarization ratio (blue line in Fig. 7) and diffuse attenuation coefficient (black line). The beam attenuation coefficients compare reasonably well with the in-water measurements (green line in Fig. 7). However, the in-water measurements were made several days after the aircraft measurements, as a frontal system moved through the region just after the July 26 flight. This time offset might be partially responsible for the differences between two measurements.

5. CONCLUSIONS

This study introduces an approach for estimating the beam attenuation coefficient, C, from lidar depolarization measurements and ocean color- or lidar-based diffuse attenuation coefficients. The concept is based on a theoretical formula established from Monte Carlo simulations that links depolarization ratio of sea water to the multiple scattering factor (the ratio of diffuse attenuation and c). The lidar-based C retrievals compare reasonably well with ocean color based estimates and in-water measurements.

In the future, the improved HSRL measurements at both 355nm and 532nm could further improve the accuracy of diffuse attenuation coefficient retrievals. Furthermore, CDOM absorption could be estimated from the dual wavelength K_d measurements [14], while dual wavelength backscatter and extinction retrievals could provide information on particle size distributions [15] and composition. By extending these developments to a space-based lidar, significant advances could be realized in quantifying ocean carbon cycle processes.

ACKNOWLEDGEMENT

The authors would like to thank the CNES mission operations team for enabling the 30° pitch maneuver and Drs. Paula Bontempi, Hal Maring and David Considine of NASA CALIPSO, ACE, and SABOR projects and NASA radiation science program for supporting this study.

REFERENCES

[1] Hu, Y., et al., 2008: "Sea surface wind speed estimation from space-based lidar measurements", Atmos. Chem. Phys., **8**, 3593-3601.

[2] Behrenfeld, M.J., et al. (2013), Space-based lidar measurements of global ocean carbon stocks, Geophys. Res. Lett., **40**, doi:10.1002/grl.50816.

[3] Lu, X., et al. (2014), Ocean subsurface studies with the CALIPSO spaceborne lidar, J.G.R., **119**, 4305–4317, doi:10.1002/2014JC009970.

[6] Behrenfeld, M. and E. Boss (2006), Beam attenuation and chlorophyll concentration as alternative optical indices of phytoplankton biomass. J. Mar. Res . **64**, 431-451.

[4] Boss, E. et al., (2009), Acceptance angle effects on the 42 beam attenuation in the ocean. Optics Express **17**(3); 1535-1550.

[5] Voss, K.J (1992), A spectral model of the beam attenuation coefficient in the ocean and coastal areas, Limn. and Ocean., 37, doi: 10.4319/lo.1992.37.3.0501.

[7] Churnside, J. (2014), Review of profiling oceanographic lidar, Optical Engineering, 53(5) 051405 doi: 10.1117/1.OE.53.5.051405.

[8] Hu, Y. et al. (2006), A simple relation between lidar multiple scattering and depolarization for water clouds, Optics Letters, **31**, 1809-1811.

[9] Hu, Y. (2007), Depolarization ratio–effective lidar ratio relation: Theoretical basis for space lidar cloud phase discrimination, Geophys. Res. Lett., **34**, L11812, doi:10.1029/ 2007GL029584.

[10] Hu, Y. et al. (2007), The depolarization-attenuated backscatter relation: CALIPSO lidar measurements vs. theory, Opt. Exp., **15**, 5327–5332.

[11] Gordon, H.R. (1989), Theoretical aspects of hydrologic optics, Limnol. Oceanogr., 34, 1389-1409.

[12] Hair, J.W., et al. (2008), Airborne High Spectral Resolution Lidar for profiling aerosol optical properties, Appl. Opt., **47**, 6734-6752

[13] Hostetler, C.A et al. (2014), Applications of lidar systems for ocean ecosystem and ocean-atmosphere studies, Oc. Opt. XXII, Portland, ME, 26-31 October.

[14] Mannino, A. et al. (2013), Algorithm development and validation of CDOM properties for estuarine and continental shelf waters along the northeastern U.S. coast, Rem. Sens. of Env., **152**, 576-602.

[15] Kostadinov, T. S., et al. (2009), "Retrieval of the particle size distribution from satellite ocean color observations." JGR: Oceans (1978–2012) 114.C9.

EPJ Web of Conferences **119**, 22004 (2016) DOI: 10.1051/epjconf/201611922004
ILRC 27

DAILY EVOLUTION OF THE INSECT BIOMASS SPECTRUM IN AN AGRICULTURAL LANDSCAPE ACCESSED WITH LIDAR

Mikkel Brydegaard [1,2,4]*, Alem Gebru[1,2,3], Carsten Kirkeby[5], Susanne Åkesson[2], Henrik Smith[2]

[1]*Lund Laser Centre (LLC), Lund University, Sölvegatan 14, 22362 Lund, Sweden, * mikkel.brydegaard@fysik.lth.se*
[2]*Centre for Animal Movement Research (CAnMove), Lund University, Sölvegatan 37, 22362 Lund, Sweden*
[3]*Laser Research Institute (LRI), Stellenbosch University, Private Bag X1, 7602 Matieland, South Africa*
[4]*Norsk Elektro Optikk AS (NEO), Prost Stabels vei 22, N-2019 Skedsmokorset, Norway*
[5]*National Veterinary Institute, Technical University of Denmark, Bülowsvej 27, 1870 Frederiksberg C, Denmark.*

ABSTRACT

We present measurements of atmospheric insect fauna intercepted by a static lidar transect over arable and pastoral land over one day. We observe nearly a quarter million of events which are calibrated to optical cross section. Biomass spectra are derived from the size distribution and presented against space and time. We discuss detection limits and instrument biasing, and we relate the insect observations to relevant ecological landscape features and land use. Future directions and improvements of the technique are also outlined.

1. BACKGROUND AND MOTIVATION

Anthropogenic exploitation of the environment, such as agricultural intensification including pesticide use, has profound effects on the atmospheric fauna in the agricultural landscape[1]. The land use may have severe negative effects on insect and natural insect predators[2,3]. The consequences of anthropogenic agricultural activity and its impact on the composition of the air-born fauna is however less known. It is expected that even specific perturbations to the population of a particular size of an organism, e.g. by using a targeted pesticide, may propagate as waves throughout the biomass spectrum, affecting populations of both smaller and larger sized organisms[4], for example through competition and the trophic chain. The effect is analogous to hole-burning spectroscopy, well known in the laser and electron population community[5].

A more heterogeneous landscape including, field borders, vegetated islands, trees, pastures used for grazing and higher variation in crop diversity is likely to have a larger abundance and diversity of the farmland insect fauna[6]. There is furthermore a current global concern about the declining pollinator fauna in the agricultural landscape[7]. Here we explore the potential of using a remote sensing technique to investigate both the diversity of the insect fauna, but also the insect occurrence in relation to field use and landscape borders. In particular, we compare the faunistic diversity between an oat field and a cattle grazed pasture at Brunslöv in south Sweden. We present the biomass spectra derived from the optical cross section (OCS) of lidar-sensed atmospheric insect fauna during one day. We define backscatter OCS as the product of diffuse reflectance and projected area, and extinction OCS as the product of ballistic opaqueness and the projected area.

2. SCENARIO AND INSTRUMENTATION

We recently developed several atmospheric CW lidars based on the Scheimpflug principle for various applications[8,9]. The lidar, shown in Fig.1, is based on telescopes on a rectangular Al crossbar - mounted on an equatorial mount with 50 kg payload capacity. The baseline separation between transmitter and receiver is 81 cm.

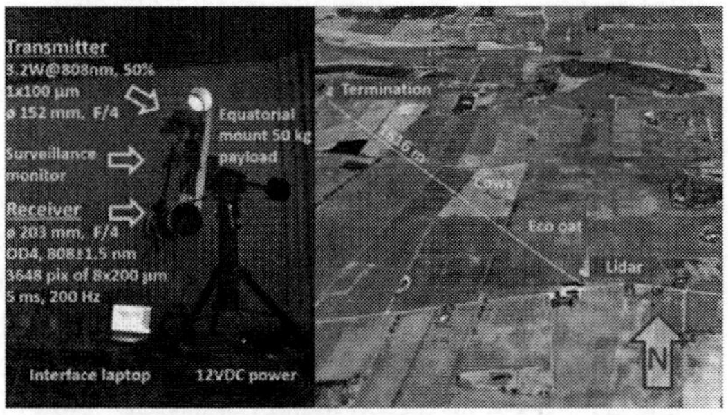

Fig.1a. *NIR photograph of the setup during field campaign.* ***1b****. The monitored transect in the vicinity of Brunslöv, southern Sweden.*

The lidar monitored a static transect of 1816 m over agricultural- and grasslands in southern Sweden. The range of interest was 30-1000 m. The lidar was positioned at 55°44'17.29"N, 13°37'48.91"E, 120 m ASL and terminated at 55°44'58.70"N, 13°36'35.25"E, 136 m ASL. The beam propagated 3-8 m over ground and the total air volume monitored is estimated to 6.2 m^3. The measurements included one week continuous monitoring in mid-July, during which the sky was clear and the atmosphere homogeneous. Here we present data recorded 16th July, 2014 starting at midnight and lasting until 21:00h the following evening, at which time the sun light entered the telescope and the experiment was terminated. At this date the winds were calm 0-4 m/s, and temperatures ranged between 289-295 K.

The transmitter is a 3.2W GaAlAs diode laser emitting at 808 nm, operated at a 50% duty cycle and expanded to ø152 mm with an F/4 refractor telescope. The source slow axis is oriented perpendicular to the lidar baseline and imaged onto the termination with a final beam size of 97x452 mm.

The receiver consists of an ø203 mm F/4 Newtonian reflector telescope. Background light is suppressed by an RG780 absorption filter and a band pass filter centered at 808 nm-with a FWHM of 3 nm. The backscattered light is imaged onto a linear Si-CCD array with 3648 pixels and a pixel size of 8x200 μm. The detector is tilted 45° according to the Scheimpflug principle to obtain infinite focal depth. The detector is operated with 5 ms exposure time at 200 Hz line rate. The data are transferred via USB2, and a strobe signal is used to alternate between the dark and bright state of the laser. After interpolating and subtracting background the effective sample rate becomes 100 Hz.

3. CALIBRATION AND SENSITIVITY

The lidar generates range-time maps in files covering 10 s each. Within this time window we made three statistical measures of the 1000 lidar curves; the min-, median- and max-curve, see Fig.2. The median gives the best approximation of the static expectation value. The discrepancy between the min- and static curve, represents system noise. The range dependent threshold is defined by any events exceeding the static signal plus twice the noise (SNR=2), these were considered as atmospheric fauna. On July 16, 2014 we identified 221.684 events, or ~1.700 events / (m^3 hour).

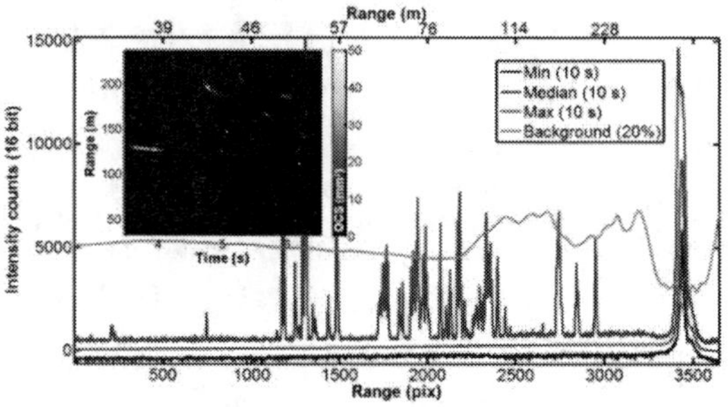

Fig.2 *Min/Median/Max/Background intensity vs. range at noon. Min indicates noise level, median static molecular air return and max atmospheric fauna. The background is rescaled to fit in the graph. The insert shows a close-up of the time-range map where insects are seen as appearing sparsely in space and time.*

The probe volume is constrained by the pixel footprints for the current experiment. To estimate the impinging laser intensity on each footprint the atmosphere is assumed to be homogeneous and insect events are divided by the static signal. Homogeneity is verified by the range-time map of the whole day. This form factor is rescaled through the known OCS, σ_{term}, for the termination; see Eq.1.

$$\sigma_{event} = \frac{\sigma_{term} r^2_{event} (I_{event(r)} - I_{static(r)})}{r^2_{term} I_{static(r)}} \qquad Eq.1$$

This calibration was verified in two ways: 1) The values relate to the apparent size derived from the opening angle (spike width) by a factor square. 2) The cross section for backscatter is one tenth of the extinction which for atmospheric insect fauna is expected to be roughly opaque with a reflectance in the order of 10% at 808 nm.

Equivalently, the noise- and CCD saturation intensity is converted into OCS providing the minimal and maximal observable OCS at any given range and time. Detection limits are 1 mm^2 at short range and 100 mm^2 at 1 km. At night time we cover a span of 1:1000 at any range, while at daytime the dynamical span is partially consumed by background and reducing to 1:100. The span within the entire range was 1:10.000. The saturation does not pose a strict limit since size may be estimated despite CCD blooming in our case.

4. STATISTICS AND BIOMASS SPECTRA

The time-range density of the aerial insect observations during the day reveals a peak activity around 13:00h which is noon in true sun time; see Fig.3. Interestingly, a rapid rush-hour is observed at the shortest range around 6:00h, lasting for only 10 minutes. As expected, activity over the cattle field is generally higher at night and peaking at dawn.

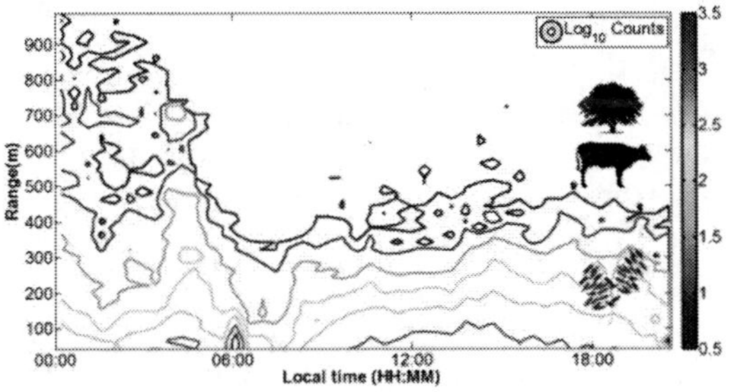

Fig.3 *Density of observations versus time and range.*

To gain further insight into the various spatio-temporal features the evolution of the biomass spectrum over the day is plotted in Fig.4. As a reference the average min and max detection limits are displayed in the background. The largest events are observed at night time, reaching sizes corresponding to birds or bats. At dawn large amounts of tiny insects are observed, and at noon the activity of intermediate sized insects peaks.

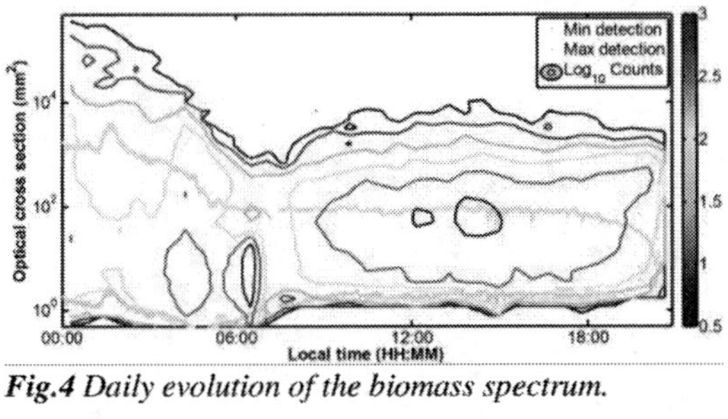

Fig.4 *Daily evolution of the biomass spectrum.*

A valid concern when comparing activity and biomass distributions at various ranges is the biasing induced by detection limits. In Fig.5 the biomass spectrum is plotted as a function of range. As a reference the average detection limits are plotted in the background. It is obvious that the biomass spectra to a large extent are associated with the observable range. However, a bimodality is observed at the shortest range, which could not be explained by instrumentation biasing. On top of this the abundance increases around 700 m range, again displaying a bimodal distribution. At this distance cattle were often resting beneath a large tree. One possible explanation for the high activity at night time and bimodal distribution over the cattle field could be blood-feeding insects or moths predated by bats. Both bats and insect are known to be attracted to solitary trees in the landscape[3].

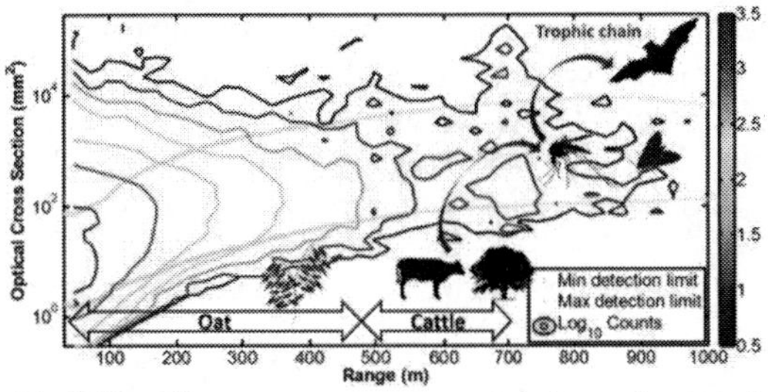

Fig.5 *The biomass spectra as a function of range is largely associated with the detection limits shown in the background. However, bimodalities are observed both in range and sizes.*

5. CONCLUSIONS AND DISCUSSION

Compared to our previous work on lidar entomology including spectral identification[10,11] and kHz modulation spectroscopy[8], the present data are crude in terms of specificity. However, it marks a record of involving a quarter million observations over a 21h period, quantities allowing for accurate distributions in several dimensions as clearly demonstrated in the figures presented in this work. Also, earlier entomological lidar work, e.g.[10,12], have not quantitatively specified cross sections and detection thresholds.

Our study shows the potential of using optical remote sensing to quantify aerial insect fauna associated with land use, farming regimes and crop type in agricultural landscapes. The method has further a great potential in identifying effects of insecticides on pollinators and pest control insects[5].

Throughout the day the temperature changes in the order of 15 K under natural field conditions, in particular when nights are clear. The lidar has weak heaters on the expander and receiver telescopes to prevent condensation in the morning, but is otherwise left drifting in temperature. This can potentially cause the laser emission wavelength to drift out of the narrow band pass filter on the receiver or the lidar structure to twist causing poor overlap of beam and FOV. Thermal stabilization, a power monitor and adaptive overlap mechanics would improve the stability.

The projected cross section is a crude measurement of the actual organism size. Minute details in the biomass spectrum would smear out due to uncertainty regarding the phase in the wingbeat cycle, interception angle and specular reflexes from the wings. We have previously elaborated on how insect scattering can be decomposed into frequency components and assigned projection of the body (DC), wing projection (fundamental tone and low harmonics) and specular reflexes (high harmonics)[13]. In this work we also alluded to how the interception angle may be derived from the harmonics content, which may reduce the spread of cross sections for identical species.

To pursue this work we are currently upgrading the lidar with an InGaAs array detector with line rates up to 10 kHz and a SWIR laser at 1.55μm. Apart from improved sizing, additional parameters of frequency, body size and wing size, fundamental tone, harmonics spectrum[14] and wing glossiness for improved species classification can be achieved. SWIR bands also implies a 10-20 fold improvement in SNR due to the reflectance spectrum of insects[15], thus resulting in lower detection limits at longer distance.

ACKNOWLEDGEMENTS

We are grateful to the Stefan Frank for permission to work at his lands. We thank Samuel Jansson, Johan Kjellman, Tianqi Li, Sune Svanberg and Annika Södermann for assistance in field. The study was financially supported from the Swedish Research Council (621-2013-4361), an infrastructural grant from the Science Faculty at Lund University and by Lund University. This project also received funding from the Swedish Research Council through Linnaeus grants to the Centre for Animal Movement Research (349-2007-8690) and the Lund Laser Centre.

REFERENCES

[1] Grüebler, M. U., Morand, M. & Naef-Daenzer, B. 2008, A predictive model of the density of airborne insects in agricultural environments. *Agriculture, ecosystems & environment* **123**, 75-80.

[2] Wickramasinghe, L. P., Harris, S., Jones, G. & Vaughan Jennings, N. 2004, Abundance and species richness of nocturnal insects on organic and conventional farms: effects of agricultural intensification on bat foraging. *Conservation Biology* **18**, 1283-1292.

[3] Fischer, J., Stott, J. & Law, B. S. 2010, The disproportionate value of scattered trees. *Biological Conservation* **143**, 1564-1567.

[4] Lawl, R., Plank, M. J., James, A. & Blanchard, J. L. 2009, Size-spectra dynamics from stochastic predation and growth of individuals. *Ecol.* **90**, 802–811.

[5] Nilsson, M., Rippe, L., Kröll, S., Klieber, R. & Suter, D. 2004, Hole-burning techniques for isolation and study of individual hyperfine transitions in inhomogeneously broadened solids demonstrated in $Pr^{3+}:Y_2SiO_5$. *Phys. Rev. B* **70**, 214116.

[6] Veres, A., Petit, S., Conord, C. & Lavigne, C. 2013, Does landscape composition affect pest abundance and their control by natural enemies? A review. *Agriculture, Ecosystems & Environment* **166**, 110-117.

[7] Potts, S. G. *et al.* 2010, Global pollinator declines: trends, impacts and drivers. *Trends Ecol. Evol.* **25**, 345-353.

[8] Brydegaard, M., Gebru, A. & Svanberg, S. 2014, Super resolution laser radar with blinking atmospheric particles – application to interacting flying insects *Progress In Electromagnetics Research* **147**, 141-151.

[9] Mei, L. & Brydegaard, M. 2014, Continous-wave differential absorption lidar. *Submitted.*

[10] Guan, Z. *et al.* 2010, Insect monitoring with fluorescence lidar techniques: Field experiments. *Appl. Opt.* **49**, 1-11.

[11] Runemark, A., Wellenreuther, M., Jayaweera, H., Svanberg, S. & Brydegaard, M. 2012, Rare events in remote dark field spectroscopy: an ecological case study of insects. *IEEE JSTQE* **18**, 1573 - 1582.

[12] Hoffman, D. S., Nehrir, A. R., Repasky, K. S., Shaw, J. A. & Carlsten, J. L. 2007, Range-resolved optical detection of honeybees by use of wing-beat modulation of scattered light for locating land mines. *Appl. Opt.* **46**, 3009-3012.

[13] Brydegaard, M. 2014. Towards quantitative entomological laser radar. in *ICO23. Santiago de Compostela, Spain.*

[14] Moore, A. & Miller, R. H. 2002, Automated identification of optically sensed aphid (*Homoptera: Aphidae*) wingbeat waveforms. *Ann. Entomol. Soc. Am.* **95**, 1-8.

[15] Brydegaard, M. 2014. Advantages of shortwave infrared LIDAR entomology. in *Imaging and Applied Optics 2014. Seattle, Washington.* LW2D.6 (Optical Society of America).

FOREST CANOPY HEIGHT ESTIMATION FROM CALIPSO LIDAR MEASUREMENT

Xiaomei Lu[1], Yongxiang Hu[2]*, Patricia L. Lucker[1], Charles Trepte[2]

[1] *Science Systems and Applications, Inc., Hampton, Virginia 23666, USA*

[2]*NASA Langley Research Center, Hampton, VA 23681, USA, *Email: Yongxiang.hu-1@nasa.gov*

ABSTRACT

The canopy height is an important parameter in aboveground biomass estimation. Lidar remote sensing from airborne or satellite platforms, has a unique capability for forestry applications. This study introduces an innovative concept to estimate canopy height using CALIOP two wavelengths lidar measurements. One main advantage is that the concept proposed here is dependent on the penetration depths at two wavelengths without making assumption about the last peak of waveform as the ground location, and it does not require the ancillary Digital Elevation Model (DEM) data in order to obtain the slope information of terrain.

Canopy penetration depths at two wavelengths indicate moderately strong relationships for estimating the canopy height. Results show that the CALIOP-derived canopy heights were highly correlated with the ICESat/GLAS-derived values with a mean RMSE of 3.4 m and correlation coefficient (R) of 0.89. Our findings present a relationship between the penetration difference and canopy height, which can be used as another metrics for canopy height estimation, except the full waveforms.

1. INTRODUCTION

Lidar remote sensing, from airborne or satellite platforms, has a unique capability for forestry applications, such as estimating forest canopy height [1], forest structural attributes [2] as well as aboveground biomass [3, 4]. Airborne small footprint full waveform lidar measurements can provide the best accuracy of canopy height, even for dense forests. However, the use of airborne lidar data is usually limited to local or regional areas. Many researchers have studied the forest canopy height globally from large footprint space-borne lidars [3-7]. For example, the Geoscience Laser Altimeter System (GLAS) on the Ice, Cloud and land Elevation Satellite (ICESat) provides forest canopy height from the full waveform data [8-11].

The methods based on the full waveform required the exact position of the top of the canopy and ground peaks in order to better and accurately estimate the canopy height. Therefore, estimating forest canopy height from large-footprint satellite lidar full waveforms is challenging due to the complex interaction between lidar waveforms, terrain, and vegetation, especially in dense tropical and equatorial forests [5]. For example, terrain slope and roughness are the main factors contributing to canopy height estimation error, since the impact of slope and roughness is to broaden the lidar waveform, and thereby introducing a bias in canopy height estimates. In addition, as canopies become more closed, forest metrics such as waveform extent, trailing and leading edge derived from full waveforms become more difficult because laser pulse returns tend to be preferentially distributed towards the outer envelope of the tree crown, with fewer pulses penetrating past the height above ground level at which canopies become closed. Generally, for the full waveforms method, the ground can be determined as the last Gaussian peak, which works best in flat areas and open canopies. However, within closed canopies, locating the ground is sometimes difficult, for example, when the last peak has low amplitude relative to another neighboring peak [6], or few lidar returns reach the ground in tropical forests. Moreover, over very sloped terrain, laser returns from both canopy and ground surfaces can occur at the same elevation, and as a result it is difficult to estimate the tree canopy height from the full waveform methods. Lidar does have certain foliage penetration capability even for densely vegetated forests. Therefore, it can offer advantage in certain detection and identification applications.

In this paper, we used the penetration difference at two wavelengths to estimate the forest canopy

 DOI: 10.1051/epjconf/201611922005

height from space-borne lidar measurements, such as CALIOP lidar.

2. TWO WAVELENGHT METHOD

CALIOP, a three-channel elastic-backscatter lidar, has been operating since June 2006 and is expected to continue for several more years. Recently, a super-resolution altimetry method has been proposed to provide improved lidar altimetry from CALIOP lidar measurements [12, 13]. The precision of the altimetry products retrieved by the method is much better than 1.4 m [8] and can be as good as GLAS [14]. Since the CALIOP lidar measurements are discontinuous with a footprint of about 70 m spaced every 333 m along ground track, the results from CALIOP data were averaged within 1° cell.

When the laser pulse cannot penetrate the thick foliage of closed forest, the exact location of peak pulse return retrieved by the super-resolution altimetry method [12] is the laser penetration depth. The penetration difference between 1064 and 532 nm was shown in Fig. 1. The penetration difference is likely due to the relative larger reflectance in infrared (>50%) than that in green (<15%) and thus there is more probability of photons penetration through forest canopy at 532 nm than that at 1064 nm. The penetration difference as shown in Fig. 1 is higher than 8 m in the tropical forest near Equator.

The GLAS has 1064- and 532-nm channels the longer wavelength of which is used for surface altimetry. Several canopy height models from GLAS waveforms have been developed in recent years depending on different parameters retrieved from full waveforms and ancillary DEM data [8-11]. The GLAS-derived canopy height within each footprint has a precision between 2 and 13 m depending on forest types and characteristics of the study site. The GLAS canopy height used in this paper is the waveform metric RH100, defined as the distance between signal beginning and the location of the ground peak. This product provides the top of the canopy height that can be downloaded from ORNL Distributed Active Archive Center for Biogeochemical Dynamics. The GLAS-derived canopy height was shown in Fig. 2. The linear relationship between penetration difference at two wavelengths and canopy height was shown in Fig. 3. Color code represents the number of occurrences. The fitted red line of Fig. 3 can be written as:

$$y = 7.0144x \quad 19.5474 \qquad (1)$$

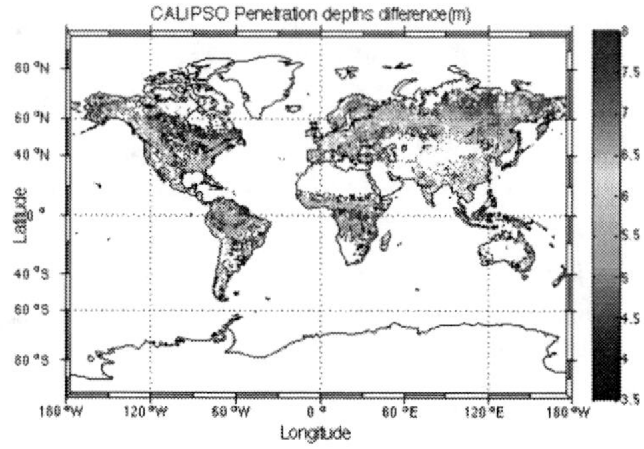

Figure 1. Global distribution (1°×1°) of the penetration differences (meter) between 1064 and 532 nm.

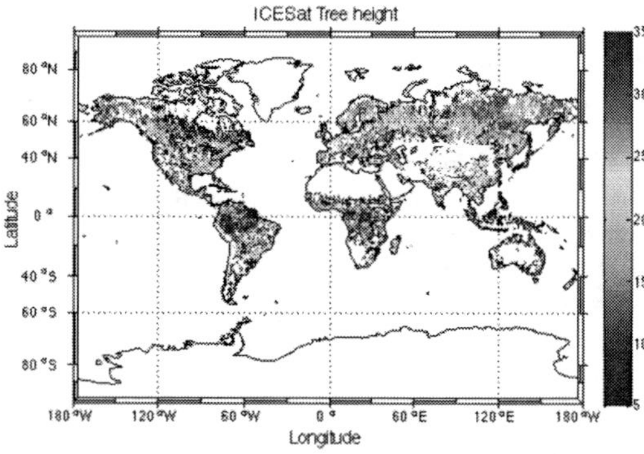

Figure 2. ICESat GLAS-derived canopy height. The unit of the height in color bar is meter.

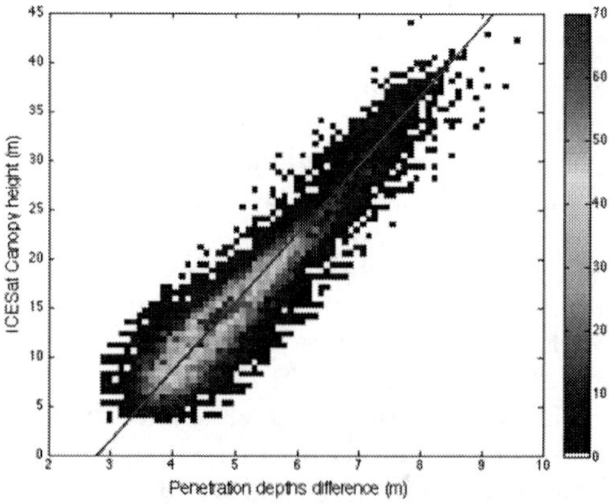

Figure 3. Relationship between penetration difference and canopy height, the red line is the

DOI: 10.1051/epjconf/201611922005

linear fit. Color code represents the number of occurrences.

3. RESULTS AND VALIDATION

The locations of the peak laser return were first retrieved by the super-resolution altimetry method at wavelengths of both 1064 and 532 nm in order to get the penetration difference between the two wavelengths. The canopy height was finally obtained based on Equation 1. The globally distributon of CALIOP-derived canopy height was shown in Fig. 4. Figure 5 shows the difference between CALIOP-derived and GLAS-derived canopy height. Results of Fig. 2, 4 and 5 show that the CALIOP-derived canopy heights were highly correlated with the ICESat/GLAS-derived values with a mean RMSE of 3.4 m and correlation coefficient (R) of 0.89.

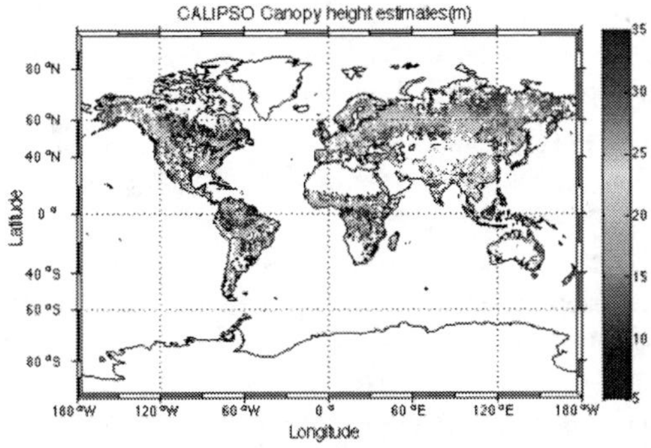

Figure 4. The CALIOP-derived canopy height (meter). The color bar represents the height.

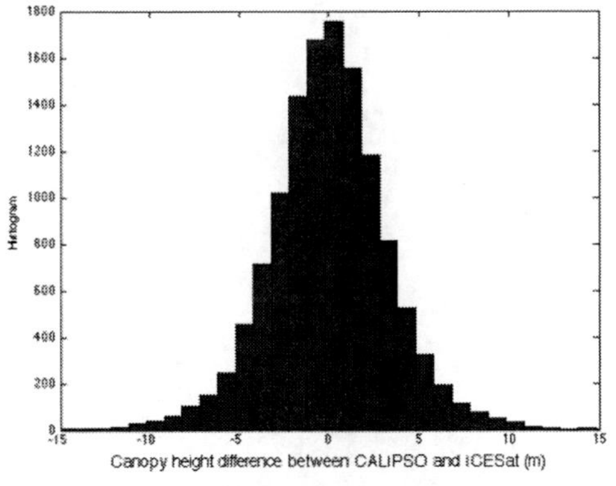

Figure 5. Canopy height difference between CALIOP and ICESat GLAS-derived ones.

NASA's Laser Vegetation Imaging Sensor (LVIS) is a scanning laser altimeter instrument that is flown by aircraft to collect data on surface topography and vegetation coverage. It was designed and developed at NASA's Goddard Space Flight Center (GSFC). LVIS operates at altitudes up to 10 km above ground, and is capable of producing a data swath up to 1000 m wide nominally with 25-m wide footprints. It has a vertical precision of about 10 cm. LVIS data with latitude from 36°N to 46° N and longitude from 67° W to 120° W were used here to valid the two-wavelength method. The LVIS data were averaged in the same 1°×1° bins.

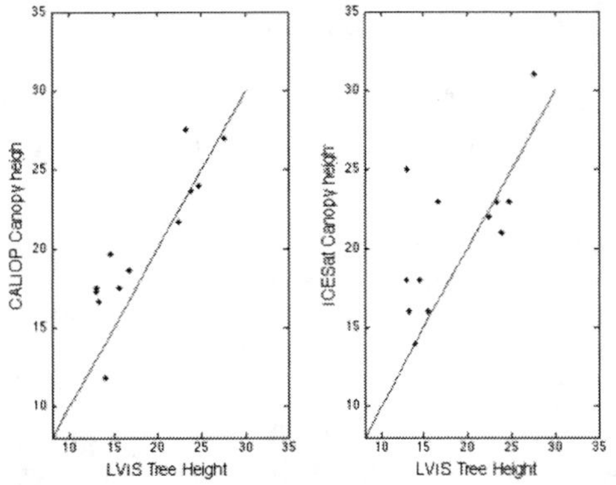

Figure 6. Canopy height (meter) comparison Left panel: CALIOP vs. LVIS; Right panel: GLAS vs. LVIS.

Figure 6 shows the canopy height comparison among CALIOP, GLAS and LVIS results. From Fig. 6, the CALIOP-derived canopy heights are in good agreement with LVIS canopy heights with standard deviation of about 2.6 m. The standard deviation of difference between GLAS-derived canopy height and LVIS height is about 4.1 m. Several factors may have caused the differences shown in Fig. 6, including not exactly footprint overlap, different measurement time and seasonal difference of canopy sturcutres. Future studies will focus on the CALIOP single laser shot and its co-located with in situ measurements in order to estimate canopy heigth with higher spatially resolution.

4. CONCLUSIONS

In this study, canopy height was estimated using CALIOP lidar measurements. This analysis was based on the different canopy penetration depths of laser pulses at wavelengths of 1064nm and 532nm. Results show that canopy penetration differences at two wavelengths indicate moderately strong relationships for estimating the canopy height, and the CALIOP-derived canopy heights were highly correlated with the ICESat/GLAS-derived canopy heights and LVIS results. The proposed method provides potential use of two-wavelength lidar for vegetation canopy height studies.

Future generations of airborne dual wavelength lidar with small footprint (meters) may provide accurate canopy height with high spatial resolution. The relationship between the penetration difference at different wavelength and canopy height will lead to significantly advance in canopy height estimation, except the full waveform method.

REFERENCES

[1] M. A. Lefsky, (2010), "A global forest canopy height map from the Moderate Resolution Imaging Spectroradiometer and the Geoscience Laser Altimeter System," *Geophysical Research Letters,* vol. 37, p. L15401.

[2] J. B. Drake, et al. (2002), "Estimation of tropical forest structural characteristics using large-footprint lidar," *Remote Sensing of Environment,* vol. 79, pp. 305-319.

[3] G. Zhang, et al. (2014), "Estimation of forest aboveground biomass in California using canopy height and leaf area index estimated from satellite data," *Remote Sensing of Environment,* vol. 151, pp. 44-56.

[4] J. B. Drake, (2001), "Estimation of tropical forest aboveground biomass using large-footprint lidar," *Doctoral Dissertation--University of Maryland, College Park,* p. 184.

[5] I. Fayad, et al. (2014), "Canopy Height Estimation in French Guiana with LiDAR ICESat/GLAS Data Using Principal Component Analysis and Random Forest Regressions," *Remote Sensing,* vol. 6, pp. 11883-11914.

[6] M. Simard, et al. (2011), "Mapping forest canopy height globally with spaceborne lidar," *Journal of Geophysical Research: Biogeosciences,* vol. 116, p. G04021.

[7] J. E. Kalshoven et al. (1993), "Remote sensing of the Earth's surface with an airborne polarized laser," *Geoscience and Remote Sensing, IEEE Transactions on,* vol. 31, pp. 438-446.

[8] M. A. Lefsky, et al., (2007), "Revised method for forest canopy height estimation from Geoscience Laser Altimeter System waveforms," *Journal of Applied Remote Sensing,* vol. 1, pp. 013537-013537-18.

[9] J. A. B. Rosette, et al., (2008), "Vegetation height estimates for a mixed temperate forest using satellite laser altimetry," *International Journal of Remote Sensing,* vol. 29, pp. 1475-1493.

[10] N. Baghdadi, et al. (2014), "Testing Different Methods of Forest Height and Aboveground Biomass Estimations From ICESat/GLAS Data in Eucalyptus Plantations in Brazil," *Selected Topics in Applied Earth Observations and Remote Sensing, IEEE Journal of,* vol. 7, pp. 290-299.

[11] M. A. Lefsky, et al. (2005), "Estimates of forest canopy height and aboveground biomass using ICESat," *Geophysical Research Letters,* vol. 32, p. L22S02.

[12] X. Lu, et al. (2014), "A Super-Resolution Laser Altimetry Concept," *Geoscience and Remote Sensing Letters, IEEE,* vol. 11, pp. 298-302.

[13] Y. Hu, et al. (2007), "Elevation information in tail (EIT) technique for lidar altimetry," Opt. Express, vol. 15, pp. 14504-14515

[14] X. Lu et al. (2014), "Accuracy of land surface elevation from CALIPSO mission data," *Optical Engineering,* vol. 54, pp. 031102-031102.

[15] X. Lu, et al. (2014), "Ocean subsurface studies with the CALIPSO spaceborne lidar," *Journal of Geophysical Research: Oceans,* vol. 119, pp. 4305-4317.

AIRBORNE UV LIDAR FOR FOREST PARAMETER RETRIEVALS

Xiaoxia Shang*, Patrick Chazette, and Julien Totems

*Laboratoire des Sciences du Climat et de l'Environnement, CEA-CNRS-UVSQ, 91191 Gif-sur-Yvette Cedex, France, *Email: xiaoxia.shang@gmail.com*

ABSTRACT

A full-waveform UV lidar performed airborne measurements over several temperate and tropical forests sites. The structural and ecological parameters (canopy height, quadratic mean canopy height and apparent foliage) were extracted from lidar backscattered profiles. The aboveground carbon and leaf area index are also evaluated from lidar measurements.

1. INTRODUCTION

Representing 80% of the continental biospheric carbon, forests play a crucial role in Earth's carbon cycle by absorbing carbon dioxide from the atmosphere and storing it into biomass. They are also important sources and sinks for several atmospheric chemical species (e.g. volatile organic compounds, ozone). Both energy and chemical component fluxes between the low troposphere and the biosphere remains however largely unknown. What is clear is that it crucially depends on forest properties such as living biomass, composition and spatial structure. Measuring the three-dimensional vegetation structure of forests is thus essential to address this issue, as such structure contains a substantial amount of information about the state of development of plant communities and their potential to interact with the low troposphere. Active remote sensing instruments, including lidar and radar, can peer through the forest canopy down to the ground level. In particular, a dedicated lidar can provide a full description of the forest vertical structure (including canopy top, tree crown base height and understory structures) [1][2]. A full-waveform lidar with a large footprint can reliably extract the vertical structure of an optically thick tropical rainforest [3].

2. METHODOLOGY

The full-waveform UV lidar ULICE (Ultraviolet LIdar for Canopy Experiment), developed in our laboratory as a spaceborne demonstrator, performed airborne measurements over varied temperate and tropical forest biomes. Its main characteristics are given in Table 1. The use of UV spectral domain (355 nm) leads to a significant reduction of the multiple scattering effects, compared to near infrared wavelengths, in the forest structures. Depending on the sampled site, ULICE has 2 operation modes: the *classic* mode for temperate forests, and *tropical* mode for tropical forest. Their footprints diameters at the ground are ~2 m or ~10 m, respectively, for a flight altitude of ~300 m above the ground level.

The lidar system was embedded on an Ultra-Light Aircraft (ULA) as shown in Figure 1. It is coupled with an ancillary instrument which provides the position and the attitude information of the ULA to derive the angle between the lidar line of sight and the nadir direction. The principle of the canopy lidar measurement is shown in Figure 2, along with a schematic lidar backscattered profile showing forest vertical structures.

Table 1. Main characteristics of ULICE (Ultraviolet LIdar for Canopy Experiment).

	Characteristics	ULICE *classic*	ULICE *tropical*
Emitter	Wavelength	355 nm	
	Mean energy per pulse	~7 mJ	
	Pulse duration	~6 ns	
	Pulse repetition frequency	1–100 Hz	
	Beam diameter	20 mm	
	Maximum beam divergence	9 mrad	40 mrad
Receiver	Reception diameter	150 mm	
	Reception optical density	adjustable	
	Filter bandwidth	0.3 nm	10 nm
	Total field of view	5 mrad	33 mrad
	Detector	Photomultiplicateur	
	Detection mode	Analog	
Acquisition	Sampling frequency	100-500 MHz	
Global	Weight of the optical head	~20 kg	
	Weight of the electronics	~15 kg	
	Electric supply (2×12V batteries)	24 V/~400 W	

 DOI: 10.1051/epjconf/201611922006

Figure 1. Photo of ULICE onboard ULA.

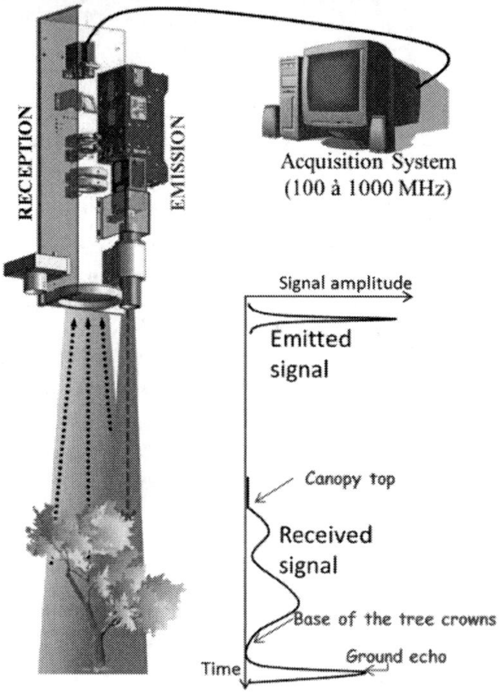

Figure 2. Principle of the canopy lidar measurement.

3. RESULTS

In order to characterize the forest biomes, the structural and ecological parameters are studied, such as the canopy height (CH), the aboveground carbon (AGC), and the leaf area index (LAI). Here we present some results for our sampling sites.

3.1 Structural parameters

Lidar backscatter profiles include a signature from vertical forest structures, which is dominant compared to the atmospheric contribution. Hence, the canopy height (CH) can be extracted as the distance between the first return at the upper surface of the vegetation and the last return at the ground surface, with a mean uncertainty of ~1.5 m. An example of lidar-derived CH is given in Figure 3. The vertical structure within the canopy is also important for forest studies, and can be extracted from lidar backscatter profiles as shown in Figure 4 for a tropical forest in the Réunion Island. The weighted height – quadratic mean canopy height (QMCH) – was used to evaluate the aboveground carbon by several authors [1][4][5], via the relationship found between the QMCH and the AGC as shown in Figure 5.

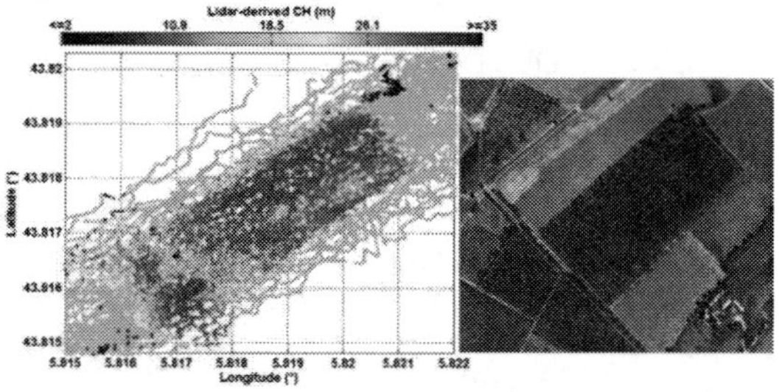

Figure 3. Canopy height (CH) of a plantation of poplar (the footprint diameter at the ground was ~2 m). The satellite view (Google Earth™) of sampled site is on the right.

Figure 4. Upper panel: Photo of the sampled tropical forest. Bottom panel: Example of lidar signals over the tropical forest (the footprint at the ground was ~10 m).

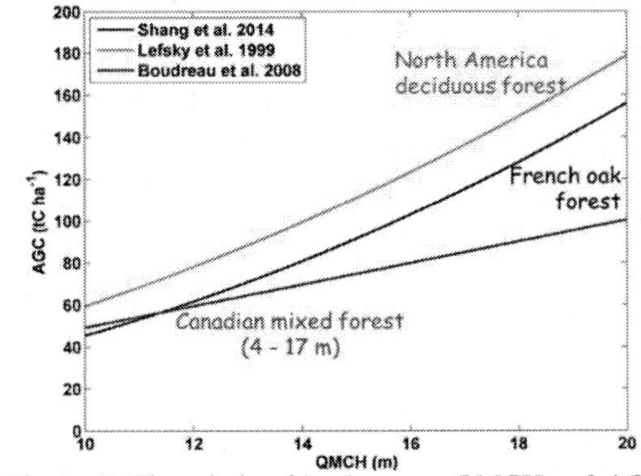

Figure 5. The relationships between QMCH and AGC published.

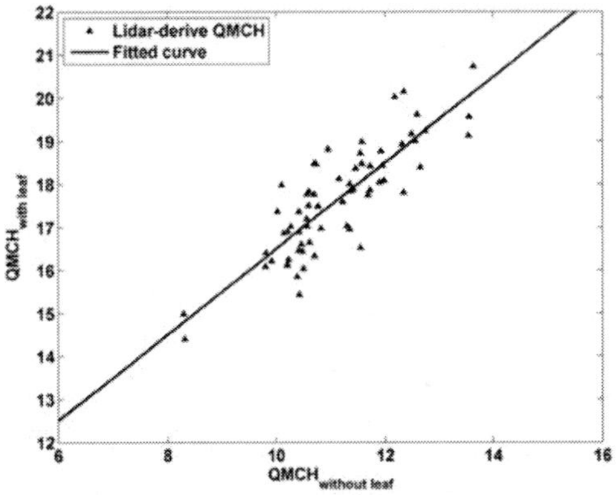

Figure 6. Comparison among the plot-level QMCHs, which are derived from experiments in summer (with leaf) and winter (without leaf). The relationship is found to be $QMCH_{with\ leaf} = 0.996 \times QMCH_{without\ leaf} + 6.5$ (r^2 ~0.7).

The effect of leaves on the AGC estimation was studied. By comparing the summer and winter lidar measurements for the same temperate forest site, we found that the different canopy structures imply a significant difference of the QMCH with a mean value ~6 m (Figure 6), which would lead to an AGC overestimation of ~50% in summer when using a calibration done in winter. This overestimation could be considered as a systematic bias and be corrected.

3.2 Ecological parameters

Lidar is also a powerful instrument to retrieve the leaf area index (LAI) [3] using the vertical profile of apparent foliage which informs on both the canopy density and the vertical distribution of leaf biomass along the profile. The lidar-derived forest optical thickness (FOT), depending on the canopy extinction coefficient, was found to get a linear relationship with the mean LAI derived from MODIS over a larger sample site of pin forest as shown in Figure 7. The lidar-derived LAI index seems to overestimate the LAI, because our lidar sampled only the forest area whereas the MODIS production integrated both forest and non-forest area for each considered pixel of 1 km × 1 km.

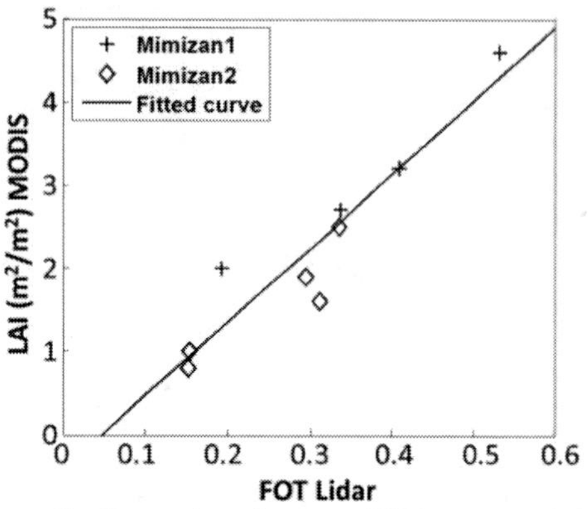

Figure 7. Comparison between FOT computed from lidar measurements and LAI estimated via MODIS for 9 pixels of 1 km × 1 km in a pin forest.

3.3 Application to forest type classification

Using the average values and variability of both CH and LAI, 7 tropical forest sites of the Réunion Island were classified.

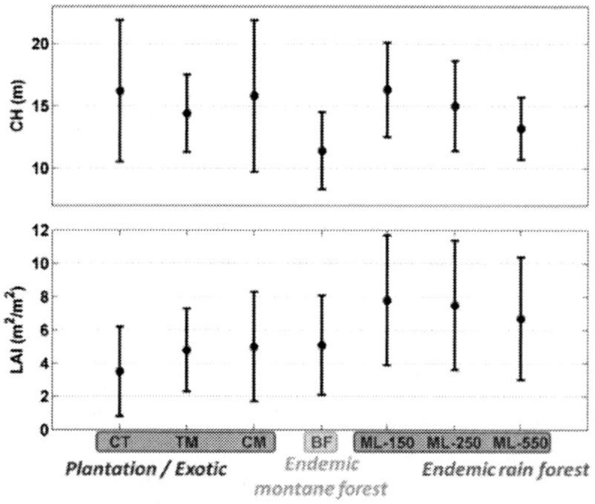

Figure 8. Mean (dot) and standard deviations (line segments) of canopy height and leaf area index for 7 tropical forest site: coastal (CT), Tamarind (TM), Cryptomeria (CM), Bélouve (BF), and Mare-Longue (ML-150, 250 and 550) sites. Measurements with CH < 5 m are not considered.

These tropical forests range from coastal to rain forest, including montane cloud forest. As shown in Figure 8 we can identify 2 groups: the endemic (montane/rain) forests and the planted/exotic forests.

4. CONCLUSIONS

Airborne lidar measurements were performed over several temperate and tropical forests sites, which allowed building a representative database of lidar vertical profiles. We extracted the structural and ecological parameters (canopy height, quadratic mean canopy height and apparent foliage) from lidar backscattered signal, and then evaluate the aboveground carbon and leaf area index. Even though UV lidar is not a good candidate for spaceborne missions due to the weak atmospheric transmission and the strong absorption by the vegetation in the UV domain, we confirm that the UV wavelength is a good candidate for airborne lidar measurements, and can be a very interesting reference for a spaceborne mission dedicated to forest studies at the global scale.

ACKNOWLEDGMENTS

The experiments have been funded by the Centre National d'Etudes Spatiales (CNES), the Commissariat à l'Energie Atomique et aux Energies Alternatives (CEA), and the Université de la Réunion through the federation Observatoire des Milieux Naturels et des Changements Globaux (OMNCG) of the Observatoire des Sciences de l'Univers de la Réunion (OSU-R). We are also grateful for the support offered by the Direction Générale de l'Armement (DGA). We finally thank our ULA pilot F. Toussaint.

REFERENCES

[1] Shang, X., and P. Chazette, 2014: Interest of a Full-Waveform Flown UV Lidar to Derive Forest Vertical Structures and Aboveground Carbon, *Forests*, **5(6)**, 1454–1480.

[2] Cuesta, J., P. Chazette, T. Allouis, P. H. Flamant, S. Durrieu, J. Sanak, P. Genau, D. Guyon, D. Loustau, and C. Flamant, 2010: Observing the Forest Canopy with a New Ultra-Violet Compact Airborne Lidar, *Sensors*, **10(8)**, 7386–7403.

[3] Tang, H., R. Dubayah, A. Swatantran, M. Hofton, S. Sheldon, D. B. Clark, and B. Blair, 2012: Retrieval of vertical LAI profiles over tropical rain forests using waveform lidar at La

Selva, Costa Rica, *Remote Sens. Environ.*, **124**, 242–250.

[4] Lefsky, M. A., D. Harding, W. B. Cohen, G. Parker, and H. H. Shugart, 1999: Surface lidar remote sensing of basal area and biomass in deciduous forests of eastern Maryland, USA, *Remote Sens. Environ.*, **67(1)**, 83–98.

[5] Boudreau, J., R. F. Nelson, H. A. Margolis, A. Beaudoin, L. Guindon, and D. S. Kimes, 2008: Regional aboveground forest biomass using airborne and spaceborne LiDAR in Québec, *Remote Sens. Environ.*, **112(10)**, 3876–3890.

EPJ Web of Conferences **119**, 23001 (2016) DOI: 10.1051/epjconf/201611923001
ILRC 27

AEROSOL BACKSCATTER AND EXTINCTION RETRIEVAL FROM AIRBORNE COHERENT DOPPLER WIND LIDAR MEASUREMENTS

F. Chouza[1], O. Reitebuch[1], S. Groß[1], S. Rahm[1], V. Freudenthaler[2], C. Toledano[3], B. Weinzierl[1,2]

[1] *Institut für Physik der Atmosphäre, Deutsches Zentrum für Luft- und Raumfahrt (DLR), Oberpfaffenhofen, Germany, *Email: fernando.chouza@dlr.de*

[2] *Meteorologisches Institut, Ludwig-Maximilians-Universität München (LMU), München, Germany*

[3] *Atmospheric Optics Group, University of Valladolid, Valladolid, Spain*

ABSTRACT

A novel method for coherent Doppler wind lidars (DWLs) calibration is shown in this work. Concurrent measurements of a ground based aerosol lidar operating at 532 nm and an airborne DWL at 2 μm are used in combination with sun photometer measurements for the retrieval of backscatter and extinction profiles.

The presented method was successfully applied to the measurements obtained during the Saharan Aerosol Long-range Transport and Aerosol-Cloud-Interaction Experiment (SALTRACE: http://www.pa.op.dlr.de/saltrace), which aimed to characterize the Saharan dust long range transport between Africa and the Caribbean.

1. INTRODUCTION

Aerosol lidars allow detailed measurements of aerosol properties, whereas wind lidars provide information on horizontal and vertical transport of aerosol layers. However, in many cases it is scientifically interesting to have aerosol and wind information synchronous from one system.

Due to the strong wavelength dependence of the atmospheric molecular backscatter and the low sensitivity of the coherent detection to spectrally broad signals, calibration methods for aerosol lidars cannot be applied to a coherent DWLs usually operating at wavelengths between 1.5 μm and 2 μm.

Up to now, different approaches where applied to retrieve calibrated backscatter coefficient from lidars to which calibration methods based on molecular return cannot be applied, including the use of ground based hard targets (Menzies and Tratt, 1994), ground return (Cutten et al., 2002), sea return (Bufton et al., 1983) and clouds (O'Connor et al., 2003).

While the main problems associated with the use of ground based hard targets are the boundary layer variability and the changes in heterodyne efficiency due to high path integrated turbulence, the sea and ground surface have either highly variable optical properties or they are limited to some specific regions (e.g. deserts).

On the other hand, the method based on cloud extinction requires presence of well characterized stratocumulus clouds.

In order to avoid the previous described problems, concurrent measurements of an airborne DWL and the ground based aerosol lidar POLIS are used in combination with sun photometer measurements to provide calibrated backscatter and extinction profiles (Chouza et al., 2005).

The proposed method was applied to the dataset obtained during the SALTRACE campaign conducted during June/July 2013, which was intended to provide a better understanding of the Saharan dust long range transport and its associated processes.

2. INTRUMENTAL DESCRIPTION

The coherent DWL deployed on the DLR Falcon 20 research aircraft during SALTRACE is based on a CLR Photonics instrument (Henderson et al., 1993) combined with a two wedge scanner and acquisition system developed at DLR (Köpp et al., 2004; Reitebuch, 2012). The system operates at a wavelength of 2.02254 μm, with a pulse full width at half maximum (FWHM) of 400 ns, a pulse energy of 1-2 mJ, and a repetition frequency of 500 Hz.

3. INSTRUMENTAL CORRECTIONS

The backscattered power $\langle P(R) \rangle$, for a given lidar line of sight $\vec{L}$ and a range gate at distance R, can be written as:

DOI: 10.1051/epjconf/201611923001

$$\langle P(R)\rangle = \frac{E_T}{R^2} k_d k_{in,\vec{L}}(R) k_\delta(R)\beta(R)T^2(R)$$

where E_T is the mean transmitted energy of the averaged laser pulses, R^2 is the squared distance of the range gate to the lidar, β is the backscatter coefficient, T the atmospheric transmission, $k_{in,\vec{L}}(R)$ groups instrumental coefficients that change with the range gate range R and the line of sight vector $\vec{L}$ (e.g. system gain as a function of the backscattered signal frequency, transmission of the transceiver optics), $k_\delta(R)$ is the detector response depending on the depolarization of the backscattered signal and k_d groups other instrumental constants (e.g. telescope area).

The transmitted energy E_T was estimated based on the pulse build up time and laboratory measurements that relates this quantity with the laser pulse energy. On the other hand, most of the different instrumental coefficients grouped in $k_{in,\vec{L}}(R)$ were estimated based on atmospheric measurements for which the other instrumental and atmospheric parameters were considered to be constant.

Finally, the received backscatter power $\langle P(R)\rangle$ is divided by the estimated instrumental constants in order to remove their effects

$$\langle P_c(R)\rangle = \frac{\langle P(R)\rangle R^2}{E_T k_{in,\vec{L}}(R)} = k_d k_\delta(R)\beta(R)T^2(R)$$

4. CALIBRATION AND RETREIVAL METHOD

The proposed calibration method relies on concurrent measurements of the airborne DWL and a ground based aerosol lidar (in our case the POLIS system based on the island of Barbados, Caribbean) (Freudenthaler et al., 2009) from which backscatter β_{532}^{POLIS}, extinction α_{532}^{POLIS} and lidar ratio $S_{532}^{POLIS}(L_n)$ can be retrieved.

Since the DWL operates at a wavelength of 2 µm and the POLIS lidar operates at 532 nm, additional sun photometer measurements relating the extinction coefficient at both wavelengths k_α are needed (Toledano et al., 2011). This relation, as well as the atmospheric depolarization, depends on the aerosol type and its vertical structure. Each layer L_n of the atmospheric model represents an aerosol type and is defined as a region in which the particle depolarization ratio, the lidar ratio and the wavelength dependency of the extinction coefficient are considered to be constant.

During the DWL overflight over the ground based lidar POLIS (Fig. 1), the same atmospheric volume is sensed by both lidar systems. For a given range gate R, the measured backscatter power by the DWL and the ground based lidar can be related trough the following equation

$$\langle P_c(R)\rangle = k_d k_\delta(L_n) k_\beta(L_n) \beta_{532}^{POLIS}(R) e^{-2\int_0^R \alpha_{532}^{POLIS}(r) k_\alpha(L_n) dr}$$

Grouping al parameter that remain constant for a given L_n in a single constant $k(L_n)$

$$\langle P_c(R)\rangle = k(L_n)\beta_{532}^{POLIS}(R) e^{-2\int_0^R \alpha_{532}^{POLIS}(r) k_\alpha(L_n) dr}$$

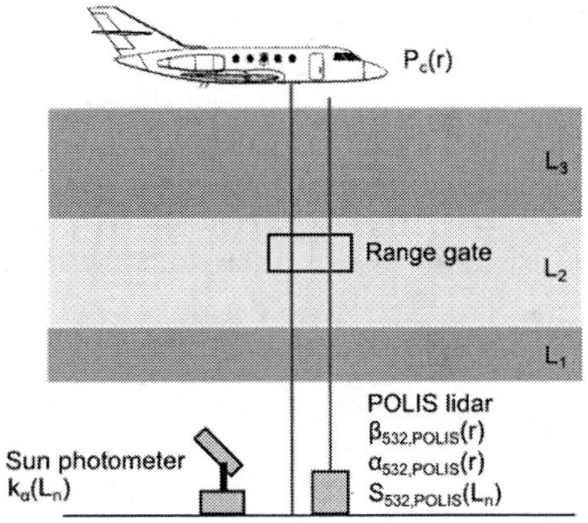

Fig. 1. Scheme of the instruments and variables involved in the calibration and retrieval process.

Finally, based on the measured extinction coefficient retrieved from POLIS and the wavelength dependency of the extinction coefficient retrieved from sun photometer measurements, a linear relation between the measured backscatter power by the DWL and the backscatter coefficient retrieved by POLIS can be established

$$\frac{\langle P_c(R)\rangle}{T^2_{2022,POLIS}(R)} = k(L_n)\beta_{532}^{POLIS}(R)$$

With

$$T^2_{2022,POLIS}(R) = e^{-2\int_0^R \alpha_{532}^{POLIS}(r) k_\alpha(L_n) dr}$$

Finally, the conversion constant $k(L_n)$ corresponding to each layer can be estimated applying a LSF (Least Squares Fit) between the backscatter coefficients β_{532}^{POLIS} measured by the ground-based lidar POLIS and the extinction corrected power measured by the DWL.

Based on the layer distribution and the conversion constants $k(L_n)$ calculated for each layer, it is possible to retrieve the backscatter coefficient at 532 nm based on the 2 μm measurements β_{532}^{DWL} through an iterative process.

$$\beta_{532,i}^{DWL}(R) = \frac{\langle P_c(R) \rangle}{T_{2022,i-1}^2(R)} \frac{1}{k(L_n)}$$

$$\alpha_{532,i}^{DWL}(R) = \beta_{532,i}^{DWL}(R) S_{532}^{POLIS}(L_n)$$

with the iteration number i and $T_{2022,0}^2(R) = 1$ as starting value.

5. RESULTS

Figure 2 shows the linear relation between the extinction corrected backscattered power measured by the airborne DWL and the backscatter coefficient retrieved by the ground based aerosol lidar for three different aerosol types based on six overflights from the flight on 26 June 2013 over Barbados.

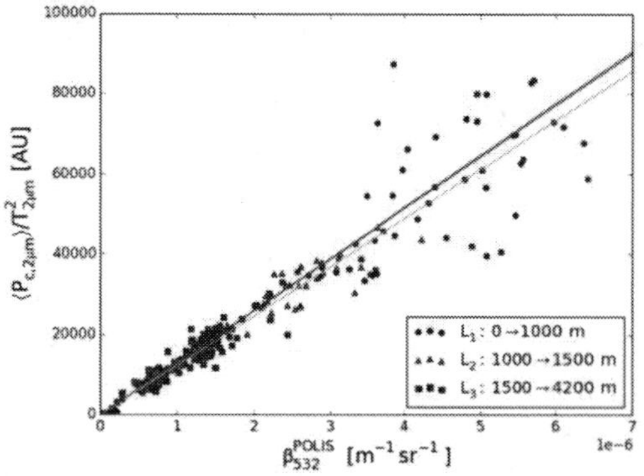

Figure 2. Correlation between the extinction corrected backscattered power of the DWL and the POLIS measured backscatter coefficient for 3 different aerosol layers: boundary layer (red dots), mixed layer (yellow triangles) and SAL (Saharan air layer) (green squares).

The retrieved calibration constants obtained from the LSF and wavelength dependency of the extinction coefficient retrieved from sun photometer measurements are presented in the following table

Layer	Calibration		
	k_α	k^{-1}	$\sigma_{k^{-1}}$
Boundary layer	0.614	$7.75 \cdot 10^{-11}$	$2.26 \cdot 10^{-12}$
Mixed layer	0.670	$8.20 \cdot 10^{-11}$	$1.80 \cdot 10^{-12}$
SAL	0.679	$7.80 \cdot 10^{-11}$	$7.85 \cdot 10^{-13}$

Figure 3 shows the retrieved backscatter coefficient for the flights on 10 July and 11 July based on the calibration constants retrieved during the calibration flight on 26 June. The retrieved backscatter coefficient from the DWL measurements are shown together with the POLIS aerosol lidar measurements in Fig. 4.

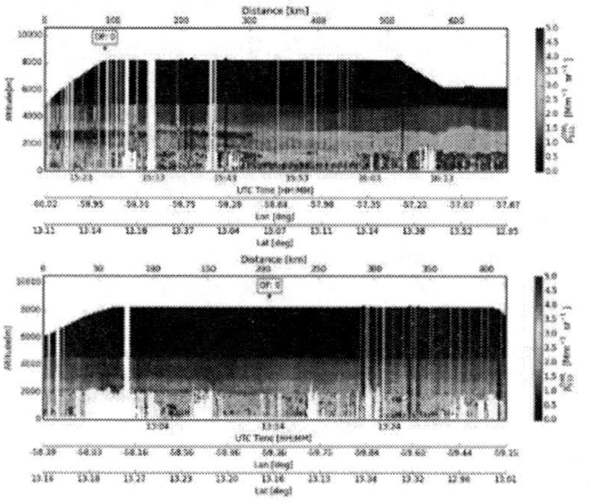

Figure 3. Overview of the retrieved backscatter coefficient for the flights on the 10 July (upper panel) and 11 July (lower panel). The labels "OF" indicate the approximated time of the overflight over POLIS lidar. The white color indicates regions were no atmospheric signal is available (e.g. below clouds, low laser energy).

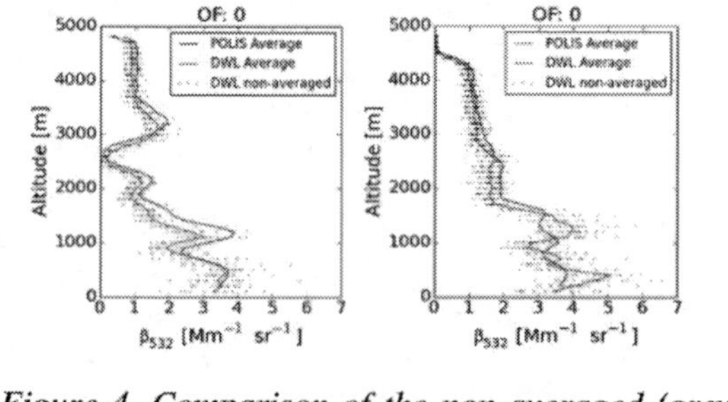

Figure 4. Comparison of the non-averaged (grey dots) and averaged backscatter coefficient profiles (green) corresponding to the retrieved

 DOI: 10.1051/epjconf/201611923001

data for the flights on the 10 July (left) and 11 July (right), and the averaged profiles measured by POLIS (blue) during the Falcon overflights (OF).

Using the calibration constant corresponding to the SAL, the attenuated backscatter coefficient from the flight on 12 June in the Dakar region was retrieved and compared with the attenuated backscatter coefficient measured by CALIPSO. The results are shown in Fig. 5.

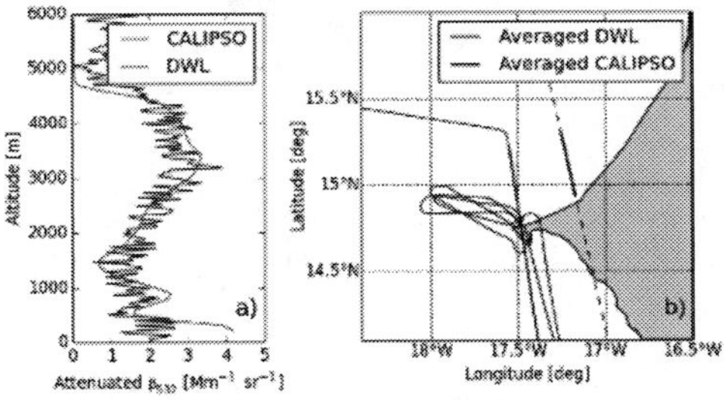

Figure 5. a) Comparison of the averaged attenuated backscatter profiles retrieved from the DWL (green) and the corresponding averaged profile measured by CALIPSO (blue) during its overpass over Dakar region on the 12 June. b) DLR Falcon (black, solid) and CALIPSO (black, dashed) tracks, together with the averaged sections (green for the DWL and blue for CALIPSO).

6. SUMMARY

In this study, we showed a method which allows the synchronous retrieval of aerosol backscatter and extinction coefficients from an (airborne) DWL giving detailed insight into aerosol transport but also aerosol properties. The retrieved calibration constants were validated using ground based aerosol lidar and CALIPSO measurements, showing an agreement within 20%.

In further studies, the sea surface return will be tested as monitor of the stability of the retrieved calibration constants.

ACKNOWLEDGEMENT

This work was funded by the Helmholtz Association under grant number VH-NG-606 (Helmholtz-Hochschul-Nachwuchsforschergruppe AerCARE). The SALTRACE campaign was mainly funded by the Helmholtz Association, DLR, LMU and TROPOS.

REFERENCES

[1] Menzies, R. T. and Tratt, D. M., 1994: Airborne CO_2 coherent lidar for measurements of atmospheric aerosol and cloud backscatter, Applied Optics, **33**, 5698-5711.

[2] Bufton, J. L., F. E. Hoge, R. N. Swift, 1983: Airborne measurements of laser backscatter from the ocean surface, Applied Optics, **22**, 2603.

[3] Cutten, D. R., J. Rothermel, M. A. Jarzembski, R. M. Hardesty, J. N. Howell, D. M. Tratt, V. Srivastava, 2002: Radiometric calibration of an airborne CO2 pulsed Doppler lidar with a natural earth surface, Applied Optics, **41**, 3530-3537.

[4] O'Connor, E. J., A. J. Illingworth, R. J.Hogan, 2004: A technique for autocalibration of cloud lidar, J. Atmos. Ocean. Tech., **21**, 777–786.

[5] Köpp, F., S. Rahm, I. Smalikho, 2004: Characterization of Aircraft Wake Vortices by 2-μm Pulsed Doppler Lidar, Journal of Atmospheric and Oceanic Technology, **21**, 194–206.

[6] Reitebuch, O., 2012: Wind Lidar for Atmospheric Research, in: Atmospheric Physics, Schumann, U., Springer, Berlin Heidelberg, 487-507.

[7] Chouza, F., O. Reitebuch, S. Groß, S. Rahm, V. Freudenthaler, C. Toledano, B. Weinzierl, 2015: Retrieval of aerosol backscatter and extinction from airborne coherent Doppler wind lidar measurements, Atmos. Meas. Tech. Discuss., **8**, 1935-1986.

[8] Freudenthaler, V., M. Esselborn, M. Wiegner, B. Heese, M. Tesche, A. Ansmann, D. Müller, D. Althause, M. Wirth, A. Fix, G. Ehret, P. Knippertz, C. Toledano, J. Gasteiger, M. Garhammer, M. Seefeldner, 2009: Depolarization ratio profiing at several wavelengths in pure Saharan dust during SAMUM 2006, Tellus B, **61**, 165-179.

[9] Toledano, C., M. Wiegner, S. Groß, V. Freudenthaler, J. Gasteiger, D. Müller, T. Müller, A. Schladitz, B. Weinzierl, B. Torres, N.T. O'Neill, 2011: Optical properties of aerosol mixtures derived from sun-sky radiometry during SAMUM-2, Tellus Series B: Chemical and Physical Meteorology, **63**, 635-648.

DOI: 10.1051/epjconf/201611923002

AEROSOL LIDAR FOR THE RELATIVE BACKSCATTER AMPLIFICATION MEASUREMENTS

Igor A. Razenkov*, Victor A. Banakh, Alexander I. Nadeev

*V.E. Zuev Institute of Atmospheric Optics, Tomsk, 634055, RUSSIA, *Email: lidaroff@iao.ru*

ABSTRACT

Backscatter amplification presents only in a turbulent atmosphere, when the laser beam is propagates twice through the same inhomogeneities. We proposed technical solution to detect backscatter amplification. An aerosol micro pulse lidar with a beam expansion via receiving telescope was built to study this effect. Our system allows simultaneous detection of two returns from the same scattering volume: exactly on the axis of the laser beam and off the axis.

1. INTRODUCTION

This amplification effect arises because of the correlation of forward and backward light waves propagating through the same inhomogeneities of a random medium [1,2]. The intensity of the scattered light on the laser beam axis should be bigger than around the laser beam. Moreover, the effect of amplification is possible in a very small angle (Fresnel zone) around the optical axis of the beam. There is one approach on the measurement backscatter amplification known [3].

We designed a lidar transceiver to measure the *relative backscatter amplification* (RBSA). It measures the average intensity of the scattered radiation exactly on the laser beam axis and the average intensity off axis, where is no amplification at all. Our design is based on laser-radar transceiver with the *beam expansion via receiving telescope* [4]. The same telescope works to transmit laser pulse to the atmosphere and to receive radiation scattered from molecules and aerosol particles. Such systems have good thermo-mechanical stability and this is important for long-term continuous observations. It is also important to get returns on-axis and off-axis at the same time. That is why our system has one transmitter channel and two receiver channels. The schematic of a lidar transceiver is shown in Fig. 1.

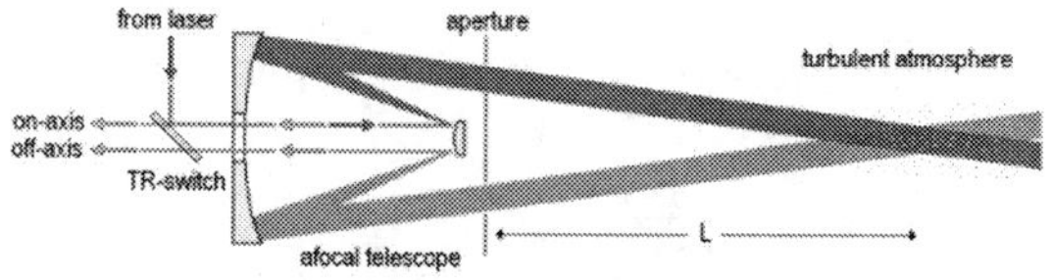

Fig. 1. Schematic drawing of the transceiver.

The main element of the system is an afocal mirror telescope. Antenna switch sends a narrow laser beam to the edge of the telescope, which is pre-focused at a distance of (L = 1 km). A screen with two round apertures is installed in front of the telescope. The laser pulse is emitted into the atmosphere through the top aperture. Scattered radiation from the atmosphere passes through both apertures to the antenna switch (transmit-receive-switch). The first on-axis beam returned to the system in exactly the same path the laser beam as the transmitted to the atmosphere beam. The second off-axis beam from the scattering volume returns through a different path. In turbulent atmosphere in the presence of RBSA on-axial return exceeds the off-axis return. In case of a weak turbulence or absence of turbulence on-axis return will be close or equal to the off-axis return.

Equations for on-axis and off-axis lidar returns are

$$P_1(z) = E_0 C_1 G_1(z) N(z) \beta_\pi(z) \frac{1}{z^2} T^2(z)\,, \qquad (1)$$

$$P_2(z) = E_0 C_2 G_2(z) \beta_\pi(z) \frac{1}{z^2} T^2(z)\,, \qquad (2)$$

where z is distance; E_0 is the pulse energy; C_1 and C_2 are hardware constants includes channel transmittance, receiving aperture, and detector efficiency; $G_1(z)$ and $G_2(z)$ - geometric factors of the receiving channels; $\beta_\pi(z)$ is the backscatter coefficient; $N(z)$ is the backscatter amplification coefficient; $T^2(z)$ is the double atmospheric transmittance.

The ratio of the lidar returns $P_1(z)$ and $P_2(z)$ is proportional to the RBSA

DOI: 10.1051/epjconf/201611923002

$$R(z) = \frac{P_1(z)}{P_2(z)} = \frac{C_1G_1(z)}{C_2G_2(z)} \times N(z) . \quad (3)$$

According to the theory [1], the amplification coefficient must be $1 \leq N(z) \leq 2$.

2. EXPERIMENTAL SETUP

Figure 2 shows an optical diagram of the lidar. A short light pulse from the laser 1 is directed to beam splitter 3 (50/50) by turning mirror 2. We use the fiber laser (λ=532 nm) with outgoing beam about 5 mm in diameter. Box 5 is energy monitor. Turning mirrors 6 and 7 direct the beam with a small offset from the axis of the Mersen telescope (mirrors 8 and 9). After the telescope, laser beam with a diameter of 45 mm is transmitted into the atmosphere through the top aperture of screen 10.

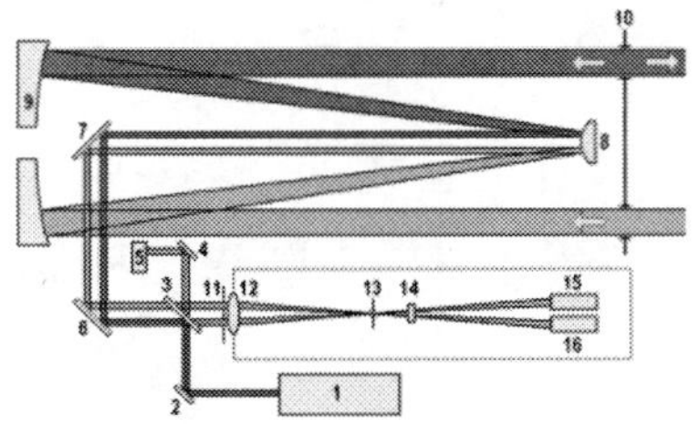

Fig. 2. Optical diagram of the RBSA lidar.

Scattered by the atmosphere radiation can be detected by the lidar only when it passes through the top and bottom holes of the screen 10. Note that the apertures' size is about equal to the Fresnel zone ($\sqrt{\lambda L}$), and the separation between them is 5 times bigger than the aperture size. Incoming beams inside the system are parallel to each other. They pass through the beam splitter 3 and the screen 11, which blocks the parasitic internally scattered radiation within the system at the moment when laser transmits the pulse.

Fig. 3. Picture of the RBSA-lidar. A white cover with two in/out ports is behind the lidar.

Lens 12 focuses the beams on the field stop 13. Size of the aperture 13, focus of lens 12 and telescope magnification ($10^{\times}$) define the field-of-view of the lidar. An interference filter 14 is installed after the field stop 13 to block the background light. Then, photodetectors 15 and 16 detect on-axis and off-axis beams respectively. The photomultiplier tubes Hamamatsu with 8-mm size of the photocathode are used. Figure 3 shows the picture of our instrument. The system parameters are listed in Table 1.

Table 1: Lidar parameters.

Transmitter	
Laser	IPG Photonics Corp. model GLPM-10-Y13
Wavelength	532 nm
Pulse duration	2 ns
Pulse repetition	50 kHz
Laser power	500 mW
Beam quality	TEM_{00}, M^2<1.3
Beam size	45 mm
Receiver	
Afocal telescope	Mersen ($10^{\times}$)
Telescope diameter	400 mm
Channels number	2
Receiving apertures	45 mm
Apertures separation	250 mm
Field-of-view	330 μrad
Space resolution	2 m
Time resolution	10 min
Detector (PMT)	Hamamatsu H10721P
PMT's efficiency	10%

3. CALIBRATION

In equation (3) for the ratio $R(z)$ has hardware constants C_1, C_2, and geometric functions $G_1(z)$, $G_2(z)$. Let's introduce the ratio

$$R_0(z) = \frac{C_1G_1(z)}{C_2G_2(z)}, \quad (4)$$

which is equal to $R(z)$ for $N(z) \equiv 1$. The ratio $R_0(z)$ is the calibration ratio, because it determines the relative sensitivity of on-axis and off-axis channels. Ratio $R_0(z)$ can be determined in a clean atmosphere with a weak wind, when the turbulence intensity is very low. Often these conditions are at dawn and at sunset. If to divide the ratio $R(z)$ (3) on the calibration ratio $R_0(z)$ (4) the normalized ratio $R'(z)$ can be obtained

$$R'(z) = \frac{R(z)}{R_0(z)} = N(z). \quad (5)$$

Ratio $R'(z)$ is equal to the relative backscatter amplification coefficient $N(z)$.

In reality, recorded returns $\tilde{P}_1(z)$ and $\tilde{P}_2(z)$ differ from the signals in the expressions (1) and (2)

$$\tilde{P}_1(z) = P_1(z) + P_{BG1} + P_{AP1}(z), \quad (6)$$

$$\tilde{P}_2(z) = P_2(z) + P_{BG2} + P_{AP2}(z), \quad (7)$$

where P_{BG1} and P_{BG2} - background noise, $P_{AP1}(z)$ and $P_{AP2}(z)$ – aftereffect (afterpulse) of photodetectors, which is caused by "blinding" of the detectors at the time of sending the laser pulse into the atmosphere and subsequent relaxation.

To obtain "clean" signals $P_1(z)$ and $P_2(z)$ is necessary to determine the response of the photodetectors $P_{AP1}(z)$ and $P_{AP2}(z)$ to the instant light at the moment of transmitting a laser pulse to the atmosphere. It is better do this at night before observation cycle. It is necessary to record the signals $\tilde{P}_1(z)$ and $\tilde{P}_2(z)$ with closed exit port to the atmosphere. Thus, the echo and background terms can be excluded from expressions (6) and (7). Let's call this procedure a "calibration-1". We usually perform calibration-1 at the beginning of the measurement cycle and at the end of it.

Figure 4 presents all signals in formulas (6) and (7). Signals from on-axis channel (#1) is shown on the left (a), and from off-axis channel (#2) is shown on the right (b). The background (BG) has the constant value. Afterpulses curves (AP) are significantly different for the two channels. This is due to much larger pulse exposure of the detector of the on-axis channel at the moment of the shot than that for the off-axis channel. First receiving channel (on-axis) is exactly opposite to the beam splitter 3 (Fig. 2), where the laser beam is directed towards the telescope. In accordance with equations (8) and (9) "net" echoes in Fig. 4 (curves P) were obtained by subtraction the afterpulse (curves AP) and background (BG) from the recorded signals (curves P+AP+BG). Data were received in October 6, 2014 at 03:40 (hereinafter, the time is local). Note the values of returns on all plots in *photons per shot*. Total number of laser shots for every profile is 3×10^7.

Signals $P_1(z)$ and $P_2(z)$ are obtained from (8) and (9) by subtracting the background (BG) and PMT's afterpulse (AP). In accordance with the formula (5), in order to determine the mutual relative sensitivity of the receiving channels a calibration ratio $R_0(z)$ in the absence of turbulence has to be determined (calibration-2). An example of this ratio is shown in Fig. 5, which uses the signals from Fig. 4.

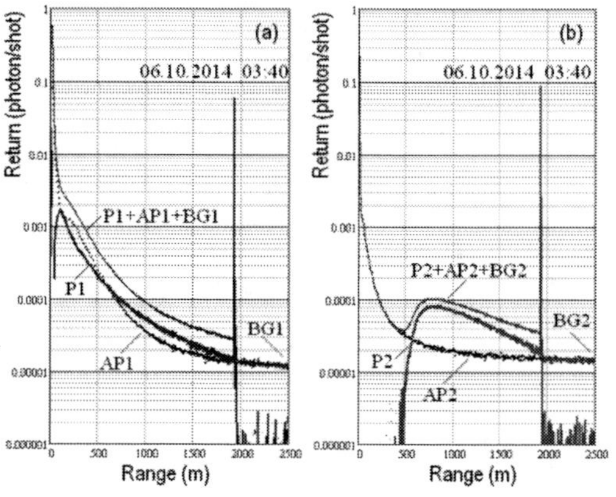

Fig. 4. Profiles of signals for (a) on-axis channel (#1) on the left and (b) for off-axis channel (#2) on the right. Background (BG), afterpulse (AP), return from the atmosphere (P+AP+BG), and "clean" return (P). The spike at 1950 m is from the solid obstacle that blocks the pulse. Data recorded on 10 October 2014 at 03:40.

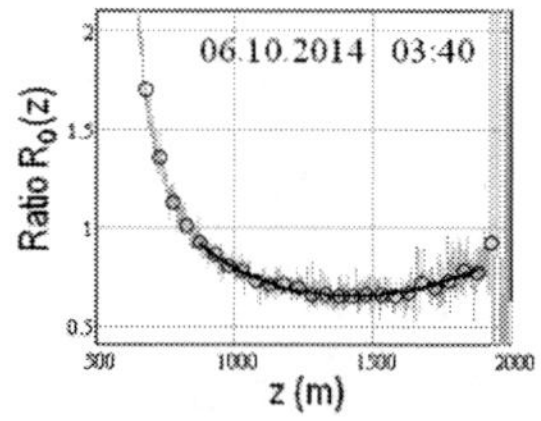

Fig. 5. The calibration ratio $R_0(z)$ (calibration-2) from data recorded on 10 October 2014 at 03:40 (see Fig. 3).

Our system has spatial resolution of 2 m. A light-gray curve in Figure 5 is a ratio $R_0(z)$ plotted with resolution of 2 m. To improve the statistical error signals were averaged over 25 points, and spatial resolution went down to 50 m. In Fig. 5

circles show $R_0(z)$ obtained by using the averaged signals. A smooth bold black curve in this plot is a 4th degree polynomial fit. This interpolation is used as calibration-2 to obtain the amplification coefficient $N(z)$.

4. DATA EXAMPLES

Ratio $R_0(z)$ (calibration-2) is the relative sensitivity of the receiving channels. So, $R_0(z)$, also can be used for comparison of returns from on-axis and off-axis channels in the analysis of experimental data. In the absence of amplification, when the ratio $N(z) = 1$, far field returns $P_1(z)$ and $R_0(z) \times P_2(z)$ should be equal. Figure 6 shows two pairs of signals received at different time. In the data example shown in Fig. 6(a), the backscatter amplification was absent and returns are nearly equal. In case shown in Fig. 6(b) the amplification occurs, because it is clear that the on-axis return $P_1(z)$ is bigger than the off-axis return $R_0(z) \times P_2(z)$.

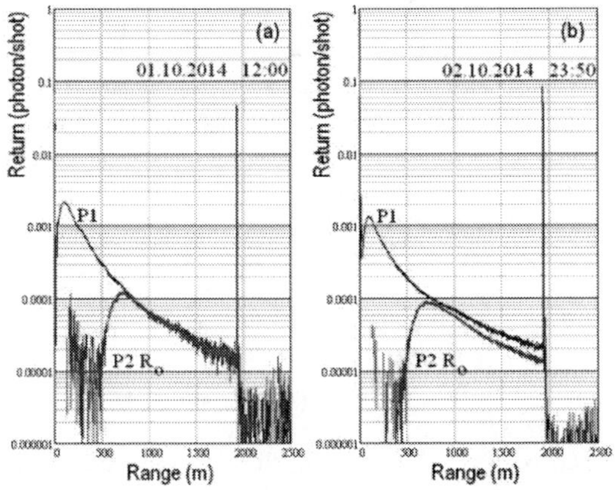

Fig. 6. On-axis (P1) and off-axis (P2) returns in (a) calm and (b) turbulent atmospheric conditions. The signal P2 is corrected on $R_0(z)$. Data recorded on 1 October 2014 at 12:00 and 2 October 2014 at 23:50.

Figure 7 presents the ratio $R'(z)$ for pairs of signals described above (see Fig. 6). Calibration ratio $R_0(z)$ retrieved on 6 October 2014 at 03:40 (see Fig. 4 and Fig. 5) is a solid gray line. Normalized ratio $R'(z)$ shown by circles. By dashed lines show area of ±5% around $N(z) = 1$, when turbulence is very low. The result of ratio $R'(z)$ has error less than 10%.

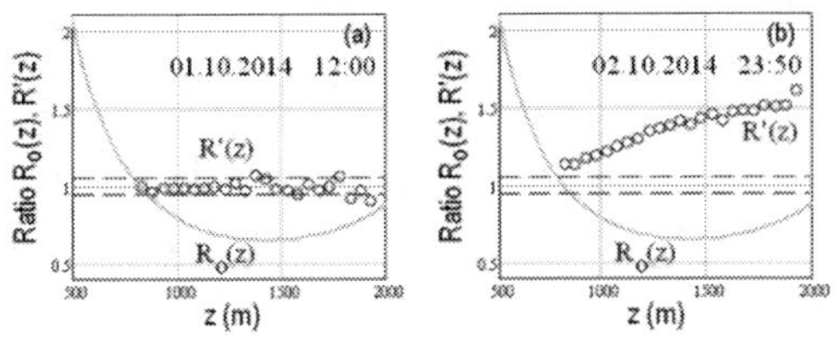

Fig. 7. Normalized ratios $R'(z)$ obtained (a) in the absence of backscatter amplification and (b) in the presence it. Ratios (a) and (b) here are corresponds to the pairs of signals (a) and (b) in Fig. 6. Calibration ratio $R_0(z)$ determined from the data on 10 October 2014 at 03:40 (see Fig. 4 and Fig. 5).

Lidar is designed for long-term and unattended operation. The system currently doesn't have a remote control, but it will be implemented in the future in the newer version of the instrument. It was possible to built instrument with the laser beam expansion via receiving telescope to detect relative backscatter amplification in the atmosphere. We hope this approach is useful for future studies of the atmospheric turbulence.

ACKNOWLEDGMENTS

The authors thank Mr. Evgenii Gordeev for his technical support. This study was supported by the Russian Scientific Foundation for Maintenance and Development (Project: 14-17-00386).

REFERENCES

[1] Vinogradov A. G., Kravtsov Yu. A., and Tatarskii V. I., "The Effect of Intensification of Back Scattering by Bodies That Are Situated in a Medium Having Random Inhomogeneities," Radiophys. Quant. Electr. **16**(7), 818–823 (1973).

[2] Banakh V. A. and Mironov V. L., *Lidar in a Turbulent Atmosphere* (Artech House, Boston and London, 1987).

[3] Gurvich A.S. // Lidar. Patent on useful model No. 116245. Registered by the State List of Useful Models of the Russian Federation of 20 may 2012 г.

[4] Razenkov, I.A., "Aerosol lidar for continuous atmospheric measurements," Atmos. Oceanic Opt. 26(1), 52-63 (2013).

DOI: 10.1051/epjconf/201611923003

CONCEPT DESIGN OF A MULTIWAVELENGTH AEROSOL LIDAR SYSTEM WITH MITIGATED DIATTENUATION EFFECTS AND DEPOLARIZATION-MEASUREMENT CAPABILITY

Adolfo Comerón [1]*, Michaël Sicard[1,2], Eric Vidal[1], Rubén Barragán[1], Constantino Muñoz[1], Alejandro Rodríguez[1], Jordi Tiana-Alsina[1], Francesc Rocadenbosch[1,2], David García-Vizcaíno[1]

[1]*RSLab, Department of Signal Theory and Communications, Universitat Politècnica de Catalunya, 08034 Barcelona, Spain, *Email:comeron@tsc.upc.edu*

[2]*Institut d'Estudis Espacials de Catalunya (IEEC), Barcelona, Spain*

ABSTRACT

It is known that the retrieval of aerosol extinction and backscatter coefficients from lidar data acquired through so-called total-power channels – intended to measure the backscattered power irrespective of the polarization – can be adversely affected by varying depolarization effects produced by the aerosol under measurement. This effect can be particularly noticeable in advanced multiwavelength systems, where different wavelengths are separated using a system of dichroic beam splitters, because in general the reflection and transmission coefficients of the beam splitters will be different for fields with polarization parallel or perpendicular to the incidence plane. Here we propose a setup for multiwavelength aerosol lidars alleviating diattenuation effects due to changing depolarization conditions while allowing measure linear depolarization.

1. INTRODUCTION

Advanced aerosol-lidar systems use several transmitted wavelengths and several receiving channels, both elastic and Raman, for the range-resolved measurement of the extinction and backscatter coefficient of atmospheric aerosols at different wavelengths. From the behavior of these optical coefficients as a function of the wavelength, microphysical properties of the particles can be inferred [1]. The addition of depolarization channels allows further microphysical characterization of the aerosols by providing information about the shape of the particles contributing to the backscatter [2]-[4]. In such advanced systems some of the channels are total-power channels, intended to receive the total power scattered from the atmosphere at a given wavelength; in some configurations, total-power channels can also be part of the arrangement for measuring depolarization [5]. To separate the different wavelengths in the lidar receiver setup, dichroic beam splitters are usually employed to direct the different incoming wavelengths to the corresponding receivers. Nevertheless the reflectance and transmittance[1] of beam splitters depend in general on the polarization of the incident radiation. Because of this, for the same amount of collected backscattered power, the power reaching the photodetector of a total power channel can vary noticeably depending on the depolarization produced by the aerosol contributing to the backscatter. The response of the beam splitters to an arbitrarily polarized wave can be expressed in terms of the responses to waves with linear polarizations respectively parallel and perpendicular to the plane of incidence (henceforth called p- and s-polarized wave respectively) [6],[7]. It has been shown [7] that, for the usual case of lidars transmitting linearly-polarized beams, when the transmission ratio (the ratio of the receiver transmittance for the p polarization to the transmittance for the s polarization) exceeds the interval (0.85,1.15) the retrieval of the aerosol backscatter coefficient in the presence of mineral dust can be biased by more than 5%.

Making the transmission ratio close to 1 to reduce the effect of varying polarization may have a big impact on the cost of the lidar receiver system, as it requires custom-coated beam splitters [4]. Nevertheless, if i) the different wavelengths transmitted all have linear polarizations parallel or

[1] For the sake of simplicity in the terminology, we will talk henceforth only about transmittance, since reflectance plays exactly the same role in the equations that give the power reaching a photodetector and has the same formal dependence on the polarization of the incident wave as the transmittance.

perpendicular between themselves, ii) the backscattering medium does not change the polarized part of the returned radiation with respect to the polarization of the transmitted beams [8] - [10], nor does it the part of the optical receiving system before the beam splitters, and iii) all the beam splitters in the receiver setup are arranged in such a way that the incidence plane is the same for all of them, the sensitivity of the receiver to the depolarization produced by the medium can be suppressed by orientating the incidence plane to be at a 45° angle with respect the polarization direction of the polarized part of the return radiation, irrespective of the transmission ratio. In this case the depolarization information after transmission or reflections is found in the 3rd and 4th components of the Stokes vector, whence it can be retrieved with a quarter-wave plate and linear polarizer arrangement.

2. PROPOSED SETUP

It can be easily shown that, if they share the same incidence plane, the Mueller matrix of a system of N cascaded beam splitters where the i-th element has a (complex) field transmittance t_{pi} in the p-plane and t_{si} in the s-plane is

$$[M]=\frac{1}{2}\begin{bmatrix} |t_p|^2+|t_s|^2 & |t_p|^2-|t_s|^2 & 0 & 0 \\ |t_p|^2-|t_s|^2 & |t_p|^2+|t_s|^2 & 0 & 0 \\ 0 & 0 & 2\,\mathrm{Re}\left[t_p^* t_s\right] & 2\,\mathrm{Im}\left[t_p^* t_s\right] \\ 0 & 0 & -2\,\mathrm{Im}\left[t_p^* t_s\right] & 2\,\mathrm{Re}\left[t_p^* t_s\right] \end{bmatrix},$$

where $t_p=\prod_{i=1}^{N} t_{pi}$ and $t_s=\prod_{i=1}^{N} t_{si}$.

If we assume that the light entering the optical setup is a combination of linearly polarized light and completely depolarized light, as would be the case if the polarization of the transmitted light is linear and the depolarizing scatterers do not have a preferred orientation [8]-[10], the Stokes vector describing the intensity and polarization state for that radiation will be [11]

$$[s_i]^T=\begin{bmatrix} I & I_{\parallel}\cos 2\gamma & I_{\parallel}\sin 2\gamma & 0 \end{bmatrix},$$

where I is the total intensity of the light, $I_{\parallel}$ is the intensity of the linearly polarized part, and γ is the angle of the linear polarization direction with respect to a reference direction perpendicular to the propagation one. If the reference direction is taken parallel to the incidence plane of the beam splitters, the Stokes vector at the setup output is found as $[s_o]=[M][s_i]$,

$$[s_o]=\frac{1}{2}\begin{bmatrix} \left(|t_p|^2+|t_s|^2\right)I+\left(|t_p|^2-|t_s|^2\right)I_{\parallel}\cos 2\gamma \\ \left(|t_p|^2-|t_s|^2\right)I+\left(|t_p|^2+|t_s|^2\right)I_{\parallel}\cos 2\gamma \\ 2\,\mathrm{Re}\left[t_p^* t_s\right]I_{\parallel}\sin 2\gamma \\ -2\,\mathrm{Im}\left[t_p^* t_s\right]I_{\parallel}\sin 2\gamma \end{bmatrix},$$

which for $\gamma=45°$ or $\gamma=135°$ reduces to

$$[s_o]=\frac{1}{2}\begin{bmatrix} \left(|t_p|^2+|t_s|^2\right)I \\ \left(|t_p|^2-|t_s|^2\right)I \\ 2\,\mathrm{Re}\left[t_p^* t_s\right]I_{\parallel} \\ -2\,\mathrm{Im}\left[t_p^* t_s\right]I_{\parallel} \end{bmatrix}. \qquad (1)$$

Eq. (1) shows that the transmitted power (1st element of the Stokes vector) has become insensitive to the depolarization ratio. The depolarization information is found in the 3rd and 4th components of the Stokes vector (the power of

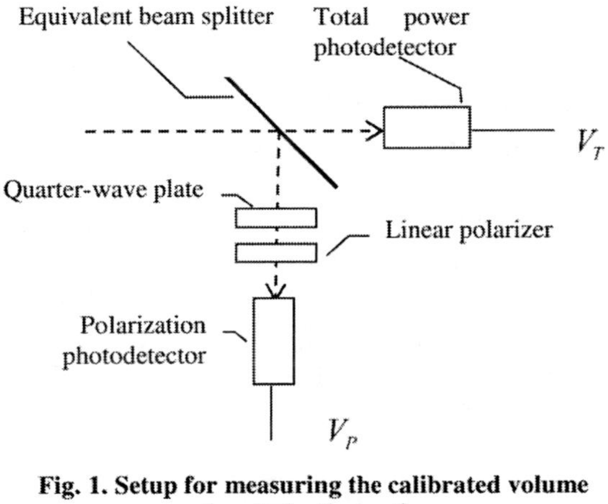

Fig. 1. Setup for measuring the calibrated volume depolarization ratio.

the depolarized light is $I_d=I-I_{\parallel}$), whence it can be extracted and calibrated using the setup outlined in fig. 1, where the equivalent beam splitter represents any combination of beam splitters with the sole condition that they share their incidence planes.

3. DEPOLARIZATION CALIBRATION

Calibration of the depolarization ratio is crucial for quantitative measurements. Assuming the (linearly) polarized part of the incident light has the polarization direction at 45° with respect to the incidence plane, the Stokes vector of the transmitted beam will have the form given by Eq. (1). As the first component of the Stokes vector corresponds to the total power of the beam and it is proportional to the total incoming power I, we will designate the detector measuring the power of the transmitted beam as the total-power photodector. At its output we will have a voltage

$$V_T = V_T^* \frac{1}{2}\left(\left|t_p\right|^2 + \left|t_s\right|^2\right) I, \qquad (2)$$

with V_T^* the voltage responsivity of the photodetector. The Stokes vector of the reflected beam, before passing through the quarter-wave plate and the linear polarizer, will have the form given by Eq. (1), with the reflection coefficients r_p, r_s substituted for the respective transmission ones.

The Mueller matrix of the cascaded quarter-wave plate and linear polarizer can be shown to be

$$[M_{e\alpha\theta}] = \frac{1}{2}\begin{bmatrix} 1 & \cos 2\alpha \cos 2(\alpha-\theta) & \sin 2\alpha \cos 2(\alpha-\theta) & \sin 2(\alpha-\theta) \\ \cos 2\theta & \cos 2\alpha \cos 2\theta \cos 2(\alpha-\theta) & \sin 2\alpha \cos 2\theta \cos 2(\alpha-\theta) & \cos 2\theta \sin 2(\alpha-\theta) \\ \sin 2\theta & \cos 2\alpha \sin 2\theta \cos 2(\alpha-\theta) & \sin 2\alpha \sin 2\theta \cos 2(\alpha-\theta) & \sin 2\theta \sin 2(\alpha-\theta) \\ 0 & 0 & 0 & 0 \end{bmatrix}$$

with α the angle between the quarter-wave plate fast axis and the p-plane and θ the angle between the polarizer transmission axis and the p-plane.

By setting $\alpha = 0$, $\theta = 0$, the Stokes vector of the light exiting the ensemble is

$$\left[s_{r00}\right]^T = \frac{1}{2}\left[\left|r_p\right|^2 I \quad \left|r_p\right|^2 I \quad 0 \quad 0\right],$$

and the polarization photodetector will yield an output voltage

$$V_{p00} = V_p^* \frac{1}{2}\left|r_p\right|^2 I,$$

with V_p^* the voltage responsivity of the polarization photodetector. By making the ratio between V_{p00} and V_T we obtain the coefficient

$$R_{00} = \frac{V_p^*}{V_T^*} \frac{\left|r_p\right|^2}{\left|t_p\right|^2 + \left|t_s\right|^2}. \qquad (3)$$

Likewise, by setting $\alpha = 0$, $\theta = 90°$, we will obtain an output voltage of the polarization photodetector that divided by V_T will result in the coefficient

$$R_{090} = \frac{V_p^*}{V_T^*} \frac{\left|r_s\right|^2}{\left|t_p\right|^2 + \left|t_s\right|^2}. \qquad (4)$$

Dividing the output of the polarization photodetector when $\alpha = 45°$, $\theta = 45°$ by V_T we obtain a coefficient

$$R_{4545} = \frac{1}{2} \frac{V_p^*\left[\left(\left|r_p\right|^2 + \left|r_s\right|^2\right) + 2\left|r_p\right|\left|r_s\right|\cos\left(\varphi_s - \varphi_p\right)\frac{I_\parallel}{I}\right]}{V_T^*\left(\left|t_p\right|^2 + \left|t_s\right|^2\right)},$$

which using Eqs. (3)and (4) can be written as

$$R_{4545} = \frac{1}{2}\left(R_{00} + R_{090}\right) + \sqrt{R_{00}R_{090}}\cos\left(\varphi_s - \varphi_p\right)\frac{I_\parallel}{I}, \qquad (5)$$

with φ_s and φ_p the phases of the complex reflection coefficients r_s and r_p respectively.

Likewise setting $\alpha = 45°$, $\theta = 0$ and dividing the output of the polarization photodetector by V_T we obtain the coefficient

$$R_{450} = \frac{1}{2}\left(R_{00} + R_{090}\right) - \sqrt{R_{00}R_{090}}\sin\left(\varphi_s - \varphi_p\right)\frac{I_\parallel}{I}. \qquad (6)$$

The diphase $\varphi_s - \varphi_p$ can be calculated from Eqs. (5) and (6):

$$\tan(\varphi_s - \varphi_p) = \frac{R_{00} + R_{090} - 2R_{450}}{2R_{4545} - R_{00} - R_{090}}, \quad (7)$$

and the quadrant uncertainty can be removed by checking the signs of the numerator and the denominator. Note that, to find $\varphi_s - \varphi_p$ from Eq. (7), the ratio $I_{\parallel} / I$ must be the same in Eqs. (5) and (6); therefore the measurements of R_{4545} and of R_{450} must be very close in time, or else they must be made on an aerosol-free region, where the depolarization ratio does not change.

Once the diphase has been found, the ratio $I_{\parallel} / I$ can be calculated from

$$\frac{I_{\parallel}}{I} = \frac{R_{4545} - \frac{1}{2}(R_{00} + R_{090})}{\sqrt{R_{00}R_{090}}\cos(\varphi_s - \varphi_p)}$$

or from

$$\frac{I_{\parallel}}{I} = \frac{\frac{1}{2}(R_{00} + R_{090}) - R_{450}}{\sqrt{R_{00}R_{090}}\sin(\varphi_s - \varphi_p)},$$

depending respectively on whether $\cos(\varphi_s - \varphi_p)$ or $\sin(\varphi_s - \varphi_p)$ has the maximum absolute value.

4. SUMMARY AND OUTLOOK

A setup has been proposed that allows alleviating the diattenuation-induced uncertainties while maintaining the depolarization measurement capability in multiwavelength aerosol lidar systems. Underlying assumptions are that all the transmitted wavelengths are linearly polarized with polarizations parallel or perpendicular to each other and that the aerosol particles are randomly oriented. Further work is necessary to study the sensitivity of the system to non-idealities such as angle tolerances and imperfect linear polarization in the transmitted beams.

ACKNOWLEDGEMENTS

This work has been supported by the following projects and grants: European Union 7th Framework Programme grant agreement No. 262254 (ACTRIS), Spanish Ministry for Economy and Competitiveness grants TEC2012-34575 and UNPC10-4E-442, and by the Catalan Agency for Support to Universities and Research (AGAUR) grant 2014 SGR 583.

REFERENCES

[1] Müller, D. et al., "Retrieval of Physical Particle Properties from Lidar Observations of Extinction and Backscatter at Multiple Wavelengths," *Appl. Opt.*, vol. 37, no. 12, p. 2260, Apr. 1998.

[2] McNeil, W. R. and A. I. Carswell, "Lidar polarization studies of the troposphere.," *Appl. Opt.*, vol. 14, no. 9, pp. 2158–68, Sep. 1975.

[3] Sassen, K., "Polarization in lidar," in *Lidar. Range-Resolved Optical Remote Sensing of the Atmosphere*, C. Weitkamp, Ed. Springer, 2005.

[4] Freudenthaler, V. et al., "Depolarization ratio profiling at several wavelengths in pure Saharan dust during SAMUM 2006," *Tellus B*, vol. 61, no. 1, Feb. 2009.

[5] De Tomasi F., and M. R. Perrone, "Multiwavelengths lidar to detect atmospheric aerosol properties," *IET Sci. Meas. Technol.*, vol. 8, no. 3, pp. 143–149, May 2014.

[6] Biele, J., G. Beyerle, and G. Baumgarten, "Polarization Lidar: Correction of instrumental effects," *Opt.Express*, vol. 7, no. 12, pp. 427–435, 2000.

[7] Mattis, I., et al., "Systematic error of lidar profiles caused by a polarization-dependent receiver transmission: quantification and error correction scheme," *Appl. Opt.*, vol. 48, no. 14, pp. 2742–2751, Oct. 2009.

[8] Mishchenko, M. I. and J. W. Hovenier, "Depolarization of light backscattered by randomly oriented nonspherical particles," *Opt.Lett.*, vol. 20, no. 12, pp. 1356–1358, Jun. 1995.

[9] Flynn, C. J. et al. "Novel polarization-sensitive micropulse lidar measurement technique," *Opt.Express*, vol. 15, no. 6, pp. 2785–2790, Mar. 2007.

[10] Gimmestad, G. G., "Reexamination of depolarization in lidar measurements," *Appl. Opt.*, vol. 47, no. 21, p. 3795, Jul. 2008.

[11] M. Born and E. Wolf, "Interference and diffraction with partially coherent light," in *Principles of Optics*, 6th ed., M. Born and E. Wolf, Eds. Pergamon Press, 1980, pp. 491–555.

RETRIEVAL OF AEROSOL PARAMETERS FROM CONTINUOUS H24 LIDAR-CEILOMETER MEASUREMENTS

D. Dionisi [1]*, F. Barnaba[1], F. Costabile[1], L. Di Liberto[1], G. P. Gobbi[1], H. Wille[2]

*[1]Institute of Atmospheric Sciences and Climate-CNR, Via del Fosso del Cavaliere 100, 00133 Roma, Italy, *Email: d.dionisi@isac.cnr.it*

[2]Jenoptik ESW, now at Lufft GmbH, 70736 Fellbach, Germany

ABSTRACT

Ceilometer technology is increasingly applied to the monitoring and the characterization of tropospheric aerosols. In this work, a method to estimate some key aerosol parameters (extinction coefficient, surface area concentration and volume concentration) from ceilometer measurements is presented. A numerical model has been set up to derive a mean functional relationships between backscatter and the above mentioned parameters based on a large set of simulated aerosol optical properties. A good agreement was found between the modeled backscatter and extinction coefficients and the ones measured by the EARLINET Raman lidars. The developed methodology has then been applied to the measurements acquired by a prototype Polarization Lidar-Ceilometer (PLC). This PLC instrument was developed within the EC- LIFE+ project "DIAPASON" as an upgrade of the commercial, single-channel Jenoptik CHM15k system. The PLC run continuously (h24) close to Rome (Italy) for a whole year (2013-2014). Retrievals of the aerosol backscatter coefficient at 1064 nm and of the relevant aerosol properties were performed using the proposed methodology. This information, coupled to some key aerosol type identification made possible by the depolarization channel, allowed a year-round characterization of the aerosol field at this site. Examples are given to show how this technology coupled to appropriate data inversion methods is potentially useful in the operational monitoring of parameters of air quality and meteorological interest.

1. INTRODUCTION

Nowadays, hundreds of automated single channel lidar ceilometers (ALCs) are in operation over Europe. Although such instruments were originally designed for cloud base detection, recent studies [1, 2] show that the ceilometer technology is now mature enough to use these systems for a quantitative evaluation of the aerosol physical properties in the lower atmosphere. The assessment of the full potential of ceilometers in the remote sensing of aerosols is a current matter of discussion in the lidar community. In Europe, such evaluation is one of the main objectives of the EU COST Action ES1303, TOPROF (Towards Operational ground-based PROFiling with ceilometers, doppler lidars and microwave radiometers).

This work presents a general model-based methodology to estimate the aerosol optical and physical properties (extinction, surface area and volume) from the aerosol backscatter profiles retrieved by ceilometers. The procedure, based on the approach previously developed by Barnaba et al. for visible and UV lidar channels [3, 4], has been upgraded and extended to the 1064 nm wavelength. The results of the model simulations were firstly compared to aerosol EARLINET (European Aerosol Research Lidar Network, [5]) database, in which the particle backscatter and the extinction coefficients are retrieved independently by Raman-lidars [6]. The model-based inversion procedure was then validated employing a year-round record retrieved by a "Polarization-sensitive Lidar Ceilometer" (PLC) prototype. The PLC was prototyped within the European EC-Life+ Project DIAPASON (Desert-dust Impact on Air quality through model-Predictions and Advanced Sensors ObservatioNs, LIFE+2010 ENV/IT/391). This instrument upgrades the commercial Jenoptik CHM15k system, now produced by the German Lufft GmbH. The PLC prototype was installed at the background coastal site of 'Castel di Guido', 20 km NW of Rome (Italy) in 2013. Here we start evaluating the capability of the developed method at quantitatively estimating the aerosol backscatter coefficients, and at deriving relevant aerosol

DOI: 10.1051/epjconf/201611923004

physical properties of a variety of aerosol types. The proposed approach could represent a valid option to extend the capabilities of these systems at providing important information for operational air quality and meteorological monitoring.

2. METHODS

2.1 THE AEROSOL MODEL

A numerical aerosol model was set up to calculate mean functional relationships between the aerosol backscatter coefficient (β_a) and some relevant aerosol properties as extinction coefficient, surface area concentration and volume concentration (α_a, S_a and V_a, respectively, e.g. [3, 4]). Following a Monte Carlo approach, the model simulates a large set of aerosol optical properties by randomly varying, within appropriate ranges, the microphysical parameters describing the aerosol size distribution and composition. A tri-modal lognormal size distribution is assumed:

$$n(r) = \frac{dN}{d\ln r} \sum_{i=1}^{3} \frac{N_i}{\sqrt{2\pi}\ln\sigma_i} \exp\left[-\frac{(\ln r - \ln r_{mi})^2}{2(\ln\sigma_i)^2}\right], \quad (1)$$

where r_{mi}, σ_i and N_i are the modal radius, the width and the particle number density of the *i*th aerosol mode, respectively. To compute β_a and α_a values, the particle refractive index (real and imaginary parts, m_r and m_i) are also randomly selected within appropriate ranges.

In a first step, we fixed the ranges of the aerosol parameters to be adopted in the simulation (Table 1), based on data available in the literature.

TABLE 1: Values ranges of the optical parameters adopted in the general model configuration

Parameter	Mode I	Mode II	Mode III
r_{mi} (μm)	0.005-0.05	0.03-0.1	0.3-0.5
σ_i	1.5-2.2	1.5-2.3	1.5-2.4
N_i/N_{tot} (%)[a]	5-55	45-95	0.001-0.1
m_r	1.3-1.6	1.3-1.6	1.4-1.6
m_i	10^{-5}-0.05	10^{-5}-0.02	10^{-5}-0.01

[a]N_{tot} varies in the range 10^3-$5\cdot10^4$ (cm^{-3}) at z = 0 km.

These ranges were intended to reproduce a 'continental aerosol type', with possible contribution of maritime or desert dust particles. For the latter, an empirical correction for non-spherical particles was introduced, based on the results of Mishchenko et al. [7]). Conversely, for non-dust species, a dependence on relative humidity of both the particle radius and refractive index was included. Simulations have been performed at 355, 532 and 1064 nm.

2.2 THE PLC INSTRUMENT

The main specifications of the DIAPASON PLC are similar to those of the Jenoptik CHM15k ('Nimbus'). In particular, the system emits laser pulses at 1064 nm (Nd:YAG-laser, class M1) with a typical pulse energy of 8 μJ and a pulse repetition rate of about 6500 Hz. The beam divergence is < 0.3 mrad; the receiver field of view is 0.46 mrad (half angle). The height of complete overlap (z_{ovl}) is about 150 m. The main difference with the standard Nimbus ceilometer is the implementation of two receiving channels for parallel and cross signals that are detected through two APDs in analog mode with a fixed sensitivity. Raw data are acquired with a temporal and vertical resolution of 15 s and 7.5 m, respectively.

2.3 INVERSION OF THE PLC H24 DATA

To derive the β_a from the PLC measurements, the backward solution of the Klett inversion algorithm [8] was applied to the data. In addition to the estimation of the molecular backscatter and extinction coefficients (β_m and α_m, respectively, calculated from climatological monthly air density profiles), the solution requires two assumptions: a boundary value at a reference height z_0 where $\beta_a(z_0) = 0$ (Rayleigh calibration) and a lidar ratio ($S_a = \alpha_a/\beta_a$). For ceilometers, the first assumption can be challenging because of their weak sensitivity to the molecular return at high altitudes. Still, appropriate averaging allows obtaining a clean molecular profile. For the PLC data, a daily calibration constant was derived applying the Rayleigh calibration to nighttime and cloud-free signal averaged over 1 h at 75 m height resolutions. The instrument gain ratio of the two polarization channels was also computed in this region. The second assumption is an inherent problem of any single-wavelength backscatter lidar that limits the accuracy of the aerosol extinction coefficients. In our case α_a is computed by the means of a mean relationship $\alpha_a = \alpha_a(\beta_a)$ derived from the numerical simulations (see Section 3). This requires an iterative inversion technique to correct the backscatter signal for

DOI: 10.1051/epjconf/201611923004

extinction losses until convergence in the Integrated Aerosol Backscatter (IAB=$\sum_0^{zcal}\beta_a(z)$) is reached. In the case of systems with polarization capabilities, as the PLC, the additional information of the depolarization signal can be integrated in the inversion algorithm. In particular, in presence of dust (highly non spherical) or marine (highly spherical) aerosol, the simulated optical properties can be constrained to the detected type of particles.

3. RESULTS

An example of the numerical simulation outcome is shown in Figure 1 (blue crosses). This presents the results in terms of model-derived α_a vs β_a values for $2x10^4$ different combinations of the aerosol microphysical parameters in Table 1.

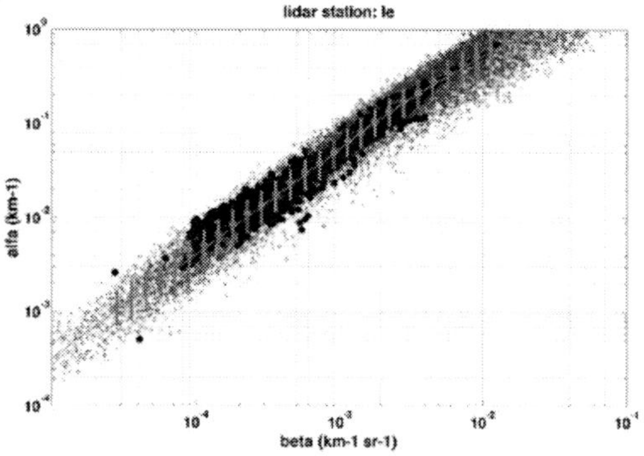

Figure 1. Model-simulated (blue crosses) and lidar measured (EARLINET Leipzig Raman lidar, black dots) aerosol α_a vs β_a. The green line and the magenta dots are respectively the fitting curve and the average values (with associated standard deviations) of the simulated aerosol ensemble.

The modeled values were compared to the backscatter and extinction coefficients measured by the EARLINET lidars. In particular we selected the coefficients retrieved at 355 nm (13 lidar stations) and at 532 nm (12 lidar stations) within the 'climatology' category in the EARLINET database). As an example, α_a and β_a values retrieved by the Leipzig Raman lidar at 355 nm (black dots) are also reported in Figure 1 showing that the 'cloud' of the modeled values well coincides with the measured α_a and β_a coefficients, and that the aerosol variability seems to be correctly reproduced by the fitting curve of the aerosol ensemble (green line).

A one-year dataset of the PLC acquired during the DIAPASON project has been analyzed through the above-described methodology. An example of the h24 PLC data is shown in Figure 2.

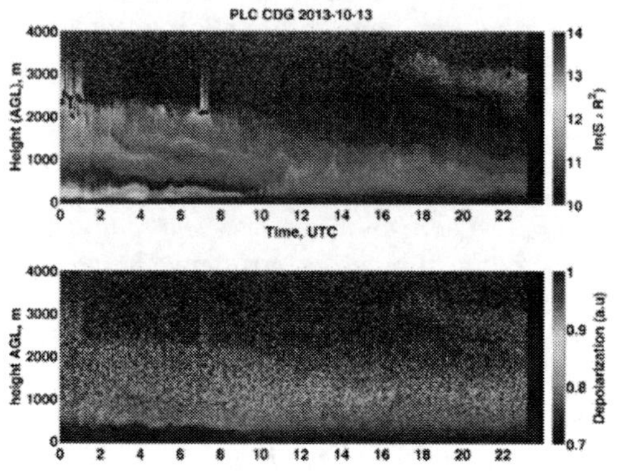

Figure 2. Upper panel: time height cross-section of the logarithm of the range-corrected signal at 1064 nm. Lower panel: time height cross-section of the ratio between the cross and parallel signals (Castel di Guido site, 13 October 2013).

This shows the time-height contour plot of the logarithm of the range corrected parallel signal (upper panel) and the ratio between the cross and parallel signals (lower panel) acquired on October 13, 2013. The time and altitude resolutions are 2 min and 15 m, respectively. These data include a Saharan dust advection event that lasted 4 days from 12 to 15 October 2013.

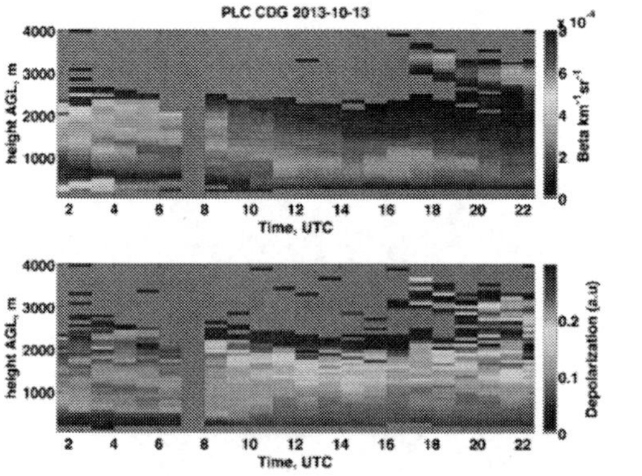

Figure 3. Upper panel: time-height cross-section of the aerosol backscatter coefficients β_a [$km^{-1}sr^{-1}$] at 1064 nm retrieved from data in Figure 2. Lower panel: corresponding time-height cross-section of the ratio between the cross and parallel signals. Each value is averaged over 60 min and 75 m. Gray areas mask not-inverted data due to either cloud presence or relative error >50%.

The corresponding retrieved values of β_a are shown in Figure 3 (upper panel) with the relevant values of depolarization (δ, lower panel). The relative error $\Delta\beta_a/\beta_a$ ranges between 0.1 and 0.5. The potential of this kind of inversion is further explored in Figure 4 showing the PLC-based retrieval of the hourly-resolved aerosol mass (µg/cm^3) at two different levels (225 and 525 m, red and blue lines, respectively) for the same date (13 October 2013). In this case the computed functional relationships $V_a = V_a(\beta_a)$ was employed to quantify the aerosol load for specific aerosol densities (ρ_a).

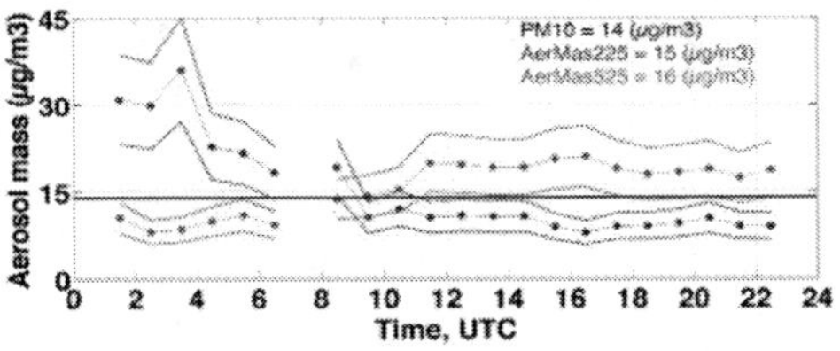

Figure 4. Hourly-resolved aerosol mass concentration estimated from PLC data at 225 m and 525 m (blue and red lines, respectively) for the 13 October 2013. Three different particle densities were assumed: 1.5 (lower line), 2 (central line) and 2.5 (upper line) g/cm^3. The horizontal black line is the in situ measured daily average PM10, the aerosol metric regulated by the relevant EC Air Quality Directive. For comparison, the daily mean values of the aerosol mass (with ρ_a= 2 g/cm^3) for the two altitudes are also reported in the figure inset.

5. CONCLUSIONS

A model-based methodology to estimate aerosol optical and physical properties from calibrated aerosol backscatter profiles retrieved by lidar-ceilometers has been introduced. The resulting functional relationships have been used to evaluate aerosol backscatter coefficient and volume out of a Polarization Lidar-Ceilometer (PLC) dataset retrieved in Rome (Italy). The PLC is a prototype developed within the European Commission LIFE+ project "DIAPASON" (www.diapason-life.eu). The relationships between β_a and α_a at 355 and 532 nm have been validated with the coefficients estimated by the EARLINET Raman lidars. The β_a coefficients derived by the PLC measurements (with a temporal and vertical resolution of 1 h and 75 m) have an associated relative error between 0.1 and 0.5 up to 3 km. The depolarization channel allowed acquiring information on the type of particles (e.g. dust-like or spherical aerosols). This approach, which will be tested on the Leosphere R-MAN super-ceilometer also employed in DIAPASON, shows that these instruments could be usefully applied to the operational monitoring of Air Quality.

ACKNOWLEDGEMENT

This work was supported by the European Commission Life+ Project "DIAPASON" (LIFE+2010 ENV/IT/391). The study also contribute to the activities of the COST Action TOPROF. The authors gratefully acknowledge the EARLINET Project for establishing and maintaining their open-access database.

REFERENCES

[1] Wiegner, M., A. Geiß, 2012: Aerosol profiling with the Jenoptik ceilometer CHM15kx, *Atmos. Meas. Tech.*, **5**, 1953–1964, doi:10.5194/amt-5-1953-2012.

[2] Wiegner, et al., 2014: What is the benefit of ceilometers for aerosol remote sensing? An answer from EARLINET, *Atmos. Meas. Tech.*, **7**, 1979–1997, doi:10.5194/amt-7-1979- 2014.

[3] Barnaba, F., G. P. Gobbi, 2001: Lidar estimation of tropospheric aerosol extinction, surface area and volume: Maritime and desert-dust cases, *J. Geophys. Res.*, **106-D3**, 3005–3018.

[4] Barnaba, F., G. P. Gobbi, 2004: Modeling the aerosol extinction versus backscatter relationship for lidar applications: maritime and continental conditions, *J. Atmos. Ocean. Technol.*, **21**, 428–442.

[5] Pappalardo, G. et al., 2014: EARLINET: towards an ad- vanced sustainable European aerosol lidar network, *Atmos. Meas. Tech.*, **7**, 2389–2409, doi:10.5194/amt-7-2389-2014.

[6] Ansmann, A., et al., 1992: Independent measurement of extinction and backscatter profiles in cirrus clouds by using a combined Raman elastic-backscatter lidar, *Appl. Optics*, **31**, 7113–7131.

[7] Mishchenko, M.I., et al., 1997: Modeling phase functions for dustlike tropospheric aerosols using a mixture of randomly oriented polydisperse spheroids. *J. Geophys. Res.*, **102**, 16831-16847, doi:10.1029/96JD02110.

[8] Klett, J. D., 1981: Stable analytical inversion solution for processing lidar returns, *Appl. Optics*, **20**, 211–220.

EPJ Web of Conferences **119**, 23005 (2016)
ILRC 27
DOI: 10.1051/epjconf/201611923005

THE SYSTEM OF THE CALIBRATION FOR VISIBILITY MEASUREMENT INSTRUMENT UNDER THE ATMOSPHERIC AEROSOL SIMULATION ENVIRONMENT

Zhifeng Shu[1,2*] , ShaoChen Yang[3] , Wenjing Xu[3]

*[1]Chinese Academy of Science, Key Laboratory of Geospace Environment, Hefei, Anhui, 230026, China, *Email:szf@ustc.edu.cn*

[2]University of Science and Technology of China, School of Earth and Space Science, Hefei, Anhui, 230026, China

3. ShenZhen Darsunlaser Tech Co.,Ltd Shenzhen Guangdong,518067,China

ABSTRACT

Visibility is one of the most important parameters for meteorological observation and numerical weather prediction (NWP).It is also an important factor in everyday life, mainly for surface and air traffic especially in the Aeronautical Meteorology. The visibility decides the taking off and landing of aircraft. If the airport visibility is lower than requirement for aircraft taking off stipulated by International Civil Aviation Administration, then the aircraft must be parked at the airport. So the accurate measurement of visibility is very important. Nowadays, many devices can be measured the visibility or meteorological optical range (MOR) such as Scatterometers, Transmissometers and visibility lidar. But there is not effective way to verify the accuracy of these devices expect the artificial visual method. We have developed a visibility testing system that can be calibration and verification these devices. The system consists of laser transmitter, optical chopper, phase-locking amplifier, the moving optic receiving system, signal detection and data acquisition system, atmospheric aerosol simulation chamber. All of them were placed in the atmosphere aerosol simulation chamber with uniform aerosol concentration. The Continuous wave laser, wavelength 550nm, has been transmitted into the collimation system then the laser beam expanded into 40mm diameter for compressing the laser divergence angle before modulated by optical chopper. The expanding beam transmitting in the atmosphere aerosol cabin received by the optic receiving system moving in the 50m length precision guide with 100mm optical aperture. The data of laser signal has been acquired by phase-locking amplifier every 5 meter range. So the 10 data points can be detected in the 50 meters guide once. The slope of the fitting curve can be obtained by linear fitting these data using the least square method. The laser extinction coefficient was calculated from the slope using the Koschmieder formula, then it been divided by 3 is MOR. The aerosol concentration in chamber can be changed by adjusting aerosol generator that producing variety of visibility atmospherical environment. The experiment has been carried out and the measurement accuracy of atmospheric transmittance is 0.3‰ Corresponding to the accuracy of MOR 4.9% at the 2km visibility environment. So this system can be calibrated and validated the other visibility measuring devices.

1. INTRODUCTION

Visibility is one of the most important parameters for meteorological observation and numerical weather prediction (NWP).It is also an important factor in everyday life, mainly for surface and air traffic especially in the Aeronautical Meteorology[1,2,3]. Visibility is the most important information for pilot of landing and taking off aircraft. One of the most complicated phases of flying an aircraft is landing approach. The ground staff provide the pilot the visibility (the runway visual range RVR). The meteorological visual range(MOR) is most important parameter for composing the RVR. So the accurate measurement of visibility is very important. Nowadays, many devices can be measured the visibility or meteorological optical range (MOR) such as Scatterometers, Transmissometers and visibility lidar. But there is not effective way to verify the accuracy of these devices expect the artificial visual method. We have developed a visibility testing system that can be calibration and verification these devices[4,5].

DOI: 10.1051/epjconf/201611923005

2. METHODOLOGY

Figure 1 shows that block diagrams of the calibration system. The main container includes a continuous wave (cw) laser with wave length 550nm and power 50mw, optical chopper with frequency 20Hz to 1kHz , detector with model thorlabs DET100A, motor control, phase-locking amplifier , high precision guide and the optical transceiver system. All of them are placed in the atmosphere aerosol simulation environment with LWH 65m╳2m╳2m. The Continuous wave laser, wavelength 550nm, has been transmitted into the collimation system then the laser beam expanded into 40mm for compressing the laser divergence angle before modulated by optical chopper. The expanding beam transmitting in the atmosphere aerosol cabin received by the optic receiving system moving in the 50m length precision guide . The data of laser signal has been acquired by phase-locking amplifier every 5 meters range. So the 10 data points can be detected in the 50 meters guide once.

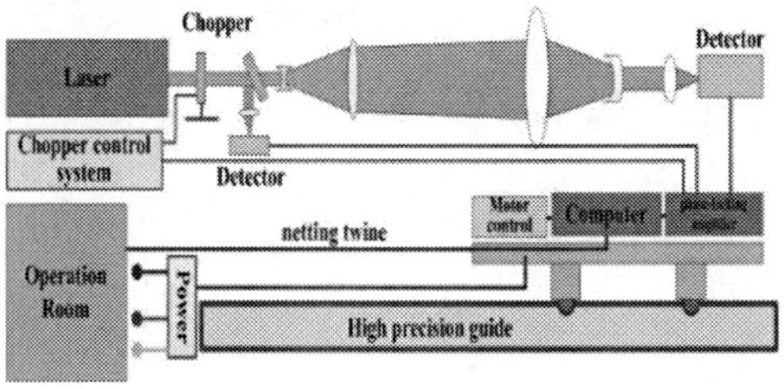

Fig1. The block diagrams of the visibility calibration system

The atmosphere aerosol simulation environment is the transparent glass cabin the all kinds of aerosol can be produced by the aerosol generator such as atomizer aerosol generator, atomizer aerosol generator, black carbon aerosol generator. These generators are placed in a room beside the glass cabin as shown in the figure 2. The aerosol can be emitted in to the cabin through the nine pipelines with uniform distribution along the guide, then it will be mixed uniformly by the 10 convective fans distributed along the guide. The aerosol concentration can be detected by the concentration to ensure the uniform aerosol distribution in the cabin. Figure 4 is the real pictures and software interface of the system.

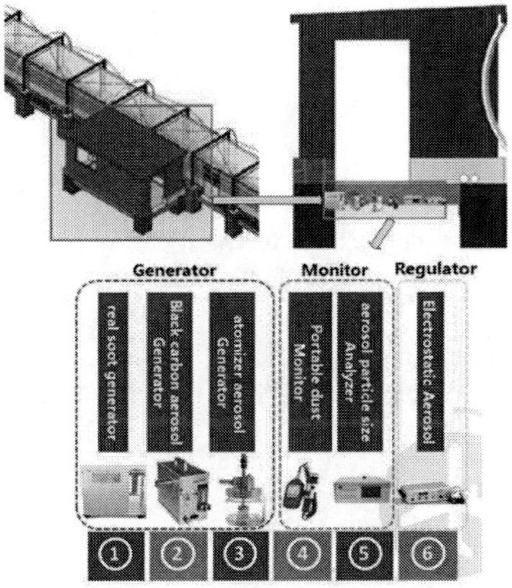

Fig2. Aerosol generation system

Fig3. The pictures of Atmospheric simulation cabin

3. RESULTS

These system of visibility calibration have been built in the tianjin province. In order to testing the performance of the system a experiment has been carried out. Laser beam has been emitted and been chopped then the sigal will be detected by the detector and data acquired by the phase-locking amplifier that they are mounted on the cart. The speed of the cart is 3m/s and it will stay 10 seconds when responsing the position sensor for detecting and acquiring of the data. There will be 10 points data for calculating the MOR. There data can be fitted linearly by the least square method and the slope can be achieved. According to the lidar equation, the slope is the laser atmosphere extinction coefficient, then it been divided by 3 is MOR using Koschmieder formula. Figure 4 is the laser energy change when the cart goes and backs along the glass cabin. Fitting these data the slope is -8.002^{-4} Corresponding to 3.6km MOR using 3 divide the slope. The standard error of the fitting is 3.11^{-5}Corresponding to the of 4.9% accuracy of MOR as shown in figure4.

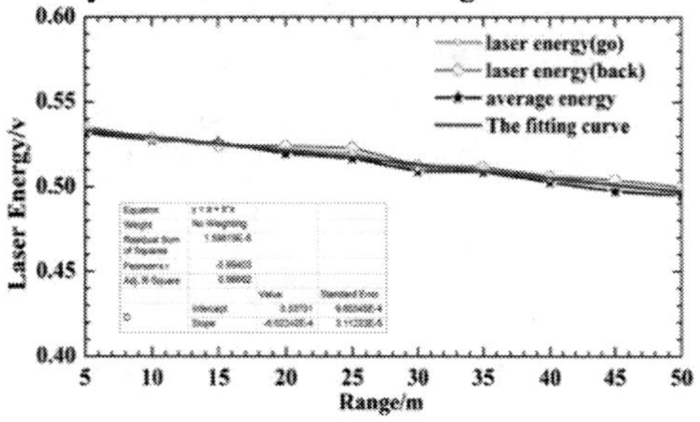

Fig4. The result of the experiment

4. CONCLUSIONS

The calibration system for visibility measurement instrument under the aerosol simulation environment has been built and the measuring MOR expriment has been carried out and the accuracy of MOR is 4.9% at 3.6km visibility. The MOR accuracy of the Civil aviation administration is ±50m@MOR 10m~500m, ±10%@MOR>500m~1500m, ±20% @ MOR > 1500m~50000m.So this system can be used to calibrate and validate the other visibility measuring devices. There system is the first visibility calibration device in the international arena .It can be do some other meteorological research under the atmosphere aerosol simulation environment.

ACKNOWLEDGEMENT

This work was funded by the National Natural Science Foundation of China (NSFC) project Nos. 41304123 and the civil aviation administration of china.

REFERENCES

[1] J. Streicher, C.Monkel, H.Borchardt, Trial of slant visual range measuring device, Journal of atmospheric and oceanic technology, Vol 10 718~724,1993.

[2] R.Boers, H.Klen baltink, H.J.Hemink, F.C.Bosveld,M.Moerman, Ground-based observation and modeling of the visibility and radar reflectivity in a radiation fog layer, Journal of atmospheric and oceanic technology, Vol 30 288~299.

[3] M.A.Vanghan, D.M.Winker and K.A.Po-well. "Part2: Feature detection and layer properties algorithms" CALIOP Algorithm Theoretical Basis Document PC-SCI-202 Part2, pp.87.

[4] Jurgen Streicher, Christian Werner. Design of a small laser ceilometers and visibili-ty measuring device for helicopter landi-ng sites[J]. Laser Radar Technology for Remote Sensing, SPIE, Vol.5240:31-41.

[5]. CHEN Min, SUN Dong-song, LI Hong-jing. Detective method and analysis of atmosphere slant visibility for lidar [J]. Infrared and Laser Engineering, Vol 35(2): 156-160..

DOI: 10.1051/epjconf/201611923006

BIOMASS BURNING AEROSOLS IN THE AMAZON BASIN, CHARACTERISED BY LIDAR, OPTICAL PARTICLE COUNTERS, AND MODELLING

Franco Marenco[1,*], Ben Johnson[1], Justin Langridge[1], Jane Mulcahy[1], Angela Benedetti[2], Samuel Remy[2], Luke Jones[2], Kate Szpek[1], and Jim Haywood[1,3]

[1]*Met Office, Exeter, United Kingdom; *Email: franco.marenco@metoffice.gov.uk*

[2]*European Centre for Medium-range Weather Forecasts, Reading, United Kingdom*

[3]*University of Exeter, United Kingdom*

1. INTRODUCTION

Biomass burning is the second largest global source of anthropogenic aerosols, and South America is one of the major source regions. In the dry season, the atmosphere of the Amazon basin features a remarkable haze, with layers containing high loadings of smoke. Aerosols with different degrees of ageing, are encountered in the boundary layer and the free troposphere. The South American Biomass Burning Analysis (SAMBBA) was an intensive observation campaign in September-October 2012 that involved measurements of the Amazonian atmosphere using the Facility for Airborne Measurements (FAAM) BAe-146 research aircraft.

2. LIDAR OBSERVATIONS

Twenty research flights were carried out from Porto Velho, Brazil, totaling 65 flying hours. A large range of conditions were sampled, from very low aerosol concentrations in pristine areas to large quantities of smoke within fresh plumes. The aircraft carried a nadir-pointing elastic backscatter lidar, operating at 355 nm. In situ probes sampled particle size distributions and gas-phase chemistry. Six flights have been selected that span a 2400 km wide area extending East-West across Brazil along a latitude of approximately 10°S. The lidar data presented here have a vertical resolution of 45 m and an integration time of 1 min (corresponding to a 9 km footprint). From these flights, 334 lidar profiles have been reviewed individually, and analysed.

Processing has undergone a double iteration, to firstly to determine the lidar ratio (extinction-to-backscatter ratio), and subsequently to estimate the aerosol extinction coefficient. The lidar ratio determination is based on iterating the retrieval method detailed in the next paragraph, until a good match to the overlying Rayleigh scattering layer is obtained (see e.g. Ref. [1]). A single value of the lidar ratio (constant with height) is thus obtained for each profile. It is then further averaged over all profiles in order to achieve a single lidar ratio for the campaign. The lidar ratio deduced from the lidar profiles using this methodology was found to be 73 ± 6 sr^{-1}, and is compatible with Ref. [2–5].

The determination of the aerosol extinction coefficient has followed Ref. [6]. This is a variant of the Fernald–Klett method [7–8], where the reference is taken within an aerosol layer. This permits using the stable (inward) solution in the unfavourable geometry represented by a nadir-looking lidar. Very large uncertainties (50-100%) exist near the surface, but they are quickly damped when moving upwards (< 20% above 2 km).

3. LIDAR RATIO ESTIMATED FROM IN SITU MEASUREMENTS

The lidar ratio obtained from the lidar profiles has been compared to estimates derived from the Mie scattering theory, using the particle size-distribution from the optical particle counters. Fig. 1(a) shows the campaign-mean size-distribution, and Fig. 1(b) shows the resulting lidar ratio as a function of the real and imaginary parts of the reftractive index.

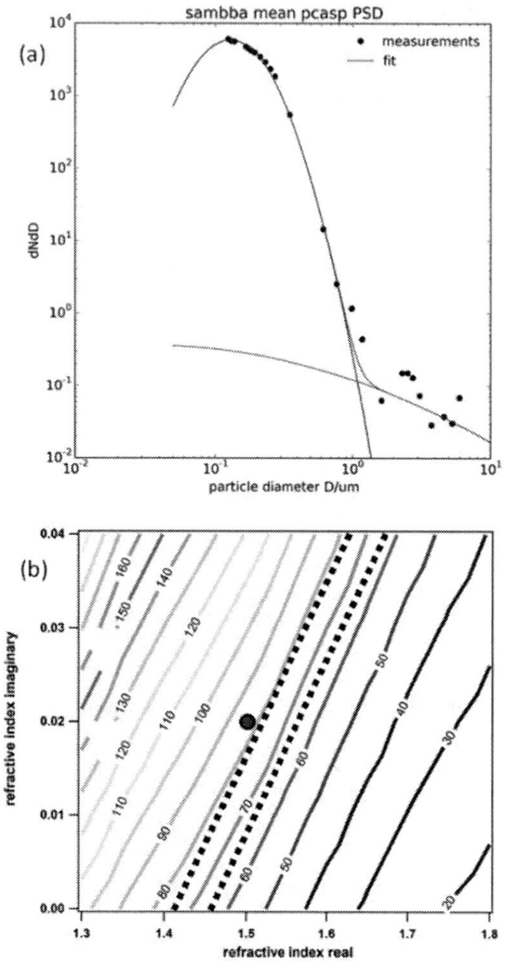

Fig. 1. (a) Campaign mean particle size-distribution from the wing-mounted PCASP optical particle counter; (b) Lidar ratio computed for the campaign mean size-distribution and different values for the real and imaginary parts of the refractive index. Red dot: SCAR-B refractive index (1.5 – 0.02i). Black dotted lines: range of lidar ratios estimated from the lidar.

4. MODEL COMPARISONS

The lidar data have been compared to high resolution (12km) simulations from a limited area model (LAM) configuration of the Met Office Unified Model (UM), which was set up over Brazil for the SAMBBA campaign. Biomass burning aerosol was simulated on-line using the CLASSIC biomass burning aerosol scheme [9], while all other aerosol species were represented by climatologies. Fire emissions were taken from the Global Fire Assimilation System, GFAS emission dataset [10]. Operationally the model used emission from the previous day due to the time lag in providing near real time data. Fig. 2 compares modeled and observed aerosol extinction from a 1500km West-East transit over Amazonia. The model represents many realistic features of the aerosol layers, although plumes from individual fires are not always captured.

The next step of the present research, will be to compare the airborne dataset with the aerosol forecasts by the European Centre for Medium-range Weather Forecasts (ECMWF), provided as part of the EU-funded projects GEMS, MACC and MACC-II [11–12]. Five types of tropospheric aerosols are considered in the model, and they are fully coupled with the meteorology. Biomass burning emissions are taken from the GFAS inventory; moreover, MODIS AOD data are routinely assimilated in a 4D-Var framework. The result of this comparison will be shown at the conference.

ACKNOWLEDGEMENTS

Airborne data was obtained using the BAe-146 Atmospheric Research Aircraft flown by Directflight Ltd. and managed by the Facility for Airborne Atmospheric Measurements (FAAM), which a joint entity of the Natural Environment Research Council (NERC) and the Met Office. SAMBBA was funded by the Met Office and NERC (grant NE/J009822/1). The National Institute for Space Research (INPE) and the University of Sao Paulo are kindly thanked for having helped make this campaign a success.

REFERENCES

[1] Marenco F., B.T. Johnson, K.F. Turnbull, S. Newman, J.M. Haywood, H. Webster, and H. Ricketts: Airborne lidar observations of the 2010 Eyjafjallajökull volcanic ash plume, J. Geophys Res 116, D00U05, doi: 10.1029/2011JD016396, 2011.

[2] Omar, A. H., Winker, D. M., Kittaka, C., Vaughan, M. A., Liu, Z., Hu, Y., Trepte, C. R., Rogers, R. R., Ferrare, R. A., Lee, K.-P., Kuehn, R. E., and Hostetler, C. A.: The CALIPSO Automated Aerosol Classification and 285 Lidar Ratio Selection Algorithm, J. Atmos. Ocean.

Technol., 26, 1994–2014, 2009.

[3] Baars, H., Ansmann, A., Althausen, D., Engelmann, R., Heese, B.,M¨uller, D., Artaxo, P., Paixao,M., Pauliquevis, T., and Souza, R.: Aerosol profiling with lidar in the Amazon Basin during the wet and dry season, J. Geophys. Res., 117, D21201, doi:10.1029/2012JD018338, 2012.

[4] Groß, S., Freudenthaler, V., Wiegner, M., Gasteiger, J., Geiß, A., and Schnell, F.: Dual-wavelength linear depolarization ratio of volcanic aerosols: Lidar measurements of the Eyjafjallaj¨okull plume over Maisach, Germany, Atmos. Environ., 48, 85–96, 2012.

[5] Lopes, F. J. S., Landulfo, E., and Vaughan, M. A.: Evaluating CALIPSO's 532 nm lidar ratio selection algorithm using AERONET sun photometers in Brazil, Atmos. Meas. Tech., 6, 3281–3299, 2013.

[6] Marenco, F.: Nadir airborne lidar observations of deep aerosol layers, Atmos.Meas. Tech., 6, 2055–2064, 2013.

[7] Fernald, F. G.: Analysis of atmospheric lidar observations: some comments, Appl. Opt., 23, 652–653, 1984.

[8] Klett, J. D., Lidar inversion with variable backscatter/extinction ratios, Appl. Opt. 24, 1638–1643, 1985.

[9] Bellouin, N., J. Rae, A. Jones, C. Johnson, J. Haywood, and O. Boucher, Aerosol forcing in the Climate Model Intercomparison Project (CMIP5) simulations by HadGEM2-ES and the role of ammonium nitrate, J. Geophys. Res., 116, D20206, doi:10.1029/2011JD016074, 2011.

[10] Kaiser, J. W., et al.: Biomass burning emissions estimated with a global fire assimilation system based on observed fire radiative power, Biogeosciences, 9, 527-554, 2012.

[11] Morcrette J-J et al, Aerosol analysis and forecast in the European centre for medium-range weather forecasts integrated forecast system: forward modeling. J Geophys Res 114, D06206. doi:10.1029/2008JD011235, 2009.

[12] Benedetti A et al, Aerosol analysis and forecast in the European Centre for Medium-Range Weather Forecasts Integrated Forecast System: 2. Data assimilation. J Geophys Res 114, D13205. doi:10.1029/2008JD011115, 2009.

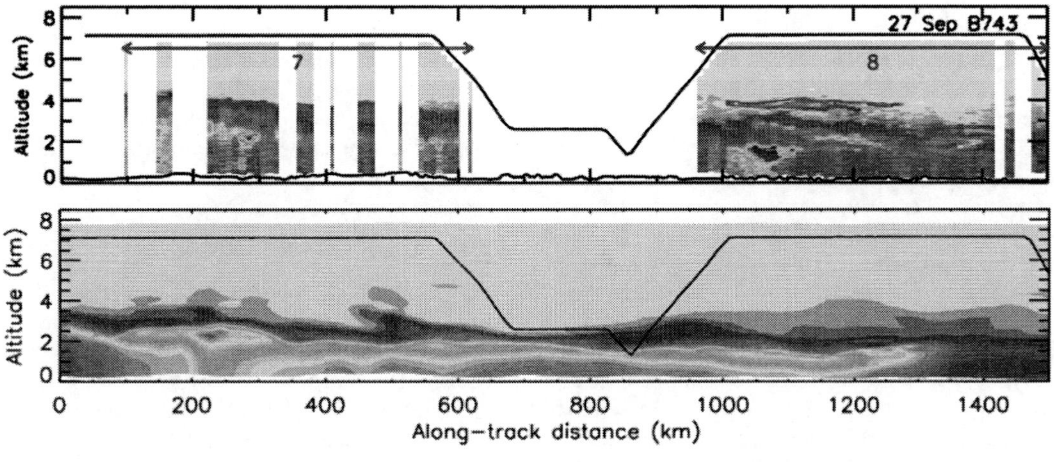

Fig. 2 Extinction coefficient evaluated by lidar for 27 September 2012 (top) and predictions with the UM.

EPJ Web of Conferences **119**, 23007 (2016) DOI: 10.1051/epjconf/201611923007
ILRC 27

CERTAIN RESULTS OF MEASUREMENTS OF CHARACTERISTICS OF STRATOSPHERIC AEROSOL LAYER AND TOTAL OZONE CONTENT AT SIBERIAN LIDAR STATION IN TOMSK

Aleksey Nevzorov*, Oleg Bazhenov, Vladimir Burlakov and Sergey Dolgii

V.E. Zuev Institute of Atmospheric Optics, Siberian Branch, Russian Academy of Sciences, Tomsk 634021, Russia, *e-mail: nevzorov@iao.ru

ABSTRACT

We consider the results of long-term remote optical monitoring, obtained at the Siberian Lidar Station of Institute of Atmospheric Optics, Siberian Branch, Russian Academy of Sciences in Tomsk (56.5°N, 85.0°E). The scattering characteristics of stratospheric aerosol layer, obtained according to data of lidar measurements since 1986, are presented. We analyze the trends of changes in the total ozone (TO) content over Tomsk for the period 1996-2013 according to data of spectrophotometric measurements with employment of Total Ozone Mapping Spectrometer (TOMS) data for the period 1979-1994. We determined the periods of elevated content of stratospheric aerosol over Tomsk after a series of explosive eruptions of volcanoes of Pacific Ring of Fire and Iceland in 2006-2011. Since the second half of 1990s, we record an increasing TO trend, equaling 0.65 DU/yr for the period 1996-2013.

1. INTRODUCTION

The most important constituents and parameters of the atmosphere, which determine its radiation-temperature regime and chemical balance, are stratospheric aerosol, ozone, gas components of ozone cycles, and temperature. The scattering and absorbing properties of the stratospheric aerosol and gas constituents should be taken into consideration in radiation codes of the models of atmospheric general circulation and climate. Since early 2000s, there are signs of stabilization and even growth of the total ozone (TO), at separate regions [1]. The reasons for the TO stabilization signs are still debatable: whether these are (1) due to the implementation of Montreal protocol and subsequent amendments, which came into force in 1989 and stated that anthropogenic emissions of ozone-depleting gases should be reduced, or they are (2) because the stratosphere from 1997 to 2006 was under the conditions of long-term volcanically quiescent period with minimal content of stratospheric aerosol (SA), which is a strong ozone-depleting factor. Therefore, the integrated remote monitoring of the state of stratospheric aerosol and ozone layers with the purpose of studying the physical mechanisms and tendencies of their changes is still urgent.

Studies of the physical mechanisms of atmospheric changes rely upon obtaining and subsequent statistical and correlation analysis of time series of long-term regular measurements. This analysis makes it possible to develop empirical models of atmospheric parameters; determine seasonal, quasibiennial, and other periodic cycles of variations; study the mechanisms that determine the dynamics and interrelations of measured parameters; reveal short-term disturbances of the atmosphere and gradually accumulated changes under the impact of natural and anthropogenic factors; and, ultimately, determine and predict the trends of atmospheric changes, together with providing estimates of degree of their effect on possible climate changes. This approach to studying the atmospheric changes on the basis of regular integrated remote optical monitoring is one of the priority directions in the present-day word practice of atmospheric studies.

2. MEASUREMENT INSTRUMENTATION

The atmospheric observatory, i.e., the **Siberian Lidar Station (SLS)**, had been created in V. E. Zuev Institute of Atmospheric Optics, Siberian Branch, Russian Academy of Sciences; it operates in Tomsk (56.5°N; 85.0°E) and comprises the set of modern means of remote laser and passive sensing for studying the aerosol and cloud fields, temperature and humidity, ozone, and gas components of ozone cycles. Aside from its broad coverage of monitored atmospheric parameters, the observatory is unique in that it permits atmospheric monitoring simultaneously in the entire significant altitude range of 0-70 km. The observatory includes the following instrumental complexes.

1. Multichannel stationary lidar complex on the basis of 0.3-, 0.5-, and 2.2-m diameter receiver mirrors for monitoring the optical and meteorological parameters and gaseous composition of the atmosphere in the altitude range of 5-70 km, thus covering the upper troposphere, stratosphere, and mesosphere.

2. Spectrophotometer complex for daily automatic measurements of total ozone content (at daytime) and twilight (morning - evening) measurements of the vertical distribution (0-50 km) and total content of NO_2.

3. RESULTS OF MONITORING THE STRATOSPHERIC AEROSOL LAYER

Lidar measurements of SA scattering characteristics have been conducted at SLS since 1986. Sensing is performed at the wavelength of 532 nm of Nd:YAG laser. Optical lidar signals are received using 0.3-m diameter mirror. Photoelectron recording of lidar signals is performed in photocurrent pulse counting mode.

The numerous studies of ozonosphere state after powerful volcanic eruptions such as El Chichon (1982), Pinatubo (1991), and others, performed abroad and at Institute of Atmospheric Optics, Siberian Branch, Russian Academy of Sciences, show the effect of elevated SA content on destruction of ozone layer, when TO depletion is observed for a few years after eruption [2-4]. The radiation-temperature and other stratospheric changes make it possible to study and evaluate field lidar observations in the periods of explosive volcanic eruptions, when sulfur-containing products are injected directly to the stratosphere, where they form additional sulfuric acid aerosol.

However, even a series of weak explosive volcanic eruptions can lead to marked radiation-temperature changes because, by scattering the solar radiation, the stratospheric aerosol layer (SAL) with increased content of volcanic aerosol reduces the solar radiative influx to the Earth's surface. After the long-term volcanically quiescent period 1997-2005 there had been a series of weak eruptions of volcanoes of Pacific Ring of Fire and Iceland, which increased the SA content [5-7]. These volcanoes could have a cooling effect on global climate [8, 9].

We identified the periods with elevated content of stratospheric aerosol over Tomsk on the basis of lidar measurements of integrated aerosol backscattering coefficient B_{π}^{a} (Table 1). In 2012-2014, we recorded background SA content over Tomsk.

Figure 1 shows the time behavior of B_{π}^{a}, determined according to data of sensing at the wavelength of 532 nm. In this figure, every point represents data, averaged over 10 days of measurements.

Table 1. Chronology of volcanic eruptions, traces of which were recorded in the stratosphere over Tomsk in 2006-2011.

Volcano	Latitude, Longitude, deg.	Date	Period of aerosol disturbance of the stratosphere
Rabaul, Papua New Guinea	4.3°S , 152.2°E.	6 Oct 2006	late October 2006 – winter of 2006/07
Okmok, Auletian Islands	53.4°N, 168.1°W	12 Jul 2008	late July – fall of 2008
Kasatochi, Aleutian Islands	52.2 N, 175.5°W	6-8 Aug 2008	
Redoubt, Alaska	60.5°N, 152.7°W	18 Mar –4 Apr 2009	April – May 2009
Sarychev Peak, Kuril Islands	48.1°N, 153.2°E	12-16 Jun 2009	July – December 2009
Eyjafjallajokull, Iceland	63.6°N, 19.6°W	14 Apr 2010	Minor disturbance of the lower stratosphere in late April 2010
Grimsvötn, Iceland	64.4°N, 17.3°W	21 May 2011	Second half-year of 2011

There are periodic bursts of increased aerosol content after every next volcanic eruption; growing trend of SA content over the volcanic activity period 2006-2011 had been $3.43602 \cdot 10^{-7} \pm 1.27198 \cdot 10^{-7}$ sr^{-1} per year (6.7±0.2% per year). Arrows indicate explosive volcanic eruptions and, in particular, those presented in the table. The maximum integrated aerosol backscattering coefficients of the stratosphere reached (4-5) 10^{-4} sr^{-1} in periods of increased SA content versus the average values $\sim 1.5 \cdot 10^{-4}$ sr^{-1} for the background period 1997-2006, which were no higher than $2.5 \cdot 10^{-4}$ sr^{-1} even in winter periods of maximal aerosol content.

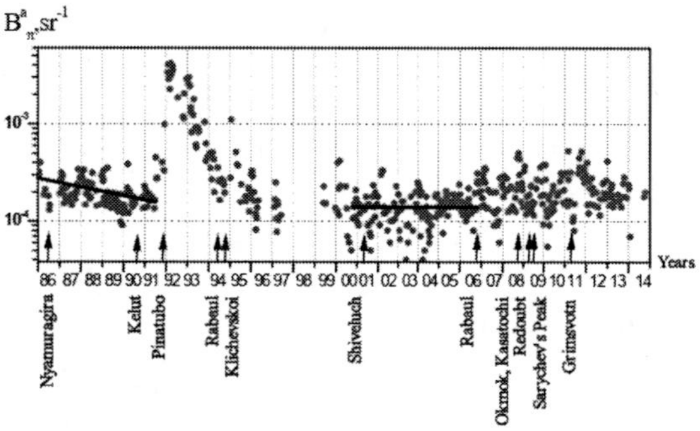

Figure 1. Dynamics of changes in the integrated aerosol backscattering coefficient at the wavelength 532 nm in the altitude range 15-30 km over Tomsk in 1986-2014. Straight lines show trend.

4. MEASUREMENTS OF THE TOTAL OZONE CONTENT

The spectrophotometric measurements of TO content have been conducted with M-124 ozonometer. Also, TOMS satellite data are employed to analyze the multiyear measurements.

The TO content, recorded by the present-day ground-based and satellite measurement facilities, had considerably depleted since early 1980s until mid-1990s. The decline until mid-1990s had conventionally and most often been attributed to the chemical ozone destruction due to the growing concentrations of ozone depleting substances (ODSs) in the stratosphere. Since the second half of 1990s to the present, the ozone layer stopped depleting almost everywhere or there are signs of increasing TO trend, most apparent at the Northern Hemisphere middle to high latitudes.

In April 2011, the ozone level had decreased by 30% as compared to multiyear average values (Fig. 2). In May 2011, the total ozone had largely returned to its seasonal behavior. However, its values remained lowered throughout 2011 and returned back only in the period of resumed wintertime ozone supply by the meridional transport from equatorial reservoir.

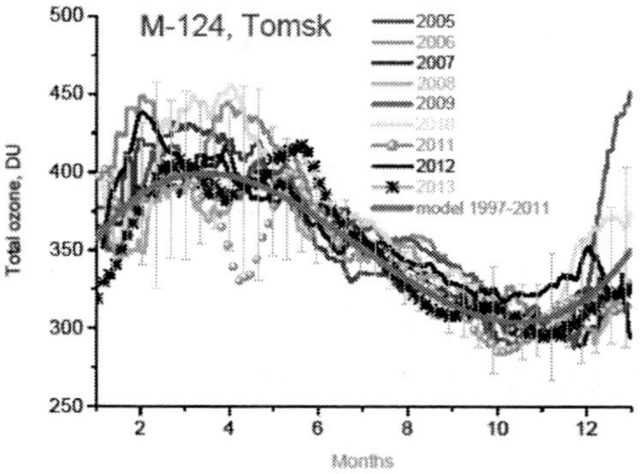

Figure 2. TO variations over Tomsk in 2005-2012 according to data of M-124 ozonometer measurements [10].

Figure 3 presents the trends of multiyear TO variations in different time intervals of period 1979-2013. Decreasing TO trend of -2.35 DU/yr is evident for the period of 1979 – 1995. Increasing TO trend, observed since 1996, was affected by the occurrence of ozone anomaly at Arctic latitudes during spring 2011. As a result of this phenomenon, the positive trend had decreased from 1.09 DU/yr for the period 1996 -2010 to 0.26 DU/yr for the period 1996-2011. In 2012-2013, the growing trend, like the ozone layer itself, started recovering and had been 0.44 DU/yr for the period 1996-2012 and 0.65 DU/yr for the period 1996-2013.

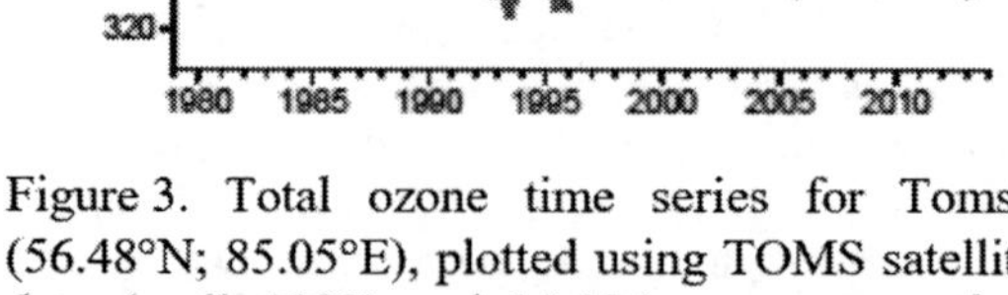

Figure 3. Total ozone time series for Tomsk (56.48°N; 85.05°E), plotted using TOMS satellite data (until 1993) and M-124 ozonometer data (since 1994). The time series was plotted using 365-point smoothing. The linear trends are plotted

for the periods 1979 – 1995, 1996 – 2010, 1996 – 2011, 1996 – 2012, and 1996 – 2013.

5. CONCLUSIONS

We determined the periods of elevated content of stratospheric aerosol over Tomsk after a series of explosive eruptions of volcanoes of Pacific Ring of Fire and Iceland in 2006-2011. The observed bursts of elevated aerosol content after every next volcanic eruption are periodic in character, and the increasing trend of stratospheric aerosol content over the volcanic activity period 2006-2011 had been $3.43602 \cdot 10^{-7} \pm 1.27198\ 10^{-7}\ sr^{-1}$ per year (6.7±0.2% per year). The maximal values of integrated aerosol backscattering coefficient of the stratosphere reached $(4\text{-}5) \cdot 10^{-4}\ sr^{-1}$ in periods of elevated SA content versus average values of $\sim 1.5 \cdot 10^{-4}\ sr^{-1}$ for background period 1997-2006.

The destruction of ozone layer has stopped and there are signs for increasing TO trend since the second half of 1990s. This can be concluded from decreasing TO trend of -2.35 DU/yr for period of 1979 – 1995 and increasing TO trend of 0.65 DU/yr for 1996 – 2013.

ACKNOWLEDGEMENT

This work was supported in part by the Ministry of Education and Science of the Russian Federation under Agreement no. 14.604.21.0100, the Russian Science Foundation under Agreement no. 14-27-00022, and the President of the Russian Federation under grant NSh-4714.2014.5 ("Laser sensing of the atmosphere and ocean") for the support of leading scientific schools.

REFERENCES

[1] Weatherhead, E. C., and S. B. Andersen. 2006. "The search for signs of recovery of the ozone layer," *Nature* 441: 39-45, doi: 10.1038.

[2] Hofmann, D. J, and S. Solomon. 1989. "Ozone destruction through heterogeneous chemistry following the eruption of El Chichon," *J. Geophys. Res.* 94 (D4): 5029-5041.

[3] Ansmann, A., F. Wagner, U. Wandinger, I. Mattis, U. Görsdorf, H.-D. Dier, and J. Reichardt. 1996. "Pinatubo aerosol and stratospheric ozone reduction: Observations over central Europe." *J. Geophys. Res.* 101 (D13): 18,775-18,785.

[4] Zuev, V. V., O. E. Bazhenov, V. D. Burlakov, M. V. Grishaev, S. I. Dolgii, and A. V. Nevzorov. 2008. "On the effect of volcanic aerosol on variations of stratospheric ozone and NO_2 according to measurements at the Siberian Lidar Station," *Atm. Oceanic Opt.* 21 (11): 825-831.

[5] Burlakov, V. D., S. I. Dolgii, and A. V. Nevzorov. 2011. "Lidar observations of the stratosphere aerosol disturbances over Tomsk (56.5 N; 85.0 E) in period of volcanic activity of 2006-2010." *Opt. Atm. Okeana*, 24 (12): 1031–1040.

[6] Vernier, J. -P., L. W. Thomason, J. -P. Pommereau, A. Bourassa, J. Pelon, A. Garnier, A. Hauchecorne, L. Blanot, C. Trepte, D. Degenstein, and F. Vargas. 2011. "Major influence of tropical volcanic eruptions on the stratospheric aerosol layer during the last decade," *Geophys. Res. Lett.* 38 (12): L12807, doi:10.1029/2011GL047563.

[7] Bourassa, A. E., A. Robock, W. J. Randel, T. Deshler, L. A. Rieger, N. D. Lloyd, E. J. Llewellyn, and D. A. Degenstein. 2012. "Large volcanic aerosol load in the stratosphere linked to Asian monsoon transport." *Science* 337: 78-81, doi:10.1126/science.1219371.

[8] Fyfe, J. C., K. Salzen, J. N. S. Cole, N. P. Gillett, and J. P. Vernier. 2013. "Surface response to stratospheric aerosol changes in a coupled atmosphere–ocean model," *Geophys. Res. Lett.* 40 (3): 584-588.

[9] Haywood, J. M., A. Jones, and G. S. Jones. 2014. "The impact of volcanic eruptions in the period 2000–2013 on global mean temperature trends evaluated in the HadGEM2-ES climate model," *Atmospheric Science Letters* 15: 92-96.

[10] Bazhenov, O. E., and A. V. Nevzorov. 2012. "Effect of ozone anomaly during spring of 2011 on long-term (1996-2011) trends of the total ozone content according to ground-based (Tomsk: 56.48°N, 85.05°E) and satellite spectrophotometric measurements." In *Proceedings of SPIE* 8696, *Eighteenth International Symposium on Atmospheric and Ocean Optics*: *Atmospheric Physics*, 86960F (November 27, 2012); doi:10.1117/12.2007478.

DOI: 10.1051/epjconf/201611923008

AEROSOL TYPING BY 3-WAVELENGTH ELASTIC LIDAR SIGNALS OVER THE CENTRAL MEDITERRANEAN

Maria Rita Perrone*, Pasquale Burlizzi

Mathematics and Physics Department, Universita' del Salento, 73100 Lecce, ITALY,
**perrone@le.infn.it*

ABSTRACT

Elastic lidar signals at 355, 532, and 1064 nm combined with aerosol optical thicknesses (AOTs) from sunphotometer measurements collocated in space and time have been used to retrieve columnar lidar ratio (LR) values at the lidar wavelengths by a constrained iterative inversion procedure. Then, the relationships of LRs with AOTs, Ångström exponents, fine mode fractions (η), and fine mode radii (R_f) have been investigated for the aerosol typing. η and R_f values have been retrieved from a graphical framework. It is shown that the implemented methodology has allowed identifying three main aerosol types over the Central Mediterranean which are designed as urban/industrial, marine-polluted, and mixed-dust. Results on the relationships of LRs with AOTs, Å, η, and R_f for each aerosol type represent main paper results.

1. INTRODUCTION

The impact of aerosol on climate is widely recognized and several efforts have been undertaken to characterize aerosol optical and microphysical properties and estimate aerosol direct and indirect radiative effects. Main difficulties in quantifying these effects arise from the very high variability in time and space of the aerosol concentration. The sensitivity of aerosol direct radiative effects and heating rates (HRs) to the vertical distribution of aerosol properties has recently been investigated by one of the authors [1]. Sensitivity tests have indicated that the uncertainties on the vertical distribution of the aerosol properties may have a large impact mainly on aerosol HR vertical profiles, both in the solar and the terrestrial domain. Lidars are now days the best tools that can allow retrieving the vertical distribution of the aerosol properties. Most of the available lidar systems can only provide elastic lidar signals during daytime. As a consequence, numerical procedures only based on elastic lidar signals have been developed to characterize the dependence on altitude of aerosol properties. Ansmann et al. have proposed the single-wavelength POLIPHON (POlarization LIdar PHOtometer Networking) technique [2]. A numerical tool (LIRIC, LIdar/Radiometer Inversion Code) to retrieve vertically resolved aerosol microphysical properties by combining backscatter coefficient measurements at 3 wavelengths and sun/sky radiance measurements is described in [3]. While, a constrained iterative inversion (CII) procedure and a graphical framework (GF) is used in [4] to estimate the dependence on altitude of the aerosol fine mode radius (R_f) and of the fine mode contribution (η) to the aerosol optical thickness (AOT).

In this study the CII-GF retrieval scheme is applied to lidar measurements performed at a coastal site of southeastern Italy with the main goal of contributing to the typing of the Central Mediterranean aerosol. Particles from the surrounding European Countries, the African Continent, and the Atlantic Sea are advected over the Mediterranean basin. As a consequence, several studies have indicated that the aerosol direct radiative effects are among the highest in the world during the Mediterranean summers. The HYbrid Single-Particle Lagrangian Integrated Trajectory (HYSPLIT) model version 4.8, from NOAA/ARL [5], has been used to compute 96-hour backward trajectories and investigate the impact of source regions on aerosol properties. A brief overview of the 3-wavelength lidar system and of the CII-GF retrieval scheme is provided in Section 2. Main results are presented and discussed in Section 3. Summary and conclusion are reported in Section 4.

2. INSTRUMENTS AND METHODOLOGY

The ground-based lidar system used in this study is located at the Mathematics and Physics Department of Universita' del Salento (Lecce,

40.33°N; 18.11°E). It is identified as UNILE (UNIversity of LEcce) lidar and operates within EARLINET since May 2000. It is nowadays it designed to derive elastically backscattered lidar profiles at 355 nm, 532 nm and 1064 nm, respectively, and it is estimated to achieve full overlap between 0.5 – 1.0 km above the ground level (agl). Elastic lidar signals suffer from a need to assume an aerosol-to-backscatter ratio LR (also referred to as the aerosol-lidar ratio) to solve the ill-posed problem of the lidar equation and extract aerosol extinction and backscattering coefficient profiles. The assumption that LR is known *a priori* is likely the largest source of systematic error within this lidar inversion procedure. However, this uncertainty can be largely reduced if an additional information, such as the AOT of the whole aerosol load, is available. A Constrained Iterative Inversion (CII) procedure which allows one to determine the aerosol extinction ($\alpha(z)$) and backscatter ($\beta(z)$) coefficient profiles by using as boundary conditions (1) the AOT of a selected altitude range and (2) the total backscattering coefficient β_T (due to molecules (β_M) and aerosol (β)) at a far-end reference height, is used in this study to reduce the uncertainties on *a priori* LR assumption. A detailed description of the CII procedure is provided in [4], where main benefits and limits of the CII procedure have also been underlined. AOT values at the lidar wavelengths have been retrieved from AERONET sun/sky photometer measurements co-located in space and time with lidar measurements. An AERONET sun/sky photometer operates at the UNILE lidar site since the year 2003 and it provides AOTs with accuracy of ± 0.01 [6]. Hence, it was required that the AOTs calculated from the retrieved aerosol extinction profiles should not exceed (within ± 0.01) the corresponding AOT values retrieved from co-located sun/sky photometer measurements. To this end, we have assumed that $\alpha(z)$ does not vary with altitude below the height where the lidar system achieves the full overlap. The capability of Ångstrom exponents (Å) and the spectral difference (ΔÅ) to be indicators of the dominant size of atmospheric aerosols has been exploited to estimate the dependence on altitude of fine mode fractions (η), and fine mode radii (R_f) from the lidar extinction profiles at 355 nm, 532 nm and 1064 nm, according to [7]. In particular, a graphical framework (GF) was developed by Gobbi et al. [7] to classify aerosol properties by the ΔÅ(z) versus Å(355,1064, z) plot. The GF provides reference points corresponding to bimodal size distributions characterized by a variety of fine mode (R_f) and coarse mode (R_c) modal radii combined in order to lead to prescribed fine mode fractions (η) at 532 nm. Angstrom coefficients at the altitude z have been calculated as follows:

$$\text{Å}(\lambda 1, \lambda 2, z) = - \ln [\alpha_{\lambda 1}(z)/(\alpha_{\lambda 2}(z)] / [\ln (\lambda 1/ \lambda 2)]$$

where $\alpha_{\lambda 1}(z)$ and $\alpha_{\lambda 2}(z)$ represent the extinction coefficient at the height z and the wavelength λ1 and λ2, respectively. $\alpha_{\lambda 1}(z)$ and $\alpha_{\lambda 2}(z)$ are calculated from the following relationship:

$$\alpha_{\lambda i}(z) = LR_{\lambda i} \times \beta_{\lambda i}(z) \qquad i = 1, 2$$

where $LR_{\lambda i}$ and $\beta_{\lambda i}(z)$ represent the lidar ratio and the backscattering coefficient at the wavelength λi, respectively. The spectral curvature is represented by the difference:

$$\Delta\text{Å}(z) = \text{Å}(355, 532, z) - \text{Å}(532, 1064,z)$$

3. RESULTS

A case study is analyzed, as an example, in this section to demonstrate the ability of the CII-GF technique on estimating the dependence on altitude of the fine mode aerosol contribution and the fine mode aerosol radius, on the basis of the Å and ΔÅ profiles retrieved from lidar measurements. Figure 1 shows the extinction coefficient profiles at 355 nm (black line), 532 nm (green line), and 1064 nm (red line) retrieved from lidar measurements performed on 5 September, 2011 from 1316 to 1343 UTC. Altitude independent LRs and AOTs calculated from the extinction coefficient profiles retrieved from the CII procedure are reported in Table 1. AOTs are in accordance within ±0.1 with the corresponding values retrieved from the AERONET (Lecce University) sunphotometer measurements at 1332 UTC. Figure 2 shows the pathways of the HYSPLIT analytical back trajectories referring to September 5, 2011 at 1400 UTC. They reveal that the air masses advected over southeastern Italy at 1.0, 3.5, and 4.5 km agl have their source region over north western Africa and as a consequence, they have likely been responsible for the advection of African dust particles up to ~5 km agl. The Å(z) profiles that are plotted in Fig. 3a for different wavelength pairs support last comment. In fact, the Å values which are smaller than 0.7 within 1-4 km agl are likely due to the contribution of coarse mode particles. The spectral curvature profile is shown in Fig. 3b.

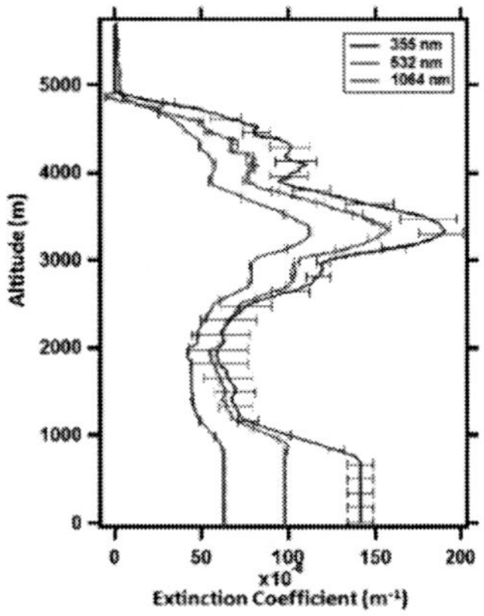

Figure 1. Extinction coefficient profiles at 355, 532, and 1064 nm.

Calculated ΔÅ(z) versus Å(355, 1064, z) values are plotted in Fig. 4 with corresponding error bars. Different colors have been used to represent ΔÅ(z)-Å(z) values referring to different z, as indicated by the color bar on the right of the figure. One observes that the aerosol load is

Table 1 Lidar ratios ± corresponding standard deviations (SD) and AOT ± uncertainties at the lidar wavelength for the September 5, 2011 study case.

λ(nm)	LR±1SD (sr)	AOT
355	56±7	0.51±0.1
532	58±2	0.41±0.1
1064	48±9	0.30±0.1

characterized by η(z) and R_f(z) values spanning the 0.50-0.70 and the 0.14-0.24 μm range, respectively. The Å(355, 1064, z) and ΔÅ(z) mean values ($Å_{av}$ and $ΔÅ_{av}$, respectively) calculated from the data points of Fig. 4 are equal to 0.4±0.1 and -0.2±0.2, respectively. These values have allowed determining from the graphical framework the η and R_f mean values ($η_{av}$ and $R_{f,av}$, respectively) which are equal to 0.54±0.04 and 0.18±0.03, respectively.

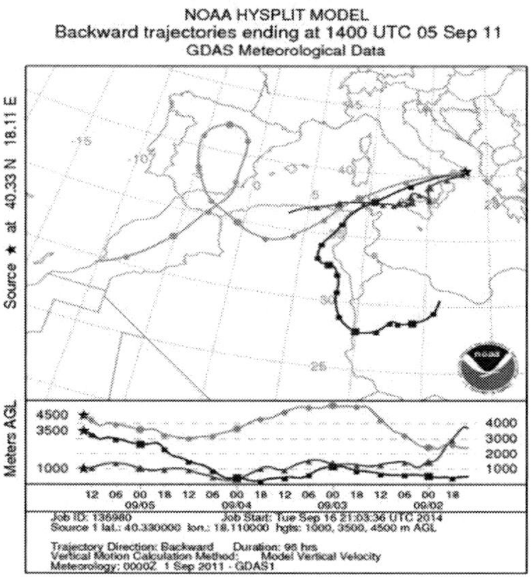

Figure 2. Pathways of analytical backward trajectories

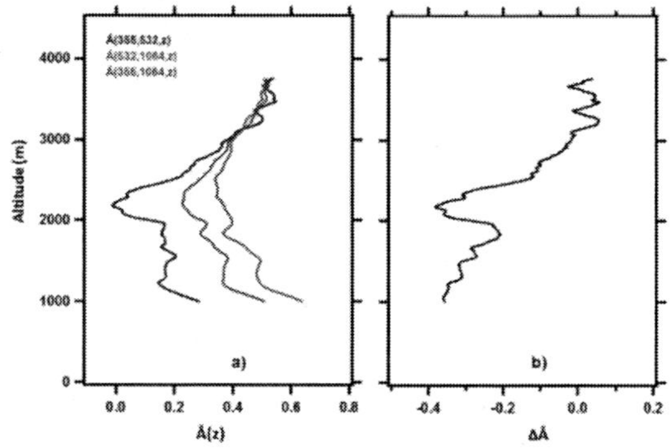

Figure 3. (a) Ångstrom exponent profiles for different wavelength pair. (b) ΔÅ(z) profile.

Lidar measurements performed on more than 20 days of the 2011-2014 years have been analyzed by the CII-GF scheme for typing Central Mediterranean aerosols by AOTs, LRs, backscatter color ratios, Ångström exponents, Ångström exponent spectral curvature, *η*, and R_f values. Then, $ΔÅ_{av}$ and $Å_{av}$ values and the corresponding $η_{av}$, and $R_{f,av}$ have been calculated for each study case. The analysis has allowed identifying 3 main aerosol types denoted as urban/industrial (UI), marine polluted (MP), and mixed dust (MD). Table 2 shows as an example, the variability range of LR (355 nm), $Å_{av}$(355,

DOI: 10.1051/epjconf/201611923008

1064nm), $\Delta Å_{av}$, η_{av}, and $R_{f,av}$ for each aerosol type. One observes that from Table 2 that UI aerosols are characterized by larger values of

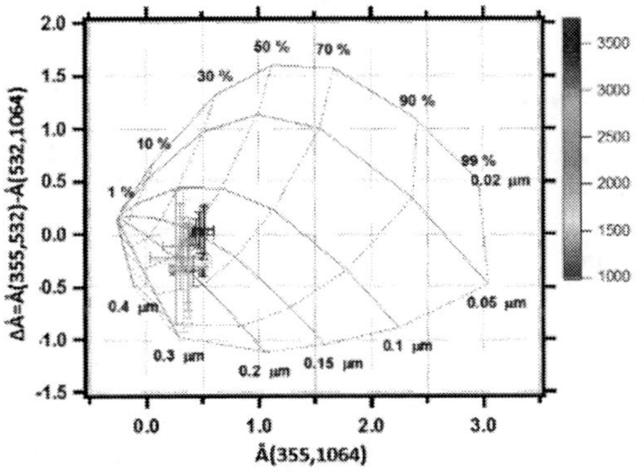

Figure 4. Graphical classification framework. Solid and dashed lines are representative of selected fine modal radii (R_f = 0.02, 0.05, 0.1, 0.15, 0.2, 0.3, and 0.4 µm) and fine mode fractions at 532 nm of (η =1, 10, 30, 50, 70, 90, 99%). Dots and errors bars represent ΔÅ(z) versus Å(355, 1064, z) with corresponding standard deviations, retrieved from the lidar measurements performed on September 5, 2011 from 1317 to 1347 UTC.

LR, $Å_{av}$, and η_{av}, for the larger presence of fine mode absorbing particles with respect to MP and MD aerosol. It has also been found that the LR and AOT spectral dependence were quite dependent on the aerosol type.

Table 2. Dependence of LR, $Å_{av}$, $\Delta Å_{av}$, η_{av}, and $R_{f,av}$ on aerosol type.

	LR(355 nm) (sr)	**$Å_{av}$ (355, 1064)**	**η_{av} (532 nm) (%)**	**$R_{f,av}$ (µm)**
UI	**55-80**	**1.4-1.6**	**80-94**	**0.10-0.13**
MP	**39-64**	**0.9-1.4**	**60-80**	**0.05-0.14**
MD	**44-64**	**0.3-0.6**	**50-60**	**0.14-0.18**

4. CONCLUSIONS

The analysis of lidar measurements performed on more than 20 monitoring days characterized by advection pattern typical of Central Mediterranean, has revealed the ability of the CII-GF scheme to characterize main aerosol types by LR, Å(λ1, λ2, z), ΔÅ = Å(355, 532, z)- Å(532, 1064, z), η(z) at 532 nm, and R_f(z). Three main aerosol types denoted as urban/industrial (UI), marine polluted (MP), and mixed dust (MD) have been identified. Therefore, paper results have contributed to the aerosol characterization over the Central Mediterranean.

ACKNOWLEDGEMENT

This study was supported by the European Community through the ACTRIS Research Infrastructure Action under the 7th Framework Programme under ACTRIS Grant Agreement n° 262254. The authors gratefully acknowledge the NOAA Air Resources Laboratory (ARL) for the provision of the HYSPLIT backtrajectories used in this publication.

REFERENCES

[1] Perrone, M. R., Tafuro, A. M., and Kinne, S., 2012: Dust layer effects on the atmospheric radiative budget and heating rate profiles, *Atmos. Environ.*, **59**, 344-354.

[2] Ansmann, A., Seifert, P., Tesche, M., and Wandinger, U. 2012: Profiling of fine and coarse particle mass: case studies of Saharan dust and Eyjafjallajokull/Grimsvötn volcanic plumes, *Atmos. Chem. Phys.*, **12**, 9399–9415.

[3] Chaikovsky, A. et al. 2012: Algorithm and software for the retrieval of vertical aerosol properties using combined lidar/radiometer data: dissemination in EARLINET, *Reviewed & Revised Papers of the 26th International Laser Radar Conference, 25-29 June, Porto Heli, Greece,* Paper SO3-09

[4] Perrone, M.R., De Tomasi, F., Gobbi G.P. 2014: Vertically resolved aerosol properties by multi wavelength lidar measurements. *Atmos. Chem. Phys.* **14**, 1185-1204.

[5] Draxler, R.R., Hess, G.D. 1998: An overview of the HYSPLIT_4 modeling system of trajectories, dispersion, and deposition. *Aust. Meteorol. Mag.* **47**, 295-308.

[6] Dubovik O., Holben B. N., Lapyonok T., Sinyuk A., Mishchenko M. I., Yang P., Slutsker I. 2002: Non-spherical aerosol retrieval method employing light scattering by spheroids. *Geophys. Res. Lett.* **29**, 1415. 2

[7] Gobbi, G. P., Kaufman, Y. J., Koren, I., Eck, T. F. 2007: Classification of aerosol properties derived from AERONET direct sun data, *Atmos. Chem. Phys.*, **7**, 453–458.

EPJ Web of Conferences **119**, 23009 (2016)
ILRC 27
DOI: 10.1051/epjconf/201611923009

LIDAR OBSERVATIONS OF POLLUTION TRANSPORT FROM LONDON TO RURAL AREAS

Hugo Ricketts [1,2]*, Geraint Vaughan[1,2], David Wareing[3]

[1]*School of Earth, Atmospheric and Environmental Sciences, University of Manchester, United Kingdom*
**Email: h.ricketts@manchester.ac.uk*

[2]*National Centre for Atmospheric Science, United Kingdom*

[3]*Institute of Mathematics, Physics and Computer Science, Aberystwyth University, United Kingdom*

ABSTRACT

The Clean Air for London (ClearfLo) Project took place in and around London, United Kingdom. The aim of the project was to learn how both atmospheric dynamics and chemistry affect air pollution in the south east of England. During the winter and summer of 2012 many different types of instrument including lidars were deployed throughout London city centre, suburbs and into rural areas. Amongst these instruments was the Boundary Layer Aerosol/Ozone Lidar owned by the National Centre for Atmospheric Sciences (NCAS) in the United Kingdom. Ozone and aerosol data are presented from data collected during July and August 2012 and compared to back trajectories to identify their origins.

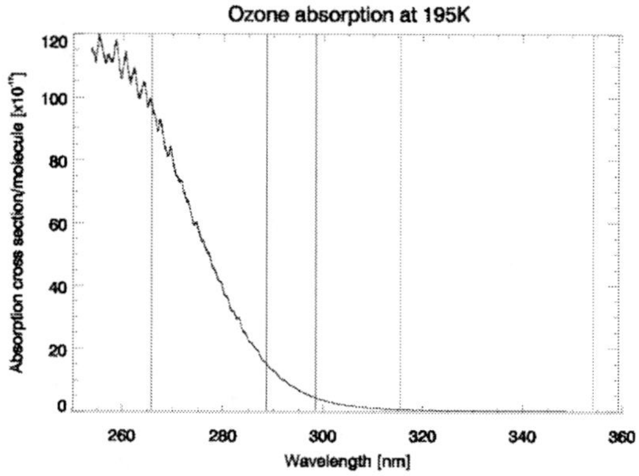

Fig. 1: Absorption cross-section of ozone at 195K [3]

1. INTRODUCTION

During the winter and summer of 2012 a field experiment called Clean air for London (ClearfLo) was carried out in the south east of England [1]. The aim of the project was to observe the dynamics and chemistry and analyse how these impact on air pollution in the Greater London area as well as further afield at more rural sites. A large amount of data was collected using many different scientific instruments including the United Kingdom's National Centre for Atmospheric Science's (NCAS) Boundary Layer Aerosol/Ozone Lidar System. This paper will discuss some of the results.

2. CLEARFLO PROJECT

As well as field sites in the Greater London area, the ClearfLo project included rural field sites to the East and West of London. During the winter and summer of 2012 two intensive observation periods (IOPs) were carried out to analyse different seasons and their effect on the chemistry and dynamics. During these IOPs, the NCAS Boundary Layer Aerosol/Ozone Lidar System was deployed to the Rutherford Appleton Laboratories (RAL) at Harwell Campus 80km west of London's City Centre as part of the field experiment. Its purpose was to detect ozone concentrations and aerosol backscatter for the atmospheric boundary layer (ABL) and lower free troposphere (FT) to provide estimates of ozone flux towards or away from London depending on wind direction. While the lidar did not produce any data in the winter IOP, it proved to be a useful tool for analyzing data from the summer IOP.

During July and August 2012 the lidar was operational most days, except during rainy periods. The synoptic situation of summer 2012 presented plenty of variety; frontal systems passing over the UK from the Atlantic Ocean generally leading to westerly winds and also some high pressure situations leading to more variable wind directions including Easterlies. Only data with wind blowing from easterly directions will be presented here as this air was advected from London over Harwell Campus. Specifically, this

DOI: 10.1051/epjconf/201611923009

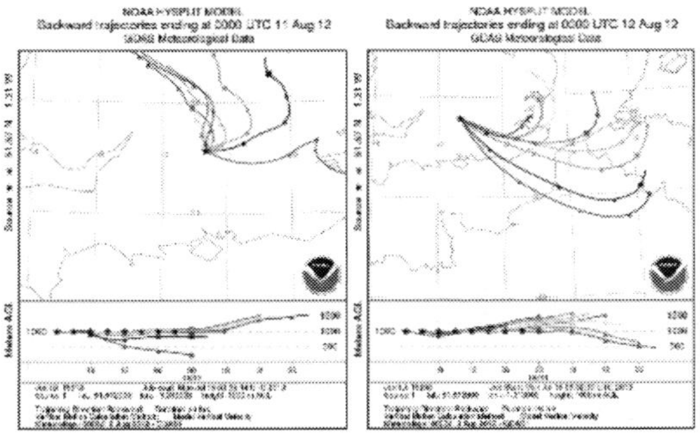

Fig. 2: Back trajectories from the HYSPLIT Model tracing the origin of the air mass seen by the lidar at 1km [2].

study will focus on the time period of 10th and 11th August.

3. NCAS AEROSOL/OZONE LIDAR

3.1. Description

The NCAS Boundary Layer Aerosol/Ozone Lidar system is mounted on a small trailer allowing it to be transported to anywhere with road access. It was originally built by Elight Systems in 2002 and has since been operational at various locations in the United Kingdom (UK). The lidar system contains a Continuum Powerlite 8020 Nd:YAG laser with a pulse repetition frequency of 20Hz capable of producing laser pulses of 355nm and 266nm through frequency-tripling and quadrupling of the original 1064nm pulse, respectively. A further three wavelengths can then be produced using Stimulated Raman Scattering, whereby the 266nm laser pulse is directed through cells containing Hydrogen, Deuterium and Argon at different ratios. The wavelengths created by these cells are 289nm, 299nm and 316nm. Finally, each of the 5 wavelengths are sequentially transmitted vertically into the atmosphere allowing a complete set of measurements to be created at a rate of 4Hz.

The backscattered light is collected using a reflector telescope with a primary mirror diameter of 40cm and the light is collected in two different focal planes to prevent saturation of the lidar detection system close to the telescope. To collect the backscattered light two Hamamatsu photomultiplier tubes (PMTs) are connected to an analogue to digital (A/D) converter designed by Licel GmbH. The data is them stored at a range resolution of 7.5 metres and a time resolution of typically 3 minutes.

3.2. Data analysis

To obtain aerosol backscatter coefficient a simple onion-peeling algorithm after Klett [4] is used and therefore a Lidar Ratio (extinction-to-backscatter ratio) needs to be assumed. For the UK, typically a Lidar Ratio of 55sr can be taken for aerosol from a continental air mass and 30sr is used for marine air masses as measured from other UK lidar systems based at Capel Dewi near Aberystwyth, Wales [5]. The wavelengths analysed using this method are 355nm and 316nm. Typically, aerosol backscatter coefficient can be obtained from 100m up to 5-8km, depending on atmospheric condition.

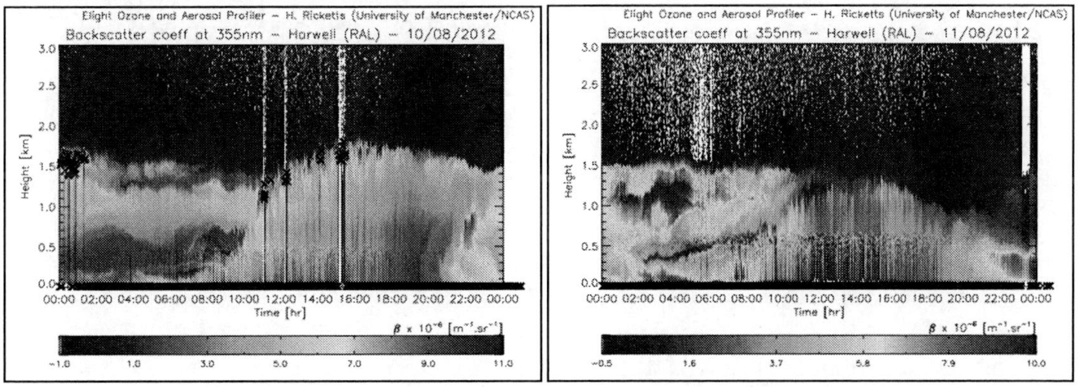

Fig. 3: Aerosol backscatter coefficient at 355nm as measured by the lidar on 10th and 11th August 2012.

DOI: 10.1051/epjconf/201611923009

To retrieve ozone mixing ratio a differential absorption lidar (DIAL) technique is used which calculates ozone mixing ratio from the slope in two of the backscatter signals. Firstly, this requires the removal of the backscatter due to aerosol and secondly, this leads to vertical averaging. The technique is also sensitive to the wavelength pairs used. Fig. 1 shows the absorption cross-section of ozone. While comparatively strong at 266nm, it decreases significantly with increasing wavelength. Therefore, in the presence of ozone, the 266nm is absorbed more rapidly than the higher wavelengths of 289nm and 299nm. This stronger absorption leads to a more reliable retrieval of ozone, but with reduced range. Therefore wavelength pairs 266nm/289nm and 266nm/299nm can be used up to about 2km with an error of about +/-2ppbv. Above that the less absorbed wavelength pair 289nm/299nm can be used up to 2.5-3.5km (depending on conditions) with an error of about +/-5ppbv, due to the much smaller difference between their respective absorption cross-sections.

4. POLLUTION ADVECTION

Fig. 2 shows back trajectory calculations calculated from the HYSPLIT Model [2]. This data shows where the air sampled by the lidar at 1 km is expected to have originated. During 10th August 2012 the wind veers from a northerly to an easterly direction and on 11th August 2012 the wind direction continues to veer to a Southeasterly. Fig. 3 is a plot of aerosol backscatter coefficient for 10th and 11th August 2012 and demonstrates the increase of aerosol backscatter in the residual layer (RL), defined as a layer of air that originates from the convective boundary layer (CBL) of a day previously. This increase at 1km during the early morning of 11th August coincides with the prediction from the back trajectory model of air that has originated from London. A similar RL can be seen on the early morning of 10th August, but this shows about a third of the values experienced the following night.

Ozone mixing ratio from the lidar and a nearby ground station are shown in Fig. 4. These show an ozone mixing ratio of up to 60ppbv at 800 metres

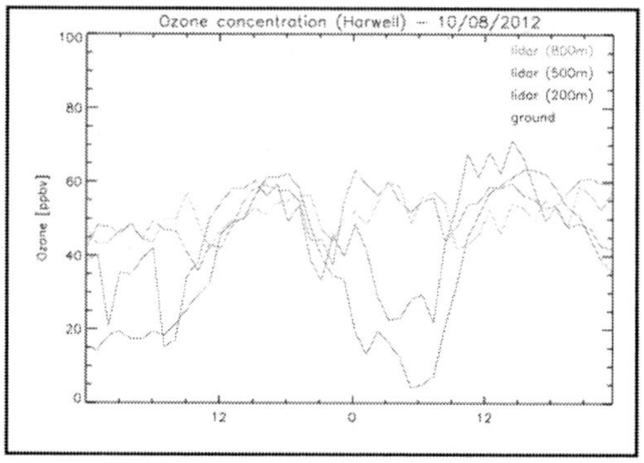

Fig. 4: Ozone mixing ratio for 10th and 11th August 2012. The ground-based ozone values are from the Harwell air quality monitoring site run by the UK's Department of Environment, Food and Rural Affairs (DEFRA) [6]

on the morning of 11th August compared to 45ppbv on the morning of 10th August.

Ozone mixing ratios are subject to chemical reactions with other gases and aerosol types in the atmosphere. This is visible in the ground-based data where ozone mixing ratio decreases to very low values of 3ppbv on the morning of 11th August. This is pending further analysis.

5. FUTURE WORK

- Further work will be carried out to analyse all of the data collected, including the other incidents of easterly winds as well as the cases of other wind directions.

- The lidar data will be compared to data from a Weather Research and Forecasting (WRF) model that includes chemistry.

- Air composition data from London and Harwell will be analysed further to understand the variations in ozone with respect to chemical reactions in the atmosphere.

- A CBL height comparison will be carried out using all available lidar data during the campaign.

ACKNOWLEDGEMENT

The authors acknowledge the Natural Environment Research Council (NERC) for the funding provided for the Clean Air for London (ClearfLo) project as well as the National Centre for Atmospheric Science's (NCAS) Atmospheric Measurement Facility (AMF) for providing the Boundary Layer Aerosol/Ozone Lidar.

REFERENCES

[1] Bohnenstengel, S. I., Belcher, S. E., Aiken, A., Allan, J. D., Allen, G., Bacak, A., Bannan, T. J., Barlow, J. F.,Beddows, D. C. S., Bloss, W. J., Booth, A. M., Chemel, C., Coceal, O., Di Marco, C. F., Dubey, M. K., Faloon, K. H., Fleming, Z. L., Furger, M., Gietl, J. K., Graves, R. R., Green, D. C., Grimmond, C. S. B., Halios, C. H.,Hamilton, J. F., Harrison, R. M., Heal, M. R., Heard, D. E., Helfter, C., Herndon, S. C., Holmes, R. E., Hopkins, J. R., Jones, A. M., Kelly, F. J., Kotthaus, S., Langford, B., Lee, J. D., Leigh, R. J., Lewis, A. C., Lidster, R. T.,Lopez-Hilfiker, F. D., McQuaid, J. B., Mohr, C., Monks, P. S., Nemitz, E., Ng, N. L., Percival, C. J., Prevot, A. S. H., Ricketts, H. M. A., Sokhi, R., Stone, D., Thornton, J. A., Tremper, A. H., Valach, A. C., Visser, S., Whalley, L. K., Williams, L. R., Xu, L., Young, D. E. and Zotter, P., 2014: Meteorology, air quality, and health in London: The ClearfLo project, *Bulletin of the American Meteorological Society*. doi:10.1175/BAMS-D-12-00245.1

[2] Draxler, R.R. and Rolph, G.D. HYSPLIT (HYbrid Single-Particle Lagrangian Integrated Trajectory) Model access via NOAA ARL READY Website (http://www.arl.noaa.gov/HYSPLIT.php). NOAA Air Resources Laboratory, College Park, MD.

[3] Bogumil, K., J. Orphal, T. Homann, S. Voigt, P. Spietz, O. C. Fleischmann, A. Vogel, M. Hartmann, H. Kromminga, H. Bovensmann, J. Frerick & J. P. Burrows 2003: Measurements of molecular absorption spectra with the sciamachy pre-flight model: instrument characterization and reference data for atmospheric remote- sensing in the 230-2380 nm region, *Journal of Photochemistry and Photobiology a-Chemistry*, **157 (2-3)**, 167–184

[4] Klett, J. D., 1981: Stable analytical inversion solution for processing lidar returns, *Applied Optics*, **20 (2)**, 211–220

[5] L. Komguem, priv. comm.

[6] Crown 2015 copyright Defra via uk-air.defra.gov.uk, licenced under the Open Government Licence (OGL)

Intense particulate pollution events observed with lidar over the Paris Megalopolis

Patrick Chazette[1], and Philippe Royer[2]

[1]*Laboratoire des Sciences du Climat et de l'Environnement, CEA-CNRS-UVSQ, 91191 Gif-sur-Yvette Cedex, France, patrick.chazette@lsce.ipsl.fr*

[2]*LEOSPHERE, 76 rue de Monceau, 75008 Paris, France,*

proyer@leosphere.fr

ABSTRACT

The great particulate pollution event that affected the Paris Megalopolis in March 2014 was due to long-range transport from the northern-northeastern Europe. Although this phenomenon has appeared as exceptional in the media, this is not an exception and similar events have already been observed by lidar measurements. Here we will briefly describe and illustrate the origin of this intense pollution obviously harmful to health.

1. INTRODUCTION

Major aerosol pollution events may occur at the beginning of spring in the Paris Megalopolis, and even more throughout the northern part of France. We sampled four of these events from lidar systems located in the south of Paris. These measurements are one-off in space and time as they generally correspond to the opportunities offered during the preparation of large experimental campaigns as the Hydrological cycle in the Mediterranean eXperiment (HyMeX) [1].

We have already conducted several research projects on pollution aerosols in Paris Megalopolis using the lidar observations [2][3]. Lidar measurements were performed from ground-based station (fix or mobile) but also from airborne facilities [4]. This new study is based on lidar measurements acquired at different times since 2010 and recently during the particle pollution event of March 2014. We will not be exhaustive here, but we will mainly illustrate our work by the no less important pollution event occurring from 21 to 25 March 2011.

2. EXPERIMENTAL SET-UP

In all cases, two lidar have been used, the first one for aerosol optical properties and the second one for the wind intensity and direction. They were located ~30 km South of Paris (~48°42' N, 2°10' W). They worked during several days for sampling the major aerosol pollution event of April 2010, March 2011 (2 events) and March 2014. The Rayleigh-Mie lidar was similar to the ALS450® lidar commercialized by Leosphere Company. The wind-lidar WindCube WLS70 is also a commercial Doppler coherent wind lidar developed and manufactured by Leosphere for meteorogical applications. The main characteristics of the lidar are given in Table 1.

Table 1. Main characteristics of ground-based aerosol (ALS 450) and wind (WLS 70) lidars.

	ALS 450	WLS 70
Laser	Nd:YAG, flash-pumped, Q-switched Quantel Ultra	Master Oscillator + Erbium Doped Fiber Amplifier
Pulse length	5 ns	400 ns
Energy	16 mJ	50 µJ
Frequency	20 Hz	10 kHz
Reception channels	// 355 nm ⊥ 355 nm	1543 nm
Reception diameter	15 cm	10 cm
Field-of-view	~ 4 mrad	collimated
Detector	Photomultiplier tubes	Avalanche photodiode
Filter bandwidth	0.3 nm	-
Vertical sampling	15 m	50 m

3. GENERAL DESCRIPTION OF EVENTS

For the 4 major aerosol pollution events observed from our lidars, the origin is very similar. For instance, the situations as observed by MODIS are presented in Figure 1 for the events of 21-25 March 2011 and 12-15 March 2014. We have selected the most impacted days (23 and 14 March) where the pollution air mass advection is very well highlighted on the MODIS aerosol optical thickness (AOT) fields.

For these two cases, the aerosol plume is spread from East to West and advected from East. It traps particulate pollutants throughout its route passing over different aerosol sources that will be discussed hereafter. The

DOI: 10.1051/epjconf/201611923010

AOT at 550 nm is significantly larger than the mean value usually recorded throughout the year. In the aerosol plume it is larger than 0.3-0.4 on a large area compared with the background value of 0.15.

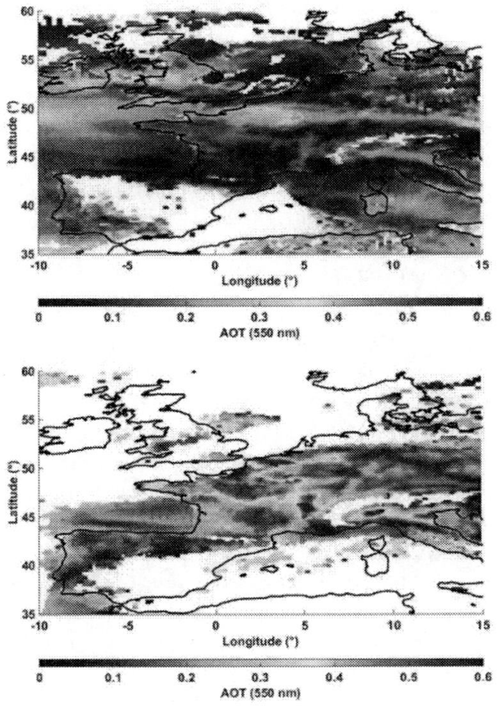

Figure 1. MODIS aerosol optical thickness (AOT) for 23 March 2011 (top panel) and 14 March 2014 (bottom panel).

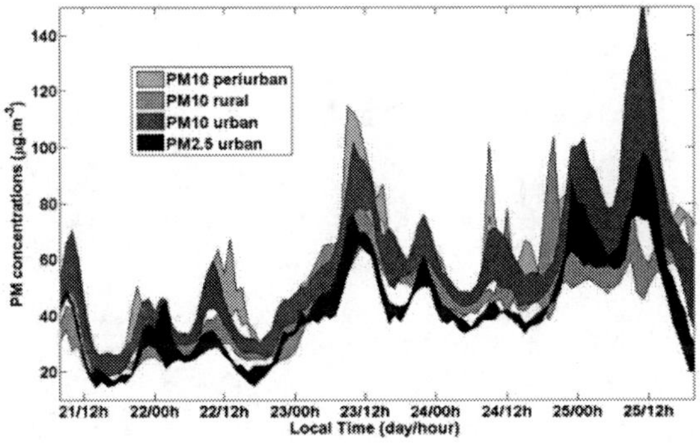

Figure 2. Example of temporal evolution of PM10 and PM2.5 concentrations observed by ground-based AIRPARIF network between 21 and 25 March 2011.

For the entire ground-based air quality network stations of AIRPARIF (http://www.airparif.asso.fr/) the mass concentrations of both PM2.5 and PM10 (Figure 2) significantly increased during all the considered events (PMx = mass concentration of particle with aerodynamic diameter less than x µm). The levels of particulate pollution are similar with PM10 reaching values larger than 100 µg m^{-3}, which is twice the health tolerance required for outside air.

4. OVERLAP FUNCTION

For such pollution studies, the Rayleigh-Mie lidar must have an overlap function close to 1 at the lowest possible altitude. This is because the aerosol pollutants are mainly trapped in the planetary boundary layer (PBL) and the lower free troposphere (LFT). The overlap function is given in Figure 3. They are close to 1 at a distance of ~150 m from the emitter. A correction is possible for lidar data between 120 and 150 m. The lidar may be considered blind below 120 m.

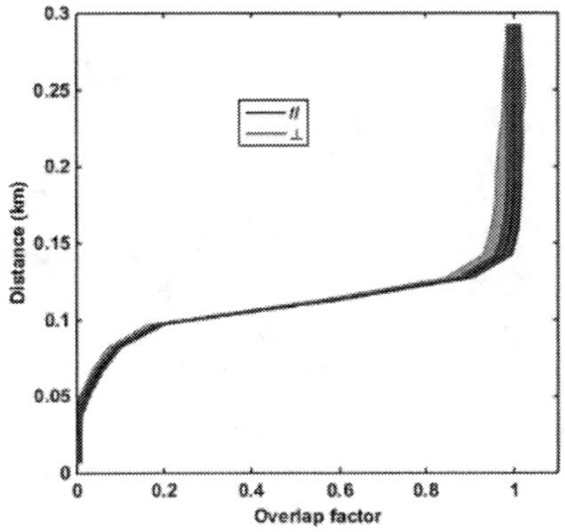

Figure 3. Overlap function of the Rayleigh-Mie lidar for the two polarization channels (parallel // and perpendicular⊥).

5. EXAMPLE OF LIDAR MEASUREMENTS

As for the PM10 recorded at the ground level, the lidar shows a significant increase of the aerosol load in the PBL and LFT from 23 March 2011 (Figure 4). The AOT at 355 nm changes from ~0.2 to ~0.7. For the other aerosol pollution events we note similar AOT, sometime slightly higher (~0.9 for March 2014).

The inversion of lidar profile was performed using the sunphotometer as a reference during cloud-free daytime [2]. Hence, an equivalent backscatter to extinction ratio (inverse of the lidar ratio) has been retrieved, which can be compared with the sunphotometer-derived Angstrom exponent (Figure 5). The higher it is, the less the Angstrom is. In general, the larger aerosols are associated with the smaller Angstrom exponent. It may indeed be a contribution of terrigenous aerosols related to the convection that develops during anticyclonic periods associated with these large particulate pollution events of early spring.

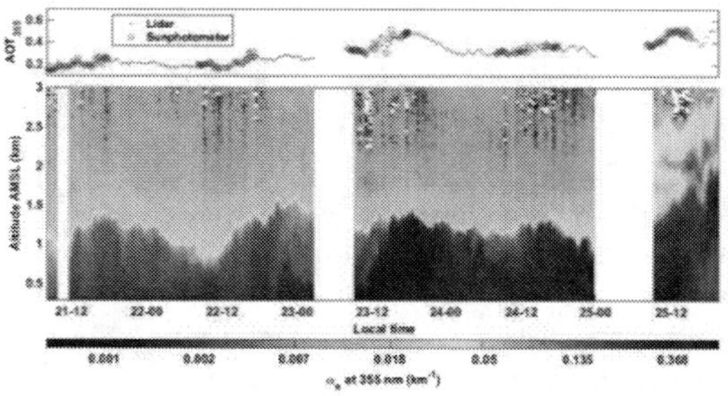

Figure 4. Time-height evolution of aerosol extinction coefficient at 355 nm (α_e) between 21 and 25 March 2011 (bottom panel). The aerosol optical thickness (AOT) is also given from the lidar (blue) and the sunphotometer (red) (top panel).

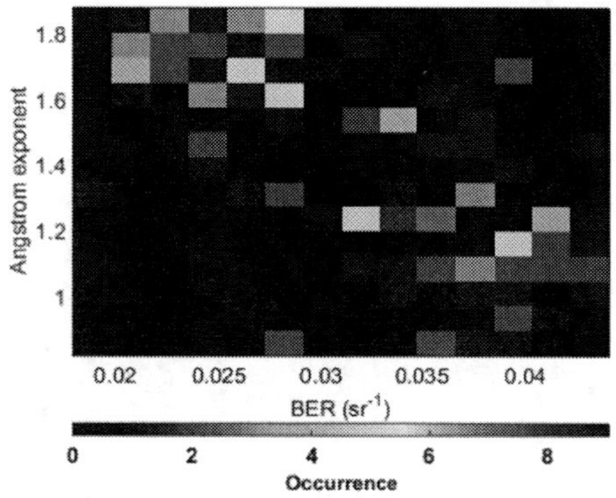

Figure 5. Angstrom exponent derived from the sunphotometer of the AERONET station of Palaiseau against the backscatter to extinction ratio (BER) from 21 to 25 March 2011.

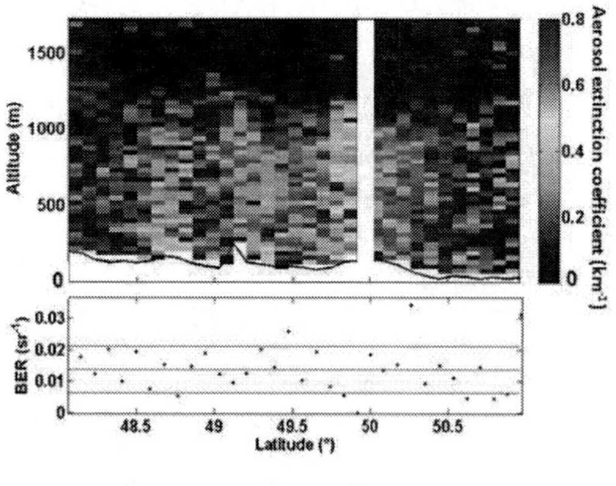

Figure 6. CALIOP-derived aerosol extinction coefficient and BER at 532 nm for 23 March 2011 1212 UTC.

We have also inverted the CALIOP profiles when the ground-track of the CALIPSO platform is close to Paris (Figure 6). The inversion was performed using MODIS-derived AOT as a reference [5]. Although the wavelengths are different (355 nm for the ground-based lidar and 532 nm for CALIOP), the comparisons between the BER seems not conclusive for the moment. In instance for the 23 March 2011, we found 0.029±0.005 sr^{-1} and 0.013±0.005 sr^{-1} at 532 and 355 nm, respectively. Note that the nearest CALIPSO orbit crosses the aerosol plume but this orbit is situated at 200 km east from Paris.

6. CONCLUSIVE DISCUSSION

The wind lidar very well highlights the origin of air masses crossing the Paris Megalopolis during the major aerosol pollution events. The two horizontal components of the wind are given in Figure 7 for the event of 21-25 March 2011. They are mainly recorder in the PBL due to the limited altitude range of the lidar. The aerosol plume is well established (Figure 1) with a horizontal wind from sector East between 3 and 7 ms^{-1}. After the 25 June, the wind veers and is from south-southwest below 1 km. This cannot sufficiently describe the air mass advection. Hence, we used backtrajectories computed from the Hysplit model (http://www.arl.noaa.gov/HYSPLIT.php).

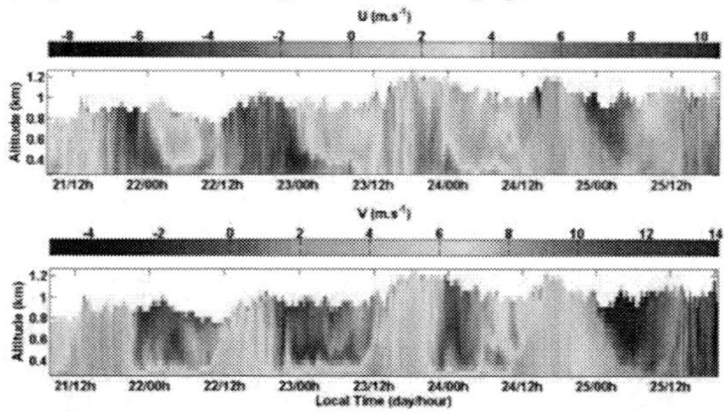

Figure 7. Temporal evolution of zonal (U, positive from North) and meridianal (V, positive from east) wind derived from wind lidar (WLS70) between 21 and 25 March 2011.

The trajectories have been compared with the emission area of PM2.5 as defined from the EMEP data base (http://www.emep.int/). By this way, we can clearly check the origin of aerosol pollutants as shown Figure 8.

The events of high particulate pollution observed in France in March and April are the conditions of i) favorable weather to an east-west transport, ii) the emissions of industrial areas of the Ruhr area and Benelux, and also iii) the use of fertilizers from animal droppings widely used in agriculture. The period from February to April is indeed favorable for mineralization of nitrogen in manure that is in the form of ammonia nitrogen. The fields thus become an important source of secondary aerosols that add to pollution from the East of France.

All these aerosols generally lead to a small particulate depolarization ratio (*PDR*<2%). In our cases study, this ratio is observed as higher and can be close to 10%, showing sometime the contribution of terrigenous particles from resuspension over dry soils. We have noted the dispersion in the PBL of dust-like aerosols is efficient during the afternoons with mean vertical speed velocity larger than 1 ms^{-1}.

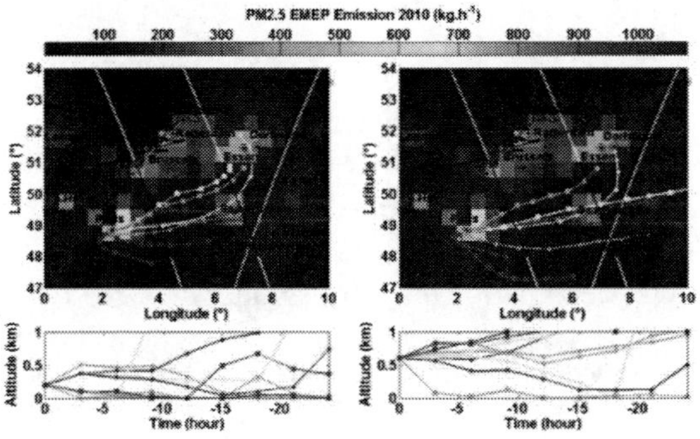

Figure 8. PM2.5 EMEP emission inventory for 2010 with a resolution of 0.5x0.5°. One-day backward trajectory from Saclay starting at 200 m agl. (left column) and 800 m agl (right column) for 21 (red), 22 (green), 23 (brown), 24 (black), 25 (yellow) March 2011 at 0h LT (squares) and 12h LT (circles). Available CALIOP orbits are indicated in white and the portion of the orbit considered is represented in black. Two daytime orbits are drawn for the 23 and 25 March, and one nighttime orbit for the 24 March.

This study has to be continued and the aerosol optical properties retrieved by the lidar measurements in terms of PDR and BER should be compared to those we met within London [6] and during the experiment conducted from Paris (France) to the lake Baikal (Russia) in 2013 [7].

ACKNOWLEDGMENTS

The MODIS and CALIOP data were obtained from the NASA Langley Research Center Atmospheric Science Data Center. The Hysplit model was used with the courtesy of NOAA Air Resources Laboratory (http://www.arl.noaa.gov).

REFERENCES

[1] Chazette, P., F. Marnas, and J. Totems, 2014: The mobile Water vapor Aerosol Raman LIdar and its implication in the framework of the HyMeX and ChArMEx programs: application to a dust transport process, Atmos. Meas. Tech., 7, 1629-1647, doi:10.5194/amt-7-1629-2014.

[2] Royer, P., Chazette, P., Sartelet, K., Zhang, Q. J., Beekmann, M., and Raut, J.-C. , 2011: Lidar-derived PM10 and comparison with regional modeling in the frame of the MEGAPOLI Paris summer campaign, Atmos. Chem. Phys., 11, 11861-11909, doi:10.5194/acp-11-11861-2011.

[3] Raut J.-C. and P. Chazette, 2009: Assessment of vertically-resolved PM10 from mobile lidar observations, Atmos. Chem. Phys., 9, 8617-8638.

[4] Chazette P., H. Randriamiarisoa, J. Sanak, P. Couvert and C. Flamant, 2005: Optical properties of urban aerosol from airborne and ground-based in situ measurements performed during the ESQUIF program, J. Geophys. Res, Vol. 110, No. D2, D0220610.1029/2004JD004810.

[5] Royer, P., J.-C. Raut, G. Ajello, S. Berthier, and P. Chazette, 2010: Synergy between CALIOP and MODIS instruments for aerosol monitoring: application to the Po Valley. Atmospheric Measurement Techniques 3, 1323-1359.

[6] McMeeking, G. R., M. Bart, P. Chazette, J. M. Haywood, J. R. Hopkins, J. B. McQuaid, W. T. Morgan, J.-C. Raut, C. L. Ryder, N. Savage, K. Turnbull, and H. Coe, 2012: Airborne measurements of trace gases and aerosols over the London metropolitan region, Atmos. Chem. Phys., 12, 5163-5187, doi:10.5194/acp-12-5163-2012.

[7] Dieudonné, E., P. Chazette, F Marnas, J. Totems, and X. Shang, 2015: Lidar profiling of aerosol optical properties from Paris to Lake Baikal (Siberia), Atmos. Chem. Phys., 15, 5007-5026, doi:10.5194/acp-15-5007-2015.

LONG-TERM VARIATION OF STRATOSPHERIC AEROSOLS OBSERVED WITH LIDAR FROM 1982 TO 2014 OVER TSUKUBA, JAPAN

Tetsu Sakai [1]*, Osamu Uchino[1], Tomohiro Nagai[1], Toshifumi Fujimoto[2], Isao Tabata[1]

[1]*Meteorological Research Institute, Japan Meteorological Agency, 1-1 Nagamine, Tsukuba, Ibaraki, 305-0052, Japan, *Email: tetsu@mri-jma.go.jp*

[2] *Fukuoka Regional Headquarters, Japan Meteorological Agency, 1-2-36 Ohori, Chuo-ku, Fukuoka 810-0052, Japan*

ABSTRACT

The vertical distribution of stratospheric aerosols has been measured with lidars at the Meteorological Research Institute (MRI) over Tsukuba since 1982. After two major volcanic eruptions (Mt. El Chichón in 1982 and Mt. Pinatubo in 1991), stratospheric aerosol loading increased about 50–100 times compared with the background level which was observed for 1997-2000. From 2000 to 2012, a slight increase (5.3% year–1) was observed by some volcanic eruptions. This long-term lidar data have been used for assessing of impact of the stratospheric aerosols on climate and the ozone layer.

1. INTRODUCTION

Stratospheric aerosols are submicrometer-size particles present persistently at altitude ranging from 10 to 30 km. The amount of the stratospheric aerosols increases more than tenth of times after the large volcanic eruptions that inject SO2 gas into the stratosphere. Recently, several researchers [1, 2] have reported the increase of the stratospheric aerosol loading after 2000. The influence of the increase of the stratospheric aerosols is non-negligible on the Earth's radiation budget because they reflect solar light to space and reduce the incoming solar radiation [3, 4]. The stratospheric aerosols also affect ozone by serving reaction surface that result in ozone loss [5]. To understand the influences of stratospheric aerosol on the climate, we are necessary to monitor the vertical distribution of aerosols.

The MRI has been measuring the vertical distribution of the stratospheric aerosols over Tsukuba (36.1°N, 140.1°E) with lidar [6, 7]. This paper reports the result of the observation over 30 years and studies the long-term variation of the stratospheric aerosols

2. METHODOLOGY

We have used six lidar systems for monitoring the stratospheric aerosol over Tsukuba since 1982. After the El Chichón volcano erupted on April 1982, we have started the measurement with a ruby lidar at a wavelength of 694.3 nm. It was operated until 1987 when the influence of the El Chichón volcano on the stratospheric aerosols almost ceased [8]. After that time, we have employed Nd:YAG lidar systems operating at a wavelength of 532 nm since 1988. Since April 2002 we have employed unmanned system that continuously measures the aerosols in the stratosphere and troposphere for 24 hours.

We compute the backscattering ratio by employing the method of Fernald [9]. We assumed the extinction-to-backscatter ratio (lidar ratio) to 35–50 sr depending on the period according to the balloon borne measurements [10, 11]. The atmospheric density was derived from radiosonde observation launched 250 m northeast of the lidar (Tateno Aerological observatory). At the altitude range where radiosonde data was unavailable (usually above 35 km) we used U.S. Standard Atmosphere (1976). In this study, we reanalyzed the data by taking absorption by ozone into account. The vertical profiles of ozone were obtained with ozonesonde launched from Tateno Aeorological Observatory every week.

3. RESULTS

Figure 1 shows the temporal variation of the vertically integrated aerosol backscattering coefficient (IBC) over Tsukuba from 1982 to 2012. The altitude ranges of the integrations are from 16.5 km and 30.5 km. For the data taken with the ruby lidar (694 nm) from 1982 to 1987, we converted the IBC values to those at 532 nm by using the conversion factor proposed by Jäger and Hofmann [12]. The orange tick marks on the top of Fig. 1 show the period of large volcanic eruptions with Volcanic Explosivity Index (VEI) ≥ 4.

The values of IBC increased rapidly after the eruptions of El Chichón (17.3°N, 93.2°W) on April 1982 and Mt. Pinatubo (15.1°N, 120.4°E) on June 1992. The maximum values of IBC were 3.1×10^{-3} sr^{-1} on July 2 in 1982 and 5.8×10^{-3} sr^{-1} at February 22 in 1992. These values are 62 and 116 times larger than the minimum value (5.01×10^{-5} sr^{-1}) during the period from October 1997 to September 2001 when the influence of volcanic eruptions was at minimum (annual average of blue line in Fig. 1). The corresponding aerosol optical thicknesses were 0.16 and 0.29, respectively, assuming the aerosol lidar ratio as 50 sr. After the eruptions the IBC value decreased to the background level over 6 years.

4. DISCUSSION: Increase of IBC from 2000 to 2012

The IBC value slightly increased from 2000 to 2012. The rate of increase was 1.06×10^{-5} $sr^{-1}year^{-1}$ (5.3% $year^{-1}$). The increase of IBC after 2000 has been observed at Mauna Loa and Boulder in the United States [1] and Lauder, New Zealand [6]. Vernier et al. [2] suggested that the major reason for the increase was volcanic eruptions with VEI ≥ 4. We notice that the IBC between the tropopause and 16.5 km (not shown), which was unavailable from satellite observation, increased significantly during the period. The increase of the stratospheric aerosols cancel the global warming by the increase of anthropogenic carbon dioxide [3, 4]. Thus, it is important to continue monitoring the stratospheric aerosol and studying the trend.

REFERENCES

[1] Hofmann, D. J., et al., 2009: Increase in background stratospheric aerosol observed with lidar at Mauna Loa Observatory and Boulder, Colorado, *Geophys. Res. Lett.*, **36**, L15808, doi:10.1029/2009GL039008.

[2] Vernier, J. P., et al., 2011: Major influence of tropical volcanic eruptions on the stratospheric aerosol layer during the last decade, *Geophys. Res. Lett.*, **38**, L12807, doi:10.1029/2011GL047563.

[3] Solomon, S., et al., 2011: The persistently variable "background" stratospheric aerosol layer and global climate change, *Science*, **333**, 866–870.

[4] Ridley, D. A., et al., 2014: Total volcanic stratospheric aerosol optical depths and implications for global climate change, *Geophys. Res. Lett.*, **41**, 7763–7769, doi:10.1002/2014GL061541.

[5] Kondo, Y., et al., 1995, Stratospheric ozone changes at 43°N and 36°N over Japan between 1991 and 1994, *Geophys. Res. Lett.*, **22**, 3223–3226, doi:10.1029/95GL03180.

[6] Uchino, O., et al., 1995: Extensive Lidar observations of the Pinatubo aerosol layers at Tsukuba (36.1°N), Japan and Lauder (45.0°S), New Zealand, *Geophys. Res. Lett.*, **22**, 57–60.

[7] Nagai, T., et al., 2010: Post Pinatubo evolution and subsequent trend of the stratospheric aerosol layer observed by mid-latitude lidars in both hemispheres, *SOLA*, **6**, 69–72, doi:10.2151/sola.2010-018.

[8] Uchino, O., et al., 1988: Five-year lidar observational results and effects of the El Chichón particles on Umkehr ozone data, *J. Meteor. Soc. Japan*, **66**, 635–643.

[9] Fernald, F. G., 1984, Analysis of atmospheric lidar observations: Some comments, *Appl. Opt.*, **23**, 652–653.

[10] Jäger H. and Deshler T., 2002: Lidar backscatter to extinction, mass and area conversions for stratospheric aerosols based on midlatitude balloonborne size distribution measurements, *Geophys. Res. Lett.*, **29**, 1929, doi:10.1029/2002GL015609.

[11] Jäger, H., and T. Deshler, 2003: Correction to "Lidar backscatter to extinction, mass and area conversions for stratospheric aerosols based on midlatitude balloonborne size distribution measurements", *Geophys. Res. Lett.*, **30**, 1382, doi:10.1029/2003GL017189.

[12] Jäger, H, and D. Hofmann, 1991: Midlatitude lidar backscatter to mass, area, and extinction conversion model based on in situ aerosol measurements from 1980 to 1987, *App. Opt.*, **30**, 127–138.

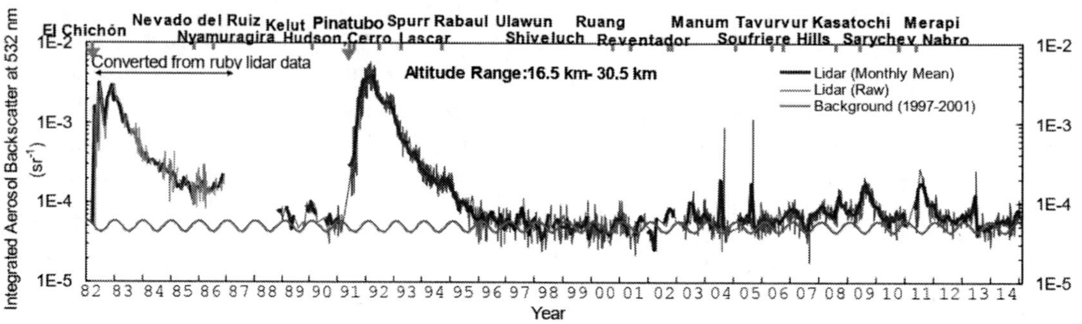

Fig. 1: Long-term variation of vertically integrated aerosol backscattering coefficient (IBC) at 532 nm between altitudes of 16.5 and 30.5 km over Tsukuba from 1982 to 2014. Thick black and thin red lines show the monthly mean and raw IBC values observed with the lidar. Blue line shows background values estimated by fitting a sinusoidal function to the observed values between 1997 and 2002. Orange tick marks on the top show the period of large volcanic eruptions.

AEROSOL OPTICAL PROPERTIES IN THE LOWER TROPOSPHERE DURING SUMMER OVER NEW DELHI

Radhakrishnan Soman Radha*, Bhuwan Chandra Arya, Sumith Kumar Misra, Chhemendra Sharma, Arun Kumar, Devesh Kumar Shukla and Jaswanth

*CSIR-National Physical Laboratory, Dr.K.S.Krishnan Road, New Delhi, India, *E-mail: rkrishnan@nplindia.org*

ABSTRACT

This work reports the variation in optical properties of aerosol in the boundary layer over the Delhi region during the summer season of the year 2014. The layering of aerosol particles in the shallow residual layer is observed in the night preceded by the hot days. The monitoring of this kind of layers in the lower troposphere is very important to study its long range transport.

1. INTRODUCTION

Aerosols play an important role in the radiation budget of Earth's atmospheric system through direct, indirect and semi direct effects. Altitude distribution of aerosols is an important factor which influences in many way the weather and climate of the Earth-atmosphere system. Atmospheric residence time as well as long-range transport of aerosols is primarily governed by the altitude distribution. The microphysical and optical properties and vertical distribution of aerosols are needed to be measured to understand their influence on radiative forcing [1].

The aerosol loading over the north Indian region is found to be increasing compared to the southern part [2]. The mega city Delhi situated in the western part of Indo Gangetic Basin (IGB) in the north Indian region experiences extreme weather conditions every year. Also it experiences the dust storm events during summer, haze and fog during winter, anthropogenic loading of aerosols associated with festivals and biomass burning etc. The increasing industrialization/urbanization and population growth are also one of the causes of loading of aerosols. Dubey et al., [3] reported vertical distribution of aerosol and clouds over New Delhi using Micropulse lidar observations.

During summer the surface heating and the result of increase in convection over land enhances the aerosol concentration in the boundary layer. The frequent dust storm activities also contribute largely to the increase in coarse mode particles during the summer season. This work reports the variation in boundary layer aerosol optical properties over the urban tropical region Delhi during summer 2014. The variation in optical properties of aerosols in the boundary layer during the May-June (summer) period of the year 2014 is studied.

2. DETAILS OF THE LIDAR SYSTEM AND METHOD OF ANALYSIS

A dual polarization micropulse lidar system is developed indigenously to study the boundary layer aerosols and clouds. This lidar uses the diode pumped Nd:YAG microchip laser as source that emits a pulse of duration of 750 ps at 532 nm wavelength with energy of 3 μJ and repetition rate of 7.1 kz. The backscattered signal is collected using a 20cm Cassegrain telescope and split into co-polarized (p) and cross polarized (s) signals using beam splitter. A single photomultiplier is using for near simultaneous measurement of the **p** and **s** signals with a stepper motor controlled mirror system. Number of photons are counted using photon counter and stored in the memory of the computer for further analysis. All the steps of the data acquisition, online display of the data etc. are controlled through the indigenously developed control software. The details of the lidar system can be seen elsewhere [3]. For the current study each lidar profile obtained during night time with a range resolution of 24 m and integrated for 10,000 laser pulses is used. Then, spatial and temporal averaging is done for better signal-to-noise ratio and then noise removed and range corrected for further analysis. Fernald's [4] algorithm is used to obtain the aerosol backscatter coefficient from the lidar signal. The lidar data obtained during night time for the period of May-June of the year 2014 is used for this study.

3. RESULTS AND DISCUSSION

During the months of May and June of the year 2014 Delhi experiences record heat and a severe storm event. The daily maximum and minimum temperature, relative humidity, wind velocity and average rainfall during May and June are shown in Figure 1.

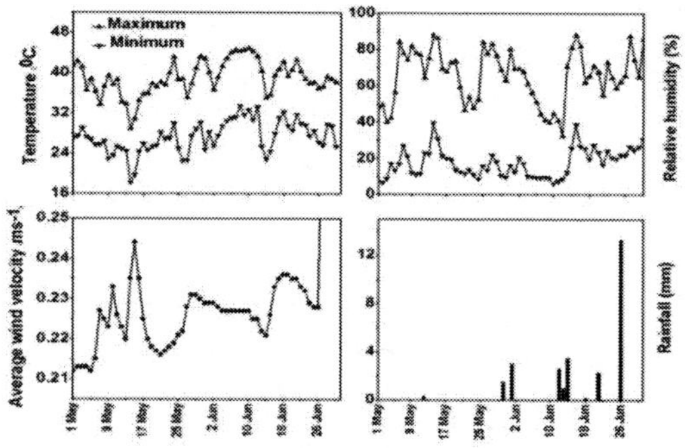

Figure 1: Daily Maximum and minimum temperature, relative humidity, wind velocity and average rainfall during May and June of the year 2014.

In most of the days during the month of June the maximum temperature was above 37 ^{0}C and in the month of June it was above 40 ^{0}C. The minimum temperature also was very high (>24 ^{0}C) during May and June. The relative humidity varies between 25 and 45 % during the observation period. The wind velocity was moderate during the observation period except the dust storm day. Some days of June there was showers. But not much change in the surface temperature was observed on or after the shower.

The variations in the boundary layer aerosol optical properties are derived from the backscattered lidar signal for the months of May and June. For convenience ground (0.072 km) to 2 km altitude range is divided into two slab thicknesses namely slab1 (0.072 to 1km) and slab2 (1km to 2km). The integrated aerosol backscatter coefficient (IABC) corresponds to a slab is obtained by integrating the aerosol backscatter coefficient from base to top of that slab.

The daily mean values of the IABC for the 0.072 to 1 km and 1 to 2 km range obtained during the observation period is shown in Figure 2.

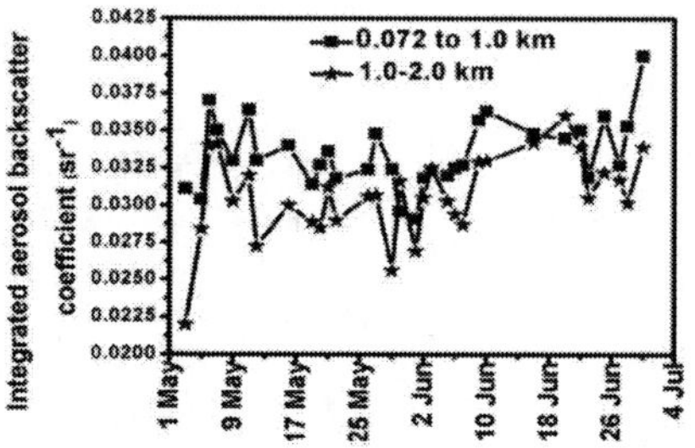

Figure 2: Integrated aerosol backscatter coefficient (sr^{-1}) during the May-June period of the year 2014.

In most of the days during the observation period the value of IABC in the 0.072 to 1 km (slab1) is comparatively higher than its value in the 1 to 2km (slab2) range. In few cases IABC in the 1-2 km slightly higher or very close to its values in the 0.072 to 1km. This was observed during the month of June in which the maximum surface temperature was above 40 ^{0}C. During summer, the strong surface convection due to high temperature and low pressure results in deep atmospheric boundary layer (ABL) and accumulation of aerosol particles in the ABL. In the evening the surface cools due to radiative cooling and a stable nocturnal boundary layer forms. The layering of aerosol particles in the shallow residual layer formed in the night time may be the reason for the observed higher aerosol backscatter coefficient during the hot days.

On 30th May 2014 Delhi was hit by a huge dust storm around 1700 hrs. The day time temperature was very high (43^{0}C) and atmosphere was hazy. Figure 3 shows the aerosol backscatter coefficient (derived by lidar) and Microtop derived aerosol optical depth during May 29 to June 2, 2014. Microtop derived AOD values in all wavelength are high on May 30th.

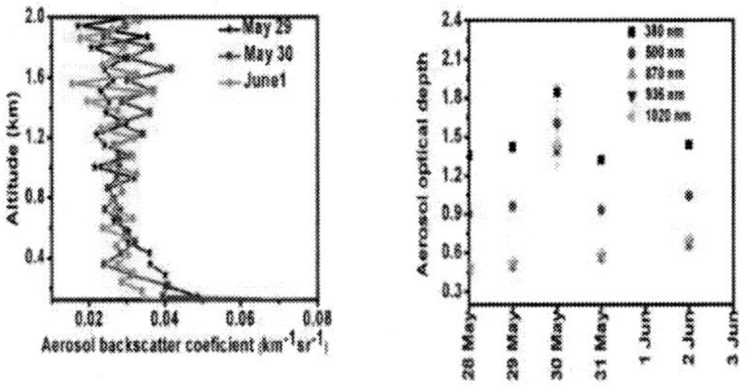

Figure 3: Altitude profile of aerosol backscatter coefficient ($km^{-1}sr^{-1}$) and Microtop derived aerosol optical depth during May 29 to June 2 of the year 2014.

This may be due to the presence of course mode particles present during this period and it is confirmed by the low value of α (an indicator of aerosol particle size). The lidar derived aerosol backscatter profile shows high values on May 29th as compared to May 30th. After the dust storm there was rain and the aerosol particle concentration might have reduced due to washout. Also the aerosol particles in the lower altitude might have transported away from the observation site due to the heavy wind.

4. CONCLUSION

The aerosol variability in the 0.072 to 2 km altitude range over the Delhi region in the summer season of the year 2014 is studied using lidar. The residual layer of aerosol contributes the enhanced backscatter coefficient in the 1-2km altitude range. The monitoring of the vertical profiles of aerosols in the lower troposphere is very useful to study its long range transport.

ACKNOWLEDGEMENT

The authors are thankful to Director NPL and Head-RASD for their consistent support to the ongoing work. Authors also acknowledge CSIR Network Project AIM_IGPHim (PSC-0112) for the financial support.

REFERENCES

[1] Haywood, J.M., and Shine, K.P., 1997: Multi-spectral calculations of the radiative forcing of tropospheric sulphate and soot aerosols using a column model, Q.J.R.Meteorol. Soc. **123**, 1907-1930.

[2] Lodhi, N.K., Beegum, S.N., Singh, S., and Kumar, K., 2013: Aerosol climatology at Delhi in the western Indo-Gangetic Plain: Microphysics, long-term trends, and source strengths. J. Geophys. Res. **118**, 1361-1375.

[3] Dubey, P. K., Jain, S.L., Arya, B.C., Ahammed, Y.N., Arun, K., Shukla, D.K., Pavan S. K., 2011: Indigenous design and development of a micro-pulse lidar for atmospheric Studies. Int. J. Remote. Sens.**32 (2)**, 337-351.

[4] Fernald F.D., 1984: Analysis of atmospheric lidar observations: some comments. Appl. Opt. **23**, 652- 653

OBSERVATIONS IN LIDAR STATION OF ST. PETERSBURG STATE UNIVERSITY FOR ECOLOGICAL SAFETY STUDYIES

Vladislav Donchenko, Irina Melnikova, Dmitriy Samulenkov, Maksim Sapunov

St. Petersburg State University, Universitetskaya nab., 7/9, 199034, St. Petersburg, Russia, Email: mikecf@mail.ru

ABSTRACT

The solution of many problems associated with the air pollution, radiative regime of the earth's surface and atmosphere, global and local environmental changes and climate, facing humanity in the early 21st century, require detailed and regular information on atmospheric aerosol and gaseous pollutants in the atmosphere. For monitoring atmospheric pollutants especially effective were the methods of laser sounding of the atmosphere, which provide a vertical profile of aerosol parameters to a height of 20 km In this regard, at the beginning of the 21st century created a continental networks of lidar sounding stations. Over Europe there is a network EARLINET. Laser station, built on the basis of St. Petersburg State University has become the first Russian station that acceded to the European research network. The article briefly presents the technical features of the equipment and demonstrates the first results of the observations.

1. INTRODUCTION

The creation of the Observatory of ecological safety of St. Petersburg state University was due to the need of research, creation of new techniques, the elaboration of advanced algorithms for regular monitoring of atmospheric pollution in the city with the use of high-tech lidar instruments [1]. Equipping the Observatory with laser systems provides environmental monitoring, the functioning of the laser station in the framework of the European laser network (EARLINET). In early 2014, the Station was associated in EARLINET, and now it is performing calibration and studying the accuracy of measurements and accomplishing observations according to the network schedule.

The Environmental Safety Observatory Resource Center of the St. Petersburg State University (SPbSU) located in the center of St. Petersburg on Vasilievsky Island, 10th Line, 33/35 at the roof of the university building: height of 35 m, geographical coordinates: Lat 59.943N, Lon 30.273E. There is a mobile lidar complex mounted in a car for monitoring air quality around the city and suburbs. The main objectives of the SPbSU Observatory of Environmental Safety Resource Center are as follows [1]:

- Formation of models of the impact on the environment of potentially hazardous industries;
- Estimation of transboundary transport of atmospheric pollutants;
- Creation of forecast models of an unfavorable situation in the environment and approaches for the assessment of possible economic damage;
- Preparation of the synthesized maps of suspected areas and sites of threats to environmental and high ecological risk areas and sites;

Because the lidar station of the SPbSU Observatory of Environmental Safety Resource Center is now part of the European lidar network (EARLINET), the EARLINET networking enables the use of the information provided by lidar monitoring stations for modeling the transboundary pollution transport over Europe. At the same time, the data from atmospheric measurements presents an interest for solution of not only ecological problems but also of the problems of atmospheric physics and meteorology (the concentrations of trace gases, aerosol layers in the atmosphere).

Another important task accomplished by the SPbSU Resource Center is the development of training programs, including lecture courses and laboratory workshop to provide students with the skills they need to operate modern remote sensing facilities.

2. METHODOLOGY

Stationary and mobile multiwave lidar complexes provide information on the concentration and physical nature of atmospheric aerosol, chemical composition of the gas phase of the atmosphere, wind and turbulence at a distance of up to 25 km

DOI: 10.1051/epjconf/201611923013

for vertical and slant paths in all directions. The lidar complex includes components and state of the art technology, available from both Russian and foreign manufacturers [2]. Aerosol lidar allows remote sensing in three elastic spectral channels: 355, 532, 1064 nm and Raman channels: 387, 408, 608 nm. Polarization channel is at 355 nm. Laser pulses of ~10 nsec duration, propagating in the atmosphere, interact with its various components and undergo partial backscattering to the gun. The backscattered signals from the pulse beam emitted by each laser into the atmosphere are collected by the same telescope and sent to the appropriate receiver. The signal is transmitted to the control computer, where it is processed in real time. The computer provides the processing of raw signal (background exposure accounting, smoothing over a group of pulses) [3]. Time scanning of the signal provides information on the points in space along the lidar beam propagation path.

The mobile complex is additionally equipped with a shortwave differential absorption lidar based on a tunable multiwave Ti:Sph laser operating in three wavelength ranges: 700...960, 350...480, 230...310 nm [2] for ozone, chlorine, oxides of nitrogen and sulfur indication in the atmosphere.

Further step consist in solving the inverse problem of lidar sensing, the retrieval of the concentrations of the substances detected. Special algorithms were developed for the solution of the entire set of inversions, which use different methods and approaches to the regularization of solutions for the different components monitored. The output appears as both the smoothed lidar backscattered signal and the quantities of polluting substances, in particular, against the corresponding threshold levels, and the direction and speed of their transport, i.e., in fact, one deals with the first stage of the inverse problem of environmental safety [1].

3. RESULTS

Figure 1 presents an example of the vertical profile of the volume extinction coefficient $\alpha(z)$ in the height (m) – time (hour) coordinates, derived from 532- and 355-nm channel measurements in the zenith direction on March 5, 2015, during 2.5 hours. A "pollution cap" above the city up to the height of 1.0 km can be clearly seen in the figure. Another characteristic of aerosol particles is their

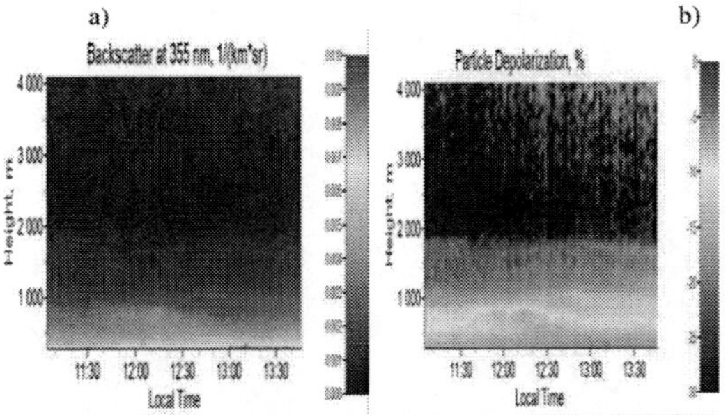

Fig.1. Dynamics of vertical profiles of a) backscattering coefficient at 532 nm and b) particle depolarization from observation 5 March, 2015

shape. Typically, spherical (or nearly spherical) and non-spherical (crystals, chains, etc.) particles are distinguished. The shape of the particles is determined by their origin and may indicate the source of aerosols. Figure 1b demonstrates the particle depolarization for the same time. It is seen that there is a lot of dust not spherical shape in the lower layer.

Figure 2 shows the profile of the atmospheric aerosols concentration.

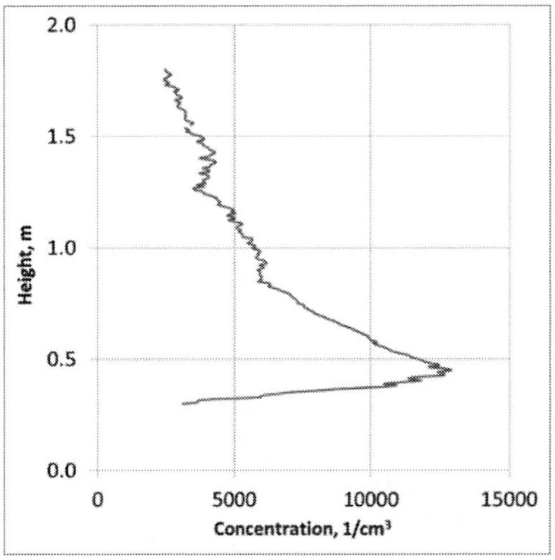

Fig.2. Vertical profiles of particle concentratione index from observation 5 March, 2015

The peak of 12000 $1/cm^3$ at the height 0.5 km is clear distinguished.

There are profiles of real and imagine parts of the refractive index on the Fig.3 those are not varied significantly and pointed to dust component in the height ranges 0.3-2 km.

DOI: 10.1051/epjconf/201611923013

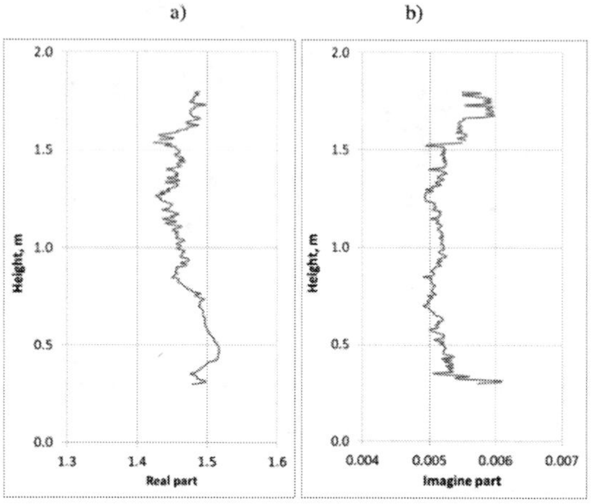

Fig.3. Vertical profiles of a) real part of refractive index, b) imagine part of refractive index from observation 5 March, 2015

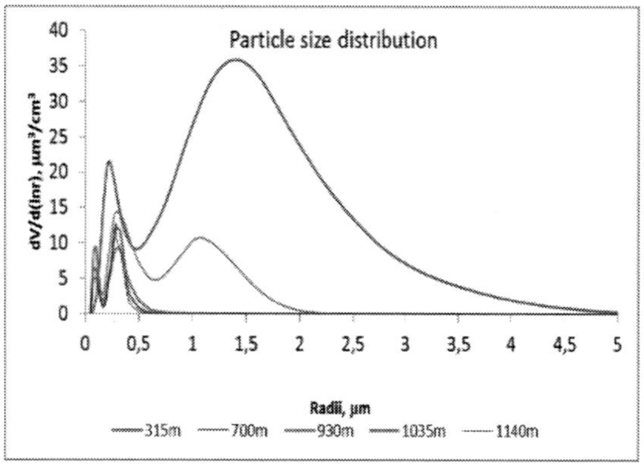

Fig.4. Particle size distributions at 5 altitudes, pointed in figure

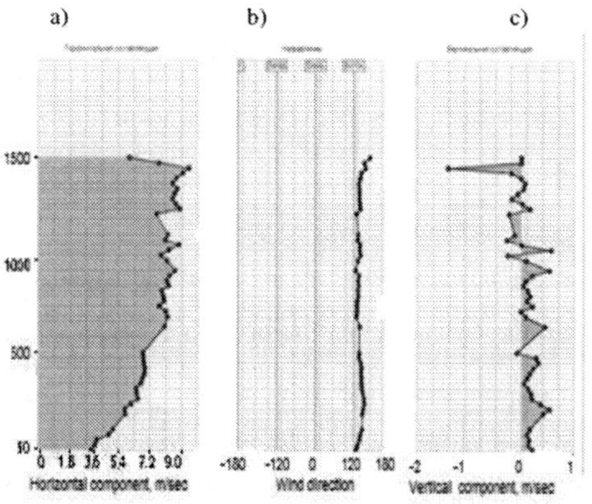

Fig.5. Vertical profiles of the horizontal and vertical velocity and wind direction from observation 5 March, 2015. a) horizontal component, b) – direction and c) vertical component (positive values correspond to upward motions).

The Fig. 4 demonstrates the particle size distribution at 5 altitudes 315 (red line), 700 (blue line), 930 (green line), 1035 (violet line) and 1140m (black line). It is clearly seen the transformation particle sizes with height: larger particle are in lower heights, and very small particles are higher 1 km.

The wind Doppler lidar [4] provides registration of the vertical profile of horizontal and vertical components of the wind velocity and the wind direction that are demonstrated in the Fig. 5. Fig 5a corresponds to horizontal wind velocity, 5b – to wind direction, and 5c – to vertical velocity component. The positive values point to upward air motion. It is detectable the wind shift at height 1.4 km and the downward motion at the same height. The zeroth vertical velocity at the height 0.5 km might explain the particle concentration peak.

4. ATMOSPHERIC POLLUTION IMPACT ON THE PUBLIC HELTH

As it was mentioned in [5, 6] the role of air pollutants in triggering asthma exacerbations in young and adult asthma patients is established. Several studies [7, 8] support the role of air pollution in the development of asthma in childhood. According to [9] long-term exposure to particulate air pollution (PM_{10}) was associated with nonaccidental mortality, mortality from respiratory diseases, and lung cancer mortality in our study population of 7 million adults. Furthermore, PM_{10} was associated with cardiovascular mortality. These points demonstrate importance of monitoring air conditions under big city conditions. The lidar instruments are very suitable for this purpose.

5. CONCLUSION

Studying atmospheric pollution with the use of stationary lidar complex in big cities, like, the case study of St. Petersburg, provides insight into the spatial distribution and temporal dynamics of the air pollution in a city. The obtained characteristics can be represented as the risks of high level of air pollution, whose impact on the health of the population can be assessed. The situational-information center could accomplish processing of the monitoring data in real time or operatively and forecast the ecological situation dynamics. Hence, the active optical methods of ground-based atmosphere remote sensing allow determining not only optical parameters of the atmosphere and atmospheric aerosols but also the environmental parameters: environmental risks of hazardous events associated with the condition of the atmosphere on the local scale.

The activities undertaken by the St. Petersburg State University Resource Center in the frames of the European network are aimed at effective exchange of data, algorithms and methods of data processing, which is currently in progress.

ACKNOWLEDGEMENT

State contract RF number 16.740.11.0619 of 31.05.2011 “Development of methods for studying the physical processes of the formation of atmospheric pollution by smoke and smog"
Agreement No. 14.B37.21.1528 from 01.10.2012 "Modeling of the formation and dynamics fields of atmospheric aerosol, formed as a result of forest fires, for operational forecasting of the atmosphere and study of regional climate"
Agreement No. 14.574.21.0088 from the 16.07.2014 (topic: Conducting applied research on the development of an automated information system for monitoring and forecasting the balance for urban storm water drainage systems") in the framework of realization of the Federal target program "Research and development on priority directions of development of scientific-technological complex of Russia for 2014 - 2020" (a Unique identifier for applied scientific research (project) **RFMEFI57414X0088**).

REFERENCES

[1] Melnikova I., Donchenko V., Boreisho A., Morozov A. 2010: Laser Complexes for the Solution of the Inverse Problem of Ecological Monitoring, *Proceedings of the 25th International Laser Radar Conference*, 2010. 131-133

[2] Boreysho A.S., Chakchir S.Ya., Konyaev M.A., Konovalov K.A., Savin A.V. 2005: Optical Heterodyning in Differential Tunable CO2 Laser Systems. *American Institute of Aeronautics and Astronautics*, Toronto, Chapter DOI: 10.2514/6.2005-5367

[3] Veselovskii I., Whiteman D.N., Kolgotin A., Andrews E., Korenskii M. 2009: Demonstration of aerosol property profiling by multi-wavelength lidar under varying relative humidity conditions *J. Atmos. Ocean Tech.* **26,** 1543-1557.

[4] Akhmetianov B.R., Vasilyev D.N., Konyaev M.A., Mishina O.A., Penkin M.S., Petrov G.A., Tezadov Ya.A., Shatalov I.V., Shiryaev I.F. 2013: Approaches and algorithms for data processing of wind coherent Doppler lidar with cone scanning. *The Journal of radio electronics*, No. 10, 11-23

[5] Peel JL, Tolbert PE, Klein M, Metzger KB, Flanders WD, Todd K, et al. 2005. Ambient air pollution and respiratory emergency department visits. *Epidemiology* 16:164-174.

[6] Sunyer J, Spix C, Quenel P, Ponce-de-Leon A, Ponka A, Barumandzadeh T, et al. 1997. Urban air pollution and emergency admissions for asthma in four european cities: The APHEA project. *Thorax* 52:760-765

[7] Anderson HR, Favarato G, Atkinson RW. 2013 Long-term exposure to air pollution and the incidence of asthma: meta-analysis of cohort studies. *Air Qual Atmos Health* 6:47-56.

[8] McConnell R, Islam T, Shankardass K, Jerrett M, Lurmann F, Gilliland F, et al. 2010. Childhood incident asthma and traffic-related air pollution at home and school. *Environ Health Perspect* 118:1021-1026.

[9] Fischer Paul H., Marra Marten, Ameling Caroline B., Hoek Gerard, Beelen Rob, Kees de Hoogh, Breugelmans Oscar, Kruize Hanneke, Janssen Nicole A.H., Houthuijs Danny. Air Pollution and Mortality in Seven Million Adults: The Dutch Environmental Longitudinal Study (DUELS). *Environ Health Perspect.* Advance Publication: 11 March 2015.

Comparison of aerosol optical and microphysical retrievals from HSRL-2 and in-situ measurements during DISCOVER-AQ 2013 (California and Texas)

Patricia Sawamura[1,2,*], Detlef Müller[3], Sharon Burton[1], Eduard Chemyakin[1,4], Chris Hostetler[1], Richard Ferrare[1], Alexei Kolgotin[5], Luke Ziemba[1], Andreas Beyersdorf [1], Bruce Anderson[1]

*[1]NASA Langley Research Center, Hampton, VA, USA, *Email: patricia.sawamura@nasa.gov, [2]Oak Ridge Associated Universities, TN, USA, [3]University of Hertfordshire, UK, [4]Science System and Applications Inc., Hampton, VA, USA, [5]Physics Instrumentation Center, Troitsk, Russia*

ABSTRACT

The combination of backscatter coefficients measured at 355, 532 and 1064 nm and extinction coefficients at 355 and 532 nm (i.e. $3\beta+2\alpha$) can be used to retrieve profiles of optical and micro-physical properties of aerosols, such as effective radius, total volume concentration and total number concentration. NASA LaRC HSRL-2 is an airborne multi-wavelength high spectral resolution lidar in operation that provides the full $3\beta+2\alpha$ dataset. HSRL-2 was deployed during DISCOVER-AQ along with other airborne and ground-based instruments that also measured many aerosol parameters in close proximity to the HSRL-2 system, allowing us to evaluate the performance of an automated and unsupervised retrieval algorithm that has been recently developed. We present the results from California (Jan/Feb 2013) and Texas (Sep 2013) DISCOVER-AQ.

1. INTRODUCTION

Since 2012, the NASA LaRC airborne HSRL-2 has been measuring backscatter coefficients at 355, 532, and 1064 nm and extinction coefficients at 355 and 532 nm (i.e. $3\beta+2\alpha$) on field missions. This lidar is an airborne prototype of the multiwavelength HSRL envisioned for the NASA Aerosol-Clouds-Ecosystem (ACE) satellite mission being formulated in response to the NRC Decadal Study recommendation.

The $3\beta+2\alpha$ lidar technique allows for the retrieval of microphysical properties of aerosols, such as particle size distributions and complex refractive index [1]. In order to handle the inversion of extensive satellite datasets, an automated and unsupervised microphysics retrieval algorithm has been recently developed [2]. The results from HSRL-2 first deployment during TCAP 2012 showed good agreement with in-situ measurements [2].

HSRL-2 was also deployed during DISCOVER-AQ (Deriving Information on Surface Conditions from Column and VERtically Resolved Observations Relevant to Air Quality) onboard NASA Langley's B-200 aircraft in 2013 (Jan/Feb: California, Aug/Sep: Texas) and 2014 (Jul/Aug: Colorado). During DISCOVER-AQ in-situ instruments were also deployed onboard NASA Wallops' P3B aircraft that flew in ascending/descending spirals over a number of locations collecting vertical profile measurements of optical, physical and chemical properties of aerosols and trace gases.

These coincident airborne in-situ measurements provide an opportunity for continued evaluation of the HSRL-2 retrievals. In this paper we examine aerosol measurements and retrievals obtained from in-situ and HSRL-2 measurements during the two deployments of DISCOVER-AQ 2013 (California and Texas).

2. METHODOLOGY

The HSRL-2 and in-situ measurements were obtained onboard two different aircraft. Therefore, in order to compare the HSRL-2 measurements/ retrievals to the in-situ measurements, a radius of colocation was set to 20 and 10 km for California and Texas, respectively. The longest period of time allowed between lidar and in-situ measurements was 30 minutes for both locations.

From the in-situ set of measurements we used: the scattering coefficients measured at 450, 550, and 700 nm with a TSI nephelometer, absorption coefficient at 532 nm measured with a Particle Soot Absorption Photometer (PSAP), and size distributions obtained with an Ultra-High Sensitivity Aerosol Spectrometer (UHSAS) that measures particles in the 60 nm – 1 μm optical diameter range. Most measurements were obtained for low relative humidity (RH) conditions except for the scattering at 550 nm, which was also measured at higher RH conditions (RH ~ 80-85%). The two (dry and wet) scattering coefficients measurements at 550 nm along with measurements of the ambient RH allow for the estimation of the scattering coefficient at ambient RH [3].

The UHSAS instrument, like most optical counters, is calibrated using polystyrene latex (PSL) spheres that are known to have a refractive index (real part) of ~ 1.59 at the 500-600 nm range. This value is higher than for most particles present in the atmosphere (except for soot). For this reason, a correction was applied in order to account for more realistic refractive index values.

After this first correction to the size distributions, the real and imaginary parts of the refractive index (mR, mI, respectively) at low RH conditions (i.e. dry) are estimated from the combination of dry size distributions, dry scattering coefficients at 450, 550 and 700 nm and dry absorption at 532 nm. Using a grid of mR and mI values we search for pairs that, along with the size distributions in a Mie code, allow the reproduction of the dry scattering and absorption coefficients within 20% and 2 Mm^{-1} accuracy, respectively. It should be noted that a wavelength-independent refractive index is assumed for this step.

Lidar systems measure aerosol backscatter and/or extinction coefficients at ambient RH conditions. In order to compare the HSRL-2 aerosol measurements and retrievals to the in-situ measurements, the latter had to be adjusted to account for water uptake effects (i.e. hygroscopicity) as these measurements were originally obtained in dry conditions.

An iterative algorithm is then used to estimate diameter-independent growth factors that are used to correct the dry refractive index and size distributions for hygroscopicity effects in a manner that allows the reproduction of the measured ambient scattering coefficients at 550 nm within 1% accuracy. The methodology used to correct the refractive index and size distribution with growth factors is similar to the one described in [4].

At the end of this process we are able to estimate vertical profiles of complex index of refraction, surface-area, volume and effective radius at ambient RH from the in-situ measurements.

As a last step, the ambient refractive indices and size distributions are used to calculate the backscatter coefficients at 355, 532 and 1064 nm, and extinction coefficients at 355, and 532 nm in order to be compared with the HSRL-2 measurements.

Details on the automated, unsupervised microphysical retrieval algorithm that was used for the HSRL-2 data are presented in [2].

3. RESULTS

After applying all the necessary corrections to the in-situ size distributions we compare number, surface-area, and volume concentrations, as well as effective radius to the HSRL-2 fine mode retrievals. Figure 1 shows the results for DISCOVER-AQ Texas and Figure 2 shows the results for DISCOVER-AQ California. The colors in Figure 1 represent density of points due to the much higher number of points obtained in Texas. DISCOVER-AQ California happened during wintertime when the planetary boundary layer is usually very shallow, yielding very short profiles.

We find good agreement between HSRL-2 fine mode retrievals and humidity-corrected in-situ measurements of number, surface-area and volume concentrations for Texas and California. These results agree with the results of a case study presented in [2].

The comparison of optical data (i.e. 3β+2α) measured by the HSRL-2 to the reconstructed optical data from the in-situ measurements showed some disagreements which are currently being investigated. In general, the extinction coefficients show better agreement than the backscatter coefficients. Some of the possible causes for the disagreements are:

- Limitation of the UHSAS instrument to submicron particles;

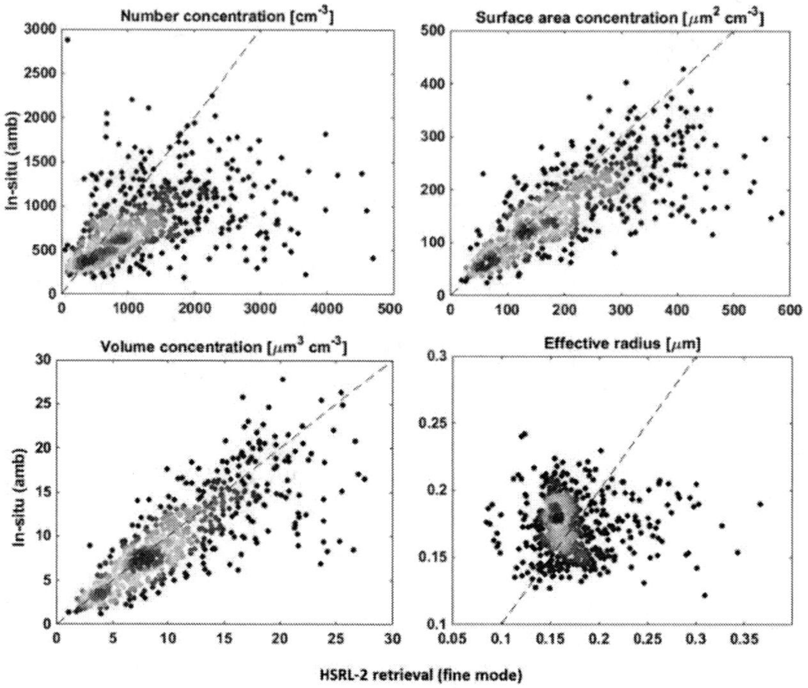

Figure 1. Comparison of size parameters retrieved from HSRL-2 (x-axis) and in-situ measurements (y-axis) from coincident profiles obtained during DISCOVER-AQ Texas (Aug-Sep 2013). Due to the large number of data points (658), a density scatter plot is presented, where warmer colors represent higher density of points.

- Estimation of imaginary refractive index (which relies strongly on the absorption coefficient measurements at one wavelength) and its correction for hygroscopic effects;
- Assumption of wavelength-independent refractive index

4. CONCLUSIONS

DISCOVER-AQ provided us with a great opportunity and datasets to evaluate the automated aerosol microphysics retrieval algorithm that was recently developed for the HSRL-2.

Given the different nature of remote sensing and in-situ measurements, a careful analysis is required when comparing such measurements/ retrievals. We attempted to account for differences originating from sampling method, instrument and retrieval limitations.

Compared to in-situ measurements, the current algorithm has shown good performance in retrieving size parameters such as number, surface-area, and volume concentrations, as well as effective radius.

More work is ongoing to improve the accuracy of refractive index and single-scattering albedo retrievals (not shown).

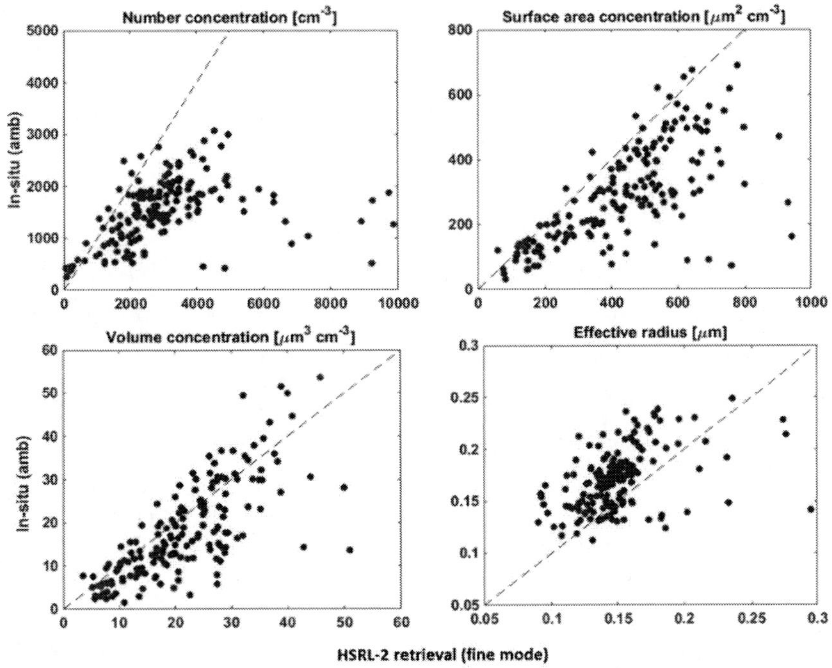

Figure 2. Comparison of size parameters retrieved from HSRL-2 (x-axis) and in-situ measurements (y-axis) from coincident profiles obtained during DISCOVER-AQ California (Jan-Feb 2013). Number of data points: 173.

ACKNOWLEDGEMENT

This work was funded by DISCOVER-AQ and NASA Postdoctoral Program (through Oak Ridge Associated Universities). The authors would like to thank the flight crew from both NASA B200 and P3B.

REFERENCES

[1] Müller, D., Wandinger, U., Althausen, D., Mattis, I., and Ansmann, A.: Retrieval of physical particle properties from lidar observations of extinction and backscatter at multiple wavelengths, Appl. Opt., 37, 2260–2263, 1998.

[2] Müller, D., Hostetler, C. A., Ferrare, R. A., Burton, S. P., Chemyakin, E., Kolgotin, A., Hair, J. W., Cook, A. L., Harper, D. B., Rogers, R. R.. Hare, R. W., Cleckner, C. S., Obland, M. D., Tomlinson, J., Berg, L. K., and Schmid, B.: Airborne Multiwavelength High Spectral Resolution Lidar (HSRL-2) observations during TCAP 2012: vertical profiles of optical and microphysical properties of a smoke/urban haze plume over the northeastern coast of the US, Atmos. Meas. Tech, 7, 3487-3496, 2014.

[3] Ziemba, L. D., Thornhill, K. L., Ferrare, R., Barrick, J., Beyersdorf, A. J., Chen, G., Crumeyrolle, S. N., Hair, J., Hostetler, C., Hudgins, C., Obland, M., Rogers, R., Scarino, A. J., Winstead, E. L., Anderson, B. E., Geophys. Res. Lett., 40, 417-422, 2013.

[4] Zieger, P., Fierz-Schmidhauser, R., Gysel, M., Ström, J., Henne, S., Yttri, K. E., Baltensperger, U., and Weingartner, E.: Effects of relative humidity on aerosol light scattering in the Arctic, Atmos. Chem. Phys., 10, 3875-3890, 2010.

EPJ Web of Conferences **119**, 23015 (2016) DOI: 10.1051/epjconf/201611923015
ILRC 27

FREE TROPOSPHERIC AEROSOLS OVER SOUTH AFRICA

Giannakaki Elina[1]*, Anne Pfüller[1], Kimmo Korhonen[1,2], Tero Mielonen[1], Lauri Laakso[3,4], Ville Vakkari[3], Holger Baars[5], Ronny Engelmann[5], Johan P. Beukes[4], Pieter G. Van Zyl[4], Miroslav Josipovic[4], Petri Tiitta[4,6], Kgaugelo Chiloane[4], Stuart Piketh[4], Heikki Lihavainen[3], Kari Lehtinen[1,2]

[1]*Finnish Meteorological Institute, FI-70211, Kuopio, Finland, *Email: eleni.giannakaki@fmi.fi*

[2]*Department of Applied Physics, University of Eastern Finland, FI-70211 Kuopio, Finland*

[3]*Finnish Meteorological Institute, FI-00101, Helsinki, Finland*

[4]*Unit for Environmental Sciences and Management, North-West University, Potchefstroom, South Africa*

[5]*Leibniz Institute for Tropospheric Research, Permoserstrasse 15, D-04318, Leipzig, Germany*

[6]*Department of Environmental Sciences, University of Eastern Finland, FI-70211 Kuopio, Finland*

ABSTRACT

Raman lidar data of one year was been analyzed to obtain information relating aerosol layers in the free troposphere over South Africa, Elandsfontein. In total, 375 layers were observed above the boundary layer during the period 30th January 2010 – 31st January 2011. The seasonal behavior of aerosol layer geometrical characteristics as well as intensive and extensive optical properties were studied. In general, layers were observed at higher altitudes during spring (2520 ± 970 m) while the geometrical layer depth did not show any significant seasonal dependence. The variations of most of the intensive and extensive optical properties analyzed were high during all seasons. Layers were observed at mean altitude of 2100 m ± 1000 m with lidar ratio at 355 nm of 67 ± 25 and extinction-related Ångström exponent between 355 and 532 nm of 1.9 ± 0.8.

1. INTRODUCTION

Atmospheric aerosols have a strong impact on radiative forcing, chemical processes in the atmosphere, cloud properties and air quality [1]. Despite their importance in atmospheric physics, significant gaps in the scientific knowledge about aerosols still exist. In particular, the vertical distribution of aerosols is not well known, but of essential relevance to understanding aerosol effects on climate [2]. Thus, systematic observations of aerosol optical properties are needed.

In South Africa, only few measurements of aerosol vertical profiles have been carried out, and most of the available data is from short field campaigns. Previous studies have indicated that South Africa is one of the most affected countries in the world with regard to aerosol load, due to various natural and anthropogenic activities [e.g. 3].

Considering the above mentioned, systematic observations of the optical properties of aerosol vertical profiles at acceptable accuracy are needed.

The measurements presented in this paper were performed within the framework of the EUCAARI project [4]. One year of Raman lidar observations over South Africa are presented in terms of geometrical and optical aerosol characteristics.

2. METHODOLOGY

2.1 Research site

The lidar measurement site was located on a hill top at Elandsfontein (26°15´S, 29°26´ E, 1745 m a.s.l.) in the Highveld region of South Africa. The station was located approximately 150 km east from the Johannesburg-Pretoria megacity, which is the largest metropolitan area in South Africa with a population of over 10 million people.

One significant source of atmospheric pollutants in the region is biomass burning through wild and agricultural fires. In addition, the burning of solid fuels such as coal, wood, dung and residual crops for domestic cooking and space heating is also an important source of emissions into the atmosphere in southern Africa. These emissions (from wild fires and domestic burning) are difficult to

DOI: 10.1051/epjconf/201611923015

distinguish from one another when they are transported above the boundary layer.

Another source of aerosols in the region investigated is anthropogenic emissions from industrialized regions in South Africa. The major industrial pollution sources within a 50 km radius of the measurement site include ten coal-fired power plants, a petrochemical plant and some metallurgical plants [5].

Dust from Botswana and Namibia might also be possible sources of aerosols in this region. The shortest distance to the Indian Ocean is approximately 350 km, therefore marine air masses can also have an impact on atmospheric aerosols at Elandsfontein.

2.2 Multi-wavelength lidar Polly^{XT}

The transportable aerosol Raman lidar Polly^{XT} was operated remotely at Elandsfontain in South Africa.

The range-resolved backscatter signal of elastic lidar contains information that can be used to derive the height of aerosol layers. The gradient method was applied to determine the bottom and top layer heights of the aerosols in the free troposphere. Planetary boundary layer top heights were retrieved from the lidar backscatter signal at 1064 nm using the Wavelet Covariance Transform method.

Extinction and the backscatter coefficient profiles at 355 and 532 nm have been obtained with the Raman method [6], while the backscatter at 1064 nm has been determined using the Klett method [7].

3. RESULTS

The percentage of the night-time measurements performed each month in terms of hourly averaged profiles every three hours is presented in Figure 1(a) (blue bars), and varies between 3 and 72 % for all the months. Measurement analysis was not performed during the presence of low clouds and rain (shaded bars) or due to the scheduled shutdowns (grey bars). The percentage of the measurements in which at least one free-tropospheric aerosol layer was observed (green bars) is shown in Figure 1(b). In total, 375 aerosol layers were observed above the boundary layer during the period studied.

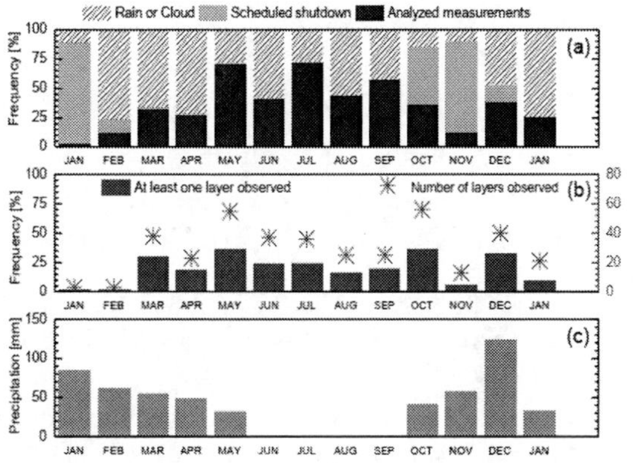

Figure 1. (a) Percentage of night-time analyzed measurements, scheduled shutdown and unanalyzed measurements due to weather conditions; (b) the percentage of measurements in which free tropospheric aerosol layers were observed (green bars) and the total number of the observed layers (red stars) per month; (c) monthly accumulated precipitation at Elandsfontein.

The total number of the observed layers per month is presented with red stars in the same figure.

Approximately 90% of the annual precipitation falls during the wet season (October - March) [5], as indicated in Figure 1 (c). Fewer measurements were performed during the wet period. The seasonal precipitation cycle should not only affect the number of measurements that could be performed but also the total number of layers observed. Therefore it is expected that the fraction of measurements of free-tropospheric aerosol layers in relation to the total number of observations should be less during the wet season due to wet scavenging.

However, our results indicate that during the wet period a relatively large number of layers are still observed in the free troposphere. The general anticyclonic nature of atmospheric circulation over the region, results in the formation of stable layers of air throughout the year [8].

Our complete dataset of lidar measurements in terms of geometrical characteristics is shown in Figure 2.

DOI: 10.1051/epjconf/201611923015

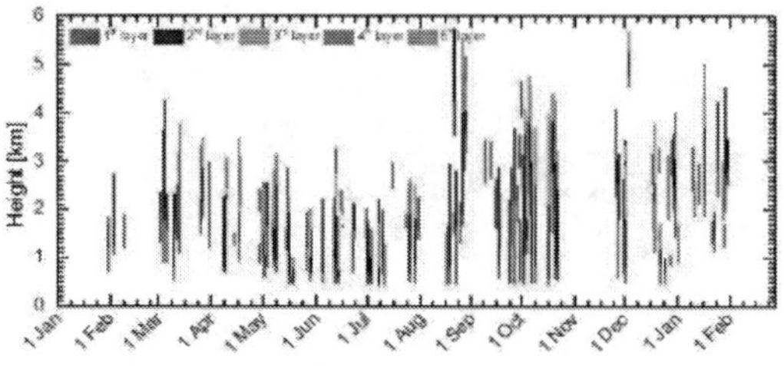

Figure 2. Geometrical boundaries of free tropospheric aerosol layers observed between 30th January 2010 and 31st January 2011.

The vertical lines present the observed height ranges of free tropospheric aerosol layers. Out of 429 hourly analyzed lidar observation, we observe free tropospheric pollution events on 223 hours (i.e. 52 %). In total 375 free tropospheric aerosol layers were observed. The geometrical depth of the free tropospheric layers varies from a few hundred meters to several kilometers throughout the period under investigation. Pollution events are observed as single layer on 52% of the cases, while they are often characterized by two (32 %), three (13 %) or more particle layers (3%).

The seasonal frequency distribution plots for the lidar ratio at 355 nm and the Ångström exponent related to extinction between 355 and 532 nm are presented in Figure 3 and 4, respectively. An average lidar ratio of 67 ± 25 sr at 355 nm and a mean extinction-related Ångström exponent of 1.9 ± 0.8 between 355 and 532 nm are measured during the entire sampling period in South Africa.

Between late winter (Jun. – Aug.) and throughout spring (Sep. – Nov.) high lidar ratios are observed. During this period a significant contribution of aerosol layers originates from wild and controlled fires. Although these fire plumes occur at ground level, they are lifted higher in the atmosphere due to the increased heat and may be transported above Elandsfontein. Domestic biomass burning in informal settlements for cooking and heating can also contribute, but these emissions are exclusively emitted at ground level. It is therefore unlikely that such emission close to Elandsfontein will have a significant impact to free tropospheric aerosol load measured. However, household emissions are common across the entire southern Africa, which implies that regional transport of such emission further away from Elandsfontein may results in a contribution to the free tropospheric load measured over Elandsfontein. The absorbing smoke aerosols, originating from all the afore-mentioned combustion sources, can explain the large lidar ratio values observed.

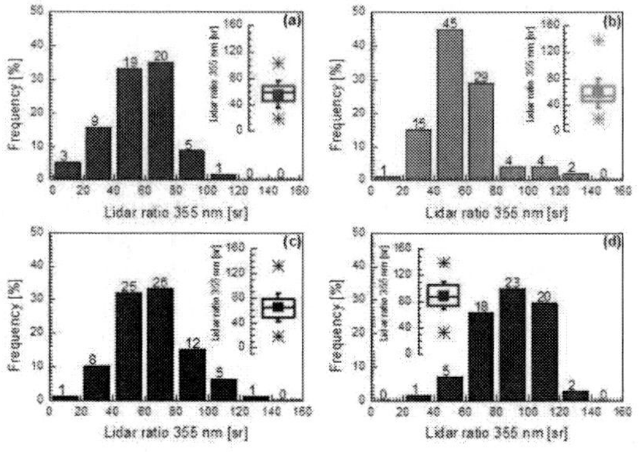

Figure 3. Frequency distribution of lidar ratio at 355 nm of free tropospheric aerosol layers in South Africa between 30th January 2010 and 31st January 2011, for (a) summer, (b) autumn, (c) winter and (d) spring. Box and whisker plots are also presented for each of the seasons: filled square is the mean value, horizontal line is the median, boxes are the 25 and 75 % percentiles respectively, whiskers represent the one standard deviation and stars the minimum and maximum values.

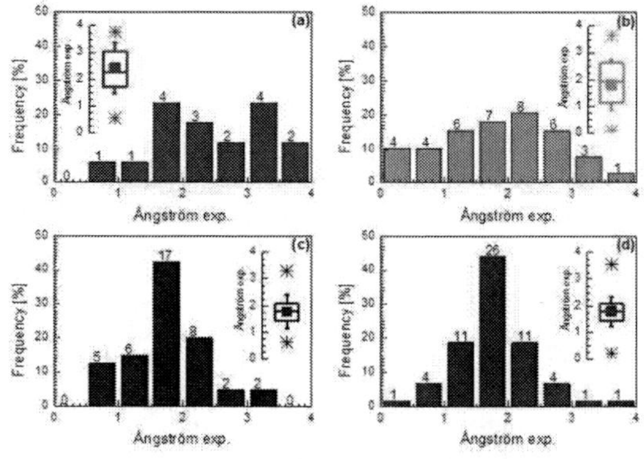

Figure 4. The same as Figure 3 but for Ångström exponent.

Larger Ångström exponents were observed during summer (Dec. – Feb.). One possible explanation is the removal of larger aerosols through wet-scavenging during these wetter months. Most of the free-tropospheric aerosol layers observed during this period were associated with an easterly air mass flow. The major industrial pollution sources are located primarily either to the north, west or southwest of the lidar at Elandsfontein (Laakso et al., 2012). Therefore, the easterly flow that prevails during summer results in a less polluted free troposphere, compared to other seasons. Aerosols measured during this period could also include maritime aerosols with air masses moving from the Indian Ocean.

4. CONCLUSIONS

Geometrical characteristics and optical properties of elevated aerosol layers at a continental site in South Africa are presented. During winter nights layers are observed up to 3000 m. These aerosol layers are probably trapped between the inversion surface layer and an absolutely stable layer that is formed in 90% of days during winter [8]. Aerosol layers are more frequently observed during biomass burning period and are determined to be at higher heights.

Except for the intensive biomass burning period the lidar ratios and Ångström exponents are within the range of previous observations for urban/industrial aerosols [9,10]. Considering that the Elandsfontein measurement station is located in the midst of large industrial plants this is reasonable. However, we observed large variability in both intensive and extensive aerosol properties of free tropospheric aerosol layers. From August to October, the lidar ratio of elevated layers is high, on average 83 sr, which is comparable to biomass burning smoke [10,11]. Mean Ångström exponent related to extinction was found to be 1.8 both for dry and biomass burning period indicating the same size of particles for both biomass burning and (industry originated) polluted aerosols.

ACKNOWLEDGEMENT

This work has been partly supported by the European Commission 6th Framework program under the EUCAARI. Elina Giannakaki acknowledges the support of the Academy of Finland. The authors acknowledge the staff of the North-West University for valuable assistance and routine maintenance of the lidar. We also acknowledge Eskom and Sasol for their logistical support for measurements at Elandsfontein.

REFERENCES

[1] Ramanathan V., P. J. Crutzen, J.T. Kiehl, D. Rosenfeld, 2001: Aerosols, climate, and the hydrological cycle, *Science*, **294**, pp. 2119-2124.

[2] Kaufman Y. J., O. Tanre, O. Boucher, 2002: A satellite view of aerosols in the climate system, *Nature*, **419**, pp. 215-223.

[3] Piketh S. J., R.J. Swap, W. Maenhaut, H.J. Annegarn, P. Formenti, 2002: Chemical evidence of long-range atmospheric transport over southern Africa, *J. Geophys. Res.*, **107**, pp. 4817.

[4] Kulmala, M., *et al.*, 2011 : General overview: European Integrated project on Aerosol Cloud Climate and Air Quality interactions (EUCAARI) – integrating aerosol research from nano to global scales, *Atmos. Chem. Phys.*, **11**, pp. 13061-13143.

[5] Laakso L., V. Vakkari, A. Virkkula, *et al.* 2012: South African EUCAARI measurements: seasonal variation of trace gases and aerosol optical properties, *Atmos. Chem. Phys.*, **12**, pp. 1847-1864.

[6] Ansmann A., U. Wandinger, M. Riebesell, C. Weitkamp, W. Michaelis, 1992: Independent measurement of extinction and backscatter profiles in cirrus clouds by using a combined Raman elastic-backscatter lidar, *Appl. Optics*, **31**, pp. 7113-7131.

[7] Klett J.D., 1981: Stable analytical inversion solution for processing lidar returns, *Appl. Optics*, **20**, pp. 211-220.

[8] Cosijn, C., P. D Tyson, 1996: Stable discontinuities in the atmosphere over South Africa, *S. Afr. J. Sci.*, **92**, pp. 381–386.

[9] Müller, D., A. Ansmann, I. Mattis, M. Tesche, U. Wandinger, D. Althausen, G. Pisani, 2007: Aerosol-type-dependent lidar ratios observed with Raman lidar, *J. Geophys. Res.*, **112**, D16202.

[10] Giannakaki, E., D. S. Balis, V. Amiridis, C. Zerefos, 2010: Optical properties of different aerosol types: seven years of combined Raman-elastic backscatter lidar measurements in Thessaloniki, Greece, *Atmos. Meas. Tech.*, **3**, pp. 569-578.

[11] Amiridis V., D.S. Balis, E. Giannakaki, A.Stohl,, S. Kazadzis, M.E. Koukouli, P. Zanis, 2009: Optical characteristics of biomass burning aerosols over Southeastern Europe determined from UV-Raman lidar measurements, *Atmos. Chem. Phys.*, **9**, pp. 2431-2440.

TROPOSPHERIC VERTICAL PROFILES OF AEROSOL OPTICAL, MICROPHYSICAL AND CONCENTRATION PROPERTIES IN THE FRAME OF THE HYGRA-CD CAMPAIGN (ATHENS, GREECE 2014): A CASE STUDY OF LONG-RANGE TRANSPORT OF MIXED AEROSOLS

Alexandros Papayannis[1]*, Athina Argyrouli[1], Detlef Müller[2], Georgios Tsaknakis[1], Panayotis Kokkalis[1], Ioannis Binietoglou[3], Stelios Kazadzis[4], Stavros Solomos[5] and Vassilis Amiridis[5]

[1]*Laser Remote Sensing Laboratory, National Technical University of Athens, 15780 Athens, Greece*
**Email: apdlidar@central.ntua.gr*

[2]*University of Hertfordshire, Hatfield, Hertfordshire, AL10 9AB, UK*

[3]*INOE, National R&D Institute for Optoelectronics, Magurele, Romania*

[4]*Physikalisch-Meteorologisches Observatorium Davos, World Radiation Center, Switzerland*

[5]*IAASARS, National Observatory of Athens, Greece*

ABSTRACT

Combined multi-wavelength aerosol Raman lidar and sun photometry measurements were performed during the HYGRA-CD campaign over Athens, Greece during May-June 2014. The retrieved aerosol optical properties (3 aerosol backscatter at 355-532-1064 nm and 2 aerosol extinction profiles at 355-532 nm) were used as input to an inversion code to retrieve the aerosol microphysical properties (effective radius r_{eff} and number concentration N) using regularization techniques. Additionally, the volume concentration profile was derived for fine particles using the LIRIC code. In this paper we selected a complex case study of long-range transport of mixed aerosols (biomass burning particles mixed with dust) arriving over Athens between 10-12 June 2014 in the 1.5-4 km height. Between 2-3 km height we measured mean lidar ratios (LR) ranging from 45 to 58 sr (at 355 and 532 nm), while the Ångström exponent (AE) aerosol extinction-related values (355nm/532nm) ranged between 0.8-1.3. The retrieved values of r_{eff} and N ranged from 0.19±0.07 to 0.22±0.07 μm and 460±230 to 2200±2800 cm^{-3}, respectively. The aerosol linear depolarization ratio (δ) at 532 nm was lower than 5-7% (except for the Saharan dust cases, where δ~10-15%).

1. INTRODUCTION

The HygrA-CD campaign was an international field campaign which took place in Athens, Greece between 15 May to 22 June, 2014, in the frame of the Initial Training on Atmospheric Remote Sensing (ITARS) project. HYGRA-CD (http://hygracd.impworks.gr) brought together a suite of different instruments and expertise aiming to enhance our understanding on the impact of aerosols and clouds on weather and climate. It was based on the synergy between remote sensing and in-situ instrumentation, making also use of numerical weather prediction and atmospheric modeling.

It is well established that tropospheric aerosols play a crucial role in climate change through scattering (cooling effect) and absorbing (warming effect) incoming solar and outgoing thermal radiation [1]. Despite recent progress documented in the latest Intergovernmental Panel for Climate Change (IPCC), the uncertainty about the current level of radiative forcing due to aerosols (0.5 Wm^{-2}) is still relatively large compared to that of global warming gases (0.25 Wm^{-2}). In order to clarify the mechanisms of aerosol radiative forcing and reduce the respective uncertainties, detailed knowledge of the vertical profiles of the particle optical (backscatter and extinction coefficients, AE and LR), micro-physical (r_{eff}, surface-area concentration, N, single-scattering albedo (ω) and mean complex refractive index) and chemical properties (water content, dry chemical composition), as well as their mass concentrations are required [1,2].

Raman lidars have proven to be the most adapted tools in aerosol characterization experiments since

DOI: 10.1051/epjconf/201611923016

they can provide vertical profiles of the aerosol optical properties (aerosol extinction and backscatter coefficients), with high spatial and temporal resolution [3,5 and references therein]. Based on the above measurements the aerosol microphysical properties can be calculated using the regularization technique [4].

2. METHODOLOGY

In this study a synergy of instruments was used (Raman depolarization lidar and sun photometry) to derive the aerosol optical and microphysical properties aloft. The Athens Raman lidar system (EOLE) [5] was run in parallel with the AIAS depolarization lidar; both are part of the ARIADNE Greek lidar network [6] and are based in the Laser Remote Sensing Unit (LRSU) of the National Technical University of Athens (NTUA) (37.9°N, 23.6°E, 200 m a.s.l.). EOLE is able to perform independent and simultaneous measurements of the vertical profiles of the aerosol backscatter (at 355, 532 and 1064 nm) and extinction coefficient (at 355 and 532 nm) and of the water vapor to dry air mixing ratio in the troposphere (using the H_2O Raman channel at 407 nm). AIAS provides the aerosol linear depolarization ratio (δ) at 532 nm.

The EOLE Raman lidar data were also used to derive the vertical profiles of the intensive (lidar ratio, Ångström exponent) optical properties of the aerosols in the ultraviolet, visible and infrared. Due to the weak Raman signals, the complete set of aerosol products can be derived for tropospheric layers only during nighttime. The data processing was based on the Single Calculus Chain (SCC) developed within EARLINET [7].

Sun photometry data were provided by an AERONET CIMEL sun photometer located at the National Observatory of Athens (NOA) (www.aeronet.gsfc.nasa.gov).

The microphysical particle properties within specific atmospheric layers were retrieved using the inversion code provided by Müller et al. (1999 and 2011) [4 and 8, respectively]. A detailed description of the original version of the algorithm, which assumes spherical shape of the investigated particles in the retrieval procedures can be found in Müller et al. (1999) [4]. The full set of the aerosol backscatter coefficient at three wavelengths (355, 532 and 1064 nm) and of the extinction coefficient at two wavelengths (355 and 532 nm) is required for the retrieval of the aerosol microphysical properties. The inversion model uses as input the mean values of the optical properties of the aerosols calculated within specific aerosol layers. The aerosol microphysical properties which were derived are the effective radius and the number concentration. Finally, we used LIRIC [9] to derive the volume concentration profile for fine particles.

3. RESULTS

In this paper we will focus on a case study of long-range transport of mixed aerosols arriving over Athens between 10-12 June 2014, in the height range between 1 and 5 km height a.s.l.

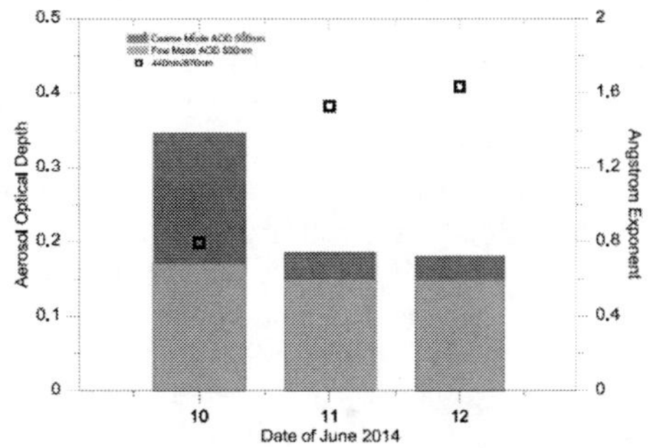

Figure 1. Evolution of the daily mean aerosol fine/coarse-mode optical depth (AOD) obtained over Athens at 500 nm as derived by the CIMEL sun photometer of NOA (left vertical axis) and the respective Ångström exponent values derived from 440 and 870 nm (right vertical axis) of the aerosol column for the period between 10 and 12 June 2014.

At first, in Figure 1 (left hand side axis) we present the evolution of the daily mean value of the aerosol optical depth (AOD) obtained over Athens at 500 nm as derived by the CIMEL sun photometer of NOA located 1.8 km from our lidar station. In the same figure (right hand side vertical axis) we present, for the same time period, the respective Ångström exponent values on 440 nm/870 nm (right vertical axis) of the aerosol column. From this figure we can see that on 10 June the AOD at 500 nm is quite high (0.35) and decreases by 30% the following 2 days, mainly due to the strong decrease of the coarse mode particles. However, for all these days, the AOD due to the fine mode particles remains nearly constant (~0.15-0.17). On the other hand the Ångström exponent values (440nm/870nm) from CIMEL rise from 0.8 (10 June) to 1.62 (12 June) (cf. Fig. 1, right hand side vertical axis). This means that on 10 June the atmospheric column over Athens was dominated by coarse particles, while 1-2 days later the fine particles clearly dominate.

8-day air mass back trajectories calculated by HYSPLIT model showed that on 10 June (night) the air masses observed between 2-3 km height over Athens, originated from the Ukraine-S. Russia and Kazakhstan regions, where biomass burning and Asian desert dust dominated, respectively, after spending 1-2 days over the Balkans. On 11 June the situation changed and after 18:00 UTC some air masses from Sahara, mixed with continental European ones, were observed at 2-3 km height. This effect was intensified more between 22:00-24:00 UTC. The situation changed completely on 12 June, where the aerosol load dropped (similarly as on the morning hours of 11 June), as the air masses sampled over Athens originated from the Balkans-Ukraine, S. Russia and Kazakhstan regions, where biomass burning and Asian desert dust dominated.

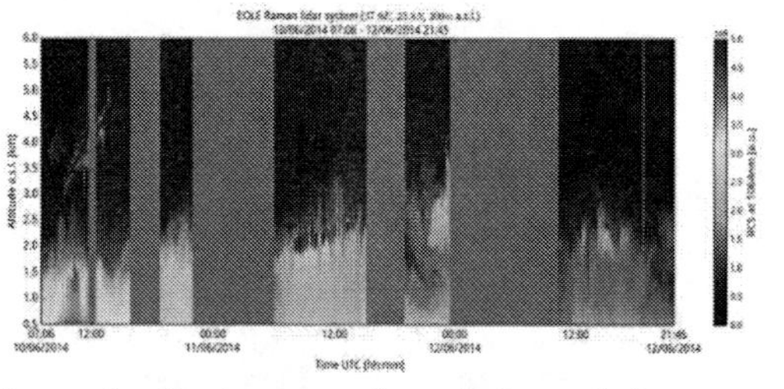

Figure 2: Spatio-temporal evolution of the range-corrected backscattered lidar signals (RCS) obtained by the EOLE lidar over Athens at 1064 nm between 10 (07:06 UTC) and 12 (21:45 UTC) June 2014.

The corresponding Raman lidar measurements, expressed in terms of the temporal evolution of the range-corrected lidar signals (RCS), between 10-12 June (Fig. 2) showed discrete aerosol layers between 2-3 km height, extending up to 3-4.5 km height on 11 June. Moreover, higher aerosol load content is observed on 10 June, compared to ones of the following 2 days.

In Figure 3 we present the aerosol optical properties (backscatter and extinction coefficients, lidar ratio and Ångström backscatter and extinction-related exponent) retrieved by EOLE lidar between 0.8 and 6 km height on 11 June 2014, between 19:00 and 20:00 UTC aerosol backscatter coefficient at 355-532-1064 nm, aerosol extinction and lidar ratio at 355-532 nm, and Ångström backscatter (355nm/532nm and 532nm/1064nm) and extinction-related (355nm/532nm) exponent.

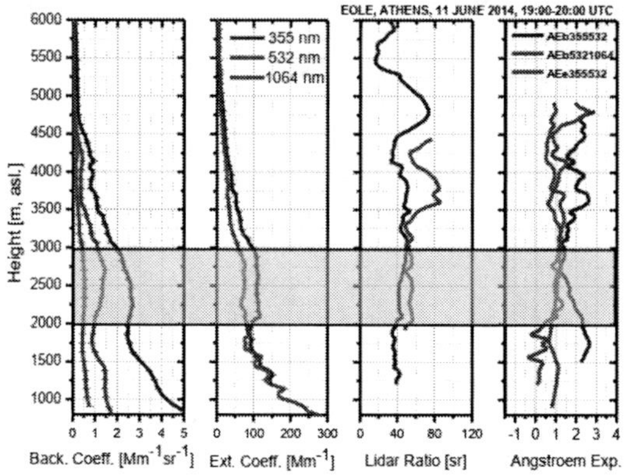

Figure 3. Vertical profiles of the aerosol optical parameters (backscatter and extinction coefficients, lidar ratio, Ångström exponent) retrieved from EOLE Raman lidar measurements on 11 June 2014 (19:00-20:00 UTC) over Athens.

In Fig. 3 we can see a strong aerosol layer between 2-3 km extending up to nearly 4.5 km height. In the 2-3 km height (shaded region) the lidar ratio values are of the order of 45-58 sr (at 355 and 532 nm), while the Ångström exponent aerosol extinction-related values (355nm/532nm) range between 0.8-1.3, close to 1.55 (440nm/870nm) measured during daytime by CIMEL in the atmospheric column. All these values compare well with those measured under similar atmospheric conditions of mixed continental aerosols with dust [10]. Using the inversion code provided by Müller et al. [8] we retrieved, within the shaded of Fig. 3, the mean r_{eff} ranging from 0.19±0.07 to 0.22±0.07 μm, and N from 460±230 to 2200±2800 cm^{-3}. Our r_{eff} values compare well with those of Mamouri et al. (2012) for a mixed dust event over Athens [3].

4. CONCLUSIONS

In this paper we selected a case study of long-range transport of mixed aerosols (biomass burning particles mixed with dust) arriving over Athens between 10-12 June 2014 in the 1.5-4 km height region. In the 2-3 km height region we measured mean lidar ratios (LR) ranging from 45 to 58 sr (at 355 and 532 nm), AE aerosol extinction-related values (355nm/532nm) ranging between 0.8-1.3; the retrieved values of r_{eff} and N ranged from 0.19±0.07 to 0.22±0.07 μm and 460±230 to 2200±2800 cm^{-3}, respectively. The aerosol linear depolarization ratio δ at 532 nm was, in most cases (except for the Saharan dust ones δ~10-15%), lower than 5-7%.

ACKNOWLEDGEMENTS

The work of AP, GT, PK was supported by the MACAVE research project which is implemented within the framework of the action Supporting of Postdoctoral Researchers of the Operational Program Education and Lifelong Learning (action's beneficiary: General Secretariat for Research and Technology), and is co-financed by the European Social Fund (ESF) and the Greek State. The work of AA and IB has received funding from the European Union Seventh Framework Programme (FP7/2007-2013): People, ITN Marie Curie Actions Programme (2012-2016) in the frame of ITaRS under grant agreement n° 289923. SS, SK and VA were supported by the European Union Seventh Framework Programme (FP7-REGPOT-2012-2013-1) in the framework of the project BEYOND: Building Capacity for a Centre of Excellence for EO-based monitoring of Natural Disasters, under grant agreement n° 316210. The authors gratefully acknowledge the NOAA Air Resources Laboratory (ARL) for the provision of the HYSPLIT transport and dispersion model and READY website (http://www.ready.noaa.gov) used in this publication. The Biomedical Research Foundation of the Academy of Athens (BRFAA) is acknowledged for the provision of the lidar mobile platform.

REFERENCES

[1] Intergovernmental Panel on Climate Change: Climate Change 2013, 2013: Synthesis Report.

[2] Bierwirth, E., et al., 2009: Spectral surface albedo over Morocco and its impact on radiative forcing of Saharan dust, *Tellus,* **B61**, 252–269.

[3] Mamouri, R.E., et al., 2012: Multi-wavelength Raman lidar, sun photometric and aircraft measurements in combination with inversion models for the estimation of the aerosol optical and physico-chemical properties over Athens, Greece, *Atmos. Meas. Tech.*, **5**, 1793-1808. http://dx.doi.org/10.5194/amt-5-1793-2012.

[4] Müller, D. et al., 1999: Microphysical particle parameters from extinction and backscatter lidar data by inversion with regularization: Theory, *Appl. Opt.*, **38**, 2346-2357.

[5] Kokkalis, P., et al., 2012: The EOLE lidar system of the National Technical University of Athens, *Proc. 26th International Laser Radar Conference,* Porto Heli, Greece, 629-632.

[6] Papayannis, A., et al., 2012: ARIADNE: The Greek Lidar Network, *Proc. 26th International Laser Radar Conference,* Porto Heli, Greece, 349-352.

[7] d'Amico, G., et al., 2012: EARLINET Single Calculus Chain for Automatic Lidar Data Processing: First Tests on Optical Products, *Proc. 26th International Laser Radar Conference*, Porto Heli, Greece, 331-334.

[8] Müller, D., et al., 2011: Vertical profiles of microphysical particle properties derived from inversion with two-dimensional regularization of multi-wavelength Raman lidar data: experiment, *Appl. Opt.,* **50**, 2069-2079.

[9] Wagner, J., et al., 2013: Evaluation of the Lidar/Radiometer Inversion Code (LIRIC) to determine microphysical properties of volcanic and desert dust, *Atmos. Meas. Tech.,* **6**, 1707-1724.

[10] Müller, D., et al., 2007: Aerosol-type-dependent lidar ratios observed with Raman lidar, *J. Geophys. Res.*, **112**, D16202, doi:10.1029/2006JD008292.

EPJ Web of Conferences **119**, 23017 (2016) DOI: 10.1051/epjconf/201611923017
ILRC 27

LIDAR RATIO DERIVED FOR PURE DUST AEROSOLS: MULTI-YEAR MICRO PULSE LIDAR OBSERVATIONS IN A SAHARAN DUST-INFLUENCED REGION

Carmen Córdoba-Jabonero[1*], José Antonio Adame[1], James R. Campbell[2], Emilio Cuevas[3], Juan Pedro Díaz[4], Francisco Expósito[4], Manuel Gil-Ojeda[1]

[1]*Instituto Nacional de Técnica Aeroespacial (INTA), Atmospheric Research and Instrumentation Branch, Torrejón de Ardoz, 28850-Madrid, Spain, *Email: cordobajc@inta.es*

[2]*Aerosol and Radiation Sciences Section, Marine Meteorology Division, Naval Research Laboratory, Monterey, CA, USA*

[3]*Agencia Estatal de Meteorología (AEMET), Izaña Atmospheric Research Centre, Santa Cruz de Tenerife, Spain*

[4]*Universidad de La Laguna (ULL), Department of Fundamental Physics, La Laguna, Spain*

ABSTRACT

A seasonal distribution of the Lidar Ratio (LR, extinction-to-backscattering coefficient ratio) for pure Saharan dust particles has been achieved. Simultaneous MPLNET/Micro Pulse lidar measurements in synergy with AERONET sun-photometer data were collected in the Tenerife area, a Saharan dust-influenced region, from June 2007 to November 2009. Dusty cases were mostly observed in summertime (71.4 % of total dusty cases). No differences were found among the LR values derived for spring, summertime and autumn times (a rather consistent seasonally averaged LR value of 57 sr is found). In wintertime, however, a higher mean LR is derived (65 sr), associated likely with a potential contamination from fine biomass burning particles coming from Sahel area during wintertime deforestation fires period. Results, obtained from a free-tropospheric pristine station (AEMET/Izaña Observatory) under Saharan dust intrusion occurrence, provide a more realistic perspective about LR values to be used in elastic lidar-derived AOD inversion for Saharan pure dust particles, and hence in improving CALIPSO AOD retrievals.

1. INTRODUCTION

The estimation of the extinction-to-backscattering coefficient ratio (Lidar Ratio: LR) from lidar measurements is a crucial task in aerosol research. In contrast to the height-resolved LR retrieval from Raman lidars, an a-priori constant LR value must be introduced in the inversion algorithms applied to elastic lidar measurements in obtaining the aerosol extinction. However, the LR is an aerosol-type dependent parameter. Hence, the elastic approach must be carefully applied to realistic aerosol scenarios with mixture of aerosols. Usually, LR is determined by elastic inversion in an iterative procedure by constraining the extinction profile to the columnar-integrated Aerosol Optical Depth (AOD) [1, 2], as obtained from ancillary instrumentation.

Several aerosol-type LR climatologies have been reported to date, with a special emphasis devoted to dust [i.e., 3-9, among others]. In particular, a single LR value is assumed by CALIOP (the current space-borne elastic lidar) aboard CALIPSO (www-calipso.larc.nasa.gov) for the retrieval of extinction products; although, the CALIPSO aerosol scheme differentiates between two dust categories: 'pure dust' and 'polluted dust' with associated LR values of 40 sr and 65 sr, respectively [10]. However, in the particular case of 'pure dust' aerosols, the LR value is under discussion. Instead, different lidar-derived values are proposed depending on the dust source region, either Sahara desert [i.e., 7] or Arabian sources [i.e., 11]. In addition, based on AERONET (AErosol Robotic NETwork, aeronet.gfsc.nasa.gov) retrievals [6], mean LR values of 55 sr and 43 sr are obtained for dust particles, respectively, from Sahara and Middle East (ME) areas.

The aim of this work is to contribute to a deeper understanding of dust LR variance globally, by

DOI: 10.1051/epjconf/201611923017

proposing a more realistic LR value for pure dust particles coming from the Sahara desert region. Our study is focused on the detection of Saharan dust over the Izaña Observatory (Tenerife, Canary Islands). Dusty profiles observed from June 2007 to November 2009 with synergetic lidar/sun-photometry measurements are analyzed.

2. METHODOLOGY

2.1 Observational area

Measurements were performed at two aerosol stations that are frequently influenced by Saharan dust intrusions during spring and summer months. They are located in the Tenerife Island (Canary Islands) at two distant altitudes:

- the Santa Cruz de Tenerife Observatory (SCO, 28.5°N 16.2°W, 52 m a.s.l.), a subtropical coastal site, belonging to several aerosol networks: NASA/AERONET and NASA/MPLNET (mplnet.gsfc.nasa.gov); and
- the Izaña Observatory (IZO, 28.3°N 16.5°W), sited in a pristine free-tropospheric environment at 2367 m a.s.l., which is a NASA/AERONET station, as well.

Both sites are managed by the Spanish Meteorological Agency (AEMET).

2.2 Measurements and instrumentation

Simultaneous measurements of both range-resolved lidar profiling and passive columnar-integrated sun-photometry observations were performed, respectively, by using a MPLNET Micropulse Lidar v.3 (MPL-3) located at SCO, and an AERONET Cimel sun-photometer (C-318) at IZO.

The MPL-3 is an elastic lidar, operated by the Instituto Nacional de Técnica Aeroespacial (INTA), with the following primary features: a highly-pulsed (2500 Hz) and low-energy (10 μJ, max.) laser at 523 nm, operational in full-time continuous mode (24 h day^{-1}/365 days $year^{-1}$), at 75-m vertical resolution and 1-minute integrating time. MPL instrumental details are discussed further [12]. The AERONET C-318 sun-photometer is operated by AEMET, and Level 1.5 (cloud screened) and 2.0 (quality-assured) inversion retrieval products used, including the spectrally-resolved Aerosol Optical Depth at 500 nm (AOD) and Ångström Exponent (AEx).

2.3 LR retrieval for pure dust

The inversion procedure used for retrieving LR at IZO station is shown in more detail in [13]. In short, the elastic inversion is applied to MPL-3 measurements using the IZO/AOD constraint for iterative method convergence, retrieving the LR for pure dust particles. Given the conditions experienced at IZO, being above the marine boundary layer, aerosol intrusions are mostly dominated by Saharan dust particles.

MPL-3 profiles and sun-photometry data are hourly-averaged. LR inversions are performed using AERONET sun-photometer AOD constraint for total signal transmission. Therefore, only day-time lidar profiles are examined.

3. RESULTS

3.1 Selection of the dusty cases

Dusty conditions are defined, as adopted from [14], based on two points: 1) AERONET AOD/AEx data reflect coarse particle presence including high-moderate AOD at 500 nm (AOD > 0.2) and low AEx for the 440/675 nm wavelength pair (AEx < 0.5) over > 40% of the day-time period, and 2) HYSPLIT backtrajectory air masses analysis indicating a likelihood for Saharan origin (North-Africa latitudes > 20°N). Five-day back-trajectories calculated at four selected altitudes (a.g.l.): 500 m (surface), 1500 m (max. BL top), 3000 m (FT heights) and 5000 m (above IZO station) are examined.

Dusty cases are selected by using these AOD and AEx data at the SCO site and confirming the Saharan origin of air masses. AERONET data from the La Laguna site (also located in Tenerife Island, ~ 8 km from SCO) were used instead from 08 May to 14 August 2007 due to the unavailability of AERONET SCO data during this period. Table 1 shows the seasonal distribution of the arrival of Saharan dust intrusions over Tenerife area from June 2007 to November 2009.

Among 189 dusty days identified using the Saharan dust criterion, 109 coincident with MPL-3 observations were analyzed. Their seasonal distribution shows that Saharan dust intrusions

DOI: 10.1051/epjconf/201611923017

mostly arrive during summer-time (51.7% of total) (see Table 1).

Table 1. Statistical analysis of the Saharan dust intrusions over Tenerife for the June 2007 – November 2009 period.

Season	Summer	Autumn	Winter	Spring
Seasonal periods	JJA 2007-2009	SON 2007-2009	DJF 2007/08-2008/09	MAM 2008-2009
Dusty days: total / MPL	105 / 72	33 / 16	16 / 4	35 / 17
Frequency (%)	51.7	17.7	8.8	21.8
Nº profiles analyzed	555	92	13	117

3.2. Evaluation of the LR derived for Saharan pure dust particles

LR derived for the Saharan dust sample are assessed seasonally. A total of 777 dusty lidar profiles were examined, divided into 117, 555, 92 and 13 profiles for spring (MAM), summer (JJA), autumn (SON) and winter (DJF) times, respectively. In order to evaluate the seasonal LR distribution, retrieved values were detached for each season time into five LR ranges: 30-35 sr, 35-45 sr, 45-55 sr, 55-65 sr and 65-70 sr. In addition, an extra LR range is included (70-75 sr) for considering a potential contamination of fine-mode biomass burning particles. Results of the seasonal LR distribution obtained for the dusty sample are illustrated in Figure 1.

71.4% of dusty cases correspond to summertime (JJA), less than 20% to spring and autumn, and less than 2% to winter. Results indicate that predominance of a given LR range is not observed. LR are mostly in the ranges of 55-65 sr (15.6%) and 70-75 sr (19.0%) in summertime. Meanwhile, similar values within all the LR ranges are observed for all other periods. The mean LR obtained for each season is 57.4 ± 14.6 sr, 56.6 ± 14.7 sr, 56.6 ± 13.2 sr and 64.9 ± 7.2 sr, respectively, for MAM, JJA, SON and DJF.

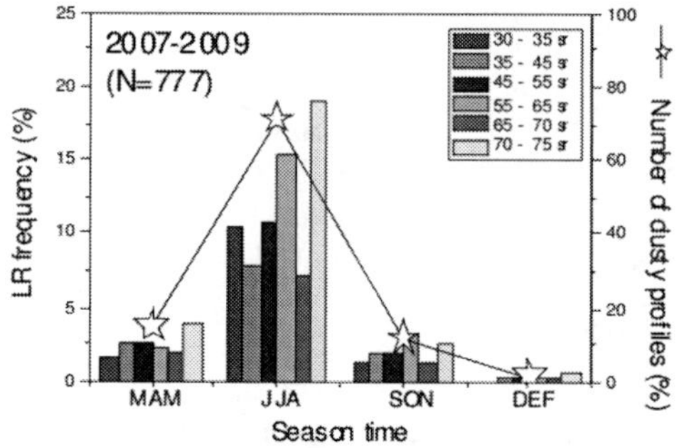

Fig. 1. (Left axis) Seasonal frequency of the retrieved LR for Saharan pure dust particles, as detached in several LR ranges (color bars). (Right axis) Number of dusty profiles examined (in %, star symbols).

In comparison with LR reported for Saharan dust in the literature, those derived in this work are mostly similar. For instance, our results match (except for wintertime) those AERONET-derived mean LR (55 sr) over Sahara [6]. They are closer to that LR (58 sr) proposed by EARLINET (European Aerosol Lidar NETwork, www.earlinet.org), on the basis of multi-year lidar measurements of Saharan dust intrusions over Europe [7]. However, they are higher than that LR of 40 sr assumed by CALIPSO for 'pure dust' aerosols in Level 2 extinction retrievals [10]. Higher LR observed in wintertime likely reflect contamination of fine-mode biomass burning particles coming from Sahel region, where seasonal deforestation is common [15].

4. CONCLUSIONS

Collocated MPL-3 lidar profiling and AERONET sun-photometry data were collected in Tenerife, a Saharan dust-influenced region, from June 2007 to November 2009 in order to retrieve the LR for pure dust particles coming from Saharan sources.

Those obtained LR have been analyzed depending on season. No differences are observed in average for spring, summer and autumn times. A relatively constant seasonally-averaged mean LR value of 57 sr is found. Sample contamination from fine-mode biomass burning particles coming from the Sahel area during wintertime deforestation fires period are likely influencing higher LR retrieved during wintertime. Outside of this latter period,

DOI: 10.1051/epjconf/201611923017

our findings are consistent with recent reports of dust LR.

In general, results obtained from Tenerife/IZO measurements during Saharan dust intrusion occurrence deepens our understanding of LR values and seasonal variance so as to be used in elastic lidar-derived AOD inversions for dusty conditions. Consequently, they can be also addressed to improve the AOD retrievals from CALIPSO measurements over Saharan areas.

ACKNOWLEDGEMENTS

This work has been supported by the Spanish Ministerio de Economía y Competitividad (MINECO) under grant CGL2011-24891 (AMISOC project). We acknowledge NOAA Air Resources Laboratory (ARL) for the provision of the HYSPLIT model and ECMWF for the input meteorological files. Authors also thank the staff of both AEMET and the University of La Laguna stations for providing maintenance and support of all the instrumentation involved in this work.

REFERENCES

[1] Fernald, 1984: Analysis of atmospheric lidar observations: some comments, *Appl. Optics,* **23**, 652–653.

[2] Klett, 1985: Stable analytic inversion solution for processing lidar returns, *Appl. Optics,* **20**, 211–220.

[3] Cattrall et al., 2005: Variability of aerosol and spectral lidar and backscatter and extinction ratios of key aerosol types derived from selected Aerosol Robotic Network locations, *J. Geophys. Res.,* **110**, D10S11, doi:10.1029/2004JD005124.

[4] Mona et al., 2006: Saharan dust intrusions in the Mediterranean area: Three years of Raman lidar measurements, *J. Geophys. Res.,* **111**, D16203, doi:10.1029/2005JD006569.

[5] Müller et al., 2007: Aerosol-type-dependent lidar ratios observed with Raman lidar, *J. Geophys. Res.,* **112**, D16202, doi:10.1029/2006JD008292.

[6] Schuster et al., 2012: Comparison of CALIPSO aerosol optical depth retrievals to AERONET measurements, and a climatology for the lidar ratio of dust, *Atmos. Chem. Phys.,* **12**, 7431–7452, doi:10.5194/acp-12-7431-2012.

[7] Amiridis et al., 2013: Optimizing CALIPSO Saharan dust retrievals, *Atmos. Chem. Phys.,* **13**, 12089-12106, doi:10.5194/acp-13-12089-2013.

[8] Omar et al., 2013: CALIOP and AERONET aerosol optical depth comparisons: One size fits none, *J. Geophys. Res. Atmos.,* **118**, 1–19, doi:10.1002/jgrd.50330.

[9] Tesche et al., 2013: Ground-based validation of CALIPSO observations of dust and smoke in the Cape Verde region, *J. Geophys. Res. Atmos.,* **118**, 2889–2902, doi:10.1002/jgrd.50248.

[10] Omar et al., 2009: The CALIPSO Automated Aerosol Classification and Lidar Ratio Selection Algorithm, *J. Atmos. Ocean. Tech.,* **26**, 1994-2014, doi:10.1175/2009JTECHA1231.1.

[11] Mamouri et al., 2013: Low Arabian dust extinction-to-backscatter ratio, *Geophys. Res. Lett.,* **40**, 4762–4766, doi:10.1002/grl.50898.

[12] Campbell et al., 2002: Full-time, eye-safe cloud and aerosol lidar observation at Atmospheric Radiation Measurement Program sites: Instruments and data processing, *J. Atmos. Oceanic Technol.,* **19**, 431– 442, 2002.

[13] Córdoba-Jabonero et al., 2014: Lidar Ratio discrimination retrieval in a two-layer aerosol system from elastic lidar measurements in synergy with sun-photometry data, *Proc. International Conference in Atmospheric Dust (DUST 2014), ProScience,* **1**, 243-248, published by Digilabs, doi:10.14644/dust.2014.040.

[14] Córdoba-Jabonero et al., 2011: Synergetic monitoring of Saharan dust plumes and potential impact on surface: a case study of dust transport from Canary Islands to Iberian Peninsula, *Atmos. Chem. Phys.,* **11**, 3067-3091, doi:10.5194/acp-11-3067-2011.

[15] Liousse et al., 2010: Updated African biomass burning emission inventories in the framework of the AMMA-IDAF program, with an evaluation of combustion aerosols, *Atmos. Chem. Phys.,* **10**, 9631-9646, doi:10.5194/acp-10-9631-2010.

AEROSOL PROPERTIES OVER SOUTHEASTERN CHINA FROM MULTI-WAVELENGTH RAMAN AND DEPOLARIZATION LIDAR MEASUREMENTS

Birgit Heese[1]*, Dietrich Althausen[1], Holger Baars[1], Stephanie Bohlmann[1], Ruru Deng[2]

*[1]Leibniz Institute for Tropospheric Research (TROPOS), Permoserstr. 15, 04318 Leipzig, Germany, *Email: heese@tropos.de*

[2]School of Geography and Planning, Sun Yat-sen University, Guangzhou, China

ABSTRACT

A dataset of particle optical properties of highly polluted urban aerosol over the Pearl River Delta, Guangzhou, China is presented. The data were derived from multi-wavelengths Raman and depolarization lidar Polly^{XT} and AERONET sun photometer measurements. The measurement campaign was conducted from Nov 2011 to June 2012. High aerosol optical depth was observed in the polluted atmosphere over this megacity, with a mean value of 0.54 ± 0.33 and a peak value of even 1.9. For the particle characterization the lidar ratio and the linear particle depolarization ratio, both at 532 nm, were used. The mean values of these properties are 48.0 sr ± 10.7 sr for the lidar ratio and 4%+-4% for the particle depolarization ratio, which means most depolarization measurements stayed below 10%. So far, most of these results indicate urban pollution particles mixed with particles arisen from biomass and industrial burning.

1. INTRODUCTION

The Southeast of China, i.e. the Pearl River Delta (PRD) is one of the largest urbanized areas in the world. High population density and a very strong economy leads to high aerosol load in the whole area around Guangzhou (PRD). The consequences for the geographical development, people's health, and atmospheric pollution were studied in the frame of the German project "Megacities – Megachallenges". The atmosphere over the PRD is influenced by high urban and industrial activity but is also affected by the vicinity of the sea. Hence the predominant atmospheric particles expected in this area may be a mixture of different aerosol types like urban haze, pollution, dust, and sea-salt particles.

The atmospheric conditions of the PRD are controlled by subtropical to tropical climate with an annual average temperature of about 22 °C. The weather is influenced by the Asian monsoon circulation. The main wind direction turns with the change of the location of the Inter Tropical Convergence Zone from north and north-easterly wind directions during the winter months to southern directions during the summer months.

Several studies concerning the vertical aerosol structure in PRD have been performed. First lidar measurements in the PRD by TROPOS were carried out during a one-month intensive field campaign in Xinken in 2004. The first prototype of the mobile single-wavelengths Raman lidar Polly with two detection channels was used. The observed aerosol was mainly dominated by local pollution [1].

As part of the Megacities project, a first short field campaign took place in November and December 2009. During this time, two events of moderate and hazy pollution were observed by the Raman lidar of the Anhui Institute of Optics and Fine Mechanics, CAS, Hefei, China Hefei Optics Institute and a sun photometer from TROPOS [2].

For the investigation of the different seasonal aerosol conditions a longer lasting field campaign was necessary. For this reason, a second field campaign was performed in the PRD in 2011/12. At that time, the multi-wavelength Raman lidar Polly^{XT} [3] with depolarization capabilities was used for the characterization of the aerosol types over the PRD. In the next chapter the deployed instruments and their products are described. The results are presented in chapter 3 and the conclusions given in chapter 4.

DOI: 10.1051/epjconf/201611923018

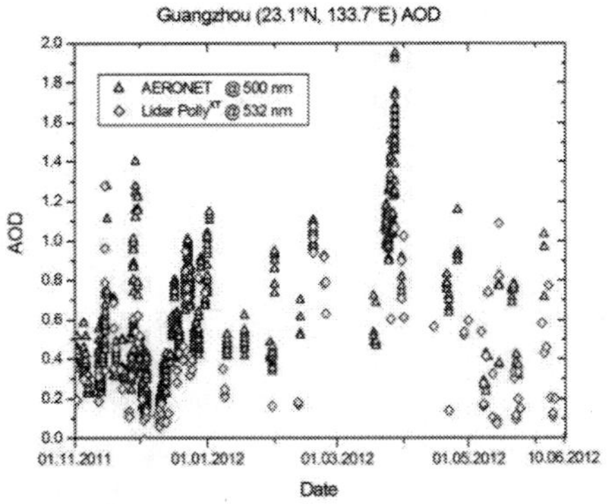

Figure 1: AOD derived from Polly^{XT} Raman extinction measurements at 532 nm and AERONET level 2 AOD derived from sun photometer measurements at 500 nm. Lidar particle extinction profiles were extrapolated as shown in Fig 2.

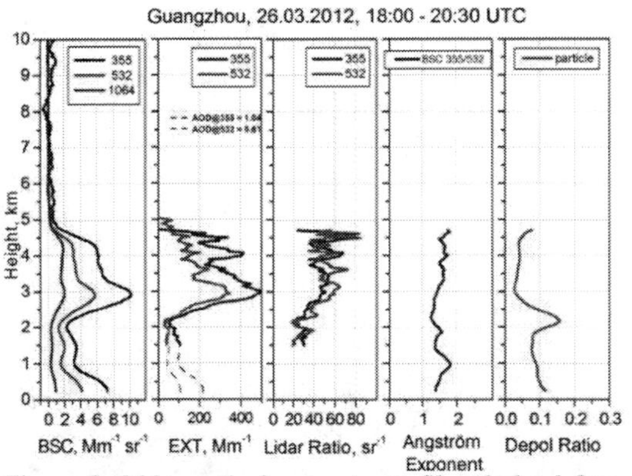

Figure 2: Lidar optical property profiles derived from Polly^{XT} measurement on March 26, 2012 (local time is +8 h, 02:00 to 04:30 LT)). Vertical smoothing was 450 m for the profiles shown here.

2. METHODOLOGY

Together with the multi-wavelength Raman lidar Polly^{XT} of TROPOS a dual-polar sun photometer of type CE-318dp was deployed on the roof top of a building on the East-Campus of the Sun Yat-sen University of Guangzhou. The measurements started in the end of October 2011 and lasted until the beginning of July 2012. The optical properties measured by the lidar are the particle backscatter coefficient at 355 nm, 532 nm and 1064 nm, the particle extinction coefficient at 355 nm and 532 nm, and the linear depolarization ratio at 532 nm. For the deter-mination of the particle backscatter coefficient and particle extinction coefficient the Raman method has been applied. Further derived optical properties are the particle extinction-to-backscatter-coefficient ratio or lidar ratio, the linear particle depolarization ratio, the AOD, and the respective backscatter and extinction related Ångström exponents.

The sun-photometer measures the direct and indirect sun radiation at eight wavelength from 340 nm up to 1640 nm. In addition, with this dual-polar instrument, the radiances at three different depolarization directions are measured by means of a second filter wheel. The data presented here are mainly the AOD derived from the direct sun measurements.

3. RESULTS

An overview over the typical aerosol conditions during the observation period is given by the aerosol optical depth (AOD). Fig. 1 shows AERONET level 2 AOD values at 500 nm measured by the sun photometer and the AOD values derived from Polly^{XT} Raman extinction profiles. During November and December 2011, sun photometer AOD values are available on 17 and 22 days, respectively. After this period, a lot of rain and cloudy weather was present over Guangzhou. Thus, in 2012, level 2 sun photometer data are only available on 32 days in total. However, these data still provide an overview over the development of the aerosol load throughout the observation period. In the beginning, in November and December 2011, most AOD values range between 0.2 and 0.6 with some peak AOD periods with high values up to 1.4.

The monthly mean AOD are still moderate with values of AOD=0.45/0.46 (SPM/Lidar) in November and AOD=0.49/0.33 in December. In January (AOD=0.57/0.24) and in February (AOD=0.93/0.68) the mean values are larger but only a few observations were available. Unusual heavy rainfall was observed which may be triggered by a La Niña event during that winter. In March, a period with very high AOD values was observed. The maximum value of 1.95 was

DOI: 10.1051/epjconf/201611923018

measured on March 28, 2012, so that the highest monthly mean AOD of 1.16 was reached.

Towards the summer season, the mean AOD decreases to 0.82/0.81 in April and 0.49/0.35 in May. In the beginning of June the mean AOD value was 0.32/0.37. For the whole observation period, the mean AOD value was 0.54±0.33, which indicates a generally high mean background level of aerosol in the atmosphere above the PRD. For comparison, in Leipzig, Germany, a continental, central European site, the yearly mean AOD at 500 nm lies around 0.2.

The mean AOD for the whole observation period derived from the lidar measurements is 0.43±0.30. These AOD values were calculated by integrating the particle extinction profiles from the uppermost level of the observed aerosol layers to the overlap height. The overlap height mostly lies around 1 km. Below this height, the remaining profile was extrapolated by fitting the particle backscatter profiles to the particle extinction profile. This is illustrated in the second graph of Fig. 2. Finally, the last few hundred meters to the ground were extrapolated using the lowermost profile value. This procedure obviously underestimates the AOD by about 12%, so that the lidar AOD values often are below the AERONET values, as can be seen in Figure 1.

Although the AOD measurements by the sun photometer already show seasonal differences in the aerosol, these measurements are still column integrated. Thus, a detailed analysis was performed using the vertical information obtained by the lidar measurements.

One example of the obtained lidar profiles is presented in Fig. 2. It shows the profiles of the particle backscatter coefficient, the particle extinction coefficient, and the resulting lidar ratio, the backscatter related Ångström exponent, and the linear particle depolarization ratio from a measurements on March 26, 2012 from 18:00 to 20:30 UTC. Here, a lofted layer at altitude ranges between 2 and 5 km with particle properties that differ from those in the planetary boundary layer (PBL) was observed. This is an example with very high aerosol content. The AOD on this day was among the highest values during the observation period: 1.04 at 355 nm and 0.61 at 532 nm. However, this type of particle layer structure was

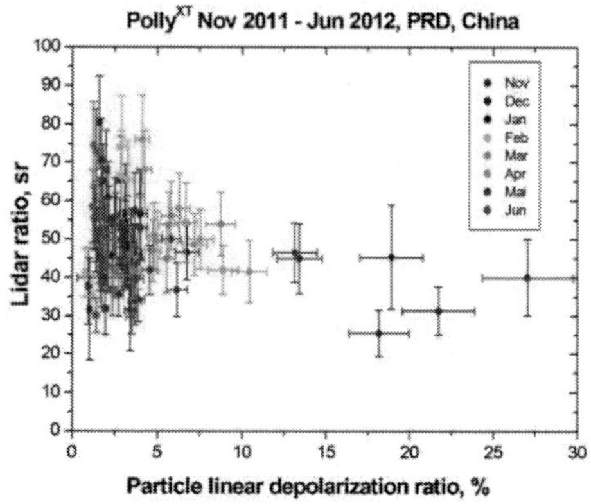

Figure 3: Lidar ratio versus linear particle depolarization ratio 532 nm for elevated layers period November 2011 to June 2012.

observed often above the measurements site. The upper boundary of these upper layers lies between 2.5 km and up to even 5 km, as in this example. In most cases, the PBL was observed below 1.5 km, as on this day. Since no overlap correction could be performed, the PBL is not captured by the Raman particle extinction profiles.

Therefore, the PBL is not included in the following discussion and the focus is laid on the lofted layers.

For all observed lofted layers above 1.5 km the lidar ratio has been calculated and has been plotted against the linear particle depolarization ratio for 532 nm (Fig. 3). For each data point the lidar measurements were integrated over at least 2 hours and vertically smoothed over a window of 750 m. Each month has been coded by a color for easier identification. The lidar ratio is derived from Raman retrievals of the particle backscatter and extinction coefficient profiles. The linear depolarization ratio was calculated using the 90-degree calibration method described in detail in [4]. Fig. 3 shows that most data points range between a lidar ratio of 30 sr and 80 sr. The linear depolarization ratio lies in 90% of all cases below 10% and 80% even below 5%. Only a few cases of high linear depolarization ratio > 20 % were observed. This was on December 2 and 3, 2011 during the only dust advection event [5]. The monthly mean values for the lidar ratios range from 43.9 sr to 53.2 sr with standard deviations of 5 to 10 sr and even 16.7 sr in one case. The monthly mean linear depolarization ratio ranges from 1.9% to 8.7%. The mean lidar ratio value for the whole observation period is 48.0 ± 10.7 sr. Here, the standard deviation represents the atmospheric variability. These lidar ratios indicate a wide variety of particles that were observed. The total mean value linear particle depolarization ratio is 4 % ± 4%, indicating mainly particles with low depolarization properties such as urban pollution particles arising from traffic and fuel and other burning processes.

4. CONCLUSIONS

A data set obtained by a multi-wavelengths Raman and depolarization lidar and dual-polar sun photometer taken in Guangzhou, PRD, China, has been presented. In this study, the upper layers that were frequently observed have been studied. The derived particle optical properties AOD, lidar ratio, and linear particle depolarization ratio allows the conclusion, that mostly urban, industrial, and possibly biomass burning pollution is frequently present in the atmosphere over the PRD. Further analysis will be done to investigate different advection regimes and particle properties inside the lower PBL.

ACKNOWLEDGEMENT

The measurements were conducted in the frame of the special priority program “Megacities-Megachallenge - Informal Dynamics of Global Change” (SPP 1233) funded by the German Research Foundation (DFG).

We thank our colleagues from the Sun Yat-sen University, Guangzhou, for their support and help throughout the measurement campaign.

REFERENCES

[1] Ansmann A., R. Engelmann, D. Althausen, U. Wandinger, M. Hu, Y. Zhang, and Q. He, 2005: High aerosol load over the Pearl River Delta, China, observed with Raman lidar and Sun photometer, *Geophys. Res. Lett.* **32**, L13815, doi:10.1029/2005GL023094.

[2] Chen, Z., W. Liu, B. Heese, D. Althausen, H. Baars, T. Cheng, X. Shu, and T. Zhang, 2014, Aerosol optical properties observed by combined Raman-elastic backscatter lidar in winter 2009 in Pearl River Delta, south China, *J. Geophys. Res.* **119**, 2496–2510, doi:10.1002/2013JD020200.

[3] Althausen, D., R. Engelmann, H. Baars, B. Heese, A. Ansmann, D. Müller, M. Komppula, 2009: Portable Raman Lidar Polly^{XT} for Automated Profiling of Aerosol Backscatter, Extinction, and Depolarization. *J. Atmos. Oceanic Technol.*, **47**, pp. 2366-2378.

[4] Freudenthaler, V., Esselborn, M., Wiegner, M., Heese, B., Tesche, M., Ansmann, A., Müller, D., Althausen, D., Wirth, M., Fix, A., Ehret, G., Knippertz, P., Toledano, C., Gasteiger, J., Garhammer, M., Seefeldner, M. 2009: Depolarization ratio profiling at several wavelengths in pure Saharan dust during SAMUM 2006. *Tellus*, **61B**, 165-179.

[5] Heese B., Althausen, D., Bauditz, M., Bao, R., Deng, R., Li, Z., 2012, Lidar depolarization and sun photometer polarization measurements for particle characterization over Guangzhou, China, Proceedings of the 26th ILRC, Porto Heli, Greece, 25-29 June, 505-509.

EPJ Web of Conferences **119**, 23019 (2016) DOI: 10.1051/epjconf/201611923019
ILRC 27

SENSITIVITY OF PARTICLE EXTINCTION AND BACKSCATTERING CALCULATION FROM MIE-RAMAN LIDAR MEASUREMENTS TO THE CHOICE OF ÅNGSTRÖM EXPONENT

Anastasia Suvorina[1], Igor Veselovskii[1], David N. Whiteman[2], Michael Korenskiy[1]

[1]Physics Instrumentation Center, Troitsk, Moscow Region, 142190, Russia, E-mail:
E-mail: igorv@pic.troitsk.ru
[2]Mesoscale Atmospheric Processes Laboratory, NASA Goddard Space Flight Center, Greenbelt, MD 20771, USA,
E-mail: david.n.whiteman@nasa.gov

ABSTRACT

Vibrational Raman scattering from nitrogen is commonly used in Mie-Raman lidars for evaluation of particle backscattering (β) and extinction (α) coefficients. However, vibrational scattering is characterized by significant frequency shift of the Raman component, so for the calculation of α and β the assumption about the extinction Ångström exponent is needed. Simulation results presented in this study demonstrate that ambiguity in the choice of this exponent can be the significant source of uncertainty in the calculation of backscattering coefficients when optically thick aerosol layers are considered. Examples of lidar measurements and optical data calculated for different values of Ångström exponent are given.

1. INTRODUCTION

The particle backscattering (β) and extinction (α) coefficients measured at multiple wavelengths with a Raman lidar can be inverted to particle microphysics. The system most often used for such measurements, due to its relative simplicity, is a Raman lidar based on a tripled Nd:YAG laser, which provides three backscattering (355 nm, 532 nm, 1064 nm) and two extinction (355 nm, 532 nm) coefficients. To get this so called 3β+2α data set the elastic backscatters are measured simultaneously with nitrogen vibrational Raman signal at 387 nm and 608 nm. However the vibrational Raman scattering is characterized by significant wavelength shift of the nitrogen Raman component with respect to the laser line. As a result, information about the spectral dependence of aerosol extinction is needed when α and β are computed, which can be an additional source of uncertainty. In our presentations we report the results of numerical simulations and analysis of experimental measurements aimed to estimate corresponding uncertainties.

2. NUMERICAL SIMULATIONS

The possibility of independent calculation of aerosol backscattering and extinction coefficients from simultaneously measured elastic and Raman backscatters was first demonstrated by Ansmann et al. [1]. Following their approach the backscattering and extinction coefficients at height z can be calculated as

$$\alpha_L^a(z)=\frac{\frac{d}{dz}\ln\left(\frac{N_R(z)}{P_R(z)\cdot z^2}\right)-\alpha_L^m(z)-\alpha_R^m(z)}{1+\left(\frac{\lambda_R}{\lambda_L}\right)^A} \quad (1)$$

$$\beta_L^a(z)=-\beta_L^m(z)+\beta_L^m(z_0)\frac{P_R(z_0)P_L(z)N_R(z)}{P_L(z_0)P_R(z)N_R(z_0)}\times$$
$$\times\exp\left\{\int_{z_0}^{z}\left[\alpha_L\left(1-\left(\frac{\lambda_L}{\lambda_R}\right)^A\right)+\left(\alpha_L^m-\alpha_R^m\right)\right]dz'\right\} \quad (2)$$

Where $P_{L,R}(z)$ is the power of elastic and Raman backscatters, $\lambda_{L,R}$ are corresponding wavelengths and $N_R(z)$ is the density of Raman scatters. The reference height z_0 is usually chosen in the free troposphere where $\beta_L^a(z_0) << \beta_L^m(z_0)$. Indices "a" and "m" here correspond to aerosol and molecular components of backscattering and extinction coefficients.

To account for spectral dependence of α^a the extinction Ångström exponent (EAE) *A* is used:

$$\frac{\alpha^a(\lambda_L)}{\alpha^a(\lambda_R)}=\left(\frac{\lambda_R}{\lambda_L}\right)^A \quad (3)$$

The value of EAE is unknown, besides it can be height and spectrally dependent, however in computations usually the fixed value of *A* is assumed. Thus the ambiguity in choice of Ångström exponent

DOI: 10.1051/epjconf/201611923019

is an additional source of uncertainties in β^a and α^a calculations.

As it follows from (1) the uncertainty in computation of aerosol extinction ε_α due to the choice of EAE depends on the ratio $\left(\frac{\lambda_R}{\lambda_L}\right)^A$ and for constant A this uncertainty is range independent. The effect of the EAE on backscattering coefficient is described by the exponent term in (2), which for height independent A can be written as

$$\exp\left\{\int_{z_0}^{z}\alpha_L(z')\left[1-\left(\frac{\lambda_L}{\lambda_R}\right)^A\right]dz'\right\}=\exp\left\{D(z)\left[1-\left(\frac{\lambda_L}{\lambda_R}\right)^A\right]\right\} \qquad (4)$$

Where $D(z)=\int_{z_0}^{z}\alpha_L(z')dz'$ is the aerosol optical depth for height interval [z_0, z]. Thus the uncertainties ε_β of backscattering coefficient calculation are integrated over distance z_0-z and are the most significant at low heights. If ΔA is deviation of EAE from "true" value, ε_β can be calculated as

$$\varepsilon_\beta=\frac{\beta^a(A+\Delta A)-\beta^a(A)}{\beta^a(A)}=\frac{\beta^T(A+\Delta A)-\beta^T(A)}{\beta^T(A)-\beta^m}= \qquad (5)$$
$$=\frac{R}{R-1}\left(\exp\left\{D\left(\frac{\lambda_L}{\lambda_R}\right)^A\left(1-\left(\frac{\lambda_L}{\lambda_R}\right)^{\Delta A}\right)\right\}-1\right)$$

Here $\beta^T=\beta^a+\beta^m$ and R is the scattering ratio $\frac{\beta^T}{\beta^m}$.

For high aerosol loading, when $\beta_L^a>>\beta_L^m$ the ratio $\frac{R}{R-1}\approx 1$ and ε_β rises quickly with increasing optical depth D and ΔA. However in the UV and visible spectral ranges the magnitude of molecular and aerosol backscattering outside of clouds can be comparable, so the influence of molecular scattering on ε_β must be considered.

For estimating the uncertainty ε_β for different laser wavelengths and aerosol loadings, numerical simulation was performed. The particle bimodal size distribution was represented as:

$$\frac{dn(r)}{d\ln(r)}=\sum_{i=f,c}\frac{N_i}{(2\pi)^{1/2}\cdot\ln\sigma_i}\cdot\exp\left[\frac{(\ln r-\ln r_{0i})^2}{2(\ln\sigma_i)^2}\right] \qquad (6)$$

Here N_i is the total particle number of the i-th mode, r_i describes the mode radius and $\ln\sigma_i$ is the dispersion. The index $i = f, c$ corresponds to the fine and coarse mode, respectively. For all computations the values $r_{0f}=0.1\mu m$, $r_{0c}=1.2\mu m$ and $\ln\sigma_{f,c}=0.4$ were used and the complex refractive index (CRI) was assumed to be the same for both modes. The concentration of particles in the find mode N_f was kept constant through all simulations, while the concentration in the coarse mode varied. The spectral dependence of the EAE is influenced by the relative contributions of the fine and the coarse modes, so to evaluate this effect three types of particle size distribution (PSD) were considered. For the first type (T1) with $\frac{N_f}{N_c}=10^4$ the main particle volume is attributed to the fine mode. For the second type (T2) $\frac{N_f}{N_c}=10^3$ the total volume of particles in each mode is comparable. And finally, in the distribution T3 with $\frac{N_f}{N_c}=10^2$ the coarse mode is predominant.

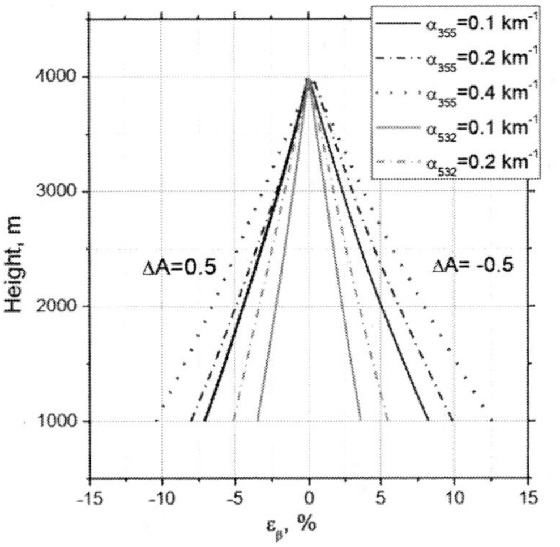

Fig.1. Profiles of backscattering coefficient uncertainties at 355 nm (blue) and 532 nm (green) for extinction coefficient 0.1 km^{-1} (solid), 0.2 km^{-1} (dash-dot), 0.4 km^{-1} (dot). Results are given for $\Delta A=\pm 0.5$.

A Raman lidar based on a tripled Nd:YAG laser can measure the EAE at 355-532 nm wavelengths ($A_{355\text{-}532}$). However, when nitrogen vibrational Raman backscatters at 387 nm and 608 nm are used to get backscattering coefficients at 355 nm and 532 nm, the Ångström exponents $A_{355-387}$ and $A_{532-608}$ are needed and these may differ from $A_{355-532}$. Table 1 shows the Ångström exponents $A_{355-532}$, $A_{355-387}$, $A_{532-608}$ for the size distributions T1-T3 and refractive indices m=1.35-i0.005; 1.45-i0.005; 1.6-i0.005; 1.6-i0.05. These values of CRI correspond to variety of particles, such as water aerosol, biomass burning products and desert dust (Dubovik et al., 2002). The spectral dependence of the EAE is the most significant for particles with a fine mode predominance. For the T1 distribution, the uncertainty in the choice of the Ångström exponent ΔA may exceed 0.5 when $A_{355\text{-}532}$ is used for the computation of both β_{355} and β_{532}. For particles with a coarse mode predominance (T3), the spectral dependence of the Ångström exponent is weaker and $\Delta A < 0.2$ when $A_{355\text{-}532}$ is used.

Table 1. Extinction Ångström exponents for different size distributions and different refractive indices.

Aerosol type	Refractive index	Ångström exponent		
		$A_{355\text{-}532}$	$A_{355\text{-}387}$	$A_{532\text{-}608}$
T1	1.35-i0.005	1.62	1.39	1.98
	1.45-i0.005	1.32	1.03	1.79
	1.6-i0.005	0.87	0.52	1.47
	1.6-i0.05	0.78	0.46	1.30
T2	1.35-i0.005	1.40	1.24	1.58
	1.45-i0.005	1.15	0.92	1.46
	1.6-i0.005	0.81	0.48	1.33
	1.6-i0.05	0.72	0.43	1.18
T3	1.35-i0.005	0.58	0.59	0.48
	1.45-i0.005	0.426	0.41	0.59
	1.6-i0.005	0.44	0.28	0.64
	1.6-i0.05	0.39	0.24	0.57

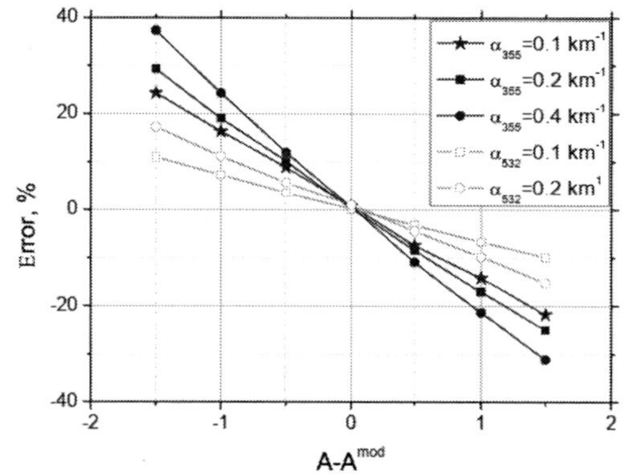

Fig.2. The errors of backscattering coefficient calculation at 355 nm (blue) and 532 nm (green) as a function of deviation from model value of Angstrom exponent.

To estimate how sensitive the backscattering coefficient is to the choice of the Ångström exponent, we considered an aerosol layer extending from the ground up to 4000 m with a height independent particle concentration. As mentioned, the spectral dependence of *A* has the most influence for PSD with a predominant fine mode, so the simulation results are presented for the T1 distribution. The uncertainty of the backscattering calculation depends on aerosol loading, so in the simulations different particle concentrations were considered in order to obtain extinction coefficients of α^a=0.1 km^{-1}, 0.2 km^{-1}, 0.4 km^{-1} at 355 nm and 532 nm. The refractive index used was m=1.45-i.005, and corresponding model values of $A_{355-387}$, $A_{532-608}$ and $A_{355-532}$ were taken from Table 1. The backscattering and extinction coefficients for the chosen size distribution were used to generate noise-free synthetic lidar signals. Then, the values of α and β were calculated using Ångström exponents *A* which differed from the model values A^{mod}. The systematic uncertainty for backscattering was determined as $\varepsilon_\beta = \frac{\beta^a(A)-\beta^a(A^{mod})}{\beta^a(A^{mod})}$. The systematic uncertainty in extinction was defined similarly.

Table 2. Uncertainties of extinction ε_α at 355 and 532 nm for different values of Ångström exponent *A* chosen in computations. Results are shown for m=1.45-i0.005.

A	355 nm		532 nm	
	A-A^{mod}	ε_α	A-A^{mod}	ε_α
0	-1	4.5%	-1.8	10.5%
0.5	-0.5	2%	-1.3	7.5%
1	0	0.15%	-0.8	4.5%
1.5	0.5	2%	-0.3	1.5%
2	1	4%	0.2	1.2%

The results of the simulations for particle extinction are summarized in Table 2, showing ε_α at 355 and 532 nm for different assumed values of *A* in range [0, 2.0]. As mentioned, the uncertainty of extinction ε_α is independent of range or aerosol loading and in our computations it doesn't exceed 2% for both wavelengths, when $A_{355\text{-}532}$ is used.

In contrast to extinction the uncertainties ε_β depend on range since the differential transmission between the two Raman wavelengths is involved and errors in this calculation accumulate with range. The height profiles of ε_β at 355 nm for the particle extinction coefficients 0.1 km^{-1}, 0.2 km^{-1} and 0.4 km^{-1} are given in Fig.1 for $\Delta A = A - A^{mod} = \pm 0.5$. For comparison, the same figure also shows results at 532 nm for α=0.1 km^{-1} and 0.2 km^{-1}. The uncertainty is negligible near the reference height (z_0=4000 m) and it rises with increasing z-z_0. At z=1000 m the uncertainties can be as high as 8% and 10% at 355 nm for α=0.1 and 0.2 km^{-1} respectively, when ΔA=-0.5. For 532 nm these uncertainties are 4% and 6%, which is significantly lower than at 355 nm. An increase of extinction up to α=0.4 km^{-1} leads to a rapid increase in ε_β and the correct choice of the

Ångström exponent becomes more important. Fig.2 illustrates the dependence of $\varepsilon_{\beta}(\Delta A)$ at z=1000 m for different values of extinction coefficient. This dependence is close to linear so an increase of ΔA to ΔA =1 doubles the uncertainty, when compared with Fig.1.

3. ANALYSIS OF LIDAR MEASUREMENTS

The observations performed with multiwavelength GSFC Raman lidar in August 2013 [2] were used to analyze sensitivity of α and β calculation to the choice of the Ångström exponent. During the period considered a strong fire smoke event occurred and the aerosol layers extended up to 4000 m height. The Ångström exponent $A_{355\text{-}532}$ varied with height from 0.5 to 2.0, as shown in Fig.3. $A_{355\text{-}532}$ was computed for values of Ångström exponent in the range 0.5-2.0 assuming AEA to be height independent. The profile of $A_{355\text{-}532}$ demonstrates low sensitivity to the choice of EAE, while sensitivity of backscattering to AEA is high, as shown in Fig.4. At 1000 m height the deviation of β_{355} from mean value may achieve 10% depending on EAE chosen. For extinction the uncertainty is below 5%.

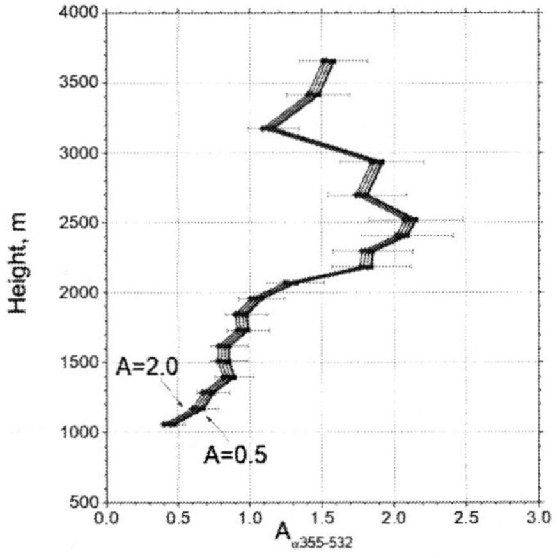

Fig.3. Vertical profile of $A_{355\text{-}532}$ computed for values of Ångström exponent in the range 0.5-2.0.

4. CONCLUSION

The simulations performed here demonstrate that the ambiguity in the choice of the Ångström exponent can be a source of significant uncertainties in backscattering calculation when the aerosol layers with high optical depth are profiled and that this effect is especially pronounced at 355 nm. In real measurements, the situation can be even more complicated because of the vertical variability of the Ångström exponent. These uncertainties can be eliminated through the use of rotational Raman scattering instead of vibrational one, thus decreasing the frequency shift of the inelastically scattered component. The first results of rotational Raman channel operation will be reported in accompanying presentation.

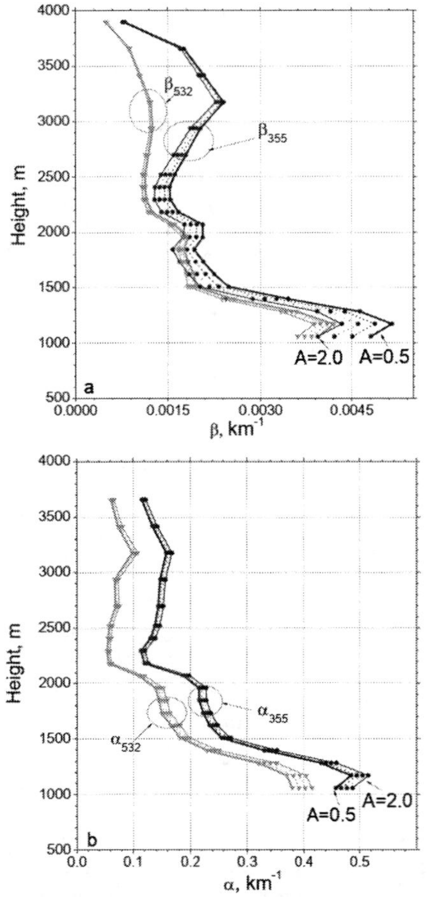

Fig.4. Vertical profiles of (a) backscattering and (b) extinction coefficient computed for values of Ångström exponent in the range 0.5-2.0.

REFERENCES

[1] Ansmann, et al.: Combined Raman elastic-backscatter lidar for vertical profiling of moisture, aerosols extinction, backscatter, and lidar ratio, Appl.Phys.B, *55*, 18-28, 1992.

[2] I. Veselovskii, et al. "Characterization of forest fire smoke event near Washington, D.C. in Summer 2013 with multi-wavelength lidar", Atmos. Chem. Phys. 15, 1647–1660, 2015.

EPJ Web of Conferences **119**, 23020 (2016) DOI: 10.1051/epjconf/201611923020
ILRC 27

GRADIENT CORRELATION METHOD FOR THE STABILIZATION OF INVERSION RESULTS OF AEROSOL MICROPHYSICAL PROPERTIES RETRIEVED FROM PROFILES OF OPTICAL DATA

Alexei Kolgotin[1], Detlef Müller[2,3], Anton Romanov[4], Eduard Chemyakin[2]

[(1)]Physics Instrumentation Center, Troitsk, Moscow Region, 142190, Russia, alexeift@yahoo.com
[(2)]Science Systems and Applications, Inc. (SSAI), NASA LaRC,
1 Enterprise Parkway, Hampton, VA, 23666, USA,
[(3)]University of Hertfordshire, Hatfield, Hertfordshire, AL10 9AB, UK, d.mueller@herts.ac.uk
[(4)]The National University of Science and Technology, Leninskii av. 4, Moscow, 119049

ABSTRACT

Correlation relationships between aerosol microphysical parameters and optical data are investigated. The results show that surface-area concentrations and extinction coefficients are linearly correlated with a correlation coefficient above 0.99 *for arbitrary particle size distribution*. The correlation relationships that we obtained can be used as constraints in our inversion of optical lidar data. Simulation studies demonstrate a significant stabilization of aerosol microphysical data products if we apply the gradient correlation method in our traditional regularization technique.

1. INTRODUCTION

Several approaches for the retrieval of particle microphysical parameters from multiwavelength lidar measurements were developed in the past years [1-3]. Since we deal with an ill-posed, ill-conditioned problem the derived solutions oscillate, they are unstable, and we may obtain non-physical results. Even the use of advanced mathematical methods such as regularization, e.g. Tikhonov's regularization does not guarantee that we find physically meaningful solutions unless we introduce additional constraints in the solution spaces. We showed that we can stabilize the solution space if we use whole sections of optical data profiles rather than optical data sets of individual height bins as input in our inversion scheme [4].

If the profiles of the optical data (OD) do not change, in other words the profile gradient is close to 0, we can expect that profiles of particle microphysical parameters (PMP) do not vary significantly either. On the contrary if the OD profiles show variations, i.e., the "gradient" profile differs from 0, we can expect that the PMP vary with height, too. In this contribution we investigated which correlations between lidar optical data and particle microphysical parameters exist and if these correlations can be used in our inversion methodology for improving the microphysical data products.

In section 2 we present the correlation relationships we found between PMP and OD. We describe how they can be used in our gradient correlation method. In section 3 we show a numerical example with synthetic OD. Retrieval results of different approaches are compared in section 4. Section 5 summarizes our results.

2. METHODOLOGY

Lidar measurements deliver the OD which can be used for the retrieval of bulk PMP (p) such as mean radius (r_{mean}) and effective radius (r_{eff}), mean width, i.e. geometrical standard deviation (σ) of a particle size distribution, number (n), surface-area (s) and volume (v) concentrations. This problem is related to solving the Fredholm integral equation of the 1st kind

$$\int_{r_{min}^{(l)}}^{r_{max}^{(l)}} K_g(\lambda, m^{(l)}, r) f^{(l)}(r) dr = g^{(l)}(\lambda), \qquad (1)$$

$$l=1,..,N_L,\ g=\alpha,\beta$$

The optical data $g^{(l)}(\lambda)$ are measured with lidar at wavelength λ. The unknown function $f^{(l)}(r)$ describes the particle size distribution (PSD), and r describe the particle radius on the domain $[r_{min}^{(l)}, r_{max}^{(l)}]$. Here we consider $r_{max}^{(l)}>0.05$ µm. The kernel functions $K_g(\lambda,m,r)$ can be computed on the basis of Mie-scattering theory in the case of spherical particle shape. The parameter $m^{(l)}=m_R^{(l)}-im_I^{(l)}$ describes the particle complex refractive index. Extinction coefficients are denoted as $g=\alpha$, backscatter coefficients are denoted as $g=\beta$. The superscript l indicates the number of the height bin we use to describe the OD profiles. The total number of height bins is N_L. The PMP can be easily estimated from the solution f.

Eq. (1) can be solved, for example, with regularization which allows us to find a solution space $F^{(l)}=\{f_1^{(l)}(r),\ldots, f_{Nsol}^{(l)}(r)\}$ at each height bin l. If we know $F^{(l)}$ we can also obtain the solution spaces for any bulk parameter $P^{(l)}=\{p_1^{(l)},\ldots, p_{Nsol}^{(l)}\}$ at each height bin l, where $P^{(l)}= R_{mean}^{(l)}$, $R_{eff}^{(l)}$, $\Sigma^{(l)}$, $N^{(l)}$, $S^{(l)}$, or $V^{(l)}$ denotes the solution space of r_{mean}, r_{eff}, σ, n, s, or v respectively [1]. The collection of the solution spaces for all N_L heights forms the profiles of the PMP. The final solution p_{av} at height l is defined by averaging all individual parameters $p_i^{(l)}$, $i \leq N_{sol}$, over some interval of the discrepancy range $[\rho_{min}^{(l)},\rho_{av}^{(l)}]$ [2,4]. In order to find physically meaningful solutions we need to apply different constraints, as for example, the discrepancy range $[\rho_{min}^{(l)},\rho_{av}^{(l)}]$, the radius range $[r_{min}^{(l)},r_{max}^{(l)}]$, and the complex refractive index $m^{(l)}$. In that way we can exclude individual solutions $p_i^{(l)}$ which for example deviate too much from the average value p_{av}. In that case we insert a threshold δp to take into consideration only those individual solutions that fulfill the condition

$$| p_i^{(l)} - p_{av} |< \delta p. \quad (2)$$

In our previous strategy of data inversion we used this averaging procedure for each solution space (in each height bin) independently, i.e., we did not take into consideration that results of successive height bins may be correlated to each other. The profile of the OD contains information regarding the variation of the profile of the PMP. In other words we can predict the PMP behavior if we know the law(s) (or correlations) that describes the interdependence of OD and PMP.

The interdependence that is generally accepted is that effective radius r_{eff} is inversely proportional to the extinction Ångström exponent (EAE) γ:

$$\gamma=\ln[\alpha(\lambda_1)/\alpha(\lambda_2)]/\ln(\lambda_2/\lambda_1). \quad (3)$$

For small variations $\delta\gamma$ we obtain

$$r_{eff} =a_r\gamma+b_r. \quad (4)$$

The constants a_r and b_r are regression coefficients (RC). The linear correlation (4) can be used as extra constraint if we want to find the solution space for $R_{eff}^{(l)}$. In fact if we measure $\gamma^{(l)}$ the correlation (4) allows us to estimate $r_{eff}^{(l)*}$ according to

$$r_{eff}^{(l)*} =a_r\gamma^{(l)}+b_r \quad (5)$$

If we introduce the threshold δr_{eff} we can rewrite condition (2) as

$$|r_{eff,i}^{(l)} - r_{eff}^{(l)*}|<\delta r_{eff} \quad (6)$$

If we approximate the kernel K_α and the PSD f by the parabolic and rectangle functions, respectively, one can show on the basis of the average theorem that there are additional linear correlations between extinction (in Mm^{-1}) and other PMP and combinations of several PMP (in Mm^{-1}) with correlation coefficients R^2 >0.99:

$$s=a_s\alpha(\lambda)+b_s, \quad (7)$$

$$v/r_{eff}= a_v\alpha(\lambda)+b_v=[a_s\alpha(\lambda)+b_s]/3, \quad (8)$$

$$nr_{eff}^2=a_n\alpha(\lambda)+b_n \quad (9)$$

The correlation (9) is valid only for a fixed standard deviation σ.

The linear correlations (7)-(9) and relationship (4) can be used as extra constraints to identify the solution spaces. We stress that any combination of correlations (4) and (8), (9), is acceptable, for example

$$v / (a_r\gamma+b_r) \approx [a_s\alpha(\lambda)+b_s]/3. \quad (10)$$

Again, introducing the threshold δp the condition (6) and the correlations (4), (8), (9) can be rewritten in a more general form as

$$|p_i^{(l)} - p^{(l)*}|<\delta p \quad p= r_{eff}, s, v/r_{eff}, nr_{eff}^2 \quad (11)$$

and

$$p^{(l)*}= a_p x^{(l)}+b_p \qquad x=\alpha,\gamma \quad (12)$$

The use of the constraints (11) in the averaging procedure [2,4] is at the heart of the gradient correlation method (GCM).

3. NUMERICAL SIMULATION

3.1. Synthetic optical data

We used synthetic optical data (SOD) to identify the correlation relationships. We generated the synthetic optical data $3\beta+2\alpha$ at 355, 532 and 1064 nm for backscatter coefficients and at 355 and 532 nm for extinction coefficients [1]. We took into account Eq. (1). We pre-defined values for f and m. We used the lognormal law for describing f. We selected a wide range of values for the mean radius, i.e. r_{mean}=20, 60, 100, 140, 180, 220, 260, and 300 nm. With regard to the mode width we used σ=1.5, 1.7, 1.9, 2.1, 2.3, and 2.5. We selected the following values for the real and imaginary parts: m_R=1.4, 1.5, 1.6, 1.7, and m_I =0, 1e-4, 1e-3, 2.5e-3, 5e-3, 7.5e-3, 1e-2, 1.5e-2, 2e-2, 2.5e-2, 3e-2, 3.5e-2, 4e-2, 4.5e-2, and 5e-2. We used all possible combinations (r_{mean}, σ, m_R, m_I) and in that way we created 2880 sets of $3\beta+2\alpha$ data for the SOD bank.

Fig. 1 presents the statistical analysis of the SOD bank. We see that r_{eff} is nearly inversely proportional to γ (Fig. 1a). The range of EAE varies from -0.5 to 4. For example, if γ > 2, which means that δr_{eff} <0.1 µm, we find for the RC a_r=-0.02 and b_r =0.1. The analysis of all entries of the SOD bank shows that these two parameters vary in the diapasons:

$$a_r\in[-0.02; -1.7], \qquad b_r\in[0.1; 1.5] \quad (13)$$

We find linear correlations of R^2=0.998 for s versus α(355) [see Fig. 1b]. In the following we denote IP as parameters we investigated with regard to useful correlation properties. If we use all values stored in the SOD bank we find for the RC:

a_s=1.7 and b_s=-0.08. A more thorough analysis of the SOD bank shows the following variations

$$a_s \in [1.33;\ 1.75], \qquad b_s \in [-0.05;\ 0.03] \quad (14)$$

These values depend on σ. The strong correlation guarantees that surface-area concentration can be directly estimated from the extinction with an uncertainty that is less than the measurement error.

Fig. 1b shows that v/r_{eff} and α are linearly correlated as well. It can be shown that Eq. (7) and (8) are true for any amount of modes in PSDs.

We also find a linear correlation of R^2=0.999 for IP nr_{eff}^2, but in that case σ needs to be kept constant (see Fig. 1c). Depending on $\sigma \in [1.5;\ 2.5]$ we find the following RCs:

$$a_n \in [0.17;\ 1.73], \qquad b_n \in [-0.05; 0.004] \quad (15)$$

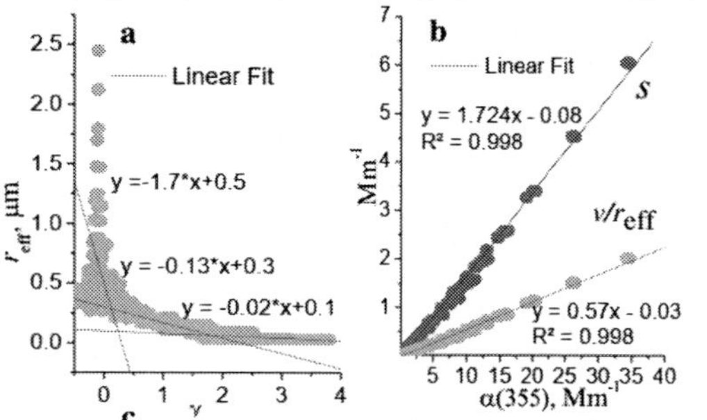

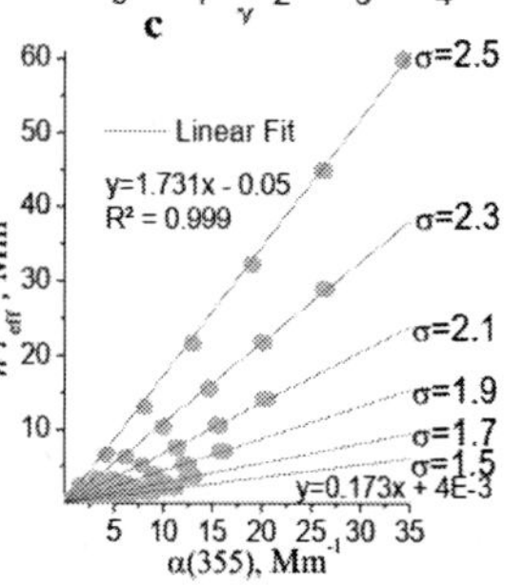

Fig. 1. Statistics (gray solid circle) for the parameters s (b), v/r_{eff} (b), and nr_{eff}^2 for different σ (c) versus α(355), for r_{eff} (a) versus γ. The solid lines describe the correlation trends according to $y=ax+b$ with correlation coefficient R^2.

3.2. Retrieval example for synthetic data

We selected N_L=6 OD sets of 3β+2α data from the SOD bank and we constructed OD profiles (see Fig. 2a). The true PMP are shown in Fig. 2b and 2c as thick lines. The real part does not change with height. It is equal to 1.5. The imaginary part increases with height, i.e. m_I=0, 0.01, 0.03, and 0.05 (not shown). In the inversion problem the true OD profile are distorted with 15% extreme error.

In the next step we identified the solution space $F^{(l)}$, l=1,…,6, with our traditional regularization technique [2]. Each solution space $F^{(l)}$ contains about 6×10^4 individual solutions which are retrieved for 100 inversion windows that slip inside the radius range from 0.03 to 8 μm. We used 20 equidistant values of the real part, i.e., $m_R \in [1.325;1.8]$ and 30 equidistant values of the imaginary part, i.e., $m_I \in [0.0;0.1]$.

Finally we post-processed the solution spaces with GCM when the RCs a_p are known (True RC) and when they are distorted up to 25% (Dist.RC). In that way we could test the stability of our novel approach. In addition we post-processed the solution spaces in the automated mode [5] without using GCM and without using any constraints of the parameters r_{eff} , n, $r \in [0.03;8]$, $m_R \in [1.325;1.8]$ and $m_I \in [0.0;0.1]$ (NoGCM).

All retrieval results are shown in Fig. 2b and Fig. 2c. We see that in the case of NoGCM (triangle) there are large retrieval errors of effective radius in 3 of the 6 height bins, and number concentration shows errors up to 200% in the height bin *l*#4. The solution can be significantly improved if we require that the correlation relationships (4), (9), and (10) are fulfilled. In the case of True CC (square) and Dist.CC (star) the retrieval errors of number concentration do not exceed 50 %, the error of effective radius is even less.

In order to obtain the GCM results we used the RCs shown in the table:

p	a_p	b_p	δp	a_p	b_p	δp
	True RC			Dist.RC		
r_{eff}	-1.0	0.76	0.25	-0.8	0.8	0.25
$v/(a_r\gamma+b_r)$	0.7	-0.05	0.1	0.56	-0.05	0.1
nr_{eff}^2	0.7	-0.04	0.1	0.5	-0.04	0.1

We present the statistics for all results derived with NoGCM (triangle), TrueRC (square) and DistRC (star) in Fig. 3. The statistics for the true data obtained from the SOD bank are plotted as circles in Fig. 3. Fig. 3 also shows the (actual) RCs, which we obtained after post-processing the solution space $F^{(l)}$, for all three cases together with RCs for the true data.

We see that the parameter s that is retrieved with NoGCM, TrueRC, and DistRC is linearly correlated with extinction (Fig. 3b). We obtain R^2=1 and a_s =1.7 which is equal to the true value. This is the reason why we do not use Eq. (7) in GCM (see table).

We find outliers in the profiles of r_{eff} and n if we use NoGCM (see Fig. 2). As a result the actual RCs a_r=1.82 (Fig. 3a, triangle) and a_n=2.37 (Fig. 3c, triangle) are out of their ranges (13) and (15), respectively.

If we implement in GCM the relationships (4) and (9) and we use the RCs from the table, we find for the actual RCs: a_r=1.14 (R^2=0.95) and a_n=0.68 (R^2=0.999) for TrueRC (Fig. 3a, 3c, square). If we use GCM with DistRC we obtain

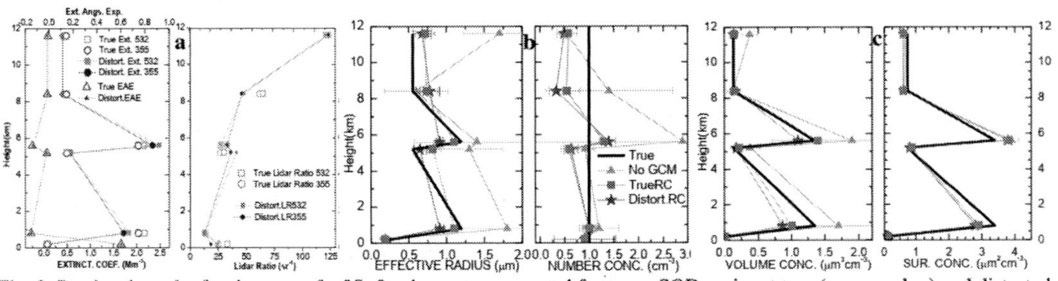

Fig. 2. Retrieval results for the case of a 3β+2α data set constructed from our SOD. **a**: input true (open marker) and distorted (close marker+line) EAE, extinctions, lidar ratios at 355 and 532 nm. **b**, **c**: true (thick line) and retrieved (line+marker) PMP. Retrieval results were obtained when GCM is not used (triangle), GCM is used with true RC (square), GCM is used with RC distorted up to 25%.

RCs a_r=0.93 (R^2=0.97), and a_n=0.49 (R^2=0.999) [see Fig. 3a, 3c, star]. Notably the actual and true RCs converge.

4. CONCLUSION

We presented the novel method of GCM for the stabilization of PMP retrieved from profiles of optical data taken with lidar. GCM uses correlation relationships between particle bulk parameters and measured optical information.

We found that the linear correlation between particle surface-area concentration and extinction coefficient is approximately $R^2 \approx 1$. The IP v/r_{eff} and extinction coefficient are linearly correlated with the same correlation coefficient. We find a linear correlation between the product nr_{eff}^2 and particle extinction. The correlation is $R^2 \approx 1$ for fixed mode width σ.

These relationships allow us to use additional constraints during the post-processing of the solution spaces. As a result the surface-area concentration can be estimated with an uncertainty that is less than the measurement error of the extinction coefficient.

Our comparisons between the results obtained with GCM and with our traditional regularization technique show a significant stabilization of the retrieved profiles of aerosol microphysical properties. We will continue with numerical simulations and the analysis of case studies in order to further assess the potential of this new approach.

REFERENCES

1. Müller D., et al., 1999: Microphysical particle parameters from extinction and backscatter lidar data by inversion with regularization: theory, *Appl. Opt.*, 38, pp. 2346 – 2357.

2. Veselovskii I., et al., 2002: Inversion with regularization for the retrieval of tropospheric aerosol parameters from multiwavelength lidar sounding, *Appl. Opt.*, 41, pp. 3685 – 3699.

3. Böckmann C., et al, 2005: Microphysical aerosol parameters from multiwavelength lidar, *Journal of Optical Society of America*, A, 22, pp. 518–528.

4. A. Kolgotin and D. Müller, 2008: Theory of inversion with two-dimensional regularization: profiles of microphysical particle properties derived from multiwavelength lidar measurements, *Appl. Opt.*, 47, pp. 4472-4490.

5. D. Müller, et al, 2014: Airborne Multiwavelength High Spectral Resolution Lidar (HSRL-2) observations during TCAP 2012: vertical profiles of optical and microphysical properties of a smoke/urban haze plume over the northeastern coast of the US, *Atmos. Meas. Tech.* 7, 3487–3496, 2014.

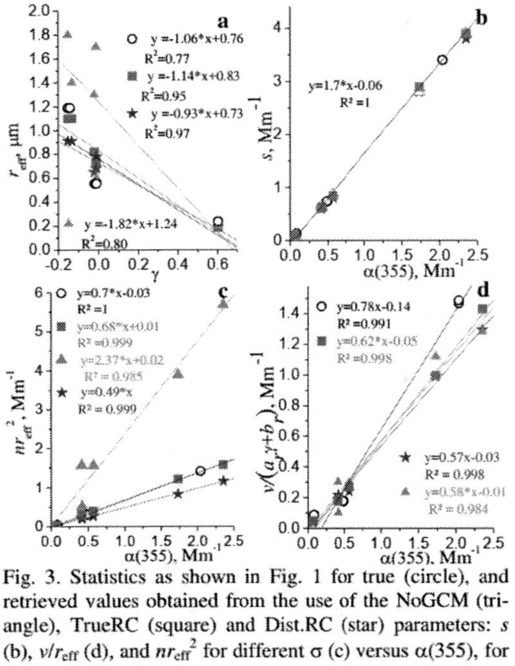

Fig. 3. Statistics as shown in Fig. 1 for true (circle), and retrieved values obtained from the use of the NoGCM (triangle), TrueRC (square) and Dist.RC (star) parameters: *s* (b), v/r_{eff} (d), and nr_{eff}^2 for different σ (c) versus α(355), for r_{eff} (a) versus γ. The solid lines describe correlation trends according to $y=ax+b$ and R^2.

APPLICATION OF THE GARRLIC ALGORITHM FOR THE CHARACTERIZATION OF DUST AND MARINE PARTICLES UTILIZING THE LIDAR-SUNPHOTOMETER SYNERGY

Alexandra Tsekeri [1]*, Vassilis Amiridis[1], Anton Lopatin[2], Eleni Marinou[1], Panos Kokkalis[1], Stavros Solomos[1], Ronny Engelmann[3], Holger Baars[3], Ulla Wandinger[3], Albert Ansmann[3], Dirk Schüttemeyer[4] and Oleg Dubovik[2]

*[1]IAASARS, National Observatory of Athens, Athens, Greece, *Email: atsekeri@noa.gr*

[2]Laboratoire d' Optique Atmosphérique, Universite de Lille, Lille, France

[3]Leibniz Institute for Tropospheric Research, Leipzig, Germany

[4]European Space Agency, Netherlands

ABSTRACT

The importance of studying the vertical distribution of aerosol plumes is prominent in regional and climate studies. The new Generalized Aerosol Retrieval from Radiometer and Lidar Combined data algorithm (GARRLiC) provides this opportunity combining active and passive ground-based remote sensing from lidar and sunphotometer measurements. Here, we utilize GARRLiC capabilities for the characterization of Saharan dust and marine particles at the Eastern Mediterranean region during the Characterization of Aerosol mixtures of Dust And Marine origin Experiment (CHARADMExp). Two different case studies are presented, a dust-dominated case which we managed to characterize successfully in terms of the particle microphysical properties and their vertical distribution and a case of two separate layers of marine and dust particles for which the characterization proved to be more challenging.

1. INTRODUCTION

Information about aerosol characterization is becoming an important topic in the context of current and future Earth Observation missions (e.g Flex, CarbonSat, EARTHCARE, ADM-Aeolus missions of the European Space Agency (ESA)). The CHARADMExp campaign (http://charadmexp.gr) was funded by ESA and was held in June and July 2014 to study the optical, microphysical and chemical properties of the Saharan dust and marine particles, as the most abundant aerosols of Eastern Mediterranean. The campaign was located in the remote Finokalia station, which belongs to the Aerosols, Clouds, and Trace gases Research InfraStructure Network (ACTRIS - http://www.actris.net/), and its location provides favorable conditions for unobstructed monitoring of marine, dust and smoke particles along with their mixtures.

The GARRLiC algorithm [1] was used for the analysis of the campaign data. Although the synergy of active with passive remote sensing has been extensively tried mainly using the sunphotometer data as ancillary information for the lidar retrieval (e.g. [2; 3]), with GARRLiC algorithm it is the first time that sunphotometer and lidar measurements are both incorporated in the retrieval of the particle microphysical properties. In [1] it is shown that GARRLiC manages to characterize successfully cases of dust mixed with urban or smoke aerosols.

2. METHODOLOGY

GARRLiC algorithm retrieves the aerosol concentration profiles along with their columnar microphysical properties, building on the AERONET operational retrieval [4] by ingesting vertically-resolved elastic lidar measurements in the retrieval scheme. This way it utilizes both the columnar information of the sun and sky measurements at four wavelengths and 35 angles from the sunphotometer, together with the vertically-resolved lidar measurements of the elastic backscatter at three wavelengths. GARRLiC input and output is shown in Figure 1.

The lidar measurements during the CHARADMExp campaign were acquired with the Polly^{XT} OCEANET lidar, provided by the TROPOS Institute (http://www.tropos.de/). The sunphotometer measurements were provided by the Cimel CE318 sunphotometer.

DOI: 10.1051/epjconf/201611923021

GARRLiC algorithm	
INPUT	**OUTPUT**
lidar elastic backscatter signals (range-resolved at 355, 532, 1064 nm)	**for fine and coarse particles** • concentration profile • size distribution • real part of refractive index • imaginary part of refractive index • spherical particle fraction } constant along the column
sunphotometer • AOD • total scattered radiances (at 440, 670, 870, 1020 nm and multiple angles (≤35))	

Figure 1: GARRLiC algorithm input and output parameters.

Furthermore, in order to identify the origin of the aerosol layers we employed source – receptor computations derived with dispersion modeling tools. These tools provided the source apportionment of the particles in the layer. The modeling simulations were performed with the atmospheric dispersion model FLEXPART-WRF [5], offline coupled with the atmospheric model WRF_ARW [6].

3. RESULTS

Dust case

On June 26 the Polly^{XT} OCEANET measurements of volume depolarization ratio at 532 nm showed a plume of non-spherical particles extending from close to the ground up to 5-6 km (Figure 2).

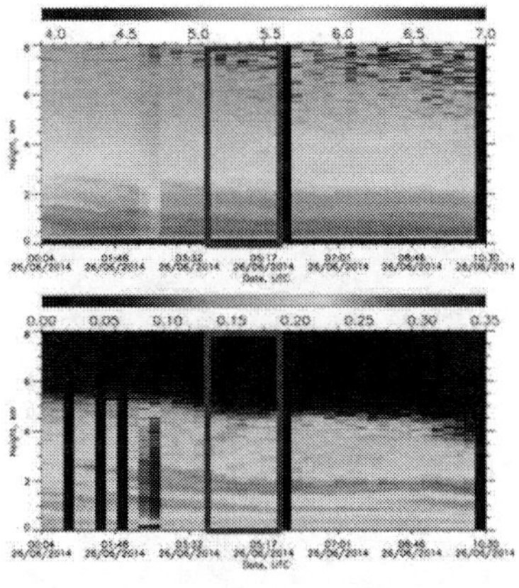

Figure 2: Atenuated backscatter at 1064 nm in arbitrary units (top) and volume depolarization ratio at 532 nm (bottom) from Polly^{XT} OCEANET lidar, on June 26. The red rectangle indicates the time range of the measurements used for GARRLiC retrieval (04:00-06:00 UTC).

The AERONET products indicate dust particles, with aerosol optical depth (AOD) at 550 nm of ~0.4, Ångstrom exponent at 440/870 nm of ~0.1 and a coarse-mode dominated size distribution. Dust transport simulations from the NMMB/BSC-Dust model [7] show Saharan dust transport in Finokalia, which is also supported from our source apportionment modelling in Figure 3: from the ground up to 6 km the particles are originated from west Sahara with some mixing with particles from the Eastern Mediterranean region, the Balkans and Greece up to 1 km.

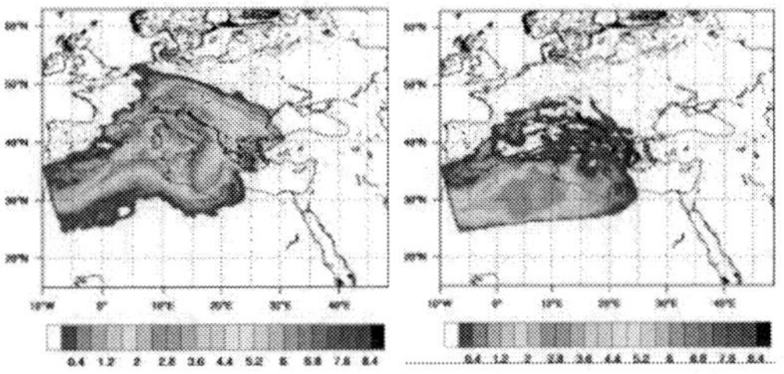

Figure 3: FLEXPART-WRF emission sensitivity (residence time) [log$\cdot$s$\cdot$m^3$\cdot$kg^{-1}], at 0-1 km (left) and 3-6 km (right), at Finokalia station, on June 26.

GARRLiC retrieved successfully the dust lidar ratio at 532 nm, as seen from the comparison with the Raman retrieval in Figure 4. Moreover, the retrieved size distribution and spectral refractive index for coarse particles agree well with climatological values for dust (e.g. [8]), as well as with AERONET (Figure 5). The AERONET retrieval is closer to the coarse particle characteristics, since this is the dominant mode along the atmospheric column.

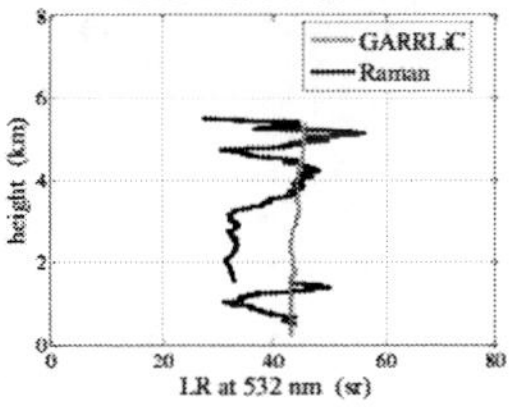

Figure 4: GARRLiC (green line, at 04:00-06:00 UTC) compared with Raman (black line, at 00:00-02:00 UTC) retrieval for the lidar ratio at 532 nm, at Finokalia station, on June 26.

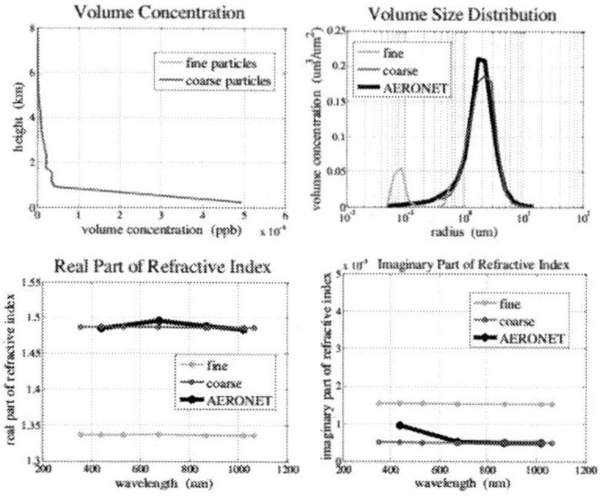

Figure 5: GARRLiC retrieval of particle concentration profiles (up-left), size distribution (up-right), spectral real and imaginary part of the refractive index (bottom –left and right), for fine (cyan) and coarse (pink) particles, at Finokalia station, on June 26, 04:00-06:00 UTC. The black line shows the AERONET retrieval at 04:54 UTC.

Dust and marine case

On July 6 two very distinct aerosol plumes were measured at Finokalia: a depolarizing dust plume at 2-6 km and a non-depolarizing, possibly containing marine and inland-transported particles plume, extending from the ground up to 1 km (Figure 6).

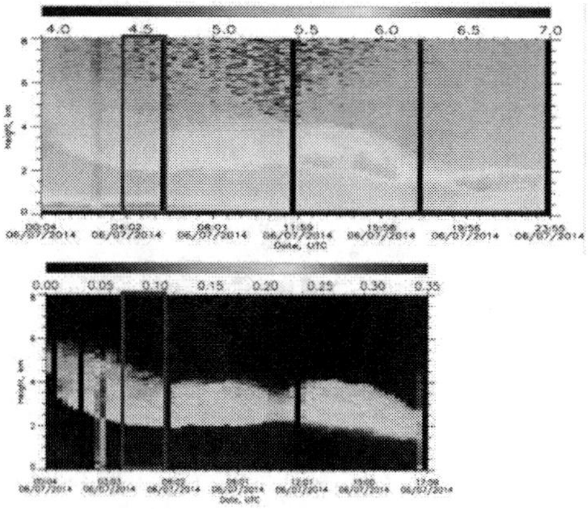

Figure 6: As in Figure 2 for the PollyXT OCEANET lidar measurements on July 6.

The NMMB/BSC-Dust model shows Saharan dust transport over Finokalia, and our source apportionment calculations show mostly Saharan dust particles at 3-4 km and marine particles (mostly from Aegean Sea) along with particles from the Balkans region at 0-1 km (Figure 7).

Based on the above we expect a dust plume aloft and a plume containing (mostly) marine particles close to the ground. GARRLiC retrieves these two coarse particle dominated layers (see concentration profiles in Figure 9) but it cannot characterize them effectively in terms of their refractive indices, since it is able to retrieve only one refractive index for the coarse mode. Thus what we get from GARRLiC is possibly closer to the average of the refractive indices of dust and marine particles.

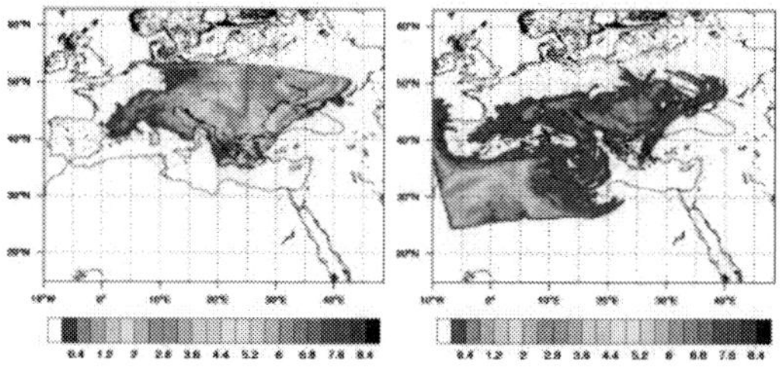

Figure 7: As in Figure 3 for FLEXPART-WRF emission sensitivity on July 6.

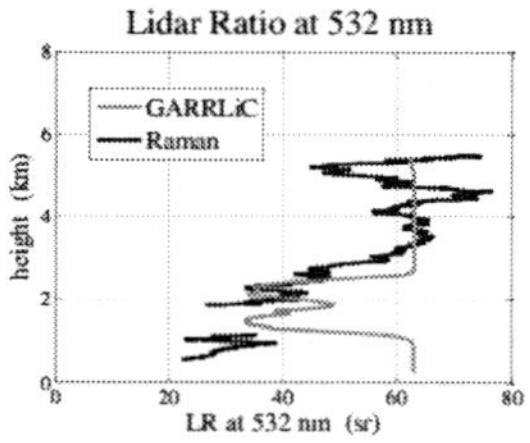

Figure 8: As in Figure 4 for the LR retrievals on July 6.

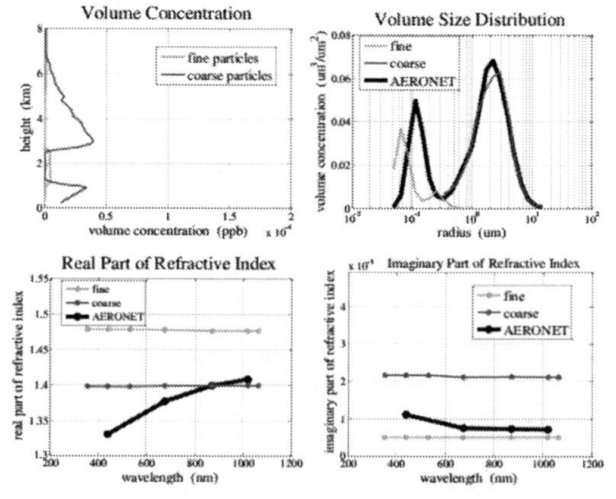

Figure 9: As in Figure 5, for GARRLiC retrieval on July 6, 04:00-06:00 UTC. The black line shows the AERONET retrieval at 05:49 UTC.

This is what we see in Figure 8 as well, showing that both layers are characterized with the same lidar ratio at 532 nm from GARRLiC. For the aloft dust plume it matches well the lidar ratio from Raman technique, but for the layer close to the surface, the Raman lidar ratio is much lower, as expected for marine particles.

4. CONCLUSIONS and FUTURE WORK

GARRLiC algorithm provides the great innovation of retrieving the vertical profile of aerosol particles utilizing the elastic backscatter lidar measurements together with the sunphotometer measurements in the retrieval. This way it can effectively characterize the particles of fine and coarse modes in terms of their size, shape and refractive index.

In this study we used GARRLiC algorithm to characterize two cases from the CHARADMExp campaign, one with pure dust and one with dust and marine particles. For the first case GARRLiC achieves a successful characterization of the dust vertical distribution and microphysichal properties that seems to agree well with climatological values for dust and Raman retrievals. For the second case the retrieval was not as successful, since GARRLiC is able to discern the properties of fine and coarse particles but not of two different coarse modes.

The retrieval of multiple modes would be possibly feasible in the future with the incorporation of polarimetric measurements from the sunphotometer or Raman signals from the lidar. We aim to continue investigating the GARRLiC potential for aerosol characterization and follow related improvements in the framework of the ACTRIS-2 project and the experimental campaigns that are dedicated to that objective.

ACKNOWLEDGEMENT

This work has been developed under the auspices of the ESA-ESTEC project "Characterization of Aerosol mixtures of Dust And Marine origin" contract no. IPL-PSO/FF/lf/14.489. The publication was supported by the European Union Seventh Framework Programme (FP7-REGPOT-2012-2013-1), in the framework of the project BEYOND, under Grant Agreement no. 316210 (BEYOND – Building Capacity for a Centre of Excellence for EO-based monitoring of Natural Disasters). The research leading to these results has received funding from the European Union Seventh Framework Programme (FP7/2007-2013) under grant agreement no. 262254 (ACTRIS)

REFERENCES

[1] Lopatin, A., et al., 2013: Enhancement of aerosol characterization using synergy of lidar and sun-photometer coincident observations: the GARRLiC algorithm, *Atmos. Meas. Tech.*, **6**, 2065-2088, doi:10.5194/amt-6-2065-2013.

[2] Fernald, F. G., et al., 1972: Determination of aerosol height distributions by lidar, *J. Appl. Meteorol.*, **11**, 482–489, 1972.

[3] Chaikovsky, A., et al., 2012: Combined lidar and radiometric sounding of atmospheric aerosol: algorithm of data processing, software, dessimination, Proceedings of XVIII International symposium "Atmospheric and ocean optics. Atmosphere physics", 2–6 July 2012, Irkutsk, Russian Federation, C1–C4.

[4] Dubovik, O. and King, M, 2000: A flexible inversion algorithm for retrieval of aerosol optical properties from Sun and sky radiance measurements, *J. Geophys. Res.*, **105**, 20673–20696, doi:10.1029/2000JD900282.

[5] Brioude, J., et al., 2013: The Lagrangian particle dispersion model FLEXPART-WRF version 3.1, Geosci. Model Dev., 6, 1889-1904, doi:10.5194/gmd-6-1889-2013.

[6] Skamarock, W. C., et al., 2008: A description of the Advanced Research WRF version 3, NCAR Technical Note 475.

[7] Basart, S., et al., 2012: Development and evaluation of the BSC-DREAM8b dust regional model over Northern Africa, the Mediterranean and the Middle East, *Tellus B*, **64**, 18539, doi:10.3402/tellusb.v64i0.18539.

[8] Dubovik, O., et al., 2002: Variability of absorption and optical properties of key aerosol types observed in worldwide locations, *J. Atmos. Sci.*, **59**, 590–608.

EPJ Web of Conferences **119**, 23022 (2016)
ILRC 27
DOI: 10.1051/epjconf/201611923022

Combined sphere-spheroid particle model for the retrieval of the microphysical aerosol parameters via regularized inversion of lidar data

Stefanos Samaras[1], **Christine Böckmann**[1], **Doina Nicolae**[2]

[1] *University of Potsdam, Am Neuen Palais 10, 14469 Potsdam, Germany, samaras@uni-potsdam.de, Germany, bockmann@rz.uni-potsdam.de*

[2] *Laser Remote Sensing Department, National Institute of R&D for Optoelectronics, 409 Atomistilor Str., Magurele, Ilfov, Romania, nnicol@inoe.ro*

Abstract

In this work we propose a two-step advancement of the Mie spherical-particle model accounting for particle non-sphericity. First, a naturally two-dimensional (2D) generalized model (GM) is made, which further triggers analogous 2D re-definitions of microphysical parameters. We consider a spheroidal-particle approach where the size distribution is additionally dependent on aspect ratio. Second, we incorporate the notion of a sphere-spheroid particle mixture (PM) weighted by a non-sphericity percentage. The efficiency of these two models is investigated running synthetic data retrievals with two different regularization methods to account for the inherent instability of the inversion procedure. Our preliminary studies show that a retrieval with the PM model improves the fitting errors and the microphysical parameter retrieval and it has at least the same efficiency as the GM. While the general trend of the initial size distributions is captured in our numerical experiments, the reconstructions are subject to artifacts. Finally, our approach is applied to a measurement case yielding acceptable results.

1. Introduction

A central topic in understanding the climate processes is the investigation of the spatial and temporal variability of the microphysical properties of aerosol particles, such as the aerosol size distribution, the complex refractive index and the mean particle radius. Recent development of continental-scale lidar networks provides the opportunity to obtain quality assured optical profiles on a large temporal and spatial scale, e.g. the European Aerosol Research Lidar Network (EARLINET) involvement. In particular, Raman-elastic multi-wavelength lidar data is used [1] to measure the aerosol optical properties and to retrieve the size distribution through the inversion of the Mie forward model.

Regularization methods are a widely preferable choice for an efficient inversion, e.g. [2,9]. While the spherical-particle (Mie) model offers a reasonable approximation for spherically modelled particles, e.g. biomass burning aerosols, it no longer provides a viable description for the more realistic non-spherical case such as mineral dust; particle shape is known to have substantial effects for the scattering in side- and backward direction [8]. Nevertheless the first non-spherical model approximation, i.e. spheroids, seems to reproduce the optical properties of realistic aerosol particles significantly better than spheres [6,8]. Steps towards particle non-sphericity incorporation for the purpose of a microphysical retrieval are a result of very recent work mostly considering spheroidal-particle models, e.g. [3,4,13].

In the present work, we use the generalization of the Mie forward model (GM) in two dimensions (2D) and perform simulations with two regularization methods to study the retrieval behavior and compare. We further adopt the concept of a mixture of spherical and spheroidal particles, quantitatively dictated by a non-sphericity percentage, introduced in [7]. The particle-mixture model (PM) is then compared to the (GM) with respect to the most appropriate of the regularization methods used previously. Our algorithm is subsequently applied to a measurement case from the Bucharest lidar station. This study is a part of our first attempts to familiarize with potential retrieval patterns of our newly developed software.

2. Mathematical standpoint: two-dimensional forward models and ill-posedness

The aerosol volume size distribution $v(r,a)$ is related to the extinction- and backscatter coefficients, $\Gamma(\lambda)$ through a generalization of Mie forward model (GM) in two dimensions and the so-called Fredholm integral equation of the first kind:

$$\Gamma(\lambda) = \int_{a_{\min}}^{a_{\max}} \int_{r_{\min}}^{r_{\max}} K_{\alpha/\beta}(r,\lambda,a;m) v(r,a)\, dr da, \quad (1)$$

where $K_{\alpha/\beta} = \frac{3S}{16\pi r^3} Q_{\alpha/\beta}$, are the extinction- ($\alpha$) or backscatter kernels (β), summarizing the relation of the scattering efficiencies $Q_{\alpha/\beta}$ with the spheroidal surface-area S. The dependency on the volume-equivalent radius and the aspect ratio is denoted by r and a, respectively, and m is the refractive index. In this work the wavelength λ is restricted to the discrete values $\lambda_i = \{355, 532, 1064\}$ µm, $i = \{1,2,3\}$. The ideal optical data setup here $(3\beta + 2\alpha + 3d)$ consists of 3 backscatter coefficients horizontally polarized $\beta_{\lambda_{1,2,3}}$, 2 extinction coefficients $\alpha_{\lambda_{1,2}}$ and 3 vertically polarized backscatter coefficients $d_{\lambda_{1,2,3}}$. The computational effort for the calculation of the kernel function using T-matrix method [8] is very high for real-time retrievals. Instead we use a precalculated database

DOI: 10.1051/epjconf/201611923022

produced by Mieschka software tool, see [12]. The refractive index is commonly considered wavelength independent, see e.g. [2]. Moreover, a refractive index grid is predefined, upon which the solution space is built. For the retrieval process, the model is ran for every different value of the grid and solution is picked by means of least residual error.

Eq. 1 belongs to the class of inverse ill-posed problems, see [11]. This means that potential noisy components in the input data $\Gamma(\lambda)$ will result in oscillatory solutions, regardless of the noise level, see e.g. [2, 10]. The degree of ill-posedness is linked to the level of smoothness of the scattering efficiencies, thus their qualitative behavior indicates where the problem arises. Fig. 1 shows a multiple plot of the total backscatter efficiencies with respect to the size parameter $s = \frac{2\pi r}{\lambda}$ for different real part of refractive index (RRI) and constant imaginary part of refractive index (IRI) and a fixed aspect ratio; the oscillations are suppressed with decreasing RRI. In Fig. 2, similar trend can be observed for increasing IRI (higher absorption levels) and constant RRI. In particular, looking along the aspect ratio axis of the lower 3D plots in Fig. 2 we observe that extinction efficiencies cannot reveal any shape information for a large IRI in the case of larger size parameters.

Eq. 1 is discretized with spline collocation using variable number (n) and degree (d) of B-splines $\phi_{j,k}^{(d)}$, using the size distribution approximation $v_{pq} = \sum_{j=1}^{p}\sum_{k=1}^{q} b_{jk}\phi_{jk}^{(d)}(r,a)$. Then, an one-dimensional index reordering $L = F(j,k)$ is used to build the matrix $\sum_{L=1}^{pq} A_{iL}\, b_L = \Gamma_i,\, i = 1,\ldots,8$. We solve the resulting linear system with the following regularization methods:

1. TSVD with the Discrepancy Principle (TSVD),
2. Tikhonov with the L-curve method (TLC).

There are well-known essential differences between these methods, but the main one here is that only method 1 makes use of a known error level.

Next, we define the particle mixture model (PM) as follows:

$$\Gamma(\lambda) = (1-\eta)\int_{r_{\min}}^{r_{\max}} K_{\alpha/\beta}^{\rm sph}(r,\lambda,m)\,v(r)\mathrm{d}r + \\ +\eta\int_{a_{\min}}^{a_{\max}}\int_{r_{\min}}^{r_{\max}} K_{\alpha/\beta}^{\rm nsph}(r,\lambda,m,a)\,v(r,a)\mathrm{d}r\mathrm{d}a, \tag{2}$$

where the superscript sph/nsph refers to the spherical ($a = 1$) or non-spherical ($a \neq 1$) kernel functions. The assumption here is that the aerosol particles consist of independent spherical and non-spherical (spheroidal) parts, a property which further passes on the optical data through a non-sphericity percentage we call asphericity (η). The solution calculation is slightly modified here. We define a range of viable asphericity options and run (PM) for every value within it and for every refractive index within the grid to finally bring out one least residual error solution.

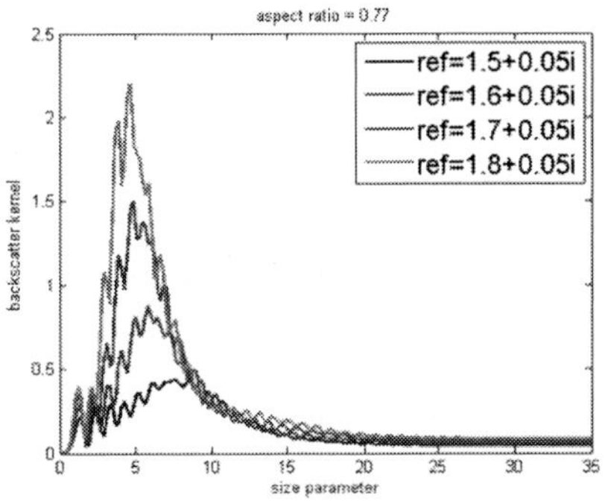

Fig. 1. Total backscatter efficiencies with respect to size parameter for fixed aspect ratio and different RRI.

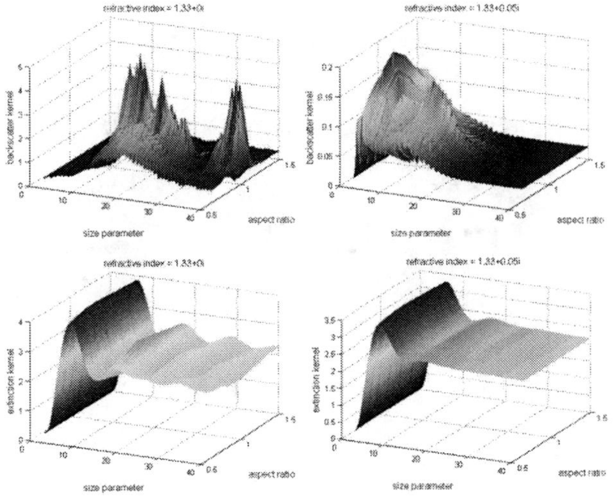

Fig. 2. Total backscatter- (upper panel) and extinction efficiencies (lower panel) for increasing IRI from left to right.

3. Synthetic data results

In all the simulations that follow we assume the same log-normally distributed radius range [0.03,1.24] with mean radius $\mu = 0.25$, width $\sigma = 1.2$ and number concentration $n_t = 1$ combined with an aspect ratio distribution of a purely oblate particle sample. In this experiment we would like to see the retrieval behavior of the methods 1-2 for the size distributions. The synthetic data were produced with a forward calculation and 3% random errors were added. We reconstructed $v(r,a)$ with each regularization method using the same spline number and degree (9,6) and a fixed refractive index $m = 1.4 + 0.01i$. Fig. 3 shows the solution reconstructions with the two methods for the case of an oblate spheroidal sample. Both methods seem to capture the general trend with almost the same efficiency. The artifacts in the case of TSVD are more prominent, and the fitting error (not shown here)

Table 1. Comparison of the models GM and PM with respect to TLC-retrieved microphysical parameters

Parameters	Synthetic values	Solution 1 (GM)	Solution 2 (GM)	Solution 1 (PM)	Solution 2 (PM)
m	1.33 + 0.1i	1.33 +0.03i	1.33+0.03i	1.33+0.05i	1.33+0.05i
n_t	0.076	0.20	0.20	0.05	0.06
a_t	0.015	0.015	0.015	0.002	0.003
v_t	0.006	0.005	0.005	0.0008	0.001
r_{eff}	1.09	0.95	0.97	0.94	0.92
$\mu_a^{(v)}$	0.84	0.90	0.90	0.86	0.89
η	100%	100%	100%	47%	56%
splines no./degree	-	22/4	21/4	22/4	21/4
reconstruction error	-	6%	7%	1%	1%

is larger than in the case of TLC. A closer look to the regularization parameter of TSVD, the cut-off level (cl), see [5], for this case reveals that such a behavior is to be expected since the noise is dominating and causes an underestimation of the cl.

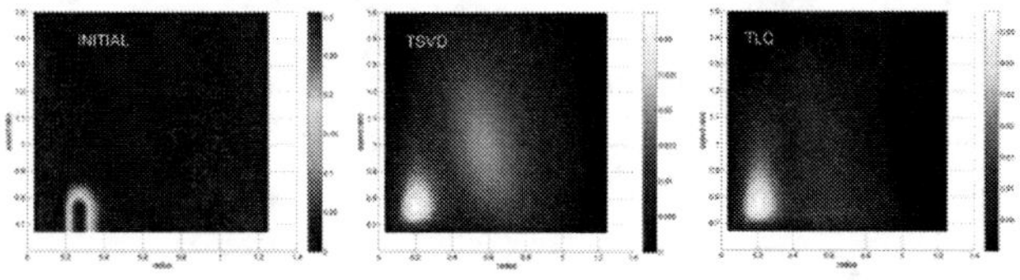

Fig. 3. From left to right: Volume size distribution reconstructions with TSVD and TLC method compared to the initial one (leftmost).

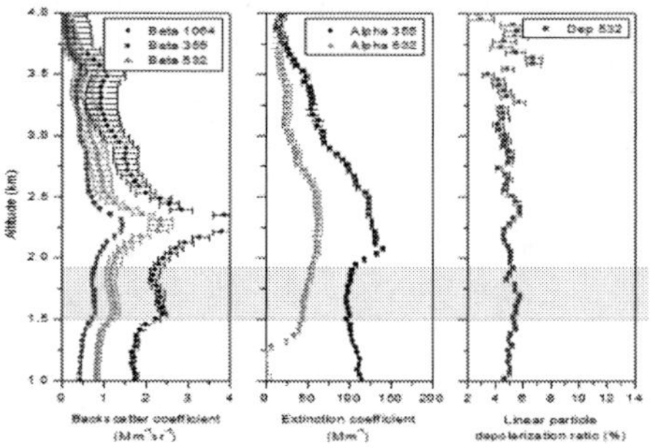

Fig. 4. From left to right: Extinction coefficient, backscatter coefficient and linear depolarization ratio profiles for the measurement case captured by the lidar station in Magurele, Bucharest. Gray-shaded areas highlight the layer values used for retrieving the microphysical properties shown in Table 2.

Table 2. Microphysical parameters derived with the method from [2] and the TLC method with the (PM) model for the measurement case captured by the lidar station in Magurele, Bucharest.

Parameters	Method from [2]	TLC - PM
m	1.35+0.002i	1.33+0.005i
n_t	0.0006	0.00017
a_t	0.0001	8.06e-06
v_t	1.36e-05	2.35e-06
r_{eff}	0.28	0.86
$\mu_a^{(v)}$	-	0.77
η	-	22.7%
splines no./degree	7/3	9/3
reconstruction error	1.87e-05	9.1e-07

4. Microphysical parameter retrieval

First we need some definitions which follow the same concept as the 2D generalization of Eq. 1: the number concentration:

$$n_t = \int_{a_{\min}}^{a_{\max}} \int_{r_{\min}}^{r_{\max}} \frac{3}{4\pi r^3} v(n,a)drda,$$

the total volume concentration:

$$v_t = \int_{a_{\min}}^{a_{\max}} \int_{r_{\min}}^{r_{\max}} v(r,a)drda,$$

the surface-area concentration

$$a_t = \int_{a_{\min}}^{a_{\max}} \int_{r_{\min}}^{r_{\max}} \frac{3}{\pi r^3} G(r,a)v(r,a)drda,$$

where $G(r,a)$ is the geometrical cross section of the particle. The effective radius can as usual be defined by $r_{\text{eff}} = \frac{3v_t}{a_t}$. Moreover, we define the effective aspect ratio:

$$\mu_a^{(v)} = \frac{\int_{a_{\min}}^{a_{\max}} a \int_{r_{\min}}^{r_{\max}} v(r,a)drda}{v_t}.$$

In this experiment we retrieve the microphysical parameters $m, n_t, v_t, a_t, r_{\text{eff}}$, and $\mu_a^{(v)}$ using the models GM (Eq. 1) and PM (Eq. 2) with TLC method and the same test-configuration used in section 3 except that now the refractive index resides within the grid $\Re(m) \times \Im(m)$, where $\Re(m) = \{1.33, 1.4, 1.5, 1.6, 1.7, 1.8\}$, $\Im(m) = \{0, 0.001, 0.005, 0.01, 0.03, 0.05, 0.1\}$, see [12].

The results are in Table 1; we include two least-residual-error solutions from each run for comparison. We can see that the new model (PM) improves the fitting errors, reproducing the initial optical data better than the (GM). The parameters m, n_t and $\mu_a^{(v)}$ are better retrieved in (PM), r_{eff}, are equivalent with both models, while a_t and v_t are underestimated and better retrieved with (GM). The (PM) model offers additional knowledge on the non-sphericity predicting a percentage of 47% and 56% for the two solutions shown. Although there is room for improvements evidently, we will consider (PM) as an enhancement of (GM) model and use this for the measurement case in section 5.

5. Retrieval from a measurement case

In this section we apply our model (PM) to a measurement case captured by the multiwavelength depolarization Raman lidar in Magurele, 6 km south of Bucharest (lat. 44.348 N, long. 26.029 E, above sea level), on June, 27th 2012 between 19:30-20:30 UTC, see Fig. 4. The HYSPLIT model and MODIS fire maps confirmed that this is probably a biomass burning aerosol case, which is in accordance with the low depolarization value of 5%, suggesting almost spherical particles. In Table 2 we show the microphysical parameters derived in two ways; (1) with an older solution algorithm of our group based on the Mie forward model, presented in [2], and (2) the TLC-(PM) method. A direct comparison requires a deeper analysis, but here we will focus on the orders of magnitude for the retrieved values using TLC-(PM), which appear reasonable. The (PM) predicts an atmospheric scenario of particles acting as an oblate spheroidal sample ($\mu_a^{(v)} = 0.77$) with a non-sphericity percentage of 22.7%.

6. Summary

We investigated a natural generalization of the Mie-forward model (GM) considering a size distribution depending both on the radius and the aspect ratio and subsequently employed both spherical and spheroidal efficiencies and the notion of a non-sphericity percentage to a new model (PM). TLC-retrieved size distributions seem to be less prune to artifacts and yield smaller fitting errors compared to TSVD ones. PM along with the use of TLC indicates that it is a correction to the GM model, offering better fitting errors, and improving the key microphysical parameters, refractive index and number concentration. Finally, the results from the application to a measurement case fall within the order-of-magnitude tolerance and the limitations of this approach.

Acknowledgements

The research leading to these results has received funding mainly from the European Unions Seventh Framework Programme for research, technological development and demonstration under grant agreement no 289923 - ITaRS.

References

1. Ansmann, A., M. Riebesell, U. Wandinger, C. Weitkamp, E. Voss, W. Lahmann, W. Michaelis, 1992: Combined Raman elastic-backscatter lidar for vertical profiling of moisture, aerosol extinction, backscatter, and lidar ratio, *Applied Physics B*, **55(1)**, 18–28.
2. Böckmann C., 2001: Hybrid regularization method for the ill-posed inversion of multiwavelength lidar data in the retrieval of aerosol size distributions, *Applied Optics*, **40(9)**, 1329–1342.
3. Böckmann C., L. Osterloh, 2014: Runge-kutta type regularization method for inversion of spheroidal particle distribution from limited optical data, *Inverse Problems in Science and Engineering*, **22**, 150–165.
4. Dubovik O., B. N. Holben, T. Lapyonok, A. Sinyuk, M. I. Mishchenko, P. Yang, I. Slutsker, 2002: Non-spherical aerosol retrieval method employing light scattering by spheroids, *Geophysical Research Letters*, **29(10)**, 1415–1418.
5. Hansen P. C., 2010: Discrete inverse problems: insight and algorithms, *SIAM*, **7**.
6. Kahnert M., A. Kylling: Radiance and flux simulations for mineral dust aerosols: Assessing the error due to using spherical or spheroidal model particles, *Journal of Geophysical Research: Atmospheres*, **109(D9)**, 2156–2202.
7. Lopatin A., O. Dubovik, A. Chaikovsky, P. Goloub, T. Lapyonok, D. Tanré, P. Litvinov: Enhancement of aerosol characterization using synergy of lidar and sun-photometer coincident observations: the GARRLiC algorithm, *Atmospheric Measurement Techniques*, **6(8)**, 2065–2088.
8. Mishchenko M. I., L. D. Travis, D. W. Mackowski: T-matrix computations of light scattering by nonspherical particles: A review, *Journal of Quantitative Spectroscopy and Radiative Transfer*, **55(5)**, 535–575.
9. Müller D., U. Wandinger, and A. Ansmann: Microphysical particle parameters from extinction and backscatter lidar data by inversion with regularization: simulation, *Applied Optics*, **38(12)**, 2358–2368.
10. Osterloh L., C. Pérez, D. Böhme, J. M. Baldasano, C. Böckmann, L. Schneidenbach, D. Vicente, 2009: Parallel software for retrieval of aerosol distribution from lidar data in the framework of EARLINET-ASOS, *Computer Physics Communications*, **180(11)**, 2095–2102.
11. Rieder A., 2003: Keine Probleme mit Inversen Problemen - Eine Einfhrung in ihre stabile Lösung, *Vieweg-Teubner Verlag*.
12. Rother T., 2009: Electromagnetic Wave Scattering on Non- spherical Particles: Basic Methodology and Simulations, *Springer*.
13. Veselovskii I., O. Dubovik, A. Kolgotin, T. Lapyonok, P. Di Girolamo, D. Summa, D. N. Whiteman, M. Mishchenko, D. Tanr, 2010: Application of randomly oriented spheroids for retrieval of dust particle parameters from multiwavelength lidar measurements, *Journal of Geophysical Research: Atmospheres*, **115(D21)**.

EPJ Web of Conferences **119**, 23023 (2016) DOI: 10.1051/epjconf/201611923023
ILRC 27

STANDARDS – AN IMPORTANT STEP FOR THE (PUBLIC) USE OF LIDARS

Dietrich Althausen [1]*, Stefan Emeis [2], Harald Flentje [3], Josef Guttenberger [4], Simon Jäckel [5], Volker Lehmann [6], Ina Mattis [3], Christoph Münkel [7], Gerhard Peters [8], Christoph Ritter [9], Matthias Wiegner [10], Holger Wille [11]

[1]*Leibniz Institute for Troposheric Research, 04318 Leipzig, Germany, *Email: dietrich@tropos.de*

[2]*Karlsruhe Institute of Technology (KIT), 82467 Garmisch-Partenkirchen, Germany*

[3]*Deutscher Wetterdienst, 82383 Hohenpreissenberg, Germany*

[4]*RSC GmbH, 92355 Velburg, Germany*

[5]*Commission on Air Pollution Prevention (KRdL) of VDI and DIN, 40468 Düsseldorf*

[6]*Deutscher Wetterdienst, 15848 Tauche OT Lindenberg, Germany*

[7]*Vaisala GmbH, 22607 Hamburg, Germany*

[8]*METEK Meteorologische Messtechnik GmbH, 25337 Elmshorn, Germany*

[9]*Alfred Wegener Institute, Helmholtz Centre for Polar and Marine Research, 14401 Potsdam, Germany*

[10]*Ludwig-Maximilians-Universität, 80333 München, Germany*

[11]*G. Lufft Mess- und Regeltechnik GmbH, 12489 Berlin, Germany*

ABSTRACT

Lidar standards are needed to ensure quality and lidar product control at the interface between lidar manufacturers and lidar users. Meanwhile three lidar standards have been published by German and international standardization organizations. This paper describes the cooperation between the lidar technique inventors, lidar instrument constructors, and lidar product users to establish useful standards. Presently a backscatter lidar standard is elaborated in Germany. Key points of this standard are presented here. Two German standards were already accepted as international standards by the International Organization for Standardization (ISO). Hence, German and international organizations for the establishment of lidar standards are introduced to encourage a cooperative work on lidar standards by lidar scientists.

1. INTRODUCTION

Aerosol lidar instruments get more and more applied for public and broad-scale purposes (compare for instance the ceilometer networks and the GAW Aerosol Lidar Observations Network GALION: [1, 2]). This can be understood as a step out of the lab by the researchers to combine synergistically measurements within lidar networks (e.g. [3]). With establishing lidar networks, the quality assurance and quality control and hence the standards of the systems need to be regarded comprehensively.

There is a steady need of improving the public health by taking into account research results and the recognition of their application. This was especially recognized during the eruption of the Eyjafjallajökull volcano in 2010 when only lidar measurements could provide information about the vertical extend of the plume and its ash concentrations (compare e.g. results from these measurements in [4]). Such events increase the awareness of the public to implement lidar systems that have been manufactured according to internationally accepted standards. Only then, measurements can be combined for continental or even global surveillance.

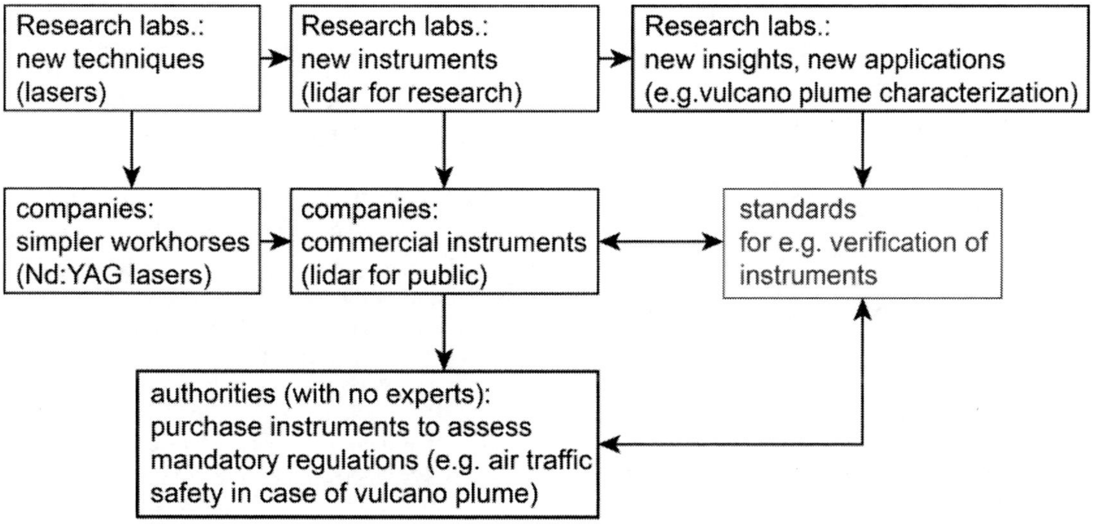

Fig.1: Schematic sketch about the cooperation of lidar scientists, lidar technicians, and lidar users that yields to the need of standards about lidars. The volcano case is mentioned as an example that has forced the work on standards.

So there is a need for standards that can be used by the manufacturers and users to understand each other.

Meanwhile three standards for lidar applications have been published by German and international standardization organizations [5-7]. Presently the responsible board in Germany deals with a standard regarding backscatter lidar.

This paper explains the cooperation that yields to standards in chapter 2. The current status of the standard regarding backscatter lidar is explained in chapter 3. The standard developing work is done in selected expert groups that are introduced in chapter 4. At the end of the paper the main ideas are summarized and an outlook is given.

2. COOPERATION OF LIDAR TECHNIQUE INVENTORS, LIDAR INSTRUMENT CONSTRUCTORS, AND LIDAR PRODUCT USERS

Figure 1 shows a schematic sketch of the cooperation between lidar scientists, lidar technicians, and lidar users for establishing standards.

Looking back, lidar systems were developed and after this (independent of this) the laser was invented. The application of lasers in lidars was a big step forward for the development and improvement of different lidar types (e.g. backscatter lidar, differential absorption lidar, Raman lidar, high-spectral-resolution lidar) and their application for atmospheric research purposes by research institutes. One recent example of such new atmospheric insights is the estimation of mass concentrations in the plume advected over Europe from the Icelandic volcano Eyjafjallajökull in 2010 (e.g. [8]).

Parallel to these research activities companies were founded for manufacturing lidar systems. At the beginning, the products of these lidar companies were only used by research institutes that have experts and know how to handle the systems. With time, smaller and "plug and play" lidar systems are offered by these companies due to the development of e.g. smaller and cheaper lasers.

With the research results, authorities realized the successful application of lidar systems to monitor specific atmospheric events (e.g., plumes from volcanic eruptions) and the benefit to their work for e.g. regulations of air quality. As a consequence, national weather services started to purchase smaller lidars, too.

This development yields directly to define standards about lidar systems and their products / output for public use.

The interest of the researchers to contribute to the development of these standards is that they often work in governmental funded institutes. Also helping writing they get applied their results.

The interest of the public (authorities) having such standards is that they can use the lidar systems for their purposes as for instance a better knowledge about the air quality. By this they can reduce or even avoid risks and severe situations.

The interest of the companies on these standards is that they get a document that can be referred to when offering their products.

These standards help the lidar manufacturers to offer their systems to authorities and help the authorities to purchase such systems since the standards can be used for e.g. system verification. So the standards define a quality and help the quality control, make systems comparable, and show the minimum the systems have to provide.

3. EXAMPLE: STANDARD REGARDING BACKSCATTER LIDAR

Presently the standard regarding the backscatter lidar is compiled in Germany. The goal is to publish the first draft beginning 2016. So all information about this standard is still preliminary.

The standard consists of 11 chapters and 1 appendix. The topics include area of application, references to other standards, basics of backscatter lidar, target quantities and operation properties, system components, planning of a measurement and requirements to the measurement site, influencing quantities and considerations to uncertainties, quality assurance and system control, and examples of measurements. Appendix A summarizes some of the information the manufactures have to provide to the data users for the data retrieval algorithms. It is not possible to describe all chapters in detail within this paper. So, only remarks to 3 chapters will be presented below.

Chapter 1 summarizes possible application areas of a backscatter lidar. This shows also possible customers of the standard. These application areas are for instance detection of atmospheric particle layers and/or cloud boundaries, derivation of profiles of the particle backscatter coefficient, estimates of size and/or shape classification of particles, monitoring of the air quality, and contributions to air traffic safety, weather forecast, satellite remote sensing.

The theoretical basis of an (elastically) backscatter lidar is described in chapter 5. The lidar equation is given after the main features / system parts of a backscatter lidar. Several considerations are made: choice of wavelength, attenuated backscatter coefficient, particle backscatter coefficient, retrieval method by Klett and Fernald, determination of reference values and/or boundary values, retrievals using system constants, particle extinction coefficient, depolarization ratio, Ångström exponent, and color ratio.

An overview of the primary and derived quantities is given in chapter 6. The range-corrected backscatter signal, the attenuated backscatter coefficient and the particle backscatter coefficient are defined as the primary target quantities of a backscatter lidar. Derived quantities are boundaries of atmospheric particle layers, cloud heights and vertical extensions, and the shape classification of particle layers (with the linear depolarization ratio). Auxiliary quantities as the air density required to apply the Klett-Fernald formalism are regarded in chapter 6, too. The operation properties are explained by describing quantities like signal-to-noise ratio, temporal resolution, range resolution, range averaging, overlap function, range of measurement, data availability, and range of coverage for certain conditions. Especially the range of coverage makes a comparison of different systems possible.

4. NATIONAL AND INTERNATIONAL BOARDS FOR LIDAR STANDARDS

In Germany two institutions are responsible for the standardization in the field of meteorology: DIN (German Institute for Standardization; Deutsches Institut für Normung e. V.; [9]) and VDI (The Association of German Engineers; Verein Deutscher Ingenieure e.V.; [10]). They work together in the Commission on Air Pollution Prevention (KRdL; [11]). One department of KRdL is responsible for the elaboration of standards on environmental meteorology, which includes meteorological measurements.

DIN is the German representative in the International Organization for Standardization (ISO). ISO is an independent, non-governmental organization of members from the national standardization bodies of 163 countries [12]. Each technical committee (TC) of ISO – e.g. ISO/TC 146 "Air quality" – has a mirror board at DIN. The subcommittee (SC) 5 of TC 146 deals with meteorology. Presently there are 17 participating and 10 observing countries in ISO/TC 146/SC 5. The steering committee of KRdL department on environmental meteorology is the mirror board for ISO standards regarding meteorological measurements. This structure is organized in the same way in all participating countries of SC5.

The standardization projects of ISO/TC 146/SC 5 are elaborated in cooperation with the World Meteorological Organization (WMO) through its Technical Commission on Instruments and Methods of Observation (CIMO).

5. CONCLUSION AND OUTLOOK

An increasing number of commercial lidars systems is available and applied in lidar networks. Most likely, lidars will soon be used by institutions without lidar experts. Hence, standards for the systems and their possible products need to be elaborated.

The provision of a standard for the most basic lidar concept, the backscatter lidar, shows that efforts are necessary to extend the standardization work to more complex lidar techniques as for instance the Raman technique.

The lidar technique is an international technique and the atmosphere is per se a common good. So there is a strong need that lidar scientists work on international lidar standards.

REFERENCES

[1] http://www.dwd.de/ceilomap (05 June 2015).

[2] ftp://ftp.wmo.int/Documents/PublicWeb/arep/gaw/gaw178-galion-27-Oct.pdf (05 June 2015).

[3] Pappalardo, G., A. Amodeo, A. Apituley, A. Comeron, V. Freudenthaler, H. Linné, A. Ansmann, J. Bösenberg, G. D'Amico, I. Mattis, L. Mona, U. Wandinger, V. Amiridis, L. Alados-Arboledas, D. Nicolae, and M. Wiegner, 2014: EARLINET: towards an advanced sustainable European aerosol lidar network, *Atmos. Meas. Tech.*, **7**, 2389-2409, doi:10.5194/amt-7-2389-2014

[4] Ansmann, A., M. Tesche, P. Seifert, S. Groß, V. Freudenthaler, A. Apituley, K. M. Wilson, I. Serikov, H. Linné, B. Heinold, A. Hiebsch, F. Schnell, J. Schmidt, I. Mattis, U. Wandinger, and M. Wiegner, 2011: Ash and fine-mode particle mass profiles from EARLINET-AERONET observations over central Europe after the eruptions of the Eyjafjallajökull volcano in 2010, *J. Geophys. Res.*, 116, D00U02, doi:10.1029/2010JD015567.

[5] VDI 4210 Part 1, 1999: Remote sensing – Atmospheric measurements with LIDAR – Measuring gaseous air pollution with DAS LIDAR. German/English. Beuth Verlag, Berlin, 47 p.

[6] ISO 28902-1, 2012: Air quality – Environmental meteorology – Part 1: Ground-based remote sensing of visual range by lidar.

[7] ISO/DIS 28902-2, 2015: Air quality – Environmental meteorology – Part 2: Ground-based remote sensing of wind by heterodyne pulsed Doppler lidar.

[8] Wiegner, M., J. Gasteiger, S. Groß, F. Schnell, V. Freudenthaler, and R. Forkel, 2012: Characterization of the Eyjafjallajökull ash-plume, 2012: Potential of lidar remote sensing, J Phys. Chem. Earth, 45-46, p 79-86, doi: 10.1016/j.pce.2011.01.006.

[9] http://www.din.com (05 June 2015).

[10] http://www.vdi.eu (05 June 2015).

[11] http://www.vdi.eu/krdl (05 June 2015).

[12] http://www.iso.org (05 June 2015).

DOI: 10.1051/epjconf/201611923024

COMPARISON OF AEROSOL BACKSCATTER AND EXTINCTION PROFILES BASED ON THE EARLINET DATABASE AND THE SINGLE CALCULUS CHAIN FOR THESSALONIKI GREECE (2001-2014)

K. Voudouri[1], N. Siomos[1], E. Giannakaki[1], V. Amiridis[1], G. d'Amico[2] and D. S. Balis[1]

[1]*Laboratory of atmospheric physics, Physics Department, Aristotle University of Thessaloniki, Greece*

[2]*Consiglio Nazionale delle Ricerche, Istituto di Metodologie per l'Analisi Ambientale (CNR-IMAA), Tito Scalo, Potenza, Italy*

ABSTRACT

Aerosol backscatter and extinction coefficient profiles derived by the Single Calculus Chain (SCC) algorithm, which was developed within the European Aerosol Research Lidar Network (EARLINET) are compared with profiles derived by the operational inversion algorithm of Thessaloniki. Measurements performed during the period 2001-2014, that have already been uploaded in the EARLINET database, are considered in this study. The objective of this study is to verify, for the case of Thessaloniki, the consistency of the climatology of the aerosol profiles based on SCC and the EARLINET database data respectively. In this paper we show example comparisons for each lidar product submitted in the official database.

1. INTRODUCTION

Measurements within the period 2001-2014 have been performed at Thessaloniki, Greece (40.5N, 22.EE) with five different lidar setups. The initial setup for the lidar system of Thessaloniki included two elastic backscatter channels at 355nm and 532nm and a Raman channel at 387nm. A second Raman channel at 607nm was added in 2008. In 2011, two channels for the measurement of the cross and parallel polarized signal at 532nm were added. Finally, a third elastic backscatter channel at 1064nm was added in 2012.

All the measurements processed within the period 2001-2014 are available in the EARLINET database (doi:10.1594/WDCC/EN_all_measurements_2000-2010). The total number of uploaded files for the period is 1234, out of which 412 files correspond to elastic backscatter profiles at 355nm, 577 files to elastic backscatter profiles at 532nm, 22 files to elastic backscatter profiles at 1064nm, 187 files to Raman extinction and backscatter profiles at 355nm and 36 files to Raman extinction and backscatter profiles at 532nm.

2. SINGLE CALCULUS CHAIN

The SCC algorithm is designed to provide high quality standardized aerosol optical products in near real time along all the EARLINET network (*D'Amico et al.*, 2012). In brief, SCC uses raw lidar data as input in a common netCDF format for all EARLINET stations. The lidar signals are initially processed by the Preprocessor module which applies system-dependent corrections to the raw signals. After that, the aerosol backscatter and extinction coefficient profiles are derived by the optical processing module (ELDA: Earlinet Lidar Data Analysis). The software is installed on a centralized server and a web interface is available to upload the data and to start the analysis. The output data is in the EARLINET database netCDF format.

3. THESSALONIKI OPERATIONAL ALGORITHM

The inversion algorithm developed for the Thessaloniki lidar has been designed to provide aerosol optical products in operational processing. It has been validated in two algorithm intercomparisons that took place during EARLINET [*Böckmann et al.*, 2004; *Pappalardo et al.*, 2004].

DOI: 10.1051/epjconf/201611923024

For the retrieval of the elastic backscatter profiles, the Fernald-Klett retrieval method [*Klett*, 1981, 1985; *Fernald et al.*, 1972; *Fernald*, 1984] is implemented whereas for the retrieval of the Raman backscatter and extinction profiles the Raman retrieval method [*Ansmann et al.*, 1992] is being utilized. More on the retrievals of the operational algorithm can be found in *Amiridis et al.*, 2005 and *Giannakaki et al.*, 2010.

4. RESULTS

The optimal configuration of the Single Calculus Chain algorithm for each of the lidar setup of the Thessaloniki Station has been investigated in order to reduce the discrepancies in the comparisons with the EARLINET database data.

Two cases will be presented here. The first case (06-06-2005) has been processed with the Fernald-Klett method in order to derive the elastic backscatter coefficient at 355nm and at 532nm [Figure 1]. There is a good agreement between the two algorithms for both wavelengths. Differences due to noise are attributed to different smoothing techniques in the two algorithms.

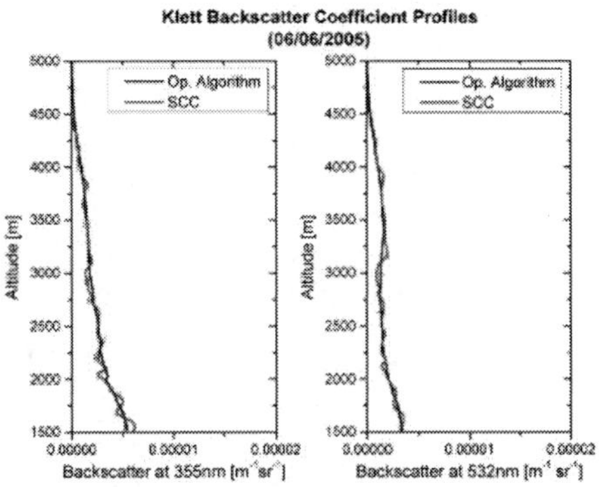

Figure 1. Comparison of the aerosol Klett backscatter coefficient profiles at 355nm 355nm (left) and 532nm (right) between the two algorithms for the case on 06/06/2005.

The second case (20-05-2013) has been processed with the Raman method in order to derive the Raman backscatter and extinction coefficients at 355nm and at 532nm [Figures 2 and 3]. The two algorithms produce quite similar Raman backscatter profiles for both wavelengths [Figure 2]. The extinction profiles show a very good agreement especially below 2km. The SCC products seem somewhat underestimated especially above 2km and below 1.3km [Figure 3]. Differences due to noise are attributed to different smoothing techniques in the two algorithms.

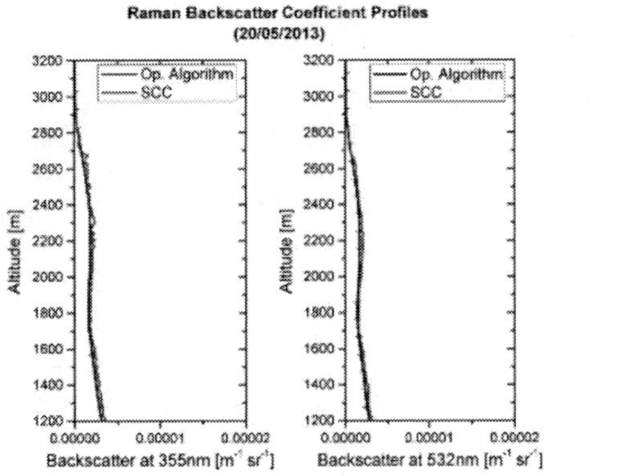

Figure 2. Comparison of the aerosol Raman backscatter coefficient profiles at 355nm (left) and 532nm (right) between the two algorithms for the case on 20/05/2013.

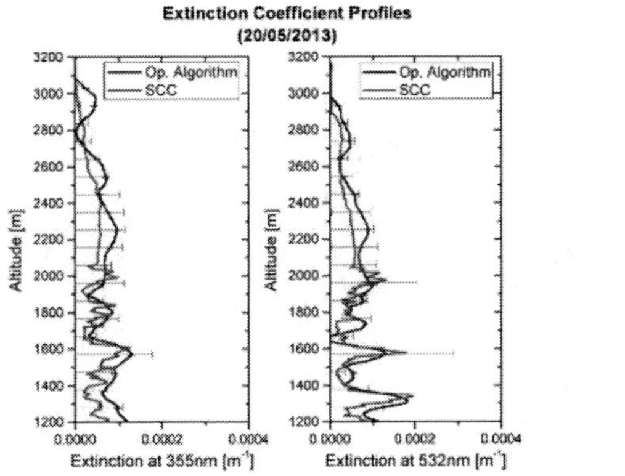

Figure 3. Comparison of the aerosol extinction coefficient profiles at 355nm (left) and 532nm (right) between the two algorithms for the case on 20/05/2013.

5. FUTURE PLANS

A complete comparison of the measurements that are processed with the operational algorithm of Thessaloniki and are uploaded in the

EARLINET database is in progress for the period 2001-2014. The objective is to verify the long-term consistency of the aerosol profiles based on SCC and the EARLINET database data respectively, in order to adopt the SCC for operational processing, without introducing inhomogeneities in the time series.

ACKNOWLEDGEMENT

The financial support by the European Commission grants RICA-025991 EARLINET-ASOS and 262254 ACTRIS is gratefully acknowledged.

REFERENCES

Amiridis, V., Balis, D. S., Kazadzis, S., Bais, A., Giannakaki, E., Papayannis, A., and Zerefos, C.: *Four-year aerosol observations with a Raman lidar at Thessaloniki, Greece, in the framework of European Aerosol Research Lidar Network (EARLINET)*, J. Geophys. Res., 110, 1–12, 2005.

Ansmann, A., Wandinger, U., Riebesell, M., Weitkamp, C., and Michaelis, W.: *Independent measurement of extinction and backscatter profiles in cirrus clouds by using a combined Raman elastic backscatter lidar*, Appl. Optics, 31, 7113–7131, 1992.

Böckmann, C., Wandinger, U., Ansmann, A., et al.: *Aerosol lidar intercomparison in the framework of the EARLINET project. 2. Aerosol backscatter algorithms*, Appl. Optics, 43, 977–989, 2004.

Fernald, F. G. (1984), *Analysis of atmospheric lidar observations: Some comments,* Applied Optics, 23:652–653

Fernald, F. G., B. M. Herman, and J. A. Reagan (1972), *Determination of aerosol height distributions by lidar*, J. Appl. Meteorol., 11, 482 – 489.

Giannakaki, E. , Balis, D. S., Amiridis V., and Zerefos, C.: *Optical properties of different aerosol types: seven years of combined Raman-elastic backscatter lidar measurements in Thessaloniki, Greece*, Atmos. Meas. Tech., 3, 569–578, 2010.

Klett, J. D. (1981), *Stable analytical inversion solution for processing lidar returns,* Applied Optics, 20:211–220.

Klett, J. D. (1985), *Lidar inversion with variable backscatter to extinction ratios*, Appl. Opt., 24, 1638 – 1643.

Pappalardo, G., Amodeo, A., Pandolfi, M., et al.: *Aerosol lidar intercomparison in the framework of the EARLINET project. 3. Raman lidar algorithm for aerosol extinction, backscatter, and lidar ratio*, Appl. Optics, 43, 5370–5385, 2004.

D'Amico, G., Binietoglou, I., Amodeo, A., Pappalardo, G., Baars, H., Engelmann, R., Wandinger, U., Mattis, I., Freudenthaler, V., Wiegner, M., Nicolae, D., Chaikovsky, A., Apituley, A., Adam, M., *Earlinet Single Calculus Chain for automatic lidar data processing: First tests on optical products*, Paper presented at the *26th International Laser Radar Conference, Porto Heli, Peloponnesus, Greece on June 25-29, 2012*

EPJ Web of Conferences **119**, 23025 (2016)
ILRC 27
DOI: 10.1051/epjconf/201611923025

THE POTENTIAL OF THE SYNERGY OF SUNPHOTOMETER AND LIDAR DATA TO VALIDATE VERTICAL PROFILES OF THE AEROSOL MASS CONCENTRATION ESTIMATED BY AN AIR QUALITY MODEL

N. Siomos[1]*, M. Filioglou[1,2], A. Poupkou[1], N. Liora[1], S. Dimopoulos[1], D. Melas[1], A. Chaikovsky[3] and D. S. Balis[1]

[1]*Laboratory of atmospheric physics, Physics Department, Aristotle University of Thessaloniki, Greece*

[2]*Finnish Meteorological Institute, Kuopio Unit, Finland*

[3]*Institute of Physics, National Academy of Science, Minsk, Belarus*

**Email: nsiomos@physics.auth.gr*

ABSTRACT

Vertical profiles of the aerosol mass concentration derived by the Lidar/Radiometer Inversion Code (LIRIC), that uses combined sunphotometer and lidar data, were used in order to validate the aerosol mass concentration profiles estimated by the air quality model CAMx. Lidar and CIMEL measurements performed at the Laboratory of Atmospheric Physics of the Aristotle University of Thessaloniki, Greece (40.5N, 22.9E) from the period 2013-2014 were used in this study.

1. ALGORITHMS AND MODELS

CAMx is an air quality model (ENVIRON, 2010) that can provide, among others, the vertical mass concentration profiles of PM2.5 and PM10 aerosol. Input data include meteorological fields from the WRF model, natural and anthropogenic emission data, landuse and photochemistry data. Natural emissions include windblown dust from soils, sea salt aerosol and vegetation non-methane volatile organic compounds (NMVOCs) formating secondary organic aerosols (SOA), while anthropogenic emissions include fossil fuel burning procedures like heating, industry, transportation and energy production at power plants. Waste treatment and disposal, agricultural activities like biomass burning and fertilization and extraction and distribution of fossil fuels and geothermal energy are also considered. It is important to mention that the emissions due to dust resuspension during agricultural activities and road traffic as well as the Saharan dust emissions are not currently included. Saharan dust emissions are taken into account only indirectly from the model's chemical boundary conditions taken from the chemistry transport model IFS-MOZART. A nesting technique is being used to increase accuracy near the area of interest and thus the spatial resolution is 30x30km in the European domain, 10x10km in the East Mediterranean domain and 2x2km around Athens and Thessaloniki. More on the simulations of CAMx for Thessaloniki can be found in *Poupkou et al.*, 2012.

LIRIC algorithm utilizes both sunphotometer data that have been processed by AERONET algorithms and elastic backscatter lidar data at three wavelengths (355nm, 532nm and 1064nm) in order to estimate the aerosol concentration profiles for fine and coarse particles. The algorithmic inversion is described in detail in *Chaikovski et al.* 2012.

2. METHODOLOGY

The aerosol mass concentration profiles of the fine and coarse mode derived by CAMx were compared with the respective profiles derived by the retrieval algorithm. For the coarse mode particles, forecasts of the Saharan dust transportation model BSC-DREAM8bV2 were also taken into account. Each of the retrieval algorithm's profiles was matched to the models' profile with the best agreement within a time window of four hours before and after the central measurement. Since LIRIC derived profiles are in ppb units it was necessary to convert them in μgr/m³ in order to compare them with the models' profiles (see e.g. *Kokkalis et al.*, 2013). We assumed a mean density of 1.5 μgr/m³ for the fine mode and 2.6 μgr/m³ for the coarse mode. These values are typical for the fine and coarse mode particles respectively (see e.g. *Kokkalis et al.*, 2013).

Mean concentration profiles of CAMx and LIRIC were derived for both modes. For the calculation of the mean bias error, the root mean square error and the temporal correlation coefficient it was necessary to interpolate LIRIC and CAMx profiles to the same vertical resolution.

Optical Properties of Aerosols and Clouds (OPAC), a software that can provide optical properties of aerosol mixtures [*Hess et al.*, 1998], was also employed in order to calculate the Angstrom exponent values for 355nm/532nm and 532nm/1064nm for each of the model's profiles, aiming for a comparison with the Angstrom exponent values derived by the retrieval algorithm for each measurement.

3. RESULTS

The mean concentration profiles for the fine mode particles and the respective mean bias error are presented below.

The comparison of the aerosol fine mode concentration profiles presented in Figure 1 shows good agreement between CAMx and the retrieval algorithm, especially above 1.5km. The mean bias error shown in Figure 2, is slightly above 5 μgr/m³ below 1km and ranges between 0-5 μgr/m³ in the layer between 1 and 5km. Only nine cases were included since the data availability of concurrent Lidar and CIMEL measurements and CAMx forecasts was low.

As far as the aerosol coarse mode concentration profiles is considered, both CAMx and BSC-DREAM8bV2 values are severely underestimated (not shown here), although, in cases of Saharan dust transportation events there is an agreement between the profiles of BSC-DREAM8bV2 model and the retrieval algorithm. A typical case is presented in Figure 3.

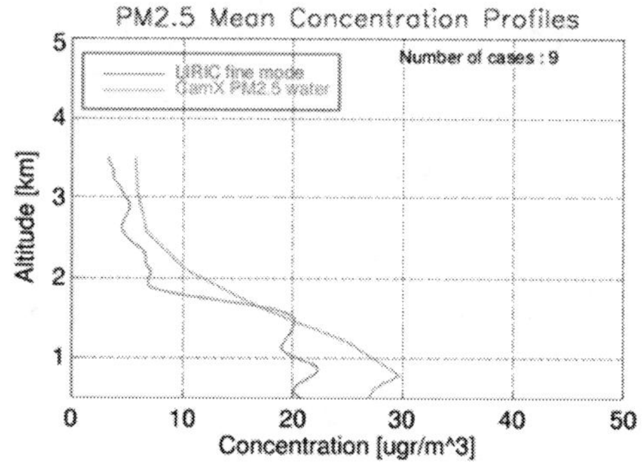

Figure 1. Comparison of the mean concentration profiles derived by CAMx and LIRIC for the fine mode particles.

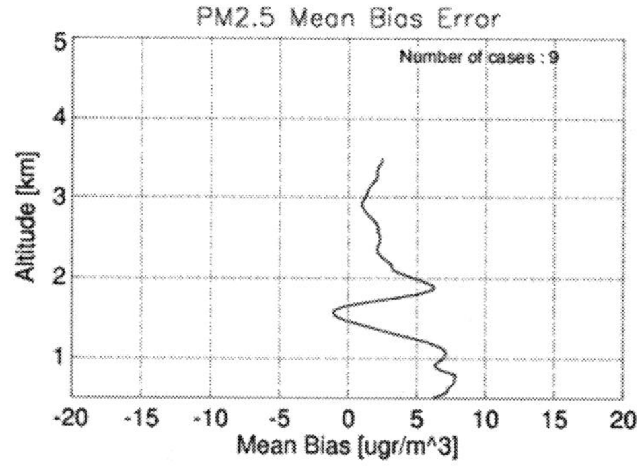

Figure 2. Mean bias error profile between CAMx and LIRIC for the fine mode particles.

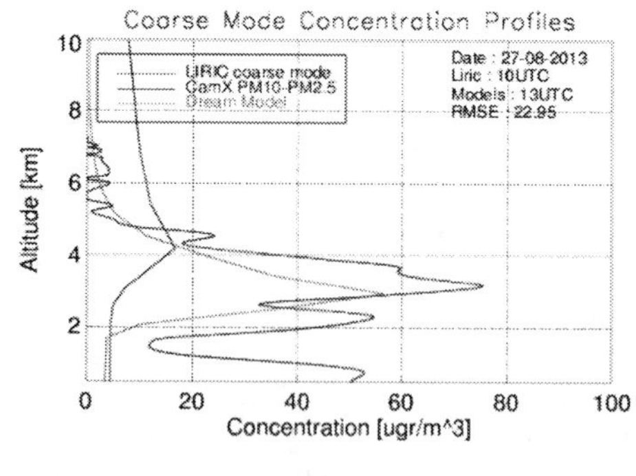

Figure 3. Comparison of the aerosol coarse mode concentration profiles between LIRIC [black], CAMx [blue] and BSC-DREAM8bV2 [red].

DOI: 10.1051/epjconf/201611923025

The aerosol Angstrom exponent comparison resulted in an underestimation of the values derived by OPAC for both 355nm-532nm and 532nm-1064nm spectral regions. The scatterplots of OPAC and LIRIC derived Angstrom exponent values are presented in Figures 4 and 5.

There is some linear behavior in both graphs. The Pearson correlation coefficient is 0.49 in the 355nm-532nm region and 0.60 in the 532nm-1064nm region. There are indications that CAMx is underestimating the sea salt emission. Greater sea salt values could lead to OPAC calculating smaller Angstrom values since sea salt consist mainly of coarse mode particles.

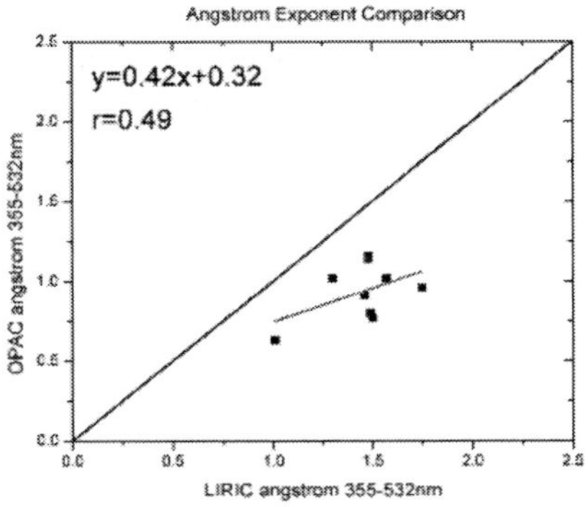

Figure 4. Scatterplot of the Aerosol Angstrom Exponent 355nm-532nm derived by LIRIC and OPAC.

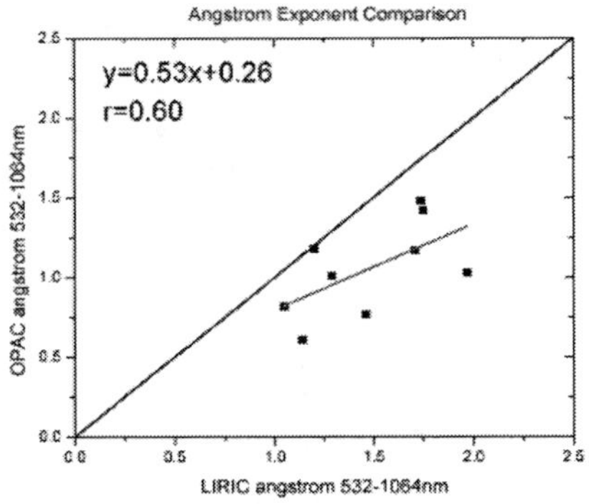

Figure 5. Scatterplot of the Aerosol Angstrom Exponent 532nm-1064nm derived by LIRIC and OPAC.

4. CONCLUSIONS

The comparison of the aerosol fine mode concentration profiles resulted in a good agreement between the model and the retrieval estimates, with the vertical mean bias mostly ranging between 0 and 5 μgr/m^3. As far as the aerosol coarse mode concentration profiles is considered, CAMx severely underestimates the aerosol concentration This probably occurs because aerosol emission sources outside the domain of CAMx, like Sahara desert, are not directly included in the model's emissions. In addition emissions from local dust sources are underestimated in CAMx. In cases of Saharan dust transport events there is a reasonable agreement between BSC-DREAM8bV2 model and the retrieval algorithm. The aerosol Angstrom exponent comparison resulted in an underestimation of the values derived by OPAC for both 355-532nm and 532-1064nm spectral regions, showing however a significant correlation of 0.5 to 0.6

5. FUTURE PLANS

Processing and comparison of additional cases in order to produce more accurate results is under progress. Furthermore the validation with a new run of CAMx is planned, which will use updated emission inventories.

ACKNOWLEDGEMENT

The financial support by the European Commission through ACTRIS project (EU INFRA-2010-1.1.16-262254) is gratefully acknowledged. CIMEL Calibration was performed at the AERONET-EUROPE calibration center, supported by ACTRIS (European Union Seventh Framework Program; FP7/2007-2013) under grant agreement no. 262254.

The IFS-MOZART simulations were performed in the framework of the MACC-III project (Monitoring Atmospheric Composition and Climate -III), Horizon 2020, under grant agreement no: 633080

DOI: 10.1051/epjconf/201611923025

REFERENCES

Chaikovski, A., Dubovik, O., Pappalardo, G., Wandinger, U., Apituley, A., Alados Arbroledas, L., Pappayannis, A., Pietruczuk, A. (2012), *Algorithm and Software for the retrieval of Vertical Aerosol Properties using combined Lidar/Radiometer data : Dissemination in EARLINET*, Reviewed and Revised Papers of the 26th International Laser Radar Conference, Porto Helli, Greece

ENVIRON International Corporation (2010), *CAMx - User's Guide - Comprehensive Air Quality Model with Extensions Version 5.30*, Novato, California

Hess, M. and Koepke, P. and Schult, I. (1998), *Optical Properties of Aerosols and Clouds: The Software Package OPAC*, American Meteorological Society

Kokkalis P., A. Papayannis, V. Amiridis, R. E. Mamouri, I. Veselovskii, A. Kolgotin, G. Tsaknakis, N. I. Kristiansen, A. Stohl5, and L. Mona (2013): *Optical, microphysical, mass and geometrical properties of aged 72volcanic particles observed over Athens, Greece, during the Eyjafjallajökull eruption in April 2010through synergy of Raman Lidar and sunphotometer measurements,* Atmos. Chem. Phys., 13, 9303–9320, 2013.

Poupkou, A., Markakis, K., Liora, N., Giannaros, T. and Melas, D., 2012. "Particulate matter source apportionment for the urban center of Thessaloniki (Greece) using the WRF-CAMx modeling system". *19th International Transport and Air Pollution Conference 2012*, 26-27 November 2012, Thessaloniki, Greece.

ARRANGE AND AVERAGE ALGORITHM FOR MICROPHYSICAL RETRIEVALS WITH A "3β+3α" LIDAR CONFIGURATION

Eduard Chemyakin[1]*, Detlef Müller[1,2], Sharon Burton[3], Chris Hostetler[3], Richard Ferrare[3]

[1]*Science Systems and Applications, Inc., NASA Langley Research Center, Mail Stop 475, Hampton, Virginia 23681-2199, USA, *Email: eduard.v.chemyakin@nasa.gov*

[2]*University of Hertfordshire, College Lane, Hatfield, AL10 9AB Hertfordshire, UK*

[3]*NASA Langley Research Center, Mail Stop 401 A, Hampton, Virginia 23681-2199, USA*

ABSTRACT

We present the results of a comparison study in which a simple, automated, and unsupervised algorithm, which we call the arrange and average algorithm, was used to infer microphysical parameters (complex refractive index (CRI), effective radius, total number, surface area, and volume concentrations) of atmospheric aerosol particles. The algorithm normally uses backscatter coefficients (β) at 355, 532, and 1064 nm and extinction coefficients (α) at 355 and 532 nm as input information. We compared the performance of the algorithm for the existing "3β+2α" and potential "3β+3α" configurations of a multiwavelength aerosol Raman lidar or high-spectral-resolution lidar (HSRL). The "3β+3α" configuration uses an extra extinction coefficient at 1064 nm. Testing of the algorithm is based on synthetic optical data that are computed from prescribed CRIs and monomodal logarithmically normal particle size distributions that represent spherical, primarily fine mode aerosols. We investigated the degree to which the microphysical results retrieved by this algorithm benefits from the increased number of input extinction coefficients.

1. INTRODUCTION

Aerosol particles affect the radiative energy balance in the atmosphere and thus influence Earth's climate [1]. Different types of aerosols, e.g., sea salt, desert dust, smoke from biomass burning, and emissions from the burning of fossil fuel, influence regional and global climate through the direct and indirect radiative effect, which may result in net cooling or net warming of the air, changes of the large-scale atmospheric circulation, cloud lifetime and occurrence, and intensity of precipitation [1]. In this context, light detection and ranging (lidar) instruments will play an important role as only these instruments provide information of aerosol properties on a comparably high vertical resolution. There are several fundamental approaches that allow one to retrieve microphysical parameters of aerosols from combinations of backscatter and extinction coefficients. The main techniques are Tikhonov's inversion with regularization [2–6], principal components analysis [7; 8], linear estimation [9], and recently we presented the arrange and average algorithm [10]. In our paper we considered four existing lidar configurations [10]. In this study we investigate if a "3β+3α" configuration has potential advantage over a "3β+2α" configuration. We have not seen yet any publication stating that the extinction coefficient at 1064 nm was successfully measured using Raman or HSRL technology but the "3β+3α" configuration is still possible via combination of lidar and column measurements of the atmospheric optical depth at 1064 nm. We investigated if this extra extinction coefficient could have a significant effect on the retrieval accuracy of microphysical properties.

2. METHODOLOGY

The general structure of the arrange and average algorithm is shown in study [10]. Few minor changes were done to adapt the algorithm for the needs of the "3β+3α" configuration. We increased the number of employed input optical data ratios from 11 for the "3β+2α" to 15 for the "3β+3α" lidar configuration [10]. The updated list contains six normalized input optical data

$$\beta^*_{355}=\frac{\beta_{355}}{\beta_3},\quad \beta^*_{532}=\frac{\beta_{532}}{\beta_3},\quad \beta^*_{1064}=\frac{\beta_{1064}}{\beta_3},$$

where $\beta_3=\sqrt{\beta^2_{355}+\beta^2_{532}+\beta^2_{1064}}$, and

$$\alpha^*_{355}=\frac{\alpha_{355}}{\alpha_3},\quad \alpha^*_{532}=\frac{\alpha_{532}}{\alpha_3},\quad \alpha^*_{1064}=\frac{\alpha_{1064}}{\alpha_3},$$

where $\alpha_3=\sqrt{\alpha_{355}^2+\alpha_{532}^2+\alpha_{1064}^2}$, and also we use nine extinction-to-backscatter ratios:

$$c_{355}^{355}=\frac{\alpha_{355}}{\beta_{355}},\quad c_{532}^{355}=\frac{\alpha_{355}}{\beta_{532}},\quad c_{1064}^{355}=\frac{\alpha_{355}}{\beta_{1064}},$$

$$c_{355}^{532}=\frac{\alpha_{532}}{\beta_{355}},\quad c_{532}^{532}=\frac{\alpha_{532}}{\beta_{532}},\quad c_{1064}^{532}=\frac{\alpha_{532}}{\beta_{1064}},$$

$$c_{355}^{1064}=\frac{\alpha_{1064}}{\beta_{355}},\quad c_{532}^{1064}=\frac{\alpha_{1064}}{\beta_{532}},\quad c_{1064}^{1064}=\frac{\alpha_{1064}}{\beta_{1064}}.$$

Thus, we kept the concept of using ratios of backscatter and extinction coefficients which allows to compare the pre-calculated and arbitrary optical data sets [10]. As the result of calibration simulations [10], the value of the reduction coefficient was increased from $w_{3\beta+2\alpha}$=0.375 to $w_{3\beta+3\alpha}$=0.476. The core of the arrange and average algorithm, i. e., the look-up table was used without any changes [10].

3. RESULTS

We performed numerical simulations using noise-free optical data. Figures 1–6 present the results. The figures display the statistical distribution of precision, which is achievable for the "$3\beta+3\alpha$" and "$3\beta+2\alpha$" lidar configurations. The left vertical axis of each plot shows the number of cases (the total number was 2880 optical data sets) that were tested [10]. The histogram displays the statistical distribution of errors of the retrieved microphysical parameters. The right vertical axis of each plot refers to the cumulative probability distribution. The cumulative probabilities are shown for the positive and for the negative retrieval errors. The left and right axes on the plots are scaled to each other. The maximum value of 61% shown on the right axis ("Cumulative probability, %") corresponds to the value of 1750 shown on the left axis ("Number of cases"). Positive values of the horizontal axes mean that the retrieval results overestimated the true values. Negative values show that the retrieval results underestimated the true values. Figures 1 and 2 show the retrieval statistics for the real and imaginary parts of the CRI. The horizontal axis of Figure 1 is split in increments of 0.05. We chose the stepsize of 0.05 because our goal is to achieve a precision of 0.05 for the retrieved real part of the CRI. The horizontal axis of Figure 2 is incremented in steps of 0.005, which we target regarding the retrieval precision of the imaginary part. Figures 3–6 show the statistics of the retrieved value of effective radius, total number, surface area, and volume concentrations. The stepsize in each case is 10%. In contrast to the CRI we show the results in terms of relative error. We consider 10% relative precision to be an ambitious goal for any algorithm that is used for the retrieval of these parameters. Table 1 summarizes the statistics of the parameters retrieved with the automated, unsupervised arrange and average algorithm for all 2880 optical data for both lidar configurations. The precision of the total number concentration estimation is worse than for any other size property. Considering all simulated cases, the effective radius has the smallest values of relative error (see the "100%" subcolumn) and has the same order of magnitude as the true effective radius in the majority of cases. The total surface area concentration has the smallest retrieval uncertainties with regard to the "68.2%" subcolumn (see the well-pronounced central peaks in Figure 5). The usage of an extra extinction coefficient by "$3\beta+3\alpha$" configuration allows to reduce the maximal value of retrieval error for all size properties (see the "100%" subcolumn).

4. CONCLUSIONS

Using our novel arrange and average algorithm we estimated its performance for one extra lidar configuration. The "$3\beta+3\alpha$" option was skipped in our key publication [10]. The comparison between the most advanced of existing "$3\beta+2\alpha$" and potential "$3\beta+3\alpha$" lidar configurations was done by using only noiseless input data. We obtained a quick-look impression about benefits of the increase of the number of extinction coefficients being used for the retrievals. The best overall performance of the arrange and average algorithm in this study was achieved with the "$3\beta+3\alpha$" configuration, however the benefit of this extra extinction coefficient does not seem to be significant. The "$3\beta+2\alpha$" results statistically are slightly worse. Thus, we confirmed a similar trend that was noticed earlier, i. e., the increase of number of used optical coefficients slowly improves the precision of microphysical products [10].

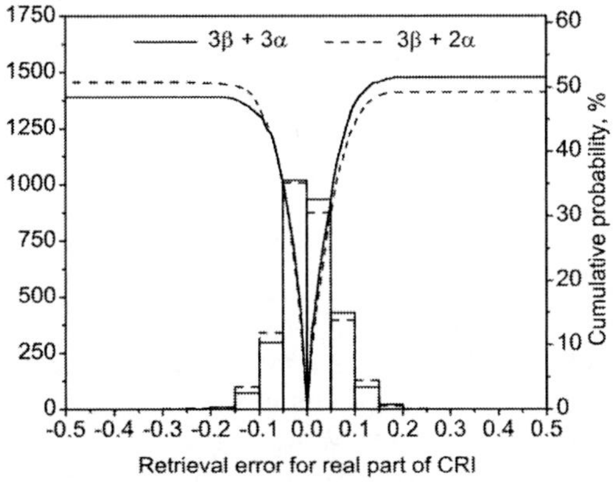

Figure 1: Retrieval of the real part of the CRI for the "3β+3α" lidar configuration (solid line) and the "3β+2α" configuration (dashed line). Negative values indicate overestimation in the retrievals whereas positive values mean underestimation.

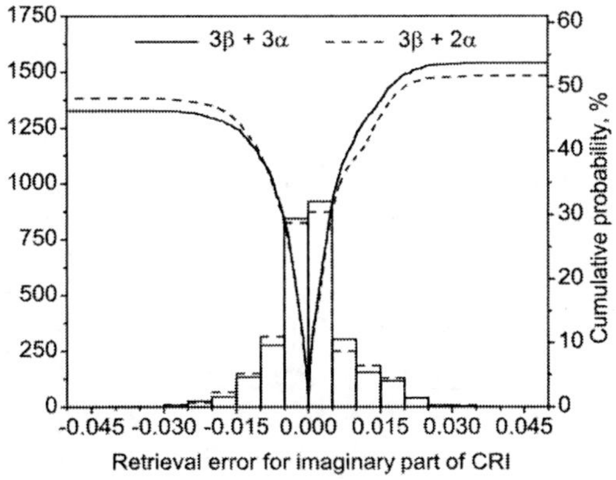

Figure 2: Retrieval of the imaginary part of the CRI. The notation is the same as in Figure 1.

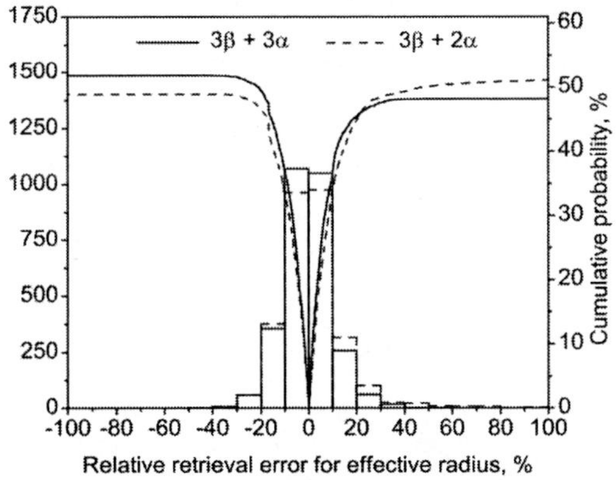

Figure 3: Retrieval of the effective radius. The notation is the same as in Figure 1.

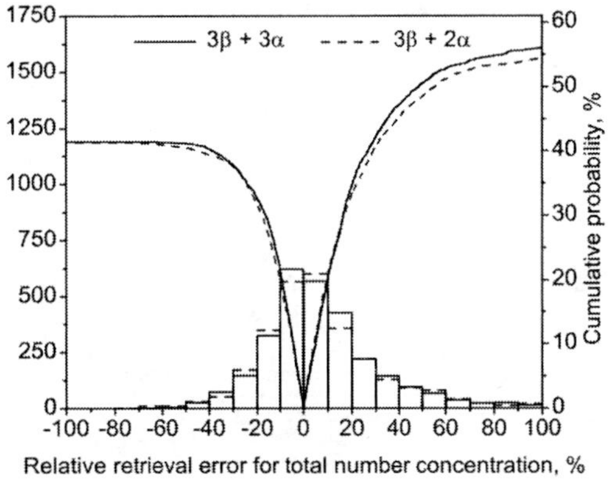

Figure 4: Retrieval of the total number concentration. The notation is the same as in Figure 1.

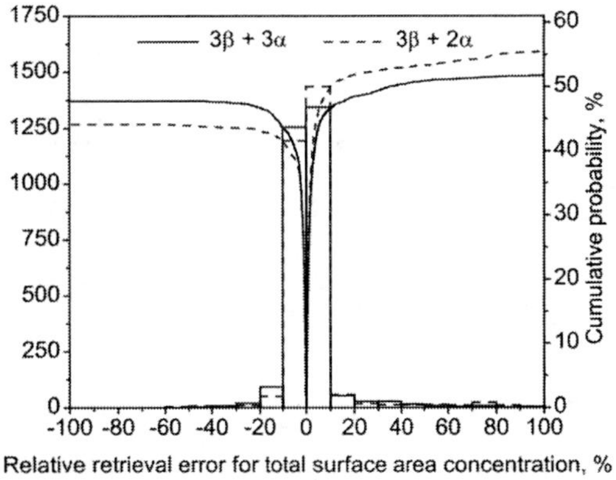

Figure 5: Retrieval of the total surface area concentration. The notation is the same as in Figure 1.

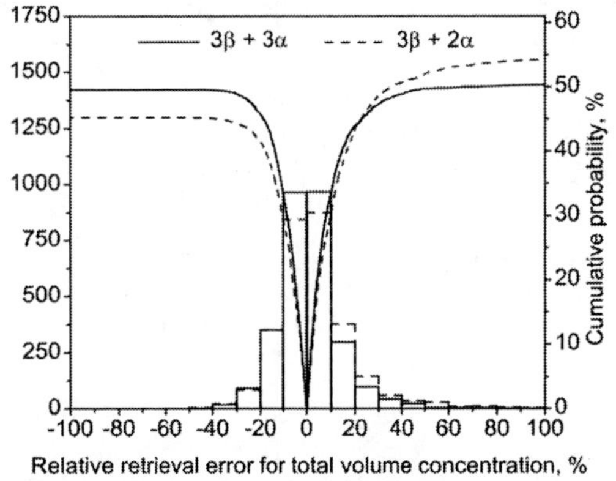

Figure 6: Retrieval of the total volume concentration. The notation is the same as in Figure 1.

Table 1: Retrieval Errors for the Microphysical Parameters[a]

Lidar Configuration	68.2%	95.4%	100%
Retrieval error for real part of CRI			
"$3\beta+3\alpha$"	0.051	0.114	0.184
"$3\beta+2\alpha$"	0.054	0.116	0.256
Retrieval error for imaginary part of CRI			
"$3\beta+3\alpha$"	0.0063	0.0182	0.0374
"$3\beta+2\alpha$"	0.0067	0.0182	0.0313
Relative retrieval error for effective radius, %			
"$3\beta+3\alpha$"	9	21	54
"$3\beta+2\alpha$"	10	25	107
Relative retrieval error for total number concentration, %			
"$3\beta+3\alpha$"	20	70	340
"$3\beta+2\alpha$"	22	94	2218
Relative retrieval error for total surface area concentration, %			
"$3\beta+3\alpha$"	2	21	186
"$3\beta+2\alpha$"	2	21	253
Relative retrieval error for total volume concentration, %			
"$3\beta+3\alpha$"	10	28	125
"$3\beta+2\alpha$"	12	38	173

[a]Shown are the precisions being achieved for 68.2% (analog of one-standard deviation), 95.4% (analog of two-standard deviations), and 100% out of 2880 of optical data.

REFERENCES

1. Solomon, S., D. Qin, M. Manning, M. Marquis, K. Averyt, M. Tignor, H. Miller, Z. Chen, 2007: The Physical Science Basis. Contribution of Working Group I to the Fourth Assessment Report of the Intergovernmental Panel on Climate Change, *Cambridge University Press, Cambridge.*

2. Müller, D., U. Wandinger, A. Ansmann, 1999: Microphysical particle parameters from extinction and backscatter lidar data by inversion with regularization: theory, *Appl. Opt.*, **38**, 2346-2357.

3. Müller, D., U. Wandinger, A. Ansmann, 1999: Microphysical particle parameters from extinction and backscatter lidar data by inversion with regularization: simulation, *Appl. Opt.*, **38**, 2358-2368.

4. Veselovskii, I., A. Kolgotin, V. Griaznov, D. Müller, U. Wandinger, D. N. Whiteman, 2002: Inversion with regularization for the retrieval of tropospheric aerosol parameters from multiwavelength lidar sounding, *Appl. Opt.*, **41**, 3685-3699.

5. Veselovskii, I., A. Kolgotin, V. Griaznov, D. Müller, K. Franke, D. N. Whiteman, 2004: Inversion of multiwavelength Raman lidar data for retrieval of bimodal aerosol size distribution, *Appl. Opt.*, **43**, 1180-1195.

6. Kolgotin, A., D. Müller, 2008: Theory of inversion with two-dimensional regularization: profiles of microphysical particle properties derived from multiwavelength lidar measurements, *Appl. Opt.*, **47**, 4472-4490.

7. Donovan, D. P., A. I. Carswell, 1997: Principle component analysis applied to multiwavelength lidar aerosol backscatter and extinction measurements, *Appl. Opt.*, **36**, 9406-9424.

8. De Graaf, M., A. Apituley, D. Donovan, 2013: Feasibility study of integral property retrieval for tropospheric aerosol from Raman lidar data using principal component analysis, *Appl. Opt.*, **52**, 2173-2186.

9. Veselovskii, I., O. Dubovik, A. Kolgotin, M. Korenskiy, D. N. Whiteman, K. Allakhverdiev, F. Huseyinoglu, 2012: Linear estimation of particle bulk parameters from multiwavelength lidar measurements, *Atmos. Meas. Tech.*, **5**, 1135-1145.

10. Chemyakin, E., D. Müller, S. Burton, A. Kolgotin, C. Hostetler, R. Ferrare, 2014: Arrange and average algorithm for the retrieval of aerosol parameters from multiwavelength high-spectral-resolution lidar / Raman lidar data, *Appl. Opt.*, **53**, 7252-7266.

A SENSITIVITY STUDY OF LIRIC ALGORITHM TO USER-DEFINED INPUT PARAMETERS, USING SELECTED CASES FROM THESSALONIKI'S MEASUREMENTS

Filioglou M. [1, 2]*, Balis D. [1], Siomos, N. [1], Poupkou, A.[1], Dimopoulos, S. [1], Chaikovsky, A. [3]

[1] *Laboratory of Atmospheric Physics, Aristotle University of Thessaloniki, Greece*

[2] *Finnish Meteorological Institute, Atmospheric Research Center of Eastern Finland, Kuopio, Finland*

[3] *Institute of Physics, National Academy of Science, Minsk, Belarus*

**Email: mfiliogl@physics.auth.gr, maria.filioglou@fmi.fi*

ABSTRACT

A targeted sensitivity study of the LIRIC algorithm was considered necessary to estimate the uncertainty introduced to the volume concentration profiles, due to the arbitrary selection of user-defined input parameters. For this purpose three different tests were performed using Thessaloniki's Lidar data. Overall, tests in the selection of the regularization parameters, an upper and a lower limit test were performed. The different sensitivity tests were applied on two cases with different predominant aerosol types, a dust episode and a typical urban case.

1. INTRODUCTION

The LIdar Radiometer Inversion Code (LIRIC) has been developed within ACTRIS by the Institute of Physics in Minsk (Belarus) in collaboration with the Laboratoire d'Optique Atmosphérique in Lille, (France). LIRIC combines active and passive remote sensing techniques from two different instruments, a Lidar and a Cimel sunphotometer. In this work, measurements from the multiwavelength Raman-Lidar and the Cimel sunphotometer which are located at the Laboratory of Atmospheric Physics (LAP) in Aristotle University of Thessaloniki (AUTH), (Greece), (40.6° N, 22.9° E, 60 m a.s.l) are being used. The three elastic backscatter channels (355, 532, 1064 nm) along with the column volume concentration derived by the sunphotometer paired to produce profiles of fine and coarse mode concentrations. If a depolarization channel is added, a further distinction in spheroid and spherical mode is possible. A detailed description of LIRIC can be found here [2] [4] [6].

The sensitivity tests were applied on two different cases. The first one is a typical urban pollution case detected at Thessaloniki on the 14th of January 2014 and the second corresponds to a Saharan dust episode captured on the 20th of January 2014 by the Lidar system.

2. METHODOLOGY

The main outputs of the algorithm are concentration profiles of fine and coarse mode expressed in parts per billion (ppbv). These concentrations were converted to $\mu gr/m^3$ so as to obtain comparable to the models' output data [5].

Firstly, a Regularization parameters test was performed for each case so as to determine the deviations on the retrieved quantities and also to conclude to a proper set of parameters for the rest of the retrievals. Once concluded to a proper set of parameters, an Upper limit test was performed continued by a Lower limit test.

The regularization parameters consist of 3 different sets referring to Lidar, sunphotometer and smoothness restrains. Smoothness parameters were remained unchanged while the others varied according to Table 1. More than 1800 possible combinations between these orders of magnitude where performed so as to evaluate the uncertainties introduced to the results. A succeeded retrieval was considered as the one where each signal's AOD output did not differ more than 5% of the AERONET's AOD and in addition the column integrated volume concentrations did not differ more than 5% of the

DOI: 10.1051/epjconf/201611923027

AERONET's one. The Upper limit test was performed by cutting the signal at three different heights above the ref. height (~500 m above) while in the Lower limit test a step of ± 200 m was applied above/below the height of the full overlap. Each test was done separately so as to determine correctly the deviations occurred. Errors were calculated by averaging the arising signals.

The photochemical model CAMx [3] has been used for the vertical concentration profiles of fine mode and the BSC-DREAM8b v2.0 [1] web based model for the comparison of the coarse mode.

Table 1: Range of Regularization Parameters.

	Parameter	*Range*
Lidar	k_{355}	from 10^{-6} to 10^{-3}
	k_{532}	from 10^{-6} to 10^{-3}
	k_{1064}	from 10^{-6} to 10^{-3}
Cimel	f_{fine}	from 10^{-1} to 10^{3}
	f_{coarse}	from 10^{-1} to 10^{3}
Smooth	d_{fine}, d_{coarse}	2

3. RESULTS

Table 2: Overview of input parameters and tests performed

	Urban case	*Dust case*
z_L **(m) :**	975	1005
z_{RH} **(m) :**	3450	4950
z_U **(m) :**	3600	5100
k_{355} **:**	$9 * 10^{-5}$	$2 * 10^{-4}$
k_{532} **:**	$5 * 10^{-4}$	$2 * 10^{-4}$
k_{1064} **:**	$4 * 10^{-4}$	$9 * 10^{-5}$
f_{fine} **:**	5	1
f_{coarse} **:**	0.7	5

The overall input parameters for the two cases are summarized in Table 2. Each of the following figures show the mean profile with one standard deviation which was calculated by averaging the arising available signals (e.g. for the upper/lower limit tests there were three arising signals (three different cuts as described above)).

Case I: 13 January 2014

a) Regularization parameters test

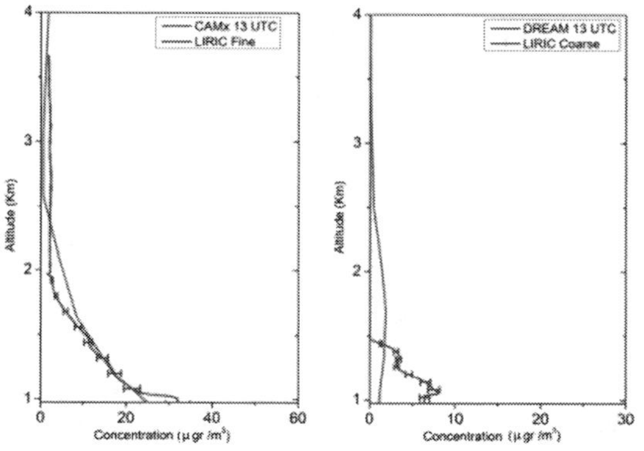

Figure 1: Mean profiles and standard deviations of fine and coarse mode derived by LIRIC (cyan) compared with models (dark yellow) for 13th January 2014.

b) Upper Limit test

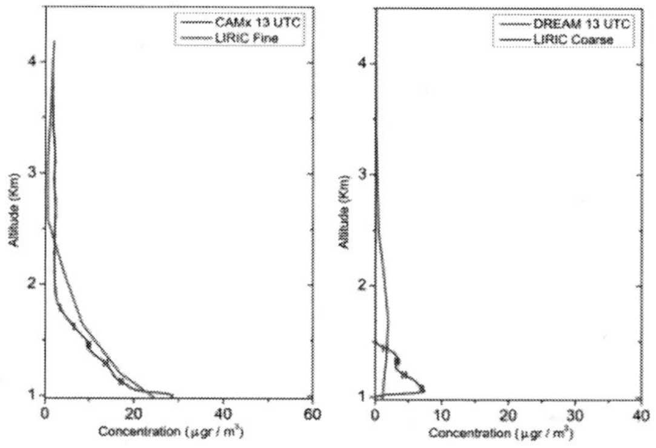

Figure 2: Respective results when varying the Upper limit height for 13th January case.

DOI: 10.1051/epjconf/201611923027

c) Lower limit test

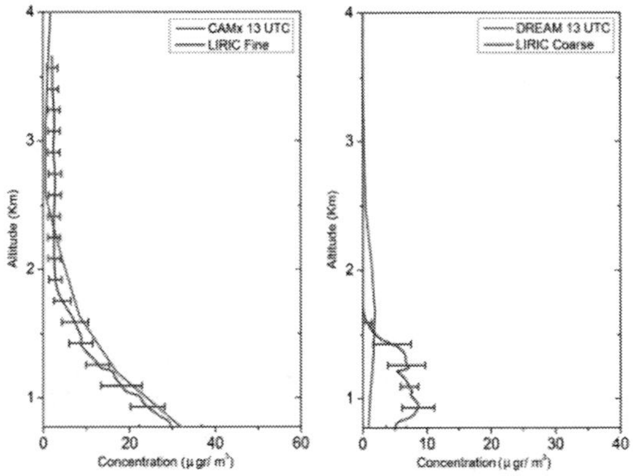

Figure 3: Respective results when varying the Lower limit height for 13th January case.

Case II: 20 January 2014

a) Regularization parameters test

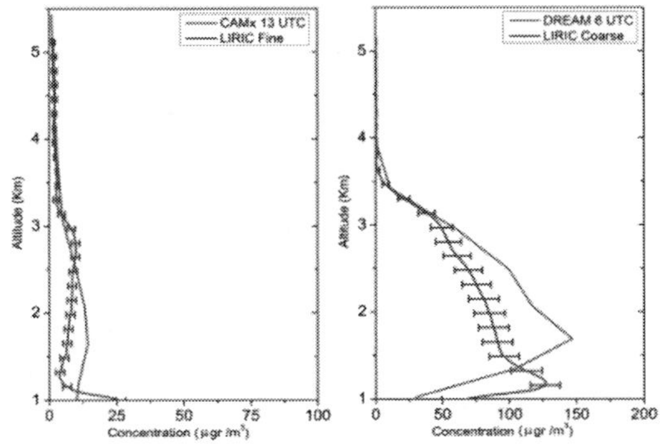

Figure 4: Respective results for the dust event occurred at 20 Jan. 2014.

b) Upper Limit test

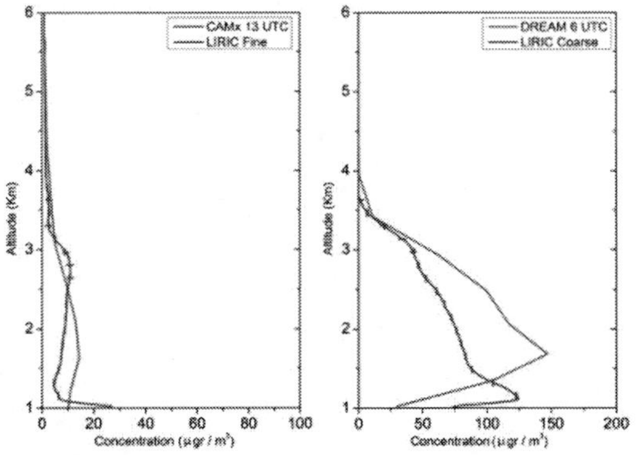

Figure 5: Respective results when varying the Upper Limit height for 20th January case.

c) Lower limit test

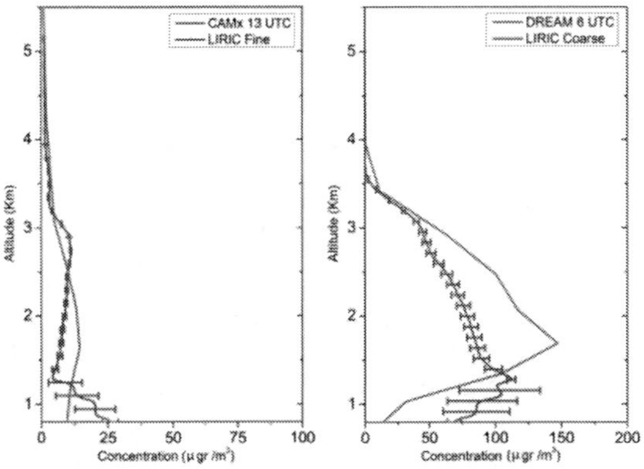

Figure 6: Respective results when varying the Lower Limit height for 20th January case.

4. CONCLUSIONS

The arbitrary selection of the regularization parameters test introduces an uncertainty of 10% in the lower part and 2.5 % in the upper part of the concentration profiles, when examining a typical urban aerosol case. The uncertainty due to the choice of the Lower limit ranged between 25-30% in the lower part for both modes and 40% for the fine mode in the upper part.

For the dust case, the uncertainty introduced by the choice of the regularization parameters is larger and ranged between 10 – 30 % below 3 km and decreases further above. The uncertainties due to the lower limit selection ranged above 30% in the lower part for both modes (at some heights reached 80%).

The upper limit selection introduced 1% variation at both cases.

Good structural agreement between CAMx and LIRIC was found in both cases for the fine mode. Good structural agreement for DREAM and LIRIC was found for the dust case, though there are differences in the height of the maximum concentration.

ACKNOWLEDGEMENTS

Special thanks to A. Chaikovsky and the associate team for providing LIRIC publicly. The algorithm has been developed within ACTRIS project. The ground-based EARLINET LIDAR database has been downloaded from www.earlinet.org .

REFERENCES

[1] Basart, S., Pérez, C., Nickovic, S., Cuevas, E. & Baldasano, J.M.: Development and evaluation of the BSC-DREAM8b dust regional model over Northern Africa, the Mediterranean and the Middle East. Tellus B, 64, 1-23, 2012.

[2] Chaikovsky A., Dubovik O., Goloub P., Tanré D., Pappalardo G., Wandinger U., Chaikovskaya L., Denisov S., Grudo Y., Lopatsin A., Karol Y., Lapyonok T., Korol M., Osipenko F., Savitski D., Slesar A., Apituley, A., Arboledas L. A., Binietoglou I., Kokkalis P., Granados Muñoz M. J., Papayannis A., Perrone M. R., Pietruczuk A., Pisani G., Rocadenbosch F., Sicard M., De Tomasi F., Wagner, J., and Wang X.: Algorithm and software for the retrieval of vertical aerosol properties using combined lidar/radiometer data: Dissemination in EARLINET, Reviewed and revised papers presented at the 26th International Laser Radar Conference, Porto Heli, Greece, 399–402, 2012.

[3] ENVIRON. User's Guide to the Comprehensive Air Quality Model with Extensions (CAMx) Version 6.10. ENVIRON International Corporation, 101 Rowland Way, Suite 220, Novato, California 94945-5010, 2014.

[4] Granados-Muñoz M. J., Guerrero-Rascado J. L., Bravo-Aranda J. A., Navas-Guzmá, F. , Valenzuela A., Lyamani H., Chaikovsky A., Wandinger U., Ansmann A., Dubovik O., Grudo J. O., Alados-Arboledas L.: Retrieving aerosol microphysical properties by LIdar-Radiometer Inversion Code (LIRIC) for different aerosol types. Journal of Geophysical Research Atmospheres 04/2014; 119(8), 2014.

[5] Kokkalis P., A. Papayannis, V. Amiridis, R. E. Mamouri, I. Veselovskii, A. Kolgotin, G. Tsaknakis, N. I. Kristiansen, A. Stohl, and L. Mona: Optical, microphysical, mass and geometrical properties of aged volcanic particles observed over Athens, Greece, during the Eyjafjallajökull eruption in April 2010 through synergy of Raman Lidar and sunphotometer measurements. Atmos. Chem. Phys., 13, 9303–9320, 2013.

[6] Wagner, J., Ansmann, A., Wandinger, U., Seifert, P., Schwarz, A., Tesche, M., Chaikovsky, A., and Dubovik, O.: Evaluation of the Lidar/Radiometer Inversion Code (LIRIC) to determine microphysical properties of volcanic and desert dust, Atmos. Meas. tech., 6, 1707–1724, doi:10.5194/amt-6-1707-2013, 2013.

DOI: 10.1051/epjconf/201611923028

CAN ULTRA-VIOLET MIE LIDAR BE AN EFFECTIVE INSTRUMENT DURING HIGH POLLUTION EPISODE?

Qiaojun Liu[1]*, Zhangjun Wang[1], Libin Du[1], Andrew Yuk Sun Cheng[1], Xiangqian Meng[1], Chao Chen[1], Xianxin Li[1], Xingtao Liu[1]

[1]Shandong Academy of Sciences Institute of Oceanographic Instrumentation, Qingdao266001, China,
**Email: qiaojunliu@163.com*

ABSTRACT

Vertical profiles of particulates were measured by a 355 nm Mie scattering lidar during a dust-storm event. A high energy pulse laser was employed as the light source to detect the extinction coefficient in the altitude up to 6 km in the day and 9 km at night. The extinction profiles showed layers of high aerosol concentrations in good agreement with ground-based pollution measurements, which indicated that such ultra-violet lidar is a very useful remote sensing instrument for monitoring extinction profiles during extreme high aerosol loading and low visibility atmospheric conditions when low energy lidar systems cannot obtain effective results.

1. INTRODUCTION

Air quality is always of great concerns. The particulate concentration keeps increasing in China over the past 20 years with the rapid economic development. Numerous ground stations make continuous round-the-clock point measurements of a range of pollutants, but they seldom provide the pollutants vertical profile which is important for modelling the transport of pollution. With good spatial and temporal resolution, real-time measurements can now be obtained from remote sensing technique such as lidar. Lidar has been widely employed to study aerosol spatial distribution [1] and explore the transport of aerosols over larger areas [2]. During episodes of dust event, lidars are important monitoring instruments [3, 4].

There is an internal relation between aerosol extinction and mass concentration. Without information on the particle size distribution and complex refractive index, the aerosol mass concentration can be retrieved from Mie lidar data using statistical correlations [5]. It is possible to extract aerosol mass concentrations within a reasonable accuracy of 20-30% without supplementary information on the aerosols [6, 7]. With varying refractive index, Del Guasta [8] obtained aerosol mass concentrations 40 m above the ground from lidar measurements.

In this paper, rough proportionality between accumulation-mode aerosol mass concentration and lidar extinction is reported. The current work aims to infer vertical profiles of PM_{10} using a single channel lidar. Mass concentrations are inferred by establishing empirical correlations between PM_{10}, visibility and extinction coefficient. In Section 2, the lidar instrumentation and data analysis method are described. Then, in Section 3, some results and discussions from dust case measurement are presented. And Section 4 is the conclusion.

2. METHODOLOGY

The lidar system used for the study, is a single wavelength Mie lidar located on top of the Taipa Grande in Macao at 112 m above sea level. The third harmonics of Nd:YAG laser is chosen as the emitter laser source with a pulse width of 5.7 ns at 50 Hz. The maximum pulse energy of the laser is 160 mJ, which is set to at most 50% of the maximum capacity in routine aerosol monitoring.

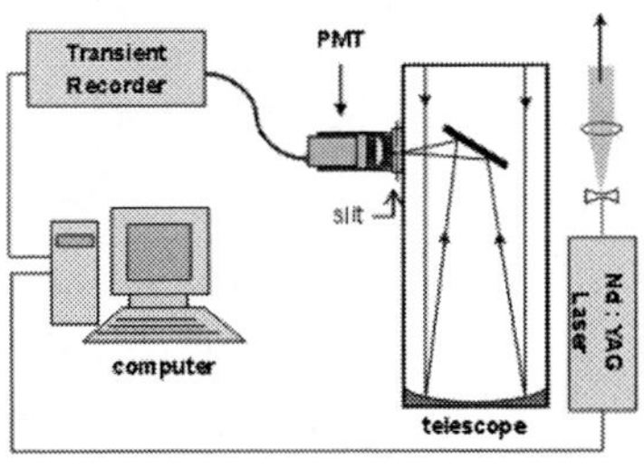

Figure 1. Configuration of 355 nm bi-static aerosol lidar system

The system layout is depicted in Figure 1. In the daytime, the sky background radiation in ultraviolet wavelength range is much weaker than that in the visible wavelength range. Under extremely high atmospheric extinction conditions this lidar can still obtain aerosol profiles up to 6 km in the day and 9 km at night.

The Fernald's method [9] is used to retrieve the extinction coefficient. The molecular Rayleigh scattering values referred were taken from the US Standard Atmosphere (1976), the lidar ratio was assumed to be 50 sr for all data presented here, based on data from a Raman lidar operated in the Pear River Delta region near Hong Kong [10]. The boundary condition was the assumption that the lidar could transmit far enough and the far range aerosol extinction was zero. In general, aerosol concentrations are higher and vary sharply near the surface, therefore, capturing values of the near range is important for air quality studies. Near range correction of the lidar signal followed the methods proposed by A.Y.S Cheng et al. [11] and Liu et al. [12].

3. RESULTS

Figure 2 presents the PM_{10} and $PM_{2.5}$ concentration in the dust episode in March, 2010. A significant increase of PM_{10} and $PM_{2.5}$ concentration was found after March 21st. Before that day, the PM_{10} concentration was much lower, and the average value was only 60.40μg/m³. PM_{10} concentration had a weak growth on 21st, however, at 18:00 on 21st, it increased rapidly. At 5:30 on 22nd it was up to the maximum value 616.83 μg/m³, which was ten times more than the average value before. Then the high concentration maintained for a period of time and slowly decreased subsequently. Data showed the concentration decreased to 30.45 μg/m³ on 24th. By comparing the two lines in figure 2, it can be found that during the dust period, particles of large size (2.5 ~ 10μm) are three times more than the small particles (2.5μm or less).

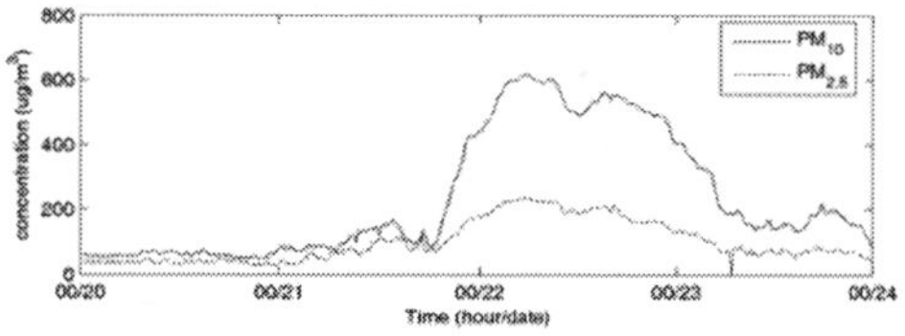

Figure 2. Particulates concentration between March 20th and March 24th

The extinction coefficients inverted from Mie lidar *in suit* measurement are present in Fig.3. Each of these profiles is the average result in an hour, with the starting time on the top. It can be seen that on 22nd, when the pollution is heavy, the particles congregation in the low altitude was evident. The extinction coefficients near the ground exceeded 0.7 km⁻¹, which decreased rapidly from the maximum value about 0.9 km⁻¹ at the low height to no larger than 0.3 km⁻¹ at about 0.9 km. However, on 26th, the moderate pollution day, the downtrend of extinction was not as sharp, and the near-surface extinction coefficients on 22nd was larger than that on 26th.

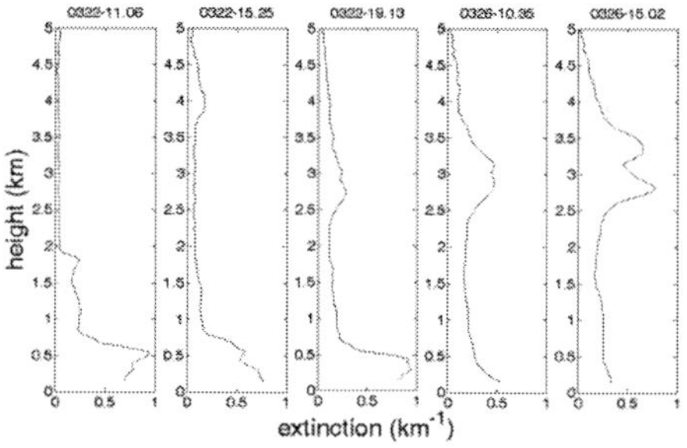

Figure 3. Extinction coefficients inverted from Mie lidar

Comparing the extinction coefficients with PM_{10} concentrations in heavy polluted episode with that of moderate polluted episode, we find an interesting phenomenon. That is, though the PM_{10} concentration has fallen from high pollution episode value (average value of 527.49μg/m³ on the 22nd) to a normal level (average value of 97.25μg/m³ on 26th, which is less than one-fifth of the high pollution episode) before the dust-storm event, the near-surface extinction coefficients from lidar measurements didn't show such difference. Lidar measurements showed that the impact by dust particles had not yet been over on 26th. There are two main reasons for explaining the disagreement between PM_{10} concentrations and extinction coefficients. On one hand, as is known, the variation of return signals caused by different particle size-distribution will impact the extinction coefficient retrieval from lidar measurement. In the heavy polluted episode, the bigger particles are the major component of aerosol. But they have a weaker influence on extinction in our back-scattering Mie lidar system, for their scattering energy is mainly concentrated in the forward. On the other hand, compared to the rapidly dropped bigger particles, the smaller particles can stay longer in the atmosphere (about one week). So extinction coefficients on 26th are still at a high level.

Based on these two reasons, extinction coefficients at 355nm are not sufficient to show the impact of dust particles on atmosphere. Figure 4 shows the near-surface extinction coefficients at 1064nm converted from 355nm results [13] and PM_{10} concentration values at the same time. The correlation is helpful for inferring the vertical profiles of PM_{10} concentration in the dust episode and tracing the possible aerosol sources.

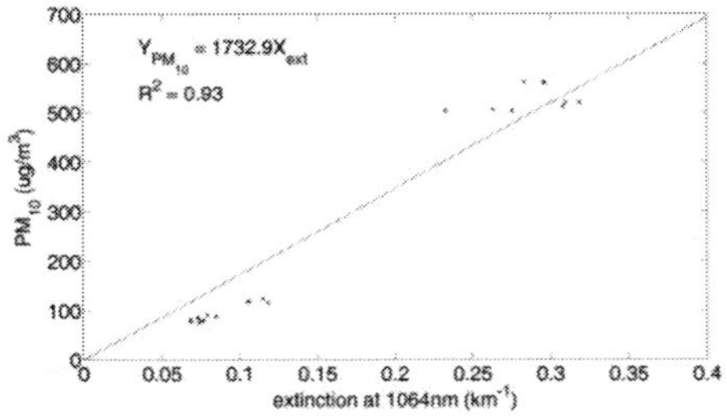

Figure 4. The agreement between extinction coefficient inverted from lidar and PM_{10} concentration

4. CONCLUSIONS

Aerosol extinction coefficients were measured by using the ultra-violet Mie lidar in a high pollution episode, which is the first study of aerosol extinction vertical profiles in a dust-storm event using lidar in Macao. The results show that the ultra-violet lidar has obtained good vertical profiles during the high pollution environment when other low energy lidar systems are ineffective.

ACKNOWLEDGEMENT

The work described in this paper is supported by the Shandong province excellent young and middle-aged scientists research award fund with grant No. BS2014HZ009, the Progress Program of Science and Technology of Shandong with grant No. 2014GSF117006, the Public science and technology research funds projects of ocean with grant No. 201205036, the International Science & Technology Cooperation Program of China with grant No. 2011DRF20090 and the Shandong province excellent young and middle-aged scientists research award fund with grant No. BS2012HZ014.

REFERENCES

[1]Pelon, J., C. Flamant, P. Chazette, J.-F. Leon, D. Tanre, M. Sicard, and S. K. Satheesh, 2002: Characterization of aerosol spatial distribution and optical properties over the Indian Ocean from airborne LIDAR and radiometry during INDOEX'99, *J. Geophys. Res.*, **107**(D19), 8029.

[2]Strawbridge, K. B., Snyder, B. J., 2004: Daytime and nighttime aircraft measurements showing evidence of particulate matter transport into the northeastern valleys of the Lower Fraser Valley BC, *Atmospheric Environment*, **38**, 5873-5886.

[3]G. P. Gobbi, F. Barnaba, R. Giorgi, and A. Santacasa, 2000: Altitude-resolved properties of a Saharan dust event over the Mediterranean, *Atmospheric Environment* ,**34**, 5119-5127

[4]Chenbo Xie, Tomoki Nishizawa, Nobuo Sugimoto, Ichiro Matsui, and Zifa Wang, 2008: Characteristics of aerosol optical properties in pollution and Asian dust episodes over Beijing, China. *Applied Optics*, **47(27)**, 4945-4951.

[5]Tam, K., et al, 2008: Studies of Urban Aerosols in Macao using a Horizontal Mie Lidar, *Journal of Light Scattering*, **20(4)**, 375-378.

[6]Kent, G.S., 1978: Deduction of aerosol concentrations from 1.06μm lidar measurements, *Applied Optics*, **17(23)**, 3763-3773.

[7]Del Guasta, M., Marini, S., 2000: On the retrieval of urban aerosol mass concentration by a 532 and 1064nm lidar, *Journal of Aerosol Science*, **31**, 1469-1488.

[8]Del Guasta, M., 2002: Daily cycles in urban aerosol observed in Florence (Italy) by means of an automatic 532-1064nm lidar, *Atmospheric Environment*, **36**, 2853-2865.

[9]Fernald, F.G., Herman, B.M., Reagan, J.A., 1972: Determination of aerosol height distribution by lidar, *Journal of Applied Meteorology*, **11**, 482-489.

[10]Ansmann, A., Engelmann R., Althausen, D., Wandinger, U., Hu M., Zhang, Y., He, Q., 2005:

High aerosol load over the Pearl River Delta, South China, observed with Raman lidar and Sun photometer, *Geophysical Research Letters*, **32(13)**, L13815.

[11]Andrew Y. S. Cheng, Andrew Walton and Johnny C. L. Chan, 2003: Urban aerosols spatial variations using a slant angle scanning Mie Lidar system. In the SPIE's 3rd International Asia-Pacific Symposium on Remote Sensing of the Atmosphere, Environment and Space, October 23-27, 2002, Proc. SPIE Volume **4893**, 17-24.

[12]Liu Qiao-jun, Yang Lin, Wang Jie-Yu, Zuo Hao-yi, Luo Shi-Rong, Andrew Y. S. Cheng, 2009: Calculation of the overlap factor and correction of near-field signal of the off-axis lidar based on the Gaussian mode of laser beam, *Acta Physica Sinica*, **58(10)**, 7376-7381.

[13]Inaba, H., 1976: Detection of atoms and molecules by Raman scattering and resonance fluorescence, in Laser Monitoring of the Atmosphere, chap. 5, Ed. Hinkley E.D., Springer-Verlag.

URBAN AEROSOL OPTICAL PROPERTIES MEASUREMENT BY ELASTIC COUNTER-LOOK LIDAR

X. Wang[1,2*], A. Boselli[2,3], Y. He[4], A. Sannino[5], C. Song[5], N. Spinelli[2,5]

[1]*CNR-SPIN, Napoli, Italy, *Email:wang@na.infn.it*

[2]*Advanced Lidar Applications srl, Napoli, Italy*

[3]*CNR-IMAA, Potenza, Italy*

[4]*School of Electronic and Information Engineering, Beihang University, Beijing, China*

[5]*CNISM and Dipartimento di Fisica, Università di Napoli "Federico II", Napoli, Italy*

ABSTRACT

The new developed elastic lidar system utilizes two identical elastic lidars, in counter-look configuration, to measure aerosol backscattering and extinction coefficients without any hypotheses. Compared to elastic-Raman lidar and high spectral resolution lidar, the proposed counter-look elastic lidar can use low power eye-safe laser and all available wavelengths. With this prototype lidar system, urban aerosol optical properties and their spatial distribution have been directly measured, including backscatter coefficient, extinction coefficient and lidar ratio. The preliminary results show that the low cost and eye-safe counter-look configured elastic lidar system can be used to measure the aerosol optical properties distribution and give the hint of aerosol type.

1. INTRODUCTION

Lidar systems are widely employed to characterize atmospheric aerosol optical properties. From Lidar signals, it is possible to determine the particles optical and microphysical parameters through the application of inversion algorithms when measurements of both aerosol extinction and backscattering coefficients are available at different wavelengths [1, 2]. Aerosol extinction and backscattering coefficients can be independently measured by Raman lidar or high spectral resolution lidar system [3]. Nevertheless, the Raman and high spectral resolution lidar technique are hardly to apply to the near infrared, due to the dependence from λ^{-4} of the cross section of the process. Furthermore, the low absolute value of the Raman cross section involves the use of high power laser beams, with resulting economical, technical and safety problems, which add up a not negligible complexity of data processing. Therefore, it is difficult to apply conventional lidars to routine measurements in the control of fine particles in urban areas, to the control of emissions of industrial plants, energy production plants, incinerators etc., so that their usage is general limited to scientific research purpose.

With the aim to characterize the fine particles in the low part of atmospheric, an innovative instrument, patent of CNR (National Council of the Research) and CNISM (National Interuniversity Consortium for the Physical Sciences of Matter), has been proposed [4]. This application can be realized with the simultaneous employment of different wavelengths (from the ultraviolet to the infrared) and the use of low power and cheap laser sources.

2. METHODOLOGY

The principle of counter-look lidar system has been descripted in references [5]. The briefly discussion of the method is as follows.

Two identical elastic lidars locate in two separation positions, usually apart in several kilometers. The orientation of lidars are counter look each other. In this way, two laser beams pass the same atmosphere area.

The two range corrected signals (RCS) of lidar A and lidar B can be written as

$$RCS_A(z) = K_A \cdot \beta(z) \cdot \exp(-2\tau_{0z})$$

$$RCS_B(z) = K_B \cdot \beta(z) \cdot \exp(-2\tau_{zd})$$

Where $\tau_{0z} = \int_{o}^{z} \alpha(\zeta) d\zeta$ and $\tau_{zd} = \int_{z}^{d} \alpha(\zeta) d\zeta$ are the optical depths from lidar A to z and from z to lidar B. K_A and K_B are two instrument constants. d is the distance from lidar A to lidar B.

From RCS_A and RCS_B, we can get backscattering and extinction coefficients, respectively.

$$\beta(z) = \sqrt{K \cdot RCS_A(z) \cdot RCS_B(z)}$$

$$\alpha(z) = \frac{1}{4}\frac{d}{dz}\ln\left(\frac{RCS_B(z)}{RCS_A(z)}\right)$$

Where K is a system constant. We can get it from the system calibration.

The aerosol backscattering and extinction can be calculated from total backscattering, $\beta(z)$, and extinction, $\alpha(z)$, subtracting the molecular contribution.

3. RESULTS

The first prototype of counter-look lidar system was implemented only for single wavelength elastic backscattering lidar. The main parameters of two elastic lidars are report in table 1.

Table 1. The main parameters of counter-look lidar system prototype

Laser		Telescope	
Energy per pulse (μJ)	5	Radius (cm)	10
Repetition frequency (Hz)	2000	Focal length (m)	0.5
Divergence (mrad) (with Beam Expander)	<0.2	Diaphragm opening (mm)	0.5
Wavelength (nm)	532	Angular opening (mrad)	1
Pulse duration (ns)	< 20		

The two lidar were located in two different places to across an urban area of Naples, Italy, shown in Figure 1.

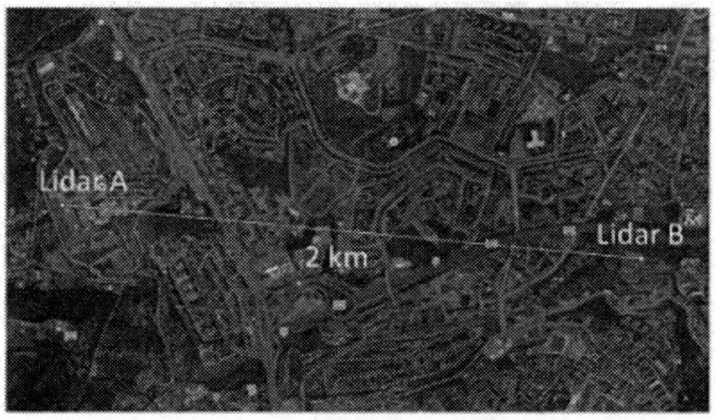

Figure 1. The location (40.838099°, 14.183132°) of counter-look lidar system prototype.

Before the real measurement, the counter-look lidar system prototype has been calibrated with our Raman lidar, which is one of the EARLINET station [6], as shown in Figure 2. From the calibration, the system constant has been evaluated, in order to perform the backscattering coefficient calculation.

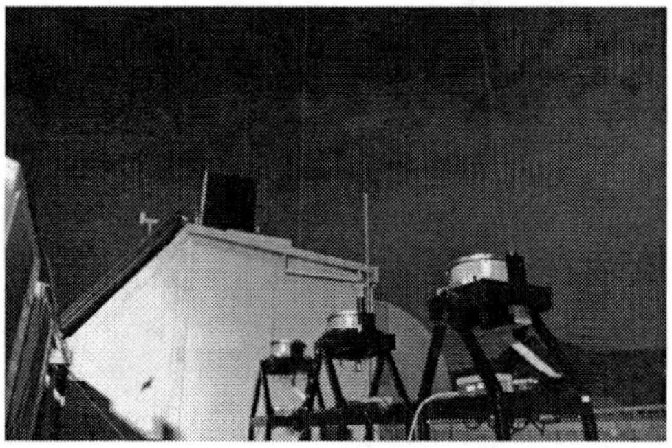

Figure 2. System calibration with the Raman Lidar of the Earlinet Naples station.

Lidar return signal has been acquired in both day and night time. 10 minutes integrated range corrected signals from lidar A and B are shown in Figure 3, as example.

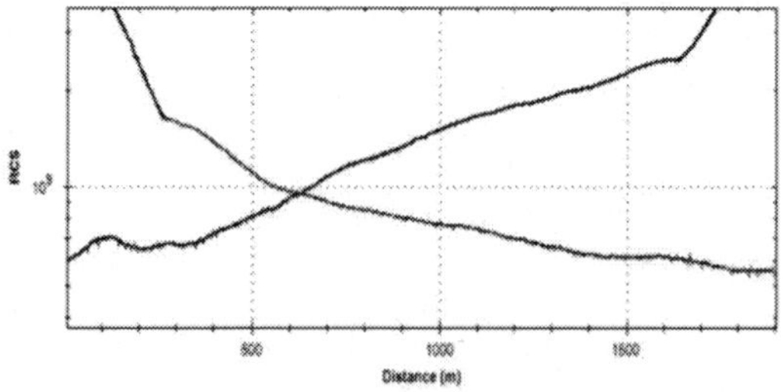

Figure 3. The 10 minutes integrated range corrected signals from two lidars measured on Dec 15, 2014

From above mentioned formula, aerosol backscattering and extinction coefficients have been retrieved without any Lidar Ratio hypothesis. Retrieved profiles are shown in Figure 4.

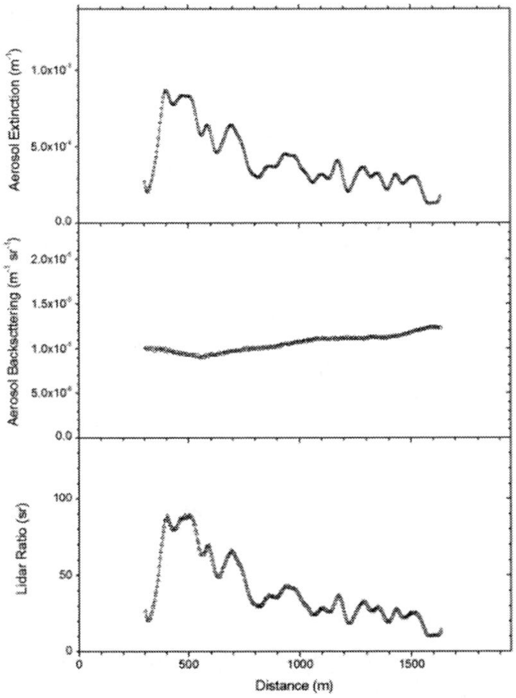

Figure 4. Retrieved aerosol backscattering and extinction coefficients distribution, measured in Dec 15, 2014.

There are not data in the first and the last 300 m. This is because in these range the lidar beam and the field of view are not total overlapped. The further evaluation of overlap function work should be done in future.

From the results in Figure 4, we can see that along the lidar path, even the aerosol backscattering coefficient dose not change very much, about 20%, the aerosol extinction coefficient vary a lot. The aerosol lidar ratio values show that there are two different aerosol types along the lidar path. The high lidar ratio, about 90 sr, around 500 m from lidar A, corresponds a high polluted urban aerosol. While beyond the 1000 m, the aerosol lidar ratio value is 20-30 sr. This is a typical natural aerosol. In fact, from the map, we can see that in that area there is little human activity.

4. CONCLUSIONS

The counter-look elastic lidar system is low cost and eye safe. It can be used in both day time and night time. With a simple lidar configuration and simple algorithm, we can independently retrieve aerosol backscattering and extinction coefficients. The preliminary result show that from the high precise lidar ratio, we can estimate the aerosol nature.

ACKNOWLEDGEMENT

The authors thank the Italian National Research Council (CNR) and National Interuniversity Consortium for the Physical Sciences of Matter (CNISM) to allow free use the patent.

The prototype of the system was funded by European Union project PONa3-00363 "I-AMICA".

REFERENCES

[1] C. Böckmann, Hybrid regularization method for the ill-posed inversion of multiwavelength lidar data to determine aerosol size distribution, *Appl. Opt.,* **40** (2001) 1329-1342.

[2] Müller, D., Wandinger, U., e Ansmann, A. Microphysical particulate parameters from extinction and backscatter lidar data by inversion with regularization: simulation, *Appl. Opt.,* **38** (12):2358–2368, (1999).

[3] Ansmann et al., Combined Raman Elastic Backscatter LIDAR for vertical profiling of moisture, aerosol extinction, backscatter and lidar ratio, *Appl. Phys.B*, **55**, 18-28. (1992)

[4] Patent application No. EP2144088 and US2010026281

[5] X. Wang et al., An Innovative Lidar Instrument For Monitoring The Atmospheric Particulate, ILRC 26

[6] Bösenberg J et al., A European Aerosol Research Lidar Network to Establish an Aerosol Climatology. *Max Planck Institut für Meteorologie - Technical Report* n. **348**. (2003)

Research in Depolarization and Extinction Coefficient of Particles in Tibetan Plateau by Lidar

Guangyao DAI, Xiaoquan SONG, Xiaochun ZHAI, Songhua WU*

[1]*Ocean University of China, Qingdao 266100, China, *Email: wush@ouc.edu.cn*

ABSTRACT

Vertical profiles of the depolarization ratio and the extinction coefficient of atmospheric particles in Tibetan Plateau were measured with the OUC Water Vapor, Cloud and Aerosol Lidar during the 3rd Tibetan Plateau Atmospheric Expedition Experiment Campaign in 2013 and 2014. The cloud types and phases, the spatial temporal distribution of the aerosols and the boundary layer height in the Tibetan Plateau were obtained using polarization lidar technique. In this paper, the depolarization ratio was validated with CALIOP polarization simultaneous data, and the extinction coefficient was retrieved by the Fernald method. The result implied that the atmosphere in the Tibetan Plateau was quite clean with low aerosol load and serious pollution. The ice-water mixed cumulus, water cumulus or stratus clouds in Litang and Nagqu were occurred and classified, respectively. The boundary layer height in Nagqu at average altitude over 4600 m was obtained at around 200 m-300 m, which was commonly lower than that in other observed sites.

1. INTRODUCTION

Atmospheric aerosol particles have a complicated influence on the earth climate by directly absorbing and scattering the atmospheric radiation and by indirectly serving as cloud condensation nuclei [1]. The Tibetan Plateau lies at a critical and sensitive area, which influences the atmosphere in East Asia area and even the whole northern hemisphere. So it is significant to study the impact of aerosol on atmospheric condition and composition in Tibetan Plateau. The polarization lidar technique [2] has been an excellent method of atmospheric probing, which plays an important role in the detection of spatial and temporal distribution of aerosol and cloud phase. In general, it bases on the particle polarization effect, that is, the polarization orientation of backscatter signal of the spherical particles is the same as the linearly polarized laser pulse while the polarization orientation of the non-spherical particle is changed and consists of perpendicular and parallel signals. The ratio of these two signals corrected by the calibration factor is referred to as the particle linear depolarization ratio. Liu et al.[3] indicated the feasibility of using backscatter coefficients and backscatter color ratios to distinguish aerosols from clouds, which showed clouds have larger backscatter coefficient (≥ 0.01) and higher color ratio around 1 than mostly aerosol types except for optically thin cloud, and the cumulus or stratus clouds generally have higher backscatter coefficients than cirrus. Sassen et al.[4] stated that different cloud phases correspond to differentiable depolarization ratio range, which can be an essential basis to distinguish cloud phases. With the help of particle linear depolarization ratio and extinction coefficient (or backscatter coefficient), the cloud types and phases, the spatial temporal distribution of the aerosols in observed site can be determined, supplying useful and synergistic data to other kinds of aerosol lidars.

2. METHODOLOGY

2.1 POLARIZATION CHANNEL SETUP DESCRIPTION

The OUC Water Vapor, Cloud and Aerosol Lidar (WVCAL) is multifunctional, and the detailed introduction can refer to a parallel paper in this conference. Fig.1 presents the schematic diagram of the polarization channel setup. The laser transmitter is a Nd:YAG laser, which produces a laser pulse at 532nm with energy per pulse of 120mJ and a repetition rate of 30Hz.

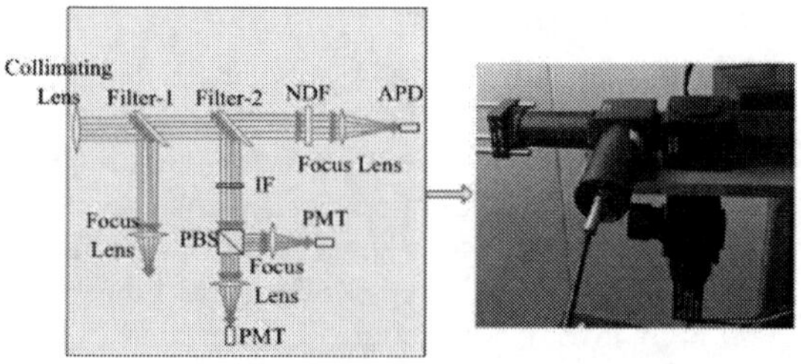

Fig.1. Schematic diagram of the polarization channel of WVCAL system at 532 nm.

The beam divergence angle is 0.05 mrad. The telescope has an aperture of 308mm with field-of-view of 1.3mrad. After the backscatter signals are collected by telescope and split by spectrophotometer system, the 532nm signal is transmitted to the polarizing beam splitter (PBS) and is separated to $P_{//}$ and $P_{\perp}$.

2.2 RETRIEVAL OF DEPOLARIZATION RATIO AND EXTINCTION COEFFICIENT

The linear volume depolarization ratio δ^{v} is commonly defined as the ratio of the total perpendicular-polarized backscatter power ($P_{\perp}$) to total parallel-polarized backscatter power ($P_{//}$), measured with polarization lidar by [5]:

$$\delta^{v}=\frac{\beta_{\perp}}{\beta_{//}}=\frac{P_{\perp}}{P_{//}} \tag{1}$$

According to the lidar equation (2):

$$P=\frac{C\Delta r}{r^{2}}(\beta_{a}+\beta_{m})\exp\{-2\int_{0}^{r}[\alpha_{m}^{\lambda}(z)+\alpha_{a}^{\lambda}(z)]dz\} \tag{2}$$

Where C is the system constant, β is the backscatter coefficient, α is the extinction coefficient, the subscripts a and m represent aerosol and molecular, respectively, and Δr is the range resolution. The $P_{//}$ and $P_{\perp}$ can be described as (3) and (4), respectively:

$$P_{//}=\frac{C_{//}\Delta r}{r^{2}}(\beta_{a}^{//}+\beta_{m}^{//})\exp\{-2\int_{0}^{r}[\alpha_{m}^{//}(z)+\alpha_{a}^{//}(z)]dz\} \tag{3}$$

$$P_{\perp}=\frac{C_{\perp}\Delta r}{r^{2}}(\beta_{a}^{\perp}+\beta_{m}^{\perp})\exp\{-2\int_{0}^{r}[\alpha_{m}^{\perp}(z)+\alpha_{a}^{\perp}(z)]dz\} \tag{4}$$

Where $C_{//}$ and $C_{\perp}$ are system constants for parallel-polarized channel and perpendicular-polarized channel, respectively.

$C_{//}$ and $C_{\perp}$ represent the different receiver efficiency on the two channels including the PBS cross-talk effect, and the lidar system depolarization effects including non-linear polarized laser source and optical device depolarization effect. As a result, δ^{v} cannot be used directly to classify the aerosol types and cloud phases. To retrieve the depolarization properties of particles, the particle linear depolarization ratio δ^{p} was defined, and the relationship between δ^{p} and δ^{v} can be described as (5) :

$$\delta^{P}=a*\delta^{v}+b \tag{5}$$

The calibration factors a and b, which depend on the system itself, were obtained by WVCAL and CALIPSO validation experiment.[6]

Based on the signal of the polarization lidar at 532nm, the aerosol extinction coefficient can be calculated by the Fernald method [7], and can be written as equation (6):

$$\alpha_{a}(z)+\frac{S_{a}(z)}{S_{m}}\alpha_{m}(z)=\frac{S_{a}(z)P(z)z^{2}\exp\{-2\int_{z_0}^{z}[\frac{S_{a}(\zeta)}{S_{m}}-1]\alpha_{m}(\zeta)d\zeta\}}{\frac{S_{a}(z_0)P(z_0)z_0^{2}}{\alpha_{a}(z_0)+\frac{S_{a}(z_0)}{S_{m}}\alpha_{m}(z_0)}-2\int_{z_0}^{z}S_{a}(\zeta)P(\zeta)\zeta^{2}\exp\{-2\int_{z_0}^{\zeta}[\frac{S_{a}(\xi)}{S_{m}}-1]\alpha_{m}(\xi)d\xi\}d\zeta} \tag{6}$$

Where $S_{m}=8\pi/3$ and $S_{a}(z)$ are the extinction to backscatter ratios for Rayleigh and Mie scattering, respectively. $S_{a}(z)$ was assumed as a constant and was independent of the range.

The reference height z_0 was chosen where the atmospheric layer was relative clean with rare aerosols. In this paper, the Minimum Value Method was used to find out z_0, that is, z_0 was the height which minimizes $P(z)z^{2}/\beta_{m}(z)$. An iteration was began with $\beta_{a,input}(z_0)=4\times10^{-10}km^{-1}sr^{-1}$ after the reference height z_0 had been fixed. The $\beta_{a,input}(z_0)$ increased by 10% for each iteration

 DOI: 10.1051/epjconf/201611923030

step until the relative difference between $\beta_{a,input}(z_0)$ and $\beta_{a,cal}(z_0)$ was less than 5% [8].

The extinction coefficients of the lidar were compared with the CALIOP polarization data. One study case measured on September 30, 2013 was shown in Fig.2. The black dashed line represents the lidar data and the other color lines represent CALIOP polarization data, which shows good consistency.

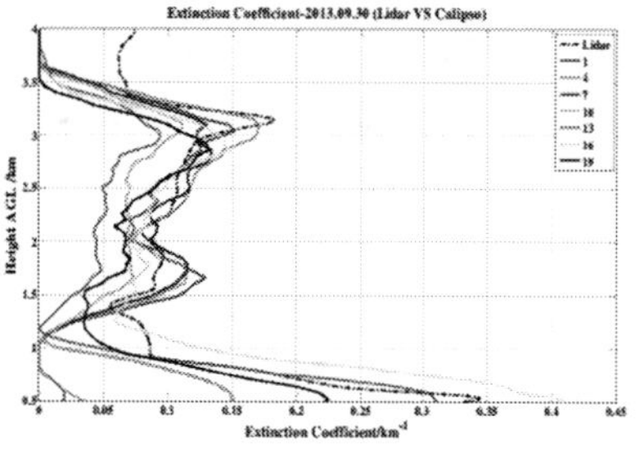

Fig.2. Comparison of extinction coefficient measured by lidar and CALIPSO on Sep 30, 2013

3. RESULTS

In this section, some cases including information about particle liner depolarization ratio, extinction coefficient and boundary layer height is discussed.

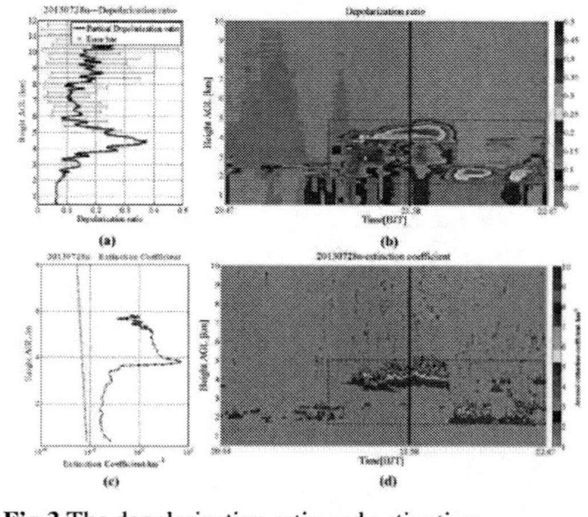

Fig.3. The depolarization ratio and extinction coefficient measured on July 28, 2013 (In Litang). (a) Profile of particle depolarization ratio (black line) and the error bar (pink line). (b) Temporal development of particle depolarization ratio. (c) Profile of α_a (black dashed line) and α_m (pink dashed line). (d) Observation of aerosol extinction coefficient α_a

In Fig.3, the δ^P at the height of 2-5 km above ground level (in red dashed box) was in the range of 0.1 to 0.35. Meanwhile, α_a was bigger than 5 km^{-1}. It was implied that clouds existed and can be classified as mixed phase cumulus clouds[9]. Furthermore, the δ^P near the ground was less than 0.1 and even less than 0.05, and α_a was about 0.05km^{-1}. So it was confirmed that the aerosol loading near the ground in Litang was rare, and the aerosols were free of biomass burning aerosols and pure dust.

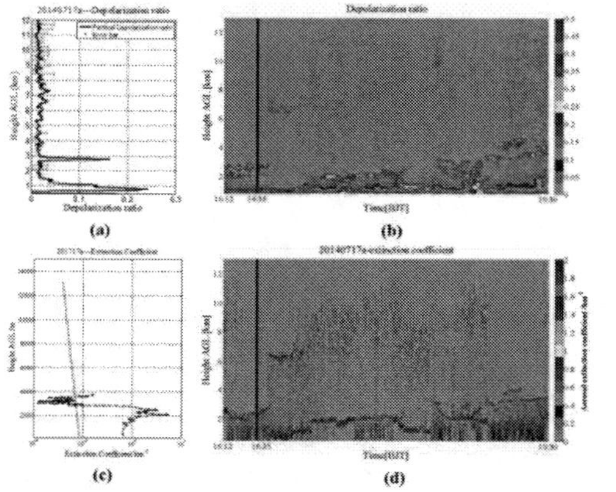

Fig.4. The depolarization ratio and extinction coefficient measured on July 17, 2014 (In Nagqu) (a)-(d) are same as Figure 3 (a)-(d)

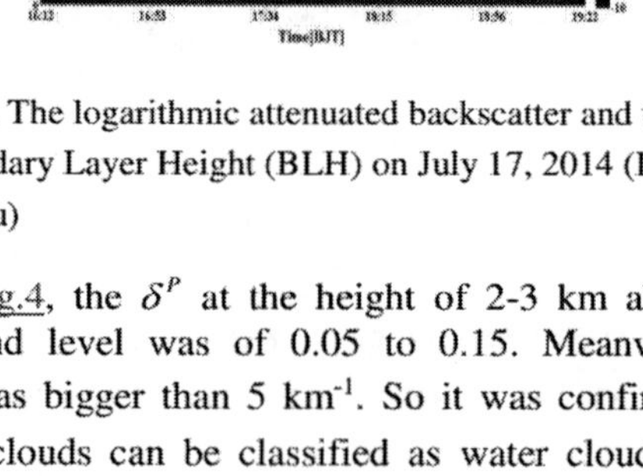

Fig.5. The logarithmic attenuated backscatter and the Boundary Layer Height (BLH) on July 17, 2014 (In Nagqu)

In Fig.4, the δ^P at the height of 2-3 km above ground level was of 0.05 to 0.15. Meanwhile α_a was bigger than 5 km^{-1}. So it was confirmed that clouds can be classified as water clouds or mixed phase clouds dominated by water clouds, and the cloud type can be classified as cumulus or stratus clouds [10]. Except the clouds, the atmosphere above 1 km was very clean and the depolarization ratios δ^p were close to zero. The air above 1 km was free or contains tiny of the non-spherical particles. Furthermore, the δ^p near the ground was less than 0.1 and even less than 0.05 and α_a was about 0.5km^{-1}. So it was implied that the air conditions in Nagqu were quite clean and had no other types of aerosols and serious pollution. Fig.5 showed the boundary layer height in Nagqu at that period was about 200m-300m, which was commonly lower than that in other observed sites.

4. CONCLUSIONS

The key conclusions of the study were listed below:

The extinction coefficients of WVCAL were compared with CALIOP polarization data, which showed good consistency. In the process of clouds classification, α_a can generally distinguish aerosols from clouds, and further classify cloud types. δ^p can then judge the cloud phases. In the Tibetan plateau experien, the ice-water mixed cumulus clouds, water cumulus or stratus clouds in Litang and Nagqu were observed and classified. Most of δ^p near ground in Tibetan Plateau were less than 0.1 and even 0.05, showing that the aerosol loading near the ground was rare and the aerosols were free of biomass burning aerosols and pure dust. Moreover, the atmosphere above 1 km was very clean and the δ^p was close to zero. The air above 1 km was free or contains tiny of the non-spherical particles. The boundary layer height in Tibetan Plateau was about 200m-300m, which was commonly lower than that in other observed sites.

ACKNOWLEDGEMENT

The observation in Tibetan Plateau was completed by the cooperation of CAMS/ LAWS(Chinese Academy of Meteorological Science/Laboratory of Severe Weather) and Ocean University of China (OUC). We thank researcher Liping Liu from CAMS for his help and support. And we thank Meteorological Administrations of Litang and Nagqu for their support.

REFERENCES

[1] Charlson, R.J., Schwartz, S.E., Hales, J.M., Cess, D., Coakley, J.A., Hansen, J.E., 1992. Climate forcing by anthropogenic aerosols. Science **255**, 423–430.

[2] Sassen K., 2005: Polarization in lidar. *Springer New York.*

[3] Liu, Zhaoyan, et al., 2004:Use of probability distribution functions for discriminating between cloud and aerosol in Lidar backscatter data, *Journal of Geophysical Research: Atmospheres (1984–2012)*, **D15** ,109.

[4] Sassen K, Zhao H., 1993: Super cooled liquid water clouds in Utah winter mountain storms: Cloud-seeding implications of a remote-sensing dataset, *Journal of Applied Meteorology*, **32(9)**, 1548-1558

[5] Freudenthaler V, Esselborn M, Wiegner M, et al., 2009: Depolarization ratio profiling at several wavelengths in pure Saharan dust during SAMUM 2006, *Tellus B*, **61(1)**, 165-179.

[6] Dai G, Wu S, Song X, et al., 2014: Research in depolarization of particles in Tibetan Plateau and coastal area by lidar, *SPIE Asia Pacific Remote Sensing. International Society for Optics and Photonics*, 92620I-92620I-12.

[7] Fernald, Frederick G., 1984: Analysis of atmospheric Lidar observations: some comments. *Applied optics*, **23.5**, 652-653

[8] Althausen D, Engelmann R, Baars H, et al., 2009: Portable Raman lidar PollyXT for automated profiling of aerosol backscatter, extinction, and depolarization, *Journal of Atmospheric and Oceanic Technology*, **26(11)**: 2366-2378.

[9] Sassen K, Zhao H., 1993: Super cooled liquid water clouds in Utah winter mountain storms: Cloud-seeding implications of a remote-sensing dataset. *Journal of Applied Meteorology*, **32(9)** 1548-1558.

[10] Toon, O., Tabazadeh, A., Browell, E. and Jordan, J., 2000: Analysis of Lidar observations of Arctic polar stratospheric clouds during January 1989, *J. Geophys. Res.* **105**, 20 589–20 615

MOBILE MULTIWAVELENGTH POLARIZATION RAMAN LIDAR FOR WATER VAPOR, CLOUD AND AEROSOL MEASUREMENT

Songhua Wu[1]*, Xiaoquan Song[1], Bingyi Liu[1], Guangyao Dai[1], Kailin Zhang[1], Shengguang Qin[1], Fei Gao[2], Dengxin Hua[2]

1. Ocean Remote Sensing Institute, Ocean University of China, Qingdao 266100, China.
2. Xi'an University of Technology, Xi'an 710048, China.
**Email: wush@ouc.edu.cn*

ABSTRACT

Aiming at the detection of water vapor mixing ratio, particle linear depolarization ratio, extinction coefficient and cloud information, the Water vapor, Cloud and Aerosol Lidar (WVCAL) was developed by the lidar group at Ocean University of China. The Lidar consists of transmitting subsystem, receiving subsystem, data acquisition and controlling subsystem and auxiliary subsystem. These parts were presented and described in this paper. For the measurement of various physical properties, three channels including Raman channel, polarization channel and infrared channel are integrated in this Lidar system. In this paper, the integration and working principle of these channels is introduced in details. Finally, a measurement example which was operated in coastal area-Qingdao, Shandong province, during 2014 is provided.

1. INTRODUCTION

Water vapor has a great impact on the determination of weather and climate due to its fundamental role in the radiative energy transfer, hydrological cycle, and atmospheric chemistry processes [1]. The complexity of atmospheric aerosols resulted from their highly variable particle number concentrations, multimodal size distributions, variable shape characteristics, complex chemical composition and mixing behavior. And the correspondingly large temporal and spatial variability in the aerosol characteristics are the main reasons for the high uncertainties in our quantitative understanding of the role of atmospheric aerosol in environmental, weather, and climate-related processes [2]. The Fourth IPCC Assessment Report [3] has identified aerosol radiative forcing and the impact of aerosols on cloud and precipitation processes as one of the major unknowns in our understanding of climate change.

For the detection of the water vapor, two Lidar techniques method are applied: the Differential Absorption Lidar (DIAL) and the Raman Lidar technique. In term of the DIAL, two laser pulses at different wavelengths, called "on-line" and "off-line", respectively, are emitted to the atmosphere [4] for the measurement. In this paper, the Lidar system applied Raman technique. This technique was firstly used by Melfi [5] and Cooney [6] and the profiles of water vapor mixing ratio was retrieved and provided. Rely on its simplicity and ability to measure the profile of water vapor field in the troposphere with high vertical and temporal resolution, the Raman lidar technique has been widely used in meteorology, boundary layer studies, and in climatology [1].

For the detection of the aerosol and cloud, the polarization Lidar technique [7] is a well-established method. Ansmann et al. set up a polarization Lidar to distinguish ice clouds from water clouds and to identify layers with ice crystals in mixed–phase clouds [8]. In order to study the evolution of contrails Freudenthaler et al. also applied a scanning polarization Lidar in 1996 [9]. The technique has been used to identify the type of polar stratospheric clouds [7] and volcanic ash in the troposphere and stratosphere [10]. The polarization Lidar can also be used to do some research on aerosol profiling [11] and help to classify desert dust from other aerosols such as biomass burning aerosols, fresh smoke, etc. [12].

For the purpose of combined meteorological research and aiming at the detection of water vapor mixing ratio, depolarization ratio extinction and cloud information, the multifunctional WVCAL was developed by the Ocean Remote Sensing Institute at Ocean University of China, Qingdao. The Lidar system was developed from

June 2012 and was deployed to a field test on July 2013. This Lidar is set up on an optical platform and housed in a weather-proof cabinet. Because of the independence of the entire system, it can be transported and installed easily in the field. The schematic diagram of the whole system is provided in Fig. 1.

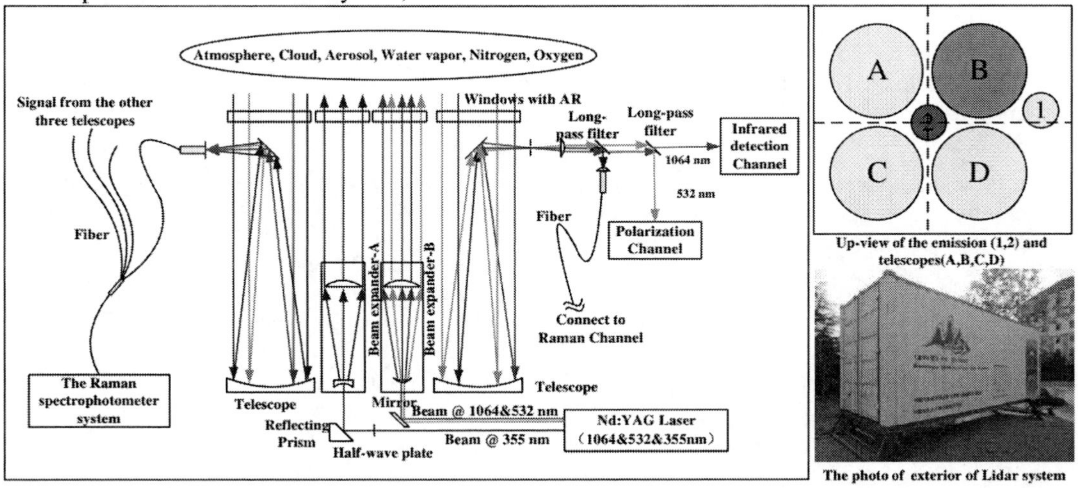

Fig.1. Schematic diagram and photos of WVCAL.

2. WATER VAPOR, CLOUD AND AEROSOL LIDAR

2.1 TRANSMITTER

The laser transmitter, Continuum Powerlite 9030, is a high peak power lamp pump Nd:YAG laser with three wavelengths of 354.7 nm, 532 nm and 1064 nm. And the single pulse energy was 410mJ, 120mJ and 700mJ, respectively. The wavelength of 354.7 nm was used for exciting Raman signal of nitrogen and water vapor. Meanwhile the wavelengths of 532 nm and 1064 nm were utilized for the detection of Mie and Rayleigh signal. In this system, for purpose of compressing divergence angle and reducing energy cost, two beam expanders were designed. As shown in Fig. 1, the transmitter system consists of laser transmitter, half-wave plate, reflecting prism, mirror, beam expanders and windows with anti-reflective coating. The expanded beams with 90-mm diameter transmitted into the atmosphere on an axis closed to the receiver axis.

2.2 RECEIVER

After a laser pulse is transmitted to the atmosphere, molecules and particles scatter the light in all directions. A portion of the light is scattered backwards to the lidar. The light is collected by telescopes and transmitted to detection systems. In order to improve the efficiency of receiving, this system deployed four telescopes A, B, C and D (shown in Fig.1.) with the diameter of 300 mm and the focal length of 1524mm. The 4 telescopes assembly serve as an equivalent receiver aperture of about 610 mm.

2.2.1 POLYCHROMATOR OF RAMAN CHANNEL

The optical setup of the polychromator of Raman channel was provided in Fig. 2.

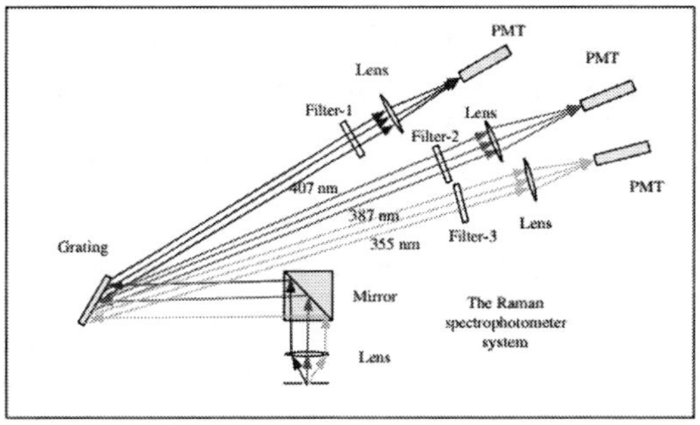

Fig.2. The optical setup of the Raman channel

In order to avoid the interference of the elastic signal, band-pass filter was used. The center wavelength of the filter is 390 nm and the FWHM is 44.6 nm. The transmission between 370 nm and

410 nm is bigger than 93% and the OD is bigger than 5 for light at the wavelength of 354.7 nm and 532 nm. Four fibers were mounted after the filters to couple the backscatter signal. The core of the fibers was 200 microns and the numerical aperture was 0.22. After the coupling of fiber, the Raman signal was delivered to the spectrophotometer system and separated as Nitrogen Raman signal and Water vapor Raman signal.

When the signal is coupled to spectrophotometer system, the light is dispersed and then collimated by the convex lens with the focal length of 50.0mm. After the reflection of the prism, the parallel light arrives at the grating. The groove density of the grating is 1302 line/mm and the blaze wavelength is 400 nm. So far, the elastic signal at 354.7 nm and Raman signal at 386.7 nm and 407.5 nm are separated. Then the filters were used to ensure the purity of each signal. The center wavelengths of the filter-1, filter-2 and filter-3 are 407.5 nm$\pm$0.1nm, 386.7 nm$\pm$0.1nm and 354.7 nm$\pm$0.08nm, respectively. The FWHM of all filters is 0.5 nm $\pm$ 0.10nm. Peak transmittance is bigger than 50% and OD is 5 for out of band blocking. After the filtration of filters, the signal was then focused by plano-convex lens with a focal length of 100mm. Finally, the signals were acquired by the photomultiplier tubes which were mounted at the focal point of plano-convex lens.

2.2.2 POLARIZATION AND INFRARED CHANNELS

Considering the high power laser and the strong Rayleigh and Mie scattering, only one telescope B is utilized for the collection of signal in polarization channel and infrared channel. The optical setup is described in Fig. 3.

One adjustable aperture was installed at the focal plane. The field of view can be changed by adjusting the aperture. In this system, the aperture is fixed as 2 mm and the FOV is determined as 1.3mrad. By utilizing the dichroic long-pass filter-1, the elastic signal at 354.7 nm and Raman signal is reflected and the signal of wavelength longer than 425 nm transmits through the filter. Similarly, the elastic signal of 532 nm and 1064 nm is separated by the dichroic long-pass filter-2. The signal at 1064 nm is detected by the APD. The signal at 532 nm is used to measure the depolarization ratio. The parallel-polarized and cross-polarized signal are separated by the polarizing beam splitter and detected by PMTs respectively.

2.3 DATA ACQUISITION

Data acquisition and controlling subsystem consists of PMT, APD, transient recorder and computer. Using PMT and APD, the optical signal are transferred to electrical signal and stored by a Licel transient recorder. The recorder has 6 channels and 2 trigger interface. The channel C1, C2 and C3 were used for Raman channel. C4 and C5 were utilized for Polarization channel and C6 was for Infrared channel. The trigger signal is from a PIN detector to sense the laser pulse at the laser head.

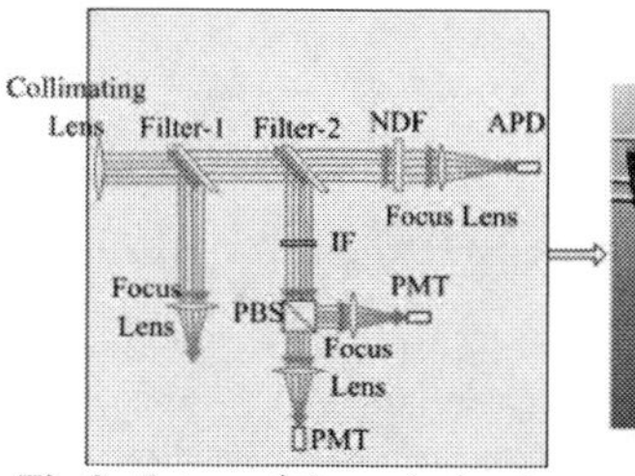

Fig.3. Setup of the polarization and infrared channels

3. RESULTS / CASE STUDY

WVCAL has been installed in coastal area and Tibetan plateau for the measurements of atmosphere since 2013. Fig. 4 and Fig. 5 show the result measured on 31 May 2014 in Qingdao.

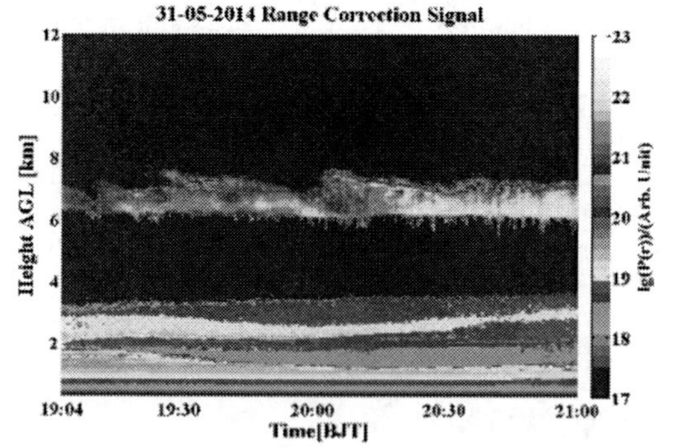

Fig.4. Temporal development of the range-corrected backscatter signal at 532 nm wavelength

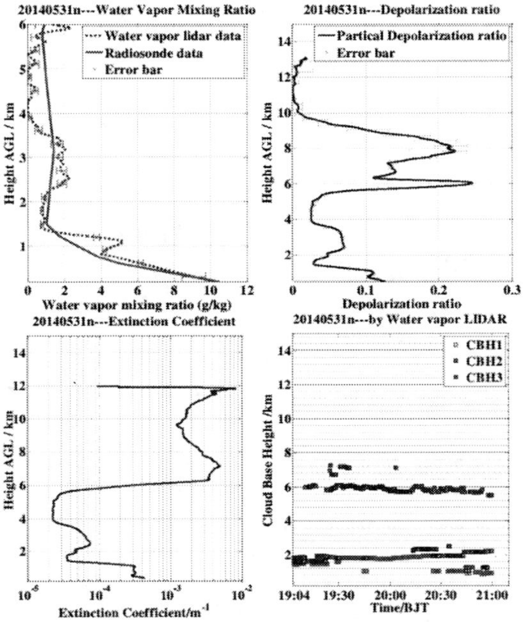

Fig.5. Measurements of water vapor mixing ratio, particle linear depolarization ration, extinction coefficient and cloud base height

ACKNOWLEDGEMENT

This work was funded by CAMS/LAWS (Chinese Academy of Meteorological Science/Laboratory of Severe Weather).

REFERENCES

[1] Dinoev, T., 2009: Automated Raman lidar for day and night operational observation of tropospheric water vapor for meteorological applications(Doctoral dissertation, ÉCOLE POLYTECHNIQUE FÉDÉRALE DE LAUSANNE).

[2] Althausen, D., Engelmann, R., Baars, H., Heese, B., Ansmann, A., Müller, D., & Komppula, M., 2009: Portable Raman lidar PollyXT for automated profiling of aerosol backscatter, extinction, and depolarization, *Journal of Atmospheric and Oceanic Technology*, **26(11)**, 2366-2378.

[3] Forster, P., and Coauthors, 2007: Changes in atmospheric constituents and in radiative forcing, *Climate Change 2007: The Physical Science Basis, S. Solomon et al., Eds., Cambridge University Press*, 129–234.

[4] Bruneau, D., Quaglia, P., Flamant, C., Meissonnier, M., & Pelon, J., 2001: Airborne lidar LEANDRE II for water-vapor profiling in the troposphere. I. System description, *Applied optics*, **40(21)**, 3450-3461.

[5] Melfi, S. H., Lawrence Jr, J. D., & McCormick, M. P., 1969: Observation of Raman scattering by water vapor in the atmosphere, *Applied Physics Letters*, **15(9)**, 295-297.

[6] Cooney, J., 1970: Remote measurements of atmospheric water vapor profiles using the Raman component of laser backscatter, *Journal of Applied Meteorology*, **9(1)**, 182-184.

[7] Sassen, K., 2005: Polarization in lidar, *Springer New York*, 19-42.

[8] Ansmann, A., Tesche, M., Althausen, D., Müller, D., Seifert, P., Freudenthaler, V., & Dubovik, O., 2008: Influence of Saharan dust on cloud glaciation in southern Morocco during the Saharan Mineral Dust Experiment, *Journal of Geophysical Research: Atmospheres (1984–2012)*, **113(D4)**.

[9] Freudenthaler, V., Homburg, F., & Jäger, H., 1996: Optical parameters of contrails from lidar measurements: linear depolarization, *Geophysical research letters*, **23(25)**, 3715-3718.

[10] Sassen, K., Zhu, J., Webley, P., Dean, K., & Cobb, P., 2007: Volcanic ash plume identification using polarization lidar: Augustine eruption, Alaska, *Geophysical Research Letters*, **34(8)**.

[11] Cairo, F., Di Donfrancesco, G., Adriani, A., Pulvirenti, L., & Fierli, F., 1999: Comparison of various linear depolarization parameters measured by lidar, *Applied Optics*, **38(21)**, 4425-4432.

[12] Murayama, T., Müller, D., Wada, K., Shimizu, A., Sekiguchi, M., & Tsukamoto, T., 2004: Characterization of Asian dust and Siberian smoke with multi- wavelength Raman lidar over Tokyo, Japan in spring 2003.*Geophysical Research Letters*, **31(23)**.

EPJ Web of Conferences **119**, 23032 (2016) DOI: 10.1051/epjconf/201611923032
ILRC 27

AEROSOL LAYERING CHARACTERIZATION NEAR THE GOBI DESERT BY A DOUBLE POLARIZATION LIDAR SYSTEM

Y. Zhao[1], A. Boselli[2], A.Sannino[3], C. Song[1,3], N. Spinelli[3], X. Wang[4,*]

[1] *BRIT, Beijing Research Institute for Telemetry, Beijing, P.R. China*

[2] *CNISM and IMAA-CNR - C.da S. Loja 85050 Tito Scalo Potenza, Italy*

[3] *CNISM and Dipartimento di Fisica – Università di Napoli "Federico II" – 80126 Napoli, Italy*

[4] *CNISM and CNR-SPIN, – 80126 Napoli, Italy, *Email:wang@na.infn.it*

ABSTRACT

In order to carry out 4-D (space and time) analysis of the atmospheric aerosol distribution and to make a characterization of their properties and time evolution, a transportable multi-wavelength, Elastic/Raman scanning lidar system with angular scanning capability has been realized. The system uses a diode pumped Nd:YAG laser source, specifically designed for this device, and a receiving systems able to detect elastic signals at 355, 532 and 1064 nm and Raman signals at 386, 407 and 607 nm. It also allows to perform aerosol depolarization measurements at both 355nm and 532nm.

A first measurement campaign has been carried out in Dunhuang, North-West of China, in the region of the Gobi desert with the aims to study and characterize desert dust at source.

Optical properties of aerosol layers developing in the atmosphere have been analyzed and lidar data are discussed in terms of profiles of aerosol backscatter coefficient at 355nm, 532nm, aerosol extinction coefficient at 355nm, aerosol depolarization ratio at 355nm and 532nm and water vapor mixing ratio. Depolarization ratio measured simultaneously at two wavelengths allowed also to study its dependence on the wavelength.

1. INTRODUCTION

Air pollution is a problem that affects for a long time the industrialized countries of Europe and more recently, with the increasing economical and industrial development, China's cities. Those last are characterized by a reduction of the visibility also in absence of special weather condition (e.g fog and rain) depending on high concentrations of particulate matter, often exceeding healthy limit values prescribed by the normative.

The pollutants resulting from human activities (related to vehicular traffic, domestic heating and industrial activity) and those related to natural sources (volcanic eruptions, forest fires, long range transport of mineral particles from desert regions) change the composition and physical-chemical characteristics of the atmosphere, thereby generating a clear reduction of visibility.

In large urban areas, atmospheric pollution is a serious problem recognized by the World Health Organization that highlighted the need to use resources and energy in the development of new methodologies for real time monitoring of atmospheric particles, in order to analyze its effect on air quality.

Advanced systems for environmental monitoring that use the lidar technology allow the real-time monitoring of suspended particles in the atmosphere and a characterization of their geometrical, optical and microphysical properties in space and time.

Lidar systems combining elastic and Raman observations are already successfully used in the aerosols tropospheric and stratospheric studies in the Asian region [1,2], especially during dust transport events, with the purpose to reveal the spatial and temporal distributions of dust and to study the ambient air quality.

In this paper, we describe the results of a measurement campaign performed in Dunhuang on 19-20 of August 2013 with the aim to study and characterize mineral aerosol directly at source.

DOI: 10.1051/epjconf/201611923032

Measurements have been performed with a new lidar apparatus designed and implemented by CNISM (National Consortium of Italian Universities for the Physical Science of the Matter) and BRIT (Beijing Research Institute of Telemetry) in the frame of the AMPLE (Aerosol Multi-wavelength Polarization Lidar Experiment) project.

2. METHODOLOGY

The AMPLE lidar system has been designed to perform volume scanning of the atmosphere and to retrieve high quality 3D map of particulate optical properties and their time evolution.

The system uses a Nd:YAG diode-pumped laser source, specifically designed for this device, doubled and tripled in frequency and with a repetition rate of 1KHz. The high repetition rate of the laser source can increase the detectable signal dynamic range allowing to monitor optically dense atmospheric layers (as cloud, and fog) and to make measurement with high speed scanning to follow rapidly variable phenomena.

The receiving system is able to detect both the elastic lidar returns at 355nm and 532nm, the N2 Raman lidar echoes at 386nm and 607nm and the H_2O Raman lidar echoes at 407nm. Moreover, polarization purity of laser line allows polarization measurements at both 355nm and 532nm.

Each detected signal is acquired with a raw spatial resolution varying from 30cm to 30m.

The lidar system has been calibrated through horizontal measurement method and iterative method for overlap function calibration. Multi-wavelength channel calibration method has been implemented by using cirrus cloud, while an achromatic depolarizer has been used to calibrate the depolarization measurements [3].

Lidar profiles of aerosol backscattering coefficient at 355nm and 532nm, aerosol extinction coefficient of 355nm, water vapor mixing ratio and aerosol depolarization ratio at 355nm and 532nm have been retrieved from lidar signals.

Their spatial and temporal distributions have been analyzed to study change in the aerosol optical properties measured in the observed layers. The Color Index (CI) expressed as:

$$CI=(Log\ (\beta_{532}/\beta_{355}))/(Log\ (355/532))$$

has been evaluated in order to study the wavelength dependence of the backscatter coefficient.

Moreover, calibrated depolarization ratio simultaneously measured at two wavelengths allowed to study the aerosol depolarization dependence on the wavelength in addition to discriminate aerosol shape identifying the presence of non-spherical particles in the observed layers.

3. RESULTS

In Fig.1, the 24 hours continuous measurements performed in Dunhuang on 19-20 of August 2013 are reported in a color map of the Range Corrected Signal (RCS) that highlights several aerosol layer located below 4 km of altitude and a cirrus cloud in the range from 9 to 11 km.

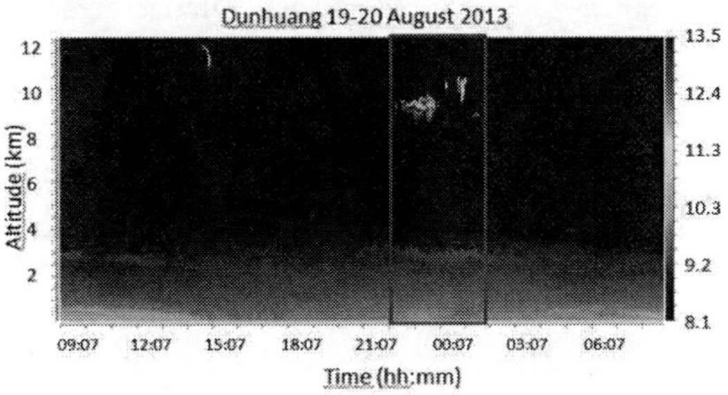

Fig:1 *RCS color map of the 24 hours continuous measurements performed in Dunhuang on 19-20 of August 2013*

A preliminary quantitative analysis of the data relative to the time interval between 22:00 UTC of 19 of August to 01:40 UTC of 20 (red line in the map) has been performed.

A very interesting feature develops after 22:00 UTC. It is spectacularly visible in the depolarization map of figure 2.

It consists in an evident dynamical effects of rising of mineral particles from the soil giving rise to a stable layer at about 2km of altitude. A layered structure of aerosol is also present up to about 5 kilometers of altitude.

DOI: 10.1051/epjconf/201611923032

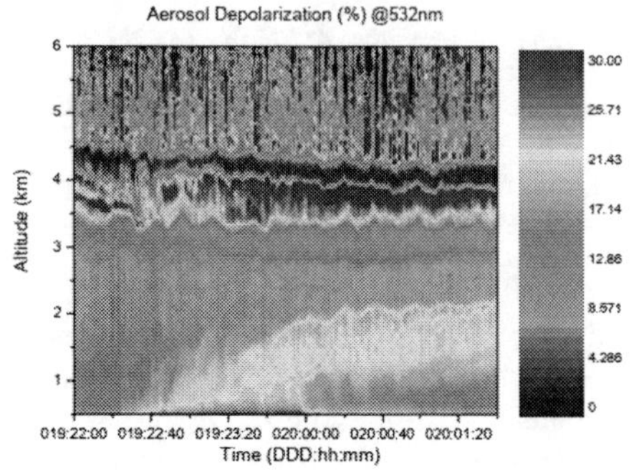

Fig:2 *dynamical effects of rising of mineral particles from the soil giving rise to a stable layer at about 2km of altitude*

Data with a time resolution of 1 minute have been analyzed in term of aerosol backscattering coefficient (β) and aerosol depolarization ratio (δ) at 355nm and 532nm. The measured values of β and δ resulted very close at the two wavelengths for cirrus, as expected from large particles, but they differs considerably for low altitude layers as a function of the different physical-chemical properties of particles in these layers.

This behavior is more clear in the scatter-plot of figure 3 (a-b) showing the depolarization ratio at the two wavelengths. The values have been averaged in different atmospheric layers a) 0.5÷2 km, d) 8÷10.5 km relating to the observed aerosol signature in the lidar maps.

In figure 3a the values obtained in the first layer are shown. We observed a different dependence on the wavelength of the aerosol depolarization with time. In fact red point of the figure are related to the first 50 minutes of the measurement, before the observed rising of mineral particles. The different dimension of the particles could explain the observed different trend.

In the upper layer (figure 3b) where cirrus is present, data are clearly correlated and fitted by a line with slope 0.98, indicating very close values between the two wavelength observed in the cloud.

Of course, a more detailed analysis of the optical and microphysical aerosol properties in these layers is required to better explain the observed results.

Data analysis of aerosol extinction and Lidar Ratio (LR) has been performed with a time resolution of 10min, in order to reduce statistical errors. Therefore, LR and Color Index (CI) have been analyzed in order to characterize aerosol and to define a particles classifications in each layer.

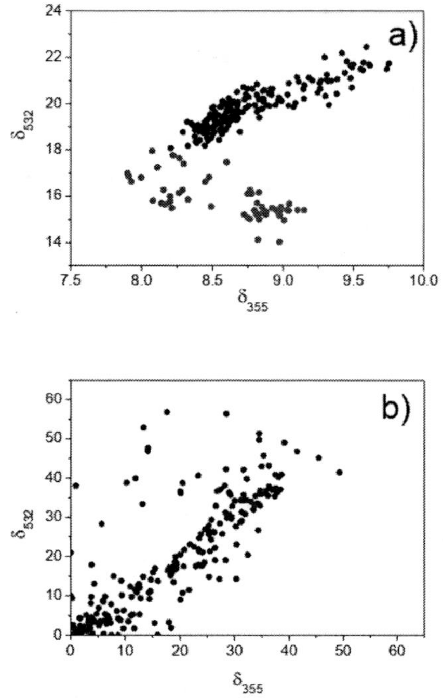

Fig:3 *depolarization ratio at the two wavelengths in the atmospheric layers: a)0.5-2 km and b) 8-10.5 km*

For cirrus we obtain values close to 0 for CI while a larger spread of LR values has been observed, indicating a change in the cloud properties with time.

Below 4 km LR decreases with the altitude as the CI increases. A comparative analysis of the obtained values allowed identifying aerosol of different nature in each layer.

Water vapor mixing ratio showed the presence of a stable moist layer extending up to 4km, with a maximum at around 3km. Therefore, physical property of particles in this layer could be strongly influenced by this.

4. CONCLUSIONS

An innovative lidar system has been developed and a first measurement campaign has been carried out in Dunhuang, North-West of China, in the region of the Gobi desert in order to study and characterize desert dust at source.

Elastic lidar signals have been analyzed in term of aerosol backscattering coefficient and aerosol depolarization ratio at 355nm and 532nm. Moreover, from the Raman analysis the lidar ratio and the water vapour mixing ratio have been also determined. The results showed a layering with aerosol of different nature in each layer. Optical and microphysical properties of the aerosol have been studied.

ACKNOWLEDGEMENT

The financial support by the European Community through the ACTRIS Research Infrastructure Action under the 7th Framework Program, ACTRIS Grant Agreement n. 262254 is gratefully acknowledged.

REFERENCES

[1] J. Liu, Y. Zheng, Z. Li, C. Flynn, E.J. Welton , M. Cribb, Transport, vertical structure and radiative properties of dust events in southeast China determined from ground and space sensors, *Atmospheric Environment* **45** (2011) 6469-6480.

[2] Z. Chen, W. Liu, B. Heese, D. Althausen, H. Baars, T. Cheng, X. Shu, and T. Zhang, Aerosol optical properties observed by combined Raman-elastic backscatter lidar in winter 2009 in Pearl River Delta, south China, *J. Geophys. Res. Atmos.*, (2013), **119**, 2496–2510.

[3] X. Wang, A. Boselli, A. Sannino, C. Song, N. Spinelli, Y. Zhao, and C. Pan (2014) Calibration of Multi-wavelength Raman Polarization Lidar, EPJ Web of Conference, **89**, 01002 (2015) .

MULTIYEAR AEROSOL STUDY BASED ON LIDAR&SUNPHOTOMETER MEASUREMENTS IN ROMANIA

Anca Nemuc [1]*, Ioannis Binietoglou[1], Simona Andrei[1], Alexandru Dandocsi [1,2], Horatiu Stefanie[1,3,4]

[1]*National Institute of Research and Development for Optoelectronics INOE, Bucharest-Magurele, Romania, *Email: anca@inoe.ro*
[2]*University Politehnica of Bucharest, Faculty of Applied Sciences, Bucharest, Romania*
[3]*"Babeş-Bolyai" University of Cluj-Napoca, Faculty of Environmental Science and Engineering, Cluj-Napoca, Romania*
[4]*University of Bucharest, Faculty of Physics, P.O.BOX MG-11, Magurele, Bucharest, Romania*

ABSTRACT

This observational study focused on three-years time-averaged data set (January 2012–2015). An investigation of long-term trends was performed on two different data sets derived from active and passive remote sensing measurements in Magurele, Romania.

Measurements of sun photometer aerosol optical depth (AOD) at 500 nm and 340 nm show the mean values of 0.230 ±0 .118 and 0.398 ± 0.185, respectively. The lidar AOD at 532 and 355nm has a mean of 0.271 ±.0.164 and 0.472 ± 0.165 respectively. The highest seasonal mean value was measured by the lidar during the summer of 2014 while the lowest seasonal value was measured by the sunphotometer in February 2012.

The origin of atmospheric aerosols has been analyzed using both backtajectories of Hysplit and Circulation Type Classification (CTCs) methods.

1. INTRODUCTION

Despite of the GCMs (General Circulation Model) improvements in the last years, aerosols are still not well captured, therefore long term observations are essential to address the questions of global climate changes.

EARLINET (European Aerosol Lidar Network) started in 2000 a coordinated research strategy to quantitatively address the complex aerosol–climate problem. Many climatological studies, Saharan dust outbreaks, volcanic eruptions, biomass burning layers have been documented for model evaluation and satellite data validation and integration [1]. The EARLINET procedures ensure instrument standardization and consistent retrievals within the network [2].

Detailed analyses of high-quality observations of aerosol microphysical and optical properties from long-term AERONET (Aerosol Robotic Network) (http://aeronet.gsfc.nasa.gov/) monitoring sites can yield a complete characterization of aerosol optical effects for a wide range of applications climatology of atmospheric aerosols [3]. In order to establish the origin of atmospheric aerosols two classical approaches (Lagrangian and Eulerian) are generally used. One is the Lagrangian analysis of back trajectories using Hysplit model [4]. Second, the Eulerian approach can be used via Circulation Type Classification methods based on synoptic fields at one time instant [5].

The methodology used for analysis of lidar measurements and sunphotometer during 2012-January 2015 near Bucharest, Romania, along with an estimation of the aerosol types influence during this time is provided in the next sections.

2. METHODOLOGY

The multiwavelenth Raman Lidar (RALI) at INOE has been submitting data to EARLINET since 2010, but major upgrades to include calibrated depolarization were only performed since 2012. Technical details are described in [6]. The climatology data used in this study was measured in the frame of EARLINET, between January 2012 and 2015, containing 162 lidar measurements of aerosol extinction and backscatter coefficients at 355 and 532 nm.

The sun-photometer is operational in the AERONET since July 2007 and provides direct solar irradiance measurements at 8 different wavelengths from UV to near IR (340, 380, 440, 500, 675, 870, 1020 and 1640 nm) and further the aerosol optical depth (AOD) at each wavelength from the surface to the top of the atmosphere [7]. Monthly variation of the AOD is described based on the sunphotometer data Bucharest_Inoe site and on RALI measurements performed every Monday and Thursday sunset time, with one hour average. Both instruments are located in Magurele, 5km South of Bucharest (45.35 N, 26.03 E). Standard deviation values are reported as errors of the mean values to provide information about the typical aerosol load and its variability.

To identify the connections between large scale circulation patterns and the short–to–long-range aerosol transport, the Wetter Lagen Klassifikation (WLKC733) developed within "COST733Action" [8] has been used. This is an automated threshold-based catalog derived from the "Objektive Wetterlagenklassifikation-OWLK" classification developed at German Weather Service [9]. It uses information from three tropospheric levels: 925, 700 and 500 hPa, and information regarding water content over the entire tropospheric column. Geopotential fields at 925 hPa and 500 hPa are used to establish the cyclonicity or anticyclonicity, while the wind at 700 hPa is used to establish the dominant flow direction. Because of the increased number of inputs the catalog generates 40 circulation types.

According to this catalog, for each time step (day in this study) - a certain type of circulation was assigned, as the most predominant ones are: NE (Nord East-Moldova, Ukraine, Russia), SE (Black Sea-South East, Turkey, Caucasus Region), SW (South West- from the Balkans, Mediterranean area, Sahara), NW (North West-Central Europe).

3. RESULTS

Seasonal variation of extinction profiles

The occurrence of strong, elevated aerosol layers can best be seen in the summer profiles (Fig.1 panels b and f). Large extinction coefficients occur frequently up to 4-5 km in summer and autumn (Fig.1 panels b,c,f,g). In contrast, the winter profiles mostly show aerosol below 2 km; only in a few profiles we observe values large values above this height (Fig.1 panels d and h).

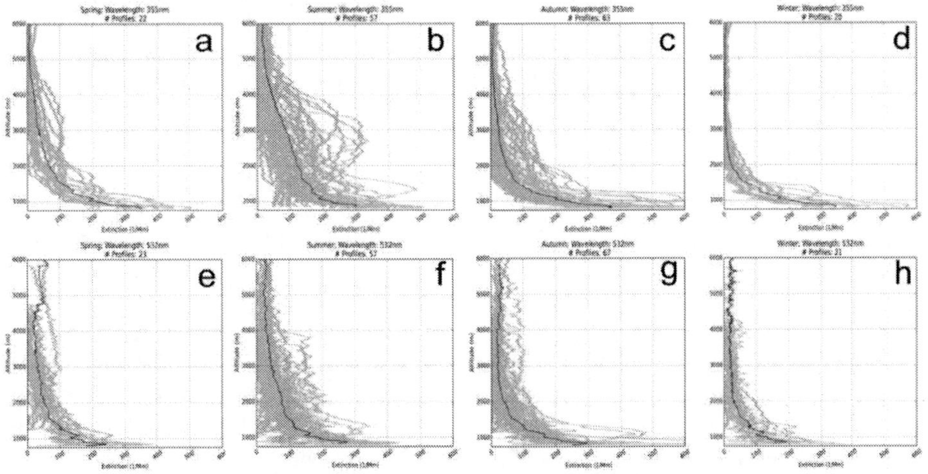

*Fig. 1 Extinction coefficient profiles in grey, grouped into season (**a,e** -spring; **b,f**- summer; **c,g** -fall; **d,h**-winter), at 355 nm (upper panel), 532 nm (lower panel), mean profiles (dark blue) derived from RALI measurements in Magurele (93m a.s.l.). Used are all EALINET climatology data from 2012 to January 2015*

Occurrences of aerosol types

In Fig.2 is presented the distribution of the occurrence of different aerosol types as detected in different layers by RALI and is confirmed by complementary tools (HYSPLIT, MODIS fire maps), separated for the four seasons.

The assessment, carried out for the entire data set (multi-annual evaluation) and seasonal evaluation (Fig.2), emphasized the presence of aerosols within three main layers. The first one, between 0.5 and about 2 km, can be assimilated with the PBL (Planetary Boundary Layer). Its height varies seasonally, the upper limit being at an altitude of 1.36 km during wintertime and 2.09 km during summertime. The second layer (between 2 and 4 km) has been detected in 44% of the analyzed cases, and the third layer (between 4 and 6 km) was identified only in 8% of cases.

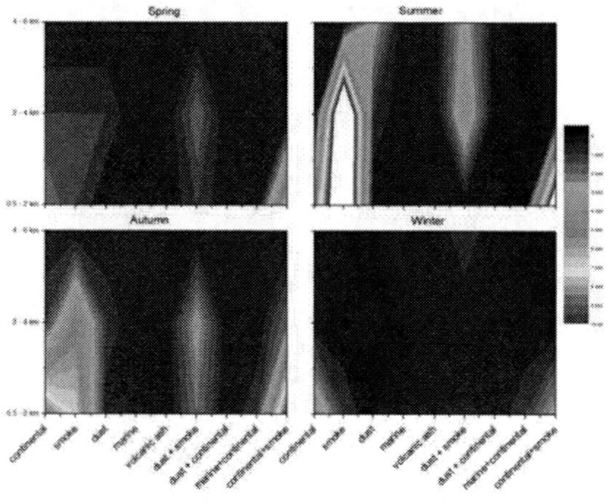

Fig. 2 Multiannual and seasonal distribution of aerosol types in different layer(0.5 -2Km;2-4 Km;4-6km) as measured at Magurele

Occasionally (1case), the presence of a fourth layer has been noticed at heights of over 10 km. This particular event was due to trans-continental & trans-oceanic transport of mineral particles from desert areas of Asia (Gobi desert in our case), following intense storms ten days earlier.

Observations of atmospheric aerosol columnar

AOD values from extinction profiles (at 532nm and 355 nm) have been calculated from the measured ones starting at 850m while for the lowest PBL part an exponential decay have been considered in order to eliminate the bias related to different altitude ranges measured by the two instruments. The AOD wavelength dependency of the two RALI wavelengths 532 nm and 355 nm can be seen in the average values (0.271±0.164, and 0.472±0.165 respectively) as well as in the monthly and seasonal variation of the AOD (Fig. 3). The AOD decreases with increasing wavelength.

Optical properties were analyzed from January 2012 to January 2015. During this period 162 days of measurements were retrieved from AERONET level 2.0 (01.2012-03.2014) and level 1.5 data (03.2014-01.2015). There are gaps in the data of about one month yearly due to the system calibrations and upgrade.

AOD monthly values range from 0.03 to 0.6 (500 nm sunphotometer, 532nm lidar) and from 0.12 to 0.45 (340nm sunphotometer, 355nm lidar) respectively. The values from May to September (spring to late summer) are slightly higher than the winter months (October to February). The lowest AOD occurs in February, while the maximum is reached in August.

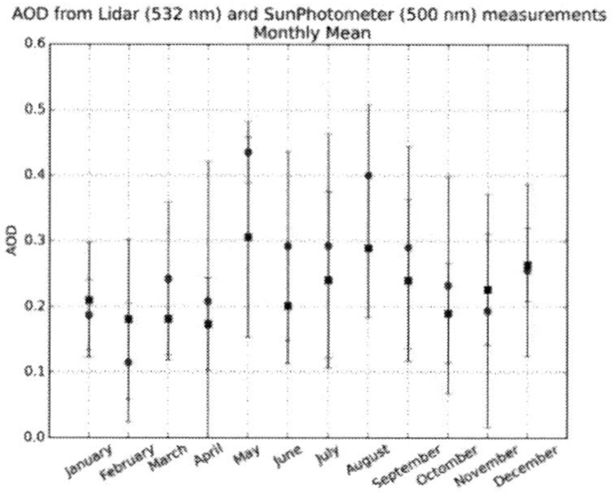

Fig. 3 Monthly mean and uncertainty of AODs from lidar in red (532 nm) and sunphotometer in blue (500nm). The error bars show the variability within the mean

The monthly means of the two instruments match well with each other, with highest values in August, and the minimum AOD of both instruments in winter, indicating that these values are credible. However, on a monthly scale some differences can be seen: some, such as that AERONET encounters absolute minima in April while RALI has it in February. Large differences between the two instruments in May, June and August can be explained by the large uncertainties. This data points to singular high AOD events moving the mean AOD to larger or lower values. These differences could also arise from the different sampling times (during night lidar and daytime-sunphotometer).

Circulation patterns

The predominant sectors identified as the origin of the air masses monthly averages are represented in Fig. 4. It can be noted the predominance of the circulation from the NW (bringing mainly continental polluted aerosol from Central Europe) during December and February and also the frequent SW circulation from June to October (mostly linked with Saharan dust intrusions and biomass burning influence).

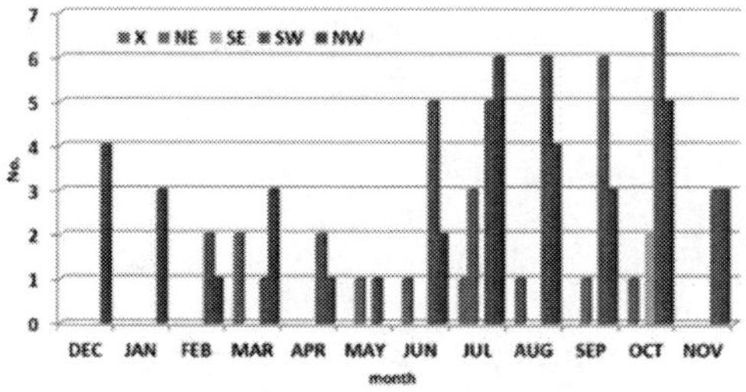

Fig. 4 Monthly distribution of Circulation types- main predominant directions January 2012 - 2015

4. CONCLUSIONS

To characterize the three years aerosol columnar properties and to quantify the contribution of different aerosol types we used lidar and sunphotometer data along with CTCs analysis.

AODs at 355 (340) nm and 532 (500) nm are characterized by a wide range of values from 0.03 to 0.6 and from 0.15 to 0.45 respectively, as shown by the measurements above Magurele, Romania. Although some discrepancies exist between the two datasets monthly averages of AOD evidence a seasonal behavior: higher AOD values are mainly observed during summer months while the minimum ones are in winter.

We observed elevated aerosol layers during summer and fall as shown by the measured lidar profiles with high extinction coefficients occurring frequently up to 4-5 km. The averaged seasonal CTCs highlighted the long-range transport of mixed aerosol: dust from desert regions and biomass burning from Ukraine and Greece (summer and fall). The local sources seem to be the only ones contributing to the aerosol load during the winter since we could see mostly aerosol below 1.2 km.

ACKNOWLEDGEMENT

This work has been supported by grants of the STAR–ESA Programme Project No. 38/2012- CAPESA and Project No.55/2013- CARESSE.

MODIS active fire products were supplied by Fire Information for Resource Management System FIRMS (NASA Information for Research Management System) at the University of Maryland. The authors gratefully acknowledge the NOAA Air Resources Laboratory (ARL) for the provision of the HYSPLIT transport model and/or READY website (http://www.arl.noaa.gov/ready.html).

REFERENCES

[1]Pappalardo, G., et. al. , 2014: EARLINET: towards an advanced sustainable European aerosol lidar network, *Atmos. Meas. Tech.*, 7, 2389-2409, doi:10.5194/amt-7-2389-2014.

[2]Böckmann, C. et al., 2004: "Aerosol lidar intercomparison in the framework of the EARLINET project. 2. Aerosol backscatter algorithms", *Applied Optics*, **43**, pp. 977-989.

[3]Schafer, J. S. et al., 2008: Characterization of the optical properties of atmospheric aerosols in Amazônia from long-term AERONET monitoring (1993–1995 and 1999–2006), *J. Geophys. Res.*, **113** D04204.

[4]Toledano, C. et al., 2009: Airmass Classification and Analysis of Aerosol Types at El Arenosillo (Spain). J. *Appl. Meteor. Climatol.*, **48**, 962–981.

[5]Cahynová M. and Huth R., 2010: Circulation vs. climatic changes over the Czech Republic: a comprehensive study based on the COST733 database of atmospheric circulation classifications, *Phys. Chem. Earth*, **35**, 422–428.

[6] Nemuc, A. et al., 2013: Assessment of aerosol's mass concentrations from measured linear particle depolarization ratio (vertically resolved) and simulations, *Atmos. Meas. Tech.*, **6**, 3243-3255.

[7]Dubovik, O. et al., 2000: Accuracy assessments of aerosol optical properties retrieved from Aerosol Robotic Network (AERONET) Sun and sky radiance measurements, *J. Geophys. Res.*, 105, 9791–9806.

[8]Philipp, A. et al., 2014: Development and comparison of circulation type classifications using the COST 733 dataset and software, *Int. J. Climatol.*, DOI: 10.1002/joc.3920.

[9]Bissolli D. and Dittmann E., 2001: The objective weather type classification of the German Weather Service and its possibilities of application to environmental and meteorological investigations, *Meteorologische Zeitschrift*, **10**(4), 253-260.

STUDY OF THE INFLUENCE OF THE STRATOSPHERIC WARMING IN JANUARY, 2013 ON THE VERTICAL STRUCTURE OF OZONE AND TEMPERATURE IN THE MIDDLE ATMOSPHERE OVER TOMSK USING MICROWAVE AND LIDAR DIAGNOSTICS

G.G. Matvienko[1], Y.Y. Kulikov[2], V.N. Marichev[1], D.A.Bochkovsky[1], A.A. Krasilnikov[2], V.G. Ryskin[2].

[1]V.E. Zuev Institute of Atmospheric Optics, Siberian Branch of the Russian Academy of Sciences, Academician Zuev Square, Tomsk 634021, Russia, E-mail: mgg@iao.ru

[2]Institute of Applied Physics, Russian Academy of Sciences, 46 Ulyanov St., Nizhny Novgorod 603950, Russia, E-mail: yuyukul@appl.sci-nnov.ru

1. INTRODUCTION

The study of response of the middle atmosphere (stratosphere and mesosphere) to any external impact (variations of sunlight, flux of cosmic ray particles, eruptions of volcanoes, and anthropogenesis factors) is an urgent physical problem. Ozone and temperature are important atmospheric parameters. The correlation of these parameters is of significant interest from the viewpoint of thermal balance of the middle atmosphere. The basic heating of the stratosphere (altitude range of 20-50 km) is caused by absorption of solar ultraviolet radiation by ozone molecules. Thermal changes influence the rate of ozone formation and destruction. Other types of wave processes can redistribute the structures of vertical profiles of ozone and temperature in the middle atmosphere. Sudden stratospheric warming affects a lot of widely known atmospheric wave processes. Microwave radiometry and Rayleigh lidar allow investigation of ozone and temperature variations during large-scale wave disturbances in the middle atmosphere, such as, for example, stratospheric warming [1-3].

2. METHODOLOGY

Experiment was arranged in the following manner. Measurements were conducted with a mobile microwave ozonemeter [4]. A microwave instrument consisted of an uncooled millimeter wave receiver and a multichannel spectrometer. Device was tuned at the frequency of ozone rotational transition at 110836.04 MHz. A receiver converts millimeter wave signals at its input to lower intermediate frequencies (IF) through heterodyning with a local oscillator. The characteristics of the microwave instrument (mobile ozonemeter) were the following: single sideband noise temperature of 2500 K, analysis bandwidth of 240 MHz, spectral resolution of 1 to 10 MHz, and number of spectral channels of 31. The ozonemeter allowed measurement of atmospheric ozone line with a time resolution 20 min and an accuracy of ~ 2%. Figure 1 shows the general view of the microwave equipment.

Figure 1. Mobile microwave ozonemeter

Figure 2. Rayleigh lidar

The ozone emission spectra were measured by the method of calibration at hot and cold reference loads. To retrieve ozone vertical profiles from measured spectra, we used the model-fitting method [6] with model vertical distributions of pressure and temperature [7]. The uncertainty of retrieval depended on errors of radio measurement and systematic artifacts of the ozone spectrum. Generally, it was about 20% in the altitude range from 20 up to 60 km.

The vertical temperature distribution was measured by the lidar with a receiving mirror 1 m in diameter and a transmitter at an Nd: AIG laser with a wavelength of 532 nm, pulse energy of 200 mJ, pulse repetition frequency of 10 Hz, signal accumulation time of 2 h, spatial resolution of 192 m, and signal receiving in the single-electron photo pulse counting mode. The backscattered radiation was received from altitudes of 15–60 km. The relative error was 2% at an altitude of 30–40 km. Figure 2 shows the general view of the Rayleigh lidar. The detailed description of joint measurements (technology and methods) can be found in [5]. The observed results were compared with MLS/AURA [8] data and Upperair sounding data from stations Kolpasevo (58.31N, 82.95E) and Novosibirsk (54.96N, 82.95E) [9].

3. RESULTS

Joint observations of temperature (lidar) and ozone (microwave ozonemeter) of the middle atmosphere above Tomsk (56.5N, 85E) were conducted between December 14, 2012 and January 29, 2013. Observations were carried out every day before and after sunset, weather conditions permitting. Figure 3 depicts variations of total ozone (OMI/AURA), ozone column in the 20-50 km layer (satellite microwave measurements MLS/AURA), and ozone column at altitudes above 22 km (ground-based microwave measurements) during the observations. Instruments measuring the ozone content observed its decrease and further growth connected with stratospheric warming. The stratospheric warming, during which the lidar and microwave observations were conducted, was of the major type. As a result, the air mass circulation in the upper stratosphere over Western Siberia has changed from westward transport to eastward transport. During winter anomalous events, marked variations of the ozone concentration and temperature in the stratosphere have been observed.

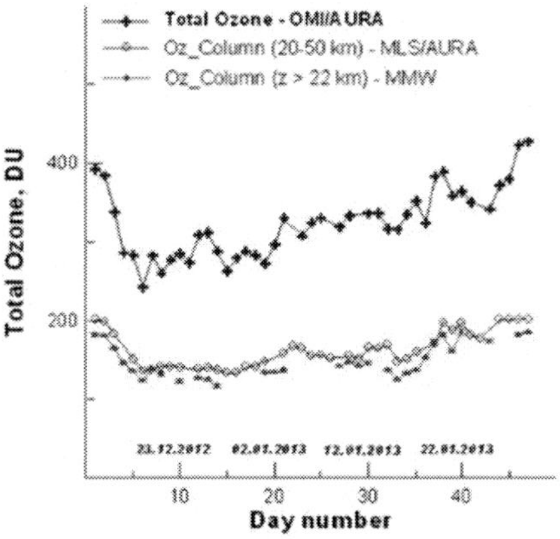

Figure 3. Total ozone (OMI/AURA), ozone content MLS/AURA and MMW (ground-based microwave ozonemeter) versus time during winter 2012/2013 over Tomsk (56.5N, 85E)

Figures 4, 5, 6 illustrate the behavior of ozone density and temperature at altitudes of 25, 40 and 60 km during stratospheric warming. The fact of warming was confirmed by the data of lidar sounding of the high-altitude temperature structure above Tomsk.

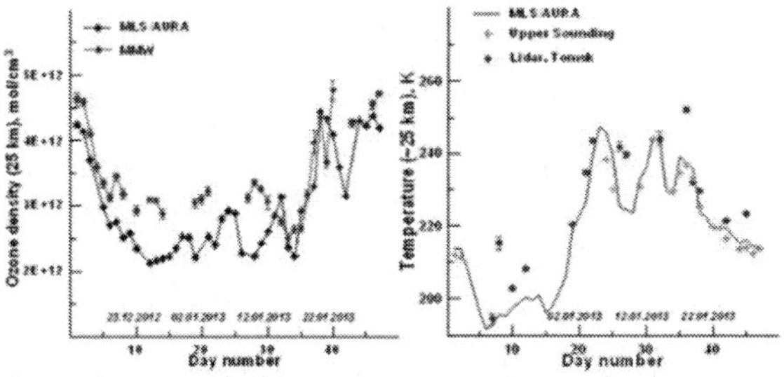

Figure 4. Variations of ozone density and temperature on altitude of 25 km during stratospheric warming in the winter 2012/2013

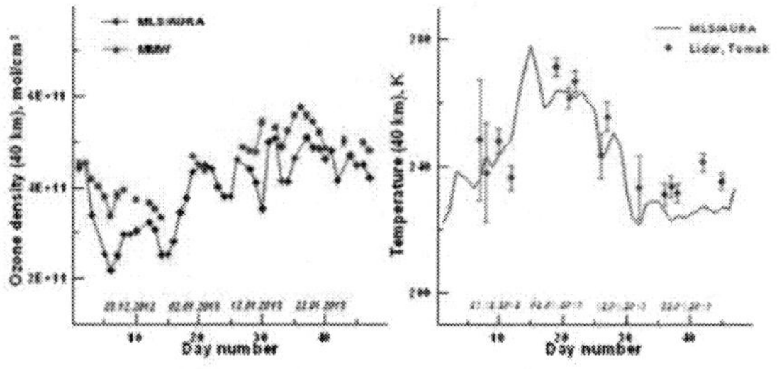

Figure 5. Variations of ozone density and temperature on altitude of 40 km during stratospheric warming in the winter 2012/2013

The ozone density at altitude levels from 25 to 60 km varied by 1.5 to 2 times, and the amplitude of the variations has increased significantly. It should be mentioned that the diurnal variation (due to sunset and sunrise) of the mesospheric ozone density (60 km) additionally obtained in our ground-based microwave measurements was, on average, 30%. The peak positive deviation (70 K) of temperature from its monthly average value was achieved at an altitude of 30 km.

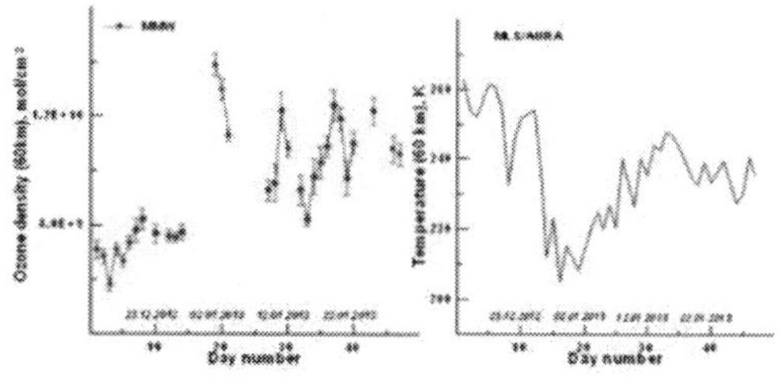

Figure 6. Variations of ozone density and temperature on altitude of 60 km during stratospheric warming in the winter 2012/2013

Correlations between temperature and ozone variations were found at both the natural and disturbed states of the middle atmosphere. Character of ozone variability depends on the thermal regime of the stratosphere, which, in its turn, is determined by the activity of wave processes in the atmosphere of mid-latitudes.

4. CONCLUSIONS

From the results of microwave and optical observations over Tomsk in the period of December—January 2012-2013, the following conclusions can be drawn:

1. The stratospheric warming corresponded to the major type, at which the air mass circulation over the Western Siberia alternated from the western transport to the eastern one.
2. During the anomalous winter events, significant variations of the ozone concentration from 1.5 to 2 times at altitudes from 25 to 60 km and the considerable growth of the amplitude of these variations were observed. The maximal positive deviation of the temperature from its monthly mean value achieved ~ 70 K at an altitude of 30 km.

ACKNOWLEDGEMENT

This work was supported in part by the Program of basic research of Department of physical sciences of the Russian Academy of Sciences "Atmospheric physics: electrical processes, radiophysical methods of study", Integration Project No. 106 of the Siberian Branch of the Russian Academy of Sciences, the Russian Foundation for Basic Researches (Project 13-05-01036a), the President of the Russian Federation (grant NSh-4714.2014.5) and the Project of the Russian Science Foundation № 14-27-00022.

REFERENCES

[1] Y.Y. Kulikov, A.A. Krasilnikov, V.G. Ryskin, Microwave studies of the structure of the polar-latitude ozone layer during winter anomalous warming events in the stratosphere, Atmos. Oceanic Phys., Vol. **38**, pp. 158-166, 2002.

[2] G.G. Matvienko, V.A. Banakh, S.M. Bobrovnikov, V.D. Burlakov, V.V. Veretennikov, B.V. Kaul, G.M. Krekov, V.N. Marichev, The development of atmospheric laser remote sensing technologies, Atmospheric and Ocean Optics, Vol. **22**, No. 10, pp. 915-930, 2009.

[3] V.N. Marichev, Lidar investigations of stratospheric warming over Tomsk in 2008-2010, Atmospheric and Ocean Optics, Vol. **24**, No. 5, pp. 386-391, 2011.

[4] Y.Y. Kulikov, A.A. Krasilnikov, A.M. Shchitov, New mobile ground-based microwave instrument for research of stratospheric ozone (some results of observation), The Sixth International Kharkov Symposium on Physics and Engineering of Microwaves, Millimeter, and Submillimeter Waves (MSMW'07) Proceedings, Kharkov, Ukraine, June 25 – 30, Vol. **1**. pp. 62 – 66. 2007.

[5] V.N. Marichev, G.G. Matvienko, A.A. Lisenko, V.Y. Iljushik, Y.Y. Kulikov, A.A. Krasilnikov, V.G. Ryskin, V.V. Byuchkov, First results of complex experiment on sounding the middle atmosphere in optical and millimeter waves (above Tomsk), Atmospheric and Ocean Optics, Vol. **25**, No. 12, pp. 1091-1095, 2012.

[6] J. De la Noe, A. Baudry, M. Perault, P. Dierich, N. Monnanteuil, M. Colmont, Measurements of the vertical distribution of ozone by ground-based microwave techniques at the Bordeaux observatory during the June 1981 Intercomparison campaign, Planet., Space Sci., Vol. 16, No. **7**, pp. 737-741, 1983.

[7] J.J. Barnett, M. Corney, Middle atmosphere reference model derived from satellite data, Handbook for MAP, Vol. **16**, pp. 47-85, 1985.

[8] http://mirador.gsfc.nasa.gov

[9]http://weather.uwyo.edu/upperair/sounding.html

DOI: 10.1051/epjconf/201611924003

COMPLEX AEROSOL EXPERIMENT IN WESTERN SIBERIA (APRIL – OCTOBER 2013)

G.G. Matvienko, B.D. Belan, M.V. Panchenko, O.A. Romanovskii[a*], S.M. Sakerin, D.M. Kabanov, S.A. Turchinovich, Yu.S. Turchinovich, T.A. Eremina, V.S. Kozlov, S.A. Terpugova, V.V. Pol'kin, E.P. Yausheva, D.G. Chernov, T.B. Zuravleva, T.V. Bedareva, S.L. Odintsov, V.D. Burlakov, M.Yu. Arshinov, G.A. Ivlev, D.E. Savkin, A.V. Fofonov, V.A. Gladkikh, A.P. Kamardin, Yu.S. Balin, G.P. Kokhanenko, I.E. Penner, S.V. Samoilova, P.N. Antokhin, V.G. Arshinova, D.K. Davydov, A.V. Kozlov, D.A. Pestunov, T.M. Rasskazchikova, D.V. Simonenkov, T.K. Sklyadneva, G.N. Tolmachev, S.B. Belan, V.P. Shmargunov

V.E. Zuev Institute of Atmospheric Optics SB RAS, 1 Academician Zuev sq., 634021, Tomsk, Russia, E-mail: roa@iao.ru

ABSTRACT

The primary project objective was to accomplish the Complex Aerosol Experiment, during which the aerosol properties should be measured in the near-ground layer and free atmosphere. Three measurement cycles were performed during the project implementation: in spring period (April), when the maximum of aerosol generation is observed; in summer (July), when atmospheric boundary layer height and mixing layer height are maximal; and in late summer – early autumn (October), when the secondary particle nucleation period is recorded. Numerical calculations were compared with measurements of fluxes of downward solar radiation. It was shown that the relative differences between model and experimental values of fluxes of direct and total radiation, on the average, do not exceed 1% and 3% respectively.

1. INTRODUCTION

The climate change observed for already more than one decade is recognized by the whole global community. Despite the common recognition of this problem, there is no consensus on the role of human activity in the global climate change. Thus, the studies of Antarctic ice cores have shown that for the last 650 000 years there were several periods of global warming accompanied by an increase in the concentration of major greenhouse gases. It was found that the increase of temperature in the Antarctic region started several centuries before the increase of the CO_2 concentration. The permanent development and improvement of climatic models requires the increasingly greater amount of data of field measurements conducted in different regions of the world. The territory of Siberia occupying about 10% of the world's land is practically not covered by the modern observation network. Taking into account the fact that every of the applied methods for investigation of aerosol characteristics is informative only in a certain wavelength range, only the combined analysis of all available experimental data can provide the correct reconstruction of the complete optical pattern of aerosol in a particular region. The objective of this study was to conduct simultaneous measurements of microphysical, chemical, and optical properties of aerosol particles in the atmospheric surface layer and in the free atmosphere with the unique set of ground-based, airborne, and spaceborne instruments in order to draw a complete pattern of the composition and state of the atmosphere over the territory of Western Siberia.

2. MEASUREMENT SYSTEM

The developed measurement system provides laser sensing of the aerosol content, sun-photometric measurements of the aerosol optical depth, acoustic sounding of the boundary layer, measurement of the gas composition, measurement of meteorological parameters of air. The composition and purpose of individual elements of the measurement system are summarized in Table 1.

Remote laser sensing of aerosol of the troposphere was carried out with the stationary LOZA lidar at three laser wavelengths of 355, 532, and 1064 nm. The lidar system detects not only echo signals of elastic backscattering, but also signals of Raman scattering in night time. The multifrequency lidar provides reconstruction of high-quality information about the vertical distribution of optical and microphysical properties of aerosol.

Table 1. Composition and purpose of individual elements of the measurement system

Element	Purpose	URL
Siberian Lidar Station	High-altitude sensing of the atmosphere	http://www.iao.ru/en/resources/equip/sls/
LOZA Lidar	For determination of aerosol profile at a distance of 1-10 km	http://www.iao.ru/en/resources/equip/lidars/loza/
Ty-134 Optik flying laboratory	Measurement of meteorological parameters, gas and aerosol composition of air, optical characteristics	http://www.iao.ru/en/resources/equip/plane/
TOR station	Monitoring of meteorological parameters, gas and aerosol composition of air, and solar radiation in the surface layer	http://lop.iao.ru/eng/index.php/about-tor-station/aero-tor
Basic Experimental Complex	Monitoring of meteorological parameters, gas and aerosol composition of air, and solar radiation in the surface layer	http://lop.iao.ru/eng/index.php/about-bec/aerosols
Fonovaya (Background) Station	Monitoring of meteorological parameters, gas and aerosol composition of air, and solar radiation in the surface layer	http://lop.iao.ru/eng/index.php/about-fonovyi/aerosols-2
DigiCora 3 system for radiosounding of the atmosphere	For determination of vertical distribution of meteorological parameters by contact tools	http://www.vaisala.ru/Vaisala%20Documents/Brochures%20and%20Datasheets/RS92SGP-Datasheet-B210358RU-E-LoRes.pdf
SP-9 sun photometer	Measurement of aerosol optical thickness and water column of the atmosphere	http://link.springer.com/article/10.1134/S102485601304012X
Aerosol Station	Measurements of characteristics of surface aerosol	http://aerosol1.iao.ru/index.php?lang=eng
2.4 XLB Direct Broadcast Reception System	Reception of satellite data from AVHRR, MODIS, and other optoelectronic devices	http://www.orbitalsystems.com/antenna-products/systems/eosdb-systems/
Meteo-2 Ultrasonic Meteostation	Measurement of mean values and characteristics of wind turbulence and air temperature	http://www.iao.ru/en/resources/equip/dev/meteo2/
Volna Sodar	Measurement of vertical profiles of wind vector components in the atmospheric boundary layer	http://www.iao.ru/en/resources/equip/dev/sodar/

We will first consider the systems used for measurements in the surface air and then describe the devices for determination of the integral content of sought parameters and their vertical distributions. Continuous measurements of the concentrations of minor atmospheric constituents were carried out at three stations for monitoring of the atmospheric composition owned by IAO SB RAS: Fonovaya Observatory, TOR station, and Basic Experimental Complex (BEC) [1]. To study the vertical distribution of the climatically significant components of the troposphere, we used the analytical equipment installed aboard the Tu-134 Optik flying laboratory [2]. The temperature-wind sounding was carried out with the Vaisala DigiCORA® MW31 radiosonde. Aerosol measurements at the stationary IAO Aerosol Station are conducted with the use of the methods for investigation of aerosol properties. The aureole photometer measures the directed scattering coefficient and is used for estimation of the mass content of coarse particles in the size spectrum. For monitoring of temperature, humidity, ozone, aerosol, and clouds the data of MODIS multichannel spectroradiometer installed on Terra and Aqua platforms were mostly used. The developed measurement system allows us to reconstruct the whole set of microphysical, chemical, and optical characteristics of aerosol from measured data.

3. MEASURING SITES

Permanent measurements of the concentration of minor constituents were carried out at the three stations for monitoring of the atmospheric composition belonging to IAO SB RAS (Fig. 1): Fonovaya Observatory, TOR Station, and Basic Experimental Complex.

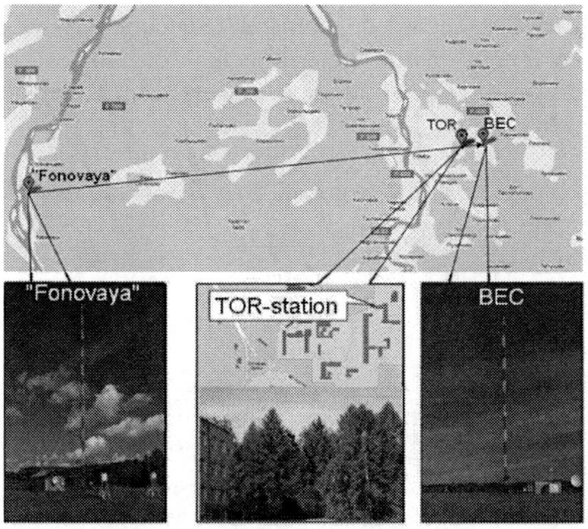

Fig. 1. Arrangement of stations for atmospheric monitoring: (left) Fonovaya Observatory, (center) TOR station, and (right) BEK.

The Fonovaya Observatory is located 60 km to the west of Tomsk, Siberian Lidar Station and the TOR station is situated at the northeastern end of the Tomsk Scientific Center (Akademgorodok), and BEC lies in suburbs 3 km to the east from Akademgorodok. This arrangement of the stations

 DOI: 10.1051/epjconf/201611924003

(nearly in line) in the case of the west-eastern transport of air mass allows us to estimate the anthropogenic contribution of the city of Tomsk to the formation of the field of atmospheric pollutants. The vertical distribution of the climatically significant components of the troposphere was studied with the Tu-134 Optik flying laboratory. The Tu-134 Optik flying laboratory carried out a sounding flight over the territory of Ob-Tom interfluve southwest of Tomsk (on Kozhevnikovo – Kurlek traverse) on April 9, 2013. Regular monitoring of the aerosol optical thickness (AOT) of the atmosphere was carried out at the two observation sites: in Tomsk (IAO SB RAS) and at the Fonovaya Observatory. At the stationary Aerosol Station of IAO SB RAS located at the southeastern end of Tomsk, characteristics of the surface aerosol were monitored every hour round-the-clock. The Complex Aerosol Experiment on 2013 included a cycle of simultaneous two-site measurements at the IAO Aerosol Station in Tomsk and at the mobile Aerosol Station based at the Fonovaya Observatory in the forest zone 70 km to the west from Tomsk. The mass concentrations of aerosol and black carbon were measured every hour round-the-clock at the two measurement sites. Aerosol fields of the vertical depth of the troposphere were sensed with the aid of the LOZA-S stationary multifrequency lidar. The sensing in 2013 was conducted continuously starting from 06:00 local time to 20:00. In 2013, the two-site radiative experiment was conducted in Tomsk and at the Fonovaya Observatory. This experiment invoked the data of MODIS multichannel spectroradiometer installed on Terra and Aqua platforms of the EOS satellite system.

4. RESULTS OF COMPLEX AEROSOL EXPERIMENT

A series of experiments on laser sensing of the troposphere was performed in period from April 2 to 11, 2013, and radiosondes were launched in some days. Figure 2 presents two examples of comparison of the aerosol spatial distribution and radiosonde data. On a color scale, the figures present the profiles of backscattering coefficient in units of scattering ratio, defined as the ratio of total to molecular scattering. The meteorological data were obtained at ultrasonic meteorological station, located near lidar at height of 3 m above the sea surface. Based on balloon-sonde data, we presented the relative humidity r, the wind direction D, and the equivalent-potential temperature (EPT) in degrees Kelvin. As can be seen from Figure 2, the meteorological situation on April 3 was characterized by the absence of clouds and considerable surface heating at preserved negative air temperatures. The balloon-sonde profiles for 08:00 LT show decrease in relative humidity and veering of wind above the boundary of the boundary layer.

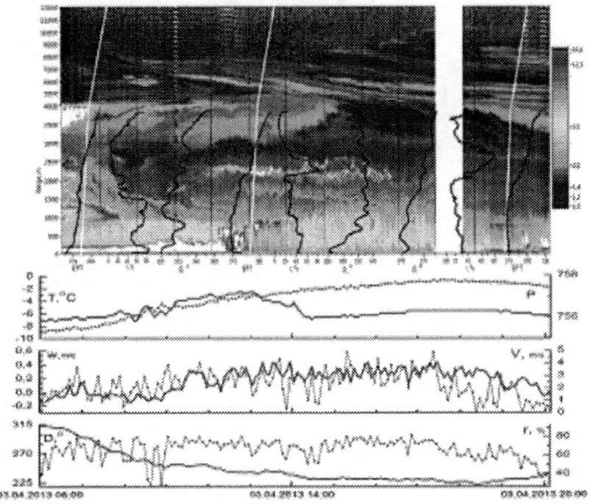

Figure 2 – The aerosol structure on April 3, 2013. White lines show the profiles of sonde ascents.

The vertical aerosol and black carbon profiles were measured onboard aircraft on pril 9 during flights in the region of Tomsk (Fig. 3).

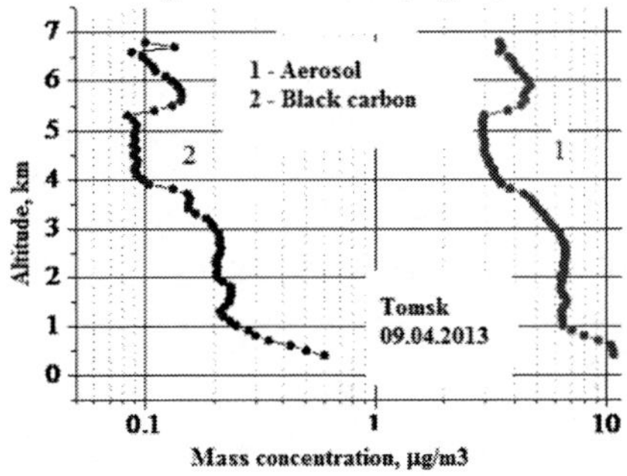

Figure 3 – Vertical aerosol and black carbon profiles according to data from flights of flying laboratory in the region of Tomsk on April 9.

Analysis of near-ground measurements at IAO aerosol station showed that aerosol characteristics exhibited qualitatively similar variations to those following from aircraft observations. The analysis was performed using diurnally average behaviors of dry bases of submicron aerosol and black carbon, normalized by their diurnally average values (Fig. 4). The diurnal behavior of condensation activity parameter γ exhibits anticorrelation with behaviors of mass concentrations of submicron aerosol and black carbon (Fig. 5).

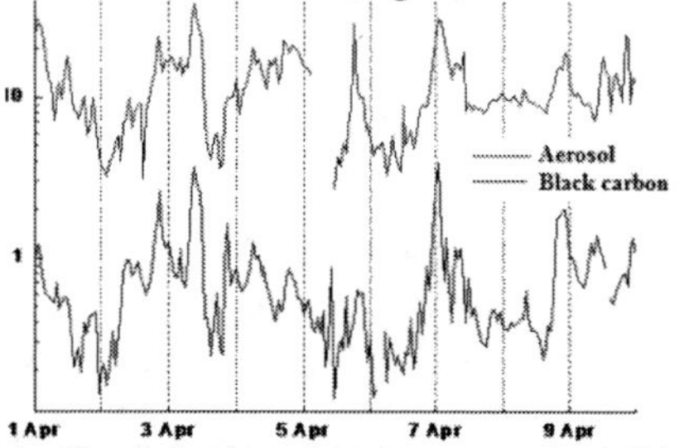

Figure 4 – Time behaviors of mass concentrations of dry base of submicron aerosol and black carbon on April 1-9, 2013.

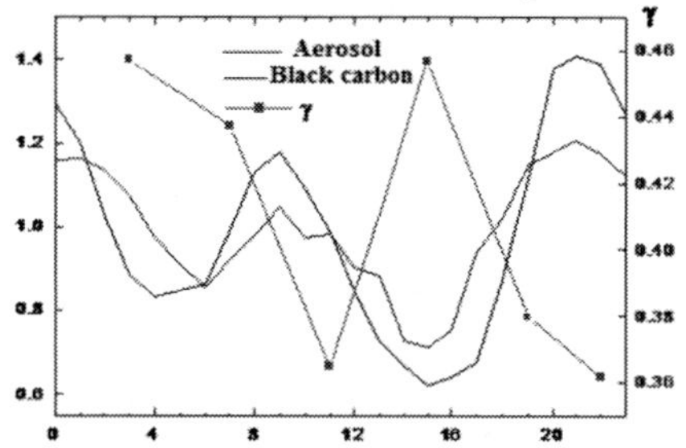

Figure 5 – Diurnally average normalized behaviors of mass concentrations of dry base of submicron aerosol and black carbon and diurnally average behavior of condensation activity parameter on April 1-9, 2013.

To date, the dataset of observations of size distribution of aerosol number concentration at TOR station and at Fonovaya observatory includes over 1900000 and 410000 measurements respectively. The decadally average values of the total number concentration varied in the ranges of $1.3{\cdot}10^3$-$1.2{\cdot}10^4$ cm^{-3} at the TOR station and $1.4{\cdot}10^3$-$5.8{\cdot}10^3$ cm^{-3} in the region of observatory. The average number concentration of aerosol particles over the entire period of observations was 4890 cm^{-3} (±4905 cm^{-3}, median= 3610 cm^{-3}) at TOR station and 3200 cm^{-3} (±3244 cm^{-3}, median= 2561 cm^{-3}) at Fonovaya station.

Comparison of numerical calculations and measurements of downward fluxes of the solar radiation at the Earth's surface. It is shown that relative differences between model and experimental values of the direct and total radiation, on the average, do not exceed 1% and 3% respectively when instrumental errors and determination uncertainties of atmospheric characteristics are taken into consideration. It is shown that under summer conditions the radiative cooling rates have local maximum at the height of approximately 3 km, with the magnitude of the maximum being approximately 3.5 *K*/day.

5. CONCLUSIONS

The Complex Aerosol Experiment on measurement of various atmospheric parameters in the Siberian Region in 2013 has become a first step on the way of joining the efforts of different research groups aimed at obtaining the as complete as possible information about the state of the atmosphere under conditions of changing climate. Despite the short duration of the experiment, it has demonstrated the potential capabilities of available measuring facilities for longer measurements in the future. Nevertheless, the data obtained during the Complex Aerosol Experiment are of great value both for validation, correction, and calibration of satellite measurements and for validation and improvement of climatic models.

ACKNOWLEDGEMENTS

This work was supported in part by the Ministry of Science and Education of the Russian Federation (Agreements No. 14.613.21.0013 (RFMEFI61314X0013) and No. 14.604.21.0100 (RFMEFI60414X0100)), the Russian Science Foundation (Agreement no. 14-27-00022), the Russian Foundation for Basic Research (Grant No. 13-05-98074 - r_sibir-a), The President of the Russian Federation (Grant NS-4714.2014.5).

REFERENCES

[1] Arshinov, M.Yu., Belan, B.D., Davydov, D.K., Ivlev, G.A., Kozlov, A.V., Pestunov, D.A., Pokrovskii, E.V., Tolmachev, G.N., Fofonov, A.V., 2007. -Sites for monitoring of greenhouse gases and gases oxidizing the atmosphere. *Atmospheric and oceanic optics,* **20**(1), 45-53.

[2] Anokhin, G.G., Antokhin, P.N., Arshinov, M.Yu., Barsuk, V.E., Belan, B.D., Belan, S.B., Davydov, D.K., Ivlev, G.A., Kozlov, A.V., Kozlov, V.S., Morozov, M.V., Panchenko, M.V., Penner, I.E., Pestunov, D.A., Sikov, G.P., Simonenkov, D.V., Sinitsyn, D.S., Tolmachev, G.N., Filipov, D.V., Fofonov, A.V., Chernov, D.G., Shamanaev, V.S., Shmargunov, V.P., 2011, - OPTIK Tu-134 aircraft laboratory. *Atmospheric and oceanic optics*. **24**(9), 805-816.

EPJ Web of Conferences **119**, 24004 (2016) DOI: 10.1051/epjconf/201611924004
ILRC 27

VARIABILITY OF BIOMASS BURNING AEROSOLS LAYERS AND NEAR GROUND

Jeni Vasilescu[1]*, Livio Belegante[1], Luminita Marmureanu[1], Flori Toanca[1,2]
[1]*National Institute of R&D for Optoelectronics, Magurele, 077125, Romania, *Email: jeni@inoe.ro*
[2]*University of Bucharest, Faculty of Physics, P.O.BOX MG-11, Magurele - Bucharest, 077125, Romania*

ABSTRACT

The aim of this study is to characterize aerosols from both chemical and optical point of view and to explore the conditions to sense the same particles in elevated layers and at the ground. Three days of continuous measurements using a multi-wavelength depolarization lidar(RALI) and a C-ToF-AMS aerosol mass spectrometer are analyzed. The presence of smoke particles was assessed in low level layers from RALI measurements. Chemical composition of submicronic volatile/semi-volatile aerosols at ground level was monitored by the CTOF AMS

Several episodes of biomass burning aerosols have been identified by both techniques due to the presence of specific markers (*f60*, linear particle depolarization ratio, Ängström exponent).

1. INTRODUCTION

Aerosol size distributions, chemical composition, optical and microphysical retrievals provide essential information regarding local and long range transported aerosols [1]. Frequently, the atmospheric layers contain combination of different aerosol types picked up during transport, which can be dry deposited on the ground or remain suspended at higher altitudes [2].

Numerous studies are discussing independently the chemical and physical properties of aerosols, as well as their origin and local component [3][4][5]. The lidar observations and aerosol mass spectrometer ground based measurements can be combined in order to assess the contribution of long range transported aerosols at ground level and to retrieve their chemical and optical properties, including the level of oxidation, dimension, shape, lidar ratio, Angstrom exponents, linear particle depolarization ratio etc. [6].

Multi-wavelength Raman lidars are suitable to track the levels of aerosols layers, their thickness and variability in time, as well as type dependent aerosols properties. Linear particle depolarization ratio can be used to differentiate between high- and low-depolarizing particles, to assess a possible dust intrusion or particles from wild fires [2]. The variability of smoke particles properties and vertical distribution have been intensively studied all over the world, discrepancies being observed even though sometimes the studies focused on the same origin and loads of aerosols [1][3][5]. Several studies already demonstrated that during the aging process Ängström exponent of the biomass burning aerosols decreases from around 2 down to 1, highlighting an increase of the particle size [1].

Organic ions loadings and proportions of smoke markers can be used to demonstrate the presence of biomass burning aerosols at ground level [7][6]. Besides organic component these aerosols contain high proportions of sulfate, nitrate and small amounts of ammonium and chloride fractions. Generally m/z 43, 44, 60 and 73 ions are used to characterize both the biomass burning presence and their aging process. The mechanism of aging process of biomass burning aerosols are still poorly described, but gives important insights of the chemical and physical evolution of organic aerosols. Depending on the level of oxidation the secondary aerosols, the oxygenated organic aerosols (OOA), could be separated into low-volatility OOA (LV-OOA - the more oxidized component) and semi-volatile OOA (SV-OOA). In this paper several cases of biomass burning

DOI: 10.1051/epjconf/201611924004

aerosols are identified and analyzed using lidar and mass spectrometry techniques.

2. METHODOLOGY

Collocated instruments have been used to sense biomass burning aerosols at Magurele site (44.35 N, 26.03 E) near Bucharest, Romania. Three days of continuous measurements of the multi-wavelength depolarization lidar [8] and C-ToF aerosol mass spectrometer have been evaluated. Frequently during summer the Romanian atmosphere has influences of biomass-burning and dust long range transported aerosols as a result of complex dynamics of the air masses [2].

A Compact Time of Flight aerosol mass spectrometer model (C-ToF AMS), based on the Aerodyne quadrupole mass spectrometer measured the chemical composition and size resolved mass concentration of aerosols [7]. The submicronic aerosols were sampled with a 2.5μm cyclone, aerodynamically focused and vaporized at 600°C. Several papers describe the operation principal and application of this instrument to ambient aerosol sampling [9][7]. The mass spectrum and time of flight modes are used to study both chemical composition on five important fractions and chemical resolved size distribution. The average concentration of organics, sulfate, nitrate, ammonium, chloride fractions are available with high time resolution, giving the possibility to sense loadings variations. The most important ions used to assess the fragments are: organics (43, 44, 55 and 57), sulphate (48, 64), ammonium (15, 16) and nitrate (30, 46) [9].

The multi-wavelength Raman lidar system is frequently used in the frame of EARLINET (European Aerosol Lidar Network) for tropospheric measurements [2]. The lidar signal, at 3.75m vertical resolution is collected at 1064, 532p, 532s, 355, 607, 387 and 408 nm, most of the channels having both analog and photon counting detection. The backscatter coefficients (1064, 532, 355 nm), extinction coefficients (532, 355 nm) and linear particle depolarization ratio (532 nm) of aerosols between 0.8 and 10 km altitude are usually used to assess the aerosols presence and their optical properties [6].

The data analysis for the three days continuous measurements started with detection of aerosol layers near ground and those with low linear particle depolarization ratios. Hysplit (Hybrid Single Particle Lagrangian Integrated Trajectory) model have been used to check the possible regional source of the aerosols, while the MODIS (Moderate Resolution Imaging Spectro-radiometer) global fire maps to assess a possible smoke particles presence. From C-ToF AMS the m/z 43, 44, 60 and 73 ions evolutions were considered to identify the biomass burning aerosols presence at ground and their oxidation level.

3. RESULTS

During the intensive lidar measurements the layers were generally present above 2 km and below 6 km altitude, but the descent of air masses from the free troposphere to the ground was favorable. The lofted layers reached the ground after downward mixing in the planetary boundary layer for several hours during the continuous lidar measurement. Three predominant layers near ground were typically sensed (altitude lower than 1.5 km, between 1.5-2.5 km and above 2.5km). The tendency of the first two aerosols lidar layers to mix with the planetary boundary layer can be observed in the time series of the average height of the layers (***Figure 1***).

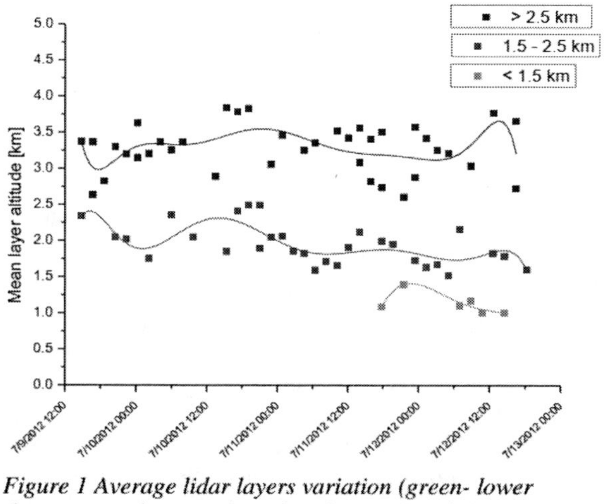

Figure 1 Average lidar layers variation (green- lower layer, near planetary boundary layer, altitude lower than 1.5KM; red - second layer, 1.5-2.5 km; blue – third layer, altitude higher than 2.5 km)

More spherical particle were observed in higher layers, while particles with increased depolarization values were evidenced for lower layers above measurement site during the investigated period (***Figure 2***). The higher linear depolarization ratio (~9%) could corresponds to smoke mixed with urban aerosols, while smaller depolarization values (~3%) for lofted layers to smoke or industrial type. On 11th of July 2012 conversion from small to higher values of depolarization, from higher to lower layers can be seen. This could be a case of downward mixing into the planetary boundary layer.

The Ängström backscatter-related coefficient values are higher than 1.2 for more than 50% of the three days data, which also indicate the presence of small particles; usually the presence of smoke. Two periods of Ängström backscatter-related coefficient values higher than 1.2 were noticed: in the afternoon of 10th and from 11th to 13th of July.

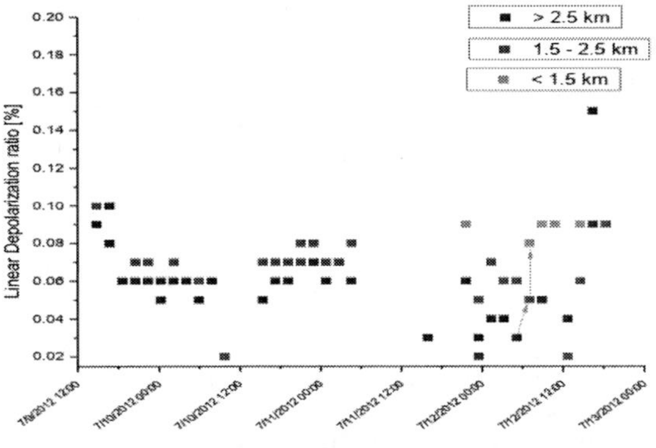

Figure 2 Average linear particle depolarization at 532nm of the three Lidar layers, arrows indicate the conversion from low to higher depolarizing particles

Organics, sulfate, ammonium and nitrate fractions presented high loads during the three days continuous measurements. The organics species represented more than 50% from the total mass of the non-refractory submicronic aerosols (***Figure 3***). Higher concentrations were evidenced during night time due to the low planetary boundary layer height.

The m/z 60 fragment represent a marker for "levoglucosan like" species - based organics, indicating the presence of biomass burning aerosols. Its concentration varied same as m/z 73, high values of *f60* (ratio of m/z 60 to total organic fragment loadings) and *f73* being retrieved for the entire campaign. The *f60* is present for all AMS measurements, a threshold value indicates the biomass burning influence or non-biomass burning sources. During our intensive measurements period the *f60* exceeded very often the background level (0.003) found by Cubison paper [4], indicating the presence of biomass burning aerosols. Doubled than the background level *f60* values were noticed ~~for~~ between 11th and 13th of July, when Hysplit back-trajectories showed air masses travelling from the West of Europe [10] and Balkans where MODIS fire map revealed a high density of forest fires.

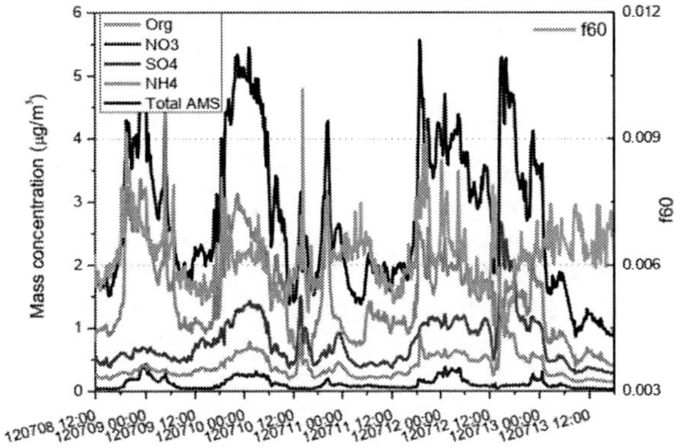

Figure 3 PM1 chemical fractions and total concentration time series as measured by CTOF AMS

The *f44* versus *f60* plot suggested the increasing oxidation state of the organic aerosols (***Figure 4***). The regional LV-OOA represents the more oxidized component, higher *f 44* (ratio of m/z 44 to total organic fragment loadings) values is present for 70% of the data. The less oxidized organic aerosols component (SV-OOA), characterized by lower *f44,* is highlighted mostly for 9th-10th of July and few hours on the 12th. The aging process of aerosols can be also identified using the Cubison *f44* versus *f60* plot, the higher *f44* and lower *f60* values corresponding to aged biomass burning aerosols.

The AMS biomass burning marker showed a strong variability of aerosol properties well correlated with Lidar mean layers altitude for the second part of the measurements period. During the aging process the biomass burning aerosols

Ängström exponent decreased from 2 to 1, while the *f60* ranged between 0.009 and 0.007.

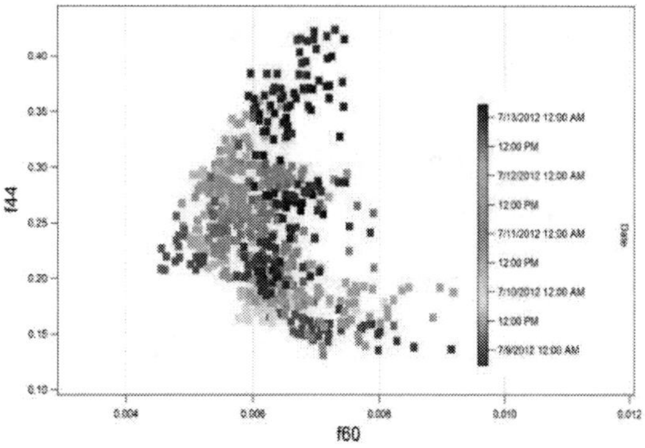

Figure 4 f44 versus f60 for biomass burning aerosols

4. CONCLUSIONS

The concentration variations of few specific organic ions have been used to identify biomass burning aerosols and to evidence their evolution. Significant chemical markers of biomass burning-like particles have been noticed for around 40% of the measurements. Also the Ängström backscatter related coefficient and linear particle depolarization ratio values from lidar continuous measurements pin pointed the presence of low depolarizing and small diameter particles.

The 11th -13th July period has been characterized by highest *f60* values when also Hysplit back-trajectories and MODIS fire indicate biomass burning presence. Correlations of aerosols mean layer height and mass spectrometry biomass burning markers are noticed for half of the measurement period, when conversion from small to higher values of depolarization, from higher to lower layers can be seen. The higher *f44* and lower *f60* values corresponding to aged biomass burning aerosols are identified for first hours of 12th of July.

ACKNOWLEDGEMENT

This work was supported by the Programme, Partnerships in S&T priority domain - PNII, MEN – UEFISCDI, project number 309/2014 (MOBBE) and 144/2014 (LIRA), and by the Romanian – Swiss Research Programme, project no IZERZO_ 142146/1. The work of the author F. Toanca was supported by the strategic grant POSDRU/159/1.5/S/137750, "Project Doctoral and Postdoctoral programs support for increased competitiveness in Exact Sciences research" cofinanced by the European Social Found within the Sectorial Operational Program Human Resources Development 2007-2013.

We acknowledge the use of data products or imagery from the Land, Atmosphere Near real-time Capability for EOS (LANCE) system operated by the NASA/GSFC/Earth Science Data and Information System (ESDIS) with funding provided by NASA/HQ.

REFERENCES

[1] Müller, D. et al., 2007: Multi-wavelength Raman lidar observations of particle growth during long-range transport of forest-fire smoke in the free troposphere, *Geophys. Res. Lett.*, **34**, L05803.

[2] Nemuc, A. et al., 2013: Assessment of aerosol's mass concentrations from measured linear particle depolarization ratio (vertically resolved) and simulations, *Atmos. Meas. Tech.*, **6**, 3243-3255.

[3] Amiridis, V. et al., 2009: Optical characteristics of biomass burning aerosols over Southeastern Europe determined from UVRaman lidar measurements, *Atmos. Chem. Phys.*, **9**, 2431–2440.

[4] Cubison, M J et al., 2011: Effects of aging on organic aerosol from open biomass burning smoke in aircraft and laboratory studies, *Atmos. Chem. Phys.* **11**, 12049–12064

[5] Alados-Arboledas L. et al., 2011: Optical and microphysical properties of fresh biomass burning aerosol retrieved by Raman lidar, and star-and sun-photometry, *Geophys. Res. Lett.*, **38**(1), L01807.

[6] Nicolae, D. et al., 2013: Characterization of fresh and aged biomass burning events using multiwavelength Raman lidar and mass spectrometry, *J. Geophys. Res. Atmos.*, **118**, 2956–2965.

[7] Adler G. et al., 2011: Chemical, physical, and optical evolution of biomass burning aerosols: a case study, *Atmos. Chem. Phys.*, **11**, 1491-1503.

[8] Wang, Y. et al., 2014: Assimilation of lidar signals: application to aerosol forecasting in the western Mediterranean basin, *Atmos. Chem. Phys.*, **14**, 12031-12053.

[9] DeCarlo, P. F. et al., 2006: Field-deployable, high-resolution, time-of-flight aerosol mass spectrometer, *Anal. Chem.*, **78**, 8281–8289.

[10] Draxler, R.R. and Rolph, G.D., 2012: HYSPLIT (HYbrid Single-Particle Lagrangian Integrated Trajectory) Model access via NOAA ARL READY Website

EPJ Web of Conferences **119**, 24005 (2016) DOI: 10.1051/epjconf/201611924005
ILRC 27

COMPARING WATER VAPOR MIXING RATIO PROFILES AND CLOUD VERTICAL STRUCTURE FROM MULTIWAVELENGTH RAMAN LIDAR RETRIEVALS AND RADIOSOUNDING MEASUREMENTS

Montserrat Costa-Surós[1]*, Iwona S.Stachlewska[1], Krzysztof Markowicz[1]

[1]*Institute of Geophysics, Faculty of Physics, University of Warsaw, Warsaw 02-093, Poland,*
**Email: montsecs@igf.fuw.edu.pl*

ABSTRACT

A study of comparison of water vapor mixing ratio profiles, relative humidity profiles, and cloud vertical structures using two different instruments, a multiwavelength Aerosol-Depolarization-Raman lidar and radiosoundings, is presented. The observations were taken by the lidar located in Warsaw center and the radiosoundings located about 30km to the North in Legionowo (Poland). We compared the ground-based remote sensing technology with in-situ method in order to improve knowledge about water content thought the atmosphere and cloud formation. The method used for retrieving the cloud vertical structure can be improved comparing the radiosonde results with the lidar observations, which show promising results.

1. INTRODUCTION

Clouds are key factor that rule climate. Improving methods for automatic and continuous description of clouds has a huge importance in order to determine the role of clouds in climate and their contribution to climate change [1]. The water vapor mixing ratio (WVMR), the relative humidity (RH) and the cloud vertical structure (CVS) in the atmosphere are important characteristics for describing the impact that clouds have in the atmosphere.

In the past, researchers undertook the approach for retrieval cloud layer structures only by using radiosounding (RS) data [2, 3]. There is ongoing work being done to improve these methods and compare them with the most advanced ground-based remote-sensing technologies, such as the Atmospheric Radiation Measurement (ARM) Program value added product, the Active Remote Sensing of Clouds (ARSCL), as well as with satellite data [4]. In this work the comparison was done in order to find a better approximation to the CVS profile, and to reveal the strengths and weaknesses of both ground-based instruments and in-situ radiosoundings. Now, the present study aim is to make similar comparisons between radiosounding methods and advanced multiwavelength lidar measurements in Poland. The idea is to see if the results obtained with ARSCL data product (which is a combination of measurements taken with cloud radar, micro-pulse lidar, and ceilometer) are comparable with those obtained with measurements taken by a multiwavelength Aerosol-Depolarization-Raman (ADR) lidar. Moreover, apart from CVS comparison, the WVMR and RH profiles retrieved from the RS are also compared, since the multiwavelength lidar provide profiles in Raman channel for water.

2. METHODOLOGY

The lidar measurements were taken in the Radiative Transfer Laboratory (RT-Lab) of the Institute of Geophysics, Faculty of Physics, University of Warsaw (Poland). The RT-Lab site located at central Warsaw (52.21°, N 20.98° E, 96m asl) joined the EARLINET [5] in 2015, contributing to the network activities with the 8-channel ($2\alpha+3\beta+2\delta+VW$) ADR lidar, that is a NeXT generation Polly XT system developed in a scientific cooperation with TROPOS, Germany [6]. Its detection channels are at 1064, 532 and 355nm (elastic), at 607 and 387nm (vibrational Raman for N_2), 407nm (Raman channel for H_2O), and 532 and 355 nm (elastic cross channels), using photon counting detection for all channels. Moreover, RT-Lab is part of the Poland AOD network (http://www.polandaod.pl).

The radiosoundings were launched in Legionowo station, which is the nearest radio sounding station (code: 12374), located about 30km to the North from the lidar at RT-Lab, and which frequency of radio sounding launches is 12h.

For the study we focused on one hand on the retrieval of WVMR and RH, and on the other, on the CVS. The main idea of our retrieval scheme is given below.

2.1. Water vapor mixing ratio and relative humidity

The mixing ratio of water vapor to dry air is obtained from nighttime observations. The available lidar wavelength for water vapor detection is not in the solar-blind region and thus the daytime measurements are roughly affected by background light noise. The WVMR profile (*m(z)*) is obtained from the measurement of the water-vapor to the reference signal ratio, where the reference gas is nitrogen [7] (eq. 1).

$$m(z)=K_m\frac{P_{\lambda_{H_2O}}(z)}{P_{\lambda_{Ref}}(z)}\times\frac{\exp\{-\int_0^z[\alpha^{aer}_{\lambda_{Ref}}(\xi)+\alpha^{mol}_{\lambda_{Ref}}(\xi)]d\xi\}}{\exp\{-\int_0^z[\alpha^{aer}_{\lambda_{H_2O}}(\xi)+\alpha^{mol}_{\lambda_{H_2O}}(\xi)]d\xi\}} \quad (1)$$

The $P_{\lambda H2O}(z)$ and $P_{\lambda Ref}(z)$ are the return signals from distance z at the laser wavelength for the water-vapor (at 407nm) and the Raman reference wavelength (at 387nm, a vibrational Raman channel for N_2). The α^{aer} and α^{mol} describe the extinction of light (either in the water-vapor or the reference wavelength), by aerosol particles and air molecules. Extinction coefficients (α) are derived from the Raman signal profile of the reference gas (eq. 2) [7].

$$\alpha^{aer}_{\lambda_0}(z)=\frac{d/dz\left\{\ln\left[N_{Ref}(z)/P_{\lambda_{Ref}}(z)z^2\right]\right\}-\alpha^{mol}_{\lambda_0}(z)-\alpha^{mol}_{\lambda_{Ref}}(z)}{1+(\lambda_0/\lambda_{Ref})^k} \quad (2)$$

The $N_{Ref}(z)$ is the molecular number density of the reference gas. The λ^{-k} describes the wavelength dependence of particle scattering; k=1 is used except for ice clouds for which we set k=0.

The overall system constant $K_m= K_{\lambda Ref} / K_{\lambda H2O}$ can in principle be deduced from the known Raman cross sections and the measured properties of the detector used. In practice it was determined from a comparison of the lidar measurement with critically evaluated data from a radiosonde ascent [7].

Furthermore, relative humidity profiles are obtained as the ratio between the MWVR (from RS and lidar respectively) and the saturation vapor pressure (calculated with Clasius-Clapeyron equations).

2.2. Cloud vertical structure

The CVS is determined from RS using the method by [8] conveniently modified as explained in [4], hereafter entitled as method COS14. Specifically, the method for determining cloud tops and bases is based in RH thresholds applied to the RS profile. As a first step, the RH with respect to liquid water is converted to RH with respect to ice when the temperature is below 0°C. Then, moist layers are identified by applying some conditions, which are related to a minimum RH threshold (min-RH) and to a minimum thickness (400 m). Subsequently, moist layers are defined as cloud layers through some additional steps: (a) if the maximum RH within the layer is greater than the corresponding maximum RH threshold (max-RH) for the base of this moist layer, (b) the base of cloud layers is set at least at 280m above ground level, and (c) two contiguous layers are considered as a single-layer cloud if the distance between these two layers is less than 300m or the minimum RH within this distance is greater than the corresponding inter-RH threshold value. Note that instead of single-humidity thresholds to define a cloud as in earlier works [2, 3], this method is based on altitude-dependent thresholds, which vary between 70% and 95% depending on the specific threshold (min-RH, inter-RH, and max-RH) and on the altitude (greater values of humidity thresholds correspond to lower atmospheric layers) [4, 8].

First approximation to the CVS from multiwavelength lidar is taken from visual inspection of the plotted signals at 1064nm. Then, to determine CVS more accurately, thresholds are applied to the profiles of backscatter coefficients (at 355, 532 and 1064nm; eq. 3), extinction coefficients (at 355 and 532 nm; eq. 2), as well as depolarization ratios (at 355 and 532; which are determined by the ratio between the cross polarized signal to the total signal).

$$\beta^{aer}_{\lambda_0}(z)=[\beta^{aer}_{\lambda_0}(z_0)+\beta^{mol}_{\lambda_0}(z_0)] \times\frac{P_{\lambda_{Ref}}(z_0)P_{\lambda_0}(z)N_{Ref}(z)}{P_{\lambda_0}(z_0)P_{\lambda_{Ref}}(z)N_{Ref}(z_0)} \times\frac{\exp\{-\int_{z_0}^{z}[\alpha^{aer}_{\lambda_{Ref}}(\xi)+\alpha^{mol}_{\lambda_{Ref}}(\xi)]d\xi\}}{\exp\{-\int_{z_0}^{z}[\alpha^{aer}_{\lambda_0}(\xi)+\alpha^{mol}_{\lambda_0}(\xi)]d\xi\}}-\beta^{mol}_{\lambda_0}(z) \quad (3)$$

3. RESULTS

Examples of lidar signals are shown in Figs. 1a and b. Both observations are at 1064nm over Warsaw since we want to see cloud particles. Fig. 1a shows lidar range-corrected signals during 29/07/14 from 18 to 24h UTC, where aerosol complex structures in boundary layer are visible from the overlap level up to 2500 m approximately, whereas a thinner aerosol layer is discernible at 4000 m. Fig. 1b shows range-corrected signals on 30/07/14 from 12 to 17h UTC. This period was partially cloudy: discontinuous cloud layer with the base at 1000 m, and other higher clouds between 2000 and 4000 m are distinguished. Note that cloud signatures are clearly distinguished from aerosol layers since the signal behind their bases are very attenuated.

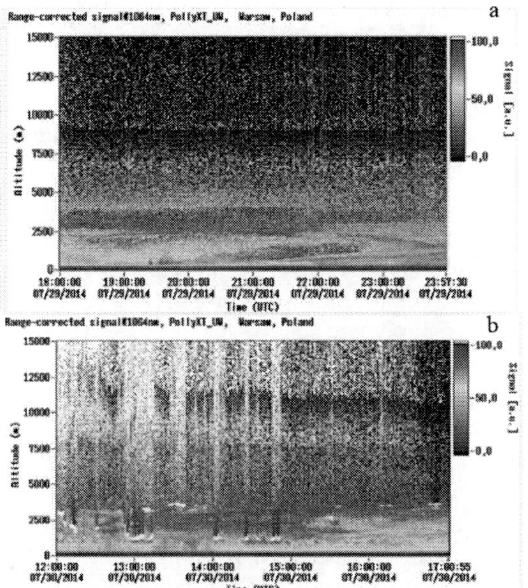

Fig. 1. a) Lidar signals at 1064nm over Warsaw on 29/07/14 from 18 to 24h, and b) on 30/07/14 from 12 to 17h UTC. These figures are available at http://polly.rsd.tropos.de/.

3.1. Water vapor mixing ratio

WVMR profile was calculated from Raman multiwavelength lidar signals over RT-Lab (Warsaw) from 23 to 24 on 29/07/14. According to lidar range corrected signals inspection there were no clouds during this period (Fig. 1a). Overall system constant (K_m) has been determined as 5. In Fig. 2a the MWVR lidar profile is plotted along with the MWVR profile obtained by the RS. The two profiles agree well from the overlap level (540 m) to 3 km, and from 4.6 to 8 km, although at higher heights the MWVR profile obtained by the lidar become noisier.

In Fig. 2b the RH from 3 sources is plotted. The first one is directly taken from the RS over Legionowo on 30/07/14 00UTC, secondly the RH derived from the MWVR from the same RS, and finally the RH derived from multiwavelength lidar signals from 23 to 24h UTC on 29/07/14 over central Warsaw. The figure shows good agreement between the RH calculated from lidar signals with the RH calculated from the RS up to 3 km, and between 4.3 and 6 km. The differences found between 3 to 4.3 km and above 6 km could be due to the distance between the lidar site and the RS launching station, which is about 30 km.

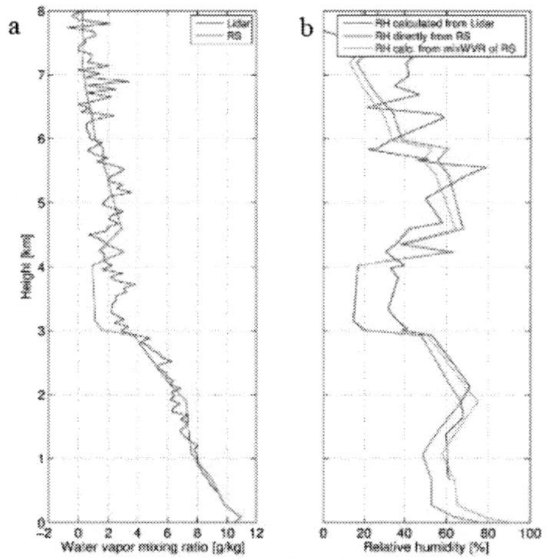

Fig. 2. a) MRWV to dry air profile from RS (in red), derived from Raman multiwavelength lidar signals (in blue); b) RH humidity profile obtained directly from RS (in red), derived from the WVMR from the RS (in green) and derived from Raman lidar signals (in blue). RS launched on 30/07/14 at 00UTC over Legionowo and lidar signals evaluated from 23 to 24 on 29/07/14 over RT-Lab (Warsaw).

3.2. Cloud vertical structure

The comparison between the CVS from lidar signals and from the COS14 RS methodology [4] has been made for two different RS launchings. The first RS evaluated with COS14 method was on 30/07/14 at 00UTC and it did not reveal any cloud layer, which agrees with the visual inspection of the IR lidar signals (Fig. 1a).

The second RS evaluated is on 30/07/14 at 12UTC and it reveals a low level cloud layer (from 1.3km to almost 3km) (Fig. 4 right), which agrees with the three layers seen on lidar signals for the same time period over Warsaw (Fig. 1b). Although the COS14 RS method only retrieves 1 layer (Fig. 4 left), it seems a good approximation to the 3 layers structures seen on lidar signals.

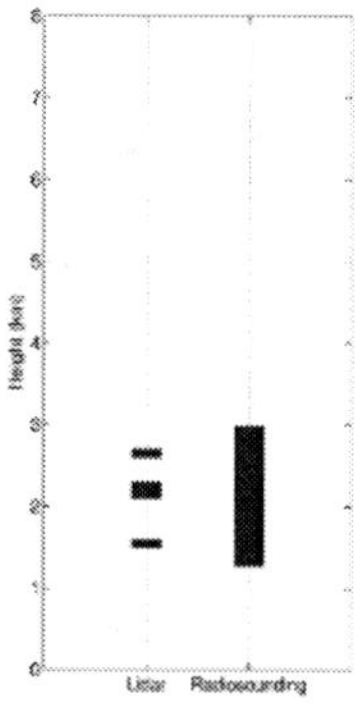

Fig. 4. CVS derived from the visual inspection of lidar signals at 1064nm over Warsaw (left), and cloud layer derived from COS14 radiosounding method from Legionowo (right), both on 30/07/14 at 12h UTC.

4. CONCLUSIONS

A comparison of the WVMR and RH between multiwavelength ground-based lidar (from 23 to 24h on 29/07/14) based on Warsaw, and RS measurement (at 00h UTC on 30/07/14) launched in Legionowo, were retrieved up to 8km successfully, and revealed a good agreement between the two methodologies at most heights, although the WVMR lidar derived showed noise. The differences in some parts of the WVMR and RH profiles could be due to geographical distance between the instruments. The comparison between CVS derived from visual inspection of ground-based multiwavelength ADR lidar and RS for two study cases produce good agreement. Despite on the second case, COS14 method only retrieves 1 cloud layer where the lidar derived 3, the cloudy height range is good determined by the lidar.

Combining lidar measurements and their retrievals with other technologies, such as RS, is very important in order to get a better knowledge about water content thought the atmosphere and CVS. Such studies have potential to improve methods for automatic and continuous description of clouds [4] and will very useful to provide better inputs for modelers. In future, more comparisons will be performed for other case studies since it will be very useful for testing and eventually improving RS methods. Moreover, comparing RS methods retrievals with different multiwavelength systems operating around the world opens a huge range of possibilities.

ACKNOWLEDGEMENT

Part of the research was done within the SONATA BIS Project "Integrated study of climate processes involving absorbing aerosols" funded by NCN, Poland (Grant No. 2012/05/E/ST10/01578).

The lidar development was financed by FNITP, Poland (Grant No.519/FNITP/115/2010).

REFERENCES

[1] Boucher, O., Randall, D., et al., 2013: Clouds and Aerosols. In: *Climate Change 2013: The Phisical Science Basis. IPCC,* 571–657.

[2] Wang, J., & Rossow, W., 1995: Determination of Cloud Vertical Structure from Upper-Air Observations. *J. Appl. Meteorol.*, ***34*(10)**, 2243–2258.

[3] Poore, K., Wang, J., & Rossow, W., 1995: Cloud Layer Thicknesses from a Combination of Surface and Upper-Air Observations. *J. Clim.*, ***8*(3)**, 550–568.

[4] Costa-Surós, M., Calbó, J., González, J. A., Long, C. N., 2014: Comparing the cloud vertical structure derived from several methods based on radiosonde profiles and ground-based remote sensing measurements. *Atmos. Meas. Tech.*, **7(8)**, 2757–2773.

[5] Pappalardo, G., et al., 2014: EARLINET towards an advanced sustainable European aerosol lidar network, *Atmos. Meas. Tech.*, **7**, 2389–2409.

[6] Althausen, D., et al., 2009: Portable Raman Lidar PollyXT for Automated Profiling of Aerosol Backscatter, Extinction, and Depolarization. *J. Atmos. Oceanic Technol.*, **47**, 2366-2378.

[7] Ansmann, A., Riebesell, et al., 1992: Combined Raman Elastic-Backscatter LIDAR for Vertical Profiling of Moisture, Aerosol Extinction, Backscatter, and LIDAR Ratio, *Applied Physics B*, **55**, 18–28.

[8] Zhang, J., et al., 2010: Analysis of cloud layer structure in Shouxian, China using RS92 radiosonde aided by 95 GHz cloud radar. *J. Geophys. Res.*, **115**, 1–13.

DOI: 10.1051/epjconf/201611924006

SPATIO-TEMPORAL VARIABILITY OF WATER VAPOR IN THE FREE TROPOSPHERE INVESTIGATED BY DIAL AND FTIR VERTICAL SOUNDINGS

H. Vogelmann, R. Sussmann, T. Trickl, A. Reichert

Karlsruhe Institute of Technology, IMK-IFU (Garmisch-Partenkirchen, Germany),
E-mail: hannes.vogelmann@kit.edu

ABSTRACT

We report on the free tropospheric spatio-temporal variability of water vapor investigated by the analysis of a five-year period of water vapor vertical soundings above Mt. Zugspitze (2962 m a.s.l., Germany). Our results are obtained from a combination of measurements of vertically integrated water vapor (IWV), recorded with a solar Fourier Transform InfraRed (FTIR) spectrometer and of water vapor profiles recorded with the nearby differential absorption lidar (DIAL). The special geometrical arrangement of one zenith-viewing and one sun-pointing instrument and the temporal resolution of both optical instruments allow for an investigation of the spatio-temporal variability of IWV on a spatial scale of less than one kilometer and on a time scale of less than one hour. We investigated the short-term variability of both IWV and water vapor profiles from statistical analyses. The latter was also examined by case studies with a clear assignment to certain atmospheric processes as local convection or long-range transport. This study is described in great detail in our recent publication [1].

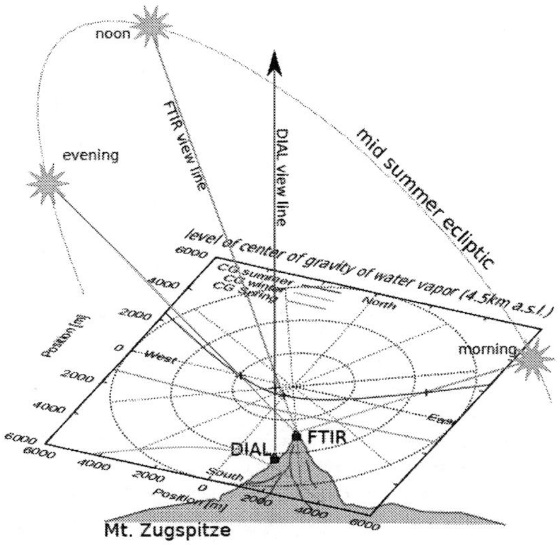

Figure 1: Geographical collocation of the water-vapor DIAL and the FTIR instrument on Mt. Zugspitze

1. INSTRUMENTATION AND GEOGRAPHICAL ARRANGEMENT

The Zugspitze (47.42°N, 10.98°E, 2962 m a.s.l.) is by far the highest mountain on the northern rim of the Alps. The free troposphere above this site is representative of Central Europe. The mountain is above the moist boundary layer for most of the year. Due to reduced absorption losses this site is ideal for sensitive spectroscopic measurements of water vapor throughout the free troposphere. While the FTIR instrument is located on the summit of Mt. Zugspitze the DIAL instrument is located at the Schneefernerhaus research station (UFS) on the steep southern slope of Mt. Zugspitze at an altitude of 2675 m a.s.l., 680 m southwest of the FTIR instrument (Fig. 1).

The solar FTIR instrument uses direct radiation from the sun in the mid-infrared as light source (Table). The instrument provides water vapor columns integrated along its slanted viewline towards the sun with a precision better than 0.05 mm. It is based on a Bruker IFS125HR interferometer and is described in detail by [2].

The Zugspitze DIAL is operated with single absorption lines in the 817-nm band of H_2O (see table) for ground-based water vapor profiling in the free troposphere. In order to keep a balanced signal-to-noise ratio a vertical resolution of 50 m to 300 m is adapted dynamically to the vertical range from 2.95 km to roughly 12 km a.s.l., respectively. Thus, statistical measurement uncertainties are kept below about 5% related to a mean humidity profile to about 8 km. The sensitivity limit is roughly 18 ppm at 10 km a.s.l.. The DIAL system at UFS and the retrieval of water vapor profiles are described in more detail by [3]. Water-vapor profiles from the Zugspitze DIAL allow for retrieving IWV with a precision better than 0.1 mm [4].

	FTIR	DIAL
Geographical	E 10°59′8.7″	E 10°58′46.8″
Coordinates	N 47°25′15.6″	N 47°25′0″
Altitude a.s.l.	2964 m	2675 m
Range a.s.l.	above 2.96 km	2.95 km - 12 km
Int. time	13.3 min	17 min
Spectral	micro windows	ν_{on}
range [cm^{-1}]	839.5 – 840.5	12236.560
	849.0 – 850.2	12237.466
	852.0 – 853.1	12243.537

Combining the sun-pointing geometry of the FTIR instrument and the fixed zenith-pointing geometry of the DIAL is ideal for studies of the variations of IWV values measured by both instruments with a defined spatial and temporal matching. For simplicity, it is assumed that

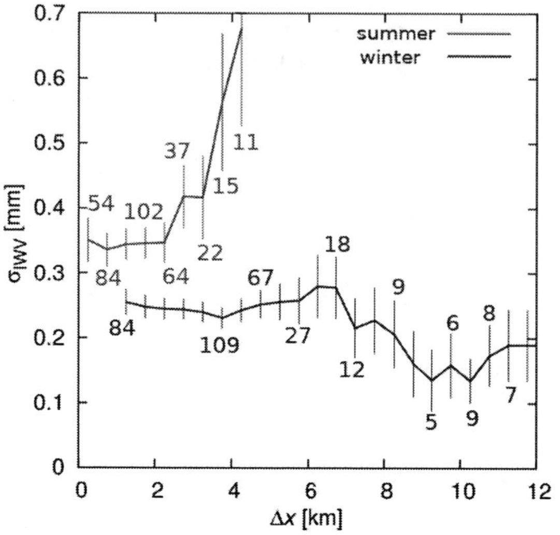

Figure 2: σ_{IWV} as a function of the horizontal distance Δx between the center of gravity of FTIR IWV and DIAL IWV in the summer season (red) and in the winter season (blue). The coincidence time interval Δt is 60 min for the blue curve and 30 min for the red curve. For geometrical reasons, the shortest distance in the winter season is 1 km. The number of measurement pairs from which σ_{IWV} was calculated is indicated by the numbers near the curves (not for all nodes). The uncertainties ($\pm\sigma$) are indicated by the error bars (for calculation see text).

the FTIR IWV is horizontally located at the point where the viewing direction of the instrument meets the altitude level of the center of gravity of the IWV distribution. From this a rough estimate of the varying horizontal position of the IWV measured by the FTIR instrument is possible. The zenith angle of the sun defines the horizontal distance from the instrument, which may vary from less than 1 km around noon in midsummer to more than 10 km at very low sun positions. The azimuth of the FTIR IWV position is equal to the azimuth of the sun position which depends on daytime and season. In contrast to this, the horizontal position of IWV measured with the DIAL is always fixed to the location of the instrument, i.e., 680 m southwest of the FTIR site.

2. VARIABILITY OF INTEGRATED WATER VAPOR IN SPACE AND TIME

Of more than 350 lidar profiles recorded in the years 2007-2009 more than 250 profiles were measured during daytime (i.e. between 5h00 and 19h00 local time). In the same period, more than 3500 column measurements were made by the FTIR instrument. The systems operate with a typical integration time of 13 min (FTIR) and 17 min (DIAL). In order to obtain a quantitative measure of the water vapor variability, we analyzed certain measurement samples recorded by the two different instruments under certain spatio-temporal matching criteria for Δx and Δt. The center of the integration time of FTIR and DIAL each was used to determine the temporal matching. We retrieved σ_{IWV} by calculating the standard deviation of the differences of IWV values from a linear model $y = a \cdot x + b$:

$$\sigma_{\mathrm{IWV}} = \sqrt{\frac{1}{n-2}\sum_{i=1}^{k}(y_i - (ax_i + b))^2},$$

whereat y_i and x_i are the IWV-values from the DIAL and the FTIR, respectively, within one sample and n is the sample size. a and b were calculated by a regression analysis using the method of least squares. Thus,

$$\sigma_{\mathrm{IWV}} = \sqrt{\frac{1}{n-2}\sum_{i=1}^{n}\left(y_i - \bar{y} - \frac{(x_i-\bar{x})\sum_j (x_j-\bar{x})y_j}{\sum_j (x_j-\bar{x})^2}\right)^2}.$$

Spatial variability

Spatial and temporal variabilities are analyzed separately for summer and winter because of two counteracting effects:

1. The special observation geometry in this study implies that the spatial overlap Δx of both soundings depends on both daytime and season. The best spatial matching ($\Delta x < 1$ km) is achieved around midsummer in the early afternoon only (between 12 h and 14 h UTC), while Δx is always larger during the winter season.

2. Due to heat-driven convective dynamics in complex mountain surroundings, spatial and temporal variabilities of IWV are expected to be higher during summer season. The convection above alpine terrain can reach an altitude of about 1.5 km above the mean summit levels in summer. During all other seasons, the convection usually does not even reach the Zugspitze summit and our measurement range.

Figure 2 (red curve) shows σ_{IWV} as a function of the horizontal distance of the probed volumes in the summer season. While σ_{IWV} constantly remains around 0.35 mm for $\Delta x < 2$ km, it rises to values of more than 0.65 m at a distance of $\Delta x = 4$ km. This result shows that the variability depends on the spatio-temporal matching. Up to $\Delta x = 2$ km, the temporal variability within the selected time interval ($\Delta t = 30$min) predominates. For larger distances, the contribution of spatial variability becomes significant. In contrast to this, σ_{IWV} is not increasing with Δx in the winter season (Fig. 2, blue curve). This is in agreement with the assumption that local convection does not reach the vertical measurement range during the winter season and that the IWV variability is probably dominated by horizontal advection of filamentary structures in the free troposphere from very different source regions. Consequently, the observed variability during winter is due to larger spatial scale processes (compared to local convection in summer), which would explain the absence of an increase with Δx in Fig. 2. Note that because IWV is much lower in winter than in summer, the relative variabilities (i.e. if σ_{IWV} were given in per cent) would be larger for the blue curve in Fig. 2. This means that advection of filaments (winter) leads to larger relative changes of IWV than local convection in summer. We will discuss this finding in more detail within the context of the variability of the vertical water vapor profile in Sect. 3. Figure 2 also indicates that σ_{IWV} even shows a trend to lower values for distances above 6 km. We explain this by the fact that measurements with large horizontal mismatch ($\Delta x > 6$ km) require extraordinarily calm and clear weather conditions, because the FTIR instrument needs a cloudless field of view while the sun position is close to the horizon. While the mean horizontal distance is low during the summer season ($\Delta x < 2$ km) due to high sun positions and stronger cloud perturbance, it reaches up to almost 10 km around midwinter. The variability over the entire field of horizontal distances within a certain time interval (e.g. 20 min) reaches its maximum of almost 0.6 mm when the temperature peaks around the end of July. We assume that this is a direct effect of the heat-driven local convection, which can reach altitudes of 4.5 km at Zugspitze site during the summer season. As expected, the minimum variability of about 0.15 mm is observed around midwinter, although the mean horizontal mismatch of both instruments is largest at this time of the year. However, as stated above, relative short-term variations are larger during the winter season.

Temporal variability

For the analysis of temporal variability, we calculated the standard deviation of differences σ_{IWV} between IWV values from both instruments as a function of temporal coincidence. This was repeated for varied spatial matching criteria. When using all IWV values from both instruments without applying any geometrical matching criteria, σ_{IWV} shows a flat minimum around a coincidence interval of $\Delta t = 20$ min (red curve in Fig. 3). The minimum is assigned to the fact the most FTIR and lidar measurements were carried out in the morning, because there are still few clouds. As a consequence, most of the pairs with the shortest coincidence intervals are found in the morning when the spatial matching is worst. This slightly increases σ_{IWV} on the very left hand side of the red curve in Fig. 3. Secondly, many pairs with good spatial matching can be found around noon, even for somewhat larger temporal coincidence intervals. This explains the decrease of σ_{IWV} towards the minimum. When considering measurement pairs with an FTIR sun azimuth close to the position of the DIAL instrument ($210° \pm 6°$) only, σ_{IWV} is much smaller in general and has its minimum at the shortest coincidence intervals (green curve in Fig. 3). For time intervals on the minutes scale, we find $\sigma_{IWV} = 0.05$ mm, which agrees with the validated (combined) precision of our instruments [4]. The temporal variability of IWV can also be estimated from the standard deviation of differences of measurements recorded by the same instrument within certain time intervals. This was done with data from the FTIR instrument (blue curve in Fig. 3). They begin to deviate increasingly from the green curve beyond 30 minutes and converge towards the red curve for larger time intervals. This corresponds

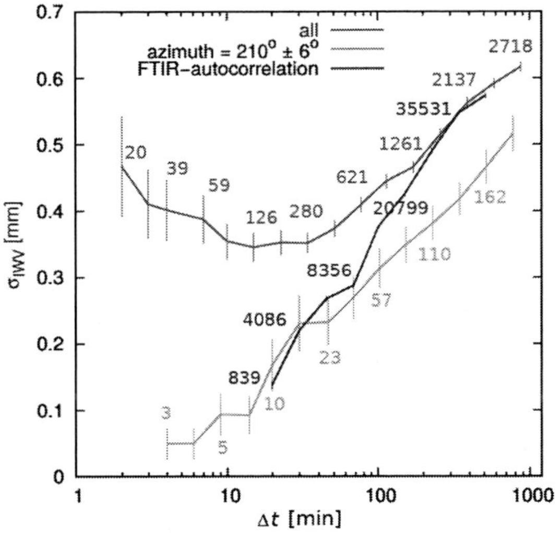

Figure 3: Variability as a function of the length of the time interval. The red curve shows σ_{IWV} from all measurements with no geometrical restrictions as a function of the length of the time interval in which data were taken into account. The green curve only includes measurements recorded in the early afternoon when the volume matching peaks with a sun azimuth of $210° \pm 6°$. The blue curve only shows σ_{IWV} of IWV values from the FTIR instrument. The quantity of measurement pairs from which σ_{IWV} was calculated is indicated by the numbers near the curves (not for all nodes). The uncertainties ($\pm\sigma$) are indicated by the error bars (for calculation see text).

to the fact that we observe a superposition of temporal and spatial variability with the solar FTIR due to its sun-pointing geometry.

3. PROFILE VARIABILITY

The variability of the vertical water vapor distribution on time scales of $\Delta t \leq 5$ h was derived from profiles retrieved from the DIAL measurements alone (Fig. 4). For the lowest layer (i.e. 3 - 4 km), where most of the entire column above Zugspitze site is located, we find equal relative variabilities as for IWV. In contrast to this, the relative variability increases with altitude above 5 km. This can be explained by the increasing wind speed at higher altitudes in the troposphere. The temporal variability of the water vapor density in the free troposphere at a certain altitude primarily features a horizontal variability combined with a horizontal wind velocity at this altitude. Depending on the pathway of the jet stream or the polar vortex, maximum wind velocities of more than 100 m/s occur. Considering a time interval of 60 min, this means a mean horizontal spread of about 80 km around 10 km altitude with a potential increase to more than 360 km in the jet stream regime. For different time intervals (1 h–5 h), the relative variance behaves quite similarly in agreement

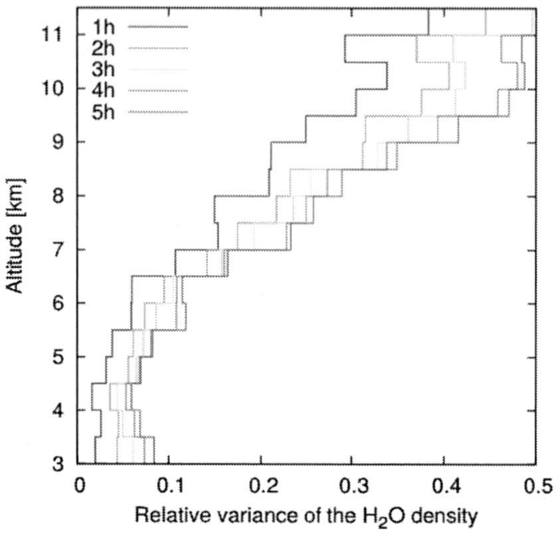

Figure 4: The short-term variability of the vertical water vapor profile is illustrated by the plot of the relative variance as a function of altitude within different time intervals.

with our results of IWV variability analysis, according to which longer time intervals lead to larger variabilities. In comparison to the 1-h profile, we see a more significant maximum at the lower edge at 3 km and a significant minimum at 4.5 km for longer time intervals. This enhanced increase between 3 km and 4 km is, to our understanding, induced by the diurnally varying upper edge of the boundary layer during the warm season.

The interconnectivity of the short-term variability of water vapor at different altitudes is analyzed using the covariance matrix of the vertical profile variability. A weak coupling between different layers is observed for lower altitudes which is in agreement with the assumption of local convection and turbulence being the dominant sources of variability in the lower part of the examined altitude range. In the upper troposphere, on the other hand, varying air masses are more coherently exchanged within the upper air flow, as a result of which layers of a wider vertical spread are affected.

4. MECHANISMS DRIVING THE OBSERVED VARIABILITY

Lidar sounding allows for clearly distinguishing between local convection and long-range transport as sources of variability, though also a superposition of both mechanisms can be observed. From these incidents we learned that the water-vapor density at a certain altitude can vary by a factor of more than 30 within a few hours. Thus, the short-term variability of water vapor induced by long-range transport and the advection of very inhomogeneous layer structures can exceed the impact of local convection by roughly one order of magnitude. This is particularly pronounced for stratospheric intrusions that descend from the Arctic to Central Europe. These intrusion layers occasionally become the main source of short-term variability of water vapor in the altitude range between 3 km and 5 km [5]. However, such events occur predominantly during the winter season and are accompanied with non-convective weather conditions. Case studies based on lidar observations will be presented at the ILRC-meeting.

5. CONCLUSIONS

Our results for the first time provide a quantitative description of the free tropospheric spatio-temporal variability of water vapor in the free troposphere on the scales of minutes and kilometers (horizontal) for IWV and on the scales of hours and 500 m (vertical) for profiles. This information can be useful for the parameterization of humidity in atmospheric models as well as for estimating the influence of the atmospheric variability of water vapor on the significance of water vapor measurements performed with a given integration time. In a related sense our results also provide the information necessary for evaluating intercomparison studies of not perfectly co-located or synchronized instruments.

ACKNOWLEDGMENTS

We thank Hans-Peter Schmid (KIT/IMK-IFU) for his continuous interest in this work and M. Rettinger (KIT/IMK-IFU) for executing the FITR measurements at Mt. Zugspitze. We also acknowledge the assistance by the team of the Schneefernerhaus research station (UFS) and the Bavarian Ministry of Environment and Consumer Protection for funding our work within the ALOMAR cooperation.

REFERENCES

1. Vogelmann, H., Sussmann, R., Trickl, T., and Reichert, A. Spatiotemporal variability of water vapor investigated using lidar and ftir vertical soundings above the zugspitze. *Atmos. Chem. Phys.*, 15:3135–3148, 2015.
2. Sussmann, R. and Schäfer, K. Infrared spectroscopy of tropospheric trace gases: combined analysis of horizontal and vertical column abundances. *Appl. Opt.*, 36:735–741, 1997.
3. Vogelmann, H. and Trickl, T. Wide Range Sounding of Free Tropospheric Water Vapor with a Differential Absorption Lidar (DIAL) at a High Altitude Station. *Appl. Opt.*, 47(12):2116–2132, 2008.
4. Vogelmann, H., Sussmann, R., Trickl, T., and Borsdorff, T. Intercomparison of atmospheric water vapor soundings from the differential absorption lidar (DIAL) and the solar FTIR system on Mt. Zugspitze. *Atmos. Meas. Tech.*, 4(5):835–841, 2011.
5. Trickl, T., Vogelmann, H., Giehl, H., Scheel, H.-E., Sprenger, M., and Stohl, A. How stratospheric are deep stratospheric intrusions? *Atmos. Chem. Phys.*, 14(18):9941–9961, 2014.

EPJ Web of Conferences 119, 24007 (2016) DOI: 10.1051/epjconf/201611924007
ILRC 27

LOCAL - AIR PROJECT: TROPOSPHERIC AEROSOL MONITORING BY CALIPSO LIDAR SATELLITE AND GROUND-BASED OBSERVATIONS

V. Sarli[1]*, S. Trippetta[1], P.Bitonto[2], N. Papagiannopoulos[1], R. Caggiano[1], A. Donvito[2] and L. Mona[1]

*[1]Consiglio Nazionale delle Ricerche, Istituto di Metodologie per l'Analisi Ambientale (CNR –IMAA), Tito Scalo (PZ), 85050, Italy, *Email: valentina.sarli@imaa.cnr.it*

[2]Digimat S.r.l., Matera, 75100, Italy

ABSTRACT

A new method for the detection of the Planetary Boundary Layer (PBL) height from CALIPSO space-borne lidar data was developed and the possibility to infer the sub-micrometric aerosol particle (i.e., PM_1) concentrations at ground level from CALIPSO observations was also explored. The comparison with ground-based lidar measurements from an EARLINET (European Aerosol Research LIdar Network) station showed the reliability of the developed method for the PBL. Moreover, empirical relationships between integrated backscatter values from CALIPSO and PM_1 concentrations were found thanks to the combined use of the retrieved PBL heights, CALIPSO aerosol profiles and typing and PM_1 in-situ measurements.

1. INTRODUCTION

Aerosol monitoring is usually carried out through ground-based monitoring networks that measure the concentration of particulate matter (PM). The main limitation of these networks is their inability to provide wide spatial coverage because of the local character of the measurements. On the other side, satellite remote sensing provides typically a larger spatial coverage but for columnar aerosol parameters. Since 2006, CALIPSO (Cloud-Aerosol Lidar and Infrared Pathfinder Satellite Observations) has been providing the vertical distribution of aerosol at global scale. In principle, this could offer the possibility to globally investigate into detail the PBL (Planetary Boundary Layer) region, but limits and uncertainties should be carefully assessed.

In this context, the LOCAL AIR (LOCAL Aerosol monitoring combining In-situ and Remote sensing observations) project aims at the development of a methodology for a synergistic use of data at different resolutions from several platforms for the characterization of tropospheric aerosols on a local scale.

2. DATA AND METHODS

2.1 Satellite observations

Lidar data from CALIPSO were used for taking advantage from its capability of providing vertical aerosol profiles at global scale [1]. CALIPSO is a 2-wavelength elastic backscatter lidar with polarization capability. It provides aerosol and clouds layering and aerosol backscatter and depolarization profiles. Aerosol extinction profiles are also provided through lidar ratio assumptions. In this project, the aerosol optical profiles (Level2 Profile products) and layering and typing information (level2 VFM products) were used.

2.2 Ground-based observations

The study was carried out in correspondence of CALIPSO overpasses at CIAO (CNR-IMAA Atmospheric Observatory) [2], the most advanced atmospheric observatory in the Mediterranean region equipped with active and passive sensors for aerosol, clouds and water vapor observations. In particular, EARLINET (European Aerosol Lidar Network) [3] observations performed at CIAO were found fitting with the scope of the project. EARLINET systems at CIAO are able to provide quality assured profiles of aerosol extinction and backscatter at 2 and 3 wavelengths respectively plus depolarization ratio profiles at 532nm.

In addition, daily concentrations and chemical composition of sub-micrometric aerosol particle (i.e., PM_1) measured at CNR-IMAA from April 2006 to March 2007 were also used [4].

2.3 Methodological approach

The main concept of the project is to use CALIPSO data for providing information about local air quality. To this aim, a specific methodology has to be developed and its performance has to be assessed using ground-based measurements as a reference. A first step in this direction was the determination of the PBL top from CALIPSO lidar data. Concerning this, only two methods have been reported in literature up today [5, 6, 7], but one method [5, 6] is limited to cloud free conditions and the other one [7] is unreliable in case of dust advection. Therefore, a new method was developed and investigated. The values obtained were used as input for determining the aerosol backscatter integrated value (IB) into the PBL, an important parameter (under well mixed situation assumption) for the air quality at the surface. Then, the relationship among IB and PM_1 was studied in order to find empirical relationships for inferring PM_1 concentration values from CALIPSO lidar data.

3. RESULTS

For the investigation of PBL top, the study was focused on 2 years, from June 2006-May 2008. During this period, a total of 163 CALIPSO overpasses (daytime and nighttime) occurred within 100 km from CIAO observatory (Figure 1).

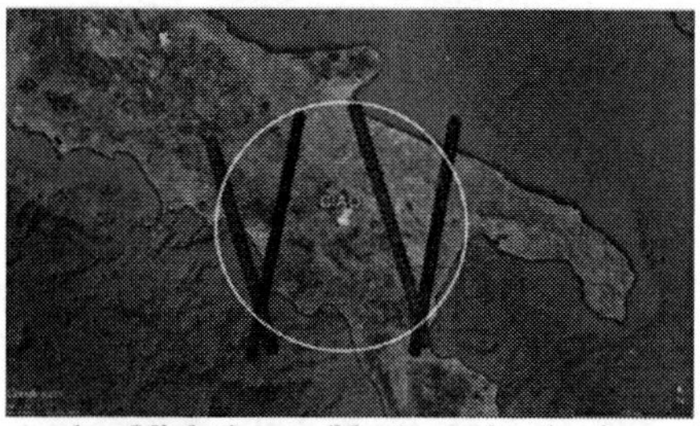

Figure 1. Nighttime (blue) and daytime (red) CALIPSO overpass traces within 100 km from CIAO observatory (white circle). Aerial photography courtesy of Google Earth (http://earth.google.com/).

In order to estimate the PBL top, the maximum variance technique was borrowed from [5, 6] and applied to the profile of the total aerosol backscatter at 532 nm reported in CALIPSO Level 2 Aerosol Profile (APro) products. This allowed to overcomes the problem of cloud presence in [5, 6] method. On the other hand, searching for the maximum variance, the new method could also overcome the problem of unreliable PBL values in [7] for dust intrusion. The new method (indicated as APro from the data product on which it is applied) was compared with the existing ones (TAB [5, 6] and ALay [7]) and with ground-based collocated measurements for checking its performances.

As Table 1 shows, the application of the maximum variance technique to TAB and APro allowed the PBL detection in most of the cases (i.e., 108 and 103 out of 163 cases, respectively).

Method	Night	Day	Total
TAB	75	33	108
ALay	43	33	76
APro	57	46	103

Table 1. Number of PBL heights detected by applying the three methods to CALIPSO aerosol data products.

The retrieved PBL heights over the nearest point of CALIPSO overpasses from CIAO were compared with the corresponding PBL heights as determined by Potenza EARLINET station (Figure 2).

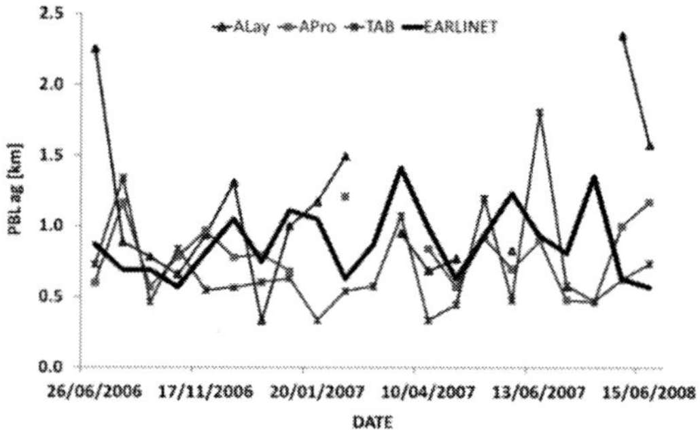

Figure 2. Temporal series of CALIPSO retrieved and EARLINET derived PBL heights.

As Figure 2 shows, the PBL heights detected by applying the maximum variance technique to APro were found to be more consistent with the ground-lidar derived ones with an associated Root Mean Square Error (RMSE) of 0.45 compared to 0.55 and 0.65 for TAB and ALay methods, respectively.

The analyses of the daily and seasonal variability of the PBL heights retrieved using the three methods confirmed the higher capability of the algorithm developed in this study to reconstruct the expected daily and seasonal variations of the

PBL top over the different areas observed (figures not shown).

In order to find empirical relationships able to infer PM_1 concentration values from CALIPSO lidar data, the existing correlations between CALIPSO observations and in-situ PM_1 data were explored. To this aim, at the nearest point of CALIPSO overpasses to CIAO, the IB values were calculated by integrating the CALIPSO total backscatter coefficient at 532 nm profiles within the PBL as estimated by the method developed in this study. Then, the resulting IB values were compared with the PM_1 daily concentrations measured on the same day of the satellite acquisition.

Regression analysis results showed no significant correlation when the entire dataset was considered. This could be due to the fact that different aerosol types have different scattering efficiency, so that the PM-IB relation could be strongly typing dependent.

Low correlation values were observed ($R^2<0.5$) also by classifying the dataset on the basis of the CALIPSO derived aerosol typing (i.e., Clean Marine - CM, Dust - D, Polluted Continental - PC, Clean Continental - CC, Polluted Dust - PD and Smoke - S.) The low correlations found could be related to some limitations in the CALIPSO aerosol typing, especially in the low troposphere [8].

A significant improvement in the correlations was observed when the dataset was classified according to the ratio between natural (N) and anthropogenic (A) contributions (i.e., N/A) to the PM_1 measured at ground level. In particular, these contributions were calculated starting from the PM_1 trace element composition by applying a methodological approach based on the combined use of the enrichment factor and the chemical mass closure techniques [9].

As Figure 3 shows, two linear relationships between PM_1 concentrations and IB values were found. The first (f1) corresponds to N/A≥0.3 and is characterized by R^2 value of 0.6. The second (f2) corresponds to N/A<0.3 and is characterized by R^2 value of 0.7.

When the CALIPSO derived aerosol type was associated to each IB-PM_1 couple, it was observed that only Polluted Continental (PC) and Polluted Dust (PD) cases corresponded both to N/A≥0.3 and N/A<0.3 (Table 2).

N/A	CM	D	PC	CC	PD	S
TOT	1	2	7	-	10	3
≥ 0.3	0%	100%	43%	-	50%	100%
< 0.3	100%	0%	57%	-	50%	0%

Table 2. Correspondence between data grouped by N/A ratio and CALIPSO derived aerosol types.

In particular, the points corresponding to PD and PC followed the regression line of the first relationship if $PM_1 > 8.5$ μgm^{-3} and the second relationship if $PM_1 < 4.5$ μgm^{-3} (Figure 3).

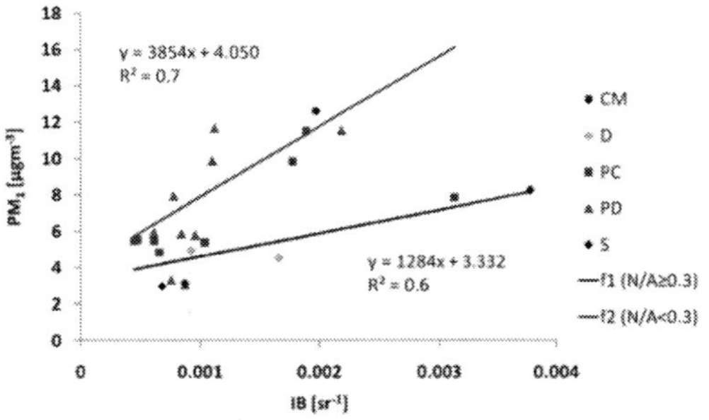

Figure 3. Scatter plot between PM_1 daily concentrations and IB values. The regression lines (f1 and f2) obtained by dividing the IB-PM_1 dataset as a function of the N/A ratio are also reported. Moreover, each IB-PM_1 couple is classified according to the corresponding CALIPSO aerosol type.

Therefore, the reconstruction of the PM_1 concentrations from CALIPSO IB values was obtained as follows. If CALIPSO derived aerosol types were dust or smoke, then f1 was applied. If, polluted dust or polluted continental types prevailed, a further analysis was performed. In particular, PM_1 concentrations were estimated by CALIPSO IB values using both f1 and f2. If the estimated value of the PM_1 concentration by f1 exceeded the value of 8.5 μgm^{-3} or the estimated value of the PM_1 concentration by f2 was lower than 4.5 μgm^{-3}, than the obtained estimate was retained. Otherwise, the PM_1 concentrations were estimated by applying f1 and f2 and averaging the obtained values. For clean marine cases, the same procedure used for polluted dust or polluted continental types was adopted because of the well-known problems for the typing of this category [10]. As Figure 4 shows, a good agreement between retrieved and measured PM_1 concentrations was observed in most of the cases.

DOI: 10.1051/epjconf/201611924007

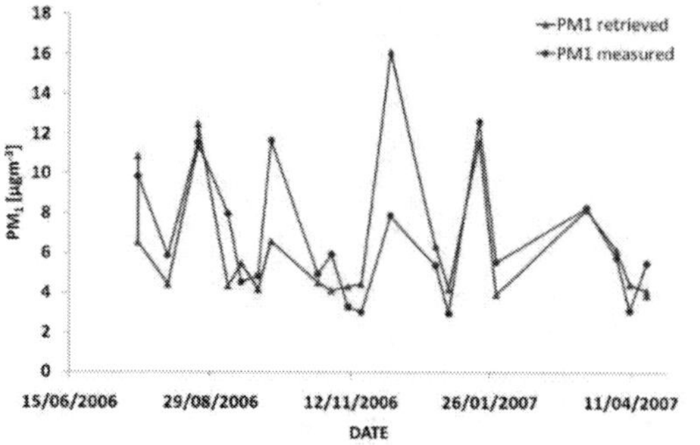

Figure 4. Comparison between retrieved and measured PM_1 concentrations.

4. CONCLUSIONS

A new algorithm for the PBL height detection from CALIPSO space-borne lidar data was developed. The comparison of the retrieved PBL heights with the the EARLINET reference point at CIAO corresponding ones as well as daily and seasonal variation analyses showed its reliability on the investigated area, also compared with methods reported in literature. The possibility to infer PM_1 values at ground from CALIPSO observations was also explored. Thanks to the combined use of retrieved PBL heights, CALIPSO aerosol profiles and typing and PM_1 in-situ measurements, empirical relationships between IB and PM_1 concentrations were found, representing the starting point for the development of a new methodology for providing PM_1 in-situ concentrations from CALIPSO aerosol data. Further work needs to be conducted in order to better understand the empirical relationships obtained and to evaluate the possibility of their transfer to different locations.

ACKNOWLEDGEMENTS

This work has been supported by PO FSE Basilicata 2007-2013 Azione n. 45/AP/05/2013/REG –CUP: G53G13000300009.

This work has been partially supported by the ACTRIS Research Infrastructure Project by the European Union under grant agreement n. 262254 in the 7th Framework Programme (FP7/2007-2013).

REFERENCES

[1] Winker, D. M., W. H. Hunt, M. J. McGill, 2007: Initial performance assessment of CALIOP, *Geophys. Res. Lett.*, **34**, L19803, doi:10.1029/2007GL030135.

[2] Madonna, F., et al., 2011: CIAO: the CNR-IMAA advanced observatory for atmospheric research, *Atmos. Meas. Tech.*, **4**, 1191-1208, doi: 10.5194/amt-4-1191-2011.

[3] Pappalardo, G., et al., 2014: EARLINET: towards an advanced sustainable European aerosol lidar network, *Atmos. Meas. Tech.*, **7**, 2389-2409, doi:10.5194/amt-7-2389-2014.

[4] Caggiano, R., et al., 2010: Levels, chemical composition and sources of fine aerosol particles (PM_1) in an area of the Mediterranean basin, *Sci. Total Environ.*, **408**, 884–895.

[5] Jordan, N. S., R. M. Hoff, J. T. Bacmeister, 2010: Validation of Goddard Earth Observing System-version 5 MERRA planetary boundary layer heights using CALIPSO, *J. Geophys. Res.*, **115**, D24218, doi:10.1029/2009JD013777.

[6] McGrath-Spangler, E. L., A. S. Denning, 2012: Estimates of North American summertime planetary boundary layer depths derived from space-borne lidar, *J. Geophys. Res.*, **117**, D15101, doi:10.1029/2012JD017615.

[7] Leventidou, E., et al., 2013: Factors affecting the comparisons of planetary boundary layer height retrievals from CALIPSO, ECMWF and radiosondes over Thessaloniki, Greece, *Atmos. Environ.*, **74**. 360-366.

[8] Papagiannopoulos, N., 2015: Climatological study of aerosol optical properties over Europe by ground-based and satellite-borne remote sensing, *Ph.D. Thesis*, 201 pp.

[9] Jaafar, M., et al., 2014: Dust episodes in Beirut and their effect on the chemical composition of coarse and fine particulate matter, *Sci. Total Environ.*, **496**, 75–83.

[10] Kanitz T., et al., 2014: Surface matters: limitations of CALIPSO V3 aerosol typing in coastal regions, *Atm. Meas. Tech.*, **7**, 2061-2072.

EPJ Web of Conferences 119, 24008 (2016) DOI: 10.1051/epjconf/201611924008
ILRC 27

Using Lidar, in-situ measurements and Trajectory Analysis to observe air pollution in Beijing, 2014

Chen Zhenyi[1]*, Liu Wenqing[1], Liu Jianguo[1], Zhang Tianshu[1] and Dong Yunsheng[1]

[1]*Key Laboratory of Environmental Optical and Technology, Anhui Institute of Optics and Fine Mechanics, Chinese Academy of Sciences, Hefei, 230031, China, *Email: zychen@aiofm.ac.cn*

ABSTRACT

We present combined Mie lidar, ozone lidar and wide-range particle spectrometer observations that were carried out in Beijing, north China during two periods—one haze period before the Asia-Pacific Economic Cooperation (APEC) meeting and one moderate pollution period during the meeting in 2014. High extinction coefficient, moderate ozone concentration and variable particle number concentration were obtained throughout the first haze observation period. The mean extinction coefficients in the two pollution periods were 0.52 km-1 and 0.23 km^{-1}, respectively, at 532 nm. The ozone concentration during the first haze phase was more various with higher average value of 49 ppb compared to that in the second pollution observations (32 ppb). The comparison of aerosols and ozone in different heights indicate different pollution sources and complicated ozone process of generation and disappearance. The four-day back trajectories from a HYSPLIT model indicate that the air masses in the lower boundary layer were advected from the densely populated south regions of China and the long pollution transportation passing through northern China.

1. INTRODUCTION

Aerosol particles can absorb and scatter radiation efficiently, thereby affecting the balance of the atmospheric radiation system of the Earth [1-2].And many studies indicated that both direct emitted particles from human activities and secondary aerosols lead to significant reduction in visibility and the increase of haze occurrence [3-5]. Especially fine particles (PM2.5) suspended in the air may easily enter the respiratory and blood circulation systems and thus undermine human health [6]. Ozone is an important source of the hydroxyl radical, an oxidant that breaks down most pollutants and some greenhouse gases in the troposphere. Much of the present tropospheric ozone burden is also a consequence of anthropogenic emissions of ozone precursors resulting in widespread increases in ozone concentrations [7-8]. Thus it is important get understanding of aerosol properties and ozone concentration variation.

Thus far, some extensive studies using different methods have been conducted in the North China, with regard to aerosol pollution [9-11]. For example, *Tesche et al.* (2007) presented lidar observations of particle optical properties (backscatter coefficient, extinction coefficient and lidar ratio) in Beijing in North China. *Xue et al.* (2009) reports the results of the measurements of sulfur dioxide (SO_2) and aerosol scattering coefficient obtained from several aircraft campaigns during summer and autumn 2007 in the north-eastern regions of China. *Pan et al* (2013) measured trace elements of size-resolved aerosols at a rural site in North China and found the metallic episodes were attributable to the long-range transportation. Recently *Guo et al* (2014) illustrated the similarity and difference in particulate matter formation between Beijing and other world regions. However, knowledge on the combined optical properties of aerosol particles and ozone over northern China is particularly sparse, at least in the vertical scale. Therefore, experimental data in northern are strongly needed to assess the role of aerosols in future scenarios of global change and the impact on human health.

The paper is organized as follows: In Section 2 we describe the experimental setup (Mie lidar, ozone lidar and Wide-range Particle Spectrometer (WPS)). In Section 3, the metrological condition and in-situ measurements are presented. And aerosol optical properties and ozone profiles are shown along with the data interpretation. Furthermore the comparison between aerosol and ozone are investigated. Back trajectory analysis is used to track the origin of the air mass arriving at our observation site and the aerosol sources. In Section 4 we provide conclusions based on the results in this experiment.

2. METHODOLOGY

2.1 532 Mie lidar

The 532Mie-backscattered lidar is made up of three sub-systems—emitting system, receiving system and signal analogue system. The laser source used is a pulsed Nd: YAG laser emitting short pulses at 532. The receiving telescope of the lidar system is based on a Cassegrainian design. The backscatter signal are then collected and focused on the telescope's focal point. After separating and passing the respective interference filters, the photons elastically backscattered at the 532 nm wavelengths are detected with photomultiplier tubes. After data preprocessing, the particle backscatter coefficient and the particle extinction coefficient at 532 nm could be retrieved by the Klett method.

2.2 Ozone DIAL

DIAL measurements of ozone in our system are made in the Hartley band with the on-line nominally set near 289 nm and the off-line set near 299 nm. And the pulsed light at wavelengths of 289 nm, 299 nm and 316 nm is generated by stimulated Raman scattering. Two lenses in front of and after the tube constitute a Raman con-focal system. The backscattered radiation from the atmosphere is collected by a 40 cm diameter telescope with an effective focal length of 400 cm. After passing through the field stop, the UV DIAL returns are reflected to a four-wavelength grating spectrometer. Both reflected beams are independently collimated to be compatible with interference filters and PMT areas. The resolution of ozone DIAL is 15 m under the height of 3km.

2.3 WPS

Particle Number Size Distribution (PNSD) in the size of 10 nm to 10 μm is measured by a Wide-range Particle Spectrometer (WPS^{TM} model 1000xp, MSP Corporation, USA). The WPS system consists of a Differential Mobility Analyzer (DMA), a Condensation Particle Counter (CPC) and a Laser Particle Spectrometer (LPS). CPC and LPS are utilized to measure PNSD ranged from 10-500 nm and 350 nm-10 μm respectively. The number concentration has a ± 10% error as indicated by the manufacturer. The instrument's time resolution was set at 5 min with 48 channels for each size for the DMA and 24 channels for the LPS.

3. RESULTS

3.1 In-situ Measurements

Figure 1 illustrates the time sequence plots of the aerosol extinction at 532nm measured by lidar during the observation period. Because of the incomplete overlap between the laser beam and the receiver field of view, only lidar measurements above 100m were subsequently used for the inversion. The lidar signals showed the enhanced pollution from October 28 to October 30 and from November 4 to November 5. More specifically, the median PM2.5 particle concentration reaches its maximum on October 30 and November 4 by the end of the afternoon (18–20 UTC) with the respective value of 100 ug/m^3 and 150 ug/m^3 (Figure 2). The highest NO concentration suggested the peak traffic in rush hours. This is due to the titration effect of ozone by freshly emitted NO from traffic in the city. And NO_2 mean concentration is high with value of 30 ppb and 50 ppb from October 29 to October 31 and from November 4 to November 5 respectively. Large NO_2 enhancements were observed in the noon of October 30 and November 4. Because the average value of the RH at ground level did not exceed 85% to reach saturation, we conclude that the two pollution periods: one haze pollution period from October 28 to October 30 and one moderate pollution period from November 4 to November 5.

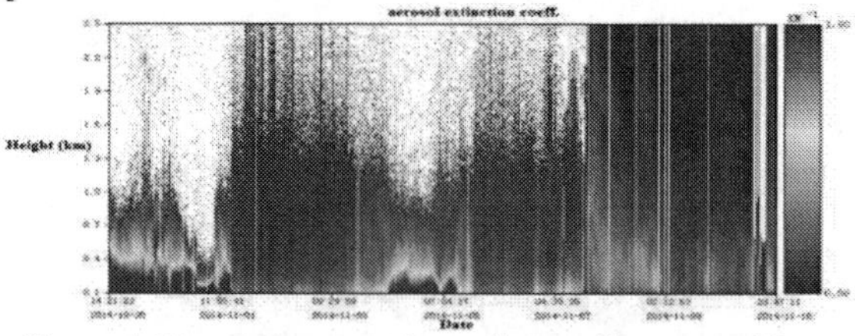

Figure 1 Time sequence plots of aerosol measured by lidar at 532 nm from October 30 to November 10, 2014

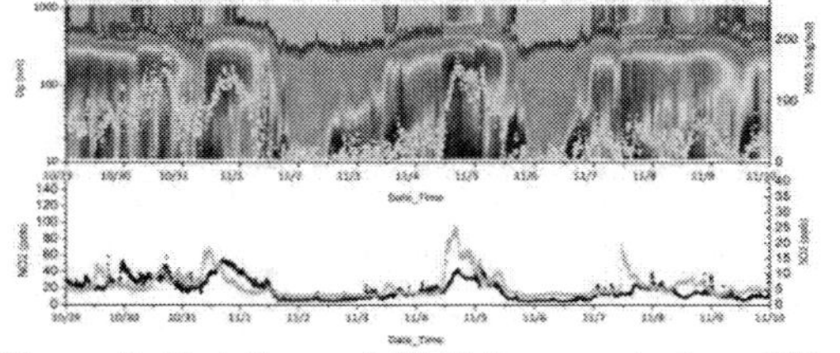

Figure 2 Variations of PM2.5 concentration, NO_2, SO_2 and PNSD from October 29 to November 10, 2014

3.2 Aerosols optical properties and ozone

The aerosol-ozone evolution and the aerosol loading impacts on ozone concentration are represented in Figure 5. During the first pollution period, the DIAL observations show pronounced a high ozone layer between 1 km and 2 km on October 29, with levels exceeding 60 ppb. The aerosol retrieval at 2 km (Figure 3(a)) indicates turbid air conditions below 2.5 km during the daytime and the average value of extinction coefficient is 0.52 km^{-1}. In the moderate pollution period, ozone concentration profile is less turbulent and maintains the average value of 32 ppb from 0.5 km to 2 km on November 4(Figure 3(b)).

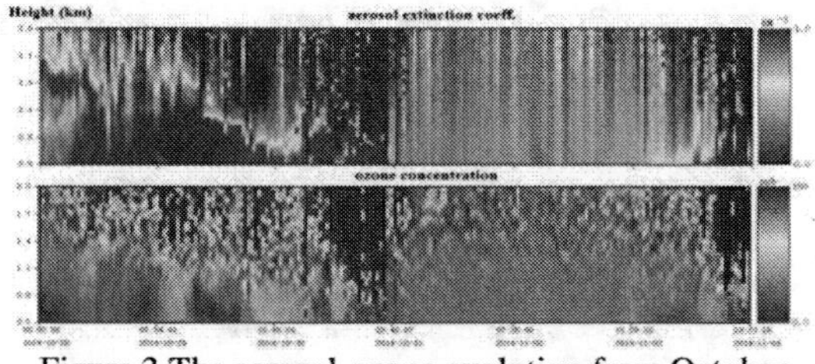

Figure 3 The aerosol-ozone evolution from October 28 to November 4, 2014

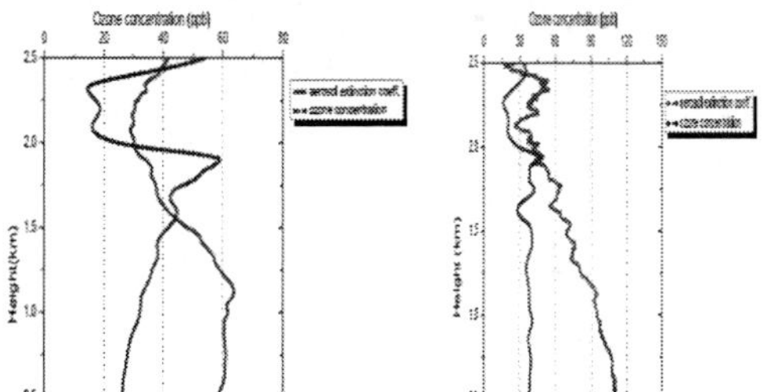

(a) (b)

Figure 4 Comparison of aerosol and ozone profiles on October 29 and November 4, 2014

In the two pollution episodes, aerosols show negative impacts on ozone concentrations on near ground level (from 100m to 200m) at all aerosol loading, which corresponds to the negative influence on photolysis rates by aerosols. However, positive impacts ozone concentrations in the lower troposphere (from 0.5 km to 2.5 km) are observed at small loadings for aerosols. Because in the upper troposphere where NOx is limited and net ozone destruction condition exists, a positive impact on photolysis rates causes a larger enhancement in ozone photo-dissociation rather than NO_2 photo-dissociation. *Lefohn et al.* [12] previously noted an increase in the low troposphere the probability distribution of ozone concentrations and attributed it to a decrease in NO emissions resulting in less effective titration. This positive correlation between ozone and aerosol in the troposphere on October 29 and November 4 suggests the same source is responsible for the enhanced ozone and aerosols.

3.3 Sources

During the first haze period (Figure 5(a)), the airflows from different directions at different heights began to converge in the J^3A since October 26. The air mass below 0.5 km was mainly from the densely populated southeast regions of China and the long pollution transportation passing through northern and northeast China. Straw burning and incomplete combustion of carbonaceous fuels during the heating period in winter is common and may cause more absorbing aerosols . The airflow dead zone made the diffusion more difficult. Thus, pollution accumulated and haze began to prevail.

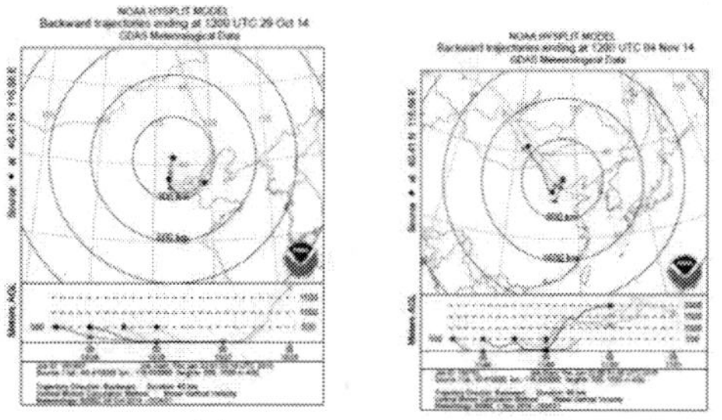

(a) (b)

Figure 5 Backward trajectories from HYSPLIT ensembles at 0.5km ending at Beijing, North China at 12:00 UTC on October 29 and November 4 2014

DOI: 10.1051/epjconf/201611924008

The air mass at 0.5km during the second pollution period was a bit different from that in the first haze period (Figure 9(b)); it came from southwest regions and northwest of China. The trajectories at 0.5 km showed the moderate pollution was due to the presence of more aerosol particles from adjacent region of southwest Beijing and the growth of aerosol under high relative humidity.

4. CONCLUSIONS

The results show the retrieval of aerosol extinction and ozone based on Mie lidar and ozone DIAL lidar measurements. Particle extinction coefficients were as large as 0.4 - 0.6 km-1 in the lower parts of the troposphere layer in the end of October 2014 in Beijing. And the ozone concentration maintained the high mean value of 49 ppb. Compared to the first pollution episode, the aerosol extinction coefficient in the second pollution period was smaller with decreased average value of 0.31km-1 and the ozone concentration was lower. The different aerosol and ozone peaks at different heights showed the different pollution sources, ambient meteorological conditions, and the resulting aerosol accumulation. And the backward trajectory indicates that both the short range and long range polluted air masses accumulated in the J^3A Beijing during the two pollution episodes.

ACKNOWLEDGEMENT

This work is supported by the project "Haze Observation Project Especially for Jing-Jin-Ji Area（HOPE-J3A， Grant No. KJZD-EW-TZ-G06-01）" of CAS. The authors also gratefully acknowledge the NOAA Air Resources Laboratory (ARL) for providing the HYSPLIT transport and dispersion model used in this research.

REFERENCES

[1] Bardouki H., Liakakou H., Economou C., Scaire J., Smolik J., V. Zdimal, 2003,Chemical composition of size resolved atmospheric aerosols in the Eastern Mediterranean during summer and winter, *Atmos. Environ.*, **37**, 195–208

[2] Ramanathan V. and Feng Y., 2009, Air pollution, greenhouse gases and climate change: Global and regional perspectives, *Atmos. Environ.*, **43**, 37–50

[3] Harrison R.G., Aerosol-induced correlation between visibility and atmospheric electricity, 2012, *J. Aerosol Sci.*, **52**, 121–126

[4] David Y.H. P., Chen S.C. and Zuo Z.L., 2014, PM2.5 in China: Measurements, sources, visibility and health effects, and mitigation, *Particuology*, **13**, 1-26

[5] Kessner A.L., Wang J., Levy R.C., Colarco P.R., 2013, Remote sensing of surface visibility from space: A look at the United States East Coast, *Atmos. Environ.*, **81**, 136–147

[6] Pope C.A., Thun M.J., Namboodiri M.M., Dockery D.W., Evans J.S., Speizer F.E., 1995, Particulate air pollution as a predictor of mortality in a prospective study of U.S. adults, *AM J RESP CRITCARE,* **151**, 669–674

[7] Shi C.Z., Wang S.S., Liu R., Zhou R., Li D.H, Wang W.X., Li Z.Q., Cheng T.T. and Zhou B., 2015, A study of aerosol optical properties during ozone pollution episodes in 2013 over Shanghai, China, *Atmos. Res.* , **153**, 235–249

[8] Hoshika Y., Watanabe W., Kitao M., Häberle K.H., Grams T. E.E., Koike T. and Matyssek R., 2015, Ozone induces stomatal narrowing in European and Siebold's beeches: A comparison between two experiments of free-air ozone exposure, *Environ. Pollut.*, **196**, 527-533

[9] Xue L.K., Ding A. J., Gao J., Wang T. , Wang W.X., Wang X.Z., Lei H.C., Jin D.J and Qi Y.B., 2010,Aircraft measurements of the vertical distribution of sulfur dioxide and aerosol scattering coefficient in China, *Atmos. Environ.*, **44**, 2, 278–282

[10] Pan Y.P., Wang Y.S., Sun Y., Tian S.L. and Cheng M.T, 2013, Size-resolved aerosol trace elements at a rural mountainous site in Northern China: Importance of regional transport, *Sci. Total Environ.,*461–462

[11] Guo S., Hu M., Zamora M. L., Peng J.F., Shang D.J., Zheng J., Du Z.F., Wu Z.J., Shao M., Zeng L.M., Molina M. J. and Zhang R.Y., 2014, Elucidating severe urban haze formation in China, *PNAS,*111 ,**49** , 17373–17378

[12] Lefohn, A. S., Shadwick, D. and Oltmans, S. J., 2008,Characterizing long-term changes in surface ozone levels in the United States (1980–2005). *Atmos. Environ.*, **42**, 8252–8262

COMPARISON OF REMOTE SPECTROPHOTOMETRIC AND LIDAR MEASUREMENTS OF O_3, NO_2, AND TEMPERATURE WITH DATA OF SATELLITE MEASUREMENTS

Bazhenov O.E., Burlakov V.D., Grishaev M.V., Gridnev Yu.V., Dolgii S.I., Makeev A.P, Nevzorov A.V., Salnikova N.S.

V.E. Zuev Institute of Atmospheric Optics, Siberian Branch, Russian Academy of Sciences

Russia, 634021, Tomsk, Academician Zuev square, 1, E-mail: dolgii@iao.ru

ABSTRACT

We consider the results of remote spectrophotometric and lidar measurements of the total ozone and nitrogen dioxide contents and temperature, obtained at the Siberian Lidar Station (SLS) of V.E. Zuev Institute of Atmospheric Optics, Siberian Branch, Russian Academy of Sciences (Tomsk: 56.5°N; 85.0°E) in comparison with the results of analogous satellite measurements.

The ground-based measurements of the total ozone (TO) content are performed with the help of M-124 ozonometer; and the measurements of the nitrogen dioxide (NO_2) content are carried out with automatic spectrophotometer. The ground-based lidar measurements of temperature are conducted on the basis of SLS measurement complex. These measurements are compared with data of balloon-sonde and satellite measurements. The satellite measurements are performed by the TOMS and IASI instrumentation.

1. TO MEASUREMENTS

TO measurements have been performed with the help of ozonometer M-124 since 1993. To compare our data, we used satellite measurements of TO, which are retrieved with the help of two algorithms: OMI-TOMS and OMI-DOAS [1]. The first algorithm is based on the version TOMS V8. The measurements are performed in four discrete spectral regions centered at 313, 318, 331, and 360 nm. The second algorithm uses the hyperspectral feature of Ozone Monitoring Instrument (OMI). It is based on the principle of Differential Optical Absorption Spectroscopy (DOAS). The retrievals are performed using measurements in the wavelength range 331.1-336.6 nm. The key difference between the two algorithms is that DOAS removes the effects of aerosol, clouds, sulfur dioxide, and underlying surface by spectral fitting; while TOMS algorithm does that through empirical correction. In addition, TOMS algorithm uses a cloud height climatology, derived with the help of infrared satellite data; while the DOAS algorithm uses cloud information derived from OMI measurements in the O_2-O_2 absorption band at wavelength of 470 nm. A single well-established algorithm of the satellite data processing has not yet been commonly accepted. Qualitative behavior of the long-term TO variations differs little between different algorithms used.

Figure 1 shows the observation time series for the first half of 2013, performed with M-124 device and processed by the OMI-DOAS algorithm. **Figure 2** compares the M-124 ozonometer measurements for 2005 – 2013 with satellite data, processed with different algorithms. We present the relative differences between M-124 values and results of satellite measurements; the differences are smoothed using sliding average over 365 days. At the Aura Validation Data Center (AVDC) website, the OMI data processed by the OMI-TOMS algorithm are abbreviated as OMTO, and data based on the OMI-DOAS algorithm are abbreviated as OMDO.

The relative discrepancy between ground-based and satellite data, retrieved with the help of OMI-DOAS algorithm, and averaged over the year, has been 13 DU (3.5% of the annual average) in 2010 and 16 DU (4.7%) in 2011. Recall that 2012 was very dry, with many fires at summertime, which probably influenced the accuracy of measurements. As a result, the discrepancy between the satellite and ground-based measurements grew from the middle to the end of 2012. Seemingly, until August 2012 this was due to the increased turbidity and smoke pollution of the atmosphere. In late 2012, the satellite measurements indicated the decreased total ozone content; while ground-based observations reported the usual seasonal TO growth. As a consequence, the discrepancy had been 18 DU in the first half and 23 DU in the second half of 2012. In 2013, the disagreement between ground-based and satellite data decreased and became 16 DU over the first half-year.

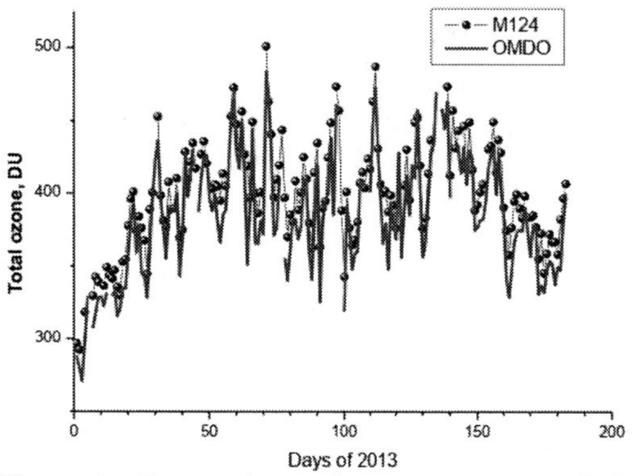

Figure 1. Comparison of measurements of the total ozone content with the help of M-124 ozonometer and TOMS instrumentation (satellite measurements are retrieved with the use of the OMI-DOAS algorithm) for the first half of 2013.

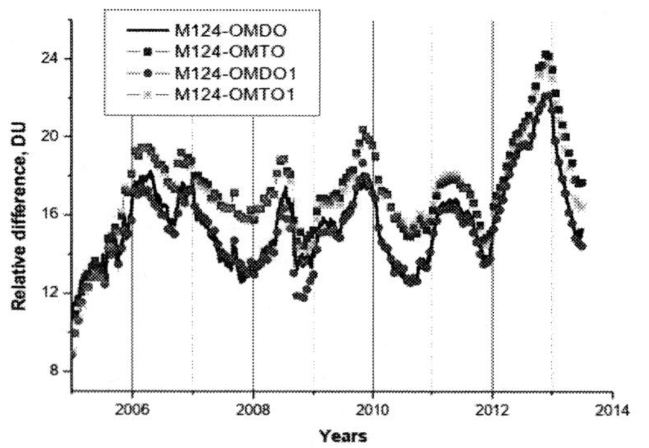

Figure 2. Relative discrepancies between measurements of the total ozone content with the help of M-124 ozonometer and on the basis of TOMS instrumentation for period of 2005 – 2013.
The satellite data are retrieved with the help of two algorithms for two closely lying geographic points: Tomsk (OMDO, OMTO) and Tomsk 1 (OMDO 1, OMTO 1).

2. MEASUREMENTS OF NO_2 CONTENT

The NO_2 content is measured with automatic spectrophotometer, which records the spectrum of zenith-scattered solar radiation in the wavelength range of 430 – 450 nm with the spectral resolution of 0.9 nm. The deformation of the spectrum is used to determine the nitrogen dioxide content in the slant atmospheric column as a function of the solar zenith angle. The measurements are performed at twilight time of the day, when the solar zenith angle varies from 83° to 96°. By solving the inverse problem, the data on the slant nitrogen dioxide content in the atmosphere are used to retrieve the vertical distribution of nitrogen dioxide in 10 layers, each 5 km thick, in the altitude range from 0 to 50 km [2]. For determining the NO_2 content, the satellite instrument OMI uses the DOAS method, by recording the spectrum of solar radiation, scattered by the Earth's atmosphere, in the wide wavelength range in the ultraviolet and visible parts of the spectrum with the spectral resolution of 0.5 nm and spatial resolution of 13x24 km^2 at nadir [3]. The wavelength range of 415 – 465 nm is used to determine the NO_2 content in the vertical atmospheric column. In the report we use OMI data for period from October 2004 to December 2012. For comparison, we took the data sample of OMI measurements of the total NO_2 content in the stratospheric column over Tomsk; these data are publicly available at AVDC website [4].

The ground-based measurements are performed at twilight time of the day; therefore, interpolation method was used to reduce them to the time of satellite measurement, taking into consideration the photochemical processes. The satellite measurements over Tomsk were performed twice daily at different times; therefore, the results of both measurements were used for comparison purposes, while the ground-based measurements were interpolated to the times of satellite measurements.

Figure 3 presents the values of the total NO_2 content in the stratospheric column according to measurements from satellite (OMI instrument) and ground-based measurements, interpolated to the times of observations from satellite. As can be seen from the figure, these satellite measurements coincide quite exactly with data of ground-based observations. The time series show annual behavior, with maximal values in summer and minimal values in winter.

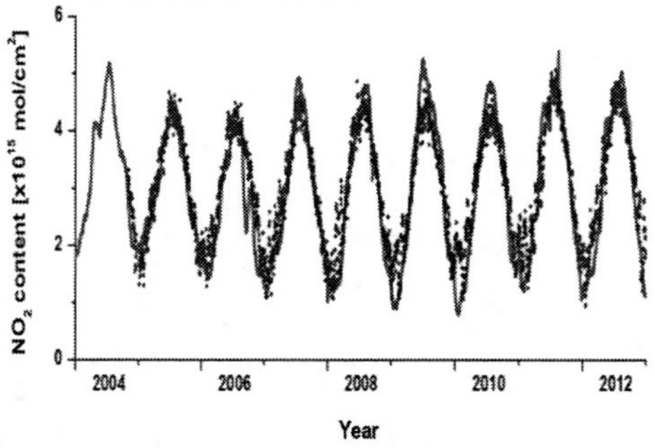

Figure 3. The NO_2 content in the vertical column of the stratosphere (10 – 50 km) over Tomsk according to measurements with the help of satellite (OMI instrument) and ground-based measurements, interpolated to the time of observation from satellite

Figure **4** presents the difference between ground-based measurements and satellite observations (grey circles). The standard deviation of the values is $0.53x10^{15}$ mol/cm^2. The average difference is positive and is equal to (0.19 ±0.01) mol/cm^2. The values of NO_2 content in the stratosphere according to data of ground-based observations are somewhat higher than satellite measurements; however, the average difference is low and is within the error of ground-based and satellite measurements. The sliding average over 30 days (thick line) demonstrates a pronounced seasonal behavior. The correlation coefficient for this observation period is 0.91.

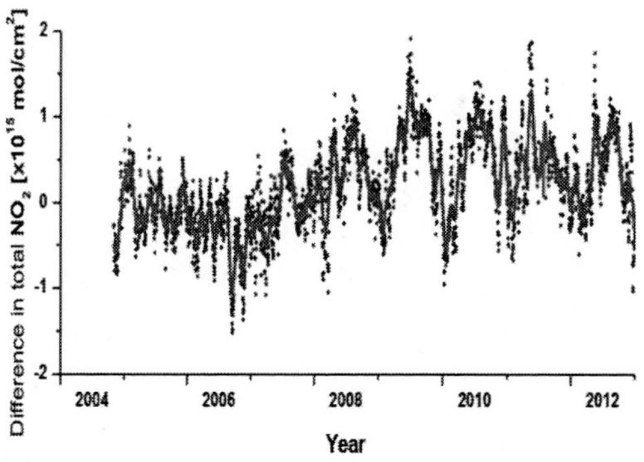

Figure 4. Difference in the diurnal values of NO_2 content in the stratosphere between ground-based and satellite measurements, performed over Tomsk.

3. TEMPERATURE MEASUREMENTS.

At heights larger than 30 km, where aerosol is practically absent (under the background conditions), the intensity of Rayleigh lidar signal (at the wavelength of the sensing radiation 532 nm) is determined by purely molecular scattering, thus making it possible to retrieve the molecular density and temperature of the middle atmosphere from these signals, because the density depends linearly on the temperature, and because Raman scattering (RS) signals from molecular nitrogen (at the wavelength of 607 nm) also have no direct contribution from aerosol scattering, i.e., the number of recorded scattered photons is directly proportional to the molecular density of the atmosphere, making it possible to retrieve the profiles of the vertical distribution of temperature.

We compared the results of lidar measurements of temperature with the simultaneous balloon-sonde measurements in Tomsk and with measurements of altitudinal temperature profiles by Infrared Atmospheric Sounding Interferometer (IASI) on MetOP satellite, ESA. The measurements and processing of IASI data can be used to obtain the profiles of temperature, humidity, and content of certain atmospheric gases.

The temperature profiles are retrieved on the basis of standard EUMETSAT and NASA methods [5]. The data on the simultaneous lidar, balloon-sonde, and satellite measurements of temperature are presented in **Fig. 5.**

From the figure, we can see that the profiles of the vertical distribution of temperature in the lower stratosphere from data of lidar measurements according to Raman scattering signals from molecular nitrogen at the wavelength of 607 nm and balloon-sonde data nearly coincide. It is noteworthy that a well-defined tropopause is observed at height of about 11 km. The satellite measurements of temperature fix the tropopause not as sharply and exceed the balloon-sonde observations. The reasons for this are both due to differences in the spatial resolutions of satellite, lidar, and balloon-sonde measurements, and because these measurements have different uncertainties.

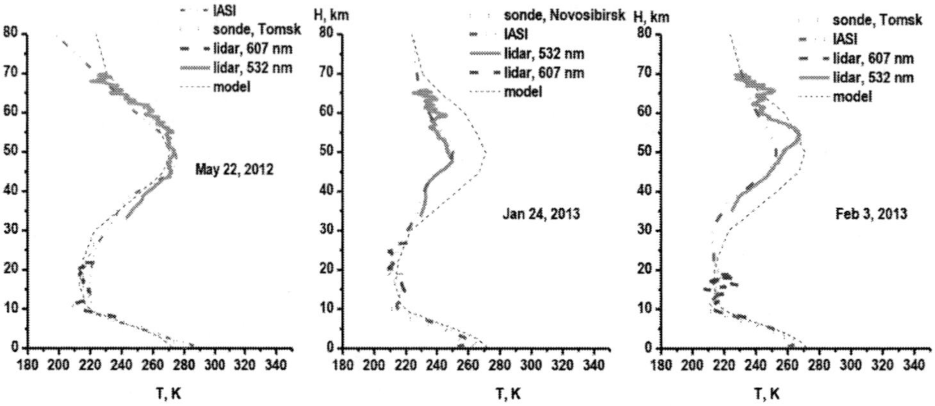

Figure 5. Examples of simultaneous lidar, balloon-sonde, and satellite measurements of temperature

4. CONCLUSIONS

When TO data are compared, M-124 ozonometer measurements best agree with data on the basis of OMDO algorithm. As a rule, the M-124 measurements exceed the satellite data.

Comparison of ground-based measurements of NO_2 at SLS with satellite (OMI instrumentation) observations showed that the wintertime (summertime) values of NO_2 content according to satellite observations are larger (smaller) than those obtained from the ground-based observations. However, the statistically average discrepancy for the stratosphere is small and is within the measurement errors.

In the upper stratosphere – mesosphere, the profiles of temperature, obtained from data of satellite and lidar measurements, do not totally coincide. The discrepancy reaches 7-8 K, probably due to the uncertainties of both instruments and spatiotemporal separation of measurements. For instance, the lidar and balloon-sonde measurements of temperature on January 24, 2012 in Tomsk (56.5°N; 85.0°E) were performed from 12:20 to 13:00 GMT (balloon-sonde was released at the location of lidar), and satellite measurements were performed at 13:53 GMT at the point with coordinates 56.447°N; 84.608°E.

ACKNOWLEDGEMENT

We thank M.U. Arshinov, B.D. Belan and G.A. Ivlev, LKAS IAO SB RAS, for kindly providing balloon-sonde data in Tomsk.

This work was supported in part by the integrated project 131a, Siberian Branch, Russian Academy of Sciences; Russian Fund for Basic Research (under grant no. 11-01-00673_a), and by the grant of the President of the Russian Federation for the support of leading scientific schools NSh-4714.2014.5.

REFERENCES

1. http://avdc.gsfc.nasa.gov/
2. McKenzie R.L., Johnston P.V., McElroy CT., Kerr J.B., Solomon J.B. Altitude distributions of stratospheric constituents from ground-based measurements at twilight // *J Geophys. Res.* 1991. V. 96. № 8. P. 15499-15511
3. Boersma K.F., Busela E.J., Brinksma E.J., Gleason J.F. OMI algorithm theoretical basis document. Volume IV OMI trace gas algorithms //ATBD-OMI-02. NASA Goddard Space Flight Centr/ Ed. K. Chance. Greenbelt, Md., 2002. P. 15-36.
4. http://avdc.gsfc.nasa.gov/pub/data/satellite/Aura/OMI/V03L2OVP/OMNO2/aura_omi_l2ovp_omno2_v03_tomsk.txt.
5. August, T., et al, IASI on Metop-A: Operational Level 2 retrievals after five years in orbit// *Journal of Quantitative Spectroscopy and Radiative Transfer.* July 2012. V 113, Issue 11, p. 1340-1371.

APPLICATION OF THE BACKSCATTER NEAR-END SOLUTION FOR THE INVERSION OF SCANNING LIDAR DATA

Vladimir Kovalev*, Cyle Wold, Alexander Petkov, and Wei Min Hao

*Missoula Fire Sciences Laboratory, MT 59808, USA, *E-mail: vladimirakovalev@fs.fed.us*

ABSTRACT

The significant issue of the classic multiangle data-processing technique is that the upper boundary height, up to which the multiangle data processing technique allows reliable extraction of optical parameters of the searched atmosphere, is always significantly less than the operative range of the scanning lidar. The existing inversion methodology yields poor accuracy when inverting the far-end data points of the signals measured in and close to zenith. In the report, the data processing technique is considered which allows using the zenith-measured signal to increase the maximal heights of profiling of the atmosphere. Simulated and experimental data are presented that illustrate the specifics of such a technique.

1. INTRODUCTION

The classical Kano-Hamilton multiangle method allows extracting the vertical extinction-coefficient profile from elastic lidar data without an a priori assumption on the vertical profile of the lidar ratio [1, 2]. However, the method requires fulfilling two conditions. First, the total backscatter coefficient should be invariable along the horizontal direction, that is,

$$\beta_{\pi,\Sigma}(h) = const. \qquad (1)$$

The requirement in Eq. (1) means that the total backscatter coefficient, $\beta_{\pi,\Sigma}(h)$ at the height h, which is the sum of the molecular and particulate components, $\beta_{\pi,m}(h)$ and $\beta_{\pi,p}(h)$ respectively, should be the same for any slope direction used during scanning. The second condition is the validity of the unanimous correlation between the total optical depth, $\tau_\varphi(0, h)$, measured along the elevation angle, φ, and the vertical optical depth, $\tau_{vert}(0, h)$. The dependence can be written as,

$$\tau_\varphi(0,h)\sin\varphi = \tau_{vert}(0,h), \qquad (2)$$

where $\tau_{vert}(0, h)$ is the total (molecular and particulate) optical depth of the layer from ground level to the height h in the vertical direction.

To extract the optical parameters of the searched atmosphere, the classic Kano-Hamilton inversion method requires determining the set of functions $y_\varphi(h) = \ln\left[P_\varphi(h)(h/\sin\varphi)^2\right]$, calculated from the lidar signals, $P_\varphi(h)$, measured in different slope directions φ. If the conditions in Eqs. (1) and (2) are met, these functions can be written as,

$$y_\varphi(x,h) = \ln\left[C\beta_{\pi,\Sigma}(h)\right] - 2\tau_{vert}(0,h)x, \qquad (3)$$

where C is the lidar equation constant and $x = 1/\sin\varphi$. The function $[C\beta_{\pi,\Sigma}(h)]$, which is used in the below inversion method, is found after determining the intercept of the linear fit of the functions $y_\varphi(x,h)$ versus x with the vertical axis.

A significant issue of the Kano-Hamilton method, as compared with the one-directional zenith profiling, is the restricted altitude range where required inversion accuracy can be achieved [3]. The maximal height, h_{up}, of the atmospheric layer up to which the acceptable inversion accuracy can be achieved is always less than the maximum ranges of the lidar signals measured in and close to zenith (Fig. 1). In the conventional multiangle

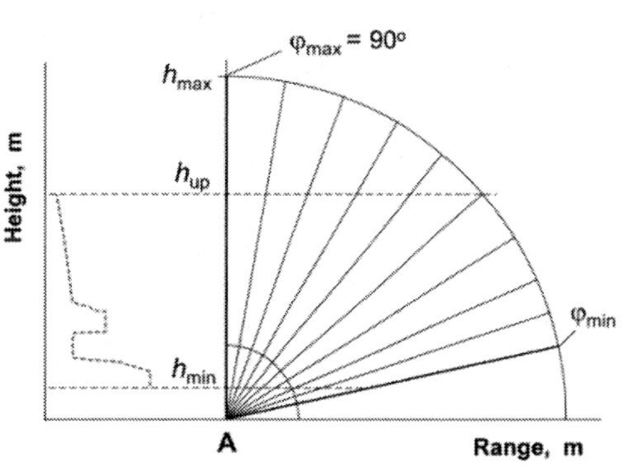

Fig. 1. Right side: Schematic of the scanning lidar operation. Left side: The vertical profile of the function $[C\beta_{\pi,\Sigma}(h)]$ derived from the Kano-Hamilton solution.

inversion method, the backscatter signals, $P_\varphi(h)$, measured within the altitude range from h_{up} to h_{max} do not provide accurate determination of the optical parameters of the atmosphere within this altitude range. The only possibility to increase the profiling height is to combine the multiangle and one-directional data processing methodologies.

Below, the methodology for processing the multiangle data is discussed that allows increasing the maximum height of the profiling of the atmosphere as compared with the conventional Kano-Hamilton method.

2. METHODOLOGY

The algorithms for the backscatter near-end solution for a zenith-directed lidar data were considered in the study by Sassen and Cho [4]. In our study, an advanced version of this methodology is adopted for multiangle measurements. The first step of the processing procedure, after performing the initial Kano-Hamilton inversion, is the selection of the height h_{up} and determination of the product $[C\beta_{\pi,\Sigma}(h_{up})]$ at this height. This quantity, same as the particulate extinction coefficient, $\kappa_p(h_{up})$, found from the initial inversion, will be used as the input parameters for the next one-directional backscatter near-end solution. The vertical profile of the extinction coefficient, $\kappa_p(h)$, for the heights $h > h_{up}$ will be extracted from the square-range-corrected zenith signal, $P(h)h^2$, measured at these heights. Thus, the height h_{up} is the upper height for the Kano-Hamilton solution, and at the same time, the near-end boundary point for the backscatter solution. The inversion procedure is similar to that in Ref. [4]. Initially the ratio of the square-range-corrected lidar signals within the altitude range from h_{up} to h_{max} is determined,

$$F_{sign}(h) = \frac{P(h)h^2}{P(h_{up})h_{up}^2}. \tag{4}$$

The function $F_{sign}(h)$ is then transformed into another function, $Z(h)$, with the following structure,

$$Z(h) = [C\beta_{\pi,\Sigma}(h)]\exp\left[-2B\int_{h_{up}}^{h}\beta_{\pi,\Sigma}(h')dh'\right], \tag{5}$$

where B is a constant.

Such a transformation of $F_{sign}(h)$ into $Z(h)$ is achieved by multiplying the function $F_{sign}(h)$ by an auxiliary function $Y(h)$. The issue is that the profile of the required $Y(h)$ is a function of the unknown particulate lidar ratio, $S_p(h)$, within the altitude range from h_{up} to h_{max}. Therefore, the solution in the form in Eq. (5) can only be obtained using the assumption of invariable lidar ratio, $S_p(h) = S_p = \text{const.}$, for all the heights $h > h_{up}$. After taking the condition $B = S_p$, as proposed in [4], the function $Y(h)$ can be written in the form,

$$Y(h) = [C\beta_{\pi,\Sigma}(h_{up})]\exp\left\{-2(S_p - S_m)\int_{h_{up}}^{h}\beta_{\pi,m}(h')dh'\right\}, \tag{6}$$

where $S_m = 8\pi/3$ is the molecular lidar ratio.

The ratio of $[C\beta_{\pi,\Sigma}(h)]$ to $[C\beta_{\pi,\Sigma}(h_{up})]$ for the height interval $h_{up} - h_{max}$ can be written as,

$$F_\beta(h) = \frac{Z(h)}{Z(h_{up}) - 2[\kappa_p(h_{up}) + S_p\beta_{\pi,m}(h_{up})]\int_{h_{up}}^{h}Z(h')dh'}. \tag{7}$$

The particulate extinction coefficient for the heights $h > h_{up}$ can be obtained in the form,

$$\kappa_p(h) = S_p\left\{F_\beta(h)\frac{[C\beta_{\pi,\Sigma}(h_1)]}{C} - \beta_{\pi,m}(h)\right\}. \tag{8}$$

Thus, two issues aggravate obtaining the reliable profile of $\kappa_p(h)$ for the heights $h > h_{up}$. First, the solution may be accurate enough only over the height intervals where the lidar ratio, S_p, is constant. Second, for obtaining the most accurate inversion result, the constant C should be someway determined. Note that if the quantity $\beta_{\pi,\Sigma}(h_{up})$ can be determined accurately from the initial Kano-Hamilton inversion results, the constant can be found as $C = Z(h_{up})/\ \beta_{\pi,\Sigma}(h_{up})$.

If the constant C cannot be determined, the solution for $\kappa_p(h)$ can be found with the formula,

$$\kappa_p(h) = F_\beta(h)\kappa_p(h_{up})\left\{\frac{1+\frac{S_p\,\kappa_m(h_{up})}{S_m\,\kappa_p(h_{up})}}{1+\frac{S_p\,\kappa_m(h)}{S_m\,\kappa_p(h)}}\right\}. \tag{9}$$

When the molecular component is significantly less than the particulate one, the term in braces is close to unit, accordingly,

$$\kappa_p(h) \approx F_\beta(h)\kappa_p(h_{up}). \tag{10}$$

DOI: 10.1051/epjconf/201611925001

With the increase of the relative level of the molecular component, the simplified solution in Eq. (10) yields less accurate inversion results.

3. ANALYSES OF SIMULATED AND EXPERIENTAL DATA

In Fig. 2, the dashed curve is the test profile $[C\beta_{\pi,\Sigma}(h)]$ used in the below simulations; the open circles show the bottom part of this profile within the altitude range from h_{min} to h_{up}, which is obtained from the data of artificial scanning lidar. The lidar operates at the wavelength 355 nm in the synthetic atmosphere where, for the heights $h > h_{up}$, the lidar ratio S_p is equal to 30 sr; the maximum range of the synthetic lidar signal measured in the zenith direction is $r_{max} = 8000$ m.

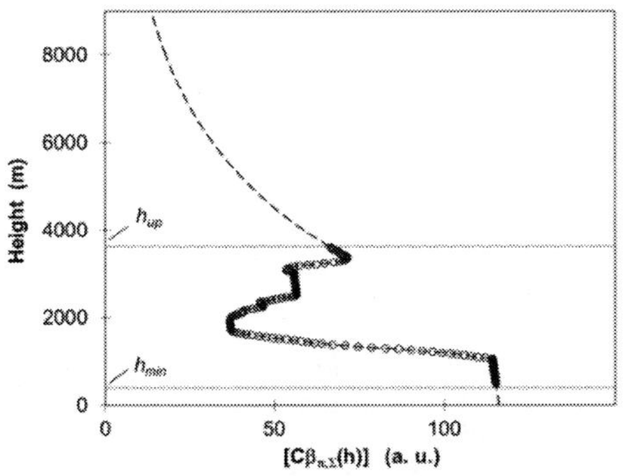

Fig. 2. Test profile $[C\beta_{\pi,\Sigma}(h)]$ used in the simulations (dashed curve) and the bottom part of that (open circles) obtained from the artificial scanning lidar; the height $h_{up} = 3600$ m.

The results of determining the profile of the extinction coefficient $\kappa_p(h)$ with Eq. (8) when the constant C is known are shown in Fig. 3. The model extinction coefficient is shown as the dashed curve; the profiles of the restored $\kappa_p(h)$ when assuming the lidar ratio equal to 20, 30, and 40 sr are shown as dotted, thick and thin solid curves, respectively. One can see that when the constant C is known, the selection of an incorrect S_p significantly shifts the location of the derived $\kappa_p(h_{up})$ from its starting point at h_{up} shown in the figure as the filled circle. Thus, the knowledge of the constant C significantly improves the accuracy of determining the profile of the extinction coefficient for the heights $h > h_{up}$.

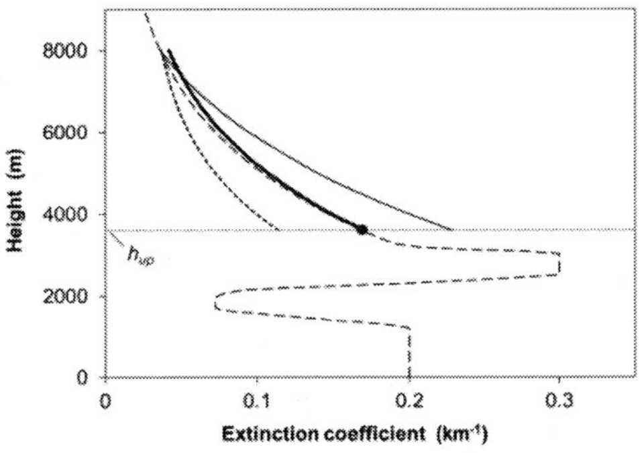

Fig. 3. Test profile $\kappa_p(h)$ used in the simulations (dashed curve) and the extinction coefficient profiles derived with Eq. (8) within the altitude range from h_{up} to $h = 8000$ m

The alternative profiles of the extinction coefficient for the same synthetic atmosphere but obtained with Eq. (10) are shown in Figs. 4 and 5.

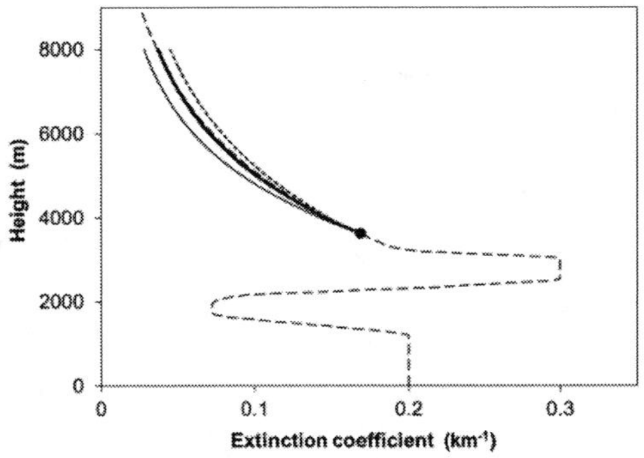

Fig. 4. Dashed curve is the test profile $\kappa_p(h)$ at 532 nm. The profiles derived with S_p equal to 10, 30, and 60 sr are shown as the dotted, thick and thin solid curves, respectively.

The profiles are calculated for the lidar signals, measured at the wavelengths 532 and 355 nm, respectively. The profiles of the extinction coefficient derived at the wavelength 532 nm, where the molecular component is relatively small, do not depend dramatically on the selected lidar ratio S_p. At the wavelength 355 nm, where the level of the molecular component is higher, the unbiased profiles of the extinction coefficient can be extracted only using an extremely restricted scope of the lidar ratios; in particular,

DOI: 10.1051/epjconf/201611925001

the retrieval can be made using only small values of the lidar ratio, $S_p < 15$ sr. The use of the higher values yields unphysical profiles of the corresponding particulate transmittance.

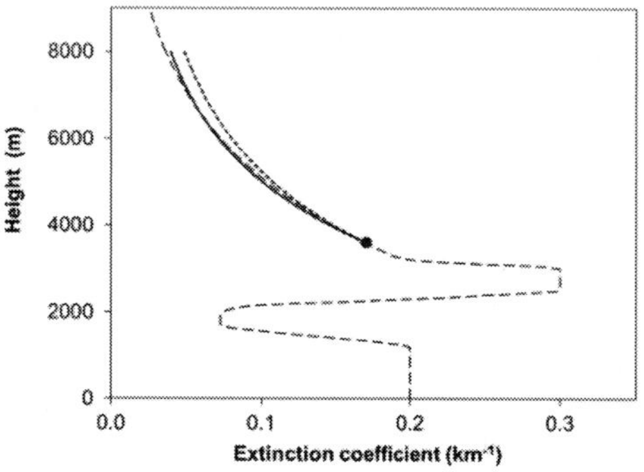

Fig. 5. Same as in Fig. 4 but calculated for the wavelength 355 nm. The profiles derived with S_p equal to 10 and 14 sr are shown as the dotted and solid curves, respectively.

In Figs. 6 and 7, the real lidar signal measured at 355 nm in clear-sky conditions and its inversion result is shown. For deriving the profiles $\kappa_p(h)$ at the heights $h > h_{up}$, the signal measured at $\varphi = 80°$ was used. One can see that the square-range-corrected signal, shown in Fig. 6, monotonically decreases with height, so that the assumption $S_p \approx$ const. for the altitude interval from 4000 to 10,000 m looks sensible. The inversion results of the scanning lidar data combined with the

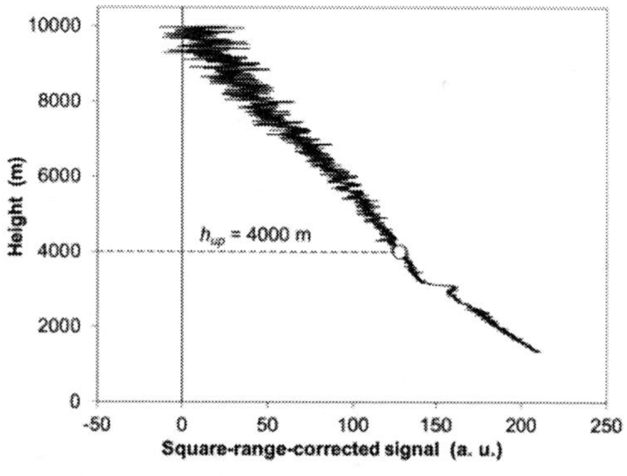

Fig. 6. Square-range-corrected signal measured with scanning lidar close to zenith direction.

backscatter near-end solution are shown in Fig. 7. The profile of $\kappa_p(h)$ derived from the Kano-Hamilton solution is shown as the dashed curve.

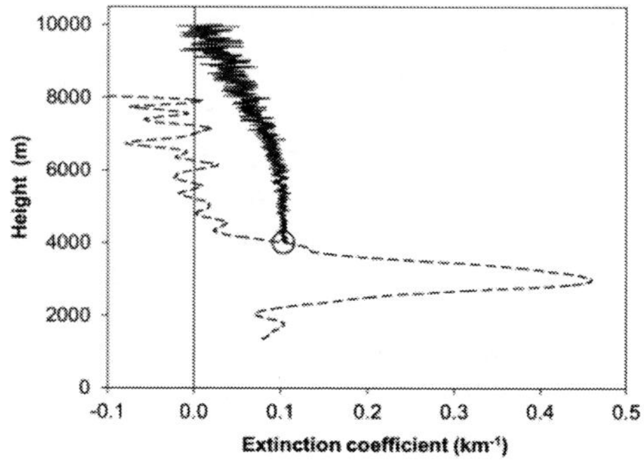

Fig. 7. Inversion results of the scanning lidar data.

The inversion result obtained with the backscatter near-end solution is shown as the solid curve. The analysis shows that the selected $S_p = 15$ sr is most adequate to the optical situation that took place during the lidar signal recordings.

4. SUMMARY

The new measurement methodology is presented which allows using the far-end zenith measured signal of scanning lidar to increase the maximal heights of determining vertical parameters of the searched atmosphere as compared to the classical Kano-Hamilton solution.

REFERENCES

[1] Kano, M., 1968: On the determination of backscattering and extinction coefficient of the atmosphere by using laser radar, *Papers Meteorol. and Geophys.* **19**, 121-129.

[2] Hamilton, P. M., 1969: Lidar measurement of backscatter and attenuation of atmospheric aerosol, *Atmos. Environ.* **3**, 221-223.

[3] Kovalev, V. A., 2015: *Solutions in Lidar Profiling of the Atmosphere*, J. Wiley & Sons Inc., pp. 195-199.

[4] Sassen, K. and Cho, B. S. 1992: Subvisual-thin cirrus lidar dataset for satellite verification and climatological research, *J. Appl. Meteorol.*, **31**, 1275-1285.

EPJ Web of Conferences **119**, 25002 (2016) DOI: 10.1051/epjconf/201611925002
ILRC 27

OVERLAP CORRECTION FUNCTION FOR AN AIRBORNE BASED LIDAR

Mariana Adam[1]*, Franco Marenco[2]

[1]*Observations branch, Met Office, Exeter, Devon EX1 3PB, United Kingdom, *Email: mariana.adam@metoffice.gov.uk*

[2]*Observations Based Research, Met Office, Exeter, Devon EX1 3PB, United Kingdom*

ABSTRACT

The present research envisages the estimation of the overlap correction function for an airborne nadir-mounted lidar using multi-angle measurements. We have scanned a series of off-nadir angles and after data processing we have been able to determine the instrument's overlap function down to 95m from the lidar. This function can be used for the correction of lidar profiles and hence reduce the near-range uncertainty of lidar measurements. To our knowledge, the estimation of the overlap function using multi-angle method for a nadir pointing lidar is a première.

1. INTRODUCTION

The Met Office is one of the main users of the FAAM BAe-146 research aircraft, alongside scientists working in universities funded by the Natural Environment Research Council [1], and is responsible for maintaining the on-board elastic backscatter lidar. Moreover, as a response to the eruption of Eyjafjallajökull in 2010, the Met Office has set up a second aircraft, the Met Office Civil Contingencies Aircraft (MOCCA), a Cessna 421 which is aimed at giving a rapid response in case of natural hazards (especially volcanic ash) [2].

A Leosphere ALS450 UV Lidar (with depolarization capability) is set up onboard both aircrafts [3]. Besides Lidar, several instruments perform in-situ measurements to characterize the microphysical and optical properties of the aerosol particles and clouds, atmospheric turbulence and gas chemistry [1-2].

The current research evaluates the overlap function of the lidar on-board the FAAM aircraft, using multi-angle measurements, which is based on the method developed by Adam et al. [4] for ground-based lidars.

2. METHODOLOGY

For ground-based Lidar, the multi-angle methods are based on measurements taken by scanning several elevation angles; the main assumption being that the atmosphere is horizontally homogeneous. This assumption is most probably fulfilled for airborne based lidars, provided a suitable altitude is selected, as we expect more homogeneity above the boundary layer (very close to a molecular atmosphere). The main steps of the method are the following. For each height, the logarithm of range corrected signals (*RCS*) (expressed as a function of height) is plotted versus $1/\cos(\varphi)$, where φ the off-nadir pointing angle. The slope of the linear regression provides the optical depth. Once we retrieve the optical depth $\tau(0,h)$ and the intercept $A^*(h)$ from the linear regression, a synthetic signal for *RCS* is calculated as in [4]:

$$Z_j^*(h) = \exp\left[A^*(h)\right]\exp\left[-2\tau(0,h)/\cos\varphi_j\right] \quad (1)$$

where j is the angle index. The individual overlap function is then computed as the ratio of the measured *RCS* and synthetic *RCS* as a function of height h.

$$q_j(h) = P_j(h)\left(h/\cos\varphi_j\right)^2 / Z_j^*(h) \quad (2)$$

The final steps are the conversion of individual overlap function (eq. 2) in terms of range and then compute the average overlap function [4].

The method was first tested over synthetic data. Sensitivity studies were employed to find out a comfortable set of bank angles capable of yielding to a good estimate of the overlap function. The main constraint is given by the capability of the aircraft to provide specific measurements at high bank angles. The Lidar is installed with a fixed pointing angle with respect to the aircraft and thus, we need to roll the aircraft in a turn in order to change the bank angle. As the aircraft spins around 360 degrees with a special bank angle we have an orbit. The steepest orbit available with the

DOI: 10.1051/epjconf/201611925002

BAe-146 aircraft is one with a 60 degrees bank angle, and it impresses on cabin occupants an acceleration of 2g. The aircraft pitch and roll are recorded every second by the on-board navigation system, and can thus be easily linked to the lidar data.

3. EXPERIMENT

The first flight performing orbits according to the scientific needs for the overlap study, took place on 17th of June 2014 (as part of a broader mission, along with other scientific objectives). The orbits were performed on the nominal bank angles: 30°, 40°, 50° and 60°. The Lidar did not perform optimally during this flight (laser energy was low and thus SNR was low). The cruise altitude was ~3.9km. However, the method was tested and the result was promising even if it was relatively noisy. Thus, during a subsequent flight, on 10th of September 2014, the specific flying strategy for the overlap study was repeated. In addition to the initial bank angles, the 10° and 20° bank angles were added. However, the cruise altitude was different (~4.8km). The flight took place over the Ocean, near the coast of Cornwall. Note that the 30°, 40° and 60° orbit was repeated twice in order to increase the number of available data points (a 60° orbit lasts less than 1 minute). The data quality was superior and thus we chose this set of data for final analyses. The flight track is shown in Fig. 1. In addition to the nine orbits, two runs were performed at ~0° and we consider these data as well.

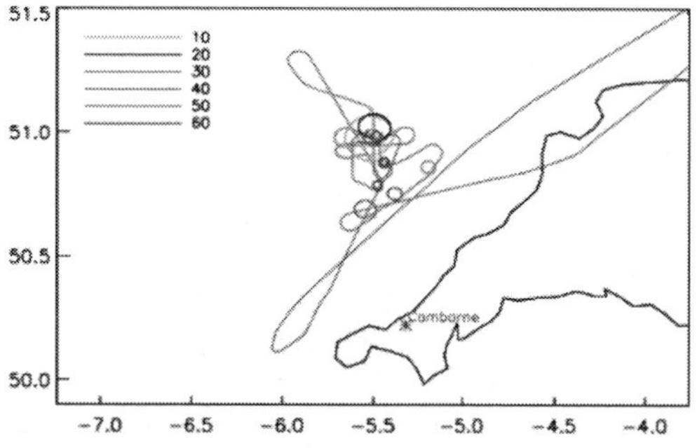

Fig. 1. Flight track.

The measurements are taken with a range resolution of 1.5 m and an integration time of 2 s. In the first stage, the number of measurements for each nominal bank angle was calculated, according to the flight summary. The cruise altitudes and the bank angles for the entire flight are shown in Fig. 2.

Figure 3 shows the bank angles used in the overlap experiment. As seen in both figures, in reality, there is a slight deviation from the nominal bank angles. The mean altitudes (along with STD) corresponding to the mean bank angles are shown in Table 1 and Fig. 3. The number of measurements for each bank angle is shown in the third column in Table 1. The overall mean cruise altitude is 4837.6 km while the STD is 13.76m.

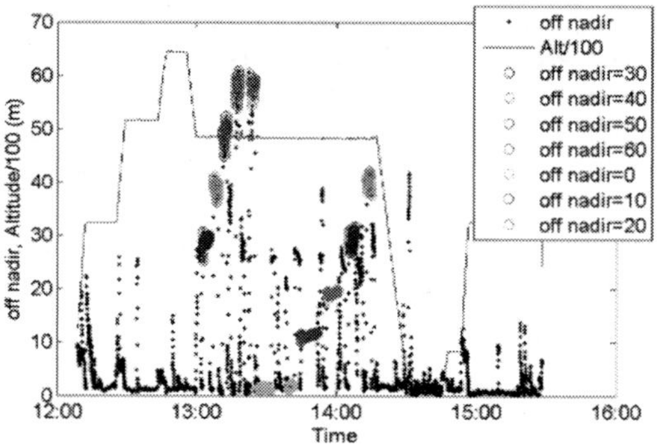

Fig. 2. Cruise altitudes and bank angles (off-nadir) for entire flight.

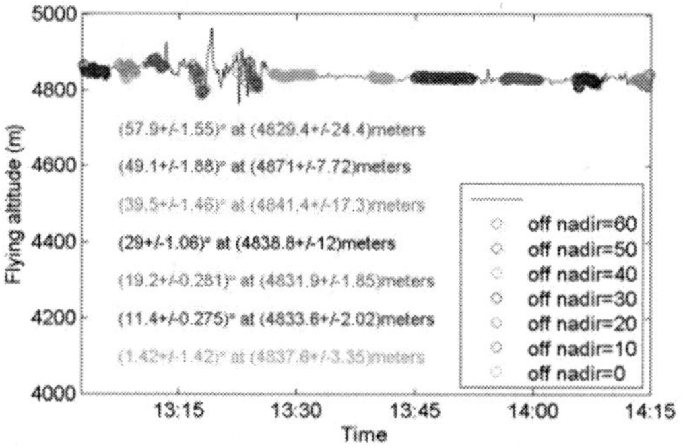

Fig. 3. The cruise altitudes for all the orbits performed for the planned bank angles (off-nadir) of 10°, 20°, 30°, 40°, 50° and 60° and runs at 0°. The mean bank angle and the corresponding mean altitude are shown.

Table 1. Mean bank angle and altitude cruise

Mean bank angle	Mean cruise altitude	# points
11.4±0.275	4833.6±2.02	203
19.2±0.281	4831.9±1.85	123

 DOI: 10.1051/epjconf/201611925002

29±1.06	4838.8±12	153
39.5±1.46	4841.4±17.3	110
49.1±1.88	4871±7.72	48
57.9±1.55	4829.4±24.4	79

4. RESULTS

According to the lidar specifications, the full overlap should be around 300m. Time-altitude *RCS* profiles for the total signal (as a combination of parallel and perpendicular signals) is shown in Fig. 4 a). PBL height was ~1 km at Camborne and ~1.3 km at Exeter (as seen in ceilometers). While at Camborne it was cloud free atmosphere, at Exeter there were clouds in the PBL. As seen in Fig. 4b, there is an increase in the *RCS* below 1.5km which may be due to the presence of a cloud in the PBL. However, the region below 1.5km altitude will not affect the overlap retrieval (which is of interest in the first few hundred meters from the aircraft which is situated more than 2000m above the PBL). For each bank angle, the mean (total) *RCS* is calculated [Fig. 4 b)].

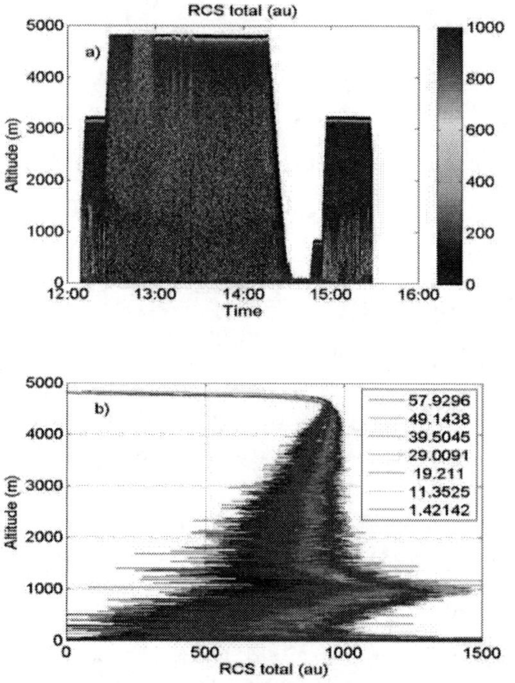

Fig. 4. Total *RCS* over the entire flight (a). Mean (total) *RCS* for each bank angle (b).

The flight took place about ~100 km north of Camborne site (Fig. 1). In the paper by Adam [4], for zenith pointing angles Lidar, there are few criteria involved such that we use the correct altitude interval to perform the linear regression ("optimal heights interval") without introducing systematic errors. Those criteria should be adapted for nadir pointing Lidar as the signals look different as compared with zenith pointing lidars. As described in [4], the use of data from incomplete overlap introduces systematic distortions in the retrieval the intercept (and further in the overlap). In the first step we did not restrict the region over the incomplete overlap and thus, all the measurements were taken into consideration. The optical depth and the intercept retrieved in this case are shown in Fig. 5 (down to 1500 m altitude). Also shown, the molecular optical depth, which is calculated using the pressure and temperature profile as given by the radiosonde launched at 12:00 at Camborne. As seen, the optical depth obtained over the incomplete overlap is negative (from the cruise altitude down to ~4500m).

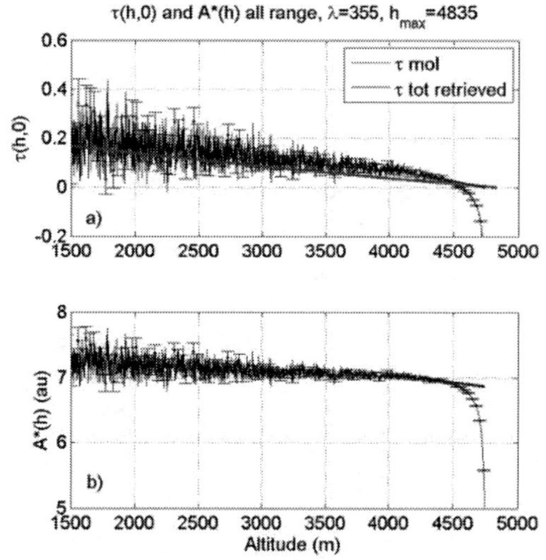

Fig. 5. Optical depth (a) and the intercept (b).

In order to avoid the use of data from incomplete overlap, we have chosen to the following option. We extrapolated the optical depth and the intercept from 4400m towards the Lidar using a linear fit over the region 4200m-4400m. The extrapolation is shown as blue line in Fig. 5. Further we have dismissed the first 95m from the

Lidar. Now we proceed to compute the synthetic *RCS* [eq. (1)] and the individual overlap functions [eq. (2)]. The overlap function, calculated as the average of the individual functions, is shown in Fig. 6. The complete overlap is shown to be around 400m (it is ~0.98 at 300m, the range of the nominal full overlap according to specifications). The overlap function looks very good in the far field up to 5000m. Note that the errors were computed using the error propagation with input error the error on the mean for signals at each bank angle.

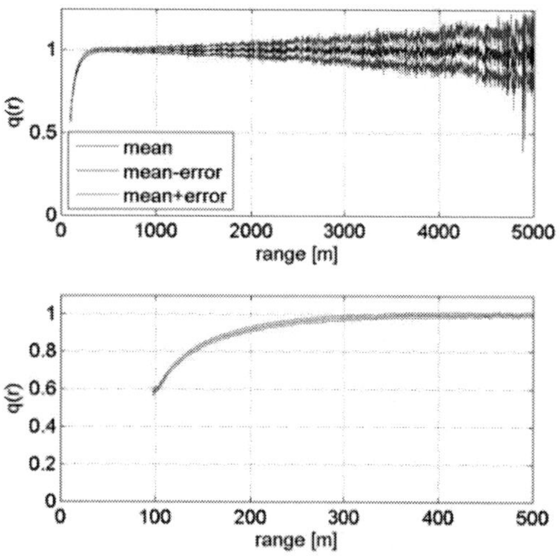

Fig. 6. Overlap function. The lower plot is zoomed region over 500m.

Note that if we consider from the beginning that signals from the first 300m should not be considered, we arrive to the situation in which it is not possible to estimate the overlap function very close to the lidar. The highest the bank angles the closest to the lidar the overlap function can be retrieved. For more details please see [4], where the smallest the elevation angles the closest to the lidar the overlap function goes. Thus, technically, for 60° bank angle, 300m range corresponds to 150m on altitude. The linear regression is performed if there are at least three points available (corresponding to three different bank angles). Thus, for the highest bank angles (40°, 50° and 60°) the corresponding heights for a range of 300m are 229.81m, 192.84m and 150m. Therefore the linear regression can be first performed at 229.81m away from the lidar.

As seen in Fig. 5 a) the optical depth is slightly larger than the molecular optical depth and thus, the comparison of the measured *RCS* with synthetic molecular *RCS* is not appropriate.

A thorough presentation of the method and a complete picture of various steps will be shown during ILRC.

5. CONCLUSIONS

We have demonstrated an innovative method to determine the overlap function for an airborne Lidar. The main constraint in retrieving the overlap function very close to the Lidar is the availability of the measurements taken at high bank angles, which is possible up to an aircraft-dependent limit (60 degrees for the BAe-146).

For the system on-board the FAAM aircraft, in the present study we estimate full overlap to be reached at ~400 m. This has to be compared with a nominal overlap of 300m; note that at this range we estimate the overlap function to be equal to 0.98 The overlap function was retrieved down to ~95m from the Lidar.

ACKNOWLEDGEMENT

Airborne data was obtained using the BAe-146 Atmospheric Research Aircraft flown by Directflight Ltd. and managed by the Facility for Airborne Atmospheric Measurements (FAAM), which a joint entity of the Natural Environment Research Council (NERC) and the Met Office.

REFERENCES

[1] http://www.faam.ac.uk/

[2] Marenco, F., J. Kent, M. Adam, J. Buxmann, P. Francis and J. Haywood, 2015, Remote Sensing of Volcanic Ash at the Met Office, *27 ILRC* – Hazard Monitoring.

[3] Marenco, F., et al., 2011: Airborne lidar observations of the 2010 Eyjafjallajökull volcanic ash plume, *J. Geophys. Res.*, **116**, D00U05, doi:10.1029/2011JD016396.

[4] Adam, M., et al., 2007: Application of the Kano–Hamilton Multiangle Inversion Method in Clear Atmospheres, *J. Atmos. Oceanic Technol.*, **24**, 2014-2028.

DATA FLOW OF A MULTIPLE INSTRUMENT ON-DEMAND PROCESSING ENGINE WITH PYTHON AND DPLKIT

Joseph P. Garcia[1]*, Edwin Eloranta[1], Raymond K. Garcia[1]

[1]The University of Wisconsin – Madison, 1225 W. Dayton St., Madison, WI 53706, USA,

**Email: jpgarcia@lidar.ssec.wisc.edu*

ABSTRACT

The University of Wisconsin LIDAR Group's High Spectral Resolution LIDAR on-demand data website and processing codebase uses Python to explore coding techniques which facilitate a flexible codebase that is reusable for various outputs, cooperative multi-instrument products, and retains stability and maintainability without hindering dynamic experimentation.

1. INTRODUCTION

The University of Wisconsin LIDAR Group has implemented the High Spectral Resolution LIDAR on-demand data website and processing codebase using Python to be more easily deployable, extensible, and reusable. The processing engine uses a data flow programming toolkit and conventions, known as DPLKit, developed together with other atmospheric research programmers at the University of Wisconsin - Madison. These techniques introduce a natural data flow end-to-end, moving incrementally toward encapsulated data actors with simple interfaces which can be independently tested. The objects created can be rearranged and reused as needed, and allow a researcher to focus on coding just the module he/she wants, unencumbered by data intake or outflow concerns. The system already includes shared input code layers for ease in supporting multiple instrument sources in different formats, resampling filters for gridding multiple sources onto a single coordinate system, image creation, processed file creation, and can be plugged into an interactive UI for active data sources. These demonstrate ways to create modules to replace what would otherwise be single-use ancillary code. The website, hsrl.ssec.wisc.edu, uses networks of actors to process data on-demand from HSRL deployments since 2003. It also includes an expanding codebase for advancements in instrumentation and processing, such as integrating products from co-located instruments in order to create new innovative measurements.

2. UNDERSTANDING FRAME STREAMS.

Many traditional scientific processing engines operate on files directly, neglecting proper handling of file borders and I/O optimization until it has already infiltrated the science code. In order to separate these tasks while addressing both real-time and retrospective batch processing, a more generally versatile approach to the task interfaces is necessary. The fundamental idea of data flow with DPLKit (DPL stands for "Data Pot-Luck", after pot-luck parties at which guests each bring a dish to share) involves thinking of data as a unidirectional flow of measurement frames in time. I/O levels need only retrieve data in chronological order, presenting every time step once and only once. File borders disappear from functional code, leaving only more approachable and repeatable framing boundaries. The specific content of the frames themselves is flexible, and only relevant to the narrow requirements of the task consuming them. Any repeatable task operating on these frames may be broken out into its own "actor," which also permits the frames be from any actor capable of making compatible structure, whether or not the two actors are familiar to each other. Each actor in the pipeline advertises what its output frame will contain as a dictionary attribute, available to the immediate downstream actor at initialization time. In adapting existing code, the manner in which actors are chained together may start out as fixed, but will evolve into a more shared interface as new actors are dynamically added that plug in anywhere along the pipeline, without other actors' awareness.

The only necessary actors for any new data source are one to discover and catalog data URIs from a storage source, referred to as a Librarian, and one to convert URIs into frames of ordered data,

 DOI: 10.1051/epjconf/201611925003

referred to as a Zookeeper. The separation of these two tasks simplify both, since it allows them to focus on the filesystem and file content separately, and either can change without impacting the other.

3. USING PYTHON: ACTORS AS PYTHONIC GENERATORS

The natural implementation of repetitive processing is a for-loop, which for our purposes means using iterable objects. The fundamental idea here is that the object will read or synthesize a frame with each step, tracking its own state along the way. Far more intriguing, Python generators add elegance by creating an iterable object with little code overhead in what deceivingly looks like a stateless function. In Python, the simplest actor will simply iterate upon its source (upstream) actor, complete its task on each frame, and "yield" the product for the downstream actor. All state of the loop is maintained by simply being a generator.

Any parameters needed by the actor should be provided at initialization time. When an actor is initialized, its parameters and source actors' frame descriptions allow it to configure itself completely, as well as present expected errors at that moment instead of mid-operation. All runtime state is maintained within the actor as necessary, or within the actor's __iter__ function scope when operating as a generator.

Because iterators are allowed to operate in a disjoint rhythm, more complex actors may accumulate multiple frames to operate on each frame in the context of those in close proximity in time and space. These actors would appear to yield a time-deferred product when compared to the input frames, but this is not at all a problem so long as chronological order is maintained. Furthermore, an actor can accumulate and assemble multiple frames into a compound-frame or a single high-level observation as its product. For example, a scanning instrument can have one stream of native resolution data frames, usable for status content, while also feeding into a separate actor that stitches together whole scans into complete images.

4. MODULAR DESIGN BENEFITS

By separating tasks into different actors that flow data in one direction, operating in its own context, we have also effectively created an efficient computational pipeline. Minimization of side effects from cross talk and backwards flow between actors makes it more feasible to parallelize their tasks. Output of one actor can be diverted to another thread or process, or entirely separated within an actor. By integrating a method of transport and resource management, this becomes a clear direction for improved efficiency.

With each actor doing a very deliberate and finite task, generalized modules can be written with full focus on the purpose of that module. This avoids duplication and allows you to benefit from robust implementations, having only to provide the proper initialization to get correct results.

Building upon this, the actor can also operate by calling any separately developed code. Heritage code, library functions, and scientific code written by an expert can be interfaced with, as Python already has many community libraries and language bridges. Actors can serve as a sandbox for any environment or scientific workspace.

5. STEERING AND CONTAINING DIVERGENT CODE EVOLUTION

In the process of adapting and separating existing code into DPL constructs, discrete tasks will have a tendency to evolve in place. When these segments are found to be useful to more than one role, they can be extracted into library functions and used where needed. This allows existing modules to maintain the same stable and mature functionality, while new development code can reuse these functions, extending the API minimally to be more use-agnostic. This approach keeps dramatic changes localized without compromising core functionality. Properly allowing development and experimentation to independently diverge from stable code produces a much more approachable and versatile system as a whole. In time, the experimental code can stabilize and mature to contribute its own reusable functionality back into the stable system.

6. SEPARATION OF TASKS IN PRACTICE: HSRL

HSRL calibration routines evolved into four distinct actors: instrumentation constants, instrument calibration tables, atmospheric profiles, and computed calibration vectors. Instrumentation constants consist of simple values that describe the instrument, which can be necessary for proper handling of both raw and inverted HSRL data. Calibration tables are more complex tables generated from properties of the instrument. These are combined with the measured atmospheric profiles to compute the calibration vectors necessary for processing HSRL raw data. Atmospheric profiles initially came from RAOB radiosonde files, which have a specific actor for reading them into a stream of individual profiles. Later, a virtual radiosonde system (VRS) using the GFS archive and forecast data became available, so an actor that streams and adapts these profiles to resemble the radiosonde source was written. For a mobile platform HSRL, the VRS profiles need to track the instrument location. This location information is stored in the HSRL raw data files, so the HSRL's raw data stream actors are reused here, adapted to serve as location input for the VRS stream. From this, the code generating the calibration vectors gets a stream of atmospheric profiles at semi-regular intervals, determined by the VRS from either significant time or position changes, without any modifications in the calibration synthesis code.

In a lab environment, overriding the production default calibration task for experimental purposes is sometimes necessary. To facilitate this, an additional actor can be injected on demand between the deployable instrumentation calibration tables and the calibration vector computation. This actor can be configured to alter or completely replace tables in any way the scientist desires. Because this filter is simply omitted from the production data flow, the system doesn't suffer any additional complexity from unused non-production code, while allowing any means of modification useful to the scientist.

Downstream from the calibration vector calculation, an actor receives these vectors and resampled raw HSRL data, and is solely responsible for the inversion task.

After the HSRL processing itself, output frames can be streamed into an actor designed for batch plotting, storage, or a live UI. For housekeeping plots, only the raw data is needed, so processing actors can be omitted entirely in that case. The same actors can also be used for custom extraction tasks, selecting specific fields to be exported to a CSV file for one-off analysis, for example. With a live updating UI, the actors can be configured to continuously loop instead of terminate on the last slice. The processing code can operate on simply the newest slice of data, allowing it to be combined in another actor tasked with caching old data. The result is a rolling window of compound data, which can be connected to a constantly updating display. Accuracy of the content is ensured since all methods use the same common tested backend.

7. COOPERATIVE PRODUCTS

With multiple HSRLs deployed in different platforms, often with other instruments, it becomes extremely appealing to be able to integrate other datasets into the data flow. Already, the processing platform includes the ability to read processed data from several co-located radar systems for the purpose of re-gridding to a common set of axes, and creating convenient runtime observations that don't already exist in the source files (radar backscatter, for example). These can be used in combination with HSRL data to produce cooperative observations. The computation tasks are compartmentalized into actors that focus on only that task, without complications from disk access, data flow, resampling, or making any assumption about the where the data is to be used down-stream.

Currently, actors exist that combine HSRL and radar data to calculate particle measurements and rainfall, as well as actors that compare these results to measured observations from other sources. As more datasets are introduced to the processing system, additional actors can be written to produce new cooperative results in much the same style.

8. GRAPH DISPATCH ENGINE

As processing engines grow in size and complexity, connecting actors in an ad-hoc fashion becomes inefficient. Structural duplication of actors and resource use become a serious problem to scalability. To help control this, a reusable many-to-many graph construct is being

DOI: 10.1051/epjconf/201611925003

developed, where actors can be added and reused via branching in order to avoid multiple instances of the same actor. With further research, this implementation can use its awareness of the entire graph structure to configure, optimize, and dispatch a multi-core pipeline, allowing actors to operate in parallel, with data-flow between processing contexts managed intelligently.

In the process of adapting our actors to work robustly a graph solution, we did discover that actors should adhere to a set of expectations if the output is to be reliable. The most important is that frames and actor attributes must be immutable. What this means is that once an object is created, the exposed content and structure stays the same for the life of the object. Immutable frames allow any single frame object to be shared across many actors and contexts, without the ability of another actor to alter it unexpectedly. Immutable actor attributes are useful to actors downstream, as it allows them to acquire some parameters from upstream actors, making their initialization simpler and consistent. For example, our cooperative actor that operates with both HSRL and radar needs system parameters and constants that aren't part of the data stream itself. Actors tasked to each instrument exposes these parameters as attributes, becoming available down stream to any actor. Hypothetically, if these attributes are allowed to change, a parallel processing system would require complex synchronization between actors or suffer from unpredictability. Any actor's attributes that do need to change as the task progresses must be provided within its frame stream, or via a separate sparse frame stream and actor. The latter is more appealing, as the attributes usually constitute useful meta-data that shouldn't depend upon other complex tasks to acquire. For HSRL, this is why the instrumentation constants became it's own stream, available to the raw data and the calibration tables, without depending upon either. With this layout, dynamic parameters are available to any actor, allowing the parameters to be used independently, or synchronized properly to another dependent data stream.

ACKNOWLEDGEMENT

This work was partially supported by NSF Grant# ARC-0946359.

A NEW APPROACH TO INVERTING AND DE-NOISING BACKSCATTER FROM LIDAR OBSERVATIONS

Willem Marais[1*], Yu Hen Hu[1], Robert Holz[2], Edwin Eloranta[3]
[1] *Department of Electrical and Computer Engineering, University of Wisconsin-Madison, USA*
[2] *Cooperative Institute for Meteorological Satellite Studies, University of Wisconsin-Madison, USA*
[3] *Space Science Engineering Center, University of Wisconsin-Madison, USA*
* *Email:* `Willem.Marais@ssec.wisc.edu`

ABSTRACT
Atmospheric lidar observations provide a unique capability to directly observe the vertical profile of cloud and aerosol scattering properties and have proven to be an important capability for the atmospheric science community. For this reason NASA and ESA have put a major emphasis on developing both space and ground based lidar instruments. Measurement noise (solar background and detector noise) has proven to be a significant limitation and is typically reduced by temporal and vertical averaging. This approach has significant limitations as it results in significant reduction in the spatial information and can introduce biases due to the non-linear relationship between the signal and the retrieved scattering properties. This paper investigates a new approach to de-noising and retrieving cloud and aerosol backscatter properties from lidar observations that leverages a technique developed for medical imaging to de-blur and de-noise images; the accuracy is defined as the error between the true and inverted photon rates. Hence non-linear bias errors can be mitigated and spatial information can be preserved.

1. INTRODUCTION

The scatter properties of cloud and aerosol particulates can be inverted through an atmospheric lidar system, and are of great use to the atmospheric science community. NASA has invested in several lidar instruments, eg. CALIPSO, CPL and CATS. And in the near future ESA will launch EarthCare, which will have an onboard HSRL (High Spectral Resolution Lidar) instrument. This paper mainly focusses on ground based photon counting HSRL observations, with the intent of extending the presented technique for both single and multi channel ground and space based lidar instruments.

The inversion of the desired scatter properties are hindered by solar background and detector noise. Hence accurate backscatter estimates are important when these are used to constrain the inversion of the scattering volume coefficient and the backscatter phase function. The backscatter volume coefficient - or backscatter in short, is one of the scatter properties that can be inverted from HSRL photon rate observations, without making a priori assumptions about the backscatter phase function. Photon counting systems inadvertently introduce noise in the measured photon rates. This is due to the stochastic properties of photon counting detectors, such as PMTs and APDs, and the noise can be described by a Poisson point process.

The most common method to de-noise lidar observations is to simply average, which reduces the noise variance. Averaging is well suited for scenes where aerosol and cloud particulates are homogenous over horizontal layers. But there are two significant drawbacks to this approach. First, nonlinear bias errors are introduced into the retrievals whenever averaging is performed over a region of a cloud which is non-homogenous. This is especially true when the scattering volume coefficient change several orders in magnitude over a short period of time. The second drawback is the loss of spatial information. In this paper we investigate a new approach, a PMLE (Penalized Maximum Likelihood Estimator) technique, which has being used in de-noising and de-blurring medical images.

The image to the left of figure 1 shows a noisy (1 profile) and averaged (40 profiles) retrieved backscatter of a cloudy scene. Between ranges 5 and 6 km the averaged retrieval overestimate the backscatter, and again at 7.2 km. The image to the right of figure 1 shows the molecular and combined

This work was funded by NASA under grant number NNX13AO09G.

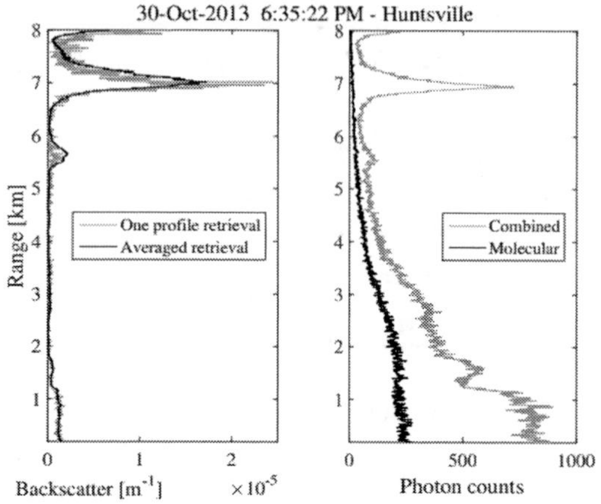

Fig. 1: The leftmost image demonstrates potential biases in the averaged backscatter volume coefficient retrieval, by comparing it to a single profile retrieval. The rightmost image shows the photon rates of the combined and molecular channels that were used for the one profile retrieval.

channel photon counts that were used to retrieve the single profile backscatter; refer to §2 for more information about these two channels. The observations were made with the Madison-WI HSRL system [2].

The PMLE technique has been used in medical imaging systems, such as confocal microscopy imaging, to de-blur and de-noise Poisson-noisy medical images [3]. As with photon counting lidar systems, some medical imaging systems also use PMTs or APDs to measure incoming photon rates. Statistical inversion techniques developed for medical imaging can also be employed in the context of photon counting lidar systems.

This paper makes a contribution by demonstrating that an accurate backscatter profile, relative to the standard retrieval, can be inverted from HSRL measured photon rates. The error criteria which is used to claim the improved accuracy is defined in §3. Both qualitative and quantitative validation approaches are used in this paper. For the qualitative approach a homogenous scene is used to estimate the backscatter, and it is compared against an averaged retrieval of the backscatter. And for the quantitative approach a simulation is used to quantify the performance of the backscatter estimates.

2. Problem formulation

A HSRL measure the photon rates with two distinct channels, which can be used to estimate the unknown backscatter [1]. With the molecular channel a HSRL takes advantage of molecular doppler broadening by attenuating the particulate contribution, and in effect measure the photon rate due to the molecular backscatter, with the molecular and particulate transmittances. Let $\beta \in \mathbb{R}_+^n$ denote the particulate scatter volume coefficient vector, where each vector component corresponds to a range bin; the symbol $\mathbb{R}_+$ denotes non-negative real numbers. Let $\alpha \in [0,1]^n$ denote the backscatter phase function vector. The discretized single scatter lidar equation for the molecular channel is

$$S_m = x \cdot (\alpha \cdot \beta \cdot \theta_m + \phi_m) \cdot \exp(-2Q\beta) + b_m,$$

where $b_m \in \mathbb{R}_+^n$ is the background photon rate, $Q \in \mathbb{R}_+^{n \times n}$ is a discrete integral operator, $\phi_m \in \mathbb{R}_+^n$ represent both the molecular backscatter and its calibration term. The calibration term is due to - for example - the iodine filter spectral shape and other optical devices [1]. The factor $\theta_m \in [0,1]$ represent the attenuation due to the iodine filter. The vector $x \in \mathbb{R}_+^n$ includes all the other parameters such as the reciprocal of the squared range, the molecular transmittance, the area of the telescope, the number of photons transmitted, etc. The symbol $\cdot$ is the point-wise multiplication of the vectors.

The photon rate of the second channel, called the combined channel, is where the particulate backscatter is not attenuated. Its single scatter lidar equation similar to that of the molecular channel.

To estimate the backscatter we can rewrite the single scatter lidar equation in a different form. Let $\iota \in \{c, m\}$ be an index holder for the molecular and combined channels. An alternative form of the single scatter lidar equation is

$$S_\iota = x \cdot (\exp(\chi) \cdot \theta_\iota + \exp(\psi) \cdot \phi_\iota) + b_\iota,$$

where $\chi \in \mathbb{R}^n$ and $\psi \in \mathbb{R}^n$. We have that

$$\exp(\chi) = \alpha \cdot \beta \cdot \exp(-2Q\beta) \text{ and } \psi = -2Q\beta.$$

The backscatter can be recovered by $\exp(\chi - \psi) = \alpha \cdot \beta$.

2.1 THE PMLE TECHNIQUE

The noise can be modeled by a Poisson PMF (Probability Mass Function), and with this we can use the MLE technique in conjunction with the single scatter lidar equations to infer the particulate backscatter $(\alpha \cdot \beta)$. This amounts to minimizing a loss function, where the minimizer of the loss function represent the estimate which best "fits" the observe photon rates. The MLE technique - in our case - is statistically ill-posed and the estimation has to be constrained; this is achieved by introducing a penalty term. For example with "optimal estimation" the loss function is the negative logarithm of the Gaussian probability density function, and the penalty term is the Euclidean norm of the parameter being estimated.

Using the rearranged single scatter lidar equation, the estimation of χ and ψ is formulated as

$$(1) \quad (\hat{\chi}_\tau, \hat{\psi}_\tau) = \underset{(\chi,\psi)\in\mathcal{B}}{\operatorname{argmin}}\, l(\chi,\psi) + \tau \mathrm{TV}(\chi,\psi),$$

where $\mathcal{B} \subset \mathbb{R}^n \times \mathbb{R}^n$ is a closed set. The function $l(\chi,\psi)$ is the MLE loss function; it is a function of the photon counts (not shown) and the parameters we want to estimate. The function TV (Total Variation) is the penalty function and the parameter $\tau \in \mathbb{R}_+$ is the regularizer parameter; these are discussed in the next paragraph. The estimates of χ and ψ are expressed as $\hat{\chi}_\tau$ and $\hat{\psi}_\tau$, where each one is function of regularizer parameter τ.

A reasonable penalty function, that has been used in medical image de-blurring and de-noising, is the so called TV penalty [3]. It is the absolute value norm of the discrete derivative. With a large regularizer parameter ($\tau \gg 1$) the discrete derivative is constrained to be small, hence the parameter being penalized is constrained to be smooth. If the regularizer parameter is close to being zero, the discrete derivative is not constrained to be small and penalized parameter is therefore not necessarily smooth. The advantages of using the TV penalty, is 1) that allows discontinuities in the parameter being estimated and 2) it is convex.

To thus far we found - empirically - that the loss function with the rearranged model has one solution, with both simulated data and actual HSRL observations.

3. METHODOLOGY

A variant of the SpaRSA algorithm called SPIRAL is used to find $(\hat{\chi}_\tau, \hat{\psi}_\tau)$ via (1) [3]. The algorithm uses sequential quadratic approximations of the loss function to find a minimizer. In each sequence of the quadratic approximation a sub-optimization problem is solved. The sub-optimization problem is the linear combination of the 2nd order Taylor series approximate of the loss function and the TV penalty function. The parameters are constrained on closed set, thus the algorithm will converge to a minimizer.

A model selection technique is be used to choose the optimum regularizer parameter τ [3, p. 162]. This is achieved by splitting the photon counts profile into two profiles using a technique called "Poisson thinning". The first profile is used to find $(\hat{\chi}_\tau, \hat{\psi}_\tau)$ for every τ, and the second profile is used to find the τ by minimizing a test loss function. It can be shown experimentally that with the described technique minimizes the error criterion

$$(2) \quad \left\| \sqrt{\hat{S}_c} - \sqrt{S_c^*} \right\|_2^2 + \left\| \sqrt{\hat{S}_m} - \sqrt{S_m^*} \right\|_2^2,$$

where S_c^* and S_m^* are the true photon rates of the corresponding channels. The error criteria (2) gives an indication of how "close" the inverted photon rates are relative to the true photon rates.

4. RESULTS AND CONCLUSION

Both qualitative and quantitative approaches are used to validate backscatter volume coefficient estimates. The qualitative approach is done using HSRL observation of the Madison-WI HSRL system. The quantitative approach is based on simulated HSRL observations.

For the qualitative validation approach two cases are considered, where a homogenous daytime scene is used. In the first case a single profile is used to compare the backscatter retrieval and inversion. And in the second case 100 accumulated profiles are used to compare the backscatter retrieval and inversion. For the second case intuitively the backscatter retrieval should be accurate, since the noise variance decrease relatively to the number of samples using for averaging.

The leftmost image of figure 2 shows the inverted and retrieve backscatter of the first case. The chosen scene has fairly homogenous horizontal layers between ranges 0 and 3 km. The rightmost image of figure 2 shows the inverted and retrieved backscatter of the second case. In this case both

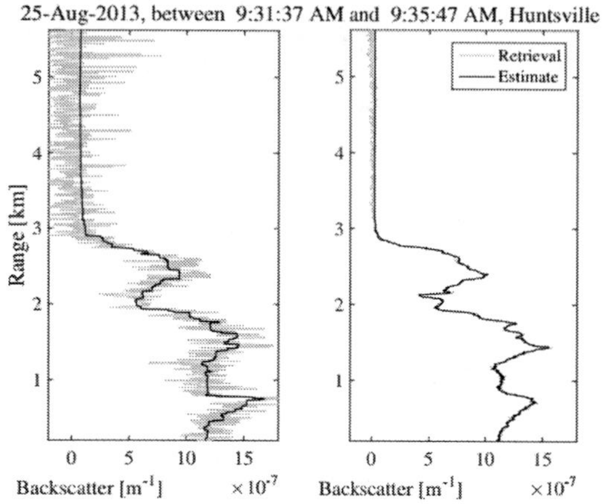

Fig. 2: The leftmost image compares the backscatter retrieval and estimate of one profile. In the rightmost image shows the retrieval and estimate of 100 accumulated profiles.

estimates of the backscatter are nearly in full agreement between ranges 0 and 3 km, while the inverted backscatter is overestimating the backscatter between ranges 3 and 5.6 km.

For the quantitative validation approach the simulated HSRL profile has an aerosol layer similar to figure 2. The dotted curves in figure 3 shows the true backscatter that is used in the simulated HSRL observations. The continuous curve of the left images shows an example of inverted backscatter profile, and the continuous curve of the right image is a retrieved backscatter profile. In order to evaluate the bias and variance of the backscatter estimates, the simulated HSRL profile is recreated several times and of each the backscatter is inverted and retrieved. The light filled curve in the left most image of figure 3 shows the upper and lower bounds of the inverted backscatter; 18 recreated HSRL simulations are used in this case. The image to the right of figure 3 shows the corresponding upper and lower bounds of the retrieved backscatter.

From figure 3 we deduce that the inverted

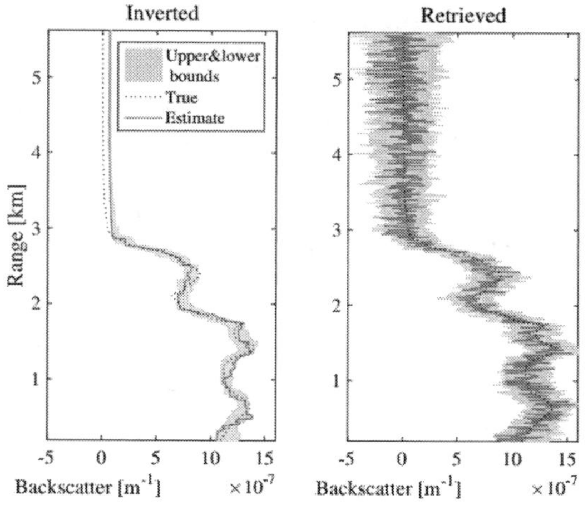

Fig. 3: The leftmost image shows inverted backscatter profile, in conjunction with the true backscatter profile. The rightmost image shows correpsonding retrieved backscatter profile.

backscatter profile is somewhat biased in comparison to the retrieved backscatter profile. But variance of inverted backscatter profile is significantly smaller than retrieved backscatter profile variance. The mean error in the simulations, according to the definition (2), of the proposed estimation procedure is 54.48. And mean error using standard averaging approach is 769.45. The results indicate that the potential exists for inverting the backscatter volume coefficient reliably, using one or very few profiles, with HSRL observations. Further research is required to 1) derive confidence bounds of the backscatter estimate and 2) determine how to extend the technique for single channel lidar systems.

References

[1] E. Eloranta, "High Spectral Resolution Lidar Measurements of Atmospheric Extinction: Progress and Challenges," in *Aerospace Conference, 2014 IEEE*. IEEE, 2014, pp. 1–6.

[2] ——. (2014, 10) http://hsrl.ssec.wisc.edu. [Online]. Available: http://hsrl.ssec.wisc.edu

[3] Z. T. Harmany, "Computational optical imaging systems: Sensing strategies, optimization methods, and performance bounds," Ph.D. dissertation, Duke University, 2012.

EPJ Web of Conferences 119, 25005 (2016)
ILRC 27
DOI: 10.1051/epjconf/201611925005

A micropulse eye-safe all-fiber molecular backscatter coherent temperature lidar

*Cyrus F. Abari[1,3,4], Xinzhao Chu[1,2], Jakob Mann[4], and Scott Spuler[3]

[1] Cooperative Institute for Research in Environmental Sciences, University of Colorado, Boulder, Colorado, USA
[2] Department of Aerospace Engineering Sciences, University of Colorado Boulder, Boulder, Colorado, USA
[3] National Center for Atmospheric Research, Earth Observing Lab, Boulder, CO, USA
[4] Department of Wind Energy, Technical University of Denmark, Roskilde, Denmark
* Email: cyrus.abari@gmail.com

Abstract—**In this paper, we analyze the performance of an all-fiber, micropulse, 1.5 μm coherent lidar for remote sensing of atmospheric temperature. The proposed system benefits from the recent advances in optics/electronics technology, especially an all-fiber image-reject homodyne receiver, where a high resolution spectrum in the baseband can be acquired. Due to the presence of a structured spectra resulting from the spontaneous Rayleigh-Brillouine scattering, associated with the relevant operating regimes, an accurate estimation of the temperature can be carried out. One of the main advantages of this system is the removal of the contaminating Mie backscatter signal by electronic filters at the baseband (before signal conditioning and amplification). The paper presents the basic concepts as well as a Monte-Carlo system simulation as the proof of concept.**

I. INTRODUCTION

Continuous high-resolution observation of atmospheric temperature profile in the troposphere is crucial for improved weather forecasting at the mesoscale [1]. In wind energy industry, where the latest turbines have reached heights beyond 200 m, remote sensing of temperature has become compelling. At these heights, wind shear, veer, and turbulence are highly sensitive to the stratification due to temperature [2]. Furthermore, profiling horizontal variations of temperature is important for the flow characterization on varying terrain.

Lidars can provide high spatial and temporal resolution monitoring of thermodynamic variables in the atmosphere. The majority of existing temperature measurement lidars benefit from the direct detection principle. In these systems sub-micron wavelengths are employed to take advantage of a stronger Rayleigh backscatter, where $\beta \propto \lambda^{-4}$ (β is the molecular backscatter cross-section and λ is the wavelength). There are a few disadvantages associated with such systems, e.g., they require high-power lasers which can be costly, bulky, and require high-maintenance. The transceivers in such systems can also be more costly and complicated to design and implement.

Coherent Doppler lidars (CDL), relying on Mie scattering, have been successfully used for remote sensing of wind [3], [4], [5]. Recent advances in fiber-optic technology and electronics have paved the way for a wide range of robust, compact, and cost-effective lasers and fiber-optic components. Successful realization of all-fiber CDLs operating at 1.5 μm wavelength has already been well established, e.g., see [4], [6]. Although the existing CDLs rely on Mie scattering to capture the wind-induced Doppler spectrum, modified systems should be capable of recording the thermally broadened (Doppler) spectrum from the molecular backscatter signal [7].

Historically, a few pitfalls have prevented the realization of a molecular coherent temperature lidar (CTL). The Rayleigh backscatter signal is weak and has a thermally broadened Doppler spectrum with a relatively wide bandwidth (BW), when compared to the narrowband Mie backscatter. The BW scales inversely with λ and is outside the detection BW of the available photodetector (PD) technology for sub-micron wavelengths. Only recently, PDs and fast analogue to digital converters (ADC) have become available with detection BWs that are comparable with the BW of the Rayleigh backscatter Doppler signal (for longer wavelengths in the infrared region such as 1.5 μm). Optical image-reject down-conversion of Doppler signals to baseband, essential for resolving positive and negative frequency shifts, has not been practically demonstrated, viz., not until recently [8]. This is an essential part of the proposed CTL to circumvent the restrictions imposed by the available BW of ADCs and PDs. The alternative, and conventionally employed, down-conversion to intermediate frequency (IF) band, places unrealistic requirements on the BW of the PDs and ADCs. A successful prototyping of an all-fiber image-reject homodyne receiver demonstrated by Abari et al. [8] facilitates the realization of a molecular backscatter CTL as suggested in this paper.

DOI: 10.1051/epjconf/201611925005

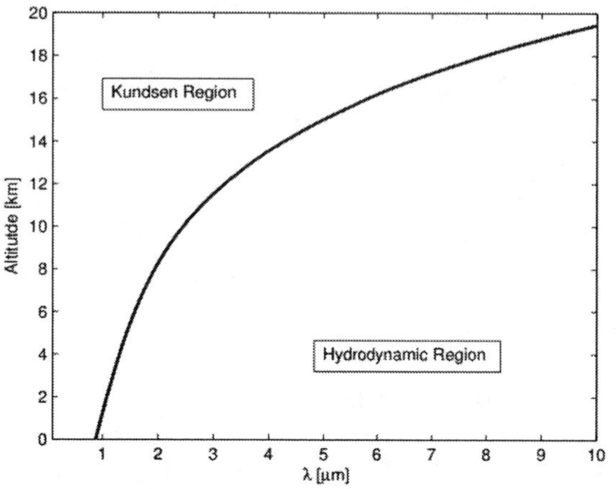

Fig. 1: Transition from the Kundsen to hydrodynamic region [7]. The region below the curve represents the area where the wavelength is longer than the mean free-path in the atmosphere, hence, SRB backscattering.

II. Spontaneous Rayleigh-Brillouin (SRB) Backscattering and System Description

A theoretical feasibility study of a molecular backscatter CTL was carried out by Rye [7] for a few wavelengths (i.e., 350, 2100, and 10600 nm). The results indicate that the longer wavelengths, in the infrared (IR) spectrum, may operate in the hydrodynamic region, where λ is larger than the mean free path in the air [9], [7]. In this region the return signal is not purely Rayleigh; The return spectra has contributions from the Rayleigh and spontaneous Brillouin backscatter signal, known as spontaneous Rayleigh-Brillouin (SRB), resulting in a triplet of peaks in the spectra. Fig. 1 shows the transition from the Kundsen (where the backscatter signal is largely dominated by the Rayeligh backscatter) to hydrodynamic region for different wavelengths and altitudes. As a result, due to the presence of a structured SRB backscatter Doppler spectrum in the hydrodynamic region [10] as well as the inherent CTL detection sensitivity, accurate estimation of temperature from a weak molecular backscatter is possible; The presence of narrowband Brillouin peaks provide a reliable measure of the atmospheric temperature [7]. As we will see in Sec. III, we have resorted to a least squares (LS) curve fitting between the estimated and expected spectra to estimate the temperature. The added benefit of LS curve fitting is the potential for extracting other parameters (e.g., atmospheric pressure) from the estimated spectra [7].

Following Tenti's S6 theory [9], the spectra associated with the SRB backscatter for $\lambda = 1560$ nm (at two different heights in reference to sea level) has been calculated and shown in Fig. 2. As the atmospheric pressure decreases (i.e., the altitude increases) the contribution from the Brillouin effect is reduced (as the mean free-path between the molecules becomes comparable with the laser wavelength) and the Rayleigh backscatter becomes more dominant. This is why the Brillouin peaks associated with the estimated spectra at an altitude of $h = 4$ km shown in Fig 2 are less pronounced when compared to the spectra estimated at $h = 1$ km.

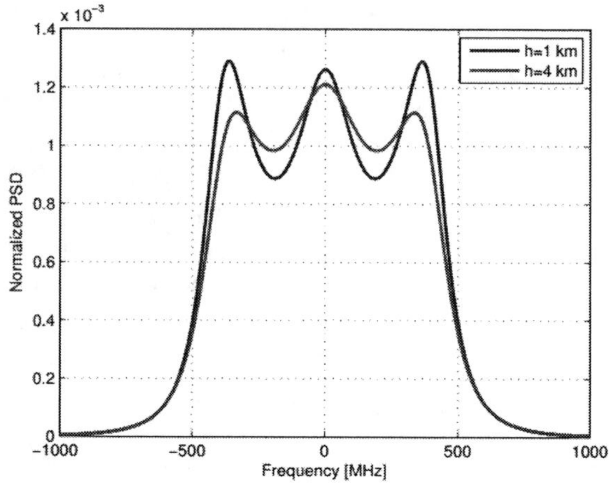

Fig. 2: SRB spectrum for $\lambda = 1560$ nm at two different altitudes in reference to sea level. The data is taken from the U.S. Standard Atmosphere Model.

In light of Rye's [7] results, we have shown that it is possible to use a micropulse 1.5μm CTL to carry out an accurate remote sensing of temperature in the lower atmosphere. Although the micropulse system cannot rival the solid-state or gas-based IR laser, in terms of the maximum available pulse energy, it provides comparable average output powers. In other words, the integration time is competitive with the high power IR systems analyzed in [7]. High-quality all-fiber 1.5 μm pulsed erbium doped fiber amplifiers (EDFA) with a PRR of 10-20 kHz and pulse energy of 110-220 μJ (average optical power of 2.2 W) are available as off-the-shelf components [11] and have been successfully integrated in commercially available all-fiber wind CDLs [6]. More powerful multi-stage EDFAs, benefiting from large core optical fibers, can be developed and integrated into the micropulse system to improve the range and required integration time.

Fig. 3 shows the schematic of the proposed all-fiber CTL. This system benefits from an all-fiber image-reject homodyne receiver [8] which allows the translation of the frequency-shifted signals into the baseband. Due to a relatively narrow BW associated with the 1560

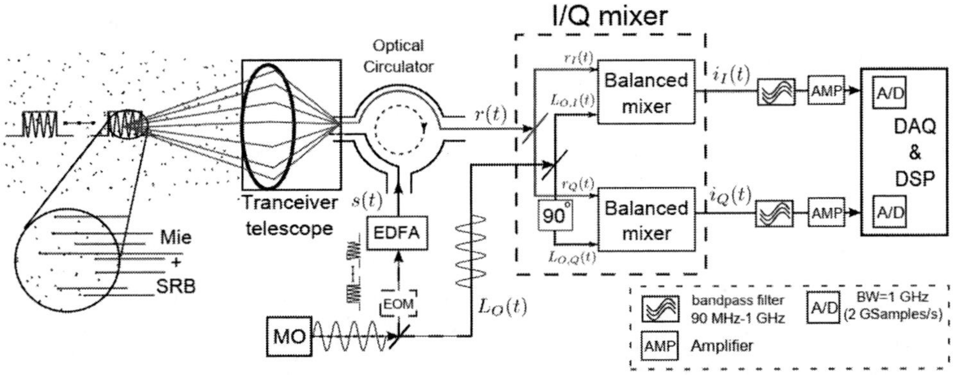

Fig. 3: An all-fiber pulsed image-reject homodybe CTL. The master oscillator (MO) provides a highly stable and narrow linewidth laser which is split into (by a beam splitter) a local oscillator (LO) and input signal to the electro-optic modulator (EOM). The EOM can be programmed to generate the desired pulse train which is amplified by the EDFA. The optical circulator isolates the transmit and receive signals. The in-phase/quadrature-phase (I/Q) mixer translates the spectral information into baseband for further processing. For a detailed analysis of the image-reject homodyne receiver please refer to Abari, et al. [8]. The bandpass filters, designed for a maximum expected mean wind speed, filter out the Mie signal.

nm molecular backscatter signal, a PD with a BW of 1 GHz and two ADCs with a sampling rate of 2 GSamples/s provide the opportunity to resolve the spectral components, which upon further processing and digital signal processing allow the estimation of the relevant parameters such as temperature. A common problem associated with the detection and characterization of the molecular backscatter signal in the lower atmosphere is a strong contamination resulting from the Mie backscattering which dominates the return signal. In direct detection techniques, such as high spectral resolution lidars (HSRL), optical filters, only available for a limited number of wavelengths, are employed to filter out the Mie component from the return signal. In the proposed system, we suggest the removal of the Mie backscatter directly at the output of the PD. If the dominant Mie component is not removed immediately before amplification and conditioning of the detected signal, it will either saturate or occupy the dynamic range of the following stages, i.e., the electronic components adopted for signal amplification, conditioning, and sampling. One of the main merits of this system is to remove the Mie component through reliable, cost-effective, and accurate electronic filters. In this system the electronic filter BW can be designed for the expected maximum mean wind speed. Moreover, due to the nature of coherent detection, background light removal, necessary in direct detection systems, is not required in the proposed system.

III. Simulation Results

The performance of the system suggested in this paper is evaluated through modeling and Monte-Carlo

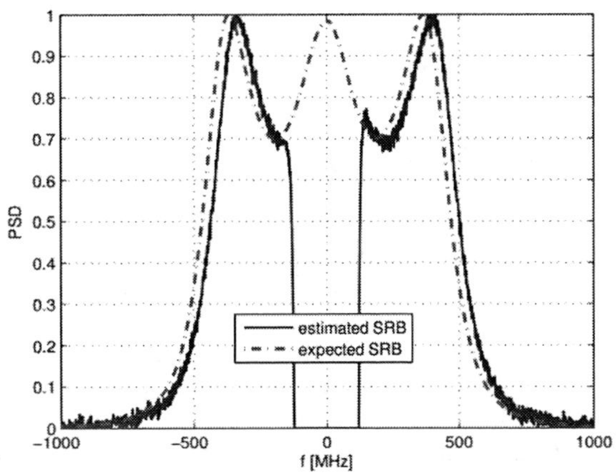

Fig. 4: The simulated return spectrum (target range of 1 km, please see Table I). The observed mean frequency shift in the blue curve is due to a presumed mean wind speed of 25 ms^{-1}.

simulations performed in Matlab. In our simulation the combined effects of Mie and molecular backscattering are taken into account. Since the aerosol type and density in the atmosphere is hard to predict and model, we have assumed a moderate loading of aerosols [12] with uniform scattering throughout the measurement range. The Rayleigh backscatter parameters have been adopted from the U.S. Standard Atmospheric Model. Furthermore, the Monte-Carlo simulations have been done for a CTL located 1 km above the sea level performing a horizontal scan. The simulation parameters are listed in Table I. Please note that β_{Mie} and D represent Mie volume backscatter coefficient and the telescope diameter, respectively.

TABLE I: Simulation parameters

E_t [μJ]	220	PRR [kHz]	10
PD BW [GHz]	1	f_s [GHz]	2
pulse length [ns]	400	resolution [m]	95
integration time [s]	60	N [FFT points]	1024
D [cm]	20	$\beta_{\text{Mie}}[\text{m}^{-1}\text{sr}^{-1}]$	1.83×10^{-7}

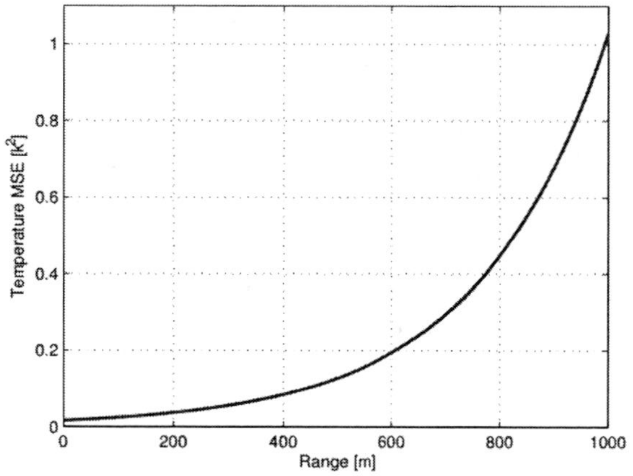

Fig. 5: MSE (variance) of the temperature estimator (for a temporal resolution of 60 s) versus target range.

For the estimation of the temperature we have resorted to an LS estimator in this paper. The estimator employs a LS fitting of the acquired spectra; An iterative procedure where the SRB curve resulting in the LS error between the acquired spectra and the curve is selected as the best estimate. For the performance characterization of our system, we have adopted the mean square error (MSE) of the temperature estimator for different ranges; The MSE can be viewed as the variance of the estimator. Fig. 5 shows the MSE of the temperature estimator adopted in this paper for different ranges. As we can see the estimator provides an estimate of the temperature with a variance better than 1 K for a measurement range of up to 1 km. It can be shown that the measurement range can be improved by increasing the pulse energy or integration time. Due to the availability of high speed computers and digital signal processors, more advanced signal processing can be adopted to improve the temperature estimation accuracy which can result in a longer measurement range for a given system configuration and atmospheric condition.

IV. CONCLUSIONS

In this paper, we have shown that in light of recent advances in fiber-optic and electronic technology it is feasible to build a compact and robust coherent lidar for remote sensing of atmospheric temperature. Compared to other lidars, the proposed system should be low-cost, low-maintenance, and can be deployed as an eye-safe system (in the same fashion 1.5μm wind Doppler lidars with similar optical output powers operate, e.g., a scanning CTL). Due to it's form factor the system can be deployed as ground-base or airborne. Furthermore, the system has the potential to provide simultaneous measurements of wind and pressure (besides temperature); This will be investigated in our future work. We have adopted only one simple estimator, i.e., LS, to demonstrate the feasibility of the proposed CTL. However, we believe there is room for improved estimation algorithms where the Cramer-Rao lower bound (CRLB), computationally calculated in [7], can be achieved.

REFERENCES

[1] Committee on Developing Mesoscale Meteorological Observational Capabilities to Meet Multiple Needs, National Research Council, *Observing Weather and Climate from the Ground Up: A Nationwide Network of Networks*. Washington, D.C.: The National Academies Press, 2009.

[2] A. Sathe, J. Mann, T. Barlas, W. Bierbooms, and G. van Bussel, "Influence of atmospheric stability on wind turbine loads," *Wind Energy*, vol. 16, no. 7, pp. 1013–1032, 2013.

[3] F. Hall, R. Huffaker, R. Hardesty, M. Jackson, T. Lawrence, M. Post, R. Richter, and B. Weber, "Wind measurement accuracy of the NOAA pulsed infrared Doppler lidar," *Appl. Opt.*, vol. 23, no. 15, pp. 2503–2506, 1984.

[4] S. M. Spuler, D. Richter, M. P. Spowart, and K. Rieken, "Optical fiber-based laser remote sensor for airborne measurement of wind velocity and turbulence," *Appl. Opt.*, vol. 50, pp. 842–851, 2011.

[5] R. M. Huffaker and R. M. Hardesty, "Remote sensing of atmospheric wind velocities using solid-state and CO_2 coherent laser systems," *Proc. IEEE*, vol. 84, pp. 181–204, 1996.

[6] WINDCUBE V2, a 200m vertical wind Doppler lidar. [Online]. Available: http://www.leosphere.com/products/vertical-profiling/windcube-v2. [Accessed Feb. 6, 2015].

[7] B. J. Rye, "Molecular backscatter heterodyne lidar: a computational evaluation," *Appl. Opt.*, vol. 37, no. 27, pp. 6321–6328, 1998.

[8] C. F. Abari, A. T. Pedersen, and J. Mann, "An all-fiber image-reject homodyne coherent Doppler wind lidar," *Opt. Express*, vol. 22, pp. 25 880–25 894, 2014.

[9] C. D. Boley, R. C. Desai, and G. Tenti, "Kinetic models and Brillouin scattering in a Molecular Gas," *Canadian Journal of Physics*, vol. 50, no. 18, pp. 2158–2173, 1972.

[10] Z. Gu, B. Witschas, W. van de Water, and W. Ubachs, "Rayleigh-Brillouin scattering profiles of air at different temperatures and pressures," *Appl. Opt.*, vol. 52, no. 19, pp. 4640–4651, 2013.

[11] 1.5 μm LONG PULSE FIBER AMPLIFIER, (Keopsys, 2015). [Online]. Available: http://www.keopsys.com

[12] V. Srivastava, J. Rothermel, A. D. Clarke, J. D. Spinhirne, R. T. Menzies, D. R. Cutten, M. A. Jarzembski, D. A. Bowdle, and E. W. McCaul, "Wavelength dependence of backscatter by use of aerosol microphysics and lidar data sets: application to 2.1-μm wavelength for space-based and airborne lidars," *Appl. Opt.*, vol. 40, pp. 4759–4769, 2001.

CHARACTERIZATION OF WATER VAPOR FLUXES BY THE RAMAN LIDAR SYSTEM BASIL AND THE UNIVERISTY OF COLOGNE WIND LIDAR IN THE FRAME OF THE HD(CP)^2 OBSERVATIONAL PROTOTYPE EXPERIMENT – HOPE

Paolo Di Girolamo[1]*, Donato Summa[1], Dario Stelitano[1], Marco Cacciani[2], Andrea Scoccione[2], Jan H. Schween[3]

[1]*Scuola di Ingegneria, Università degli Studi della Basilicata, Viale dell'Ateneo Lucano n. 10, 85100 Potenza – Italy , *Email: digirolamo@unibas.it*

[2]*Dipartimento di Fisica, Università di Roma "Sapienza", Piazzale Aldo Moro, n. 2, 00100 Roma – Italy*

[3]*Institut fuer Geophysik und Meteorologie, Universität zu Köln, Pohligstrasse 3, 50969 Köln – Germany*

ABSTRACT

Measurements carried out by the Raman lidar system *BASIL* and the University of Cologne wind lidar are reported to demonstrate the capability of these instruments to characterize water vapour fluxes within the Convective Boundary Layer (CBL). In order to determine the water vapour flux vertical profiles, high resolution water vapour and vertical wind speed measurements, with a temporal resolution of 1 sec and a vertical resolution of 15-90, are considered.

Measurements of water vapour flux profiles are based on the application of covariance approach to the water vapour mixing ratio and vertical wind speed time series. The algorithms are applied to a case study (IOP 11, 04 May 2013) from the HD(CP)2 Observational Prototype Experiment (HOPE), held in Central Germany in the spring 2013. For this case study, the water vapour flux profile is characterized by increasing values throughout the CBL with lager values (around 0.1 g/kg m/s) in the entrainment region. The noise errors are demonstrated to be small enough to allow the derivation of water vapour flux profiles with sufficient accuracy.

1. INTRODUCTION

Measurements of water vapour flux profiles provide unique and essential information for the characterization of turbulent processes within the convective boundary layer (CBL). Evaporation and horizontal/vertical humidity transport are key components of the hydrological cycle which are still poorly modelled and observed. Most models parameterize convective transport at sub-grid scale, while operational water vapour observations used for this purpose are inaccurate due to instrumental limitations [1].

Additionally, ground-based *in situ* measurements of water vapour and latent heat fluxes, while being difficult to perform, also have a limited significance because they are representative for a given area. Additionally, ground-based *in situ* measurements observe only the lowest part of the CBL and are influenced by local surface properties.

Lidar systems, based on their capability to provide high space and time resolution and accurate measurements of atmospheric water vapour and vertical wind speed, have nowadays reached the level of maturity needed to measure vertical profiles of water vapour and latent heat fluxes [2][3][4]. The major advantage of the lidar techniques is represented by its capability to characterize turbulent variables from the proximity of the surface up to interfacial layer and above.

In the present work we report measurements of the water vapour flux profiles within the Convective Boundary Layer (CBL) obtained from the Raman lidar system *BASIL* and the University of Cologne wind lidar.

2. METHODOLOGY

BASIL is a ground-based Raman Lidar hosted in a transportable sea-tainer. The major feature of *BASIL* is represented by its capability to perform high-resolution and accurate measurements of atmospheric temperature and water vapour, both in daytime and night-time, based on the application of the rotational and vibrational Raman lidar techniques in the UV [5],[6]. Besides temperature and water vapour, *BASIL* is also capable of providing measurements of particle backscatter at 355, 532 and 1064 nm, particle extinction at 355 and 532 nm and particle depolarization at 355 and 532 nm ([7]).

The University of Cologne wind lidar is 1.5-µm Doppler lidar produced by HALO Photonics [8] capable of measuring along-beam wind speeds in any arbitrary direction typically within the range of the atmospheric boundary layer (starting 75 m above the ground). Profiles of horizontal wind speed and direction are determined every hour by means of a VAD scan, the rest of the time is used for vertical wind (w) profiling.

The accuracies of the measurements performed by BASIL and the University of Cologne wind lidar are high enough to allow the retrieval of water vapour flux profiles throughout the atmospheric CBL.

Water vapour flux profiles Φ are obtained as the covariance of the vertical velocity and water vapour mixing ratio fluctuations:

$$\Phi = \langle w'.q' \rangle$$

with w' being the vertical velocity fluctuations and q' being the water vapour mixing ratio fluctuations.

3. RESULTS

In the selection of the case study to be included to this paper, attention was paid on identifying weather conditions characterized by the presence of a consolidated, well-mixed and quasi-stationary CBL. Typically time segments with a duration of 1-2 hours are considered as in fact for longer periods the CBL can no longer be considered as being quasi-stationary, while the consideration of shorter periods would reduce the number of sampled thermals and thus increase the sampling errors.

Figure 1 illustrates the time-height cross section of the vertical wind speed (upper panel) and water vapour mixing ratio (lower panel) in the time interval 14-18 UTC on 04 May 2013. The figure reveals the alternation of updrafts associated with thermals of warm air rising from the ground and downdrafts associated with thermals of cool air sinking from the free troposphere, with the stronger updrafts in the time interval 14:15-15:30 UTC and weaker updrafts afterwards. The largest variability is observed in the interfacial layer, which is characterized by the penetration of the warm humid air rising from the ground and the entrainment of cool dry air sinking from the free troposphere. The lower panel of the figure also reveals an increase in water vapour mixing ratio after 16:30 UTC.

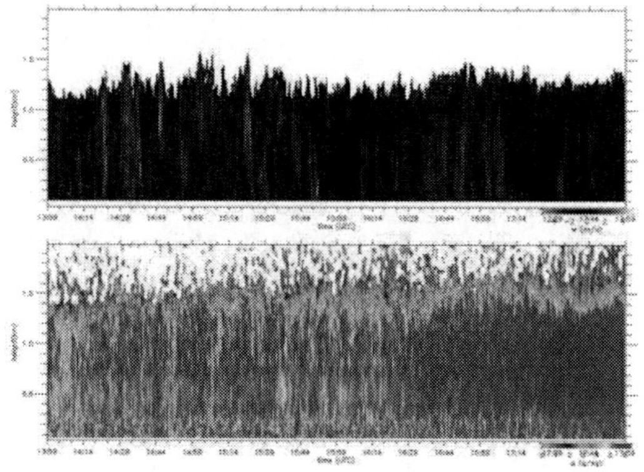

Figure 1: Time-height cross-section of the vertical wind speed (upper panel) and water vapour mixing ratio (lower panel) in the time interval 14-18 UTC on 04 May 2013.

For the purpose of the application of the covariance approach to estimate the water vapour flux profile, we then focused our attention on the time interval 14:15-16:15 UTC on this day.

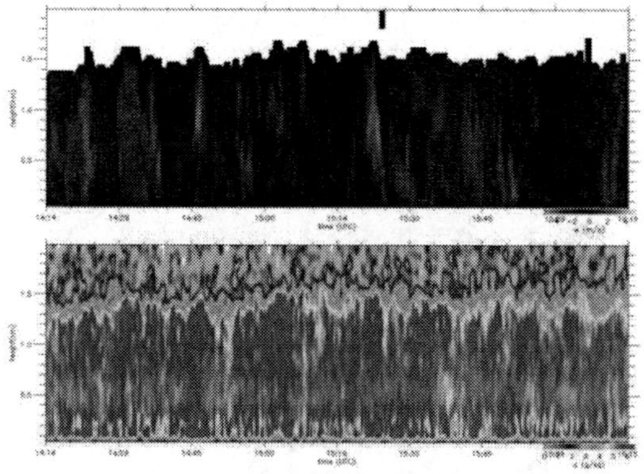

Figure 2: Time-height cross-section of the vertical wind speed (upper panel) and water vapour mixing ratio (lower panel) in the time interval 14:15-16:15 UTC on 04 May 2013.

Figure 2 illustrates the time-height cross section of the vertical wind speed (upper panel) and water vapour mixing ratio (lower panel) in the time interval 14:15-16:15. The data in figure 2 have been smoothed with a gaussian kernel with

L=90m and T=30sec. The integral scale of vertical wind speed and water vapour mixing ratio fluctuations are in the ranges 40-80 s and 40-90 s, respectively. Considering an average wind speed of ~9m/s, length scales for vertical wind speed and water vapour mixing ratio are in the range 360-720 m and 360-800m, respectively. Before applying the covariance approach, specific algorithms for the removal of linear trends and outliers were applied to the wind speed and water vapour mixing ratio time series.

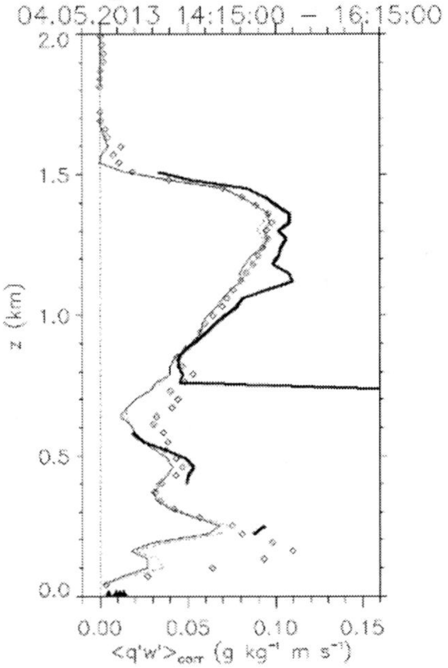

Figure 3: Water vapour flux profiles corresponding to the time interval 14:15-16:15 UTC on 04 May 2013. Thick line: profile corrected for separation, i.e. extrapolation to zero spatial shift; diamonds: profile corresponding to the largest covariance; solid line: profile with zero temporal shift; triangles: independent surface flux measurements.

Figure 3 illustrates the water vapour flux profiles corresponding to the time interval 14:15-16:15 UTC on 04 May 2013. Different profiles are shown for different time shifts. More specifically, the thick line represents the profile corrected for separation, i.e. extrapolation to zero spatial shift, the diamonds identify the profile corresponding to the largest covariance, the solid line represent the profile with zero temporal shift, while the triangles at surface level are independent surface flux measurements.

The flux profile is characterized by values increasing with height, this behaviour being compatible with the presence of a turbulent flux divergence having the same value at all heights within the CBL. In the absence of horizontal advection, this flux profile structure should produce a humidity loss to free troposphere of 8% per hour. However, this humidity loss does not appear in the water vapour lidar data. Consequently, we assume that this loss is compensated by local advection. Arguments is support of this hypothesis will be illustrated and discussed at the Conference.

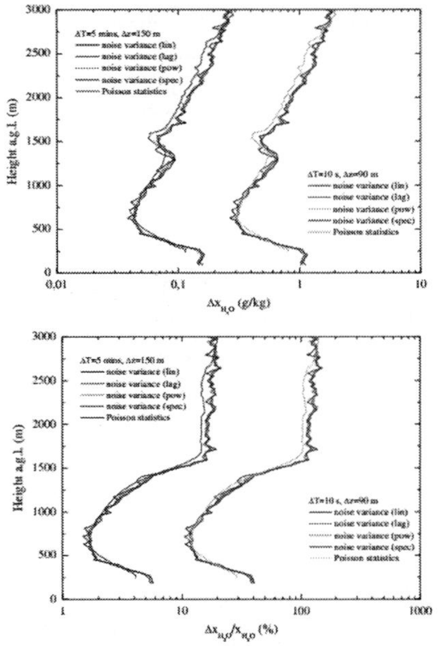

Figure 4: Profiles of noise error affecting water vapour mixing ratio measurements (upper panel in g/kg and lower panel in %).

In order to characterize the quality of water vapour mixing ratio measurements which are considered in the determination of the water vapour flux profile, an accurate assessment of noise error profiles is necessary. Profiles of noise error affecting water vapour mixing ratio measurements are illustrated in figure 4 (as estimated at 11:30 and 13:30 UTC on 20 April 2013), this being quantified as the root-square of the noise variance. For the selection of the vertical and temporal resolution considered for the flux measurements, i.e. 90 m and 10 s, respectively,

DOI: 10.1051/epjconf/201611925006

the statistical error affecting water vapour mixing ratio measurements is smaller than 0.06 g/kg or 50 % up to 1.4 km.

4. CONCLUSIONS

This research effort demonstrates that water vapour flux profiles can be determined throughout the CBL with sufficient accuracy based on the use of state-of-art water vapour Raman and Doppler lidars. For this purpose measurements from the HD(CP)2 Observational Prototype Experiment (HOPE), held in Central Germany in the spring 2013, are considered.

REFERENCES

[1] Weckwerth, T. M, D. B. Parsons, S. E. Koch, J. A. Moore, M. A. LeMone, B. B. Demoz, C. Flamant, B. Geerts, J. Wang, W. F. Feltz, 2004: An overview of the International H2O Project (IHOP 2002) and some preliminary Highlights, *Bull. Am. Meteorol. Soc.,* **85**, 253–277.

[2] Senff C, J. Bösenberg, G. Peters, 1994: Measurement of water vapor flux profiles in the convective boundary layer with lidar and radar-RASS, *J. Atmos. Oceanic Technol.*, **11**, 85–93.

[3] Wulfmeyer, V., 1999: Investigation of turbulent processes in the lower troposphere with water-vapour DIAL and radar-RASS, *J. Atm. Sci.*, **56**, 1055–1076.

[4] Kiemle C, M. Wirth, A. Fix, S. Rahm, U. Corsmeier, P. Di Girolamo, 2011: Latent heat flux measurements over complex terrain by airborne water vapour and wind lidars, *Q. J. R. Meteorol. Soc.*, **137**, 190–203.

[5] Di Girolamo, P., R. Marchese, D. N. Whiteman, B. B. Demoz, 2004: Rotational Raman Lidar measurements of atmospheric temperature in the UV. *Geophys. Res. Lett.*, **31**, doi: 10.1029/2003GL018342.

[6] Di Girolamo, P., D. Summa, R. Ferretti, 2009: Multiparameter Raman Lidar Measurements for the Characterization of a Dry Stratospheric Intrusion Event, *J. Atmos. Ocean. Tech.*, **26**, 1742-1762.

[7] Di Girolamo, P., D. Summa, R. Bhawar, T. Di Iorio, M. Cacciani, I. Veselovskii, O. Dubovik, A. Kolgotin: 2012, Raman lidar observations of a Saharan dust outbreak event: Characterization of the dust optical properties and determination of particle size and microphysical parameters, *Atm. Env.*, **50**, 66-78.

[8] Pearson, G., F. Davies, C. Collier, 2009: An Analysis of the Performance of the UFAM Pulsed Doppler Lidar for Observing the Boundary Layer. *J. Atmos. Ocean. Tech.*, **26**, 240-250.

EPJ Web of Conferences 119, 25007 (2016)
ILRC 27
DOI: 10.1051/epjconf/201611925007

INDEPENDENT CALIBRATION OF WATER VAPOR RAMAN LIDAR BY USING ADDITIONAL ELASTIC MEASUREMENTS AT WATER VAPOR RAMAN WAVELENGTH

Decheng Wu [1, 2*], Zhien Wang [1], Dong Liu[2], Chenbo Xie[2], Yingjian Wang[2]

[1]*Department of Atmospheric Sciences, University of Wyoming, WY 82070, USA,*

[2] *Hefei Institutes of Physical Science, Chinese Academy of Sciences, Anhui 230031, China*

**Email: dwu5@uwyo.edu*

ABSTRACT

A new method for independent calibration of water vapor Raman lidar by using additional measurements of elastic signals at water vapor Raman wavelength was introduced. The water vapor mixing ratio calibration constant was derived from nitrogen mixing ratio calibration constant and Rayleigh to water vapor Raman backscattering cross section ratio. The nitrogen mixing ratio calibration constant with uncertainties of 1~2% can be determined from the nitrogen mixing ratio measurement in an aerosol-free zone in the upper troposphere with the additional elastic measurement and nitrogen Raman measurement, the Rayleigh to water vapor Raman backscattering cross section ratio can be calculated with uncertainties of about 8%, or can be determined experimentally with uncertainties of less than 5%. Using the experimental cross section ratio, a total uncertainty of the water vapor calibration less than 6% is possible to achieve. Because this calibration method utilizes actual lidar signals in nitrogen and water vapor detection channels, the calibration can fully capture lidar state changes.

1. INTRODUCTION

Water vapor mixing ratio (WVMR) calibration is an important issue for water vapor Raman lidar (WVRL), in particular for climate monitoring applications [1]. The calibration usually relies on an external, independent measurement of water vapor, such as radiosonde or microwave radiometer. The calibration accuracy relies on accuracies of these independent measurements and synchronization of the measurements [2~3]. As the changes of the WVRL state, such as changes in overall system efficiency and detector sensitivity due to ageing, will influence the calibration constant, a regular calibration is needed for a WVRL [1, 3]. Therefore, it is difficult to obtain long-term regularly independent water vapor measurements at most of WVRL sites.

In general, WVRL calibrations need to determine the difference of system detection efficiencies between the nitrogen and water vapor channels and nitrogen to water vapor Raman backscattering cross section ratio. Using a calibrated tungsten lamp, the ratio of lidar channel detection efficiencies could be calibrated with errors of about 1% [4]. The backscattering cross section ratio of nitrogen to water vapor is needed in the lamp independent calibration, the cross section ratio can be calculated with uncertainties of about ±10%, or can be determined experimentally with an error of less than 5% based on a well-calibrated WVRL [4, 5]. But the cross section ratio is system dependent due to different filters used. However, there are several serious limitations in the lamp independent calibration due to not able to fully capture lidar state changes [6].

In this paper, a new independent calibration method for WVRL based on additional elastic measurement at water vapor Raman wavelength was introduced and discussed.

2. METHODOLOGY

The measured Raman signals from nitrogen and water vapor molecules can be written as below.

$$V_N(z,\lambda_N) = E_L \frac{c}{2} A\eta_N \frac{1}{z^2} N_N(z)\sigma_{N,\pi}(\lambda_L,\lambda_N) T(z,z_0,\lambda_L)T(z,z_0,\lambda_N) \quad (1)$$

$$V_{wv}(z,\lambda_{wv}) = E_L \frac{c}{2} A\eta_{wv} \frac{1}{z^2} N_{wv}(z)\sigma_{wv,\pi}(\lambda_L,\lambda_{wv}) T(z,z_0,\lambda_L)T(z,z_0,\lambda_{wv}) \quad (2)$$

where, z is the altitude of scattering volume; z_0 is the altitude of lidar; $V_N(z, \lambda_N)$ and $V_{wv}(z, \lambda_{wv})$ are

DOI: 10.1051/epjconf/201611925007

the voltage values of nitrogen and water vapor Raman signals without background signals in nitrogen and water vapor channels; λ_L is the laser wavelength; λ_N and λ_{wv} are the nitrogen and water vapor Raman shifted wavelength; E_L is the laser pulse energy; c is the light speed; A is the effective telescope receiving area; η_N and η_{wv} are the total efficiencies of nitrogen and water vapor Raman channels, including optical efficiency, quantum efficiency, and electrical efficiency; $N_N(z)$ and $N_{wv}(z)$ are the nitrogen and water vapor molecular number densities at z; $\sigma_{N,\pi}(\lambda_L, \lambda_N)$ and $\sigma_{wv,\pi}(\lambda_L, \lambda_{wv})$ are nitrogen and water vapor Raman backscattering cross sections; $T(z, z_0, \lambda_N)$ and $T(z, z_0, \lambda_{wv})$ are the atmospheric transmissions from z_0 to z at the wavelength of λ_N and λ_{wv}.

Temperature influence on the backscattering cross section was not considered in lidar equations in this section, it will be discussed in section 3. Lidar signals in the range of full overlap were selected for calibration to avoid uncertainties caused by the overlap function. Therefore, the overlap functions were neglected in lidar equations in this paper.

Water vapor mixing ratio w(z) is derived from the ratio of water vapor to nitrogen Raman signals,

$$w(z) = C_{wv}\frac{V_{wv}(z,\lambda_{wv})}{V_N(z,\lambda_N)}\frac{T(z,z_0,\lambda_N)}{T(z,z_0,\lambda_{wv})} \quad (3)$$

where, C_{wv} is the WVMR calibration constant.

$$C_{wv} = 0.485\frac{\eta_N}{\eta_{wv}}\frac{\sigma_{N,\pi}(\lambda_L,\lambda_N)}{\sigma_{wv,\pi}(\lambda_L,\lambda_{wv})} \quad (4)$$

After WVMR measurement, extra laser pulses at a wavelength λ_C, which is very close to λ_{wv}, were emitted into the atmosphere. If the difference between λ_C and λ_{wv} is less than the half of the filter bandwidth in water vapor channel $\Delta\lambda_{wv}$, $|\lambda_C-\lambda_{wv}|<\Delta\lambda_{wv}/2$, the elastic signal from the atmosphere at the wavelength λ_C can be measured by using the water vapor channel.

$$V_C(z,\lambda_C) = E_C\frac{c}{2}A\eta_{wv}\varepsilon(\lambda_C,\lambda_{wv})\frac{1}{z^2} \left[\beta_a(z,\lambda_C) + N_{air}(z)\sigma_{m,\pi}(\lambda_C)\right] [T(z,z_0,\lambda_C)]^2 \quad (5)$$

where, $V_C(z, \lambda_C)$ is the voltage value of elastic signal without background signal received by the water vapor channel; $\varepsilon(\lambda_C, \lambda_{wv})$ is the ratio of total efficiency of the water vapor channel at λ_C to the one at λ_{wv}, the difference between λ_C and λ_{wv} is so little that $\varepsilon(\lambda_C, \lambda_{wv})$ is mainly determined by the narrow bandwidth filter in the water vapor channel; $\beta_a(z, \lambda_C)$ is the aerosol backscattering coefficient at λ_C; $N_{air}(z)$ is the air molecular number density; $\sigma_{m,\pi}(\lambda_C)$ is Rayleigh backscattering cross section at λ_C; $T(z, z_0, \lambda_C)$ is the atmospheric transmission from z_0 to z at λ_C, $T(z, z_0, \lambda_C) \approx T(z, z_0, \lambda_{wv})$.

In the free troposphere, an aerosol-free zone, where the contribution of aerosol backscattering to the elastic signal could be neglected, can be often found. In the aerosol-free zone, the elastic signal measured by using water vapor channel could be written as below,

$$V_C(z,\lambda_C) = E_C\frac{c}{2}A\eta_{wv}\varepsilon(\lambda_C,\lambda_{wv})\frac{1}{z^2}N_{air}(z)\sigma_{m,\pi}(\lambda_C) [T(z,z_0,\lambda_C)]^2 \quad (6)$$

In the aerosol-free zone, nitrogen mixing ratio (NMR), which is a constant in the atmosphere below about 100 km, could be derived from the ratio of nitrogen Raman signal in Eq. (1) to Rayleigh signal in Eq. (6).

$$NMR(z) = 0.755 = C_N\frac{E_C}{E_L}\frac{V_N(z,\lambda_N)}{V_C(z,\lambda_C)}\frac{[T(z,z_0,\lambda_C)]^2}{T(z,z_0,\lambda_L)T(z,z_0,\lambda_N)} \quad (7)$$

where, C_N is the NMR calibration constant.

$$C_N = 0.967\varepsilon(\lambda_C,\lambda_{wv})\frac{\eta_{wv}}{\eta_N}\frac{\sigma_{m,\pi}(\lambda_C)}{\sigma_{N,\pi}(\lambda_L,\lambda_N)} \quad (8)$$

Using Eq. (7), C_N could be determined in the aerosol-free zone. C_{wv} could be derived from the determined C_N and Rayleigh to water vapor backscattering cross section ratio by using Eq. (9) shown below.

$$C_{wv} = 0.469\frac{1}{C_N}\varepsilon(\lambda_C,\lambda_{wv})\frac{\sigma_{m,\pi}(\lambda_C)}{\sigma_{wv,\pi}(\lambda_L,\lambda_{wv})} \quad (9)$$

3. DISCUSSIONS

3.1 Rayleigh Backscattering Signal

Measuring Rayleigh backscattering signals by using the water vapor channel is the key in the proposed method.

Firstly, an extra laser source is needed, and its output wavelength should be very close to the filter central wavelength (CWL) in the water vapor channel. For the most of WVRLs, the filter CWL in the water vapor channel is about 407.5

DOI: 10.1051/epjconf/201611925007

nm. OPO based on a type II BBO crystal can generate narrow linewidth nanosecond laser pulse at the wavelength of 407.5 nm. The nanosecond laser pulse at 354.7 nm employed in the most of WVRLs, can be used to pump the OPO to get the needed wavelength.

Secondly, an aerosol-free zone should be found. When the efficiency of the OPO is 5%, the intensity of the Rayleigh signal at the height of 10 km approximately equals the water vapor Raman signal at the height of 1 km with a WVMR of 10 g/kg under a fair atmosphere condition. Therefore, the elastic signal from the upper troposphere with the OPO laser could be detected by the water vapor channel under the cloud-free atmosphere condition. Because a high pulse energy laser and a large telescope are employed in ground-based WVRLs and temporal averaging is normally used in the data processing, the averaged nitrogen Raman signal is quite smooth in the upper troposphere, even up to the tropopause. In summary, an aerosol-free zone could be measured by the proposed system in the upper troposphere.

In the aerosol-free zone, backscattering ratio at 407.5 nm should be less than 1.01, and the error cause by the aerosol-free assumption will be less than 1%. To reduce the noise level, the calibration would be carried out during nighttime. Due to temporal averaging in data processing, the error of laser energy correction could be neglected.

3.2 Backscattering Cross Sections and Temperature Dependence

Nitrogen and water vapor Raman backscattering cross sections, as well as Rayleigh backscattering cross section, are involved in this method.

With filter bandwidths of 0.3 nm, calculated temperature-dependent effective nitrogen Raman backscattering cross section at 386.7 nm and Rayleigh backscattering cross section at 407.5 nm, as well as the ratio of Rayleigh to nitrogen Raman backscattering cross section, were shown in figure 1. The relative difference between Rayleigh to nitrogen Raman effective backscattering cross section ratios at 320 K and 200 K is about 0.26%, therefore, the temperature influence could be neglected in the determination of C_N.

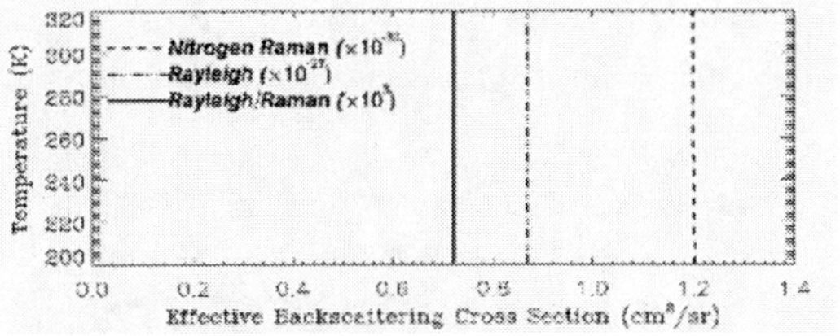

Figure 1. Calculations of temperature-dependent effective nitrogen Raman backscattering cross section at 386.7 nm (dash), Rayleigh backscattering cross section at 407.5 nm (dash dot), and the ratio of Rayleigh to nitrogen Raman effective backscattering cross section (solid).

In the determination of C_{wv} using Eq. (9), temperature-dependent Rayleigh to water vapor Raman backscattering cross section ratio is needed. The error of theoretical calculation of Rayleigh scattering cross section is less than 1% [7], and the error of calculated water vapor absolute Raman differential scattering cross section of the Q branch of ν1 is less than 8% [8]. However, the cross section ratio could be determined experimentally if the lidar could point horizontally to provide C_{wv} by comparing lidar measurements with accurate in situ measurements on the laser path. Then with Eq. (9), the cross section ratio can be determined with accuracy better than 5% for future vertical pointing measurements. The temperature-dependence of the cross section ratio could be calculated with uncertainties of about 1% [4].

3.3 Atmospheric Transmission Correction

As shown in Eq. (7), two-way transmission of the atmosphere between lidar and the aerosol-free zone needs to be corrected in the determination of C_N. The two-way transmission corrections calculated by using different aerosol Angstrom exponents k under different aerosol optical depth (AOD) conditions were shown in figure 2. From figure 2, it can be seen that the error of atmospheric transmission correction caused by the uncertainty of k increases along with the increasing AOD, so the calibration should be carried out under low aerosol loading conditions. AOD and k can be retrieved from the elastic signals at 354.7 nm and 407.5 nm and Raman signal at 386.7 nm measured by the lidar during the calibration experiment. Even for AOD at 355 nm equals 0.2, 20% uncertainty in k yields about 0.6% uncertainty in transmission correction.

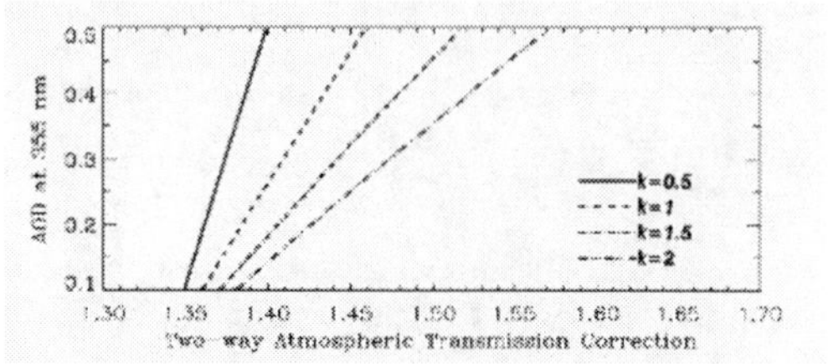

Figure 2. Calculated two-way atmospheric transmission corrections in NMR in Eq. (7) as a function of AOD by using different Angstrom exponent k.

4. CONCLUSIONS

In this paper, a new independent calibration method for WVRL was proposed based on additional elastic measurements at water vapor Raman wavelength. The water vapor mixing ratio calibration constant was derived from the nitrogen mixing ratio calibration constant and Rayleigh to water vapor Raman backscattering cross section ratio.

Nitrogen mixing ratio calibration constant indicates difference of total efficiencies between the nitrogen and water vapor channels. The error of the determined nitrogen mixing ratio calibration constant is about 1~2%. The Rayleigh to water vapor Raman backscattering cross section ratio can be calculated with uncertainties of about 8%. The cross section ratio also can be determined experimentally from a well-calibrated WVRL with uncertainties of less than 5%. Using the experimental cross section ratio, the total error of the water vapor calibration will be less than 6%.

Compared with the lamp calibration method, this new approach can fully capture the changes of lidar state. Furthermore, after the initial design and integration, the calibration can be done routinely and automatically.

ACKNOWLEDGEMENTS

This research work is supported by NSFC under grant No.41405032 and partly supported by NSF under AGS-1337599.

REFERENCES

[1] Commission for instruments and methods of observation, 2013: Expert team on new technologies and testbeds first section final report, Geneva, Switzerland, 4-7 March.

[2] Davide Dionisi, Fernando Congeduti, Gian Luigi Liberti, Francesco Cardillo, 2010: Calibration of a multichannel water vapor Raman lidar through noncollocated operational soundings: optimization and characterization of accuracy and variability, *J. Atmos. Oceanic Technol.*, **27**, 108-121.

[3] T. Leblanc, I. S. McDermid, 2008: Accuracy of Raman lidar water vapor calibration and its applicability to long-term measurements, *Appl. Opt.*, **47(30)**, 5592–5603.

[4] Whiteman, D. N, T. Berkoff, D. D. Turner, T. P. Tooman, R. A. Ferrare, and L. A. Heilman, 2000: Research efforts in the absolute calibration of a Raman water vapor lidar, *Proc. of the 20th International Laser Radar Conference*, Vichy, France.

[5] D. D. Venable, D. N. Whiteman, M. N. Calhoun, A. O. Dirisu, R. N. Connell, and E. Landulfo, 2011: A lamp mapping technique for independent determination of the water vapor mixing ratio calibration factor for a Raman lidar system, *Appl. Opt.*, **50(23)**, 4622–4632.

[6] D. N. Whiteman, D. D. Venable, and E. Landulfo, 2011: Comments on: Accuracy of Raman lidar water vapor calibration and its applicability to long-term measurements, *Appl. Opt.*, **50(15)**, 2170–2176.

[7] R. Thalman, K. J. Zarzana, M. A. TOLBERT, R. Volkamer, 2014: Rayleigh scattering cross-section measurements of nitrogen, argon, oxygen and air, *JQSRT*, **147**, 171-177.

[8] G. Avila, J. M. Fernández, G. Tejeda, S. Montero, 2004: The Raman spectra and cross-sections of HO, DO, and HDO in the OH/OD stretching regions, *J. Mol. Spectrosc.*, **228**, 38-65.

PERFORMANCES OF A HGCDTE APD BASED DETECTOR WITH ELECTRIC COOLING FOR 2-μm DIAL/IPDA APPLICATIONS

A. Dumas[1]*, J. Rothman[2], F. Gibert[1], G. Lasfargues[2], J-P. Zanatta[2] and D. Edouart[1].

[1] ***LMD, École Polytechnique/CNRS, France, *Email: arnaud.dumas@lmd.polytechnique.edu***

[2] ***LETI, CEA - Campus Minatech, France***

1. ABSTRACT

In this work we report on design and testing of an HgCdTe Avalanche Photodiode (APD) detector assembly for lidar applications in the Short Wavelength Infrared Region (SWIR : 1,5 - 2 μm). This detector consists in a set of diodes set in parallel -making a 200 μm large sensitive area- and connected to a custom high gain TransImpedance Amplifier (TIA). A commercial four stages Peltier cooler is used to reach an operating temperature of 185K. Crucial performances for lidar use are investigated : linearity, dynamic range, spatial homogeneity, noise and resistance to intense illumination.

2. INTRODUCTION

Within the framework of greenhouse gases (GHG) monitoring with lidar, the Laboratoire de Météorologie Dynamique (LMD, IPSL, Paris) has developed a 2μm instrument dedicated to Differential Absorption Lidar (DIAL) and Integrated Path Differential Absorption (IPDA) techniques. Results on the laser emitter were reported in [1] and first CO_2 DIAL measurements using coherent detection are to be published. The main reason for using coherent detection at the time was the lack of a very highly-sensitive detector in SWIR. However, recent advances in the field of HgCdTe APD may have shifted the balance of power between direct and coherent detection.

Indeed, the remarkable properties for amplification of HgCdTe APD (large gain, low dark current and close to unity excess noise factor) makes this technology especially suitable for very low intensity signal detection in SWIR : on the one hand photomultiplier tube technology is not available because the energy per photon is not large enough to use external photoeffect [2], on the other hand, InGaAs APD have intrinsic limitation when it comes to amplification because of higher dark current and large excess noise factor. Today HgCdTe APD technology has reached some level of maturity and applications are currently investigated : space laser-based telecommunications, spectroscopy, lidar, etc.

The challenging issue for lidar use is to reach sufficiently large sensitive area while keeping a low level of noise. One proposal has been carried out at NASA Goddard Space Flight Center (GSFC) in collaboration with DRS technologies to develop a multi-element HgCdTe APD detector. This latter consists in an array of 16 independant pixels whose individual size is 80 μm. Each pixel is a set of four diodes set in parallel that inherit the p around n junction architecture. A Noise Equivalent Power (NEP) as low as 4 fW/$\sqrt{\text{Hz}}$ is reported when operated at 77K [3]. On the other hand, we have developed a single element detector from existing Focal Plane Array (FPA) matrix by connecting 37 diodes in parallel. This detector was manufactured at the Centre à l'Energie Atomique -Laboratoire d'Electronique et Techniques de l'Information (CEA-LETI) in Grenoble (France) and therefore inherits planar architecture for diode junctions. Regarding cooling, we made the decision to use Thermo-Electric Cooling (TEC) because it requires less energy, which is an appreciable asset for any space equipment.

In this study we first describe this detector design. Previous communication with wider scope on the development of this technology at CEA-LETI can be found here [4]. Then we report on test results for the detector figures of merit that are especially relevant for lidar use. Namely : dark current, spatial homogeneity, linearity, gain, NEP and resistance to intense illumination.

3. DESIGN OF LIDAR DETECTOR

A. Assembly overview

A side view of the detector assembly is presented in figure 1. The APD and the TIA are set up side by side within a cryostat. Distance between the APD and the TIA PCB is minimized to avoid parasite capacitance.

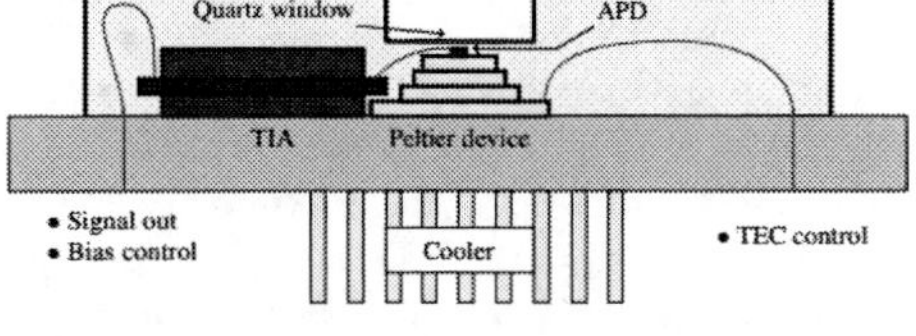

Figure 1: Sideview of the detector assembly that was developed at CEA-LETI.

A quartz window is used to enable IR illumination of the APD. This latter was set up as close to the APD as possible in order to use short focal lenses outside the cryostat. Commercial references and main characteristics of the commercial Peltier device and TIA are provided in table 1.

Table 1: References and characteristics for Peltier device and TIA.

Petier device :	TEC Marlow NL4040
	$\Delta T = 120$K
TIA :	FirstLightImaging
	Gain : 6,97·10^5 V/A (1% accuracy)
	Bandwidth : 20,5 MHz
	Input noise current density : 0,9 pA/$\sqrt{\text{Hz}}$

B. Requirements

For the setup presented in last section, the Noise Equivalent Power (NEP$_W$) in Watts can be expressed as [2]:

$$\frac{1}{M}\cdot\sqrt{<i^2_{\text{input TIA}}>_T + 2q(i_d+i_s)M^2F\cdot\frac{1}{2T}}\cdot\frac{1}{\eta}\cdot\frac{hc}{q\lambda} \quad (1)$$

where M is the APD gain, $< i^2_{\text{input TIA}} >_T$ the variance of the input TIA current over a time interval with duration T, q the elementary charge, i_d the mean value of the dark current, i_s the mean value of the current converted photon signal, T the caracteristic time (defined as $1/2B$, where B is the bandwidth of the TIA), F the excess noise factor, h the Planck constant, c the speed of light and λ the wavelength.

As the NEP$_W$ is often normalized per unit $\sqrt{\text{Hz}}$, we rewrite it NEP in W/$\sqrt{\text{Hz}}$ as follows :

$$\text{NEP} = \frac{1}{M}\sqrt{i^2_{TIA} + 2q(i_d+i_s)M^2F}\cdot\frac{hc}{q\lambda}\cdot\frac{1}{\eta} \quad (2)$$

where i_{TIA} is the noise density at the input of the TIA (in A/$\sqrt{Hz}$). As long as the noise due to the TIA is dominant over the shot noise of the signal and the dark current, it is worth to increase the gain. Once the amplified shot noise or the amplified dark current dominates, any increase of the gain would decrease the signal to noise ratio because of the excess noise factor (see [2]). In particular, equation 2 shows that for a certain gain, a balance is obtained between dark current, shot noise amplification and TIA noise reduction.

A numerical model of APD gain as a function of reverse bias was used to compute the expectable gain at the temperature of 185K . The maximum value for reverse bias is limited to approximately 14V. Beyond this value, tunnel current is not negligible anymore. At 12 V reverse bias and 185 K temperature, a gain of approximately 40 is expected. For a NEP of 50 fW/$\sqrt{\text{Hz}}$ for instance, this means that the input referred current noise of the TIA should be lower than 2 pA/$\sqrt{\text{Hz}}$.

C. Macro-diode geometry

Usually in lidar applications, the telescope collecting mirror, which serves as pupil, is imaged onto the sensitive area of the detector. Indeed in such a configuration, the enlightened area of the detector is always the same, which is suitable to avoid potential errors due to spatial inhomogeneities of detector response. On the one hand, the larger the sensitive area, the easier it is to make the image of the pupil on it. On the other hand, a large sensitive area limits the performances of the diode itself, especially in terms of noise. As a result, the size of 200 μm stands as a trade off between these constraints.

The architecture of the macro photodiode is inherited from standard production of FPAs at CEA-LETI (see figure 2).

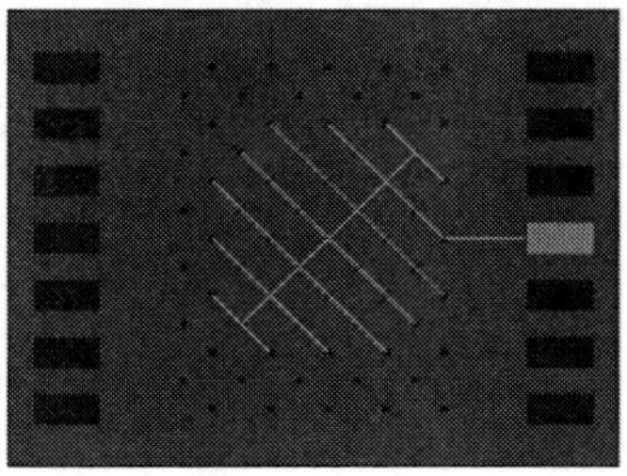

Figure 2: The macro APD structure is inherited from FPA production. 37 diodes are connected in parallel. Horizontal step between two diodes is roughly 40 μm.

Planar architecture is used and the 37 diodes are bonded in parallel on the front side. As a consequence, and because the detector is in this case back-side illuminated, remarkable homogeneity in photons collection is achieved (see experimental evaluation).

4. HGCDTE APD TEST RESULTS

Having presented the detector setup, we focus here on experimental tests that have been carried out to assess the macro-diode performances : dark current, spatial homogeneity of the response and response to intense flux. In these experiments the macro diode was isolated from the TIA described above and cooled with liquid nitrogen. This also means that the cryostat was different. Experimental conditions are described for each subsection.

A. Dark current

First the dark current has been evaluated to roughly 0.2 nA at 185 K(see figure 3). A cryostat with cold shield was used.

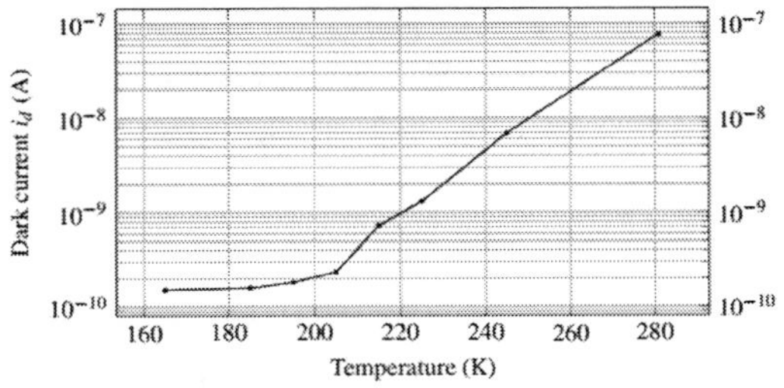

Figure 3: Dark current as a function of APD temperature

It should be noted that the observed current would rather be called residual current. Indeed as any object at ambient temperature emits in the SWIR region and despite of cold shield, some undesirable photons are always collected. Therefore the distinction between residual photonic noise and dark current is hard to grasp.

B. Spatial homogeneity

A so-called spotscan experiment has been performed to assess spatial response of the APD. A SWIR source is used together with a pupil and a imaging lens to form a 10 μm large spot. This spot can be moved to every position in a square of roughly 220 μm edge by use of commanded mirrors. In our configuration the APD was cooled to 77 K with liquid nitrogen and connected to a high-gain transimpedance amplifier. Let (x, y) be the coordinates of the spot within the square. For any (x, y) the intensity at the output of the APD is recorded and we obtain that way the APD spatial response.

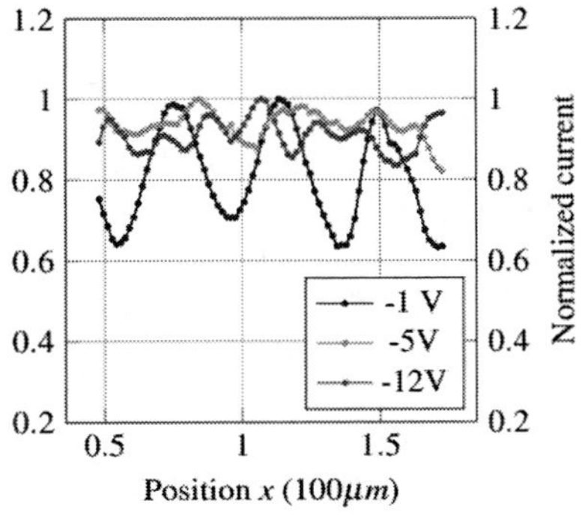

Figure 4: Relative variation of the normalized current along the central row for different reverse bias

One should be aware that the intensity figure obtained is the product of an optical intensity transfer function $\mathcal{H}(x, y)$ by the detector spatial response $\mathcal{R}(x, y)$. $\mathcal{H}$ accounts for any disturbing effect that modify the spot intensity at the position (x, y). It can be due to inhomogeneity of the source itself or to dust on optics for example. In the following paragraph we describe how we have managed to obtain an estimation of relative variation of $\mathcal{R}$.

We introduce $(\mathcal{G}_i)$ as the set of figures produced with a series of spotscan experiments during which the detector is steadily translated with respect to the square mentioned before. Namely, for any i, we have

$$\mathcal{G}_i(x, y) = \mathcal{R}(x + i\delta x, y)\mathcal{H}(x, y) \tag{3}$$

Therefore we can write

$$[\mathcal{G}_i(x, y)]_{i\in[n,m]} = \frac{\mathcal{H}(x, y)}{\mathcal{H}(x + \delta x, y)} \cdot [\mathcal{G}_i(x + \delta x, y)]_{i\in[n-1,m-1]} \tag{4}$$

From this set of measurements it is not possible to derive an absolute value of $\mathcal{R}$ but it is possible to express the relative difference of pixels along a row. From equation 4 we get relative variation of $\mathcal{H}$ along a row, and then variation of $\mathcal{R}$ along the x axis for different reverse bias. Figure 4 presents the results of this analysis. For 1 V reverse bias, the depletion area is not extended enough to collect efficiently photons that are incident far from the diode center. On the other hand, for reverse bias greater than 5 V, collection of photons is almost homogeneous. Slight heterogeneity of spatial response comes from manufacturing choices and flaws in experimental procedure. For a reverse bias of 5V, a standard deviation of 3.5 % for the set $\mathcal{R}(\cdot, y)$ where y corresponds to the central row of the APD. At a reverse bias of 12V, standard deviation is 3.8 %.

C. Behaviour under intense illumination

Previous experiences with space lidar have shown that because of intense reflexion on clouds, the detector might be intensely illuminated by such an echo. We have reproduced these conditions in order to estimate a damage threshold in terms of power as well as a recovery time. Therefore incident optical power was gradually increased as well as reverse bias. We went up to 20V reverse bias and 10 μW incident average power without noticing any damage. Depolarization of the diode was observed as a result of the mitigation voltage induced by the current in the series resistance.

5. DETECTOR ASSEMBLY TEST RESULTS

The results presented in this section concern the detector assembly, ie the APD with thermo-electric cooling and the TIA presented in first section. Therefore they account for the global behaviour of the macro APD plus the TIA.

A. Thermo-electric cooling efficiency

For ambient temperature of approximately 23 degrees and a pressure inside the cryostat of $5{\cdot}10^{-7}$ mbar, the TEC device presented in section 1 enables to reach an operating temperature of 184 K on the long term.

B. Linearity

To assess detector linearity, a 2-μm laser diode Nanoplus has been used in combination with a wheel of calibrated densities. The laser diode is set up in front of the quartz window of the detector and illuminates widely the APD. Densities are added between the quartz window and the laser diode to provide a large set of transmittances (see figure 5).

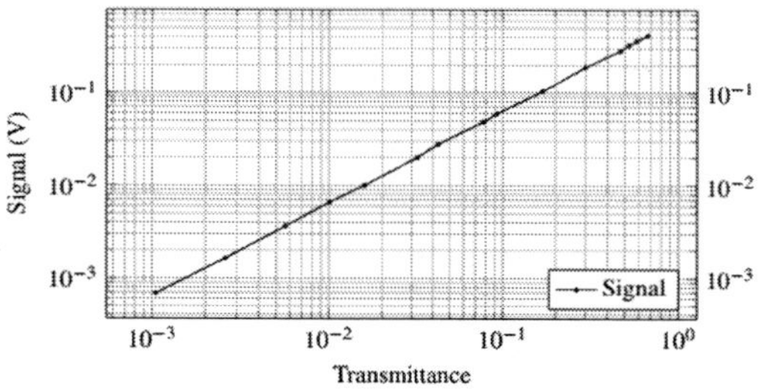

Figure 5: Signal intensity versus transmittance at 2 μm. Transmittance is accurately known thanks to FTIR calibrated values of the optical densities. Laser intensity drift is taken into account to correct the signal. Reverse bias is set to 12V and operating temperature is 185 K.

The incident power with transmittance $T = 1$ was roughly 2 μW. Hence the detector was proved to achieve close to linear reponse over 3 orders of magnitude (from nW to μW). The experimental setup shows a root mean square error of 1,8 % from ideal linear response.

C. Gain and NEP as a function of revers bias

We confront here the numerical prediction presented in first section with experimental data (see figure 6).

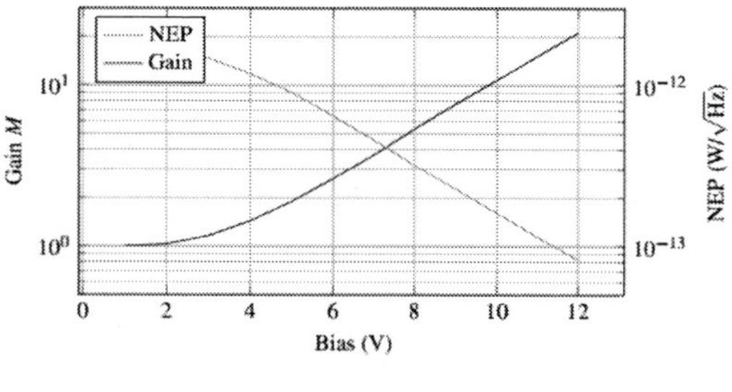

Figure 6: Experimental gain and NEP records as a function of reverse bias. Operating temperature is 185K.

For a reverse bias of 12V, a gain of 21 is achieved. Corresponding NEP is about 80 fW/$\sqrt{\text{Hz}}$.

D. Noise

Finally we investigate the noise properties with Allan variance tool (see figure 7). This mathematical tool allows to distinguish between white noise, flicker noise, etc. We use the simple estimator $\text{std}(u,\tau) = \sqrt{\frac{1}{m}\sum_{i=1}^{m}(u_{i+1} - u_i)^2}$ where u_i is the averaged voltage value over the duration τ. The $\tau^{-1/2}$ slope corresponds to white noise.

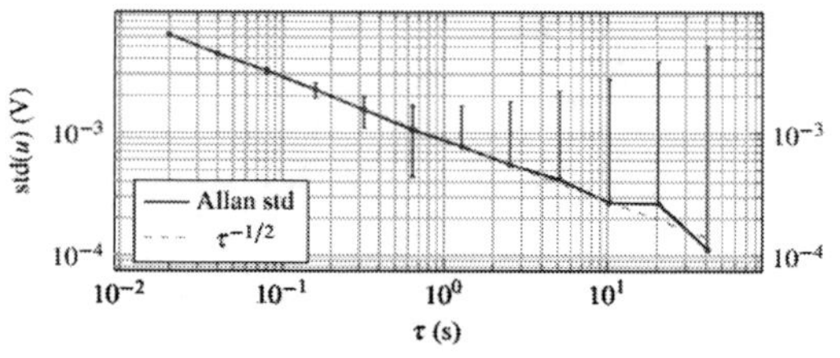

Figure 7: Allan standard deviations computed for a series of 5000 samples with 0,02 s as time step.

Results presented in figure 7 show that there is no evidence of other noises than white noise at the time scale of a few seconds, which was highly suitable to avoid any potential error while averaging over such time gate.

6. CONCLUSION

A HgCdTe APD-based detector for lidar use at 2 μm has been characterized. Key figures of merit are a NEP of 80 fW/$\sqrt{\text{Hz}}$ and a gain of 22 at 12 V reverse bias. At the same reverse bias, spatial response variability is about 2 % and the close to linear dynamic range covers at least three orders of magnitude. Overall bandwidth is 20,5 MHz.

AKNOWLEDGMENTS

The authors thank the french national space agency CNES for supporting this work.

REFERENCES

[1] F. Gibert, D. Edouart, C. Cenac, and F. L. Mounier, ``2-μm high-power multiple-frequency single-mode Q-switched Ho:YLF laser for DIAL application," *Applied Physics B*, 2014.

[2] B. Saleh and M. Teich, *Fundamentals of photonics, p645.* Wiley series, 1991.

[3] J. Beck, T. Welsch, P. Mitra, K. Reiff, X. Sun, and J. Abshire, ``A Highly Sensitive Multi-element HgCdTe e-APD Detector for IPDA Lidar Applications," *Journal of Electronic Materials*, 2014.

[4] K. Foubert, G. Lasfargues, L.Mathieu, and S. Benahmed, ``Developement of hgcdte single-element apds based detectors for low flux short wave infrared applications," *Proceedings SPIE*, 2013.

DOI: 10.1051/epjconf/201611925009

ABOUT THE POTENTIAL OF LIDARS WITH DIFFERENT PHOTODETECTORS UNDER DAYTIME SKY RADIATION

Ravil Agishev[1], Adolfo Comeron[2], Alexander Gilerson[3]
[1] Kazan State Power Engineering University, 51, Krasnoselskaya St., Kazan 420066, Russia
[2] Polytechnic University of Catalonia, 1–3, Jordi Girona, Barcelona 08034, Spain
[3] City College of the City University of New York, 140 St. & Convent Ave., New York, NY 10031, USA

ABSTRACT
Results of theoretical analysis and experimental developments implemented as advanced methods and means to improve a noise-immunity of lidar systems for practical applications are discussed. A particular attention is paid to assessing the developed methods and technical solutions effectiveness and their comparison with existing lidar systems and real receivers.

1. INTRODUCTION
We discuss a renewed approach to the generalized methodology for atmospheric lidar assessment, which uses the dimensionless-parameterization as a core component [1-9]. It allows predicting its work under the presence of the sky background radiation using three different photodetectors - traditional PMTs, SiPMTs and APDs.
The renewed approach can be widely used to evaluate a broad range of lidar system capabilities for a variety of lidar remote sensing applications, as well as to serve as a basis for selection of appropriate lidar system parameters for a specific application.

2. METHODOLOGY
Such a modernized methodology provides generalized, uniform and objective approach for the evaluation of a broad range of lidar types and systems (aerosol, Raman, DIAL), operating on different targets (backscatter or topographic) and under intense sky background conditions. It can be used within the lidar community to compare different lidar instruments.
The signal to noise ratio at the lidar photodetector input Ψ_X can be decomposed as a product of five independent dimensionless parameters, each of which follows from a different source [8]:

$$\Psi_X = V\, Q_X\, W^2 U^{-1} r^{-2} \qquad (1)$$

where V is the ratio of the echo-signal power P_{sref} received from the reference range R_{ref} of the reference molecular atmosphere to the photo-detector threshold power P_{t0} in the absence of background noise; Q_X is the backscatter efficiency of the arbitrary species to the molecular reference; W is the normalized atmospheric component determined by transparencies ratio of the current atmosphere state and the standard molecular atmosphere; U is the ratio of photodetector' s threshold powers taken in presence and in absence of background radiation; r is the normalized range-factor.

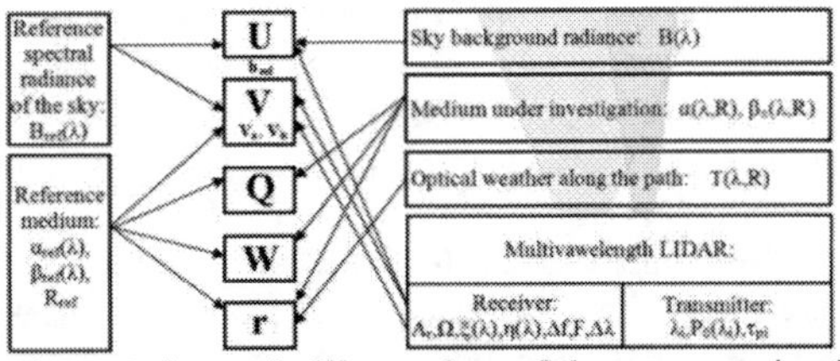

Fig.1. Schematic illustration of the parametrization principle.

power P_{t0} in the absence of background noise; Q_X is the backscatter efficiency of the arbitrary species to the molecular reference; W is the normalized atmospheric component determined by transparencies ratio of the current atmosphere state and the standard molecular atmosphere; U is the ratio of photodetector' s threshold powers taken in presence and in absence of background radiation; r is the normalized range-factor.
When analyzing the potential of any lidar system and in order to have a clear idea about the particular lidar system response at standard atmospheric conditions, in [9] it was proposed to adapt the optical parameters of sounded medium to parameters of the molecular atmosphere with pre-agreed parameters. This allows you to predict more effectively the behavior of lidar at certain atmospheric conditions.

3. RESULTS
As noted, the previously entered reference medium [1-9] has enabled to adapt the arbitrary optical "weather" to standardized parameters of molecular atmosphere. The modified universal lidar parame-

ter for each sounding wavelength can be obtained by using the following three facts: (a) by introduction of standardized conditions of external environment's background radiation in the form of its brightness according to Eq. (2); (b) by taking into account the background brightness in Eq. (4) for threshold power of lidar signal; and (c) by subsequent joint account of reference environment and reference background brightness in analytical expressions for V-parameter calculations.

So when comparing different types of lidars it is useful to identify the V-parameter as a universal dimensionless constant of multi-wavelength system at the specific wavelength $\lambda=\lambda_i$ (i.e. V_i-parameter at λ_i) that characterizes the lidar's capability to remote sensing [7-9] and describes not only the energetic potential, but the lidar instrument's immunity to background radiation as well. Numerically, the V-parameter can be redefined [1] as a ratio of the echo-signal power P_{srefi} received from the reference range R_{ref} of the reference atmosphere (for example, at molecular atmosphere conditions) to the photodetector's threshold power $P_{tBref}(\lambda_i)$ at reference background conditions (for example, at conditions of «averaged» sky background brightness on the middle latitudes):

$$V_i \equiv P_{s0}(\lambda_i) / P_{tBref}(\lambda_i) = \sqrt{PRF \cdot T_0} \cdot P_{stref} / P_{t0} \quad (2)$$

The signal-to-noise ratio at the photodetector output of a lidar receiver channel can be presented in a simple and translucent form [9] with no pulse inte-gration (i.e. $(RFT \times T_0)^{1/2}=1$) as follow:

$$SNR = P_s / [P_q(P_s+P_b)+P_n^2]^{1/2} \quad (3)$$

where P_s and P_b are respectively the optical signal power and the background-radiation power reaching the active surface of the photodetector. The parameter P_q, with power units, is given by

$$P_q = 2\,h\,c\,F\,\Delta f / \eta\,\lambda \quad (4)$$

where h is Planck's constant, c *is* the speed of light, F *is* the photodetector excess noise ratio, Δf

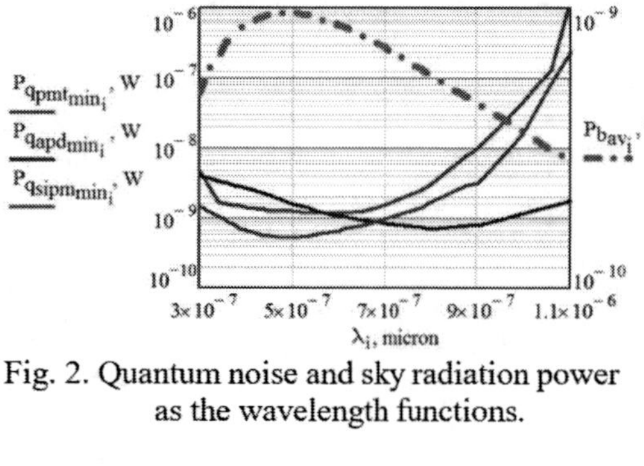

Fig. 2. Quantum noise and sky radiation power as the wavelength functions.

is the photoreceiver's electrical bandwidth, η is the photodetector's quantum efficiency, and λ is the radiation wavelength.

In turn, the parameter P_n is given by

$$P_n = NEP \cdot \Delta f^{1/2} \quad (5)$$

with *NEP* the noise equivalent power of the photoreceiver.

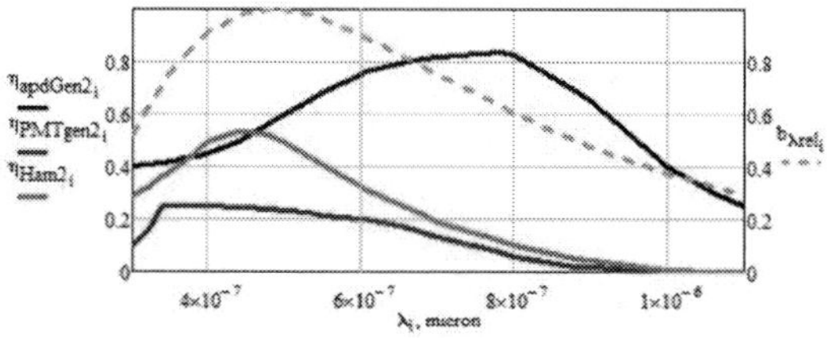

Fig. 3. Relations between three photodetectors quantum efficiency and the wavelength.

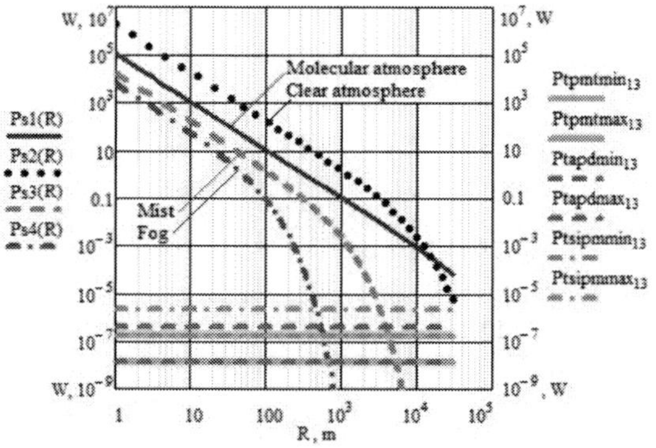

Fig. 4. Lidar returns and threshold signals of different photodetectors.

Under no background radiation conditions ($P_b=0$) we will call the signal-to-noise ratio SNR_D, where the subscript D stands for "dark" as far as background radiation is concerned. A doubly logarithmic graph of as a function of can be approximated very well by two asymptotic straight lines, one of slope 1, given by [2]

$$SNR_D = P_s / P_n \quad (6)$$

for $P_s<P_n^2/P_q$, and another of slope 1/2, given by

$$SNR_D = (P_s / P_n)^{1/2} \quad (7)$$

for $P_s>P_n^2/P_q$.

We can define the «knee» signal power at which the transition from slope 1 to slope 1/2 occurs as

$$P_{sk} = P_n^2 / P_q \quad (8)$$

Accepting the asymptotic approximation, which will overestimate the output signal-to-noise ratio in the neighborhood of the «knee», with a maximum overestimation of 3 dB just when $P_s = P_{sk}$, the out-

DOI: 10.1051/epjconf/201611925009

put SNR_{out} can be easily calculated as a function of the dimensionless parameters V_a and V_k, and of the power decrease V′ [2] with respect to the reference input signal level P_{sref}. Working in logarithmic units, if $V'_k < V_k$ we have

$$SNR_D = \rho_0 + 1/2(V_a - V') \quad (9)$$

whereas if $V'_k > V_k$ we will have

$$SNR_D = \rho_0 + 1/2(V_a - V') + 1/2(V_k - V') \quad (10)$$

where ρ_0 is the reference signal-to-noise ratio.
Eqs. (9) and (10) can be merged [2] in a single one using Heaviside's step function u(x):

$$SNR_D = \rho_0 + 1/2(V_a - V') + 1/2(V_k - V')\, u(V' - V_k) \quad (11)$$

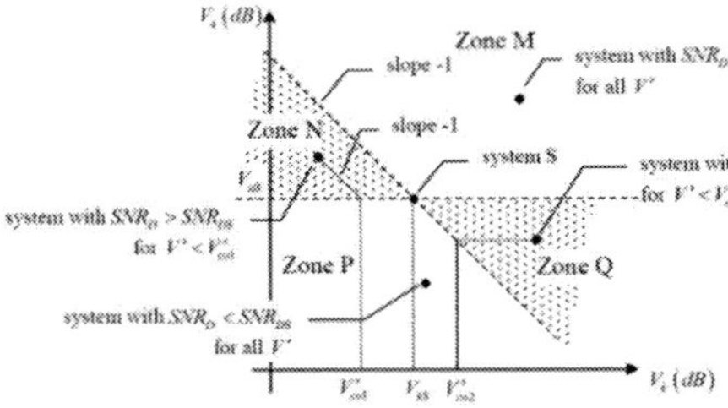

Fig. 5. Graphical comparison of systems.

The physical meaning of the earlier introduced and used U-parameter [1-9] is an extent of the lidar photo-detector's detection threshold degradation under intense daytime sky background.

$$U_i \equiv P_t^B(\lambda_i) / P_t^{Bref}(\lambda_i) \approx \quad (12)$$

$$\approx \left[1 + \sqrt{1 + \frac{4}{SNR_{out}^2} \frac{P_b(\lambda_i)}{P_q(\lambda_i)}}\right] / \left[1 + \sqrt{1 + \frac{4}{SNR_{out}^2} \frac{P_{bref}(\lambda_i)}{P_q(\lambda_i)}}\right]$$

Since it is true for SiPM, PMT and APD detectors: $P_n << P_q$, the formula (9) can be simplified. Then it is easy to see that under intense daytime sky background with $P_b >> P_q >> P_n$, the U-parameter as a degree of detection threshold increasing under the intense background radiation, for $\rho_{out} = 1$ becomes as follows [1]:

$$U_i(\lambda_i) \approx \sqrt{\frac{P_b(\lambda_i)}{P_{bref}(\lambda_i)}} \quad (13)$$

Since we have chosen the reference background brightness in order to provide the reference background power to be the same order of magnitude as the photodetector's quantum noise, i.e. $P_{bref} \sim P_q$, then Eq. (13) can be rewritten in a simpler form as follows:

$$U_i = \frac{2\sqrt{B_{\lambda i} A_r \xi(\lambda_i) \Omega \Delta\lambda \lambda_i \eta(\lambda_i) / \mathrm{h\,c\,F}\,\Delta f}}{1 + \sqrt{1 + 2B_{ref\lambda i} A_r \xi(\lambda_i) \Omega \Delta\lambda \lambda_i \eta(\lambda_i) / \mathrm{h\,c\,F}\,\Delta f}} \quad (14)$$

where $B_{\lambda i}$ is the spectral radiance of the sky zone falling in the receiver field of view at $\lambda = \lambda_i$.

As we are interested to know what is the extent of operation range reduction under strong sky background for different detectors, by following [7] we can obtain the expression:

$$r_b \equiv \frac{R_B}{R_{ref}} \approx \sqrt{\left[\sqrt{1 + \frac{4}{\rho_{out}^2} \frac{P_n^2}{P_q^2} \left(1 + \frac{P_b P_q}{P_n^2}\right)} - 1\right] / \left[\left(1 + \frac{P_b P_q}{P_n^2}\right)\left(\sqrt{1 + \frac{4}{\rho_{out}^2} \frac{P_n^2}{P_q^2}} - 1\right)\right]} \quad (15)$$

If we apply (15) to lidars with different photodetectors, when inequalities $P_b >> P_q >> P_n$ are satisfied, and then use U_i from (10), we can obtain:

$$r_b = U_i^{-1/2} \quad (16)$$

and for specific receiver parameters and certain sky background

$$r_b = \sqrt[4]{2hc\Delta f / \lambda} \cdot \sqrt[4]{F / \eta(\lambda)} / \sqrt[4]{P_b} \quad (17)$$

By applying all the specific features of the lidar system and assessments necessary to the silicon photodetectors under consideration, we obtained results shown in Fig. 6.

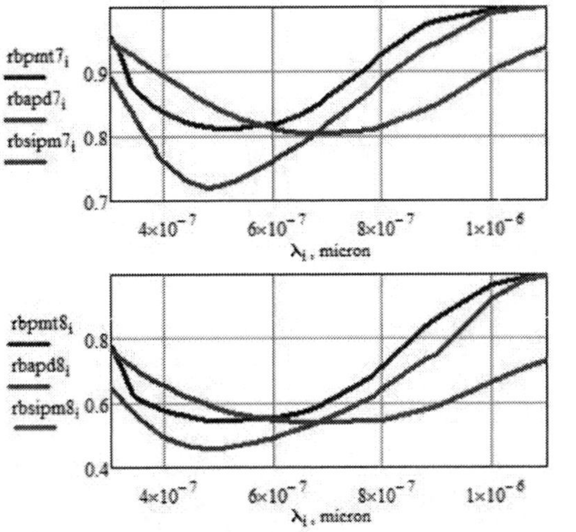

Fig. 6. r_b as a function of λ for $B_\lambda = 10^7$ and $B_\lambda = 10^8$ $W \cdot m^{-2} sr^{-1} m^{-1}$ at λ=483 nm.

CONCLUSIONS

In the development of our previous approaches [1-9], we considered the practical applicability of the V- and U- parameters by extending it to all wavelengths of multi-wavelength sensing lidar and by adapting it to the reference level of the daytime sky back-ground on the basis of use of a more general approach to estimate the threshold power. Using the reference brightness of background radiation within its usual range of changes allows to evaluate the vulnerability of a particular system of impact of the background noise of moderate intensity and to

predict reasonably the degree of deterioration of the lidar' s detection threshold for the impact of sky background of arbitrary intensity.
These parameters allow predicting with a good approximation the signal-to-noise ratio at the photoreceiver output as a function of the signal variation with respect to a reference level. They also provide a quick way of comparing the performance of different systems.

REFERENCES

[1] R.Agishev, A.Comerón, A.Rodriguez, M.Sicard. Dimensionless parameterization of LIDAR for laser remote sensing of the atmosphere and its application to systems with SiPM and PMT detectors, *Applied Optics,* 2014, **53**, № 15, 3164-3175.

[2] A.Comerón, R.Agishev. Dimensionless parameters for lidar performance characterization / in: Remote Sensing of Clouds and the Atmosphere XIX; and Optics in Atmospheric Propagation and Adaptive Systems XVII, *SPIE Proceedings*, 2014, **9242**, pp.1-10.

[3] R.Agishev, A.Comerón, J.Bach, A.Rodriguez, M.Sicard, J.Riu, S.Royo. Lidar with SiPM: Some capabilities and limitations in real environment, *Optics & Laser Technology,* 2013, **49**, 86-90.

[4] R.Agishev. Introduction to SiPMs as novel and advanced detectors for Lidar remote sensing: Operation principles, main parameters, and state-of-the-art. Analytical Review, *UPC Publishing House, Barcelona, Spain*, 2012, 115 pp.

[5] R.Agishev. Advanced methods and means to improve atmospheric lidar stability against sky background clutter, in: Remote sensing of clouds and the atmosphere, *SPIE Proceedings,* 2011, **8177**, 177-183.

[6] R.Agishev. Lidar monitoring of the atmosphere: Monograph (in Russian), *PhysMathGIz, Moscow*, 2009. 316 pp.

[7] R.Agishev, B.Gross, F.Moshary, A.Gilerson, S.Ahmed. Simple approach to predict APD/PMT lidar detector performance under sky background using dimensionless parameterization, *Optics and Lasers in Engineering,* 2006, **44**, No.8, 779-796.

[8] R.Agishev, B.Gross, F.Moshary, A.Gilerson, S.Ahmed, Development of a SNR parameterization scheme for general lidar assessment, *Applied Physics B: Lasers and Optics*, 2005, vol. 80, No. 6, pp. 765-776.

[9] R.Agishev, A.Comeron, B.Gross, F.Moshary, S.Ahmed, A.Gilerson, V.Vlasov, Application of the method of decomposition of lidar signal-to-noise ratio to the assessment of laser instruments for gaseous pollution detection, *Applied Physics B,* 2004, **79(2)**, 255–264 (2004).

EPJ Web of Conferences 119, 25010 (2016) DOI: 10.1051/epjconf/201611925010
ILRC 27

PURE ROTATIONAL RAMAN LIDAR FOR TEMPERATURE MEASUREMENTS FROM 5-40 KM OVER WUHAN, CHINA

Yajuan Li[1,2], Shalei Song[1], Yong Yang[1,3*], Faquan Li[1],Xuewu Cheng[1], Zhenwei Chen[1], Linmei Liu[1], M. Patrick McCormick[3], Shunsheng Gong[1]

[1] *Wuhan Institute of Physics and Mathematics, The Chinese Academy of Sciences, State Key Laboratory of Magnetic Resonance and Atomic and Molecular Physics , Wuhan 430071, China*

**Email:* yangyong2019@sina.com

[2] *University of Chinese Academy of Sciences, Beijing 100049, China*

[3] *Department of Atmospheric and Planetary Sciences, Center for Atmospheric Sciences, Hampton University, Hampton, Virginia 23668, USA.*

ABSTRACT

In this paper a pure rotational Raman lidar (PRR) was established for the atmospheric temperature measurements from 5 km to 40 km over Wuhan, China (30.5°N, 114.5°E). To extract the expected PRR signals and simultaneously suppress the elastically backscattered light, a high-spectral resolution polychromator for light splitting and filtering was designed. Observational results revealed that the temperature difference measured by PRR lidar and the local radiosonde below 30 km was less than 3.0 K. The good agreement validated the reliability of the PRR lidar. With the 1-h integration and 150-m spatial resolution, the statistical temperature error for PRR lidar increases from 0.4 K at 10 km up to 4 K at altitudes of about 30 km. In addition, the whole night temperature profiles were obtained for study of the long-term observation of atmospheric fluctuations.

1. INTRODUCTION

The temperature structure of the atmosphere plays a key role in understanding the dynamical, radiative and coupling processes of different atmospheric layers [1]. To study atmospheric energy transport and the impact on the environment from anthropogenic behaviors, it is of great significance to understand the connective detection of the Earth's atmospheric temperature over a wide vertical range.

To date, versatile lidar techniques have been developed for routine atmospheric temperature measurements with high precision, and high temporal and spatial resolution. Based on different methodologies, temperature profiles can be retrieved from the ground to the mesosphere and lower thermosphere (MLT). Based on the Rayleigh integration technique, absolute temperature above ~ 35 km can be accurately deduced from the relative air density, assuming that the pure atmosphere is in hydrostatic equilibrium and obeys the ideal gas law [2]. When Mie scattering from aerosols gradually contaminate Rayleigh signals, a vibrational Raman lidar that depends on a specific frequency shift can be used instead to extend the temperature detection to lower altitudes [3]. The weaker cross section of vibrational Raman and the interference of the aerosols, clouds and ozone concentration present considerable experimental difficulty for high altitude and high precision detection. With a relatively stronger cross section than vibrational Raman, rotational Raman (RR) lidar is a candidate for high precision temperature measurements from the ground to the upper stratosphere, even in the presence of aerosols and optically-thick cloud layers [4]. With progress in technologies of spectral extraction and weak signal detection, the temperature detection altitude for RR lidar can be extended upward to the stratosphere [5,6]. However, it is still a challenge for the RR technique to provide temperature measurements up to above 30 km with high precision. A large power-aperture product and sufficient suppression of the stronger elastic backscattering (about 6 - 8 orders of magnitude) are important steps to obtain signal-to-noise ratios (SNR) and precision that are required.

In this paper, a PRR lidar system with a high-spectral resolution polychromator was built for high precision temperature measurements from 5-

DOI: 10.1051/epjconf/201611925010

40 km over Wuhan. Temperature measurements up to 40 km by PRR lidar showed the great potential for the further combination with Rayleigh lidar that would provide for study of the middle to lower atmosphere.

2. METHODOLOGY

The principle for PRR temperature measurements is mainly based on the fact that the intensities of molecular rotational Raman lines exhibit different dependencies on temperature [7] as a result of the Maxwell-Boltzmann distribution. To determine the temperature, two domains (RR1 & RR2) with opposite temperature sensitivities are extracted from the anti-Stokes branch of the RR spectrum. To suppress the elastic backscatter returns, state-of-the art interference filters are selected to realize the signal separation with high spectral resolution.

In our simulation, the central wavelengths (CWL) and the full-width-at-half-maximum bandwidths (FWHM) of the filters are the primary parameters taken into account. Two opposite temperature dependence regions are distinguished at the wavelength of ~530.5 nm. In terms of the minimum statistical temperature error, the optimal CWLs for the two RR channels are selected at 531.4 nm and 528.7 nm separately for the FWHMs of 0.7 nm and 1.1 nm [8]. The ratio of the detected signal in RR1 and RR2 channels is sensitive to temperature as shown in equation (1).

$$Q(T) = \frac{S_{RR2}(T)}{S_{RR1}(T)} = \exp\left(\frac{a}{T^2} + \frac{b}{T} + c\right) \tag{1}$$

$$T = \frac{-2a}{b \pm \sqrt{b^2 - 4a(c - \ln Q)}} \tag{2}$$

$$\Delta T = \frac{\partial T}{\partial Q} Q \cdot \sqrt{\frac{1}{S_{RR1}} + \frac{1}{S_{RR2}}} \tag{3}$$

In Equations (1-3), T is the temperature; Q is the signal ratio; S_{RR1} and S_{RR2} are the detected signals in RR1 and RR2 channels; a, b, c are the calibration coefficients; ΔT is the statistical error of RR temperature measurements. With the calibration function in Eq. (1) and the local radiosonde temperature data, the calibration state will be determined for a stable-working lidar system.

Fig. 1 shows the schematic overview of the PRR lidar system whose main characteristics are presented in Table 1. A frequency-doubled and seeded-injection Nd:YAG laser is used as a transmitter to emit light at 532.1 nm with a repetition rate of 30 Hz. After expansion the laser beam with a divergence of ~170 μrad is directed vertically into the atmosphere.

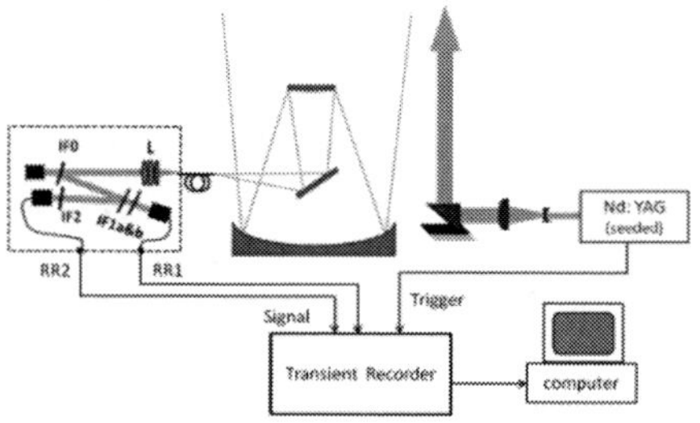

Fig.1 System setup of the rotational Raman lidar

Table1. Technical data of the rotational Raman lidar

Laser	Nd:YAG (seeded)
Wavelength	532.1 nm (in air)
Pulse energy	~300 mJ
Divergence	~170 μrad
Telescope	Cassegrain
Field of View	~300 μrad

Signals backscattered from the atmosphere are collected by a Cassegrain telescope and transferred into the beam collimating and splitting system. Finally two RR signals are captured by a transient reorder (TR20-160, manufactured by Licel) and stored in a computer for the later data processing. Fig.2 is a photo of the transmitter and receiver scene of the lidar system.

Fig.2 Transmitter and receiver of the rotational Raman lidar system

DOI: 10.1051/epjconf/201611925010

Since filter performance changes with the angle of incidence, a higher level of collimation is demanded for the 1.5 m Z-shaped separating light path. Optical simulation shows that the collected backscattering can be coupled efficiently into the fiber. With a set of combined lenses for collimation, the beam spot size is confined within the area of the tilted interference filters. The transmission curves of the narrowband interference filters (IF0, IF1a, IF1b, and IF2) under small angles are shown in Fig. 3.

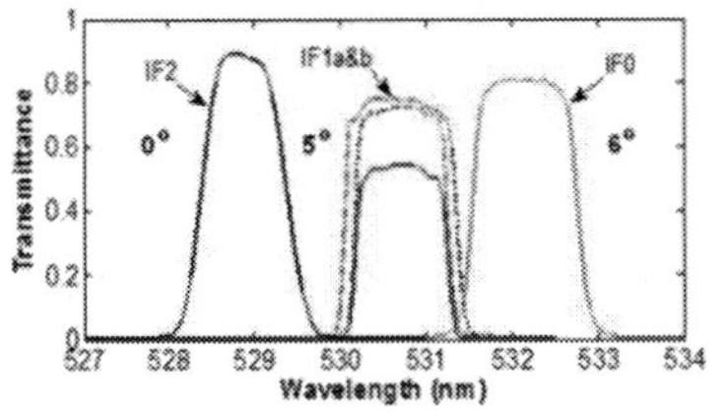

Fig. 3 Transmission curves of the interference filters

3. RESULTS

The PRR temperature measurement lidar has been operated for routine observations in the Wuhan institute of physics and mathematics (30.5°N, 114.4°E) since March 2014. Up to the end of 2014, a total of ~ 400 hours of temperature data were collected from more than 50 lidar observational nights.

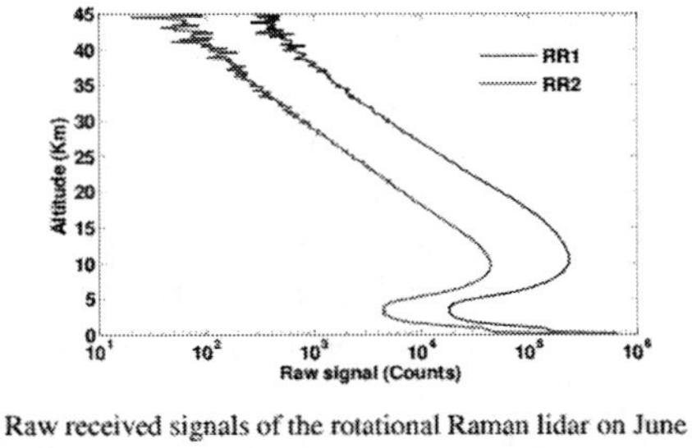

Fig. 4 Raw received signals of the rotational Raman lidar on June 4, 2014.

Fig.4 exhibits the 1h-integrated signal profiles (background-corrected) received separately from the RR1 and RR2 channels with a range resolution of 150 m. The local time (LT) is 20:00–21:00, June 4, 2014. With the bi-axial configuration of the laser emitter and telescope receiver, the stronger signals at lower altitudes are effectively restrained, which ensures the detection of the weaker signals at higher altitudes.

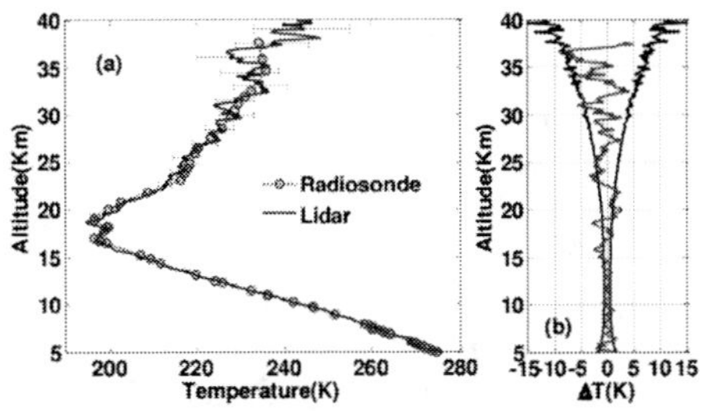

Fig.5 (a) Temperature profiles (5-40 km) measured by the rotational Raman lidar and local radiosonde. (b) Statistical error and the temperature deviation between lidar and radiosonde on June 4, 2014.

Based on the received signals shown in Fig.4, the temperature profile measured by PRR lidar from 5-40 km is plotted in Fig. 5 (a) together with the temperature profile measured by the local radiosonde. The launch time for the local radiosonde is 20:00 LT, June 4, 2014, which is close to the lidar measurements in space and time. Fig. 5 (b) shows the statistical temperature error of the lidar measurements as well as the temperature deviation between lidar and radiosonde. The maximum deviation is not more than 3.0 K for altitudes below 30 km. With the 1-h integration and 150-m spatial resolution, the smallest statistical temperature error is 0.4 K at ~10 km where the signal-to-noise level is highest. Below 10 km there is partial geometric overlap between the laser divergence and the telescope field-of-view, thus the statistical error becomes larger at lower altitudes where the geometric factor (GF) decreases. While the statistical error increases up to 4 K at 30 km, and above that altitude the error becomes larger as well when the signal-to-noise level lowers. With nearly the same GF, temperature retrieved by two RR channels is obtainable over altitudes of 5-40 km.

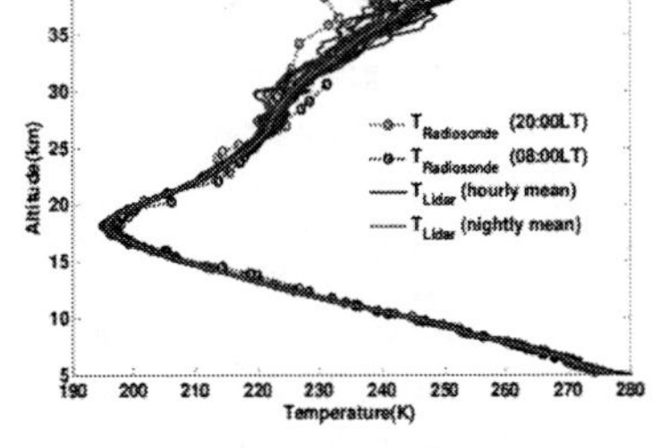

Fig.6 The whole night temperature profiles measured by the rotational Raman lidar and the two local-time radiosondes on August 4-5, 2014.

To achieve smaller temperature error at high altitudes, the temporal or spatial resolution must be reduced. Thus a trade-off exists between the measurement precision and the temporal and spatial resolution. Fig. 6 shows the whole night temperature profiles observed on August 4-5, 2014 with 300-m resolution, 1-h integration and 15-min time shift. As the lidar system operates stably, the temperature is obtained with the initial calibration on July 22, 2014. By comparing the lidar-derived temperature with the radiosondes', the good agreement demonstrates that PRR lidar can work continuously and steadily with long term dependability. Besides the given hourly mean temperature profiles (blue lines), the nightly mean temperature shown with the magenta line is also presented. Estimated by subtracting the nightly mean temperature from the individual hour to hour profiles, temperature fluctuations will be observed to study the dynamic processes of the atmosphere.

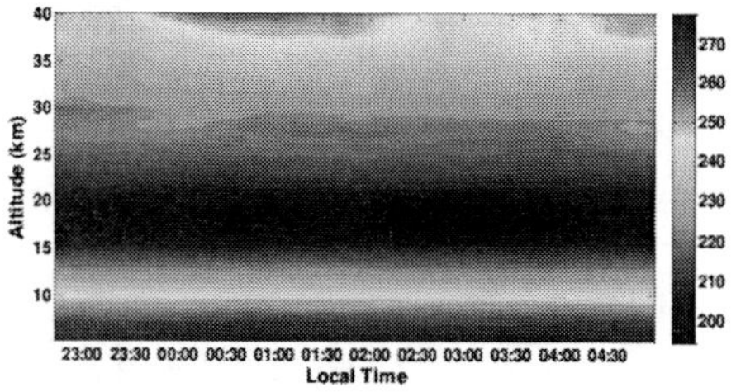

Fig.7 Time-height contour of the temperature profiles observed on August 4-5, 2014.

As observed from the time height contour of the temperature profiles during the night shown in Fig. 7, the thermal structure in the lower atmosphere from 5-40 km is evident with the coldest temperature, ~195 K, at about 18 km.

4. CONCLUSIONS

This paper described the system design and observational results of a PRR lidar for the temperature measurements from 5-40 km over Wuhan. With the high-spectral resolution polychromator for light splitting and filtering, RR scattering returns were detected by weak signal detection technology. Observational results were presented to investigate the overall lidar performance. The temperature profiles simultaneously measured by PRR lidar and the local meteorological radiosonde were compared. The deviation below 30 km was less than 3.0 K, which validated the reliability of the PRR lidar system. The entire night's temperature profiles demonstrate the long-term capability of the PRR lidar system. With high precision, high temporal and spatial resolution temperature measurements, a more detailed analysis of the fluctuations will allow study of atmospheric dynamical processes. Temperature measurements using PRR lidar combined with Rayleigh lidar will allow an even greater lidar capability.

ACKNOWLEDGEMENT

We acknowledge the Wuhan Weather Station for providing the radiosonde data. This work was supported by the National Natural Science Foundation of China, under the Grant nos. 41127901, 41101334, 11403085.

REFERENCES

[1] Alpers, M., R. Eixmann, C. Fricke-Begemann, et al., 2004: Temperature lidar measurements from 1 to 105 km altitude using resonance, Rayleigh, and Rotational Raman scattering, *Atmos. Chem. Phys.*, **4**: 793–800.

[2] Hauchecorne, A., M. L. Chanin, 1980: Density and temperature profiles obtained by lidar between 35 and 70 km, *Geophys. Res. Lett.*, **7**: 565–568.

[3] Keckhut, P., M. L.Chanin, A. Hauchecorne, 1990: Stratosphere temperature measurement using Raman Lidar, *Appl. Opt.*, **29(34)**: 5182–5186.

[4] Behrendt, A., T. Nakamura, T.Tsuda, 2004: Combined temperature lidar for measurements in the troposphere, stratosphere, and mesosphere, *Appl. Opt.*, **43(14)**: 2930-2939.

[5] Achtert, P., M. Khaplanov, F.Khosrawi et al., 2013: Pure rotational-Raman channels of the Esrange lidar for temperature and particle extinction measurements in the troposphere and lower stratosphere. *Atmos. Meas. Tech.*, **6**: 91–98.

[6] Behrendt, A., 2005: Temperature Measurements with Lidar, Lidar: Range-Resolved Optical Remote Sensing of the Atmosphere. C. E. Weitkamp,ed., *Springer*, **102**: 273-305.

[7] Cooney, J., 1972: Measurement of atmospheric temperature profiles by Raman backscatter, *J. Appl. Meteorol.*, **11**: 108–112.

[8] Li, Y., Song, S., Li, F. et al., 2015: High Precision Measurements of lower atmospher ic temperature based on Pure Rotational Raman Li dar, *submitted to Chinese J. Geophys.*

EPJ Web of Conferences **119**, 25011 (2016)
ILRC 27

DOI: 10.1051/epjconf/201611925011

DEVELOPMENT OF MULTI-WAVELENGTH RAMAN LIDAR AND ITS APPLICATION ON AEROSOL AND CLOUD RESEARCH

Dong Liu[1], Yingjian Wang[1, 2], Zhenzhu Wang[1], Zongming Tao[3], Decheng Wu[1,4], Bangxin Wang[1], Zhiqing Zhong[1] and Chenbo Xie[1]

[1]Key Laboratory of Atmospheric Composition and Optical Radiation, Anhui Institute of Optics and Fine Mechanics, Chinese Academy of Sciences, Hefei, Anhui 230031, China
[2]University of Science and Technology of China, Hefei, Anhui 230031, China
[3]Department of Basic Sciences, New Star Institute of Applied Technology, Hefei, Anhui 230031, China
[4]Department of Atmospheric Sciences, University of Wyoming, WY 82070, USA
**Email: dliu@aiofm.ac.cn*

ABSTRACT

A movable multi-wavelength Raman lidar (TMPRL) was built in Hefei, China. Emitting with three wavelengths at 1064, 532, and 355nm, receiving three above Mie scattering signals and two nitrogen Raman signals at 386 and 607nm, and depolarization signal at 532nm, TMPRL has the capacity to investigate the height resolved optical and microphysical properties of aerosol and cloud. The retrieval algorithms of optical parameters base on Mie-Raman technique and the microphysical parameters based on Bayesian optimization method were also developed and applied to observed lidar data. Designing to make unattended operation and 24/7 continuous working, TMPRL has joined several field campaigns to study on the aerosol, cloud and their interaction researches. Some observed results of aerosol and cloud optical properties and the first attempt to validate the vertical aerosol size distribution retrieved by TMPRL and *in-situ* measurement by airplane are presented and discussed.

1. INTRODUCTION

Aerosol and cloud play an important role in modulating the balance of the radiation budget between the earth and its atmosphere directly and indirectly. They still show a big uncertainty on radiative forcing and climate studies[1]. Though variable means have been carried out to make observation of their optical and other properties, the vertical structures are still lack especial for their microphysics as well as the optical properties. Taking advantage of the profiling tool, a moveable multi-wavelength Raman lidar (TMPRL) was built to investigate the height resolved optical and microphysical properties of aerosol and cloud. In this paper, the overall structures and the main specifications the lidar system are introduced. The retrieval algorithms of optical and microphysical parameters of aerosol and cloud are described. The observed results and the airplane validation experiment is presented and discussed. More results will be shown during the conference.

2. METHODOLOGY

2.1 The TMPRL lidar system

TMPRL is a powerful and continuous working lidar installed in a standard container with a window on the roof for easy transportation. A three-wavelength Nd:YAG laser was equipped as the transmitter. Three Mie scattering, two Raman scattering and one depolarization signals were collected by the telescope simultaneously, totaled in six receiving channels. A glass with a self-designed heater was covered on the roof window to get rid of the dew especially before the sunrise to ensure it could work under all weather conditions. The optical and mechanical structures of the lidar system were elaborated designed and installed to keep stable to meet the transportation request.

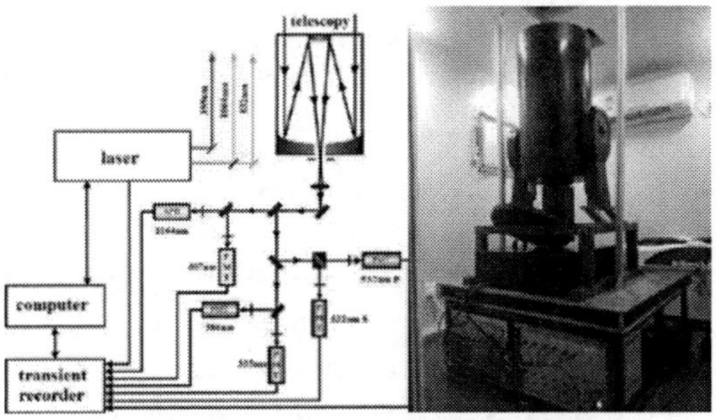

Fig.1 The diagram and photo of TMPRL

DOI: 10.1051/epjconf/201611925011

Fig.1 gives the system diagram and the photo of the TMPRL. Table 1 lists the main specifications of this lidar system. The Range-corrected signals of the six channels are shown in fig.2.

Table 1 Main specifications of TMPRL

Laser	Nd:YAG（Quantel Brilliant B）
Wavelength	1064 nm ,532 nm ,355 nm
Pulse energy	280 mJ , 260 mJ ,160 mJ
Repetition rate	10 Hz
Divergence	0.5 mrad
Telescope	Cassegrain LX400-ACF-14"
Detectors	APD for 1064nm, PMT for others
Acquisition	Licel TR-20-160

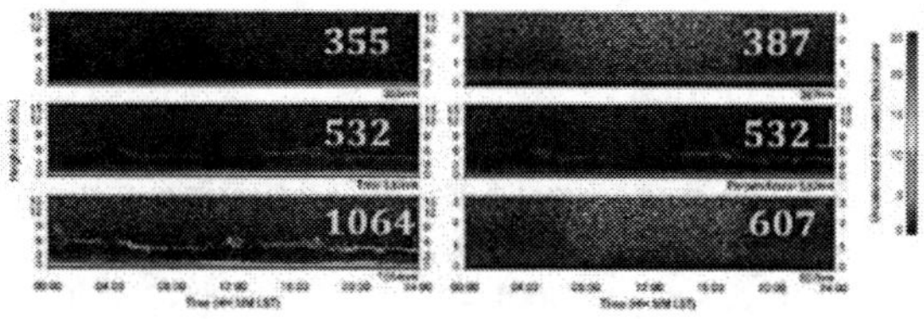

Fig.2 Range-corrected signals for 6 receiving channels

From fig.2, one can see TMPRL worked continuously for the whole day. The Mie signals including the perpendicular signal at 532nm wavelength didn't show any day/night difference which indicated the signal-to-noise ratio of the Mie scattering are pretty good. For the Raman scattering signal is not the case due to their weaker cross section of the nitrogen stimulated by the 355 and 532nm wavelength laser pulse.

2.2 Optical parameters retrieval

The common two component Mie scattering lidar equation can be expressed as:

$$P(r)=P_t k r^{-2}[\beta_{mol}(r)+\beta_{aer}(r)] \times \exp\{-2\int_0^r[\alpha_{mol}(r')+\alpha_{aer}(r')]dr'\} \quad (1)$$

P_t stands for emitted laser power, P stands for received backscatter power in distance r , α and β stands for extinction and backscattering coefficient, the subscript mol and aer stand for molecule and aerosol, k is the system constant.

The backward iteration solution developed by Fernald(1984)[2] is selected for retrieving the extinction coefficient as shown in equation (2).

$$\alpha_a(r)=-\frac{S_a}{S_m}\cdot\alpha_m(r)+\frac{P(r)r^2\cdot\exp[2(\frac{S_a}{S_m}-1)\int_r^{r_c}\alpha_m(r')dr']}{\frac{P(r_c)r^2}{\alpha_a(r_c)+\frac{S_a}{S_m}\alpha_m(r_c)}+2\int_r^{r_c}P(r')r'^2\exp[2(\frac{S_a}{S_m}-1)\int_{r'}^{r_c}\alpha_m(r'')dr'']dr'} \quad (2)$$

For the Mie-Raman combined method, equation (3) and (4) gives the solution of extinction and backscattering coefficient[3].

$$\alpha_{\lambda_0}^{aer}(z)=\frac{\frac{d}{dz}\ln\frac{N_R(z)}{P_{\lambda_R}z^2}-a_{\lambda_0}^{mol}(z)-a_{\lambda_R}^{mol}(z)}{1+(\frac{\lambda_0}{\lambda_R})^{k(z)}} \quad (3)$$

$$\beta_{\lambda_0}^{aer}(z)=[\beta_{\lambda_0}^{aer}(z_0)+\beta_{\lambda_0}^{mol}(z_0)]\frac{p_{\lambda_R}(z_0)p_{\lambda_0}(z)N_R(z)}{p_{\lambda_0}(z_0)p_{\lambda_R}(z)N_R(z_0)} \times\frac{\exp(-\int_{r_0}^{z}[\alpha_{\lambda_R}^{aer}(z')+\alpha_{\lambda_R}^{mol}(z')]dz')}{\exp(-\int_{r_0}^{z}[\alpha_{\lambda_0}^{aer}(z')+\alpha_{\lambda_0}^{mol}(z')]dz')}-\beta_{\lambda_0}^{mol}(z) \quad (4)$$

2.3 Aerosol microphysical parameters retrieval

The aerosol optical properties such as backscattering and extinction coefficient are closely related to their microphysics as shown in equation(5). Q_{ext} and Q_π stands for extinction and backscattering efficiency. m stands for the complex refractive index. n stands for the size distribution.

$$\alpha(\lambda,z)=\int_{r_{min}}^{r_{max}}\pi r^2 Q_{ext}(r,\lambda;m)n(r,z)dr \quad (5a)$$

$$\beta(\lambda,z)=\int_{r_{min}}^{r_{max}}\pi r^2 Q_\pi(r,\lambda;m)n(r,z)dr \quad (5b)$$

The Bayesian optimization method is applied to estimate the aerosol size distribution and refractive index as shown in equation (6).

$$J=\sum_{i=1}^{n_y}\frac{[y_i-H(\mathbf{x})]^2}{\sigma_{y_i}^2}+\sum_{i=1}^{n_x}\frac{[x_i-b_i]^2}{\sigma_{b_i}^2} \quad (6)$$

J is the cost function. The forward model $H(\mathbf{x})$ predicts the observations from the state vector $\mathbf{x}$. Each observation y_i is weighted by the inverse of its error variance $\sigma_{y_i}^2$. For the initial guess b_i, some elements of $\mathbf{X}$ are constrained by an a priori estimate. In the forward model, a three modes aerosol size distribution[4] is adopted. Each mode assumes a lognormal particle size distribution with two parameters.

3. RESULTS

3.1 Aerosol optical properties

Aerosol extinction coefficient profile retrieval is straightforward as described above. Fig.3 shows an example for the retrieved extinction coefficient for 532nm and 355nm, respectively and compared with two different retrieval methods.

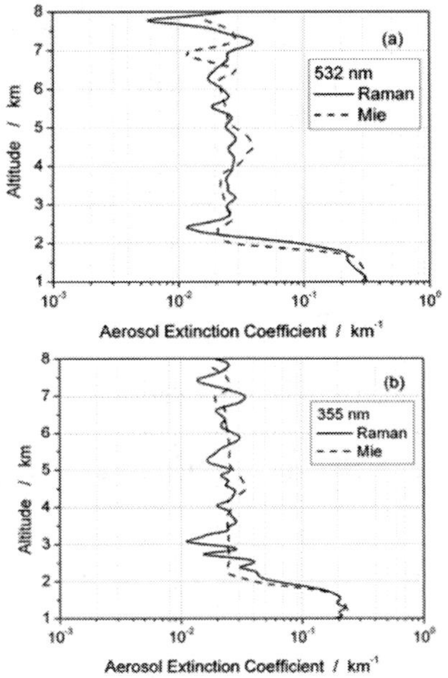

Fig 3 Aerosol extinction coefficient retrieved compared by Mie scattering and combined Raman technique (a) for 532nm wavelength (b) for 355nm wavelength

From the fig.3, one can see these two profiles coincided well, for each wavelength which indicated these two methods both worked correctly. Due to the differential process when applied the Raman technique the variation of the retrieved extinction coefficients is bigger than the Fernald algorithm.

3.2 Cloud optical properties

For the optical thin cirrus clouds which the lidar can penetrated through, applied the optical depth constrained method[5], the lidar ratio cloud be estimate precisely. The color ratio of different wavelength pairs could also been calculated. Based on two-year dataset obtained by TMPRL in Hefei, the backscattering color ratio was calculated and statistically analysed as shown in Fig.4. One can see the maximum occurrence is 0.9, 0.7 and 0.6 for 1064/532, 532/355 and 1064/355 wavelength pairs, respectively.

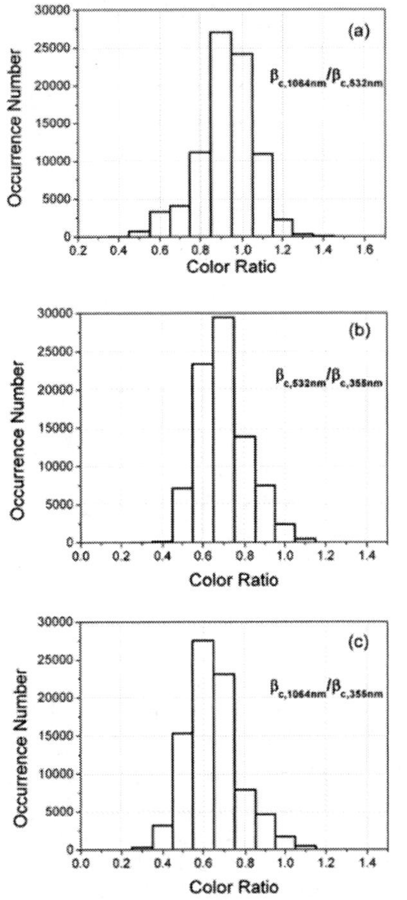

Fig.4 Statistics of the backscatter color ratio for three wavelength pairs (a) 1064/532nm (b) 532/355nm (c) 1064/355nm

For estimating the shape of the ice crystal of these cirrus clouds, the ray tracing method is applied to simulate the backscatter color ratio of six types[6], i.e., aggregate, hollow column, plate, bullet rosette, dendrite and solid column. Three of them is shown in the fig.5.

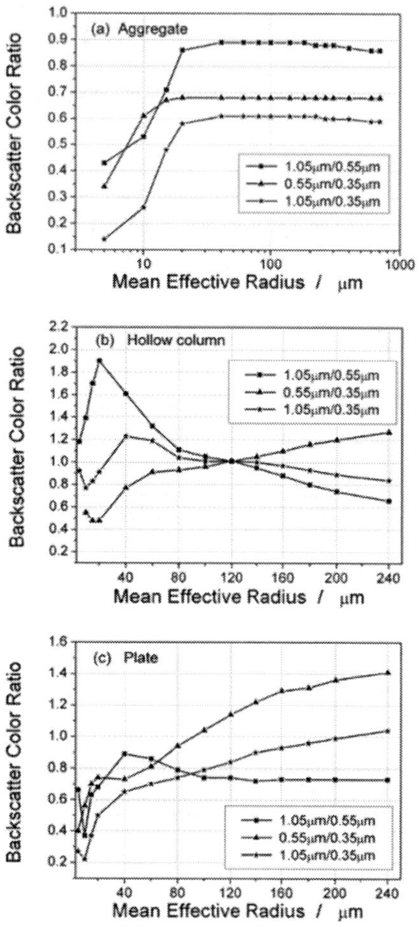

Fig.5 Simulated backscatter color ratio of three wavelength pairs for different shape of ice crystal (a) aggregate (b) hollow column (c) plate

From fig.5(a), one can see the backscatter color ratio kept stable on 0.6, 0.7 and 0.9 when the mean effective radius greater than 20um for 1.05/0.35, 0.55/0.35 and 1.05/0.55um wavelength pairs, respectively. It could be infer that most ice crystal observed in fig.4 seemed to be aggregate shape with mean effective radius greater than 20um.

3.3 Aerosol microphysical properties and validation

In August 2013, TMPRL was shipped to Wenshui city in Shanxi province which was about 1000km far from Hefei to join the field campaign to study the aerosol and cloud vertical properties and their interaction. A PCASP instrument was carried on an airplane to measure the vertical structure of the aerosol size distribution over the location of TMPRL. The retrieved results are shown in fig.6. The trend of these two profiles looks reasonable. The effective radiuses measured by PCASP were larger than the retrieval by TMPRL due to the PCASP measuring the aerosol radius greater than 0.2um. The TMPRL algorithm needs be revised to match the size range of PCASP. More detail results will be done and shown in the conference.

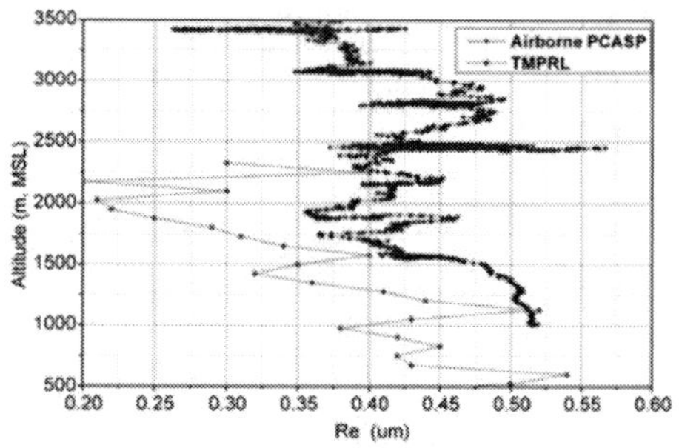

Fig.6 Aerosol effective radius retrieved by TMPRL compared with Airborne measurement

4. CONCLUSIONS

A movable multi-wavelength Raman lidar was built to study both optical and microphysical properties of aerosol and cloud. The first attempt to validate the aerosol vertical size distribution with airborne measurement was done and shown promising. More detail works need to be elaborated in the future, including the robust retrieval algorithms and more validation field campaign.

ACKNOWLEDGEMENTS

This work was supported by the National Natural Science Foundation of China under Grant No. 41075016 and National Basic Research Program of China under Grant No. 2013CB955802.

REFERENCES

[1] IPCC AR5, Summary for Policymakers. (2013)
[2] F. G. Fernald, Appl. Opt. 23, 652-653 (1984)
[3] A. Ansmann, et al. Applied Physics B 55(1): 18-28 (1992)
[4] G. Chen, et.al. Atmospheric Chemistry and Physics, Volume 10, Issue 5, pp.13445-13493 (2011)
[5] S. A. Young, Appl. Opt., 34, 7019–7031 (1995)
[6] Z. Tao, et al., Chinese Optics Letters, 10(5), p050101, (2012)

EPJ Web of Conferences **119**, 25012 (2016) DOI: 10.1051/epjconf/201611925012
ILRC 27

COMPACT RAMAN LIDAR MEASUREMENT OF LIQUID AND VAPOR PHASE WATER UNDER THE INFLUENCE OF IONIZING RADIATION

Tatsuo Shiina[1]*, Tomoyuki Chigira[1], Hayato Saito[2], Naohiro Manago[2], Hiroaki Kuze[2], Toshinori Hanyu[3], Fumihiko Kanayama[3], Mineo Fukushima[3]

[1] *Graduate School of Advanced Integration Science, Chiba University, Chiba, 263-8522, Japan, shiina@faculty.chiba-u.jp*
[2] *Center for Environmental Remote Sensing, Chiba University, Chiba, 263-8522, Japan*
[3] *Japan Atomic Energy Agency, Ibaraki, 319-1194, Japan*

ABSTRACT

A compact Raman lidar has been developed for studying phase changes of water in the atmosphere under the influence of ionization radiation. The Raman lidar is operated at the wavelength of 349 nm and backscattered Raman signals of liquid and vapor phase water are detected at 396 and 400 nm, respectively. Alpha particles emitted from ^{241}Am of 9 MBq ionize air molecules in a scattering chamber, and the resulting ions lead to the formation of liquid water droplets. From the analysis of Raman signal intensities, it has been found that the increase in the liquid water Raman channel is approximately 3 times as much as the decrease in the vapor phase water Raman channel, which is consistent with the theoretical prediction based on the Raman cross-sections. In addition, the radius of the water droplet is estimated to be 0.2 μm.

1. INTRODUCTION

Monitoring of radioactive substances is indispensable for ensuring safety for the ongoing dismantling operation of the nuclear reactors damaged in the wake of the Great East Japan Earthquake. In addition to conventional proximity detectors such as a Geiger counter, stand-off detectors such as a gamma camera has been developed and applied at the site. For obtaining early warning, however, stand-off detection of aerosol particles that are produced under the influence of ionizing radiation is highly desirable.

In this study, we describe a compact Raman lidar system developed for studying phase changes of water in the atmosphere under the influence of ionization radiation. This is a modified version of the system originally developed for hydrogen leak detection: it has attained the detection of 1% hydrogen in the air with the stand-off distance of less than 50 m [1]. Radiation from radioactive materials ionizes molecules in the ambient atmosphere. The reaction of these ions or ion clusters with water vapor leads to the formation of hydrates. Overall, it is expected that water droplets are produced as a result of ionization. In this paper, the effect is studied in a chamber equipped with 9 MBq ^{241}Am sources that produce alpha particles (5.4 MeV). The signals observed for water vapor Raman and liquid water Raman channels are compared for monitoring the phase changes of water.

2. COMPACT RAMAN LIDAR

The compact Raman lidar has an in-line configuration, in which the same optics is employed as both transmitter and receiver. The 349 nm laser output (~100 μJ/pulse, 1 kHz) is generated as the third harmonic of a Nd:YLF laser. As shown in Fig. 1, the Raman backscattering signals of liquid water (396 nm) and water vapor (400 nm) are detected in addition to the hydrogen (380 nm) and nitrogen (408 nm) Raman signals. A filter wheel is used to switch the filters (3 nm FWHM) for 396 and 400 nm. The measurement is conducted with photon counting mode. A narrow field-of-view of 2 mrad is attained with the in-line optics. The maximum observation ranges for atmospheric nitrogen and water vapor detection are 70 and 50 m, respectively. The count of water vapor Raman channel is corrected for the contribution of liquid water Raman, which exhibits long-wavelength tail within the filter transmittance centered at 400 nm.

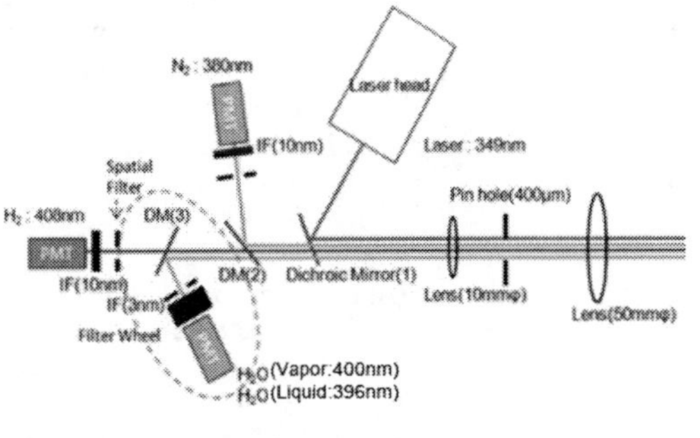

Fig. 1 Improved compact Raman lidar for water phase change observation.

3. RAMAN MEASUREMENT UNDER RADIATION

The radioactivity and radiation dose of the ^{241}Am source are 3 MBq and 50 mSv/h, respectively. Three pieces of sealed ^{241}Am source are employed for investigating the influence of radioactivity. Figure 2 shows the experimental setup of the radiation chamber, which can be humidified with wet towels soaked in a water-filled vessel. Under room temperature, the humidity inside the chamber reaches 90% after 30 - 60 min. The laser beam is passed through fused-silica glass windows, with its position located around 15 mm beneath the surface of the Am sources, in consideration of the relatively short range of alpha particles (a few cm).

4. RESULTS AND DISCUSSION

The experimental results obtained with the humidity of 90% are summarized in Fig. 3. Figure 3(a) and (b) show the Raman signal counts of liquid water and water vapor observed at 396 and 400 nm, respectively, after subtracting the background due to the atmospheric signals outside the chamber. As compared with the original condition in which all the Am sources were shielded, the liquid water counts increased by 16.8% whereas the water vapor counts decreased by 5.1% when all the Am sources were opened (9 MBq in total). The experiments were repeated under different levels of humidity with different numbers of Am sources opened.

If the increase of liquid water Raman counts was caused by water cluster, the decrease of water vapor Raman counts is considered as the reflection of the ionization effect, leading to the formation of water droplets. The Raman lidar signal expressed in terms of the number of photons can be written as

$$N(R) = n \cdot l \cdot s \cdot \frac{A_r}{R^2} \cdot P(R) \cdot f / \frac{hc}{\lambda}, \qquad (1)$$

where n: number density, s: differential cross-section of scattering, l: optical path length, A_r: receiver aperture, R: distance, $P(R)$: laser intensity, f: laser repetition frequency, λ: laser wavelength, h: Planck's coefficient, and c: speed of light. The differential cross-sections of liquid water and water vapor are 5.6×10^{-29} and 2.0×10^{-29} cm^2/sr, respectively [2-4]. Equation (1) indicates that the counted photon number is proportional to the value of ns. If we assume that the effective number densities of water vapor and liquid water molecules are the same, the net ionization effect as detected by the change in Raman signal intensities should be equal to the ratio of the differential cross-sections, which is calculated to be 2.8. This value is fairly close to the experimental result of the ratio of 3.3. Thus, the present results suggest that the changes observed in the two Raman channels are caused by the ionization process under the influence of alpha particles from the Am sources. Moreover, if we assume that each water cluster contributes to Raman scattering, the radius of water cluster is estimated to be 0.2 μm, a reasonable size for a water droplet.

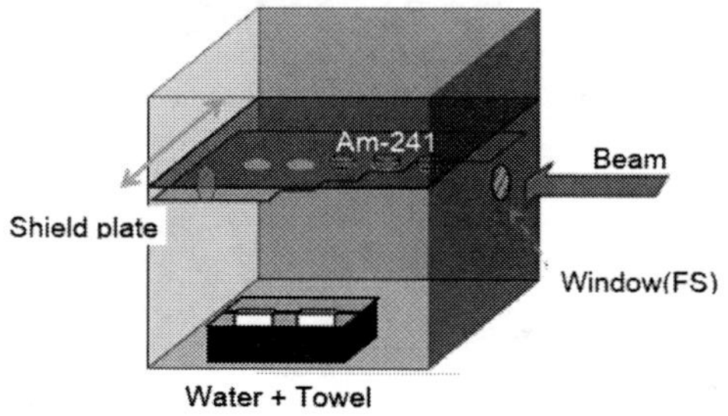

Fig. 2 Radiation chamber with humidifier.

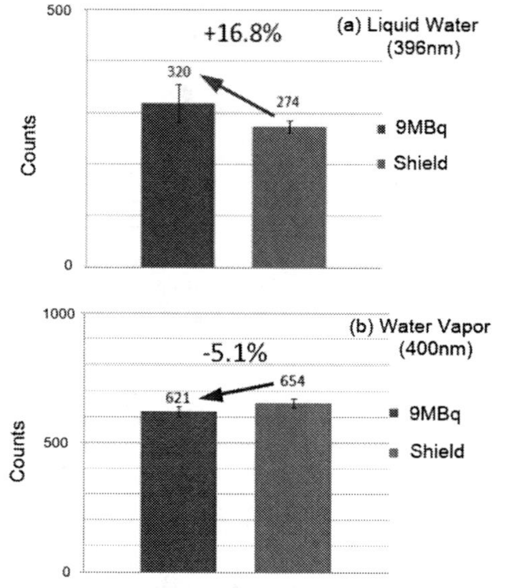

Fig. 3 Phase change measurement with 90% humidity.

REFERENCES

[1] H. Miya, T. Shiina, T. Kato, K. Noguchi, T. Fukuchi, I. Asahi, S. Sugimoto, H. Ninomiya, Y. Shimamoto, Compact Raman lidar for hydrogen gas leak detection, CLEO/Pacific Rim, 2009. ISBN: 978-1-4244-3830-3

[2] S.H. Melfi, K.D. Evans, J. Li, D. Whiteman, R. Ferrare, G. Schwemmer, Observation of Raman scattering by cloud droplets in the atmosphere, 1997, Appl. Opt., 36(15), pp. 3551-3559.

[3] D. Turner and D. Whiteman Eds., 2002: *Handbook of Vibrational Spectroscopy*, Chichester, Wiley & Sons.

[4] B. Li and A. Myers, 1990: Absolute Raman cross sections for cyclohexane, acetonitrile, and water in the far-ultraviolet region, J. Phys. Chem. 94(10), pp.4051-4054.

EPJ Web of Conferences 119, 25013 (2016) DOI: 10.1051/epjconf/201611925013
ILRC 27

ACCURACY OF LINEAR DEPOLARISATION RATIOS IN CLEAN AIR RANGES MEASURED WITH POLIS-6 AT 355 AND 532 NM

Volker Freudenthaler[1]*, Meinhard Seefeldner[1], Silke Groß[2], Ulla Wandinger[3]

[1] *Meteorologisches Institut, LMU, 80333 Munich, Germany, *Email: volker.freudenthaler@lmu.de*

[2] *Insitut für Physik der Atmosphäre, DLR, 82234 Oberpaffenhofen, Germany*

[3] *Leibniz Institute for Tropospheric Research, TROPOS, 04318 Leipzig, Germany*

ABSTRACT

Linear depolarization ratios in clean air ranges were measured with POLIS-6 at 355 and 532 nm. The mean deviation from the theoretical values, including the rotational Raman lines within the filter bandwidths, amounts to 0.0005 at 355 nm and to 0.0012 at 532 nm. The mean uncertainty of the measured linear depolarization ratio of clean air is about 0.0005 at 355 nm and about 0.0006 at 532 nm.

1. INTRODUCTION

Large errors of the linear depolarization ratio (LDR, δ) result from the use of uncertain atmospheric reference values in the relative calibration of the two polarization channels [1]. Several instrumental calibration techniques have been proposed to overcome that, which still can have large uncertainties due to an unknown state of polarization of the emitted laser beam, polarizing optics, rotational misalignment of the optics and of the calibration device, and cross talk of the polarizing beam splitter [2] . In this work we show how those uncertainties have been minimized in the design of POLIS-6, which is a six-channel upgrade of POLIS [3], developed and manufactured by the Meteorological Institute of the LMU in Munich, Germany, in 2013. Furthermore, we show the validation of the measurement accuracy by means of a comparison with theoretically well known LDRs from air molecules.

2. LIDAR SETUP AND SPECIFICATIONS

POLIS-6 is a truly portable lidar system (see Fig. 1) with specifications shown in Tab. 1. It is operated on a tripod and can be pointed to any direction. The desired low distance of full overlap of 70 m requires a large field of view of ±2.5 mrad (tilted slit [4]) and an optimized optical design to keep the optical paths in the receiving optics short. The laser is directly mounted on the rigid telescope tube (see Fig. 1) without any emitter optics, which avoids possible elliptical polarization of the emitted laser beam. The pointing is achieved with a high precision and very stable two-axis tilt mount and controlled with a camera module in place of one of the detector modules.

Table 1 Specifications of POLIS-6

Laser	Nd:YAG Litron LG-250-10
Emitted wavelengths [nm]	355, pol. vertical / 532, pol. horiz.
SHG/THG	KTP II / BBO
Emitted pulse energy	50 / 27 mJ*
Repetition rate	10 Hz
Puls length	4 - 6 ns
Pointing stability	< 70 µrad fw
Beam divergence	<0.5 mrad**
Telescope	Dall-Kirkham
Effective diameter	175 mm
Focal length	1200 mm
Field of view [mrad]	variable, typ. ±2.5
Detection channels	355s, 355p, 387, 532s, 532p, 607
Filter bandwidths: CWL, BW fwhm [nm]	354.6 s&p, 1.1 386.7, 0.52 532.04 s&p, 0.97 607.54, 1.38
Additional polarization-filters: WL (nm), type, extinction ratio	ITOS: 355, XP-38R, 6.4e-4 532, XP-40HT, 2e-4
Data acquisition	6x Licel TR 40-160
Range resolution	3.75 m

* with internal attenuation
** full width (fw) at 90% of output energy

The receiving optics has been selected to yield minimal diattenuation $|D_O|$ with total values (before the polarizing beam splitter cubes (PBC) at 355 and 532 nm, and before the detectors at 387 and 607 nm) smaller than [0.002, 0.032, 0.022, 0.033] at [355, 387, 532, 607] nm, respectively. Additional polarization filters behind the PBCs eliminate all cross-talk. The Δ90°-calibration with mechanical rotation before the receiving optics [2, 3] (see Fig. 1) is used for the relative calibration of the polarization channels. The retardation of the optics can be neglected because the rotational alignment of the receiving optics module can be determined and adjusted with better than 0.5° precision with respect to the laser polarization (see day-to-day variability after 25.06. in Fig. 5).

Figure 1 POLIS-6 with the receiving optics rotated at ±45° (left, right) for the Δ90°-calibration, and at 0° (middle) for atmospheric measurements.

Figure 2 Inside of the receiving optics module of POLIS-6 with the top cover removed. The telescope mount is at the image bottom, indicated by the red circle, which also provides the rotation with the fixed ±45° and 0° positions of the module with respect to the laser/telescope assembly (see Fig.1). The green circle indicates the rotation mount for exact 0° adjustment. At the right and top the detector modules are protruding.

3. THEORY

For the comparison of the measured LDR of clean air ranges with the theoretical values, the rotational Raman lines (RRL) within the bandwidth of the interference filters have to be considered for the calculation of the theoretical molecular (air) LDR. This depends on the bandwidth (BW) and center wavelength (CWL) of the interference filter (IFF) and on the clean air temperature, but also on the exact laser wavelengths. The latter are usually not measured by the laser manufacturers. The fundamental laser wavelength depends, among others, on the temperature of the Nd:YAG rod [5], which we assume for a first estimation to be in the range between 25°C and 85°C. Assuming that the wavelength in air is 1064.15 nm at a rod temperature of 300 K [6] and shifts about +0.005nm/K [5], the second and third harmonics are between 532.07 and 532.22, and 354.71 and 354.81 nm, respectively. While we have measured transmission values for the used 355 nm IFF, we use the specified CWL and BW of the 532 nm IFF (Tab. 1) to calculate the typical two-cavity IFF transmission shape (Fig. 3).

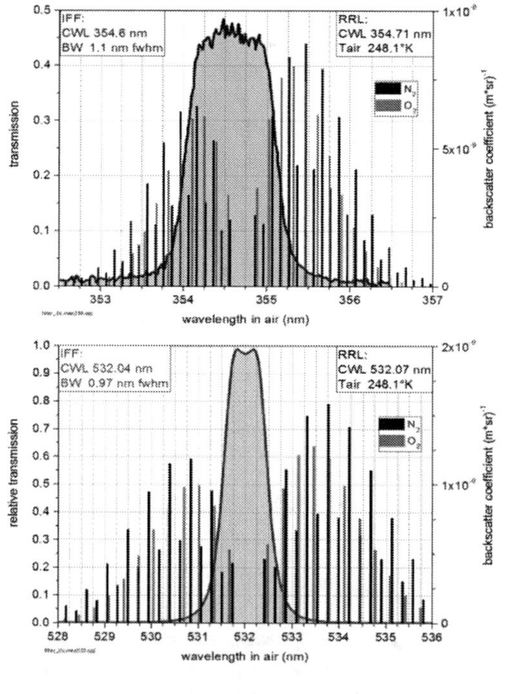

Figure 3 Rotational Raman lines (RRL, backscatter coefficient, right scale) of N_2 and O_2 at 248 K air temperature (central line omitted) and the transmission of the used interference filters (IFF) (see Table 1 for BW and CWL).

We calculate the backscatter coefficients of individual RRLs of N_2 and O_2 (Fig. 3) at different air temperatures according to [7, 8] and weight them with the IFF transmissions for the sum of their LDRs (see Fig. 4).

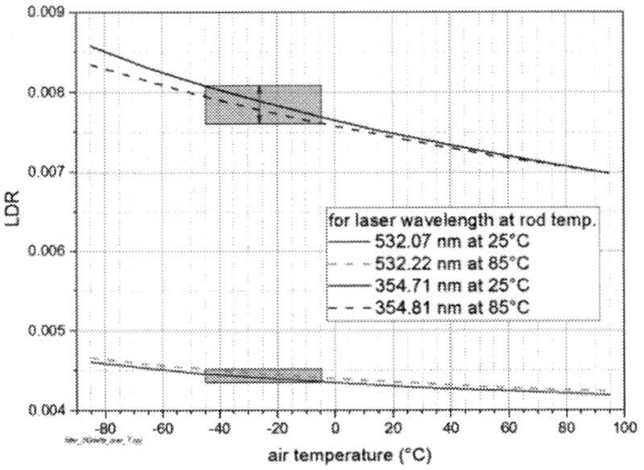

Figure 4 Theoretical LDRs of clean air (with 385 ppmv CO_2 and 0% RH) over air temperature including the rotational Raman lines of O_2 and N_2 within the used IFF bandwidths and considering laser wavelength ranges for rod temperatures between 25°C and 85°C. The red rectangles show the considered variability of air temperature and laser wavelength, and the red arrow the resulting uncertainty of the theoretical LDR.

4. MEASUREMENTS

POLIS-6 was first deployed during a one-month field campaign in 2013. At several nights, measurements of assumed clean air ranges above the aerosol layers were possible with sufficient high signal-to-noise ratios after temporal averaging. With temporal close Δ90°-calibration measurements it was possible to determine the calibration factor and from that the calibrated signal ratio δ^* [2, 3] with an accuracy better than ±2%, and to estimate the rotation ε between the plane of polarization of the laser beam and the receiving optics [2] (see Fig. 5). The corrected LDRs (δ) calculated from Eq. (1) [2] are shown in Fig. 6 together with the error bars, derived with Eq. (2).

$$\delta = \frac{1+D_O}{1-D_O}\frac{\delta^* - \tan^2\varepsilon}{1-\delta^*\tan^2\varepsilon} \qquad (1)$$

$$|\Delta\delta| \approx \frac{1+D_O}{1-D_O}\left(|\Delta\delta^*| + |2\varepsilon\Delta\varepsilon|\right) \qquad (2)$$

After initial system adjustments until 25.06.13, the uncertainty due to Δε becomes negligible compared to Δδ*; the latter mainly stems from signal noise. For the period after 25.06. the mean LDRs are δ = [0.0083, 0.0054] and the mean uncertainties Δδ = [4.8e-4, 5.9e-4] at [355, 532] nm, respectively. The theoretical LDRs (gray areas in Figs. 4 and 6) range between 0.00761 and 0.00808 at 355 nm, and between 0.00436 and 0.00452 at 532 nm, which means that the measured values are about 0.0005 and 0.0012 higher at 355 and 532 nm, respectively.

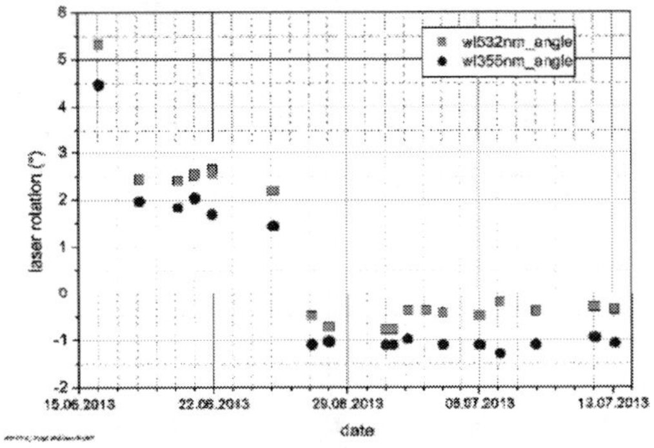

Figure 5 Rotation between the plane of polarization of the laser beam and the incidence plane of the receiving optics (laser rotation) determined with the Δ90°-calibration.

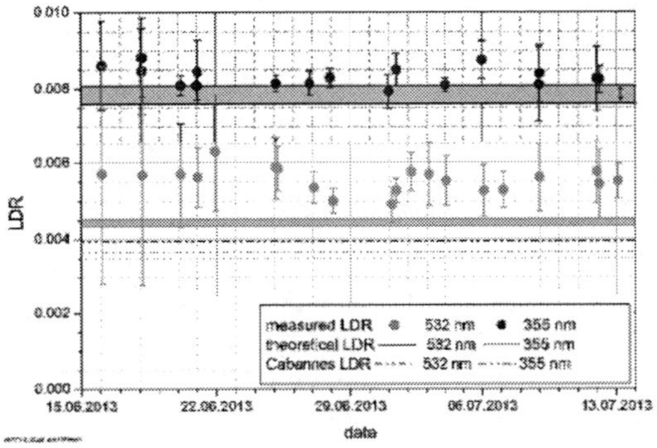

Figure 6 Linear depolarization ratios (LDR) of presumable clean air ranges measured during SALTRACE with POLIS-6 at 355 and 532 nm (dots). The dash-dotted lines show the theoretical LDR if only the central Cabannes line passes the IFF, and the gray areas between solid lines show the theoretical ranges from Fig. 4.

It is remarkable in Fig. 5 that the difference

between the laser rotation at 355 and 532 is about 0.5° over the whole campaign, with a variability well explainable by signal noise.

5. CONCLUSIONS

We show that it is possible to measure the linear depolarization ratio of clean air with a lidar with high accuracy, if polarizing optics in the emitter and receiver (before the polarizing splitter) are avoided, if the cross talk of the polarizing splitter is suppressed, if the simple but high accurate Δ90°-calibration with mechanical rotation is used, and if the plane of the laser polarization is well aligned with the incidence plane of the receiving optics. At this accuracy level it becomes necessary to include in the calculation of the molecular linear depolarization ratio the influence of the rotational Raman lines within the interference filter bandwidth and to account for the uncertainty of the laser wavelengths due to the unknown laser rod temperature. Furthermore, we find a difference in the rotation of the plane of polarization of about 0.5° between the second and third harmonic of the employed laser, which might be caused by the third harmonic generator, because this is the only optical component between the second harmonic generator and the Δ90°-calibrator. While the difference between the measured and theoretical LDR at 355 nm is only about 0.0005 with a mean measurement uncertainty in the same range, the difference is 0.0012 at 532 nm with an uncertainty of about 0.0006. Possible reasons are residual aerosol in the assumed clean air range, or/and elliptical polarization of the 532 nm due to the third harmonic generator.

REFERENCES

[1] Reichardt, J., R. Baumgart, T. McGee, 2003: Three-Signal Method for Accurate Measurements of Depolarization Ratio with Lidar, *Appl. Opt.*, **42 (24)**, 4909-4913.

[2] Freudenthaler, V., 2016: About the effects of polarising optics on lidar signals and the Δ90-calibration, *AMTD*, doi:10.5194/amt-2015-338, in review, 2016.

[3] Freudenthaler, V., M. Esselborn, M. Wiegner, B. Heese, M. Tesche, A. Ansmann, D. Müller, D. Althausen, M. Wirth, A. Fix, G. Ehret, P. Knippertz, C. Toledano, J. Gasteiger, M. Garhammer, M. Seefeldner, 2009: Depolarization ratio profiling at several wavelengths in pure Saharan dust during SAMUM 2006, *Tellus B*, **61 (1)**, 165-179.

[4] Freudenthaler, V., 2003: Optimized background suppression in near field lidar telescopes, http://epub.ub.uni-muenchen.de/12957/.

[5] Kushida, T., 1969: Linewidths and Thermal Shifts of Spectral Lines in Neodymium-Doped Yttrium Aluminum Garnet and Calcium Fluorophosphate, *Phys. Rev.*, **185** , 500-508.

[6] Kaminskii, A., 1990: Laser Crystals, *Springer Berlin Heidelberg*.

[7] Behrendt, A., 2005: Temperature Measurements with Lidar, in Weitkamp, C. (ed.): Lidar, *Springer New York*, 2005, 273-305.

[8] Wandinger, U., 2005: Raman Lidar, in Weitkamp, C. (ed.): Lidar, *Springer New York*, 2005, 241-271.

ENTRAINMENT HEAT FLUX COMPUTED WITH LIDAR AND WAVELET TECHNIQUE IN BUENOS AIRES DURING LAST CHAITÉN VOLCANO ERUPTION

Ezequiel Eduardo Pawelko[1], Jacobo Omar Salvador[1], Pablo Roberto Ristori[1], Juan Vicente Pallotta[1], Lidia Ana Otero[1], Eduardo Jaime Quel[1]

[1] *CEILAP (CITEDEF-CONICET), UMI-IFAECI-CNRS 3351, Juan B. de La Salle 4397, B1603ALO, Villa Martelli, Buenos Aires, Argentina, E-mail: epawelko@citedef.gob.ar*

ABSTRACT

At Lidar Division of CEILAP (CITEDEF-CONICET) a multiwavelength Raman-Rayleigh lidar optimized to measure the atmospheric boundary layer is being operated. This instrument is used for monitoring important aerosol intrusion events in Buenos Aires, such as the arrival of volcanic ashes from the Chaitén volcano eruption on May 2008. That was the first monitoring of volcanic ash with lidar in Argentina. In this event several volcanic ash plumes with high aerosol optical thickness were detected in the free atmosphere, affecting the visibility, surface radiation and therefore, the ABL evolution. In this work, the impact of ashes in entrainment flux ratio is studied. This parameter is obtained from the atmospheric boundary layer height and entrainment zone thickness using algorithms based on covariance wavelet transform.

1. INTRODUCTION

The Atmospheric Boundary Layer (ABL) is the lowest layer that is directly influenced by the Earth´s surface. It responds to surface forcing by frictional drag, evaporation and transpiration, and sensible heat transfer with a timescale of one hour or less [1]. The ABL height is variable in time and space, ranging typically from a few hundred meters up to 1 – 3 km, depending on the nature of the surface and on the meteorological conditions [2]. So, the ABL acts as a reservoir of chemical species where the solar radiation triggers photochemical processes [3].
The lidar (LIght Detection And Ranging) technique is a powerful tool to obtain the boundary layer structures and its evolution with high spatial and time scale resolution. Lidars are able to identify several dynamic parameters of the ABL, using aerosols as tracers of atmospheric dynamics, such as Mixing Layer Depth (MLD) and Entrainment Zone Thickness (EZT). The MLD is the height of the ABL in which pollutants are dispersed by the action of turbulence [4]. Measurements, parameterizations and predictions of the MLD have many theoretical and practical applications such as the forecast of pollutant concentrations and surface temperature, turbulence scaling measurements or the treatment of the mixed layers in numerical weather prediction and climate models [5].
The MLD can be resolved by different methods, however algorithms based on Covariance Wavelet Transform (CWT) [6] [7] [8] are preferred because of their simplicity, robustness and low computational cost.
The EZT is the region at the top of the mixed layer where the free atmosphere above is entrained downward into the mixed layer, and thermals overshoot upward of the mixed layer [9]. This is an important variable that has received little attention due to the difficult measurement. Currently, an iterative algorithm based on CWT to identify these transition layers is used [10].
The effectiveness of EZT and MLD recognition algorithms based on CWT has been tested on lidar measurements with boundary layers, well defined and free of clouds and pollution plumes.
Entrainment Heat Flux (EHF) represents the amount of energy that is supplied to the ABL due to the entrainment processes. It is another important parameter that could be easily computed using the EZT and MLD results. The EHF calculated with lidar profiles has high temporal resolution, unlike the radiosonde data obtained [11].
In Argentina, a lidar network to monitor atmospheric aerosol events, such as those produced by volcanic eruptions is being built [12]. The ABL monitoring behavior is one of the main objectives [13]. A prominent ABL study of aerosol intrusion occurred on May 9th and 10th, 2008, in Buenos Aires. In this event the aerosol lidar of CEILAP (34°33' S; 58°30' W; 17 m asl) registered volcanic ash plumes from the Chaitén volcano (42°50'S; 72°39'W; 962 m asl), that was erupting at that moment [14]. The measurements showed dense aerosol plumes that affected the surface radiation, visibility, ABL and therefore the typical EHF evolution. The study of this case is important because it was the first time that a volcanic eruption was measured and studied with lidar techniques in Argentina.

2. METHOD AND RESULTS

2.a. Instrumentation

The lidar system located at the Lidar Division was designed to monitor the ABL evolution. On May 2008, the instrument was operating with a biaxial overlap configuration. The distance between the emission and the reception was about 30 cm, therefore the blind zone was approximately 300 m. The transmitter was a Nd:YAG laser model Continuum Surelite III P-IV operating at 10 Hz repetition rate and 650 mJ @ 1064 nm. The backscattered light was collected with a Newtonian telescope of 0.5 m diameter and 1 m focal length. In the focus of the telescope there was a 1 mm diameter optical fiber. The data acquisition system had a 20 MHz sampling rate and a 7.5 m vertical resolution [15].

2.b.- Identification of boundary layers

The detection of the ABL structures was done using a Covariance Wavelet Transform and Haar function based algorithm. A combination of the ABL height recognition based on wavelet variance technique and the EZT retrieving based on the iterative wavelet method are synergistically applied to enhance the recognition. First of all, clouds and aerosol plumes are identified and removed in free troposphere of the lidar profiles. This is done automatically using the minimum wavelet dilation with the algorithm proposed by [16]. Then, wavelet variance technique is used in the clean profile to detect the ABL height region. This result is used as a pointer to identify the area where the entrainment zone has to be calculated. EZT is correctly identified with the iterative algorithm based on CWT whenever the upper and lower limits around the pointer are previously calculated. Finally, the MLD is the altitude in the middle of EZT.

This algorithm reduces detection errors because of clouds and aerosol plumes along the lidar profile and it can be applied to extract diurnal and nocturnal boundary layers.

Figure 1 shows a 1064 nm lidar backscatter on May 9th and 10th 2008 with volcanic ash in the free atmosphere which affects the typical ABL evolution. This measurement shows that the ABL has a lower MLD in the second day than the one during the first day.

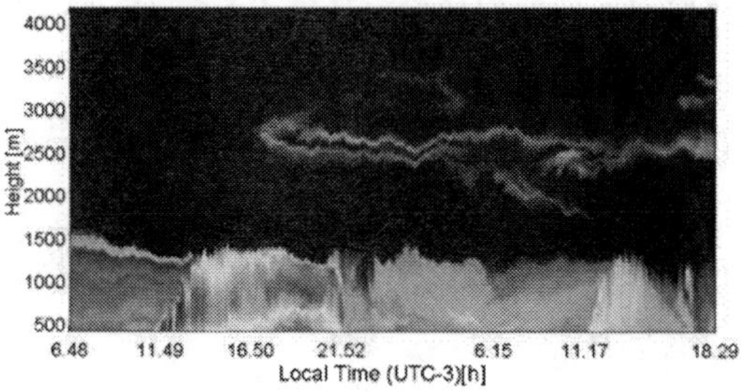

Figure 1. Lidar measurement at 1064 nm (one minute temporal averaged) on May 9th and 10th 2008 in Buenos Aires. Volcanic ashes of Chaitén volcano in eruption at that moment can be seen in the free atmosphere.

Figure 2 shows the ABL height evolution calculated with wavelet variance method.

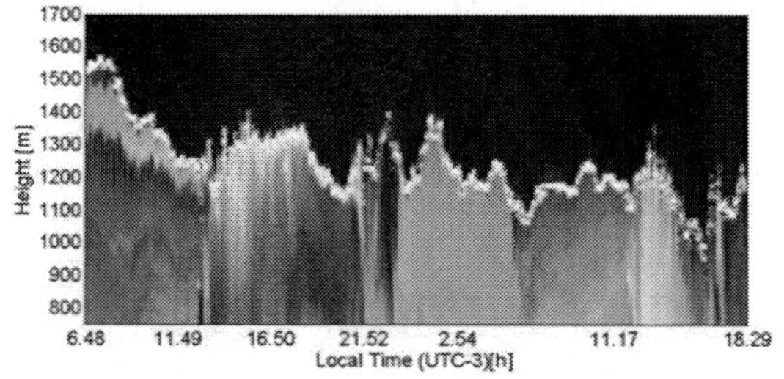

Figure 2. Daytime and nocturnal (residual) ABL height evolution retrieved with wavelet variance technique.

Figure 3 shows the result of the combined method used over the entire matrix of lidar measurements. Figure 4 shows the evolution of ABL height and upper and lower limits of the transition zone for convective ABL on May 9th and 10th. The MLD is the mean value between the upper and lower limits of the EZT.

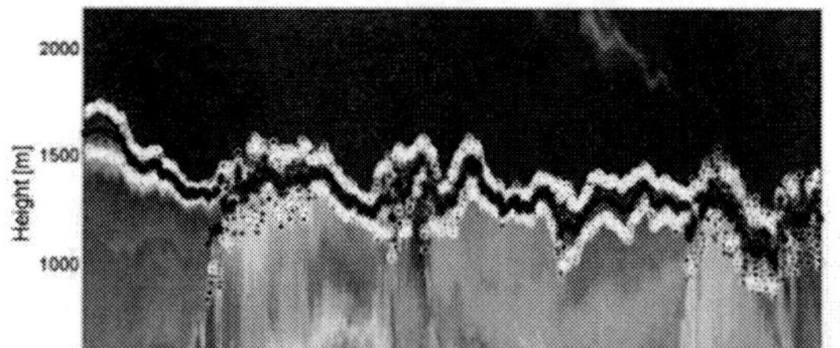

Figure 3. Boundary layers on May 9th and 10th, obtained with covariance wavelet transform. The black line represents the ABL height and the white lines show the upper and lower limits of the transition zone.

DOI: 10.1051/epjconf/201611925014

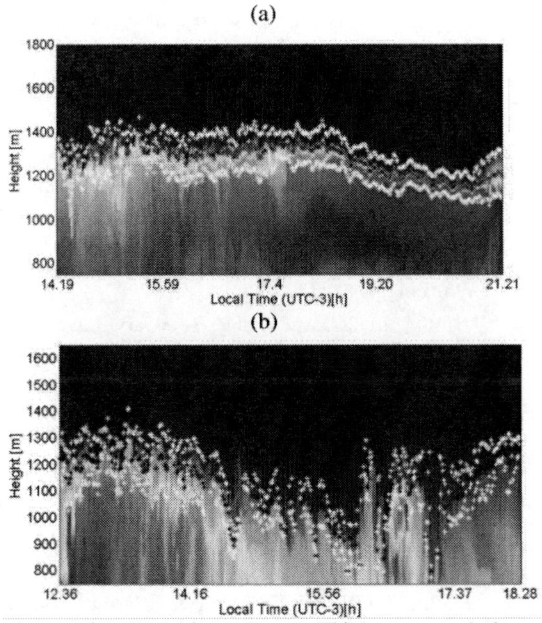

Figure 4. Transition Zone Thickness evolution, obtained with wavelet algorithm, during the monitoring of Chaitén volcano eruption on May 2008 in Buenos Aires. MLD on (a) May 9th and (b) May 10th.

Table I presents statistical analysis for MLD and EZT on May 9th and 10th. The h_2 and h_0 parameters are the upper and lower height of EZT.

Table I. Atmospheric boundary layer geometric parameters on May 9th and 10th, 2008.

Date	Parameter	Mean [m]	Standard deviation [m]	Maximum [m]
May 9th.	MLD	1223	156	1391
	Lower limit of EZT (h_0)	1158	138	1350
	Upper limit of EZT (h_2)	1289	177	1470
May 10th.	MLD	1076	164	1357
	Lower limit of EZT (h_0)	1018	144	1290
	Upper limit of EZT (h_2)	1133	188	1410

2.c.- Heat flux entrainment evolution

The Entrainment Heat Flux ratio (A_r) is calculated with the upper (h_2) and lower (h_0) limits of EZT. This ratio quantifies the amount of energy that is supplied to the ABL by entrainment process. The A_r expression is shown in equation 1.

$$A_r = \frac{h_2 - h_0}{2h_0} \qquad (1)$$

The A_r parameter takes values between 0.1 and 0.4 in atmospheres under free convectivity [17]. Figure 5-a shows the A_r evolution on May 9th between 2.19 and 10.21 pm (local time). Figure 5-b shows the A_r evolution on May 10th between 12.36 and 6.5 pm (local time). Both figures show that the dispersion exceeds the free convectivity limits with a value of A_r smaller than 0.1. On May 10th most of measurements have a value of A_r less than 0.1.

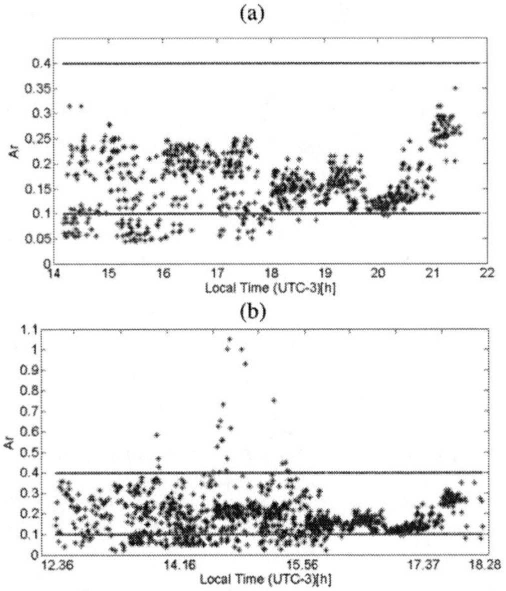

Figure 5. Entrainment Heat Flux (A_r) evolution. (a) May 9th and (b) May 10th.

The A_r on May 9th has a mean value of 0.1623 (with a standard deviation of 0.0576) and on May 10th it has a mean value of 0.2407 (with a standard deviation of 0.1540).

3. CONCLUSIONS

An application of the wavelet technique was used to study the Buenos Aires ABL affected by the presence of volcanic ash. The method allows to extract from the lidar profiles the lower and upper height of the transition zone in order to calculate the entrainment heat flux. On May 9th, It is registered an maximum ABL height of 1391 m and a mean entrainment heat flux (A_r) of 0.1623 (with one standard deviation of 0.0576). On May 10th, It is registered a maximum ABL height of 1357 m and a mean A_r of 0.2407 (with one standard deviation of 0.1540). Finally, A_r results are found below 0.1 on both days studied. This trend was more significant on May 10th, when volcanic ash plumes with greater aerosol optical thickness were recorded.

ACKNOWLEDGMENTS

Authors would like to thank JICA (Japan International Cooperation Agency) and JST through the Science and Technology Research Partnership for Sustainable Development (SATREPS) for supporting this work under the project: "Development of Atmospheric Environmental Risk Management System in South America".

REFERENCES

[1] Stull R. 1998: "*An Introduction to Boundary Layer Meteorology*", Kluwer Academic Publishers, Dordrecht, Boston, London.

[2] Loaec S., S. Lolli, L. Sauvage, M. Boquet, I. Xueref-Remy, 2009: An automatic Planetary Boundary Layer height retrieval method with compact EZ backscattering Lidar in the frame of ICOS campaign, American Geophysical Union, Meeting.

[3] Pallotta J., E. Pawelko, L. Otero, P. Ristori, R. D'Elia, F. Gonzalez, J. Dworniczak, O. Vilar. E. Quel, 2009: Monitoring Aerosol Optical Properties in the ABL, Using Lidar System and Sunphotometer in Buenos Aires, Argentina, Proceedings of the International Radiation Symposium. AIP Conference Proceedings, **1100**, pp. 315-318.

[4] Sicard, M., C. Perez, F. Rocadenbosch, J. Baldasano, D. Garcia-Vizcaino, 2006: Mixed-layer depth determination in the Barcelona coastal area from regular lidar measurements: Methods, results and limitations, Boundary-Layer Meteorology **119**, pp. 135-157.

[5] Seibert P., P. Kromp-Kolb, A. Kasper, M. Kalina, H. Puxbaum, D. Jost, M. Schwikowski, U. Baltensperger, 1998: Transport of polluted boundary layer air from the Fohn Valley to high-alpine sites, Atmos. Environ., **32**, pp. 3953-3965.

[6] Gamage N. and C. Hagelberg, 1993: Detection and analysis of microfronts and associated coherent events using localized transforms, J. Atmos. Sci., **50**, pp. 750–756.

[7] Davis K, D. Lenschow, S. Oncley, C. Kiemle, G. Ehret, A. Giez, 1997: Role of entrainment in surface-atmosphere interactions over a boreal forest", J. Geophys. Res., **102**, pp. 29219-29230.

[8] Davis K., N. Gamage, C. Hagelberg, C. Kiemle, D. Lenschow, P. Sullivan, 2000: An objective method for deriving atmospheric structure from airborne lidar observations, J. Atmos. Oceanic. Technol., **17**, pp. 1455-1468.

[9] Nemuc A., C. Talianu, L. Belegante, R. Ngo, C. Derognat, 2012: The boundary layer height and entrainment zone assessment from lidar, meteorological and forecast model data, 9th International Symposium on Tropospheric Profiling, L'Aquila, Italy.

[10] Brooks I., 2003: Finding Boundary Layer Top: Application of a wavelet covariance transform to lidar backscatter profiles, J. Atmos. Ocean. Tech., **20**, pp. 1092–1105.

[11] Flamant C., J. Pelon, P. Flamant, P. Durand, 1997, Lidar determination of the entrainment zone thickness at the top of the unstable marine atmospheric boundary layer, Journal Boundary-Layer Meteorology, **83**, pp. 247-284.

[12] Ristori P., L. Otero, E. Pawelko, J. Pallotta, R. D'Elía, F. Chouza, F Gonzalez, J. Dworniczak, A. Pereyra, M. Fernández, S. Lema, N. Sugimoto, E. Quel, 2012: Development of an Argentinean lidar nework to monitor the volcanic plume and dust in Patagonia, Proceeding of the 26th International Laser Radar Conference, Porto Heli, Grecia.

[13] Pawelko E., P. Ristori, I. Otero, R. D'Elia, A. Pereyra, O. Vilar, F. Chouza, J. Pallotta, F. González, M. Fernandez, M. Lema, N. Sugimoto , E. Quel, 2012: Multiwavelength Raman lidar construction to monitor volcanic ash and aerosols in Bariloche international airport, Argentina; Proceeding of the 26th International Laser Radar Conference, Porto Heli, Grecia,

[14] Pawelko E., L. Otero, P. Ristori, E. Quel, 2008: Chaitén Volcanic Aerosol Transport Study During May 2008, Proceeding of the 25th International Laser Radar Conference, St. Petersburg, Rusia, **1**, pp. 215-218.

[15] Otero L., P. Ristori, E. Pawelko, J. Pallotta, E. Quel, 2011: Six-Year Evolution of Multiwavelength Lidar System at CEILAP, V Workshop on Lidar Measurements in Latin America, Opt. Pura Apl., **44**, pp. 13-18.

[16] Baars H. , 2007: "*Thesis: Continuous monitoring of the planetary-boundary-layer depth with lidar*", University of Leipzig, Germany.

[17] Deardorff, J., G. Willis, B. Stckton, 1980: Laboratory Studies of the Entrainment Zone of a Convectively Mixed Layer, J. Fluid Mech, **100**, pp. 41-46.

 DOI: 10.1051/epjconf/201611925015

THE ONLINE SYSTEM FOR LIDAR DATA HANDLING AND REAL TIME MONITORING OF LIDAR OPERATIONS AT ALO-USU

Luis A. Navarro [1]*, Vincent B. Wickwar[1], Jose Gamboa[2], Marco Milla[2]

[1]*Physics and Center for Atmospheric and Space Science, Utah State University, Logan, UT 84322, USA,*
**Email: luis.navarro.dominguez@gmail.com*

[2]*Jicamarca Radio Observatory, Instituto Geofisico del Peru, Lima, Peru*

ABSTRACT

It is no longer sufficient to use lidar, such as the Rayleigh lidar at the Atmospheric Lidar Observatory (ALO) at Utah State University (USU), to observe the middle atmosphere and reduce the data to geophysical parameters. Extended operations, with inevitable equipment, data reduction, and analysis improvements, require us to keep careful track of all these changes and how they affect the scientific products. Furthermore, many of the funding agencies and the journals now require us to do, at least, some of this. We have built three interconnected data structures to organize and manage the different hardware and software set-ups from the Internet as well as to keep track of the products generated by these. These structures were implemented as a database, which is particularly important for groups with a large volume of information like the Rayleigh Group at ALO-USU.

1. INTRODUCTION

The Rayleigh-scatter Lidar Group in the Center for Atmospheric and Space Sciences (CASS) at Utah State University has been developing and operating Rayleigh lidars since 1992 to study the middle atmosphere. During this period, different hardware and software have been developed and implemented to improve the lidar's capabilities and the scientific products. More generally, researchers in many fields have faced a similar challenge keeping track of the constantly improving hardware and software in their systems and their effects. To help, they use several tools such as Evernote, TeamViewer, Google Drive, as well as the traditional logbooks and notebooks. However, few of these tools have been designed for scientific instrumentation, let alone for lidar systems.

In addition, to make it easier to verify scientific conclusions and to make scientific results more available to tax payers, funding agencies such as NSF and NIH have added a requirement for proposals to include a Data Management Plan. These plans are focused, in general, on the management, preservation and sharing of data and, more specifically, on sharing the scientific products generated by the proposed research.

So far, different institutions and communities have developed their own solutions for sharing and presenting results. Among others, this is the case of the DMPTool [1] developed by the University of California, San Diego. This is an online platform that helps produce the required Data Management Plan document. This platform uses different document templates, depending on the funding agency, and lets one institution share them with other institutions doing similar research. This last characteristic is particularly important for promoting networking and collaboration among groups.

Another example is the OpenMadrigal Initiative [2]. It is a platform, developed in Python that supports an on-line database for atmospheric data produced by a variety of instruments and models. It enables anyone to examine and download the results of scientific measurements from the web. It includes some lidar data. Contents of the database can be downloaded in several possible formats, such as the CEDAR and HDF5 formats.

Aware of this necessity, the Optical Instrumentation Department from the Jicamarca Radio Observatory started a development for optical data handling that was subsequently specialized and improved for lidar instrumentation at USU.

In the following section, a description of the data structures for the Rayleigh Group at ALO-USU is presented. Section 3 shows some of the online features available from the Internet. Finally, Section 4 indicates how it is being used at USU for lidar and correlative data.

2. METHODOLOGY

The system is based on three different structures designed to save different types of information. These are interconnected to give the user the ability to save and retrieve information in a quick and ordered way.

2.1. Structure for the Instrumentation

This structure is based on past and present experience operating optical instrumentation. It systematically organizes the relationships between the instrumentation and the observations.

It starts by defining a Type of Instrument, which in turn is used to define many Instruments, each of which are mainly described by a Type, Geographical Location and a Person to Contact. It also has a Code to match with other classifications such as the instrument code in the Madrigal data model [2]. For each instrument, we can define several Instrument Models, which are meant to describe the instrument hardware through a set of parameters and their values. The Rayleigh Group at ALO-USU used a design from 1992 to 2004, called "Old design". Since 2012, it has used another design, called "New Design". Many of these terms, and examples, are shown in Figure 1.

Following up, several Operating modes might be available due to the hardware specifications for each Instrument Model. Each Operating Mode describes how the hardware parameters interact through a set of software settings that control how the hardware functions.

As an example, the Instrument Model "New Design" has servomotors to position the optical fiber at the focal point of the telescope mirror [4]. This is an Operating Mode called "Calibration". In addition, this Instrument Model is able to collect data continuously throughout the night in an Operating Mode called "Extended Sounding". This is possible through the Value of Setting for the number of laser pulses to integrate over [4]. These can also be seen in Figure 1.

In the same sense, due to the Value of Settings, each Operating Mode opens up the capability of performing different experiments. An Experiment can also be reached by many Operating Modes. Then, an experiment Trial represents the execution of the Experiment for a specified date or time interval. In Figure 1, the experiment Coupling Temperatures could measure temperatures and relative densities from the stratosphere to the lower thermosphere.

In addition, Trials are related to a Campaign, which is a label given to an interval of time that usually characterizes the motivation or purpose of an Experiment or set of observations. This is particularly important when different groups run different instruments. Examples are described for a series of optical campaigns (IMPAC) [3].

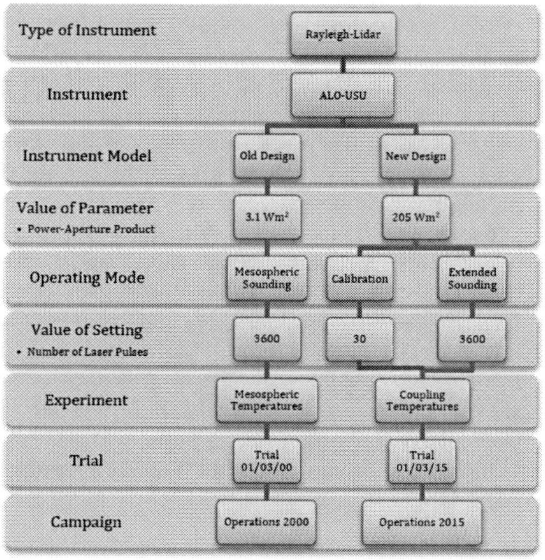

Figure 1. Configuration of the instrumentation structure for Type of Instrumemt, Rayleigh Lidar, at USU.

2.2. Structure for Products

This structure is intended to manage the Type of Product Set. A Product is, in general, any type of information generated by a Trial of an Experiment or by a Campaign such as files or data depending on what information is required to be handed directly from the database. These products can be classified in Types of Product Sets like raw data, processed data, published data or any other type defined by the user. Also, even when the Generators of Products are meant to be a Trial or a Campaign, other types of Generators can be setup by the user.

For ALO-USU, four different Types of Product Sets are defined: "Calibration Data", "Raw Data", "Temperature Data" and "Cedar Data". The

"Raw Data", shown in Figure 2, is designed for the data recorded from a Multi-Channel Scaler in normal operation, whereas the Type of Product Set "Calibration Data" is used for data recorded during telescope alignment. In the same sense, the Type of Product Set "Temperature Data" is for data generated by software that processes the Raw Data to produce mesospheric temperatures. "Cedar Data" is used for the processed data that is transformed into the Cedar data format.

On the other hand, the user can process a Product Set to generate more information. Usually, the routines, scripts or programs that process information need initial conditions or, in general, specific setups that the software will use to perform the task.

For the case of Raw Data generated by the experiment "Coupling Temperatures", a task was created to reduce these data and to save the resulting temperatures in the database as products with type "Temperature Data". To fulfill this task, to determine the temperatures, we need an initial temperature [6], which was setup as a condition.

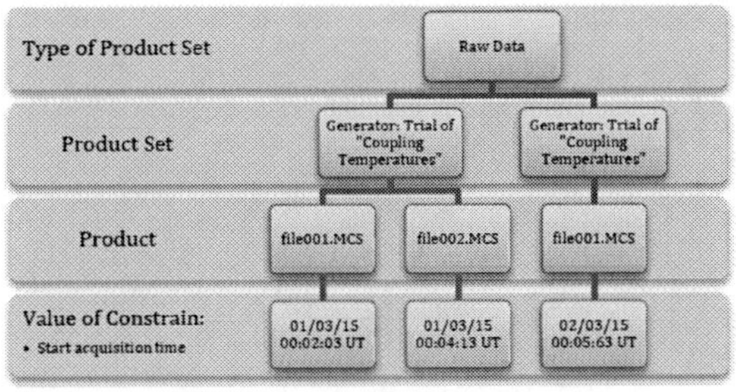

Figure 2. Configuration for Type of Product Set "Raw data" for ALO-USU.

2.3. Structure for Supervisory Devices

To supervise the operation of the lidar system, we define an entity called the Supervisory Device (SD). It is a web server running on the computer controlling the lidar operation. It receives and executes commands to inform the lidar operator about the lidar status or to send back data files to be saved.

A basic set of commands were implemented in the SD so as to perform an initial registration of the SD (see Figure 3), list directories, send files, start a watchdog thread to look over any newly generated file and send it to the server, start a ping thread to find the average communication time to check the network connection quality (e.g., Figure 4), send a status command and send a shutdown command to the SD.

For the instrument Rayleigh lidar at ALO-USU, this was implemented with the HTTPServer class of the Python core library. The results, state and variables handled by this communication are saved in the database following this structure.

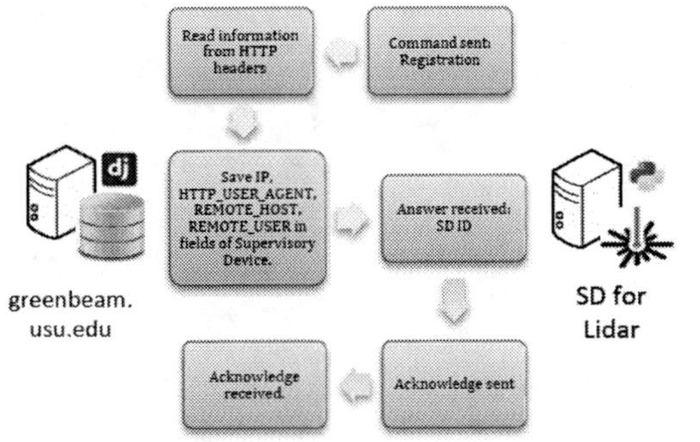

Figure 3. Procedure for registering the SD into the Supervisory Device structure.

3. RESULTS

For the Rayleigh Lidar Group at ALO-USU, the site https://greenbeam.usu.edu was implemented to monitor and handle the lidar data (see Figure 5). The database engine MySQL and the Python web framework Django are used to implement the structures discussed in Section 2. The scheduling of tasks is done by the distributed task queue system Celery. The Google API is used to handle user authentication and the data are backed up into the cloud storage Box.

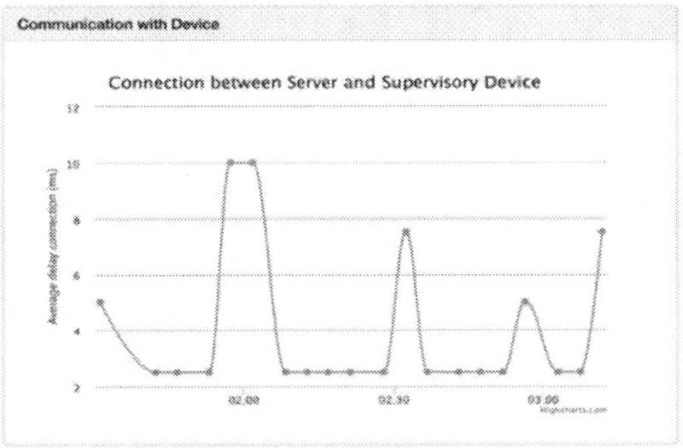

Figure 4. Example of online chart showing the quality of the connection to the Supervisory Device.

DOI: 10.1051/epjconf/201611925015

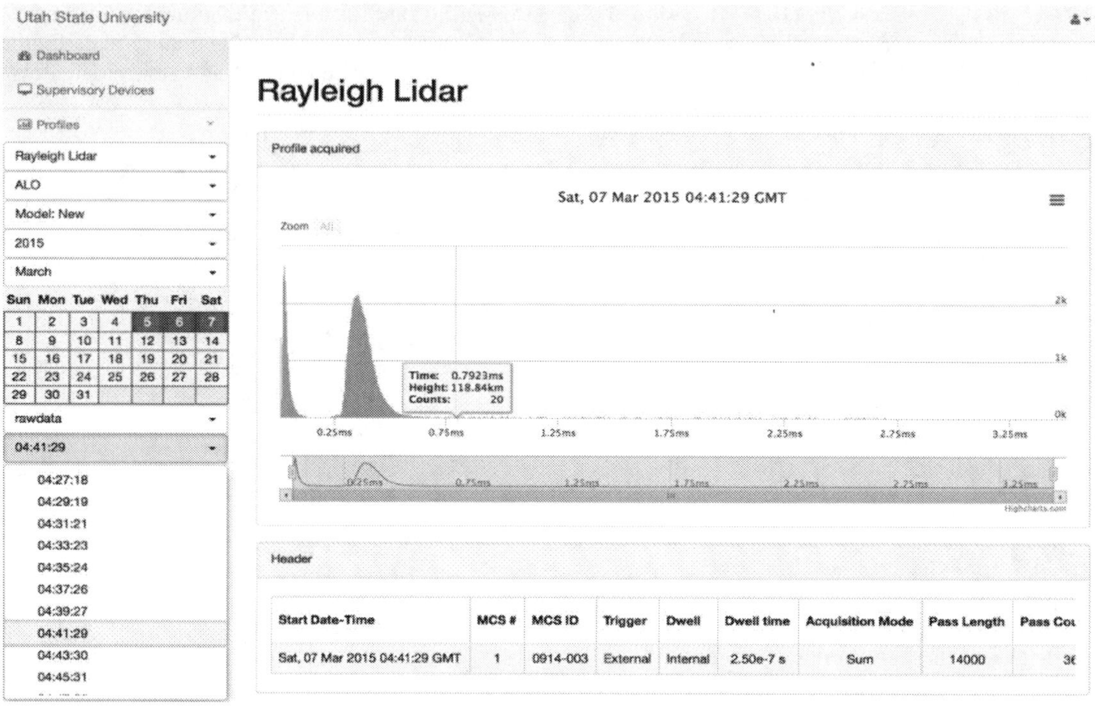

Figure 5. Example of online viewer of Products for the Rayleigh Lidar at ALO-USU.

4. CONCLUSIONS

The interconnected design of the structures and the use of informatic tools like MySQL, Django, and Celery (see Figure 6), make it possible to handle a large volume of data. While this was developed for lidar at ALO-USU, much of this can be adapted to the needs of other lidar groups.

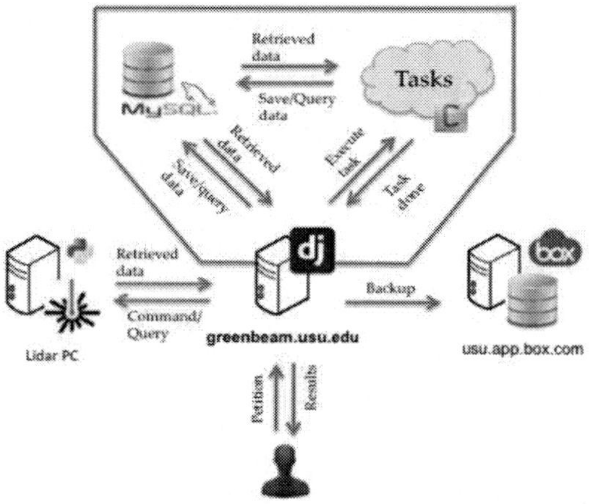

Figure 6. Data flow design for the Rayleigh Lidar at ALO-USU located in https://greenbeam.usu.edu.

REFERENCES

[1] https://dmptool.org

[2] http://www.openmadrigal.org

[3] Wu, Q., W. Wang, M. G. Conde, A. L. Arulian, R. Niciejewski, and J. W. Meriwether, 2009: Ionosphere Mesosphere Polar Aeronomy Campaign (IMPAC). AGU Fall Meeting 2009, San Francisco, CA.

[4] Wickwar, V. B., L. Sox, D. L. Barton, J. P. Herron, J., and M. T. Emerick. June 22, 2014: Extremely Sensitive Rayleigh-Scatter Lidar at USU. Presented at the 2014 CEDAR Workshop, University of Washington, Seattle, WA.

[5] Peterson, L. W., 2011: *Lidar User's Manual*, Undergraduate Honors Theses, 20 pp, Utah State University, Logan, UT.

[6] Wickwar, V. B., and L. Sox, June 22, 2014: Interpretation (or is it Calibration?) of Rayleigh-Scatter Lidar Signals, CEDAR Workshop, University of Washington, Seattle, WA.

EPJ Web of Conferences **119**, 25016 (2016) DOI: 10.1051/epjconf/201611925016
ILRC 27

APPLICATIONS OF KHZ-CW LIDAR IN ECOLOGICAL ENTOMOLOGY

Elin Malmqvist [1,2]*, Mikkel Brydegaard[1,2]

[1]*Lunds Laser Centre (LLC), Lund University, Lund, Sweden, *Email: elin.malmqvist@forbrf.lth.se*

[2]*Centre of Animal Movement Research (CAnMove), Lund University, Sweden*

ABSTRACT

The benefits of kHz lidar in ecological entomology are explained. Results from kHz-measurements on insects, carried out with a CW-lidar system, employing the Scheimpflug principle to obtain range resolution, are presented. A method to extract insect events and analyze the large amount of lidar data is also described.

1. INTRODUCTION

Insects are a very dominant group of animals on Earth and they are essential in most ecosystems. The ability to systematically and efficiently detect, identify and monitor insects is of great interest from an ecological, as well as an economic and social, point of view. For instance, pollinators and insect pests have great impact on agriculture, and therefore also on food production. Many insects, such as malaria mosquitos, also function as disease vectors [1].

Over the last several decades, lidar methods have been used and developed for different varieties of atmospheric monitoring, see e.g. [2]. Our research group at Lund University have in recent years explored opportunities for applications in ecological entomology [3]. Lidar measurements can provide continuous and extensive monitoring of insects, which is impossible for manual methods such as collection in sweep nets or traps. Information about insect properties such as size, velocity and wing beat frequency can be extracted from lidar data. These properties span a parameter space with a certain dimensionality. A higher dimensionally of this space increases the likelihood of successful separation between different species, sexes and age groups.

Due to their natural wing beat cycle, flying insects are seen as blinking scattering particles, meaning that the backscattered lidar signal, defined in terms of an optical cross section (OCS) with unit cm^2, is modulated. The fundamental frequency of insect wing beats can vary between 10-1000 Hz [4], and we have seen that the signal can contain up to 20 harmonics due to specular reflexes. When the Nyquist frequency is considered, it is clear that a lidar system with a kHz sampling rate is required to obtain the full harmonic content of the signal from insect wing beats. The frequency content of wing beats is related to factors such as body mass, wing span and wing area through fundamental aerodynamic constraints (se e.g. [5]) and it is thus of high value for species classification purposes, see e.g. [6].

Resolved wing-beat modulation have previously been used to discriminate lidar signal of honeybees from the signal of the surrounding environment, see e.g. [7]. While pulsed Nd:YAG lasers with kHz-repetition rates were utilized in those cases, our research group has developed a kHz-continuous wave (CW) lidar system, relaying on cheap and small diode lasers. The system has been used in several field campaigns, aimed at insect detection. A method for reducing time-range data, obtained with kHz lidar, into relevant parameterized observations, such as wing beat frequency and size has also been developed. Examples of observed insect events from the kHz-lidar data, and their parameterized properties will be presented here.

2. METHODOLOGY

Figure 1 displays a schematic setup of the CW lidar system. The beam is expanded and transmitted by a refractor telescope, and the beam is terminated on a hard, preferably black, target. The light is collected using another refractor telescope. Range resolution is obtained by sharply imaging the laser beam onto a line-detector. This is achieved by arranging the planes of the laser beam, the detector, and the collection lens according to the Scheimpflug principle. It states that to achieve focus, the three planes must intersect along a common line, see upper panel of Fig. 1. For more information about the Scheimpflug setup see [3].

The specific data set utilized in this work was acquired in Brunnslov, Sweden, in the summer of 2013, using the Lund University Mobile Biosphere Observatory (LUMBO). LUMBO is a moveable research platform intended for quantitative monitoring of atmospheric fauna and it contains, among other instrumentation, a CW-lidar system. An 808-nm laser beam was transmitted into the atmosphere, above various patches of farmland and fields, by a refractor telescope and it was terminated on a treetop, 1966 m from the transmitter. Figure 1 display a map of the beam line at Brunnslov. The backscatter from the beam was imaged on a line scan camera with a sampling rate of 2 kHz. The samples were stored as image files, covering a 4 second window.

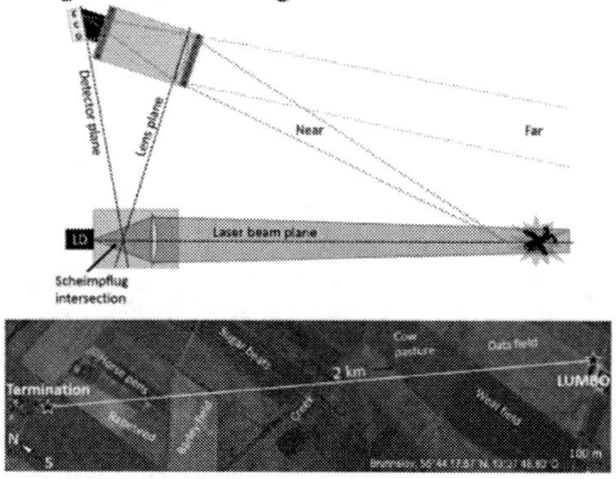

Figure 1. Scheimpflug-lidar setup and a map of the beam line at Brunnslov.

3. DATA PROCESSING

To be able to handle the analysis of the large volume of image files acquired during a measurement campaign in an efficient and robust manner, an algorithm which automatically identifies and extracts insect events from the files has been developed. The algorithm analyzes the insect events and parameterizes properties such as size and wing beat frequency. Identification and extraction of insect events from the data are done by defining OCS thresholds above the static OCS level, σ_{stat}. σ_{stat} is the laser induced signal when no insect, or other large particle, is present in the beam, and it is a result of the laser backscattering from the beam termination and the air in the atmosphere. It is defined as the median intensity value over a 4 second time window. Each range has a specific threshold value and it is obtained by exploiting the asymmetry of the distribution in an intensity histogram over the time window (see Fig 2). When no rare events occurs in the time window the histogram has a Gaussian distribution (parabola with log scale), around σ_{stat}, where the width of the distribution represent the noise level. An insect event will cause the Gaussian distribution to become skewed towards higher values, which can be seen in Fig 2. The threshold is defined by folding the minimum OCS value over the static median value, and multiplying with a factor corresponding to the minimum signal-to-noise ratio (SNR) to be considered. Hereafter we identify rare events by considering all cases where the OCS exceeds the threshold.

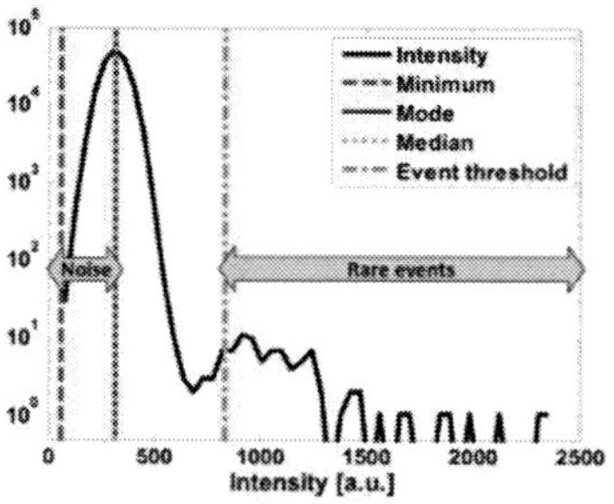

Figure 2 Histogram of intensity values in a 4 second window. The asymmetry of the distribution, is used to define a threshold for an insect event.

A typical lidar signal from an insect is modulated due to its wing beat cycle which makes it sparse in both time and space. Figure 3a presents two, modulated, lidar signals from insects in time range maps, extracted from the Brunnslov data. The time series, σ_t, of the insect events are displayed in Fig 3b. They consists of a non-oscillatory, DC contribution from the insect body during the transit time through the laser beam (Δt), and an oscillatory contribution resulting from the wing beat cycle. When the insect propels its wings, the projected cross section area which backscatters light changes and the detected signal is therefore modulated. This modulation is directly related to the wing beat cycle. Beside this incoherent modulation, the oscillatory signal from insects with glossy wings may also include a specular (coherent) part. Specular reflections show up as narrow spikes on the signal in the time domain as seen in Fig 3b, or as high harmonics in the frequency domain. The algorithm determines the size of the body and the wings of the insect by evaluating their contributions to the time series.

DOI: 10.1051/epjconf/201611925016

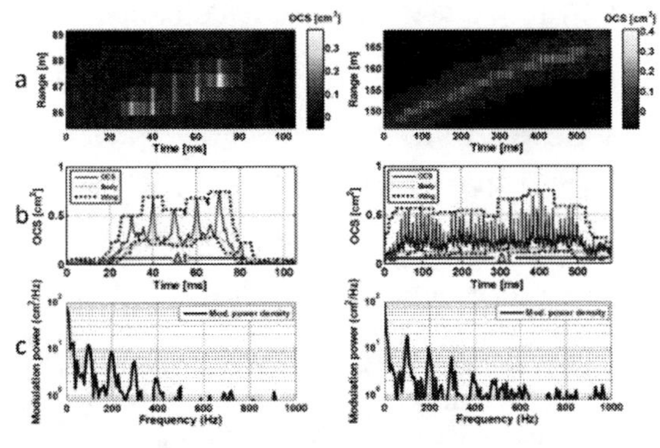

Figure 3 a) Two insect events and b) their OCS time series, in which the contributions from the body and wings are marked. c) Their power spectra.

The algorithm also perform frequency analysis on the time series to estimate the fundamental frequency of the wing beats, ω_0. A power spectrum of σ_t is obtained with a fast Fourier transform. The power spectra for our two case events are presented in Fig. 3c. These power spectra contain three contributions; a DC component, corresponding to the insect flying in and out of the beam, the fundamental tone of the wing beats and its harmonics, and finally noise. The power spectrum is used to make an inital guess of the fundamental frequency of the wing beats. This guess is later used to find the combination of wing beat harmonics (multiples of ω_0) that best represents the original insect event. The initial fundamental tone estimation has great impact on the final parameterized ω_0, and the best method to obtain a robust estimation for all events is still being developed.

The first step to find ω_0 is to express the wing contribution of the event as a linear combination of harmonic basis functions, σ_{wing}, and using these to acquire the absolute strength and phase of the different harmonics signal. The strength and phase are then fitted on top of the harmonic functions to recreate the oscillating signal contribution for the wing beats. This process corresponds to projecting σ_{wing} the harmonic basis functions. To determine the final ω_0 and its harmonics, frequencies within 30% of the initial guess are used in the harmonic functions and the resulting signal is compared to the original event, using a least square fit. The ω_0, which gives the best fit, is set as the fundamental wing beat frequency and this is used to obtain the final parameterized event. The parameterized fundamental frequency and harmonics obtained for the case events are represented in Fig 4. For the event in Fig 4a, the strength of ω_0 is lower than the strength of $2\omega_0$. It is our understanding that this may occur when the detector sees the insect from the side, since it will appear big twice during one wing beat cycle. If the detector instead sees the insect from the back or the front it only appears large once per cycle and ω_0 should dominate (as is the case for the event in Fig 4b). By this reasoning, it is probable that the insect in Fig 4a flew across the beam, while the movement of the insect in Fig 4b was more along the beam direction. This conclusion also correlates with the duration of the two events.

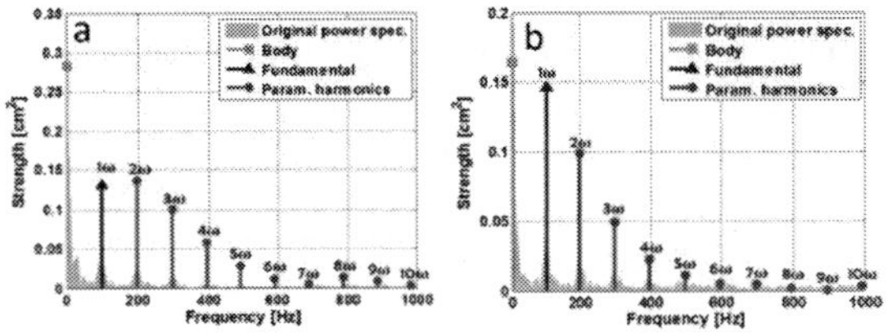

Figure 4 The strength of the fundamental frequency and the harmonics for the two case events.

4. STATISTICS

Some examples of statistical results the lidar system and data analysis method can deliver to researchers are presented below. Figure 5 displays a scatter plot of all parameterized events during one night of measurements. It shows body size, wing size, range, time and ω_0 of each event. The scatterplot contains 2154 parameterized events. A majority of the insects detected close to the detector are small in size while the insects detected at longer ranges are large. This is due to a bias in the system which is a result of a higher system sensitivity closer to the detector. The best way to determine and compensated for the system sensitivity is still being investigated. Figure 6 presents a 2-D contour plots of ω_0 against time of night. It is clear from both the scatter plot and the contour plot that there are two peaks in insect activity during the night. Both occurs around twilight (dawn and dusk). This is not very surprising, since many insects are crepuscular, meaning that they are primarily active during twilight. In Fig 6, it is clear that the insects have a wide range of wing beat frequency with peaks in an interval around 100 Hz. The peaks at dusk and dawn are at roughly the same frequencies which might indicate that it is the same species (which have the same ω_0) that are active at these times.

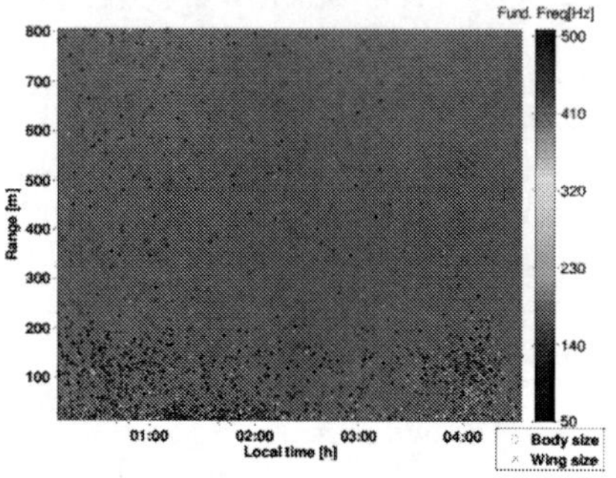

Figure 5 Scatter plot of all the parameterized insect events during one night of measurements at Brunnslov.

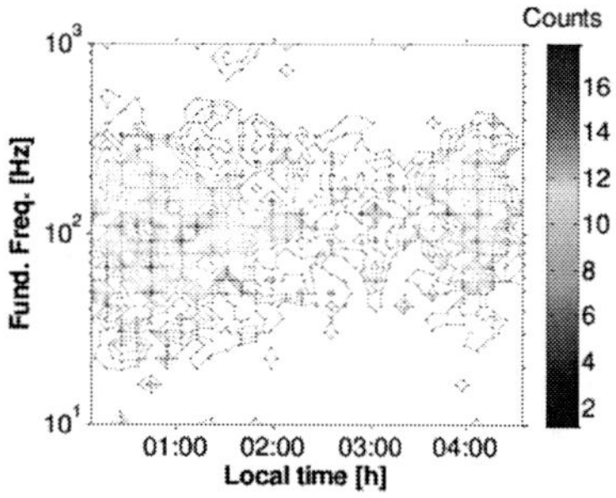

Figure 6. 2-D contour plot of all insects during the measurements period. It displays the fundamental frequency vs time of night.

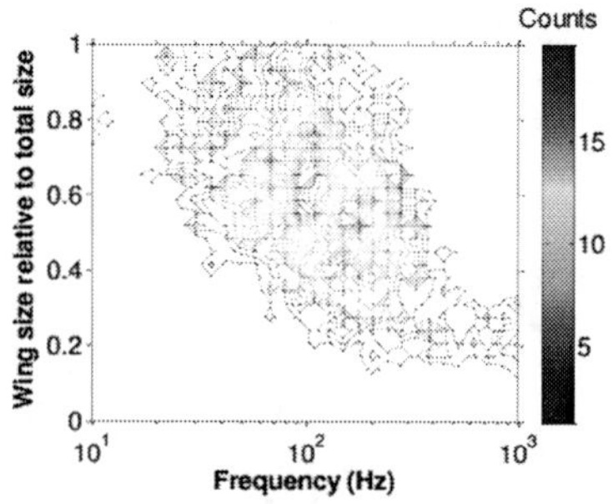

Figure 7. 2-D contour plot of all insects during the measurement period. It displays the relative wing size vs fundamental frequency.

Figure 7 displays a 2-D contour plot representing the wing size relative to total size and ω_0 of all the parameterized insects. The distribution indicates that insects with large wings compare to their body have a low wing beat frequency and insects with small wings have a high wing beat frequency.

5. CONCLUSIONS

Our kHz-lidar system and data analysis routine are still being developed. Aspects, such as system sensitivity at different ranges and fundamental tone estimation must be further investigated and improved. Nevertheless, the system already displays great potential to provide entomological biologists and ecologists with extensive and time resolved data on insect activity and insect properties. Such properties can be used by the researchers for species identification and study of the behavior and movements of insects, which in turn can be used to understand spreading of insect borne diseases, pests and pollinators.

One future aim is to create a database which relates insect properties, such as wing beat frequency content, to species and possibly sex and age group. Such a tool would make it possible for us to provide ecologists with very valuable information in the form of species, age and sex distributions in time and space.

REFERENCES

[1] Gillott, C. 2005: Entomology, 3rd Edition, Dordrecht: Springer.

[2] Kovalev, V. A., W. E. Eichinger, 2004: Elastic lidar: theory, practice, and analysis methods, John Wiley & Sons.

[3] Brydegaard, M., A. Gebru, S. Svanberg, 2014: Super resolution laser radar with blinking atmospheric particles – application to interacting flying insects, *Prog. in Electromagnetics Research*, **147**, 141-151.

[4] Sotavalta, O., 1952: Flight-Tone and Wing-Stroke Frequency of Insects and the Dynamics of Insect Flight, *Nature*, **170**, 1057-1058, 12/20/print.

[5] Fry, S. N., R. Sayaman, M. H. Dickinson, 2005: The aerodynamics of hovering flight in Drosophila, J. Exp. Biol., **208**, 2303-2318.

[6] Moore, A., R. H. Miller, 2002: Automated identification of optically sensed aphid (Homoptera: Aphidea) wingbeat waveforms, Ann. Entomol. Soc. Am. **95(1)**, 1-8.

[7] Hoffman, D. S., A. R. Nehrir, K. S., Repasky, J. A. Shaw, J. L., Carlsten, 2007: Range-resolved optical detection of honeybees by use of wing-beat modulation of scattered light for locating land mines, Appl. Opt, **46 (15)**, 3007-3012.

ACKNOWLEDGMENTS

Financial support from the Swedish Research Council, Susanne Åkesson and CAnMove. Thanks to Sandra Török, Alem Gebru, Carsten Kirkeby and Annika Söderman for assistance in the field.

Raman Spectroscopic Measurements of CO_2 Dissolved in Seawater for Laser Remote Sensing in Water

Toshihiro Somekawa [1]*, Masayuki Fujita[1,2]

[1]*Institute for Laser Technology, 2-6 Yamadaoka, Suita, Osaka 565-0871, Japan,*

**Email: somekawat@ilt.or.jp*

[2]*Institute of Laser Engineering, Osaka Univ., 2-6 Yamadaoka, Suita, Osaka 565-0871, Japan*

ABSTRACT

We examined the applicability of Raman lidar technique as a laser remote sensing tool in water. The Raman technique has already been used successfully for measurements of CO_2 gas dissolved in water and bubbles. Here, the effect of seawater on CO_2 Raman spectra has been evaluated. A frequency doubled Q-switched Nd:YAG laser (532 nm) was irradiated to CO_2 gas dissolved in a standard seawater. In seawater, the Raman signals at 984 and 1060-1180 cm^{-1} from SO_4^{2-} were detected, which shows no spectral interference caused by Raman signals derived from CO_2.

1. INTRODUCTION

Lidar (light detection and ranging) is a very promising technique for monitoring the vast seafloor, and it is very useful for exploring mineral and natural gas deposits on a seafloor, monitoring the environmental change and working condition within submarine industrial facilities [1-3]. However, its development is limited almost to atmospheric remote measurements [4], because water is a strong light absorber showing relatively high transmission only in the shorter wavelength region from UV to Green spectral range [5]. For example, CO_2 gas is usually detected by infrared absorption methods [6,7], but conventional infrared absorption spectroscopy is not suited for sensing gasses in water due to strong IR absorption [8]. Therefore, we proposed Raman lidar using green laser with relatively high transmission in water.

We reported on progress toward developing a technique for the underwater remote detection of gases in water using Raman lidar in our previous papers [1-3]. We used CO_2 gas to demonstrate our approach because it is easy to handle. Firstly, its effectiveness was demonstrated with remote identification of CO_2 dissolved in water in a glass bottle located 20 m away [1]. Next, applications of our approach to the quantitative analysis of the CO_2 concentration in water were presented for comparison with CO_2 dissolved in water and CO_2 bubbles [2, 3].

In this paper, we report measurements of CO_2 dissolved in seawater by laser Raman spectroscopy. Seawater Raman spectrum has characteristic peaks at 984 and 1060-1180 cm^{-1} from SO_4^{2-}, which does not cause interference by CO_2 Raman measurements. Hence, submarine Raman lidar could be a useful technique for 3D laser mapping of CO_2 leakages in water.

2. EXPERIMENT

A schematic diagram of the Raman spectroscopy is given in Fig. 1. We use a standard Q-switched Nd:YAG laser (Continuum, Surelite) operating at its second harmonic wavelength of 532 nm. This laser generates a pulse-width of 10 ns and a pulse energy of 100 mJ at a 10 Hz repetition rate. The wavelength of the laser was chosen at 532 nm due to lower attenuation in water.

The gas-water mixtures were realized by bubbling

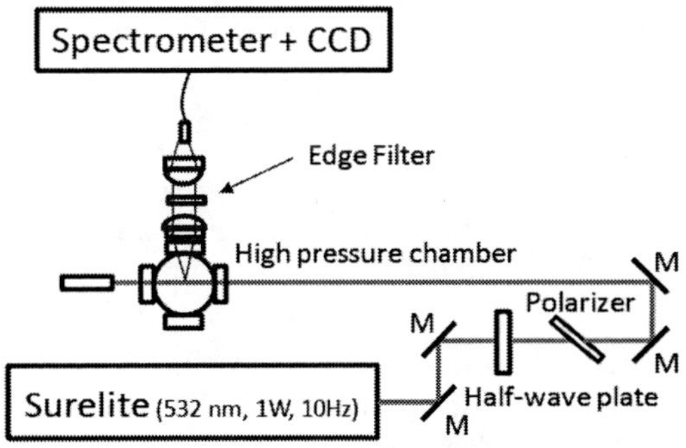

Figure 1. Schematic diagram of Raman spectroscopy experimental setup.

DOI: 10.1051/epjconf/201611925017

the CO_2 gas into the water in a high pressure chamber. This chamber is designed to withstand 10 MPa pressure, which is equivalent to a depth of 1000 m in the ocean. This chamber has four sapphire crystal windows and a rotation blade. The high-speed agitation by the rotation blade has efficiently achieved the equilibrium between the pressure in the water and the pressure in CO_2 gas. The pressure is monitored with both analog and digital pressure gauges. Also, a thermocouple located in water provided a measure of the water temperature.

The chamber is initially filled with about 180 mL deionized water or IAPSO (International Association for the Physical Sciences of the Ocean) standard seawater (OSIL). After degassing in vacuum for 30 minutes, CO_2 gas is introduced from a gas cylinder into the water. Each of the equilibrium steps is carried out for 5 minutes at about 22.5 degrees Celsius with rapid agitation (1000 rpm).

The Raman signal from CO_2 in water is collected using an achromatic lens at an angle of 90° to the laser beam. After passing through the edge filter at 532 nm, the Raman signal is coupled into an optical fiber bundle by using an achromatic lens. The collected Raman signal was delivered to a spectrometer (Acton, SpectraPro-2300i) equipped with a liquid nitrogen-cooled charge-coupled device (CCD) camera (Princeton Instruments, SPEC-10).

3. RESULTS AND DISCUSSION

Figure 2 shows typical spectra of CO_2 dissolved in water and seawater. These spectra have been normalized at ~1645 cm^{-1} peak value after subtraction of the background, and offset vertically for better visualization. The exposure time was 90 ms. The accumulation number was 3000 and it was repeated 5 times in each case. The entrance slit width of the spectrometer was set to 20 μm and the spectral resolution of this system was estimated to be 0.23 nm.

The broad spectrum at ~1645 cm^{-1} is mainly the H-O-H bending band detected in both samples. The relatively sharp bands at ~1277 and ~1383 cm^{-1} are CO_2 Raman signals. The relatively strong Raman signal at ~1383 cm^{-1} is a good candidate for doing Raman lidar experiment in water [1]. Also, the water Raman signal at ~1645 cm^{-1} is utilized as an internal standard for evaluating the dissolved CO_2 concentration [2]. On the other hand, the spectrum of the standard seawater has characteristic peaks associated with the SO_4^{2-} of the S-O symmetric stretching mode at ~984 cm^{-1} and anti-symmetric stretching modes from 1060 to 1180 cm^{-1}[9], which do not interfere with the CO_2 Raman spectrum. The dissolved CO_2

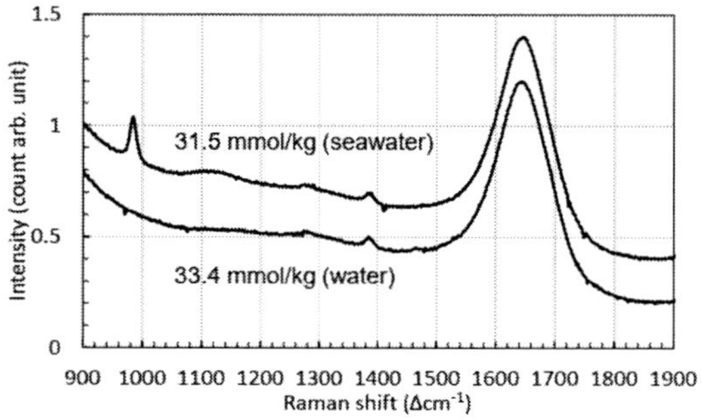

Figure 2. Raman spectra of CO_2 dissolved in water and seawater.

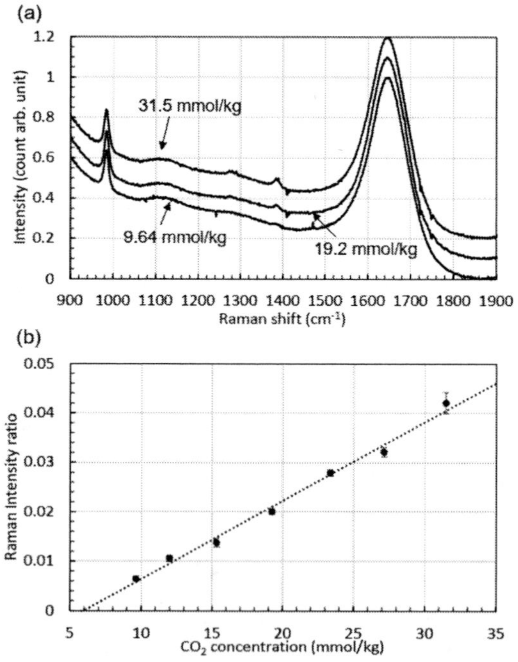

Figure 3. (a) Raman spectra of CO_2 dissolved in seawater at different CO_2 concentrations and (b) dependence of dissolved CO_2 concentration versus CO_2/H_2O Raman bands peak ratio.

concentrations were calculated from the pressure of the chamber using Henry's law [10].

Figure 3(a) shows Raman spectra of CO_2 dissolved in seawater at different dissolved CO_2 concentrations. Raman peak intensity at ~1383 cm^{-1} increased linearly versus increasing dissolved CO_2 concentration. Intensity ratios of CO_2/H_2O Raman signals at ~1383 cm^{-1} and ~1645 cm^{-1} for different CO_2 concentrations are plotted in Fig. 3(b). The error bars were evaluated using the standard deviation of 5 consecutive spectra. Using our current detection system, the spectrum of the dissolved CO_2 concentration at about 10 mmol/kg could be detected. These ratios can be reasonably well fitted by a straight line, however, the fitting line does not pass through the "0" origin. This may be due to the saturated vapor pressure of the gas-water mixture system.

4. CONCLUSIONS

We have demonstrated the applicability of Raman lidar technique for 3D laser mapping in water by measuring CO_2 gas dissolved in seawater. In seawater, the Raman signals at 984 and 1060-1180 cm^{-1} from SO_4^{2-} were detected, showing no spectral interference caused by Raman signals derived from CO_2. The results show that our technique can be applied to an efficient underwater ocean monitoring system.

ACKNOWLEDGEMENT

This work was supported by JSPS KAKENHI Grant Number 25871083.

REFERENCES

[1] Somekawa, T., A. Tani, M. Fujita, 2011: Remote detection and identification of CO_2 dissolved in water using a Raman Lidar system, *Appl. Phys. Express,* **4**, 112401.

[2] Somekawa, T., M. Fujita, 2012: Development of Raman Lidar for Water Dissolved CO2 Detection, *Proc. of ILRC26*, S10-06.

[3] Somekawa, T., T. Takeuchi, C. Yamanaka, M. Fujita, 2014: Raman spectroscopy measurements of CO_2 dissolved in water and CO_2 bubbles for laser remote sensing in water, *Proc. of SPIE*, **9240**, 92400J.

[4]Weitkamp C., 2005: *Lidar Range-Resolved Optical Remote Sensing of the Atmosphere*, Springer.

[5] Smith R. C. and K. S. Baker, 1981: Optical properties of the clearest natural waters (200-800 nm), *Appl. Opt.*, **20**, 177-184.

[6] Somekawa T., M. Fujita, Y. Izawa, 2010: Direct Absorption Spectroscopy of CO2 Using a Coherent White Light Continuum, *Appl. Phys. Express*, **3**, 082401.

[7] Somekawa T., N. Manago, H. Kuze, M. Fujita, 2011: Differential optical absorption spectroscopy measurement of CO2 using a nanosecond white light continuum, *Opt. Lett.*, **36**, 4782-4784.

[8] Friedman D., 1969: Infrared Characteristics of Ocean Water (1.5-15μ), *Appl. Opt.*, **8**, 2073-2078.

[9] Mabrouk, K. B., T. H. Kauffmann, H. Aroui, M. D. Fontana, 2013: Raman study of cation effect on sulfate vibration modes in solid state and in aqueous solutions, *J. Raman Spectrosc.,* **44**, 1603-1608.

[10] Fogg P. and J. Stangster, 2003: *Chemicals in the Atmosphere: Solubility, Sources and Reactivity*, John Wiley and Sons, Hoboken, NJ.

MASSIVELY PARALLEL COMPUTATION OF SOIL SURFACE ROUGHNESS PARAMETERS ON A FERMI GPU

Xiaojie Li [1,3]*, Changhe Song[2]

*[1]Northeast Institute of Geography and Agroecology, Chinese Academy of Sciences, Changchun, 130102, China, *Email: lixiaojie@iga.ac.cn*

[2] State Key Laboratory of Integrated Service Networks, Xidian University, Xi'an 710071, China

[3]Space Science and Engineering Center, University of Wisconsin, Madison, WI, 53706, USA

ABSTRACT

Surface roughness is description of the surface micro topography of randomness or irregular. The standard deviation of surface height and the surface correlation length describe the statistical variation for the random component of a surface height relative to a reference surface. When the number of data points is large, calculation of surface roughness parameters is time-consuming. With the advent of Graphics Processing Unit (GPU) architectures, inherently parallel problem can be effectively solved using GPUs. In this paper we propose a GPU-based massively parallel computing method for 2D bare soil surface roughness estimation. This method was applied to the data collected by the surface roughness tester based on the laser triangulation principle during the field experiment in April 2012. The total number of data points was 52,040. It took 47 seconds on a Fermi GTX 590 GPU whereas its serial CPU version took 5422 seconds, leading to a significant 115x speedup.

Keywords: Soil surface roughness; Correlation length; Graphics Processing Unit (GPU); High Performance Computing (HPC)

1. INTRODUCTION

Surface roughness is a description of the randomness or irregularity of the microtopography. The standard deviation of surface height (σ) and the surface correlation length (*l*) describe the statistical variation for the random component of a surface height relative to a reference surface. Surface roughness forms a key factor to soil microwave radiation, while observing the earth using a microwave radiometer [1-2].

Several researchers focused on determination of surface roughness parameters. Benjamin [3] and Wang [4] studied satellite images for extraction of soil roughness parameters. A method using a roller chain was developed by Ali Saleh to measure soil roughness [5]. Oh [6] developed an inversion method of soil moisture and surface roughness from backscatter measurements of vegetation canopy. Tang [7] obtained soil roughness parameters through digital image processing method. McDonald developed a stereo vision system capable of measuring soil surfaces [8]. Also, the contact method was used by García [9] and Šařec [10]. Lidar technology is also using in the field of testing soil roughness [11-12]. Moreover, our lab presented a surface roughness testing that features rapid testing speed and high testing precision, requires no manual work and obtains 3D parameters for the surface [13]. It has the similar principle with Lidar, but easier and quicker. So they can share the data processing method. The total data points in the range of 40,000~100,000 can be obtained by a 500mm*600mm area. However, calculations of the soil roughness parameters based on these data points present a huge work and require a large computing time.

Graphics Processing Units (GPUs) possess hundreds of parallel processor cores and are capable of executing tens of thousands of threads in parallel. GPUs have been used very successfully for many different computational problems [14-15]. In this paper we propose a GPU-based massively parallel computing method for 2D bare soil surface roughness estimation. We reached a speedup factor of 115x as compared to the single threaded CPU version when GPU I/O transfer is taken into account.

2. METHODOLOGY

2.1 Formulas of soil surface roughness

The standard deviation of a surface height (σ) and the surface correlation length (l) are defined as follows [16]:

a). Standard Deviation of a Surface Height:
Consider a surface in the *x-y* plane whose height at a point *(x, y)* is *z(x, y)* above the *x-y* plane. For a statistically representative segment of the surface, of dimensions Lx and Ly, the standard deviation σ for the discrete one-dimensional case is given by

$$\sigma = (\overline{z^2} - \bar{z}^2)^{1/2} = [\frac{1}{N-1}(\sum_{i=1}^{N}(z_i)^2 - N(\bar{z})^2]^{1/2} \quad (1)$$

where $\bar{z} = \frac{1}{N}\sum_{i=1}^{N} z_i$ and N is the number of samples.

b). Surface Correlation Length
The normalized autocorrelation function for a one-dimensional surface profile z(x) is defined as

$$\rho(x') = \frac{\int_{-L_x/2}^{L_x/2} z(x)z(x+x')dx}{\int_{-L_y/2}^{L_y/2} z^2(x)dx} \quad (2)$$

In our program, because the result produced by the microwave radiometer and radar detector, the scattering values are spread over a 2D area. So, we want to calculate the 2D correlation coefficient. First we need to calculate the distance between each pair of data points as follows,

$$r = \sqrt{(x_i - x_k)^2 + (y_i - y_k)^2} \quad i,k = 1,2,\ldots N \quad (3)$$

where $x' = (j-1)\Delta x$, j is an integer larger than 1. Take a Δx, with each value of j, find x' . Calculate Equation (2) to get all the points (the total is M) where $r = x'$, x' being the spatial distance between two points. Considering the normalized data point, equation becomes

$$\rho(j) = \frac{\sum_{i=1,k=1}^{M}(z_i - \bar{z}_i)(z_k - \bar{z}_k)|_{r=(j-1)\Delta x}}{\sum_{i=1}^{M}(z_i - \bar{z}_i)^2|_{r=(j-1)\Delta x}} \quad (4)$$

The surface correlation length l usually is defined as the displacement x' for which $\rho(x')$ is equal to 1/e, i.e. $\rho(l) = 1/e$. The unit of l is mm.

2.2 GPU Hardware Specification and CUDA Computing Basics

We use CUDA to implement a GPU-based surface roughness parameter computation scheme because of its easy deployment and high computing performance [17]. Moreover, it is best suited to NVIDIA GPU implementations [17-18] as compared to the other computing techniques such as OpenCL and Direct Compute. The hardware specifications of one NVIDIA GTX 590 GPU employed in our study are depicted below

Table 1. Device parameters of the GPU and CPU

	CPU	GPU
Model	Intel i7 970	NVIDIA GTX 590
Number of cores	6	512 stream processors
Frequency	3.2GHz	1.215GHz
L1 cache	6x64 KBytes	48 KBytes or 16 KBytes
L2 cache	6x256 KBytes	768 KBytes
L3 cache	12 MBytes	n/a
Shared memory	n/a	16 KBytes or 48 KBytes

2.3 CUDA Implementation

As 3D tester, the data includes 3 parts, x-coordinate, y-coordinate and z-coordinate. First of all, we need write C codes. Once correct C codes were obtained, in the conversion from C codes to CUDA codes, some caution need to be taken.
In our implementation, 300 different distances are thus chosen. As the CUDA architecture automatically schedules the execution in different blocks, the number of threads in one block, i.e. the block size, is however optimized experimentally. For the influence of the thread block size on the computing performance, we carry out a test to vary the block size. The results are presented in Figure 1.

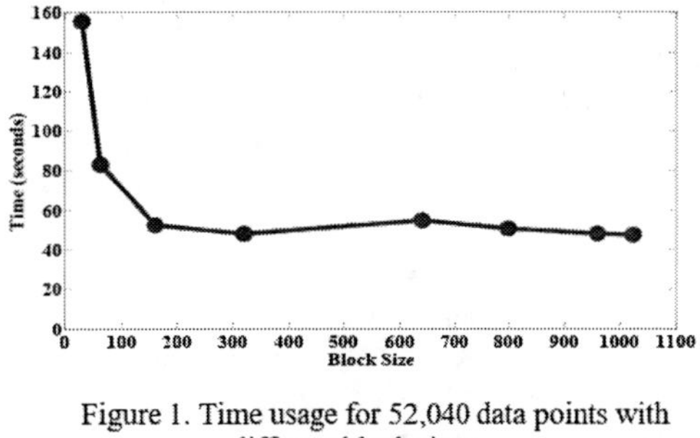

Figure 1. Time usage for 52,040 data points with different block sizes

From Figure 1, we can observe that the best performance is achieved when the block size is 1024. Hence, we set the block size to be 1024. The shared memory usage is consequently determined.

A command "cudaFuncCachePreferL1" was launched in our CUDA codes, which means that the usage of L1 cache is larger than shared memory.

2.4 The other Considerations

1). Memory Coalescing

The input data is a 3D data array stored in a text file. The first column is x-coordinate, the second column is y-coordinate, and third is z-coordinate. For coalesced memory access, the array is stored column wise in the memory, which means the x-coordinates of all points are stored together in the memory, followed by all the y-coordinates and then all the z-coordinates of all points, that is shown in Figure 2.

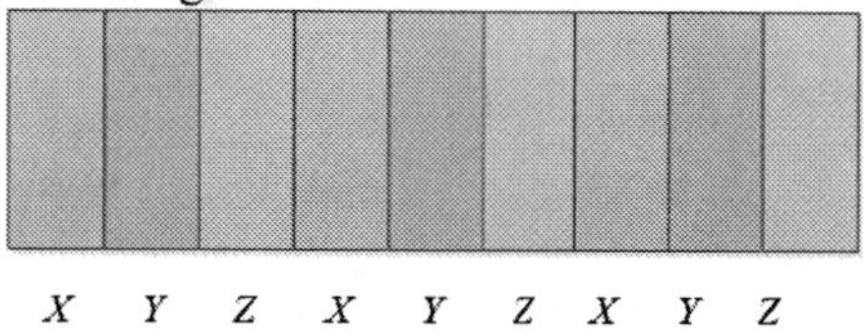

(a) no coalesced memory

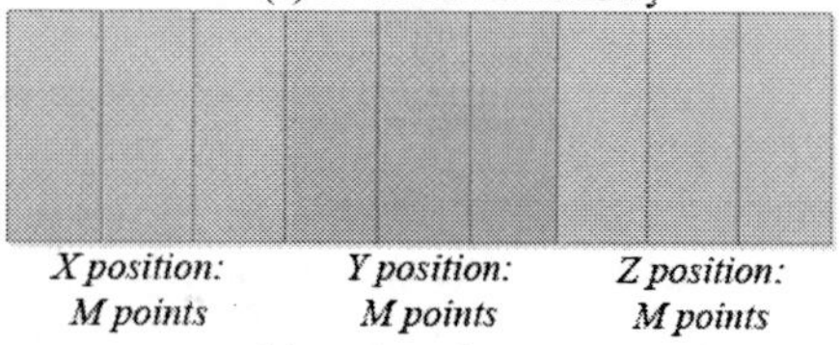

(b) coalesced memory

Figure2. A pictorial layout of the data structure.

2). Coalesced memory access

The coalesced memory access technique combines several memory accesses together so as to save the memory bandwidth. In our study, contiguous data will be accessed by neighboring threads due to our designed data layout. This access pattern makes coalesced memory access possible. However, in the case of two-dimensional data, the length of the first dimension is required to be a multiple of warp size in order to achieve full efficiency during coalesced memory access. Therefore, we pad the number of points to be a multiple of the warp size.

3). Shared Memory

We used shared memory to save the local sum of two local variables. When all threads are finished, the local sums are buffered to the shared memory. Then the local sums in the shared memory are accessed and summed up again by each of these threads.

3. RESULTS

In order to validate this method, we used the field test date to calculate, which was carried out at the Jilin Agricultural University, China in April 19, 2012. The pictures of bare soil and the original data by the surface roughness tester are below.

Figure 3. (a) Bare soil picture.

Figure 3. (b) The original data obtained by 3D tester

The computation is originally implemented in C programming language using gcc as the compiler. We used gcc -lstdc++ correlation.c -O3 for C codes. The CUDA codes were compiled using nvcc (NVIDIA CUDA compiler) version of 5.5 and executed on NVIDIA GTX590 GPU with compute capability of 2.0. The compiler options thus are –arch=compute_20, code=sm_20 -fmad=false -ftz=true. The timing results calculated at different data points using CPU and GPU are shown in Table 2 along with the speedup for the GPU based implementation. The correlation length of 49.09 mm is obtained in this case. The standard deviation of the surface height is 12.70 mm.

Table 2. Timing results for the CPU and GPU

Data points	Time usage (unit:s)		Speedup
	CPU	GPU	
52040	5422	47	115

4. CONCLUSIONS

In this paper, we developed a GPU-based parallel computing method for 2D surface roughness estimation. It was written in CUDA and was executed on an NVIDIA GeForce GTX 590 GPU. Calculating the distance between any two points and estimating whether this distance is an assignment value can be carried out by threads. In our implementation, we designed the computations for different appointed distances such that they are performed in different blocks. This method was applied to the data collected by the surface roughness tester on April 2012 based on laser triangulation principle. For a total of 52,040 data points, the serial CPU version of the implementation took 5422 seconds whereas the GPU implementation took 47 seconds, leading to a significant 115x speedup.

ACKNOWLEDGEMENT

This work was supported by the National Natural Science Foundation of China Nos. 41201335 at Space Science and Engineering Center, University of Wisconsin-Madison USA. The authors thank Professor Bormin Huang for his help.

REFERENCES

[1] Sungwook, H, 2010: Surface roughness and polarization ratio in microwave remote sensing, *International Journal of Remote Sensing*, **31(10)**, 2709-2716.

[2]Zheng, X. M., Zhao, K, 2010: A method for surface roughness parameter estimation in passive microwave remote sensing. *Chinese Geographical Science*, **20(4)**, 345-352.

[3]Benjamin, S., Leonie, D. F., Jo-Ann, H.; Felix, W., 2010: Fast derivation of soil surface roughness parameters using multi-band SAR imagery and the Integral Equation Model. *2010 International Conference on Pattern Recognition*, IEEE computer society, 3931-3934.

[4]Wang, S. G., Li, X, Han, X. J., Jin, R., 2011: Estimation of surface soil moisture and roughness from multi-angular ASAR imagery in the Watershed Allied Telemetry Experimental Research (WATER). *Hydrology and Earth System Sciences*, **15**, 1415-1426.

[5]Ali Saleh., 1993: Soil roughness measurement: Chain method. *Journal of Soil and Water Conservation*. **48(6)**, 527-529.

[6]Oh, Y., Hong, J. Y., Jung, S. G., 2007: Retrieval of Soil Moisture and Surface Roughness from Backscatter Measurements of Vegetation Canopy. *Asia-Pacific Microwave Conference*, 1-3.

[7]Tang, X. J., Xiao, H., Ding, H., Liu, J. H., 2009: Surface roughness measurement based on image processing and image recognition. *Computers and Simulation in Modern Science*. 91-96.

[8]McDonald, A. J., Crossley, S., Bennett, J. C, Cookmartin, G., Morrison, K., Quegan, S., 1999: Stereo vision measurements of soil surfaces and their utility in testing microwave backscattering models. *Geoscience and Remote Sensing Symposium*, **5**, 2634 - 2636.

[9] Garcia Morenoa, R., Diaz Alvarezb, M.C., Saa Requejob, A., Valencia Delfac, A.M., 2010: Tarquisb Multiscaling analysis of soil roughness variability. *Complexity and Nonlinearity in Soils*.**160(1)**, 22–30.

[10] Šařec, P., Šařec, O., Prošek, V., Čížková, K., 2007: Laser profilometer testing by laboratory measurements. *RES. AGR. ENG.*, **53 (1)**, 1-7.

[11] Russell T., Rocco, P., Mihai A. Tanase, Kim L., Jorg M. Hacker, Jeffrey P. Walker. 2014: Estimation of soil surface roughness of agricultural soils using airborne LiDAR. *Remote Sensing of Environment*, **140**, 107-117.

[12] Shi, X., Girod, L., Long, R., DeKett, R., Philippe, J., Burke, Tom. 2012: A comparison of LiDAR-based DEMs and USGS-sourced DEMs in terrain analysis for knowledge-based digital soil mapping. *Geoderma*, **170**, 217-226.

[13] Li, X. J., Zhao, K., Zheng, X. M., 2012: Development of surface roughness tester based on laser triangulation method. *Transactions of the Chinese Society of Agricultural Engineering (Transactions of the CSAE)*, **28(8)**, 116-121.

[14] Sui, H.G., Peng, F. F., Xu, C., Sun, K.M., Gong, J.Y., 2012: GPU-accelerated MRF segmentation algorithm for SAR images. *Computers & Geosciences*, **43**,159-166.

[15] Devrim, A., Pakize, E.□ 2015: GPU accelerated training of image convolution filter weights using genetic algorithms. *Applied Soft Computing*, **30**, 585-594.

[16] Ulaby, F. T., Moore, R. K., Fung, A. K., 1987: Microwave remote sensing active and passive. *Radar remote sensing and surface scattering and emission theory*, **2**, 819-833.

[17] Mielikainen, J., Huang, B. M., Huang, H. L. A, Goldberg, M. D., 2012: Improved GPU/CUDA Based Parallel Weather and Research Forecast (WRF) Single Moment 5-Class (WSM5) Cloud Microphysics. *Applied Earth Observations and Remote Sensing*, **5(4)**, 1256-1265.

A NEW AIRBORNE LIDAR FOR REMOTE SENSING OF CANOPY FLUORESCENCE AND VERTICAL PROFILE

Ounis A.[1*], Bach J.[1], Mahjoub A.[2], Daumard F.[3], Moya I.[1], Goulas Y.[1]

[1]*Laboratoire de Météorologie Dynamique, École Polytechnique, F91128 Palaiseau (France)*
[2]*Laboratoire de Physique des Plasmas, École Polytechnique, F91128 Palaiseau (France)*
[3] *INRA-EGC, Thiverval-Grignon (France)*
[*]*Email: ounis@lmd.polytechnique.fr*

ABSTRACT

We report the development of a new lidar system for airborne remote sensing of chlorophyll fluorescence (ChlF) and vertical profile of canopies. By combining laser-induced fluorescence (LIF), sun-induced fluorescence (SIF) and canopy height distribution, the new instrument will allow the simultaneous assessment of gross primary production (GPP), photosynthesis efficiency and above ground carbon stocks. Technical issues of the lidar development are discussed and expected performances are presented.

1. INTRODUCTION

Terrestrial vegetation plays a fundamental role in the global carbon cycle; however, the coupling between climate change and plant eco-physiological functions still remains poorly understood. Experimental investigations at the leaf, canopy, ecosystem and regional scales are needed in order to improve climate predictions and mitigation strategies.

In this framework, current remote sensing techniques only deliver information on biophysical variables of the canopy, not on eco-physiological functions. ChlF is emitted by the photosynthetic pigments pool shortly after light absorption. It is considered as a promising remote observable of vegetation because of its close link with photosynthesis (for a review see [1]). Recently, a strong correlation was evidenced between SIF measured from space and GPP of various biomes derived from eddy covariance measurements at ground level [2]. However, GPP depends on both: (i) APAR, which itself depends on canopy structure, leaf biochemical composition and on incident radiation level, and (ii) the light use efficiency (LUE), which reflects the eco-physiological state of the plant [3]. On the other hand, SIF depends also on APAR and on the fluorescence quantum yield (FQY), which is linked to photosynthesis efficiency. Hence, disentangling APAR effects from LUE effects in GPP is possible if we are able to measure FQY independently. Assessment of relative FQY changes is feasible at canopy level by laser-induced fluorescence (LIF) if all measuring conditions are kept constant.

In this paper, we describe a new lidar concept for measurements of FQY and vertical profile (VP) of canopies (LASVEG, LASer VEGetation) combined to simultaneous SIF measurements from a single airborne platform. We discuss development issues and new possibilities that are open up by the future measurements.

2. DERIVATION OF CANOPY PHOTOSYNTHESIS

Our instrument aims to fill the gap between eddy fluxes towers and space observation:
- SIF gives access to GPP, a major component of carbon fluxes between surface and atmosphere [2] [4],
- LIF, as a proxy of FQY, allows assessment of fluorescence quenching, thus giving access to photosynthetic efficiency [5],
- lidar waveform maps above ground carbon stocks through the use of allometric relationships [6].

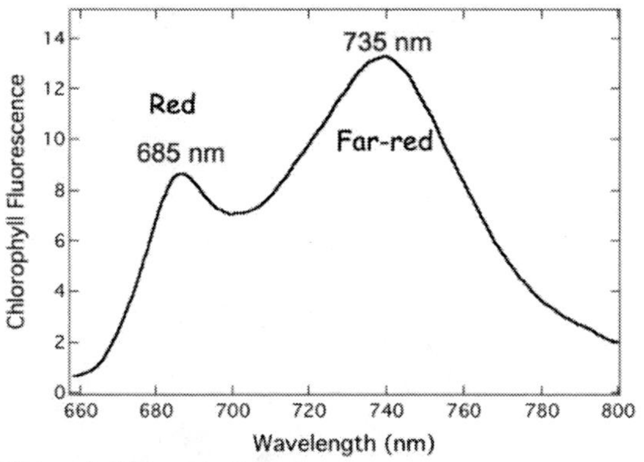

Figure 1: Chlorophyll fluorescence spectrum.

SIF and GPP share APAR as a common factor [4]. GPP is expressed as:

$$GPP = APAR \times LUE \qquad (1)$$

where LUE, as seen above, represents the fraction of absorbed PAR that is used for carbon fixation. On the other hand, SIF can be written as:

$$SIF = APAR \times FQY \times \tau_F \qquad (2)$$

where τ_F is the fraction of emitted fluorescence that escapes from the canopy. FQY can be assessed from LIF exitance considering Equation 3:

$$LIF = I_{laser} \times fAPAR_{laser} \times FQY \times \tau_{Flaser} \qquad (3)$$

I_{laser} being the laser intensity, $fAPAR_{laser}$ the fraction of laser light absorbed by vegetation and τ_{Flaser} the fraction of laser induced fluorescence that escapes from the canopy. As illumination geometry is constant in the case of LIF, these later factors are constant and LIF can be regarded as a proxy for FQY. In the far-red (see Figure 1), τ_F is close to one because fluorescence is not reabsorbed at these wavelengths. Hence, combining Equations 1 to 3 gives the possibility to investigate the relationship between LUE and FQY and to improve GPP modeling using fluorescence.

This instrument will also estimate the ground carbon stocks through the measurement of canopy vertical structure by lidar waveform analysis [6, 7]. Biophysical variables such as LAI, FAPAR and chlorophyll content will also be derived from the spectral reflectance [8].

3. THE LIDAR SYSTEM

3.1 Signal to noise ratio

Because of the low fluorescence yield (<1%) of *in vivo* chlorophyll, the instrumental design needed to perform high-quality data become challenging. Therefore, we setup a ground test to investigate the overall detectability of the LIF. Using laser pulses of 9 mJ/pulse at 532 nm on a tree canopy situated at a distance of 42 m, we achieved fluorescence measurements with SNR=100. By increasing the laser pulse energy up to 30 mJ, and the telescope diameter from 15 cm to 35 cm, and by improving the detector response time (+50%) and the stray light rejection (+50%), a SNR of 100 at a flying altitude of 300 m should be achieved.

3.2 The LASVEG excitation source

The LASVEG system uses a compact and rugged lamp-pumped and frequency doubled Nd:YAG laser pulsed at a frequency of 20 Hz with 30 mJ energy per pulse (Ultra 50 from Quantel).

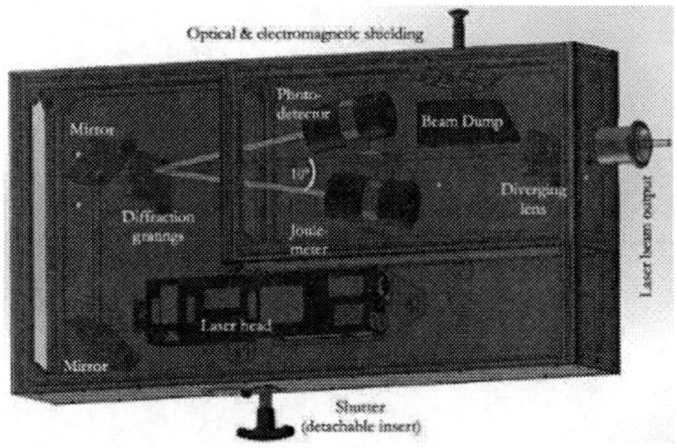

Figure 2 : Schematic drawing of LASVEG excitation source.

Figure 2 shows a schematic drawing of LASVEG excitation source. The laser beam is expanded with a negative lens so that the illuminated area (foot print) is around 5-m diameter for a typical operational flight-altitude of 300 m. Such a beam-size is required to operate in eye-safety conditions on the ground level. Energy pulses are recorded with a pyro-electric joule-meter (ED 200, Gentec) in order to obtain normalized fluorescence and reflectance ratios. To record the energy of each pulse and trigger the data acquisition system we

DOI: 10.1051/epjconf/201611925019

use the first order beams transmitted by a diffraction grating: at 532 nm, these beams have a sampling ratio of 1.06% at an angle of 10°. A photo-detector is used to detect the presence of laser pulses.

3.3 Eye safety experimental condition

To avoid eye injury during operation, the energy received at eye level must be far under the Maximum Permissible Exposure. MPEs are defined by the international standard IEC 60825, and depend on wavelength, pulses frequency and exposure time. This later parameter depends on spot size at ground level and aircraft speed. Figure 3 gives MPE as a function of illuminated spot diameter at ground level, compared to the exposure (1.5 mJ m^{-2}) induced by the LASVEG source at 300 m. It shows that eye-safety is achievable under good operational conditions.

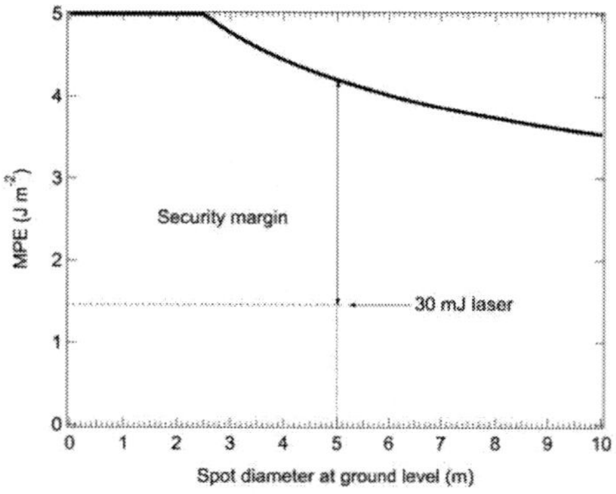

Figure 3 : Maximum Permissible Exposure as a function of illuminated spot size at ground level with aircraft speed at 50 m s^{-1}.

3.4 The LASVEG telescope

Photon collection is performed with a Fresnel lens of 33 cm clear-aperture (Figure 4). Fresnel lenses have several benefits such as large aperture, large numerical aperture, reduced weight and low cost. In the other hand, commercial available telescopes are not suited for this kind of detection due to their typical numerical aperture. A set of 4 baffles is used to minimize stray-light. Fluorescence and back reflected-light are split into three channels by two dichroic mirrors: one channel at 532 nm for waveform analysis and two fluorescence channels at 685 nm and 740 nm (see Figures 4 & 5).

Figure 4: Schematic cross-section drawing of the LASVEG telescope.

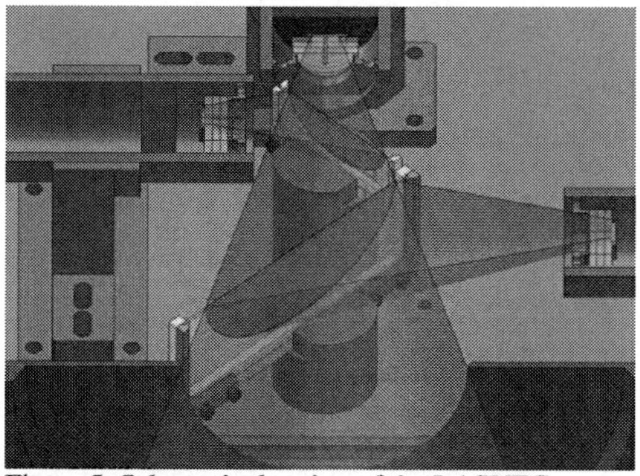

Figure 5: Schematic drawing of the LASVEG detector chamber with dichroïc mirrors.

The whole lidar system is mounted above the aircraft aperture through silent blocs.

3.5 LASVEG detectors

Three high-speed PIN photodiodes are used for fluorescence and waveform detection. Specific trans-impedance amplifiers have been developed to selectively amplify the pulsed signal and providing a high rejection of the background illumination (≈ 50 dB). The three signals are sampled at high rate (250 MSample/s) with a fast analog to digital card

(NI 5761 coupled to a FPGA Module, National Instruments) and recorded on a hard disk (250 Go) for post processing. Peak values of fluorescence channels are divided by the pulse energy to give apparent fluorescence yield. The vertical structure of the canopy is deduced from waveform analysis of the reflected laser beam.

4. PASSIVE INSTRUMENTS

The SIF sensor is based on an improved version of the AIRFLEX sensor previously used in airborne campaigns [9]. AIRFLEX is a six-channels high performance radiometer dedicated to measurements in the oxygen bands for fluorescence retrieval. Passive instruments include also a wide-band (400-1000 nm) spectrometer (from Ocean Optics) to record canopy reflectance. A synchronized video camera is used to record images of the scene that will help for target identification.

5. CONCLUSIONS AND PERSPECTIVES

We presented a new instrumental concept that combines measurements of LIF, SIF and vertical profile of canopies. This multi-parameter instrument can simultaneously provide photosynthetic efficiency, gross instantaneous carbon flux and above ground carbon stock, which are major eco-physiological variables of canopy functioning. It is expected to resolve major issues in the improvement of GPP models based on fluorescence, with benefits for fluorescence space missions like GOSAT, OCO-2 or FLEX.

ACKNOWLEDGMENTS

This instrument is developed in the framework of the CALSIF project with the support of the French national agency ANR and the French space agency CNES.

REFERENCES

[1] G. Papageorgiou and Govindjee, *Chlorophyll a fluorescence. A signature of photosynthesis*. Dordrecht, The Netherlands: Springer, 2004.

[2] C. Frankenberg, J. B. Fisher, J. Worden, G. Badgley, S. S. Saatchi, J.-E. Lee, G. C. Toon, A. Butz, M. Jung, A. Kuze, and T. Yokota, "New global observations of the terrestrial carbon cycle from GOSAT: Patterns of plant fluorescence with gross primary productivity," *Geophys. Res. Lett.,* **38(17)**, L17706, 2011.

[3] J. L. Monteith, "Solar radiation and productivity in tropical ecosystems," *Journal of Applied Ecology,* **9(3)**, 747-766, 1972.

[4] M. Rossini, M. Meroni, M. Migliavacca, G. Manca, S. Cogliati, L. Busetto, V. Picchi, A. Cescatti, G. Seufert, and R. Colombo, "High resolution field spectroscopy measurements for estimating gross ecosystem production in a rice field," *Agricultural and Forest Meteorology,* **150**, 1283-1296, 2010.

[5] J. Flexas, J. M. Escalona, S. Evain, J. Gulias, I. Moya, C. B. Osmond, and H. Medrano, "Steady-state chlorophyll fluorescence (Fs) measurements as a tool to follow variations of net CO2 assimilation and stomatal conductance during water-stress in C-3 plants," *Physiologia Plantarum,* **114(2)**, 231-240, Feb 2002.

[6] G. P. Asner, J. Mascaro, H. C. Muller-Landau, G. Vieilledent, R. Vaudry, M. Rasamoelina, J. S. Hall, and M. van Breugel, "A universal airborne LiDAR approach for tropical forest carbon mapping," *Oecologia,* **168(4)**, 1147-1160, Apr 2012.

[7] J. Mascaro, M. Detto, G. P. Asner, and H. C. Muller-Landau, "Evaluating uncertainty in mapping forest carbon with airborne LiDAR," *Remote Sensing of Environment,* **115(12)**, 3770-3774, 2011.

[8] J. Verrelst, L. Alonso, G. Camps-Valls, J. Delegido, and J. Moreno, "Retrieval of Vegetation Biophysical Parameters Using Gaussian Process Techniques," *IEEE Transactions on Geoscience and Remote Sensing,* **50(5)**, 1832-1843, May 2012.

[9] I. Moya, F. Daumard, N. Moise, A. Ounis, and Y. Goulas, "First airborne multiwavelength passive chlorophyll fluorescence measurements over La Mancha (Spain) fields," in *2nd International Symposium on Recent Advances in Quantitative Remote Sensing: RAQRS'II, 25-29th September 2006,* Torrent (Valencia)-Spain, 2006.

EPJ Web of Conferences **119**, 25020 (2016) DOI: 10.1051/epjconf/201611925020
ILRC 27

EXPERIMENTAL DETERMINATION OF DUAL-WAVELENGTH MIE LIDAR GEOMETRIC FORM FACTOR COMBINING SIDE-SCATTER AND BACK-SCATTER SIGNALS

Zhenzhu Wang*, Zongming Tao, Dong Liu, Chenbo Xie, and Yingjian Wang

*Key laboratory of Atmospheric Composition and Optical Radiation, Anhui Institute of Optics and Fine Mechanics, Chinese Academy of Sciences, Hefei, Anhui, 230031 China, *Email: zzwang@aiofm.ac.cn*

ABSTRACT

In theory, lidar overlap factor can be derived from the difference between the particle backscatter coefficient retrieved from lidar elastic signal without overlap correction and the actual particle backscatter coefficient, which can be obtained by other measured techniques. The side-scatter signal using a CCD camera is testified to be a powerful tool to detect the particle backscatter coefficient in near ground layer during night time. In experiment, by combining side-scatter and back-scatter signals the geometric form factor for vertically-pointing Mie lidar in 532 nm channel is determined successfully, which is corrected by an iteration algorithm combining the retrieved particle backscatter coefficient using CCD side-scatter method and Fernald method. In this study, the method will be expanded to 1064 nm channel in dual-wavelength Mie lidar during routine campaigns. The experimental results in different atmosphere conditions demonstrated that the method present in this study is available in practice.

1. INTRODUCTION

The incomplete overlap between laser beam and receiver field of view significantly leads to lidar signal loss in the near-field range, and affects the lidar measurement in this range. However, lidar measurement in the near-field range (e.g. particles in the boundary layer) is very essential. The fraction of the laser beam intersecting the receiver field of view is described by geometric form factor (named GFF), and the influence of the incomplete lidar signal can be corrected by GFF. Theoretical and experimental methods have been explored to determine GFF [1-6]. In theoretical methods, GFF is determined by application of a ray-tracing model, and some parameters of the lidar must be known with a sufficient accuracy [1-2], which is not the best idea in practice. In experimental methods, a trustworthy determination of GFF is possible under homogenous atmosphere [3], and this method only can be used when the instrument is oriented horizontally. However, some lidars are only oriented vertically because of complex structure or mobile carrier, and the atmosphere along the laser path is generally inhomogeneous. In order to solve this problem, Ulla Wandinger [4] developed a method based on combined Raman-Mie lidar, and Hu [5] proposed a method based on nitrogen Raman signal and nitrogen molecule density profile along the laser path. For these two methods, some inherent hypotheses limit their application, as well as some lidars with only Mie channels. Wang [6] introduced an experiment method to determine GFF for vertically-pointing lidar by iterative algorithm based on the CCD side-scatter technique. In this paper, the method will be expanded to 1064 nm channel in dual-wavelength Mie lidar combining side-scatter and back-scatter signals. The Mie lidar system consists of laser, telescope, PMT, APD, CCD camera, ICC and related optical and mechanical parts as shown in figure 1, which will not be introduced in detail here.

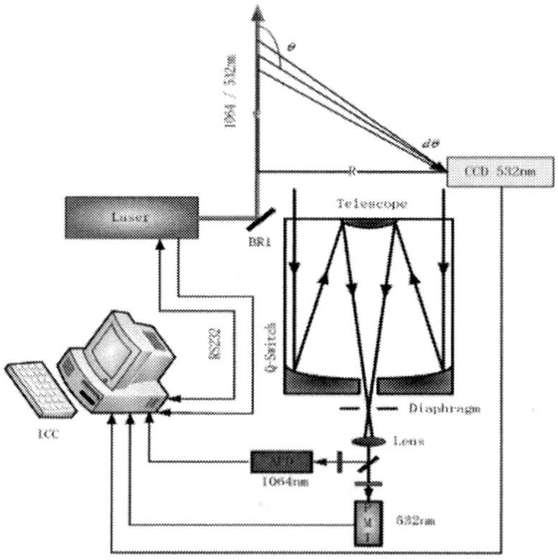

Figure 1. The diagram of lidar system

2. METHODOLOGY

The CCD side-scatter technique is a developing technique in which the CCD camera is spatially separated to look the laser beam at a certain distance. The side-scatter lidar has no overlap problem as in a backscatter lidar, and is especially suitable to detect actual particle backscatter coefficient in the near range during nighttime [7]. Particle backscatter coefficient can be retrieved from elastic signal using Fernald method [8]. The backscattering coefficient can also be retrieved from side-scatter technique by numerical inversion method [9]. An iterative algorithm is developed to determine GFF from lidar elastic signal in 532 nm when the profile of particle backscatter coefficient had been known correctly. This method can also be used to get GFF for 1064 nm channel. As introduced above, the actual profile of particle backscatter coefficient in 1064 nm has to be known firstly. Note here, the color ratio of 532/1064 for aerosol backscatter coefficient above the height z0 (GFF = 1) can be used to obtain the "actual profile" in 1064 nm with the assumption of a fair homogenous atmosphere condition.

3. RESULTS

Measurement was made on Sep. 4, 2014 using our Mie lidar with side-scatter technique as shown in figure 2. In fig. 2(a), the aerosol profiles were obtained at 19:00 local time on September 04, 2014 when when the atmosphere condition is relative homogenous below the height of 1.0 km with an integrated AOD value 0.65 on the assumption of S1 equal 50 for 532 nm. In the height of 0.8~1.5 km, the aerosol backscatter coefficients from these two lidar techniques are almost the same. Below 0.8 km, aerosol backscatter coefficients by Mie lidar are affected by the overlap factor. Through iteration algorithm, the inversed aerosol profiles by Mie lidar (blue line) and CCD side-scatter lidar (red line) are matched together perfectly. After that, the GFF for Mie lidar signal is obtained and shown in sub-figure. In fig. 2(b), the aerosol profiles in 1064 nm are obtained comined Mie lidar and CCD side-scatter lidar through a homogenous atmosphere assumption (i.e., using the calculated color ratio and inversed backscatter coefficient for 532 nm by CCD lidar). Then the AOD value is 0.28 using S1 = 40 for 1064 nm. The incomplete part affected by GFF is also given after iteration calculation as shown in sub-figure of Fig. 2(b). As to 1064 nm channel, the input parameters would induce uncertainty in determining GFF, receiver field of view and detector for two techniques are different, when there are particle layers on the laser paths, the measurements given by two techniques are probably different, which may be caused by the multi-scattering and different electronic response at the layer. Thus, the measurements should be performed under fair and homogenous atmosphere condition.

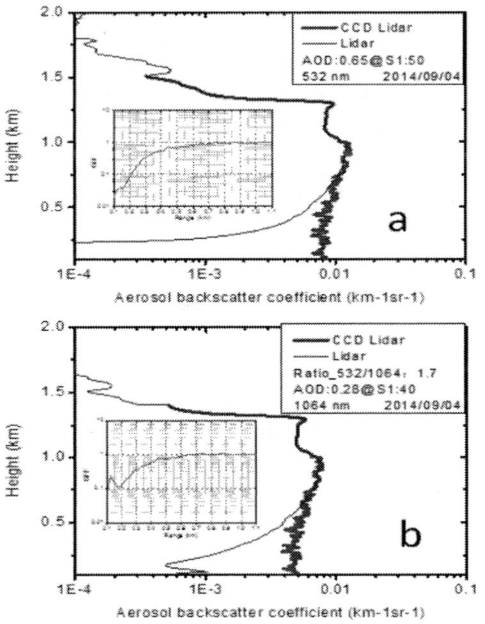

Figure 2. Particle backscatter coefficients obtained from CCD lidar (read line) and Mie lidar (blue line), and corresponding obtained GFF in sub-figure for 532 nm (a) and 1064 nm (b) channels on Sep. 04, 2014 over Hefei, China

4. CONCLUSIONS

In summary, a new experiment approach for determining overlap factor for vertically-pointing lidar is presented by iteration algorithm combining the retrieved particle backscatter coefficient using CCD side-scatter method and Fernald method. The effect of overlap factor on Mie lidar is corrected successfully in 532 and 1064 nm channels. This method is available in practice and GFF measurements for lidars can be determined easily in cloud-free night time by adding a CCD detector with a proper way. But for 1064 nm, some assumption will be done, and multi-scattering may be another problem, further investigation will be carried out in the near future.

ACKNOWLEDGEMENT

This research is supported by the National Natural Science Foundation of China (No. 41305022), the Strategic Priority Research Program of the Chinese Academy of Sciences (XDA05100302), and the Anhui Provincial Natural Science Foundation (No. 1308085MD53).

REFERENCES

[1]T. Halldórsson and J. Langerholc, "Geometrical form factors for the lidar function," Appl. Opt. 17, 240–244 (1978).

[2]K. Sassen and G.C. Dodd, "Lidar crossover function and misalignment effects," Appl. Opt. 21, 3162-3165 (1982).

[3]Y. Sasano, H. Shimizu, N. Takeuchi, and M. Okuda, "Geometrical form factor in the laser radar equation: an experimental determination," Appl. Opt. 18, 2886-2890 (1979).

[4]Ulla Wandinger and Albert Ansmann, "Experimental determination of the lidar overlap profile with Raman lidar," Appl. Opt. 41, 511-514 (2002).

[5]Shunxing Hu, Xiaobin Wang, Yonghua Wu, Chen Li, and Huanling Hu, "Geometrical form factor determination with Raman backscattering signals," Opt. Let. 30, 1879-1881 (2005).

[6] Z. Wang, Z. Tao, D. Liu, D. Wu, C. Xie, Y. Wang, A new experiment method for lidar overlap factor using a CCD side-scatter technique, Opt. Let. 40, (2015).

[7]J. E. Barnes, N. C. Parikh Sharma, and T. B. Kaplan, Atmospheric aerosol profiling with a bistatic imaging lidar system, Appl. Opt. 46(15), 2922-2929 (2007).

[8]F.G. Fernald, "Analysis of atmospheric lidar observations: some comments," Appl. Opt. 23, 652-653 (1984).

[9]Z. Tao, D. Liu, Z. Wang, X. Ma, Q. Zhang, C. Xie, S. Hu, Y. Wang, Measurements of aerosol phase function and vertical backscattering coefficient using a charge-couple device side-scatter lidar, Opt. Express 1127 DOI: 10.1364/OE.22.001127, (2014)

GLUING LIDAR SIGNALS DETECTED IN ANALOG-TO-DIGITAL AND PHOTON COUNTING MODES

Chang-Zhong Feng [1], Bing-Yi Liu [1]*, Jin-Tao Liu [1], Song-Hua Wu [1]

[1]*Ocean University of China, 238 Songling Road, Qingdao 266100, China, *Email: liubingyi@ouc.edu.cn*

ABSTRACT

Lidar is one of the most effective tools for atmospheric remote sensing. For a ground-based lidar system, the backscattered light usually has large dynamic range. Photon-counting mode has the capability to measure weak signal from high altitude, while Analog-to-Digital mode with better linearity is good at measuring strong signal at low altitude. In some lidar systems, atmospheric return signal is measured in both Analog-to-Digital and Photon Counting modes and combined into an entire profile by using a gluing algorithm. A method for gluing atmospheric return signal is developed and tested. For the Photon Counting signal, the saturation characteristics are analyzed to calculate the coefficients for correction. Then the Analog-to-Digital and Photon Counting signals are glued by a weighted average process. Results show the glued signal is reliable at both low and high altitudes.

1. INTRODUCTION

Photoelectric detection is an important issue in the receiving subsystem of lidar. Since the dynamic range of received signal is usually very large for a ground-based lidar, both Analog-to-Digital (AD) mode and photon counting (PC) mode are used to cover the whole dynamic range in many systems. For near-field signal, AD can convert analog voltage signal into digital signal with good linearity, but it cannot efficiently detect weak signal from far-field. On the contrary, PC mode can properly detect the far-field signal by counting the pulses generated by single photon. However, the intensity of near-field signal can easily exceed the maximum counting rate of PC, which results in saturation of photon counting rate. In that case, nonlinear error introduced from pulse pile-up appears [1]. Therefore, AD and PC can detect the near-filed and far-field signal separately and two data streams should be combined into one signal profile before further processing.

The gluing algorithm is one of the essential issues to be considered for improving measurement accuracy of various atmospheric parameters [2-5]. One method is to find regressions coefficient from two measurement modes to convert AD data to PC data and replace the saturated PC data with the converted AD data [6] . The other method is to calculate the variance of the data with Poisson's distribution to determine coefficients between AD and PC data [7]. In addition, Newsom et al., pointed that PC data should be the independent variable in linear regression calculation between AD data and PC data [6]. This method is also used in this paper. However, there is still some space to improve the gluing algorithm. Firstly, since the solar background light has influence on the selection of gluing region, it should not be subtracted from the return signal before applying the gluing process. Secondly, the non-linear relationship of AD and PC data calculated by curve-fitting is useful to determine the gluing region and perform the correction.

This paper develops a new gluing method to solve the problems mentioned above. Firstly, the relationship between original data of AD and PC is analyzed and the linear region is calculated. Secondly, the unit of AD data is converted into photon counting rates and is used to correct the saturated PC data according to the relationship obtained by curve-fitting. At last, the gluing signal is calculated by the weighted average method.

2. METHODOLOGY

In lidar system, PMT (photomultiplier tube) is illustrated by backscattered light and outputs analog voltage signal which is converted into digital signal by AD converter or into photon counts by photon counter. The process of gluing AD and PC data is shown in Fig. 1. The glued signal is calculated from the converted AD and PC data by linear-fitting and curve-fitting method.

The original AD data and PC data are compared and shown in Fig. 2. It shows that, when the PC value is less than about 100, there is a good linear relationship between PC and AD data, which means that the measured PC and AD value are

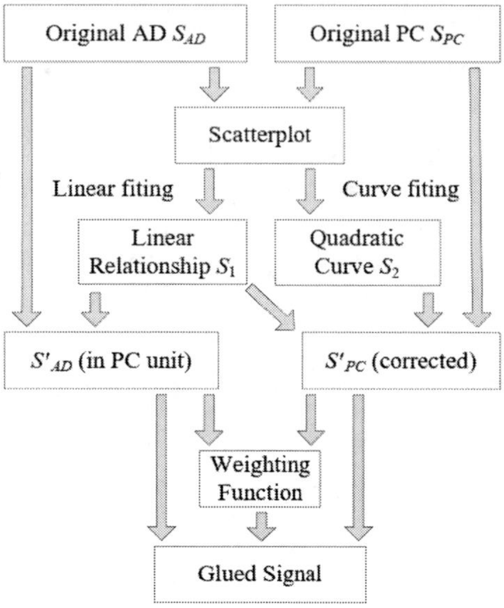

Fig. 1 Flow chart of gluing PC data and AD data

both correct. However, when the PC value is more than 200, there is obvious non-linear relationship between AD and PC data which is in saturated state. As shown in Fig. 2, the black curve is the result by curve fitting method, which is expressed as

$$E = \sum_{i=0}^{m} E_i^2 = \sum_{i=0}^{m} [S_2(x_i) - y_i]^2, \qquad (1)$$

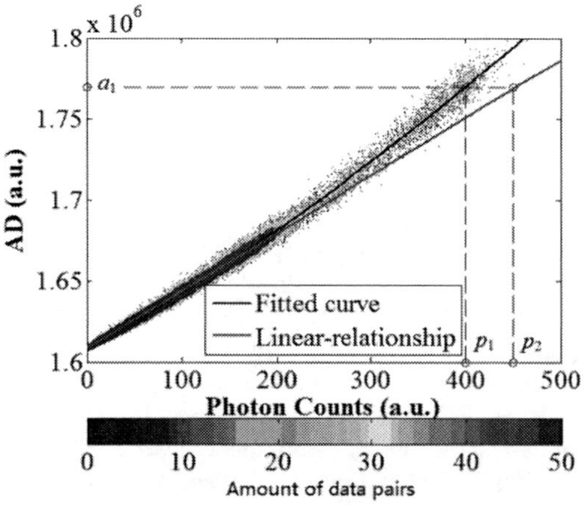

Fig. 2. Scatterplot of PC data and AD data. Red line is the linear-relationship between AD and PC data, black curve is the fitted curve, and colored points represent the amount of data pairs.

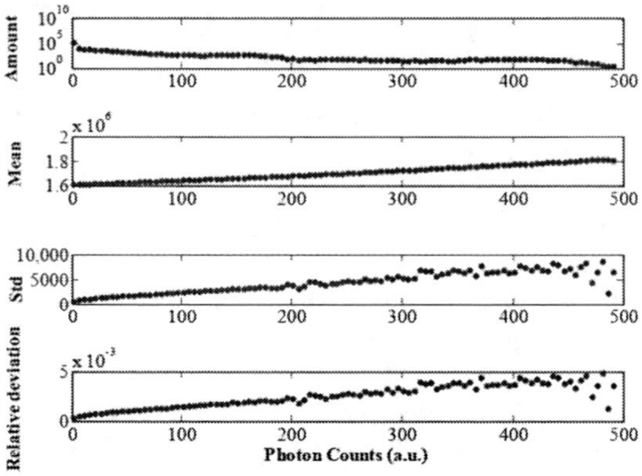

Fig. 3. Analysis of PC and AD data showing the amount of data pairs, mean, standard deviation and relative deviation in corresponding bin of photon counts from top to bottom.

$$S_2(x) = ax^2 + bx + c, \qquad (2)$$

where $S_2(x)$ is the expression of quadratic curve, y_i is the data before fitting, E is the total residual. a, b, c are the coefficients of the quadratic curve. The solution of S_2 can be obtained by minimizing the total residual E.

To find a good linear region between AD and PC data, the PC axis is divided into photon counts bins to calculate the amount of data pairs, mean, standard deviation, the relative deviation (ratio of standard deviation and mean) in each bin. As shown in Fig. 3, the relative deviation increases with PC value. The linear region is determined by the threshold of the relative deviation, which is selected to be about 0.003. The data within this region is used for fitting linear relationship shown by the red line in Fig. 2. The linear fitting method is expressed as

$$E = \sum_{i=0}^{m} E_i^2 = \sum_{i=0}^{m} [S_1(x_i) - y_i]^2, \qquad (3)$$

$$S_1(x) = ax + b, \qquad (4)$$

where $S_1(x)$ is the expression of linear relationship, y_i is the data before fitting, E is the total residual. a and b are the slope and offset of the line. The solution of S_1 can be obtained by minimizing the total residual E.

An example of the process of gluing method is shown in Fig. 2. In the first step, for any PC value p_1, the corresponding AD value a_1 can be

DOI: 10.1051/epjconf/201611925021

calculated according to the fitted curve shown by the black curve in Fig. 2, which can correct the saturated PC data. In the second step, to convert the AD data into PC unit, p_2 is calculated from the AD value a_1 according to the linear relationship shown by the red line in Fig. 2. These two steps carry out the correction of PC data. As shown in Fig. 2, when original PC value is about 400, the corrected PC value is about 450, corresponding to a saturation amount of about 13%. When PC value is beyond 400, the saturation amount is more than 13%. To convert any original AD data to PC unit, just the second step is needed.

Finally, the gluing signal is calculated by weighted average method expressed as

$$S_{Glued} = (1-W)S'_{PC} + WS'_{AD}, \qquad (5)$$

$$W = \begin{cases} 0, (S'_{PC} \le C_{\min}) \\ \dfrac{S'_{PC} - C_{\min}}{C_{\max} - C_{\min}}, (C_{\min} < S'_{PC} < C_{\max}), \\ 1, (S'_{PC} \ge C_{\max}) \end{cases} \qquad (6)$$

where S_{Glued} is the glued data, S'_{PC} is the corrected PC data, S'_{AD} is the converted AD data, W is weighting function, $C_{\max}$ and $C_{\min}$ are the maximum and minimum value in the gluing region, respectively.

3. RESULTS

The lidar system [8-10] developed by Ocean University of China (OUC) is used to acquire data for testing the proposed gluing method. Specifications of the lidar are shown in Tab. 1. Atmospheric signal was measured on January 17, 2015 from 20:08 to 20:12 when the weather was

Tab. 1 Specifications of the lidar system

Parameters	Specifications
Wavelength	532 nm
Reputation rates	100 Hz
Pulse energy	0.12J
Pulse duration	10ns
Linewidth	100MHz
Spectrum purity	> 99.9%
Analog-to-Digital converters	400MHz 14bit
Photon counters	400MHz

cloudy. Data of every 100 pulses (about a second) is saved in a data file and a scatterplot of all data pairs is shown in Fig. 2.

After subtracting background, the AD, PC and glued data is shown in Fig. 4. The signal shows a cloud layer at altitude of about 4.7km, which accords with weather condition. Fig. 4(b)(c)(d) show the details at different altitudes. As shown in Fig. 4(a) the saturation of PC signal is obvious when the data is beyond 300. As shown in Fig. 4(c) and (d), the glued signal is consistent with AD, and PC data. However, it is shown in Fig. 4(b) that the PC is saturated and the glued signal corrects the saturation.

Fig. 5 shows the deviation of AD and PC data compared to the glued data. The difference between PC data, AD data and glued data is obviously at the height of about 1km, especially the absolute deviation of PC data due to saturation. However, absolute deviation of AD data is much smaller than PC data because AD data becomes noisy in the case of weak signal with small

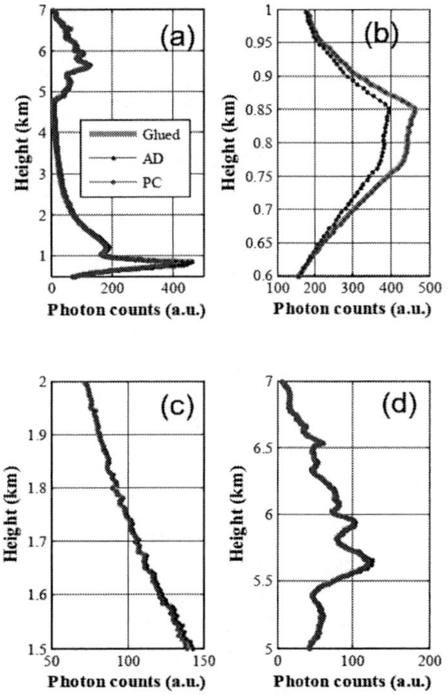

Fig. 4 Signals after background subtraction in the height of (a) 0.5km to 7km, (b) 0.6km to 1km, (c) 1.5km to 2km and (d) 5km to 7km.

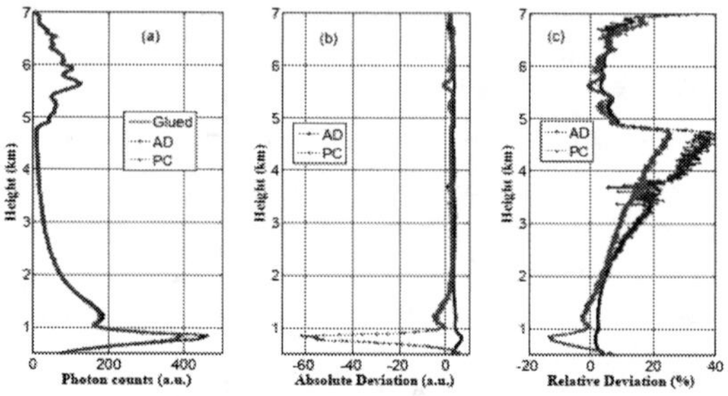

Fig. 5 Deviation of AD and PC data compared to the glued data. (a) glued data, AD data and PC data, (b) absolute deviation of AD (AD data minus glued data) and PC (PC data minus glued data), (c) relative deviation (absolute deviation divided by glued data) of AD and PC.

absolute value. In addition, relative deviation increases below cloud layer. However, in the cloud region at the height of 5km to 7km, relative deviation decreases.

4. CONCLUSIONS

A gluing method is proposed and tested in this paper. Original AD data and PC data are used to calculate the relationship by curve-fitting and to correct PC data. A linear region is determined to obtain a linear relationship to change AD data to PC unit. Finally, the glued signal is calculated by weighted average. Results show the gluing method can correct the saturated PC data. Comparison in cloud region indicates that strong signal keeps the reliability after gluing. This method can be used for gluing PC data and AD data in lidar systems for measuring various atmospheric parameters.

ACKNOWLEDGEMENT

This work was supported by NSFC under grant 41375016.

REFERENCES

[1] Whiteman D N, Demoz B, Rush K, et al, 2006: Raman lidar measurements during the International H2O Project. Part I: Instrumentation and analysis techniques. *Journal of Atmospheric and Oceanic Technology*, **23(2)**, 157-169.

[2] Zhi-Shen Liu, Zhi-Gang Li, Bing-Yi Liu, et al, 2009: Analysis of saturation signal correction of the troposphere lidar. *Chinese Optics Letters*, **7(11)**, 1051-1054.

[3] D. Petty and D. D. Turner, 2006: "Combine analog-to-digital and photon counting detection utilized for continuous Raman lidar measurements," *presented at the 23rd International Laser Radar Conference, Nara, Japan.*

[4] R. K. Newsom, D. D. Turner, M. Clayton, and R.Ferrare, 2008: "Progress on the use of combined analog and photon counting detection for Raman lidar," *presented at the 24th International Laser Radar Conference, Boulder, Colorado, USA.*

[5] D. D. Turner and J. E. M. Goldsmith, 1999: "Twenty-four-hour Raman lidar water vapor measurements during the Atmospheric Radiation Measurement Program's 1996 and 1997 water vapor intensive observation periods," *J. Atmos. Ocean. Technol.***16**, 1062–1076.

[6] Newsom R K, Turner D D, Mielke B, et al, 2009: Simultaneous analog and photon counting detection for Raman lidar. *Applied optics*, 2009, **48(20)**, 3903-3914.

[7] Fei Gao, 2012: Study of Process in Atmospheric Boundary Layer over Land-Sea T ransition Interface Using Scanning Lidar. *Nova Gorica.*

[8] Zhi-Shen Liu, Bing-Yi Liu, Zhi-Gang Li, et al, 2007: Wind measurements with incoherent Doppler lidar based on iodine filters at night and day. *Applied Physics B*, **88(2)**, 327-335.

[9] Bing-Yi Liu, 2008: Performance optimization and wind field retrieval for mobile wind lidar. *Ocean University of China.*

[10] Zhi-Shen Liu, Bing-Yi Liu, Song-Hua Wu, et al, 2008: High spatial and temporal resolution mobile incoherent Doppler lidar for sea surface wind measurements. *Optics letters*, **33(13)**, 1485-1487.

MERLIN (METHANE REMOTE SENSING LIDAR MISSION): AN OVERVIEW

C. Pierangelo[*1] B. Millet[*1], F. Esteve[1], M. Alpers[3], G. Ehret[4], P. Flamant[5], S. Berthier[5] , F. Gibert[5], O. Chomette[5], D. Edouart[5], C. Deniel[2], P. Bousquet[6], F. Chevallier[6]

[1]*Centre National d'Etudes Spatiales (CNES), 18 Av. E. Belin, 31401 Toulouse, FRANCE*
**Email: clemence.pierangelo@cnes.fr, bruno.millet@cnes.fr*
[2]*Centre National d'Etudes Spatiales (CNES), 2 place Maurice Quentin, 75039 Paris, FRANCE*
[3]*Deutsches Zentrum für Luft-und Raumfahrt (DLR), Königswinterer Str. 522-524, 53227 Bonn, GERMANY*
[4] *Deutsches Zentrum für Luft-und Raumfahrt (DLR), Münchener Straße 20, 82234 Weßling, GERMANY*
[5]*Laboratoire de Météorologie Dynamique (LMD), Ecole Polytechnique, 91128 Palaiseau, FRANCE*
[6]*Laboratoire des Sciences du Climat et de l'Environnement (LSCE) 91198 Gif-sur-Yvette, FRANCE*

ABSTRACT

The Methane Remote Sensing Lidar Mission (MERLIN), currently in phase B, is a joint cooperation between France and Germany on the development, launch and operation of a methane (CH4) monitoring satellite. MERLIN is focused on global measurements of the spatial and temporal gradients of atmospheric CH4, the second most anthropogenic gas, with a precision and accuracy sufficient to constrain Methane fluxes significantly better than with the current observation network.

For the first time, measurements of atmospheric composition will be performed from space thanks to an IPDA (Integrated Path Differential Absorption) LIDAR (Light Detecting And Ranging). This payload is under the responsibility of the German space agency (DLR), while the platform (MYRIADE Evolutions product line) is developed by the French space agency (CNES). The IPDA technique relies on DIAL (Differential Absorption LIDAR) measurements using a pulsed laser emitting at two wavelengths, one wavelength accurately locked on a spectral feature of the methane absorption line, and the other wavelength free from absorption to be used as reference. This technique enables measurements in all seasons, at all latitudes. It also guarantees almost no contamination by aerosols or water vapour cross-sensitivity, and thus has the advantage of an extremely low level of systematic error on the dry-air column mixing ratio of CH4.

1. INTRODUCTION

Carbon dioxide (CO_2), methane (CH_4) and nitrous oxide (N_2O) have been recognized by the International Panel of Climate Change (IPCC) [1] as the major greenhouse gases whose concentration has been directly modified by human activities.

The Methane Remote Sensing Lidar Mission (MERLIN), aims at global measurements of the spatial and temporal gradients of atmospheric CH4, the second most anthropogenic gas, with a precision and accuracy sufficient to constrain Methane fluxes significantly better than with the current observation network.

We show here an overview of the MERLIN mission: its science objectives, the measurement principle and the related expected performances, next the status of the project, including details on the space segment, the ground segment and the orbit.

2. SCIENCE OBJECTIVES

The major anthropogenic sources of CH_4 are the emissions from energy production, landfills, waste treatment, cattle, rice fields, and incomplete biomass burning. In addition there are significant natural CH_4 sources from wetlands. The observing Key regions are the Arctic regions (boreal forest, permafrost), Eurasia (anthropogenic emissions), the tropical regions (forest, wetlands).

For the regional quantification of greenhouse gas sources and sinks, the "top-down" approach relies on atmospheric gas concentration measurements and the use of an "inverted" model of atmospheric transport. At present the top-down approach is severely limited by the small number of observing stations. The goal of MERLIN is to improve this situation significantly by enhancing the CH_4 observations to the point where regional to continental scale CH_4 fluxes can be estimated to generally within less than 20% of their mean values. This requires high-precision and high accuracy observations, the ability to make measurements globally (even in the high-latitudes, during the winter months), and to avoid geographical biases.

To meet these objectives, MERLIN will provide column-weighted dry-air mixing ratios of CH_4, referred to as XCH_4, over the satellite sub-track. The IPDA-lidar instrument considered for the MERLIN mission is not dependent on sunlight and will allow global coverage. Besides, the small footprint and the relative insensitivity to thin cirrus clouds and aerosol layers will allow observations in tropical regions and other

DOI: 10.1051/epjconf/201611926001

areas that are frequently cloud covered or contaminated by biomass burning (e.g. boreal forests). Moreover, a considerable amount of CH_4 is due to biomass burning of boreal forests which calls for a sensor that is immune to aerosol scattering. The MERLIN instrument shall operate in a range-gated mode and therefore will not be prone to biases that arise from aerosol scattering or scattering from thin cirrus layers.

3. MEASUREMENT CONCEPT

The core element of the MERLIN payload is the nadir-viewing CH_4 IPDA LIDAR system. This instrument measures the light scattered and reflected from the Earth's surface and cloud tops which are illuminated by laser pulses having slightly different wavelengths denoted as λ_{on} and λ_{off} as illustrated in Figure 1. The online wavelength λ_{on} is accurately positioned in the trough of one of the CH4 absorption line manifold in the 1.64 µm near-infrared spectral region. The measurement at λ_{off} serves as the reference measurement with negligible absorption by the CH4 molecules in the path. The on- and off-line wavelengths are spectrally close enough to consider the atmospheric and surface interaction processes to be identical with the exception of CH_4 absorption. Because the CH_4 line is pressure broadened throughout the entire troposphere, a significant detuning of the on-line wavelength with respect to the CH_4 absorption line center permits to put a higher weight on the contributions from the lower troposphere [2].

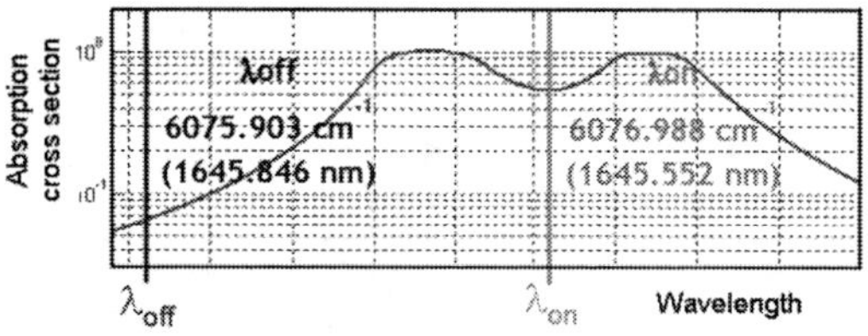

Figure 1: selected wavelength for MERLIN

The nadir viewing LIDAR instrument measures the reflected signals from ground along the track in a quasi-continuous operation mode. The distance on ground between successive on-/off-line laser pulses is about 350m and depends on the pulse-pair repetition frequency (baseline: 20 Hz).

Since the return signals are very weak, it is necessary to accumulate and average several single measurements of the return signals along the track in order to achieve the required measurement sensitivity for the CH_4-column.

From the backscattered signals at both wavelengths, Pon, Poff,, a direct measurement of the differential absorption optical depth (DAOD) of methane is possible

$$DAOD = \frac{1}{2}\ln\left(\frac{P_{off}}{P_{on}}\right) \tag{1}$$

The DAOD is converted into vertically-integrated dry mixing ratio of methane, XCH_4, which is the measurement used for assimilation in atmospheric transport models:

$$XCH_4 = \frac{DAOD}{\int_0^{P_{surf}} \mathrm{WF}(p)dp} \tag{2}$$

where the weighting function (WF) is integrated vertically from the surface pressure P_{surf} to the top-of-atmosphere.

It can be demonstrated that the weighting function, WF, is:

$$\mathrm{WF} = \frac{(\sigma_{on} - \sigma_{off})}{g\left(m_{dry} + m_{H2O}x_{H2O}\right)} \tag{3}$$

where σ is the methane absorption cross-section for the "on" or "off" wavelength, m denotes the molecular mass of dry air or of water vapour, g is the gravity acceleration, and x_{H2O} is the water vapour mixing ratio.

The cross-sections are computed with an absorption line model using spectroscopic parameters and information on meteorological profiles (pressure, temperature and humidity) The surface pressure $\mathbf{P_{surf}}$ in Eq. 2 is calculated using **auxiliary data** from NWP models (pressure and temperature) and information on the altitude of the scattering surface) where the lidar beam hits the ground.

4. MISSION PRODUCTS

The MERLIN mission will provide the following data products:

- Level 0: for each individual laser shot the range-gated backscatter signal data together with auxiliary data such as ranging information, laser pulse energy, time, wavelength in a chronological data sequence.
- Level 1a: Level 0 data with all calibrations computed and appended.
- Level 1b: integrated signal at echo for off- and on-wavelengths, slant-path DAOD, shot-by-shot and averaged, geo-location and time.
- Level 2: column-weighted dry-air mixing ratios of methane (XCH4).
- Level 3: visualizations of level 2 data such as XCH4 maps.

Aside from these data products, directly delivered by the MERLIN mission, level 4 data products are methane surface fluxes at various temporal and spatial scales, obtained through assimilation in transport model. Level 4 data products will not be part of the official MERLIN mission but provided by scientists at various scientific research centers using lower level MERLIN data.

In the context of the overall mission objective, additional data products can be derived from MERLIN. MERLIN secondary products can provide information about the surface (topography, estimates of the vegetation height and occasionally on the vertical structure, lidar retro-reflectance) and the atmosphere (cloud boundaries, including cloud base for small to moderate cloud optical thickness).

5. EXPECTED PERFORMANCES

MERLIN aims at providing global measurements of the spatial and temporal gradients of atmospheric methane (CH_4) with a precision and accuracy sufficient to significantly constrain Methane surface fluxes (Table 1). There are requirements on the random part of the error (high frequency, typically > 1 Hz) and the systematic part (slowly varying component, e.g. orbital variations, or scene dependent errors). The very low level of systematic error aims at avoiding geographical biases in the XCH4 fields that could lead to uncertainties in fluxes.

Table 1: User Requirements on performances

parameter	threshold	breakthrough	target
XCH4 random error	36 ppb	18 ppb	8 ppb
XCH4 systematic error	3 ppb	2 ppb	1 ppb
Spatial coverage	Global	Global	Global
Resolution	Horizontal: 50 km, vertical: total column		

From the analysis of the IPDA equations, an error budget on XCH_4 has been established, both for random error and systematic error.

The main contributor for the random error on XCH_4 is the error on the DAOD, due mostly to the Signal-to-Noise Ratio of the instrument. In order to improve the random performance, averaging of the data of several shot pairs is foreseen, on distances of typically 10 to 50 km.

Conversely to the random error, dominated by DAOD uncertainty, several elements almost equally contribute to the systematic errors: (i) the DAOD measured by the payload, (ii) the error on the surface pressure, caused by both the uncertainty on NWP products and the altitude (with platform and lidar ranging contributors), (iii) the error on the weighting function. Regarding this last contributor, the cross-section knowledge depends on external parameters (temperature profile from NWP, spectroscopic parameters), but also on the lidar itself, through the knowledge of the emitted frequency. As shown in Figure 1, the "on" frequency has been chosen in a local minimum of methane manifold in order to reduce the sensitivity to laser stability. Finally, processing itself may contribute to the systematic error through the introduction of scene-dependent biases, for example in the horizontal averaging step.

6. STATUS OF THE PROJECT

6.1 SPACE SEGMENT

The MERLIN satellite is based on a platform and a payload developed respectively by France and Germany.

The MERLIN platform, developed by Airbus Defence and Space-France under CNES contract, is based on the new MYRIADE product line so-named MYRIADE Evolutions. The MYRIADE Evolutions program offers: a new structure in order to reach 400 kg for satellite mass, an increase of the solar array capacity and the power distribution, a new adapted propulsion system, an improved Attitude and Orbital Control Systems (AOCS), an increase of the Payload data storage, a compatibility with low flight altitudes (Atomic Oxygen concerns).

Table 2. MERLIN Platform characteristics

Satellite Mass	kg	400
Platform Mass	kg	260
Dimensions (body)	mm^3	570 × 940 × 940
Power consumption	W	180

The instrument is developed by Airbus Defence and Space-Germany under a DLR contract. Its main components are the instrument control unit, laser transmitter, a frequency reference unit, an energy calibration unit, receiver chain, receiving telescope and transmitter telescope. One of the main challenges of the MERLIN mission is to implement a complex LIDAR instrument on a small satellite platform with limited resources (power consumption, mass and volume allocations, thermal behavior, and radiation hardness). The main parameters of the MERLIN instrument can be found in table 3.

Until now, no direct pumped laser at this wavelength is available, a laser concept based on a Nd:YAG pumped optical parametric oscillator (OPO) was selected. The laser will be operated in a pressurized housing. The stability of the laser and the actual knowledge of the laser frequency are crucial for this kind of instrument and therefore a compact and precise frequency reference is necessary: an etalon and a CCD-row for fringe measurements, combined with absolute wavelength calibration using a CH_4 gas cell.

Aside from the laser, the receiver telescope is a challenging component of MERLIN. Its Field-of-View

 DOI: 10.1051/epjconf/201611926001

has to be large enough so that it "sees" all of the laser spot on ground.The other limiting factor is the size of the detector which should be small enough to limit the detector noise These factors finally result in a complex optical design. Furthermore, the thermal design of the telescope has to be in such a way that a purely passive thermal control system is sufficient. Due to the limited pulse energy and telescope size, the receiver and detector noise are both critical. More details on the instrument design and sub-components can be found in [3].

Table 3. MERLIN Payload characteristics

Payload Mass	kg	119
Payload Volume	mm^3	820 × 830 × 1010
Payload Power	W	150
On-line λon	nm	1645.552
Off-line λoff	nm	1645.846
Pulse energy	mJ	9
Repetition rate	Hz	20 (double pulse)
Telescope diameter	mm	690

6.2 GROUND SEGMENT

The MERLIN Command and Control Centre (CCC) is built from the future generic ISIS CCC (ISIS: Initiative for Space Innovation Standard). Its main tasks are to control and monitor the satellite and ensure its health over the mission lifetime. The CNES earth terminal network, based on stations implemented in Europe, Africa and America, will be used for commanding and receiving housekeeping data (S-band terminals), and for downloading of the Science data (X-band terminals).

The main functions of the payload ground segment are:

- Payload support functions: generation of commands to control the payload, monitoring of the payload housekeeping data for long term analysis, calibration and management of the payload on-board configuration.
- Payload data processing functionality for science data and products: data processing, archiving and dissemination up-to level 3.

6.3 ORBIT DEFINITION

For MERLIN, a near-polar sun-synchronous orbit (SSO) with an orbit height of about 500 km, a Local Time Ascending Node (LTAN) of 06:00 or 18:00 (as shown in figure 9) and a repeat cycle of about 28 days is foreseen. The selection of the orbit is based on the following considerations:

- to get measurements on a global scale: selection of a near-polar sun-synchronous orbit
- the orbit should be as low as possible in order to get a return signal with an appropriate signal-to-noise ratio. The lower limit here is set by the platform (atomic oxygen and atmospheric drag).
- The LTAN selection is mainly based on satellite power issues.
- The repeat cycle determines the spacing between two subsequent revolutions and the re-visit time on certain areas on the globe. 28 day cycle leads to spacing between two subsequent tracks of about 100 km at the Equator.

Table 4. MERLIN orbit characteristics

Orbit type		Sun-synchronous, dusk_dawn
Semi major axis	km	≈ 6884
Altitude	km	≈ 500
Inclination	°	≈ 97.4
Eccentricity		$\approx 10^{-3}$
Repeat cycle	days	28

6.4 SCHEDULE

The MERLIN project is currently in phase B: preliminary design. The MERLIN feasibility was demonstrated in the phase A, it was shown that it is possible to operate a complex IPDA LIDAR instrument on a MYRIADE Evolutions small satellite platform. The major milestones foreseen for the rest of the development are a Preliminary Design Reviews for satellite and payload in 2015, Critical Design Reviews in 2017 and a Launch date in December 2019.

REFERENCES

[1] Intergovernmental Panel on Climate Change, Climate Change 2007: The Physical Science Basis: Contribution of Working Group I to the Fourth Assessment Report of the IPCC, edited by S. Solomon et al., 996 pp., *Cambridge Univ. Press*, Cambridge, UK

[2] Kiemle, C., Quatrevalet, M., Ehret, G., Amediek, A., Fix, A., and Wirth, M.: Sensitivity studies for a space-based methane lidar mission, Atmos. Meas. Tech., 4, 2195-2211, doi:10.5194/amt-4-2195-2011, 2011.

[3] C. Stephan, et al., 2011: MERLIN – a Space-based Methane Monitor, Proceedings of the SPIE, Volume 8159, 815908-15.

INTEGRATED PATH DETECTION OF CO_2 AND CH_4 USING A WAVEFORM DRIVEN ELECTRO-OPTIC SINGLE SIDEBAND LASER SOURCE

Gerd Wagner[1], Stephen Maxwell[2], David Plusquellic[1*]

[1]*Quantum Electronics and Photonics Division, Physical Measurement Laboratory, National Institute of Standards and Technology, Boulder, CO 80305, USA, *Email: david.plusquellic@nist.gov*
[2]*Sensor Science Division, Physical Measurement Laboratory, National Institute of Standards and Technology, Gaithersburg, MD 20899, USA*

ABSTRACT

Integrated path concentrations of ambient levels of carbon dioxide and methane have been measured during nighttime periods at NIST, Boulder (CO, USA), using a ground-based, eye-safe laser system. In this contribution, we describe the transmitter and receiver system, demonstrate measurements of CO_2 and CH_4 in comparison with an *in situ* point sensor measurement using a commercial cavity ring-down instrument, and demonstrate a speckle noise reduction method.

1. INTRODUCTION

Differential absorption lidar (DIAL) is an active remote sensing technique for measuring range-resolved concentrations of trace gases, e.g., CO_2, CH_4, and H_2O in the atmosphere. The integrated path differential absorption (IPDA) method is similar to DIAL except that it uses natural or mirror surface reflections from hard targets to determine the column concentrations of trace gases in the atmosphere. IPDA can achieve much higher signal-to-noise (SNR) ratio compared to DIAL due to the 100% duty cycle and large return signals from surface reflection. DIAL and IPDA are advantageous compared to passive remote sensing instruments using Fourier-transform infrared (FTIR) and microwave (MW) methods. First, active remote sensing achieves a higher SNR compared to passive remote sensing methods. Second, aspects of DIAL and IPDA instruments are self-calibrating due to their measurement principle. Third, given sufficient power, active remote sensing measurements can be performed during night and day times.
The paper is structured as follows. Section 2 introduces the experimental setup of the NIST IPDA system and its specifications. Section 3 describes the theoretical modeling of the absorption line and the nonlinear fitting routine of the experimental data. IPDA measurements of $CO_2 + H_2O$ and $CH_4 + H_2O$ in comparison with an *in situ* point sensor are presented in Section 4. A speckle noise reduction method is introduced in section 5. A summary with an outlook is given in section 6.

2. NIST IPDA SYSTEM SETUP [1]

The experimental setup of the IPDA system is depicted in Fig. 1 and is based on [2]. A single mode external cavity diode laser (ECDL) tunable from 1585 nm to 1646 nm is launched into a polarization maintaining (PM) single mode fiber, coupled through two isolators to minimize optical feedback and split into two paths using a fiber splitter with a 90% : 10% ratio to provide signals for the scan leg and lock leg, respectively.
The lock leg is coupled to an electro-optic phase modulator (EOM) and then to an acousto-optic modulator (AOM). The EOM is driven at 15 MHz to provide sidebands for stabilization of the ECDL to the filter cavity. The AOM is driven at 251.73 MHz (250 MHz center frequency) with a Rubidium (Rb) referenced radio frequency (RF) source to provide a frequency offset for sideband selection of the scan leg. The output is coupled through a circulator and free space launched through a polarizing beam splitter (PBS) and mode matched to the filter cavity. The signal from a 150 MHz bandwidth photodiode that monitors the reflected beam from the circulator (CIR) is mixed with the 15 MHz reference and demodulated for stabilization (Pound-Drever-Hall (PDH) stabilization technique) [3].
The scan leg is amplified to 20 mW using a semiconductor optical amplifier, and coupled to a broadband EOM (EOM2) to provide sidebands for frequency tuning. EOM2 is driven by an arbitrary

DOI: 10.1051/epjconf/201611926002

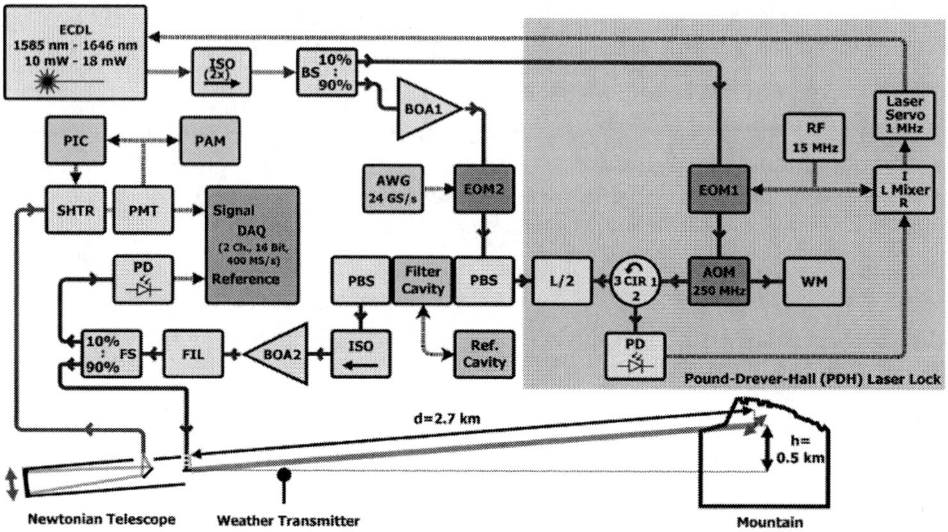

Figure 1. NIST IPDA system setup (PAM, picoammeter; PIC, PMT protection circuit; WM, HeNe reference).

waveform generator (AWG) having 10 GHz of bandwidth (interleaved clock rate of 24 GigaSamples/s (GS/s)). The PDH error signal is conditioned in a 10 MHz band-pass proportional-integral-differential (PID) gain controller and the PID output is limited to < 1 MHz bandwidth to drive the diode laser current for stabilization to the filter cavity. The lock stability of the laser relative to the filter cavity is estimated at < 200 kHz from the noise content of the error signal.

The filter cavity is a linear confocal resonator with a transverse mode spacing of 150 MHz (cavity length is 0.5 m) and a cavity finesse of 43 (mirror reflectivity R=93%). For the transmission and rapid tuning of the 2nd order EOM sideband, the AOM frequency determines the MW tuning frequencies and filter properties of the cavity. An AOM frequency near 250 MHz is used to give sideband and carrier rejection of the scan leg through 3rd order. After mode matching, the resonant contribution of the 4th order sideband is typically less than 0.5% of the 2nd order power.

An absolute frequency reference for the diode laser was achieved by locking the zero-order beam from the AOM (with 15 MHz sidebands) to an optical reference (transfer) cavity. The reference cavity is actively locked to a polarization-stabilized HeNe laser using a piezo-electric transducer (PZT) and heater with a long term absolute frequency stability of better than ± 0.5 MHz. Since the laser is frequency locked to the filter cavity, slow feedback corrections (with a 0.1 s time constant) to the laser frequency are indirectly enforced by controlling the filter cavity length using the PZT driven cavity mirror.

The free space output of the filter cavity is coupled back into a PM fiber, fed through an optical isolator and amplified using a booster amplifier (BOA2). BOA2 is driven near saturation with up to 27.5 mW of continuous wave output power. Due to amplified spontaneous emission (ASE), the output of BOA2 is filtered with an optical bandpass (FIL, 12 nm FWHM, 70% transmission) and split using a 90% : 10% fiber splitter. The 10% leg is used for the initial power leveling at each of the 123 step frequencies, and for the long-term power normalization of the remote sensing signals. The power of the 90% leg is < 9.5 mW (an eye-safe level) and is coupled to the transmitter mounted on the telescope receiver. The transmitter unit consists of a fiber collimator and a (1:5) beam expander resulting in an output beam diameter of 1.9 cm (1/e^2 beam diameter).

A 25 cm Newtonian–type telescope (currently being updated to a 28 cm Schmidt-Cassegrain-type telescope) is used as a receiver. The IPDA return signal is collected at the telescope focus using a multimode fiber with diameter, d = 550 μm or 910 μm, numerical aperture, NA = 0.22, and coupled to a photomultiplier tube (PMT) with a quantum efficiency of 3% at 1600 nm, and 0.3% at 1645 nm.

For photon counting, a preamplifier-amplifier-discriminator setup is used to generate a 5.0 ns output pulse for each electron pulse from the PMT above a discriminator threshold. A 16-bit analog-to-digital converter (ADC) is used to digitize two channels at a combined rate and throughput of 400 MegaSamples/sec (MS/s). Channel 1 is used to monitor the discriminator output and to count photons (signal channel), and channel 2 is used to digitize a 100 MHz InGaAs photodiode used to monitor the reference power.

Figure 2 shows the signal and reference channel with the 123 frequencies (pulse width is 600 ns at each frequency). Each scan consists of 123 frequencies and takes 0.1 ms to acquire (scan repetition rate is 10 kHz).

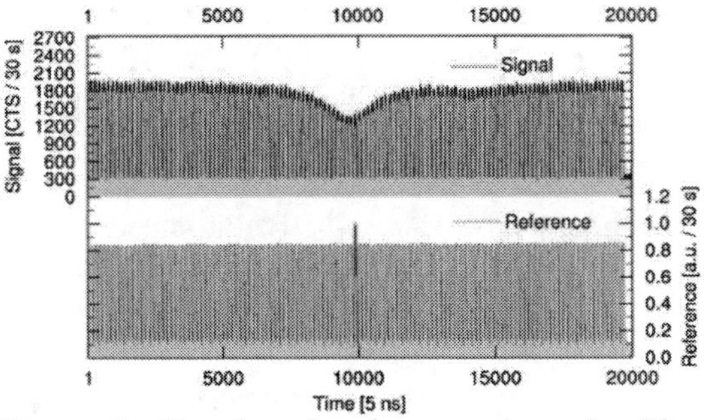

Figure 2. Signal and reference channels. The central dip in signal channel is due to ambient CO_2 and the dip on the right shoulder is water vapor.

3. ABSORPTION LINE MODELING

The atmospheric transmission model to retrieve the column-integrated trace gas concentrations is based on the sum of Voigt line shapes generated from the information contained in the HITRAN 2008 [4] for CH_4 and HITRAN 2012 [5] database for CO_2 and H_2O. A weather transmitter located near the start of the optical path provides initial environmental conditions for the model. Since the optical path is extended over an elevation of 500 m (see Section 4), the line shape model that is fit to the data is first integrated over a set of temperature and pressure range bins. The experimental absorption data is fit to the model line using the Levenberg–Marquardt algorithm [6].

4. IPDA MEASUREMENTS OF CO_2 AND CH_4

IPDA nighttime measurements of CO_2 and CH_4 have been performed from January to March 2015 at NIST, Boulder (CO, USA). The IPDA system was located inside a NIST lab and was pointing towards a rock formation (a Flatiron mountain face) at a distance of 2.7 km and over a difference in altitude of 500 m. Figures 3, 4 show NIST IPDA measurements of CO_2 and CH_4, respectively, in comparison with the *in situ* cavity-ring down measurements (Picarro G2301). For most of these nighttime runs, water vapor was also fit as a shoulder feature but not shown here. Corrections for dry air concentrations are underway.

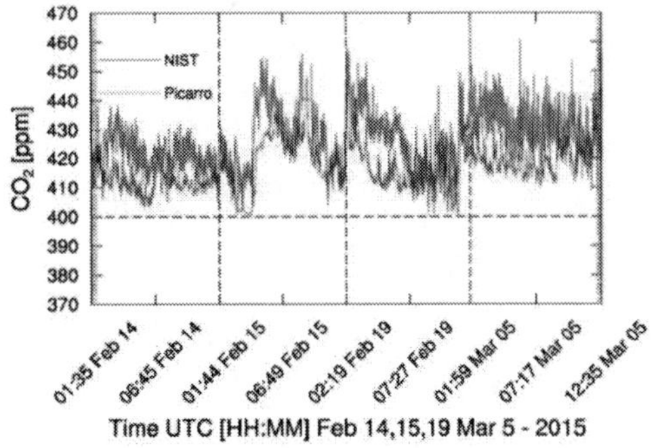

Figure 3. IPDA nighttime data of CO_2 at 1602 nm.

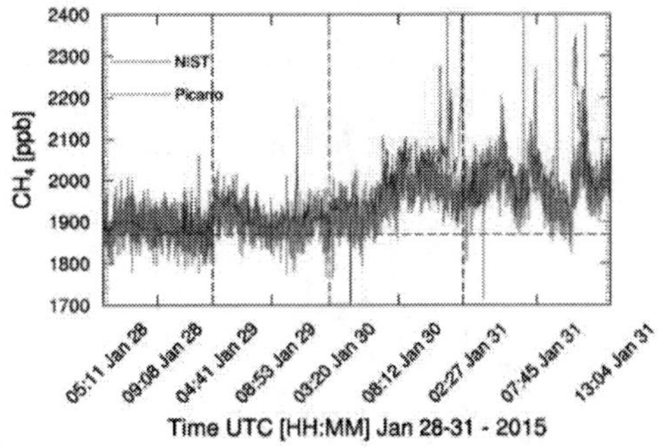

Figure 4. IPDA nighttime data of CH_4 at 1645 nm.

5. SPECKLE NOISE REDUCTION

Speckle is the main source of noise in the IPDA measurement signals [7]. A method of reducing speckle noise is achieved by spatial averaging to remove the temporal coherence over a large number of speckle cells (currently > 30,000). During IPDA signal acquisition, we rapidly modify the pointing position of the telescope. The telescope is slewed repeatedly by ± 1 deg. over a 40 m x 0.4 m area of the target every 4 sec. A galvanometer scanner also moves the transmitted beam within the field of view of the telescope perpendicular to the slewing direction of the telescope at a repetition rate of 50 Hz.

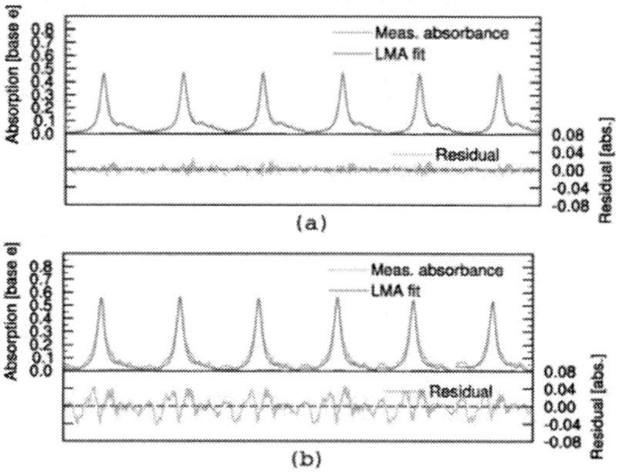

Figure 5. (a) Speckle reduction by moving telescope, (b) speckle with stationary telescope.

Figure 5 shows a comparison of an IPDA absorption measurement of CO_2 at 1602 nm with (a) and without (b) speckle noise reduction, respectively. The signals are shown for 6 consecutive absorption intervals, each averaged for a 30 s period (300,000 scans). As evident in the residuals, this method removes a significant fraction of the systematic error associated with the baseline features. For CO_2 data in Figure 5a, the statistical uncertainty in fitted concentration per interval is near 0.5 %, which is still more than a factor of 5 larger than the quantum (Poisson) noise floor.

6. SUMMARY AND OUTLOOK

In this contribution, we have presented the setup of the NIST IPDA system and showed IPDA measurements of CO_2 and CH_4 in comparison with *in situ* results from a commercial cavity-ring down instrument. We demonstrated a significantly reduce systematic error associated with speckle. For CO_2 retrievals, the addition of a fiber amplifier is planned to permit day time data acquisition. The IPDA system will be used for comparison of CO_2 and CH_4 retrievals obtained using range-resolved DIAL that is under development.

ACKNOWLEDGEMENT

The NIST Greenhouse Gas Measurement Initiative provided financial support. The Picarro G2301 analyzer was kindly loaned by Colm Sweeney and Anna Karion (NOAA Earth System Research Laboratory, Global Monitoring Division, Boulder CO). We also wish to thank Alan Brewer (NOAA Chemical Sciences Division, Boulder CO) for assistance with the transmitter alignment.

REFERENCES

[1] Certain equipment, instruments or materials are identified in this paper in order to adequately specify the experimental details. Such identification does not imply recommendation by the National Institute of Standards and Technology nor does it imply the materials are necessarily the best available for the purpose.

[2] Douglass, K. O., S. E. Maxwell, G.-W. Truong, R. D. van Zee, J. R. Whetstone, J. T. Hodges, D. A. Long, D. F. Plusquellic, 2013: Rapid scan absorption spectroscopy using a waveform-driven electro-optic phase modulator in the 1.6-1.65 μm region, JOSA B, **30**, 2696-2703.

[3] Drever, R. W. P., J. L. Hall, F. V. Kowalski, J. Hough, G. M. Ford, A. J. Munley, H. Ward, 1983: Laser phase and frequency stabilization using an optical resonator, Appl. Phys. B, **31**, 97-105.

[4] Rothman, L. S., and 41 co-authors, 2009: The HITRAN 2008 molecular spectroscopic database, J. Quant. Spectrosc. Ra., **110**, 533-572.

[5] Rothman, L. S., and 48 co-authors, 2013: The HITRAN 2012 molecular spectroscopic database, J. Quant. Spectrosc. Ra., **130**, 4-50.

[6] Moré, J. J., 1978: The Levenberg-Marquardt algorithm: implementation and theory, Numerical Analysis, Lecture Notes in Math., **630**, 105-116.

[7] Ehret, G., C. Kiemle, M. Wirth, A. Amediek, A. Fix, S. Houweling, 2008: Space-borne remote sensing of CO_2, CH_4, and N_2O by integrated path differential absorption lidar: a sensitivity analysis, Appl. Phys. B, **90**, 593-608.

GREENHOUSE GAS LASER IMAGING TOMOGRAPHY EXPERIMENT (GreenLITE)

Jeremy Dobler [1]*, Nathan Blume[1], Michael Braun[1], T. Scott Zaccheo[2], Tim Pernini[2], and Chris Botos[2]

[1]*Exelis Inc., Fort Wayne, IN 46818, USA, *Email: jeremy.dobler@exelisinc.com*

[2]*Atmospheric and Environmental Research Inc., Lexington, MA, USA*

ABSTRACT

Exelis has recently developed a novel laser-based instrument to aid in the autonomous real-time monitoring and mapping of CO_2 concentration over a two-dimensional area. The Greenhouse gas Laser Imaging Tomography Experiment (GreenLITE) instrument uses two transceivers and a series of retroreflectors to continuously measure the differential transmission over a number of overlapping lines of sight or "chords", forming a plane. By inverting the differential transmission measurements along with locally measured temperature (T), pressure (P) and relative humidity (RH) the average concentration of CO_2 along each chord can be determined and, based on the overlap between chords, a 2D map of CO_2 concentration over the measurement plane can be estimated. The GreenLITE system was deployed to the Zero Emissions Research and Technology (ZERT) center in Bozeman, Montana, in Aug-Sept 2014, where more than 200 hours of data were collected over a wide range of environmental conditions, while utilizing a controlled release of CO_2 into a segmented underground pipe [1]. The system demonstrated the ability to identify persistent CO_2 sources at the test facility and showed strong correlation with an independent measurement using a LI-COR based system. Here we describe the measurement approach, instrument design, and results from the deployment to the ZERT site.

1. INTRODUCTION

Under a cooperative agreement with the National Energy Technology Laboratory (NETL), Exelis and our partners at Atmospheric and Environmental Research (AER) have been developing a system with the focus of supporting the identification of possible leak sources for pilot Ground Carbon Sequestration (GCS) sites. The resulting GreenLITE system provides continuous, quantified, 2D CO_2 concentrations and flux over a large area with high accuracy. It is based on proven technology utilized in a new and innovative way. The system uses two laser sensors, located 2-3 m off the ground, nominally preprogrammed to point to a series of 20 retroreflector targets each, as notionally depicted in Figure 1. This sensor target system provides an interwoven series of high accuracy, integrated path differential absorption measurements that are subsequently ingested into a tomographic retrieval algorithm used to generate a 2D reconstruction of the CO_2 concentration across an entire carbon storage field. In addition the system was designed to operate remotely with minimal interaction, and all of the data processing and display is handled through a custom open web interface.

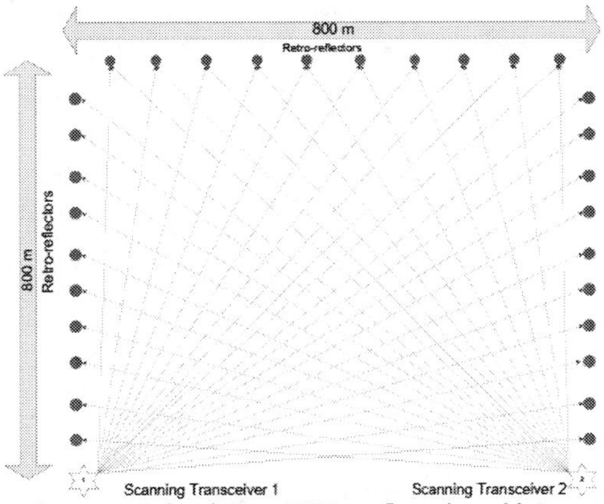

Figure 1: Notional GreenLITE configuration with 2 transceivers, 30 retroreflectors, and 20 chords per transceiver.

The laser sensor technology is based on more than a decade's worth of investments and experience developing and testing the Exelis Multifunctional Fiber-Laser Lidar (MFLL), an airborne instrument that has demonstrated measurements of column-integrated CO_2 from 3 different aircraft over a wide range of conditions. The Exelis-built system has been under evaluation by NASA Langley Research Center since 2005, and recent flights (2010, 2011, 2013 & 2014) have demonstrated accuracies for the remote measurement of better than 0.67 ppmv, on average, with a precision of 1.7 ppmv when compared to an on-board in situ monitor traceable to World Meteorological Organization (WMO) standards [2,3].

In addition to MFLL, Exelis has developed a novel bi-static lidar approach which has ground-to-ground and geostationary orbit-to-ground capability. This system is referred to as the Laser Atmospheric Transmitter Receiver-Network (LAnTeRN), and a proof of concept

This document is not subject to the controls of the International Traffic in Arms Regulations (ITAR) or the Export Administration Regulations (EAR).

DOI: 10.1051/epjconf/201611926003

was recently (2012) demonstrated through Exelis internal research and development funding which supported the modeled performance estimates at extremely low signal levels [4]. Both systems, MFLL and LAnTeRN, rely on a unique single beam amplitude modulation approach and a digital lock-in amplifier to separate and measure the small signals out of a large background. The system being developed in this work takes advantage of the best features of both instruments.

2. METHODOLOGY

The measurement concept is based on Integrated Path Differential Absorption (IPDA) Laser Absorption Spectroscopy (LAS) [5,6]. In this approach two or more narrow-line (~4 MHz) laser wavelengths are positioned on a vibrational-rotational absorption feature of the desired molecule, such that at least one of the lines is in a highly absorbed region and at least one line is in a region of little or no absorption due to molecular vibration/rotation. The wavelengths are spaced very closely (50 – 100 pm) and thus experience nearly identical extinction as they traverse an atmospheric path, except for the extinction due to absorption by the molecule of interest.

Using the Beer-Lambert law, if we know the transmitted power and the received power for a given wavelength we can get the Optical Depth (τ) through the relation $I/I_o=\exp(-2\tau)$, where I is the received power, I_0 is the transmitted power, τ is the one way optical depth over the given path due to scattering and absorption, and the 2 accounts for the round trip in this configuration. By taking the ratio of the optical depth at the two closely spaced wavelengths, all terms in the τ fall out, or become negligible, other than the absorption difference ($d\tau$).

A unique and enabling aspect of the Exelis implementation is the simultaneous transmission and reception of multiple laser wavelengths using a Continuous Wave (CW) Intensity Modulated (IM) approach. In this approach each laser has a unique IM applied, and they are then combined in optical fiber before being transmitted to the atmosphere and subsequently received. The individual wavelength components are then separated after digitization using a custom digital lock-in amplifier or matched filter approach. This simultaneity causes several noise sources due to the atmosphere and instrument to become common-mode and fall out in the ratio, unlike other common pulsed or frequency swept approaches.

If the cross-section of the molecule is known, the difference in absorption over the path between the two wavelengths is proportional to the total number of molecules along the path, or chord. Additionally if the surface pressure and the path length are known, the average concentration of the molecule in the path can be determined based on an iterative approach that compares the differences between the observed values of τ with those computed via the Line-by-Line Radiative Transfer Model (LBLRTM) [7] developed by AER. The IPDA method results in an integrated column molecular number density value between the transmitter and the receiver [5,6].

For the GreenLITE concept a large number of these density values are obtained over multiple intersecting chords, defined by the line between a transceiver and a retroreflector, as illustrated in Figure 1. An estimate of the 2D spatial concentration of CO_2 within the field of regard can be determined using the grid of average concentration per chord and an analytic model that describes the underlying field distribution [8]. This is essentially a sparsely sampled tomographic problem. Our current approach assumes a limited number of Gaussian plumes, which can have different amplitudes and be stretched or rotated along a given axis. The solution is then iterated using a maximum descent approach to arrive at the solution that results in the smallest residual error between the observed differential transmission and retrieved concentrations.

3. SYSTEM DESCRIPTION

The transceivers consist of two main sections: a "hotbox" and the optics head. The hotbox houses the electronics, lasers, modulators, computers and communications equipment; is water tight; and maintains everything at an ambient temperature of 38 °C. The optics head, which houses the transmit/receive optics and a camera, is also sealed and heated.

Figure 2: GreenLITE transceiver hardware at Exelis Farm test site in New Haven, IN.

The optics head is attached to a mechanical scanner which points the transceiver at the specified retroreflector based on a map programmed at setup. The scanner is capable of being reprogrammed remotely. The optics head and the hotbox are connected by a watertight hose containing the electrical cables and optical fibers. An additional component is the calibration target which allows the system to take a zero path measurement, where the ratio of the channels should be one, in order to allow monitoring of the instrument stability over time.

The system is designed for continuous remote operation. A list of the key system parameters is included in Table 1.

Table 1: Key system parameters of GreenLITE system.

Online Wavelength	1571.1129 nm
Offline Wavelength	1571.0629 nm
Optical Transmit Power	5 mW (2.5 mW per ch.)
Modulation Rate	19.2 – 24.6 kHz
Modulation Waveform	sinusoidal
Sampling Rate	1.0 MHz
Sampling Resolution	16 bits
Transmitter Optics	25 mm
Receiver Optics	25 mm
Optics Configuration	biaxial, fiber-coupled
Detector	InGaAs PIN
TIA Gain	10^5 or 10^6 (programmed)
Retroreflectors	50 mm
Lockin Period	10 sec (adjustable)
Power	110 V, 60 Hz, 3 A

An additional key component of GreenLITE, in terms of performance enhancements, is the real time web-based processing and user interface. These web-based tools allow remote autonomous deployment of the system and handle all data storage, processing and dissemination of the data to the user. The user can explore the 2D reconstruction and individual chord data in real time, or scroll through a select period of time. Additionally, users can get information on the raw data and the instrument performance and health,

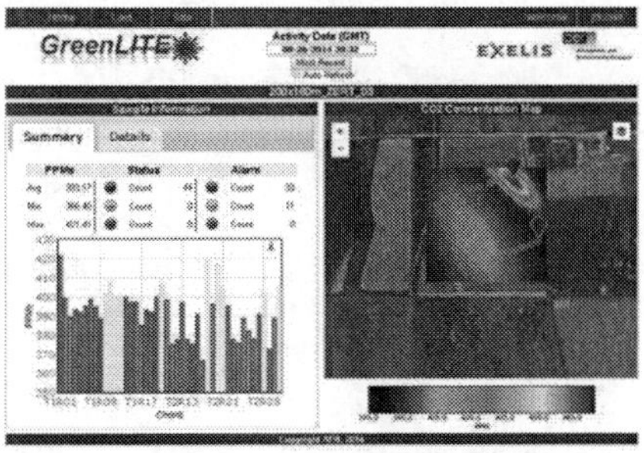

Figure 3: Example of GreenLITE web-based interface showing live individual chord concentration on left and the associated 2D reconstruction with geolocation on the right.

either in real time or at selected times in the past.

4. RESULTS

The GreenLITE system provides a direct measurement of the CO_2 spatial distribution across a large area, which can be used to identify and locate areas of persistent higher concentrations indicative of a source, such as a leak from a GCS site. This was seen in the ZERT testing of the GreenLITE system where higher concentration areas were repeatedly identified over an underground pipe used for the controlled release and near a large manure pile just off the edge of the site. The manure pile was measured independently using a LI-COR based system and was found to be producing spikes that were 10X larger than the concentration spikes being generated by the controlled release of 0.3T/day at the height above ground used for the GreenLITE measurements (~6 ft).

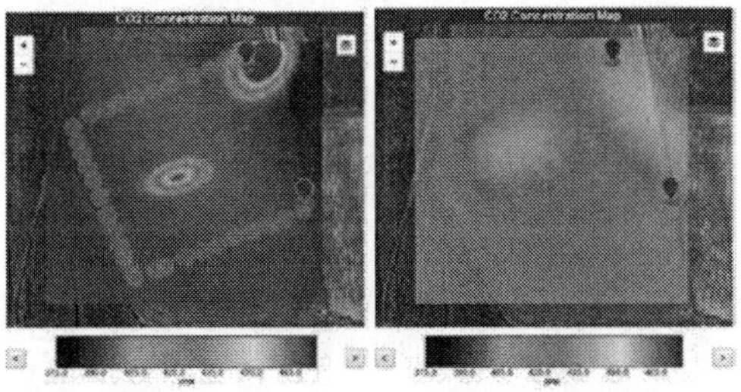

Figure 4: Measurements at the ZERT site on 27 Aug 2014, showing signature from ground release and from an adjacent manure pile (upper right corner of images). Scale goes from 375 ppm to 475 ppm. Left image was taken in the afternoon and right image was taken at 7 am showing the difference in background due to vegetation drawdown and respiration, respectively.

The LI-COR system was a West system designed for chamber measurements. During the ZERT testing, the

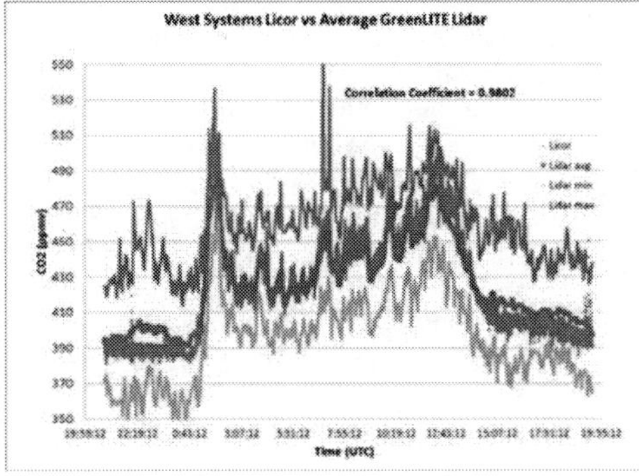

Figure 5: Comparison of average of all GreenLITE chords (red) to an independent LI-COR based measurement system (blue) over a 23 hour period with max and min chord values in purple and green, respectively.

DOI: 10.1051/epjconf/201611926003

chamber was placed on a rebar post away from the underground pipe, in order to provide an independent measurement of the variability of the background CO_2. Figure 5 shows the comparison of the LI-COR data to the lidar data. The correlation coefficient over this 23 hour period was 0.98. Several other comparisons were made and showed correlations of >0.94 with the lidar data.

Evaluation of the instrument ratio using 10 s averages showed SNR values typically > 3000 in the differential absorption ratio. Measured values over a 1400 m path were > 4000 which is equivalent to < 1 ppm precision. Realized setups at the ZERT site and local testing at the Exelis Farm test site required the use of path lengths much shorter than the design range of 1.2 km, resulting in low absorption over the path and precisions ranging from 2-10 ppm per 10 s measurement.

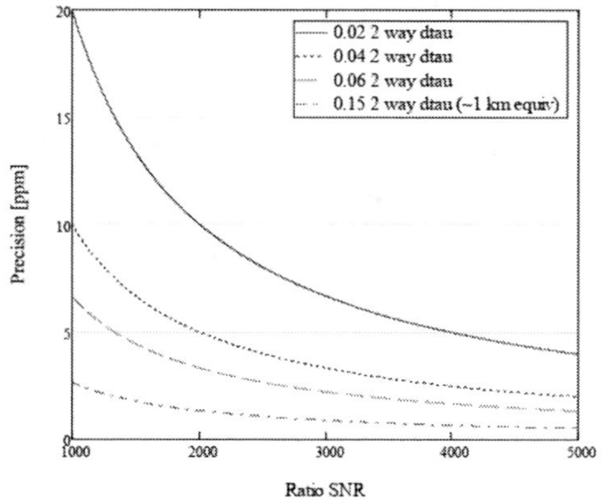

Figure 6: Trade between low differential optical depth (dτ) for short paths and SNR versus precision

4. CONCLUSIONS

The GreenLITE system has demonstrated the ability to generate 2D spatial concentration maps of CO_2 to aid in the identification and location of likely sources. The system was tested at the Exelis Farm test site in New Haven, IN, and is exceeding the design requirements. It has also been tested using a quantifiable underground leak source at the ZERT site in Bozeman, MT, and was able to identify persistent elevated CO_2 sources even in the presence of large background fluctuations. To further test the system's utility and robustness, it has recently been deployed to an industrial GCS site in Illinois, where it is currently running autonomously, uploading measurement data to the web-based processing, and displaying system data and 2D reconstructions in near real-time.

Exelis and AER are in the process of scaling the measurement capability to 5 km scales to enhance the applicability to a wider range of GCS sites, and are evaluating the technology for other applications such as urban areas, or facilities monitoring.

ACKNOWLEDGEMENT

Funding for this work has been partially provided by the National Energy Technology Laboratory under the Department of Energy cooperative agreement # DE-FE0012574. We would also like to acknowledge Dr. Zachary Levine for assistance and discussion regarding more advanced approaches for estimating the 2D reconstructions.

REFERENCES

[1] Spangler, L. E., et al, 2009: A shallow subsurface controlled release facility in Bozeman, Montana, USA, for testing near surface CO2 detection techniques and transport models, *Environmental Earth Sciences*, **60 (2)**, 227-239.

[2] Browell, E. V., Dobler, J., Kooi, S., Choi, Y., Harrison, F. W., Moore, B., & Zaccheo, T. S., 2010: Airborne validation of laser remote measurements of atmospheric carbon dioxide, *25th International Laser Radar Conference*, St. Petersburg, Russia, July 5-9.

[3] Dobler, J. T., Harrison, F. W.; Browell, E. V.; Lin, B.; McGregor, D. P.; Kooi, S. A.; Choi, Y.; Ismail, S., 2013: Atmospheric CO2 column measurements with an airborne intensity-modulated continuous wave 1.57 μm fiber laser lidar, *Appl. Opt.*, **52**, 2874–2892.

[4] Dobler, J., Braun, M., Blume, N., Zaccheo, T. S., 2013: A new laser based approach for measuring atmospheric greenhouse gases, *Remote Sens.*, **5**, 6284-6304.

[5] Korb, C. L., Weng, C. Y., 1983: Differential absorption lidar technique for measurement of the atmospheric profile, *Applied Optics*, **22(23)**, 3759-3770.

[6] Measures, R., 1992: Laser remote sensing: fundamentals and applications, *Krieger Publishing Company*, New York, 243-247 and 305-310.

[7] Clough, S. A.; Shephard, M. W.; Mlawer, E. J.; Delamere, J. S.; Iacono, M. J.; Cady-Pereira, K.; Boukabara, S.; Brown, P. D., 2005: Atmospheric radiative transfer modeling: a summary of the AER codes, Short Communication. *J. Quant. Spectrosc. Radiat. Transfer*, ***91***, 233-244.

[8] Giuli, D., Toccafondi, A., Biffi Gentili, G., Freni, A., 1991: Tomographic reconstruction of rainfall fields through microwave attenuation measurements, *J. Appl. Meteorol.*, **30**, 1323–1340.

EPJ Web of Conferences **119**, 26004 (2016) DOI: 10.1051/epjconf/201611926004
ILRC 27

STANDOFF STACK EMISSIONS MONITORING USING SHORT RANGE LIDAR

Jean-Francois Y. Gravel, Francois Babin*, Martin Allard

*INO Québec, QC, G1P 4S4, Canada.*Email: francois.babin@ino.ca*

ABSTRACT

There are well documented methods for stack emissions monitoring. These are all based on stack sampling through sampling ports in well defined conditions. Once sampled, the molecules are quantified in instruments that often use optical techniques. Unfortunately sampling ports are not found on all stacks/ducts or the use of the sampling ports cannot be planned efficiently because of operational constraints or the emissions monitoring equipment cannot be driven to a remote stack/duct. Emissions monitoring using many of the same optical techniques, but at a standoff distance, through the atmosphere, using short range high spatial resolution lidar techniques was thus attempted. Standoff absorption and Raman will be discussed and results from a field campaign will be presented along with short descriptions of the apparatus. In the first phase of these tests, the molecules that were targeted were NO and O_2. Spatially resolved optical measurements allow for standoff identification and quantification of molecules, much like the standardized methods, except for the fact that it is not done in the stack, but in the plume formed by the emissions from the stack. The pros and cons will also be discussed, and in particular the problem of mass emission estimates that require the knowledge of the flow rate and the distribution of molecular concentration in the plane of measurement.

1. INTRODUCTION

The standoff monitoring for NO_x , SO_x, CO_x, particulates and O_2 emissions from compressor stations is of great interest for pipeline operators. New regulations for more testing is a strong motivator behind this work.

The well documented methods for stack emissions monitoring are all based on stack sampling through sampling ports in well defined conditions. Once sampled, the exhaust gases are analyzed by the usual laboratory instruments that are often based on optical techniques.

Following a first test phase on a TransCanada compression station where NO, CO_2 and O_2 were successfully measured, the work presented here mainly focuses on standoff absorption and Raman for NO and O_2 measurements. Results obtained from intermediate scale simulated stacks will be presented along with description of the apparatus and procedures. Spatially resolved optical measurements allow for standoff identification and quantification of molecules, much like the standardized methods. However this it is not done using in stack sampling, but directly in the plume formed by the emissions from the stack.

2. METHODOLOGY

To evaluate the potential of the approach in a simple but controlled way, two different systems were used, one using a propane burner, the other one a PVC tube. A versatile spectroscopic LiDAR breadboard system based on a pulsed tunable laser source is used for the measurements. Installed in a mobile laboratory located about 45 m away from the mock-up stacks, measurements are performed outdoors in a parking lot.

The first mock-up stack is based on a propane gas burner commonly used for heating. The high temperature exhaust generated by this burner is of particular interest because it creates a depleted oxygen region. Oxygen analysis is then performed by UV-Raman LiDAR.

The other system is based on a 4 inches PVC tube connected to a commercial blower. This represents the smallest stack size expected to be encountered on a compression station plant. The blower is used to create a high velocity gas output at one end of the tube. An injection port is also added between the blower and the PVC tube exit. A controlled NO gas mixture can therefore be introduced within the flowing air simulating a NO emission source. Nitric oxide analysis is then performed using an absorption method close to UV-DiAL.

In both cases, the probing laser is aimed directly over the mock-up stack exit.

For Raman measurements, the 355 nm output of the Nd:YAG source is used for excitation.

For absorption measurements, the third harmonic of the Nd:YAG laser is used as the pumping source for a tunable Optical Parametric Oscillator which is then frequency doubled to generate deep UV light in the 225-230 nm range. The wavelength is precisely selected to match an absorption feature of NO molecule.

Signal collection is achieved through a 20 cm aperture telescope routed either to a filter based 2 channel wavelength selector with PMTs or to a compact monochromator also equipped with a PMT. Both options allow for full waveform measurements of the LiDAR returns.

Absorption measurements

By opposition to traditional DiAL methods that require high stability of the wavelength output, the wavelength of the laser is not alternated between ON and OFF transition wavelengths. Making the hypothesis that the NO emission in the stack exhaust is approximately constant within a few minutes required for the measurement, the wavelength is slowly scanned across an absorption feature. The absorption cross section of NO is at the same time monitored thanks to a laser pick up that goes through a NO reference cell followed by a photodetector.

All the LiDAR curves obtained at each wavelength are normalized to a reference wavelength lidar curve, the wavelength being the one for which there is minimum absorption in the reference cell. Once these normalized curves have been obtained, an average is made over a pre-determined number of points of the lidar curve. For each LiDAR return curve (for each wavelength), the ratio of the averaged values after the stack over the averaged values before the stack is computed. Finally, the resulting spectrum is re-normalize with respect to the highest transmission point in the spectrum, in order to have only transmissions lower than 1, including noise. The reference transmission spectrum from the reference cell is also re-normalized to 1, and these two spectra, along with the concentration in the reference cell, are used as inputs for the DOASIS software.

Raman measurements

For the analysis of O_2 in the propane burner exhaust, Raman returns for oxygen and nitrogen (for normalization) are collected. Ratio of O_2 signal over N_2 signal, indicates the variation in O_2 concentration as N_2 is expected to be constant within the atmosphere as well as within the exhaust gas.

3. RESULTS

A portion of the NO absorption features can be found in Figure 1. This spectrum has been measured in the NO reference cell. The absorption is measured throughout all of the experiments which means that at any wavelength used within a scan, the absorption cross-section is known.

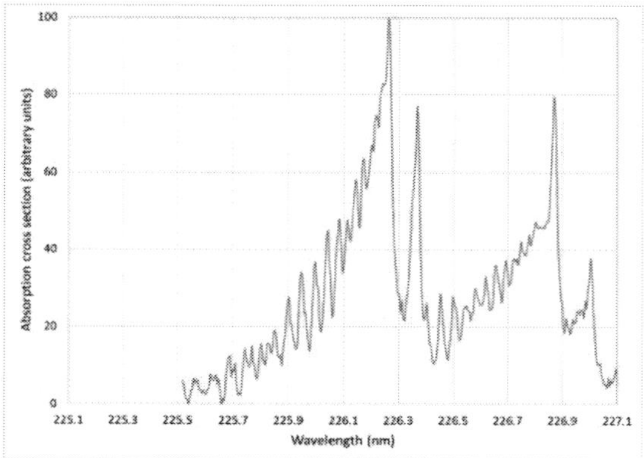

Figure 1. NO absorption spectrum as measured with the INO UV-DiAL platform in a reference cell uncorrected for baseline effects.

Scanning the laser wavelength across the absorption feature results in LiDAR traces used to build an absorption spectrum that reflects the presence of NO. In the case of Figure 2, it is obvious that the blue dotted line is returned by a wavelength that is closer to an absorption maximum than the wavelength represented by the orange line. It is also clear that the absorbing region (stack simulator) is located approximately 46 m from the LiDAR.

DOI: 10.1051/epjconf/201611926004

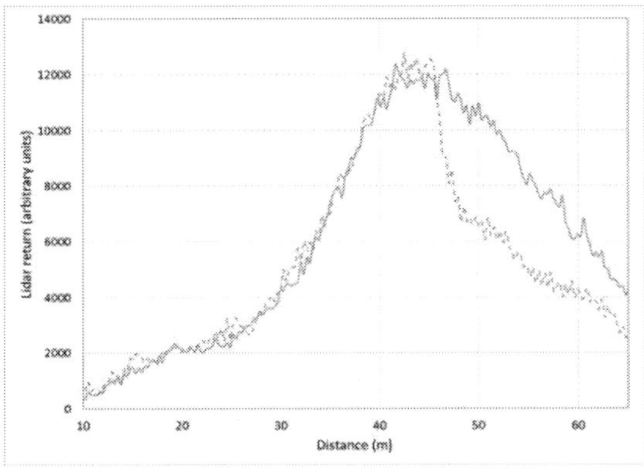

Figure 2. Lidar return curves for NO in the exhaust stack with and without absorption.

Eleven different wavelengths were used for scanning of the absorption line from the stack plume. Using the reference cell for cross-section evaluation, the results were inputted to the DOASIS software which computed the best fit to retrieve a concentration in the stack output plume. Figure 3 shows the absorption results computed for both the reference cell and the exhaust stack.

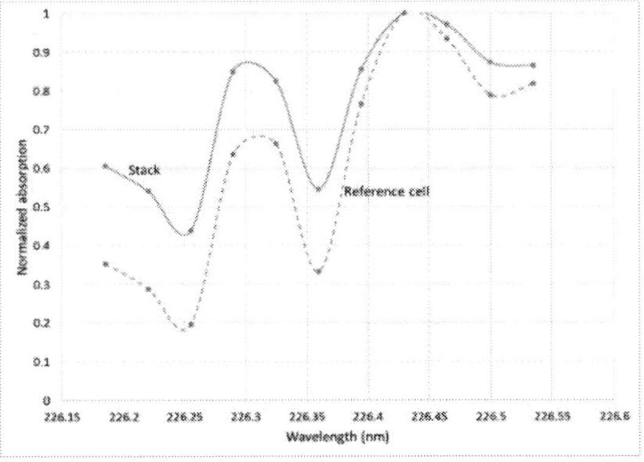

Figure 3. Normalized UV-DiAL and reference cell transmission spectra of NO.

Tests were performed at 3 different nitric oxide concentrations and 3 replicates. Detailed results are expressed in Table 1.

Table 1. Determined NO concentrations vs expected values.

	Expected (ppm-m)	Measured average (ppm-m)	Standard deviation
Concentration 1	116	111.5	15.4
Concentration 2	52	40.4	3.5
Concentration 3	23	20.3	3.8

It is to note that there is a large uncertainty in the expected values that is related to the velocity measurements and blower stability. More experiments need to be done to properly evaluate the accuracy and repeatability of the approach. However, a rough estimate of the limit of detection can be made using the standard deviation of the measurements times 3 thus giving 11 ppm-m for a 3min40s measurement.

For Raman measurements of oxygen, the propane burner exhaust was used. Full waveform return of both O_2 and N_2 signals analysis showed a clear decrease in signal at approximately 44 m. This dip is mostly related to the lower density of the hot exhaust gases; around 600C (see Figure 4).

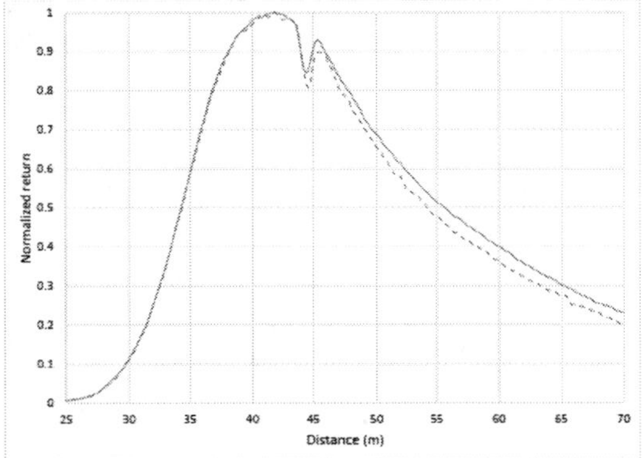

Figure 4. Lidar return curves of O_2 (blue dotted line) and N_2 (orange line) with propane burner « ON ».

Taking each point of the O_2 curve and normalizing it by N_2 curve allow to determine the relative decrease in oxygen concentration related to the combustion process of the burner. Even at lower air density, N_2 concentration should remain constant.

In this case, the drop in concentration is ~ 9% of the normal 20.9% O_2 in air which correspond to about 19% in air. Here again using the 3 standard deviation approximation to estimate the limit of detection gives 0.4% O_2 within a 13 minutes measurement.

4. CONCLUSIONS

It is not easy to simulate real stack exhaust conditions at a small but representative scale. However two different approaches were proposed and two different spectroscopic LiDAR approach for monitoring of the NO emission and determination of O_2 concentration were tested. It has been demonstrated that it is possible to achieve stack emission monitoring from a distance using spectroscopic LiDAR to analyze the exhaust gases. The quantitative potential of the approaches have been demonstrated and limit of detection for NO was estimated to 11 ppm-m for 3 min 40 s measurement time. For O_2, it was determined that the lowest variation in concentration that could be detected would be roughly 0.4% on a 13 min basis. Parts of the mock-up would need to be improved to allow better evaluation of the accuracy and precision of the methods used above. An important hardware limitation of the approaches was related to the lack of spatial resolution that would be needed for very small diameter stacks. For example, the 6 ns pulse length laser would need to be replaced by a shorter one to get better spatial resolution.

Although this work is already interesting from a scientific point of view, it is important that the approaches be accepted by the industry and even more so by regulators. For this, at least two major aspects need to be addressed. First, the measurement is done in the plume just above the stack. This way of doing a measurement has not been yet documented in the standard monitoring methods, although there is a German standard for lidar plume monitoring. A clear assessment that the open path sampling through exhaust gases just above the stack is equivalent to an upstream sampling done from a port located inside the stack needs to be made. Second, it is clear that the concentration value found in the plume needs to be related to a mass emission rate. To achieve this, one needs the size and aspect of the stack as well as the velocity of output gases. The accepted method asked for the velocity to be measured directly in the stack. However, it is clear that it makes no sense to do remote monitoring if someone still need to access the sampling ports of the stack to measure velocity. A project is ongoing for evaluating possible approaches to do this measurement by another standoff method.

LIDAR MEASUREMENTS OF INDUSTRIAL BENZENE EMISSIONS

A.J.C Berkhout[1]*, G.R. van der Hoff[1], L.F.L. Gast[1]

[1]*National Institute of Public Health and the Environment RIVM, Bilthoven, the Netherlands, *Email: stijn.berkhout@rivm.nl*

ABSTRACT

The ability to measure benzene concentrations was added to the RIVM mobile DIAL system. In a ten-days campaign, it was used to measure benzene emissions in the Rijnmond, a heavily industrialised area in the South-west of the Netherlands with petrochemical industry, petrochemical products storage and the port of Rotterdam. On two of the ten days, benzene emissions were found. Combined with measurements of wind speed and wind direction, the Lidar measurements indicated the possible origins of these emissions. This makes the Lidar a valuable tool, augmenting the data collected at fixed monitoring stations.

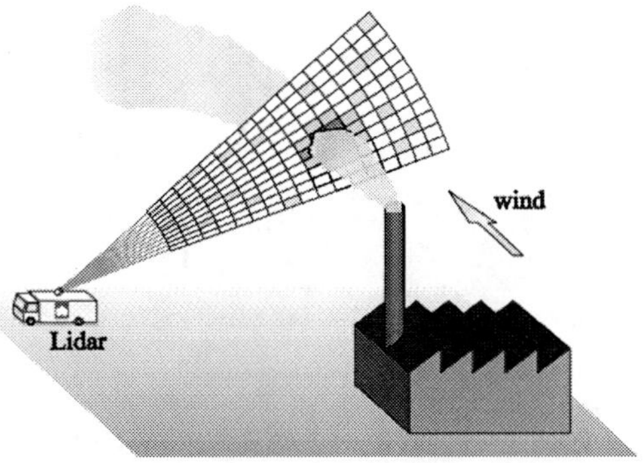

Figure 1. Measuring the emission from an industrial source. A mapping of the emitted plume is combined with measurements of wind speed and wind direction to determine an emission rate.

1. INTRODUCTION

Benzene is a widely used but hazardous and carcinogenic substance. During production, processing or transhipment, some benzene often escapes into the air. In 2012, the total emission of benzene to the air in the Netherlands amounted to slightly more than two million kg. Exposure to benzene should be avoided as much as possible. This makes it important to know where sources of benzene are located. For this reason, the DCMR, the environmental protection agency of the local authorities in the Rijnmond area, ordered RIVM to conduct a Lidar measurement campaign, to map benzene sources. This campaign was in addition to the air quality monitoring network that is present in the area.

2. METHODOLOGY

The measurement campaign was conducted with the RIVM mobile Lidar system. This uses the DIAL technique to make two- and three-dimensional scans of the benzene concentration in the air. This can be used to map benzene plumes (see Figure 1). When combined with a measurement of wind speed and wind direction, an emission rate in kg/h can be determined.

Figure 2 shows a schematic representation of the mobile DIAL system. The light source is a pump laser – dye laser combination of a Spectra-Physics Quanta-Ray Pro 250 pulsed Nd:YAG laser pumping a Sirah PrecisionScan PRSC-D-30 dye laser. For benzene measurements, the dye Coumarin 307 is being used. This generates green light with a wavelength of 506 nm, which is frequency doubled to 253 nm. This UV light is separated off the residual green light and is used for the measurements. It has an energy of about 5 mJ/pulse, a linewidth of 0.0012 nm and a pulse duration of about 10 ns. The laser repetition rate is 30 Hz, pulses are tuned between the *on* and *off* wavelengths on each consecutive pulse. The UV laser beam has a divergence of about 0.5 mrad. The wavelength of each pulse is checked with a Cluster LM007 wavemeter. If the wavelength has drifted too much, the dye laser is tuned to the proper wavelength. The spectroscopy branch shown in Figure 2 is used to determine the effective molecular cross section of the gas being measured.

DOI: 10.1051/epjconf/201611926005

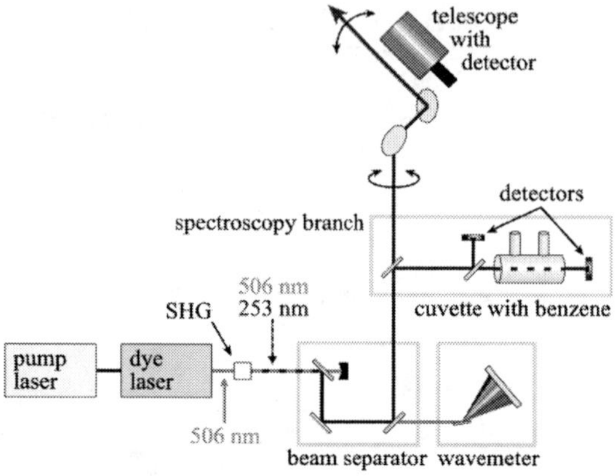

Figure 2. Schematic representation of the mobile Lidar DIAL system.

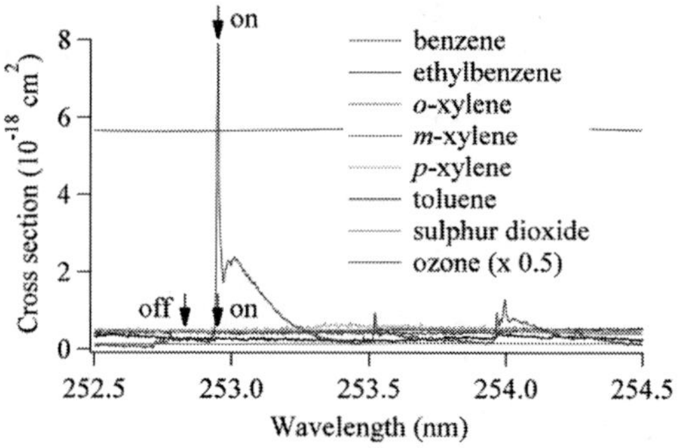

Figure 3. Absorption spectra for benzene and some possible interfering gases. Data sources: [1] (benzene, toluene, xylenes), [2] (ethylbenzene), [3] (ozone) and [4] (sulphur dioxide). The wavelength scale has been shifted by 0.076 nm to show wavelengths in vacuum.

The wavelengths selected for the measurements were 252.95052 nm as *on* wavelength and 252.83175 nm as *off* wavelength (both wavelengths in vacuum). A short literature study was performed to determine possible interference by a number of gases that may be present in the industrial environment where the measurements took place. Their spectra are shown in Figure 3, their relative sensitivities in Table 1. As can be seen, the interference by these compounds is negligible.

Table 1. Relative sensitivities of some possible interfering gases. Benzene = 1.

gas	**relative sensitivity**
benzene	1
ethylbenzene	-0.0012
ortho-xylene	-0.0004
meta-xylene	-0.0029
para-xylene	0.0001
toluene	0.0012
sulphur dioxide	$1.8 \cdot 10^{-5}$
ozone	0.0028

The instrument is housed in a fully self-supporting measurement vehicle (Figure 4). It is able to measure sulphur dioxide [5], nitrogen dioxide [6] and ammonia [7]. For the campaign described here, the ability to benzene was added to the instrument.

Figure 4. The Lidar in an industrial area.

Measuring a single two-dimensional concentration slice took 52 to 90 s, depending on the number of directions (the blue lines in Figure 1). During this campaign, vertical wind profiles were measured using a SODAR, Sound Detection And Ranging. The SODAR used employed the Doppler shift of reflected sound pulses to construct vertical wind speed and wind direction profiles. See [8] for a discussion of the SODAR used in this campaign.

The current campaign encompassed ten measurement days, all in the Rijnmond area (Figure 5). This is a heavily industrialised area, with the extensive facilities of the port of Rotterdam, petrochemical and other industry, storage of petrochemical products and the associated transport movements. Residential areas

DOI: 10.1051/epjconf/201611926005

are often in close proximity to industrial activities, especially in the east of the Rijnmond area.

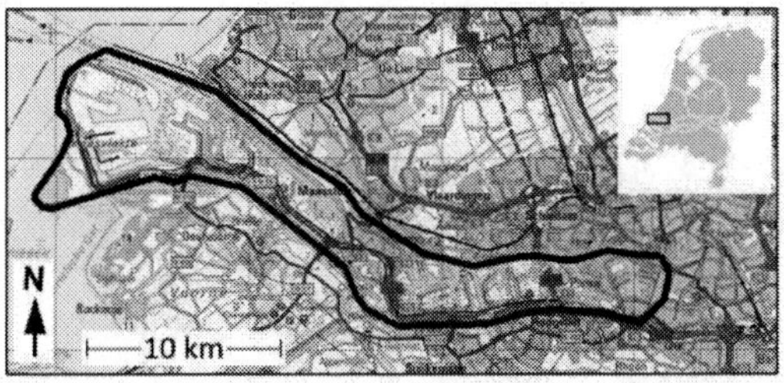

Figure 5. Map of the Rijnmond area and its position in the Netherlands. The most heavily industrialised area is outlined in black.

Measurements were in most cases conducted from public roads, but on a number of occasions the Lidar was positioned within the perimeter fence of an industrial site. Figure 6 shows an example of a measurement location. In this case, two measurement directions are used. Any benzene found in direction 1 but not in direction 2 is assumed to originate in the area between the two directions, as the wind will carry any benzene from the source through the measurement plane.

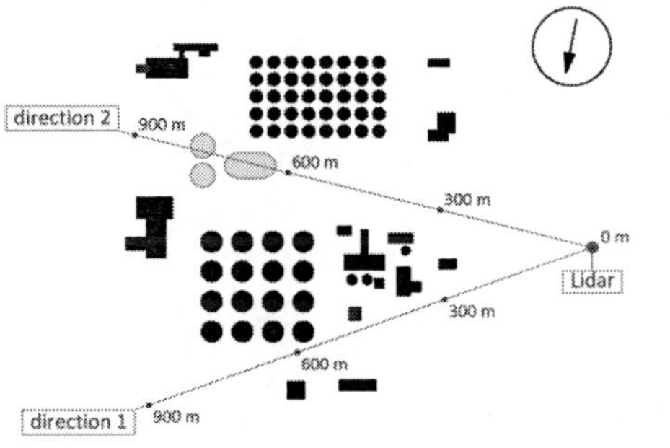

Figure 6. Example location. The red lines indicate the measurement directions. The arrow shows the wind direction during the measurements.

3. RESULTS

In Figure 7, an example of a benzene plume is shown. The Lidar measures concentrations between 300 and 900 m distance. The detection limit of the instrument was estimated to be 12 μg/m^3. The highest concentration measured during the campaign was 3400 μg/m^3.

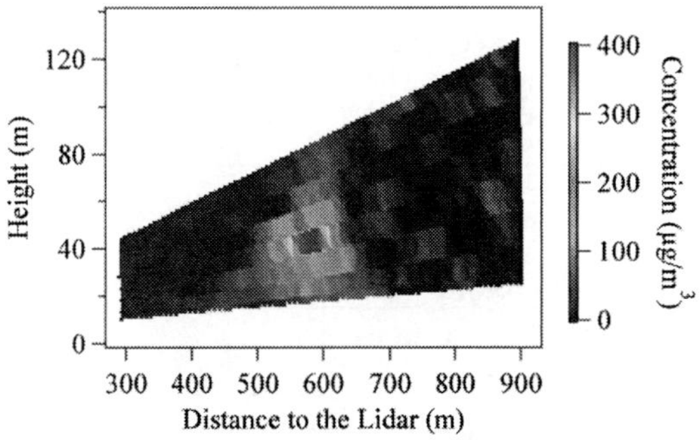

Figure 7. Example of a benzene plume.

In Figure 8, an example is shown of an emission event. This event lasted only 25 minutes, other events lasted longer. Emission events were found on two of the ten measurement locations. The detection limit of the instrument for benzene emissions was estimated to be 1.6 kg/h. The highest emission measured was 25 kg/h.

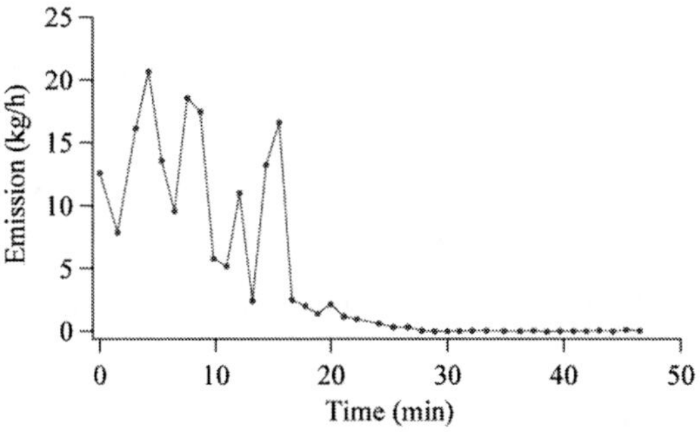

Figure 8. Example of the development of an emission event. Time is in minutes after the start of the event.

The total uncertainty of the emission rate was estimated to be 40% (1 sigma). Table 2 lists a number of contributions to this uncertainty. The largest source is the uncertainty in the representativity of the wind measurement, i.e. the observation that the wind field is disturbed by structures in the vicinity of the measurement slice. This may lead to a difference between the wind profile being measured and the wind profile at the position where the benzene concentration are measured. This contibution is difficult to quantify, it is recommended to conduct further research to reduce this source of uncertainty.

Table 2. Various sources of uncertainty and their contributions to the total uncertainty in the measured emission rate.

description	uncertainty
wind speed	10%
molecular cross section	2.8%
change in cross section with temperature	2%
Lidar elevation angle	2%
wavelength of emitted light	underestimation of 1.1%
Lidar azimuth angle	0.004% to 0.3%
wind direction	0.14%
SODAR azimuth angle	0.14%
saturation of light absorption by the benzene plume	negligible, except at very high concentrations
interference by other gases	negligible
representativity wind profile	difficult to quantify, estimate: <20%
total	**ca. 40%**

4. CONCLUSIONS

The Lidar was successfully deployed to measure benzene emissions. Combined with measurements of wind speed and wind direction, the Lidar measurements indicated the possible origins of these emissions. This makes the Lidar a valuable tool, augmenting the data collected at fixed monitoring stations.

The measured emissions were high, the highest instantaneous emission rate observed amounted to 10% of the average total emission rate in the Netherlands. However, because of their high variability in time, these emission rates cannot be extrapolated to a daily, monthly or yearly emission.

ACKNOWLEDGEMENT

The measurements described in this paper were commissioned by DCMR, the joint environmental protection agency of the province of South Holland and 16 municipalities in the Rijnmond area.

REFERENCES

[1] Fally, S., M. Carleer, and A.C. Vandaele, *UV Fourier transform absorption cross sections of benzene, toluene, meta-, ortho-, and para-xylene.* Journal of Quantitative Spectroscopy and Radiative Transfer, 2009. **110**(9-10): p. 766-782.

[2] Etzkorn, T., B. Klotz, *et al.*, *Gas-phase absorption cross sections of 24 monocyclic aromatic hydrocarbons in the UV and IR spectral ranges.* Atmospheric Environment, 1999. **33**(4): p. 525-540.

[3] Burrows, J.P., A. Richter, *et al.*, *Atmospheric remote-sensing reference data from GOME - 2. Temperature-dependent absorption cross sections of O_3 in the 231-794 nm range.* Journal of Quantitative Spectroscopy and Radiative Transfer, 1999. **61**(4): p. 509-517.

[4] Bogumil, K., J. Orphal, *et al.*, *Measurements of molecular absorption spectra with the SCIAMACHY pre-flight model: Instrument characterization and reference data for atmospheric remote-sensing in the 230-2380 nm region.* Journal of Photochemistry and Photobiology A: Chemistry, 2003. **157**(2-3): p. 167-184.

[5] Berkhout, A.J.C., D.P.J. Swart, G.R. van der Hoff, and J.B. Bergwerff, *Sulphur dioxide emissions of oceangoing vessels measured remotely with Lidar.* RIVM, Bilthoven, Research report 609021119, 61 pp., 2012.

[6] Brinksma, E.J., G. Pinardi, *et al.*, *The 2005 and 2006 DANDELIONS NO_2 and aerosol intercomparison campaigns.* Journal of Geophysical Research D: Atmospheres, 2008. **113**(16).

[7] Berkhout, A.J.C., G.R. van der Hoff, *et al.*, *Measuring ammonia emissions from manured fields.* RIVM, Bilthoven, Research report 680150003, 80 pp., 2008.

[8] Hartogensis, O.K., D. van Dinther, and A.A.M. Holtslag, *WindVisions: First Phase Final Report.* WUR, Wageningen, Research report KfC 56/2012, 2012.

EPJ Web of Conferences **119**, 26006 (2016)
ILRC 27
DOI: 10.1051/epjconf/201611926006

Water vapor measurements by mobile Raman lidar over the Mediterranean Sea in the framework of HyMeX: application to multi-platform validation of moisture profiles

Julien Totems[1], Patrick Chazette[1], Xiaoxia Shang[1], Cyrille Flamant[2], Jean-Christophe Raut[2], Alexis Doerenbecher[3], Véronique Ducrocq[3], Olivier Bock[4] and Fabien Marnas[1]

[1]*IPSL/LSCE, UMR 8212 CEA-CNRS-UVSQ, 91191 Gif-sur-Yvette Cedex, France, julien.totems@lsce.ipsl.fr*
[2]*IPSL/LATMOS, UMR 8190 CNRS-UVSQ, Sorbonne Universités, UPMC (Paris 06), 75005 Paris*
[3]*CNRM/GAME, UMR 3589 Météo-France-CNRS, 31057 Toulouse Cedex*
[4]*IGN/LAREG, Sorbonne Paris Cité, Université Paris Diderot (Paris 07), 75013 Paris*

ABSTRACT

The Water Aerosol Lidar (WALI) system, deployed for 14 weeks during 2012 & 2013 on the island of Menorca, provided the Hydrological cycle in the Mediterranean eXperiment (HyMeX) with an opportunity to perform a multi-platform comparison on moisture retrievals at the timescales relevant for extreme precipitation events in the West Mediterranean basin. After calibration, the WALI lidar yields night-time profiles of water vapor with ~7% accuracy from the ground up to 7 km, and daytime coverage of the lower layers, alongside common aerosol retrievals. It is used to characterize the water vapor profile product given by the IASI instrument on-board MetOp-B, and the fields simulated by the Météo-France AROME-WMED model and the open-source WRF model. IASI is found to be reliable above 1 km altitude, and the two models obtain similar high scores in the middle troposphere; WRF beneficiates from a more accurate modelling of the planetary boundary layer.

1. INTRODUCTION

Because of high orography and intense sea evaporation, the western Mediterranean coastline concentrates major weather hazards. Heavy precipitation and flash-flooding, occurring mainly in the fall, can cause extreme events such as the Gard disaster in France (700 mm daily precipitation in Sep. 2002 [1]). An hourly characterization of water vapor variability across the Mediterranean basin is essential to elucidate the life-cycle of high impact weather systems and improve forecasts of numerical weather prediction (NWP) models. The dedicated Hydrological cycle in the Mediterranean eXperiment (HyMeX), conducted in 2012, opened the opportunity for a multi-instrument (lidar, satellites, balloons...), multi-model comparison to properly apprehend the water vapor variability at the relevant timescales.

The Water vapor Aerosol Lidar (WALI) system developed by LSCE [2] was deployed in Sep.-Nov. 2012 and June-July 2013 on the island of Menorca (shown on Fig. 1). The lidar was embedded in LSCE's Mobile Atmospheric Station (MAS) laboratory van. Its aim during HyMeX was to provide WVMR profiles during the intense observation periods preceding forecasts of extreme precipitation events in Spain and France, to investigate the reliability of moisture fields given by state-of-the-art NWP models and remote sensing instruments.

In this paper, after a description of the instruments and models involved, we present the results of the lidar calibration and validation, and proceed to study the accuracy of high resolution water vapor products by the AROME-WMED and WRF NWP models, as well as the IASI spectrometer on board ESA's MetOp satellite.

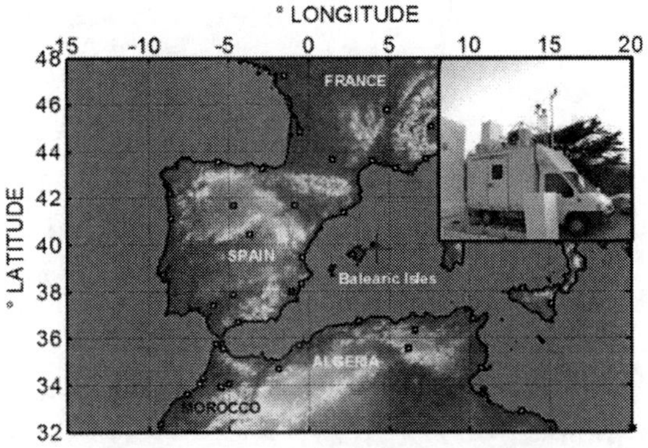

Figure 1. Location of the WALI lidar on Menorca Island near the center of the western Mediterranean, and image of the MAS laboratory van in which WALI is embedded.

2. INSTRUMENTS AND MODELS

The WALI is a transportable and modular lidar instrument dedicated to atmospheric research activities [2,3]. Emitting at 354.7 nm wavelength, it is designed to fulfill eye safety conditions. Its emitter is a pulsed Nd:YAG (Quantel Brilliant) laser, with ~60 mJ UV pulse energy and 20 Hz pulse repetition frequency. Its wide field-of-view ensures a full-overlap of the transmission and reception paths beyond 150 to 200 m. Its characteristics are summarized in Table 1.

Using analog and photon count detection merging, and variable photomultiplier gain as permitted by skylight-controlled high voltage values, the dynamic range of the receiver is sufficient to allow both nighttime measurements up to ~7 km altitude, and daytime measurements up to 1 km altitude.

Table 1. Characteristics of the mobile WALI Raman lidar.

Laser	**Nd:YAG (freq.tripled)**
Energy	60 mJ at 355 nm
Frequency	20 Hz
Reception channels	Elastic // 355 nm Elastic ⊥355 nm Raman-N_2 387 nm Raman-H_2O 407 nm
Reception diameters	3 x 150 mm
Field of view	2.3 mrad
Full overlap	~200 m
Detector	Photomultiplier tubes
Filter bandwidth	0.2 nm @355&387 nm 0.3 nm @407 nm
Vertical sampling	1.5 m (analog) 15 m (photon counting)
Vertical resolution	~30 m
Acquisition system	PXi techno. at 200 MHz
Lidar head size	~ 90x60x18 cm^3
Total weight	150 kg

In this work, we use lidar measurements from Ciutadella de Menorca (40°01′N, 3°49′E) and Cap d'en Font (39°50′N, 4°12′E) to qualify the moisture fields provided by a space-borne instrument (MetOp/IASI), as well as two high-resolution NWP models (AROME-WMED & WRF):

- The IASI spectrometer [4] was launched in 2012 onboard the polar orbiting meteorological satellite MetOp-B (Operational Meteorology), which forms the space segment of the overall EUMETSAT Polar System. In particular, low resolution vertical profiles of WVMR are retrieved from infrared radiances at the global scale [5]. Here, we use the IASI-derived WVMR operational Level-2 products available via EUMETSAT. It was only validated punctually against radiosoundings and models before, and not in the low troposphere [6].
- AROME-WMED [7] is a dedicated high resolution (2.5 km) operational modelling system covering the western Mediterranean, based on an extension of the adiabatic equations of the ALADIN NWP System [8]. For the HyMeX program it has been run by Météo-France in real-time from Sep. 2012 to March 2013. The data has been horizontally interpolated at the Menorca site, on the 60 model hybrid levels. We use 3-hourly operational analysis, as well as hourly 48-h forecasts.
- We have specifically run the open-source regional non-hydrostatic Weather Research and Forecasting (WRF) model (version 3.5, [9]) for simulations in Sep. and Oct. 2012. Boundary conditions are taken from the 1°×1° Final Global Analysis. Nudging has been applied every 6 hours to wind, temperature and humidity, above the planetary boundary layer (PBL). The horizontal resolution is set to 10 km, and there are 50 pressure levels along the vertical column. Specific models represent land surface, PBL, surface layer, microphysics, subgrid-scale convective, shortwave and longwave radiation schemes.

3. LIDAR CALIBRATION

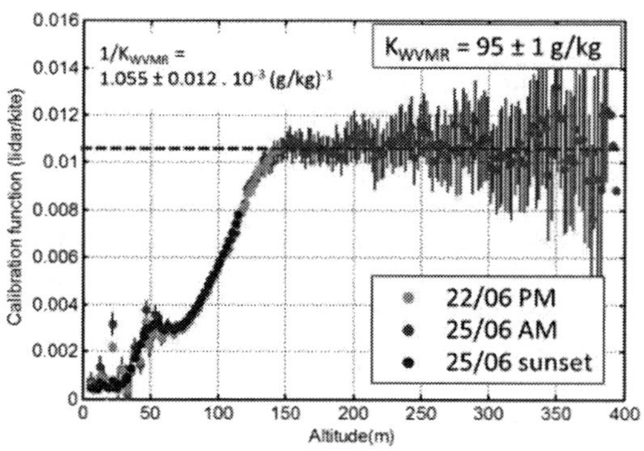

Figure 2. Lidar calibration function as estimated by simultaneous co-located kite-borne water vapor measurements (ratio of lidar/kite WVMRs).

The WVMR is proportional to the ratio between the received signals at 407 nm (water vapor) and 387 nm (nitrogen). The lidar signals are corrected from the differential absorption of the atmosphere at these wavelengths. The coefficient of proportionality, comprising the system constant of the channels and their overlap factors, has to be carefully calibrated. Numerous methods have been proposed to perform this calibration [10,11], but comparison to co-located sounding on site is preferable in the case of a mobile lidar with separate channels. This ensures more precise measurements in the lower troposphere by minimizing the calibration bias.

For both deployments of the WALI lidar, a Vaisala sonde measuring pressure, temperature and relative humidity was flown in the near vicinity of the laser beam, either by plane or by kite directly from the lidar site. In the latter case, a very precise assessment of the calibration function (<2% uncertainty) was possible down to the ground, as shown in Fig. 2, during 3 separate flights of the kite.

To validate the calibration and the daytime WVMR retrieval, Figure 3 shows a daytime inter-comparison between the lidar-derived WVMR, measurements by an airborne sonde and nearby rawindsoundings at Mahon, as well as profiles from the considered meteorological models. In spite of excess noise due to sunlight, WALI's mean uncertainty is below 10% under 1 km during daytime.

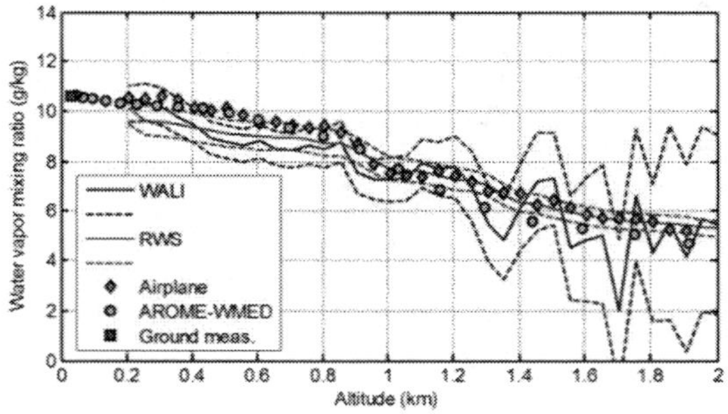

Figure 3. Vertical profiles of WVMR obtained by the calibrated WALI lidar and various measurements (rawindsounding RWS, sonde onboard a plane above the lidar, AROME NWP model) on October 17th, 2013 (daytime, explaining the strong noise on WALI). Dashed lines represent standard deviations.

4. WVMR RETRIEVALS

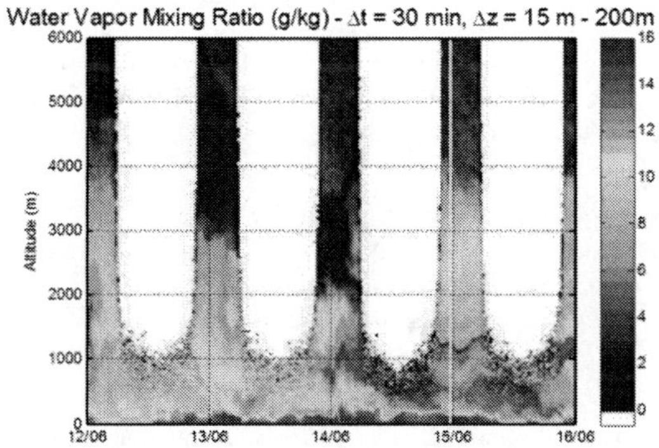

Figure 4. Lidar-derived WVMR from June 12th to June 16th 2013, showing the daytime range limit and the strong variability of atmospheric moisture below 4 km altitude.

In Fig. 4, we show time-altitude cross-sections of WVMR as retrieved by the WALI lidar. The lower profile between the ground and 90 m (first retrievable signal) is completed by a humidity sonde on the lidar van, allowing the observation of the full PBL. The maximum range of the lidar retrieval is artificially limited to Raman signal-to-noise ratios above 5, highlighting the daytime range limit (due to sunlight). During night-time, the accuracy of the WVMR retrieval is around 7% at 5 km [2].

Fig. 4 also highlights a dry layer, from 2 to 6 km altitude on June 12th & 13th, coming from the north Atlantic, as well as three wet plumes, coming from Spain, which have been identified as polluted airmasses using both their optical properties and back-trajectories. This situation is a good example of the great variability of tropospheric water vapor in the region.

5. INTER-COMPARISONS

30 night-time coincidences have been identified across the two campaigns between MetOp-B/IASI and the WALI at Menorca. The Correlation (COR) profile of IASI versus WALI over these cases is plotted in Fig. 5.

Overall, we find a good agreement (COR > 0.75), but only above 1.5 km altitude. By studying individual cases, we find that the remaining deviation is almost always due to strong variability of WVMR with altitude (sharp transitions between wet/dry layers). The weight function of IASI, badly resolved in the low troposphere, is responsible for these errors. Nevertheless integrated water contents provided by IASI are highly reliable [3].

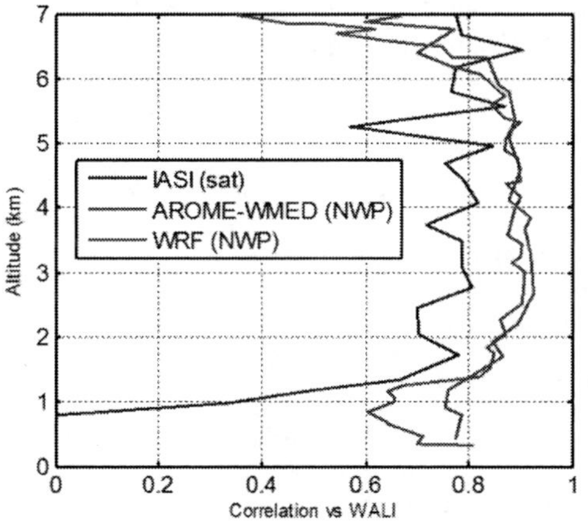

Figure 5. Profiles of correlation scores between the IASI (satellite) WVMR retrievals (blue) and WVRM profiles from the NWP models (AROME, green and WRF, red), versus measurements by the WALI lidar, as a function of altitude.

Also plotted on Fig. 5 are the correlation scores of NWP models AROME-WMED and WRF, compared to WALI over Sep. and Oct. 2012. They are equivalently very good between 2 and 6 km altitude (COR > 0.85), but WRF is better in the PBL thanks to its specific planetary boundary layer scheme. It fares worse in the

DOI: 10.1051/epjconf/201611926006

higher layers due to an underestimation of moisture, which origin remains to be investigated.

6. CONCLUSION

The WALI system, developed by LSCE and deployed for 14 weeks during 2012 & 2013 on the island of Menorca provided the HyMeX program. It was an opportunity to validate the capacity of space-borne instruments and numerical weather models to apprehend moisture variability at the relevant timescales for extreme precipitation events in the West Mediterranean basin. After in-situ calibration, the WALI lidar yielded night-time profiles of water vapor with ~7% accuracy from the ground up to 7 km, and daytime coverage of the lower layers, alongside common aerosol retrievals (extinction, depolarization, lidar ratio…).

WALI was used to characterize the water vapor profile product given by the IASI instrument on-board MetOp-B, and the fields simulated by the Météo-France AROME-WMED model and the open-source WRF model. IASI is found to be quantitatively reliable at low resolution above 1 km altitude, and the two models obtain similarly high scores in the middle troposphere; WRF beneficiates from a more accurate modelling of the moisture trapped in the planetary boundary layer.

Future works within HyMeX will study the accuracy of AROME-WMED forecasts and their potential improvement by the assimilation of lidar data.

ACKNOWLEDGEMENT

This work was supported by the French Agence Nationale de la Recherche (ANR) via the IODA-MED Grant ANR-11-BS56-0005, the MISTRALS/HyMeX programme the French space agency (CNES) and the Commissariat à l'Energie Atomique (CEA).

REFERENCES

[1] Ducrocq, V., Lebeaupin, C., Thouvenin, T. , Giordani, H., Chancibault, K., Anquetin, S., and Saulnier, G.-M.: The 8-9 September 2002 extreme flash-flood : Meteorological description and mesoscale simulations. La Houille Blanche, 6, 86–92, http://hal-sde.archives-ouvertes.fr/halsde-00441440 (Accessed August 4, 2014), 2004.

[2] Chazette, P., Marnas F., and Totems J.: The mobile Water vapour Aerosol Raman LIdar and its implication in the framework of the HyMeX and ChArMEx programs: application to a dust transport process. Atmos. Meas. Tech., 7, 1629–1647, doi:10.5194/amt-7-1629-2014, 2014.

[3] Chazette, P., Marnas, F., Totems, J., and Shang, X.: Comparison of IASI water vapor retrieval with H2O-Raman lidar in the framework of the Mediterranean HyMeX and ChArMEx programs, Atmos. Chem. Phys., 14, 9583-9596, doi:10.5194/acp-14-9583-2014, 2014.

[4] Zhou, D. K., Smith, W. L., Larar, A. M., Liu, X., Taylor, J. P., Schlüssel, P., Strow, L. L., and Mango, S. A.: All weather IASI single field-of-view retrievals: case study – validation with JAIVEx data, Atmos. Chem. Phys., 9, 2241-2255, doi:10.5194/acp-9-2241-2009, 2009.

[5] Amato, U., Antoniadis, A., De Feis, I., Masiello, G., Matricardi, M., and Serio, C.: Technical Note: Functional sliced inverse regression to infer temperature, water vapour and ozone from IASI data, Atmos. Chem. Phys., 9, 5321-5330, doi:10.5194/acp-9-5321-2009, 2009.

[6] Pougatchev, N., August, T., Calbet, X., Hultberg, T., Oduleye, O., Schlüssel, P., Stiller, B., Germain, K. St., and Bingham, G.: IASI temperature and water vapor retrievals – error assessment and validation, Atmos. Chem. Phys., 9, 6453-6458, doi:10.5194/acp-9-6453-2009, 2009.

[7] Fourrié, N., Bresson, É., Nuret, M., Jany, C., Brousseau, P., Doerenbecher, A., Kreitz, M., Nuissier, O., Sevault, E., Bénichou, H., Amodei, M., and Pouponneau, F.: AROME-WMED, a real-time mesoscale model designed for the HyMeX Special Observation Periods, Geosci. Model Dev. Discuss., 8, 1801-1856, doi:10.5194/gmdd-8-1801-2015, 2015.

[8] Bubnová, R., G. Hello, P. Bénard, and J.-F. Geleyn, 1995: Integration of the Fully Elastic Equations Cast in the Hydrostatic Pressure Terrain-Following Coordinate in the Framework of the ARPEGE/Aladin NWP System. Mon. Weather Rev., 123, 515–535, doi:10.1175/1520-0493, 1995.

[9] Skamarock, W., J. Klemp, J. Dudhia, D. Gill, D. Barker, W. Wang, X. Huang, and M. Duda: A Description of the Advanced Research WRF Version 3. http://opensky.library.ucar.edu/collections/TECH-NOTE-000-000-000-855 (Accessed February 19, 2015).

[10] Whiteman, D., Venable, D., and Landulfo, E., Comments on "Accuracy of Raman lidar water vapor calibration and its applicability to long-term measurements", Appl. Opt. 50, 2170-2176, doi: 10.1364/AO.50.002170, 2011.

[11] Sherlock, V., Hauchecorne, A. and Lenoble, J.: Methodology for the independent calibration of Raman backscatter water-vapor lidar systems., Appl. Opt. 38, 5816–5837, doi:10.1364/AO.38.005816, 1999

EPJ Web of Conferences **119**, 27001 (2016)
ILRC 27

DOI: 10.1051/epjconf/201611927001

LIDAR DETECTION OF EXPLOSIVES TRACES

Sergei M. Bobrovnikov[1,2]*, Evgeny V. Gorlov[1,2], Victor I. Zharkov[1], Yury N. Panchenko[3]

[1] *V.E. Zuev Institute of Atmospheric Optics SB RAS, Tomsk, Russia, *Email: bsm@iao.ru*

[2] *National Research Tomsk State University, Tomsk, Russia*

[3] *Institute of High Current Electronics SB RAS, Tomsk, Russia*

ABSTRACT

The possibility of remote detection of *traces* of explosives using laser fragmentation/laser-induced fluorescence (LF/LIF) is studied. Experimental data on the remote visualization of traces of trinitrotoluene (TNT), hexogen (RDX), trotyl-hexogen (Comp B), octogen (HMX), and tetryl with a scanning lidar detector of *traces* of nitrogen-containing explosives at a distance of 5 m are presented.

1. INTRODUCTION

In previous studies [1–4], we experimentally demonstrated the ability to remotely detect the vapors of explosive nitrocompounds in the atmosphere using the laser fragmentation/laser-induced fluorescence method (LF/LIF). The detection range was more than 10 meters at the detection sensitivity of about 10^{-12} g/cm^3.

Further works on exploring the possibilities of the LF/LIF method allowed to prove the applicability of this technology for the detection of *traces* of explosives and formulate the research directions for solving the problem of creating a remote detector of traces of explosives.

In the period from 2012 to 2013, a working model of the lidar for the remote detection of traces of explosives was designed and built by the efforts of the three institutes of the Siberian Branch of the Russian Academy of Sciences (V.E. Zuev Institute of Atmospheric Optics, Institute of High Current Electronics, and Institute for Problems of Chemical and Energetic Technologies).

The operating principle of the lidar is based on the combination of the effects of laser fragmentation of explosive molecules and laser-induced fluorescence of NO-fragments excited from the second vibrational level (v"= 2) of the ground electronic state $X^2\Pi$ to a zero vibrational level (v'= 0) of the excited electronic state $A^2\Sigma^+$.

In this case, the formation of both fragments presumably occurs both in the solid phase of the trace and in the vapors over the surface resulting from the laser desorption of the trace substance.

2. EXPERIMENTAL

General view of the lidar is shown in Fig. 1.

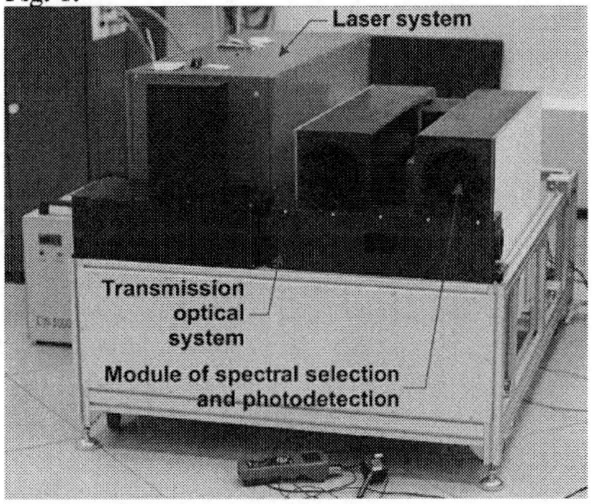

Fig. 1. General view of the lidar.

Functionally, the lidar consists of an excitation radiation source, a spectral filtering system of the probe radiation, a system of the laser beam formation, a system of receiving and spectral selection of the optical signal, a 64-channel system of photodetection, and a system of collecting and processing the information.

In the lidar narrow line excimer KrF-laser, tunable in the frequency range of the $A^2\Sigma^+$ (v' = 0) ← $X^2\Pi$ (v" = 2) transition of the NO molecule is used [5]. For desorption and fragmentation of the explosives traces the same frequency is applied.

To suppress the stray light from the discharge glow of laser chamber and from the stimulated anti-Stokes emission of optical elements of the laser cavity, a system of spectral cleaning of the output radiation is provided.

DOI: 10.1051/epjconf/201611927001

The system of formation of the laser beam is intended to form the illumination line in the scanning area.

The system of receiving and spectral selection of the optical signal consists of a double monochromator designed for the spectral selection of the fluorescence band of the NO molecule ($A^2\Sigma^+$ (v' = 0) → $X^2\Pi$ (v'' = 0)) at a wavelength of 226.5 nm and for the suppression of the unshifted scattering line of the exciting radiation.

A multichannel photodetection system made on the basis of the combination of the two 32-anode photomultipliers H7260-04 (Hamamatsu Photonics) provides a 64 pixel of the image of the illumination line, which corresponds to the spatial resolution along the row in the scanning area of about 1 cm.

Photodetection process in each of the pixels detection channel is carried out in the photon counting mode with a time resolution of 5 ns.

The 64-channel data acquisition and processing system receives data from the photodetection system and sends them to the storage. The program of the traces visualization recognizes the lidar response pattern in the scanning area according to detection criterion.

The process of scanning of the object surface is carried out by sequential moving of the horizontal illumination line from upper start position in the downward direction.

During tests of the lidar, a series of experiments was performed to investigate the possibility of remote visualization of traces of explosives on the surface of an object.

Experiments were carried out according to the scheme shown in Fig. 2. In the detection zone at a distance of 5 m from the lidar, the sample with traces of explosives was fixed. Then, a sequential scanning of the examined area was made.

The traces of explosives were deposited on the surface of the aluminum foil in the form of an abbreviation of the corresponding explosive name: trinitrotoluene - "TNT", hexogen - "RDX", trotyl-hexogen - "C-B", octogen - "HMX", and tetryl - "TET".

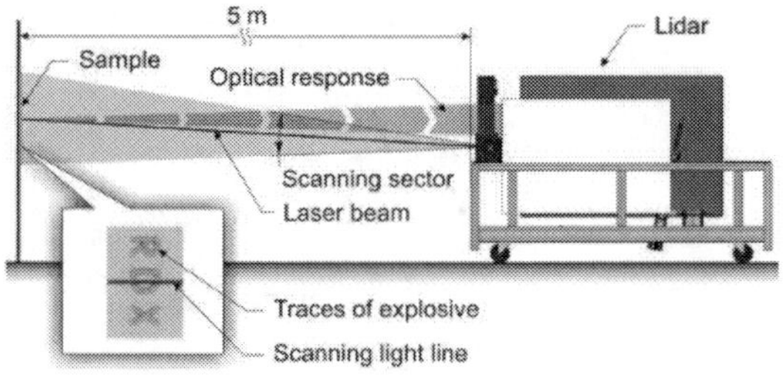

Fig. 2. The scheme of experiments on the remote visualization of traces of explosives.

The surface concentration of particles of explosives in traces was 1 μg/cm^2. The inscription had a size of 7cm×22cm and a vertical direction (Fig. 3).

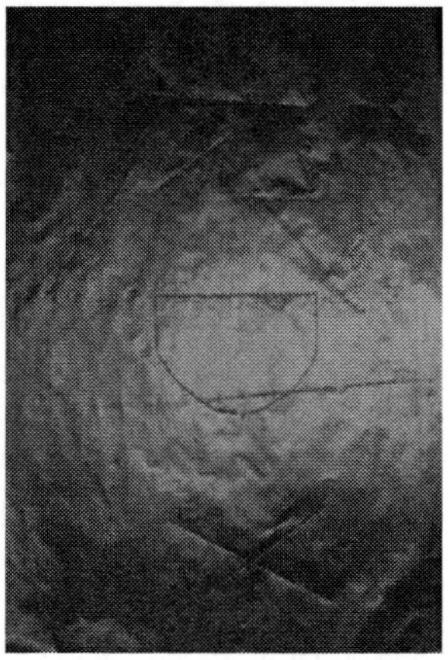

Fig. 3. The sample with traces of hexogen deposited in the form of the abbreviation "RDX".

Detection of the optical response was carried out in the photon counting mode. For the signal accumulation volume over 50 laser pulses per row, number of rows 40, and a laser pulse repetition rate of 50 Hz, the examination time of the sample area was about 30 s.

All the measurements were performed at a laser pulse energy surface density about 13 mJ/cm^2.

3. RESULTS

The results of experiments are shown in Fig. 4. In the figure, the variable "*Pixel #*" denotes the corresponding number of the channel of the multi-anode photomultiplier and the variable "*Row #*" denotes the illumination row number. In the right-hand side of figures, the gray scale legend is given that shows the values of the recorded signals in the number of photon counts.

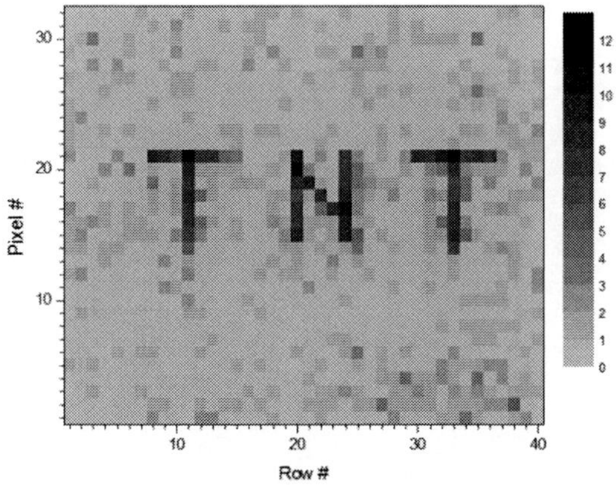

a)

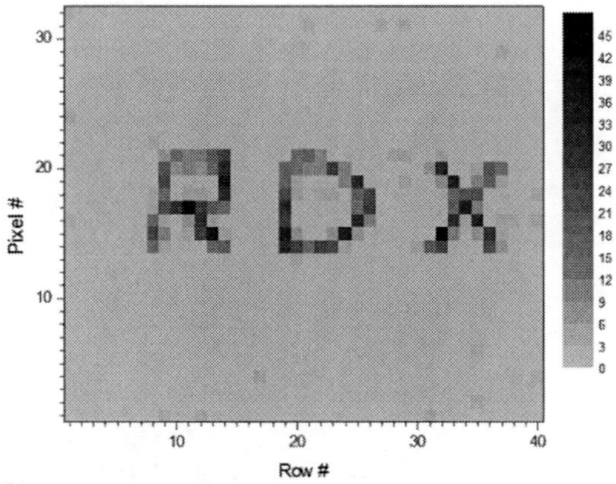

b)

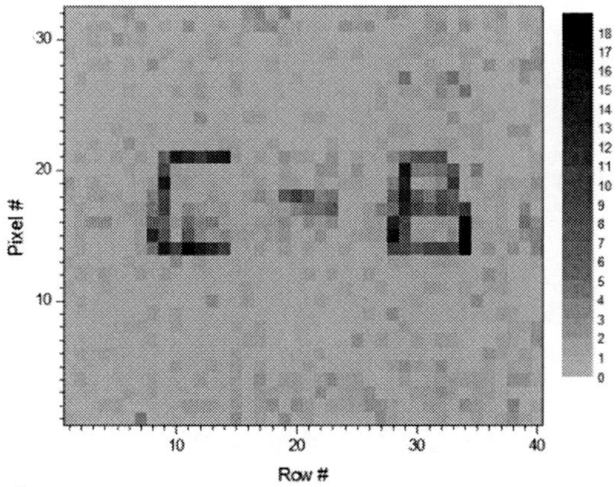

c)

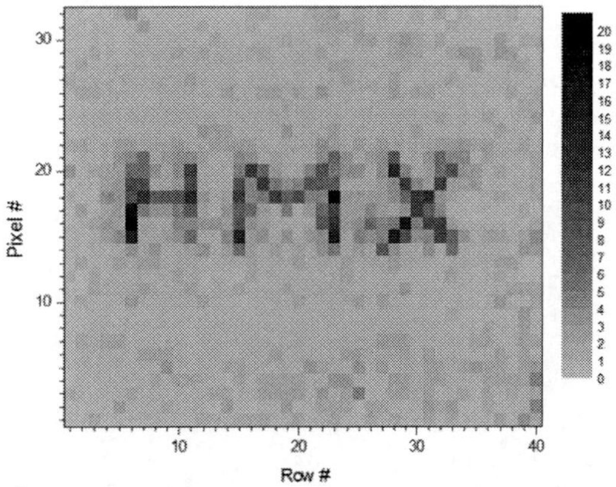

d)

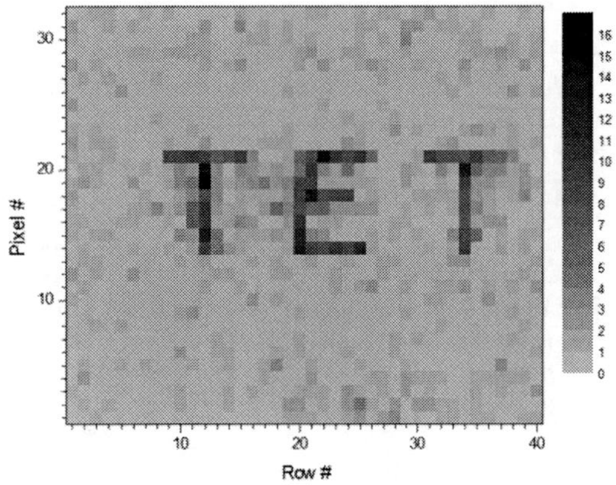

e)

Fig. 4. The results of remote visualization of the explosives traces on the aluminum foil: a) trinitrotoluene, b) cyclonite, c) composition B, d) octogen, e) tetryl.

4. CONCLUSIONS

The investigations have shown the prospects of the LF/LIF method for the fabrication of detectors that enable remotely, at a distance of several meters, to detect trace amounts of explosives with the ability to visualize the places of their localization on the surfaces of objects.

ACKNOWLEDGEMENT

This work was supported in part by the RFBR under Grant No. 13-05-00096a and No. 13-08-98013, the RF President grant No. NSH-4714.2014.5 and No. MK-6286.2015.10, and the Russian Science Foundation (Agreement No. 14-27-00022).

DOI: 10.1051/epjconf/201611927001

REFERENCES

[1] Bobrovnikov, S.M. and Gorlov, E.V., 2011: Lidar method for remote detection of vapors of explosives in the atmosphere, *J. Atm. Ocean. Opt.*, **24** (**3**), 235-241.

[2] Bobrovinikov, S.M., Gorlov, E.V., Sakovich, G.V., Vorozhtsov, A.B., Reznev, A.A., Maksimov, E.M., Perederiy, A.N. and Stepanov, I.V., 2011: Stand off detection of explosives vapors in the atmosphere, *Proc. 1st EUCDE*, 28.

[3] Bobrovnikov, S.M., Gorlov, E.V., Zharkov, V.I. and Panchenko, Yu.N., 2012: Scanning UV lidar for explosives detection, *Proc. 26-ILRC*, 257-260.

[4] Bobrovnikov, S., Gorlov, E., Zharkov, V., Panchenko, Yu., Sakovich, G. and Vorozhtsov, A., 2013: Lidar detector of explosive vapors, *Proc. 2nd EUCDE*, 63-64.

[5] Panchenko, Y.N., Andreev, M.V., Dudarev, V.V., Ivanov, N.G., Pavlinskii, A.V., Puchikin, A.V., Bobrovnikov, S.M., Gorlov, E.V., Zharkov, V.I. and Losev, V.F., 2012: Narrow-band Tunable Laser for a Lidar Facility, *Russ. Phys. J.*, **55** (**6**), 609-615.

EXPLOITATION OF MULTI-BAND LIDAR FOR THE CLASSIFICATION OF FREE-FLYING MIGRATORY BIRDS: A PILOT STUDY OVER ATHENS, GREECE

Samuel Jansson[1*], Alexandros Papayannis[2], Susanne Åkesson[3], Georgios Tsaknakis[2], Mikkel Brydegaard[1]

[1]Lund Laser Centre, Lund University, P.O. Box 118, SE-221 00 Lund, Sweden
[2]Laser Remote Sensing Laboratory, National Technical University of Athens, 15780 Athens, Greece
[3]Centre for Animal Movement Research, Lund University, Ecology Building, SE-223 62 Lund, Sweden
**Email: jansson.samuel@gmail.com*

ABSTRACT

A multi-wavelength lidar system was used to detect free-flying birds passing over Athens, Greece. The location is strategically located in one of the important migratory corridors for birds migrating between Europe and Africa. Multi-wavelength aerosol lidars are operated regularly across Europe in the frame of EARLINET. Here, the feasibility of using this existing infrastructure for assessing fluxes of migratory birds is explored. The backscattered lidar signals were detected at three elastic bands and one Raman band. The monitoring was extended over a period of three months covering predominantly the summer and early autumn period during which approximately 100 hours of lidar data was gathered.

1. INTRODUCTION AND BACKGROUND

Bird migration is a global seasonal phenomenon, engaging billions of birds transporting themselves between distant locations used for breeding and wintering [1]. The long-distance migration has evolved repeatedly in birds and routes often follow topographical features that are concentrating in coastal areas before sea crossings [2]. Different species, age classes and populations depart at different times, and flight parameters such as timing, routes, and distances travelled differ substantially between them. It is therefore desirable to be able to identify migrants in flight. Furthermore, their movements can be affected by climate change [3], and as such the study of bird migration is of interest not only from a migration ecological point of view, but can also shed light on other large-scale phenomena.

Since a major part of the migration flights take place at night and at altitudes of several km, bird migration is a challenging subject to study in detail. Whereas most of the atmospheric lidars pursue the monitoring of gases [4] and aerosols [5] our group has recently started exploring lidar monitoring of the atmospheric fauna, such as insects [6] and birds [7].

The EARLINET comprises 27 regularly operated lidar systems across Europe, each with at least three elastic bands, and a majority with several Raman channels [8]. Coincidentally, many of these systems are located inside or close to common migratory corridors of birds [1], and it is therefore of interest to investigate the potential of utilizing these existing lidar systems parasitically for the classification of migratory birds.

2. SITE AND INSTRUMENTATION

The present study is based on data acquired from the EOLE Raman lidar system [9], located at the National Technical University of Athens, Greece (37.96°N, 23.78°E, 220 m ASL). The measurements were performed between May 13 and July 25 in 2013, with a total of 100 hours of lidar data acquired during this period.

EOLE is based on a pulsed Nd:YAG laser with a 10 Hz repetition rate, emitting coaxially at 1064, 532 and 355 nm. The optical receiver is a ø300 mm modified Cassegrainian reflecting telescope, f = 600 mm, coupled through an optical fiber and divided optically into six spectral bands. The laser beam and telescope field-of-view overlap entirely above ~500 m above ground level. The elastic lidar bands at 532 and 355 nm and the H_20 Raman band at 407 nm were recorded by photomultipliers, while the one at 1064 nm was recorded by an avalanche photodiode, with spectral bandwidths ranging between 0.4-1.5 nm FWHM.

The conventional LICEL GmbH electronics used with the system averaged 100-1000 shots for aerosol purposes (time resolution of 1.5 min). Birds, however, would only appear in a single

 DOI: 10.1051/epjconf/201611927002

shot. Therefore a 5 GS/s 4-channel digital oscilloscope with 300 MHz bandwidth (PicoScope 6400-series) was used parasitically on EOLE and used to sample the 355, 407, 532 and 1064 nm bands in single laser shot mode.

3. DATA PROCESSING

The data was recorded and stored in files, each file containing 4 bands with 20000 time-of-flight/altitude bins, the first 1000 being pre-pulse and the remaining 19000 distributed over the lowest 4 km of the atmosphere, and 600 time slots, each file thereby spanning 1 minute in time. The occurrence of a bird being intercepted by a laser pulse can be considered scarce, and with close to 6000 files, each containing $48*10^6$ 16-bit integer values, efficient algorithms for scanning through the data and extracting such bird events were needed. A false-color time-range map of a file containing a bird event is shown in Fig. 1.

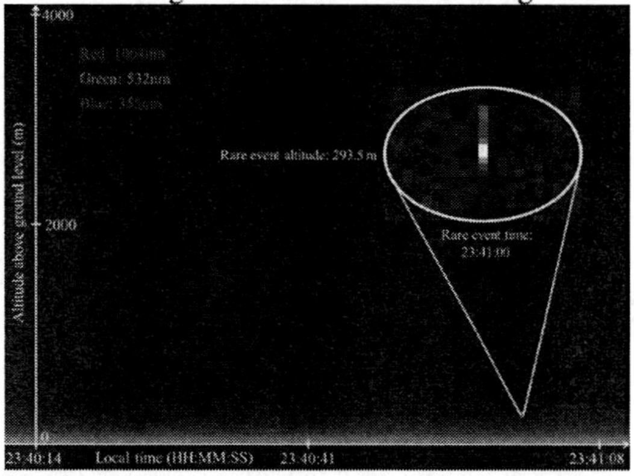

Figure 1: False-color representation of the three elastic bands of a data-file recorded at 27-05-2013 in Athens, Greece. The file contains a bird event at an altitude of 293.5 m above ground level at 23:41:00 LT.

To prevent false identifications, files recorded in inhomogeneous atmospheric conditions were excluded. This was done by fitting an analytical lidar form factor [10] against the static return signal in the 1064 nm channel. The files were rejected according to the residuals of the fit.

When a file was deemed to having been recorded in homogeneous atmospheric conditions it was scoured for bird events. First a detection threshold was set using the properties of the statistical intensity distribution in the elastic bands. In clean air the signal can be assumed to be normally distributed around a static value. The static return and the minimum signal were obtained, and the difference between the two was used to represent the noise amplitude. Events exceeding the static return plus twice the noise amplitude (SNR>2) were considered, and a threshold was set accordingly. The asymmetric intensity histograms for the three elastic bands, together with static return, minimum signal and threshold can be seen in Fig. 2. Lidar signals for the three elastic bands, together with the same three statistical parameters, are shown in Fig. 3.

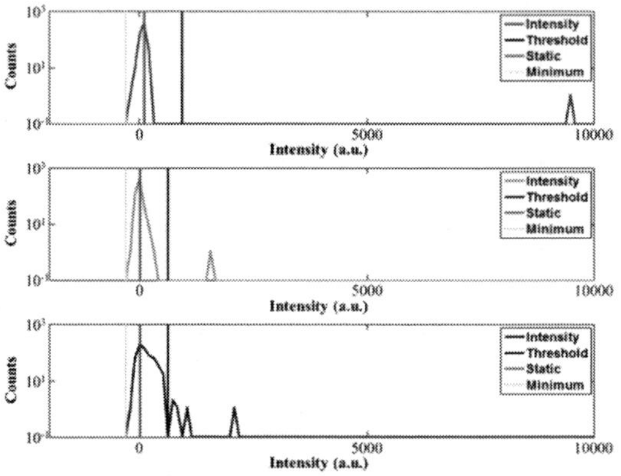

Figure 2: Intensity histograms for the three elastic bands at an altitude of 293.5 m. The intensity distribution of the NIR, Green and UV bands is shown in the upper, middle and lower panels respectively. The static return, minimum signal and resulting threshold are shown in all three histograms. The event exceeds the threshold in all three bands.

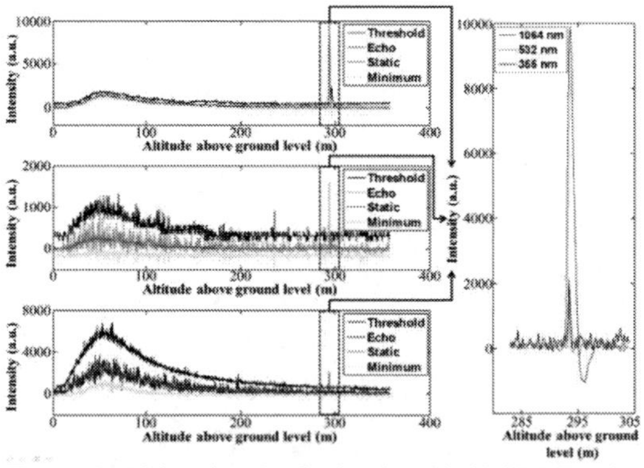

Figure 3: Lidar signals obtained at 23:41:00 LT using the same file as illustrated in Fig. 1. The upper, middle and lower panels show the intensities at 1064, 532 and 355 nm respectively, together with static return, minimum signal and threshold. The rightmost panel shows a close-up of the signals in the marked regions, exceeding the threshold in all bands simultaneously.

While cosmic particles and single photons may cause spikes in the detectors, spikes exceeding the range dependent threshold in all three bands simultaneously were considered to arise from atmospheric fauna. Events fulfilling this criterion were considered as bird events and extracted for further analysis.

4. PROPERTIES OF BIRD EVENTS

Using the methods described in the previous section a total of 1735 events were found in homogeneous atmospheric conditions. The optical properties of these events were then further analyzed.

One of the properties analyzed was the lidar signal over the events in the 407 nm H_2O Raman band. Due to high atmospheric background this band was only active at night. Bird plumages consist of a keratin-air matrix with embedded melanin granules or other pigments [11] and can be expected to become fluorescent when excited by a 355 nm laser pulse [7]. The red-shifted light could then be detected in the 407 nm band. An example of the 407 nm lidar signal over a low altitude event can be seen in Fig. 4, together with the backscattered lidar signal at the three elastic bands.

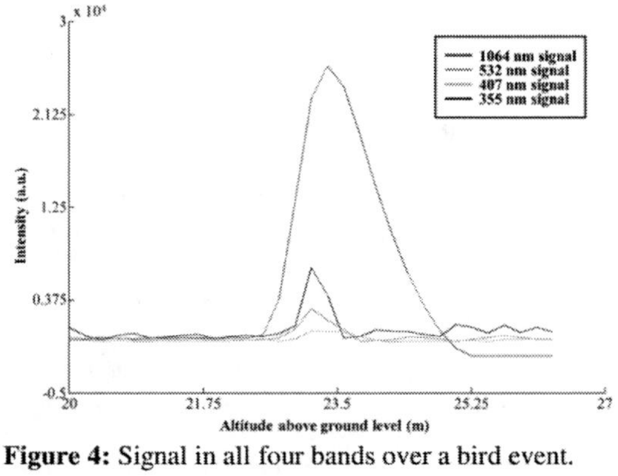

Figure 4: Signal in all four bands over a bird event.

Beam extinction was also analyzed, for the purpose of estimating bird-beam overlap. When part of the laser beam is blocked by an object the signal from the subsequent air volume will decrease, which could be used to assess the absolute reflectance of birds. An example of beam extinction after an event is shown in Fig. 5.

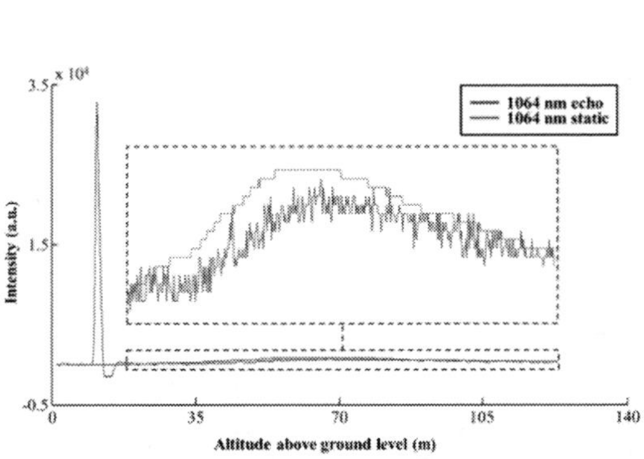

Figure 5: Laser beam extinction at 1064 nm.

Beam extinction after events could, except for in a few cases, not be seen clearly. In the example above some 23% of the beam seems to have been blocked by the low altitude bird event and it is further observed that the signal recovers towards the static return at higher altitudes.

With an elastic backscatter signal in three bands, 3-point scatter spectra were constructed for all events. This was done by integrating the signal intensity over the event after subtracting and dividing with the static return, thus obtaining echo energies. These energies were then divided by the sum of all energies for the same bird event, thereby obtaining chromatic values. Fig. 6 shows a 2D histogram of the chromatic echo energies in the UV and Green bands.

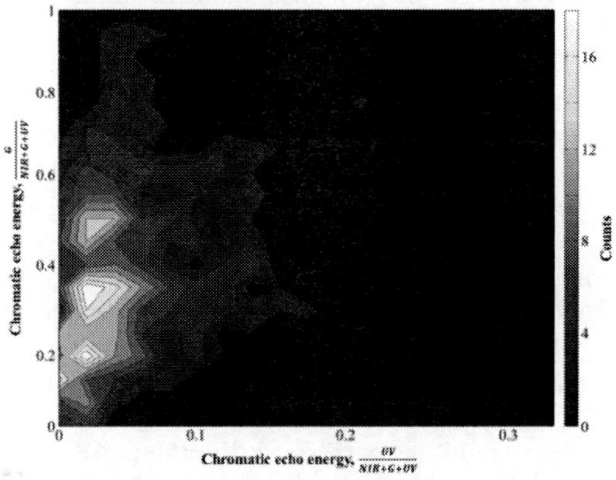

Figure 6: 2D histogram of the chromatic echo energies in the UV and Green bands. The congregation spots in this histogram could be associated with different observed bird species.

5. CONCLUSIONS

A few aspects of the monitoring of free-flying birds using an aerosol lidar system have been illustrated in this study. The repetition rate of 10 Hz requires single shot observation and is somewhat inefficient for bird detection. Random noise spikes exceeding the threshold, specifically in the bands using photomultipliers could be circumvented by imposing that all bands spike simultaneously.

With fluorescence being a broadband phenomenon and the 407 nm band operating with a 0.41 nm filter bandwidth, any attempt to measure fluorescence induced in bird plumages can be assumed to be severely hampered. Though some spikes were observed at typical fluorescence timings they were too small to be significantly distinguished from noise. The inelastic band may be improved making it broader but would still only work at night.

The intent was to use beam extinction as a means of recalibrating the echo intensity into absolute reflectance, but since extinction could scarcely be detected this was not possible. It may, however, be possible at other concentric lidar nodes. In the presented example the bird event occurs before overlap between laser beam and telescope field of view is achieved, which is why the signal recovers at higher altitudes.

The peaks in the 2D histogram represent the most common reflectance spectra. These histogram regions with high event concentrations indicate a multimodality that could be related to different species of birds flying through the beam.

In this study it has been shown that it is feasible to monitor bird activity using the EARLINET and that in-flight species identification could be possible. Should wider filters be installed and the system run during migration seasons at night, many interesting studies could be performed. Furthermore, since the EARLINET encompasses 27 lidar systems, many of which are located in prime locations for the study of bird migration, this offers the opportunity of monitoring bird migration on a continental scale.

ACKNOWLEDGEMENTS

The study was financially supported by grants to S. Å. from the Swedish Research Council (621-2013-4361) and Lund University. This project is a contribution from the Centre for Animal Movement Research financed by a Linnaeus grant from the Swedish Research Council (349-2007-8690) and Lund University. The financial support by the European Community through the ACTRIS Research Infrastructure Action under the 7th Framework Programme under Grant Agreement no 262254 is gratefully acknowledged.

REFERENCES

[1] Newton, I., 2008: The Migration Ecology of Birds, Academic.

[2] Alerstam, T., Hedenström, A., and Åkesson, S., 2003: Long-distance migration: evolution and determinants, *Oikos* **103**, 247–260.

[3] Cox, G., 2010: Bird migration and Global Change, Island Press.

[4] Papayannis, A., et al., 1990: Multiwavelength lidar for ozone measurements in the troposphere and the lower stratosphere, *Appl. Opt.,* **29**, 467-476.

[5] Papayannis, A., et al., 2008: Systematic lidar observations of Saharan dust over Europe in the fram of EARLINET (2000-2002), J. *Geophys. Res.,* **113**, D10204, doi:10.1029/2007JD009028.

[6] Guan, Z., et al., 2010: Insect monitoring with fluorescence lidar techniques: field experiments, *Appl. Opt.*, **49**, 5133-5142.

[7] Lundin, P., et al., 2011: Remote nocturnal bird classification by spectroscopy in extended wavelength ranges, *Appl. Opt.*, **50**, 3396-3411.

[8] Pappalardo, G., et al., 2014: EARLINET: Towards an advanced sustainable European aerosol lidar network, *Atmos. Meas. Tech.,* **7**, 2389-2409.

[9] Kokkalis, P., et al., 2012: The EOLE lidar system of the National Technical University of Athens, *Proc. 26th Intern. Laser Radar Conf.*, 629-632, Porto Heli, Greece, A. Papayannis and D. Balis and V. Amiridis (Eds), 25-29 June 2012.

[10] Kuze, H., et al., 1998: Field-of-view dependence of lidar signals by use of Newtonian and Cassegrainian telescopes, *Appl. Opt.,* **37**, , 3128-3132.

[11] Hill, G. E., and McGraw, K. J., 2006: Bird coloration vol. 1., Mechanisms and measurements, Cambridge, Massachusetts: Harvard University Press.

Observation of Arabian and Saharan Dust in Cyprus with a New Generation of the Smart Raman Lidar Polly

Ronny Engelmann[1]*, Albert Ansmann[1], Johannes Bühl[1], Birgit Heese[1], Holger Baars[1], Dietrich Althausen[1], Eleni Marinou[2], Vassilis Amiridis[2], Rodanthi-Elisavet Mamouri[3], Mihalis Vrekoussis[4]

[1]*Leibniz Institute of Tropospheric Research (TROPOS), Leipzig, Germany, *Email: ronny@tropos.de*

[2]*National Observatory of Athens (NOA), Greece*

[3]*Cyprus University of Technology (CUT), Limassol, Cyprus*

[4]*Cyprus Institute (CyI), Nicosia, Cyprus*

ABSTRACT

The atmospheric science community demands for autonomous and quality-assured vertically resolved measurements of aerosol and cloud properties. Aiming this goal, TROPOS developed the fully automated multiwavelength polarization Raman lidar Polly since over 10 years [1, 2]. In cooperation with different partner research institutes the system was improved continuously. Our latest lidar developments include aside the "3+2" measurements also a near-range receiver to measure aerosol extinction and backscatter down to 120 m above the lidar, a water-vapor channel, and measurements of the linear depolarization at two wavelengths.

The latest system was built in cooperation with the National Observatory of Athens (NOA). Its first campaign however was performed at the Cyprus Institute of Nicosia from March to April 2015, aiming specifically at the observation of ice nuclei with in-situ and lidar remote sensing techniques in the framework of BACCHUS [3, 4].

1. INTRODUCTION

Lidar profiling of atmospheric aerosol and cloud layers have become important for climate research during the last decades. More recently, the volcanic eruption hazards of Eyjafjallajökull and Grimsvötn for aircraft safety have shown the need for a height–resolved monitoring of the aerosol concentration on continental scales. Aside from spaceborne observations (e.g. CALIPSO, ATLID, EARTHCare) continental networks of lidars such as EARLINET, MPLNET, AD-Net, LALINET, and CORALNET are a great tool for such observations. Nowadays, and with the technical and methodological developments of the recent years it is possible not only to observe aerosol layers in terms of an optical characterizations but also to infer on many microphysical aerosol properties such as the type, the shape, the concentration, and even to separate aerosol mixtures in terms of smoke and dust concentration as well as fine and coarse mode fractions of dust [5, 6].

Such observations require fully automated and advanced lidar systems because manpower for the continuous operation is sparse. At TROPOS meanwhile we developed nine of our automated lidars in the framework of the Polly (POrtabLe Lidar sYstem) idea. Advanced Raman and depolarization lidar techniques combined in a cabinet for the installation basically anywhere around the globe. The requirements for the operation are power, an internet connection, and occasional maintenance (e.g. window cleaning, flash lamp replacement in the laser, and replacement of neutral density filters if required).

We follow the idea of a PollyNET [7], where all users have a platform to upload the data, receive immediate quicklooks, and exchange ideas, or solve technical problems. Many technical details of the different systems are similar and allow network observations in such a harmonized way (laser system, signal acquisition, data structure), whereas constant improvement is possible when needed (automated calibration routines, higher photon count rates, additional near-range capabilities, aircraft-safety radar). However, becoming part of PollyNET and sharing data and knowledge is a mandatory prerequisite in our "philosophy".

Our latest systems include a full set of measurement channels always guided by the needs, recommendations, and techniques that have

DOI: 10.1051/epjconf/201611927003

been developed within the community of the European Aerosol Research Lidar Network (EARLINET). Nowadays, a Polly can measure the aerosol backscatter and extinction coefficients at 3 and 2 wavelengths, respectively. Also the linear depolarization is measured at two wavelengths. Measurements of the water-vapor mixing ratio and near-range capabilities are another new feature.

In the framework of the BACCHUS campaign in March 2015 we installed the new NOA system on the roof of the Cyprus Institute in Nicosia, Cyprus. In this contribution we will summarize the latest technical developments of our smart Polly systems (section 2) and give an overview about the scientifically unique location of Cyprus (section 3). We applied the automated system over four weeks continuously in Cyprus and present first results in section 4.

2. THE LIDAR POLLY

Figure 1 shows the setup of the Polly-XT NOA at the Cyprus Institute. The typical cabinet-like shape together with a water chiller for the laser cooling shows the apparent simplicity of the system. Equipped with a precipitation sensor an automated roof shuts down the measurements if required by the meteorological conditions. It has a weight of 500 kg but can be easily maneuvered on four wheels.

Figure 1: The Polly lidar on the roof of the Cyprus Institute, Nicosia, together with an AERONET photometer and a Doppler lidar.

In terms of the optical setup the lidar is completely build on a carbon fiber board which gives a light-weight but sturdy framework for the components. In case of major maintenance requirements the board can even be removed from the system and realigned in an optical laboratory. Figure 2 shows the schematically setup of the optical parts. The flash-lamp pumped Nd:YAG laser (E1) emits radiation of 500 mJ, 20 Hz at 1064 nm wavelength which is then frequency doubled and tripled at (E2) and "post"-polarized (E3). After two beam steering mirrors an external safety shutter (E4) is implemented before the beam is expanded 8X (E5) and transmitted into the atmosphere. The receiving telescopes are R1 (50-mm near-range telescope) and R2 (300-mm main telescope). An automated depolarization calibration unit (R3) is placed in front of the field stop. By use of the Δ90° method [8] absolute calibration is possible on a regular basis. Various beam splitters are then used to separate all paths for the detection channels seen in Fig. 2.

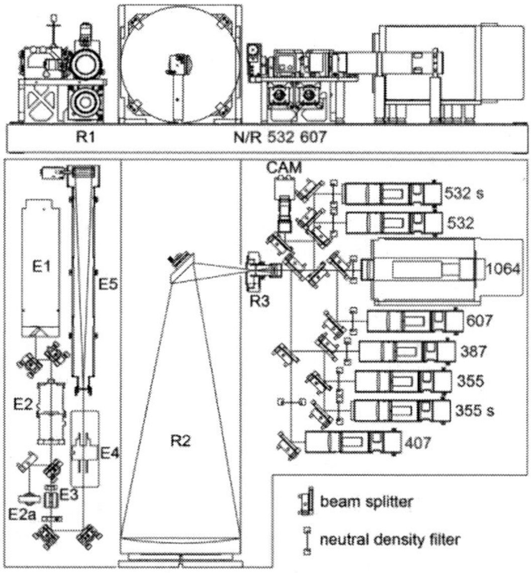

Figure 2: The optical setup of Polly-XT NOA. The upper part displays a top view and the lower part the front view of the system.

For simplicity (e.g., statistical error calculation) and stability (no induced signal noise, no interference with laser discharges …) all detectors operate in photon counting mode. We employ a data acquisition that has been developed by Holger Linné of the Max-Planck Institute, Hamburg, Germany. With this acquisition system

DOI: 10.1051/epjconf/201611927003

we obtain dead-times of approximately 2.3 ns. Furthermore in order to better correct for dead-time effects we measure this effect with a defined ramping light source for each photomultiplier separately in a special laboratory setup and store the correction coefficients in the final data files.

3. THE CYPRUS SITE

The island of Cyprus is a unique spot in the Eastern Mediterranean Sea where different and very complex aerosol mixtures occur [9]. One the one hand, Saharan dust and marine aerosol reach the island quite frequently. One the other hand, Arabian dust mixed with anthropogenic pollution from the Middle East or Europe can be observed, too, as well as clean maritime conditions with Aerosol optical depth <0.05 at 500 nm.

In the framework of a BACCHUS campaign (Impact of Biogenic versus Anthropogenic emissions on Clouds and Climate: towards a Holistic UnderStanding) together with the Raman lidar Polly we installed a SUN/SKY/LUNAR AERONET photometer and a Doppler wind lidar (cf. Fig. 1) at Nicosia (35.14°N, 33.38°E). In this way we obtained an observational area spanning from the very south of Cyprus with the CUT lidar at Limassol to the northernmost area of the Republic of Cyprus also to observe the local island effects.

4. FIRST RESULTS

Figure 3 shows the measurements of the first week of the campaign. Except for a few power breaks during the start of the campaign the lidar and the photometer were operating continuously. During the first days from 4 - 6 March clean, almost maritime, conditions with aerosol optical thicknesses (AOT, 550 nm) of 0.05 - 0.1 and Angström exponents (AE, 380 - 870 nm) of 1.2 – 1.7 prevailed. Local dust (e.g. slightly increased volume depolarization) was detected during the diurnal boundary-layer development up to a height of 0.8 km. At midnight of 7 March a plume of Arabian dust arrived at Nicosia, as indicated by backward trajectories and different by dust models (DREAM8b v2.0, SKIRON). This dust layer extended from 0.8 – 2.8 km height above ground and the AOT increased towards 0.2 – 0.3 while the AE was observed to be around 0.4. At noon of 8 March the Arabian air layer was subsequently replaced by Saharan air and Saharan dust. This Saharan dust was observed to heights up to 5 km with a maximum AOT of 0.38. The AE values were around 0.5 indicating not pure Saharan dust conditions but moreover a mixture with pollution. In the upper part of the Saharan layer we already observed heterogeneous cloud formation in the first week of this BACCHUS campaign. Further and more detailed investigations will be performed after the end of the campaign.

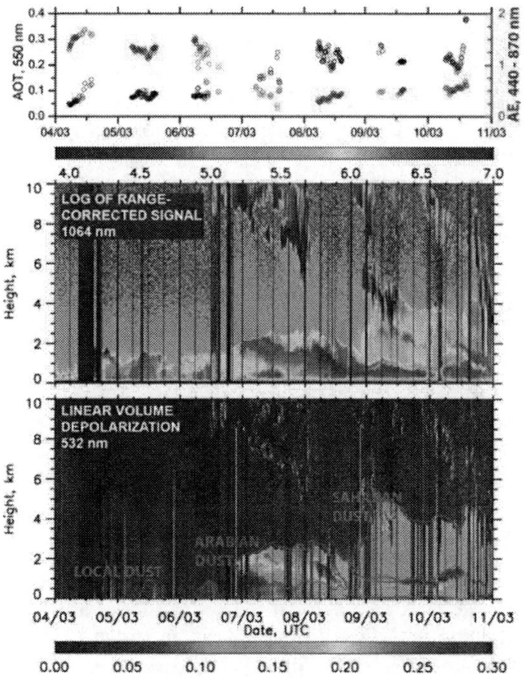

Figure 3: AERONET optical thickness and Angström exponent, Range-corrected lidar signal and volume linear depolarization ratio measured with Polly at Nicosia from 4-10 March 2015.

5. OUTLOOK

At the conference we will present more detailed case studies of profiles of backscatter and extinction, lidar ratio, and particle linear depolarization ratios contrasting Arabian and Saharan dust. We will also compare the profiles with the ones obtained from the Limassol lidar station located much closer to the sea. Additionally in-situ INC observations were performed during the campaign aboard Unmanned Aerial Vehicles (UAV) from the Cyprus institute. Our goal is to show first comparisons between in-situ and lidar derived ice nuclei concentrations.

ACKNOWLEDGEMENT

We gratefully thank Holger Linné for his effort in the developments of the new data acquisition system and for his eagerness to share the hard- and software among the community. The same yields for the outstanding collaboration with Volker Freudenthaler, who spends endless working hours to unify and improve the quality of lidar systems in EARLINET and ACTRIS. We also want to thank our collaboration partners for their confidence in our work. The research leading to these results has received partly funding from the European Union Seventh Framework Programme (FP7/2007-2013) under grant agreement n° 262254. The lidar setup received funding from the European Union's Seventh Framework Programme for research, technological development and demonstration (in the framework of BEYOND funded under: FP7-REGPOT-2012-2013-1) under grant agreement n° 316210.

REFERENCES

[1] Althausen, D., R. Engelmann, H. Baars, B. Heese, A. Ansmann, D. Müller, and M. Komppula, 2009: Portable Raman Lidar PollyXT for Automated Profiling of Aerosol Backscatter, Extinction, and Depolarization, *J. Atmos. Oceanic Technol.*, **26**, 2366–2378..

[2] Engelmann, R., et al., 2015: EARLINET Raman Lidar PollyXT: The NeXT Generation, *Atmos. Meas. Tech.*, to be submitted

[3] http://www.bacchus-env.eu/ (30 March 2015)

[4] Mamouri, R. E. and A. Ansmann, 2015: Estimated desert-dust ice nuclei profiles from polarization lidar: methodology and case studies, *Atmos. Chem. Phys.*, **15**, 3463-3477

[5] Tesche, M., A. Ansmann, D. Müller, D. Althausen, R. Engelmann, V. Freudenthaler, and S. Groß, 2009: Vertically resolved separation of dust and smoke over Cape Verde using multiwavelength Raman and polarization lidars during Saharan Mineral Dust Experiment 2008, *Journal of Geophysical Research*, **114**, D13202,

[6] Mamouri, R. and A. Ansmann, 2014: Fine and coarse dust separation with polarization lidar, *Atmos. Meas. Tech.*, **7**, 3717–3735.

[7] http://polly.tropos.de (30 March 2015)

[8] Freudenthaler, V., M. Esselborn, M. Wiegner, B. Heese, M. Tesche, A. Ansmann, D. Müller, D. Althausen, M. Wirth, A. Fix, G. Ehret, P. Knippertz, C. Toledano, J. Gasteiger, M. Garhammer, and M. Seefeldner, 2009: Depolarization ratio profiling at several wavelengths in pure Saharan dust during SAMUM 2006, *Tellus Series B*, **61**, 165–179.

[9] Mamouri, R. E., Ansmann, A., Nisantzi, A., Kokkalis, P., Schwarz, A., and Hadjimitsis, D., 2013: Low Arabian dust extinction-to-backscatter ratio, *Geophys. Res. Lett.*, **40**, 4762-4766.

THE MEASUREMENT OF TROPOSPHERIC TEMPERATURE PROFILES USING RAYLEIGH-BRILLOUIN SCATTERING: RESULTS FROM LABORATORY AND ATMOSPHERIC STUDIES

Benjamin Witschas[1]*, Oliver Reitebuch[1], Christian Lemmerz[1], Pau Gomez Kableka[1], Sergey Kondratyev[2], Ziyu Gu[3], Wim Ubachs[3]

[1]*German Aerospace Center (DLR), Institute of Atmospheric Physics, 82234 Oberpfaffenhofen, Germany*

[2]*Angstrom Ltd., 630090 Inzhenernaya16, Novosibirsk, Russia*

[3]*Department of Physics and Astronomy, LaserLaB, VU University, 1081 HV Amsterdam, Netherlands*

**Email: Benjamin.Witschas@dlr.de*

ABSTRACT

In this letter, we suggest a new method for measuring tropospheric temperature profiles using Rayleigh-Brillouin (RB) scattering. We report on laboratory RB scattering measurements in air, demonstrating that temperature can be retrieved from RB spectra with an absolute accuracy of better than 2 K. In addition, we show temperature profiles from 2 km to 15.3 km derived from RB spectra, measured with a high spectral resolution lidar during daytime. A comparison with radiosonde temperature measurements shows reasonable agreement. In cloud-free conditions, the temperature difference reaches up to 5 K within the boundary layer, and is smaller than 2.5 K above. The statistical error of the derived temperatures is between 0.15 K and 1.5 K.

1. INTRODUCTION

Temperature is a key parameter of the state of the atmosphere and hence, temperature data play an important role in atmospheric dynamics, climatology, meteorology, and chemistry. For this reason, observational data on atmospheric temperature profiles are important inputs to models predicting the atmospheric state and to retrieve details of other atmospheric properties, such as wind, pressure, relative humidity, or other trace gas concentrations.

At present, lidar instruments enable the measurement of temperature with high accuracy (≈1 K), high vertical resolution (≈100 m), and long range (from ground up to 105 km) [1]. In order to achieve this performance, it is common to use integration technique lidar systems, fluorescence lidars, or rotational Raman lidars. The former two are only applicable to stratospheric or mesospheric measurements because they can only be applied in air masses that do not contain aerosols, or where appropriate fluorescing metal atoms exist. In contrast, Raman lidars can be applied for tropospheric measurements [2]. Nevertheless, it has to be mentioned that the Raman scattering cross section is quite low. Thus, powerful lasers, sophisticated background filters, or night-time operation are required to obtain reliable results. In particular, the rotational Raman differential backscattering cross section (considering Stokes and anti-Stokes branches) is about a factor of 50 smaller than the one of Rayleigh scattering [3]. Regarding this, it is a preferred method to retrieve temperatures from Rayleigh–Brillouin (RB) scattered light, as it can be done by high spectral resolution lidars (HSRL). In particular, atmospheric temperature can be derived by resolving the RB spectrum with high spectral resolution filters, such as atomic vapor cells [4] or Fabry–Pérot interferometers [5], and relating the measured spectrum to an appropriate line shape model. Thus, apart from the challenging optical setup, such an approach is depending on the detailed knowledge of the RB scattering process and on the accuracy of line shape models that describe the spectral distribution of RB scattered light. In 1974, Tenti et al. [6] developed a model - the so-called Tenti S6 model - for describing RB spectra of light scattered in molecular gases of single species, which is since then considered as the best model available. Recently performed laboratory studies [7-8] showed that the Tenti S6 model describes the measured RB spectra in air with deviations smaller than 2% if modifications on the model calculations are performed taking

DOI: 10.1051/epjconf/201611927004

into account the mixed nature of air as scattering medium. In order to retrieve temperature from measured RB profiles, they have to be analyzed with the Tenti S6 model in a certain optimization procedure as for instance a least-squares fit with temperature as free fitting parameter. As for such method it is an important issue how the 2% model deviation transfers into a temperature error.

So as to deal with this issue, RB scattering measurements in air at a wavelength of 403 nm, at a scattering angle of 91.7°, for temperatures from 257 K to 330 K and pressures from 871 hPa to 1013 hPa were performed and used to verify a temperature retrieval algorithm. The results from this study are discussed in section 2. In addition, atmospheric daytime measurements of tropospheric temperature profiles (2–15 km) using RB scattering were performed demonstrating that temperatures can be derived from RB profiles with an accuracy better than 2 K (see section 3). Based on the results from the laboratory and atmospheric studies, a novel lidar temperature receiver is suggested and is currently under development at the German Aerospace center DLR (see section 4). It is based on a Fizeau interferometer used to resolve the RB spectrum without any scanning procedures.

2. LABORATORY MEASUREMENTS

In order to investigate the principal accuracy that can be reached by deriving temperature from RB spectra in air, highly accurate laboratory measurements were performed at the LaserLaB of VU University Amsterdam. Details about the experimental setup and the applied retrieval algorithm can be found in [9]. An example of measured RB profiles at different temperatures and pressures can be seen in Figure 1 (top). The spectra illustrate the high signal-to-noise ratio on the measured RB spectra obtained, as well as the line shape broadening with increasing temperature.

In Figure 1 (bottom), the retrieved temperatures T_{model} are shown as a function of the reference temperature T_{Pt100}. It is obvious that there is very good accordance between retrieved- and reference temperature. In particular, the absolute difference is less than 2 K for all measurements. A linear correlation (Figure 1, blue dashed lines) yields a slope of 1.01 ± 0.02, and thus clearly demonstrates the possibility of retrieving absolute temperatures from RB profiles with high accuracy.

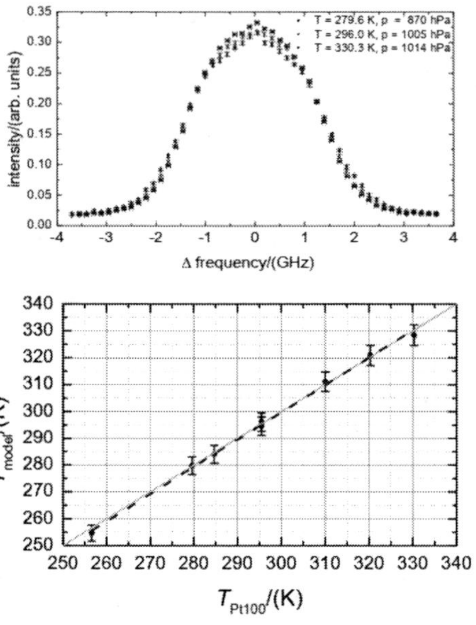

Figure 1: (top): RB spectra measured at 403 nm, a scattering angle of 91.7° and different temperatures (colors). (bottom): Temperatures derived from measured spectra by using the Tenti S6 model (solid blue dots) and the corresponding linear fit (dashed blue line). The gray line represents the $T_{model} = T_{Pt100}$ line.

3. ATMOSPHERIC MEASUREMENTS

In addition to the laboratory measurements, atmospheric measurements have been performed. The lidar used for this study was not originally built for this purpose, but is actually a Doppler wind lidar, used for validating the instrument of the ADM-Aeolus satellite mission initiated by ESA [10]. However, as it is shown in [11], the instrument is also suited for measuring temperature by means of RB scattering.

To record RB profiles, the frequency of the laser is changed in defined steps Δf (e.g., Δf = 125 MHz), over a range of 12 GHz (at 354.89 nm) and a Fabry-Perot interferometer (Full Width Half Maximum (FWHM) ~ 1.7 GHz, Free Spectral Range (FSR) ~ 11 GHz) is used as frequency discriminator. Considering 90 steps and duration of 18 s per step, the sampling of the entire RB line shape requires 28 min. It is worth mentioning that sampling the RB spectrum with only a few points (≈15) would be enough to reach maximum accuracy for the temperature retrieval (according to simulations). However, as the measurement uncertainty is inversely proportional to the signal level, reducing the sample number or rather measurement time increases the uncertainty of the derived temperatures. Furthermore, it is beneficial to have more samples to resolve the peak resulting from particle scattering in aerosol-rich regions. An example of measured RB line shapes (Δf = 250 MHz) at different distances (6.4 km–12.7 km) is shown in Figure 2 (top). It is pointed out that the signal from bin 21 (10.2 km–11.5 km) shows a remarkable contribution from significantly narrower particle scattering on a cirrus cloud, which also has to be considered for temperature retrieval.

A generic temperature profile retrieved from the measured RB spectra is shown in Figure 2 (bottom, left, red dots). For this measurement, the corresponding RB profile was sampled with 93 points separated by 125 MHz, leading to a frequency range of about 11.6 GHz and a measurement time of 28 min. The range gate size was vertically adapted between 0.3 km and 2.2 km to keep the number of received photons large enough to minimize the statistical error. The measured temperature profile is compared with radiosonde temperatures (Figure 2, bottom, left, black line), launched during the lidar measurement about 27 km away from the lidar. The statistical error of the obtained temperature values (Figure 2, bottom, left, error bars) is calculated by applying a maximum likelihood estimator and considering solely Poisson noise on the measured data points (details can be found in [11]).

Although the temperature measurements were performed during noon, they show good agreement to radiosonde measurements. The profiles range from 2 to 15.3 km, and temperature difference reaches up to 5 K within the boundary layer and is smaller than 2.5 K above. The inversion layer at 3 km is obvious. The statistical error varies from 0.15 K (at 2 km) to 1.5 K (at 15 km). However, it has to be mentioned that other measurements showed contamination by particle scattering that led to temperature errors of several Kelvin [14]. Although the influence of particle scattering was considered within the temperature retrieval, uncertainties are caused by temporal variations in the particle load during line shape sampling. Simulations confirm that discrepancies of larger than 20 K can occur if the particle scattering is not constant within the sampling process. In addition, it has to be pointed out that the limited FPI resolution of 1.7 GHz is not perfectly suited to resolving the influence of particle scattering on the RB spectrum.

Thus, it is concluded that, although very promising for clear air conditions, the suggested approach still suffers from the influence of particle scattering.

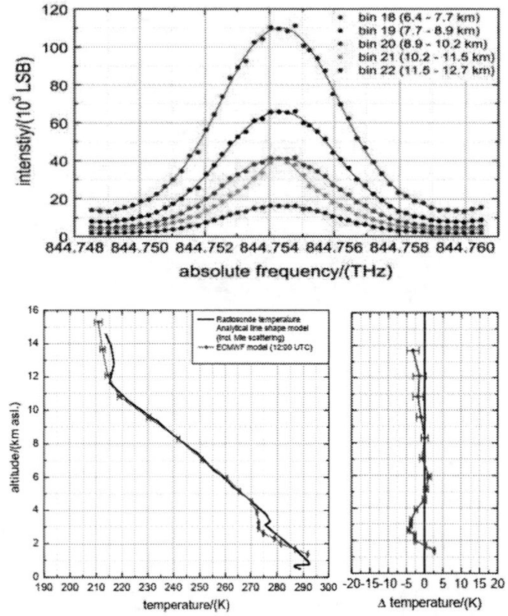

Figure 2: (top): Measured RB line shapes for different distances from the lidar (dots) and best-fits according to the Tenti S6 model (lines). (bottom, left): Temperature profiles derived from lidar measurements (red dots) compared with temperatures measured by radiosonde (black line). (bottom, right): Difference between radiosonde and lidar temperature values.

4. OUTLOOK

To further reduce the measurement time for one temperature profile, an instrument that is able to sample the entire RB line shape simultaneously without any scanning procedures is proposed. This can be realized by imaging techniques such as, for instance, resolving the RB spectrum with a

Fizeau interferometer and imaging the throughput on a multichannel PMT array. By using such a detector, even a varying aerosol load would not negatively influence the temperature retrieval. Furthermore, by designing the Fizeau interferometer with sufficiently high resolution, the aerosol peak can be better distinguished from the RB spectrum, which will additionally improve the temperature retrieval in regions of larger particle scattering. In addition to that, it is feasible to design a hybrid system to measure temperature profiles that uses both Raman scattering in cloudy conditions and RB scattering during daytime and cloudless conditions.

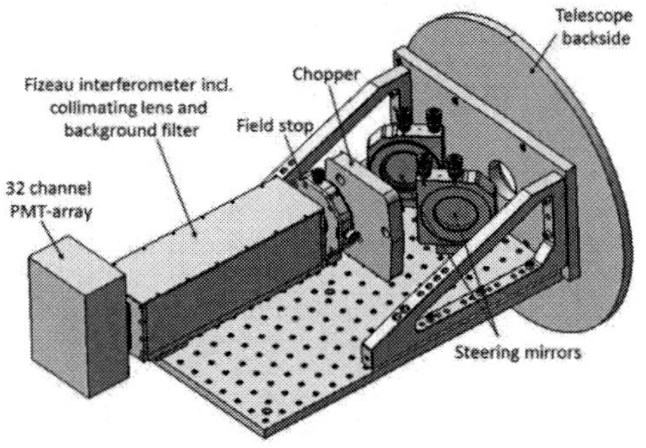

Figure 3: Sketch of a new lidar temperature receiver currently under development at DLR

Such kind of novel temperature receiver is currently under development at DLR and first atmospheric measurements are foreseen for mid of 2015. A sketch of the temperature receiver is shown in Figure 3.

ACKNOWLEDGEMENT

This work was partly funded by DLR and ESA within the Network Partnering Initiative, Contract No. 4000104130. The valuable suggestions of A.-G. Straume (ESA) and the technical support of E. Nagel (DLR) are highly appreciated.

REFERENCES

[1] Alpers M., R. Eixmann, C. Fricke-Begemann, M. Greding, and J. Höffner, 2004: Temperature lidar measurements from 1 to 105 km altitude using resonance, Rayleigh, and Rotational Raman scattering, Atmos. Chem. Phys. **4**, 793-800.

[2] Behrendt A., T. Nakamura, T. Tsuda, 2004: Combined temperature lidar for measurements in the troposphere, stratosphere, and mesosphere, Appl. Opt. **43**, 2930-2939.

[3] Witschas B. 2012: Light Scattering on Molecules in the Atmosphere, in *Atmospheric Physics: Background - Methods - Trends*, U. Schumann, ed. (Springer Berlin Heidelberg, 2012), pp. 69-83.

[4] Hair J., L. Caldwell, D. Krueger, and C. She, 2001: High-Spectral-Resolution Lidar with Iodine-Vapor Filters: Measurement of Atmospheric-State and Aerosol Profiles, Appl. Opt. **40**, 5280-5294.

[5] Hua D., M. Uchida, and T. Kobayashi, 2005: Ultraviolet Rayleigh-Mie lidar for daytime-temperature profiling of the troposphere, Appl. Opt. **44**, 1315-1322.

[6] Tenti G., C. Boley, and R. Desai, 1974: On the kinetic model description of Rayleigh-Brillouin scattering from molecular gases, Can. J. Phys. **52**, 285-290.

[7] Witschas B., M. O. Vieitez, E.-J. van Duijn, O. Reitebuch, W. van de Water, and W. Ubachs, 2010: Spontaneous Rayleigh-Brillouin scattering of ultraviolet light in nitrogen, dry air, and moist air, Appl. Opt. **49**, 4217-4227.

[8] Gu Z., B. Witschas, W. van de Water, W. Ubachs, 2013: Rayleigh–Brillouin scattering profiles of air at different temperatures and pressures, Appl. Opt. **52**, 4640-4651.

[9] Witschas B., Z. Gu, W. Ubachs, 2014: Temperature retrieval from Rayleigh-Brillouin scattering profiles measured in air, Opt. Exp. **22**, 29655-29667.

[10] Reitebuch O., C. Lemmerz, E. Nagel, and U. Paffrath, 2009: The Airborne Demonstrator for the Direct-Detection Doppler Wind Lidar ALADIN on ADM-Aeolus. Part I: Instrument Design and Comparison to Satellite Instrument, J. Atmos. Oceanic Technol. **26**, 2501-2515.

[11] Witschas B., C. Lemmerz, O. Reitebuch, 2014: Daytime measurements of atmospheric temperature profiles (2–15 km) by lidar utilizing Rayleigh–Brillouin scattering, Opt. Lett. **39**, 1972-1975.

EPJ Web of Conferences **119**, 27005 (2016) DOI: 10.1051/epjconf/201611927005
ILRC 27

DEVELOPMENT OF A SCHEIMPFLUG LIDAR SYSTEM FOR ATMOSPHERIC AEROSOL MONITORING

Liang Mei [1,2]*, Mikkel Brydegaard[1,2]

[1] *Norsk Elektro Optikk AS, Lørenskog N-1471, Norway, *Email: liang@neo.no*

[2]*Department of Physics, Lund University, Lund SE-221 00, Sweden*

ABSTRACT

This work presents a Scheimpflug lidar system which was employed for atmospheric aerosol monitoring in southern Sweden. Atmospheric aerosol fluctuation was observed around rush-hour. The extinction coefficient over 6 km was retrieved, i.e., 0.15 km^{-1}, by employing the slop-method during the time when the atmosphere was relatively homogenous. The measurements successfully demonstrate the potential of using a Scheimpflug lidar technique for atmospheric aerosol monitoring applications.

1. INTRODUCTION

Atmospheric aerosol plays a critical role in global radiation budget, and their chemical/physical properties and distribution are the utmost important parameters for earth climatology [1]. Further, as the unprecedented rapid development of industrialization and urbanization during recent decades, aerosol particles, produced by anthropogenic activities, have an significant impact on the global climate and may be harmful to humans and animals [2]. It thus important to study and monitor aerosol particles in atmosphere in terms of climatology and public health.

Although point-monitoring suction devices are widely available in the market, the distribution of the aerosol particles is of great importance in many applications. One prevailing method for atmospheric aerosol remote sensing is to employ the lidar technique which detects backscattering echoes of a nanosecond–pulsed light source from atmospheric molecules and aerosol particles [3]. From the backscattering echoes, the spatial distribution as well as temporal evolution of atmospheric aerosol can be retrieved in a high resolution. The aerosol type, size, and concentration can also be analyzed with the help of sophisticated Raman lidar, multi–wavelength lidar [4,5], etc. However, the high cost and complexity of conventional pulsed lidar systems limit the use of general applications. Even for the simplest single wavelength elastic lidar – micro-pulse lidar (MPL) [6,7], there are still limited installations.

In this article, we will demonstrate a new lidar technique for atmospheric aerosol monitoring, i.e., Scheimpflug Lidar which is based on the Scheimpflug principle that describes the relationship between the image- and object-planes when they are not parallel in an optical imaging system. The Scheimpflug principle has actually been utilized for short-scale range sensing since 1990s. However it was until quite recently that the Scheimpflug principle, for the first time, was employed for atmospheric fauna sounding [8] and atmospheric oxygen molecules detection [9] over kilometer range, benefit by the high power continuous-wave laser diodes available in recent years. The Scheimpflug lidar system is a monostatic lidar system and the range-resolved backscattering echo is recorded by employing either a line scan or an area camera, while infinite focal depth can still be achieved with large aperture.

This paper presents the system schematic of the Scheimpflug lidar system as well as the aerosol monitoring results performed in southern Sweden during March, 2015. In the following sections, we will first discuss the principle and the system schematic of the Scheimpflug technique in Sections 2, and then present the atmospheric aerosol monitoring measurements with the Scheimpflug lidar system in Section 3. A short conclusion is given in Section 4.

2. METHODOLOGY

2.1 Scheimpflug lidar principle

According to the Scheimpflug principle, in the situation that the object plane is nonparallel to the lens plane, sharp focus can still be achieved by employing a tilt image sensor when the corresponding image plane intersects both with the object and lens plane.

DOI: 10.1051/epjconf/201611927005

When implementing the Scheimpflug principle for atmospheric remote sensing, the backscattering echoes of the laser beam transmitted into atmosphere can be in focus simultaneously on a tilt image plane which intersects both with the object plane (the plane of laser beam) and the lens plane, as shown in **Figure 1**. In other words, infinite focal depth is achieved while employing large optical aperture. The main advantage of the Scheimpflug lidar technique is that continuous-wave lasers can be employed for range-resolved remote sensing instead of utilizing ns-scale pulsed light source, which significantly reduces the complexity and cost of building up an atmospheric lidar system.

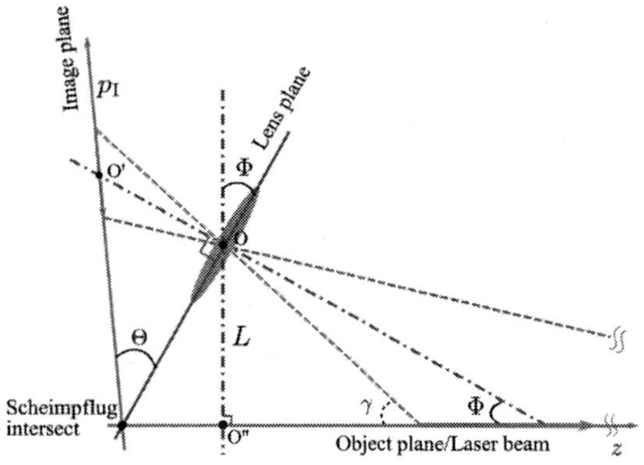

Figure 1 Principle of the Scheimpflug lidar. f – the focal length of the receiving telescope, L – the distance of the lens from the object plane, L_{IL} – the distance from the center of the image plane to the lens plane, p_I – the pixel position of the image sensor on the image plane, z – measurement distance along the object plane, Φ – the swing angle of lens, Θ – the tilt angle of the image plane to the lens plane, γ – the sampling angle for each pixel. The pixel number is counted in the opposite direction of p_I.

When implementing the Scheimpflug principle for atmospheric remote sensing, the lens swing angle (Φ) is rather small, typically less than 1°. Thus, the origo O″ coincides with the Scheimpflug intersect. The relationship between the pixel and range can be derived from the lens equation and trigonometric formulas [9]:

$$z=\frac{L\left[p_I(\sin\Theta-\cos\Theta\tan\Phi)+L_{IL}\right]}{p_I(\cos\Theta+\sin\Theta\tan\Phi)+L_{IL}\tan\Phi} \quad (1)$$

The swing angle of the lens (Φ) and the value of L_{IL} can be retrieved from a reference measurement on a hard target with known distance. The range-pixel relationship can then be derived from Eq. (1).

The atmospheric backscattering echo for elastic Scheimpflug lidar also follows the lidar equation used in a conventional pulsed lidar system:

$$P_S(\lambda,z)=P_0(\lambda)C\frac{O(z)dz}{z^2}\beta(\lambda,z)\exp\left[-2\int_0^z\alpha(\lambda,z')dz'\right] \quad (2)$$

Here λ is the laser center wavelength, C is the system constant, $P_0(\lambda)$ is the output power of the laser, z is the measurement range, $\beta(\lambda,z)$ and $\alpha(\lambda,z)$ are the backscattering and extinction coefficients at range z, respectively, dz is the length of the sampling volume at range z, $O(z)$ is the overlap function between pixel foot prints and the laser beam.

It can be deduced from Eq. (1) that $dz\propto z^2$, thus the lidar signal does not decrease over the square of range and the z^2 factor is cancelled out in the lidar equation. This implies improved exploitation of the dynamic range of the Scheimpflug lidar over pulsed systems. When employing an area 2D camera, the whole laser beam in the field of view of the telescope can be imaged without truncation. Thus, the overlap function is equal to 1 for the whole detection range. The backscattering intensity for each measurement distance can be retrieved from the intensity summation of all the pixels perpendicular to the direction of laser beam image. The Scheimpflug lidar equation can then be simplified as

$$P_S(\lambda,z)=K\beta(\lambda,z)\exp\left[-2\int_0^z\alpha(\lambda,z')dz'\right] \quad (3)$$

Here K is a new system constant. Equation (3) implies that the atmospheric backscattering echo detected by a Scheimpflug lidar system does not need any range-correction.

2.2 Scheimpflug lidar system

Figure 2 shows a typical system schematic of a single-band Scheimpflug lidar system, employing a wide-stripe (multiple transverse modes) laser diode running in the near infrared region (808 nm). The output power of the laser diode is ~3.2 W, 50% duty cycle. The laser beam is transmitted into atmosphere by an achromatic doublet (F/5, f=750 mm, ø150 mm). The backscattering echo is collected by a Newtonian telescope (F/4, f=800 mm, ø200 mm). A 2D CMOS camera (Lumenera, 2088×1088 pixels, 5.5-μm pixel width) is employed to record the image of the laser beam. For the 808-nm single-band operation, an interference filter (808 nm, 10 nm FWHM) and an absorption filter (RG780) are utilized. A Johnson counter is triggered by the strobe out signal from the CMOS camera to synchronize the laser diodes for single-band operation. A dark-time slot (CH0), when the laser diode is turned off, is also taken for dynamic background subtraction.

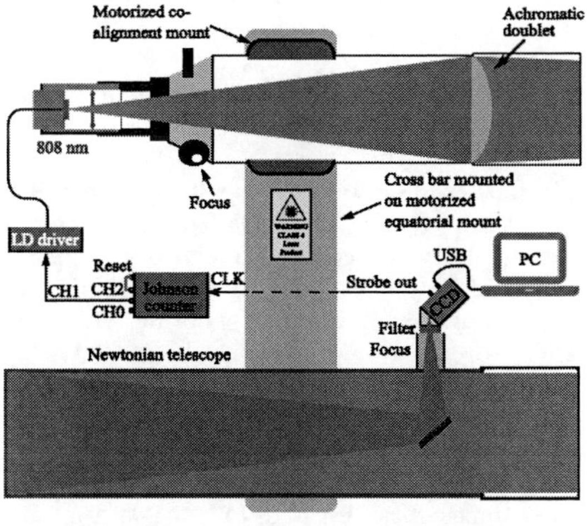

Figure 2 System schematic of Scheimpflug.

3. RESULTS

To demonstrate the potential of using Scheimpflug lidar for aerosol monitoring, single-band (808 nm) lidar measurements were performed on March 19th in the morning during the rush hour from 07:29 to 10:18, in Lund, southern Sweden (**Figure 3**). The laser beam was transmitted into atmosphere with an elevation angle of 4°. The integration time for the CMOS camera was 20 ms, and 200-time averaging gives a single lidar recording, corresponding to 4 s. Since the atmospheric background signal was taken dynamically, the total measurement time for each recording is about 10 s, including the data transfer time. The measurement distance is from 84 to 6000 meters, covering both urban and rural area.

Figure 4 shows the recorded backscattering intensities at different times, while the corresponding range-time map is given in **Figure 5**. As can be seen, substantial aerosol emission activities were observed in the urban area, while the aerosol distribution in the rural area did not change too much. The mixture of aerosol particles in the urban-rural area can be observed. After the rush-hour, atmosphere along the measurement path became relatively homogeneous. According to the slope-method, the extinction coefficient can be retrieved by the slop of the log scale backscattering intensity, i.e., $\alpha = 0.15\,\mathrm{km}^{-1}$ at 09:37, which implies very good atmosphere visibility.

Figure 3 Experimental site (Lund, southern Sweden), the elevation angle is about 4°. The mean wind speed measured by a weather station nearby the lidar system (60 m away) was approximately 1 m/s from east.

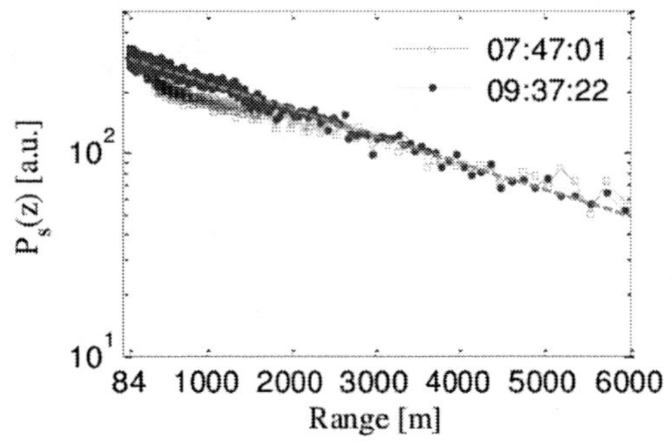

Figure 4 Backscattering intensity recorded over 6 km at different times during the measurements. The dash curve is the linear fit of the log scale backscattering intensity recorded at 09:37:22. The main contribution of the noise in the lidar signal is the readout noise of the CMOS camera.

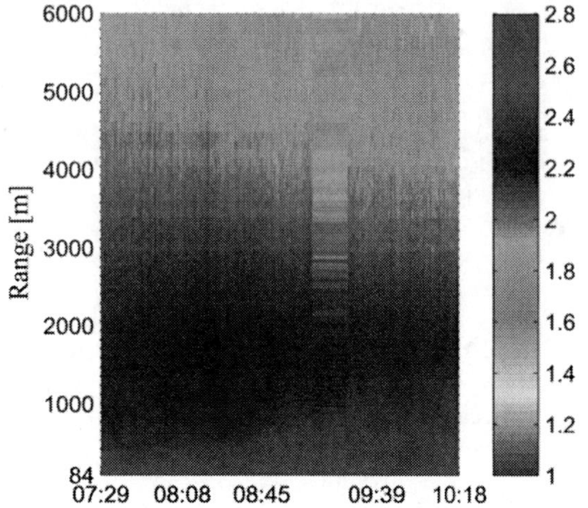

Figure 5 Range-time backscattering intensity map from 07:29 to 10:18 on March 19th in Lund, southern Sweden. 0-1 km – urban area, 1-2.5 km – rural-urban area, over 2.5 km is rural area. The outdoor temperature varied from 2.9° to 13.2° during the measurements, while the relative humidity decreased from 68% to 36%. Note: the electric power was suddenly off at around 09:15 for about 15 mins during the recording.

4. CONCLUSIONS

The aerosol backscattering intensity over three hours was successfully recorded by the present single-band Scheimpflug lidar system. Completely overlap between the field of view of the telescope and the laser beam was achieved by employing a 2D camera. The closest measurement range is about 84 meter, while the maximum detection range is over 6 km with a measurement time of 10 s in a clean atmosphere. The extinction coefficient was found to be 0.15 km^{-1} in a relatively homogeneous atmosphere based on the slope-method. This work presents a great potential of using the Scheimpflug lidar technique for many atmospheric monitoring application, especially considering a multi-band system employing several cheap, compact and high-power laser diodes.

REFERENCES

[1] Jacqueline Lenoble, Lorraine Remer, and Didier Tanre, (Springer, Verlag Berlin Heidelberg, 2013).

[2] A. J. McMichael, R. E. Woodruff, and S. Hales, 2006: Climate change and human health: present and future risks, Lancet, **367(9513)**, 859-869.

[3] Vladimir A. Kovalev and William E. Eichinger, Elastic Lidar: Theory, Practice, and Analysis Methods. (John Wiley & Sons, Hoboken, New Jersey, 2004).

[4] A. Papayannis, R. E. Mamouri, V. Amiridis, E. Remoundaki, G. Tsaknakis, P. Kokkalis, I. Veselovskii, A. Kolgotin, A. Nenes, and C. Fountoukis, 2012: Optical-microphysical properties of Saharan dust aerosols and composition relationship using a multi-wavelength Raman lidar, in situ sensors and modelling: a case study analysis, Atmos. Chem. Phys., **12(9)**, 4011-4032.

[5] M. R. Perrone, F. De Tomasi, and G. P. Gobbi, 2014: Vertically resolved aerosol properties by multi-wavelength lidar measurements, Atmos. Chem. Phys., **14(3)**, 1185-1204.

[6] J.D. Spinhirne, 1993: Micro pulse lidar, IEEE Trans. Geoscience and Remote Sensing, **31(1)**, 48-55.

[7] E. J. Welton, J. R. Campbell, J. D. Spinhirne, and V. S. Scott, 2001: Global monitoring of clouds and aerosols using a network of micro-pulse lidar systems, Lidar Remote Sensing for Industry and Environment Monitoring, **4153**, 151-158.

[8] M. Brydegaard, A. Gebru, and S. Svanberg, 2014: Super Resolution Laser Radar with Blinking Atmospheric Particles - Application to Interacting Flying Insects, Progress In Electromagnetics Research, **147**, 141-151.

[9] Liang Mei and Mikkel Brydegaard, 2014: Continuous-wave differential absorption lidar, Laser and Photonics Reviews, Submitted.

CEILOMETER AEROSOL PROFILING VERSUS RAMAN LIDAR IN THE FRAME OF INTERACT CAMPAIGN OF ACTRIS

F. Madonna[1], F. Amato[1], M. Rosoldi[1], J. Vande Hey[2], G. Pappalardo[1]

1) Consiglio Nazionale delle Ricerche, Istituto di Metodologie per l'Analisi Ambientale (CNR-IMAA), C.da S. Loja, 85050 Tito Scalo, Potenza, Italy.
2) University of Leicester, Department of Physics and Astronomy, Earth Observation Science Group, University Road, Leicester, LE1 7RH, United Kingdom.
**Corresponding author: fabio.madonna@imaa.cnr.it*

ABSTRACT

In this paper, multi-wavelength Raman lidar measurements are used to investigate the capability of ceilometers to provide reliable information about atmospheric aerosol properties through the INTERACT (INTERcomparison of Aerosol and Cloud Tracking) campaign carried out at the CNR-IMAA Atmospheric Observatory (760 m a.s.l., 40.60 N, 15.72 E), in the framework of ACTRIS (Aerosol Clouds Trace gases Research InfraStructure) FP7 project. This work is the first time that three different commercial ceilometers with an advanced Raman lidar are compared over a period of six month. The comparison of the attenuated backscatter coefficient profiles from a multi-wavelength Raman lidar and three ceilometers (CHM15k, CS135s, CT25K) reveals differences due to the expected discrepancy in the SNR but also due to effect of changes in the ambient temperature on the stability of ceilometer calibration over short and mid-term. Technological improvements of ceilometers towards their operational use in the monitoring of the atmospheric aerosol in the low and free troposphere are likely needed.

1. INTRODUCTION

For the study of both climate, air pollution and its influence on health, knowledge of vertical distributions of aerosols is a key factor. This scenario has pushed the demand for continuous aerosol measurements provided by high resolution networks of ground-based instruments also to validate and improve aerosol and pollution forecasting. In order to achieve broad, high resolution coverage, low-cost and low-maintenance instruments are needed. Despite of their differences from more advanced and more powerful lidars, low construction and operation cost of ceilometer, originally designed for cloud base height monitoring, have fostered their use for the quantitative study of aerosol properties. The large number of ceilometers available worldwide represent a strong motivation to investigate to which extent they can be used to fill the geographical gaps between advanced lidar stations and how their continuous data flow can be linked to existing networks of the advanced lidars, like EARLINET (European Aerosol research LIdar NETwork – Pappalardo et al., 2014).

In order to make the best use of existing and future ceilometer deployments, ceilometer must be better characterized. This is the purpose of the INTERACT campaign (Madonna et al., 2014) carried out at the CNR-IMAA Atmospheric Observatory (760 m a.s.l., 40.60N, 15.72E), named CIAO (Madonna et al., 2011), in the framework of ACTRIS (Aerosol Clouds Trace gases Research InfraStructure) FP7 project.

This work is the first time that three different commercial ceilometers with an advanced Raman lidar, MUSA (Multi-wavelength System for Aerosol) are compared over a period of six months. An overview of the results achieved during the campaign is provided.

2. METHODOLOGY

The INTERACT campaign, was held at CIAO in Potenza, Italy from 1 July 2013 to 12 January 2014, with the main scientific objective to evaluate the stability, sensitivity, and uncertainties of ceilometer aerosol backscatter profiles. Here, three commercial ceilometers (CHM15k, CT25K, CS135s) from different manufactures are compared with an advanced multi-wavelength Raman lidar (MUSA), and their aerosol detection sensitivity and stability are assessed using a dataset collected over a period of more than six months.

MUSA data (aerosol extinction and backscatter coefficients) are processed using the automatic Single Calculus Chain (SCC) of EARLINET (Pappalardo et al., 2014). The SCC is able to pre-processing lidar signals to provide aerosol optical and geometrical properties (e.g. layering) using Raman and elastic algorithms.

CHM15k data are collected using the JO-Dataclient software provided by the manufacturer, while the attenuated backscatter coefficient profiles are obtained by normalizing the ceilometer range-corrected signals to the corresponding MUSA attenuated backscatter coefficient profiles. Normalization was first attempted using a region 1-2 km wide, located at 6-7 km a.g.l. and identified as an aerosol free region identified from the quicklooks of the lidar measurements time series. This choice, however, tended to underestimate the normalization factor because of the very poor SNR of the ceilometer at those altitude levels. Throughout the campaign the ceilometer proved to be able to detect values of the attenuated backscatter coefficient larger than 1.0 10^{-7} m-1 sr-1 at altitude levels lower than 4 km a.g.l. with a vertical resolution of 30 m and a time resolution ranging from 45 to 120 minutes. Therefore, the normalization has been performed over a vertical range of 1 km, below the altitude level where this threshold value is detected. This typically occurs around 4 km and a detailed visual inspection has been also performed.

CS135's raw signals have been collected using a terminal emulator: attenuated backscatter coefficient profiles are obtained upon normalization on the corresponding MUSA attenuated backscatter coefficient profiles, using the same procedure followed for the CHM15k. Since signals in the upper troposphere were affected by electronic distortion, the normalization region has been selected typically immediately below 2.0-2.5 km in order to have a sufficient SNR to obtain a stable normalization over the lidar profiles.

The CT25K does not provide any raw signal but the so-called normalized sensitivity backscatter coefficient (that is proportional to the range corrected signal) through the manufacturer's software; therefore, attenuated backscatter coefficient profiles can only be obtained using the stratocumulus calibration typically performed within Cloudnet (www.cloud-net.org).

The time and vertical resolutions have been selected in order to keep the ceilometer SNR to sufficient levels to allow them to be compared to and calibrated by MUSA. In addition, for the CHM15k (to avoid problems with the sudden change of the internal gain automatically selected by the ceilometer itself) only signals corresponding to a value of the "base" (daylight correction factor) parameter less than 0.0015 (low background light level) are considered.

A comparison between the attenuated backscatter coefficient profiles provided by the four instruments (MUSA, CHM15k, CT25K, CS135S) is reported in Figure 1 and it is related to the observations collected on 15 July 2013 from in the time interval between 01:18 and 02:54. CHM15k profile is not corrected for the overlap function.

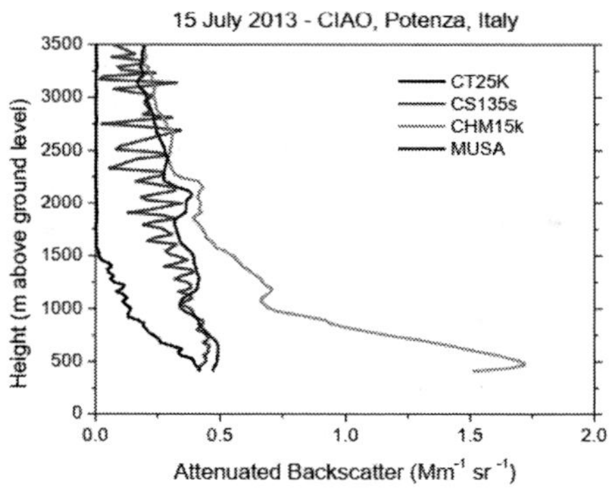

Figure 1: comparison between the attenuated backscatter coefficient profiles provided by the four instruments (MUSA, CHM15k, CT25K, CS135S) on 15 July 2013 from in the time interval between 01:18 and 02:54. CHM15k is not corrected for the overlap function.

3. RESULTS

In this section, a statistical comparison of the MUSA and the ceilometer attenuated backscatter coefficient is discussed. This is performed by comparing the probability density functions (pdfs) of the ß' retrieved for the simultaneous observations performed by the four instruments. Results of this comparison are summarized in Figure 2 that show the probability density functions (pdfs) of ß' measured by MUSA and each of the ceilometers calculated for the whole INTERACT campaign from 405 m a.g.l. to 10000 m a.g.l. MUSA pdf's are considered as the truth/reference. The number of cases available for each ceilometer and MUSA simultaneously is not the same due to the use of the selection criteria mainly affecting the CHM15k selected data. Moreover, for the CHM15k only data above 1.3 km above the ground have been considered to remove the region of incomplete overlap. This data selection is the reason for the difference among the MUSA pdf reported in the different panels of Figure 2. Under ideal conditions, the pdf's of the ceilometers and MUSA should be the same. Calibration error and a low SNR can largely affect the comparison. In the case of a calibration error, the pdf could show much higher or much lower values for the ceilometer with respect to the MUSA pdf, though the effect might compensate over the whole dataset. A low SNR, however, can show very high positive and very low negative values affecting, respectively, the values of the pdf higher than the maximum value observed by MUSA and values lower than 1.0 10^{-10} m^{-1} sr^{-1}.

The comparison of the pdfs shows that CHM15k agrees closely with MUSA; CT25K underestimates in a more significant way the values of ß' measured by MUSA. CS135s is in very good agreement with MUSA for values lower than 1.7 10^{-6} m^{-1} sr^{-1}, but few larger values of ß' are measured by CS135s probably because of the distortion affecting the signal. This indicates that the suppression of the electronic distortion might strongly improve the CS135s performance. Moreover, CT25K shows several very low values of ß' (<1.0-10 m^{-1} sr^{-1}) corresponding to much larger values of ß' for MUSA and CHM15k. The other deviations for both the CT25K and CS135s are mainly due their lower SNR than MUSA.

The relationship between the 355 nm aerosol extinction coefficient (α) provided by MUSA and the attenuated backscatter coefficient ß' obtained at 1064 nm by MUSA and by the three ceilometers, respectively, have been compared to further investigate the ceilometers' performance and their sensitivity to different aerosol types, i.e. different extinction coefficients. The parameter is calculated over the same time window as ß' but using a lower effective vertical resolution (typically within 480 m) in order to reduce the uncertainty and the related oscillation affecting the extinction profile calculated using the Raman lidar signal. The profile is output at 30 m vertical resolution to match the backscatter coefficient vertical resolution.

CHM15k shows a very good agreement with MUSA (the regression coefficient of the two attenuated backscatter coefficient is R=0.95), though a bit larger dispersion than MUSA in the relationship between and the ß' is observed. With an increasing value of the difference becomes larger and the value of the ß' is overestimated. This is particularly evident for values of α larger than 0.5 10^{-4} m^{-1}. Values of α larger than 0.5 10^{-4} m^{-1} are mainly located in the atmospheric region below 3 km above the ground.

In the case of the CS135s, three clusters of data are observed: the first corresponds to values of ß' higher than 5.0 10^{-7} m^{-1} sr^{-1} and values of lower than 0.5 10^{-4} m^{-1} where the values of ß' are largely overestimated by the CS135s because of the signal distortion; a second cluster corresponds to values of ß' lower than 5.0 10^{-7} m^{-1} sr^{-1} and values of lower than 0.5 10^{-4} m^{-1}, where the relationship looks well estimated but the noise affecting the CS135s is much larger than MUSA; finally, a third cluster corresponding to values of higher than 5.0 10^{-4} m^{-1}, where a small systematic effect seems to increase the values of ß' with respect to those measured by MUSA, which is probably related to the effect of environmental temperature, described above, on the CS135s hardware.

Finally, the CT25k show values in fair agreement with MUSA for the values of lower than 0.5 10^{-4} m^{-1}. Above this value, in agreement with the analysis reported in section 4.2, a systematic effect appears to increase the values of ß' with respect to those measured by MUSA. Values of above 0.5 10^{-4} m^{-1} the values are also strongly affected by the decrease of the SNR.

Differences between MUSA and the ceilometers are expected due to the large difference in the SNR due to their different laser sources. Nevertheless, all the plots show a difference between MUSA and ceilometers that looks proportional to the value of ß' and α. At CIAO, higher values of aerosol optical thickness are typically observed in summer than in fall and winter (Mona et al., 2006; Boselli et al., 2012). Therefore, the sensitivity issues ceilometers face in higher aerosol optical thicknesses are compounded by the larger discrepancies between ceilometers and MUSA at higher temperatures and, for the 905 nm instruments, by the higher water vapor content in the summer. In particular, warmer temperatures can decrease the efficiency of the ceilometer hardware and increase the bias of ceilometer attenuated backscatter coefficient profiles if calibration is not performed frequently. At large values of both of β' and α also the insufficient dynamic ranges of the systems significantly affects the comparison.

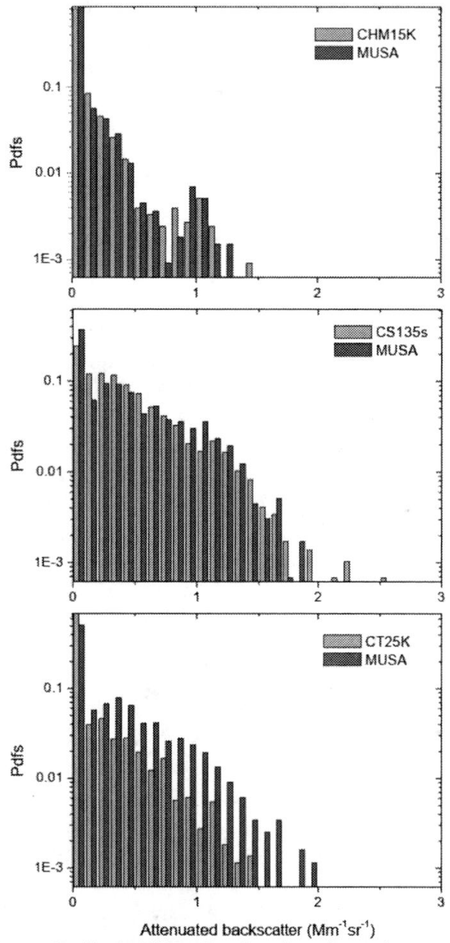

Figure 2: Probability density function of attenuated backscatter coefficient values retrieved from simultaneous observations performed by CHM15k and MUSA (upper panels), CS135s and MUSA (middle panels), CT25K and MUSA (lower panels).

4. CONCLUSIONS

The outcome of INTERACT campaign how ceilometers, though rugged instruments, are quite sensitive to the large changes in external temperature and collected background levels that occur on daily or seasonal bases; this generates adjustments of system parameters that affect the stability of the instrument response over the short and mid-term time period. Therefore, the use of a forward approach to calibrate a ceilometer using lidar observations or the use of a different calibration method should be frequently re-evaluated. Differences in the value of aerosol attenuated backscatter (β') among the MUSA Raman lidar and the ceilometers look proportional to the aerosol extinction coefficient (α). Changes in the values of α can be associated to seasonal changes affecting the ambient temperatures and, therefore, the ceilometer stability.

In conclusion, further technological improvements of ceilometers towards their operational use in the monitoring of the atmospheric aerosol are likely needed. It is planned to study historical data available at CIAO to confirm or improve the outcome of the INTERACT campaign.

ACKNOWLEDGEMENT

The financial support for EARLINET by the European Union under grant RICA 025991 in the Sixth Framework Programme is gratefully acknowledged. Since 2011 EARLINET is integrated in the ACTRIS Research Infrastructure Project supported by the European Union Seventh Framework Programme (FP7/2007-2013) under grant agreement n. 262254. The authors would also like to thank Campbell Scientific, Ltd., for contributing the CS135s for this campaign.

REFERENCES

[1] Madonna, F. and co-authors.: CIAO: the CNR-IMAA advanced observatory for atmospheric research, Atmos. Meas. Tech., 4, 1191-1208, doi:10.5194/amt-4-1191-2011, 2011.

[2] Madonna, F., Amato, F., Vande Hey, J., and Pappalardo, G.: Ceilometer aerosol profiling vs. Raman lidar in the frame of INTERACT campaign of ACTRIS, Atmos. Meas. Tech. Discuss., 7, 12407-12447, doi:10.5194/amtd-7-12407-2014, 2014.

[3] Pappalardo, G. and co-authors: EARLINET: towards an advanced sustainable European aerosol lidar network, Atmos. Meas. Tech., 7, 2389-2409, doi:10.5194/amt-7-2389-2014, 2014.

EPJ Web of Conferences **119**, 27007 (2016) DOI: 10.1051/epjconf/201611927007
ILRC 27

FROM OPERATIONAL CEILOMETER NETWORK TO OPERATIONAL LIDAR NETWORK

Mariana Adam[1]*, Myles Turp[1], Andrew Horseman[1], Carlos Ordóñez[2], Joelle Buxmann[1], Jacqueline Sugier[1]

[1]*Observations branch, Met Office, Exeter, Devon EX1 3PB, United Kingdom,*
**Email:mariana.adam@metoffice.gov.uk*

[2] *Weather Science branch, Met Office, Exeter, Devon EX1 3PB, United Kingdom*

ABSTRACT

During the eruption of Eyjafjallajökull in 2010, the Met Office ceilometers (Laser Cloud Based Recorders - LCBR) provided reasonable information about volcanic ash plumes over the United Kingdom [1]. This capability triggered the development of an operational system to provide quick looks of the range corrected signals (*RCS*) in near-real-time (NRT). Moreover, the Met Office acquired eleven Jenoptik ceilometers to supplement the operational ceilometer network. The combined network became operational in 2012 and currently comprises a total of 43 ceilometers reporting backscatter profiles in NRT. In 2013, Civil Aviation Authority (CAA) and the Department for transport (DfT) sponsored the acquisition of 9 fixed lidars and one mobile unit (each accompanied by a sunphotometer), to further improve the quantitative monitoring of volcanic ash. The current status of both ceilometer and lidar/sun-photometer networks is discussed and further developments are proposed.

1. INTRODUCTION

The Met Office hosts and runs the London Volcanic Ash Advisory Centre (VAAC) as part of the nine VAACs worldwide. The VAAC forecasters produce volcanic ash advisories and guidance products using a combination of various observation of volcanic plume (i.e. ground-based, satellite-based and aircraft-based observations as well as weather forecast and dispersion models) [2]. The ground-based observations in the UK are currently provided by the ceilometer network and by the end of 2016 additionally by the VA network (Lidar and sun-photometer network for Volcanic Ash observations). A full description of the remote sensing capabilities for volcanic ash observation at the Met Office is presented by Marenco et al. [3].

The ultimate goal for the VA network is to provide estimates of the volcanic ash concentrations within the volcanic plumes for VAAC. In order to reduce uncertainties in such estimations, a thorough quality control of the input data is performed.

2. MET OFFICE CEILOMETER NETWORK

The Met Office ceilometer network includes 32 Vaisala (25 of CL31 and 7 of CT25k) and 11 Jenoptik (Nimbus) ceilometers. Most of CT25k were replaced with CL31 during 2014 while the remaining seven will be upgraded this year. NRT plots of *RCS*, overlapped with cloud base height (*CBH*) are available internally to Met Office as well as externally within few hours delay (http://www.dwd.de/ceilomap). Note that some sites have both Vaisala and Jenoptik instruments while two sites are located overseas (Gibraltar and St Helena). The *NRT* plots are updated every 15 min. More recently, a further six stations from Ireland were added to this ceilometer network. Currently, the plots are updated each six hours (to be improved to hourly data soon).

The data flow for the ceilometer network (and in future for the VA network) is shown in Fig. 1. Data from ceilometers C_1, C_2,... C_n arrive to Met Office ftp server and then they are directed towards HERMES which is the Observations central processing system. The ceilometers data are temporally stored in HERMES database. The data are processed and then stored in MetDB (Met Office database). Once processed, the raw data are moved to MASS (Managed Archive Storage System). The website plots are generated using processed data stored in MetDB. Various errors are monitored and reported to the network manager.

The Met Office ceilometer network is shown in Fig. 2. The user selects a location and the image of *RCS* over the latest 24h is shown (Fig. 3). The option of plotting up to 15 km altitude range is available for Jenoptik ceilometers. The temporal resolution is 30s for all instruments. The spatial resolution is different for the three instruments (15m for Vaisala CT25 and Jenoptik and 10m or 20m for Vaisala CL31). Plots with the latest three days are available internally for all the stations.

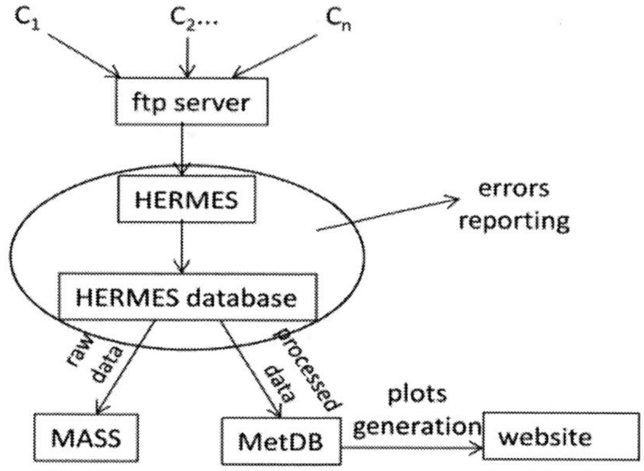

Fig. 1. Data flow chart for ceilometer network.

Fig. 2. The Met Office ceilometer network. Green dots represent Jenoptik ceilometers while blue and red dots represent Vaisala CL31 and CT25 ceilometers respectively.

Besides the primary use of the operational ceilometer network as ground-based observation for VAAC, this network can also provide valuable information for monitoring pollution events and validating air quality model parameters.

Fig. 3. Example of *RCS* from the Jenoptik ceilometer at Exeter (logarithmic scale).

Figure 4 shows an example of Saharan dust outbreak over UK in March 2014. Comparisons with AQUM (Air Quality Unified Model) are investigated. AQUM is a limited area configuration of the Unified model used to provide operational air quality forecast at the regional scale. The model includes relevant tropospheric chemistry and aerosol processes. At present time the model domain covers part of Western Europe, including the UK, at a horizontal resolution of 12 km x 12 km [4]. The model revealed enhanced particulate matter (PM) concentrations over the end of March, especially on 30th (Fig. 5). Unusual high Planetary Boundary Layer is observed over Exeter (Figs. 4-5).

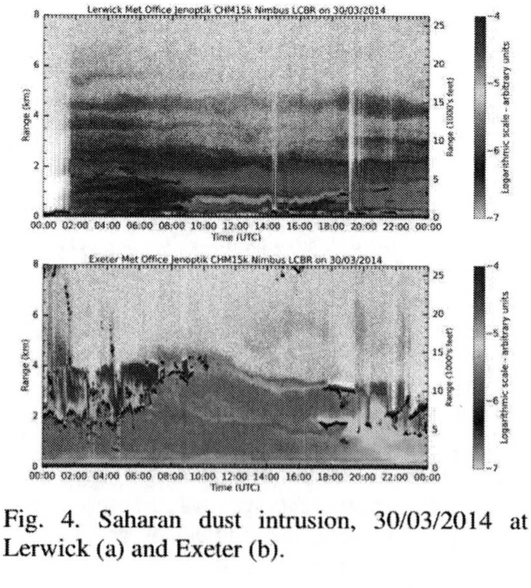

Fig. 4. Saharan dust intrusion, 30/03/2014 at Lerwick (a) and Exeter (b).

Other pollution events (e.g. smoke fires) will be shown during ILRC presentation.

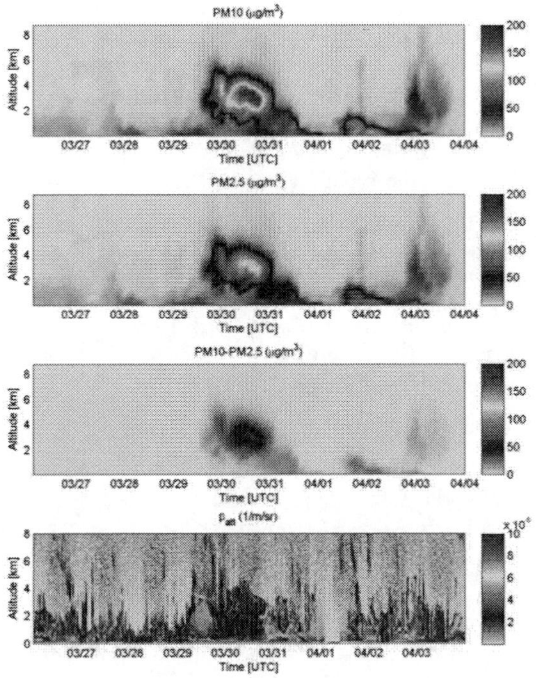

Fig. 5. Concentrations of PM10, PM2.5 and PM10 – PM2.5 (representing coarse aerosols) from AQUM as well as particle backscatter coefficient retrieved from Jenoptik ceilometer, in Exeter.

3. MET OFFICE VA NEWORK

The procurement of the lidars was initiated in 2013 and by the end of March 2014 the first two systems were ordered. The VA network is also equipped with sunphotometers. The sunphotometers were all acquired during 2013 from Cimel (CE318 NE DPS9). The lidars' manufacturer is Raymetrics. The Lidar is an UV Raman Lidar with depolarization capability (polar and co-polar). The Lidar was designed according to several specific requirements as appropriate for operational use. The lidar should work unattended and self-recover after various faults. For safety reasons, the laser output is eye-safe at the emission (after beam expansion) for 10s exposure. Thus, the energy per pulse is ~50mJ/pulse while the pulse repetition rate is 20Hz. The laser is a CFR200 from Quantel. The telescope has a 30cm diameter. The full overlap is ~250m which corresponds to a ~2.2 mrad FOV. The laser beam divergence after collimation and expansion is ~0.21mrad. The envisaged resolutions for operational use are 1 min and 15 m. However, the particles optical properties will be retrieved to a coarse resolution (at least 10min temporal resolution). Additional safety features include four interlocks which stop the laser operating during specific conditions. More details on the technical characteristics of the Lidar will be given during ILRC presentation.

The first two systems were delivered by the end of November 2014. One system was installed at Camborne, in SW England (Fig. 6) while the other system was installed in a van (mobile unit). Factory Acceptance Tests were performed at the manufacturer site in Athens in October 2014, while the Site Acceptance Tests were performed between December 2014 and February 2015. The most important requirement over the data quality was the provision of good signal to noise ratio (*SNR*) profiles over certain altitude ranges, as required for each channel.

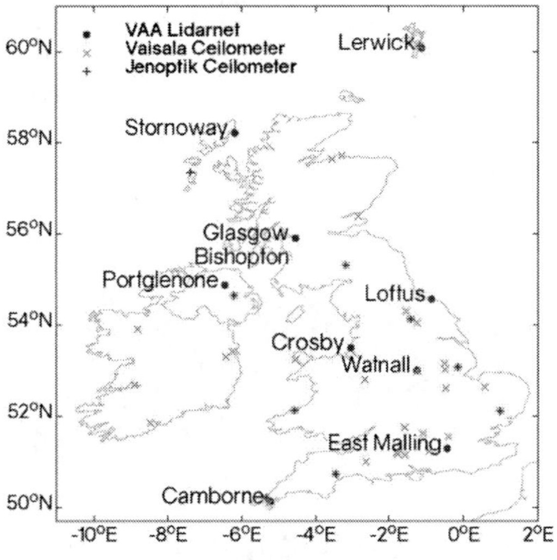

Fig. 6. The Met Office lidar network. Also shown, ceilometer network.

At the moment, the procurement of the remaining 8 systems is in progress. The project aims to have the network installed and operational by April 2016. The locations of the Lidar network are shown in Fig. 6.

An example of the *SNR* is shown in Fig. 7 for the

mobile Lidar for 10min temporal resolution profiles for both day (7:30-9:20) (a) and night (00:00-06:00) (b) conditions. In general, for both systems, a *SNR*=10 is found above 10km in most of the analyzed periods (day and night) for elastic channels. For Raman channel, *SNR*=10 exceeds 10km during night and 2-5km during day. An example of *RCS* and *VDR* is given in Fig. 8.

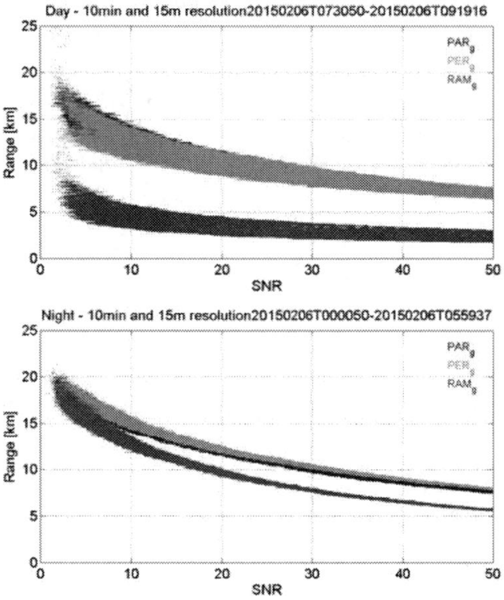

Fig. 7. SNR for all receiving channels (co-polar PARg, cross-polar PERg and N_2 Raman RAMg). "g" stands for glued signal.

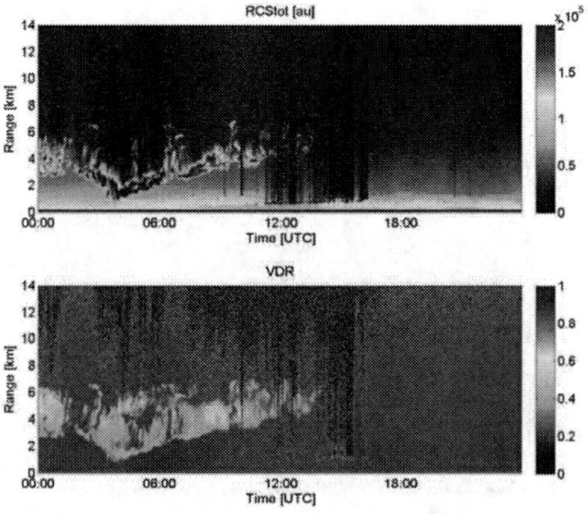

Fig. 8. *RCS* and *VDR* recorded on 03/02/2015 at Exeter.

Currently, various aspects of the data quality control (QC) are under development. In an operational context these QC algorithm have to be applied in real-time and therefore robust procedures have to be developed and implemented. This will require the statistical studies of the gluing coefficients, the depolarization constant and of various techniques to derive the Lidar calibration.

NRT plots will be available for *RCS* and *VDR* along with aerosol optical depth and Ångström coefficient provided by sunphotometer.

A complete illustration of the Lidar capabilities (including the retrieval of the particles optical properties) will be given during ILRC.

4. CONCLUSIONS

Coordinated measurements by both the ceilometer network and lidar network will improve the spatial coverage over UK for volcanic ash monitoring as well as other pollution events.

Synergetic measurements by the aircraft as well as satellites retrievals and dispersion models output will bring a more comprehensive knowledge of the pollutants, from their dynamics to their microphysical and optical properties.

ACKNOWLEDGEMENT

The ground based Lidar network is funded by United Kingdom Civil Aviation Authority and Department for Transport.

REFERENCES

[1] Lee, D. S. et al., 2011: The UK Met Office observations programme response to the Eyjafjallajökull eruption of Spring 2010, *AMS, Special International Application Session: The Eyjafjallajökull Volcanic Erruption of 2010, (https://ams.confex.com/ams/91Annual/webprogram/Paper184471.html)*.

[2] http://www.metoffice.gov.uk/aviation/vaac

[3] Marenco, F. et al., 2015: Remote Sensing of Volcanic Ash at the Met Office, *27 ILRC* – Hazard Monitoring.

[4] Savage, N. H. et al., 2013: Air quality modelling using the Met Office Unified Model (AQUM OS24-26): model description and initial evaluation, *Geosci. Model Dev.*, **6**, 353–372.

AUTHOR INDEX

AUTHOR INDEX

AUTHOR INDEX

AUTHOR INDEX

AUTHOR INDEX

AUTHOR INDEX

AUTHOR INDEX

AUTHOR INDEX

AUTHOR INDEX

AUTHOR INDEX